SALAS AND HILLE'S
CALCULUS

ONE AND SEVERAL VARIABLES

SEVENTH EDITION

REVISED BY

GARRET J. ETGEN

JOHN WILEY & SONS, INC.

New York Chichester Brisbane Toronto Singapore

In fond remembrance of
EINAR HILLE

ACQUISITIONS EDITOR	Ruth Baruth
DEVELOPMENTAL EDITOR	Nancy Perry
MARKETING MANAGER	Susan Elbe
PRODUCTION EDITOR	Charlotte Hyland
DESIGNER	Ann Marie Renzi
MANUFACTURING MANAGER	Susan Stetzer
ILLUSTRATION EDITOR	Sigmund Malinowski
ELECTRONIC ILLUSTRATION	Fine Line
CHAPTER TITLE DESIGN	Carol Grobe
COVER PHOTOGRAPH	Paul Silverman

This book was set in New Times Roman by Progressive Information Technologies and printed and bound by Von Hoffman Press. The cover was printed by Phoenix Color Corp.

Recognizing the importance of preserving what has been written, it is a policy of John Wiley & Sons, Inc. to have books of enduring value published in the United States printed on acid-free paper, and we exert our best efforts to that end.

The paper in this book was manufactured by a mill whose forest management programs include sustained yield harvesting of its timberlands. Sustained yield harvesting principles ensure that the number of trees cut each year does not exceed the amount of new growth.

Library of Congress Cataloging in Publication Data:
Salas, Saturnino L.
    Salas and Hille's calculus: one and several variables/revised by Garret J. Etgen. — 7th ed.
    p.   cm.
    Rev. ed. of: Calculus: one and several variables. 6th ed. 1990.
    Includes index.
    ISBN 0-471-58719-2
    1. Calculus. I. Hille, Einar (deceased) II. Salas, Saturnino L. III. Etgen, Garret J., Calculus: one and several variables. IV. Title.
    QA303.S17   1995b
    515 — dc20
                                                                94-30557
                                                                CIP
Printed in the United States of America

10 9 8 7 6 5 4 3

PREFACE

Above all, this is a text on mathematics. The subject is calculus, and the emphasis is on the three basic concepts: limit, derivative, and integral.

This text is designed for a standard introductory single and multivariable calculus sequence. Our fundamental goal in preparing the Seventh Edition has been to preserve and enhance the notable strengths that characterized previous editions, including:

- An emphasis on the mathematical exposition—an accurate, understandable treatment of the topics.
- A clear, concise approach. Basic ideas and important points are not obscured by excess verbiage.
- An appropriate level of rigor. Mathematical statements are careful and precise, and all important theorems are proved. This formality is presented in a way that is completely accessible to the beginning calculus student.
- A balance of theory and applications, illustrated by many examples and exercises.

At the same time, we recognize that with the rapid advances in computer technology and the current scrutiny of mathematics education at all levels, the teaching of calculus is undergoing a serious examination. Thus, an equally important and parallel goal of the Seventh Edition has been to incorporate modern technology and current trends without sacrificing the acknowledged strengths of the text.

FEATURES OF THE SEVENTH EDITION

Problem-Solving Skills and Real-World Applications

Over 2000 new problems have been added to the Seventh Edition.

- In order to develop students' problem-solving skills, we have significantly increased the number of problems at all levels. A large number of challenging and routine problems are now available in all exercise sets. Many additional medium-level problems are included to assist students in developing the understanding necessary to attack the challenging problems. In some problems, students are called upon to interpret and justify their answers to improve their analytical and communication skills.

- An even wider variety of real-world applications motivates students' study of mathematical topics.
- More illustrations have been added to exercise sets to provide students with visual support as they devise their problem-solving strategies.

Technology

Because the use of graphing calculators and/or computer algebra systems has increased in calculus courses, we have considerably expanded the application of technology in the text. We do not attempt to teach any particular technology and so use a generic approach. Technology problems are clearly designated with an icon (⊙) and may be skipped by instructors who prefer that their students not use calculators or computers.

- New technology-based examples appear within the chapter discussions of the material. These support the numerous exercises requiring the use of a graphics calculator or other graphing software located in the end-of-section problems sets.
- "Projects and Explorations Using Technology," a set of problems that requires a combination of approaches involving both analytical and technology skills, ends each chapter. As their title suggests, these problems are also suitable for use by students working in groups. A few of the problems introduce concepts to be developed later in the text, while others explore realistic applications of topics that have already been studied.

Increased Emphasis on Visualization

We recognize the importance of visualization in developing students' understanding of mathematical concepts. For that reason:

- All the artwork from the previous edition has been redrawn for increased clarity and understanding.
- Over 130 new figures have been added.
- Representations in three dimensions are now in full color for increased geometric understanding and include many new computer-generated figures of curves and surfaces in space.

Early Introduction of Differential Equations

A subsection on differential equations (separable equations) has been added to the Exponential Growth and Decay section of Chapter 7, allowing exercises on differential equations to be used throughout the rest of the book.

CONTENT AND ORGANIZATION CHANGES
IN THE SEVENTH EDITION

In response to the evolutionary state of the current calculus curriculum, many changes have been made in organization and content to meet the needs of today's students and instructors.

Precalculus Review (Chapter 1)

- This material, which provides a brief yet comprehensive review of the precalculus topics basic to the study of calculus, has been rewritten and expanded.
- The real number system and the real line are now discussed; the section covering methods for solving inequalities in one variable has been reorganized; and the treatment of analytic geometry—the Cartesian coordinate system, straight lines, and the conic sections—has been expanded.
- The treatment of the function concept, the elementary functions, and graphing (including technology) has been reorganized and expanded.
- The treatment of one-to-one functions and inverses has been moved to Chapter 7, where it serves to connect logarithmic and exponential functions.
- The brief section on proofs now includes mathematical induction with examples and exercises.

Limits and Continuity (Chapter 2)

- To improve students' understanding of these critical topics, a few of the discussions have been expanded and some figures have been added.
- The section on limits now includes a technology approach, illustrated by examples and new exercises.
- The derivative in various forms is introduced in the exercises.

Differentiation and Applications of the Derivative (Chapters 3 and 4)

- The interpretations of the derivative as a rate of change are treated in one section rather than two, and the coverage includes examples from economics.
- Differentiation of inverse functions has been moved to Chapter 7.
- The derivative of rational powers is now approached through implicit differentiation.
- The treatment of applications of the derivative has been reorganized slightly to emphasize the two objectives of Chapter 4: optimization and curve sketching.

Integration and Applications of the Integral (Chapters 5 and 6)

- The interpretation of the definite integral as "area" has been expanded; a subsection on "signed area" has been added.
- The treatment of u-substitutions has been streamlined; for example, trigonometric integrals are now incorporated in the change of variables section rather than as a separate section.
- The introduction of Riemann sums serves to introduce and motivate the approach taken in the applications chapter.
- The treatments of the various applications of the definite integral have been expanded.

The Transcendental Functions (Chapter 7)

- Inverse functions and the calculus of inverse functions are treated here rather than in separate sections earlier in the text.
- Integration by parts has been moved to Chapter 8, Techniques of Integration, and simple harmonic motion has been moved to Chapter 18, Differential Equations.
- The treatment of applications of exponential and logarithmic functions now includes a subsection on differential equations (separable equations).
- The treatment of the inverse trig functions now includes the inverse secant function.

Techniques of Integration (Chapter 8)

- Integration by parts has been moved from Chapter 7 to Section 8.2, and the section on partial fractions now follows trig substitutions.
- Almost all the treatments have been expanded, with many new examples, increased coverage of reduction formulas, and a discussion of hyperbolic substitutions.

Conic Sections; Polar Coordinates and Parametric Equations (Chapter 9)

- This new chapter combines material from two chapters in the previous edition. The conic sections are treated in one section; translation of axes and rotation of axes have been moved to Chapter 1 or the Appendices.
- The treatment of arc length has been softened a little: the proof based on the least upper bound axiom has been replaced by an intuitive argument.

Sequences and Series (Chapters 10 and 11)

- The least upper bound axiom now serves as a prelude to sequences, and there is more emphasis on boundedness in the treatment of sequences.
- The treatment of indeterminate forms has been modified: The "other" indeterminate forms—differences, products, exponential forms—are now treated in a separate subsection rather than integrating them with the $0/0$ and ∞/∞ forms.
- The treatment of power series has been expanded slightly; there are some new examples and figures, and the Lagrange form of the remainder is stated explicitly and used to derive bounds on the remainder.

Multivariable Calculus (Chapters 12–17)

- Substantial changes were not necessary in the treatment of these chapters. The major effort in this edition was to upgrade the illustrations and the exercises.
- A large number of computer-generated figures illustrating curves and surfaces in space have been added, and full color has been used where it is most helpful to students' understanding—in three-dimensional figures.

Differential Equations (Chapter 18)

- The material on differential equations has been thoroughly updated and revised to include numerous examples and applications throughout the chapter.
- A new introductory section familiarizes students with the basic terminology and concepts of differential equations.

FEATURES OF THE BOOK

Concise exposition The concepts of calculus are presented clearly and accurately without hand waving.

Theorems and proofs Highlighted theorems direct students to accurate mathematical statements. Most proofs are included to provide a high level of precision.

■ 2.5 THE PINCHING THEOREM; TRIGONOMETRIC LIMITS

Figure 2.5.1 shows the graphs of three functions f, g, h. Suppose that, as suggested by the figure, for x close to c, f is trapped between g and h. (The values of these functions at c itself are irrelevant.) If, as x tends to c, both $g(x)$ and $h(x)$ tend to the same limit L, then $f(x)$ also tends to L. This idea is made precise in what we call *the pinching theorem*.

Figure 2.5.1

THEOREM 2.5.1 THE PINCHING THEOREM

Let $p > 0$. Suppose that, for all x such that $0 < |x - c| < p$,
$$h(x) \leq f(x) \leq g(x).$$
If
$$\lim_{x \to c} h(x) = L \quad \text{and} \quad \lim_{x \to c} g(x) = L,$$
then
$$\lim_{x \to c} f(x) = L.$$

PROOF Let $\epsilon > 0$. Let $p > 0$ be such that

if $0 < |x - c| < p$, then $h(x) \leq f(x) \leq g(x)$.

Choose $\delta_1 > 0$ such that

if $0 < |x - c| < \delta_1$, then $L - \epsilon < h(x) < L + \epsilon$.

Choose $\delta_2 > 0$ such that

$< |x - c| < \delta_2$, then $L - \epsilon < g(x) < L + \epsilon$.

$, \delta_2\}$. For x satisfying $0 < |x - c| < \delta$, we have

$L - \epsilon < h(x) \leq f(x) \leq g(x) < L + \epsilon$,

$|f(x) - L| < \epsilon.$ ❑

Example 3 Figure 4.7.10 is a computer-generated graph of the function
$$f(x) = \frac{\cos x}{x}.$$
As $x \to 0^-$, $f(x) \to -\infty$; as $x \to 0^+$, $f(x) \to \infty$. The line $x = 0$ (the y-axis) is a vertical asymptote.

$f(x) = \frac{\cos x}{x}$

Figure 4.7.10

As $x \to \pm\infty$,
$$f(x) = \frac{\cos x}{x} \to 0.$$
This follows from the fact that
$$\left|\frac{\cos x}{x}\right| \leq \frac{1}{|x|} \quad \text{for all} \quad x$$
and $1/|x| \to 0$ as $x \to \pm\infty$. Thus, the line $y = 0$ (the x-axis) is a horizontal asymptote. Note that f is an odd function [$f(-x) = -f(x)$] so its graph is symmetric with respect to the origin. ❑

Example 4 Find the vertical and horizontal asymptotes, if any, of the function
$$g(x) = \frac{x + 1 - \sqrt{x}}{x^2 - 2x + 1} = \frac{x + 1 - \sqrt{x}}{(x - 1)^2}.$$

SOLUTION The domain of g is $0 \leq x < \infty$, $x \neq 1$. As $x \to 1$, $g(x) \to \infty$. Thus, the line $x = 1$ is a vertical asymptote. The behavior of g as $x \to \infty$ can be made more apparent by writing
$$g(x) = \frac{x + 1 - \sqrt{x}}{x^2 - 2x + 1} = \frac{x\left(1 + \frac{1}{x} - \frac{1}{\sqrt{x}}\right)}{x^2\left(1 - \frac{2}{x} + \frac{1}{x^2}\right)} = \frac{1 + \frac{1}{x} - \frac{1}{\sqrt{x}}}{x\left(1 - \frac{2}{x} + \frac{1}{x^2}\right)}.$$
Now, it is easy to see that $g(x) \to 0$ as $x \to \infty$. The line $y = 0$ (the x-axis) is a horizontal asymptote. ❑

New examples To facilitate students' understanding, many examples have been revised and new examples have been added.

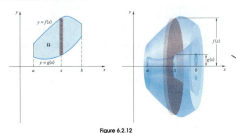

Figure 6.2.12

PROOF The cross section with coordinate x is a *washer* of outer radius $f(x)$, inner radius $g(x)$, and area

$$A(x) = \pi[f(x)]^2 - \pi[g(x)]^2 = \pi([f(x)]^2 - [g(x)]^2).$$

We can get the volume of the solid by integrating this function from a to b. ❑

Suppose now that the boundaries are functions of y rather than x (see Figure 6.2.13). By revolving Ω *about the y-axis,* we obtain a solid. It is clear from (6.2.4) that in this case

(6.2.6) $$V = \int_c^d \pi([F(y)]^2 - [G(y)]^2)\, dy.$$ (washer method about y-axis)

Figure 6.2.13

Improved visualization A completely new and expanded illustration program, including three-dimensional illustrations in full color, provides a better visual representation of concepts.

Example 2 A metal plate in the form of a trapezoid is affixed to a vertical dam as in Figure 6.6.5. The dimensions shown are given in meters; the weight density of water in the metric system is approximately 9800 newtons per cubic meter. Find the force on the plate.

Figure 6.6.5

SOLUTION First we find the width of the plate x meters below the water level. By similar triangles (see Figure 6.6.6)

$$t = \tfrac{1}{2}(8 - x) \qquad \text{so that} \qquad w(x) = 8 + 2t = 16 - x.$$

Figure 6.6.6

The force against the plate is

$$\int_4^8 9800x(16 - x)\, dx = 9800 \int_4^8 (16x - x^2)\, dx$$
$$= 9800 \left[8x^2 - \tfrac{1}{3}x^3 \right]_4^8 \approx 2,300,000 \text{ newtons.} \quad ❑$$

Real-world applications Students see how the concepts and methods of calculus connect with important problems in science and engineering.

New exercises The exercise sets have been revised and over 2000 new problems added, resulting in an improved balance between drill problems and more challenging exercises involving either theory or applications.

Technology problems Problems marked by the icon ▷ encourage students to use technology as a tool to enhance understanding and problem-solving skills.

26. $f(x) = x + \cos 2x, \quad 0 < x < \pi$.

27. $f(x) = \sin^2 x - \sqrt{3} \sin x, \quad 0 < x < \pi$.

28. $f(x) = \sin^2 x, \quad 0 < x < 2\pi$.

29. $f(x) = \sin x \cos x - 3 \sin x + 2x, \quad 0 < x < 2\pi$.

30. $f(x) = 2 \sin^3 x - 3 \sin x, \quad 0 < x < \pi$.

31. Prove Theorem 4.3.4 by applying Theorem 4.2.3.

32. Prove the validity of the second-derivative test in the case that $f''(c) < 0$.

33. Find the critical numbers and the local extreme values of the polynomial

$$P(x) = x^4 - 8x^3 + 22x^2 - 24x + 4.$$

Then show that the equation $P(x) = 0$ has exactly two real roots, both positive.

34. A function f has derivative f' given by

$$f'(x) = x^3(x-1)^2(x+1)(x-2).$$

At what numbers x, if any, does f have a local maximum? A local minimum?

35. A polynomial function $p(x) = a_n x^n + a_{n-1} x^{n-1} + \cdots + a_1 x + a_0$ has critical numbers at $x = -1, 1, 2,$ and 3, and corresponding values $p(-1) = 6, \ p(1) = 1, \ p(2) = 3,$ and $p(3) = 1.$ Sketch a possible graph for p if: (a) n is odd, (b) n is even.

36. The quadratic function $f(x) = Ax^2 + Bx + C$ has a local minimum at $x = 2$ and passes through the points $(-1, 3)$ and $(3, -1)$. Find $A, B,$ and C.

37. Determine a and b such that the function $f(x) = ax/(x^2 + b^2)$ has a local minimum at $x = -2$ and $f'(0) = 1$.

38. Let $f(x) = x^p(1-x)^q$, where $p \geq 2$ and $q \geq 2$ are integers.
(a) Show that the critical numbers of f are $x = 0$, $p/(p+q)$, and 1.
(b) Show that if p is even, then f has a local minimum at 0.
(c) Show that if q is even, then f has a local minimum at 1.
(d) Show that f has a local maximum at $p/(p+q)$ for all p and q.

39. Let

$$f(x) = \begin{cases} x^2 \sin(1/x), & x \neq 0 \\ 0, & x = 0. \end{cases}$$

In Exercise 67, Section 3.1, we saw that f is differentiable at 0 and that $f'(0) = 0$. Show that f has neither a local maximum nor a local minimum at 0.

40. Suppose that $C(x)$, $R(x)$, and $P(x)$ are the cost, revenue, [and profit] corresponding to the production and [sale of x] certain item. Suppose, also, that C and [R are] functions. Then, since $P = R - C$, it [is dif]ferentiable. Prove that if it is possible

to maximize the profit by producing and selling x_0 items, then $C'(x_0) = R'(x_0)$. That is, the marginal cost equals the marginal revenue when the profit is maximized.

41. Let $y = f(x)$ be differentiable and suppose that the graph of f does not pass through the origin. Then the distance D from the origin to a point $P(x, f(x))$ on the graph is given by

$$D = \sqrt{x^2 + [f(x)]^2}.$$

Show that if D has a local extreme value at c, then the line through $(0, 0)$ and $(c, f(c))$ is perpendicular to the tangent line to the graph of f at c.

42. Prove that a polynomial of degree n has at most $n - 1$ local extreme values.

▷ 43. Let $f(x) = x^4 - 2x^2 - 3x + 2$.
(a) Show that f has exactly one critical number c in the interval $(1, 2)$.
(b) Use the bisection method (see Section 2.6) to approximate c to within $\frac{1}{16}$. Does f have a local maximum, a local minimum, or neither a maximum nor a minimum at c?

▷ 44. Let $f(x) = 2 + 20x + 4x^2 - x^4$.
(a) Show that f has exactly one critical number in the interval $(2, 3)$.
(b) Use the bisection method to approximate c to within $\frac{1}{16}$. Does f have a local maximum, a local minimum, or neither a maximum nor a minimum at c?

▷ 45. Let $f(x) = x^4 - 7x^2 + 2x - 3$.
(a) Show that f has exactly one critical number c in the interval $(2, 3)$.
(b) Use the Newton-Raphson method to approximate c; calculate x_3 and round your answer to four decimal places. Does f have a local maximum, a local minimum, or neither a maximum nor a minimum at c?

▷ 46. Let $f(x) = x \cos x$.
(a) Show that f has exactly one critical number in the interval $(0, \pi/2)$.
(b) Use the Newton-Raphson method to approximate c; calculate x_3 and round your answer to four decimal places. Does f have a local maximum, a local minimum, or neither a maximum nor a minimum at c?

▷ 47. Let $f(x) = \sin x + (x^2/2) - 2x$.
(a) Show that f has exactly one critical number in the interval $[2, 3]$.
(b) Use the Newton-Raphson method to approximate c; calculate x_3 and round your answer to four decimal places. Does f have a local maximum, a local minimum, or neither a maximum nor a minimum at c?

▷ In Exercises 48–51, use a graphing utility to graph the function f on the indicated interval. (a) Use the graph to estimate the critical numbers and the local extreme values; and (b) estimate the intervals on which f increases and the intervals on which f decreases. Round off your estimates to three decimal places.

Chapter Highlights End-of-chapter lists stress important terms, ideas, and theorems.

The inverse secant, $y = \sec^{-1} x$, is the inverse of $y = \sec x, \ x \in [0, \frac{1}{2}\pi) \cup (\frac{1}{2}\pi, \pi]$.

graph of $y = \sin^{-1} x$ (p. 462)   graph of $y = \tan^{-1} x$ (p. 465)
graph of $y = \sec^{-1} x$ (p. 468)

$$\frac{d}{dx}(\sin^{-1} x) = \frac{1}{\sqrt{1-x^2}} \qquad \int \frac{dx}{\sqrt{a^2 - x^2}} = \sin^{-1}\left(\frac{x}{a}\right) + C \quad (a > 0)$$

$$\frac{d}{dx}(\tan^{-1} x) = \frac{1}{1+x^2} \qquad \int \frac{dx}{a^2 + x^2} = \frac{1}{a}\tan^{-1}\left(\frac{x}{a}\right) + C \quad (a \neq 0)$$

$$\frac{d}{dx}(\sec^{-1} x) = \frac{1}{|x|\sqrt{x^2 - 1}} \qquad \int \frac{dx}{x\sqrt{x^2 - a^2}} = \frac{1}{a}\sec^{-1}\left(\frac{|x|}{a}\right) + C \quad (a > 0)$$

definition of the remaining inverse trigonometric functions (p. 471)

7.9   The Hyperbolic Sine and Cosine

$$\sinh x = \frac{1}{2}(e^x - e^{-x}), \qquad \cosh x = \frac{1}{2}(e^x + e^{-x}),$$

$$\frac{d}{dx}(\sinh x) = \cosh x, \qquad \frac{d}{dx}(\cosh x) = \sinh x.$$

graphs (pp. 475–476)   basic identities (p. 477)

*7.10   The Other Hyperbolic Functions

$$\tanh x = \frac{\sinh x}{\cosh x}, \qquad \coth x = \frac{\cosh x}{\sinh x},$$

$$\operatorname{sech} x = \frac{1}{\cosh x}, \qquad \operatorname{csch} x = \frac{1}{\sinh x}.$$

derivatives (p. 479)   hyperbolic inverses (p. 481)
derivatives of hyperbolic inverses (p. 482)

■ **PROJECTS AND EXPLORATIONS USING TECHNOLOGY**

To do these exercises you will need a graphics calculator or a computer with graphing capability. The majority of these problems are open-ended so different approaches may be used to solve them. You should be aware that different approaches can result in slight variations in the answers. Round your numerical answers to at least four decimal places. The rounding method that your calculator or computer uses may also cause variations in answers.

7.1 The functions $f(x) = a \ln x$, where a is a constant, have a number of applications, one of which will be considered in a later exercise.
(a) Find the values of a for which the graph of f is tangent to the line $y = x$.
(b) For each real number a, how many solutions will there be of $f(x) = x$? What is the value of f' at each solution of $f(x) = x$?
(c) How many solutions are there to $f[f(x)] = x$? What is the value of f' at each of these solutions?
(d) Represent f as a logarithm function in another base.

7.2 Let $A(t)$ denote the area of the rectangle of width $2t$ that has its lower vertices on the x-axis and its upper vertices on the graph of

$$f(x) = e^{(1-x^2)(2+x^2)}.$$

See the figure.

Projects and Explorations Using Technology Special problem sets encourage deeper investigation of the material and can be used for cooperative learning activities.

SUPPLEMENTS

Student Aids

Answers to Odd-Numbered Exercises Answers to all the odd-numbered exercises are included at the back of the text.

Student Solutions Manual, Prepared by Garret J. Etgen, University of Houston This manual contains worked-out solutions to all the odd-numbered exercises and is available through your bookstore.

Instructor Aids

Instructor's Manual, by Garret J. Etgen and Sylvain Laroche This manual contains solutions to all the problems in the text.

Test Bank, by Sylvain Laroche A wide range of problems and their solutions are keyed to the text material and exercise sets.

Computerized Test Bank Available in both IBM and Macintosh formats, the Computerized Test Bank allows instructors to create, customize, and print a test containing any combination of questions from the test bank. Instructors can also edit the questions or add their own.

Technology Manuals

Discovering Calculus with Derive, by Jerry Johnson, University of Nevada-Reno, and Benny Evans, Oklahoma State University

- Derive instructions and tutorials
- Solved problems
- Practice problems
- Laboratory exercises

Discovering Calculus with Mathematica, by Cecilia A. Knoll, Florida Institute of Technology, Michael D. Shaw, Florida Institute of Technology, Jerry Johnson, and Benny Evans

- Mathematica introduction and commands
- Solved problems
- Exercises
- Laboratory projects

Discovering Calculus with Maple, by Kent Harris, Western Illinois University, and Robert J. Lopez, Rose-Hulman Institute of Technology

- Maple commands
- Example problems and step-by-step solutions
- Exercises

Discovering Calculus with Graphing Calculators, by Joan McCarter, Arizona State University

- Introductions to various calculators currently on the market (this manual is calculator nonspecific)
- Projects

- Additional exercises
- Critical thinking questions

ACKNOWLEDGMENTS

The revision of a text of this magnitude and stature requires a lot of encouragement and a lot of help. I was fortunate to have an ample supply of both from a variety of sources.

Each edition of this text was developed from those that preceded it. The present book owes much to the people who contributed to the first six editions. I am deeply indebted to them.

The reviewers of the seventh edition supplied detailed criticisms and valuable suggestions. I offer my sincere appreciation to the following individuals:

Linda Becerra
University of Houston

Jay Bourland
Colorado State University

Gary Crown
Wichita State University

Steve Davis
Davidson College

Anthony Dooley
University of New South Wales

William Fox
U.S. Military Academy — West Point

Barbara Gale
Prince Georges Community College

Pamela Gorkin
Bucknell University

Gary Itzkowitz
Rowan College of New Jersey

Harold Jacobs
East Stroudsburg University of Pennsylvania

Adam Lutoborski
Syracuse University

Douglas Mackenzie
University of New South Wales

Katherine Murphy
University of North Carolina-Chapel Hill

Jean E. Rubin
Purdue University

Stuart Smith
University of Toronto

I am especially grateful to Richard D. Byrd, University of Houston, who read the first draft of the manuscript letter by letter, made comments to match, read all the reviewers' comments, and then read the revised manuscript. This was a truly noble effort.

I am equally grateful to John Oman, University of Wisconsin — Oshkosh, who created the Projects and Explorations Using Technology problems.

I also want to thank Jun Ma, University of Houston, who provided detailed solutions for all the new exercises; there were more than 2000 of them.

I am deeply indebted to the editorial staff at John Wiley & Sons: Ruth Baruth, Mathematics Editor, who invited me to undertake this project in the first place and who provided encouragement when I needed encouragement and prodding when I needed prodding; Joan Carrafiello and Nancy Perry, Development Editors — Joan offered invaluable help in the beginning stages of this revision and Nancy continued it, along with exceptional organization and support; Charlotte Hyland, Production Editor, who was patient and understanding as she guided the project through the production stages; Sigmund Malinowski, who carefully directed the entirely new art program; and

Ann Marie Renzi, whose creativity produced the attractive interior design as well as the cover.   This is an efficient, professional (and patient!) group.

Finally, I want to acknowledge the contributions of my wife, Charlotte.   She supplied the sustenance and support that I had to have to complete this work.

Garret J. Etgen

CONTENTS

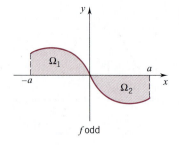

CHAPTER 6 SOME APPLICATIONS OF THE INTEGRAL   347

CHAPTER 7 THE TRANSCENDENTAL FUNCTIONS   395

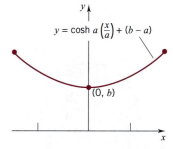

CHAPTER 8 TECHNIQUES OF INTEGRATION   489

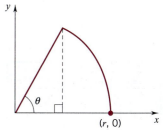

---

* Denotes optional section.

CHAPTER 16   DOUBLE AND TRIPLE INTEGRALS   1033

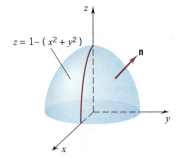

$z = 1 - (x^2 + y^2)$

CHAPTER 17   LINE INTEGRALS AND SURFACE INTEGRALS   1115

CHAPTER 18 ELEMENTARY DIFFERENTIAL EQUATIONS 1195

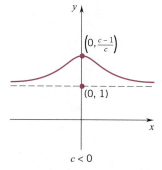

APPENDIX A SOME ADDITIONAL TOPICS A-1

APPENDIX B SOME ADDITIONAL PROOFS A-11

SALAS AND HILLE'S
CALCULUS

ONE VARIABLE

INTRODUCTION

CHAPTER 1

■ **1.1 WHAT IS CALCULUS?**

To a Roman in the days of the empire a "calculus" was a pebble used in counting and in gambling. Centuries later "calculare" came to mean "to compute," "to reckon," "to figure out." To the mathematician, physical scientist, and social scientist of today calculus is elementary mathematics (algebra, geometry, trigonometry) enhanced by *the limit process.*

Calculus takes ideas from elementary mathematics and extends them to a more general situation. Here are some examples. On the left-hand side you will find an idea from elementary mathematics; on the right, this same idea as extended by calculus.

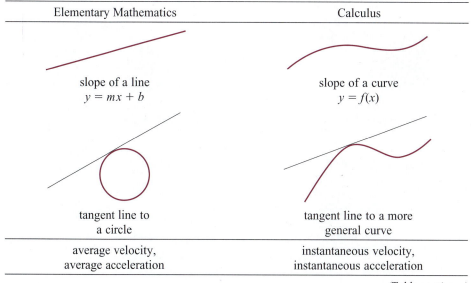

Elementary Mathematics	Calculus
slope of a line $y = mx + b$	slope of a curve $y = f(x)$
tangent line to a circle	tangent line to a more general curve
average velocity, average acceleration	instantaneous velocity, instantaneous acceleration

(Table continues)

1

distance moved under a constant velocity	distance moved under varying velocity
area of a region bounded by line segments	area of a region bounded by curves
sum of a finite collection of numbers $a_1 + a_2 + \cdots + a_n$	sum of an infinite series $a_1 + a_2 + \cdots + a_n + \cdots$
average of a finite collection of numbers	average value of a function on an interval
length of a line segment	length of a curve
center of a circle	centroid of a region
volume of a rectangular solid	volume of a solid with a curved boundary
surface area of a cylinder	surface area of a more general solid

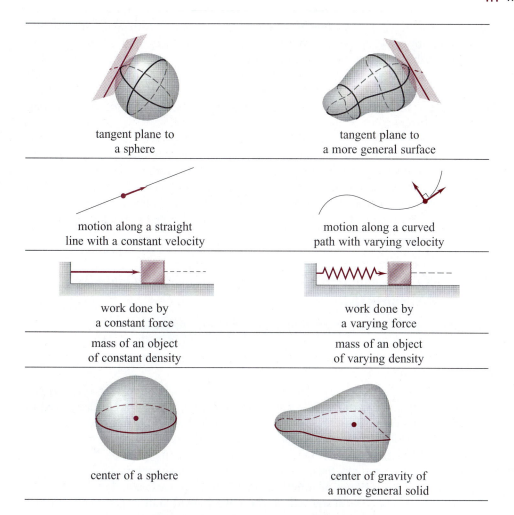

tangent plane to
a sphere

tangent plane to
a more general surface

motion along a straight
line with a constant velocity

motion along a curved
path with varying velocity

work done by
a constant force

work done by
a varying force

mass of an object
of constant density

mass of an object
of varying density

center of a sphere

center of gravity of
a more general solid

It is fitting to say something about the history of calculus. The origins can be traced back to ancient Greece. The ancient Greeks raised many questions (often paradoxical) about tangents, motion, area, the infinitely small, the infinitely large— questions that today are clarified and answered by calculus. Here and there the Greeks themselves provided answers (some very elegant), but mostly they provided only questions.

After the Greeks, progress was slow. Communication was limited, and each scholar was obliged to start almost from scratch. Over the centuries some ingenious solutions to calculus-type problems were devised, but no general techniques were put forth. Progress was impeded by the lack of a convenient notation. Algebra, founded in the ninth century by Arab scholars, was not fully systemized until the sixteenth century. Then, in the seventeenth century, Descartes established analytic geometry, and the stage was set.

The actual invention of calculus is credited to Sir Isaac Newton (1642–1727) and to Gottfried Wilhelm Leibniz (1646–1716), an Englishman and a German. Newton's invention is one of the few good turns that the great plague did mankind. The plague forced the closing of Cambridge University in 1665 and young Isaac Newton of Trinity College returned to his home in Lincolnshire for eighteen months of meditation, out of which grew his *method of fluxions,* his *theory of gravitation,* and his *theory of light*. The method of fluxions is what concerns us here. A treatise with this title was written by Newton in 1672, but it remained unpublished until 1736, nine years after his death. The new method (calculus to us) was first announced in 1687, but in vague general terms without symbolism, formulas, or applications. Newton himself seemed reluctant to publish anything tangible about his new method, and it is not surprising that the development on the Continent, in spite of a late start, soon overtook Newton and went beyond him.

Leibniz started his work in 1673, eight years after Newton. In 1675 he initiated the basic modern notation: dx and $\int$. His first publications appeared in 1684 and 1686. These made little stir in Germany, but the two brothers Bernoulli of Basel (Switzerland) took up the ideas and added profusely to them. From 1690 onward calculus grew rapidly and reached roughly its present state in about a hundred years. Certain theoretical subtleties were not fully resolved until the twentieth century.

■ 1.2 NOTATIONS AND FORMULAS FROM ELEMENTARY MATHEMATICS

The following outline is presented for review and easy reference.

Sets

A set is a well-defined collection of distinct objects. The objects in a set are called the *elements* or *members* of the set. Capital letters A, B, C, . . . , are usually used to denote sets and lowercase letters a, b, c, . . . , to denote the elements of a set.

For a collection S of objects to be a set it must be *well-defined* in the sense that given any object x whatsoever, it must be possible to determine whether or not x is an element of S. For example, the collection of capital letters on this page, the collection of states in the United States of America, and the collection of natural numbers are all examples of sets. On the other hand, suppose we wanted to discuss the collection of beautiful cities in England? Different people might have different collections. Thus, the collection of beautiful cities in England is not a set. Collections based on subjective judgments such as "all good football players," or "all intelligent adults" are not sets.

Notation

The object x is in the set A	$x \in A$
The object x is not in the set A	$x \notin A$
The set of all x for which property P holds	$\{x: P\}$
(For example, $A = \{x: x$ is a vowel$\} = \{a, e, i, o, u\}$)	
A is a subset of B (A is contained in B)	$A \subseteq B$
The union of A and B	$A \cup B$
($A \cup B = \{x: x \in A$ or $x \in B\}$)	
The intersection of A and B	$A \cap B$
($A \cap B = \{x: x \in A$ and $x \in B\}$)	
The empty set	$\varnothing$

These are the basic notions from set theory that you will need for this text.

Real Numbers

Our study of calculus is based on the real number system. The real number system consists of the set of real numbers together with the familiar arithmetic operations— addition, subtraction, multiplication, division—and certain other properties which are reviewed briefly here.

Classification

Natural numbers (*or positive integers*)　　1, 2, 3, ...

Integers　　0, 1, -1, 2, -2, ...

Rational numbers　　$\{x: x = p/q$, where p, q are integers and $q \neq 0\}$†

Examples　　$\frac{2}{5}$, $\frac{-19}{7}$, $\frac{4}{1}$ $(=4)$.

Irrational numbers　　real numbers that are not rational numbers.

Examples　　$\sqrt{2}$, $\sqrt[3]{7}$, π, the solutions of the equation $3x^2 - 5 = 0$.

Decimal Representation

Every real number can be represented by a decimal. If $r = p/q$ is a rational number, then its decimal representation is found by dividing the denominator q into the numerator p. The resulting decimal expansion will either *terminate* (for example, $3/5 = 0.6$, $27/20 = 1.35$) or *repeat* (for example, $2/3 = 0.666666 \ldots = 0.\overline{6}$, $15/11 = 1.363636 \ldots = 1.\overline{36}$, $19/7 = 2.714285714285 \ldots = 2.\overline{714285}$; the bar over the sequence of digits indicates that the sequence repeats indefinitely). The converse is also true: every repeating or terminating decimal represents a rational number.

The decimal expansion of an irrational number neither terminates nor repeats. For example,

$$\sqrt{2} = 1.414213562 \ldots, \quad \pi = 3.141592653 \ldots, \text{ and}$$
$$r = 0.10110111011110 \ldots$$

are irrational numbers.

If we stop the decimal expansion of a given number at a certain decimal place, then the result is a rational number approximation for the number. For instance, $1.414 = 1414/1000$ and $3.14 = 314/100$ are common rational number approximations for $\sqrt{2}$ and π, respectively. More accurate approximations can be obtained by taking more decimal places in the expansions.

Geometric Representation

A fundamental concept in mathematics which connects the abstract notion of *real number* with the geometric notion of *point* is the representation of the real numbers as points on a straight line. This is done by choosing an arbitrary point O on a line to represent the number 0, and another point U (usually taken to the right of O) to represent the number 1. See Figure 1.2.1. The point O is called the *origin* and the distance between O and U determines a scale (a unit length). With O and U specified, each real number can be represented as a point on the line and, conversely,

Figure 1.2.1

---

† Recall that division by 0 is undefined.

each point on the line represents a real number. A line representing the real numbers is called a *number line* or *real line*; the number associated with a point P on the line is called the *coordinate* of P. The *positive* numbers are identified with the points on the right side of O and the *negative* numbers with the points to the left of O. The point representing the number x ($x \neq 0$) is x units from O if x is positive and $-x$ units from O if x is negative. In this context, we will frequently refer to real numbers as "points," and by this we will mean "points on a number line." Figure 1.2.2 shows some numbers plotted as points on a number line.

Figure 1.2.2

Order Properties

If a and b are real numbers, then *a is less than b*, denoted $a < b$, if $b - a$ is a positive number. This is equivalent to saying that *b is greater than a*, which is denoted $b > a$. Geometrically, $a < b$ if the point a lies to the left of the point b on a number line. The notation $a \leq b$ means that either $a < b$ or $a = b$ (equivalently, $b \geq a$).

The real numbers are *ordered* in the sense that if a and b are real numbers, then exactly one of the following holds:

$$a < b, \qquad a = b, \qquad a > b \qquad \text{(trichotomy).}$$

The symbols $<, >, \leq, \geq$ are called *inequalities*. Inequalities satisfy the following properties:

(i) If $a < b$ and $b < c$, then $a < c$ (transitive property).

(ii) If $a < b$, then $a + c < b + c$ for all real numbers c.

(iii) If $a < b$ and $c < d$, then $a + c < b + d$.

(iv) If $a < b$ and $c > 0$, then $ac < bc$.

(v) If $a < b$ and $c < 0$, then $ac > bc$.

The corresponding properties hold for $>, \leq, \geq$. A crucial point is Property (v): If an inequality is multiplied by a negative quantity, then the "direction" of the inequality is reversed. Techniques for solving inequalities use these properties and are reviewed in Section 1.3.

Density

Between any two real numbers there are infinitely many rational numbers and infinitely many irrational numbers. In particular, *there is no smallest positive real number.*

Absolute Value

Two important properties of a real number a are its *sign* and its *size, or magnitude*. Geometrically, the sign of a tells us whether the point a is on the right or left of 0 on a number line. The magnitude of a is the distance between the point a and 0; 0 itself does not have a sign and its magnitude is 0. The magnitude of a is more commonly called *absolute value of a*, denoted $|a|$. The absolute value of a is also given by

$$|a| = \begin{cases} a & \text{if } a \geq 0 \\ -a & \text{if } a < 0. \end{cases}$$

Other characterizations $|a| = \max\{a, -a\}$; $|a| = \sqrt{a^2}$.
Geometric interpretations $|a| = $ distance between a and 0;

$$|a - c| = \text{distance between } a \text{ and } c.$$

Other properties   (i) $|a| = 0$ iff $a = 0$.†
                                   (ii) $|-a| = |a|$.
                                   (iii) $|ab| = |a||b|$.
                                   (iv) $|a + b| \leq |a| + |b|$ (the triangle inequality).††
                                   (v) $\big||a| - |b|\big| \leq |a - b|$ (a variant of the triangle inequality).

Techniques for solving inequalities involving absolute values are also reviewed in Section 1.3.

Intervals

Suppose that $a < b$. The *open interval* (a, b) is the set of all numbers between a and b:

$$(a, b) = \{x: a < x < b\}.$$

The *closed interval* $[a, b]$ is the open interval (a, b) together with the endpoints a and b:

$$[a, b] = \{x: a \leq x \leq b\}.$$

There are seven other types of intervals:

$$(a, b] = \{x: a < x \leq b\}.$$
$$[a, b) = \{x: a \leq x < b\}.$$
$$(a, \infty) = \{x: a < x\}.$$
$$[a, \infty) = \{x: a \leq x\}.$$
$$(-\infty, b) = \{x: x < b\}.$$
$$(-\infty, b] = \{x: x \leq b\}.$$
$$(-\infty, \infty) = \text{set of real numbers}.$$

Interval notation is easy to remember: we use a square bracket to include an endpoint and a parenthesis when the endpoint is not included. On a number line, the inclusion or exclusion of an endpoint is signified with a solid "dot" or an open "dot," respectively. The symbols ∞ and $-\infty$, read "infinity" and "negative infinity" (or

---

† By "iff" we mean "if and only if." This expression is used so often in mathematics that it's convenient to have an abbreviation for it.

†† The absolute value of the sum of two numbers cannot exceed the sum of their absolute values. This is analogous to the fact that in a triangle the length of one side cannot exceed the sum of the lengths of the other two sides.

"minus infinity") are not real numbers. The ∞ symbol in the intervals just given indicates that the interval extends indefinitely in the positive direction; the $-\infty$ indicates that the interval extends indefinitely in the negative direction. Since ∞ and $-\infty$ are not real numbers, we do not have intervals of the form $[a, \infty]$ or $[-\infty, b)$, and so forth.

Boundedness

A set S of real numbers is said to be:

  (i) *Bounded above* iff there exists a real number M such that

$$x \leq M \qquad \text{for all } x \in S;$$

    M is called an *upper bound* for S.

 (ii) *Bounded below* iff there exists a real number m such that

$$x \geq m \qquad \text{for all } x \in S;$$

    m is called a *lower bound* for S.

(iii) *Bounded* iff it is bounded above and below.

For example, the intervals $(-\infty, 2]$ and $(-\infty, 2)$ are bounded above but not below. The set of positive integers is bounded below but not above. The intervals $[1, 5]$, $(-3, 7)$, and $(-5, \sqrt{2}]$ are bounded (these intervals are bounded above and below).

Algebra

Basic Factoring Formulas

$$(a + b)^2 = a^2 + 2ab + b^2$$
$$(a - b)^2 = a^2 - 2ab + b^2$$
$$(a + b)^3 = a^3 + 3a^2b + 3ab^2 + b^3$$
$$(a - b)^3 = a^3 - 3a^2b + 3ab^2 - b^3$$
$$a^2 - b^2 = (a - b)(a + b)$$
$$a^3 - b^3 = (a - b)(a^2 + ab + b^2)$$
$$a^3 + b^3 = (a + b)(a^2 - ab + b^2)$$

Polynomials and Rational Expressions

A *variable* (also called an *unknown*) is a symbol used to represent an arbitrary element in a given set. Lowercase letters near the end of the alphabet such as $t, u, x, y, \ldots$, will normally be used to denote variables. In this text, the sets under consideration are usually sets of real numbers, and so variables are often referred to as *real variables*.

    A polynomial in one variable is an expression of the form

$$P(x) = a_n x^n + a_{n-1} x^{n-1} + \cdots + a_2 x^2 + a_1 x + a_0$$

where $a_0, a_1, \ldots, a_n \, (a_n \neq 0)$ are real numbers called the *coefficients* of P, and n is a nonnegative integer called the *degree* of P. For example,

$$P(x) = 2x^4 - 5x^3 + \pi x - \sqrt{2}$$

is a polynomial of degree 4;

$$P(x) = 3x + 1 \qquad \text{and} \qquad P(x) = 6$$

are polynomials of degrees 1 and 0, respectively. In general, each nonzero real number is a polynomial of degree 0; 0 is also a polynomial, but no degree is assigned to it. The expression

$$F(x) = x^3 + 4x^{3/2} + \frac{1}{x^2}$$

is not a polynomial.

A number r is a *root* or *solution* of the polynomial equation

$$P(x) = a_n x^n + a_{n-1} x^{n-1} + \cdots + a_2 x^2 + a_1 x + a_0 = 0$$

iff

$$P(r) = a_n r^n + a_{n-1} r^{n-1} + \cdots + a_2 r^2 + a_1 r + a_0 = 0.$$

The number r is also called a *zero* of the polynomial P.

Recall the *factor theorem* which states that $x - r$ is a factor of a polynomial P iff r is a zero of P, that is

$$P(x) = (x - r)Q(x) \qquad \text{for some polynomial } Q$$

iff

$$P(r) = 0.$$

A polynomial of degree 1 has the form $P(x) = ax + b$, $a \neq 0$, and is called a *linear polynomial*. The number $r = -b/a$ is the (unique) zero of P.

A polynomial of degree 2 has the form $P(x) = ax^2 + bx + c$, $a \neq 0$, and is called a *quadratic polynomial.* The quadratic polynomial P has two zeros which are given by the *quadratic formula*:

$$r_1 = \frac{-b + \sqrt{b^2 - 4ac}}{2a}, \qquad r_2 = \frac{-b - \sqrt{b^2 - 4ac}}{2a},$$

and P can be written in terms of r_1 and r_2 as

$$P(x) = a(x - r_1)(x - r_2) \qquad \qquad \text{(verify this).}$$

This expression is called the *factored form* of P.

Quadratic polynomials can be used to illustrate a limitation of the set of real numbers, namely the existence of polynomials with real coefficients that do not have real zeros. In particular, if $b^2 - 4ac < 0$, then the zeroes of the quadratic polynomial P are not real numbers. This situation is remedied by introducing a new "number" i defined by $i^2 = -1$ (equivalently, i is a zero of the quadratic polynomial $P(x) = x^2 + 1$). The *complex numbers* consist of all numbers of the form $a + bi$, where a and b are real numbers. For example, $3 + 4i$, $-2 - 3i$, $0 + 6i = 6i$, and $\frac{1}{2} + 0i = \frac{1}{2}$ are complex numbers. The real numbers can be considered to be a subset of the complex numbers since any real number a can be written $a = a + 0i$.

The *Fundamental Theorem of Algebra* states that if $P(z)$ is a polynomial of degree n, $n \geq 1$, with complex coefficients, then P has a (complex) zero. Based on this fact, and using the factor theorem, it is easy to show that P has exactly n zeros. Thus, a polynomial of degree $n \geq 1$ with real coefficients has exactly n zeros, but some (or even all) of the zeros may be complex numbers. In Section 8.6, we will need to use the fact that a polynomial of degree $n \geq 1$ with real coefficients can be factored into a product of linear and quadratic factors corresponding to its real and complex zeros.

In this text we will be concerned almost exclusively with the set of real numbers, and so if the term "number" is used without qualification, it will be interpreted to mean "real number." You should be aware, however, that complex numbers can arise in the course of solving equations.

An expression of the form $P(x)/Q(x)$, where P and Q are polynomials, is called a *rational expression* (in one variable). For example,

$$\frac{1}{x-1}, \qquad \frac{2x^3 + 3x}{x^2 - 9}, \qquad \text{and} \qquad \frac{x}{x^2 + 1}$$

are rational expressions, while

$$\frac{x^{2/3} - 2x}{\sqrt{x^3 + 2x - 1}}$$

is not a rational expression. It is important to note that a rational expression $R(x) = P(x)/Q(x)$ is not defined at the points (numbers) x where $Q(x) = 0$. Also, $R(x) = 0$ at the points x where $P(x) = 0$.

Factorials

Let n be a positive integer. Then n *factorial*, denoted $n!$, is the product of the integers from 1 to n. Thus,

$$1! = 1, \quad 2! = 1 \cdot 2, \quad 3! = 1 \cdot 2 \cdot 3, \quad \cdots \quad n! = 1 \cdot 2 \cdot 3 \cdot \; \ldots \; \cdot (n-1) \cdot n.$$

Note that

$$1! = 1, \qquad 2! = 2, \qquad 3! = 6.$$

You can verify that $4! = 24$, $5! = 120$, and so on. Finally, it is convenient to define $0! = 1$.

Geometry

Elementary Figures

Triangle	Equilateral Triangle
area $= \frac{1}{2}bh$	area $= \frac{1}{4}\sqrt{3}\, s^2$

Rectangle	Rectangular Solid

area $= lw$

perimeter $= 2l + 2w$

diagonal $= \sqrt{l^2 + w^2}$

volume $= lwh$

surface area $= 2lw + 2lh + 2wh$

Square	Cube

 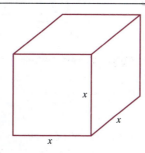

area $= x^2$

perimeter $= 4x$

diagonal $= x\sqrt{2}$

volume $= x^3$

surface area $= 6x^2$

Circle	Sphere

 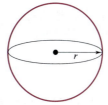

area $= \pi r^2$

circumference $= 2\pi r$

volume $= \frac{4}{3}\pi r^3$

surface area $= 4\pi r^2$

Sector of a Circle: radius r, central angle θ measured in radians (see Section 1.6).

arc length $= r\theta$

area $= \frac{1}{2}r^2\theta$

(Table continues)

Right Circular Cylinder	Right Circular Cone

volume $= \frac{1}{3}\pi r^2 h$
slant height $= \sqrt{r^2 + h^2}$

volume $= \pi r^2 h$
lateral area $= 2\pi rh$
total surface area $= 2\pi r^2 + 2\pi rh$

lateral area $= \pi r\sqrt{r^2 + h^2}$
total surface area $= \pi r^2 + \pi r\sqrt{r^2 + h^2}$

EXERCISES 1.2

In Exercises 1–12, use the terms *integer, rational, irrational, complex* to classify the given number; use all the terms that apply.

1. $\frac{17}{7}$.

2. -6.

3. $2.131313 \ldots = 2.\overline{13}$.

4. $\sqrt{2} - 3$.

5. 0.

6. $\pi - 2$.

7. $\sqrt[3]{8}$.

8. 0.125.

9. $-\sqrt{9}$.

10. $0.9999 \ldots = 0.\overline{9}$.

11. $0.21211211121111 \ldots$.

12. $13\frac{2}{7}$.

A method for writing the repeating decimal $3.\overline{135}$ in the form p/q, p, q integers, $q \neq 0$, is as follows:

Set $x = 3.135135 \ldots$.

Then $1000x = 3135.135135 \ldots$.

Now subtract:

$1000x - x = 3135.135135 \ldots - 3.135135 \ldots = 3132$.

Thus

$$999x = 3132,$$

and

$$x = \frac{3132}{999} = \frac{116}{37}.$$

Use this method in Exercises 13–17 to express the given repeating decimal in the rational number form p/q.

13. $13.\overline{201}$.

14. $2.777 \ldots$.   HINT: Multiply by 10.

15. $0.2323 \ldots$.

16. $4.16\overline{3}$.

17. $5.252252 \ldots$.

18. Show that the repeating decimal $0.9999 \ldots$ represents the number 1. Note that $1 = 1.0$ also represents the number 1. Thus, it follows that a rational number may have more than one decimal representation.

In Exercises 19–24, replace the symbol $*$ by $<$, $>$, or $=$ to make the statement true.

19. $\frac{3}{4} * 0.75$.

20. $0.33 * \frac{1}{3}$.

21. $\sqrt{2} * 1.414$.

22. $4 * \sqrt{16}$.

23. $-\frac{2}{7} * -0.285714$.

24. $\pi * \frac{22}{7}$.

In Exercises 25–31, write the given number without absolute value.

25. $|6|$.

26. $|-4|$.

27. $|3 - 7|$.

28. $|-5| - |8|$.

29. $|-5| + |-8|$.

30. $|2 - \pi|$.

31. $|5 - \sqrt{5}|$.

In Exercises 32–41, indicate on a number line all the values of x that satisfy the given condition.

32. $x \geq 3$.

33. $x \leq -\frac{3}{2}$.

34. $-2 \leq x \leq 3$.

35. $x^2 < 16$.

36. $x^2 \geq 16$.

37. $|x| \leq 0$.

38. $x^2 \geq 0$.

39. $|x - 4| \leq 2$.

40. $|x + 1| > 3$.

41. $|x + 3| \leq 0$.

In Exercises 42–50, sketch the given set on a number line.

42. $[3, \infty)$.

43. $(-\infty, 2)$.

44. $(-4, 3]$.

45. $[-2, 3] \cup [1, 5]$.

46. $(1, 4] \cup (4, 8)$.

47. $[-2, 2] \cap [0, 4]$.

48. $[-3, \frac{3}{2}] \cap (\frac{3}{2}, \frac{5}{2}]$. **49.** $(-\infty, -1) \cup (-2, \infty)$.

50. $(-\infty, 2) \cap [3, \infty)$.

In Exercises 51–57, determine whether the given set is bounded above, bounded below, or bounded. If a set is bounded above, give an upper bound; if it is bounded below, give a lower bound; if it is bounded, give an upper and a lower bound.

51. $\{0, 1, 2, 3, 4\}$. **52.** $\{0, -1, -2, -3, \ldots\}$.

53. S is the set of even integers.

54. $S = \{x: x \leq 4\}$. **55.** $S = \{x: x^2 > 3\}$.

56. $S = \left\{\dfrac{n-1}{n}: n = 1, 2, 3, \ldots\right\}$.

57. S is the set of rational numbers less than $\sqrt{2}$.

In Exercises 58–63, rewrite the given expression in factored form.

58. $9x^2 - 4$. **59.** $x^2 - 10x + 25$.

60. $27x^3 - 8$. **61.** $8x^6 + 64$.

62. $4x^4 + 4x^2 + 1$. **63.** $4x^2 + 12x + 9$.

In Exercises 64–70, find the real zeros of the given quadratic polynomial.

64. $P(x) = x^2 - 9$. **65.** $P(x) = x^2 - x - 2$.

66. $P(x) = 2x^2 - 5x - 3$. **67.** $P(x) = x^2 - 6x + 9$.

68. $P(x) = x^2 + 8x + 16$. **69.** $P(x) = x^2 - 2x + 2$.

70. $P(x) = x^2 - 2x + 5$.

In Exercises 71–75, determine all real numbers x for which the given rational expression $R(x)$ does not represent a real number. Also, determine all real numbers x for which $R(x) = 0$.

71. $R(x) = \dfrac{x-1}{x^2}$. **72.** $R(x) = \dfrac{x^2 - 5x + 6}{x^2 - 4}$.

73. $R(x) = \dfrac{x^2 - 1}{x^2 + 1}$. **74.** $R(x) = \dfrac{1}{x^2 - 2x + 8}$.

75. $R(x) = \dfrac{2x + 3}{x^2 + 2x + 5}$.

In Exercises 76–79, determine the value of the given expression.

76. $5!$. **77.** $\dfrac{5!}{8!}$.

78. $\dfrac{8!}{3!5!}$. **79.** $\dfrac{9!}{3!6!}$.

80. Prove that the sum of two rational numbers is a rational number.

81. Prove that the sum of a rational number and an irrational number is irrational.

82. Prove that the product of two rational numbers is a rational number.

83. What can you say about the product of a rational number and an irrational number?

84. Show by examples that the sum or product of two irrational numbers can either be a rational number or an irrational number.

85. Prove that $\sqrt{2}$ is an irrational number. HINT: Assume that $\sqrt{2} = p/q$ and that p and q have no common factors (other than 1 and -1). Square both sides of this equation and argue that both p and q must be divisible by 2. This contradicts the assumption.

86. Prove that $\sqrt{3}$ is irrational. HINT: Use the method of Exercise 85.

87. Let R be the set of all rectangles that have the same perimeter P. Show that the square (whose side necessarily has length $P/4$) is the element of R that has the largest area.

88. Show that if a circle and a square have the same perimeter, then the circle has a larger area than the square. Suppose a circle and an arbitrary rectangle have equal perimeters. Which has the larger area?

■ 1.3 INEQUALITIES

(All our work with inequalities is based on the order properties of real numbers as given on page 6.) In this section we work with the sort of inequalities that abound in calculus, inequalities that involve a variable.

To solve an equation in x is to find the set of numbers x for which the equation holds. To solve an inequality in x is to find the set of numbers x for which the inequality holds.

The way we solve an inequality is very similar to the way we solve an equation, but there is one important difference. We can maintain an inequality by adding the same number to both sides, or by subtracting the same number from both sides, or by multiplying or dividing both sides by the same *positive* number. But if we multiply or divide by a *negative* number, then the inequality is *reversed*:

$$x - 2 < 4 \quad \text{gives} \quad x < 6, \qquad x + 2 < 4 \quad \text{gives} \quad x < 2,$$

$$\tfrac{1}{2}x < 4 \quad \text{gives} \quad x < 8,$$

but

$$-\tfrac{1}{2}x < 4 \quad \text{gives} \quad x > -8.$$

            note

Example 1 Solve the inequality

$$-3(4 - x) < 12.$$

SOLUTION Multiplying both sides of the inequality by $-\tfrac{1}{3}$, we have

$$4 - x > -4 \qquad \text{(the inequality has been reversed).}$$

Subtracting 4, we get

$$-x > -8.$$

To isolate x, we multiply by -1. This gives

$$x < 8 \qquad \text{(the inequality has been reversed again).}$$

The solution set is the interval $(-\infty, 8)$. ❑

There are generally several ways to solve a given inequality. For example, the last inequality could have been solved this way:

$$-3(4 - x) < 12,$$

$$-12 + 3x < 12,$$

$$3x < 24 \qquad \text{(we added 12),}$$

$$x < 8 \qquad \text{(we divided by 3).}$$

The solution set, of course, is once again $(-\infty, 8)$.

The usual way to solve a quadratic inequality is to factor the quadratic.

Example 2 Solve the inequality

$$\tfrac{1}{5}(x^2 - 4x + 3) < 0.$$

SOLUTION First we eliminate the outside factor $\tfrac{1}{5}$ by multiplying through by 5: this gives

$$x^2 - 4x + 3 < 0.$$

Factoring the quadratic, we have

$$(x - 1)(x - 3) < 0.$$

The product $(x - 1)(x - 3)$ is zero at 1 and 3. We mark these points on a number line (Figure 1.3.1). The points 1 and 3 separate three intervals:

$$(-\infty, 1), \qquad (1, 3), \qquad (3, \infty).$$

$$+ + + + + + + + + + + + 0 \,\text{---------------}\, 0 + + + + + + + + + + +$$

$$1 \qquad\qquad\qquad\qquad 3$$

Figure 1.3.1

On each of these intervals the product $(x - 1)(x - 3)$ keeps a constant sign:

on $(-\infty, 1)$ [to the left of 1] sign of $(x - 1)(x - 3) = (-)(-) = +$;

on $(1, 3)$ [between 1 and 3] sign of $(x - 1)(x - 3) = (+)(-) = -$;

on $(3, \infty)$ [to the right of 3] sign of $(x - 1)(x - 3) = (+)(+) = +$.

The product $(x - 1)(x - 3)$ is negative only on $(1, 3)$. The solution set is the open interval $(1, 3)$. ❑

More generally, consider an expression of the form

$$(x - a_1)(x - a_2) \cdots (x - a_n) \qquad \text{with } a_1 < a_2 < \cdots < a_n.$$

Such an expression is zero at $a_1, a_2, \ldots, a_n$. It is positive on those intervals where an even number of terms are negative, and it is negative on those intervals where an odd number of terms are negative.

As an example, take the expression

$$(x + 2)(x - 1)(x - 3).$$

This product is zero at $-2, 1, 3$. It is

negative on $(-\infty, -2)$,	(3 negative terms)
positive on $(-2, 1)$,	(2 negative terms)
negative on $(1, 3)$,	(1 negative term)
positive on $(3, \infty)$.	(0 negative terms)

See Figure 1.3.2.

$$\text{-----} 0 + 0 \,\text{--------------}\, 0 + + + +$$

$$-2 \qquad\qquad\qquad\qquad 1 \qquad\qquad\qquad 3$$

Figure 1.3.2

Example 3 Solve the inequality

$$(x + 3)^5(x - 1)(x - 4)^2 < 0.$$

SOLUTION We view $(x + 3)^5(x - 1)(x - 4)^2$ as the product of three factors: $(x + 3)^5$, $(x - 1)$, $(x - 4)^2$. The product is zero at -3, 1, and 4. These points separate the intervals

$$(-\infty, -3), \qquad (-3, 1), \qquad (1, 4), \qquad (4, \infty).$$

On each of these intervals the product keeps a constant sign. It is

positive on $(-\infty, -3)$,   (2 negative factors)

negative on $(-3, 1)$,   (1 negative factor)

positive on $(1, 4)$,   (0 negative factors)

positive on $(4, \infty)$.   (0 negative factors)

See Figure 1.3.3.

$$+++++++0 - - - - - - - - - - - -0+++++++++0+++++++$$

-3   1   4

Figure 1.3.3

The solution set is the open interval $(-3, 1)$. ❑

-3   1

Example 4 Solve the inequality

$$x^2 + 4x - 2 \le 0.$$

SOLUTION To factor the quadratic we first complete the square by adding and sub-tracting 4:

$$x^2 + 4x + 4 - 4 - 2 \le 0$$

$$(x^2 + 4x + 4) - 6 \le 0$$

$$(x + 2)^2 - 6 \le 0.$$

We can factor this quadratic as the difference of two squares:

$$(x + 2 + \sqrt{6})(x + 2 - \sqrt{6}) \le 0.$$

The product is 0 at $-2 - \sqrt{6}$ and $-2 + \sqrt{6}$. It is negative in between. The solution set is the closed interval $[-2 - \sqrt{6}, -2 + \sqrt{6}]$. ❑

$-2 - \sqrt{6}$ $-2 + \sqrt{6}$

Remark The roots of the quadratic $x^2 + 4x - 2 = 0$ could have been obtained from the quadratic formula. ❑

In solving inequalities that involve quotients we use the fact that

(1.3.1) $\dfrac{a}{b} > 0$ iff $ab > 0$ and $\dfrac{a}{b} < 0$ iff $ab < 0.$

Example 5 Solve the inequality

$$\frac{x + 2}{1 - x} < 1, \qquad (x \ne 1).$$

SOLUTION   Note that we cannot multiply the given inequality by $1 - x$ since we don't know its sign.   Subtracting 1 from both sides, we have

$$\frac{x + 2}{1 - x} - 1 < 0,$$

$$\frac{x + 2 - (1 - x)}{1 - x} < 0,$$

$$\frac{2x + 1}{1 - x} < 0,$$

$$(2x + 1)(1 - x) < 0, \qquad\qquad \text{[by (1.3.1)]}$$

$$(2x + 1)(x - 1) > 0 \qquad\qquad \text{(we multiplied by } -1),$$

$$(x + \tfrac{1}{2})(x - 1) > 0 \qquad\qquad \text{(we divided by 2).}$$

The product $(x + \tfrac{1}{2})(x - 1)$ is zero at $-\tfrac{1}{2}$ and 1.   As you can check, it is positive on the open intervals $(-\infty, -\tfrac{1}{2})$ and $(1, \infty)$.   Thus the solution set of our inequality is the set $(-\infty, -\tfrac{1}{2}) \cup (1, \infty)$; its graph is shown in the following figure.   ❑

Example 6   Solve the inequality

$$\frac{(x + 3)^5(x - 1)}{(x - 4)^2} < 0.$$

SOLUTION   By (1.3.1) this inequality has the same solution as the inequality

$$(x + 3)^5(x - 1)(x - 4)^2 < 0,$$

which we solved in Example 3.   ❑

Inequalities and Absolute Value

Now we take up some inequalities that involve absolute values.   With an eye toward Chapter 2 we introduce two Greek letters: δ (delta) and ϵ (epsilon).

Recall that for each real number a

(1.3.2)
$$|a| = \begin{cases} a, & \text{if } a \geq 0 \\ -a, & \text{if } a < 0, \end{cases} \qquad |a| = \max\{a, -a\}, \qquad |a| = \sqrt{a^2}.$$

We begin with the inequality

$$|x| < \delta$$

where δ is some positive number. To say that $|x| < \delta$ is to say that x lies within δ units of 0, or, equivalently, that x lies between $-\delta$ and δ. Thus

(1.3.3)
$$|x| < \delta \quad \text{iff} \quad -\delta < x < \delta.$$

To say that $|x - c| < \delta$ is to say that x lies within δ units of c, or, equivalently, that x lies between $c - \delta$ and $c + \delta$. Thus

(1.3.4)
$$|x - c| < \delta \quad \text{iff} \quad c - \delta < x < c + \delta.$$

Somewhat more delicate is the inequality

$$0 < |x - c| < \delta.$$

Here we have $|x - c| < \delta$ with the additional requirement that $x \neq c$. Consequently

(1.3.5)
$$0 < |x - c| < \delta \quad \text{iff} \quad c - \delta < x < c \quad \text{or} \quad c < x < c + \delta.$$

Thus, for example,

$$|x| < \tfrac{1}{2} \quad \text{iff} \quad -\tfrac{1}{2} < x < \tfrac{1}{2}; \qquad\qquad\qquad\qquad \textit{Solution: } (-\tfrac{1}{2}, \tfrac{1}{2})$$

$$|x - 5| < 1 \quad \text{iff} \quad 4 < x < 6; \qquad\qquad\qquad\qquad\qquad \textit{Solution: } (4, 6)$$

$$0 < |x - 5| < 1 \quad \text{iff} \quad 4 < x < 5 \quad \text{or} \quad 5 < x < 6. \qquad \textit{Solution: } (4, 5) \cup (5, 6). \quad \square$$

Example 7 Solve the inequality

$$|x + 2| < 3.$$

SOLUTION The inequality $|x + 2| < 3$ holds iff

$$|x - (-2)| < 3 \quad \text{iff} \quad -2 - 3 < x < -2 + 3 \quad \text{iff} \quad -5 < x < 1.$$

The solution set is the open interval $(-5, 1)$. $\quad \square$

Example 8 Solve the inequality

$$|3x - 4| < 2.$$

SOLUTION Since

$$|3x - 4| = |3(x - \tfrac{4}{3})| = |3||x - \tfrac{4}{3}| = 3|x - \tfrac{4}{3}|,$$

the inequality can be rewritten

$$3|x - \tfrac{4}{3}| < 2.$$

This gives $|x - \tfrac{4}{3}| < \tfrac{2}{3}$. Therefore

$$\tfrac{4}{3} - \tfrac{2}{3} < x < \tfrac{4}{3} + \tfrac{2}{3},$$

$$\tfrac{2}{3} < x < 2.$$

The solution set is the open interval $(\tfrac{2}{3}, 2)$.

ALTERNATIVE SOLUTION The inequality

$$|3x - 4| < 2$$

is equivalent to

$$-2 < 3x - 4 < 2$$

by (1.3.3). Therefore

$$2 < 3x < 6 \qquad\qquad \text{(add 4 to both inequalities)}$$

and

$$\tfrac{2}{3} < x < 2 \qquad\qquad \text{(divide through by 3)}$$

as before. ❑

Let $\epsilon > 0$. If you think of $|a|$ as the distance between a and 0, then obviously

(1.3.6) $\qquad |a| > \epsilon \quad \text{iff} \quad a > \epsilon \quad \text{or} \quad a < -\epsilon.$

Example 9 Solve the inequality

$$|2x + 3| > 5.$$

SOLUTION As you saw, in general

$$|a| > \epsilon \qquad \text{iff} \qquad a > \epsilon \quad \text{or} \quad a < -\epsilon.$$

So here

$$2x + 3 > 5 \quad \text{or} \quad 2x + 3 < -5.$$

The first possibility gives $2x > 2$ and thus

$$x > 1.$$

The second possibility gives $2x < -8$ and thus

$$x < -4.$$

The total solution is therefore the union

$$(-\infty, -4) \cup (1, \infty). \quad \square$$

We come now to one of the fundamental inequalities of calculus: for all real numbers a and b,

(1.3.7)

$$|a + b| \leq |a| + |b|.$$

This is called the *triangle inequality* in analogy with the geometric maxim "in a triangle the length of each side is less than or equal to the sum of the lengths of the other two sides."

PROOF OF THE TRIANGLE INEQUALITY   The key here is to think of $|x|$ as $\sqrt{x^2}$. Note first that

$$(a + b)^2 = a^2 + 2ab + b^2 \leq |a|^2 + 2|a||b| + |b|^2 = (|a| + |b|)^2.$$

Comparing the extremes of the inequality and taking square roots, we have

$$\sqrt{(a + b)^2} \leq |a| + |b|. \qquad \text{(Exercise 69)}$$

The result follows from observing that

$$\sqrt{(a + b)^2} = |a + b|. \quad \square$$

Here is a variant of the triangle inequality that also comes up in calculus: for all real numbers a and b,

(1.3.8)

$$\big||a| - |b|\big| \leq |a - b|.$$

The proof is left to you as an exercise.

EXERCISES 1.3

In Exercises 1–36, solve the given inequality and graph the solution set on a number line.

1. $2 + 3x < 5$.

2. $\frac{1}{2}(2x + 3) < 6$.

3. $16x + 64 \leq 16$.

4. $3x + 5 > \frac{1}{4}(x - 2)$.

5. $\frac{1}{2}(1 + x) < \frac{1}{3}(1 - x)$.

6. $3x - 2 \leq 1 + 6x$.

7. $x^2 - 1 < 0$.

8. $x^2 + 9x + 20 < 0$.

9. $x(x - 1)(x - 2) > 0$.

10. $x(2x - 1)(3x - 5) \leq 0$.

11. $x^3 - 2x^2 + x \geq 0$.

12. $x^2 - 4x + 4 \leq 0$.

13. $x^2 + 1 > 4x$.

14. $2x^2 + 9x + 6 \geq x + 2$.

15. $1 - 3x^2 < \frac{1}{2}(2 - x^2)$.

16. $6x^2 + 2x \leq (x - 1)^2$.

17. $\dfrac{1}{x} < x$.

18. $x + \dfrac{1}{x} \geq 0$.

19. $\dfrac{x}{x - 5} \geq 0$.

20. $\dfrac{x}{x + 5} < 0$.

21. $\dfrac{x}{x - 5} > \dfrac{1}{4}$.

22. $\dfrac{1}{3x - 5} < 2$.

23. $\dfrac{x^2 - 9}{x + 1} > 0$.

24. $\dfrac{x^2}{x^2 - 4} < 0$.

25. $x^3(x - 2)(x + 3)^2 < 0$.

26. $x^2(x - 3)(x + 4)^2 > 0$.

27. $x^2(x - 2)(x + 6) > 0$.

28. $7x(x - 4)^2 < 0$.

29. $\dfrac{2x}{x^2 - 4} > 0.$

30. $\dfrac{x^2 - 9}{3x} > 0.$

31. $\dfrac{1}{x - 1} + \dfrac{4}{x - 6} > 0.$

32. $\dfrac{3}{x - 2} - \dfrac{5}{x - 6} < 0.$

33. $\dfrac{2x - 6}{x^2 - 6x + 5} < 0.$

34. $\dfrac{2x + 8}{x^2 + 8x + 7} > 0.$

35. $\dfrac{x^2 - 4x + 3}{x^2} > 0.$

36. $\dfrac{x^2 - 4x}{x + 2} > 0.$

In Exercises 37–53, solve the given inequality and express the solution set in terms of intervals.

37. $|x| < 2.$

38. $|x| \geq 1.$

39. $|x| > 3.$

40. $|x - 1| < 1.$

41. $|x - 2| < \frac{1}{2}.$

42. $|x - \frac{1}{2}| < 2.$

43. $0 < |x| < 1.$

44. $0 < |x| < \frac{1}{2}.$

45. $0 < |x - 2| < \frac{1}{2}.$

46. $0 < |x - \frac{1}{2}| < 2.$

47. $0 < |x - 3| < 8.$

48. $|3x - 5| < 3.$

49. $|2x + 1| < \frac{1}{4}.$

50. $|5x - 3| < \frac{1}{2}.$

51. $|2x + 5| > 3.$

52. $|3x + 1| > 5.$

53. $|5x - 1| > 9.$

In Exercises 54–59, find an inequality of the form $|x - c| < \delta$, the solution of which is the given open interval.

54. $(-2, 2).$

55. $(-3, 3).$

56. $(0, 4).$

57. $(-3, 7).$

58. $(-4, 0).$

59. $(-7, 3).$

In Exercises 60–63, determine all values of $A > 0$ for which the statement is true.

60. If $|x - 2| < 1$, then $|2x - 4| < A.$

61. If $|x - 2| < A$, then $|2x - 4| < 3.$

62. If $|x + 1| < A$, then $|3x + 3| < 4.$

63. If $|x + 1| < 2$, then $|3x + 3| < A.$

64. Arrange the following in order: $1, x, \sqrt{x}, 1/x, 1/\sqrt{x}$, given that $x > 1.$

65. Redo Exercise 64 for the case $0 < x < 1.$

66. Compare

$$\sqrt{\dfrac{x}{x + 1}} \quad \text{and} \quad \sqrt{\dfrac{x + 1}{x + 2}}$$

given that $x > 0.$

67. Let a and b have the same sign. If $a < b$, show that $(1/b) < (1/a).$

68. Let a and b be nonnegative numbers. Show that if

$$a^2 \leq b^2 \quad \text{then} \quad a \leq b.$$

69. Let a and b be nonnegative numbers. Show that if

$$a \leq b \quad \text{then} \quad \sqrt{a} \leq \sqrt{b}.$$

70. Prove that for all real numbers a and b.

$$|a - b| \leq |a| + |b|.$$

71. Prove that for all real numbers a and b

$$\big||a| - |b|\big| \leq |a - b|.$$

HINT: Calculate $\big||a| - |b|\big|^2$ and use the fact that $ab \leq |a||b|.$

72. Show that $|a + b| = |a| + |b|$ iff $ab \geq 0.$

73. Show that if

$$0 \leq a \leq b \quad \text{then} \quad \dfrac{a}{1 + a} \leq \dfrac{b}{1 + b}.$$

74. Let a, b, c be nonnegative numbers. Show that

$$\text{if } a \leq b + c, \quad \text{then} \quad \dfrac{a}{1 + a} \leq \dfrac{b}{1 + b} + \dfrac{c}{1 + c}.$$

75. Prove that if a and b are real numbers and $a < b$, then $a < (a + b)/2 < b.$ The number $(a + b)/2$ is called the *arithmetic mean* of a and b. How are the three numbers, $a, (a + b)/2, b$, related on the number line?

76. Let a and b be nonnegative numbers with $a \leq b$. Prove that

$$a \leq \sqrt{ab} \leq \dfrac{a + b}{2} \leq b.$$

The number $\sqrt{ab}$ is called the *geometric mean* of a and b.

■ 1.4 COORDINATE PLANE; ANALYTIC GEOMETRY

Rectangular Coordinates

The relation between the real numbers and points on a straight line can be used to construct a coordinate system for the plane. In the plane, draw two number lines which are perpendicular to each other and intersect at their origins; let O be the point of intersection. It is customary to have one of the lines be horizontal with the positive numbers to the right of O and the other vertical with the positive numbers above O. The point O is called the *origin* and the number lines are called the *coordinate axes*. The horizontal axis is usually labeled the *x-axis* and the vertical axis is usually labeled the *y-axis*. The coordinate axes separate the plane into four regions called *quadrants*. The quadrants are numbered in the counterclockwise direction starting with the upper right. See Figure 1.4.1.

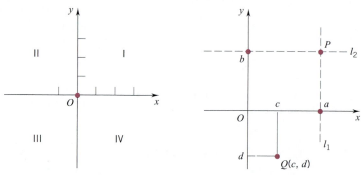

| Figure 1.4.1 | Figure 1.4.2 |

Choose any point P, $P \neq O$, in the plane. Let l_1 be the line through P parallel to the *y*-axis and l_2 the line through P parallel to the *x*-axis. Let a be the coordinate of the point of intersection of l_1 and the *x*-axis, and let b be the coordinate of the point of intersection of l_2 and the *y*-axis (Figure 1.4.2). Then P is identified with the *ordered pair* of real numbers (a, b). The ordered pair associated with the origin is $(0, 0)$. Conversely, starting with an ordered pair (c, d), we can reverse these steps and associate with it a unique point Q. If $P(a, b)$ is a point in the plane, then a and b are the *coordinates* of P with a the *x-coordinate* or *abscissa* and b the *y-coordinate* or *ordinate*. This coordinate system for the plane is called the *rectangular* or *Cartesian* coordinate system.

Distance and Midpoint Formulas

Let $P_0(x_0, y_0)$ and $P_1(x_1, y_1)$ be points in the plane. The distance $d(P_0, P_1)$ between P_0 and P_1 can be computed using the *Pythagorean Theorem*:

Distance (Figure 1.4.3): $d(P_0, P_1) = \sqrt{(x_1 - x_0)^2 + (y_1 - y_0)^2}$

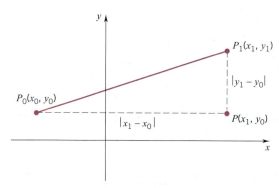

Figure 1.4.3

Let $M(x, y)$ be the midpoint of the line segment joining P_0 and P_1. Using the fact that the two triangles in Figure 1.4.4 are congruent, it can be shown that

$$x = \frac{x_0 + x_1}{2} \quad \text{and} \quad y = \frac{y_0 + y_1}{2}.$$

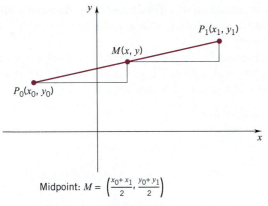

Midpoint: $M = \left(\frac{x_0 + x_1}{2}, \frac{y_0 + y_1}{2}\right)$

Figure 1.4.4

Lines

Let l be a straight line in the plane, and let $P_0(x_0, y_0)$ and $P_1(x_1, y_1)$ be any two distinct points on l.

(i) Slope The *slope* of l is given by

$$m = \frac{y_1 - y_0}{x_1 - x_0} \qquad \text{provided } x_1 \neq x_0.$$

Also,

$$m = \tan \theta,\dagger$$

where θ, called the *inclination* of l, is the angle between the positively directed x-axis and l measured in the counterclockwise direction (Figure 1.4.5). If $x_1 = x_0$, then l is a vertical line; its slope is undefined and $\theta = \pi/2 = 90°$.

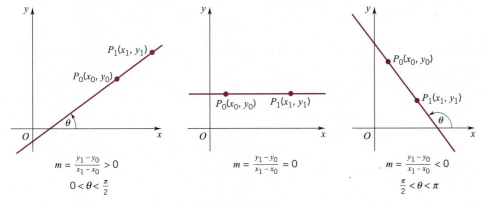

Figure 1.4.5

(ii) Intercepts If l is a nonvertical line, then it intersects the y-axis at a point $(0, b)$. The number b is called the *y-intercept* of l. Similarly, if l is not horizontal, then it will

† The trigonometric functions are reviewed in the Section 1.6.

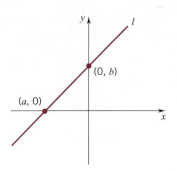

Figure 1.4.6

intersect the x-axis at a point $(a, 0)$. The number a is the *x-intercept* of l. These ideas are illustrated in Figure 1.4.6.

(iii) Equations An equation for l is an equation which involves two variables, say x and y, and has the property that a point $P(x_0, y_0)$ lies on l iff its coordinates, x_0 and y_0, satisfy the equation. The standard forms are:

$$\begin{aligned}
\textit{vertical line} \quad & x = a. \\
\textit{horizontal line} \quad & y = b. \\
\textit{point-slope form} \quad & y - y_0 = m(x - x_0). \\
\textit{two-point form} \quad & y - y_0 = \frac{y_1 - y_0}{x_1 - x_0}(x - x_0). \\
\textit{slope-intercept form} \quad & y = mx + b \quad (y = b \text{ when } x = 0). \\
\textit{two intercept form} \quad & \frac{x}{a} + \frac{y}{b} = 1. \\
\textit{general form} \quad & Ax + By + C = 0 \quad (A \text{ and } B \text{ not both } 0).
\end{aligned}$$

(iv) Parallel and Perpendicular Lines Given two nonvertical lines l_1 and l_2 with slopes m_1 and m_2, respectively, then l_1 and l_2 are:

$$\textit{parallel} \quad \text{iff} \quad m_1 = m_2,$$
$$\textit{perpendicular} \quad \text{iff} \quad m_1 m_2 = -1.$$

Example 1 Find the slope and y-intercept of each of the following lines:

$$l_1 : 20x - 24y - 30 = 0, \qquad l_2 : 2x - 3 = 0, \qquad l_3 : 4y + 5 = 0.$$

SOLUTION The equation of l_1 can be written

$$y = \tfrac{5}{6}x - \tfrac{5}{4}.$$

This is in the form $y = mx + b$. The slope is $\tfrac{5}{6}$, and the y-intercept is $-\tfrac{5}{4}$.
  The equation of l_2 can be written

$$x = \tfrac{3}{2}.$$

The line is vertical and the slope is not defined. Since the line does not cross the y-axis, it has no y-intercept.
  The third equation can be written

$$y = -\tfrac{5}{4}.$$

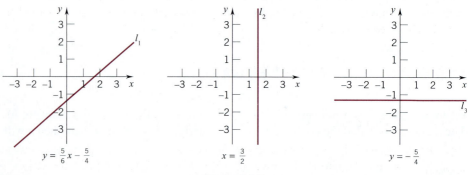

Figure 1.4.7

The line is horizontal. Its slope is 0 and the y-intercept is $-\frac{5}{4}$. The three lines are drawn in Figure 1.4.7. ❑

Example 2 Write an equation for the line l_2 that is parallel to

$$l_1 : 3x - 5y + 8 = 0$$

and passes through the point $P(-3, 2)$.

SOLUTION We can rewrite the equation of l_1 as

$$y = \tfrac{3}{5}x + \tfrac{8}{5}.$$

The slope of l_1 is $\frac{3}{5}$. The slope of l_2 must also be $\frac{3}{5}$. (*Remember*: For nonvertical parallel lines, $m_1 = m_2$.)
   Since l_2 passes through $(-3, 2)$ with slope $\frac{3}{5}$, we can use the point-slope formula and write the equation as

$$y - 2 = \tfrac{3}{5}(x + 3). ❑$$

Example 3 Write an equation for the line that is perpendicular to

$$l_1 : x - 4y + 8 = 0$$

and passes through the point $P(2, -4)$.

SOLUTION The equation for l_1 can be written

$$y = \tfrac{1}{4}x + 2.$$

The slope of l_1 is $\frac{1}{4}$. The slope of l_2 is therefore -4. (*Remember:* For nonvertical perpendicular lines, $m_1 m_2 = -1$.)
   Since l_2 passes through $(2, -4)$ with slope -4, we can use the point-slope formula and write the equation as

$$y + 4 = -4(x - 2). ❑$$

The Angle Between Two Lines

Figure 1.4.8 shows two intersecting lines l_1, l_2 with inclinations θ_1, θ_2. In the figure, $\theta_1 < \theta_2$. These two lines form angles, $\theta_2 - \theta_1 = \alpha$ and $180° - (\theta_2 - \theta_1) = 180° - \alpha$. If these two angles are equal, then $\alpha = 90°$ and the two lines are perpendicular. If α and $180° - \alpha$ are not equal, then the smaller of the two (the one less than 90°) is called the *angle between l_1 and l_2*.

Figure 1.4.8

Example 4 Find the point where the lines

$$3x - 4y + 8 = 0 \quad \text{and} \quad 12x - 5y - 12 = 0$$

intersect, and then determine the angle between the lines.

SOLUTION To find the point of intersection, we solve the two equations simultaneously. As you can check, this gives $x = 8/3$, $y = 4$. The point of intersection is $P(8/3, 4)$.

To find the angle between the lines, we find the inclination of each line:

$$3x - 4y + 8 = 0 \qquad\qquad 12x - 5y - 12 = 0$$

$$4y = 3x + 8 \qquad\qquad 5y = 12x - 12$$

$$y = \tfrac{3}{4}x + 2. \qquad\qquad y = \tfrac{12}{5}x - \tfrac{12}{5}.$$

$$\tan \theta = \tfrac{3}{4} = 0.75 \qquad\qquad \tan \theta = \tfrac{12}{5} = 2.4$$

$$\theta \cong 37°. \qquad\qquad \theta \cong 67°.$$

The angle between the lines is approximately $67° - 37° = 30°$. See Figure 1.4.9. ❏

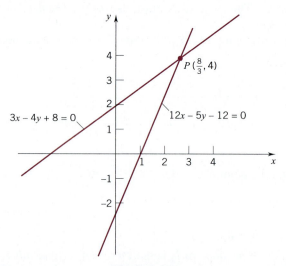

Figure 1.4.9

Distance Between a Point and a Line

Let l be a line and let P be a point which is not on l. It is easy to see that the point Q on l that is closest to P is the foot of the perpendicular from P to l. See Figure 1.4.10. The *distance between P and l*, denoted $d(P, l)$, is defined to be $d(P, Q)$.

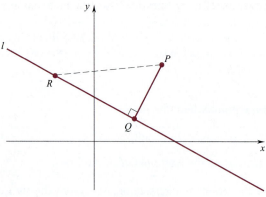

Figure 1.4.10

Example 5 Let l be the line with equation $Ax + By + C = 0$, with $A \neq 0$ and $B \neq 0$. Show that

$$d(O, l) = \frac{|C|}{\sqrt{A^2 + B^2}}.$$

SOLUTION Since l has slope $-A/B$, the line through the origin perpendicular to l has slope B/A and equation

$$y = \frac{B}{A}x \qquad \text{or} \qquad Bx - Ay = 0.$$

Solving the equations

$$Ax + By + C = 0$$

$$Bx - Ay \qquad = 0$$

simultaneously gives

$$x = \frac{-AC}{A^2 + B^2}, \qquad y = \frac{-BC}{A^2 + B^2}.$$

Thus the foot of the perpendicular from the origin to l is the point

$$Q\left(\frac{-AC}{A^2 + B^2}, \frac{-BC}{A^2 + B^2}\right)$$

and

$$d(O, l) = d(O, Q) = \sqrt{\left(\frac{-AC}{A^2 + B^2}\right)^2 + \left(\frac{-BC}{A^2 + B^2}\right)^2}$$

$$= \frac{1}{A^2 + B^2}\sqrt{A^2C^2 + B^2C^2} = \frac{|C|}{\sqrt{A^2 + B^2}}. \qquad \square$$

Translations

In Figure 1.4.11 we have drawn a rectangular coordinate system and marked a point $O'(x_0, y_0)$. Think of the Oxy system as a rigid frame and in your mind slide it along the plane, without letting it turn, so that the origin O falls on the point O' (Figure 1.4.12). Such a move, called a *translation,* produces a new coordinate system $O'XY$.

Figure 1.4.11

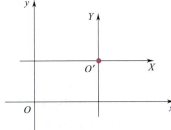

Figure 1.4.12

A point P now has two pairs of coordinates: a pair (x, y) with respect to the Oxy system and a pair (X, Y) with respect to the $O'XY$ system (Figure 1.4.13). To see the relation between these coordinates, note that, starting at O, we can reach P by first going to O' and then going on to P; thus

$$x = x_0 + X, \qquad y = y_0 + Y.$$

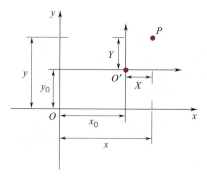

Figure 1.4.13

Translations are often used to simplify geometric arguments. In Example 5 we found the distance between the origin and a line l. Now we will derive the formula for the distance between a line and an arbitrary point by translating the coordinate system

DISTANCE BETWEEN A POINT AND A LINE

The distance between the line l: $Ax + By + C = 0$ and the point $P_1(x_1, y_1)$ is given by the formula

$$d(P_1, l) = \frac{|Ax_1 + By_1 + C|}{\sqrt{A^2 + B^2}}.$$

PROOF We translate the Oxy coordinate system to obtain a new system $O'XY$ with O' falling on the point P_1 (see Figure 1.4.14). The new coordinates are now related to the old coordinates as follows:

$$x = x_1 + X, \qquad y = y_1 + Y.$$

In the xy-system, l has equation

$$Ax + By + C = 0.$$

In the XY-system, l has equation

$$A(x_1 + X) + B(y_1 + Y) + C = 0.$$

We can write this last equation as

$$AX + BY + K = 0 \qquad \text{with} \quad K = Ax_1 + By_1 + C.$$

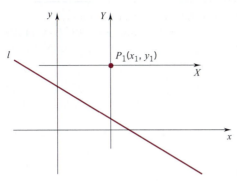

Figure 1.4.14

The distance we want is the distance between the line $AX + BY + K = 0$ and the new origin O'. From Example 5, this distance is

$$\frac{|K|}{\sqrt{A^2 + B^2}}.$$

Since $K = Ax_1 + By_1 + C$, we have

$$d(P_1, l) = \frac{|Ax_1 + By_1 + C|}{\sqrt{A^2 + B^2}}. \quad ❏$$

Example 6 Find the distance between the line

$$l: 3x + 4y - 5 = 0$$

and (a) the origin, (b) the point $P(-6, 2)$.

SOLUTION

(a) $d(O, l) = \dfrac{|C|}{\sqrt{A^2 + B^2}} = \dfrac{|-5|}{\sqrt{3^2 + 4^2}} = \dfrac{5}{\sqrt{25}} = 1.$

(b) $d(P, l) = \dfrac{|Ax_1 + By_1 + C|}{\sqrt{A^2 + B^2}} = \dfrac{|3(-6) + 4(2) - 5|}{\sqrt{3^2 + 4^2}} = \dfrac{15}{\sqrt{25}} = 3. \quad ❏$

Conic Sections

We now consider equations of the form

$$Ax^2 + By^2 + Cx + Dy + E = 0,$$

where A, B, C, D, E are constants, and A and B are not both 0. The graph of such an equation is, in general, a conic section; that is, a circle, ellipse, parabola, or hyperbola. We say "in general" because there are degenerate cases in which the graph is not a conic section. For example, the graph of $x^2 + y^2 = 0$ is a point, $(0, 0)$. In fact, the graph might not exist at all; there are no ordered pairs (x, y) which satisfy the equation $x^2 + y^2 + 1 = 0$.

A more formal treatment of the conic sections is given in Section 9.1. For purposes of review here, we list the standard forms for the equations of the conics and indicate their graphs.

Circle: $A = B, A \neq 0$; $Ax^2 + Ay^2 + Cx + Dy + E = 0$ or
$$x^2 + y^2 + C_1 x + D_1 y + E_1 = 0 \text{ (after dividing by } A).$$

Standard forms: $x^2 + y^2 = r^2$ *circle of radius r centered at the origin; (Figure 1.4.15a)*

$(x - h)^2 + (y - k)^2 = r^2$ *circle of radius r centered at P(h, k). (Figure 1.4.15b)*

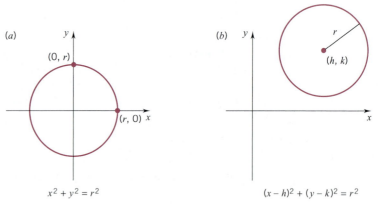

Figure 1.4.15

Ellipse: A, B have the same sign, $A \neq B$; $Ax^2 + By^2 + Cx + Dy + E = 0$.

Standard form: $\dfrac{(x - h)^2}{a^2} + \dfrac{(y - k)^2}{b^2} = 1$ *ellipse centered at the point P(h, k). (Figure 1.4.16)*

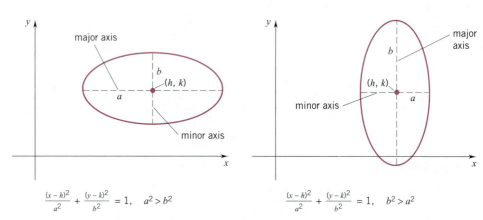

Figure 1.4.16

Parabola: Either $A = 0$ or $B = 0$; $By^2 + Cx + Dy + E = 0$ or
$$Ax^2 + Cx + Dy + E = 0.$$

Standard forms:

$(x - h)^2 = 4c(y - k)$ or *parabola, vertex at the point P(h, k) and axis vertical; (Figure 1.4.17a)*

$(y - k)^2 = 4c(x - h)$ *parabola, vertex at the point P(h, k) and axis horizontal. (Figure 1.4.17b)*

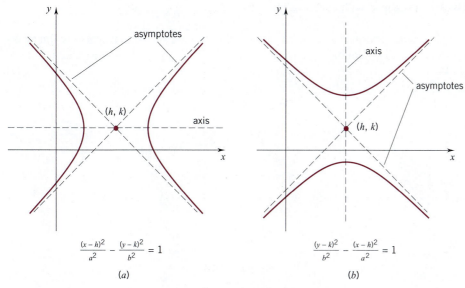

$(x - h)^2 = 4c(y - k), \quad c > 0$

(a)

$(y - k)^2 = 4c(x - k), \quad c > 0$

(b)

Figure 1.4.17

Hyperbola: A, B have opposite sign: $\quad Ax^2 + By^2 + Cx + Dy + E = 0$.

Standard forms:

$$\frac{(x - h)^2}{a^2} - \frac{(y - k)^2}{b^2} = 1$$ *hyperbola, center at the point $P(h, k)$ and axis horizontal;* (Figure 1.4.18a)

$$\frac{(y - k)^2}{b^2} - \frac{(x - h)^2}{a^2} = 1$$ *hyperbola, center at the point $P(h, k)$ and axis vertical.* (Figure 1.4.18b)

$\dfrac{(x-h)^2}{a^2} - \dfrac{(y-k)^2}{b^2} = 1$

(a)

$\dfrac{(y-k)^2}{b^2} - \dfrac{(x-h)^2}{a^2} = 1$

(b)

Figure 1.4.18

Standard forms are obtained from the general form $Ax^2 + By^2 + Cx + Dy + E = 0$ by completing the square; the general form is obtained from the standard forms by expanding the $(x - h)^2$ and $(y - k)^2$ terms and "clearing fractions."

For example, consider the hyperbola

$$\frac{(x - 2)^2}{16} - \frac{(y + 3)^2}{4} = 1.$$

This equation can be written

$$4(x - 2)^2 - 16(y + 3)^2 = 64,$$

from which we get the general form

$$4x^2 - 16y^2 - 16x - 96y - 192 = 0$$

or

$$x^2 - 4y^2 - 4x - 24y - 48 = 0.$$

Example 7 Identify the conic $9x^2 + 4y^2 - 18x + 16y - 11 = 0$ and rewrite the equation in standard form.

SOLUTION Since $A = 9$ and $B = 4$ have the same sign and are unequal, the conic is (we expect) an ellipse. We group the x and y terms

$$9x^2 - 18x + 4y^2 + 16y - 11 = 0,$$

complete the square

$$9(x^2 - 2x + 1) + 4(y^2 + 4y + 4) - 11 - 9 - 16 = 0$$

$$9(x - 1)^2 + 4(y + 2)^2 = 36$$

and get the standard form

$$\frac{(x - 1)^2}{4} + \frac{(y + 2)^2}{9} = 1,$$

which is an ellipse with center at $(1, -2)$. ❑

The conic sections can also be generated in three-dimensional space by intersecting a plane with a "double right circular cone." This approach is illustrated in Figure 1.4.19.

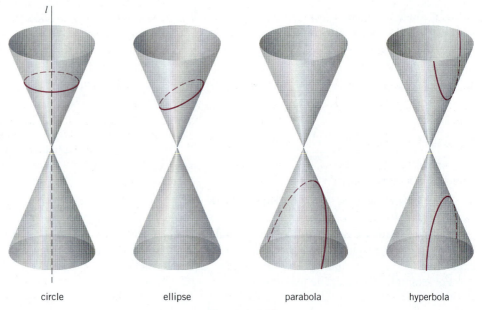

circle   ellipse   parabola   hyperbola

Figure 1.4.19

EXERCISES 1.4

In Exercises 1–6, find the distance between the given points.

1. $P_0(0, 5)$, $P_1(6, -3)$.   **2.** $P_0(2, 2)$, $P_1(5, 5)$.

3. $P_0(5, -2)$, $P_1(-3, 2)$.   **4.** $P_0(-5, 12)$, $O(0, 0)$.

5. $P_0(2, -2)$, $P_1(2, 4)$.   **6.** $P_0(2, 7)$, $P_1(-4, 7)$.

In Exercises 7–12, find the midpoint of the line segment $\overline{P_0P_1}$.

7. $P_0(2, 4)$, $P_1(6, 8)$.   **8.** $P_0(3, -1)$, $P_1(-1, 5)$.

9. $P_0(2, -3)$, $P_1(7, -3)$.   **10.** $P_0(5, -1)$, $P_1(5, 6)$.

11. $P_0(\sqrt{3}, 0)$, $P_1(0, \sqrt{3})$.   **12.** $P_0(a, 3)$, $P_1(3, a)$.

In Exercises 13–20, find the slope of the line through the given points.

13. $P_0(-2, 5)$, $P_1(4, 1)$.

14. $P_0(4, -3)$, $P_1(-2, -7)$.

15. $P_0(4, 2)$, $P_1(-3, 2)$.   **16.** $P_0(5, 0)$, $P_1(0, 5)$.

17. $P(a, b)$, $Q(b, a)$.   **18.** $P(4, -1)$, $Q(-3, -1)$.

19. $P(x_0, 0)$, $Q(0, y_0)$.   **20.** $O(0, 0)$, $P(x_0, y_0)$.

In Exercises 21–28, find the slope and y-intercept.

21. $y = 2x - 4$.   **22.** $6 - 5x = 0$.

23. $3y = x + 6$.   **24.** $y = 4x - 2$.

25. $4x = 1$.   **26.** $6y - 3x + 8 = 0$.

27. $7x - 3y + 4 = 0$.   **28.** $4y = 3$.

In Exercises 29–32, write an equation for the line with

29. slope 5 and y-intercept 2.

30. slope 5 and y-intercept -2.

31. slope -5 and y-intercept 2.

32. slope -5 and y-intercept -2.

In Exercises 33–34, write an equation for the horizontal line 3 units

33. above the x-axis.   **34.** below the x-axis.

In Exercises 35–36, write an equation for the vertical line 3 units

35. to the left of the y-axis.   **36.** to the right of the y-axis.

In Exercises 37–42, find an equation for the line that passes through the point $P(2, 7)$ and is

37. parallel to the x-axis.   **38.** parallel to the y-axis.

39. parallel to the line $3y - 2x + 6 = 0$.

40. perpendicular to the line $y - 2x + 5 = 0$.

41. perpendicular to the line $3y - 2x + 6 = 0$.

42. parallel to the line $y - 2x + 5 = 0$.

In Exercises 43–48, find the inclination of the line.

43. $x - y + 2 = 0$.   **44.** $6y + 5 = 0$.

45. $2x - 3 = 0$.   **46.** $x + y - 3 = 0$.

47. $3x + 4y - 12 = 0$.   **48.** $9x + 10y + 4 = 0$.

In Exercises 49–52, write an equation for the line with

49. inclination 30°, y-intercept 2.

50. inclination 60°, y-intercept 2.

51. inclination 120°, y-intercept 3.

52. inclination 135°, y-intercept 3.

In Exercises 53–56, determine the point(s) where the line intersects the circle.

53. $y = x$, $x^2 + y^2 = 1$.   **54.** $y = mx$, $x^2 + y^2 = 4$.

55. $4x + 3y = 24$, $x^2 + y^2 = 25$.

56. $y = mx + b$, $x^2 + y^2 = b^2$.

In Exercises 57–60, find the point where the lines intersect and determine the angle between the lines.

57. $l_1: 4x - y - 3 = 0$, $l_2: 3x - 4y + 1 = 0$.

58. $l_1: 3x + y - 5 = 0$, $l_2: 7x - 10y + 27 = 0$.

59. $l_1: 4x - y + 2 = 0$, $l_2: 19x + y = 0$.

60. $l_1: 5x - 6y + 1 = 0$, $l_2: 8x + 5y + 2 = 0$.

61. Find the distance between the line $5x + 12y + 2 = 0$ and
(a) the origin.   (b) the point $P(1, -3)$.

62. Find the distance between the line $2x - 3y + 1 = 0$ and
(a) the origin.   (b) the point $P(-2, 5)$.

63. Which of the points $(0, 1)$, $(1, 0)$, and $(-1, 1)$ is closest to $l: 8x + 7y - 6 = 0$? Which is farthest from l?

64. Consider the triangle with vertices $A(2, 0)$, $B(4, 3)$, $C(5, -1)$. Which of these vertices is farthest from the opposite side?

65. Find the area of the triangle with vertices $(1, -2)$, $(-1, 3)$, $(2, 4)$.

66. Find the area of the triangle with vertices $(-1, 1)$, $(3, \sqrt{2})$, $(\sqrt{2}, -1)$.

In Exercises 67–74, identify the given equation as a conic section. If it is a circle, give the center and radius; if it is an ellipse, give the center; if it is a parabola, give the vertex; if it is a hyperbola, give the center.

67. $y^2 + 2y - x = 0$.

68. $x^2 + y^2 - 4x + 6y - 3 = 0$.

69. $2x^2 + 3y^2 - 8x + 6y + 5 = 0$.

70. $2x^2 + 2y^2 - 4x + 8y = -10$.

71. $y^2 - 4y - 4x^2 + 8x = 4$.

72. $4x^2 + 16x - 16y = 32$.

73. $4x^2 - y^2 - 24x - 4y + 16 = 0$.

74. $4x^2 + y^2 - 8x + 4y + 4 = 0$.

75. Determine the slope of the line that intersects the circle $x^2 + y^2 = 169$ only at the point (5, 12).

76. Find an equation for the line which is tangent to the circle $x^2 + y^2 - 2x + 6y - 15 = 0$ at the point (4, 1) on the circle.   HINT: A line is tangent to a circle at a point P iff it is perpendicular to the radius at P.

77. The point $P(1, -1)$ is on the circle whose center is at $C(-1, 3)$. Find an equation for the tangent line to the circle at P.

The *perpendicular bisector* of the line segment $\overline{PQ}$ is the line which is perpendicular to $\overline{PQ}$ and passes through the midpoint of $\overline{PQ}$. In Exercises 78 and 79, find an equation for the perpendicular bisector of the line segment joining the given points.

78. $P(-1, 3)$, $Q(3, -4)$.   **79.** $P(1, -4)$, $Q(4, 9)$.

In Exercises 80–83, the given points are the vertices of a triangle. State whether the triangle is *isoceles* (two sides equal), a right triangle, both of these, or neither of these.

80. $P_0(-4, 3)$, $P_1(-4, -1)$, $P_2(2, 1)$.
81. $P_0(-2, 5)$, $P_1(1, 3)$, $P_2(-1, 0)$.
82. $P_0(-2, -1)$, $P_1(0, 7)$, $P_2(3, 2)$.
83. $P_0(3, 4)$, $P_1(1, 1)$, $P_2(-2, 3)$.

84. An *equilateral triangle* is a triangle in which the three sides have equal length. If two of the vertices of an equilateral triangle are (0, 0) and (4, 3), find the third vertex. How many such triangles are there?

85. Show that the midpoint M of the hypotenuse of a right triangle is equidistant from the three vertices of the triangle. HINT: Introduce a coordinate system in which the sides of the triangle are on the coordinate axes; see the figure.

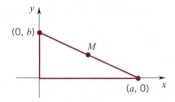

86. A *median* of a triangle is a line segment from a vertex to the midpoint of the opposite side. Find the lengths of the medians of the triangle with vertices $(-1, -2)$, (2, 1), $(4, -3)$.

87. The vertices of a triangle are (1, 0), (3, 4), and $(-1, 6)$. Find the point(s) where the medians of this triangle intersect.

88. Show that the medians of a triangle intersect in a single point (called the *centroid* of the triangle). HINT: Introduce a coordinate system such that one vertex is at the origin and one side is on the positive x-axis; see the figure.

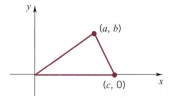

89. Prove that the diagonals of a parallelogram bisect each other. HINT: Introduce a coordinate system such that one vertex of the parallelogram is at the origin and one side is on the positive x-axis.

90. Let $P_1(x_1, y_1)$, $P_2(x_2, y_2)$, $P_3(x_3, y_3)$, $P_4(x_4, y_4)$, be the vertices of a quadrilateral. Show that the quadrilateral formed by joining the midpoints of adjacent sides is a parallelogram.

91. Temperature is normally measured either in degrees Fahrenheit (F) or in degrees Celsius (C). The relation between F and C is linear. The freezing point of water is 0°C or 32°F, and the boiling point of water is 100°C or 212°F. Find an equation giving the Fahrenheit temperature F in terms of the Celsius temperature C. Is there a temperature at which the Fahrenheit and Celsius temperatures are equal? If so, find it.

92. The relation between Fahrenheit temperature F and absolute (or Kelvin) temperature K is linear. If K = 273° when F = 32°, and K = 373° when F = 212°, express K in terms of F. Also, use Exercise 91 to determine the relation between Celsius temperature and absolute temperature. Is this relation linear?

■ **1.5 FUNCTIONS**

The fundamental processes of calculus (called *differentiation* and *integration*) are processes applied to functions. To understand these processes and to be able to carry them out, you have to be thoroughly familiar with functions. Here we review some of the basic notions.

Function; Domain and Range

First we need a working definition for the term *function*. Let D and R be sets. A *function from D into R* is a rule that assigns to each element of D a unique element of R. The set D is called the *domain* of the function. Functions are usually denoted by letters such as f, g, F, G, and so on. If f is a function from D into R, then the element $y \in R$ that is assigned to the element $x \in D$ by f is denoted $f(x)$ (read "f of x"), and is called the *value of f at x*, or the *image of x under f*. The elements $x \in D$ and $y \in R$ are called *variables*, with x the *independent variable* and y the *dependent variable*. The set of values of f, that is $\{y \in R: y = f(x) \text{ for } x \in D\}$, is called the *range* of f. Throughout most of this text, D and R will be sets of real numbers, and functions from D into R are said to be *real-valued functions of a real variable*. In general, the rules for functions will be specified by equations, although there will be occasions when some other means, such as a set of equations, are used to define a function.

Here are some examples. We begin with the squaring function

$$f(x) = x^2, \qquad \text{for all real numbers } x.$$

The domain of f, denoted dom(f), is given explicitly as the set of all real numbers. Values of f can be found by substituting for x in the equation. For example,

$$f(4) = 4^2 = 16, \qquad f(-3) = (-3)^2 = 9, \qquad f(0) = 0^2 = 0.$$

As x runs through the real numbers, x^2 runs through all the nonnegative numbers. Therefore, the range of f is $[0, \infty)$. In abbreviated form we write

$$\text{dom}(f) = (-\infty, \infty) \qquad \text{and} \qquad \text{range}(f) = [0, \infty),$$

and say that f *maps* $(-\infty, \infty)$ *onto* $[0, \infty)$.

Now consider the function

$$g(x) = \sqrt{2x + 4}, \qquad x \in [0, 6].$$

The domain of g is given as the closed interval $[0, 6]$. At $x = 0$, g takes on the value 2:

$$g(0) = \sqrt{2 \cdot 0 + 4} = \sqrt{4} = 2,$$

and at $x = 6$, g has the value 4:

$$g(6) = \sqrt{2 \cdot 6 + 4} = \sqrt{16} = 4.$$

As x runs through the numbers from 0 to 6, $g(x)$ runs through the numbers from 2 to 4. Thus, the range of g is the closed interval $[2, 4]$. The function g maps $[0, 6]$ onto $[2, 4]$.

As suggested previously, the rule for a function might be specified by a set of equations rather than just one equation. For example, consider the function

$$h(x) = \begin{cases} x^2 & \text{if } x \geq 0 \\ 2x + 1 & \text{if } x < 0. \end{cases}$$

This function is said to be *piecewise defined*. The definition of h indicates that its domain is all real numbers. It can be verified that range(h) is all real numbers as well, although the vertification here is not as straightforward as in the first two examples. The function h maps $(-\infty, \infty)$ onto $(-\infty, \infty)$.

The *absolute value function* defined by

$$|x| = \begin{cases} x & \text{if } x \geq 0 \\ -x & \text{if } x < 0. \end{cases}$$

is a familiar example of a piecewise defined function. This function has domain $(-\infty, \infty)$ and range $[0, \infty)$.

Graphs of Functions

If f is a function with domain D, then the *graph* of f is the set of all points $P(x, f(x))$ in the plane, where $x \in D$. That is, the graph of f is the graph of the equation $y = f(x)$:

$$\text{graph of } f = \{(x, y): x \in D \text{ and } y = f(x)\}.$$

An obvious method for sketching the graph of a function f is to plot some points $P(x, f(x))$, where $x \in \text{dom}(f)$. We plot enough points so that we can "see" what the graph is and then, typically, we connect the points with a "curve." Of course, if we can identify the curve in advance (for example, a straight line, a parabola, and so on), then it is much easier to draw its graph.

The graph of the squaring function

$$f(x) = x^2, \qquad x \in (-\infty, \infty)$$

is the parabola shown in Figure 1.5.1. The points that we plotted are indicated in the table and on the graph. The graph of the function

$$g(x) = \sqrt{2x + 4}, \qquad x \in [0, 6]$$

is the arc shown in Figure 1.5.2.

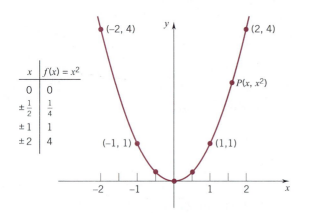

x	$f(x) = x^2$
0	0
$\pm\frac{1}{2}$	$\frac{1}{4}$
± 1	1
± 2	4

Figure 1.5.1

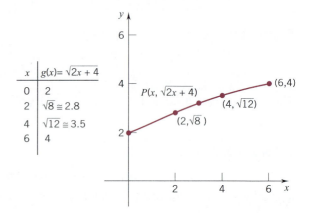

x	$g(x) = \sqrt{2x+4}$
0	2
2	$\sqrt{8} \cong 2.8$
4	$\sqrt{12} \cong 3.5$
6	4

Figure 1.5.2

The graph of the piecewise defined function

$$h(x) = \begin{cases} x^2, & x \geq 0 \\ 2x + 1, & x < 0 \end{cases}$$

and the graph of the absolute value function are given in Figures 1.5.3 and 1.5.4.

Graphing calculators and computer software packages (that is, graphing utilities) can be valuable aids to sketching the graphs of functions, provided they are used properly. However, these technologies are not a substitute for learning the basic methods. We will not attempt to teach you how to use a graphing calculator or computer software in this text, but technology-based examples and exercises will occur throughout. You are encouraged to use these to enhance your understanding of the mathematical concepts and methods.

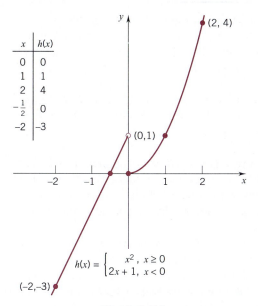

$$h(x) = \begin{cases} x^2, & x \geq 0 \\ 2x+1, & x < 0 \end{cases}$$

Figure 1.5.3

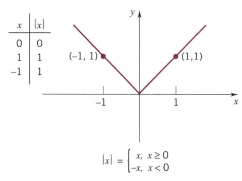

$$|x| = \begin{cases} x, & x \geq 0 \\ -x, & x < 0 \end{cases}$$

Figure 1.5.4

The next example illustrates why you must be careful when using a graphing utility to draw a graph.

Example 1 Let

$$f(x) = \tfrac{1}{2}x^3 - 2x + \tfrac{1}{2}.$$

The graph of f "looks like" the curve shown to the right.

A graphing utility was used to obtain the graphs of f shown in Figure 1.5.5(a)–(c).

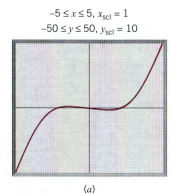

$-5 \leq x \leq 5$, $x_{scl} = 1$
$-50 \leq y \leq 50$, $y_{scl} = 10$

(a)

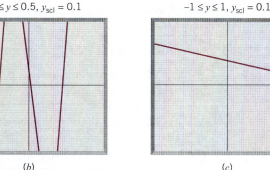

$-5 \leq x \leq 5$, $x_{scl} = 1$
$-0.5 \leq y \leq 0.5$, $y_{scl} = 0.1$

(b)

$-0.2 \leq x \leq 0.2$, $x_{scl} = 0.05$
$-1 \leq y \leq 1$, $y_{scl} = 0.1$

(c)

Figure 1.5.5

$-5 \le x \le 5, x_{scl} = 1$
$-10 \le y \le 10, Y_{scl} = 1$

Figure 1.5.6

An accurate sketch of the graph of f is shown in Figure 1.5.6.

The difficulty with the graph in Figure 1.5.5(a) is that the y-scale, $[-50, 50]$, is too large; the "oscillation" near the origin is obscured by the large scale. On the other hand, it was reasonable to use this scale since the minimum and maximum values of f on the x-interval $[-5, 5]$ are approximately -50 and 50, respectively.

In Figure 1.5.5(b), the y-scale, $[-0.5, 0.5]$, is too small; the behavior of the graph for larger values of x and y has been eliminated. In Figure 1.5.5(c), the x-scale is too small; not enough of the graph is shown. The scaling in Figure 1.5.6 provides an accurate sketch of the graph.

This example illustrates that a basic knowledge of graphing, the general shapes of graphs of various functions, and so forth, is a prerequisite to the effective use of a graphing utility. The example also illustrates that you might have to "experiment" with several views of a graph before obtaining one with the desired accuracy. ❑

In general, the graph of a function is a "curve" in the plane; it is a geometric representation of the function. A graph provides a useful picture of the behavior of a function and, among other things, it can be used to determine the range. For example, it is clear from the graph of the function h in Figure 1.5.3 that range(h) $= (-\infty, \infty)$.

This geometric representation of a function raises a converse question, namely: Under what conditions is a given curve in the plane the graph of a function? By the so-called *vertical line test*, a curve in the plane is the graph of a function iff no vertical line intersects the curve in more than one point. Thus, we can conclude that the graphs of the circles, ellipses, and hyperbolas given at the end of Section 1.4 are not the graphs of functions. The curve shown in Figure 1.5.7 is the graph of a function, while the curve shown in Figure 1.5.8 is not.

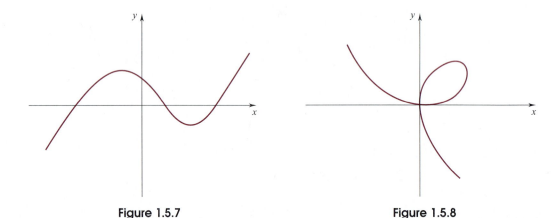

Figure 1.5.7   **Figure 1.5.8**

Even and Odd Functions; Symmetry

A function f is said to be *even* iff

$$f(-x) = f(x) \qquad \text{for all } x \in \text{dom}(f).$$

A function f is said to be *odd* iff

$$f(-x) = -f(x) \qquad \text{for all } x \in \text{dom}(f).$$

The graph of an even function is *symmetric about the y-axis* (Figure 1.5.9), and the graph of an odd function is *symmetric about the origin* (Figure 1.5.10).

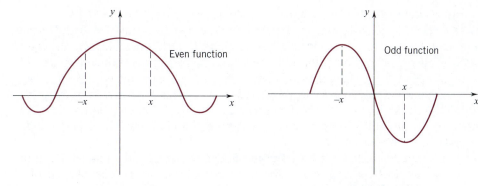

Figure 1.5.9   **Figure 1.5.10**

The squaring function $f(x) = x^2$ is even since

$$f(-x) = (-x)^2 = x^2 = f(x);$$

its graph is symmetric about the y-axis (see Figure 1.5.1). The function $f(x) = 4x - x^3$ is odd since

$$f(-x) = 4(-x) - (-x)^3 = -4x + x^3 = -(4x - x^3) = -f(x)$$

The graph of f is given in Figure 1.5.11. Note that it is symmetric with respect to the origin.

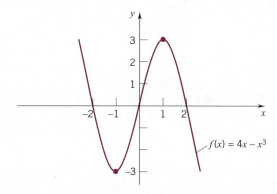

Figure 1.5.11

Convention on Domains

It will often be the case that the domain of a function will not be given explicitly. For example, you might be asked to consider

$$f(x) = x^3 + 1, \quad \text{or} \quad f(x) = \frac{1}{x-2}, \quad \text{or} \quad f(x) = \sqrt{x} \qquad \text{(see Figure 1.5.12)}$$

without further explanation. The *convention* in such cases is to take as the domain the largest set of real numbers x for which $f(x)$ is itself a real number. The first function is defined for all real numbers x, and so we would take $\operatorname{dom}(f) = (-\infty, \infty)$. The second function is defined for all real numbers except $x = 2$, and so $\operatorname{dom}(f) = \{x : x \neq 2\} = (-\infty, 2) \cup (2, \infty)$. For the square-root function, $\operatorname{dom}(f) = [0, \infty)$.

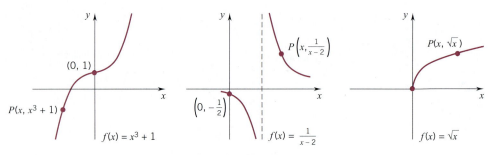

Figure 1.5.12

Example 2   Find the domain of each of the following functions:

$$\textbf{(a)} \ f(x) = \frac{x + 1}{x^2 + x - 6}. \qquad \textbf{(b)} \ g(x) = \frac{\sqrt{4 - x^2}}{x - 1}.$$

SOLUTION   **(a)** The function f is defined for all real numbers x such that $x^2 + x - 6 = (x + 3)(x - 2) \neq 0$.   Thus, $\text{dom}(f) = \{x: x \neq -3, \ 2\} = (-\infty, \ -3) \cup (-3, \ 2) \cup (2, \infty)$.

   (b) Since $\sqrt{4 - x^2}$ is a real number iff $4 - x^2 \geq 0$, we must have $-2 \leq x \leq 2$. Also, g is not defined when $x - 1 = 0$, that is when $x = 1$.   Thus,

$$\text{dom}(g) = \{x: -2 \leq x \leq 2 \quad \text{and} \quad x \neq 1\} = [-2, 1) \cup (1, 2]. \ \ \Box$$

Example 3   Find the domain and range of the function:

$$f(x) = \frac{1}{\sqrt{2 - x}} + 5.$$

SOLUTION   First we look for the domain.   Since $\sqrt{2 - x}$ is real iff $2 - x \geq 0$, we need $x \leq 2$.   But at $x = 2$, $\sqrt{2 - x} = 0$ and its reciprocal is not defined.   We must therefore restrict x to $x < 2$.   The domain is $(-\infty, 2)$.

   Now we look for the range.   As x runs through $(-\infty, 2)$, $\sqrt{2 - x}$ takes on all positive values and so does its reciprocal.   The range of f is therefore $(5, \infty)$. $\ \Box$

Examples of Applications

Functions are used in a wide variety of disciplines as mathematical models which describe how variable quantities are related.   In such cases, the application itself will usually dictate a domain for the function serving as the model.

Example 4   A candy manufacturer wants to make a tray by cutting four squares from the corners of a 12-by-20-inch rectangular sheet of cardboard and folding up the edges (see Figure 1.5.13).   Express the volume V of the box as a function of the side x of the square and give the domain of the function.

SOLUTION   From the figure, the dimensions of the tray are length $l = 20 - 2x$, width $w = 12 - 2x$, and height $h = x$.   Thus,

$$V(x) = l \times w \times h = (20 - 2x)(12 - 2x)x = 240x - 64x^2 + 4x^3 \qquad \text{(cubic inches)}.$$

While this cubic is defined for all values of x, the context of this problem dictates that $x > 0$, $12 - 2x > 0$, and $20 - 2x > 0$ (length, width, and height are all positive quantities).   It follows from these inequalities that $0 < x < 6$.   Therefore, the domain of this function is the interval $(0, 6)$. $\ \Box$

Figure 1.5.13

Example 5 A rectangle has an area of $k(>0)$ square units. Express the perimeter P of the rectangle as a function of the length x of one of its sides and give the domain of the function.

SOLUTION See Figure 1.5.14. Since the area $xy = k$, we have $y = k/x$. Now, the perimeter P of the rectangle is given by

$$P = 2x + 2y,$$

and so it follows that

$$P(x) = 2x + \frac{2k}{x}.$$

Figure 1.5.14

Since the length of a side of a rectangle cannot be negative, and since the area of the given rectangle is positive, we can conclude that $x > 0$. Also, there is no upper bound on the length of x. Thus, x satisfies $0 < x < \infty$ and dom$(P) = (0, \infty)$. ❏

EXERCISES 1.5

In Exercises 1–6, calculate (a) $f(0)$, (b) $f(1)$, (c) $f(-2)$, (d) $f(3/2)$.

1. $f(x) = 2x^2 - 3x + 2.$

2. $f(x) = \dfrac{2x - 1}{x^2 + 4}.$

3. $f(x) = \sqrt{x^2 + 2x}.$

4. $f(x) = |x + 3| - 5x.$

5. $f(x) = \dfrac{2x}{|x + 2| + x^2}.$

6. $f(x) = 1 - \dfrac{1}{(x + 1)^2}.$

In Exercises 7–10, calculate (a) $f(-x)$, (b) $f(1/x)$, (c) $f(a + b)$.

7. $f(x) = x^2 - 2x.$

8. $f(x) = \dfrac{x}{x^2 + 1}.$

9. $f(x) = \sqrt{1 + x^2}.$

10. $f(x) = \dfrac{x}{|x^2 - 1|}.$

In Exercises 11 and 12, calculate (a) $f(a + h)$ and (b) $[f(a + h) - f(a)]/h, \ h \neq 0.$

11. $f(x) = 2x^2 - 3x.$

12. $f(x) = \dfrac{1}{x - 2}.$

In Exercises 13–18, find the number(s) x, if any, where f takes on the value 1.

13. $f(x) = |2 - x|.$

14. $f(x) = \sqrt{1 + x}.$

15. $f(x) = x^2 + 4x + 5.$

16. $f(x) = 4 + 10x - x^2.$

17. $f(x) = \dfrac{2}{\sqrt{x^2 - 5}}.$

18. $f(x) = \dfrac{x}{|x|}.$

In Exercises 19–30, find the domain and range of the given function.

19. $f(x) = |x|.$

20. $g(x) = x^2 - 1.$

21. $f(x) = 2x - 3.$

22. $g(x) = \sqrt{x} + 5.$

23. $f(x) = \dfrac{1}{x^2}.$

24. $g(x) = \dfrac{4}{x}.$

25. $f(x) = \sqrt{1 - x}$.

26. $g(x) = \sqrt{x - 3}$.

27. $f(x) = \sqrt{7 - x} - 1$.

28. $g(x) = \sqrt{x - 1} - 1$.

29. $f(x) = \dfrac{1}{\sqrt{2 - x}}$.

30. $g(x) = \dfrac{1}{\sqrt{4 - x^2}}$.

In Exercises 31–44, find the domain and sketch the graph of the given function.

31. $f(x) = 1$.

32. $f(x) = -1$.

33. $f(x) = 2x$.

34. $f(x) = 2x + 1$.

35. $f(x) = \frac{1}{2}x$.

36. $f(x) = -\frac{1}{2}x$.

37. $f(x) = \frac{1}{2}x + 2$.

38. $f(x) = -\frac{1}{2}x - 3$.

39. $f(x) = \sqrt{4 - x^2}$.

40. $f(x) = \sqrt{9 - x^2}$.

41. $f(x) = x^2 - x - 6$.

42. $f(x) = |x - 1|$.

43. $f(x) = |x| + x$.

44. $f(x) = \dfrac{x^2 - 1}{x - 1}$.

In Exercises 45–48, sketch the graph and specify the domain and range of the given function.

45. $f(x) = \begin{cases} -1 & x < 0 \\ 1, & x > 0. \end{cases}$

46. $f(x) = \begin{cases} x^2, & x \le 0 \\ 1 - x, & x > 0. \end{cases}$

47. $f(x) = \begin{cases} 1 + x, & 0 \le x \le 1 \\ x, & 1 < x < 2 \\ \frac{1}{2}x + 1, & 2 \le x. \end{cases}$

48. $f(x) = \begin{cases} x^2, & x < 0 \\ -1, & 0 < x < 2 \\ x, & 2 < x. \end{cases}$

In Exercises 49–52, state whether the given curve is the graph of a function. If it is, find the domain and range.

49.

50.

51.

52.

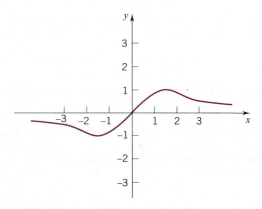

In Exercises 53–58, determine whether the function is odd, even, or neither.

53. $f(x) = x^3$.

54. $f(x) = x^2 + 1$.

55. $g(x) = x(x - 1)$.   **56.** $g(x) = x(x^2 + 1)$.

57. $f(x) = \dfrac{x^2}{1 - |x|}$.   **58.** $F(x) = x + \dfrac{1}{x}$.

59. The graph of $f(x) = \frac{1}{3}x^3 + \frac{1}{2}x^2 - 12x - 6$ looks like

(a) Use a graphing utility to sketch an accurate graph of f.
(b) Find the zero(s) of f accurate to three decimal places.
(c) Find the coordinates of the points marked A and B, accurate to three decimal places.

60. The graph of $f(x) = -x^4 + 8x^2 + x - 1$ looks like

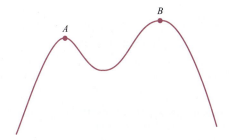

(a) Use a graphing utility to sketch an accurate graph of f.
(b) Find the zero(s) of f, if any. Use three decimal place accuracy.
(c) Find the coordinates of the points marked A and B, accurate to three decimal places.

In Exercises 61 and 62, use a graphing utility to draw several views of the graph of the function f. Select one that most accurately shows the important features of the graph and include the ranges on the variables x and y in your answer.

61. $f(x) = |x^3 - 3x^2 - 24x + 4|$.

62. $f(x) = \sqrt{x^3 - 8}$.

In Exercises 63–80, determine a formula for the function that is described and give its domain.

63. Express the area of a circle as a function of its circumference.

64. Express the volume of a sphere as a function of its surface area.

65. Express the volume of a cube as a function of the area of one of its faces.

66. Express the volume of a cube as a function of its total surface area.

67. Express the surface area of a cube as a function of the length of a diagonal of one face.

68. Express the volume of a cube as a function of one of its diagonals.

69. Express the area of an equilateral triangle as a function of the length of a side.

70. A right triangle with hypotenuse c is rotated around one of its legs to form a cone (see the figure). If x is the length of the other leg, express the volume of the cone as a function of x.

71. U.S. Post Office regulations require that the length plus the girth of a package for mailing cannot exceed 108 inches. A rectangular box with a square end whose edge length is x is designed to exactly meet the regulation (see the figure). Express the volume of the box as a function of x.

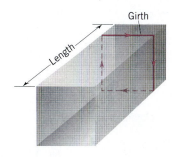

72. Suppose that a cylindrical mailing container exactly meets the U.S. Post Office regulations given in Exercise

71 (see the figure). Express the volume of the container as a function of the radius of the end.

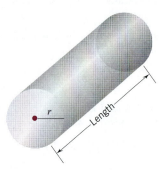

73. A Norman window is a window in the shape of a rectangle surmounted by a semicircle (see the figure). If the perimeter of the window is 15 feet, express the area as a function of its width x.

74. A window has the shape of a rectangle surmounted by an equilateral triangle. If the perimeter of the window is 15 feet, express the area as a function of one side of the equilateral triangle.

75. A rectangular beam of width x is cut from a circular log of diameter d (see the figure). Express the cross sectional area of the beam as a function of x.

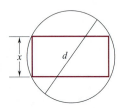

76. A rectangular sheet of metal with a perimeter of 30 inches is rolled into a cylinder of height h. Express the volume of the cylinder as a function of h.

77. Express the area of the rectangle shown in the figure as a function of the x-coordinate of the point P.

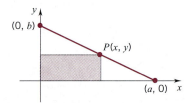

78. A right triangle is formed by the coordinate axes and a line through the point $(2, 5)$ (see the figure). Express the area of the triangle as a function of the x-intercept.

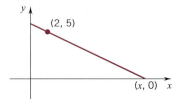

79. A string 28 inches long is to be cut into two pieces, one piece to form a square and the other to form a circle. Express the total area enclosed by the square and circle as a function of the perimeter of the square.

80. A tank in the shape of an inverted cone is being filled with water (see the figure). Express the volume of water in the tank as a function of its height.

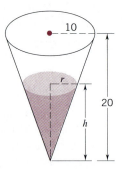

■ 1.6 THE ELEMENTARY FUNCTIONS

The basic types of functions that will be used throughout this text are polynomial functions, rational functions, trigonometric functions, and exponential and logarithmic functions. These functions are commonly called the *elementary functions*. We

review polynomial, rational, and trigonometric functions here. Exponential and logarithmic functions are defined and studied in Chapter 7.

Polynomial Functions

Polynomials were introduced in Section 1.2. A *polynomial function* is a function
defined by an equation of the form

$$P(x) = a_n x^n + a_{n-1} x^{n-1} + \cdots + a_1 x + a_0,$$

where the coefficients $a_n, a_{n-1}, \ldots, a_1, a_0$ are real numbers, $a_n \neq 0$, and n, the
degree of P, is a nonnegative integer. The function $P(x) = 0$ is also a polynomial
function, but no degree is assigned to it.

The domain of a polynomial function P is $(-\infty, \infty)$. It can be shown that the
range of P is $(-\infty, \infty)$ if its degree is odd. If $n > 1$ is even, then the range of P is an
interval of the form $[b, \infty)$ when $a_n > 0$, and an interval of the form $(-\infty, b]$ when
$a_n < 0$. Finally, if $P(x) = c$ is constant, then the range of P is a point.

Polynomial functions of degree 0 are the nonzero constant functions $P(x) = c$
(constant), $c \neq 0$. Their graphs are horizontal lines $|c|$ units above or below the
x-axis, depending on the sign of c (Figure 1.6.1). The x-axis is the graph of $P(x) = 0$.

Polynomial functions of degree 1 have the form $P(x) = ax + b$, $a \neq 0$, and are
called *linear polynomials* since their graphs are straight lines with (nonzero) slope a
and y-intercept b (Figure 1.6.2).

$P(x) = c \quad (c > 0)$

Figure 1.6.1

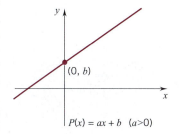

$P(x) = ax + b \quad (a > 0)$

Figure 1.6.2

$a > 0$ $\qquad$ $a < 0$

Figure 1.6.3

Polynomial functions of degree 2 have the form $P(x) = ax^2 + bx + c$, $a \neq 0$, and
are often called *quadratic functions*. The graph of a quadratic function is a parabola
which opens up if $a > 0$ and down if $a < 0$ (Figure 1.6.3). The location of the graph
with respect to the x- and y-axes depends upon the coefficients a, b, and c. For
example, if $a > 0$, then three possibilities for the location of the graph of $P(x) =
ax^2 + bx + c$ with respect to the x-axis exist. These possibilities are indicated in
Figure 1.6.4.

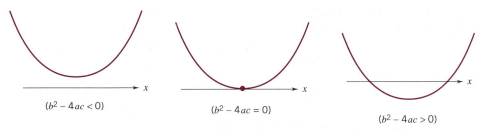

$(b^2 - 4ac < 0)$ $\qquad$ $(b^2 - 4ac = 0)$ $\qquad$ $(b^2 - 4ac > 0)$

Figure 1.6.4

Polynomial functions of degree 3 have the form $P(x) = ax^3 + bx^2 + cx + d$,
$a \neq 0$. These functions are also called *cubics*. In general, the graph of a cubic

polynomial has one of the following two shapes, again determined by the sign of a (Figure 1.6.5). Also, we have not tried to locate these graphs with respect to the coordinate axes because there are so many possibilities. Our purpose is simply to indicate the two typical shapes.

$a > 0$ $a < 0$

Figure 1.6.5

Polynomial functions become more complicated as their degrees increase. In Chapter 4 we show how to use techniques of calculus to analyze polynomial functions and to draw accurate graphs.

Rational Functions

Rational expressions were also defined in Section 1.2. A *rational function* is simply a function of the form

$$R(x) = \frac{P(x)}{Q(x)},$$

where P and Q are polynomials. Note that every polynomial P is a rational function because $P(x) = P(x)/1$ is the quotient of two polynomials. Since division by 0 is not allowed, a rational function $R = P/Q$ is not defined at those points x (if any) where $Q(x) = 0$; R is defined at all other points. Thus, $\text{dom}(R) = \{x : Q(x) \neq 0\}$.

Rational functions are more difficult to analyze and graph than polynomials. For example, it is important to analyze the behavior of a rational function $R = P/Q$ in the vicinity of a zero of Q, as well as the behavior of R for large values of x, both positive and negative. If P and Q have no common factors (P/Q is in "lowest terms"), then the zeros of Q correspond to *vertical asymptotes* of the graph of R; the existence of *horizontal asymptotes* depends on the behavior of R for large positive and large negative values of x (that is, as $x \to \pm\infty$). Vertical and horizontal asymptotes are defined and treated in detail in Chapter 4.

A sketch of the graph of

$$R(x) = \frac{1}{x^2 - 4x + 4} = \frac{1}{(x-2)^2}$$

is shown in Figure 1.6.6. Note that $x = 2$ is a vertical asymptote and the x-axis is a horizontal asymptote.

The graph of

$$R(x) = \frac{x^2}{x^2 - 1} = \frac{x^2}{(x-1)(x+1)}$$

is sketched in Figure 1.6.7. The lines $x = 1$ and $x = -1$ are vertical asymptotes of the graph and $y = 1$ is a horizontal asymptote.

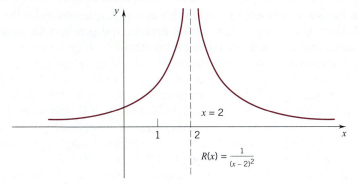

Figure 1.6.6

$R(x) = \dfrac{1}{(x-2)^2}$

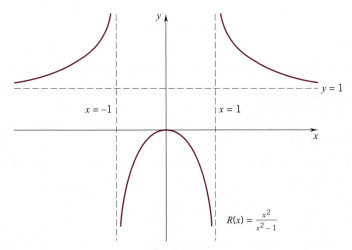

Figure 1.6.7

$R(x) = \dfrac{x^2}{x^2-1}$

Trigonometric Functions

We assume that the reader is already familiar with trigonometry, and so we present here a brief review of the definitions, basic properties, and graphs of the trigonometric functions.

Angles An angle in the plane is generated by rotating a ray (half-line) about its endpoint. The starting position of the ray is called the *initial side* of the angle, and the final position of the ray is called the *terminal side*. The point of intersection of the initial and terminal sides is called the *vertex*. An angle is said to be *positive* if it is generated by a counterclockwise rotation and *negative* if the rotation is clockwise (see Figure 1.6.8).

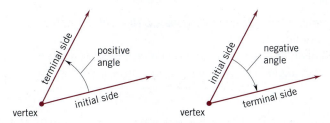

Figure 1.6.8

Degree measure, traditionally used to measure angles, has a serious drawback. It is artificial; there is no intrinsic connection between a degree and the geometry of a circle. Why choose 360° for one complete revolution? Why not 100°? or 400°?

There is another way of measuring angles that is more natural and lends itself better to the methods of calculus; measuring angles in *radians*. Let θ be an angle in the plane. Introduce a coordinate system such that the vertex of θ is at the origin and the initial side is the positive x-axis. In this context, θ is said to be in *standard position*. Let $P(1, 0)$ be the point of intersection of the initial side with the unit circle: $x^2 + y^2 = 1$. As the initial side is rotated onto the terminal side of θ, the point P traces an arc on the unit circle. Let s be the length of that arc. The *radian measure* of θ is s if θ is a positive angle and it is $-s$ if θ is a negative angle. Thus, the radian measure of θ is the signed length of the arc that θ (in standard position) subtends on the unit circle (see Figure 1.6.9). If there is no rotation, then the radian measure of θ is 0.

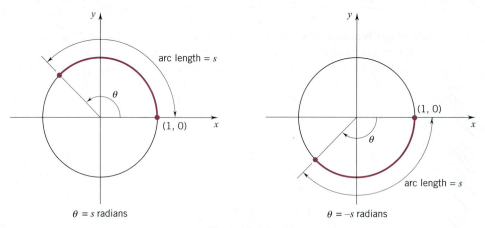

Figure 1.6.9

Since the circumference of the unit circle is 2π, the measure of the angle corresponding to one complete revolution in the counterclockwise direction is 2π radians; half a revolution (a straight angle) comprises π radians; and a quarter revolution (a right angle) comprises $\pi/2$ radians (see Figure 1.6.10).

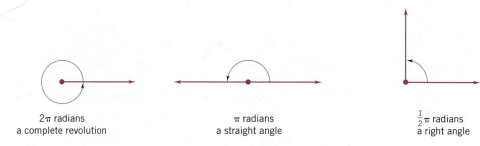

Figure 1.6.10

The conversion of degrees to radians and vice versa is made by noting that if A is the measure of an angle in degrees and x is the measure of the same angle in radians, then

$$\frac{A}{360} = \frac{x}{2\pi}.$$

In particular,

$$1 \text{ radian} = \frac{360}{2\pi} \text{ degrees} \cong 57.30°$$

and

$$1° = \frac{2\pi}{360} \text{ radian} \cong 0.0175 \text{ radian}$$

The following table gives some common angles measured in both degrees and radians.

degrees	0°	30°	45°	60°	90°	120°	135°	150°	180°	270°	360°
radians	0	$\pi/6$	$\pi/4$	$\pi/3$	$\pi/2$	$2\pi/3$	$3\pi/4$	$5\pi/6$	π	$3\pi/2$	2π

Definitions of the Trigonometric Functions As mentioned at the beginning of this section, we are concerned with real-valued functions of a real variable and so we use the unit circle and the radian measure of angles as opposed to right triangles and degree measure to define the trigonometric functions. Let θ be an angle in standard position and let $P(x, y)$ be the point of intersection of the terminal side of θ and the unit circle (see Figure 1.6.11).

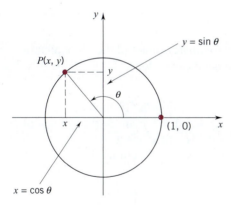

Figure 1.6.11

The six trigonometric functions are defined as follows:

sine: $\sin \theta = y$

cosine: $\cos \theta = x$

tangent: $\tan \theta = \dfrac{y}{x} = \dfrac{\sin \theta}{\cos \theta}, \qquad x \neq 0$

cosecant: $\csc \theta = \dfrac{1}{y} = \dfrac{1}{\sin \theta}, \qquad y \neq 0$

secant: $\sec \theta = \dfrac{1}{x} = \dfrac{1}{\cos \theta}, \qquad x \neq 0$

cotangent: $\cot \theta = \dfrac{x}{y} = \dfrac{\cos \theta}{\sin \theta}, \qquad y \neq 0.$

The sine and cosine functions are defined for all real numbers. That is, both sine and cosine have domain $(-\infty, \infty)$. The domains of the tangent and secant functions are all real numbers except for those values of θ for which $x = 0$, namely $\theta = (2n + 1)\pi/2$, for any integer n. The domains of the cosecant and cotangent functions are all real numbers except for those values of θ for which $y = 0$, namely $\theta = n\pi$, for any integer n.

Values We concentrate on the sine, cosine, and tangent functions since cosecant, secant, and cotangent are simply the reciprocals of these functions. The values of the trigonometric functions for the angles 0 ($=0°$), $\pi/6$ ($=30°$), $\pi/4$ ($=45°$), $\pi/3$ ($=60°$), $\pi/2$ ($=90°$), and selected multiples up to 2π, are given in the following table. You should already have these values memorized.

	0	$\pi/6$	$\pi/4$	$\pi/3$	$\pi/2$	$2\pi/3$	$3\pi/4$	$5\pi/6$	π	$3\pi/2$	2π
$\sin\theta$	0	$\dfrac{1}{2}$	$\dfrac{\sqrt{2}}{2}$	$\dfrac{\sqrt{3}}{2}$	1	$\dfrac{\sqrt{3}}{2}$	$\dfrac{\sqrt{2}}{2}$	$\dfrac{1}{2}$	0	-1	0
$\cos\theta$	1	$\dfrac{\sqrt{3}}{2}$	$\dfrac{\sqrt{2}}{2}$	$\dfrac{1}{2}$	0	$-\dfrac{1}{2}$	$-\dfrac{\sqrt{2}}{2}$	$-\dfrac{\sqrt{3}}{2}$	-1	0	1
$\tan\theta$	0	$\dfrac{\sqrt{3}}{3}$	1	$\sqrt{3}$	Undefined	$-\sqrt{3}$	-1	$-\dfrac{\sqrt{3}}{3}$	0	Undefined	0

In general, the (approximate) values of the trigonometric functions for any angle θ can be obtained with a hand calculator or from a table of values.

Periodicity A function f is *periodic* if there is a number p, $p \neq 0$, such that $f(x + p) = f(x)$ whenever x and $x + p$ are in the domain of f. The smallest positive number with this property is called the *period* of f. It is easy to verify from the definitions of the sine and cosine functions that they are periodic with period 2π.

sine

$y = \sin x$
period 2π

cosine

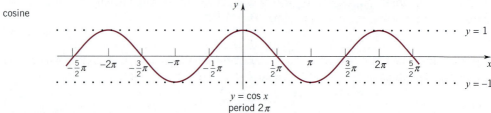

$y = \cos x$
period 2π

Figure 1.6.12

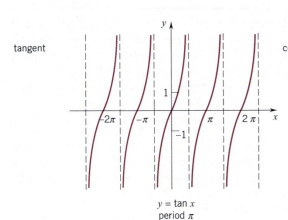

tangent

$y = \tan x$
period π

vertical asymptotes $x = (n + \frac{1}{2})\pi$, n an integer

cotangent

$y = \cot x$
period π

vertical asymptotes $x = n\pi$, n an integer

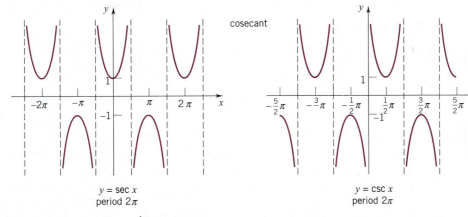

secant

$y = \sec x$
period 2π

vertical asymptotes $x = (n + \frac{1}{2})\pi$, n an integer

cosecant

$y = \csc x$
period 2π

vertical asymptotes $x = n\pi$, n an integer

Figure 1.6.13

That is,

$$\sin(\theta + 2\pi) = \sin\theta \quad \text{and} \quad \cos(\theta + 2\pi) = \cos\theta \quad \text{for all } \theta.$$

It follows that the secant and cosecant functions are also periodic with period 2π. It can be verified that the tangent and cotangent functions have period π.

Graphs The graphs of the six trigonometric functions are shown in Figures 1.6.12 and 1.6.13. Since we normally use x and y to represent the independent and dependent variables when drawing a graph, we follow that convention here, replacing θ by x and letting $y = \sin x$, $y = \cos x$, and so forth.

Important Identities We assume that you are familiar with the common trigonometric identities. The most important of these are listed here for reference.

(i) *unit circle*

$$\sin^2\theta + \cos^2\theta = 1, \quad \tan^2\theta + 1 = \sec^2\theta, \quad 1 + \cot^2\theta = \csc^2\theta.$$

(ii) *addition formulas*

$$\sin(\alpha + \beta) = \sin\alpha\cos\beta + \cos\alpha\sin\beta,$$
$$\sin(\alpha - \beta) = \sin\alpha\cos\beta - \cos\alpha\sin\beta,$$
$$\cos(\alpha + \beta) = \cos\alpha\cos\beta - \sin\alpha\sin\beta,$$
$$\cos(\alpha - \beta) = \cos\alpha\cos\beta + \sin\alpha\sin\beta.$$

(iii) *even/odd functions*

$$\sin(-\theta) = -\sin\theta$$
$$\cos(-\theta) = \cos\theta.$$

Thus, the sine function is an odd function; its graph is symmetric with respect to the origin. The cosine function is an even function; its graph is symmetric with respect to the y-axis. See Figure 1.6.12.

(iv) *double-angle formulas*

$$\sin 2\theta = 2\sin\theta\cos\theta,$$
$$\cos 2\theta = \cos^2\theta - \sin^2\theta = 2\cos^2\theta - 1 = 1 - 2\sin^2\theta.$$

(v) *half-angle formulas*

$$\sin^2\theta = \frac{1}{2}(1 - \cos 2\theta), \qquad \cos^2\theta = \frac{1}{2}(1 + \cos 2\theta).$$

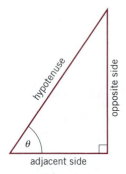

Figure 1.6.14

In Terms of a Right Triangle For angles θ between 0 and $\pi/2$, the trigonometric functions can also be defined as ratios of the sides of a right triangle (see Figure 1.6.14).

$$\sin\theta = \frac{\text{opposite side}}{\text{hypotenuse}}, \qquad \csc\theta = \frac{\text{hypotenuse}}{\text{opposite side}},$$

$$\cos\theta = \frac{\text{adjacent side}}{\text{hypotenuse}}, \qquad \sec\theta = \frac{\text{hypotenuse}}{\text{adjacent side}},$$

$$\tan\theta = \frac{\text{opposite side}}{\text{adjacent side}}, \qquad \cot\theta = \frac{\text{adjacent side}}{\text{opposite side}}.$$

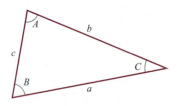

Figure 1.6.15

Arbitrary Triangles Let a, b, and c be the sides of an arbitrary triangle and let A, B and C be the corresponding angles. (See Figure 1.6.15.)

$$\text{area:} \qquad \frac{1}{2}ab\sin C = \frac{1}{2}ac\sin B = \frac{1}{2}bc\sin A.$$

$$\text{law of sines:} \qquad \frac{\sin A}{a} = \frac{\sin B}{b} = \frac{\sin C}{c}.$$

$$\text{law of cosines:} \qquad a^2 = b^2 + c^2 - 2bc\cos A.$$
$$b^2 = a^2 + c^2 - 2ac\cos B.$$
$$c^2 = a^2 + b^2 - 2ab\cos C.$$

EXERCISES 1.6

In Exercises 1–10, state whether the given function is a polynomial function, a rational function (but not a polynomial), or neither a polynomial function nor a rational function. If the function is a polynomial, state its degree.

1. $f(x) = 3$.

2. $f(x) = 1 + \frac{1}{2}x$.

3. $g(x) = \frac{1}{x}$.

4. $h(x) = \frac{x^2 - 4}{\sqrt{2}}$.

5. $F(x) = \frac{x^3 - 3x^{3/2} + 2x}{x^2 - 1}$.

6. $f(x) = 5x^4 - \pi x^2 + \frac{1}{2}$.

7. $f(x) = \sqrt{x}(\sqrt{x} + 1)$.

8. $g(x) = \frac{x^2 - 2x - 8}{x + 2}$.

9. $f(x) = \frac{\sqrt{x^2 + 1}}{x^2 - 1}$.

10. $h(x) = \frac{\sin^2 x + \cos^2 x}{x^3 + 8}$.

In Exercises 11–16, determine the domain of the given function and sketch its graph.

11. $f(x) = 3x - \frac{1}{2}$.

12. $f(x) = \frac{1}{x + 1}$.

13. $g(x) = x^2 - x - 6$.

14. $F(x) = x^3 - x$.

15. $f(x) = \frac{1}{x^2 - 4}$.

16. $g(x) = x + \frac{1}{x}$.

In Exercises 17–22, convert the given degree measure into radian measure.

17. $225°$.

18. $-210°$.

19. $-300°$.

20. $450°$.

21. $15°$.

22. $3°$.

In Exercises 23–28, convert the given radian measure into degree measure.

23. $-\frac{3\pi}{2}$.

24. $\frac{5\pi}{4}$.

25. $\frac{5\pi}{3}$.

26. $-\frac{11\pi}{6}$.

27. 2.

28. $-\sqrt{3}$.

In Exercises 29–36, determine the exact value(s) of x in the interval $[0, 2\pi)$ which satisfy the given equation.

29. $\sin x = \frac{1}{2}$.

30. $\cos x = -\frac{1}{2}$.

31. $\tan \frac{1}{2}x = 1$.

32. $\sqrt{\sin^2 x} = 1$.

33. $\cos x = \frac{\sqrt{2}}{2}$.

34. $\sin 2x = -\frac{\sqrt{3}}{2}$.

35. $\cos 2x = 0$.

36. $\tan x = -\sqrt{3}$.

In Exercises 37–46, find the approximate values of the given expression. Use four decimal place accuracy.

37. $\sin 51°$.

38. $\cos 17°$.

39. $\sin(2.352)$.

40. $\cos(-13.461)$.

41. $\tan 72.4°$.

42. $\cot(-13.5°)$.

43. $\tan(11.249)$.

44. $\cot(7.311)$.

45. $\sec(4.360)$.

46. $\csc(-9.725)$.

In Exercises 47–52, find the solutions of the given equation that are in the interval $[0, 2\pi)$. Express your answers in radians and use four decimal place accuracy.

47. $\sin x = 0.5231$.

48. $\cos x = -0.8243$.

49. $\tan x = 6.7192$.

50. $\cot x = -3.0649$.

51. $\sec x = -4.4073$.

52. $\csc x = 10.260$.

In Exercises 53–58, determine the domain and range of the given function.

53. $f(x) = |\sin x|$.

54. $g(x) = \sin^2 x + \cos^2 x$.

55. $f(x) = 2 \cos 3x$.

56. $F(x) = 1 + \sin x$.

57. $f(x) = 1 + \tan^2 x$.

58. $h(x) = \sqrt{\cos^2 x}$.

In Exercises 59–64, sketch the graph of the given function.

59. $f(x) = 3 \sin 2x$.

60. $f(x) = 1 + \sin x$.

61. $g(x) = 1 - \cos x$.

62. $F(x) = \tan \frac{1}{2}x$.

63. $f(x) = \sqrt{\sin^2 x}$.

64. $g(x) = -2 \cos x$.

In Exercises 65–70, determine whether the given function is odd, even, or neither.

65. $f(x) = \sin 3x$.

66. $g(x) = \tan x$.

67. $f(x) = 1 + \cos 2x$.

68. $g(x) = \sec x$.

69. $f(x) = x^3 + \sin x$.

70. $h(x) = \frac{\cos x}{x^2 + 1}$.

71. Suppose that l_1 and l_2 are two nonvertical lines. If $m_1 m_2 = -1$, then l_1 and l_2 intersect at right angles. Show that if l_1 and l_2 do not intersect at right angles, then the angle α between l_1 and l_2 is given by the formula:

$$\tan \alpha = \left| \frac{m_2 - m_1}{1 + m_1 m_2} \right|.$$

HINT: Use the identity

$$\tan(\theta - \phi) = \frac{\tan \theta - \tan \phi}{1 + \tan \theta \tan \phi}.$$

72. For angles θ between 0 and $\pi/2$, verify that the definition of the trigonometric functions in terms of the unit circle and the definitions in terms of a right triangle are equivalent. HINT: Introduce a coordinate system with one of the angles of a right triangle in standard position. See the figure.

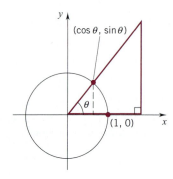

In Exercises 73 and 74, consider a general triangle with angles A, B, C and corresponding sides a, b, c. For example, see Figure 1.6.15.

73. Prove the law of sines:

$$\frac{\sin A}{a} = \frac{\sin B}{b} = \frac{\sin C}{c}.$$

HINT: Drop a perpendicular from one vertex to the opposite side and use the two right triangles that are formed.

74. Prove the law of cosines:

$$a^2 = b^2 + c^2 - 2bc \cos A.$$

HINT: Drop a perpendicular from angle B to side b, and use trigonometry and the Pythagorean theorem.

75. Prove that the area A of an isosceles triangle is given by $A = \frac{1}{2}a^2\sin\theta$, where a is the length of the equal sides and θ is the angle included between them.

76. (a) Use a graphing utility to graph the polynomial functions

$$f(x) = x^4 + 2x^3 - 5x^2 - 3x + 1 \quad \text{and}$$
$$g(x) = -x^4 + x^3 + 4x^2 - 3x + 2.$$

(b) Based on your graphs in part (a), make a conjecture as to the general shape of the graph of a polynomial function of degree 4 (a *quartic* polynomial).

(c) Now graph the quartic polynomials

$$f(x) = x^4 - 4x^2 + 4x + 2 \quad \text{and} \quad g(x) = -x^4$$

and compare these graphs with your conjectures in part (b). The methods of calculus will help us distinguish between the various possibilities for the graph of a quartic polynomial. Can you determine

one property that the graphs of all quartic polynomials of the form

$$P(x) = x^4 + ax^3 + bx^2 + cx + d$$

share? What about quartics of the form $Q(x) = -x^4 + ax^3 + bx^2 + cx + d$?

77. (a) Use a graphing utility to graph the polynomial functions

$$f(x) = x^5 - 7x^3 + 6x + 2 \quad \text{and}$$
$$g(x) = -x^5 + 5x^3 - 3x - 3.$$

(b) Based on your graphs in part (a), make a conjecture as to the general shape of the graph of a polynomial function of degree 5 (a *quintic* polynomial).

(c) Now graph

$$P(x) = x^5 + ax^4 + bx^3 + cx^2 + dx + e$$

for several choices of a, b, c, d, and e. (For example, try $a = b = c = d = e = 0$.) How do these graphs compare with your graph of f from part (a)?

78. (a) Use a graphing utility to graph $f_A(x) = A\cos x$ for several values of A; use both positive and negative values. Compare your graphs with the graph of $f(x) = \cos x$.

(b) Now graph $f_B(x) = \cos Bx$ for several values of B. Since the cosine function is even, it is sufficient to use only positive values for B. Use some values between 0 and 1 and some values greater than 1. Again, compare your graphs with the graph of $f(x) = \cos x$.

(c) Describe the effects that the coefficients A and B have on the graph of the cosine function.

79. Let $f_n(x) = x^n$, $n = 1, 2, 3, \ldots$.

(a) Using a graphing utility, draw the graphs of f_n for $n = 2, 4, 6$, in one coordinate system, and in another coordinate system draw the graphs of f_n for $n = 1, 3, 5$.

(b) Based on your results in part (a), make a sketch of the graph of f_n when n is even, and when n is odd.

(c) For a given positive integer k, compare the graphs of f_k and f_{k+1} on $[0, 1]$ and on $[1, \infty)$.

80. (a) Use a graphing utility to draw the graphs of

$$f(x) = \frac{1}{x}\sin x \quad \text{and}$$
$$g(x) = x\sin\left(\frac{1}{x}\right) \quad \text{for} -\frac{\pi}{2} \le x \le \frac{\pi}{2}, \quad x \ne 0.$$

(b) Use the "zoom in" feature to examine the behavior of the graphs of f and g at the origin. What can you say about the values of $f(x)$ and $g(x)$ when x is close to 0?

■ 1.7 COMBINATIONS OF FUNCTIONS

As we will see in our study of calculus, functions often arise as combinations of other functions. In this section we review the basic methods for combining two or more functions.

Algebraic Combinations of Functions

Let f and g be real-valued functions of a real variable. Then the sum, difference, product, and quotient of f and g are defined in a natural way in terms of the sum, difference, product, and quotient of real numbers. These algebraic combinations of f and g are defined as follows:

$$\textit{sum:} \quad [f + g](x) = f(x) + g(x).$$
$$\textit{difference:} \quad [f - g](x) = f(x) - g(x).$$
$$\textit{product:} \quad [f \cdot g](x) = f(x) \cdot g(x).$$

$$\textit{quotient:} \quad \left[\frac{f}{g}\right](x) = \frac{f(x)}{g(x)}.$$

According to these definitions, $(f + g)(x)$, $(f - g)(x)$, and $(f \cdot g)(x)$ exist only when both $f(x)$ and $g(x)$ exist. Thus, $f + g$, $f - g$, and $f \cdot g$ each have domain $\text{dom}(f) \cap \text{dom}(g)$. The definition of f/g requires the additional condition that $g(x) \neq 0$.

Example 1 Let

$$f(x) = x^3 + x - 2 \quad \text{and} \quad g(x) = \tfrac{1}{2}x - \sqrt{x}.$$

Determine (a) $f + g$, (b) $f \cdot g$, and (c) f/g, and give the domain in each case.

SOLUTION Since f is a polynomial, $\text{dom}(f) = (-\infty, \infty)$. You can verify that $\text{dom}(g) = [0, \infty)$.

(a) $(f + g)(x) = f(x) + g(x) = (x^3 + x - 2) + (\tfrac{1}{2}x - \sqrt{x}) = x^3 + \tfrac{3}{2}x - 2 - \sqrt{x}$

and

$$\text{dom}(f + g) = (-\infty, \infty) \cap [0, \infty) = [0, \infty).$$

(b)
$$(f \cdot g)(x) = f(x) \cdot g(x) = (x^3 + x - 2)(\tfrac{1}{2}x - \sqrt{x})$$
$$= \tfrac{1}{2}x^4 + \tfrac{1}{2}x^2 - x - x^3\sqrt{x} - x\sqrt{x} + 2\sqrt{x}$$

and $\text{dom}(f \cdot g) = [0, \infty)$. Note that this product can also be written:

$$(f \cdot g)(x) = \tfrac{1}{2}x^4 + \tfrac{1}{2}x^2 - x - x^{7/2} - x^{3/2} + 2x^{1/2}.$$

(c)
$$\left(\frac{f}{g}\right)(x) = \frac{f(x)}{g(x)} = \frac{x^3 + x - 2}{\tfrac{1}{2}x - \sqrt{x}}.$$

To find the domain of f/g we start with $\text{dom}(f) \cap \text{dom}(g) = [0, \infty)$ and impose the additional restriction that $g(x) \neq 0$. Now $g(x) = 0$ implies

$$\tfrac{1}{2}x = \sqrt{x} \quad \text{or} \quad \tfrac{1}{4}x^2 = x.$$

It follows from the second equation that $g(x) = 0$ for $x = 0$ and $x = 4$. Thus

$$\text{dom}(f/g) = \{x: x \in [0, \infty) \text{ and } x \neq 0, 4\} = (0, 4) \cup (4, \infty). \quad \square$$

With α and β real numbers, we can also form *scalar multiples* of functions

$$[\alpha f](x) = \alpha \cdot f(x)$$

and *linear combinations* of functions

$$[\alpha f + \beta g](x) = \alpha \cdot f(x) + \beta \cdot g(x).$$

It should be clear that $\mathrm{dom}(\alpha f) = \mathrm{dom}(f)$ and $\mathrm{dom}(\alpha f + \beta g) = \mathrm{dom}(f) \cap \mathrm{dom}(g)$.

Example 2   Let

$$f(x) = x^2 - 3x + \frac{2}{x} \quad \text{and} \quad g(x) = x - \frac{1}{x} + \sqrt{x + 1}.$$

Then

$$\mathrm{dom}(f) = \{x : x \neq 0\} \quad \text{and} \quad \mathrm{dom}(g) = \{x : x \geq -1 \text{ and } x \neq 0\}.$$

Now

$$(3f)(x) = 3\left(x^2 - 3x + \frac{2}{x}\right) = 3x^2 - 9x + \frac{6}{x}$$

with $\mathrm{dom}(3f) = \{x : x \neq 0\}$ and

$$(4f - 3g)(x) = 4\left(x^2 - 3x + \frac{2}{x}\right) - 3\left(x - \frac{1}{x} + \sqrt{x + 1}\right)$$

$$= 4x^2 - 15x + \frac{11}{x} - 3\sqrt{x + 1}$$

with $\mathrm{dom}(4f - 3g) = \mathrm{dom}(f) \cap \mathrm{dom}(g) = [-1, 0) \cup (0, \infty)$ (verify this).   ❏

Composition of Functions

We have seen how to combine functions algebraically.   There is another way to combine functions called *composition*.   To describe it, we begin with two functions f and g, and a number x in the domain of g.   By applying g to x we get the number $g(x)$. If $g(x)$ is in the domain of f, then we can apply f to $g(x)$ and thereby obtain the number $f(g(x))$.

What is $f(g(x))$?   It is the result of first applying g to x and then applying f to $g(x)$.   The idea is illustrated in Figure 1.7.1.   This new function—it takes x in the domain of g such that $g(x)$ is in the domain of f, and assigns to it the value $f(g(x))$—is called the *composition* of f and g and is denoted by the symbol $f \circ g$ (see Figure 1.7.2).

Figure 1.7.1

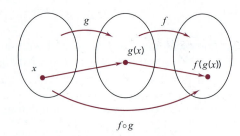

Figure 1.7.2

DEFINITION 1.7.1 COMPOSITION

Let f and g be functions and let $D = \{x \in \text{dom}(g): g(x) \in \text{dom}(f)\}$. The *composition* $f \circ g$ (read "f circle g") is the function defined on D by

$$(f \circ g)(x) = f(g(x)).$$

Example 3 Suppose that

$$g(x) = x^2 \qquad \text{(the squaring function)}$$

and

$$f(x) = x + 3. \qquad \text{(the function that adds 3)}$$

Then

$$(f \circ g)(x) = f(g(x)) = g(x) + 3 = x^2 + 3.$$

In other words, $f \circ g$ is the function that *first* squares and *then* adds 3. Since f and g each have domain the set of all real numbers, the domain of $f \circ g$ is all real numbers.

On the other hand,

$$(g \circ f)(x) = g(f(x)) = [f(x)]^2 = (x + 3)^2$$

Thus, $g \circ f$ is the function that *first* adds 3 and *then* squares; $\text{dom}(g \circ f)$ is all real numbers. Notice that $g \circ f$ is *not* the same as $f \circ g$. ❑

Example 4 Let

$$g(x) = \sqrt{3 - x}, \qquad \text{dom}(g) = (-\infty, 3]$$

and

$$f(x) = x^2 - 1, \qquad \text{dom}(f) = (-\infty, \infty).$$

Then

$$(f \circ g)(x) = f(g(x)) = \left(\sqrt{3 - x}\right)^2 - 1 = (3 - x) - 1 = 2 - x$$

and $\text{dom}(f \circ g) = \{x \in \text{dom}(g): g(x) \in \text{dom}(f)\} = \text{dom}(g)$ since $\text{dom}(f)$ is all real numbers. Thus, $\text{dom}(f \circ g) = (-\infty, 3]$.

Also,

$$(g \circ f)(x) = g(f(x)) = \sqrt{3 - (x^2 - 1)} = \sqrt{4 - x^2}$$

and $\text{dom}(g \circ f) = \{x \in \text{dom}(f): f(x) \in (-\infty, 3]\} = \{x: x^2 - 1 \leq 3\} = [-2, 2]$. ❑

It is possible to form the composition of more than two functions. For example, the triple composition $f \circ g \circ h$ consists of first h, then g, and then f:

$$(f \circ g \circ h)(x) = f[g(h(x))].$$

We can go on in this manner with as many functions as we like.

Example 5   If

$$f(x) = \frac{1}{x}, \qquad g(x) = x^2 + 1, \qquad h(x) = \cos x,$$

then

$$(f \circ g \circ h)(x) = f[g(h(x))] = \frac{1}{g(h(x))} = \frac{1}{[h(x)]^2 + 1} = \frac{1}{\cos^2 x + 1}. \quad \square$$

Example 6   Find functions f and g such that $f \circ g = F$ if
$$F(x) = (x + 1)^5.$$

A SOLUTION   The function consists of first adding 1 and then taking the fifth power. We can therefore set

$$g(x) = x + 1 \qquad \qquad \text{(adding 1)}$$

and

$$f(x) = x^5. \qquad \qquad \text{(taking the fifth power)}$$

As you can see,

$$(f \circ g)(x) = f(g(x)) = [g(x)]^5 = (x + 1)^5. \quad \square$$

Example 7   Find three functions f, g, h such that $f \circ g \circ h = F$ if

$$F(x) = \frac{1}{|x| + 3}.$$

A SOLUTION F takes the absolute value, adds 3, and then inverts.   Let h take the absolute value:

$$\text{set} \quad h(x) = |x|.$$

Let g add 3:

$$\text{set} \quad g(x) = x + 3.$$

Let f do the inverting:

$$\text{set} \quad f(x) = \frac{1}{x}.$$

With this choice of f, g, h, we have

$$(f \circ g \circ h)(x) = f(g(h(x))) = \frac{1}{g(h(x))} = \frac{1}{h(x) + 3} = \frac{1}{|x| + 3}. \quad \square$$

EXERCISES 1.7

In Exercises 1–8, let $f(x) = 2x^2 - 3x + 1$ and $g(x) = x^2 + 1/x$.   Determine the indicated values.

1. $(f + g)(2)$.

2. $(f - g)(-1)$.

3. $(fg)(-2)$.

4. $\left(\dfrac{f}{g}\right)(1)$.

5. $(2f - 3g)\left(\dfrac{1}{2}\right)$.

6. $\left(\dfrac{f + 2g}{f}\right)(-1)$.

7. $(f \circ g)(1)$.

8. $(g \circ f)(1)$.

In Exercises 9–12, determine $f + g$, $f - g$, fg, and f/g, and give the domain of each.

9. $f(x) = 2x - 3$, $g(x) = 2 - x$.

10. $f(x) = x^2 - 1$, $g(x) = x + \dfrac{1}{x}$.

11. $f(x) = \sqrt{x - 1}$, $g(x) = x - \sqrt{x + 1}$.

12. $f(x) = \sin^2 x$, $g(x) = \cos 2x$.

13. Given that $f(x) = x + 1/\sqrt{x}$ and $g(x) = \sqrt{x} - 2/\sqrt{x}$, find (a) $6f + 3g$, (b) $f - g$, (c) f/g.

14. Given that

$$f(x) = \begin{cases} 1 - x, & x \le 1 \\ 2x - 1, & x > 1 \end{cases}, \quad g(x) = \begin{cases} 0, & x < 2 \\ -1, & x \ge 2, \end{cases}$$

find $f + g$, $f - g$, and fg.   HINT: Break up the domains of the two functions in the same manner.

In Exercises 15–22, sketch the graph with f and g as shown in the figure.

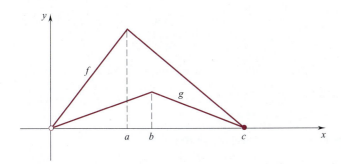

15. $2g$.

16. $\frac{1}{2}f$.

17. $-f$.

18. $0 \cdot g$.

19. $-2g$.

20. $f + g$.

21. $f - g$.

22. $f + 2g$.

In Exercises 23–32, form the composition $f \circ g$ and give the domain.

23. $f(x) = 2x + 5$, $g(x) = x^2$.

24. $f(x) = x^2$, $g(x) = 2x + 5$.

25. $f(x) = \sqrt{x}$, $g(x) = x^2 + 5$.

26. $f(x) = x^2 + x$, $g(x) = \sqrt{x}$.

27. $f(x) = \dfrac{1}{x}$, $g(x) = \dfrac{x - 2}{x}$.

28. $f(x) = \dfrac{1}{x - 1}$, $g(x) = x^2$.

29. $f(x) = \dfrac{1}{\sqrt{x} - 3}$, $g(x) = (x^2 - 1)^2$.

30. $f(x) = \dfrac{1}{x} - \dfrac{1}{x + 1}$, $g(x) = \dfrac{x}{x^2 - 1}$.

31. $f(x) = \sqrt{1 - x^2}$, $g(x) = \cos 2x$.

32. $f(x) = \sqrt{1 - x}$, $g(x) = 2 \cos x$ for $x \in [0, 2\pi]$.

In Exercises 33–36, form the composition $f \circ g \circ h$ and give the domain.

33. $f(x) = 4x$, $g(x) = x - 1$, $h(x) = x^2$.

34. $f(x) = x - 1$, $g(x) = 4x$, $h(x) = x^2$.

35. $f(x) = \dfrac{1}{x}$, $g(x) = \dfrac{1}{2x + 1}$, $h(x) = x^2$.

36. $f(x) = \dfrac{x + 1}{x}$, $g(x) = \dfrac{1}{2x + 1}$, $h(x) = x^2$.

In Exercises 37–40, find f such that $f \circ g = F$ given that

37. $g(x) = \dfrac{1 + x^2}{1 + x^4}$, $F(x) = \dfrac{1 + x^4}{1 + x^2}$.

38. $g(x) = x^2$, $F(x) = ax^2 + b$.

39. $g(x) = 3x$, $F(x) = 2 \sin 3x$.

40. $g(x) = -x^2$, $F(x) = \sqrt{a^2 + x^2}$.

In Exercises 41–44, find g such that $f \circ g = F$ given that

41. $f(x) = x^3$, $F(x) = \left(1 - \dfrac{1}{x^4}\right)^2$.

42. $f(x) = x + \dfrac{1}{x}$, $F(x) = a^2 x^2 + \dfrac{1}{a^2 x^2}$.

43. $f(x) = x^2 + 1$, $F(x) = (2x^3 - 1)^2 + 1$.

44. $f(x) = \sin x$, $F(x) = \sin \dfrac{1}{x}$.

In Exercises 45–49, find $f \circ g$ and $g \circ f$.

45. $f(x) = \sqrt{x}$, $g(x) = x^2$.

46. $f(x) = 3x + 1$, $g(x) = x^2$.

47. $f(x) = 1 - x^2$, $g(x) = \sin x$.

48. $f(x) = 2x$, $g(x) = \frac{1}{2}$.

49. $f(x) = x^3 + 1$, $g(x) = \sqrt[3]{x - 1}$.

50. Suppose that f and g are odd functions. What can you conclude about fg? Justify your answer.

51. Suppose that f and g are even functions. What can you conclude about fg? Justify your answer.

52. Suppose that f is an even function and g is an odd function. What can you conclude about fg? Justify your answer.

53. Suppose that

$$f(x) = \begin{cases} x, & 0 \le x \le 1 \\ 1, & x > 1. \end{cases}$$

How is f defined for $x < 0$ if (a) f is even? (b) f is odd?

54. Suppose that $f(x) = x^2 - x$ for $x \ge 0$. How is f defined for $x < 0$ if (a) f is even? (b) f is odd?

55. Given that the function f is defined for all real numbers, show that the function $g(x) = f(x) + f(-x)$ is an even function.

56. Show that every function that is defined for all real numbers can be written as the sum of an even function and an odd function.

57. For $x \neq 0, 1$, define

$$f_1(x) = x, \qquad f_2(x) = \frac{1}{x}, \qquad f_3(x) = 1 - x,$$

$$f_4(x) = \frac{1}{1-x}, \qquad f_5(x) = \frac{x-1}{x}, \qquad f_6(x) = \frac{x}{x-1}.$$

This family of functions is *closed* under composition; that is, the composition of any two of these functions is again one of these functions. Tabulate the results of composing these functions one with the other by filling in the table shown in the figure. To indicate that $f_i \circ f_j = f_k$, write "f_k" in the ith row, jth column. We have already made two entries in the table. Check out these two entries and then fill in the rest of the table.

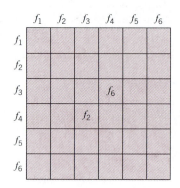

	f_1	f_2	f_3	f_4	f_5	f_6
f_1						
f_2						
f_3				f_6		
f_4			f_2			
f_5						
f_6						

58. Find the minimal family of functions that contains the given functions and is closed under composition.
   (a) $f(x) = 1 - x$.   (b) $f(x) = 1 + x$.   (c) $f(x) = -x^3$.

Two functions f and g are *inverses* of each other if

$$(f \circ g)(x) = x \qquad \text{for all } x \in \text{dom}(g)$$

and

$$(g \circ f)(x) = x \qquad \text{for all } x \in \text{dom}(f).$$

For example, $f(x) = 2x + 3$ and $g(x) = \frac{1}{2}(x - 3)$ are inverses of each other since

$$(f \circ g)(x) = f(g(x)) = 2[\tfrac{1}{2}(x - 3)] + 3 = x$$

$$(g \circ f)(x) = g(f(x)) = \tfrac{1}{2}([2x + 3] - 3) = x.$$

In Exercises 59–62, verify that the given functions are inverses of each other.

59. $f(x) = 3x - 5, \quad g(x) = \frac{1}{3}(x + 5)$.

60. $f(x) = x^3, \quad g(x) = \sqrt[3]{x}$.

61. $f(x) = \dfrac{1}{x+1}, \quad g(x) = \dfrac{1-x}{x}$.

62. $f(x) = \sqrt[5]{x+1}, \quad g(x) = x^5 - 1$.

▷ In Exercises 63 and 64, use the family of functions $F(x) = (x - a)^2 + b$.

63. (a) Choose a value for b and, using a graphing utility, graph F for several different values of a. Be sure to choose both positive and negative values. Compare your graphs with the graph of $f(x) = x^2$, and describe the effect that varying a has on the graph of F.
   (b) Now fix a value of a and graph F for several values of b; again, use both positive and negative values. Compare your graphs with the graph of $f(x) = x^2$, and describe the effect on that varying b has on the graph of F.

64. For any values of a and b, the graph of F is a parabola which opens up. Find values for a and b such that the parabola will have x-intercepts at $-\frac{3}{2}$ and 2. Verify your result algebraically.

▷ In Exercises 65 and 66, let $f(x) = x^3 - 3x + 1$.

65. (a) Using a graphing utility, graph $f(x - a)$ for $a = -2, -1, 0, 1, 2$. Describe the effect that varying a has on the graph of f.
   (b) Graph $f(bx)$ for $b = -2, -1, 0, 1, 2, 3$. Describe the effect that varying b has on the graph of f.
   (c) Graph $f(x) + c$ for $c = -2, -1, 0, 1, 2$. Describe the effect that varying c has on the graph of f.

66. Use your graphing utility to find values for a, b, and c such that the graph of $f(b[x - a]) + c$ has x-intercepts at $-2, \frac{1}{2}$, and 3. Verify your result algebraically.

▷ In Exercises 67 and 68, let $f(x) = \sin x$

67. (a) Using a graphing utility, graph Af for $A = -3, -2, -1, 2, 3$. Compare your graphs with the graph of f.
   (b) Now graph $f(Bx)$ for $B = -3, -2, -\frac{1}{2}, \frac{1}{3}, \frac{1}{2}, 2, 3$. Compare your graphs with the graph of f.

68. (a) Using a graphing utility, graph $f(x - C)$ for $C = -\frac{\pi}{2}, -\frac{\pi}{4}, \frac{\pi}{3}, \frac{\pi}{2}, \pi, 2\pi$. Compare your graphs with the graph of f.
   (b) Now graph $Af(Bx - C)$ for several values of A, B, and C. Describe the effects of A, B, and C.

■ 1.8 A NOTE ON MATHEMATICAL PROOF; MATHEMATICAL INDUCTION

The notion of proof goes back to Euclid's *Elements,* and the rules of proof have changed little since they were formulated by Aristotle. We work in a deductive system where truth is argued on the basis of assumptions, definitions, and previously proved results. We cannot claim that such and such is true without clearly stating the basis on which we make that claim.

A theorem is an implication; it consists of a hypothesis and a conclusion:

$$\text{if (hypothesis)} \ldots, \text{then (conclusion)} \ldots.$$

Here is an example:

$$\text{if } a \text{ and } b \text{ are positive numbers, then } ab \text{ is positive.}$$

A common mistake is to ignore the hypothesis and persist with the conclusion: to insist, for example, that $ab > 0$ just because a and b are numbers.

Another common mistake is to confuse a theorem

$$\text{if } A, \text{ then } B$$

with its converse

$$\text{if } B, \text{ then } A.$$

The fact that a theorem is true does not mean that its converse is true: While it is true that

$$\text{if } a \text{ and } b \text{ are positive numbers, then } ab \text{ is positive,}$$

it is *not* true that

$$\text{if } ab \text{ is positive, then } a \text{ and } b \text{ are positive numbers;}$$

$[(-2)(-3)$ is positive, but -2 and -3 are not positive$]$.

A third, more subtle mistake is to presume that the hypothesis of a theorem represents the only condition under which the conclusion is true. There may well be other conditions under which the conclusion is true. Thus, for example, not only is it true that

$$\text{if } a \text{ and } b \text{ are positive numbers, then } ab \text{ is positive}$$

but it is also true that

$$\text{if } a \text{ and } b \text{ are negative numbers, then } ab \text{ is positive.}$$

In the event that a theorem

$$\text{if } A, \text{ then } B$$

and its converse

$$\text{if } B, \text{ then } A$$

are both true, then we can write

$$A \text{ if and only if } B \qquad \text{or more briefly} \qquad A \text{ iff } B.$$

We know, for example, that

$$\text{if} \quad x \geq 0, \quad \text{then} \quad |x| = x;$$

we also know that

$$\text{if} \quad |x| = x, \qquad \text{then} \quad x \geq 0.$$

We can summarize by writing

$$x \geq 0 \qquad \text{iff} \qquad |x| = x.$$

A final point.   One way of proving

$$\text{if } A, \text{ then } B$$

is to assume that

(1) A holds and B does not hold

and then arrive at a contradiction.   The contradiction is taken to indicate that (1) is false and therefore

$$\text{if } A \text{ holds, then } B \text{ must hold.}$$

Some of the theorems of calculus are proved by this method.

Calculus provides procedures for solving a wide range of problems in the physical and social sciences.   The fact that these procedures give us answers that seem to make sense is comforting, but it is only because we can prove our theorems that we can be confident in the results.   Accordingly, your study of calculus should include the study of some proofs.

Mathematical Induction

Mathematical induction is a method of proof which can be used to show that certain propositions are true for all positive integers n.   The method is based upon the following axiom:

1.8.1 AXIOM OF INDUCTION

Let S be a set of positive integers.   If

(A) $1 \in S$, and

(B) $k \in S$ implies that $k + 1 \in S$,

then all the positive integers are in S.

You can think of the axiom of induction as a kind of "domino theory."   If the first domino falls (Figure 1.8.1), and if each domino that falls causes the next one to fall, then, according to the axiom of induction, each domino will fall.

While we cannot prove that this axiom is valid (axioms are by their very nature assumptions and therefore not subject to proof), we can argue that it is *plausible*.

Let's assume that we have a set S that satisfies conditions (A) and (B).   Now let's choose a positive integer m and "argue" that $m \in S$.

From (A) we know that $1 \in S$.   Since $1 \in S$, we know from (B) that $1 + 1 \in S$, and thus that $(1 + 1) + 1 \in S$, and so on.   Since m can be obtained from 1 by adding 1 successively $(m - 1)$ times, it *seems clear* that $m \in S$.   ❏

domino theory

Figure 1.8.1

To prove that a given proposition is true for *all* positive integers n, we let S be the set of positive integers for which the proposition is true. We prove first that $1 \in S$, that is, that the proposition is true for $n = 1$. Next we assume that the proposition is true for some positive integer k, and prove that it is also true for $k + 1$, that is, $k \in S$ implies that $k + 1 \in S$. Then, by the axiom of induction, it follows that S contains the set of positive integers and so the proposition is true for all positive integers.

Example 1 Show that

$$1 + 2 + 3 + \cdots + n = \frac{n(n + 1)}{2}$$

for all positive integers n.

SOLUTION Let S be the set of positive integers n for which

$$1 + 2 + 3 + \cdots + n = \frac{n(n + 1)}{2}.$$

Then $1 \in S$ since

$$1 = \frac{1(1 + 1)}{2}.$$

Next, assume that the positive integer $k \in S$, that is, assume

$$1 + 2 + 3 + \cdots + k = \frac{k(k + 1)}{2}.$$

Now consider the sum of the first $k + 1$ integers:

$$
\begin{aligned}
1 + 2 + 3 + \cdots + k + (k + 1) &= [1 + 2 + 3 + \cdots k] + (k + 1) \\
&= \frac{k(k + 1)}{2} + (k + 1) \quad \text{(by the induction hypothesis)} \\
&= \frac{k(k + 1) + 2(k + 1)}{2} \\
&= \frac{(k + 1)(k + 2)}{2},
\end{aligned}
$$

and so $k + 1 \in S$. Thus, by the axiom of induction, we can conclude that all positive integers are in S, that is

$$1 + 2 + 3 + \cdots + n = \frac{n(n + 1)}{2}$$

for all positive integers n. ❑

Example 2 Show that, if $x \geq -1$, then

$$(1 + x)^n \geq 1 + nx \qquad \text{for all positive integers } n.$$

SOLUTION Take $x \geq -1$ and let S be the set of positive integers n for which

$$(1 + x)^n \geq 1 + nx.$$

Since

$$(1 + x)^1 \geq 1 + 1 \cdot x,$$

you can see that $1 \in S$.
    Assume now that $k \in S$. By the definition of S,

$$(1 + x)^k \geq 1 + kx.$$

Since

$$(1 + x)^{k+1} = (1 + x)^k(1 + x) \geq (1 + kx)(1 + x) \qquad \text{(explain)}$$

and

$$(1 + kx)(1 + x) = 1 + (k + 1)x + kx^2 \geq 1 + (k + 1)x,$$

it follows that

$$(1 + x)^{k+1} \geq 1 + (k + 1)x$$

and thus that $k + 1 \in S$.
    We have shown that

$$1 \in S \qquad \text{and that} \qquad k \in S \quad \text{implies} \quad k + 1 \in S.$$

By the axiom of induction, all the positive integers are in S. ❑

EXERCISES 1.8

In Exercises 1–10, show that the statement holds for all positive integers n.

1. $2n \leq 2^n$.

2. $1 + 2n \leq 3^n$.

3. $n(n + 1)$ is divisible by 2.
   HINT: $(k + 1)(k + 2) = k(k + 1) + 2(k + 1)$.

4. $1 + 3 + 5 + \cdots + (2n - 1) = n^2$.

5. $1^2 + 2^2 + 3^2 + \cdots + n^2 = \frac{1}{6}n(n + 1)(2n + 1)$.

6. $1^3 + 2^3 + 3^3 + \cdots + n^3 = (1 + 2 + 3 + \cdots + n)^2$.
   HINT: Use Example 1.

7. $1^3 + 2^3 + \cdots + (n - 1)^3 < \frac{1}{4}n^4 < 1^3 + 2^3 + \cdots + n^3$.

8. $1^2 + 2^2 + \cdots + (n - 1)^2 < \frac{1}{3}n^3 < 1^2 + 2^2 + \cdots + n^2$.

9. $\dfrac{1}{\sqrt{1}} + \dfrac{1}{\sqrt{2}} + \dfrac{1}{\sqrt{3}} + \cdots + \dfrac{1}{\sqrt{n}} > \sqrt{n}$.

10. $\dfrac{1}{1 \cdot 2} + \dfrac{1}{2 \cdot 3} + \dfrac{1}{3 \cdot 4} + \cdots + \dfrac{1}{n(n + 1)} = \dfrac{n}{n + 1}$.

11. For what integers n is $3^{2n+1} + 2^{n+2}$ divisible by 7? Prove that your answer is correct.

12. For what integers n is $9^n - 8n - 1$ divisible by 64? Prove that your answer is correct.

13. Find a simplifying expression for the product

$$\left(1 - \frac{1}{2}\right)\left(1 - \frac{1}{3}\right)\cdots\left(1 - \frac{1}{n}\right)$$

and verify its validity for all integers $n \geq 2$.

14. Find a simplifying expression for the product

$$\left(1 - \frac{1}{2^2}\right)\left(1 - \frac{1}{3^2}\right)\cdots\left(1 - \frac{1}{n^2}\right)$$

and verify its validity for all integers $n \geq 2$.

15. Prove that an N-sided convex polygon has $\frac{1}{2}N(N - 3)$ diagonals, $N > 3$.

16. Prove that the sum of the angles in an N-sided convex polygon is $(N - 2)180°$, $N > 2$.

17. Prove that all sets with n elements have 2^n subsets. Count the empty set $\varnothing$ and the whole set as subsets.

■ CHAPTER HIGHLIGHTS

1.1 What Is Calculus?

Calculus as elementary mathematics (algebra, geometry, trigonometry enhanced by the limit process).

> Sir Isaac Newton (p. 3)   Gottfried Wilhelm Leibniz (p. 3)

1.2 Notations and Formulas from Elementary Mathematics

sets
> notation (p. 4)

real numbers
> real line (p. 6)   order properties (p. 6)   absolute value (p. 7)
> intervals (p. 7)   boundedness (p. 8)

algebra
> factoring formulas (p. 8)
> polynomials and rational expressions (p. 8)

geometry
> elementary figures (pp. 10–12)

1.3 Inequalities

To solve an inequality is to find the set of numbers for which the inequality holds.

We can maintain an inequality by adding the same quantity to both sides, or by subtracting the same quantity from both sides, or by multiplying or dividing both sides by the same *positive* quantity. But if we multiply or divide both sides by a *negative* quantity, then the inequality is *reversed*.

Inequalities and Absolute Value

$$|x - c| < \delta \quad \text{iff} \quad c - \delta < x < c + \delta$$

$$0 < |x - c| < \delta \quad \text{iff} \quad c - \delta < x < c \quad \text{or} \quad c < x < c + \delta$$

$$|x| > \epsilon \quad \text{iff} \quad x > \epsilon \quad \text{or} \quad x < -\epsilon$$

$$|a + b| \leq |a| + |b|, \; \big||a| - |b|\big| \leq |a - b|$$

1.4 Coordinate Plane; Analytic Geometry

rectangular coordinates (p. 22)
distance and midpoint formulas (p. 22)
lines
> slope (p. 23)   intercepts (p. 23)   equations (p. 24)
> parallel and perpendicular lines (p. 24)   angle between two lines (p. 25)

distance between a point and a line (p. 28)

■ PROJECTS AND EXPLORATIONS USING TECHNOLOGY

To do these exercises you will need a graphics calculator or a computer with graphing capability. The majority of these problems are open-ended so different approaches may be used to solve them. You should be aware that different approaches can result in slight variations in the answers. Round your numerical answers to at least four decimal places. The rounding method that your calculator or computer uses also may cause variations in answers.

1.1 When economic analysts study such systems as the stock market or the seasonal sales of some product, they often try to smooth out short-term fluctuations in order to determine long-term trends. One method for doing this is to study a *moving average* that averages the current value with the value from the previous period. For example, if $f(x)$ denotes the sales of a certain product for the current period, then the function g, defined by

(1) $$g(x) = \frac{f(x-1) + f(x)}{2},$$

is such a moving average.

(a) Draw the graphs of the functions f and g for each of the following functions:

$$f(x) = x^4 - 3x^2 + 0.1x + 1 \qquad \text{and} \qquad f(x) = \frac{x^3 - 6}{x^4 + 3}$$

(b) For each function f in part (a), find and compare the domains and ranges of f and g.

(c) For each function f in part (a), find the points of intersection of the graphs of f and g. What characteristic do the points of intersection have?

A function f is *one-to-one* if $f(x_1) \neq f(x_2)$ whenever x_1, x_2 are in the domain of f and $x_1 \neq x_2$. Geometrically, a function f is one-to-one if it has the property that no horizontal line intersects its graph in more than one point.

(d) Based on the examples just given, would you conjecture that g will be one-to-one if f is one-to-one?

1.2 Sometimes it might be preferable to weight a moving average more heavily in favor of recent values. For example, the function

(2)
$$h(x) = \frac{f(x-1) + 2f(x)}{3}$$

gives twice as much weight to the current value versus the value for the preceding period.

(a) Let g and h be the moving averages defined by (1) and (2). Draw the graphs of f, g, and h for the two functions given in Exercise 1.

(b) For each of the functions in Exercise 1, compare the intersections of h and f to those of g and f.

(c) Answer questions (b) and (d) in Exercise 1 for h and the two functions given in Exercise 1.

(d) Suppose that the sales of a new product are modeled by the function

$$f(x) = 2x + 0.1x \sin(\pi x),$$

where $f(x)$ is measured in thousands of units and x is in months since the introduction of the product. Draw the graphs of f, g, and h.

(e) For a given problem, it is not always obvious which moving average is better. Explain why (1) might be better if the time period is ''one day,'' while (2) might be better if the time period is ''one year.''

1.3 Another method for analyzing functions is to study the *change function*

$$c(x) = f(x) - f(x-1).$$

(a) Graph the functions $f(x)$ and $c(x)$ for each of the following functions:

$$f(x) = \sqrt{1+x} \qquad \text{and} \qquad f(x) = |x^2 - 7x + 2|.$$

(b) What is the meaning in terms of f if, for all x, $c(x) > 0$? $c(x) < 0$? $c(x) = 0$?

(c) Explain why the change function c may be one-to-one even when f is not one-to-one.

(d) Give an example of a function f such that f is even and its change function c is odd.

1.4 Suppose that some oil has been spilled in water and has formed a circular oil slick. Suppose, in addition, that one minute after the spill, the radius of the slick is 2 meters and 3 minutes after the spill, the radius is 6 meters.

(a) Express the radius r of the spill as a function of time t if it is observed that the radius is increasing at a constant rate. Was the radius 0 at time $t = 0$?

(b) Using composition of functions, express the area A and circumference C of the spill as functions of time.

(c) Apply the change function defined in Exercise 3 to each of the functions r, A, and C. What do they tell us about the spill?

(d) Suppose that the problem is modified to consider a gas that has been released in the atmosphere and has a spherical shape. Use the given data to answer parts (a)–(c) for the radius r, the volume V, and the surface area S of the sphere.

LIMITS AND CONTINUITY

CHAPTER

2

◼ 2.1 THE IDEA OF LIMIT

Introduction

We could begin by saying that limits are important in calculus, but that would be a major understatement. *Without limits calculus simply does not exist. Every single notion of calculus is a limit in one sense or another.*

What is instantaneous velocity? It is the limit of average velocities.
What is the slope of a curve? It is the limit of slopes of secant lines. See Figure 2.1.1.
What is the length of a curve? It is the limit of the lengths of polygonal paths.
What is the sum of an infinite series? It is the limit of finite sums.
What is the area of a region bounded by a curve? It is the limit of areas of regions bounded by line segments. See Figure 2.1.2

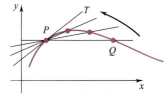

Figure 2.1.1

The Idea of Limit

We begin with a number L and a function f defined *near the number c but not necessarily defined at c itself.* A rough translation of

$$\lim_{x \to c} f(x) = L \qquad \text{(the limit of } f(x) \text{ as } x \text{ tends to } c \text{ is } L\text{)}$$

might read

as x approaches c, f(x) approaches L

or, equivalently,

when x is close to c but different from c, f(x) is close to L.

region R

1 approximating
rectangle

3 approximating
rectangles

8 approximating
rectangles

14 approximating
rectangles

Figure 2.1.2

Figure 2.1.3

Figure 2.1.4

We illustrate the idea in Figure 2.1.3. The curve represents the graph of a function f. The number c appears on the x-axis, the limit L on the y-axis. As x approaches c along the x-axis, $f(x)$ approaches L along the y-axis.

In taking the limit as x approaches c, it does not matter whether f is defined at c and if so, how it is defined there. The only thing that matters is the values that f takes on when x is *near* c. For example, in Figure 2.1.4 the graph of f is a broken curve defined peculiarly at c, and yet

$$\lim_{x \to c} f(x) = L$$

because, as suggested in Figure 2.1.5,

as x approaches c, $f(x)$ approaches L.

Numbers x near c fall into two natural categories: those that lie to the left of c and those that lie to the right of c. We write

$$\lim_{x \to c^-} f(x) = L \quad \text{(the left-hand limit of } f(x) \text{ as } x \text{ tends to } c \text{ is } L)$$

to indicate that

as x approaches c from the left, $f(x)$ approaches L.

Similarly,

$$\lim_{x \to c^+} f(x) = L \quad \text{(the right-hand limit of } f(x) \text{ as } x \text{ tends to } c \text{ is } L)$$

indicates that

as x approaches c from the right, $f(x)$ approaches L.†

___

† The left-hand limit is sometimes written $\lim_{x \uparrow c} f(x)$ and the right-hand limit, $\lim_{x \downarrow c} f(x)$.

For an example, see the graph of the function f in Figure 2.1.6. As x approaches 5 from the left, $f(x)$ approaches 2. Thus,

$$\lim_{x \to 5^-} f(x) = 2.$$

As x approaches 5 from the right, $f(x)$ approaches 4, so

$$\lim_{x \to 5^+} f(x) = 4.$$

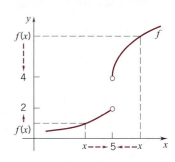

Figure 2.1.6

Note that $\lim_{x \to 5} f(x)$ does not exist since it is not true that there is a single number L with the property that $f(x)$ is close to L *whenever* x is close to 5 ($f(x)$ is close to 2 when x is close to 5 and is less than 5; $f(x)$ is close to 4 when x is close to 5 and is greater than 5).

The left- and right-hand limits are called *one-sided limits.* The relationship between the *limit of f as x approaches c* and the two *one-sided limits of f as x approaches c* should be intuitively clear, namely

$$\lim_{x \to c} f(x) = L$$

if and only if both

$$\lim_{x \to c^-} f(x) = L \qquad and \qquad \lim_{x \to c^+} f(x) = L.$$

Example 1 For the function f graphed in Figure 2.1.7

$$\lim_{x \to (-2)^-} f(x) = 5 \qquad and \qquad \lim_{x \to (-2)^+} f(x) = 5$$

and therefore

$$\lim_{x \to -2} f(x) = 5.$$

For the limit, it does not matter that $f(-2) = 3$.

Examining the graph of f near $x = 4$, we find that

$$\lim_{x \to 4^-} f(x) = 7 \qquad whereas \qquad \lim_{x \to 4^+} f(x) = 2.$$

Since these one-sided limits are different,

$$\lim_{x \to 4} f(x) \quad \text{does not exist.} \quad \square$$

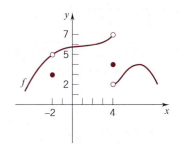

Figure 2.1.7

Example 2 Consider the function f graphed in Figure 2.1.8. As x approaches 3 from either side, $f(x)$ becomes arbitrarily large.† Becoming arbitrarily large, $f(x)$ does not approach any fixed number L. Therefore

$$\lim_{x \to 3} f(x) \quad \text{does not exist.}$$

Now let's focus on what happens near 7. As x approaches 7 from the left, $f(x)$ becomes arbitrarily large negative (i.e., less than any preassigned number).†† Under these circumstances $f(x)$ can't possibly approach a fixed number. Thus

$$\lim_{x \to 7^-} f(x) \quad \text{does not exist.}$$

Figure 2.1.8

---

† This particular behavior is symbolized by writing $f(x) \to \infty$ as $x \to 3$; the line $x = 3$ is a *vertical asymptote* of the function. Vertical and horizontal asymptotes are treated in Section 4.7.

†† This behavior is symbolized by writing $f(x) \to -\infty$ as $x \to 7^-$; the line $x = 7$ is a vertical asymptote of f.

Since the left-hand limit at $x = 7$ does not exist,

$$\lim_{x \to 7} f(x) \quad \text{does not exist.}$$

(Note that the right-hand limit of f at $x = 7$ does exist; $\lim_{x \to 7^+} f(x) = 2$.) ❏

Example 3 Let f be the function defined by $f(x) = \dfrac{|x|}{x}$ for $x \neq 0$. This function can also be written equivalently as

$$f(x) = \begin{cases} 1, & \text{if } x > 0 \\ -1, & \text{if } x < 0. \end{cases} \qquad \text{(Figure 2.1.9)}$$

We consider $\lim_{x \to c} f(x)$ for several values of c.

For $c = 0$,

$$\lim_{x \to 0^-} f(x) = \lim_{x \to 0^-} (-1) = -1$$

$f(x) = \frac{|x|}{x}$

Figure 2.1.9

and

$$\lim_{x \to 0^+} f(x) = \lim_{x \to 0^+} (1) = 1.$$

Thus,

$$\lim_{x \to 0} \frac{|x|}{x} \quad \text{does not exist.}$$

If $c = c_1$ is any positive number, then $f(x)$ "is close to 1" (in fact, $f(x) = 1$) when x is close to c_1. Therefore, when $c_1 > 0$

$$\lim_{x \to c_1} \frac{|x|}{x} = 1.$$

A similar analysis shows that when $c = c_2$ is a negative number, then

$$\lim_{x \to c_2} \frac{|x|}{x} = -1. \quad ❏$$

Our next example is somewhat exotic. It is an adaptation of an example known as the Dirichlet function (P. G. Lejeune Dirichlet, 1805–1859).

Example 4 Let f be the function defined as follows:

$$f(x) = \begin{cases} 7, & x \text{ rational} \\ 4, & x \text{ irrational.} \end{cases} \qquad \text{(Figure 2.1.10)}$$

Figure 2.1.10

Now let c be any real number. As x approaches c, x passes through both rational and irrational numbers. As this happens, $f(x)$ jumps wildly back and forth between 7 and 4, and thus cannot stay close to any fixed number L. Therefore, $\lim_{x \to c} f(x)$ does not exist. ❏

We go on with examples. Where a graph has not been provided, you may find it useful to sketch one.

Example 5

$$\lim_{x \to 3} (2x + 5) = 11.$$

As x approaches 3, $2x$ approaches 6, and $2x + 5$ approaches 11. ❏

Example 6

$$\lim_{x \to 2} (2x^2 - 3x + 5) = 7.$$

When x is close to 2, $2x^2$ is close to 8, $3x$ is close to 6, and $2x^2 - 3x + 5$ is close to $8 - 6 + 5 = 7$. ❏

Example 7

$$\lim_{x \to 3} \frac{1}{x - 1} = \frac{1}{2}.$$

As x approaches 3, $x - 1$ approaches 2, and $1/(x - 1)$ approaches $\frac{1}{2}$. ❏

Example 8

$$\lim_{x \to 2} \frac{1}{x - 2} \quad \text{does not exist.}$$

Here the function $f(x) = 1/(x - 2)$ is not defined at $x = 2$. However, if x is close to 2 and is greater than 2, then $1/(x - 2)$ is a large positive number ($f(x) \to \infty$ as $x \to 2^+$); if x is close to 2 and is less than 2, then $1/(x - 2)$ is a large negative number ($f(x) \to -\infty$ as $x \to 2^-$). See the table of values and Figure 2.1.11.

x	1.5	1.9	1.99	1.999	2	2.001	2.01	2.1	2.5
$f(x)$	-2	-10	-100	-1000		1000	100	10	2

❏

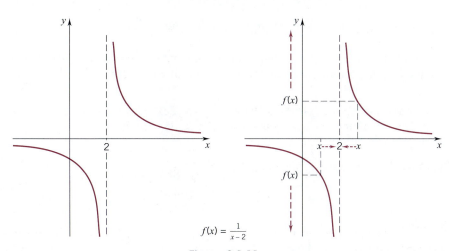

Figure 2.1.11

Before going on to the next examples, remember that, in taking the limit as x approaches a given number c, it does not matter whether the function is defined at the number c and if so, how it is defined there. The only thing that matters is how the function is defined *near* the number c.

Example 9

$$\lim_{x \to 4} \frac{x^2 - 2x - 8}{x - 4} = 6.$$

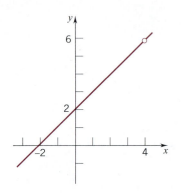

Figure 2.1.12

At $x = 4$, the function is undefined: both numerator and denominator are 0 and 0/0 makes no sense. But that doesn't matter. For all $x \neq 4$, and therefore *for all x near* 4,

$$\frac{x^2 - 2x - 8}{x - 4} = \frac{(x + 2)(x - 4)}{x - 4} = x + 2.$$

Therefore,

$$\lim_{x \to 4} \frac{x^2 - 2x - 8}{x - 4} = \lim_{x \to 4} (x + 2) = 6.$$

The graph of $f(x) = (x^2 - 2x - 8)/(x - 4)$ is shown in Figure 2.1.12. ❏

Example 10

$$\lim_{x \to -3} \frac{2x + 6}{x + 3} = 2.$$

At $x = -3$, the function is undefined: both numerator and denominator are 0. But, as we said before, that doesn't matter. For all $x \neq -3$, and therefore for all x near -3,

$$\frac{2x + 6}{x + 3} = \frac{2(x + 3)}{x + 3} = 2 \qquad \text{so that} \qquad \lim_{x \to -3} \frac{2x + 6}{x + 3} = \lim_{x \to -3} 2 = 2.$$

Sketch the graph of $f(x) = \dfrac{2x + 6}{x + 3}$. ❏

Example 11

$$\lim_{x \to 2} \frac{x - 2}{x^2 - 4x + 4} \qquad \text{does not exist.}$$

This result does not follow from the fact that the function is not defined at 2, but rather from the fact that for all $x \neq 2$, and therefore for all x near 2,

$$\frac{x - 2}{x^2 - 4x + 4} = \frac{x - 2}{(x - 2)^2} = \frac{1}{x - 2},$$

and, as you saw before,

$$\lim_{x \to 2} \frac{1}{x - 2} \qquad \text{does not exist.} \quad ❏$$

Example 12

$$\text{If } f(x) = \begin{cases} 3x - 1, & x \neq 2 \\ 45, & x = 2, \end{cases} \qquad \text{then} \quad \lim_{x \to 2} f(x) = 5.$$

It doesn't matter to us that $f(2) = 45$. For $x \neq 2$, and thus for all x near 2,

$$f(x) = 3x - 1 \qquad \text{so that} \qquad \lim_{x \to 2} f(x) = \lim_{x \to 2} (3x - 1) = 5. \quad ❏$$

Example 13 Let

$$f(x) = \begin{cases} 1 - x^2, & x < 1 \\ 2x, & x > 1. \end{cases}$$

Then

$$\lim_{x \to 1} f(x) \quad \text{does not exist}$$

and

$$\lim_{x \to 1.05} f(x) = 2.10.$$

The first limit does not exist since the one-sided limits are different:

$$\lim_{x \to 1^-} f(x) = \lim_{x \to 1^-} (1 - x^2) = 0 \quad \text{and} \quad \lim_{x \to 1^+} f(x) = \lim_{x \to 1^+} (2x) = 2.$$

The second limit does exist because for values of x sufficiently near 1.05, the values of the function are computed using the rule $2x$ whether x is to the left or the right of 1.05, and

$$\lim_{x \to 1.05} 2x = 2.10. \quad \square$$

The preceding examples illustrate two intuitive methods for evaluating the limit of a function f at a number c: (1) using the graph of f (see Examples 1–4), and (2) "analyzing" the behavior of f for values of x close to c (Examples 5–13). We conclude this section with a third method which might be termed *numerical* or *experimental*. This method can be used when the intuitive approaches are difficult (or perhaps impossible) to apply. Typical numerical problems will require a calculator or a computer; calculators/computers with graphing capabilities can be used to support numerical results.

Example 14 Let $f(x) = (\sin x)/x$. If we try to evaluate f at 0, we get the meaningless ratio $0/0$. However, f is defined for all $x \neq 0$, and so we can consider

$$\lim_{x \to 0} \frac{\sin x}{x}.$$

We select a sequence of numbers which approaches 0 from the right, and a sequence of numbers which approaches 0 from the left. Using a calculator, we evaluate f at these numbers. The results are tabulated in Table 2.1.1.

■ **Table 2.1.1**

(Left side)		(Right side)	
x (radians)	$\dfrac{\sin x}{x}$	x (radians)	$\dfrac{\sin x}{x}$
-1	0.84147	1	0.84147
-0.5	0.95885	0.5	0.95885
-0.1	0.99833	0.1	0.99833
-0.01	0.99998	0.01	0.99998
-0.001	0.99999	0.001	0.99999

Based on these calculations, it appears that

$$\lim_{x \to 0^+} \frac{\sin x}{x} = 1 \qquad \text{and} \qquad \lim_{x \to 0^-} \frac{\sin x}{x} = 1.$$

Thus, we are led to conclude

$$\lim_{x \to 0} \frac{\sin x}{x} = 1.$$

The graphs shown in Figure 2.1.13 support this conclusion. First, we graphed the function with the x-interval $-10 \le x \le 10$ (part a). Then we zoomed in to obtain the graphs in 2.1.13(b) and 2.1.13(c). The fact that this limit is 1 is established rigorously in Section 2.5. ❏

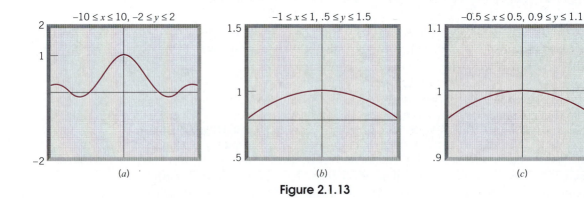

Figure 2.1.13

Example 15 In the function $f(x) = 2^x$, the independent variable x appears as an exponent. This is an example of an *exponential function*. At this stage, an expression like 2^x is defined only when x is a rational number; *logarithms* are needed to define irrational powers.† However, we can use a limit to try to estimate a number like $2^{\sqrt{2}}$. We take a sequence of rational numbers $x_1, x_2, x_3, \ldots$ which approaches $\sqrt{2}$ and evaluate $2^{x_1}, 2^{x_2}, 2^{x_3}, \ldots$ Then $2^{\sqrt{2}}$ is defined to be the "limit," if it exists. Recall that

$$\sqrt{2} = 1.414213562 \ldots .$$

Using the values $x = 1.4, 1.41, 1.414, \ldots$, we get the results shown in Table 2.1.2.

■ **Table 2.1.2**

x	1	1.4	1.41	1.414	1.4142	1.41421	1.414213
2^x	2	2.639016	2.657372	2.664750	2.665119	2.665138	2.665143

It appears, therefore, that $2^{\sqrt{2}} = \lim\limits_{x \to (\sqrt{2})^-} 2^x \cong 2.6651$ (to four decimal places). The value of $2^{\sqrt{2}}$ given by a hand calculator is 2.665144143. The graph of $f(x) = 2^x$,

---

† Logarithmic and exponential functions are defined and studied in Chapter 7.

$0 \le x \le 2$, shown in Figure 2.14(a), was obtained on a graphics calculator. The estimates found in Table 2.1.2 can be confirmed by repeatedly zooming in on $\sqrt{2}$. See Figure 2.14(b). ❑

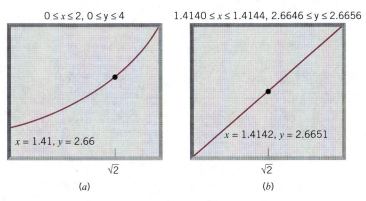

$0 \le x \le 2, 0 \le y \le 4$ $1.4140 \le x \le 1.4144, 2.6646 \le y \le 2.6656$

$x = 1.41, y = 2.66$ $x = 1.4142, y = 2.6651$

$\sqrt{2}$ $\sqrt{2}$

(a)   (b)

Figure 2.1.14

If you have found all this very imprecise, you are absolutely right. It is imprecise, but it need not remain so. One of the great triumphs of calculus has been its capacity to formulate limit statements with precision, but for this you have to wait until Section 2.2.

EXERCISES 2.1

In Exercises 1–12 you are given a number c and the graph of a function f. Use the graph of f to find

(a) $\lim\limits_{x \to c^-} f(x)$.   (b) $\lim\limits_{x \to c^+} f(x)$.   (c) $\lim\limits_{x \to c} f(x)$.   (d) $f(c)$

1. $c = 2$.

2. $c = 3$.

3. $c = 3$.

4. $c = 4$.

5. $c = -2$.

6. $c = 1$.

7. $c = 1$.

8. $c = -1$.

9. $c = 3.$

10. $c = 3.$

11. $c = 2.$

12. $c = -2.$

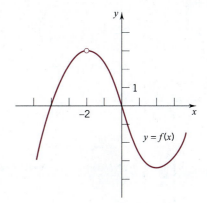

In Exercises 13 and 14, state the values of c for which $\lim_{x \to c} f(x)$ does not exist.

13.

14.

In Exercises 15–50, decide on intuitive grounds whether or not the indicated limit exists, and evaluate the limit if it does exist.

15. $\lim\limits_{x \to 0} (2x - 1)$.

16. $\lim\limits_{x \to 1} (2 - 5x)$.

17. $\lim\limits_{x \to -2} x^2$.

18. $\lim\limits_{x \to 4} \sqrt{x}$.

19. $\lim\limits_{x \to -3} (|x| - 2)$.

20. $\lim\limits_{x \to 0} \dfrac{1}{|x|}$.

21. $\lim\limits_{x \to 1} \dfrac{3}{x + 1}$.

22. $\lim\limits_{x \to -1} \dfrac{4}{x + 1}$.

23. $\lim\limits_{x \to -1} \dfrac{2}{x + 1}$.

24. $\lim\limits_{x \to 2} \dfrac{1}{3x - 6}$.

25. $\lim\limits_{x \to 3} \dfrac{2x - 6}{x - 3}$.

26. $\lim\limits_{x \to 3} \dfrac{x^2 - 6x + 9}{x - 3}$.

27. $\lim\limits_{x \to 3} \dfrac{x - 3}{x^2 - 6x + 9}$.

28. $\lim\limits_{x \to 2} \dfrac{x^2 - 3x + 2}{x - 2}$.

29. $\lim\limits_{x \to 2} \dfrac{x - 2}{x^2 - 3x + 2}$.

30. $\lim\limits_{x \to 1} \dfrac{x - 2}{x^2 - 3x + 2}$.

31. $\lim\limits_{x \to 0} \left(x + \dfrac{1}{x} \right)$.

32. $\lim\limits_{x \to 1} \left(x + \dfrac{1}{x} \right)$.

33. $\lim\limits_{x \to 0} \dfrac{2x - 5x^2}{x}$.

34. $\lim\limits_{x \to 3} \dfrac{x - 3}{6 - 2x}$.

35. $\lim\limits_{x \to 1} \dfrac{x^2 - 1}{x - 1}$.

36. $\lim\limits_{x \to 1} \dfrac{x^3 - 1}{x - 1}$.

37. $\lim\limits_{x \to 1} \dfrac{x^3 - 1}{x + 1}$.

38. $\lim\limits_{x \to 1} \dfrac{x^2 + 1}{x^2 - 1}$.

39. $\lim\limits_{x \to 0} f(x)$; $f(x) = \begin{cases} 1, & x \neq 0 \\ 3, & x = 0. \end{cases}$

40. $\lim\limits_{x \to 1} f(x)$; $f(x) = \begin{cases} 3x, & x < 1 \\ 3, & x > 1. \end{cases}$

41. $\lim\limits_{x \to 4} f(x)$; $f(x) = \begin{cases} x^2, & x \neq 4 \\ 0, & x = 4. \end{cases}$

42. $\lim\limits_{x \to 0} f(x)$; $f(x) = \begin{cases} -x^2, & x < 0 \\ x^2, & x > 0. \end{cases}$

43. $\lim\limits_{x \to 0} f(x)$; $f(x) = \begin{cases} x^2, & x < 0 \\ 1 + x, & x > 0. \end{cases}$

44. $\lim\limits_{x \to 1} f(x)$; $f(x) = \begin{cases} 2x, & x < 1 \\ x^2 + 1, & x > 1. \end{cases}$

45. $\lim\limits_{x \to 0} f(x)$; $f(x) = \begin{cases} 2, & x \text{ rational} \\ -2, & x \text{ irrational}. \end{cases}$

46. $\lim\limits_{x \to 1} f(x)$; $f(x) = \begin{cases} 2x, & x \text{ rational} \\ 2, & x \text{ irrational}. \end{cases}$

47. $\lim\limits_{x \to 2} f(x)$; $f(x) = \begin{cases} 3x, & x < 1 \\ x + 2, & x \geq 1. \end{cases}$

48. $\lim\limits_{x \to 0} f(x)$; $f(x) = \begin{cases} 2x, & x \leq 1 \\ x + 1, & x > 1. \end{cases}$

49. $\lim\limits_{x \to 1} \dfrac{\sqrt{x^2 + 1} - \sqrt{2}}{x - 1}$.

50. $\lim\limits_{x \to 5} \dfrac{\sqrt{x^2 + 5} - \sqrt{30}}{x - 5}$.

Consider function f defined on an interval I and a point $a \in I$. Let $h \neq 0$ be a number such that $a + h \in I$. The line determined by the points $(a, f(a))$ and $(a + h, f(a + h))$ is called a *secant line* (see the figure).

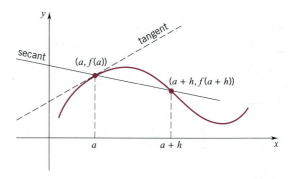

The slope of the secant line is given by

$$m_{\sec} = \frac{f(a + h) - f(a)}{h}.$$

If we let h approach 0, then the secant line will approach the dashed line as a limiting position. The *tangent line* to the graph of f at the point $(a, f(a))$ is defined to be the line through $(a, f(a))$ with slope

$$\lim_{h \to 0} \frac{f(a + h) - f(a)}{h} \qquad \text{(provided this limit exists)}.$$

In Exercises 51–57, calculate $(f(a + h) - f(a))/h$, determine whether $\lim\limits_{h \to 0} (f(a + h) - f(a))/h$ exists, and give an equation for the tangent line to the graph of f at $(a, f(a))$ when the limit does exist.

51. $f(x) = x^2$, $a = 2$.

52. $f(x) = x^3 + 1$, $a = 1$.

53. $f(x) = 1 - 2x + x^2$, $a = -1$.

54. $f(x) = 1/x$, $a = 2$.

55. $f(x) = \sqrt{x}$, $a = 1$.
    HINT: Rationalize the numerator.

56. $f(x) = |x|$, $a = 0$.
    HINT: Consider the cases $h > 0$ and $h < 0$ separately.

57. $f(x) = \sqrt[3]{x}$, $a = 0$.

Exercises 58 and 59 are concerned with the function $f(x) = \sin (1/x)$ and its limit as x approaches 0.

58. (a) Show that $f(1/(n\pi)) = 0$ for every nonzero integer n.
    (b) Show that $f(1/(2n\pi + \pi/2)) = 1$ for every integer n.

(c) Show that $f(1/(2n\pi + 3\pi/2)) = -1$ for every integer n.

59. (a) Plot the points $(x, f(x))$ on the graph of f with $x = \dfrac{1}{\pi}$,

$$\frac{1}{2\pi}, \frac{1}{3\pi}, \frac{1}{4\pi}, \frac{1}{(\pi/2)}, \frac{1}{(5\pi/2)}, \frac{1}{(9\pi/2)}, \frac{1}{(3\pi/2)},$$
$$\frac{1}{(7\pi/2)}, \frac{1}{(11\pi/2)}.$$

(b) Does $\lim\limits_{x \to 0^+} f(x)$ exist? What would you conjecture about $\lim\limits_{x \to 0^-} f(x)$?

(c) Sketch the graph of f for $-\pi \le x \le \pi$.

▶ In Exercises 60–65, estimate $\lim\limits_{x \to a} f(x)$ by creating a table of values. Then use a graphing utility to zoom in on the graph of f near $x = a$ to justify or improve your guess.

60. Estimate

$$\lim_{x \to 0} \frac{\cos x - 1}{x} \qquad \text{(radian measure)}$$

after evaluating the quotient at $x = \pm 1, \pm 0.1, \pm 0.01, \pm 0.001$.

61. Estimate

$$\lim_{x \to 0} \frac{\tan 2x}{x} \qquad \text{(radian measure)}$$

after evaluating the quotient at $x = \pm 1, \pm 0.1, \pm 0.01, \pm 0.001$.

62. Estimate

$$\lim_{x \to 0} \frac{x - \sin x}{x^3} \qquad \text{(radian measure)}$$

after evaluating the quotient at $x = \pm 1, \pm 0.1, \pm 0.01, \pm 0.001, \pm 0.0001$.

63. Estimate

$$\lim_{x \to 1} \frac{x^{3/2} - 1}{x - 1}$$

after evaluating the quotient at $x = 0.9, 0.99, 0.999, 0.9999$ and $x = 1.1, 1.01, 1.001, 1.0001$.

64. Estimate

$$\lim_{x \to 0} \frac{2^x - 1}{x}.$$

after evaluating the quotient at $x = \pm 1, \pm 0.1, \pm 0.01, \pm 0.001, \pm 0.0001$.

65. Estimate

$$\lim_{x \to 0^+} (1 + x)^{1/x}$$

by evaluating $(1 + x)^{1/x}$ at $x = 1, 0.1, 0.001, 0.0001, 0.00001, 0.000001, 0.0000001$.

▶ 66. Estimate the value of 3^π by taking a sequence $x_1, x_2, x_3, \dots,$ of rational numbers which approach π and evaluating $3^{x_1}, 3^{x_2}, 3^{x_3}, \dots$. Recall $\pi \cong 3.14159265$.

■ 2.2 DEFINITION OF LIMIT

Our work with limits in Section 2.1 was informal. It is time to be more precise.

Let f be a function and let c be a real number. We do not require that f be defined at c, but we do require that f be defined at least on a set of the form $(c - p, c) \cup (c, c + p)$ with $p > 0$. (This guarantees that we can form $f(x)$ for all $x \neq c$ that are "sufficiently close" to c.)

To say that

$$\lim_{x \to c} f(x) = L$$

means we can make $f(x)$ be as close to L as we want simply by choosing x close enough to c, $x \neq c$. That is:

Informal Statement	**ϵ, δ Statement**
$\|f(x) - L\|$ can be made arbitrarily small	you pick $\epsilon > 0$; $\|f(x) - L\|$ can be made less than ϵ
simply by requiring that	simply by requiring that
$\|x - c\|$ be sufficiently small but different from zero.	$\|x - c\|$ satisfy an inequality of the form $0 < \|x - c\| < \delta$ for δ sufficiently small.

Putting the various pieces together in compact form we have the following precise definition.

DEFINITION 2.2.1 THE LIMIT OF A FUNCTION

Let f be a function defined at least on some set of the form

$$(c - p, c) \cup (c, c + p), \text{ with } p > 0.$$

Then

$$\lim_{x \to c} f(x) = L \quad \text{iff}$$

for each $\epsilon > 0$, there exists $\delta > 0$ such that

$$\text{if} \quad 0 < |x - c| < \delta, \quad \text{then} \quad |f(x) - L| < \epsilon.$$

Figures 2.2.1 and 2.2.2 illustrate this definition.

In general, the choice of δ depends upon the previous choice of ϵ. We do not require that there exists a number δ which "works" for *all* ϵ, but rather, that for *each* ϵ there exists a δ which "works" for it.

Figure 2.2.1

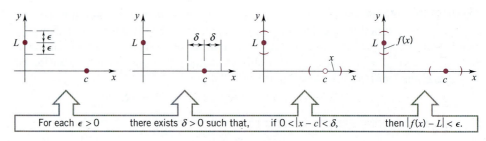

For each $\epsilon > 0$ there exists $\delta > 0$ such that,   if $0 < |x - c| < \delta$,   then $|f(x) - L| < \epsilon$.

Figure 2.2.2

In Figure 2.2.3, we give two choices of ϵ and for each we display a suitable δ. For a δ to be suitable, all the points within δ of c (with the possible exception of c itself) must be taken by the function f to within ϵ of L. In part (b) of the figure we began with a smaller ϵ and had to use a smaller δ.

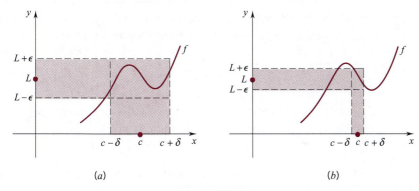

Figure 2.2.3

The δ of Figure 2.2.4 is too large for the given ϵ. In particular the points marked x_1 and x_2 in the figure are not taken by the function f to within ϵ of L.

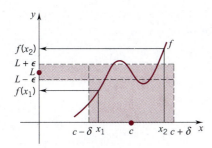

Figure 2.2.4

Next we apply the ϵ, δ definition of limit to a variety of functions. If you have never run across ϵ, δ arguments before, you may find them confusing at first. It usually takes a little while for the ϵ, δ idea to take hold.

Example 1 Show that

$$\lim_{x \to 2} (2x - 1) = 3. \qquad \text{(Figure 2.2.5)}$$

Finding a δ. Let $\epsilon > 0$. We seek a number $\delta > 0$ such that,

$$\text{if} \quad 0 < |x - 2| < \delta, \quad \text{then} \quad |(2x - 1) - 3| < \epsilon.$$

What we have to do first is establish a connection between

$$|(2x - 1) - 3| \qquad \text{and} \qquad |x - 2|.$$

The connection is simple:

$$|(2x - 1) - 3| = |2x - 4|$$

so that

$$(*) \qquad\qquad |(2x - 1) - 3| = 2|x - 2|.$$

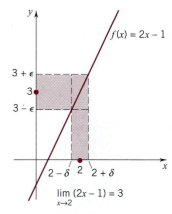

$$\lim_{x \to 2} (2x - 1) = 3$$

Figure 2.2.5

To make $|(2x - 1) - 3|$ less than ϵ, we need to make $2|x - 2| < \epsilon$, which can be accomplished by making $|x - 2| < \epsilon/2$. This suggests that we choose $\delta = \frac{1}{2}\epsilon$.

Showing that the δ "works." If $0 < |x - 2| < \frac{1}{2}\epsilon$, then $2|x - 2| < \epsilon$ and, by (*), $|(2x - 1) - 3| < \epsilon$. ❑

Remark In Example 1 we chose $\delta = \frac{1}{2}\epsilon$, but we could have chosen *any* positive number δ such that $\delta \leq \frac{1}{2}\epsilon$. In general, if a certain δ^* "works" for a given ϵ, then any $\delta < \delta^*$ will also work. ❑

Example 2 Show that

$$\lim_{x \to -1} (2 - 3x) = 5. \qquad \text{(Figure 2.2.6)}$$

Finding a δ. Let $\epsilon > 0$. We seek a number $\delta > 0$ such that

$$\text{if} \quad 0 < |x - (-1)| < \delta, \qquad \text{then} \quad |(2 - 3x) - 5| < \epsilon.$$

To find a connection between

$$|x - (-1)| \qquad \text{and} \qquad |(2 - 3x) - 5|,$$

we simplify both expressions:

$$|x - (-1)| = |x + 1|$$

and

$$|(2 - 3x) - 5| = |-3x - 3| = |-3||x + 1| = 3|x + 1|.$$

We can conclude that

$$(**) \qquad\qquad |(2 - 3x) - 5| = 3|x - (-1)|.$$

We can make the expression on the left less than ϵ by making $|x - (-1)|$ less than $\epsilon/3$. This suggests that we set $\delta = \frac{1}{3}\epsilon$.

Showing that the δ "works." If $0 < |x - (-1)| < \frac{1}{3}\epsilon$, then $3|x - (-1)| < \epsilon$ and, by (**), $|(2 - 3x) - 5| < \epsilon$. ❑

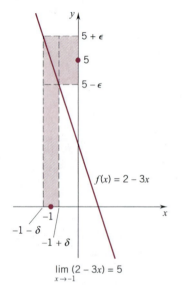

$$\lim_{x \to -1} (2 - 3x) = 5$$

Figure 2.2.6

Example 3 Let c be any number. Then

(2.2.2)

$$\lim_{x \to c} x = c. \qquad \text{(Figure 2.2.7)}$$

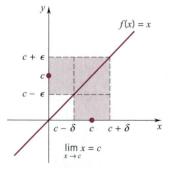

$$\lim_{x \to c} x = c$$

Figure 2.2.7

PROOF Let $\epsilon > 0$. We must find $\delta > 0$ such that

$$\text{if} \quad 0 < |x - c| < \delta, \qquad \text{then} \quad |x - c| < \epsilon.$$

Obviously we can choose $\delta = \epsilon$. ❑

Example 4 Let c be any number. Then

(2.2.3)

$$\lim_{x \to c} |x| = |c|.$$

(Figure 2.2.8)

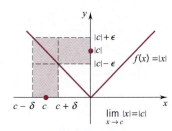

Figure 2.2.8

PROOF Let $\epsilon > 0$. We seek $\delta > 0$ such that

$$\text{if} \quad 0 < |x - c| < \delta, \qquad \text{then} \quad \big||x| - |c|\big| < \epsilon.$$

Since

$$\big||x| - |c|\big| \leq |x - c|,$$

(Recall: $\big||a| - |b|\big| \leq |a - b|$)

we can choose $\delta = \epsilon$; for

$$\text{if} \quad 0 < |x - c| < \epsilon, \qquad \text{then} \quad \big||x| - |c|\big| < \epsilon. \quad ❑$$

Example 5

(2.2.4)

$$\lim_{x \to c} a = a.$$

(Figure 2.2.9)

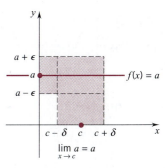

Figure 2.2.9

PROOF Here we are dealing with the constant function

$$f(x) = a.$$

Let $\epsilon > 0$. We must find $\delta > 0$ such that

$$\text{if} \quad 0 < |x - c| < \delta, \qquad \text{then} \quad |a - a| < \epsilon.$$

Since $|a - a| = 0$, we always have

$$|a - a| < \epsilon$$

no matter how δ is chosen; in short, any positive number will do for δ. ❑

Usually ϵ, δ arguments are carried out in two stages. First we do a little algebraic scratchwork, labeled "finding a δ" in some of the preceding examples. This scratchwork involves working backward from $|f(x) - L| < \epsilon$ to find a $\delta > 0$ sufficiently small so that we can begin with the inequality $0 < |x - c| < \delta$ to arrive at $|f(x) - L| < \epsilon$. This first stage is just preliminary, but it shows us how to proceed in the second stage. The second stage consists of showing that the δ "works" by verifying that for our choice of δ the implication

$$\text{if} \quad 0 < |x - c| < \delta, \qquad \text{then} \quad |f(x) - L| < \epsilon$$

is true. The next two examples are more complicated and therefore may give you a better feeling for this idea of working backward to find a δ.

Figure 2.2.10

Example 6

$$\lim_{x \to 4} \sqrt{x} = 2.$$ (Figure 2.2.10)

Finding a δ. Let $\epsilon > 0$. We seek $\delta > 0$ such that

$$\text{if} \quad 0 < |x - 4| < \delta, \quad \text{then} \quad |\sqrt{x} - 2| < \epsilon.$$

First we want a relation between $|x - 4|$ and $|\sqrt{x} - 2|$. To be able to form $\sqrt{x}$ at all we need $x \geq 0$. To ensure this we must have $\delta \leq 4$. (Why?)

Remembering that we must have $\delta \leq 4$, let's go on. If $x \geq 0$, then we can form $\sqrt{x}$ and write

$$x - 4 = (\sqrt{x})^2 - 2^2 = (\sqrt{x} + 2)(\sqrt{x} - 2).$$

Taking absolute values, we have

$$|x - 4| = |\sqrt{x} + 2||\sqrt{x} - 2|.$$

Since $|\sqrt{x} + 2| > 1$, we have

$$|\sqrt{x} - 2| < |x - 4|.$$

This last inequality suggests that we simply set $\delta = \epsilon$. But remember the previous requirement $\delta \leq 4$. We can meet all requirements by setting $\delta = $ minimum of 4 and ϵ.

Showing that the δ *"works."* Let $\epsilon > 0$. Choose $\delta = \min\{4, \epsilon\}$ and assume that

$$0 < |x - 4| < \delta.$$

Since $\delta \leq 4$, we have $x \geq 0$ and so $\sqrt{x}$ is defined. Now, we can write

$$|x - 4| = |\sqrt{x} + 2||\sqrt{x} - 2|.$$

Since $|\sqrt{x} + 2| > 1$, we can conclude that

$$|\sqrt{x} - 2| < |x - 4|.$$

Since $|x - 4| < \delta$ and $\delta \leq \epsilon$, it does follow that

$$|\sqrt{x} - 2| < \epsilon. \quad \square$$

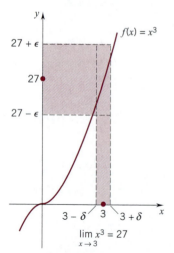

Figure 2.2.11

Example 7

$$\lim_{x \to 3} x^3 = 27.$$ (Figure 2.2.11)

Finding a δ. Let $\epsilon > 0$. We seek $\delta > 0$ such that

$$\text{if} \quad 0 < |x - 3| < \delta, \quad \text{then} \quad |x^3 - 27| < \epsilon.$$

The needed connection between $|x - 3|$ and $|x^3 - 27|$ is found by factoring:

$$x^3 - 27 = (x - 3)(x^2 + 3x + 9)$$

and thus

(∗) $$|x^3 - 27| = |x - 3||x^2 + 3x + 9|.$$

At this juncture we need to get a handle on the size of $|x^2 + 3x + 9|$ for x close to 3. For convenience let's take x within one unit of 3.

If $|x - 3| < 1$, then $2 < x < 4$ and

$$|x^2 + 3x + 9| \leq |x^2| + |3x| + |9|$$
$$= |x|^2 + 3|x| + 9$$
$$< 16 + 12 + 9 = 37.$$

It follows by (∗) that

(∗∗) if $|x - 3| < 1,$ then $|x^3 - 27| < 37|x - 3|.$

If, in addition, $|x - 3| < \epsilon/37$, then it will follow that

$$|x^3 - 27| < 37(\epsilon/37) = \epsilon.$$

This means that we can take $\delta = $ minimum of 1 and $\epsilon/37$.

Showing that the δ "works." Let $\epsilon > 0$. Choose $\delta = \min\{1, \epsilon/37\}$ and assume that

$$0 < |x - 3| < \delta.$$

Then

$$|x - 3| < 1 \quad\text{and}\quad |x - 3| < \epsilon/37.$$

By (∗∗)

$$|x^3 - 27| < 37|x - 3|,$$

and, since $|x - 3| < \epsilon/37$, we have

$$|x^3 - 27| < 37(\epsilon/37) = \epsilon. \quad \square$$

There are many different ways of formulating the same limit statement. Sometimes one formulation is more convenient, sometimes another. In any case, it is useful to recognize that the following are equivalent:

(2.2.5)

(i) $\lim\limits_{x \to c} f(x) = L$ (ii) $\lim\limits_{h \to 0} f(c + h) = L$

(iii) $\lim\limits_{x \to c} (f(x) - L) = 0$ (iv) $\lim\limits_{x \to c} |f(x) - L| = 0.$

The equivalence of these four statements is obvious. For example, the equivalence of (i) and (ii) can be seen in Figure 2.2.12: simply think of h as being the signed distance from c to x. Then $x = c + h$, and x approaches c iff h approaches 0. Also, it is a good exercise in ϵ, δ technique to prove that (i) is equivalent to (ii).

Example 8 For $f(x) = x^3$, we have

$$\lim\limits_{x \to 2} x^3 = 8, \qquad \lim\limits_{h \to 0} (2 + h)^3 = 8,$$

$$\lim\limits_{x \to 2} (x^3 - 8) = 0, \qquad \lim\limits_{x \to 2} |x^3 - 8| = 0. \quad \square$$

Figure 2.2.12

We come now to the ϵ, δ definitions of one-sided limits. These are just the usual ϵ, δ statement, except that for a left-hand limit, the δ has to "work" only for x to the left of c, and for a right-hand limit, the δ has to "work" only for x to the right of c.

DEFINITION 2.2.6 LEFT-HAND LIMIT

Let f be a function defined at least on an interval of the form $(c - p, c)$, with $p > 0$. Then

$$\lim_{x \to c^-} f(x) = L \quad \text{iff}$$

for each $\epsilon > 0$ there exists $\delta > 0$ such that

$$\text{if} \quad c - \delta < x < c, \quad \text{then} \quad |f(x) - L| < \epsilon.$$

DEFINITION 2.2.7 RIGHT-HAND LIMIT

Let f be a function defined at least on an interval of the form $(c + p)$, with $p > 0$. Then

$$\lim_{x \to c^+} f(x) = L \quad \text{iff}$$

for each $\epsilon > 0$ there exists $\delta > 0$ such that

$$\text{if} \quad c < x < c + \delta, \quad \text{then} \quad |f(x) - L| < \epsilon.$$

As indicated in Section 2.1, one-sided limits give us a simple way of determining whether or not a (two-sided) limit exists:

(2.2.8)

$$\lim_{x \to c} f(x) = L \quad \text{iff} \quad \lim_{x \to c^-} f(x) = L \quad \text{and} \quad \lim_{x \to c^+} f(x) = L.$$

The proof is left as an exercise. The proof is easy because any δ that "works" for the limit will work for both one-sided limits, and any δ that "works" for both one-sided limits will work for the limit.

You should also understand that one-sided limits arise naturally; that is, independent of (two-sided) limits. For example, suppose the domain of a function f is the interval (a, b). Then the only limit that we can consider at $x = a$ is the right-hand limit, $\lim_{x \to a^+} f(x)$. Similarly, at $x = b$, the only limit we can consider is the left-hand limit, $\lim_{x \to b^-} f(x)$. For the function $f(x) = \sqrt{x - 4}$, the only limit at $x = 4$ that makes sense is the right-hand limit ($\sqrt{x - 4}$ is not defined for $x < 4$), and

$$\lim_{x \to 4^+} \sqrt{x - 4} = 0.$$

Example 9 If f is the function defined by

$$f(x) = \begin{cases} 2x + 1, & x \le 0 \\ x^2 - x, & x > 0, \end{cases} \qquad \text{(Figure 2.2.13)}$$

then $\lim\limits_{x \to 0} f(x)$ does not exist.

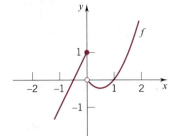

Figure 2.2.13

PROOF The left- and right-hand limits at 0 are

$$\lim_{x \to 0^-} f(x) = \lim_{x \to 0^-} (2x + 1) = 1 \qquad \text{and} \qquad \lim_{x \to 0^+} f(x) = \lim_{x \to 0^+} (x^2 - x) = 0.$$

Since these one-sided limits are different, $\lim\limits_{x \to 0} f(x)$ does not exist. ❑

Example 10 If g is the function defined by

$$g(x) = \begin{cases} 1 + x^3, & x < 1 \\ 3, & x = 1 \\ 4 - 2x, & x > 1, \end{cases} \qquad \text{(Figure 2.2.14)}$$

then $\lim\limits_{x \to 1} g(x) = 2$.

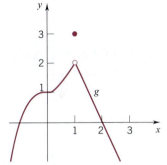

Figure 2.2.14

PROOF The left- and right-hand limits at 1 are

$$\lim_{x \to 1^-} g(x) = \lim_{x \to 1^-} (1 + x^2) = 2 \qquad \text{and} \qquad \lim_{x \to 1^+} g(x) = \lim_{x \to 1^+} (4 - 2x) = 2.$$

Thus, $\lim\limits_{x \to 1} g(x) = 2$.   NOTE: it does not matter that $g(1) \ne 2$. ❑

EXERCISES 2.2

In Exercises 1–22, decide in the manner of Section 2.1 whether or not the indicated limit exists. Evaluate the limits that do exist.

1. $\lim\limits_{x \to 1} \dfrac{x}{x + 1}$.

2. $\lim\limits_{x \to 0} \dfrac{x^2(1 + x)}{2x}$.

3. $\lim\limits_{x \to 0} \dfrac{x(1 + x)}{2x^2}$.

4. $\lim\limits_{x \to 4} \dfrac{x}{\sqrt{x + 1}}$.

5. $\lim\limits_{x \to 4} \dfrac{x}{\sqrt{x + 1}}$.

6. $\lim\limits_{x \to -1} \dfrac{1 - x}{x + 1}$.

7. $\lim\limits_{x \to 1} \dfrac{x^4 - 1}{x - 1}$.

8. $\lim\limits_{x \to 2} \dfrac{x}{|x|}$.

9. $\lim\limits_{x \to 0} \dfrac{x}{|x|}$.

10. $\lim\limits_{x \to 1} \dfrac{x^2 - 1}{x^2 - 2x + 1}$.

11. $\lim\limits_{x \to -2} \dfrac{|x|}{x}$.

12. $\lim\limits_{x \to 9} \dfrac{x - 3}{\sqrt{x} - 3}$.

13. $\lim\limits_{x \to 3^+} \dfrac{x + 3}{x^2 - 7x + 12}$.

14. $\lim\limits_{x \to 0^-} \dfrac{x}{|x|}$.

15. $\lim\limits_{x \to 1^+} \dfrac{\sqrt{x} - 1}{x}$.

16. $\lim\limits_{x \to 3^-} \sqrt{9 - x^2}$.

17. $\lim\limits_{x \to 2^+} f(x)$ if $f(x) = \begin{cases} 2x - 1, & x \le 2 \\ x^2 - x, & x > 2. \end{cases}$

18. $\lim\limits_{x \to -1^-} f(x)$ if $f(x) = \begin{cases} 1, & x \le -1 \\ x + 2, & x > -1. \end{cases}$

19. $\lim\limits_{x \to 2} f(x)$ if $f(x) = \begin{cases} 3, & x \text{ an integer} \\ 1, & \text{otherwise.} \end{cases}$

20. $\lim\limits_{x \to 3} f(x)$ if $f(x) = \begin{cases} x^2, & x < 3 \\ 7, & x = 3 \\ 2x + 3, & x > 3. \end{cases}$

21. $\lim\limits_{x \to \sqrt{2}} f(x)$ if $f(x) = \begin{cases} 3, & x \text{ an integer} \\ 1, & \text{otherwise.} \end{cases}$

22. $\lim\limits_{x \to 2} f(x)$ if $f(x) = \begin{cases} x^2, & x \le 1 \\ 5x, & x > 1. \end{cases}$

23. Which of the δ's displayed in the figure "works" for the given ϵ?

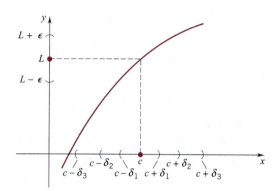

24. For which of the ϵ's given in the figure does the specified δ work?

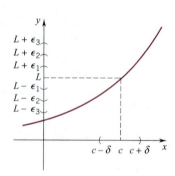

For the limits in Exercises 25–28, find the largest δ that "works" for the given ϵ.

25. $\lim\limits_{x \to 1} 2x = 2$; $\epsilon = 0.1$.

26. $\lim\limits_{x \to 4} 5x = 20$; $\epsilon = 0.5$.

27. $\lim\limits_{x \to 2} \frac{1}{2}x = 1$; $\epsilon = 0.01$.

28. $\lim\limits_{x \to 2} \frac{1}{5}x = \frac{2}{5}$; $\epsilon = 0.1$.

In Exercises 29–34, give an ϵ, δ proof for the following limits.

29. $\lim\limits_{x \to 4} (2x - 5) = 3$.

30. $\lim\limits_{x \to 2} (3x - 1) = 5$.

31. $\lim\limits_{x \to 3} (6x - 7) = 11$.

32. $\lim\limits_{x \to 0} (2 - 5x) = 2$.

33. $\lim\limits_{x \to 2} |1 - 3x| = 5$.

34. $\lim\limits_{x \to 2} |x - 2| = 0$.

35. Let f be some function of which you know only that

if $\quad 0 < |x - 3| < 1, \quad$ then $\quad |f(x) - 5| < 0.1$.

Which of the following statements are necessarily true?

(a) If $|x - 3| < 1$, then $|f(x) - 5| < 0.1$.
(b) If $|x - 2.5| < 0.3$, then $|f(x) - 5| < 0.1$.
(c) $\lim\limits_{x \to 3} f(x) = 5$.
(d) If $0 < |x - 3| < 2$, then $|f(x) - 5| < 0.1$.
(e) If $0 < |x - 3| < 0.5$, then $|f(x) - 5| < 0.1$.
(f) If $0 < |x - 3| < \frac{1}{4}$, then $|f(x) - 5| < \frac{1}{4}(0.1)$.
(g) If $0 < |x - 3| < 1$, then $|f(x) - 5| < 0.2$.
(h) If $0 < |x - 3| < 1$, then $|f(x) - 4.95| < 0.05$.
(i) If $\lim\limits_{x \to 3} f(x) = L$, then $4.9 \le L \le 5.1$.

36. Suppose that $|A - B| < \epsilon$ for each $\epsilon > 0$. Prove that $A = B$. HINT: Consider what happens if $A \ne B$ and $\epsilon = \frac{1}{2}|A - B|$.

In Exercises 37–38, give the four limit statements displayed in 2.2.5, taking

37. $f(x) = \dfrac{1}{x - 1}$, $\quad c = 3$. **38.** $f(x) = \dfrac{x}{x^2 + 2}$, $\quad c = 1$.

In Exercises 39–44, assume that x is a fixed number and find

$$\lim\limits_{h \to 0} \frac{f(x + h) - f(x)}{h}$$

for the given function f.

39. $f(x) = 5x + 2$.

40. $f(x) = 2x^2 - 3x$.

41. $f(x) = 4x + 5x^2$.

42. $f(x) = x^3 + 2x + 4$.

43. $f(x) = \dfrac{1}{x + 1}$.

44. $f(x) = \sqrt{x + 2}$. HINT: rationalize the numerator.

45. Prove that

(2.2.9) $\qquad \lim\limits_{x \to c} f(x) = 0 \quad$ iff $\quad \lim\limits_{x \to c} |f(x)| = 0$.

46. (a) Prove that

if $\lim\limits_{x \to c} f(x) = L \quad$ then $\quad \lim\limits_{x \to c} |f(x)| = |L|$.

(b) Show that the converse is false. Give an example where

$\lim\limits_{x \to c} |f(x)| = |L| \qquad$ and $\qquad \lim\limits_{x \to c} f(x) = M \ne L$.

and then give an example where

$\lim\limits_{x \to c} |f(x)| \quad$ exists $\quad$ but $\quad \lim\limits_{x \to c} f(x) \quad$ does not exist.

▶ In Exercises 47–50, $\lim\limits_{x \to a} f(x) = L$ and ϵ are given. Use a graphing utility to graph f, then zoom in at $x = a$ to find a δ

such that $|f(x) - a| < \epsilon$ whenever $|x - a| < \delta$.
HINT: graph f, $L + \epsilon$ and $L - \epsilon$.

47. $\lim\limits_{x \to 2} (x^2 + x - 3) = 3$; $\epsilon = 0.01$.

48. $\lim\limits_{x \to 1} \dfrac{x - 1}{\sqrt{x} - 1} = 2$; $\epsilon = 0.1$.

49. $\lim\limits_{x \to 0} \dfrac{\sin 3x}{x} = 3$; $\epsilon = 0.01$.

50. $\lim\limits_{x \to 1} \tan\left(\dfrac{\pi}{4}x\right) = 1$; $\epsilon = 0.1$.

51. Give an ϵ, δ proof that statement (i) of 2.2.5 is equivalent to (ii).

52. Give an ϵ, δ proof of (2.2.8).

53. (a) Show that $\lim\limits_{x \to c} \sqrt{x} = \sqrt{c}$ for $c > 0$.

  HINT: If x and c are positive, then

$$0 \le \left|\sqrt{x} - \sqrt{c}\right| = \frac{|x - c|}{\sqrt{x} + \sqrt{c}} < \frac{1}{\sqrt{c}}|x - c|.$$

  (b) Show that $\lim\limits_{x \to 0^+} \sqrt{x} = 0$.

In Exercises 54–57, give an ϵ, δ proof for the following limits.

54. $\lim\limits_{x \to 2} x^2 = 4$.

55. $\lim\limits_{x \to 1} x^3 = 1$.

56. $\lim\limits_{x \to 3} \sqrt{x + 1} = 2$.

57. $\lim\limits_{x \to 3^-} \sqrt{3 - x} = 0$.

58. Prove that if

$$g(x) = \begin{cases} x, & x \text{ rational} \\ 0, & x \text{ irrational,} \end{cases}$$

  then $\lim\limits_{x \to 0} g(x) = 0$.

59. The function f defined by

$$f(x) = \begin{cases} 1, & x \text{ rational} \\ 0, & x \text{ irrational} \end{cases}$$

  is called the *Dirichlet function*. Prove that for no number c does $\lim\limits_{x \to c} f(x)$ exist.

In Exercises 60–62, prove the given limit statement.

60. $\lim\limits_{x \to c^-} f(x) = L$ iff $\lim\limits_{h \to 0} f(c - |h|) = L$.

61. $\lim\limits_{x \to c^+} f(x) = L$ iff $\lim\limits_{h \to 0} f(c + |h|) = L$.

62. $\lim\limits_{x \to c} f(x) = L$ iff $\lim\limits_{x \to c^-} f(x) = \lim\limits_{x \to c^+} f(x) = L$.

63. Suppose that $\lim\limits_{x \to c} f(x) = L$.

(a) Prove that if $L > 0$, then $f(x) > 0$ for all $x \neq c$ in an interval of the form $(c - \gamma, c + \gamma)$, $\gamma > 0$.
HINT: Use an ϵ, δ argument, setting $\epsilon = L$.

(b) Prove that if $L < 0$, then $f(x) < 0$ for all $x \neq c$ in an interval of the form $(c - \gamma, c + \gamma)$, $\gamma > 0$.

64. Prove or give a counterexample: If $f(c) > 0$ and $\lim\limits_{x \to c} f(x)$ exists, then $f(x) > 0$ for all x in an interval of the form $(c - \gamma, c + \gamma)$, $\gamma > 0$.

65. Suppose that $f(x) \le g(x)$ for all $x \in (c - p, c + p)$, $p > 0$, except possibly at c itself.

(a) Prove that $\lim\limits_{x \to c} f(x) \le \lim\limits_{x \to c} g(x)$, provided each of these limits exist.

(b) If $f(x) < g(x)$ on $(c - p, c + p)$, except possibly at c itself, does it follow that $\lim\limits_{x \to c} f(x) < \lim\limits_{x \to c} g(x)$?

66. Prove that if $\lim\limits_{x \to c} f(x) = L$, then there is a number $\delta > 0$ and a number B such that $|f(x)| < B$ if $0 < |x - c| < \delta$.
HINT: It suffices to prove that $L - 1 < f(x) < L + 1$ for $0 < |x - c| < \delta$.

In Exercises 2.1, the slope of the tangent line to the graph of a function f at the point $(a, f(a))$ was defined to be

$$\lim\limits_{h \to 0} \frac{f(a + h) - f(a)}{h} \qquad \text{provided this limit exists.}$$

The equivalence of (i) and (ii) in (2.2.5) implies that the slope of the tangent line is also given by

$$\lim\limits_{x \to a} \frac{f(x) - f(a)}{x - a} \qquad \text{provided this limit exists.}$$

In Exercises 67–70, determine if $\lim\limits_{x \to a} (f(x) - f(a))/(x - a)$ exists, and give an equation for the tangent line to the graph of f at $(a, f(a))$ when the limit does exist.

67. $f(x) = 2x^2 - 3x$, $a = 2$.

68. $f(x) = x^3 + 1$, $a = -1$.

69. $f(x) = \sqrt{x}$, $a = 4$.

70. $f(x) = 1/(x + 1)$, $a = 1$.

▶ In Exercises 71–74, estimate $\lim\limits_{x \to a} (f(x) - f(a))/(x - a)$ numerically by evaluating the quotient at $a \pm 0.1$, $a \pm 0.01$, $a \pm 0.001$. Use a graphing utility to support or improve your numerical result.

71. $f(x) = \cos \pi x$, $a = \frac{1}{3}$.

72. $f(x) = \sqrt{x}$, $a = 1$.

73. $f(x) = 3^x$, $a = 0$.

74. $f(x) = 2^x$, $a = 1$.

■ 2.3 SOME LIMIT THEOREMS

As you saw in the last section, it can be rather tedious to apply the ϵ, δ limit test to individual functions. By proving some general theorems about limits we can avoid some of this repetitive work. The theorems themselves, of course (at least the first ones), will have to be proved by ϵ, δ methods.

We begin by showing that, if a limit exists, it is unique.

THEOREM 2.3.1 THE UNIQUENESS OF A LIMIT

If $\lim\limits_{x \to c} f(x) = L$ and $\lim\limits_{x \to c} f(x) = M$, then $L = M$.

PROOF We show that $L = M$ by proving that the assumption that $L \neq M$ leads to the false conclusion

$$|L - M| < |L - M|.$$

Let's assume that $L \neq M$. Then $|L - M|/2 > 0$. Since

$$\lim_{x \to c} f(x) = L,$$

we know that there exists $\delta_1 > 0$ such that

(1) $\qquad$ if $\;\; 0 < |x - c| < \delta_1,\qquad$ then $\;\; |f(x) - L| < \dfrac{|L - M|}{2}.$

$\hfill$ (Here we are using $|L - M|/2$ as ϵ.)

Also, since

$$\lim_{x \to c} f(x) = M,$$

we know that there exists $\delta_2 > 0$ such that

(2) $\qquad$ if $\;\; 0 < |x - c| < \delta_2,\qquad$ then $\;\; |f(x) - M| < \dfrac{|L - M|}{2}.$

$\hfill$ (Again, using $|L - M|/2$ as ϵ.)

Now let x_1 be a number that satisfies the inequality

$$0 < |x_1 - c| < \text{minimum of } \delta_1 \text{ and } \delta_2.$$

Then by (1) and (2)

$$|f(x_1) - L| < \frac{|L - M|}{2} \qquad \text{and} \qquad |f(x_1) - M| < \frac{|L - M|}{2}.$$

It follows that

$$|L - M| = |[L - f(x_1)] + [f(x_1) - M]|$$
$$\leq |L - f(x_1)| + |f(x_1) - M|$$

by the triangle $\underline{\qquad}\uparrow$
inequality

$$= |f(x_1) - L| + |f(x_1) - M| < \frac{|L - M|}{2} + \frac{|L - M|}{2} = |L - M|. \quad \square$$

$|a| = |-a| \underline{\qquad}\uparrow$

THEOREM 2.3.2

If $\lim\limits_{x \to c} f(x) = L$ and $\lim\limits_{x \to c} g(x) = M$, then

$$\textbf{(i)} \quad \lim_{x \to c} [f(x) + g(x)] = L + M,$$

$$\textbf{(ii)} \quad \lim_{x \to c} [\alpha f(x)] = \alpha L \qquad \text{for each real } \alpha,$$

$$\textbf{(iii)} \quad \lim_{x \to c} [f(x)g(x)] = LM.$$

PROOF Let $\epsilon > 0$. To prove (i) we must show that there exists $\delta > 0$ such that

$$\text{if} \quad 0 < |x - c| < \delta, \quad \text{then} \quad |[f(x) + g(x)] - [L + M]| < \epsilon.$$

Note that

$$(*) \qquad |[f(x) + g(x)] - [L + M]| = |[f(x) - L] + [g(x) - M]|$$
$$\leq |f(x) - L| + |g(x) - M|.$$

We make $|[f(x) + g(x)] - [L + M]|$ less than ϵ by making $|f(x) - L|$ and $|g(x) - M|$ each less than $\frac{1}{2}\epsilon$. Since $\epsilon > 0$, we know that $\frac{1}{2}\epsilon > 0$. Since

$$\lim_{x \to c} f(x) = L \qquad \text{and} \qquad \lim_{x \to c} g(x) = M,$$

we know that there exist positive numbers δ_1 and δ_2 such that

$$\text{if} \quad 0 < |x - c| < \delta_1, \quad \text{then} \quad |f(x) - L| < \tfrac{1}{2}\epsilon$$

and

$$\text{if} \quad 0 < |x - c| < \delta_2, \quad \text{then} \quad |g(x) - M| < \tfrac{1}{2}\epsilon.$$

Now we set $\delta = $ minimum of δ_1 and δ_2 and note that, if $0 < |x - c| < \delta$, then

$$|f(x) - L| < \tfrac{1}{2}\epsilon \qquad \text{and} \qquad |g(x) - M| < \tfrac{1}{2}\epsilon,$$

and thus by $(*)$

$$|[f(x) + g(x)] - [L + M]| < \epsilon.$$

In summary, by setting $\delta = \min\{\delta_1, \delta_2\}$, we found that,

$$\text{if} \quad 0 < |x - c| < \delta, \quad \text{then} \quad |[f(x) + g(x)] - [L + M]| < \epsilon.$$

Thus (i) is proved. For proofs of (ii) and (iii), see the supplement at the end of this section. ❏

If you are wondering about $\lim\limits_{x \to c} [f(x) - g(x)]$, note that $f(x) - g(x) = f(x) + (-1)g(x)$, and so the result

(2.3.3)
$$\lim_{x \to c} [f(x) - g(x)] = L - M$$

follows from (i) and (ii).

The results of Theorem 2.3.2 are easily extended (by mathematical induction) to any finite collection of functions; namely, if

$$\lim_{x \to c} f_1(x) = L_1, \qquad \lim_{x \to c} f_2(x) = L_2, \ldots, \lim_{x \to c} f_n(x) = L_n,$$

then

(2.3.4)
$$\lim_{x \to c} [\alpha_1 f_1(x) + \alpha_2 f_2(x) + \cdots + \alpha_n f_n(x)] = \alpha_1 L_1 + \alpha_2 L_2 + \cdots + \alpha_n L_n$$

and

(2.3.5)
$$\lim_{x \to c} [f_1(x) f_2(x) \cdots f_n(x)] = L_1 L_2 \cdots L_n.$$

Let $P(x) = a_n x^n + \cdots + a_1 x + a_0$ be a polynomial and let c be any real number. Then it follows that

(2.3.6)
$$\lim_{x \to c} P(x) = P(c).$$

PROOF   We already know that

$$\lim_{x \to c} x = c. \qquad \text{(see (2.2.2) on page 84)}$$

Applying (2.3.5) to $f(x) = x$ multiplied k times by itself, we have

$$\lim_{x \to c} x^k = c^k \qquad \text{for each positive integer } k.$$

We also know from (2.2.4) on page 85 that $\lim_{x \to c} a_0 = a_0$. It follows now from (2.3.4) that

$$\lim_{x \to c} (a_n x^n + \cdots + a_1 x + a_0) = a_n c^n + \cdots + a_1 c + a_0;$$

that is,

$$\lim_{x \to c} P(x) = P(c). \quad \square$$

In words, this result says that the limit of a polynomial $P(x)$ as x approaches c not only exists, but is actually the value of P at c. Functions that have this property are said to be *continuous at* c. Thus, polynomial functions are continuous at $x = c$ for each number c. Continuity will be treated in detail in the next section.

Examples

$$\lim_{x \to 1} (5x^2 - 12x + 2) = 5(1)^2 - 12(1) + 2 = -5,$$

$$\lim_{x \to 0} (14x^5 - 7x^2 + 2x + 8) = 14(0)^5 - 7(0)^2 + 2(0) + 8 = 8,$$

$$\lim_{x \to -1} (2x^3 + x^2 - 2x - 3) = 2(-1)^3 + (-1)^2 - 2(-1) - 3 = -2. \quad \square$$

We come now to reciprocals and quotients.

THEOREM 2.3.7

$$\text{If } \lim_{x \to c} g(x) = M \quad \text{with} \quad M \neq 0, \qquad \text{then} \quad \lim_{x \to c} \frac{1}{g(x)} = \frac{1}{M}.$$

PROOF See the supplement at the end of this section. $\square$

Examples

$$\lim_{x \to 4} \frac{1}{x^2} = \frac{1}{16}, \qquad \lim_{x \to 2} \frac{1}{x^3 - 1} = \frac{1}{7}, \qquad \lim_{x \to -3} \frac{1}{|x|} = \frac{1}{|-3|} = \frac{1}{3}. \quad \square$$

Once you know that reciprocals present no trouble, quotients become easy to handle.

THEOREM 2.3.8

$$\text{If } \lim_{x \to c} f(x) = L \text{ and } \lim_{x \to c} g(x) = M \text{ with } M \neq 0, \quad \text{then} \quad \lim_{x \to c} \frac{f(x)}{g(x)} = \frac{L}{M}.$$

PROOF The key here is to observe that the quotient can be written as a product:

$$\frac{f(x)}{g(x)} = f(x) \frac{1}{g(x)}.$$

With

$$\lim_{x \to c} f(x) = L \qquad \text{and} \qquad \lim_{x \to c} \frac{1}{g(x)} = \frac{1}{M},$$

the product rule (part (iii) of Theorem 2.3.2) gives

$$\lim_{x \to c} \frac{f(x)}{g(x)} = L \frac{1}{M} = \frac{L}{M}. \quad \square$$

As an immediate consequence of this theorem on quotients you can see that, if $R = P/Q$ is a rational function (quotient of polynomials) and $Q(c) \neq 0$, then

(2.3.9)

$$\lim_{x \to c} R(x) = \lim_{x \to c} \frac{P(x)}{Q(x)} = \frac{P(c)}{Q(c)} = R(c).$$

This result says that a rational function is *continuous* at the numbers c where $Q(c) \neq 0$.

Examples

$$\lim_{x \to 2} \frac{3x - 5}{x^2 + 1} = \frac{6 - 5}{4 + 1} = \frac{1}{5}, \qquad \lim_{x \to 3} \frac{x^3 - 3x^2}{1 - x^2} = \frac{27 - 27}{1 - 9} = 0. \quad ❑$$

There is no point looking for a limit that does not exist. The next theorem gives a condition under which a quotient does not have a limit.

THEOREM 2.3.10

If $\lim\limits_{x \to c} f(x) = L$ with $L \neq 0$ and $\lim\limits_{x \to c} g(x) = 0$, then $\lim\limits_{x \to c} \dfrac{f(x)}{g(x)}$ does not exist.

PROOF Suppose, on the contrary, that there exists a real number K such that

$$\lim_{x \to c} \frac{f(x)}{g(x)} = K.$$

Then

$$L = \lim_{x \to c} f(x) = \lim_{x \to c} \left[g(x) \cdot \frac{f(x)}{g(x)} \right] = \lim_{x \to c} g(x) \cdot \lim_{x \to c} \frac{f(x)}{g(x)} = 0 \cdot K = 0,$$

which contradicts our assumption that $L \neq 0$. ❑

Examples From Theorem 2.3.10 you can see that

$$\lim_{x \to 1} \frac{x^2}{x - 1}, \qquad \lim_{x \to 2} \frac{3x - 7}{x^2 - 4}, \qquad \text{and} \qquad \lim_{x \to 0} \frac{5}{x}$$

all fail to exist. ❑

Suppose $\lim\limits_{x \to c} f(x) = L$ and $\lim\limits_{x \to c} g(x) = M$. Combining Theorems 2.3.8 and 2.3.10, we have

$$\lim_{x \to c} \frac{f(x)}{g(x)} = \begin{cases} L/M & \text{if } M \neq 0 \\ \text{does not exist} & \text{if } M = 0 \text{ and } L \neq 0. \end{cases}$$

This raises the question: What happens if $L = M = 0$? This is the most difficult and the most interesting case. A limit of a quotient in which both the numerator and the denominator approach 0 is called an *indeterminate form of type 0/0*; as you will see in the examples below, the limit may or may not exist. At this stage, we will use algebraic methods to analyze limits of this type. We will learn other, more powerful, methods later.

Example 1 Evaluate the limits that exist:

(a) $\displaystyle\lim_{x\to 3}\frac{x^2 - x - 6}{x - 3},$ **(b)** $\displaystyle\lim_{x\to 4}\frac{(x^2 - 3x - 4)^2}{x - 4},$ **(c)** $\displaystyle\lim_{x\to -1}\frac{x + 1}{(2x^2 + 7x + 5)^2}.$

SOLUTION In each case both numerator and denominator tend to zero, and so we have to be careful.

(a) First we factor the numerator:

$$\frac{x^2 - x - 6}{x - 3} = \frac{(x + 2)(x - 3)}{x - 3}.$$

For $x \neq 3$,

$$\frac{x^2 - x - 6}{x - 3} = x + 2.$$

Thus

$$\lim_{x\to 3}\frac{x^2 - x - 6}{x - 3} = \lim_{x\to 3}(x + 2) = 5.$$

(b) Note that

$$\frac{(x^2 - 3x - 4)^2}{x - 4} = \frac{[(x + 1)(x - 4)]^2}{x - 4} = \frac{(x + 1)^2(x - 4)^2}{x - 4}$$

so that, for $x \neq 4$,

$$\frac{(x^2 - 3x - 4)^2}{x - 4} = (x + 1)^2(x - 4).$$

It follows then that

$$\lim_{x\to 4}\frac{(x^2 - 3x - 4)^2}{x - 4} = \lim_{x\to 4}(x + 1)^2(x - 4) = 0.$$

(c) Since

$$\frac{x + 1}{(2x^2 + 7x + 5)^2} = \frac{x + 1}{[(2x + 5)(x + 1)]^2} = \frac{x + 1}{(2x + 5)^2(x + 1)^2},$$

you can see that, for $x \neq -1$,

$$\frac{x + 1}{(2x^2 + 7x + 5)^2} = \frac{1}{(2x + 5)^2(x + 1)}.$$

By Theorem 2.3.10

$$\lim_{x\to -1}\frac{1}{(2x + 5)^2(x + 1)} \quad \text{does not exist,}$$

and therefore

$$\lim_{x\to -1}\frac{x + 1}{(2x^2 + 7x + 5)^2} \quad \text{does not exist either.} \quad \square$$

Example 2 Evaluate the following limits:

(a) $\displaystyle\lim_{x\to 2}\frac{1/x - 1/2}{x - 2}$ **(b)** $\displaystyle\lim_{x\to 9}\frac{x - 9}{\sqrt{x} - 3}$.

SOLUTION

(a) For $x \neq 2$,

$$\frac{1/x - 1/2}{x - 2} = \frac{\dfrac{2 - x}{2x}}{x - 2} = \frac{-(x - 2)}{2x(x - 2)} = \frac{-1}{2x}.$$

Thus,

$$\lim_{x\to 2}\frac{1/x - 1/2}{x - 2} = \lim_{x\to 2}\left[\frac{-1}{2x}\right] = -\frac{1}{4}.$$

(b) First, we "rationalize" the denominator:

$$\frac{x - 9}{\sqrt{x} - 3} = \frac{x - 9}{\sqrt{x} - 3} \cdot \frac{\sqrt{x} + 3}{\sqrt{x} + 3} = \frac{(x - 9)(\sqrt{x} + 3)}{x - 9} = \sqrt{x} + 3 \qquad (x \neq 9).$$

Alternatively, we could have factored the numerator:

$$\frac{x - 9}{\sqrt{x} - 3} = \frac{(\sqrt{x} - 3)(\sqrt{x} + 3)}{\sqrt{x} - 3} = \sqrt{x} + 3 \qquad (x \neq 9)$$

and arrived at the same result. Either way,

$$\lim_{x\to 9}\frac{x - 9}{\sqrt{x} - 3} = \lim_{x\to 9}[\sqrt{x} + 3] = 6. \quad \square$$

Remark In this section we have phrased everything in terms of two-sided limits. Although we won't stop here to prove it, *all these results carry over to one-sided limits.* ❏

EXERCISES 2.3

1. Given that

$$\lim_{x\to c} f(x) = 2, \qquad \lim_{x\to c} g(x) = -1, \qquad \lim_{x\to c} h(x) = 0,$$

evaluate the limits that exist. If the limit does not exist, explain why.

(a) $\displaystyle\lim_{x\to c} [f(x) - g(x)]$.   (b) $\displaystyle\lim_{x\to c} [f(x)]^2$.

(c) $\displaystyle\lim_{x\to c} \frac{f(x)}{g(x)}$.   (d) $\displaystyle\lim_{x\to c} \frac{h(x)}{f(x)}$.

(e) $\displaystyle\lim_{x\to c} \frac{f(x)}{h(x)}$.   (f) $\displaystyle\lim_{x\to c} \frac{1}{f(x) - g(x)}$.

2. Given that

$$\lim_{x\to c} f(x) = 3, \qquad \lim_{x\to c} g(x) = 0, \qquad \lim_{x\to c} h(x) = -2,$$

evaluate the limits that exist. If the limit does not exist, explain why.

(a) $\displaystyle\lim_{x\to c} [3f(x) + 2h(x)]$.   (b) $\displaystyle\lim_{x\to c} [h(x)]^3$.

(c) $\displaystyle\lim_{x\to c} \frac{h(x)}{x - c}$.   (d) $\displaystyle\lim_{x\to c} \frac{g(x)}{h(x)}$.

(e) $\displaystyle\lim_{x\to c} \frac{4}{f(x) - h(x)}$.   (f) $\displaystyle\lim_{x\to c} [3 + g(x)]^2$.

3. When asked to evaluate

$$\lim_{x\to 4}\left(\frac{1}{x} - \frac{1}{4}\right)\left(\frac{1}{x - 4}\right),$$

I reply that the limit is zero since $\displaystyle\lim_{x\to 4} [1/x - \tfrac{1}{4}] = 0$ and cite Theorem 2.3.2 (limit of a product) as justification. Verify that the limit is actually $-\tfrac{1}{16}$ and identify my error.

4. When asked to evaluate

$$\lim_{x \to 3} \frac{x^2 + x - 12}{x - 3},$$

I say that the limit does not exist since $\lim_{x \to 3} (x - 3) = 0$ and cite Theorem 2.3.10 (limit of a quotient) as justification. Verify that the limit is actually 7 and identify my error.

In Exercises 5–38, evaluate the limits that exist.

5. $\lim_{x \to 2} 3$.

6. $\lim_{x \to 3} (5 - 4x)^2$.

7. $\lim_{x \to -4} (x^2 + 3x - 7)$.

8. $\lim_{x \to -2} 3|x - 1|$.

9. $\lim_{x \to \sqrt{3}} |x^2 - 8|$.

10. $\lim_{x \to -1} \frac{x^2 + 1}{3x^5 + 4}$.

11. $\lim_{x \to 0} \left(x - \frac{4}{x} \right)$.

12. $\lim_{x \to 5} \frac{2 - x^2}{4x}$.

13. $\lim_{x \to 0} \frac{x^2 + 1}{x - 1}$.

14. $\lim_{x \to 0} \frac{x^2}{x^2 + 1}$.

15. $\lim_{x \to 2} \frac{x}{x^2 - 4}$.

16. $\lim_{h \to 0} h \left(1 - \frac{1}{h} \right)$.

17. $\lim_{h \to 0} h \left(1 + \frac{1}{h} \right)$.

18. $\lim_{x \to 2} \frac{x - 2}{x^2 - 4}$.

19. $\lim_{x \to 2} \frac{x^2 - 4}{x - 2}$.

20. $\lim_{x \to -2} \frac{(x^2 - x - 6)^2}{x + 2}$.

21. $\lim_{x \to 4} \frac{\sqrt{x} - 2}{x - 4}$.

22. $\lim_{x \to 1} \frac{x - 1}{\sqrt{x} - 1}$.

23. $\lim_{x \to 1} \frac{x^2 - x - 6}{(x + 2)^2}$.

24. $\lim_{x \to -2} \frac{x^2 - x - 6}{(x + 2)^2}$.

25. $\lim_{h \to 0} \frac{1 - 1/h^2}{1 - 1/h}$.

26. $\lim_{h \to 0} \frac{1 - 1/h^2}{1 + 1/h^2}$.

27. $\lim_{h \to 0} \frac{1 - 1/h}{1 + 1/h}$.

28. $\lim_{h \to 0} \frac{1 + 1/h}{1 + 1/h^2}$.

29. $\lim_{t \to -1} \frac{t^2 + 6t + 5}{t^2 + 3t + 2}$.

30. $\lim_{x \to 2^+} \frac{\sqrt{x^2 - 4}}{x - 2}$.

31. $\lim_{t \to 0} \frac{t + a/t}{t + b/t}$.

32. $\lim_{x \to 1} \frac{x^2 - 1}{x^3 - 1}$.

33. $\lim_{x \to 1} \frac{x^5 - 1}{x^4 - 1}$.

34. $\lim_{h \to 0} h^2 \left(1 + \frac{1}{h} \right)$.

35. $\lim_{h \to 0} h \left(1 + \frac{1}{h^2} \right)$.

36. $\lim_{x \to -4} \left(\frac{3x}{x + 4} + \frac{8}{x + 4} \right)$.

37. $\lim_{x \to -4} \left(\frac{2x}{x + 4} + \frac{8}{x + 4} \right)$.

38. $\lim_{x \to -4} \left(\frac{2x}{x + 4} - \frac{8}{x + 4} \right)$.

39. Evaluate the limits that exist.

(a) $\lim_{x \to 4} \left(\frac{1}{x} - \frac{1}{4} \right)$.

(b) $\lim_{x \to 4} \left[\left(\frac{1}{x} - \frac{1}{4} \right) \left(\frac{1}{x - 4} \right) \right]$.

(c) $\lim_{x \to 4} \left[\left(\frac{1}{x} - \frac{1}{4} \right) (x - 2) \right]$.

(d) $\lim_{x \to 4} \left[\left(\frac{1}{x} - \frac{1}{4} \right) \left(\frac{1}{x - 4} \right)^2 \right]$.

40. Evaluate the limits that exist.

(a) $\lim_{x \to 3} \frac{x^2 + x + 12}{x - 3}$.

(b) $\lim_{x \to 3} \frac{x^2 + x - 12}{x - 3}$.

(c) $\lim_{x \to 3} \frac{(x^2 + x - 12)^2}{x - 3}$.

(d) $\lim_{x \to 3} \frac{x^2 + x - 12}{(x - 3)^2}$.

41. Given that $f(x) = x^2 - 4x$, evaluate the limits that exist.

(a) $\lim_{x \to 4} \frac{f(x) - f(4)}{x - 4}$.

(b) $\lim_{x \to 1} \frac{f(x) - f(1)}{x - 1}$.

(c) $\lim_{x \to 3} \frac{f(x) - f(1)}{x - 3}$.

(d) $\lim_{x \to 3} \frac{f(x) - f(2)}{x - 3}$.

42. Given that $f(x) = x^3$, evaluate the limits that exist.

(a) $\lim_{x \to 3} \frac{f(x) - f(3)}{x - 3}$.

(b) $\lim_{x \to 3} \frac{f(x) - f(2)}{x - 3}$.

(c) $\lim_{x \to 3} \frac{f(x) - f(3)}{x - 2}$.

(d) $\lim_{x \to 1} \frac{f(x) - f(1)}{x - 1}$.

43. Show by example that $\lim_{x \to c} [f(x) + g(x)]$ can exist even if $\lim_{x \to c} f(x)$ and $\lim_{x \to c} g(x)$ do not exist.

44. Show by example that $\lim_{x \to c} [f(x)g(x)]$ can exist even if $\lim_{x \to c} f(x)$ and $\lim_{x \to c} g(x)$ do not exist.

Exercises 45–51: True or false? Justify your answers.

45. If $\lim_{x \to c} [f(x) + g(x)]$ exists but $\lim_{x \to c} f(x)$ does not exist, then $\lim_{x \to c} g(x)$ does not exist.

46. If $\lim_{x \to c} [f(x) + g(x)]$ and $\lim_{x \to c} f(x)$ exist, then it can happen that $\lim_{x \to c} g(x)$ does not exist.

47. If $\lim_{x \to c} \sqrt{f(x)}$ exists, then $\lim_{x \to c} f(x)$ exists.

48. If $\lim_{x \to c} f(x)$ exists, then $\lim_{x \to c} \sqrt{f(x)}$ exists.

49. If $\lim_{x \to c} f(x)$ exists, then $\lim_{x \to c} \frac{1}{f(x)}$ exists.

50. If $f(x) \leq g(x)$ for all $x \neq c$, then $\lim\limits_{x \to c} f(x) \leq \lim\limits_{x \to c} g(x)$.

51. If $f(x) < g(x)$ for all $x \neq c$, then $\lim\limits_{x \to c} f(x) < \lim\limits_{x \to c} g(x)$.

52. (a) Verify that
$$\max \{f(x), g(x)\} = \tfrac{1}{2}\{[f(x) + g(x)] + |f(x) - g(x)|\}.$$
(b) Find a similar expression for $\min \{f(x), g(x)\}$.

53. Let $h(x) = \min \{f(x), g(x)\}$ and $H(x) = \max \{f(x), g(x)\}$. Show that

$$\text{if } \lim\limits_{x \to c} f(x) = L \quad \text{and} \quad \lim\limits_{x \to c} g(x) = L,$$

$$\text{then } \lim\limits_{x \to c} h(x) = L \quad \text{and} \quad \lim\limits_{x \to c} H(x) = L.$$

HINT: Use Exercise 52.

54. (*The stability of limit*) Given a function f defined on an interval $I = (c - p, c + p), p > 0$. Suppose the function g is also defined on I and $g(x) = f(x)$ except, possibly, at a finite set of points $x_1, x_2, \ldots, x_n$ in I.
(a) Show that if $\lim\limits_{x \to c} f(x) = L$, then $\lim\limits_{x \to c} g(x) = L$.
(b) Show that if $\lim\limits_{x \to c} f(x)$ does not exist, then $\lim\limits_{x \to c} g(x)$ does not exist.

55. (a) Suppose that $\lim\limits_{x \to c} f(x) = 0$ and $\lim\limits_{x \to c} [f(x)g(x)] = 1$.

Prove that $\lim\limits_{x \to c} g(x)$ does not exist.

(b) Suppose that
$$\lim\limits_{x \to c} f(x) = L \neq 0 \text{ and } \lim\limits_{x \to c} [f(x)g(x)] = 1.$$
Does $\lim\limits_{x \to c} g(x)$ exist, and if so, what is it?

56. Suppose f is a function with the following property: $\lim\limits_{x \to c} [f(x) + g(x)]$ does not exist whenever $\lim\limits_{x \to c} g(x)$ does not exist. Prove that $\lim\limits_{x \to c} f(x)$ *does* exist.

57. Fix a number x and evaluate

$$\lim_{h \to 0} \frac{f(x + h) - f(x)}{h}$$

for
(a) $f(x) = x$.   (b) $f(x) = x^2$.
(c) $f(x) = x^3$.   (d) $f(x) = x^4$.
(e) Guess the limit for $f(x) = x^n$, where n is any positive integer.

58. Fix a number $x \neq 0$ and repeat Exercise 57 for
(a) $f(x) = \dfrac{1}{x}$.   (b) $f(x) = \dfrac{1}{x^2}$.   (c) $f(x) = \dfrac{1}{x^3}$.
(d) Guess the limit for $f(x) = 1/x^n$, where n is any positive integer.
(e) What will this limit be for $f(x) = x^n$, where n is any nonzero integer? Does your "formula" also hold for $n = 0$?

---

*SUPPLEMENT TO SECTION 2.3

PROOF OF THEOREM 2.3.2 (ii)

We consider two cases: $\alpha \neq 0$ and $\alpha = 0$. If $\alpha \neq 0$, then $\epsilon/|\alpha| > 0$ and, since

$$\lim_{x \to c} f(x) = L$$

we know that there exists $\delta > 0$ such that,

$$\text{if } \ 0 < |x - c| < \delta, \qquad \text{then} \quad |f(x) - L| < \frac{\epsilon}{|\alpha|}.$$

From the last inequality we obtain

$$|\alpha||f(x) - L| < \epsilon \quad \text{and thus} \quad |\alpha f(x) - \alpha L| < \epsilon.$$

The case $\alpha = 0$ was treated before in (2.2.4). ❏

PROOF OF THEOREM 2.3.2 (iii)

We begin with a little algebra:
$$\begin{aligned}
|f(x)g(x) - LM| &= |[f(x)g(x) - f(x)M] + [f(x)M - LM]| \\
&\leq |f(x)g(x) - f(x)M| + |f(x)M - LM| \\
&= |f(x)||g(x) - M| + |M||f(x) - L| \\
&\leq |f(x)||g(x) - M| + (1 + |M|)|f(x) - L|.
\end{aligned}$$

Now let $\epsilon < 0$. Since $\lim_{x \to c} f(x) = L$ and $\lim_{x \to c} g(x) = M$, we know:

(1) That there exists $\delta_1 > 0$ such that, if $0 < |x - c| < \delta_1$, then

$$|f(x) - L| < 1 \qquad \text{and thus} \qquad |f(x)| < 1 + |L|.$$

(2) That there exists $\delta_2 > 0$ such that

$$\text{if } 0 < |x - c| < \delta_2, \qquad \text{then} \qquad |g(x) - M| < \left(\frac{\frac{1}{2}\epsilon}{1 + |L|} \right).$$

(3) That there exists $\delta_3 > 0$ such that

$$\text{if } \quad 0 < |x - c| < \delta_3, \qquad \text{then} \qquad |f(x) - L| < \left(\frac{\frac{1}{2}\epsilon}{1 + |M|} \right).$$

We now set $\delta = \min \{\delta_1, \delta_2, \delta_3\}$ and observe that, if $0 < |x - c| < \delta$, then

$$|f(x)g(x) - LM| < |f(x)| \, |g(x) - M| + (1 + |M|)|f(x) - L|$$

$$< (1 + |L|)\left(\frac{\frac{1}{2}\epsilon}{1 + |L|} \right) + (1 + |M|)\left(\frac{\frac{1}{2}\epsilon}{1 + |M|} \right) = \epsilon. \quad \square$$

by (1) ⎯⎯↑   ↑⎯⎯ by (2)   ↑⎯⎯ by (3)

PROOF OF THEOREM 2.3.7

For $g(x) \neq 0$.

$$\left| \frac{1}{g(x)} - \frac{1}{M} \right| = \frac{|g(x) - M|}{|g(x)| |M|}.$$

Choose $\delta_1 > 0$ such that

$$\text{if } \quad 0 < |x - c| < \delta_1, \qquad \text{then} \quad |g(x) - M| < \frac{|M|}{2}.$$

For such x,

$$|g(x)| > \frac{|M|}{2} \quad \text{so that} \quad \frac{1}{|g(x)|} < \frac{2}{|M|}$$

and thus

$$\left| \frac{1}{g(x)} - \frac{1}{M} \right| = \frac{|g(x) - M|}{|g(x)| |M|} \leq \frac{2}{|M|^2} |g(x) - M|.$$

Now let $\epsilon > 0$ and choose $\delta_2 > 0$ such that

$$\text{if } \quad 0 < |x - c| < \delta_2, \qquad \text{then} \quad |g(x) - M| < \frac{|M|^2}{2} \epsilon.$$

Setting $\delta = \min \{\delta_1, \delta_2\}$, we find that

$$\text{if } \quad 0 < |x - c| < \delta, \qquad \text{then} \quad \left| \frac{1}{g(x)} - \frac{1}{M} \right| < \epsilon. \quad \square$$

■ 2.4 CONTINUITY

In ordinary language, to say that a certain process is "continuous" is to say that it goes on without interruption and without abrupt changes. In mathematics the word "continuous" has much the same meaning.

The idea of continuity is so important to calculus and its applications that we discuss it with some care. First we treat the idea of *continuity at a point* (or number) *c*, and then we discuss *continuity on an interval*.

Continuity at a Point

DEFINITION 2.4.1

Let f be a function defined at least on an open interval $(c - p, c + p)$ with $p > 0$. Then f is *continuous at c* iff

$$\lim_{x \to c} f(x) = f(c).$$

Remark Recall that in the definition of "limit of f at c" (Definition 2.2.1) we did not require that f be defined at c itself. In contrast, the definition of "continuity at c" requires f to be defined at c. Thus, according to this definition, a function f is continuous at a point c iff:

(i) f is defined at c,

(ii) $\lim_{x \to c} f(x)$ exists,

(iii) $\lim_{x \to c} f(x) = f(c)$.

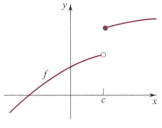

Figure 2.4.1

The function f is said to be *discontinuous at c* if it is not continuous there. ❏

If the domain of f contains an interval $(c - p, c + p)$ (so that f is defined at c), then f can fail to be continuous at c only for one of two reasons: either $f(x)$ does not have a limit as x tends to c, or it does have a limit, but the limit is not $f(c)$. In the latter case, f is said to have a *removable discontinuity at c*; the discontinuity can be removed by redefining f at c. If the limit is L, define f at c to be L.

The function depicted in Figure 2.4.1 is discontinuous at c because it does not have a limit at c. Note that $\lim_{x \to c^-} f(x)$ and $\lim_{x \to c^+} f(x)$ each exist, but they are not equal. A discontinuity of this particular type is called a *jump discontinuity*.

The function depicted in Figure 2.4.2 does have a limit at c. It is discontinuous at c only because its limit at c is not its value at c. The discontinuity is removable; it can be removed by lowering the dot into place (by redefining f at c).

Figure 2.4.2

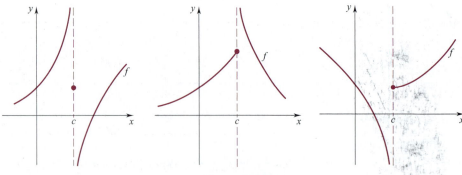

Figure 2.4.3

Additional examples of functions that fail to be continuous at a point c are shown in Figure 2.4.3. In each of these examples, at least one of the one-sided limits

$$\lim_{x \to c^-} f(x) \quad \text{or} \quad \lim_{x \to c^+} f(x)$$

fails to exist.

In Figure 2.4.4, we have tried to suggest the Dirichlet function

$$f(x) = \begin{cases} 1, & x \text{ rational} \\ 0, & x \text{ irrational.} \end{cases}$$

At no point c does f have a limit. It is therefore everywhere discontinuous.

Most of the functions that you have encountered so far are continuous at each point of their domains. In particular, this is true for polynomials P,

$$\lim_{x \to c} P(x) = P(c), \qquad \text{(see 2.3.6)}$$

for rational functions (quotients of polynomials) $R = P/Q$,

$$\lim_{x \to c} R(x) = \lim_{x \to c} \frac{P(x)}{Q(x)} = \frac{P(c)}{Q(c)} = R(c) \qquad \text{provided} \quad Q(c) \neq 0, \quad \text{(see 2.3.9)}$$

and for the absolute-value function,

$$\lim_{x \to c} |x| = |c|. \qquad \text{(see 2.2.3)}$$

Figure 2.4.4

In Exercise 53, Section 2.2, you were asked to show that

$$\lim_{x \to c} \sqrt{x} = \sqrt{c} \qquad \text{for each } c > 0.$$

This makes the square-root function continuous at each positive number. We discuss later what happens for $c = 0$.

With f and g continuous at c, we have

$$\lim_{x \to c} f(x) = f(c), \qquad \lim_{x \to c} g(x) = g(c)$$

and thus, by the limit theorems,

$$\lim_{x \to c} [f(x) + g(x)] = f(c) + g(c), \qquad \lim_{x \to c} [\alpha f(x)] = \alpha f(c) \quad \text{for each real } \alpha$$

$$\lim_{x \to c} [f(x)g(x)] = f(c)g(c), \qquad \text{and, if } g(c) \neq 0, \lim_{x \to c} f(x)/g(x) = f(c)/g(c).$$

In summary, we have the following theorem:

THEOREM 2.4.2

If f and g are continuous at c, then

  (i) $f + g$ is continuous at c **(ii)** αf is continuous at c for each real α,
  (iii) fg is continuous at c,   **(iv)** f/g is continuous at c provided $g(c) \neq 0$.

Parts (i) and (iii) can be extended to any finite number of functions. Also, since $f - g = f + (-1)g$, it follows from (i) and (ii) that $f - g$ is continuous at c.

Example 1   The function

$$F(x) = 3|x| + \frac{x^3 - x}{x^2 - 5x + 6} + 4$$

is continuous at all real numbers other than 2 and 3.   You can see this by noting that

$$F = 3f + g/h + k$$

where

$$f(x) = |x|, \ g(x) = x^3 - x, \ h(x) = x^2 - 5x + 6, \ \text{and} \ k(x) = 4.$$

Since f, g, h, and k are everywhere continuous, F is continuous except at 2 and 3, the numbers where h takes on the value 0 (that is, the numbers where F is not defined).   ❏

Since continuity at a point c is defined in terms of a limit, there is an ϵ, δ version of the definition.   A direct translation of

$$\lim_{x \to c} f(x) = f(c)$$

into ϵ, δ terms reads like this: for each $\epsilon > 0$ there exists $\delta > 0$ such that

$$\text{if} \quad 0 < |x - c| < \delta \quad \text{then} \quad |f(x) - f(c)| < \epsilon.$$

In the case of continuity at c, the restriction $0 < |x - c|$ is unnecessary.   We can allow $|x - c| = 0$ because then $x = c$, $f(x) = f(c)$, and thus $|f(x) - f(c)| = 0$.   Being 0, $|f(x) - f(c)|$ is certainly less than ϵ.

Thus, an ϵ, δ characterization of continuity at c reads as follows:

(2.4.3) f is continuous at c iff $\begin{cases} \text{for each } \epsilon > 0 \text{ there exists } \delta > 0 \text{ such that} \\ \text{if} \quad |x - c| < \delta \quad \text{then} \quad |f(x) - f(c)| < \epsilon. \end{cases}$

In simple intuitive language

f is continuous at c iff   for x close to c, $f(x)$ is close to $f(c)$.

We are now ready to take up the continuity of composite functions.   Remember the defining formula: $(f \circ g)(x) = f(g(x))$.

THEOREM 2.4.4

If g is continuous at c and f is continuous at $g(c)$, then the composition $f \circ g$ is continuous at c.

The idea here is simple: with g continuous at c, we know that

for x close to c, $g(x)$ is close to $g(c)$;

from the continuity of f at $g(c)$, we know that

with $g(x)$ close to $g(c)$, $f(g(x))$ is close to $f(g(c))$.

In summary,

with x close to c, $f(g(x))$ is close to $f(g(c))$.   ❏

The argument we just gave is too vague to be a proof. Here in contrast is a proof. We begin with $\epsilon > 0$. We must show that there exists some number $\delta > 0$ such that

$$\text{if} \quad |x - c| < \delta, \quad \text{then} \quad |f(g(x)) - f(g(c))| < \epsilon.$$

In the first place, we observe that, since f is continuous at $g(c)$, there does exist a number $\delta_1 > 0$ such that

(1) $$\text{if} \quad |t - g(c)| < \delta_1, \quad \text{then} \quad |f(t) - f(g(c))| < \epsilon.$$

With $\delta_1 > 0$, we know from the continuity of g at c that there exists a number $\delta > 0$ such that

(2) $$\text{if} \quad |x - c| < \delta, \quad \text{then} \quad |g(x) - g(c)| < \delta_1.$$

Combining (2) and (1) we have what we want: by (2)

$$\text{if} \quad |x - c| < \delta, \quad \text{then} \quad |g(x) - g(c)| < \delta_1$$

so that by (1)

$$|f(g(x)) - f(g(c))| < \epsilon. \quad \Box$$

This proof is illustrated in Figure 2.4.5. The numbers within δ of c are taken by g to within δ_1 of $g(c)$, and then by f to within ϵ of $f(g(c))$.

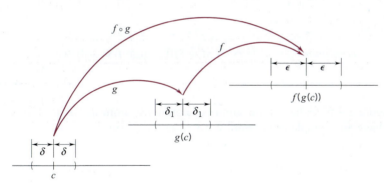

Figure 2.4.5

It is time to look at some examples.

Example 2   The function

$$F(x) = \sqrt{\frac{x^2 + 1}{(x - 8)^3}}$$

is continuous at all numbers greater than 8. To see this, note that $F = f \circ g$, where

$$f(x) = \sqrt{x} \quad \text{and} \quad g(x) = \frac{x^2 + 1}{(x - 8)^3}.$$

Now, take any $c > 8$. Since g is a rational function and g is defined at c, g is continuous at c. Also, since $g(c)$ is positive and f is continuous at each positive number, f is continuous at $g(c)$. By Theorem 2.4.4, F is continuous at c. $\quad \Box$

The continuity of composites holds for any finite number of functions. The only requirement is that each function be continuous *where it is applied.*

Example 3 The function

$$F(x) = \frac{1}{5 - \sqrt{x^2 + 16}}$$

is continuous everywhere except at $x = \pm 3$, where it is not defined. To see this, note that $F = f \circ g \circ k \circ h$, where

$$f(x) = \frac{1}{x}, \qquad g(x) = 5 - x, \qquad k(x) = \sqrt{x}, \qquad h(x) = x^2 + 16$$

and observe that each of these functions is being evaluated only where it is continuous. In particular, g and h are continuous everywhere, f is being evaluated only at nonzero numbers, and k is being evaluated only at positive numbers. ❑

Just as we considered one-sided limits, we can consider one-sided continuity.

DEFINITION 2.4.5 ONE-SIDED CONTINUITY

A function f is called

continuous from the left at c   iff $\lim\limits_{x \to c^-} f(x) = f(c)$.

It is called

continuous from the right at c   iff $\lim\limits_{x \to c^+} f(x) = f(c)$.

In Figure 2.4.6 we have an example of continuity from the right at 0; in Figure 2.4.7 we have an example of continuity from the left at 1.

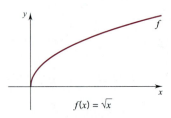

$$f(x) = \sqrt{x}$$

Figure 2.4.6

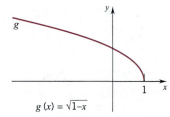

$$g(x) = \sqrt{1 - x}$$

Figure 2.4.7

It is clear from (2.2.8) that a function f is continuous at c iff it is continuous from both sides at c. Thus

(2.4.6)

f is continuous at c iff $f(c)$, $\lim\limits_{x \to c^-} f(x)$, $\lim\limits_{x \to c^+} f(x)$ all exist and are equal.

We apply this result to piecewise-defined functions in the following examples.

Example 4 Determine the discontinuities, if any, of the following function:

$$f(x) = \begin{cases} 2x + 1, & x \le 0 \\ 1, & 0 < x \le 1 \\ x^2 + 1, & x > 1. \end{cases} \qquad \text{(Figure 2.4.8)}$$

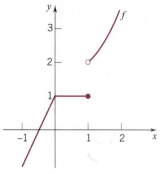

SOLUTION Clearly, f is continuous at each point in the open intervals $(-\infty, 0)$, $(0, 1)$, and $(1, \infty)$ since f is a polynomial in each of these intervals. Thus, we only have to check the behavior of f at $x = 0$ and $x = 1$. The figure suggests that f is continuous at 0 and discontinuous at 1. Indeed, at $x = 0$, $f(0) = 1$,

$$\lim_{x \to 0^-} f(x) = \lim_{x \to 0^-} (2x + 1) = 1 \qquad \text{and} \qquad \lim_{x \to 0^+} f(x) = \lim_{x \to 0^+} (1) = 1,$$

so f is continuous at 0. At $x = 1$,

$$\lim_{x \to 1^-} f(x) = \lim_{x \to 1^-} (1) = 1 \qquad \text{and} \qquad \lim_{x \to 1^+} f(x) = \lim_{x \to 1^+} (x^2 + 1) = 2.$$

Thus, f is discontinuous at 1; in fact, f has a jump discontinuity at this point. ❏

Figure 2.4.8

Example 5 Determine the discontinuities of the following function:

$$f(x) = \begin{cases} x^3, & x \le -1 \\ x^2 - 2, & -1 < x < 0 \\ 3 - x, & 0 \le x < 2 \\ \dfrac{4x - 1}{x - 1}, & 2 \le x < 4 \\ \dfrac{15}{7 - x}, & 4 < x < 7 \\ 5x + 2, & 7 \le x. \end{cases}$$

SOLUTION It should be clear that f is continuous at each point in the open intervals $(-\infty, -1)$, $(-1, 0)$, $(0, 2)$, $(2, 4)$, $(4, 7)$, and $(7, \infty)$. All we have to check is the behavior of f at $x = -1, 0, 2, 4,$ and 7. To do so, we apply (2.4.6).

At $x = -1$, f is continuous since $f(-1) = (-1)^3 = -1$,

$$\lim_{x \to -1^-} f(x) = \lim_{x \to -1^-} x^3 = -1, \qquad \text{and} \qquad \lim_{x \to -1^+} f(x) = \lim_{x \to -1^+} (x^2 - 2) = -1.$$

Our findings at the other four points are displayed in the chart below. Try to verify each entry.

c	$f(c)$	$\displaystyle\lim_{x \to c^-} f(x)$	$\displaystyle\lim_{x \to c^+} f(x)$	Conclusion
0	3	-2	3	Discontinuous
2	7	1	7	Discontinuous
4	Not defined	5	5	Discontinuous
7	37	Does not exist	37	Discontinuous

Note that the discontinuity at 4 is removable; if we define $f(4) = 5$, then f is continuous at 4. The discontinuities at $x = 0$ and $x = 2$ are jump discontinuities. ❏

Continuity on Intervals

Let (a, b) be an open interval. A function f is *continuous on* (a, b) if it is continuous at each point $c \in (a, b)$. For a function defined on a closed interval $[a, b]$, the most continuity that we can expect is:

1. Continuity on the open interval (a, b),
2. Continuity from the right at a,
3. Continuity from the left at b.

A function f that fulfills these requirements is said to be *continuous on* $[a, b]$. These definitions can be extended in an obvious way to functions defined on half-open intervals and on infinite intervals.

Functions that are continuous on a closed interval $[a, b]$ have some important special properties not shared by functions in general. We discuss two such properties in Section 2.6.

The concept of *continuity on an interval* also has a geometric interpretation, namely a function f defined on an interval I is continuous on that interval if its graph has no "holes" or "jumps." That is, f is continuous on I if its graph is a "connected" curve. In practical terms, this means that you can draw the graph of f without lifting your pencil from the paper. Look back over the examples in this section to verify this.

$f(x) = \sqrt{1 - x^2}$

Figure 2.4.9

Example 6   Consider the function

$$f(x) = \sqrt{1 - x^2}.$$

The graph of f is the semicircle shown in Figure 2.4.9. The function is continuous on $[-1, 1]$ because it is continuous on the open interval $(-1, 1)$, continuous from the right at -1, and continuous from the left at 1.   ❏

Example 7   Let g be the function defined by

$$g(x) = \sqrt{x^2 - 4}.$$

Note that $\text{dom}(g) = \{x : x^2 - 4 \geq 0\}$. You can verify that this is the same as the set $(-\infty, -2] \cup [2, \infty)$. The function g is continuous on each of these intervals since it is continuous on each of the open intervals $(-\infty, -2)$ and $(2, \infty)$, continuous from the left at -2, and continuous from the right at 2.   ❏

EXERCISES 2.4

1. The graph of a function f is given in the figure.
   (a) At which points is f discontinuous?
   (b) For each point of discontinuity found in (a), determine whether f is continuous from the right, from the left, or neither.
   (c) Which, if any, of the points of discontinuity found in (a) is removable? Which, if any, is a jump discontinuity?

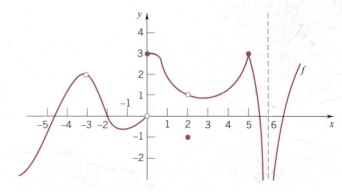

2. The graph of a function g is given in the figure. Determine the intervals on which g is continuous.

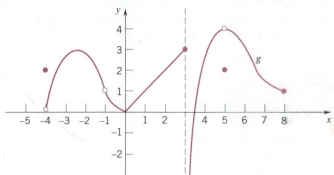

In Exercises 3–16, determine whether or not the function is continuous at the indicated point. If not, determine whether the discontinuity is a removable discontinuity, a jump discontinuity, or neither.

3. $f(x) = x^3 - 5x + 1$; $x = 2$.

4. $g(x) = \sqrt{(x-1)^2 + 5}$; $x = 1$.

5. $f(x) = \sqrt{x^2 + 9}$; $x = 3$.

6. $f(x) = |4 - x^2|$; $x = 2$.

7. $f(x) = \begin{cases} x^2 + 4, & x < 2 \\ x^3, & x \geq 2; \end{cases}$ $x = 2$.

8. $h(x) = \begin{cases} x^2 + 5, & x < 2 \\ x^3, & x \geq 2; \end{cases}$ $x = 2$.

9. $g(x) = \begin{cases} x^2 + 4, & x < 2 \\ 5, & x = 2 \\ x^3, & x > 2; \end{cases}$ $x = 2$.

10. $g(x) = \begin{cases} x^2 + 5, & x < 2 \\ 10, & x = 2 \\ 1 + x^3, & x > 2; \end{cases}$ $x = 2$.

11. $f(x) = \begin{cases} -1, & x < 0 \\ 0, & x = 0 \\ 1, & x > 0; \end{cases}$ $x = 0$.

12. $f(x) = \begin{cases} 1 - x, & x < 1 \\ 1, & x = 1 \\ x^2 - 1, & x > 1; \end{cases}$ $x = 1$.

13. $h(x) = \begin{cases} \dfrac{x^2 - 1}{x + 1}, & x \neq -1 \\ -2, & x = -1; \end{cases}$ $x = -1$.

14. $g(x) = \begin{cases} \dfrac{1}{x + 1}, & x \neq -1 \\ 0, & x = -1; \end{cases}$ $x = -1$.

15. $f(x) = \begin{cases} -x^2, & x < 0 \\ 1 - \sqrt{x}, & x \geq 0; \end{cases}$ $x = 0$.

16. $g(x) = \dfrac{x(x + 1)(x + 2)}{\sqrt{(x-1)(x-2)}}$; $x = -2$.

In Exercises 17–30, sketch the graph and classify the discontinuities (if any) as being a removable discontinuity, a jump discontinuity, or neither.

17. $f(x) = \dfrac{x^2 - 4}{x - 2}$.

18. $f(x) = \dfrac{x - 3}{x^2 - 9}$.

19. $f(x) = |x - 1|$.

20. $h(x) = |x^2 - 1|$.

21. $g(x) = \begin{cases} x - 1, & x < 1 \\ 0, & x = 1 \\ x^2, & x > 1. \end{cases}$

22. $g(x) = \begin{cases} 2x - 1, & x < 1 \\ 0, & x = 1 \\ x^2, & x > 1. \end{cases}$

23. $f(x) = \max \{x, x^2\}$

24. $f(x) = \min \{x, x^2\}$.

25. $f(x) = \begin{cases} -1, & x < -1 \\ x^3, & -1 \leq x \leq 1 \\ 1, & 1 < x. \end{cases}$

26. $g(x) = \begin{cases} 1, & x \leq -2 \\ \frac{1}{2}x, & -2 < x < 4 \\ \sqrt{x}, & 4 \leq x. \end{cases}$

27. $h(x) = \begin{cases} 1, & x \leq 0 \\ x^2, & 0 < x < 1 \\ 1, & 1 \leq x < 2 \\ x, & 2 \leq x. \end{cases}$

28. $g(x) = \begin{cases} -x^2, & x < -1 \\ 3, & x = -1 \\ 2 - x, & -1 < x \leq 1 \\ x^2, & 1 < x. \end{cases}$

29. $f(x) = \begin{cases} 2x + 9, & x < -2 \\ x^2 + 1, & -2 < x \leq 1 \\ 3x - 1, & 1 < x < 3 \\ x + 6, & 3 < x. \end{cases}$

30. $g(x) = \begin{cases} x + 7, & x < -3 \\ |x - 2|, & -3 < x < -1 \\ x^2 - 2x, & -1 < x < 3 \\ 2x - 3, & 3 \leq x. \end{cases}$

Each of the functions f in Exercises 31–34 is defined everywhere except at $x = 1$. Where possible, define f at 1 so that it becomes continuous at 1.

31. $f(x) = \dfrac{x^2 - 1}{x - 1}$.

32. $f(x) = \dfrac{1}{x - 1}$.

33. $f(x) = \dfrac{x - 1}{|x - 1|}$.

34. $f(x) = \dfrac{(x - 1)^2}{|x - 1|}$.

35. Let $f(x) = \begin{cases} x^2, & x < 1 \\ Ax - 3, & x \geq 1. \end{cases}$ Find A given that f is continuous at 1.

36. Let $f(x) = \begin{cases} A^2 x^2, & x \leq 2 \\ (1 - A)x, & x > 2. \end{cases}$ For what values of A is f continuous at 2?

37. Give necessary and sufficient conditions on A and B for the function

$$f(x) = \begin{cases} Ax - B, & x \leq 1 \\ 3x, & 1 < x < 2 \\ Bx^2 - A, & 2 \leq x \end{cases}$$

to be continuous at $x = 1$ but discontinuous at $x = 2$.

38. Give necessary and sufficient conditions on A and B for the function in Exercise 37 to be continuous at $x = 2$ but discontinuous at $x = 1$.

In Exercises 39–42, define the function at 5 so that it becomes continuous at 5.

39. $f(x) = \dfrac{\sqrt{x + 4} - 3}{x - 5}$.

40. $f(x) = \dfrac{\sqrt{x + 4} - 3}{\sqrt{x} - 5}$.

41. $f(x) = \dfrac{\sqrt{2x - 1} - 3}{x - 5}$.

42. $f(x) = \dfrac{\sqrt{x^2 - 7x + 16} - \sqrt{6}}{(x - 5)\sqrt{x + 1}}$.

In Exercises 43–45, at what points (if any) is the given function continuous?

43. $f(x) = \begin{cases} 1, & x \text{ rational} \\ 0, & x \text{ irrational.} \end{cases}$

44. $g(x) = \begin{cases} x, & x \text{ rational} \\ 0, & x \text{ irrational.} \end{cases}$

45. $g(x) = \begin{cases} 2x, & x \text{ an integer} \\ x^2, & \text{otherwise.} \end{cases}$

46. Three important functions in science and engineering are:

1. The *Heaviside function* $H_c(x) = \begin{cases} 0, & x < c \\ 1, & x \geq c, \end{cases}$

   where c is a given number.

2. The *unit pulse function*

$$P_{\epsilon,c}(x) = \tfrac{1}{\epsilon}[H_c(x) - H_{c+\epsilon}(x)]$$

3. The *Dirac delta function* $\delta_c(x) = \lim\limits_{\epsilon \to 0} P_{\epsilon,c}(x)$.

   (a) Graph the functions H_c, $P_{\epsilon,c}$, and δ_c.
   (b) Determine where each of the functions is continuous.
   (c) Find $\lim\limits_{x \to c^-} H_c(x)$ and $\lim\limits_{x \to c^+} H_c(x)$. What can you say about $\lim\limits_{x \to c} H(x)$?

47. (*Important*) Prove that

$$f \text{ is continuous at } c \quad \text{iff} \quad \lim_{h \to 0} f(c + h) = f(c).$$

48. (*Important*) Let f and g be continuous at c. Prove that if:
   (a) $f(c) > 0$, then there exists $\delta > 0$ such that
       $f(x) > 0$ for all $x \in (c - \delta, c + \delta)$.
   (b) $f(c) < 0$, then there exists $\delta > 0$ such that
       $f(x) < 0$ for all $x \in (c - \delta, c + \delta)$.
   (c) $f(c) < g(c)$, then there exists $\delta > 0$ such that
       $f(x) < g(x)$ for all $x \in (c - \delta, c + \delta)$.

49. Suppose that f has a discontinuity at c which is not removable. Let g be a function such that $g(x) = f(x)$ except, possibly, at a finite set of points x_1, x_2, $\ldots$, x_n. Show that g also has a nonremovable discontinuity at c.

50. (a) Prove that if f is continuous everywhere, then $|f|$ is continuous everywhere.
   (b) Give an example to show that the continuity of $|f|$ does not imply the continuity of f.
   (c) Give an example of a function f such that f is continuous nowhere, but $|f|$ is continuous everywhere.

51. Suppose the function f has the property that there exists a number B such that

$$|f(x) - f(c)| \leq B|x - c|$$

for all x in the interval $(c - p, c + p)$, $p > 0$. Prove that f is continuous at c.

52. Suppose the function f has the property that

$$|f(x) - f(t)| \leq |x - t|$$

for each pair of points x, t in the interval (a, b). Prove that f is continuous on (a, b).

53. Prove that if

$$\lim_{h \to 0} \frac{f(c + h) - f(c)}{h} = L$$

exists, then f is continuous at c.

54. Suppose that the function f is continuous on $(-\infty, \infty)$. Show that f can be written

$$f = f_e + f_o,$$

where f_e is an even function which is continuous on $(-\infty, \infty)$ and f_o is an odd function which is continuous on $(-\infty, \infty)$. HINT: See Exercise 56, Section 1.7.

■ 2.5 THE PINCHING THEOREM; TRIGONOMETRIC LIMITS

Figure 2.5.1 shows the graphs of three functions f, g, h. Suppose that, as suggested by the figure, for x close to c, f is trapped between g and h. (The values of these functions at c itself are irrelevant.) If, as x tends to c, both $g(x)$ and $h(x)$ tend to the same limit L, then $f(x)$ also tends to L. This idea is made precise in what we call *the pinching theorem.*

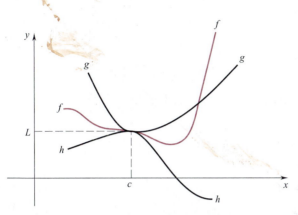

Figure 2.5.1

THEOREM 2.5.1   THE PINCHING THEOREM

Let $p > 0$. Suppose that, for all x such that $0 < |x - c| < p$,

$$h(x) \leq f(x) \leq g(x).$$

If

$$\lim_{x \to c} h(x) = L \quad \text{and} \quad \lim_{x \to c} g(x) = L,$$

then

$$\lim_{x \to c} f(x) = L.$$

PROOF   Let $\epsilon > 0$. Let $p > 0$ be such that

$$\text{if} \quad 0 < |x - c| < p, \quad \text{then} \quad h(x) \leq f(x) \leq g(x).$$

Choose $\delta_1 > 0$ such that

$$\text{if} \quad 0 < |x - c| < \delta_1, \quad \text{then} \quad L - \epsilon < h(x) < L + \epsilon.$$

Choose $\delta_2 > 0$ such that

$$\text{if} \quad 0 < |x - c| < \delta_2, \quad \text{then} \quad L - \epsilon < g(x) < L + \epsilon.$$

Let $\delta = \min \{p, \delta_1, \delta_2\}$. For x satisfying $0 < |x - c| < \delta$, we have

$$L - \epsilon < h(x) \leq f(x) \leq g(x) < L + \epsilon,$$

and thus

$$|f(x) - L| < \epsilon. \quad \square$$

$(x-1)(x+1)$

Remark With obvious modifications, the pinching theorem also holds for one-sided limits. We do not spell out the details here because we will be working with two-sided limits throughout this section. ❏

We come now to some trigonometric limits. All calculations are based on radian measure.

As our first application of the pinching theorem we prove that

(2.5.2)

$$\lim_{x \to 0} \sin x = 0.$$

PROOF To follow the argument, see Figure 2.5.2.†

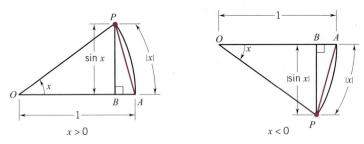

Figure 2.5.2

For small $x \neq 0$

$$0 < |\sin x| = \text{length of } \overline{BP} < \text{length of } \overline{AP} < \text{length of } \overparen{AP} = |x|.$$

Thus, for such x

$$0 < |\sin x| < |x|.$$

Since

$$\lim_{x \to 0} 0 = 0 \qquad \text{and} \qquad \lim_{x \to 0} |x| = 0.$$

we know from the pinching theorem that

$$\lim_{x \to 0} |\sin x| = 0 \qquad \text{and therefore} \qquad \lim_{x \to 0} \sin x = 0. \quad ❏$$

From this it follows readily that

(2.5.3)

$$\lim_{x \to 0} \cos x = 1.$$

PROOF In general, $\cos^2 x + \sin^2 x = 1$. For x close to 0, the cosine is positive and we have

$$\cos x = \sqrt{1 - \sin^2 x}.$$

As x tends to 0, $\sin x$ tends to 0, $\sin^2 x$ tends to 0, and therefore $\cos x$ tends to 1. ❏

---

† Recall that in a circle of radius 1, a central angle of x radians subtends an arc of length $|x|$.

Next we show that the sine and cosine are everywhere continuous; which is to say, for all real numbers c,

(2.5.4)

$$\lim_{x \to c} \sin x = \sin c \quad \text{and} \quad \lim_{x \to c} \cos x = \cos c.$$

PROOF Take any real number c. By (2.2.5) we can write

$$\lim_{x \to c} \sin x \quad \text{as} \quad \lim_{h \to 0} \sin (c + h).$$

This form of the limit suggests that we use the addition formula

$$\sin (c + h) = \sin c \cos h + \cos c \sin h.$$

Since $\sin c$ and $\cos c$ are constants, we have

$$\lim_{h \to 0} \sin (c + h) = (\sin c)(\lim_{h \to 0} \cos h) + (\cos c)(\lim_{h \to 0} \sin h)$$
$$= (\sin c)(1) + (\cos c)(0) = \sin c.$$

The proof that $\lim_{x \to c} \cos x = \cos c$ is left to you. ❏

The remaining trigonometric functions

$$\tan x = \frac{\sin x}{\cos x}, \quad \cot x = \frac{\cos x}{\sin x}, \quad \sec x = \frac{1}{\cos x}, \quad \csc x = \frac{1}{\sin x}$$

are all continuous where defined. Justification? They are all quotients of continuous functions.

We turn now to two limits, the importance of which will become clear in Chapter 3:

(2.5.5)

$$\lim_{x \to 0} \frac{\sin x}{x} = 1 \quad \text{and} \quad \lim_{x \to 0} \frac{1 - \cos x}{x} = 0.$$

Remark These results were suggested in Section 2.1 using a numerical approach. See Example 14 and Exercise 60 in Section 2.1. ❏

PROOF We show that

$$\lim_{x \to 0} \frac{\sin x}{x} = 1$$

by using some simple geometry and the pinching theorem. For any x satisfying $0 < x < \pi/2$ (see Figure 2.5.3), length of $\overline{PB} = \sin x$, length of $\overline{OB} = \cos x$, and length $\overline{OA} = 1$. Also, since triangle OAQ is a right triangle, we have $\tan x = \overline{QA}/1$ or

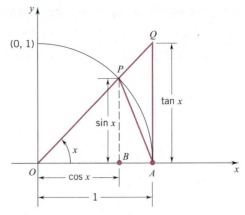

Figure 2.5.3

$\overline{QA} = \tan x$. Now

$$\text{Area of triangle } OAP = \tfrac{1}{2}(1) \sin x = \tfrac{1}{2} \sin x$$

$$\text{Area of sector } OAP = \tfrac{1}{2}(1)^2 x = \tfrac{1}{2}x$$

$$\text{Area of triangle } OAQ = \tfrac{1}{2}(1) \tan x = \tfrac{1}{2} \frac{\sin x}{\cos x}.$$

Since triangle $OAP \subseteq$ sector $OAP \subseteq$ triangle OAQ (and these are all proper containments), we have

$$\tfrac{1}{2} \sin x < \tfrac{1}{2}x < \tfrac{1}{2} \frac{\sin x}{\cos x}$$

$$1 < \frac{x}{\sin x} < \frac{1}{\cos x}$$

$$\cos x < \frac{\sin x}{x} < 1.$$

(taking reciprocals, see Exercise 67, Section 1.3)

This inequality was derived for $x > 0$, but since

$$\cos(-x) = \cos x \qquad \text{and} \qquad \frac{\sin(-x)}{-x} = \frac{-\sin x}{-x} = \frac{\sin x}{x},$$

this inequality also holds for $x < 0$.

We can now apply the pinching theorem.   Since

$$\lim_{x \to 0} \cos x = 1 \qquad \text{and} \qquad \lim_{x \to 0} 1 = 1,$$

we can conclude that

$$\lim_{x \to 0} \frac{\sin x}{x} = 1.$$

We can now show that

$$\lim_{x \to 0} \frac{1 - \cos x}{x} = 0.$$

For small $x \neq 0$, $\cos x$ is close to 1 and so $\cos x \neq -1$. Therefore, we can write

$$\frac{1 - \cos x}{x} = \left(\frac{1 - \cos x}{x}\right)\left(\frac{1 + \cos x}{1 + \cos x}\right)^{\dagger}$$

$$= \frac{1 - \cos^2 x}{x(1 + \cos x)}$$

$$= \frac{\sin^2 x}{x(1 + \cos x)}$$

$$= \left(\frac{\sin x}{x}\right)\left(\frac{\sin x}{1 + \cos x}\right).$$

Now, since

$$\lim_{x \to 0} \frac{\sin x}{x} = 1 \quad \text{and} \quad \lim_{x \to 0} \frac{\sin x}{1 + \cos x} = 0,$$

it follows that

$$\lim_{x \to 0} \frac{1 - \cos x}{x} = 0. \quad \blacksquare$$

We are now in a position to evaluate a variety of trigonometric limits.

Example 1 Find

$$\lim_{x \to 0} \frac{\sin 4x}{x}.$$

SOLUTION You know that

$$\lim_{x \to 0} \frac{\sin x}{x} = 1.$$

From this it follows that

$$\lim_{x \to 0} \frac{\sin 4x}{4x} = 1. \qquad \text{(see Exercise 34)}$$

Now,

$$\lim_{x \to 0} \frac{\sin 4x}{x} = \lim_{x \to 0} 4\left(\frac{\sin 4x}{4x}\right) = 4 \lim_{x \to 0} \frac{\sin 4x}{4x} = 4. \quad \square$$

Example 2 Find

$$\lim_{x \to 0} x \cot 3x.$$

SOLUTION We begin by writing

$$x \cot 3x = x \frac{\cos 3x}{\sin 3x}.$$

---

† This "trick" is a fairly common procedure with trigonometric expressions. It is much like using "conjugates" to revise algebraic expressions:

$$\frac{3}{4 + \sqrt{2}} = \frac{3}{4 + \sqrt{2}} \cdot \frac{4 - \sqrt{2}}{4 - \sqrt{2}} = \frac{3(4 - \sqrt{2})}{14}.$$

Since

$$\lim_{x \to 0} \frac{\sin x}{x} = 1 \qquad \text{and equivalently} \qquad \lim_{x \to 0} \frac{x}{\sin x} = 1,$$

we would like to "pair off" $\sin 3x$ with $3x$. Thus, we write

$$x \frac{\cos 3x}{\sin 3x} = \frac{1}{3} \left(\frac{3x}{\sin 3x} \right) (\cos 3x).$$

Now we take the limit:

$$\lim_{x \to 0} x \cot 3x = \frac{1}{3} \left(\lim_{x \to 0} \frac{3x}{\sin 3x} \right) \left(\lim_{x \to 0} \cos 3x \right) = \frac{1}{3} (1)(1) = \frac{1}{3}. \qquad \square$$

Example 3 Find

$$\lim_{x \to \pi/4} \frac{\sin (x - \frac{1}{4}\pi)}{(x - \frac{1}{4}\pi)^2}.$$

SOLUTION

$$\frac{\sin (x - \frac{1}{4}\pi)}{(x - \frac{1}{4}\pi)^2} = \left[\frac{\sin (x - \frac{1}{4}\pi)}{(x - \frac{1}{4}\pi)} \right] \cdot \frac{1}{x - \frac{1}{4}\pi}.$$

We know that

$$\lim_{x \to \pi/4} \frac{\sin (x - \frac{1}{4}\pi)}{x - \frac{1}{4}\pi} = 1.$$

Since $\lim_{x \to \pi/4} (x - \frac{1}{4}\pi) = 0$, you can see by Theorem 2.3.10 that

$$\lim_{x \to \pi/4} \frac{\sin (x - \frac{1}{4}\pi)}{(x - \frac{1}{4}\pi)^2} \qquad \text{does not exist.} \qquad \square$$

Example 4 Find

$$\lim_{x \to 0} \frac{x^2}{\sec x - 1}.$$

SOLUTION The evaluation of this limit requires a little imagination. Since both the numerator and denominator approach zero as x approaches zero, it is not clear what happens to the fraction; the limit is an indeterminate form of type $0/0$. However, we can rewrite the fraction in a more amenable form by using the trick just introduced:

$$\frac{x^2}{\sec x - 1} = \frac{x^2}{\sec x - 1} \left(\frac{\sec x + 1}{\sec x + 1} \right)$$

$$= \frac{x^2(\sec x + 1)}{\sec^2 x - 1} = \frac{x^2(\sec x + 1)}{\tan^2 x}$$

$$= \frac{x^2 \cos^2 x \, (\sec x + 1)}{\sin^2 x}$$

$$= \left(\frac{x}{\sin x} \right)^2 (\cos^2 x)(\sec x + 1).$$

Since each of these expressions has a limit as x tends to 0, the fraction we began with has a limit:

$$\lim_{x \to 0} \frac{x^2}{\sec x - 1} = \lim_{x \to 0} \left(\frac{x}{\sin x} \right)^2 \cdot \lim_{x \to 0} \cos^2 x \cdot \lim_{x \to 0} (\sec x + 1)$$

$$= (1)(1)(2) = 2. \quad \square$$

EXERCISES 2.5

In Exercises 1–32, evaluate the limits that exist.

1. $\lim\limits_{x \to 0} \dfrac{\sin 3x}{x}$.

2. $\lim\limits_{x \to 0} \dfrac{2x}{\sin x}$.

3. $\lim\limits_{x \to 0} \dfrac{3x}{\sin 5x}$.

4. $\lim\limits_{x \to 0} \dfrac{\sin 3x}{2x}$.

5. $\lim\limits_{x \to 0} \dfrac{\sin 4x}{\sin 2x}$.

6. $\lim\limits_{x \to 0} \dfrac{\sin 3x}{5x}$.

7. $\lim\limits_{x \to 0} \dfrac{\sin x^2}{x}$.

8. $\lim\limits_{x \to 0} \dfrac{\sin x^2}{x^2}$.

9. $\lim\limits_{x \to 0} \dfrac{\sin x}{x^2}$.

10. $\lim\limits_{x \to 0} \dfrac{\sin^2 x^2}{x^2}$.

11. $\lim\limits_{x \to 0} \dfrac{\sin^2 3x}{5x^2}$.

12. $\lim\limits_{x \to 0} \dfrac{\tan^2 3x}{4x^2}$.

13. $\lim\limits_{x \to 0} \dfrac{2x}{\tan 3x}$.

14. $\lim\limits_{x \to 0} \dfrac{4x}{\cot 3x}$.

15. $\lim\limits_{x \to 0} x \csc x$.

16. $\lim\limits_{x \to 0} \dfrac{\cos x - 1}{2x}$.

17. $\lim\limits_{x \to 0} \dfrac{x^2}{1 - \cos 2x}$.

18. $\lim\limits_{x \to 0} \dfrac{x^2 - 2x}{\sin 3x}$.

19. $\lim\limits_{x \to 0} \dfrac{1 - \sec^2 2x}{x^2}$.

20. $\lim\limits_{x \to 0} \dfrac{1}{2x \csc x}$.

21. $\lim\limits_{x \to 0} \dfrac{2x^2 + x}{\sin x}$.

22. $\lim\limits_{x \to 0} \dfrac{1 - \cos 4x}{9x^2}$.

23. $\lim\limits_{x \to 0} \dfrac{\tan 3x}{2x^2 + 5x}$.

24. $\lim\limits_{x \to 0} x^2(1 + \cot^2 3x)$.

25. $\lim\limits_{x \to 0} \dfrac{\sec x - 1}{x \sec x}$.

26. $\lim\limits_{x \to \pi/4} \dfrac{1 - \cos x}{x}$.

27. $\lim\limits_{x \to \pi/4} \dfrac{\sin x}{x}$.

28. $\lim\limits_{x \to 0} \dfrac{\sin^2 x}{x(1 - \cos x)}$.

29. $\lim\limits_{x \to \pi/2} \dfrac{\cos x}{x - \frac{1}{2}\pi}$.

30. $\lim\limits_{x \to \pi} \dfrac{\sin x}{x - \pi}$.

31. $\lim\limits_{x \to \pi/4} \dfrac{\sin (x + \frac{1}{4}\pi) - 1}{x - \frac{1}{4}\pi}$.

32. $\lim\limits_{x \to \pi/6} \dfrac{\sin (x + \frac{1}{3}\pi) - 1}{x - \frac{1}{6}\pi}$.

33. Show that $\lim\limits_{x \to c} \cos x = \cos c$ for all real numbers c.

34. Show that

if $\lim\limits_{x \to 0} f(x) = L$ then $\lim\limits_{x \to 0} f(ax) = L$ for all $a \neq 0$.

HINT: Let $\epsilon > 0$. If $\delta_1 > 0$ "works" for the first limit, then $\delta = \delta_1/|a|$ "works" for the second limit.

In Exercises 35–38, determine whether $\lim\limits_{h \to 0} (f(a + h) - f(a))/h$ exists, and write an equation for the tangent line to the graph of f at $(a, f(a))$ when the limit does exist.

35. $f(x) = \sin x$, $a = \pi/4$. HINT: Use the addition formula for the sine function.

36. $f(x) = \cos x$, $a = \pi/3$.

37. $f(x) = \cos 2x$, $a = \pi/6$.

38. $f(x) = \sin 3x$, $a = \pi/2$.

39. Prove that $\lim\limits_{x \to 0} x \sin (1/x) = 0$. HINT: Use the pinching theorem.

40. Prove that $\lim\limits_{x \to \pi} (x - \pi) \cos^2[1/(x - \pi)] = 0$.

41. Prove that $\lim\limits_{x \to 1} |x - 1| \sin x = 0$.

42. Let f be the Dirichlet function

$$f(x) = \begin{cases} 1 & x \text{ rational} \\ 0 & x \text{ irrational.} \end{cases}$$

Prove that $\lim\limits_{x \to 0} x f(x) = 0$.

43. Prove that if there is a number B such that $|f(x)| \leq B$ for all $x \neq 0$, then $\lim\limits_{x \to 0} x f(x) = 0$. NOTE: Exercises 39 and 42 are special cases of this general result.

44. Prove that if there is a number B such that $|f(x)/x| \leq B$ for all $x \neq 0$, then $\lim\limits_{x \to 0} f(x) = 0$.

45. Prove that if there is a number B such that $|f(x) - L|/|x - c| \leq B$ for all $x \neq c$, then $\lim\limits_{x \to c} f(x) = L$.

■ 2.6 TWO BASIC THEOREMS

A function that is continuous on an interval does not "skip" any values, and thus its graph is an "unbroken curve." There are no "holes" in it and no "jumps." This is the idea behind the *intermediate-value theorem*.

> ### THEOREM 2.6.1 THE INTERMEDIATE-VALUE THEOREM
>
> If f is continuous on $[a, b]$ and C is a number between $f(a)$ and $f(b)$, then there is at least one number c in the interval $[a, b]$ for which $f(c) = C$.

Figure 2.6.1

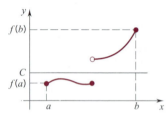

Figure 2.6.2

We illustrate the theorem in Figure 2.6.1. What can happen in the discontinuous case is illustrated in Figure 2.6.2. There the number C has been "skipped." You can find a proof of the intermediate-value theorem in Appendix B. We will assume the result and use it.

We apply the intermediate-value theorem to the important problem of approximating the zeros of a function. In particular, suppose that f is continuous on $[a, b]$ and that either

$$f(a) < 0 < f(b) \qquad \text{or} \qquad f(b) < 0 < f(a). \qquad \text{(Figure 2.6.3)}$$

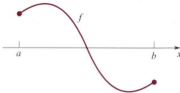

Figure 2.6.3

Then, by the intermediate-value theorem, we know that the equation $f(x) = 0$ has at least one root in $[a, b]$. For simplicity let's assume that there is only one such root and call it c. How can we estimate the location of c? The intermediate-value theorem itself gives us no clue. The simplest method of finding the approximate location of c is called the *bisection method*.

The Bisection Method

The bisection method is an iterative process. A basic step is iterated (carried out repeatedly) until the goal is reached, in this case an approximation for c that is as accurate as we wish.

It is standard practice to label the elements of successive iterations by using as subscripts $n = 1, 2, 3$, and so forth. We begin by setting $u_1 = a$ and $v_1 = b$. We now bisect $[u_1, v_1]$. If c is the midpoint of $[u_1, v_1]$, our search is over. If not, then c lies in one of the halves of $[u_1, v_1]$. We call that half $[u_2, v_2]$. If c is the midpoint of $[u_2, v_2]$, then our search is over. If not, then c lies in one of the halves of $[u_2, v_2]$. We call that half $[u_3, v_3]$ and continue. The first three iterations for a particular function are depicted in Figure 2.6.4.

After n bisections we are examining the midpoint m_n of the interval $[u_n, v_n]$. We can thus be certain that

$$|c - m_n| \le \frac{1}{2}(v_n - u_n) = \frac{1}{2}\left(\frac{v_{n-1} - u_{n-1}}{2}\right) = \cdots = \frac{b - a}{2^n}.$$

Thus m_n approximates c within $(b - a)/2^n$. If we want m_n to approximate c within a

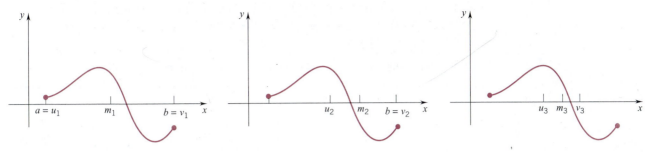

Figure 2.6.4

given ϵ, then we must carry out the iteration to the point where

$$\frac{b - a}{2^n} < \epsilon.$$

Example 1 The function $f(x) = x^2 - 2$ is zero at $x = \sqrt{2}$. Since $f(1) < 0$ and $f(2) > 0$, we can estimate $\sqrt{2}$ by applying the bisection method to f on the interval $[1, 2]$. Let's say we want a numerical estimate accurate to within 0.001. To achieve this degree of accuracy we must carry out the iteration to the point where

$$\frac{2 - 1}{2^n} < 0.001,$$

or, equivalently, to the point where $2^n > 1000$. As you can check, this condition is met by taking $n \geq 10$. Thus 10 bisections will suffice.

$u_1 = 1, v_1 = 2, m_1 = 1.5:$
    $f(u_1) < 0, \quad f(v_1) > 0, \quad f(m_1) = (1.5)^2 - 2 = 0.25 > 0,$
$u_2 = 1, v_2 = 1.5, m_2 = 1.25:$
    $f(u_2) < 0, \quad f(v_2) > 0, \quad f(m_2) = (1.25)^2 - 2 = -0.4375 < 0,$
$u_3 = 1.25, v_3 = 1.5, m_3 = 1.375:$
    $f(u_3) < 0, \quad f(v_3) > 0, \quad f(m_3) = (1.375)^2 - 2 = -0.109375 < 0,$
$u_4 = 1.375, v_4 = 1.5, m_4 = 1.4375:$
    and so on.

The complete computations are summarized in Table 2.6.1. To increase readability, all the entries in the table have been rounded off to 5 decimal places.

■ **Table 2.6.1**

n	u_n	v_n	m_n	$f(m_n)$
1	1.00000	2.00000	1.50000	0.25000
2	1.00000	1.50000	1.25000	-0.43750
3	1.25000	1.50000	1.37500	-0.10938
4	1.37500	1.50000	1.43750	0.06641
5	1.37500	1.43750	1.40625	-0.02246
6	1.40625	1.43750	1.42187	0.02171
7	1.40625	1.42187	1.41406	-0.00043
8	1.41406	1.42187	1.41797	0.01064
9	1.41406	1.41797	1.41601	0.00508
10	1.41406	1.41601	1.41503	0.00231

We can conclude that $m_{10} = 1.41503$ approximates $\sqrt{2}$ within 0.001. ❑

Example 2 Let $f(x) = \frac{1}{2}x^4 - 3x^2 + 4x + 4$. Use the intermediate-value theorem to show that f has at least two zeros in the interval $[-3, 0]$. Sketch the graph of f and find approximations for the two zeros.

$-4 \le x \le 1, -10 \le y \le 10$

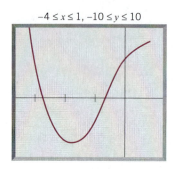

Figure 2.6.5

SOLUTION The values of f at the integers from -3 to 0 are as follows:

x	-3	-2	-1	0
$f(x)$	5.5	-8	-2.5	4

Thus f has a zero r_1 in the interval $[-3, -2]$ and a zero r_2 in the interval $[-1, 0]$.

The graph of f shown in Figure 2.6.5 was obtained with a graphics calculator. Using the zoom feature, we found that approximate values for these zeros are: $r_1 \cong -2.80$ and $r_2 \cong -0.68$. ❑

Boundedness Properties

Suppose that f is a function which is defined on an interval I. Then f is said to be *bounded on I* if there exist numbers k and K such that

$$k \le f(x) \le K \quad \text{for all} \quad x \in I.$$

Equivalently, f is bounded on I if there exists a number B such that $|f(x)| \le B$ for all $x \in I$. The function f is *unbounded on I* if it is not bounded.

The function f has a *maximum value* on I if there is a number $d \in I$ such that $f(d) \ge f(x)$ for all $x \in I$. The value $f(d) = M$ is called the *maximum value of f on I*. Similarly, f has a *minimum value* on I if there is a number $c \in I$ such that $f(c) \le f(x)$ for all $x \in I$; the value $f(c) = m$ is called the *minimum value of f on I*. The maximum and minimum values of f, if they exist, are called the *extreme values of f on I*.

The trigonometric functions sine and cosine are bounded on $(-\infty, \infty)$ since

$$-1 \le \sin x \le 1 \qquad \text{and} \qquad -1 \le \cos x \le 1$$

for all x. Also, 1 is the maximum value of both sine and cosine, and -1 is the minimum value of both functions: $\sin(\pi/2) = 1$, $\cos(0) = 1$, $\sin(3\pi/2) = -1$, and $\cos(\pi) = -1$. The sine and cosine functions also illustrate the fact that the maximum and/or minimum values of a function may be attained at more than one point on the interval I; $\sin x$ and $\cos x$ attain their extreme values at infinitely many points on $(-\infty, \infty)$. See Figure 2.6.6.

The tangent function, $\tan x$, is unbounded on $(-\pi/2, \pi/2)$ and is bounded on $(-\pi/4, \pi/4)$. See Figure 2.6.7.

If f has a maximum value M and a minimum value m on I, then clearly f is bounded on I since

$$m \le f(x) \le M$$

for all $x \in I$. On the other hand, a function can be bounded on an interval I without having a maximum value or a minimum value on the interval. For example, $\tan x$ has neither a maximum nor a minimum value on $(-\pi/4, \pi/4)$ (see Figure 2.6.7).

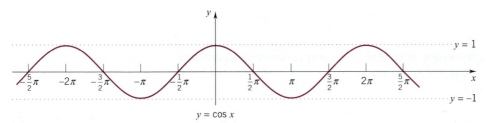

$y = \cos x$

Figure 2.6.6

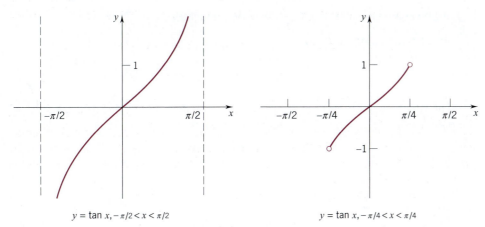

$y = \tan x, -\pi/2 < x < \pi/2$

$y = \tan x, -\pi/4 < x < \pi/4$

Figure 2.6.7

Example 3   Let

$$g(x) = \begin{cases} 1/x^2, & x > 0 \\ 0, & x = 0. \end{cases}$$
(Figure 2.6.8)

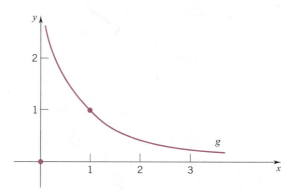

Figure 2.6.8

It is clear from the graph that g is unbounded on $[0, \infty)$. On the other hand, g is bounded on $[1, \infty)$, since

$$0 < g(x) \le 1$$

for each x on this interval. Note that g has a maximum value on $[1, \infty)$, $g(1) = 1$, but no minimum value. Why? ❏

THEOREM 2.6.2   BOUNDEDNESS: EXTREME-VALUE THEOREM

Let f be continuous on the closed, bounded interval $[a, b]$.   Then

(1) f is bounded on $[a, b]$,

(2) f attains both its maximum value M and its minimum value m on $[a, b]$.

The situation is illustrated in Figure 2.6.9.   The maximum value M is taken on at the point marked d, and the minimum value m is taken on at the point marked c.

Figure 2.6.9

It is interesting to note that the full hypothesis of this theorem is needed.   That is, if either the continuity requirement on the function f, or the requirement that the interval be closed and bounded, is dropped, then the conclusion of the theorem does not follow. As an example, we can take the function

$$f(x) = \begin{cases} 3, & x = 1 \\ x, & 1 < x < 5 \\ 3, & x = 5. \end{cases}$$

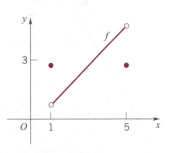

Figure 2.6.10

The graph of f is pictured in Figure 2.6.10. The function is defined on $[1, 5]$, but it takes on neither a maximum nor a minimum value.

If, instead of dropping the continuity requirement, we drop the requirement that the interval be of the form $[a, b]$, then again the result fails. For example, as previously illustrated, the tangent function, $\tan x$, is continuous and unbounded on $(-\pi/2, \pi/2)$; on the other hand, $\tan x$ is continuous and bounded on $(-\pi/4, \pi/4)$, but has neither a maximum value nor a minimum value on this interval (see Figure 2.6.7). The function g in Example 3 has a maximum value on $[1, \infty)$, but no minimum value.

A proof of the maximum-minimum theorem appears in Appendix B. Techniques for determining the maximum and minimum values of functions are described in Chapter 4. These techniques require an understanding of "differentiation," the subject to which we devote Chapter 3.

EXERCISES 2.6

In Exercises 1–16, sketch the graph of a function f that is defined on $[0, 1]$ and meets the given conditions (if possible).

1. f is continuous on $[0, 1]$, minimum value 0, maximum value $\frac{1}{2}$.

2. f is continuous on $[0, 1)$, minimum value 0, no maximum value.

3. f is continuous on $(0, 1)$, takes on the values 0 and 1, but does not take on the value $\frac{1}{2}$.

4. f is continuous on $[0, 1]$, takes on the values -1 and 1, but does not take on the value 0.

5. f is continuous on $[0, 1]$, maximum value 1, minimum value 1.

6. f is continuous on $[0, 1]$ and nonconstant, takes on no integer values.

7. f is continuous on $[0, 1]$, takes on no rational values.

8. f is not continuous on $[0, 1]$, takes on both a maximum value and a minimum value and every value in between.

9. f is continuous on $[0, 1]$, takes on only two distinct values.

10. f is continuous on $(0, 1)$, takes on only three distinct values.

11. f is not continuous on $(0, 1)$ and the range of f is an open, bounded interval.

12. f is not continuous on $(0, 1)$ and the range of f is a closed, bounded interval.

13. f is continuous on $(0, 1)$ and the range of f is an unbounded interval.

14. f is continuous on $[0, 1]$ and the range of f is an unbounded interval.

15. f is not continuous on $[0, 1]$ and the range of f is $[0, \infty)$.

16. f is continuous on $[0, 1)$ and the range of f is a closed, bounded interval.

In Exercises 17–19 use the intermediate-value theorem to show that there is a root of the given equation in the indicated interval.

17. $2x^3 - 4x^2 + 5x - 4 = 0$, $[1, 2]$.

18. $2\cos x - x^2 = 0$, $[0, \pi/2]$.

19. $\dfrac{1}{x-1} + \dfrac{1}{x-4} = 0$, $(1, 4)$.

20. Given that $\lim_{x \to c} f(x) = 0$ and that g is bounded on $(c - p, c + p)$ with $p > 0$, prove that

$$\lim_{x \to c} f(x)g(x) = 0.$$

21. *(Fixed-point property)* Show that if f is continuous and $0 \le f(x) \le 1$ for all $x \in [0, 1]$, then there exists at least one point c in $[0, 1]$ for which $f(c) = c$.
HINT: Apply the intermediate-value theorem to the function $g(x) = x - f(x)$.

22. Let n be a positive integer.
   (a) Prove that if $0 \leq a < b$, then $a^n < b^n$.
      HINT: Use mathematical induction.
   (b) Prove that every nonnegative real number x has a unique nonnegative nth root $x^{1/n}$. HINT: The existence of $x^{1/n}$ can be seen by applying the intermediate-value theorem to the function $f(t) = t^n$ for $t \geq 0$. The uniqueness follows from part (a).

23. Suppose that f is bounded on the interval $(-p, p)$ with $p > 0$. Define g on $(-p, p)$ by

$$g(x) = \begin{cases} xf(x), & x \neq 0 \\ 0, & x = 0. \end{cases}$$

Prove that g is continuous at 0.

24. Given that f and g are continuous functions on $[a, b]$, and that $f(a) < g(a)$ and $g(b) < f(b)$, show that there exists at least one number $c \in (a, b)$ such that $f(c) = g(c)$. HINT: Consider $f(x) - g(x)$.

25. The intermediate-value theorem can be used to prove that each polynomial equation of odd degree

$$x^n + a_{n-1}x^{n-1} + \cdots + a_1 x + a_0 \quad \text{with } n \text{ odd}$$

has at least one real root. Prove that the cubic equation

$$x^3 + ax^2 + bx + c = 0$$

has at least one real root.

26. Assume that at any given instant in time, the temperature on the earth's surface varies continuously with position. Prove that there is at least one pair of points diametrically opposite each other on the equator where the temperature is the same. HINT: Form a function that relates the temperature at diametrically opposite points on the equator.

27. Let $\mathscr{C}$ denote the set of all circles whose radii are less than or equal to 10 inches. Prove that there is at least one member of $\mathscr{C}$ whose area is exactly 250 square inches.

28. Fix a positive number P. Let $\mathscr{R}$ denote the set of all rectangles having perimeter P. Prove that there is a member of $\mathscr{R}$ that has maximum area. What are the dimensions of the rectangle that has maximum area? HINT: Express the area of an arbitrary element of R as a function of the length of one of its sides.

29. Given a circle C of radius R. Let $\mathscr{F}$ denote the set of all rectangles that can be inscribed in C. Prove that there is a member of $\mathscr{F}$ that has maximum area.

▷ In Exercises 30–33, use the bisection method to approximate the root of $f(x) = 0$ on the given interval to within 0.001.

30. $f(x) = 5 - x^2$; $[-3, -2]$.

31. $f(x) = x^2 - 3$; $[1, 2]$.

32. $f(x) = x^3 + x - 10$; $[1, 4]$.

33. $f(x) = \cos x - x$; $[0, 1]$.

▷ In Exercises 34–37, use the intermediate-value theorem to locate the zeros of the given function. Then use a graphing utility to approximate the zeros to within 0.001.

34. $f(x) = 2x^3 + 5x - 7$. **35.** $f(x) = x^3 - 5x + 3$.

36. $f(x) = x^5 - 3x + 1$.

37. $f(x) = x^3 - 2 \sin x + \frac{1}{2}$.

▷ In Exercises 38–41, use a graphing utility to graph f on the given interval. Is f bounded? If so, estimate its maximum and minimum values.

38. $f(x) = \dfrac{x^3 - 8x + 6}{4x + 1}$; $[0, 5]$.

39. $f(x) = \dfrac{2x}{1 + x^2}$; $[-2, 2]$.

40. $f(x) = \dfrac{\sin 2x}{x^2}$; $[-\pi/2, \pi/2]$.

41. $f(x) = \dfrac{1 - \cos x}{x^2}$; $[-2, 2]$.

■ CHAPTER HIGHLIGHTS

2.1 The Idea of Limit

Intuitive Interpretation of $\lim_{x \to c} f(x) = L$:

as x approaches c, $f(x)$ approaches L

or, equivalently,

for x close to c but different from c, $f(x)$ is close to L.

Restricting our attention to values of x to one side of c, we have *one-sided limits*:

$$\lim_{x \to c^-} f(x) \quad \text{and} \quad \lim_{x \to c^+} f(x).$$

2.2 Definition of Limit

limit of a function (p. 82)   left-hand limit (p. 88)   right-hand limit (p. 88)
four formulations of the limit statement (p. 87)

2.3 Some Limit Theorems

uniqueness of limit (p. 92)   stability of limit (p. 101)

The "arithmetic of limits": if $\lim\limits_{x \to c} f(x) = L$ and $\lim\limits_{x \to c} g(x) = M$, then

$$\lim\limits_{x \to c} [f(x) + g(x)] = L + M, \qquad \lim\limits_{x \to c} [\alpha f(x)] = \alpha L \quad \text{for each real } \alpha,$$

$$\lim\limits_{x \to c} [f(x)g(x)] = LM, \qquad \lim\limits_{x \to c} \frac{f(x)}{g(x)} = \frac{L}{M} \quad \text{provided } M \neq 0.$$

If $M = 0$ and $L \neq 0$, then $\lim\limits_{x \to c} f(x)/g(x)$ does not exist.

The same results hold for one-sided limits.

2.4 Continuity

continuity at c (p. 102)   removable discontinuity, jump discontinuity (p. 102)
one-sided continuity (p. 106)   continuity on intervals (p. 108)

The sum, difference, product, quotient, and composition of continuous functions are continuous where defined. Polynomials and the absolute value function are everywhere continuous. Rational functions are continuous on their domains. The square-root function is continuous at all positive numbers and continuous from the right at 0.

2.5 The Pinching Theorem; Trigonometric Limits

$$\text{pinching theorem (p. 111)} \qquad \lim\limits_{x \to 0} \frac{\sin x}{x} = 1, \qquad \lim\limits_{x \to 0} \frac{1 - \cos x}{x} = 0.$$

Each trigonometric function is continuous on its domain.

2.6 Two Basic Theorems

intermediate-value theorem (p. 118)   bisection method (p. 118)
bounded functions (p. 120)   maximum value (p. 120)
minimum value (p. 120)   extreme values (p. 120)
boundedness; extreme-value theorem (p. 122)

■ PROJECTS AND EXPLORATIONS USING TECHNOLOGY

To do these exercises you will need a graphics calculator or a computer with graphing capability. The majority of these problems are open-ended so different approaches may be used to solve them. You should be aware that different approaches can result in slight variations in the answers. Round your numerical answers to at least four decimal places. The rounding method that your calculator or computer uses also may cause variations in answers.

2.1 Let

$$F(x) = \frac{\sin(\tan x) - \tan(\sin x)}{x^7},$$

and consider the limit

$$\lim\limits_{x \to 0} F(x).$$

(a) Explain why F is defined for values of x near 0.   Then explain why

$$\lim_{x \to 0} [\sin(\tan x) - \tan(\sin x)] = 0.$$

(b) Use your calculator to evaluate F for values of x near 0 (be sure to use both positive and negative values) and approximate the limit.

(c) Graph F and approximate the limit.   Zoom in as close as you can and note the "hole" in the graph at $x = 0$.

(d) If you zoom in too close, the picture becomes distorted.   Explain why your calculator or computer has trouble with this graph.

2.2 The *greatest integer function,* denoted by $[\![x]\!]$, has the set of all real numbers for its domain and, for each real number x, $[\![x]\!]$ is the greatest integer that is less than or equal to x.   For example,

$$[\![1.75]\!] = 1, \quad [\![4]\!] = 4, \quad [\![-2.3]\!] = -3, \quad [\![-\tfrac{1}{2}]\!] = -1, \quad \text{and} \quad [\![-6]\!] = -6.$$

(a) Find the points of discontinuity of $f(x) = [\![x]\!]$.   Classify the discontinuities of f as being removable, jump, or neither.   Suppose that f is discontinuous at the number a.   Is f continuous from the right at a?   From the left?

(b) Let

$$g(x) = [\![3 \sin(x^2)]\!], \quad x \in [-\pi, \pi].$$

Find the points of discontinuity of g.   Then use the theorems on the composition of continuous functions to verify your answers.

(c) Let

$$h(x) = 3 \sin([\![x^2]\!]), \quad x \in [-\pi, \pi].$$

Find the points of discontinuity of h.   Again, use the theorems on the composition of continuous functions to verify your answers.

2.3 A number x is a *fixed point* of a function f if $f(x) = x$.

(a) Let $f(x) = \cos x$.   Use the continuity of f and the intermediate value theorem to explain why f must have a fixed point on the interval $[0, 1]$.   HINT: Consider $F(x) = f(x) - x$.

(b) Show that f has exactly one fixed point on $[0, 1]$ and use the bisection method to approximate it, accurate to two decimal places.   HINT: Verify first that f has a fixed point on the interval $[\tfrac{1}{2}, 1]$.

(c) Starting with $x_1 = f(1)$, generate the values $x_2 = f(x_1)$, $x_3 = f(x_2)$, . . . , $x_n = f(x_{n-1})$, ...   Note that these values appear to approach the fixed point found in part (b).

(d) Repeat parts (a)–(c) with the function

$$g(x) = \frac{3}{x^2 + 1}, \quad x \in [1, 2].$$

This time, observe that the values $x_1 = g(2)$, $x_2 = g(x_1)$, $x_3 = g(x_2)$, . . . , do not approach the fixed point.

(e) Draw the graph of each of these functions near their respective fixed points.   The graph of $f(x) = \cos x$ intersects the line $y = x$ at a "flatter" angle than $g(x) = 3/(x^2 + 1)$ does.   This is one of the properties of functions that will be studied in the following chapters.

2.4 The following modification of the bisection method for finding the zeros of a function is sometimes used.   The line connecting the endpoints of the graph of the function on the interval is calculated and its x-intercept is used instead of the midpoint. For example, if

$$f(x) = x^3 + x - 9,$$

then f has a zero in the interval $[1, 2]$. The line joining the endpoints $(1, -7)$ and $(2, 1)$ is $y = 8x - 15$ and its x-intercept is $15/8$. See the figure.

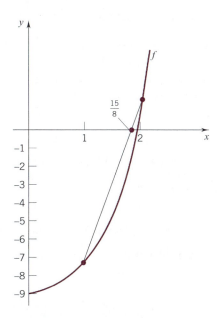

(a) Continuing with this example, calculate the next three x-intercepts. Notice that these values approach the zero of f very quickly, and that only one endpoint of the interval changes. Can you explain this from the "shape" of the graph of f near the zero? A difference between this method and the bisection method is that the lengths of the intervals for the bisection method must approach zero while the lengths of the intervals approach a positive constant with this method.

(b) While this method requires more work per step, it has the advantage of providing better approximations in fewer steps. How many bisection steps would be required to get an answer as accurate as that found in part (a)?

(c) A hybrid method uses the following modification of the method introduced in this exercise. If this method changes the same endpoint twice in succession, then one step of the midpoint method is done. Try the hybrid method on the function f just given. The hybrid method takes advantage of the good feature of the bisection method (the lengths of the intervals go to zero) and the modified bisection method (fewer steps are required).

(d) Suppose that a mistake is made and you start the hybrid method on the interval $[2, 3]$ instead of $[1, 2]$. Show that the procedure can still be used.

(e) Find an example of a function that has zeros in an interval but where the calculated line is horizontal so that the method terminates. Explain under what conditions (if any) the bisection method might terminate.

DIFFERENTIATION

CHAPTER

3

■ 3.1 THE DERIVATIVE

Introduction: The Tangent to a Curve

We begin with a function f, and on its graph we choose a point $(x, f(x))$. See Figure 3.1.1. What line, if any, should be called tangent to the graph at that point?

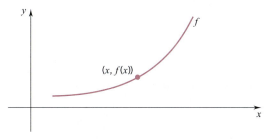

Figure 3.1.1

To answer this question, we choose a small number $h \neq 0$ and on the graph mark the point $(x + h, f(x + h))$. Now we draw the secant line that passes through these two points. The situation is pictured in Figure 3.1.2, first with $h > 0$ and then with $h < 0$.

As h tends to zero from the right (see the figure), the secant line tends to the limiting position indicated by the dashed line, and it tends to the same limiting position as h tends to zero from the left. The line at this limiting position is what we call "the tangent to the graph at the point $(x, f(x))$."

129

Figure 3.1.2

Since the approximating secant lines have slopes

(∗)
$$\frac{f(x + h) - f(x)}{h}$$
(verify this)

you can expect the tangent line, the limiting position of these secants, to have slope

(∗∗)
$$\lim_{h \to 0} \frac{f(x + h) - f(x)}{h}.$$

While (∗) measures the steepness of the line that passes through the points $(x, f(x))$ and $(x + h, f(x + h))$, (∗∗) measures the steepness of the graph at $(x, f(x))$ and is called the "slope of the graph" at that point.

Derivatives and Differentiation

In the introduction we spoke informally about tangent lines and gave a geometric interpretation to limits of the form

$$\lim_{h \to 0} \frac{f(x + h) - f(x)}{h}.$$

Here we begin the systematic study of such limits, what mathematicians call the theory of *differentiation*.

DEFINITION 3.1.1

A function f is said to be *differentiable at x* iff

$$\lim_{h \to 0} \frac{f(x + h) - f(x)}{h} \quad \text{exists.}$$

If this limit exists, it is called the *derivative of f at x*, and is denoted by $f'(x)$.†
The function f is a *differentiable function* if it is differentiable at each x in its domain.

Geometrically,

$$f'(x) = \lim_{h \to 0} \frac{f(x + h) - f(x)}{h}$$

---

† This prime notation goes back to the French mathematician Joseph-Louis Lagrange (1736–1813). Other notations are introduced later.

is the *slope of the graph at the point* $(x, f(x))$. The line that passes through the point $(x, f(x))$ with slope $f'(x)$ is called the *tangent line to the graph of f at* $(x, f(x))$. This is the line that best approximates the graph of *f* near the point $(x, f(x))$.

It is time to consider some examples.

Example 1   We begin with the quadratic function

$$f(x) = x^2. \qquad\qquad \text{(Figure 3.1.3)}$$

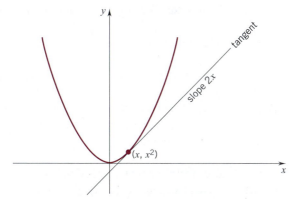

$$\left(x^2 + 2xh + h^2\right)(x+h)$$
$$x^3 + x^2h + h^2x$$

Figure 3.1.3

The domain of *f* is all real numbers.   Fix any number *x*.   To find $f'(x)$ we form the *difference quotient*

$$\frac{f(x + h) - f(x)}{h} = \frac{(x + h)^2 - x^2}{h}, \qquad h \neq 0.$$

Since

$$\frac{(x + h)^2 - x^2}{h} = \frac{x^2 + 2xh + h^2 - x^2}{h} = \frac{2xh + h^2}{h} = 2x + h,$$

we have

$$\frac{f(x + h) - f(x)}{h} = 2x + h,$$

and therefore

$$f'(x) = \lim_{h \to 0} \frac{f(x + h) - f(x)}{h} = \lim_{h \to 0} (2x + h) = 2x. \quad \square$$

Example 2   The domain of a linear function

$$f(x) = mx + b$$

is all real numbers.   For any number *x*,

$$f'(x) = m.$$

In other words, the slope is constantly *m*.   To verify this, note that for $h \neq 0$

$$\frac{f(x + h) - f(x)}{h} = \frac{[m(x + h) + b] - [mx + b]}{h} = \frac{mh}{h} = m.$$

It follows that

$$f'(x) = \lim_{h \to 0} \frac{f(x + h) - f(x)}{h} = \lim_{h \to 0} m = m.$$

This result can also be verified geometrically: $f'(x)$ is the slope of the tangent line to the graph of the linear function $f(x) = mx + b$ at the point $(x, f(x))$; but the tangent line to the graph of a linear function is simply the line itself (you should verify this). Thus, it follows that $f'(x) = m$. ❏

It is important to note that in order to apply the definition of the derivative

$$f'(x) = \lim_{h \to 0} \frac{f(x + h) - f(x)}{h},$$

x and $x + h$ must be in the domain of f and, since the limit involved is a two-sided limit, f must be defined on some open interval containing x. Also, in calculating the limit, x is fixed and the "variable" is h.

In our next example we will be dealing with the square-root function. Although this function is defined for all $x \geq 0$, you can expect a derivative only for $x > 0$.

Example 3 The square-root function

$$f(x) = \sqrt{x}, \qquad x \geq 0 \qquad\qquad \text{(Figure 3.1.4)}$$

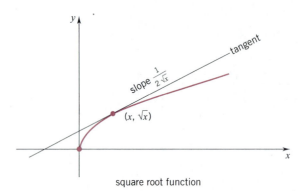

square root function

Figure 3.1.4

has derivative

$$f'(x) = \frac{1}{2\sqrt{x}}, \qquad \text{for } x > 0.$$

To verify this, fix an $x > 0$ and form the difference quotient

$$\frac{f(x + h) - f(x)}{h} = \frac{\sqrt{x + h} - \sqrt{x}}{h}, \qquad h \neq 0.$$

To simplify this expression, we multiply both the numerator and the denominator by $\sqrt{x + h} + \sqrt{x}$. This gives

$$\frac{f(x + h) - f(x)}{h} = \left(\frac{\sqrt{x + h} - \sqrt{x}}{h} \right) \left(\frac{\sqrt{x + h} + \sqrt{x}}{\sqrt{x + h} + \sqrt{x}} \right)$$

$$= \frac{(x + h) - x}{h(\sqrt{x + h} + \sqrt{x})} = \frac{1}{\sqrt{x + h} + \sqrt{x}}.$$

Thus,

$$f'(x) = \lim_{h \to 0} \frac{f(x+h) - f(x)}{h} = \frac{1}{2\sqrt{x}} \qquad \text{for } x > 0. \quad \square$$

To *differentiate* a function is to find its derivative. As illustrated by the preceding examples, when we differentiate a function f we get a new function f' called the *derivative* of f. The domain of f' is necessarily a subset of the domain of f. In particular,

$$\text{dom}(f') = \left\{ x \in \text{dom}(f): \lim_{h \to 0} \frac{f(x+h) - f(x)}{h} \text{ exists} \right\}.$$

Example 4 Here we differentiate the function

$$f(x) = \frac{1}{x}, \qquad x \neq 0.$$

(See Figure 3.1.5.) We form the difference quotient

$$\frac{f(x+h) - f(x)}{h} = \frac{\dfrac{1}{x+h} - \dfrac{1}{x}}{h}$$

and simplify:

$$\frac{\dfrac{1}{x+h} - \dfrac{1}{x}}{h} = \frac{\dfrac{x}{x(x+h)} - \dfrac{x+h}{x(x+h)}}{h} = \frac{\dfrac{-h}{x(x+h)}}{h} = \frac{-1}{x(x+h)}.$$

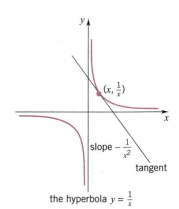

slope $-\dfrac{1}{x^2}$

tangent

the hyperbola $y = \frac{1}{x}$

Figure 3.1.5

Thus

$$f'(x) = \lim_{h \to 0} \frac{f(x+h) - f(x)}{h} = \lim_{h \to 0} \frac{-1}{x(x+h)} = -\frac{1}{x^2}. \quad \square$$

Evaluating Derivatives

Example 5 Find $f'(-2)$ given that

$$f(x) = 2 + 3x - x^2.$$

SOLUTION We let $x = -2$ in the definition:

$$f'(-2) = \lim_{h \to 0} \frac{f(-2+h) - f(-2)}{h}$$

$$= \lim_{h \to 0} \frac{[2 + 3(-2+h) - (-2+h)^2] - [2 + 3(-2) - (-2)^2]}{h}$$

$$= \lim_{h \to 0} \frac{[2 - 6 + 3h - (4 - 4h + h^2)] - (-8)]}{h}$$

$$= \lim_{h \to 0} \frac{7h - h^2}{h} = \lim_{h \to 0} (7 - h) = 7.$$

Alternatively, we can calculate $f'(x)$ for an arbitrary x:

$$f'(x) = \lim_{h \to 0} \frac{f(x + h) - f(x)}{h}$$

$$= \lim_{h \to 0} \frac{[2 + 3(x + h) - (x + h)^2] - [2 + 3x - x^2]}{h}$$

$$= \lim_{h \to 0} \frac{[2 + 3x + 3h - (x^2 + 2xh + h^2)] - [2 + 3x - x^2]}{h}$$

$$= \lim_{h \to 0} \frac{[3h - 2xh - h^2]}{h} = \lim_{h \to 0} (3 - 2x - h) = 3 - 2x.$$

Then, substituting -2 for x, we have

$$f'(-2) = 3 - 2(-2) = 7. \quad \square$$

You might be wondering why we showed you the alternate solution in Example 5. Suppose that, in addition to calculating $f'(-2)$, we also wanted to find $f'(0), f'(\frac{1}{2})$, and $f'(4)$? Obviously, it would be tedious to form each of the difference quotients and calculate the limits individually. On the other hand, since

$$f'(x) = 3 - 2x,$$

we have $f'(0) = 3 - 2(0) = 3$, $f'(\frac{1}{2}) = 3 - 2(\frac{1}{2}) = 2$, and $f'(4) = 3 - 2(4) = -5$. The advantage of having a "general" expression for f' should now be clear.

Example 6 Find $f'(-3)$ and $f'(1)$ given that

$$f(x) = \begin{cases} x^2, & x \le 1 \\ 2x - 1, & x > 1. \end{cases}$$ (see Figure 3.1.6)

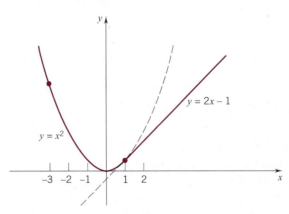

Figure 3.1.6

SOLUTION By definition

$$f'(-3) = \lim_{h \to 0} \frac{f(-3 + h) - f(-3)}{h}.$$

However, for all x sufficiently close to -3, $f(x) = x^2$, and from Example 1, we know that $f'(x) = 2x$. Thus, $f'(-3) = 2(-3) = -6$.

Now let's find

$$f'(1) = \lim_{h \to 0} \frac{f(1 + h) - f(1)}{h}.$$

Since f is not defined by the same formula on both sides of 1, we will evaluate this limit by taking one-sided limits. Note that $f(1) = 1^2 = 1$.

To the left of 1, $f(x) = x^2$. Thus

$$\lim_{h \to 0^-} \frac{f(1 + h) - f(1)}{h} = \lim_{h \to 0^-} \frac{(1 + h)^2 - 1}{h}$$

$$= \lim_{h \to 0^-} \frac{(1 + 2h + h^2) - 1}{h} = \lim_{h \to 0^-} (2 + h) = 2.$$

To the right of 1, $f(x) = 2x - 1$. Thus

$$\lim_{h \to 0^+} \frac{f(1 + h) - f(1)}{h} = \lim_{h \to 0^+} \frac{[2(1 + h) - 1] - 1}{h} = \lim_{h \to 0^+} 2 = 2.$$

The limit of the difference quotient exists and is 2:

$$f'(1) = \lim_{h \to 0} \frac{f(1 + h) - f(1)}{h} = 2. \quad \square$$

Tangent Lines and Normal Lines

We begin with a function f and choose a point (x_0, y_0) on the graph. If f is differentiable at x_0, then, as you know, the tangent line through this point has slope $f'(x_0)$. To get an equation for this tangent line, we use the point-slope formula

$$y - y_0 = m(x - x_0).$$

In this case $m = f'(x_0)$ and the equation becomes

(3.1.2)
$$y - y_0 = f'(x_0)(x - x_0).$$

The line through (x_0, y_0) that is perpendicular to the tangent is called the *normal line*. Since the slope of the tangent is $f'(x_0)$, the slope of the normal is $-1/f'(x_0)$, provided of course that $f'(x_0) \neq 0$. (*Remember*: For perpendicular lines, neither of which is vertical, $m_1 m_2 = -1$.) As an equation for this normal line we have

(3.1.3)
$$y - y_0 = -\frac{1}{f'(x_0)}(x - x_0), \quad \text{provided } f'(x_0) \neq 0.$$

Some tangents and normals are pictured in Figure 3.1.7.

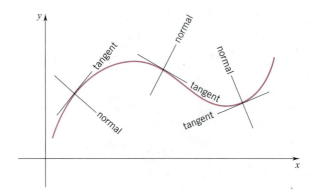

Figure 3.1.7

Example 7 Find equations for the tangent and normal to the graph of

$$f(x) = x^2$$

at the point $(-3, f(-3)) = (-3, 9)$.

SOLUTION From Example 1, we know that

$$f'(x) = 2x.$$

At the point $(-3, 9)$ the slope is $f'(-3) = -6$. As an equation for the tangent we have

$$y - 9 = -6(x - (-3)), \quad \text{which simplifies to} \quad y - 9 = -6(x + 3).$$

An equation for the normal can be written

$$y - 9 = \tfrac{1}{6}(x + 3). \quad \square$$

If $f'(x_0) = 0$, Equation (3.1.3) does not apply. If $f'(x_0) = 0$, the tangent line at (x_0, y_0) is horizontal (equation $y = y_0$) and the normal line is vertical (equation $x = x_0$). In Figure 3.1.8 you can see several instances of a horizontal tangent. In the first two cases the point of tangency is the origin, the tangent line is the x-axis ($y = 0$), and the normal line is the y-axis ($x = 0$).

Figure 3.1.8

A Note on Vertical Tangents

It is also possible for the graph of a function to have a vertical tangent. Figure 3.1.9 shows the graph of the cube-root function $f(x) = x^{1/3}$. The difference quotient at $x = 0$,

$$\frac{f(0 + h) - f(0)}{h} = \frac{h^{1/3} - 0}{h} = \frac{1}{h^{2/3}},$$

increases without bound as h tends to zero. The graph has a vertical tangent at the origin (the line $x = 0$) and it has a horizontal normal (the line $y = 0$).

Vertical tangents are mentioned here only in passing. They are discussed in detail in Section 4.7.

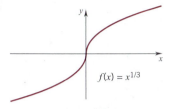

Figure 3.1.9

Differentiability and Continuity

Differentiability, like continuity, is a so-called "point concept": a function f is differentiable at the point x iff

$$\lim_{h \to 0} \frac{f(x + h) - f(x)}{h} \quad \text{exists.}$$

Also like continuity, we say that f is differentiable on the open interval (a, b) if it is differentiable at each $x \in (a, b)$. Similarly, f is differentiable on $(-\infty, b)$, (a, ∞), or $(-\infty, \infty)$ if it is differentiable at each x in the indicated interval.

For example, $f(x) = x^2$ and $f(x) = mx + b$ are each differentiable on $(-\infty, \infty)$, while $f(x) = \sqrt{x}$ is differentiable on $(0, \infty)$. The function $f(x) = 1/x$ is differentiable on $(-\infty, 0)$ and on $(0, \infty)$.

We can extend the definition of differentiability to include either of the endpoints of an interval [a, b] by introducing one-sided derivatives. This is developed in the Exercises. Finally, as stated in Definition 3.1.1, a function f is said to be a *differentiable function* if it is differentiable at each point in its domain.

A function can be continuous at some number x without being differentiable there. For example, the absolute-value function $f(x) = |x|$ is continuous at 0 (it is everywhere continuous), but it is not differentiable at 0:

$$\frac{f(0 + h) - f(0)}{h} = \frac{|0 + h| - |0|}{h} = \frac{|h|}{h} = \begin{cases} -1, & h < 0 \\ 1, & h > 0 \end{cases}$$

so that

$$\lim_{h \to 0^-} \frac{f(0 + h) - f(0)}{h} = -1, \qquad \lim_{h \to 0^+} \frac{f(0 + h) - f(0)}{h} = 1$$

and thus

$$\lim_{h \to 0} \frac{f(0 + h) - f(0)}{h} \quad \text{does not exist.} \quad ❏$$

In Figure 3.1.10 we have drawn the graph of the absolute-value function. The lack of differentiability at 0 is evident geometrically. At the point $(0, 0) = (0, f(0))$ the graph changes direction abruptly and there is no single tangent line. You should also note that the absolute-value function *is* differentiable on $(-\infty, 0)$, $f'(x) = -1$ on this interval; and it *is* differentiable on $(0, \infty)$ with $f'(x) = 1$.

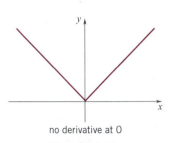

no derivative at 0

Figure 3.1.10

You can see a similar change of direction in the graph of

$$f(x) = \begin{cases} x^2, & x \le 1 \\ \frac{1}{2}x + \frac{1}{2}, & x > 1 \end{cases}$$ (Figure 3.1.11)

at the point $(1, 1)$. Once again, f is everywhere continuous (verify this), but it is not differentiable at 1:

$$\lim_{h \to 0^-} \frac{f(1+h) - f(1)}{h} = \lim_{h \to 0^-} \frac{(1+h)^2 - 1}{h} = \lim_{h \to 0^-} \frac{h^2 + 2h}{h} = \lim_{h \to 0^-} (h + 2) = 2,$$

$$\lim_{h \to 0^+} \frac{f(1+h) - f(1)}{h} = \lim_{h \to 0^+} \frac{\frac{1}{2}(1+h) + \frac{1}{2} - 1}{h} = \lim_{h \to 0^+} \left(\frac{1}{2}\right) = \frac{1}{2}.$$

Since these one-sided limits are different, the two-sided limit

$$\lim_{h \to 0} \frac{f(1+h) - f(1)}{h} \quad \text{does not exist.} \quad \square$$

no derivative at 1

Figure 3.1.11

Contrast these two illustrations with Example 6.

A graphing utility was used to obtain the graph of

$$f(x) = |x^3 - 6x^2 + 8x| + 3$$

shown in Figure 3.1.12. It appears that f is differentiable except, possibly, at $x = 0$, $x = 2$, and $x = 4$ where abrupt changes in direction seem to occur. By zooming in near the point $(2, f(2))$ you can confirm that the left-hand and right-hand limits of the difference quotient both exist at $x = 2$, but they are not equal. See Figure 3.1.13. A similar situation occurs at $x = 0$ and $x = 4$. Thus, f fails to be differentiable at $x = 0$, 2, and 4.

$-1 \le x \le 5, 0 \le y \le 10$

Figure 3.1.12

$1.997 \le x \le 2.003, 2.996 \le y \le 3.006$

Figure 3.1.13

Although not every continuous function is differentiable, every differentiable function is continuous.

THEOREM 3.1.4

If f is differentiable at x, then f is continuous at x.

PROOF For $h \ne 0$ and $x + h$ in the domain of f,

$$f(x + h) - f(x) = \frac{f(x+h) - f(x)}{h} \cdot h.$$

With f differentiable at x,

$$\lim_{h \to 0} \frac{f(x + h) - f(x)}{h} = f'(x).$$

Since $\lim_{h \to 0} h = 0$, we have

$$\lim_{h \to 0} [f(x + h) - f(x)] = \left[\lim_{h \to 0} \frac{f(x + h) - f(x)}{h} \right] \cdot \left[\lim_{h \to 0} h \right] = f'(x) \cdot 0 = 0.$$

It follows that

$$\lim_{h \to 0} f(x + h) = f(x) \qquad \text{(explain)}$$

and thus (by Exercise 47, Section 2.4) f is continuous at x. See, also, Exercise 53, Section 2.4. ❏

EXERCISES 3.1

In Exercises 1–12, differentiate the given function by forming a difference quotient

$$\frac{f(x + h) - f(x)}{h}$$

and taking the limit as h tends to 0.

1. $f(x) = 4$.
2. $f(x) = c$.
3. $f(x) = 2 - 3x$.
4. $f(x) = 4x + 1$.
5. $f(x) = 5x - x^2$.
6. $f(x) = 2x^3 + 1$.
7. $f(x) = x^4$.
8. $f(x) = 1/(x + 3)$.
9. $f(x) = \sqrt{x - 1}$.
10. $f(x) = x^3 - 4x$.
11. $f(x) = 1/x^2$.
12. $f(x) = 1/\sqrt{x}$.

In Exercises 13–18, find $f'(2)$ by forming the difference quotient

$$\frac{f(2 + h) - f(2)}{h}$$

and taking the limit as $h \to 0$.

13. $f(x) = (3x - 7)^2$.
14. $f(x) = 7x - x^2$.
15. $f(x) = 9/(x + 4)$.
16. $f(x) = 5 - x^4$.
17. $f(x) = x + \sqrt{2x}$.
18. $f(x) = \sqrt{6 - x}$.

In Exercises 19–24, find equations for the tangent and normal to the graph of f at the point $(a, f(a))$.

19. $f(x) = x^2$; $a = 2$.
20. $f(x) = \sqrt{x}$; $a = 4$.
21. $f(x) = 5x - x^2$; $a = 4$.
22. $f(x) = 5 - x^3$; $a = 2$.
23. $f(x) = 1/(x + 2)$; $a = -3$.
24. $f(x) = 1/x^2$; $a = -2$.

25. The graph of a function f is shown in the figure below.

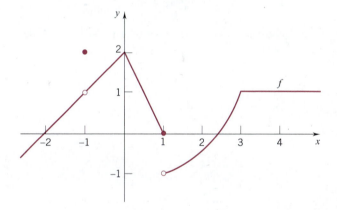

(a) For which numbers c does f fail to be continuous? For each discontinuity, state whether it is a removable discontinuity, a jump discontinuity, or neither.
(b) At which numbers c is f continuous but not differentiable?

26. Repeat Exercise 25 for the function g whose graph is shown in the figure below.

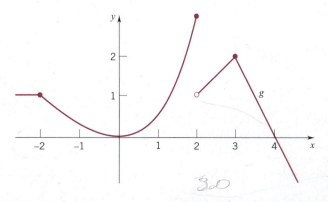

In Exercises 27–32, draw the graph of each of the following functions and indicate where it is not differentiable.

27. $f(x) = |x + 1|$.

28. $f(x) = |2x - 5|$.

29. $f(x) = \sqrt{|x|}$.

30. $f(x) = |x^2 - 4|$.

31. $f(x) = \begin{cases} x^2, & x \le 1 \\ 2 - x, & x > 1. \end{cases}$

32. $f(x) = \begin{cases} x^2 - 1, & x \le 2 \\ 3, & x > 2. \end{cases}$

In Exercises 33–36, find $f'(c)$ if it exists.

33. $f(x) = \begin{cases} 4x, & x < 1 \\ 2x^2 + 2, & x \ge 1; \end{cases} \quad c = 1.$

34. $f(x) = \begin{cases} 3x^2, & x \le 1 \\ 2x^3 + 1, & x > 1; \end{cases} \quad c = 1.$

35. $f(x) = \begin{cases} x + 1, & x \le -1 \\ (x + 1)^2, & x > -1; \end{cases} \quad c = -1.$

36. $f(x) = \begin{cases} -\frac{1}{2}x^2, & x < 3 \\ -3x, & x \ge 3; \end{cases} \quad c = 3.$

Sketch the graph of the derivative of the function with the given graph.

37.

$(2, -2)$

38.

39.

$(2, 2)$

$(-2, -2)$

40.

-2

41.

1

-1

42.

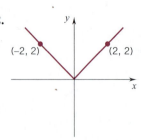

$(-2, 2)$ $(2, 2)$

In Exercises 43–48, each limit represents the derivative of a function f at a value c. Determine f and c in each case.

43. $\lim_{h \to 0} \dfrac{(1 + h)^2 - 1}{h}$.

44. $\lim_{h \to 0} \dfrac{(-2 + h)^3 + 8}{h}$.

45. $\lim_{h \to 0} \dfrac{\sqrt{4 + h} - 2}{h}$.

46. $\lim_{h \to 0} \dfrac{(8 + h)^{1/3} - 2}{h}$.

47. $\lim_{h \to 0} \dfrac{\cos(\pi + h) + 1}{h}$.

48. $\lim_{h \to 0} \dfrac{\sin(\frac{\pi}{6} + h) - \frac{1}{2}}{h}$.

49. Show that

$$f(x) = \begin{cases} x^2, & x \le 1 \\ 2x, & x > 1 \end{cases}$$

is not differentiable at $x = 1$. HINT: Use Theorem 3.1.4.

50. Find A and B given that the function

$$f(x) = \begin{cases} x^3, & x \le 1 \\ Ax + B, & x > 1 \end{cases}$$

is differentiable at $x = 1$.

51. Let

$$f(x) = \begin{cases} (x + 1)^2, & x \le 0 \\ (x - 1)^2, & x > 0. \end{cases}$$

(a) Compute $f'(x)$ for $x \ne 0$.
(b) Show that f is not differentiable at $x = 0$.

52. Given that f is differentiable at c, let g be the function defined by

$$g(x) = \begin{cases} f(x), & x \le c \\ f'(c)(x - c) + f(c), & x > c. \end{cases}$$

(a) Show that g is differentiable at c. What is $g'(c)$?
(b) Suppose that the graph of f is as shown in the figure below. Sketch the graph of g. Note that this exercise is a generalization of Exercise 50.

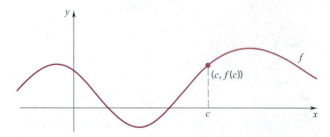

$(c, f(c))$

c

f

In Exercises 53–58, give an example of a function f that is defined for all real numbers and satisfies the given conditions.

53. $f'(x) = 0$ for all real x.

54. $f'(x) = 0$ for all $x \ne 0$; $f'(0)$ does not exist.

55. $f'(x)$ exists for all $x \ne -1$; $f'(-1)$ does not exist.

56. $f'(x)$ exists for all $x \neq \pm 1$; neither $f'(1)$ nor $f'(-1)$ exists.

57. $f'(1) = 2$ and $f(1) = 7$.

58. $f'(x) = 1$ for $x < 0$ and $f'(x) = -1$ for $x > 0$.

59. Suppose that f is an odd function ($f(-x) = -f(x)$). Prove that if f is differentiable, then f' is an even function.

60. Suppose that f is an even function. Prove that if f is differentiable, then f' is an odd function.

(*Important*) Given a function f and a value x in the domain of f. The *left-hand derivative of f at x* and the *right-hand derivative of f at x*, denoted $f'_-(x)$ and $f'_+(x)$, respectively, are defined by

$$f'_-(x) = \lim_{h \to 0^-} \frac{f(x+h) - f(x)}{h}$$

and

$$f'_+(x) = \lim_{h \to 0^+} \frac{f(x+h) - f(x)}{h}$$

provided these limits exist.

61. Let $f(x) = \begin{cases} x^2 - x & x \leq 2 \\ 2x - 2 & x > 2. \end{cases}$
   (a) Show that f is continuous at 2.
   (b) Find $f'_-(2)$ and $f'_+(2)$.
   (c) Is f differentiable at 2?

62. Let $f(x) = x\sqrt{x}$, $x \geq 0$.
   (a) Calculate $f'(x)$ for any $x > 0$.
   (b) Calculate $f'_+(0)$.

63. Let $f(x) = \sqrt{1-x}$ for $0 \leq x \leq 1$.
   (a) Calculate $f'(x)$ for any $x \in (0, 1)$.
   (b) Find $f'_+(0)$, if it exists.
   (c) Find $f'_-(1)$, if it exists.

64. Let $f(x) = \begin{cases} 1 - x^2 & x \leq 0 \\ x^2 & x > 0. \end{cases}$
   (a) Find $f'_-(0)$, if it exists.
   (b) Find $f'_+(0)$, if it exists.
   (c) Is f differentiable at 0?   HINT: Sketch the graph of f.

65. Given a function f and a number c in the domain of f, prove that if $f'_-(c)$ and $f'_+(c)$ both exist and are equal, then f is differentiable at c.

66. Let $f(x) = \begin{cases} x & x \text{ rational} \\ 0 & x \text{ irrational} \end{cases}$ and

$g(x) = \begin{cases} x^2 & x \text{ rational} \\ 0 & x \text{ irrational}. \end{cases}$

Both functions are continuous at 0 and are discontinuous at any $x \neq 0$. (See Exercise 44, Section 2.4.)
   (a) Can either function be differentiable at a value x where $x \neq 0$? Explain.
   (b) Show that f is not differentiable at 0.
   (c) Show that g is differentiable at 0 and give $g'(0)$.

67. Let $f(x) = \begin{cases} x \sin(1/x), & x \neq 0 \\ 0, & x = 0 \end{cases}$ and

$g(x) = \begin{cases} x^2\sin(1/x), & x \neq 0 \\ 0, & x = 0. \end{cases}$

The graphs of f and g are indicated in the figures below.

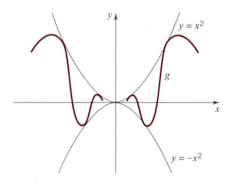

   (a) Show that f and g are both continuous at 0.
   (b) Show that f is not differentiable at 0.
   (c) Show that g is differentiable at 0 and give $g'(0)$.

(*Important*) According to Definition 3.1.1, the derivative of f at c is given by

$$f'(c) = \lim_{h \to 0} \frac{f(c+h) - f(c)}{h}$$

provided this limit exists. Setting $x = c + h$, we can write

(3.1.5) $$f'(c) = \lim_{x \to c} \frac{f(x) - f(c)}{x - c}$$

(see (2.2.5), Section 2.2). This alternative definition of the derivative has advantages in certain situations. Convince yourself of the equivalence of both definitions by calculating $f'(c)$ by both methods in Exercises 68–72.

68. $f(x) = x^3 + 1$; $c = 2$.

69. $f(x) = x^2 - 3x$; $c = 1$.

70. $f(x) = \sqrt{1 + x}$; $c = 3$.

71. $f(x) = x^{1/3}$; $c = -1$.

72. $f(x) = \dfrac{1}{x + 2}$; $c = 0$.

▶ **73.** Let $f(x) = x^{5/2}$ and consider the difference quotient

$$D(h) = \frac{f(2 + h) - f(2)}{h}.$$

(a) Use a graphing utility to graph D for $-1 \le h \le 1$.
(b) Use the zoom feature to obtain successive magnifications of the graph of D near $h = 0$. Estimate $f'(2)$ accurate to three decimal places.
(c) Calculate D at $h = \pm 0.001$ and compare your result with your estimate in (b).

▶ **74.** Repeat Exercise 73 for $f(x) = x^{2/3}$.

▶ **75.** Let $f(x) = 4x - x^3$.
(a) Find an equation for the tangent line T to the graph of f at $x_0 = \frac{3}{2}$; write your equation in the form $T(x) = f'(\frac{3}{2})(x - \frac{3}{2}) + f(\frac{3}{2})$.
(b) Use a graphing utility to graph f and T together.
(c) Notice that the tangent line is a good approximation to the graph of f for values of x close to $x = \frac{3}{2}$. Determine an interval on which $|f(x) - T(x)| < 0.01$.

▶ **76.** Let $f(x) = \sqrt{x - 1}$ and $x_0 = 2$. Repeat Exercise 75.

■ **3.2 SOME DIFFERENTIATION FORMULAS**

Calculating the derivative of

$$f(x) = (x^3 + 2x - 3)(4x^2 + 1) \qquad \text{or} \qquad f(x) = \frac{6x^2 - 1}{x^4 + 5x + 1}$$

by forming the proper difference quotient

$$\frac{f(x + h) - f(x)}{h}$$

and then taking the limit as h tends to 0 is somewhat laborious. Here we derive some general formulas that enable us to calculate such derivatives quite quickly and easily. We begin by pointing out that constant functions have derivative identically 0:

(3.2.1)

> if $f(x) = \alpha$, $\quad \alpha$ any constant, $\quad$ then $\quad f'(x) = 0$ $\quad$ for all real x

and the identity function $f(x) = x$ has constant derivative 1:

(3.2.2)

> if $\quad f(x) = x$, $\quad$ then $\quad f'(x) = 1$ $\quad$ for all real x.

PROOF For $f(x) = \alpha$,

$$f'(x) = \lim_{h \to 0} \frac{f(x + h) - f(x)}{h} = \lim_{h \to 0} \frac{\alpha - \alpha}{h} = \lim_{h \to 0} 0 = 0.$$

For $f(x) = x$,

$$f'(x) = \lim_{h \to 0} \frac{f(x + h) - f(x)}{h} = \lim_{h \to 0} \frac{(x + h) - x}{h} = \lim_{h \to 0} \frac{h}{h} = \lim_{h \to 0} 1 = 1. \quad ❑$$

Remark These results can also be verified geometrically. The graph of a constant function $f(x) = \alpha$ is a horizontal line and the slope of a horizontal line is 0. The graph of the identity function $f(x) = x$ is the graph of the line $y = x$, which has slope 1. ❑

THEOREM 3.2.3 DERIVATIVES OF SUMS AND SCALAR MULTIPLES

Let α be a real number. If f and g are differentiable at x, then $f + g$ and αf are differentiable at x. Moreover,

$$(f + g)'(x) = f'(x) + g'(x) \qquad \text{and} \qquad (\alpha f)'(x) = \alpha f'(x).$$

PROOF To verify the first formula note that

$$\frac{(f + g)(x + h) - (f + g)(x)}{h} = \frac{[f(x + h) + g(x + h)] - [f(x) + g(x)]}{h}$$

$$= \frac{f(x + h) - f(x)}{h} + \frac{g(x + h) - g(x)}{h}.$$

By definition

$$\lim_{h \to 0} \frac{f(x + h) - f(x)}{h} = f'(x), \qquad \lim_{h \to 0} \frac{g(x + h) - g(x)}{h} = g'(x).$$

Thus

$$\lim_{h \to 0} \frac{(f + g)(x + h) - (f + g)(x)}{h} = f'(x) + g'(x),$$

which means that

$$(f + g)'(x) = f'(x) + g'(x).$$

To verify the second formula we must show that

$$\lim_{h \to 0} \frac{(\alpha f)(x + h) - (\alpha f)(x)}{h} = \alpha f'(x).$$

This follows directly from the fact that

$$\frac{(\alpha f)(x + h) - (\alpha f)(x)}{h} = \frac{\alpha f(x + h) - \alpha f(x)}{h} = \alpha \left[\frac{f(x + h) - f(x)}{h} \right]. \quad \square$$

Remark In this section and the next few sections we will be presenting rules and formulas for calculating derivatives. It will be to your advantage to memorize them. To do so, it sometimes helps to verbalize the rule or formula. For example, Theorem 3.2.3 says that

"the derivative of the sum of two functions is the sum of the derivatives," and

"the derivative of a constant times a function is that constant times the derivative of the function." ❏

Since $f - g = f + (-1)g$, it follows that if f and g are differentiable at x, then $f - g$ is differentiable at x, and

$$(f - g)'(x) = f'(x) - g'(x).$$

"The derivative of the difference of two functions is the difference of the derivatives."

The results of Theorem 3.2.3 are easily extended, by mathematical induction, to any finite collection of functions. That is, if $f_1, f_2, \ldots, f_n$ are each differentiable at x, and $\alpha_1, \alpha_2, \ldots, \alpha_n$ are real numbers, then $\alpha_1 f_1 + \alpha_2 f_2 + \cdots + \alpha_n f_n$ is differentiable at x and

(3.2.4) $\qquad (\alpha_1 f_1 + \alpha_2 f_2 + \cdots + \alpha_n f_n)'(x) = \alpha_1 f_1'(x) + \alpha_2 f_2'(x) + \cdots + \alpha_n f_n'(x).$

THEOREM 3.2.5 THE PRODUCT RULE

If f and g are differentiable at x, then so is their product, and

$$(fg)'(x) = f(x)g'(x) + g(x)f'(x).$$

"The derivative of the product of two functions is the first function times the derivative of the second plus the second function times the derivative of the first."

PROOF We form the difference quotient

$$\frac{(fg)(x + h) - (fg)(x)}{h} = \frac{f(x + h)g(x + h) - f(x)g(x)}{h}$$

and rewrite it as

$$f(x + h)\left[\frac{g(x + h) - g(x)}{h}\right] + g(x)\left[\frac{f(x + h) - f(x)}{h}\right].$$

(Here we have added and subtracted $f(x + h)g(x)$ in the numerator and then regrouped the terms so as to display difference quotients for f and g.) Since f is differentiable at x, we know that f is continuous at x (Theorem 3.1.4) and thus

$$\lim_{h \to 0} f(x + h) = f(x).$$

Since

$$\lim_{h \to 0} \frac{g(x + h) - g(x)}{h} = g'(x) \qquad \text{and} \qquad \lim_{h \to 0} \frac{f(x + h) - f(x)}{h} = f'(x),$$

we obtain

$$\lim_{h \to 0} \frac{(fg)(x + h) - (fg)(x)}{h} = f(x)g'(x) + g(x)f'(x). \quad \square$$

Using the product rule, it is not hard to prove that

(3.2.6)

$$\text{for each positive integer } n$$
$$p(x) = x^n \quad \text{has derivative} \quad p'(x) = nx^{n-1}.$$

In particular,

$$p(x) = x \quad \text{has derivative} \quad p'(x) = 1 \cdot x^0 = 1,$$
$$p(x) = x^2 \quad \text{has derivative} \quad p'(x) = 2x,$$
$$p(x) = x^3 \quad \text{has derivative} \quad p'(x) = 3x^2,$$
$$p(x) = x^4 \quad \text{has derivative} \quad p'(x) = 4x^3,$$

and so on.

PROOF OF (3.2.6) We proceed by mathematical induction on n. If $n = 1$, then we have the identity function

$$p(x) = x,$$

which we know satisfies

$$p'(x) = 1 = 1 \cdot x^0.$$

This means that the formula holds for $n = 1$.

We suppose now that the result holds for $n = k$, that is, we assume that if $p(x) = x^k$, then $p'(x) = kx^{k-1}$, and go on to show that it holds for $n = k + 1$. We let

$$p(x) = x^{k+1}$$

and note that

$$p(x) = x \cdot x^k.$$

Applying the product rule (3.2.5) and our induction hypothesis, we obtain

$$p'(x) = x \cdot kx^{k-1} + x^k \cdot 1 = (k + 1)x^k.$$

This shows that the formula holds for $n = k + 1$.

By the axiom of induction the formula holds for all positive integers n. ❑

The formula for differentiating polynomials follows from (3.2.4) and (3.2.6):

(3.2.7)

If	$P(x) = a_n x^n + a_{n-1} x^{n-1} + \cdots + a_2 x^2 + a_1 x + a_0,$
then	$P'(x) = na_n x^{n-1} + (n-1)a_{n-1} x^{n-2} + \cdots + 2a_2 x + a_1.$

For example,

$$P(x) = 12x^3 - 6x - 2 \quad \text{has derivative} \quad P'(x) = 36x^2 - 6$$

and

$$Q(x) = \tfrac{1}{4}x^4 - 2x^2 + x + 5 \quad \text{has derivative} \quad Q'(x) = x^3 - 4x + 1.$$

Example 1 Differentiate

$$F(x) = (x^3 + 2x - 3)(4x^2 + 1)$$

and find $F'(-1)$.

SOLUTION We have a product $F(x) = f(x)g(x)$ with

$$f(x) = x^3 - 2x + 3 \quad \text{and} \quad g(x) = 4x^2 + 1.$$

$$3x^2 - 2$$

The product rule gives

$$F'(x) = f(x)g'(x) + g(x)f'(x)$$
$$= (x^3 - 2x + 3)(8x) + (4x^2 + 1)(3x^2 - 2)$$
$$= 8x^4 - 16x^2 + 24x + 12x^4 - 5x^2 - 2$$
$$= 20x^4 - 21x^2 + 24x - 2.$$

Setting $x = -1$, we have

$$F'(-1) = 20(-1)^4 - 21(-1)^2 + 24(-1) - 2 = 20 - 21 - 24 - 2 = -27. \quad ❑$$

Example 2   Differentiate

$$F(x) = (ax + b)(cx + d), \qquad \text{(where } a, b, c, d \text{ are constants).}$$

SOLUTION   We have a product $F(x) = f(x)g(x)$ with

$$f(x) = ax + b \qquad \text{and} \qquad g(x) = cx + d.$$

Again we use the product rule

$$F'(x) = f(x)g'(x) + g(x)f'(x).$$

In this case

$$F'(x) = (ax + b)c + (cx + d)a = 2acx + bc + ad.$$

We can also do the problem without using the product rule by first carrying out the multiplication:

$$F(x) = acx^2 + bcx + adx + bd$$

and then differentiating:

$$F'(x) = 2acx + bc + ad.$$

The result is the same.   ❑

We come now to reciprocals.

THEOREM 3.2.8   THE RECIPROCAL RULE

If g is differentiable at x and $g(x) \neq 0$, then $1/g$ is differentiable at x and

$$\left(\frac{1}{g}\right)'(x) = -\frac{g'(x)}{[g(x)]^2}.$$

PROOF   Since g is differentiable at x, g is continuous at x (Theorem 3.1.4).   Since $g(x) \neq 0$, we know that $1/g$ is continuous at x, and thus that

$$\lim_{h \to 0} \frac{1}{g(x + h)} = \frac{1}{g(x)}.$$

For h different from 0 and sufficiently small, $g(x + h) \neq 0$ (the continuity of g at x and the fact that $g(x) \neq 0$ guarantee this—see Exercise 48, Section 2.4). Now the difference quotient is:

$$\frac{1}{h}\left[\frac{1}{g(x+h)} - \frac{1}{g(x)}\right] = \frac{1}{h}\left[\frac{g(x) - g(x+h)}{g(x+h)g(x)}\right]$$

$$= -\left[\frac{g(x+h) - g(x)}{h}\right]\frac{1}{g(x+h)}\frac{1}{g(x)}.$$

As h tends to zero, the right-hand side (and thus the left) tends to

$$-\frac{g'(x)}{[g(x)]^2}. \quad \square$$

From this last result it is easy to see that the formula for the derivative of a positive integer power, x^n, also applies to negative integer powers; namely,

(3.2.9)

> for each negative integer n
>
> $p(x) = x^n$ has derivative $p'(x) = nx^{n-1}$.

This formula holds for all x except, of course, at $x = 0$ where no negative power is even defined. In particular, for $x \neq 0$,

$$p(x) = x^{-1} \quad \text{has derivative} \quad p'(x) = (-1)x^{-2} = -x^{-2},$$
$$p(x) = x^{-2} \quad \text{has derivative} \quad p'(x) = -2x^{-3},$$
$$p(x) = x^{-3} \quad \text{has derivative} \quad p'(x) = -3x^{-4},$$

and so on.

PROOF OF (3.2.9) Note that

$$p(x) = \frac{1}{g(x)} \quad \text{where} \quad g(x) = x^{-n} \text{ and } -n \text{ is a positive integer.}$$

The rule for reciprocals gives

$$p'(x) = -\frac{g'(x)}{[g(x)]^2} = -\frac{(-nx^{-n-1})}{x^{-2n}} = (nx^{-n-1})x^{2n} = nx^{n-1}. \quad \square$$

Remark An obvious question now is: What about the derivative of $p(x) = x^0$? Since $x^0 = 1$ for all $x \neq 0$ (0^0 is another indeterminate form; it will be treated in Chapter 10), it follows that $p'(x) = 0$ for all $x \neq 0$. But,

$$0 = 0 \cdot x^{-1} \quad \text{for all} \quad x \neq 0,$$

and so the formula also holds for $n = 0$. Thus,

> for *any* integer n
>
> $p(x) = x^n$ has derivative $p'(x) = nx^{n-1}$.

$\square$

Example 3 Differentiate

$$f(x) = \frac{5}{x^2} - \frac{6}{x}$$

and find $f'(\frac{1}{2})$.

SOLUTION To apply (3.2.9) we write

$$f(x) = 5x^{-2} - 6x^{-1}.$$

Differentiation gives

$$f'(x) = -10x^{-3} + 6x^{-2}.$$

Back in fractional notation

$$f'(x) = -\frac{10}{x^3} + \frac{6}{x^2}.$$

Setting $x = \frac{1}{2}$, we have

$$f'(\tfrac{1}{2}) = -\frac{10}{(\frac{1}{2})^3} + \frac{6}{(\frac{1}{2})^2} = -80 + 24 = -56. \quad \square$$

Example 4 Differentiate

$$f(x) = \frac{1}{ax^2 + bx + c}, \qquad \text{(where } a, b, c \text{ are constants).}$$

SOLUTION Here we have a reciprocal $f(x) = 1/g(x)$ with

$$g(x) = ax^2 + bx + c.$$

The reciprocal rule (3.2.8) gives

$$f'(x) = -\frac{g'(x)}{[g(x)]^2} = -\frac{2ax + b}{[ax^2 + bx + c]^2}. \quad \square$$

Finally we come to quotients in general.

THEOREM 3.2.10 THE QUOTIENT RULE

If f and g are differentiable at x and $g(x) \neq 0$, then the quotient f/g is differentiable at x and

$$\left(\frac{f}{g}\right)'(x) = \frac{g(x)f'(x) - f(x)g'(x)}{[g(x)]^2}$$

"The derivative of a quotient is the denominator times the derivative of the numerator minus the numerator times the derivative of the denominator all divided by the square of the denominator."

The proof of the quotient rule is left to you as an exercise. Also, you should note that the reciprocal rule is a special case of the quotient rule (take $f(x) = 1$). Therefore it is not necessary for you to memorize the reciprocal rule.

From the quotient rule you can see that all rational functions (quotients of polynomials) are differentiable wherever they are defined.

Example 5 Differentiate

$$F(x) = \frac{ax + b}{cx + d} \qquad \text{(where } a, b, c, d \text{ are constants).}$$

SOLUTION We are dealing with a quotient $F(x) = f(x)/g(x)$. The quotient rule,

$$F'(x) = \frac{g(x)f'(x) - f(x)g'(x)}{[g(x)]^2},$$

gives

$$F'(x) = \frac{(cx + d) \cdot a - (ax + b) \cdot c}{(cx + d)^2} = \frac{ad - bc}{(cx + d)^2}. \quad \square$$

Example 6 Differentiate

$$F(x) = \frac{6x^2 - 1}{x^4 + 5x + 1}.$$

SOLUTION Again we are dealing with a quotient $F(x) = f(x)/g(x)$. The quotient rule gives

$$F'(x) = \frac{(x^4 + 5x + 1)(12x) - (6x^2 - 1)(4x^3 + 5)}{(x^4 + 5x + 1)^2}$$

$$= \frac{-12x^5 + 4x^3 + 30x^2 + 12x + 5}{(x^4 + 5x + 1)^2}. \quad \square$$

Example 7 Find $f'(0), f'(1),$ and $f'(2)$ for the function

$$f(x) = \frac{5x}{1 + x}.$$

SOLUTION First we find a general expression for $f'(x)$, and then we evaluate it at 0, 1, and 2. Using the quotient rule, we get

$$f'(x) = \frac{(1 + x)5 - 5x(1)}{(1 + x)^2} = \frac{5}{(1 + x)^2}.$$

This gives

$$f'(0) = \frac{5}{(1 + 0)^2} = 5, \qquad f'(1) = \frac{5}{(1 + 1)^2} = \frac{5}{4}, \qquad f'(2) = \frac{5}{(1 + 2)^2} = \frac{5}{9}. \quad \square$$

Example 8 Find $f'(-1)$ for

$$f(x) = \frac{x^2}{ax^2 + b} \qquad \text{(where } a \text{ and } b \text{ are constants).}$$

SOLUTION We first find $f'(x)$ in general by the quotient rule:

$$f'(x) = \frac{(ax^2 + b)2x - x^2(2ax)}{(ax^2 + b)^2} = \frac{2bx}{(ax^2 + b)^2}.$$

Now we evaluate f' at -1:

$$f'(-1) = -\frac{2b}{(a+b)^2}. \quad \square$$

Remark Some expressions are easier to differentiate if we rewrite them in more convenient form. For example, we can differentiate

$$f(x) = \frac{x^5 - 2x}{x^2} = \frac{x^4 - 2}{x}, \qquad x \neq 0,$$

by the quotient rule, or we can write

$$f(x) = (x^4 - 2)x^{-1}$$

and use the product rule; even better, we can notice that

$$f(x) = x^3 - 2x^{-1}$$

and proceed from there:

$$f'(x) = 3x^2 + 2x^{-2}. \quad \square$$

EXERCISES 3.2

In Exercises 1–20, find the derivative of the given function.

1. $F(x) = 1 - x$.

2. $F(x) = 2(1 + x)$.

3. $F(x) = 11x^5 - 6x^3 + 8$.

4. $F(x) = \dfrac{3}{x^2}$.

5. $F(x) = ax^2 + bx + c$; a, b, c constant.

6. $F(x) = \dfrac{x^4}{4} - \dfrac{x^3}{3} + \dfrac{x^2}{2} - \dfrac{x}{1}$.

7. $F(x) = -\dfrac{1}{x^2}$.

8. $F(x) = \dfrac{(x^2 + 2)}{x^3}$.

9. $G(x) = (x^2 - 1)(x - 3)$.

10. $F(x) = x - \dfrac{1}{x}$.

11. $G(x) = \dfrac{x^3}{1 - x}$.

12. $F(x) = \dfrac{ax - b}{cx - d}$; a, b, c, d constant.

13. $G(x) = \dfrac{x^2 - 1}{2x + 3}$.

14. $G(x) = \dfrac{7x^4 + 11}{x + 1}$.

15. $G(x) = (x - 1)(x - 2)$.

16. $G(x) = \dfrac{2x^2 + 1}{x + 2}$.

17. $G(x) = \dfrac{6 - 1/x}{x - 2}$.

18. $G(x) = \dfrac{1 + x^4}{x^2}$.

19. $G(x) = (9x^8 - 8x^9)\left(x + \dfrac{1}{x}\right)$.

20. $G(x) = \left(1 + \dfrac{1}{x}\right)\left(1 + \dfrac{1}{x^2}\right)$.

In Exercises 21–26, find $f'(0)$ and $f'(1)$.

21. $f(x) = \dfrac{1}{x - 2}$.

22. $f(x) = x^2(x + 1)$.

23. $f(x) = \dfrac{1 - x^2}{1 + x^2}$.

24. $f(x) = \dfrac{2x^2 + x + 1}{x^2 + 2x + 1}$.

25. $f(x) = \dfrac{ax + b}{cx + d}$; a, b, c, d constant.

26. $f(x) = \dfrac{ax^2 + bx + c}{cx^2 + bx + a}$; a, b, c constant.

In Exercises 27–30, find $f'(0)$ given that $h(0) = 3$ and $h'(0) = 2$.

27. $f(x) = xh(x)$.

28. $f(x) = 3x^2h(x) - 5x$.

29. $f(x) = h(x) - \dfrac{1}{h(x)}$.

30. $f(x) = h(x) + \dfrac{x}{h(x)}$.

In Exercises 31–36, find an equation for the tangent to the graph of f at the point $(a, f(a))$.

31. $f(x) = \dfrac{x}{x + 2}$; $a = -4$.

32. $f(x) = (x^3 - 2x + 1)(4x - 5)$; $a = 2$.

33. $f(x) = (x^2 - 3)(5x - x^3)$; $a = 1$.

34. $f(x) = x^2 - \dfrac{10}{x}$; $a = -2$.

35. $f(x) = 6/x^2$; $a = 3$.

36. $f(x) = \dfrac{3x + 5}{7x - 3}$; $a = 1$.

In Exercises 37–42, find the points where the tangent to the curve is horizontal.

37. $f(x) = (x - 2)(x^2 - x - 11)$.

38. $f(x) = x^2 - \dfrac{16}{x}$.

39. $f(x) = \dfrac{5x}{x^2 + 1}$.

40. $f(x) = (x + 2)(x^2 - 2x - 8)$.

41. $f(x) = x + \dfrac{4}{x^2}$.

42. $f(x) = \dfrac{x^2 - 2x + 4}{x^2 + 4}$.

In Exercises 43–46, find the points where the tangent to the curve

43. $f(x) = -x^2 - 6$ is parallel to the line $y = 4x - 1$.

44. $f(x) = x^3 - 3x$ is perpendicular to the line $5y - 3x - 8 = 0$.

45. $f(x) = x^3 - x^2$ is perpendicular to the line $5y + x + 2 = 0$.

46. $f(x) = 4x - x^2$ is parallel to the line $2y = 3x - 5$.

47. Find the area of the triangle formed by the x-axis and the lines tangent and normal to the curve $f(x) = 6x - x^2$ at the point $(5, 5)$.

48. Find the area of the triangle formed by the x-axis and the lines tangent and normal to the curve $f(x) = 9 - x^2$ at the point $(2, 5)$.

49. Determine the coefficients A, B, C so that the curve $f(x) = Ax^2 + Bx + C$ will pass through the point $(1, 3)$ and be tangent to the line $4x + y = 8$ at the point $(2, 0)$.

50. Determine A, B, C, D so that the curve $f(x) = Ax^3 + Bx^2 + Cx + D$ will be tangent to the line $y = 3x - 3$ at the point $(1, 0)$ and tangent to the line $y = 18x - 27$ at the point $(2, 9)$.

51. Find the value(s) of x where the tangent line to the graph of the quadratic function $f(x) = ax^2 + bx + c$ has a horizontal tangent line.   NOTE: This gives a way to find the vertex of the parabola $y = ax^2 + bx + c$.

52. Find conditions on a, b, c, and d that will guarantee that the graph of the cubic polynomial $p(x) = ax^3 + bx^2 + cx + d$ has:
(a) Exactly two horizontal tangents.
(b) Exactly one horizontal tangent.
(c) No horizontal tangents.

53. Find the value(s) of c, if any, where the tangent line to the graph of $f(x) = x^3 - x$ at $(c, f(c))$ is parallel to the secant line through $(-1, f(-1))$ and $(2, f(2))$.

54. Find the value(s) of c, if any, where the tangent line to the graph of $f(x) = x/(x + 1)$ at $(c, f(c))$ is parallel to the secant line through $(1, f(1))$ and $(3, f(3))$.

55. Let $f(x) = 1/x$, $x > 0$. Show that the triangle that is formed by *any* tangent line to the graph of f and the coordinate axes has an area of 2 square units.

56. Let $f(x) = x^3$.
(a) Determine an equation for the tangent line to the graph of f at $x = c$, where $c \neq 0$.
(b) Determine whether the tangent line found in (a) intersects the graph of f at a point other than (c, c^3). If it does, determine the x-coordinate of the point of intersection.

57. Given two functions f and g, show that if f and $f + g$ are differentiable, then g is differentiable. Give an example to show that the differentiability of $f + g$ does not imply that f and g are each differentiable.

58. Given two functions f and g, if f and $f \cdot g$ are differentiable, does it follow that g is differentiable? If not, find a condition that will imply that g is differentiable when both f and $f \cdot g$ are differentiable.

59. Prove the validity of the quotient rule.
HINT: $f/g = f \cdot (1/g)$.

60. Verify that, if f, g, h are differentiable, then

$$(fgh)'(x) = f'(x)g(x)h(x) + f(x)g'(x)h(x) + f(x)g(x)h'(x).$$

HINT: Apply the product rule to $[f(x)g(x)]h(x)$.

61. Use the result in Exercise 60 to find the derivative of $F(x) = (x^2 + 1)(1 + (1/x))(2x^3 - x + 1)$.

62. Use the result in Exercise 60 to find the derivative of $G(x) = \sqrt{x}[1/(1 + 2x)](x^2 + x - 1)$.

63. Use the product rule (3.2.5) to show that, if f is differentiable, then

$$g(x) = [f(x)]^2 \quad \text{has derivative} \quad g'(x) = 2f(x)f'(x).$$

64. Show that, if f is differentiable, then

$$g(x) = [f(x)]^n \text{ has derivative } g'(x) = n[f(x)]^{n-1}f'(x)$$

for each nonzero integer n.   HINT: Mimic the inductive proof of (3.2.6).

65. Use the result in Exercise 64 to find the derivative of $g(x) = (x^3 - 2x^2 + x + 2)^3$.

66. Use the result in Exercise 64 to find the derivative of $g(x) = [x^2/(1 + 2x)]^{10}$.

67. Let $f(x) = x^2 - 4x + 2$ on $[-1, 3]$, and let $g(x) = x^2 - 4x + 2$ on $(-\infty, \infty)$.
(a) Find $f'_+(-1)$ and $f'_-(3)$ (see the definition in Exercises 3.1).
(b) Find $g'(x)$ for any x, and evaluate g' at -1 and at 3. How do these results compare with the results in part (a)?

68. Suppose that g is a differentiable function on an open interval I. Let $[a, b]$ be a closed interval contained in I and let f be the function defined on $[a, b]$ by

$$f(x) = g(x) \qquad \text{for all} \quad x \in [a, b].$$

Prove that $f'_+(a) = g'(a)$ and $f'_-(b) = g'(b)$. This result provides an alternative method for calculating one-sided derivatives. When it can be applied, it is usually easier to use than the definition in Exercises 3.1.

69. Find A and B given that the derivative of

$$f(x) = \begin{cases} Ax^3 + Bx + 2, & x \le 2 \\ Bx^2 - A, & x > 2 \end{cases}$$

is continuous for all real x.   HINT: f itself must be continuous.

70. Find A and B given that the derivative of

$$f(x) = \begin{cases} Ax^2 + B, & x < -1 \\ Bx^5 + Ax + 4, & x \ge -1 \end{cases}$$

is continuous for all real x.

▷ 71. Let $f(x) = \sin x$.

(a) Approximate $f'(x)$ at $x = 0$, $x = \pi/6$, $x = \pi/4$, $x = \pi/3$, and $x = \pi/2$ using the difference quotient

$$\frac{f(x + h) - f(x)}{h}$$

with $h = \pm 0.001$.

(b) Compare the approximate values of $f'(x)$ found in (a) with the values of $\cos x$ at each of these points.

(c) Use the results in (b) to guess a formula for the derivative of the sine function.

▷ 72. Repeat Exercise 71 with $f(x) = \cos x$, using $\sin x$ in part (b) and the cosine function in part (c).

▷ 73. Let $f(x) = 2^x$.

(a) Approximate $f'(x)$ at $x = 0$, $x = 1$, $x = 2$, and $x = 3$ using the difference quotient

$$\frac{f(x + h) - f(x)}{h}$$

with $h = \pm 0.001$.

(b) Calculate the ratio $f'(x)/f(x)$ for each of the values of x in (a).

(c) Use the results in (b) to guess a formula for the derivative of $f(x) = 2^x$.

▷ 74. Let $f(x) = x^4 + x^3 - 5x^2 + 2$.

(a) Use a graphing utility to graph f on the interval $[-4, 4]$ and estimate the x-coordinates of the points where the tangent line to the graph of f is horizontal.

(b) Use a graphing utility to graph $|f|$. Are there any points where f is not differentiable? If so, estimate where f fails to be differentiable.

■ 3.3 THE d/dx NOTATION; DERIVATIVES OF HIGHER ORDER

The d/dx Notation

So far we have indicated the derivative by a prime, but there are other notations that are widely used, particularly in science and engineering. The most popular of these is the double-d notation of Leibniz. In the Leibniz notation the derivative of a function y is indicated by writing

$$\frac{dy}{dx}, \qquad \frac{dy}{dt}, \qquad \text{or} \qquad \frac{dy}{dz}, \qquad \text{and so forth,}$$

depending upon whether the letter x, t, or z, and so on, is being used for the elements of the domain of y. For instance, if y is initially defined by

$$y(x) = x^3,$$

then the Leibniz notation gives

$$\frac{dy(x)}{dx} = 3x^2.$$

Usually writers drop the (x) and simply write

$$y = x^3 \qquad \text{and} \qquad \frac{dy}{dx} = 3x^2.$$

The symbols

$$\frac{d}{dx}, \qquad \frac{d}{dt}, \qquad \frac{d}{dz}, \qquad \text{and so forth,}$$

are also used as prefixes before expressions to be differentiated.   For example,

$$\frac{d}{dx}(x^3 - 4x) = 3x^2 - 4, \qquad \frac{d}{dt}(t^2 + 3t + 1) = 2t + 3, \qquad \frac{d}{dz}(z^5 - 1) = 5z^4.$$

In the Leibniz notation the differentiation formulas are:

$$\frac{d}{dx}[f(x) + g(x)] = \frac{d}{dx}[f(x)] + \frac{d}{dx}[g(x)], \qquad \frac{d}{dx}[\alpha f(x)] = \alpha \frac{d}{dx}[f(x)],$$

$$\frac{d}{dx}[f(x)g(x)] = f(x)\frac{d}{dx}[g(x)] + g(x)\frac{d}{dx}[f(x)],$$

$$\frac{d}{dx}\left[\frac{1}{g(x)}\right] = -\frac{1}{[g(x)]^2}\frac{d}{dx}[g(x)],$$

$$\frac{d}{dx}\left[\frac{f(x)}{g(x)}\right] = \frac{g(x)\dfrac{d}{dx}[f(x)] - f(x)\dfrac{d}{dx}[g(x)]}{[g(x)]^2}.$$

Often functions *f* and *g* are replaced by *u* and *v* and the *x* is left out altogether.   Then the formulas look like this:

$$\frac{d}{dx}(u + v) = \frac{du}{dx} + \frac{dv}{dx}, \qquad \frac{d}{dx}(\alpha u) = \alpha\frac{du}{dx},$$

$$\frac{d}{dx}(uv) = u\frac{dv}{dx} + v\frac{du}{dx},$$

$$\frac{d}{dx}\left(\frac{1}{v}\right) = -\frac{1}{v^2}\frac{dv}{dx}, \qquad \frac{d}{dx}\left(\frac{u}{v}\right) = \frac{v\dfrac{du}{dx} - u\dfrac{dv}{dx}}{v^2}.$$

The only way to develop a feeling for this notation is to use it.   Below we work out some examples.

Example 1   Find

$$\frac{dy}{dx} \quad \text{for} \quad y = \frac{3x - 1}{5x + 2}.$$

SOLUTION   We use the quotient rule:

$$\frac{dy}{dx} = \frac{(5x + 2)\dfrac{d}{dx}(3x - 1) - (3x - 1)\dfrac{d}{dx}(5x + 2)}{(5x + 2)^2}$$

$$= \frac{(5x + 2)(3) - (3x - 1)(5)}{(5x + 2)^2} = \frac{11}{(5x + 2)^2}. \quad ❏$$

$15x + 6 - 15 - 5$

Example 2   Find

$$\frac{dy}{dx} \quad \text{for} \quad y = (x^3 + 1)(3x^5 + 2x - 1).$$

SOLUTION Here we use the product rule:

$$\frac{dy}{dx} = (x^3 + 1)\frac{d}{dx}(3x^5 + 2x - 1) + (3x^5 + 2x - 1)\frac{d}{dx}(x^3 + 1)$$

$$= (x^3 + 1)(15x^4 + 2) + (3x^5 + 2x - 1)(3x^2)$$
$$= (15x^7 + 15x^4 + 2x^3 + 2) + (9x^7 + 6x^3 - 3x^2)$$
$$= 24x^7 + 15x^4 + 8x^3 - 3x^2 + 2. \quad ❏$$

Example 3 Find

$$\frac{d}{dt}\left(t^3 - \frac{t}{t^2 - 1}\right).$$

SOLUTION

$$\frac{d}{dt}\left(t^3 - \frac{t}{t^2 - 1}\right) = \frac{d}{dt}(t^3) - \frac{d}{dt}\left(\frac{t}{t^2 - 1}\right)$$

$$= 3t^2 - \left[\frac{(t^2 - 1)(1) - t(2t)}{(t^2 - 1)^2}\right] = 3t^2 + \frac{t^2 + 1}{(t^2 - 1)^2}. \quad ❏$$

Example 4 Find

$$\frac{du}{dx} \quad \text{for} \quad u = x(x + 1)(x + 2).$$

SOLUTION You can think of u as

$$[x(x + 1)](x + 2) \quad \text{or as} \quad x[(x + 1)(x + 2)].$$

From the first point of view,

$$\frac{du}{dx} = [x(x + 1)](1) + (x + 2)\frac{d}{dx}[x(x + 1)]$$

$$= x(x + 1) + (x + 2)[x(1) + (x + 1)(1)]$$
(∗) $$= x(x + 1) + (x + 2)(2x + 1).$$

From the second point of view,

$$\frac{du}{dx} = x\frac{d}{dx}[(x + 1)(x + 2)] + (x + 1)(x + 2)(1)$$

$$= x[(x + 1)(1) + (x + 2)(1)] + (x + 1)(x + 2)$$
(∗∗) $$= x(2x + 3) + (x + 1)(x + 2).$$

Both (∗) and (∗∗) can be multiplied out to give

$$\frac{du}{dx} = 3x^2 + 6x + 2.$$

Alternatively, this same result can be obtained by first carrying out the multiplication and then differentiating:

$$u = x(x + 1)(x + 2) = x(x^2 + 3x + 2) = x^3 + 3x^2 + 2x$$

so that

$$\frac{du}{dx} = 3x^2 + 6x + 2. \quad ❏$$

Example 5   Evaluate dy/dx at $x = 0$ and $x = 1$ given that

$$y = \frac{x^2}{x^2 - 4}.$$

SOLUTION

$$\frac{dy}{dx} = \frac{(x^2 - 4)2x - x^2(2x)}{(x^2 - 4)^2} = -\frac{8x}{(x^2 - 4)^2}.$$

At $x = 0$,

$$\frac{dy}{dx} = -\frac{8 \cdot 0}{(0^2 - 4)^2} = 0; \quad \text{at } x = 1, \quad \frac{dy}{dx} = -\frac{8 \cdot 1}{(1^2 - 4)^2} = -\frac{8}{9}. \quad \square$$

Remark   The notations

$$\frac{dy}{dx}\bigg|_{x=a} \quad \text{and} \quad \frac{dy}{dx}(a)$$

are sometimes used to emphasize the fact that we are evaluating the derivative dy/dx at $x = a$. Thus, in Example 5, we have

$$\frac{dy}{dx}\bigg|_{x=0} = 0 \quad \text{or} \quad \frac{dy}{dx}(0) = 0,$$

and

$$\frac{dy}{dx}\bigg|_{x=1} = -\tfrac{8}{9} \quad \text{or} \quad \frac{dy}{dx}(1) = -\tfrac{8}{9}. \quad \square$$

Derivatives of Higher Order

As we noted in Section 3.1, when we differentiate a function f we get a new function f', the derivative of f, whose domain is a subset of the domain of f. Now suppose that f' can be differentiated. If we calculate $(f')'$, we get the *second derivative of f*, which is denoted f''. So long as we have differentiability, we can continue in this manner, forming the *third derivative of f*, denoted f''', and so on. However, the prime notation is not used beyond the third derivative. For the *fourth derivative of f*, we write $f^{(4)}$ and more generally, the nth derivative of f is denoted $f^{(n)}$. The functions f', f'', f''', $f^{(4)}$, . . . , $f^{(n)}$, are called the derivatives of f of *orders* 1, 2, 3, 4, . . . , n, respectively. For example, if $f(x) = x^5$, then

$$f'(x) = 5x^4, \quad f''(x) = 20x^3, \quad f'''(x) = 60x^2, \quad f^{(4)}(x) = 120x, \quad f^{(5)}(x) = 120.$$

In this case, all higher derivatives are identically zero. As a variant of this notation you can write $y = x^5$ and then

$$y' = 5x^4, \quad y'' = 20x^3, \quad y''' = 60x^2, \quad \text{and so on.}$$

Since each polynomial P has a derivative P' that is in turn a polynomial, and each rational function Q has a derivative Q' that is in turn a rational function, polynomials and rational functions have derivatives of all orders. In the case of a polynomial of degree n, derivatives of order greater than n are all identically zero. (Explain.)

In the Leibniz notation the derivatives of higher order are written

$$\frac{d^2y}{dx^2} = \frac{d}{dx}\left(\frac{dy}{dx}\right), \frac{d^3y}{dx^3} = \frac{d}{dx}\left(\frac{d^2y}{dx^2}\right), \quad \ldots \quad , \frac{d^ny}{dx^n} = \frac{d}{dx}\left(\frac{d^{n-1}y}{dx^{n-1}}\right), \quad \ldots$$

or

$$\frac{d^2}{dx^2}[f(x)] = \frac{d}{dx}\left[\frac{d}{dx}[f(x)]\right], \qquad \frac{d^3}{dx^3}[f(x)] = \frac{d}{dx}\left[\frac{d^2}{dx^2}[f(x)]\right],$$

$$\ldots, \qquad \frac{d^n}{dx^n}[f(x)] = \frac{d}{dx}\left[\frac{d^{n-1}}{dx^{n-1}}[f(x)]\right], \ldots$$

Below we work out some examples.

Example 6 If

$$f(x) = x^4 - 3x^{-1} + 5,$$

then

$$f'(x) = 4x^3 + 3x^{-2} \qquad \text{and} \qquad f''(x) = 12x^2 - 6x^{-3}. \quad \square$$

Example 7

$$\frac{d}{dx}(x^5 - 4x^3 + 7x) = 5x^4 - 12x^2 + 7$$

so that

$$\frac{d^2}{dx^2}(x^5 - 4x^3 + 7x) = \frac{d}{dx}(5x^4 - 12x^2 + 7) = 20x^3 - 24x$$

and

$$\frac{d^3}{dx^3}(x^5 - 4x^3 + 7x) = \frac{d}{dx}(20x^3 - 24x) = 60x^2 - 24. \quad \square$$

Example 8 Finally we consider $y = x^{-1}$. In the Leibniz notation

$$\frac{dy}{dx} = -x^{-2}, \qquad \frac{d^2y}{dx^2} = 2x^{-3}, \qquad \frac{d^3y}{dx^3} = -6x^{-4}, \qquad \frac{d^4y}{dx^4} = 24x^{-5}, \ldots$$

Based on these calculations, we are led to the general result

$$\frac{d^ny}{dx^n} = (-1)^n n! \, x^{-n-1} \qquad \text{[Recall that } n! = n(n-1)(n-2)\cdots 3\cdot 2\cdot 1.]$$

In the prime notation

$$y' = -x^{-2}, \qquad y'' = 2x^{-3}, \qquad y''' = -6x^{-4}, \qquad y^{(4)} = 24x^{-5}$$

and

$$y^{(n)} = (-1)^n n! \, x^{-n-1}. \quad \square$$

EXERCISES 3.3

In Exercises 1–10, find dy/dx.

1. $y = 3x^4 - x^2 + 1$.

2. $y = x^2 + 2x^{-4}$.

3. $y = x - \dfrac{1}{x}$.

4. $y = \dfrac{2x}{1-x}$.

5. $y = \dfrac{x}{1+x^2}$.

6. $y = x(x-2)(x+1)$.

7. $y = \dfrac{x^2}{1-x}$.

8. $y = \left(\dfrac{x}{1+x}\right)\left(\dfrac{2-x}{3}\right)$.

9. $y = \dfrac{x^3+1}{x^3-1}$. **10.** $y = \dfrac{x^2}{(1+x)}$.

In Exercises 11–26, find the indicated derivative.

11. $\dfrac{d}{dx}(2x-5)$. **12.** $\dfrac{d}{dx}(5x+2)$.

13. $\dfrac{d}{dx}[(3x^2-x^{-1})(2x+5)]$.

14. $\dfrac{d}{dx}[(2x^2+3x^{-1})(2x-3x^{-2})]$.

15. $\dfrac{d}{dt}\left(\dfrac{t^2+1}{t^2-1}\right)$. **16.** $\dfrac{d}{dt}\left(\dfrac{2t^3+1}{t^4}\right)$.

17. $\dfrac{d}{dt}\left(\dfrac{t^4}{2t^3-1}\right)$. **18.** $\dfrac{d}{dt}\left[\dfrac{t}{(1+t)^2}\right]$.

19. $\dfrac{d}{du}\left(\dfrac{2u}{1-2u}\right)$. **20.** $\dfrac{d}{du}\left(\dfrac{u^2}{u^3+1}\right)$.

21. $\dfrac{d}{du}\left(\dfrac{u}{u-1}-\dfrac{u}{u+1}\right)$.

22. $\dfrac{d}{du}[u^2(1-u^2)(1-u^3)]$.

23. $\dfrac{d}{dx}\left(\dfrac{x^2}{1-x^2}-\dfrac{1-x^2}{x^2}\right)$.

24. $\dfrac{d}{dx}\left(\dfrac{3x^4+2x+1}{x^4+x-1}\right)$.

25. $\dfrac{d}{dx}\left(\dfrac{x^3+x^2+x+1}{x^3-x^2+x-1}\right)$.

26. $\dfrac{d}{dx}\left(\dfrac{x^3+x^2+x-1}{x^3-x^2+x+1}\right)$.

In Exercises 27–30, evaluate dy/dx at $x=2$.

27. $y = (x+1)(x+2)(x+3)$.

28. $y = (x+1)(x^2+2)(x^3+3)$.

29. $y = \dfrac{(x-1)(x-2)}{(x+2)}$. **30.** $y = \dfrac{(x^2+1)(x^2-2)}{x^2+2}$.

In Exercises 31–36, find the second derivative.

31. $f(x) = 7x^3 - 6x^5$.

32. $f(x) = 2x^5 - 6x^4 + 2x - 1$.

33. $f(x) = \dfrac{x^2-3}{x}$. **34.** $f(x) = x^2 - \dfrac{1}{x^2}$.

35. $f(x) = (x^2-2)(x^{-2}+2)$.

36. $f(x) = (2x-3)\left(\dfrac{2x+3}{x}\right)$.

In Exercises 37–42, find d^3y/dx^3.

37. $y = \frac{1}{3}x^3 + \frac{1}{2}x^2 + x + 1$.

38. $y = (1+5x)^2$. **39.** $y = (2x-5)^2$.

40. $y = \frac{1}{6}x^3 - \frac{1}{4}x^2 + x - 3$.

41. $y = x^3 - \dfrac{1}{x^3}$. **42.** $y = \dfrac{x^4+2}{x}$.

In Exercises 43–48, find the indicated derivatives.

43. $\dfrac{d}{dx}\left[x\dfrac{d}{dx}(x-x^2)\right]$.

44. $\dfrac{d^2}{dx^2}\left[(x^2-3x)\dfrac{d}{dx}(x+x^{-1})\right]$.

45. $\dfrac{d^4}{dx^4}[3x-x^4]$.

46. $\dfrac{d^5}{dx^5}[ax^4+bx^3+cx^2+dx+e]$, $a,\ldots,e$ constant.

47. $\dfrac{d^2}{dx^2}\left[(1+2x)\dfrac{d^2}{dx^2}(5-x^3)\right]$.

48. $\dfrac{d^3}{dx^3}\left[\dfrac{1}{x}\dfrac{d^2}{dx^2}(x^4-5x^2)\right]$.

49. Find a quadratic polynomial p such that $p(1)=3$, $p'(1)=-2$, and $p''(1)=4$.

50. Find a cubic polynomial p such that $p(-1)=0$, $p'(-1)=3$, $p''(-1)=-2$, and $p'''(-1)=6$.

51. Let $f(x)=x^n$, where n is a positive integer.
 (a) Find $f^{(k)}(x)$ for $k=n$.
 (b) Find $f^{(k)}(x)$ when $k>n$.
 (c) Find $f^{(k)}(x)$ when $k<n$.

52. Given the polynomial function
$p(x) = a_n x^n + a_{n-1}x^{n-1} + \cdots + a_1 x + a_0$:
 (a) Find $(d^n/dx^n)[p(x)]$
 (b) What is $(d^k/dx^k)[p(x)]$ when $k>n$?

53. Let $f(x) = \begin{cases} x^2, & x \geq 0 \\ 0, & x < 0. \end{cases}$
 (a) Show that f is differentiable at 0 and give $f'(0)$.
 (b) Determine $f'(x)$ for all x.
 (c) Show that $f''(0)$ does not exist.
 (d) Sketch the graph of f.

54. Let $g(x) = \begin{cases} x^3, & x \geq 0 \\ 0, & x < 0. \end{cases}$
 (a) Show $g'(0)$ and $g''(0)$ both exist and give their values.
 (b) Determine $g'(x)$ and $g''(x)$ for all x.
 (c) Show that $g'''(0)$ does not exist.
 (d) Sketch the graph of g.

55. Show that in general

$$(fg)''(x) \neq f(x)g''(x) + f''(x)g(x).$$

56. Verify the identity

$$f(x)g''(x) - f''(x)g(x) = \frac{d}{dx}[f(x)g'(x) - f'(x)g(x)].$$

In Exercises 57–60, determine the values of x for which
(a) $f''(x) = 0$, (b) $f''(x) > 0$, (c) $f''(x) < 0$.

57. $f(x) = x^3$.   **58.** $f(x) = x^4$.

59. $f(x) = x^4 + 2x^3 - 12x^2 + 1$.

60. $f(x) = x^4 + 3x^3 - 6x^2 - x$.

61. Prove by mathematical induction that

$$\text{if} \quad y = x^{-1}, \quad \text{then} \quad \frac{d^n y}{dx^n} = (-1)^n n! \, x^{-n-1}.$$

62. Calculate y', y'', y''' for $y = 1/x^2$. Use these results to guess a formula for $y^{(n)}$ for any positive integer n, and then prove your conjecture using mathematical induction.

63. If f and g are differentiable functions, then

$$(f \cdot g)' = f'g + fg'$$

by the product rule.
 (a) Show that if f and g are twice differentiable, then

$$(f \cdot g)'' = f''g + 2f'g' + fg''.$$

 (b) Let f and g have derivatives of order 3. Show that

$$(f \cdot g)''' = f'''g + 3f''g' + 3f'g'' + fg'''.$$

64. Let f and g have derivatives of order n. The results in Exercise 63 suggest that the expansion of $(f \cdot g)^n$ parallels the expansion of $(a + b)^n$ by the binomial theorem. That is,

$$(*) \quad (f \cdot g)^{(n)} = f^{(n)}g + nf^{(n-1)}g' + \cdots + \binom{n}{k}f^{(n-k)}g^{(k)}$$
$$+ \cdots + nf'g^{(n-1)} + fg^{(n)}$$

where $\binom{n}{k} = n!/(k!(n - k)!)$ is the coefficient of $a^{n-k}b^k$ in the expansion of $(a + b)^n$, called the kth binomial coefficient. Formula $(*)$ is known as *Leibniz's rule*. Use mathematical induction to prove this formula.

65. Let u, v, w be differentiable functions of x. Express the derivative of the product uvw in terms of the functions u, v, w, and their derivatives.

66. Let $f(x) = \frac{1}{2}x^3 - 3x^2 + 3x + 3$.
 (a) Calculate $f'(x)$.
 (b) Use a graphing utility to graph f and f' together.
 (c) Find the x-coordinates of the points where the tangent line to the graph of f is horizontal by finding the zeros of f'. Approximate the zeros of f' with three decimal place accuracy.

67. Let $f(x) = x^3 + x^2 - 4x + 1$.
 (a) Calculate $f'(x)$.
 (b) Use a graphing utility to graph f and f' together.
 (c) What can you say about the graph of f when $f'(x) < 0$? What can you say about the graph of f when $f'(x) > 0$?

68. Let $f(x) = x^3 - x$.
 (a) Find an equation for the tangent line to the graph of f at the point $(1, 0)$.
 (b) Use a graphing utility to graph f and the tangent line together. Notice that the graph and the tangent line intersect at another point (a, b).
 (c) Find the point (a, b).

69. Let $f(x) = \frac{1}{2}x^3 - 3x^2 + 4x + 1$.
 (a) Find an equation for the tangent line to the graph of f at the point $(0, 1)$.
 (b) Use a graphing utility to graph f and the tangent line together. Show that the graph and the tangent line intersect at another point (a, b).
 (c) Find the point (a, b).

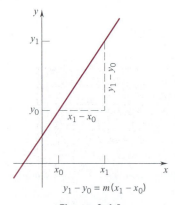

Figure 3.4.1

■ 3.4 THE DERIVATIVE AS A RATE OF CHANGE

In the case of a linear function $y = mx + b$ the graph is a straight line and the slope m measures the steepness of the line by giving the rate of climb of the line, *the rate of change of y with respect to x.*

As x changes from x_0 to x_1, y changes m times as much:

$$y_1 - y_0 = m(x_1 - x_0). \tag{Figure 3.4.1}$$

Thus the slope $m = (y_1 - y_0)/(x_1 - x_0)$ gives the change in y per unit change in x.

In the more general case of a differentiable function $y = f(x)$, the difference quotient

$$\frac{f(x + h) - f(x)}{(x + h) - h} = \frac{f(x + h) - f(x)}{h}, \quad h \neq 0,$$

gives the *average rate of change of y (or f) with respect to x.* Geometrically, this ratio is the slope of the secant line through $(x, f(x))$ and $(x + h, f(x + h))$. The limit as h approaches zero is the derivative $dy/dx = f'(x)$, which can be interpreted as the *instantaneous rate of change of y with respect to x,* or as the slope of the tangent line to the graph of f at the point $(x, f(x))$. Since the graph is a curve, the rate of change of y with respect to x can vary from point to point. At $x = x_1$ (see Figure 3.4.2) the rate of change of y with respect to x is $f'(x_1)$; the steepness of the graph is that of a line of slope $f'(x_1)$. At $x = x_2$ the rate of change of y with respect to x is $f'(x_2)$; the steepness of the graph is that of a line of slope $f'(x_2)$. At $x = x_3$ the rate of change of y with respect to x is $f'(x_3)$; the steepness of the graph is that of a line of slope $f'(x_3)$.

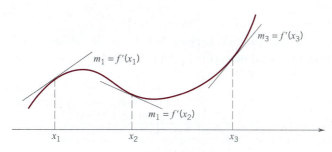

Figure 3.4.2

From your understanding of slope it should be apparent to you when examining Figure 3.4.2 that $f'(x_1) > 0, f'(x_2) < 0,$ and $f'(x_3) > 0$, and that, in general, a function increases on any interval where the derivative remains positive and decreases on any interval where the derivative remains negative. We will take up this matter carefully in Chapter 4. Right now we look at rates of change in the context of some simple geometric figures.

Example 1 A square of side s has area $A = s^2$. Now suppose that s increases. As s increases from 1 to 2, A increases by 3 square units:

$$2^2 - 1^2 = 4 - 1 = 3. \qquad \text{(See Figure 3.4.3.)}$$

As s increases from 2 to 3, A increases by 5 square units:

$$3^2 - 2^2 = 9 - 4 = 5.$$

As s increases from 3 to 4, A increases by 7 square units:

$$4^2 - 3^2 = 16 - 9 = 7,$$

and so on. Continued unit changes in s cause ever-increasing changes in A.

Figure 3.4.3

Figure 3.4.4

Figure 3.4.5

In Figure 3.4.4 we have plotted A against s. The instantaneous rate of change of A with respect to s is the derivative

$$\frac{dA}{ds} = 2s.$$

This appears in Figure 3.4.4 as the slope of the tangent line. At $s = 1$, $dA/ds = 2$; thus, at $s = 1$, A is increasing at the rate of a linear function of slope 2. At $s = 2$, $dA/ds = 4$; thus, at $s = 2$, A is increasing at the rate of a linear function of slope 4, and so on. At $s = k$, $dA/ds = 2k$ and A is increasing at the rate of a linear function of slope $2k$. ❑

Example 2 Suppose that we have a right circular cylinder of changing dimensions (Figure 3.4.5). When the base radius is r and the height is h, the cylinder has volume

$$V = \pi r^2 h.$$

If r remains constant while h changes, then V can be viewed as a function of h. The rate of change of V with respect to h is the derivative

$$\frac{dV}{dh} = \pi r^2.$$

If h remains constant while r changes, then V can be viewed as a function of r. The rate of change of V with respect to r is the derivative

$$\frac{dV}{dr} = 2\pi r h.$$

Suppose now that r is changing but V is being kept constant. How is h changing with respect to r? To answer this we express h in terms of r and V:

$$h = \frac{V}{\pi r^2} = \frac{V}{\pi} r^{-2}.$$

Since V is being held constant, h is now a function of r. The rate of change of h with respect to r is the derivative

$$\frac{dh}{dr} = -\frac{2V}{\pi} r^{-3} = -\frac{2(\pi r^2 h)}{\pi} r^{-3} = -\frac{2h}{r}. \quad ❑$$

The derivative as a rate of change is one of the fundamental ideas of calculus. We will come back to it again and again.

Velocity and Acceleration

Suppose that an object is moving along a straight line and that, for each time t during a certain time interval, the object has position (coordinate) $x(t)$. Then, at time $t + h$, the position of the object is $x(t + h)$, and $x(t + h) - x(t)$ is the distance that the object traveled during the time period t to $t + h$. The ratio

$$\frac{x(t + h) - x(t)}{(t + h) - t} = \frac{x(t + h) - x(t)}{h}$$

gives the *average velocity* of the object during this time period. If

$$\lim_{h \to 0} \frac{x(t + h) - x(t)}{h} = x'(t)$$

exists, then $x'(t)$ gives the (*instantaneous*) *rate of change of position with respect to time*. This rate of change of position is called the *velocity* of the object; in symbols

(3.4.1)
$$v(t) = x'(t).$$

If the velocity function is itself differentiable, then its rate of change with respect to time is called the *acceleration*; in symbols,

(3.4.2)
$$a(t) = v'(t) = x''(t).$$

In the Leibniz notation,

(3.4.3)
$$v = \frac{dx}{dt} \quad \text{and} \quad a = \frac{dv}{dt} = \frac{d^2x}{dt^2}.$$

The *speed* is by definition the absolute value of the velocity:

(3.4.4)
$$\text{speed at time } t = |v(t)|.$$

1. Positive velocity indicates motion in the positive direction (x is increasing). Negative velocity indicates motion in the negative direction (x is decreasing).
2. Positive acceleration indicates increasing velocity (increasing speed in the positive direction or decreasing speed in the negative direction). Negative acceleration indicates decreasing velocity (decreasing speed in the positive direction or increasing speed in the negative direction).
3. It follows from (2) that, if the velocity and acceleration have the same sign, the object is speeding up, but if the velocity and acceleration have opposite signs, the object is slowing down.

Example 3 An object moves along the x-axis, its position at each time t given by the function

$$x(t) = t^3 - 12t^2 + 36t - 27.$$

Let's study the motion from time $t = 0$ to time $t = 9$.

The object starts out 27 units to the left of the origin:

$$x(0) = 0^3 - 12(0)^2 + 36(0) - 27 = -27$$

and ends up 54 units to the right of the origin:

$$x(9) = 9^3 - 12(9)^2 + 36(9) - 27 = 54.$$

We can find the velocity function by differentiating the position function:

$$v(t) = x'(t) = 3t^2 - 24t + 36 = 3(t - 2)(t - 6).$$

It is clear that

$$v(t) \text{ is } \begin{cases} \text{positive,} & \text{for } 0 \leq t < 2 \\ 0, & \text{at } t = 2 \\ \text{negative,} & \text{for } 2 < t < 6 \\ 0, & \text{at } t = 6 \\ \text{positive,} & \text{for } 6 < t \leq 9. \end{cases}$$

We can interpret all this as follows: The object begins by moving to the right [$v(t)$ is positive for $0 \leq t < 2$]; it comes to a stop at time $t = 2$ [$v(2) = 0$]; it then moves left [$v(t)$ is negative for $2 < t < 6$]; it stops at time $t = 6$ [$v(6) = 0$]; it then moves right and keeps going right [$v(t) > 0$ for $6 < t \leq 9$].

We can find the acceleration by differentiating the velocity:

$$a(t) = v'(t) = 6t - 24 = 6(t - 4).$$

Obviously

$$a(t) \text{ is } \begin{cases} \text{negative,} & \text{for } 0 \leq t < 4 \\ 0, & \text{at } t = 4 \\ \text{positive,} & \text{for } 4 < t \leq 9. \end{cases}$$

At the beginning the velocity decreases, reaching a minimum at time $t = 4$. Then the velocity starts to increase and continues to increase.

Figure 3.4.6 shows a diagram for the sign of the velocity and a comparable diagram for the sign of the acceleration. Combining the two diagrams we have a brief description of the motion in convenient form.

The direction of the motion at each time $t \in [0, 9]$ is represented schematically in Figure 3.4.7.

Figure 3.4.6

Figure 3.4.7

A better way to represent the motion is to graph x as a function of t, as we did in Figure 3.4.8. The velocity $v(t) = x'(t)$ then appears as the slope of the tangent to the curve. Note, for example, that at $t = 2$ and $t = 6$, the tangent is horizontal and the slope is 0. This reflects the fact that $v(2) = 0$ and $v(6) = 0$.   ❑

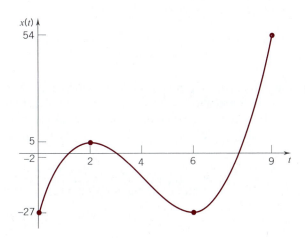

Figure 3.4.8

Before going on, a few words about units. The units of velocity and acceleration depend upon the units used to measure distance and the units used to measure time. The units of velocity are units of distance per unit time:

feet per second,   meters per second,   miles per hour,   and so forth.

The units of acceleration are units of distance per unit time per unit time:

feet per second per second,   meters per second per second,
miles per hour per hour,   and so forth.

Free Fall (Near the Surface of the Earth)

Imagine an object (for example, a rock or an apple) falling to the ground (Figure 3.4.9). We will assume that the object is in *free fall*: namely, that the gravitational pull on the object is constant throughout the fall and that there is no air resistance.†

Figure 3.4.9

---

† In practice, neither of these conditions is ever fully met. Gravitational attraction near the surface of the earth does vary somewhat with altitude, and there is always some air resistance. Nevertheless, in the setting within which we will be working the results that we obtain are correct within close approximation.

Galileo's formula for free fall gives the height of the object at each time t of the fall:

(3.4.5)
$$y(t) = -\tfrac{1}{2}gt^2 + v_0 t + y_0.\dagger$$

Let's examine this formula. First, the formula assumes that the positive y direction is up. Next, since $y(0) = y_0$, the constant y_0 represents the height of the object at time $t = 0$. This is called the *initial position*. Differentiation gives

$$y'(t) = -gt + v_0.$$

Since $y'(0) = v_0$, the constant v_0 gives the velocity of the object at time $t = 0$. This is called *the initial velocity*. A second differentiation gives

$$y''(t) = -g.$$

This indicates that the object falls with constant negative acceleration $-g$. (Why negative?)

There is a minor technicality that we should mention here. Although not explicitly stated, there is the implication that the height function $y(t)$ given by (3.4.5) is defined only for $t \geq 0$. Therefore, to be precise, the initial velocity $v(0) = y'(0)$ and the initial acceleration $a(0) = y''(0)$ should be defined as right-handed derivatives (see Exercises 3.1 for the definition of one-sided derivatives). However, it is easy to verify that the right-hand derivative of y at $t = 0$ is v_0 and the right-hand derivative of y' at $t = 0$ is $-g$, and so the values given here as ordinary (two-sided) derivatives are correct.

The constant g is a *gravitational constant*. If time is measured in seconds and distance in feet, then g is approximately 32 feet per second per second; if time is measured in seconds and distance in meters, then g is approximately 9.8 meters per second per second.$\dagger\dagger$ In making numerical calculations, we will take g as 32 feet per second per second or as 9.8 meters per second per second. Equation (3.4.5) then reads

$$y(t) = -16t^2 + v_0 t + y_0 \qquad \text{(distance in feet)}$$

or

$$y(t) = -4.9t^2 + v_0 t + y_0 \qquad \text{(distance in meters)}.$$

Example 4 A stone is dropped from a height of 98 meters. In how many seconds does it hit the ground? What is the speed at the instant of impact?

SOLUTION Here $y_0 = 98$ and $v_0 = 0$. Consequently, by (3.4.5),

$$y(t) = -4.9t^2 + 98.$$

To find t at the moment of impact, we set $y(t) = 0$. This gives

$$-4.9t^2 + 98 = 0, \qquad t^2 = 20, \qquad t = \pm\sqrt{20} = \pm2\sqrt{5}.$$

$\dagger$ Galileo Galilei (1564–1642), a great Italian astronomer and mathematician, is popularly known today for his early experiments with falling objects. His astronomical observations led him to support the Copernican view of the solar system. For this he was brought before the Inquisition.

$\dagger\dagger$ The value of this constant varies with latitude and elevation. It is approximately 32 feet per second per second at the equator at elevation zero. In Greenland it is about 32.23.

We can disregard the negative value because the stone was not dropped before time $t = 0$. We conclude that it takes $2\sqrt{5} \cong 4.47$ seconds for the stone to hit the ground.

The velocity at impact is the velocity at time $t = 2\sqrt{5}$. Since

$$v(t) = y'(t) = -9.8t,$$

we have

$$v(2\sqrt{5}) = -(19.6)\sqrt{5}.$$

(One might argue that we should use the left-hand derivative here; you can verify that the left-hand derivative of y at $t = 2\sqrt{5}$ is $-(19.6)\sqrt{5}$. If we were to use one-sided derivatives, what do you think the right-hand derivative of y at $t = 2\sqrt{5}$ would be?) The speed at impact is $|v(2\sqrt{5})| = (19.6)\sqrt{5} \cong 43.8$ meters per second. ❏

Example 5 An explosion causes debris to rise vertically with an initial velocity of 72 feet per second.

(a) In how many seconds does it attain maximum height?

(b) What is this maximum height?

(c) What is the speed of the debris as it reaches a height of 32 feet (i) going up? (ii) coming back down?

SOLUTION The basic equation in this case is

$$y(t) = -16t^2 + v_0 t + y_0.$$

Here $y_0 = 0$ (it starts at ground level) and $v_0 = 72$ (the initial velocity is 72 feet per second). The equation of motion is therefore

$$y(t) = -16t^2 + 72t.$$

Differentiation gives

$$v(t) = y'(t) = -32t + 72.$$

The maximum height is attained when the velocity is 0. This occurs at time $t = \frac{72}{32} = \frac{9}{4}$. Since $y(\frac{9}{4}) = 81$, the maximum height attained is 81 feet.

To answer part (c) we must first find those times t for which $y(t) = 32$. Since

$$y(t) = -16t^2 + 72t,$$

the condition $y(t) = 32$ gives

$$16t^2 - 72t + 32 = 0.$$

This quadratic has two solutions, $t = \frac{1}{2}$ and $t = 4$. Since $v(\frac{1}{2}) = 56$ and $v(4) = -56$, the velocity going up is 56 feet per second and the velocity coming down is -56 feet per second. In each case the speed is 56 feet per second. ❏

Economics

In business and economics one is often interested in how changes in such variables as production, supply, or price will affect other variables such as cost, revenue, or profit. If f is a function that describes the relationship between a pair of these variables, then the term *marginal* is used to specify the derivative of f.

For example, suppose $C = C(x)$ represents the cost of producing x units of a certain commodity. Although in reality x is a nonnegative integer, in theory and practice it is convenient to assume that C is defined for all x in some interval, and that C is differentiable. The derivative $C'(x)$ is called the *marginal cost*.

Originally, economists defined the marginal cost at a production level x to be $C(x + 1) - C(x)$, which is the cost of producing one additional unit of the commodity. Since

$$C(x + 1) - C(x) = \frac{C(x + 1) - C(x)}{1} \cong \lim_{h \to 0} \frac{C(x + h) - C(x)}{h} = C'(x),$$

it follows that the marginal cost $C'(x)$ at the production level x is approximately the cost of producing the $(x + 1)$-st unit.

Similarly, if $R = R(x)$ is the revenue received for selling x units of the commodity, then $R'(x)$ is called the *marginal revenue*. The marginal revenue at a sales level x is approximately the revenue obtained by selling one additional unit.

If $C = C(x)$ and $R = R(x)$ are the cost and revenue functions associated with producing and selling x units of the commodity, then

$$P(x) = R(x) - C(x)$$

is called the *profit function*. The values of x (if any) at which $C(x) = R(x)$, that is "cost" = "revenue," are called *break-even points*. The derivative, P', is the *marginal profit*. In Chapter 4 we show that the maximum profit occurs when marginal cost equals marginal revenue.

Example 6 A manufacturer of computer components determines that the total cost C of producing x components per week is

$$C(x) = 2000 + 50x - \frac{x^2}{20} \text{ (dollars).}$$

What is the marginal cost at the production level of 20 units? What is the exact cost of producing the 21st component?

SOLUTION The marginal cost at a production level x is given by the derivative

$$C'(x) = 50 - \frac{x}{10}.$$

Thus, the marginal cost at the production level of 20 components is

$$C'(20) = 50 - \frac{20}{10} = 48 \text{ (dollars).}$$

The exact cost of producing the 21st component is

$$C(21) - C(20) = 2000 + 50(21) - \frac{(21)^2}{20} - \left[2000 + 50(20) - \frac{(20)^2}{20} \right]$$

$$= 3027.95 - 2980 = 47.95 \text{ (dollars).} \quad \square$$

Example 7 A manufacturer of digital watches determines that the cost and revenue functions involved in producing and selling x watches are:

$$C(x) = 1200 + 13x$$

and

$$R(x) = 75x - \frac{x^2}{2},$$

respectively. Find the profit function and determine the break-even points. Find the marginal profit and determine the production/sales level at which the marginal profit is zero.

SOLUTION The profit function P is given by

$$P(x) = R(x) - C(x) = 75x - \frac{x^2}{2} - (1200 + 13x) = 62x - \frac{x^2}{2} - 1200.$$

To find the break-even points, set $P(x) = 0$:

$$62x - \frac{x^2}{2} - 1200 = 0$$

$$x^2 - 124x + 2400 = 0$$

$$(x - 24)(x - 100) = 0.$$

Thus, the break-even points are $x = 24$ and $x = 100$.

The marginal profit is given by the derivative:

$$P'(x) = 62 - x$$

and $P'(x) = 0$ when $x = 62$.

Figure 3.4.10 gives a graphical representation of the cost and revenue functions, the break-even points, and indicates the regions of profit and loss. ❑

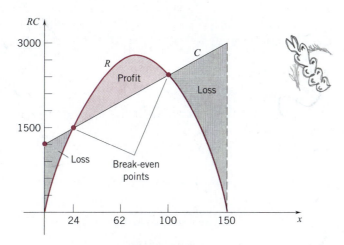

Figure 3.4.10

EXERCISES 3.4

1. Find the rate of change of the area of a circle with respect to the radius r. What is the rate when $r = 2$?

2. Find the rate of change of the volume of a cube with respect to the length s of a side. What is the rate when $s = 4$?

3. Find the rate of change of the area of a square with respect to the length z of a diagonal. What is the rate when $z = 4$?

4. Find the rate of change of $y = 1/x$ with respect to x at $x = -1$.

5. Find the rate of change of $y = [x(x + 1)]^{-1}$ with respect to x at $x = 2$.

6. Find the values of x for which the rate of change of $y = x^3 - 12x^2 + 45x - 1$ with respect to x is zero.

7. Find the rate of change of the volume of a sphere with respect to the radius r.

8. Find the rate of change of the surface area of a sphere with respect to the radius r. What is this rate of change when $r = r_0$? How must r_0 be chosen so that the rate of change is 1?

9. Find x_0 given that the rate of change of $y = 2x^2 + x - 1$ with respect to x at $x = x_0$ is 4.

10. Find the rate of change of the area A of a circle with respect to
    (a) the diameter d.   (b) the circumference C.

11. Find the rate of change of the volume V of a cube with respect to
    (a) the length w of a diagonal on one of the faces.
    (b) the length z of one of the diagonals of the cube.

12. The dimensions of a rectangle are changing in such a way that the area of the rectangle remains constant. Find the rate of change of the height h with respect to the base b.

13. The area of a sector in a circle is given by the formula $A = \frac{1}{2}r^2\theta$ where r is the radius and θ is the central angle measured in radians.
    (a) Find the rate of change of A with respect to θ if r remains constant.
    (b) Find the rate of change of A with respect to r if θ remains constant.
    (c) Find the rate of change of θ with respect to r if A remains constant.

14. The total surface area of a right circular cylinder is given by the formula $A = 2\pi r(r + h)$ where r is the radius and h is the height.
    (a) Find the rate of change of A with respect to h if r remains constant.
    (b) Find the rate of change of A with respect to r if h remains constant.
    (c) Find the rate of change of h with respect to r if A remains constant.

15. For what value of x is the rate of change of
    $$y = ax^2 + bx + c \quad \text{with respect to } x$$
    the same as the rate of change of
    $$z = bx^2 + ax + c \quad \text{with respect to } x?$$
    Assume that a, b, c are constant with $a \neq b$.

16. Find the rate of change of the product $f(x)g(x)h(x)$ with respect to x at $x = 1$ given that
    $$f(1) = 0, \quad g(1) = 2, \quad h(1) = -2,$$
    $$f'(1) = 1, \quad g'(1) = -1, \quad h'(1) = 0.$$

In Exercises 17–24, an object moves along a coordinate line, its position at each time $t \geq 0$ given by $x(t)$. Find the position, velocity, acceleration, and speed at time t_0.

17. $x(t) = 4 + 3t - t^2$; $t_0 = 5$.

18. $x(t) = 5t - t^3$; $t_0 = 3$.

19. $x(t) = t^3 - 6t$; $t_0 = 2$.

20. $x(t) = 4t^2 - 3t + 6$; $t_0 = 1$.

21. $x(t) = \dfrac{18}{t + 2}$; $t_0 = 1$.

22. $x(t) = \dfrac{2t}{t + 3}$; $t_0 = 3$.

23. $x(t) = (t^2 + 5t)(t^2 + t - 2)$; $t_0 = 1$.

24. $x(t) = (t^2 - 3t)(t^2 + 3t)$; $t_0 = 2$.

In Exercises 25–30, an object moves along a coordinate line, its position at each time $t \geq 0$ given by $x(t)$. Determine when, if ever, the object changes direction.

25. $x(t) = t^3 - 3t^2 + 3t$.   26. $x(t) = t + \dfrac{3}{t + 1}$.

27. $x(t) = t + \dfrac{5}{t + 2}$.

28. $x(t) = t^4 - 4t^3 + 4t^2$.

29. $x(t) = \dfrac{t}{t^2 + 8}$.

30. $x(t) = 3t^5 + 10t^3 + 15t$.

Objects A, B, C move along the x-axis. Their positions from time $t = 0$ to time $t = t_3$ have been graphed in the figure below.

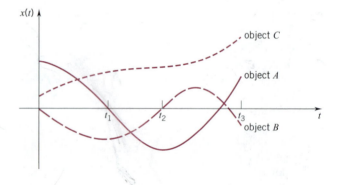

In Exercises 31–40, find:

31. Which object begins furthest to the right? A

32. Which object finishes furthest to the right? C

33. Which object has the greatest speed at time t_1? A

34. Which object maintains the same direction during the time interval $[t_1, t_3]$? C

35. Which object begins by moving left? B

36. Which object finishes moving left? B

37. Which object changes direction at time t_2? A

38. Which object speeds up during the time interval $[0, t_1]$?

39. Which object slows down during the time interval $[t_1, t_2]$?

40. Which object changes direction during the time interval $[t_2, t_3]$? B

An object moves along the x-axis, its position at each time $t \geq 0$ given by $x(t)$. In Exercises 41–48, determine the time interval(s), if any, during which the object satisfies the given condition.

41. $x(t) = t^4 - 12t^3 + 28t^2$; moving right.

42. $x(t) = t^3 - 12t^2 + 21t$; moving left.

43. $x(t) = 5t^4 - t^5$; speeding up.

44. $x(t) = 6t^2 - t^4$; slowing down.

45. $x(t) = t^3 - 6t^2 - 15t$; moving left and slowing down.

46. $x(t) = t^3 - 6t^2 - 15t$; moving right and slowing down.

47. $x(t) = t^4 - 8t^3 + 16t^2$; moving right and speeding up.

48. $x(t) = t^4 - 8t^3 + 16t^2$; moving left and speeding up.

In Exercises 49–60, neglect air resistance. For the numerical calculations, take g as 32 feet per second per second or as 9.8 meters per second per second.

49. An object is dropped and hits the ground 6 seconds later. From what height was it dropped?

50. Supplies are dropped from a helicopter and seconds later hit the ground at 98 meters per second. How high was the helicopter?

51. An object is projected vertically upward from ground level with velocity v_0. What is the height in meters attained by the object?

52. An object projected vertically upward from ground level returns to earth in 8 seconds. What was the initial velocity in feet per second?

53. An object projected vertically upward passes every height less than the maximum twice, once on the way up, once on the way down. Show that the speed is the same in each direction. Assume that height is measured in feet.

54. An object is projected vertically upward from the ground. Show that it takes the object the same amount of time to reach its maximum height as it takes for it to drop from that height back to the ground. Assume that height is measured in meters.

55. A rubber ball is thrown straight down from a height of 224 feet at a speed of 80 feet per second. If the ball always rebounds with one-fourth of its impact speed, what will be the speed of the ball the third time it hits the ground?

56. A ball is thrown straight up from ground level. How high will the ball go if it reaches a height of 64 feet in 2 seconds?

57. A stone is thrown upward from ground level. The initial speed is 32 feet per second. (a) In how many seconds will the stone hit the ground? (b) How high will it go? (c) With what minimum speed should the stone be thrown so as to reach a height of 36 feet?

58. To estimate the height of a bridge a man drops a stone into the water below. How high is the bridge (a) if the stone

hits the water 3 seconds later? (b) if the man hears the splash 3 seconds later? (Use 1080 feet per second as the speed of sound.)

59. A falling stone is at a certain instant 100 feet above the ground. Two seconds later it is only 16 feet above the ground. (a) From what height was it dropped? (b) If it was thrown down with an initial speed of 5 feet per second, from what height was it thrown? (c) If it was thrown upward with an initial speed of 10 feet per second, from what height was it thrown?

60. A rubber ball is thrown straight down from a height of 4 feet. If the ball rebounds with one-half of its impact speed and returns exactly to its original height before falling again, how fast was it thrown originally?

In Exercises 61–64, a cost function for a certain commodity is given. Find the marginal cost function at a production level of 100 units and compare that with the actual cost of producing the 101st unit.

61. $C(x) = 200 + 0.02x + 0.0001x^2$, $x > 0$.

62. $C(x) = 1000 + 2x + 0.02x^2 + 0.0001x^3$, $x \geq 0$.

63. $C(x) = 200 + 0.01x + \dfrac{100}{x}$, $x > 0$.

64. $C(x) = 2000 + 2\sqrt{x}$, $x \geq 0$.

65. A manufacturer of electric motors estimates that the cost of producing x motors per day is given by

$$C(x) = 1000 + 25x - \frac{x^2}{10}, \quad 0 \leq x \leq 200.$$

Find the marginal cost of producing 10 motors and compare it with the exact cost of producing the 11th motor.

66. The cost and revenue functions for the production of x units of a certain commodity are

$C(x) = 4x + 1400$ and

$$R(x) = 20x - \frac{x^2}{50} \quad \text{for} \quad 0 \leq x \leq 1000.$$

(a) Find the profit function and determine the break-even points.

(b) Find the marginal profit and determine the production level at which the marginal profit is zero.

(c) Sketch the cost and revenue functions in the same coordinate system and indicate the regions of profit and loss. What production level do you think will produce the maximum profit?

Economists often study costs, revenues, and profits in terms of averages of these quantities. The average cost $\overline{C}$ (cost per unit), average revenue $\overline{R}$ (revenue per unit), and average profit $\overline{P}$ (profit per unit) are given by

$$\overline{C}(x) = \frac{C(x)}{x}, \qquad \overline{R}(x) = \frac{R(x)}{x}, \qquad \text{and} \qquad \overline{P}(x) = \frac{P(x)}{x},$$

where C, R, and P are the cost, revenue, and profit functions. The marginal average cost, marginal average revenue, and marginal average profit functions are the corresponding derivatives of $\overline{C}$, $\overline{R}$, and $\overline{P}$.

67. Calculate the marginal average cost functions for the given cost functions.

(a) $C(x) = 200 + 0.02x + 0.0001x^2$.

(b) $C(x) = 200 + 0.01x + \dfrac{100}{x}$.

68. Calculate the marginal average cost, marginal average revenue, and marginal average profit functions corresponding to the cost and revenue functions given in Exercise 66.

▶ **69.** An object moves along the x-axis. Its position at time t is given by

$$x(t) = t^3 - 7t^2 + 10t + 5, \quad 0 \le t \le 5.$$

(a) Determine the velocity function v. Use a graphing utility to graph v.

(b) Use the graph to estimate when the object is moving to the right and when it is moving to the left.

(c) Use the graphing utility to graph the speed $|v|$ of the object. Estimate when the object stops. Estimate the maximum speed for $1 \le t \le 4$.

(d) Determine the acceleration a and use the graphing utility to graph it. Estimate when the object is speeding up and when it is slowing down.

▶ **70.** The cost and revenue functions for the production of a certain commodity are

$$C(x) = 4 + 0.75x \qquad \text{and}$$

$$R(x) = \frac{10x}{1 + 0.25x^2} \qquad \text{for } x \ge 0,$$

where x is measured in hundreds of units.

(a) Using a graphing utility, graph the cost and revenue functions together. Estimate the break-even points accurate to two decimal places.

(b) Graph the profit function and estimate the production level that will yield the maximum profit. Use two decimal place accuracy. How many units should be produced to maximize the profit?

■ 3.5 THE CHAIN RULE

In this section we take up the differentiation of composite functions. Until we get to Theorem 3.5.7, our approach is completely intuitive—no real definitions, no proofs, just informal discussion. Our purpose is to give you some experience with the standard computational procedures and some insight into why these procedures work. Theorem 3.5.7 puts it all on a sound footing.

Suppose that y is a differentiable function of u:

$$y = f(u)$$

and u in turn is a differentiable function of x:

$$u = g(x).$$

Then y is a composite function of x:

$$y = f(u) = f(g(x)) = (f \circ g)(x).$$

Does y have a derivative with respect to x? Yes it does, and it's given by a formula that is easy to remember:

(3.5.1)
$$\frac{dy}{dx} = \frac{dy}{du}\frac{du}{dx}.$$

This formula, known as the *chain rule*, says that "the rate of change of y with respect to x is the rate of change of y with respect to u times the rate of change of u with respect to x." (Plausible as all this sounds, remember that we have proved nothing. All we have done is asserted that the composition of differentiable functions is differentiable and given you a formula, a formula that needs justification and is justified at the end of this section.)

Before using the chain rule in elaborate computations, let's confirm its validity in some simple instances.

If $y = 2u$ and $u = 3x$, then $y = 6x$. Clearly

$$\frac{dy}{dx} = 6 = 2 \cdot 3 = \frac{dy}{du}\frac{du}{dx},$$

and so, in this case certainly, the chain rule is confirmed:

$$\frac{dy}{dx} = \frac{dy}{du}\frac{du}{dx}.$$

If $y = u^3$ and $u = x^2$, then $y = (x^2)^3 = x^6$. This time

$$\frac{dy}{dx} = 6x^5, \qquad \frac{dy}{du} = 3u^2 = 3x^4, \qquad \frac{du}{dx} = 2x$$

and once again

$$\frac{dy}{dx} = 6x^5 = 3x^4 \cdot 2x = \frac{dy}{du}\frac{du}{dx}.$$

Example 1 Find dy/dx by the chain rule given that

$$y = \frac{u-1}{u+1} \qquad \text{and} \qquad u = x^2.$$

SOLUTION

$$\frac{dy}{du} = \frac{(u+1)1 - (u-1)1}{(u+1)^2} = \frac{2}{(u+1)^2} \qquad \text{and} \qquad \frac{du}{dx} = 2x$$

so that

$$\frac{dy}{dx} = \frac{dy}{du}\frac{du}{dx} = \left[\frac{2}{(u+1)^2}\right]2x = \frac{4x}{(x^2+1)^2}. \qquad ❑$$

Remark We would have obtained the same result without the chain rule by first writing y as a function of x and then differentiating:

$$\text{with} \quad y = \frac{u-1}{u+1} \qquad \text{and} \qquad u = x^2 \quad \text{we have} \quad y = \frac{x^2-1}{x^2+1}$$

and therefore

$$\frac{dy}{dx} = \frac{(x^2+1)2x - (x^2-1)2x}{(x^2+1)^2} = \frac{4x}{(x^2+1)^2}. \qquad ❑$$

If f is a differentiable function of u and u is a differentiable function of x, then, according to (3.5.1),

(3.5.2)

$$\frac{d}{dx}[f(u)] = \frac{d}{du}[f(u)]\frac{du}{dx}.$$

(All we have done here is written y as $f(u)$.) This is the formulation of the chain rule that we will use most frequently in later work.

Suppose now that we were asked to calculate

$$\frac{d}{dx}[(x^2 - 1)^{100}].$$

We could expand $(x^2 - 1)^{100}$ into polynomial form by using the binomial theorem or, if we were masochistic, by repeated multiplication, but we would have a terrible mess on our hands; for example, $(x^2 - 1)^{100}$ has 101 terms. With (3.5.2) we can derive a formula that will render such calculations almost trivial.

Assuming (3.5.2), we can prove that, if u is a differentiable function of x and n is an integer, then

(3.5.3)

$$\boxed{\frac{d}{dx}(u^n) = nu^{n-1}\frac{du}{dx}.}$$

PROOF Set $f(u) = u^n$. Then

$$\frac{d}{dx}(u^n) = \underset{\underset{(3.5.2)}{\uparrow}}{\frac{d}{du}(u^n)}\frac{du}{dx} = nu^{n-1}\frac{du}{dx}. \quad ❑$$

To calculate

$$\frac{d}{dx}[(x^2 - 1)^{100}]$$

we set $u = x^2 - 1$. Then by our formula

$$\frac{d}{dx}[(x^2 - 1)^{100}] = 100(x^2 - 1)^{99}\frac{d}{dx}(x^2 - 1) = 100(x^2 - 1)^{99}2x = 200x(x^2 - 1)^{99}.$$

Remark While it is clear that the chain rule is the only practical way to calculate the derivative of $f(x) = (x^2 - 1)^{100}$, you do have a choice in differentiating a similar but simpler function such as $g(x) = (x^2 - 1)^5$. Calculating the derivative of g by the chain rule, we get

$$\frac{d}{dx}[(x^2 - 1)^5] = 5(x^2 - 1)^4\frac{d}{dx}(x^2 - 1) = 5(x^2 - 1)^4 2x = 10x(x^2 - 1)^4.$$

On the other hand, if we were to first expand the expression $(x^2 - 1)^5$, we would get

$$g(x) = x^{10} - 5x^8 + 10x^6 - 10x^4 + 5x^2 - 1$$

and then

$$g'(x) = 10x^9 - 40x^7 + 60x^5 - 40x^3 + 10x.$$

As a final answer, this is correct, but it is somewhat unwieldy. To simplify, notice that $10x$ is a factor of g':

$$g'(x) = 10x(x^8 - 4x^6 + 6x^4 - 4x^2 + 1),$$

but note that the expression in parentheses is the binomial expansion for $(x^2 - 1)^4$ so that

$$g'(x) = 10x(x^2 - 1)^4$$

as we saw earlier. The point is, the chain rule gave us the neat, compact result immediately, without the need for expansions and factoring. ❏

Here are more examples of a similar sort.

Example 2

$$\frac{d}{dx}\left[\left(x + \frac{1}{x}\right)^{-3}\right] = -3\left(x + \frac{1}{x}\right)^{-4}\frac{d}{dx}\left(x + \frac{1}{x}\right) = -3\left(x + \frac{1}{x}\right)^{-4}\left(1 - \frac{1}{x^2}\right).$$

❏

Example 3

$$\frac{d}{dx}[1 - (2 + 3x)^2]^3 = 3[1 - (2 + 3x)^2]^2\frac{d}{dx}[1 - (2 + 3x)^2].$$

Since

$$\frac{d}{dx}[1 - (2 + 3x)^2] = -2(2 + 3x)\frac{d}{dx}(2 + 3x) = -2(2 + 3x)3 = -6(2 + 3x),$$

we have

$$\frac{d}{dx}[1 - (2 + 3x)^2]^3 = 3[1 - (2 + 3x)^2]^2[-6(2 + 3x)]$$

$$= -18(2 + 3x)[1 - (2 + 3x)^2]^2. \quad ❏$$

Example 4 Calculate the derivative of

$$f(x) = 4x(x^2 + 3)^3.$$

SOLUTION Here we need to use the product rule and the chain rule:

$$\frac{d}{dx}[4x(x^2 + 3)^3] = 4x\frac{d}{dx}[(x^2 + 3)^3] + (x^2 + 3)^3\frac{d}{dx}(4x)$$

$$= 4x[3(x^2 + 3)^2\, 2x] + 4(x^2 + 3)^3$$

$$= 24x^2(x^2 + 3)^2 + 4(x^2 + 3)^3$$

$$= (x^2 + 3)^2(28x^2 + 12) = 4(x^2 + 3)^2(7x^2 + 3). \quad ❏$$

The formula

$$\frac{dy}{dx} = \frac{dy}{du}\frac{du}{dx}$$

can easily be extended to more variables. For example, if x itself depends on s, then we have

(3.5.4)

$$\boxed{\frac{dy}{ds} = \frac{dy}{du}\frac{du}{dx}\frac{dx}{ds}.}$$

If, in addition, s depends on t, then

(3.5.5)

$$\frac{dy}{dt} = \frac{dy}{du}\frac{du}{dx}\frac{dx}{ds}\frac{ds}{dt},$$

and so on.   Each new dependence adds a new link to the chain.

Example 5   Find dy/ds given that

$$y = 3u + 1, \qquad u = x^{-2}, \qquad x = 1 - s.$$

SOLUTION

$$\frac{dy}{du} = 3, \qquad \frac{du}{dx} = -2x^{-3}, \qquad \frac{dx}{ds} = -1.$$

Therefore

$$\frac{dy}{ds} = \frac{dy}{du}\frac{du}{dx}\frac{dx}{ds} = (3)(-2x^{-3})(-1) = 6x^{-3} = 6(1-s)^{-3}. \quad \square$$

Example 6   Find dy/dt at $t = 9$ given that

$$y = \frac{u + 2}{u - 1}, \qquad u = (3s - 7)^2, \qquad s = \sqrt{t}.$$

SOLUTION   As you can check, $2(3s-7)3$

$$\frac{dy}{du} = -\frac{3}{(u-1)^2}, \qquad \frac{du}{ds} = 6(3s - 7), \qquad \frac{ds}{dt} = \frac{1}{2\sqrt{t}}.†$$

At $t = 9$, we have $s = 3$ and $u = 4$ so that

$$\frac{dy}{du} = -\frac{3}{(4-1)^2} = -\frac{1}{3}, \qquad \frac{du}{ds} = 6(9 - 7) = 12, \qquad \frac{ds}{dt} = \frac{1}{2\sqrt{9}} = \frac{1}{6}.$$

Thus, at $t = 9$,

$$\frac{dy}{dt} = \frac{dy}{du}\frac{du}{ds}\frac{ds}{dt} = \left(-\frac{1}{3}\right)(12)\left(\frac{1}{6}\right) = -\frac{2}{3}. \quad \square$$

So far we have worked entirely in Leibniz's notation.   What does the chain rule look like in prime notation?

Let's go back to the beginning.   Once again let y be a differentiable function of u:

$$y = f(u),$$

and let u be a differentiable function of x:

$$u = g(x).$$

---

† It was shown in Section 3.1 that

$$\frac{d}{dx}(\sqrt{x}) = \frac{1}{2\sqrt{x}}.$$

Then

$$y = f(u) = f(g(x)) = (f \circ g)(x)$$

and, according to the chain rule (as yet unproved),

$$\frac{dy}{dx} = \frac{dy}{du}\frac{du}{dx}.$$

Since

$$\frac{dy}{dx} = \frac{d}{dx}[(f \circ g)(x)] = (f \circ g)'(x), \qquad \frac{dy}{du} = f'(u) = f'(g(x)), \qquad \frac{du}{dx} = g'(x),$$

the chain rule can be written

(3.5.6)
$$(f \circ g)'(x) = f'(g(x))g'(x).$$

The chain rule in prime notation says that

> "the derivative of a composition $f \circ g$ at x is the derivative of f at $g(x)$ times the derivative of g at x."

In Leibniz's notation the chain rule *appears* seductively simple, to some even obvious. "After all, to prove it, all you have to do is cancel the du's":

$$\frac{dy}{dx} = \frac{dy}{d\!\!\!/u}\frac{d\!\!\!/u}{dx}.$$

Of course, this is just nonsense. What would one cancel from

$$(f \circ g)'(x) = f'(g(x))g'(x)?$$

Although Leibniz's notation is useful for routine calculations, mathematicians generally turn to the prime notation when precision is required.

It is time for us to be precise. How do we know that the composition of differentiable functions is differentiable? What assumptions do we need? Precisely under what circumstances is it true that

$$(f \circ g)'(x) = f'(g(x))g'(x)?$$

The following theorem provides the definitive answer.

THEOREM 3.5.7   THE CHAIN-RULE THEOREM

If g is differentiable at x and f is differentiable at $g(x)$, then the composition $f \circ g$ is differentiable at x and

$$(f \circ g)'(x) = f'(g(x))g'(x).$$

A proof of this theorem appears in the supplement to this section. The argument is not as easy as "canceling" the du's.

We conclude this section with an application that involves the chain rule.

Example 7   Gravel is being poured by a conveyor onto a conical pile at the constant rate of 60π cubic feet per minute.   Frictional forces within the pile are such that the height is always two-thirds of the radius. How fast is the radius of the pile changing at the instant it is 5 feet?

SOLUTION   The formula for the volume V of a right circular cone of radius r and height h is

$$V = \tfrac{1}{3}\pi r^2 h.$$

However, in this case we are told that $h = \tfrac{2}{3}r$, and so we have

(∗) $$V = \tfrac{2}{9}\pi r^3.$$

Since gravel is being poured onto the pile, the volume, and hence the radius, are functions of time t.   We are given that $dV/dt = 60\pi$ and we want to find dr/dt at the instant when $r = 5$.   Differentiating (∗) with respect to t using the chain rule, we get

$$\frac{dV}{dt} = \frac{dV}{dr} \cdot \frac{dr}{dt} = (\tfrac{2}{3}\pi r^2)\frac{dr}{dt}.$$

Solving for dr/dt and using the fact that $dV/dt = 60\pi$, we find that

$$\frac{dr}{dt} = \frac{180\pi}{2\pi r^2} = \frac{90}{r^2}.$$

Now, when $r = 5$,

$$\frac{dr}{dt} = \frac{90}{(5)^2} = \frac{90}{25} = 3.6.$$

Thus, the radius is increasing at the rate of 3.6 feet per minute at the instant the radius is 5 feet.   ❑

EXERCISES 3.5

In Exercises 1–6, differentiate the given function: (a) by expanding before differentiation, (b) by using the chain rule. Then reconcile the results.

1. $f(x) = (x^2 + 1)^2$.

2. $f(x) = (x^3 - 1)^2$.

3. $f(x) = (2x + 1)^3$.

4. $f(x) = (x^2 + 1)^3$.

5. $f(x) = (x + x^{-1})^2$.

6. $f(x) = (3x^2 - 2x)^2$.

In Exercises 7–26, differentiate the given function.

7. $f(x) = (1 - 2x)^{-1}$.

8. $f(x) = (1 + 2x)^5$.

9. $f(x) = (x^5 - x^{10})^{20}$.

10. $f(x) = \left(x^2 + \dfrac{1}{x^2}\right)^3$.

11. $f(x) = \left(x - \dfrac{1}{x}\right)^4$.

12. $f(x) = \left(x + \dfrac{1}{x}\right)^3$.

13. $f(x) = (x - x^3 - x^5)^4$.

14. $f(t) = \left(\dfrac{1}{1 + t}\right)^4$.

15. $f(t) = (t^2 - 1)^{100}$.

16. $f(t) = (t - t^2)^3$.

17. $f(t) = (t^{-1} + t^{-2})^4$.

18. $f(x) = \left(\dfrac{4x + 3}{5x - 2}\right)^3$.

19. $f(x) = \left(\dfrac{3x}{x^2 + 1}\right)^4$.

20. $f(x) = [(2x + 1)^2 + (x + 1)^2]^3$.

21. $f(x) = (x^4 + x^2 + x)^2$.

22. $f(x) = (x^2 + 2x + 1)^3$.

23. $f(x) = \left(\dfrac{x^3}{3} + \dfrac{x^2}{2} + \dfrac{x}{1}\right)^{-1}$.

24. $f(x) = [(6x + x^5)^{-1} + x]^2$.

25. $f(x) = [(x + x^{-1})^2 - (x^2 + x^{-2})^{-1}]^3$.

26. $f(x) = [(x^{-1} + 2x^{-2})^3 + 3x^{-3}]^4$.

In Exercises 27–30, find dy/dx at $x = 0$.

27. $y = \dfrac{1}{1 + u^2}$, $u = 2x + 1$.

28. $y = u + \dfrac{1}{u}$, $u = (3x + 1)^4$.

29. $y = \dfrac{2u}{1 - 4u}$, $u = (5x^2 + 1)^4$.

30. $y = u^3 - u + 1$, $\quad u = \dfrac{1 - x}{1 + x}$.

In Exercises 31 and 32, find dy/dt.

31. $y = \dfrac{1 - 7u}{1 + u^2}$, $\quad u = 1 + x^2$, $\quad x = 2t - 5$.

32. $y = 1 + u^2$, $\quad u = \dfrac{1 - 7x}{1 + x^2}$, $\quad x = 5t + 2$.

In Exercises 33 and 34, find dy/dx at $x = 2$.

33. $y = (s + 3)^2$, $\quad s = \sqrt{t - 3}$, $\quad t = x^2$.

34. $y = \dfrac{1 + s}{1 - s}$, $\quad s = t - \dfrac{1}{t}$, $\quad t = \sqrt{x}$.

Given that

$$f(0) = 1, \quad f'(0) = 2, \quad f(1) = 0, \quad f'(1) = 1,$$
$$f(2) = 1, \quad f'(2) = 1,$$

$$g(0) = 2, \quad g'(0) = 1, \quad g(1) = 1, \quad g'(1) = 0,$$
$$g(2) = 2, \quad g'(2) = 1,$$

$$h(0) = 1, \quad h'(0) = 2, \quad h(1) = 2, \quad h'(1) = 1,$$
$$h(2) = 0, \quad h'(2) = 2,$$

evaluate the following.

35. $(f \circ g)'(0)$.

36. $(f \circ g)'(1)$.

37. $(f \circ g)'(2)$.

38. $(g \circ f)'(0)$.

39. $(g \circ f)'(1)$.

40. $(g \circ f)'(2)$.

41. $(f \circ h)'(0)$.

42. $(f \circ h \circ g)'(1)$.

43. $(g \circ f \circ h)'(2)$.

44. $(g \circ h \circ f)'(0)$.

In Exercises 45–48, find $f''(x)$.

45. $f(x) = (x^3 + x)^4$

46. $f(x) = (x^2 - 5x + 2)^{10}$.

47. $f(x) = \left(\dfrac{x}{1 - x}\right)^3$.

48. $f(x) = \sqrt{x^2 + 1}$ $\left(\text{Recall that } \dfrac{d}{dx}[\sqrt{x}] = \dfrac{1}{2\sqrt{x}}\right)$.

In Exercises 49–52, find the indicated derivative.

49. $\dfrac{d}{dx}[f(x^2 + 1)]$.

50. $\dfrac{d}{dx}\left[f\left(\dfrac{x - 1}{x + 1}\right)\right]$.

51. $\dfrac{d}{dx}[[f(x)]^2 + 1]$.

52. $\dfrac{d}{dx}\left[\dfrac{f(x) - 1}{f(x) + 1}\right]$.

In Exercises 53–56, determine the values of x for which
(a) $f'(x) = 0$.   (b) $f'(x) > 0$.   (c) $f'(x) < 0$.

53. $f(x) = (1 + x^2)^{-2}$.

54. $f(x) = (1 - x^2)^2$.

55. $f(x) = x(1 + x^2)^{-1}$.

56. $f(x) = x(1 - x^2)^3$.

An object moves along a coordinate line, its position at each time $t \geq 0$ given by $x(t)$. In Exercises 57–60, determine when the object changes direction.

57. $x(t) = (t + 1)^2(t - 9)^3$.

58. $x(t) = t(t - 8)^3$.

59. $x(t) = (t^3 - 12t)^4$.

60. $x(t) = (t^2 - 8t + 15)^3$.

61. A function L has the property that $L'(x) = 1/x$ for $x \neq 0$. Determine the derivative of $L(x^2 + 1)$.

62. Let f and g be differentiable functions such that $f'(x) = g(x)$ and $g'(x) = f(x)$, and let

$$H(x) = [f(x)]^2 - [g(x)]^2.$$

Find $H'(x)$.

63. Let f and g be differentiable functions such that $f'(x) = g(x)$ and $g'(x) = -f(x)$, and let

$$T(x) = [f(x)]^2 + [g(x)]^2.$$

Find $T'(x)$.

64. Let f be a differentiable function. Use the chain rule to show that:
  (a) if f is even, then f' is odd.
  (b) if f is odd, then f' is even.
    (Also see Exercises 59 and 60, Section 3.1.)

65. The number a is a *double zero* (or a zero of *multiplicity* 2) of the polynomial function $p(x)$ if

$$p(x) = (x - a)^2 q(x) \quad \text{and} \quad q(a) \neq 0.$$

Prove that a is a double zero of $p(x)$ iff a is a zero of both $p(x)$ and $p'(x)$ and $p''(a) \neq 0$.   HINT: Recall the factor theorem; see Section 1.2.

66. The number a is a *triple zero* (or a zero of *multiplicity* 3) of the polynomial function $p(x)$ if

$$p(x) = (x - a)^3 q(x) \quad \text{and} \quad q(a) \neq 0.$$

Prove that a is a triple zero of $p(x)$ iff a is a zero of $p(x)$, $p'(x)$, and $p''(x)$, and $p'''(a) \neq 0$.

67. The number a is a *zero of multiplicity* k of the polynomial function $p(x)$ if

$$p(x) = (x - a)^k q(x) \quad \text{and} \quad q(a) \neq 0.$$

Use the results in Exercises 65 and 66 to state a theorem about a zero of multiplicity k.

68. Two functions f and g are said to be *inverse functions* if

$$(f \circ g)(x) = x \quad \text{and} \quad (g \circ f)(x) = x.$$

For example, $f(x) = 3x + 2$ and $g(x) = \frac{1}{3}x - \frac{2}{3}$ are inverse functions. Let f and g be differentiable. Show that if f and g are inverse functions, then

$$f'(y) = \dfrac{1}{g'(x)},$$

where $y = g(x)$, provided $g'(x) \neq 0$.

69. An object is moving along the curve $y = x^3 - 3x + 5$ so that its x-coordinate at time t is $x = 2t^2 - t + 2$, $t \geq 0$. At what rate is the y-coordinate changing when $t = 2$?

70. An equilateral triangle of side length x and altitude h has area A given by

$$A = \dfrac{\sqrt{3}}{4}x^2 \quad \text{where} \quad x = \dfrac{2\sqrt{3}}{3}h.$$

Find the rate of change of A with respect to h and determine the rate of change of A when $h = 2\sqrt{3}$.

71. Air is being pumped into a spherical balloon in such a way that its radius is increasing at the constant rate of 2 centimeters per second. What is the rate of change of the balloon volume at the instant the radius is 10 centimeters? (The volume V of a sphere of radius r is $V = \frac{4}{3}\pi r^3$.)

72. Air is being pumped into a spherical balloon at the constant rate of 200 cubic centimeters per second. How fast is the surface area of the balloon changing at the instant the radius is 5 centimeters? (The surface area S of a sphere of radius r is $S = 4\pi r^2$.)

73. If an object of mass m has speed v, then its *kinetic energy*, KE, is given by

$$\text{KE} = \tfrac{1}{2}mv^2.$$

Suppose that v is a function of time. What is the rate of change of KE with respect to t?

74. Newton's Law of Gravitational Attraction states that if two bodies are at a distance r apart, then the force F exerted by one body on the other is given by

$$F(r) = \frac{k}{r^2},$$

where k is a positive constant. Suppose that, as a function of time, the distance between the two bodies is given by

$$r(t) = 49t - 4.9t^2, \qquad 0 \le t \le 10.$$

(a) Find the rate of change of F with respect to t.
(b) Show that $(F \circ r)'(3) = -(F \circ r)'(7)$.

---

*SUPPLEMENT TO SECTION 3.5

To prove Theorem 3.5.7 it is convenient to use the formulation of derivative given in (3.1.5), Exercises 3.1.

THEOREM 3.5.8

The function f is differentiable at x iff

$$\lim_{t \to x} \frac{f(t) - f(x)}{t - x} \quad \text{exists.}$$

If this limit exists, it is $f'(x)$.

Proof

For each t in the domain of f, $t \ne x$, define

$$G(t) = \frac{f(t) - f(x)}{t - x}.$$

The theorem follows from observing that f is differentiable at x iff

$$\lim_{h \to 0} G(x + h) \quad \text{exists,}$$

and noting that

$$\lim_{h \to 0} G(x + h) = L \quad \text{iff} \quad \lim_{t \to x} G(t) = L. \quad \square$$

Proof of Theorem 3.5.7

By Theorem 3.5.8 it is enough to show that

$$\lim_{t \to x} \frac{f(g(t)) - f(g(x))}{t - x} = f'(g(x))g'(x).$$

We begin by defining an auxiliary function F on the domain of f by setting

$$F(y) = \begin{cases} \dfrac{f(y) - f(g(x))}{y - g(x)}, & y \ne g(x) \\ f'(g(x)), & y = g(x). \end{cases}$$

F is continuous at $g(x)$ since

$$\lim_{y \to g(x)} F(y) = \lim_{y \to g(x)} \frac{f(y) - f(g(x))}{y - g(x)},$$

and the right-hand side is (by Theorem 3.5.8) $f'(g(x))$, which is the value of F at $g(x)$. For $t \neq x$,

(1)
$$\frac{f(g(t)) - f(g(x))}{t - x} = F(g(t)) \left[\frac{g(t) - g(x)}{t - x} \right].$$

To see this we note that, if $g(t) = g(x)$, then both sides are 0. If $g(t) \neq g(x)$, then

$$F(g(t)) = \frac{f(g(t)) - f(g(x))}{g(t) - g(x)},$$

so that again we have equality.

Since g, being differentiable at x, is continuous at x and since F is continuous at $g(x)$, we know that the composition $F \circ g$ is continuous at x. Thus

$$\lim_{t \to x} F(g(t)) = F(g(x)) = f'(g(x)).$$

————— by our definition of F

This, together with Equation (1), gives

$$\lim_{t \to x} \frac{f(g(t)) - f(g(x))}{t - x} = f'(g(x))g'(x). \quad \square$$

■ 3.6 DIFFERENTIATING THE TRIGONOMETRIC FUNCTIONS

An outline review of trigonometry—definitions, identities, and graphs—appears in Chapter 1. As discussed there, the calculus of the trigonometric functions is simplified by the use of radian measure. Therefore, we use radian measure throughout our work and refer to degree measure only in passing.

The derivative of the sine function is the cosine function:

(3.6.1)
$$\frac{d}{dx}(\sin x) = \cos x.$$

PROOF Fix any number x. For $h \neq 0$

$$\frac{\sin(x + h) - \sin x}{h} = \frac{[\sin x \cos h + \cos x \sin h] - [\sin x]}{h}$$

$$= \sin x \frac{\cos h - 1}{h} + \cos x \frac{\sin h}{h}.$$

Now, as shown in Section 2.5, (2.5.5),

$$\lim_{h \to 0} \frac{\cos h - 1}{h} = 0 \quad \text{and} \quad \lim_{h \to 0} \frac{\sin h}{h} = 1.$$

Since $\sin x$ and $\cos x$ remain constant as h approaches zero, it follows that

$$\lim_{h \to 0} \frac{\sin(x+h) - \sin x}{h} = \lim_{h \to 0} \left(\sin x \frac{\cos h - 1}{h} + \cos x \frac{\sin h}{h} \right)$$

$$= \sin x \left(\lim_{h \to 0} \frac{\cos h - 1}{h} \right) + \cos x \left(\lim_{h \to 0} \frac{\sin h}{h} \right).$$

Thus

$$\lim_{h \to 0} \frac{\sin(x+h) - \sin x}{h} = (\sin x)(0) + (\cos x)(1) = \cos x. \quad \square$$

The derivative of the cosine function is the negative of the sine function:

(3.6.2)

$$\frac{d}{dx}(\cos x) = -\sin x.$$

PROOF   Fix any number x.   For $h \neq 0$

$$\cos(x+h) = \cos x \cos h - \sin x \sin h.$$

Therefore

$$\lim_{h \to 0} \frac{\cos(x+h) - \cos x}{h} = \lim_{h \to 0} \frac{[\cos x \cos h - \sin x \sin h] - [\cos x]}{h}$$

$$= \cos x \left(\lim_{h \to 0} \frac{\cos h - 1}{h} \right) - \sin x \left(\lim_{h \to 0} \frac{\sin h}{h} \right)$$

$$= -\sin x. \quad \square$$

Example 1   To differentiate

$$f(x) = \cos x \sin x$$

we use the product rule:

$$f'(x) = \cos x \frac{d}{dx}(\sin x) + \sin x \frac{d}{dx}(\cos x)$$

$$= \cos x \, (\cos x) + \sin x \, (-\sin x) = \cos^2 x - \sin^2 x. \quad \square$$

We come now to the tangent function.   Since $\tan x = \sin x / \cos x$, we have

$$\frac{d}{dx}(\tan x) = \frac{\cos x \dfrac{d}{dx}(\sin x) - \sin x \dfrac{d}{dx}(\cos x)}{\cos^2 x} = \frac{\cos^2 x + \sin^2 x}{\cos^2 x} = \frac{1}{\cos^2 x} = \sec^2 x.$$

The derivative of the tangent function is the secant squared:

(3.6.3)

$$\frac{d}{dx}(\tan x) = \sec^2 x.$$

The derivatives of the other trigonometric functions are as follows:

(3.6.4)

$$\frac{d}{dx}(\cot x) = -\csc^2 x,$$

$$\frac{d}{dx}(\sec x) = \sec x \tan x,$$

$$\frac{d}{dx}(\csc x) = -\csc x \cot x.$$

The verification of these formulas is left as an exercise.

It is time for some sample problems.

Example 2   Find $f'(\pi/4)$ for

$$f(x) = x \cot x.$$

SOLUTION   We first find $f'(x)$ by the product rule:

$$f'(x) = x\frac{d}{dx}(\cot x) + \cot x\frac{d}{dx}(x) = -x \csc^2 x + \cot x.$$

Now we evaluate f' at $\pi/4$:

$$f'(\pi/4) = -\frac{\pi}{4}(\sqrt{2})^2 + 1 = 1 - \frac{\pi}{2}. \quad \square$$

Example 3   Find

$$\frac{d}{dx}\left[\frac{1 - \sec x}{\tan x}\right].$$

SOLUTION   By the quotient rule,

$$\frac{d}{dx}\left[\frac{1 - \sec x}{\tan x}\right] = \frac{\tan x \dfrac{d}{dx}(1 - \sec x) - (1 - \sec x)\dfrac{d}{dx}(\tan x)}{\tan^2 x}$$

$$= \frac{\tan x\,(-\sec x \tan x) - (1 - \sec x)(\sec^2 x)}{\tan^2 x}$$

$$= \frac{\sec x\,(\sec^2 x - \tan^2 x) - \sec^2 x}{\tan^2 x}$$

$$= \frac{\sec x - \sec^2 x}{\tan^2 x} = \frac{\sec x\,(1 - \sec x)}{\tan^2 x}. \quad \square$$

(recall $\sec^2 x - \tan^2 x = 1$) _____↑

Example 4   Find an equation for the tangent to the curve

$$y = \cos x$$

at the point where $x = \pi/3$.

SOLUTION Since $\cos \pi/3 = \frac{1}{2}$, the point of tangency is $(\pi/3, \frac{1}{2})$. To find the slope of the tangent, we evaluate the derivative

$$\frac{dy}{dx} = -\sin x$$

at $x = \pi/3$. This gives $m = -\sqrt{3}/2$. An equation for the tangent can be written

$$y - \frac{1}{2} = -\frac{\sqrt{3}}{2}\left(x - \frac{\pi}{3}\right). \quad \square$$

Example 5 An object moves along the x-axis, its position at each time t given by the function

$$x(t) = t + 2 \sin t.$$

Determine those times t from 0 to 2π when the object is moving to the left.

SOLUTION The object is moving to the left only when its velocity $v(t) < 0$. Since

$$v(t) = x'(t) = 1 + 2 \cos t,$$

the object is moving to the left only when $\cos t < -\frac{1}{2}$. As you can check, the only t in $[0, 2\pi]$ for which $\cos t < -\frac{1}{2}$ are those t that lie between $2\pi/3$ and $4\pi/3$. Thus from time $t = 0$ to $t = 2\pi$, the object is moving left only during the time interval $(2\pi/3, 4\pi/3)$. $\square$

The Chain Rule and the Trigonometric Functions

If f is a differentiable function of u and u is a differentiable function of x, then, as you saw in Section 3.5,

$$\frac{d}{dx}[f(u)] = \frac{d}{du}[f(u)]\frac{du}{dx}.$$

Written in this form, the derivatives of the six trigonometric functions appear as follows:

(3.6.5)

$$\frac{d}{dx}(\sin u) = \cos u \frac{du}{dx}, \qquad \frac{d}{dx}(\cos u) = -\sin u \frac{du}{dx},$$

$$\frac{d}{dx}(\tan u) = \sec^2 u \frac{du}{dx}, \qquad \frac{d}{dx}(\cot u) = -\csc^2 u \frac{du}{dx},$$

$$\frac{d}{dx}(\sec u) = \sec u \tan u \frac{du}{dx}, \qquad \frac{d}{dx}(\csc u) = -\csc u \cot u \frac{du}{dx}.$$

Example 6

$$\frac{d}{dx}(\cos 2x) = -\sin 2x \frac{d}{dx}(2x) = -2 \sin 2x. \quad \square$$

Example 7

$$\frac{d}{dx}[\sec(x^2 + 1)] = \sec(x^2 + 1)\tan(x^2 + 1)\frac{d}{dx}(x^2 + 1)$$

$$= 2x\sec(x^2 + 1)\tan(x^2 + 1). \quad \square$$

Example 8

$$\frac{d}{dx}(\sin^3 \pi x) = \frac{d}{dx}(\sin \pi x)^3$$

$$= 3(\sin \pi x)^2 \frac{d}{dx}(\sin \pi x)$$

$$= 3(\sin \pi x)^2 \cos \pi x \frac{d}{dx}(\pi x)$$

$$= 3(\sin \pi x)^2 \cos \pi x\, (\pi) = 3\pi \sin^2 \pi x \cos \pi x. \quad \square$$

Our treatment of the trigonometric functions has been based entirely on radian measure. When degrees are used, the derivatives of the trigonometric functions contain the extra factor $\frac{1}{180}\pi \cong 0.0175$.

Example 9   Find

$$\frac{d}{dx}(\sin x°).$$

SOLUTION   Since $x° = \frac{1}{180}\pi x$ radians,

$$\frac{d}{dx}(\sin x°) = \frac{d}{dx}\left(\sin \frac{1}{180}\pi x\right) = \frac{1}{180}\pi \cos \frac{1}{180}\pi x = \frac{1}{180}\pi \cos x°. \quad \square$$

The extra factor $\frac{1}{180}\pi$ is a disadvantage, particularly in problems where it occurs repeatedly. This tends to discourage the use of degree measure in theoretical work.

EXERCISES 3.6

In Exercises 1–12, differentiate the given function.

1. $y = 3\cos x - 4\sec x$.
2. $y = x^2 \sec x$.
3. $y = x^3 \csc x$.
4. $y = \sin^2 x$.
5. $y = \cos^2 t$.
6. $y = 3t^2 \tan t$.
7. $y = \sin^4 \sqrt{u}$.
8. $y = u\csc u^2$.
9. $y = \tan x^2$.
10. $y = \cos \sqrt{x}$.
11. $y = [x + \cot \pi x]^4$.
12. $y = [x^2 - \sec 2x]^3$.

In Exercises 13–24, find the second derivative.

13. $y = \sin x$.
14. $y = \cos x$.
15. $y = \dfrac{\cos x}{1 + \sin x}$.
16. $y = \tan^3 2\pi x$.
17. $y = \cos^3 2u$.
18. $y = \sin^5 3t$.
19. $y = \tan 2t$.
20. $y = \cot 4u$.

21. $y = x^2 \sin 3x$.
22. $y = \dfrac{\sin x}{1 - \cos x}$.
23. $y = \sin^2 x + \cos^2 x$.
24. $y = \sec^2 x - \tan^2 x$.

In Exercises 25–30, find the indicated derivative.

25. $\dfrac{d^4}{dx^4}(\sin x)$.
26. $\dfrac{d^4}{dx^4}(\cos x)$.
27. $\dfrac{d}{dt}\left[t^2 \dfrac{d^2}{dt^2}(t\cos 3t)\right]$.
28. $\dfrac{d}{dt}\left[t\dfrac{d}{dt}(\cos t^2)\right]$.
29. $\dfrac{d}{dx}[f(\sin 3x)]$.
30. $\dfrac{d}{dx}[\sin(f(3x))]$.

In Exercises 31–36, find an equation for the tangent to the curve at $x = a$.

31. $y = \sin x;\quad a = 0$.
32. $y = \tan x;\quad a = \pi/6$.
33. $y = \cot x;\quad a = \pi/6$.
34. $y = \cos x;\quad a = 0$.

35. $y = \sec x$; $a = \pi/4$.   **36.** $y = \csc x$; $a = \pi/3$.

In Exercises 37–46, determine the numbers x between 0 and 2π where the tangent to the curve is horizontal.

37. $y = \cos x$.

38. $y = \sin x$.

39. $y = \sin x + \sqrt{3} \cos x$.

40. $y = \cos x - \sqrt{3} \sin x$.

41. $y = \sin^2 x$.

42. $y = \cos^2 x$.

43. $y = \tan x - 2x$.

44. $y = 3 \cot x + 4x$.

45. $y = 2 \sec x + \tan x$.

46. $y = \cot x - 2 \csc x$.

An object moves along the x-axis, its position at each time t given by $x(t)$. In Exercises 47–52, determine those times from $t = 0$ to $t = 2\pi$ when the object is moving to the right with increasing speed.

47. $x(t) = \sin 3t$.

48. $x(t) = \cos 2t$.

49. $x(t) = \sin t - \cos t$.

50. $x(t) = \sin t + \cos t$.

51. $x(t) = t + 2 \cos t$.

52. $x(t) = t - \sqrt{2} \sin t$.

In Exercises 53–56, find dy/dt (a) by using (3.5.4) and (b) by writing y as a function of t and then differentiating.

53. $y = u^2 - 1$, $u = \sec x$, $x = \pi t$.

54. $y = [\frac{1}{2}(1 + u)]^3$, $u = \cos x$, $x = 2t$.

55. $y = [\frac{1}{2}(1 - u)]^4$, $u = \cos x$, $x = 2t$.

56. $y = 1 - u^2$, $u = \csc x$, $x = 3t$.

57. It can be shown by induction that the nth derivative of the sine function is given by the formula

$$\frac{d^n}{dx^n}(\sin x) = \begin{cases} (-1)^{(n-1)/2} \cos x, & n \text{ odd} \\ (-1)^{n/2} \sin x, & n \text{ even}. \end{cases}$$

Persuade yourself that this formula is correct and obtain a similar formula for the nth derivative of the cosine function.

58. Verify the following differentiation formulas.

(a) $\dfrac{d}{dx}(\cot x) = -\csc^2 x$.

(b) $\dfrac{d}{dx}(\sec x) = \sec x \tan x$.

(c) $\dfrac{d}{dx}(\csc x) = -\csc x \cot x$.

59. Let $f(x) = \begin{cases} x \sin\left(\dfrac{1}{x}\right), & x \neq 0 \\ 0, & x = 0 \end{cases}$ and

$g(x) = \begin{cases} x^2 \sin\left(\dfrac{1}{x}\right), & x \neq 0 \\ 0, & x = 0. \end{cases}$

In Exercise 67, Section 3.1, you were asked to show that f is continuous at 0, but not differentiable there, and that g is differentiable at 0. Both f and g are differentiable for any $x \neq 0$.

(a) Find $f'(x)$ and $g'(x)$ for $x \neq 0$.

(b) Show that $g'(x)$ is not continuous at 0.

60. Let $f(x) = \begin{cases} \cos x, & x \geq 0 \\ ax + b, & x < 0. \end{cases}$

(a) Determine a and b so that f is differentiable at 0.

(b) Using the values found in part (a), sketch the graph of f.

Let $g(x) = \begin{cases} \sin x, & 0 \leq x \leq 2\pi/3 \\ ax + b, & 2\pi/3 < x \leq 2\pi. \end{cases}$

(c) Determine a and b so that g is differentiable at $2\pi/3$.

(d) Using the values found in part (c), sketch the graph of g.

61. Let $y = A \sin \omega t + B \cos \omega t$, where A, B, and ω are constants. Show that y satisfies the equation

$$\frac{d^2 y}{dx^2} + \omega^2 y = 0.$$

62. A simple pendulum consists of a mass m swinging at the end of a (massless) rod or wire of length L (see the figure). The angular displacement θ of the pendulum at time t is given by

$$\theta(t) = a \cos(\omega t + \phi),$$

where a, ω, and ϕ are constants.

(a) Show that θ satisfies the equation $d^2\theta/dt^2 + \omega^2\theta = 0$. (Compare this result with the result in Exercise 61.)

(b) Show that θ can be written in the form

$$\theta(t) = A \sin \omega t + B \cos \omega t,$$

where A, B, and ω are constants.

63. An isosceles triangle has equal sides of length c, and the angle between them is x radians. Express the area A of the triangle as a function of x and find the rate of change of A with respect to x.

64. A triangle has sides of length a and b, and the angle between them is x radians. If a and b are constants, find the rate of change of the third side c with respect to x. HINT: Use the law of cosines.

65. A 13-foot ladder leans against a vertical wall. If the bottom of the ladder is being pulled away from the wall at the rate of 2 feet per second, how fast is the angle between the ladder and the ground changing at the instant the bottom of the ladder is 5 feet from the wall?

66. A lighthouse is on an island 1 mile from a straight shoreline. Suppose that the light is turning at the rate of 2

revolutions per minute in the counterclockwise direction.
(a) How fast is the light beam moving along the shore at the instant it makes an angle of 45° with the shore-line?
(b) How fast is the light beam moving at the instant the angle is 90°?

67. A man starts at a point A and walks 40 feet north. He then turns and walks due east at 4 feet per second. If a searchlight placed at A follows him, at what rate is the light turning at the end of 6 seconds?

▶68. Let θ be measured in degrees.
(a) Estimate

$$\lim_{\theta \to 0} \frac{\sin\theta}{\theta}$$

by letting $\theta = 5, 1, 0.1, 0.01,$ and 0.001 degrees.
(b) Find the value of $\pi/180$ and compare the result with the limit found in part (a). This gives another demonstration of the reason for using radian measure rather than degree measure.

▶69. Let $f(x) = \cos x^2$. Determine

$$f'(0) = \lim_{h \to 0} \frac{f(h) - f(0)}{h}$$

numerically by letting $h = \pm 0.1, \pm 0.01,$ and ± 0.001. Compare your answer to what you obtain by the chain rule.

▶70. Let $f(x) = \cos^2 x$. Determine

$$f'(\tfrac{1}{4}\pi) = \lim_{h \to 0} \frac{f(\tfrac{1}{4}\pi + h) - f(\tfrac{1}{4}\pi)}{h}$$

numerically by letting $h = \pm 0.1, \pm 0.01,$ and ± 0.001. Compare your answer to what you obtain by the chain rule.

▶71. Let $f(x) = x^2 - \sin 3x$.
(a) Use a graphing utility to graph f for $-\pi \le x \le \pi$.
(b) Estimate the zeros of f with two decimal place accuracy.
(c) Estimate with two decimal place accuracy the x-coordinates of the points where the tangent line to the graph of f is horizontal.

▶72. Let $f(x) = \sin x, -\pi \le x \le \pi$.
(a) Using a graphing utility to graph f and $p_1(x) = x$ together. Notice that $p_1 = x$ is a good approximation for $\sin x$ near 0.
(b) Compare the graphs of f and $p_3(x) = x - \frac{1}{6}x^3$.
(c) Compare the graphs of f and
$$p_5(x) = x - \tfrac{1}{6}x^3 + \tfrac{1}{120}x^5.$$

▶73. Let $f(x) = \cos x, -\pi \le x \le \pi$. Repeat Exercise 72 for f and $p_0(x) = 1, p_2(x) = 1 - \frac{1}{2}x^2,$ and $p_4(x) = 1 - \frac{1}{2}x^2 + \frac{1}{24}x^4$. The use of polynomials to approximate the trigonometric functions (and other functions) is one of the main topics in Chapter 11.

■ 3.7 IMPLICIT DIFFERENTIATION; RATIONAL POWERS

The functions that we have been studying up to this point have been defined by expressing the dependent variable *explicitly* in terms of the independent variable. For instance

$$y = (x^2 - 1)^3, \qquad y = x^2\sec x, \qquad y = \begin{cases} x^2 + 1, & x \le 0 \\ 1 - x, & x > 0, \end{cases}$$

or, in general, $y = f(x)$. In this section we consider functional relationships between two variables in which the dependent variable is not given explicitly in terms of the independent variable.

For example, suppose you know that y is a function of x that satisfies the equation

$$3x^3y - 4y - 2x + 1 = 0.$$

We say that this equation defines y *implicitly* as a function of x. We want to develop a method for finding the derivative of y with respect to x when the functional relationship between x and y is given implicitly. Continuing with this example, one way to find dy/dx is to solve first for y:

$$(3x^3 - 4)y - 2x + 1 = 0,$$

$$(3x^3 - 4)y = 2x - 1,$$

$$y = \frac{2x - 1}{3x^3 - 4},$$

and then differentiate:

$$\frac{dy}{dx} = \frac{(3x^3 - 4)2 - (2x - 1)(9x^2)}{(3x^3 - 4)^2} = \frac{6x^3 - 8 - 18x^3 + 9x^2}{(3x^3 - 4)^2}$$

$$(1) \qquad = -\frac{12x^3 - 9x^2 + 8}{(3x^3 - 4)^2}.$$

It is also possible to find dy/dx without first solving the equation for y. The technique is called *implicit differentiation.*

Returning to the original equation

$$3x^3y - 4y - 2x + 1 = 0,$$

we differentiate both sides of the equation with respect to x (remembering that y is a function of x). We have

$$\frac{d}{dx}(3x^3y - 4y - 2x + 1) = \frac{d}{dx}(0)$$

$$\underbrace{\left(3x^3\frac{dy}{dx} + 9x^2y\right)} - 4\frac{dy}{dx} - 2 = 0$$

by the product rule ⟶

$$(3x^3 - 4)\frac{dy}{dx} = 2 - 9x^2y,$$

$$(2) \qquad \frac{dy}{dx} = \frac{2 - 9x^2y}{3x^3 - 4}.$$

The answer looks different from what we obtained before because this time y appears on the right-hand side. To satisfy yourself that the two answers are really the same all you have to do is substitute

$$y = \frac{2x - 1}{3x^3 - 4}$$

in (2). This gives

$$\frac{dy}{dx} = \frac{2 - 9x^2\left(\dfrac{2x - 1}{3x^3 - 4}\right)}{3x^3 - 4} = \frac{6x^3 - 8 - 18x^3 + 9x^2}{(3x^3 - 4)^2} = -\frac{12x^3 - 9x^2 + 8}{(3x^3 - 4)^2}$$

which is the answer (1) we obtained before.

Implicit differentiation is particularly useful where it is inconvenient (or impossible) to first solve the given equation for y.

Example 1 Use implicit differentiation to find dy/dx.

(a) $2xy - y^3 + 1 = x + 2y$. **(b)** $\cos(x - y) = (2x + 1)^2y$.

SOLUTION **(a)** Differentiating implicitly with respect to x, we have

$$\underbrace{2x\frac{dy}{dx} + 2y} - \underbrace{3y^2\frac{dy}{dx}} = 1 + 2\frac{dy}{dx}$$

(by the product rule) ⟶ ⟶ (by the chain rule)

$$[2x - 3y^2 - 2]\frac{dy}{dx} = 1 - 2y$$

and

$$\frac{dy}{dx} = \frac{1 - 2y}{2x - 3y^2 - 2}.$$

(b) Differentiate implicitly with respect to x:

$$-\sin(x - y)\left[1 - \frac{dy}{dx}\right] = (2x + 1)^2\frac{dy}{dx} + 2(2x + 1)(2)y$$

(by the chain rule) _____↑

$$[\sin(x - y) - (2x + 1)^2]\frac{dy}{dx} = 4(2x + 1)y + \sin(x - y).$$

Thus,

$$\frac{dy}{dx} = \frac{4(2x + 1)y + \sin(x - y)}{\sin(x - y) - (2x + 1)^2}. \quad ❑$$

Example 2   Find the slope of the curve

$$x^3 + y^3 = 1 + 3xy^2 \qquad \text{at} \quad (2, -1).$$

SOLUTION   Differentiating this equation implicitly with respect to x gives

$$3x^2 + 3y^2\frac{dy}{dx} = 3x\left(2y\frac{dy}{dx}\right) + 3y^2$$

or

$$x^2 + y^2\frac{dy}{dx} = 2xy\frac{dy}{dx} + y^2.$$

Solving for dy/dx, we have

$$\frac{dy}{dx} = \frac{y^2 - x^2}{y^2 - 2xy}.$$

At $x = 2$ and $y = -1$, we find that

$$\frac{dy}{dx} = \frac{(-1)^2 - 2^2}{(-1)^2 - 2(2)(-1)} = -\frac{3}{5}.$$

The slope is $-\frac{3}{5}$.   ❑

We can also find higher derivatives by implicit differentiation.

Example 3   Find d^2y/dx^2 given that

$$y^3 - x^2 = 4.$$

SOLUTION   Differentiation with respect to x gives

$$3y^2\frac{dy}{dx} - 2x = 0$$

and

(*) $$\frac{dy}{dx} = \frac{2x}{3y^2}.$$

Differentiating again with respect to x, using the quotient rule, we get

$$\frac{d^2y}{dx^2} = \frac{d(dy/dx)}{dx} = \frac{3y^2(2) - 2x(6y(dy/dx))}{(3y^2)^2} = \frac{6y^2 - 12xy(dy/dx)}{9y^4}.$$

Now, substituting (∗) into this equation gives

$$\frac{d^2y}{dx^2} = \frac{6y^2 - 12xy(2x/3y^2)}{9y^4}$$

which, as you can check, simplifies to

$$\frac{d^2y}{dx^2} = \frac{6y^3 - 8x^2}{9y^5}. \quad \square$$

A Theoretical Aside

In the preceding discussion and examples, we assumed that the given equations determined y as a differentiable function of x. However, this is not always true; proper interpretations have to be made.

For example, if we differentiate $x^2 + y^2 = -9$ implicitly, we find that

$$2x + 2y\frac{dy}{dx} = 0 \qquad \text{and therefore} \qquad \frac{dy}{dx} = -\frac{x}{y}.$$

However, the result is meaningless. It is meaningless because there is no real-valued function y of x that satisfies the equation $x^2 + y^2 = -9$. Implicit differentiation can be applied meaningfully to an equation in x and y only if there is a differentiable function y of x that satisfies the equation.

Now consider the equation $x^2 + y^2 = 9$. The graph is a circle of radius 3 centered at the origin (Figure 3.7.1). Clearly this graph does not define y as a function of x (remember the vertical line test). If we solve this equation for y, we get

$$y = \pm\sqrt{9 - x^2} \qquad -3 \leq x \leq 3.$$

Although this equation does not define y as a function of x,

$$y = \sqrt{9 - x^2} \qquad \text{(Figure 3.7.2)} \qquad \text{and} \qquad y = -\sqrt{9 - x^2} \qquad \text{(Figure 3.7.3)}$$

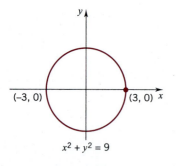

$x^2 + y^2 = 9$

Figure 3.7.1

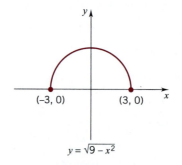

$y = \sqrt{9 - x^2}$

Figure 3.7.2

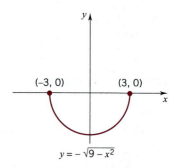

$y = -\sqrt{9 - x^2}$

Figure 3.7.3

are functions of x on $[-3, 3]$ that satisfy the original equation. In fact, there are an infinite number of functions $y = f(x)$ that satisfy the given equation:

$$y = -\sqrt{9 - x^2} \quad \text{for} \quad 0 \leq x \leq 3 \quad \text{and} \quad y = \begin{cases} -\sqrt{9 - x^2}, & -3 \leq x < 0 \\ 0\sqrt{9 - x^2}, & \leq x \leq 3 \end{cases}$$

are other examples. However, the functions $y = \sqrt{9 - x^2}$ and $y = -\sqrt{9 - x^2}$ are the only two that are continuous on $[-3, 3]$ and differentiable on $(-3, 3)$.

The point of this discussion is that an equation involving x and y does not always define y implicitly as a function of x, and, on the other hand, such an equation may give rise to more than one function. Also, as noted before, implicit differentiation can be applied to an equation in x and y only if there is at least one differentiable function y of x that satisfies the equation.

Rational Powers

We have shown that the differentiation formula

$$\frac{d(x^n)}{dx} = nx^{n-1}$$

holds for all integers n. Here we show that this formula also holds for rational exponents p/q.

(3.7.1)

$$\boxed{\frac{d(x^{p/q})}{dx} = \frac{p}{q} x^{(p/q) - 1}.}$$

PROOF Let $y = x^{p/q}$, where p and q are nonzero integers. Then $y^q = x^p$. Differentiating this equation implicitly with respect to x, we have

$$qy^{q-1}\frac{dy}{dx} = px^{p-1}$$

and

$$\frac{dy}{dx} = \frac{p}{q}\frac{x^{p-1}}{y^{q-1}} = \frac{p}{q} x^{p-1}y^{1-q}.$$

Now, replacing y by $x^{p/q}$, we have

$$\frac{dy}{dx} = \frac{p}{q} x^{p-1}(x^{p/q})^{(1-q)} = \frac{p}{q} x^{p-1}x^{(p/q)-p}$$

$$= \frac{p}{q} x^{(p/q)-1}. \quad \square$$

Here are some simple examples:

$$\frac{d}{dx}(x^{2/3}) = \frac{2}{3}x^{-1/3}, \qquad \frac{d}{dx}(x^{5/2}) = \frac{5}{2}x^{3/2}, \qquad \frac{d}{dx}(x^{-7/9}) = -\frac{7}{9}x^{-16/9}.$$

If u is a differentiable function of x, then, by the chain rule,

(3.7.2)

$$\frac{d}{dx}(u^{p/q}) = \frac{p}{q}\, u^{(p/q)-1}\frac{du}{dx}.$$

The result holds on every open x-interval where $u^{(p/q)-1}$ is defined.

Example 4

(a) $\dfrac{d}{dx}[(1 + x^2)^{1/5}] = \frac{1}{5}(1 + x^2)^{-4/5}(2x) = \frac{2}{5}x(1 + x^2)^{-4/5}.$

(b) $\dfrac{d}{dx}[(1 - x^2)^{1/5}] = \frac{1}{5}(1 - x^2)^{-4/5}(- 2x) = - \frac{2}{5}x(1 - x^2)^{-4/5}.$

(c) $\dfrac{d}{dx}[(1 - x^2)^{1/4}] = \frac{1}{4}(1 - x^2)^{-3/4}(- 2x) = - \frac{1}{2}x(1 - x^2)^{-3/4}.$

You should verify that the first statement holds for all real x, the second for all $x \neq \pm 1$, the third only for $x \in (-1, 1)$. ❏

Example 5

$$\frac{d}{dx}\left[\left(\frac{x}{1 + x^2}\right)^{1/2}\right] = \frac{1}{2}\left(\frac{x}{1 + x^2}\right)^{-1/2}\frac{d}{dx}\left(\frac{x}{1 + x^2}\right)$$

$$= \frac{1}{2}\left(\frac{x}{1 + x^2}\right)^{-1/2}\frac{(1 + x^2)(1) - x(2x)}{(1 + x^2)^2}$$

$$= \frac{1}{2}\left(\frac{1 + x^2}{x}\right)^{1/2}\frac{1 - x^2}{(1 + x^2)^2}$$

$$= \frac{1 - x^2}{2x^{1/2}(1 + x^2)^{3/2}}.$$

The result holds for all $x > 0$. ❏

EXERCISES 3.7

In Exercises 1–10, use implicit differentiation to obtain dy/dx in terms of x and y.

1. $x^2 + y^2 = 4.$

2. $x^3 + y^3 - 3xy = 0.$

3. $4x^2 + 9y^2 = 36.$

4. $\sqrt{x} + \sqrt{y} = 4.$

5. $x^4 + 4x^3y + y^4 = 1.$

6. $x^2 - x^2y + xy^2 + y^2 = 1.$

7. $(x - y)^2 - y = 0.$

8. $(y + 3x)^2 - 4x = 0.$

9. $\sin(x + y) = xy.$

10. $\tan xy = xy.$

In Exercises 11–16, express d^2y/dx^2 in terms of x and y.

11. $y^2 + 2xy = 16.$

12. $x^2 - 2xy + 4y^2 = 3.$

13. $y^2 + xy - x^2 = 9.$

14. $x^2 - 3xy = 18.$

15. $4 \tan y = x^3.$

16. $\sin^2 x + \cos^2 y = 1.$

In Exercises 17–20, find dy/dx and d^2y/dx^2 at the given point.

17. $x^2 - 4y^2 = 9;$ $(5, 2).$

18. $x^2 + 4xy + y^3 + 5 = 0;$ $(2, -1).$

19. $\cos(x + 2y) = 0;$ $(\pi/6, \pi/6).$

20. $x = \sin^2 y;$ $(\frac{1}{2}, \pi/4).$

In Exercises 21–26, find equations for the tangent and normal at the indicated point.

21. $2x + 3y = 5;$ $(-2, 3).$

22. $9x^2 + 4y^2 = 72;$ $(2, 3).$

23. $x^2 + xy + 2y^2 = 28;$ $(-2, -3)$.

24. $x^3 - axy + 3ay^2 = 3a^3;$ (a, a).

25. $x = \cos y;$ $\left(\dfrac{1}{2}, \dfrac{\pi}{3}\right)$

26. $\tan xy = x;$ $\left(1, \dfrac{\pi}{4}\right)$.

Find dy/dx in Exercises 27–34.

27. $y = (x^3 + 1)^{1/2}$.

28. $y = (x + 1)^{1/3}$.

29. $y = x\sqrt{x^2 + 1}$.

30. $y = x^2\sqrt{x^2 + 1}$.

31. $y = \sqrt[4]{2x^2 + 1}$.

32. $y = (x + 1)^{1/3}(x + 2)^{2/3}$.

33. $y = \sqrt{2 - x^2}\sqrt{3 - x^2}$.

34. $y = \sqrt{(x^4 - x + 1)^3}$.

Compute the indicated derivative in Exercises 35–40.

35. $\dfrac{d}{dx}\left(\sqrt{x} + \dfrac{1}{\sqrt{x}}\right)$.

36. $\dfrac{d}{dx}\left(\sqrt{\dfrac{3x + 1}{2x + 5}}\right)$.

37. $\dfrac{d}{dx}\left(\dfrac{x}{\sqrt{x^2 + 1}}\right)$.

38. $\dfrac{d}{dx}\left(\dfrac{\sqrt{x^2 + 1}}{x}\right)$

39. $\dfrac{d}{dx}\left(\sqrt[3]{x} + \dfrac{1}{\sqrt[3]{x}}\right)$.

40. $\dfrac{d}{dx}\left(\sqrt{\dfrac{ax + b}{cx + d}}\right)$, a, b, c, d constant.

41. (*Important*) Show the general form of the graph.
(a) $f(x) = x^{1/n}$, n a positive even integer.
(b) $f(x) = x^{1/n}$, n a positive odd integer.
(c) $f(x) = x^{2/n}$, n an odd integer greater than 1.

Find the second derivative in Exercises 42–46.

42. $y = \sqrt{a^2 + x^2}$.

43. $y = \sqrt[3]{a + bx}$.

44. $y = x\sqrt{a^2 - x^2}$.

45. $y = \sqrt{x}\tan\sqrt{x}$.

46. $y = \sqrt{x}\sin\sqrt{x}$.

47. Show that all normals to the circle $x^2 + y^2 = r^2$ pass through the center of the circle.

48. Determine the x-intercept of the tangent to the parabola $y^2 = x$ at the point where $x = a$.

The angle between two curves is the angle between their tangents at the point of intersection. If the slopes are m_1 and m_2, then the angle of intersection α can be obtained from the formula

$$\tan \alpha = \left|\frac{m_2 - m_1}{1 + m_1 m_2}\right|. \quad \text{(Exercise 71, Section 1.6)}$$

49. At what angles do the parabolas $y^2 = 2px + p^2$ and $y^2 = p^2 - 2px$ intersect?

50. At what angles does the line $y = 2x$ intersect the curve $x^2 - xy + 2y^2 = 28$?

51. The curves $y = x^2$ and $x = y^3$ intersect at the points $(1, 1)$ and $(0, 0)$. Find the angle between the curves at each of these points.

52. Find the angles at which the circles $(x - 1)^2 + y^2 = 10$ and $x^2 + (y - 2)^2 = 5$ intersect.

Two curves are said to be *orthogonal* if at each point of intersection, the angle between them is a right angle. In Exercises 53 and 54, show that the given curves are orthogonal.

53. The hyperbola $x^2 - y^2 = 5$ and the ellipse $4x^2 + 9y^2 = 72$.

54. The ellipse $3x^2 + 2y^2 = 5$ and $y^3 = x^2$.
HINT: The curves intersect at $(1, 1)$ and $(-1, 1)$.

Two families of curves are said to be *orthogonal trajectories* (of each other) if each member of one family is orthogonal to each member of the other family. In Exercises 55 and 56, show that the given families of curves are orthogonal trajectories.

55. The family of circles $x^2 + y^2 = r^2$ and the family of lines $y = mx$.

56. The family of parabolas $x = ay^2$ and the family of ellipses $x^2 + \frac{1}{2}y^2 = b$.

57. Find equations for the tangents to the ellipse $4x^2 + y^2 = 72$ that are perpendicular to the line $x + 2y + 3 = 0$.

58. Find equations for the normals to the hyperbola $4x^2 - y^2 = 36$ that are parallel to the line $2x + 5y - 4 = 0$.

59. The curve defined by the equation $(x^3 + y^2)^2 = x^2 - y^2$ is called a *lemniscate*. The graph is shown in the figure. Find the four points on the curve at which the tangent line is horizontal.

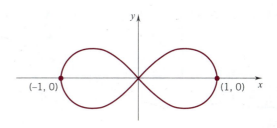

60. The curve defined by the equation $x^{2/3} + y^{2/3} = a^{2/3}$ is called an *astroid*. The graph is shown in the figure.

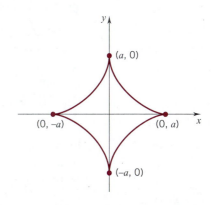

(a) Find the slope of the tangent line to the graph at an arbitrary point (x_1, y_1), $x_1 \neq 0$.

(b) At what points on the curve is the slope of the tangent line 0, 1, or -1?

61. Prove that the sum of the x- and y-intercepts of any tangent line to the graph of

$$x^{1/2} + y^{1/2} = c^{1/2}$$

is constant and equal to c.

62. A circle of radius 1 with center on the y-axis is inscribed in the parabola $y = 2x^2$. See the figure. Find the points of intersection.   HINT: The circle has equation $x^2 + (y - a)^2 = 1$, and the circle and parabola have the same tangent lines at the points of intersection.

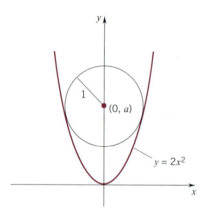

63. The function y defined by

$$y = \begin{cases} \sqrt{1 - x^2} & -1 \leq x < -\frac{1}{2} \\ -\sqrt{1 - x^2} & -\frac{1}{2} \leq x \leq \frac{1}{2} \\ \sqrt{1 - x^2} & \frac{1}{2} < x \leq 1. \end{cases}$$

satisfies the equation $x^2 + y^2 = 1$

(a) Sketch the graph of this function.

(b) Use implicit differentiation to find the slope of the tangent line to the graph of y at the points $(-\sqrt{\frac{3}{2}}, \frac{1}{2})$, $(\sqrt{\frac{3}{2}}, \frac{1}{2})$ and $(0, -1)$.

(c) Find $y'_+(-\frac{1}{2})$, $y'_-(\frac{1}{2})$.

▷ 64. Determine

$$\lim_{h \to 0} \frac{(1 + h)^{98.6} - 1}{h}$$

numerically by letting $h = \pm 0.1, \pm 0.01, \pm 0.001$. Confirm your answer by other means.

▷ 65. Let $f(x) = x^{3/4}$. Determine

$$f'(16) = \lim_{h \to 0} \frac{f(16 + h) - f(16)}{h}$$

numerically by letting $h = \pm 0.1, \pm 0.01, \pm 0.001$. Compare your answer to what you obtain by the methods of this section.

▷ 66. Determine

$$\lim_{x \to 32} \frac{x^{2/5} - 4}{x^{4/5} - 16}$$

numerically. Confirm your answer by other means.

▷ 67. A graphing utility in parametric mode can be used to graph some equations in x and y. Sketch the graph of the equation $x^2 + y^2 = 4$ by first setting $x = t$, $y = \sqrt{4 - t^2}$ and then setting $x = t$, $y = -\sqrt{4 - t^2}$.

▷ 68. Use a graphing utility to sketch the graph of $(2 - x)y^2 = x^3$. This curve is called a *cissoid*.

▷ 69. (a) Use a graphing utility to sketch the graph of $x^4 = x^2 - y^2$. This curve is called a *figure eight curve*.

(b) Find the x-coordinates of the points on the graph where the tangent line is horizontal.

■ 3.8 RATES OF CHANGE PER UNIT TIME

Earlier you saw that velocity is the rate of change of position with respect to time t, and acceleration the rate of change of velocity with respect to time t. In this section we work with other quantities that vary with time. The fundamental point is this: *If Q is a quantity that varies with time, then the derivative dQ/dt gives the rate of change of that quantity with respect to time.*

Example 1   A spherical balloon is expanding. If the radius is increasing at the rate of 2 inches per minute, at what rate is the volume increasing when the radius is 5 inches?

SOLUTION

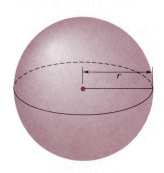

Find $\dfrac{dV}{dt}$ when $r = 5$ inches

given that $\dfrac{dr}{dt} = 2$ in./min.

$V = \frac{4}{3}\pi r^3$.   (volume of a sphere of radius r)

Since r and V are functions of t, we differentiate the equation $V = \frac{4}{3}\pi r^3$ implicitly with respect to t. We have

$$\frac{dV}{dt} = 4\pi r^2 \frac{dr}{dt}.$$

Substituting $r = 5$ and $dr/dt = 2$, we find that

$$\frac{dV}{dt} = 4\pi(5)^2(2) = 200\pi.$$

When the radius is 5 inches, the volume is increasing at the rate of 200π cubic inches per minute. ❑

Example 2 An object is moving in the clockwise direction around the unit circle $x^2 + y^2 = 1$. As it passes through the point $(\frac{1}{2}, \frac{1}{2}\sqrt{3})$, its y-coordinate is decreasing at the rate of 3 units per second. At what rate is the x-coordinate changing at this point?

SOLUTION

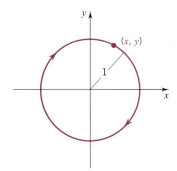

Find $\dfrac{dx}{dt}$ when $x = \frac{1}{2}$ and $y = \frac{1}{2}\sqrt{3}$

given that $\dfrac{dy}{dt} = -3$ units/sec.

$x^2 + y^2 = 1$.   (equation of circle)

Differentiating $x^2 + y^2 = 1$ with respect to t, we have

$$2x\frac{dx}{dt} + 2y\frac{dy}{dt} = 0 \quad \text{and thus} \quad x\frac{dx}{dt} + y\frac{dy}{dt} = 0.$$

Substitution of $x = \frac{1}{2}$, $y = \frac{1}{2}\sqrt{3}$, and $dy/dt = -3$ gives

$$\frac{1}{2}\frac{dx}{dt} + \frac{1}{2}\sqrt{3}(-3) = 0 \quad \text{so that} \quad \frac{dx}{dt} = 3\sqrt{3}.$$

As the particle passes through the point $(\frac{1}{2}, \frac{1}{2}\sqrt{3})$, the x-coordinate is increasing at the rate of $3\sqrt{3}$ units per second. ❑

As you must have noticed, the first two sample problems were solved by the same general method, a method that we recommend to you for solving problems of this type:

Step 1. Draw a diagram, where relevant, and indicate the quantities that vary.

Step 2. Specify in mathematical form the rate of change you are looking for, and record all given information.

Step 3. Find an equation involving the variable whose rate of change is to be found.

Step 4. Differentiate with respect to time t the equation found in step 3.

Step 5. State the final answer in coherent form, specifying the units that you are using.

Example 3   A water trough with vertical cross section in the shape of an equilateral triangle is being filled at a rate of 4 cubic feet per minute. Given that the trough is 12 feet long, how fast is the level of the water rising at the instant that the water reaches a depth of $1\frac{1}{2}$ feet?

SOLUTION   Let x be the depth of the water in feet and V the volume of water in cubic feet.

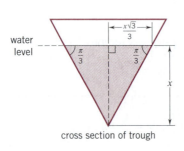

cross section of trough

Find $\dfrac{dx}{dt}$ when $x = \frac{3}{2}$

given that $\dfrac{dV}{dt} = 4$ ft³/min.

Area of cross section $= \dfrac{1}{2}\left(\dfrac{2x}{\sqrt{3}}\right)x = \dfrac{\sqrt{3}}{3}x^2.$

Volume of water $= 12\left(\dfrac{\sqrt{3}}{3}x^2\right) = 4\sqrt{3}x^2.$

Differentiation of $V = 4\sqrt{3}x^2$ with respect to t gives

$$\frac{dV}{dt} = 8\sqrt{3}x\frac{dx}{dt}.$$

Substituting $x = \frac{3}{2}$ and $dV/dt = 4$, we have

$$4 = 8\sqrt{3}\left(\frac{3}{2}\right)\frac{dx}{dt} \qquad \text{and thus} \qquad \frac{dx}{dt} = \frac{1}{9}\sqrt{3}.$$

At the instant that the water reaches a depth of $1\frac{1}{2}$ feet, the water level is rising at the rate of $\frac{1}{9}\sqrt{3}$ feet per second (about 0.19 feet per second).   ❑

Example 4   Two ships, one heading west and the other east, approach each other on parallel courses 8 nautical miles apart.† Given that each ship is cruising at 20 nautical miles per hour (knots), at what rate is the distance between them diminishing when they are 10 nautical miles apart?

---

† The international nautical mile measures 6080 feet. Nautical speed is measured in knots—nautical miles per hour.

SOLUTION Let y be the distance between the ships measured in nautical miles. Since the ships are moving in opposite directions at the rate of 20 knots each, the distance x (see figure) is decreasing at the rate of 40 knots. Thus, we want to

find $\dfrac{dy}{dt}$ when $y = 10$

given that $\dfrac{dx}{dt} = -40$ knots.

(Note that dx/dt is taken as negative since x is decreasing.)

$$x^2 + 8^2 = y^2. \qquad \text{(Pythagorean theorem)}$$

Differentiating $x^2 + 8^2 = y^2$ with respect to t, we find that

$$2x\frac{dx}{dt} + 0 = 2y\frac{dy}{dt} \qquad \text{and consequently} \qquad x\frac{dx}{dt} = y\frac{dy}{dt}.$$

When $y = 10$, $x = 6$. (Why?) Substituting $x = 6$, $y = 10$, and $dx/dt = -40$, we have

$$6(-40) = 10\frac{dy}{dt} \quad \text{so that} \quad \frac{dy}{dt} = -24.$$

(Note that dy/dt is negative since y is decreasing.) When the two ships are 10 miles apart, the distance between them is diminishing at the rate of 24 knots. ❑

Example 5 A conical paper cup 8 inches across the top and 6 inches deep is full of water. The cup springs a leak at the bottom and loses water at the rate of 2 cubic inches per minute. How fast is the water level dropping at the instant when the water is exactly 3 inches deep?

SOLUTION We begin with a diagram that represents the situation after the cup has been leaking for a while (Figure 3.8.1). We label the radius and height of the remaining "cone of water" r and h. We can relate r and h by similar triangles (Figure 3.8.2). Let h be the depth of the water measured in inches.

$$\text{Find } \frac{dh}{dt} \text{ when } h = 3 \quad \text{given that } \frac{dV}{dt} = -2 \text{ in.}^3/\text{min.}$$

$$V = \frac{1}{3}\pi r^2 h \quad \text{(volume of a cone)} \qquad \text{and} \qquad \frac{r}{h} = \frac{4}{6} \quad \text{(similar triangles)}.$$

Using the second equation to eliminate r from the first equation, we have

$$V = \frac{1}{3}\pi \left(\frac{2h}{3}\right)^2 h = \frac{4}{27}\pi h^3.$$

Differentiation with respect to t gives

$$\frac{dV}{dt} = \frac{4}{9}\pi h^2 \frac{dh}{dt}.$$

Substituting $h = 3$ and $dV/dt = -2$, we have

$$-2 = \frac{4}{9}\pi(3)^2 \frac{dh}{dt} \qquad \text{and thus} \qquad \frac{dh}{dt} = -\frac{1}{2\pi}.$$

Figure 3.8.1

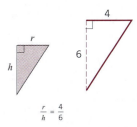

$$\frac{r}{h} = \frac{4}{6}$$

Figure 3.8.2

At the instant when the water is exactly 3 inches deep, the water level is dropping at the rate of $1/2\pi$ inches per minute (about 0.16 inches per minute). ❑

Example 6 A balloon leaves the ground 500 feet away from an observer and rises vertically at the rate of 140 feet per minute. At what rate is the angle of inclination of the observer's line of sight increasing at the instant when the balloon is exactly 500 feet above the ground?

SOLUTION Let x be the altitude of the balloon and θ the angle of inclination of the observer's line of sight.

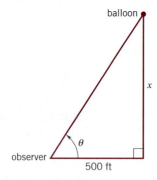

Find $\dfrac{d\theta}{dt}$ when $x = 500$ feet,

given that $\dfrac{dx}{dt} = 140$ ft/min.

$$\tan \theta = \frac{x}{500}.$$

Differentiation with respect to t gives

$$\sec^2 \theta \, \frac{d\theta}{dt} = \frac{1}{500} \frac{dx}{dt}.$$

When $x = 500$, the triangle is isosceles, and therefore $\sec \theta = \sqrt{2}$. Substituting $\sec \theta = \sqrt{2}$ and $dx/dt = 140$, we have

$$(\sqrt{2})^2 \frac{d\theta}{dt} = \frac{1}{500}(140) \qquad \text{and therefore} \qquad \frac{d\theta}{dt} = 0.14.$$

At the instant when the balloon is exactly 500 feet above the ground, the inclination of the observer's line of sight is increasing at the rate of 0.14 radians per minute (about 8 degrees per minute). ❑

Example 7 A ladder 13 feet long is leaning against the side of a building. If the foot of the ladder is pulled away from the building at the rate of 0.1 foot per second, how fast is the angle formed by the ladder and the ground changing at the instant when the top of the ladder is 12 feet above the ground?

SOLUTION

Find $\dfrac{d\theta}{dt}$ when $y = 12$ feet

given that $\dfrac{dx}{dt} = 0.1$ ft/sec.

$$\cos \theta = \frac{x}{13}.$$

Differentiation with respect to t gives

$$- \sin \theta \frac{d\theta}{dt} = \frac{1}{13} \frac{dx}{dt}.$$

When $y = 12$, $\sin \theta = \frac{12}{13}$. Substituting $\sin \theta = \frac{12}{13}$ and $dx/dt = 0.1$, we have

$$-\left(\frac{12}{13}\right) \frac{d\theta}{dt} = \frac{1}{13}(0.1) \qquad \text{and thus} \qquad \frac{d\theta}{dt} = -\frac{1}{120}.$$

At the instant that the top of the ladder is 12 feet above the ground, the angle formed by the ladder and the ground is decreasing at the rate of $\frac{1}{120}$ radians per second (about half a degree per second).   ❑

EXERCISES 3.8

1. A point moves along the straight line $x + 2y = 2$. Find (a) the rate of change of the y-coordinate, given that the x-coordinate is increasing at a rate of 4 units per second; (b) the rate of change of the x-coordinate, given that the y-coordinate is decreasing at a rate of 2 units per second.

2. A particle is moving in circular orbit $x^2 + y^2 = 25$. As it passes through the point $(3, 4)$, its y-coordinate is decreasing at the rate of 2 units per second. How is the x-coordinate changing?

3. A particle is moving along the parabola $y^2 = 4(x + 2)$. As it passes through the point $(7, 6)$ its y-coordinate is increasing at the rate of 3 units per second. How fast is the x-coordinate changing at this instant?

4. A particle is moving along the parabola $4y = (x + 2)^2$ in such a way that its x-coordinate is increasing at the constant rate of 2 units per second. How fast is the particle's distance to the point $(-2, 0)$ changing at the instant it is at the point $(2, 4)$?

5. A particle is moving around the ellipse $4x^2 + 16y^2 = 64$. At any time t its x- and y- coordinates are given by $x = 4 \cos t$, $y = 2 \sin t$. At what rate is the particle's distance to the origin changing at an arbitrary time t? At what rate is the distance to the origin changing when $t = \pi/4$?

6. A particle is moving along the curve $y = x\sqrt{x}$, $x \geq 0$. At time t its coordinates are $(x(t), y(t))$. Find the points on the curve, if any, at which both coordinates are changing at the same rate.

7. A heap of rubbish in the shape of a cube is being compacted. Given that the volume decreases at a rate of 2 cubic meters per minute, find the rate of change of an edge of the cube when the volume is exactly 27 cubic meters. What is the rate of change of the surface area of the cube at that instant?

8. At a certain instant the dimensions of a rectangle are a and b. These dimensions are changing at the rates m, n, respectively. Find the rate at which the area is changing.

9. The volume of a spherical balloon is increasing at a constant rate of 8 cubic feet per minute. How fast is the radius increasing when the radius is exactly 10 feet? How fast is the surface area increasing at that instant?

10. A spherical balloon has an initial volume of V_0 cubic meters of helium. Suppose that the gas is being released from the balloon at the rate of $1/(1 + t^2)$ cubic meters per second, $t \geq 0$.
   (a) What is the rate of change of the radius of the balloon at any time t? At what rate is the radius changing when $t = 3$?
   (b) What is the rate of change of the surface area of the balloon at any time t? At what rate is the surface area changing when $t = 3$?

11. An isosceles triangle has equal sides of length 10 centimeters. Let θ denote the angle included between the two equal sides.
   (a) Express the area A of the triangle as a function of θ (radians).
   (b) Suppose that θ is increasing at the rate of $10°$ per minute. How fast is A changing at the instant $\theta = 60°$?
   (c) At what value of θ will the triangle have maximum area?

12. At a certain instant the side of an equilateral triangle is α centimeters long and increasing at the rate of k centimeters per minute. How fast is the area increasing?

13. The perimeter of a rectangle is fixed at 24 centimeters. If the length l of the rectangle is increasing at the rate of 1 centimeter per second, for what value of l does the area of the rectangle start to decrease?

14. The dimensions of a rectangle are changing in such a way that the perimeter is always 24 inches. Show that, at the instant when the area is 32 square inches, the area is either increasing or decreasing 4 times as fast as the length is increasing.

15. A rectangle is inscribed in a circle of radius 5 inches. If the length of the rectangle is decreasing at the rate of 2

inches per second, how fast is the area changing at the instant when the length is 6 inches? HINT: A diagonal of the rectangle is a diameter of the circle.

16. A boat is held by a bow line that is wound about a bollard 6 feet higher than the bow of the boat. If the boat is drifting away at the rate of 8 feet per minute, how fast is the line unwinding when the bow is exactly 30 feet from the bollard?

17. Two boats are racing with constant speed toward a finish marker, boat A sailing from the south at 13 mph and boat B approaching from the east. When equidistant from the marker the boats are 16 miles apart and the distance between them is decreasing at the rate of 17 mph. Which boat will win the race?

18. A ladder 13 feet long is leaning against a wall. If the foot of the ladder is pulled away from the wall at the rate of 0.5 foot per second, how fast will the top of the ladder be dropping at the instant when the base is 5 feet from the wall?

19. A tank contains 1000 cubic feet of natural gas at a pressure of 5 pounds per square inch. Find the rate of change of the volume if the pressure decreases at a rate of 0.05 pound per square inch per hour. (Assume Boyle's law: *pressure* × *volume* = *constant*.)

20. The adiabatic law for the expansion of air is $PV^{1.4} = C$. At a given instant the volume is 10 cubic feet and the pressure 50 pounds per square inch. At what rate is the pressure changing if the volume is decreasing at a rate of 1 cubic foot per second?

21. A man standing 3 feet from the base of a lamppost casts a shadow 4 feet long. If the man is 6 feet tall and walks away from the lamppost at a speed of 400 feet per minute, at what rate will his shadow lengthen? How fast is the tip of his shadow moving?

22. A light is attached to the wall of a building 64 feet above the ground. A ball is dropped from the same height, but 20 feet away from the side of the building. The height y of the ball at time t is given by $y(t) = 64 - 16t^2$. How fast is the shadow of the ball moving along the ground after 1 second? After 3 seconds? After 4 seconds? (Explain.)

23. An object that weighs 150 pounds on the surface of the earth will weigh $150(1 + \frac{1}{4000} r)^{-2}$ pounds when it is r miles above the surface. If the object's altitude is increasing at the rate of 10 miles per second, how fast is its weight decreasing at the instant it is 400 miles above the surface?

24. In the special theory of relativity the mass of a particle moving at velocity v is

$$\frac{m}{\sqrt{1 - v^2/c^2}}$$

where m is the mass at rest and c is the speed of light. At what rate is the mass changing when the particle's velocity is $\frac{1}{2}c$ and the rate of change of the velocity is $0.01\,c$ per second?

25. Water is dripping through the bottom of a conical cup 4 inches across and 6 inches deep. Given that the cup loses half a cubic inch of water per minute, how fast is the water level dropping when the water is 3 inches deep?

26. Water flows from a faucet into a hemispherical basin 14 inches in diameter at a rate of 2 cubic inches per second. How fast does the water rise (a) when the water is exactly halfway to the top? (b) just as it runs over? (The volume of a spherical segment is given by $\pi r h^2 - \frac{1}{3}\pi h^3$ where r is the radius of the sphere and h is the depth of the segment.)

27. The base of an isosceles triangle is 6 feet. If the altitude is 4 feet and increasing at the rate of 2 inches per minute, at what rate is the vertex angle changing?

28. As a boy winds up the cord, his kite is moving horizontally at a height of 60 feet with a speed of 10 feet per minute. How fast is the inclination of the cord changing when its length is 100 feet?

29. A revolving searchlight $\frac{1}{2}$ mile from shore makes 1 revolution per minute. How fast is the light traveling along the straight beach at the instant it passes over a shorepoint 1 mile away from the shorepoint nearest to the searchlight?

30. Two cars, car A traveling east at 30 mph and car B traveling north at 22.5 mph, are heading toward an intersection I. At what rate is angle IAB changing at the instant when cars A and B are 300 feet and 400 feet, respectively, from the intersection?

31. A rope 32 feet long is attached to a weight and passed over a pulley 16 feet above the ground. The other end of the rope is pulled away along the ground at the rate of 3 feet per second. At what rate is the angle between the rope and the ground changing at the instant when the weight is exactly 4 feet off the ground?

32. A slingshot is made by fastening the two ends of a 10-inch rubber strip 6 inches apart. If the midpoint of the strip is drawn back at the rate of 1 inch per second, at what rate is the angle between the segments of the strip changing 8 seconds later?

33. A balloon is released 500 feet away from an observer. If the balloon rises vertically at the rate of 100 feet per minute and at the same time the wind is carrying it horizontally away from the observer at the rate of 75 feet per minute, at what rate is the angle of inclination of the observer's line of sight changing 6 minutes after the balloon has been released?

34. A searchlight is trained on a plane that flies directly above the light at an altitude of 2 miles at a speed of 400 miles per hour. How fast must the light be turning 2 seconds after the plane passes directly overhead?

(*Important*) Angular Velocity; Uniform Circular Motion
As a particle moves along a circle of radius r, it effects a change in the central angle marked θ in the figure. The angle θ is measured in radians. The time rate of change of θ, $\omega = d\theta/dt$, is called the *angular velocity* of the particle. Circular motion with constant positive angular velocity is called *uniform circular motion*.

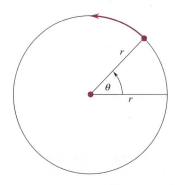

35. A particle in uniform circular motion traces out a circular arc. The time rate of change of the length of that arc is called the *speed* of the particle. What is the speed of a particle that moves along a circle of radius r with constant positive angular velocity ω?

36. The kinetic energy KE of a particle of mass m is given by

$$\text{KE} = \tfrac{1}{2}mv^2$$

where v is the speed. What is the kinetic energy of the particle in Exercise 35?

37. A point P moves uniformly along the circle $x^2 + y^2 = r^2$ with angular velocity ω. Find the xy-coordinates of P at time t given that the motion starts at time $t = 0$ with $\theta = \theta_0$.

38. With P as in Exercise 37 find the velocity and acceleration of the projection of P onto†
(a) the x-axis.   (b) the y-axis.

39. The figure shows a sector in a circle of radius r. The sector is the union of triangle T and segment S. Suppose that the radius vector rotates counterclockwise with a constant angular velocity of ω radians per second. Show that the area of the sector changes at a constant rate, but the area of T and the area of S do not change at a constant rate.

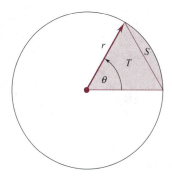

40. Take S and T as in Exercise 39. In general the area of S and the area of T change at different rates. There is, however, one value of θ between 0 and π at which both areas have the same instantaneous rate of change. Find this value of θ.

■ 3.9 DIFFERENTIALS; NEWTON-RAPHSON APPROXIMATIONS

Differentials

Let f be a differentiable function. In Figure 3.9.1, you can see the graph of f and below it the graph of the tangent line at the point $(x, f(x))$. As the figure suggests, for small h, $h \neq 0$, $f(x + h) - f(x)$, the change in f from x to $x + h$, can be approximated by the product $f'(x)h$:

(3.9.1)
$$f(x + h) - f(x) \cong f'(x)h.$$

---

† The *projection* of $P(x, y)$ onto the x-axis is the point $Q(x, 0)$; the projection of $P(x, y)$ onto the y-axis is the point $R(0, y)$.

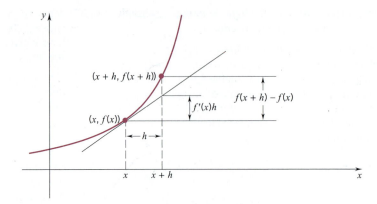

Figure 3.9.1

How good is this approximation? It is good in the sense that, for small h, the difference between the two quantities,

$$[f(x + h) - f(x)] - f'(x)h,$$

is small compared to h. How small? Small enough that the ratio

$$\frac{[f(x + h) - f(x)] - f'(x)h}{h}$$

tends to 0 as h tends to 0:

$$\lim_{h \to 0} \frac{[f(x + h) - f(x)] - f'(x)h}{h} = \lim_{h \to 0} \frac{f(x + h) - f(x)}{h} - \lim_{h \to 0} \frac{f'(x)h}{h}$$

$$= f'(x) - f'(x) = 0.$$

The quantities $f(x + h) - f(x)$ and $f'(x)h$ have names:

DEFINITION 3.9.2

Let $h \neq 0$. The difference $f(x + h) - f(x)$ is called the *increment of f from x to x + h*, and is denoted by Δf:

$$\Delta f = f(x + h) - f(x).\dagger$$

The product $f'(x)h$ is called the *differential of f at x with increment h*, and is denoted by df:

$$df = f'(x)h.$$

Display (3.9.1) says that, for small h, Δf, and df are approximately equal:

$$\boxed{\Delta f \cong df.}$$

---

† The symbol Δ is the capital of the Greek letter δ, which corresponds to the English letter d; Δf is read "delta f."

How approximately equal are they? Enough so that the ratio

$$\frac{\Delta f - df}{h}$$

tends to 0 as h tends to 0.

Let's see what all this amounts to in a very simple case. The area of a square of side x is given by the function

$$f(x) = x^2, \qquad x > 0.$$

If the length of each side is increased from x to $x + h$, the area increases from $f(x)$ to $f(x + h)$. The change in area is the increment Δf:

$$
\begin{aligned}
\Delta f &= f(x + h) - f(x) \\
&= (x + h)^2 - x^2 \\
&= (x^2 + 2xh + h^2) - x^2 \\
&= 2xh + h^2.
\end{aligned}
$$

As an estimate for this change we can use the differential

$$df = f'(x)h = 2xh. \qquad \text{(Figure 3.9.2)}$$

The error of this estimate, the difference between the actual change Δf and the estimated change df, is the difference

$$\Delta f - df = h^2.$$

As promised, the error is small compared to h in the sense that

$$\frac{\Delta f - df}{h} = \frac{h^2}{h} = h$$

tends to 0 as h tends to 0.

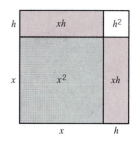

Figure 3.9.2

Example 1 Use a differential to estimate the change in $f(x) = x^{2/5}$ if:

(a) x is increased from 32 to 34.   **(b)** x is decreased from 1 to $\frac{9}{10}$.

SOLUTION Since $f'(x) = \frac{2}{5}x^{-3/5} = 2/(5x^{3/5})$, we have

$$df = f'(x)h = \frac{2}{5x^{3/5}}h.$$

For part (a) we set $x = 32$ and $h = 2$. The differential then becomes

$$df = \frac{2}{5(32)^{3/5}}(2) = \frac{4}{40} = 0.10.$$

A change in x from 32 to 34 increases the value of f by approximately 0.10. Checking this result with a hand calculator, we have

$$\Delta f = f(34) - f(32) = (34)^{2/5} - (32)^{2/5} \cong 4.0982 - 4 = 0.0982.$$

For part (b) we set $x = 1$ and $h = -\frac{1}{10}$. In this case, the differential is

$$df = \frac{2}{5(1)^{3/5}}\left(-\frac{1}{10}\right) = -\frac{2}{50} = -0.04.$$

A change in x from 1 to $\frac{9}{10}$ decreases the value of f by approximately 0.04. Checking this result with a hand calculator, we have

$$\Delta f = f(0.9) - f(1) = (0.9)^{2/5} - (1)^{2/5} \cong 0.9587 - 1 = -0.0413. \quad \square$$

Example 2 Use a differential to estimate $\sqrt{104}$.

SOLUTION We know $\sqrt{100}$. What we need is an estimate for the increase of

$$f(x) = \sqrt{x}$$

when x increases from 100 to 104. Here

$$f'(x) = \frac{1}{2\sqrt{x}} \quad \text{and thus} \quad df = f'(x)h = \frac{h}{2\sqrt{x}}.$$

With $x = 100$ and $h = 4$, df becomes

$$\frac{4}{2\sqrt{100}} = \frac{1}{5} = 0.2.$$

A change in x from 100 to 104 increases the value of the square root by approximately 0.2. It follows that

$$\sqrt{104} \cong \sqrt{100} + 0.2 = 10 + 0.2 = 10.2.$$

As you can check, $(10.2)^2 = 104.04$, so that we are not far off. $\quad \square$

Example 3 Use a differential to approximate $\cos 40°$.

SOLUTION Let $f(x) = \cos x$. We know that $\cos 45° = \cos \pi/4 = \sqrt{2}/2$. Converting $40°$ to radians, we have

$$40° = 45° - 5° = \frac{\pi}{4} - \left(\frac{\pi}{180}\right)5 = \frac{\pi}{4} - \frac{\pi}{36} \text{ radians.}$$

We will use a differential to estimate the change in $\cos x$ when x decreases from $\pi/4$ to $(\pi/4) - (\pi/36)$. Now,

$$f'(x) = -\sin x \quad \text{and} \quad df = f'(x)h = -h \sin x.$$

With $x = \pi/4$ and $h = -\pi/36$, df is given by

$$df = -\left(-\frac{\pi}{36}\right)\sin\left(\frac{\pi}{4}\right) = \frac{\pi}{36}\frac{\sqrt{2}}{2} = \frac{\pi\sqrt{2}}{72} \cong 0.0617.$$

A decrease in x from $\pi/4$ to $(\pi/4) - (\pi/36)$ increases the value of the cosine function by approximately 0.0617. Therefore

$$\cos 40° \cong \cos 45° + 0.0617 \cong 0.7071 + 0.0617 = 0.7688.$$

Checking this result with a hand calculator, we find that $\cos 40° \cong 0.7660.$ $\quad \square$

Newton-Raphson Method

Figure 3.9.3 shows the graph of a function f. Since the graph of f crosses the x-axis at $x = c$, the number c is a solution (root) of the equation $f(x) = 0$. In the setup of Figure 3.9.3, we can approximate c as follows: Start at a point x_1 (see the figure). The tangent line at $(x_1, f(x_1))$ intersects the x-axis at a point x_2, which is closer to c than

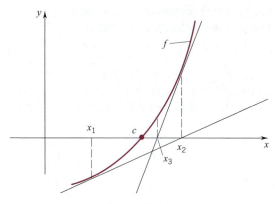

Figure 3.9.3

x_1. The tangent line at $(x_2, f(x_2))$ intersects the x-axis at a point x_3, which is closer to c than x_2. Continuing in this manner, we will obtain better and better approximations $x_4, x_5, \ldots , x_n$ to the root c.

There is an algebraic connection between x_n and x_{n+1} that we now develop. The tangent line at $(x_n, f(x_n))$ has equation

$$y - f(x_n) = f'(x_n)(x - x_n).$$

The x-intercept of this line, x_{n+1}, can be found by setting $y = 0$:

$$0 - f(x_n) = f'(x_n)(x_{n+1} - x_n).$$

Solving this equation for x_{n+1}, we have

(3.9.3)

$$x_{n+1} = x_n - \frac{f(x_n)}{f'(x_n)}.$$

This method for locating a root of an equation $f(x) = 0$ is called the *Newton-Raphson method*. The method does not work in all cases. First, there are some conditions that must be placed on the function f. Clearly, f must be differentiable on some interval that contains the root c. Also, note that if $f'(x_n) = 0$ for some n, then the tangent line at $(x_n, f(x_n))$ is horizontal and the next approximation x_{n+1} cannot be calculated. See Figure 3.9.4. Thus, we assume that $f'(x) \neq 0$ in some interval containing c.

Figure 3.9.4

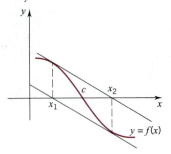

Figure 3.9.5

The method can also fail if proper care is not taken in choosing the first approximation x_1. For example, it can happen that the first approximation x_1 produces a second approximation x_2 which, in turn, gives the same x_1 as the third approximation, and so on—the approximations simply alternate between x_1 and x_2. See Figure 3.9.5. Another type of difficulty can arise if $f'(x_1)$ is close to zero. In this case, the second approximation x_2 could be worse than x_1, the third approximation x_3 could be worse than x_2, and so forth. See Figure 3.9.6.

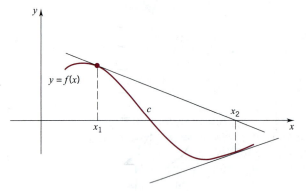

Figure 3.9.6

There is a condition that guarantees that the Newton-Raphson method will work. Suppose that f is twice differentiable and that $f(x)f''(x) > 0$ on the open interval I joining c and x_1. If $f(x) > 0$ on I, then the graph of f is concave up† on I and we have the situation pictured in Figure 3.9.7. On the other hand, if $f(x) < 0$ on I, then the graph of f is concave down on I and we have the situation pictured in Figure 3.9.8. In either case, the sequence of approximations $x_1, x_2, x_3, \ldots$ will "converge" to the root c.

Figure 3.9.7

Figure 3.9.8

Example 4 The number $\sqrt{3}$ is a root of the equation $x^2 - 3 = 0$. We will estimate $\sqrt{3}$ by applying the Newton-Raphson method to the function $f(x) = x^2 - 3$ starting at $x_1 = 2$. (As you can check, $f(x)f''(x) > 0$ on $(\sqrt{3}, 2)$ and therefore we can be sure that the method applies.) Since $f'(x) = 2x$, the Newton-Raphson formula gives

$$x_{n+1} = x_n - \left(\frac{x_n^2 - 3}{2x_n}\right) = \frac{x_n^2 + 3}{2x_n}.$$

---

† The concavity of a graph and the role of the second derivative are treated in Chapter 4.

Successive calculations with this formula (using a calculator) are given in the following table:

n	x_n	$x_{n+1} = \dfrac{x_n^2 + 3}{2x_n}$
1	2	1.75000
2	1.75000	1.73214
3	1.73214	1.73205

Since $(1.73205)^2 = 2.999997$, the method has generated a very accurate estimate of $\sqrt{3}$ in only three steps. ❑

EXERCISES 3.9

1. Use a differential to estimate the change in the volume of a cube caused by an increase h in the length of each side. Interpret geometrically the error of your estimate $\Delta V - dV$.

2. Use a differential to estimate the area of a ring of inner radius r and width h. What is the exact area?

▶ In Exercises 3–10, use differentials to estimate the value of the indicated expression. Then compare your estimate with the result given by a calculator.

3. $\sqrt[3]{1010}$.
4. $\sqrt{125}$.
5. $\sqrt[4]{15}$.
6. $1/\sqrt{24}$.
7. $\sqrt[5]{30}$.
8. $(26)^{2/3}$.
9. $(33)^{3/5}$.
10. $(33)^{-1/5}$.

▶ In Exercises 11–14, use differentials to estimate the indicated expression. (Remember to convert to radian measure.) Compare your estimate with the result given by a calculator.

11. $\sin 46°$.
12. $\cos 62°$.
13. $\tan 28°$.
14. $\sin 43°$.

15. Estimate $f(2.8)$ given that $f(3) = 2$ and $f'(x) = (x^3 + 5)^{1/5}$.

16. Estimate $f(5.4)$ given that $f(5) = 1$ and $f'(x) = \sqrt[3]{x^2 + 2}$.

17. Find the approximate volume of a thin cylindrical sheet with open ends given that the inner radius is r, the height is h, and the thickness is t.

18. The diameter of a steel ball is measured to be 16 centimeters, with a maximum error of 0.3 centimeters. Estimate by differentials the maximum error in (a) the surface area when calculated by the formula $S = 4\pi r^2$; (b) the volume when calculated by the formula $V = \frac{4}{3}\pi r^3$.

19. A box is to be constructed in the form of a cube to hold 1000 cubic feet. Use a differential to estimate how accurately the inner edge must be made so that the volume will be correct to within 3 cubic feet.

20. Use differentials to estimate the values of x for which
(a) $\sqrt{x+1} - \sqrt{x} < 0.01$.
(b) $\sqrt[4]{x+1} - \sqrt[4]{x} < 0.002$.

21. A hemispherical dome with radius 50 feet will be given a coat of paint 0.01 inch thick. The contractor for the job wants to estimate the number of gallons of paint that will be needed. Use differentials to obtain an estimate (there are 231 cubic inches in a gallon). HINT: Approximate the change in the volume of the hemisphere corresponding to an increase of 0.01 inch in the radius.

22. Assume that the earth is a sphere of radius 4000 miles. The volume of ice at the north and south poles is estimated to be 8 million cubic miles. Suppose that this ice melts and the water produced distributes uniformly over the surface of the earth. Estimate the depth of the added water at any point on the earth.

23. For oscillations of small amplitude, the relationship between the period P of one complete oscillation and the length L of a simple pendulum (see Exercise 62, Section 3.6) is given by the equation

$$P = 2\pi\sqrt{\frac{L}{g}},$$

where g is the (constant) acceleration of gravity. Show that a small change dL in the length of a pendulum produces a change dP in the period that satisfies

$$\frac{dP}{P} = \frac{1}{2}\frac{dL}{L}.$$

24. Suppose that the pendulum in a pendulum clock has length 90 centimeters. Use the result in Exercise 23 to determine how the length of the pendulum should be adjusted if the clock is losing 15 seconds per hour?

25. A pendulum has length 3.26 feet and goes through one complete oscillation in 2 seconds. Use Exercise 23 to find the approximate change in P if the pendulum is lengthened by 0.01 foot.

26. As a metal cube is heated, the length of each edge increases $\frac{1}{10}\%$ per degree increase in temperature. Show by differentials that the surface area increases about $\frac{2}{10}\%$ per degree and the volume increases about $\frac{3}{10}\%$ per degree.

27. We are trying to determine the area of a circle by measuring the diameter. How accurately must we measure the diameter if our estimate is to be correct within 1%?

28. Estimate by differentials how precisely x must be determined if (a) x^n is to be accurate within 1%; (b) $x^{1/n}$ is to be accurate within 1%.

▶ In Exercises 29–36, use the Newton-Raphson method to estimate a root of the equation $f(x) = 0$ starting at the indicated value of x: (a) Express x_{n+1} in terms of x_n. (b) Give x_4 rounded off to five decimal places and evaluate f at that approximation.

$$x^2 - 24 = 0$$
$$x^2 = 24$$
$$\sqrt[n]{16} = 2\sqrt{6}$$

29. $f(x) = x^2 - 24$; $x_1 = 5$.

30. $f(x) = x^2 - 17$; $x_1 = 5$.

31. $f(x) = x^3 - 25$; $x_1 = 3$.

32. $f(x) = x^5 - 30$; $x_1 = 2$.

33. $f(x) = \cos x - x$; $x_1 = 1$.

34. $f(x) = \sin x - x^2$; $x_1 = 1$.

35. $f(x) = 2 \sin x - x$; $x_1 = 2$.

36. $f(x) = x^3 - 4x + 1$; $x_1 = 2$.

37. The function $f(x) = x^{1/3}$ is 0 at $x = 0$. Show that the Newton-Raphson method applied to f starting at *any* value $x_1 \neq 0$ fails to generate values that approach the solution $x = 0$. Describe the sequence $x_1, x_2, x_3,$ . . . , that the method generates.

38. Let $f(x) = 2x^3 - 3x^2 - 1$.
  (a) Prove that the equation $f(x) = 0$ has a root in the interval $(1, 2)$. HINT: Use the intermediate-value theorem.
  (b) Show that the Newton-Raphson method with $x_1 = 1$ fails to generate values that approach the root in $(1, 2)$.
  (c) Estimate the root by starting at $x_1 = 2$. Determine x_4 rounded off to five decimal places, and evaluate $f(x_4)$.

39. The function $f(x) = x^4 - 2x^2 - \frac{17}{16}$ has two zeros, one at a, where $0 < a < 2$, and the other at $-a$ since f is an even function.
  (a) Show that the Newton-Raphson methods fails if $x_1 = \frac{1}{2}$ is used as the initial estimate for finding the root a. Describe the sequence $x_1, x_2, x_3,$ . . . that the method generates.
  (b) Estimate the root a by starting at $x_1 = 2$. Determine x_4 rounded off to five decimal places, and evaluate $f(x_4)$.

▶ **40.** Let $f(x) = x^2 - a$, where $a > 0$. The roots of the equation $f(x) = 0$ are $\pm\sqrt{a}$.

$= 29.93747916$

(a) Show that if $x_1 > 0$ is any initial estimate for $\sqrt{a}$, then the Newton-Raphson method gives the iteration formula

$$x_{n+1} = \frac{1}{2}\left(x_n + \frac{a}{x_n}\right), \qquad n \geq 1.$$

(b) Let $a = 5$. Beginning with $x_1 = 2$, use the formula in part (a) to calculate x_4 rounded off to five decimal places and evaluate $f(x_4)$.

▶ **41.** Let $f(x) = x^k - a$, where k is a positive integer and $a > 0$. The number $a^{1/k}$ is a root of the equation $f(x) = 0$.
  (a) Show that if $x_1 > 0$ is any initial estimate for $a^{1/k}$, then the Newton-Raphson method gives the iteration formula

$$x_{n+1} = \frac{1}{k}\left[(k-1)x_n + \frac{a}{x_n^{k-1}}\right].$$

  Note that for $k = 2$, this formula reduces to the formula given in part (a) of Exercise 40.
  (b) Use the formula in part (a) to approximate $\sqrt[3]{23}$. Begin with $x_1 = 3$ and calculate x_4 rounded off to five decimal places. Evaluate $f(x_4)$.

Let f be a function that is defined on an interval I. A number $c \in I$ is a *fixed point* of f if $f(c) = c$ (see Exercise 21, Section 2.6). A given function may or may not have a fixed point. If f is differentiable, then the Newton-Raphson method applied to the function $F(x) = f(x) - x$ can be used to approximate the fixed points of f.

▶ **42.** Let $f(x) = 2x^3 - 4x - 3$.
  (a) Show that f has a fixed point on $[1, 2]$. HINT: Let $F(x) = f(x) - x$ and use the intermediate-value theorem.
  (b) Use the Newton-Raphson method to approximate the fixed point of f. Begin with $x_1 = 2$ and calculate x_4 rounded off to four decimal places; evaluate $f(x_4)$.

▶ **43.** Let $f(x) = \frac{1}{2} \cos x$.
  (a) Show that f has a fixed point on $[0, \pi/2]$.
  (b) Use the Newton-Raphson method to approximate the fixed point of f. Begin with $x_1 = 0$ and calculate x_4 rounded off to four decimal places; evaluate $f(x_4)$.

▶ **44.** Let $f(x) = \frac{2}{3} \sin x + 1$.
  (a) Show that f has a fixed point on $[0, 2]$.
  (b) Use the Newton-Raphson method to approximate the fixed point of f. Begin with $x_1 = 2$ and calculate x_4 rounded off to four decimal places; evaluate $f(x_4)$.

Let g be a function defined at least on some open interval containing the number 0. We say that g is of *smaller order* *than* h, or that $g(h)$ is *little-o(h)*, and write $g(h) = o(h)$, iff $g(h)$ is small enough compared with h that

$$\lim_{h \to 0} \frac{g(h)}{h} = 0, \qquad \text{or equivalently,} \qquad \lim_{h \to 0} \frac{g(h)}{|h|} = 0.$$

45. Which of the following statements is true?

(a) $h^3 = o(h)$.　(b) $\dfrac{h^2}{h-1} = o(h)$.　(c) $h^{1/3} = o(h)$.

46. Show that, if $g(h) = o(h)$, then $\lim\limits_{h \to 0} g(h) = 0$.

47. Show that, if $g_1(h) = o(h)$ and $g_2(h) = o(h)$, then

$$g_1(h) + g_2(h) = o(h) \qquad \text{and} \qquad g_1(h)g_2(h) = o(h).$$

48. The figure to the right shows the graph of a differentiable function f and a line with slope m that passes through the point $(x, f(x))$. The vertical separation at $x + h$ between the line with slope m and the graph of f has been labeled $g(h)$.
(a) Calculate $g(h)$.
(b) Show that, of all lines that pass through $(x, f(x))$, the tangent line is the line that best approximates the

graph of f near the point $(x, f(x))$ by showing that

$$g(h) = o(h) \qquad \text{iff} \qquad m = f'(x).$$

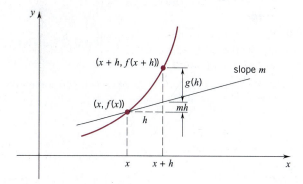

■ CHAPTER HIGHLIGHTS

3.1 The Derivative

derivative of f at x: $\lim\limits_{h \to 0} \dfrac{f(x+h) - f(x)}{h}$ (p. 130)

tangent line (p. 135)　normal line (p. 135)　vertical tangent (p. 137)

If f is differentiable at x, then f is continuous at x (p. 138). The converse is false.

left-hand derivative of f at x: $\lim\limits_{h \to 0^-} \dfrac{f(x+h) - f(x)}{h}$.

right-hand derivative of f at x: $\lim\limits_{h \to 0^+} \dfrac{f(x+h) - f(x)}{h}$.

alternative definition of the derivative at c: $\lim\limits_{h \to 0} \dfrac{f(x) - f(c)}{x - c}$.

3.2 Some Differentiation Formulas

If $f(x) = \alpha$, α a constant, then $f'(x) = 0$. If $f(x) = x$, then $f'(x) = 1$.

$(f + g)'(x) = f'(x) + g'(x),$　　　$(\alpha f)'(x) = \alpha f'(x),$

$(fg)'(x) = f(x)g'(x) + g(x)f'(x),$　　$\left(\dfrac{f}{g}\right)'(x) = \dfrac{g(x)f'(x) - f(x)g'(x)}{[g(x)]^2}$　　$(g(x) \neq 0).$

For any integer n, $p(x) = x^n$ has derivative $p'(x) = nx^{n-1}$.

If $p(x) = a_n x^n + a_{n-1} x^{n-1} + \cdots + a_2 x^2 + a_1 x + a_0$ is a polynomial, then

$$p'(x) = na_n x^{n-1} + (n-1)a_{n-1} x^{n-2} + \cdots + 2a_2 x + a_1.$$

3.3 The d/dx Notation; Derivatives of Higher Order

Another notation for the derivative is the *double-d* notation of Leibniz: if $y = f(x)$, then

$$\frac{dy}{dx} = f'(x).$$

Higher derivatives are calculated by repeated differentiation. For example, the second derivative of $y = f(x)$ is the derivative of the derivative:

$$f''(x) = [f'(x)]' \qquad \text{or} \qquad \frac{d^2y}{dx^2} = \frac{d}{dx}\left(\frac{dy}{dx}\right).$$

3.4 The Derivative as a Rate of Change

dy/dx gives the rate of change of y with respect to x

Velocity and Acceleration

velocity, acceleration, speed (p. 161)
free fall (p. 163)

If velocity and acceleration have the same sign, the object is *speeding up.* If the velocity and acceleration have opposite signs, the object is *slowing down.*

Free Fall (Near the Surface of the Earth)

Economics

marginal cost, marginal revenue, marginal profit, break-even points.

3.5 The Chain Rule

Various forms of the chain rule:

$$\frac{dy}{dx} = \frac{dy}{du}\frac{du}{dx}, \qquad \frac{d}{dx}[f(u)] = \frac{d}{du}[f(u)]\frac{du}{dx}, \qquad (f \circ g)'(x) = f'(g(x))g'(x).$$

If u is a differentiable function of x and n is an integer, then

$$\frac{d}{dx}(u^n) = nu^{n-1}\frac{du}{dx}.$$

3.6 Differentiating the Trigonometric Functions

$$\frac{d}{dx}(\sin u) = \cos u \frac{du}{dx}, \qquad \frac{d}{dx}(\cos u) = -\sin u \frac{du}{dx},$$

$$\frac{d}{dx}(\tan u) = \sec^2 u \frac{du}{dx}, \qquad \frac{d}{dx}(\cot u) = -\csc^2 u \frac{du}{dx},$$

$$\frac{d}{dx}(\sec u) = \sec u \tan u \frac{du}{dx}, \qquad \frac{d}{dx}(\csc u) = -\csc u \cot u \frac{du}{dx}.$$

3.7 Implicit Differentiation; Rational Powers

finding dy/dx from an equation in x and y without first solving the equation for y

If u is a differentiable function of x and r is a rational number, then

$$\frac{d}{dx}(u^r) = ru^{r-1}\frac{du}{dx}.$$

3.8 Rates of Change per Unit Time

If a quantity Q varies with time, then the derivative dQ/dt gives the rate of change of that quantity with respect to time.

A 5-step procedure for solving *related rate problems* is outlined on p. 194.

angular velocity; uniform circular motion.

3.9 Differentials; Newton-Raphson Approximations

increment: $\Delta f = f(x + h) - f(x)$ differential: $df = f'(x)h$

$\Delta f \cong df$ in the sense that $\dfrac{\Delta f - df}{h}$ tends to 0 as $h \to 0$

Newton-Raphson method (p. 202).

■ PROJECTS AND EXPLORATIONS USING TECHNOLOGY

To do these exercises you will need a graphics calculator or a computer with graphing capability. The majority of these problems are open-ended so different approaches may be used to solve them. You should be aware that different approaches can result in slight variations in the answers. Round your numerical answers to at least four decimal places. The rounding method that your calculator or computer uses also may cause variations in answers.

3.1 When designing the curved shape of a car fender, the following type of problem must be solved: Find two cubic polynomials p and q, with p defined on $[1, 3]$, q defined on $[3, 9]$, and such that:

$$p(1) = 3, \quad p(3) = 7, \quad q(3) = 7, \quad q(9) = -2, \quad p'(3) = q'(3),$$
$$p''(3) = q''(3), \quad p''(1) = 0 \quad \text{and} \quad q''(9) = 0.$$

The conditions $p(1) = 3$, $p(3) = 7$, $q(3) = 7$, and $q(9) = -2$ state that the curved fender passes through the points $(1, 3)$, $(3, 7)$, and $(9, -2)$. The condition $p'(3) = q'(3)$ ensures that curve is "smooth" at the point where the graphs of p and q meet. The condition $p''(3) = q''(3)$ is also a smoothness condition and, further, it ensures that certain forces on the fender will not be troublesome. The last conditions $p''(1) = 0$ and $q''(9) = 0$ help in the manufacture of the fender; they imply that the fender will be "straight" at the ends. A curve that satisfies these conditons is called a *cubic spline.*

(a) Let $p(x) = ax^3 + bx^2 + cx + d$ and $q(x) = \alpha x^3 + \beta x^2 + \gamma x + \delta$. Write the system of linear equations generated by the specified conditions and find the solutions of the system.

(b) Draw the graphs of p and q on their respective intervals and verify visually that the resulting curve on $[1, 9]$ is continuous and smooth.

(c) Let

$$F(x) = \begin{cases} p(x), & 1 \leq x \leq 3 \\ q(x), & 3 < x \leq 9. \end{cases}$$

Show that F, F', and F'' are continuous on $[1, 9]$.

3.2 Let $g(a)$ be the largest (rightmost) zero of

$$f_a(x) = 20x^3 - 40x^2 + 25x - 5 + a.$$

For example, the graph of $f_0(x)$ is shown in the figure and $g(0) = 1$.

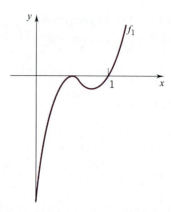

(a) Draw the graphs of f for $a = 0.1, 0.2, 0.3, \ldots, 0.9, 1$. Use Newton-Raphson iteration to approximate the values of $g(a)$ for these values of a.

(b) The function g has a discontinuity on $[0, 1]$. Based on your results in part (a), find the point of discontinuity.

(c) Using difference quotients, approximate the graph of g' on $[0, 1]$.

(d) By calculating additional values of g near the point of discontinuity, determine whether the left- and right-hand limits of g' exist at the point of discontinuity.

(e) Use difference quotients on g' to approximate g'' on $[0, 1]$.

(f) Make a table of values of g' and $f' \circ g$. Can you determine a relation between these two functions?

3.3 Suppose that we want to draw the graph of the equation

$$(x^2 + y^2 + 3x)^2 = 2(x^2 + y^2).$$

(a) Use implicit differentiation to find dy/dx.

(b) Assign some specific values to y, and for each such value graph the function

$$f(x) = (x^2 + y^2 + 3x)^2 - 2(x^2 + y^2).$$

Then record the approximate values of the zeros of f.

(c) Assign some specific values to x, and for each such value graph the function

$$g(y) = (x^2 + y^2 + 3x)^2 - 2(x^2 + y^2).$$

Then record the approximate values of the zeros of g.

(d) Use the information from parts (b) and (c) to find values a, b, c, and d such that the graph of the equation lies inside the rectangle R given by

$$R: a \leq x \leq b, \quad c \leq y \leq d.$$

(e) In how many points can a vertical line intersect the graph of the equation? In how many points can a horizontal line intersect the graph?

(f) Use Newton-Raphson iteration and the values from parts (b) and (c) to calculate and plot some points on the graph.

(g) Evaluate dy/dx at each of the points calculated in part (f) and sketch a short segment of the tangent line at each of these points. Use this information to sketch the graph of the equation (if necessary find additional points and tangent lines).

(h) Use a graphing utility to graph the equation and compare with your sketch from part (g).

THE MEAN-VALUE THEOREM AND APPLICATIONS

■ 4.1 THE MEAN-VALUE THEOREM

In this section we prove a result known as *the mean-value theorem.* First stated by the French mathematician Joseph-Louis Lagrange† (1736–1813), this theorem has come to permeate the theoretical structure of all calculus.

THEOREM 4.1.1   THE MEAN-VALUE THEOREM

If f is differentiable on the open interval (a, b) and continuous on the closed interval $[a, b]$ then there is at least one number c in (a, b) for which

$$f'(c) = \frac{f(b) - f(a)}{b - a}$$

or, equivalently,

$$f(b) - f(a) = f'(c)(b - a).$$

The number

$$\frac{f(b) - f(a)}{b - a}$$

is the slope of the line l that passes through the points $(a, f(a))$ and $(b, f(b))$. To say that there is at least one number c for which

$$f'(c) = \frac{f(b) - f(a)}{b - a}$$

---

† Lagrange, whom you encountered earlier in connection with the prime notation for differentiation, was born in Turin, Italy. He spent twenty years as mathematician in residence at the court of Frederick the Great. Later he taught at the renowned École Polytechnique in France.

is to say the the graph of f has at least one point $(c, f(c))$ at which the tangent is parallel to l. See Figure 4.1.1.

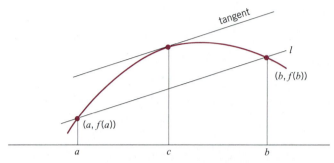

Figure 4.1.1

We can also give a "physical" interpretation of the mean-value theorem. Suppose that an object is moving in one direction along a straight line and that, during a certain time interval $t = a$ to $t = b$, its position at time t is denoted by $x(t)$ (see Section 3.4). Assume that the position function $x(t)$ is continuous on $[a, b]$ and differentiable on (a, b). Now, $x(b) - x(a)$ is the distance that the object travels during the time interval $[a, b]$ and

$$\frac{x(b) - x(a)}{b - a}$$

is the average velocity of the object over this interval. By the mean-value theorem, there is at least one time $t = c$ for which

$$x'(c) = \frac{x(b) - x(a)}{b - a}.$$

But, $x'(c) = v(c)$ is the instantaneous velocity of the object at $t = c$. Thus, the mean-value theorem tells us that there is at least one time $t = c$ at which the velocity at that instant equals the average velocity over the whole interval. Think about this when you are driving your car. For example, if you drive 240 miles in 4 hours, then your average velocity is 60 miles per hour and, by the mean-value theorem, your speedometer must have registered 60 miles per hour at least once during your trip.

We will prove the mean-value theorem in steps. First we will show that if a function f has a nonzero derivative at some point x_0, then, close to x_0, $f(x)$ is greater than $f(x_0)$ on one side of x_0 and less than $f(x_0)$ on the other side of x_0.

THEOREM 4.1.2

Let f be differentiable at x_0. If $f'(x_0) > 0$, then

$$f(x_0 - h) < f(x_0) < f(x_0 + h)$$

for all positive h sufficiently small. If $f'(x_0) < 0$, then

$$f(x_0 + h) < f(x_0) < f(x_0 - h)$$

for all positive h sufficiently small.

PROOF   We take the case $f'(x_0) > 0$ and leave the other case to you. By the definition of the derivative,

$$\lim_{k \to 0} \frac{f(x_0 + k) - f(x_0)}{k} = f'(x_0).$$

With $f'(x_0) > 0$ we can use $f'(x_0)$ itself as ϵ and conclude that there exists $\delta > 0$ such that

$$\text{if} \quad 0 < |k| < \delta, \qquad \text{then} \quad \left| \frac{f(x_0 + k) - f(x_0)}{k} - f'(x_0) \right| < f'(x_0).$$

For such k we have

$$-f'(x_0) < \frac{f(x_0 + k) - f(x_0)}{k} - f'(x_0) < f'(x_0)$$

and it follows that

$$\frac{f(x_0 + k) - f(x_0)}{k} > 0. \qquad\qquad \text{(why?)}$$

Now let $0 < h < \delta$. Replacing k with h, we have

$$\frac{f(x_0 + h) - f(x_0)}{h} > 0 \qquad \text{and so} \qquad f(x_0) < f(x_0 + h).$$

Replacing k with $-h$, we have

$$\frac{f(x_0 - h) - f(x_0)}{-h} > 0 \qquad \text{and so} \qquad f(x_0 - h) < f(x_0). \quad \square$$

Next we prove a special case of the mean-value theorem, known as Rolle's theorem (after the French mathematician Michel Rolle (1652–1719) who first announced the result in 1691). In Rolle's theorem we make the additional assumption that $f(a)$ and $f(b)$ are both 0. (See Figure 4.1.2.) In this case the line joining $(a, f(a))$ to $(b, f(b))$ is horizontal. (It is the x-axis.) The conclusion is that there is a point $(c, f(c))$ at which the tangent is horizontal.

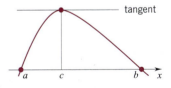

Figure 4.1.2

THEOREM 4.1.3   ROLLE'S THEOREM

Let f be differentiable on the open interval (a, b) and continuous on the closed interval $[a, b]$. If $f(a)$ and $f(b)$ are both 0, then there is at least one number c in (a, b) for which

$$f'(c) = 0.$$

PROOF   If f is constantly 0 on $[a, b]$, the result is obvious. If f is not constantly 0 on $[a, b]$, then either f takes on some positive values or some negative values. We assume the former and leave the other case to you.

Since f is continuous on $[a, b]$, f must take on a maximum value at some point c of $[a, b]$. (Theorem 2.6.2.) This maximum value, $f(c)$, must be positive. Since $f(a)$ and $f(b)$ are both 0, c cannot be a and it cannot be b. This means that c must lie in the open interval (a, b) and therefore $f'(c)$ exists. Now $f'(c)$ cannot be greater than 0 and it cannot be less than 0 because each of these conditions would imply that f takes on values greater than $f(c)$. (This follows from Theorem 4.1.2.) We conclude therefore that $f'(c) = 0$. ❑

Remark Rolle's theorem is sometimes stated as:

Let g be differentiable on the open interval (a, b) and continuous on the closed interval $[a, b]$. If $g(a) = g(b)$, then there is at least one number c in (a, b) for which

$$g'(c) = 0.$$

In Exercise 38, you are asked to show that this version of Rolle's theorem is an immediate consequence of the given version. ❑

Rolle's theorem has some useful applications independent of the mean-value theorem.

Example 1 Show that the polynomial function $p(x) = 2x^3 + 5x - 1$ has exactly one real zero.

SOLUTION From Exercise 25, Section 2.6, we know that p has at least one real zero (by the Fundamental Theorem of Algebra [see Section 1.2], p has at most three real zeros). Suppose that p has more than one real zero. In particular, suppose that $p(a) = p(b) = 0$, where a and b are real numbers and $a \neq b$. Without loss of generality, assume that $a < b$. Since a polynomial is differentiable everywhere, p is differentiable on (a, b) and continuous on $[a, b]$. Thus, by Rolle's theorem, there is a number c in (a, b) such that $p'(c) = 0$. But

$$p'(x) = 6x^2 + 5 \geq 5 \qquad \text{for all} \quad x,$$

and so $p'(x) \neq 0$ for all x. Thus, we have a contradiction and we can conclude that p has exactly one real zero. ❑

We are now ready to give a proof of the mean-value theorem.

PROOF OF THE MEAN-VALUE THEOREM. We will create a function g that satisfies the conditions of Rolle's theorem and is so related to f that the conclusion $g'(c) = 0$ leads to the conclusion

$$f'(c) = \frac{f(b) - f(a)}{b - a}.$$

It is not hard to see that

$$g(x) = f(x) - \left[\frac{f(b) - f(a)}{b - a} (x - a) + f(a) \right]$$

is exactly such a function. Geometrically $g(x)$ is represented in Figure 4.1.3. The line that passes through $(a, f(a))$ and $(b, f(b))$ has the equation

$$y = \frac{f(b) - f(a)}{b - a}(x - a) + f(a).$$

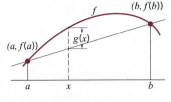

Figure 4.1.3

[This is not hard to verify. The slope is right, and, when $x = a$, $y = f(a)$.] The difference

$$g(x) = f(x) - \left[\frac{f(b) - f(a)}{b - a}(x - a) + f(a) \right]$$

is simply the vertical separation between the graph of f and the line in question.

If f is differentiable on (a, b) and continuous on $[a, b]$, then so is g. As you can check, $g(a)$ and $g(b)$ are both 0. Therefore, by Rolle's theorem, there is at least one number c in (a, b) for which $g'(c) = 0$. Since in general

$$g'(x) = f'(x) - \frac{f(b) - f(a)}{b - a},$$

in particular

$$g'(c) = f'(c) - \frac{f(b) - f(a)}{b - a}.$$

With $g'(c) = 0$, we have

$$f'(c) = \frac{f(b) - f(a)}{b - a}. \quad \square$$

The hypothesis of the mean-value theorem requires f to be differentiable on the open interval (a, b) and continuous on the closed interval $[a, b]$. As you would expect, if either of these conditions fails to hold, then the conclusion of the theorem may not hold. This is illustrated in the Exercises.

EXERCISES 4.1

In Exercises 1–4, show that f satisfies the conditions of Rolle's theorem on the indicated interval and find all numbers c on the interval such that $f'(c) = 0$.

1. $f(x) = x^3 - x$; [0, 1].

2. $f(x) = x^4 - 2x^2 - 8$; [−2, 2].

3. $f(x) = \sin 2x$; [0, 2π].

4. $f(x) = x^{2/3} - 2x^{1/3}$; [0, 8].

In Exercises 5–10, verify that f satisfies the conditions of the mean-value theorem on the indicated interval and find all numbers c that satisfy the conclusion of the theorem.

5. $f(x) = x^2$; [1, 2].

6. $f(x) = 3\sqrt{x} - 4x$; [1, 4].

7. $f(x) = x^3$; [1, 3].

8. $f(x) = x^{2/3}$; [1, 8].

9. $f(x) = \sqrt{1 - x^2}$; [0, 1].

10. $f(x) = x^3 - 3x$; [−1, 1].

11. Determine whether the function $f(x) = \sqrt{1 - x^2}/(3 + x^2)$ satisfies the conditions of Rolle's theorem on the interval [−1, 1]. If so, find the values of c such that $f(c) = 0$.

12. The function $f(x) = x^{2/3} - 1$ takes on the value 0 at $x = -1$ and at $x = 1$.

(a) Show that $f'(x) \neq 0$ for all $x \in (-1, 1)$.

(b) Explain why this function does not contradict Rolle's theorem.

13. Does there exist a differentiable function f that satisfies $f(0) = 2$, $f(2) = 5$, and $f'(x) \leq 1$ on (0, 2)? If not, why not?

14. Does there exist a differentiable function f that has the value 1 only when $x = 0, 2$, and 3, and $f'(x) = 0$ only when $x = -1, 3/4$, and 3/2? If not, why not?

15. Sketch the graph of

$$f(x) = \begin{cases} 2x + 2, & x \le -1 \\ x^3 - x, & x > -1 \end{cases}$$

and compute the derivative. Determine whether f satisfies the conditions of the mean-value theorem on the interval $[-3, 2]$ and, if so, find the values of c guaranteed by the theorem.

16. Sketch the graph of

$$f(x) = \begin{cases} 2 + x^3, & x \le 1 \\ 3x, & x > 1 \end{cases}$$

and compute the derivative. Determine whether f satisfies the conditions of the mean-value theorem on the interval $[-1, 2]$ and, if so, find the values of c guaranteed by the theorem.

17. Consider the quadratic function $f(x) = Ax^2 + Bx + C$. Show that, for any interval $[a, b]$, the value of c that satisfies the conclusion of the mean-value theorem is $(a + b)/2$, the midpoint of the interval.

18. Set $f(x) = x^{-1}$, $a = -1$, $b = 1$. Verify that there is no number c for which

$$f'(c) = \frac{f(b) - f(a)}{b - a}.$$

Explain how this does not violate the mean-value theorem.

19. Exercise 18 with $f(x) = |x|$.

20. Graph the function $f(x) = |2x - 1| - 3$ and compute the derivative. Verify that $f(-1) = 0 = f(2)$ and yet $f'(x)$ is never 0. Explain how this does not violate Rolle's theorem.

21. Show that the equation $6x^4 - 7x + 1 = 0$ does not have more than two distinct real roots. (Use Rolle's theorem.)

22. Show that the equation $6x^5 + 13x + 1 = 0$ has exactly one real root. (Use Rolle's theorem and the intermediate-value theorem.)

23. Show that the equation $x^3 + 9x^2 + 33x - 8 = 0$ has exactly one real root.

24. (a) Let f be differentiable on (a, b). Prove that if $f'(x) \ne 0$ for all $x \in (a, b)$, then f has at most one zero in (a, b).
(b) Let f be twice differentiable on (a, b). Prove that if $f''(x) \ne 0$ for all $x \in (a, b)$, then f has at most two zeros in (a, b).

25. Let $P(x) = a_n x^n + \cdots + a_1 x + a_0$ be a nonconstant polynomial. Show that between any two consecutive roots of the equation $P'(x) = 0$ there is at most one root of the equation $P(x) = 0$.

26. Let f be twice differentiable. Show that, if the equation $f(x) = 0$ has n distinct real roots, then the equation

$f'(x) = 0$ has at least $n - 1$ distinct real roots and the equation $f''(x) = 0$ has at least $n - 2$ distinct real roots.

27. Recall that a number a is a fixed point of a function f if $f(a) = a$ (see Exercises 3.9). Prove that if f is differentiable on an interval I and $f'(x) < 1$ for all $x \in I$, then f has at most one fixed point in I. HINT: Assume that f has two fixed points in I and let $g(x) = f(x) - x$.

28. Show that the equation $x^3 + ax + b = 0$ has exactly one real root if $a \ge 0$ and at most one real root between $-\frac{1}{3}\sqrt{3}|a|$ and $\frac{1}{3}\sqrt{3}|a|$ if $a < 0$.

29. Given that $|f'(x)| \le 1$ for all real numbers x, show that $|f(x_1) - f(x_2)| \le |x_1 - x_2|$ for all real numbers x_1 and x_2.

30. Let f be differentiable on an open interval I. Prove that, if $f'(x) = 0$ for all x in I, then f is constant on I.

31. Let f be differentiable on (a, b) with $f(a) = f(b) = 0$ and $f'(c) = 0$ for some c in (a, b). Show by example that f need not be continuous on $[a, b]$.

32. Prove that for all real x and y
(a) $|\cos x - \cos y| \le |x - y|$.
(b) $|\sin x - \sin y| \le |x - y|$.

33. Let f be differentiable on (a, b) and continuous on $[a, b]$.
(a) Prove that if there is a constant M such that $f'(x) \le M$ for all $x \in (a, b)$, then

$$f(b) \le f(a) + M(b - a).$$

(b) Prove that if there is a constant m such that $f'(x) \ge m$ for all $x \in (a, b)$, then

$$f(b) \ge f(a) + m(b - a).$$

(c) Parts (a) and (b) imply that if there exists a constant L such that $|f'(x)| \le L$ on (a, b), then

$$f(a) - L(b - a) \le f(b) \le f(a) + L(b - a).$$

Verify this result.

34. Let f and g be differentiable functions that satisfy the condition $f(x)g'(x) - g(x)f'(x) \ne 0$ for all x in some interval I. Suppose that $f(a) = f(b) = 0$, where $a, b \in I$, $a < b$, and $f(x) \ne 0$ for all $x \in (a, b)$. Prove that if $g(a) \ne 0$ and $g(b) \ne 0$, then $g(x) = 0$ for exactly one $x \in (a, b)$. HINT: Suppose $g(x) \ne 0$ for all $x \in (a, b)$ and consider $h = f/g$. Then reverse the roles of f and g.

35. Let $f(x) = \cos x$ and $g(x) = \sin x$ on $I = (-\infty, \infty)$. Prove that the zeros of f and g separate each other on I; that is, prove that between two consecutive zeros of $\cos x$ there is exactly one zero of $\sin x$ and conversely. HINT: Use Exercise 34.

36. (*Important*) Use the mean-value theorem to show that, if f is continuous at x and $x + h$ and differentiable in between, then

$$f(x + h) - f(x) = f'(x + \theta h)h$$

for some number θ between 0 and 1. (In some texts this is how the mean-value theorem is stated.)

37. Let $h > 0$. Suppose that f is continuous on $[x_0 - h, x_0 + h]$ and differentiable on $(x_0 - h, x_0) \cup (x_0, x_0 + h)$. Show that if

$$\lim_{x \to x_0^-} f'(x) = \lim_{x \to x_0^+} f'(x) = L,$$

then f is differentiable at x_0 and $f'(x_0) = L$. HINT: Use Exercise 36.

38. Let f be continuous on $[a, b]$ and differentiable on (a, b). Without assuming the mean-value theorem, prove that if $f(a) = f(b)$, then there is at least one number $c \in (a, b)$ for which $f'(c) = 0$. HINT: Suppose $f(a) = f(b) = k$ and let $g(x) = f(x) - k$.

39. An object moves along a coordinate line, its position at time t given by $f(t)$ where f is a differentiable function. (a) What does Rolle's theorem say about the motion? (b) What does the mean-value theorem say about the motion?

40. A certain tollway is 120 miles long and the speed limit is 65 miles per hour. If a driver's entry ticket at one end of the tollway is stamped 12 noon and she exits at the other end at 1:45 P.M., should she be given a speeding ticket? Explain.

41. The results of an investigation of a car accident showed that the driver applied his brakes and skidded 280 feet in 6 seconds. If the speed limit on the street where the accident occurred was 30 miles per hour, was the driver exceeding the speed limit at the instant he applied his brakes? Explain. HINT: 30 miles per hour = 44 feet per second.

42. Let f be a differentiable function. Apply the mean-value theorem to f with $a = x$ and $b = x + \Delta x$ to show that

$$\Delta f = f(x + \Delta x) - f(x) = f'(c)\Delta x,$$

where $x < c < x + \Delta x$. Compare this result with the result in Section 3.9. If f' is continuous and Δx is small, then $f'(c) \cong f'(x)$ and we have

$$\Delta f \cong f'(x)\Delta x$$

which is the result in Section 3.9.

43. Use the mean-value theorem to estimate $\sqrt{65}$. HINT: Let $f(x) = \sqrt{x}$, $a = 64$, $b = 65$, and use Exercise 42.

44. Let $f(x) = x^3 - 3x^2 + 6x - 12$.
(a) Show that the equation $f(x) = 0$ has exactly one solution.
(b) Find an interval that contains the solution.
(c) Use the Newton-Raphson method to approximate the solution. Calculate x_3 and round off to four decimal places.

45. Repeat Exercise 44 for the function
$$f(x) = 1 + 4x - 2\cos x.$$

In Exercises 46–47, show that the given function satisfies the hypotheses of Rolle's theorem on the indicated interval. Use a graphing utility to graph f' and estimate the number(s) c where $f'(c) = 0$. Round off your estimates to three decimal places.

46. $f(x) = 2x^3 + 3x^2 - 3x - 2$; $[-2, 1]$.

47. $f(x) = 1 - x^3 - \cos(\pi x/2)$; $[0, 1]$.

Suppose that the function f satisfies the hypotheses of the mean-value theorem on an interval $[a, b]$. We can find the numbers c that satisfy the conclusion of the mean-value theorem by finding the zeros of the function g given by

$$g(x) = f'(x) - \frac{f(b) - f(a)}{b - a}.$$

In Exercises 48–50 use a graphing utility to graph the function g corresponding to the given function f on the indicated interval. Estimate the zeros of g accurate to three decimal places. For each zero c in the interval, graph the tangent line to the graph of f at $(c, f(c))$, and graph the line through $(a, f(a))$ and $(b, f(b))$. Verify that they are parallel.

48. $f(x) = x^2 + 1/x$; $[2, 4]$.

49. $f(x) = x^4 - 7x^2 + 2$; $[1, 3]$.

50. $f(x) = x \cos x + 4 \sin x$; $[-\pi/2, \pi/2]$.

Figure 4.2.1

Figure 4.2.2

Figure 4.2.3

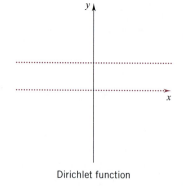

Dirichlet function

Figure 4.2.4

■ **4.2 INCREASING AND DECREASING FUNCTIONS**

To place our discussion on a solid footing we begin with a definition.

DEFINITION 4.2.1

A function f is said to:

(i) *increase* on the interval I iff for every two numbers x_1, x_2 in I,

$$x_1 < x_2 \quad \text{implies} \quad f(x_1) < f(x_2);$$

(ii) *decrease* on the interval I iff for every two numbers x_1, x_2 in I,

$$x_1 < x_2 \quad \text{implies} \quad f(x_1) > f(x_2).$$

Preliminary Examples

(a) The quadratic function

$$f(x) = x^2 \qquad \text{(Figure 4.2.1)}$$

decreases on $(-\infty, 0]$ and increases on $[0, \infty)$.

(b) The function

$$f(x) = \begin{cases} 1, & x < 0 \\ x, & x \geq 0 \end{cases} \qquad \text{(Figure 4.2.2)}$$

is constant on $(-\infty, 0)$, it neither increases nor decreases; f increases on $[0, \infty)$.

(c) The cubic function

$$f(x) = x^3 \qquad \text{(Figure 4.2.3)}$$

is everywhere increasing.

(d) In the case of the Dirichlet function

$$g(x) = \begin{cases} 1, & x \text{ rational} \\ 0, & x \text{ irrational} \end{cases} \qquad \text{(Figure 4.2.4)}$$

there is no interval on which the function increases and no interval on which the function decreases. On every interval the function jumps back and forth between 0 and 1 an infinite number of times. ❏

If f is a differentiable function, then we can determine the intervals on which f increases and the intervals on which f decreases by examining the sign of the first derivative.

THEOREM 4.2.2

Let f be differentiable on an open interval I.

(i) If $f'(x) > 0$ for all x in I, then f increases on I.
(ii) If $f'(x) < 0$ for all x in I, then f decreases on I.
(iii) If $f'(x) = 0$ for all x in I, then f is constant on I.

PROOF Choose any two points x_1 and x_2 in I with $x_1 < x_2$. Since f is differentiable on I, it is continuous on I. Therefore we know that f is differentiable on (x_1, x_2) and continuous on $[x_1, x_2]$. By the mean-value theorem there is a number c in (x_1, x_2) for which

$$f'(c) = \frac{f(x_2) - f(x_1)}{x_2 - x_1}.$$

In (i), $f'(c) > 0$ and we have

$$\frac{f(x_2) - f(x_1)}{x_2 - x_1} > 0.$$

Thus $f(x_1) < f(x_2)$. In (ii), $f'(c) < 0$ and we have

$$\frac{f(x_2) - f(x_1)}{x_2 - x_1} < 0.$$

Thus $f(x_1) > f(x_2)$. In (iii), $f'(c) = 0$ and we have

$$\frac{f(x_2) - f(x_1)}{x_2 - x_1} = 0.$$

Thus $f(x_1) = f(x_2)$. ❏

Remark In Section 3.2 we showed that if $f(x) = \alpha$, constant, on an open interval I, then $f'(x) = 0$ for all $x \in I$. Part (iii) of Theorem 4.2.2 gives the converse: if $f'(x) = 0$ for all x in an open interval I, then $f(x) = \alpha$ for some constant α. Combining these two statements, we have:

Suppose that f is differentiable on an open interval I. Then

$$f(x) = \alpha \ (\alpha \text{ a constant}) \text{ for all } x \in I \text{ iff}$$

$$f'(x) = 0 \text{ for all } x \in I.$$

❏

Theorem 4.2.2 is useful, but it doesn't tell the complete story. Look, for example, at the function $f(x) = x^2$. The derivative $f'(x) = 2x$ is negative for x in $(-\infty, 0)$, zero at $x = 0$, and positive for x in $(0, \infty)$. Theorem 4.2.2 assures us that

f decreases on $(-\infty, 0)$ and increases on $(0, \infty)$, but actually

$$f \text{ decreases on } (-\infty, 0] \text{ and increases on } [0, \infty).$$

To get these stronger results we need a theorem that works for closed intervals too.

To extend Theorem 4.2.2 so that it works for an arbitrary interval I, the only additional condition we need is continuity at the endpoint(s).

THEOREM 4.2.3

Let f be continuous on an arbitrary interval I and differentiable on the interior of I.

  (i) If $f'(x) > 0$ for all x in the interior of I, then f increases on all of I.
  (ii) If $f'(x) < 0$ for all x in the interior of I, then f decreases on all of I.
  (iii) If $f'(x) = 0$ for all x in the interior of I, then f is constant on all of I.

A proof of this result is not hard to construct. As you can verify, a word-for-word repetition of our proof of Theorem 4.2.2 also works here.

It is time for some examples.

Example 1 The function

$$f(x) = \sqrt{1 - x^2}$$

has derivative

$$f'(x) = -\frac{x}{\sqrt{1 - x^2}}.$$

Since $f'(x) > 0$ for all x in $(-1, 0)$ and f is continuous on $[-1, 0]$, f increases on $[-1, 0]$. Since $f'(x) < 0$ for all x in $(0, 1)$ and f is continuous on $[0, 1]$, f decreases on $[0, 1]$. The graph of f is a semicircle. (Figure 4.2.5) ❑

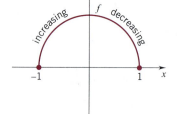

Figure 4.2.5

Example 2 The function

$$f(x) = \frac{1}{x}$$

is defined for all $x \neq 0$. The derivative

$$f'(x) = -\frac{1}{x^2}$$

is negative for all $x \neq 0$. Thus the function f decreases both on $(-\infty, 0)$ and on $(0, \infty)$. (Figure 4.2.6) ❑

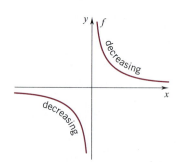

Figure 4.2.6

Example 3 The function

$$g(x) = \tfrac{4}{5}x^5 - 3x^4 - 4x^3 + 22x^2 - 24x + 6$$

is a polynomial. It is therefore everywhere continuous and everywhere differentiable. Differentiation gives

$$g'(x) = 4x^4 - 12x^3 - 12x^2 + 44x - 24$$
$$= 4(x^4 - 3x^3 - 3x^2 + 11x - 6)$$
$$= 4(x + 2)(x - 1)^2(x - 3).$$

The derivative g' takes on the value 0 at -2, at 1, and at 3. These numbers determine four intervals on which g' keeps a constant sign (see Section 1.3):

$$(-\infty, -2), \qquad (-2, 1), \qquad (1, 3), \qquad (3, \infty).$$

The sign of g' on these intervals and the consequences for g are as follows:

Since g is everywhere continuous, g increases on $(-\infty, -2]$, decreases on $[-2, 3]$, and increases on $[3, \infty)$. (Figure 4.2.7) ❑

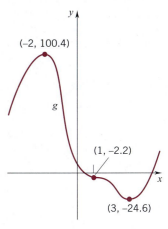

Figure 4.2.7

Example 4 Let $f(x) = x - 2 \sin x$, $0 \leq x \leq 2\pi$. Find the intervals on which f increases and the intervals on which f decreases.

SOLUTION The derivative of f is

$$f'(x) = 1 - 2 \cos x.$$

Setting $f'(x) = 0$, we have

$$1 - 2 \cos x = 0$$

$$\cos x = \frac{1}{2}.$$

The solutions are $x = \pi/3$ and $x = 5\pi/3$. Thus, f' has constant sign on the intervals $(0, \pi/3)$, $(\pi/3, 5\pi/3)$, and $(5\pi/3, 2\pi)$. The sign of f' and the behavior of f are recorded below.

Since f is continuous on $[0, 2\pi]$, f decreases on $[0, \pi/3]$, increases on $[\pi/3, 5\pi/3]$, and decreases on $[5\pi/3, 2\pi]$. (Figure 4.2.8) ❑

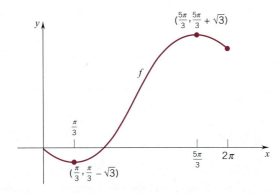

Figure 4.2.8

While Theorems 4.2.2 and 4.2.3 have wide applicability, there are also some limitations. If, for example, f is discontinuous at some point in its domain, then Theorems 4.2.2 and 4.2.3 do not tell the whole story.

Example 5 The function

$$f(x) = \begin{cases} x^3, & x < 1 \\ \frac{1}{2}x + 2, & x \geq 1 \end{cases}$$

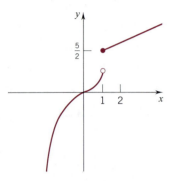

is graphed in Figure 4.2.9. Obviously there is a discontinuity at $x = 1$. Differentiation gives

$$f'(x) = \begin{cases} 3x^2, & x < 1 \\ \text{does not exist}, & x = 1 \\ \frac{1}{2}, & x > 1. \end{cases}$$

Since $f'(x) > 0$ on $(-\infty, 0) \cup (0, 1)$, we know from Theorem 4.2.2 that f increases on $(-\infty, 1)$. Since $f'(x) > 0$ on $(1, \infty)$ and f is continuous on $[1, \infty)$, we know from Theorem 4.2.3 that f increases on $[1, \infty)$. The obvious fact that f increases on all of $(-\infty, \infty)$ is not derivable from the theorems.

Figure 4.2.9

Now consider the function

$$g(x) = \begin{cases} \frac{1}{2}x + 2, & x < 1 \\ x^3, & x \geq 1, \end{cases}$$

which is graphed in Figure 4.2.10. Again, there is a discontinuity at $x = 1$. The derivative of g is

$$g'(x) = \begin{cases} \frac{1}{2}, & x < 1 \\ \text{does not exist}, & x = 1 \\ 3x^2, & x > 1, \end{cases}$$

so g increases on $(-\infty, 1)$ and on $[1, \infty)$ but, clearly, g does not increase on $(-\infty, \infty)$. ❑

Figure 4.2.10

Equality of Derivatives

If two differentiable functions differ by a constant,

$$f(x) = g(x) + C,$$

then their derivatives are equal:

$$f'(x) = g'(x).$$

The converse is also true. In fact, we have the following theorem.

THEOREM 4.2.4

(i) Let I be an open interval. If $f'(x) = g'(x)$ for all x in I, then f and g differ by a constant on I.

(ii) Let I be an arbitrary interval. If $f'(x) = g'(x)$ for all x in the interior of I and f and g are continuous on I, then f and g differ by a constant on I.

PROOF Set $H = f - g$. For the first assertion apply (iii) of Theorem 4.2.2 to H. For the second assertion apply (iii) of Theorem 4.2.3 to H. We leave the details as an exercise. ❏

We illustrate the theorem in Figure 4.2.11. At points with the same x-coordinate the slopes are equal, and thus the curves have the same steepness. The separation between the curves remains constant; the curves are "parallel."

Example 6 Find f given that

$$f'(x) = 6x^2 - 7x - 5 \quad \text{for all real } x \quad \text{and} \quad f(2) = 1.$$

SOLUTION It is not hard to find a function that has $6x^2 - 7x - 5$ as its derivative:

$$\frac{d}{dx}\left(2x^3 - \frac{7}{2}x^2 - 5x\right) = 6x^2 - 7x - 5.$$

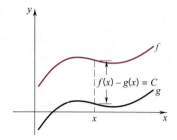

Figure 4.2.11

By Theorem 4.2.4 we know that $f(x)$ differs from $2x^3 - \frac{7}{2}x^2 - 5x$ only by some constant C. Thus we can write

$$f(x) = 2x^3 - \frac{7}{2}x^2 - 5x + C.$$

To evaluate C we use the fact that $f(2) = 1$. Since $f(2) = 1$ and also

$$f(2) = 2(2)^3 - \frac{7}{2}(2)^2 - 5(2) + C = 16 - 14 - 10 + C = -8 + C,$$

we have $-8 + C = 1$, and therefore $C = 9$. Thus

$$f(x) = 2x^3 - \frac{7}{2}x^2 - 5x + 9. \quad ❏$$

EXERCISES 4.2

In Exercises 1–27, find the intervals on which f increases and the intervals on which f decreases.

1. $f(x) = x^3 - 3x + 2$.

2. $f(x) = x^3 - 3x^2 + 6$.

3. $f(x) = x + \dfrac{1}{x}$.

4. $f(x) = (x - 3)^3$.

5. $f(x) = x^3(1 + x)$.

6. $f(x) = x(x + 1)(x + 2)$.

7. $f(x) = (x + 1)^4$.

8. $f(x) = 2x - \dfrac{1}{x^2}$.

9. $f(x) = \dfrac{1}{|x - 2|}$.

10. $f(x) = \dfrac{x}{1 + x^2}$.

11. $f(x) = \dfrac{x^2 + 1}{x^2 - 1}$.

12. $f(x) = \dfrac{x^2}{x^2 + 1}$.

13. $f(x) = |x^2 - 5|$.

14. $f(x) = x^2(1 + x)^2$.

15. $f(x) = \dfrac{x - 1}{x + 1}$.

16. $f(x) = x^2 + \dfrac{16}{x^2}$.

17. $f(x) = \left(\dfrac{1 - \sqrt{x}}{1 + \sqrt{x}}\right)^7$.

18. $f(x) = \sqrt{\dfrac{2 + x}{1 + x}}$.

19. $f(x) = \sqrt{\dfrac{1 + x^2}{2 + x^2}}$.

20. $f(x) = |x + 1||x - 2|$.

21. $f(x) = \sqrt{\dfrac{3 - x}{x}}$.

22. $f(x) = x + \sin x, \quad 0 \le x \le 2\pi$.

23. $f(x) = x - \cos x, \quad 0 \le x \le 2\pi$.

24. $f(x) = \cos^2 x, \quad 0 \le x \le \pi$.

25. $f(x) = \cos 2x + 2 \cos x, \quad 0 \le x \le \pi$.

26. $f(x) = \sin^2 x - \sqrt{3} \sin x, \quad 0 \le x \le \pi$.

27. $f(x) = \sqrt{3} x - \cos 2x, \quad 0 \le x \le \pi$.

In Exercises 28–36, find f given the following information.

28. $f'(x) = x^2 - 1$ for all real $x, \quad f(0) = 1$.

29. $f'(x) = x^2 - 1$ for all real $x, \quad f(1) = 2$.

30. $f'(x) = 2x - 5$ for all real $x, \quad f(2) = 4$.

31. $f'(x) = 5x^4 + 4x^3 + 3x^2 + 2x + 1$ for all real x, $f(0) = 5$.

32. $f'(x) = 4x^{-3}$ for $x > 0, \quad f(1) = 0$.

33. $f'(x) = x^{1/3} - x^{1/2}$ for $x > 0$, $f(0) = 1$.

34. $f'(x) = x^{-5} - 5x^{-1/5}$ for $x > 0$, $f(1) = 0$.

35. $f'(x) = 2 + \sin x$ for all real x, $f(0) = 3$.

36. $f'(x) = 4x + \cos x$ for all real x, $f(0) = 1$.

In Exercises 37–40, find the intervals on which f increases and the intervals on which f decreases.

37. $f(x) = \begin{cases} x + 7, & x < -3 \\ |x + 1|, & -3 \le x < 1 \\ 5 - 2x, & 1 \le x. \end{cases}$

38. $f(x) = \begin{cases} (x - 1)^2, & x < 1 \\ 5 - x, & 1 \le x < 3 \\ 7 - 2x, & 3 \le x. \end{cases}$

39. $f(x) = \begin{cases} 4 - x^2, & x < 1 \\ 7 - 2x, & 1 \le x < 3 \\ 3x - 10, & 3 \le x. \end{cases}$

40. $f(x) = \begin{cases} x + 2, & x < 0 \\ (x - 1)^2, & 0 < x < 3 \\ 8 - x, & 3 < x < 7 \\ 2x - 9, & 7 < x \\ 6, & x = 0, 3, 7. \end{cases}$

In Exercises 41–44, f is a differentiable function and the graph of its derivative f' is given. If $f(0) = 1$, give a rough sketch of the graph of f.

41.

42.

43.

44.

In Exercises 45 and 46, the graph of a continuous function f is given. Sketch the graph of f'.

45.

46.

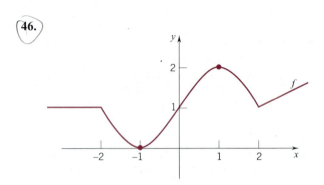

In Exercises 47–50, sketch the graph of a differentiable function f that satisfies the given conditions, if possible. If it is not possible, explain why.

47. $f(x) > 0$ for all x, $f(0) = 1$, and $f'(x) < 0$ for all x.

48. $f(1) = -1$, $f'(x) < 0$ for all $x \ne 1$, and $f'(1) = 0$.

49. $f(-1) = 4$, $f(2) = 2$, and $f'(x) > 0$ for all x.

50. f has x-intercepts only at $x = 1$ and $x = 2$, $f(3) = 4$, and $f(5) = -1$.

In Exercises 51–54, an object is moving along a straight line with its position at time t given by $x(t)$. Determine the time interval(s) when the object is moving to the right, determine the time interval(s) when it is moving to the left, and find the time(s) when it changes direction. Also, find the time interval(s) when the velocity of the object is increasing and when it is decreasing. Draw a figure that illustrates the motion of the object. See Example 3, Section 3.4.

51. $x(t) = t^3 - 6t^2 + 9t + 2$.

52. $x(t) = (2t - 1)(t - 1)^2$.

53. $x(t) = 2 \sin 3t$, $t \in [0, \pi]$.

54. $x(t) = 3 \cos (2t + \frac{\pi}{4})$, $t \in [0, 2\pi]$.

55. Suppose that the function f increases on (a, b) and on (b, c). Suppose further that $\lim\limits_{x \to b^-} f(x) = M$, $\lim\limits_{x \to b^+} f(x) = N$ and $f(b) = L$. For which values of L, if any, can you conclude that f increases on (a, c) if (a) $M < N$? (b) $M > N$? (c) $M = N$?

56. Suppose that the function f increases on (a, b) and decreases on (b, c). Under what conditions can you conclude that f increases on $(a, b]$ and decreases on $[b, c)$?

57. Give an example of a function f that satisfies all of the following conditions: (i) f is defined for all real x, (ii) f' is positive wherever it exists, (iii) there is no interval on which f increases.

58. Prove Theorem 4.2.4.

59. Let $f(x) = \sec^2 x$ and $g(x) = \tan^2 x$ on the interval $I = (-\pi/2, \pi/2)$.
(a) Show that $f'(x) = g'(x)$ for all $x \in I$.
(b) The result in part (a) implies, by Theorem 4.2.4, that $f - g = C$, constant, on $(-\pi/2, \pi/2)$. Find the value of C.

60. Let L be a differentiable function on $(0, \infty)$ such that $L'(x) = 1/x$ and $L(1) = 0$. This function is studied in detail in Chapter 7. Prove that for any two positive numbers a and b, $L(ab) = L(a) + L(b)$. HINT: Begin by showing that, for any $x > 0$, $L(ax)$ and $L(x)$ have the same derivative.

61. Let f and g be differentiable functions such that $f'(x) = -g(x)$ and $g'(x) = f(x)$ for all x.
(a) Prove that $f^2(x) + g^2(x) = C$, constant.
(b) Suppose there is a number a such that $f(a) = 1$ and $g(a) = 0$. Use this information to find the value of C. Can you name a pair of functions f, g that have these properties?

62. Prove that if f is differentiable on an interval I and $f'(x) \geq 0$ (≤ 0) on I with $f'(x) = 0$ for at most finitely many values of $x \in I$, then f increases (decreases) on I.

63. Let $f(x) = x - \sin x$.
(a) Prove that f is increasing on $(-\infty, \infty)$.
(b) Use the result in part (a) to show that $\sin x < x$ on $(0, \infty)$ and $\sin x > x$ on $(-\infty, 0)$.

64. Let f and g be differentiable functions on the interval $(-c, c)$ such that $f(0) = g(0)$.
(a) Prove that if $f'(x) > g'(x)$ for all $x \in (0, c)$, then $f(x) > g(x)$ for all $x \in (0, c)$. HINT: Set $F = f - g$ and use Theorem 4.2.2.
(b) Prove that if $f'(x) > g'(x)$ for all $x \in (-c, 0)$, then $f(x) < g(x)$ for all $x \in (-c, 0)$.

65. Prove that $\tan x > x$ for all $x \in (0, \pi/2)$. HINT: Use Exercise 64(a).

66. Prove that $1 - x^2/2 < \cos x$ for all $x \in (0, \infty)$. HINT: Use Exercises 63 and 64.

67. Let $n > 1$ be an integer. Prove that $(1 + x)^n > 1 + nx$ for all $x \in (0, \infty)$.

68. Prove that $x - x^3/6 < \sin x$ for all $x \in (0, \infty)$.

▷ 69. It follows from Exercises 63 and 68 that
$$x - \frac{1}{6}x^3 < \sin x < x \qquad \text{for } x > 0.$$
Use this result to estimate $\sin 4°$.

▷ In Exercises 70–73, use a graphing utility to graph the function f and its derivative f' on the indicated interval. Estimate the zeros of f' correct to three decimal places. Then estimate the subintervals on which f increases and the subintervals on which f decreases.

70. $f(x) = 2x^3 - x^2 - 13x - 6$; $[-3, 4]$.
71. $f(x) = 3x^4 - 10x^3 - 4x^2 + 10x + 9$; $[-2, 5]$.
72. $f(x) = x^4 + 3x^3 - 2x^2 + 4x + 4$; $[-5, 3]$.
73. $f(x) = x \cos x - 3 \sin 2x$; $[0, 6]$.

Energy of a Falling Body (Near the Surface of the Earth)

If we lift an object, we counteract the force of gravity. In so doing, we increase what physicists call the *gravitational potential energy* of the object. The gravitational potential energy, GPE, of an object is defined by

$$\text{GPE} = \text{weight} \times \text{height}.$$

Since the weight of an object of mass m is mg (we take this from physics), where g is the gravitational constant, we can write

$$\text{GPE} = mgy,$$

where y is the height.

If we lift an object and release it, the object drops. As it drops, it loses height and therefore loses gravitational potential energy, but it picks up speed. The speed with which the object falls gives the object another form of energy called *kinetic energy*, the energy of motion. The kinetic energy KE of an object in straight-line motion is given by

$$\text{KE} = \frac{1}{2} mv^2$$

where v is the velocity of the object.

74. Prove the Law of Conservation of Energy: GPE + KE = C, constant. HINT: Differentiate the expression GPE + KE and use the fact that $dv/dt = -g$.

75. An object initially at rest falls freely from height y_0. Show that the speed of the object at height y is given by
$$|v| = \sqrt{2g(y_0 - y)}.$$

76. According to the results in Section 3.4, the position of an object that falls from rest from a height y_0 is given by
$$y(t) = -\tfrac{1}{2}gt^2 + y_0.$$
Calculate the speed of the object starting from this equation and show that the result is equivalent to the result found in Exercise 75.

77. A bobsled run descends from a height of 150 meters from its starting point to the bottom of the run. Suppose that a bobsled starts from rest at the top and slides down the run without friction. Find the speed of the bobsled at the bottom of the run (use $g = 9.8$ meters per second per second). (NOTE: To solve this problem by calculating the forces and acceleration would require a detailed knowledge of the path of the run down the hill. The Law of Conservation of Energy provides a simple method for finding the speed at the bottom.)

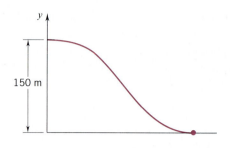

■ 4.3 LOCAL EXTREME VALUES

In many problems of economics, engineering, and physics it is important to determine how large or how small a certain quantity can be. If the problem admits a mathematical formulation, it is often reducible to the problem of finding the maximum or minimum value of some function.

In this section we consider maximum and minimum values for functions defined on an *open interval* or on a *union of open intervals*. We will take up functions defined on closed or half-closed intervals in the next section. We begin with a definition.

DEFINITION 4.3.1 LOCAL EXTREME VALUES

A function f is said to have a *local maximum at c* iff

$$f(c) \geq f(x) \qquad \text{for all } x \text{ sufficiently close to } c.\dagger$$

A function f is said to have a *local minimum at c* iff

$$f(c) \leq f(x) \qquad \text{for all } x \text{ sufficiently close to } c.$$

The local maxima and local minima of f are called the *local extreme values of f.*

We illustrate these notions in Figure 4.3.1. A careful look at the figure suggests that local maxima and minima occur only at points where the tangent is horizontal $[f'(c) = 0]$ or where there is no tangent line $[f'(c)$ does not exist]. This is indeed the case.

---

† What do we mean by saying that "such and such is true *for all x sufficiently close to c*"? We mean that it is true for all x in some open interval $(c - \delta, c + \delta)$ centered at c.

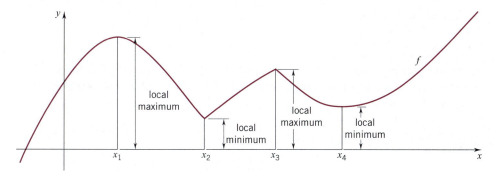

Figure 4.3.1

THEOREM 4.3.2

If f has a local maximum or minimum at c, then either

$$f'(c) = 0 \quad \text{or} \quad f'(c) \text{ does not exist.}$$

PROOF We are given that f has a local extreme value at c. Suppose that $f'(c)$ exists. If $f'(c) > 0$ or $f'(c) < 0$, then, by Theorem 4.1.2, there must be numbers x_1 and x_2 that are arbitrarily close to c which satisfy

$$f(x_1) < f(c) < f(x_2).$$

This makes it impossible for a local maximum or a local minimum to occur at c. Therefore, if $f'(c)$ exists, then it must have the value 0. The only other possibility is that $f'(c)$ does not exist. ❑

Based on this result, we have the following important definition:

DEFINITION 4.3.3

Given a function f. The numbers c in the domain of f for which either

$$f'(c) = 0 \quad \text{or} \quad f'(c) \text{ does not exist}$$

are called *critical numbers of f.*†

In view of Theorem 4.3.2, when searching for the local maxima and minima of a function f, the only numbers that we need to consider are the critical numbers of f.

We illustrate the technique for finding local maxima and minima by some examples. In each example the first step will be to find the critical numbers. We begin with very simple cases.

Example 1 For

$$f(x) = 3 - x^2 \qquad \text{(Figure 4.3.2)}$$

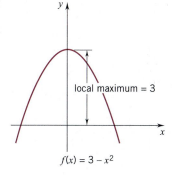

$f(x) = 3 - x^2$

Figure 4.3.2

---

† The terms *critical values* and *critical points* are also used.

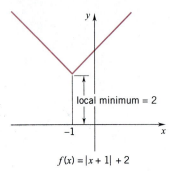

$f(x) = |x + 1| + 2$

Figure 4.3.3

the derivative

$$f'(x) = -2x$$

exists everywhere. Since $f'(x) = 0$ only at $x = 0$, 0 is the only critical number. The number $f(0) = 3$ is obviously a local maximum since the graph of f is a parabola that opens down. ❑

Example 2 In the case of

$$f(x) = |x + 1| + 2 = \begin{cases} -x + 1, & x < -1 \\ x + 3, & x \geq -1, \end{cases} \qquad \text{(Figure 4.3.3)}$$

differentiation gives

$$f'(x) = \begin{cases} -1, & x < -1 \\ 1, & x > -1. \end{cases}$$

This derivative is never 0. It fails to exist only at -1. The number -1 is the only critical number. The value $f(-1) = 2$ is a local minimum. ❑

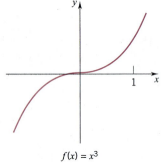

$f(x) = \frac{1}{x-1}$

Figure 4.3.4

Example 3 Figure 4.3.4 shows the graph of the function

$$f(x) = \frac{1}{x - 1}.$$

The domain is $(-\infty, 1) \cup (1, \infty)$. The derivative

$$f'(x) = -\frac{1}{(x - 1)^2}$$

exists throughout the domain of f and is never 0. Thus there are no critical numbers. In particular, 1 is not a critical number of f because 1 is not in the domain of f. Since f has no critical numbers, there are no local extreme values. ❑

CAUTION The fact that c is a critical number of f does not guarantee that $f(c)$ is a local extreme value. This is illustrated in the next two examples.

$f(x) = x^3$

Figure 4.3.5

Example 4 In the case of the function

$$f(x) = x^3, \qquad \text{(Figure 4.3.5)}$$

the derivative $f'(x) = 3x^2$ is 0 at 0, but $f(0) = 0$ is not a local extreme value. This function is everywhere increasing. ❑

Example 5 The function

$$f(x) = \begin{cases} -2x + 5 & x < 2 \\ -\frac{1}{2}x + 2 & x \geq 2 \end{cases} \qquad \text{(Figure 4.3.6)}$$

is everywhere decreasing. Although 2 is a critical number ($f'(2)$ does not exist), $f(2) = 1$ is not a local extreme value. ❑

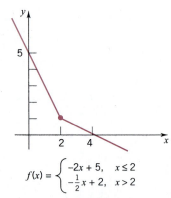

$f(x) = \begin{cases} -2x + 5, & x \leq 2 \\ -\frac{1}{2}x + 2, & x > 2 \end{cases}$

Figure 4.3.6

There are two widely used tests for determining the behavior of a function at a critical number. The first test (given in Theorem 4.3.4) requires that we examine the sign of the first derivative on both sides of the critical number. The second test (given in Theorem 4.3.5) requires that we examine the sign of the second derivative at the critical number itself.

THEOREM 4.3.4   THE FIRST DERIVATIVE TEST

Suppose that c is a critical number of f and that f is continuous at c. If there is a positive number δ such that:

(i) $f'(x) > 0$ for all x in $(c - \delta, c)$ and $f'(x) < 0$ for all x in $(c, c + \delta)$, then $f(c)$ is a local maximum.   (Figures 4.3.7 and 4.3.8.)

(ii) $f'(x) < 0$ for all x in $(c - \delta, c)$ and $f'(x) > 0$ for all x in $(c, c + \delta)$, then $f(c)$ is a local minimum.   (Figures 4.3.9 and 4.3.10.)

(iii) $f'(x)$ keeps constant sign on $(c - \delta, c) \cup (c, c + \delta)$, then $f(c)$ is not a local extreme value.

Figure 4.3.7

Figure 4.3.8

Figure 4.3.9

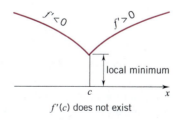

Figure 4.3.10

PROOF   The result is a direct consequence of Theorem 4.2.3. The details of the proof are left to you as Exercise 31.   ❑

Example 6   The function

$$f(x) = x^4 - 2x^3, \qquad \text{all real } x$$

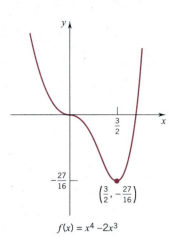

$$f(x) = x^4 - 2x^3$$

Figure 4.3.11

has derivative

$$f'(x) = 4x^3 - 6x^2 = 2x^2(2x - 3), \qquad \text{all real } x.$$

The only critical numbers are 0 and $\frac{3}{2}$. The sign of f' is recorded below.

Since f' keeps the same sign on both sides of 0, $f(0) = 0$ is not a local extreme value. However, $f(\frac{3}{2}) = -\frac{27}{16}$ is a local minimum. The graph of f appears in Figure 4.3.11. ❑

Example 7 The function

$$f(x) = 2x^{5/3} + 5x^{2/3},$$

is defined for all real x. The derivative of f is given by

$$f'(x) = \frac{10}{3}x^{2/3} + \frac{10}{3}x^{-1/3} = \frac{10}{3}x^{-1/3}(x + 1), \qquad x \neq 0.$$

Since $f'(-1) = 0$ and $f'(0)$ does not exist, the critical numbers are -1 and 0. The sign of f' is recorded below. (To save space in the diagram we write "dne" for "does not exist.")

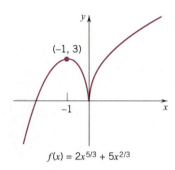

$$f(x) = 2x^{5/3} + 5x^{2/3}$$

Figure 4.3.12

In this case $f(-1) = 3$ is a local maximum and $f(0) = 0$ is a local minimum. The graph appears in Figure 4.3.12. ❑

Remark The requirement that f be continuous at c in the hypothesis of Theorem 4.3.4 is necessary. For example, 1 is a critical number of the function

$$f(x) = \begin{cases} 1 + 2x, & x \leq 1 \\ 5 - x, & x > 1 \end{cases} \qquad \text{(Figure 4.3.13)}$$

since $f'(1)$ does not exist. Note that $f'(x) = 2 > 0$ to the left of 1 and $f'(x) = -1 < 0$ to the right of 1, but $f(1) = 3$ is not a local maximum of f. ❑

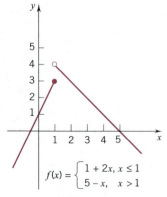

$$f(x) = \begin{cases} 1 + 2x, & x \leq 1 \\ 5 - x, & x > 1 \end{cases}$$

Figure 4.3.13

In some situations it might be difficult to determine the sign of the first derivative on one side or the other of a critical number c. For example, $x = \frac{2}{3}$ is a critical number of the polynomial function

$$p(x) = \tfrac{3}{5}x^5 + x^4 - \tfrac{1}{3}x^3 - \tfrac{1}{2}x^2 - \tfrac{2}{3}x + 1$$

since

$$p'(x) = 3x^4 + 4x^3 - x^2 - x - \tfrac{2}{3} \qquad \text{and} \qquad p'(\tfrac{2}{3}) = 0.$$

Examining the sign of p' on either side of $\frac{2}{3}$ will be difficult, especially since we do not know what the other critical numbers of p are.

If c is a critical number for a function f and if f is twice differentiable at c, then the following test is sometimes easier to apply.

THEOREM 4.3.5 **THE SECOND-DERIVATIVE TEST**

Suppose that $f'(c) = 0$ and that $f''(c)$ exists.

If $f''(c) > 0$, then $f(c)$ is a local minimum value.
If $f''(c) < 0$, then $f(c)$ is a local maximum value.

PROOF   We handle the case $f''(c) > 0$.   The other is left as an exercise.   Since f'' is the derivative of f', we see from Theorem 4.1.2 that there exists $\delta > 0$ such that, if

$$c - \delta < x_1 < c < x_2 < c + \delta,$$

then

$$f'(x_1) < f'(c) < f'(x_2).$$

Since $f'(c) = 0$, we have

$$f'(x) < 0 \quad \text{for } x \text{ in } (c - \delta, c) \qquad \text{and} \qquad f'(x) > 0 \quad \text{for } x \text{ in } (c, c + \delta).$$

By the first-derivative test $f(c)$ is a local minimum.   ❑

Continuing with the illustration begun earlier,

$$p''(x) = 12x^3 + 12x^2 - 2x - 1$$

and

$$p''(\tfrac{2}{3}) = 12(\tfrac{2}{3})^3 + 12(\tfrac{2}{3})^2 - 2(\tfrac{2}{3}) - 1 = \tfrac{59}{9} > 0.$$

Thus p has a local minimum at $x = \tfrac{2}{3}$.

Example 8   For

$$f(x) = 2x^3 - 3x^2 - 12x + 5$$

we have

$$f'(x) = 6x^2 - 6x - 12 = 6(x^2 - x - 2) = 6(x - 2)(x + 1)$$

and

$$f''(x) = 12x - 6.$$

The critical numbers are 2 and -1; the first derivative is 0 at each of these points. Since $f''(2) = 18 > 0$ and $f''(-1) = -18 < 0$, we can conclude from the second-derivative test that $f(2) = -15$ is a local minimum and $f(-1) = 12$ is a local maximum.   ❑

We can apply the first-derivative test even at points where the function is not differentiable, provided that it is continuous there.   On the other hand, the second-derivative test can only be applied at points where the function is twice differentiable and then only if the second derivative is different from zero.   This is a drawback.

Take, for example, the function $f(x) = x^{4/3}$. We have $f'(x) = \frac{4}{3}x^{1/3}$ so that

$$f'(0) = 0, \qquad f'(x) < 0 \quad \text{for } x < 0, \qquad f'(x) > 0 \quad \text{for } x > 0.$$

By the first-derivative test, $f(0) = 0$ is a local minimum. We cannot get this information from the second-derivative test because $f''(x) = \frac{4}{9}x^{-2/3}$ does not exist at $x = 0$.

To show what happens if the second derivative equals zero at a critical number c, consider the functions

$$f(x) = x^3, \qquad g(x) = x^4 \quad \text{and} \quad h(x) = -x^4. \qquad \text{(Figure 4.3.14)}$$

In the case of the function $f(x) = x^3$, we already know that $x = 0$ is a critical point and that f has neither a local maximum nor a local minimum value at 0 (see Example 4). Note that $f''(x) = 6x$ and $f''(0) = 0$. In the case of $g(x) = x^4$, we have $g'(x) = 4x^3$ so that

$$g'(0) = 0, \qquad g'(x) < 0 \quad \text{for } x < 0, \qquad g'(x) > 0 \quad \text{for } x > 0.$$

By the first-derivative test, $g(0) = 0$ is a local minimum. But here again the second-derivative test is of no avail: $g''(x) = 12x^2$ is 0 at $x = 0$.

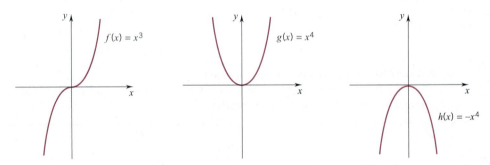

Figure 4.3.14

Finally, since the function h is the negative of the function g, it is clear that $h'(0) = h''(0) = 0$ and h has a local maximum at $x = 0$. ❑

EXERCISES 4.3

In Exercises 1–30, find the critical numbers of f and the local extreme values.

1. $f(x) = x^3 + 3x - 2.$

2. $f(x) = 2x^4 - 4x^2 + 6.$

3. $f(x) = x + \dfrac{1}{x}.$

4. $f(x) = x^2 - \dfrac{3}{x^2}.$

5. $f(x) = x^2(1 - x).$

6. $f(x) = (1 - x)^2(1 + x).$

7. $f(x) = \dfrac{1 + x}{1 - x}.$

8. $f(x) = \dfrac{2 - 3x}{2 + x}.$

9. $f(x) = \dfrac{2}{x(x + 1)}.$

10. $f(x) = |x^2 - 16|.$

11. $f(x) = x^3(1 - x)^2.$

12. $f(x) = \left(\dfrac{x - 2}{x + 2}\right)^3.$

13. $f(x) = (1 - 2x)(x - 1)^3.$

14. $f(x) = (1 - x)(1 + x)^3.$

15. $f(x) = \dfrac{x^2}{1 + x}.$

16. $f(x) = \dfrac{|x|}{1 + |x|}.$

17. $f(x) = |x - 1||x + 2|.$

18. $f(x) = x\sqrt[3]{1 - x}.$

19. $f(x) = x^2\sqrt[3]{2 + x}.$

20. $f(x) = \dfrac{1}{x + 1} - \dfrac{1}{x - 2}.$

21. $f(x) = |x - 3| + |2x + 1|.$

22. $f(x) = x^{7/3} - 7x^{1/3}.$

23. $f(x) = x^{2/3} + 2x^{-1/3}.$

24. $f(x) = \dfrac{x^3}{x + 1}.$

25. $f(x) = \sin x + \cos x, \quad 0 < x < 2\pi.$

26. $f(x) = x + \cos 2x, \quad 0 < x < \pi$.

27. $f(x) = \sin^2 x - \sqrt{3} \sin x, \quad 0 < x < \pi$.

28. $f(x) = \sin^2 x, \quad 0 < x < 2\pi$.

29. $f(x) = \sin x \cos x - 3 \sin x + 2x, \quad 0 < x < 2\pi$

30. $f(x) = 2 \sin^3 x - 3 \sin x, \quad 0 < x < \pi$.

31. Prove Theorem 4.3.4 by applying Theorem 4.2.3.

32. Prove the validity of the second-derivative test in the case that $f''(c) < 0$.

33. Find the critical numbers and the local extreme values of the polynomial

$$P(x) = x^4 - 8x^3 + 22x^2 - 24x + 4.$$

Then show that the equation $P(x) = 0$ has exactly two real roots, both positive.

34. A function f has derivative f' given by

$$f'(x) = x^3(x - 1)^2(x + 1)(x - 2).$$

At what numbers x, if any, does f have a local maximum? A local minimum?

35. A polynomial function $p(x) = a_n x^n + a_{n-1} x^{n-1} + \cdots + a_1 x + a_0$ has critical numbers at $x = -1, 1, 2,$ and 3, and corresponding values $p(-1) = 6$, $p(1) = 1$, $p(2) = 3$, and $p(3) = 1$. Sketch a possible graph for p if:
(a) n is odd, (b) n is even.

36. The quadratic function $f(x) = Ax^2 + Bx + C$ has a local minimum at $x = 2$ and passes through the points $(-1, 3)$ and $(3, -1)$. Find A, B, and C.

37. Determine a and b such that the function $f(x) = ax/(x^2 + b^2)$ has a local minimum at $x = -2$ and $f'(0) = 1$.

38. Let $f(x) = x^p(1 - x)^q$, where $p \geq 2$ and $q \geq 2$ are integers.
(a) Show that the critical numbers of f are $x = 0$, $p/(p + q)$, and 1.
(b) Show that if p is even, then f has a local minimum at 0.
(c) Show that if q is even, then f has a local minimum at 1.
(d) Show that f has a local maximum at $p/(p + q)$ for all p and q.

39. Let

$$f(x) = \begin{cases} x^2 \sin(1/x), & x \neq 0 \\ 0, & x = 0. \end{cases}$$

In Exercise 67, Section 3.1, we saw that f is differentiable at 0 and that $f'(0) = 0$. Show that f has neither a local maximum nor a local minimum at 0.

40. Suppose that $C(x)$, $R(x)$, and $P(x)$ are the cost, revenue, and profit functions corresponding to the production and sale of x units of a certain item. Suppose, also, that C and R are differentiable functions. Then, since $P = R - C$, it follows that P is differentiable. Prove that if it is possible

to maximize the profit by producing and selling x_0 items, then $C'(x_0) = R'(x_0)$. That is, the marginal cost equals the marginal revenue when the profit is maximized.

41. Let $y = f(x)$ be differentiable and suppose that the graph of f does not pass through the origin. Then the distance D from the origin to a point $P(x, f(x))$ on the graph is given by

$$D = \sqrt{x^2 + [f(x)]^2}.$$

Show that if D has a local extreme value at c, then the line through $(0, 0)$ and $(c, f(c))$ is perpendicular to the tangent line to the graph of f at c.

42. Prove that a polynomial of degree n has at most $n - 1$ local extreme values.

▶ 43. Let $f(x) = x^4 - 2x^2 - 3x + 2$.
(a) Show that f has exactly one critical number c in the interval $(1, 2)$.
(b) Use the bisection method (see Section 2.6) to approximate c to within $\frac{1}{16}$. Does f have a local maximum, a local minimum, or neither a maximum nor a minimum at c?

▶ 44. Let $f(x) = 2 + 20x + 4x^2 - x^4$.
(a) Show that f has exactly one critical number in the interval $(2, 3)$.
(b) Use the bisection method to approximate c to within $\frac{1}{16}$. Does f have a local maximum, a local minimum, or neither a maximum nor a minimum at c?

▶ 45. Let $f(x) = x^4 - 7x^2 + 2x - 3$.
(a) Show that f has exactly one critical number c in the interval $(1, 2)$.
(b) Use the Newton-Raphson method to approximate c; calculate x_3 and round your answer to four decimal places. Does f have a local maximum, a local minimum, or neither a maximum nor a minimum at c?

▶ 46. Let $f(x) = x \cos x$.
(a) Show that f has exactly one critical number in the interval $(0, \pi/2)$.
(b) Use the Newton-Raphson method to approximate c; calculate x_3 and round your answer to four decimal places. Does f have a local maximum, a local minimum, or neither a maximum nor a minimum at c?

▶ 47. Let $f(x) = \sin x + (x^2/2) - 2x$.
(a) Show that f has exactly one critical number in the interval $[2, 3]$.
(b) Use the Newton-Raphson method to approximate c; calculate x_3 and round your answer to four decimal places. Does f have a local maximum, a local minimum, or neither a maximum nor a minimum at c?

▶ In Exercises 48–51, use a graphing utility to graph the function f on the indicated interval. (a) Use the graph to estimate the critical numbers and the local extreme values; and (b) estimate the intervals on which f increases and the intervals on which f decreases. Round off your estimates to three decimal places.

48. $f(x) = 3x^3 - 7x^2 - 14x + 24$; $[-3, 4]$.

49. $f(x) = |3x^3 + x^2 - 10x + 2| + 3x$; $[-4, 4]$.

50. $f(x) = x^4 - 6x^2 + 9x + 3$; $[-5, 5]$.

51. $f(x) = \dfrac{8 \sin 2x}{1 + \frac{1}{2}x^2}$; $[-3, 3]$.

▶ In Exercises 52 and 53, the derivative f' of a function f is given. Use a graphing utility to graph f' on the indicated

interval. Estimate the critical numbers of f and determine whether f has a local maximum, a local minimum, or neither a maximum nor a minimum at each. Round off your estimates to three decimal places.

52. $f'(x) = 2x^3 + x^2 - 4x + 3$; $[-4, 4]$.

53. $f'(x) = \sin^2 x + 2 \sin 2x$; $[-2, 2]$.

■ 4.4 ENDPOINT AND ABSOLUTE EXTREME VALUES

Endpoint Maxima and Minima

For functions defined on an open interval or on the union of open intervals the critical numbers are those where the derivative is 0 or the derivative does not exist. For functions defined on a closed or half-closed interval

$$[a, b], \quad [a, b), \quad (a, b], \quad [a, \infty), \quad \text{or} \quad (-\infty, b]$$

or on a union of such intervals, the *endpoints* of the domain (a and b in the case of $[a, b]$, a in the case of $[a, b)$ or $[a, \infty)$, and so forth) are also called *critical numbers*.

Endpoints can give rise to what are called *endpoint maxima* and *endpoint minima*. See, for example, Figures 4.4.1, 4.4.2, 4.4.3, and 4.4.4.

Figure 4.4.1

Figure 4.4.2

Figure 4.4.3

Figure 4.4.4

DEFINITION 4.4.1 ENDPOINT EXTREME VALUES

If c is an endpoint of the domain of f, then f is said to have an *endpoint maximum* at c iff

$$f(c) \geq f(x) \qquad \text{for all } x \text{ in the domain of } f \text{ sufficiently close to } c.$$

It is said to have an *endpoint minimum* at c iff

$$f(c) \leq f(x) \qquad \text{for all } x \text{ in the domain of } f \text{ sufficiently close to } c.$$

Endpoints are usually tested either by calculating a one-sided derivative at the endpoint, or by examining the sign of the derivative at nearby points. Suppose, for example, that a is the left endpoint and f is continuous from the right at a. If $f'_+(a) < 0$, or if $f'(x) < 0$ for all x sufficiently close to a, then f decreases on an interval of the form $[a, a + \delta)$ and therefore $f(a)$ must be an endpoint maximum. (Figure 4.4.1.) On the other hand, if $f'_+(a) > 0$, or if $f'(x) > 0$ for all x sufficiently close to a, then f increases on an interval of the form $[a, a + \delta)$ and so $f(a)$ must be an endpoint minimum. (Figure 4.4.2.) Similar reasoning can be applied to right endpoints.

Absolute Maxima and Minima

Whether or not a function f has a local or endpoint extreme value at some point depends entirely on the behavior of f for x close to that point. Absolute extreme values, which we define below, depend on the behavior of the function on its entire domain.

We begin with a number d in the domain of f. Here d can be an interior point or an endpoint.

DEFINITION 4.4.2 ABSOLUTE EXTREME VALUES

A function f is said to have an *absolute maximum at* d iff

$$f(d) \geq f(x) \qquad \text{for all } x \text{ in the domain of } f;$$

f is said to have an *absolute minimum at* d iff

$$f(d) \leq f(x) \qquad \text{for all } x \text{ in the domain of } f.$$

These values are called the *absolute extreme values of* f.

A function can be continuous on an interval (or even differentiable there) without taking on an absolute maximum or an absolute minimum. (The functions depicted in Figures 4.4.2 and 4.4.3 have no absolute maximum. The function depicted in Figure 4.4.4 has no absolute minimum.) All we can say in general is that, if f takes on an absolute extreme value, then it does so at a critical point.

There are some special cases, however, where the continuity of f does guarantee the existence of absolute extreme values. Recall Theorem 2.6.2 which states:

> If f is continuous on a closed, bounded interval [a, b], then
>
> **(i)** f is bounded on $[a, b]$, and
>
> **(ii)** f attains its (absolute) maximum value M and its (absolute) minimum value m on $[a, b]$.

In this case the absolute extreme values of f on $[a, b]$ can be determined as follows:

1. Find the critical numbers $c_1, c_2, \ldots, c_n$ of f in the open interval (a, b).
2. Calculate $f(a), f(c_1), f(c_2), \ldots, f(b)$.
3. The largest of the numbers found in step 2 is the absolute maximum of f and the smallest is the absolute minimum of f.

Example 1 Find the critical numbers and classify all extreme values of the function f defined on $[-\frac{1}{2}, \frac{3}{2}]$ by

$$f(x) = x^2 - 2|x| + 2 = \begin{cases} x^2 + 2x + 2, & -\frac{1}{2} \leq x < 0 \\ x^2 - 2x + 2, & 0 \leq x < \frac{3}{2}. \end{cases}$$

SOLUTION Since f is continuous on a closed, bounded interval, we know that it has an absolute maximum and an absolute minimum. This function is differentiable on the open interval $(-\frac{1}{2}, \frac{3}{2})$, except at $x = 0$:

$$f'(x) = \begin{cases} 2x + 2, & -\frac{1}{2} < x < 0 \\ 2x - 2, & 0 < x < \frac{3}{2}. \end{cases} \qquad \text{(verify this)}$$

This makes 0 a critical number. Since $f'(x) = 0$ at $x = 1$, 1 is a critical number. The endpoints $-\frac{1}{2}$ and $\frac{3}{2}$ are also critical numbers. (Note that $2x + 2 = 0$ when $x = -1$, but we have ignored this value since it is not in the domain of f.)

The sign of f' and behavior of f are as follows:

sign of f': $\quad +++++++$dne $---------------$ 0$++++++++$

behavior of f: $\quad -\frac{1}{2}$ increases 0 $\qquad$ decreases $\qquad$ 1 increases $\frac{3}{2}$ $\qquad x$

Therefore

$$f(-\tfrac{1}{2}) = \tfrac{1}{4} - 1 + 2 = \tfrac{5}{4} \quad \text{is an endpoint minimum;}$$

$$f(0) = 2 \quad \text{is a local maximum;}$$

$$f(1) = 1 - 2 + 2 = 1 \quad \text{is a local minimum;}$$

$$f(\tfrac{3}{2}) = \tfrac{9}{4} - 3 + 2 = \tfrac{5}{4} \quad \text{is an endpoint maximum.}$$

Also $f(1) = 1$ is the absolute minimum and $f(0) = 2$ is the absolute maximum. The graph of the function is shown in Figure 4.4.5. ❑

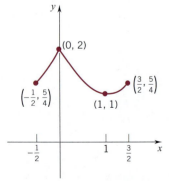

Figure 4.4.5

Behavior of f as $x \to \pm\infty$†

We now state four definitions. Once you grasp the first one, the others become transparent.

To say that

$$\text{as} \quad x \to \infty, \quad f(x) \to \infty$$

is to say that, *as x increases without bound, f(x) becomes arbitrarily large.* In formal terms, the definition is: Given any positive number M, there exists a positive number K such that

$$\text{if} \quad x \geq K, \quad \text{then} \quad f(x) \geq M.$$

Thus, for example, as $x \to \infty$,

$$x^2 \to \infty, \qquad \sqrt{1 + x^2} \to \infty, \qquad \tan\left(\frac{\pi}{2} - \frac{2}{x^2}\right) \to \infty. \quad \square$$

To say that

$$\text{as} \quad x \to \infty, \quad f(x) \to -\infty$$

is to say that, *as x increases without bound, f(x) becomes arbitrarily large negative.* Again in formal terms, the definition is: Given any negative number M, there exists a positive number K such that

$$\text{if} \quad x \geq K, \quad \text{then} \quad f(x) \leq M.$$

Thus, for example, as $x \to \infty$,

$$-x^4 \to -\infty, \qquad 1 - \sqrt{x} \to -\infty, \qquad \tan\left(\frac{1}{x^2} - \frac{\pi}{2}\right) \to -\infty. \quad \square$$

To say that

$$\text{as} \quad x \to -\infty, \quad f(x) \to \infty$$

is to say that, *as x decreases without bound, f(x) becomes arbitrarily large:* given any positive number M, there exists a negative number K such that,

$$\text{if} \quad x \leq K, \quad \text{then} \quad f(x) \geq M.$$

Thus, for example, as $x \to -\infty$,

$$x^2 \to \infty, \qquad \sqrt{1 - x} \to \infty, \qquad \tan\left(\frac{\pi}{2} - \frac{1}{x^2}\right) \to \infty. \quad \square$$

Finally, to say that

$$\text{as} \quad x \to -\infty, \quad f(x) \to -\infty$$

is to say that, *as x decreases without bound, f(x) becomes arbitrarily large negative:* given any negative number M, there exists a negative number K such that,

$$\text{if} \quad x \leq K, \quad \text{then} \quad f(x) \leq M.$$

---

† We will use the notation $x \to \pm\infty$ to mean $x \to \infty$ or $x \to -\infty$.

Thus, for example, as $x \to -\infty$,

$$x^3 \to -\infty, \qquad -\sqrt{1-x} \to -\infty, \qquad \tan\left(\frac{1}{x^2} - \frac{\pi}{2}\right) \to -\infty. \quad \square$$

Remark As you can readily see, $f(x) \to -\infty$ iff $-f(x) \to \infty$. $\quad \square$

Suppose now that P is a nonconstant polynomial:

$$P(x) = a_n x^n + a_{n-1} x^{n-1} + \cdots + a_1 x + a_0 \qquad (a_n \neq 0, \, n \geq 1).$$

When $|x|$ is large—that is, when x is either a large positive number or a large negative number—the leading term $a_n x^n$ dominates. Thus, what happens to $P(x)$ as $x \to \pm\infty$ depends entirely on what happens to $a_n x^n$. (For confirmation, do Exercise 41.)

Example 2

(a) As $x \to \infty$, $3x^4 \to \infty$, and therefore $3x^4 - 100x^3 + 2x - 5 \to \infty$.

(b) As $x \to -\infty$, $5x^3 \to -\infty$, and therefore $5x^3 + 12x^2 + 80 \to -\infty$. $\quad \square$

Finally, we point out that, if $f(x) \to \infty$, then f cannot have an absolute maximum value and, if $f(x) \to -\infty$, then f cannot have an absolute minimum value.

The following theorem is sometimes helpful in finding absolute extreme values of a continuous function defined on an arbitrary interval I.

THEOREM 4.4.3

Let f be continuous on an interval I and suppose that f has exactly one local extreme value in the interior of I, say at $x = c$.

If $f(c)$ is a local maximum, then $f(c)$ is the absolute maximum of f on I.
If $f(c)$ is a local minimum, then $f(c)$ is the absolute minimum of f on I.

Figure 4.4.6

An intuitive proof of this theorem goes as follows: Suppose that f has a local maximum at $x = c$. If $f(c)$ is not the absolute maximum of f, then there is a point d such that $f(d) \geq f(c)$. But this would imply that there exists a local minimum of f between $x = c$ and $x = d$. See Figure 4.4.6.

Example 3 Find the critical numbers and classify all the extreme values of

$$f(x) = 6\sqrt{x} - x\sqrt{x}.$$

SOLUTION The domain is $[0, \infty)$. Therefore the endpoint 0 is a critical number.
This function will be easier to differentiate if we rewrite it using fractional exponents:

$$f(x) = 6x^{1/2} - x^{3/2}.$$

Now, on $(0, \infty)$,

$$f'(x) = 3x^{-1/2} - \tfrac{3}{2}x^{1/2} = \frac{3(2-x)}{2\sqrt{x}}. \qquad \text{(verify this)}$$

Since $f'(x) = 0$ at $x = 2$, we see that 2 is an interior critical number.

The sign of f' and the behavior of f are as follows:

sign of f': $+ + + + + + + + 0 - - - - - - - -$

0   2 x

behavior of f: increases   decreases

Therefore,

$$f(0) = 0 \text{ is an endpoint minimum;}$$

$$f(2) = 6\sqrt{2} - 2\sqrt{2} = 4\sqrt{2} \text{ is a local maximum.}$$

Since $f(x) = \sqrt{x}(6 - x)$, it is easy to see that $f(x) \to -\infty$ as $x \to \infty$. Thus the function has no absolute minimum value. Also, since f has exactly one local extreme value on $(0, \infty)$, $f(2) = 4\sqrt{2}$ is the absolute maximum value of f by Theorem 4.4.3. The graph of f appears in Figure 4.4.7. ❑

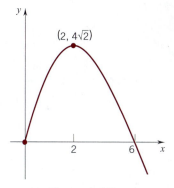

$(2, 4\sqrt{2})$

Figure 4.4.7

A Summary on Finding All the Extreme Values (Local, Endpoint, and Absolute) of a Continuous Function f

1. Find the critical numbers. These are the endpoints of the domain (if any) and the interior numbers c at which $f'(c) = 0$ or $f'(c)$ does not exist.

2. Test each endpoint of the domain by examining the sign of the first derivative nearby.

3. Test each interior critical number c by examining the sign of the first derivative on both sides of c (first-derivative test) or by checking the sign of the second derivative at c itself (second-derivative test).

4. If f is defined on (a, ∞) for some number a or on $(-\infty, b)$ for some number b, determine the behavior of f as $x \to \infty$ or as $x \to -\infty$.

5. Determine whether any of the endpoint extremes and local extremes are absolute extremes; use Theorems 2.6.2 and Theorem 4.4.3 when applicable.

Example 4 Find the critical numbers and classify all the extreme values of

$$f(x) = \tfrac{1}{4}(x^3 - \tfrac{3}{2}x^2 - 6x + 2), \qquad x \in [-2, \infty).$$

SOLUTION The left endpoint -2 is a critical number. To find the interior critical numbers, we differentiate:

$$f'(x) = \tfrac{1}{4}(3x^2 - 3x - 6) = \tfrac{3}{4}(x + 1)(x - 2).$$

Since $f'(x) = 0$ at $x = -1$ and $x = 2$, the numbers -1 and 2 are interior critical numbers.

The sign of f' and the behavior of f are:

sign of f': $+ + + + + + + + 0 - - - - - - - - - - - - - - - 0 + + + + + +$

behavior of f: -2 increases -1 decreases   2   increases x

We can see from the sign of f' that

$$f(-2) = \tfrac{1}{4}(-8 - 6 + 12 + 2) = 0 \quad \text{is an endpoint minimum;}$$

$$f(-1) = \tfrac{1}{4}(-1 - \tfrac{3}{2} + 6 + 2) = \tfrac{11}{8} \quad \text{is a local maximum;}$$

$$f(2) = \tfrac{1}{4}(8 - 6 - 12 + 2) = -2 \quad \text{is a local minimum.}$$

The function takes on no absolute maximum value since $f(x) \to \infty$ as $x \to \infty$; $f(2) = -2$ is the absolute minimum value of f. The graph of f is shown in Figure 4.4.8. ❏

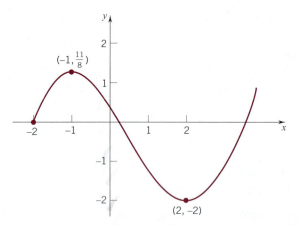

Figure 4.4.8

Example 5 Find the critical numbers and classify all the extreme values of

$$f(x) = \sin x - \sin^2 x, \qquad x \in [0, 2\pi].$$

SOLUTION The endpoints 0 and 2π are critical numbers. On $(0, 2\pi)$

$$\begin{aligned} f'(x) &= \cos x - 2 \sin x \cos x \\ &= \cos x (1 - 2 \sin x). \end{aligned}$$

Setting $f'(x) = 0$, we have

$$\cos x (1 - 2 \sin x) = 0.$$

This equation is satisfied when $\cos x = 0$, which gives $x = \pi/2$ and $x = 3\pi/2$; and when $\sin x = \tfrac{1}{2}$, which gives $x = \pi/6$ and $x = 5\pi/6$. Thus, the numbers $\pi/6$, $\pi/2$, $5\pi/6$, and $3\pi/2$ are interior critical numbers.

The sign of f' and the behavior of f are as follows:

Therefore,

$$f(0) = 0 \quad \text{is an endpoint minimum;}$$

$$f(\pi/6) = \tfrac{1}{4} \quad \text{is a local maximum;}$$

$$f(\pi/2) = 0 \quad \text{is a local minimum;}$$

$$f(5\pi/6) = \tfrac{1}{4} \quad \text{is local maximum;}$$

$$f(3\pi/2) = -2 \quad \text{is a local minimum;}$$

$$f(2\pi) = 0 \quad \text{is an endpoint maximum}$$

Also, $f(\pi/6) = f(5\pi/6) = \frac{1}{4}$ is the absolute maximum of f and $f(3\pi/2) = -2$ is the absolute minimum. The graph of the function is shown in Figure 4.4.9. ❑

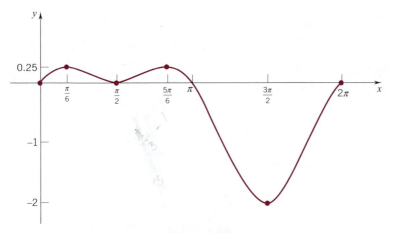

Figure 4.4.9

In the next section we will illustrate some practical applications of the theory that we have developed up to this point.

EXERCISES 4.4

In Exercises 1–30, find the critical numbers and classify the extreme values.

1. $f(x) = \sqrt{x + 2}$.

2. $f(x) = (x - 1)(x - 2)$.

3. $f(x) = x^2 - 4x + 1$, $x \in [0, 3]$.

4. $f(x) = 2x^2 + 5x - 1$, $x \in [-2, 0]$.

5. $f(x) = x^2 + \dfrac{1}{x}$.

6. $f(x) = x + \dfrac{1}{x^2}$.

7. $f(x) = x^2 + \dfrac{1}{x}$, $x \in [\frac{1}{10}, 2]$.

8. $f(x) = x + \dfrac{1}{x^2}$, $x \in [1, \sqrt{2}]$.

9. $f(x) = (x - 1)(x - 2)$, $x \in [0, 2]$.

10. $f(x) = (x - 1)^2(x - 2)^2$, $x \in [0, 4]$.

11. $f(x) = \dfrac{x}{4 + x^2}$, $x \in [-3, 1]$.

12. $f(x) = \dfrac{x^2}{1 + x^2}$, $x \in [-1, 2]$.

13. $f(x) = (x - \sqrt{x})^2$.

14. $f(x) = x\sqrt{4 - x^2}$.

15. $f(x) = x\sqrt{3 - x}$.

16. $f(x) = \sqrt{x} - \dfrac{1}{\sqrt{x}}$.

17. $f(x) = 1 - \sqrt[3]{x - 1}$.

18. $f(x) = (4x - 1)^{1/3}(2x - 1)^{2/3}$.

19. $f(x) = \sin^2 x - \sqrt{3}\cos x$, $0 \le x \le \pi$.

20. $f(x) = \cot x + x$, $0 < x \le \dfrac{2\pi}{3}$.

21. $f(x) = 2\cos^3 x + 3\cos x$, $0 \le x \le \pi$.

22. $f(x) = \sin 2x - x$, $0 \le x \le \pi$.

23. $f(x) = \tan x - x$, $-\dfrac{\pi}{3} \le x < \dfrac{\pi}{2}$.

24. $f(x) = \sin^4 x - \sin^2 x$, $0 \le x \le \dfrac{2\pi}{3}$.

25. $f(x) = \begin{cases} -2x, & 0 \le x < 1 \\ x - 3, & 1 \le x \le 4 \\ 5 - x, & 4 < x \le 7. \end{cases}$

26. $f(x) = \begin{cases} x + 9, & -8 \le x < -3 \\ x^2 + x, & -3 \le x \le 2 \\ 5x - 4, & 2 < x < 5. \end{cases}$

27. $f(x) = \begin{cases} x^2 + 1, & -2 \le x < -1 \\ 5 + 2x - x^2, & -1 \le x \le 3 \\ x - 1, & 3 < x < 6. \end{cases}$

28. $f(x) = \begin{cases} 2 - 2x - x^2, & -2 \le x \le 0 \\ |x - 2|, & 0 < x < 3 \\ \frac{1}{3}(x - 2)^3, & 3 \le x \le 4. \end{cases}$

29. $f(x) = \begin{cases} |x + 1|, & -3 \le x < 0 \\ x^2 - 4x + 2, & 0 \le x < 3 \\ 2x - 7, & 3 \le x < 4. \end{cases}$

30. $f(x) = \begin{cases} -x^2, & 0 \le x < 1 \\ -2x, & 1 < x < 2 \\ -\frac{1}{2}x^2, & 2 \le x \le 3. \end{cases}$

In Exercises 31–34, sketch the graph of a differentiable function satisfying the given conditions, if possible. If it is not possible, state why.

31. Domain is $(-\infty, \infty)$, local maximum at -1, local minimum at 1, $f(3) = 6$ is the absolute maximum, no absolute minimum.

32. Domain is $[0, \infty)$, $f(0) = 1$ is the absolute minimum, local maximum at 4, local minimum at 7, no absolute maximum.

33. Domain is $[-3, 3]$, $f(-3) = f(3) = 0$, $f(x) \neq 0$ for all $x \in (-3, 3)$, $f'(-1) = f'(1) = 0$, $f'(x) > 0$ if $|x| > 1$, $f'(x) < 0$ if $|x| < 1$.

34. Domain is $(-1, 3]$, local minimum at 2, $f(1) = 5$ is the absolute maximum, no absolute minimum.

35. Show that the cubic polynomial $p(x) = x^3 + ax^2 + bx + c$ has no extreme values iff $a^2 \leq 3b$.

36. Let r be a rational number, $r > 1$, and let f be the function defined by

$$f(x) = (1 + x)^r - (1 + rx) \qquad \text{for} \quad x \geq -1.$$

Show that 0 is a critical number of f and show that $f(0) = 0$ is the absolute minimum of f.

37. Suppose that c is a critical number for f and $f'(x) > 0$ for $x \neq c$. Prove that, if $f(c)$ is a local maximum, then f is not continuous at c.

38. What can be said about the function f continuous on $[a, b]$, if for some c in (a, b), $f(c)$ is both a local maximum and a local minimum?

39. Suppose that f is continuous on its domain $[a, b]$ and $f(a) = f(b)$. Prove that f has at least one critical point in (a, b).

40. Suppose that $c_1 < c_2$ and that both $f(c_1)$ and $f(c_2)$ are local maxima. Prove that, if f is continuous on $[c_1, c_2]$, then, for some c in (c_1, c_2), $f(c)$ is a local minimum.

41. Let P be a nonconstant polynomial with positive leading coefficient:

$$P(x) = a_n x^n + a_{n-1} x^{n-1} + \cdots + a_1 x + a_0$$
$$(n \geq 1, a_n > 0).$$

Clearly, as $x \to \infty$, $a_n x^n \to \infty$. Show that, as $x \to \infty$, $P(x) \to \infty$ by showing that, given any positive number M, there exists a positive number K such that, if $x \geq K$, then $f(x) \geq M$.

42. An object moving along a straight line with position function $x(t)$ given by

$$x(t) = A \sin(\omega t + \phi_0),$$

where A, ω, and ϕ_0 are constants, is said to be in *simple harmonic motion*. For example, consider the motion of a mass m suspended by a spring and oscillating about its equilibrium position.

(a) Show that the position function $x(t)$ satisfies

$$x''(t) + \omega^2 x(t) = 0.$$

(b) Find the absolute maximum and the absolute minimum values of $x(t)$.

43. Of all rectangles with a given diagonal of length c, prove that the square has the largest area.

44. The sum of two numbers is 16. Find the numbers if the sum of their cubes is an absolute minimum.

45. A piece of wire of length L will be cut into two pieces, one piece to form a square and the other piece to form an equilateral triangle. How should the wire be cut so as to:
(a) Maximize the sum of the areas of the square and triangle?
(b) Minimize the sum of the areas of the square and triangle?

▷ In Exercises 46–49, use a graphing utility to graph the function f on the indicated interval. Estimate the critical numbers of f and classify all the extreme values. Round off your estimates to three decimal places.

46. $f(x) = x^4 - 7x^2 + 10x + 3$; $[-3, 3]$.

47. $f(x) = x^3 - 4x + 2x \sin x$; $[-2.5, 3]$.

48. $f(x) = 5x^{2/3} + 3x^{5/3} + 1$; $[-3, 1]$.

49. $f(x) = x \cos 2x - \cos^2 x$; $[-\pi, \pi]$.

■ 4.5 SOME MAX-MIN PROBLEMS

The techniques of the last two sections can be brought to bear on a wide variety of max-min problems. The *key* idea in solving such problems is to express the quantity to be maximized or minimized as a function of one variable. If the function is differentiable, we can then differentiate and analyze the results.

Example 1   An architect wants to design a window in the shape of a rectangle capped by a semicircle. If the perimeter of the window is constrained to be 24 feet, what dimensions should the architect choose for the window in order to admit the greatest amount of light?

SOLUTION   The maximum amount of light will be admitted when the area of the window is a maximum. As indicated in Figure 4.5.1, we let x be the radius of the semicircular part and y the height of the rectangular part. Then the area A of the window is given by

$$A = \tfrac{1}{2}\pi x^2 + 2xy.$$

In order to apply the methods of the preceding sections, we need to express A as a function of one variable only.

Since the perimeter is 24 feet, we have

$$2x + 2y + \pi x = 24$$

and so

$$y = \frac{24 - (2 + \pi)x}{2} = 12 - \frac{(2 + \pi)}{2}\,x.$$

Now the area can be expressed in terms of x alone:

$$A(x) = \tfrac{1}{2}\pi x^2 + 2x\left[12 - \frac{(2 + \pi)}{2}\,x\right]$$
$$= \tfrac{1}{2}\pi x^2 + 24x - (2 + \pi)x^2 = 24x - \left(2 + \tfrac{1}{2}\pi\right)x^2.$$

Also, since x and y represent lengths and A is an area, these variables cannot be negative. Therefore, it follows that $0 \le x \le 24/(2 + \pi)$.

Our problem can now be formulated as: Find the absolute maximum of the function

$$A(x) = 24x - \left(2 + \tfrac{1}{2}\pi\right)x^2, \qquad x \in \left[0, \frac{24}{2 + \pi}\right].$$

The derivative

$$A'(x) = 24 - 2\left(2 + \tfrac{1}{2}\pi\right)x = 24 - (4 + \pi)x$$

is 0 only at $x = 24/(4 + \pi)$. Since $A'(x) > 0$ for $0 < x < 24/(4 + \pi)$ and $A'(x) < 0$ for $24/(4 + \pi) < x < 24/(2 + \pi)$, it follows that $A[24/(4 + \pi)]$ is a local maximum. (NOTE: We could just as easily have used the second-derivative test to find that A has a local maximum at this value.) Also, since A has only one local extreme value on the interval, we can conclude, by Theorem 4.4.3, that $A[24/(4 + \pi)]$ is the absolute maximum.

The dimensions of the window that will admit the greatest amount of light are:

$$\text{radius of semicircular part} \qquad x = \frac{24}{4 + \pi} \cong 3.36 \text{ feet,}$$

$$\text{height of rectangular part} \qquad y = 12 - \left(\frac{2 + \pi}{2}\right)\left(\frac{24}{4 + \pi}\right) = \frac{24}{4 + \pi},$$

the height of the rectangle equals the radius of the semicircle.   ❏

Figure 4.5.1

This example illustrates a basic strategy for solving max-min problems.

Strategy

1. Draw a representative figure and assign labels to the relevant quantities involved in the problem.
2. Identify the quantity that is to be maximized or minimized and find a formula for it.
3. Express the quantity to be maximized or minimized as a function of one variable only; use the conditions given in the problem to eliminate variables.
4. Determine the domain of the function found in (3).
5. Apply the techniques of the preceding sections to find the extreme value(s).

Steps 3 and 4 are the key steps in this procedure.

Figure 4.5.2

Example 2   A soft-drink manufacturer wants to fabricate cylindrical cans for its product.   The can is to have a volume of 12 fluid ounces, which is approximately 22 cubic inches.   Find the dimensions of the can that will require the least amount of material.   See Figure 4.5.2.

Let r be the radius of the can and h its height.   The total surface area (top, bottom, lateral area) of a right circular cylinder of radius r and height h is

$$S = 2\pi r^2 + 2\pi rh,$$

and this is the quantity that we want to minimize.

Now, since the volume $V = \pi r^2 h$ has to be 22 cubic inches, we have

$$\pi r^2 h = 22$$

and

$$h = \frac{22}{\pi r^2}.$$

Also, since r and h must both be positive, it follows that $0 < r < \infty$.

Thus, we want to minimize the function

$$S(r) = 2\pi r^2 + 2\pi r \left(\frac{22}{\pi r^2}\right)$$

$$= 2\pi r^2 + \frac{44}{r}, \qquad r \in (0, \infty).$$

(Key steps completed.)

The derivative of S is

$$\frac{dS}{dr} = 4\pi r - \frac{44}{r^2}.$$

To find the critical numbers of S, set $dS/dr = 0$:

$$4\pi r - \frac{44}{r^2} = 0$$

$$r^3 = \frac{11}{\pi}$$

and

$$r = \sqrt[3]{\frac{11}{\pi}}$$

is the only critical number. Since

$$\frac{d^2S}{dr^2} = 4\pi + \frac{88}{r^3} > 0$$

at $r = \sqrt[3]{11/\pi}$, S has a local minimum at this point. By Theorem 4.4.3, $S(\sqrt[3]{11/\pi})$ is the absolute minimum value of S. Therefore, the dimensions of the can that will require the least amount of material are:

radius $\qquad r = \sqrt[3]{\frac{11}{\pi}} \cong 1.5$ inches,

height $\qquad h = \frac{22}{\pi(11/\pi)^{2/3}} = 2\sqrt[3]{\frac{11}{\pi}} \cong 3$ inches;

the height of the cylindrical can should be twice the radius. ❏

Example 3 An isosceles triangle has base 6 and height 12. Find the maximum possible area of a rectangle that can be placed inside the triangle with one side on the base of the triangle.

SOLUTION Figure 4.5.3 shows the isosceles triangle with a typical rectangle placed inside. In Figure 4.5.4, we have introduced a rectangular coordinate system. With x and y as indicated in Figure 4.5.4, we want to maximize the area A, which is given by

$$A = 2xy.$$

Now, since the point (x, y) lies on the line through $(0, 12)$ and $(3, 0)$, it can be shown that $y = 12 - 4x$. Also, since x and y are both nonnegative, we have $0 \leq x \leq 3$. Thus, our problem is:

Maximize the function

$$A(x) = 2x(12 - 4x) = 24x - 8x^2, \qquad x \in [0, 3].$$

(Key steps completed.)

The derivative

$$A'(x) = 24 - 16x$$

is 0 only at $x = 24/16 = 3/2$. Since $A'(x) > 0$ for $0 < x < 3/2$ and $A'(x) < 0$ for $3/2 < x < 3$, it follows that A has a local maximum and hence an absolute maximum at $x = 3/2$. If $x = 3/2$, then $y = 6$ and the maximum possible area is $A = 2(3/2)6 = 18$ square units. ❏

Example 4 A state highway department plans to construct a new road between town A and town B. Town A lies on an abandoned road that runs east-west. Town B is 3 miles north of the point on that road that is 5 miles east of A. The engineering division proposes that the road be constructed by restoring a section of the old road from A up to a point P and joining it to a new road that connects P and B. If the cost of restoring the old road is $200,000 per mile and the cost of the new road is $400,000 per mile, how much of the old road should be restored in order to minimize the department's costs?

Figure 4.5.3

Figure 4.5.4

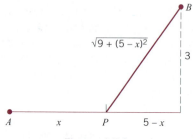

Figure 4.5.5

SOLUTION Figure 4.5.5 shows the geometry of the problem. Notice that we have chosen a straight line joining P and B rather than some curved path (recall that the shortest distance between two points is a straight line). We let x be the amount of old road that will be restored. Then

$$\sqrt{9 + (5 - x)^2} = \sqrt{34 - 10x + x^2}$$

is the length of the new road. The total cost of constructing the two sections of the road is

$$C(x) = 200{,}000x + 400{,}000[34 - 10x + x^2]^{1/2}, \qquad 0 \le x \le 5.$$

We want to find the value of x that minimizes this function.

(*Key steps completed.*)

Differentiation gives

$$C'(x) = 200{,}000 + 400{,}000 \left(\tfrac{1}{2}\right) [34 - 10x + x^2]^{-1/2}(2x - 10)$$

$$= 200{,}000 + \frac{400{,}000(x - 5)}{[34 - 10x + x^2]^{1/2}}, \qquad 0 < x < 5.$$

Setting $C'(x) = 0$, we find that

$$1 + \frac{2(x - 5)}{[34 - 10x + x^2]^{1/2}} = 0$$

$$2(x - 5) = -[34 - 10x + x^2]^{1/2}$$

$$4(x^2 - 10x + 25) = 34 - 10x + x^2$$

$$3x^2 - 30x + 66 = 0.$$

$$x^2 - 10x + 22 = 0.$$

Now, by the quadratic formula, we have

$$x = \frac{10 \pm \sqrt{100 - 4(22)}}{2} = 5 \pm \sqrt{3}.$$

The value $x = 5 + \sqrt{3}$ is not in the domain of our function; the value we want is $x = 5 - \sqrt{3}$. You can verify that the sign of C' is:

Thus $x = 5 - \sqrt{3} \cong 3.27$ gives the absolute minimum value for C. The highway department will minimize its costs by restoring 3.27 miles of the old road.

You can check that the minimum cost is $200,000(5 + 3\sqrt{3}) \cong \$2,039,230$. Using all of the old road, the cost would be exactly $2,200,000, while using none it would cost $400,000\sqrt{34} \cong \$2,332,400$. The most expensive option costs almost $300,000 more than the cheapest option. ❏

Example 5 A right circular cylinder is inscribed in a cone of height H and radius R. Determine the dimensions of the cylinder with the largest possible volume. What is the maximum volume?

SOLUTION Let r be the radius of the cylinder and h its height. We want to find the maximum value of

$$V = \pi r^2 h.$$

Using similar triangles (see Figure 4.5.6), we see that

$$\frac{H}{R} = \frac{H - h}{r},$$

and so

$$h = H - \frac{Hr}{R}.$$

Since r and h are both nonnegative, it follows that $0 \leq r \leq R$. Thus, we want to maximize the function

$$V(r) = \pi r^2 \left[H - \frac{Hr}{R} \right] = \pi r^2 H - \frac{\pi r^3 H}{R}, \qquad r \in [0, R].$$

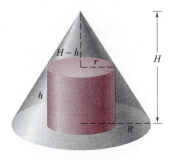

Figure 4.5.6

(*Key steps completed.*)

The endpoints, 0 and R, are critical numbers of V, and it is easy to see that $V(0) = V(R) = 0$ is the absolute minimum of V. The derivative of V on $(0, R)$ is

$$\frac{dV}{dr} = 2\pi rH - \frac{3\pi r^2 H}{R}.$$

Setting $dV/dr = 0$, we have

$$\pi rH \left(2 - \frac{3r}{R} \right) = 0$$

so

$$r = 0 \qquad \text{and} \qquad r = \frac{2R}{3}.$$

The only critical number in $(0, R)$ is $r = 2R/3$. By examining the sign of the first derivative, or by calculating the second derivative and evaluating at $r = 2R/3$, you can verify that V has a local maximum and hence an absolute maximum at this value. The dimensions of the cylinder with the largest possible volume are $r = 2R/3$ and $h = H/3$; the maximum volume is $V = 4\pi R^2 H/27$. ❏

Figure 4.5.7

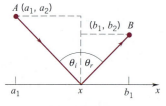

Figure 4.5.8

Example 6 (*The angle of incidence equals the angle of reflection.*) Figure 4.5.7 depicts light from a point A reflected to a point B by a mirror. Two angles have been marked: the *angle of incidence*, θ_i, and the *angle of reflection*, θ_r. Experiment shows that $\theta_i = \theta_r$. Derive this result by postulating that the light that travels from A to the mirror and then to B follows the shortest possible path.†

SOLUTION We can write the length of the path as a function of x: in the setup of Figure 4.5.8,

$$L(x) = \sqrt{(x - a_1)^2 + a_2^2} + \sqrt{(x - b_1)^2 + b_2^2}, \qquad x \in [a_1, b_1].$$

Differentiation gives

$$L'(x) = \frac{x - a_1}{\sqrt{(x - a_1)^2 + a_2^2}} + \frac{x - b_1}{\sqrt{(x - b_1)^2 + b_2^2}}.$$

We can therefore see that

$$L'(x) = 0 \quad \text{iff} \quad \frac{x - a_1}{\sqrt{(x - a_1)^2 + a_2^2}} = \frac{b_1 - x}{\sqrt{(x - b_1)^2 + b_2^2}}$$

$$\text{iff} \quad \sin \theta_i = \sin \theta_r \qquad \text{(see the figure)}$$
$$\text{iff} \quad \theta_i = \theta_r.$$

That $L(x)$ is minimal when $\theta_i = \theta_r$ can be seen from noting that $L''(x)$ is always positive:

$$L''(x) = \frac{a_2^2}{[(x - a_1)^2 + a_2^2]^{3/2}} + \frac{b_2^2}{[(x - b_1)^2 + b_2^2]^{3/2}} > 0. \quad ❑$$

We must admit that there is a simpler way to do Example 6 that requires no calculus. Can you find it?

Now we will work out a problem in which the function to be maximized is defined, not on an interval or union of intervals, but on a discrete set of points, in this case a finite collection of integers.

Example 7 A manufacturing plant has a capacity of 25 articles per week. Experience has shown that n articles per week can be sold at a price of p dollars each where $p = 110 - 2n$ and the cost of producing n articles is $600 + 10n + n^2$ dollars. How many articles should be made each week to give the largest profit?

SOLUTION The profit (P dollars) on the sale of n articles is

$$P = \text{revenue} - \text{cost} = np - (600 + 10n + n^2).$$

With $p = 110 - 2n$, this simplifies to

$$P = 100n - 600 - 3n^2.$$

In this problem n must be an integer, and thus it makes no sense to differentiate P with respect to n. The formula shows that P is negative if n is less than 8 or greater than

---

† This is a special case of Fermat's *Principle of Least Time*, which says that, of all (neighboring) paths, light chooses the one that demands the least time. If light passes from one medium to another, the geometrically shortest path is not necessarily the path of least time.

25. By direct calculation we construct Table 4.5.1. The table shows that the largest profit comes from setting production at 17 articles per week.

■ **Table 4.5.1**

n	P	n	P	n	P
8	8	14	212	20	200
9	57	15	225	21	177
10	100	16	232	22	148
11	137	17	233	23	113
12	168	18	228	24	72
13	193	19	217	25	25

We can avoid such massive computation by considering the function

$$f(x) = 100x - 600 - 3x^2, \qquad 8 \le x \le 25.$$

This function is differentiable with respect to x, and for integral values of x it agrees with P. Differentiation of f gives

$$f'(x) = 100 - 6x.$$

Obviously, $f'(x) = 0$ at $x = \frac{100}{6} = 16\frac{2}{3}$. Since $f'(x) > 0$ on $(8, 16\frac{2}{3})$, f increases on $[8, 16\frac{2}{3}]$. Since $f'(x) < 0$ on $(16\frac{2}{3}, 25)$, f decreases on $[16\frac{2}{3}, 25]$. The largest value of f corresponding to an integer value of x will therefore occur at $x = 16$ or $x = 17$. Direct calculation of $f(16)$ and $f(17)$ shows that the choice of $x = 17$ is correct. ❏

EXERCISES 4.5

1. Find the greatest possible value of xy given that x and y are both positive and $x + y = 40$.

2. Find the dimensions of the rectangle of perimeter 24 that has the largest area.

3. A rectangular garden 200 square feet in area is to be fenced off against rabbits. Find the dimensions that will require the least amount of fencing if one side of the garden is already protected by a barn.

4. Find the largest possible area for a rectangle with base on the x-axis and upper vertices on the curve $y = 4 - x^2$.

5. Find the largest possible area for a rectangle inscribed in a circle of radius 4.

6. The cross section of a beam is in the form of a rectangle of length l and width w. (See the figure.) Assuming that the strength of the beam varies directly with w^2l, what are the dimensions of the strongest beam that can be sawed from a round log of diameter 3 feet?

7. A rectangular playground is to be fenced off and divided in two by another fence parallel to one side of the playground. Six hundred feet of fencing is used. Find the dimensions of the playground that will enclose the greatest total area.

8. A rectangular warehouse will have 5000 square feet of floor space and will be separated into two rectangular rooms by an interior wall. The cost of the exterior walls is $150 per linear foot and the cost of the interior wall is

$100 per linear foot. Find the dimensions that will minimize the cost of building the warehouse.

9. Rework Example 1, this time assuming that the semicircular portion of the window admits only one-third as much light per square foot as does the rectangular portion of the window.

10. One side of a rectangular field is bounded by a straight river. The other three sides are bounded by straight fences. The total length of the fence is 800 feet. Determine the dimensions of the field given that its area is a maximum.

11. Find the coordinates of P that maximize the area of the rectangle shown in the figure below.

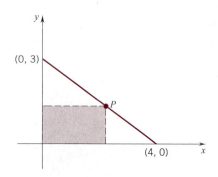

12. The base of a triangle is on the x-axis, one side lies along the line $y = 3x$, and the third side passes through the point $(1, 1)$. What is the slope of the third side if the area of the triangle is to be a minimum?

13. A triangle is formed by the coordinate axes and a line through the point $(2, 5)$ as in the figure below. Determine the slope of this line if the area of the triangle is to be a minimum.

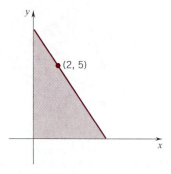

14. In the setting of Exercise 13 determine the slope of the line if the area is to be a maximum.

15. What are the dimensions of the base of the rectangular box of greatest volume that can be constructed from 100 square inches of cardboard if the base is to be twice as long as it is wide? Assume that the box has a top.

16. Exercise 15 under the assumption that the box has no top.

17. Find the dimensions of the isosceles triangle of largest area with perimeter 12.

18. Find the point(s) on the parabola $y = \frac{1}{8}x^2$ closest to the point $(0, 6)$.

19. Find the point(s) on the parabola $x = y^2$ closest to the point $(0, 3)$.

20. Find A and B given that the function $y = Ax^{-1/2} + Bx^{1/2}$ has a minimum value of 6 at $x = 9$.

21. A pentagon with a perimeter of 30 inches is to be constructed by adjoining an equilateral triangle to a rectangle. Find the dimensions of the rectangle and triangle that will maximize the area of the pentagon.

22. A 10-foot section of gutter is made from a 12-inch-wide strip of sheet metal by folding up 4-inch strips on each side so that they make the same angle with the bottom of the gutter. Determine the depth of the gutter that has the greatest carrying capacity. *Caution:* there are two ways to sketch the trapezoidal cross section, as shown in the figure below.

23. From a rectangular piece of cardboard of dimensions 8×15, four congruent squares are to be cut out, one at each corner. (See the figure below.) The remaining crosslike piece is then to be folded into an open box. What size squares should be cut out if the volume of the resulting box is to be a maximum?

24. A page is to contain 81 square centimeters of print. The margins at the top and bottom are to be 3 centimeters each and, at the sides, 2 centimeters each. Find the most economical dimensions given that the cost of a page varies directly with the perimeter of the page.

25. Let ABC be a triangle with vertices $A = (-3, 0)$, $B = (0, 6)$, $C = (3, 0)$. Let P be a point on the line segment that joins B to the origin. Find the position of P that

minimizes the sum of the distances between P and the vertices.

26. Solve Exercise 25 with $A = (-6, 0)$, $B = (0, 3)$, $C = (6, 0)$.

27. An 8-foot-high fence is located 1 foot from a building. Determine the length of the shortest ladder that can be leaned against the building and touch the top of the fence.

28. Two hallways, one 8 feet wide and the other 6 feet wide, meet at right angles. Determine the length of the longest ladder that can be carried horizontally from one hallway into the other.

29. A rectangular banner has a red border and a white center. The width of the border at top and bottom is 8 inches and along the sides it is 6 inches. The total area is 27 square feet. What should be the dimensions of the banner if the area of the white center is to be a maximum?

30. Find the absolute maximum value of $y = x(r^2 + x^2)^{-3/2}$.

31. A string 28 inches long is to be cut into two pieces, one piece to form a square and the other to form a circle. How should the string be cut so as to (a) maximize the sum of the two areas? (b) minimize the sum of the two areas?

32. What is the maximum volume for a rectangular box (square base, no top) made from 12 square feet of cardboard?

33. The figure below shows a cylinder inscribed in a right circular cone of height 8 and base radius 5. Find the dimensions of the cylinder if its volume is to be a maximum.

34. A variant of Exercise 33. This time find the dimensions of the cylinder if the area of its curved surface is to be a maximum.

35. A rectangular box with square base and top is to be made to contain 1250 cubic feet. The material for the base costs 35 cents per square foot, for the top 15 cents per square foot, and for the sides 20 cents per square foot. Find the dimensions that will minimize the cost of the box.

36. What is the largest possible area for a parallelogram in-

scribed in a triangle ABC in the manner of the figure below?

37. Find the dimensions of the isosceles triangle of least area that circumscribes a circle of radius r.

38. What is the maximal possible area for a triangle inscribed in a circle of radius r?

39. The figure below shows a right circular cylinder inscribed in a sphere of radius R. Find the dimensions of the cylinder if its volume is to be a maximum.

40. A variant of Exercise 39. This time find the dimensions of the right circular cylinder if the area of its curved surface is to be a maximum.

41. A right circular cone is inscribed in a sphere of radius R as in the figure below. Find the dimensions of the cone if its volume is to be a maximum.

42. What is the largest possible volume for a right circular cone of slant height a?

43. A power line is needed to connect a power station on the shore of a river to an island 4 kilometers downstream and 1 kilometer offshore. Find the minimum cost for such a line given that it costs $50,000 per kilometer to lay wire under water and $30,000 per kilometer to lay wire under ground.

44. A tapestry 7 feet high hangs on a wall. The lower edge is 9 feet above an observer's eye. How far from the wall

should the observer stand to obtain the most favorable view? Namely, what distance from the wall maximizes the visual angle of the observer? HINT: Use the formula for $\tan(A - B)$.

45. A body of weight W is dragged along a horizontal plane by means of a force P whose line of action makes an angle θ with the plane. The magnitude of the force is given by the equation

$$P = \frac{mW}{m \sin \theta + \cos \theta}$$

where m denotes the coefficient of friction. For what value of θ is the pull a minimum?

46. If a projectile is fired from O so as to strike an inclined plane that makes a constant angle α with the horizontal, its range is given by the formula

$$R = \frac{2v^2 \cos \theta \sin (\theta - \alpha)}{g \cos^2 \alpha},$$

where v and g are constants and θ is the angle of elevation. Calculate θ for maximum range.

47. The lower edge of a movie theater screen 30 feet high is 6 feet above an observer's eye. How far from the screen should the observer sit to obtain the most favorable view? Namely, what distance from the screen maximizes the visual angle of the observer?

48. A local bus company offers charter trips to Blue Mountain Museum at a fare of $37 per person if 16 to 35 passengers sign up for the trip. The company does not charter trips for fewer than 16 passengers. The bus has 48 seats. If more than 35 passengers sign up, then the fare for every passenger is reduced by 50 cents for each passenger in excess of 35 that signs up. Determine the number of passengers that generates the greatest revenue for the bus company.

49. The Hotwheels Rent-A-Car Company derives an average net profit of $12 per customer if it services 50 customers or less. If it services over 50 customers, then the average net profit is decreased by 6 cents for each customer over 50. What number of customers produces the greatest total net profit for the company?

■ 4.6 CONCAVITY AND POINTS OF INFLECTION

We begin with a picture of the graph of a function f, Figure 4.6.1. To the left of c_1 and between c_2 and c_3, the graph "curves up" (we call it *concave up*); between c_1 and c_2, and to the right of c_3, the graph "curves down" (we call it *concave down*). These terms deserve a precise definition.

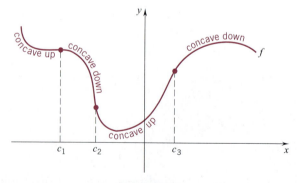

Figure 4.6.1

DEFINITION 4.6.1 CONCAVITY

Let the function f be differentiable on an open interval I. The graph of f is said to be *concave up* on I iff f' increases on I; it is said to be *concave down* on I iff f' decreases on I.

In other words, the graph is concave up on an open interval where the slope increases and concave down on an open interval where the slope decreases.

There is also a geometric interpretation of concavity: The graph of f is concave up on an interval I if the tangent lines to the graph are *below* the curve; the graph of f is concave down on I if the tangent lines to the graph are *above* the curve. You can verify this interpretation using Figure 4.6.1.

Points that join arcs of opposite concavity are called *points of inflection*. The graph in Figure 4.6.1 has three of them: $(c_1, f(c_1))$, $(c_2, f(c_2))$, $(c_3, f(c_3))$. Here is a formal definition.

DEFINITION 4.6.2 POINT OF INFLECTION

Let the function f be continuous at $x = c$. The point $(c, f(c))$ is called a *point of inflection* iff there exists $\delta > 0$ such that the graph of f is concave in one sense on $(c - \delta, c)$ and concave in the opposite sense on $(c, c + \delta)$.

Example 1   The graph of the quadratic function $f(x) = x^2 - 4x + 3$ is concave up everywhere since the derivative $f'(x) = 2x - 4$ is everywhere increasing. (See Figure 4.6.2.) ❏

Example 2   Consider the cubic function $f(x) = x^3$. You can verify that

$$f'(x) = 3x^2 \qquad \text{decreases on } (-\infty, 0] \text{ and increases on } [0, \infty).$$

Thus, the graph of f is concave down on $(-\infty, 0)$ and concave up on $(0, \infty)$. The origin, $(0, f(0)) = (0, 0)$, is a point of inflection. (See Figure 4.6.3.) ❏

If f is twice differentiable, then, remembering that $f'' = (f')'$, we can determine the concavity of the graph by looking at the sign of the second derivative.

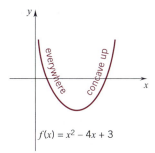

$f(x) = x^2 - 4x + 3$

Figure 4.6.2

THEOREM 4.6.3

Let f be twice differentiable on an open interval I.

(i) If $f''(x) > 0$ for all x in I, then f' increases on I and the graph of f is concave up.

(ii) If $f''(x) < 0$ for all x in I, then f' decreases on I and the graph of f is concave down.

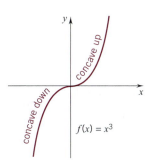

$f(x) = x^3$

Figure 4.6.3

PROOF   Apply the proof of Theorem 4.2.2 to the function f'. ❏

The following result gives us a way of identifying possible points of inflection.

> **THEOREM 4.6.4**
>
> If the point $(c, f(c))$ is a point of inflection, then either
> $$f''(c) = 0 \quad \text{or} \quad f''(c) \quad \text{does not exist.}$$

PROOF Suppose that $(c, f(c))$ is a point of inflection. Let's assume that the graph of f is concave up to the left of c and concave down to the right of c. The other case can be handled in a similar manner.

In this situation f' increases on an interval $(c - \delta, c)$ and decreases on an interval $(c, c + \delta)$.

Suppose now that $f''(c)$ exists. Then f' is continuous at c. It follows that f' increases on the half-open interval $(c - \delta, c]$ and decreases on the half-open interval $[c, c + \delta)$.† This says that f' has a local maximum at c. Since, by assumption, $f''(c)$ exists, $f''(c) = 0$. (Theorem 4.3.2 applied to f'.)

We have shown that if $f''(c)$ exists, then $f''(c) = 0$. The other possibility, of course, is that $f''(c)$ does not exist (see Example 4 below). ❑

Example 3 For the function
$$f(x) = x^3 - 6x^2 + 9x + 1 \qquad \text{(Figure 4.6.4)}$$
we have
$$f'(x) = 3x^2 - 12x + 9 = 3(x^2 - 4x + 3)$$
and
$$f''(x) = 6x - 12 = 6(x - 2).$$
Since $f''(x) = 0$ only at $x = 2$, f'' has constant sign on $(-\infty, 2)$ and on $(2, \infty)$. The sign of f'' on these intervals and the consequences for the graph of f are as follows:

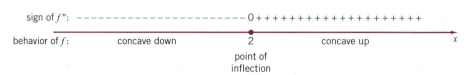

The point $(2, f(2)) = (2, 3)$ is a point of inflection. ❑

y

5 •

(2, 3) •

point of
inflection

1 2 3 x

$f(x) = x^3 - 6x^2 + 9x + 1$

Figure 4.6.4

Example 4 For
$$f(x) = 3x^{5/3} - 5x \qquad \text{(Figure 4.6.5)}$$
we have
$$f'(x) = 5x^{2/3} - 5 \qquad \text{and} \qquad f''(x) = \tfrac{10}{3} x^{-1/3}.$$
The second derivative does not exist at $x = 0$. Since
$$f''(x) \text{ is } \begin{cases} \text{negative,} & \text{for } x < 0 \\ \text{positive,} & \text{for } x > 0, \end{cases}$$

---
† See Exercise 56, Section 4.2

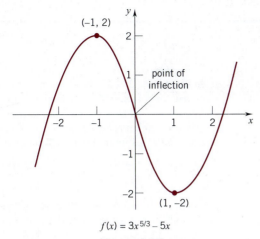

$$f(x) = 3x^{5/3} - 5x$$

Figure 4.6.5

the graph is concave down on $(-\infty, 0)$ and concave up on $(0, \infty)$. Since f is continuous at 0, the point $(0, f(0)) = (0, 0)$ is a point of inflection. ❑

CAUTION The fact that

$$f''(c) = 0 \quad \text{or} \quad f''(c) \text{ does not exist}$$

does not guarantee that $(c, f(c))$ is a point of inflection. As you can verify, the function $f(x) = x^4$ satisfies $f''(0) = 0$, but the graph is always concave up and there are no points of inflection. If f is discontinuous at c, then $f''(c)$ does not exist, but $(c, f(c))$ cannot be a point of inflection. A point of inflection occurs at c iff f is continuous at c and the point $(c, f(c))$ joins arcs of opposite concavity. ❑

Example 5 Determine the concavity and find the points of inflection (if any) of the graph of

$$f(x) = x + \cos x, \qquad x \in [0, 2\pi].$$

SOLUTION For $x \in (0, 2\pi)$, we have

$$f'(x) = 1 - \sin x$$

and

$$f''(x) = -\cos x.$$

Since $f''(x) = 0$ at $x = \pi/2$ and $3\pi/2$, f'' has constant sign on $(0, \pi/2)$, $(\pi/2, 3\pi/2)$, and $(3\pi/2, 2\pi)$. The sign of f'' on these intervals and the consequences for f are as follows:

points of inflection

$$\left(\tfrac{3\pi}{2}, \tfrac{3\pi}{2}\right)$$

$$\left(\tfrac{\pi}{2}, \tfrac{\pi}{2}\right)$$

$$f(x) = x + \cos x$$

Figure 4.6.6

The points $(\pi/2, f(\pi/2)) = (\pi/2, \pi/2)$ and $(3\pi/2, f(3\pi/2)) = (3\pi/2, 3\pi/2)$ are points of inflection. The graph of f is shown in Figure 4.6.6. ❑

EXERCISES 4.6

1. The graph of a function f is given in the figure below. (a) Determine the intervals on which f increases and the intervals on which f decreases; (b) determine the intervals on which the graph of f is concave up, the intervals on which the graph is concave down, and give the x-coordinates of the points of inflection.

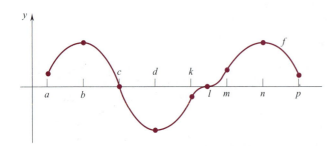

2. Repeat Exercise 1 for the function g whose graph is given in the figure below.

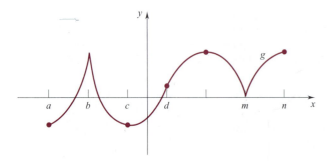

In Exercises 3–20, describe the concavity of the graph of f and find the points of inflection (if any).

3. $f(x) = \dfrac{1}{x}$.

4. $f(x) = x + \dfrac{1}{x}$.

5. $f(x) = x^3 - 3x + 2$.

6. $f(x) = 2x^2 - 5x + 2$.

7. $f(x) = \dfrac{1}{4}x^4 - \dfrac{1}{2}x^2$.

8. $f(x) = x^3(1 - x)$.

9. $f(x) = \dfrac{x}{x^2 - 1}$.

10. $f(x) = \dfrac{x + 2}{x - 2}$.

11. $f(x) = (1 - x)^2(1 + x)^2$.

12. $f(x) = \dfrac{6x}{x^2 + 1}$.

13. $f(x) = \dfrac{1 - \sqrt{x}}{1 + \sqrt{x}}$.

14. $f(x) = (x - 3)^{1/5}$.

15. $f(x) = (x + 2)^{5/3}$.

16. $f(x) = x\sqrt{4 - x^2}$.

17. $f(x) = \sin^2 x, \quad x \in [0, \pi]$.

18. $f(x) = 2\cos^2 x - x^2, \quad x \in [0, \pi]$.

19. $f(x) = x^2 + \sin 2x, \quad x \in [0, \pi]$.

20. $f(x) = \sin^4 x, \quad x \in [0, \pi]$.

In Exercises 21–28, find: (a) the intervals on which f increases or decreases; (b) the local maxima and minima; (c) the intervals on which the graph of f is concave up and the intervals on which it is concave down; and (d) the points of inflection. Use this information to sketch the graph of f.

21. $f(x) = x^3 - 9x$.

22. $f(x) = 3x^4 + 4x^3 + 1$.

23. $f(x) = \dfrac{2x}{x^2 + 1}$.

24. $f(x) = x^{1/3}(x - 6)^{2/3}$.

25. $f(x) = x + \sin x, \quad x \in [-\pi, \pi]$.

26. $f(x) = \sin x + \cos x, \quad x \in [0, 2\pi]$.

27. $f(x) = \begin{cases} x^3, & x < 1 \\ 3x - 2, & x \ge 1. \end{cases}$

28. $f(x) = \begin{cases} 2x + 4, & x \le -1 \\ 3 - x^2, & x > -1. \end{cases}$

In Exercises 29–32, sketch the graph of a continuous function f that satisfies the given conditions.

29. $f''(x) > 0$ if $x < 0$ or if $x > 2$, $f''(x) < 0$ if $0 < x < 2$; $f'(0) = f'(4) = 0, f'(x) < 0$ if $x < 4$ and $x \ne 0, f'(x) > 0$ if $x > 4$; $f(0) = 0, f(2) = -2, f(4) = -5$.

30. $f''(x) > 0$ if $|x| > 2$, $f''(x) < 0$ if $|x| < 2$; $f'(0) = 0$, $f'(x) > 0$ if $x < 0, f'(x) < 0$ if $x > 0; f(0) = 1, f(-2) = f(2) = \frac{1}{2}, f(x) > 0$ for all x and f is an even function.

31. $f''(x) < 0$ if $x < 0, f''(x) > 0$ if $x > 0; f'(-1) = f'(1) = 0, f'(0)$ does not exist, $f'(x) > 0$ if $|x| > 1, f'(x) < 0$ if $|x| < 1 \ (x \ne 0); f(-1) = 2, f(1) = -2, f$ is an odd function.

32. $f''(x) < 0$ if $x < -3$ or if $0 < x < 4$; $f''(x) > 0$ if $-3 < x < 0$ or if $x > 4; f'(-1) = f'(2) = 0, f'(x) < 0$ if $x < -1$ or if $x > 2, f'(x) > 0$ if $-1 < x < 2; f(-1) = -2, f(0) = 0, f(2) = 2, f(x) < 0$ if $x < 0, f(x) > 0$ if $x > 0$.

33. Find d given that $(d, f(d))$ is a point of inflection of the graph of
$$f(x) = (x - a)(x - b)(x - c).$$

34. Find c given that the graph of $f(x) = cx^2 + x^{-2}$ has a point of inflection at $(1, f(1))$.

35. Find a and b given that the graph of $f(x) = ax^3 + bx^2$ passes through $(-1, 1)$ and has a point of inflection at $x = \frac{1}{3}$.

36. Determine A and B so that the curve
$$y = Ax^{1/2} + Bx^{-1/2}$$
will have a point of inflection at $(1, 4)$.

37. Determine A and B so that the curve

$$y = A \cos 2x + B \sin 3x$$

will have a point of inflection at $(\pi/6, 5)$

38. Find necessary and sufficient conditions on A and B for $f(x) = Ax^2 + Bx + C$:
  (a) to decrease between A and B with graph concave up.
  (b) to increase between A and B with graph concave down.

39. Find a function f such that $f'(x) = 3x^2 - 6x + 3$ and $(1, -2)$ is a point of inflection of the graph of f.

40. Prove that the graph of a polynomial of degree n has at most $n-2$ points of inflection.

41. Given the cubic polynomial $p(x) = x^3 + ax^2 + bx + c$.
  (a) Prove that the graph of p has exactly one point of inflection.
  (b) Show that p has local extreme values iff $a^2 \geq 3b$.

42. Prove that if the cubic polynomial $p(x) = x^3 + ax^2 + bx + c$ has a local maximum and a local minimum, then the point of inflection is the midpoint of the line segment connecting the local extrema on the graph.

In Exercises 43–46, use a graphing utility to graph the function f on the indicated interval. (a) Estimate the intervals where the graph of f is concave up and the intervals where it is concave down. (b) Estimate the x-coordinate of each point of inflection. Round off your estimates to three decimal places.

43. $f(x) = x^4 - 5x^2 + 3; [-4, 4]$.

44. $f(x) = x \sin x; [-2\pi, 2\pi]$.

45. $f(x) = 1 + x^2 - 2x \cos x; [-\pi, \pi]$.

46. $f(x) = x^{2/3}(x^2 - 4); [-5, 5]$.

In Exercises 47–48, the second derivative of a function f is given. Use a graphing utility to graph f'' on the indicated interval. (a) Estimate the intervals where the graph of f is concave up and the intervals where it is concave down. (b) Estimate the x-coordinate of each point of inflection. Round off your estimates to three decimal places.

47. $f''(x) = x^3 + \tfrac{1}{2}x^2 - \tfrac{16}{3}x + 2; [-3, 2.5]$.

48. $f''(x) = 2x \sin x - 2; [-\pi, \pi]$.

■ 4.7 VERTICAL AND HORIZONTAL ASYMPTOTES; VERTICAL TANGENTS AND CUSPS

Vertical and Horizontal Asymptotes

In Figure 4.7.1 you can see the graph of the function

$$f(x) = \frac{1}{|x - c|} \qquad \text{for } x \text{ close to } c.$$

As $x \to c$, $f(x) \to \infty$, that is, given any positive number M there exists a positive number δ such that

$$\text{if} \quad 0 < |x - c| < \delta, \qquad \text{then} \quad f(x) \geq M.$$

The line $x = c$ is called a *vertical asymptote*. Figure 4.7.2 shows the graph of

$$g(x) = -\frac{1}{|x - c|} \qquad \text{for } x \text{ close to } c.$$

In this case, we see that $g(x) \to -\infty$ as $x \to c$. Again the line $x = c$ is called a *vertical asymptote*.

Figure 4.7.1

Figure 4.7.2

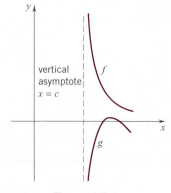

Figure 4.7.3

Figure 4.7.4

Vertical asymptotes can also arise from one-sided behavior. With f and g as in Figure 4.7.3, we write

$$\text{as} \quad x \to c^-, \quad f(x) \to \infty \quad \text{and} \quad g(x) \to -\infty.$$

With f and g as in Figure 4.7.4, we write

$$\text{as} \quad x \to c^+, \quad f(x) \to \infty \quad \text{and} \quad g(x) \to -\infty.$$

In each case the vertical line $x = c$ is a vertical asymptote for both functions. To summarize, the line $x = c$ is a vertical asymptote for a function f if any one of the following conditions holds: $f(x) \to \infty$ or $-\infty$ as $x \to c$; $f(x) \to \infty$ or $-\infty$ as $x \to c^-$; or $f(x) \to \infty$ or $-\infty$ as $x \to c^+$. Typically, to locate the vertical asymptotes for a function f, find the values $x = c$ at which f is discontinuous and examine the behavior of f as x approaches c.

Example 1 The function

$$f(x) = \frac{3x + 6}{x^2 - 2x - 8} = \frac{3(x + 2)}{(x + 2)(x - 4)}$$

is defined and continuous everywhere except $x = 4$ and $x = -2$. As $x \to 4^+$, $f(x) \to \infty$; as $x \to 4^-$, $f(x) \to -\infty$. Thus, the line $x = 4$ is a vertical asymptote. Since $\lim_{x \to -2} f(x) = -\frac{1}{2}$ exists, f does not have a vertical asymptote at $x = -2$. See Figure 4.7.5. ❑

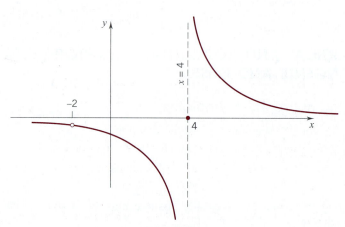

Figure 4.7.5

Recall that $f(x) = \tan x \to \infty$ as $x \to (\pi/2)^-$ and $\tan x \to -\infty$ as $x \to (\pi/2)^+$. Thus the line $x = \pi/2$ is a vertical asymptote for the tangent function. Indeed, the lines $x = (2n + 1)\pi/2$, $n = 0, \pm 1, \pm 2, \ldots$, are all vertical asymptotes for the tangent function (Figure 4.7.6).

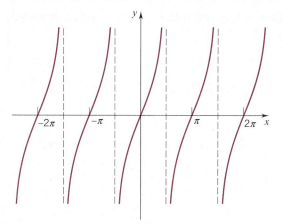

Figure 4.7.6

It is also possible for a function to have a *horizontal asymptote.* If there is a number L such that $f(x) \to L$ as $x \to \infty$, or if $f(x) \to L$ as $x \to -\infty$, then the horizontal line $y = L$ is called a *horizontal asymptote.* See Figures 4.7.7. and 4.7.8. To be precise, $f(x) \to L$ as $x \to \infty$ means that given a positive number ϵ there is a positive number K such that if $x \geq K$, then $|f(x) - L| < \epsilon$. Similarly, $f(x) \to L$ as $x \to -\infty$ means that given a positive number ϵ there is a negative number M such that if $x \leq M$, then $|f(x) - L| < \epsilon$.

Figure 4.7.7

Figure 4.7.8

Example 2 Figure 4.7.9 shows the graph of the function

$$f(x) = \frac{x}{x - 2}.$$

As $x \to 2^-$, $f(x) \to -\infty$; as $x \to 2^+$, $f(x) \to \infty$. The line $x = 2$ is a vertical asymptote.
As $x \to \pm\infty$,†

$$f(x) = \frac{x}{x - 2} = \frac{x}{x\left(1 - \dfrac{2}{x}\right)} = \frac{1}{1 - \dfrac{2}{x}} \to 1.$$

The line $y = 1$ is a horizontal asymptote. ❑

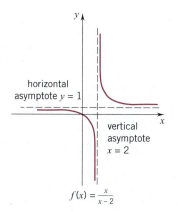

$f(x) = \frac{x}{x-2}$

Figure 4.7.9

† Recall that $x \to \pm\infty$ means $x \to \infty$ or $x \to -\infty$.

Example 3 Figure 4.7.10 is a computer-generated graph of the function

$$f(x) = \frac{\cos x}{x}.$$

As $x \to 0^-$, $f(x) \to -\infty$; as $x \to 0^+$, $f(x) \to \infty$. The line $x = 0$ (the y-axis) is a vertical asymptote.

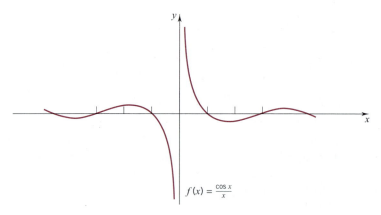

$f(x) = \frac{\cos x}{x}$

Figure 4.7.10

As $x \to \pm\infty$,

$$f(x) = \frac{\cos x}{x} \to 0.$$

This follows from the fact that

$$\left| \frac{\cos x}{x} \right| \leq \frac{1}{|x|} \qquad \text{for all} \quad x$$

and $1/|x| \to 0$ as $x \to \pm\infty$. Thus, the line $y = 0$ (the x-axis) is a horizontal asymptote. Note that f is an odd function $[\, f(-x) = -f(x)]$ so its graph is symmetric with respect to the origin. ❑

Example 4 Find the vertical and horizontal asymptotes, if any, of the function

$$g(x) = \frac{x + 1 - \sqrt{x}}{x^2 - 2x + 1} = \frac{x + 1 - \sqrt{x}}{(x - 1)^2}.$$

SOLUTION The domain of g is $0 \leq x < \infty$, $x \neq 1$. As $x \to 1$, $g(x) \to \infty$. Thus, the line $x = 1$ is a vertical asymptote. The behavior of g as $x \to \infty$ can be made more apparent by writing

$$g(x) = \frac{x + 1 - \sqrt{x}}{x^2 - 2x + 1} = \frac{x \left(1 + \dfrac{1}{x} - \dfrac{1}{\sqrt{x}} \right)}{x^2 \left(1 - \dfrac{2}{x} + \dfrac{1}{x^2} \right)} = \frac{1 + \dfrac{1}{x} - \dfrac{1}{\sqrt{x}}}{x \left(1 - \dfrac{2}{x} + \dfrac{1}{x^2} \right)}.$$

Now, it is easy to see that $g(x) \to 0$ as $x \to \infty$. The line $y = 0$ (the x-axis) is a horizontal asymptote. ❑

The technique suggested in Examples 2 and 4—factoring out the highest power of x from the numerator and the denominator—can be used to prove the following general result concerning the behavior of a rational function as $x \to \pm\infty$. Let

$$R(x) = \frac{p(x)}{q(x)} = \frac{a_n x^n + \cdots + a_1 x + a_0}{b_k x^k + \cdots + b_1 x + b_0}, \quad a_n \neq 0, b_k \neq 0,$$

be a rational function. Then,

(4.7.1)
$$\text{as} \quad x \to \pm\infty, \begin{cases} R(x) \to 0, & \text{if } n < k \\ R(x) \to \dfrac{a_n}{b_n}, & \text{if } n = k \\ R(x) \to \pm\infty, & \text{if } n > k. \end{cases}$$

Example 5 The function

$$f(x) = \frac{5 - 3x^2}{1 - x^2}$$

is continuous everywhere except $x = \pm 1$.

The line $x = 1$ is a vertical asymptote:

$$\text{as} \quad x \to 1, \quad 5 - 3x^2 \to 2 \quad \text{and} \quad 1 - x^2 \to 0.$$

In particular:

$$\text{as} \quad x \to 1^-, \quad 1 - x^2 \text{ is positive so } f(x) \to \infty;$$
$$\text{as} \quad x \to 1^+, \quad 1 - x^2 \text{ is negative so } f(x) \to -\infty.$$

The line $x = -1$ is a vertical asymptote:

$$\text{as} \quad x \to -1, \quad 5 - 3x^2 \to 2 \quad \text{and} \quad 1 - x^2 \to 0.$$

In particular:

$$\text{as} \quad x \to -1^+, \quad 1 - x^2 \text{ is positive so } f(x) \to \infty;$$
$$\text{as} \quad x \to -1^-, \quad 1 - x^2 \text{ is negative so } f(x) \to -\infty.$$

Since f is a rational function, we can use (4.7.1) to investigate the behavior as $x \to \pm\infty$. We have $n = k = 2$ and so $f(x) \to -3/-1 = 3$ as $x \to \pm\infty$. Thus, the line $y = 3$ is a horizontal asymptote. The graph of f is shown in Figure 4.7.11. Note that f is an even function $[f(-x) = f(x)]$; its graph is symmetric with respect to the y-axis. ❑

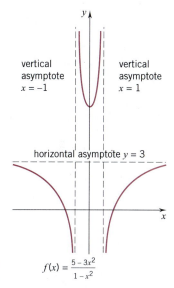

$$f(x) = \frac{5 - 3x^2}{1 - x^2}$$

Figure 4.7.11

Vertical Tangents; Vertical Cusps

(For the remainder of this section assume that f is continuous at $x = c$ and differentiable for $x \neq c$.)

We say that the graph of f has a *vertical tangent* at the point $(c, f(c))$ iff

$$\text{as} \quad x \to c, \quad f'(x) \to \infty \quad \text{or} \quad f'(x) \to -\infty.$$

Examples

(a) The function $f(x) = x^{1/3}$ has a vertical tangent at the point $(0,0)$ since

$$f'(x) = \frac{1}{3}x^{-2/3} \to \infty \quad \text{as} \quad x \to 0 \qquad \text{(Figure 4.7.12a)}.$$

(b) The function $g(x) = (2 - x)^{1/5}$ has a vertical tangent at the point $(2, 0)$ since

$$g'(x) = -\frac{1}{5}(2 - x)^{-4/5} \to -\infty \quad \text{as} \quad x \to 2 \qquad \text{(Figure 4.7.12b)}.$$

❏

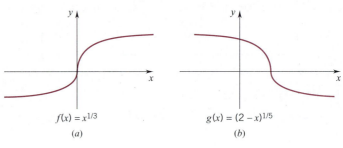

$f(x) = x^{1/3}$ $g(x) = (2 - x)^{1/5}$

(a)   (b)

Figure 4.7.12

On occasion you will see a graph become almost vertical and then virtually double back on itself. Such a pattern signals the presence of what is known as a "vertical cusp." We say that the graph of f has a *vertical cusp* at $(c, f(c))$ iff

$$\text{as} \quad x \to c^-, \quad f'(x) \to -\infty \qquad \text{and} \qquad \text{as} \quad x \to c^+, \quad f'(x) \to \infty,$$

or

$$\text{as} \quad x \to c^-, \quad f'(x) \to \infty \qquad \text{and} \qquad \text{as} \quad x \to c^+, \quad f'(x) \to -\infty.$$

Examples

(a) The function $f(x) = x^{2/3}$ has a vertical cusp at $(0, 0)$: $f'(x) = \frac{2}{3}x^{-1/3}$ and

$$\text{as} \quad x \to 0^-, \quad f'(x) \to -\infty, \qquad \text{as} \quad x \to 0^+, \quad f'(x) \to \infty \quad \text{(Figure 4.7.13a)}.$$

(b) The function $g(x) = 2 - (x - 1)^{2/5}$ has a vertical cusp at $(1, 2)$: $g'(x) = -\frac{2}{5}(x - 1)^{-3/5}$ and

$$\text{as} \quad x \to 1^-, \quad g'(x) \to \infty, \qquad \text{as} \quad x \to 1^+, \quad g'(x) \to -\infty. \quad \text{(Figure 4.7.13b)}.$$

❏

$f(x) = x^{2/3}$ $g(x) = 2 - (x - 1)^{2/5}$

(a)   (b)

Figure 4.7.13

The fact that $f'(c)$ does not exist does *not* mean that the graph of f has either a vertical tangent or a vertical cusp at $(c, f(c))$. Unless the conditions spelled out earlier are met, the graph of f can simply be making a "corner" at $(c, f(c))$. For example, the function

$$f(x) = |x^3 - 1|$$

has derivative

$$f'(x) = \begin{cases} -3x^2, & x < 1 \\ 3x^2, & x > 1. \end{cases}$$

At $x = 1$, $f'(x)$ does not exist. As $x \to 1^-$, $f'(x) \to -3$, and as $x \to 1^+$, $f'(x) \to 3$. There is no vertical tangent here and no vertical cusp. For the graph, see Figure 4.7.14.

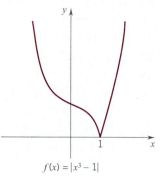

$f(x) = |x^3 - 1|$

Figure 4.7.14

EXERCISES 4.7

1. The graph of a function f is given in the figure below.

(a) As $x \to -1$, $f(x) \to ?$ (b) As $x \to 1^-$, $f(x) \to ?$
(c) As $x \to 1^+$, $f(x) \to ?$ (d) As $x \to \infty$, $f(x) \to ?$
(e) As $x \to -\infty$, $f(x) \to ?$
(f) Give the equations of the vertical asymptotes, if any.
(g) Give the equations of the horizontal asymptotes, if any.

2. The graph of a function g is given in the figure below.

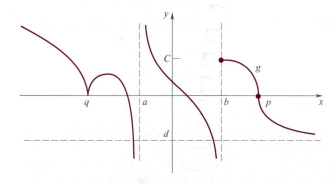

(a) As $x \to \infty$, $g(x) \to ?$ (b) As $x \to b^+$, $g(x) \to ?$
(c) Give the equations of the vertical asymptotes, if any.

(d) Give the equations of the horizontal asymptotes, if any.
(e) Give the numbers c, if any, at which the graph of g has a vertical tangent line.
(f) Give the numbers c, if any, at which the graph of g has a vertical cusp.

In Exercises 3–20, find the vertical and horizontal asymptotes.

3. $f(x) = \dfrac{x}{3x - 1}$.

4. $f(x) = \dfrac{x^3}{x + 2}$.

5. $f(x) = \dfrac{x^2}{x - 2}$.

6. $f(x) = \dfrac{4x}{x^2 + 1}$.

7. $f(x) = \dfrac{2x}{x^2 - 9}$.

8. $f(x) = \dfrac{\sqrt{x}}{4\sqrt{x} - x}$.

9. $f(x) = \left(\dfrac{2x - 1}{4 + 3x}\right)^2$.

10. $f(x) = \dfrac{4x^2}{(3x - 1)^2}$.

11. $f(x) = \dfrac{3x}{(2x - 5)^2}$.

12. $f(x) = \left(\dfrac{x}{1 - 2x}\right)^3$.

13. $f(x) = \dfrac{3x}{\sqrt{4x^2 + 1}}$.

14. $f(x) = \dfrac{x^{1/3}}{x^{2/3} - 4}$.

15. $f(x) = \dfrac{\sqrt{x}}{2\sqrt{x} - x - 1}$.

16. $f(x) = \dfrac{2x}{\sqrt{x^2 - 1}}$.

17. $f(x) = \sqrt{x + 4} - \sqrt{x}$.

18. $f(x) = \sqrt{x} - \sqrt{x - 2}$.

19. $f(x) = \dfrac{\sin x}{\sin x - 1}$.

20. $f(x) = \dfrac{1}{\sec x - 1}$.

In Exercises 21–34, determine whether or not the graph of f has a vertical tangent or a vertical cusp at c.

21. $f(x) = (x + 3)^{4/3}$; $c = -3$.

22. $f(x) = 3 + x^{2/5}$; $c = 0$.

23. $f(x) = (2 - x)^{4/5}$; $c = 2$.

24. $f(x) = (x + 1)^{-1/3}$; $c = -1$.

25. $f(x) = 2x^{3/5} - x^{6/5}; c = 0.$

26. $f(x) = (x - 5)^{7/5}; c = 5.$

27. $f(x) = (x + 2)^{-2/3}; c = -2.$

28. $f(x) = 4 - (2 - x)^{3/7}; c = 2.$

29. $f(x) = \sqrt{|x - 1|}; c = 1.$

30. $f(x) = x(x - 1)^{1/3}; c = 1.$

31. $f(x) = |(x + 8)^{1/3}|; c = -8.$

32. $f(x) = \sqrt{4 - x^2}; c = 2.$

33. $f(x) = \begin{cases} x^{1/3} + 2, & x \le 0 \\ 1 - x^{1/5}, & x > 0; \end{cases} c = 0.$

34. $f(x) = \begin{cases} 1 + \sqrt{-x}, & x \le 0 \\ (4x - x^2)^{1/3}, & x > 0; \end{cases} c = 0.$

In Exercises 35–38, sketch the graph of the given function, showing all asymptotes.

35. $f(x) = \dfrac{x + 1}{x - 2}.$

36. $f(x) = \dfrac{1}{(x + 1)^2}.$

37. $f(x) = \dfrac{x}{1 + x^2}.$

38. $f(x) = \dfrac{x - 2}{x^2 - 5x + 6}.$

▶C In Exercises 39–44, find (a) the intervals on which f increases or decreases, and (b) the intervals of which the graph of f is concave up and the intervals on which it is concave down. Also, determine whether the graph of f has vertical tangents or vertical cusps. Confirm your results using a graphing utility. Then sketch the graph of f.

39. $f(x) = x - 3x^{1/3}.$

40. $f(x) = x^{2/3} - x^{1/3}.$

41. $f(x) = \frac{3}{5}x^{5/3} - 3x^{2/3}.$

42. $f(x) = \sqrt{|x|}.$

43. $f(x) = \dfrac{x^{2/3} - 1}{|x^{1/3} - 1|}.$

44. $f(x) = |(4 + x)^{3/5}|.$

In Exercises 45–50, sketch the graph of a continuous function f that satisfies the given conditions. Indicate whether the graph of f has any horizontal or vertical asymptotes, and whether the graph has any vertical tangent lines or vertical cusps. If no such function exists, explain why.

45. $f(3) = 0, f(0) = 4, f(-1) = 0, f(-2) = -3, f(x) \to \infty$ as $x \to 1^-$, $f(x) \to -\infty$ as $x \to 1^+$, $f(x) \to 2$ as $x \to \infty$, $f(x) \to 0$ as $x \to -\infty$; $f'(x) < 0$ if $x < -2$, $f'(x) > 0$ if $x > -2$, $x \ne 1$; $f''(x) < 0$ if $x > 1$ or if $x < -4$, $f''(x) > 0$ if $-4 < x < 1$.

46. $f(0) = 0,\ f(3) = f(-3) = 0,\ f(x) \to -\infty$ as $x \to 1$, $f(x) \to -\infty$ as $x \to -1$, $f(x) \to 1$ as $x \to \infty$, $f(x) \to 1$ as $x \to -\infty$; $f''(x) < 0$ for all $x, x \ne \pm 1$.

47. $f(x) \ge 1$ for all x, $f(0) = 1$; $f''(x) < 0$ for all $x \ne 0$; $f'(x) \to \infty$ as $x \to 0^+$, $f'(x) \to -\infty$ as $x \to 0^-$.

48. $f(0) = 1,\ f(x) \to 4$ as $x \to \infty,\ f(x) \to -\infty$ as $x \to -\infty$; $f'(x) \to \infty$ as $x \to 0$; $f''(x) > 0$ if $x < 0$, $f''(x) < 0$ if $x > 0$.

49. $f(0) = 0,\ f(x) \to -1$ as $x \to \infty,\ f(x) \to 1$ as $x \to -\infty$; $f'(x) \to -\infty$ as $x \to 0$; $f''(x) < 0$ if $x < 0$, $f''(x) > 0$ if $x > 0$; f is an odd function.

50. $f(0) = 1, f(2) = 0; f'(x) \to -\infty$ as $x \to 0^-, f'(x) \to 0$ as $x \to 0^+; f''(x) < 0$ for all $x \ne 0$.

51. Let p and q be positive integers with q odd and $p < q$. Let $f(x) = x^{p/q}$. Specify conditions on p and q so that:
  (a) The graph of f will have a vertical tangent line at $x = 0$.
  (b) The graph of f will have a vertical cusp at $x = 0$.

52. Let $r(x) = [p(x)/q(x)]$ be a rational function and suppose that (degree of p) = (degree of q) + 1.
  (a) Show that r can be written in the form
$$r(x) = ax + b + \frac{Q(x)}{q(x)},$$
  where (degree Q) < (degree q).   HINT: divide q into p.
  (b) Prove that $[r(x) - (ax + b)] \to 0$ as $x \to \infty$ and $[r(x) - (ax + b)] \to 0$ as $x \to -\infty$. Thus the graph of f "approaches" the line $y = ax + b$ as $x \to \pm\infty$. The line $y = ax + b$ is called an *oblique asymptote* of the graph of r.

In Exercises 53–56, sketch the graph of the given function showing all vertical and oblique asymptotes.

53. $f(x) = \dfrac{x^2 - 4}{x}.$

54. $f(x) = \dfrac{2x^2 + 3x - 2}{x + 1}.$

55. $f(x) = \dfrac{x^3}{(x - 1)^2}.$

56. $f(x) = \dfrac{1 + x - 3x^2}{x}.$

■ 4.8 SOME CURVE SKETCHING

During the course of the last few sections you have learned how to find the extreme values of a function, the intervals where it increases and the intervals where it decreases; you have seen how to determine the concavity of the graph and how to find the points of inflection; and in the preceding section you learned how to investigate the asymptotic properties of a function. This knowledge makes it possible to sketch an accurate graph of a somewhat complicated function without having to plot point, after point, after point.

Before attempting to sketch the graph of a function f, we try to gather the necessary information and record it in an organized form. Here is an outline of the procedure we will follow.

1. Domain of f: Determine the domain of f; identify endpoints; find the vertical asymptotes of f; determine the behavior of f as $x \to \pm\infty$, find the horizontal asymptotes.

2. Intercepts: Determine the x- and y-intercepts of the graph; the y-intercept is the value $f(0)$; the x-intercepts are the solutions of the equation $f(x) = 0$.

3. Symmetry and periodicity: If f is an *even function* $[\,f(-x) = f(x)\,]$, then the graph is symmetric with respect to the y-axis; if f is an *odd function* $[\,f(-x) = -f(x)\,]$, then the graph is symmetric with respect to the origin. A function f is *periodic* if there exists a positive number p such that $f(x + p) = f(x)$ whenever x and $x + p$ are in the domain of f.

4. Calculate f': Determine the critical numbers of f; examine the sign of f' to determine the intervals on which f increases and the intervals on which f decreases; determine vertical tangents and cusps.

5. Calculate f'': Examine the sign of f'' to determine the intervals on which the graph is concave up and the intervals on which the graph is concave down; determine the points of inflection.

6. Plot the points of interest in a preliminary sketch: intercepts, "extreme" points (i.e., local and absolute extrema), and points of inflection.

7. Sketch the graph by connecting the points in the preliminary sketch; make sure that the graph "rises," "falls," and "bends" in the proper way.

Figure 4.8.1 shows some of the elements that we might include in a preliminary sketch.

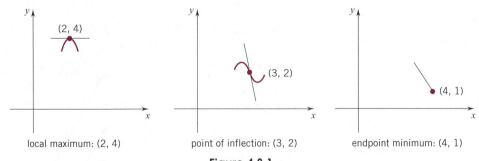

local maximum: (2, 4)   point of inflection: (3, 2)   endpoint minimum: (4, 1)

Figure 4.8.1

Example 1 Sketch the graph of the function

$$f(x) = \frac{1}{4}x^4 - 2x^2 + \frac{7}{4}.$$

SOLUTION

(1) Domain: This is a polynomial function, so its domain is the set of all real numbers. Since the leading term is $\frac{1}{4}x^4$, $f(x) \to \infty$ as $x \to \pm\infty$. There are no asymptotes.

(2) Intercepts: The y-intercept is $f(0) = \frac{7}{4}$. To find the x-intercepts, we solve the equation

$$f(x) = \tfrac{1}{4}x^4 - 2x^2 + \tfrac{7}{4} = 0$$

$$x^4 - 8x^2 + 7 = 0$$

$$(x^2 - 1)(x^2 - 7) = 0$$

$$(x + 1)(x - 1)(x + \sqrt{7})(x - \sqrt{7}) = 0.$$

Thus, the x-intercepts are $x = \pm 1$ and $x = \pm \sqrt{7}$.

(3) Symmetry/periodicity: Since
$f(-x) = \frac{1}{4}(-x)^4 - 2(-x)^2 + \frac{7}{4} = \frac{1}{4}x^4 - 2x^2 + \frac{7}{4} = f(x)$, f is an even function and its graph is symmetric with respect to the y-axis; f is not a periodic function.

(4) First derivative:

$$f'(x) = x^3 - 4x = x(x^2 - 4) = x(x + 2)(x - 2).$$

The critical numbers of f are $x = 0, x = \pm 2$. The sign of f' and behavior of f are:

(5) Second derivative:

$$f''(x) = 3x^2 - 4 = 3\left(x - \frac{2}{\sqrt{3}}\right)\left(x + \frac{2}{\sqrt{3}}\right).$$

The sign of f'' and the behavior of the graph of f are:

(6) Points of interest: (See Figure 4.8.2 for a preliminary sketch.)

$(0, \frac{7}{4})$: y-intercept $\frac{7}{4}$.

$(-1, 0), (1, 0), (-\sqrt{7}, 0), (\sqrt{7}, 0)$: x-intercepts.

$(0, \frac{7}{4})$: $f(0) = \frac{7}{4}$ is a local maximum.

$(-2, -\frac{9}{4}), (2, -\frac{9}{4})$: $f(-2) = f(2) = -\frac{9}{4}$ are local (and absolute) minimums.

$(-2/\sqrt{3}, -17/36), 2/\sqrt{3}, -17/36)$:   points of inflection; slopes $\pm 16/(3\sqrt{3}) \cong \pm 3.1$.

(7) Sketch the graph: Of course, since the graph is symmetric with respect to the y-axis, we can sketch the graph for $x \geq 0$, and then the graph for $x > 0$ is the mirror image of that in the y-axis. See Figure 4.8.3. ❏

Figure 4.8.2

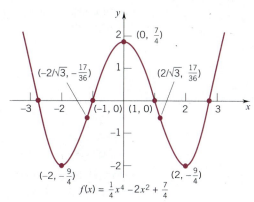

$$f(x) = \tfrac{1}{4}x^4 - 2x^2 + \tfrac{7}{4}$$

Figure 4.8.3

Example 2 Sketch the graph of the function

$$f(x) = x^4 - 4x^3 + 1 \qquad x \in [-1, 5).$$

SOLUTION

(1) Domain: The domain is given as $[-1, 5)$; -1 is the left endpoint; and 5 is a "missing right endpoint." There are no asymptotes and we do not consider the behavior of f as $x \to \pm\infty$ since f is defined only on $[-1, 5)$.

(2) Intercepts: The y-intercept is $f(0) = 1$. To find the x-intercepts, we must solve the equation

$$x^4 - 4x^3 + 1 = 0.$$

We cannot do this exactly, but you can verify that $f(0) > 0$ and $f(1) < 0$, and that $f(3) < 0$ and $f(4) > 0$. Thus there are x-intercepts on the intervals $(0, 1)$ and $(3, 4)$. We could find approximate values for these intercepts using the Newton–Raphson method but we do not need to do this since we are only trying to sketch the graph.

(3) Symmetry/periodicity: The graph is not symmetric with respect to the y-axis or the origin $[f(-x) \neq \pm f(x)]$; f is not periodic.

(4) First derivative: For $x \in (-1, 5)$

$$f'(x) = 4x^3 - 12x^2 = 4x^2(x - 3).$$

The critical numbers of f are $x = 0$ and $x = 3$.

(5) Second derivative:

$$f''(x) = 12x^2 - 24x = 12x(x - 2).$$

sign of f'': $+ + + + + + + 0 - - - - - - - - - - - - - - 0 +$

behavior of graph: -1 0   2   5 x

concave up   point of inflection   concave down   point of inflection   concave up

Figure 4.8.4

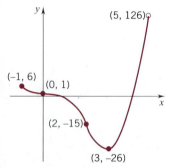

Figure 4.8.5

(6) Points of interest: (See Figure 4.8.4 for a preliminary sketch.)

$(0, 1)$: y-intercept; point of inflection with horizontal tangent.

$(-1, 6)$: $f(-1) = 6$ is an endpoint maximum.

$(2, -15)$: point of inflection; slope -16.

$(3, -26)$: $f(3) = -26$ is a local (and absolute) minimum.

As x approaches the missing endpoint 5 from the left, $f(x)$ increases toward a value of 126.

(7) Sketch the graph. Since the range of f makes a scale drawing impractical, we must be content with a rough sketch as in Figure 4.8.5. In cases like this, it is particularly important to give the coordinates of the points of interest. ❏

Example 3 Sketch the graph of the function

$$f(x) = \frac{x^2 - 3}{x^3}.$$

SOLUTION

(1) Domain: The domain of f is $\{x: x \neq 0\} = (-\infty, 0) \cup (0, \infty)$. The y-axis (the line $x = 0$) is a vertical asymptote since $f(x) \to \infty$ as $x \to 0^-$ and $f(x) \to -\infty$ as $x \to 0^+$. The x-axis (the line $y = 0$) is a horizontal asymptote: $f(x) \to 0$ as $x \to \pm\infty$.

(2) Intercepts: There is no y-intercept since $f(0)$ is not defined. The x-intercepts are $x = \pm\sqrt{3}$.

(3) Symmetry: Since

$$f(-x) = \frac{(-x)^2 - 3}{(-x)^3} = -\frac{x^2 - 3}{x^3} = -f(x),$$

the graph is symmetric with respect to the origin; f is not periodic.

(4) First derivative: It is easier to calculate f' if we first rewrite f as

$$f(x) = \frac{x^2 - 3}{x^3} = x^{-1} - 3x^{-3}.$$

Now,

$$f'(x) = -x^{-2} + 9x^{-4} = \frac{9 - x^2}{x^4}.$$

The critical numbers of f are $x = \pm 3$. (NOTE: $x = 0$ is not a critical number since 0 is not in the domain of f.)

The sign of f' and the behavior of f are:

(5) Second derivative:

$$f''(x) = 2x^{-3} - 36x^{-5} = \frac{2(x^2 - 18)}{x^5} = \frac{2(x - 3\sqrt{2})(x + 3\sqrt{2})}{x^5}.$$

The sign of f'' and the behavior of the graph of f are:

sign of f': $------0+++++++++++$ dne $----------0+++++++++$ x

behavior of graph:

$\frac{-3}{\sqrt{2}}$ 0 $\frac{3}{\sqrt{2}}$

concave   point of   concave   concave   point of   concave
down   inflection   up   down   inflection   up

(6) Points of interest: (See Figure 4.8.6 for a preliminary sketch.)

$(-\sqrt{3}, 0), (\sqrt{3}, 0)$: x-intercepts.

$(-3, -2/9)$:   local minimum.

$(3, 2/9)$:   local maximum.

$(-3\sqrt{2}, -5\sqrt{2}/36), (3\sqrt{2}, 5\sqrt{2}/36)$:   points of inflection.

Figure 4.8.6

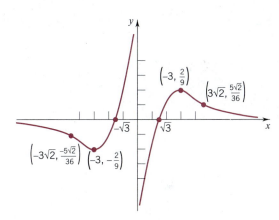

Figure 4.8.7

(7) Sketch the graph: see Figure 4.8.7. ❑

Example 4   Sketch the graph of the function

$$f(x) = \frac{3}{5}x^{5/3} - 3x^{2/3}.$$

SOLUTION

(1) Domain: The domain of f is the set of real numbers. Since we can write $f(x) = x^{2/3}(\frac{3}{5}x - 3)$, it is clear that, as $x \to \infty$, $f(x) \to \infty$, and as $x \to -\infty$, $f(x) \to -\infty$. There are no asymptotes.

(2) Intercepts: $f(0) = 0$ is both the y-intercept and an x-intercept; $x = 5$ is also an x-intercept.

(3) Symmetry/periodicity: There is no symmetry; f is not periodic.

(4) First derivative:

$$f'(x) = x^{2/3} - 2x^{-1/3} = \frac{x - 2}{x^{1/3}}.$$

The critical numbers of f are $x = 0$ and $x = 2$. The sign of f' and the behavior of f are:

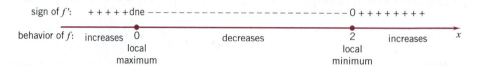

Note that, as $x \to 0^-$, $f'(x) \to \infty$, and as $x \to 0^+$, $f'(x) \to -\infty$. Thus the graph of f has a vertical cusp at $x = 0$.

(5) Second derivative:

$$f''(x) = \frac{2}{3}x^{-1/3} + \frac{2}{3}x^{-4/3} = \frac{2}{3}x^{-4/3}(x + 1).$$

The sign of f'' and the behavior of the graph of f are:

(6) Points of interest: (See Figure 4.8.8 for a preliminary sketch.)

    $(0, 0)$: y-intercept and a local maximum; vertical cusp at $(0, 0)$.

    $(0, 0), (5, 0)$: x-intercepts.

    $(2, -9\sqrt[3]{4}/5)$: $f(2) = -9\sqrt[3]{4}/5 \approx -2.9$ is a local minimum.

    $(-1, -18/5)$:   point of inflection; slope 3.

(7) Sketch the graph: See Figure 4.8.9. ❏

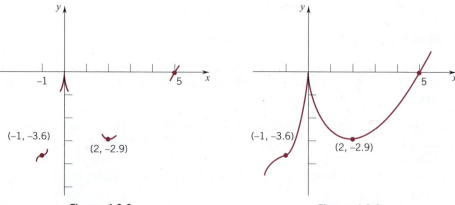

Figure 4.8.8   Figure 4.8.9

Example 5   Sketch the graph of

$$f(x) = \sin 2x - 2 \sin x.$$

SOLUTION

(1) Domain: The domain of f is the set of all real numbers.   There are no asymptotes and you should verify that the graph of f oscillates between $\frac{3}{2}\sqrt{3}$ and $-\frac{3}{2}\sqrt{3}$ as $x \to \pm\infty$.

(2) Intercepts: The y-intercept is $f(0) = 0$.   To find the x-intercepts, we set $f(x) = 0$:

$$\sin 2x - 2 \sin x = 2 \sin x \cos x - 2 \sin x$$
$$= 2 \sin x(\cos x - 1) = 0.$$

Now, $\sin x = 0$ at $x = n\pi$, n an integer; and $\cos x = 1$ at $x = 2n\pi$, n an integer. Thus, the x-intercepts are $x = n\pi$, $n = 0, \pm 1, \pm 2, \ldots$.

(3) Symmetry/periodicity: Since
$f(-x) = \sin 2(-x) - 2 \sin(-x) = -\sin 2x + 2 \sin x = -f(x)$, f is an odd function and the graph is symmetric with respect to the origin.   Also, f is periodic with period 2π.   Based on these two properties, it would be sufficient to sketch the graph of f on the interval $[0, \pi]$.   The result could then be extended to the interval $[-\pi, 0]$ using the symmetry, and then to $(-\infty, \infty)$ using the periodicity.   However, for purposes of illustration here, we will consider f and its derivatives on $[-\pi, \pi]$.

(4) First derivative:

$$f'(x) = 2 \cos 2x - 2 \cos x$$
$$= 2(2 \cos^2 x - 1) - 2 \cos x$$
$$= 4 \cos^2 x - 2 \cos x - 2$$
$$= 2(2 \cos x + 1)(\cos x - 1).$$

The critical points of f (on $[-\pi, \pi]$) are $x = -2\pi/3$, $2\pi/3$, and $x = 0$.

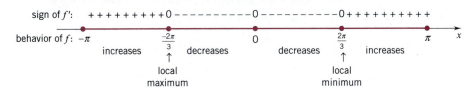

(5) Second derivative:

$$f''(x) = -4 \sin 2x + 2 \sin x$$
$$= -8 \sin x \cos x + 2 \sin x$$
$$= 2 \sin x(-4 \cos x + 1).$$

Now $f''(x) = 0$ at $x = -\pi, 0, \pi$, and at the values of x where $\cos x = \frac{1}{4}$, which yields $x \cong \pm 1.3$. The sign of f'' and the behavior of the graph (on $[-\pi, \pi]$) are:

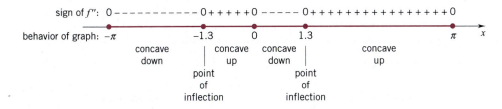

(6) Points of interest: (See Figure 4.8.10 for a preliminary sketch.)

$(0, 0)$: y-intercept.

$(-\pi, 0), (0, 0), (\pi, 0)$: x-intercepts; these are also points of inflection.

$(-\tfrac{2}{3}\pi, \tfrac{3}{2}\sqrt{3})$: $f(-2\pi/3) = \tfrac{3}{2}\sqrt{3} \cong -2.6$ is a local (and absolute) maximum.

$(\tfrac{2}{3}\pi, -\tfrac{3}{2}\sqrt{3})$: $f(2\pi/3) = -\tfrac{3}{2}\sqrt{3} \cong -2.6$ is a local (and absolute) minimum.

$(-1.3, 1.4), (1.3, -1.4)$:   points of inflection (approximately), in addition to those noted above.

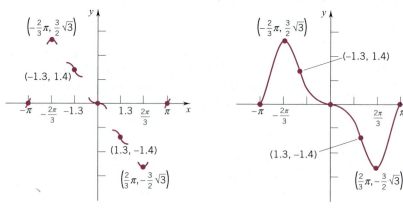

Figure 4.8.10   Figure 4.8.11

(7) Sketch the graph: The graph of f on the interval $[-\pi, \pi]$ is shown in Figure 4.8.11. An indication of the complete graph is shown in Figure 4.8.12

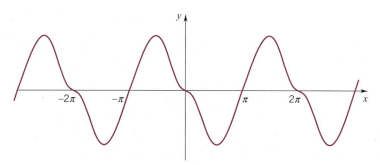

Figure 4.8.12

EXERCISES 4.8

In Exercises 1–54, sketch the graph of the function f using the approach presented in this section.

1. $f(x) = (x - 2)^2$.

2. $f(x) = 1 - (x - 2)^2$.

3. $f(x) = x^3 - 2x^2 + x + 1$.

4. $f(x) = x^3 - 9x^2 + 24x - 7$.

5. $f(x) = x^3 + 6x^2$, $x \in [-4, 4]$.

6. $f(x) = x^4 - 8x^2$, $x \in [0, \infty)$.

7. $f(x) = \tfrac{2}{3}x^3 - \tfrac{1}{2}x^2 - 10x - 1$.

8. $f(x) = x(x^2 + 4)^2$.

9. $f(x) = x^2 + \dfrac{2}{x}$.

10. $f(x) = x - \dfrac{1}{x}$.

11. $f(x) = \dfrac{x - 4}{x^2}$.

12. $f(x) = \dfrac{x + 2}{x^3}$.

13. $f(x) = 2\sqrt{x} - x, \quad x \in [0, 4]$.

14. $f(x) = \frac{1}{4}x - \sqrt{x}, \quad x \in [0, 9]$.

15. $f(x) = 2 + (x + 1)^{6/5}$.

16. $f(x) = 2 + (x + 1)^{7/5}$.

17. $f(x) = 3x^5 + 5x^3$.

18. $f(x) = 3x^4 + 4x^3$.

19. $f(x) = 1 + (x - 2)^{5/3}$.

20. $f(x) = 1 + (x - 2)^{4/3}$.

21. $f(x) = \dfrac{2x}{4x - 3}$.

22. $f(x) = \dfrac{2x^2}{x + 1}$.

23. $f(x) = \dfrac{x}{(x + 3)^2}$.

24. $f(x) = \dfrac{x}{x^2 + 1}$.

25. $f(x) = \dfrac{x^2}{x^2 - 4}$.

26. $f(x) = \dfrac{2x}{x - 4}$.

27. $f(x) = x\sqrt{1 - x}$.

28. $f(x) = (x - 1)^4 - 2(x - 1)^2$.

29. $f(x) = x + \sin 2x, \quad x \in [0, \pi]$.

30. $f(x) = \cos^3 x + 6 \cos x, \quad x \in [0, \pi]$.

31. $f(x) = \cos^4 x, \quad x \in [0, \pi]$.

32. $f(x) = \sqrt{3}x - \cos 2x, \quad x \in [0, \pi]$.

33. $f(x) = 2\sin^3 x + 3\sin x, \quad x \in [0, \pi]$.

34. $f(x) = \sin^4 x, \quad x \in [0, \pi]$.

35. $f(x) = (x + 1)^3 - 3(x + 1)^2 + 3(x + 1)$.

36. $f(x) = x^3(x + 5)^2$.

37. $f(x) = x^2(5 - x)^3$.

38. $f(x) = 4 - |2x - x^2|$.

39. $f(x) = 3 - |x^2 - 1|$.

40. $f(x) = x - x^{1/3}$.

41. $f(x) = x(x - 1)^{1/5}$.

42. $f(x) = x^2(x - 7)^{1/3}$.

43. $f(x) = x^2 - 6x^{1/3}$.

44. $f(x) = \dfrac{2x}{\sqrt{x^2 + 1}}$.

45. $f(x) = \sqrt{\dfrac{x}{x - 2}}$.

46. $f(x) = \sqrt{\dfrac{x}{x + 4}}$.

47. $f(x) = \dfrac{x^2}{\sqrt{x^2 - 2}}$.

48. $f(x) = 3\cos 4x, \quad x \in [0, \pi]$.

49. $f(x) = 2\sin 3x, \quad x \in [0, \pi]$.

50. $f(x) = 3 + 2\cot x + \csc^2 x, \quad x \in (0, \frac{1}{2}\pi)$.

51. $f(x) = 2\tan x - \sec^2 x, \quad x \in (0, \frac{1}{2}\pi)$.

52. $f(x) = 2\cos x + \sin^2 x$.

53. $f(x) = \dfrac{\sin x}{1 - \sin x}, \quad x \in (-\pi, \pi)$.

54. $f(x) = \dfrac{1}{1 - \cos x}, \quad x \in (-\pi, \pi)$.

55. A function f is continuous on $(-\infty, \infty)$, differentiable for all $x \neq 0$, and $f(0) = 0$. The graph of the derivative of f is shown in the figure below.

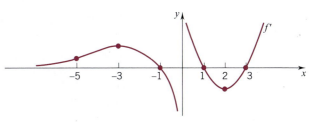

(a) Determine the intervals on which f increases and the intervals on which it decreases; find the critical numbers of f.

(b) Sketch the graph of $f''(x)$ and determine where the graph of f is concave up and where it is concave down.

(c) Does the graph of f have any horizontal asymptotes? Sketch the graph of f.

56. Set

$$F(x) = \begin{cases} \sin(1/x), & x \neq 0 \\ 0, & x = 0, \end{cases}$$

$$G(x) = \begin{cases} x \sin(1/x), & x \neq 0 \\ 0, & x = 0, \end{cases}$$

$$H(x) = \begin{cases} x^2\sin(1/x), & x \neq 0 \\ 0, & x = 0. \end{cases}$$

(a) Sketch a figure displaying the general nature of the graph of F.

(b) Sketch a figure displaying the general nature of the graph of G.

(c) Sketch a figure displaying the general nature of the graph of H.

(d) Which of these functions is continuous at 0?

(e) Which of these functions is differentiable at 0?

57. Show that the lines $y = (b/a)x$ and $y = -(b/a)x$ are oblique asymptotes of the hyperbola

$$\frac{x^2}{a^2} - \frac{y^2}{b^2} = 1.$$

■ CHAPTER HIGHLIGHTS

4.1 The Mean-Value Theorem

mean-value theorem (p. 211)   Rolle's theorem (p. 213)

4.2 Increasing and Decreasing Functions

f increases on an interval I (p. 218) f decreases on an interval I (p. 218)

One can find the intervals on which a differentiable function increases or decreases or is constant by examining the sign of the derivative.

If two functions have the same derivative on an interval, then they differ by a constant on that interval (p. 222).

4.3 Local Extreme Values

local extreme: local maximum, local minimum (p. 226)
critical number (p. 227)   first-derivative test (p. 229)
second-derivative test (p. 231)

If f has a local extreme value at c, then either $f'(c) = 0$ or $f'(c)$ does not exist; the converse is false.

The first-derivative test requires that we examine the sign of the first derivative on both sides of the critical number; the test is conclusive provided the function is continuous at the critical number. The second-derivative test requires that we examine the sign of the second derivative at the critical number itself; the test is inconclusive if the second derivative is 0 at the critical number.

4.4 Endpoint and Absolute Extreme Values

endpoint maximum and endpoint minimum (p. 235)
absolute maximum and absolute minimum (p. 235)
$f(x) \to \pm\infty$ as $x \to \pm\infty$ (p. 237)

4.5 Optimization Problems

strategy for solving optimization problems (p. 244)

The *key steps* are to express the quantity to be maximized or minimized as a function of one variable and to specify the domain of the function.

4.6 Concavity and Points of Inflection

concave up and concave down (p. 252)
point of inflection (p. 253)

If $(c, f(c))$ is a point of inflection, then either $f''(c) = 0$ or $f''(c)$ does not exist; the converse is false.

4.7 Vertical and Horizontal Asymptotes; Vertical Tangents and Cusps

$f(x) \to \pm\infty$ as $x \to c$ (p. 257) $f(x) \to L$ as $x \to \pm\infty$ (p. 259)
vertical asymptote (p. 257)   horizontal asymptote (p. 259)
vertical tangent (p. 261)   vertical cusp (p. 262)

4.8 Curve Sketching

systematic procedure for sketching graphs (p. 265)

■ PROJECTS AND EXPLORATIONS USING TECHNOLOGY

To do these exercises you will need a graphics calculator or a computer with graphing capability. The majority of these problems are open-ended so different approaches may be used to solve them. You should be aware that different approaches can result in slight variations in the answers. Round your numerical answers to at least four decimal places. The rounding method that your calculator or computer uses also may cause variations in answers.

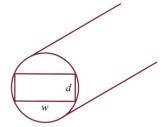

Figure A

4.1 A rectangular beam is to be cut from a cylindrical log of radius r (see Figure A).

(a) Show that the beam that has maximal cross-sectional area is a square beam.
(b) If rectangular planks are to be cut from the four pieces that remain after the square beam has been cut, find the ratio of the width to the depth that will give the plank of maximal cross-sectional area. See Figure B.
(c) Suppose that the strength of a beam is proportional to the product of the width and the square of the depth (see Figure A). Find the ratio of width to depth of the beam of maximum strength.
(d) If rectangular planks are to be cut from the four pieces that remain, find the ratio of width to depth that will give the plank of maximal strength.
(e) Suppose the log is elliptical instead of circular (Figure C). In particular, let the equation of the ellipse be

$$x^2 + 2y^2 = r^2.$$

Answer parts (a)–(d) for this log.

Figure B

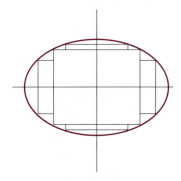

Figure C

4.2 In the Projects and Explorations section of Chapter 1 we introduced moving averages as a possible means of analyzing the stock market or seasonal sales figures. In this problem we consider a slight generalization of that moving average. Fix $a > 0$. For a given function f, define the moving average function g_a by

$$g_a(x) = \frac{f(x - a) + f(x)}{2}.$$

(a) Let f be a differentiable function. Show that the derivative of g_a is the moving average of f'.
(b) Set

$$f(x) = \frac{x^2 + 3x - 5}{x^3 - 6x^2 - 7}.$$

Let $x = 1$ and consider $g_a(1)$ as a function of a. Find the derivative of $g_a(1)$ with respect to a. Now fix any number x and use the chain rule to find the derivative with respect to a of $g_a(x)$.
(c) Suppose that f is an increasing function on an interval I. Prove that $g_a(x)$ is increasing on I.
(d) If $f'(b) = 0$, then we might conjecture that $g_a'[(b - a)/2] = 0$. Show that this conjecture is false. However, give an example of a function for which this conjecture is true.
(e) Suppose that f is twice differentiable and that its graph is concave up on an interval I. Prove that the graph of $g_a(x)$ is concave up on I.

4.3 Calculators and computers often approximate derivatives by using finite difference quotients. Fix $h > 0$. The *left quotient* is defined by

$$D_L(x) = \frac{f(x) - f(x - h)}{h};$$

the *right quotient* by

$$D_R(x) = \frac{f(x + h) - f(x)}{h};$$

and the *symmetric difference* by

$$D_C(x) = \frac{f(x + h) - f(x - h)}{2h}.$$

(a) Let $f(x) = x^3$ and set $x = 2$. Calculate $D_L(2)$, $D_R(2)$, and $D_C(2)$ for $h = 0.1, 0.01, 0.001$, and 0.0005. Which gives the best estimate for $f'(2)$?
(b) We would expect that the estimates will improve by taking smaller and smaller values of h. However, round-off error can cause the estimates to get worse for very small values of h. For your calculator, find some values of h that give large errors.

(c) Based on the concavity of $f(x) = x^3$, explain why the answers you obtained in part (a) were too small or too large.

(d) Consider the function

$$E(h) = D_L(2) - f'(2).$$

If you calculate the algebraic formula for $E(h)$, it follows that its graph is concave up. However, when you graph E (with h small) using your graphing utility this may not be as evident. Compare what the graph tells you with what the algebra tells you.

4.4 Let

$$f(x) = \frac{x^2 + x + 10}{x^4 + 7x^2 + 1}.$$

It is easy to show that f is defined and positive for all x, and that the x-axis is a horizontal asymptote of the graph of f. We want to find the rectangle of maximum area that has its upper vertices on the graph of f and its lower vertices on the x-axis. See the figure.

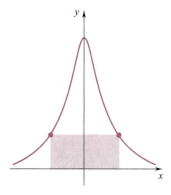

Solve this problem by using the following sequence of steps:

(a) Find f' and f''. Determine where f is increasing and where it is decreasing. Show that f has an absolute maximum value and find it. Determine where the graph of f is concave up and where it is concave down. Find the points of inflection.

(b) Verify that the x-axis is a horizontal asymptote of the graph of f and show that the graph is concave up when $|x|$ is large.

(c) Use the information from parts (a) and (b) to sketch the graph of f. Then verify that a horizontal line will either (1) not intersect the graph at all, (2) intersect the graph in exactly one point (namely, when the line passes through the absolute maximum point on the graph), or (3) intersect the graph in exactly two points.

(d) For several y values between 0 and the absolute maximum value of f, use a graphing utility to approximate the points of intersection of the horizontal lines and the graph.

(e) Using the points found in part (d), drop vertical lines to the x-axis to obtain inscribed rectangles. Calculate the areas of each. From these calculations, does it appear that the areas of the rectangles approach 0 as $y \to 0$ or as y approaches the absolute maximum value of f?

(f) Using the values found in part (e), make some new guesses of horizontal lines to use to find inscribed rectangles of larger area.

(g) After you have found your approximation to the rectangle of maximum area, evaluate f' at each of the upper vertices of the rectangle. What do you notice?

INTEGRATION

■ 5.1 AN AREA PROBLEM; A SPEED–DISTANCE PROBLEM

An Area Problem

You are already familiar with the formulas for the area of such regular geometric figures as a triangle, rectangle, square and circle (Figure 5.1.1): In Figure 5.1.2 you can see a region Ω bounded above by the graph of a continuous nonnegative function f, bounded below by the x-axis, bounded to the left by $x = a$, and bounded to the right by

area $= \frac{1}{2} bh$

Figure 5.1.1A

area $= lw$

Figure 5.1.1B

area $= x^2$

Figure 5.1.1C

area $= \pi r^2$

Figure 5.1.1D

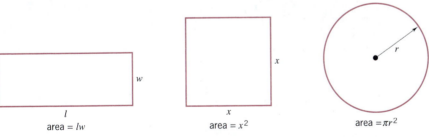

$y = f(x)$

Ω

Figure 5.1.2

$x = b$. The question before us is this: What number, if any, should measure the area of Ω?

To begin to answer this question, we split up the interval $[a, b]$ into a finite number of subintervals

$$[x_0, x_1], [x_1, x_2], \ldots, [x_{n-1}, x_n] \quad \text{with} \quad a = x_0 < x_1 < \cdots < x_n = b.$$

This breaks up the region Ω into n subregions:

$$\Omega_1, \Omega_2, \ldots, \Omega_n. \qquad \text{(Figure 5.1.3)}$$

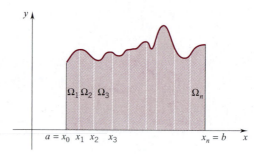

Figure 5.1.3

We can estimate the total area of Ω by estimating the area of each subregion Ω_i and adding up the results. Let's denote by M_i the maximum value of f on $[x_{i-1}, x_i]$ and by m_i the minimum value. (Recall Theorem 2.6.2.) Consider now the rectangles r_i and R_i of Figure 5.1.4. Since

$$r_i \subseteq \Omega_i \subseteq R_i,$$

Figure 5.1.4

we must have

$$\text{area of } r_i \leq \text{area of } \Omega_i \leq \text{area of } R_i.$$

Since the area of a rectangle is the length times the width,

$$m_i(x_i - x_{i-1}) \leq \text{area of } \Omega_i \leq M_i(x_i - x_{i-1}).$$

Setting $x_i - x_{i-1} = \Delta x_i$ we have

$$m_i \Delta x_i \leq \text{area of } \Omega_i \leq M_i \Delta x_i.$$

This inequality holds for $i = 1, i = 2, \ldots, i = n$. Adding up these inequalities we get on the one hand

(5.1.1) $$m_1 \Delta x_1 + m_2 \Delta x_2 + \cdots + m_n \Delta x_n \leq \text{area of } \Omega$$

and on the other hand

(5.1.2)   area of $\Omega \leq M_1 \Delta x_1 + M_2 \Delta x_2 + \cdots + M_n \Delta x_n$.

A sum of the form

$$m_1 \Delta x_1 + m_2 \Delta x_2 + \cdots + m_n \Delta x_n \qquad \text{(Figure 5.1.5)}$$

is called a *lower sum for f.*   A sum of the form

$$M_1 \Delta x_1 + M_2 \Delta x_2 + \cdots + M_n \Delta x_n \qquad \text{(Figure 5.1.6)}$$

is called an *upper sum for f.*

area of shaded region is a
lower sum for f

Figure 5.1.5

area of shaded region is an
upper sum for f

Figure 5.1.6

Inequalities (5.1.1) and (5.1.2) together tell us that for a number to be a candidate for the title "area of Ω" it must be greater than or equal to every lower sum for f, and it must be less than or equal to every upper sum.   By an argument that we omit here, it can be proved that with f continuous on $[a, b]$ there is one and only one such number. This number we call *the area of Ω.*

Later we will return to the subject of area.   At this point we turn to a speed–distance problem.   As you will see, this new problem can be approached by the same technique that we just applied to the area problem.

A Speed–Distance Problem

If an object moves at a constant speed for a given period of time, then the total distance traveled is given by the familiar formula

$$\text{distance} = \text{speed} \times \text{time}.$$

Suppose now that during the course of the motion the speed does not remain constant but instead varies continuously.   How can the total distance traveled be computed then?

To answer this question, we suppose that the motion begins at time a, ends at time b, and that during the time interval $[a, b]$ the speed varies continuously.

As in the case of the area problem we begin by breaking up the interval $[a, b]$ into a finite number of subintervals:

$$[t_0, t_1], [t_1, t_2], \ldots , [t_{n-1}, t_n], \quad \text{with} \quad a = t_0 < t_1 < \cdots < t_n = b.$$

On each subinterval $[t_{i-1}, t_i]$ the object attains a certain maximum speed M_i and a certain minimum speed m_i.   (How do we know this?)   If throughout the time interval $[t_{i-1}, t_i]$ the object were to move constantly at its minimum speed, m_i, then it would cover a distance of $m_i \Delta t_i$ units.   If instead it were to move constantly at its maximum speed, M_i, then it would cover a distance of $M_i \Delta t_i$ units.   As it is, the actual distance traveled, call it s_i, must lie somewhere in between; namely, we must have

$$m_i \Delta t_i \leq s_i \leq M_i \Delta t_i.$$

The total distance traveled during the time interval $[a, b]$, call it s, must be the sum of the distances traveled during the subintervals $[t_{i-1}, t_i]$. In other words we must have

$$s = s_1 + s_2 + \cdots + s_n.$$

Since

$$m_1 \Delta t_1 \leq s_1 \leq M_1 \Delta t_1$$
$$m_2 \Delta t_2 \leq s_2 \leq M_2 \Delta t_2$$
$$\vdots$$
$$m_n \Delta t_n \leq s_n \leq M_n \Delta t_n,$$

it follows by the addition of these inequalities that

$$m_1 \Delta t_1 + m_2 \Delta t_2 + \cdots + m_n \Delta t_n \leq s \leq M_1 \Delta t_1 + M_2 \Delta t_2 + \cdots + M_n \Delta t_n.$$

A sum of the form

$$m_1 \Delta t_1 + m_2 \Delta t_2 + \cdots + m_n \Delta t_n$$

is called a *lower sum* for the speed function. A sum of the form

$$M_1 \Delta t_1 + M_2 \Delta t_2 + \cdots + M_n \Delta t_n$$

is called an *upper sum* for the speed function. The inequality we just obtained for s tells us that s must be greater than or equal to every lower sum for the speed function, and it must be less than or equal to every upper sum. As in the case of the area problem, it turns out that there is one and only one such number, and this is the total distance traveled.

■ 5.2 THE DEFINITE INTEGRAL OF A CONTINUOUS FUNCTION

The process we used to approach the two problems of Section 5.1 is called *integration,* and the end results of this process are called *definite integrals.* Our purpose here is to establish these notions more precisely.

(5.2.1)

> By a *partition* of the closed interval $[a, b]$ we mean a finite subset of $[a, b]$ which contains the points a and b.

We index the elements of a partition according to their natural order. Thus, if we write

$$P = \{x_0, x_1, \ldots, x_n\} \text{ is a partition of } [a, b],$$

you can conclude that

$$a = x_0 < x_1 < \cdots < x_n = b.$$

Example 1 The sets

$$\{0, 1\}, \quad \{0, \tfrac{1}{2}, 1\}, \quad \{0, \tfrac{1}{4}, \tfrac{1}{2}, 1\}, \quad \{0, \tfrac{1}{4}, \tfrac{1}{3}, \tfrac{1}{2}, \tfrac{5}{8}, 1\}$$

are all partitions of the interval $[0, 1]$. ❑

If $P = \{x_0, x_1, \ldots, x_n\}$ is a partition of $[a, b]$, then P breaks up $[a, b]$ into a finite number of subintervals

$$[x_0, x_1], [x_1, x_2], \ldots, [x_{n-1}, x_n]$$

of lengths $\Delta x_1, \Delta x_2, \ldots, \Delta x_n$, respectively.

Suppose now that f is continuous on $[a, b]$. Then on each interval $[x_{i-1}, x_i]$ the function f takes on a maximum value, M_i, and a minimum value, m_i.

(5.2.2)

The number

$$U_f(P) = M_1 \Delta x_1 + M_2 \Delta x_2 + \cdots + M_n \Delta x_n$$

is called *the P upper sum for f*, and the number

$$L_f(P) = m_1 \Delta x_1 + m_2 \Delta x_2 + \cdots + m_n \Delta x_n$$

is called *the P lower sum for f*.

[handwritten margin note:] Def. Let f be continuous an [a, b]. the unique num. I that satisfies the inequality $L_f(P) \le I \le U_f(P)$

Example 2 The quadratic function

$$f(x) = x^2$$

is continuous on $[0, 1]$. The partition $P = \{0, \frac{1}{4}, \frac{1}{2}, 1\}$ breaks up $[0, 1]$ into three subintervals:

$$[x_0, x_1] = [0, \tfrac{1}{4}], \qquad [x_1, x_2] = [\tfrac{1}{4}, \tfrac{1}{2}], \qquad [x_2, x_3] = [\tfrac{1}{2}, 1]$$

of lengths

$$\Delta x_1 = \tfrac{1}{4} - 0 = \tfrac{1}{4}, \qquad \Delta x_2 = \tfrac{1}{2} - \tfrac{1}{4} = \tfrac{1}{4}, \qquad \Delta x_3 = 1 - \tfrac{1}{2} = \tfrac{1}{2},$$

respectively. Since f is increasing on $[0, 1]$, it takes on its maximum value at the right endpoint of each subinterval:

$$M_1 = f(\tfrac{1}{4}) = \tfrac{1}{16}, \qquad M_2 = f(\tfrac{1}{2}) = \tfrac{1}{4}, \qquad M_3 = f(1) = 1.$$

The minimum values of f are taken on at the left endpoints of each subinterval:

$$m_1 = f(0) = 0, \qquad m_2 = f(\tfrac{1}{4}) = \tfrac{1}{16}, \qquad m_3 = f(\tfrac{1}{2}) = \tfrac{1}{4}.$$

Thus,

$$U_f(P) = M_1 \Delta x_1 + M_2 \Delta x_2 + M_3 \Delta x_3 = \tfrac{1}{16}(\tfrac{1}{4}) + \tfrac{1}{4}(\tfrac{1}{4}) + 1(\tfrac{1}{2}) = \tfrac{37}{64}$$

(see Figure 5.2.1a)

and

$$L_f(P) = m_1 \Delta x_1 + m_2 \Delta x_2 + m_3 \Delta x_3 = 0(\tfrac{1}{4}) + \tfrac{1}{16}(\tfrac{1}{4}) + \tfrac{1}{4}(\tfrac{1}{2}) = \tfrac{9}{64}. \quad \square$$

(see Figure 5.2.1b)

Example 3 Now consider the function

$$f(x) = -2x - 1$$

on the closed interval $[-1, \frac{1}{2}]$. The partition $P = \{-1, -\frac{3}{4}, -\frac{1}{4}, \frac{1}{4}, \frac{1}{2}\}$ breaks up $[-1, \frac{1}{2}]$ into four subintervals: $[-1, -\frac{3}{4}]$, $[-\frac{3}{4}, -\frac{1}{4}]$, $[-\frac{1}{4}, \frac{1}{4}]$, and $[\frac{1}{4}, \frac{1}{2}]$. See Figure 5.2.2. Since f is a decreasing function, the maximum values of f will occur at the left

Figure 5.2.1

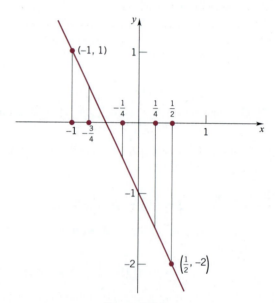

Figure 5.2.2

endpoints of the subintervals, and the minimum values will occur at the right endpoints. As you can check

$$U_f(P) = 1(\tfrac{1}{4}) + \tfrac{1}{2}(\tfrac{1}{2}) + (-\tfrac{1}{2})(\tfrac{1}{2}) + (-\tfrac{3}{2})(\tfrac{1}{4}) = -\tfrac{1}{8}$$

and

$$L_f(P) = \tfrac{1}{2}(\tfrac{1}{4}) + (-\tfrac{1}{2})(\tfrac{1}{2}) + (-\tfrac{3}{2})(\tfrac{1}{2}) + (-2)(\tfrac{1}{4}) = -\tfrac{11}{8}. \quad \square$$

By an argument that we omit here (it appears in Appendix B.4) it can be proved that, with f continuous on $[a, b]$, there is one and only one number I that satisfies the inequality

$$L_f(P) \le I \le U_f(P) \qquad \text{for } all \text{ partitions } P \text{ of } [a, b].$$

This is the number we want.

DEFINITION 5.2.3 THE DEFINITE INTEGRAL

Let f be continuous on $[a, b]$. The unique number I that satisfies the inequality

$$L_f(P) \leq I \leq U_f(P) \qquad \text{for all partitions } P \text{ of } [a, b]$$

is called the *definite integral* (or more simply *the integral*) of f from a to b and is denoted by

$$\int_a^b f(x)\, dx.$$

The symbol $\int$ dates back to Leibniz and is called an *integral sign*. It is really an elongated S—as in *Sum*. The numbers a and b are called *the limits of integration*, a is the *lower limit* and b is the *upper limit*, and we'll speak of *integrating* a function f from a to b. The function f being integrated is called *the integrand*. This is not the only notation. Some mathematicians omit the dx and write simply $\int_a^b f$. We will keep the dx. As we go on, you will see that it does serve a useful purpose.

In the expression

$$\int_a^b f(x)\, dx$$

the letter x is a "dummy variable"; in other words, it can be replaced by any other letter not already engaged. Thus, for example,

$$\int_a^b f(x)\, dx, \qquad \int_a^b f(t)\, dt, \qquad \text{and} \qquad \int_a^b f(z)\, dz.$$

All denote exactly the same quantity, the definite integral of f from a to b.

Section 5.1 gives two immediate applications of the definite integral:

I. If f is nonnegative on $[a, b]$, then

$$A = \int_a^b f(x)\, dx$$

gives the area below the graph of f. See Figure 5.1.2.

II. If $|v(t)|$ is the speed of an object at time t, then

$$s = \int_a^b |v(t)|\, dt$$

gives the distance traveled from time a to time b.

We will come back to these ideas later. Right now we do some simple computations.

Example 4 If $f(x) = 3$ for all x in $[a, b]$ (see Figure 5.2.3),

$$\int_a^b f(x)\, dx = 3(b - a).$$

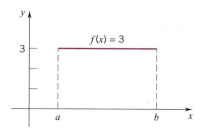

Figure 5.2.3

To see this, take $P = \{x_0, x_1, \ldots, x_n\}$ as an arbitrary partition of $[a, b]$. Since f is constantly 3 on $[a, b]$, it is constantly 3 on each subinterval $[x_{i-1}\ x_i]$. Thus, M_i and m_i are both 3. It follows that

$$U_f(P) = 3\Delta x_1 + 3\Delta x_2 + \cdots + 3\Delta x_n$$
$$= 3(\Delta x_1 + \Delta x_2 + \cdots + \Delta x_n)$$
$$= 3[(x_1 - x_0) + (x_2 - x_1) + (x_3 - x_2) + \cdots + (x_{n-1} - x_{n-2}) + (x_n - x_{n-1})]$$
$$= 3(b - a) \qquad (\text{recall } x_0 = a \text{ and } x_n = b)$$

and

$$L_f(P) = 3\Delta x_1 + 3\Delta x_2 + \cdots + 3\Delta x_n$$
$$= 3(\Delta x_1 + \Delta x_2 + \cdots + \Delta x_n) = 3(b - a).$$

Obviously then

$$L_f(P) \leq 3(b - a) \leq U_f(P)$$

Since this inequality holds for all partitions P of $[a, b]$, we can conclude that

$$\int_a^b f(x)\,dx = 3(b - a). \quad \square$$

Clearly, this same argument will hold with any constant k in place of 3. Thus, we have

(5.2.4)
$$\int_a^b k\,dx = k(b - a), \qquad \text{where } k \text{ is any constant.}$$

For example

$$\int_{-1}^1 3\,dx = 3[1 - (-1)] = 3(2) = 6 \qquad \text{and}$$

$$\int_4^{10} -2\,dx = -2[10 - 4] = -2(6) = -12.$$

If $k > 0$, the region between the graph and the x-axis is a rectangle of height k erected on the interval $[a, b]$ (Figure 5.2.4). The integral gives the area of the rectangle.

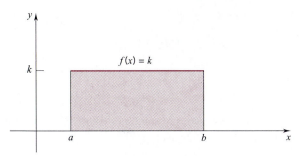

Figure 5.2.4

Example 5 Let $f(x) = x$ for all $x \in [a, b]$ (see Figure 5.2.5). Then

(5.2.5)
$$\int_a^b x \, dx = \tfrac{1}{2}(b^2 - a^2).$$

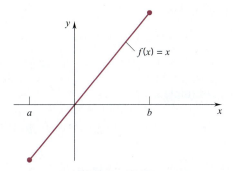

Figure 5.2.5

To see this, take $P = \{x_0, x_1, \ldots, x_n\}$ as an arbitrary partition of $[a, b]$. On each subinterval $[x_{i-1}, x_i]$ the function $f(x) = x$ has a maximum value M_i and a minimum value m_i. Since f is an increasing function, the maximum value $M_i = x_i$ occurs at the right endpoint of the subinterval and the minimum value $m_i = x_{i-1}$ occurs at the left endpoint. If follows that

$$U_f(P) = x_1 \Delta x_1 + x_2 \Delta x_2 + \cdots + x_n \Delta x_n$$

and

$$L_f(P) = x_0 \Delta x_1 + x_1 \Delta x_2 + \cdots + x_{n-1} \Delta x_n.$$

For each index i

(*) $$x_{i-i} \leq \tfrac{1}{2}(x_i + x_{i-1}) \leq x_i. \qquad \text{(explain)}$$

Multiplication by $\Delta x_i = x_i - x_{i-1}$ gives

$$x_{i-1} \Delta x_i \leq \tfrac{1}{2}(x_i^2 - x_{i-1}^2) \leq x_i \Delta x_i.$$

Summing from $i = 1$ to $i = n$, we find that

(**) $$L_f(P) \leq \tfrac{1}{2}(x_1^2 - x_0^2) + \tfrac{1}{2}(x_2^2 - x_1^2) + \cdots + \tfrac{1}{2}(x_n^2 - x_{n-1}^2) \leq U_f(P).$$

The sum in the middle collapses to

$$\tfrac{1}{2}(x_n^2 - x_0^2) = \tfrac{1}{2}(b^2 - a^2).$$

Consequently we have

$$L_f(P) \le \tfrac{1}{2}(b^2 - a^2) \le U_f(P).$$

Since P was chosen arbitrarily, we can conclude that this inequality holds for all partitions P of $[a, b]$. It follows then that

$$\int_a^b x \, dx = \tfrac{1}{2}(b^2 - a^2). \quad \square$$

For example

$$\int_{-1}^3 x \, dx = \tfrac{1}{2}[3^2 - (-1)^2] = \tfrac{1}{2}(8) = 4 \qquad \text{and} \qquad \int_{-2}^2 x \, dx = \tfrac{1}{2}[2^2 - (-2)^2] = 0.$$

Remark   You might be wondering why we chose the inequality ($*$) in the preceding example instead of some other possible inequality, such as

$$x_{i-1} \le \tfrac{2}{3}x_{i-1} + \tfrac{1}{3}x_i \le x_i.$$

The reason is that the choice of ($*$) resulted in the "collapsing" sum ($**$). If we had not been clever in this way, the problem would have been very difficult. You will see this idea again in the next example. $\square$

If the interval $[a, b]$ lies to the right of the origin, then the region below the graph of

$$f(x) = x, \qquad x \in [a, b]$$

is the trapezoid of Figure 5.2.6. The integral

$$\int_a^b x \, dx$$

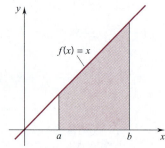

area of shaded region: $\int_a^b x \, dx = \tfrac{1}{2}(b^2 - a^2)$

Figure 5.2.6

gives the area of this trapezoid: $A = (b - a)[\tfrac{1}{2}(a + b)] = \tfrac{1}{2}(b^2 - a^2).$

Example 6

$$\int_0^1 x^2 \, dx = \tfrac{1}{3}. \qquad \text{(Figure 5.2.7)}$$

This time take $P = \{x_0, x_1, \ldots, x_n\}$ as a partition of $[0, 1]$. On each subinterval $[x_{i-1}, x_i]$ the function $f(x) = x^2$ has a maximum $M_i = x_i^2$ and a minimum $m_i = x_{i-1}^2$. It follows that

$$U_f(P) = x_1^2 \Delta x_1 + \cdots + x_n^2 \Delta x_n$$

and

$$L_f(P) = x_0^2 \Delta x_1 + \cdots + x_{n-1}^2 \Delta x_n.$$

For each index i,

$$x_{i-1}^2 \le \tfrac{1}{3}(x_{i-1}^2 + x_{i-1} x_i + x_i^2) \le x_i^2. \qquad \text{(verify this)}$$

$f(x) = x^2$

area of the shaded region: $\int_0^1 x^2 \, dx = \frac{1}{3}$

Figure 5.2.7

(As we will see later in this chapter, $\tfrac{1}{3}(x_{i-1}^2 + x_{i-1} x_i + x_i^2)$ is the average value of x^2 on $[x_{i-1}, x_i]$.) We now multiply this inequality by $\Delta x_i = x_i - x_{i-1}$. The middle term is

$$\tfrac{1}{3}(x_{i-1}^2 + x_{i-1} x_i + x_i^2)(x_i - x_{i-1}) = \tfrac{1}{3}(x_i^3 - x_{i-1}^3),$$

showing us that

$$x_{i-1}^2 \Delta x_i \le \tfrac{1}{3}(x_i^3 - x_{i-1}^3) \le x_i^2 \Delta x_i.$$

The sum of the terms on the left is $L_f(P)$. The sum of all the middle terms collapses to $\tfrac{1}{3}$:

$$\tfrac{1}{3}(x_1^3 - x_0^3 + x_2^3 - x_1^3 + \cdots + x_n^3 - x_{n-1}^3) = \tfrac{1}{3}(x_n^3 - x_0^3) = \tfrac{1}{3}(1^3 - 0^3) = \tfrac{1}{3}.$$

The sum of the terms on the right is $U_f(P)$. Clearly then

$$L_f(P) \le \tfrac{1}{3} \le U_f(P).$$

Since P was chosen arbitrarily, we can conclude that this inequality holds for all partitions P of $[0, 1]$. It follows therefore that

$$\int_0^1 x^2 \, dx = \tfrac{1}{3}. \quad \square$$

EXERCISES 5.2

In Exercises 1–12, find $L_f(P)$ and $U_f(P)$.

1. $f(x) = 2x$, $x \in [0, 1]$; $P = \{0, \tfrac{1}{4}, \tfrac{1}{2}, 1\}$.
2. $f(x) = 1 - x$, $x \in [0, 2]$; $P = \{0, \tfrac{1}{3}, \tfrac{3}{4}, 1, 2\}$.
3. $f(x) = x^2$, $x \in [-1, 0]$; $P = \{-1, -\tfrac{1}{2}, -\tfrac{1}{4}, 0\}$.
4. $f(x) = 1 - x^2$, $x \in [0, 1]$; $P = \{0, \tfrac{1}{4}, \tfrac{1}{2}, 1\}$.
5. $f(x) = 1 + x^3$, $x \in [0, 1]$; $P = \{0, \tfrac{1}{2}, 1\}$.
6. $f(x) = \sqrt{x}$, $x \in [0, 1]$; $P = \{0, \tfrac{1}{25}, \tfrac{4}{25}, \tfrac{9}{25}, \tfrac{16}{25}, 1\}$.
7. $f(x) = |x|$, $x \in [-1, 1]$; $P = \{-1, -\tfrac{1}{2}, 0, \tfrac{1}{4}, 1\}$.
8. $f(x) = |x|$, $x \in [-1, 1]$; $P = \{-1, -\tfrac{1}{2}, -\tfrac{1}{4}, \tfrac{1}{2}, 1\}$.

9. $f(x) = x^2$, $x \in [-1, 1]$; $P = \{-1, -\tfrac{1}{4}, \tfrac{1}{4}, \tfrac{1}{2}, 1\}$.
10. $f(x) = x^2$, $x \in [-1, 1]$; $P = \{-1, -\tfrac{3}{4}, -\tfrac{1}{4}, \tfrac{1}{4}, \tfrac{1}{2}, 1\}$.
11. $f(x) = \sin x$, $x \in [0, \pi]$; $P = \{0, \tfrac{1}{6}\pi, \tfrac{1}{2}\pi, \pi\}$.
12. $f(x) = \cos x$, $x \in [0, \pi]$; $P = \{0, \tfrac{1}{3}\pi, \tfrac{1}{2}\pi, \pi\}$.
13. Let f be a continuous function on $[-1, 1]$. Explain why each of the following three statements is false. Take P as a partition of $[-1, 1]$.
    (a) $L_f(P) = 3$ and $U_f(P) = 2$.

    (b) $L_f(P) = 3$, $U_f(P) = 6$, and $\displaystyle\int_{-1}^1 f(x) \, dx = 2$.

(c) $L_f(P) = 3$, $U_f(P) = 6$, and $\int_{-1}^{1} f(x)\, dx = 10$.

14. (a) Given that $P = \{x_0, x_1, \ldots, x_n\}$ is an arbitrary partition of $[a, b]$, find $L_f(P)$ and $U_f(P)$ for $f(x) = x + 3$.
    (b) Use your answers to part (a) to evaluate
$$\int_a^b f(x)\, dx.$$

15. Repeat Exercise 14 with $f(x) = -3x$.

16. Repeat Exercise 14 with $f(x) = 1 + 2x$.

17. Evaluate
$$\int_0^1 x^3\, dx$$
by the methods of this section.   HINT: $b^4 - a^4 = (b^3 + b^2 a + b a^2 + a^3)(b - a)$.

18. Evaluate
$$\int_0^1 x^4\, dx$$
by the methods of this section.

19. A partition $P = \{x_0, x_1, x_2, \ldots, x_{n-1}, x_n\}$ of the interval $[a, b]$ is *regular* if the subintervals $[x_{i-1}, x_i]$, $i = 1, 2, \ldots, n$, all have the same length, namely $\Delta x = (b - a)/n$. Suppose that $P = \{x_0, x_1, \ldots, x_{n-1}, x_n\}$ is a regular partition of the interval $[a, b]$. Show that if f is continuous and increasing on $[a, b]$, then
$$U_f(P) - L_f(P) = [f(b) - f(a)]\, \Delta x.$$

20. Let $P = \{x_0, x_1, x_2, \ldots, x_{n-1}, x_n\}$ be a regular partition of the interval $[a, b]$. Show that if f is continuous and decreasing on $[a, b]$, then
$$U_f(P) - L_f(P) = [f(a) - f(b)]\, \Delta x.$$

21. Let f be continuous on the interval $[a, b]$ and let $I = \int_a^b f(x)\, dx$. Let $P = \{x_0, x_1, x_2, \ldots, x_{n-1}, x_n\}$ be an arbitrary partition of the interval $[a, b]$.
    (a) Show that $I - L_f(P) \le U_f(P) - L_f(P)$. The difference $I - L_f(P)$ is sometimes called the *error* in using $L_f(P)$ to approximate I.
    (b) Show that $U_f(P) - I \le U_f(P) - L_f(P)$ ($U_f(P) - I$ is the error in using $U_f(P)$ to approximate I).

22. Let $P = \{x_0, x_1, x_2, \ldots, x_{n-1}, x_n\}$ be a regular partition of the interval $[a, b]$ and let f be continuous. Show that if f is increasing on $[a, b]$, or if f is decreasing on $[a, b]$, then
$$I - L_f(P) \le |f(b) - f(a)|\, \Delta x \qquad \text{and}$$
$$U_f(P) - I \le |f(b) - f(a)|\, \Delta x.$$

▶ 23. Let $f(x) = \sqrt{1 + x^2}$.
    (a) Verify that f is increasing on $[0, 2]$.
    (b) Use Exercise 22 to determine a value of n such that if $P = \{x_0, x_1, x_2, \ldots, x_n\}$ is a regular partition of $[0, 2]$, then
$$\int_0^2 f(x)\, dx - L_f(P) \le 0.1.$$
    (c) Use a programmable calculator or computer to calculate $\int_0^2 f(x)\, dx$ with an error of less than 0.1.

▶ 24. Let $f(x) = 1/(1 + x^2)$.
    (a) Verify that f is decreasing on $[0, 1]$.
    (b) Use Exercise 22 to determine a value of n such that if $P = \{x_0, x_1, x_2, \ldots, x_n\}$ is a regular partition of $[0, 1]$, then
$$U_f(P) - \int_0^1 f(x)\, dx \le 0.05.$$
    (c) Use a programmable calculator or computer to calculate $\int_0^1 f(x)\, dx$ with an error of less than 0.05.
    NOTE: You will see in Chapter 7 that the exact value of this integral is $\pi/4$.

25. Use mathematical induction to show that the sum of the first k positive integers is $\frac{1}{2} k(k + 1)$. That is,
$$1 + 2 + 3 + \cdots + k = \frac{k(k + 1)}{2},$$
where k is any positive integer.

26. Use mathematical induction to show that the sum of the squares of the first k positive integers is $\frac{1}{6} k(k + 1)(2k + 1)$. That is,
$$1^2 + 2^2 + 3^2 + \cdots + k^2 = \frac{k(k + 1)(2k + 1)}{6},$$
where k is any positive integer.

27. Let $P = \{x_0, x_1, x_2, \ldots, x_{n-1}, x_n\}$ be a regular partition of the interval $[0, b]$, and let $f(x) = x$.
    (a) Show that
$$L_f(P) = \frac{b^2}{n^2}\, [0 + 1 + 2 + 3 + \cdots + (n - 1)].$$
    (b) Show that
$$U_f(P) = \frac{b^2}{n^2}\, [1 + 2 + 3 + \cdots + n].$$
    (c) Use Exercise 25 to show that $L_f(P)$, $U_f(P) \to b^2/2$ as $n \to \infty$. Thus, $\int_0^b x\, dx = \frac{1}{2} b^2$. Compare with Example 5.

28. Let $P = \{x_0, x_1, x_2, \ldots, x_{n-1}, x_n\}$ be a regular partition of $[0, b]$, and let $f(x) = x^2$.
   (a) Show that
$$L_f(P) = \frac{b^3}{n^3}[0^2 + 1^2 + 2^2 + \cdots + (n-1)^2].$$

   (b) Show that
$$U_f(P) = \frac{b^3}{n^3}[1^2 + 2^2 + 3^2 + \cdots + n^2].$$

   (c) Use Exercise 26 to show that $L_f(P), U_f(P) \to$ $(b^3/3)$ as $n \to \infty$. Thus, $\int_0^b x^2\, dx = \frac{1}{3}b^3$. Compare with Example 6.

29. (*Important*) The definition of integral that we have given (Definition 5.2.3) can be applied to functions with a finite number of discontinuities. Suppose that f is continuous on $[a, b]$. If g is defined on $[a, b]$ and differs from f only at a finite number of points, then g can be integrated on $[a, b]$ and
$$\int_a^b g(x)\, dx = \int_a^b f(x)\, dx.$$

For example, the function
$$g(x) = \begin{cases} 2, & x \in [0, 3) \cup (3, 4) \\ 7, & x = 3 \end{cases}$$

differs from the constant function
$$f(x) = 2, \qquad x \in [0, 4]$$

only at $x = 3$. Clearly
$$\int_0^4 f(x)\, dx = 8.$$

Show that
$$\int_0^4 g(x)\, dx = 8$$

by showing that 8 is the unique number I that satisfies the inequality
$$L_g(P) \le I \le U_g(P) \qquad \text{for all partitions } P \text{ of } [0, 4].$$

30. (*Important*) A function f is *piecewise continuous* on $[a, b]$ if it is continuous except, possibly, for a finite set of points at which it has a jump discontinuity. It can be shown that if f is piecewise continuous on $[a, b]$, then $\int_a^b f(x)\, dx$ exists. Let
$$f(x) = \begin{cases} x^2 & 0 \le x \le 2 \\ x & 2 < x \le 5. \end{cases}$$

   (a) Sketch the graph of f.
   (b) Find the area of the region bounded by the graph of f and the x-axis between $x = 0$ and $x = 5$. HINT: $\int_0^5 f(x)\, dx = \int_0^2 f(x)\, dx + \int_2^5 f(x)\, dx$, then use Exercise 28 and Example 5.

31. Consider the function f on $[2, 10]$:
$$f(x) = \begin{cases} 7 & \text{for } x \text{ rational} \\ 4 & \text{for } x \text{ irrational.} \end{cases}$$

   (a) Show that for all partitions P of $[2, 10]$, $L_f(P) \le 40 \le U_f(P)$.
   (b) Explain why we can *not* conclude that
$$\int_2^{10} f(x)\, dx = 40.$$

   (c) Show that, for all partitions P of $[2, 10]$, $L_f(P) = 32$ and $U_f(P) = 56$. Hence we can conclude that $\int_2^{10} f(x)\, dx$ does not exist.

■ **5.3 THE FUNCTION** $F(x) = \int_a^x f(t)\, dt$

The evaluation of a definite integral
$$\int_a^b f(x)\, dx$$

directly from its definition as the unique number I satisfying the inequality $L_f(P) \le I \le U_f(P)$ for all partitions P of $[a, b]$ is usually a laborious and difficult process. Try, for example, to evaluate
$$\int_2^5 \left(x^3 + x^{5/2} - \frac{2x}{1 - x^2} \right) dx \qquad \text{or} \qquad \int_{-1/2}^{1/4} \frac{x}{1 - x^2}\, dx$$

by this method. Theorem 5.4.2, called *the fundamental theorem of integral calculus,* gives us another way to evaluate such integrals. This other way depends on a connection between integration and differentiation described in Theorem 5.3.5. The main purpose of this section is to prove Theorem 5.3.5. Along the way we will pick up some information that is of interest in itself.

THEOREM 5.3.1

Let f be continuous on $[a, b]$ and let P and Q be partitions of the interval $[a, b]$. If $P \subseteq Q$, then

$$L_f(P) \leq L_f(Q) \qquad \text{and} \qquad U_f(Q) \leq U_f(P).$$

This result is easy to see. By adding points to a partition we make the subintervals $[x_{i-1}, x_i]$ smaller. This tends to make the minima, m_i, larger and the maxima, M_i, smaller. Thus the lower sums are made bigger, and the upper sums are made smaller. The idea is illustrated (for a positive function) in Figures 5.3.1 and 5.3.2.

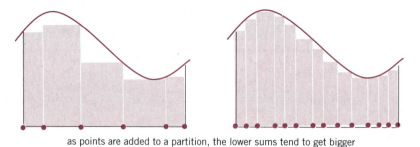

as points are added to a partition, the lower sums tend to get bigger

Figure 5.3.1

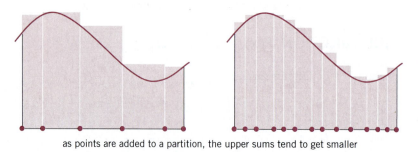

as points are added to a partition, the upper sums tend to get smaller

Figure 5.3.2

The next theorem says that the integral is *additive* on intervals.

THEOREM 5.3.2

If f is continuous on $[a, b]$ and $a < c < b$, then

$$\int_a^c f(t)\ dt + \int_c^b f(t)\ dt = \int_a^b f(t)\ dt.$$

For nonnegative functions f this theorem is easily understood in terms of area. The area of part I in Figure 5.3.3 is given by

$$\int_a^c f(t)\ dt;$$

the area of part II by

$$\int_c^b f(t)\ dt;$$

the area of the entire region by

$$\int_a^b f(t)\ dt.$$

The theorem says that

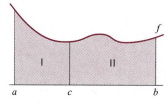

Figure 5.3.3

---

the area of part I + the area of part II = the area of the entire region.

Theorem 5.3.2 can also be viewed in terms of speed and distance. With $f(t) \geq 0$, we can interpret $f(t)$ as the speed of an object at time t. With this interpretation,

$$\int_a^c f(t)\ dt = \text{distance traveled from time } a \text{ to time } c,$$

$$\int_c^b f(t)\ dt = \text{distance traveled from time } c \text{ to time } b,$$

$$\int_a^b f(t)\ dt = \text{distance traveled from time } a \text{ to time } b.$$

It is not surprising that the sum of the first two distances should equal the third.

The fact that the additivity theorem is so easy to understand does not relieve us of the necessity to prove it. Here is a proof.

PROOF OF THEOREM 5.3.2. To prove the theorem we need only show that for each partition P of $[a, b]$

$$L_f(P) \leq \int_a^c f(t)\ dt + \int_c^b f(t)\ dt \leq U_f(P) \qquad \text{(Why?)}$$

We begin with an arbitrary partition of $[a, b]$:

$$P = \{x_0, x_1, \ldots, x_n\}.$$

Since the partition $Q = P \cup \{c\}$ contains P, we know from Theorem 5.3.1 that

(1) $$L_f(P) \leq L_f(Q) \quad \text{and} \quad U_f(Q) \leq U_f(P).$$

The sets

$$Q_1 = Q \cap [a, c] \quad \text{and} \quad Q_2 = Q \cap [c, b]$$

are partitions of $[a, c]$ and $[c, b]$, respectively. Moreover

$$L_f(Q_1) + L_f(Q_2) = L_f(Q) \quad \text{and} \quad U_f(Q_1) + U_f(Q_2) = U_f(Q).$$

Since

$$L_f(Q_1) \leq \int_a^c f(t) \, dt \leq U_f(Q_1) \quad \text{and} \quad L_f(Q_2) \leq \int_c^b f(t) \, dt \leq U_f(Q_2),$$

we have

$$L_f(Q_1) + L_f(Q_2) \leq \int_a^c f(t) \, dt + \int_c^b f(t) \, dt \leq U_f(Q_1) + U_f(Q_2),$$

which is the same as

$$L_f(Q) \leq \int_a^c f(t) \, dt + \int_c^b f(t) \, dt \leq U_f(Q).$$

Therefore, by (1),

$$L_f(P) \leq \int_a^c f(t) \, dt + \int_c^b f(t) \, dt \leq U_f(P). \quad \square$$

Until now we have integrated only from left to right: from a number a to a number b greater than a. We integrate in the other direction by defining

(5.3.3)
$$\int_b^a f(t) \, dt = -\int_a^b f(t) \, dt.$$

The integral from any number to itself is defined to be zero:

(5.3.4)
$$\int_c^c f(t) \, dt = 0$$

for all real c. With these additional conventions, the additivity condition

$$\int_a^c f(t) \, dt + \int_c^b f(t) \, dt = \int_a^b f(t) \, dt$$

holds for all choices of a, b, c from an interval on which f is continuous, no matter what the order of a, b, c happens to be. We have left the proof of this to you as an exercise.

We are now ready to state the all-important connection that exists between integration and differentiation. Our first step is to point out that, if f is continuous on $[a, b]$, then for each x in $[a, b]$, the integral

$$\int_a^x f(t)\, dt$$

is a number, and consequently we can define a function F on $[a, b]$ by setting

$$F(x) = \int_a^x f(t)\, dt.$$

THEOREM 5.3.5

Let f be continuous on $[a, b]$. The function F defined on $[a, b]$ by

$$F(x) = \int_a^x f(t)\, dt$$

is continuous on $[a, b]$, differentiable on (a, b), and has derivative

$$F'(x) = f(x) \qquad \text{for all} \quad x \text{ in } (a, b).$$

PROOF We begin with x in the half-open interval $[a, b)$ and show that

$$\lim_{h \to 0^+} \frac{F(x + h) - F(x)}{h} = f(x).$$

(For a pictorial outline of the proof for f nonnegative, see Figure 5.3.4.)
If $x < x + h \leq b$, then

$$F(x + h) - F(x) = \int_a^{x+h} f(t)\, dt - \int_a^x f(t)\, dt.$$

It follows that

$$F(x + h) - F(x) = \int_x^{x+h} f(t)\, dt. \qquad \text{(justify this step)}$$

Now set

$$M_h = \text{maximum value of } f \text{ on } [x, x + h]$$

and

$$m_h = \text{minimum value of } f \text{ on } [x, x + h].$$

See Figure 5.3.5. Since

$$M_h[(x + h) - x] = M_h \cdot h$$

is an upper sum for f on $[x, x + h]$ and

$$m_h[(x + h) - x] = m_h \cdot h$$

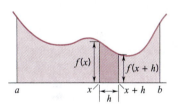

$F(x + h) = $ area from a to $x + h$
and
$F(x) = $ area from a to x.

Therefore

$$F(x + h) - F(x)$$
$= $ area from x to $x + h \cong f(x)h$ if h is small

$$\frac{F(x + h) - F(x)}{h} = \frac{\text{area from } x \text{ to } x + h}{h} \cong f(x)$$

Figure 5.3.4

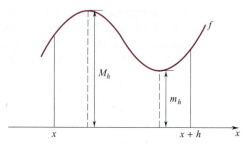

Figure 5.3.5

is a lower sum for f on $[x, x + h]$, we have

$$m_h \cdot h \le \int_x^{x+h} f(t)\, dt \le M_h \cdot h.$$

Now, using (1) and the fact that $h > 0$, it follows that

$$m_h \le \frac{F(x + h) - F(x)}{h} \le M_h.$$

Also, since f is continuous on $[x, x + h]$,

(2) $$\lim_{h \to 0^+} m_h = f(x) = \lim_{h \to 0^+} M_h$$

and thus

(3) $$\lim_{h \to 0^+} \frac{F(x + h) - F(x)}{h} = f(x)$$

by the "pinching theorem," Theorem 2.5.1.

In a similar manner we can prove that, for x in the half-open interval $(a, b]$,

(4) $$\lim_{h \to 0^-} \frac{F(x + h) - F(x)}{h} = f(x).$$

For x in the open interval (a, b), both (3) and (4) hold, and we have

$$F'(x) = \lim_{h \to 0} \frac{F(x + h) - F(x)}{h} = f(x).$$

This proves that F is differentiable on (a, b) and has derivative $F'(x) = f(x)$.

All that remains to be shown is that F is continuous from the right at a and continuous from the left at b. Limit (3) at $x = a$ gives

$$\lim_{h \to 0^+} \frac{F(a + h) - F(a)}{h} = f(a).$$

Now, for $h > 0$,

$$F(a + h) - F(a) = \frac{F(a + h) - F(a)}{h} \cdot h$$

and so

$$\lim_{h \to 0^+} [F(a + h) - F(a)] = \lim_{h \to 0^+} \left[\frac{F(a + h) - F(a)}{h} \cdot h \right] = f(a) \cdot \lim_{h \to 0^+} h = 0.$$

Therefore

$$\lim_{h \to 0^+} F(a + h) = F(a).$$

This shows that F is continuous from the right at a. The continuity of F from the left at b can be shown by applying limit (4) at $x = b$. ❏

Example 1 If F is defined by

$$F(x) = \int_{-1}^{x} (2t + t^2)\, dt \qquad \text{for } -1 \le x \le 5,$$

then

$$F'(x) = 2x + x^2 \qquad \text{on } (-1, 5). \quad ❏$$

Example 2 If F is defined by

$$F(x) = \int_0^x \sin \pi t \, dt,$$

determine $F'(3/4)$.

SOLUTION By Theorem 5.3.5,

$$F'(x) = \sin \pi x.$$

Thus, $F'(3/4) = \sin(3\pi/4) = \sqrt{2}/2$. ❏

Example 3 Let F be defined by

$$F(x) = \int_0^x \frac{1}{1 + t^2}\, dt, \qquad \text{where } x \text{ is any real number.}$$

(a) Find the critical numbers of F and determine the intervals on which F is increasing and the intervals on which F is decreasing.

(b) Determine the concavity of the graph of F and find the points of inflection (if any).

(c) Sketch the graph of F.

SOLUTION

(a) To find where F is increasing or decreasing, we need to examine the first derivative. By Theorem 5.3.5,

$$F'(x) = \frac{1}{1 + x^2}.$$

Since $F'(x) > 0$ for all x, F is increasing on $(-\infty, \infty)$; there are no critical numbers.

(b) We use the second derivative to determine the concavity of the graph and to find the points of inflection:

$$F''(x) = \frac{-2x}{(1 + x^2)^2}.$$

The sign of F'' and the behavior of the graph of F are:

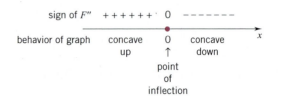

(c) Since $F(0) = 0$ and $F'(0) = 1$, the graph goes through the origin with slope 1. A sketch of the graph is shown in Figure 5.3.6. Although it is not at all clear now, you will see in Chapter 7 that $y = \pi/2$ and $y = -\pi/2$ are horizontal asymptotes of the graph. ❑

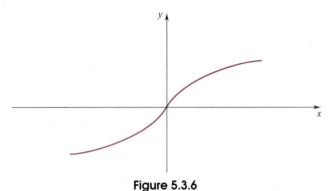

Figure 5.3.6

EXERCISES 5.3

1. Given that

$$\int_0^1 f(x)\, dx = 6, \qquad \int_0^2 f(x)\, dx = 4, \qquad \int_2^5 f(x)\, dx = 1,$$

find each of the following:

(a) $\displaystyle\int_0^5 f(x)\, dx.$ (b) $\displaystyle\int_1^2 f(x)\, dx.$ (c) $\displaystyle\int_1^5 f(x)\, dx.$

(d) $\displaystyle\int_0^0 f(x)\, dx.$ (e) $\displaystyle\int_2^0 f(x)\, dx.$ (f) $\displaystyle\int_5^1 f(x)\, dx.$

2. Given that

$$\int_1^4 f(x)\, dx = 5, \qquad \int_3^4 f(x)\, dx = 7, \qquad \int_1^8 f(x)\, dx = 11,$$

find each of the following:

(a) $\displaystyle\int_4^8 f(x)\, dx.$ (b) $\displaystyle\int_4^3 f(x)\, dx.$ (c) $\displaystyle\int_1^3 f(x)\, dx.$

(d) $\displaystyle\int_3^8 f(x)\, dx.$ (e) $\displaystyle\int_8^4 f(x)\, dx.$ (f) $\displaystyle\int_4^4 f(x)\, dx.$

3. Use upper and lower sums to show that

$$0.5 < \int_1^2 \frac{dx}{x} < 1.$$

4. Use upper and lower sums to show that

$$0.6 < \int_0^1 \frac{dx}{1 + x^2} < 1.$$

5. For $x > -1$ set $F(x) = \displaystyle\int_0^x t\sqrt{t + 1}\, dt.$

(a) Find $F(0)$.   (b) Find $F'(x)$.   (c) Find $F'(2)$.

(d) Express $F(2)$ as an integral of $t\sqrt{t + 1}$.

(e) Express $-F(x)$ as an integral of $t\sqrt{t + 1}$.

6. Let $F(x) = \displaystyle\int_\pi^x t \sin t\, dt.$

(a) Find $F(\pi)$.   (b) Find $F'(x)$.   (c) Find $F'(\pi/2)$.

(d) Express $F(2\pi)$ as an integral of $t \sin t$.

(e) Express $-F(x)$ as an integral of $t \sin t$.

7. For $x > 0$, set $F(x) = \displaystyle\int_1^x (1/t)\, dt.$

(a) Find the critical numbers of F and determine the intervals on which F is increasing and the intervals on which F is decreasing.

(b) Determine the concavity of the graph of F and find the points of inflection, if any.

(c) Sketch the graph of F.

8. Let $F(x) = \displaystyle\int_0^x t(t - 3)^2\, dt.$

(a) Find the critical numbers of F and determine the intervals on which F is increasing and the intervals on which F is decreasing.

(b) Determine the concavity of the graph of F and find the points of inflection, if any.

(c) Sketch the graph of F.

For the given function F in Exercises 9–14, compute the following:

(a) $F'(-1)$.   (b) $F'(0)$.   (c) $F'(\frac{1}{2})$.   (d) $F''(x)$.

9. $F(x) = \int_0^x \dfrac{dt}{t^2 + 9}$.

10. $F(x) = \int_x^0 \sqrt{t^2 + 1}\, dt$.

11. $F(x) = \int_x^1 \sqrt{t^2 + 1}\, dt$.

12. $F(x) = \int_1^x \sin \pi t\, dt$.

13. $F(x) = \int_1^x \cos \pi t\, dt$.

14. $F(x) = \int_2^x (t + 1)^3\, dt$.

15. Explain why each of the following statements must be false.

(a) $U_f(P_1) = 4$ for the partition $P_1 = \{0, 1, \frac{3}{2}, 2\}$ and
$U_f(P_2) = 5$ for the partition $P_2 = \{0, \frac{1}{4}, 1, \frac{3}{2}, 2\}$.

(b) $L_f(P_1) = 5$ for the partition $P_1 = \{0, 1, \frac{3}{2}, 2\}$ and
$L_f(P_2) = 4$ for the partition $P_2 = \{0, \frac{1}{4}, 1, \frac{3}{2}, 2\}$.

16. Show that, if f is continuous on an interval I, then

$$\int_a^c f(t)\, dt + \int_c^b f(t)\, dt = \int_a^b f(t)\, dt$$

for *every* choice of a, b, c from I.   HINT: Assume $a < b$ and consider the four cases: $c = a$, $c = b$, $c < a$, $b < c$. Then consider what happens if $a > b$ or $a = b$.

In Exercises 17–22, find the derivative of the function F.

17. $F(x) = \int_0^{x^3} t \cos t\, dt$ HINT: Let $u = x^3$ and use the chain rule.

18. $F(x) = \int_1^{\cos x} \sqrt{1 - t^2}\, dt$.

19. $F(x) = \int_{x^2}^1 (t - \sin^2 t)\, dt$.

20. $F(x) = \int_0^{\sqrt{x}} \dfrac{t^2}{1 + t^4}\, dt$.

21. $F(x) = \int_{x^2}^{x^4} \sin \sqrt{t}\, dt$

HINT: $\displaystyle\int_{x^2}^{x^4} f(t)\, dt = \int_{x^2}^0 f(t)\, dt + \int_0^{x^4} f(t)\, dt$.

22. $F(x) = \int_{\sin x}^{3x+1} t(1 + t^2)\, dt$.

23. Assume that f is a continuous function and that

$$\int_0^x f(t)\, dt = \dfrac{2x}{4 + x^2}.$$

(a) Determine $f(0)$.

(b) Find the zeros of f, if any.

24. Assume that f is a continuous function and that

$$\int_0^x t f(t)\, dt = \sin x - x \cos x.$$

(a) Determine $f(\pi/2)$.

(b) Find $f'(x)$.

25. (*A Mean-Value Theorem for Integrals*)   Show that, if f is continuous on $[a, b]$, then there is at least one number c in (a, b) for which

$$\int_a^b f(x)\, dx = f(c)(b - a).$$

HINT: Apply the mean-value theorem to the function

$$F(x) = \int_a^x f(t)\, dt \quad \text{on } [a, b].$$

26. Show the validity of equation (2) in the proof of Theorem 5.3.5.

27. Extend Theorem 5.3.5 by showing that, if f is continuous on $[a, b]$ and c is *any* number in $[a, b]$, then the function

$$F(x) = \int_c^x f(t)\, dt$$

is continuous on $[a, b]$, differentiable on (a, b), and satisfies

$$F'(x) = f(x) \qquad \text{for all } x \text{ in } (a, b).$$

HINT: $\displaystyle\int_c^x f(t)\, dt = \int_c^a f(t)\, dt + \int_a^x f(t)\, dt$.

28. Complete the proof of Theorem 5.3.5 by showing that

$$\lim_{h \to 0^-} \dfrac{F(x + h) - F(x)}{h} = f(x) \qquad \text{for all } x \text{ in } (a, b].$$

■ 5.4 THE FUNDAMENTAL THEOREM OF INTEGRAL CALCULUS

DEFINITION 5.4.1   ANTIDERIVATIVE

Let f be continuous on $[a, b]$. A function G is called an *antiderivative* for f on $[a, b]$ iff

    G is continuous on $[a, b]$ and $G'(x) = f(x)$ for all $x \in (a, b)$.

Theorem 5.3.5 says that if f is continuous on $[a, b]$, then

$$F(x) = \int_a^x f(t)\, dt$$

is an antiderivative for f on $[a, b]$. This gives us a prescription for constructing antiderivatives. It tells us that we can construct an antiderivative for f by integrating f.

The so-called "fundamental theorem" goes the other way. It gives us a prescription, not for finding antiderivatives, but for evaluating integrals. It tells us that we can evaluate

$$\int_a^b f(t)\, dt$$

by finding an antiderivative for f.

THEOREM 5.4.2 THE FUNDAMENTAL THEOREM OF INTEGRAL CALCULUS

Let f be continuous on $[a, b]$. If G is any antiderivative for f on $[a, b]$, then

$$\int_a^b f(t)\, dt = G(b) - G(a).$$

PROOF   From Theorem 5.3.5 we know that the function

$$F(x) = \int_a^x f(t)\, dt$$

is an antiderivative for f on $[a, b]$. If G is also an antiderivative for f on $[a, b]$, then both F and G are continuous on $[a, b]$ and satisfy $F'(x) = G'(x)$ for all x in (a, b). From Theorem 4.2.4 we know that there exists a constant C such that

$$F(x) = G(x) + C \qquad \text{for all } x \text{ in } [a, b].$$

Since $F(a) = 0$,

$$G(a) + C = 0 \quad \text{and thus} \quad C = -G(a).$$

It follows that

$$F(x) = G(x) - G(a) \qquad \text{for all } x \text{ in } [a, b].$$

In particular,

$$\int_a^b f(t)\, dt = F(b) = G(b) - G(a). \quad \square$$

We now evaluate some integrals by applying the fundamental theorem. In each case we use the simplest antiderivative we can think of.

Example 1 Evaluate

$$\int_1^4 x^2 \, dx.$$

SOLUTION As an antiderivative for $f(x) = x^2$, we can use the function

$$G(x) = \tfrac{1}{3}x^3. \qquad\qquad \text{(verify this)}$$

By the fundamental theorem

$$\int_1^4 x^2 \, dx = G(4) - G(1) = \tfrac{1}{3}(4)^3 - \tfrac{1}{3}(1)^3 = \tfrac{64}{3} - \tfrac{1}{3} = 21.$$

NOTE: Any other antiderivative of $f(x) = x^2$ has the form $H(x) = \tfrac{1}{3}x^3 + C$ for some constant C. If we had chosen such an H instead of G, then

$$\int_1^4 x^2 \, dx = H(4) - H(1) = [\tfrac{1}{3}(4)^3 + C] - [\tfrac{1}{3}(1)^3 + C] = \tfrac{64}{3} + C - \tfrac{1}{3} - C = 21;$$

the C "cancels out."   ❑

Example 2 Evaluate

$$\int_0^{\pi/2} \sin x \, dx.$$

SOLUTION Here we use the antiderivative $G(x) = -\cos x$:

$$\int_0^{\pi/2} \sin x \, dx = G(\pi/2) - G(0) = -\cos(\pi/2) - [-\cos(0)] = 0 - (-1) = 1.$$

Expressions of the form $G(b) - G(a)$ are conveniently written

$$\Big[G(x) \Big]_a^b.$$

In this notation

$$\int_1^4 x^2 \, dx = \Big[\tfrac{1}{3}x^3 \Big]_1^4 = \tfrac{1}{3}(4)^3 - \tfrac{1}{3}(1)^3 = 21$$

and

$$\int_0^{\pi/2} \sin x \, dx = \Big[-\cos x \Big]_0^{\pi/2} = -\cos(\pi/2) - [-\cos(0)] = 1. ❑$$

Obviously the essential step in applying the fundamental theorem is the determination of an antiderivative G for the integrand f. Thus, we need to have a supply of antiderivatives of functions. From our study of differentiation, we know that

$$\frac{d}{dx}(x^r) = rx^{r-1}, \qquad r \text{ any rational number.}$$

It follows immediately from this formula that

$$G(x) = \frac{x^{r+1}}{r+1}, \qquad (r \neq -1)$$

is an antiderivative of $f(x) = x^r$. Similarly, the differentiation formula

$$\frac{d}{dx}(\sin x) = \cos x$$

implies that $G(x) = \sin x$ is an antiderivative of $f(x) = \cos x$. In general, each differentiation formula yields an immediate antidifferentiation formula. Table 5.4.3 lists some particular antiderivatives for our use in this section.

■ **Table 5.4.3**

Function	Antiderivative
x^r	$\dfrac{x^{r+1}}{r+1}$ (r a rational number $\neq -1$)
$\sin x$	$-\cos x$
$\cos x$	$\sin x$
$\sec^2 x$	$\tan x$
$\sec x \tan x$	$\sec x$
$\csc^2 x$	$-\cot x$
$\csc x \cot x$	$-\csc x$

You can verify that the derivative of each function in the right column is the corresponding function in the left column.

The following examples make use of the entries in Table 5.4.3.

$$\int_1^2 \frac{dx}{x^2} = \int_1^2 x^{-2}\, dx = \left[-x^{-1}\right]_1^2 = \left[-\frac{1}{x}\right]_1^2 = -\tfrac{1}{2} - (-1) = \tfrac{1}{2},$$

$$\int_0^1 t^{5/3}\, dt = \left[\tfrac{3}{8}t^{8/3}\right]_0^1 = \tfrac{3}{8}(1)^{8/3} - \tfrac{3}{8}(0)^{8/3} = \tfrac{3}{8}.$$

$$\int_{-\pi/4}^{\pi/3} \sec^2 x\, dx = \left[\tan x\right]_{-\pi/4}^{\pi/3} = \tan\frac{\pi}{3} - \tan\frac{-\pi}{4} = \sqrt{3} - (-1) = \sqrt{3} + 1.$$

$$\int_{\pi/6}^{\pi/2} \csc x \cot x\, dx = \left[-\csc x\right]_{\pi/6}^{\pi/2} = -\csc\frac{\pi}{2} - \left[-\csc\frac{\pi}{6}\right] = -1 - (-2)$$
$$= 1. \quad \square$$

Example 3 Evaluate

$$\int_0^1 (2x - 6x^4 + 5)\, dx.$$

SOLUTION As an antiderivative we use $G(x) = x^2 - \frac{6}{5}x^5 + 5x$:

$$\int_0^1 (2x - 6x^4 + 5)\, dx = \left[x^2 - \tfrac{6}{5}x^5 + 5x\right]_0^1 = 1 - \tfrac{6}{5} + 5 = 4\tfrac{4}{5}. \quad \square$$

Example 4 Evaluate

$$\int_{-1}^1 (x - 1)(x + 2)\, dx.$$

SOLUTION First we carry out the indicated multiplication:

$$(x - 1)(x + 2) = x^2 + x - 2.$$

As an antiderivative we use $G(x) = \frac{1}{3}x^3 + \frac{1}{2}x^2 - 2x$:

$$\int_{-1}^{1} (x - 1)(x + 2)\, dx = \left[\frac{1}{3}x^3 + \frac{1}{2}x^2 - 2x \right]_{-1}^{1} = -\frac{10}{3}. \quad \square$$

We now give some slightly more complicated examples. The essential step in each case is the determination of an antiderivative. Check each computation in detail.

$$\int_{1}^{2} \frac{x^4 + 1}{x^2}\, dx = \int_{1}^{2} (x^2 + x^{-2})\, dx = \left[\frac{1}{3}x^3 - x^{-1} \right]_{1}^{2} = \frac{17}{6}.$$

$$\int_{1}^{5} \sqrt{x - 1}\, dx = \int_{1}^{5} (x - 1)^{1/2}\, dx = \left[\frac{2}{3}(x - 1)^{3/2} \right]_{1}^{5} = \frac{16}{3}.$$

$$\int_{0}^{1} (4 - \sqrt{x})^2\, dx = \int_{0}^{1} (16 - 8\sqrt{x} + x)\, dx = \left[16x - \frac{16}{3}x^{3/2} + \frac{1}{2}x^2 \right]_{0}^{1} = \frac{67}{6}.$$

$$\int_{1}^{2} -\frac{dt}{(t + 2)^2} = \int_{1}^{2} -(t + 2)^{-2}\, dt = \left[(t + 2)^{-1} \right]_{1}^{2} = -\frac{1}{12}. \quad \square$$

The Linearity of the Integral

The preceding examples suggest some simple properties of the integral that are used regularly in computations. Throughout, take f and g as continuous functions and α and β as constants.

I. Constants may be factored through the integral sign:

(5.4.4)
$$\int_{a}^{b} \alpha f(x)\, dx = \alpha \int_{a}^{b} f(x)\, dx.$$

For example,

$$\int_{1}^{4} \frac{3}{7}\sqrt{x}\, dx = \frac{3}{7} \int_{1}^{4} x^{1/2}\, dx = \frac{3}{7} \left[\frac{x^{3/2}}{3/2} \right]_{1}^{4} = \frac{2}{7}[(4)^{3/2} - (1)^{3/2}] = \frac{2}{7}[8 - 1]$$
$$= 2.$$

$$\int_{0}^{\pi/4} 2\cos x\, dx = 2 \int_{0}^{\pi/4} \cos x\, dx = 2 \left[\sin x \right]_{0}^{\pi/4} = 2 \left[\sin\frac{\pi}{4} - \sin(0) \right]$$
$$= 2\frac{\sqrt{2}}{2} = \sqrt{2}. \quad \square$$

Remark We want to emphasize that only constants can be factored through the integral sign; an expression containing a variable cannot! For example,

$$\int_{0}^{\pi/2} x \cos x\, dx \neq x \int_{0}^{\pi/2} \cos x\, dx.$$

You can verify that an antiderivative for $h(x) = x \cos x$ is $H(x) = x \sin x + \cos x$, and so

$$\int_0^{\pi/2} x \cos x \, dx = \left[x \sin x + \cos x \right]_0^{\pi/2} = \frac{\pi}{2} - 1.$$

On the other hand,

$$x \int_0^{\pi/2} \cos x \, dx = x \left[\sin x \right]_0^{\pi/2} = x. \quad \square$$

II. The integral of a sum is the sum of the integrals:

$$(5.4.5) \qquad \int_a^b [f(x) + g(x)] \, dx = \int_a^b f(x) \, dx + \int_a^b g(x) \, dx.$$

For example,

$$\int_1^2 \left[(x - 1)^2 + \frac{1}{(x + 2)^2} \right] dx = \int_1^2 (x - 1)^2 \, dx + \int_1^2 \frac{dx}{(x + 2)^2}$$

$$= \left[\frac{1}{3}(x - 1)^3 \right]_1^2 + \left[-(x + 2)^{-1} \right]_1^2$$

$$= \frac{1}{3} - \frac{1}{4} + \frac{1}{3} = \frac{5}{12}. \quad \square$$

III. The integral of a linear combination is the linear combination of the integrals:

$$(5.4.6) \qquad \int_a^b [\alpha f(x) + \beta g(x)] \, dx = \alpha \int_a^b f(x) \, dx + \beta \int_a^b g(x) \, dx.$$

For example,

$$\int_0^1 (2x - 6x^4 + 5) \, dx = 2 \int_0^1 x \, dx - 6 \int_0^1 x^4 \, dx + \int_0^1 5 \, dx$$

$$= 2 \left[\frac{x^2}{2} \right]_0^1 - 6 \left[\frac{x^5}{5} \right]_0^1 + \left[5x \right]_0^1 = 1 - \frac{6}{5} + 5 = 4\frac{4}{5}$$

as we saw in Example 3.

Properties I and II are particular instances of Property III. To prove III, let F be an antiderivative for f and let G be an antiderivative for g. Then, since

$$[\alpha F(x) + \beta G(x)]' = \alpha F'(x) + \beta G'(x) = \alpha f(x) + \beta g(x),$$

it follows that $\alpha F + \beta G$ is an antiderivative for $\alpha f + \beta g$. Therefore,

$$\int_a^b [\alpha f(x) + \beta g(x)] \, dx = \left[\alpha F(x) + \beta G(x) \right]_a^b$$

$$= [\alpha F(b) + \beta G(b)] - [\alpha F(a) + \beta G(a)]$$
$$= \alpha [F(b) - F(a)] + \beta [G(b) - G(a)]$$
$$= \alpha \int_a^b f(x) \, dx + \beta \int_a^b g(x) \, dx. \quad \square$$

Example 5 Evaluate

$$\int_0^{\pi/4} \sec x \, [2 \tan x - 5 \sec x] \, dx.$$

SOLUTION

$$\int_0^{\pi/4} \sec x \, [2 \tan x - 5 \sec x] \, dx = \int_0^{\pi/4} [2 \sec x \tan x - 5 \sec^2 x] \, dx$$

$$= 2 \int_0^{\pi/4} \sec x \tan x \, dx - 5 \int_0^{\pi/4} \sec^2 x \, dx$$

$$= 2 \left[\sec x \right]_0^{\pi/4} - 5 \left[\tan x \right]_0^{\pi/4}$$

$$= 2 \left[\sec \frac{\pi}{4} - \sec(0) \right] - 5 \left[\tan \frac{\pi}{4} - \tan(0) \right]$$

$$= 2[\sqrt{2} - 1] - 5[1 - 0] = 2\sqrt{2} - 7. \quad \square$$

EXERCISES 5.4

In Exercises 1–40, evaluate the definite integral.

1. $\int_0^1 (2x - 3) \, dx.$

2. $\int_0^1 (3x + 2) \, dx.$

3. $\int_{-1}^0 5x^4 \, dx.$

4. $\int_1^2 (2x + x^2) \, dx.$

5. $\int_1^4 2\sqrt{x} \, dx.$

6. $\int_0^4 \sqrt[3]{x} \, dx.$

7. $\int_1^5 2\sqrt{x - 1} \, dx.$

8. $\int_1^2 \left(\frac{3}{x^3} + 5x \right) dx.$

9. $\int_{-2}^0 (x + 1)(x - 2) \, dx.$

10. $\int_2^0 \frac{dx}{(x + 1)^2}.$

11. $\int_3^3 \sqrt{x} \, dx.$

12. $\int_{-1}^0 (t - 2)(t + 1) \, dt.$

13. $\int_0^1 \frac{dt}{(t + 2)^3}.$

14. $\int_1^1 (t^3 + t^2) \, dt.$

15. $\int_1^2 \left(3t + \frac{4}{t^2} \right) dt.$

16. $\int_{-1}^{-1} 7x^6 \, dx.$

17. $\int_0^1 (x^{3/2} - x^{1/2}) \, dx.$

18. $\int_0^1 (x^{3/4} - 2x^{1/2}) \, dx.$

19. $\int_0^1 (x + 1)^{17} \, dx.$

20. $\int_0^a (a^2 x - x^3) \, dx.$

21. $\int_0^a (\sqrt{a} - \sqrt{x})^2 \, dx.$

22. $\int_{-1}^1 (x - 2)^2 \, dx.$

23. $\int_1^2 \frac{6 - t}{t^3} \, dt.$

24. $\int_1^2 \frac{2 - t}{t^3} \, dt.$

25. $\int_0^1 x^2(x - 1) \, dx.$

26. $\int_1^3 \left(x^2 - \frac{1}{x^2} \right) dx.$

27. $\int_1^2 2x(x^2 + 1) \, dx.$

28. $\int_0^1 3x^2(x^3 + 1) \, dx.$

29. $\int_0^{\pi/2} \cos x \, dx.$

30. $\int_0^\pi 3 \sin x \, dx.$

31. $\int_0^{\pi/4} 2 \sec^2 x \, dx.$

32. $\int_{\pi/6}^{\pi/3} \sec x \tan x \, dx.$

33. $\int_{\pi/6}^{\pi/4} \csc x \cot x \, dx.$

34. $\int_{\pi/4}^{\pi/3} -\csc^2 x \, dx.$

35. $\int_0^{2\pi} \sin x \, dx.$

36. $\int_0^\pi \frac{1}{2} \cos x \, dx.$

37. $\int_0^{\pi/3} \left(\dfrac{2}{\pi}x - 2\sec^2 x \right) dx.$

38. $\int_{\pi/4}^{\pi/2} \csc x \, (\cot x - 3 \csc x) \, dx.$

39. $\int_0^3 \left[\dfrac{d}{dx}(\sqrt{4 + x^2}) \right] dx.$

40. $\int_0^{\pi/2} \left[\dfrac{d}{dx}(\sin^3 x) \right] dx.$

41. Define a function F on $[1, 8]$ such that $F'(x) = 1/x$ and (a) $F(2) = 0$; (b) $F(2) = -3$.

42. Define a function F on $[0, 4]$ such that $F'(x) = \sqrt{1 + x^2}$ and (a) $F(3) = 0$; (b) $F(3) = 1$.

In Exercises 43–46, verify that the function f is nonnegative on the given interval, and then calculate the area between the graph of f and the x-axis on that interval.

43. $f(x) = 4x - x^2$; $[0, 4]$.

44. $f(x) = x\sqrt{x} + 1$; $[1, 9]$.

45. $f(x) = 2 \cos x$; $[-\pi/2, \pi/4]$.

46. $f(x) = \sec x \tan x$; $[0, \pi/3]$.

In Exercises 47–50, evaluate the given integrals.

47. (a) $\int_2^5 (x - 3) \, dx.$ (b) $\int_2^5 |x - 3| \, dx.$

48. (a) $\int_{-4}^2 (2x + 3) \, dx.$ (b) $\int_{-4}^2 |2x + 3| \, dx.$

49. (a) $\int_{-2}^2 (x^2 - 1) \, dx.$ (b) $\int_{-2}^2 |x^2 - 1| \, dx.$

50. (a) $\int_{-\pi/2}^{\pi} \cos x \, dx.$ (b) $\int_{-\pi/2}^{\pi} |\cos x| \, dx.$

51. An object starts at the origin and moves along the x-axis with velocity

$$v(t) = 10t - t^2, \qquad 0 \le t \le 10.$$

(a) What is the position of the object at any time t, $0 \le t \le 10$?

(b) When is the object's velocity a maximum and what is its position at that time?

52. The velocity of a weight suspended on a spring is given by

$$v(t) = 3 \sin t + 4 \cos t, \qquad t \ge 0.$$

At time $t = 0$, the weight is one unit below the equilibrium position (see the figure).

(a) Determine the position of the weight at any time $t \ge 0$.

(b) What is the weight's maximum displacement from the equilibrium position?

(NOTE: The motion of the weight is called *simple harmonic motion*.   Also see Exercises 61 and 62, Section 3.6.)

0 ⊢ equilibrium position

1 ⊢

In Exercises 53–56, evaluate the given definite integral.

53. $\int_0^4 f(x) \, dx,$ where $f(x) = \begin{cases} 2x + 1, & 0 \le x \le 1 \\ 4 - x, & 1 < x \le 4. \end{cases}$

54. $\int_{-2}^4 f(x) \, dx,$ where $f(x) = \begin{cases} x^2, & -2 \le x < 0 \\ \frac{1}{2}x + 2, & 0 \le x \le 4. \end{cases}$

55. $\int_{-\pi/2}^{\pi} f(x) \, dx,$ where $f(x) = \begin{cases} \cos x, & -\pi/2 \le x \le \pi/ \\ (3/\pi)\,x + 1, & \pi/3 < x \le \pi. \end{cases}$

56. $\int_0^{3\pi/2} f(x) \, dx,$ where $f(x) = \begin{cases} 2 \sin x, & 0 \le x \le \pi/2 \\ \frac{1}{2} \cos x, & \pi/2 < x \le 3\pi/2. \end{cases}$

57. Let

$$f(x) = \begin{cases} x + 2, & -2 \le x \le 0 \\ 2, & 0 < x \le 1 \\ 4 - 2x, & 1 < x \le 2 \end{cases}$$

and let $g(x) = \displaystyle\int_{-2}^x f(t) \, dt.$

(a) Find an expression for $g(x)$.

(b) Sketch the graphs of f and g.

(c) Where is f continuous? Where is f differentiable? Where is g differentiable?

58. Let

$$f(x) = \begin{cases} 1 - x^2, & -1 \le x \le 1 \\ 1, & 1 < x < 3 \\ 2x - 5, & 3 \le x \le 5 \end{cases}$$

and let $g(x) = \displaystyle\int_{-1}^x f(t) \, dt.$

(a) Find an expression for $g(x)$.

(b) Sketch the graphs of f and g.

(c) Where is f continuous? Where is f differentiable? Where is g differentiable?

59. Compare $\dfrac{d}{dx}\left[\displaystyle\int_a^x f(t)\,dt\right]$ to $\displaystyle\int_a^x \dfrac{d}{dt}[f(t)]\,dt$.

60. Let $f_1, f_2, \ldots, f_n$ be continuous functions on $[a, b]$ and let $\alpha_1, \alpha_2, \ldots, \alpha_n$ be constants. Use mathematical induction to prove that

$$\int_a^b [\alpha_1 f_1(x) + \alpha_2 f_2(x) + \cdots + \alpha_n f_n(x)]\,dx$$

$$= \alpha_1 \int_a^b f_1(x)\,dx + \alpha_2 \int_a^b f_2(x)\,dx + \cdots$$

$$+ \alpha_n \int_a^b f_n(x)\,dx.$$

■ 5.5 SOME AREA PROBLEMS

In Section 5.2 we noted that, if f is nonnegative and continuous on $[a, b]$, then the integral of f from a to b gives the area of the region below the graph of f, that is, the area of the region bounded by the graph of f and the x-axis between $x = a$ and $x = b$. With Ω as in Figure 5.5.1

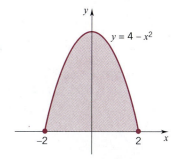

(5.5.1)

$$\text{Area of } \Omega = \int_a^b f(x)\,dx.$$

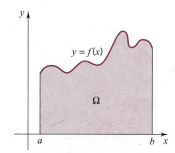

Figure 5.5.1

Example 1 Find the area below the graph of the square-root function from $x = 0$ to $x = 1$.

SOLUTION The graph is pictured in Figure 5.5.2. The area below the graph is $\frac{2}{3}$:

$$\int_0^1 \sqrt{x}\,dx = \int_0^1 x^{1/2}\,dx = \left[\tfrac{2}{3}x^{3/2}\right]_0^1 = \tfrac{2}{3}. \quad \square$$

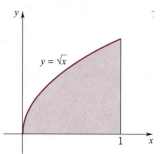

Example 2 Find the area of the region bounded above by the curve $y = 4 - x^2$ and below by the x-axis.

Figure 5.5.2

SOLUTION The curve intersects the x-axis at $x = -2$ and $x = 2$. See Figure 5.5.3. The area of the region is $\frac{32}{3}$:

$$\int_{-2}^2 (4 - x^2)\,dx = \left[4x - \tfrac{1}{3}x^3\right]_{-2}^2 = \tfrac{32}{3}.$$

NOTE: The region is symmetric with respect to the y-axis. Therefore, the area of the region is also $2\displaystyle\int_0^2 (4 - x^2)\,dx$, and this integral is a little easier to compute:

$$2\int_0^2 (4 - x^2)\,dx = 2\left[4x - \tfrac{1}{3}x^3\right]_0^2 = 2(8 - \tfrac{8}{3}) = 2(\tfrac{16}{3}) = \tfrac{32}{3}.$$

We will have more to say about symmetry in integration in Section 5.8. □

Figure 5.5.3

Now we calculate the areas of somewhat more complicated regions such as region Ω shown in Figure 5.5.4. To avoid excessive repetition, let's agree at the outset that throughout this section the symbols f, g, h represent continuous functions.

Look at the region Ω shown in Figure 5.5.4. The upper boundary of Ω is the graph of a nonnegative function f and the lower boundary is the graph of a nonnegative area of Ω = area of Ω_1 − area of Ω_2

Figure 5.5.4

function g. We can obtain the area of Ω by calculating the area of Ω_1 and subtracting off the area of Ω_2. Since

$$\text{Area of } \Omega_1 = \int_a^b f(x)\, dx \quad \text{and} \quad \text{Area of } \Omega_2 = \int_a^b g(x)\, dx,$$

we have

$$\text{Area of } \Omega = \int_a^b f(x)\, dx - \int_a^b g(x)\, dx.$$

We can combine the two integrals [by (5.4.6)] and write

(5.5.2)
$$\text{Area of } \Omega = \int_a^b [f(x) - g(x)]\, dx.$$

Example 3 Find the area of the region bounded above by $y = x + 2$ and below by $y = x^2$.

SOLUTION The region bounded by the graphs of the two equations is shown in Figure 5.5.5. Our first step is to find the points of intersection of the two curves:

$$x + 2 = x^2 \quad \text{iff} \quad x^2 - x - 2 = 0$$
$$\text{iff} \quad (x + 1)(x - 2) = 0,$$

and so $x = -1, 2$. The curves intersect at $(-1, 1)$ and $(2, 4)$.

Now, the area of the region is:

$$\int_{-1}^2 [(x + 2) - x^2]\, dx = \left[\tfrac{1}{2}x^2 + 2x - \tfrac{1}{3}x^3 \right]_{-1}^2$$
$$= (2 + 4 - \tfrac{8}{3}) - (\tfrac{1}{2} - 2 + \tfrac{1}{3})$$
$$= \tfrac{9}{2}. \quad \square$$

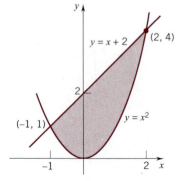

Figure 5.5.5

We derived Formula 5.5.2 under the assumption that f and g were both nonnegative, but that assumption is unnecessary. The formula holds for any region Ω that has

an upper boundary of the form $\quad y = f(x), \quad x \in [a, b]$

and

$$\text{a lower boundary of the form} \quad y = g(x), \quad x \in [a, b].$$

To see this, take Ω as in Figure 5.5.6. Obviously, Ω is congruent to the region marked Ω' in Figure 5.5.7; Ω' is Ω raised C units. Since Ω' lies entirely above the x-axis, the area of Ω' is given by the integral

$$\int_a^b \{[f(x) + C] - [g(x) + C]\} \, dx = \int_a^b [f(x) - g(x)] \, dx.$$

Since area of Ω = area of Ω',

$$\text{Area of } \Omega = \int_a^b [f(x) - g(x)] \, dx$$

as asserted.

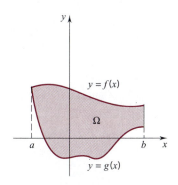

Figure 5.5.6

Example 4 Find the area of the region shown in Figure 5.5.8.

Figure 5.5.8

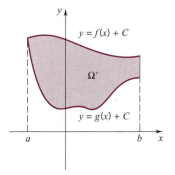

Figure 5.5.7

SOLUTION

$$\begin{aligned}
\text{Area of } \Omega &= \int_{\pi/4}^{5\pi/4} [\sin x - \cos x] \, dx \\
&= \Big[-\cos x - \sin x \Big]_{\pi/4}^{5\pi/4} \\
&= 2\sqrt{2}. \quad \square
\end{aligned}$$

Example 5 Find the area between the curves

$$y = 4x \qquad \text{and} \qquad y = x^3$$

from $x = -2$ to $x = 2$. See Figure 5.5.9.

SOLUTION Notice that $y = x^3$ is the upper boundary from $x = -2$ to $x = 0$, but it is the lower boundary from $x = 0$ to $x = 2$. Thus

$$\begin{aligned}
\text{Area} &= \int_{-2}^0 [x^3 - 4x] \, dx + \int_0^2 [4x - x^3] \, dx \\
&= \Big[\tfrac{1}{4}x^4 - 2x^2 \Big]_{-2}^0 + \Big[2x^2 - \tfrac{1}{4}x^4 \Big]_0^2 \\
&= [0 - (-4)] + [4 - 0] = 8. \quad \square
\end{aligned}$$

Figure 5.5.9

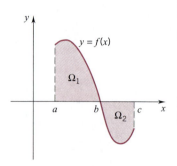

Figure 5.5.10

Example 6 Use integrals to represent the area of the region $\Omega = \Omega_1 \cup \Omega_2$ shaded in Figure 5.5.10.

SOLUTION From $x = a$ to $x = b$, the curve $y = f(x)$ is above the x-axis. Therefore

$$\text{Area of } \Omega_1 = \int_a^b f(x)\, dx.$$

From $x = b$ to $x = c$, the curve $y = f(x)$ is below the x-axis. The upper boundary for Ω_2 is the curve $y = 0$ (the x-axis) and the lower boundary is the curve $y = f(x)$. Thus

$$\text{Area of } \Omega_2 = \int_b^c [0 - f(x)]\, dx = -\int_b^c f(x)\, dx.$$

The area of Ω is the sum of these two areas:

$$\text{Area of } \Omega = \int_a^b f(x)\, dx - \int_b^c f(x)\, dx. \quad \square$$

Signed Area

The solution of Example 6 suggests a general geometric interpretation of the definite integral of a (continuous) function f. For example, if f is as denoted in Figure 5.5.10, then $\int_a^c f(x)\, dx$ is a number. But what does this number represent in terms of area?

Since

$$\int_a^c f(x)\, dx = \int_a^b f(x)\, dx + \int_b^c f(x)\, dx = \text{area } \Omega_1 - \text{area } \Omega_2,$$

it follows that $\int_a^c f(x)\, dx$ is the area of the region above the x-axis *minus* the area of the region below the x-axis. This result is true in general: If f changes sign on the interval $[a, b]$, then $\int_a^b f(x)\, dx$ can be interpreted as the *difference* of two areas—the total area of the regions that lie above the x-axis *minus* the total area of the regions that lie below the x-axis. Consider the function f shown in Figure 5.5.11:

$$\int_a^b f(x)\, dx = \int_a^c f(x)\, dx + \int_c^d f(x)\, dx + \int_d^e f(x)\, dx + \int_e^b f(x)\, dx$$

$$= \text{area of } \Omega_1 - \text{area of } \Omega_2 + \text{area of } \Omega_3 - \text{area of } \Omega_4$$

$$= [\text{area of } \Omega_1 + \text{area of } \Omega_3] - [\text{area of } \Omega_2 + \text{area of } \Omega_4].$$

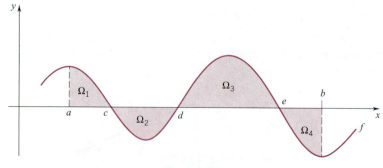

Figure 5.5.11

This value is sometimes referred to as the *signed area*; $\int_a^b f(x)\,dx$ can be interpreted geometrically as the signed area of the region bounded by the graph of f and the x-axis between $x = a$ and $x = b$.

Example 7 Evaluate

$$\int_{-1}^{3} (x^2 - 2x)\,dx$$

and interpret the result in terms of areas. Also, determine the area of the region bounded by the graph of $f(x) = x^2 - 2x$ and the x-axis between $x = -1$ and $x = 3$.

SOLUTION The graph of $f(x) = x^2 - 2x$ is shown in Figure 5.5.12. Now

$$\int_{-1}^{3} (x^2 - 2x)\,dx = \left[\tfrac{1}{3}x^3 - x^2 \right]_{-1}^{3}$$

$$= \left[\tfrac{1}{3}(3)^3 - (3)^2 \right] - \left[\tfrac{1}{3}(-1)^3 - (-1)^2 \right] = \tfrac{4}{3}.$$

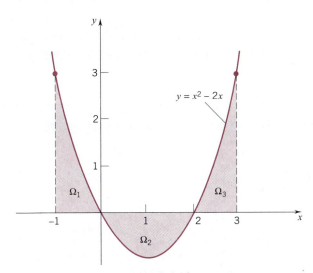

$$y = x^2 - 2x$$

Figure 5.5.12

Thus, area of Ω_1 + area of Ω_3 − area of $\Omega_2 = \tfrac{4}{3}$. The area A of the region bounded by the graph of f and the x-axis from $x = -1$ to $x = 3$ is given by

$$A = \int_{-1}^{0} (x^2 - 2x)\,dx + \int_{0}^{2} (2x - x^2)\,dx + \int_{2}^{3} (x^2 - 2x)\,dx$$

$$= \left[\tfrac{1}{3}x^3 - x^2 \right]_{-1}^{0} + \left[x^2 - \tfrac{1}{3}x^3 \right]_{0}^{2} + \left[\tfrac{1}{3}x^3 - x^2 \right]_{2}^{3}$$

$$= \tfrac{4}{3} + \tfrac{4}{3} + \tfrac{4}{3} = 4. \quad \square$$

Example 8 Use integrals to represent the area of the region shaded in Figure 5.5.13.

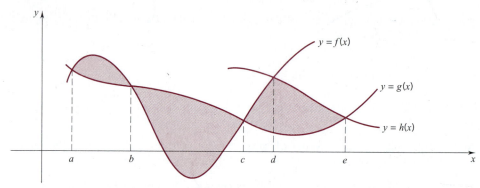

Figure 5.5.13

SOLUTION

$$\text{Area} = \int_a^b [f(x) - g(x)]\, dx + \int_b^c [g(x) - f(x)]\, dx$$

$$+ \int_c^d [f(x) - g(x)]\, dx + \int_d^e [h(x) - g(x)]\, dx. \quad \square$$

Remark The examples in this section also demonstrate the importance of having an accurate figure to illustrate the problem being considered. Drawing an accurate figure is usually the first and crucial step in effective problem solving. ❑

EXERCISES 5.5

In Exercises 1–10, find the area between the graph of f and the x-axis.

1. $f(x) = 2 + x^3$, $x \in [0, 1]$.

2. $f(x) = (x + 2)^{-2}$, $x \in [0, 2]$.

3. $f(x) = \sqrt{x + 1}$, $x \in [3, 8]$.

4. $f(x) = x^2(3 + x)$, $x \in [0, 8]$.

5. $f(x) = (2x^2 + 1)^2$, $x \in [0, 1]$.

6. $f(x) = \frac{1}{2}(x + 1)^{-1/2}$, $x \in [0, 8]$.

7. $f(x) = x^2 - 4$, $x \in [1, 2]$.

8. $f(x) = \cos x$, $x \in [\frac{1}{6}\pi, \frac{1}{3}\pi]$.

9. $f(x) = \sin x$, $x \in [\frac{1}{3}\pi, \frac{1}{2}\pi]$.

10. $f(x) = x^3 + 1$, $x \in [-2, -1]$.

In Exercises 11–26, sketch the region bounded by the curves and find its area.

11. $y = \sqrt{x}$, $y = x^2$.

12. $y = 6x - x^2$, $y = 2x$.

13. $y = 5 - x^2$, $y = 3 - x$.

14. $y = 8$, $y = x^2 + 2x$.

15. $y = 8 - x^2$, $y = x^2$.

16. $y = \sqrt{x}$, $y = \frac{1}{4}x$.

17. $x^3 - 10y^2 = 0$, $x - y = 0$.

18. $y^2 - 27x = 0$, $x + y = 0$.

19. $x - y^2 + 3 = 0$, $x - 2y = 0$.

20. $y^2 = 2x$, $x - y = 4$.

21. $y = x$, $y = 2x$, $y = 4$.

22. $y = x^2$, $y = -\sqrt{x}$, $x = 4$.

23. $y = \cos x$, $y = 4x^2 - \pi^2$.

24. $y = \sin x$, $y = \pi x - x^2$.

25. $y = x$, $y = \sin x$, $x = \pi/2$.

26. $y = x + 1$, $y = \cos x$, $x = \pi$.

27. The graph of $f(x) = x^2 - x - 6$ is shown in the figure.

(a) Evaluate $\int_{-3}^{4} f(x)\, dx$ and interpret the result in terms of areas.

(b) Find the area of the region bounded by the graph of f and the x-axis for $x \in [-3, 4]$.

(c) Find the area of the region bounded by the graph of f and the x-axis for $x \in [-2, 3]$.

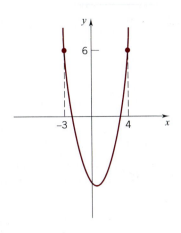

28. The graph of $f(x) = 2 \sin x$, $x \in [-\pi/2, 3\pi/4]$ is shown in the figure below.

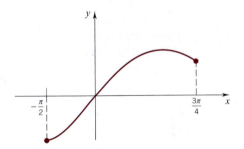

(a) Evaluate $\int_{-\pi/2}^{3\pi/4} f(x)\, dx$ and interpret the result in terms of areas.

(b) Find the area of the region bounded by the graph of f and the x-axis for $x \in [-\pi/2, 3\pi/4]$.

(c) Find the area of the region bounded by the graph of f and the x-axis for $x \in [-\pi/2, 0]$.

29. Let $f(x) = x^3 - x$.

(a) Evaluate $\int_{-2}^{2} f(x)\, dx$.

(b) Sketch the graph of f and find the area of the region bounded by the graph and the x-axis for $x \in [-2, 2]$.

30. Let $f(x) = \cos x + \sin x$, $x \in [-\pi, \pi]$.

(a) Evaluate $\int_{-\pi}^{\pi} f(x)\, dx$.

(b) Sketch the graph of f and find the area of the region bounded by the graph and the x-axis for $x \in [-\pi, \pi]$.

31. Let $f(x) = x^3 - 4x + 2$.

(a) Evaluate $\int_{-2}^{3} f(x)\, dx$.

(b) Use a graphing utility to graph f and estimate the area bounded by the graph and the x-axis for $x \in [-2, 3]$. Use two-decimal place accuracy in your approximations.

32. Let $f(x) = 3x^2 - 2 \cos x$.

(a) Evaluate $\int_{-\pi/2}^{\pi/2} f(x)\, dx$.

(b) Use a graphing utility to graph f and estimate the area bounded by the graph and the x-axis for $x \in [-\pi/2, \pi/2]$. Use two-decimal place accuracy in your approximations.

33. Let
$$f(x) = \begin{cases} x^2 + 1, & 0 \le x \le 1 \\ 3 - x, & 1 < x \le 3. \end{cases}$$

Sketch the graph of f and find the area of the region bounded by the graph and the x-axis.

34. Let
$$f(x) = \begin{cases} 3\sqrt{x}, & 0 \le x \le 1 \\ 4 - x^2, & 1 < x \le 2. \end{cases}$$

Sketch the graph of f and find the area of the region bounded by the graph and the x-axis.

35. Sketch the region bounded by the x-axis the curves $y = \sin x$ and $y = \cos x$, $x \in [0, \pi/2]$, and find its area.

36. Sketch the region bounded by the curves $y = 1$ and $y = 1 + \cos x$ for $x \in [0, \pi]$ and find its area.

37. Use a graphing utility to sketch the region bounded by the curves $y = x^3 + 2x$ and $y = 3x + 1$ for $x \in [0, 2]$, and estimate its area. Use two-decimal place accuracy in your approximations.

38. Use a graphing utility to sketch the region bounded by the curves $y = x^4 - 2x^2$ and $y = 4 - x^2$ for $x \in [-2, 2]$, and estimate its area. Use two-decimal place accuracy in your approximations.

(a) Evaluate $\int_{-\pi}^{\pi} f(x)\, dx$.

(b) Sketch the graph of f and find the area of the region bounded by the graph and the x-axis for $x \in [-\pi, \pi]$.

■ 5.6 INDEFINITE INTEGRALS

We begin with a continuous function f. If F is an antiderivative for f on $[a, b]$, then

(1)
$$\int_a^b f(x)\, dx = \left[F(x) \right]_a^b.$$

If C is a constant, then

$$\left[F(x) + C \right]_a^b = [F(b) + C] - [F(a) + C] = F(b) - F(a) = \left[F(x) \right]_a^b.$$

Thus we can replace (1) by writing

$$\int_a^b f(x)\, dx = \left[F(x) + C \right]_a^b.$$

If we have no particular interest in the interval $[a, b]$, but wish instead to emphasize that F is an antiderivative for f on *some* interval, then we can omit the a and the b and simply write

$$\int f(x)\, dx = F(x) + C.$$

Antiderivatives expressed in this manner are called *indefinite integrals.* The constant C is called the *constant of integration;* it is an *arbitrary* constant since it can be assigned any real value.

The indefinite integral of a function f is actually a *family* of functions; a specific member of the family is determined by assigning a particular value to the constant of integration. This family has the property that each of its members is an antiderivative of f and, conversely, every antiderivative of f is a member of the family. The latter fact follows from Theorem 4.2.4.

Thus, for example,

$$\int x^2\, dx = \tfrac{1}{3}x^3 + C \qquad \text{and} \qquad \int \sqrt{s}\, ds = \tfrac{2}{3}s^{3/2} + C.$$

The graphs of some specific members of these families are sketched in Figures 5.6.1 and 5.6.2, respectively.

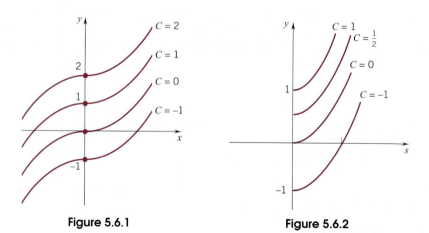

Figure 5.6.1   **Figure 5.6.2**

The Table of Antiderivatives (5.4.3) can be restated in terms of indefinite integrals:

5.6.1 TABLE OF INDEFINITE INTEGRALS

$$\int x^r \, dx = \frac{x^{r+1}}{r+1} + C \quad (r \text{ rational} \neq -1). \qquad \int \sec x \tan x \, dx = \sec x + C.$$

$$\int \sin x \, dx = -\cos x + C. \qquad \int \csc^2 x \, dx = -\cot x + C.$$

$$\int \cos x \, dx = \sin x + C. \qquad \int \csc x \cot x \, dx = -\csc x + C.$$

$$\int \sec^2 x \, dx = \tan x + C.$$

The linearity properties of definite integrals (5.4.4), (5.4.5), and (5.4.6) also hold for indefinite integrals.

(5.6.2)
$$\int \alpha f(x) \, dx = \alpha \int f(x) \, dx, \quad \alpha \text{ a constant.}$$

(5.6.3)
$$\int [f(x) + g(x)] \, dx = \int f(x) \, dx + \int g(x) \, dx.$$

and, in general,

(5.6.4)
$$\int [\alpha f(x) + \beta g(x)] \, dx = \alpha \int f(x) \, dx + \beta \int g(x) \, dx,$$
$$\text{where } \alpha \text{ and } \beta \text{ are constants.}$$

Example 1 Calculate
$$\int [5x^{3/2} - 2 \csc^2 x] \, dx.$$

SOLUTION

$$\int [5x^{3/2} - 2 \csc^2 x] \, dx = 5 \int x^{3/2} \, dx - 2 \int \csc^2 x \, dx$$
$$= 5\left(\tfrac{2}{5}\right) x^{5/2} + C_1 - 2(-\cot x) + C_2$$
$$= 2x^{5/2} + 2 \cot x + C,$$

where $C = C_1 + C_2$. (We will routinely combine the constants of integration in this manner.) ❏

Example 2   Find f given that

$$f'(x) = x^3 + 2 \qquad \text{and} \qquad f(0) = 1.$$

SOLUTION   Since f' is the derivative of f, f is an antiderivative for f'.   Thus

$$f(x) = \int (x^3 + 2)\, dx = \tfrac{1}{4}x^4 + 2x + C,$$

for some value of the constant C.   To evaluate C we use the fact that $f(0) = 1$.   Since

$$f(0) = 1 \qquad \text{and} \qquad f(0) = \tfrac{1}{4}(0)^4 + 2(0) + C = C,$$

we see that $C = 1$.   Therefore

$$f(x) = \tfrac{1}{4}x^4 + 2x + 1.$$

Some of the members of the family of curves $F(x) = \tfrac{1}{4}x^4 + 2x + C$ are shown in Figure 5.6.3.   The graph of f is highlighted. ❏

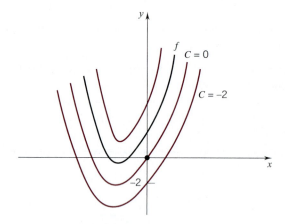

Figure 5.6.3

Example 3   Find f given that

$$f''(x) = 6x - 2, \qquad f'(1) = -5, \qquad \text{and} \qquad f(1) = 3.$$

SOLUTION   First we get f' by integrating f'':

$$f'(x) = \int (6x - 2)\, dx = 3x^2 - 2x + C.$$

Since

$$f'(1) = -5 \qquad \text{and} \qquad f'(1) = 3(1)^2 - 2(1) + C = 1 + C,$$

we have

$$-5 = 1 + C \quad \text{and thus} \quad C = -6.$$

Therefore

$$f'(x) = 3x^2 - 2x - 6.$$

Now we get to f by integrating f':

$$f(x) = \int (3x^2 - 2x - 6)\, dx = x^3 - x^2 - 6x + K.$$

(We are writing the constant of integration as K because we used C before and it would be confusing to have C with two different values in the same problem.)   Since

$$f(1) = 3 \quad \text{and} \quad f(1) = (1)^3 - (1)^2 - 6(1) + K = -6 + K,$$

we have

$$3 = -6 + K \quad \text{and thus} \quad K = 9.$$

Therefore

$$f(x) = x^3 - x^2 - 6x + 9. \quad \square$$

Some Motion Examples

Example 4   An object moves along a coordinate line with velocity

$$v(t) = 2 - 3t + t^2 \quad \text{units per second.}$$

Its initial position (position at time $t = 0$) is 2 units to the right of the origin.   Find the position of the object 4 seconds later.

SOLUTION   Let $x(t)$ be the position (coordinate) of the object at time t.   We are given that $x(0) = 2$.   Since $x'(t) = v(t)$,

$$x(t) = \int v(t)\, dt = \int (2 - 3t + t^2)\, dt = 2t - \tfrac{3}{2}t^2 + \tfrac{1}{3}t^3 + C.$$

Since $x(0) = 2$ and $x(0) = 2(0) - \tfrac{3}{2}(0)^2 + \tfrac{1}{3}(0)^3 + C = C$, we have $C = 2$ and

$$x(t) = 2t - \tfrac{3}{2}t^2 + \tfrac{1}{3}t^3 + 2.$$

The position of the object at time $t = 4$ is the value of this function at $t = 4$:

$$x(4) = 2(4) - \tfrac{3}{2}(4)^2 + \tfrac{1}{3}(4)^3 + 2 = 7\tfrac{1}{3}.$$

At the end of 4 seconds the object is $7\tfrac{1}{3}$ units to the right of the origin.
     The motion of the object is represented schematically in Figure 5.6.4. $\square$

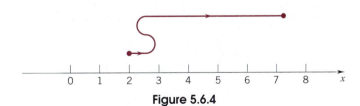

Figure 5.6.4

Recall that speed is the absolute value of velocity:

$$\text{Speed at time } t = |v(t)|$$

and the integral of the speed function gives the distance traveled:

$$\int_a^b |v(t)|\, dt = \text{distance traveled from time } t = a \text{ to time } t = b.$$

Example 5 An object moves along the x-axis with acceleration $a(t) = 2t - 2$ units per second per second. Its initial position (position at time $t = 0$) is 5 units to the right of the origin. One second later the object is moving left at the rate of 4 units per second.

(a) Find the position of the object at time $t = 4$ seconds.

(b) How far does the object travel during these 4 seconds?

SOLUTION (a) Let $x(t)$ and $v(t)$ denote the position and velocity of the object at time t. We are given that $x(0) = 5$ and $v(1) = -4$. Since $v'(t) = a(t)$,

$$v(t) = \int a(t)\, dt = \int (2t - 2)\, dt = t^2 - 2t + C.$$

Since

$$v(1) = -4 \qquad \text{and} \qquad v(1) = (1)^2 - 2(1) + C = -1 + C,$$

we have $C = -3$ and

$$v(t) = t^2 - 2t - 3.$$

Since $x'(t) = v(t)$,

$$x(t) = \int v(t)\, dt = \int (t^2 - 2t - 3)\, dt = \tfrac{1}{3}t^3 - t^2 - 3t + K.$$

Since

$$x(0) = 5 \qquad \text{and} \qquad x(0) = \tfrac{1}{3}(0)^3 - (0)^2 - 3(0) + K = K,$$

we have $K = 5$. Therefore

$$x(t) = \tfrac{1}{3}t^3 - t^2 - 3t + 5.$$

As you can check, $x(4) = -\tfrac{5}{3}$. At time $t = 4$ the object is $\tfrac{5}{3}$ units to the left of the origin.

(b) The distance traveled from time $t = 0$ to $t = 4$ is given by the integral

$$s = \int_0^4 |v(t)|\, dt = \int_0^4 |t^2 - 2t - 3|\, dt.$$

To evaluate this integral we first remove the absolute-value sign. As you can verify,

$$|t^2 - 2t - 3| = \begin{cases} -(t^2 - 2t - 3), & 0 \le t < 3 \\ t^2 - 2t - 3, & 3 \le t \le 4. \end{cases}$$

Thus

$$s = \int_0^3 (3 + 2t - t^2)\, dt + \int_3^4 (t^2 - 2t - 3)\, dt$$

$$= \left[3t + t^2 - \tfrac{1}{3}t^3 \right]_0^3 + \left[\tfrac{1}{3}t^3 - t^2 - 3t \right]_3^4 = \tfrac{34}{3}.$$

During the 4 seconds the object travels a distance of $\tfrac{34}{3}$ units.
The motion of the object is represented schematically in Figure 5.6.5. ❏

Figure 5.6.5

QUESTION: The object in Example 5 leaves $x = 5$ at time $t = 0$ and arrives at $x = -\frac{5}{3}$ at time $t = 4$. The separation between $x = 5$ and $x = -\frac{5}{3}$ is only $|5 - (-\frac{5}{3})| = \frac{20}{3}$. How is it possible that the object travels a distance of $\frac{34}{3}$ units?

ANSWER: The object does not maintain a fixed direction. It changes direction at time $t = 3$. You can see this by noting that the velocity function

$$v(t) = t^2 - 2t - 3 = (t - 3)(t + 1)$$

changes signs at $t = 3$. ❑

Example 6 Find the equation of motion for an object that moves along a straight line with constant acceleration a from an initial position x_0 with initial velocity v_0.

SOLUTION Call the line of the motion the x-axis. Here $a(t) = a$ at all times t. To find the velocity we integrate the acceleration:

$$v(t) = \int a \, dt = at + C.$$

The constant C is the initial velocity v_0:

$$v_0 = v(0) = a \cdot 0 + C = C.$$

We see therefore that

$$v(t) = at + v_0.$$

To find the position function we integrate the velocity:

$$x(t) = \int v(t) \, dt = \int (at + v_0) \, dt = \tfrac{1}{2}at^2 + v_0 t + K.$$

The constant K is the initial position x_0:

$$x_0 = x(0) = \tfrac{1}{2}a \cdot 0^2 + v_0 \cdot 0 + K = K.$$

The equation of motion is

(5.6.5)
$$\boxed{x(t) = \tfrac{1}{2}at^2 + v_0 t + x_0.}^{\dagger} \quad ❑$$

---

† In the case of a free-falling body, $a = -g$ and we have Galileo's equation for free fall. See (3.4.5).

EXERCISES 5.6

In Exercises 1–18, calculate the given indefinite integral.

1. $\displaystyle\int \frac{dx}{x^4}.$

2. $\displaystyle\int (x-1)^2\,dx.$

3. $\displaystyle\int (ax+b)\,dx.$

4. $\displaystyle\int (ax^2+b)\,dx.$

5. $\displaystyle\int \frac{dx}{\sqrt{1+x}}.$

6. $\displaystyle\int \left(\frac{x^3+1}{x^5}\right)dx.$

7. $\displaystyle\int \left(\frac{x^3-1}{x^2}\right)dx.$

8. $\displaystyle\int \left(\sqrt{x}-\frac{1}{\sqrt{x}}\right)dx.$

9. $\displaystyle\int (t-a)(t-b)\,dt.$

10. $\displaystyle\int (t^2-a)(t^2-b)\,dt.$

11. $\displaystyle\int \frac{(t^2-a)(t^2-b)}{\sqrt{t}}\,dt.$

12. $\displaystyle\int (2-\sqrt{x})(2+\sqrt{x})\,dx.$

13. $\displaystyle\int g(x)g'(x)\,dx.$

14. $\displaystyle\int \sin x \cos x\,dx.$

15. $\displaystyle\int \tan x \sec^2 x\,dx.$

16. $\displaystyle\int \frac{g'(x)}{[g(x)]^2}\,dx.$

17. $\displaystyle\int \frac{4}{(4x+1)^2}\,dx.$

18. $\displaystyle\int \frac{3x^2}{(x^3+1)^2}\,dx.$

In Exercises 19–32, find f from the information given.

19. $f'(x) = 2x - 1, \quad f(3) = 4.$

20. $f'(x) = 3 - 4x, \quad f(1) = 6.$

21. $f'(x) = ax + b, \quad f(2) = 0.$

22. $f'(x) = ax^2 + bx + c, \quad f(0) = 0.$

23. $f'(x) = \sin x, \quad f(0) = 2.$

24. $f'(x) = \cos x, \quad f(\pi) = 3.$

25. $f''(x) = 6x - 2, \quad f'(0) = 1, \quad f(0) = 2.$

26. $f''(x) = -12x^2, \quad f'(0) = 1, \quad f(0) = 2.$

27. $f''(x) = x^2 - x, \quad f'(1) = 0, \quad f(1) = 2.$

28. $f''(x) = 1 - x, \quad f'(2) = 1, \quad f(2) = 0.$

29. $f''(x) = \cos x, \quad f'(0) = 1, \quad f(0) = 2.$

30. $f''(x) = \sin x, \quad f'(0) = -2, \quad f(0) = 1.$

31. $f''(x) = 2x - 3, \quad f(2) = -1, \quad f(0) = 3.$

32. $f''(x) = 5 - 4x, \quad f(1) = 1, \quad f(0) = -2.$

33. Compare $\dfrac{d}{dx}\left[\displaystyle\int f(x)\,dx\right]$ to $\displaystyle\int \frac{d}{dx}[f(x)]\,dx.$

34. Calculate

$$\int [f(x)g''(x) - g(x)f''(x)]\,dx.$$

35. An object moves along a coordinate line with velocity $v(t) = 6t^2 - 6$ units per second. Its initial position (position at time $t = 0$) is 2 units to the left of the origin. (a) Find the position of the object 3 seconds later. (b) Find the total distance traveled by the object during those 3 seconds.

36. An object moves along a coordinate line with acceleration $a(t) = (t + 2)^3$ units per second per second. (a) Find the velocity function given that the initial velocity is 3 units per second. (b) Find the position function given that the initial velocity is 3 units per second and the initial position is the origin.

37. An object moves along a coordinate line with acceleration $a(t) = (t + 1)^{-1/2}$ units per second per second. (a) Find the velocity function given that the initial velocity is 1 unit per second. (b) Find the position function given that the initial velocity is 1 unit per second and the initial position is the origin.

38. An object moves along a coordinate line with velocity $v(t) = t(1 - t)$ units per second. Its initial position is 2 units to the left of the origin. (a) Find the position of the object 10 seconds later. (b) Find the total distance traveled by the object during those 10 seconds.

39. A car traveling at 60 mph decelerates at 20 feet per second per second. (a) How long does it take for the car to come to a complete stop? (b) What distance is required to bring the car to a complete stop?

40. An object moves along the x-axis with constant acceleration. Express the position $x(t)$ in terms of the initial position x_0, the initial velocity v_0, the velocity $v(t)$, and the elapsed time t.

41. An object moves along the x-axis with constant acceleration a. Verify that

$$[v(t)]^2 = v_0^2 + 2a[x(t) - x_0].$$

42. A bobsled moving at 60 mph decelerates at a constant rate to 40 mph over a distance of 264 feet and continues to decelerate at that same rate until it comes to a full stop. (a) What is the acceleration of the sled in feet per second per second? (b) How long does it take to reduce the speed to 40 mph? (c) How long does it take to bring the sled to a complete stop from 60 mph? (d) Over what distance does the sled come to a complete stop from 60 mph?

43. In the AB-run, minicars start from a standstill at point A, race along a straight track, and come to a full stop at point B one-half mile away. Given that the cars can accelerate uniformly to a maximum speed of 60 mph in 20 seconds and can brake at a maximum rate of 22 feet per second per second, what is the best possible time for the completion of the AB-run?

In Exercises 44–46, find the general law of motion of an object that moves in a straight line with acceleration $a(t)$. Write x_0 for initial position and v_0 for initial velocity.

44. $a(t) = \sin t$.   **45.** $a(t) = 2A + 6Bt$.

46. $a(t) = \cos t$.

47. As a particle moves about the plane, its x-coordinate changes at the rate of $t^2 - 5$ units per second and its y-coordinate changes at the rate of $3t$ units per second. If the particle is at the point $(4, 2)$ when $t = 2$ seconds, where is the particle 4 seconds later?

48. As a particle moves about the plane, its x-coordinate changes at the rate of $t - 2$ units per second and its y-coordinate changes at the rate of $\sqrt{t}$ units per second. If the particle is at the point $(3, 1)$ when $t = 4$ seconds, where is the particle 5 seconds later?

49. A particle moves along the x-axis with velocity $v(t) = At + B$. Determine A and B given that the initial velocity of the particle is 2 units per second and the position of the particle after 2 seconds of motion is 1 unit to the left of the initial position.

50. A particle moves along the x-axis with velocity $v(t) = At^2 + 1$. Determine A given that $x(1) = x(0)$. Compute the total distance traveled by the particle during the first second.

51. An object moves along a coordinate line with velocity $v(t) = \sin t$ units per second. The object passes through the origin at time $t = \pi/6$ seconds. When is the next time: (a) that the object passes through the origin? (b) that the object passes through the origin moving from left to right?

52. Repeat Exercise 51 with $v(t) = \cos t$.

53. An automobile with varying velocity $v(t)$ moves in a fixed direction for 5 minutes and covers a distance of 4 miles. What theorem would you invoke to argue that for at least one instant the speedometer must have read 48 miles per hour?

54. A speeding motorcyclist sees his way blocked by a haywagon some distance s ahead and slams on his brakes. Given that the brakes impart to the motorcycle a constant negative acceleration a and that the haywagon is moving with speed v_1 in the same direction as the motorcycle, show that the motorcyclist can avoid collision only if he is traveling at a speed less than $v_1 + \sqrt{2|a|s}$.

55. Find the velocity $v(t)$ given that $a(t) = 2[v(t)]^2$ and $v_0 \neq 0$.

■ **5.7 THE *u*-SUBSTITUTION; CHANGE OF VARIABLES**

When we differentiate a composite function, we do so by the chain rule. In trying to calculate an indefinite integral, we are often called upon to apply the chain rule in reverse. This process can be facilitated by making what we call a "*u*-substitution."

An integral of the form

$$\int f(g(x)) \, g'(x) \, dx$$

can be written

$$\int f(u) \, du$$

by setting

$$u = g(x), \qquad du = g'(x) \, dx.†$$

If F is an antiderivative for f, then

$$[F(g(x))]' = F'(g(x))g'(x) = f(g(x))g'(x)$$

by the chain rule _____↑   ↑_____ $F' = f$

_____

† Think of $du = g'(x) \, dx$ as a "formal differential," writing dx for h.

and so

$$\int f(g(x))g'(x)\,dx = \int [F(g(x))]'\,dx = F(g(x)) + C.$$

As illustrated below, we can obtain the same result by calculating

$$\int f(u)\,du$$

and then substituting $g(x)$ back in for u:

$$\int f(u)\,du = F(u) + C = F(g(x)) + C.$$

Example 1 Calculate

$$\int (x^2 - 1)^4\,2x\,dx$$

and then check the result by differentiation.

SOLUTION Set

$$u = x^2 - 1 \qquad \text{so that} \qquad du = 2x\,dx.$$

Then

$$\int (x^2 - 1)^4\,2x\,dx = \int u^4\,du = \tfrac{1}{5}u^5 + C = \tfrac{1}{5}(x^2 - 1)^5 + C.$$

CHECKING

$$\frac{d}{dx}\left[\frac{1}{5}(x^2 - 1)^5 + C\right] = \frac{5}{5}(x^2 - 1)^4 \frac{d}{dx}(x^2 - 1) \overset{\checkmark}{=} (x^2 - 1)^4\,2x. \quad \square$$

Example 2 Calculate

$$\int \sin x \cos x\,dx.$$

and then check the result by differentiation.

SOLUTION Set

$$u = \sin x, \qquad du = \cos x\,dx.$$

Then

$$\int \sin x \cos x\,dx = \int u\,du = \tfrac{1}{2}u^2 + C = \tfrac{1}{2}\sin^2 x + C.$$

CHECKING

$$\frac{d}{dx}[\tfrac{1}{2}\sin^2 x + C] = \tfrac{1}{2}(2)\sin x \frac{d}{dx}[\sin x] \overset{\checkmark}{=} \sin x \cos x.$$

ALTERNATE SOLUTION   Since $\sin x \cos x = \frac{1}{2} \sin 2x$,

$$\int \sin x \cos x \, dx = \frac{1}{2} \int \sin 2x \, dx.$$

Set

$$u = 2x, \quad du = 2 \, dx \quad \text{or} \quad dx = \frac{1}{2} \, du.$$

Then

$$\frac{1}{2} \int \sin 2x \, dx = \frac{1}{2} \int \sin u \left(\frac{1}{2} \, du \right) = \frac{1}{4} \int \sin u \, du = -\frac{1}{4} \cos u + C$$

$$= -\frac{1}{4} \cos 2x + C.$$

We leave it to you to check this result and to reconcile it with the result given in the first solution.   ❏

Example 3   Calculate

$$\int \frac{dx}{(3 + 5x)^2}.$$

and then check the result by differentiation.

SOLUTION   Set

$$u = 3 + 5x \qquad \text{so that} \qquad du = 5 \, dx.$$

Then

$$\frac{dx}{(3 + 5x)^2} = \frac{\frac{1}{5} \, du}{u^2} = \frac{1}{5} \frac{du}{u^2}$$

and

$$\int \frac{dx}{(3 + 5x)^2} = \frac{1}{5} \int \frac{du}{u^2} = -\frac{1}{5u} + C\dagger = -\frac{1}{5(3 + 5x)} + C.$$

CHECKING

$$\frac{d}{dx}\left[-\frac{1}{5(3 + 5x)} + C \right] = \frac{d}{dx}\left[-\frac{1}{5}(3 + 5x)^{-1} \right]$$

$$= \left(-\frac{1}{5} \right)(-1)(3 + 5x)^{-2}(5) \overset{\checkmark}{=} \frac{1}{(3 + 5x)^2}. \quad ❏$$

In the remaining examples we leave the checking to you.

---

† We could write

$$\frac{1}{5} \int \frac{du}{u^2} = -\frac{1}{5}\left[\frac{1}{u} + C \right] = -\frac{1}{5u} - \frac{C}{5},$$

but, since C is arbitrary, $-C/5$ is arbitrary, and we can therefore write C instead.

Example 4   Calculate

$$\int x^2 \sqrt{4 + x^3}\ dx.$$

SOLUTION   Set

$$u = 4 + x^3, \qquad du = 3x^2\ dx.$$

Then

$$x^2 \sqrt{4 + x^3}\ dx = \underbrace{\sqrt{4 + x^3}}_{\sqrt{u}}\ \underbrace{x^2\ dx}_{\frac{1}{3}\,du} = \tfrac{1}{3} \sqrt{u}\ du$$

and

$$\int x^2 \sqrt{4 + x^3}\ dx = \tfrac{1}{3} \int \sqrt{u}\ du$$

$$= \tfrac{2}{9}\,u^{3/2} + C = \tfrac{2}{9}\,(4 + x^3)^{3/2} + C. \quad ❑$$

Example 5   Find

$$\int \sec^3 x\ \tan x\ dx.$$

SOLUTION   We can write the integrand as $\sec^2 x \sec x \tan x\ dx$. Then, setting

$$u = \sec x, \qquad du = \sec x \tan x\ dx,$$

We have

$$\sec^3 x \tan x\ dx = \underbrace{\sec^2 x}_{u^2}\ \underbrace{(\sec x \tan x)\ dx}_{du} = u^2\ du$$

and

$$\int \sec^3 x \tan x\ dx = \int u^2 du = \tfrac{1}{3}\,u^3 + C = \tfrac{1}{3} \sec^3 x + C. \quad ❑$$

The key step in making a u-substitution is to find a substitution $u = g(x)$ such that the expression $du = g'(x)\ dx$ appears in the original integral (at least up to a constant factor) and the new integral

$$\int f(u)\ du$$

is easier to calculate than the original integral.   In most cases the form of the original integrand will suggest a good choice for u.

Example 6   Find

$$\int x \cos \pi x^2\ dx.$$

SOLUTION   Set

$$u = \pi x^2, \qquad du = 2 \pi x\ dx.$$

Then

$$x \cos \pi x^2 \, dx = \underbrace{\cos \pi x^2}_{\cos u} \underbrace{x \, dx}_{\frac{1}{2\pi} du} = \frac{1}{2\pi} \cos u \, du$$

and

$$\int x \cos \pi x^2 \, dx = \frac{1}{2\pi} \int \cos u \, du = \frac{1}{2\pi} \sin u + C = \frac{1}{2\pi} \sin \pi x^2 + C. \quad \square$$

Example 7 Evaluate the definite integral

$$\int_0^2 (x^2 - 1)(x^3 - 3x + 2)^3 \, dx.$$

SOLUTION To evaluate this definite integral, we need to find an antiderivative for the integrand. The indefinite integral

$$\int (x^2 - 1)(x^3 - 3x + 2)^3 \, dx$$

gives the set of all antiderivatives, and so we will calculate it first. Set

$$u = x^3 - 3x + 2, \qquad du = (3x^2 - 3) \, dx = 3(x^2 - 1) \, dx.$$

Then

$$(x^2 - 1)(x^3 - 3x + 2)^3 \, dx = \underbrace{(x^3 - 3x + 2)^3}_{u^3} \underbrace{(x^2 - 1) \, dx}_{\frac{1}{3} du} = \tfrac{1}{3} u^3 \, du$$

and

$$\int (x^2 - 1)(x^3 - 3x + 2)^3 \, dx = \tfrac{1}{3} \int u^3 \, du = \tfrac{1}{12} u^4 + C = \tfrac{1}{12} (x^3 - 3x + 2)^4 + C.$$

To evaluate the given definite integral, we need only one antiderivative, and so we will choose the one with $C = 0$. This gives

$$\int_0^2 (x^2 - 1)(x^3 - 3x + 2)^3 \, dx = \left[\tfrac{1}{12} (x^3 - 3x + 2)^4 \right]_0^2 = 20. \quad \square$$

Later in this section we give another, more formal, approach to evaluating a definite integral when a substitution is involved in the integration step.

Example 8 Find

$$\int x^2 \csc^2 x^3 \cot^4 x^3 \, dx.$$

SOLUTION We can effect some simplification by setting

$$u = x^3, \qquad du = 3x^2 \, dx.$$

Then

$$x^2 \csc^2 x^3 \cot^4 x^3 \, dx = \underbrace{\cot^4 x^3}_{\cot^4 u} \underbrace{\csc^2 x^3}_{\csc^2 u} \underbrace{x^2 \, dx}_{\frac{1}{3} du}$$

$$= \tfrac{1}{3} \cot^4 u \, \csc^2 u \, du$$

and

$$\int x^2 \csc^2 x^3 \cot^4 x^3 \, dx = \tfrac{1}{3} \int \cot^4 u \, \csc^2 u \, du.$$

We can calculate the integral on the right by setting

$$t = \cot u, \qquad dt = -\csc^2 u \, du.$$

Then

$$\tfrac{1}{3} \int \cot^4 u \, \csc^2 u \, du = -\tfrac{1}{3} \int t^4 \, dt = -\tfrac{1}{15} t^5 + C = -\tfrac{1}{15} \cot^5 u + C,$$

and thus

$$\int x^2 \csc^2 x^3 \cot^4 x^3 \, dx = -\tfrac{1}{15} \cot^5 u + C = -\tfrac{1}{15} \cot^5 x^3 + C.$$

We arrived at this by making two consecutive substitutions. First we set $u = x^3$ and then we set $t = \cot u$. We could have saved ourselves some work by setting $u = \cot x^3$ at the very beginning. With

$$u = \cot x^3, \qquad du = -\csc^2 x^3 \cdot 3x^2 \, dx$$

we have

$$x^2 \csc^2 x^3 \cot^4 x^3 \, dx = \underbrace{\cot^4 x^3}_{u^4} \underbrace{\csc^2 x^3 \, x^2 \, dx}_{-\frac{1}{3} du}$$

$$= -\tfrac{1}{3} u^4 \, du$$

and

$$\int x^2 \csc^2 x^3 \cot^4 x^3 \, dx = -\tfrac{1}{3} \int u^4 \, du = -\tfrac{1}{15} u^5 + C = -\tfrac{1}{15} \cot^5 x^3 + C. \quad ❑$$

Remark Thus far, all the integrals that we have calculated using a u-substitution can be calculated without substitution. All that is required is a good sense of the chain rule. For example:

$\displaystyle\int \sin x \cos x \, dx$ (Example 2). The cosine is the derivative of the sine. Thus

$$\int \sin x \cos x \, dx = \int \sin x \, \frac{d}{dx} (\sin x) \, dx = \tfrac{1}{2} \sin^2 x + C.$$

$\displaystyle\int x^2 \sqrt{4 + x^3} \, dx$ (Example 4). The derivative of $4 + x^3$ is $3x^2$. Therefore

$$\frac{d}{dx} (4 + x^3) = 3x^2 \qquad \text{and} \qquad x^2 = \tfrac{1}{3} \frac{d}{dx} (4 + x^3).$$

Thus,

$$\int x^2\sqrt{4 + x^3}\,dx = \tfrac{1}{3}\int (4 + x^3)^{1/2}\frac{d}{dx}(4 + x^3)\,dx$$

$$= \tfrac{1}{3}\cdot\tfrac{2}{3}(4 + x^3)^{3/2} + C = \tfrac{2}{9}(4 + x^3)^{3/2} + C.$$

$\int \sec^3 x\,\tan x\,dx$ (Example 5).   Write the integrand as

$$\sec^2 x(\sec x\,\tan x) = \sec^2 x\frac{d}{dx}(\sec x).$$

Then

$$\int \sec^3 x\,\tan x\,dx = \int \sec^2 x\frac{d}{dx}(\sec x)\,dx = \tfrac{1}{3}\sec^3 x + C.$$

$\int x^2\,\csc^2 x^3\,\cot^4 x^3\,dx$ (Example 8).   This integral may look complicated, but looked at in the right way, it becomes very simple.   You know that the derivative of the cotangent is the negative of the cosecant squared.   Therefore, by the chain rule,

$$\frac{d}{dx}(\cot x^3) = -\csc^2 x^3 \cdot 3x^2$$

and

$$x^2\,\csc^2 x^3 = -\frac{1}{3}\frac{d}{dx}(\cot x^3).$$

Thus

$$\int x^2\,\csc^2 x^3\,\cot^4 x^3\,dx = -\frac{1}{3}\int \cot^4 x^3 \cdot \frac{d}{dx}(\cot x^3)\,dx = -\tfrac{1}{15}\cot^5 x^3 + C.$$

There is nothing wrong with calculating integrals by substitution.   All we are saying is that, with some experience, you will be able to calculate many integrals without it.   ❑

Substitution and Definite Integrals

In Example 7 we evaluated a definite integral by first calculating a corresponding indefinite integral.   Then we chose one of the antiderivatives for the evaluation step. Here we present an alternative approach in which the limits of integration are changed along with the variable.   This approach uses the *change of variables formula*:

(5.7.1)

$$\int_a^b f(g(x))g'(x)\,dx = \int_{g(a)}^{g(b)} f(u)\,du.$$

The formula holds provided that f and g' are both continuous.   More precisely, g' must be continuous on $[a, b]$, and f must be continuous on the set of values taken on by g.

PROOF   Let F be an antiderivative for f.   Then $F' = f$ and

$$\int_a^b f(g(x))g'(x)\, dx = \int_a^b F'(g(x))g'(x)\, dx$$

$$= \Big[F(g(x)) \Big]_a^b = F(g(b)) - F(g(a)) = \int_{g(a)}^{g(b)} f(u)\, du. \quad \square$$

We redo Example 7 using the change of variables formula.

Example 9   Evaluate

$$\int_0^2 (x^2 - 1)(x^3 - 3x + 2)^3\, dx.$$

SOLUTION   As before, set

$$u = x^3 - 3x + 2, \qquad du = 3(x^2 - 1)\, dx.$$

Then

$$(x^2 - 1)(x^3 - 3x + 2)^3\, dx = \tfrac{1}{3}u^3\, du.$$

Now, at $x = 0$, $u = 2$. At $x = 2$, $u = 4$.   Therefore,

$$\int_0^2 (x^2 - 1)(x^3 - 3x + 2)^3\, dx = \tfrac{1}{3}\int_2^4 u^3\, du = \Big[\tfrac{1}{12}u^4 \Big]_2^4 = \tfrac{1}{12}(4)^4 - \tfrac{1}{12}(2)^4 = 20. \quad \square$$

Example 10   Evaluate

$$\int_0^{1/2} \cos^3 \pi x \sin \pi x\, dx.$$

$$\frac{du}{dx} = -\pi \sin \pi x$$

$$du = -\pi \sin \pi x\, dx$$

SOLUTION   Set

$$u = \cos \pi x, \qquad du = -\pi \sin \pi x\, dx.$$

Then

$$\cos^3 \pi x \sin \pi x\, dx = \underbrace{\cos^3 \pi x}_{u^3}\ \underbrace{\sin \pi x\, dx}_{-\frac{1}{\pi}du} = -\frac{1}{\pi}u^3\, du.$$

At $x = 0$, $u = 1$. At $x = 1/2$, $u = 0$.   Therefore,

$$\int_0^{1/2} \cos^3 \pi x \sin \pi x\, dx = -\frac{1}{\pi}\int_1^0 u^3 du = \frac{1}{\pi}\int_0^1 u^3\, du = \frac{1}{\pi}\Big[\tfrac{1}{4}u^4 \Big]_0^1 = \frac{1}{4\pi}. \quad \square$$

Up to this point our focus has been on substitutions connected with the chain rule. However, as we shall illustrate here and later in Chapter 8, "substitution" is a very general method of integration. Choosing the proper substitution is a matter of practice and experience. At this early stage, we will provide hints and guidance.

Example 11 Calculate

$$\int x(x - 3)^5 \, dx.$$

SOLUTION Set

$$u = x - 3, \qquad du = dx.$$

Then

$$x(x - 3)^5 \, dx = (u + 3)u^5 \, du = (u^6 + 3u^5) \, du$$

and

$$\int x(x - 3)^5 \, dx = \int (u^6 + 3u^5) \, du$$

$$= \tfrac{1}{7}u^7 + \tfrac{1}{2}u^6 + C = \tfrac{1}{7}(x - 3)^7 + \tfrac{1}{2}(x - 3)^6 + C. \quad \square$$

Example 12 Evalulate

$$\int_0^{\sqrt{3}} x^5\sqrt{x^2 + 1} \, dx.$$

SOLUTION Set

$$u = x^2 + 1, \qquad du = 2x \, dx.$$

$x = (u - 1)^{1/2}$

Then

$$x^5\sqrt{x^2 + 1} \, dx = \underbrace{x^4}_{(u-1)^2} \underbrace{\sqrt{x^2 + 1}}_{\sqrt{u}} \underbrace{x \, dx}_{\tfrac{1}{2}du} = \tfrac{1}{2}(u - 1)^2\sqrt{u} \, du.$$

At $x = 0, u = 1$. At $x = \sqrt{3}, u = 4$. Thus

$$\int_0^{\sqrt{3}} x^5\sqrt{x^2 + 1} \, dx = \tfrac{1}{2}\int_1^4 (u - 1)^2\sqrt{u} \, du$$

$$= \tfrac{1}{2}\int_1^4 (u^{5/2} - 2u^{3/2} + u^{1/2}) \, du$$

$$= \tfrac{1}{2}\left[\tfrac{2}{7}u^{7/2} - \tfrac{4}{5}u^{5/2} + \tfrac{2}{3}u^{3/2} \right]_1^4$$

$$= \left[u^{3/2}(\tfrac{1}{7}u^2 - \tfrac{2}{5}u + \tfrac{1}{3}) \right]_1^4 = \tfrac{848}{105}. \quad \square$$

EXERCISES 5.7

In Exercises 1–24, evaluate the integral by a *u*-substitution.

1. $\displaystyle\int \frac{dx}{(2 - 3x)^2}$.

2. $\displaystyle\int \frac{dx}{\sqrt{2x + 1}}$.

3. $\displaystyle\int \sqrt{2x + 1} \, dx$

4. $\displaystyle\int \sqrt{ax + b} \, dx$.

5. $\displaystyle\int (ax + b)^{3/4} \, dx$.

6. $\displaystyle\int 2ax(ax^2 + b)^4 \, dx$.

7. $\displaystyle\int \frac{t}{(4t^2 + 9)^2} \, dt$.

8. $\displaystyle\int \frac{3t}{(t^2 + 1)^2} \, dt$.

9. $\displaystyle\int t^2(5t^3 + 9)^4 \, dt$

10. $\displaystyle\int t(1 + t^2)^3 \, dt$.

11. $\displaystyle\int x^2(1 + x^3)^{1/4}\, dx.$

12. $\displaystyle\int x^{n-1}\sqrt{a + bx^n}\, dx.$

13. $\displaystyle\int \frac{s}{(1 + s^2)^3}\, ds.$

14. $\displaystyle\int \frac{2s}{\sqrt[3]{6 - 5s^2}}\, ds.$

15. $\displaystyle\int \frac{x}{\sqrt{x^2 + 1}}\, dx.$

16. $\displaystyle\int \frac{3ax^2 - 2bx}{\sqrt{ax^3 - bx^2}}\, dx.$

17. $\displaystyle\int x^2(1 - x^3)^{2/3}\, dx.$

18. $\displaystyle\int \frac{x^2}{(1 - x^3)^{2/3}}\, dx.$

19. $\displaystyle\int 5x(x^2 + 1)^{-3}\, dx.$

20. $\displaystyle\int 2x^3(1 - x^4)^{-1/4}\, dx.$

21. $\displaystyle\int x^{-3/4}(x^{1/4} + 1)^{-2}\, dx.$

22. $\displaystyle\int \frac{4x + 6}{\sqrt{x^2 + 3x + 1}}\, dx.$

23. $\displaystyle\int \frac{b^3x^3}{\sqrt{1 - a^4x^4}}\, dx.$

24. $\displaystyle\int \frac{x^{n-1}}{\sqrt{a + bx^n}}\, dx.$

In Exercises 25–32, evaluate the definite integral by a u-substitution.

25. $\displaystyle\int_0^1 x(x^2 + 1)^3\, dx.$

26. $\displaystyle\int_{-1}^0 3x^2(4 + 2x^3)^2\, dx.$

27. $\displaystyle\int_0^1 5x(1 + x^2)^4\, dx.$

28. $\displaystyle\int_1^2 (6 - x)^{-3}\, dx.$

29. $\displaystyle\int_{-1}^1 \frac{r}{(1 + r^2)^4}\, dr.$

30. $\displaystyle\int_0^3 \frac{r}{\sqrt{r^2 + 16}}\, dr.$

31. $\displaystyle\int_0^a y\sqrt{a^2 - y^2}\, dy.$

32. $\displaystyle\int_{-a}^0 y^2\left(1 - \frac{y^3}{a^2}\right)^{-2}\, dy.$

In Exercises 33–44, evaluate the integral by a u-substitution.

33. $\displaystyle\int x\sqrt{x + 1}\, dx.$ [set $u = x + 1$]

34. $\displaystyle\int 2x\sqrt{x - 1}\, dx.$ [set $u = x - 1$]

35. $\displaystyle\int x\sqrt{2x - 1}\, dx.$

36. $\displaystyle\int x^2\sqrt{x + 1}\, dx.$

37. $\displaystyle\int y(y + 1)^{12}\, dy.$

38. $\displaystyle\int t(2t + 3)^8\, dt.$

39. $\displaystyle\int t^2(t - 2)^{-5}\, dt.$

40. $\displaystyle\int x^2(2x - 1)^{-2/3}\, dx.$

41. $\displaystyle\int_0^1 \frac{x + 3}{\sqrt{x + 1}}\, dx.$

42. $\displaystyle\int_0^1 \frac{x^2}{\sqrt{x + 1}}\, dx.$

43. $\displaystyle\int_{-1}^0 x^3(x^2 + 1)^6\, dx.$ [set $u = x^2 + 1$]

44. $\displaystyle\int_1^{\sqrt2} x^3(x^2 - 1)^7\, dx$

In Exercises 45–64, calculate the indefinite integral.

45. $\displaystyle\int \cos(3x - 1)\, dx.$

46. $\displaystyle\int \sin 2\pi x\, dx.$

47. $\displaystyle\int \csc^2 \pi x\, dx.$

48. $\displaystyle\int \sec 2x \tan 2x\, dx.$

49. $\displaystyle\int \sin(3 - 2x)\, dx.$

50. $\displaystyle\int \sin^2 x \cos x\, dx.$

51. $\displaystyle\int \cos^4 x \sin x\, dx.$

52. $\displaystyle\int x \sec^2 x^2\, dx.$

53. $\displaystyle\int x^{-1/2}\sin x^{1/2}\, dx.$

54. $\displaystyle\int \csc(1 - 2x) \cot(1 - 2x)\, dx.$

55. $\displaystyle\int \sqrt{1 + \sin x} \cos x\, dx.$

56. $\displaystyle\int \frac{\sin x}{\sqrt{1 + \cos x}}\, dx.$

57. $\displaystyle\int \frac{1}{\cos^2 x}\, dx.$

58. $\displaystyle\int (1 + \tan^2 x) \sec^2 x\, dx.$

59. $\displaystyle\int x \sin^3 x^2 \cos x^2\, dx.$

60. $\displaystyle\int \sqrt{1 + \cot x} \csc^2 x\, dx.$

61. $\displaystyle\int (1 + \cot^2 x) \csc^2 x\, dx.$

62. $\displaystyle\int \frac{1}{\sin^2 x}\, dx.$

63. $\displaystyle\int \frac{\sec^2 x}{\sqrt{1 + \tan x}}\, dx.$

64. $\displaystyle\int x^2\sin(4x^3 - 7)\, dx.$

Evaluate the integral in Exercises 65–71.

65. $\displaystyle\int_{-\pi}^\pi \sin x\, dx.$

66. $\displaystyle\int_{-\pi/3}^{\pi/3} \sec x \tan x\, dx.$

67. $\displaystyle\int_{1/4}^{1/3} \sec^2 \pi x\, dx.$

68. $\displaystyle\int_0^1 \cos^2\frac{\pi}{2}x \sin\frac{\pi}{2}x\, dx.$

69. $\displaystyle\int_0^{\pi/2} \sin^3 x \cos x\, dx.$

70. $\displaystyle\int_0^\pi x \cos x^2\, dx$

71. $\displaystyle\int_{\pi/6}^{\pi/4} \csc x \cot x\, dx.$

72. Derive the formula

$$\int \cos^2 x\, dx = \tfrac{1}{2}x + \tfrac{1}{4}\sin 2x + C.$$

HINT: Recall the half-angle formula $\cos^2\theta = \tfrac{1}{2}(1 + \cos 2\theta)$.

73. Derive the formula

$$\int \sin^2 x\, dx = \tfrac{1}{2}x - \tfrac{1}{4}\sin 2x + C.$$

HINT: Use the half-angle formula for the sine function.

Evaluate the integrals in Exercises 74–77.

74. $\displaystyle\int \sin^2 3x\, dx.$

75. $\displaystyle\int \cos^2 5x\, dx.$

76. $\displaystyle\int_0^{2\pi} \sin^2 x\, dx.$

77. $\displaystyle\int_0^{\pi/2} \cos^2 2x\, dx.$

In Exercises 78–81, find the area bounded by the following curves.

78. $y = \cos \pi x$, $y = \sin \pi x$, $x = 0$, $x = \frac{1}{4}$.

79. $y = \cos^2 \pi x$, $y = \sin^2 \pi x$, $x = 0$, $x = \frac{1}{4}$.
HINT: Use Exercises 72 and 73.

80. $y = \cos^2 \pi x$, $y = -\sin^2 \pi x$, $x = 0$, $x = \frac{1}{4}$.

81. $y = \csc^2 \pi x$, $y = \sec^2 \pi x$, $x = \frac{1}{6}$, $x = \frac{1}{4}$.

82. In Example 2 we found that

$$\int \sin x \cos x \, dx = \tfrac{1}{2} \sin^2 x + C$$

by setting $u = \sin x$. Calculate the integral by setting $u = \cos x$ and then reconcile the two answers.

83. Calculate

$$\int \sec^2 x \tan x \, dx$$

(a) Setting $u = \sec x$.   (b) Setting $u = \tan x$.
(c) Reconcile your answers to (a) and (b).

84. (*The area of a circular region*) The circle $x^2 + y^2 = r^2$ encloses a circular disc of radius r. Justify the familiar formula $A = \pi r^2$ by integration. HINT: The quarter disc in the first quadrant is the region below the curve $y = \sqrt{r^2 - x^2}$, $x \in [0, r]$. Therefore

$$A = 4 \int_0^r \sqrt{r^2 - x^2} \, dx.$$

Evaluate the integral by setting $x = r \sin u$, $dx = r \cos u \, du$ and use Exercise 72.

85. Find the area enclosed by the ellipse $b^2 x^2 + a^2 y^2 = a^2 b^2$.

■ 5.8 SOME FURTHER PROPERTIES OF THE DEFINITE INTEGRAL

In this section we feature some important general properties of the integral. The proofs are left mostly to you. You can assume throughout that the functions involved are continuous and that $a < b$.

I. The integral of a nonnegative function is nonnegative:

$$\frac{1}{4} - 1$$

(5.8.1)   if $f(x) \geq 0$ for all $x \in [a, b]$,   then $\int_a^b f(x) \, dx \geq 0$.

The integral of a positive function is positive:

(5.8.2)   if $f(x) > 0$ for all $x \in [a, b]$,   then $\int_a^b f(x) \, dx > 0$.

The next property is an immediate consequence of Property I and linearity (5.4.6).

II. The integral is order-preserving:

(5.8.3)   if $f(x) \leq g(x)$ for all $x \in [a, b]$,   then $\int_a^b f(x) \, dx \leq \int_a^b g(x) \, dx$

and

(5.8.4)

$$\text{if } f(x) < g(x) \text{ for all } x \in [a, b], \quad \text{then} \quad \int_a^b f(x)\, dx < \int_a^b g(x)\, dx.$$

PROOF OF (5.8.3) Since $f(x) \le g(x)$, it follows that $g(x) - f(x) \ge 0$. Therefore, using the linearity of the integral and (5.8.1), we have

$$\int_a^b g(x)\, dx - \int_a^b f(x)\, dx = \int_a^b [g(x) - f(x)]\, dx \ge 0.$$

and so

$$\int_a^b f(x)\, dx \le \int_a^b g(x)\, dx.$$

The proof of (5.8.4) follows in exactly the same manner. ❏

III. Just as the absolute value of a sum is less than or equal to the sum of the absolute values,

$$|x_1 + x_2 + \cdots + x_n| \le |x_1| + |x_2| + \cdots + |x_n|,$$

the absolute value of an integral is less than or equal to the integral of the absolute value:

(5.8.5)

$$\left| \int_a^b f(x)\, dx \right| \le \int_a^b |f(x)|\, dx.$$

HINT FOR PROOF OF (5.8.5) Show that

$$\int_a^b f(x)\, dx \quad \text{and} \quad -\int_a^b f(x)\, dx$$

are both less than or equal to

$$\int_a^b |f(x)|\, dx. \quad ❏$$

You should already be familiar with the next property.

IV. If m is the minimum value of f on $[a, b]$ and M is the maximum value, then

(5.8.6)

$$m(b - a) \le \int_a^b f(x)\, dx \le M(b - a).$$

Recall Theorem (5.3.5), which states that

$$\frac{d}{dx}\left(\int_a^x f(t)\,dt\right) = f(x).$$

Our next result is a generalization of this theorem. Its importance will become apparent in Chapter 7.

V. If u is a differentiable function of x and f is continuous, then

(5.8.7)
$$\frac{d}{dx}\left(\int_a^u f(t)\,dt\right) = f(u)\frac{du}{dx}.$$

PROOF The function

$$F(u) = \int_a^u f(t)\,dt$$

is differentiable with respect to u and

$$\frac{d}{du}[F(u)] = f(u). \qquad\qquad \text{(Theorem 5.3.5)}$$

Thus

$$\frac{d}{dx}\left(\int_a^u f(t)\,dt\right) = \frac{d}{dx}[F(u)] = \frac{d}{du}[F(u)]\frac{du}{dx} = f(u)\frac{du}{dx}. \qquad \square$$

$$\text{chain rule} \longrightarrow$$

Example 1 Find

$$\frac{d}{dx}\left(\int_0^{x^3} \frac{dt}{1+t}\right).$$

SOLUTION At this stage you would be hard put to carry out the integration; it requires the natural logarithm function, which is not introduced in this text until Chapter 7. But, for our purposes, that doesn't matter. We have $f(t) = 1/(1+t)$ and $u(x) = x^3$. By (5.8.7), you know that

$$\frac{d}{dx}\left(\int_0^{x^3} \frac{dt}{1+t}\right) = f(u)\frac{du}{dx} = \frac{1}{1+x^3}\,3x^2 = \frac{3x^2}{1+x^3}$$

without carrying out the integration. ❑

Example 2 Find

$$\frac{d}{dx}\left(\int_x^{2x} \frac{dt}{1+t^2}\right).$$

SOLUTION The idea is to express the integral in terms of integrals that have constant lower limits of integration. Once we have done that, we can apply (5.8.7).

In this case, we choose 0 as a convenient lower limit. Then, by the additivity of the integral,

$$\int_0^x \frac{dt}{1 + t^2} + \int_x^{2x} \frac{dt}{1 + t^2} = \int_0^{2x} \frac{dt}{1 + t^2}.$$

Thus

$$\int_x^{2x} \frac{dt}{1 + t^2} = \int_0^{2x} \frac{dt}{1 + t^2} - \int_0^x \frac{dt}{1 + t^2}.$$

Differentiation gives

$$\frac{d}{dx} \left(\int_x^{2x} \frac{dt}{1 + t^2} \right) = \frac{d}{dx} \left(\int_0^{2x} \frac{dt}{1 + t^2} \right) - \frac{d}{dx} \left(\int_0^x \frac{dt}{1 + t^2} \right)$$

$$= \frac{1}{1 + (2x)^2} (2) - \frac{1}{1 + x^2} (1) = \frac{2}{1 + 4x^2} - \frac{1}{1 + x^2}. \quad ❏$$

(5.8.7)

VI. Now a few words about the role of symmetry in integration. Suppose that f is a continuous function defined on some interval of the form $[-a, a]$, some closed interval that is symmetric about the origin.

(5.8.8)

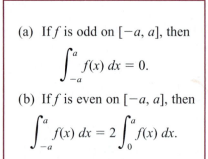

(a) If f is odd on $[-a, a]$, then

$$\int_{-a}^a f(x)\, dx = 0.$$

(b) If f is even on $[-a, a]$, then

$$\int_{-a}^a f(x)\, dx = 2 \int_0^a f(x)\, dx.$$

These assertions can be verified by a simple change of variables (Exercise 40). Here we look at these assertions from the standpoint of area, referring to Figures 5.8.1 and 5.8.2.

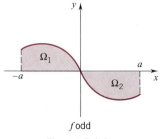

f odd

Figure 5.8.1

For the odd function,

$$\int_{-a}^a f(x)\, dx = \int_{-a}^0 f(x)\, dx + \int_0^a f(x)\, dx = \text{area of } \Omega_1 - \text{area of } \Omega_2 = 0.$$

For the even function,

$$\int_{-a}^a f(x)\, dx = \text{area of } \Omega_1 + \text{area of } \Omega_2 = 2(\text{area of } \Omega_2) = 2 \int_0^a f(x)\, dx.$$

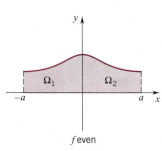

f even

Figure 5.8.2

Suppose we were asked to evaluate

$$\int_{-\pi}^\pi (\sin x - x \cos x)^3\, dx.$$

A laborious calculation would show that this integral is zero. We don't have to carry out that calculation. The integrand is an odd function, and thus we can tell immediately that the integral is zero.

EXERCISES 5.8

Assume: f and g continuous, $a < b$, and

$$\int_a^b f(x)\, dx > \int_a^b g(x)\, dx.$$

Answer questions 1–6, giving supporting reasons.

1. Does it necessarily follow that $\int_a^b [f(x) - g(x)]\, dx > 0$?

2. Does it necessarily follow that $f(x) > g(x)$ for all $x \in [a, b]$?

3. Does it necessarily follow that $f(x) > g(x)$ for at least some $x \in [a, b]$?

4. Does it necessarily follow that $\left| \int_a^b f(x)\, dx \right| > \left| \int_a^b g(x)\, dx \right|$?

5. Does it necessarily follow that $\int_a^b |f(x)|\, dx > \int_a^b |g(x)|\, dx$?

6. Does it necessarily follow that $\int_a^b |f(x)|\, dx > \int_a^b g(x)\, dx$?

Assume f continuous, $a < b$, and

$$\int_a^b f(x)\, dx = 0.$$

Answer questions 7–15, giving supporting reasons.

7. Does it necessarily follow that $f(x) = 0$ for all $x \in [a, b]$?

8. Does it necessarily follow that $f(x) = 0$ for at least some $x \in [a, b]$?

9. Does it necessarily follow that $\int_a^b |f(x)|\, dx = 0$?

10. Does it necessarily follow that $\left| \int_a^b f(x)\, dx \right| = 0$?

11. Must all upper sums $U_f(P)$ be nonnegative?

12. Must all upper sums $U_f(P)$ be positive?

13. Can a lower sum, $L_f(P)$, be positive?

14. Does it necessarily follow that $\int_a^b [f(x)]^2\, dx = 0$?

15. Does it necessarily follow that $\int_a^b [f(x) + 1]\, dx = b - a$?

16. Let u be a differentiable function of x and let f be continuous. Derive a formula for

$$\frac{d}{dx}\left(\int_u^b f(t)\, dt \right).$$

In Exercises 17–23, calculate the derivative.

17. $\dfrac{d}{dx}\left(\displaystyle\int_0^{1+x^2} \dfrac{dt}{\sqrt{2t+5}} \right).$

18. $\dfrac{d}{dx}\left(\displaystyle\int_1^{x^2} \dfrac{dt}{t} \right).$

19. $\dfrac{d}{dx}\left(\displaystyle\int_x^a f(t)\, dt \right).$

20. $\dfrac{d}{dx}\left(\displaystyle\int_0^{x^3} \dfrac{dt}{\sqrt{1+t^2}} \right).$

21. $\dfrac{d}{dx}\left(\displaystyle\int_{x^2}^3 \dfrac{\sin t}{t}\, dt \right).$

22. $\dfrac{d}{dx}\left(\displaystyle\int_{\tan x}^4 \sin(t^2)\, dt \right).$

23. $\dfrac{d}{dx}\left(\displaystyle\int_1^{\sqrt{x}} \dfrac{t^2}{1+t^2}\, dt \right).$

24. Show that, if u and v are differentiable functions of x and f is continuous, then

$$\frac{d}{dx}\left(\int_u^v f(t)\, dt \right) = f(v)\frac{dv}{dx} - f(u)\frac{du}{dx}.$$

HINT: Take a number a from the domain of f. Express the integral as the difference of two integrals, each with lower limit a.

In Exercises 25–32, calculate the derivative using Exercise 24.

25. $\dfrac{d}{dx}\left(\displaystyle\int_x^{x^2} \dfrac{dt}{t} \right).$

26. $\dfrac{d}{dx}\left(\displaystyle\int_{1-x}^{1+x} \dfrac{t-1}{t}\, dt \right).$

27. $\dfrac{d}{dx}\left(\displaystyle\int_{x^{1/3}}^{2+3x} \dfrac{dt}{1+t^{3/2}} \right).$

28. $\dfrac{d}{dx}\left(\displaystyle\int_{\sqrt{x}}^{x^2+x} \dfrac{dt}{2+\sqrt{t}} \right).$

29. $\dfrac{d}{dx}\left(\displaystyle\int_{\sqrt{x}}^{x^2} t\cos t^2\, dt \right).$

30. $\dfrac{d}{dx}\left(\displaystyle\int_{3x}^{1/x} \cos 2t\, dt \right).$

31. $\dfrac{d}{dx}\left(\displaystyle\int_{\tan x}^{2x} t\sqrt{1+t^2}\, dt \right).$

32. $\dfrac{d}{dx}\left(\displaystyle\int_{x^2}^{\sin x} \dfrac{t}{4+t^2}\, dt \right).$

33. Verify (5.8.1): (a) by considering lower sums $L_f(P)$; (b) by using an antiderivative.

34. Verify (5.8.4).   **35.** Verify (5.8.5).

36. (*Important*) Prove that, if f is continuous on $[a, b]$ and

$$\int_a^b |f(x)|\, dx = 0,$$

then $f(x) = 0$ for all x in $[a, b]$.   HINT: Exercise 48, Section 2.4.

37. Find $H'(2)$ given that

$$H(x) = \int_{2x}^{x^3-4} \frac{t}{1+\sqrt{t}}\, dt.$$

38. Find $H'(3)$ given that

$$H(x) = \frac{1}{x} \int_3^x [2t - 3H'(t)]\, dt.$$

39. (a) Let f be continuous on $[-a, 0]$. Use a change of variable to prove that

$$\int_{-a}^0 f(x)\, dx = \int_0^a f(-x)\, dx.$$

(b) Let f be continuous on $[-a, a]$. Prove that

$$\int_{-a}^a f(x)\, dx = \int_0^a [f(x) + f(-x)]\, dx.$$

40. Let f be a continuous function on $[-a, a]$. Use Exercise 39 to prove that:

(a) $\displaystyle\int_{-a}^a f(x)\, dx = 0$ if f is odd.

(b) $\displaystyle\int_{-a}^a f(x)\, dx = 2\int_0^a f(x)\, dx$ if f is even.

In Exercises 41–44, use Exercise 40 to evaluate the given integral.

41. $\displaystyle\int_{-\pi/4}^{\pi/4} (x + \sin 2x)\, dx.$ **42.** $\displaystyle\int_{-3}^3 \frac{t^3}{1+t^2}\, dt.$

43. $\displaystyle\int_{-\pi/3}^{\pi/3} (1 + x^2 - \cos x)\, dx.$

44. $\displaystyle\int_{-\pi/4}^{\pi/4} (x^2 - 2x + \sin x + \cos 2x)\, dx.$

■ 5.9 MEAN-VALUE THEOREMS FOR INTEGRALS; AVERAGE VALUES

We begin with a result that we asked you to prove in Exercise 25, Section 5.3.

THEOREM 5.9.1   **THE FIRST MEAN-VALUE THEOREM FOR INTEGRALS**

If f is continuous on $[a, b]$, then there is at least one number c in (a, b) for which

$$\int_a^b f(x)\, dx = f(c)(b - a).$$

This number $f(c)$ is called *the average* (or *mean*) *value of f on $[a, b]$*.

We then have the following identity:

(5.9.2) $$\int_a^b f(x)\, dx = (\textit{the average value of } f \textit{ on } [a, b]) \cdot (b - a).$$

This identity provides a powerful and intuitive way of viewing the definite integral. Think for a moment about area. If f is constant and positive on $[a, b]$, then Ω, the region below the graph, is a rectangle. Its area is given by the formula

Area of Ω = (the constant value of f on $[a, b]$) $\cdot$ $(b - a)$.   (Figure 5.9.1)

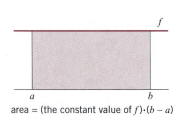

area = (the constant value of f)·$(b - a)$

Figure 5.9.1

If f is now allowed to vary continuously on $[a, b]$, then we have

$$\text{Area of } \Omega = \int_a^b f(x) \, dx,$$

and the area formula reads

Area of Ω = *(the average value of* f *on* $[a, b]$*)* $\cdot$ $(b - a)$.   (Figure 5.9.2)

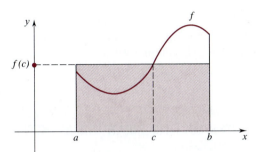

Figure 5.9.2

Think now about motion. If an object moves along a line with constant speed $|v|$ during the time interval $[a, b]$, then

The distance traveled = (the constant value of $|v|$ on $[a, b]$) $\cdot$ $(b - a)$.

If the speed $|v|$ varies, then we have

$$\text{Distance traveled} = \int_a^b |v(t)| \, dt$$

and the formula reads

Distance traveled = *(the average value of* $|v|$ *on* $[a, b]$*)* $\cdot$ $(b - a)$.

We take an interval $[a, b]$ and calculate the average value of the simplest possible functions on that interval. For convenience, let f_{avg} denote the average value of f on $[a, b]$. Solving identity (5.9.2) for f_{avg}, we have

$$f_{\text{avg}} = \frac{1}{b - a} \int_a^b f(x) \, dx.$$

The average value of a constant function $f(x) = k$ is, of course, k:

$$f_{\text{avg}} = \frac{1}{b - a} \int_a^b k \, dx = \frac{k}{b - a} \left[x \right]_a^b = \frac{k}{b - a}(b - a) = k.$$

The average value of $f(x) = x$ on $[a, b]$ is $\frac{1}{2}(b + a)$, the midpoint of the interval:

$$f_{avg} = \frac{1}{b-a} \int_a^b x \, dx = \frac{1}{b-a} \left[\frac{x^2}{2} \right]_a^b = \frac{1}{b-a} \left(\frac{b^2 - a^2}{2} \right) = \frac{1}{2}(b + a).$$

What is the average value of $f(x) = x^2$?

$$f_{avg} = \frac{1}{b-a} \int_a^b x^2 \, dx = \frac{1}{b-a} \left[\frac{x^3}{3} \right]_a^b = \frac{1}{b-a} \left(\frac{b^3 - a^3}{3} \right)$$

$$= \frac{1}{b-a} \left[\frac{(b^2 + ab + a^2)(b - a)}{3} \right] = \frac{1}{3}(b^2 + ab + a^2).$$

Thus, the average value of $f(x) = x^2$ on $[a, b]$ is $\frac{1}{3}(b^2 + ab + a^2)$ (see Example 6 in Section 5.2.). On $[1, 3]$ the values of x^2 range from 1 to 9; the average value is $\frac{13}{3}$.

There is an extension of Theorem 5.9.1 that, as you will see, is useful in applications.

THEOREM 5.9.3 THE SECOND MEAN-VALUE THEOREM FOR INTEGRALS

If f and g are continuous on $[a, b]$ and g is nonnegative, then there is a number c in (a, b) for which

$$\int_a^b f(x)g(x) \, dx = f(c) \int_a^b g(x) \, dx.$$

This number $f(c)$ is called *the g-weighted average of f on* $[a, b]$.

We will prove this theorem (and thereby have a proof of Theorem 5.9.1) at the end of the section. First a brief excursion into physics.

The Mass of a Rod Imagine a thin rod (a straight material wire of negligible thickness) lying on the x-axis from $x = a$ to $x = b$. If the *mass density* of the rod (the mass per unit length) is constant, then the total mass M of the rod is simply the density λ times the length of the rod: $M = \lambda(b - a)$.† If the density λ varies continuously from point to point, say $\lambda = \lambda(x)$, then the mass of the rod is the average density of the rod times the length of the rod:

$$M = (\text{average density}) \cdot (\text{length}).$$

This is an integral:

(5.9.4)

$$M = \int_a^b \lambda(x) \, dx.$$

---

† The symbol λ is the Greek letter "lambda."

The Center of Mass of a Rod Continue with that same rod. If the rod is homogeneous (constant density), then the center of mass x_M of the rod is simply the midpoint of the rod:

$$x_M = \tfrac{1}{2}(a + b). \qquad \text{(the average of } x \text{ from } a \text{ to } b)$$

If the rod is not homogeneous, the center of mass is still an average, but now a weighted average, *the density-weighted average of x from a to b*; namely, x_M is the point for which

$$x_M \int_a^b \lambda(x)\, dx = \int_a^b x\lambda(x)\, dx.$$

Since the integral on the left is simply M, we have

(5.9.5)

$$x_M M = \int_a^b x\lambda(x)\, dx.$$

Example 1 A rod of length L is placed on the x-axis from $x = 0$ to $x = L$. Find the mass of the rod and the center of mass if the density of the rod varies directly as the distance from the $x = 0$ endpoint of the rod.

SOLUTION Here $\lambda(x) = kx$ where k is some positive constant. Therefore

$$M = \int_0^L kx\, dx = \left[\tfrac{1}{2}kx^2 \right]_0^L = \tfrac{1}{2}kL^2$$

and

$$x_M M = \int_0^L x(kx)\, dx = \int_0^L kx^2\, dx = \left[\tfrac{1}{3}kx^3 \right]_0^L = \tfrac{1}{3}kL^3.$$

Division by M gives x_M gives $x_M = \tfrac{2}{3}L$.

In this instance the center of mass is to the right of the midpoint. This makes sense. After all, the density increases from left to right. Thus there is more mass to the right of the midpoint than there is to the left of it. ❑

We go back now to Theorem 5.9.3 and prove it. [There is no reason to worry about Theorem 5.9.1. That is just Theorem 5.9.3 with $g(x)$ identically 1.]

PROOF OF THEOREM 5.9.3 Since f is continuous on $[a, b]$, f takes on a minimum value m on $[a, b]$ and a maximum value M. Since g is nonnegative on $[a, b]$,

$$mg(x) \le f(x)g(x) \le Mg(x) \qquad \text{for all } x \text{ in } [a, b].$$

Therefore

$$\int_a^b mg(x)\, dx \le \int_a^b f(x)g(x)\, dx \le \int_a^b Mg(x)\, dx$$

and

$$m \int_a^b g(x)\, dx \le \int_a^b f(x)g(x)\, dx \le M \int_a^b g(x)\, dx.$$

We know that $\displaystyle\int_a^b g(x)\,dx \geq 0$. If $\displaystyle\int_a^b g(x)\,dx = 0$, then $\displaystyle\int_a^b f(x)g(x)\,dx = 0$ and the theorem holds for all choices of c in (a, b). If $\displaystyle\int_a^b g(x)\,dx > 0$, then

$$m \leq \frac{\displaystyle\int_a^b f(x)g(x)\,dx}{\displaystyle\int_a^b g(x)\,dx} \leq M$$

and by the intermediate-value theorem (Theorem 2.6.1) there exists c in (a, b) for which

$$f(c) = \frac{\displaystyle\int_a^b f(x)g(x)\,dx}{\displaystyle\int_a^b g(x)\,dx}.$$

Obviously then

$$f(c)\int_a^b g(x)\,dx = \int_a^b f(x)g(x)\,dx. \quad \square$$

EXERCISES 5.9

In Exercises 1–12, determine the average value on the indicated interval and find a point in this interval at which the function takes on this average value.

1. $f(x) = mx + b$, $x \in [0, c]$.

2. $f(x) = x^2$, $x \in [-1, 1]$.

3. $f(x) = x^3$, $x \in [-1, 1]$.

4. $f(x) = x^{-2}$, $x \in [1, 4]$.

5. $f(x) = |x|$, $x \in [-2, 2]$.

6. $f(x) = x^{1/3}$, $x \in [-8, 8]$.

7. $f(x) = 2x - x^2$, $x \in [0, 2]$.

8. $f(x) = 3 - 2x$, $x \in [0, 3]$.

9. $f(x) = \sqrt{x}$, $x \in [0, 9]$.

10. $f(x) = 4 - x^2$, $x \in [-2, 2]$.

11. $f(x) = \sin x$, $x \in [0, 2\pi]$.

12. $f(x) = \cos x$, $x \in [0, \pi]$.

▶ 13. Let $f(x) = \sqrt{x}$ for $x \in [1, 9]$.
  (a) Find the average value of f on this interval.
  (b) Estimate with three decimal place accuracy a value c in the interval at which f takes on its average value.
  (c) Use a graphing utility to illustrate your results with a figure similar to Figure 5.9.2.

▶ 14. Let $f(x) = x^3 - x + 1$ for $x \in [-1, 2]$.
  (a) Find the average value of f on this interval.
  (b) Estimate with three decimal place accuracy a value of c in the interval at which f takes on its average value.
  (c) Use a graphing utility to illustrate your results with a figure similar to Figure 5.9.2.

▶ 15. Find $f(x) = \sin x$ for $x \in [0, \pi]$.
  (a) Find the average value of f on this interval.
  (b) Estimate with three decimal place accuracy the smallest value on the interval at which the average value occurs.
  (c) Use a graphing utility to illustrate your results with a figure similar to Figure 5.9.2.

▶ 16. Let $f(x) = 2 \cos 2x$ for $x \in [-\pi/4, \pi/6]$
  (a) Find the average value of f on this interval.
  (b) Estimate with three decimal place accuracy the smallest value on the interval at which the average value occurs.
  (c) Use a graphing utility to illustrate your results with a figure similar to Figure 5.9.2.

17. Solve the following equation for A:

$$\int_a^b [f(x) - A]\,dx = 0.$$

18. Given that f is continuous on $[a, b]$, compare

$$f(b)(b - a) \quad \text{and} \quad \int_a^b f(x) \, dx$$

if f is: (a) constant on $[a, b]$; (b) increasing on $[a, b]$; (c) decreasing on $[a, b]$.

19. In Chapter 3, we viewed $[f(b) - f(a)]/(b - a)$ as the average rate of change of f on $[a, b]$ and $f'(t)$ as the instantaneous rate of change at time t. If our new sense of average is to be consistent with the old one, we must have

$$\frac{f(b) - f(a)}{b - a} = \text{average of } f' \text{ on } [a, b].$$

Prove that this is the case.

20. Determine whether the assertion is true or false.

(a) $\left(\begin{array}{c} \text{the average of } f + g \\ \text{on } [a, b] \end{array} \right)$
$= \left(\begin{array}{c} \text{the average of } f \\ \text{on } [a, b] \end{array} \right) + \left(\begin{array}{c} \text{the average of } g \\ \text{on } [a, b] \end{array} \right).$

(b) $\left(\begin{array}{c} \text{the average of } \alpha f \\ \text{on } [a, b] \end{array} \right)$
$= \alpha \left(\begin{array}{c} \text{the average of } f \\ \text{on } [a, b] \end{array} \right), \alpha \text{ a constant.}$

(c) $\left(\begin{array}{c} \text{the average of } fg \\ \text{on } [a, b] \end{array} \right)$
$= \left(\begin{array}{c} \text{the average of } f \\ \text{on } [a, b] \end{array} \right) \left(\begin{array}{c} \text{the average of } g \\ \text{on } [a, b] \end{array} \right).$

21. Find the average distance of the parabolic arc

$$y = x^2, \quad x \in [0, \sqrt{3}]$$

from: (a) the x-axis; (b) the y-axis; (c) the origin.

22. Find the average distance of the line segment

$$y = mx, \quad x \in [0, 1]$$

from: (a) the x-axis; (b) the y-axis; (c) the origin.

23. A stone falls from rest in a vacuum for t seconds (see Section 3.4). (a) Compare its terminal velocity to its average velocity; (b) compare its average velocity during the first $\frac{1}{2}t$ seconds to its average velocity during the next $\frac{1}{2}t$ seconds.

24. Let f be continuous. Show that, if f is an odd function, then its average value on every interval of the form $[-a, a]$ is zero.

25. An object starts from rest at the point x_0 and moves along the x-axis with constant acceleration a.
(a) Derive formulas for the velocity and position of the object at any time $t \geq 0$.
(b) Show that the average velocity over any time interval $[t_1, t_2]$ is the average of the initial and final velocities on that interval.

26. A rod 6 meters long is placed on the x-axis from $x = 0$ to $x = 6$. The mass density is $12/\sqrt{x + 1}$ kilograms per meter.
(a) Find the mass of the rod and the center of mass.
(b) Find the average mass density of the rod.

27. A rod of length L is placed on the x-axis from $x = 0$ to $x = L$. Find the mass of the rod and the center of mass if the mass density of the rod varies directly: (a) as the square root of the distance from $x = 0$; (b) as the square of the distance from $x = L$.

28. A rod of varying density, mass M, and center of mass x_M extends from $x = a$ to $x = b$. A partition $P = \{x_0, x_1, \ldots, x_n\}$ of $[a, b]$ decomposes the rod into n pieces in the obvious way. Show that, if the n pieces have masses $M_1, M_2, \ldots, M_n$ and centers of mass x_{M_1}, $x_{M_2}, \ldots, x_{M_n}$, then

$$x_M M = x_{M_1} M_1 + x_{M_2} M_2 + \cdots + x_{M_n} M_n.$$

29. A rod that has mass M and extends from $x = 0$ to $x = L$ consists of two pieces with masses M_1, M_2. Given that the center of mass of the entire rod is at $x = \frac{1}{4}L$ and the center of mass of the first piece is at $x = \frac{1}{8}L$, determine the center of mass of the second piece.

30. A rod that has mass M and extends from $x = 0$ to $x = L$ consists of two pieces. Find the mass of each piece given that the center of mass of the entire rod is at $x = \frac{2}{3}L$, the center of mass of the first piece is at $x = \frac{1}{4}L$, and the center of mass of the second piece is at $x = \frac{7}{8}L$.

31. Prove Theorem 5.9.1 by applying the mean-value theorem of differential calculus to the function

$$F(x) = \int_a^x f(t) \, dt, \quad x \in [a, b].$$

(This exercise was given before.)

32. Let f be continuous on $[a, b]$. Let $a < c < b$. Prove that

$$f(c) = \lim_{h \to 0^+} (\text{average value of } f \text{ on } [c - h, c + h]).$$

33. Prove that two distinct continuous functions cannot have the same average on every interval.

■ **5.10 THE INTEGRAL AS THE LIMIT OF RIEMANN SUMS**

Let f be some function continuous on a closed interval $[a, b]$. With our approach to integration (at this point we ask you to review Section 5.2) the definite integral

$$\int_a^b f(x)\, dx$$

is the unique number that satisfies the inequality

$$L_f(P) \le \int_a^b f(x)\, dx \le U_f(P)$$

for all partitions P of $[a, b]$. This method of obtaining the definite integral by *squeezing* toward it with upper and lower sums is called the *Darboux method*.†

There is another way to obtain the integral that is frequently used. Take a partition $P = \{x_0, x_1, \ldots, x_n\}$ of $[a, b]$. Then P breaks up $[a, b]$ into n subintervals

$$[x_0, x_1], [x_1, x_2], \ldots, [x_{n-1}, x_n]$$

of lengths

$$\Delta x_1, \Delta x_2, \ldots, \Delta x_n.$$

Now pick a point x_1^* from $[x_0, x_1]$ and form the product $f(x_1^*)\, \Delta x_1$; pick a point x_2^* from $[x_1, x_2]$ and form the product $f(x_2^*)\, \Delta x_2$; go on in this manner until you have formed the products

$$f(x_1^*)\, \Delta x_1, f(x_2^*)\, \Delta x_2, \ldots, f(x_n^*)\, \Delta x_n.$$

The sum of these products

$$S^*(P) = f(x_1^*)\, \Delta x_1 + f(x_2^*)\, \Delta x_2 + \cdots + f(x_n^*)\, \Delta x_n$$

is called a *Riemann sum*.†† As an exercise, you are asked to show that if P is any partition of the interval $[a, b]$ and $S^*(P)$ is any corresponding Riemann sum, then

$$L_f(P) \le S^*(P) \le U_f(P).$$

Example 1 Let $f(x) = 1 - x^2$, $x \in [-1, 2]$. Take $P = \{-1, 0, \frac{1}{2}, \frac{3}{2}, 2\}$ and let $x_1^* = -\frac{1}{2}$, $x_2^* = \frac{1}{2}$, $x_3^* = \frac{5}{4}$, $x_4^* = \frac{7}{4}$. See Figure 5.10.1. Then $\Delta x_1 = 1$, $\Delta x_2 = \frac{1}{2}$, $\Delta x_3 = 1$, $\Delta x_4 = \frac{1}{2}$, and

$$S^*(P) = f(-\tfrac{1}{2}) \cdot 1 + f(\tfrac{1}{2}) \cdot \tfrac{1}{2} + f(\tfrac{5}{4}) \cdot 1 + f(\tfrac{7}{4}) \cdot \tfrac{1}{2}$$
$$= \tfrac{3}{4}(1) + \tfrac{3}{4}(\tfrac{1}{2}) + (-\tfrac{9}{16})(1) + (-\tfrac{33}{16})(\tfrac{1}{2})$$
$$= -\tfrac{15}{32}.$$

You can verify that

$$L_f(P) = f(-1) \cdot 1 + f(\tfrac{1}{2}) \cdot \tfrac{1}{2} + f(\tfrac{3}{2}) \cdot 1 + f(2) \cdot \tfrac{1}{2} = \tfrac{-19}{8}$$

and

$$U_f(P) = f(0) \cdot 1 + f(0) \cdot \tfrac{1}{2} + f(\tfrac{1}{2}) \cdot 1 + f(\tfrac{3}{2}) \cdot \tfrac{1}{2} = \tfrac{13}{8}. \quad \square$$

Let f be continuous on the interval $[a, b]$. The definite integral of f can be viewed as the *limit* of Riemann sums in the following sense. For any partition P of $[a, b]$, define $\|P\|$, the *norm* of P, by setting

$$\|P\| = \max \Delta x_i, \qquad i = 1, 2, \ldots, n.$$

---

† After the French mathematician J.-G. Darboux (1842–1917).
†† After the German mathematician G. F. B. Riemann (1826–1866).

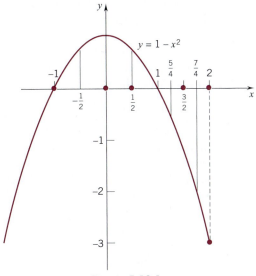

Figure 5.10.1

Then, given any $\epsilon > 0$, there exists a $\delta > 0$ such that

$$\text{if} \quad \|P\| < \delta \quad \text{then} \quad \left| S^*(P) - \int_a^b f(x)\, dx \right| < \epsilon,$$

no matter how the x_i^* are chosen within $[x_{i-1}, x_i]$.

In symbols we write

(5.10.1) $$\int_a^b f(x)\, dx = \lim_{\|P\| \to 0} [f(x_1^*)\, \Delta x_1 + f(x_2^*)\, \Delta x_2 + \cdots + f(x_n^*)\, \Delta x_n].$$

A proof of this assertion appears in Appendix B.5. Figure 5.10.2 illustrates the idea. Here the base interval is broken up into 8 subintervals. The point x_1^* is chosen from $[x_0, x_1]$, x_2^* from $[x_1, x_2]$, and so on. While the integral represents the area under the curve, the Riemann sum represents the sum of the areas of the shaded rectangles. The

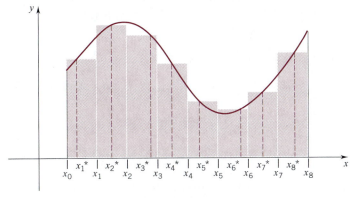

Figure 5.10.2

difference between the two can be made as small as we wish (less than ϵ) simply by making the maximum length of the base subintervals sufficiently small—that is, by making $\|P\|$ sufficiently small. ❑

You have now seen two approaches to the definite integral of a continuous function, through upper and lower sums, and through Riemann sums. Since

$$L_f(P) \leq S^*(P) \leq U_f(P)$$

for any partition P, it follows that the definite integral of f exists in the sense of upper and lower sums iff it exists in the sense of Riemann sums.

We have also pointed out that the definite integrals of certain discontinuous functions also exist. See, for example, Exercises 29 and 30 in Section 5.2. In general, a function f defined on the interval $[a, b]$ is *integrable on* $[a, b]$ if its definite integral

$$\int_a^b f(x)\, dx$$

exists either in the sense of upper and lower sums, or in the Riemann sense. Determining whether a given function is integrable is generally a very difficult problem. In addition to continuous functions, however, a function f is integrable on an interval $[a, b]$ if:

1. f is increasing on $[a, b]$, or f is decreasing on $[a, b]$.

2. f is bounded on $[a, b]$ and has at most a finite number of discontinuities.

EXERCISES 5.10

1. Let Ω be the region below the graph of $f(x) = x^2$, $x \in [0, 1]$. Draw a figure showing the Riemann sum $S^*(P)$ as an estimate for this area. Take

$$P = \{0, \tfrac{1}{4}, \tfrac{1}{2}, \tfrac{3}{4}, 1\} \quad \text{and}$$

$$x_1^* = \tfrac{1}{8}, \quad x_2^* = \tfrac{3}{8}, \quad x_3^* = \tfrac{5}{8}, \quad x_4^* = \tfrac{7}{8}.$$

2. Let Ω be the region below the graph of $f(x) = \tfrac{3}{2}x + 1$, $x \in [0, 2]$. Draw a figure showing the Riemann sum $S^*(P)$ as an estimate for this area. Take

$$P = \{0, \tfrac{1}{4}, \tfrac{3}{4}, 1, \tfrac{3}{2}, 2\} \quad \text{and let the } x_i^* \text{ be the}$$
midpoints of the subintervals.

3. Let $f(x) = 2x$, $x \in [0, 1]$. Take $P = \{0, \tfrac{1}{8}, \tfrac{1}{4}, \tfrac{1}{2}, \tfrac{3}{4}, 1\}$ and set

$$x_1^* = \tfrac{1}{16}, \quad x_2^* = \tfrac{3}{16}, \quad x_3^* = \tfrac{3}{8}, \quad x_4^* = \tfrac{5}{8}, \quad x_5^* = \tfrac{3}{4}.$$

Calculate the following:
(a) $\Delta x_1, \Delta x_2, \Delta x_3, \Delta x_4, \Delta x_5$.
(b) $\|P\|$.
(c) m_1, m_2, m_3, m_4, m_5.
(d) $f(x_1^*), f(x_2^*), f(x_3^*), f(x_4^*), f(x_5^*)$.
(e) M_1, M_2, M_3, M_4, M_5.
(f) $L_f(P)$.   (g) $S^*(P)$.

(h) $U_f(P)$.   (i) $\displaystyle\int_a^b f(x)\, dx$.

4. Let f be continuous on $[a, b]$, let $P = \{x_0, x_1, \ldots, x_n\}$ be a partition of $[a, b]$, and let $S^*(P)$ be any Riemann sum generated by P. Show that

$$L_f(P) \leq S^*(P) \leq U_f(P).$$

In numerical computations the base interval $[a, b]$ is usually broken up into n subintervals each of length $(b - a)/n$. The Riemann sums then take the form

$$S_n^* = \frac{b - a}{n}[f(x_1^*) + f(x_2^*) + \cdots + f(x_n^*)].$$

In Exercises 5–7, choose $\epsilon > 0$. Break up the interval $[0, 1]$ into n subintervals each of length $1/n$. Take $x_1^* = 1/n$, $x_2^* = 2/n, \ldots, x_n^* = n/n$.

5. (a) Determine the Riemann sum S_n^* for

$$\int_0^1 x\, dx. \quad \text{(See Exercise 27, Section 5.2.)}$$

(b) Show that

$$\left| S_n^* - \int_0^1 x\, dx \right| < \epsilon \qquad \text{if } n > 1/\epsilon.$$

HINT: $1 + 2 + \cdots + n = \tfrac{1}{2}n(n + 1)$.

6. (a) Determine the Riemann sum S_n^* for

$$\int_0^1 x^2\, dx. \quad \text{(See Exercise 28, Section 5.2.)}$$

(b) Show that

$$\left| S_n^* - \int_0^1 x^2\, dx \right| < \epsilon \quad \text{if } n > 1/\epsilon.$$

HINT: $1^2 + 2^2 + \cdots + n^2 = \frac{1}{6}n(n+1)(2n+1)$.

7. (a) Determine the Riemann sum S_n^* for

$$\int_0^1 x^3\, dx.$$

(b) Show that

$$\left| S_n^* - \int_0^1 x^3\, dx \right| < \epsilon \quad \text{if } n > 1/\epsilon.$$

HINT: $1^3 + 2^3 + \cdots + n^3 = (1 + 2 + \cdots + n)^2$.

8. Let f be a function continuous on $[a, b]$. Show that, if P is a partition of $[a, b]$, then $L_f(P)$, $U_f(P)$, and $\frac{1}{2}[L_f(P) + U_f(P)]$ are all Riemann sums.

9. Set $f(x) = x \cos x^2$, $x \in [0, 2]$. Take $P = \{0, \frac{1}{3}, \frac{2}{3}, 1, \frac{4}{3}, \frac{5}{3}, 2\}$ and set

$$x_1^* = \tfrac{1}{6},\ x_2^* = \tfrac{3}{6},\ x_3^* = \tfrac{5}{6},\ x_4^* = \tfrac{7}{6},\ x_5^* = \tfrac{9}{6},\ x_6^* = \tfrac{11}{6}.$$

Calculate $S^*(P)$ and compare this to the value of the integral

$$\int_0^2 x \cos x^2\, dx.$$

10. Set $f(x) = \sec^2 x$, $x \in [0, 1]$. Take $P = \{0, \frac{2}{10}, \frac{4}{10}, \frac{6}{10}, \frac{8}{10}, 1\}$ and set

$$x_1^* = \tfrac{1}{10},\ x_2^* = \tfrac{3}{10},\ x_3^* = \tfrac{5}{10},\ x_4^* = \tfrac{7}{10},\ x_5^* = \tfrac{9}{10}.$$

Calculate $S^*(P)$ and compare this to the value of the integral

$$\int_0^1 \sec^2 x\, dx.$$

■ CHAPTER HIGHLIGHTS

5.1 An Area Problem; A Speed–Distance Problem

5.2 The Definite Integral of a Continuous Function

partition (p. 280) — upper sum; lower sum (p. 281)
definite integral (p. 283) — limits of integration; integrand (p. 283)

5.3 The Function $F(x) = \int_a^x f(t)\, dt$

additivity of the integral (p. 291)

$$\frac{d}{dx}\left[\int_a^x f(t)\, dt\right] = f(x) \text{ provided } f \text{ is continuous (p. 293)}$$

5.4 The Fundamental Theorem of Integral Calculus

antiderivative (p. 297) — fundamental theorem (p. 298)
Table of Antiderivatives (p. 300)
linearity of the integral (p. 301)

5.5 Some Area Problems

If f and g are continuous and $f(x) \geq g(x)$ for all x in $[a, b]$, then

$$\int_a^b [f(x) - g(x)]\, dx$$

gives the area between the graph of f and the graph of g over $[a, b]$.

signed area (p. 308)

5.6 Indefinite Integrals

indefinite integral; constant of integration (p. 312)
Table of Indefinite integrals (p. 313)
some motion problems (p. 315)

For an object that moves along a coordinate line with velocity $v(t)$

$$\int_a^b |v(t)| \, dt = \text{distance traveled from time } t = a \text{ to time } t = b$$

whereas

$$\int_a^b v(t) \, dt = \text{net displacement from time } t = a \text{ to time } t = b.$$

equation for linear motion with constant acceleration (p. 317)

5.7 The *u*-Substitution; Change of Variables

An integral of the form $\int f(g(x))g'(x) \, dx$ can be written $\int f(u) \, du$ by setting

$$u = g(x), \qquad du = g'(x) \, dx.$$

For definite integrals

$$\int_a^b f(g(x))g'(x) \, dx = \int_{g(a)}^{g(b)} f(u) \, du.$$

5.8 Some Further Properties of the Definite Integral

The integral of a nonnegative function is nonnegative; the integral of a positive function is positive; the integral is order-preserving (pp. 329–330)

$$\left| \int_a^b f(x) \, dx \right| \leq \int_a^b |f(x)| \, dx, \qquad \frac{d}{dx}\left[\int_a^u f(t) \, dt \right] = f(u)\frac{du}{dx}.$$

5.9 Mean-Value Theorems for Integrals; Average Values

first mean-value theorem for integrals (p. 334) average value (p. 334)
second mean-value theorem for integrals (p. 336) weighted average (p. 336)
mass of a rod (p. 336) center of mass of a rod (p. 337)

5.10 The Integral as the Limit of Riemann Sums

Darboux method (p. 340) Riemann sum (p. 340)

$$\int_a^b f(x) \, dx = \lim_{\|P\| \to 0} [f(x_1^*) \, \Delta x_1 + f(x_2^*) \, \Delta x_2 + \cdots + f(x_n^*) \, \Delta x_n].$$

■ PROJECTS AND EXPLORATIONS USING TECHNOLOGY

To do these exercises you will need a graphics calculator or a computer with graphing capability. The majority of these problems are open-ended so different approaches may be used to solve them. You should be aware that different approaches can result in slight variations in the answers. Round your numerical answers to at least four decimal places. The rounding method that your calculator or computer uses also may cause variations in answers.

5.1 There is no simple way to represent the function

$$F(x) = \int_0^x \sin^2(t^2) \, dt.$$

For example, you cannot represent it in terms of a sum, difference, product, quotient or composition of elementary functions. However, using the methods of this chapter and the preceding chapters, we can still obtain a lot of information about F.

(a) Give the domain of F.

(b) Find F' and F''.

(c) Explain why F is continuous.

(d) Show that F is a nondecreasing function. Is F one-to-one?

(e) Discuss the symmetry of the graph of F.

(f) Discuss the concavity of the graph of F, and find the points of inflection.

(g) Based on the information in parts (a)–(f), sketch the graph of F. Then use a calculator or graphing utility to check your graph.

5.2 Consider the family of functions given by

$$f_p(x) = 40x^6 - 24x^5 - 45x^4 + 50x^3 - 15x^2 + 3(2p - 1)x + 6,$$

where p is an arbitrary constant. The graph of $f_1(x)$ is shown in the figure.

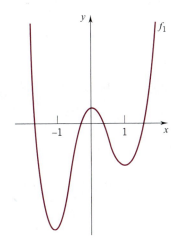

(a) Graph f_p for several values of p to get an idea about the graphs of the members of this family.

Let $g(p)$ be the x-coordinate of the largest (rightmost) local minimum of f_p. For example $g(1) \cong 1.02$. See the figure.

(b) Estimate $g(p)$ for the values of p used in part (a). HINT: Consider the zeros of f_p'.

(c) Based on the information obtained in parts (a) and (b), give the domain of g and determine where g is continuous.

Now consider the integral

$$G(z) = \int_0^z g(p)\, dp.$$

(d) Based on the theorems in this chapter, where does $G'(z) = g(z)$?

(e) Using the values of g calculated in part (b), approximate the values of G using Riemann sums. Estimate where G is increasing and where it is decreasing.

5.3 Assume that the velocity of a racing car is given by

$$v(t) = 40t - t^2,$$

where v is measured in feet per second and t is measured in seconds from the start of the race.

(a) Find T, the maximum value of t for which this formula is reasonable physically.
(b) Find the distance traveled from $t = 0$ to $t = T$.
(c) Find the average velocity from $t = 2$ to $t = 5.5$, that is, find

$$\frac{1}{3.5} \int_2^{5.5} (40t - t^2) \, dt.$$

(d) Find the instantaneous velocity of the car at $t = 2$ and at $t = 5.5$, and calculate the average of these two velocities.
(e) Find the instantaneous velocity at each half second starting at $t = 2$ and ending at $t = 5.5$. Then calculate the average of these velocities.
(f) Find the instantaneous velocity at $t = 2$, $t = 2.1$, $t = 2.2$, $t = 2.3$, . . . , $t = 5.4$, and $t = 5.5$. Then calculate the average of these velocities.
(g) Compare the results from parts (c)–(f) and explain their ordering.
(h) Approximate the average value of the velocity by generating 100 random values of t in the interval $[2, 5.5]$ and averaging the instantaneous velocities at each of these times.

5.4 In previous chapters we considered moving averages of a function f as a means of smoothing data to find long-term trends. The examples given in the Projects and Explorations section of Chapter 1 averaged two values. More generally, we could average n values:

$$A(x) = \frac{1}{n} \sum_{i=1}^n f(x_i) = \frac{1}{a} \left[\frac{a}{n} \sum_{i=1}^n f(x_i) \right],$$

where $x_i = x - a + ia/n$, $i = 1, 2, \ldots, n$. Since the expression on the right is a Riemann sum for A, we can take the limit as $n \to \infty$ to obtain the moving average F of f given by

$$F(x) = \frac{1}{a} \int_{x-a}^x f(t) \, dt.$$

In the problems that follow, let $f(x) = \sin(3x + 2)$ and $a = \pi$.
(a) Graph $F(x)$. Also, graph

$$g(x) = \frac{f(x - \pi) + f(x)}{2}.$$

Analysts look for long-term "trends" rather than short-term "random fluctuations." Moving averages are one method for doing such analyses. Generally, a curve is considered to be "smooth" if it is differentiable and has few relative extrema or points of inflection. Which of these moving averages seems to do the best job of smoothing the data?

(b) Find $F'(x)$.
(c) Now find the moving average of f'. That is, find

$$F(x) = \frac{1}{\pi} \int_{x-\pi}^x f'(t) \, dt.$$

Is the moving average of f' equal to F'?
(d) Can you generalize this example to other functions?
(e) Repeat parts (a)–(c) for the greatest integer function $f(x) = [\![x]\!]$ with $a = 1$, and then with $a = 0.5$. How well does the moving average smooth a discontinuous function?

SOME APPLICATIONS OF THE INTEGRAL

CHAPTER

6

■ 6.1 MORE ON AREA

Representative Rectangles

You have seen that the definite integral can be viewed as the limit of Riemann sums:

$$(1) \qquad \int_a^b f(x)\,dx = \lim_{\|P\| \to 0} [\, f(x_1^*)\,\Delta x_1 + f(x_2^*)\,\Delta x_2 + \cdots + f(x_n^*)\,\Delta x_n\,].$$

With x_i^* chosen arbitrarily from $[x_{i-1}, x_i]$, you can think of $f(x_i^*)$ as a *representative value* of f for that interval. If f is positive, then the product

$$f(x_i^*)\,\Delta x_i$$

gives the area of the *representative rectangle* shown in Figure 6.1.1. Formula (1) tells us that we can approximate the area under the curve as closely as we wish by adding up the areas of representative rectangles (Figure 6.1.2).

Figure 6.1.1

Figure 6.1.2

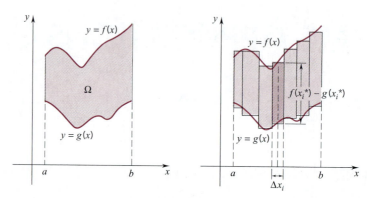

Figure 6.1.3

Figure 6.1.3 shows a region Ω bounded above by the graph of a function f and bounded below by the graph of a function g. As you have seen before, we can calculate the area of Ω by integrating with respect to x the *vertical separation*

$$f(x) - g(x)$$

from $x = a$ to $x = b$:

$$\text{area } (\Omega) = \int_a^b [f(x) - g(x)] \, dx.$$

In this case the approximating Riemann sums are of the form

$$[f(x_1^*) - g(x_1^*)] \, \Delta x_1 + [f(x_2^*) - g(x_2^*)] \, \Delta x_2 + \cdots + [f(x_n^*) - g(x_n^*)] \, \Delta x_n.$$

The dimensions of a representative rectangle are now

$$\text{"height"} = f(x_i^*) - g(x_i^*) \qquad \text{and} \qquad \text{"width"} \, \Delta x_i.$$

Example 1 Find the area of the shaded region shown in Figure 6.1.4.

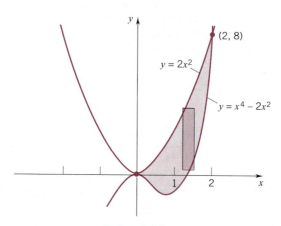

Figure 6.1.4

SOLUTION The area of the representative rectangle is $[2x^2 - (x^4 - 2x^2)] \Delta x$, $0 \le x \le 2$. Thus, the area of the shaded region is

$$A = \int_0^2 [2x^2 - (x^4 - 2x^2)] \, dx = \int_0^2 (4x^2 - x^4) \, dx$$

$$= \left[\tfrac{4}{3}x^3 - \tfrac{1}{5}x^5 \right]_0^2 = \tfrac{64}{15}. \quad \square$$

Areas by Integration with Respect to y

In Figure 6.1.5 you can see a region, the boundaries of which are not functions of x but functions of y instead. In such a case we set the representative rectangles horizontally and we calculate the area of the region as the limit of Riemann sums of the form

$$[F(y_1^*) - G(y_1^*)] \Delta y_1 + [F(y_2^*) - G(y_2^*)] \Delta y_2 + \cdots + [F(y_n^*) - G(y_n^*)] \Delta y_n.$$

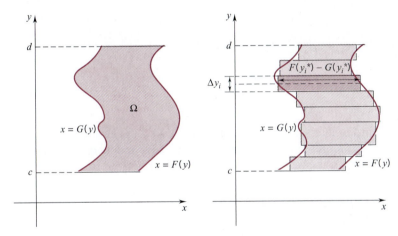

Figure 6.1.5

The area of the region is thus given by the integral

$$\int_c^d [F(y) - G(y)] \, dy.$$

Here we are integrating the *horizontal separation*

$$F(y) - G(y)$$

with respect to y.

Example 2 Find the area of the region bounded on the left by $x = y^2$ and on the right by $x = 3 - 2y^2$.

SOLUTION The region is sketched in Figure 6.1.6. The points of intersection are found by solving the two equations simultaneously:

$$x = y^2 \quad \text{and} \quad x = 3 - 2y^2$$

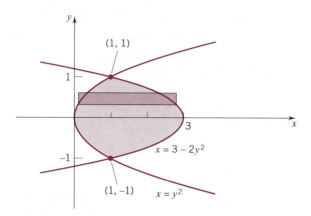

Figure 6.1.6

implies

$$y^2 = 3 - 2y^2$$
$$3y^2 = 3$$

and

$$y = \pm 1.$$

Thus the points of intersection are $(1, 1)$ and $(1, -1)$. The easiest way to calculate the area is to set our representative rectangles horizontally and integrate with respect to y. We find the area of the region by integrating the horizontal separation

$$(3 - 2y^2) - y^2 = 3 - 3y^2$$

from $y = -1$ to $y = 1$:

$$A = \int_{-1}^{1} (3 - 3y^2) \, dy = \left[3y - y^3 \right]_{-1}^{1} = 4.$$

NOTE: In our solution we did not take advantage of the symmetry. The region is symmetric with respect to the x-axis (the integrand is an even function of y) and so the area is also given by

$$A = 2\int_{0}^{1} (3 - 3y^2) \, dy = 2\left[3y - y^3 \right]_{0}^{1} = 4. \quad \square$$

Example 3 Calculate the area of the region bounded by the curves $x = y^2$ and $x - y = 2$ by integrating (a) with respect to x, (b) with respect to y.

SOLUTION You can verify that the two curves intersect at the points $(1, -1)$ and $(4, 2)$.

(a) To integrate with respect to x we set the representative rectangles vertically, and we solve the equations for y. Solving $x = y^2$ for y we get $y = \pm\sqrt{x}$; $y = \sqrt{x}$ is the upper half of the parabola and $y = -\sqrt{x}$ is the lower half. The equation of the line can be written $y = x - 2$. See Figure 6.1.7.

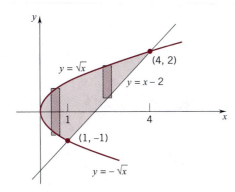

Figure 6.1.7

The upper boundary of the region is the curve $y = \sqrt{x}$. However, the lower boundary is described by two different equations: $y = -\sqrt{x}$ from $x = 0$ to $x = 1$, and $y = x - 2$ from $x = 1$ to $x = 4$. Thus, we need two integrals:

$$A = \int_0^1 [\sqrt{x} - (-\sqrt{x})]\,dx + \int_1^4 [\sqrt{x} - (x-2)]\,dx$$

$$= 2 \int_0^1 \sqrt{x}\,dx + \int_1^4 (\sqrt{x} - x + 2)\,dx = \left[\tfrac{4}{3}x^{3/2}\right]_0^1 + \left[\tfrac{2}{3}x^{3/2} - \tfrac{1}{2}x^2 + 2x\right]_1^4 = \tfrac{9}{2}.$$

(b) See Figure 6.1.8. To integrate with respect to y we set the representative rectangles horizontally. Now the right boundary is the line $x = y + 2$ and the left boundary is the curve $x = y^2$. Since y ranges from -1 to 2,

$$A = \int_{-1}^2 [(y+2) - y^2]\,dy = \left[\tfrac{1}{2}y^2 + 2y - \tfrac{1}{3}y^3\right]_{-1}^2 = \tfrac{9}{2}. \quad \square$$

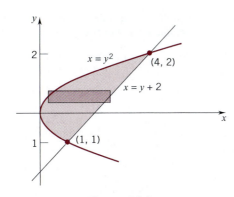

Figure 6.1.8

EXERCISES 6.1

In Exercises 1–14, sketch the region bounded by the curves. Represent the area of the region by one or more integrals (a) in terms of x; (b) in terms of y.

1. $y = x^2$, $y = x + 2$. **2.** $y = x^2$, $y = -4x$.

3. $y = x^3$, $y = 2x^2$. **4.** $y = \sqrt{x}$, $y = x^3$.

5. $y = -\sqrt{x}$, $y = x - 6$, $y = 0$.

6. $x = y^3$, $x = 3y + 2$. **7.** $y = |x|$, $3y - x = 8$.

8. $y = x$, $y = 2x$, $y = 3$.

9. $x + 4 = y^2$, $x = 5$. **10.** $x = |y|$, $x = 2$.

11. $y = 2x$, $x + y = 9$, $y = x - 1$.

12. $y = x^3$, $y = x^2 + x - 1$.

13. $y = x^{1/3}, \quad y = x^2 + x - 1$.

14. $y = x + 1, \quad y + 3x = 13, \quad 3y + x + 1 = 0$.

In Exercises 15–24, sketch the region bounded by the curves and find its area.

15. $4x = 4y - y^2, \quad 4x - y = 0$.

16. $x + y^2 - 4 = 0, \quad x + y = 2$.

17. $x = y^2, \quad x = 3 - 2y^2$.

18. $x + y = 2y^2, \quad y = x^3$.

19. $x + y - y^3 = 0, \quad x - y + y^2 = 0$.

20. $8x = y^3, \quad 8x = 2y^3 + y^2 - 2y$.

21. $y = \cos x, \quad y = \sec^2 x, \quad x \in [-\pi/4, \pi/4]$.

22. $y = \sin^2 x, \quad y = \tan^2 x, \quad x \in [-\pi/4, \pi/4]$.
HINT: See Exercise 73, Section 5.7.

23. $y = 2 \cos x, \quad y = \sin 2x, \quad [-\pi, \pi]$.

24. $y = \sin x, \quad y = \sin 2x, \quad x \in [0, \pi/2]$.

In Exercises 25 and 26, use integration to find the area of the triangle with the given vertices.

25. $(0, 0), (1, 3), (3, 1)$.   **26.** $(0, 1), (2, 0), (3, 4)$.

27. Use integration to find the area of the trapezoid with vertices $(-2, -2), (1, 1), (5, 1), (7, -2)$.

28. Sketch the region bounded by the three curves $y = x^3$, $y = -x$, $y = 1$, and find its area.

29. Sketch the region bounded by the three curves $y = 6 - x^2$, $y = x$ $(x < 0)$, and $y = -x$ $(x > 0)$, and find its area.

30. Find the area of the region bounded by the parabolas $x^2 = 4py$ and $y^2 = 4px$, p a positive constant.

31. Sketch the region bounded by $y = x^2$ and $y = 4$. This region is divided into two subregions of equal area by the line $y = c$. Find c.

32. The region between $y = \cos x$ and the x-axis for $x \in [0, \pi/2]$ is divided into two subregions of equal area by the line $x = c$. Find c.

◄**C 33.** Use a graphing utility to sketch the region bounded by the curves $y = x^4 - 2x^2$ and $y = x + 2$. Then find (approximately) the area of the region.

◄**C 34.** Use a graphing utility to sketch the region bounded by the curves $y = \sin x$ and $y = |x - 1|$. Then find (approximately) the area of the region.

35. A section of rain gutter is 8 feet long. Vertical cross sections of the gutter are in the shape of the parabolic region bounded by $y = \frac{4}{9}x^2$ and $y = 4$, where these measurements are in inches. What is t̅.   ̅ ̅me of the rain gutter? HINT: $V = $ (cross-sectic ̅   × length.

Distribution of Income. Economis̅ ̅̅ ̅ ̅ ̅ ̅ ̅ ̅ ̅ ̅ julative distribution called a *Lorenz curve* to measure̅ ̅ ̅ distribution of income between households in a given country. Typically, a

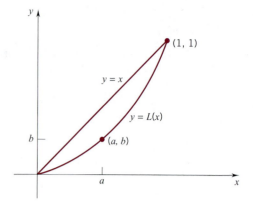

Lorenz curve is defined on $[0, 1]$, is continuous, increasing and concave up, and passes through $(0, 0)$ and $(1, 1)$. See the figure. The points on the curve are determined by ranking all households by income and then computing the percentage of households whose total income is less than or equal to a given percentage of the total income of the country. For example, the point (a, b) on the curve represents the fact that the bottom $a\%$ of the households receive less than or equal to $b\%$ of the total income. *Absolute equality* of income distribution would occur if the bottom $a\%$ of the households received $a\%$ of the total income. The Lorenz curve in this case would be the line $y = x$. The area of the region between the line $y = x$ and the Lorenz curve measures the extent to which the income distribution differs from absolute equality. The *coefficient of inequality* is the ratio of the area of the region between $y = x$ and the Lorenz curve to the area under $y = x$.

36. Let $y = L(x)$ be a Lorenz curve. Show that the coefficient of inequality C is given by

$$C = 2 \int_0^1 [x - L(x)] \, dx.$$

37. The income distribution of a certain country is represented by the Lorenz curve

$$L(x) = \tfrac{7}{12}x^2 + \tfrac{5}{12}x.$$

(a) What is the percentage of total income received by the bottom 50% of the households.

(b) Determine the coefficient of inequality.

◄**C 38.** Let

$$L(x) = \frac{5x^3}{4 + x^2}.$$

(a) Show that L has the characteristics of a Lorenz curve.

(b) Estimate the coefficient of inequality by calculating the Riemann sum for $f(x) = x - L(x)$ with the partition $P = \{0, 0.2, 0.4, 0.6, 0.8, 1\}$ and $x_1^* = 0.1, x_2^* = 0.3, x_3^* = 0.5, x_4^* = 0.7, x_5^* = 0.9$.

■ 6.2 VOLUME BY PARALLEL CROSS SECTIONS; DISCS AND WASHERS

In this section we use definite integrals to find the volumes of certain three-dimensional solids. To begin, let Ω be a plane region. A *right cylinder with cross section* Ω is a solid that is formed by translating Ω along a line, or *axis*, that is perpendicular to it. See Figure 6.2.1. Let A be the area of Ω. If a right cylinder is formed by translating the region Ω through a distance h, then the volume of the cylinder is given by

$$V = A \cdot h.$$

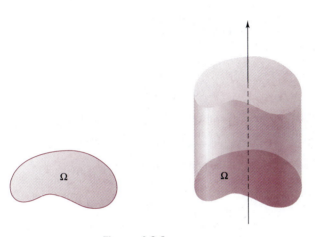

Figure 6.2.1

Some familiar examples are the right circular cylinder of radius r and height h, and a rectangular box of length l, width w, and height h (Figure 6.2.2).

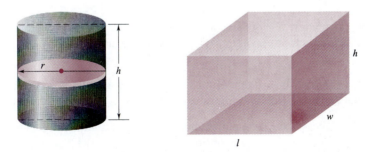

$V = \pi r^2 h = $ (cross-sectional area) · height $V = l \cdot w \cdot h = $ (cross-sectional area) · height

Figure 6.2.2

area $A(x)$

Figure 6.2.3

To calculate the volume of a more general solid, we introduce a coordinate axis and then examine the cross sections of the solid that are perpendicular to that axis. In Figure 6.2.3 we picture a solid and a coordinate axis that we label the x-axis. As in the figure, we suppose that the solid lies entirely between $x = a$ and $x = b$. The figure shows an arbitrary cross section perpendicular to the x-axis. We denote by $A(x)$ the area of the cross section that has coordinate x.

If the cross-sectional area, $A(x)$, varies continuously with x, then we can find the volume of the solid by integrating $A(x)$ from a to b:

(6.2.1)

$$V = \int_a^b A(x)\, dx.$$

To see this, let $P = \{x_0, x_1, \ldots, x_n\}$ be a partition of $[a, b]$. The solid from x_{i-1} to x_i is approximated by a cylinder whose cross section has area $A(x_i^*)$, where x_i^* is chosen arbitrarily from $[x_{i-1}, x_i]$, and whose thickness is Δx_i. The volume of this cylinder is

$$A(x_i^*)\, \Delta x_i. \qquad \text{(Figure 6.2.4)}$$

The volume of the entire solid can therefore be approximated by sums of the form

$$A(x_1^*)\, \Delta x_1 + A(x_2^*)\, \Delta x_2 + \cdots + A(x_n^*)\, \Delta x_n.$$

These are Riemann sums which, as $\|P\| \to 0$, converge to

$$\int_a^b A(x)\, dx. \qquad \square$$

$x_0 = a \qquad x_{i-1}\ x_i \qquad x_n = b$

Figure 6.2.4

Remark By the mean-value theorem for integrals (see 5.9.2), the formula for the volume of a solid with varying cross-sectional area $A = A(x)$ can also be written

(6.2.2) | $V = $ (the average cross-sectional area) $\cdot (b - a) = A_{\text{avg}}(b - a)$. | $\square$

The calculation of volumes of solids of arbitrary shape is left for Chapter 16. Here we restrict our attention to solids with simple cross sections.

Example 1 Find the volume of a pyramid whose altitude is h and whose base is a square with sides of length r (Figure 6.2.5a).

SOLUTION Locate the x-axis perpendicular to the base, with origin at the center of the base. The cross section with coordinate x is a square. Let s denote the length of the side of that square. Then, by similar triangles (see Figure 6.2.5b), we have

$$\frac{\frac{1}{2}s}{h - x} = \frac{\frac{1}{2}r}{h},$$

and so

$$s = \frac{r}{h}(h - x).$$

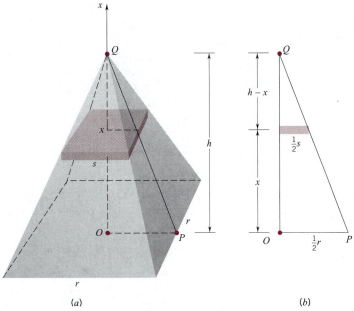

Figure 6.2.5

Now the area of the square at coordinate x is $A(x) = s^2 = (r^2/h^2)(h - x)^2$ and

$$V = \int_0^h A(x)\, dx = \frac{r^2}{h^2} \int_0^h (h - x)^2\, dx$$

$$= \frac{r^2}{h^2} \left[-\frac{(h - x)^3}{3} \right]_0^h = \tfrac{1}{3} r^2 h. \quad \square$$

Example 2 The base of a solid is the region bounded by the ellipse

$$\frac{x^2}{a^2} + \frac{y^2}{b^2} = 1.$$

Find the volume of the solid given that each cross section perpendicular to the x-axis is an isosceles triangle with base in the region and altitude equal to one-half the base.

SOLUTION Take x as in Figure 6.2.6. The cross section with coordinate x is an isosceles triangle with base $\overline{PQ}$ and altitude $\tfrac{1}{2}\overline{PQ}$. The equation of the ellipse can be written

$$y^2 = \frac{b^2}{a^2}(a^2 - x^2).$$

Since

$$\text{length of } \overline{PQ} = 2y = \frac{2b}{a}\sqrt{a^2 - x^2},$$

the isosceles triangle has area

$$A(x) = \tfrac{1}{2}bh = \tfrac{1}{2}\left(\frac{2b}{a}\sqrt{a^2 - x^2} \right)\left(\frac{b}{a}\sqrt{a^2 - x^2} \right) = \frac{b^2}{a^2}(a^2 - x^2).$$

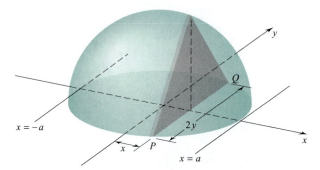

Figure 6.2.6

We can find the volume of the solid by integrating $A(x)$ from $x = -a$ to $x = a$:

$$V = \int_{-a}^{a} A(x)\,dx = 2 \int_{0}^{a} A(x)\,dx$$

$\underset{\text{by symmetry}}{\uparrow}$

$$= \frac{2b^2}{a^2} \int_{0}^{a} (a^2 - x^2)\,dx$$

$$= \frac{2b^2}{a^2} \left[a^2 x - \frac{x^3}{3} \right]_{0}^{a} = \tfrac{4}{3}ab^2. \quad \square$$

Example 3 The base of a solid is the region between the parabolas

$$x = y^2 \qquad \text{and} \qquad x = 3 - 2y^2. \quad \text{(See Example 2, Section 6.1)}$$

Find the volume of the solid given that the cross sections perpendicular to the x-axis are squares.

SOLUTION The two parabolas intersect at $(1, 1)$ and $(1, -1)$ (Figure 6.2.7). From $x = 0$ to $x = 1$, the cross section with coordinate x has area

$$A(x) = (2y)^2 = 4y^2 = 4x.$$

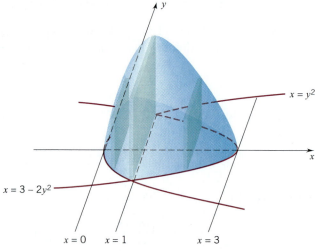

Figure 6.2.7

(Here we are measuring the span across the first parabola $x = y^2$.) The volume of the solid from $x = 0$ to $x = 1$ is

$$V_1 = \int_0^1 4x \, dx = \left[2x^2 \right]_0^1 = 2.$$

From $x = 1$ to $x = 3$, the cross section with coordinate x has area

$$A(x) = (2y)^2 = 4y^2 = 2(3 - x) = 6 - 2x.$$

(Here we are measuring the span across the second parabola $x = 3 - 2y^2$.) The volume of the solid from $x = 1$ to $x = 3$ is

$$V_2 = \int_1^3 (6 - 2x) \, dx = \left[6x - x^2 \right]_1^3 = 4.$$

The total volume is

$$V_1 + V_2 = 6. \quad ❑$$

Solids of Revolution: Disc Method

Suppose that f is nonnegative and continuous on $[a, b]$. (See Figure 6.2.8.) If we revolve the region bounded by the graph of f and the x-axis about the x-axis, we obtain a solid. The volume of this solid is given by the formula

(6.2.3)

$$V = \int_a^b \pi [f(x)]^2 \, dx.$$

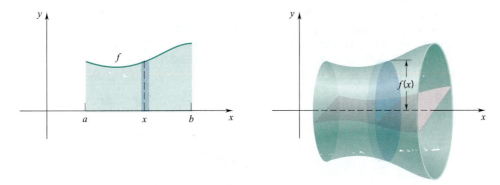

Figure 6.2.8

PROOF The cross section with coordinate x is a circular *disc* of radius $f(x)$. The cross-sectional area is thus $\pi [f(x)]^2$. ❑

Among the simplest solids of revolution are the cone and the sphere.

Example 4 We can generate a cone of base radius r and height h by revolving about the x-axis the region below the graph of

$$f(x) = \frac{r}{h} x, \qquad 0 \le x \le h. \qquad \text{(Figure 6.2.9)}$$

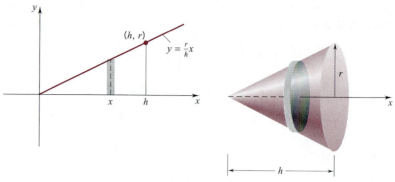

Figure 6.2.9

By Formula (6.2.3),

$$\text{volume of cone} = \int_0^h \pi \left[\frac{r}{h}x\right]^2 dx = \frac{\pi r^2}{h^2} \int_0^h x^2 \, dx = \frac{\pi r^2}{h^2} \left[\frac{x^3}{3}\right]_0^h = \frac{1}{3}\pi r^2 h. \quad \square$$

Example 5   A sphere of radius r can be obtained by revolving about the x-axis the region below the graph of

$$f(x) = \sqrt{r^2 - x^2}, \qquad -r \le x \le r. \qquad \text{(draw a figure)}$$

Therefore

$$\text{volume of sphere} = \int_{-r}^r \pi(r^2 - x^2) \, dx = \pi \left[r^2 x - \frac{1}{3}x^3\right]_{-r}^r = \frac{4}{3}\pi r^3.$$

This result was obtained by Archimedes in the third century B.C. $\square$

Now suppose that $x = g(y)$ is continuous and nonnegative for $c \le y \le d$ (see Figure 6.2.10). If we revolve the region bounded by the graph of g and the y-axis about the y-axis, we obtain a solid such that each cross section perpendicular to the y-axis is a disc with area $A(y) = \pi[g(y)]^2$. Thus the volume of this solid

Figure 6.2.10

is given by

(6.2.4)

$$V = \int_c^d \pi[g(y)]^2 \, dy.$$

Example 6 Find the volume of the solid generated by revolving the region bounded by $y = x^{2/3} + 1$, $0 \leq x \leq 8$, about the y-axis.

SOLUTION See Figure 6.2.11. We first solve $y = x^{2/3} + 1$, $0 \leq x \leq 8$, for x as a function of y:

$$x^{2/3} = y - 1$$
$$x = (y - 1)^{3/2}, \qquad 1 \leq y \leq 5.$$

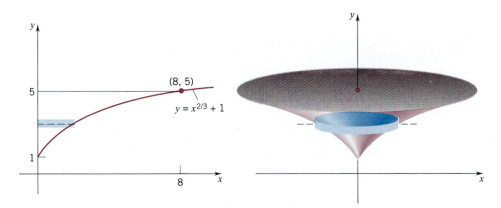

Figure 6.2.11

Now, applying (6.2.4), we have

$$V = \int_1^5 \pi[g(y)]^2 \, dy = \pi \int_1^5 [(y - 1)^{3/2}]^2 \, dy$$
$$= \pi \int_1^5 (y - 1)^3 \, dy = \pi \left[\frac{(y - 1)^4}{4} \right]_1^5 = 64\pi. \quad \square$$

Solids of Revolution: Washer Method

The washer method is a slight generalization of the disc method. Suppose that f and g are nonnegative continuous functions with $g(x) \leq f(x)$ for all x in $[a, b]$ (see Figure 6.2.12). If we revolve the region Ω about the x-axis, we obtain a solid. The volume of this solid is given by the formula

(6.2.5)

$$V = \int_a^b \pi([f(x)]^2 - [g(x)]^2) \, dx.$$

(washer method about x-axis)

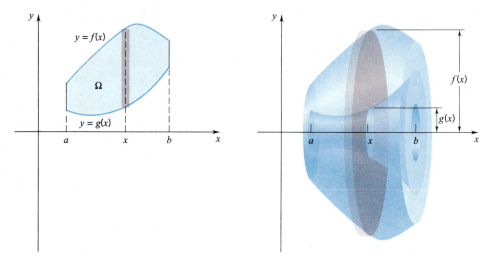

Figure 6.2.12

PROOF The cross section with coordinate x is a *washer* of outer radius $f(x)$, inner radius $g(x)$, and area

$$A(x) = \pi[f(x)]^2 - \pi[g(x)]^2 = \pi([f(x)]^2 - [g(x)]^2).$$

We can get the volume of the solid by integrating this function from a to b. ❑

Suppose now that the boundaries are functions of y rather than x (see Figure 6.2.13). By revolving Ω *about the y-axis,* we obtain a solid. It is clear from (6.2.4) that in this case

(6.2.6)
$$V = \int_c^d \pi([F(y)]^2 - [G(y)]^2)\, dy.$$

(washer method about y-axis)

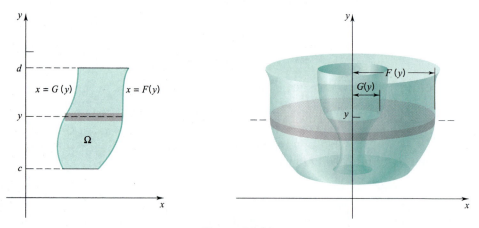

Figure 6.2.13

Example 7 Find the volume of the solid generated by revolving the region between $y = x^2$ and $y = 2x$ about the x-axis.

SOLUTION Setting $x^2 = 2x$, we get $x = 0, 2$. Thus, the curves intersect at the points $(0, 0)$ and $(2, 4)$. See Figure 6.2.14. For each x from 0 to 2, the x cross section is a washer of outer radius $2x$ and inner radius x^2. By (6.2.5)

$$V = \int_0^2 \pi[(2x)^2 - (x^2)^2] \, dx = \pi \int_0^2 (4x^2 - x^4) \, dx = \pi\left[\frac{4}{3}x^3 - \frac{1}{5}x^5\right]_0^2 = \frac{64}{15}\pi. \quad \square$$

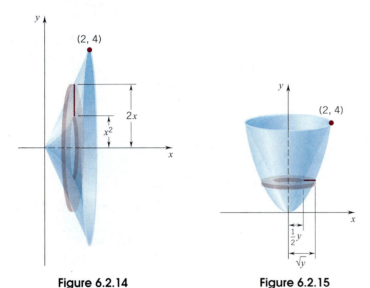

Figure 6.2.14   Figure 6.2.15

Example 8 Find the volume of the solid generated by revolving the region between $y = x^2$ and $y = 2x$ about the y-axis.

SOLUTION The solid is depicted in Figure 6.2.15. For each y from 0 to 4, the y cross section is a washer of outer radius $\sqrt{y}$ and inner radius $\frac{1}{2}y$. By (6.2.6)

$$V = \int_0^4 \pi[(\sqrt{y})^2 - (\tfrac{1}{2}y)^2] \, dy = \pi \int_0^4 (y - \tfrac{1}{4}y^2) \, dy$$
$$= \pi[\tfrac{1}{2}y^2 - \tfrac{1}{12}y^3]_0^4 = \tfrac{8}{3}\pi. \quad \square$$

Remark These last two examples concerned solids generated by revolving the *same* region about *different* axes. Notice that the solids are different and have different volumes. ❑

EXERCISES 6.2

In Exercises 1–16, sketch the region Ω bounded by the curves and find the volume of the solid generated by revolving this region about the x-axis.

1. $y = x$, $y = 0$, $x = 1$.
2. $x + y = 3$, $y = 0$, $x = 0$.
3. $y = x^2$, $y = 9$.
4. $y = x^3$, $y = 8$, $x = 0$.
5. $y = \sqrt{x}$, $y = x^3$.
6. $y = x^2$, $y = x^{1/3}$.
7. $y = x^3$, $x + y = 10$, $y = 1$.
8. $y = \sqrt{x}$, $x + y = 6$, $y = 1$.
9. $y = x^2$, $y = x + 2$.
10. $y = x^2$, $y = 2 - x$.
11. $y = \sqrt{4 - x^2}$, $y = 0$.
12. $y = 1 - |x|$, $y = 0$.
13. $y = \sec x$, $x = 0$, $x = \frac{1}{4}\pi$, $y = 0$.
14. $y = \csc x$, $x = \frac{1}{4}\pi$, $x = \frac{3}{4}\pi$, $y = 0$.
15. $y = \cos x$, $y = x + 1$, $x = \frac{1}{2}\pi$.
16. $y = \sin x$, $x = \frac{1}{4}\pi$, $x = \frac{1}{2}\pi$, $y = 0$.

In Exercises 17–26, sketch the region Ω bounded by the curves and find the volume of the solid generated by revolving this region about the y-axis.

17. $y = 2x$, $y = 4$, $x = 0$.
18. $x + 3y = 6$, $x = 0$, $y = 0$.
19. $x = y^3$, $x = 8$, $y = 0$.
20. $x = y^2$, $x = 4$.
21. $y = \sqrt{x}$, $y = x^3$.
22. $y = x^2$, $y = x^{1/3}$.
23. $y = x$, $y = 2x$, $x = 4$.
24. $x + y = 3$, $2x + y = 6$, $x = 0$.
25. $x = y^2$, $x = 2 - y^2$.
26. $x = \sqrt{9 - y^2}$, $x = 0$.

27. The base of a solid is the circle $x^2 + y^2 = r^2$. Find the volume of the solid given that the cross sections perpendicular to the x-axis are: (a) squares; (b) equilateral triangles.

28. The base of a solid is the region bounded by the ellipse $b^2x^2 + a^2y^2 = a^2b^2$. Find the volume of the solid given that the cross sections perpendicular to the x-axis are: (a) isosceles right triangles each with hypotenuse on the xy-plane; (b) squares; (c) isosceles triangles of height 2.

29. The base of a solid is the region bounded by $y = x^2$ and $y = 4$. Find the volume of the solid given that the cross sections perpendicular to the x-axis are: (a) squares; (b) semicircles; (c) equilateral triangles.

30. The base of a solid is the region between the parabolas $x = y^2$ and $x = 3 - 2y^2$. Find the volume of the solid given that the cross sections perpendicular to the x-axis are: (a) rectangles of height h; (b) equilateral triangles; (c) isosceles right triangles each with hypotenuse on the xy-plane.

31. Exercise 29 with the cross sections perpendicular to the y-axis.

32. Exercise 30 with the cross sections perpendicular to the y-axis.

33. Find the volume of the solid generated by revolving the ellipse $b^2x^2 + a^2y^2 = a^2b^2$ about the x-axis.

34. Exercise 33 with the ellipse revolved about the y-axis.

35. Derive a formula for the volume of the frustum of a right circular cone in terms of the height h, the lower base radius R, and the upper base radius r. (See the figure.)

36. Find the volume enclosed by the surface that is generated by revolving the equilateral triangle with vertices $(0, 0)$, $(a, 0)$, $(\frac{1}{2}a, \frac{1}{2}\sqrt{3}a)$ about the x-axis.

37. A hemispheric basin of radius r feet is being used to store water. To what percent of capacity is it filled when the water is: (a) $\frac{1}{2}r$ feet deep? (b) $\frac{1}{3}r$ feet deep?

38. A sphere of radius r is cut by two parallel planes: one, a units above the equator; the other, b units above the equator. Find the volume of the portion of the sphere that lies between the two planes. Assume that $a < b$.

39. The region shown in the figure is revolved around the y-axis to form a parabolic-shaped container. The indi-

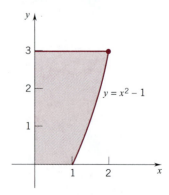

cated units are in feet. If a liquid is poured into the container at the rate of two cubic feet per minute, how fast is the level of the liquid rising at the instant its depth in the container is one foot? Two feet?

Rotation about Lines Parallel to a Coordinate Axis

In Exercises 40 and 41, let f be a continuous, nonnegative function on $[a, b]$ and let Ω be the region between the graph of f and the x-axis.

40. Let k be a positive constant. Prove that if the region Ω is revolved around the line $y = -k$ (Figure A), then the volume of the solid of revolution (Figure B) is given by

$$V = \int_a^b \pi([\,f(x) + k]^2 - k^2)\, dx.$$

Figure A

Figure B

41. Let k be a positive constant satisfying $k - f(x) \geq 0$ on $[a, b]$. Prove that if the region Ω is revolved around the line $y = k$ (Figure C), then the volume of the solid of revolution (Figure D) is given by

$$V = \int_a^b \pi(k^2 - [k - f(x)]^2)\, dx.$$

Figure C

Figure D

42. Let f and g be continuous functions on $[a, b]$ and let k be a constant.

(a) Suppose that $f(x) \geq g(x) \geq k$ on $[a, b]$. Let Ω be the region between the graphs of f and g. Show that if Ω is revolved around the line $y = k$, then the volume of the resulting solid of revolution is given by

$$V = \int_a^b \pi([\,f(x) - k]^2 - [g(x) - k]^2)\, dx.$$

(b) Suppose that $k \geq f(x) \geq g(x)$ on $[a, b]$, and let Ω be the region between the graphs of f and g. Derive a formula for the volume of the solid which is generated by revolving Ω around the line $y = k$.

NOTE: Analogs of the formulas in Exercises 40–42 hold for functions $x = h(y)$ and regions revolved around vertical lines $x = l$.

43. The region between the graph of $f(x) = \sqrt{x}$ and the x-axis for $0 \leq x \leq 4$ is revolved around the line $y = 2$. Find the volume of the solid that is generated.

44. The region bounded by the curves $y = (x - 1)^2$ and $y = x + 1$ is revolved around the line $y = -1$. Find the volume of the solid that is generated.

45. The region between the graph of $y = \sin x$ and the x-axis, $0 \leq x \leq \pi$, is revolved around the line $y = 1$. Find the volume of the solid that is generated. HINT: See Exercise 73, Section 5.7.

46. The region bounded by $y = \sin x$ and $y = \cos x$, $\pi/4 \leq x \leq \pi$, is revolved around the line $y = 1$. Find the volume of the solid that is generated. HINT: Use a trigonometric identity.

47. The region bounded by the curves $y = x^2 - 2x$ and $y = 3x$ is revolved around the line $y = -1$. Find the volume of the solid that is generated.

48. Find the volume of the solid that is generated by revolving the region bounded by $y = x^2$ and $x = y^2$ around: (a) the line $x = -2$; (b) the line $x = 3$.

49. Find the volume of the solid generated by revolving the region bounded by $y^2 = 4x$ and $y = x$ around: (a) the x-axis; (b) the line $x = 4$.

50. Find the volume of the solid that is generated by revolving the region bounded by $y = x^2$ and $y = 4x$ around: (a) the line $x = 5$; (b) the line $x = -1$.

51. Find the volume when the region OAB in Figure E is revolved about:
(a) the x-axis; (b) the line AB; (c) the line CA; (d) the y-axis.

Figure E

52. Find the volume when the region OAC in Figure E is revolved about:
(a) the y-axis; (b) the line CA; (c) the line AB; (d) the x-axis.

■ 6.3 VOLUME BY THE SHELL METHOD

To describe the shell method of computing volumes, we begin with a solid cylinder of radius R and height h, and from it we cut out a cylindrical core of radius r (Figure 6.3.1).

Since the original cylinder had volume $\pi R^2 h$ and the piece removed had volume $\pi r^2 h$, the solid cylindrical shell that remains has volume

Figure 6.3.1

(6.3.1)
$$\pi R^2 h - \pi r^2 h = \pi h (R + r)(R - r).$$

We will use this shortly.

Now let $[a, b]$ be an interval that does not contain 0 in its interior, and let f be a nonnegative continuous function on $[a, b]$. For convenience, we assume that $a \geq 0$. If the region bounded by the graph of f and the x-axis is revolved around the y-axis, then a solid T is generated (Figure 6.3.2). The volume of this solid is given by the *shell method formula*:

(6.3.2)
$$V = \int_a^b 2\pi x f(x) \, dx.$$

To see this, we take a partition $P = \{x_0, x_1, \ldots, x_n\}$ of $[a, b]$ and concentrate on what's happening on the ith subinterval $[x_{i-1}, x_i]$. Recall that, when we form a Riemann sum, we are free to choose x_i^* as any point from $[x_{i-1}, x_i]$. For convenience

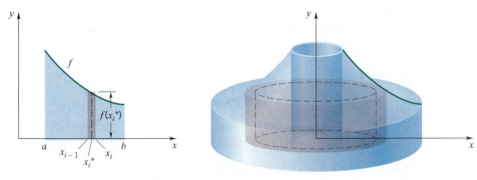

Figure 6.3.2

we take x_i^* as the midpoint $\frac{1}{2}(x_{i-1} + x_i)$. The representative rectangle of height $f(x_i^*)$ and base Δx_i (see Figure 6.3.2) generates a cylindrical shell of height $h = f(x_i^*)$, inner radius $r = x_{i-1}$, and outer radius $R = x_i$. We can calculate the volume of this shell by (6.3.1). Since

$$h = f(x_i^*) \quad \text{and} \quad R + r = x_i + x_{i-1} = 2x_i^* \quad \text{and} \quad R - r = \Delta x_i,$$

the volume of this shell is

$$\pi h(R + r)(R - r) = 2\pi x_i^* f(x_i^*)\Delta x_i.$$

The volume of the entire solid can be approximated by adding up the volumes of these shells:

$$V \cong 2\pi x_1^* f(x_1^*)\,\Delta x_1 + 2\pi x_2^* f(x_2^*)\,\Delta x_2 + \cdots + 2\pi x_n^* f(x_n^*)\,\Delta x_n.$$

The sum on the right is a Riemann sum which, as $\|P\| \to 0$, converge to

$$\int_a^b 2\pi x f(x)\, dx. \quad \square$$

For a simple interpretation of Formula 6.3.2 we refer to Figure 6.3.3.

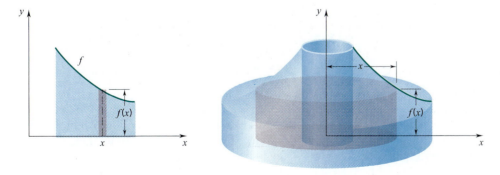

Figure 6.3.3

As the region below the graph of f is revolved about the y-axis, the line segment x units from the y-axis generates a cylinder of radius x, height $f(x)$, and lateral area

$2\pi x f(x)$. The shell method formula expresses the volume of a solid of revolution as the average lateral area of these cylinders times the length of the base interval:

(6.3.3)
$$V = \text{(average lateral area of the cylinders)} \cdot (b - a).$$

Example 1 The region bounded by $f(x) = 4x - x^2$ and the x-axis between $x = 1$ and $x = 4$ is rotated about the y-axis. Find the volume of the solid that is generated.

SOLUTION See Figure 6.3.4. The line segment x units from the y-axis, $1 \le x \le 4$, generates a cylinder of radius x, height $f(x)$, and lateral area $2\pi x f(x)$. Thus, by (6.3.2)

$$V = \int_1^4 2\pi x(4x - x^2)\, dx = 2\pi \int_1^4 (4x^2 - x^3)\, dx = 2\pi \left[\frac{4x^3}{3} - \frac{x^4}{4} \right]_1^4$$

$$= 2\pi \left[\frac{256}{3} - 64 \right] - 2\pi \left[\frac{4}{3} - \frac{1}{4} \right] = \frac{81\pi}{2}. \quad \square$$

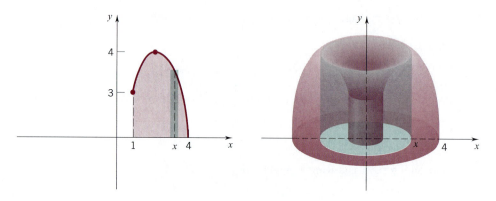

Figure 6.3.4

The shell method formula can be generalized. With Ω as in Figure 6.3.5, the volume generated by revolving Ω about the y-axis is given by the formula

(6.3.4)
$$V = \int_a^b 2\pi x[\, f(x) - g(x)]\, dx. \qquad \text{(shell method about } y\text{-axis)}$$

The integrand $2\pi x[\, f(x) - g(x)]$ is the lateral area of the cylinder in Figure 6.3.5.

We can also apply the shell method to solids generated by revolving a region about the x-axis. See Figure 6.3.6. In this instance the curved boundaries are functions of y rather than x, and we have

(6.3.5)
$$V = \int_c^d 2\pi y[F(y) - G(y)]\, dy. \qquad \text{(shell method about } x\text{-axis)}$$

Figure 6.3.5

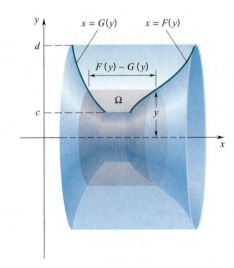

Figure 6.3.6

The integrand $2\pi y[F(y) - G(y)]$ is the lateral area of the cylinder in Figure 6.3.6.

Example 2 Find the volume of the solid generated by revolving the region between

$$y = x^2 \quad \text{and} \quad y = 2x$$

about the y-axis.

SOLUTION We refer to Figure 6.3.7. For each x from 0 to 2 the line segment x units from the y-axis generates a cylinder of radius x, height $(2x - x^2)$, and lateral area $2\pi x(2x - x^2)$. By (6.3.4)

$$V = \int_0^2 2\pi x(2x - x^2)\, dx = 2\pi \int_0^2 (2x^2 - x^3)\, dx = 2\pi \left[\tfrac{2}{3}x^3 - \tfrac{1}{4}x^4 \right]_0^2 = \tfrac{8}{3}\pi. \quad \square$$

Figure 6.3.7

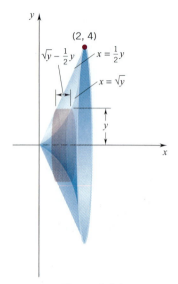

Figure 6.3.8

Example 3 Find the volume of the solid generated by revolving the region be-tween

$$y = x^2 \quad \text{and} \quad y = 2x$$

about the x-axis.

SOLUTION We begin by expressing these boundaries as functions of y. We write $x = \sqrt{y}$ for the right boundary and $x = \frac{1}{2}y$ for the left boundary (see Figure 6.3.8). The shell of radius y has height $(\sqrt{y} - \frac{1}{2}y)$. Thus, by (6.3.5)

$$V = \int_0^4 2\pi y(\sqrt{y} - \tfrac{1}{2}y)\, dy = \pi \int_0^4 (2y^{3/2} - y^2)\, dy$$

$$= \pi \left[\tfrac{4}{5}y^{5/2} - \tfrac{1}{3}y^3 \right]_0^4 = \tfrac{64}{15}\pi. \quad \square$$

Remark In Section 6.2 we calculated the volumes of the same solids (and got the same answers) by the washer method. ❑

Remark Examples 2 and 3 raise an obvious question: Which of the three methods —disc, washer, or shell—should you use to solve a particular problem? As you probably have guessed, the "best" method depends upon the problem. Since a disc is a special case of a washer, you will typically have to decide between the shell method and the washer or disc method. For the solids in Examples 2 and 3, the shell and washer methods were essentially equivalent. On the other hand, if you try to use the washer method to calculate the volume of the solid in Example 1, you will be led to the expression

$$V = \int_0^3 (\pi[g_1(y)]^2 - \pi[1]^2)\, dy + \int_3^4 (\pi[g_1(y)]^2 - \pi[g_2(y)]^2)\, dy, \quad \text{(see Figure 6.3.4)}$$

where $x = g_1(y)$ and $x = g_2(y)$ are the functions obtained by solving $y = 4x - x^2$ for x. While the two functions g_1 and g_2 can be determined and the integrals can be evaluated, it is not trivial. Try it! In general, it might not be possible to solve an equation $y = f(x)$ for x, or an equation $x = g(y)$ for y. The selection of the method to use on a particular problem will depend upon the region that is being revolved, the axis of revolution, and the flexibility you have in expressing the functions involved in terms of either x or y, whichever is required. ❑

Example 4 A round hole of radius r is drilled through the center of a hemisphere of radius a $(r < a)$. Find the volume of the portion of the hemisphere that remains.

SOLUTION A hemisphere of radius a can be generated by revolving the region in the first quadrant bounded by $x^2 + y^2 = a^2$ around the y-axis. See Figure 6.3.9a. The portion of the hemisphere that remains after the hole is drilled is the solid generated by revolving the shaded region around the y-axis (Figure 6.3.9b).

(a) Washer method. Refer to Figure 6.3.10.

$$V = \int_0^{\sqrt{a^2 - r^2}} \left(\pi \left[\sqrt{a^2 - y^2} \right]^2 - \pi r^2 \right) dy = \pi \int_0^{\sqrt{a^2 - r^2}} (a^2 - r^2 - y^2)\, dy$$

$$= \pi \left[(a^2 - r^2)y - \frac{y^3}{3} \right]_0^{\sqrt{a^2 - r^2}} = \tfrac{2}{3}\pi(a^2 - r^2)^{3/2}.$$

(a) (b)

Figure 6.3.9

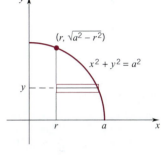

Figure 6.3.10

(b) Shell method. Refer to Figure 6.3.11.

$$V = \int_r^a 2\pi x \sqrt{a^2 - x^2} \, dx.$$

Let $u = a^2 - x^2$. Then $du = -2x \, dx$, and $u = a^2 - r^2$ at $x = r$, 0 at $x = a$.
Thus

$$V = \int_r^a 2\pi x \sqrt{a^2 - x^2} \, dx = -\pi \int_{a^2 - r^2}^0 u^{1/2} \, du = \pi \int_0^{a^2 - r^2} u^{1/2} \, du$$

$$= \pi \left[\tfrac{2}{3} u^{3/2} \right]_0^{a^2 - r^2} = \tfrac{2}{3} \pi (a^2 - r^2)^{3/2}.$$

Note that if we let $r = 0$, then $V = \tfrac{2}{3} \pi a^3$, the volume of the hemisphere. ❑

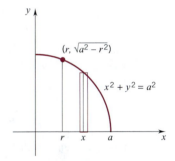

Figure 6.3.11

We conclude this section with an example in which we revolve a region about a
line parallel to the y-axis.

Example 5   The region Ω between $y = \sqrt{x}$ and $y = x^2$, $0 \leq x \leq 1$, is revolved around
the line $x = -2$ (see Figure 6.3.12). Find the volume of the solid which is generated.

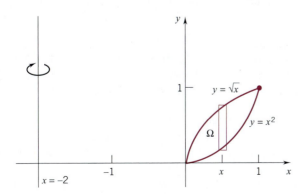

Figure 6.3.12

SOLUTION Using the shell method, for each x in $[0, 1]$, the line segment at x generates a cylinder of radius $x + 2$, height $\sqrt{x} - x^2$, and lateral area $2\pi(x + 2)(\sqrt{x} - x^2)$. Thus, the volume of the solid of revolution is

$$V = \int_0^1 2\pi(x + 2)(\sqrt{x} - x^2)\, dx$$

$$= 2\pi \int_0^1 (x^{3/2} + 2x^{1/2} - x^3 - 2x^2)\, dx$$

$$= 2\pi \left[\tfrac{2}{5}x^{5/2} + \tfrac{4}{3}x^{3/2} - \tfrac{1}{4}x^4 - \tfrac{2}{3}x^3 \right]_0^1 = \tfrac{49}{30}\pi.$$

See Exercise 48, Section 6.2. ❑

EXERCISES 6.3

In Exercises 1–12, sketch the region Ω bounded by the curves and use the shell method to find the volume of the solid generated by revolving Ω about the y-axis.

1. $y = x$, $y = 0$, $x = 1$.

2. $x + y = 3$, $y = 0$, $x = 0$.

3. $y = \sqrt{x}$, $x = 4$, $y = 0$.

4. $y = x^3$, $x = 2$, $y = 0$.

5. $y = \sqrt{x}$, $y = x^3$.   **6.** $y = x^2$, $y = x^{1/3}$.

7. $y = x$, $y = 2x$, $y = 4$.

8. $y = x$, $y = 1$, $x + y = 6$.

9. $x = y^2$, $x = y + 2$.   **10.** $x = y^2$, $x = 2 - y$.

11. $x = \sqrt{9 - y^2}$, $x = 0$.   **12.** $x = |y|$, $x = 2 - y^2$.

In Exercises 13–24, sketch the region Ω bounded by the curves and use the shell method to find the volume of the solid generated by revolving Ω about the x-axis.

13. $x + 3y = 6$, $y = 0$, $x = 0$.

14. $y = x$, $y = 5$, $x = 0$.   **15.** $y = x^2$, $y = 9$.

16. $y = x^3$, $y = 8$, $x = 0$.

17. $y = \sqrt{x}$, $y = x^3$.   **18.** $y = x^2$, $y = x^{1/3}$.

19. $y = x^2$, $y = x + 2$.   **20.** $y = x^2$, $y = 2 - x$.

21. $y = x$, $y = 2x$, $x = 4$.

22. $y = x$, $x + y = 8$, $x = 1$.

23. $y = \sqrt{1 - x^2}$, $x + y = 1$.

24. $y = x^2$, $y = 2 - |x|$.

The curves $y = x^2$, $y = \sqrt{x}$, $x = 0$, $x = 1$, $y = 0$, and $y = 1$ determine the three regions indicated in the figure. In Exercises 25–30, express the volume of the solid that is generated when the indicated region is revolved around the given line by: (a) a definite integral in which you will be integrating with respect to x; (b) a definite integral in which you will be

integrating with respect to y. Then choose one of the two integrals and calculate the volume of the solid.

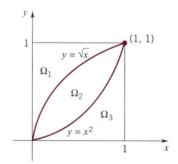

25. Ω_1 is revolved around the y-axis.

26. Ω_1 is revolved around the line $y = 2$.

27. Ω_2 is revolved around the x-axis.

28. Ω_2 is revolved around the line $x = -3$.

29. Ω_3 is revolved around the y-axis.

30. Ω_3 is revolved around the line $y = -1$.

31. Use the shell method to find the volume enclosed by the surface obtained by revolving the ellipse $b^2x^2 + a^2y^2 = a^2b^2$ about the y-axis.

32. Exercise 31 with the ellipse revolved about the x-axis.

33. Find the volume enclosed by the surface generated by revolving the equilateral triangle with vertices $(0, 0)$, $(a, 0)$, $(\tfrac{1}{2}a, \tfrac{1}{2}\sqrt{3}a)$ about the y-axis.

34. A ball of radius r is cut into two pieces by a horizontal plane a units above the center of the ball. Determine the volume of the upper piece by using the shell method.

35. Carry out Exercise 51 of Section 6.2, this time using the shell method.

36. Carry out Exercise 52 of Section 6.2, this time using the shell method.

37. (a) Verify that $F(x) = x \sin x + \cos x$ is an antiderivative of $f(x) = x \cos x$.
   (b) Use the shell method to find the volume of the solid that is generated by revolving the region between $y = \cos x$ and the x-axis, $0 \le x \le \pi/2$, around the y-axis.

38. (a) Sketch the region in the right half-plane that is outside the parabola $y = x^2$ and is between the lines $y = x + 2$ and $y = 2x - 2$.
   (b) The region in (a) is revolved around the y-axis. Use the method that you think is most practical to calculate the volume of the solid that is generated.

In Exercises 39–42, let

$$f(x) = \begin{cases} \sqrt{3}x & 0 \le x < 1 \\ \sqrt{4 - x^2} & 1 \le x \le 2, \end{cases}$$

and let Ω be the region between the graph of f and the x-axis (see the figure).

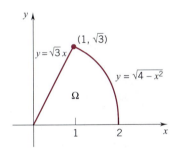

39. Revolve Ω around the y-axis. Express the volume of the solid that is generated as:
   (a) A definite integral in which you will be integrating with respect to x.
   (b) A definite integral in which you will be integrating with respect to y.
   (c) Choose one of the two integrals and determine the volume of the solid.

40. Revolve Ω around the x-axis and repeat Exercise 39.

41. Revolve Ω around the line $x = 2$ and repeat parts (a) and (b) of Exercise 39.

42. Revolve Ω around the line $y = -1$ and repeat parts (a) and (b) of Exercise 39.

43. Let Ω be the circular disc $(x - b)^2 + y^2 \le a^2$, where $b > a > 0$. The doughnut-shaped region that is generated when Ω is revolved around the y-axis is called a *torus*. Express the volume V of the torus as:
   (a) A definite integral in which you will be integrating with respect to x.
   (b) A definite integral in which you will be integrating with respect to y.
   (c) Show that $V = 2\pi^2 a^2 b$ by calculating one of the two definite integrals.

44. The circular disc $x^2 + y^2 \le a^2$, $a > 0$, is revolved around the line $x = a$. Find the volume of the solid that is generated.

■ 6.4 THE CENTROID OF A REGION; PAPPUS'S THEOREM ON VOLUMES

The Centroid of a Region

In Section 5.9 you saw how to calculate the center of mass of a thin rod. Suppose now that we have a thin distribution of matter, a *plate,* laid out in the xy-plane in the shape of some region Ω (Figure 6.4.1). If the mass density of the plate varies from point to point, then the determination of the center of mass of the plate requires the evaluation of a double integral (Chapter 16). If, however, the mass density of the plate is constant throughout Ω, then the center of mass of the plate depends only on the shape of Ω and falls on a point $(\bar{x}, \bar{y})$ that we call the *centroid*. Unless Ω has a very complicated shape, we can calculate its centroid by ordinary one-variable integration.

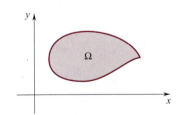

Figure 6.4.1

We will use two guiding principles to find the centroid of a plane region Ω. The first is obvious. The second we take from physics; this result is easily justified by double integration.

Principle 1: Symmetry If the region has an axis of symmetry, then the centroid $(\bar{x}, \bar{y})$ lies somewhere along that axis. In particular, if the region has a center, then that center is the centroid.

Principle 2: Additivity If the region, having area A, consists of a finite number of pieces with areas $A_1, \ldots, A_n$ and centroids $(\bar{x}_1, \bar{y}_1), \ldots, (\bar{x}_n, \bar{y}_n)$, then

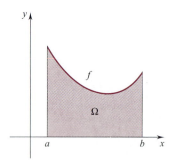

(6.4.1) $\quad \bar{x}A = \bar{x}_1 A_1 + \cdots + \bar{x}_n A_n \quad \text{and} \quad \bar{y}A = \bar{y}_1 A_1 + \cdots + \bar{y}_n A_n.$

We are now ready to bring the techniques of calculus into play. Figure 6.4.2 shows the region Ω under the graph of a continuous function f. Let's denote the area of Ω by A. The centroid $(\bar{x}, \bar{y})$ of Ω can be obtained from the following formulas:

Figure 6.4.2

(6.4.2) $\quad \bar{x}A = \displaystyle\int_a^b xf(x)\,dx, \qquad \bar{y}A = \int_a^b \tfrac{1}{2}[f(x)]^2\,dx.$

To derive these formulas we choose a partition $P = \{x_0, x_1, \ldots, x_n\}$ of $[a, b]$. This breaks up $[a, b]$ into n subintervals $[x_{i-1}, x_i]$. Choosing x_i^* as the midpoint of $[x_{i-1}, x_i]$, we form the midpoint rectangles R_i shown in Figure 6.4.3. The area of R_i is $f(x_i^*)\,\Delta x_i$ and the centroid of R_i is its center $(x_i^*, \tfrac{1}{2}f(x_i^*))$. By (6.4.1), the centroid $(\bar{x}_P, \bar{y}_P)$ of the union of all these rectangles satisfies the following equations:

$$\bar{x}_P A_P = x_1^* f(x_1^*)\,\Delta x_1 + \cdots + x_n^* f(x_n^*)\,\Delta x_n,$$
$$\bar{y}_P A_P = \tfrac{1}{2}[f(x_1^*)]^2\,\Delta x_1 + \cdots + \tfrac{1}{2}[f(x_n^*)]^2\,\Delta x_n.$$

Figure 6.4.3

(Here A_P represents the area of the union of the n rectangles.) As $\|P\| \to 0$, the union of rectangles tends to the shape of Ω and the equations we just derived tend to the formulas given in (6.4.2). ❑

Example 1 Find the centroid of the quarter-disc shown in Figure 6.4.4.

SOLUTION The quarter-disc is symmetric about the line $x = y$. We know therefore that $\bar{x} = \bar{y}$. Here

$$\bar{y}A = \int_0^r \tfrac{1}{2}[f(x)]^2\,dx = \int_0^r \tfrac{1}{2}(r^2 - x^2)\,dx = \tfrac{1}{2}\left[r^2 x - \frac{x^3}{3} \right]_0^r = \tfrac{1}{3}r^3.$$

$$f(x) = \sqrt{r^2 - x^2} \longrightarrow$$

Figure 6.4.4

Since $A = \tfrac{1}{4}\pi r^2$,

$$\bar{y} = \frac{\tfrac{1}{3}r^3}{\tfrac{1}{4}\pi r^2} = \frac{4r}{3\pi}.$$

The centroid of the quarter-disc is the point

$$\left(\frac{4r}{3\pi}, \frac{4r}{3\pi}\right).$$

NOTE: It is almost as easy to calculate $\bar{x}A$:

$$\bar{x}A = \int_0^r xf(x)\,dx = \int_0^r x\sqrt{r^2 - x^2}\,dx$$

$$= -\tfrac{1}{2}\int_{r^2}^0 u^{1/2}\,du \qquad\qquad [u = (r^2 - x^2),\, du = -2x\,dx]$$

$$= -\tfrac{1}{2}\left[\tfrac{2}{3}u^{3/2}\right]_{r^2}^0 = \tfrac{1}{3}r^3. \quad \square$$

Example 2 Find the centroid of the right triangle shown in Figure 6.4.5

SOLUTION There is no symmetry that we can use here. The hypotenuse lies on the line

$$y = -\frac{h}{b}x + h.$$

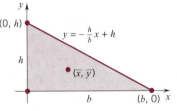

Figure 6.4.5

Hence

$$\bar{x}A = \int_0^b xf(x)\,dx = \int_0^b \left(-\frac{h}{b}x^2 + hx\right)dx = \tfrac{1}{6}b^2h$$

and

$$\bar{y}A = \int_0^b \tfrac{1}{2}[f(x)]^2\,dx = \tfrac{1}{2}\int_0^b \left(\frac{h^2}{b^2}x^2 - \frac{2h^2}{b}x + h^2\right)dx = \tfrac{1}{6}bh^2.$$

Since $A = \tfrac{1}{2}bh$, we have

$$\bar{x} = \frac{\tfrac{1}{6}b^2h}{\tfrac{1}{2}bh} = \tfrac{1}{3}b \qquad \text{and} \qquad \bar{y} = \frac{\tfrac{1}{6}bh^2}{\tfrac{1}{2}bh} = \tfrac{1}{3}h.$$

The centroid is the point $(\tfrac{1}{3}b, \tfrac{1}{3}h)$. $\quad \square$

Figure 6.4.6 shows the region Ω between the graphs of two continuous functions f and g. In this case, if Ω has area A and centroid $(\bar{x}, \bar{y})$, then

(6.4.3) $\quad \boxed{\bar{x}A = \int_a^b x[f(x) - g(x)]\,dx, \qquad \bar{y}A = \int_a^b \tfrac{1}{2}([f(x)]^2 - [g(x)]^2)\,dx.}$

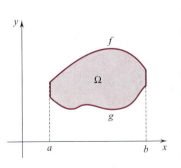

Figure 6.4.6

PROOF Let A_f be the area below the graph of f and let A_g be the area below the graph of g. Then in obvious notation

$$\bar{x}A + \bar{x}_g A_g = \bar{x}_f A_f \qquad \text{and} \qquad \bar{y}A + \bar{y}_g A_g = \bar{y}_f A_f. \qquad (6.4.1)$$

Therefore

$$\bar{x}A = \bar{x}_f A_f - \bar{x}_g A_g = \int_a^b xf(x)\,dx - \int_a^b xg(x)\,dx = \int_a^b x[f(x) - g(x)]\,dx$$

and

$$\bar{y}A = \bar{y}_f A_f - \bar{y}_g A_g = \int_a^b \tfrac{1}{2}[f(x)]^2\,dx - \int_a^b \tfrac{1}{2}[g(x)]^2\,dx$$

$$= \int_a^b \tfrac{1}{2}([f(x)]^2 - [g(x)]^2)\,dx. \quad \square$$

Example 3   Find the centroid of the region shown in Figure 6.4.7.

SOLUTION   Here there is no symmetry we can appeal to. We must carry out the calculations.

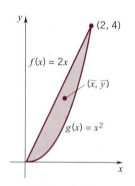

Figure 6.4.7

$$A = \int_0^2 [f(x) - g(x)]\,dx = \int_0^2 (2x - x^2)\,dx = \left[x^2 - \tfrac{1}{3}x^3\right]_0^2 = \tfrac{4}{3},$$

$$\bar{x}A = \int_0^2 x[f(x) - g(x)]\,dx = \int_0^2 (2x^2 - x^3)\,dx = \left[\tfrac{2}{3}x^3 - \tfrac{1}{4}x^4\right]_0^2 = \tfrac{4}{3},$$

$$\bar{y}A = \int_0^2 \tfrac{1}{2}([f(x)]^2 - [g(x)]^2)\,dx = \tfrac{1}{2}\int_0^2 (4x^2 - x^4)\,dx = \tfrac{1}{2}\left[\tfrac{4}{3}x^3 - \tfrac{1}{5}x^5\right]_0^2 = \tfrac{32}{15}.$$

Therefore $\bar{x} = \tfrac{4/3}{4/3} = 1$ and $\bar{y} = \tfrac{32/15}{4/3} = \tfrac{8}{5}$. $\square$

Pappus's Theorem on Volumes

All the formulas that we have derived for volumes of solids of revolution are simple corollaries to an observation made by a brilliant ancient Greek, Pappus of Alexandria (circa 300 A.D.).

THEOREM 6.4.4   PAPPUS'S THEOREM ON VOLUMES†

A plane region is revolved about an axis that lies in its plane. If the region does not cross the axis, then the volume of the resulting solid of revolution is the area of the region multiplied by the circumference of the circle described by the centroid of the region:

$$V = 2\pi \bar{R} A$$

where A is the area of the region and $\bar{R}$ is the distance from the axis to the centroid of the region. See Figure 6.4.8.

---

† This theorem is found in Book VII of Pappus's "Mathematical Collection," largely a survey of ancient geometry to which Pappus made many original contributions (among them this theorem). Much of what we know today of Greek geometry we owe to Pappus.

Figure 6.4.8

Basically we have derived only two formulas for the volumes of solids of revolution:

1. *The Washer Method Formula.* If the region Ω of Figure 6.4.6 is revolved about the x-axis, the resulting solid has volume

$$V_x = \int_a^b \pi([f(x)]^2 - [g(x)]^2) \, dx.$$

2. *The Shell Method Formula.* If the region Ω of Figure 6.4.6 is revolved about the y-axis, the resulting solid has volume

$$V_y = \int_a^b 2\pi x[f(x) - g(x)] \, dx.$$

Note that

$$V_x = \int_a^b \pi([f(x)]^2 - [g(x)]^2) \, dx = 2\pi \int_a^b \tfrac{1}{2}([f(x)]^2 - [g(x)]^2) \, dx = 2\pi \bar{y} A = 2\pi \bar{R} A$$

and

$$V_y = \int_a^b 2\pi x[f(x) - g(x)] \, dx = 2\pi \bar{x} A = 2\pi \bar{R} A,$$

exactly as predicted by Pappus.

Remark In stating Pappus's theorem we assumed a complete revolution. If Ω is only partially revolved about the given axis, then the volume of the resulting solid is simply the area of Ω multiplied by the length of the circular arc described by the centroid of Ω. ❏

Applications of Pappus's Theorem

Example 4 Earlier we saw that the region shown in Figure 6.4.7 has area $\tfrac{4}{3}$ and centroid $(1, \tfrac{8}{5})$. Find the volumes of the solids formed by revolving this region about: (a) the y-axis, and (b) the line $y = 5$.

SOLUTION (a) We have already calculated this volume two ways, by the washer method and by the shell method. (See Example 8, Section 6.2, and Example 2,

Section 6.3.) The result was $V = \frac{8}{3}\pi$. Now we will use Pappus's theorem. We have $\overline{R} = 1$ and $A = \frac{4}{3}$. Thus

$$V = 2\pi(1)(\tfrac{4}{3}) = \tfrac{8}{3}\pi.$$

(b) Here $\overline{R} = 5 - \frac{8}{5} = \frac{17}{5}$ and $A = \frac{4}{3}$. Hence

$$V = 2\pi(\tfrac{17}{5})(\tfrac{4}{3}) = \tfrac{136}{15}\pi \quad \square$$

Example 5 Find the volume of the torus generated by revolving the circular disc

$$(x - h)^2 + (y - k)^2 \leq c^2, \qquad h, k \geq c > 0 \qquad \text{(Figure 6.4.9)}$$

about (a) the x-axis, (b) the y-axis.

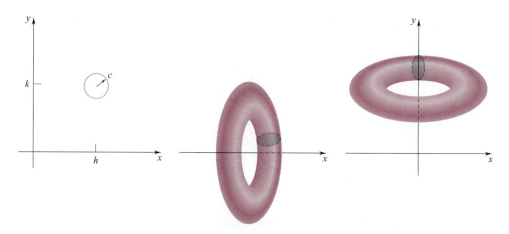

Figure 6.4.9

SOLUTION The centroid of the disc is the center (h, k). This lies k units from the x-axis and h units from the y-axis. The area of the disc is πc^2. Therefore
(a) $V_x = 2\pi(k)(\pi c^2) = 2\pi^2 k c^2,$ (b) $V_y = 2\pi(h)(\pi c^2) = 2\pi^2 h c^2.$ ❑

Example 6 Find the centroid of the half-disc

$$x^2 + y^2 \leq r^2, \qquad y \geq 0$$

by appealing to Pappus's theorem.

SOLUTION Since the half-disc is symmetric about the y-axis, we know that $\overline{x} = 0$. All we need to find is $\overline{y}$.

If we revolve the half-disc about the x-axis, we obtain a solid ball of volume $\frac{4}{3}\pi r^3$. The area of the half-disc is $\frac{1}{2}\pi r^2$. By Pappus's theorem

$$\tfrac{4}{3}\pi r^3 = 2\pi\overline{y}(\tfrac{1}{2}\pi r^2).$$

Simple division gives $\overline{y} = 4r/3\pi$. ❑

Remark Centroids of solids of revolution are discussed in the exercises. ❑

EXERCISES 6.4

In Exercises 1–14, sketch the region bounded by the curves. Determine the centroid of the region and the volume generated by revolving the region about each of the coordinate axes.

1. $y = \sqrt{x},\ y = 0,\ x = 4$.

2. $y = x^3,\ y = 0,\ x = 2$.

3. $y = x^2,\ y = x^{1/3}$.

4. $y = x^3,\ y = \sqrt{x}$.

5. $y = 2x,\ y = 2,\ x = 3$.

6. $y = 3x,\ y = 6,\ x = 1$.

7. $y = x^2 + 2,\ y = 6,\ x = 0$.

8. $y = x^2 + 1,\ y = 1,\ x = 3$.

9. $\sqrt{x} + \sqrt{y} = 1,\ x + y = 1$.

10. $y = \sqrt{1 - x^2},\ x + y = 1$.

11. $y = x^2,\ y = 0,\ x = 1,\ x = 2$.

12. $y = x^{1/3},\ y = 1,\ x = 8$.

13. $y = x,\ x + y = 6,\ y = 1$.

14. $y = x,\ y = 2x,\ x = 3$.

In Exercises 15–24, find the centroid of the bounded region determined by the following curves.

15. $y = 6x - x^2,\ y = x$.

16. $y = 4x - x^2,\ y = 2x - 3$.

17. $x^2 = 4y,\ x - 2y + 4 = 0$.

18. $y = x^2,\ 2x - y + 3 = 0$.

19. $y^3 = x^2,\ 2y = x$.

20. $y^2 = 2x,\ y = x - x^2$.

21. $y = x^2 - 2x,\ y = 6x - x^2$.

22. $y = 6x - x^2,\ x + y = 6$.

23. $x + 1 = 0,\ x + y^2 = 0$.

24. $\sqrt{x} + \sqrt{y} = \sqrt{a},\ x = 0,\ y = 0$.

25. Let Ω be the annular region (ring) formed by the circles

$$x^2 + y^2 = \tfrac{1}{4} \quad \text{and} \quad x^2 + y^2 = 4.$$

(a) Locate the centroid of Ω. (b) Locate the centroid of the first-quadrant part of Ω. (c) Locate the centroid of the upper half of Ω.

26. The ellipse $b^2x^2 + a^2y^2 = a^2b^2$ encloses a region of area πab. Locate the centroid of the upper half of the region.

27. The rectangle in the figure is revolved about the line marked l. Find the volume of the resulting solid.

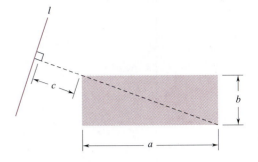

28. In Example 2 of this section you saw that the centroid of the triangle in Figure 6.4.5 is the point $(\tfrac{1}{3}b, \tfrac{1}{3}h)$.
(a) Verify that the line segments that join the centroid to the vertices divide the triangle into three triangles of equal area.
(b) Find the distance d from the centroid of the triangle to the hypotenuse.
(c) Find the volume generated by revolving the triangle about the hypotenuse.

29. The triangular region in the figure is the union of two right triangles $\Omega_1,\ \Omega_2$. Find the centroid: (a) of Ω_1; (b) of Ω_2; (c) of the entire region.

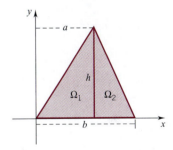

30. Find the volume of the solid generated by revolving the entire triangular region in the figure in Exercise 29 about: (a) the x-axis; (b) the y-axis.

31. (a) Find the volume of the ice-cream cone shown in the figure (a right circular cone topped by a solid hemisphere).

(b) Find $\bar{x}$ for the region Ω in the figure.

32. The region Ω in the figure consists of a square S of side $2r$ and four semidiscs of radius r. Find the centroid of each of the following.
   (a) Ω.   (b) Ω_1.   (c) $S \cup \Omega_1$.   (d) $S \cup \Omega_3$.
   (e) $S \cup \Omega_1 \cup \Omega_3$.   (f) $S \cup \Omega_1 \cup \Omega_2$.
   (g) $S \cup \Omega_1 \cup \Omega_2 \cup \Omega_3$.

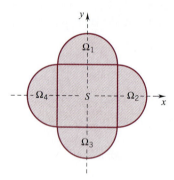

33. (a) In Section 5.9 we stated that the mass of a rod that extends from $x = a$ to $x = b$ is given by the formula

$$M = \int_a^b \lambda(x)\, dx$$

   where λ is the mass density function. Derive this formula by taking an arbitrary partition $P = \{x_0, x_1, \ldots, x_n\}$ of $[a, b]$, estimating the mass contributed by $[x_{i-1}, x_i]$, adding up these contributions, and then taking the limit as $\|P\| \to 0$.

   (b) In Section 5.9 we also gave the center of mass formula

$$x_M M = \int_a^b x\, \lambda(x)\, dx.$$

   This formula can be derived from the physical premise that, if the rod is broken up into a finite number of

pieces of masses $M_1, \ldots, M_n$ and centers of mass $x_{M_1}, \ldots, x_{M_n}$, then

$$x_M M = x_{M_1} M_1 + \cdots + x_{M_n} M_n.$$

Carry out the derivation.

34. (*The Centroid of a Solid of Revolution.*) If a solid is *homogeneous* (constant mass density), then the center of mass of the solid depends only on the shape of the solid and is called the *centroid*. The determination of the centroid of a solid of arbitrary shape requires triple integration (Chapter 16). However, if the solid is a solid of revolution, then the centroid can be found by ordinary one-variable integration.

   (a) Let Ω be the region of Figure 6.4.2 and let T be the solid generated by revolving Ω about the x-axis. By symmetry the centroid of T lies on the x-axis. Thus the centroid of T is completely determined by its x-coordinate $\bar{x}$. Show that

(**6.4.5**) $$\bar{x} V_x = \int_a^b \pi x [f(x)]^2\, dx$$

   basing your argument on the following additivity principle: If a solid of volume V consists of a finite number of pieces with volumes $V_1, \ldots, V_n$ and the centroids of the pieces have x-coordinates $\bar{x}_1, \ldots, \bar{x}_n$, then

$$\bar{x} V = \bar{x}_1 V_1 + \cdots + \bar{x}_n V_n.$$

   (b) Let Ω be the region of Figure 6.4.2 and let T be the solid generated by revolving Ω about the y-axis. By symmetry the centroid of T lies on the y-axis. Thus the centroid of T is completely determined by its y-coordinate $\bar{y}$. Show that

(**6.4.6**) $$\bar{y} V_y = \int_a^b \pi x [f(x)]^2\, dx$$

   basing your argument on the following additivity principle: If a solid of volume V consists of a finite number of pieces with volumes $V_1, \ldots, V_n$ and the centroids of the pieces have y-coordinates $\bar{y}_1, \ldots, \bar{y}_n$, then

$$\bar{y} V = \bar{y}_1 V_1 + \cdots + \bar{y}_n V_n.$$

In Exercises 35–40, locate the centroid of each of the given configurations.

35. A solid cone of base radius r and height h.

36. A solid hemisphere of radius r.

37. The solid generated by revolving the region below the graph of $f(x) = \sqrt{x}$, $x \in [0, 1]$ about: (a) the x-axis; (b) the y-axis.

38. The solid generated by revolving the region below the graph of $f(x) = 4 - x^2$, $x \in [0, 2]$ about: (a) the x-axis; (b) the y-axis.

39. The solid generated by revolving about the x-axis the first-quadrant part of the region enclosed by the ellipse $b^2x^2 + a^2y^2 = a^2b^2$.

40. A segment of height h cut from a ball of radius r. Take $h < r$.

■ 6.5 THE NOTION OF WORK

We begin with a constant force F that acts along some line that we call the x-axis. By convention F is positive if it acts in the direction of increasing x and negative if it acts in the direction of decreasing x (Figure 6.5.1).

Figure 6.5.1

Suppose now that an object moves along the x-axis from $x = a$ to $x = b$ subject to this constant force F. The *work* done by F during the displacement is by definition the *force times the displacement*:

(6.5.1)
$$W = F \times (b - a).$$

It is not hard to see that, if F acts in the direction of the motion, then $W > 0$, but, if F acts against the motion, then $W < 0$. Thus, for example, if an object slides off a table and falls to the floor, then the work done by gravity is positive (after all, earth's gravity points down). But if an object is lifted from the floor and raised to tabletop level then the work done by gravity is by definition negative. However, the work done by the hand that lifts the object is positive.

To repeat, if an object moves from $x = a$ to $x = b$ subject to a constant force F, then the work done by F is the constant value of F times $b - a$. What is the work done by F if F does not remain constant but instead varies continuously as a function of x? As you would expect, we then define the work done by F as the *average value* of F times $b - a$:

(6.5.2)
$$W = \int_a^b F(x)\, dx.$$

(Figure 6.5.2)

constant force
$$W = F \times (b - a)$$

variable force
$$W = \int_a^b F(x)\, dx$$

Figure 6.5.2

Hooke's Law

You can sense a variable force in the action of a steel spring. Stretch a spring within its elastic limit and you feel a pull in the opposite direction. The greater the stretching, the harder the pull of the spring. Compress a spring within its elastic limit and you feel a push against you. The greater the compression, the harder the push. According to Hooke's law (Robert Hooke, 1635–1703) the force exerted by the spring can be written

$$F(x) = -kx$$

where k is a positive number, called *the spring constant,* and x is the displacement from the equilibrium position. The minus sign indicates that the spring force always acts in the direction opposite to the direction in which the spring has been deformed (the force always acts so as to restore the spring to its equilibrium state).

Remark Hooke's law is only an approximation, but it is a good approximation for small displacements. In the problems that follow we assume that the restoring force of the spring is given by Hooke's law. ❏

Example 1 A spring of natural length L compressed to length $\frac{7}{8}L$ exerts a force F_0. Find the work done by the spring in restoring itself to natural length.

SOLUTION Place the spring on the x-axis so that the equilibrium point falls at the origin. View compression as a move to the left. See Figure 6.5.3.

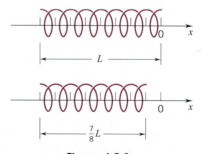

Figure 6.5.3

Compressed $\frac{1}{8}L$ units to the left, the spring exerts a force F_0. Thus by Hooke's law

$$F_0 = F(-\tfrac{1}{8}L) = -k(-\tfrac{1}{8}L) = \tfrac{1}{8}kL.$$

This tells us that $k = 8F_0/L$. Therefore the force law for this spring reads

$$F(x) = -\left(\frac{8F_0}{L}\right)x.$$

To find the work done by this spring in restoring itself to equilibrium, we integrate $F(x)$ from $x = -\frac{1}{8}L$ to $x = 0$:

$$W = \int_{-L/8}^{0} F(x)\, dx = \int_{-L/8}^{0} -\left(\frac{8F_0}{L}\right) x\, dx = -\frac{8F_0}{L}\left[\frac{x^2}{2}\right]_{-L/8}^{0} = \frac{LF_0}{16}. \quad \square$$

Example 2 For the spring in Example 1, what work must we do to stretch the spring to length $\frac{11}{10}L$?

SOLUTION Refer to Figure 6.5.4. To stretch the spring, we must counteract the force of the spring. The force exerted by the spring when stretched x units is

$$F(x) = -\left(\frac{8F_0}{L}\right) x. \qquad\qquad \text{(Example 1)}$$

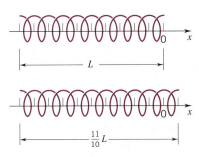

Figure 6.5.4

To counter this force we must apply the opposite force

$$-F(x) = \left(\frac{8F_0}{L}\right) x.$$

The work we must do to stretch the spring to length $\frac{11}{10}L$ can be found by integrating $-F(x)$ from $x = 0$ to $x = \frac{1}{10}L$:

$$W = \int_{0}^{L/10} -F(x)\, dx = \int_{0}^{L/10} \left(\frac{8F_0}{L}\right) x\, dx = \frac{8F_0}{L}\left[\frac{x^2}{2}\right]_{0}^{L/10} = \frac{LF_0}{25}. \quad \square$$

Remark about Units. One unit of work is the work done by a unit force in moving an object a unit distance. If force is measured in pounds and distance in feet, then the units of work are *foot-pounds*. In the metric system, the standard units of force are *newtons* (one newton is the force required to give a mass of 1 kilogram an acceleration of 1 meter per second per second) and *dynes* (one dyne is the force required to give a mass of 1 gram an acceleration of 1 centimeter per second per second). The units of work in the metric system are *newton-meters*, also called *joules*, and *dyne-centimeters*, also called *ergs*. ❑

Example 3 Stretched $\frac{1}{3}$ meter beyond its natural length, a certain spring exerts a restoring force of 10 newtons. What work must we do to stretch the spring another $\frac{1}{3}$ meter?

SOLUTION Place the spring on the x-axis so that the equilibrium point falls at the origin. View stretching as moving to the right and assume Hooke's law: $F(x) = -kx$.

When the spring is stretched $\frac{1}{3}$ meter, it exerts a force of -10 newtons (10 newtons to the left). Therefore, $-10 = -k(\frac{1}{3})$ and $k = 30$ (that is, 30 newtons per meter).

To find the work we must do to stretch the spring another $\frac{1}{3}$ meter, we integrate the opposite force $-F(x) = 30x$ from $x = \frac{1}{3}$ to $x = \frac{2}{3}$:

$$W = \int_{1/3}^{2/3} 30x\,dx = 30\left[\frac{1}{2}x^2\right]_{1/3}^{2/3} = 5 \text{ joules.} \quad \square$$

Pumping Out a Tank

To lift an object we must counteract the force of gravity. Consequently, the work done in lifting an object is given by the formula

Work = (weight of the object) × (distance lifted).

If we lift a leaking sand bag or pump out a water tank from above, the calculation of work is more complicated. In the first instance the weight varies during the motion. (There is less sand in the bag as we keep lifting.) In the second instance the distance varies. (Water at the top of the tank does not have to be pumped as far as water at the bottom.) The sand bag problem is left to the exercises. Here we take up the second problem.

Figure 6.5.5 depicts a storage tank filled to within a feet of the top with some liquid. Assume that the liquid is homogeneous and weighs σ† pounds per cubic foot. Suppose now that this storage tank is pumped out from above until the level of the liquid drops to b feet below the top of the tank. How much work has been done?

Figure 6.5.5

We can answer this question by the methods of integral calculus. For each $x \in [a, b]$, we let

$A(x)$ = cross-sectional area x feet below the top of the tank,

$s(x)$ = distance that the x-level must be lifted.

We let $P = \{x_0, x_1, \ldots, x_n\}$ be an arbitrary partition of $[a, b]$ and focus our attention on the ith subinterval $[x_{i-1}, x_i]$ (Figure 6.5.6). Taking x_i^* as an arbitrary point in the ith subinterval, we have

---

† The symbol σ is the lower case Greek letter "sigma."

$A(x_i^*) \, \Delta x_i$ = approximate volume of the ith layer of liquid,

$\sigma A(x_i^*) \, \Delta x_i$ = approximate weight of this volume,

$s(x_i^*)$ = approximate distance this weight is to be moved.

Figure 6.5.6

Therefore

$\sigma s(x_i^*)A(x_i^*) \, \Delta x_i$ = approximate work (weight × distance) required to pump this layer of liquid to the top of the tank.

The work required to pump out all the liquid can be approximated by adding up all these last terms:

$$W \cong \sigma s(x_1^*)A(x_1^*) \, \Delta x_1 + \sigma s(x_2^*)A(x_2^*) \, \Delta x_2 + \cdots + \sigma s(x_n^*)A(x_n^*) \, \Delta x_n.$$

The sum on the right is a Riemann sum which, as $\|P\| \to 0$, converge to give

(6.5.3)

$$W = \int_a^b \sigma s(x)A(x) \, dx. \qquad \square$$

We use this formula in the next problem.

Example 4 A hemispherical water tank of radius 10 feet is being pumped out (Figure 6.5.7). Find the work done in lowering the water level from 2 feet below the top of the tank to 4 feet below the top of the tank given that the pump is placed (a) at the top of the tank, (b) 3 feet above the top of the tank.

SOLUTION Take 62.5 pounds per cubic foot as the weight of water. It is not hard to see that the cross section x feet from the top of the tank is a disc of radius $\sqrt{100 - x^2}$. Its area is therefore

$$A(x) = \pi(100 - x^2).$$

Figure 6.5.7

For part (a) we have $s(x) = x$, so that

$$W = \int_2^4 62.5\pi x(100 - x^2) \, dx = 33{,}750\pi \cong 106{,}029 \text{ foot-pounds.}$$

For part (b) we have $s(x) = x + 3$, so that

$$W = \int_2^4 62.5\pi(x + 3)(100 - x^2)\, dx = 67{,}750\pi \cong 212{,}843 \text{ foot-pounds.} \quad \square$$

Example 5   A 40-foot chain which weighs 4 pounds per foot hangs from a beam which projects from the top of a 50-foot building (Figure 6.5.8). How much work is done in pulling the chain to the top of the building?

SOLUTION   A partition $P = \{x_0, x_1, x_2, \dots, x_n\}$ of the interval $[0, 40]$ will partition the chain into n pieces of length Δx_i, $i = 1, 2, \dots, n$. For each integer i, $1 \le i \le n$, let x_i^* be an arbitrary point in the ith subinterval (Figure 6.5.9).

Figure 6.5.8   Figure 6.5.9

Now, the ith piece of the chain weighs $4\,\Delta x_i$ and is approximately x_i^* feet from the top of the building. Thus, the work done in pulling this piece to the top is $x_i^* (4\,\Delta x_i) = 4\,x_i^*\,\Delta x_i$ foot-pounds. Adding the work done to pull each piece to the top, we have

$$4x_1^*\,\Delta x_1 + 4x_2^*\,\Delta x_2 + \cdots + 4x_n^*\,\Delta x_n \text{ foot-pounds.} \quad \square$$

This is a Riemann sum which, as $\|P\| \to 0$, converges to the definite integral $\int_0^{40} 4x\, dx$. Thus, the work done in pulling the chain to the top of the building is

$$\int_0^{40} 4x\, dx = \left[2x^2 \right]_0^{40} = 3200 \text{ foot-pounds.}$$

EXERCISES 6.5

In Exercises 1 and 2, an object is moving along the x-axis under the action of a force of $F(x)$ pounds when it is x feet from the origin. Find the work done by F in moving the object from $x = a$ to $x = b$.

1. $F(x) = x(x^2 + 1)^2$; $\quad a = 1, b = 4$.

2. $F(x) = 2x\sqrt{x + 1}$; $\quad a = 3, b = 8$.

In Exercises 3 and 4, an object is moving along the x-axis under the action of a force of $F(x)$ newtons when it is x meters from the origin. Find the work done by F in moving the object from $x = a$ to $x = b$.

3. $F(x) = x + \sin 2x$; $\quad a = \dfrac{\pi}{6}, b = \pi$.

4. $F(x) = \dfrac{\cos 2x}{\sqrt{2 + \sin 2x}}; \quad a = 0, b = \dfrac{\pi}{2}.$

5. A 600-pound force will compress a 10-inch automobile coil spring 1 inch. How much work must be done to compress the spring to 5 inches?

6. Five foot-pounds of work is needed to stretch a certain spring from 1 foot to 3 feet beyond its natural length. How far beyond its natural length will a 6-pound weight stretch the spring?

7. Stretched 4 feet beyond its natural length, a certain spring exerts a restoring force of 200 pounds. What work must we do to stretch the spring: (a) 1 foot beyond its natural length? (b) $1\frac{1}{2}$ feet beyond its natural length?

8. A certain spring has natural length L. Given that W is the work done in stretching the spring from L feet to $L + a$ feet, find the work done in stretching the spring: (a) from L feet to $L + 2a$ feet; (b) from L feet to $L + na$ feet; (c) from $L + a$ feet to $L + 2a$ feet; (d) from $L + a$ feet to $L + na$ feet.

9. Find the natural length of a heavy metal spring, given that the work done in stretching it from a length of 2 feet to a length of 2.1 feet is one-half of the work done in stretching it from a length of 2.1 feet to a length of 2.2 feet.

10. A vertical cylindrical tank of radius 2 feet and height 6 feet is full of water. Find the work done in pumping out the water: (a) to an outlet at the top of the tank; (b) to a level 5 feet above the top of the tank. (Assume that the water weighs 62.5 pounds per cubic foot.)

11. A horizontal cylindrical tank of radius 3 feet and length 8 feet is half full of oil weighing 60 pounds per cubic foot.
   (a) Show that the work done in pumping out the oil to the top of the tank is given by the integral

   $$960 \int_0^3 (x + 3) \sqrt{9 - x^2}\, dx.$$

   Evaluate this integral by evaluating the integrals

   $$\int_0^3 x\sqrt{9 - x^2}\, dx \quad \text{and} \quad \int_0^3 \sqrt{9 - x^2}\, dx$$

   separately. HINT: Verify that the second integral represents one-quarter of the area of a circle of radius 3.
   (b) What is the work done in pumping out the oil to a level 4 feet above the top of the tank?

12. What is the work done by gravity if the tank of Exercise 11 is completely drained through an opening at the bottom?

13. A conical container (vertex down) of radius r feet and height h feet is full of a liquid weighing σ pounds per cubic foot. Find the work done in pumping out the top $\frac{1}{2}h$ feet of liquid: (a) to the top of the tank; (b) to a level k feet above the top of the tank.

14. What is the work done by gravity if the tank of Exercise 13 is completely drained through an opening at the bottom?

15. The force of gravity exerted by the earth on a mass m at a distance r from the center of the earth is given by Newton's formula

   $$F = -G\frac{mM}{r^2},$$

   where M is the mass of the earth and G is the gravitational constant. Find the work done by gravity in pulling a mass m from $r = r_1$ to $r = r_2$.

16. A box that weights w pounds is dropped to the floor from a height of d feet. (a) What is the work done by gravity? (b) Show that the work is the same if the box slides to the floor along a smooth inclined plane.

17. A chain that weighs 15 pounds per foot is hanging from the top of an 80-foot building to the ground. How much work is done in pulling the chain to the top of the building?

18. An 8-pound monkey is attached to the end of a 20-foot rope that is hanging from a tree. The monkey climbs to the top of the rope. How much work has it done if the rope weighs 0.25 pound per foot?

19. A bucket of sand weighing 200 pounds is hoisted at a constant rate by a chain from the ground to the top of a building 100 feet high.
   (a) How much work is required to lift the bucket if the weight of the chain is negligible?
   (b) How much work is required to lift the bucket if the chain weighs 2 pounds per foot?

20. Suppose that the bucket in Exercise 19 has a hole in the bottom and the sand leaks out at a constant rate so that only 150 pounds of sand are left when the bucket reaches the top.
   (a) How much work is required to lift the bucket if the weight of the chain is negligible?
   (b) How much work is required to lift the bucket if the chain weighs 2 pounds per foot?

21. A 100-pound bag of sand is lifted for 2 seconds at the rate of 4 feet per second. Find the work done in lifting the bag if the sand leaks out at the rate of $1\frac{1}{2}$ pounds per second.

22. A water container initially weighing w pounds is hoisted by a crane at the rate of n feet per second. What is the work done if the tank is raised m feet and the water leaks out constantly at the rate of p gallons per second? (Assume that the water weighs 8.3 pounds per gallon.)

23. A rope of length l feet that weights σ pounds per foot is lying on the ground. What is the work done in lifting the rope so that it hangs from a beam: (a) l feet high; (b) $2l$ feet high?

24. A load of weight w is lifted from the bottom of a shaft h feet deep. Find the work done given that the rope used to hoist the load weighs σ pounds per foot.

25. An 800-pound steel beam hangs from a 50-foot cable which weighs 6 pounds per foot. Find the work done in winding 20 feet of the cable about a steel drum.

26. Two electrons repel each other with a force that is inversely proportional to the distance between them. Suppose one electron is located at a point A and the other is at a point B which is 8 centimeters from A. How much work is done against the repelling force if the electron at B is moved along the line connecting A and B to a point which is 4 centimeters from A? Express your answer in terms of the constant of proportionality.

27. An object of mass m is moving along the x-axis. At $x = a$, it has velocity v_a and at $x = b$ it has velocity v_b. Use Newton's second law of motion, $F = ma = m(dv/dt)$ to show that

$$W = \int_a^b F(x)\,dx = \tfrac{1}{2}mv_b^2 - \tfrac{1}{2}mv_a^2.$$

HINT: $\dfrac{dv}{dt} = \dfrac{dv}{dx}\dfrac{dx}{dt} = v\dfrac{dv}{dx}.$

Power. Power measures the rate at which work is done. Suppose an object starts at $x = a$ and moves along the x-axis under the action of a force F. Then the work done by F in moving the object to the position x is

$$W(x) = \int_a^x F(u)\,du.$$

If the position of the object is a function of time, that is, if $x = x(t)$, then

$$p = \frac{dW}{dt} = F[x(t)]\frac{dx}{dt} = F[x(t)]v(t)$$

is called the *power input* of the force F. The units of measurement of power are foot-pounds per second if force is measured in pounds, distance in feet, and time in seconds; joules per second, or *watts,* if force is measured in newtons, distance in meters, and time in seconds.

$$1 \text{ horsepower} = 550 \text{ foot-pounds per second}$$
$$= 746 \text{ watts.}$$

28. (a) Assuming constant acceleration, what horsepower must an engine produce to accelerate a 3000-pound truck from 0 to 60 miles per hour (88 feet per second) in 15 seconds along a level road?
  (b) What horsepower must the engine produce if the road rises 4 feet for every 100 feet of road? HINT: Integration is not required to answer these questions.

29. A vertical cylindrical tank with a height of 10 feet and a radius of 5 feet is half filled with water. If a 1 horsepower pump can do 550 foot-pounds of work per second, how long will it take a $\tfrac{1}{2}$-horsepower pump to:
  (a) Pump the water out of the top of the tank?
  (b) Pump the water to a point 5 feet above the top of the tank?

■ 6.6 FLUID PRESSURE AND FLUID FORCES

When an object is submerged in a liquid, it experiences a force caused by *pressure* from the liquid around it. By pressure we mean force per unit area. This force is perpendicular to the object at each of its points. The pressure P at a depth h below the surface is given by

$$P = \sigma h,$$

where σ is the *weight density* of the liquid (the weight per unit volume). Weight densities of liquids are determined experimentally.

When a horizontal surface is submerged in a liquid, the force against it is the total weight of the fluid above it (see Figure 6.6.1). If the horizontal surface has area A, if the depth is h, and if the weight density of the fluid is σ, then the force is given by the formula

(6.6.1)

$$F = \sigma hA = PA.$$

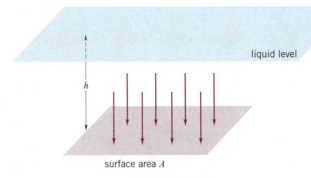

liquid level

h

surface area A

Figure 6.6.1

Remark about Units. If area is measured in square feet, depth in feet, and weight density in pounds per cubic foot, then the fluid force is measured in pounds. If, in the metric system, area is measured in square meters, depth in meters, and weight density in newtons per cubic meter, then the fluid force is measured in newtons. ❏

For example, the weight density of water is approximately 62.5 pounds per cubic foot, so the water pressure on a plate suspended horizontally in water at a depth of 6 feet is $(62.5)(6) = 375$ pounds per square foot. If the plate has an area of 4 square feet, then the force exerted on one side of the plate is $(375)(4) = 1500$ pounds.

Now suppose that a plate is suspended vertically in a liquid. The force on the plate is more difficult to determine in this case because the force exerted by the liquid at the bottom of the plate is greater than the force at the top. In Figure 6.6.2 we have depicted a vertical wall standing against a body of liquid. (Think of it as a dam or part of a container.) We want to calculate the force exerted by the liquid on this wall.

Figure 6.6.2   **Figure 6.6.3**

As in the figure we assume that the liquid sits from depth a to depth b, and we let $w(x)$ denote the width of the wall at depth x. A partition $P = \{x_0, x_1, \ldots, x_n\}$ of $[a, b]$ of small norm subdivides the wall into n narrow horizontal strips (Figure 6.6.3).

We can estimate the force on the ith strip by taking x_i^* as the midpoint of $[x_{i-1}, x_i]$. Then

$$w(x_i^*) = \text{the approximate width of the } i\text{th strip}$$

and

$$w(x_i^*) \, \Delta x_i = \text{the approximate area of the } i\text{th strip.}$$

Since this strip is narrow, all the points of the strip are approximately at depth x_i^*. Thus, using (6.6.1), we can estimate the force on the ith strip by the product

$$\sigma x_i^* w(x_i^*) \, \Delta x_i.$$

Adding up all these estimates we have an estimate for the force on the entire wall:

$$F \cong \sigma x_1^* w(x_1^*) \, \Delta x_1 + \sigma x_2^* w(x_2^*) \, \Delta x_2 + \cdots + \sigma x_n^* w(x_n^*) \, \Delta x_n.$$

The sum on the right is a Riemann sum for the integral

$$\int_a^b \sigma x w(x) \, dx$$

and as such converges to that integral as $\|P\| \to 0$. Thus we have

(6.6.2)

$$\text{force against wall} = \int_a^b \sigma x w(x) \, dx.$$

Example 1 A horizontal circular water main (Figure 6.6.4) 6 feet in diameter is capped when half full of water. What is the force of the water against the cap?

Figure 6.6.4

SOLUTION Here

$$w(x) = 2\sqrt{9 - x^2} \quad \text{and} \quad \sigma = 62.5 \text{ pounds per cubic foot.}$$

The force exerted against the cap is

$$F = \int_0^3 (62.5)x(2\sqrt{9 - x^2}) \, dx = 62.5 \int_0^3 2x\sqrt{9 - x^2} \, dx = 1125 \text{ pounds.} \quad \square$$

Example 2 A metal plate in the form of a trapezoid is affixed to a vertical dam as in Figure 6.6.5. The dimensions shown are given in meters; the weight density of water in the metric system is approximately 9800 newtons per cubic meter. Find the force on the plate.

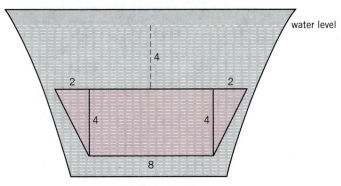

Figure 6.6.5

SOLUTION First we find the width of the plate x meters below the water level. By similar triangles (see Figure 6.6.6)

$$t = \tfrac{1}{2}(8 - x) \qquad \text{so that} \qquad w(x) = 8 + 2t = 16 - x.$$

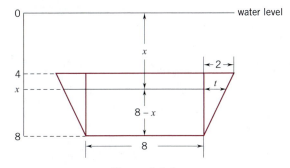

Figure 6.6.6

The force against the plate is

$$\int_{4}^{8} 9800x(16 - x)\, dx = 9800 \int_{4}^{8} (16x - x^2)\, dx$$

$$= 9800 \left[8x^2 - \tfrac{1}{3}x^3 \right]_{4}^{8} \cong 2{,}300{,}000 \text{ newtons.} \quad \square$$

EXERCISES 6.6

1. A rectangular plate 8 feet wide by 6 feet deep is submerged vertically in a tank of water with the upper edge (that is, the 8-foot edge) at the surface of the water. Find the force of the water on one side of the plate.

2. A 6-foot by 6-foot square plate is submerged vertically in a tank of water with upper edge parallel to the surface. The center of the plate is 4 feet below the surface. Find the force of the water on one side of the plate.

3. A vertical dam in a river is in the shape of an isosceles trapezoid 100 meters across at the surface of the river, 60 meters across the bottom. If the river is 20 meters deep, find the force of the water on the dam.

4. A 5-meter by 5-meter square gate is at the bottom of the dam in Exercise 3. What is the force of the water on the gate?

5. A gate in the shape of an isosceles trapezoid 4 meters at the top, 6 meters at the bottom, and 3 meters high has its upper edge 10 meters below the top of the dam in Exercise 3. Find the force of the water on this gate.

6. A vertical dam in the shape of a rectangle is 1000 feet wide and 100 feet high. Calculate the force on the dam when:
   (a) the water level is 75 feet above the bottom.
   (b) the water level is 50 feet above the bottom.

7. Each end of a horizontal oil tank is an ellipse with horizontal axis 12 feet long and vertical axis 6 feet long. Calculate the force on one end when the tank is half full of oil weighing 60 pounds per cubic foot.

8. Each vertical end of a vat is a segment of a parabola (vertex down) 8 feet across the top and 16 feet deep. Calculate the force on an end when the vat is full of liquid weighing 70 pounds per cubic foot.

9. The vertical ends of a water trough are isosceles right triangles with the 90° angle at the bottom. Calculate the force on each triangle when the trough is full of water given that the legs of the triangle are 8 feet long.

10. The vertical ends of a water trough are isosceles triangles 5 feet across the top and 5 feet deep. Calculate the force on an end when the trough is full of water.

11. The ends of a water trough are semicircular discs with radius 2 feet. Find the force of the water on one end if the trough is full of water.

12. The ends of a water trough have the shape of the parabolic region bounded by $y = x^2 - 4$ and $y = 0$, where the measurements are in feet. Assuming that the trough is full of water, set up a definite integral whose value is the force of the water on one end.

13. A horizontal cylindrical tank of diameter 8 feet is half full of oil weighing 60 pounds per cubic foot. Calculate the force on one end.

14. Calculate the force on one end if the tank of Exercise 13 is full.

15. A rectangular metal plate 10 feet by 6 feet is affixed to a vertical dam with the center of the plate 11 feet below water level. Find the force on the plate if: (a) the 10-foot sides are horizontal; (b) the 6-foot sides are horizontal.

16. A vertical cylindrical tank of diameter 30 feet and height 50 feet is full of oil weighing 60 pounds per cubic foot. Find the force on the curved surface.

17. A swimming pool is 8 meters wide and 14 meters long. The pool is 1 meter deep at the shallow end and 3 meters deep at the other end; the depth increases linearly from the shallow to the deep end. If the pool is full of water, find:
   (a) the force of the water on each of the sides;
   (b) the force of the water on each of the ends.

18. Prove that if a plate is submerged in a liquid at an angle of θ with the vertical, $0 < \theta < \pi/2$, then the force on the plate is given by

$$F = \int_a^b \sigma x w(x) \sec \theta \, dx,$$

where σ is the weight density of the liquid, and x, $w(x)$, a, b are as shown in the figure below.

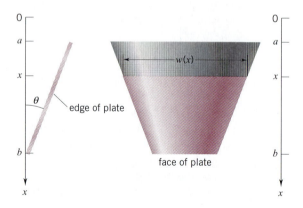

19. Find the force of the water on the bottom of the swimming pool in Exercise 17.

20. The face of a rectangular dam in a river is 1000 feet wide by 100 feet deep, and it makes an angle of 30° with the vertical. Find the force of the water on the face of the dam if:
   (a) the water level is at the top of the dam;
   (b) the water level is 75 feet from the bottom of the river.

21. Relate the force on a dam to the centroid of the submerged surface of the dam.

22. Two identical metal plates are affixed to a vertical dam. The centroid of the first plate is at depth h_1, the centroid of the second plate is at depth h_2. Compare the forces on the two plates.   HINT: Exercise 21.

■ CHAPTER HIGHLIGHTS

6.1   More on Area

representative rectangle (p. 347)
area by integration with respect to y (p. 349)

6.2   Volume by Parallel Cross Sections; Discs and Washers

volume by parallel cross sections (p. 354)
disc method (p. 357)
washer method (x-axis) (p. 359)
washer method (y-axis) (p. 360)

6.3   Volume by the Shell Method

shell method (y-axis) (p. 364)   shell method (x-axis) (p. 366)

6.4   The Centroid of a Region; Pappus's Theorem on Volumes

principles for finding the centroid of a region (p. 372)

$$\bar{x}A = \int_a^b x[\,f(x) - g(x)]\,dx, \qquad \bar{y}A = \int_a^b \tfrac{1}{2}([\,f(x)]^2 - [g(x)]^2)\,dx$$

Pappus's theorem on volumes (p. 374)

centroid of a solid of revolution: $\displaystyle \bar{x}V_x = \int_a^b \pi x[\,f(x)]^2\,dx, \quad \bar{y}V_y = \int_a^b \pi x[\,f(x)]^2\,dx$

6.5   The Notion of Work

Hooke's law (p. 380)   pumping out a tank (p. 382)

6.6   The Force Against a Dam

force against a vertical wall $= \displaystyle \int_a^b \sigma x w(x)\,dx$

■ PROJECTS AND EXPLORATIONS USING TECHNOLOGY

To do these exercises you will need a graphics calculator or a computer with graphing capability   The majority of these problems are open-ended so different approaches may be used to solve them.   You should be aware that different approaches can result in slight variations in the answers.   Round your numerical answers to at least four decimal places.   The rounding method that your calculator or computer uses also may cause variations in answers.

6.1 Let

$$f(x) = \frac{x^2 + 3x + 9}{x^6 - 2x^3 + x + 20},$$

and for each real number p, let

$$g_p(x) = px^2.$$

(a) Find the values of p such that f and g_p intersect at two points to form a bounded region.
(b) For each value of p found in part (a), let $A(p)$ be the area of the corresponding region. Evaluate $A(p)$ for several values of p.
(c) Find the values of p for which $A(p)$ is continuous.
(d) Calculate values of $A(p)$ to approximate $A'(p)$.

6.2 We are going to calculate areas and volumes in yet one more way.

(a) To find the area of a circle of radius r, subdivide it into a set of concentric rings as shown in the figure.   Write a Riemann sum to approximate this area.
HINT: The area of a "ring" with inner radius r_1 and outer radius r_2 is given by $\pi(r_2^2 - r_1^2)$, which can be written $2\pi[\frac{1}{2}(r_1 + r_2)](r_2 - r_1)$.

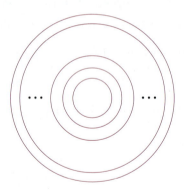

(b) Convert the Riemann sum from part (a) into a definite integral and evaluate it to find the area of the circle.   When viewed in this manner, note that the derivative of the area with respect to the radius r is the circumference of the circle.
(c) Suppose that the density of a circular oil spill is $3/\sqrt{r^2 + 1}$ grams per square centimeter (we are ignoring the thickness) at a distance r from the center.   Find the mass of the oil spill when the radius is 3 kilometers.
(d) Find a Riemann sum for approximating the volume of a sphere of radius r by subdividing it into spherical shells.
(e) Convert the Riemann sum from part (d) into a definite integral and evaluate it to find the volume of the sphere.
(f) If the density of the sphere is a function $\lambda(r)$, write the definite integral for finding the mass of the sphere.   Evaluate this definite integral in the case where

$$\lambda(r) = \frac{1}{(r^3 + 3)^2} \quad \text{grams per square centimeter}$$

and the radius of the sphere is 3 centimeters.

6.3 The Great Pyramid in Giza, Egypt, has a square base 750 feet on a side, and a height of 400 feet. See Figure A.
(a) Find its volume.
(b) Suppose that the pyramid is solid stone, having a density of 175 pounds per cubic foot. Find the weight of the pyramid.
(c) Assume that the blocks were lifted into place from ground level. Find the work required to build the pyramid.
(d) Assume that the pyramid has a facade that is 10 feet thick and that the density of this stone is 225 pounds per cubic foot. Find the weight of the facade and the work required to construct it. Assume that once the side length near the top of the pyramid becomes less than 20 feet, the rest of the pyramid is all facade.

Figure A

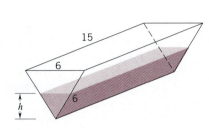

Figure B

6.4 Consider a water trough that is 15 feet long and has ends that are equilateral triangles of side length 6 feet.
(a) Find the function $V(h)$ that represents the volume of the trough from the bottom to a height of h feet. See Figure B.
(b) If water weighs 62.5 pounds per cubic foot, find the force of the water on an end of the trough if the height of the water in the trough is h feet.
(c) If water is running into the trough at the rate of 2 cubic feet per minute, find the rate at which the water level is rising.
(d) Find the rate at which the force on an end is increasing.

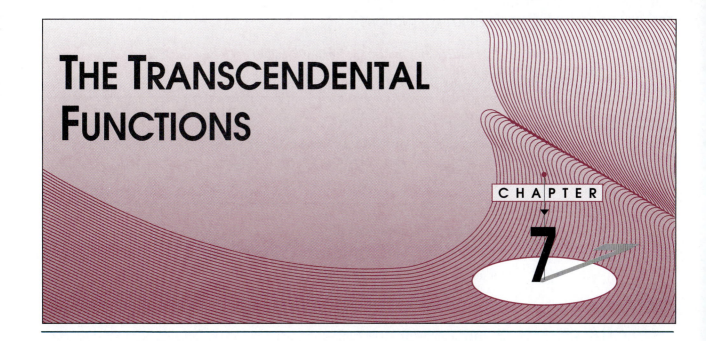

THE TRANSCENDENTAL FUNCTIONS

Some real numbers satisfy polynomial equations with integer coefficients:

$$\tfrac{3}{5} \text{ satisfies the equation } 5x - 3 = 0,$$
$$\sqrt{2} \text{ satisfies the equation } x^2 - 2 = 0.$$

Such numbers are called *algebraic*. There are, however, numbers that are not algebraic, among them π. Such numbers are called *transcendental*.

Some functions f satisfy polynomial equations with polynomial coefficients:

$$f(x) = \frac{x}{\pi x + \sqrt{2}} \quad \text{satisfies the equation} \quad (\pi x + \sqrt{2})f(x) - x = 0,$$

$$f(x) = 2\sqrt{x} - 3x^2 \quad \text{satisfies the equation} \quad [f(x)]^2 + 6x^2 f(x) + (9x^4 - 4x) = 0.$$

Such functions are called *algebraic*. There are, however, functions that are not algebraic; these are called *transcendental*. The functions that we study in this chapter (the logarithm, the exponential, and the trigonometric functions and their inverses) are all transcendental.

■ 7.1 ONE-TO-ONE FUNCTIONS; INVERSES

One-to-One Functions

A function can take on the same value at different points of its domain. Constant functions, for example, take on the same value at all points of their domains. The quadratic function $f(x) = x^2$ takes on the same value at $-c$ as it does at c; so does the absolute value function $g(x) = |x|$. The function

$$f(x) = 1 + (x - 3)(x - 5)$$

takes on the same value at $x = 5$ as it does at $x = 3$:

$$f(3) = 1, \qquad f(5) = 1.$$

Functions for which this kind of repetition *does not* occur are called *one-to-one* functions.

DEFINITION 7.1.1

A function f with domain D is said to be *one-to-one* iff no two distinct points in D have the same image under f, that is,

$$f(x_1) \neq f(x_2) \quad \text{whenever} \quad x_1 \neq x_2, \quad x_1, x_2 \in D.$$

Equivalently, f is one-to-one iff

$$f(x_1) = f(x_2) \quad \text{implies} \quad x_1 = x_2$$

for any pair of points x_1, x_2 in the domain of f.

The functions

$$f(x) = x^3 \quad \text{and} \quad g(x) = \sqrt{x}$$

are each one-to-one. The cubing function is one-to-one because no two distinct numbers have the same cube. The square-root function is one-to-one because no two distinct numbers have the same square root.

There is a simple geometric test to determine whether a function is one-to-one. Look at the graph of the function. If some horizontal line intersects the graph of the function more than once, then the function is not one-to-one (Figure 7.1.1). If, on the other hand, no horizontal line intersects the graph more than once, then the function is one-to-one (Figure 7.1.2).

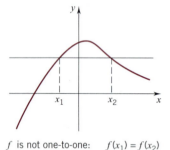

f is not one-to-one: $\quad f(x_1) = f(x_2)$

Figure 7.1.1

Inverses

We begin with a theorem about one-to-one functions.

THEOREM 7.1.2

If f is a one-to-one function, then there is one and only one function g that is defined on the range of f and satisfies the equation

$$f(g(x)) = x \qquad \text{for all } x \text{ in the range of } f.$$

---

f is one-to-one:

Figure 7.1.2

PROOF The proof is easy. If x is in the range of f, then f must take on the value x at some number. Since f is one-to-one, there can be only one such number. We have called that number $g(x)$. ❑

The function that we have named g in the theorem is called the *inverse* of f and is usually denoted f^{-1}.

DEFINITION 7.1.3 INVERSE FUNCTION

Let f be a one-to-one function. The *inverse* of f, denoted by f^{-1}, is the unique function that is defined on the range of f and satisfies the equation

$$f(f^{-1}(x)) = x \qquad \text{for all } x \text{ in the range of } f.$$

Remark   The notation f^{-1} for the inverse function of f is standard.   Unfortunately, there is the danger of confusing f^{-1} with the reciprocal of f, that is, with $1/f(x)$.   The "-1" in the notation for the inverse of f is *not an exponent*; $f^{-1}(x)$ *does not mean* $1/f(x)$.   On those occasions when we want to express $1/f(x)$ using the exponent -1, we will write $[\,f(x)]^{-1}$.

Example 1   You have seen that the cubing function

$$f(x) = x^3$$

is one-to-one.   Find the inverse.

SOLUTION   We set $t = f^{-1}(x)$ and solve the equation $f(t) = x$ for t:

$$f(t) = x$$
$$t^3 = x \qquad\qquad (f \text{ is the cubing function})$$
$$t = x^{1/3}.$$

Substituting $f^{-1}(x)$ back in for t, we have

$$f^{-1}(x) = x^{1/3}.$$

The inverse of the cubing function is the cube-root function.   The graphs of $f(x) = x^3$ and $f^{-1}(x) = x^{1/3}$ are shown in Figure 7.1.3.   ❑

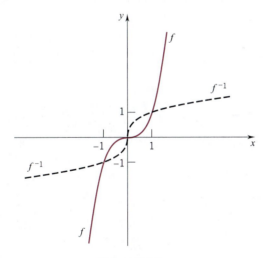

Figure 7.1.3

Remark   We substitute t for $f^{-1}(x)$ merely to simplify the calculations.   It is easier to work with the symbol t than with the string of symbols $f^{-1}(x)$.   ❑

Example 2   Show that the linear function

$$f(x) = 3x - 5$$

is one-to-one.   Then find the inverse.

SOLUTION   To show that f is one-to-one, let's suppose that

$$f(x_1) = f(x_2).$$

Then

$$3x_1 - 5 = 3x_2 - 5$$

$$3x_1 = 3x_2$$

$$x_1 = x_2.$$

The function is one-to-one since

$$f(x_1) = f(x_2) \quad \text{implies} \quad x_1 = x_2.$$

(Another way to see that this function is one-to-one is to note that the graph is a line of slope 3 and as such cannot be intersected by a horizontal line more than once.)

Now let's find the inverse. To do this, we set $t = f^{-1}(x)$ and solve the equation $f(t) = x$ for t:

$$f(t) = x$$

$$3t - 5 = x$$

$$3t = x + 5$$

$$t = \tfrac{1}{3}x + \tfrac{5}{3}.$$

Substituting $f^{-1}(x)$ back in for t, we have

$$f^{-1}(x) = \tfrac{1}{3}x + \tfrac{5}{3}.$$

The graphs of f and f^{-1} are shown in Figure 7.1.4. ❏

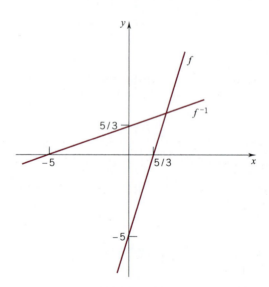

Figure 7.1.4

Example 3 Show that the function

$$f(x) = (1 - x^3)^{1/5} + 2$$

is one-to-one. What is the inverse?

SOLUTION First we show that f is one-to-one. Suppose that

$$f(x_1) = f(x_2).$$

Then

$$(1 - x_1^3)^{1/5} + 2 = (1 - x_2^3)^{1/5} + 2$$

$$(1 - x_1^3)^{1/5} = (1 - x_2^3)^{1/5}$$

$$1 - x_1^3 = 1 - x_2^3$$

$$x_1^3 = x_2^3$$

$$x_1 = x_2.$$

The function is one-to-one since

$$f(x_1) = f(x_2) \quad \text{implies} \quad x_1 = x_2.$$

To find the inverse we set $t = f^{-1}(x)$ and solve the equation $f(t) = x$ for t:

$$f(t) = x$$

$$(1 - t^3)^{1/5} + 2 = x$$

$$(1 - t^3)^{1/5} = x - 2$$

$$1 - t^3 = (x - 2)^5$$

$$t^3 = 1 - (x - 2)^5$$

$$t = [1 - (x - 2)^5]^{1/3}.$$

Substituting $f^{-1}(x)$ back in for t, we have

$$f^{-1}(x) = [1 - (x - 2)^5]^{1/3}. \quad ❑$$

In Chapter 4, Section 4.2, we studied increasing and decreasing functions: A function f is *increasing* if $x_1 < x_2$ implies $f(x_1) < f(x_2)$; f is *decreasing* if $x_1 < x_2$ implies $f(x_1) > f(x_2)$; x_1, x_2 in the domain of f.

THEOREM 7.1.4

If f is either an increasing function or a decreasing function, then f is one-to-one and hence has an inverse.

PROOF Suppose that f is increasing, and let x_1 and x_2 be points in the domain of f with $x_1 \neq x_2$. If $x_1 < x_2$, then $f(x_1) < f(x_2)$; if $x_1 > x_2$, then $f(x_1) > f(x_2)$. In either case, $f(x_1) \neq f(x_2)$ and so f is one-to-one. The proof for f as a decreasing function can be done in the same way. ❑

Theorem 7.1.4 has an important application to differentiable functions whose domains are intervals. By Theorem 4.2.3, if f is continuous on an interval I, and if $f'(x) > 0 (f'(x) < 0)$ on the interior of I, then f is increasing (decreasing) on I. In Exercise 62, Section 4.2, you were asked to show that the result of Theorem 4.2.3 still holds if $f'(x) \geq 0 (f'(x) \leq 0)$ on the interior of I with $f'(x) = 0$ for at most finitely many values of x.

Example 4 Each of the following functions has an inverse.

(a) $f(x) = 3x - 5$ $\qquad\qquad$ $f'(x) = 3 > 0.$

(b) $g(x) = 4 - 2x^3$ $\qquad\qquad$ $\begin{cases} g'(x) = -6x^2 \leq 0; \\ g'(x) = 0 \text{ iff } x = 0. \end{cases}$

(c) $F(x) = x^5 + 2x^3 + 3x - 4$ $\qquad$ $F'(x) = 5x^4 + 6x^2 + 3 > 0.$

(d) $G(x) = \cos x,$ $\quad$ for $\quad 0 \leq x \leq \pi$ $\qquad$ $G'(x) = -\sin x < 0$ $\quad$ for $\quad 0 < x < \pi.$

We found the inverse of f in Example 2. You can verify that $g^{-1}(x) = [(4 - x)/2]^{1/3}.$ To find the inverse of F we would have to solve the equation

$$t^5 + 2t^3 + 3t - 4 = x$$

for t. It is not clear that this can be done. Thus, we have the situation that we know F^{-1} exists, but we may not be able to find a "formula" for it. The same is true for G. The inverse trigonometric functions are defined and studied in Section 7.8. ❏

Suppose that the function f has an inverse. Then, by definition, f^{-1} satisfies the equation

(7.1.5) $\qquad\qquad$ $\boxed{f(f^{-1}(x)) = x \quad \text{for all } x \text{ in the range of } f.}$

It is also true that

(7.1.6) $\qquad\qquad$ $\boxed{f^{-1}(f(x)) = x \quad \text{for all } x \text{ in the domain of } f.}$

PROOF Take x in the domain of f and set $y = f(x)$. Since y is in the range of f,

$$f(f^{-1}(y)) = y.$$

This means that

$$f(f^{-1}(f(x))) = f(x)$$

and tells us that f takes on the same value at $f^{-1}(f(x))$ as it does at x. With f one-to-one, this can only happen if

$$f^{-1}(f(x)) = x. \quad ❏$$

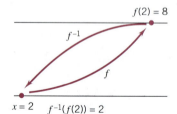

Figure 7.1.5

Equation (7.1.6) tells us that f^{-1} "undoes" the action of f:

if $\;f$ takes x to $f(x),\qquad$ then $\;f^{-1}$ takes $f(x)$ back to x.

Figure 7.1.5 illustrates this relationship in the case of $f(x) = x^3$, $f^{-1}(x) = x^{1/3}$ and $x = 2$.

Equation (7.1.5) tells us that f "undoes" the work of f^{-1}:

if $\;f^{-1}$ takes x to $f^{-1}(x),\qquad$ then $\;f$ takes $f^{-1}(x)$ back to x.

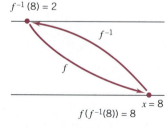

Figure 7.1.6

This is illustrated in Figure 7.1.6 with $f(x) = x^3$, $f^{-1}(x) = x^{1/3}$ and $x = 8$.

Finally, we emphasize that Equations (7.1.5) and (7.1.6) tell us that

$$\text{domain of } f^{-1} = \text{range of } f$$
$$\text{range of } f^{-1} = \text{domain of } f.$$

The Graphs of f and f^{-1}

There is an important relation between the graph of a one-to-one function f and the graph of f^{-1} (see Figure 7.1.7). The graph of f consists of points of the form $P(x, f(x))$. Since f^{-1} takes on the value x at $f(x)$, the graph of f^{-1} consists of points of the form $Q(f(x), x)$.

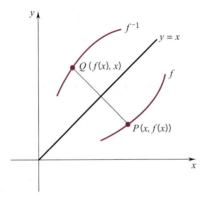

Figure 7.1.7

Now suppose that the scale on the y-axis is the same as that on the x-axis. Then, for each x, the points $P(x, f(x))$ and $Q(f(x), x)$ are symmetric with respect to the line $y = x$. Hence the graph of f^{-1} is the graph of f reflected in the line $y = x$.

Example 5   Given the graph of f in Figure 7.1.8, sketch the graph of f^{-1}.

SOLUTION   First draw in the line $y = x$. Then reflect the graph of f in that line. See Figure 7.1.9.   ❑

Continuity and Differentiability of Inverses

Let f be a one-to-one function. Then f has an inverse, f^{-1}. Suppose, in addition, that f is continuous. Since the graph of f has no "holes" or "gaps," and since the graph of f^{-1} is simply the reflection of the graph of f in the line $y = x$, we can conclude that the graph of f^{-1} has no holes or gaps. That is, f^{-1} must also be continuous. We state this result formally; a proof is given in Appendix B3.

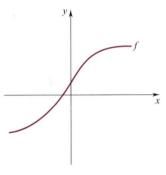

Figure 7.1.8

THEOREM 7.1.7

Let f be a one-to-one function defined on an interval I. If f is continuous, then its inverse f^{-1} is also continuous.

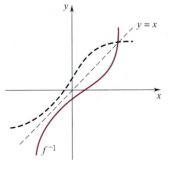

Figure 7.1.9

Now suppose that f is differentiable. Is f^{-1} necessarily differentiable? The answer is yes, provided f' does not take on the value 0. We will assume this result for now and go on to describe how to calculate the derivative of f^{-1}.

To simplify the notation, we set $f^{-1} = g$. Then

$$f(g(x)) = x \qquad \text{for all} \quad x \text{ in the range of } f.$$

Differentiation gives

$$\frac{d}{dx}[f(g(x))] = 1,$$

and so

$$f'(g(x))g'(x) = 1 \qquad \text{(by the chain rule; we are assuming that } f \text{ and } g \text{ are differentiable).}$$

Therefore,

$$g'(x) = \frac{1}{f'(g(x))} \qquad \text{(we are assuming that } f' \text{ is never 0).}$$

Substituting f^{-1} back in for g, we have

(7.1.8)
$$(f^{-1})'(x) = \frac{1}{f'[f^{-1}(x)]}.$$

This is the standard formula for the derivative of an inverse.

Before illustrating this formula, we restate it in a more specific form:

(7.1.9)
Suppose that f has an inverse and is differentiable. Let a be a point in the domain of f and let $b = f(a)$. If $f'(a) \neq 0$, then $(f^{-1})'(b)$ exists and is given by

$$(f^{-1})'(b) = \frac{1}{f'(a)}.$$

Remark Be sure to notice that $(f^{-1})'(b)$ is obtained by evaluating f' at a, not at b. Notice, also, that $a = f^{-1}(b)$, so that

$$(f^{-1})'(b) = \frac{1}{f'(a)} = \frac{1}{f'[f^{-1}(b)]},$$

which is simply (7.1.8) with $x = b$. ❏

Example 6 Let $f(x) = x^3 + \frac{1}{2}x$. (a) Show that f has an inverse. (b) Show that 9 is in the range of f and calculate $(f^{-1})'(9)$.

SOLUTION (a) The domain of f is $(-\infty, \infty)$ and

$$f'(x) = 3x^2 + \frac{1}{2} > 0$$

for all x. Thus f has an inverse by Theorem 7.1.4.

(b) It is easy to verify that the range of f is also $(-\infty, \infty)$. Thus $9 \in \text{range}(f)$. To calculate $(f^{-1})'(9)$ we must first find the value a such that $f(a) = 9$. The solution of

$$a^3 + \tfrac{1}{2}a = 9$$

is $a = 2$. Thus

$$(f^{-1})'(9) = \frac{1}{f'(2)} = \frac{1}{3(2)^2 + \frac{1}{2}} = \frac{2}{25}.$$

The graphs of f and f^{-1} are shown in Figure 7.1.10. (These were obtained on a graphing calculator in parametric mode.) ❏

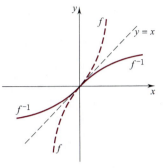

Figure 7.1.10

Formula (7.1.8) takes a simple form when it is expressed in Leibniz's notation. Suppose that the function f has an inverse and is differentiable. Let $y = f(x)$ so that $x = f^{-1}(y)$. Then

$$\frac{dx}{dy} = (f^{-1})'(y) \qquad \text{and} \qquad \frac{dy}{dx} = f'(x) = f'(f^{-1}(y)).$$

Substituting these expressions in (7.1.8), we get

(7.1.10)

$$\boxed{\dfrac{dx}{dy} = \dfrac{1}{dy/dx}.}$$

If $y = f(x)$, then the rate of change of x with respect to y is the reciprocal of the rate of change of y with respect to x. Of course, we have to be precise about where these derivatives are evaluated. The analog of (7.1.9) is

(7.1.11)

> Suppose that f has an inverse and is differentiable, and set $y = f(x)$. Let a be a number in the domain of f and let $b = f(a)$. If
>
> $$\frac{dy}{dx}(a) \neq 0, \qquad \text{then} \qquad \frac{dx}{dy}(b) \text{ exists and is given by}$$
>
> $$\frac{dx}{dy}(b) = \frac{1}{\dfrac{dy}{dx}(a)}$$

Example 7 The function defined by $y = x - \cos x$, for $-\pi \leq x \leq \pi$, is differentiable. **(a)** Show that this function has an inverse. **(b)** Calculate the derivative of the inverse function using Formula (7.1.10). **(c)** Evaluate the derivative of the inverse function at $y = -1$.

SOLUTION **(a)** Since $dy/dx = 1 + \sin x \geq 0$ for $-\pi < x < \pi$ and $1 + \sin x = 0$ only at $x = -\pi/2$, it follows that this function has an inverse.
(b) By Formula (7.1.10)

$$\frac{dx}{dy} = \frac{1}{dy/dx} = \frac{1}{1 + \sin x}.$$

(c) To calculate

$$\frac{dx}{dy}(-1)$$

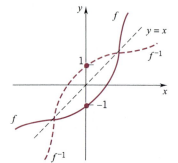

Figure 7.1.11

we must first find the value a such that $y = -1$ when $x = a$. Note that if we let $x = 0$, then $y = -1$. Thus

$$\frac{dx}{dy}(-1) = \frac{1}{\dfrac{dy}{dx}(0)} = \frac{1}{1 + \sin 0} = 1.$$

The graphs of $y = x - \cos x$ and its inverse are shown in Figure 7.1.11. (As in Example 6, these graphs were obtained on a graphing calculator in parametric mode.) ❑

For some geometric reassurance of the validity of Formula (7.1.8) we refer you to Figure 7.1.12. The graphs of f and f^{-1} are reflections of one another in the line $y = x$. The tangent lines l_1 and l_2 are also reflections of one another. From the figure

$$(f^{-1})'(x) = \text{slope of } l_1 = \frac{f^{-1}(x) - b}{x - b}, \qquad f'(f^{-1}(x)) = \text{slope of } l_2 = \frac{x - b}{f^{-1}(x) - b};$$

namely, $(f^{-1})'(x)$ and $f'(f^{-1}(x))$ are indeed reciprocals.

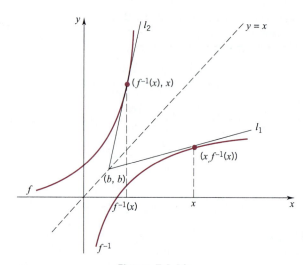

Figure 7.1.12

The figure also suggests the difficulty caused by having $f'(z) = 0$ at some point z. For if $f'(z) = 0$, then the graph of f has a horizontal tangent line at $(z, f(z))$. The reflection of a horizontal tangent in the line $y = x$ will be a vertical tangent line to the graph of f^{-1} at the point $(f(z), z)$ and this will be a point of nondifferentiability for f^{-1}.

EXERCISES 7.1

In Exercises 1–26, determine whether or not the given function is one-to-one and, if so, find its inverse.

1. $f(x) = 5x + 3$.

2. $f(x) = 3x + 5$.

3. $f(x) = 4x - 7$.

4. $f(x) = 7x - 4$.

5. $f(x) = 1 - x^2$.

6. $f(x) = x^5$.

7. $f(x) = x^5 + 1$.

8. $f(x) = x^2 - 3x + 2$.

9. $f(x) = 1 + 3x^3$.

10. $f(x) = x^3 - 1$.

11. $f(x) = (1 - x)^3$.

12. $f(x) = (1 - x)^4$.

13. $f(x) = (x + 1)^3 + 2$.

14. $f(x) = (4x - 1)^3$.

15. $f(x) = x^{3/5}$.

16. $f(x) = 1 - (x - 2)^{1/3}$.

17. $f(x) = (2 - 3x)^3$.

18. $f(x) = (2 - 3x^2)^3$.

19. $f(x) = \dfrac{1}{x}$.

20. $f(x) = \dfrac{1}{1 - x}$.

21. $f(x) = x + \dfrac{1}{x}$.

22. $f(x) = \dfrac{x}{|x|}$.

23. $f(x) = \dfrac{1}{x^3 + 1}$.

24. $f(x) = \dfrac{1}{1 - x} - 1$.

25. $f(x) = \dfrac{x + 2}{x + 1}$.

26. $\dfrac{1}{(x + 1)^{2/3}}$.

27. What is the relation between f and $(f^{-1})^{-1}$?

In Exercises 28–33, sketch the graph of f^{-1} given the graph of f.

28.

29.

30.

31.

32.

33.

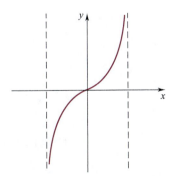

34. (a) Show that the composition of two one-to-one functions, f and g, is one-to-one.

(b) Express $(f \circ g)^{-1}$ in terms of f^{-1} and g^{-1}.

In Exercises 35–42, the given function f is differentiable. Verify that f has an inverse and find $(f^{-1})'(c)$.

35. $f(x) = x^3 + 1$; $c = 9$.

36. $f(x) = 1 - 2x - x^3$; $c = 4$.

37. $f(x) = x + 2\sqrt{x}$, $x > 0$; $c = 8$.

38. $f(x) = x + \sin x$; $c = 0$.

39. $f(x) = 2x + \cos x$; $c = \pi$.

40. $f(x) = \dfrac{x+3}{x-1}$, $x > 1$; $c = 3$.

41. $f(x) = \tan x$, $-\frac{\pi}{2} < x < \frac{\pi}{2}$; $c = \sqrt{3}$.

42. $f(x) = x^5 + 2x^3 + 2x$; $c = -5$.

In Exercises 43–45, find a formula for $(f^{-1})'(x)$ given that f is one-to-one and its derivative satisfies the indicated equation.

43. $f'(x) = f(x)$.   **44.** $f'(x) = 1 + [f(x)]^2$.

45. $f'(x) = \sqrt{1 - [f(x)]^2}$.

46. Let

$$f(x) = \begin{cases} x^3 - 1, & x < 0 \\ x^2, & x \geq 0. \end{cases}$$

(a) Sketch the graph of f and verify that f is one-to-one.

(b) Find f^{-1}.

47. Let

$$f(x) = \frac{ax + b}{cx + d}, \quad x \neq -\frac{d}{c}.$$

(a) Show that f is one-to-one iff $ad - bc \neq 0$.

(b) Suppose that $ad - bc \neq 0$. Find f^{-1}.

48. Let

$$f(x) = \int_2^x \sqrt{1 + t^2}\, dt.$$

(a) Prove that f has an inverse.

(b) Find $(f^{-1})'(0)$.

49. Let

$$f(x) = \int_1^{2x} \sqrt{16 + t^4}\, dt.$$

(a) Prove that f has an inverse.

(b) Find $(f^{-1})'(0)$.

50. Suppose that the function f has an inverse and fix a number $c \neq 0$.

(a) Let $g(x) = f(x + c)$ for all x such that $x + c \in \mathrm{dom}(f)$. Prove that g has an inverse and find it.

(b) Let $h(x) = f(cx)$ for all x such that $cx \in \mathrm{dom}(f)$. Prove that h has an inverse and find it.

51. Let f be a one-to-one, twice differentiable function and let $g = f^{-1}$.

(a) Show that

$$g''(x) = -\frac{f''[g(x)]}{(f'[g(x)])^3}.$$

(b) Suppose that the graph of f is concave upward (downward) on an interval I. What can you say about the graph of f^{-1} on I?

52. Let $P(x) = a_n x^n + a_{n-1}x^{n-1} + \cdots + a_1 x + a_0$ be a polynomial function.

(a) Can P have an inverse if its degree n is even? Explain why or why not.

(b) Can P have an inverse if n is odd? If so, give an example. Also, give an example of a polynomial function of odd degree that does not have an inverse.

53. Suppose that f is differentiable and has an inverse. If we let $y = f^{-1}(x)$, then $f(y) = x$ and we can find dy/dx by differentiating implicitly. This is how we derived (7.1.8). The function $f(x) = \sin x$, $-\frac{\pi}{2} \leq x \leq \frac{\pi}{2}$, is one-to-one and differentiable. Let $y = f^{-1}(x)$ and find dy/dx. Express your result as a function of x.

54. Repeat Exercise 53 for the function $f(x) = \tan x$, $-\frac{\pi}{2} < x < \frac{\pi}{2}$.

▶ In Exercises 55–58, use a graphing utility to draw the graph of f and verify that f is one-to-one. Then sketch the graph of f^{-1}.

55. $f(x) = x^3 + 3x + 2$.   **56.** $f(x) = x^{3/5} - 1$.

57. $f(x) = 4 \sin 2x$, $-\frac{\pi}{4} \leq x \leq \frac{\pi}{4}$.

58. $f(x) = 2 - \cos 3x$, $0 \leq x \leq \frac{\pi}{3}$.

■ 7.2 THE LOGARITHM FUNCTION, PART I

If B is a positive number different from 1, the logarithm to the base B is defined in elementary mathematics by setting

$$C = \log_B A \quad \text{iff} \quad B^C = A.$$

Historically, the base 10 was chosen because our number system is based on the powers of 10. The defining relation then becomes:

$$C = \log_{10} A \quad \text{iff} \quad 10^C = A.$$

The basic properties of $\log_{10}$ can then be summarized as follows: with $X,\ Y > 0$,

$$\log_{10}(XY) = \log_{10}X + \log_{10}Y, \qquad \log_{10}1 = 0,$$

$$\log_{10}(1/Y) = -\log_{10}Y, \qquad \log_{10}(X/Y) = \log_{10}X - \log_{10}Y,$$

$$\log_{10}X^Y = Y\log_{10}X, \qquad \log_{10}10 = 1.$$

This elementary notion of logarithm is inadequate for calculus. It is unclear: What is meant by 10^C if C is irrational? It does not lend itself well to the methods of calculus: How would you differentiate $Y = \log_{10}X$ knowing only that $10^Y = X$?

Here we take an entirely different approach to logarithms. Instead of trying to tamper with the elementary definition, we discard it altogether. From our point of view the fundamental property of logarithms is that they transform multiplication into addition:

The log of a product = the sum of the logs.

Taking this as the central idea we are led to a general notion of logarithm that encompasses the elementary notion, lends itself well to the methods of calculus, and leads us naturally to a choice of base that simplifies many calculations.

DEFINITION 7.2.1

A *logarithm* function is a nonconstant differentiable function f defined on the set of positive numbers such that for all $x > 0$ and $y > 0$

$$f(xy) = f(x) + f(y).$$

Let's assume for the time being that such logarithm functions exist, and let's see what we can find out about them. In the first place, if f is such a function, then

$$f(1) = f(1 \cdot 1) = f(1) + f(1) = 2f(1) \quad \text{and so} \quad f(1) = 0.$$

Taking $y > 0$, we have

$$0 = f(1) = f(y \cdot 1/y) = f(y) + f(1/y)$$

and therefore

$$f(1/y) = -f(y).$$

Taking $x > 0$ and $y > 0$, we have

$$f(x/y) = f(x \cdot 1/y) = f(x) + f(1/y),$$

which, in view of the previous result, means that

$$f(x/y) = f(x) - f(y).$$

We are now ready to look for the derivative. (Remember, we are *assuming* that f is differentiable.) We begin by forming the difference quotient

$$\frac{f(x+h) - f(x)}{h},$$

where $x > 0$ is fixed. From what we have discovered about f,

$$f(x + h) - f(x) = f\left(\frac{x + h}{x}\right) = f(1 + h/x),$$

and therefore

$$\frac{f(x + h) - f(x)}{h} = \frac{f(1 + h/x)}{h}.$$

Multiplying the denominator by x/x and using the fact that $f(1) = 0$, we can write the difference quotient as

$$\frac{f(x + h) - f(x)}{h} = \frac{1}{x}\left[\frac{f(1 + h/x) - f(1)}{h/x}\right].$$

Now let $k = h/x$. Then we have

$$\frac{f(x + h) - f(x)}{h} = \frac{1}{x}\left[\frac{f(1 + k) - f(1)}{k}\right].$$

Since $k \to 0$ iff $h \to 0$ (remember, x is fixed), it follows that

$$f'(x) = \lim_{h \to 0}\frac{f(x + h) - f(x)}{h} = \lim_{k \to 0}\frac{1}{x}\left[\frac{f(1 + k) - f(1)}{k}\right]$$

$$= \frac{1}{x}\lim_{k \to 0}\left[\frac{f(1 + k) - f(1)}{k}\right] = \frac{1}{x}f'(1).$$

In short,

(7.2.2)

$$f'(x) = \frac{1}{x}f'(1).$$

Thus we have shown that if f is a logarithm function and x is any positive number, then

$$f(1) = 0 \qquad \text{and} \qquad f'(x) = \frac{1}{x}f'(1).$$

We can't have $f'(1) = 0$, for that would make f constant. (Explain.) The most natural choice, the one that will keep calculations as simple as possible, is to set $f'(1) = 1.$† The derivative is then $1/x$.

This function, which takes on the value 0 at 1 and has derivative $1/x$ for $x > 0$, must, by the fundamental theorem of calculus, take the form

$$\int_1^x \frac{dt}{t}.$$

(verify this)

---

† This, as you will see later, is tantamount to a choice of base.

> **DEFINITION 7.2.3**
>
> The function
>
> $$L(x) = \int_1^x \frac{dt}{t}, \qquad x > 0$$
>
> is called the (*natural*) *logarithm function*.

Here are some obvious properties of L:

(1) L is defined on $(0, \infty)$ with derivative

$$L'(x) = \frac{1}{x} \qquad \text{for all } x > 0.$$

L' is positive on $(0, \infty)$, and therefore L is an increasing function.

(2) L is continuous on $(0, \infty)$ since it is differentiable there.

(3) For $x > 1$, $L(x)$ gives the area of the region shaded in Figure 7.2.1.

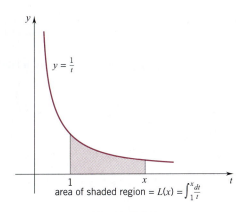

area of shaded region $= L(x) = \int_1^x \frac{dt}{t}$

Figure 7.2.1

(4) $L(x)$ is negative if $0 < x < 1$, $L(1) = 0$, $L(x)$ is positive if $x > 1$.

The following result is fundamental; it establishes that L *is* a logarithm function by showing that it satisfies the equation in Definition 7.2.1.

> **THEOREM 7.2.4**
>
> If x and y are positive, then
>
> $$L(xy) = L(x) + L(y).$$

PROOF Fix any positive number y. Then, since

$$\frac{d}{dx}[L(x)] = \frac{1}{x} \quad \text{and} \quad \frac{d}{dx}[L(xy)] = \frac{y}{xy} = \frac{1}{x},$$

$$\underset{\text{chain rule}}{\underbrace{\qquad}}$$

we can be sure that

$$L(xy) = L(x) + C,$$

for some constant C. We can evaluate the constant by taking $x = 1$:

$$L(y) = L(1 \cdot y) = L(1) + C = C.$$

$$L(1) = 0 \underset{\qquad}{\overset{\uparrow}{}}$$

Therefore $L(xy) = L(x) + L(y)$ as asserted. ❑

We come now to another important result.

THEOREM 7.2.5

If x is positive and p/q is rational, then

$$L(x^{p/q}) = \frac{p}{q}L(x).$$

PROOF You have seen that $d/dx[L(x)] = 1/x$. By the chain rule

$$\frac{d}{dx}[L(x^{p/q})] = \frac{1}{x^{p/q}}\frac{d}{dx}(x^{p/q}) = \frac{1}{x^{p/q}}\left(\frac{p}{q}\right)x^{(p/q)-1} = \frac{p}{q}\left(\frac{1}{x}\right) = \frac{d}{dx}\left[\frac{p}{q}L(x)\right].$$

$$\underset{(3.7.1)}{\overset{\uparrow}{}}$$

Since both functions have the same derivative, they differ by a constant:

$$L(x^{p/q}) = \frac{p}{q}L(x) + C.$$

Since both functions are zero at $x = 1$, $C = 0$ and the theorem is proved. ❑

The domain of L is $(0, \infty)$. What is the range of L?

THEOREM 7.2.6

The range of L is $(-\infty, \infty)$.

PROOF Since L is continuous on $(0, \infty)$, we know from the intermediate-value theorem that it "skips" no values. Thus, its range is an interval. To show that the interval is $(-\infty, \infty)$, we need only show that it is unbounded above and unbounded below. We can do this by taking M as an arbitrary positive number and showing that L takes on values greater than M and values less than $-M$.

Since

$$L(2) = \int_1^2 \frac{dt}{t}$$

is positive (explain), we know that some positive multiple of $L(2)$ must be greater than M; that is, since $L(2) > 0$, $nL(2) \to \infty$ as $n \to \infty$, and so we can choose a positive integer n big enough such that

$$nL(2) > M.$$

Multiplying this equation by -1 we have

$$-nL(2) < -M.$$

Since

$$nL(2) = L(2^n) \quad \text{and} \quad -nL(2) = L(2^{-n}),$$

we have

$$L(2^n) > M \quad \text{and} \quad L(2^{-n}) < -M.$$

This proves the unboundedness. ❑

The Number e

Since the range of L is $(-\infty, \infty)$ and L is an increasing function, we know that L takes on every value and does so only once. In particular, there is one and only one number at which the function L takes on the value 1. *This unique number is denoted by the letter e.*†

The number e can also be defined geometrically: e is the unique number with the property that the area under the graph of $f(t) = 1/t$ from $t = 1$ to $t = e$ is 1. See Figure 7.2.2.

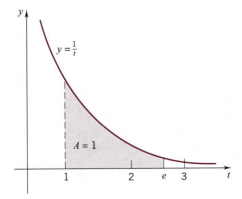

Figure 7.2.2

---

† After the Swiss mathematician Leonhard Euler (1707–1783), considered by many the greatest mathematician of the eighteenth century.

Since

(7.2.7)

$$L(e) = \int_1^e \frac{dt}{t} = 1,$$

it follows from Theorem 7.2.5 that

(7.2.8)

$$L(e^{p/q}) = \frac{p}{q} \qquad \text{for all rational numbers } \frac{p}{q}.$$

Because of this relation, we call L *the logarithm to the base e* and sometimes write

$$L(x) = \log_e x.$$

The number e arises naturally in many settings. Accordingly, we call $L(x)$ the natural logarithm and write

(7.2.9)

$$L(x) = \ln x. \quad †$$

Here are the basic properties we have established for $\ln x$:

(7.2.10)

$$\ln 1 = 0, \qquad \ln e = 1.$$
$$\ln xy = \ln x + \ln y. \qquad\qquad (x > 0, y > 0)$$
$$\ln 1/x = -\ln x. \qquad\qquad (x > 0)$$
$$\ln x/y = \ln x - \ln y. \qquad\qquad (x > 0, y > 0)$$
$$\ln x^r = r \ln x. \qquad\qquad (x > 0, r \text{ rational})$$

Notice how closely these rules parallel the familiar rules for common logarithms (base 10). Later we will show that the last of these rules also holds for irrational exponents.

---

† Logarithms to bases other than e will be taken up later [they arise by other choices of $f'(1)$], but by far the most important logarithm in calculus is the logarithm to the base e. So much so, that when we speak of the logarithm of a number x and don't specify the base, you can be sure that we are talking about the *natural logarithm* $\ln x$.

The Graph of the Logarithm Function

You know that the logarithm function

$$\ln x = \int_1^x \frac{dt}{t}$$

has domain $(0, \infty)$, range $(-\infty, \infty)$, and derivative

$$\frac{d}{dx}(\ln x) = \frac{1}{x} > 0.$$

For small x the derivative is large (near 0 the curve is steep); for large x the derivative is small (far out the curve flattens out). At $x = 1$ the logarithm is 0 and its derivative $1/x$ is 1. (The graph crosses the x-axis at the point $(1, 0)$, and the tangent line at that point is parallel to the line $y = x$.) The second derivative

$$\frac{d^2}{dx^2}(\ln x) = -\frac{1}{x^2}$$

is negative on $(0, \infty)$. (The graph is concave down throughout.) We have sketched the graph in Figure 7.2.3. The y-axis is a vertical asymptote:

$$\text{as } x \to 0^+, \quad \ln x \to -\infty.$$

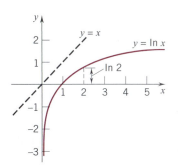

Figure 7.2.3

Example 1 Using upper and lower sums, estimate

$$\ln 2 = \int_1^2 \frac{dt}{t} \qquad \text{(Figure 7.2.4)}$$

from the partition

$$P = \{1 = \tfrac{10}{10}, \tfrac{11}{10}, \tfrac{12}{10}, \tfrac{13}{10}, \tfrac{14}{10}, \tfrac{15}{10}, \tfrac{16}{10}, \tfrac{17}{10}, \tfrac{18}{10}, \tfrac{19}{10}, \tfrac{20}{10} = 2\}.$$

SOLUTION Using a calculator we find that

$$L_f(P) = \tfrac{1}{10}(\tfrac{10}{11} + \tfrac{10}{12} + \tfrac{10}{13} + \tfrac{10}{14} + \tfrac{10}{15} + \tfrac{10}{16} + \tfrac{10}{17} + \tfrac{10}{18} + \tfrac{10}{19} + \tfrac{10}{20})$$
$$= \tfrac{1}{11} + \tfrac{1}{12} + \tfrac{1}{13} + \tfrac{1}{14} + \tfrac{1}{15} + \tfrac{1}{16} + \tfrac{1}{17} + \tfrac{1}{18} + \tfrac{1}{19} + \tfrac{1}{20} > 0.668$$

and

$$U_f(P) = \tfrac{1}{10}(\tfrac{10}{10} + \tfrac{10}{11} + \tfrac{10}{12} + \tfrac{10}{13} + \tfrac{10}{14} + \tfrac{10}{15} + \tfrac{10}{16} + \tfrac{10}{17} + \tfrac{10}{18} + \tfrac{10}{19})$$
$$= \tfrac{1}{10} + \tfrac{1}{11} + \tfrac{1}{12} + \tfrac{1}{13} + \tfrac{1}{14} + \tfrac{1}{15} + \tfrac{1}{16} + \tfrac{1}{17} + \tfrac{1}{18} + \tfrac{1}{19} < 0.719.$$

Thus we have

$$0.668 < L_f(P) \le \ln 2 \le U_f(P) < 0.719.$$

The average of these two estimates is

$$\tfrac{1}{2}(0.668 + 0.719) = 0.6935.$$

Rounded off to four decimal places, the value of $\ln 2$ given on a calculator is 0.6931, so we are not far off. ❑

Table 7.2.1 gives the natural logarithms of the integers 1 through 10 rounded off to the nearest hundredth.

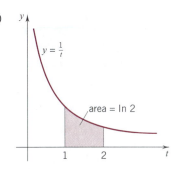

Figure 7.2.4

■ **Table 7.2.1**

n	$\ln n$	n	$\ln n$
1	0.00	6	1.79
2	0.69	7	1.95
3	1.10	8	2.08
4	1.39	9	2.20
5	1.61	10	2.30

Example 2 Use the properties of logarithms and Table 7.2.1 to estimate the following logarithms.

(a) ln 0.2.   **(b)** ln 0.25.   **(c)** ln 2.4.   **(d)** ln 90.

SOLUTION

(a) $\ln 0.2 = \ln \frac{1}{5} = -\ln 5 \cong -1.61$.   **(b)** $\ln 0.25 = \ln \frac{1}{4} = -\ln 4 \cong -1.39$.

(c) $\ln 2.4 = \ln \dfrac{12}{5} = \ln \dfrac{(3)(4)}{5} = \ln 3 + \ln 4 - \ln 5 \cong 0.88$.

(d) $\ln 90 = \ln[(9)(10)] = \ln 9 + \ln 10 \cong 4.50$.   ❏

Example 3 Estimate e on the basis of Table 7.2.1.

SOLUTION So far all we know is that $\ln e = 1$. From the table you can see that

$$3 \ln 3 - \ln 10 \cong 1.$$

The expression on the left can be written

$$\ln 3^3 - \ln 10 = \ln 27 - \ln 10 = \ln \tfrac{27}{10} = \ln 2.7.$$

Thus $\ln 2.7 \cong 1$, and so $e \cong 2.7$. It can be shown that e is an irrational number; its decimal expansion to twelve decimal places is

$$e \cong 2.718281828459.$$ ❏

EXERCISES 7.2

In Exercises 1–12, estimate the given natural logarithm on the basis of Table 7.2.1; check your results on a calculator.

1. ln 20.

2. ln 16.

3. ln 1.6.

4. ln 3^4.

5. ln 0.1.

6. ln 2.5.

7. ln 7.2.

8. ln $\sqrt{630}$.

9. ln $\sqrt{2}$.

10. ln 0.4.

11. ln 2^5.

12. ln 1.8.

13. Interpret the equation $\ln n = \ln mn - \ln m$ in terms of area under the curve $y = 1/x$. Draw a figure.

14. Given that $0 < x < 1$, express as a logarithm the area under the curve $y = 1/t$ from $t = x$ to $t = 1$.

15. Estimate

$$\ln 1.5 = \int_1^{1.5} \frac{dt}{t}$$

using the approximation $\frac{1}{2}[L_f(P) + U_f(P)]$ with $P = \{1 = \frac{8}{8}, \frac{9}{8}, \frac{10}{8}, \frac{11}{8}, \frac{12}{8} = 1.5\}$.

16. Estimate

$$\ln 2.5 = \int_1^{2.5} \frac{dt}{t}$$

using the approximation $\frac{1}{2}[L_f(P) + U_f(P)]$ with $P = \{1 = \frac{4}{4}, \frac{5}{4}, \frac{6}{4}, \frac{7}{4}, \frac{8}{4}, \frac{9}{4}, \frac{10}{4} = \frac{5}{2}\}$.

17. Taking $\ln 5 \cong 1.61$, use differentials to estimate
(a) ln 5.2.   (b) ln 4.8.   (c) ln 5.5.

18. Taking $\ln 10 \cong 2.30$, use differentials to estimate
(a) ln 10.3.   (b) ln 9.6.   (c) ln 11.

In Exercises 19–24, solve the given equation for x.

19. $\ln x = 2$.

20. $\ln x = -1$.

21. $(2 - \ln x) \ln x = 0$.

22. $\frac{1}{2} \ln x = \ln(2x - 1)$.

23. $\ln[(2x + 1)(x + 2)] = 2 \ln(x + 2)$.

24. $2 \ln(x + 2) - \frac{1}{2} \ln x^4 = 1$.

25. Show that

$$\lim_{x \to 1} \frac{\ln x}{x - 1} = 1.$$

HINT: $\dfrac{\ln x}{x - 1} = \dfrac{\ln x - \ln 1}{x - 1}$; interpret the limit in terms of the derivative of $\ln x$.

26. (a) Use the mean-value theorem to show that

$$\frac{x - 1}{x} < \ln x < x - 1 \quad \text{for all } x > 0.$$

HINT: Consider the cases $x > 1$ and $0 < x < 1$ separately.

(b) Use the result in part (a) to give an alternative proof that

$$\lim_{x \to 1} \frac{\ln x}{x - 1} = 1.$$

27. (a) Show that for $n \geq 2$,

$$\frac{1}{2} + \frac{1}{3} + \cdots + \frac{1}{n} < \ln n < 1 + \frac{1}{2} + \frac{1}{3} + \cdots + \frac{1}{n - 1}.$$

HINT: See the figure.

(b) Show that the area of the shaded region in the figure is given by

$$1 + \frac{1}{2} + \frac{1}{3} + \cdots + \frac{1}{n - 1} - \ln n.$$

As $n \to \infty$, this area approaches the number γ known as *Euler's constant*.

(c) Use geometric reasoning to show that $\frac{1}{2} < \gamma < 1$. (To three decimal places, $\gamma \cong 0.577$).

 In Exercises 28–30, a function g is given. (a) Show that there is a number r in the indicated interval such that $\ln r = g(r)$. HINT: Use the intermediate value theorem. (b) Use a graphing utility to graph $\ln x$ and g together. Then use your graphs to find r accurate to four decimal places.

28. $g(x) = 2x - 3$; $[1, 2]$.

29. $g(x) = \sin x$; $[2, 3]$. **30.** $g(x) = \dfrac{1}{x^2}$; $[1, 2]$.

 In Exercises 31–33, estimate $\lim_{x \to a} f(x)$ numerically by evaluating f at the indicated values. Then use a graphing utility to zoom in on the graph near $x = a$ to justify your estimate.

31. $\lim_{x \to 1} \dfrac{\ln x}{x - 1}$; $\quad x = 1 \pm 0.5, 1 \pm 0.1, 1 \pm 0.01,$ $1 \pm 0.001, 1 \pm 0.0001.$

32. $\lim_{x \to 0} x \ln |x|$; $\quad x = \pm 0.5, \pm 0.1, \pm 0.01, \pm 0.001,$ $\pm 0.0001.$

33. $\lim_{x \to 0^+} \sqrt{x} \ln x$; $\quad x = 0.5, 0.1, 0.01, 0.001, 0.0001.$

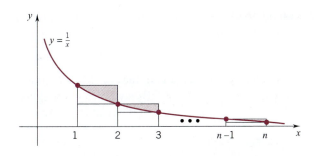

■ 7.3 THE LOGARITHM FUNCTION, PART II

Differentiation and Graphing

We know that for $x > 0$

$$\frac{d}{dx}(\ln x) = \frac{1}{x}.$$

If u is a positive differentiable function of x, then by the chain rule

$$\frac{d}{dx}(\ln u) = \frac{d}{du}(\ln u)\frac{du}{dx} = \frac{1}{u}\frac{du}{dx}.$$

Thus, for example,

$$\frac{d}{dx}[\ln(1 + x^2)] = \frac{1}{1 + x^2} \cdot 2x = \frac{2x}{1 + x^2} \quad \text{for all real } x$$

and

$$\frac{d}{dx}[\ln(1+3x)] = \frac{1}{1+3x} \cdot 3 = \frac{3}{1+3x} \qquad \text{for all } x > -\tfrac{1}{3}.$$

Example 1 Find the domain of f and find $f'(x)$ if

$$f(x) = \ln(x\sqrt{1+3x}).$$

SOLUTION For x to be in the domain of f, we must have $x\sqrt{1+3x} > 0$, and consequently, $x > 0$. The domain is the set of positive numbers.

Before differentiating f, we make use of the special properties of the logarithm:

$$f(x) = \ln(x\sqrt{1+3x}) = \ln x + \ln[(1+3x)^{1/2}] = \ln x + \tfrac{1}{2}\ln(1+3x).$$

From this we have

$$f'(x) = \frac{1}{x} + \frac{1}{2}\left(\frac{3}{1+3x}\right) = \frac{9x+2}{6x^2+2x}. \qquad \square$$

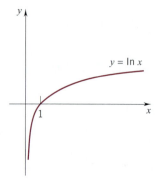

$y = \ln x$

Figure 7.3.1

Example 2 Sketch the graph of

$$f(x) = \ln|x|.$$

SOLUTION The domain of f is all numbers $x \neq 0$, and f is an even function: $f(-x) = f(x)$ for all $x \neq 0$. Thus, the graph has two branches:

$$y = \ln(-x), \qquad x < 0 \qquad \text{and} \qquad y = \ln x, \quad x > 0,$$

and is symmetric with respect to the y-axis. As we saw in the preceding section, the graph of $y = \ln x$, $x > 0$, is as shown in Figure 7.3.1. By symmetry, the graph of $y = \ln(-x)$, $x < 0$, is the mirror image of $y = \ln x$, $x > 0$, in the y-axis. Figure 7.3.2 shows the graph of $y = \ln|x|$. $\square$

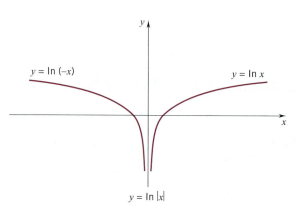

$y = \ln(-x)$ $y = \ln x$

$y = \ln|x|$

Figure 7.3.2

Example 3 Show that

(7.3.1)
$$\frac{d}{dx}(\ln|x|) = \frac{1}{x} \qquad \text{for all } x \neq 0.$$

SOLUTION For $x > 0$,

$$\frac{d}{dx}(\ln |x|) = \frac{d}{dx}(\ln x) = \frac{1}{x}.$$

For $x < 0$, we have $|x| = -x > 0$, so that

$$\frac{d}{dx}(\ln |x|) = \frac{d}{dx}[\ln(-x)] = \frac{1}{-x}\frac{d}{dx}(-x) = \left(\frac{1}{-x}\right)(-1) = \frac{1}{x}. \quad \square$$

Example 4 Find the domain of f and find $f'(x)$ given that

$$f(x) = (\ln x^2)^3.$$

SOLUTION Since the logarithm function is defined only for positive numbers, we must have $x^2 > 0$. Thus, the domain of f consists of all $x \neq 0$. Before differentiating, note that

$$f(x) = (\ln x^2)^3 = (2 \ln |x|)^3 = 8(\ln |x|)^3.$$

It follows that

$$f'(x) = 24(\ln |x|)^2 \frac{d}{dx}(\ln |x|) = 24(\ln |x|)^2 \frac{1}{x} = \frac{24}{x}(\ln |x|)^2. \quad \square$$

Example 5 Let $f(x) = x \ln x$.

(a) Specify the domain of f and find the intercepts, if any. **(b)** On what intervals does f increase? Decrease? **(c)** Find the extreme values of f. **(d)** Determine the concavity of the graph and find the points of inflection. **(e)** Sketch the graph of f.

SOLUTION Since the logarithm function is defined only for positive numbers, the domain of f is $(0, \infty)$ and there is no y-intercept. Since $f(1) = 1 \cdot \ln 1 = 0$, 1 is an x-intercept.

Differentiating f, we have

$$f'(x) = x \cdot \frac{1}{x} + \ln x = 1 + \ln x.$$

To find the critical numbers of f, we set $f'(x) = 0$:

$$1 + \ln x = 0, \qquad \ln x = -1, \qquad x = \frac{1}{e} \qquad \text{(verify this)}.$$

Recalling that the logarithm function is increasing on $(0, \infty)$, and that $\ln x \to -\infty$ as $x \to 0^+$ and $\ln x \to \infty$ as $x \to \infty$, we have

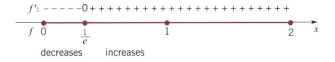

Therefore, f decreases on $(0, 1/e)$ and increases on $(1/e, \infty)$. By the first derivative test,

$$f(1/e) = \frac{1}{e}\ln\left(\frac{1}{e}\right) = -\frac{1}{e} \cong -0.368$$

is a local and absolute minimum of f.

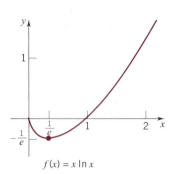

$f(x) = x \ln x$

Figure 7.3.3

Since $f''(x) = \dfrac{1}{x} > 0$ for $x > 0$, the graph of f is concave up on $(0, \infty)$; there are no points of inflection.

You can verify numerically that $\displaystyle\lim_{x \to 0^+} x \ln x = 0$ (see Exercise 32, Section 7.2), and it is clear that $x \ln x \to \infty$ as $x \to \infty$. Also, $\displaystyle\lim_{x \to 0^+} f'(x) = \lim_{x \to 0^+} (1 + \ln x) = -\infty$.

A sketch of the graph of f is shown in Figure 7.3.3. ❏

Example 6 Show that, if u is a differentiable function of x, then for $u \neq 0$

(7.3.2)

$$\frac{d}{dx}(\ln |u|) = \frac{1}{u}\frac{du}{dx}.$$

SOLUTION

$$\frac{d}{dx}(\ln |u|) = \frac{d}{du}(\ln |u|)\frac{du}{dx} = \frac{1}{u}\frac{du}{dx}. ❏$$

Here are two examples:

$$\frac{d}{dx}(\ln |1 - x^3|) = \frac{1}{1 - x^3}\frac{d}{dx}(1 - x^3) = \frac{-3x^2}{1 - x^3} = \frac{3x^2}{x^3 - 1}.$$

$$\frac{d}{dx}\left(\ln \left|\frac{x - 1}{x - 2}\right|\right) = \frac{d}{dx}(\ln |x - 1|) - \frac{d}{dx}(\ln |x - 2|) = \frac{1}{x - 1} - \frac{1}{x - 2}. ❏$$

Example 7 Let

$$f(x) = \ln \left(\frac{x^4}{x - 1}\right).$$

(a) Specify the domain of f. **(b)** On what intervals does f increase? Decrease? **(c)** Find the extreme values of f. **(d)** Determine the concavity of the graph and find the points of inflection. **(e)** Sketch the graph specifying the asymptotes if any.

SOLUTION Since the logarithm function is defined only for positive numbers, the domain of f is the open interval $(1, \infty)$.

Making use of the special properties of the logarithm we write

$$f(x) = \ln x^4 - \ln(x - 1) = 4 \ln x - \ln(x - 1).$$

Differentiation gives

$$f'(x) = \frac{4}{x} - \frac{1}{x - 1} = \frac{3x - 4}{x(x - 1)}$$

and

$$f''(x) = -\frac{4}{x^2} + \frac{1}{(x - 1)^2} = -\frac{(x - 2)(3x - 2)}{x^2(x - 1)^2}.$$

Since the domain of f is $(1, \infty)$, we consider only the values of x greater than 1.

It is not hard to see that

$$f'(x) \text{ is } \begin{cases} \text{negative,} & \text{for} & 1 < x < \frac{4}{3} \\ 0, & \text{at} & x = \frac{4}{3} \\ \text{positive,} & \text{for} & x > \frac{4}{3}. \end{cases}$$

Thus f decreases on $(1, \frac{4}{3}]$ and increases on $[\frac{4}{3}, \infty)$. By the first derivative test

$$f(\tfrac{4}{3}) = 4 \ln 4 - 3 \ln 3 \cong 2.25$$

is a local and absolute minimum. There are no other extreme values.

Since

$$f''(x) \text{ is } \begin{cases} \text{positive,} & \text{for} & 1 < x < 2 \\ 0, & \text{at} & x = 2 \\ \text{negative,} & \text{for} & x > 2, \end{cases}$$

the graph is concave up on $(1, 2)$ and concave down on $(2, \infty)$. The point

$$(2, f(2)) = (2, 4 \ln 2) \cong (2, 2.77)$$

is the only point of inflection.

Before sketching the graph, note that the derivative

$$f'(x) = \frac{4}{x} - \frac{1}{x-1}$$

is very large negative for x close to 1 and very close to 0 for x large. This means that the graph is very steep for x close to 1 and very flat for x large. See Figure 7.3.4. The line $x = 1$ is a vertical asymptote: as $x \to 1^+$, $f(x) \to \infty$. ❑

Figure 7.3.4

Integration

The integral counterpart of (7.3.1) takes the form

(7.3.3)

$$\int \frac{dx}{x} = \ln |x| + C, \qquad x \neq 0.$$

In practice

$$\int \frac{g'(x)}{g(x)} dx \quad \text{is reduced to} \quad \int \frac{du}{u}$$

by setting

$$u = g(x), \qquad du = g'(x)\, dx.$$

Thus, we have the important generalization of (7.3.3)

(7.3.4)
$$\int \frac{g'(x)}{g(x)} dx = \ln |g(x)| + C, \qquad g(x) \neq 0.$$

Example 8 Calculate

$$\int \frac{x^2}{1 - 4x^3} dx.$$

SOLUTION Set

$$u = 1 - 4x^3, \qquad du = -12x^2\, dx.$$

$$\int \frac{x^2}{1 - 4x^3} dx = -\frac{1}{12} \int \frac{du}{u} = -\frac{1}{12} \ln |u| + C = -\frac{1}{12} \ln |1 - 4x^3| + C. \quad \square$$

Example 9 Calculate

$$\int \frac{\ln x}{x} dx.$$

SOLUTION Set

$$u = \ln x, \qquad du = \frac{1}{x} dx.$$

$$\int \frac{\ln x}{x} dx = \int u\, du = \tfrac{1}{2} u^2 + C = \tfrac{1}{2}(\ln x)^2 + C. \quad \square$$

Example 10 Evaluate

$$\int_1^2 \frac{6x^2 - 2}{x^3 - x + 1} dx.$$

SOLUTION Set

$$u = x^3 - x + 1, \qquad du = (3x^2 - 1)\, dx.$$

At $x = 1$, $u = 1$; at $x = 2$, $u = 7$.

$$\int_1^2 \frac{6x^2 - 2}{x^3 - x + 1} dx = 2 \int_1^7 \frac{du}{u} = 2 \left[\ln |u| \right]_1^7 = 2(\ln 7 - \ln 1) = 2 \ln 7. \quad \square$$

Logarithmic Differentiation

We can differentiate a lengthy product

$$g(x) = g_1(x)g_2(x) \cdots g_n(x)$$

by first writing

$$\ln|g(x)| = \ln(|g_1(x)||g_2(x)| \cdots |g_n(x)|)$$
$$= \ln|g_1(x)| + \ln|g_2(x)| + \cdots + \ln|g_n(x)|$$

and then differentiating:

$$\frac{g'(x)}{g(x)} = \frac{g_1'(x)}{g_1(x)} + \frac{g_2'(x)}{g_2(x)} + \cdots + \frac{g_n'(x)}{g_n(x)}.$$

Multiplication by $g(x)$ gives

(7.3.5)
$$g'(x) = g(x)\left(\frac{g_1'(x)}{g_1(x)} + \frac{g_2'(x)}{g_2(x)} + \cdots + \frac{g_n'(x)}{g_n(x)}\right).$$

The process by which $g'(x)$ was obtained is called *logarithmic differentiation*. Logarithmic differentiation is valid at all points x where $g(x) \neq 0$. At points x where $g(x) = 0$, none of it makes sense.

A product of n factors

$$g(x) = g_1(x)g_2(x) \cdots g_n(x)$$

can, of course, also be differentiated by repeated applications of the product rule (3.2.5). The great advantage of logarithmic differentiation is that it gives us an explicit formula for the derivative, a formula that's easy to remember and easy to work with.

Example 11 Given that

$$g(x) = x(x - 1)(x - 2)(x - 3)$$

find $g'(x)$ for $x \neq 0, 1, 2, 3$.

SOLUTION We can write down $g'(x)$ directly from Formula (7.3.5),

$$g'(x) = x(x - 1)(x - 2)(x - 3)\left(\frac{1}{x} + \frac{1}{x - 1} + \frac{1}{x - 2} + \frac{1}{x - 3}\right)$$

or we can go through the process by which we derived Formula (7.3.5):

$$\ln|g(x)| = \ln|x| + \ln|x - 1| + \ln|x - 2| + \ln|x - 3|,$$

$$\frac{g'(x)}{g(x)} = \frac{1}{x} + \frac{1}{x - 1} + \frac{1}{x - 2} + \frac{1}{x - 3},$$

$$g'(x) = x(x - 1)(x - 2)(x - 3)\left(\frac{1}{x} + \frac{1}{x - 1} + \frac{1}{x - 2} + \frac{1}{x - 3}\right). \quad \square$$

Logarithmic differentiation can also be used with quotients.

Example 12 Given that

$$g(x) = \frac{(x^2 + 1)^3(2x - 5)^2}{(x^2 + 5)^2},$$

find $g'(x)$ for $x \neq \frac{5}{2}$.

SOLUTION Our first step is to write

$$g(x) = (x^2 + 1)^3(2x - 5)^2(x^2 + 5)^{-2}.$$

Then according to Formula (7.3.5)

$$g'(x) = \frac{(x^2 + 1)^3(2x - 5)^2}{(x^2 + 5)^2} \left[\frac{3(x^2 + 1)^2(2x)}{(x^2 + 1)^3} + \frac{2(2x - 5)(2)}{(2x - 5)^2} + \frac{(-2)(x^2 + 5)^{-3}(2x)}{(x^2 + 5)^{-2}} \right]$$

$$= \frac{(x^2 + 1)^3(2x - 5)^2}{(x^2 + 5)^2} \left(\frac{6x}{x^2 + 1} + \frac{4}{2x - 5} - \frac{4x}{x^2 + 5} \right).$$

Equivalently, using the basic properties listed in (7.2.10), we have

$$\ln|g(x)| = \ln(x^2 + 1)^3 + \ln(2x - 5)^2 - \ln(x^2 + 5)^2$$
$$= 3 \ln(x^2 + 1) + 2 \ln|2x - 5| - 2 \ln(x^2 + 5).$$

(We have omitted absolute value in the first and third terms since $x^2 + 1$ and $x^2 + 5$ are positive for all x.) Now

$$\frac{g'(x)}{g(x)} = \frac{3(2x)}{x^2 + 1} + \frac{2(2)}{2x - 5} - \frac{2(2x)}{x^2 + 5}$$

and so

$$g'(x) = g(x) \left[\frac{6x}{x^2 + 1} + \frac{4}{2x - 5} - \frac{4x}{x^2 + 5} \right]$$

as we saw above. ❏

EXERCISES 7.3

In Exercises 1–20, determine the domain and find the derivative.

1. $f(x) = \ln 4x$.

2. $f(x) = \ln(2x + 1)$.

3. $f(x) = \ln(x^3 + 1)$.

4. $f(x) = \ln[(x + 1)^3]$.

5. $f(x) = \ln\sqrt{1 + x^2}$.

6. $f(x) = (\ln x)^3$.

7. $f(x) = \ln|x^4 - 1|$.

8. $f(x) = \ln(\ln x)$.

9. $f(x) = x^2 \ln x$.

10. $f(x) = \ln \left| \dfrac{x + 2}{x^3 - 1} \right|$.

11. $f(x) = \dfrac{1}{\ln x}$.

12. $f(x) = \ln \sqrt[4]{x^2 + 1}$.

13. $f(x) = \dfrac{\ln(x + 1)}{x + 1}$.

14. $f(x) = \ln\sqrt{\dfrac{1 - x}{2 - x}}$.

15. $f(x) = \sin(\ln x)$.

16. $f(x) = \ln(\sin x)$.

17. $f(x) = \ln|\cos x|$.

18. $f(x) = \cos(\ln x)$.

19. $f(x) = \ln|\sec x + \tan x|$.

20. $f(x) = \ln|\csc x - \cot x|$.

In Exercises 21–38, calculate the given integral.

21. $\displaystyle\int \frac{dx}{x + 1}$.

22. $\displaystyle\int \frac{dx}{3 - x}$.

23. $\displaystyle\int \frac{x}{3 - x^2} dx$.

24. $\displaystyle\int \frac{x + 1}{x^2} dx$.

25. $\displaystyle\int \frac{x}{(3 - x^2)^2} dx$.

26. $\displaystyle\int \frac{\ln(x + a)}{x + a} dx$.

27. $\displaystyle\int \frac{\sin x}{2 + \cos x} dx$.

28. $\displaystyle\int \cot x \, dx$.

29. $\displaystyle\int \tan x \, dx$.

30. $\displaystyle\int \frac{\sec^2 2x}{4 - \tan 2x} dx$.

31. $\displaystyle\int \left(\frac{1}{x + 2} - \frac{1}{x - 2} \right) dx$.

32. $\int \dfrac{x^2}{2x^3 - 1} dx.$

33. $\int \dfrac{dx}{x(\ln x)^2}.$

34. $\int \left(\dfrac{1}{x - a} - \dfrac{1}{x - b} \right) dx.$

35. $\int \dfrac{\sin x - \cos x}{\sin x + \cos x} dx.$

36. $\int \dfrac{1}{\sqrt{x}(1 + \sqrt{x})} dx$ HINT: Let $u = \sqrt{x}$.

37. $\int \dfrac{\sqrt{x}}{1 + x\sqrt{x}} dx.$

38. $\int x \left(\dfrac{1}{x^2 - a^2} - \dfrac{1}{x^2 - b^2} \right) dx.$

In Exercises 39–46, evaluate the definite integral.

39. $\int_1^e \dfrac{dx}{x}.$

40. $\int_1^{e^2} \dfrac{dx}{x}.$

41. $\int_e^{e^2} \dfrac{dx}{x}.$

42. $\int_3^4 \dfrac{dx}{1 - x}.$

43. $\int_4^5 \dfrac{x}{x^2 - 1} dx.$

44. $\int_1^e \dfrac{\ln x}{x} dx.$

45. $\int_0^1 \dfrac{\ln(x + 1)}{x + 1} dx.$

46. $\int_0^1 \left(\dfrac{1}{x + 1} - \dfrac{1}{x + 2} \right) dx.$

In Exercises 47–52, calculate the derivative by logarithmic differentiation.

47. $g(x) = (x^2 + 1)^2 (x - 1)^5 x^3.$

48. $g(x) = x(x + a)(x + b)(x + c).$

49. $g(x) = \dfrac{x^4(x - 1)}{(x + 2)(x^2 + 1)}.$

50. $g(x) = \dfrac{(1 + x)(2 + x)x}{(4 + x)(2 - x)}.$

51. $g(x) = \sqrt{\dfrac{(x - 1)(x - 2)}{(x - 3)(x - 4)}}.$

52. $g(x) = \dfrac{x^2(x^2 + 1)(x^2 + 2)}{(x^2 - 1)(x^2 - 5)}.$

53. The region bounded by the graph of $f(x) = 1/\sqrt{1 + x}$ and the x-axis for $0 \leq x \leq 8$ is revolved around the x-axis. Find the volume of the solid that is generated.

54. The region bounded by the graph of $f(x) = 3/(1 + x^2)$ and the x-axis for $0 \leq x \leq 3$ is revolved around the y-axis. Find the volume of the solid that is generated.

In Exercises 55 and 56, find the area of the part of the first quadrant that lies between

55. $x + 4y - 5 = 0$ and $xy = 1.$

56. $x + y - 3 = 0$ and $xy = 2.$

57. A particle moves along a line with acceleration $a(t) = -(t + 1)^{-2}$ feet per second per second. Find the distance traveled by the particle during the time interval $[0, 4]$, given that the initial velocity $v(0)$ is 1 foot per second.

58. Exercise 57 with $v(0)$ as 2 feet per second.

In Exercises 59–62, find a formula for the nth derivative.

59. $\dfrac{d^n}{dx^n}(\ln x).$

60. $\dfrac{d^n}{dx^n}[\ln(1 - x)].$

61. $\dfrac{d^n}{dx^n}(\ln 2x).$

62. $\dfrac{d^n}{dx^n}\left(\ln \dfrac{1}{x} \right).$

63. (a) Verify that

$$\ln x = \int_1^x \dfrac{dt}{t} < \int_1^x \dfrac{dt}{\sqrt{t}} \qquad \text{for } x > 1.$$

(b) Use (a) to show that

$$0 < \ln x < 2(\sqrt{x} - 1) \qquad \text{for } x > 1.$$

(c) Use (b) to show that

$$2x\left(1 - \dfrac{1}{\sqrt{x}} \right) < x \ln x < 0 \qquad \text{for } 0 < x < 1.$$

(d) Use (c) to show that

$$\lim_{x \to 0^+} (x \ln x) = 0.$$

64. (a) Show that for $n = 2$ Formula (7.3.5) reduces to the product rule (3.2.5).

(b) Show that Formula (7.3.5) applied to

$$g(x) = \dfrac{g_1(x)}{g_2(x)}$$

reduces to the quotient rule (3.2.10).

In Exercises 65–71, (i) find the domain of f, (ii) find the intervals where the function increases and the intervals where it decreases, (iii) find the extreme values, (iv) determine the concavity of the graph and find the points of inflection, and finally, (v) sketch the graph, indicating asymptotes.

65. $f(x) = \ln 2x.$

66. $f(x) = x - \ln x.$

67. $f(x) = \ln(4 - x).$

68. $f(x) = \ln(4 - x^2).$

69. $f(x) = x^2 \ln x.$

70. $f(x) = \ln(8x - x^2).$

71. $f(x) = \ln \left(\dfrac{x}{1 + x^2} \right).$

72. (a) Show that $f(x) = \ln 2x$ and $g(x) = \ln 3x$ have the same derivative.
   (b) Calculate the derivative of $F(x) = \ln kx$, where k is any positive number.
   (c) Explain these results in terms of the properties of logarithms.

► In Exercises 73–76, use a graphing utility to graph f on the indicated interval. Estimate the x-intercepts of the graph of f and the values of x where f has either a local or absolute extremum (if any). Use four decimal place accuracy in your answers.

73. $f(x) = \sqrt{x} \ln x$; $[0, 10]$.

74. $f(x) = x^2 \ln x$; $[0, 2]$.

75. $f(x) = \sin(\ln x)$; $[1, 100]$.

76. $f(x) = x^2 \ln(\sin x)$; $[0, 2]$.

► **77.** A particle moves along a line with acceleration $a(t) = 4 - 2(t + 1) + 3/(t + 1)$ feet per second per second for $0 \le t \le 3$.
   (a) Find the velocity v of the particle at any time t during the time interval $[0, 3]$, given that the initial velocity $v(0) = 2$.
   (b) Use a graphing utility to graph v and a together.
   (c) Estimate the time $t \in [0, 3]$ at which the particle has maximum velocity and the time at which it has minimum velocity. Use four decimal place accuracy.

► **78.** Repeat Exercise 77 for a particle whose acceleration is $a(t) = 2 \cos 2(t + 1) + 2/(t + 1)$ feet per second per second for $0 \le t \le 7$.

■ 7.4 THE EXPONENTIAL FUNCTION

Rational powers of e already have an established meaning: by $e^{p/q}$ we mean the qth root of e raised to the pth power. But what is meant by $e^{\sqrt{2}}$ or e^π?

Earlier we proved that each rational power $e^{p/q}$ had logarithm p/q:

(7.4.1)
$$\ln e^{p/q} = \frac{p}{q}.$$

The definition of e^z for z irrational is patterned after this relation.

DEFINITION 7.4.2

If z is irrational, then by e^z we mean the unique number which has logarithm z:

$$\ln e^z = z.$$

What is $e^{\sqrt{2}}$? It is the unique number that has logarithm $\sqrt{2}$. What is e^π? It is the unique number that has logarithm π. Note that e^x now has meaning for every real value of x: it is the unique number with logarithm x.

DEFINITION 7.4.3

The function

$$E(x) = e^x \qquad \text{for all real } x$$

is called the *exponential function.*

Some properties of the exponential function follow.

(1) In the first place

(7.4.4)

$$\ln e^x = x \qquad \text{for all real } x.$$

Writing $L(x) = \ln x$ and $E(x) = e^x$, we have

$$L(E(x)) = x \qquad \text{for all real } x.$$

This says that *the exponential function is the inverse of the logarithm function.*

(2) The graph of the exponential function appears in Figure 7.4.1. It can be obtained from the graph of the logarithm by reflection in the line $y = x$.

(3) Since the graph of the logarithm lies to the right of the y-axis, the graph of the exponential function lies above the x-axis:

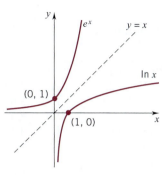

Figure 7.4.1

(7.4.5)

$$e^x > 0 \qquad \text{for all real } x.$$

(4) Since the graph of the logarithm crosses the x-axis at $(1, 0)$, the graph of the exponential function crosses the y-axis at $(0, 1)$:

$$\ln 1 = 0 \quad \text{gives} \quad e^0 = 1.$$

(5) Since the y-axis is a vertical asymptote for the graph of the logarithm function, the x-axis is a horizontal asymptote for the graph of the exponential function:

$$\text{as } x \to -\infty, \quad e^x \to 0.$$

(6) Since the exponential function is the inverse of the logarithm function, the logarithm function is the inverse of the exponential function; namely, we must have

(7.4.6)

$$e^{\ln x} = x \qquad \text{for all } x > 0.$$

You can verify this equation directly by observing that both sides have the same logarithm:

$$\ln(e^{\ln x}) = \ln x$$

since, for all real t, $\ln e^t = t$. ❑

You know that for rational exponents

$$e^{(p/q + r/s)} = e^{p/q} \cdot e^{r/s}.$$

This property holds for all exponents, including irrational exponents.

THEOREM 7.4.7

$$e^{x+y} = e^x \cdot e^y \qquad \text{for all real } x \text{ and } y.$$

PROOF

$$\ln(e^x \cdot e^y) = \ln e^x + \ln e^y = x + y = \ln e^{x+y}.$$

Since the logarithm is one-to-one, we must have

$$e^{x+y} = e^x \cdot e^y. \quad \square$$

We leave it to you to show that

■ Table 7.4.1

x	e^x	x	e^x
0.1	1.11	1.1	3.00
0.2	1.22	1.2	3.32
0.3	1.35	1.3	3.67
0.4	1.49	1.4	4.06
0.5	1.65	1.5	4.48
0.6	1.82	1.6	4.95
0.7	2.01	1.7	5.47
0.8	2.23	1.8	6.05
0.9	2.46	1.9	6.68
1.0	2.72	2.0	7.39

(7.4.8)

$$e^{-y} = \frac{1}{e^y} \quad \text{and} \quad e^{x-y} = \frac{e^x}{e^y}.$$

Table 7.4.1 gives some values of the exponential function rounded off to the nearest hundredth.

Example 1 Estimate the following powers of e using the approximations in Table 7.4.1.
(a) $e^{-0.2}$. **(b)** $e^{2.4}$. **(c)** $e^{3.1}$.

SOLUTION The idea, of course, is to use the laws of exponents.

(a) $e^{-0.2} = 1/e^{0.2} \cong 1/1.22 \cong 0.82$.

(b) $e^{2.4} = e^{2+0.4} = (e^2)(e^{0.4}) \cong (7.39)(1.49) \cong 11.01$.

(c) $e^{3.1} = e^{1.7+1.4} = (e^{1.7})(e^{1.4}) \cong (5.47)(4.06) \cong 22.21$ $\square$

We come now to one of the most important results in calculus. It is marvelously simple.

THEOREM 7.4.9

The exponential function is its own derivative: for all real x

$$\frac{d}{dx}(e^x) = e^x.$$

PROOF The logarithm function is differentiable, and its derivative is never 0. It follows from our discussion in Section 7.1 that its inverse, the exponential function, is also differentiable. Knowing this, we can show that

$$\frac{d}{dx}(e^x) = e^x$$

by differentiating the identity

$$\ln e^x = x.$$

On the left-hand side, the chain rule gives

$$\frac{d}{dx}(\ln e^x) = \frac{1}{e^x}\frac{d}{dx}(e^x).$$

On the right-hand side, the derivative is 1:

$$\frac{d}{dx}(x) = 1.$$

Equating these derivatives, we have

$$\frac{1}{e^x}\frac{d}{dx}(e^x) = 1 \quad \text{and thus} \quad \frac{d}{dx}(e^x) = e^x. \quad \square$$

We frequently run across expressions of the form e^u where u is a function of x. If u is differentiable, then the chain rule gives

(7.4.10)

$$\boxed{\frac{d}{dx}(e^u) = e^u\frac{du}{dx}.}$$

PROOF

$$\frac{d}{dx}(e^u) = \frac{d}{du}(e^u)\frac{du}{dx} = e^u\frac{du}{dx}. \quad \square$$

Example 2

(a) $\dfrac{d}{dx}(e^{kx}) = e^{kx} \cdot k = ke^{kx}.$

(b) $\dfrac{d}{dx}(e^{\sqrt{x}}) = e^{\sqrt{x}}\dfrac{d}{dx}(\sqrt{x}) = e^{\sqrt{x}}\left(\dfrac{1}{2\sqrt{x}}\right) = \dfrac{1}{2\sqrt{x}}e^{\sqrt{x}}.$

(c) $\dfrac{d}{dx}(e^{-x^2}) = e^{-x^2}\dfrac{d}{dx}(-x^2) = e^{-x^2}(-2x) = -2xe^{-x^2}. \quad \square$

The relation

$$\frac{d}{dx}(e^x) = e^x \quad \text{and its corollary} \quad \frac{d}{dx}(e^{kx}) = ke^{kx}$$

have important applications to engineering, physics, chemistry, biology, and economics. We discuss some of these applications in Section 7.6.

Example 3 Let

$$f(x) = xe^{-x} \qquad \text{for all real } x.$$

(a) On what intervals does f increase? Decrease? **(b)** Find the extreme values of f. **(c)** Determine the concavity of the graph and find the points of inflection. **(d)** Sketch the graph indicating the asymptotes if any.

SOLUTION   We have

$$f(x) = xe^{-x},$$
$$f'(x) = xe^{-x}(-1) + e^{-x} = (1 - x)e^{-x},$$
$$f''(x) = (1 - x)e^{-x}(-1) - e^{-x} = (x - 2)e^{-x}.$$

Since $e^{-x} > 0$ for all real x,

$$f'(x) \text{ is } \begin{cases} \text{positive,} & \text{for} \quad x < 1 \\ 0, & \text{at} \quad x = 1 \\ \text{negative,} & \text{for} \quad x > 1. \end{cases}$$

The function f increases on $(-\infty, 1]$ and decreases on $[1, \infty)$. The number

$$f(1) = \frac{1}{e} \cong \frac{1}{2.72} \cong 0.368$$

is a local and absolute maximum. There are no other extreme values. Since

$$f''(x) \text{ is } \begin{cases} \text{negative,} & \text{for} \quad x < 2 \\ 0, & \text{at} \quad x = 2 \\ \text{positive,} & \text{for} \quad x > 2, \end{cases}$$

the graph is concave down on $(-\infty, 2)$ and concave up on $(2, \infty)$. The point

$$(2, f(2)) = (2, 2e^{-2}) \cong \left(2, \frac{2}{(2.72)^2}\right) \cong (2, 0.27)$$

is the only point of inflection. In Section 10.6, we prove that $f(x) = x/e^x \to 0$ as $x \to \infty$. Accepting this result for now, it follows that the x-axis is a horizontal asymptote. The graph is given in Figure 7.4.2.

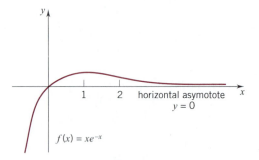

Figure 7.4.2

Example 4   Let

$$f(x) = e^{-x^2/2} \qquad \text{for all real } x.$$

(a) Determine the symmetry of the graph and the asymptotes, if any. **(b)** On what intervals does f increase? Decrease? **(c)** Find the extreme values. **(d)** Determine the concavity of the graph and find the points of inflection. **(e)** Sketch the graph.

NOTE.   This function places a very important role in the mathematical fields of probability and statistics. As you will see after we complete (a)–(e), its graph is the familiar bell-shaped curve.

SOLUTION   Since $f(-x) = e^{-(-x)^2/2} = e^{-x^2/2} = f(x)$, f is an even function so its graph is symmetric with respect to the y-axis. As $x \to \pm\infty$, $e^{-x^2/2} \to 0$. Therefore, the x-axis is a horizontal asymptote. There are no vertical asymptotes.

Differentiating f, we have

$$f'(x) = e^{-x^2/2}(-x) = -xe^{-x^2/2}$$

$$f''(x) = -x(-xe^{-x^2/2}) - e^{-x^2/2} = (x^2 - 1)e^{-x^2/2}.$$

Since $e^{-x^2/2} > 0$ for all x,

$$f'(x) \text{ is } \begin{cases} \text{positive,} & \text{for} \quad x < 0 \\ 0, & \text{at} \quad x = 0 \\ \text{negative,} & \text{for} \quad x > 0. \end{cases}$$

Thus, f is increasing on $(-\infty, 0]$ and decreasing on $[0, \infty)$. The value

$$f(0) = e^0 = 1$$

is a local and absolute maximum. There are no other extreme values.

Now consider $f''(x) = (x^2 - 1)e^{-x^2/2}$. The sign of f'' and the behavior of the graph of f are:

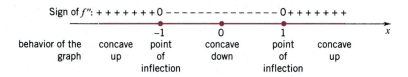

Thus, the graph of f is concave up on $(-\infty, -1)$ and $(1, \infty)$, the graph is concave down on $(-1, 1)$, and $(-1, e^{-1/2})$ and $(1, e^{-1/2})$ are points of inflection.

The graph of f is shown in Figure 7.4.3. ❑

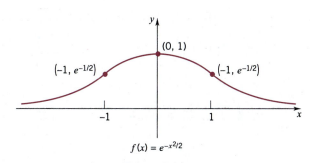

Figure 7.4.3

The integral counterpart of (7.4.9) takes the form

(7.4.11)
$$\int e^x \, dx = e^x + C.$$

In practice

$$\int e^{g(x)}g'(x) \, dx \quad \text{is reduced to} \quad \int e^u \, du$$

by setting

$$u = g(x), \quad du = g'(x) \, dx.$$

Thus, we have the important generalization of (7.4.11)

(7.4.12)

$$\int e^{g(x)}g'(x)\,dx = e^{g(x)} + C.$$

Example 5 Find

$$\int 9e^{3x}\,dx.$$

SOLUTION Set

$$u = 3x, \qquad du = 3\,dx.$$

$$\int 9e^{3x}\,dx = 3\int e^u\,du = 3e^u + C = 3e^{3x} + C.$$

If you recognize at the very beginning that

$$3e^{3x} = \frac{d}{dx}(e^{3x}),$$

then you can dispense with the u-substitution and simply write

$$\int 9e^{3x}\,dx = 3\int 3e^{3x}\,dx = 3e^{3x} + C. \quad \square$$

Example 6 Find

$$\int \frac{e^{\sqrt{x}}}{\sqrt{x}}\,dx.$$

SOLUTION Set

$$u = \sqrt{x}, \qquad du = \frac{1}{2\sqrt{x}}\,dx.$$

$$\int \frac{e^{\sqrt{x}}}{\sqrt{x}}\,dx = 2\int e^u\,du = 2e^u + C = 2e^{\sqrt{x}} + C.$$

If you recognize that

$$\tfrac{1}{2}\left(\frac{e^{\sqrt{x}}}{\sqrt{x}}\right) = \frac{d}{dx}(e^{\sqrt{x}}),$$

then you can dispense with the u-substitution and integrate directly:

$$\int \frac{e^{\sqrt{x}}}{\sqrt{x}}\,dx = 2\int \frac{1}{2}\left(\frac{e^{\sqrt{x}}}{\sqrt{x}}\right)dx = 2e^{\sqrt{x}} + C. \quad \square$$

Example 7 Find

$$\int \frac{e^{3x}}{e^{3x} + 1} \, dx.$$

SOLUTION We can put this integral in the form

$$\int \frac{du}{u}$$

by setting

$$u = e^{3x} + 1, \qquad du = 3e^{3x} \, dx.$$

Then

$$\int \frac{e^{3x}}{e^{3x} + 1} \, dx = \frac{1}{3} \int \frac{du}{u} = \frac{1}{3} \ln |u| + C = \frac{1}{3} \ln(e^{3x} + 1) + C. \quad ❑$$

Example 8 Integrals involving $xe^{-x^2/2}$ also play an important role in probability and statistics. Evaluate

$$\int_0^{\sqrt{2 \ln 3}} xe^{-x^2/2} \, dx.$$

SOLUTION Set

$$u = -\tfrac{1}{2}x^2, \qquad du = -x \, dx.$$

At $x = 0$, $u = 0$; at $x = \sqrt{2 \ln 3}$, $u = -\ln 3$. Thus

$$\int_0^{\sqrt{2 \ln 3}} xe^{-x^2/2} \, dx = -\int_0^{-\ln 3} e^u \, du = -\left[e^u \right]_0^{-\ln 3} = 1 - e^{-\ln 3} = 1 - \tfrac{1}{3} = \tfrac{2}{3}. \quad ❑$$

Example 9 Evaluate

$$\int_0^1 e^x(e^x + 1)^{1/5} \, dx.$$

SOLUTION Set

$$u = e^x + 1, \qquad du = e^x \, dx.$$

At $x = 0$, $u = 2$; at $x = 1$, $u = e + 1$. Thus

$$\int_0^1 e^x(e^x + 1)^{1/5} \, dx = \int_2^{e+1} u^{1/5} \, du = \left[\tfrac{5}{6}u^{6/5} \right]_2^{e+1} = \tfrac{5}{6}[(e + 1)^{6/5} - 2^{6/5}]. \quad ❑$$

Remark The u-substitution simplifies many calculations, but you will find that in many cases you can carry out the integration more quickly without it. ❑

EXERCISES 7.4

In Exercises 1–30, differentiate the given function.

1. $y = e^{-2x}$.

2. $y = 3e^{2x+1}$.

3. $y = e^{x^2-1}$.

4. $y = 2e^{-4x}$.

5. $y = e^x \ln x$.

6. $y = x^2 e^x$.

7. $y = x^{-1} e^{-x}$.

8. $y = e^{\sqrt{x+1}}$.

9. $y = \frac{1}{2}(e^x + e^{-x})$.

10. $y = \frac{1}{2}(e^x - e^{-x})$.

11. $y = e^{\sqrt{x}} \ln \sqrt{x}$.

12. $y = (1 - e^{4x})^2$.

13. $y = (e^x + e^{-x})^2$.

14. $y = (3 - 2e^{-x})^3$.

15. $y = (e^{x^2} + 1)^2$.

16. $y = (e^{2x} - e^{-2x})^2$.

17. $y = (x^2 - 2x + 2)e^x$.

18. $y = x^2 e^x - xe^{x^2}$.

19. $y = \dfrac{e^x - 1}{e^x + 1}$.

20. $y = \dfrac{e^{2x} - 1}{e^{2x} + 1}$.

21. $y = \dfrac{e^{ax} - e^{bx}}{e^{ax} + e^{bx}}$.

22. $y = \ln e^{3x}$.

23. $y = e^{4\ln x}$.

24. $y = e^{\sqrt{1-x^2}}$.

25. $f(x) = \sin(e^{2x})$.

26. $f(x) = e^{\sin 2x}$.

27. $f(x) = e^{-2x} \cos x$.

28. $f(x) = \ln(\cos e^{2x})$.

29. $f(x) = \tan\sqrt{e^{-3x}}$.

30. $f(x) = \sec(e^{\tan x})$.

In Exercises 31–52, calculate the indefinite integral.

31. $\displaystyle\int e^{2x}\, dx$.

32. $\displaystyle\int e^{-2x}\, dx$.

33. $\displaystyle\int e^{kx}\, dx$.

34. $\displaystyle\int e^{ax+b}\, dx$.

35. $\displaystyle\int xe^{x^2}\, dx$.

36. $\displaystyle\int xe^{-x^2}\, dx$.

37. $\displaystyle\int \frac{e^{1/x}}{x^2}\, dx$.

38. $\displaystyle\int \frac{e^{2\sqrt{x}}}{\sqrt{x}}\, dx$.

39. $\displaystyle\int (e^x + e^{-x})^2\, dx$.

40. $\displaystyle\int e^{\ln x}\, dx$.

41. $\displaystyle\int \ln e^x\, dx$.

42. $\displaystyle\int (e^{-x} - 1)^2\, dx$.

43. $\displaystyle\int \frac{4}{\sqrt{e^x}}\, dx$.

44. $\displaystyle\int \frac{e^x}{e^x + 1}\, dx$.

45. $\displaystyle\int \frac{e^x}{\sqrt{e^x + 1}}\, dx$.

46. $\displaystyle\int \frac{2e^x}{\sqrt[3]{e^x + 1}}\, dx$.

47. $\displaystyle\int \frac{e^{2x}}{2e^{2x} + 3}\, dx$.

48. $\displaystyle\int \frac{xe^{ax^2}}{e^{ax^2} + 1}\, dx$.

49. $\displaystyle\int \cos x\, e^{\sin x}\, dx$.

50. $\displaystyle\int \frac{\sin(e^{-2x})}{e^{2x}}\, dx$.

51. $\displaystyle\int e^{-x}[1 + \cos(e^{-x})]\, dx$.

52. $\displaystyle\int \sec^2 2x\, e^{\tan 2x}\, dx$.

In Exercises 53–64, evaluate the definite integral.

53. $\displaystyle\int_0^1 e^x\, dx$.

54. $\displaystyle\int_0^1 e^{-kx}\, dx$.

55. $\displaystyle\int_0^{\ln \pi} e^{-6x}\, dx$.

56. $\displaystyle\int_0^1 xe^{-x^2}\, dx$.

57. $\displaystyle\int_0^1 \frac{e^x + 1}{e^x}\, dx$.

58. $\displaystyle\int_0^1 \frac{4 - e^x}{e^x}\, dx$.

59. $\displaystyle\int_0^{\ln 2} \frac{e^x}{e^x + 1}\, dx$.

60. $\displaystyle\int_0^1 \frac{e^x + e^{-x}}{2}\, dx$.

61. $\displaystyle\int_{\ln 2}^{\ln 3} (e^x - e^{-x})^2\, dx$.

62. $\displaystyle\int_0^1 \frac{e^x}{4 - e^x}\, dx$.

63. $\displaystyle\int_0^1 x(e^{x^2} + 2)\, dx$.

64. $\displaystyle\int_1^2 (2 - e^{-x})^2\, dx$.

In Exercises 65–68, estimate on the basis of Table 7.4.1.

65. $e^{-0.4}$.

66. $e^{2.6}$.

67. $e^{2.8}$.

68. $e^{-2.1}$.

69. A particle moves on a coordinate line with its position at time t given by the function $x(t) = Ae^{ct} + Be^{-ct}$. Show that the acceleration of the particle is proportional to its position.

In Exercises 70–73, sketch the region bounded by the curves and find its area.

70. $x = e^{2y}$, $x = e^{-y}$, $x = 4$.

71. $y = e^x$, $y = e^{2x}$, $y = e^4$.

72. $y = e^x$, $y = e$, $y = x$, $x = 0$.

73. $x = e^y$, $y = 1$, $y = 2$, $x = 2$.

For each of the functions in Exercises 74–78, (i) find the domain, (ii) find the intervals where the function increases and the intervals where it decreases, (iii) find the extreme values, (iv) determine the concavity and find the points of inflection, and finally, (v) sketch the graph, indicating all asymptotes.

74. $f(x) = \frac{1}{2}(e^x - e^{-x})$.

75. $f(x) = \frac{1}{2}(e^x + e^{-x})$.

76. $f(x) = (1 - x)e^x$.

77. $f(x) = e^{(1/x)^2}$.

78. $f(x) = x^2 e^{-x}$.

79. Take $a > 0$ and refer to the figure.

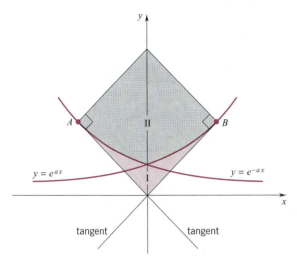

(a) Find the points of tangency, marked A and B.
(b) Find the area of region I.
(c) Find the area of region II.

80. Prove that for all $x > 0$ and all positive integers n

$$e^x > 1 + x + \frac{x^2}{2!} + \frac{x^3}{3!} + \cdots + \frac{x^n}{n!}.$$

Recall, $n! = n(n-1)(n-2) \cdots 3 \cdot 2 \cdot 1$.

HINT: $\quad e^x = 1 + \displaystyle\int_0^x e^t \, dt > 1 + \int_0^x dt = 1 + x,$

$$e^x = 1 + \int_0^x e^t \, dt > 1 + \int_0^x (1+t) \, dt$$

$$= 1 + x + \frac{x^2}{2}, \quad \text{and so on.}$$

81. Prove that, if n is a positive integer, then

$$e^x > x^n \qquad \text{for all } x \text{ sufficiently large.}$$

HINT: Use Exercise 80.

▶ In Exercises 82–84, evaluate the limit numerically and then justify your answers using a graphing utility.

82. $\displaystyle\lim_{x \to 0} \frac{e^{10x} - 1}{x}.$

83. $\displaystyle\lim_{x \to 1} \frac{e^{x^3} - e}{x - 1}.$

84. $\displaystyle\lim_{x \to 1} \frac{e^x - e}{\ln x}.$

HINT: Remember the alternative definition of derivative

$$f'(c) = \lim_{x \to c} \frac{f(x) - f(c)}{x - c}.$$

▶ **85.** Let $f(x) = e^x$ and $g(x) = 4 - x^2$.
(a) Use a graphing utility to sketch the graphs of f and g.
(b) Estimate the x-coordinates of the points of intersection of the two curves. Use four decimal place accuracy.
(c) Estimate the area of the region bounded by the two curves.

▶ **86.** Repeat Exercise 85 for the functions $f(x) = e^{-x^2}$ and $g(x) = x^2$.

▶ In Exercises 87–90, use a graphing utility to graph the functions f and g. Your graphs should suggest that f and g are inverses of each other. Confirm this by using the methods in Section 7.1.

87. $f(x) = e^{2x}$; $\quad g(x) = \ln \sqrt{x}.$

88. $f(x) = e^{x/2}$; $\quad g(x) = \ln x^2.$

89. $f(x) = e^{x^2}$; $\quad g(x) = \sqrt{\ln x}.$

90. $f(x) = e^{x-2}$; $\quad g(x) = 2 + \ln x.$

■ 7.5 ARBITRARY POWERS; OTHER BASES; ESTIMATING e

Arbitrary Powers: The Function $f(x) = x^r$

The elementary notion of exponent applies only to rational numbers. Expressions such as

$$10^5, \quad 2^{1/3}, \quad 7^{-4/5}, \quad \pi^{-1/2}$$

make sense, but so far we have attached no meaning to expressions such as

$$10^{\sqrt{2}}, \quad 2^\pi, \quad 7^{-\sqrt{3}}, \quad \pi^e.$$

The extension of our sense of exponent to allow for irrational exponents is conveniently done by making use of the logarithm function and the exponential function. The heart of the matter is to observe that for $x > 0$ and p/q rational

$$x^{p/q} = e^{(p/q) \ln x}.$$

(To verify this, take the log of both sides.)   We *define* x^z for irrational z by setting

$$x^z = e^{z \ln x}.$$

We then have the following result:

(7.5.1)

> if $x > 0$, then
>
> $$x^r = e^{r \ln x} \quad \text{for all real numbers } r.$$

In particular

$$10^{\sqrt{2}} = e^{\sqrt{2} \ln 10}, \quad 2^{\pi} = e^{\pi \ln 2}, \quad 7^{-\sqrt{3}} = e^{-\sqrt{3} \ln 7}, \quad \pi^e = e^{e \ln \pi}. \quad \square$$

With this extended sense of exponent the usual laws of exponents

(7.5.2)

> $$x^{r+s} = x^r x^s, \qquad x^{r-s} = \frac{x^r}{x^s}, \qquad (x^r)^s = x^{rs}$$

still hold:

$$x^{r+s} = e^{(r+s) \ln x} = e^{r \ln x} \cdot e^{s \ln x} = x^r x^s,$$

$$x^{r-s} = e^{(r-s) \ln x} = e^{r \ln x} \cdot e^{-s \ln x} = \frac{e^{r \ln x}}{e^{s \ln x}} = \frac{x^r}{x^s},$$

$$(x^r)^s = e^{s \ln x^r} = e^{rs \ln x} = x^{rs}. \quad \square$$

In Chapter 3, Section 3.7, we proved that $d(x^p)/dx = px^{p-1}$ for any rational number p.   We can now extend this result to arbitrary powers.   For *any real number r:*

(7.5.3)

> $$\frac{d}{dx}(x^r) = rx^{r-1} \qquad \text{for all } x > 0.$$

PROOF

$$\frac{d}{dx}(x^r) = \frac{d}{dx}(e^{r \ln x}) = e^{r \ln x}\frac{d}{dx}(r \ln x) = x^r\frac{r}{x} = rx^{r-1}.$$

You can also write $f(x) = x^r$ and use logarithmic differentiation:

$$\ln f(x) = r \ln x$$

$$\frac{f'(x)}{f(x)} = \frac{r}{x}$$

$$f'(x) = \frac{rf(x)}{x} = \frac{rx^r}{x} = rx^{r-1}. \quad \square$$

Thus

$$\frac{d}{dx}(x^{\sqrt{2}}) = \sqrt{2}x^{\sqrt{2}-1}, \qquad \frac{d}{dx}(x^{\pi}) = \pi x^{\pi-1}.$$

If u is a positive differentiable function of x and r is any real number, then, by the chain rule,

(7.5.4)

$$\frac{d}{dx}(u^r) = ru^{r-1}\frac{du}{dx}.$$

PROOF

$$\frac{d}{dx}(u^r) = \frac{d}{du}(u^r)\frac{du}{dx} = ru^{r-1}\frac{du}{dx}. \quad \square$$

For example,

$$\frac{d}{dx}[(x^2 + 5)^{\sqrt{3}}] = \sqrt{3}(x^2 + 5)^{\sqrt{3}-1}(2x) = 2\sqrt{3}x(x^2 + 5)^{\sqrt{3}-1}.$$

Each derivative formula gives rise to a companion integral formula. The integral form of (7.5.3) is:

(7.5.5)

$$\int x^r \, dx = \frac{x^{r+1}}{r+1} + C, \qquad \text{for } r \neq -1.$$

Example 1 Find

$$\int \frac{x^3}{(2x^4 + 1)^{\pi}} dx.$$

SOLUTION Set

$$u = 2x^4 + 1, \qquad du = 8x^3 \, dx.$$

$$\int \frac{x^3}{(2x^4 + 1)^{\pi}} dx = \frac{1}{8}\int \frac{du}{u^{\pi}} = \frac{1}{8}\left(\frac{u^{1-\pi}}{1-\pi}\right) + C = \frac{(2x^4 + 1)^{1-\pi}}{8(1-\pi)} + C. \quad \square$$

Example 2 Find

$$\frac{d}{dx}[(x^2 + 1)^{3x}].$$

SOLUTION One way to find the derivative is to observe that $(x^2 + 1)^{3x} = e^{3x\ln(x^2+1)}$ and then differentiate

$$\frac{d}{dx}[(x^2 + 1)^{3x}] = \frac{d}{dx}[e^{3x\ln(x^2+1)}] = e^{3x\ln(x^2+1)}\left[3x \cdot \frac{2x}{x^2 + 1} + \{\ln(x^2 + 1)\}(3)\right]$$

$$= (x^2 + 1)^{3x}\left[\frac{6x^2}{x^2 + 1} + 3\ln(x^2 + 1)\right].$$

Another way to find the derivative of $(x^2 + 1)^{3x}$ is to set $f(x) = (x^2 + 1)^{3x}$ and use logarithmic differentiation:

$$\ln f(x) = 3x \cdot \ln(x^2 + 1)$$

$$\frac{f'(x)}{f(x)} = 3x \cdot \frac{2x}{x^2 + 1} + [\ln(x^2 + 1)](3) = \frac{6x^2}{x^2 + 1} + 3\ln(x^2 + 1)$$

$$f'(x) = f(x)\left[\frac{6x^2}{x^2 + 1} + 3\ln(x^2 + 1)\right] = (x^2 + 1)^{3x}\left[\frac{6x^2}{x^2 + 1} + 3\ln(x^2 + 1)\right]. \quad \square$$

Base p: The Function $f(x) = p^x$

To form the function $f(x) = x^r$ we take a positive variable x and raise it to a constant power r. To form the function $f(x) = p^x$ we take a positive constant p and raise it to a variable power x. Since $1^x = 1$ for all x, the function is of interest only if $p \neq 1$.

Functions of the form $f(x) = p^x$ are called *exponential functions with base p*. The high status enjoyed by Euler's number e comes from the fact that

$$\frac{d}{dx}(e^x) = e^x.$$

For other bases the derivative has an extra factor:

(7.5.6)

$$\frac{d}{dx}(p^x) = p^x \ln p.$$

PROOF

$$\frac{d}{dx}(p^x) = \frac{d}{dx}(e^{x\ln p}) = e^{x\ln p}\ln p = p^x \ln p. \quad \square$$

Thus, for example,

$$\frac{d}{dx}(2^x) = 2^x \ln 2 \qquad \text{and} \qquad \frac{d}{dx}(10^x) = 10^x \ln 10.$$

If u is a differentiable function of x, then, by the chain rule,

(7.5.7)

$$\frac{d}{dx}(p^u) = p^u \ln p \frac{du}{dx}.$$

PROOF

$$\frac{d}{dx}(p^u) = \frac{d}{du}(p^u)\frac{du}{dx} = p^u \ln p \frac{du}{dx}. \quad \square$$

For example,

$$\frac{d}{dx}(2^{3x^2}) = 2^{3x^2}(\ln 2)(6x) = 6x\,2^{3x^2}\ln 2.$$

The integral form of (7.5.6) is

(7.5.8)
$$\int p^x \, dx = \frac{1}{\ln p} p^x + C, \qquad p > 0, \quad p \neq 1.$$

For example,

$$\int 2^x \, dx = \frac{1}{\ln 2} 2^x + C.$$

Example 3 Find

$$\int x 5^{-x^2} \, dx.$$

SOLUTION Set

$$u = -x^2, \qquad du = -2x \, dx.$$

$$\int x 5^{-x^2} \, dx = -\tfrac{1}{2} \int 5^u \, du = -\tfrac{1}{2} \left(\frac{1}{\ln 5} \right) 5^u + C$$

$$= \frac{-1}{2 \ln 5} 5^{-x^2} + C. \quad \square$$

Example 4 Evaluate

$$\int_1^2 3^{2x-1} \, dx.$$

SOLUTION Set

$$u = 2x - 1, \qquad du = 2 \, dx.$$

At $x = 1$, $u = 1$; at $x = 2$, $u = 3$. Thus

$$\int_1^2 3^{2x-1} \, dx = \tfrac{1}{2} \int_1^3 3^u \, du = \tfrac{1}{2} \left[\frac{1}{\ln 3} \cdot 3^u \right]_1^3 = \frac{12}{\ln 3} \cong 10.923. \quad \square$$

Base *p*: The Function $f(x) = \log_p x$

If $p > 0$, then

$$\ln p^t = t \ln p \qquad \text{for all } t.$$

If p is also different from 1, then $\ln p \neq 0$, and we have

$$\frac{\ln p^t}{\ln p} = t.$$

This indicates that the function

$$f(x) = \frac{\ln x}{\ln p}, \qquad x > 0,$$

satisfies the relation

$$f(p^t) = t \qquad \text{for all real } t.$$

In view of this we call

$$\frac{\ln x}{\ln p}$$

the logarithm of x to the base p and write:

(7.5.9)

$$\boxed{\begin{array}{l} \text{For } p > 0, p \neq 1, \\[2mm] \log_p x = \dfrac{\ln x}{\ln p}, \qquad x > 0. \end{array}}$$

For example,

$$\log_2 32 = \frac{\ln 32}{\ln 2} = \frac{\ln 2^5}{\ln 2} = \frac{5 \ln 2}{\ln 2} = 5$$

and

$$\log_{100}(\tfrac{1}{10}) = \frac{\ln(\tfrac{1}{10})}{\ln 100} = \frac{\ln 10^{-1}}{\ln 10^2} = \frac{-\ln 10}{2 \ln 10} = -\frac{1}{2}. \quad \square$$

We can obtain these same results more directly from the relation

(7.5.10)

$$\boxed{\log_p p^t = t.}$$

Accordingly

$$\log_2 32 = \log_2 2^5 = 5 \qquad \text{and} \qquad \log_{100}(\tfrac{1}{10}) = \log_{100}(100^{-1/2}) = -\tfrac{1}{2}. \quad \square$$

Differentiating (7.5.9) we have

(7.5.11)

$$\boxed{\frac{d}{dx}(\log_p x) = \frac{1}{x \ln p}.\dagger}$$

---

† The function $f(x) = \log_p x$ satisfies

$$f'(x) = \frac{1}{x \ln p}, \qquad f'(1) = \frac{1}{\ln p}.$$

This means that in general

$$f'(x) = \frac{1}{x} f'(1).$$

We predicted this from general considerations in Section 7.2. (See Formula (7.2.2).)

When p is e, the factor $\ln p$ is 1 and we have

$$\frac{d}{dx}(\log_e x) = \frac{1}{x}.$$

The logarithm to the base e, $\ln = \log_e$, is called the "*natural logarithm*" because it is the logarithm with the simplest derivative.

If u is a positive, differentiable function of x, then, by the chain rule

(7.5.12)
$$\boxed{\frac{d}{dx}(\log_p u) = \frac{1}{u \ln p} \cdot \frac{du}{dx}.}$$

Example 5 Find

(a) $\dfrac{d}{dx}[\log_2(3x^2 + 1)]$ and **(b)** $\dfrac{d}{dx}(\log_5 |x|)$.

SOLUTION

(a) Using (7.5.12), we have

$$\frac{d}{dx}[\log_2(3x^2 + 1)] = \frac{1}{(3x^2 + 1)\ln 2} \cdot \frac{du}{dx} = \frac{6x}{(3x^2 + 1)\ln 2}.$$

Alternatively, we could have used the definition of $\log_p x$ given by (7.5.9):

$$\frac{d}{dx}[\log_2(3x^2 + 1)] = \frac{d}{dx}\left[\frac{\ln(3x^2 + 1)}{\ln 2}\right] = \frac{1}{\ln 2} \cdot \frac{1}{(3x^2 + 1)} \cdot 6x$$

$$= \frac{6x}{(3x^2 + 1)\ln 2}$$

(b) It is easier to use (7.5.9) here:

$$\frac{d}{dx}(\log_5 |x|) = \frac{d}{dx}\left[\frac{\ln|x|}{\ln 5}\right] = \frac{1}{x \ln 5}. \quad ❑$$

There is no need to consider the companion integral form of (7.5.11) since the integral

$$\int \frac{1}{x \ln p}\,dx = \frac{1}{\ln p}\int \frac{1}{x}\,dx = \frac{\ln|x|}{\ln p} + C,$$

which can be expressed in terms of $\log_p$ by using (7.5.9) if we wished to do so. That is,

$$\frac{\ln|x|}{\ln p} + C = \log_p |x| + C.$$

Estimating the Number *e*

We defined the logarithm function by setting

$$\ln x = \int_1^x \frac{dt}{t}, \qquad x > 0.$$

We can derive a numerical estimate for e from this integral representation of the logarithm.

Since e is irrational, we cannot hope to express e as a repeating or terminating decimal. What we can do is describe a simple technique by which we can compute the value of e to any desired degree of accuracy.

THEOREM 7.5.13

For each positive integer n,

$$\left(1 + \frac{1}{n}\right)^n \leq e \leq \left(1 + \frac{1}{n}\right)^{n+1}.$$

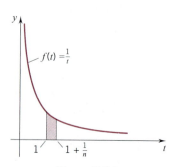

$f(t) = \frac{1}{t}$

Figure 7.5.1

PROOF Refer throughout to Figure 7.5.1.

$$\ln\left(1 + \frac{1}{n}\right) = \int_1^{1+1/n} \frac{dt}{t} \leq \int_1^{1+1/n} 1 \, dt = \frac{1}{n}.$$

since $\frac{1}{t} \leq 1$ throughout the interval of integration

$$\ln\left(1 + \frac{1}{n}\right) = \int_1^{1+1/n} \frac{dt}{t} \geq \int_1^{1+1/n} \frac{dt}{1 + 1/n} = \frac{1}{1 + 1/n} \cdot \frac{1}{n} = \frac{1}{n + 1}.$$

since $\frac{1}{t} \geq \frac{1}{1 + 1/n}$ throughout the interval of integration

We have now shown that

$$\frac{1}{n + 1} \leq \ln\left(1 + \frac{1}{n}\right) \leq \frac{1}{n}.$$

From the inequality on the right

$$1 + \frac{1}{n} \leq e^{1/n} \qquad \text{and thus} \qquad \left(1 + \frac{1}{n}\right)^n \leq e.$$

From the inequality on the left

$$e^{1/(n+1)} \leq 1 + \frac{1}{n} \qquad \text{and thus} \qquad e \leq \left(1 + \frac{1}{n}\right)^{n+1}. \quad ❑$$

The inequality

$$\left(1 + \frac{1}{n}\right)^n \leq e \leq \left(1 + \frac{1}{n}\right)^{n+1}$$

is an elegant characterization of e, but not a very efficient tool for calculating e. Reading from our calculator we find that

$$(1 + \tfrac{1}{100})^{100} \cong 2.7048138 \qquad \text{and} \qquad (1 + \tfrac{1}{100})^{101} \cong 2.7318619.$$

This gives e rounded off to one decimal place: $e \cong 2.7$. Going on to $n = 1000$, we find that

$$(1 + \tfrac{1}{1000})^{1000} \cong 2.7169239 \qquad \text{and} \qquad (1 + \tfrac{1}{1000})^{1001} \cong 2.7196409.$$

This gives a two-place estimate for e: $e \cong 2.72$. To get a five-place estimate for e, we have to go to about $n = 1,000,000$:

$$(1 + \tfrac{1}{1,000,000})^{1,000,000} \cong 2.7182805 \quad\text{and}\quad (1 + \tfrac{1}{1,000,000})^{1,000,001} \cong 2.7182832$$

and therefore, to five decimal places, $e \cong 2.71828$.

Based on these calculations, we are led to the conclusion that $(1 + 1/n)^n \to e$ as $n \to \infty$. This result is confirmed in Exercise 90. A more efficient way of estimating e is provided by Taylor series (Chapter 11). ❑

EXERCISES 7.5

In Exercises 1–12, find the indicated quantity.

1. $\log_2 64$.

2. $\log_2 \frac{1}{64}$.

3. $\log_{64} \frac{1}{2}$.

4. $\log_{10} 0.01$.

5. $\log_5 1$.

6. $\log_5 0.2$.

7. $\log_5 125$.

8. $\log_2 4^3$.

9. $\log_{32} 8$.

10. $\log_{100} 10^{-4/5}$.

11. $\log_{10} 100^{-4/5}$.

12. $\log_9 \sqrt{3}$.

In Exercises 13–16, show that the given identity holds.

13. $\log_p xy = \log_p x + \log_p y$.

14. $\log_p \dfrac{1}{x} = -\log_p x$.

15. $\log_p x^y = y \log_p x$.

16. $\log_p \dfrac{x}{y} = \log_p x - \log_p y$.

In Exercises 17–22, find those numbers x, if any, which satisfy the given equation.

17. $10^x = e^x$.

18. $\log_5 x = 0.04$.

19. $\log_x 10 = \log_4 100$.

20. $\log_x 2 = \log_3 x$.

21. $\log_2 x = \displaystyle\int_2^x \dfrac{dt}{t}$.

22. $\log_x 10 = \log_2(\tfrac{1}{10})$.

23. Estimate $\ln a$ given that $e^{t_1} < a < e^{t_2}$.

24. Estimate e^b given that $\ln x_1 < b < \ln x_2$.

In Exercises 25–36, find the derivative of the given function.

25. $f(x) = 3^{2x}$.

26. $g(x) = 4^{3x^2}$.

27. $f(x) = 2^{5x}\, 3^{\ln x}$.

28. $F(x) = 5^{-2x^2 + x}$.

29. $g(x) = \sqrt{\log_3 x}$.

30. $h(x) = 7^{\sin x^2}$.

31. $f(x) = \tan(\log_5 x)$.

32. $g(x) = \dfrac{\log_{10} x}{x^2}$.

33. $F(x) = \cos(2^x + 2^{-x})$.

34. $f(x) = \log_5\left(\dfrac{x}{x+1}\right)$.

35. $g(x) = \log_2[\log_4(2x + 1)]$.

36. $h(x) = a^{-x}\cos bx$.

In Exercises 37–45, calculate the following integrals.

37. $\displaystyle\int 3^x\, dx$.

38. $\displaystyle\int 2^{-x}\, dx$.

39. $\displaystyle\int (x^3 + 3^{-x})\, dx$.

40. $\displaystyle\int x 10^{-x^2}\, dx$.

41. $\displaystyle\int \dfrac{dx}{x \ln 5}$.

42. $\displaystyle\int \dfrac{\log_5 x}{x}\, dx$.

43. $\displaystyle\int \dfrac{\log_2 x^3}{x}\, dx$.

44. $\displaystyle\int \dfrac{\log_3 \sqrt{x} - 1}{x}\, dx$.

45. $\displaystyle\int x 10^{x^2}\, dx$.

46. Show that, if a, b, c are positive, then

$$\log_a c = \log_a b \, \log_b c$$

provided that a and b are both different from 1.

In Exercises 47–50, find $f'(e)$.

47. $f(x) = \log_3 x$.

48. $f(x) = x \log_3 x$.

49. $f(x) = \ln(\ln x)$.

50. $f(x) = \log_3(\log_2 x)$.

In Exercises 51 and 52, derive the formula for $f'(x)$ by logarithmic differentiation.

51. $f(x) = p^x$.

52. $f(x) = p^{g(x)}$.

In Exercises 53–64, find the derivative by logarithmic differentiation.

53. $\dfrac{d}{dx}[(x + 1)^x]$.

54. $\dfrac{d}{dx}[(\ln x)^x]$.

55. $\dfrac{d}{dx}[(\ln x)^{\ln x}]$.

56. $\dfrac{d}{dx}\left[\left(\dfrac{1}{x}\right)^x\right]$.

57. $\dfrac{d}{dx}[(x^2 + 2)^{\ln x}]$.

58. $\dfrac{d}{dx}[(\ln x)^{x^2 + 2}]$.

59. $\dfrac{d}{dx}[x^{\sin x}]$.

60. $\dfrac{d}{dx}[(\cos x)^{(x^2 + 1)}]$.

61. $\dfrac{d}{dx}[(\sin x)^{\cos x}]$.

62. $\dfrac{d}{dx}[x^{x^2}]$.

63. $\dfrac{d}{dx}[x^{2^x}]$.

64. $\dfrac{d}{dx}[(\tan x)^{\sec x}]$.

In Exercises 65–72, sketch figures in which you compare the following pairs of graphs.

65. $f(x) = e^x$ and $g(x) = 2^x$.

66. $f(x) = e^x$ and $g(x) = 3^x$.

67. $f(x) = e^x$ and $g(x) = e^{-x}$.

68. $f(x) = 2^x$ and $g(x) = 2^{-x}$.

69. $f(x) = \ln x$ and $g(x) = \log_3 x$.

70. $f(x) = \ln x$ and $g(x) = \log_2 x$.

71. $f(x) = 2^x$ and $g(x) = \log_2 x$.

72. $f(x) = 10^x$ and $g(x) = \log_{10} x$.

In Exercises 73–76, (i) specify the domain of f, (ii) find the intervals where f increases and those where it decreases, (iii) find the extreme values of f.

73. $f(x) = 10^{1-x^2}$.

74. $f(x) = 10^{1/(1-x^2)}$.

75. $f(x) = 10^{\sqrt{1-x^2}}$.

76. $f(x) = \log_{10}\sqrt{1-x^2}$.

In Exercises 77–85, evaluate the given integral.

77. $\displaystyle\int_1^2 2^{-x}\, dx$.

78. $\displaystyle\int_0^1 4^x\, dx$.

79. $\displaystyle\int_1^4 \dfrac{dx}{x \ln 2}$.

80. $\displaystyle\int_0^2 p^{x/2}\, dx$.

81. $\displaystyle\int_0^1 x 10^{1+x^2}\, dx$.

82. $\displaystyle\int_1^3 \dfrac{\log_3 x}{x}\, dx$.

83. $\displaystyle\int_{10}^{100} \dfrac{dx}{x \log_{10} x}$.

84. $\displaystyle\int_0^1 \dfrac{5p^{\sqrt{x+1}}}{\sqrt{x+1}}\, dx$.

85. $\displaystyle\int_0^1 (2^x + x^2)\, dx$.

In Exercises 86–89, estimate the logarithm on the basis of Table 7.2.1.

86. $\log_{10} 4$.

87. $\log_{10} 7$.

88. $\log_{10} 12$.

89. $\log_{10} 45$.

90. Prove that
$$\lim_{h \to 0} (1 + h)^{1/h} = e.$$

HINT: At $x = 1$ the logarithm function has derivative 1:
$$\lim_{h \to 0} \dfrac{\ln(1 + h) - \ln 1}{h} = 1.$$

▷ In Exercises 91–93, evaluate and then explain the result.

91. $5^{(\ln 17)/(\ln 5)}$.

92. $7^{1/\ln 7}$.

93. $16^{1/\ln 2}$.

■ 7.6 EXPONENTIAL GROWTH AND DECAY

We begin by comparing exponential change to linear change. Let y be a quantity that changes with time: $y = y(t)$.

If y is a linear function,
$$y(t) = kt + C, \qquad k, C \text{ constants,}$$
then y changes by the *same amount during all periods of the same duration*:
$$y(t + \Delta t) = k[t + \Delta t] + C = (kt + C) + k\,\Delta t = y(t) + k\,\Delta t.$$

During every period of length Δt, y changes by the amount $k\,\Delta t$.

If, on the other hand, y is an exponential function,
$$y(t) = Ce^{kt},$$
then y changes by *the same factor during all periods of the same duration*:
$$y(t + \Delta t) = Ce^{k[t + \Delta t]} = Ce^{kt + k\Delta t} = Ce^{kt} \cdot e^{k\Delta t} = e^{k\Delta t} y(t).$$

During every period of length Δt, y changes by a factor of $e^{k\Delta t}$.

The exponential function
$$f(t) = Ce^{kt}$$
has the property that its derivative $f'(t)$ is proportional to $f(t)$:
$$f'(t) = Cke^{kt} = kCe^{kt} = kf(t).$$

Moreover it is the only such function:

THEOREM 7.6.1

If

$$f'(t) = kf(t) \qquad \text{for all } t \text{ in some interval } I,$$

then f is an exponential function of the form

$$f(t) = Ce^{kt} \qquad \text{for all } t \text{ in } I \text{ and some constant } C.$$

PROOF   We are given that

$$f'(t) = kf(t)$$

or

$$f'(t) - kf(t) = 0.$$

Multiplying this equation by e^{-kt}, we have

$$(*) \qquad\qquad e^{-kt}f'(t) - ke^{-kt}f(t) = 0.$$

Now observe that the left-hand side of this equation is

$$\frac{d}{dt}[e^{-kt}f(t)] \qquad\qquad \text{(verify this).}$$

Therefore, equation $(*)$ can be written

$$\frac{d}{dt}[e^{-kt}f(t)] = 0.$$

It now follows that

$$e^{-kt}f(t) = C \qquad \text{for some constant } C,$$

and

$$f(t) = Ce^{kt}. \quad \square$$

Remark   According to Theorem 7.6.1, every solution of the equation $f'(t) = kf(t)$ has the form $f(t) = Ce^{kt}$ for some constant C. The value of C is often determined by evaluating f at 0 (assuming that $0 \in I$):

$$f(0) = Ce^0 = C \quad \text{and} \quad f(t) = f(0)e^{kt}.$$

The value $f(0)$ is called the *initial value of f*. $\square$

Example 1   Find $f(t)$ given that $f'(t) = 2f(t)$ for all t and $f(0) = \sqrt{2}$.

SOLUTION   The fact that $f'(t) = 2f(t)$ tells us that $f(t) = Ce^{2t}$ where C is some constant. The fact that $f(0) = \sqrt{2}$ tells us that $C = \sqrt{2}$. Thus $f(t) = \sqrt{2}\,e^{2t}$. $\square$

Population Growth

Under ideal conditions (for example, unlimited space, adequate food supply, freedom from disease, and so on), the rate of increase of a population P at time t is proportional to the size of the population at time t. That is,

$$P'(t) = kP(t),$$

where $k > 0$ is a constant, called the *growth constant*.† Thus, applying Theorem 7.6.1, the size of the population at any time t is given by

$$P(t) = P(0)e^{kt},$$

and the population is said to grow *exponentially*. This is a model of uninhibited growth. In reality, the rate of increase of a population will not continue to be proportional to the size of the population. After some time has passed, factors such as limitations on space or food supply, diseases, and so forth, will affect the growth rate of the population.

Example 2 The number of bacteria in a certain culture is increasing at a rate proportional to the number present. Suppose that there are 1000 bacteria present initially, and that 1500 are present after 2 hours.

(a) Determine the number of bacteria in the culture at any time t. How many bacteria are in the culture after 5 hours?

(b) How long will it take for the number of bacteria in the culture to double?

SOLUTION Let $P(t)$ be the number of bacteria in the culture at time t. The basic equation $P'(t) = kP(t)$ gives

$$P(t) = P(0)e^{kt}.$$

Since $P(0) = 1000$, we have

(1)
$$P(t) = 1000e^{kt}.$$

We can use the fact that $P(2) = 1500$ to eliminate k from (1):

$$1500 = 1000e^{2k}, \qquad e^{2k} = \frac{1500}{1000} = \frac{3}{2}, \qquad 2k = \ln\left[\frac{3}{2}\right],$$

and so

(2)
$$k = \tfrac{1}{2} \ln\left[\frac{3}{2}\right] \cong 0.203.$$

(a) The number of bacteria in the culture at any time t is (approximately):

$$P(t) = 1000e^{0.203t}.$$

After 5 hours, the number of bacteria in the culture is $P(5) = 1000e^{0.203(5)} = 1000e^{1.015} \cong 2759$.

---

† The growth constant of a population is often expressed as a percentage. In such instances, the percentage must be converted to its decimal equivalent for use in the formula. For example, if the growth constant of a certain population is given as 1.7%, then $k = 0.017$.

(b) To find out how long it will take for the number of bacteria to double, we want to find the value of t for which $P(t) = 2000$.† Thus we set

$$1000e^{0.203t} = 2000,$$

$$e^{0.203t} = 2, \qquad 0.203t = \ln[2], \qquad \text{and} \qquad t = \frac{\ln[2]}{0.203} \cong 3.4 \text{ hours.} \quad \square$$

Remark There is an alternative way of expressing P which uses the exact value of k. From (2), we have $k = \frac{1}{2} \ln[\frac{3}{2}]$. Thus,

$$P(t) = 1000e^{(t/2)\ln[3/2]} = 1000e^{\ln[3/2]^{t/2}}$$

and

$$P(t) = 1000 \left(\frac{3}{2} \right)^{t/2}.$$

Now, for example, to find the doubling time, we solve

$$1000 \left(\frac{3}{2} \right)^{t/2} = 2000$$

for t:

$$\left(\frac{3}{2} \right)^{t/2} = 2, \qquad \left(\frac{t}{2} \right) \ln \left(\frac{3}{2} \right) = \ln(2), \qquad t = \frac{2 \ln(2)}{\ln(3/2)} \cong 3.4. \quad \square$$

Example 3 The world population in 1960 was approximately 3 billion, and in 1980 it was approximately 4.5 billion. Assume that the world population is increasing at a rate proportional to the size of the population.

(a) Determine the growth constant and the (approximate) world population at any time t.

(b) How long will it take for the world population to reach 12 billion (quadruple the 1960 population)?

(c) The current (1994) world population is estimated to be 5.6. billion. What does the model determined in part (a) predict for the world population in 1994?

SOLUTION Let $P(t)$ be the world population (in billions) t years after 1960. The basic equation $P'(t) = kP(t)$, with $P(0) = 3$, gives

$$P(t) = 3e^{kt}.$$

(a) Since $P(20) = 4.5$, we have

$$3e^{20k} = 4.5, \qquad 20k = \ln \left(\frac{4.5}{3} \right), \qquad k = \frac{\ln(1.5)}{20} \cong 0.0203.$$

Thus the growth constant is approximately 0.0203 or 2.03%. The population at any time t is given (approximately) by

$$P(t) = 3e^{0.0203t}.$$

---

† The length of time that it takes for a population to double in size is called the *doubling time*.

(b) To find the value of t for which $P(t) = 12$, we set

$$3e^{0.0203t} = 12$$

$$e^{0.0203t} = 4, \qquad 0.0203t = \ln[4], \qquad t = \frac{\ln[4]}{0.0203} \cong 68.3.$$

Thus, based on the given data, the world population will reach 12 billion in just over 68 years (the year 2028).

(c) According to our model, $P(34) = 3e^{0.0203(34)} = 3e^{0.69} \cong 5.98$ billion. This is 380 million more than the current estimate.

Radioactive Decay

Experimental evidence shows that radioactive substances decay at a rate which is proportional to the amount of such substance present. Therefore, if $A(t)$ is the amount of a radioactive substance present at time t, then

$$A'(t) = kA(t),$$

where $k < 0$ is a constant, called the *decay constant*. Thus,

$$A(t) = A(0)e^{kt},$$

where $A(0)$ is the amount of the substance present initially.

   The *half-life* of a radioactive substance is the length of time that it takes for half the initial amount to decay. If we let T denote the half-life and k the decay constant of a radioactive substance, then the relation between T and k is given by

(7.6.2)

$$\boxed{kT = -\ln 2.}$$

PROOF We have $A(T) = \frac{1}{2}A(0)$. Thus,

$$\tfrac{1}{2}A(0) = A(0)e^{kT}$$

$$e^{kT} = \tfrac{1}{2}$$

$$kT = \ln[\tfrac{1}{2}] = -\ln 2. \quad \square$$

Example 4 One-third of a radioactive substance decays every 5 years. Today we have $A(0) = A_0$ grams of the substance.

(a) Find the decay constant and determine how much will be left t years from now.
(b) What is the half-life of the substance?

SOLUTION Let $A(t)$ be the amount of the radioactive substance present at time t. Since $A'(t) = kA(t)$, we know that $A(t) = A_0 e^{kt}$.

(a) At the end of 5 years, one-third of the substance will have decayed and therefore two-thirds of A_0 will remain:

$$A(5) = \tfrac{2}{3}A_0.$$

Now

$$\tfrac{2}{3}A_0 = A_0 e^{5k}$$

$$e^{5k} = \tfrac{2}{3}$$

$$5k = \ln[\tfrac{2}{3}]$$

and

$$k = \frac{\ln[\tfrac{2}{3}]}{5} \cong -0.081.$$

The amount of the substance that remains after t years is (approximately):

$$A(t) = A_0 e^{-0.081t}.$$

(b) We use Equation (7.6.2) to find the half-life of the substance:

$$T = \frac{-\ln 2}{k} = \frac{-\ln 2}{-0.081} \cong 8.56.$$

Thus, the half-life of the substance is approximately 8.56 years. ❑

Remark You can verify that the exact value of A in Example 4 is $A(t) = A_0 (\tfrac{2}{3})^{t/5}$ and the exact value of the half-life is $T = (5 \ln 2)/(\ln 1.5)$. ❑

Example 5 Cobalt-60 is a radioactive substance that is used extensively in medical radiology. It has a half-life of 5.3 years. Suppose an initial sample of cobalt-60 has a mass of 100 grams.

(a) Find the decay constant and determine an expression for the amount of the sample that will remain t years from now.

(b) How long will it take for 90% of the sample to decay?

SOLUTION **(a)** The decay constant k is given by

$$k = \frac{-\ln 2}{T} = \frac{-\ln 2}{5.3} \cong -0.131.$$

With $A(0) = 100$, the amount of material that will remain after t years is

$$A(t) = 100 e^{-0.131t}.$$

(b) If 90% of the sample decays, then 10%, or 10 grams, remains. Thus, we need to solve the equation

$$100 e^{-0.131t} = 10$$

for t. We have

$$e^{-0.131t} = 0.1, \qquad -0.131t = \ln[0.1], \qquad t = \frac{\ln[0.1]}{-0.131} \cong 17.6.$$

It will take approximately 17.6 years for 90% of the sample to decay. ❑

Compound Interest

Consider money invested at interest rate r. If the accumulated interest is credited once a year, then the interest is said to be compounded annually; if twice a year, then semiannually; if four times a year, then quarterly. The idea can be pursued further. Interest can be credited every day, every hour, every second, every half-second, and so on. In the limiting case, interest is credited instantaneously. Economists call this *continuous compounding*.

The economists' formula for continuous compounding is a simple exponential:

(7.6.3)
$$A(t) = A_0 e^{rt}.$$

Here t is measured in years,

$$A(t) = \text{the principal in dollars at time } t,$$

$$A_0 = A(0) = \text{the initial investment,}$$

$$r = \text{the annual interest rate.}$$

The rate r is also called the *nominal* interest rate and it is conventional to express this rate as a percentage. However, as in the case of the growth and decay constants, the decimal equivalent of the percentage must be used in all calculations.

A DERIVATION OF THE COMPOUND INTEREST FORMULA Fix t and take h as a small time increment. Then

$$A(t + h) - A(t) = \text{interest earned from time } t \text{ to time } t + h.$$

Had the principal remained $A(t)$ from time t to time $t + h$, the interest earned during this time period would have been

$$rhA(t).$$

Had the principal been $A(t + h)$ throughout the time interval, the interest earned would have been

$$rhA(t + h).$$

The actual interest earned must be somewhere in between:

$$rhA(t) \leq A(t + h) - A(t) \leq rhA(t + h).$$

Dividing by h, we get

$$rA(t) \leq \frac{A(t + h) - A(t)}{h} \leq rA(t + h).$$

If A varies continuously, then, as h tends to zero, $rA(t + h)$ tends to $rA(t)$ and (by the pinching theorem) the difference quotient in the middle must also tend to $rA(t)$:

$$\lim_{h \to 0} \frac{A(t + h) - A(t)}{h} = rA(t).$$

This says that

$$A'(t) = rA(t),$$

that is, with continuous compounding, the principal increases at a rate proportional to the amount present and the growth constant is the nominal rate r. Now, it follows that

$$A(t) = Ce^{rt}.$$

With A_0 as the initial investment, we have $C = A_0$ and

$$A(t) = A_0e^{rt}. \quad \square$$

Remark It is interesting to compare the accumulation of principal at a fixed nominal rate r using different compounding periods. For example, a principal of $1000 invested at 6% compounded:

(a) annually (once per year) will have the value $A(1) = 1000(1 + 0.06) = \$1060$ at the end of one year.

(b) quarterly (4 times per year) will have the value $A(1) = 1000(1 + (0.06/4))^4 \cong \$1061.36$ at the end of one year.

(c) monthly (12 times per year) will have the value $A(1) = 1000(1 + (0.06/12))^{12} \cong \$1061.67$ at the end of one year.

(d) continuously will have the value $A(1) = 1000e^{0.06} \cong \$1061.84$ at the end of one year. $\square$

Example 6 Suppose that $1000 is deposited in a bank that pays 5% compounded continuously. How much money will be in the account after 5 years and what is the interest earned during this period?

SOLUTION Here we have $A_0 = 1000$ and $r = 0.05$. Thus, the amount of money in the account at any time t is given by

$$A(t) = 1000e^{0.05t}.$$

The amount of money in the account after 5 years is

$$A(5) = 1000e^{0.05(5)} = 1000e^{0.25} \cong \$1284.03,$$

and the interest earned during this period is $284.03. $\square$

Example 7 How long does it take for a sum of money to double at 6.5% compounded continuously?

SOLUTION In general

$$A(t) = A_0e^{0.065t}.$$

We set

$$2A_0 = A_0e^{0.065t}$$

and solve for t:

$$2 = e^{0.065t}, \qquad \ln 2 = 0.065t, \qquad t = \frac{\ln 2}{0.065} \cong 10.66.$$

It takes about 10 years and 8 months. $\square$

Newton's Law of Cooling

Newton's law of cooling states that the rate of change of the temperature T of an object is proportional to the difference between T and the (constant) temperature τ of the surrounding medium, called the *ambient temperature*. Assuming that the temperature of the object is a differentiable function, the mathematical formulation of Newton's law is

$$\frac{dT}{dt} = m(T - \tau), \qquad \text{where } m \text{ is a constant.}$$

Remark The constant m in this model must be negative; for if the object is warmer than the ambient temperature ($T - \tau > 0$), then its temperature will decrease ($dT/dt < 0$), which implies $m < 0$; if the object is colder than the ambient temperature ($T - \tau < 0$), its temperature will increase ($dT/dt > 0$), which again implies $m < 0$. ❏

To emphasize the sign of the constant of proportionality, we write Newton's law of cooling as

(7.6.4)
$$\frac{dT}{dt} = -k(T - \tau), \qquad \text{where } k \text{ is a positive constant.}$$

Although this equation is not of the same form as the models for exponential growth or decay, it can be converted into that form by a change of variable. Let $y(t) = T(t) - \tau$. Then

$$\frac{dy}{dt} = \frac{d}{dt}(T - \tau) = \frac{dT}{dt} \qquad \text{since } \tau \text{ is a constant.}$$

With this change of variable, (7.6.4) becomes

$$\frac{dy}{dt} = -ky,$$

and, by Theorem 7.6.1,

$$y(t) = Ce^{-kt}.$$

Therefore

$$T(t) - \tau = Ce^{-kt},$$

and so

$$T(t) = \tau + Ce^{-kt}.$$

The constant C is determined by the initial temperature:

$$T(0) = \tau + C \qquad \text{and so} \qquad C = T(0) - \tau.$$

Thus, the solutions are given by

(7.6.5)
$$T(t) = \tau + [T(0) - \tau]e^{-kt}.$$

Example 8 A turkey at 325°F is removed from the oven and allowed to cool in a room of constant temperature 72°F. Three minutes later, the temperature of the turkey is 275°. How long will it take for the turkey to cool down to a normal eating temperature of 110°?

SOLUTION We know that $T(0) = 325$ and $\tau = 72$. Therefore, by Formula (7.6.5), $T(t) = 72 + 253e^{-kt}$. Since $T(3) = 275$, we have

$$275 = 72 + 253e^{-3k}$$

$$e^{-3k} = \frac{203}{253}$$

and

$$k = \frac{\ln[203/253]}{-3} \cong 0.073.$$

Thus

$$T(t) = 72 + 253e^{-0.073t}.$$

To find out how long it will take for the turkey to cool down to 110°, we solve the equation

$$72 + 253e^{-0.073t} = 110$$

for t:

$$253e^{-0.073t} = 38, \qquad -0.073t = \ln\left[\frac{38}{253}\right], \qquad t = \frac{\ln[38/253]}{-0.073} \cong 25.97.$$

It will take approximately 26 minutes for the temperature of the turkey to reach 110°. ❑

A Mixing Problem

In our approach to mixing problems of the sort presented below we assume that uniform concentration is maintained by some kind of "stirring."

Example 9 Water from a polluted reservoir (pollution level p_0 grams per liter) is being drawn off at the rate of n liters per hour and being replaced by less polluted water (pollution level p_1 grams per liter). Given that the reservoir contains M liters, how long will it take to reduce the pollution to p grams per liter, $p_1 < p < p_0$?

SOLUTION Let $A(t)$ be the total number of grams of pollution in the reservoir at time t. We want to find the time t at which

$$\frac{A(t)}{M} = p.$$

At time t, pollutants leave the reservoir at the rate of

$$n\frac{A(t)}{M} \quad \text{grams per hour}$$

and enter the reservoir at the rate of

$$np_1 \quad \text{grams per hour.}$$

It follows that

$$A'(t) = \text{rate in} - \text{rate out} = np_1 - n\frac{A(t)}{M}$$

and we have

$$A'(t) = -\frac{n}{M}[A(t) - Mp_1].$$

This equation has the same general form as the model for Newton's law of cooling. Thus, if we let $y = A(t) - Mp_1$, we have

$$\frac{dy}{dt} = -\frac{n}{M}y,$$

and so

$$y(t) = A(t) - Mp_1 = Ce^{-nt/M}.$$

Thus

$$\frac{A(t)}{M} = p_1 + \frac{C}{M}e^{-nt/M}.$$

The constant C is determined by the initial pollution level p_0:

$$p_0 = \frac{A(0)}{M} = p_1 + \frac{C}{M} \qquad \text{so that} \qquad C = M(p_0 - p_1).$$

Using this value for C, we have

$$\frac{A(t)}{M} = p_1 + (p_0 - p_1)e^{-nt/M}.$$

This equation gives the pollution level at each time t. To find the time at which the pollution level has been reduced to p, we set

$$p_1 + (p_0 - p_1)e^{-nt/M} = p$$

and solve for t:

$$e^{-nt/M} = \frac{p - p_1}{p_0 - p_1}, \qquad -\frac{n}{M}t = \ln\frac{p - p_1}{p_0 - p_1}, \qquad t = \frac{M}{n}\ln\frac{p_0 - p_1}{p - p_1}. \quad \square$$

Example 10   In the problem just considered, how long would it take to reduce the pollution by 50% if the reservoir contained 50,000,000 liters, the water was replaced at the rate of 25,000 liters per hour, and the replacement water was completely pollution free?

SOLUTION   In general

$$t = \frac{M}{n}\ln\frac{p_0 - p_1}{p - p_1}.$$

Here

$$M = 50{,}000{,}000, \quad n = 25{,}000, \quad p_1 = 0, \quad p = \tfrac{1}{2}p_0,$$

so that

$$t = 2000 \ln 2 \cong 1386.$$

It would take about 1386 hours, or about $8\frac{1}{4}$ weeks. ❑

Differential Equations

An equation that involves an unknown function and one or more of its derivatives is called a *differential equation*. The *order* of a differential equation is the order of the highest derivative that appears in the equation. Thus, the equations that we have used in this section to model population growth, radioactive decay, and continuous compounding of interest,

$$(1) \qquad \frac{dy}{dt} = ky, \qquad k \text{ constant},$$

are examples of first-order differential equations. The motion of a simple pendulum or a weight suspended by a spring (simple harmonic motion) is described by the second-order differential equation

$$(2) \qquad \frac{d^2y}{dx^2} + \omega^2 y = 0, \qquad \omega \text{ constant}.$$

See Exercises 61 and 62, Section 3.6.

A *solution* of a differential equation is simply a function f that satisfies the equation. As we have seen, the functions $f(t) = Ce^{kt}$, where C is a constant, are solutions of (1). The functions $y = A \sin \omega x + B \cos \omega x$, where A and B are constants, are solutions of (2).

Differential equations serve as mathematical models for a wide variety of phenomena in the physical and life sciences, engineering, the social sciences, and business. Chapter 18 is an introduction to the study of differential equations and their applications.

Equation (1) is a specific example of a class of first-order differential equations known as *separable equations*. In general, a separable equation is a first-order differential equation which can be written in the form

$$\frac{dy}{dx} = \frac{f(x)}{g(y)}.$$

An informal approach to solving this equation is as follows: Using differentials, the equation can be written

$$g(y) \, dy = f(x) \, dx. \qquad \text{(separate the variables)}$$

Integrating both sides, we have

$$\int g(y) \, dy = \int f(x) \, dx.$$

Now, if G is an antiderivative of g and F is an antiderivative of f, then

$$G(y) = F(x) + C, \qquad \text{where } C \text{ is an arbitrary constant},$$

gives a relation between x and y which satisfies the differential equation.

Example 11   Solve the differential equation

$$\frac{dy}{dx} = r(y - a), \qquad \text{where } r \text{ and } a \text{ are constants,}$$

using the approach outlined above.   Note that Newton's law of cooling and mixing problems led to this type of equation.

SOLUTION   Separating the variables and using differentials, we can write the equation as

$$\frac{1}{y - a}\, dy = r\, dx \qquad (y \neq a).$$

Integrating both sides, we get

$$\int \frac{1}{y - a}\, dy = \int r\, dx$$

$$\ln|y - a| = rx + C, \qquad (C \text{ an arbitrary constant})$$

which gives y as an implicit function of x.   In this case, we can solve for y explicitly as a function of x by taking the exponential of both sides.   In Exercise 47 you are asked to show that the solutions of the differential equation can be written

$$y = a + Be^{rx} \qquad (B \text{ an arbitrary constant}).$$

You should verify that this result is consistent with the results given previously for Newton's law of cooling and the mixing problems.   ❏

Again, our approach here has been informal.   A more precise treatment is given in Section 18.3.

EXERCISES 7.6

Note: Most of these exercises require a calculator. Exercises 38 and 48 require a graphics calculator or a computer with graphing software.

1. Find the amount of interest earned by $500 compounded continuously for 10 years (a) at 6%, (b) at 8%, (c) at 10%.

2. How long does it take for a sum of money to double when compounded continuously (a) at 6%? (b) at 8%? (c) at 10%?

3. At what rate r of continuous compounding does a sum of money triple in 20 years?

4. At what rate r of continuous compounding does a sum of money double in 10 years?

5. A certain species of viral bacteria is being grown in a culture.   It is observed that the rate of growth of the bac-

teria population is proportional to the number present.   If there were 1000 bacteria in the initial population and the number doubled after the first 30 minutes, how many bacteria will be present after 2 hours?

6. In a bacteria growing experiment, a biologist observes that the number of bacteria in a certain culture triples every 4 hours.   After 12 hours, it is estimated that there are 1 million bacteria in the culture.
   (a) How many bacteria were present initially?
   (b) What is the doubling time for the bacteria population?

7. At a certain moment a 100-gallon mixing tank is full of brine containing 0.25 pound of salt per gallon.   Find the amount of salt present t minutes later if the brine is being continuously drawn off at the rate of 3 gallons per minute and replaced by brine containing 0.2 pound of salt per gallon.

8. Water is pumped into a tank to dilute a saline solution. The volume of the solution, call it V, is kept constant by continuous outflow. The amount of salt in the tank, call it s, depends on the amount of water that has been pumped in, call this x. Given that

$$\frac{ds}{dx} = -\frac{s}{V},$$

find the amount of water that must be pumped into the tank to eliminate 50% of the salt. Take V as 10,000 gallons.

9. A 200-liter tank initially full of water develops a leak at the bottom. Given that 20% of the water leaks out in the first 5 minutes, find the amount of water left in the tank t minutes after the leak develops if the water drains off at a rate that is proportional to the product of the time elapsed and the amount of water present.

10. In Exercise 9 assume that the water drains off at a rate that is proportional to the amount of water present.

In Exercises 11–16 remember that the rate of decay of a radioactive substance is proportional to the amount of substance present.

11. What is the half-life of a radioactive substance if it takes 4 years for a quarter of the substance to decay?

12. A year ago there were 4 grams of a radioactive substance. Now there are 3 grams. How much was there 10 years ago?

13. Two years ago there were 5 grams of a radioactive substance. Now there are 4 grams. How much will remain 3 years from now?

14. What is the half-life of a radioactive substance if it takes 5 years for a third of the substance to decay?

15. Suppose the half-life of a radioactive substance is n years. What percentage of the substance present at the start of a year will decay during the ensuing year?

16. A radioactive substance weighed n grams at time $t = 0$. Today, 5 years later, the substance weighs m grams. How much will it weigh 5 years from now?

17. (*The power of exponential growth*) Imagine two racers competing on the x-axis (which has been calibrated in meters), a linear racer LIN [position function of the form $x_1(t) = kt + C$] and an exponential racer EXP [position function of the form $x_2(t) = e^{kt} + C$]. Suppose that both racers start out simultaneously from the origin, LIN at one million meters per second, EXP at only one meter per second. In the early stages of the race, fast-starting LIN will move far ahead of EXP, but in time EXP will catch up to LIN, pass her, and leave her hopelessly behind. Show that this is true as follows:
(a) Express the position of each racer as a function of time, measuring t in seconds.

(b) Show that LIN's lead over EXP starts to decline about 13.8 seconds into the race.
(c) Show that LIN is still ahead of EXP some 15 seconds into the race but far behind 3 seconds later. (Use $e^3 \cong 20$.)
(d) Show that, once EXP passes LIN, LIN can never catch up.

18. (*The weakness of logarithmic growth*) Having been soundly beaten in the race in Exercise 17, LIN finds an opponent she can beat, LOG, the logarithmic racer [position function $x_3(t) = k \ln(t + 1) + C$]. Once again the racetrack is the x-axis calibrated in meters. Both racers start out at the origin, LOG at one million meters per second, LIN at only one meter per second. (LIN is tired from the previous race.) In this race LOG will shoot ahead and remain ahead for a long time, but eventually LIN will catch up to LOG, pass her, and leave her permanently behind. Show that this is true as follows:
(a) Express the position of each racer as a function of time, measuring t in seconds.
(b) Show that LOG's lead over LIN starts to decline $10^6 - 1$ seconds into the race.
(c) Show that LOG is still ahead of LIN $10^7 - 1$ seconds into the race but behind LIN $10^8 - 1$ seconds into the race.
(d) Show that, once LIN passes LOG, LOG can never catch up.

19. Atmospheric pressure p varies with altitude h according to the equation

$$\frac{dp}{dh} = kp, \qquad \text{where } k \text{ is a constant.}$$

Given that p is 15 pounds per square inch at sea level and 10 pounds per square inch at 10,000 feet, find p at: (a) 5000 feet; (b) 15,000 feet.

20. According to the compound interest formula (7.6.3), if P dollars are deposited in an account now at an interest rate r compounded continuously, then the amount of money in the account t years from now will be

$$Q = Pe^{rt}.$$

The quantity Q is sometimes called the *future value* of P (at the interest rate r). Solving this equation for P, we get

$$P = Qe^{-rt}.$$

In this formulation, the quantity P is called the *present value* of Q. Find the present value of $20,000 at 6% compounded continuously for 4 years.

21. Find the interest rate r that is needed to have $6000 be the present value of $10,000 over an 8-year period.

22. You are 45 years old and are looking forward to an annual pension of $50,000 per year at age 65. What is the present-day purchasing power (present value) of your pension if money can be invested over this period at a continuously compounded interest rate of (a) 4%? (b) 6%? (c) 8%?

23. The cost of the tuition, fees, room, and board at XYZ College is currently $16,000 per year. What would you expect to pay 3 years from now if the costs at XYZ are rising at the continuously compounded rate of (a) 5%? (b) 8%? (c) 12%?

24. A boat moving in still water is subject to a retardation proportional to its velocity. Show that the velocity t seconds after the power is shut off is given by the formula $v = ce^{-kt}$ where c is the velocity at the instant the power is shut off.

25. A boat is drifting in still water at 4 miles per hour; 1 minute later, at 2 miles per hour. How far has the boat drifted in that 1 minute? (See Exercise 24.)

26. Population tends to grow with time at a rate roughly proportional to the population present. According to the Bureau of the Census, the population of the United States in 1970 was approximately 203 million and in 1980, 227 million. Use this information to estimate the population of 1960. (The actual figure was about 179 million.)

27. Use the data of Exercise 26 to predict the population for the year 2000.

28. Use the data of Exercise 26 to estimate how long it takes for the population to double.

29. The method of *carbon dating* makes use of the fact that all living organisms contain two isotopes of carbon, carbon-12, denoted ^{12}C, (a stable isotope) and carbon-14, denoted ^{14}C (a radioactive isotope). The ratio of the amount of ^{14}C to the amount of ^{12}C is essentially constant (approximately 1/10,000). When an organism dies, the amount of ^{12}C present remains unchanged, but the ^{14}C decays at a rate proportional to the amount present with a half-life of approximately 5700 years. This change in the amount of ^{14}C relative to the amount of ^{12}C makes it possible to estimate the time at which the organism lived. A fossil found in an archaeological dig was found to contain 25% of the original amount of ^{14}C. What is the approximate age of the fossil?

30. The Dead Sea Scrolls are approximately 2000 years old. How much of the original ^{14}C remains in them?

31. A lumber company finds from experience that the value of its standing timber increases with time according to the formula

$$V(t) = V_0 (\tfrac{3}{2})^{\sqrt{t}}.$$

Here t is measured in years and V_0 is the value at planting time. How long should the company wait before cutting the timber? Neglect costs and assume continuous compounding at 5%.

32. Determine the time period during which $y = Ce^{kt}$ changes by a factor of q.

33. During the process of inversion, the amount A of raw sugar present decreases at a rate proportional to A. During the first 10 hours, 1000 pounds of raw sugar have been reduced to 800 pounds. How many pounds will remain after 10 more hours of inversion?

34. The number of bacteria present in a given culture increases at a rate proportional to the number present. When first observed, the culture contained n_0 bacteria, an hour later, n_1.
    (a) Find the number present t hours after the observations began.
    (b) How long did it take for the number of bacteria to double?

35. Let k_1, k_2 be constants, $k_1 \neq 0$. Show that

    if $f'(t) = k_1 f(t) + k_2$ then $f(t) = Ce^{k_1 t} - k_2/k_1$

    where C is an arbitrary constant.

36. Find $f(t)$ given that $f'(t) = k[2 - f(t)]$ and $f(0) = 0$.

37. An object falling from rest in air is subject to not only gravitational force but also air resistance. If we assume that the air resistance is proportional to the velocity and acts in a direction opposite to the motion, then the velocity of the object at time t satisfies

    $$v'(t) + K v(t) = 32 \quad \text{and} \quad v(0) = 0.$$

    (a) Find $v(t)$.
    (b) Show that $v(t)$ cannot exceed $32/K$ and that $\lim_{t \to \infty} v(t) = 32/K$. The value $32/K$ is called the *terminal velocity* of the object.
    (c) Sketch the graph of v.

38. The solutions of the differential equation that model Newton's law of cooling are given by (7.6.5):

    $$T(t) = \tau + [T(0) - \tau]e^{-kt},$$

    where τ is a constant representing the ambient temperature and $k > 0$ is the constant of proportionality.
    (a) Show that $\lim_{t \to \infty} T(t) = \tau$.
    (b) Set $\tau = 20$ and $k = 0.1$. Use a graphing utility to graph T in the cases where
        (i) $T(0) = 1$ and (ii) $T(0) = 40$.
    (c) For each of the cases in part (b), estimate from your graph how long it takes for T to be within 1 unit of 20. Use four decimal place accuracy.

39. A cup of coffee is served to you at 185°F in a room where the temperature is 65°F. Two minutes later the temperature of the coffee has dropped to 155°F. How many more minutes would you expect to wait for the coffee to cool to 105°F?

40. A thermometer that reads 72°F is taken outside where the air temperature is 32°F. Three minutes later, the thermometer reading is 40°F.
  (a) What will the thermometer read 5 minutes after it is brought outside?
  (b) How long will it take for the thermometer to be within $\frac{1}{2}$°F of the air temperature?

41. A small metal bar with an initial temperature of 20°C is dropped into a large container of boiling water (100°C). The water continues to boil and 20 seconds later, the temperature of the bar is 30°C.
  (a) What will the temperature of the bar be after 1 minute?
  (b) How long will it take for the temperature of the bar to exceed 98°C?

42. The current i in an electric circuit varies with time t according to the formula

$$L\frac{di}{dt} + Ri = E$$

where E (the voltage), L (the inductance), and R (the resistance) are positive constants. Measure time in seconds, current in amperes, and suppose that the initial current is 0.
  (a) Find a formula for the current at each subsequent time t.
  (b) What upper limit does the current approach as t increases?
  (c) In how many seconds will the current reach 90% of its upper limit?

In Exercises 43–46, show that the given equation is a separable differential equation and then find the solutions.

43. $\dfrac{dy}{dx} = \dfrac{y}{x}$.

44. $\dfrac{dy}{dx} + y \cos x = 0$.

45. $\dfrac{dy}{dx} = e^y \cos x$.

46. $\dfrac{dy}{dx} = \dfrac{3x^2}{2y + \sin 2y}$.

47. Show that the solutions of the differential equation

$$\frac{dy}{dx} = r(y - a)$$

in Example 11 can be written in the form $y = a + Be^{rx}$, where B is an arbitrary constant.

48. The differential equation

$$\frac{dP}{dt} = (k \cos t)P, \quad k > 0 \text{ a constant,}$$

models a population that experiences periodic growth and decline.
  (a) Find $P(t)$ if $P(0) = P_0$.
  (b) Suppose that the population is measured in thousands, and $P(0) = 3$ and $k = 2$. Use a graphing utility to sketch the graph of P.
  (c) Estimate the absolute maximum value and the absolute minimum value of P from your graph. Use four decimal place accuracy in your answers.

■ 7.7 MORE ON THE INTEGRATION OF THE TRIGONOMETRIC FUNCTIONS

In Section 5.6 you saw that

(7.7.1)

$$\int \sin x \, dx = -\cos x + C. \qquad \int \sec x \tan x \, dx = \sec x + C.$$

$$\int \cos x \, dx = \sin x + C. \qquad \int \csc^2 x \, dx = -\cot x + C.$$

$$\int \sec^2 x \, dx = \tan x + C. \qquad \int \csc x \cot x \, dx = -\cot x + C.$$

Now that you are familiar with the natural logarithm function, we can add four more basic formulas to the list

(7.7.2)

$$
\textbf{(i)} \int \tan x \, dx = \ln|\sec x| + C.
$$

$$
\textbf{(ii)} \int \cot x \, dx = \ln|\sin x| + C.
$$

$$
\textbf{(iii)} \int \sec x \, dx = \ln|\sec x + \tan x| + C.
$$

$$
\textbf{(iv)} \int \csc x \, dx = \ln|\csc x - \cot x| + C.
$$

Each of these formulas can be derived as follows:

$$
\int \frac{g'(x)}{g(x)} \, dx = \int \frac{du}{u} = \ln|u| + C = \ln|g(x)| + C.
$$

$$
\text{set } u = g(x), \quad du = g'(x) \, dx
$$

Derivation of Formulas (i) – (iii)

$$
\textbf{(i)} \int \tan x \, dx = \int \frac{\sin x}{\cos x} \, dx \qquad\qquad [\text{set } u = \cos x, \quad du = -\sin x \, dx]
$$

$$
= -\int \frac{du}{u} = -\ln|u| + C
$$

$$
= -\ln|\cos x| + C = \ln \left| \frac{1}{\cos x} \right| + C
$$

$$
= \ln|\sec x| + C. \quad \square
$$

$$
\textbf{(ii)} \int \cot x \, dx = \int \frac{\cos x}{\sin x} \, dx \qquad\qquad [\text{set } u = \sin x, \quad du = \cos x \, dx]
$$

$$
= \int \frac{du}{u} = \ln|u| + C = \ln|\sin x| + C. \quad \square
$$

$$
\textbf{(iii)} \int \sec x \, dx \overset{\dagger}{=} \int \sec x \frac{\sec x + \tan x}{\sec x + \tan x} \, dx
$$

$$
= \int \frac{\sec x \tan x + \sec^2 x}{\sec x + \tan x} \, dx
$$

$$
[\text{set } u = \sec x + \tan x, \quad du = (\sec x \tan x + \sec^2 x) \, dx]
$$

$$
= \int \frac{du}{u} = \ln|u| + C = \ln|\sec x + \tan x| + C. \quad \square
$$

The derivation of formula (iv) is left to you.

---

† Only experience prompts us to multiply numerator and denominator by $\sec x + \tan x$.

Example 1 Calculate

$$\int \cot \pi x \, dx.$$

SOLUTION Set

$$u = \pi x, \qquad du = \pi \, dx$$

$$\int \cot \pi x \, dx = \frac{1}{\pi} \int \cot u \, du = \frac{1}{\pi} \ln|\sin u| + C = \frac{1}{\pi} \ln|\sin \pi x| + C. \quad \square$$

Remark The u-substitution simplifies many calculations, but you will find with experience that you can carry out many integrations without it. ❏

Example 2 Evaluate

$$\int_0^{\pi/8} \sec 2x \, dx.$$

SOLUTION It is easy to check that $\frac{1}{2} \ln|\sec 2x + \tan 2x|$ is an antiderivative of $\sec 2x$. Thus

$$\int_0^{\pi/8} \sec 2x \, dx = \frac{1}{2} \left[\ln|\sec 2x + \tan 2x| \right]_0^{\pi/8}$$

$$= \frac{1}{2} [\ln(\sqrt{2} + 1) - \ln 1] = \frac{1}{2} \ln(\sqrt{2} + 1) \cong 0.44. \quad \square$$

When the integrand is a quotient, it is worth checking to see if the integral can be written in the form

$$\int \frac{du}{u}.$$

Example 3 Calculate

$$\int \frac{\sec^2 3x}{1 + \tan 3x} \, dx.$$

SOLUTION Set

$$u = 1 + \tan 3x, \qquad du = 3 \sec^2 3x \, dx.$$

$$\int \frac{\sec^2 3x}{1 + \tan 3x} \, dx = \frac{1}{3} \int \frac{du}{u} = \frac{1}{3} \ln|u| + C = \frac{1}{3} \ln|1 + \tan 3x| + C. \quad \square$$

Example 4 Calculate the area of the region bounded by the graph of

$$f(x) = \frac{\sin x}{2 + \cos x}$$

and the x-axis for $x \in [0, 2\pi/3]$.

SOLUTION The graph of f is shown in Figure 7.7.1. The area A is given by

$$A = \int_0^{2\pi/3} \frac{\sin x}{2 + \cos x} \, dx.$$

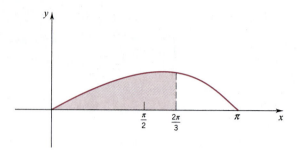

Figure 7.7.1

Set

$$u = 2 + \cos x, \qquad du = -\sin x \, dx.$$

At $x = 0$, $u = 3$, and at $x = 2\pi/3$, $u = 3/2$. Thus,

$$A = \int_0^{2\pi/3} \frac{\sin x}{2 + \cos x} \, dx = \int_3^{3/2} \frac{-du}{u} = \int_{3/2}^3 \frac{du}{u}$$

$$= \left[\ln|u| \right]_{3/2}^3 = \ln(3) - \ln(3/2) = \ln(2) \cong 0.69. \quad \square$$

EXERCISES 7.7

In Exercises 1–20, calculate the indefinite integral.

1. $\int \tan 3x \, dx.$

2. $\int \sec \tfrac{1}{2}\pi x \, dx.$

3. $\int \csc \pi x \, dx.$

4. $\int \cot(\pi - x) \, dx.$

5. $\int e^x \cot e^x \, dx.$

6. $\int \frac{\csc^2 x}{2 + \cot x} \, dx.$

7. $\int \frac{\sin 2x}{3 - 2\cos 2x} \, dx.$

8. $\int e^{\csc x} \csc x \cot x \, dx.$

9. $\int e^{\tan 3x} \sec^2 3x \, dx.$

10. $\int e^x \cos e^x \, dx.$

11. $\int x \sec x^2 \, dx.$

12. $\int \frac{\sec e^{-2x}}{e^{2x}} \, dx.$

13. $\int \cot x \ln(\sin x) \, dx.$

14. $\int \frac{\tan(\ln x)}{x} \, dx.$

15. $\int (1 + \sec x)^2 \, dx.$

16. $\int \tan x \ln(\sec x) \, dx.$

17. $\int \left(\frac{\csc x}{1 + \cot x} \right)^2 dx.$

18. $\int (3 - \csc x)^2 \, dx.$

19. $\int \frac{\sec x}{\sqrt{\ln|\sec x + \tan x|}} \, dx.$

20. $\int \frac{\sec 2x \tan 2x}{1 + \sec 2x} \, dx.$

In Exercises 21–26, evaluate the definite integral.

21. $\int_{\pi/6}^{\pi/2} \frac{\cos x}{1 + \sin x} \, dx.$

22. $\int_{\pi/4}^{\pi/2} (1 + \csc x)^2 \, dx.$

23. $\int_{\pi/4}^{\pi/2} \cot x \, dx.$

24. $\int_{1/4}^{1/3} \tan \pi x \, dx.$

25. $\int_0^{\ln \pi/4} e^x \sec e^x \, dx.$

26. $\int_{\pi/4}^{\pi/2} \frac{\csc^2 x}{3 + \cot x} \, dx.$

In Exercises 27–32, sketch the region bounded by the curves and find its area.

27. $y = \sec x$, $y = 2$, $x = 0$, $x = \pi/6$.

28. $y = \csc \tfrac{1}{2}\pi x$, $y = x$, $x = \tfrac{1}{2}$.

29. $y = \tan x$, $y = 1$, $x = 0$.

30. $y = \sec x$, $y = \cos x$, $x = 0, x = \dfrac{\pi}{4}$.

31. $y = \tan 2x$, $y = 0$, $x = -\pi/8$, $x = \pi/6$.

32. $y = \sec^2 x$, $y = \tfrac{1}{2}$, $x = 0$, $x = \pi/3$.

33. The region bounded by the graph of $f(x) = \sqrt{\sec x}$ and the x-axis for $-\pi/3 \le x \le \pi/3$ is revolved around the x-axis. Find the volume of the solid that is generated.

34. The region bounded by the graph of $f(x) = \tan x$ and the x-axis for $0 \le x \le \pi/4$ is revolved around the x-axis. Find the volume of the solid that is generated.

35. Find the average value of $f(x) = \dfrac{\sin x}{1 + 2 \cos x}$ on $[0, \pi/2]$.

36. Find the average value of $f(x) = (1 + \tan x)^2$ on $[0, \pi/4]$.

37. Show that $\displaystyle \int \csc x \, dx = \ln|\csc x - \cot x| + C$ using the methods of this section.

38. (a) Explain why $\displaystyle \int_0^{\pi} \sqrt{1 - \sin^2 x} \, dx > 0$.
   (b) Evaluate the integral in part (a).

■ 7.8 THE INVERSE TRIGONOMETRIC FUNCTIONS

Since none of the trigonometric functions are one-to-one, none of them have inverses. What then are the inverse trigonometric functions?

The Inverse Sine

The graph of $y = \sin x$ is shown in Figure 7.8.1. Since every horizontal line between -1 and 1 intersects the graph at infinitely many points, the sine function does not have an inverse. However, observe that if we restrict the domain to the interval $[-\frac{1}{2}\pi, \frac{1}{2}\pi]$ (the solid portion of the graph in Figure 7.8.1), then $y = \sin x$ is one-to-one, and on that interval it takes on as a value every number in $[-1, 1]$. Thus, if $x \in [-1, 1]$, there is one and only one number in the interval $[-\frac{1}{2}\pi, \frac{1}{2}\pi]$ at which the sine function has the value x. This number is called the *inverse sine of x*, or *the angle whose sine is x*, and is written $\sin^{-1} x$.

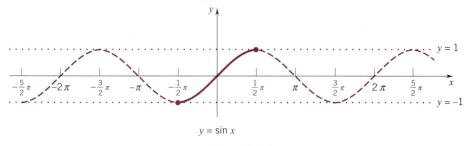

$y = \sin x$

Figure 7.8.1

Remarks Another common notation for the inverse sine function is arcsin x, read "arc sine of x." For consistency in the treatment here, we use $\sin^{-1} x$ throughout, but we use both notations in the Exercises. Remember that the "-1" is not an exponent; do not confuse $\sin^{-1} x$ with the reciprocal $1/\sin x$. ❑

The *inverse sine function*

$$y = \sin^{-1} x, \qquad \text{domain: } [-1, 1], \qquad \text{range: } [-\tfrac{1}{2}\pi, \tfrac{1}{2}\pi]$$

is the inverse of the function

$$y = \sin x, \qquad \text{domain: } [-\tfrac{1}{2}\pi, \tfrac{1}{2}\pi], \qquad \text{range: } [-1, 1].$$

The graphs of these functions are shown in Figure 7.8.2. Each curve is the reflection of the other in the line $y = x$.

Figure 7.8.2

Because these functions are inverses,

■ **Table 7.8.1**

x	$\sin x$
$-\frac{1}{2}\pi$	-1
$-\frac{1}{3}\pi$	$-\frac{1}{2}\sqrt{3}$
$-\frac{1}{4}\pi$	$-\frac{1}{2}\sqrt{2}$
$-\frac{1}{6}\pi$	$-\frac{1}{2}$
0	0
$\frac{1}{6}\pi$	$\frac{1}{2}$
$\frac{1}{4}\pi$	$\frac{1}{2}\sqrt{2}$
$\frac{1}{3}\pi$	$\frac{1}{2}\sqrt{3}$
$\frac{1}{2}\pi$	1

(7.8.1)

$$\text{for all } x \in [-1, 1], \qquad \sin(\sin^{-1}x) = x$$

and

(7.8.2)

$$\text{for all } x \in [-\tfrac{1}{2}\pi, \tfrac{1}{2}\pi], \qquad \sin^{-1}(\sin x) = x.$$

Table 7.8.1 gives some representative values of the sine function from $x = -\frac{1}{2}\pi$ to $x = \frac{1}{2}\pi$. Reversing the order of the columns we have a table for the inverse sine (Table 7.8.2).

On the basis of Table 7.8.2 one could guess that for all $x \in [-1, 1]$

$$\sin^{-1}(-x) = -\sin^{-1}x.$$

This is indeed the case. Being the inverse of an odd function ($\sin(-x) = -\sin x$ for all $x \in [-\frac{1}{2}\pi, \frac{1}{2}\pi]$), the inverse sine is itself an odd function. (Verify this.)

■ **Table 7.8.2**

x	$\sin^{-1} x$
-1	$-\frac{1}{2}\pi$
$-\frac{1}{2}\sqrt{3}$	$-\frac{1}{3}\pi$
$-\frac{1}{2}\sqrt{2}$	$-\frac{1}{4}\pi$
$-\frac{1}{2}$	$-\frac{1}{6}\pi$
0	0
$\frac{1}{2}$	$\frac{1}{6}\pi$
$\frac{1}{2}\sqrt{2}$	$\frac{1}{4}\pi$
$\frac{1}{2}\sqrt{3}$	$\frac{1}{3}\pi$
1	$\frac{1}{2}\pi$

Example 1 Calculate if defined:

(a) $\sin^{-1}(\sin \frac{1}{16}\pi)$.   (b) $\sin^{-1}(\sin \frac{8}{3}\pi)$.   (c) $\sin(\sin^{-1}\frac{1}{3})$.

(d) $\sin^{-1}(\sin \frac{9}{5}\pi)$.   (e) $\sin(\sin^{-1}2)$.

SOLUTION

(a) Since $\frac{1}{16}\pi$ is within the interval $[-\frac{1}{2}\pi, \frac{1}{2}\pi]$, we know by (7.8.2) that

$$\sin^{-1}(\sin \tfrac{1}{16}\pi) = \tfrac{1}{16}\pi.$$

(b) Since $\frac{8}{3}\pi$ is not in the interval $[-\pi/2, \pi/2]$, we cannot apply (7.8.2) directly. However, $\frac{8}{3}\pi = \frac{2}{3}\pi + 2\pi$ and $\sin(\frac{2}{3}\pi + 2\pi) = \sin(\frac{2}{3}\pi)$ (recall that the sine function is periodic with period 2π). Thus

$$\sin^{-1}(\sin \tfrac{8}{3}\pi) = \sin^{-1}(\sin[\tfrac{2}{3}\pi + 2\pi]) = \sin^{-1}(\sin \tfrac{2}{3}\pi) = \tfrac{2}{3}\pi.$$

by (7.8.2)

(c) By (7.8.1)

$$\sin(\sin^{-1}\tfrac{1}{3}) = \tfrac{1}{3}.$$

(d) Since $\tfrac{9}{5}\pi$ is not within the interval $[-\tfrac{1}{2}\pi, \tfrac{1}{2}\pi]$, we cannot apply (7.8.2) directly. However, $\tfrac{9}{5}\pi = 2\pi - \tfrac{1}{5}\pi$. Thus

$$\sin^{-1}(\sin \tfrac{9}{5}\pi) = \sin^{-1}(\sin(2\pi - \tfrac{1}{5}\pi)) = \sin^{-1}(\sin(-\tfrac{1}{5}\pi)) = -\tfrac{1}{5}\pi.$$
$$\text{by (7.8.2)} \underline{}\uparrow$$

(e) The expression $\sin(\sin^{-1}2)$ makes no sense since 2 is not in the domain of the inverse sine; there is *no* angle whose sine is 2. The inverse sine is defined only on $[-1, 1]$. ❏

If $0 < x < 1$, then $\sin^{-1}x$ is the radian measure of the acute angle whose sine is x. We can construct an angle of radian measure $\sin^{-1}x$ by drawing a right triangle with a leg of length x and a hypotenuse of length 1. See Figure 7.8.3.

Reading from the figure we have

$$\sin(\sin^{-1}x) = x, \qquad\qquad \cos(\sin^{-1}x) = \sqrt{1-x^2}$$

$$\tan(\sin^{-1}x) = \frac{x}{\sqrt{1-x^2}}, \qquad \cot(\sin^{-1}x) = \frac{\sqrt{1-x^2}}{x}$$

$$\sec(\sin^{-1}x) = \frac{1}{\sqrt{1-x^2}}, \qquad \csc(\sin^{-1}x) = \frac{1}{x}.$$

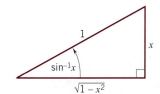

Figure 7.8.3

Since the derivative of the sine function.,

$$\frac{d}{dx}(\sin x) = \cos x,$$

does not take on the value 0 on the *open* interval $(-\tfrac{1}{2}\pi, \tfrac{1}{2}\pi)$ (recall that $\cos x > 0$ for $-\tfrac{1}{2}\pi < x < \tfrac{1}{2}\pi$), the inverse sine function is differentiable on the open interval $(-1, 1)$.† We can find the derivative as follows:

$$y = \sin^{-1}x,$$

$$\sin y = x,$$

$$\cos y \frac{dy}{dx} = 1,$$

$$\frac{dy}{dx} = \frac{1}{\cos y} = \frac{1}{\sqrt{1-x^2}} \quad \text{(see the figure).} \quad ❏$$

In short

(7.8.3)

$$\boxed{\frac{d}{dx}(\sin^{-1}x) = \frac{1}{\sqrt{1-x^2}}.}$$

---

† See Section 7.1.

Example 2 Find

$$\frac{d}{dx}[\sin^{-1}(3x^2)].$$

NOTE: Although we have not made a point of it, we continue with the convention that if the domain of a function f is not specified explicitly, then it is understood to be the largest set of real numbers x for which $f(x)$ is a real number. In this case, the domain is the set of all real numbers x such that $-1 \le 3x^2 \le 1$, or $|x| \le 1/\sqrt{3}$.

SOLUTION In general, by the chain rule,

$$\frac{d}{dx}[\sin^{-1}u] = \frac{d}{du}[\sin^{-1}u]\frac{du}{dx} = \frac{1}{\sqrt{1-u^2}}\frac{du}{dx}.$$

Thus

$$\frac{d}{dx}[\sin^{-1}(3x^2)] = \frac{1}{\sqrt{1-(3x^2)^2}}\frac{d}{dx}(3x^2) = \frac{6x}{\sqrt{1-9x^4}}. \quad \square$$

Example 3 Show that for $a > 0$

(7.8.4)
$$\boxed{\int \frac{dx}{\sqrt{a^2-x^2}} = \sin^{-1}\left(\frac{x}{a}\right) + C.}$$

SOLUTION We change variables so that the a^2 becomes 1 and we can use (7.8.3). We set

$$au = x, \qquad a\,du = dx.$$

Then

$$\int \frac{dx}{\sqrt{a^2-x^2}} = \int \frac{a\,du}{\sqrt{a^2-a^2u^2}} = \int \frac{du}{\sqrt{1-u^2}} = \sin^{-1}u + C = \sin^{-1}\left(\frac{x}{a}\right) + C. \quad \square$$

since $a > 0$ ——↑

Example 4 Evaluate

$$\int_0^{\sqrt{3}} \frac{dx}{\sqrt{4-x^2}}.$$

SOLUTION By (7.8.4)

$$\int \frac{dx}{\sqrt{4-x^2}} = \sin^{-1}\left(\frac{x}{2}\right) + C.$$

It follows that

$$\int_0^{\sqrt{3}} \frac{dx}{\sqrt{4-x^2}} = \left[\sin^{-1}\left(\frac{x}{2}\right)\right]_0^{\sqrt{3}} = \sin^{-1}\left(\frac{\sqrt{3}}{2}\right) - \sin^{-1}0 = \frac{\pi}{3} - 0 = \frac{\pi}{3}. \quad \square$$

The Inverse Tangent

Although not one-to-one on its full domain, the tangent function is one-to-one on the open interval $(-\frac{1}{2}\pi, \frac{1}{2}\pi)$ and on that interval it takes on as a value every real number (see Figure 7.8.4). Thus, for any real number x, there is one and only one number in the open interval $(-\frac{1}{2}\pi, \frac{1}{2}\pi)$ at which the tangent function has the value x. This number is called the *inverse tangent of x*, or *the angle whose tangent is x*, and is written $\tan^{-1}x$. Like the inverse sine function, the notation arctan x is also commonly used.

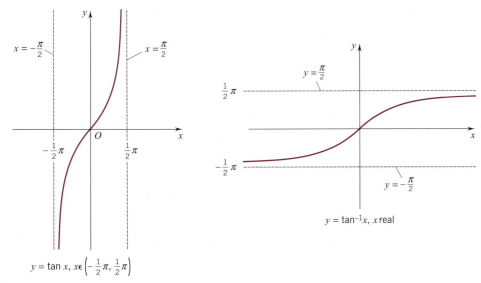

Figure 7.8.4

The *inverse tangent function*

$$y = \tan^{-1}x, \qquad \text{domain: } (-\infty, \infty), \qquad \text{range: } (-\tfrac{1}{2}\pi, \tfrac{1}{2}\pi)$$

is the inverse of the function

$$y = \tan x, \qquad \text{domain: } (-\tfrac{1}{2}\pi, \tfrac{1}{2}\pi), \qquad \text{range: } (-\infty, \infty).$$

The graphs of these two functions are given in Figure 7.8.4. Each curve is a reflection of the other in the line $y = x$. While the tangent has vertical asymptotes, the inverse tangent has horizontal asymptotes. Both functions are odd functions.

Because these functions are inverses,

(7.8.5)

$$\text{for all real numbers } x, \qquad \tan(\tan^{-1}x) = x$$

and

(7.8.6)

$$\text{for all } x \in (-\tfrac{1}{2}\pi, \tfrac{1}{2}\pi), \qquad \tan^{-1}(\tan x) = x.$$

It is hard to make a mistake with the first relation since it applies for all real numbers. The second relation requires the usual care:

$$\tan^{-1}(\tan \tfrac{1}{4}\pi) = \tfrac{1}{4}\pi \qquad \text{but} \qquad \tan^{-1}(\tan \tfrac{7}{5}\pi) \neq \tfrac{7}{5}\pi.$$

We can calculate $\tan^{-1}(\tan \tfrac{7}{5}\pi)$ as follows: $\tfrac{7}{5}\pi = \tfrac{2}{5}\pi + \pi$ and $\tan(\tfrac{2}{5}\pi + \pi) = \tan(\tfrac{2}{5}\pi)$ (recall that the tangent function is periodic with period π). The relation $\tan^{-1}(\tan \tfrac{2}{5}\pi) = \tfrac{2}{5}\pi$ is valid because $\tfrac{2}{5}\pi$ is within the interval $(-\tfrac{1}{2}\pi, \tfrac{1}{2}\pi)$.

If $x > 0$, then $\tan^{-1}x$ is the radian measure of the acute angle that has tangent x. We can construct an angle of radian measure $\tan^{-1}x$ by drawing a right triangle with legs of length x and 1 (Figure 7.8.5). Reading from the triangle, we have

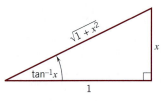

Figure 7.8.5

$$\tan(\tan^{-1}x) = x \qquad\qquad \cot(\tan^{-1}x) = \frac{1}{x}$$

$$\sin(\tan^{-1}x) = \frac{x}{\sqrt{1 + x^2}} \qquad \cos(\tan^{-1}x) = \frac{1}{\sqrt{1 + x^2}}$$

$$\sec(\tan^{-1}x) = \sqrt{1 + x^2} \qquad \csc(\tan^{-1}x) = \frac{\sqrt{1 + x^2}}{x}.$$

Since the derivative of the tangent function

$$\frac{d}{dx}(\tan x) = \sec^2 x = \frac{1}{\cos^2 x},$$

is never 0, the inverse tangent function is everywhere differentiable (Section 7.1). We can find the derivative as we did for the inverse sine:

$$y = \tan^{-1}x,$$

$$\tan y = x,$$

$$\sec^2 y \frac{dy}{dx} = 1,$$

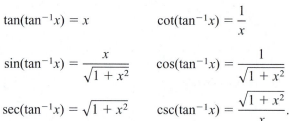

$$\frac{dy}{dx} = \frac{1}{\sec^2 y} = \cos^2 y = \frac{1}{1 + x^2} \quad \text{(see the figure).}$$

We have found that

(7.8.7)

$$\frac{d}{dx}(\tan^{-1}x) = \frac{1}{1 + x^2}.$$

Example 5 Calculate

$$\frac{d}{dx}[\tan^{-1}(ax^2 + bx + c)].$$

SOLUTION In general, by the chain rule,

$$\frac{d}{dx}[\tan^{-1}u] = \frac{d}{du}[\tan^{-1}u]\frac{du}{dx} = \frac{1}{1 + u^2}\frac{du}{dx}.$$

Thus

$$\frac{d}{dx}[\tan^{-1}(ax^2 + bx + c)] = \frac{1}{1 + (ax^2 + bx + c)^2} \frac{d}{dx}(ax^2 + bx + c)$$

$$= \frac{2ax + b}{1 + (ax^2 + bx + c)^2}. \quad \square$$

Example 6 Show that, for $a \neq 0$,

(7.8.8)
$$\int \frac{dx}{a^2 + x^2} = \frac{1}{a}\tan^{-1}\left(\frac{x}{a}\right) + C.$$

SOLUTION We change variables so that a^2 is replaced by 1 and we can use (7.8.7). We set

$$au = x, \qquad a\, du = dx.$$

Then

$$\int \frac{dx}{a^2 + x^2} = \int \frac{a\, du}{a^2 + a^2 u^2} = \frac{1}{a}\int \frac{du}{1 + u^2}$$

$$= \frac{1}{a}\tan^{-1}u + C = \frac{1}{a}\tan^{-1}\left(\frac{x}{a}\right) + C. \quad \square$$

$$(7.8.7)\text{---}\uparrow$$

Example 7 Evaluate

$$\int_0^2 \frac{dx}{4 + x^2}.$$

SOLUTION By (7.8.8)

$$\int \frac{dx}{4 + x^2} = \int \frac{dx}{2^2 + x^2} = \frac{1}{2}\tan^{-1}\left(\frac{x}{2}\right) + C$$

so that

$$\int_0^2 \frac{dx}{4 + x^2} = \left[\frac{1}{2}\tan^{-1}\left(\frac{x}{2}\right)\right]_0^2 = \frac{1}{2}\tan^{-1}1 - \frac{1}{2}\tan^{-1}0 = \frac{\pi}{8}. \quad \square$$

Inverse Secant

The procedure for defining the inverse secant function is the same as the procedure for the inverse sine and inverse tangent functions. The graph of $y = \sec x$ is indicated in Figure 7.8.6. Note that $|\sec x| \geq 1$ for all x in the domain. If we restrict the domain to the set $[0, \frac{1}{2}\pi) \cup (\frac{1}{2}\pi, \pi]$ (the solid portion of the graph in Figure 7.8.6), then the secant function is one-to-one, and for every real number x such that $|x| \geq 1$ there is one and only one number in $[0, \frac{1}{2}\pi) \cup (\frac{1}{2}\pi, \pi]$ at which the secant function has the value x. This number is called the *inverse secant of x*, or *the angle whose secant is x*, and is written $\sec^{-1}x$ or arcsec x.

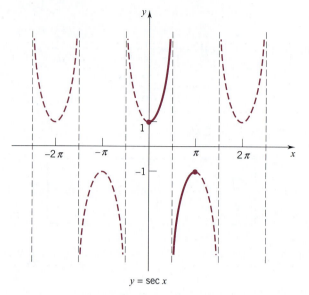

$y = \sec x$

Figure 7.8.6

The *inverse secant function*

$$y = \sec^{-1}x, \qquad \text{domain: } (-\infty, -1] \cup [1, \infty), \qquad \text{range: } [0, \tfrac{1}{2}\pi) \cup (\tfrac{1}{2}\pi, \pi]$$

is the inverse of the function

$$y = \sec x, \qquad \text{domain:} [0, \tfrac{1}{2}\pi) \cup (\tfrac{1}{2}\pi, \pi], \qquad \text{range: } (-\infty, -1] \cup [1, \infty).$$

The graphs of these two functions are given in Figure 7.8.7.

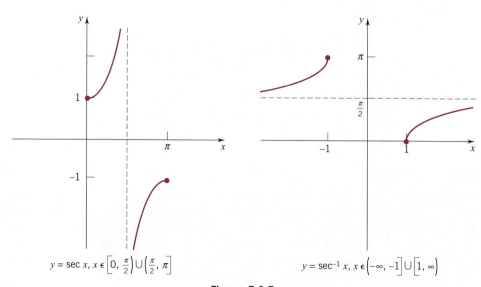

$y = \sec x, \; x \in \left[0, \dfrac{\pi}{2}\right) \cup \left(\dfrac{\pi}{2}, \pi\right]$

$y = \sec^{-1} x, \; x \in \left(-\infty, -1\right] \cup \left[1, \infty\right)$

Figure 7.8.7

The function–inverse function relationships for $\sec x$ and $\sec^{-1}x$ are:

(7.8.9)

> for all such that $|x| \geq 1$, $\quad \sec(\sec^{-1}x) = x$

(7.8.10)

> for all $x \in [0, \frac{1}{2}\pi) \cup (\frac{1}{2}\pi, \pi]$, $\quad \sec^{-1}(\sec x) = x.$

As we have seen previously, the second relation requires careful attention:

$$\sec^{-1}(\sec \tfrac{1}{3}\pi) = \tfrac{1}{3}\pi \quad \text{but} \quad \sec^{-1}(\sec \tfrac{7}{4}\pi) \neq \tfrac{7}{4}\pi.$$

We calculate $\sec^{-1}(\sec \frac{7}{4}\pi)$ as follows:

$$\sec^{-1}(\sec \tfrac{7}{4}\pi) = \sec^{-1}(\sec[2\pi - \tfrac{1}{4}\pi]) = \sec^{-1}(\sec \tfrac{1}{4}\pi) = \tfrac{1}{4}\pi.$$

If $x > 1$, then $\sec^{-1}x$ is the radian measure of the acute angle that has secant x. We can construct an angle of radian measure $\sec^{-1}x$ by drawing a right triangle with hypotenuse of length x and a side of length 1 (Figure 7.8.8). The values

$$\sec(\sec^{-1}x) = x \qquad \csc(\sec^{-1}x) = \frac{x}{\sqrt{x^2-1}}$$

$$\sin(\sec^{-1}x) = \frac{\sqrt{x^2-1}}{x} \qquad \cos(\sec^{-1}x) = \frac{1}{x}$$

$$\tan(\sec^{-1}x) = \sqrt{x^2-1} \qquad \cot(\sec^{-1}x) = \frac{1}{\sqrt{x^2-1}}$$

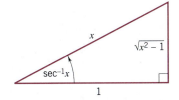

Figure 7.8.8

can all be read from this triangle.

The derivative of the secant function,

$$\frac{d}{dx}(\sec x) = \sec x \tan x,$$

is nonzero on $(0, \frac{1}{2}\pi) \cup (\frac{1}{2}\pi, \pi)$. Therefore, the inverse secant function is differentiable for $|x| > 1$. We can find the derivative as follows:

$$y = \sec^{-1}x,$$

$$\sec y = x,$$

$$\sec y \tan y \frac{dy}{dx} = 1,$$

$$\frac{dy}{dx} = \frac{1}{\sec y \tan y}.$$

To complete the calculation, we have to express the product $\sec y \tan y$ in terms of x. Clearly, $\sec y = x$, and from Figure 7.8.9 we see that $\tan y = \pm\sqrt{x^2-1}$. Now, if $x > 1$, then $0 < y < \frac{1}{2}\pi$ and $\tan y = \sqrt{x^2-1}$; if $x < -1$, then $\frac{1}{2}\pi < y < \pi$ and

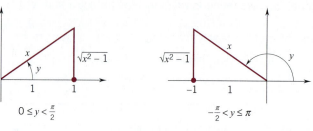

Figure 7.8.9

$\tan y = -\sqrt{x^2 - 1}$. In either case, the product $\sec y \tan y$ is *positive*. Therefore, we have

(7.8.11)
$$\frac{d}{dx}(\sec^{-1}x) = \frac{1}{|x|\sqrt{x^2 - 1}}.$$

Note that $dy/dx > 0$ on $(-\infty, -1)$ and on $(1, \infty)$, which implies that $y = \sec^{-1}x$ is an increasing function on each of these intervals. This is consistent with the graph shown in Figure 7.8.7.

Example 8 Find
$$\frac{d}{dx}[\sec^{-1}(2 \ln x)].$$

SOLUTION By the chain rule, we have
$$\frac{d}{dx}[\sec^{-1}u] = \frac{d}{du}[\sec^{-1}u]\frac{du}{dx} = \frac{1}{|u|\sqrt{u^2 - 1}}\frac{du}{dx}.$$

Thus
$$\frac{d}{dx}[\sec^{-1}(2 \ln x)] = \frac{1}{|2 \ln x|\sqrt{(2 \ln x)^2 - 1}}\frac{d}{dx}(2 \ln x) = \frac{1}{2|\ln x|\sqrt{4(\ln x)^2 - 1}} \cdot \frac{2}{x}$$
$$= \frac{1}{x|\ln x|\sqrt{4(\ln x)^2 - 1}}. \quad \square$$

Example 9 Show that
$$\int \frac{dx}{x\sqrt{x^2 - 1}} = \sec^{-1}|x| + C.$$

SOLUTION If $x > 1$, $\sec^{-1}|x| = \sec^{-1}x$ and
$$\frac{d}{dx}(\sec^{-1}x) = \frac{1}{|x|\sqrt{x^2 - 1}} = \frac{1}{x\sqrt{x^2 - 1}}.$$

If $x < -1$, $\sec^{-1}|x| = \sec^{-1}(-x)$ and

$$\frac{d}{dx}[\sec^{-1}(-x)] = \frac{1}{|-x|\sqrt{x^2 - 1}}(-1) = \frac{-1}{-x\sqrt{x^2 - 1}} = \frac{1}{x\sqrt{x^2 - 1}}.$$

Thus, it follows that $F(x) = \sec^{-1}|x|$ is an antiderivative for $f(x) = \dfrac{1}{x\sqrt{x^2 - 1}}$. ❏

We leave it as an exercise to show that if $a > 0$ is a constant, then

(7.8.12)
$$\int \frac{1}{x\sqrt{x^2 - a^2}}\,dx = \frac{1}{a}\sec^{-1}\frac{|x|}{a} + C.$$

Example 10 Evaluate

$$\int_{2\sqrt{2}}^{4} \frac{dx}{x\sqrt{x^2 - 4}}.$$

SOLUTION   Use (7.8.12) with $a = 2$:

$$\int_{2\sqrt{2}}^{4} \frac{dx}{x\sqrt{x^2 - 4}} = \left[\frac{1}{2}\sec^{-1}\frac{|x|}{2}\right]_{2\sqrt{2}}^{4} = \frac{1}{2}\sec^{-1}2 - \frac{1}{2}\sec^{-1}\sqrt{2}$$

$$= \frac{1}{2}\left(\frac{\pi}{3} - \frac{\pi}{4}\right) = \frac{\pi}{24}. \quad ❏$$

The Other Trigonometric Inverses

There are three other trigonometric inverses:

the *inverse cosine,* $y = \cos^{-1}x$, is the inverse of $y = \cos x$, $x \in [0, \pi]$;

the *inverse cotangent,* $y = \cot^{-1}x$, is the inverse of $y = \cot x$, $x \in (0, \pi)$;

the *inverse cosecant,* $y = \csc^{-1}x$, is the inverse of $y = \csc x$, $x \in [-\tfrac{1}{2}\pi, 0) \cup (0, \tfrac{1}{2}\pi]$.

Figure 7.8.10 illustrates each of these inverses between 0 and $\tfrac{1}{2}\pi$ in terms of right triangles.

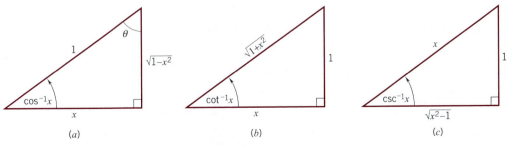

(a) (b) (c)

Figure 7.8.10

The differentiation formulas for these functions are:

$$\frac{d}{dx}(\cos^{-1}x) = \frac{-1}{\sqrt{1-x^2}} = -\frac{d}{dx}(\sin^{-1}x)$$

$$\frac{d}{dx}(\cot^{-1}x) = \frac{-1}{1+x^2} = -\frac{d}{dx}(\tan^{-1}x)$$

$$\frac{d}{dx}(\csc^{-1}x) = \frac{-1}{|x|\sqrt{x^2-1}} = -\frac{d}{dx}(\sec^{-1}x).$$

You are asked to verify these formulas in the Exercises. Since the derivatives of these functions differ from the derivatives of their corresponding cofunctions by a constant factor, they are not needed for finding antiderivatives.

Remark   In Figure 7.8.10a, we let θ denote the other acute angle of the right triangle. Since the two acute angles in a right triangle are complementary, we have

$$\theta + \cos^{-1}x = \frac{\pi}{2}.$$

But note that $\sin\theta = x$, which means that $\theta = \sin^{-1}x$. Therefore,

$$\sin^{-1}x + \cos^{-1}x = \frac{\pi}{2} \quad \text{or} \quad \cos^{-1}x = \frac{\pi}{2} - \sin^{-1}x.$$

You can verify that the corresponding relations hold for the pairs $\{\tan x, \cot x\}$ and $\{\sec x, \csc x\}$:

$$\cot^{-1}x = \frac{\pi}{2} - \tan^{-1}x \quad \text{and} \quad \csc^{-1}x = \frac{\pi}{2} - \sec^{-1}x.$$

This is another reason why it is sufficient to restrict our attention to $\sin^{-1}x$, $\tan^{-1}x$, and $\sec^{-1}x$.   ❑

EXERCISES 7.8

In Exercises 1–12, determine the exact value of the given expression.

1. $\tan^{-1}0$.

2. $\sin^{-1}(-\sqrt{3}/2)$.

3. $\text{arcsec } 2$.

4. $\tan^{-1}\sqrt{3}$.

5. $\cos^{-1}(-\frac{1}{2})$.

6. $\sec^{-1}(-\sqrt{2})$.

7. $\arcsin(\sin 7\pi/4)$.

8. $\arctan(\tan 11\pi/4)$.

9. $\sec[\sec^{-1}(-2/\sqrt{3})]$.

10. $\sin[\arccos(-\frac{1}{2})]$.

11. $\cos(\sec^{-1}2)$.

12. $\arctan(\sec 0)$.

▶ In Exercises 13–18, find the approximate value by using your calculator. Use four decimal place accuracy.

13. $\sin^{-1}0.918$.

14. $\arcsin(-0.795)$.

15. $\tan^{-1}(-0.493)$.

16. $\arctan 3.111$.

17. $\text{arcsec }(2.761)$.

18. $\cos^{-1}(-0.142)$.

In Exercises 19–24, take $x > 0$ and calculate the following from Figure 7.8.5.

19. $\cos(\tan^{-1}x)$.

20. $\sin(\tan^{-1}x)$.

21. $\tan(\tan^{-1}x)$.

22. $\cot(\tan^{-1}x)$.

23. $\sec(\tan^{-1}x)$.

24. $\csc(\tan^{-1}x)$.

In Exercises 25–30, take $0 < x < 1$ and calculate the following from an appropriate right triangle.

25. $\sin(\cos^{-1}x)$.

26. $\cos(\cot^{-1}x)$.

27. $\sec(\cot^{-1}x)$.

28. $\tan(\cos^{-1}x)$.

29. $\cot(\cos^{-1}x)$.

30. $\csc(\cot^{-1}x)$.

In Exercises 31–54, differentiate the given function.

31. $y = \tan^{-1}(x+1)$.

32. $y = \tan^{-1}\sqrt{x}$.

33. $f(x) = \sec^{-1}(2x^2)$.

34. $f(x) = e^x\sin^{-1}x$.

35. $f(x) = x \sin^{-1} 2x.$

36. $f(x) = e^{\tan^{-1} x}.$

37. $u = (\sin^{-1} x)^2.$

38. $v = \tan^{-1}(e^x).$

39. $y = \dfrac{\tan^{-1} x}{x}.$

40. $y = \sec^{-1}\sqrt{x^2 + 2}.$

41. $f(x) = \sqrt{\tan^{-1} 2x}.$

42. $f(x) = \ln(\tan^{-1} x).$

43. $y = \tan^{-1}(\ln x).$

44. $g(x) = \sec^{-1}(\cos x + 2).$

45. $\theta = \sin^{-1}(\sqrt{1 - r^2}).$

46. $\theta = \sin^{-1}\left(\dfrac{r}{r + 1}\right).$

47. $g(x) = x^2 \sec^{-1}\left(\dfrac{1}{x}\right).$

48. $\theta = \tan^{-1}\left(\dfrac{1}{1 + r^2}\right).$

49. $y = \sin[\sec^{-1}(\ln x)].$

50. $f(x) = e^{\sec^{-1} x}.$

51. $f(x) = \sqrt{c^2 - x^2} + c \sin^{-1}\left(\dfrac{x}{c}\right), c > 0.$

52. $f(x) = \frac{1}{3}\sin^{-1}(3x - 4x^2).$

53. $y = \dfrac{x}{\sqrt{c^2 - x^2}} - \sin^{-1}\left(\dfrac{x}{c}\right), c > 0.$

54. $y = x\sqrt{c^2 - x^2} + c^2 \sin^{-1}\left(\dfrac{x}{c}\right), c > 0.$

55. Show that for $a > 0$

(7.8.13)
$$\int \frac{dx}{\sqrt{a^2 - (x + b)^2}} = \sin^{-1}\left(\frac{x + b}{a}\right) + C.$$

56. Show that for $a \neq 0$

(7.8.14)
$$\int \frac{dx}{a^2 + (x + b)^2} = \frac{1}{a}\tan^{-1}\left(\frac{x + b}{a}\right) + C.$$

57. (a) Verify (7.8.12).
(b) Show that for $a > 0$
$$\int \frac{dx}{(x + b)\sqrt{(x + b)^2 - a^2}} = \frac{1}{a}\sec^{-1}\frac{|x + b|}{a} + C$$

In Exercises 58–60, give the domain and range of each function, and verify the differentiation formula.

58. $f(x) = \cos^{-1} x; \quad \dfrac{d}{dx}(\cos^{-1} x) = \dfrac{-1}{\sqrt{1 - x^2}}.$

59. $f(x) = \cot^{-1} x; \quad \dfrac{d}{dx}(\cot^{-1} x) = \dfrac{-1}{1 + x^2}.$

60. $f(x) = \csc^{-1} x; \quad \dfrac{d}{dx}(\csc^{-1} x) = \dfrac{-1}{|x|\sqrt{x^2 - 1}}.$

In Exercises 61–74, evaluate the given integral.

61. $\displaystyle\int_0^1 \frac{dx}{1 + x^2}.$

62. $\displaystyle\int_{-1}^1 \frac{dx}{1 + x^2}.$

63. $\displaystyle\int_0^{1/\sqrt{2}} \frac{dx}{\sqrt{1 - x^2}}.$

64. $\displaystyle\int_0^1 \frac{dx}{\sqrt{4 - x^2}}.$

65. $\displaystyle\int_0^5 \frac{dx}{25 + x^2}.$

66. $\displaystyle\int_1^2 \frac{dx}{x\sqrt{x^2 - 16}}.$

67. $\displaystyle\int_0^{3/2} \frac{dx}{9 + 4x^2}.$

68. $\displaystyle\int_2^5 \frac{dx}{9 + (x - 2)^2}.$

69. $\displaystyle\int_{3/4}^3 \frac{dx}{x\sqrt{16x^2 - 9}}.$

70. $\displaystyle\int_4^6 \frac{dx}{(x - 3)\sqrt{x^2 - 6x + 8}}.$

71. $\displaystyle\int_{-3}^{-2} \frac{dx}{\sqrt{4 - (x + 3)^2}}.$

72. $\displaystyle\int_{\ln 2}^{\ln 3} \frac{e^{-x}}{\sqrt{1 - e^{-2x}}}dx.$

73. $\displaystyle\int_0^{\ln 2} \frac{e^x}{1 + e^{2x}}dx.$

74. $\displaystyle\int_0^{1/2} \frac{1}{\sqrt{3 - 4x^2}}dx.$

In Exercises 75–84, calculate the indefinite integral.

75. $\displaystyle\int \frac{x}{\sqrt{1 - x^4}}dx.$

76. $\displaystyle\int \frac{\sec^2 x}{\sqrt{9 - \tan^2 x}}dx.$

77. $\displaystyle\int \frac{x}{1 + x^4}dx.$

78. $\displaystyle\int \frac{dx}{\sqrt{4x - x^2}}.$

79. $\displaystyle\int \frac{\sec^2 x}{9 + \tan^2 x}dx.$

80. $\displaystyle\int \frac{\cos x}{3 + \sin^2 x}dx.$

81. $\displaystyle\int \frac{\sin^{-1} x}{\sqrt{1 - x^2}}dx.$

82. $\displaystyle\int \frac{\tan^{-1} x}{1 + x^2}dx.$

83. $\displaystyle\int \frac{dx}{x\sqrt{1 - (\ln x)^2}}.$

84. $\displaystyle\int \frac{dx}{x[1 + (\ln x)^2]}.$

85. Find the area of the region bounded by graph of $y = 1/\sqrt{4 - x^2}$ and the x-axis between $x = -1$ and $x = 1$.

86. Find the area of the region bounded by the graph of $y = 3/(9 + x^2)$ and the x-axis between $x = -3$ and $x = 3$.

87. Sketch the region bounded by the graphs of $4y = x^2$ and $y = 8/(x^2 + 4)$ and find its area.

88. The region bounded by the graph of $y = 1/(x^2\sqrt{x^2 - 9})$ and the x-axis between $x = 2\sqrt{3}$ and $x = 6$ is revolved around the y-axis. Find the volume of the solid that is generated.

89. A billboard k feet wide is perpendicular to a straight road and is s feet from the road. At what point on the road would a motorist have the best view of the billboard; that is, at what point on the road is the angle θ subtended by the billboard a maximum (see the figure)?

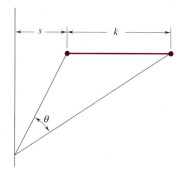

90. A person walking along a straight path at the rate of 6 feet per second is followed by a spotlight that is located 30 feet from the path. How fast is the spotlight turning at the instant the person is 50 feet past the point on the path that is closest to the spotlight?

 91. Let

$$f(x) = \tan^{-1}\left(\frac{2 + x}{1 - 2x}\right), \quad x \neq \tfrac{1}{2}.$$

(a) Use a graphing utility to graph f and $g(x) = \tan^{-1}x$. What do you notice about these two graphs?

(b) Estimate the limits

$$\lim_{x \to (1/2)^+} f(x) \quad \text{and} \quad \lim_{x \to (1/2)^-} f(x).$$

(c) Show that $f'(x) = \dfrac{1}{1 + x^2}$, $x \neq \tfrac{1}{2}$.

(d) Show that there is no constant C such that $f(x) = \tan^{-1}x + C$ for all $x \neq \tfrac{1}{2}$.

(e) Find constants C_1 and C_2 such that

$$f(x) = \tan^{-1}x + C_1 \quad \text{for } x < 1/2$$
$$f(x) = \tan^{-1}x + C_2 \quad \text{for } x > 1/2.$$

92. Let

$$f(x) = \tan^{-1}\left(\frac{a + x}{1 - ax}\right), \quad x \neq 1/a.$$

(a) Show that $f'(x) = \dfrac{1}{1 + x^2}$, $x \neq 1/a$.

(b) Show that there is no constant C such that $f(x) = \tan^{-1}x + C$ for all $x \neq 1/a$.

(c) Find constants C_1 and C_2 such that

$$f(x) = \tan^{-1}x + C_1 \quad \text{for } x < 1/a$$
$$f(x) = \tan^{-1}x + C_2 \quad \text{for } x > 1/a.$$

93. Let $f(x) = \tan^{-1}(1/x)$, $x \neq 0$.

(a) Use a graphing utility to graph f and $g(x) = \tan^{-1}x$. What do you notice about these two graphs?

(b) Use your graph of f to estimate the limits

$$\lim_{x \to 0^+} f(x) \quad \text{and} \quad \lim_{x \to 0^-} f(x)$$

(c) Show that $f'(x) = \dfrac{-1}{1 + x^2}$, $x \neq 0$.

(d) Show that there does not exist a constant C such that $f(x) + \tan^{-1}x = C$ for all x.

(e) Find constants C_1 and C_2 such that

$$f(x) + \tan^{-1}x = C_1 \quad \text{for } x > 0$$
$$f(x) + \tan^{-1}x = C_2 \quad \text{for } x < 0.$$

94. Evaluate

$$\lim_{x \to 0} \frac{\sin^{-1}x}{x}$$

numerically. Justify your answer by other means.

95. Estimate the integral

$$\int_0^{0.5} \frac{1}{\sqrt{1 - x^2}}\, dx$$

by using the partition $\{0, 0.1, 0.2, 0.3, 0.4, 0.5\}$ and the intermediate points

$$x_1^* = 0.05, \quad x_2^* = 0.15, \quad x_3^* = 0.25,$$
$$x_4^* = 0.35, \quad x_5^* = 0.45.$$

Note that the sine of your estimate is close to 0.5. Explain the reason for this.

■ 7.9 THE HYPERBOLIC SINE AND COSINE

Certain combinations of the exponential functions e^x and e^{-x} occur so frequently in mathematical applications that they are given special names. The *hyperbolic sine* (sinh) and *hyperbolic cosine* (cosh) are the functions defined by:

(7.9.1)

$$\sinh x = \tfrac{1}{2}(e^x - e^{-x}), \qquad \cosh x = \tfrac{1}{2}(e^x + e^{-x}).$$

The reasons for these names will become apparent as we go on.

Since

$$\frac{d}{dx}(\sinh x) = \frac{d}{dx}[\tfrac{1}{2}(e^x - e^{-x})] = \tfrac{1}{2}(e^x + e^{-x})$$

and

$$\frac{d}{dx}(\cosh x) = \frac{d}{dx}[\tfrac{1}{2}(e^x + e^{-x})] = \tfrac{1}{2}(e^x - e^{-x}),$$

we have

(7.9.2)
$$\frac{d}{dx}(\sinh x) = \cosh x, \qquad \frac{d}{dx}(\cosh x) = \sinh x.$$

In short, each of these functions is the derivative of the other.

The Graphs

We begin with the hyperbolic sine. Since

$$\sinh(-x) = \tfrac{1}{2}(e^{-x} - e^x) = -\tfrac{1}{2}(e^x - e^{-x}) = -\sinh x,$$

the hyperbolic sine is an odd function. The graph is therefore symmetric about the origin. Since

$$\frac{d}{dx}(\sinh x) = \cosh x = \tfrac{1}{2}(e^x + e^{-x}) > 0 \qquad \text{for all real } x,$$

the hyperbolic sine increases everywhere. Since

$$\frac{d^2}{dx^2}(\sinh x) = \frac{d}{dx}(\cosh x) = \sinh x = \tfrac{1}{2}(e^x - e^{-x}),$$

you can see that

$$\frac{d^2}{dx^2}(\sinh x) \quad \text{is} \quad \begin{cases} \text{negative,} & \text{for} \quad x < 0 \\ 0, & \text{at} \quad x = 0 \\ \text{positive,} & \text{for} \quad x > 0. \end{cases}$$

The graph is therefore concave down on $(-\infty, 0)$ and concave up on $(0, \infty)$. The point $(0, \sinh 0) = (0, 0)$ is the only point of inflection. The slope at the origin is $\cosh 0 = 1$. A sketch of the graph appears in Figure 7.9.1. ❑

We turn now to the hyperbolic cosine. Since

$$\cosh(-x) = \tfrac{1}{2}(e^{-x} + e^x) = \tfrac{1}{2}(e^x + e^{-x}) = \cosh x,$$

the hyperbolic cosine is an even function. The graph is therefore symmetric about the y-axis. Since

$$\frac{d}{dx}(\cosh x) = \sinh x,$$

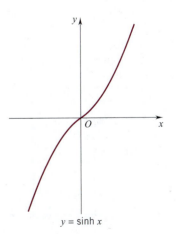

$y = \sinh x$

Figure 7.9.1

you can see that

$$\frac{d}{dx}(\cosh x) \quad \text{is} \quad \begin{cases} \text{negative}, & \text{for} \quad x < 0 \\ \quad\quad 0, & \text{at} \quad x = 0 \\ \text{positive}, & \text{for} \quad x > 0. \end{cases}$$

The function therefore decreases on $(-\infty, 0]$ and increases on $[0, \infty)$. The number

$$\cosh 0 = \tfrac{1}{2}(e^0 + e^{-0}) = \tfrac{1}{2}(1 + 1) = 1$$

is a local and absolute minimum. There are no other extreme values. Since

$$\frac{d^2}{dx^2}(\cosh x) = \frac{d}{dx}(\sinh x) = \cosh x > 0 \qquad \text{for all real } x,$$

the graph is everywhere concave up. (Figure 7.9.2) ❑

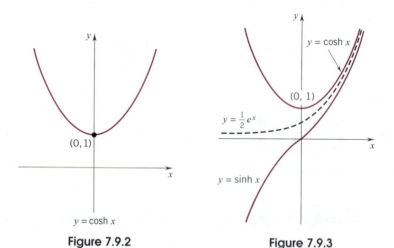

Figure 7.9.2   **Figure 7.9.3**

Figure 7.9.3 shows the graphs of three functions

$$y = \sinh x = \tfrac{1}{2}(e^x - e^{-x}), \qquad y = \tfrac{1}{2}e^x, \qquad y = \cosh x = \tfrac{1}{2}(e^x + e^{-x}).$$

For all real x

$$\sinh x < \tfrac{1}{2}e^x < \cosh x. \qquad\qquad (e^{-x} > 0)$$

Although markedly different for negative x, these functions are almost indistinguishable for large positive x. This follows from the fact that $e^{-x} \to 0$ as $x \to \infty$.

Figure 7.9.4

Applications

The hyperbolic functions have a variety of applications in science and engineering. Perhaps the best known application is the use of the hyperbolic cosine function to describe the shape of a flexible chain or cable that is suspended between two points. For example, think of a telephone wire or power line that sags under its own weight (Figure 7.9.4).

Suppose we have a flexible cable of uniform density that is suspended between two points of equal height. If we introduce an x-y coordinate system so that the lowest point of the cable is on the y-axis (see Figure 7.9.5), then it can be shown that the shape of the curve $y = f(x)$ must satisfy the differential equation

$$\frac{d^2y}{dx^2} = \frac{1}{a}\sqrt{1 + \left(\frac{dy}{dx}\right)^2},$$

where a is a constant that depends on the density of the cable and the tension or horizontal force at its two ends. In the Exercises you are asked to show that

$$y = a \cosh\left(\frac{x}{a}\right) + C$$

is a solution of this differential equation. The graph of this solution is called a *catenary*.

The Gateway to the West Arch in St. Louis, Missouri, is in the shape of an inverted catenary (see Figure 7.9.6). This arch is 630 feet high at its center, and it measures 630 across its base. The value of the constant a for this arch is approximately 127.7 and its equation is

$$y = -127.7 \cosh(x/127.7) + 757.7.$$

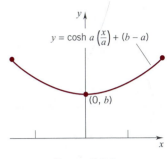

$y = \cosh a\left(\frac{x}{a}\right) + (b - a)$

$(0, b)$

Figure 7.9.5

Figure 7.9.6

Identities

The hyperbolic sine and cosine functions satisfy identities similar to those satisfied by the "circular" sine and cosine.

(7.9.3)

$$\cosh^2 t - \sinh^2 t = 1,$$
$$\sinh(t + s) = \sinh t \cosh s + \cosh t \sinh s,$$
$$\cosh(t + s) = \cosh t \cosh s + \sinh t \sinh s,$$
$$\sinh 2t = 2 \sinh t \cosh t,$$
$$\cosh 2t = \cosh^2 t + \sinh^2 t.$$

The verification of these identities is left to you as an exercise.

Relation to the Hyperbola $x^2 - y^2 = 1$

The hyperbolic sine and cosine are related to the hyperbola $x^2 - y^2 = 1$ much as the "circular" sine and cosine are related to the circle $x^2 + y^2 = 1$:

1. For each real t

$$\cos^2 t + \sin^2 t = 1,$$

and thus the point $(\cos t, \sin t)$ lies on the circle $x^2 + y^2 = 1$. For each real t

$$\cosh^2 t - \sinh^2 t = 1,$$

and thus the point $(\cosh t, \sinh t)$ lies on the hyperbola $x^2 - y^2 = 1$.

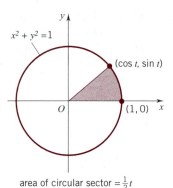

$x^2 + y^2 = 1$

$(\cos t, \sin t)$

O $(1, 0)$ x

area of circular sector $= \frac{1}{2}t$

Figure 7.9.7

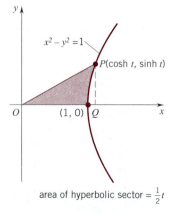

$x^2 - y^2 = 1$

$P(\cosh t, \sinh t)$

O $(1, 0)$ Q x

area of hyperbolic sector $= \frac{1}{2}t$

Figure 7.9.8

2. For each t in $[0, 2\pi]$ (see Figure 7.9.7), the number $\frac{1}{2}t$ gives the area of the circular sector generated by the circular arc that begins at $(1, 0)$ and ends at $(\cos t, \sin t)$. As we prove below, for each $t > 0$ (see Figure 7.9.8), the number $\frac{1}{2}t$ gives the area of the hyperbolic sector generated by the hyperbolic arc that begins at $(1, 0)$ and ends at $(\cosh t, \sinh t)$.

PROOF Let's call the area of the hyperbolic sector $A(t)$. It is not hard to see that

$$A(t) = \tfrac{1}{2}\cosh t \sinh t - \int_1^{\cosh t} \sqrt{x^2 - 1}\ dx.$$

The first term, $\frac{1}{2}\cosh t \sinh t$, gives the area of the triangle OPQ, and the integral

$$\int_1^{\cosh t} \sqrt{x^2 - 1}\ dx$$

gives the area of the unshaded portion of the triangle. We wish to show that

$$A(t) = \tfrac{1}{2}t \qquad \text{for all } t \geq 0.$$

We will do so by showing that

$$A'(t) = \tfrac{1}{2} \quad \text{for all } t > 0 \qquad \text{and} \qquad A(0) = 0.$$

Differentiating $A(t)$, we have

$$A'(t) = \tfrac{1}{2}\left[\cosh t \frac{d}{dt}(\sinh t) + \sinh t \frac{d}{dt}(\cosh t) \right] - \frac{d}{dt}\left(\int_1^{\cosh t} \sqrt{x^2 - 1}\ dx \right)$$

and therefore

(1) $$A'(t) = \tfrac{1}{2}(\cosh^2 t + \sinh^2 t) - \frac{d}{dt}\left(\int_1^{\cosh t} \sqrt{x^2 - 1}\ dx \right).$$

Now we differentiate the integral:

$$\frac{d}{dt}\left(\int_1^{\cosh t} \sqrt{x^2 - 1}\ dx \right) = \sqrt{\cosh^2 t - 1}\,\frac{d}{dt}(\cosh t) = \sinh t \cdot \sinh t = \sinh^2 t.$$

$\underset{\text{(5.8.7)}}{\uparrow}$

Substituting this last expression into Equation (1), we have

$$A'(t) = \tfrac{1}{2}(\cosh^2 t + \sinh^2 t) - \sinh^2 t = \tfrac{1}{2}(\cosh^2 t - \sinh^2 t) = \tfrac{1}{2}.$$

It is not hard to see that $A(0) = 0$:

$$A(0) = \tfrac{1}{2}\cosh 0 \sinh 0 - \int_1^{\cosh 0} \sqrt{x^2 - 1}\ dx = \tfrac{1}{2}(1)(0) - \int_1^1 \sqrt{x^2 - 1}\ dx = 0. \quad \square$$

EXERCISES 7.9

In Exercises 1–18, differentiate the given function.

1. $y = \sinh x^2$.

2. $y = \cosh(x + a)$.

3. $y = \sqrt{\cosh ax}$.

4. $y = (\sinh ax)(\cosh ax)$.

5. $y = \dfrac{\sinh x}{\cosh x - 1}$.

6. $y = \dfrac{\sinh x}{x}$.

7. $y = a \sinh bx - b \cosh ax$.

8. $y = e^x(\cosh x + \sinh x)$.

9. $y = \ln|\sinh ax|$.

10. $y = \ln|1 - \cosh ax|$.

11. $y = \sinh(e^{2x})$.

12. $y = \cosh(\ln x^3)$.

13. $y = e^{-x}\cosh 2x$.

14. $y = \tan^{-1}(\sinh x)$.

15. $y = \ln(\cosh x)$.

16. $y = \ln(\sinh x)$.

17. $y = (\sinh x)^x$.

18. $y = x^{\cosh x}$.

In Exercises 19–25, verify the given identity

19. $\cosh^2 t - \sinh^2 t = 1$.

20. $\sinh(t + s) = \sinh t \cosh s + \cosh t \sinh s$.

21. $\cosh(t + s) = \cosh t \cosh s + \sinh t \sinh s$.

22. $\sinh 2t = 2 \sinh t \cosh t$.

23. $\cosh 2t = \cosh^2 t + \sinh^2 t = 2 \cosh^2 t - 1$
$= 2 \sinh^2 t + 1$.

24. $\cosh(-t) = \cosh t$;   the hyperbolic cosine function is even.

25. $\sinh(-t) = -\sinh t$;   the hyperbolic sine function is odd.

In Exercises 26–28, find the absolute extreme values.

26. $y = 5 \cosh x + 4 \sinh x$.

27. $y = -5 \cosh x + 4 \sinh x$.

28. $y = 4 \cosh x + 5 \sinh x$.

29. Show that for each positive integer n

$$(\cosh x + \sinh x)^n = \cosh nx + \sinh nx.$$

30. Verify that $y = A \cosh cx + B \sinh cx$ satisfies the differential equation $y'' - c^2 y = 0$.

31. Determine A, B, and c so that $y = A \cosh cx + B \sinh cx$ satisfies the conditions $y'' - 9y = 0$, $y(0) = 2$, $y'(0) = 1$. Take $c > 0$.

32. Determine A, B, and c so that $y = A \cosh cx + B \sinh cx$ satisfies the conditions $4y'' - y = 0$, $y(0) = 1$, $y'(0) = 2$. Take $c > 0$.

In Exercises 33–44, calculate the given indefinite integral.

33. $\displaystyle\int \cosh ax \, dx$.

34. $\displaystyle\int \sinh ax \, dx$.

35. $\displaystyle\int \sinh^2 ax \cosh ax \, dx$.

36. $\displaystyle\int \sinh ax \cosh^2 ax \, dx$.

37. $\displaystyle\int \frac{\sinh ax}{\cosh ax} \, dx$.

38. $\displaystyle\int \frac{\cosh ax}{\sinh ax} \, dx$.

39. $\displaystyle\int \frac{\sinh ax}{\cosh^2 ax} \, dx$.

40. $\displaystyle\int \sinh^2 x \, dx$.

41. $\displaystyle\int \cosh^2 x \, dx$.

42. $\displaystyle\int \sinh 2x \, e^{\cosh 2x} \, dx$.

43. $\displaystyle\int \frac{\sinh \sqrt{x}}{\sqrt{x}} \, dx$.

44. $\displaystyle\int \frac{\sinh x}{1 + \cosh x} \, dx$.

In Exercises 45 and 46, find the average value of the function on the indicated interval.

45. $f(x) = \cosh x$, $x \in [-1, 1]$.

46. $f(x) = \sinh 2x$, $x \in [0, 4]$.

47. Find the volume of the solid generated by revolving the region bounded by $y = \cosh x$ and $y = \sinh x$ between $x = 0$ and $x = 1$ about the x-axis.

48. (a) Evaluate the limit

$$\lim_{x \to \infty} \frac{\sinh x}{e^x}.$$

(b) Evaluate the limit

$$\lim_{x \to \infty} \frac{\cosh x}{e^{ax}}$$

when $0 < a < 1$ and when $a > 1$.

*■ 7.10 THE OTHER HYPERBOLIC FUNCTIONS

The hyperbolic tangent is defined by setting

$$\tanh x = \frac{\sinh x}{\cosh x} = \frac{e^x - e^{-x}}{e^x + e^{-x}}.$$

There is also a *hyperbolic cotangent*, a *hyperbolic secant*, and a *hyperbolic cosecant*:

$$\coth x = \frac{\cosh x}{\sinh x}, \qquad \operatorname{sech} x = \frac{1}{\cosh x}, \qquad \operatorname{csch} x = \frac{1}{\sinh x}.$$

The derivatives are as follows:

(7.10.1)

$$\frac{d}{dx}(\tanh x) = \operatorname{sech}^2 x, \qquad \frac{d}{dx}(\coth x) = -\operatorname{csch}^2 x,$$

$$\frac{d}{dx}(\operatorname{sech} x) = -\operatorname{sech} x \tanh x, \qquad \frac{d}{dx}(\operatorname{csch} x) = -\operatorname{csch} x \coth x.$$

These formulas are easy to verify. For instance,

$$\frac{d}{dx}(\tanh x) = \frac{d}{dx}\left(\frac{\sinh x}{\cosh x}\right) = \frac{\cosh x \dfrac{d}{dx}(\sinh x) - \sinh x \dfrac{d}{dx}(\cosh x)}{\cosh^2 x}$$

$$= \frac{\cosh^2 x - \sinh^2 x}{\cosh^2 x} = \frac{1}{\cosh^2 x} = \operatorname{sech}^2 x.$$

We leave it to you to verify the other formulas. ❏

Let's examine the hyperbolic tangent a little further. Since

$$\tanh(-x) = \frac{\sinh(-x)}{\cosh(-x)} = \frac{-\sinh x}{\cosh x} = -\tanh x,$$

the hyperbolic tangent is an odd function and thus the graph is symmetric about the origin. Since

$$\frac{d}{dx}(\tanh x) = \operatorname{sech}^2 x > 0 \qquad \text{for all real } x,$$

the function is everywhere increasing. From the relation

$$\tanh x = \frac{e^x - e^{-x}}{e^x + e^{-x}} = \frac{e^x - e^{-x}}{e^x + e^{-x}} \cdot \frac{e^x}{e^x} = \frac{e^{2x} - 1}{e^{2x} + 1} = \frac{e^{2x} + 1 - 2}{e^{2x} + 1} = 1 - \frac{2}{e^{2x} + 1},$$

you can see that $\tanh x$ always remains between -1 and 1. Moreover,

$$\text{as } x \to \infty, \quad \tanh x \to 1 \qquad \text{and} \qquad \text{as } x \to -\infty, \quad \tanh x \to -1.$$

The lines $y = 1$ and $y = -1$ are horizontal asymptotes. To check on the concavity of the graph, we take the second derivative:

$$\frac{d^2}{dx^2}(\tanh x) = \frac{d}{dx}(\operatorname{sech}^2 x) = 2 \operatorname{sech} x \frac{d}{dx}(\operatorname{sech} x)$$

$$= 2 \operatorname{sech} x (-\operatorname{sech} x \tanh x)$$
$$= -2 \operatorname{sech}^2 x \tanh x.$$

Since

$$\tanh x = \frac{e^x - e^{-x}}{e^x + e^{-x}} \quad \text{is} \quad \begin{cases} \text{negative}, & \text{for} \quad x < 0 \\ 0, & \text{at} \quad x = 0 \\ \text{positive}, & \text{for} \quad x > 0, \end{cases}$$

you can see that

$$\frac{d^2}{dx^2}(\tanh x) \quad \text{is} \quad \begin{cases} \text{positive}, & \text{for} \quad x < 0 \\ 0, & \text{at} \quad x = 0 \\ \text{negative}, & \text{for} \quad x > 0. \end{cases}$$

The graph is therefore concave up on $(-\infty, 0)$ and concave down on $(0, \infty)$. The point $(0, \tanh 0) = (0, 0)$ is a point of inflection. At the origin the slope is

$$\operatorname{sech}^2 0 = \frac{1}{\cosh^2 0} = 1.$$

The graph is shown in Figure 7.10.1.

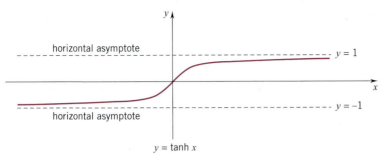

Figure 7.10.1

The Hyperbolic Inverses

Of the six hyperbolic functions, only the hyperbolic cosine and hyperbolic secant are not one-to-one (refer to the graphs of $y = \sinh x$, $y = \cosh x$, and $y = \tanh x$). Thus, the hyperbolic sine, hyperbolic tangent, hyperbolic cosecant, and hyperbolic cotangent functions all have inverses. If we restrict the domains of the hyperbolic cosine and hyperbolic secant functions to $x \geq 0$, then these functions will also have inverses. The hyperbolic inverses that are important to us are the *inverse hyperbolic sine*, the *inverse hyperbolic cosine*, and the *inverse hyperbolic tangent*. These functions

$$y = \sinh^{-1}x, \qquad y = \cosh^{-1}x, \qquad y = \tanh^{-1}x$$

are the inverses of

$$y = \sinh x, \quad y = \cosh x \quad (x \geq 0), \qquad y = \tanh x,$$

respectively.

THEOREM 7.10.2

   (i) $\sinh^{-1}x = \ln (x + \sqrt{x^2 + 1})$, x real

   (ii) $\cosh^{-1}x = \ln (x + \sqrt{x^2 - 1})$, $x \geq 1$

   (iii) $\tanh^{-1}x = \dfrac{1}{2}\ln \left(\dfrac{1 + x}{1 - x}\right)$, $-1 < x < 1$.

PROOF To prove (i), we set $y = \sinh^{-1}x$ and note that

$$\sinh y = x.$$

This gives in sequence

$$\tfrac{1}{2}(e^y - e^{-y}) = x, \qquad e^y - e^{-y} = 2x, \qquad e^y - 2x - e^{-y} = 0, \qquad e^{2y} - 2xe^y - 1 = 0.$$

This last equation is a quadratic equation in e^y. From the quadratic formula we find that

$$e^y = \tfrac{1}{2}(2x \pm \sqrt{4x^2 + 4}) = x \pm \sqrt{x^2 + 1}.$$

Since $e^y > 0$, the minus sign on the right is impossible. Consequently, we have

$$e^y = x + \sqrt{x^2 + 1}$$

and, taking the natural log of both sides,

$$y = \ln (x + \sqrt{x^2 + 1}).$$

To prove (ii), we set

$$y = \cosh^{-1}x, \qquad x \geq 1$$

and note that

$$\cosh y = x \quad \text{and} \quad y \geq 0.$$

This gives in sequence

$$\tfrac{1}{2}(e^y + e^{-y}) = x, \qquad e^y + e^{-y} = 2x, \qquad e^{2y} - 2xe^y + 1 = 0.$$

Again we have a quadratic in e^y. Here the general quadratic formula gives

$$e^y = \tfrac{1}{2}(2x \pm \sqrt{4x^2 - 4}) = x \pm \sqrt{x^2 - 1}.$$

Since y is nonnegative,

$$e^y = x \pm \sqrt{x^2 - 1}$$

cannot be less than 1. This renders the negative sign impossible (check this out) and leaves

$$e^y = x + \sqrt{x^2 - 1}$$

as the only possibility. Taking the natural log of both sides, we get

$$y = \ln(x + \sqrt{x^2 - 1}).$$

The proof of (iii) is left as an exercise. ❏

EXERCISES *7.10

In Exercises 1–10, differentiate the given function.

1. $y = \tanh^2 x$.

2. $y = \tanh^2 3x$.

3. $y = \ln(\tanh x)$.

4. $y = \tanh(\ln x)$.

5. $y = \sinh(\tan^{-1} e^{2x})$.

6. $y = \operatorname{sech}(3x^2 + 1)$.

7. $y = \coth(\sqrt{x^2 + 1})$.

8. $y = \ln(\operatorname{sech} x)$.

9. $y = \dfrac{\operatorname{sech} x}{1 + \cosh x}$.

10. $y = \dfrac{\cosh x}{1 + \operatorname{sech} x}$.

In Exercises 11–13, verify the given differentiation formula.

11. $\dfrac{d}{dx}(\coth x) = -\operatorname{csch}^2 x$.

12. $\dfrac{d}{dx}(\operatorname{sech} x) = -\operatorname{sech} x \tanh x$.

13. $\dfrac{d}{dx}(\operatorname{csch} x) = -\operatorname{csch} x \coth x$.

14. Show that

$$\tanh(t + s) = \frac{\tanh t + \tanh s}{1 + \tanh t \tanh s}.$$

15. Given that $\tanh x_0 = \tfrac{4}{5}$, find (a) $\operatorname{sech} x_0$.
HINT: $1 - \tanh^2 x = \operatorname{sech}^2 x$. Then find (b) $\cosh x_0$, (c) $\sinh x_0$, (d) $\coth x_0$, (e) $\operatorname{csch} x_0$.

16. Given that $\tanh t_0 = -\tfrac{5}{12}$, evaluate the remaining hyperbolic functions at t_0.

17. Show that, if $x^2 \geq 1$, then $x - \sqrt{x^2 - 1} \leq 1$.

18. Show that

$$\tanh^{-1} x = \frac{1}{2} \ln\left(\frac{1 + x}{1 - x}\right), \qquad -1 < x < 1.$$

19. Show that

(7.10.3) $\quad \dfrac{d}{dx}(\sinh^{-1} x) = \dfrac{1}{\sqrt{x^2 + 1}}, \qquad x \text{ real.}$

20. Show that

(7.10.4) $\quad \dfrac{d}{dx}(\cosh^{-1} x) = \dfrac{1}{\sqrt{x^2 - 1}}, \qquad x > 1.$

21. Show that

(7.10.5) $\quad \dfrac{d}{dx}(\tanh^{-1} x) = \dfrac{1}{1 - x^2}, \qquad -1 < x < 1.$

22. Show that

$$\frac{d}{dx}(\operatorname{sech}^{-1} x) = \frac{-1}{x\sqrt{1 - x^2}}, \qquad 0 < x < 1.$$

23. Show that

$$\frac{d}{dx}(\operatorname{csch}^{-1}x) = \frac{-1}{|x|\sqrt{1+x^2}}, \quad x \neq 0.$$

24. Show that

$$\frac{d}{dx}(\coth^{-1}x) = \frac{1}{1-x^2}, \quad |x| > 1.$$

25. Sketch the graph of $y = \operatorname{sech} x$, finding: (a) the extreme values; (b) the points of inflection; and (c) the concavity.

26. Sketch the graphs of (a) $y = \coth x$, (b) $y = \operatorname{csch} x$.

27. Graph $y = \sinh x$ and $y = \sinh^{-1}x$ in the same coordinate system. Find all points of inflection.

28. Sketch the graphs of (a) $y = \cosh^{-1}x$. (b) $y = \tanh^{-1}x$.

29. Given that $\tan \phi = \sinh x$, show that

(a) $\dfrac{d\phi}{dx} = \operatorname{sech} x$.

(b) $x = \ln (\sec \phi + \tan \phi)$.

(c) $\dfrac{dx}{d\phi} = \sec \phi$.

30. The region bounded by the graph of $y = \operatorname{sech} x$ between $x = -1$ and $x = 1$ is revolved around the x-axis. Find the volume of the solid that is generated.

In Exercises 31–40, calculate the given integral.

31. $\displaystyle\int \tanh x \, dx.$

32. $\displaystyle\int \coth x \, dx.$

33. $\displaystyle\int \operatorname{sech} x \, dx.$

34. $\displaystyle\int \operatorname{csch} x \, dx.$

35. $\displaystyle\int \operatorname{sech}^3 x \tanh x \, dx.$

36. $\displaystyle\int x \operatorname{sech}^2 x^2 \, dx.$

37. $\displaystyle\int \tanh x \ln(\cosh x) \, dx.$

38. $\displaystyle\int \frac{1 + \tanh x}{\cosh^2 x} \, dx.$

39. $\displaystyle\int \frac{\operatorname{sech}^2 x}{1 + \tanh x} \, dx.$

40. $\displaystyle\int \tanh^5 x \operatorname{sech}^2 x \, dx.$

In Exercises 41–43, verify the given integration formula. In each case, assume that $a > 0$.

41. $\displaystyle\int \frac{1}{\sqrt{a^2 + x^2}} \, dx = \sinh^{-1}\left(\frac{x}{a}\right) + C.$

42. $\displaystyle\int \frac{1}{\sqrt{x^2 - a^2}} \, dx = \cosh^{-1}\left(\frac{x}{a}\right) + C.$

43. $\displaystyle\int \frac{1}{a^2 - x^2} \, dx = \begin{cases} \dfrac{1}{a}\tanh^{-1}\left(\dfrac{x}{a}\right) + C & \text{if } |x| < a. \\[2mm] \dfrac{1}{a}\coth^{-1}\left(\dfrac{x}{a}\right) + C & \text{if } |x| > a. \end{cases}$

44. If a body of mass m falling from rest under the action of gravity encounters air resistance that is proportional to the square of its velocity, then the velocity $v(t)$ of the body at time t satisfies the differential equation

$$m\frac{dv}{dt} = mg - kv^2$$

where $k > 0$ is the constant of proportionality and g is the gravitational constant.

(a) Show that

$$v(t) = \sqrt{\frac{mg}{k}}\tanh\left(\sqrt{\frac{gk}{m}}\,t\right)$$

is a solution of the differential equation which satisfies $v(0) = 0$.

(b) Find

$$\lim_{t\to\infty} v(t).$$

This limit is called the *terminal velocity* of the body.

■ CHAPTER HIGHLIGHTS

7.1 Inverse Functions

one-to-one function; inverse function (p. 396)
one-to-one and increasing/decreasing functions; derivatives (p. 399)
relation between graph of f and the graph of f^{-1} (p. 401)
continuity and differentiability of inverse functions (p. 401)
derivative of an inverse (p. 402)

7.2 The Logarithm Function, Part I

definition of a logarithm function (p. 407)

natural logarithm: $\ln x = \displaystyle\int_1^x \frac{dt}{t}, \quad x > 0;$ domain $(0, \infty)$, range $(-\infty, \infty)$

$$\ln 1 = 0, \quad \ln e = 1$$

for $x, y > 0$ and r rational

$$\ln xy = \ln x + \ln y, \quad \ln\frac{1}{x} = -\ln x, \quad \ln\frac{x}{y} = \ln x - \ln y, \quad \ln x^r = r \ln x.$$

graph of $y = \ln x$ (p. 413)

7.3 The Logarithm Function, Part II

$$\frac{d}{dx}(\ln|u|) = \frac{1}{u}\frac{du}{dx} \qquad \int \frac{g'(x)}{g(x)}dx = \ln|g(x)| + C$$

logarithmic differentiation (p. 421)

7.4 The Exponential Function

The exponential function $y = e^x$ is the inverse of the logarithm function $y = \ln x$.

graph of $y = e^x$ (p. 425);   domain $(-\infty, \infty)$, range $(0, \infty)$

$$\ln(e^x) = x, \qquad e^{\ln x} = x, \qquad e^0 = 1, \qquad e^{-x} = \frac{1}{e^x}, \qquad e^{x+y} = e^x e^y, \qquad e^{x-y} = \frac{e^x}{e^y}$$

$$\frac{d}{dx}(e^u) = e^u\frac{du}{dx} \qquad \int e^{g(x)}g'(x)\,dx = e^{g(x)} + C$$

7.5 Arbitrary Powers; Other Bases; Estimating e

$$x^r = e^{r\ln x} \quad \text{for all } x > 0, \text{ all real } r$$

$$\frac{d}{dx}(p^u) = p^u \ln p \frac{du}{dx} \quad (p \text{ a positive constant}) \qquad \log_p x = \frac{\ln x}{\ln p}$$

$$\frac{d}{dx}(\log_p u) = \frac{1}{u \ln p}\frac{du}{dx}$$

$$\left(1 + \frac{1}{n}\right)^n \le e \le \left(1 + \frac{1}{n}\right)^{n+1} \qquad e \cong 2.71828$$

7.6 Exponential Growth and Decay

Linear functions of time change by the same *amount* during all periods of the same duration; exponential functions of time change by the same *factor* during all periods of the same duration.

All the functions that satisfy the equation $f'(t) = kf(t)$ are of the form $f(t) = Ce^{kt}$.

population growth (p. 444)
radioactive decay, half-life (p. 446)
compound interest, continuous compounding (p. 448)
Newton's Law of Cooling (p. 450)

7.7 More on the Integration of the Trigonometric Functions

$$\int \tan x \, dx = \ln|\sec x| + C, \qquad \int \sec x \, dx = \ln|\sec x + \tan x| + C,$$

$$\int \cot x \, dx = \ln|\sin x| + C, \qquad \int \csc x \, dx = \ln|\csc x - \cot x| + C.$$

7.8 The Inverse Trigonometric Functions

The inverse sine, $y = \sin^{-1} x$, is the inverse of $y = \sin x$, $x \in [-\frac{1}{2}\pi, \frac{1}{2}\pi]$.
The inverse tangent, $y = \tan^{-1} x$, is the inverse of $y = \tan x$, $x \in (-\frac{1}{2}\pi, \frac{1}{2}\pi)$.

The inverse secant, $y = \sec^{-1}x$, is the inverse of $y = \sec x$, $x \in [0, \frac{1}{2}\pi) \cup (\frac{1}{2}\pi, \pi]$.

graph of $y = \sin^{-1} x$ (p. 462)   graph of $y = \tan^{-1}x$ (p. 465)
graph of $y = \sec^{-1}x$ (p. 468)

$$\frac{d}{dx}(\sin^{-1}x) = \frac{1}{\sqrt{1 - x^2}} \qquad \int \frac{dx}{\sqrt{a^2 - x^2}} = \sin^{-1}\left(\frac{x}{a}\right) + C \quad (a > 0)$$

$$\frac{d}{dx}(\tan^{-1}x) = \frac{1}{1 + x^2} \qquad \int \frac{dx}{a^2 + x^2} = \frac{1}{a}\tan^{-1}\left(\frac{x}{a}\right) + C \quad (a \neq 0)$$

$$\frac{d}{dx}(\sec^{-1}x) = \frac{1}{|x|\sqrt{x^2 - 1}} \qquad \int \frac{dx}{x\sqrt{x^2 - a^2}} = \frac{1}{a}\sec^{-1}\left(\frac{|x|}{a}\right) + C \quad (a > 0)$$

definition of the remaining inverse trigonometric functions (p. 471)

7.9 The Hyperbolic Sine and Cosine

$$\sinh x = \tfrac{1}{2}(e^x - e^{-x}), \qquad \cosh x = \tfrac{1}{2}(e^x + e^{-x}),$$

$$\frac{d}{dx}(\sinh x) = \cosh x, \qquad \frac{d}{dx}(\cosh x) = \sinh x.$$

graphs (pp. 475–476)   basic identities (p. 477)

*7.10 The Other Hyperbolic Functions

$$\tanh x = \frac{\sinh x}{\cosh x}, \qquad \coth x = \frac{\cosh x}{\sinh x},$$

$$\text{sech } x = \frac{1}{\cosh x}, \qquad \text{csch } x = \frac{1}{\sinh x}.$$

derivatives (p. 479)   hyperbolic inverses (p. 481)
derivatives of hyperbolic inverses (p. 482)

■ PROJECTS AND EXPLORATIONS USING TECHNOLOGY

To do these exercises you will need a graphics calculator or a computer with graphing capability. The majority of these problems are open-ended so different approaches may be used to solve them. You should be aware that different approaches can result in slight variations in the answers. Round your numerical answers to at least four decimal places. The rounding method that your calculator or computer uses also may cause variations in answers.

7.1 The functions $f(x) = a \ln x$, where a is a constant, have a number of applications, one of which will be considered in a later exercise.
   (a) Find the values of a for which the graph of f is tangent to the line $y = x$.
   (b) For each real number a, how many solutions will there be of $f(x) = x$? What is the value of f' at each solution of $f(x) = x$?
   (c) How many solutions are there to $f[f(x)] = x$? What is the value of f' at each of these solutions?
   (d) Represent f as a logarithm function in another base.

7.2 Let $A(t)$ denote the area of the rectangle of width $2t$ that has its lower vertices on the x-axis and its upper vertices on the graph of

$$f(x) = e^{(1 - x^4)/(2 + x^2)}.$$

See the figure.

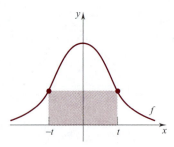

(a) Express A as a function of t and sketch its graph.
(b) Estimate the size of the rectangle of maximal area. Describe the properties of f at the value(s) of t where $A(t)$ is maximal.
(c) Discuss the concavity of the graph of A and find its points of inflection.
(d) Relate the values at which the graphs of A or f have points of inflection. For example, show that if f has a point of inflection at x_0, then the graph of A is concave down at x_0.

7.3 When we approximate the derivative of a function f on a calculator by the difference quotient

$$D(x, h) = \frac{f(x + h) - f(x)}{h},$$

we know that if h is "too small," then the calculator rounds off digits and the difference quotient fails to give accurate estimates. We will look at an unusual version of the difference quotient that will allow us to analyze different types of errors more effectively. Let

$$D_x(h) = \frac{f(x + e^{-h}) - f(x)}{e^{-h}}.$$

(a) Let f be a given function and fix a value of x. Explain why $y = f'(x)$ is a horizontal asymptote of the graph of D_x (as a function of h).
(b) Use your graphing technology to verify this for $f(x) = \log_5(x)$ at $x = 5$, $x = 7$, and $x = 100$.
(c) Fix x and let $E(h) = D_x(h) - f'(x)$; $E(h)$ is the error for this value of x at h. Is E strictly decreasing to 0 for all functions f?
(d) Find the smallest value of n such that

$$\lim_{h \to \infty} E(h)e^{nh}$$

is nonzero. Be careful about graphic and numeric solutions because there may be roundoff errors when using computer arithmetic. The importance of n is that it is a measure of how good $D(h)$ approximates $f'(x)$. That is, the bigger n is, the better the approximation.

7.4 The functions $\sinh x$, $\cosh x$, $\tanh x$ are used in the study of non-Euclidean geometry.
(a) Use the tangent line approximation:

$$f(x) \cong f(0) + f'(0)x$$

to show that

$$\sinh x \cong x, \qquad \cosh x \cong 1, \qquad \text{and} \qquad \tanh x \cong x,$$

for x small enough. From the concavity of these functions, determine whether these approximations are too small or too large.

Triangles are labeled by capital letters at the vertices and corresponding lower case letters for the sides opposite the angles (see the figure).

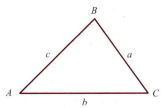

(b) Use the results in part (a) to show that the hyperbolic law of sines is approximately equal to the Euclidean law of sines for small side lengths. That is, show that

$$\frac{\sin A}{\sinh a} = \frac{\sin B}{\sinh b} = \frac{\sin C}{\sinh c} \quad \text{is approximately} \quad \frac{\sin A}{a} = \frac{\sin B}{b} = \frac{\sin C}{c}.$$

(c) A better approximation for $\cosh x$ is

$$\cosh x \cong 1 + \tfrac{1}{2}x^2,$$

for x small enough. Use this approximation to show that the hyperbolic law of cosines:

$$\cosh c = \cosh a \cosh b - \sinh a \sinh b \cos C$$

reduces to the Euclidean law of cosines for triangles with small sides.

(d) In Euclidean geometry if we fix the length of sides a and b, with $b > a$, and vary the angle C between them, then the angle A increases from 0 to a maximum value and then decreases back to 0. A way to find the maximum value is: Use the law of cosines to show that

(1) $$\cos A = \frac{b - a \cos C}{\sqrt{a^2 + b^2 - 2ab \cos C}}$$

and then differentiate implicitly with respect to C to find dA/dC.

(e) In hyperbolic geometry, the equation corresponding to (1) is

$$\cos A = \frac{\sinh b \cosh a - \sinh a \cos C}{\sqrt{[\cosh a \cosh b - \sinh a \sinh b \cos C]^2 - 1}}.$$

Find the value of C for which A is maximal.

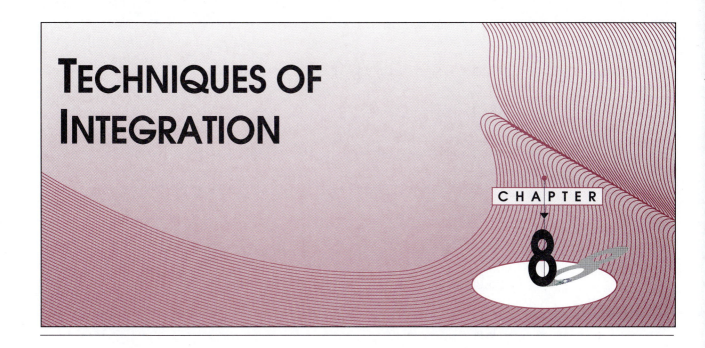

TECHNIQUES OF INTEGRATION

■ 8.1 REVIEW

We begin by listing the more important integrals with which you are already familiar.
An extended table of integrals appears on the inside covers of the book.

1. $\displaystyle\int k\,dx = kx + C, \quad k$ constant.

2. $\displaystyle\int x^r\,dx = \frac{1}{r+1}x^{r+1} + C, \quad r \neq -1.$

3. $\displaystyle\int \frac{dx}{x} = \ln|x| + C.$ **4.** $\displaystyle\int e^x\,dx = e^x + C.$

5. $\displaystyle\int p^x\,dx = \frac{p^x}{\ln p} + C, \quad p > 0$ constant, $p \neq 1.$

6. $\displaystyle\int \sin x\,dx = -\cos x + C.$ **7.** $\displaystyle\int \cos x\,dx = \sin x + C.$

8. $\displaystyle\int \tan x\,dx = \ln|\sec x| + C.$ **9.** $\displaystyle\int \cot x\,dx = \ln|\sin x| + C.$

10. $\displaystyle\int \sec x\,dx = \ln|\sec x + \tan x| + C.$ **11.** $\displaystyle\int \csc x\,dx = \ln|\csc x - \cot x| + C.$

12. $\displaystyle\int \sec x \tan x\,dx = \sec x + C.$ **13.** $\displaystyle\int \csc x \cot x\,dx = -\csc x + C.$

14. $\displaystyle\int \sec^2 x\,dx = \tan x + C.$ **15.** $\displaystyle\int \csc^2 x\,dx = -\cot x + C.$

16. $\displaystyle\int \frac{dx}{\sqrt{a^2 - x^2}} = \sin^{-1}\left(\frac{x}{a}\right) + C, \quad a \neq 0$ constant.

17. $\displaystyle\int \frac{dx}{a^2 + x^2} = \frac{1}{a}\tan^{-1}\left(\frac{x}{a}\right) + C,$ $a \neq 0$ constant.

18. $\displaystyle\int \frac{dx}{x\sqrt{x^2 - a^2}} = \frac{1}{a}\sec^{-1}\left|\frac{x}{a}\right| + C,$ $a \neq 0$ constant.

19. $\displaystyle\int \sinh x\, dx = \cosh x + C.$ **20.** $\displaystyle\int \cosh x\, dx = \sinh x + C.$

For review we work out a few integrals involving u-substitutions.

Example 1   Find

$$\int x \tan x^2\, dx.$$

SOLUTION   Set

$$u = x^2, \qquad du = 2x\, dx.$$

Then

$$\int x \tan x^2\, dx = \tfrac{1}{2}\int \tan u\, du = \tfrac{1}{2}\ln|\sec u| + C = \tfrac{1}{2}\ln|\sec x^2| + C. \quad \square$$

$\uparrow$ ———— Formula 8

Example 2   Compute

$$\int_0^1 \frac{e^x}{e^x + 2}\, dx.$$

SOLUTION   Set

$$u = e^x + 2, \qquad du = e^x\, dx.$$

At $x = 0$, $u = 3$; at $x = 1$, $u = e + 2$. Thus

$$\int_0^1 \frac{e^x}{e^x + 2}\, dx = \int_3^{e+2} \frac{du}{u} = \left[\ln|u|\right]_3^{e+2}$$

Formula 3 ———————

$$= \ln(e + 2) - \ln 3 = \ln\left[\tfrac{1}{3}(e + 2)\right] \cong 0.45. \quad \square$$

Example 3   Find

$$\int \frac{\cos 2x}{(2 + \sin 2x)^{1/3}}\, dx.$$

SOLUTION   Set

$$u = 2 + \sin 2x, \qquad du = 2\cos 2x\, dx.$$

Then

$$\int \frac{\cos 2x}{(2 + \sin 2x)^{1/3}}\, dx = \frac{1}{2}\int \frac{1}{u^{1/3}}\, du = \frac{1}{2}\int u^{-1/3}\, du = \frac{1}{2}\left(\frac{3}{2}\right)u^{2/3} + C$$

Formula 2 —————————

$$= \frac{3}{4}(2 + \sin 2x)^{2/3} + C. \quad \square$$

The final example involves a little algebra.

Example 4 Find

$$P = \int \frac{dx}{x^2 + 2x + 5}.$$

SOLUTION First we complete the square in the denominator:

$$P = \int \frac{dx}{x^2 + 2x + 5} = \int \frac{dx}{(x^2 + 2x + 1) + 4} = \int \frac{dx}{(x + 1)^2 + 2^2}.$$

We know that

$$\int \frac{du}{u^2 + a^2} = \frac{1}{a} \tan^{-1}\left(\frac{u}{a}\right) + C.$$

Setting

$$u = x + 1, \qquad du = dx, \qquad a = 2,$$

we have

$$P = \int \frac{du}{u^2 + 2^2} = \frac{1}{2}\tan^{-1}\left(\frac{u}{2}\right) + C = \frac{1}{2}\tan^{-1}\left(\frac{x + 1}{2}\right) + C. \quad \square$$

EXERCISES 8.1

In Exercises 1–48, evaluate the given integral.

1. $\displaystyle\int e^{2-x}\,dx.$

2. $\displaystyle\int \cos \tfrac{2}{3}x\,dx.$

3. $\displaystyle\int_0^1 \sin \pi x\,dx.$

4. $\displaystyle\int_0^t \sec \pi x \tan \pi x\,dx.$

5. $\displaystyle\int \sec^2(1 - x)\,dx.$

6. $\displaystyle\int \frac{dx}{5^x}.$

7. $\displaystyle\int_{\pi/6}^{\pi/3} \cot x\,dx.$

8. $\displaystyle\int_0^1 \frac{x^3}{1 + x^4}\,dx.$

9. $\displaystyle\int \frac{x}{\sqrt{1 - x^2}}\,dx.$

10. $\displaystyle\int_{-\pi/4}^{\pi/4} \frac{dx}{\cos^2 x}.$

11. $\displaystyle\int_{-\pi/4}^{\pi/4} \frac{\sin x}{\cos^2 x}\,dx.$

12. $\displaystyle\int \frac{e^{\sqrt{x}}}{\sqrt{x}}\,dx.$

13. $\displaystyle\int_1^2 \frac{e^{1/x}}{x^2}\,dx.$

14. $\displaystyle\int_{\pi/6}^{\pi/3} \csc x\,dx.$

15. $\displaystyle\int \frac{x}{x^2 + 1}\,dx.$

16. $\displaystyle\int \frac{x^3}{\sqrt{1 - x^4}}\,dx.$

17. $\displaystyle\int_0^c \frac{dx}{x^2 + c^2}.$

18. $\displaystyle\int a^x e^x\,dx.$

19. $\displaystyle\int \frac{\sec^2\theta}{\sqrt{3\tan\theta + 1}}\,d\theta.$

20. $\displaystyle\int \frac{\sin \phi}{3 - 2\cos \phi}\,d\phi.$

21. $\displaystyle\int \frac{e^x}{ae^x - b}\,dx.$

22. $\displaystyle\int_0^{\pi/4} \frac{\sec^2 x \tan x}{\sqrt{2 + \sec^2 x}}\,dx.$

23. $\displaystyle\int \frac{1 + \cos 2x}{\sin^2 2x}\,dx.$

24. $\displaystyle\int \frac{dx}{x^2 - 4x + 13}.$

25. $\displaystyle\int \frac{x}{(x + 1)^2 + 4}\,dx.$

26. $\displaystyle\int \frac{\ln x}{x}\,dx.$

27. $\displaystyle\int \frac{x}{\sqrt{1 - x^4}}\,dx.$

28. $\displaystyle\int \frac{e^x}{1 + e^{2x}}\,dx.$

29. $\displaystyle\int \frac{dx}{x^2 + 6x + 10}.$

30. $\displaystyle\int e^x \tan e^x\,dx.$

31. $\displaystyle\int x \sin x^2\,dx.$

32. $\displaystyle\int \frac{x}{9 + x^4}\,dx.$

33. $\displaystyle\int \tan^2 x\,dx.$

34. $\displaystyle\int \cosh 2x \sinh^3 2x\,dx.$

35. $\displaystyle\int_{-1}^2 \frac{x}{x^2 + 4}\,dx.$

36. $\displaystyle\int_0^\pi \sin\left(\tfrac{\pi}{2}x\right)\,dx.$

37. $\displaystyle\int_1^e \frac{\ln x^3}{x}\,dx.$

38. $\displaystyle\int_0^{\pi/4} \frac{\tan^{-1}x}{1 + x^2}\,dx.$

39. $\int \dfrac{\sin^{-1}x}{\sqrt{1-x^2}}\,dx.$

40. $\int (x+1)\cosh(x^2+2x+1)\,dx.$

41. $\int \dfrac{e^{2x}}{\sqrt{1-e^{4x}}}\,dx.$

42. $\int \tanh 2x\,dx.$

43. $\int \dfrac{1}{x\,\ln x}\,dx.$

44. $\int e^x\cosh(2-e^x)\,dx.$

45. $\int_1^4 \dfrac{\sqrt{\ln x}}{x}\,dx.$

46. $\int_{-1}^1 \dfrac{x^2}{x^2+1}\,dx.$ HINT: divide the denominator into the numerator.

47. $\int_0^{\pi/4} \dfrac{1+\sin x}{\cos^2x}\,dx.$ HINT: write the integrand as two fractions.

48. $\int_0^{1/2} \dfrac{1+x}{\sqrt{1-x^2}}\,dx.$ (See the hint for Exercise 47.)

49. Evaluate $\int_0^\pi \sqrt{1+\cos x}\,dx.$
HINT: $\cos x = 2\cos^2(x/2)-1.$

50. Evaluate $\int \sec^2x\tan x\,dx$ in two ways.

(a) Let $u=\tan x$ and verify that the result is:
$$\int \sec^2x\tan x\,dx = \tfrac{1}{2}\tan^2x + C_1.$$

(b) Let $u=\sec x$ and verify that the result is:
$$\int \sec^2x\tan x\,dx = \tfrac{1}{2}\sec^2x + C_2.$$

(c) Reconcile the results in parts (a) and (b).

51. Verify that, for any positive integer n:

(a) $\int_0^\pi \sin^2nx\,dx = \tfrac{1}{2}\pi.$
HINT: $\sin^2\theta = \dfrac{1-\cos 2\theta}{2}.$

(b) $\int_0^\pi \sin nx\cos nx\,dx = 0.$

(c) $\int_0^{\pi/n} \sin nx\cos nx\,dx = 0.$

52. (a) Evaluate $\int \sin^3x\,dx.$ HINT: $\sin^2x = 1-\cos^2x.$

(b) Evaluate $\int \sin^5x\,dx.$

(c) Explain how to evaluate $\int \sin^{2k+1}x\,dx$ for any positive integer k.

53. (a) Evaluate $\int \tan^3x\,dx.$ HINT: $\tan^2x = \sec^2x-1.$

(b) Evaluate $\int \tan^5x\,dx.$

(c) Evaluate $\int \tan^7x\,dx.$

(d) Explain how to evaluate $\int \tan^{2k+1}x\,dx$ for any positive integer k.

54. (a) Sketch the region bounded on the left by the line $x=\pi/6$, and by the curves $y=\csc x$ and $y=\sin x$ for $\pi/6 \le x \le \pi/2$.

(b) Calculate the area of the region in part (a).

(c) The region in part (a) is rotated around the x-axis. Find the volume of the solid that is generated.

▶ 55. (a) Use a graphing utility to sketch the graph of
$$f(x) = \frac{1}{\sin x + \cos x} \qquad \text{for} \qquad 0 \le x \le \frac{\pi}{2}.$$

(b) Find A and B such that
$\sin x + \cos x = A\sin(x+B).$

(c) Find the area of the region bounded by the graph of f and the x-axis.   HINT: Use the result in part (b).

▶ 56. (a) Use a graphing utility to sketch the graph of $f(x) = e^{-x^2}.$

(b) Fix a number $a>0$ and rotate the region bounded by the graph of f and the y-axis on the interval $[0,a]$ around the y-axis. Find the volume of the solid that is generated.

(c) Find a such that the solid in part (b) has volume 2 cubic units.

▶ 57. (a) Use a graphing utility to sketch the graphs of
$$f(x) = \frac{x^2+1}{x+1}, \quad x>-1, \quad \text{and} \quad x+2y=16$$
in the same coordinate system.

(b) The two graphs intersect at two points and determine a bounded region R. Estimate the x-coordinates of the two points of intersection accurate to two decimal places.

(c) Determine the approximate area of the region R.

▶ 58. (a) Use a graphing utility to sketch the graph of
$$y^2 = x^2(1-x)$$

(b) Your sketch in part (a) should show that the curve forms a loop for $0 \le x \le 1$. Calculate the area of the loop.   HINT: Use the symmetry of the curve.

■ 8.2 **INTEGRATION BY PARTS**

We begin with the formula for the derivative of a product

$$f(x)g'(x) + f'(x)g(x) = (f \cdot g)'(x).$$

Integrating both sides, we get

$$\int f(x)g'(x)\, dx + \int f'(x)g(x)\, dx = \int (f \cdot g)'(x)\, dx$$

Since

$$\int (f \cdot g)'(x)\, dx = f(x)g(x) + C,$$

we have

$$\int f(x)g'(x)\, dx + \int f'(x)g(x)\, dx = f(x)g(x) + C$$

and therefore

$$\int f(x)g'(x)\, dx = f(x)g(x) - \int f'(x)g(x)\, dx + C.$$

Since the computation of

$$\int f'(x)g(x)\, dx$$

will yield its own arbitrary constant, there is no reason to keep the constant C. We therefore drop it and write

(8.2.1)
$$\int f(x)g'(x)\, dx = f(x)g(x) - \int f'(x)g(x)\, dx.$$

This formula, called the formula for *integration by parts,* enables us to find

$$\int f(x)g'(x)\, dx$$

by computing

$$\int f'(x)g(x)\, dx$$

instead. Of course, it is of practical use only if the second integral is easier to compute than the first.

   In practice we usually set

$$u = f(x), \qquad dv = g'(x)\, dx.$$

Then

$$du = f'(x)\, dx, \qquad v = g(x).$$

Now, with these substitutions, the formula for integration by parts can be written

(8.2.2)

$$\int u \, dv = uv - \int v \, du.$$

Success with this formula depends on choosing u and dv so that

$$\int v \, du \quad \text{is easier to calculate than} \quad \int u \, dv.$$

Example 1 Calculate

$$\int xe^x \, dx.$$

SOLUTION We want to separate x from e^x. Setting

$$u = x, \qquad dv = e^x \, dx,$$

we have

$$du = dx, \qquad v = e^x.$$

Accordingly,

$$\int xe^x \, dx = \int u \, dv = uv - \int v \, du = xe^x - \int e^x \, dx = xe^x - e^x + C.$$

Our choice of u and dv worked out very well. However, if we had set

$$u = e^x, \qquad dv = x \, dx,$$

then we would have had

$$du = e^x \, dx, \qquad v = \tfrac{1}{2}x^2.$$

Integration by parts would then have given

$$\int xe^x \, dx = \int u \, dv = uv - \int v \, du = \tfrac{1}{2}x^2 e^x - \tfrac{1}{2} \int x^2 e^x \, dx,$$

an integral more complicated than the one we started with. A good choice of u and dv is crucial. ❑

Example 2 Calculate

$$\int x \sin 2x \, dx.$$

SOLUTION Setting

$$u = x, \qquad dv = \sin 2x \, dx,$$

we have

$$du = dx, \qquad v = -\tfrac{1}{2}\cos 2x.$$

Therefore,

$$\int x \sin 2x \, dx = -\tfrac{1}{2}x \cos 2x - \int -\tfrac{1}{2} \cos 2x \, dx = -\tfrac{1}{2}x \cos 2x + \tfrac{1}{4} \sin 2x + C.$$

You can verify that if we had set

$$u = \sin 2x, \qquad dv = x \, dx,$$

then we would have run into the same kind of difficulty that we saw at the end of Example 1. ❏

In Examples 1 and 2 there was only one effective choice for u and dv. As the next example shows, however, there can be more than one way to choose u and dv.

Example 3   Calculate

$$\int x \ln x \, dx.$$

SOLUTION   Setting

$$u = \ln x, \qquad dv = x \, dx,$$

we have

$$du = \frac{1}{x}dx, \qquad v = \frac{x^2}{2}.$$

Then

$$\int x \ln x \, dx = \int u \, dv = uv - \int v \, du$$

$$= \frac{x^2}{2}\ln x - \int \frac{1}{x}\frac{x^2}{2}dx = \tfrac{1}{2}x^2\ln x - \tfrac{1}{4}x^2 + C.$$

ALTERNATE SOLUTION   This time we set

$$u = x \ln x, \qquad dv = dx,$$

so that

$$du = (1 + \ln x) \, dx, \qquad v = x.$$

Substituting these selections in

$$\int u \, dv = uv - \int v \, du,$$

we find that

(1) $$\int x \ln x \, dx = x^2 \ln x - \int x(1 + \ln x) \, dx.$$

It may seem that the integral on the right is more complicated than the one we started with.   However, we can rewrite equation (1) as

$$\int x \ln x \, dx = x^2 \ln x - \int x \, dx - \int x \ln x \, dx.$$

If we let $P = \int x \ln x \, dx$, then this equation is

$$P = x^2 \ln x - \int x \, dx - P.$$

Solving for P, we have

$$2P = x^2 \ln x - \int x \, dx$$

and

$$P = \tfrac{1}{2}x^2 \ln x - \tfrac{1}{2} \int x \, dx = \tfrac{1}{2}x^2 \ln x - \tfrac{1}{4}x^2 + C.$$

Thus

$$\int x \ln x \, dx = \tfrac{1}{2}x^2 \ln x - \tfrac{1}{4}x^2 + C$$

as before. ❏

Remark   Given the two successful approaches in Example 3, you might also be tempted to try setting

$$u = x, \qquad dv = \ln x \, dx.$$

This won't work, however, because you can't calculate v, that is, you don't know an antiderivative for $\ln x$.   Actually, as you will see at the end of this section, $\int \ln x \, dx$ is, itself, calculated using integration by parts.   ❏

Integration by parts is often used to calculate integrals where the integrand is a mixture of function types; for example, polynomials and exponentials, or polynomials and trigonometric functions, and so forth.   Some integrands, however, are better left as mixtures; for example, it is easy to see that

$$\int 2xe^{x^2} \, dx = e^{x^2} + C \qquad \text{and} \qquad \int 3x^2 \cos x^3 \, dx = \sin x^3 + C.$$

Any attempt to separate these integrands for integration by parts is counterproductive.   The mixtures in these integrands arise from the chain rule, and we need these mixtures to calculate the integrals.

Example 4   Calculate

$$\int x^5 e^{x^3} \, dx.$$

SOLUTION   To integrate e^{x^3} we need an x^2 factor.   So we will keep x^2 together with e^{x^3}.   Setting

$$u = x^3, \qquad dv = x^2 e^{x^3} \, dx,$$

we have

$$du = 3x^2 \, dx, \qquad v = \tfrac{1}{3} e^{x^3}.$$

Thus

$$\int x^5 e^{x^3} dx = \int u \, dv = uv - \int v \, du$$

$$= \tfrac{1}{3} x^3 e^{x^3} - \int x^2 e^{x^3} dx$$

$$= \tfrac{1}{3} x^3 e^{x^3} - \tfrac{1}{3} e^{x^3} + C = \tfrac{1}{3}(x^3 - 1)e^{x^3} + C. \quad \square$$

Integration by parts can also be used in connection with definite integrals. Suppose that f' and g' are continuous on the interval $[a, b]$. Integrating the formula for the derivative of a product, we have

$$\int_a^b f(x)g'(x) \, dx + \int_a^b f'(x)g(x) \, dx = \int_a^b [f(x) \cdot g(x)]' \, dx.$$

However, by the fundamental theorem of calculus, Theorem 5.4.2,

$$\int_a^b [f(x) \cdot g(x)]' \, dx = \left[f(x) \cdot g(x) \right]_a^b.$$

Thus

$$\int_a^b f(x)g'(x) \, dx + \int_a^b f'(x)g(x) \, dx = \left[f(x) \cdot g(x) \right]_a^b$$

and

(8.2.3)
$$\int_a^b f(x)g'(x) \, dx = \left[f(x) \cdot g(x) \right]_a^b - \int_a^b f'(x)g(x) \, dx,$$

which is the definite integral version of Formula (8.2.1).

Example 5 Evaluate

$$\int_1^2 x^3 \ln x \, dx.$$

SOLUTION Setting

$$u = \ln x, \qquad dv = x^3 \, dx,$$

we have

$$du = \frac{1}{x} \, dx, \qquad v = \tfrac{1}{4} x^4.$$

Substituting into (8.2.3) gives

$$\int_1^2 x^3 \ln x \, dx = \left[\tfrac{1}{4} x^4 \ln x \right]_1^2 - \tfrac{1}{4} \int_1^2 x^3 \, dx$$

$$= 4 \ln 2 - \tfrac{1}{16} \left[x^4 \right]_1^2 = 4 \ln 2 - \tfrac{15}{16}. \quad \square$$

To calculate some integrals you may have to integrate by parts more than once.

Example 6 Evaluate

$$\int_0^1 x^2 e^x \, dx.$$

SOLUTION First we calculate the indefinite integral

$$\int x^2 e^x \, dx$$

by separating x^2 from e^x. Setting

$$u = x^2, \qquad dv = e^x \, dx,$$

we have

$$du = 2x \, dx, \qquad v = e^x,$$

and thus

$$\int x^2 e^x \, dx = \int u \, dv = uv - \int v \, du = x^2 e^x - \int 2x \, e^x \, dx.$$

We now compute the integral on the right again by parts. This time we set

$$u = 2x, \qquad dv = e^x \, dx.$$

This gives

$$du = 2 \, dx, \qquad v = e^x$$

and thus

$$\int 2x e^x \, dx = \int u \, dv = uv - \int v \, du = 2x e^x - \int 2e^x \, dx = 2x e^x - 2e^x + C.$$

This together with our earlier calculation gives

$$\int x^2 e^x \, dx = x^2 e^x - 2x e^x + 2e^x + C.$$

It follows now that

$$\int_0^1 x^2 e^x \, dx = \left[x^2 e^x - 2x e^x + 2e^x \right]_0^1 = (e - 2e + 2e) - 2 = e - 2. \quad ❏$$

Remark Note that we did not use Formula (8.2.3) in evaluating the definite integral in Example 6. Instead, we calculated the corresponding indefinite integral and then substituted the limits of integration at the end. This approach is sometimes easier than using (8.2.3), especially if several steps are involved. ❏

Example 7 Find

$$\int e^x \cos x \, dx.$$

SOLUTION Here we integrate by parts twice. First we write

$$u = e^x, \qquad dv = \cos x \, dx,$$

$$du = e^x \, dx, \qquad v = \sin x.$$

This gives

(1) $$\int e^x \cos x \, dx = \int u \, dv = uv - \int v \, du = e^x \sin x - \int e^x \sin x \, dx.$$

Now we work with the integral on the right.   Setting

$$u = e^x, \qquad dv = \sin x \, dx,$$
$$du = e^x \, dx, \qquad v = -\cos x,$$

we have

(2) $$\int e^x \sin x \, dx = \int u \, dv = uv - \int v \, du = -e^x \cos x + \int e^x \cos x \, dx.$$

Substituting (2) into (1), we get

$$\int e^x \cos x \, dx = e^x \sin x + e^x \cos x - \int e^x \cos x \, dx,$$

and we have a situation similar to the one we had in the alternate solution in Example 3.   "Solving" this equation for $\int e^x \cos x \, dx$ gives

$$2 \int e^x \cos x \, dx = e^x(\sin x + \cos x),$$

$$\int e^x \cos x \, dx = \tfrac{1}{2} e^x(\sin x + \cos x).$$

Since this integral is an indefinite integral, we add an arbitrary constant C:

$$\int e^x \cos x \, dx = \tfrac{1}{2} e^x(\sin x + \cos x) + C. \quad \square$$

Remark   You should verify that if you switch the roles of u and dv in Example 7 by letting

$$u = \cos x, \qquad dv = e^x \, dx,$$
$$du = -\sin x \, dx, \qquad v = e^x,$$

then you will get precisely the same result in precisely the same way.   ❑

Finally, the techniques of integration by parts enable us to integrate the logarithm function and the inverse trigonometric functions:

(8.2.4)

$$\int \ln x \, dx = x \ln x - x + C.$$

(8.2.5)

$$\int \sin^{-1}x \, dx = x \sin^{-1}x + \sqrt{1 - x^2} + C.$$

(8.2.6)
$$\int \tan^{-1}x \, dx = x\tan^{-1}x - \tfrac{1}{2}\ln(1 + x^2) + C.$$

(8.2.7)
$$\int \sec^{-1}x \, dx = x\sec^{-1}x - \ln|x + \sqrt{x^2 - 1}| + C.$$

We will derive (8.2.5) and leave the others to you as exercises. Set

$$u = \sin^{-1}x, \qquad dv = dx.$$

Then

$$du = \frac{dx}{\sqrt{1 - x^2}}, \qquad v = x,$$

and

$$\int \sin^{-1}x \, dx = x\sin^{-1}x - \int \frac{x}{\sqrt{1 - x^2}} \, dx.$$

The new integral is easy to evaluate by means of the substitution

$$w = 1 - x^2, \qquad dw = -2x \, dx,$$

which gives

$$\int \frac{x}{\sqrt{1 - x^2}} \, dx = -\tfrac{1}{2}\int \frac{dw}{\sqrt{w}} = -\sqrt{w} = -\sqrt{1 - x^2}.$$

Therefore,

$$\int \sin^{-1}x \, dx = x\sin^{-1}x + \sqrt{1 - x^2} + C. \quad \square$$

EXERCISES 8.2

In Exercises 1–46, evaluate the given integral.

1. $\displaystyle\int xe^{-x} \, dx.$

2. $\displaystyle\int \ln x \, dx.$ (8.2.4)

3. $\displaystyle\int_1^e x^2 \ln x \, dx.$

4. $\displaystyle\int_0^2 x\, 2^x \, dx.$

5. $\displaystyle\int x^2 e^{-x^3} \, dx.$

6. $\displaystyle\int x \ln x^2 \, dx.$

7. $\displaystyle\int_0^1 x^2 e^{-x} \, dx.$

8. $\displaystyle\int x^3 e^{-x^2} \, dx.$

9. $\displaystyle\int \frac{x^2}{\sqrt{1 - x}} \, dx.$

10. $\displaystyle\int \frac{dx}{x(\ln x)^3}.$

11. $\displaystyle\int_1^{e^2} x \ln \sqrt{x} \, dx.$

12. $\displaystyle\int_0^3 x\sqrt{x + 1} \, dx.$

13. $\displaystyle\int \frac{\ln(x + 1)}{\sqrt{x + 1}} \, dx.$

14. $\displaystyle\int x^2(e^x - 1) \, dx.$

15. $\displaystyle\int (\ln x)^2 \, dx.$

16. $\displaystyle\int x(x + 5)^{-14} \, dx.$

17. $\displaystyle\int x^3 3^x \, dx.$

18. $\displaystyle\int \sqrt{x} \ln x \, dx.$

19. $\int x(x + 5)^{14}\, dx.$

20. $\int (2^x + x^2)^2\, dx.$

21. $\int_0^{1/2} x \cos \pi x\, dx.$

22. $\int_0^{\pi/2} x^2 \sin x\, dx.$

23. $\int x^2(x + 1)^9\, dx.$

24. $\int x^2(2x - 1)^{-7}\, dx.$

25. $\int e^x \sin x\, dx.$

26. $\int (e^x + 2x)^2\, dx.$

27. $\int_0^1 \ln(1 + x^2)\, dx.$

28. $\int x \ln(x + 1)\, dx.$

29. $\int x^n \ln x\, dx. \quad (n \neq -1)$

30. $\int e^{3x} \cos 2x\, dx.$

31. $\int x^3 \sin x^2\, dx.$

32. $\int x^3 \sin x\, dx.$

33. $\int x^4 e^x\, dx.$

34. $\int \dfrac{\sin^{-1} 2x}{\sqrt{1 - 4x^2}}\, dx.$

35. $\int_0^{1/4} \sin^{-1} 2x\, dx.$

36. $\int \tan^{-1} x\, dx. \quad (8.2.6)$

37. $\int \sec^{-1} x\, dx. \quad (8.2.7)$

38. $\int \cos \sqrt{x}\, dx.$ HINT: Let $u = \sqrt{x}$.

39. $\int_0^1 x \tan^{-1}(x^2)\, dx.$

40. $\int_{-1}^1 x \sinh(2x^2)\, dx.$

41. $\int x^2 \cosh 2x\, dx.$

42. $\int \cos(\ln x)\, dx.$ HINT: Integrate by parts twice.

43. $\int \sin(\ln x)\, dx.$

44. $\int \cos x \tan^{-1}(\sin x)\, dx.$

45. $\int \dfrac{1}{x} \sin^{-1}(\ln x)\, dx.$

46. $\int_1^{2e} x^2(\ln x)^2\, dx.$

In Exercises 47 and 48, find the area of the region bounded by the graph of f and the x-axis.

47. $f(x) = \sin^{-1} x, \quad x \in [0, \tfrac{1}{2}].$

48. $f(x) = xe^{-2x}, \quad x \in [0, 2].$

49. Let Ω be the region bounded by the graph of $f(x) = \ln x$ and the x-axis between $x = 1$ and $x = e$. (a) Find the area of Ω. (b) Find the centroid of Ω. (c) Find the volume of the solids generated by revolving Ω about each of the coordinate axes.

50. Let $f(x) = \dfrac{\ln x}{x}$ on $[1, 2e]$.

(a) Find the area of the region R bounded by the graph of f and the x-axis.

(b) Find the volume of the solid generated by revolving R around the x-axis.

In Exercises 51–54, find the centroid of the region under the graph.

51. $f(x) = e^x, \quad x \in [0, 1].$

52. $f(x) = e^{-x}, \quad x \in [0, 1].$

53. $f(x) = \sin x, \quad x \in [0, \pi].$

54. $f(x) = \cos x, \quad x \in [0, \tfrac{1}{2}\pi].$

55. The mass density of a rod that extends from $x = 0$ to $x = 1$ is given by the function $\lambda(x) = e^{kx}$, where k is a constant. (a) Calculate the mass of the rod. (b) Find the center of mass of the rod.

56. The mass density of a rod that extends from $x = 2$ to $x = 3$ is given by the logarithm function $f(x) = \ln x$. (a) Calculate the mass of the rod. (b) Find the center of mass of the rod.

In Exercises 57–62, find the volume generated by revolving the region under the graph about the y-axis.

57. $f(x) = e^{ax}, \quad x \in [0, 1].$

58. $f(x) = \sin \pi x, \quad x \in [0, 1].$

59. $f(x) = \cos \tfrac{1}{2}\pi x, \quad x \in [0, 1].$

60. $f(x) = x \sin x, \quad x \in [0, \pi].$

61. $f(x) = xe^x, \quad x \in [0, 1].$

62. $f(x) = x \cos x, \quad x \in [0, \tfrac{1}{2}\pi].$

63. Let Ω be the region under the curve $y = e^x$, $x \in [0, 1]$. Find the centroid of the solid generated by revolving Ω about the x-axis. (6.4.5)

64. Let Ω be the region under the curve $y = \sin x$, $x \in [0, \tfrac{1}{2}\pi]$. Find the centroid of the solid generated by revolving Ω about the x-axis. (6.4.5)

65. Let Ω be the region bounded by the graph of $y = \cosh x$ and the x-axis between $x = 0$ and $x = 1$. Find the area of Ω and determine the centroid.

66. Let Ω be the region given in Exercise 65. Find the centroid of the solid generated by revolving Ω about: (a) the x-axis; (b) the y-axis.

67. Let n be a positive integer. Use integration by parts to derive the formula

$$\int x^n e^{ax}\, dx = \frac{x^n e^{ax}}{a} - \frac{n}{a} \int x^{n-1} e^{ax}\, dx, \ a \neq 0.$$

Note: This formula is known as a *reduction formula* since the exponent n in the integrand has been reduced.

68. Let n be a positive integer. Use integration by parts to derive the reduction formula

$$\int (\ln x)^n \, dx = x \, (\ln x)^n - n \int (\ln x)^{n-1} \, dx.$$

HINT: Let $u = (\ln x)^n$.

In Exercises 69–72, use the reduction formulas in Exercises 67 and 68 to evaluate the given integral.

69. $\displaystyle\int x^3 e^{2x} \, dx.$

70. $\displaystyle\int x^2 e^{-x} \, dx.$

71. $\displaystyle\int (\ln x)^3 \, dx.$

72. $\displaystyle\int (\ln x)^4 \, dx.$

73. Let f and g have continuous second derivatives. Show that if $f(a) = g(a) = f(b) = g(b) = 0$, then

$$\int_a^b f(x)g''(x) \, dx = \int_a^b g(x)f''(x) \, dx.$$

74. Consider the identity

$$f(b) - f(a) = \int_a^b f'(x) \, dx.$$

(a) Assume that f has a continuous second derivative. Use integration by parts to derive the identity

$$f(b) - f(a) = f'(a)(b - a) - \int_a^b f''(x)(x - b) \, dx.$$

(b) Assume that f has a continuous third derivative. Use the result in part (a) and integration by parts to derive the identity

$$f(b) - f(a) = f'(a)(b - a) + \frac{f''(a)}{2}(b - a)^2$$

$$+ \int_a^b \frac{f'''(x)}{2}(x - b)^2 \, dx.$$

These results generalize the mean-value theorem. The identities begun here will be extended and used in Chapter 11.

Revenue Streams

In Section 7.6, we considered continuous compounding and saw that the present value of A dollars t years from now is given by the formula

$$P.V. = Ae^{-rt},$$

where r is the annual rate of continuous compounding.

75. Suppose that revenue flows continuously at a constant rate of R dollars per year for n years. Show that the present value of such a revenue stream is given by

$$P.V. = \int_0^n Re^{-rt} \, dt.$$

76. Given a constant revenue stream of $R = \$1000$ per year, what is the present value of the first four years of revenue if the annual rate of continuous compounding is: (a) 4%? (b) 8%?

Revenue streams usually do not flow at a constant rate R; rather they normally flow at a time-dependent rate $R(t)$. In general, $R(t)$ tends to increase when business is good and tends to decrease when business is poor. "Growth companies" are those that have a continually increasing flow rate. The so-called "cyclical companies" owe their name to a fluctuating flow rate. If we postulate a time dependent flow rate $R(t)$, then the present value of an n-year stream is given by the formula

$$P.V. = \int_0^n R(t)e^{-rt} \, dt.$$

In Exercises 77 and 78, assume a time-dependent revenue stream with $R(t) = 1000 + 60t$ dollars per year.

77. What is the present value of the first two years of revenue if the annual rate of continuous compounding is: (a) 5%? (b) 10%?

78. What is the present value of the third year of revenue if the annual rate of continuous compounding is: (a) 5%? (b) 10%?

■ 8.3 POWERS AND PRODUCTS OF SINES AND COSINES

I. We begin by explaining how to calculate integrals of the form

$$\int \sin^m x \cos^n x \, dx \qquad \text{with } m \text{ or } n \text{ an odd positive integer.}$$

Suppose that n is odd. If $n = 1$, we have

(1) $$\int \sin^m x \cos x \, dx = \frac{1}{m + 1} \sin^{m+1} x + C, \qquad m \neq -1.$$

If $n > 1$, write

$$\cos^n x = \cos^{n-1} x \cos x.$$

Since $n - 1$ is even, $\cos^{n-1} x$ can be expressed in powers of $\sin^2 x$ by noting that $\cos^2 x = 1 - \sin^2 x$. The integral then takes the form

$$\int (\text{sum of powers of } \sin x) \cdot \cos x \, dx,$$

which can be broken up into integrals of the form (1).

Similarly if m is odd, write

$$\sin^m x = \sin^{m-1} x \sin x$$

and use the substitution $\sin^2 x = 1 - \cos^2 x$.

Example 1

$$\int \sin^2 x \cos^5 x \, dx = \int \sin^2 x \cos^4 x \cos x \, dx$$

$$= \int \sin^2 x \, (1 - \sin^2 x)^2 \cos x \, dx$$

$$= \int (\sin^2 x - 2 \sin^4 x + \sin^6 x) \cos x \, dx$$

$$= \int \sin^2 x \cos x \, dx - 2 \int \sin^4 x \cos x \, dx + \int \sin^6 x \cos x \, dx$$

$$= \tfrac{1}{3} \sin^3 x - \tfrac{2}{5} \sin^5 x + \tfrac{1}{7} \sin^7 x + C. \quad \square$$

Example 2

$$\int \sin^5 x \, dx = \int \sin^4 x \sin x \, dx$$

$$= \int (1 - \cos^2 x)^2 \sin x \, dx$$

$$= \int (1 - 2 \cos^2 x + \cos^4 x) \sin x \, dx$$

$$= \int \sin x \, dx - 2 \int \cos^2 x \sin x \, dx + \int \cos^4 x \sin x \, dx$$

$$= -\cos x + \tfrac{2}{3} \cos^3 x - \tfrac{1}{5} \cos^5 x + C. \quad \square$$

II. To calculate integrals of the form

$$\int \sin^m x \cos^n x \, dx \qquad \text{with } m \text{ and } n \text{ both even positive integers}$$

use the following trigonometric identities:

$$\sin x \cos x = \tfrac{1}{2} \sin 2x, \qquad \sin^2 x = \tfrac{1}{2} - \tfrac{1}{2} \cos 2x, \qquad \cos^2 x = \tfrac{1}{2} + \tfrac{1}{2} \cos 2x.$$

The first of these identities is derived from the double-angle formula for the sine function. The other two are the half-angle formulas for sine and cosine.

Example 3   (See Exercise 72, Section 5.7.)

$$\int \cos^2 x \, dx = \int (\tfrac{1}{2} + \tfrac{1}{2} \cos 2x) \, dx$$

$$= \tfrac{1}{2} \int dx + \tfrac{1}{2} \int \cos 2x \, dx = \tfrac{1}{2} x + \tfrac{1}{4} \sin 2x + C. \quad \square$$

Example 4

$$\int \sin^2 x \cos^2 x \, dx = \tfrac{1}{4} \int \sin^2 2x \, dx$$

$$= \tfrac{1}{4} \int (\tfrac{1}{2} - \tfrac{1}{2} \cos 4x) \, dx$$

$$= \tfrac{1}{8} \int dx - \tfrac{1}{8} \int \cos 4x \, dx = \tfrac{1}{8} x - \tfrac{1}{32} \sin 4x + C. \quad \square$$

Example 5

$$\int \sin^4 x \cos^2 x \, dx = \int (\sin x \cos x)^2 \sin^2 x \, dx$$

$$= \int \tfrac{1}{4} \sin^2 2x \, (\tfrac{1}{2} - \tfrac{1}{2} \cos 2x) \, dx$$

$$= \tfrac{1}{8} \int \sin^2 2x \, dx - \tfrac{1}{8} \int \sin^2 2x \cos 2x \, dx$$

$$= \tfrac{1}{8} \int (\tfrac{1}{2} - \tfrac{1}{2} \cos 4x) \, dx - \tfrac{1}{8} \int \sin^2 2x \cos 2x \, dx$$

$$= \tfrac{1}{16} x - \tfrac{1}{64} \sin 4x - \tfrac{1}{48} \sin^3 2x + C. \quad \square$$

III. Next we derive *reduction formulas* for integrals of the form

$$\int \sin^n x \, dx \quad \text{and} \quad \int \cos^n x \, dx, \qquad \text{where } n \text{ is a positive integer.}$$

We will use integration by parts to show that

(8.3.1)
$$\int \sin^n x \, dx = -\frac{1}{n} \sin^{n-1} x \cos x + \frac{n-1}{n} \int \sin^{n-2} x \, dx.$$

The corresponding formula for $\int \cos^n x \, dx$ is

(8.3.2)
$$\int \cos^n x \, dx = \frac{1}{n} \cos^{n-1} x \sin x + \frac{n-1}{n} \int \cos^{n-2} x \, dx.$$

The verification of this formula is left as an exercise.

Formulas such as (8.3.1) and (8.3.2) are called reduction formulas because the exponent is reduced (from n to $n - 2$ in each of these cases).

To establish (8.3.1), we write $\sin^n x$ as $\sin^{n-1} x \sin x$ and set

$$u = \sin^{n-1} x, \qquad dv = \sin x \, dx.$$

Then

$$du = (n - 1) \sin^{n-2} x \cos x \, dx, \qquad v = -\cos x,$$

and

$$\int \sin^n x \, dx = -\sin^{n-1} x \cos x + (n - 1) \int \sin^{n-2} x \cos^2 x \, dx$$

$$= -\sin^{n-1} x \cos x + (n - 1) \int \sin^{n-2} x \, (1 - \sin^2 x) \, dx.$$

Therefore,

$$\int \sin^n x \, dx = -\sin^{n-1} x \cos x + (n - 1) \int \sin^{n-2} x \, dx - (n - 1) \int \sin^n x \, dx.$$

Now, adding $(n - 1) \int \sin^n x \, dx$ to both sides, we get

$$n \int \sin^n x \, dx = -\sin^{n-1} x \cos x + (n - 1) \int \sin^{n-2} x \, dx$$

and (8.3.1) follows. ❑

Repeated applications of the formulas (8.3.1) or (8.3.2) will reduce a power of the sine or cosine to 0 or 1, depending upon whether n is even or odd.

Example 6

$$\int \sin^5 x \, dx = -\tfrac{1}{5} \sin^4 x \cos x + \tfrac{4}{5} \int \sin^3 x \, dx$$

$$= -\tfrac{1}{5} \sin^4 x \cos x + \tfrac{4}{5} \left[-\tfrac{1}{3} \sin^2 x \cos x + \tfrac{2}{3} \int \sin x \, dx \right]$$

$$= -\tfrac{1}{5} \sin^4 x \cos x - \tfrac{4}{15} \sin^2 x \cos x - \tfrac{8}{15} \cos x + C. ❑$$

You should verify that this result is equivalent to the result in Example 2.

IV. Finally we come to integrals of the form

$$\int \sin mx \cos nx \, dx, \qquad \int \sin mx \sin nx \, dx, \qquad \int \cos mx \cos nx \, dx.$$

If $m = n$, there is no difficulty. For $m \neq n$ use the identities

$$\sin A \cos B = \tfrac{1}{2}[\sin (A - B) + \sin (A + B)],$$

$$\sin A \sin B = \tfrac{1}{2}[\cos (A - B) - \cos (A + B)],$$

$$\cos A \cos B = \tfrac{1}{2}[\cos (A - B) + \cos (A + B)].$$

These identities follow readily from the familiar addition formulas:

$$\sin (A + B) = \sin A \cos B + \cos A \sin B,$$
$$\sin (A - B) = \sin A \cos B - \cos A \sin B,$$
$$\cos (A + B) = \cos A \cos B - \sin A \sin B,$$
$$\cos (A - B) = \cos A \cos B + \sin A \sin B.$$

Example 7

$$\int \sin 5x \sin 3x \, dx = \int \tfrac{1}{2}[\cos(5x - 3x) - \cos(5x + 3x)] \, dx$$

$$= \tfrac{1}{2} \int [\cos 2x - \cos 8x] \, dx = \tfrac{1}{4} \sin 2x - \tfrac{1}{16} \sin 8x + C. \quad ❑$$

EXERCISES 8.3

In Exercises 1–32, evaluate the given integral.

1. $\displaystyle\int \sin^3 x \, dx.$

2. $\displaystyle\int_0^{\pi/8} \cos^2 4x \, dx.$

3. $\displaystyle\int_0^{\pi/6} \sin^2 3x \, dx.$

4. $\displaystyle\int \cos^3 x \, dx.$

5. $\displaystyle\int \cos^4 x \sin^3 x \, dx.$

6. $\displaystyle\int \sin^3 x \cos^2 x \, dx.$

7. $\displaystyle\int \sin^3 x \cos^3 x \, dx.$

8. $\displaystyle\int \sin^2 x \cos^4 x \, dx.$

9. $\displaystyle\int \sin^2 x \cos^3 x \, dx.$

10. $\displaystyle\int \sin^4 x \cos^3 x \, dx.$

11. $\displaystyle\int_0^{\pi} \sin^4 x \, dx.$

12. $\displaystyle\int \cos^3 x \cos 2x \, dx.$

13. $\displaystyle\int \sin 2x \cos 3x \, dx.$

14. $\displaystyle\int_0^{\pi/2} \cos 2x \sin 3x \, dx.$

15. $\displaystyle\int \sin^2 x \sin 2x \, dx.$

16. $\displaystyle\int_0^{\pi/2} \cos^4 x \, dx.$

17. $\displaystyle\int \sin^4 x \cos^4 x \, dx.$

18. $\displaystyle\int \sin^7 x \, dx.$

19. $\displaystyle\int \sin^6 x \, dx.$

20. $\displaystyle\int \cos^5 x \sin^5 x \, dx.$

21. $\displaystyle\int \cos^7 x \, dx.$

22. $\displaystyle\int \cos^6 x \, dx.$

23. $\displaystyle\int_0^{\pi/2} \cos 3x \cos 2x \, dx.$

24. $\displaystyle\int \sin 3x \sin 2x \, dx.$

25. $\displaystyle\int \sin 5x \sin 2x \, dx.$

26. $\displaystyle\int_0^{\pi/4} \sin 5x \cos 2x \, dx.$

27. $\displaystyle\int_{-1/6}^{1/3} \sin^4 3\pi x \cos^3 3\pi x \, dx.$

28. $\displaystyle\int_0^{1/2} \cos \pi x \cos (\tfrac{\pi}{2} x) \, dx.$

29. $\displaystyle\int_0^{\pi/4} \cos 4x \sin 2x \, dx.$

30. $\displaystyle\int (\sin 3x - \sin x)^2 \, dx.$

31. $\displaystyle\int \sin(\tfrac{1}{2} x) \cos 2x \, dx.$

32. $\displaystyle\int_0^{2\pi} \sin^2 ax \, dx, \quad a \neq 0.$

In Exercises 33–36, calculate the integral by the indicated trigonometric substitution. Note: trigonometric substitutions will be studied in more detail in Section 8.5.

33. $\displaystyle\int \frac{dx}{(x^2 + 1)^3}.$ [set $x = \tan u$]

34. $\displaystyle\int \frac{dx}{(x^2 + 4)^3}.$ [set $x = 2 \tan u$]

35. $\displaystyle\int \frac{dx}{[(x + 1)^2 + 1]^2}.$ [set $x + 1 = \tan u$]

36. $\displaystyle\int \frac{dx}{[(2x + 1)^2 + 9]^2}.$ [set $2x + 1 = 3 \tan u$]

37. Find the area of the region R bounded by the graph of $f(x) = \sin^2 x$ and the x-axis, $x \in [0, \pi]$.

38. The region bounded by the graph of $f(x) = \cos x$ and the x-axis, $x \in [-\pi/2, \pi/2]$, is revolved around the x-axis. Find the volume of the solid that is generated.

39. The region R in Exercise 37 is revolved around the x-axis. Find the volume of the solid that is generated.

40. The region bounded by the y-axis and the graphs of $y = \sin x$ and $y = \cos x$, $0 \le x \le \pi/4$, is revolved around the x-axis. Find the volume of the solid that is generated.

41. Let m and n be positive integers. Prove that

$$\int \sin mx \sin nx \, dx$$

$$= \begin{cases} \dfrac{\sin(m - n)x}{2(m - n)} - \dfrac{\sin(m + n)x}{2(m + n)} + C & \text{if } m \neq n \\[2mm] \dfrac{x}{2} - \dfrac{\sin 2mx}{4m} + C & \text{if } m = n. \end{cases}$$

42. Derive formulas corresponding to the formula in Exercise 41 for:

(a) $\displaystyle\int \sin mx \cos nx \, dx.$

(b) $\displaystyle\int \cos mx \cos nx \, dx.$

43. Use the result in Exercise 41 to prove that

$$\int_{-\pi}^{\pi} \sin mx \sin nx \, dx = \begin{cases} 0 & \text{if } m \neq n \\ \pi & \text{if } m = n. \end{cases}$$

44. (a) Evaluate

$$\int_{-\pi}^{\pi} \sin mx \cos nx \, dx.$$

(b) Evaluate

$$\int_{-\pi}^{\pi} \cos mx \cos nx \, dx.$$

Note: The formulas in Exercises 43 and 44 are important in applied mathematics. They arise in Fourier analysis where the approximation of functions by sums of sines and cosines is studied.

45. Verify the reduction formula (8.3.2).

46. (a) Use the reduction formula (8.3.1) to show that

$$\int_0^{\pi/2} \sin^n x \, dx = \frac{n - 1}{n} \int_0^{\pi/2} \sin^{n-2} x \, dx.$$

(b) Show that

$$\int_0^{\pi/2} \sin^n x \, dx$$

$$= \begin{cases} \left(\dfrac{(n - 1) \cdots 5 \cdot 3 \cdot 1}{n \cdots 6 \cdot 4 \cdot 2} \right) \dfrac{\pi}{2}, & n \text{ even, } n \geq 2 \\[4mm] \dfrac{(n - 1) \cdots 4 \cdot 2}{n \cdots 5 \cdot 3}, & n \text{ odd, } n \geq 3. \end{cases}$$

These formulas are known as the *Wallis sine formulas.*

(c) Show that

$$\int_0^{\pi/2} \cos^n x \, dx = \int_0^{\pi/2} \sin^n x \, dx.$$

Use the results in Exercise 46 to evaluate the integrals in Exercises 47 and 48.

47. $\displaystyle\int_0^{\pi/2} \sin^7 x \, dx.$ **48.** $\displaystyle\int_0^{\pi/2} \cos^6 x \, dx.$

■ 8.4 OTHER TRIGONOMETRIC POWERS

I. First we consider integrals of the form

$$\int \tan^n x \, dx, \qquad \int \cot^n x \, dx, \qquad n \geq 2 \text{ an integer.}$$

To integrate $\tan^n x$, set

(1) $\quad \tan^n x = \tan^{n-2} x \tan^2 x = (\tan^{n-2} x)(\sec^2 x - 1) = \tan^{n-2} x \sec^2 x - \tan^{n-2} x.$

If $n - 2 \geq 2$, then repeat the reduction to get

$$\tan^n x = \tan^{n-2} x \sec^2 x - (\tan^{n-4} x \sec^2 x - \tan^{n-4} x)$$
$$= \tan^{n-2} x \sec^2 x - \tan^{n-4} x \sec^2 x + \tan^{n-4} x.$$

Continue until the final term is either $\pm \tan x$ (n odd) or ± 1 (n even). The idea is to obtain a sum of terms of the form $\tan^k x \sec^2 x$, pairing a power of $\tan x$ with $\sec^2 x$. These terms will be easy to integrate since they have the form $\int u^k \, du$, where $u = \tan x$, $du = \sec^2 x \, dx$.

Example 1

$$\int \tan^6 x \, dx = \int (\tan^4 x \sec^2 x - \tan^4 x) \, dx$$

$$= \int (\tan^4 x \sec^2 x - \tan^2 x \sec^2 x + \tan^2 x) \, dx$$

$$= \int (\tan^4 x \sec^2 x - \tan^2 x \sec^2 x + \sec^2 x - 1) \, dx$$

$$= \int \tan^4 x \sec^2 x \, dx - \int \tan^2 x \sec^2 x \, dx + \int \sec^2 x \, dx - \int dx$$

$$= \tfrac{1}{5} \tan^5 x - \tfrac{1}{3} \tan^3 x + \tan x - x + C. \quad \square$$

Remark The reduction formula

(8.4.1)
$$\int \tan^n x \, dx = \frac{1}{n-1} \tan^{n-1} x - \int \tan^{n-2} x \, dx, \qquad n \ge 2.$$

follows immediately from (1). $\square$

Powers of the cotangent function are handled in the same manner:

$$\cot^n x = \cot^{n-2} x \cot^2 x = (\cot^{n-2} x)(\csc^2 x - 1) = \cot^{n-2} x \csc^2 x - \cot^{n-2} x.$$

The reduction formula for $\int \cot^n x \, dx$ is left as an exercise.

II. Next we consider the integrals

$$\int \sec^n x \, dx, \qquad \int \csc^n x \, dx, \qquad n \ge 2 \text{ an integer.}$$

A. n even.

For even powers write:

$$\sec^n x = \sec^{n-2} x \sec^2 x = (\tan^2 x + 1)^{(n-2)/2} \sec^2 x \qquad \text{and put } u = \tan x,$$

$$\csc^n x = \csc^{n-2} x \csc^2 x = (\cot^2 x + 1)^{(n-2)/2} \csc^2 x \qquad \text{and put } u = \cot x.$$

Example 2

$$\int \sec^6 x \, dx = \int \sec^4 x \sec^2 x \, dx = \int (\tan^2 x + 1)^2 \sec^2 x \, dx$$

$$= \int (u^2 + 1)^2 \, du \qquad (u = \tan x, \ du = \sec^2 x \, dx)$$

$$= \int (u^4 + 2u^2 + 1) \, du = \tfrac{1}{5} u^5 + \tfrac{2}{3} u^3 + u + C$$

$$= \tfrac{1}{5} \tan^5 x + \tfrac{2}{3} \tan^3 x + \tan x + C. \quad \square$$

B. *n* odd.

You can use integration by parts to integrate odd powers. For $\sec^n x$, set

$$u = \sec^{n-2} x, \qquad dv = \sec^2 x \, dx,$$

and when $\tan^2 x$ appears, use the identity $\tan^2 x = \sec^2 x - 1$. You can handle $\csc^n x$ in a similar manner.

Example 3   For

$$\int \sec^3 x \, dx,$$

set

$$u = \sec x, \qquad\qquad dv = \sec^2 x \, dx,$$

$$du = \sec x \tan x \, dx, \qquad v = \tan x.$$

Then

$$\int \sec^3 x \, dx = \int \sec x \sec^2 x \, dx$$

$$= \sec x \tan x - \int \tan^2 x \sec x \, dx$$

$$= \sec x \tan x - \int (\sec^2 x - 1)\sec x \, dx$$

$$= \sec x \tan x - \int \sec^3 x \, dx + \int \sec x \, dx.$$

Therefore

$$\int \sec^3 x \, dx = \sec x \tan x - \int \sec^3 x \, dx + \ln|\sec x + \tan x|.$$

Adding $\int \sec^3 x \, dx$ to both sides, we get

$$2 \int \sec^3 x \, dx = \sec x \tan x + \ln|\sec x + \tan x|$$

and

$$\int \sec^3 x \, dx = \tfrac{1}{2} \sec x \tan x + \tfrac{1}{2} \ln|\sec x + \tan x|.$$

Now add the arbitrary constant:

$$\int \sec^3 x \, dx = \tfrac{1}{2} \sec x \tan x + \tfrac{1}{2} \ln|\sec x + \tan x| + C.$$

This integral occurs so frequently in applications that you will find it listed on the inside covers of this text.   ❏

Remark   The integration by parts method used to integrate odd powers of the secant or cosecant can be used to derive reduction formulas that incorporate both the even and odd cases simultaneously.

Set

$$u = \sec^{n-2}x, \qquad dv = \sec^2x\,dx.$$

Then

$$du = (n-2)\sec^{n-3}x\,\sec x\,\tan x\,dx = (n-2)\sec^{n-2}x\,\tan x\,dx, \qquad v = \tan x,$$

and

$$\int \sec^n x\,dx = \sec^{n-2}x\,\tan x - (n-2)\int \sec^{n-2}x\,\tan^2 x\,dx$$

$$= \sec^{n-2}x\,\tan x - (n-2)\int \sec^{n-2}x(\sec^2 x - 1)\,dx$$

$$= \sec^{n-2}x\,\tan x - (n-2)\int \sec^n x\,dx + (n-2)\int \sec^{n-2}x\,dx.$$

Now, solving for $\int \sec^n x\,dx$, we get

$$(n-1)\int \sec^n x\,dx = \sec^{n-2}x\,\tan x + (n-2)\int \sec^{n-2}x\,dx$$

from which it follows that

(8.4.2)
$$\int \sec^n x\,dx = \frac{1}{n-1}\sec^{n-2}x\,\tan x + \frac{n-2}{n-1}\int \sec^{n-2}x\,dx, \qquad n \geq 2.$$

For example,

$$\int \sec^6 x\,dx = \tfrac{1}{5}\sec^4 x\,\tan x + \tfrac{4}{5}\int \sec^4 x\,dx$$

$$= \tfrac{1}{5}\sec^4 x\,\tan x + \tfrac{4}{5}\left[\tfrac{1}{3}\sec^2 x\,\tan x + \tfrac{2}{3}\int \sec^2 x\,dx\right]$$

$$= \tfrac{1}{5}\sec^4 x\,\tan x + \tfrac{4}{15}\sec^2 x\,\tan x + \tfrac{8}{15}\tan x + C.$$

You should reconcile this result with that of Example 2. ❑

The reduction formula for $\int \csc^n x\,dx$ is left as an exercise.

III. Finally we come to integrals of the form

$$\int \tan^m x\,\sec^n x\,dx, \qquad \int \cot^m x\,\csc^n x\,dx, \qquad m,\,n \text{ positive integers.}$$

A. *n* even.
When *n* is even, write

$$\tan^m x\,\sec^n x = \tan^m x\,\sec^{n-2}x\,\sec^2 x$$

and express $\sec^{n-2}x$ entirely in terms of $\tan^2 x$ using $\sec^2 x = \tan^2 x + 1$.

Example 4

$$\int \tan^5 x \sec^4 x \, dx = \int \tan^5 x \sec^2 x \sec^2 x \, dx$$

$$= \int \tan^5 x (\tan^2 x + 1) \sec^2 x \, dx$$

$$= \int \tan^7 x \sec^2 x \, dx + \int \tan^5 x \sec^2 x \, dx$$

$$= \tfrac{1}{8} \tan^8 x + \tfrac{1}{6} \tan^6 x + C. \quad \square$$

B. n, m odd.
When n and m are both odd, write

$$\tan^m x \sec^n x = \tan^{m-1} x \sec^{n-1} x \sec x \tan x$$

and express $\tan^{m-1} x$ entirely in terms of $\sec^2 x$ using $\tan^2 x = \sec^2 x - 1$.

Example 5

$$\int \tan^5 x \sec^3 x \, dx = \int \tan^4 x \sec^2 x \sec x \tan x \, dx$$

$$= \int (\sec^2 x - 1)^2 \sec^2 x \sec x \tan x \, dx$$

$$= \int (\sec^6 x - 2 \sec^4 x + \sec^2 x) \sec x \tan x \, dx$$

$$= \tfrac{1}{7} \sec^7 x - \tfrac{2}{5} \sec^5 x + \tfrac{1}{3} \sec^3 x + C. \quad \square$$

C. n odd, m even.
Finally, if n is odd and m is even, use $\tan^2 x = \sec^2 x - 1$ to write the product as a sum of odd powers of the secant. Then you can either use integration by parts as in II just given, or the reduction formula (8.4.2).

Example 6

$$\int \tan^2 x \sec x \, dx = \int (\sec^2 x - 1) \sec x \, dx$$

$$= \int (\sec^3 x - \sec x) \, dx = \int \sec^3 x \, dx - \int \sec x \, dx.$$

We have calculated each of these integrals before:

$$\int \sec^3 x \, dx = \tfrac{1}{2} \sec x \tan x + \tfrac{1}{2} \ln|\sec x + \tan x| + C$$

and

$$\int \sec x \, dx = \ln|\sec x + \tan x| + C.$$

It follows that

$$\int \tan^2 x \sec x \, dx = \tfrac{1}{2} \sec x \tan x - \tfrac{1}{2} \ln|\sec x + \tan x| + C. \quad \square$$

You can handle integrals of $\cot^m x \csc^n x$ in a similar manner.

EXERCISES 8.4

In Exercises 1–26, evaluate the given integral.

1. $\int \tan^2 3x \, dx.$

2. $\int \cot^2 5x \, dx.$

3. $\int \sec^2 \pi x \, dx.$

4. $\int \csc^2 2x \, dx.$

5. $\int \tan^3 x \, dx.$

6. $\int \cot^3 x \, dx.$

7. $\int \tan^2 x \sec^2 x \, dx.$

8. $\int \cot^2 x \csc^2 x \, dx.$

9. $\int \csc^3 x \, dx.$

10. $\int \sec^3 \pi x \, dx.$

11. $\int \cot^4 x \, dx.$

12. $\int \tan^4 x \, dx.$

13. $\int \cot^3 x \csc^3 x \, dx.$

14. $\int \tan^3 x \sec^3 x \, dx.$

15. $\int \csc^4 2x \, dx.$

16. $\int \sec^4 3x \, dx.$

17. $\int \cot^2 x \csc x \, dx.$

18. $\int \csc^3 (\tfrac{1}{2}x) \, dx.$

19. $\int \tan^5 3x \, dx.$

20. $\int \cot^5 2x \, dx.$

21. $\int \sec^5 x \, dx.$

22. $\int \csc^5 x \, dx.$

23. $\int \tan^4 x \sec^4 x \, dx.$

24. $\int \cot^4 x \csc^4 x \, dx.$

25. $\int e^{2x} \tan^2(e^{2x}) \sec^2(e^{2x}) \, dx.$

26. $\int \tan^2 (\tfrac{1}{2}x) \sec^3 (\tfrac{1}{2}x) \, dx.$

In Exercises 27–32, evaluate the definite integral.

27. $\int_0^{\pi/4} \tan^3 x \sec^2 x \, dx.$

28. $\int_{\pi/4}^{\pi/2} \csc^3 x \cot x \, dx.$

29. $\int_0^{\pi/6} \tan^2 2x \, dx.$

30. $\int_0^{\pi/3} \tan x \sec^{3/2} x \, dx.$

31. $\int_{\pi/6}^{\pi/3} \cot^3 x \csc^3 x \, dx.$

32. $\int_{\pi/6}^{\pi/2} \cot^2 x \, dx.$

33. The region bounded by the y-axis, the line $y = 1$ and the graph of $y = \tan x$, $x \in [0, \pi/4]$, is revolved around the x-axis. Find the volume of the solid that is generated.

34. The region bounded by the graph of $y = \tan^2 x$ and the x-axis, $x \in [0, \pi/4]$, is revolved around the x-axis. Find the volume of the solid that is generated.

35. The region bounded by the graph of $y = \tan x$ and the x-axis, $x \in [0, \pi/4]$, is revolved around the line $y = -1$. Find the volume of the solid that is generated.

36. The region bounded by the graph of $y = \sec^2 x$ and the x-axis, $x \in [0, \pi/4]$, is revolved around the x-axis. Find the volume of the solid that is generated.

37. Derive the reduction formula: for $n > 1$,

$$\int \cot^n x \, dx = \frac{-\cot^{n-1} x}{n-1} - \int \cot^{n-2} x \, dx.$$

38. Use Exercise 37 to evaluate the following integrals:

(a) $\int \cot^3 x \, dx.$ (b) $\int \cot^4 x \, dx.$ (c) $\int \cot^5 2x \, dx.$

39. Derive the reduction formula: for $n > 1$,

$$\int \csc^n x \, dx = \frac{-\csc^{n-2} x \cot x}{n-1} + \frac{n-2}{n-1} \int \csc^{n-2} x \, dx.$$

40. Use Exercise 39 to evaluate the integrals:

(a) $\int \csc^3 x \, dx.$ (b) $\int \csc^4 x \, dx.$ (c) $\int \csc^5 3x \, dx.$

■ 8.5 INTEGRALS INVOLVING $\sqrt{a^2 \pm x^2}$ and $\sqrt{x^2 - a^2}$; TRIGONOMETRIC SUBSTITUTIONS

Integrals involving one of the forms $\sqrt{a^2 - x^2}$, $\sqrt{a^2 + x^2}$, or $\sqrt{x^2 - a^2}$ can often be simplified by making a trigonometric substitution. These substitutions transform the integrand into a trigonometric form like those studied in the two preceding sections. The three cases with their suggested substitutions are:

1. For $\sqrt{a^2 - x^2}$ set $a \sin u = x$.
2. For $\sqrt{a^2 + x^2}$ set $a \tan u = x$.
3. For $\sqrt{x^2 - a^2}$ set $a \sec u = x$.

In each case, take $a > 0$.

To illustrate the idea behind these substitutions, consider the first case: $\sqrt{a^2 - x^2}$. If we set $a \sin u = x$, then

$$\sqrt{a^2 - x^2} = \sqrt{a^2 - a^2\sin^2 u} = a\sqrt{1 - \sin^2 u} = a\sqrt{\cos^2 u} = a\,|\cos u|.$$

Since $a \sin u = x$, we have $u = \sin^{-1}(x/a)$, which implies that $-\pi/2 \le u \le \pi/2$ (see Section 7.8). Hence, $\cos u \ge 0$, and

$$\sqrt{a^2 - x^2} = a \cos u.$$

Thus, the substitution $a \sin u = x$ has the effect of replacing the radical $\sqrt{a^2 - x^2}$ by $a \cos u$. The other two cases can be illustrated in a similar manner.

Right triangles provide an easy way to relate the forms with their corresponding substitutions:

$\sqrt{a^2 - x^2}$ $\sin u = x/a$ or $x = a \sin u$
$\sqrt{a^2 - x^2} = a \cos u$

$\sqrt{a^2 + x^2}$ $\tan u = x/a$ or $x = a \tan u$
$\sqrt{a^2 + x^2} = a \sec u$

$\sqrt{x^2 - a^2}$ $\sec u = x/a$ or $x = a \sec u$
$\sqrt{x^2 - a^2} = a \tan u$

Example 1   Find

$$\int \frac{dx}{(9 - x^2)^{3/2}}.$$

SOLUTION   First note that

$$\int \frac{dx}{(9 - x^2)^{3/2}} = \int \frac{dx}{[\sqrt{9 - x^2}]^3};$$

the integral involves $\sqrt{9 - x^2}$. Therefore, we set

$$3 \sin u = x, \qquad 3 \cos u\, du = dx.$$

Then

$$\sqrt{9 - x^2} = 3 \cos u. \qquad \text{(Figure 8.5.1)}$$

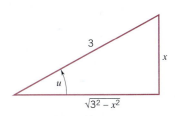

Figure 8.5.1

Now

$$\int \frac{dx}{(9-x^2)^{3/2}} = \int \frac{3\cos u \, du}{(3\cos u)^3}$$

$$= \tfrac{1}{9} \int \frac{\cos u}{\cos^3 u} \, du$$

$$= \tfrac{1}{9} \int \sec^2 u \, du$$

$$= \tfrac{1}{9} \tan u + C = \frac{x}{9\sqrt{9-x^2}} + C \quad \text{(see Figure 8.5.1).} \quad \square$$

Example 2 Find

$$\int \sqrt{a^2 + x^2} \, dx.$$

SOLUTION Set

$$a \tan u = x, \qquad a \sec^2 u \, du = dx.$$

Then

$$\sqrt{a^2 + x^2} = a \sec u. \qquad \text{(Figure 8.5.2)}$$

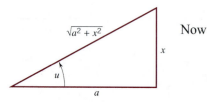

Figure 8.5.2

Now

$$\int \sqrt{a^2 + x^2} \, dx = \int (a \sec u) a \sec^2 u \, du$$

$$= a^2 \int \sec^3 u \, du$$

$$= \frac{a^2}{2} (\sec u \tan u + \ln|\sec u + \tan u|) + C$$

Example 3, Section 8.4 ⟶ $$= \frac{a^2}{2} \left[\frac{\sqrt{a^2 + x^2}}{a} \left(\frac{x}{a} \right) + \ln \left| \frac{\sqrt{a^2 + x^2}}{a} + \frac{x}{a} \right| \right] + C$$

$$= \tfrac{1}{2} x \sqrt{a^2 + x^2} + \tfrac{1}{2} a^2 \ln(x + \sqrt{a^2 + x^2}) - \tfrac{1}{2} a^2 \ln a + C$$

(see Figure 8.5.2).

We can absorb the constant $-\tfrac{1}{2} a^2 \ln a$ in C and write

(8.5.1) $$\int \sqrt{a^2 + x^2} \, dx = \tfrac{1}{2} x \sqrt{a^2 + x^2} + \tfrac{1}{2} a^2 \ln(x + \sqrt{a^2 + x^2}) + C.$$

This is a standard integration formula. □

Example 3 Find

$$\int \frac{dx}{x^2\sqrt{x^2 - 4}}.$$

SOLUTION Set

$$2 \sec u = x, \qquad 2 \sec u \tan u \, du = dx.$$

Then

$$\sqrt{x^2 - 4} = 2 \tan u. \qquad \text{(Figure 8.5.3)}$$

Now

Figure 8.5.3

$$\int \frac{dx}{x^2\sqrt{x^2 - 4}} = \int \frac{2 \sec u \tan u}{4 \sec^2 u \cdot 2 \tan u} \, du$$

$$= \tfrac{1}{4} \int \frac{1}{\sec u} \, du$$

$$= \tfrac{1}{4} \int \cos u \, du$$

$$= \tfrac{1}{4} \sin u + C = \frac{\sqrt{x^2 - 4}}{4x} + C \qquad \text{(see Figure 8.5.3).} \quad \square$$

Remark Consider the integral

$$\int \frac{x}{\sqrt{a^2 - x^2}} \, dx.$$

The appearance of $\sqrt{a^2 - x^2}$ in the integrand suggests the substitution $a \sin u = x$, and if you carry through with this substitution, you will arrive at the correct result. However, there is a much simpler approach here. Set

$$u = a^2 - x^2, \qquad du = -2x \, dx.$$

Then

$$\int \frac{x}{\sqrt{a^2 - x^2}} \, dx = -\tfrac{1}{2} \int \frac{1}{\sqrt{u}} \, du = -\sqrt{u} + C = -\sqrt{a^2 - x^2} + C.$$

You should be aware that there will sometimes be more than one way to calculate an integral. In such cases, it will obviously be to your advantage to choose the easiest method. Only experience and practice can give you the insight to do this consistently. ❑

Now a slight variation.

Example 4 Find

$$\int \frac{dx}{x\sqrt{4x^2 + 9}}.$$

SOLUTION Set

$$3 \tan u = 2x, \qquad 3 \sec^2 u \, du = 2 \, dx.$$

Then

$$\sqrt{4x^2 + 9} = 3 \sec u \qquad \text{(Figure 8.5.4)}$$

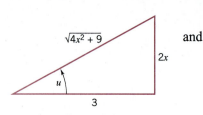

Figure 8.5.4

and

$$\int \frac{dx}{x\sqrt{4x^2 + 9}} = \int \frac{\frac{3}{2} \sec^2 u}{\frac{3}{2} \tan u \cdot 3 \sec u} \, du$$

$$= \frac{1}{3} \int \frac{\sec u}{\tan u} \, du$$

$$= \frac{1}{3} \int \csc u \, du$$

$$= \frac{1}{3} \ln|\csc u - \cot u| + C$$

$$= \frac{1}{3} \ln\left|\frac{\sqrt{4x^2 + 9} - 3}{2x}\right| + C \qquad \text{(see Figure 8.5.4).} \quad \square$$

The next example requires that we first complete the square under the radical.

Example 5 Find

$$\int \frac{x}{\sqrt{x^2 + 2x - 3}} \, dx.$$

SOLUTION First note that

$$\int \frac{x}{\sqrt{x^2 + 2x - 3}} \, dx = \int \frac{x}{\sqrt{(x + 1)^2 - 4}} \, dx.$$

Now set

$$2 \sec u = x + 1, \qquad 2 \sec u \tan u \, du = dx.$$

Then

$$\sqrt{(x + 1)^2 - 4} = 2 \tan u \qquad \text{(Figure 8.5.5)}$$

Figure 8.5.5

and

$$\int \frac{x}{\sqrt{(x + 1)^2 - 4}} \, dx = \int \frac{(2 \sec u - 1)2 \sec u \tan u}{2 \tan u} \, du$$

$$= \int (2 \sec^2 u - \sec u) \, du$$

$$= 2 \tan u - \ln|\sec u + \tan u| + C$$

$$= \sqrt{x^2 + 2x - 3} - \ln\left|\frac{x + 1 + \sqrt{x^2 + 2x - 3}}{2}\right| + C. \quad \square$$

Example 6 Find the area enclosed by the ellipse $(x^2/a^2) + (y^2/b^2) = 1$ (Figure 8.5.6).

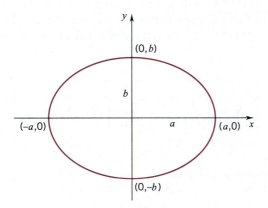

Figure 8.5.6

SOLUTION We first solve the equation of the ellipse for y:

$$y = \pm \frac{b}{a}\sqrt{a^2 - x^2}.$$

Since the graph is symmetric with respect to the x- and y-axes, the total area A enclosed by the ellipse is four times the area of the region in the first quadrant. This region is bounded by the graph of

$$y = \frac{b}{a}\sqrt{a^2 - x^2}, \qquad 0 \le x \le a.$$

Therefore

$$A = 4 \int_0^a \frac{b}{a}\sqrt{a^2 - x^2}\, dx = \frac{4b}{a} \int_0^a \sqrt{a^2 - x^2}\, dx.$$

Set

$$a \sin u = x, \qquad a \cos u\, du = dx.$$

Then

$$\sqrt{a^2 - x^2} = a \cos u.$$

Also, $u = 0$ at $x = 0$ and $u = \pi/2$ at $x = a$. Thus

$$A = \frac{4b}{a} \int_0^a \sqrt{a^2 - x^2}\, dx = \frac{4b}{a} \int_0^{\pi/2} (a \cos u) \cdot a \cos u\, du$$

$$= 4ab \int_0^{\pi/2} \cos^2 u\, du = 4ab \int_0^{\pi/2} (\tfrac{1}{2} + \tfrac{1}{2} \cos 2u)\, du$$

$$= 2ab \left[u + \tfrac{1}{2} \sin 2u \right]_0^{\pi/2} = \pi ab.$$

Hence, the area enclosed by an ellipse with semiaxes a and b is πab. Note that if $a = b$, then the ellipse is a circle of radius $r = a = b$ and the area is the familiar formula $A = \pi r^2$. ❑

Finally, we note that a trigonometric substitution may also be effective in cases where the quadratic in the integrand is not under a radical. For example, the reduction formula

(8.5.2)
$$\int \frac{dx}{(x^2 + a^2)^n} = \frac{1}{a^{2n-1}} \int \cos^{2(n-1)}u \, du$$

follows from the substitution: $a \tan u = x$, $a \sec^2 u \, du = dx$. The proof is left to you as an exercise.

Hyperbolic Substitutions

Integrals involving $\sqrt{a^2 + x^2}$ or $\sqrt{x^2 - a^2}$, $a > 0$, can also be handled using a hyperbolic substitution. The ideas here are the same as the ideas supporting trigonometric substitutions, and are based on the fundamental identity

$$\cosh^2 u - \sinh^2 u = 1, \qquad \text{for all real } u. \qquad \text{[see (7.9.3)]}$$

For an integral involving $\sqrt{a^2 + x^2}$, set

$$a \sinh u = x.$$

Then

$$\sqrt{a^2 + x^2} = \sqrt{a^2 + a^2 \sinh^2 u} = a\sqrt{1 + \sinh^2 u} = a\sqrt{\cosh^2 u} = a \cosh u$$

since $\cosh u > 0$ for all u.

For an integral involving $\sqrt{x^2 - a^2}$, where $x \geq a$, set

$$a \cosh u = x, \qquad u \geq 0.$$

Then

$$\sqrt{x^2 - a^2} = \sqrt{a^2\cosh^2 u - a^2} = a\sqrt{\cosh^2 u - 1} = a\sqrt{\sinh^2 u} = a \sinh u,$$

since $\sinh u \geq 0$ when $u \geq 0$.

Example 7 Use a hyperbolic substitution to find

$$\int \sqrt{a^2 + x^2} \, dx. \qquad \text{(see Example 2)}$$

SOLUTION Set $a \sinh u = x$. Then $dx = a \cosh u \, du$, $\sqrt{a^2 + x^2} = a \cosh u$, and

$$\int \sqrt{a^2 + x^2} \, dx = \int (a \cosh u)(a \cosh u) \, du$$

$$= a^2 \int \cosh^2 u \, du$$

$$= a^2 \int (\tfrac{1}{2} \cosh 2u + \tfrac{1}{2}) \, du \qquad \text{[see (7.9.3)]}$$

$$= \frac{a^2}{2} [\tfrac{1}{2} \sinh 2u + u] + C$$

$$= \frac{a^2}{2} [\sinh u \cosh u + u] + C \qquad\qquad \text{[see (7.9.3)]}$$

$$= \frac{a^2}{2} \left[\frac{x}{a} \cdot \frac{\sqrt{a^2 + x^2}}{a} + \sinh^{-1}\left(\frac{x}{a}\right) \right] + C$$

$$= \frac{a^2}{2} \left[\frac{x\sqrt{a^2 + x^2}}{a^2} + \ln\left(\frac{x}{a} + \frac{\sqrt{a^2 + x^2}}{a}\right) \right] + C \qquad \text{(see Theorem 7.10.2)}$$

$$= \tfrac{1}{2} x\sqrt{a^2 + x^2} + \tfrac{1}{2} a^2 \ln(x + \sqrt{a^2 + x^2}) + C$$

as we saw in Example 2. ❏

EXERCISES 8.5

In Exercises 1–44, evaluate the given integral.

1. $\displaystyle\int \frac{dx}{\sqrt{a^2 - x^2}}.$

2. $\displaystyle\int \frac{dx}{(x^2 + 2)^{3/2}}.$

3. $\displaystyle\int_0^1 \frac{dx}{(5 - x^2)^{3/2}}.$

4. $\displaystyle\int_{5/2}^4 \frac{x}{\sqrt{x^2 - 4}} \, dx.$

5. $\displaystyle\int \sqrt{x^2 - 1} \, dx.$

6. $\displaystyle\int \frac{x}{\sqrt{4 - x^2}} \, dx.$

7. $\displaystyle\int \frac{x^2}{\sqrt{4 - x^2}} \, dx.$

8. $\displaystyle\int \frac{x^2}{\sqrt{x^2 - 4}} \, dx.$

9. $\displaystyle\int \frac{x}{(1 - x^2)^{3/2}} \, dx.$

10. $\displaystyle\int \frac{x^2}{\sqrt{4 + x^2}} \, dx.$

11. $\displaystyle\int_0^{1/2} \frac{x^2}{(1 - x^2)^{3/2}} \, dx.$

12. $\displaystyle\int \frac{x}{a^2 + x^2} \, dx.$

13. $\displaystyle\int x\sqrt{4 - x^2} \, dx.$

14. $\displaystyle\int_0^2 \frac{x^3}{\sqrt{16 - x^2}} \, dx.$

15. $\displaystyle\int_0^5 x^2\sqrt{25 - x^2} \, dx.$

16. $\displaystyle\int \frac{e^x}{\sqrt{9 - e^{2x}}} \, dx.$

17. $\displaystyle\int \frac{x^2}{(x^2 + 8)^{3/2}} \, dx.$

18. $\displaystyle\int \frac{\sqrt{1 - x^2}}{x^4} \, dx.$

19. $\displaystyle\int \frac{dx}{x\sqrt{a^2 - x^2}}.$

20. $\displaystyle\int \frac{dx}{\sqrt{x^2 + a^2}}.$

21. $\displaystyle\int \frac{dx}{\sqrt{x^2 - a^2}}.$

22. $\displaystyle\int_{-a}^a \sqrt{a^2 - x^2} \, dx.$

23. $\displaystyle\int e^x\sqrt{e^{2x} - 1} \, dx.$

24. $\displaystyle\int_0^2 \frac{1}{\sqrt{4 + x^2}} \, dx.$

25. $\displaystyle\int_0^3 \frac{x^3}{\sqrt{9 + x^2}} \, dx.$

26. $\displaystyle\int \frac{\sqrt{x^2 - 1}}{x} \, dx.$

27. $\displaystyle\int \frac{dx}{x^2\sqrt{a^2 + x^2}}.$

28. $\displaystyle\int \frac{dx}{x^2\sqrt{a^2 - x^2}}.$

29. $\displaystyle\int \frac{dx}{x^2\sqrt{x^2 - a^2}}.$

30. $\displaystyle\int_4^6 \frac{1}{x\sqrt{x^2 - 4}} \, dx.$

31. $\displaystyle\int_2^{2\sqrt{2}} \frac{\sqrt{x^2 - 4}}{x} \, dx.$

32. $\displaystyle\int \frac{dx}{e^x\sqrt{4 + e^{2x}}}.$

33. $\displaystyle\int \frac{dx}{e^x\sqrt{e^{2x} - 9}}.$

34. $\displaystyle\int \frac{dx}{\sqrt{x^2 - 2x - 3}}.$

35. $\displaystyle\int \frac{dx}{(x^2 - 4x + 4)^{3/2}}.$

36. $\displaystyle\int \frac{x}{\sqrt{6x - x^2}} \, dx.$

37. $\displaystyle\int x\sqrt{6x - x^2 - 8} \, dx.$

38. $\displaystyle\int \frac{x + 2}{\sqrt{x^2 + 4x + 13}} \, dx.$

39. $\displaystyle\int \frac{x}{(x^2 + 2x + 5)^2} \, dx.$

40. $\displaystyle\int \frac{x}{\sqrt{x^2 - 2x - 3}} \, dx.$

41. $\displaystyle\int \frac{x + 3}{\sqrt{x^2 + 4x + 13}} \, dx.$

42. $\displaystyle\int x\sqrt{x^2 + 6x} \, dx.$

43. $\displaystyle\int \sqrt{6x - x^2 - 8} \, dx.$

44. $\displaystyle\int \frac{dx}{(4 - x^2)^2}.$

45. Verify the reduction formula (8.5.2).

In Exercises 46 and 47, use the reduction formula (8.5.2) to evaluate the given integral.

46. $\displaystyle\int \frac{1}{(x^2 + 1)^2} \, dx.$

47. $\displaystyle\int \frac{1}{(x^2 + 1)^3} \, dx.$

48. Calculate

$$\int x \sin^{-1}x \, dx.$$

49. (a) Use the trigonometric substitution $x = a\tan u$ to show that

$$\int \frac{1}{\sqrt{x^2 + a^2}} \, dx = \ln(x + \sqrt{x^2 + a^2}) + C_1.$$

(b) Use the hyperbolic substitution $x = a \sinh u$ to show that

$$\int \frac{1}{\sqrt{x^2 + a^2}} \, dx = \sinh^{-1}\left(\frac{x}{a}\right) + C_2.$$

(c) Justify the results in parts (a) and (b). HINT: see Section 7.10.

50. (a) Use a trigonometric substitution to show that

$$\int \sqrt{x^2 - a^2} \, dx =$$
$$\frac{1}{2}\left[x\sqrt{x^2 - a^2} - a^2\ln(x + \sqrt{x^2 - a^2}) \right] + C_1, \ x \geq a.$$

(b) Use a hyperbolic substitution to show that

$$\int \sqrt{x^2 - a^2} \, dx =$$
$$\frac{1}{2}\left[x\sqrt{x^2 - a^2} - a^2 \cosh^{-1}\left(\frac{x}{a}\right) \right] + C_2.$$

(c) Reconcile the results in parts (a) and (b).

51. The region bounded by the graph of $f(x) = 1/(1 + x^2)$ and the x-axis between $x = 0$ and $x = 1$ is revolved around the x-axis. Find the volume of the solid that is generated.

52. In a disc of radius r a chord h units from the center generates a region of the disc called a *segment* (see the figure). Find a formula for the area of the segment.

53. Derive the formula $A = \frac{1}{2}r^2\theta$ for the area of a sector of a circle of radius r and central angle θ (measured in radians). HINT: assume first that $0 < \theta < \frac{1}{2}\pi$ and subdivide the region as indicated in the figure. Then verify that the formula holds for any sector.

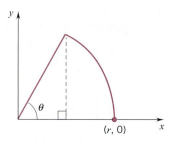

54. Find the area of the region bounded on the left and right by the two branches of the hyperbola $(x^2/a^2) - (y^2/b^2) = 1$, and above and below by the lines $y = \pm b$.

55. Calculate the mass and the center of mass of a rod that extends from $x = 0$ to $x = a > 0$ and has mass density $\lambda(x) = (x^2 + a^2)^{-1/2}$.

56. Calculate the mass and the center of mass of the rod in Exercise 55 if the mass density is given by $\lambda(x) = (x^2 + a^2)^{-3/2}$.

For Exercises 57–59, let Ω be the region under the curve $y = \sqrt{x^2 - a^2}$, $x \in [a, \sqrt{2}\, a]$.

57. Sketch Ω, find its area, and locate the centroid.

58. Find the volume of the solid generated by revolving Ω about the x-axis and determine the centroid of that solid.

59. Find the volume of the solid generated by revolving Ω about the y-axis and determine the centroid of that solid.

▶ 60. Let

$$f(x) = \frac{x^2}{\sqrt{1 - x^2}}.$$

(a) Use a graphing utility to sketch the graph of f.
(b) Find the area of the region bounded by the graph of f and the x-axis between $x = 0$ and $x = \frac{1}{2}$.
(c) Find the volume of the solid generated by revolving the region in part (b) around the y-axis.

▶ 61. Let

$$f(x) = \frac{\sqrt{x^2 - 9}}{x^2}, \quad x \geq 3.$$

(a) Use a graphing utility to sketch the graph of f.
(b) Find the area of the region bounded by the graph of f and the x-axis between $x = 3$ and $x = 6$.
(c) Find the centroid of the region.

■ 8.6 PARTIAL FRACTIONS

In this section we present a method for integrating rational functions. Recall that a rational function is, by definition, the quotient of two polynomials. For example,

$$\frac{1}{x^2 - 4}, \qquad \frac{2x^2 + 3}{x(x - 1)^2}, \qquad \frac{-2x}{(x + 1)(x^2 + 1)}, \qquad \frac{1}{x(x^2 + x + 1)},$$

$$\frac{3x^4 + x^3 + 20x^3 + 3x + 31}{(x + 1)(x^2 + 4)^2}, \qquad \frac{x^5}{x^2 - 1}$$

are all rational functions, but

$$\frac{1}{\sqrt{x}}, \qquad \ln x, \qquad \frac{|x - 2|}{x^2}$$

are not rational functions.

A rational function $R(x) = P(x)/Q(x)$ is said to be *proper* if the degree of the numerator is less than the degree of the denominator. If the degree of the numerator is greater than or equal to the degree of the denominator, then the rational function is *improper*.† We will focus our attention on *proper rational functions* because any improper rational function can be written as the sum of a polynomial and a proper rational function

$$\frac{P(x)}{Q(x)} = p(x) + \frac{r(x)}{Q(x)}.\text{††}$$

This is accomplished simply by dividing the denominator into the numerator (the polynomial p is called the quotient and the polynomial r is the remainder). Take, for example,

$$\frac{x^5 + 2}{x^2 - 1}.$$

Carrying out the suggested division, we have

$$
\begin{array}{r}
x^3 + x \\
x^2 - 1 \overline{)x^5 + 2} \\
\underline{x^5 - x^3} \\
x^3 \\
\underline{x^3 - x} \\
x + 2.
\end{array}
$$

This gives

$$\frac{x^5 + 2}{x^2 - 1} = x^3 + x + \frac{x + 2}{x^2 - 1}.$$

It is shown in algebra that every proper rational function can be written in one and only one way as a sum of fractions of the form

(8.6.1)

$$\boxed{\frac{A}{(x - \alpha)^k} \quad \text{and} \quad \frac{Bx + C}{(x^2 + \beta x + \gamma)^k}}$$

---

† These terms are taken from the familiar terms used to describe (numerical) fractions.

†† This is analogous to writing an improper fraction as a so-called *mixed number*.

with the quadratic $x^2 + \beta x + \gamma$ irreducible (that is, not factorable into linear factors with real coefficients; $\beta^2 - 4\gamma < 0$). Such fractions are called *partial fractions*. When a rational function $R(x)$ is expressed as a sum of partial fractions, the sum is called the *partial fraction decomposition of R*.

We begin by giving some examples of partial fraction decompositions. We then illustrate how these are used to calculate integrals.

Example 1 (*The denominator splits into distinct linear factors.*) The partial fraction decomposition of the rational function

$$\frac{1}{x^2 - 4} = \frac{1}{(x - 2)(x + 2)}$$

has the form

$$\frac{1}{x^2 - 4} = \frac{A}{x - 2} + \frac{B}{x + 2},$$

where A and B are constants whose values are to be determined. This is an identity that holds for all $x \neq 2, -2$. Clearing the fractions, we have

(1) $$1 = A(x + 2) + B(x - 2),$$

which is still an identity, but now it holds for all x. We illustrate two methods for finding A and B.

METHOD 1 Substitute numbers for x in (1):

Setting $x = 2$, we get $1 = 4A$, which gives $A = \frac{1}{4}$;

Setting $x = -2$, we get $1 = -4B$, which gives $B = -\frac{1}{4}$.

Thus, the desired decomposition is

$$\frac{1}{x^2 - 4} = \frac{1}{4(x - 2)} - \frac{1}{4(x + 2)}.$$

You can verify this by carrying out the subtraction on the right.

METHOD 2 Clear the parentheses on the right-hand side of (1) and rewrite the equation as

$$1 = (A + B)x + 2A - 2B.$$

Now, equate the coefficients of the corresponding powers of x to produce the system of equations

$$A + B = 0$$

$$2A - 2B = 1.$$

We can then get A and B by solving these equations simultaneously. The solutions are, of course, $A = \frac{1}{4}$, $B = -\frac{1}{4}$. ❏

In general, each distinct linear factor $x - \alpha$ in the denominator gives rise to a term of the form

$$\frac{A}{x - \alpha}.$$

Example 2 (*The denominator has a repeated linear factor.*) For

$$\frac{2x^2 + 3}{x(x - 1)^2},$$

we write

$$\frac{2x^2 + 3}{x(x - 1)^2} = \frac{A}{x} + \frac{B}{x - 1} + \frac{C}{(x - 1)^2}.$$

This leads to

$$2x^2 + 3 = A(x - 1)^2 + Bx(x - 1) + Cx.$$

To determine the three coefficients A, B, C we need to substitute three values for x. We select 0 and 1 because for those values of x several terms on the right side will drop out. As a third value of x, any other number will do; we select 2 just to keep the arithmetic simple.

Setting $x = 0$, we get $3 = A$.

Setting $x = 1$, we get $5 = C$.

Setting $x = 2$, we get $11 = A + 2B + 2C$,

which, with $A = 3$ and $C = 5$, gives $B = -1$.

The decomposition is therefore

$$\frac{2x^2 + 3}{x(x - 1)^2} = \frac{3}{x} - \frac{1}{x - 1} + \frac{5}{(x - 1)^2}. \quad ❑$$

Remark You can verify that the alternative method for finding the constants A, B, and C leads to the system of equations

$$A + B \qquad = 2$$
$$-2A - B + C = 0$$
$$A \qquad\qquad = 3.$$

We can then get A, B, and C by solving these equations simultaneously. In general, this approach involves more algebra, and so we will emphasize the method of substituting *well-chosen* values of x in the examples which follow. It is also possible to combine the two methods and, with practice, you will see that this is often a convenient approach. We will illustrate this in the next example. ❑

In general, each factor of the form $(x - \alpha)^k$ in the denominator gives rise to an expression of the form

$$\frac{A_1}{x - \alpha} + \frac{A_2}{(x - \alpha)^2} + \cdots + \frac{A_k}{(x - \alpha)^k}.$$

Example 3 (*The denominator has an irreducible quadratic factor.*) For

$$\frac{x^2 + 5x + 2}{(x + 1)(x^2 + 1)},$$

we write

$$\frac{x^2 + 5x + 2}{(x + 1)(x^2 + 1)} = \frac{A}{x + 1} + \frac{Bx + C}{x^2 + 1}$$

and obtain

$$x^2 + 5x + 2 = A(x^2 + 1) + (Bx + C)(x + 1).$$

This time we substitute $x = -1, 0,$ and 1.

Setting $x = -1$, we get $-2 = 2A$, which gives $A = -1$.

Setting $x = 0$, we get $2 = A + C$, which gives $C = 3$.

Setting $x = 1$, we get $8 = 2A + 2B + 2C$,

which, with $A = -1$ and $C = 3$, gives $B = 2$.

Alternatively, after substituting -1 and 0, we could have noted that the coefficient of x^2 on the right-hand side is $A + B$ and on the left-hand side it's 1. Thus, $A + B = 1$ and $B = 1 - (-1) = 2$. This illustrates the "combined" approach. The decomposition reads

$$\frac{x^2 + 5x + 2}{(x + 1)(x^2 + 1)} = \frac{-1}{x + 1} + \frac{2x + 3}{x^2 + 1}. \quad \square$$

Example 4 (*The denominator has an irreducible quadratic factor.*) For

$$\frac{1}{x(x^2 + x + 1)}$$

we write

$$\frac{1}{x(x^2 + x + 1)} = \frac{A}{x} + \frac{Bx + C}{x^2 + x + 1}$$

and obtain

$$1 = A(x^2 + x + 1) + (Bx + C)x.$$

Again we select values of x that produce zeros or simple arithmetic on the right side.

$$1 = A \qquad (x = 0),$$
$$1 = 3A + B + C \qquad (x = 1),$$
$$1 = A + B - C \qquad (x = -1).$$

From this we find that

$$A = 1, \qquad B = -1, \qquad C = -1,$$

and therefore

$$\frac{1}{x(x^2 + x + 1)} = \frac{1}{x} - \frac{x + 1}{x^2 + x + 1}. \quad \square$$

In general, each irreducible quadratic factor $x^2 + \beta x + \gamma$ in the denominator gives rise to a term of the form

$$\frac{Ax + B}{x^2 + \beta x + \gamma}.$$

Example 5 (*The denominator has a repeated irreducible quadratic factor.*) For

$$\frac{3x^4 + x^3 + 20x^2 + 3x + 31}{(x + 1)(x^2 + 4)^2}$$

we write

$$\frac{3x^4 + x^3 + 20x^2 + 3x + 31}{(x + 1)(x^2 + 4)^2} = \frac{A}{x + 1} + \frac{Bx + C}{x^2 + 4} + \frac{Dx + E}{(x^2 + 4)^2}.$$

This gives

$$3x^4 + x^3 + 20x^2 + 3x + 31$$
$$= A(x^2 + 4)^2 + (Bx + C)(x + 1)(x^2 + 4) + (Dx + E)(x + 1).$$

This time we use $-1, 0, 1, 2,$ and -2.

$$50 = 25A \qquad\qquad\qquad\qquad (x = -1),$$
$$31 = 16A \qquad + 4C \qquad + E \qquad (x = 0),$$
$$58 = 25A + 10B + 10C + 2D + 2E \qquad (x = 1),$$
$$173 = 64A + 48B + 24C + 6D + 3E \qquad (x = 2),$$
$$145 = 64A + 16B - 8C + 2D - E \qquad (x = -2).$$

With a little patience you can see that

$$A = 2, \qquad B = 1, \qquad C = 0, \qquad D = 0, \qquad \text{and} \qquad E = -1.$$

This gives the decomposition

$$\frac{3x^4 + x^3 + 20x^2 + 3x + 31}{(x + 1)(x^2 + 4)^2} = \frac{2}{x + 1} + \frac{x}{x^2 + 4} - \frac{1}{(x^2 + 4)^2}. \quad \Box$$

In general, each multiple irreducible quadratic factor $(x^2 + \beta x + \gamma)^k$ in the denominator gives rise to an expression of the form

$$\frac{A_1 x + B_1}{x^2 + \beta x + \gamma} + \frac{A_2 x + B_2}{(x^2 + \beta x + \gamma)^2} + \cdots + \frac{A_k x + B_k}{(x^2 + \beta x + \gamma)^k}.$$

As indicated at the beginning of this section, if the rational function is improper, then a polynomial will appear in the decomposition.

Example 6 (*The decomposition contains a polynomial.*) For the improper rational function introduced at the beginning of this section:

$$\frac{x^5 + 2}{x^2 - 1},$$

we saw that

$$\frac{x^5 + 2}{x^2 - 1} = x^3 + x + \frac{x + 2}{x^2 - 1}.$$

Since the denominator of the fraction splits into linear factors, we write

$$\frac{x + 2}{x^2 - 1} = \frac{A}{x + 1} + \frac{B}{x - 1}.$$

This gives

$$x + 2 = A(x - 1) + B(x + 1).$$

Substitution of $x = 1$ gives $B = \frac{3}{2}$; substitution of $x = -1$ gives $A = -\frac{1}{2}$. The decomposition takes the form

$$\frac{x^5 + 2}{x^2 - 1} = x^3 + x - \frac{1}{2(x + 1)} + \frac{3}{2(x - 1)}. \quad \square$$

We have been decomposing rational functions into partial fractions so as to be able to integrate them. Here we carry out the integrations, leaving some of the details to you.

Example 1′

$$\int \frac{dx}{x^2 - 4} = \frac{1}{4} \int \left(\frac{1}{x - 2} - \frac{1}{x + 2} \right) dx$$

$$= \frac{1}{4} (\ln |x - 2| - \ln |x + 2|) + C = \frac{1}{4} \ln \left| \frac{x - 2}{x + 2} \right| + C. \quad \square$$

Example 2′

$$\int \frac{2x^2 + 3}{x(x - 1)^2} dx = \int \left[\frac{3}{x} - \frac{1}{x - 1} + \frac{5}{(x - 1)^2} \right] dx$$

$$= 3 \ln |x| - \ln |x - 1| - \frac{5}{x - 1} + C$$

$$= \ln \left| \frac{x^3}{x - 1} \right| - \frac{5}{x - 1} + C \quad \square$$

Example 3′

$$\int \frac{x^2 + 5x + 2}{(x + 1)(x^2 + 1)} dx = \int \left(\frac{-1}{x + 1} + \frac{2x + 3}{x^2 + 1} \right) dx = -\int \frac{1}{x + 1} dx + \int \frac{2x + 3}{x^2 + 1} dx.$$

Since

$$-\int \frac{1}{x + 1} dx = -\ln |x + 1| + C_1$$

and

$$\int \frac{2x + 3}{x^2 + 1}\, dx = \int \frac{2x}{x^2 + 1}\, dx + 3 \int \frac{1}{x^2 + 1}\, dx = \ln(x^2 + 1) + 3 \tan^{-1}x + C_2,$$

we have

$$\int \frac{x^2 + 5x + 2}{(x + 1)(x^2 + 1)}\, dx = -\ln|x + 1| + \ln(x^2 + 1) + 3 \tan^{-1}x + C$$

$$= \ln \left| \frac{x^2 + 1}{x + 1} \right| + 3 \tan^{-1}x + C. \quad \square$$

Example 4′

$$P = \int \frac{dx}{x(x^2 + x + 1)} = \int \left(\frac{1}{x} - \frac{x + 1}{x^2 + x + 1} \right) dx = \ln|x| - \int \frac{x + 1}{x^2 + x + 1}\, dx.$$

To compute the remaining integral, note that $(d/dx)(x^2 + x + 1) = 2x + 1$, and so we manipulate the integrand to get a term of the form du/u, where $u = x^2 + x + 1$ and $du = (2x + 1)\, dx$:

$$\frac{x + 1}{x^2 + x + 1} = \frac{\frac{1}{2}[2x + 1] + \frac{1}{2}}{x^2 + x + 1} = \frac{1}{2} \frac{2x + 1}{x^2 + x + 1} + \frac{1}{2} \frac{1}{x^2 + x + 1}.$$

Therefore,

$$\int \frac{x + 1}{x^2 + x + 1}\, dx = \frac{1}{2} \int \frac{2x + 1}{x^2 + x + 1}\, dx + \frac{1}{2} \int \frac{1}{x^2 + x + 1}\, dx.$$

Now

$$\frac{1}{2} \int \frac{2x + 1}{x^2 + x + 1}\, dx = \frac{1}{2} \ln(x^2 + x + 1) + C_1. \quad \text{(NOTE: } x^2 + x + 1 > 0 \text{ for all } x.\text{)}$$

You can verify that the second integral is an inverse tangent:

$$\frac{1}{2} \int \frac{dx}{x^2 + x + 1} = \frac{1}{2} \int \frac{dx}{(x + \frac{1}{2})^2 + (\sqrt{3}/2)^2} = \frac{1}{\sqrt{3}} \tan^{-1} \left[\frac{2}{\sqrt{3}} \left(x + \frac{1}{2} \right) \right] + C_2.$$

Combining results, we have

$$P = \ln|x| - \frac{1}{2} \ln(x^2 + x + 1) - \frac{1}{\sqrt{3}} \tan^{-1} \left[\frac{2}{\sqrt{3}} \left(x + \frac{1}{2} \right) \right] + C. \quad \square$$

Example 5′

$$\int \frac{3x^4 + x^3 + 20x^2 + 3x + 31}{(x + 1)(x^2 + 4)^2}\, dx = \int \left[\frac{2}{x + 1} + \frac{x}{x^2 + 4} - \frac{1}{(x^2 + 4)^2} \right] dx.$$

The first two fractions are easy to integrate:

$$\int \frac{2}{x + 1}\, dx = 2 \ln|x + 1| + C_1,$$

$$\int \frac{x}{x^2 + 4}\, dx = \frac{1}{2} \int \frac{2x}{x^2 + 4}\, dx = \frac{1}{2} \ln(x^2 + 4) + C_2.$$

The integral of the last fraction is of the form

$$\int \frac{dx}{(x^2 + a^2)^n}.$$

As we saw in the preceding section, such integrals can be calculated by using the trigonometric substitution $a \tan u = x$ [see (8.5.2)]. Here we have

$$\int \frac{dx}{(x^2 + 4)^2} = \frac{1}{8} \int \cos^2 u \, du$$

$$\overline{} \; 2 \tan u = x$$

$$= \frac{1}{16} \int (1 + \cos 2u) \, du$$

half-angle formula

$$= \frac{1}{16} u + \frac{1}{32} \sin 2u + C_3$$

$$= \frac{1}{16} u + \frac{1}{16} \sin u \cos u + C_3$$

$\sin 2u = 2 \sin u \cos u$

$$= \frac{1}{16} \tan^{-1} \frac{x}{2} + \frac{1}{16} \left(\frac{x}{\sqrt{x^2 + 4}} \right) \left(\frac{2}{\sqrt{x^2 + 4}} \right) + C_3$$

$$= \frac{1}{16} \tan^{-1} \frac{x}{2} + \frac{1}{8} \left(\frac{x}{x^2 + 4} \right) + C_3.$$

The integral we want is therefore equal to

$$2 \ln |x + 1| + \frac{1}{2} \ln(x^2 + 4) - \frac{1}{8} \left(\frac{x}{x^2 + 4} \right) - \frac{1}{16} \tan^{-1} \frac{x}{2} + C. \quad \square$$

Example 6′

$$\int \frac{x^5 + 2}{x^2 - 1} \, dx = \int \left[x^3 + x - \frac{1}{2(x + 1)} + \frac{3}{2(x - 1)} \right] dx$$

$$= \tfrac{1}{4} x^4 + \tfrac{1}{2} x^2 - \tfrac{1}{2} \ln |x + 1| + \tfrac{3}{2} \ln |x - 1| + C. \quad \square$$

Partial fractions come up in a variety of applications. Here is an example from the life sciences.

Example 7 A theory used by scientists to describe the spread of a disease through a population states that the rate at which the disease spreads is proportional to the product of the number of individuals that are infected and the number that are not. If M denotes the total number of individuals in the population and $y(t)$ is the number of individuals that are infected at time t, then the mathematical formulation of the theory is

(8.6.2) $$\frac{dy}{dt} = ky(M - y),$$

where $k > 0$ is the constant of proportionality. Note that the term "population" being used here does not necessarily mean "people." We might want to study the spread of a disease through a population of animals, or through a population of plants. Equation (8.6.2) is known as the *logistic equation* and its solutions are called *logistic functions*. The equation is a separable differential equation, and the solutions are found by "separating the variables" and integrating (see Section 7.6):

$$\frac{dy}{y(M - y)} = k \, dt$$

$$\int \frac{dy}{y(M - y)} = \int k \, dt = kt + C, \qquad C \text{ an arbitrary constant.}$$

The method of partial fractions is needed to calculate the integral on the left-hand side:

$$\frac{1}{y(M - y)} = \frac{A}{y} + \frac{B}{M - y}$$

$$1 = A(M - y) + By.$$

Setting $y = 0$ and $y = M$, we find that $A = B = 1/M$. Thus

$$\int \frac{dy}{y(M - y)} = \frac{1}{M} \int \frac{dy}{y} + \frac{1}{M} \int \frac{dy}{M - y} = \frac{1}{M}[\ln y - \ln(M - y)] = \frac{1}{M} \ln\left(\frac{y}{M - y}\right)$$

(note: absolute values are not needed since y and $M - y$ are both positive) and

$$\frac{1}{M} \ln\left(\frac{y}{M - y}\right) = kt + C.$$

Multiplying by M, we get

$$\ln\left(\frac{y}{M - y}\right) = Mkt + MC$$

or

$$\ln\left(\frac{y}{M - y}\right) = Mkt + C_1 \qquad (C_1 = MC \text{ is arbitrary}).$$

Taking the exponential of both sides of this equation, we find that

$$\frac{y}{M - y} = e^{Mkt + C_1} = e^{C_1} e^{Mkt} = C_2 e^{Mkt}$$

If $y(0) = R$ individuals have the disease initially, then $C_2 = R/(M - R)$ and

$$\frac{y}{M - y} = \frac{R}{M - R} e^{Mkt}.$$

This equation can be solved for y. The result is

$$y(t) = \frac{MR}{R + (M - R)\, e^{-Mkt}}$$

and this gives the number of individuals that are infected at time t. The graph of y is shown in Figure 8.6.1. Note that the disease spreads rapidly in the beginning, and then "levels off." Note, also, that $y(t) \to M$ as $t \to \infty$, so that everyone will get the disease eventually unless something is introduced to stop it. ❏

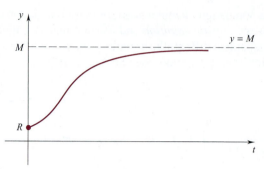

Figure 8.6.1

EXERCISES 8.6

In Exercises 1–8, determine the partial fraction decomposition of the given rational function.

1. $r(x) = \dfrac{1}{x^2 + 7x + 6}$.

2. $R(x) = \dfrac{x^2}{(x - 1)(x^2 + 4x + 5)}$.

3. $r(x) = \dfrac{x}{x^4 - 1}$. **4.** $R(x) = \dfrac{x^4}{(x - 1)^3}$.

5. $r(x) = \dfrac{x^2 - 3x - 1}{x^3 + x^2 - 2x}$.

6. $R(x) = \dfrac{x^3 + x^2 + x + 2}{x^4 + 3x^2 + 2}$.

7. $r(x) = \dfrac{2x^2 + 1}{x^3 - 6x^2 + 11x - 6}$.

8. $R(x) = \dfrac{1}{x(x^2 + 1)^2}$.

In Exercises 9–30, evaluate the given integral.

9. $\displaystyle \int \frac{7}{(x - 2)(x + 5)}\, dx$.

10. $\displaystyle \int \frac{x}{(x + 1)(x + 2)(x + 3)}\, dx$.

11. $\displaystyle \int \frac{2x^2 + 3}{x^2(x - 1)}\, dx$. **12.** $\displaystyle \int \frac{x^2 + 1}{x(x^2 - 1)}\, dx$.

13. $\displaystyle \int \frac{x^5}{(x - 2)^2}\, dx$. **14.** $\displaystyle \int \frac{x^5}{x - 2}\, dx$.

15. $\displaystyle \int \frac{x + 3}{x^2 - 3x + 2}\, dx$. **16.** $\displaystyle \int \frac{x^2 + 3}{x^2 - 3x + 2}\, dx$.

17. $\displaystyle \int \frac{dx}{(x - 1)^3}$. **18.** $\displaystyle \int \frac{dx}{x^2 + 2x + 2}$.

19. $\displaystyle \int \frac{x^2}{(x - 1)^2(x + 1)}\, dx$. **20.** $\displaystyle \int \frac{2x - 1}{(x + 1)^2(x - 2)^2}\, dx$.

21. $\displaystyle \int \frac{dx}{x^4 - 16}$. **22.** $\displaystyle \int \frac{x}{x^3 - 1}\, dx$.

23. $\displaystyle \int \frac{x^3 + 4x^2 - 4x - 1}{(x^2 + 1)^2}\, dx$.

24. $\displaystyle \int \frac{dx}{(x^2 + 16)^2}$. **25.** $\displaystyle \int \frac{dx}{x^4 + 4}$. †

26. $\displaystyle \int \frac{dx}{x^4 + 16}$. † **27.** $\displaystyle \int \frac{x - 3}{x^3 + x^2}\, dx$.

28. $\displaystyle \int \frac{1}{(x - 1)(x^2 + 1)^2}\, dx$. **29.** $\displaystyle \int \frac{x + 1}{x^3 + x^2 - 6x}\, dx$.

30. $\displaystyle \int \frac{x^3 + x^2 + x + 3}{(x^2 + 1)(x^2 + 3)}\, dx$.

In Exercises 31–36, evaluate the given definite integral.

31. $\displaystyle \int_0^2 \frac{x}{x^2 + 5x + 6}\, dx$. **32.** $\displaystyle \int_1^3 \frac{1}{x^3 + x}\, dx$.

33. $\displaystyle \int_3^6 \frac{2x}{x^3 - 2x^2 - 4x + 8}\, dx$.

34. $\displaystyle \int_2^4 \frac{x^4 - x^3 - x - 1}{x^3 - x^2}\, dx$.

35. $\displaystyle \int_1^3 \frac{x^2 - 4x + 3}{x^3 + 2x^2 + x}\, dx$. **36.** $\displaystyle \int_0^2 \frac{x^3}{(x^2 + 2)^2}\, dx$.

37. Show that if $y = \dfrac{1}{x^2 - 1}$, then

$$\frac{d^n y}{dx^n} = \frac{(-1)^n n!}{2}\left[\frac{1}{(x - 1)^{n+1}} - \frac{1}{(x + 1)^{n+1}}\right].$$

38. Calculate

$$\int x^3 \tan^{-1} x\, dx.$$

---

† HINT: With $a > 0$, $x^4 + a^2 = (x^2 + \sqrt{2a}\,x + a)(x^2 - \sqrt{2a}\,x + a)$.

39. Find the centroid of the region under the curve $y = (x^2 + 1)^{-1}$, $x \in [0, 1]$.

40. Find the centroid of the solid generated by revolving the region of Exercise 39 about: (a) the x-axis; (b) the y-axis.

 41. Let

$$f(x) = \frac{x}{x^2 + 5x + 6}.$$

(a) Use a graphing utility to sketch the graph of f.
(b) Calculate the area of the region bounded by the graph of f and the x-axis between $x = 0$ and $x = 4$.

42. (a) The region in Exercise 41 is revolved around the y-axis. Find the volume of the solid that is generated.
(b) Find the centroid of the solid in part (a).

43. Let

$$f(x) = \frac{9 - x}{(x + 3)^2}.$$

(a) Use a graphing utility to sketch the graph of f.
(b) Find the area of the region bounded by the graph of f and the x-axis between $x = -2$ and $x = 9$.

44. (a) The region in Exercise 43 is revolved around the x-axis. Find the volume of the solid that is generated.
(b) Find the centroid of the solid in part (a).

45. Suppose that a chemical A combines with a chemical B to form a compound C. In addition, suppose that the rate at which C is produced at time t varies directly with the amounts of A and B present at time t. With this model, if A_0 grams of A are mixed with B_0 grams of B, then

$$\frac{dC}{dt} = k(A_0 - C)(B_0 - C).$$

(a) Find the amount of compound C present at time t if $A_0 = B_0$.
(b) Find the amount of compound C present at time t if $A_0 \neq B_0$.

46. A mathematical model for the growth of a certain strain of bacteria is

$$\frac{dP}{dt} = 0.0020P(800 - P),$$

where $P = P(t)$ denotes the number of bacteria present at time t. There are 100 bacteria present initially.
(a) Find the population $P = P(t)$ at any time t.
(b) Use a graphing utility to sketch the graph of P and dP/dt.
(c) Approximate the time at which the bacteria culture experiences its most rapid growth rate. That is, locate the maximum value of dP/dt. Use three decimal place accuracy. What point on the graph of P corresponds to the maximum growth rate?

47. When an object of mass m is moving through air or a viscous medium, it is acted upon by a frictional force that acts in a direction opposite to its motion. This frictional force depends on the velocity of the object and (within close approximation) is given by

$$F(v) = -\alpha v - \beta v^2,$$

where α and β are positive constants.
(a) From Newton's second law, $F = ma$, we have

$$m\frac{dv}{dt} = -\alpha v - \beta v^2.$$

Solve this differential equation to find $v = v(t)$.
(b) Find v if the object has initial velocity $v(0) = v_0$.
(c) What is $\lim\limits_{t \to \infty} v(t)$?

48. A descending parachutist is acted on by two forces: a constant downward force mg and the upward force of air resistance, which (within close approximation) is of the form $-\beta v^2$ where β is a positive constant. (In this problem we are taking the downward direction as positive.)
(a) Express t in terms of the velocity v, the initial velocity v_0, and the constant $v_c = \sqrt{mg/\beta}$.
(b) Express v as a function of t.
(c) Express the acceleration a as a function of t. Verify that the acceleration never changes sign and in time tends to zero.
(d) Show that in time v tends to v_c. (This number v_c is called the *terminal velocity*.)

■ 8.7 SOME RATIONALIZING SUBSTITUTIONS

In this section we discuss two types of substitutions that lead to an integral of a rational function. Such substitutions are known as *rationalizing substitutions*.

First we consider integrals in which the integrand involves an expression of the form $\sqrt[n]{f(x)}$ for some function f. In such cases, the substitution $u = \sqrt[n]{f(x)}$, which is equivalent to $u^n = f(x)$, is sometimes effective. The idea behind this substitution is that it will replace fractional exponents with integer exponents; integer exponents are, in general, easier to handle.

Example 1 Find

$$\int \frac{dx}{1 + \sqrt{x}}.$$

SOLUTION To rationalize the integrand, we set

$$u^2 = x, \qquad 2u\, du = dx.$$

Then $u = \sqrt{x}$ and

$$\int \frac{dx}{1 + \sqrt{x}} = \int \frac{2u}{1 + u}\, du = \int \left(2 - \frac{2}{1 + u}\right) du$$

divide

$$= 2u - 2\ln|1 + u| + C$$
$$= 2\sqrt{x} - 2\ln|1 + \sqrt{x}| + C. \quad \square$$

Example 2 Find

$$\int \frac{dx}{\sqrt[3]{x} + \sqrt{x}}.$$

SOLUTION Here the integrand contains two distinct roots, $x^{1/3}$ and $x^{1/2}$. If we set $u^6 = x$, then both terms will be rationalized simultaneously:

$$u^6 = x, \qquad 6u^5\, du = dx, \qquad \text{and} \qquad x^{1/3} = u^2, \qquad x^{1/2} = u^3.$$

Thus, we have

$$\int \frac{dx}{\sqrt[3]{x} + \sqrt{x}} = \int \frac{6u^5}{u^2 + u^3}\, du = 6\int \frac{u^3}{1 + u}\, du$$

$$= 6\int \left(u^2 - u + 1 - \frac{1}{1 + u}\right) du$$

divide

$$= 6(\tfrac{1}{3}u^3 - \tfrac{1}{2}u^2 + u - \ln|1 + u|) + C$$
$$= 2\sqrt{x} - 3\sqrt[3]{x} + 6\sqrt[6]{x} - 6\ln|1 + \sqrt[6]{x}| + C. \quad \square$$

Example 3 Find

$$\int \sqrt{1 - e^x}\, dx.$$

SOLUTION To rationalize the integrand we set

$$u^2 = 1 - e^x.$$

To find dx in terms of u and du we solve the equation for x:

$$1 - u^2 = e^x, \qquad \ln(1 - u^2) = x, \qquad -\frac{2u}{1 - u^2}\, du = dx.$$

The rest is straightforward:

$$\int \sqrt{1 - e^x}\, dx = \int u \left(-\frac{2u}{1 - u^2} \right) du$$

$$= \int \frac{2u^2}{u^2 - 1}\, du = \int \left(2 + \frac{1}{u - 1} - \frac{1}{u + 1} \right) du$$

divide, then
partial fractions

$$= 2u + \ln|u - 1| - \ln|u + 1| + C$$

$$= 2u + \ln \left| \frac{u - 1}{u + 1} \right| + C$$

$$= 2\sqrt{1 - e^x} + \ln \left| \frac{\sqrt{1 - e^x} - 1}{\sqrt{1 - e^x} + 1} \right| + C. \quad \square$$

The second type of rationalizing substitution is applied to integrands which are rational expressions in sine and cosine. For example, suppose we want to calculate

$$\int \frac{\cos x}{\sin x + \sin^2 x}\, dx.$$

To convert a rational expression in sine and cosine to a rational function in u, we use the substitution

$$u = \tan \left(\frac{x}{2} \right), \qquad -\pi < x < \pi.$$

Then

$$\cos \left(\frac{x}{2} \right) = \frac{1}{\sec(x/2)} = \frac{1}{\sqrt{1 + \tan^2(x/2)}} = \frac{1}{\sqrt{1 + u^2}},$$

and

$$\sin \left(\frac{x}{2} \right) = \cos \left(\frac{x}{2} \right) \tan \left(\frac{x}{2} \right) = \frac{u}{\sqrt{1 + u^2}}.$$

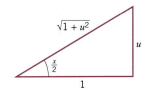

Figure 8.7.1

The right triangle in Figure 8.7.1 illustrates these relationships for $0 < x < \pi$.
Now,

$$\sin x = 2 \sin \left(\frac{x}{2} \right) \cos \left(\frac{x}{2} \right) = \frac{2u}{1 + u^2},$$

and

$$\cos x = \cos^2 \left(\frac{x}{2} \right) - \sin^2 \left(\frac{x}{2} \right) = \frac{1 - u^2}{1 + u^2}.$$

Also, since $u = \tan(x/2)$, $x = 2 \tan^{-1} u$ and so

$$dx = \frac{2}{1 + u^2}\, du.$$

To summarize, if an integrand is a rational expression in sine and cosine, then the substitution

$$\sin x = \frac{2u}{1 + u^2}, \qquad \cos x = \frac{1 - u^2}{1 + u^2}, \qquad dx = \frac{2}{1 + u^2}\, du,$$

where $u = \tan(x/2)$, $-\pi < x < \pi$, will convert the integrand into a rational function in u. The result can then be calculated using the methods of Section 8.6.

Example 4 Find

$$\int \frac{\cos x}{\sin x + \sin^2 x} dx.$$

SOLUTION If we let $u = \tan\left(\dfrac{x}{2}\right)$, then

$$\frac{\cos x}{\sin x + \sin^2 x} = \frac{(1 - u^2)/(1 + u^2)}{[2u/(1 + u^2)] + [4u^2/(1 + u^2)^2]} = \frac{(1 - u^2)(1 + u^2)}{2u^3 + 4u^2 + 2u}$$

$$= \frac{(1 - u)(1 + u^2)}{2u(1 + u)}$$

and

$$\int \frac{\cos x}{\sin x + \sin^2 x} dx = \int \frac{(1 - u)(1 + u^2)}{2u(1 + u)} \cdot \frac{2}{1 + u^2} du = \int \frac{1 - u}{u(1 + u)} du.$$

Now

$$\frac{1 - u}{u(1 + u)} = \frac{1}{u} - \frac{2}{1 + u}, \qquad \text{(partial fractions)}$$

and so

$$\int \frac{1 - u}{u(1 + u)} du = \int \frac{1}{u} du - 2 \int \frac{1}{1 + u} du$$

$$= \ln|u| - 2 \ln|1 + u| + C$$

$$= \ln\left|\frac{u}{(1 + u)^2}\right| + C$$

$$= \ln\left|\frac{\tan(x/2)}{[1 + \tan(x/2)]^2}\right| + C. \quad \square$$

EXERCISES 8.7

In Exercises 1–30, evaluate the given integral.

1. $\displaystyle\int \frac{dx}{1 - \sqrt{x}}.$

2. $\displaystyle\int \frac{\sqrt{x}}{1 + x} dx.$

3. $\displaystyle\int \sqrt{1 + e^x}\, dx.$

4. $\displaystyle\int \frac{dx}{x(x^{1/3} - 1)}.$

5. $\displaystyle\int x\sqrt{1 + x}\, dx.$ [(a) set $u^2 = 1 + x$; (b) set $u = 1 + x$]

6. $\displaystyle\int x^2\sqrt{1 + x}\, dx.$ [(a) set $u^2 = 1 + x$; (b) set $u = 1 + x$]

7. $\displaystyle\int (x + 2)\sqrt{x - 1}\, dx.$

8. $\displaystyle\int (x - 1)\sqrt{x + 2}\, dx.$

9. $\displaystyle\int \frac{x^3}{(1 + x^2)^3} dx.$

10. $\displaystyle\int x(1 + x)^{1/3}\, dx.$

11. $\displaystyle\int \frac{\sqrt{x}}{\sqrt{x} - 1} dx.$

12. $\displaystyle\int \frac{x}{\sqrt{x} + 1} dx.$

13. $\displaystyle\int \frac{\sqrt{x - 1} + 1}{\sqrt{x - 1} - 1} dx.$

14. $\displaystyle\int \frac{1 - e^x}{1 + e^x} dx.$

15. $\displaystyle\int \frac{dx}{\sqrt{1 + e^x}}.$

16. $\displaystyle\int \frac{dx}{1 + e^{-x}}.$

17. $\displaystyle\int \frac{x}{\sqrt{x + 4}} dx.$

18. $\displaystyle\int \frac{x + 1}{x\sqrt{x - 2}} dx.$

19. $\int 2x^2(4x + 1)^{-5/2} \, dx.$

20. $\int x^2\sqrt{x - 1} \, dx.$

21. $\int \dfrac{x}{(ax + b)^{3/2}} \, dx.$

22. $\int \dfrac{x}{\sqrt{ax + b}} \, dx.$

23. $\int \dfrac{1}{1 + \cos x - \sin x} \, dx.$

24. $\int \dfrac{1}{2 + \cos x} \, dx.$

25. $\int \dfrac{1}{2 + \sin x} \, dx.$

26. $\int \dfrac{\sin x}{1 + \sin^2 x} \, dx.$

27. $\int \dfrac{1}{\sin x + \tan x} \, dx.$

28. $\int \dfrac{1}{1 + \sin x + \cos x} \, dx.$

29. $\int \dfrac{1 - \cos x}{1 + \sin x} \, dx.$

30. $\int \dfrac{1}{5 + 3 \sin x} \, dx.$

In Exercises 31–36, evaluate the definite integral.

31. $\displaystyle\int_0^4 \dfrac{x^{3/2}}{x + 1} \, dx.$

32. $\displaystyle\int_0^8 \dfrac{1}{1 + \sqrt[3]{x}} \, dx.$

33. $\displaystyle\int_0^{\pi/2} \dfrac{\sin 2x}{2 + \cos x} \, dx.$

34. $\displaystyle\int_0^{\pi/2} \dfrac{1}{1 + \sin x} \, dx.$

35. $\displaystyle\int_0^{\pi/3} \dfrac{1}{\sin x - \cos x - 1} \, dx.$

36. $\displaystyle\int_0^1 \dfrac{\sqrt{x}}{1 + \sqrt{x}} \, dx.$

37. Use the method of this section to show that

$$\int \sec x \, dx = \int \dfrac{1}{\cos x} \, dx = \ln \left| \dfrac{1 + \tan(x/2)}{1 - \tan(x/2)} \right| + C.$$

38. (a) Another expression for $\int \sec x \, dx$ can be obtained as follows:

$$\int \sec x \, dx = \int \dfrac{\cos x}{\cos^2 x} \, dx = \int \dfrac{\cos x}{1 - \sin^2 x} \, dx.$$

Use the method of this section to show that

$$\int \sec x \, dx = \ln\sqrt{\dfrac{1 + \sin x}{1 - \sin x}} + C.$$

(b) Show that the result in part (a) is equivalent to the familiar formula

$$\int \sec x \, dx = \ln|\sec x + \tan x| + C.$$

39. (a) Use the approach given in Exercise 38 (a) to show that

$$\int \csc x \, dx = \ln\sqrt{\dfrac{1 - \cos x}{1 + \cos x}} + C.$$

(b) Show that the result in part (a) is equivalent to the formula

$$\int \csc x \, dx = \ln|\csc x - \cot x| + C.$$

40. The integral of a rational function of $\sinh x$ and $\cosh x$ can be transformed into a rational function of u by means of the substitution $u = \tanh(x/2)$. With this substitution, show that

$$\sinh x = \dfrac{2u}{1 - u^2}, \quad \cosh x = \dfrac{1 + u^2}{1 - u^2}, \quad dx = \dfrac{2}{1 - u^2} \, du.$$

In Exercises 41–44, use the subsitution $u = \tanh(x/2)$ to evaluate the given integral.

41. $\int \operatorname{sech} x \, dx.$

42. $\int \dfrac{1}{1 + \cosh x} \, dx.$

43. $\int \dfrac{1}{\sinh x + \cosh x} \, dx.$

44. $\int \dfrac{1 - e^x}{1 + e^x} \, dx.$

■ 8.8 NUMERICAL INTEGRATION

To evaluate a definite integral by the formula

$$\int_a^b f(x) \, dx = F(b) - F(a)$$

we must be able to find an antiderivative F and we must be able to evaluate this antiderivative both at a and at b. When this is not possible, the method fails.

The method fails even for such simple-looking integrals as

$$\int_0^1 \sqrt{x} \sin x \, dx \quad \text{and} \quad \int_0^1 e^{-x^2} \, dx.$$

There are no *elementary functions* with derivatives $\sqrt{x} \sin x$ and e^{-x^2}.

Here we take up some simple numerical methods for estimating definite integrals —methods that you can use whether or not you can find an antiderivative. All the methods we describe involve only simple arithmetic and are ideally suited to the computer.

We focus now on

$$\int_a^b f(x)\,dx.$$

We suppose that f is continuous on $[a, b]$ and, for pictorial convenience, assume that f is positive. Take a regular partition $P = \{x_0, x_1, x_2, \ldots, x_{n-1}, x_n\}$ of $[a, b]$, subdividing the interval into n subintervals each of length $(b-a)/n$:

$$[a, b] = [x_0, x_1] \cup \cdots \cup [x_{i-1}, x_i] \cup \cdots \cup [x_{n-1}, x_n],$$

with

$$\Delta x_i = \frac{b-a}{n}.$$

The region Ω_i pictured in Figure 8.8.1 can be approximated in many ways.

Figure 8.8.1

Figure 8.8.2

Figure 8.8.3

(1) By the left-endpoint rectangle (Figure 8.8.2):

$$\text{area} = f(x_{i-1})\,\Delta x_i$$
$$= f(x_{i-1})\left(\frac{b-a}{n}\right).$$

(2) By the right-endpoint rectangle (Figure 8.8.3):

$$\text{area} = f(x_i)\,\Delta x_i$$
$$= f(x_i)\left(\frac{b-a}{n}\right).$$

(3) By the midpoint rectangle (Figure 8.8.4):

$$\text{area} = f\left(\frac{x_{i-1}+x_i}{2}\right)\Delta x_i$$
$$= f\left(\frac{x_{i-1}+x_i}{2}\right)\left(\frac{b-a}{n}\right)$$

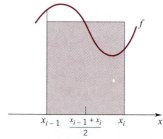

Figure 8.8.4

(4) By a trapezoid (Figure 8.8.5):

$$\text{area} = \frac{1}{2}[f(x_{i-1}) + f(x_i)]\,\Delta x_i$$

$$= \frac{1}{2}[f(x_{i-1}) + f(x_i)]\left(\frac{b-a}{n}\right).$$

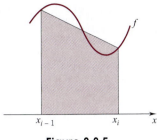

(5) By a parabolic region (Figure 8.8.6): take the parabola $y = Ax^2 + Bx + C$ that passes through the three points indicated.

$$\text{area} = \frac{1}{6}\left[f(x_{i-1}) + 4f\left(\frac{x_{i-1}+x_i}{2}\right) + f(x_i)\right]\Delta x_i$$

$$= \left[f(x_{i-1}) + 4f\left(\frac{x_{i-1}+x_i}{2}\right) + f(x_i)\right]\left(\frac{b-a}{6n}\right).$$

Figure 8.8.5

You can verify this formula for the area under the parabola by doing Exercises 11 and 12. (If the three points are collinear, the parabola degenerates to a straight line and the parabolic region becomes a trapezoid. The formula then gives the area of the trapezoid.)

The approximations to Ω_i just considered yield the following estimates for

$$\int_a^b f(x)\,dx.$$

Figure 8.8.6

(1) The left-endpoint estimate:

$$L_n = \frac{b-a}{n}[f(x_0) + f(x_1) + \cdots + f(x_{n-1})].$$

(2) The right-endpoint estimate:

$$R_n = \frac{b-a}{n}[f(x_1) + f(x_2) + \cdots + f(x_n)].$$

(3) The midpoint estimate:

$$M_n = \frac{b-a}{n}\left[f\left(\frac{x_0 + x_1}{2}\right) + \cdots + f\left(\frac{x_{n-1} + x_n}{2}\right)\right].$$

(4) The trapezoidal estimate (*trapezoidal rule*):

$$T_n = \frac{b-a}{n}\left[\frac{f(x_0) + f(x_1)}{2} + \frac{f(x_1) + f(x_2)}{2} + \cdots + \frac{f(x_{n-1}) + f(x_n)}{2}\right]$$

$$= \frac{b-a}{2n}[f(x_0) + 2f(x_1) + \cdots + 2f(x_{n-1}) + f(x_n)].$$

(5) The parabolic estimate (*Simpson's rule*):

$$S_n = \frac{b-a}{6n}\left\{f(x_0) + f(x_n) + 2[f(x_1) + \cdots + f(x_{n-1})]\right.$$

$$\left. + 4\left[f\left(\frac{x_0 + x_1}{2}\right) + \cdots + f\left(\frac{x_{n-1} + x_n}{2}\right)\right]\right\}.$$

The first three estimates, L_n, R_n, M_n, are Riemann sums (Section 5.10); T_n and S_n, although not explicitly defined as Riemann sums, can be written as Riemann sums.

(See Exercise 26.)   It follows from (5.10.1) that any one of these estimates can be used to approximate the integral as closely as we may wish.   All we have to do is take n sufficiently large.

As an example, we will find the approximate value of

$$\ln 2 = \int_1^2 \frac{dx}{x}$$

by applying each of the five estimates.   Here

$$f(x) = \frac{1}{x}, \qquad [a, b] = [1, 2].$$

Taking $n = 5$, we have

$$\frac{b - a}{n} = \frac{2 - 1}{5} = \frac{1}{5}.$$

The partition points are

$$x_0 = \tfrac{5}{5}, \quad x_1 = \tfrac{6}{5}, \quad x_2 = \tfrac{7}{5}, \quad x_3 = \tfrac{8}{5}, \quad x_4 = \tfrac{9}{5}, \quad x_5 = \tfrac{10}{5}. \qquad \text{(Figure 8.8.7)}$$

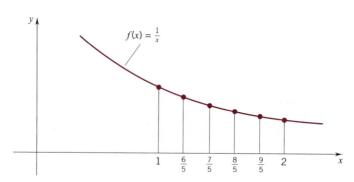

Figure 8.8.7

Using a calculator and rounding off to four decimal places, the five estimates are as follows:

$$L_5 = \tfrac{1}{5}(\tfrac{5}{5} + \tfrac{5}{6} + \tfrac{5}{7} + \tfrac{5}{8} + \tfrac{5}{9}) = (\tfrac{1}{5} + \tfrac{1}{6} + \tfrac{1}{7} + \tfrac{1}{8} + \tfrac{1}{9}) \cong 0.7457.$$

$$R_5 = \tfrac{1}{5}(\tfrac{5}{6} + \tfrac{5}{7} + \tfrac{5}{8} + \tfrac{5}{9} + \tfrac{5}{10}) = (\tfrac{1}{6} + \tfrac{1}{7} + \tfrac{1}{8} + \tfrac{1}{9} + \tfrac{1}{10}) \cong 0.6457.$$

$$M_5 = \tfrac{1}{5}(\tfrac{10}{11} + \tfrac{10}{13} + \tfrac{10}{15} + \tfrac{10}{17} + \tfrac{10}{19}) = 2(\tfrac{1}{11} + \tfrac{1}{13} + \tfrac{1}{15} + \tfrac{1}{17} + \tfrac{1}{19}) \cong 0.6919.$$

$$T_5 = \tfrac{1}{10}(\tfrac{5}{5} + \tfrac{10}{6} + \tfrac{10}{7} + \tfrac{10}{8} + \tfrac{10}{9} + \tfrac{5}{10}) = (\tfrac{1}{10} + \tfrac{1}{6} + \tfrac{1}{7} + \tfrac{1}{8} + \tfrac{1}{9} + \tfrac{1}{20}) \cong 0.6957.$$

$$S_5 = \tfrac{1}{30}[\tfrac{5}{5} + \tfrac{5}{10} + 2(\tfrac{5}{6} + \tfrac{5}{7} + \tfrac{5}{8} + \tfrac{5}{9}) + 4(\tfrac{10}{11} + \tfrac{10}{13} + \tfrac{10}{15} + \tfrac{10}{17} + \tfrac{10}{19})] \cong 0.6935.$$

Since the integrand $1/x$ decreases throughout the interval $[1, 2]$, you can expect the left-endpoint estimate, 0.7457, to be too large and you can expect the right-endpoint estimate, 0.6457, to be too small.   The other estimates should be better.

The value of $\ln 2$ given on a calculator is $\ln 2 \cong 69314718$, or 0.6931 rounded to four decimal places.   Thus S_5 is correct to the nearest thousandth.

Example 1 Find the approximate value of

$$\int_0^3 \sqrt{4 + x^3}\, dx.$$

by the trapezoid rule. Take $n = 6$.

SOLUTION Each subinterval has length

$$\frac{b - a}{n} = \frac{3 - 0}{6} = \frac{1}{2}.$$

The partition points are

$$x_0 = 0, \quad x_1 = \tfrac{1}{2}, \quad x_2 = 1, \quad x_3 = \tfrac{3}{2}, \quad x_4 = 2, \quad x_5 = \tfrac{5}{2}, \quad x_6 = 3. \qquad \text{(See Figure 8.8.8)}$$

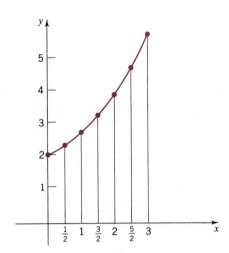

Figure 8.8.8

Now

$$T_6 = \tfrac{1}{4}[\, f(0) + 2f(\tfrac{1}{2}) + 2f(1) + 2f(\tfrac{3}{2}) + 2f(2) + 2f(\tfrac{5}{2}) + f(3)\,],$$

with $f(x) = \sqrt{4 + x^3}$. Using a calculator and rounding off to three decimal places, we have

$$f(0) = 2.000, \quad f(\tfrac{1}{2}) \cong 2.031, \quad f(1) \cong 2.236, \quad f(\tfrac{3}{2}) \cong 2.716,$$
$$f(2) \cong 3.464, \quad f(\tfrac{5}{2}) \cong 4.430, \quad f(3) \cong 5.568.$$

Thus

$$T_6 \cong \tfrac{1}{4}(2.000 + 4.062 + 4.472 + 5.432 + 6.928 + 8.860 + 5.568) \cong 9.331. \quad \square$$

Example 2 Find the approximate value of

$$\int_0^3 \sqrt{4 + x^3}\, dx$$

by Simpson's rule. Take $n = 3$.

SOLUTION There are three subintervals each of length

$$\frac{b - a}{n} = \frac{3 - 0}{3} = 1.$$

Here

$$x_0 = 0, \quad x_1 = 1, \quad x_2 = 2, \quad x_3 = 3,$$

$$\frac{x_0 + x_1}{2} = \frac{1}{2}, \quad \frac{x_1 + x_2}{2} = \frac{3}{2}, \quad \frac{x_2 + x_3}{2} = \frac{5}{2}.$$

Simpson's rule yields

$$S_3 = \tfrac{1}{6}[\,f(0) + f(3) + 2f(1) + 2f(2) + 4f(\tfrac{1}{2}) + 4f(\tfrac{3}{2}) + 4f(\tfrac{5}{2})\,],$$

with $f(x) = \sqrt{4 + x^3}$. Using the values of f from Example 1, we have

$$S_3 = \tfrac{1}{6}(2.000 + 5.568 + 4.472 + 6.928 + 8.124 + 10.864 + 17.72) \cong 9.279.$$

The value of this integral accurate to 5 decimal places is 9.27972. ❑

Error Estimates

A numerical estimate is useful only to the extent that we can gauge its accuracy. When we use any kind of approximation method, we face two forms of error: the error inherent in the method we use (we call this the *theoretical error*) and the error that accumulates from rounding off the decimals that arise during the course of computation (we call this the *round-off error*). The nature of round-off error is obvious. We will speak first about theoretical error.

We begin with a function f continuous and increasing on $[a, b]$. We subdivide $[a, b]$ into n nonoverlapping intervals, each of length $(b - a)/n$. We want to estimate

$$\int_a^b f(x)\,dx$$

by the left-endpoint method. What is the theoretical error? It should be clear from Figure 8.8.9 that the theoretical error does not exceed

$$[\,f(b) - f(a)\,]\left(\frac{b - a}{n}\right).$$

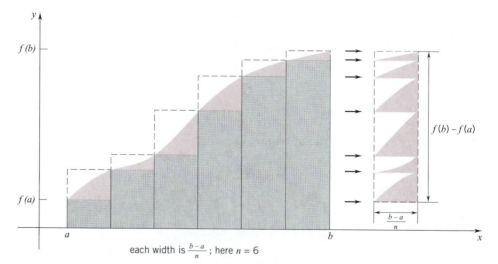

each width is $\frac{b-a}{n}$; here $n = 6$

Figure 8.8.9

The error is represented by the sum of the areas of the shaded regions. These regions, when shifted to the right, all fit together within a rectangle of height $f(b) - f(a)$ and base $(b - a)/n$.

Similar reasoning shows that, under the same circumstances, the theoretical error associated with the trapezoidal method does not exceed

$$\frac{1}{2}[f(b) - f(a)]\left(\frac{b - a}{n}\right).$$

In this setting, at least, the trapezoidal estimate does a better job than the left-endpoint estimate.

This is a crude estimate of the error in the trapezoidal method. Even without the requirement that f be increasing, it can be improved considerably. It is shown in texts on numerical analysis that if f is continuous on $[a, b]$ and twice differentiable on (a, b), then the theoretical error of the trapezoidal rule

$$E_n^T = \int_a^b f(x)\, dx - T_n,$$

can be written

(8.8.1)

$$E_n^T = -\frac{(b - a)^3}{12n^2}f''(c),$$

where c is some number between a and b. Usually we cannot pinpoint c any further. However, if $f''(x)$ is bounded on $[a, b]$, that is, if there is a constant M such that $|f''(x)| \leq M$ for $a \leq x \leq b$, then it follows that

(8.8.2)

$$|E_n^T| \leq \frac{(b - a)^3}{12n^2}M.$$

This provides a useful method for analyzing the error in a trapezoidal estimate.

Recall the trapezoidal-rule estimate of $\ln 2$ derived at the beginning of the section:

$$\ln 2 = \int_1^2 \frac{dx}{x} \cong 0.696.$$

We apply (8.8.2) to find the theoretical error. Here

$$f(x) = \frac{1}{x}, \qquad f'(x) = \frac{-1}{x^2}, \qquad f''(x) = \frac{2}{x^3}.$$

Now, it is easy to see that $|f''(x)| \leq 2$ for $1 \leq x \leq 2$. Therefore, with $a = 1$, $b = 2$, and $n = 5$, we have

$$|E_5^T| \leq \frac{(2 - 1)^3}{12 \cdot 5^2}2 = \frac{1}{150} < 0.007.$$

The estimate 0.696 is in theoretical error by less than 0.007.

Suppose, on the other hand, that we wanted an estimate for

$$\ln 2 = \int_1^2 \frac{dx}{x}$$

that is accurate to four decimal places. Then, using (8.8.2), we need

(1)
$$\frac{(b-a)^3}{12n^2} M < 0.00005.$$

Since

$$\frac{(b-a)^3}{12n^2} M \le \frac{1}{12n^2} 2 = \frac{1}{6n^2},$$

we can guarantee that (1) holds by having

$$\frac{1}{6n^2} < 0.00005,$$

which is equivalent to

$$n^2 > 3333.$$

As you can check, $n = 58$ is the smallest integer that satisfies this inequality. Thus, the trapezoid rule requires a regular partition with at least 58 points to guarantee four-decimal-place accuracy in the approximation of $\int_1^2 (1/x)\, dx$.

Simpson's rule is more effective than the trapezoidal rule. If f is continuous on $[a, b]$ and if $f^{(4)}$ exists on (a, b), then the theoretical error for Simpson's rule

$$E_n^S = \int_a^b f(x)\, dx - S_n,$$

can be written

(8.8.3)
$$E_n^S = -\frac{(b-a)^5}{2880n^4} f^{(4)}(c),$$

where, as before, c is some number between a and b. Whereas (8.8.1) varies as $1/n^2$, this quantity varies as $1/n^4$. Thus, for comparable n, we can expect greater accuracy from Simpson's rule. In addition, if we assume that $f^{(4)}(x)$ is bounded on $[a, b]$, say $|f^{(4)}(x)| \le M$ for $a \le x \le b$, then

(8.8.4)
$$|E_n^S| \le \frac{(b-a)^5}{2880n^4} M,$$

and this can be used to analyze the theoretical error for Simpson's rule in the same way that we used (8.8.2) to analyze the error for the trapezoidal rule.

For example, to achieve four-decimal-place accuracy in estimating

$$\ln 2 = \int_1^2 \frac{dx}{x},$$

using Simpson's rule, we set $b = 2$ and $a = 1$ in (8.8.4), and then try to find n so that

$$|E_n^S| \le \frac{(2-1)^5}{2880n^4} M = \frac{1}{2880n^4} M \le 0.00005,$$

where M is a bound for

$$|f^{(4)}(x)| = \frac{24}{x^5} \quad \text{on } [1, 2].$$

Clearly, $|f^{(4)}(x)| \le 24$ on $[1, 2]$, and so we want to find n such that

$$\frac{1}{2880n^4} 24 = \frac{1}{120n^4} < 0.00005.$$

This is equivalent to

$$n^4 > 167.$$

You can verify that $n = 4$ is the smallest positive integer which satisfies this inequality. Obviously this is a considerable improvement in efficiency over the trapezoidal rule.

Finally, a word about round-off error. Any numerical procedure requires careful consideration of round-off error. To illustrate this point, we rework our trapezoidal estimate for

$$\int_1^2 \frac{dx}{x},$$

again taking $n = 5$, but this time assuming that our computer or calculator can store only two significant digits. As before,

(2) $$\int_1^2 \frac{dx}{x} \cong \tfrac{1}{10}[\tfrac{1}{1} + 2(\tfrac{5}{6}) + 2(\tfrac{5}{7}) + 2(\tfrac{5}{8}) + 2(\tfrac{5}{9}) + (\tfrac{1}{2})].$$

Now our limited round-off machine goes to work:

$$\int_1^2 \frac{dx}{x} \cong (0.10)[(1.0) + 2(0.83) + 2(0.71) + 2(0.62) + 2(0.44) + (0.50)]$$

$$\text{``=''} (0.10)[(1.0) + (1.7) + (1.4) + (1.2) + (0.88) + (0.50)]$$

$$= (0.10)[6.7] = 0.67.$$

Earlier we used (8.8.1) to show that estimate (2) is in error by no more than 0.007 and found 0.70 as the approximation. Now with our limited round-off machine, we simplified (2) in a different way and obtained 0.67 as the approximation. The apparent error due to crude round off, $0.70 - 0.67 = 0.03$, exceeds the error of the approximation method itself. The lesson should be clear: round-off error is important.

EXERCISES 8.8

In Exercises 1–10, round your answers to four decimal places.

1. Estimate

$$\int_0^{12} x^2 \, dx$$

using: (a) the left-endpoint estimate, $n = 12$; (b) the right-endpoint estimate, $n = 12$; (c) the midpoint estimate, $n = 6$; (d) the trapezoidal rule, $n = 12$; (e) Simpson's rule, $n = 6$. Check your results by performing the integration.

2. Estimate

$$\int_0^1 \sin^2 \pi x \, dx$$

using: (a) the midpoint estimate, $n = 3$; (b) the trapezoidal rule, $n = 6$; (c) Simpson's rule, $n = 3$. Check your results by performing the integration.

3. Estimate

$$\int_0^3 \frac{dx}{1 + x^3}$$

using: (a) the left-endpoint estimate, $n = 6$; (b) the right-endpoint estimate, $n = 6$; (c) the midpoint estimate, $n = 3$; (d) the trapezoidal rule, $n = 6$; (e) Simpson's rule, $n = 3$.

4. Estimate

$$\int_0^\pi \frac{\sin x}{\pi + x} \, dx$$

using: (a) the trapezoidal rule, $n = 6$; (b) Simpson's rule, $n = 3$.

5. Find the approximate value of π by estimating the integral

$$\frac{\pi}{4} = \tan^{-1} 1 = \int_0^1 \frac{dx}{1 + x^2}$$

using: (a) the trapezoidal rule, $n = 4$; (b) Simpson's rule, $n = 4$.

6. Estimate

$$\int_0^2 \frac{dx}{\sqrt{4 + x^3}}$$

using: (a) the trapezoidal rule, $n = 4$; (b) Simpson's rule, $n = 2$.

7. Estimate

$$\int_{-1}^1 \cos(x^2) \, dx$$

using: (a) the midpoint estimate, $n = 4$; (b) the trapezoidal rule, $n = 8$; (c) Simpson's rule, $n = 4$.

8. Estimate

$$\int_1^2 \frac{e^x}{x} \, dx$$

using: (a) the midpoint estimate, $n = 4$; (b) the trapezoidal rule, $n = 8$; (c) Simpson's rule, $n = 4$.

9. Estimate

$$\int_0^2 e^{-x^2} \, dx$$

using: (a) the trapezoidal rule, $n = 10$; (b) Simpson's rule, $n = 5$.

10. Estimate

$$\int_2^4 \frac{1}{\ln x} \, dx$$

using: (a) the midpoint estimate, $n = 4$; (b) the trapezoidal rule, $n = 8$; (c) Simpson's rule, $n = 4$.

11. Show that there is one and only one curve of the form $y = Ax^2 + Bx + C$ through three distinct points with different x-coordinates.

12. Show that the function $g(x) = Ax^2 + Bx + C$ satisfies the condition

$$\int_a^b g(x) \, dx = \frac{b - a}{6} \left[g(a) + 4g\left(\frac{a + b}{2} \right) + g(b) \right]$$

for every interval $[a, b]$.

In Exercises 13–22, determine the values of n for which a theoretical error less than ϵ can be guaranteed if the integral is estimated using: (a) the trapezoidal rule; (b) Simpson's rule.

13. $\int_1^4 \sqrt{x} \, dx; \quad \epsilon = 0.01.$

14. $\int_1^3 x^5 \, dx; \quad \epsilon = 0.01.$

15. $\int_1^4 \sqrt{x} \, dx; \quad \epsilon = 0.00001.$

16. $\int_1^3 x^5 \, dx; \quad \epsilon = 0.00001.$

17. $\int_0^\pi \sin x \, dx; \quad \epsilon = 0.001$

18. $\int_0^\pi \cos x \, dx; \quad \epsilon = 0.001.$

19. $\int_1^3 e^x \, dx; \quad \epsilon = 0.01.$

20. $\int_1^e \ln x \, dx; \quad \epsilon = 0.01.$

21. $\int_0^2 e^{-x^2} \, dx; \quad \epsilon = 0.0001$

22. $\int_0^2 e^x \, dx; \quad \epsilon = 0.00001.$

23. Show that Simpson's rule is exact (theoretical error zero) for every polynomial of degree 3 or less.

24. Show that the trapezoidal rule is exact (theoretical error zero) if f is linear.

25. (a) Let $f(x) = x^2$ on $[0, 1]$ and let $n = 2$. Show that

$$E_n^T = \frac{(b - a)^3}{12 \, n^2} M$$

so that the inequality (8.8.2) can reduce to equality.
(b) Let $f(x) = x^4$ on $[0, 1]$ and let $n = 1$. Show that

$$E_n^S = \frac{(b - a)^5}{2880 \, n^4} M.$$

26. Show that, if f is continuous, then T_n and S_n can both be written as Riemann sums. HINT: Both

$$\frac{1}{2}[f(x_{i-1}) + f(x_i)] \quad \text{and}$$

$$\frac{1}{6}\left[f(x_{i-1}) + 4f\left(\frac{x_{i-1} + x_i}{2}\right) + f(x_i) \right]$$

lie between m_i and M_i, the minimum and maximum values of f on $[x_{i-1}, x_i]$.

27. Let f be a twice differentiable function. Show that if $f(x) > 0$ and $f''(x) > 0$ on $[a, b]$, then

$$M_n \le \int_a^b f(x) \, dx \le T_n \qquad \text{for any } n,$$

HINT: See the figure. Show that the area of the rectangle $ABCD$ equals the area of the trapezoid $AEFD$.

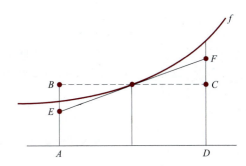

28. Show that $\frac{1}{3} T_n + \frac{2}{3} M_n = S_n$.

■ CHAPTER HIGHLIGHTS

8.1 Review

important integral formulas (p. 489)
a table of integrals appears on the inside covers

8.2 Integration by Parts

$$\int u \, dv = uv - \int v \, du$$

success with the technique depends on choosing u and dv so that $\int v \, du$ is easier to integrate than $\int u \, dv$

integration by parts is often used to calculate integrals when the integrand is a product of a mixture of functions; for example, polynomials and exponentials, or polynomials and trigonometric functions

to calculate some integrals, you may have to integrate by parts more than once

the integral of $\ln x$ and the integrals of the inverse trigonometric functions are calculated using integration by parts (p. 499)

8.3 Powers of Products of Sines and Cosines

Integrals of the form

$$\int \sin^m x \, \cos^n x \, dx$$

can be calculated by using the basic identity $\sin^2 x + \cos^2 x = 1$ and the double-angle formulas

$$\sin x \cos x = \tfrac{1}{2} \sin 2x, \qquad \sin^2 x = \tfrac{1}{2} - \tfrac{1}{2} \cos 2x, \qquad \cos^2 x = \tfrac{1}{2} + \tfrac{1}{2} \cos 2x.$$

Reduction formulas:

$$\int \sin^n x \, dx = -\frac{1}{n} \sin^{n-1} x \cos x + \frac{n-1}{n} \int \sin^{n-2} x \, dx$$

$$\int \cos^n x \, dx = \frac{1}{n} \cos^{n-1} x \sin x + \frac{n-1}{n} \int \cos^{n-2} x \, dx$$

8.4 Other Trigonometric Powers

the main tools for calculating such integrals are the identities

$$1 + \tan^2 x = \sec^2 x, \qquad 1 + \cot^2 x = \csc^2 x,$$

and integration by parts.

8.5 Integrals Involving $\sqrt{a^2 \pm x^2}$ and $\sqrt{x^2 - a^2}$; Trigonometric Substitutions

such integrals may be calculated by a trigonometric substitution:

$$\text{for } \sqrt{a^2 - x^2} \quad \text{set} \quad a \sin u = x$$
$$\text{for } \sqrt{a^2 + x^2} \quad \text{set} \quad a \tan u = x$$
$$\text{for } \sqrt{x^2 - a^2} \quad \text{set} \quad a \sec u = x$$

integrals involving $\sqrt{a^2 + x^2}$ or $\sqrt{x^2 - a^2}$ may also be calculated by a hyperbolic substitution (p. 518)

it may be necessary to complete the square under the radical before making a trigonometric or hyperbolic substitution (p. 516)

8.6 Partial Fractions

a proper rational function may be integrated by writing it as a sum of fractions of the form

$$\frac{A}{(x - \alpha)^k} \qquad \text{and} \qquad \frac{Bx + C}{(x^2 + \beta x + \gamma)^k}$$

called the *partial fraction decomposition* (p. 522)

to integrate an improper rational function, express it as a polynomial plus a proper rational function by dividing the denominator into the numerator (p. 521)

8.7 Rationalizing Substitutions

for integrals involving $\sqrt[n]{f(x)}$ for some function f, let $u^n = f(x)$, $nu^{n-1} du = f'(x) \, dx$

for rational expressions in sine and cosine, let $u = \tan(x/2)$. Then

$$\sin x = \frac{2u}{1 + u^2}, \qquad \cos x = \frac{1 - u^2}{1 + u^2}, \qquad du = \frac{2}{1 + u^2} du \qquad \text{(p. 533)}$$

8.8 Numerical Integration

left-endpoint, right-endpoint, and midpoint estimates; trapezoidal rule; Simpson's rule (p. 537)

the theoretical error in the trapezoidal rule varies as $1/n^2$

the theoretical error in Simpson's rule varies as $1/n^4$

■ PROJECTS AND EXPLORATIONS USING TECHNOLOGY

To do these exercises you will need a graphics calculator or a computer with graphing capability. The majority of these problems are open-ended so different approaches may be used to solve them. You should be aware that different approaches can result in slight variations in the answers. Round your numerical answers to at least four decimal places. The rounding method that your calculator or computer uses also may cause variations in answers.

8.1 The function

$$B(x) = \int_0^x e^{-t^2}\, dt, \qquad x \text{ real,}$$

is of fundamental importance in science and engineering. (The function $(2/\sqrt{\pi})B(x)$ is called the *error function*.) While we know that B exists, we must approximate its values numerically. Nevertheless, we can determine many of the properties of this function.
(a) Explain why B is continuous on $(-\infty, \infty)$.
(b) Find B' and show that B is increasing on $(-\infty, \infty)$.
(c) Show that the graph of B has exactly one point of inflection and find it.
(d) Show that the graph of B has a horizontal asymptote and find it.
(e) Explain why B has an inverse and find $B^{-1}(0.5)$.
(f) Sketch the graph of B^{-1}.

8.2 Consider the definite integral

$$\int_1^3 e^{-x^2}\, dx.$$

(a) Obtain 20 approximations of this integral in the following manner: Evaluate the integrand at 100 random points between 1 and 3; calculate the average of these 100 values; then multiply by the length of the interval. Note that this procedure is equivalent to calculating 20 Riemann sums.
(b) Let $X_1, X_2, X_3, \ldots, X_{20}$ be the 20 values that you obtained in part (a). Find the mean m and standard deviation σ of these 20 values. The *mean* is the average of the values and the *standard deviation* is given by

$$\sigma = \sqrt{\sum_{n=1}^{20} \frac{(X_n - m)^2}{19}}.$$

(c) For each real number x, let $D(x)$ be the percentage of the values $X_1, X_2, X_3, \ldots, X_{20}$ which are less than or equal to x. Sketch the graph of D.
(d) Using the graph of D, sketch the graph of D' and discuss its properties.

8.3 Let F be the function defined by

$$F(x) = \int_0^x \cos\,[\sin t]\, dt,$$

and suppose that we want to approximate

$$\int_1^3 F(x)\, dx$$

correct to two decimal places using the trapezoidal rule. Since F changes its concavity on the interval $[1, 3]$, it is difficult to estimate the number of subintervals necessary to get this accuracy. The following procedure suggests one way.
(a) Find the points of inflection of F and use these to subdivide the interval $[1, 3]$ into subintervals on which the graph of F does not change its concavity.
(b) Approximate $\int F$ correct to two decimal places on each of these subintervals by both the trapezoid rule and the midpoint rule using the same n.

(c) Use these values to justify why your answer is within two decimal places. (Note that on each subinterval, the integral is between the trapezoidal approximation and the midpoint approximation. Upper and lower bounds for the whole interval can be obtained by adding the upper and lower bounds on each subinterval.

8.4 Let

$$f_n(x) = \cos\left(\frac{n}{x}\right)$$

for $n = 1, 2, 3, 4, 5$. Use six decimal place accuracy in the problems that follow.

(a) For each n, approximate

$$F(n) = \int_2^{10} f_n(x)\, dx$$

using Simpson's rule with $k = 2, 4, 8, 16$ subintervals. Denote these approximations by $S(n, k)$.

(b) Show that

$$F(n) - S(n, 16) \cong \frac{[F(n) - S(n, 8)]}{16}.$$

(c) Solve the approximation given in part (b) for $F(n)$ to get a final approximation for $F(n)$, $n = 1, 2, 3, 4, 5$.

(d) For each n compute a better approximation for $F(n)$ (for example, use Simpson's rule with $k > 16$ subintervals) and verify that the formula you obtained in part (c) did improve the approximation.

(e) Now estimate

$$\int_1^5 F(n)\, dn$$

by using Simpson's rule with 1 and 2 subintervals. Find a formula similar to that used in parts (b) and (c) for 1 and 2 subintervals instead of 8 and 16 subintervals. Then use this formula to get an improved approximation.

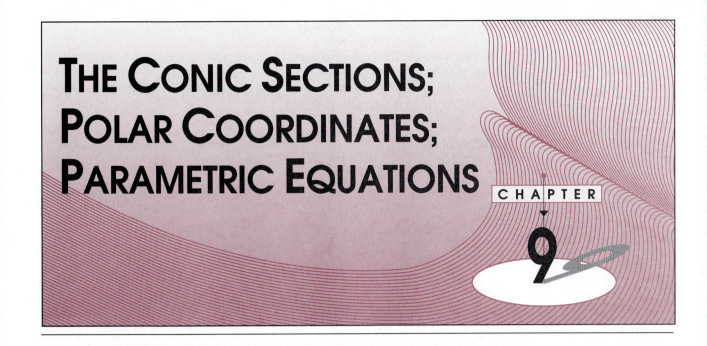

THE CONIC SECTIONS; POLAR COORDINATES; PARAMETRIC EQUATIONS

CHAPTER

9

■ 9.1 CONIC SECTIONS

Conic sections were introduced and discussed briefly in Section 1.4. We expand on that discussion here.

If a "double right circular cone" is cut by a plane, the resulting intersection is called a *conic section* or, more briefly, a *conic*. In Figure 9.1.1, we show a double cone and depict three important cases: the parabola, ellipse, and hyperbola.

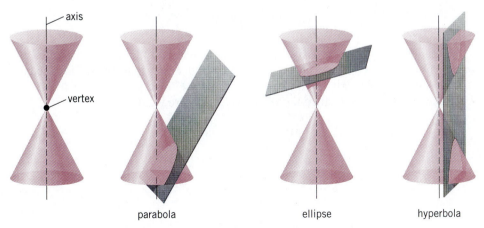

axis

vertex

parabola   ellipse   hyperbola

Figure 9.1.1

By choosing a plane perpendicular to the axis of the cone, we can obtain a circle. Other, less interesting, cases occur when the plane passes through the vertex of the cone. Depending on the orientation of the plane with respect to the axis of the cone, the intersection may either be a point, a line, or a pair of intersecting lines. Try to visualize these possibilities.

549

This three-dimensional approach to the conic sections goes back to Apollonius of Perga, a Greek of the third century B.C. He wrote eight books on the subject.

We will take a different approach. We will define parabola, ellipse, and hyperbola entirely in terms of plane geometry.

The Parabola

Figure 9.1.2 shows a line *l* and a point *F* not on *l*.

Figure 9.1.2

(9.1.1) | The set of points *P* equidistant from *F* and *l* is called a *parabola*.

See Figure 9.1.3.

Figure 9.1.3

The line *l* is called the *directrix* of the parabola, and the point *F* is called the *focus*. (You will see why later on.) The line through *F* perpendicular to *l* is called the *axis* of the parabola. (It is the axis of symmetry.) The point at which the axis intersects the parabola is called the *vertex*. (See Figure 9.1.4.)

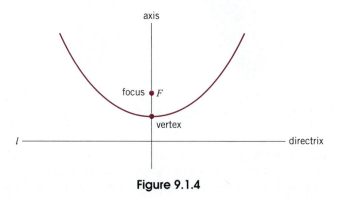

Figure 9.1.4

The equation of a parabola is particularly simple if we place the vertex at the origin and the focus on one of the coordinate axes. Suppose for the moment that the focus F is on the y-axis.   Then F has coordinates of the form $(0, c)$.   With the vertex at the origin, the directrix has equation $y = -c$.   (See Figure 9.1.5.)

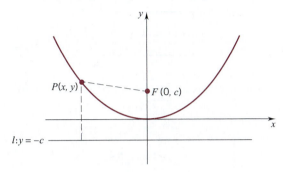

Figure 9.1.5

Every point $P(x, y)$ that lies on this parabola has the property that
$$d(P, F) = d(P, l).$$
Since
$$d(P, F) = \sqrt{x^2 + (y - c)^2} \quad \text{and} \quad d(P, l) = |y + c|,$$
you can see that
$$\sqrt{x^2 + (y - c)^2} = |y + c|,$$
$$x^2 + (y - c)^2 = |y + c|^2 = (y + c)^2,$$
$$x^2 + y^2 - 2cy + c^2 = y^2 + 2cy + c^2,$$
$$x^2 = 4cy. \quad \square$$

You have just seen that the equation

(9.1.2)
$$\boxed{x^2 = 4cy}$$
(Figure 9.1.6)

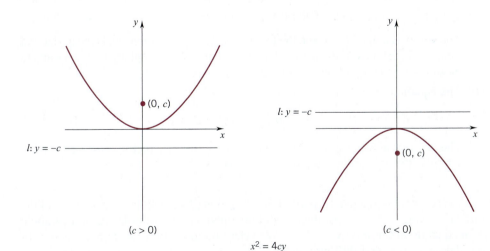

$x^2 = 4cy$

Figure 9.1.6

represents a parabola with vertex at the origin and focus at $(0, c)$. By interchanging the roles of x and y, you can see that the equation

(9.1.3)

$$y^2 = 4cx$$

(Figure 9.1.7)

represents a parabola with vertex at the origin and focus at $(c, 0)$.

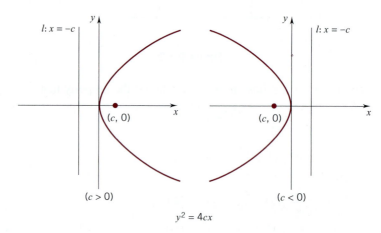

$$y^2 = 4cx$$

Figure 9.1.7

Example 1 Sketch the parabola specifying the vertex, focus, directrix, and axis:

$$\text{(a) } x^2 = -4y. \qquad \text{(b) } y^2 = 3x.$$

SOLUTION

(a) The equation $x^2 = -4y$ has the form

$$x^2 = 4cy \qquad \text{with} \quad c = -1.$$

The vertex is at the origin, and the focus is at $(0, -1)$; the directrix is the horizontal line $y = 1$; the axis of the parabola is the y-axis. We also plotted a few points to ensure the accuracy of our sketch. See Figure 9.1.8.

(b) The equation $y^2 = 3x$ has the form

$$y^2 = 4cx \qquad \text{with} \quad c = \tfrac{3}{4}.$$

(Figure 9.1.8)

The vertex is at the origin, and the focus is at $(\tfrac{3}{4}, 0)$; the directrix is the vertical line $x = -\tfrac{3}{4}$; the axis of the parabola is the x-axis. ❏

Every parabola with vertical axis is a translation of a parabola with equation of the form $x^2 = 4cy$, and every parabola with horizontal axis is a translation of a parabola with equation of the form $y^2 = 4cx$. We refer you to Section 1.4 for a discussion of translation of axes and shifting graphs.

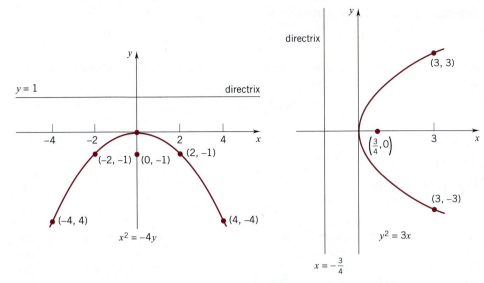

Figure 9.1.8

Example 2 Identify the curve

$$(x - 4)^2 = 8(y + 3).$$

SOLUTION The curve is the parabola

$$x^2 = 8y \qquad\qquad \text{(here } c = 2\text{)}$$

displaced 4 units right and 3 units down. The parabola $x^2 = 8y$ has vertex at the origin; the focus is at $(0, 2)$, and the directrix is the line $y = -2$. Thus, the parabola $(x - 4)^2 = 8(y + 3)$ has vertex at $(4, -3)$; the focus is at $(4, -1)$, and the directrix is the line $y = -5$. (See Figure 9.1.9.) ◻

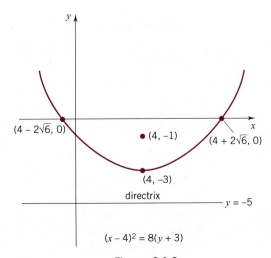

Figure 9.1.9

Example 3 Identify the curve

$$y^2 = 2y - 12x - 37.$$

SOLUTION First we gather the y terms on the left side of the equation and then complete the square:

$$y^2 - 2y = -12x - 37,$$

$$y^2 - 2y + 1 = -12x - 36,$$

$$(y - 1)^2 = -12(x + 3).$$

This is the parabola

$$y^2 = -12x \qquad \text{(here } c = -3\text{)}$$

displaced 3 units left, 1 unit up. The parabola $y^2 = -12x$ has vertex at the origin; the focus is at $(-3, 0)$ and the directrix is the vertical line $x = 3$. Thus, the parabola $(y - 1)^2 = -12(x + 3)$ has vertex at $(-3, 1)$; the focus is at $(-6, 1)$, and the directrix is the line $x = 0$ (the y-axis). See Figure 9.1.10. ❑

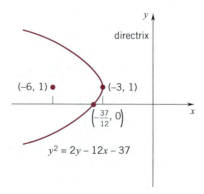

Figure 9.1.10

The Ellipse

Start with two points F_1, F_2 and a number k greater than the distance between them.

(9.1.4)

> The set of all points P for which
> $$d(P, F_1) + d(P, F_2) = k$$
> is called an *ellipse*. F_1 and F_2 are called the *foci*.

The idea is illustrated in Figure 9.1.11. A string is looped over tacks placed at the foci. The pencil placed in the loop traces out an ellipse.

Figure 9.1.12 shows an ellipse in what is called *standard position*: foci along the x-axis at equal distances from the origin. We now derive an equation for this ellipse setting $a = k/2$ or $k = 2a$. The reason for this change of constant will become apparent.

$$d(P, F_1) + d(P, F_2) = k$$

Figure 9.1.11

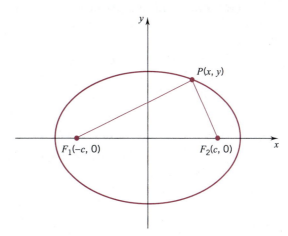

Figure 9.1.12

A point $P(x, y)$ lies on the ellipse iff

$$d(P, F_1) + d(P, F_2) = 2a.$$

With F_1 at $(-c, 0)$ and F_2 at $(c, 0)$, $c > 0$, we have

$$\sqrt{(x + c)^2 + y^2} + \sqrt{(x - c)^2 + y^2} = 2a.$$

Transferring the second term to the right-hand side and squaring both sides, we get

$$(x + c)^2 + y^2 = 4a^2 + (x - c)^2 + y^2 - 4a\sqrt{(x - c)^2 + y^2}.$$

This reduces to

$$4a\sqrt{(x - c)^2 + y^2} = 4(a^2 - cx).$$

Canceling the factor 4 and squaring again, we obtain

$$a^2(x^2 - 2cx + c^2 + y^2) = a^4 - 2a^2cx + c^2x^2.$$

This in turn reduces to

$$(a^2 - c^2)x^2 + a^2y^2 = a^2(a^2 - c^2),$$

which we write as

$$\frac{x^2}{a^2} + \frac{y^2}{a^2 - c^2} = 1.$$

Usually we set $b = \sqrt{a^2 - c^2}$ (you should verify that $a > c$). The equation for an ellipse in standard position then takes the form

(9.1.5)

$$\frac{x^2}{a^2} + \frac{y^2}{b^2} = 1 \qquad \text{with} \quad a > b. \qquad \square$$

The roles played by a, b, c can be read from Figure 9.1.13.

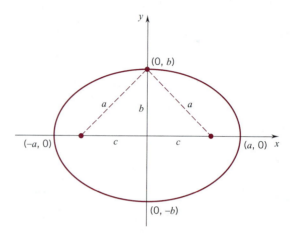

Figure 9.1.13

Every ellipse has four *vertices.* In Figure 9.1.13 these are marked $(a, 0)$, $(-a, 0)$, $(0, b)$, $(0, -b)$. The line segments that join opposite vertices are called the *axes* of the ellipse. The axis that contains the foci is called the *major axis,* the other the *minor axis.* In standard position the major axis is horizontal and has length $2a$; the minor axis is vertical and has length $2b$. The point at which the axes intersect is called the *center* of the ellipse. In standard position the center is at the origin.

Example 4   The equation $16x^2 + 25y^2 = 400$ can be written

$$\frac{x^2}{25} + \frac{y^2}{16} = 1. \qquad \text{(divide by 400)}$$

Here $a = 5$, $b = 4$, and $c = \sqrt{a^2 - b^2} = \sqrt{9} = 3$. The equation is in the form of (9.1.5). It is an ellipse in standard position with foci at $(-3, 0)$ and $(3, 0)$. The major axis has length $2a = 10$, and the minor axis has length $2b = 8$. The center is at the origin. The ellipse is sketched in Figure 9.1.14. ❏

Example 5   The equation

$$\frac{x^2}{16} + \frac{y^2}{25} = 1$$

does not represent an ellipse in standard position because $25 > 16$. This equation is the equation of Example 4 with x and y interchanged. It represents the ellipse of Example 4 reflected in the line $y = x$. (See Figure 9.1.15.) The foci are now on the y-axis, at $(0, -3)$ and $(0, 3)$. The major axis, now vertical, has length 10, and the minor axis has length 8. The center remains at the origin. ❏

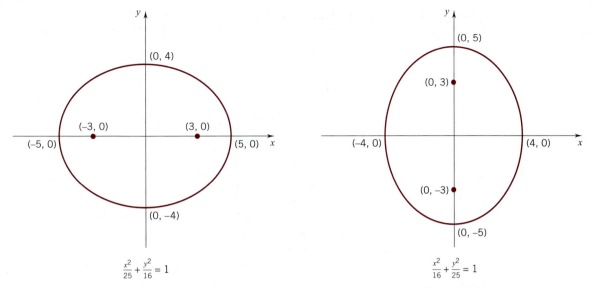

$$\frac{x^2}{25} + \frac{y^2}{16} = 1$$

Figure 9.1.14

$$\frac{x^2}{16} + \frac{y^2}{25} = 1$$

Figure 9.1.15

Example 6 Figure 9.1.16 shows two ellipses:

$$\frac{x^2}{25} + \frac{y^2}{9} = 1 \quad \text{and} \quad \frac{(x-1)^2}{25} + \frac{(y+4)^2}{9} = 1.$$

The first ellipse is in standard position. Here $a = 5$, $b = 3$, $c = \sqrt{a^2 - b^2} = 4$. The foci are at $(-4, 0)$ and $(4, 0)$. The major axis has length 10, and the minor axis has length 6.

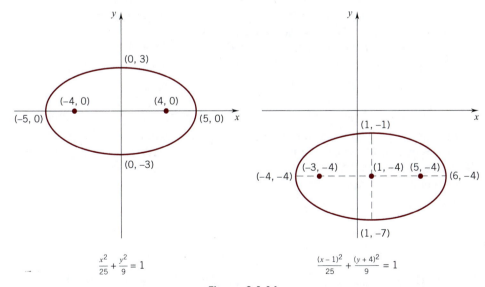

$$\frac{x^2}{25} + \frac{y^2}{9} = 1$$

$$\frac{(x-1)^2}{25} + \frac{(y+4)^2}{9} = 1$$

Figure 9.1.16

The second ellipse is the first ellipse displaced 1 unit right and 4 units down. The center is now at the point $(1, -4)$. The foci are at $(-3, -4)$ and $(5, -4)$. ❑

Example 7 To identify the curve

$$4x^2 - 8x + y^2 + 4y - 8 = 0.$$

we write

$$4(x^2 - 2x + \quad) + (y^2 + 4y + \quad) = 8,$$

and complete the squares within the parentheses. This gives

$$4(x^2 - 2x + 1) + (y^2 + 4y + 4) = 8 + 4 + 4,$$

$$4(x - 1)^2 + (y + 2)^2 = 16,$$

(*) $$\frac{(x - 1)^2}{4} + \frac{(y + 2)^2}{16} = 1.$$

This is the ellipse

(**) $$\frac{x^2}{16} + \frac{y^2}{4} = 1 \qquad (a = 4, b = 2, c = \sqrt{16 - 4} = 2\sqrt{3})$$

reflected in the line $y = x$ and then displaced 1 unit right and 2 units down. Since the foci of (**) are at $(-2\sqrt{3}, 0)$ and $(2\sqrt{3}, 0)$, the foci of (*) are at $(1, -2 - 2\sqrt{3})$ and $(1, -2 + 2\sqrt{3})$. The major axis, now vertical, has length 8; the minor axis has length 4. ❏

The Hyperbola

Start with two points F_1, F_2 and take a positive number k less than the distance between them.

(9.1.6)

> The set of all points P for which
>
> $$|d(P, F_1) - d(P, F_2)| = k$$
>
> is called a *hyperbola*. F_1 and F_2 are called the *foci*.

Figure 9.1.17 shows a hyperbola in what is called *standard position*: foci along the x-axis at equal distances from the origin. We will derive an equation for this hyperbola, again setting $a = k/2$ or $k = 2a$.

A point $P(x, y)$ lies on the hyperbola iff

$$|d(P, F_1) - d(P, F_2)| = 2a.$$

With F_1 at $(-c, 0)$ and F_2 at $(c, 0)$, $c > 0$, we have

$$\sqrt{(x + c)^2 + y^2} - \sqrt{(x - c)^2 + y^2} = \pm 2a. \qquad \text{(explain)}$$

Transferring the second term to the right and squaring both sides, we obtain

$$(x + c)^2 + y^2 = 4a^2 \pm 4a\sqrt{(x - c)^2 + y^2} + (x - c)^2 + y^2.$$

This equation reduces to

$$xc - a^2 = \pm a\sqrt{(x - c)^2 + y^2}.$$

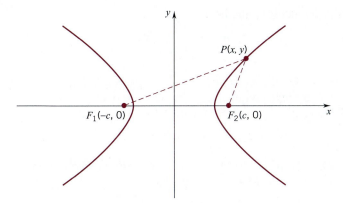

Figure 9.1.17

Squaring once more, we find that

$$x^2c^2 - 2a^2xc + a^4 = a^2(x^2 - 2xc + c^2 + y^2),$$

which reduces to

$$(c^2 - a^2)x^2 - a^2y^2 = a^2(c^2 - a^2),$$

and thus to

$$\frac{x^2}{a^2} - \frac{y^2}{c^2 - a^2} = 1.$$

Usually we set $b = \sqrt{c^2 - a^2}$ (verify that $c > a$). The equation for a hyperbola in standard position then takes the form

(9.1.7)
$$\frac{x^2}{a^2} - \frac{y^2}{b^2} = 1.$$ ❑

The roles played by a, b, c can be read from Figure 9.1.18. As the figure suggests, the

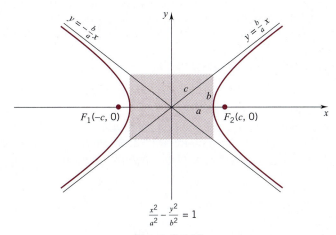

$$\frac{x^2}{a^2} - \frac{y^2}{b^2} = 1$$

Figure 9.1.18

hyperbola remains between the lines

$$y = \frac{b}{a}x \qquad \text{and} \qquad y = -\frac{b}{a}x.$$

These lines are called the *asymptotes* of the hyperbola. They can be obtained from Equation (9.1.7) by replacing the 1 on the right-hand side by 0:

$$\frac{x^2}{a^2} - \frac{y^2}{b^2} = 0 \qquad \text{gives} \qquad y = \pm\frac{b}{a}x.$$

As $x \to \pm\infty$, the vertical separation between the hyperbola and the asymptotes tends to zero. To see this, solve the equation

$$\frac{x^2}{a^2} - \frac{y^2}{b^2} = 1$$

for y. This gives

$$y = \pm\sqrt{\frac{b^2}{a^2}x^2 - b^2} = \pm\frac{b}{a}\sqrt{x^2 - a^2}.$$

In all four quadrants the vertical separation between the hyperbola and the asymptotes can be written

$$\left| \frac{b}{a}|x| - \frac{b}{a}\sqrt{x^2 - a^2} \right| = \frac{b}{a}\left| |x| - \sqrt{x^2 - a^2} \right|. \qquad \text{(check this)}$$

As $x \to \pm\infty$,

$$\frac{b}{a}\left| |x| - \sqrt{x^2 - a^2} \right| = \frac{b}{a}\left| |x| - \sqrt{x^2 - a^2} \right| \frac{|x| + \sqrt{x^2 - a^2}}{|x| + \sqrt{x^2 - a^2}}$$

$$= \frac{b}{a} \cdot \frac{x^2 - (x^2 - a^2)}{|x| + \sqrt{x^2 - a^2}} = \frac{b}{a} \cdot \frac{a^2}{|x| + \sqrt{x^2 - a^2}} \to 0.$$

The line determined by the foci of a hyperbola intersects the hyperbola at two points, called the *vertices*. The line segment that joins the vertices is called the *transverse axis*. The midpoint of the transverse axis is called the *center* of the hyperbola.

In standard position (Figure 9.1.18), the vertices are $(\pm a, 0)$, the transverse axis has length $2a$, and the center is at the origin.

Example 8 The equation

$$\frac{x^2}{1} - \frac{y^2}{3} = 1 \qquad \text{(Figure 9.1.19)}$$

represents a hyperbola in standard position; here $a = 1$, $b = \sqrt{3}$, $c = \sqrt{1 + 3} = 2$. The center is at the origin. The foci are at $(-2, 0)$ and $(2, 0)$. The vertices are at $(-1, 0)$ and $(1, 0)$. The transverse axis has length 2. We can obtain the asymptotes by setting

$$\frac{x^2}{1} - \frac{y^2}{3} = 0.$$

The asymptotes are the lines $y = \pm\sqrt{3}x$. ❑

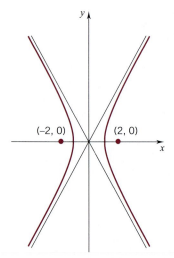

Figure 9.1.19

Example 9 The hyperbola

$$\frac{(x-4)^2}{1} - \frac{(y+5)^2}{3} = 1 \qquad \text{(Figure 9.1.20)}$$

is the hyperbola

$$\frac{x^2}{1} - \frac{y^2}{3} = 1$$

of Example 8 displaced 4 units right and 5 units down. The center of the hyperbola is now at the point $(4, -5)$. The foci are at $(2, -5)$ and $(6, -5)$. The vertices are at $(3, -5)$ and $(5, -5)$. The new asymptotes are the lines $y + 5 = \pm\sqrt{3}(x - 4)$. ❑

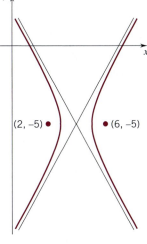

Figure 9.1.20

Example 10 The hyperbola

$$\frac{y^2}{1} - \frac{x^2}{3} = 1 \qquad \text{(Figure 9.1.21)}$$

is the hyperbola

$$\frac{x^2}{1} - \frac{y^2}{3} = 1$$

of Example 8 reflected in the line $y = x$. The center is still at the origin. The foci are now at $(0, -2)$ and $(0, 2)$. The vertices are at $(0, -1)$ and $(0, 1)$. The asymptotes are the lines $x = \pm\sqrt{3}y$. ❑

Figure 9.1.21

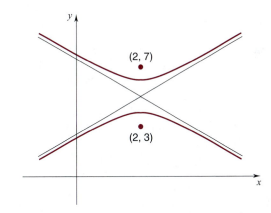

Figure 9.1.22

Example 11 The hyperbola

$$\frac{(y-5)^2}{1} - \frac{(x-2)^2}{3} = 1 \qquad \text{(Figure 9.1.22)}$$

is the hyperbola of Example 10 displaced 2 units right and 5 units up. ❑

Some Applications of Conic Sections

Parabolic Mirrors You are already familiar with the geometric principle of reflected light: that the angle of reflection equals the angle of incidence (Example 6, Section 4.5).

Now take a parabola and revolve it about its axis. This gives you a parabolic surface. A curved mirror of that form is called a *parabolic mirror*. Such mirrors are used in searchlights, automotive headlights, and reflecting telescopes. Our purpose here is to explain why.

We begin with a parabola and choose the coordinate system so that the equation takes the form $x^2 = 4cy$ with $c > 0$. We can express y in terms of x by writing

$$y = \frac{x^2}{4c}.$$

Since

$$\frac{dy}{dx} = \frac{2x}{4c} = \frac{x}{2c},$$

the tangent line at the point $P(x_0, y_0)$ has slope $m = x_0/2c$ and equation

(1) $$(y - y_0) = \frac{x_0}{2c}(x - x_0).$$

For the rest we refer to Figure 9.1.23.

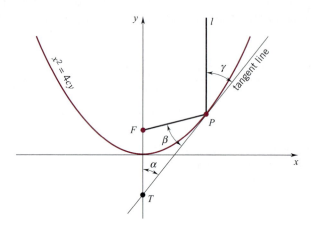

Figure 9.1.23

In the figure we have drawn a ray (a half-line) l parallel to the axis of the parabola, the y-axis. We want to show that the angles marked β and γ are equal.

Setting $x = 0$ in Equation (1), we find that

$$y = y_0 - \frac{x_0^2}{2c}.$$

Since the point (x_0, y_0) lies on the parabola, we have $x_0^2 = 4cy_0$, and thus

$$y_0 - \frac{x_0^2}{2c} = y_0 - \frac{4cy_0}{2c} = -y_0.$$

The y-coordinate of the point marked T is $-y_0$. Since the focus F is at $(0, c)$,

$$d(F, T) = y_0 + c.$$

The distance between F and P is also $y_0 + c$:

$$d(F, P) = \sqrt{x_0^2 + (y_0 - c)^2} = \sqrt{4cy_0 + (y_0 - c)^2} = \sqrt{(y_0 + c)^2} = y_0 + c.$$

$$x_0^2 = 4cy_0 \longrightarrow \qquad\qquad\qquad \longleftarrow y_0 + c > 0$$

Since $d(F, T) = d(F, P)$, the triangle TFP is isosceles and the angles marked α and β are equal. Since l is parallel to the y-axis, $\alpha = \gamma$ and thus (and this is what we wanted to show)

$$\beta = \gamma.$$

The fact that $\beta = \gamma$ has important optical consequences. It means (Figure 9.1.24) that light emitted from a source at the focus of a parabolic mirror is reflected in a beam parallel to the axis of that mirror; this is the principle of the searchlight. It also means that light coming to a parabolic mirror in a beam parallel to the axis of the mirror is reflected entirely to the focus; this is the principle of the reflecting telescope.

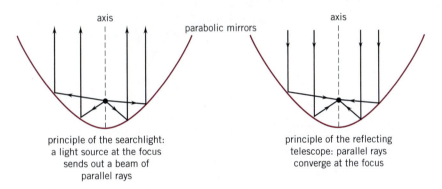

principle of the searchlight: a light source at the focus sends out a beam of parallel rays

principle of the reflecting telescope: parallel rays converge at the focus

Figure 9.1.24

Elliptical Reflectors Like the parabola, the ellipse has an interesting reflecting property. To derive it, we consider the ellipse

$$\frac{x^2}{a^2} + \frac{y^2}{b^2} = 1.$$

Differentiating implicitly with respect to x, we get

$$\frac{2x}{a^2} + \frac{2y}{b^2}\frac{dy}{dx} = 0 \qquad \text{and thus} \qquad \frac{dy}{dx} = -\frac{b^2 x}{a^2 y}.$$

The slope at the point $P(x_0, y_0)$ is therefore

$$-\frac{b^2 x_0}{a^2 y_0}, \qquad y_0 \neq 0,$$

and the tangent line has equation

$$y - y_0 = -\frac{b^2 x_0}{a^2 y_0}(x - x_0).$$

We can rewrite this last equation as

$$(b^2 x_0)x + (a^2 y_0)y - a^2 b^2 = 0.$$

We can now show the following:

(9.1.8)	At each point P of the ellipse, the focal radii $\overline{F_1 P}$ and $\overline{F_2 P}$ make equal angles with the tangent.

PROOF If P lies on the x-axis, the focal radii are coincident and there is nothing to show. To visualize the argument for a point $P = P(x_0, y_0)$ not on the x-axis, see Figure 9.1.25.

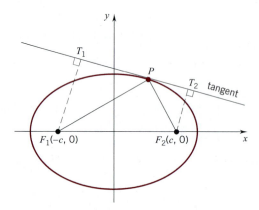

Figure 9.1.25

To show that $\overline{F_1 P}$ and $\overline{F_2 P}$ make equal angles with the tangent we need only show that the triangles $PT_1 F_1$ and $PT_2 F_2$ are similar. We can do this by showing that

$$\frac{d(T_1, F_1)}{d(F_1, P)} = \frac{d(T_2, F_2)}{d(F_2, P)}$$

or, equivalently, by showing that

$$\frac{|-b^2 x_0 c - a^2 b^2|}{\sqrt{(x_0 + c)^2 + y_0^2}} = \frac{|b^2 x_0 c - a^2 b^2|}{\sqrt{(x_0 - c)^2 + y_0^2}}. \qquad \text{(verify this)}$$

The validity of this last equation can be seen by canceling the factor b^2 and then squaring. This gives

$$\frac{(x_0 c + a^2)^2}{(x_0 + c)^2 + y_0^2} = \frac{(x_0 c - a^2)^2}{(x_0 - c)^2 + y_0^2},$$

which can be simplified to

$$(a^2 - c^2)x_0^2 + a^2 y_0^2 = a^2(a^2 - c^2) \qquad \text{and thus to} \qquad \frac{x_0^2}{a^2} + \frac{y_0^2}{b^2} = 1.$$

This last equation holds since the point $P(x_0, y_0)$ is on the ellipse. ❑

The result we just proved has the following physical consequence:

(9.1.9) An elliptical reflector takes light or sound originating at one focus and converges it at the other focus.

In elliptical rooms called "whispering chambers," a whisper at one focus, inaudible nearby, is easily heard at the other focus. You will experience this phenomenon if you visit the Statuary Room in the Capitol in Washington, D.C.

Hyperbolic Reflectors A straightforward calculation that you are asked to carry out in the exercises shows that

(9.1.10) at each point P of a hyperbola, the tangent line bisects the angle between the focal radii $\overline{F_1P}$ and $\overline{F_2P}$.

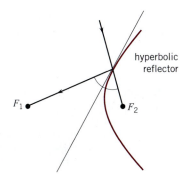

The optical consequences of this are illustrated in Figure 9.1.26. There you see the right branch of a hyperbola with foci F_1, F_2. Light or sound aimed at F_2 from any point to the left of the reflector is beamed to F_1.

Figure 9.1.26

An Application to Range Finding If observers, located at two listening posts at a known distance apart, time the firing of a cannon, the time difference multiplied by the velocity of sound gives the value of $2a$ and hence determines a hyperbola on which the cannon must be located. A third listening post gives two more hyperbolas. The cannon is found where the hyperbolas intersect.

EXERCISES 9.1

In Exercises 1–8, sketch the parabola and give an equation for it.

1. vertex $(0, 0)$, focus $(2, 0)$.

2. vertex $(0, 0)$, focus $(-2, 0)$.

3. vertex $(-1, 3)$, focus $(-1, 0)$.

4. vertex $(1, 2)$, focus $(1, 3)$.

5. focus $(1, 1)$, directrix $y = -1$.

6. focus $(2, -2)$, directrix $x = -5$.

7. focus $(1, 1)$, directrix $x = 2$.

8. focus $(2, 0)$, directrix $y = 3$.

In Exercises 9–16, find the vertex, focus, axis, and directrix; then sketch the parabola.

9. $y^2 = 2x$.

10. $x^2 = -5y$.

11. $2y = 4x^2 - 1$.

12. $y^2 = 2(x - 1)$.

13. $(x + 2)^2 = 12 - 8y$.

14. $y - 3 = 2(x - 1)^2$.

15. $x = y^2 + y + 1$.

16. $y = x^2 + x + 1$.

In Exercises 17–24, find (a) the center, (b) the foci, (c) the length of the major axis, and (d) the length of the minor axis for the given ellipse. Then sketch the graph.

17. $x^2/9 + y^2/4 = 1$.

18. $x^2/4 + y^2/9 = 1$.

19. $3x^2 + 2y^2 = 12$.

20. $3x^2 + 4y^2 - 12 = 0$.

21. $4x^2 + 9y^2 - 18y = 27$.

22. $4x^2 + y^2 - 6y + 5 = 0$.

23. $4(x - 1)^2 + y^2 = 64$.

24. $16(x - 2)^2 + 25(y - 3)^2 = 400$.

In Exercises 25–32, find an equation for the ellipse that satisfies the given conditions.

25. Foci at $(-1, 0)$, $(1, 0)$; major axis has length 6.

26. Foci at $(0, -1)$, $(0, 1)$; major axis has length 6.

27. Foci at $(1, 3)$, $(1, 9)$; minor axis has length 8.

28. Foci at $(3, 1)$, $(9, 1)$; minor axis has length 10.

29. Focus at $(1, 1)$; center at $(1, 3)$; major axis has length 10.

30. Center at $(2, 1)$; vertices at $(2, 6)$, $(1, 1)$.

31. Major axis 10; vertices at $(3, 2)$, $(3, -4)$.

32. Focus at $(6, 2)$; vertices at $(1, 7)$, $(1, -3)$.

In Exercises 33–40, find an equation for the indicated hyperbola.

33. Foci at $(-5, 0)$, $(5, 0)$; transverse axis has length 6.

34. Foci at $(-13, 0)$, $(13, 0)$; transverse axis has length 10.

35. Foci at $(0, -13)$, $(0, 13)$; transverse axis has length 10.

36. Foci at $(0, -13)$, $(0, 13)$; transverse axis has length 24.

37. Foci at $(-5, 1)$, $(5, 1)$; transverse axis has length 6.

38. Foci at $(-3, 1)$, $(7, 1)$; transverse axis has length 6.

39. Foci at $(-1, -1)$, $(-1, 1)$; transverse axis has length $\frac{1}{2}$.

40. Foci at $(2, 1)$, $(2, 5)$; transverse axis has length 3.

In Exercises 41–50, find the center, the vertices, the foci, the asymptotes, and the length of the transverse axis of the given hyperbola. Then sketch the graph.

41. $x^2 - y^2 = 1$.

42. $y^2 - x^2 = 1$.

43. $x^2/9 - y^2/16 = 1$.

44. $x^2/16 - y^2/9 = 1$.

45. $y^2/16 - x^2/9 = 1$.

46. $y^2/9 - x^2/16 = 1$.

47. $(x - 1)^2/9 - (y - 3)^2/16 = 1$.

48. $(x - 1)^2/16 - (y - 3)^2/9 = 1$.

49. $4x^2 - 8x - y^2 + 6y - 1 = 0$.

50. $-3x^2 + y^2 - 6x = 0$.

In Exercises 51–54, find an equation for the indicated parabola.

51. focus $(1, 2)$, directrix $x + y + 1 = 0$.

52. vertex $(2, 0)$, directrix $2x - y = 0$.

53. vertex $(2, 0)$, focus $(0, 2)$.

54. vertex $(3, 0)$, focus $(0, 1)$.

55. Show that every parabola has an equation of the form

$$(\alpha x + \beta y)^2 = \gamma x + \delta y + \epsilon \quad \text{with} \quad \alpha^2 + \beta^2 \neq 0.$$

HINT: Take l: $Ax + By + C = 0$ as the directrix and $F(a, b)$ as the focus.

56. A line through the focus of a parabola (different from the axis) intersects the parabola at two points P and Q. Show that the tangent line through P meets the tangent line through Q at right angles.

57. A parabola intersects a rectangle of area A at two opposite vertices. Show that, if one side of the rectangle falls on the axis of the parabola, then the parabola subdivides the rectangle into two pieces, one of area $\frac{1}{3}A$, the other of area $\frac{2}{3}A$.

58. (a) Show that every parabola with axis parallel to the y-axis has an equation of the form $y = Ax^2 + Bx + C$ with $A \neq 0$. (b) Find the vertex, the focus, and the directrix of the parabola $y = Ax^2 + Bx + C$.

59. Find equations for all the parabolas that pass through the point $(5, 6)$ and have directrix $y = 1$, axis $x = 2$.

60. Find an equation for the parabola that has horizontal axis, vertex $(-1, 1)$, and passes through the point $(-6, 13)$.

61. Find the distance between the foci of an ellipse of area A if the length of the major axis is $2a$.

62. Show that the set of all points $(a \cos t, b \sin t)$, with a, $b \neq 0$ and t real, lie on an ellipse.

63. Locate the foci of the ellipse given that the point $(3, 4)$ lies on the ellipse and the ends of the major axis are at $(0, 0)$ and $(10, 0)$.

64. Show that in an ellipse the product of the distances between the foci and a tangent to the ellipse $[d(F_1, T_1) \, d(F_2, T_2)$ in Figure 9.1.25$]$ is always the square of one-half the length of the minor axis.

The shape of an ellipse depends on its *eccentricity e*. This is half the distance between the foci divided by half the length of the major axis:

$$e = c/a.$$

For every ellipse $0 < e < 1$.

In Exercises 65–68, determine the eccentricity of the ellipse.

65. $x^2/25 + y^2/16 = 1$.

66. $x^2/16 + y^2/25 = 1$.

67. $(x - 1)^2/25 + (y + 2)^2/9 = 1$.

68. $(x + 1)^2/169 + (y - 1)^2/144 = 1$.

69. Suppose that E_1 and E_2 are both ellipses with the same major axis. Compare the shape of E_1 to the shape of E_2 if their eccentricities e_1, e_2 satisfy $e_1 < e_2$.

70. What happens to an ellipse with major axis $2a$ if its eccentricity e tends to zero?

71. What happens to an ellipse with major axis $2a$ if e tends to 1?

In Exercises 72 and 73, write an equation for the ellipse.

72. Major axis from $(-3, 0)$ to $(3, 0)$, eccentricity $\frac{1}{3}$.

73. Major axis from $(-3, 0)$ to $(3, 0)$, eccentricity $\frac{2}{3}\sqrt{2}$.

74. Let l be a line and let F be a point not on l. You have seen that the set of points P for which

$$d(F, P) = d(l, P)$$

is a parabola. Show that, if $0 < e < 1$, then the set of all points P for which

$$d(F, P) = e \, d(l, P)$$

is an ellipse of eccentricity e.   HINT: Begin by choosing a coordinate system whereby F falls on the origin and l is a vertical line $x = d$.

75. Find the center, the vertices, the foci, the asymptotes, and the length of the transverse axis of the hyperbola with equation $xy = 1$.   HINT: Define new XY-coordinates by setting $x = X + Y$ and $y = X - Y$.

For Exercises 76–78, we refer to the hyperbola in Figure 9.1.18.

76. Find functions $x = x(t)$, $y = y(t)$ such that, as t ranges over the set of real numbers, the points $(x(t), y(t))$ traverse: (a) the right branch of the hyperbola; (b) the left branch of the hyperbola.

77. Find the area of the region between the right branch of the hyperbola and the vertical line $x = 2a$.

78. Show that at each point P of the hyperbola the tangent at P bisects the angle between the focal radii $\overline{F_1 P}$ and $\overline{F_2 P}$.

The shape of a hyperbola is determined by its *eccentricity e.* This is half the distance between the foci divided by half the length of the transverse axis:

$$e = c/a.$$

For all hyperbolas $e > 1$.

In Exercises 79–82, determine the eccentricity of the hyperbola.

79. $x^2/9 - y^2/16 = 1$.   **80.** $x^2/16 - y^2/9 = 1$.

81. $x^2 - y^2 = 1$.   **82.** $x^2/25 - y^2/144 = 1$.

83. Suppose H_1 and H_2 are both hyperbolas with the same transverse axis.   Compare the shape of H_1 to the shape of H_2 if their eccentricities e_1, e_2 satisfy $e_1 < e_2$.

84. What happens to a hyperbola if e tends to 1?

85. What happens to a hyperbola if its eccentricity e increases without bound?

86. (Compare to Exercise 74)   Let l be a line and let F be a point not on l. Show that, if $e > 1$, then the set of all points P for which

$$d(F, P) = e\, d(l, P)$$

is a hyperbola of eccentricity e.   HINT: Begin by choosing a coordinate system whereby F falls on the origin and l is a vertical line $x = d$.

The line that passes through the focus of a parabola and is parallel to the directrix intersects the parabola at two points A and B.   The line segment $\overline{AB}$ is called the *latus rectum* of the parabola.   In Exercises 87–90 we work with the parabola

$x^2 = 4cy$, $c > 0$.   By Ω we mean the region bounded below by the parabola and above by the latus rectum.

87. Find the length of the latus rectum.

88. What is the slope of the parabola at the endpoints of the latus rectum?

89. Determine the area of Ω and locate the centroid.

90. Find the volume of the solid generated by revolving Ω about the y-axis and determine the centroid of the solid (6.4.6).

91. Suppose that a flexible inelastic cable (see the figure) fixed at the ends supports a horizontal load.   (Imagine a suspension bridge and think of the load on the cable as the roadway.)   Show that, if the load has constant weight per unit length, then the cable hangs in the form of a parabola.

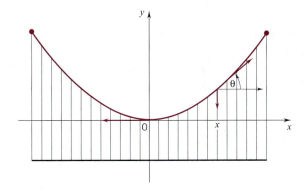

HINT: The part of the cable that supports the load from 0 to x is subject to the following forces:

(1) the weight of the load, which in this case is proportional to x

(2) the horizontal pull at 0: $p(0)$.

(3) the tangential pull at x: $p(x)$.

Balancing the vertical forces we have

$$kx = p(x) \sin \theta. \quad \text{[weight = vertical pull at x]}$$

Balancing the horizontal forces we have

$$p(0) = p(x) \cos \theta. \quad \text{[pull at 0 = horizontal pull at x]}$$

92. The parabolic mirror of a telescope gathers parallel light rays from a distant star and directs them all to the focus. Show that all the light paths to the focus are of the same length.

93. All equilateral triangles are similar; they differ only in scale.   Show that the same is true of all parabolas.

▷ **94.** A radio signal is received at the points marked P_1, P_2, P_3, P_4 in the figure on the next page.   Suppose that the signal arrives at P_1 six hundred microseconds after it arrives at P_2 and arrives at P_4 eight hundred microseconds after it arrives at P_3. Locate the source of the signal given that radio waves travel at the speed of light, 186,000 miles per second. (A microsecond is a millionth of a second.)

95. A meteor crashes somewhere in the hills that lie north of point A. The impact is heard at point A and four seconds later it is heard at point B. Two seconds still later it is heard at point C. Locate the point of impact given that A lies two miles due east of B and two miles due west of C. (Take 0.20 mile per second as the speed of sound.)

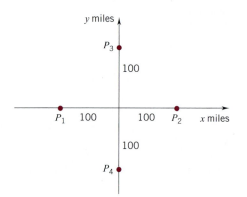

■ 9.2 POLAR COORDINATES

The purpose of coordinates is to fix position with respect to a frame of reference. When we use rectangular coordinates, our frame of reference is a pair of lines that intersect at right angles. For a *polar coordinate system,* the frame of reference is a point O that we call the *pole* and a ray that emanates from it that we call the *polar axis* (Figure 9.2.1).

Figure 9.2.1

In Figure 9.2.2 we have drawn two more rays from the pole. One lies at an angle of θ radians from the polar axis; we call it *ray θ.* The opposite ray lies at an angle of $\theta + \pi$ radians; we call it *ray $\theta + \pi$.*

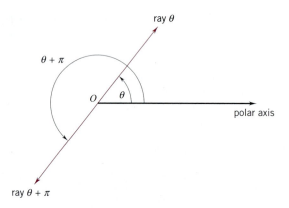

Figure 9.2.2

Figure 9.2.3 shows some points along these same rays, labeled with *polar coordinates.*

(9.2.1)

In general, a point is given *polar coordinates* $[r, \theta]$ iff it lies at a distance $|r|$ from the pole

along the ray θ, if $r \geq 0$, and along the ray $\theta + \pi$, if $r < 0$.

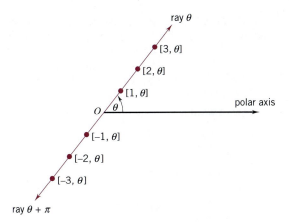

Figure 9.2.3

Figure 9.2.4 shows the point $[2, \frac{2}{3}\pi]$ at a distance of 2 units from the pole along the ray $\frac{2}{3}\pi$. The point $[-2, \frac{2}{3}\pi]$ also lies 2 units from the pole, not along the ray $\frac{2}{3}\pi$, but along the opposite ray.

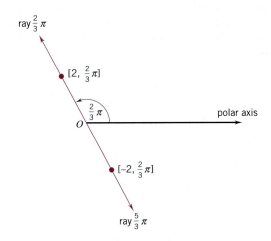

Figure 9.2.4

Polar coordinates are not unique. Many pairs $[r, \theta]$ can represent the same point.

1. If $r = 0$, it does not matter how we choose θ. The resulting point is still the pole:

(9.2.2)
$$O = [0, \theta] \qquad \text{for all } \theta.$$

2. Geometrically there is no distinction between angles that differ by an integer multiple of 2π. Consequently, as suggested in Figure 9.2.5,

(9.2.3)
$$[r, \theta] = [r, \theta + 2n\pi] \qquad \text{for all integers } n.$$

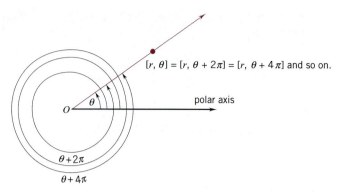

Figure 9.2.5

3. Adding π to the second coordinate is equivalent to changing the sign of the first coordinate:

(9.2.4)

$$[r, \theta + \pi] = [-r, \theta].$$

(Figure 9.2.6)

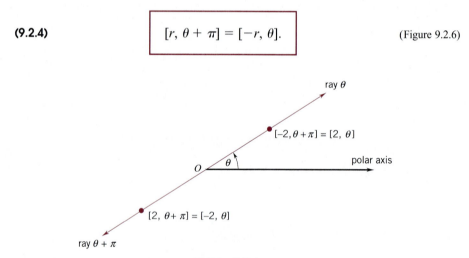

Figure 9.2.6

Remark Some authors do not allow r to take on negative values. There are some advantages to this approach. For example, the polar coordinates of a point are unique if θ is restricted to the interval $[0, 2\pi)$, or to $(-\pi, \pi]$. On the other hand, there are advantages in graphing and finding points of intersection that follow from letting r take on any real value. Since there is no convention on this issue, you should be aware that there are two approaches. ❏

Relation to Rectangular Coordinates

In Figure 9.2.7 we have superimposed a polar coordinate system on a rectangular coordinate system. We have placed the pole at the origin and the polar axis along the positive x-axis.

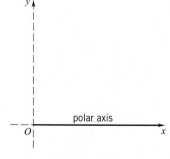

Figure 9.2.7

The relation between polar coordinates $[r, \theta]$ and rectangular coordinates (x, y) is given by the following equations:

(9.2.5)

$$x = r \cos \theta, \qquad y = r \sin \theta.$$

PROOF If $r = 0$, the formulas hold, since the point $[r, \theta]$ is then the origin and both x and y are 0:

$$0 = 0 \cos \theta, \qquad 0 = 0 \sin \theta.$$

For $r > 0$, we refer to Figure 9.2.8.† From the figure,

$$\cos \theta = \frac{x}{r}, \qquad \sin \theta = \frac{y}{r}$$

and therefore

$$x = r \cos \theta, \qquad y = r \sin \theta.$$

Suppose now that $r < 0$. Since $[r, \theta] = [-r, \theta + \pi]$ and $-r > 0$, we know from the previous case that

$$x = -r \cos (\theta + \pi), \qquad y = -r \sin (\theta + \pi).$$

Since

$$\cos (\theta + \pi) = -\cos \theta \qquad \text{and} \qquad \sin (\theta + \pi) = -\sin \theta,$$

once again we have

$$x = r \cos \theta, \qquad y = r \sin \theta. \quad \square$$

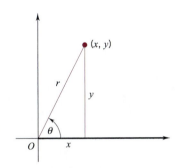

Figure 9.2.8

From the relations we just proved you can see that, unless $x = 0$,

(9.2.6)

$$\tan \theta = \frac{y}{x}$$

and, under all circumstances,

(9.2.7)

$$x^2 + y^2 = r^2. \qquad \text{(check this out)}$$

Example 1 Find the rectangular coordinates of the point P with polar coordinates $[-2, \frac{1}{3}\pi]$.

SOLUTION The relations

$$x = r \cos \theta, \qquad y = r \sin \theta$$

---

† For simplicity we have placed (x, y) in the first quadrant. A similar argument works in each of the other quadrants.

give

$$x = -2 \cos \tfrac{1}{3}\pi = -2(\tfrac{1}{2}) = -1, \qquad y = -2 \sin \tfrac{1}{3}\pi = -2(\tfrac{1}{2}\sqrt{3}) = -\sqrt{3}.$$

The point P has rectangular coordinates $(-1, -\sqrt{3})$. ❏

Example 2 Find all possible polar coordinates for the point P that has rectangular coordinates $(-2, 2\sqrt{3})$.

SOLUTION We know that

$$r \cos \theta = -2, \qquad r \sin \theta = 2\sqrt{3}.$$

We can get the possible values of r by squaring these expressions and then adding them:

$$r^2 = r^2 \cos^2 \theta + r^2 \sin^2 \theta = (-2)^2 + (2\sqrt{3})^2 = 16,$$

so that $r = \pm 4$.

Taking $r = 4$, we have

$$4 \cos \theta = -2, \qquad 4 \sin \theta = 2\sqrt{3}$$

$$\cos \theta = -\tfrac{1}{2}, \qquad \sin \theta = \tfrac{1}{2}\sqrt{3}.$$

These equations are satisfied by setting $\theta = \tfrac{2}{3}\pi$, or more generally, by setting

$$\theta = \tfrac{2}{3}\pi + 2n\pi.$$

The polar coordinates of P with first coordinate $r = 4$ are all pairs of the form

$$[4, \tfrac{2}{3}\pi + 2n\pi]$$

where n ranges over the set of all integers.

We could go through the same process again, this time taking $r = -4$, but there is no need to do so. Since $[r, \theta] = [-r, \theta + \pi]$, we know that

$$[4, \tfrac{2}{3}\pi + 2n\pi] = [-4, (\tfrac{2}{3}\pi + \pi) + 2n\pi].$$

The polar coordinates of P with first coordinate $r = -4$ are thus all pairs of the form

$$[-4, \tfrac{5}{3}\pi + 2n\pi]$$

where n again ranges over the set of all integers. ❏

Let's specify some simple sets in polar coordinates.

1. In rectangular coordinates the circle of radius a centered at the origin has equation

$$x^2 + y^2 = a^2.$$

The equation for this circle in polar coordinates is simply

$$r = a.$$

The interior of the circle is given by $0 \leq r < a$ and the exterior by $r > a$.

Figure 9.2.9

2. In rectangular coordinates the line through the origin with angle of inclination α has equation $y = mx$, where $m = \tan \alpha$ (Figure 9.2.9). The polar equation of this line is

$$\theta = \alpha. \qquad \text{(verify this)}$$

3. The vertical line $x = a$ becomes

$$r \cos \theta = a$$

and the horizontal line $y = b$ becomes

$$r \sin \theta = b.$$

4. The line $Ax + By + C = 0$ can be written

$$r(A \cos \theta + B \sin \theta) + C = 0.$$

Example 3   Find an equation in polar coordinates for the hyperbola $x^2 - y^2 = a^2$.

SOLUTION   Setting $x = r \cos \theta$ and $y = r \sin \theta$, we have

$$r^2 \cos^2 \theta - r^2 \sin^2 \theta = a^2,$$

$$r^2(\cos^2 \theta - \sin^2 \theta) = a^2,$$

$$r^2 \cos 2\theta = a^2. \quad ❏$$

Example 4   Show that the equation $r = 2a \cos \theta$, $a > 0$, represents a circle.

SOLUTION   Multiplication by r gives

$$r^2 = 2ar \cos \theta,$$

$$x^2 + y^2 = 2ax,$$

$$x^2 - 2ax + y^2 = 0,$$

$$x^2 - 2ax + a^2 + y^2 = a^2,$$

$$(x - a)^2 + y^2 = a^2.$$

This is a circle of radius a centered at the point with rectangular coordinates $(a, 0)$. See Figure 9.2.10.   ❏

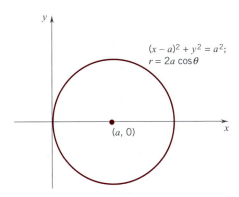

$(x - a)^2 + y^2 = a^2;$
$r = 2a \cos\theta$

$(a, 0)$

Figure 9.2.10

Symmetry

Symmetry with respect to each of the coordinate axes and with respect to the origin is illustrated in Figure 9.2.11. The coordinates marked are, of course, not the only ones possible. (The difficulties that this can cause are explained in Section 9.5.)

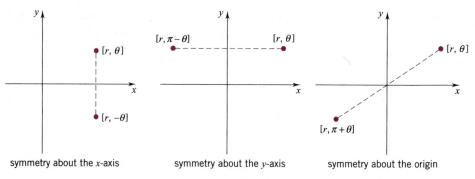

symmetry about the x-axis   symmetry about the y-axis   symmetry about the origin

Figure 9.2.11

Example 5 Test the equation $r^2 = \cos 2\theta$ for symmetry.

SOLUTION Since

$$\cos [2(-\theta)] = \cos (-2\theta) = \cos 2\theta,$$

you can see that, if $[r, \theta]$ is on the curve, then so is $[r, -\theta]$. This says that the curve is symmetric about the x-axis. Since

$$\cos [2(\pi - \theta)] = \cos (2\pi - 2\theta) = \cos (-2\theta) = \cos 2\theta,$$

you can see that, if $[r, \theta]$ is on the curve, then so is $[r, \pi - \theta]$. The curve is therefore symmetric about the y-axis.

Being symmetric about both axes, the curve must also be symmetric about the origin. You can verify this directly by noting that

$$\cos [2(\pi + \theta)] = \cos (2\pi + 2\theta) = \cos 2\theta,$$

so that, if $[r, \theta]$ lies on the curve, then so does $[r, \pi + \theta]$. A sketch of the curve, which is called a *lemniscate,* appears in Figure 9.2.12. ❑

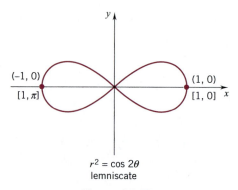

$r^2 = \cos 2\theta$
lemniscate

Figure 9.2.12

EXERCISES 9.2

In Exercises 1–8, plot the point given in polar coordinates.

1. $[1, \frac{1}{3}\pi]$.

2. $[1, \frac{1}{2}\pi]$.

3. $[-1, \frac{1}{3}\pi]$.

4. $[-1, -\frac{1}{3}\pi]$.

5. $[4, \frac{5}{4}\pi]$.

6. $[-2, 0]$.

7. $[-\frac{1}{2}, \pi]$.

8. $[\frac{1}{3}, \frac{2}{3}\pi]$.

In Exercises 9–16, find the rectangular coordinates of the given point.

9. $[3, \frac{1}{2}\pi]$.

10. $[4, \frac{1}{6}\pi]$.

11. $[-1, -\pi]$.

12. $[-1, \frac{1}{4}\pi]$.

13. $[-3, -\frac{1}{3}\pi]$.

14. $[2, 0]$.

15. $[3, -\frac{1}{2}\pi]$.

16. $[2, 3\pi]$.

In Exercises 17–24, the points are given in rectangular coordinates. Find all possible polar coordinates for each point.

17. $(0, 1)$.

18. $(1, 0)$.

19. $(-3, 0)$.

20. $(4, 4)$.

21. $(2, -2)$.

22. $(3, -3\sqrt{3})$.

23. $(4\sqrt{3}, 4)$.

24. $(\sqrt{3}, -1)$.

25. Find a formula for the distance between $[r_1, \theta_1]$ and $[r_2, \theta_2]$.

26. Show that for $r_1 > 0$, $r_2 > 0$, $|\theta_1 - \theta_2| < \pi$ the distance formula you found in Exercise 25 is just the law of cosines.

In Exercises 27–30, find the point $[r, \theta]$ symmetric to the given point about: (a) the x-axis; (b) the y-axis; (c) the origin. Express your answer with $r > 0$ and $\theta \in [0, 2\pi)$.

27. $[\frac{1}{2}, \frac{1}{6}\pi]$.

28. $[3, -\frac{5}{4}\pi]$.

29. $[-2, \frac{1}{3}\pi]$.

30. $[-3, -\frac{7}{4}\pi]$.

In Exercises 31–36, test the given curve for symmetry about the coordinate axes and the origin.

31. $r = 2 + \cos\theta$.

32. $r = \cos 2\theta$.

33. $r(\sin\theta + \cos\theta) = 1$.

34. $r\sin\theta = 1$.

35. $r^2\sin 2\theta = 1$.

36. $r^2\cos 2\theta = 1$.

In Exercises 37–48, write the equation in polar coordinates.

37. $x = 2$.

28. $y = 3$.

39. $2xy = 1$.

40. $x^2 + y^2 = 9$.

41. $x^2 + (y - 2)^2 = 4$.

42. $(x - a)^2 + y^2 = a^2$.

43. $y = x$.

44. $x^2 - y^2 = 4$.

45. $x^2 + y^2 + x = \sqrt{x^2 + y^2}$.

46. $y = mx$.

47. $(x^2 + y^2)^2 = 2xy$.

48. $(x^2 + y^2)^2 = x^2 - y^2$.

In Exercises 49–58, identify the curve and write the equation in rectangular coordinates.

49. $r\sin\theta = 4$.

50. $r\cos\theta = 4$.

51. $\theta = \frac{1}{3}\pi$.

52. $\theta^2 = \frac{1}{9}\pi^2$.

53. $r = 2(1 - \cos\theta)^{-1}$.

54. $r = 4\sin(\theta + \pi)$.

55. $r = 3\cos\theta$.

56. $\theta = -\frac{1}{2}\pi$.

57. $\tan\theta = 2$.

58. $r = 2\sin\theta$.

In Exercises 59–62, write the equation in rectangular coordinates and identify it.

59. $r = \dfrac{4}{2 - \cos\theta}$.

60. $r = \dfrac{6}{1 + 2\sin\theta}$.

61. $r = \dfrac{4}{1 - \cos\theta}$.

62. $r = \dfrac{2}{3 + 2\sin\theta}$.

63. Show that if a and b are not both zero, then the curve

$$r = a\sin\theta + b\cos\theta$$

is a circle. Find the center and the radius.

64. Find a polar equation for the set of points $P[r, \theta]$ such that the distance from P to the pole equals the distance from P to the line $x = -d$. See the figure.

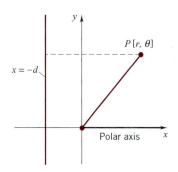

65. Find a polar equation for the set of points $P[r, \theta]$ such that the distance from P to the pole is half the distance from P to the line $x = -d$.

66. Find a polar equation for the set of points $P[r, \theta]$ such that the distance from P to the pole is twice the distance from P to the line $x = -d$.

■ 9.3 GRAPHING IN POLAR COORDINATES

We begin with the equation

$$r = \theta, \qquad \theta \geq 0.$$

The graph is a nonending spiral, part of the famous *spiral of Archimedes*. The curve is shown in detail from $\theta = 0$ to $\theta = 2\pi$ in Figure 9.3.1. At $\theta = 0$, $r = 0$; at $\theta = \frac{1}{4}\pi$, $r = \frac{1}{4}\pi$; at $\theta = \frac{1}{2}\pi$, $r = \frac{1}{2}\pi$; and so on.

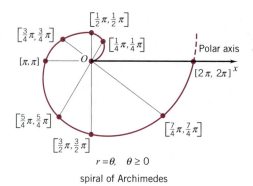

$r = \theta, \quad \theta \geq 0$

spiral of Archimedes

Figure 9.3.1

The next examples involve trigonometric functions.

Example 1   Sketch the curve

$$r = 1 - 2 \cos \theta.$$

SOLUTION   Since the cosine function is periodic, the curve $r = 1 - 2 \cos \theta$ is a closed curve. We will draw it from $\theta = 0$ to $\theta = 2\pi$. The curve just repeats itself for values of θ outside the interval $[0, 2\pi]$.

We begin by calculating a table of values:

θ	0	$\pi/4$	$\pi/3$	$\pi/2$	$2\pi/3$	$3\pi/4$	π	$5\pi/4$	$4\pi/3$	$3\pi/2$	$5\pi/3$	$7\pi/4$	2π
r	-1	-0.41	0	1	2	2.41	3	2.41	2	1	0	-0.41	-1

These points are plotted in Figure 9.3.2. The values of θ for which $r = 0$ or $|r|$ is a local maximum are:

$r = 0$ at $\theta = \tfrac{1}{3}\pi, \tfrac{5}{3}\pi$ for then $\cos \theta = \tfrac{1}{2};$ $|r|$ is a local maximum at $\theta = 0, \pi, 2\pi.$

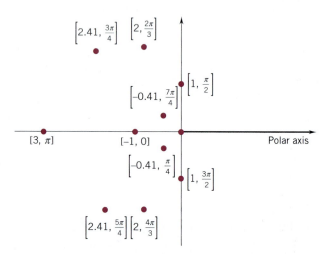

Figure 9.3.2

These five values of θ generate four intervals:

$$[0, \tfrac{1}{3}\pi], \quad [\tfrac{1}{3}\pi, \pi], \quad [\pi, \tfrac{5}{3}\pi], \quad [\tfrac{5}{3}\pi, 2\pi].$$

We sketch the curve in four stages. These stages are shown in Figure 9.3.3.

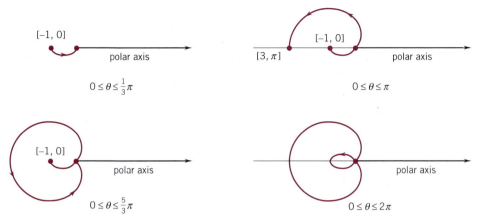

Figure 9.3.3

As θ increases from 0 to $\frac{1}{3}\pi$, $\cos\theta$ decreases from 1 to $\frac{1}{2}$ and $r = 1 - 2\cos\theta$ increases from -1 to 0.

As θ increases from $\frac{1}{3}\pi$ to π, $\cos\theta$ decreases from $\frac{1}{2}$ to -1 and r increases from 0 to 3.

As θ increases from π to $\frac{5}{3}\pi$, $\cos\theta$ increases from -1 to $\frac{1}{2}$ and r decreases from 3 to 0.

Finally, as θ increases from $\frac{5}{3}\pi$ to 2π, $\cos\theta$ increases from $\frac{1}{2}$ to 1 and r decreases from 0 to -1.

As we could have read from the equation, the curve is symmetric about the x-axis. ❏

EXAMPLE 2 Sketch the curve

$$r = \cos 2\theta, \qquad 0 \leq \theta \leq 2\pi.$$

SOLUTION As an alternative to compiling a table of values as we did in Example 1, we refer to the graph of $\cos 2\theta$ in rectangular coordinates for the values of r. See Figure 9.3.4. The values of θ for which r is zero or has an extreme value are as follows:

$$\theta = \tfrac{1}{4}n\pi, \qquad n = 0, 1, \ldots, 8.$$

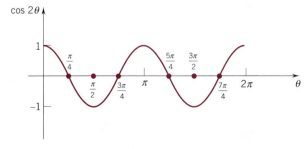

Figure 9.3.4

The curve drawn in the corresponding eight stages is shown in Figure 9.3.5.   ❑

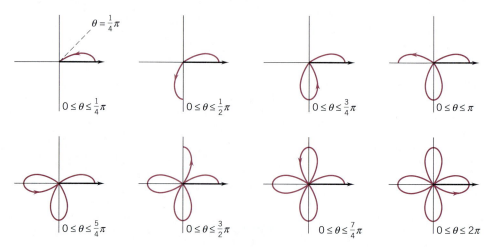

Figure 9.3.5

EXAMPLE 3   Figure 9.3.6 shows four *cardioids*, heart-shaped curves.   Rotation of $r = 1 + \cos \theta$ by $\frac{1}{2}\pi$ radians, measured in the counterclockwise direction, gives

$$r = 1 + \cos\left(\theta - \tfrac{1}{2}\pi\right) = 1 + \sin \theta.$$

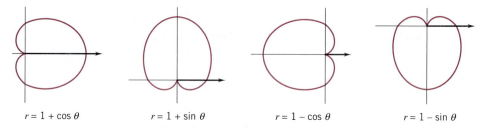

Figure 9.3.6

Rotation by another $\frac{1}{2}\pi$ radians gives

$$r = 1 + \cos\left(\theta - \pi\right) = 1 - \cos \theta.$$

Rotation by yet another $\frac{1}{2}\pi$ radians gives

$$r = 1 + \cos\left(\theta - \tfrac{3}{2}\pi\right) = 1 - \sin \theta.$$

Notice how easy it is to rotate axes in polar coordinates: each change

$$\cos \theta \longrightarrow \sin \theta \longrightarrow -\cos \theta \longrightarrow -\sin \theta$$

represents a rotation by $\frac{1}{2}\pi$ radians.   ❑

At this point we will try to give you a brief survey of some of the basic polar curves, leaving the parabola, the ellipse, and the hyperbola to Section *9.4. (The numbers a and b that appear below are to be interpreted as nonzero constants.)

Lines: $\theta = a$, $r = a \sec \theta$, $r = a \csc \theta$.   (Figure 9.3.7)

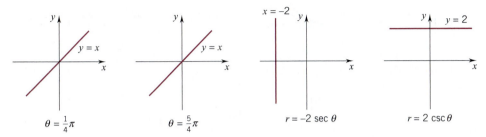

$\theta = \frac{1}{4}\pi$ $\theta = \frac{5}{4}\pi$ $r = -2 \sec \theta$ $r = 2 \csc \theta$

Figure 9.3.7

Circles: $r = a$, $r = a \sin \theta$, $r = a \cos \theta$.   (Figure 9.3.8)

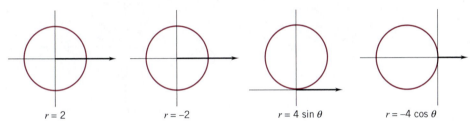

$r = 2$ $r = -2$ $r = 4 \sin \theta$ $r = -4 \cos \theta$

Figure 9.3.8

Limaçons:† $r = a + b \sin \theta$, $r = a + b \cos \theta$.   (Figure 9.3.9)

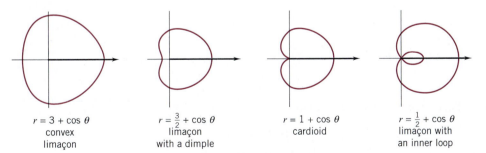

$r = 3 + \cos \theta$	$r = \frac{3}{2} + \cos \theta$	$r = 1 + \cos \theta$	$r = \frac{1}{2} + \cos \theta$
convex	limaçon	cardioid	limaçon with
limaçon	with a dimple		an inner loop

Figure 9.3.9

The general shape of the curve depends on the relative magnitudes of $|a|$ and $|b|$.

---

† From the French term for snail. The word is pronounced with a soft c.

Lemniscates:† $r^2 = a \sin 2\theta,$ $r^2 = a \cos 2\theta.$ (Figure 9.3.10)

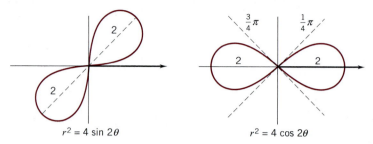

$r^2 = 4 \sin 2\theta$ $r^2 = 4 \cos 2\theta$

Figure 9.3.10

Petal Curves: $r = a \sin n\theta,$ $r = a \cos n\theta$ integer n. (Figure 9.3.11)

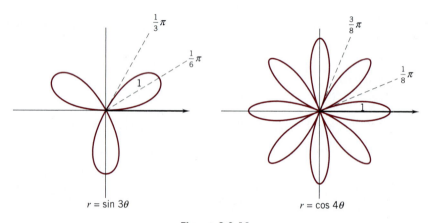

$r = \sin 3\theta$ $r = \cos 4\theta$

Figure 9.3.11

If n is *odd*, there are n petals. If n is *even*, there are $2n$ petals.

Spirals: $r = a\theta,$ spiral of Archimedes. (Figure 9.3.12)
$r = e^{a\theta},$ $\ln r = a\theta$ logarithmic spiral. (Figure 9.3.13)

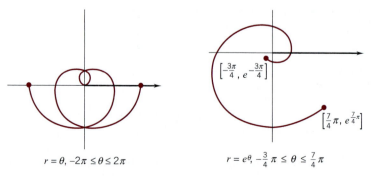

$r = \theta,\ -2\pi \le \theta \le 2\pi$ $r = e^{\theta},\ -\dfrac{3}{4}\pi \le \theta \le \dfrac{7}{4}\pi$

Figure 9.3.12 **Figure 9.3.13**

† From the Latin *lemniscatus*, meaning "adorned with pendent ribbons."

EXERCISES 9.3

In Exercises 1–34, sketch the given polar curve.

1. $\theta = -\frac{1}{4}\pi$.

2. $r = -3$.

3. $r = 4$.

4. $r = 3\cos\theta$.

5. $r = -2\sin\theta$.

6. $\theta = \frac{2}{3}\pi$.

7. $r\csc\theta = 3$.

8. $r = 1 - \cos\theta$.

9. $r = \theta$, $-\frac{1}{2}\pi \le \theta \le \pi$.

10. $r\sec\theta = -2$.

11. $r = \sin 3\theta$.

12. $r^2 = \cos 2\theta$.

13. $r^2 = \sin 2\theta$.

14. $r = \cos 2\theta$.

15. $r^2 = 4$, $0 \le \theta \le \frac{3}{4}\pi$.

16. $r = \sin\theta$.

17. $r^3 = 9r$.

18. $\theta = -\frac{1}{4}\pi$, $1 \le r < 2$.

19. $r = -1 + \sin\theta$.

20. $r^2 = 4r$.

21. $r = \sin 2\theta$.

22. $r = \cos 3\theta$, $0 \le \theta \le \frac{1}{2}\pi$.

23. $r = \cos 5\theta$, $0 \le \theta \le \frac{1}{2}\pi$.

24. $r = e^\theta$, $-\pi \le \theta \le \pi$.

25. $r = 2 + \sin\theta$.

26. $r = \cot\theta$.

27. $r = \tan\theta$.

28. $r = 2 - \cos\theta$.

29. $r = 2 + \sec\theta$.

30. $r = 3 - \csc\theta$.

31. $r = -1 + 2\cos\theta$.

32. $r = \dfrac{1}{1 + \sin\theta}$.

33. $r = \dfrac{2}{3 - 2\cos\theta}$.

34. $r = \dfrac{6}{1 - 2\cos\theta}$.

35. Suppose that you rotate the polar axis in a given polar coordinate system through an angle α. Show that if a point P has coordinates $[r, \theta]$ in the original system, then it has coordinates $[r', \theta']$ in the new system, where $r' = r$ and $\theta' = \theta - \alpha$. See the figure. Thus, the graph of a polar equation $r = f(\theta - \alpha)$ is simply the graph of the equation $r = f(\theta)$ rotated through the angle α.

36. (a) By changing to rectangular coordinates, verify that the graph of

$$r = \frac{c}{1 - \cos\theta}$$

is a parabola with its focus at the pole and directrix $x = -c$. Find the vertex and the directrix.

(b) Describe the graphs of each of the following equations:

(i) $r = \dfrac{c}{1 - \sin\theta}$,　(ii) $r = \dfrac{c}{1 + \cos\theta}$,

(iii) $r = \dfrac{c}{1 + \sin\theta}$.

HINT: Use Exercise 35.

37. (a) By changing to rectangular coordinates, verify that the graph of

$$r = \frac{ce}{1 - e\cos\theta}, \quad 0 < e < 1,$$

is an ellipse with one focus at the pole and eccentricity e. Find the center and the lengths of the major and minor axes.

(b) Describe the graphs of each of the following equations:

(i) $r = \dfrac{ce}{1 - e\sin\theta}$,　(ii) $r = \dfrac{ce}{1 + e\cos\theta}$,

(iii) $r = \dfrac{ce}{1 + e\sin\theta}$, $0 < e < 1$.

38. (a) By changing to rectangular coordinates, verify that the graph of

$$r = \frac{ce}{1 - e\cos\theta}, \quad e > 1,$$

is a hyperbola with one focus at the pole and eccentricity e. Find the center, vertices, asymptotes, and length of the transverse axis.

(b) Describe the graphs of each of the following equations:

(i) $r = \dfrac{ce}{1 - e\sin\theta}$,　(ii) $r = \dfrac{ce}{1 + e\cos\theta}$,

(iii) $r = \dfrac{ce}{1 + e\sin\theta}$, $e > 1$.

> 39. Let

$$r = \frac{ed}{1 - e\cos\theta}.$$

(Again, the symbol "e" will be representing eccentricity here.)

(a) Let $d = 1$. Use a graphing utility to graph this equation for $e = 0.5, 0.1, 0.01, 0.001$. What is the "limiting curve" as $e \to 0^+$?

(b) Continuing to let $d = 1$, graph the equation for $e = 0.9, 0.99, 0.999$. What happens as $e \to 1^-$?

(c) Now let $e = 0.5$, and graph the equation for several values of d. Use both positive and negative numbers. What effect does d have on the graph?

 40. Let

$$r = A \cos k\theta \qquad \text{or} \qquad r = A \sin k\theta$$

(a) Use a graphing utility to graph these equations for $A = 1, 2,$ and 4, and $k = 2, 3,$ and 4, respectively.
(b) What effect does the coefficient A have on the graphs? What is the difference between the "cosine" curves and the "sine" curves?

(c) Use your graphing utility to graph these equations for $A = 2$ and $k = 3/2, 5/2$. Can you predict what the graph will look like for $k = m/2$ for any odd integer m?

41. Use a graphing utility to draw the graph

$$r = 1 + \sin k\theta + \cos^2(2k\theta)$$

for $k = 1, 2, 3, 4, 5$. If you were asked to give a name to this family of curves, what would you suggest?

---

*■ 9.4 THE CONIC SECTIONS IN POLAR COORDINATES

(This material is used later in this text only in Section *13.6 where we examine planetary motion.)

In Section 9.1, we defined a parabola in terms of a focus and a directrix, but our definitions of the ellipse and hyperbola were given in terms of two foci; there was no mention of a directrix for either of these conics. In this section, we give a unified approach to the conic sections that involves a focus and directrix for all three cases.

DEFINITION 9.4.1

Let F be a fixed point (the *focus*) and ℓ a fixed line (the *directrix*) in the plane. Let e be a fixed positive number (the *eccentricity*). The set of all points P in the plane such that

$$\frac{\text{distance from } P \text{ to } F}{\text{distance from } P \text{ to } \ell} = e$$

is an ellipse if $0 < e < 1$, a parabola if $e = 1$, and a hyperbola if $e > 1$.

Figure 9.4.1 illustrates this definition. We have superimposed a polar and a rectangular coordinate system, and for convenience we refer to the pole as the origin. Without loss of generality, we have taken F at the origin and ℓ as the vertical line $x = d$.

We now have the following theorem.

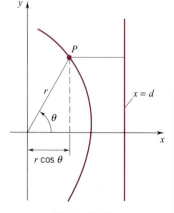

Figure 9.4.1

THEOREM 9.4.2

Let e and d be positive numbers. The set of all points P such that

$$\frac{\text{distance from } P \text{ to the origin}}{\text{distance from } P \text{ to the line } x = d} = e$$

is described by the polar equation: $r = \dfrac{ed}{1 + e \cos \theta}$

THEOREM 9.4.2 (continued)

There are three cases (see Figure 9.4.2):

I. If $0 < e < 1$, the equation represents an ellipse of eccentricity e with right focus at the origin, major axis horizontal:

$$\frac{(x + c)^2}{a^2} + \frac{y^2}{a^2 - c^2} = 1 \quad \text{with} \quad a = \frac{ed}{1 - e^2}, \quad c = ea.$$

II. If $e = 1$, the equation represents a parabola with focus at the origin and directrix $x = d$:

$$y^2 = -4\frac{d}{2}\left(x - \frac{d}{2}\right).$$

III. If $e > 1$, the equation represents a hyperbola of eccentricity e with left focus at the origin, transverse axis horizontal:

$$\frac{(x - c)^2}{a^2} - \frac{y^2}{c^2 - a^2} = 1 \quad \text{with} \quad a = \frac{ed}{e^2 - 1}, \quad c = ea.$$

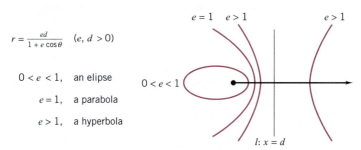

$$r = \frac{ed}{1 + e\cos\theta} \quad (e, d > 0)$$

$0 < e < 1$, an elipse

$e = 1$, a parabola

$e > 1$, a hyperbola

Figure 9.4.2

The derivation of the polar equation is left to you as an exercise.

To obtain an equation in rectangular coordinates, we first clear the denominator. Multiplication by $1 + e\cos\theta$ gives

$$r + er\cos\theta = ed.$$

Therefore

$$r = ed - er\cos\theta$$

$$r^2 = e^2d^2 - 2e^2dr\cos\theta + e^2r^2\cos^2\theta$$

(∗) $$x^2 + y^2 = e^2d^2 - 2e^2dx + e^2x^2.$$

If $0 < e < 1$, equation (∗) can be written

$$(1 - e^2)x^2 + 2e^2dx + y^2 = e^2d^2.$$

By completing the square for x we can rearrange this equation to read

$$\left(x + \frac{e^2d}{1 - e^2}\right)^2 + \frac{y^2}{1 - e^2} = \frac{e^2d^2}{(1 - e^2)^2}. \qquad \text{(carry out the details)}$$

Setting $a = ed/(1 - e^2)$ and $c = ea$, we have

$$(x + c)^2 + \frac{y^2}{1 - e^2} = a^2,$$

$$\frac{(x + c)^2}{a^2} + \frac{y^2}{(1 - e^2)a^2} = 1,$$

$$\frac{(x + c)^2}{a^2} + \frac{y^2}{a^2 - c^2} = 1.$$

This is an equation of an ellipse of eccentricity e with right focus at the origin, major axis horizontal.

If $e = 1$, equation (∗) reads $y^2 = d^2 - 2dx$. This can be written

$$y^2 = -4\frac{d}{2}\left(x - \frac{d}{2}\right).$$

As you can check, this is an equation of a parabola with focus at the origin and directrix $x = d$.

The case $e > 1$ is left to you. ❏

Example 1 The ellipse

$$r = \frac{8}{4 + 3\cos\theta}$$

has right focus at the pole, major axis horizontal. Without resorting to xy-coordinates, **(a)** find the eccentricity of the ellipse, **(b)** locate the ends of the major axis, **(c)** locate the center of the ellipse, **(d)** locate the second focus, **(e)** determine the length of the minor axis, **(f)** determine the width of the ellipse at the foci, and **(g)** sketch the ellipse.

SOLUTION

(a) Dividing numerator and denominator by 4, we have

$$r = \frac{2}{1 + \frac{3}{4}\cos\theta}.$$

The eccentricity of the ellipse is $\frac{3}{4}$.

(b) At the right end of the major axis, $\theta = 0$, $\cos\theta = 1$, and $r = \frac{8}{7}$. At the left end, $\theta = \pi$, $\cos\theta = -1$, and $r = 8$. One end of the major axis lies 8 units to the left of the pole, the other end lies $\frac{8}{7}$ units to the right of the pole.

(c) The center of the ellipse lies halfway between the endpoints of the major axis, in this case $\frac{24}{7}$ units to the left of the pole.

(d) In general the focal separation $2c$ divided by the length of the major axis $2a$ gives the eccentricity. Here $(2c) \cdot (\frac{7}{64}) = \frac{3}{4}$ and therefore $2c = \frac{48}{7}$. The second focus lies $\frac{48}{7}$ units to the left of the pole. (You can get the same result by symmetry: since the right focus lies $\frac{8}{7}$ units from the right end of the major axis, the other focus lies $\frac{8}{7}$ units from the left end of the major axis and thus $8 - \frac{8}{7} = \frac{48}{7}$ units to the left of the pole.)

(e) In general the length of the minor axis is $2b = 2\sqrt{a^2 - c^2}$. Here

$$2b = 2\sqrt{(\tfrac{32}{7})^2 - (\tfrac{24}{7})^2} = \tfrac{16}{7}\sqrt{7}.$$

The minor axis has length $\frac{16}{7}\sqrt{7} \cong 6.05$.

(f) The width of the ellipse at the foci is $2r$ where $\theta = \frac{1}{2}\pi$. The width at the foci is 4.

(g) A sketch of the ellipse appears in Figure 9.4.3. ❑

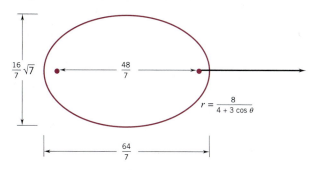

$\frac{16}{7}\sqrt{7}$ $\frac{48}{7}$ $r = \dfrac{8}{4 + 3 \cos \theta}$ $\frac{64}{7}$

Figure 9.4.3

Example 2 Sketch the ellipse

$$r = \frac{8}{4 + 3 \sin \theta}$$

SOLUTION No need to do much here. First,

$$r = \frac{8}{4 + 3 \sin \theta} = \frac{2}{1 + \frac{3}{4}\sin \theta},$$

and since $\sin \theta = \cos(\theta - \pi/2)$, this ellipse is the ellipse in Example 1 rotated by $\frac{1}{2}\pi$ radians in the counterclockwise direction. See Figure 9.4.4. ❑

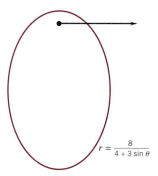

$r = \dfrac{8}{4 + 3 \sin \theta}$

Figure 9.4.4

EXERCISES *9.4

1. The parabola

$$r = \frac{1}{1 + \cos \theta}$$

has focus at the pole and directrix $x = 1$. Without resorting to xy-coordinates, (a) locate the vertex of the parabola, (b) find the length of the latus rectum (the width of the parabola at the focus), and (c) sketch the parabola.

Sketch the parabolas in Exercises 2–4.

2. $r = \dfrac{1}{1 + \sin \theta}$.

3. $r = \dfrac{1}{1 - \cos \theta}$.

4. $r = \dfrac{1}{1 - \sin \theta}$.

5. The ellipse

$$r = \frac{2}{3 + 2 \cos \theta}$$

has right focus at the pole, major axis horizontal. Without resorting to xy-coordinates, (a) find the eccentricity of the ellipse, (b) locate the ends of the major axis, (c) locate

the center of the ellipse, (d) locate the second focus, (e) determine the length of the minor axis, (f) determine the width of the ellipse at the foci, and (g) sketch the ellipse.

Sketch the ellipses in Exercises 6–8.

6. $r = \dfrac{2}{3 - 2 \sin \theta}$.

7. $r = \dfrac{2}{3 - 2 \cos \theta}$.

8. $r = \dfrac{2}{3 + 2 \sin \theta}$.

9. The hyperbola

$$r = \frac{6}{1 + 2 \cos \theta}$$

has left focus at the pole, transverse axis horizontal. Without resorting to xy-coordinates, (a) find the eccentricity of the hyperbola, (b) locate the ends of the transverse axes (for one of these you need $r < 0$), (c) locate the center of the hyperbola, (d) locate the second focus, (e) determine the width of the hyperbola at the foci, and (f) sketch the hyperbola.

Sketch the hyperbolas in Exercises 10–12.

10. $r = \dfrac{6}{1 + 2 \sin \theta}$.

11. $r = \dfrac{6}{1 - 2 \sin \theta}$.

12. $r = \dfrac{6}{1 - 2 \cos \theta}$.

In Exercises 13–15, identify the conic section and write an equation for it in rectangular coordinates.

13. $r = \dfrac{12}{2 + \sin \theta}$.

14. $r = \dfrac{4}{1 + 3 \cos \theta}$.

15. $r = \dfrac{9}{5 - 4 \sin \theta}$.

16. Let e and d be positive numbers and let ℓ be the vertical line $x = d$. Show that the equation

$$r = \frac{ed}{1 + e \cos \theta}$$

gives the set of all points P for which dist. $(P, O) = ed \cdot$ dist. (P, ℓ).

■ **9.5 THE INTERSECTION OF POLAR CURVES**

The fact that a single point has many pairs of polar coordinates can cause complications. In particular, it means that a point $[r_1, \theta_1]$ can lie on a curve given by a polar equation although the coordinates r_1 and θ_1 do not satisfy the equation. For example, the coordinates of $[2, \pi]$ do not satisfy the equation $r^2 = 4 \cos \theta$:

$$r^2 = 2^2 = 4 \quad \text{but} \quad 4 \cos \theta = 4 \cos \pi = -4.$$

Nevertheless the point $[2, \pi]$ does lie on the curve $r^2 = 4 \cos \theta$. It lies on the curve because $[2, \pi] = [-2, 0]$ and the coordinates of $[-2, 0]$ satisfy the equation:

$$r^2 = (-2)^2 = 4, \qquad 4 \cos \theta = 4 \cos 0 = 4.$$

In general, a point $P[r_1, \theta_1]$ lies on a curve given by a polar equation if it has at least one polar coordinate representation $[r, \theta]$ which satisfies the equation. The difficulties are compounded when we deal with two or more curves. Here is an example.

Example 1 Find the points where the cardioids

$$r = a(1 - \cos \theta) \qquad \text{and} \qquad r = a(1 + \cos \theta) \qquad (a > 0)$$

intersect.

SOLUTION We begin by solving the two equations simultaneously. Adding these equations, we get $2r = 2a$ and thus $r = a$. This tells us that $\cos \theta = 0$ and therefore $\theta = \frac{1}{2}\pi + n\pi$. The points $[a, \frac{1}{2}\pi + n\pi]$ all lie on both curves. Not all of these points are distinct:

for n even, $[a, \frac{1}{2}\pi + n\pi] = [a, \frac{1}{2}\pi]$;   for n odd, $[a, \frac{1}{2}\pi + n\pi] = [a, \frac{3}{2}\pi]$.

In short, by solving the two equations simultaneously we have arrived at two common points:

$$[a, \tfrac{1}{2}\pi] = (0, a) \qquad \text{and} \qquad [a, \tfrac{3}{2}\pi] = (0, -a).$$

However, by sketching the graphs of the two curves (see Figure 9.5.1), we see that there is a third point at which the curves intersect; the two curves intersect at the origin, which clearly lies on both curves:

for $r = a(1 - \cos \theta)$ take $\theta = 0, 2\pi, \ldots,$

for $r = a(1 + \cos \theta)$ take $\theta = \pi, 3\pi, \ldots.$

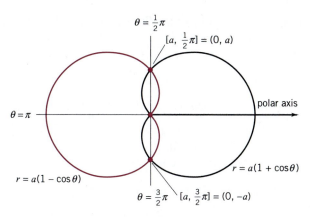

Figure 9.5.1

The reason that the origin does not appear when we solve the two equations simultaneously is that the curves do not pass through the origin "simultaneously"; that is, they do not pass through the origin for the same values of θ. Think of each of the equations

$$r = a(1 - \cos \theta) \qquad \text{and} \qquad r = a(1 + \cos \theta)$$

as giving the position of an object at time θ. At the points we found by solving the two equations simultaneously, the objects collide. (They both arrive there at the same time.) At the origin the situation is different. Both objects pass through the origin, but no collision takes place because the objects pass through the origin at *different* times. ❏

Example 2 Sketch the graphs of the polar curves

$$r = 2 \sin \theta \qquad \text{and} \qquad r = 2 \sin 2\theta,$$

and find their points of intersection, if any.

SOLUTION The graph of $r = 2 \sin \theta$ is a circle with center on the ray $\theta = \pi/2$ and radius 1. The entire circle is traced out when θ varies from 0 to π. The graph of $r = 2 \sin 2\theta$ is a petal curve with four petals. This curve is traced out when θ varies from 0 to 2π. The graphs are shown in Figure 9.5.2.

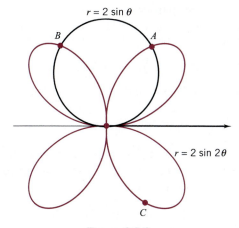

Figure 9.5.2

From the figure, we see that there are three points of intersection: the origin, the point labeled A, and the point labeled B. Solving the two equations simultaneously, we have

$$2 \sin 2\theta = 2 \sin \theta,$$

$$2 \sin \theta \cos \theta = \sin \theta,$$

and

$$\sin \theta (2 \cos \theta - 1) = 0.$$

Setting $\sin \theta = 0$, we get $\theta = n\pi$, n an integer. We can take $[0, 0]$ as the coordinates of this point of intersection.

Setting $2 \cos \theta - 1 = 0$, we get $\theta = (\pi/3) + 2n\pi$ and $\theta = (5\pi/3) + 2n\pi$. The point A clearly has coordinates $[\sqrt{3}, \pi/3]$, but the point B involves a complication. The coordinates $[\sqrt{3}, 2\pi/3]$ satisfy the equation for the circle, but they do not satisfy the equation for the petal curve: at $\theta = 2\pi/3$, $r = 2 \sin 2(2\pi/3) = -\sqrt{3}$ (the point labeled C in the figure).

You can verify that both the sets of coordinates $[\sqrt{3}, (2\pi/3) + 2n\pi]$ and $[-\sqrt{3}, (5\pi/3) + 2n\pi]$ satisfy $r = 2 \sin \theta$, while only $[-\sqrt{3}, (5\pi/3) + 2n\pi]$ satisfies $r = 2 \sin 2\theta$. Using the analogy in Example 1, if the equations describe the positions of two objects at time θ, one moving around the circle and the other around the petal curve, then at $\theta = 2\pi/3$ the object on the circle is at B and the object on the petal curve is at C. However, at time $\theta = 5\pi/3$ both objects are at the point B and they collide. The point $[-\sqrt{3}, (5\pi/3)]$ satisfies both equations simultaneously. ❑

Remark Problems of incidence (does such and such a point lie on the curve with the following polar equation?) and problems of intersection (where do such and such polar curves intersect?) can usually be analyzed by sketching the graphs. However, there are situations where such problems can be handled more readily by first changing to rectangular coordinates. For example, the rectangular equation of the curve $r^2 = 4 \cos \theta$ discussed at the beginning of this section is

$$(x^2 + y^2)^3 = 16x^2.$$

The point $P[2, \pi]$ has rectangular coordinates $(-2, 0)$, and it is easy to verify that the pair $x = -2$, $y = 0$ satisfies this equation.

In a similar manner, the symmetry properties of a curve can often be analyzed more easily in rectangular coordinates. To illustrate this, the curve C given by

$$r^2 = \sin \theta$$

is symmetric about the x-axis:

(1) $\qquad\qquad$ if $[r, \theta] \in C$, then $[r, -\theta] \in C$.

But this is not easy to see from the polar equation because, in general, if the coordinates of the first point satisfy the equation, the coordinates of the second point do not. One way to see that (1) is valid is to note that

$$[r, -\theta] \ = \ [-r, \pi - \theta]$$

and then verify that, if the coordinates of $[r, \theta]$ satisfy the equation, then so do the coordinates of $[-r, \pi - \theta]$. But all this is very cumbersome. The easiest way to see that the curve $r^2 = \sin \theta$ is symmetric about the x-axis is to write it as

$$(x^2 + y^2)^3 = y^2.$$

The other symmetries are then also clear. ❑

EXERCISES 9.5

In Exercises 1–4, determine whether the point lies on the curve.

1. $r^2 \cos \theta = 1$; $[1, \pi]$.

2. $r^2 = \cos 2\theta$; $[1, \frac{1}{4}\pi]$.

3. $r = \sin \frac{1}{3}\theta$; $[\frac{1}{2}, \frac{1}{2}\pi]$.

4. $r^2 = \sin 3\theta$; $[1, -\frac{5}{6}\pi]$.

5. Show that the point $[2, \pi]$ lies on both $r^2 = 4 \cos \theta$ and $r = 3 + \cos \theta$.

6. Show that the point $[2, \frac{1}{2}\pi]$ lies on both $r^2 \sin \theta = 4$ and $r = 2 \cos 2\theta$.

In Exercises 7–16, sketch the curves and find the points at which they intersect. Express your answers in rectangular coordinates.

7. $r = \sin \theta$, $r = -\cos \theta$.

8. $r^2 = \sin \theta$, $r = 2 - \sin \theta$.

9. $r = \cos^2 \theta$, $r = -1$.

10. $r = 2 \sin \theta$, $r = 2 \cos \theta$.

11. $r = 1 - \cos \theta$, $r = \cos \theta$.

12. $r = 1 - \cos \theta$, $r = \sin \theta$.

13. $r = \dfrac{1}{1 - \cos \theta}$, $r \sin \theta = 2$.

14. $r = \dfrac{1}{2 - \cos \theta}$, $r \cos \theta = 1$.

15. $r = \cos 3\theta$, $r = \cos \theta$.

16. $r = \sin 2\theta$, $r = \sin \theta$.

▶ **17.** (a) Use a graphing utility to draw the graphs of

$$r = 1 + \sin \theta \quad \text{and} \quad r^2 = 4 \sin 2\theta$$

in the same coordinate system.

(b) Find all the points of intersection of the two graphs and estimate their coordinates accurate to three decimal places.

▶ **18.** Repeat Exercise 17 for the pair of equations

$$r = 1 + \cos \theta \quad \text{and} \quad r = 1 + \cos (\tfrac{1}{2}\theta).$$

▶ **19.** Repeat Exercise 17 for the pair of equations

$$r = 1 - 3 \cos \theta \quad \text{and} \quad r = 2 - 5 \sin \theta.$$

▶ **20.** (a) The electrostatic charge distribution consisting of a charge q, $(q > 0)$ at the point $[r, 0]$, and a charge $-q$ at $[r, \pi]$ is called a *dipole*. The *lines of force* for the dipole are given by the equations

$$r = k \sin^2 \theta.$$

Use a graphing utility to draw the lines of force for $k = 1, 2, 3$.

(b) The *equipotential lines* (the set of points with equal electric potential) for the dipole are given by the equations

$$r^2 = m \cos \theta.$$

Use a graphing utility to draw the equipotential lines for $m = 1, 2, 3$.

(c) Graph the curves $r = 2 \sin^2 \theta$ and $r^2 = 2 \cos \theta$ in the same coordinate system and estimate the coordinates of their points of intersection. Use four decimal place accuracy.

■ 9.6 AREA IN POLAR COORDINATES

Here we develop a technique for calculating the area of a region the boundary of which is given in polar coordinates.

As a start, we suppose that α and β are two real numbers with $\alpha < \beta \leq \alpha + 2\pi$. We take ρ as a function that is continuous on $[\alpha, \beta]$ and keeps a constant sign on that interval. We want the area of the polar region Γ generated by the curve

$$r = \rho(\theta), \quad \alpha \leq \theta \leq \beta.$$

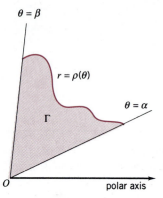

Figure 9.6.1

Such a region is portrayed in Figure 9.6.1.

In the figure $\rho(\theta)$ remains nonnegative. If $\rho(\theta)$ were negative, the region Γ would appear on the opposite side of the pole. In either case, the area of Γ is given by the formula

(9.6.1)

$$A = \int_{\alpha}^{\beta} \tfrac{1}{2}[\rho(\theta)]^2 \, d\theta.$$

PROOF We consider the case where $\rho(\theta) \geq 0$. We take $P = \{\theta_0, \theta_1, \ldots, \theta_n\}$ as a partition of $[\alpha, \beta]$ and direct our attention to what happens between θ_{i-1} and θ_i. We set

$r_i = $ min value of ρ on $[\theta_{i-1}, \theta_i]$ and $R_i = $ max value of ρ on $[\theta_{i-1}, \theta_i]$.

The part of Γ that lies between θ_{i-1} and θ_i contains a circular sector of radius r_i and central angle $\Delta \theta_i = \theta_i - \theta_{i-1}$ and is contained in a circular sector of radius R_i and central angle $\Delta \theta_i$. (See Figure 9.6.2.) Its area A_i must therefore satisfy the inequality

$$\tfrac{1}{2}r_i^2 \, \Delta\theta_i \leq A_i \leq \tfrac{1}{2}R_i^2 \, \Delta\theta_i.†$$

By summing these inequalities from $i = 1$ to $i = n$, you can see that the total area A must satisfy the inequality

(1)

$$L_f(P) \leq A \leq U_f(P)$$

where

$$f(\theta) = \tfrac{1}{2}\, [\rho(\theta)]^2.$$

Since f is continuous and (1) holds for every partition P of $[a, b]$, we must have

$$A = \int_{\alpha}^{\beta} f(\theta) \, d\theta = \int_{\alpha}^{\beta} \tfrac{1}{2}[\rho(\theta)]^2 \, d\theta. \quad \square$$

Figure 9.6.2

Example 1 Calculate the area enclosed by the cardioid

$$r = 1 - \cos \theta. \qquad \text{(Figure 9.6.3)}$$

SOLUTION The entire curve is traced out when θ increases from 0 to 2π and $1 - \cos \theta \geq 0$ on $[0, 2\pi]$. Thus, the area A is given by

$$A = \int_0^{2\pi} \tfrac{1}{2}(1 - \cos \theta)^2 \, d\theta = \tfrac{1}{2} \int_0^{2\pi} (1 - 2\cos \theta + \cos^2 \theta) \, d\theta$$

$$= \tfrac{1}{2} \int_0^{2\pi} (\tfrac{3}{2} - 2\cos \theta + \tfrac{1}{2}\cos 2\theta) \, d\theta.$$

half-angle formula: $\cos^2 \theta = \tfrac{1}{2} + \tfrac{1}{2}\cos 2\theta$

---

† The area of a circular sector of radius r and central angle α is $\tfrac{1}{2}r^2\alpha$.

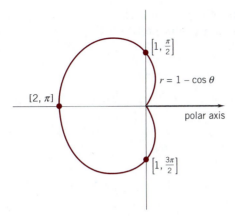

Figure 9.6.3

Since

$$\int_0^{2\pi} \cos \theta \, d\theta = 0 \qquad \text{and} \qquad \int_0^{2\pi} \cos 2\theta \, d\theta = 0,$$

we have

$$A = \tfrac{1}{2} \int_0^{2\pi} \tfrac{3}{2} \, d\theta = \tfrac{3}{4} \int_0^{2\pi} d\theta = \tfrac{3}{2}\pi. \quad \square$$

A slightly more complicated type of region is pictured in Figure 9.6.4. We approach the problem of calculating the area of the region Ω in the same way that we calculated the area between two curves in Section 5.5. That is, we calculate the area out to $r = \rho_2(\theta)$ and subtract from it the area out to $r = \rho_1(\theta)$. This gives the formula

(9.6.2)
$$\text{area of } \Omega = \int_\alpha^\beta \tfrac{1}{2}([\rho_2(\theta)]^2 - [\rho_1(\theta)]^2) \, d\theta.$$

Figure 9.6.4

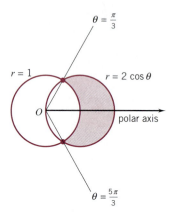

$\theta = \frac{\pi}{3}$

$r = 1$

$r = 2 \cos \theta$

O

polar axis

$\theta = \frac{5\pi}{3}$

Figure 9.6.5

Example 2 Find the area of the region that consists of all points that lie within the circle $r = 2 \cos \theta$ but outside the circle $r = 1$.

SOLUTION The region is shown in Figure 9.6.5. Our first step is to find values of θ for the two points where the circles intersect:

$$2 \cos \theta = 1, \qquad \cos \theta = \tfrac{1}{2}, \qquad \theta = \tfrac{1}{3}\pi, \tfrac{5}{3}\pi.$$

Since the region is symmetric about the polar axis, the area below the polar axis equals the area above the polar axis. Thus

$$A = 2 \int_0^{\pi/3} \tfrac{1}{2}([2 \cos \theta]^2 - [1]^2) \, d\theta.$$

If you carry out the integration, you will see that $A = \tfrac{1}{3}\pi + \tfrac{1}{2}\sqrt{3} \cong 1.91$. ❑

To find the area between two polar curves, we first determine the curves that serve as outer and inner boundaries of the region and the intervals of θ values over which these boundaries are traced out. Since the polar coordinates of a point are not unique, extra care is needed to determine these intervals of θ values.

Example 3 Find the area A of the region between the inner and outer loops of the limaçon

$$r = 1 - 2 \cos \theta. \qquad \text{(Figure 9.6.6)}$$

$r = 1 - 2 \cos\theta$

$\theta = \frac{1}{3}\pi$

$\theta = \frac{5}{3}\pi$

Figure 9.6.6

SOLUTION It is easy to verify that $r = 0$ when $\theta = \pi/3$ and when $\theta = 5\pi/3$. The outer loop is formed by having θ increase from $\pi/3$ to $5\pi/3$. Thus

$$\text{area within outer loop} = A_1 = \int_{\pi/3}^{5\pi/3} \tfrac{1}{2}[1 - 2 \cos \theta]^2 \, d\theta.$$

The lower half of the inner loop is formed when θ increases from 0 to $\pi/3$, and the upper half when θ increases from $5\pi/3$ to 2π (verify this). Therefore, we have

$$\text{area within inner loop} = A_2 = \int_0^{\pi/3} \tfrac{1}{2}[1 - 2 \cos \theta]^2 \, d\theta + \int_{5\pi/3}^{2\pi} \tfrac{1}{2}[1 - 2 \cos \theta]^2 \, d\theta.$$

Now

$$\int \tfrac{1}{2}[1 - 2 \cos \theta]^2 \, d\theta = \tfrac{1}{2} \int [1 - 4 \cos \theta + 4 \cos^2\theta] \, d\theta$$

$$= \tfrac{1}{2} \int [1 - 4 \cos \theta + 2(1 + \cos 2\theta)] \, d\theta$$

$$= \tfrac{1}{2} \int [3 - 4 \cos \theta + 2 \cos 2\theta] \, d\theta$$

$$= \tfrac{1}{2}[3\theta - 4 \sin \theta + \sin 2\theta] + C.$$

Therefore,

$$A_1 = \tfrac{1}{2} \left[3\theta - 4 \sin \theta + \sin 2\theta \right]_{\pi/3}^{5\pi/3} = 2\pi + \tfrac{3}{2}\sqrt{3}$$

and

$$A_2 = \tfrac{1}{2}\Big[3\theta - 4\sin\theta + \sin 2\theta \Big]_0^{\pi/3} + \tfrac{1}{2}\Big[3\theta - 4\sin\theta + \sin 2\theta \Big]_{5\pi/3}^{2\pi}$$
$$= \tfrac{1}{2}\pi - \tfrac{3}{4}\sqrt{3} + \tfrac{1}{2}\pi - \tfrac{3}{4}\sqrt{3} = \pi - \tfrac{3}{2}\sqrt{3}.$$

Thus,

$$A = A_1 - A_2 = 2\pi + \tfrac{3}{2}\sqrt{3} - (\pi - \tfrac{3}{2}\sqrt{3}) = \pi + 3\sqrt{3} \cong 8.34. \quad ❑$$

Remark   A more efficient approach to the solution in Example 3 would have been to use the symmetry of the region. The region is symmetric about the x-axis, and so the area A is also given by

$$A = 2\int_{\pi/3}^{\pi} \tfrac{1}{2}[1 - 2\cos\theta]^2\, d\theta - 2\int_0^{\pi/3} \tfrac{1}{2}[1 - 2\cos\theta]^2\, d\theta. \quad ❑$$

Example 4   The region Ω common to the circle $r = 2\sin\theta$ and the limaçon $r = \tfrac{3}{2} - \sin\theta$ is indicated in Figure 9.6.7. The θ coordinates of the points of intersection can be found by solving the two equations simultaneously:

$$2\sin\theta = \tfrac{3}{2} - \sin\theta, \qquad \sin\theta = \tfrac{1}{2}, \qquad \text{and} \qquad \theta = \tfrac{\pi}{6},\ \tfrac{5\pi}{6}.$$

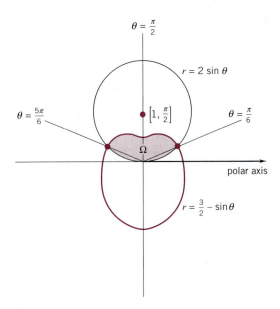

Figure 9.6.7

Thus, the area of Ω can be represented as follows:

$$\text{area of } \Omega = \int_0^{\pi/6} \tfrac{1}{2}[2\sin\theta]^2\, d\theta + \int_{\pi/6}^{5\pi/6} \tfrac{1}{2}[\tfrac{3}{2} - \sin\theta]^2\, d\theta + \int_{5\pi/6}^{\pi} \tfrac{1}{2}[2\sin\theta]^2\, d\theta$$

or, using the symmetry of the region,

$$\text{area of } \Omega = 2\int_0^{\pi/6} \tfrac{1}{2}[2\sin\theta]^2\, d\theta + 2\int_{\pi/6}^{\pi/2} \tfrac{1}{2}[\tfrac{3}{2} - \sin\theta]^2\, d\theta.$$

You can verify that the area of Ω is $\tfrac{5}{4}\pi - \tfrac{15}{8}\sqrt{3} \cong 0.68. \quad ❑$

EXERCISES 9.6

In Exercises 1–6, calculate the area enclosed by the given curve. Take $a > 0$.

1. $r = a \cos \theta$ from $\theta = -\frac{1}{2}\pi$ to $\theta = \frac{1}{2}\pi$.

2. $r = a \cos 3\theta$ from $\theta = -\frac{1}{6}\pi$ to $\theta = \frac{1}{6}\pi$.

3. $r = a\sqrt{\cos 2\theta}$ from $\theta = -\frac{1}{4}\pi$ to $\theta = \frac{1}{4}\pi$.

4. $r = a(1 + \cos 3\theta)$ from $\theta = -\frac{1}{3}\pi$ to $\theta = \frac{1}{3}\pi$.

5. $r^2 = a^2 \sin^2 \theta$. 6. $r^2 = a^2 \sin^2 2\theta$.

In Exercises 7–12, calculate the area of the given region.

7. $r = \tan 2\theta$ and the rays $\theta = 0$, $\theta = \frac{1}{8}\pi$.

8. $r = \cos \theta$, $r = \sin \theta$, and the rays $\theta = 0$, $\theta = \frac{1}{4}\pi$.

9. $r = 2 \cos \theta$, $r = \cos \theta$, and the rays $\theta = 0$, $\theta = \frac{1}{4}\pi$.

10. $r = 1 + \cos \theta$, $r = \cos \theta$, and the rays $\theta = 0$, $\theta = \frac{1}{2}\pi$.

11. $r = a(4 \cos \theta - \sec \theta)$ and the rays $\theta = 0$, $\theta = \frac{1}{4}\pi$.

12. $r = \frac{1}{2} \sec^2 \frac{1}{2}\theta$ and the vertical line through the origin.

In Exercises 13–16, find the area of the given region.

13. $r = e^\theta$, $0 \le \theta \le \pi$; $r = \theta$, $0 \le \theta \le \pi$; the rays $\theta = 0$, $\theta = \pi$.

14. $r = e^\theta$, $2\pi \le \theta \le 3\pi$; $r = \theta$, $0 \le \theta \le \pi$; the rays $\theta = 0$, $\theta = \pi$.

15. $r = e^\theta$, $0 \le \theta \le \pi$; $r = e^{\theta/2}$, $0 \le \theta \le \pi$; the rays $\theta = 2\pi$, $\theta = 3\pi$.

16. $r = e^\theta$, $0 \le \theta \le \pi$; $r = e^\theta$, $2\pi \le \theta \le 3\pi$; the rays $\theta = 0$, $\theta = \pi$.

In Exercises 17–28, represent the indicated area by one or more integrals.

17. Outside $r = 2$, but inside $r = 4 \sin \theta$.

18. Outside $r = 1 - \cos \theta$, but inside $r = 1 + \cos \theta$.

19. Inside $r = 4$, and to the right of $r = 2 \sec \theta$.

20. Inside $r = 2$, but outside $r = 4 \cos \theta$.

21. Inside $r = 4$, and between the lines $\theta = \frac{1}{2}\pi$ and $r = 2 \sec \theta$.

22. Inside the inner loop of $r = 1 - 2 \sin \theta$.

23. Inside one petal of $r = 2 \sin 3\theta$.

24. Outside $r = 1 + \cos \theta$, but inside $r = 2 - \cos \theta$.

25. Interior to both $r = 1 - \sin \theta$ and $r = \sin \theta$.

26. Inside one petal of $r = 5 \cos 6\theta$.

27. Outside $r = \cos 2\theta$, but inside $r = 1$.

28. Interior to both $r = 2a \cos \theta$ and $r = 2a \sin \theta$, $a > 0$.

29. Let n be a positive integer. Prove that the petal curves $r = a \cos 2n\theta$ and $r = a \sin 2n\theta$, $a > 0$, all enclose exactly the same area. Find the area.

30. Let n be a positive integer. Prove that the petal curves $r = a \cos([2n + 1]\theta)$ and $r = a \sin([2n + 1]\theta)$, $a > 0$, all enclose exactly the same area. Find the area.

Centroids in Polar Coordinates

Let Ω be the region bounded by the polar curve $r = f(\theta)$ between $\theta = \alpha$ and $\theta = \beta$. Since the centroid of a triangle lies on each median, two-thirds of the distance from the vertex to the opposite side, it follows that the x and y coordinates of the centroid of the "triangular" region shown in the figure are given approximately by

$$\bar{x} = \tfrac{2}{3} r \cos \theta, \qquad \bar{y} = \tfrac{2}{3} r \sin \theta.$$

These approximations improve as $\Delta \theta \to 0$.

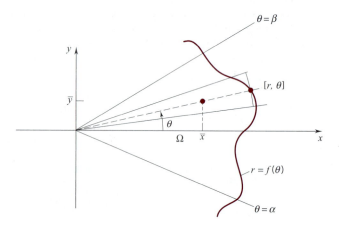

31. Following the approach used in Section 6.4, show that the rectangular coordinates of the centroid of Ω are

$$\bar{x} = \frac{\tfrac{2}{3} \displaystyle\int_\alpha^\beta r^3 \cos \theta \, d\theta}{\displaystyle\int_\alpha^\beta r^2 \, d\theta};$$

$$\bar{y} = \frac{\tfrac{2}{3} \displaystyle\int_\alpha^\beta r^3 \sin \theta \, d\theta}{\displaystyle\int_\alpha^\beta r^2 \, d\theta}.$$

In Exercises 32–34, use the result of Exercise 31 to find the rectangular coordinates of the centroid of the given region.

32. The region enclosed by $r = 4$ between $\theta = -\alpha$ and $\theta = \alpha$, $0 < \alpha < \pi/2$.

33. The region enclosed by the cardioid $r = 1 + \cos \theta$.

34. The region enclosed by $r = 2 + \sin \theta$.

35. The curve whose equation in rectangular coordinates is

$$y^2 = x^2\left(\frac{a-x}{a+x}\right), a > 0,$$

is called a *strophoid*.
(a) Show that the polar equation of this curve has the form

$$r = a \cos 2\theta \sec \theta.$$

(b) Use a graphing utility to draw the graph of the curves for $a = 1, 2$, and 4.
(c) Let $a = 2$. Find the area inside the loop.

36. The curve whose equation in rectangular coordinates is

$$(x^2 + y^2)^2 = ax^2 y, a > 0,$$

is called a *bifolium*.
(a) Show that the polar equation of this curve has the form

$$r = a \sin \theta \cos^2 \theta.$$

(b) Use a graphing utility to draw the graph of the curves for $a = 1, 2$, and 4.
(c) Let $a = 2$. Find the area inside one of the loops.

■ 9.7 CURVES GIVEN PARAMETRICALLY

We begin with a pair of functions $x = x(t), y = y(t)$ differentiable on the interior of an interval I. At the endpoints of I (if any) we require only continuity.

For each number t in I we can interpret $(x(t), y(t))$ as the point with x-coordinate $x(t)$ and y-coordinate $y(t)$. Then, as t ranges over I, the point $(x(t), y(t))$ traces out a path in the xy-plane. (Figure 9.7.1) We call such a path a *parametrized curve* and refer to t as the *parameter*.

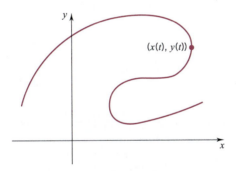

Figure 9.7.1

Example 1   Identify the curve parametrized by the functions

$$x(t) = t + 1, \quad y(t) = 2t - 5, \qquad t \text{ real.}$$

SOLUTION   We can express $y(t)$ in terms of $x(t)$:

$$y(t) = 2[x(t) - 1] - 5 = 2x(t) - 7.$$

The functions parametrize the line $y = 2x - 7$: as t ranges over the set of real numbers, the point $(x(t), y(t))$ traces out the line $y = 2x - 7$.   ❑

Example 2   Identify the curve parametrized by the functions

$$x(t) = 2t, \quad y(t) = t^2, \qquad t \in [0, \infty).$$

SOLUTION   From the first equation $t = \frac{1}{2}x(t)$, and so

$$y(t) = \frac{1}{4}[x(t)]^2.$$

The functions parametrize that part of the parabola $y = \frac{1}{4}x^2$ which lies in the right half plane: as t ranges over the interval $[0, \infty)$, the point $(x(t), y(t))$ traces out the parabola $y = \frac{1}{4}x^2$ for $x \geq 0$ (Figure 9.7.2). ❏

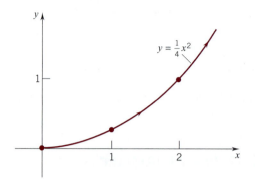

Figure 9.7.2

Example 3 Identify the curve parametrized by the functions

$$x(t) = \sin^2 t, \quad y(t) = \cos t, \qquad t \in [0, \pi].$$

SOLUTION Note first that

$$x(t) = \sin^2 t = 1 - \cos^2 t = 1 - [y(t)]^2.$$

The points $(x(t), y(t))$ all lie on the parabola

$$x = 1 - y^2. \tag{Figure 9.7.3}$$

At $t = 0$, $x = 0$ and $y = 1$; at $t = \pi$, $x = 0$ and $y = -1$. As t ranges from 0 to π, the point $(x(t), y(t))$ traverses the parabolic arc

$$x = 1 - y^2, \qquad -1 \leq y \leq 1$$

from the point $(0, 1)$ to the point $(0, -1)$. ❏

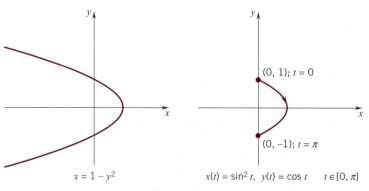

Figure 9.7.3

Remark   Changing the domain in the previous problem to all real t does *not* give us any more of the parabola.   For any given t we still have

$$0 \le x(t) \le 1 \quad \text{and} \quad -1 \le y(t) \le 1.$$

As t ranges over the set of real numbers, the point $(x(t), y(t))$ traces out that same parabolic arc back and forth an infinite number of times.   ❏

The functions

(9.7.1)
$$x(t) = x_0 + t(x_1 - x_0), \quad y(t) = y_0 + t(y_1 - y_0), \qquad t \text{ real}$$

parametrize the line that passes through (x_0, y_0) and (x_1, y_1), where $(x_0, y_0) \ne (x_1, y_1)$.

PROOF   If $x_1 = x_0$, then we have

$$x(t) = x_0, \quad y(t) = y_0 + t(y_1 - y_0). \qquad (y_0 \ne y_1)$$

As t ranges over the set of real numbers, $x(t)$ remains constantly x_0 and $y(t)$ ranges over the set of real numbers.   The functions parametrize the vertical line $x = x_0$.   Since $x_1 = x_0$, both (x_0, y_0) and (x_1, y_1) lie on this vertical line.
   If $x_1 \ne x_0$, then we can solve the first equation for t:

$$t = \frac{x(t) - x_0}{x_1 - x_0}.$$

Substituting this into the second equation we obtain the identity

$$y(t) - y_0 = \frac{y_1 - y_0}{x_1 - x_0} [x(t) - x_0].$$

The functions parametrize the line with equation

$$y - y_0 = \frac{y_1 - y_0}{x_1 - x_0}(x - x_0).$$

This is the line determined by (x_0, y_0) and (x_1, y_1).   ❏

The functions $x(t) = a \cos t$, $y(t) = b \sin t$, where $a, b > 0$, satisfy the identity

$$\frac{[x(t)]^2}{a^2} + \frac{[y(t)]^2}{b^2} = 1.$$

As t ranges over any interval of length 2π, the point $(x(t), y(t))$ traces out the ellipse

$$\frac{x^2}{a^2} + \frac{y^2}{b^2} = 1.$$

Usually we let t range from 0 to 2π and parametrize the ellipse by setting

(9.7.2)
$$x(t) = a \cos t, \quad y(t) = b \sin t, \qquad t \in [0, 2\pi].$$

If $b = a$, we have a circle. We can parametrize the circle

$$x^2 + y^2 = a^2$$

by setting

(9.7.3)

$$x(t) = a \cos t, \quad y(t) = a \sin t, \quad t \in [0, 2\pi].$$

The functions $x(t) = a \cosh t$, $y(t) = b \sinh t$, where a, $b > 0$, satisfy the identity

$$\frac{[x(t)]^2}{a^2} - \frac{[y(t)]^2}{b^2} = 1.$$

As t ranges over the real numbers, $x(t) = a \cosh t > 0$ and the point $(x(t), y(t))$ traces out the right branch of the hyperbola

$$\frac{x^2}{a^2} - \frac{y^2}{b^2} = 1. \qquad \text{(Figure 9.7.4)}$$

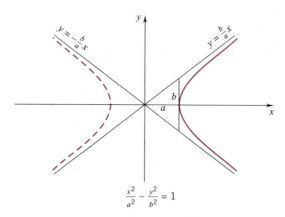

Figure 9.7.4

Thus, the right branch of the hyperbola is parametrized by

$$x(t) = a \cosh t, \quad y(t) = b \cosh t, \qquad t \text{ real.}$$

The left branch is parametrized by

$$x(t) = -a \cosh t, \quad y(t) = b \sinh t, \qquad t \text{ real.}$$

Interpreting the parameter t as time measured in seconds, we can think of a pair of parametric equations, $x = x(t)$ and $y = y(t)$, as describing the motion of a particle in the xy-plane. Different parametrizations of the same curve represent different ways of traversing that curve.

Example 4 The line that passes through the points $(1, 2)$ and $(3, 6)$ has equation $y = 2x$. The line segment that joins these same points is given by

$$y = 2x, \qquad 1 \le x \le 3.$$

We will parametrize this line segment in different ways and interpret each parametrization as the motion of a particle.

We begin by setting

$$x(t) = t, \quad y(t) = 2t, \qquad t \in [1, 3].$$

At time $t = 1$, the particle is at the point $(1, 2)$. It traverses the line segment and arrives at the point $(3, 6)$ at time $t = 3$.

Now we set

$$x(t) = t + 1, \quad y(t) = 2t + 2, \qquad t \in [0, 2].$$

At time $t = 0$, the particle is at the point $(1, 2)$. It traverses the line segment and arrives at the point $(3, 6)$ at time $t = 2$.

The equations

$$x(t) = 3 - t, \quad y(t) = 6 - 2t, \qquad t \in [0, 2]$$

represent a traversal of that same line segment in the opposite direction. At time $t = 0$, the particle is at $(3, 6)$. It arrives at $(1, 2)$ at time $t = 2$.

Set

$$x(t) = 3 - 4t, \quad y(t) = 6 - 8t, \qquad t \in [0, \tfrac{1}{2}].$$

Now the particle traverses the same line segment in only half a second. At time $t = 0$, the particle is at $(3, 6)$. It arrives at $(1, 2)$ at time $t = \tfrac{1}{2}$.

Finally we set

$$x(t) = 2 - \cos t, \quad y(t) = 4 - 2\cos t, \qquad t \in [0, 4\pi].$$

In this instance the particle begins and ends its motion at the point $(1, 2)$, having traced and retraced the line segment twice during a span of 4π seconds. See Figure 9.7.5. ❏

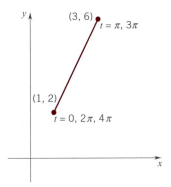

Figure 9.7.5

Remark If the path of an object is given in terms of a time parameter t and we eliminate the parameter to obtain an equation in x and y, it may be that we obtain a clearer view of the path, but we do so at considerable expense. The equation in x and y does not tell us where the particle is at any time t. The parametric equations do. ❏

Example 5 We return to the ellipse

$$\frac{x^2}{a^2} + \frac{y^2}{b^2} = 1.$$

A particle with position given by the equations

$$x(t) = a \cos t, \quad y(t) = b \sin t \qquad \text{with } t \in [0, 2\pi]$$

traverses the ellipse in a counterclockwise manner. It begins at the point $(a, 0)$ and makes a full circuit in 2π seconds. If the equations of motion are

$$x(t) = a \cos 2\pi t, \quad y(t) = -b \sin 2\pi t \qquad \text{with } t \in [0, 1],$$

the particle still travels the same ellipse, but in a different manner. Once again it starts at $(a, 0)$, but this time it moves clockwise and makes the full circuit in only one second. If the equations of motion are

$$x(t) = a \sin 4\pi t, \quad y(t) = b \cos 4\pi t \qquad \text{with } t \in [0, \infty),$$

the motion begins at $(0, b)$ and goes on in perpetuity. The motion is clockwise, a complete circuit taking place every half second. ❏

Intersections and Collisions

Example 6 Two particles start at the same instant, the first along the linear path

$$x_1(t) = \tfrac{16}{3} - \tfrac{8}{3}t, \quad y_1(t) = 4t - 5, \qquad t \geq 0$$

and the second along the elliptical path

$$x_2(t) = 2 \sin \tfrac{1}{2}\pi t, \quad y_2(t) = -3 \cos \tfrac{1}{2}\pi t, \qquad t \geq 0.$$

(a) At what points, if any, do the paths intersect?

(b) At what points, if any, do the particles collide?

SOLUTION To see where the paths intersect, we find equations for them in x and y. The linear path can be written

$$3x + 2y - 6 = 0, \qquad x \leq \tfrac{16}{3}$$

and the elliptical path

$$\frac{x^2}{4} + \frac{y^2}{9} = 1.$$

Solving the two equations simultaneously, we get

$$x = 2, \quad y = 0 \qquad \text{and} \qquad x = 0, \quad y = 3.$$

This means that the paths intersect at the points $(2, 0)$ and $(0, 3)$. This answers part (a). Now for part (b). The first particle passes through $(2, 0)$ only when

$$x_1(t) = \tfrac{16}{3} - \tfrac{8}{3}t = 2 \qquad \text{and} \qquad y_1(t) = 4t - 5 = 0.$$

As you can check, this happens only when $t = \tfrac{5}{4}$. When $t = \tfrac{5}{4}$, the second particle is elsewhere. Hence no collision takes place at $(2, 0)$. There is, however, a collision at $(0, 3)$ because both particles get there at exactly the same time, $t = 2$:

$$x_1(2) = 0 = x_2(2), \qquad y_1(2) = 3 = y_2(2). \quad \square$$

EXERCISES 9.7

In Exercises 1–12, express the curve by an equation in x and y.

1. $x(t) = t^2, \quad y(t) = 2t + 1.$

2. $x(t) = 3t - 1, \quad y(t) = 5 - 2t.$

3. $x(t) = t^2, \quad y(t) = 4t^4 + 1.$

4. $x(t) = 2t - 1, \quad y(t) = 8t^3 - 5.$

5. $x(t) = 2 \cos t, \quad y(t) = 3 \sin t.$

6. $x(t) = \sec^2 t, \quad y(t) = 2 + \tan t.$

7. $x(t) = \tan t, \quad y(t) = \sec t.$

8. $x(t) = 2 - \sin t, \quad y(t) = \cos t.$

9. $x(t) = \sin t, \quad y(t) = 1 + \cos^2 t.$

10. $x(t) = e^t, \quad y(t) = 4 - e^{2t}.$

11. $x(t) = 4 \sin t, \quad y(t) = 3 + 2 \sin t.$

12. $x(t) = \csc t, \quad y(t) = \cot t.$

In Exercises 13–21, express the curve by an equation in x and y; then sketch the curve.

13. $x(t) = e^{2t}, \quad y(t) = e^{2t} - 1, \quad t \leq 0.$

14. $x(t) = 3 \cos t, \quad y(t) = 2 - \cos t, \quad 0 \leq t \leq \pi.$

15. $x(t) = \sin t, \quad y(t) = \csc t, \quad 0 < t \leq \tfrac{1}{4}\pi.$

16. $x(t) = 1/t, \quad y(t) = 1/t^2, \quad 0 < t < 3.$

17. $x(t) = 3 + 2t, \quad y(t) = 5 - 4t, \quad -1 \leq t \leq 2.$

18. $x(t) = \sec t, \quad y(t) = \tan t, \quad 0 \leq t \leq \tfrac{1}{4}\pi.$

19. $x(t) = \sin \pi t, \quad y(t) = 2t, \quad 0 \leq t \leq 4.$

20. $x(t) = 2 \sin t, \quad y(t) = \cos t, \quad 0 \leq t \leq \frac{1}{2}\pi.$

21. $x(t) = \cot t, \quad y(t) = \csc t, \quad \frac{1}{4}\pi \leq t < \frac{1}{2}\pi.$

22. (*Important*) Parametrize: (a) the curve $y = f(x)$, $x \in [a, b]$; (b) the polar curve $r = f(\theta)$, $\theta \in [\alpha, \beta]$.

23. A particle with position given by the equations

$$x(t) = \sin 2\pi t, \quad y(t) = \cos 2\pi t, \quad t \in [0, 1],$$

starts at the point $(0, 1)$ and traverses the unit circle $x^2 + y^2 = 1$ once in a clockwise manner. Write equations of the form

$$x(t) = f(t), \quad y(t) = g(t), \quad t \in [0, 1],$$

so that the particle
(a) begins at $(0, 1)$ and traverses the circle once in a counterclockwise manner;
(b) begins at $(0, 1)$ and traverses the circle twice in a clockwise manner;
(c) traverses the quarter circle from $(1, 0)$ to $(0, 1)$;
(d) traverses the three-quarter circle from $(1, 0)$ to $(0, 1)$.

24. A particle with position given by the equations

$$x(t) = 3 \cos 2\pi t, \quad y(t) = 4 \sin 2\pi t, \quad t \in [0, 1],$$

starts at the point $(3, 0)$ and traverses the ellipse $16x^2 + 9y^2 = 144$ once in a counterclockwise manner. Write equations of the form

$$x(t) = f(t), \quad y(t) = g(t), \quad t \in [0, 1],$$

so that the particle
(a) begins at $(3, 0)$ and traverses the ellipse once in a clockwise manner;
(b) begins at $(0, 4)$ and traverses the ellipse once in a clockwise manner;
(c) begins at $(-3, 0)$ and traverses the ellipse twice in a counterclockwise manner;
(d) traverses the upper half of the ellipse from $(3, 0)$ to $(0, 3)$.

25. Find a parametrization

$$x = x(t), \quad y = y(t), \quad t \in (-1, 1),$$

for the horizontal line $y = 2$.

26. Find a parametrization

$$x(t) = \sin f(t), \quad y(t) = \cos f(t), t \in (0, 1),$$

which traces out the unit circle infinitely often.

In Exercises 27–33, find a parametrization

$$x = x(t), \quad y = y(t), \quad t \in [0, 1],$$

for the given curve.

27. The line segment from $(3, 7)$ to $(8, 5)$.

28. The line segment from $(2, 6)$ to $(6, 3)$.

29. The parabolic arc $x = 1 - y^2$ from $(0, -1)$ to $(0, 1)$.

30. The parabolic arc $x = y^2$ from $(4, 2)$ to $(0, 0)$.

31. The curve $y^2 = x^3$ from $(4, 8)$ to $(1, 1)$.

32. The curve $y^3 = x^2$ from $(1, 1)$ to $(8, 4)$.

33. The curve $y = f(x), x \in [a, b]$.

34. (*Important*) Suppose that the curve

$$C: \quad x = x(t), \quad y = y(t), \quad t \in [c, d],$$

is the graph of a nonnegative function $y = f(x)$ over an interval $[a, b]$. Suppose that $x'(t)$ and $y(t)$ are continuous, $x(c) = a$, and $x(d) = b$.
(a) (*The area under a parametrized curve*) Show that

(9.7.4)
$$\text{the area below } C = \int_c^d y(t)x'(t)\, dt.$$

HINT: Since C is the graph of f, we know that $y(t) = f(x(t))$.
(b) (*The centroid of a region under a parametrized curve*) Show that, if the region under C has area A and centroid $(\bar{x}, \bar{y})$, then

(9.7.5)
$$\bar{x}A = \int_c^d x(t)y(t)x'(t)\, dt,$$
$$\bar{y}A = \int_c^d \tfrac{1}{2}[y(t)]^2 x'(t)\, dt.$$

(c) (*The volume of the solid generated by revolving about a coordinate axis the region under a parametrized curve*) Show that

(9.7.6)
$$V_x = \int_c^d \pi[y(t)]^2 x'(t)\, dt,$$
$$V_y = \int_c^d 2\pi x(t)y(t)x'(t)\, dt.$$
provided $x(c) \geq 0$

(d) (*The centroid of the solid generated by revolving about a coordinate axis the region under a parametrized curve*) Show that

(9.7.7)
$$\bar{x}V_x = \int_c^d \pi x(t)[y(t)]^2 x'(t)\, dt,$$
$$\bar{y}V_y = \int_c^d \pi x(t)[y(t)]^2 x'(t)\, dt.$$
provided $x(c) \geq 0$

35. Sketch the curve

$$x(t) = at, \quad y(t) = a(1 - \cos t), \quad a > 0, \quad t \in [0, 2\pi]$$

and find the area below it.

36. Determine the centroid of the region under the curve in Exercise 35.

37. Find the volume generated by revolving the region in Exercise 36 about: (a) the x-axis; (b) the y-axis.

38. Find the centroid of the solid generated by revolving the region of Exercise 36 about: (a) the x-axis; (b) the y-axis.

39. Give a parametrization for the upper half of the ellipse $b^2x^2 + a^2b^2 = a^2b^2$ that satisfies the requirements of Exercise 34.

40. Use the parametrization you chose for Exercise 39 to find (a) the area of the region enclosed by the ellipse; (b) the centroid of the upper half of that region.

41. Two particles start at the same instant, the first along the ray

$$x(t) = 2t + 6, \quad y(t) = 5 - 4t, \quad t \geq 0,$$

and the second along the circular path

$$x(t) = 3 - 5 \cos \pi t, \quad y(t) = 1 + 5 \sin \pi t, \quad t \geq 0.$$

(a) At what points, if any, do these paths intersect?
(b) At what points, if any, will the particles collide?

42. Two particles start at the same instant, the first along the elliptical path

$$x_1(t) = 2 - 3 \cos \pi t, \quad y_1(t) = 3 + 7 \sin \pi t, \quad t \geq 0,$$

and the second along the parabolic path

$$x_2(t) = 3t + 2, \quad y_2(t) = -\tfrac{7}{15}(3t + 1)^2 + \tfrac{157}{15}, \quad t \geq 0.$$

(a) At what points, if any, do these paths intersect?
(b) At what points, if any, will the particles collide?

We can determine the points where a parametrized curve

$$C: \quad x = x(t), \quad y = y(t), \quad t \in I$$

intersects itself by finding the numbers r and s in I ($r \neq s$) for which

$$x(r) = x(s) \quad \text{and} \quad y(r) = y(s).$$

Use this method to find the point(s) of self-intersection for each of the curves in Exercises 43–46.

43. $x(t) = t^2 - 2t, \quad y(t) = t^3 - 3t^2 + 2t, \quad t$ real.

44. $x(t) = \cos t \, (1 - 2 \sin t), \quad y(t) = \sin t \, (1 - 2 \sin t),$ $t \in [0, \pi]$.

45. $x(t) = \sin 2\pi t, \quad y(t) = 2t - t^2, \quad t \in [0, 4]$.

46. $x(t) = t^3 - 4t, \quad y(t) = t^3 - 3t^2 - 3t^2 + 2t, \quad t$ real.

▷ 47. A *cycloid* is the curve traced out by a point P on a circle of radius a as it moves without slipping along the x-axis. See

the figure. The parametric equations of the curve are

$$x(\theta) = a(\theta - \sin \theta),$$
$$y(\theta) = a(1 - \cos \theta).$$

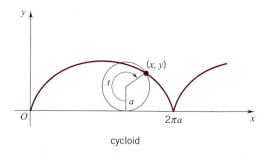

cycloid

(a) Use a graphing utility to draw the graph of the cycloid for $a = 1, 2$, and 4. What effect does the coefficient a have on the graph?
(b) The curve has a cusp at $\theta = 0, 2\pi$, and so on. Let $a = 2$. Use your graph to determine the slope of the curve for points near the cusp corresponding to $\theta = 2\pi$.
(c) Does the curve have a vertical tangent line at the cusp?

▷ 48. An *epicycloid* is the curve traced out by a point P on a circle of radius a as it rolls without slipping around the outside of a circle of radius R. See the figure. The parametric equations of the epicycloid are

$$x(\theta) = (R + a)\cos \theta - a \cos \left(\frac{R + a}{a} \right) \theta,$$

$$y(\theta) = (R + a)\sin \theta - a \sin \left(\frac{R + a}{a} \right) \theta, \quad 0 \leq \theta \leq 2\pi.$$

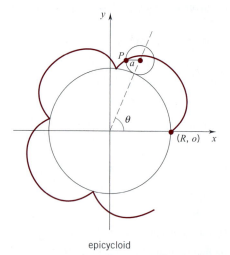

epicycloid

Use a graphing utility to draw the graph of the epicycloid for the following values of a and R:

(a) $a = 1, R = 4$.

(b) $a = 2, R = 6$.

49. A *hypocycloid* is the curve traced out by a point P on a circle of radius a as it rolls without slipping around the inside of a circle of radius R ($R > a$). See the figure. The parametric equations of the hypocycloid are

$$x(\theta) = (R - a)\cos\theta + a\cos\left(\frac{R - a}{a}\right)\theta,$$

$$y(\theta) = (R - a)\sin\theta - a\sin\left(\frac{R - a}{a}\right)\theta, \quad 0 \le \theta \le 2\pi.$$

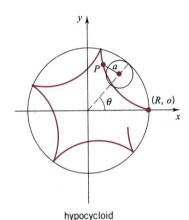

hypocycloid

Use a graphing utility to draw the graph of the hypocycloid for the following values of a and R.

(a) $a = 1, R = 4$.

(b) $a = 2, R = 6$.

(c) $a = 1, R = 8$.

50. Use a graphing utility to draw the graph of

$$x(\theta) = \cos\theta\,(a - b\sin\theta),$$

$$y(\theta) = \sin\theta\,(a - b\sin\theta), \quad 0 \le \theta \le 2\pi,$$

for the following values of a and b:

(a) $a = 1, b = 2$

(b) $a = 2, b = 2$.

(c) $a = 2, b = 1$.

(d) In general, what can you say about the graphs when $a < b$, and $a > b$?

---

*SUPPLEMENT TO SECTION 9.7

PARABOLIC TRAJECTORIES

In the early part of the seventeenth century Galileo Galilei observed the motion of stones projected from the tower of Pisa and observed that their trajectory was parabolic. By simple calculus, together with some simplifying physical assumptions, we obtain results that agree with Galileo's observations.

Consider a projectile fired at angle θ, $0 < \theta < \frac{1}{2}\pi$, from a point (x_0, y_0) with initial velocity v_0 (Figure 9.7.6). The horizontal component of v_0 is $v_0\cos\theta$, and the vertical component is $v_0\sin\theta$ (Figure 9.7.7).

Figure 9.7.6

Figure 9.7.7

We neglect air resistance and the curvature of the earth. Under these circumstances there is no horizontal acceleration:

$$x''(t) = 0.$$

The only vertical acceleration is due to gravity:

$$y''(t) = -g.$$

From the first equation, we have

$$x'(t) = C$$

and, since $x'(0) = v_0 \cos \theta$,

$$x'(t) = v_0 \cos \theta.$$

Integrating again, we have

$$x(t) = (v_0 \cos \theta)t + C$$

and, since $x(0) = x_0$,

(1) $$x(t) = (v_0 \cos \theta)t + x_0.$$

The relation $y''(t) = -g$ gives

$$y'(t) = -gt + C,$$

and, since $y'(0) = v_0 \sin \theta$,

$$y'(t) = -gt + v_0 \sin \theta.$$

Integrating again, we find that

$$y(t) = -\tfrac{1}{2}gt^2 + (v_0 \sin \theta)t + C.$$

Since $y(0) = y_0$, we have

(2) $$y(t) = -\tfrac{1}{2}gt^2 + (v_0 \sin \theta)t + y_0.$$

Thus, the parametric equations of the path of the projectile are

$$x(t) = (v_0 \cos \theta)t + x_0, \quad y(t) = \tfrac{1}{2}gt^2 + (v_0 \sin \theta)t + y_0.$$

From the first equation

$$t = \frac{1}{v_0 \cos \theta} [x(t) - x_0].$$

Substitute this expression for t in the second equation and you will find that

$$y(t) = -\frac{g}{2v_0^2} \sec^2 \theta \, [x(t) - x_0]^2 + \tan \theta \, [x(t) - x_0] + y_0.$$

The trajectory (the path followed by the projectile) is the curve

(9.7.8)

$$y = -\frac{g}{2v_0^2} \sec^2 \theta \, [x - x_0]^2 + \tan \theta \, [x - x_0] + y_0.$$

This is a quadratic in x and therefore a parabola. ❏

In Exercises 51–56, we neglect air resistance and the curvature of the earth. We measure distance in feet, time in seconds, and set $g = 32$ feet per second per second. We take O as the origin, the x-axis as ground level, and consider a projectile fired from O at an angle θ with initial velocity v_0.

51. Find an equation for the trajectory.

52. What is the maximum height attained by the projectile?

53. Find the range of the projectile.

54. How many seconds after firing does the impact take place?

55. How should θ be chosen so as to maximize the range?

56. How should θ be chosen so that the range becomes r?

■ 9.8 TANGENTS TO CURVES GIVEN PARAMETRICALLY

Let C be a curve which is parametrized by the functions

$$x = x(t), \quad y = y(t),$$

where x and y are defined on some interval I. Since C can intersect itself, it can have, at any given point:

(i) one tangent, (ii) two or more tangents, or (iii) no tangent at all.

We illustrate these possibilities in Figure 9.8.1.

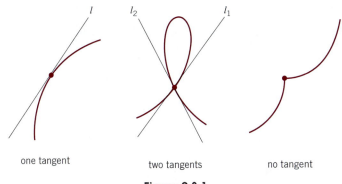

one tangent   two tangents   no tangent

Figure 9.8.1

As before, we are assuming that $x'(t)$ and $y'(t)$ exist, at least on the interior of I. To make sure that at least one tangent line exists at each point of C, we will make the additional *assumption* that

(9.8.1)
$$[x'(t)]^2 + [y'(t)]^2 \neq 0.$$

(Without this assumption most anything can happen. See Exercises 31–35.)

Now choose a point (x_0, y_0) on the curve C and a time t_0 at which

$$x(t_0) = x_0 \quad \text{and} \quad y(t_0) = y_0.$$

We want the slope of the curve as it passes through the point (x_0, y_0) at time t_0.† To find this slope, we assume that $x'(t_0) \neq 0$. With $x'(t_0) \neq 0$, we can be sure that, for h sufficiently small, $h \neq 0$,

$$x(t_0 + h) - x(t_0) \neq 0. \hspace{4em} \text{(explain)}$$

For such h we can form the quotient

$$\frac{y(t_0 + h) - y(t_0)}{x(t_0 + h) - x(t_0)}.$$

This quotient is the slope of the secant line pictured in Figure 9.8.2. The limit of this quotient as h tends to zero is the slope of the tangent line and thus the slope of the curve. Since

$$\frac{y(t_0 + h) - y(t_0)}{x(t_0 + h) - x(t_0)} = \frac{(1/h)[y(t_0 + h) - y(t_0)]}{(1/h)[x(t_0 + h) - x(t_0)]} \rightarrow \frac{y'(t_0)}{x'(t_0)} \qquad \text{as } h \rightarrow 0,$$

you can see that

(9.8.2)
$$\boxed{m = \frac{y'(t_0)}{x'(t_0)}.}$$

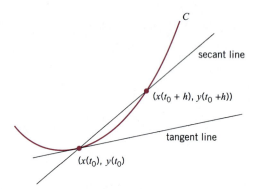

Figure 9.8.2

An equation for the tangent line can be written

$$y - y(t_0) = \frac{y'(t_0)}{x'(t_0)}[x - x(t_0)].$$

Multiplication by $x'(t_0)$ gives

$$y'(t_0)[x - x(t_0)] - x'(t_0)[y - y(t_0)] = 0,$$

---

† It could pass through the point (x_0, y_0) at other times also.

and thus

(9.8.3)

$$y'(t_0)[x - x_0] - x'(t_0)[y - y_0] = 0.$$

We derived this equation under the assumption that $x'(t_0) \neq 0$. If $x'(t_0) = 0$, Equation (9.8.3) still makes sense. It is simply $y'(t_0)[x - x_0] = 0$, which, since $y'(t_0) \neq 0,$† can be simplified to read

(9.8.4)

$$x = x_0.$$

In this instance the line is vertical, and we say that the curve has a *vertical tangent*.

Example 1 Find equation(s) for the tangent(s) to the curve

$$x(t) = t^3, \quad y(t) = 1 - t, \qquad t \text{ real},$$

at the point $(8, -1)$.

SOLUTION Since the curve passes through the point $(8, -1)$ only when $t = 2$, there can be only one tangent line at that point. With

$$x(t) = t^3, \quad y(t) = 1 - t,$$

we have

$$x'(t) = 3t^2, \quad y'(t) = -1,$$

and therefore

$$x'(2) = 12, \quad y'(2) = -1.$$

The tangent line has equation

$$(-1)[x - 8] - (12)[y - (-1)] = 0. \tag{9.8.3}$$

This reduces to

$$x + 12y + 4 = 0. \quad \square$$

Example 2 Find the points of the curve

$$x(t) = 3 - 4 \sin t, \quad y(t) = 4 + 3 \cos t, \qquad t \text{ real},$$

at which there is (i) a horizontal tangent, (ii) a vertical tangent.

SOLUTION Observe first of all that the derivatives

$$x'(t) = -4 \cos t \quad \text{and} \quad y'(t) = -3 \sin t$$

are never 0 simultaneously.

---

† We are assuming that $[x'(t)]^2 + [y'(t)]^2$ is never 0. Since $x'(t_0) = 0, y'(t_0) \neq 0$.

To find the points at which there is a horizontal tangent, we set $y'(t) = 0$. This gives $t = n\pi$, $n = 0, \pm1, \pm2, \ldots$. Horizontal tangents occur at all points of the form $(x(n\pi), y(n\pi))$. Since

$$x(n\pi) = 3 - 4 \sin n\pi = 3 \qquad \text{and} \qquad y(n\pi) = 4 + 3 \cos n\pi = \begin{cases} 7, & n \text{ even} \\ 1, & n \text{ odd,} \end{cases}$$

there are horizontal tangents at $(3, 7)$ and $(3, 1)$.

To find the vertical tangents, we set $x'(t) = 0$. This gives $t = \tfrac{1}{2}\pi + n\pi$, $n = 0, \pm1, \pm2, \ldots$. Vertical tangents occur at all points of the form $(x(\tfrac{1}{2}\pi + n\pi), y(\tfrac{1}{2}\pi + n\pi))$. Since

$$x(\tfrac{1}{2}\pi + n\pi) = 3 - 4 \sin (\tfrac{1}{2}\pi + n\pi) = \begin{cases} -1, & n \text{ even} \\ 7, & n \text{ odd} \end{cases}$$

and

$$y(\tfrac{1}{2}\pi + n\pi) = 4 + 3 \cos (\tfrac{1}{2}\pi + n\pi) = 4,$$

there are vertical tangents at $(-1, 4)$ and $(7, 4)$. ❏

Remark The curve parametrized by the equations in Example 2 is the ellipse

$$\frac{(x - 3)^2}{16} + \frac{(y - 4)^2}{9} = 1$$

with center at $(3, 4)$, major axis horizontal with length 8, and minor axis vertical with length 6. (See Figure 9.8.3). ❏

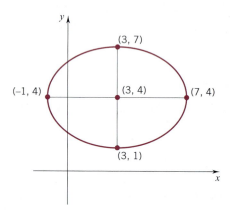

Figure 9.8.3

Example 3 The curve parametrized by

$$x(t) = \frac{1 - t^2}{1 + t^2}, \quad y(t) = \frac{t(1 - t^2)}{1 + t^2}, \qquad t \text{ real,}$$

is called a *strophoid*. Its graph is shown in Figure 9.8.4. Find equations for the tangent lines to the curve at the origin. Also, find the points at which the tangent line is horizontal.

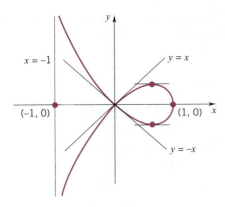

Figure 9.8.4

SOLUTION The curve passes through the origin when $t = -1$ and when $t = 1$ (verify this). Differentiating $x(t)$ and $y(t)$, we obtain

$$x'(t) = \frac{(1 + t^2)(-2t) - (1 - t^2)(2t)}{(1 + t^2)^2} = \frac{-4t}{(1 + t^2)^2},$$

and

$$y'(t) = \frac{(1 + t^2)(1 - 3t^2) - t(1 - t^2)(2t)}{(1 + t^2)^2} = \frac{1 - 4t^2 - t^4}{(1 + t^2)^2}.$$

When $t = -1$, $x'(-1) = 1$ and $y'(-1) = -1$. Thus, an equation for the tangent line is

$$-1(y - 0) - 1(x - 0) = 0 \qquad \text{or} \qquad x + y = 0. \qquad \text{(Figure 9.8.4)}$$

When $t = 1$, $x'(1) = -1$ and $y'(1) = -1$. Thus, an equation for the tangent line is

$$-1(y - 0) - (-1)(x - 0) = 0 \qquad \text{or} \qquad y - x = 0. \qquad \text{(Figure 9.8.4)}$$

To find the points at which there is a horizontal tangent, we set $y'(t) = 0$. This implies that

$$1 - 4t^2 - t^4 = 0.$$

Treating this equation as a quadratic in t^2, we have, by the quadratic formula,

$$t^2 = \frac{-4 \pm \sqrt{16 + 4}}{2} = -2 \pm \sqrt{5}.$$

Since $t^2 \geq 0$, it follows that $t^2 = \sqrt{5} - 2$, and therefore $t = \pm\sqrt{\sqrt{5} - 2}$. Now,

$$x\left(\pm\sqrt{\sqrt{5} - 2}\right) = \frac{1 - (\sqrt{5} - 2)}{1 + (\sqrt{5} - 2)} = \frac{\sqrt{5} - 1}{2} \cong 0.62$$

and

$$y\left(\pm\sqrt{\sqrt{5} - 2}\right) = \left(\pm\sqrt{\sqrt{5} - 2}\right)x = \left(\pm\sqrt{\sqrt{5} - 2}\right)\left(\frac{\sqrt{5} - 1}{2}\right) \cong \pm 0.30.$$

Thus, there is a horizontal tangent line at the points $(0.62, \pm 0.30)$ (approximately).

We can apply these ideas to a curve given in polar coordinates by an equation of the form $r = f(\theta)$. The coordinate transformations

$$x = r \cos \theta, \quad y = r \sin \theta$$

enable us to parametrize such a curve by setting

$$x(\theta) = f(\theta) \cos \theta, \quad y(\theta) = f(\theta) \sin \theta.$$

Example 4 Find the slope of the spiral

$$r = a\theta, \qquad \theta \in [0, \infty) \quad (a > 0)$$

at $\theta = \frac{1}{2}\pi$.

SOLUTION We write

$$x(\theta) = r \cos \theta = a\theta \cos \theta, \quad y(\theta) = r \sin \theta = a\theta \sin \theta.$$

Now we differentiate:

$$x'(\theta) = -a\theta \sin \theta + a \cos \theta, \quad y'(\theta) = a\theta \cos \theta + a \sin \theta.$$

Since

$$x'(\tfrac{1}{2}\pi) = -\tfrac{1}{2}\pi a \quad \text{and} \quad y'(\tfrac{1}{2}\pi) = a,$$

the slope of the curve at $\theta = \frac{1}{2}\pi$ is

$$\frac{y'(\tfrac{1}{2}\pi)}{x'(\tfrac{1}{2}\pi)} = -\frac{2}{\pi} \cong -0.64.$$

See Figure 9.8.5. ❑

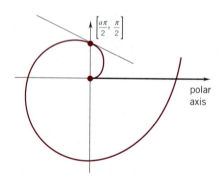

Figure 9.8.5

Example 5 Find the points of the cardioid

$$r = 1 - \cos \theta$$

at which the tangent line is vertical.

SOLUTION Since the cosine function has period 2π, we need only concern ourselves with θ in $[0, 2\pi)$. Parametrically we have

$$x(\theta) = (1 - \cos \theta) \cos \theta, \quad y(\theta) = (1 - \cos \theta) \sin \theta.$$

Differentiating and simplifying, we find that

$$x'(\theta) = (2\cos\theta - 1)\sin\theta, \quad y'(\theta) = (1 - \cos\theta)(1 + 2\cos\theta).$$

The only numbers in the interval $[0, 2\pi)$ at which x' is zero and y' is not zero are $\frac{1}{3}\pi$, π, and $\frac{5}{3}\pi$. The tangent line is vertical at

$$[\tfrac{1}{2}, \tfrac{1}{3}\pi], \quad [2, \pi], \quad [\tfrac{1}{2}, \tfrac{5}{3}\pi].$$

These points have rectangular coordinates

$(\tfrac{1}{4}, \tfrac{1}{4}\sqrt{3})$, $(-2, 0)$, $(\tfrac{1}{4}, -\tfrac{1}{4}\sqrt{3})$. ❑   (See Figure (9.8.6))

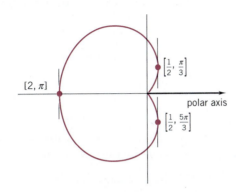

Figure 9.8 6

EXERCISES 9.8

In Exercises 1–8, find an equation in x and y for the line tangent to the curve.

1. $x(t) = t$, $y(t) = t^3 - 1$; $t = 1$.

2. $x(t) = t^2$, $y(t) = t + 5$; $t = 2$.

3. $x(t) = 2t$, $y(t) = \cos\pi t$; $t = 0$.

4. $x(t) = 2t - 1$, $y(t) = t^4$; $t = 1$.

5. $x(t) = t^2$, $y(t) = (2 - t)^2$; $t = \frac{1}{2}$.

6. $x(t) = 1/t$, $y(t) = t^2 + 1$; $t = 1$.

7. $x(t) = \cos^3 t$, $y(t) = \sin^3 t$; $t = \frac{1}{4}\pi$.

8. $x(t) = e^t$, $y(t) = 3e^{-t}$; $t = 0$.

In Exercises 9–14, find an equation in x and y for the line tangent to the polar curve.

9. $r = 4 - 2\sin\theta$, $\theta = 0$.

10. $r = 4\cos 2\theta$, $\theta = \frac{1}{2}\pi$.

11. $r = \dfrac{4}{5 - \cos\theta}$, $\theta = \frac{1}{2}\pi$.

12. $r = \dfrac{5}{4 - \cos\theta}$, $\theta = \frac{1}{6}\pi$.

13. $r = \dfrac{\sin\theta - \cos\theta}{\sin\theta + \cos\theta}$, $\theta = 0$.

14. $r = \dfrac{\sin\theta + \cos\theta}{\sin\theta - \cos\theta}$, $\theta = \frac{1}{2}\pi$.

In Exercises 15–18, parametrize the curve by a pair of differentiable functions

$$x = x(t), \quad y = y(t) \qquad \text{with} \quad [x'(t)]^2 + [y'(t)]^2 \neq 0.$$

Sketch the curve and determine the tangent line at the origin by the method of this section.

15. $y = x^3$.

16. $x = y^3$.

17. $y^5 = x^3$.

18. $y^3 = x^5$.

In Exercises 19–26, find the points (x, y) at which the curve has: (a) a horizontal tangent; (b) a vertical tangent. Then sketch the curve.

19. $x(t) = 3t - t^3$, $y(t) = t + 1$.

20. $x(t) = t^2 - 2t$, $y(t) = t^3 - 12t$.

21. $x(t) = 3 - 4\sin t$, $y(t) = 4 + 3\cos t$.

22. $x(t) = \sin 2t$, $y(t) = \sin t$.

23. $x(t) = t^2 - 2t$, $y(t) = t^3 - 3t^2 + 2t$.

24. $x(t) = 2 - 5\cos t$, $y(t) = 3 + \sin t$.

25. $x(t) = \cos t$, $y(t) = \sin 2t$.

26. $x(t) = 3 + 2\sin t$, $y(t) = 2 + 5\sin t$.

27. Find the tangent(s) to the curve

$$x(t) = -t + 2 \cos \tfrac{1}{4}\pi t, \quad y(t) = t^4 - 4t^2$$

at the point $(2, 0)$.

28. Find the tangent(s) to the curve.

$$x(t) = t^3 - t, \quad y(t) = t \sin \tfrac{1}{2}\pi t$$

at the point $(0, 1)$.

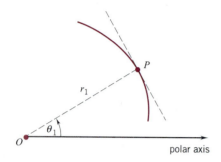

29. Let $P = [r_1, \theta_1]$ be a point on a polar curve $r = f(\theta)$ as in the figure. Show that, if $f'(\theta_1) = 0$ but $f(\theta_1) \neq 0$, then the tangent line at P is perpendicular to the line segment $\overline{OP}$.

30. If $0 < a < 1$, the polar curve $r = a - \cos \theta$ is a limaçon with an inner loop. Choose a so that the curve will intersect itself at the pole in a right angle.

In Exercises 31–35, verify that $x'(0) = y'(0) = 0$ and that the given description holds at the point where $t = 0$. Sketch the graph.

31. $x(t) = t^3, \quad y(t) = t^2$; cusp.

32. $x(t) = t^3, \quad y(t) = t^5$; horizontal tangent.

33. $x(t) = t^5, \quad y(t) = t^3$; vertical tangent.

34. $x(t) = t^3 - 1, \quad y(t) = 2t^3$; tangent with slope 2.

35. $x(t) = t^2, \quad y(t) = t^2 + 1$; no tangent line.

36. Suppose that $x = x(t)$, $y = y(t)$ are twice differentiable functions that parametrize a curve. Take a point on the curve at which $x'(t) \neq 0$ and d^2y/dx^2 exists. Show that

(9.8.5)
$$\frac{d^2y}{dx^2} = \frac{x'(t)y''(t) - y'(t)x''(t)}{[x'(t)]^3}.$$

In Exercises 37–40, calculate d^2y/dx^2 at the indicated point without eliminating the parameter.

37. $x(t) = \cos t, \quad y(t) = \sin t; \quad t = \tfrac{1}{6}\pi$.

38. $x(t) = t^3, \quad y(t) = t - 2; \quad t = 1$.

39. $x(t) = e^t, \quad y(t) = e^{-t}; \quad t = 0$.

40. $x(t) = \sin^2 t, \quad y(t) = \cos t; \quad t = \tfrac{1}{4}\pi$.

■ **9.9 ARC LENGTH AND SPEED**

Arc Length

We come now to the notion of arc length. In Figure 9.9.1 we have sketched a curve C which we assume is parametrized by a pair of *continuously differentiable* functions†

$$x = x(t), \quad y = y(t), \qquad t \in [a, b].$$

We want to determine the length of C.

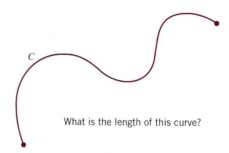

What is the length of this curve?

Figure 9.9.1

---

† By this we mean functions that have continuous first derivatives.

Here our experience in Chapter 5 can be used as a model. To decide what should be meant by the area of a region Ω, we approximated Ω by the union of a finite number of rectangles. To decide what should be meant by the length of C, we approximate C by the union of a finite number of line segments.

Each number t in $[a, b]$ gives rise to a point $P = P(x(t), y(t))$ that lies on C. By choosing a finite number of points in $[a, b]$,

$$a = t_0 < t_1 < \cdots < t_{i-1} < t_i < \cdots < t_{n-1} < t_n = b,$$

we obtain a finite number of points on C,

$$P_0, P_1, \ldots, P_{i-1}, P_i, \ldots, P_{n-1}, P_n.$$

We join these points consecutively by line segments and call the resulting path,

$$\gamma = \overline{P_0 P_1} \cup \cdots \cup \overline{P_{i-1} P_i} \cup \cdots \cup \overline{P_{n-1} P_n},$$

a *polygonal path* inscribed in C. (See Figure 9.9.2.)

a polygonal path inscribed in the curve C

Figure 9.9.2

The length of such a polygonal path is the sum of the distances between consecutive vertices:

$$\text{length of } \gamma = L(\gamma) = d(P_0, P_1) + \cdots + d(P_{i-1}, P_i) + \cdots + d(P_{n-1}, P_n).$$

Now, the ith line segment $\overline{P_{i-1} P_i}$ has length

$$d(P_{i-1}, P_i) = \sqrt{[x(t_i) - x(t_{i-1})]^2 + [y(t_i) - y(t_{i-1})]^2}$$

$$= \sqrt{\left[\frac{x(t_i) - x(t_{i-1})}{t_i - t_{i-1}}\right]^2 + \left[\frac{y(t_i) - y(t_{i-1})}{t_i - t_{i-1}}\right]^2} \, (t_i - t_{i-1}).$$

By the mean-value theorem, there exist points t_i^* and t_i^{**}, both in the interval (t_{i-1}, t_i), such that

$$\frac{x(t_i) - x(t_{i-1})}{t_i - t_{i-1}} = x'(t_i^*) \quad \text{and} \quad \frac{y(t_i) - y(t_{i-1})}{t_i - t_{i-1}} = y'(t_i^{**}).$$

Letting $\Delta t_i = t_i - t_{i-1}$, we have

$$d(P_{i-1}, P_i) = \sqrt{[x'(t_i^*)]^2 + [y'(t_i^{**})]^2} \, \Delta t_i.$$

Thus,

$$L(\gamma) = \sqrt{[x'(t_1^*)]^2 + [y'(t_1^{**})]^2} \, \Delta t_1 + \sqrt{[x'(t_2^*)]^2 + y'(t_2^{**})]^2} \, \Delta t_2 + \cdots$$
$$+ \sqrt{[x'(t_n^*)]^2 + [y'(t_n^{**})]^2} \, \Delta t_n.$$

This is not a Riemann sum because it will not be true, in general, that $t_i^* = t_i^{**}$ for each i. However, this sum is "close" to a Riemann sum, close enough so that, as $\max \Delta t_i \to 0$, $i = 1, 2, \ldots, n$, it approaches the integral

$$\int_a^b \sqrt{[x'(t)]^2 + [y'(t)]^2} \, dt.$$

Arc Length Formulas

Based on the intuitive argument just given, we have a formula for calculating the length of a parametrized curve. If C is parametrized by a pair of continuously differentiable functions

$$x = x(t), \quad y = y(t), \qquad t \in [a, b],$$

then

(9.9.1)

$$\text{the length of } C = L(C) = \int_a^b \sqrt{[x'(t)]^2 + [y'(t)]^2} \, dt.$$

Obviously, this is not something to be taken on faith. It has to be proved. We will do so, but not until Chapter 13.

There is an implicit assumption underlying the arc length formula that we need to illustrate before going on. The functions

$$x(t) = a \cos t, \quad y = a \sin t, \qquad t \in [0, 2\pi],$$

parametrize the circle C of radius a centered at the origin. Since

$$x'(t) = -a \sin t, \quad y'(t) = a \cos t,$$

Formula (9.9.1) gives

$$L(C) = \int_0^{2\pi} \sqrt{a^2 \sin^2 t + a^2 \cos^2 t} \, dt = \int_0^{2\pi} a \, dt = 2\pi a.$$

This is, of course, correct; a circle of radius a has circumference $2\pi a$.

But notice that C can also be parametrized by the functions

$$x(t) = a \cos 2t, \quad y(t) = a \sin 2t, \qquad t \in [0, 2\pi].$$

Here we have

$$x'(t) = -2a \sin 2t, \quad y'(t) = 2a \cos 2t$$

and

$$L(C) = \int_0^{2\pi} \sqrt{4a^2 \sin^2 2t + 4a^2 \cos^2 2t} \, dt = \int_0^{2\pi} 2a \, dt = 4\pi a,$$

which is obviously not the length of C. The difference between the two cases is that with the first parametrization, the point $(x(t), y(t))$ traced out the circle exactly once as t increased from 0 to 2π; while in the second case the point $(x(t), y(t))$ traced out the curve twice as t increased from 0 to 2π.

In finding the arc length of a given curve C we will always use a parametrization that traverses the curve exactly once.

Suppose now that C is the graph of a continuously differentiable function

$$y = f(x), \qquad x \in [a, b].$$

Then we can parametrize C by setting

$$x(t) = t, \quad y(t) = f(t) \qquad t \in [a, b].$$

Since

$$x'(t) = 1 \qquad \text{and} \qquad y'(t) = f'(t),$$

Formula (9.9.1) gives

$$L(C) = \int_a^b \sqrt{1 + [f'(t)]^2} \, dt.$$

Replacing t by x we can write

(9.9.2)
$$\text{the length of the graph of } f = \int_a^b \sqrt{1 + [f'(x)]^2} \, dx.$$

A direct derivation of this formula is outlined in Exercise 42.

Example 1 If

$$f(x) = \tfrac{1}{6}x^3 + \tfrac{1}{2}x^{-1},$$

then

$$f'(x) = \tfrac{1}{2}x^2 - \tfrac{1}{2}x^{-2}.$$

Therefore

$$1 + [f'(x)]^2 = 1 + (\tfrac{1}{4}x^4 - \tfrac{1}{2} + \tfrac{1}{4}x^{-4}) = \tfrac{1}{4}x^4 + \tfrac{1}{2} + \tfrac{1}{4}x^{-4} = (\tfrac{1}{2}x^2 + \tfrac{1}{2}x^{-2})^2.$$

The length of the graph from $x = 1$ to $x = 3$ is

$$\int_1^3 \sqrt{1 + [f'(x)]^2} \, dx = \int_1^3 (\tfrac{1}{2}x^2 + \tfrac{1}{2}x^{-2}) \, dx = \left[\tfrac{1}{6}x^3 - \tfrac{1}{2}x^{-1} \right]_1^3 = \tfrac{14}{3}. \quad ❑$$

Example 2 The graph of the function

$$f(x) = x^2, \qquad x \in [0, 1]$$

is a parabolic arc. The length of this arc is given by

$$\int_0^1 \sqrt{1 + [f'(x)]^2} \, dx = \int_0^1 \sqrt{1 + 4x^2} \, dx = 2 \int_0^1 \sqrt{(\tfrac{1}{2})^2 + x^2} \, dx$$

$$\underset{\underset{\text{by (8.5.1)}}{\uparrow}}{=} \left[x\sqrt{(\tfrac{1}{2})^2 + x^2} + (\tfrac{1}{2})^2 \ln (x + \sqrt{(\tfrac{1}{2})^2 + x^2}) \right]_0^1$$

$$= \tfrac{1}{2}\sqrt{5} + \tfrac{1}{4} \ln (2 + \sqrt{5}) \cong 1.48. \quad ❑$$

Suppose now that C is the graph of a polar function

$$r = \rho(\theta), \qquad \alpha \leq \theta \leq \beta,$$

where ρ is continuously differentiable.

We can parametrize C by setting

$$x(\theta) = \rho(\theta) \cos \theta, \quad y(\theta) = \rho(\theta) \sin \theta \qquad \theta \in [\alpha, \beta].$$

A straightforward calculation that we leave to you shows that

$$[x'(\theta)]^2 + [y'(\theta)]^2 = [\rho(\theta)]^2 + [\rho'(\theta)]^2.$$

The arc length formula now reads

(9.9.3)
$$L(C) = \int_\alpha^\beta \sqrt{[\rho(\theta)]^2 + [\rho'(\theta)]^2}\, d\theta.$$

Example 3 For fixed $a > 0$ the equation $r = a$ represents a circle of radius a. Here

$$\rho(\theta) = a \qquad \text{and} \qquad \rho'(\theta) = 0.$$

The circumference of this circle is

$$\int_0^{2\pi} \sqrt{[\rho(\theta)]^2 + [\rho'(\theta)]^2}\, d\theta = \int_0^{2\pi} \sqrt{a^2 + 0^2}\, d\theta = \int_0^{2\pi} a\, d\theta = 2\pi a. \quad ❑$$

Example 4 In the case of the cardioid $r = a(1 - \cos \theta)$, $a > 0$, we have

$$\rho(\theta) = a(1 - \cos \theta), \qquad \rho'(\theta) = a \sin \theta.$$

Here

$$[\rho(\theta)]^2 + [\rho'(\theta)]^2 = a^2(1 - 2\cos\theta + \cos^2\theta) + a^2\sin^2\theta = 2a^2(1 - \cos\theta).$$

The identity

$$\tfrac{1}{2}(1 - \cos \theta) = \sin^2 \tfrac{1}{2}\theta$$

gives

$$[\rho(\theta)]^2 + [\rho'(\theta)]^2 = 4a^2\sin^2 \tfrac{1}{2}\theta.$$

The length of the cardioid is $8a$:

$$\int_0^{2\pi} \sqrt{[\rho(\theta)]^2 + [\rho'(\theta)]^2}\, d\theta = \int_0^{2\pi} 2a \sin \tfrac{1}{2}\theta\, d\theta = 4a\left[-\cos\tfrac{1}{2}\theta\right]_0^{2\pi} = 8a. \quad ❑$$

The Geometric Significance of dx/ds and dy/ds

Figure 9.9.3 shows the graph of a function $y = f(x)$ which we assume to be continuously differentiable on (a, b). At the point (x, y) the tangent line has an inclination marked α_x.

The length of the graph from a to x can be written

$$s(x) = \int_a^x \sqrt{1 + [f'(t)]^2}\, dt.$$

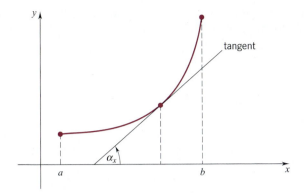

Figure 9.9.3

Differentiation with respect to x gives $s'(x) = \sqrt{1 + [f'(x)]^2}$ (Theorem 5.3.5). Using the Leibniz notation we have

$$\frac{ds}{dx} = \sqrt{1 + \left(\frac{dy}{dx}\right)^2} = \sqrt{1 + \tan^2 \alpha_x} = \sec \alpha_x.$$

$$\underset{\sec \alpha_x > 0 \text{ for } \alpha_x \in (-\tfrac{1}{2}\pi, \tfrac{1}{2}\pi)}{\uparrow}$$

By (7.1.10)

$$\frac{dx}{ds} = \frac{1}{\sec \alpha_x} = \cos \alpha_x.$$

To find dy/ds we note that

$$\tan \alpha_x = \frac{dy}{dx} = \frac{dy}{ds}\frac{ds}{dx} = \frac{dy}{ds}\sec \alpha_x.$$

$$\underset{\text{chain rule}}{\uparrow}$$

Multiplication by $\cos \alpha_x$ gives

$$\frac{dy}{ds} = \sin \alpha_x.$$

For the record

(9.9.4) $\dfrac{dx}{ds} = \cos \alpha_x$ and $\dfrac{dy}{ds} = \sin \alpha_x$ where α_x is the inclination of the tangent line at the point (x, y).

Speed Along a Plane Curve

So far we have talked about speed only in connection with straight-line motion. How can we calculate the speed of an object that moves along a curve? Imagine an object moving along some curved path. Suppose that $(x(t), y(t))$ gives the position of the object at time t. The distance traveled by the object from time zero to any later time t is simply the length of the path up to time t:

$$s(t) = \int_0^t \sqrt{[x'(u)]^2 + [y'(u)]^2}\, du.$$

The time rate of change of this distance is what we call the *speed* of the object. Denoting the speed of the object at time t by $\nu(t)$ we have

(9.9.5)

$$\nu(t) = s'(t) = \sqrt{[x'(t)]^2 + [y'(t)]^2}.$$

Example 5 The path of the projectile in Exercises 51–56 in the Supplement to Section 9.7 is given in terms of the time parameter t by the following equations:

$$x(t) = (v_0 \cos \theta)t, \qquad y(t) = -16t^2 + (v_0 \sin \theta)t.$$

From those exercises we know that the projectile impacts at time $t = \frac{1}{16} v_0 \sin \theta$. Calculate the speed at impact.

SOLUTION Since

$$x'(t) = v_0 \cos \theta \quad \text{and} \quad y'(t) = -32t + v_0 \sin \theta,$$

we have

$$\nu(t) = \sqrt{v_0^2 \cos^2\theta + (-32t + v_0 \sin \theta)^2}.$$

The speed at impact is therefore

$$\sqrt{v_0^2 \cos^2\theta + (-2v_0 \sin \theta + v_0 \sin \theta)^2} = \sqrt{v_0^2 \cos^2\theta + v_0^2 \sin^2\theta} = |v_0|,$$

which is exactly the speed with which the projectile was fired.† ❑

In the Leibniz notation the equation for speed reads

(9.9.6)

$$\nu = \frac{ds}{dt} = \sqrt{\left(\frac{dx}{dt}\right)^2 + \left(\frac{dy}{dt}\right)^2}.$$

If we know the speed of an object and we know its mass, then we can calculate its kinetic energy.

Example 6 A particle of mass m slides down a frictionless curve (see Figure 9.9.4) from a point (x_0, y_0) to a point (x_1, y_1) under the force of gravity. As discussed in Exercises 4.2, the particle has two forms of energy during the motion: gravitational potential energy mgy and kinetic energy $\frac{1}{2}mv^2$. Show that the sum of these two quantities remains constant:

$$\overset{\text{GPE}}{mgy} + \overset{\text{KE}}{\tfrac{1}{2}mv^2} = C.$$

SOLUTION The particle is subjected to a vertical force $-mg$ (a downward force of magnitude mg). Since the particle is constrained to remain on the curve, the effective force on the particle is tangential. The tangential component of the vertical force is

---

† This can be obtained from energy considerations. See Exercise 43.

Figure 9.9.4

$-mg \sin \alpha$. (See Figure 9.9.4.)   The speed of the particle is ds/dt and the tangential acceleration is d^2s/dt^2.   (It is as if the particle were moving along the tangent line.) Therefore, by Newton's law $F = ma$, we have

$$m\frac{d^2s}{dt^2} = -mg \sin \alpha = -mg\frac{dy}{ds}.$$

by (9.9.4) ⟶

We can therefore write

$$mg\frac{dy}{ds} + m\frac{d^2s}{dt^2} = 0,$$

$$mg\frac{dy}{ds}\frac{ds}{dt} + m\frac{ds}{dt}\frac{d^2s}{dt^2} = 0, \qquad \left(\text{multiplied by } \frac{ds}{dt}\right)$$

$$mg\frac{dy}{dt} + mv\frac{dv}{dt} = 0. \qquad \text{(chain rule)}$$

Integrating with respect to t we have

$$mgy + \tfrac{1}{2}mv^2 = C$$

as asserted.   ❏

EXERCISES 9.9

In Exercises 1–18, find the length of the graph and compare it to the straight-line distance between the endpoints of the graph.

1. $f(x) = 2x + 3, \quad x \in [0, 1]$.

2. $f(x) = 3x + 2, \quad x \in [0, 1]$.

3. $f(x) = (x - \tfrac{4}{9})^{3/2}, \quad x \in [1, 4]$.

4. $f(x) = x^{3/2}, \quad x \in [0, 44]$.

5. $f(x) = \tfrac{1}{3}\sqrt{x}\,(x - 3), \quad x \in [0, 3]$.

6. $f(x) = \tfrac{2}{3}(x - 1)^{3/2}, \quad x \in [1, 2]$.

7. $f(x) = \tfrac{1}{3}(x^2 + 2)^{3/2}, \quad x \in [0, 1]$.

8. $f(x) = \tfrac{1}{3}(x^2 - 2)^{3/2}, \quad x \in [2, 4]$.

9. $f(x) = \tfrac{1}{4}x^2 - \tfrac{1}{2}\ln x, \quad x \in [1, 5]$.

10. $f(x) = \tfrac{1}{8}x^2 - \ln x, \quad x \in [1, 4]$.

11. $f(x) = \tfrac{3}{8}x^{4/3} - \tfrac{3}{4}x^{2/3}, \quad x \in [1, 8]$.

12. $f(x) = \tfrac{1}{10}x^5 + \tfrac{1}{6}x^{-3}, \quad x \in [1, 2]$.

13. $f(x) = \ln (\sec x), \quad x \in [0, \tfrac{1}{4}\pi]$.

14. $f(x) = \tfrac{1}{2}x^2, \quad x \in [0, 1]$.

15. $f(x) = \tfrac{1}{2}x\sqrt{x^2 - 1} - \tfrac{1}{2}\ln (x + \sqrt{x^2 - 1}), \quad x \in [1, 2]$.

16. $f(x) = \cosh x, \quad x \in [0, \ln 2]$.

17. $f(x) = \tfrac{1}{2}x\sqrt{3 - x^2} + \tfrac{3}{2}\sin^{-1}(\tfrac{1}{3}\sqrt{3}x), \quad x \in [0, 1]$.

18. $f(x) = \ln (\sin x), \quad x \in [\tfrac{1}{6}\pi, \tfrac{1}{2}\pi]$.

In Exercises 19–24, the equations give the position of a particle at each time t during the time interval specified. Find the initial speed of the particle, the terminal speed, and the distance traveled.

19. $x(t) = t^2$, $y(t) = 2t$ from $t = 0$ to $t = \sqrt{3}$.

20. $x(t) = t - 1$, $y(t) = \frac{1}{2}t^2$ from $t = 0$ to $t = 1$.

21. $x(t) = t^2$, $y(t) = t^3$ from $t = 0$ to $t = 1$.

22. $x(t) = a\cos^3 t$, $y(t) = a\sin^3 t$ from $t = 0$ to $t = \frac{1}{2}\pi$.

23. $x(t) = e^t \sin t$, $y(t) = e^t \cos t$ from $t = 0$ to $t = \pi$.

24. $x(t) = \cos t + t \sin t$, $y(t) = \sin t - t \cos t$ from $t = 0$ to $t = \pi$.

25. The curve defined parametrically by

$$x(\theta) = a(\theta - \sin \theta), \qquad y(\theta) = a(1 - \cos \theta), \qquad a > 0,$$

for $0 \le \theta \le 2\pi$, is one arch of a cycloid (see Exercise 47, Section 9.7). Find the length of this curve.

26. Find the length of the epicycloid (see Exercise 48, Section 9.7)

$$x(\theta) = 2a \cos \theta - a \cos 2\theta,$$

$$y(\theta) = 2a \sin \theta - a \sin 2\theta,$$

where $a > 0$ and $0 \le \theta \le 2\pi$.

27. (a) Find the length of the hypocycloid (see Exercise 49, Section 9.7)

$$x(\theta) = 3a \cos \theta + a \cos 3\theta,$$

$$y(\theta) = 3a \sin \theta - a \sin 3\theta,$$

where $a > 0$ and $0 \le \theta \le 2\pi$.

(b) Show that this hypocycloid can also be represented by the parametric equations

$$x(\theta) = 4a \cos^3\theta, \quad y(\theta) = 4a \sin^3\theta, \quad 0 \le \theta \le 2\pi.$$

28. The curve defined parametrically by

$$x(\theta) = \theta \cos \theta, \qquad y(\theta) = \theta \sin \theta,$$

is called an *Archimedean spiral*. Find the length of this spiral for $0 \le \theta \le 2\pi$.

In Exercises 29–36, find the length of the polar curve.

29. $r = 1$ from $\theta = 0$ to $\theta = 2\pi$.

30. $r = 3$ from $\theta = 0$ to $\theta = \pi$.

31. $r = e^\theta$ from $\theta = 0$ to $\theta = 4\pi$. (logarithmic spiral)

32. $r = ae^\theta$, $a > 0$, from $\theta = -2\pi$ to $\theta = 2\pi$.

33. $r = e^{2\theta}$ from $\theta = 0$ to $\theta = 2\pi$.

34. $r = 1 + \cos \theta$ from $\theta = 0$ to $\theta = 2\pi$.

35. $r = 1 - \cos \theta$ from $\theta = 0$ to $\theta = \frac{1}{2}\pi$.

36. $r = 2a \sec \theta$, $a > 0$, from $\theta = 0$ to $\theta = \frac{1}{4}\pi$.

37. At time t a particle has position

$$x(t) = 1 + \tan^{-1}t, \quad y(t) = 1 - \ln \sqrt{1 + t^2}$$

Find the total distance traveled from time $t = 0$ to time $t = 1$. Give the initial speed and the terminal speed.

38. At time t a particle has position

$$x(t) = 1 - \cos t, \quad y(t) = t - \sin t.$$

Find the total distance traveled from time $t = 0$ to time $t = 2\pi$. Give the initial speed and the terminal speed.

39. Find c given that the length of the curve $y = \ln x$ from $x = 1$ to $x = e$ equals the length of the curve $y = e^x$ from $x = 0$ to $x = c$.

40. Find the length of the curve $y = x^{2/3}$, $x \in [1, 8]$. HINT: Work with the mirror image $y = x^{3/2}, x \in [1, 4]$.

41. Show that the function $f(x) = \cosh x$ has the following property: for every interval $[a, b]$ the length of the graph equals the area under the graph.

42. The figure shows the graph of a continuously differentiable function f from $x = a$ to $x = b$ together with a polygonal approximation. Show that the length of this polygonal approximation can be written as the following Riemann sum:

$$\sqrt{1 + [f'(x_1^*)]^2} \, \Delta x_1 + \sqrt{1 + [f'(x_2^*)]^2} \, \Delta x_2$$
$$+ \cdots + \sqrt{1 + [f'(x_n^*)]^2} \, \Delta x_n.$$

As $\|P\| = \max \Delta x_i$ tends to 0, such Riemann sums tend to

$$\int_a^b \sqrt{1 + [f'(x)]^2} \, dx.$$

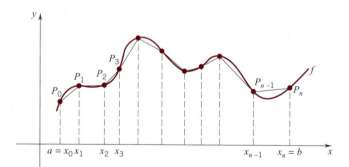

43. Rework Example 5, this time from energy considerations. (a) Verify that GPE + KE remains constant. (b) Then derive the speed at impact from an energy equation.

44. The path of the projectile discussed under the heading "Parabolic Trajectories" in Section 9.7 is given in terms of the time parameter t by the following equations:

$$x(t) = (v_0\cos \theta)t + x_0, \, y(t) = -\tfrac{1}{2}gt^2 + (v_0\sin \theta)t + y_0.$$

The projectile impacts at ground level, $y = 0$. Find the speed at impact.

45. Suppose that f is continuously differentiable from $x = a$ to $x = b$. Show that the

(9.9.7)

$$\text{Length of the graph of } f = \int_a^b |\sec [\alpha(x)]| \, dx,$$

where $\alpha(x)$ is the inclination of the tangent line at $(x, f(x))$.

46. Show that a homogeneous, flexible, inelastic rope hanging from two fixed points assumes the shape of a *catenary*:

$$f(x) = a \cosh\left(\frac{x}{a}\right) = \frac{a}{2}(e^{x/a} + e^{-x/a}), \, a > 0.$$

HINT: Refer to the figure below. The part of the rope that corresponds to the interval $[0, x]$ is subject to the following forces:

(1) its weight, which is proportional to its length;
(2) a horizontal pull at 0, $p(0)$;
(3) a tangential pull at x, $p(x)$.

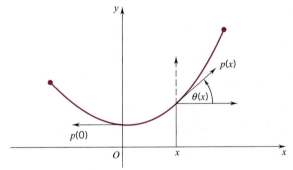

▷47. (a) Use a graphing utility to draw the graph of the curve defined parametrically by

$$x(t) = t^2, \qquad y(t) = t^3 - t, \quad -\infty < t < \infty.$$

(b) Your graph in part (a) should show that the curve has a loop. Use Simpson's rule with $n = 3$ to find the approximate length of the loop. Round your answer to four decimal places.

▷48. Sketch the graph of the limaçon $r = 2 + \sin \theta$. Use Simpson's rule with $n = 4$ to estimate the length of this curve. Round your answer to four decimal places.

▷49. Elliptic integrals.
(a) Let $a > b > 0$. Show that the length of the ellipse

$$x(t) = a \cos t, \quad y(t) = b \sin t, \quad 0 \le t \le 2\pi,$$

is given by

$$L = 4a \int_0^{\pi/2} \sqrt{1 - e \cos^2 t} \, dt,$$

where $e = \sqrt{a^2 - b^2}/a$ is the eccentricity. The integrand does not have an elementary antiderivative.
(b) Let $a = 5$ and $b = 4$, and approximate the length of the ellipse using Simpson's rule with $n = 4$. Round your answer to four decimal places.

▷50. The curve defined parametrically by

$$x(t) = \frac{3t}{t^3 + 1}, \qquad y(t) = \frac{3t^2}{t^3 + 1}, \qquad t \neq -1,$$

is called the *folium of Descartes*.
(a) Use a graphing utility to draw the graph of this curve.
(b) Your graph in part (a) should show that the curve has a loop in the first quadrant. Use Simpson's rule with $n = 3$ to estimate the length of the loop. Round your answer to four decimal places. HINT: Use symmetry.

■ **9.10 THE AREA OF A SURFACE OF REVOLUTION; THE CENTROID OF A CURVE; PAPPUS'S THEOREM ON SURFACE AREA**

The Area of a Surface of Revolution

In Figure 9.10.1 you can see the frustum of a cone; one radius is marked r, the other R, and the slant height is marked s. An interesting elementary calculation that we leave to you shows that the area of this slanted surface is given by the formula

(9.10.1)

$$A = \pi(r + R)s.$$

(Exercise 21)

Figure 9.10.1

This little formula forms the basis for all that follows.

Let C be a curve in the upper half-plane (Figure 9.10.2). The curve can meet the x-axis, but only at a finite number of points. We will assume that C is parametrized by a pair of continuously differentiable functions

$$x = x(t), \quad y = y(t), \qquad t \in [c, d].$$

Furthermore, we will assume that C is *simple*: no two values of t between c and d give rise to the same point of C, that is, the curve does not intersect itself.

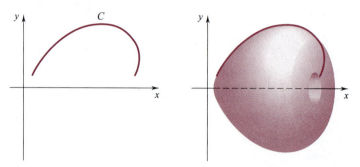

Figure 9.10.2

If we revolve C about the x-axis, we obtain a surface of revolution. The area of that surface is given by the formula

(9.10.2)
$$A = \int_c^d 2\pi y(t)\sqrt{[x'(t)]^2 + [y'(t)]^2}\, dt.$$

We will try to outline how this formula comes about. The argument is similar to the one given in Section 9.9 for the length of a curve.

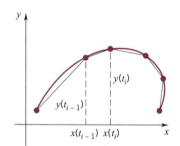

Figure 9.10.3

Each partition $P = \{c = t_0 < t_1 < \cdots < t_n = d\}$ of $[c, d]$ generates a polygonal approximation to C (Figure 9.10.3). Call this polygonal approximation C_p. By revolving C_p about the x-axis we get a surface made up of n conical frustums. The ith frustum (Figure 9.10.4) has slant height

$$s_i = \sqrt{[x(t_i) - x(t_{i-1})]^2 + [y(t_i) - y(t_{i-1})]^2}$$
$$= \sqrt{\left[\frac{x(t_i) - x(t_{i-1})}{t_i - t_{i-1}}\right]^2 + \left[\frac{y(t_i) - y(t_{i-1})}{t_i - t_{i-1}}\right]^2}\,(t_i - t_{i-1})$$

and the lateral area $\pi[y(t_{i-1}) + y(t_i)]s_i$ (see Formula (9.10.1)) can be written

$$\pi[y(t_{i-1}) + y(t_i)] \sqrt{\left[\frac{x(t_i) - x(t_{i-1})}{t_i - t_{i-1}}\right]^2 + \left[\frac{y(t_i) - y(t_{i-1})}{t_i - t_{i-1}}\right]^2}\,(t_i - t_{i-1}).$$

There exist points $t_i^*, t_i^{**}, t_i^{***}$ all in $[t_{i-1}, t_i]$ such that

$$y(t_i) + y(t_{i-1}) = 2y(t_i^*), \qquad \frac{x(t_i) - x(t_{i-1})}{t_i - t_{i-1}} = x'(t_i^{**}), \qquad \frac{y(t_i) - y(t_{i-1})}{t_i - t_{i-1}} = y'(t_i^{***}).$$

intermediate-value theorem — mean-value theorem

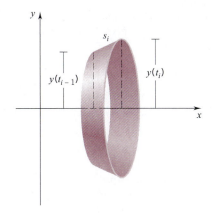

Figure 9.10.4

Let $\Delta t_i = t_i - t_{i-1}$. Then we can write the lateral area of the ith frustum as

$$2\pi y(t_i^*) \sqrt{[x'(t_i^{**})]^2 + [y'(t_i^{***})]^2}\; \Delta t_i.$$

The area generated by revolving all of C_p is the sum of these terms:

$$2\pi y(t_1^*)\sqrt{[x'(t_1^{**})]^2 + [y'(t_1^{***})]^2}\; \Delta t_1 + \cdots + 2\pi y(t_n^*)\sqrt{[x'(t_n^{**})]^2 + [y'(t_n^{***})]^2}\; \Delta t_n.$$

This is not a Riemann sum: we don't know that $t_i^* = t_i^{**} = t_i^{***}$. But it is "close" to a Riemann sum. Close enough that, as $\|P\| \to 0$, this "almost" Riemann sum tends to the integral

$$\int_c^d 2\pi y(t)\sqrt{[x'(t)]^2 + [y'(t)]^2}\; dt.$$

That this is so follows from a theorem of advanced calculus known as Duhamel's principle. We will not attempt to fill in the details. ❏

Example 1 Derive a formula for the surface area of a sphere from (9.10.2).

SOLUTION We can generate a sphere of radius r by revolving the arc

$$x(t) = r \cos t, \quad y(t) = r \sin t, \qquad t \in [0, \pi]$$

about the x-axis. Differentiation gives

$$x'(t) = -r \sin t, \quad y'(t) = r \cos t.$$

By Formula (9.10.2)

$$A = 2\pi \int_0^\pi r \sin t \sqrt{r^2(\sin^2 t + \cos^2 t)}\; dt$$

$$= 2\pi r^2 \int_0^\pi \sin t\; dt = 2\pi r^2 \left[-\cos t \right]_0^\pi = 4\pi r^2. \quad ❏$$

Example 2 Find the area of the surface generated by revolving about the x-axis the curve

$$y^2 - 2 \ln y = 4x \qquad \text{from } y = 1 \text{ to } y = 2.$$

SOLUTION We can represent the curve parametrically by setting

$$x(t) = \tfrac{1}{4}(t^2 - 2 \ln t), \quad y(t) = t, \qquad t \in [1, 2].$$

Here

$$x'(t) = \tfrac{1}{2}(t - t^{-1}), \quad y'(t) = 1$$

and

$$[x'(t)]^2 + [y'(t)]^2 = [\tfrac{1}{2}(t + t^{-1})]^2. \qquad \text{(check this)}$$

It follows that

$$A = \int_1^2 2\pi t[\tfrac{1}{2}(t + t^{-1})]\, dt = \int_1^2 \pi(t^2 + 1)\, dt = \pi\left[\tfrac{1}{3}t^3 + t\right]_1^2 = \tfrac{10}{3}\pi. \quad ❑$$

Suppose now that f is some nonnegative function defined from $x = a$ to $x = b$. If f' is continuous, then the graph of f is a continuously differentiable curve in the upper half-plane. The area of the surface generated by revolving this graph about the x-axis is given by the formula

(9.10.3)

$$A = \int_a^b 2\pi f(x)\sqrt{1 + [f'(x)]^2}\, dx.$$

This follows readily from (9.10.2). Set

$$x = t, \quad y(t) = f(t), \qquad t \in [a, b].$$

Apply (9.10.2) and then replace the dummy variable t by x.

Example 3 Find the area of the surface generated by revolving about the x-axis the graph of the sine function from $x = 0$ to $x = \tfrac{1}{2}\pi$.

SOLUTION Setting $f(x) = \sin x$, we have $f'(x) = \cos x$ and therefore

$$A = \int_0^{\pi/2} 2\pi \sin x\sqrt{1 + \cos^2 x}\, dx.$$

To calculate this integral, we set

$$u = \cos x, \qquad du = -\sin x\, dx.$$

At $x = 0$, $u = 1$; at $x = \tfrac{1}{2}\pi$, $u = 0$. Therefore

$$A = -2\pi \int_1^0 \sqrt{1 + u^2}\, du = 2\pi \int_0^1 \sqrt{1 + u^2}\, du$$

$$= 2\pi\left[\tfrac{1}{2}u\sqrt{1 + u^2} + \tfrac{1}{2}\ln(u + \sqrt{1 + u^2})\right]_0^1$$

by (8.5.1) ⟶

$$= \pi[\sqrt{2} + \ln(1 + \sqrt{2})] \cong 2.3\pi \cong 7.23. \quad ❑$$

The Centroid of a Curve

The centroid of a plane region Ω is the center of mass of a homogeneous plate in the shape of Ω. Likewise, the centroid of a solid of revolution T is the center of mass of a homogeneous solid in the shape of T. All this you know from Section 6.4.

What do we mean by the centroid of a plane curve C? Exactly what you would expect. By the *centroid* of a plane curve C, we mean the center of mass of a homogeneous wire in the shape of C. It should be noted, however, that the centroid of a plane curve does not necessarily lie on the curve. We can calculate the centroid of a curve from the following principles, which we take from physics.

Principle 1: Symmetry If a curve has an axis of symmetry, then the centroid $(\bar{x}, \bar{y})$ lies somewhere along that axis.

Principle 2: Additivity If a curve with length L is broken up into a finite number of pieces with arc lengths $\Delta s_1, \ldots, \Delta s_n$ and centroids $(\bar{x}_1, \bar{y}_1), \ldots, (\bar{x}_n, \bar{y}_n)$, then

$$\bar{x}L = \bar{x}_1 \, \Delta s_1 + \cdots + \bar{x}_n \, \Delta s_n \quad \text{and} \quad \bar{y}L = \bar{y}_1 \, \Delta s_1 + \cdots + \bar{y}_n \, \Delta s_n.$$

Figure 9.10.5 shows a curve C that begins at A and ends at B. Let's suppose that the curve is continuously differentiable and that the length of the curve is L. We want a formula for the centroid $(\bar{x}, \bar{y})$.

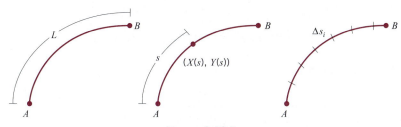

Figure 9.10.5

Let $(X(s), Y(s))$ be the point on C that is at an arc distance s from the initial point A. (What we are doing here is called *parametrizing C by arc length*.) A partition $P = \{0 = s_0 < s_1 < \cdots < s_n = L\}$ of $[0, L]$ breaks up C into n little pieces of lengths $\Delta s_1, \ldots, \Delta s_n$ and centroids $(\bar{x}_1, \bar{y}_1), \ldots, (\bar{x}_n, \bar{y}_n)$. From Principle 2 we know that

$$\bar{x}L = \bar{x}_1 \, \Delta s_1 + \cdots + \bar{x}_n \Delta s_n \quad \text{and} \quad \bar{y}L = \bar{y}_1 \, \Delta s_1 + \cdots + \bar{y}_n \, \Delta s_n.$$

Now for each i there exists s_i^* in $[s_{i-1}, s_i]$ for which $\bar{x}_i = X(s_i^*)$ and s_i^{**} in $[s_{i-1}, s_i]$ for which $\bar{y}_i = Y(s_i^{**})$. We can therefore write

$$\bar{x}L = X(s_i^*) \, \Delta s_1 + \cdots + X(s_n^*) \, \Delta s_n, \qquad \bar{y}L = Y(s_i^{**}) \, \Delta s_1 + \cdots + Y(s_n^{**}) \, \Delta s_n.$$

The sums on the right are Riemann sums tending to easily recognizable limits: letting $\|P\| \to 0$ we have

(9.10.4)
$$\bar{x}L = \int_0^L X(s) \, ds \quad \text{and} \quad \bar{y}L = \int_0^L Y(s) \, ds.$$

These formulas give the centroid of a curve in terms of the parameter arc length. It is but a short step from here to formulas of more ready applicability.

Suppose that the curve C is given parametrically by the functions

$$x = x(t), \quad y = y(t), \qquad t \in [c, d]$$

where t is now an arbitrary parameter. Then

$$s(t) = \int_c^t \sqrt{[x'(u)]^2 + [y'(u)]^2}\, du, \qquad ds = s'(t)\, dt = \sqrt{[x'(t)]^2 + [y'(t)]^2}\, dt.$$

At $s = 0$, $t = c$; at $s = L$, $t = d$. Changing variables in (9.10.4) from s to t, we have

$$\bar{x}L = \int_c^d X(s(t))s'(t)\, dt = \int_c^d X(s(t))\sqrt{[x'(t)]^2 + [y'(t)]^2}\, dt$$

and

$$\bar{y}L \int_c^d Y(s(t))s'(t)\, dt = \int_c^d Y(s(t))\sqrt{[x'(t)]^2 + [y'(t)]^2}\, dt.$$

A moment's reflection shows that

$$X(s(t)) = x(t) \quad \text{and} \quad Y(s(t)) = y(t).$$

We can then write

(9.10.5)

$$\bar{x}L = \int_c^d x(t)\sqrt{[x'(t)]^2 + [y'(t)]^2}\, dt,$$

$$\bar{y}L = \int_c^d y(t)\sqrt{[x'(t)]^2 + [y'(t)]^2}\, dt.$$

These are the centroid formulas in their most useful form.

Example 4 Find the centroid of the quarter-circle shown in Figure 9.10.6.

SOLUTION We can parametrize that quarter-circle by setting

$$x(t) = r \cos t, \quad y(t) = r \sin t, \quad t \in [0, \pi/2].$$

Since the curve is symmetric about the line $x = y$, we know that $\bar{x} = \bar{y}$. Here $x'(t) = -r \sin t$ and $y'(t) = r \cos t$. Therefore

$$\sqrt{[x'(t)]^2 + [y'(t)]^2} = \sqrt{r^2 \sin^2 t + r^2 \cos^2 t} = r.$$

By (9.10.5)

$$\bar{y}L = \int_0^{\pi/2} (r \sin t)r\, dt = r^2 \int_0^{\pi/2} \sin t\, dt = r^2 \left[-\cos t \right]_0^{\pi/2} = r^2.$$

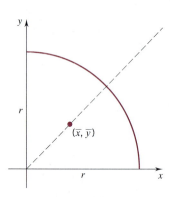

Figure 9.10.6

Note that $L = \pi r/2$. Therefore $\bar{y} = r^2/L = 2r/\pi$. The centroid of the quarter-circle is the point $(2r/\pi, 2r/\pi)$. [Note that this point is closer to the curve than the centroid of the quarter-disc. (Example 1, Section 6.4).] ❏

Example 5 Find the centroid of the cardioid $r = a(1 - \cos \theta)$, $a > 0$.

SOLUTION The curve (see Figure 9.5.1) is symmetric about the x-axis. Thus $\bar{y} = 0$. To find $\bar{x}$ we parametrize the curve as follows:

$$x(\theta) = r \cos \theta = a(1 - \cos \theta) \cos \theta,$$
$$y(\theta) = r \sin \theta = a(1 - \cos \theta) \sin \theta, \qquad \theta \in [0, 2\pi].$$

A straightforward calculation shows that

$$[x'(\theta)]^2 + [y'(\theta)]^2 = 4a^2 \sin^2 \tfrac{1}{2}\theta.$$

Applying (9.10.5) we have

$$\bar{x}L = \int_0^{2\pi} [a(1 - \cos\theta)\cos\theta][2a\sin\tfrac{1}{2}\theta]\,d\theta = -\tfrac{32}{5}a^2.$$

check this out ⎯⎯⎯⎯↑

By Example 4 in Section 9.9, $L = 8a$. Thus $\bar{x} = (-\tfrac{32}{5}a^2)/8a = -\tfrac{4}{5}a$. The centroid of the curve is the point $(-\tfrac{4}{5}a, 0)$. ❑

If C is a curve of the form

$$y = f(x), \qquad x \in [a, b],$$

where f is continuously differentiable, then the formulas in (9.10.5) give

(9.10.6)
$$\bar{x}L = \int_a^b x\sqrt{1 + [f'(x)]^2}\,dx, \qquad \bar{y}L = \int_a^b f(x)\sqrt{1 + [f'(x)]^2}\,dx.$$

The details are left to you.

Pappus's Theorem on Surface Area

That same Pappus who gave us that wonderful theorem on volumes of solids of revolution (Theorem 6.4.4) gave us the following equally marvelous result on surface area:

THEOREM 9.10.7 PAPPUS'S THEOREM ON SURFACE AREA

A plane curve is revolved about an axis that lies in its plane. The curve may meet the axis but, if so, only at a finite number of points. If the curve does not cross the axis, then the area of the resulting surface of revolution is the length of the curve multiplied by the circumference of the circle described by the centroid of the curve:

$$A = 2\pi\bar{R}L$$

where L is the length of the curve and $\bar{R}$ is the distance from the axis to the centroid of the curve.

Pappus did not have calculus to help him when he made his inspired guesses: he did his work 13 centuries before Newton or Leibniz were born. With the formulas that we have developed through calculus (through Newton and Leibniz, that is) Pappus's theorem is easily verified. Call the plane of the curve the xy-plane and call the axis of rotation the x-axis. Then $\bar{R} = \bar{y}$ and

$$A = \int_c^d 2\pi y(t)\sqrt{[x'(t)]^2 + [y'(t)]^2}\,dt$$

$$= 2\pi \int_c^d y(t)\sqrt{[x'(t)]^2 + [y'(t)]^2}\,dt = 2\pi\bar{y}L = 2\pi\bar{R}L. ❑$$

EXERCISES 9.10

In Exercises 1–10, find the length of the curve, locate the centroid, and determine the area of the surface generated by revolving the curve about the *x*-axis.

1. $f(x) = 4$, $x \in [0, 1]$. **2.** $f(x) = 2x$, $x \in [0, 1]$.
3. $y = \frac{4}{3}x$, $x \in [0, 3]$.
4. $y = -\frac{12}{5}x + 12$, $x \in [0, 5]$.
5. $x(t) = 3t$, $y(t) = 4t$; $t \in [0, 2]$.
6. $r = 5$, $\theta \in [0, \frac{1}{4}\pi]$.
7. $x(t) = 2 \cos t$, $y(t) = 2 \sin t$; $t \in [0, \frac{1}{6}\pi]$.
8. $x(t) = \cos^3 t$, $y(t) = \sin^3 t$; $t \in [0, \frac{1}{2}\pi]$.
9. $x^2 + y^2 = a^2$, $x \in [-\frac{1}{2}a, \frac{1}{2}a]$, $a > 0$.
10. $r = 1 + \cos \theta$, $\theta \in [0, \pi]$.

In Exercises 11–18, find the area of the surface generated by revolving the curve about the *x*-axis.

11. $f(x) = \frac{1}{3}x^3$, $x \in [0, 2]$. **12.** $f(x) = \sqrt{x}$, $x \in [1, 2]$.
13. $4y = x^3$, $x \in [0, 1]$. **14.** $y^2 = 9x$, $x \in [0, 4]$.
15. $y = \cos x$, $x \in [0, \frac{1}{2}\pi]$.
16. $f(x) = 2\sqrt{1 - x}$, $x \in [-1, 0]$.
17. $r = e^\theta$, $\theta \in [0, \frac{1}{2}\pi]$.
18. $y = \cosh x$, $x \in [0, \ln 2]$.
19. One arch of a cycloid is defined parametrically by

$$x(\theta) = a(\theta - \sin \theta), \qquad y(\theta) = a(1 - \cos \theta),$$

$$a > 0, \quad 0 \le \theta \le 2\pi.$$

(a) Find the area under the curve. HINT: See Exercise 47, Section 9.7.
(b) Find the area of the surface generated by revolving the arch around the *x*-axis.

 20. Given the hypocycloid

$$x(\theta) = 3a \cos \theta + a \cos 3\theta,$$

$$y(\theta) = 3a \sin \theta - a \sin 3\theta,$$

$$a > 0, \quad 0 \le \theta \le 2\pi.$$

See Exercise 49, Section 9.7.
(a) Use a graphing utility to draw the graph.
(b) Find the area enclosed by the hypocycloid.
(c) Set up a definite integral that gives the area of the surface generated by revolving the hypocycloid around the *x*-axis.

21. By cutting a cone of slant height *s* and base radius *r* along a lateral edge and laying the surface flat, we can form a sector of a circle of radius *s*. (See the figure.) Use this idea to verify Formula (9.10.1).

$$\text{area} = \tfrac{1}{2}\theta s^2, \ \theta \text{ in radians}$$

22. The figure shows a ring formed by two quarter-circles. Call the corresponding quarter-discs Ω_a and Ω_r. By Section 6.4, Ω_a has centroid $(4a/3\pi, 4a/3\pi)$ and Ω_r has centroid $(4r/3\pi, 4r/3\pi)$.

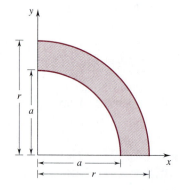

(a) Without integration calculate the centroid of the ring.
(b) Find the centroid of the outer arc from your answer to (a) by letting *a* tend to *r*.

23. (a) Find the centroid of each side of the triangle in the figure.

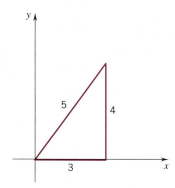

(b) Use your answers to (a) to calculate the centroid of the triangle.

(c) What is the centroid of the triangular region?

(d) What is the centroid of the curve consisting of sides 4 and 5?

(e) Use Pappus's theorem to find the slanted surface area of a cone of base radius 4 and height 3.

(f) Use Pappus's theorem to find the total surface area of the cone in (e). (This time include the base.)

24. Find the area of the surface generated by revolving about the x-axis the curve

(a) $2x = y\sqrt{y^2 - 1} + \ln|y - \sqrt{y^2 - 1}|$, $y \in [2, 5]$.

(b) $6a^2xy = y^4 + 3a^4$, $y \in [a, 3a]$.

25. Use Pappus's theorem to find the surface area of the *torus* generated by revolving about the x-axis the circle $x^2 + (y - b)^2 = a^2$. $(0 < a \le b)$

26. (a) We calculated the total surface area of a sphere from (9.10.2) not (9.10.3). Could we just as well have used (9.10.3)? Explain.

(b) Verify that Formula (9.10.2) applied to

$$C: \quad x(t) = \cos t, \quad y(t) = r, \quad \text{with } t \in [0, 2\pi],$$

gives $A = 8\pi r$. Note that the surface obtained by revolving C about the x-axis is a cylinder of base radius r and height 2, and therefore A should be $4\pi r$. What's wrong?

27. (*An interesting property of the sphere*) Slice a sphere along two parallel planes that are a fixed distance apart. Show that the surface area of the band so obtained is independent of where the cuts are made.

28. Locate the centroid of a first-quadrant circular arc

$$C: \quad x(t) = r \cos t, \quad y(t) = r \sin t, \quad t \in [\theta_1, \theta_2].$$

29. Find the surface area of the ellipsoid obtained by revolving the ellipse

$$\frac{x^2}{a^2} + \frac{y^2}{b^2} = 1 \qquad (0 < b < a)$$

(a) about its major axis; (b) about its minor axis.

The Centroid of a Surface of Revolution

If a material surface of revolution is homogeneous (constant mass density), then the center of mass of that material surface is called the *centroid*. The determination of the centroid of a surface of arbitrary shape requires surface integration (Chapter 17). However, if the surface is a surface of revolution, then the centroid can be found by ordinary one-variable integration.

30. Let C be a simple curve in the upper half-plane parametrized by a pair of continuously differentiable functions.

$$x = x(t), \quad y = y(t), \quad t \in [c, d].$$

By revolving C about the x-axis we obtain a surface of revolution, the area of which we denote by A. By symmetry the centroid of the surface lies on the x-axis. Thus the centroid is completely determined by its x-coordinate $\bar{x}$. Show that

(9.10.8) $$\bar{x}A = \int_c^d 2\pi x(t)y(t)\sqrt{[x'(t)]^2 + [y'(t)]^2}\, dt$$

by assuming the following additivity principle: If the surface is broken up into n surfaces of revolution with areas $A_1, \ldots, A_n$ and the centroids of the surfaces have x-coordinates $\bar{x}_1, \ldots, \bar{x}_n$, then

$$\bar{x}A = \bar{x}_1A_1 + \cdots + \bar{x}_nA_n.$$

31. Locate the centroid of a hemisphere of radius r.

32. Locate the centroid of a conical surface of base radius r and height h.

33. Where is the centroid of the lateral surface of the frustum of a cone of height h with base radii r and R?

34. (a) Show that the circle

$$(x - a)^2 + y^2 = b^2$$

can be parametrized by

$$x(t) = a + b \cos t, \quad y(t) = b \sin t, \quad 0 \le t \le 2\pi.$$

(b) Suppose that $0 < b < a$. The solid generated by revolving the circle around the y-axis is a torus. Find the volume of the torus.

(c) Find the surface area of the torus.

*■ 9.11 THE CYCLOID

(In this section the exercises are intertwined with the text.)

Take a wheel (a roll of tape will do) and fix your eyes on some point of the rim. Call that point P. Now roll the wheel slowly, keeping your eyes on P. The jumping-kangaroo path described by P is called a *cycloid*.

To obtain a mathematical characterization of the cycloid, call the radius of the wheel R and set the wheel on the x-axis so that the point P starts out at the origin. Figure 9.11.1 shows P after a turn of θ radians.

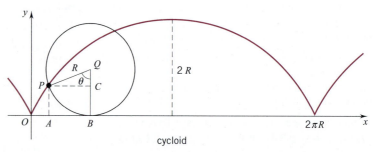

cycloid

Figure 9.11.1

1. Show that the cycloid can be parametrized by the functions

(9.11.1)
$$x(\theta) = R(\theta - \sin\theta), \quad y(\theta) = R(1 - \cos\theta).$$

HINT: Length of $\overline{OB}$ = length of $\overparen{PB}$ = $R\theta$.

2. At the end of each arch the cycloid comes to a cusp. Verify that x' and y' are both 0 at the end of each arch.

3. Show that the line tangent to the cycloid at P (a) passes through the top of the circle and (b) intersects the x-axis at an angle of $\frac{1}{2}(\pi - \theta)$ radians.

4. Express the cycloid from $\theta = 0$ to $\theta = \pi$ by an equation in x and y. (The equation is quite complicated and not easy to work with. The properties of the cycloid are much easier to fathom from the parametric equations.)

5. Find the length of an arch of the cycloid. (9.9.1)

6. Show that the area under an arch of the cycloid is three times the area of the rolling circle. (9.7.4)

7. Locate the centroid of the region under the first arch of the cycloid. (9.7.5)

8. Find the volume of the solid generated by revolving about the x-axis the region under an arch of the cycloid. (9.7.6)

9. Find the volume of the solid generated by revolving about the y-axis the region under the first arch of the cycloid. (9.7.6)

10. Determine the centroid of the first arch of the cycloid. (9.10.5)

11. Find the area of the surface generated by revolving about the x-axis an arch of the cycloid. (9.10.2)

THE INVERTED CYCLOID

12. Show that the arch of the inverted cycloid shown in Figure 9.11.2 can be parametrized by setting

$$x = R(\phi + \sin\phi), \quad y = R(1 - \cos\phi), \qquad \phi \in [-\pi, \pi].$$

Proceed as follows:
(a) Reflect the arch of Figure 9.11.1 in the x-axis and find the θ parametrization for that.
(b) Raise the arch obtained in part (a) so that the low point rests on the x-axis and find the θ parametrization for that.

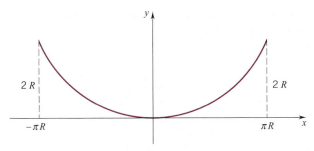

Figure 9.11.2

(c) Obtain the arch of Figure 9.11.2 by translating πR units to the left the arch obtained in part (b). Write down the θ parametrization for this arch and then set $\phi = \theta - \pi$.

13. Find the inclination α of the line tangent to the inverted arch at the point $(x(\phi), y(\phi))$.

14. Let s be the arc distance from the low point of the inverted arch to the point $(x(\phi), y(\phi))$ of that same arch. Show that $s = 4R \sin \frac{1}{2}\phi = 4R \sin \alpha$ where α is the inclination of the tangent line at $(x(\phi), y(\phi))$.

THE TAUTOCHRONE

Visualize two particles (beads if you prefer) sliding without friction down an arch of an inverted cycloid (Figure 9.11.3). If the two particles are released at the same time from different positions, which will reach the bottom first? Neither: They will both get there at exactly the same time.† Being the only curve that produces this effect, the inverted arch of a cycloid is known as *the tautochrone,* the *same-time* curve.

Figure 9.11.3

15. Verify that the inverted arch of a cycloid has the tautochrone property by taking the following steps:

(a) Show that the effective gravitational force on a particle of mass m is $- mg \sin \alpha$ where α is the inclination of the tangent line at the position of the particle. Conclude then that

(∗)
$$\frac{d^2s}{dt^2} = -g \sin \alpha.$$

(b) Combine (∗) with Exercise 14 to show that the particle is in simple harmonic motion with period

$$T = 4\pi\sqrt{R/g}.$$

(Thus, while the amplitude of the motion depends on the point of release, the frequency does not. Two particles released simultaneously from different points of the curve will reach the low point of the curve in exactly the same amount of time: $T/4 = \pi\sqrt{R/g}$.)

THE BRACHYSTOCHRONE

A particle is to descend without friction along a curve from some point A to a point B not directly below it. (See Figure 9.11.4.) What should be the shape of the curve so that the particle descends from A to B in the least possible time?

Figure 9.11.4

---

† This was understood in 1673 by Christian Huygens, the inventor of the pendulum clock. By tracking the bob of the pendulum along the arch of an inverted cycloid, Huygens was able to stabilize the frequency of the oscillations of the pendulum and thereby improve the accuracy of his clock.

This question was first formulated by Johann Bernoulli and posed by him as a challenge to the scientific community in 1696. The challenge was readily accepted and within months the answer was found—by Johann Bernoulli himself, by his brother Jacob, by Newton, by Leibniz, and by L'Hospital. The answer? Part of an inverted cycloid. Because of this, the inverted cycloid is heralded as the *brachystochrone,* the *least-time* curve.

A proof that the inverted cycloid is the least-time curve, the curve of quickest descent, is beyond our reach. The argument requires a sophisticated variant of calculus known as *the calculus of variations.* We can, however, compare the time of descent along a cycloid to the time of descent along a straight-line path.

16. You have seen that a particle descends along the inverted arch of a cycloid from $(\pi R, 2R)$ to $(0, 0)$ in time $t = T/4 = \pi\sqrt{R/g}$. What is the time of descent along a straight-line path?

■ CHAPTER HIGHLIGHTS

9.1 The Conic Sections

The Parabola
    directrix, focus, axis, vertex (p. 550)
    reflecting property (p. 562)

A parabola is the set of points equidistant from a fixed line and a fixed point not on that line.

The Ellipse
    foci, standard position (p. 554)   vertices, axes, center (p. 556)
    elliptical reflectors (p. 563)   eccentricity (p. 566)

An ellipse is the set of points the sum of whose distances from two fixed points is constant.

The Hyperbola
    foci, standard position (p. 558)   asymptotes (p. 560)
    vertices, transverse axis, center (p. 560)   hyperbolic reflectors (p. 565)
    eccentricity (p. 567)

A hyperbola is the set of points the difference of whose distances from two fixed points is constant.

9.2 Polar Coordinates

Relation between rectangular coordinates (x, y) and polar coordinates $[r, \theta]$:

$$x = r \cos \theta, \qquad y = r \sin \theta \qquad \tan \theta = \frac{y}{x}, \qquad x^2 + y^2 = r^2$$

9.3 Graphing in Polar Coordinates

    cardioids (p. 578)   lines (p. 579)   circles (p. 579)
    limaçons (p. 579)   spirals (p. 580)   lemniscates (p. 580)
    petal curves (p. 580)

*9.4 The Conic Sections in Polar Coordinates

$$r = \frac{ed}{1 + e \cos \theta} \qquad \begin{matrix} \text{ellipse} & 0 < e < 1 \\ \text{parabola} & e = 1 \\ \text{hyperbola} & e > 1 \end{matrix}$$

9.5 The Intersection of Polar Curves

If we think of each of two polar curves as giving the position of an object at time θ, then the simultaneous solution of the two equations gives the points where the objects collide. A figure will often help to identify the points where two polar curves intersect, but do not collide; conversion to rectangular coordinates may also help to locate such points.

9.6 Area in Polar Coordinates

Let ρ_1 and ρ_2 be positive continuous functions defined on a closed interval $[\alpha, \beta]$ of length 2π, at most. If $\rho_1(\theta) \le \rho_2(\theta)$ for all θ in $[\alpha, \beta]$, then the area of the region between the polar curves $r = \rho_1(\theta)$ and $r = \rho_2(\theta)$ is given by the formula

$$A = \int_\alpha^\beta \tfrac{1}{2}([\rho_2(\theta)]^2 - [\rho_1(\theta)]^2) \, d\theta.$$

9.7 Curves Given Parametrically

Let $x = x(t), y = y(t)$ be a pair of functions differentiable on the interior of some interval I. At the endpoints of I (if any) we require only continuity. For each t in I we can interpret $(x(t), y(t))$ as the point with x-coordinate $x(t)$ and y-coordinate $y(t)$. Then, as t ranges over I, the point $(x(t), y(t))$ traces a path in the xy-plane. We call such a path a *parametrized curve* and refer to t as the *parameter*.

line: $\quad x(t) = x_0 + t(x_1 - x_0), \qquad y(t) = y_0 + t(y_1 - y_0), \qquad t$ real.

circle: $\quad x(t) = a \cos t, \qquad y(t) = a \sin t, \qquad t \in [0, 2\pi].$

ellipse: $\quad x(t) = a \cos t, \qquad y(t) = b \sin t, \qquad t \in [0, 2\pi].$

9.8 Tangents to Curves Given Parametrically

Tangent line at (x_0, y_0): $y'(t_0)[x - x_0] - x'(t_0)[y - y_0] = 0$, (where $x(t_0) = x_0, y(t_0) = y_0$)
provided that $[x'(t_0)]^2 + [y'(t_0)]^2 \ne 0.$

9.9 Arc Length and Speed

polygonal path (p. 613)   definition of arc length (p. 614)
significance of $dx/ds, dy/ds$ (p. 616)

length of a parametrized curve: $\quad L = \int_a^b \sqrt{[x'(t)]^2 + [y'(t)]^2} \, dt$

length of a graph: $\quad L = \int_a^b \sqrt{1 + [f'(x)]^2} \, dx$

speed along a curve parametrized by time t: $\quad v(t) = \sqrt{[x'(t)]^2 + [y'(t)]^2}$

9.10 The Area of a Surface of Revolution; The Centroid of a Curve; Pappus's Theorem on Surface Area

revolution of a curve about x-axis: $A = \int_c^d 2\pi y(t) \sqrt{[x'(t)]^2 + [y'(t)]^2} \, dt$

revolution of a graph about x-axis: $A = \int_a^b 2\pi f(x) \sqrt{1 + [f'(x)]^2} \, dx$

parametrizing a curve by arc length (p. 625)

principles for finding the centroid of a plane curve (p. 625)

centroid of a plane curve: $\bar{x} L = \int_c^d x(t) \sqrt{[x'(t)]^2 + [y'(t)]^2} \, dt$

$$\bar{y} L = \int_c^d y(t) \sqrt{[x'(t)]^2 + [y'(t)]^2} \, dt$$

Pappus's theorem on surface area: $A = 2\pi \bar{R} L$

*9.11 The Cycloid

cycloid (p. 629)   tautochrone (p. 631)   brachystochrone (p. 631)

■ PROJECTS AND EXPLORATIONS USING TECHNOLOGY

To do these exercises you will need a graphics calculator or a computer with graphing capability. The majority of these problems are open-ended so different approaches may be used to solve them. You should be aware that different approaches can result in slight variations in the answers. Round your numerical answers to at least four decimal places. The rounding method that your calculator or computer uses also may cause variations in answers.

9.1 Consider the ellipse $x^2 + 4y^2 = 9$ and define the function $c = c(x)$ to be the cosine of the angle whose vertex is at the point $P(x, y)$ on the ellipse and whose sides pass through the foci of the ellipse. See the figure.

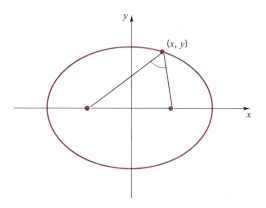

(a) Find a formula for $c(x)$.

(b) Use both the geometry and the formula that you found in part (b) to explain why c is an even function. As an even function, what can you say about the extrema of c?

(c) Suppose that the line segment extending between the two foci is a stage. Where on the ellipse should you sit to maximize your viewing angle?

(d) Repeat parts (a)–(c) for the ellipse

$$\frac{x^2}{a^2} + \frac{y^2}{b^2} = 1.$$

(e) Will the equation $c(x) = 0$ always have a solution? That is, is one of these angles always a right angle?

9.2 (*This problem requires software with an arc-length function.*) Let

$$f(x) = \sin x + x^{0.515}$$

and let $L(t)$ be the arc length of the graph of f from $x = 0$ to $x = t$.

(a) Graph L, L', and L''.

(b) Explain why L must have an inverse function and give an interpretation for it.

(c) Graph the inverse of L. What can you say about the points where L and its inverse intersect?

(d) Can you relate the local maxima, local minima, and points of inflection of f with those of L.

9.3 Consider using the summation

$$\sum_{i=1}^{n} \sqrt{(x_i - x_{i+1})^2 + (y_i - y_{i-1})^2}$$

as an approximation for arc length, where the points $P_i(x_i, y_i)$ are spaced along the curve. For convenience, either the x_i or the y_i are usually equally spaced. Not only is this method intuitively clear, but it can have advantages over approximations of appropriate integrals.

(a) Using arc length software or software to evaluate the arc length integral, find the arc length of the curve $f(x) = x^5$ on $[0, 1]$ to at least five decimal places. Then approximate the arc length using this summation with $n = 1, 2, 4, 8, 16, 32$.

(b) Using the values in part (a), plot n versus the difference between arc length calculated by software and the approximations. The differences should be positive. Why?

The summations

$$\sum_{i=1}^{n} \sqrt{1 + [f'(x_{i-1})]^2} \, \Delta x_i, \qquad \sum_{i=1}^{n} \sqrt{1 + [f'(m_i)]^2} \, \Delta x_i,$$

where $\Delta x_i = x_i - x_{i-1}$ and $m_i = (x_{i-1} + x_i)/2$ is the midpoint of $[x_{i-1}, x_i]$, can also be used to approximate arc length.

(c) Repeat part (a) using each of these summations and discuss the convergence of these approximations.

(d) Let $g(x) = \sin x$ on $[0, \pi/2]$. Use these summation formulas to approximate the arc length of the graph of g. Do you get the same type of results as you did for f?

SEQUENCES; INDETERMINATE FORMS; IMPROPER INTEGRALS

CHAPTER

10

■ 10.1 THE LEAST UPPER BOUND AXIOM

In the next few sections and in the next chapter, we will be talking about limits of sequences. To be able to do that with precision we first need to look a little deeper into the real number system.

We begin with a nonempty set S of real numbers. As indicated in Chapter 1, a number M is called an *upper bound* for S iff

$$x \leq M \qquad \text{for all } x \in S.$$

Note that if M is an upper bound for S, then any number greater than M will also be an upper bound for S. Thus, if a set has *an* upper bound, then it has, in fact, infinitely many upper bounds. Of course, not all sets of real numbers have upper bounds. Those that do are said to be *bounded above*.

For example, the number 3 is an upper bound for the set $\{x: x^2 \leq 1\} = \{x: -1 \leq x \leq 1\}$; so are $\frac{3}{2}$ and 1. This set is bounded above; 1 and every number greater than 1 are upper bounds. The set $\{2, 4, 6, 8, \ldots, 2n, \ldots\}$ is not bounded above.

It is clear that every set that has a largest element has an upper bound: if b is the largest element of S, then

$$x \leq b \qquad \text{for all } x \in S;$$

this makes b an upper bound for S. The converse is false: the sets

$$(-\infty, 0) \quad \text{and} \quad \left\{ \frac{1}{2}, \frac{2}{3}, \frac{3}{4}, \ldots, \frac{n}{n+1}, \ldots \right\}$$

both have upper bounds (2 for instance), but neither has a largest element.

Let's return to the first set, $(-\infty, 0)$. While $(-\infty, 0)$ does not have a largest element, the set of its upper bounds, $[0, \infty)$, does have a least element, 0. We call 0 the *least upper bound of* $(-\infty, 0)$.

Now let's reexamine the second set. While the set of quotients

$$\frac{n}{n+1} = 1 - \frac{1}{n+1}$$

does not have a greatest element, the set of its upper bounds, $[1, \infty)$, does have a least element, 1. We call 1 the *least upper bound* of that set of quotients.

In general, we have

DEFINITION 10.1.1 LEAST UPPER BOUND

Let S be a nonempty set of real numbers which is bounded above. A number M is the least upper bound of S iff

(i) M is an upper bound for S,

(ii) $M \leq K$, where K is any upper bound for S.

We are ready now to state explicitly one of the key *assumptions* that we make about the real number system. It is called the *least upper bound axiom*. Although we have not made an issue of it up to this point, this axiom has been an implicit assumption throughout the previous nine chapters; it underlies all of calculus.

AXIOM 10.1.2 THE LEAST UPPER BOUND AXIOM

Every nonempty set of real numbers that has an upper bound has a *least* upper bound.

Remark It is easy to see that if a set of real numbers S is bounded above, then its least upper bound is unique. For if L and M are least upper bounds of S, then $M \leq L$ and $L \leq M$ by property (ii) of Definition 10.1.1, and so $L = M$. Thus, it makes sense to use the term "*the* least upper bound" of S. ❏

To indicate the least upper bound of a set S, we will write lub S. Here are some examples:

1. lub $(-\infty, 0) = 0$,   lub $(-\infty, 0] = 0$.

2. lub $(-4, -1) = -1$,   lub $(-4, -1] = -1$.

3. lub $\left\{ \dfrac{1}{2}, \dfrac{2}{3}, \dfrac{3}{4}, \cdots, \dfrac{n}{n+1}, \cdots \right\} = 1$.

4. lub $\left\{ -\dfrac{1}{2}, -\dfrac{1}{8}, -\dfrac{1}{27}, \cdots, -\dfrac{1}{n^3}, \cdots \right\} = 0$.

5. lub $\{x : x^2 < 3\} = $ lub $\{x : -\sqrt{3} < x < \sqrt{3}\} = \sqrt{3}$. ❏

The least upper bound of a set has a special property that deserves particular attention. The idea is this: the fact that M is the least upper bound of the set S

does not tell us that M is in S (indeed, it need not be, as illustrated in the preceding examples), but it does tell us that we can approximate M as closely as we wish by elements of S.

THEOREM 10.1.3

If M is the least upper bound of the set S and ϵ is a positive number, then there is at least one number s in S such that

$$M - \epsilon < s \leq M.$$

PROOF Let $\epsilon > 0$. Since M is an upper bound for S, the condition $s \leq M$ is satisfied by all numbers s in S. All we have to show therefore is that there is some number s in S such that

$$M - \epsilon < s.$$

Suppose on the contrary that there is no such number in S. We then have

$$x \leq M - \epsilon \qquad \text{for all } x \in S.$$

This makes $M - \epsilon$ an upper bound for S. But this cannot be, for then $M - \epsilon$ is an upper bound for S that is *less* than M, and by assumption, M is the *least* upper bound. ❑

The theorem we just proved is illustrated in Figure 10.1.1. Take S as the set of points marked in the figure. If $M = \text{lub } S$, then S has at least one element in every half-open interval of the form $(M - \epsilon, M]$.

$$M - \epsilon \qquad\qquad\qquad M$$

Figure 10.1.1

Example 1

(a) Let

$$S = \left\{ \frac{1}{2}, \frac{2}{3}, \frac{3}{4}, \cdots, \frac{n}{n + 1}, \cdots \right\}$$

and take $\epsilon = 0.0001$. Since 1 is the least upper bound of S, there must be a number s in S such that

$$1 - 0.0001 < s \leq 1.$$

There is: take, for example, $s = \frac{99999}{100000}$.

(b) Let

$$S = \{1, 2, 3\}$$

and take $\epsilon = 0.00001$. It is clear that 3 is the least upper bound of S. Therefore, there must be a number $s \in S$ such that

$$3 - 0.00001 < s \leq 3.$$

There is: $s = 3$. ❑

We come now to lower bounds. In the first place, a number m is called a *lower bound* for a nonempty set S iff

$$m \leq x \quad \text{for all } x \in S.$$

Sets that have lower bounds are said to be *bounded below*. Not all sets have lower bounds; those that do have *greatest lower bounds*. Paralleling the definition of least upper bound, if a nonempty set S is bounded below, then a number m is the greatest lower bound of S if (i) m is a lower bound, and (ii) $m \geq k$, where k is any lower bound for S.

The existence of a greatest lower bound of a set S which is bounded below does not need to be taken as an axiom. We can prove it as a theorem using the least upper bound axiom.

THEOREM 10.1.4

Every nonempty set of real numbers that has a lower bound has a *greatest lower bound*.

PROOF Suppose that S is nonempty and that it has a lower bound x. Then

$$x \leq s \quad \text{for all } s \in S.$$

It follows that $-s \leq -x$ for all $s \in S$; that is,

$$\{-s : s \in S\} \quad \text{has an upper bound } -x.$$

From the least upper bound axiom we conclude that $\{-s : s \in S\}$ has a least upper bound; call it x_0. Since $-s \leq x_0$ for all $s \in S$, we can see that

$$-x_0 \leq s \quad \text{for all } s \in S,$$

and thus $-x_0$ is a lower bound for S. We now assert that $-x_0$ is the greatest lower bound of the set S. To see this, note that, if there existed a number x_1 satisfying

$$-x_0 < x_1 \leq s \quad \text{for all } s \in S,$$

then we would have

$$-s \leq -x_1 < x_0 \quad \text{for all } s \in S,$$

and thus x_0 would not be the *least* upper bound of $\{-s : s \in S\}$.† ❑

---

† We proved Theorem 10.1.4 by assuming the least upper bound axiom. We could have proceeded the other way. We could have set Theorem 10.1.4 as an axiom, and then proved the least upper bound axiom as a theorem.

As in the case of the least upper bound, the greatest lower bound of a set is unique. Also, the greatest lower bound of a set need not be in the set, but can be approximated as closely as we wish by members of the set. In short, we have the following theorem, the proof of which is left as an exercise.

THEOREM 10.1.5

If m is the greatest lower bound of the set S and ϵ is a positive number, then there is at least one number s in S such that

$$m \leq s < m + \epsilon.$$

The theorem is illustrated in Figure 10.1.2. If $m = $ glb S (that is, if m is the greatest lower bound of the set S), then S has at least one element in every half-open interval of the form $[m, m + \epsilon)$.

m $\qquad\qquad\qquad$ $m + \epsilon$

Figure 10.1.2

Remark Remember the intermediate-value theorem? It states that a continuous function skips no values. Remember the maximum-minimum theorem? It states that on a bounded closed interval a continuous function takes on both a maximum and a minimum value. We have been using these two results right along, but we have not proved them. Now that you understand least upper bounds and greatest lower bounds, you are in a position to follow proofs of both of these theorems. (See Appendix B.) Better still, try to prove the theorems yourself. ❏

EXERCISES 10.1

In Exercises 1–20, find the least upper bound (if it exists) and the greatest lower bound (if it exists) for the given set.

1. $(0, 2)$.

2. $[0, 2]$.

3. $(0, \infty)$.

4. $(-\infty, 1)$.

5. $\{x: x^2 < 4\}$.

6. $\{x: |x - 1| < 2\}$.

7. $\{x: x^3 \geq 8\}$.

8. $\{x: x^4 \leq 16\}$.

9. $\{2\frac{1}{2}, 2\frac{1}{3}, 2\frac{1}{4}, \ldots\}$.

10. $\{-1, -\frac{1}{2}, -\frac{1}{3}, -\frac{1}{4}, \ldots\}$.

11. $\{0.9, 0.99, 0.999, \ldots\}$.

12. $\{-2, 2, -2.1, 2.1, -2.11, 2.11, \ldots\}$.

13. $\{x: \ln x < 1\}$.

14. $\{x: \ln x > 0\}$.

15. $\{x: x^2 + x - 1 < 0\}$.

16. $\{x: x^2 + x + 2 \geq 0\}$.

17. $\{x: x^2 > 4\}$.

18. $\{x: |x - 1| > 2\}$.

19. $\{x: \sin x \geq -1\}$.

20. $\{x: e^x < 1\}$.

In Exercises 21–24, illustrate the validity of Theorem 10.1.5 taking S and ϵ as given.

21. $S = \{\frac{1}{11}, (\frac{1}{11})^2, (\frac{1}{11})^3, \ldots, (\frac{1}{11})^n, \ldots\}$, $\qquad \epsilon = 0.001$.

22. $S = \{1, 2, 3, 4\}$, $\qquad \epsilon = 0.0001$.

23. $S = \{\frac{1}{10}, \frac{1}{1000}, \frac{1}{100000}, \ldots, (\frac{1}{10})^{2n-1}, \ldots\}$, $\qquad \epsilon = (\frac{1}{10})^k \quad (k \geq 1)$.

24. $S = \{\frac{1}{2}, \frac{1}{4}, \frac{1}{8}, \ldots, (\frac{1}{2})^n, \ldots\}$, $\qquad \epsilon = (\frac{1}{4})^k \quad (k \geq 1)$.

25. Prove Theorem 10.1.5 by imitating the proof of Theorem 10.1.3.

26. Let $S = \{a_1, a_2, a_3, \ldots, a_n\}$ be a nonempty, finite set of real numbers.
    (a) Prove that S is bounded.
    (b) Prove that lub S and glb S are elements of S.

27. Suppose that b is an upper bound for a set S of real numbers. Prove that if $b \in S$, then $b =$ lub S.

28. Let S be a bounded set of real numbers and suppose that lub $S =$ glb S. What can you conclude about S?

29. Suppose that S is a nonempty, bounded set of real numbers and that T is a nonempty subset of S.
    (a) Prove that T is bounded.
    (b) Prove that glb $S \leq$ glb $T \leq$ lub $T \leq$ lub S.

30. Let S and T be nonempty sets of real numbers such that $x \leq y$ for all $x \in S$ and all $y \in T$.
    (a) Prove that lub $S \leq y$ for all $y \in T$.
    (b) Prove that lub $S \leq$ glb T.

31. Let c be a positive number. Prove that the set $S = \{c, 2c, 3c, \ldots, nc, \ldots\}$ is not bounded above.

32. Prove that if a and b are any two positive numbers with $a < b$, then there exists a rational number r such that $a < r < b$. HINT: Show that there is a positive integer n such that $na > 1$ and $n(b - a) > 1$. Then show that there is a positive integer m such that $m > na$ and $m - 1 \leq na$. Let $r = m/n$.

▶ 33. Let S be the set of irrational numbers

$$S = \{\sqrt{2}, \sqrt{2\sqrt{2}}, \sqrt{2\sqrt{2\sqrt{2}}}, \ldots\}.$$

That is, $S = \{a_1, a_2, a_3, \ldots, a_n, \ldots\}$ where $a_1 = \sqrt{2}$, and for each positive integer n, $a_{n+1} = \sqrt{2a_n}$.
    (a) Calculate the numbers $a_1, a_2, a_3, \ldots, a_{10}$.
    (b) Use mathematical induction to prove that $a_n < 2$ for all n.
    (c) Is 2 the least upper bound of S?
    (d) Choose a positive number other than 2 and repeat this exercise. What can you conclude?

▶ 34. Let S be the set of irrational numbers

$$S = \{\sqrt{2}, \sqrt{2 + \sqrt{2}}, \sqrt{2 + \sqrt{2 + \sqrt{2}}}, \ldots\}$$

That is, $a_1 = \sqrt{2}$, and for each positive integer n, $a_{n+1} = \sqrt{2 + a_n}$.
    (a) Calculate the numbers $a_1, a_2, a_3, \ldots, a_{10}$.
    (b) Use mathematical induction to show that $a_n < 2$ for all n.
    (c) Is 2 the least upper bound for S?

Choose a positive number other than 2 and repeat this exercise. What can you conclude?

■ 10.2 SEQUENCES OF REAL NUMBERS

So far in our study of calculus, our attention has been fixed on functions defined on an interval or on a union of intervals. Here we study functions defined on the set of positive integers.

DEFINITION 10.2.1   SEQUENCE OF REAL NUMBERS

A real-valued function defined on the set of positive integers is called a *sequence of real numbers*.

The functions defined on the set of positive integers by setting

$$a(n) = n^2, \quad b(n) = \frac{n}{n+1}, \quad c(n) = \sqrt{\ln n}, \quad d(n) = \frac{e^n}{n}, \quad \text{for } n = 1, 2, 3, \ldots$$

are examples of sequences of real numbers.

The notions developed for functions carry over to sequences. For example, if the functions a and b are sequences, then the linear combination $\alpha a + \beta b$ (α and β are real numbers) and their product ab are also sequences:

$$(\alpha a + \beta b)(n) = \alpha a(n) + \beta b(n) \quad \text{and} \quad (ab)(n) = a(n) \cdot b(n).$$

If the sequence b does not take on the value 0, then the reciprocal $1/b$ is a sequence and so is the quotient a/b:

$$\frac{1}{b}(n) = \frac{1}{b(n)} \qquad \text{and} \qquad \frac{a}{b}(n) = \frac{a(n)}{b(n)}.$$

Let a be a sequence. The numbers $a(1)$, $a(2)$, $a(3)$, . . . are called the *terms* of a. In particular, $a(n)$ is called the *nth term of a*. In discussing sequences it is conventional to use subscript notation a_n rather than functional notation $a(n)$, $n = 1, 2, 3, . . .$, to denote the terms, and the sequence itself is often written

$$\{a_1, a_2, a_3, . . .\}$$

or, even more simply,

$$\{a_n\}.$$

For example, the sequence of reciprocals defined by setting

$$a_n = 1/n \qquad \text{for all } n$$

can be written

$$\{1, \tfrac{1}{2}, \tfrac{1}{3}, . . .\} \qquad \text{or} \qquad \{1/n\}.$$

The sequence defined by setting

$$a_n = 10^{1/n} \qquad \text{for all } n$$

can be written

$$\{10, 10^{1/2}, 10^{1/3}, . . .\} \qquad \text{or} \qquad \{10^{1/n}\}.$$

In this notation, we can write

$$\alpha\{a_n\} + \beta\{b_n\} = \{\alpha a_n + \beta b_n\}, \qquad \{a_n\}\{b_n\} = \{a_n b_n\}$$

and, provided that none of the b_n are zero,

$$\frac{1}{\{b_n\}} = \{1/b_n\} \qquad \text{and} \qquad \frac{\{a_n\}}{\{b_n\}} = \{a_n/b_n\}.$$

The following assertions will serve to illustrate the notation further.

(1) The sequence $\{1/2^n\}$ multiplied by 5 is the sequence $\{5/2^n\}$:

$$5\{1/2^n\} = \{5/2^n\}.$$

We can also write

$$5\{\tfrac{1}{2}, \tfrac{1}{4}, \tfrac{1}{8}, \tfrac{1}{16}, . . .\} = \{\tfrac{5}{2}, \tfrac{5}{4}, \tfrac{5}{8}, \tfrac{5}{16}, . . .\}.$$

(2) The sequence $\{n\}$ plus the sequence $\{1/n\}$ is the sequence $\{n + 1/n\}$:

$$\{n\} + \{1/n\} = \{n + 1/n\}.$$

In expanded form

$$\{1, 2, 3, . . .\} + \{1, \tfrac{1}{2}, \tfrac{1}{3}, . . .\} = \{2, 2\tfrac{1}{2}, 3\tfrac{1}{3}, . . .\}.$$

(3) The sequence $\{n\}$ times the sequence $\{\sqrt{n}\}$ is the sequence $\{n\sqrt{n}\}$:

$$\{n\}\{\sqrt{n}\} = \{n\sqrt{n}\} = \{n^{3/2}\};$$

the sequence $\{2^n\}$ divided by the sequence $\{n\}$ is the sequence $\{2^n/n\}$:

$$\frac{\{2^n\}}{\{n\}} = \{2^n/n\}. \quad ❑$$

The concept of boundedness for functions also carries over to sequences. In particular, a sequence a is said to be *bounded above* if $\{a_1, a_2, a_3, \ldots\}$ is bounded above; a is *bounded below* if $\{a_1, a_2, a_3, \ldots\}$ is bounded below. The sequence a is *bounded* if it is bounded above *and* below.

By the least upper bound axiom (10.1.2), we know that if the sequence a is bounded above, then $\{a_1, a_2, a_3, \ldots\}$ has a least upper bound which is called the *least upper bound of the sequence*. Similarly, if a is bounded below, then it has a *greatest lower bound*. The following assertions illustrate these ideas.

(1) The sequence $\{1/n\}$ is bounded above and below:

$$0 \leq 1/n \leq 1 \qquad \text{for } n = 1, 2, 3, \ldots .$$

The number 0 is the greatest lower bound of the set $\{1, \frac{1}{2}, \frac{1}{3}, \ldots\}$ and 1 is the least upper bound.

(2) The sequence $\{2^n\}$ is bounded below. For example, 1 is a lower bound:

$$1 \leq 2^n \qquad \text{for } n = 1, 2, 3, \ldots .$$

The number 2 is the greatest lower bound for this sequence. It is not bounded above: there is no fixed number M that satisfies

$$2^n \leq M \qquad \text{for all } n.$$

(3) The sequence $a_n = (-1)^n 2^n$ can be written

$$\{-2, 4, -8, 16, -32, 64, \ldots\}.$$

It is unbounded below and unbounded above. ❑

DEFINITION 10.2.2

A sequence $\{a_n\}$ is said to be

  (i) *increasing*   iff $a_n < a_{n+1}$ for each positive integer n,

  (ii) *nondecreasing*   iff $a_n \leq a_{n+1}$ for each positive integer n,

  (iii) *decreasing*   iff $a_n > a_{n+1}$ for each positive integer n,

  (iv) *nonincreasing*   iff $a_n \geq a_{n+1}$ for each positive integer n.

If any of these four properties holds, the sequence is said to be *monotonic*.

Remark An increasing sequence is also nondecreasing. However, a nondecreasing sequence will not, in general, be increasing. The sequence $\{1, 1, 1, \ldots\}$ is nondecreasing; it is not increasing. A corresponding relationship holds for decreasing and nonincreasing sequences. ❑

The sequences

$$\{1, \tfrac{1}{2}, \tfrac{1}{3}, \ldots\}, \qquad \{2, 4, 8, 16, \ldots\}, \qquad \{2, 2, 4, 4, 8, 8, 16, 16, \ldots\}$$

are all monotonic, but the sequences

$$\{1, \tfrac{1}{2}, 1, \tfrac{1}{3}, 1, \tfrac{1}{4}, \ldots\} \qquad \text{and} \qquad \{-2, 4, -8, 16, -32, 64, \ldots\}$$

are not monotonic.

The following examples are less trivial.

Example 1 The sequence $\{a_n\}$ defined by

$$a_n = \frac{n}{n+1}$$

is increasing. It is bounded below by $\tfrac{1}{2}$ (the greatest lower bound) and above by 1 (the least upper bound).

PROOF Since

$$\frac{a_{n+1}}{a_n} = \frac{n+1}{n+2} \cdot \frac{n+1}{n} = \frac{n^2 + 2n + 1}{n^2 + 2n} > 1$$

we have $a_n < a_{n+1}$. This confirms that the sequence is increasing. Since the sequence can be written

$$\{\tfrac{1}{2}, \tfrac{2}{3}, \tfrac{3}{4}, \tfrac{4}{5}, \tfrac{5}{6}, \ldots\}$$

it is easy to see that $\tfrac{1}{2}$ is the greatest lower bound. To show that 1 is the least upper bound, note first that

$$a_n = \frac{n}{n+1} < 1 \qquad \text{for all } n.$$

Thus, the sequence is bounded above and 1 is an upper bound. Now choose any number $K < 1$ and let p be a positive integer such that

$$\frac{1}{p+1} < 1 - K.$$

Then

$$K < 1 - \frac{1}{p+1} = \frac{p}{p+1} = a_p.$$

Thus, no number less than 1 can be an upper bound for the sequence. This means that 1 is the least upper bound. ❑

Example 2 The sequence $\{a_n\}$ defined by

$$a_n = \frac{2^n}{n!} \dagger$$

is nonincreasing. Moreover $a_n > a_{n+1}$ for $n \geq 2$.

PROOF The first two terms are equal:

$$a_1 = \frac{2^1}{1!} = 2 = \frac{2^2}{2!} = a_2.$$

---

† Recall that $n! = n(n-1)(n-2) \cdots 3 \cdot 2 \cdot 1$. See Section 1.2.

For $n \geq 2$ the sequence decreases:

$$\frac{a_{n+1}}{a_n} = \frac{2^{n+1}}{(n+1)!} \cdot \frac{n!}{2^n} = \frac{2}{n+1} < 1 \qquad \text{if } n \geq 2. \quad \square$$

Remark The sequence in Example 2 is not decreasing since $a_1 = a_2 = 2$, but for $n \geq 2$, the terms *are* decreasing. This suggests an extension of Definition 10.2.2. We will say that a sequence $\{a_n\}$ is *increasing for $n \geq k$ if $a_n < a_{n+1}$ for all $n \geq k$*. The terms *nondecreasing for $n \geq k$, decreasing for $n \geq k$,* and *nonincreasing for $n \geq k$* are defined similarly. ❏

Example 3 If $c > 1$, the sequence $\{a_n\}$ defined by

$$a_n = c^n$$

increases without bound.

PROOF Choose a number $c > 1$. Then

$$\frac{a_{n+1}}{a_n} = \frac{c^{n+1}}{c^n} = c > 1.$$

This shows that the sequence increases. To show the unboundedness, we take an arbitrary positive number M and show that there exists a positive integer k for which

$$c^k \geq M.$$

A suitable k is one such that

$$k \geq \frac{\ln M}{\ln c},$$

for then

$$k \ln c \geq \ln M, \qquad \ln c^k \geq \ln M, \qquad \text{and thus} \qquad c^k \geq M. \quad \square$$

Since sequences are defined on the set of positive integers and not on an interval, they are not directly susceptible to the methods of calculus. Fortunately, we can sometimes circumvent this difficulty by dealing initially, not with the sequence itself, but with a function of a real variable x that agrees with the given sequence for all positive integers n.

Example 4 The sequence $\{a_n\}$ defined by

$$a_n = \frac{n}{e^n}$$

is decreasing. It is bounded above by $1/e$ and below by 0.

PROOF We will work with the function

$$f(x) = \frac{x}{e^x}. \qquad \text{(See Example 3, Section 7.4.)}$$

Note that $f(1) = 1/e = a_1, f(2) = 2/e^2 = a_2, f(3) = 3/e^3 = a_3$, and so on.

Differentiating f, we get

$$f'(x) = \frac{e^x - xe^x}{e^{2x}} = \frac{1-x}{e^x}.$$

Since $f'(x) < 0$ for $x > 1$, f is decreasing on $[1, \infty)$. Thus $f(1) > f(2) > f(3) > \cdots$, that is $a_1 > a_2 > a_3 > \cdots$, and $\{a_n\}$ is decreasing.

It now follows that $a_1 > a_n$ for all $n \geq 2$, and so $a_1 = 1/e$ is an upper bound for the sequence. In fact, it is the least upper bound. Since all the terms in the sequence are positive, 0 is a lower bound.

Figure 10.2.1 illustrates the graphical relationship between $f(x) = x/e^x$ and $a_n = n/e^n$. The graph of f indicates that $a_n \to 0$ as $n \to \infty$, and so 0 is, in fact, the greatest lower bound. ❏

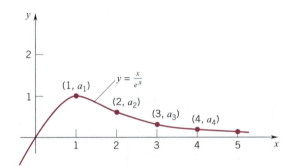

Figure 10.2.1

Example 5 The sequence $\{a_n\}$ defined by

$$a_n = n^{1/n}$$

decreases for $n \geq 3$.

PROOF We could compare a_n with a_{n+1} directly, but it is easier to consider the function

$$f(x) = x^{1/x}$$

instead. Since

$$f(x) = e^{(1/x)\ln x},$$

we have

$$f'(x) = e^{(1/x)\ln x} \frac{d}{dx}\left(\frac{1}{x}\ln x\right) = x^{1/x}\left(\frac{1 - \ln x}{x^2}\right).$$

For $x > e$, $f'(x) < 0$. This shows that f decreases on $[e, \infty)$. Since $3 > e$, f decreases on $[3, \infty)$. It follows that $\{a_n\}$ decreases for $n \geq 3$. ❏

Remark We must be careful when we examine a function of a real variable x in order to analyze the behavior of a sequence. The function $y = f(x)$ and the sequence $\{y_n\} = \{f(n)\}$ may behave differently. For example, the sequence $\{y_n\}$ defined by

$$y_n = f(n) = \frac{1}{n - 4.5}$$

is bounded (by 2 and -2) even though the function

$$f(x) = \frac{1}{x - 4.5}$$

is not bounded; the graph of f has a vertical asymptote at $x = 4.5$.   See Figure 10.2.2.   ❑

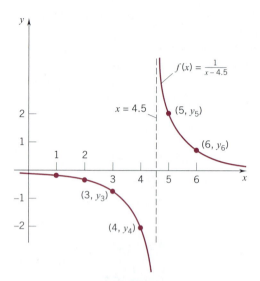

Figure 10.2.2

Clearly, the function $f(x) = \sin \pi x$ is not monotonic on $[0, \infty)$, but the sequence $\{u_n\}$ defined by $u_n = f(n) = \sin n\pi = 0$ for all n is monotonic.   Indeed, it is constant. See Figure 10.2.3.

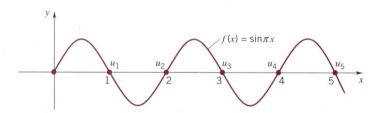

Figure 10.2.3

EXERCISES 10.2

In Exercises 1–8, the first several terms of a sequence $\{a_n\}$ are given.   Assuming that the pattern continues as indicated, find an explicit formula for a_n.

1. 2, 5, 8, 11, 14, . . .

2. 2, 0, 2, 0, 2, . . .

3. $1, -\frac{1}{3}, \frac{1}{5}, -\frac{1}{7}, \frac{1}{9}, \ldots$

4. $\frac{1}{2}, \frac{3}{4}, \frac{7}{8}, \frac{15}{16}, \frac{31}{32}, \ldots$

5. $2, \frac{5}{2}, \frac{10}{3}, \frac{17}{4}, \frac{26}{5}, \ldots$

6. $-\frac{1}{4}, \frac{2}{9}, -\frac{3}{16}, \frac{4}{25}, -\frac{5}{36}, \ldots$

7. $1, \frac{1}{2}, 3, \frac{1}{4}, 5, \frac{1}{6}, \ldots$

8. $1, 2, \frac{1}{9}, 4, \frac{1}{25}, 6, \frac{1}{49}, \ldots$

In Exercises 9–44, determine the boundedness and monotonicity of the indicated sequence.

9. $\left\{\dfrac{2}{n}\right\}$.

10. $\left\{\dfrac{(-1)^n}{n}\right\}$.

11. $\{\sqrt{n}\}$.

12. $\{(1.001)^n\}$.

13. $\left\{\dfrac{n + (-1)^n}{n}\right\}$.

14. $\left\{\dfrac{n - 1}{n}\right\}$.

15. $\{(0.9)^n\}$.

16. $\{\sqrt{n^2 + 1}\}$.

17. $\left\{\dfrac{n^2}{n+1}\right\}.$

18. $\left\{\dfrac{2^n}{4^n+1}\right\}.$

19. $\left\{\dfrac{4n}{\sqrt{4n^2+1}}\right\}.$

20. $\left\{\dfrac{n+1}{n^2}\right\}.$

21. $\left\{\dfrac{4^n}{2^n+100}\right\}.$

22. $\left\{\dfrac{n^2}{\sqrt{n^3+1}}\right\}.$

23. $\left\{\dfrac{10^{10}\sqrt{n}}{n+1}\right\}.$

24. $\left\{\dfrac{n^2+1}{3n+2}\right\}.$

25. $\left\{\ln\left(\dfrac{2n}{n+1}\right)\right\}.$

26. $\left\{\dfrac{n+2}{3^{10}\sqrt{n}}\right\}.$

27. $\left\{\dfrac{(n+1)^2}{n^2}\right\}.$

28. $\{(-1)^n\sqrt{n}\}.$

29. $\left\{\sqrt{4-\dfrac{1}{n}}\right\}.$

30. $\left\{\ln\left(\dfrac{n+1}{n}\right)\right\}.$

31. $\{(-1)^{2n+1}\sqrt{n}\}.$

32. $\left\{\dfrac{\sqrt{n+1}}{\sqrt{n}}\right\}.$

33. $\left\{\dfrac{2^n-1}{2^n}\right\}.$

34. $\left\{\dfrac{1}{2n}-\dfrac{1}{2n+3}\right\}.$

35. $\left\{\sin\dfrac{\pi}{n+1}\right\}.$

36. $\{(-\tfrac{1}{2})^n\}.$

37. $\{(1.2)^{-n}\}.$

38. $\left\{\dfrac{n+3}{\ln(n+3)}\right\}.$

39. $\left\{\dfrac{1}{n}-\dfrac{1}{n+1}\right\}.$

40. $\{\cos n\pi\}.$

41. $\left\{\dfrac{\ln(n+2)}{n+2}\right\}.$

42. $\left\{\dfrac{(-2)^n}{n^{10}}\right\}.$

43. $\left\{\dfrac{3^n}{(n+1)^2}\right\}.$

44. $\left\{\dfrac{1-(\tfrac{1}{2})^n}{(\tfrac{1}{2})^n}\right\}.$

45. Show that the sequence $\{5^n/n!\}$ decreases for $n \geq 5$. Is the sequence nonincreasing?

46. Let M be a positive integer. Show that $\{M^n/n!\}$ decreases for $n \geq M$.

47. Show that, if $0 < c < d$, then the sequence

$$a_n = (c^n + d^n)^{1/n}$$

is bounded and monotonic.

48. Show that linear combinations and products of bounded sequences are bounded.

A sequence $\{a_n\}$ is said to be defined *recursively* if, for some $k \geq 1$, the terms $a_1, a_2, \ldots, a_k$ are given and a_n is specified in terms of $a_1, a_2, \ldots, a_{n-1}$ for each $n \geq k$. The formula specifying a_n for $n \geq k$ in terms of some (or all) of its predecessors is called a *recurrence relation*. In Exercises 49–64,

write down the first six terms of the sequence and then give the general term.

49. $a_1 = 1;\quad a_{n+1} = \dfrac{1}{n+1}a_n.$

50. $a_1 = 1;\quad a_{n+1} = a_n + 3n(n+1) + 1.$

51. $a_1 = 1;\quad a_{n+1} = \tfrac{1}{2}(a_n + 1).$

52. $a_1 = 1;\quad a_{n+1} = \tfrac{1}{2}a_n + 1.$

53. $a_1 = 1;\quad a_{n+1} = a_n + 2.$

54. $a_1 = 1;\quad a_{n+1} = \dfrac{n}{n+1}a_n.$

55. $a_1 = 1;\quad a_{n+1} = 3a_n + 1.$

56. $a_1 = 1;\quad a_{n+1} = 4a_n + 3.$

57. $a_1 = 1;\quad a_{n+1} = a_n + 2n + 1.$

58. $a_1 = 1;\quad a_{n+1} = 2a_n + 1.$

59. $a_1 = 1;\quad a_{n+1} = a_n + \cdots + a_1.$

60. $a_1 = 3;\quad a_{n+1} = 4 - a_n.$

61. $a_1 = 2, a_2 = 1, a_3 = 2;\quad a_{n+1} = 6 - (a_n + a_{n-1} + a_{n-2}), n \geq 3.$

62. $a_1 = 1, a_2 = 2;\quad a_{n+1} = 2a_n - a_{n-1}, n \geq 2.$

63. $a_1 = 1, a_2 = 3;\quad a_{n+1} = 2a_n - a_{n-1}, n \geq 2.$

64. $a_1 = 1, a_2 = 3;\quad a_{n+1} = 3a_n - 2n - 1, n \geq 2.$

In Exercises 65–68, use mathematical induction to prove the following assertions for all $n \geq 1$.

65. If $a_1 = 1$ and $a_{n+1} = 2a_n + 1$, then $a_n = 2^n - 1$.

66. If $a_1 = 3$ and $a_{n+1} = a_n + 5$, then $a_n = 5n - 2$.

67. If $a_1 = 1$ and $a_{n+1} = \dfrac{n+1}{2n}a_n$, then $a_n = \dfrac{n}{2^{n-1}}$.

68. If $a_1 = 1$ and $a_{n+1} = a_n - \dfrac{1}{n(n+1)}$, then $a_n = \dfrac{1}{n}$.

69. Let r be a real number, $r \neq 0$. Define a sequence $\{S_n\}$ by

$$S_1 = 1$$
$$S_2 = 1 + r$$
$$S_3 = 1 + r + r^2$$
$$\cdot$$
$$\cdot$$
$$\cdot$$
$$S_n = 1 + r + r^2 + \cdots + r^{n-1}$$
$$\cdot$$
$$\cdot$$
$$\cdot$$

(a) Suppose $r = 1$. What is S_n for $n = 1, 2, 3, \ldots$?

(b) Suppose $r \neq 1$. Find a formula for S_n that does not involve adding up the powers of r. HINT: Calculate $S_n - rS_n$.

70. Let $a_n = \dfrac{1}{n(n+1)}$, $n = 1, 2, 3, \ldots$, and let $\{S_n\}$ be the sequence defined by

$$S_1 = a_1$$

$$S_2 = a_1 + a_2$$

$$S_3 = a_1 + a_2 + a_3$$

.
.
.

$$S_n = a_1 + a_2 + a_3 + \cdots + a_n$$

.
.
.

Find a formula for S_n, $n = 1, 2, 3, \ldots$, that does not involve adding up the terms $a_1, a_2, a_3, \ldots$. HINT: Use partial fractions to write $1/k(k+1)$ as the sum of two fractions.

71. A ball is dropped from a height of 100 feet. Each time it hits the ground it rebounds to 75% of its previous height.
(a) Let S_n be the distance that the ball travels between the nth and $(n + 1)$st bounce, $n = 1, 2, 3, \ldots$. Find a formula for S_n.

(b) Let T_n be the time that the ball is in the air between the nth and $(n + 1)$st bounce, $n = 1, 2, 3, \ldots$. Find a formula for T_n.

72. Suppose that the number of bacteria in a culture is growing exponentially (see Section 7.6) and that the number doubles every 12 hours. Find a formula for the number P_n of bacteria in the culture after n hours, given that there are 500 bacteria initially.

▷ **73.** Let $\{a_n\}$ be the sequence defined recursively by

$$a_1 = 1; \qquad a_n = 1 + \sqrt{a_{n-1}}, \quad n = 2, 3, 4, \ldots$$

Use mathematical induction to show that:
(a) $\{a_n\}$ is an increasing sequence.
(b) $\{a_n\}$ is bounded above.
(c) Calculate $a_2, a_3, a_4, \ldots, a_{15}$. Can you estimate the least upper bound for $\{a_n\}$?

▷ **74.** Let $\{a_n\}$ be the sequence defined recursively by

$$a_1 = 1; \qquad a_n = \sqrt{3a_{n-1}}, \quad n = 2, 3, 4, \ldots$$

Use mathematical induction to show that:
(a) $\{a_n\}$ is an increasing sequence.
(b) $\{a_n\}$ is bounded above.
(c) Calculate $a_1, a_2, a_3, \ldots, a_{15}$. Can you estimate the least upper bound for $\{a_n\}$?

■ 10.3 LIMIT OF A SEQUENCE

The meaning of

$$\lim_{x \to c} f(x) = L$$

is that we can make $f(x)$ as close as we wish to the number L simply by requiring that x be sufficiently close to c. The meaning of

$$\lim_{n \to \infty} a_n = L$$

(read "the limit of a_n as n tends to infinity is L") is that we can make a_n as close as we wish to the number L simply by requiring that n be sufficiently large. For another analogy, recall Section 4.7 where we considered limits of the form

$$\lim_{x \to \infty} f(x)$$

in connection with the problem of finding the horizontal asymptotes for the graph of a function f; if $f(x) \to L$ as $x \to \infty$, then the line $y = L$ is a horizontal asymptote of the graph. The statement

$$\lim_{n \to \infty} a_n = L$$

is the same as saying $a_n \to L$ as $n \to \infty$.

> **DEFINITION 10.3.1   LIMIT**
>
> $$\lim_{n \to \infty} a_n = L$$
>
> iff for each $\epsilon > 0$, there exists a positive integer K such that
>
> $$\text{if} \quad n \geq K, \quad \text{then} \quad |a_n - L| < \epsilon.$$

Example 1

$$\lim_{n \to \infty} \frac{1}{n} = 0.$$

PROOF   Let $\epsilon > 0$ and choose an integer $K > 1/\epsilon$. Then $1/K < \epsilon$. Now, for any $n \geq K$, we have

$$0 < \frac{1}{n} \leq \frac{1}{K},$$

and so

$$\left| \frac{1}{n} - 0 \right| = \left| \frac{1}{n} \right| = \frac{1}{n} \leq \frac{1}{K} < \epsilon. \quad \square$$

Example 2

$$\lim_{n \to \infty} \frac{2n - 1}{n} = 2.$$

PROOF   Let $\epsilon > 0$. We must show that there exists an integer K such that

$$\left| \frac{2n - 1}{n} - 2 \right| < \epsilon \qquad \text{for all } n \geq K.$$

Since

$$\left| \frac{2n - 1}{n} - 2 \right| = \left| \frac{2n - 1 - 2n}{n} \right| = \left| -\frac{1}{n} \right| = \frac{1}{n},$$

again we need only choose $K > 1/\epsilon$. $\square$

The next example justifies the familiar statement

$$\tfrac{1}{3} = 0.333 \ . \ . \ . \ .$$

Example 3   The decimal fractions

$$a_n = 0.\overset{n}{\overbrace{33 \ . \ . \ . \ 3}}, \qquad n = 1, 2, 3, \ . \ . \ .$$

tend to $\frac{1}{3}$ as a limit:

$$\lim_{n \to \infty} a_n = \tfrac{1}{3}.$$

PROOF   Let $\epsilon > 0$.   In the first place

$$(1) \quad \left| a_n - \frac{1}{3} \right| = \left| \overbrace{0.33 \ \ldots \ 3}^{n} - \frac{1}{3} \right| = \left| \frac{\overbrace{0.99 \cdots 9}^{n} - 1}{3} \right| = \frac{1}{3} \cdot \frac{1}{10^n} < \frac{1}{10^n}.$$

Now choose K so that $1/10^K < \epsilon$.   If $n \geq K$, then by (1)

$$\left| a_n - \frac{1}{3} \right| < \frac{1}{10^n} \leq \frac{1}{10^K} < \epsilon. \quad \square$$

Example 4   The limit

$$\lim_{n \to \infty} \frac{n}{\sqrt{n+1}}$$

does not exist.

PROOF   Note that

$$\frac{n}{\sqrt{n+1}} = \frac{1}{\dfrac{1}{n}\sqrt{n+1}} = \frac{1}{\sqrt{\dfrac{1}{n} + \dfrac{1}{n^2}}}.$$

Since $1/n \to 0$ and $1/n^2 \to 0$ as $n \to \infty$, the limit does not exist. $\square$

Remark   Following the ideas in Section 4.7, we will use the notation $a_n \to \infty$ as $n \to \infty$ to mean that corresponding to any positive number M there is a positive integer K such that $a_n > M$ for all $n \geq K$.   Thus, in Example 4, we have

$$\frac{n}{\sqrt{n+1}} \to \infty \quad \text{as } n \to \infty.$$

The notation $a_n \to -\infty$ means that for any negative number Q there is a positive integer K such that $a_n < Q$ for all $n \geq K$. $\square$

Limit Theorems

The limit process for sequences is so similar to the limit process you have already studied that you may find you can prove many of the limit theorems yourself.   In any case, try to come up with your own proofs and refer to these only if necessary.

THEOREM 10.3.2   UNIQUENESS OF LIMIT

If $\lim_{n \to \infty} a_n = L$ and $\lim_{n \to \infty} a_n = M$,   then $L = M$.

A proof, similar to the proof of Theorem 2.3.1, is given in the supplement at the end of this section.

DEFINITION 10.3.3

A sequence that has a limit is said to be *convergent*. A sequence that has no limit is said to be *divergent*.

Instead of writing

$$\lim_{n \to \infty} a_n = L,$$

we will often write

$$a_n \to L \qquad \text{(read ``}a_n \text{ converges to } L \text{'')}$$

or more fully

$$a_n \to L \qquad \text{as} \quad n \to \infty.$$

THEOREM 10.3.4

Every convergent sequence is bounded.

PROOF Assume that $a_n \to L$ and choose any positive number: 1, for instance. Using 1 as ϵ, you can see that there must exist a positive integer K such that

$$|a_n - L| < 1 \qquad \text{for all } n \geq K.$$

This means that

$$|a_n| < 1 + |L| \qquad \text{for all } n \geq K$$

and, consequently,

$$|a_n| \leq \max\{|a_1|, |a_2|, \ldots, |a_{K-1}|, 1 + |L|\} \qquad \text{for all } n.$$

This proves that $\{a_n\}$ is bounded. ❑

Since every convergent sequence is bounded, a sequence that is not bounded cannot be convergent; namely,

(10.3.5)

> every unbounded sequence is divergent.

The sequences

$$a_n = \tfrac{1}{2}n, \qquad b_n = \frac{n^2}{n+1}, \qquad c_n = n \ln n$$

are all unbounded. Each of these sequences is therefore divergent.

Boundedness does not imply convergence. As a counterexample, consider the oscillating sequence

$$\{1, 0, 1, 0, \ldots\} = \left\{ \frac{1 + (-1)^{n+1}}{2} \right\}.$$

This sequence is certainly bounded (above by 1 and below by 0), but obviously it does not converge: the limit would have to be arbitrarily close to both 0 and 1 simultaneously.

Boundedness together with monotonicity does imply convergence.

THEOREM 10.3.6

A bounded nondecreasing sequence converges to its least upper bound; a bounded nonincreasing sequence converges to its greatest lower bound.

PROOF Suppose that $\{a_n\}$ is bounded and nondecreasing. If L is the least upper bound of this sequence, then

$$a_n \leq L \qquad \text{for all } n.$$

Now let ϵ be an arbitrary positive number. By Theorem 10.1.3 there exists a_k such that

$$L - \epsilon < a_k.$$

Since the sequence is nondecreasing,

$$a_k \leq a_n \qquad \text{for all } n \geq k.$$

It follows that

$$L - \epsilon < a_n \leq L \qquad \text{for all } n \geq k.$$

This shows that

$$|a_n - L| < \epsilon \qquad \text{for all } n \geq k$$

and proves that

$$a_n \to L.$$

The nonincreasing case can be handled in a similar manner. ❑

Example 5 Take the sequence

$$\{(3^n + 4^n)^{1/n}\}.$$

Since

$$3 = (3^n)^{1/n} < (3^n + 4^n)^{1/n} < (2 \cdot 4^n)^{1/n} = 2^{1/n} \cdot 4 \leq 8,$$

the sequence is bounded. Note that

$$(3^n + 4^n)^{(n+1)/n} = (3^n + 4^n)^{1/n}(3^n + 4^n)$$
$$= (3^n + 4^n)^{1/n}3^n + (3^n + 4^n)^{1/n}4^n.$$

Since

$$(3^n + 4^n)^{1/n} > (3^n)^{1/n} = 3 \qquad \text{and} \qquad (3^n + 4^n)^{1/n} > (4^n)^{1/n} = 4,$$

it follows that

$$(3^n + 4^n)^{(n+1)/n} > 3 \cdot (3^n) + 4 \cdot (4^n) = 3^{n+1} + 4^{n+1}.$$

Taking the $(n + 1)$st root of the left and right sides of this inequality, we have

$$(3^n + 4^n)^{1/n} > (3^{n+1} + 4^{n+1})^{1/(n+1)}.$$

The sequence is decreasing. Being bounded, it must be convergent. (Later you will be asked to show that the limit is 4.) ❑

THEOREM 10.3.7

Let α be a real number. If $a_n \to L$ and $b_n \to M$, then

(i) $a_n + b_n \to L + M$,   (ii) $\alpha a_n \to \alpha L$,   (iii) $a_n b_n \to LM$.

If, in addition, $M \neq 0$ and $b_n \neq 0$ for all n, then

$$\text{(iv)} \ \frac{1}{b_n} \to \frac{1}{M} \quad \text{and} \quad \text{(v)} \ \frac{a_n}{b_n} \to \frac{L}{M}.$$

Proofs of parts (i) and (ii) are left as exercises. For proofs of parts (iii)–(v), see the supplement at the end of this section.

We are now in a position to handle any rational sequence $\{a_n\}$, where

(2) $$a_n = \frac{\alpha_k n^k + \alpha_{k-1} n^{k-1} + \cdots + \alpha_0}{\beta_j n^j + \beta_{j-1} n^{j-1} + \cdots + \beta_0}, \quad \alpha_k \neq 0, \ \beta_j \neq 0.$$

To determine the behavior of such a sequence we need only divide both numerator and denominator by the highest power of n that occurs.

Example 6

$$\frac{3n^4 - 2n^2 + 1}{n^5 - 3n^3} = \frac{3/n - 2/n^3 + 1/n^5}{1 - 3/n^2} \to \frac{0}{1} = 0. \quad ❑$$

Example 7

$$\frac{1 - 4n^7}{n^7 + 12n} = \frac{1/n^7 - 4}{1 + 12/n^6} \to \frac{-4}{1} = -4. \quad ❑$$

Example 8

$$\frac{n^4 - 3n^2 + n + 2}{n^3 + 7n} = \frac{1 - 3/n^2 + 1/n^3 + 2/n^4}{1/n + 7/n^3}.$$

Since the numerator tends to 1 and the denominator tends to 0, the sequence is unbounded. Therefore it cannot converge. ❑

In general, if a_n is given by (2), then,

$$\text{as} \quad n \to \infty, \quad a_n \to \begin{cases} 0 & \text{if } k < j \\ \alpha_k / \beta_k & \text{if } k = j \\ \pm\infty & \text{if } k > j. \end{cases}$$

(See Exercise 40.) The corresponding result for the limit of a rational function $R(x) = P(x)/Q(x)$ as $x \to \infty$ was discussed in Section 4.7.

THEOREM 10.3.8

$$a_n \to L \qquad \text{iff} \qquad a_n - L \to 0 \qquad \text{iff} \qquad |a_n - L| \to 0.$$

We leave the proof to you.

THEOREM 10.3.9 THE PINCHING THEOREM FOR SEQUENCES

Suppose that there is a positive integer K such that for all $n \geq K$

$$a_n \leq b_n \leq c_n.$$

If $a_n \to L$ and $c_n \to L$, then $b_n \to L$.

Once again the proof is left to you.

As an immediate and obvious consequence of the pinching theorem we have the following corollary.

(10.3.10)

Suppose that there is a positive integer K such that for all $n \geq K$

$$|b_n| \leq c_n.$$

If $c_n \to 0$, then $|b_n| \to 0$.

Example 9

$$\frac{\cos n}{n} \to 0 \qquad \text{since} \qquad \left| \frac{\cos n}{n} \right| \leq \frac{1}{n} \quad \text{and} \quad \frac{1}{n} \to 0. \quad \square$$

Example 10

$$\sqrt{4 + \left(\frac{1}{n}\right)^2} \to 2$$

since

$$2 \leq \sqrt{4 + \left(\frac{1}{n}\right)^2} \leq \sqrt{4 + 4\left(\frac{1}{n}\right) + \left(\frac{1}{n}\right)^2} = 2 + \frac{1}{n} \quad \text{and} \quad 2 + \frac{1}{n} \to 2. \quad \square$$

Example 11 Recall from Chapter 7 that there are several ways of defining the number e. Here we complete the argument begun in Theorem 7.5.13.

(10.3.11)
$$\lim_{n \to \infty} \left(1 + \frac{1}{n}\right)^n = e.$$

PROOF You have already seen that, for all positive integers n,

$$\left(1 + \frac{1}{n}\right)^n \le e \le \left(1 + \frac{1}{n}\right)^{n+1}. \qquad \text{(Theorem 7.5.13)}$$

Dividing the right-hand inequality by $1 + 1/n$, we have

$$\frac{e}{1 + 1/n} \le \left(1 + \frac{1}{n}\right)^n.$$

Combining this with the left-hand inequality, we can write

$$\frac{e}{1 + 1/n} \le \left(1 + \frac{1}{n}\right)^n \le e.$$

Since

$$\frac{e}{1 + 1/n} \to \frac{e}{1} = e,$$

we can conclude from the pinching theorem that

$$\left(1 + \frac{1}{n}\right)^n \to e. \quad ❑$$

The sequences

$$\left\{\cos\frac{\pi}{n}\right\}, \qquad \left\{\ln\left(\frac{n}{n+1}\right)\right\}, \qquad \{e^{1/n}\}, \qquad \left\{\tan\left(\sqrt{\frac{\pi^2 n^2 - 8}{16n^2}}\right)\right\}$$

are all of the form $\{f(c_n)\}$ with f a continuous function. Such sequences are frequently easy to deal with. The basic idea is this: When a continuous function is applied to a convergent sequence, the result is itself a convergent sequence. More precisely, we have the following theorem.

THEOREM 10.3.12

Suppose that

$$c_n \to c$$

and that, for each n, c_n is in the domain of f. If f is continuous at c, then

$$f(c_n) \to f(c).$$

PROOF   We assume that f is continuous at c and take $\epsilon > 0$. From the continuity of f at c we know that there exists $\delta > 0$ such that

$$\text{if} \quad |x - c| < \delta, \qquad \text{then} \quad |f(x) - f(c)| < \epsilon.$$

Since $c_n \to c$, we know that there exists a positive integer K such that

$$\text{if} \quad n \geq K, \qquad \text{then} \quad |c_n - c| < \delta.$$

It follows therefore that

$$\text{if} \quad n \geq K, \qquad \text{then} \quad |f(c_n) - f(c)| < \epsilon. \quad \square$$

The following examples illustrate the use of Theorem 10.3.12 in calculating limits.

Example 12   Since $\pi/n \to 0$ and the cosine function is continuous at 0,

$$\cos\left(\frac{\pi}{n}\right) \to \cos 0 = 1. \quad \square$$

Example 13   Since

$$\frac{n}{n+1} = \frac{1}{1 + 1/n} \to 1$$

and the logarithm function is continuous at 1,

$$\ln\left(\frac{n}{n+1}\right) \to \ln 1 = 0. \quad \square$$

Example 14   Since $1/n \to 0$ and the exponential function is continuous at 0,

$$e^{1/n} \to e^0 = 1. \quad \square$$

Example 15   Since

$$\frac{\pi^2 n^2 - 8}{16n^2} = \frac{\pi^2 - 8/n^2}{16} \to \frac{\pi^2}{16}$$

and the function $f(x) = \tan\sqrt{x}$ is continuous at $\pi^2/16$,

$$\tan\left(\sqrt{\frac{\pi^2 n^2 - 8}{16n^2}}\right) \to \tan\left(\sqrt{\frac{\pi^2}{16}}\right) = \tan\frac{\pi}{4} = 1. \quad \square$$

Example 16   Since

$$\frac{2n+1}{n} + \left(5 - \frac{1}{n^2}\right) \to 7$$

and the square-root function is continuous at 7,

$$\sqrt{\frac{2n+1}{n} + \left(5 - \frac{1}{n^2}\right)} \to \sqrt{7}. \quad \square$$

Example 17 Since the absolute-value function is everywhere continuous,

$$a_n \to L \quad \text{implies} \quad |a_n| \to |L|. \quad \square$$

Remark For some time now we have asked you to take on faith two fundamentals of integration: that continuous functions do have definite integrals and that these integrals can be expressed as limits of Riemann sums. We could not give you proofs of these assertions because we did not have the necessary tools. Now we do. Proofs are given in Appendix B. □

EXERCISES 10.3

In Exercises 1–36, state whether or not the sequence converges and, if it does, find the limit.

1. $\{2^n\}$.

2. $\left\{\dfrac{2}{n}\right\}$.

3. $\left\{\dfrac{(-1)^n}{n}\right\}$.

4. $\{\sqrt{n}\}$.

5. $\left\{\dfrac{n-1}{n}\right\}$.

6. $\left\{\dfrac{n+(-1)^n}{n}\right\}$.

7. $\left\{\dfrac{n+1}{n^2}\right\}$.

8. $\left\{\sin\dfrac{\pi}{2n}\right\}$.

9. $\left\{\dfrac{2^n}{4^n+1}\right\}$.

10. $\left\{\dfrac{n^2}{n+1}\right\}$.

11. $\{(-1)^n\sqrt{n}\}$.

12. $\left\{\dfrac{4n}{\sqrt{n^2+1}}\right\}$.

13. $\{(-\tfrac{1}{2})^n\}$.

14. $\left\{\dfrac{4^n}{2^n+10^6}\right\}$.

15. $\left\{\tan\dfrac{n\pi}{4n+1}\right\}$.

16. $\left\{\dfrac{10^{10}\sqrt{n}}{n+1}\right\}$.

17. $\left\{\dfrac{(2n+1)^2}{(3n-1)^2}\right\}$.

18. $\left\{\ln\left(\dfrac{2n}{n+1}\right)\right\}$.

19. $\left\{\dfrac{n^2}{\sqrt{2n^4+1}}\right\}$.

20. $\left\{\dfrac{n^4-1}{n^4+n-6}\right\}$.

21. $\{\cos n\pi\}$.

22. $\left\{\dfrac{n^5}{17n^4+12}\right\}$.

23. $\{e^{1/\sqrt{n}}\}$.

24. $\left\{\sqrt{4-\dfrac{1}{n}}\right\}$.

25. $\{(0.9)^{-n}\}$.

26. $\left\{\dfrac{2^n-1}{2^n}\right\}$.

27. $\{\ln n - \ln(n+1)\}$.

28. $\left\{\dfrac{1}{n}-\dfrac{1}{n+1}\right\}$.

29. $\left\{\dfrac{\sqrt{n+1}}{2\sqrt{n}}\right\}$.

30. $\{(0.9)^n\}$.

31. $\left\{\left(1+\dfrac{1}{n}\right)^{2n}\right\}$.

32. $\left\{\left(1+\dfrac{1}{n}\right)^{n/2}\right\}$.

33. $\left\{\dfrac{2^n}{n^2}\right\}$.

34. $\left\{\dfrac{(n+1)\cos\sqrt{n}}{n(1+\sqrt{n})}\right\}$.

35. $\left\{\dfrac{\sqrt{n}\sin(e^n\pi)}{n+1}\right\}$.

36. $\{2\ln 3n - \ln(n^2+1)\}$.

37. Prove that, if $a_n \to L$ and $b_n \to M$, then $a_n + b_n \to L + M$.

38. Let α be a real number. Prove that, if $a_n \to L$, then $\alpha a_n \to \alpha L$.

39. Prove that

$$\left(1+\dfrac{1}{n}\right)^{n+1} \to e \quad \text{given that} \quad \left(1+\dfrac{1}{n}\right)^n \to e.$$

40. Determine the convergence or divergence of a rational sequence

$$a_n = \dfrac{\alpha_k n^k + \alpha_{k-1}n^{k-1} + \cdots + \alpha_0}{\beta_j n^j + \beta_{j-1}n^{j-1} + \cdots + \beta_0}$$

with $\alpha_k \neq 0$, $\beta_j \neq 0$,

given that: (a) $k = j$; (b) $k < j$; (c) $k > j$. Justify your answers.

41. Prove that a bounded nonincreasing sequence converges to its greatest lower bound.

42. Let $\{a_n\}$ be a sequence of real numbers. Let $\{e_n\}$ be the sequence of even terms:

$$e_n = a_{2n}$$

and let $\{o_n\}$ be the sequence of odd terms:

$$o_n = a_{2n-1}.$$

Show that

$$a_n \to L \quad \text{iff} \quad e_n \to L \quad \text{and} \quad o_n \to L.$$

43. Prove the pinching theorem for sequences.

44. Let $\{a_n\}$ and $\{b_n\}$ be sequences such that $a_n \to 0$ and $\{b_n\}$ is bounded. Prove that $a_n b_n \to 0$.

45. Let $\{a_n\}$ be a convergent sequence with limit L. Prove that if $a_n \leq M$ for all n, then $L \leq M$.

46. According to Example 17, if $a_n \to L$, then $|a_n| \to |L|$. Is the converse true? That is, if $|a_n| \to |L|$, does it follow that $a_n \to L$? Prove or give a counter-example.

47. Let f be a continuous function on $(-\infty, \infty)$ and let r be a real number. Define the sequence $\{a_n\}$ as follows:

$$a_1 = r, \ a_2 = f(r), \ a_3 = f[f(r)],$$
$$a_4 = f\{f[f(r)]\}, \ \ldots \ .$$

Prove that if $a_n \to L$, then $f(L) = L$; that is, L is a fixed point of f.

48. Show that

$$\frac{2^n}{n!} \to 0.$$

HINT: First show that

$$\frac{2^n}{n!} = \frac{2}{1} \cdot \frac{2}{2} \cdot \frac{2}{3} \cdot \ldots \cdot \frac{2}{n} \leq \frac{4}{n}.$$

49. Prove that $(1/n)^{1/p} \to 0$ for all positive integers p.

50. Prove Theorem 10.3.8.

In Exercises 51–58, the sequences are defined recursively.† Determine in each case whether the sequence converges and, if so, find the limit. Start each sequence with $a_1 = 1$.

51. $a_{n+1} = \dfrac{1}{e} a_n.$

52. $a_{n+1} = 2^{n+1} a_n.$

53. $a_{n+1} = \dfrac{1}{n+1} a_n.$

54. $a_{n+1} = \dfrac{n}{n+1} a_n.$

55. $a_{n+1} = 1 - a_n.$

56. $a_{n+1} = -a_n.$

57. $a_{n+1} = \frac{1}{2} a_n + 1.$

58. $a_{n+1} = \frac{1}{3} a_n + 1.$

▶ In Exercises 59–66, evaluate numerically the limit of each sequence as $n \to \infty$. Some of these sequences converge more rapidly than others. Determine for each sequence the least value of n for which the nth term differs from the limit by less than 0.001.

59. $\left\{ \dfrac{1}{n^2} \right\}.$

60. $\left\{ \dfrac{1}{\sqrt{n}} \right\}.$

61. $\left\{ \dfrac{n}{10^n} \right\}.$

62. $\left\{ \dfrac{n^{10}}{10^n} \right\}.$

---

† The notion was introduced in Exercises 10.2.

63. $\left\{ \dfrac{1}{n!} \right\}.$

64. $\left\{ \dfrac{2^n}{n!} \right\}.$

65. $\left\{ \dfrac{\ln n}{n^2} \right\}.$

66. $\left\{ \dfrac{\ln n}{n} \right\}.$

67. (a) Find the exact value of the limit of the sequence $\{a_n\}$ given in Exercise 73, Section 10.2. HINT: Suppose $a_n \to L$, then $a_{n-1} \to L$.
(b) Find the exact value of the limit of the sequence $\{a_n\}$ given in Exercise 74, Section 10.2.

▶ **68.** Let $\{a_n\}$ be the sequence defined recursively by

$$a_1 = 1, \quad a_n = \sqrt{6 + a_{n-1}}, \quad n = 2, 3, 4, \ldots .$$

(a) Approximate a_2, a_3, a_4, a_5, a_6. Round your answers to six decimal places.
(b) Use mathematical induction to show that $a_n \leq 3$ for all n.
(c) Show that $\{a_n\}$ is an increasing sequence. HINT: $a_{n+1}^2 - a_n^2 = (3 - a_n)(2 + a_n)$.
(d) What is the limit of this sequence?

▶ **69.** Let $\{a_n\}$ be the sequence defined recursively by

$$a_1 = 1, \quad a_n = \cos a_{n-1}, \quad n = 2, 3, 4, \ldots .$$

(a) Approximate $a_2, a_3, a_4, \ldots, a_{10}$. Round your answers to six decimal places.
(b) Assuming that $a_n \to L$, approximate L to six decimal places and interpret your result geometrically. HINT: Use Exercise 47.

▶ **70.** Let $\{a_n\}$ be the sequence defined recursively by

$$a_1 = 1, \quad a_n = a_{n-1} + \cos a_{n-1}, \quad n = 2, 3, 4, \ldots .$$

(a) Approximate $a_2, a_3, a_4, \ldots, a_{10}$. Round your answers to six decimal places.
(b) Assuming that $a_n \to L$, approximate L to six decimal places and interpret your result geometrically.

▶ **71.** Let R be a positive number. Approximations to $\sqrt{R}$ are generated by the sequence defined recursively by

$$a_1 = 1, \quad a_n = \frac{1}{2}\left(a_{n-1} + \frac{R}{a_{n-1}}\right), \ n = 2, 3, 4, \ldots .$$

Let $R = 3$.

(a) Approximate $a_2, a_3, \cdots, a_8$. Round your answers to six decimal places.
(b) Assuming that $a_n \to L$, prove that $L = \sqrt{3}$.
(c) Show that the recursion relation given above is simply the sequence generated by the Newton-Raphson method applied to the function $f(x) = x^2 - R$ (see Section 3.9).

72. The Newton-Raphson method (Section 3.9) applied to a differentiable function f generates a sequence $\{x_n\}$ that, under certain conditions, converges to a zero of f. The recursion formula for the sequence is given by

$$x_{n+1} = x_n - \frac{f(x_n)}{f'(x_n)}, \quad n = 1, 2, 3, \ldots$$

Determine whether the following sequences converge, and if so, give the limit. HINT: Identify the function that generates the sequence.

(a) $x_{n+1} = x_n - \dfrac{x_n^3 - 8}{3x_n^2}$; $\quad x_1 = 1$.

(b) $x_{n+1} = x_n - \dfrac{\sin x_n - 0.5}{\cos x_n}$; $\quad x_1 = 0$.

(c) $x_{n+1} = x_n - \dfrac{\ln x_n - 1}{1/x_n} = x_n[2 - \ln x_n]$; $\quad x_1 = 1$.

---

*SUPPLEMENT TO SECTION 10.3

PROOF OF THEOREM 10.3.2

If $L \neq M$, then

$$\tfrac{1}{2}|L - M| > 0.$$

The assumption that $\lim_{n \to \infty} a_n = L$ and $\lim_{n \to \infty} a_n = M$ gives the existence of K_1 such that

$$\text{if} \quad n \geq K_1, \quad \text{then} \quad |a_n - L| < \tfrac{1}{2}|L - M|$$

and the existence of K_2 such that

$$\text{if} \quad n \geq K_2, \quad \text{then} \quad |a_n - m| < \tfrac{1}{2}|L - M|.\dagger$$

For $n \geq \max\{K_1, K_2\}$ we have

$$|a_n - L| + |a_n - M| < |L - M|.$$

By the triangle inequality we have

$$|L - M| = |(L - a_n) + (a_n - M)| \leq |L - a_n| + |a_n - M| = |a_n - L| + |a_n - M|.$$

Combining the last two statements, we have

$$|L - M| < |L - M|.$$

The hypothesis $L \neq M$ has led to an absurdity. We conclude that $L = M$. ❑

PROOF OF THEOREM 10.3.7 (iii)–(v)

To prove (iii), we set $\epsilon > 0$. For each n,

$$|a_n b_n - LM| = |(a_n b_n - a_n M) + (a_n M - LM)|$$
$$\leq |a_n||b_n - M| + |M||a_n - L|.$$

Since $\{a_n\}$ is convergent, $\{a_n\}$ is bounded; that is, there exists $Q > 0$ such that

$$|a_n| \leq Q \quad \text{for all } n.$$

Since $|M| < |M| + 1$, we have

(1) $$|a_n b_n - LM| \leq Q|b_n - M| + (|M| + 1)|a_n - L|.\dagger\dagger$$

Since $b_n \to M$, we know that there exists K_1 such that

$$\text{if} \quad n \geq K_1, \quad \text{then} \quad |b_n - M| < \frac{\epsilon}{2Q}.$$

---

† We can reach these conclusions from Definition 10.3.1 by taking $\tfrac{1}{2}|L - M|$ as ϵ.

†† Soon we will want to divide by the coefficient of $|a_n - L|$. We have replaced $|M|$ by $|M| + 1$ because $|M|$ can be zero.

Since $a_n \to L$, we know that there exists K_2 such that

$$\text{if} \quad n \geq K_2, \quad \text{then} \quad |a_n - L| < \frac{\epsilon}{2(|M| + 1)}.$$

For $n \geq \max \{K_1, K_2\}$ both conditions hold, and consequently

$$Q|b_n - M| + (|M| + 1)|a_n - L| < \frac{\epsilon}{2} + \frac{\epsilon}{2} = \epsilon.$$

In view of (1), we can conclude that

$$\text{if} \quad n \geq \max \{K_1, K_2\}, \quad \text{then} \quad |a_n b_n - LM| < \epsilon.$$

This proves that

$$a_n b_n \to LM. \quad \square$$

To prove (iv), once again we set $\epsilon > 0$. In the first place

$$\left| \frac{1}{b_n} - \frac{1}{M} \right| = \left| \frac{M - b_n}{b_n M} \right| = \frac{|b_n - M|}{|b_n||M|}.$$

Since $b_n \to M$ and $|M|/2 > 0$, there exists K_1 such that

$$\text{if} \quad n \geq K_1, \quad \text{then} \quad |b_n - M| < \frac{|M|}{2}.$$

This tells us that for $n \geq K_1$ we have

$$|b_n| > \frac{|M|}{2} \quad \text{and thus} \quad \frac{1}{|b_n|} < \frac{2}{|M|}.$$

Thus for $n \geq K_1$ we have

(2) $$\left| \frac{1}{b_n} - \frac{1}{M} \right| \leq \frac{2}{|M|^2}|b_n - M|.$$

Since $b_n \to M$ there exists K_2 such that

$$\text{if} \quad n \geq K_2, \quad \text{then} \quad |b_n - M| < \frac{\epsilon|M|^2}{2}.$$

Thus for $n \geq K_2$ we have

$$\frac{2}{|M|^2}|b_n - M| < \epsilon.$$

In view of (2), we can be sure that

$$\text{if} \quad n \geq \max\{K_1, K_2\}, \quad \text{then} \quad \left| \frac{1}{b_n} - \frac{1}{M} \right| < \epsilon.$$

This proves that

$$\frac{1}{b_n} \to \frac{1}{M}. \quad \square$$

The proof of (v) is now easy:

$$\frac{a_n}{b_n} = a_n \cdot \frac{1}{b_n} \to L \cdot \frac{1}{M} = \frac{L}{M}. \quad \square$$

■ 10.4 SOME IMPORTANT LIMITS

Our purpose here is to familiarize you with some limits that are particularly important in calculus and to give you more experience with limit arguments.

(10.4.1)

$$
\text{If } x > 0, \text{ then}
$$
$$
x^{1/n} \to 1 \qquad \text{as} \quad n \to \infty.
$$

PROOF Fix any $x > 0$. Note that

$$
\ln(x^{1/n}) = \frac{1}{n} \ln x \to 0 \qquad \text{as} \quad n \to \infty.
$$

Since the exponential function is continuous at 0, it follows from Theorem 10.3.12 that

$$
x^{1/n} = e^{(1/n)\ln x} \to e^0 = 1. \quad \square
$$

(10.4.2)

$$
\text{If } |x| < 1, \text{ then}
$$
$$
x^n \to 0 \qquad \text{as} \quad n \to \infty.
$$

PROOF The result clearly holds if $x = 0$. Fix any x with $|x| < 1$ and observe that $\{|x|^n\}$ is a decreasing sequence:

$$
|x|^{n+1} = |x|\,|x|^n < |x|^n.
$$

Now let $\epsilon > 0$. By (10.4.1)

$$
\epsilon^{1/n} \to 1 \qquad \text{as} \quad n \to \infty.
$$

Thus there exists an integer $k > 0$ such that

$$
|x| < \epsilon^{1/k}. \qquad\qquad \text{(explain)}
$$

Obviously, then, $|x|^k < \epsilon$. Since $\{|x|^n\}$ is a decreasing sequence,

$$
|x^n| = |x|^n < \epsilon \qquad \text{for all } n \geq k. \quad \square
$$

(10.4.3)

$$
\text{For each } \alpha > 0
$$
$$
\frac{1}{n^\alpha} \to 0 \qquad \text{as} \quad n \to \infty.
$$

PROOF Since $\alpha > 0$, there exists an odd positive integer p such that $1/p < \alpha$. Then

$$
0 < \frac{1}{n^\alpha} = \left(\frac{1}{n}\right)^\alpha \leq \left(\frac{1}{n}\right)^{1/p}.
$$

Since $1/n \to 0$ and $f(x) = x^{1/p}$ is continuous at 0, we have

$$\left(\frac{1}{n}\right)^{1/p} \to 0 \quad \text{and thus by the pinching theorem} \quad \frac{1}{n^\alpha} \to 0. \quad \square$$

(10.4.4)

> For each real x
>
> $$\frac{x^n}{n!} \to 0 \qquad \text{as} \quad n \to \infty.$$

PROOF Fix any real number x and choose an integer k such that $k > |x|$. For $n > k + 1$,

$$\frac{k^n}{n!} = \left(\frac{k^k}{k!}\right) \underbrace{\left[\frac{k}{k+1} \frac{k}{k+2} \cdots \frac{k}{n-1}\right]}_{\text{the middle term is less than 1}} \left(\frac{k}{n}\right) < \left(\frac{k^{k+1}}{k!}\right)\left(\frac{1}{n}\right).$$

Since $k > |x|$, we have

$$0 < \frac{|x|^n}{n!} < \frac{k^n}{n!} < \left(\frac{k^{k+1}}{k!}\right)\left(\frac{1}{n}\right).$$

Since k is fixed and $1/n \to 0$, it follows from the pinching theorem that

$$\frac{|x|^n}{n!} \to 0 \qquad \text{and thus} \qquad \frac{x^n}{n!} \to 0. \quad \square$$

(10.4.5)

> $$\frac{\ln n}{n} \to 0 \qquad \text{as} \quad n \to \infty.$$

PROOF A routine proof can be based on L'Hospital's rule (10.6.1), but that is not available to us yet. We will appeal to the pinching theorem and base our argument on the integral representation of the logarithm:

$$0 \le \frac{\ln n}{n} = \frac{1}{n} \int_1^n \frac{dt}{t} \le \frac{1}{n} \int_1^n \frac{dt}{\sqrt{t}} = \frac{2}{n}(\sqrt{n} - 1)$$

$$= 2\left(\frac{1}{\sqrt{n}} - \frac{1}{n}\right) \to 0. \quad \square$$

(10.4.6)

> $$n^{1/n} \to 1 \qquad \text{as} \quad n \to \infty.$$

PROOF We know that

$$n^{1/n} = e^{(1/n)\ln n}.$$

Since

(11.1.1) $(1/n)\ln n \to 0$ (10.4.5)

and the exponential function is continuous at 0, it follows from Theorem 10.3.12 that

$$n^{1/n} \to e^0 = 1. \quad \square$$

(10.4.7)

> For each real x
>
> $$\left(1 + \frac{x}{n}\right)^n \to e^x \quad \text{as} \quad n \to \infty.$$

PROOF For $x = 0$, the result is obvious. For $x \neq 0$,

$$\ln\left(1 + \frac{x}{n}\right)^n = n \ln\left(1 + \frac{x}{n}\right) = x\left[\frac{\ln(1 + x/n) - \ln 1}{x/n}\right].$$

The crux here is to recognize that the bracketed expression is a difference quotient for the logarithm function. Once we see this, we let $h = x/n$ and write

$$\lim_{n \to \infty}\left[\frac{\ln(1 + x/n) - \ln 1}{x/n}\right] = \lim_{h \to 0}\left[\frac{\ln(1 + h) - \ln 1}{h}\right] = 1.\dagger$$

It follows that

$$\ln\left(1 + \frac{x}{n}\right)^n \to x \quad \text{and therefore} \quad \left(1 + \frac{x}{n}\right)^n = e^{\ln(1 + x/n)^n} \to e^x. \quad \square$$

---

† For each $t > 0$

$$\lim_{h \to 0}\frac{\ln(t + h) - \ln t}{h} = \frac{d}{dt}(\ln t) = \frac{1}{t}.$$

EXERCISES 10.4

In Exercises 1–36, state whether or not the sequence converges as $n \to \infty$; if it does, find the limit.

1. $\{2^{2/n}\}$.

2. $\{e^{-\alpha/n}\}$.

3. $\left\{\left(\dfrac{2}{n}\right)^n\right\}$.

4. $\left\{\dfrac{\log_{10} n}{n}\right\}$.

5. $\left\{\dfrac{\ln(n + 1)}{n}\right\}$.

6. $\left\{\dfrac{3^n}{4^n}\right\}$.

7. $\left\{\dfrac{x^{100n}}{n!}\right\}$.

8. $\{n^{1/(n+2)}\}$.

9. $\{n^{\alpha/n}\}, \quad \alpha > 0$.

10. $\left\{\ln\left(\dfrac{n + 1}{n}\right)\right\}$.

11. $\left\{\dfrac{3^{n+1}}{4^{n-1}}\right\}$.

12. $\left\{\displaystyle\int_{-n}^{0} e^{2x}\, dx\right\}$.

13. $\{(n + 2)^{1/n}\}$.

14. $\left\{\left(1 - \dfrac{1}{n}\right)^n\right\}$.

15. $\left\{\displaystyle\int_{0}^{n} e^{-x}\, dx\right\}$.

16. $\left\{\dfrac{2^{3n-1}}{7^{n+2}}\right\}$.

17. $\left\{\displaystyle\int_{-n}^{n} \dfrac{dx}{1 + x^2}\right\}$.

18. $\left\{\displaystyle\int_{0}^{n} e^{-nx}\, dx\right\}$.

19. $\{(n + 2)^{1/(n + 2)}\}$.

20. $\{n^2 \sin n\pi\}$.

21. $\left\{\dfrac{\ln n^2}{n}\right\}$.

22. $\left\{\displaystyle\int_{-1+1/n}^{1-1/n} \dfrac{dx}{\sqrt{1 - x^2}}\right\}$.

23. $\left\{ n^2 \sin \dfrac{\pi}{n} \right\}.$

24. $\left\{ \dfrac{n!}{2n} \right\}.$

25. $\left\{ \dfrac{5^{n+1}}{4^{2n-1}} \right\}.$

26. $\left\{ \left(1 + \dfrac{x}{n} \right)^{3n} \right\}.$

27. $\left\{ \left(\dfrac{n+1}{n+2} \right)^n \right\}.$

28. $\left\{ \displaystyle\int_{1/n}^{1} \dfrac{dx}{\sqrt{x}} \right\}.$

29. $\left\{ \displaystyle\int_{n}^{n+1} e^{-x^2}\, dx \right\}.$

30. $\left\{ \left(1 + \dfrac{1}{n^2} \right)^n \right\}.$

31. $\left\{ \dfrac{n^n}{2^{n^2}} \right\}.$

32. $\left\{ \displaystyle\int_{0}^{1/n} \cos e^x\, dx \right\}.$

33. $\left\{ \left(1 + \dfrac{x}{2n} \right)^{2n} \right\}.$

34. $\left\{ \left(1 + \dfrac{1}{n} \right)^{n^2} \right\}.$

35. $\left\{ \displaystyle\int_{-1/n}^{1/n} \sin x^2\, dx \right\}.$

36. $\left\{ \left(t + \dfrac{x}{n} \right)^n \right\}, \quad t > 0, \quad x > 0.$

37. Show that $\lim\limits_{n \to \infty} (\sqrt{n+1} - \sqrt{n}) = 0.$

38. Show that $\lim\limits_{n \to \infty} (\sqrt{n^2 + n} - n) = \frac{1}{2}.$

39. (a) Show that a regular polygon of n sides inscribed in a circle of radius r has perimeter $p_n = 2rn \sin(\pi/n).$
(b) Find

$$\lim_{n \to \infty} p_n$$

and give a geometric interpretation of your result.

40. Show that

$$\text{if} \quad 0 < c < d, \quad \text{then} \quad (c^n + d^n)^{1/n} \to d.$$

In Exercises 41–43, find the indicated limit.

41. $\lim\limits_{n \to \infty} \dfrac{1 + 2 + \cdots + n}{n^2}.$

HINT: $1 + 2 + \cdots + n = \dfrac{n(n+1)}{2}.$

42. $\lim\limits_{n \to \infty} \dfrac{1^2 + 2^2 + \cdots + n^2}{(1+n)(2+n)}.$

HINT: $1^2 + 2^2 + \cdots + n^2 = \dfrac{n(n+1)(2n+1)}{6}.$

43. $\lim\limits_{n \to \infty} \dfrac{1^3 + 2^3 + \cdots + n^3}{2n^4 + n - 1}.$

HINT: $1^3 + 2^3 + \cdots + n^3 = \dfrac{n^2(n+1)^2}{4}.$

44. A sequence $\{a_n\}$ is said to be a *Cauchy sequence†* iff

(10.4.8)

> for each $\epsilon > 0$ there exists a positive integer K such that
>
> $|a_n - a_m| < \epsilon$ for all $m, n \geq K.$

Show that

(10.4.9)

> every convergent sequence is a Cauchy sequence.

It is also true that every Cauchy sequence is convergent, but this is more difficult to prove.

45. (*Arithmetic means*) For a given sequence $\{a_n\}$, let

$$m_n = \frac{1}{n}(a_1 + a_2 + \cdots + a_n).$$

(a) Prove that if $\{a_n\}$ is increasing, then $\{m_n\}$ is increasing.
(b) Prove that if $a_n \to 0$, then $m_n \to 0.$
HINT: Choose an integer $j > 0$ such that, if $n \geq j$, then $a_n < \epsilon/2$. Then for $n \geq j$,

$$|m_n| < \frac{|a_1 + a_2 + \cdots + a_j|}{n} + \frac{\epsilon}{2}\left(\frac{n-j}{n} \right).$$

46. (a) Let $\{a_n\}$ be a convergent sequence. Prove that

$$\lim_{n \to \infty} (a_n - a_{n-1}) = 0.$$

(b) What can you say about the converse? That is, suppose that $\{a_n\}$ is a sequence such that

$$\lim_{n \to \infty} (a_n - a_{n-1}) = 0.$$

Does $\{a_n\}$ necessarily converge? Prove or give a counter-example.

47. Let a and b be positive numbers with $b > a$. Define two sequences $\{a_n\}$ and $\{b_n\}$ as follows:

$$a_1 = \frac{a+b}{2} \text{ (the arithmetic mean of } a \text{ and } b),$$

$$b_1 = \sqrt{ab} \text{ (the geometric mean of } a \text{ and } b),$$

---

† After the French baron Augustin-Louis Cauchy (1789–1857), one of the most prolific mathematicians of all time.

and

$$a_n = \frac{a_{n-1} + b_{n-1}}{2},$$

$$b_n = \sqrt{a_{n-1}b_{n-1}}, \quad n = 2, 3, 4, \ldots$$

(a) Use mathematical induction to show that

$$a_{n-1} > a_n > b_n > b_{n-1} \quad \text{for } n = 2, 3, 4, \ldots$$

(b) Prove that $\{a_n\}$ and $\{b_n\}$ are convergent sequences and that $\lim_{n\to\infty} a_n = \lim_{n\to\infty} b_n$. The value of these limits is known as the *arithmetic-geometric mean* of a and b.

48. You have seen that for all real x

$$\lim_{n\to\infty} \left(1 + \frac{x}{n}\right)^n = e^x.$$

However, the rate of convergence is different for different x. Verify that at $n = 100$, $(1 + 1/n)^n$ is within 1% of its limit, while $(1 + 5/n)^n$ is still about 12% from its limit. Give comparable accuracy estimates for these two sequences at $n = 1000$.

49. Evaluate

$$\lim_{n\to\infty} \left(\sin \frac{1}{n}\right)^{1/n}$$

numerically and justify your answer by other means.

50. We have stated that

$$\lim_{n\to\infty} (\sqrt{n^2 + n} - n) = \tfrac{1}{2}. \quad \text{(Exercise 38)}$$

Evaluate numerically

$$\lim_{n\to\infty} [(n^3 + n^2)^{1/3} - n].$$

Formulate a conjecture about

$$\lim_{n\to\infty} [(n^k + n^{k-1})^{1/k} - n], \quad k = 1, 2, 3, \ldots$$

and prove that your conjecture is valid.

51. The *Fibonacci sequence* is defined recursively by

$$a_{n+2} = a_{n+1} + a_n, \quad \text{with} \quad a_1 = a_2 = 1.$$

(a) Calculate $a_3, a_4, a_5, \ldots, a_{10}$.
(b) Now define the sequence $\{r_n\}$ by

$$r_n = \frac{a_{n+1}}{a_n}$$

Calculate $r_1, r_2, \ldots, r_6$.
(c) Assuming that $r_n \to L$, find L. HINT: Show that

$$r_n = 1 + \frac{1}{r_{n-1}}.$$

52. Let $\{a_n\}$ be the sequence defined by

$$a_n = \frac{1}{n^2} + \frac{2}{n^2} + \frac{3}{n^2} + \cdots + \frac{n}{n^2}.$$

Show that a_n is a Riemann sum for $\int_0^1 x \, dx$ for each $n \geq 1$. Does the sequence $\{a_n\}$ converge? If so, to what?

■ 10.5 THE INDETERMINATE FORM (0/0)

Recall that the quotient rule for evaluating the limit of a quotient $f(x)/g(x)$ fails when $f(x)$ and $g(x)$ both tend to zero (see Section 2.3). Some important examples of this behavior were

$$\lim_{x\to 0} \frac{\sin x}{x} \quad \text{and} \quad \lim_{x\to 0} \frac{1 - \cos x}{x}.$$

Also, the definition of the derivative:

$$\lim_{h\to 0} \frac{f(x + h) - f(x)}{h}$$

typically leads to a limit of this type. We called such limits *indeterminates of the form* 0/0. In this section we present a method for handling this kind of limit when elementary methods fail or are difficult to apply.

THEOREM 10.5.1   L'HOSPITAL'S RULE (0/0)†

Suppose that f and g are differentiable functions and that

$$f(x) \to 0 \qquad \text{and} \qquad g(x) \to 0$$

as $x \to c^+, x \to c^-, x \to c, x \to \infty,$ or $x \to -\infty.$

$$\text{If} \quad \frac{f'(x)}{g'(x)} \to L, \qquad \text{then} \quad \frac{f(x)}{g(x)} \to L.$$

If $\dfrac{f'(x)}{g'(x)} \to \infty$ or $-\infty,$ then $\dfrac{f(x)}{g(x)} \to \infty$ or $-\infty,$ respectively.

We will prove the validity of L'Hospital's rule later in the section. First we demonstrate its usefulness.

Example 1 Find

$$\lim_{x \to \pi/2} \frac{\cos x}{\pi - 2x}.$$

SOLUTION As $x \to \pi/2$, both the numerator $f(x) = \cos x$ and the denominator $g(x) = \pi - 2x$ tend to zero, but it is not at all obvious what happens to the quotient

$$\frac{f(x)}{g(x)} = \frac{\cos x}{\pi - 2x}.$$

Therefore we test the quotient of derivatives:

$$\frac{f'(x)}{g'(x)} = \frac{-\sin x}{-2} = \frac{\sin x}{2} \to \frac{1}{2} \qquad \text{as} \quad x \to \frac{\pi}{2}.$$

It follows from L'Hospital's rule that

$$\frac{\cos x}{\pi - 2x} \to \frac{1}{2} \qquad \text{as} \quad x \to \frac{\pi}{2}.$$

We can express all this on just one line using $*$ to indicate the differentiation of numerator and denominator:

$$\lim_{x \to \pi/2} \frac{\cos x}{\pi - 2x} \overset{*}{=} \lim_{x \to \pi/2} \frac{-\sin x}{-2} = \lim_{x \to \pi/2} \frac{\sin x}{2} = \frac{1}{2}. \qquad \square$$

† Named after a Frenchman G. F. A. L'Hospital (1661–1704). The result was actually discovered by his teacher, Jakob Bernoulli (1654–1705).

Example 2 Find

$$\lim_{x \to 0^+} \frac{x}{\sin\sqrt{x}}.$$

SOLUTION As $x \to 0^+$, both numerator and denominator tend to 0. Since

$$\frac{f'(x)}{g'(x)} = \frac{1}{(\cos\sqrt{x})(1/2\sqrt{x})} = \frac{2\sqrt{x}}{\cos\sqrt{x}} \to 0 \qquad \text{as} \quad x \to 0^+,$$

it follows from L'Hospital's rule that

$$\frac{x}{\sin\sqrt{x}} \to 0 \qquad \text{as} \quad x \to 0^+.$$

For short we can write

$$\lim_{x \to 0^+} \frac{x}{\sin\sqrt{x}} \overset{*}{=} \lim_{x \to 0^+} \frac{2\sqrt{x}}{\cos\sqrt{x}} = 0. \quad ❏$$

Remark It is important to understand that Theorem 10.5.1 does not apply to quotients in general; you should verify first that the numerator and denominator both tend to zero. For example,

$$\lim_{x \to 0} \frac{x}{x + \cos x} = \frac{0}{1} = 0,$$

but a blind application of L'Hospital's rule would lead to

$$\lim_{x \to 0} \frac{x}{x + \cos x} \overset{*}{=} \lim_{x \to 0} \frac{1}{1 - \sin x} = 1.$$

This is, of course, incorrect. ❏

Sometimes it is necessary to differentiate numerator and denominator more than once. The next problem gives such an instance.

Example 3 Find

$$\lim_{x \to 0} \frac{e^x - x - 1}{x^2}.$$

SOLUTION As $x \to 0$, both numerator and denominator tend to 0. Here

$$\frac{f'(x)}{g'(x)} = \frac{e^x - 1}{2x}.$$

Since both numerator and denominator still tend to 0 as $x \to 0$, we differentiate again:

$$\frac{f''(x)}{g''(x)} = \frac{e^x}{2}.$$

Since this last quotient tends to $\frac{1}{2}$, we can conclude that

$$\frac{e^x - 1}{2x} \to \frac{1}{2} \qquad \text{and therefore} \qquad \frac{e^x - x - 1}{x^2} \to \frac{1}{2} \qquad \text{as} \quad x \to 0.$$

For short we can write

$$\lim_{x \to 0} \frac{e^x - x - 1}{x^2} \overset{*}{=} \lim_{x \to 0} \frac{e^x - 1}{2x} \overset{*}{=} \lim_{x \to 0} \frac{e^x}{2} = \frac{1}{2}. \quad \square$$

In the next example we use L'Hospital's rule to find the limit of a sequence.

Example 4   Given the sequence

$$\left\{ \frac{e^{2/n} - 1}{1/n} \right\}.$$

Find

$$\lim_{n \to \infty} \frac{e^{2/n} - 1}{1/n}.$$

SOLUTION   This is an indeterminate of the form $0/0$. To apply the methods of this section, we replace the integer variable n by the real variable x and examine the behavior of

$$\frac{e^{2/x} - 1}{1/x} \quad \text{as} \quad x \to \infty.$$

For an analogous example, see Example 4, Section 10.2. Applying L'Hospital's rule, we have

$$\lim_{x \to \infty} \frac{e^{2/x} - 1}{1/x} \overset{*}{=} \lim_{x \to \infty} \frac{e^{2/x}(-2/x^2)}{(-1/x^2)} = 2 \lim_{x \to \infty} e^{2/x} = 2.$$

This shows that

$$\frac{e^{2/x} - 1}{1/x} \to 2 \quad \text{and so} \quad \frac{e^{2/n} - 1}{1/n} \to 2 \quad \text{as} \quad n \to \infty. \quad \square$$

To derive L'Hospital's rule, we need a generalization of the mean-value theorem.

THEOREM 10.5.2   THE CAUCHY MEAN-VALUE THEOREM†

Suppose that f and g are differentiable on (a, b) and continuous on $[a, b]$. If g' is never 0 in (a, b), then there is a number c in (a, b) for which

$$\frac{f'(c)}{g'(c)} = \frac{f(b) - f(a)}{g(b) - g(a)}.$$

PROOF   We can prove this by applying Rolle's theorem (4.1.3) to the function

$$G(x) = [g(b) - g(a)][f(x) - f(a)] - [g(x) - g(a)][f(b) - f(a)].$$

---

† Another contribution of A. L. Cauchy, after whom Cauchy sequences were named.

Since

$$G(a) = 0 \quad \text{and} \quad G(b) = 0,$$

there exists (by Rolle's theorem) a number c in (a, b) for which

$$G'(c) = 0.$$

Since, in general,

$$G'(x) = [g(b) - g(a)]f'(x) - g'(x)[f(b) - f(a)],$$

we must have

$$[g(b) - g(a)]f'(c) - g'(c)[f(b) - f(a)] = 0,$$

and thus

$$[g(b) - g(a)]f'(c) = g'(c)[f(b) - f(a)].$$

Since g' is never 0 in (a, b),

$$g'(c) \neq 0 \quad \text{and} \quad g(b) - g(a) \neq 0.$$
$$\underset{\qquad\qquad\text{explain}}{\uparrow}$$

We can therefore divide by these numbers and obtain

$$\frac{f'(c)}{g'(c)} = \frac{f(b) - f(a)}{g(b) - g(a)}. \quad ❑$$

Now we prove L'Hospital's rule for the case $x \to c^+$. The case $x \to c^-$ is handled in the same way, and the two cases together can be used to prove the rule for the case $x \to c$.

Let f and g be defined on an interval (c, b) for some $b > c$. We assume that, as $x \to c^+$,

$$f(x) \to 0, \quad g(x) \to 0, \quad \text{and} \quad \frac{f'(x)}{g'(x)} \to L$$

and show that

$$\frac{f(x)}{g(x)} \to L.$$

PROOF   The fact that

$$\frac{f'(x)}{g'(x)} \to L \quad \text{as} \quad x \to c^+$$

assures us that both f' and g' exist on a set of the form $(c, c + h]$ and that g' is not zero there. By setting $f(c) = 0$ and $g(c) = 0$, we ensure that f and g are both continuous on $[c, c + h]$. We can now apply the Cauchy mean-value theorem and conclude that there exists a number c_h between c and $c + h$ such that

$$\frac{f'(c_h)}{g'(c_h)} = \frac{f(c + h) - f(c)}{g(c + h) - g(c)} = \frac{f(c + h)}{g(c + h)}.$$

The result is now obtained by letting $h \to 0^+$. Since the left side tends to L, the right side tends to L.   ❑

Here is a proof of L'Hospital's rule for the case $x \to \infty$.

PROOF The key here is to set $x = 1/t$:

$$\lim_{x \to \infty} \frac{f'(x)}{g'(x)} = \lim_{t \to 0^+} \frac{f'(1/t)}{g'(1/t)} = \lim_{t \to 0^+} \frac{-t^{-2} f'(1/t)}{-t^{-2} g'(1/t)} = \lim_{t \to 0^+} \frac{f(1/t)}{g(1/t)} = \lim_{x \to \infty} \frac{f(x)}{g(x)}. \quad ❏$$

↑ by L'Hospital's rule for the case $t \to 0^+$

EXERCISES 10.5

In Exercises 1–32, find the indicated limit.

1. $\displaystyle\lim_{x \to 0^+} \frac{\sin x}{\sqrt{x}}$.

2. $\displaystyle\lim_{x \to 1} \frac{\ln x}{1 - x}$.

3. $\displaystyle\lim_{x \to 0} \frac{e^x - 1}{\ln(1 + x)}$.

4. $\displaystyle\lim_{x \to 4} \frac{\sqrt{x} - 2}{x - 4}$.

5. $\displaystyle\lim_{x \to \pi/2} \frac{\cos x}{\sin 2x}$.

6. $\displaystyle\lim_{x \to a} \frac{x - a}{x^n - a^n}$.

7. $\displaystyle\lim_{x \to 0} \frac{2^x - 1}{x}$.

8. $\displaystyle\lim_{x \to 0} \frac{\tan^{-1} x}{x}$.

9. $\displaystyle\lim_{x \to 1} \frac{x^{1/2} - x^{1/4}}{x - 1}$.

10. $\displaystyle\lim_{x \to 0} \frac{e^x - 1}{x(1 + x)}$.

11. $\displaystyle\lim_{x \to 0} \frac{e^x - e^{-x}}{\sin x}$.

12. $\displaystyle\lim_{x \to 0} \frac{1 - \cos x}{3x}$.

13. $\displaystyle\lim_{x \to 0} \frac{x + \sin \pi x}{x - \sin \pi x}$.

14. $\displaystyle\lim_{x \to 0} \frac{a^x - (a + 1)^x}{x}$.

15. $\displaystyle\lim_{x \to 0} \frac{e^x + e^{-x} - 2}{1 - \cos 2x}$.

16. $\displaystyle\lim_{x \to 0} \frac{x - \ln(x + 1)}{1 - \cos 2x}$.

17. $\displaystyle\lim_{x \to 0} \frac{\tan \pi x}{e^x - 1}$.

18. $\displaystyle\lim_{x \to 0} \frac{\cos x - 1 + x^2/2}{x^4}$.

19. $\displaystyle\lim_{x \to 0} \frac{1 + x - e^x}{x(e^x - 1)}$.

20. $\displaystyle\lim_{x \to 0} \frac{\ln(\sec x)}{x^2}$.

21. $\displaystyle\lim_{x \to 0} \frac{x - \tan x}{x - \sin x}$.

22. $\displaystyle\lim_{x \to 0} \frac{x e^{nx} - x}{1 - \cos nx}$.

23. $\displaystyle\lim_{x \to 1^-} \frac{\sqrt{1 - x^2}}{\sqrt{1 - x^3}}$.

24. $\displaystyle\lim_{x \to 0} \frac{2x - \sin \pi x}{4x^2 - 1}$.

25. $\displaystyle\lim_{x \to \pi/2} \frac{\ln(\sin x)}{(\pi - 2x)^2}$.

26. $\displaystyle\lim_{x \to 0^+} \frac{\sqrt{x}}{\sqrt{x} + \sin \sqrt{x}}$.

27. $\displaystyle\lim_{x \to 0} \frac{\cos x - \cos 3x}{\sin(x^2)}$.

28. $\displaystyle\lim_{x \to 0} \frac{\sqrt{a + x} - \sqrt{a - x}}{x}$.

29. $\displaystyle\lim_{x \to \pi/4} \frac{\sec^2 x - 2 \tan x}{1 + \cos 4x}$.

30. $\displaystyle\lim_{x \to 0} \frac{x - \sin^{-1} x}{\sin^3 x}$.

31. $\displaystyle\lim_{x \to 0} \frac{\tan^{-1} x}{\tan^{-1} 2x}$.

32. $\displaystyle\lim_{x \to 0} \frac{\sin^{-1} x}{x}$.

In Exercises 33–36, find the limit of the sequence.

33. $\displaystyle\lim_{n \to \infty} \frac{(\pi/2 - \tan^{-1} n)}{1/n}$.

34. $\displaystyle\lim_{n \to \infty} \frac{\ln(1 - 1/n)}{\sin(1/n)}$.

35. $\displaystyle\lim_{n \to \infty} \frac{1}{n[\ln(n + 1) - \ln n]}$.

36. $\displaystyle\lim_{n \to \infty} \frac{\sinh \pi/n - \sin \pi/n}{\sin^3 \pi/n}$.

37. Find the fallacy:

$$\lim_{x \to 0} \frac{2 + x + \sin x}{x^3 + x - \cos x} \overset{\text{\tiny ?}}{=} \lim_{x \to 0} \frac{1 + \cos x}{3x^2 + 1 + \sin x}$$

$$\overset{\text{\tiny ?}}{=} \lim_{x \to 0} \frac{-\sin x}{6x + \cos x} = \frac{0}{1} = 0.$$

38. Show that, if $a > 0$, then

$$\lim_{n \to \infty} n(a^{1/n} - 1) = \ln a.$$

39. Find values for a and b such that

$$\lim_{x \to 0} \frac{\cos ax - b}{2x^2} = -4.$$

40. Find values for a and b so that

$$\lim_{x \to 0} \frac{\sin 2x + ax + bx^3}{x^3} = 0$$

41. Given that f is continuous, use L'Hospital's rule to determine

$$\lim_{x \to 0} \left(\frac{1}{x} \int_0^x f(t) \, dt \right).$$

42. Let f be a twice differentiable function and fix a value of x.

(a) Prove that

$$\lim_{h \to 0} \frac{f(x + h) - f(x - h)}{2h} = f'(x).$$

(b) Prove that

$$\lim_{h \to 0} \frac{f(x + h) - 2f(x) + f(x - h)}{h^2} = f''(x).$$

43. Let $A(b)$ be the area of the region bounded by the parabola $y = x^2$ and the horizontal line $y = b$ ($b > 0$), and let $T(b)$ be the area of the triangle AOB (see the figure). Find $\lim\limits_{b \to 0^+} T(b)/A(b)$.

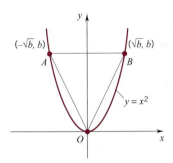

44. Choose an angle θ, $0 < \theta < \pi/2$, in standard position as shown in the figure. Let $T(\theta)$ be the area of the triangle ABC, and let $S(\theta)$ be the area of the segment of the circle formed by the chord AB. Find $\lim\limits_{\theta \to 0^+} T(\theta)/S(\theta)$.

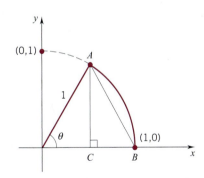

45. Let

$$f(x) = \frac{x^2 - 16}{\sqrt{x^2 + 9} - 5}.$$

(a) Use a graphing utility to graph f. What is the behavior of the graph as $x \to \infty$ and as $x \to -\infty$?

(b) What is the behavior of f as $x \to 4$? Confirm your answer using L'Hospital's rule.

46. Let

$$f(x) = \frac{x - \sin x}{x^3}.$$

(a) Use a graphing utility to graph f. What is the behavior of the graph as $x \to \infty$ and as $x \to -\infty$?

(b) What is the behavior of f as $x \to 0$? Confirm your answer using L'Hospital's rule.

47. Let $f(x) = \dfrac{2^{\sin x} - 1}{x}$.

(a) Use a graphing utility to graph f. Estimate

$$\lim_{x \to 0} f(x).$$

(b) Use L'Hospital's rule to confirm your estimate in part (a).

48. Let $g(x) = \dfrac{3^{\cos x} - 3}{x^2}$

(a) Use a graphing utility to graph g. Estimate

$$\lim_{x \to 0} g(x).$$

(b) Use L'Hospital's rule to confirm your estimate in part (a).

■ 10.6 THE INDETERMINATE FORM (∞/∞); OTHER INDETERMINATE FORMS

We come now to limits of quotients $f(x)/g(x)$ where numerator and denominator both tend to ∞. Such limits are called *indeterminates of the form ∞/∞.*

THEOREM 10.6.1   L'HOSPITAL'S RULE (∞/∞)

Suppose that f and g are differentiable, and that

$$f(x) \to \pm\infty \qquad \text{and} \qquad g(x) \to \pm\infty$$

as $x \to c^+$, $x \to c^-$, $x \to c$, $x \to \infty$, or $x \to -\infty$.

$$\text{If } \frac{f'(x)}{g'(x)} \to L, \qquad \text{then} \qquad \frac{f(x)}{g(x)} \to L.$$

$$\text{If } \frac{f'(x)}{g'(x)} \to \infty \text{ or } -\infty, \qquad \text{then} \qquad \frac{f(x)}{g(x)} \to \infty \text{ or } -\infty, \text{ respectively}$$

While the proof of L'Hospital's rule in this setting is a little more complicated than it was in the (0/0) case,† the application of the rule is much the same.

Example 1   Let α be any positive number.   Show that

(10.6.2)
$$\lim_{x \to \infty} \frac{\ln x}{x^\alpha} = 0.$$

SOLUTION   Both numerator and denominator tend to ∞ as $x \to \infty$.   L'Hospital's rule gives

$$\lim_{x \to \infty} \frac{\ln x}{x^\alpha} \overset{*}{=} \lim_{x \to \infty} \frac{1/x}{\alpha x^{\alpha-1}} = \lim_{x \to \infty} \frac{1}{\alpha x^\alpha} = 0. \quad ❑$$

For example,

$$\frac{\ln x}{x^{0.01}} \to 0 \qquad \text{and} \qquad \frac{\ln x}{x^{0.001}} \to 0$$

as $x \to \infty$.

Example 2   Let k be any positive integer.   Show that

(10.6.3)
$$\lim_{x \to \infty} \frac{x^k}{e^x} = 0.$$

SOLUTION   Here we differentiate numerator and denominator k times:

$$\lim_{x \to \infty} \frac{x^k}{e^x} \overset{*}{=} \lim_{x \to \infty} \frac{kx^{k-1}}{e^x} \overset{*}{=} \lim_{x \to \infty} \frac{k(k-1)x^{k-2}}{e^x} \overset{*}{=} \cdots \overset{*}{=} \lim_{x \to \infty} \frac{k!}{e^x} = 0. \quad ❑$$

---

† We omit the proof.

For example,

$$\frac{x^{100}}{e^x} \to 0 \quad \text{and} \quad \frac{x^{1000}}{e^x} \to 0$$

as $x \to \infty$.

Remark The limits (10.6.2) and (10.6.3) tell us that $\ln x$ tends to infinity *more slowly than* any positive power of x and that e^x tends to infinity *faster than* any positive integral power of x. In the Exercises you are asked to show that e^x tends to infinity faster than *any* positive power of x and that *any* positive power of $\ln x$ tends to infinity more slowly than x. That is, for any positive number α,

$$\lim_{x \to \infty} \frac{x^\alpha}{e^x} = 0 \quad \text{and} \quad \lim_{x \to \infty} \frac{[\ln x]^\alpha}{x} = 0.$$

Comparisons of logarithmic and exponential growth were also given in Exercises 17 and 18 in Section 7.6. ❑

Example 3 Find the limit as $n \to \infty$ of the sequence $\{a_n\}$ given by

$$a_n = \frac{2^n}{n^2}.$$

SOLUTION To use the methods of calculus, we investigate

$$\lim_{x \to \infty} \frac{2^x}{x^2}.$$

Since both numerator and denominator tend to ∞ with x, we try L'Hospital's rule:

$$\lim_{x \to \infty} \frac{2^x}{x^2} \overset{*}{=} \lim_{x \to \infty} \frac{2^x \ln 2}{2x} \overset{*}{=} \lim_{x \to \infty} \frac{2^x (\ln 2)^2}{2} = \infty.$$

Therefore, the limit of the sequence must also be ∞. ❑

Other Indeterminate Forms: $0 \cdot \infty$, $\infty - \infty$, 0^0, 1^∞, ∞^0

If f tends to 0 and g tends to ∞ (or $-\infty$) as x approaches some number c (or $\pm\infty$), then it is not clear what the product $f \cdot g$ will do. For example, as $x \to 1$,

$$(x - 1)^3 \to 0, \qquad \frac{1}{(x - 1)^2} \to \infty, \quad \text{and} \quad (x - 1)^3 \cdot \frac{1}{(x - 1)^2} = (x - 1) \to 0.$$

On the other hand,

$$\lim_{x \to 1} \left[(x - 1)^3 \cdot \frac{1}{(x - 1)^4} \right] = \lim_{x \to 1} \left[\frac{1}{x - 1} \right] \qquad \text{does not exist.}$$

A limit of this type is called an *indeterminate of the form* $0 \cdot \infty$. We handle these indeterminates by writing the product $f \cdot g$ as a quotient

$$\frac{f}{1/g} \quad \text{or} \quad \frac{g}{1/f}.$$

In the first case, the result will be an indeterminate of the form $0/0$, and in the second it will have the form ∞/∞.

Example 4 Find

$$\lim_{x \to 0^+} \sqrt{x} \ln x.$$

SOLUTION   As $x \to 0^+$, $\sqrt{x} \to 0$ and $\ln x \to -\infty$. Thus the given limit is an indeterminate of the form $0 \cdot \infty$. Rewriting the product as a quotient, we have

$$\lim_{x \to 0^+} \sqrt{x} \ln x = \lim_{x \to 0^+} \frac{\ln x}{1/\sqrt{x}} \overset{*}{=} \lim_{x \to 0^+} \frac{1/x}{-\frac{1}{2} x^{-3/2}} = \lim_{x \to 0^+} -2\sqrt{x} = 0.$$

Therefore, $\lim_{x \to 0^+} \sqrt{x} \ln x = 0$.

Of course, we could have chosen to write $\sqrt{x} \ln x$ as the quotient

$$\frac{\sqrt{x}}{1/\ln x}.$$

Try to evaluate the limit using this quotient.   ❑

If f and g both tend to ∞, or if both tend to $-\infty$, as x tends to c (or $\pm\infty$), then $\lim(f - g)$ is called an *indeterminate of the form* $\infty - \infty$. Like the preceding case, indeterminates of this type are handled by converting the difference to a quotient.

Example 5 Find

$$\lim_{x \to (\pi/2)^-} (\tan x - \sec x).$$

SOLUTION   Both $\tan x$ and $\sec x$ tend to ∞ as x tends to $\pi/2$ from the left. We first rewrite the difference as a quotient:

$$\tan x - \sec x = \frac{\sin x}{\cos x} - \frac{1}{\cos x} = \frac{\sin x - 1}{\cos x}.$$

Now

$$\lim_{x \to (\pi/2)^-} \frac{\sin x - 1}{\cos x}$$

is an indeterminate of the form $0/0$, and

$$\lim_{x \to (\pi/2)^-} \frac{\sin x - 1}{\cos x} \overset{*}{=} \lim_{x \to (\pi/2)^-} \frac{\cos x}{-\sin x} = \frac{0}{-1} = 0.$$

Thus, $\lim_{x \to (\pi/2)^-} (\tan x - \sec x) = 0.$ ❑

Limits involving exponential expressions $[f(x)]^{g(x)}$ are indeterminate when: (1) f and g both tend to 0; (2) f tends to 1 and g tends to $\pm\infty$; and (3) f tends to $\pm\infty$ and g tends to 0. These cases are called *indeterminates of the forms* 0^0, 1^∞, and ∞^0, respectively. Exponential indeterminate forms are treated by taking natural logarithms:

$$\text{if} \quad y = [f(x)]^{g(x)}, \quad \text{then} \quad \ln y = g(x)\ln[f(x)].$$

Now, $\lim \ln y = \lim g(x)\ln[f(x)]$ will be an indeterminate of the form $0 \cdot \infty$

Example 6 Show that

(10.6.4)

$$\lim_{x \to 0^+} x^x = 1.$$

SOLUTION Here we are dealing with an indeterminate of the form 0^0. Our first step is to take the logarithm of x^x. Then we apply L'Hospital's rule:

$$\lim_{x \to 0^+} \ln(x^x) = \lim_{x \to 0^+} (x \ln x) = \lim_{x \to 0^+} \frac{\ln x}{1/x} \overset{*}{=} \lim_{x \to 0^+} \frac{1/x}{-1/x^2} = \lim_{x \to 0^+} (-x) = 0.$$

Since $\ln(x^x) \to 0$ as $x \to 0^+$, $x^x = e^{\ln(x^x)} \to e^0 = 1$. ❏

Example 7 Find

$$\lim_{x \to 0^+} (1 + x)^{1/x}.$$

SOLUTION Here we are dealing with an indeterminate of the form 1^∞: as $x \to 0^+$, $1 + x \to 1$ and $1/x$ increases without bound. Taking the logarithm and then applying L'Hospital's rule, we have

$$\lim_{x \to 0^+} \ln(1 + x)^{1/x} = \lim_{x \to 0^+} \frac{\ln(1 + x)}{x} \overset{*}{=} \lim_{x \to 0^+} \frac{1}{1 + x} = 1.$$

Since $\ln(1 + x)^{1/x} \to 1$ as $x \to 0^+$, $(1 + x)^{1/x} = e^{\ln(1 + x)^{1/x}} \to e^1 = e$. Note that if we set $x = 1/n$, we have the familiar result: $[1 + (1/n)]^n \to e$ as $n \to \infty$. ❏

Example 8 Show that

$$\lim_{x \to \infty} (x^2 + 1)^{1/\ln x} = e^2.$$

SOLUTION Here we have an indeterminate of the form ∞^0. Taking the logarithm and then applying L'Hospital's rule, we find that

$$\lim_{x \to \infty} \ln[(x^2 + 1)^{1/\ln x}] = \lim_{x \to \infty} \frac{\ln(x^2 + 1)}{\ln x} \overset{*}{=} \lim_{x \to \infty} \frac{2x/(x^2 + 1)}{1/x} = \lim_{x \to \infty} \frac{2x^2}{x^2 + 1} = 2.$$

It follows that

$$\lim_{x \to \infty} (x^2 + 1)^{1/\ln x} = e^2. \qquad ❏$$

Concluding Remarks Suppose that $\lim(f/g)$ is an indeterminate form (either $0/0$ or ∞/∞). Both versions of L'Hospital's rule (Theorems 10.5.1 and 10.6.1) tell us that if

$$\lim \frac{f'}{g'} = L \text{ (or } \pm\infty), \qquad \text{then} \qquad \lim \frac{f}{g} = L \text{ (or } \pm\infty).$$

However, the rules do not provide any information when $\lim(f'/g')$ fails to exist; $\lim(f/g)$ may or may not exist. For example,

$$\lim_{x \to \infty} \frac{x + \cos x}{x}$$

is an indeterminate of the form ∞/∞. It is easy to show that this limit is 1 (divide numerator and denominator by x). On the other hand, if we try to apply Theorem 10.6.1, we consider the limit

$$\lim_{x\to\infty} \frac{f'(x)}{g'(x)} = \lim_{x\to\infty} \frac{1 - \sin x}{1},$$

and this limit does not exist.

Finally, as noted in Section 10.5, you should always check first to make sure that a given limit actually involves an indeterminate form before trying to apply the methods of these sections. In the case of a quotient, L'Hospital's rule does not apply when either the numerator or denominator has a finite nonzero limit. For example,

$$\lim_{x\to 0^+} \frac{1 + x}{\sin x} = \infty,$$

but a misapplication of L'Hospital's rule would lead to the limit

$$\lim_{x\to 0^+} \frac{1}{\cos x} = 1,$$

and the *incorrect conclusion* that

$$\lim_{x\to 0^+} \frac{1 + x}{\sin x} = 1. \quad \square$$

EXERCISES 10.6

In Exercises 1–36, find the indicated limit.

1. $\lim\limits_{x\to-\infty} \dfrac{x^2 + 1}{1 - x}$.

2. $\lim\limits_{x\to\infty} \dfrac{20x}{x^2 + 1}$.

3. $\lim\limits_{x\to\infty} \dfrac{x^3}{1 - x^3}$.

4. $\lim\limits_{x\to\infty} \dfrac{x^3 - 1}{2 - x}$.

5. $\lim\limits_{x\to\infty} \left(x^2 \sin \dfrac{1}{x} \right)$.

6. $\lim\limits_{x\to\infty} \dfrac{\ln x^k}{x}$.

7. $\lim\limits_{x\to\pi/2^-} \dfrac{\tan 5x}{\tan x}$.

8. $\lim\limits_{x\to 0} (x \ln|\sin x|)$.

9. $\lim\limits_{x\to 0^+} x^{2x}$.

10. $\lim\limits_{x\to\infty} \left(x \sin \dfrac{\pi}{x} \right)$.

11. $\lim\limits_{x\to 0} [x(\ln|x|)^2]$.

12. $\lim\limits_{x\to 0^+} \dfrac{\ln x}{\cot x}$.

13. $\lim\limits_{x\to\infty} \left(\dfrac{1}{x} \displaystyle\int_0^x e^{t^2}\, dt \right)$.

14. $\lim\limits_{x\to\infty} \dfrac{\sqrt{1 + x^2}}{x}$.

15. $\lim\limits_{x\to 0} \left[\dfrac{1}{\sin^2 x} - \dfrac{1}{x^2} \right]$.

16. $\lim\limits_{x\to 0} |\sin x|^x$.

17. $\lim\limits_{x\to 1} x^{1/(x-1)}$.

18. $\lim\limits_{x\to 0^+} x^{\sin x}$.

19. $\lim\limits_{x\to\infty} \left(\cos \dfrac{1}{x} \right)^x$.

20. $\lim\limits_{x\to\pi/2} |\sec x|^{\cos x}$.

21. $\lim\limits_{x\to 0} \left[\dfrac{1}{\ln(1 + x)} - \dfrac{1}{x} \right]$.

22. $\lim\limits_{x\to\infty} (x^2 + a^2)^{(1/x)^2}$.

23. $\lim\limits_{x\to 0} \left(\dfrac{1}{x} - \cot x \right)$.

24. $\lim\limits_{x\to\infty} \ln \left(\dfrac{x^2 - 1}{x^2 + 1} \right)^3$.

25. $\lim\limits_{x\to\infty} (\sqrt{x^2 + 2x} - x)$.

26. $\lim\limits_{x\to\infty} \dfrac{1}{x} \displaystyle\int_0^x \sin \left(\dfrac{1}{t + 1} \right) dt$.

27. $\lim\limits_{x\to\infty} (x^3 + 1)^{1/\ln x}$.

28. $\lim\limits_{x\to\infty} (e^x + 1)^{1/x}$.

29. $\lim\limits_{x\to\infty} (\cosh x)^{1/x}$.

30. $\lim\limits_{x\to\infty} (x^4 + 1)^{1/\ln x}$.

31. $\lim\limits_{x\to 0} (e^x + x)^{1/x}$.

32. $\lim\limits_{x\to\infty} \left(1 + \dfrac{1}{x} \right)^{3x}$.

33. $\lim\limits_{x\to 0} \left(\dfrac{1}{\sin x} - \dfrac{1}{x} \right)$.

34. $\lim\limits_{x\to 0} (e^x + 3x)^{1/x}$.

35. $\lim\limits_{x\to 1} \left(\dfrac{1}{\ln x} - \dfrac{x}{x - 1} \right)$.

36. $\lim\limits_{x\to 0} \left(\dfrac{1 + 2^x}{2} \right)^{1/x}$.

In Exercises 37–44, find the limit of the sequence.

37. $\lim\limits_{n\to\infty} \left(\dfrac{1}{n} \ln \dfrac{1}{n} \right)$.

38. $\lim\limits_{n\to\infty} \dfrac{n^k}{2^n}$.

39. $\lim\limits_{n\to\infty} (\ln n)^{1/n}.$

40. $\lim\limits_{n\to\infty} \dfrac{\ln n}{n^p}, \quad (p > 0).$

41. $\lim\limits_{n\to\infty} (n^2 + n)^{1/n}.$

42. $\lim\limits_{n\to\infty} n^{\sin(\pi/n)}.$

43. $\lim\limits_{n\to\infty} \dfrac{n^2\ln n}{e^n}.$

44. $\lim\limits_{n\to\infty} (\sqrt{n} - 1)^{1/\sqrt{n}}.$

In Exercises 45–50, sketch the curve, specifying all vertical and horizontal asymptotes.

45. $y = x^2 - \dfrac{1}{x^3}.$

46. $y = \sqrt{\dfrac{x}{x - 1}}.$

47. $y = xe^x.$

48. $y = xe^{-x}.$

49. $y = x^2 e^{-x}.$

50. $y = \dfrac{\ln x}{x}.$

The graphs of two functions $y = f(x)$ and $y = g(x)$ are said to be *asymptotic as $x \to \infty$* iff

$$\lim\limits_{x\to\infty} [f(x) - g(x)] = 0;$$

they are said to be *asymptotic as $x \to -\infty$* iff

$$\lim\limits_{x\to-\infty} [f(x) - g(x)] = 0.$$

51. Show that the hyperbolic arc $y = (b/a)\sqrt{x^2 - a^2}$ is asymptotic to the line $y = (b/a)x$ as $x \to \infty$.

52. Show that the graphs of $y = \cosh x$ and $y = \sinh x$ are asymptotic.

53. Give an example of a function the graph of which is asymptotic to the parabola $y = x^2$ as $x \to \infty$ and crosses the graph of the parabola twice.

54. Give an example of a function the graph of which is asymptotic to the line $y = x$ as $x \to \infty$ and crosses the graph of the line infinitely often.

55. Find the fallacy:

$$\lim\limits_{x\to0^+} \dfrac{x^2}{\sin x} \overset{*}{=} \lim\limits_{x\to0^+} \dfrac{2x}{\cos x} \overset{*}{=} \lim\limits_{x\to0^+} \dfrac{2}{-\sin x} = -\infty.$$

56. Let α be a positive number. Show that

$$\lim\limits_{x\to\infty} \dfrac{x^\alpha}{e^x} = 0.$$

57. (a) Show by induction that, for each positive integer k,

$$\lim\limits_{x\to\infty} \dfrac{(\ln x)^k}{x} = 0.$$

(b) Show that, for each positive number α,

$$\lim\limits_{x\to\infty} \dfrac{(\ln x)^\alpha}{x} = 0.$$

58. (a) Try to evaluate

$$\lim\limits_{x\to\infty} \dfrac{x}{\sqrt{x^2 + 1}}$$

using L'Hospital's rule and see what happens.

(b) Evaluate this limit by some other method.

59. (a) Try to evaluate

$$\lim\limits_{x\to0} \dfrac{e^{-1/x^2}}{x}$$

using L'Hospital's rule and see what happens. Then rewrite the quotient in an equivalent form and show that the limit is 0.

(b) Define the function f by

$$f(x) = \begin{cases} e^{-1/x^2} & \text{if } x \neq 0 \\ 0 & \text{if } x = 0. \end{cases}$$

Show that f is differentiable at 0. What is $f'(0)$?

60. The differential equation governing the velocity of an object of mass m dropped from rest under the influence of gravity with air resistance directly proportional to the velocity is

$$(*) \qquad m\dfrac{dv}{dt} + kv = mg,$$

where $k > 0$ is the constant of proportionality, g is the gravitational constant and $v(0) = 0$. See Exercise 37, Section 7.6. The velocity of the object at time t is given by

$$v(t) = (mg/k)(1 - e^{-(k/m)t}).$$

(a) Fix t and find $\lim\limits_{k\to0^+} v(t)$.

(b) Set $k = 0$ in equation $(*)$ and solve

$$m\dfrac{dv}{dt} = mg, \quad v(0) = 0.$$

Does this result agree with the result found in part (a)?

61. Let $f(x) = (1 + x)^{1/x}$ and $g(x) = (1 + x^2)^{1/x}$ on $(0, \infty)$.

(a) Use a graphing utility to graph f and g in the same coordinate system. Estimate

$$\lim\limits_{x\to0^+} g(x).$$

(b) Use L'Hospital's rule to confirm your estimate in part (a).

62. Let $f(x) = \sqrt{x^2 + 3x + 1} - x.$

(a) Use a graphing utility to graph f. Then use your graph to estimate

$$\lim\limits_{x\to\infty} f(x)$$

(b) Use L'Hospital's rule to confirm your estimate. HINT: "Rationalize."

63. Let $g(x) = \sqrt[3]{x^3 - 5x^2 + 2x + 1} - x.$

(a) Use a graphing utility to graph g. Then use your graph to estimate

$$\lim_{x \to \infty} g(x).$$

(b) Use L'Hospital's rule to confirm your estimate.

64. Exercises 62 and 63 can be generalized as follows: Let n be a positive integer and let P be the polynomial

$$P(x) = x^n + b_1 x^{n-1} + b_2 x^{n-2} + \cdots + b_{n-1} x + b_n.$$

Prove that

$$\lim_{x \to \infty} \left([P(x)]^{1/n} - x \right) = \frac{b_1}{n}.$$

■ 10.7 IMPROPER INTEGRALS

In all of our work involving the theory and applications of the definite integral

$$\int_a^b f(x) \, dx,$$

it has been assumed that the interval $[a, b]$ is finite and that the function f is bounded on $[a, b]$. In the context to be developed here, such integrals are said to be *proper*. In this section, we will use a limit process to calculate integrals in cases where the interval is infinite or where the function is unbounded. Such integrals are called *improper integrals*.

Integrals over Infinite Intervals

We begin with a function f continuous on an unbounded interval $[a, \infty)$. For each number $b > a$ we can form the definite integral

$$\int_a^b f(x) \, dx.$$

If, as b tends to ∞, this integral tends to a finite limit L,

$$\lim_{b \to \infty} \int_a^b f(x) \, dx = L,$$

then we write

$$\int_a^\infty f(x) \, dx = L$$

and say that

> the improper integral $\displaystyle\int_a^\infty f(x) \, dx$ *converges to L.*

Otherwise, we say that

> the improper integral $\displaystyle\int_a^\infty f(x) \, dx$ *diverges.*

In a similar manner, if f is continuous on the unbounded interval $(-\infty, b]$, then for each number $a < b$, we can form the definite integral

$$\int_a^b f(x) \, dx$$

and calculate

$$\lim_{a \to -\infty} \int_a^b f(x) \, dx.$$

If this limit exists and equals L, then

the improper integral $\displaystyle\int_{-\infty}^b f(x) \, dx$ converges to L;

otherwise,

the improper integral $\displaystyle\int_{-\infty}^b f(x) \, dx$ diverges.

Example 1

(a) $\displaystyle\int_0^\infty e^{-2x} \, dx = \tfrac{1}{2}.$

(b) $\displaystyle\int_1^\infty \frac{dx}{x}$ diverges.

(c) $\displaystyle\int_1^\infty \frac{dx}{x^2} = 1.$

(d) $\displaystyle\int_{-\infty}^1 \cos \pi x \, dx$ diverges.

VERIFICATION

(a) $\displaystyle\int_0^\infty e^{-2x} \, dx = \lim_{b \to \infty} \int_0^b e^{-2x} \, dx = \lim_{b \to \infty} \left[-\frac{e^{-2x}}{2} \right]_0^b = \lim_{b \to \infty} \left(\frac{1}{2} - \frac{1}{e^{2b}} \right) = \frac{1}{2}.$

(b) $\displaystyle\int_1^\infty \frac{dx}{x} = \lim_{b \to \infty} \int_1^b \frac{dx}{x} = \lim_{b \to \infty} \ln b = \infty.$

(c) $\displaystyle\int_1^\infty \frac{dx}{x^2} = \lim_{b \to \infty} \int_1^b \frac{dx}{x^2} = \lim_{b \to \infty} \left[-\frac{1}{x} \right]_1^b = \lim_{b \to \infty} \left(1 - \frac{1}{b} \right) = 1.$

(d) Note first that

$$\int_a^1 \cos \pi x \, dx = \left[\frac{1}{\pi} \sin \pi x \right]_a^1 = -\frac{1}{\pi} \sin \pi a.$$

As a tends to $-\infty$, $\sin \pi a$ oscillates between -1 and 1. Therefore the integral oscillates between $1/\pi$ and $-1/\pi$ and does not converge. ❑

The usual formulas for area and volume are extended to the unbounded case by means of improper integrals.

Example 2 Let p be a positive number. If Ω is the region between the x-axis and the graph of

$$f(x) = \frac{1}{x^p}, \qquad x \geq 1 \qquad\qquad \text{(Figure 10.7.1)}$$

then

$$\text{area of } \Omega = \begin{cases} \dfrac{1}{p-1}, & \text{if } p > 1 \\ \infty, & \text{if } p \leq 1. \end{cases}$$

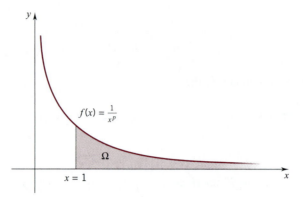

Figure 10.7.1

This comes about from setting

$$\text{area of } \Omega = \lim_{b \to \infty} \int_1^b \frac{dx}{x^p} = \int_1^\infty \frac{dx}{x^p}.$$

For $p \neq 1$,

$$\int_1^\infty \frac{dx}{x^p} = \lim_{b \to \infty} \int_1^b \frac{dx}{x^p} = \lim_{b \to \infty} \frac{1}{1-p}(b^{1-p} - 1) = \begin{cases} \dfrac{1}{p-1} & \text{if } p > 1 \\ \infty & \text{if } p < 1. \end{cases}$$

For $p = 1$,

$$\int_1^\infty \frac{dx}{x^p} = \int_1^\infty \frac{dx}{x} = \infty,$$

as you have seen already. ❑

Remark It is easy to verify that if $p \leq 0$ and Ω is the region between the x-axis and the graph of $f(x) = 1/x^p = x^{-p}$, $-p \geq 0$, then area of $\Omega = \infty$. Thus the conclusion in Example 2 actually holds for all real numbers p. ❑

Example 3 From the last example you know that the region below the graph of

$$f(x) = \frac{1}{x}, \qquad x \geq 1$$

has infinite area. Suppose that this region with infinite area is revolved about the x-axis (see Figure 10.7.2). What is the volume V of the resulting solid? It may surprise you somewhat, but the volume is not infinite. In fact, it is π. Using the disc method to calculate the volume (see Section 6.2), we have

$$V = \int_1^\infty \pi[f(x)]^2 \, dx = \pi \int_1^\infty \frac{dx}{x^2} = \pi \lim_{b \to \infty} \int_1^b \frac{dx}{x^2}$$

$$= \pi \lim_{b \to \infty} \left[\frac{-1}{x} \right]_1^b = \pi \cdot 1 = \pi. \quad ❑$$

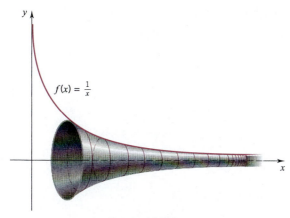

Figure 10.7.2

For future reference we record the following:

(10.7.1)

$$\int_1^\infty \frac{dx}{x^p} \text{ converges for } p > 1 \text{ and diverges for } p \le 1.$$

It is often difficult to determine the convergence or divergence of a given improper integral by direct methods, that is, by calculating the definite integral and evaluating the limit. In such cases we can sometimes gain information by comparison with integrals of known behavior.

(10.7.2)

(*A comparison test*) Suppose that f and g are continuous and

$$0 \le f(x) \le g(x) \qquad \text{for all } x \in [a, \infty).$$

(i) If $\displaystyle\int_a^\infty g(x)\, dx$ converges, then $\displaystyle\int_a^\infty f(x)\, dx$ converges.

(ii) If $\displaystyle\int_a^\infty f(x)\, dx$ diverges, then $\displaystyle\int_a^\infty g(x)\, dx$ diverges.

A similar result holds for integrals from $-\infty$ to b. Figure 10.7.3 illustrates the comparison test (10.7.2). The proof of the result is left to you as an exercise.

Example 4 The improper integral

$$\int_1^\infty \frac{dx}{\sqrt{1 + x^3}}$$

converges since

$$\frac{1}{\sqrt{1 + x^3}} < \frac{1}{x^{3/2}} \quad \text{for } x \in [1, \infty) \qquad \text{and} \qquad \int_1^\infty \frac{dx}{x^{3/2}} \text{ converges.}$$

Figure 10.7.3

In contrast, if we tried to evaluate

$$\lim_{b \to \infty} \int_1^b \frac{dx}{\sqrt{1 + x^3}}$$

directly, we would have to calculate the integral

$$\int \frac{dx}{\sqrt{1 + x^3}},$$

and this cannot be done by any of the methods we have developed so far. ❏

Example 5   The improper integral

$$\int_1^\infty \frac{dx}{\sqrt{1 + x^2}}$$

diverges since

$$\frac{1}{1 + x} \leq \frac{1}{\sqrt{1 + x^2}} \quad \text{for } x \in [1, \infty) \quad \text{and} \quad \int_1^\infty \frac{dx}{1 + x} \text{ diverges.}$$

This result can also be obtained by evaluating

$$\int_1^b \frac{dx}{\sqrt{1 + x^2}}$$

and then calculating the limit as $b \to \infty$. Try it. ❏

Suppose now that f is continuous on $(-\infty, \infty)$. The *improper integral*

$$\int_{-\infty}^\infty f(x) \, dx$$

is said to *converge* iff

$$\int_{-\infty}^0 f(x) \, dx \quad \text{and} \quad \int_0^\infty f(x) \, dx$$

both converge. We then set

$$\int_{-\infty}^\infty f(x) \, dx = L + M,$$

where

$$\int_{-\infty}^0 f(x) \, dx = L \quad \text{and} \quad \int_0^\infty f(x) \, dx = M.$$

Example 6   Determine whether the improper integral

$$\int_{-\infty}^\infty \frac{e^x}{1 + e^{2x}} \, dx$$

converges or diverges. If it converges, give its value.

SOLUTION   First consider the indefinite integral

$$\int \frac{e^x}{1 + e^{2x}} \, dx.$$

Let $u = e^x$ and $du = e^x \, dx$.   Then

$$\int \frac{e^x}{1 + e^{2x}} \, dx = \int \frac{1}{1 + u^2} \, du = \tan^{-1}u + C = \tan^{-1}(e^x) + C.$$

Now,

$$\int_{-\infty}^0 \frac{e^x}{1 + e^{2x}} \, dx = \lim_{a \to -\infty} \int_a^0 \frac{e^x}{1 + e^{2x}} \, dx = \lim_{a \to -\infty} \left[\tan^{-1}(e^x) \right]_a^0$$

$$= \tan^{-1}(1) - \lim_{a \to -\infty} \tan^{-1}(e^a) = \frac{\pi}{4} - 0 = \frac{\pi}{4}$$

and

$$\int_0^{\infty} \frac{e^x}{1 + e^{2x}} \, dx = \lim_{b \to \infty} \int_0^b \frac{e^x}{1 + e^{2x}} \, dx = \lim_{b \to \infty} \left[\tan^{-1}(e^x) \right]_0^b$$

$$= \lim_{b \to \infty} \tan^{-1}(e^b) - \tan^{-1}(1) = \frac{\pi}{2} - \frac{\pi}{4} = \frac{\pi}{4}.$$

Therefore the improper integral

$$\int_{-\infty}^{\infty} \frac{e^x}{1 + e^{2x}} \, dx$$

converges.   Its value is $\frac{1}{4}\pi + \frac{1}{4}\pi = \frac{1}{2}\pi$.   ❑

Remark   You might be wondering why we did not define

$$\int_{-\infty}^{\infty} f(x) \, dx$$

in terms of the limit

$$\lim_{b \to \infty} \int_{-b}^b f(x) \, dx.$$

In fact, it can be shown that if the integral from $-\infty$ to ∞ exists in the sense of the original definition, then the limit just given also exists and they are equal.   On the other hand, if we let $f(x) = x$, then

$$\lim_{b \to \infty} \int_{-b}^b x \, dx = \lim_{b \to \infty} \left[\frac{x^2}{2} \right]_{-b}^b = \lim_{b \to \infty} \left[\frac{b^2}{2} - \frac{b^2}{2} \right] = 0,$$

while the calculation of

$$\int_{-\infty}^{\infty} x \, dx$$

using the definition will lead to an indeterminate of the form $\infty - \infty$.   ❑

Integrals of Unbounded Functions

Improper integrals can also arise on bounded intervals. Suppose that f is continuous on the half-open interval $[a, b)$ but is unbounded there. See Figure 10.7.4. For each number $c < b$, we can form the definite integral

$$\int_a^c f(x)\, dx.$$

If

$$\lim_{c \to b^-} \int_a^c f(x)\, dx = L$$

exists, then we say that

the improper integral $\displaystyle\int_a^b f(x)\, dx$ *converges to L.*

Otherwise, *the improper integral diverges.*

In a similar manner, if f is continuous on $(a, b]$ and unbounded at a, then we consider the limit

$$\lim_{c \to a^+} \int_c^b f(x)\, dx.$$

If this limit exists and has the value L, then

the improper integral $\displaystyle\int_a^b f(x)\, dx$ *converges to L.*

Otherwise, *the improper integral diverges.*

Figure 10.7.4

Example 7

(a) $\displaystyle\int_0^1 (1 - x)^{-2/3}\, dx = 3.$ **(b)** $\displaystyle\int_0^2 \frac{dx}{x}$ diverges.

VERIFICATION

(a) $\displaystyle\int_0^1 (1 - x)^{-2/3}\, dx = \lim_{c \to 1^-} \int_0^c (1 - x)^{-2/3}\, dx$

$$= \lim_{c \to 1^-} \left[-3(1 - x)^{1/3} \right]_0^c = \lim_{c \to 1^-} [-3(1 - c)^{1/3} + 3] = 3.$$

(b) $\displaystyle\int_0^2 \frac{dx}{x} = \lim_{c \to 0^+} \int_c^2 \frac{dx}{x} = \lim_{c \to 0^+} [\ln 2 - \ln c] = \infty.$ ❑

Now suppose that f is continuous on an interval $[a, b]$ except at some point c in (a, b) where $f(x) \to \pm\infty$ as $x \to c^-$ or as $x \to c^+$. We say that the *improper integral*

$$\int_a^b f(x)\, dx$$

converges iff *both* of the integrals

$$\int_a^c f(x)\, dx \quad \text{and} \quad \int_c^b f(x)\, dx$$

converge. If

$$\int_a^c f(x)\, dx = L \quad \text{and} \quad \int_c^b f(x)\, dx = M,$$

then

$$\int_a^b f(x)\, dx = L + M.$$

Example 8 To evaluate

(*)
$$\int_1^4 \frac{dx}{(x-2)^2}$$

we need to calculate

$$\lim_{c \to 2^-} \int_1^c \frac{dx}{(x-2)^2} \quad \text{and} \quad \lim_{c \to 2^+} \int_c^4 \frac{dx}{(x-2)^2}.$$

As you can verify, neither of these limits exists and thus improper integral (*) diverges.

Notice that, if we ignore the fact that integral (*) is improper, then we are led to the *incorrect conclusion* that

$$\int_1^4 \frac{dx}{(x-2)^2} = \left[\frac{-1}{x-2}\right]_1^4 = -\frac{3}{2}. \qquad \text{(see Figure 10.7.5)} \quad \square$$

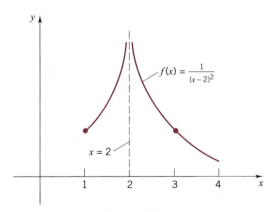

Figure 10.7.5

Example 9 Evaluate

$$\int_{-2}^1 \frac{dx}{x^{4/5}}.$$

SOLUTION Since $1/x^{4/5} \to \infty$ as $x \to 0^-$ and as $x \to 0^+$, the given integral is improper. Therefore, we need to calculate

$$\int_{-2}^0 \frac{dx}{x^{4/5}} \quad \text{and} \quad \int_0^1 \frac{dx}{x^{4/5}}.$$

Now

$$\int_{-2}^{0} \frac{dx}{x^{4/5}} = \lim_{c \to 0^-} \int_{-2}^{c} \frac{dx}{x^{4/5}} = \lim_{c \to 0^-} \left[5x^{1/5} \right]_{-2}^{c} = \lim_{c \to 0^-} [5c^{1/5} - 5(-2)^{1/5}] = 5(2^{1/5})$$

and

$$\int_{0}^{1} \frac{dx}{x^{4/5}} = \lim_{c \to 0^+} \int_{c}^{1} \frac{dx}{x^{4/5}} = \lim_{c \to 0^+} \left[5x^{1/5} \right]_{c}^{1} = \lim_{c \to 0^+} [5 - 5c^{1/5}] = 5.$$

Thus, the improper integral converges and

$$\int_{-2}^{1} \frac{dx}{x^{4/5}} = 5 + 5(2^{1/5}) \cong 10.74. \quad \square$$

EXERCISES 10.7

In Exercises 1–34, evaluate the improper integrals that converge.

1. $\int_{1}^{\infty} \frac{dx}{x^2}.$

2. $\int_{0}^{\infty} \frac{dx}{1 + x^2}.$

3. $\int_{0}^{\infty} \frac{dx}{4 + x^2}.$

4. $\int_{0}^{\infty} e^{-px} \, dx, \quad p > 0.$

5. $\int_{0}^{\infty} e^{px} \, dx, \quad p > 0.$

6. $\int_{0}^{1} \frac{dx}{\sqrt{x}}.$

7. $\int_{0}^{8} \frac{dx}{x^{2/3}}.$

8. $\int_{0}^{1} \frac{dx}{x^2}.$

9. $\int_{0}^{1} \frac{dx}{\sqrt{1 - x^2}}.$

10. $\int_{0}^{1} \frac{dx}{\sqrt{1 - x}}.$

11. $\int_{0}^{2} \frac{x}{\sqrt{4 - x^2}} \, dx.$

12. $\int_{0}^{a} \frac{dx}{\sqrt{a^2 - x^2}}.$

13. $\int_{e}^{\infty} \frac{\ln x}{x} \, dx.$

14. $\int_{e}^{\infty} \frac{dx}{x \ln x}.$

15. $\int_{0}^{1} x \ln x \, dx.$

16. $\int_{e}^{\infty} \frac{dx}{x(\ln x)^2}.$

17. $\int_{-\infty}^{\infty} \frac{dx}{1 + x^2}.$

18. $\int_{2}^{\infty} \frac{dx}{x^2 - 1}.$

19. $\int_{-\infty}^{\infty} \frac{dx}{x^2}.$

20. $\int_{1/3}^{3} \frac{dx}{\sqrt[3]{3x - 1}}.$

21. $\int_{1}^{\infty} \frac{dx}{x(x + 1)}.$

22. $\int_{-\infty}^{0} xe^x \, dx.$

23. $\int_{3}^{5} \frac{x}{\sqrt{x^2 - 9}} \, dx.$

24. $\int_{1}^{4} \frac{dx}{x^2 - 4}.$

25. $\int_{-3}^{3} \frac{dx}{x(x + 1)}.$

26. $\int_{1}^{\infty} \frac{x}{(1 + x^2)^2} \, dx.$

27. $\int_{-3}^{1} \frac{dx}{x^2 - 4}.$

28. $\int_{0}^{\infty} \sinh x \, dx.$

29. $\int_{0}^{\infty} \cosh x \, dx.$

30. $\int_{1}^{4} \frac{dx}{x^2 - 5x + 6}.$

31. $\int_{0}^{\infty} e^{-x} \sin x \, dx.$

32. $\int_{0}^{\infty} \cos^2 x \, dx.$

33. $\int_{0}^{1} \frac{e^{\sqrt{x}}}{\sqrt{x}} \, dx.$

34. $\int_{0}^{\pi/2} \frac{\cos x}{\sqrt{\sin x}} \, dx.$

35. The integral

$$\int_{0}^{1} \sin^{-1} x \, dx$$

is a "proper" definite integral. But the integration technique (integration by parts) will lead to an improper integral. Evaluate this integral.

36. (a) For what values of r is

$$\int_{0}^{\infty} x^r e^{-x} \, dx$$

convergent?
(b) Use mathematical induction to show that

$$\int_{0}^{\infty} x^n e^{-x} \, dx = n!, \quad n = 1, 2, 3, \ldots.$$

37. The integral

$$\int_{0}^{\infty} \frac{1}{\sqrt{x}(1 + x)} \, dx$$

is improper for both of the reasons discussed in this section: the interval is infinite and the integrand is unbounded. If we rewrite the integral as

$$\int_0^\infty \frac{1}{\sqrt{x}(1+x)}\,dx = \int_0^1 \frac{1}{\sqrt{x}(1+x)}\,dx + \int_1^\infty \frac{1}{\sqrt{x}(1+x)}\,dx$$

then we have two improper integrals, the first having an unbounded integrand and the second an infinite interval. If each of these integrals converges with values L_1 and L_2, respectively, then the given integral converges and has the value $L_1 + L_2$. Evaluate the given integral.

38. Evaluate

$$\int_1^\infty \frac{1}{x\sqrt{x^2-1}}\,dx$$

using the method given in Exercise 37.

39. Let Ω be the region bounded by the coordinate axes, the graph of $y = 1/\sqrt{x}$, and the line $x = 1$. (a) Sketch Ω. (b) Show Ω has finite area and find it. (c) Show that if Ω is revolved about the x-axis, the solid obtained does not have finite volume.

40. Let Ω be the region between the graph of $y = 1/(1 + x^2)$ and the x-axis, $x \geq 0$. (a) Sketch Ω. (b) Find the area of Ω. (c) Find the volume of the solid obtained by revolving Ω about the x-axis. (d) Find the volume of the solid obtained by revolving Ω about the y-axis.

41. Let Ω be the region bounded by the curve $y = e^{-x}$ and the x-axis, $x \geq 0$. (a) Sketch Ω. (b) Find the area of Ω. (c) Find the volume of the solid obtained by revolving Ω about the x-axis. (d) Find the volume obtained by revolving Ω about the y-axis. (e) Find the lateral surface area of the solid in part (c).

42. What point would you call the centroid of the region in Exercise 41? Does Pappus's theorem work in this instance?

43. Let Ω be the region bounded by the curve $y = e^{-x^2}$ and the x-axis, $x \geq 0$. (a) Show that Ω has finite area. (The area is actually $\frac{1}{2}\sqrt{\pi}$, as you will see in Chapter 16.) (b) Calculate the volume generated by revolving Ω about the y-axis.

44. Let Ω be the region bounded below by $y(x^2 + 1) = x$, above by $xy = 1$, and to the left by $x = 1$. (a) Find the area of Ω. (b) Show that the solid generated by revolving Ω about the x-axis has finite volume. (c) Calculate the volume generated by revolving Ω about the y-axis.

45. Let Ω be the region bounded by the curve $y = x^{-1/4}$ and the x-axis, $0 < x \leq 1$. (a) Sketch Ω. (b) Find the area of Ω. (c) Find the volume of the solid obtained by revolving Ω about the x-axis. (d) Find the volume of the solid obtained by revolving Ω about the y-axis.

46. Prove the validity of the comparison test (10.7.2).

In Exercises 47–52, use the comparison test (10.7.2) to determine which of the integrals converge.

47. $\displaystyle\int_1^\infty \frac{x}{\sqrt{1 + x^5}}\,dx.$

48. $\displaystyle\int_1^\infty 2^{-x^2}\,dx.$

49. $\displaystyle\int_0^\infty (1 + x^5)^{-1/6}\,dx.$

50. $\displaystyle\int_\pi^\infty \frac{\sin^2 2x}{x^2}\,dx.$

51. $\displaystyle\int_1^\infty \frac{\ln x}{x^2}\,dx.$

52. $\displaystyle\int_e^\infty \frac{dx}{\sqrt{x+1}\,\ln x}.$

53. Calculate the arc distance from the origin to the point $(x(\theta_1),\ y(\theta_1))$ along the exponential spiral $r = ae^{c\theta}$. (Take $a > 0, c > 0$.)

54. The function

$$f(x) = \frac{1}{\sqrt{2\pi}} \int_{-\infty}^x e^{-t^2/2}\,dt$$

is important in statistics. Prove that the integral on the right converges for all real x.

Exercises 55–58: **Laplace transforms.** Let f be continuous on $[0, \infty)$. The *Laplace transform* of f is the function F defined by

$$F(s) = \int_0^\infty e^{-sx}f(x)\,dx.$$

The domain of F is the set of all real numbers s such that the improper integral converges. Find the Laplace transform F of each of the following functions and give the domain of F.

55. $f(x) = 1.$

56. $f(x) = x.$

57. $f(x) = \cos 2x.$

58. $f(x) = e^{ax}.$

Exercises 59–62: **Probability density functions**. A nonnegative function f defined on $(-\infty, \infty)$ is a *probability density function* if

$$\int_{-\infty}^\infty f(x)\,dx = 1.$$

59. Show that the function f defined by

$$f(x) = \begin{cases} 6x/(1 + 3x^2)^2 & x \geq 0 \\ 0 & x < 0 \end{cases}$$

is a probability density function.

60. Show that the function f defined by

$$f(x) = \begin{cases} ke^{-kx} & x \geq 0 \\ 0 & x < 0, \quad k > 0, \end{cases}$$

is a probability density function. It is called the *exponential density function.*

61. If f is a probability density function, then its *mean* μ is given by

$$\mu = \int_{-\infty}^\infty x f(x)\,dx.$$

Calculate the mean for the exponential density function.

62. If f is a probability density function, then its *standard deviation* σ is given by

$$\sigma = \left[\int_{-\infty}^{\infty} (x - \mu)^2 f(x) \, dx \right]^{1/2}$$

where μ is the mean. Calculate the standard deviation for the exponential density function.

63. (*Useful later*) Let f be a continuous, positive, decreasing function on $[1, \infty)$. Show that

$$\int_{1}^{\infty} f(x) \, dx \quad \text{converges} \quad \text{iff} \left\{ \int_{1}^{n} f(x) \, dx \right\} \text{ converges.}$$

■ CHAPTER HIGHLIGHTS

10.1 The Least Upper Bound Axiom

Upper bound, bounded above, least upper bound (p. 637)
Least upper bound axiom (p. 638)
Lower bound, bounded below, greatest lower bound (p. 640)

10.2 Sequences of Real Numbers

Sequence (p. 642)
Bounded above, bounded below, bounded (p. 644)
Increasing, nondecreasing, decreasing, nonincreasing (p. 644)
Recurrence relation (p. 649)

It is sometimes possible to obtain useful information about a sequence $y_n = f(n)$ by applying the techniques of calculus to the function $y = f(x)$.

10.3 Limit of a Sequence

Limit of a sequence (p. 651)   Uniqueness of the limit (p. 652)
Convergent, divergent (p. 653)   Pinching theorem (p. 656)

Every convergent sequence is bounded (p. 653); thus, every unbounded sequence is divergent.

A bounded, monotonic sequence converges. (p. 654)

Suppose that $c_n \to c$ as $n \to \infty$, and all the c_n are in the domain of f. If f is continuous at c, then $f(c_n) \to f(c)$. (p. 657)

10.4 Some Important Limits

For $x > 0$, $\lim_{n \to \infty} x^{1/n} = 1$

For $|x| < 1$, $\lim_{n \to \infty} x^n = 0$

For each $\alpha > 0$, $\lim_{n \to \infty} \dfrac{1}{n^\alpha} = 0$

For each real x, $\lim_{n \to \infty} \dfrac{x^n}{n!} = 0$

$\lim_{n \to \infty} \dfrac{\ln n}{n} = 0$

$\lim_{n \to \infty} n^{1/n} = 1$

For each real x, $\lim_{n \to \infty} \left(1 + \dfrac{x}{n} \right)^n = e^x$ Cauchy sequence (p. 680)

10.5 The Indeterminate Form (0/0)

L'Hospital's rule (0/0) (p. 668)   Cauchy mean-value theorem (p. 670)

10.6 The Indeterminate Form (∞/∞)

L'Hospital's rule (∞/∞) (p. 674)

$\lim_{x \to \infty} \dfrac{\ln x}{x^\alpha} = 0$ $\lim_{x \to \infty} \dfrac{x^k}{e^x} = 0$ $\lim_{x \to \infty} x^x = 1$

Other indeterminate forms: $0 \cdot \infty$, $\infty - \infty$, 0^0, 1^∞, ∞^0 (p. 675)

10.7 Improper Integrals

Integrals over infinite intervals (p. 680)   convergent, divergent (p. 680)

$$\int_1^\infty \frac{dx}{x^p} \text{ converges for } p > 1 \text{ and diverges for } p \le 1.$$

A comparison test (p. 683)

Integrals of unbounded functions (p. 686)   convergent, divergent (p. 686)

■ PROJECTS AND EXPLORATIONS USING TECHNOLOGY

*To do these exercises you will need a graphics calculator or a computer with graphing capabil-
ity. The majority of these problems are open-ended so different approaches may be used to
solve them. You should be aware that different approaches can result in slight variations in the
answers. Round your numerical answers to at least four decimal places. The rounding
method that your calculator or computer uses also may cause variations in answers.*

10.1 It is easy to show using the comparison test that the improper integral

$$\int_1^\infty \frac{1}{x^3 + 1}\, dx$$

converges. However, computer programs may have trouble evaluating this integral.
(a) Use the comparison test to show that this integral converges.
(b) Use the method of partial fractions to evaluate this integral.
(c) The function

$$F(x) = \int_1^x \frac{1}{t^3 + 1}\, dt$$

is an increasing function on $[1, \infty)$. Explain why.
(d) Evaluate F at a sequence of x values (for example, $x = 10$, $x = 100$, $x = 1000$, $x = 10{,}000$, and so on) using a computer program. Does F start to decrease as x increases? Can you explain why?
(e) Set up a procedure for checking the accuracy of the values that you compute.

10.2 This problem investigates the convergence of the improper integral

$$\int_1^\infty 1000xe^{-0.1\sqrt{x}}\, dx.$$

(a) Let $\{I_k\}$ be the sequence defined by

$$I_k = \int_1^k 1000xe^{-0.1\sqrt{x}}\, dx.$$

Calculate I_k for $k = 10,\ 100,\ 1000,\ 10{,}000$, and so on. Can you determine $L = \lim\limits_{k \to \infty} I_k$?
(b) The difference $E(k) = |L - I_k|$ is the "error." The limit L can also be calculated using integration by parts or by software (be careful that the software gives reasonable answers). Sketch the graph of E to see how quickly the values I_k approach L.
(c) Let

$$a_k = \int_{10^{k-1}}^{10^k} 1000xe^{-0.1\sqrt{x}}\, dx.$$

Show that

$$I_{10^k} = a_0 + a_1 + a_2 + \cdots + a_k.$$

and compare the values for I_k obtained this way with the values calculated directly in part (a).

(d) Explain why a_k might have advantages over I_k in terms of approximating the integral.

10.3 When calculating compound interest, we work with the functions

$$f_r(n) = \left(1 + \frac{r}{n}\right)^n,$$

where r is the annual interest rate and n is the number of compounding periods per year.

(a) Let $r = 0.1$. Find and discuss the meaning of $f_r(4)$, $f_r(12)$, $f_r(365)$, $f_r(8760)$, $f_r(525,600)$, and $f_r(31,536,600)$.

(b) Repeat the calculations in part (a) using $r = 0.01, 0.05, 0.075$, and 0.12.

(c) Mathematically, f_r can be considered for all real numbers and not just for positive integers n. Based on your results in parts (a) and (b), does it appear that

$$\lim_{x \to \infty} f_r(x)$$

exists when r is a given number?

(d) Now let

$$g_n(r) = \left(1 + \frac{r}{n}\right)^n.$$

For $n = 4, 12, 365, 8760, 525,600$, and $31,536,000$, respectively, approximate the following limits

$$\lim_{x \to 0^+} g_n(x).$$

What is the meaning of these limits?

(e) For the values of n and r given in parts (a) and (b), approximate $f_r'(n)$ and $g_n'(r)$. What conclusions can you draw? What are the relationships between these derivatives for a given n and r?

(f) Use a graphing utility to graph the functions $f_r(x)$ and $g_n(x)$ for the values of n and r given in parts (a) and (b). Are these functions increasing or decreasing? Do they have inverse functions?

INFINITE SERIES

■ 11.1 SIGMA NOTATION

In Section 10.2 we defined a sequence as a real-valued function whose domain is the set of positive integers. For example, to indicate the sequence

$$\{1, \tfrac{1}{2}, \tfrac{1}{4}, \tfrac{1}{8}, \ldots\}$$

we would set $a_n = (\tfrac{1}{2})^{n-1}$, $n = 1, 2, 3, \ldots$, and write

$$\{a_1, a_2, a_3, a_4, \ldots\}.$$

In this chapter, however, it will often be convenient to begin a sequence with an index other than 1. So, continuing with the example, we can also set $b_n = (\tfrac{1}{2})^n$, $n = 0, 1, 2, \ldots$, and write

$$\{b_0, b_1, b_2, b_3, \ldots\},$$

thereby beginning with the index 0. In general, we can set $c_n = (\tfrac{1}{2})^{n-p}$, where p is an integer and $n = p, p + 1, p + 2, \ldots$, and write

$$\{c_p, c_{p+1}, c_{p+2}, c_{p+3}, \ldots\},$$

so that we begin with the index p.

The symbol Σ is the capital Greek letter "sigma." We write

(1) $$\sum_{k=0}^{n} a_k$$

(read "the sum of the a sub k from k equals 0 to k equals n") to indicate the sum

$$a_0 + a_1 + \cdots + a_n.$$

693

More generally, if $n \geq m$, we write

(2)
$$\sum_{k=m}^{n} a_k$$

to indicate the sum

$$a_m + a_{m+1} + \cdots + a_n.$$

In (1) and (2) the letter "k" is being used as a "dummy" variable. That is, it can be replaced by any letter not already engaged. For instance,

$$\sum_{i=3}^{7} a_i, \qquad \sum_{j=3}^{7} a_j, \qquad \sum_{k=3}^{7} a_k$$

can all be used to indicate the sum

$$a_3 + a_4 + a_5 + a_6 + a_7.$$

Translating

$$(a_0 + \cdots + a_n) + (b_0 + \cdots + b_n) = (a_0 + b_0) + \cdots + (a_n + b_n),$$

$$\alpha(a_0 + \cdots + a_n) = \alpha a_0 + \cdots + \alpha a_n,$$

$$(a_0 + \cdots + a_m) + (a_{m+1} + \cdots + a_n) = a_0 + \cdots + a_n$$

into the Σ-notation, we have

$$\sum_{k=0}^{n} a_k + \sum_{k=0}^{n} b_k = \sum_{k=0}^{n} (a_k + b_k), \qquad \alpha \sum_{k=0}^{n} a_k = \sum_{k=0}^{n} \alpha a_k,$$

$$\sum_{k=0}^{m} a_k + \sum_{k=m+1}^{n} a_k = \sum_{k=0}^{n} a_k.$$

At times it is convenient to change indices. In this connection note that

$$\sum_{k=j}^{n} a_k = \sum_{i=0}^{n-j} a_{i+j}. \qquad \text{(set } i = k - j\text{)}$$

Both expressions are abbreviations for $a_j + a_{j+1} + \cdots + a_n$.

You can familiarize yourself further with this notation by doing the exercises below, but first one more remark. If all the a_k are equal to some fixed number x, then

$$\sum_{k=0}^{n} a_k \quad \text{can be written} \quad \sum_{k=0}^{n} x.$$

Obviously then

$$\sum_{k=0}^{n} x = \overbrace{x + x + \cdots + x}^{n+1} = (n + 1)x.$$

In particular

$$\sum_{k=0}^{n} 1 = n + 1.$$

EXERCISES 11.1

In Exercises 1–12, evaluate the given expression.

1. $\displaystyle\sum_{k=0}^{2} (3k + 1)$.

2. $\displaystyle\sum_{k=1}^{4} (3k - 1)$.

3. $\displaystyle\sum_{k=0}^{3} 2^k$.

4. $\displaystyle\sum_{k=0}^{3} (-1)^k 2^k$.

5. $\displaystyle\sum_{k=0}^{3} (-1)^k 2^{k+1}$.

6. $\displaystyle\sum_{k=2}^{5} (-1)^{k+1} 2^{k-1}$.

7. $\displaystyle\sum_{k=1}^{4} \frac{1}{2^k}$.

8. $\displaystyle\sum_{k=2}^{5} \frac{1}{k!}$.

9. $\displaystyle\sum_{k=3}^{5} \frac{(-1)^k}{k!}$.

10. $\displaystyle\sum_{k=2}^{4} \frac{1}{3^{k-1}}$.

11. $\displaystyle\sum_{k=0}^{3} (\tfrac{1}{2})^{2k}$.

12. $\displaystyle\sum_{k=1}^{3} (-1)^{k+1}(\tfrac{1}{2})^{2k-1}$.

In Exercises 13–28, express in sigma notation.

13. $1 + 3 + 5 + 7 + \cdots + 21$.

14. $1 - 3 + 5 - 7 + \cdots - 19$.

15. $2 \cdot 1 + 2 \cdot 2 + 2 \cdot 3 + \cdots + 2 \cdot 25$.

16. $1 \cdot 2 + 2 \cdot 3 + 3 \cdot 4 + \cdots + 35 \cdot 36$.

17. $1 - \sqrt{2} + \sqrt{3} - 2 + \sqrt{5} - \cdots + 9$.

18. $\dfrac{\tan 1}{2} + \dfrac{\tan 2}{5} + \dfrac{\tan 3}{10} + \dfrac{\tan 4}{17} + \cdots + \dfrac{\tan 10}{101}$.

19. The lower sum $m_1 \Delta x_1 + m_2 \Delta x_2 + \cdots + m_n \Delta x_n$.

20. The upper sum $M_1 \Delta x_1 + M_2 \Delta x_2 + \cdots + M_n \Delta x_n$.

21. The Riemann sum
$$f(x_1^*)\Delta x_1 + f(x_2^*)\Delta x_2 + \cdots + f(x_n^*)\Delta x_n.$$

22. $a^5 + a^4 b + a^3 b^2 + a^2 b^3 + ab^4 + b^5$.

23. $a^5 - a^4 b + a^3 b^2 - a^2 b^3 + ab^4 - b^5$.

24. $a^n + a^{n-1} b + \cdots + ab^{n-1} + b^n$.

25. $a_0 x^4 + a_1 x^3 + a_2 x^2 + a_3 x + a_4$.

26. $a_0 x^n + a_1 x^{n-1} + \cdots + a_{n-1} x + a_n$.

27. $1 - 2x + 3x^2 - 4x^3 + 5x^4$.

28. $3x - 4x^2 + 5x^3 - 6x^4$.

In Exercises 29–32, write the given sums as $\displaystyle\sum_{k=3}^{10} a_k$ and $\displaystyle\sum_{i=0}^{7} a_{i+3}$:

29. $\dfrac{1}{2^3} + \dfrac{1}{2^4} + \cdots + \dfrac{1}{2^{10}}$.

30. $\dfrac{3^3}{3!} + \dfrac{4^4}{4!} + \cdots + \dfrac{10^{10}}{10!}$.

31. $\dfrac{3}{4} - \dfrac{4}{5} + \cdots - \dfrac{10}{11}$.

32. $\dfrac{1}{3} + \dfrac{1}{5} + \dfrac{1}{7} + \cdots + \dfrac{1}{17}$.

In Exercises 33–36, verify by a change of indices that the two sums are identical.

33. $\displaystyle\sum_{k=2}^{10} \frac{k}{k^2 + 1}$; $\displaystyle\sum_{n=-1}^{7} \frac{n + 3}{n^2 + 6n + 10}$.

34. $\displaystyle\sum_{n=2}^{12} \frac{(-1)^n}{n - 1}$; $\displaystyle\sum_{k=1}^{11} \frac{(-1)^{k+1}}{k}$.

35. $\displaystyle\sum_{k=4}^{25} \frac{1}{k^2 - 9}$; $\displaystyle\sum_{n=7}^{28} \frac{1}{n^2 - 6n}$.

36. $\displaystyle\sum_{k=0}^{15} \frac{3^{2k}}{k!}$; $81 \displaystyle\sum_{n=-2}^{13} \frac{3^{2n}}{(n + 2)!}$.

37. (a) (*Important*) Show that for $x \neq 1$
$$\sum_{k=0}^{n} x^k = \frac{1 - x^{n+1}}{1 - x}.$$

  (b) Determine whether the sequence $a_n = \displaystyle\sum_{k=0}^{n} \frac{1}{3^k}$ converges and, if it does, find the limit.

38. Express $\displaystyle\sum_{k=1}^{n} \frac{a_k}{10^k}$ as a decimal fraction, given that each a_k is an integer from 0 to 9.

39. Let p be a positive integer. Show that, as $n \to \infty$,
$$a_n \to L \qquad \text{iff} \qquad a_{n-p} \to L.$$

40. Show that
$$\sum_{k=1}^{n} \frac{1}{\sqrt{k}} \geq \sqrt{n}.$$

In Exercises 41–44, verify by induction.

41. $\displaystyle\sum_{k=1}^{n} k = \tfrac{1}{2}(n)(n + 1)$.

42. $\displaystyle\sum_{k=1}^{n} (2k - 1) = n^2$.

43. $\displaystyle\sum_{k=1}^{n} k^2 = \tfrac{1}{6}(n)(n + 1)(2n + 1)$.

44. $\displaystyle\sum_{k=1}^{n} k^3 = \left(\sum_{k=1}^{n} k \right)^2$.

In Exercises 45–48, evaluate the sum.

45. $\displaystyle\sum_{k=1}^{10} (2k + 3)$.

46. $\displaystyle\sum_{k=1}^{10} (2k^2 + 3k)$.

47. $\displaystyle\sum_{k=1}^{8} (2k - 1)^2$.

48. $\displaystyle\sum_{k=1}^{n} k(k^2 - 5)$.

■ 11.2 INFINITE SERIES

Introduction; Definitions

While it is possible to add two numbers, three numbers, a hundred numbers, or even a million numbers, how can we attach meaning to the sum of an infinite number of numbers? The theory of infinite series arose from attempts to answer this question. As you might expect, a limit process is involved.

To form an infinite series we begin with an infinite sequence of real numbers: $a_0, a_1, a_2, \ldots$. We can't form the sum of all the a_k (there is an infinite number of them), but we can form the *partial sums*

$$s_0 = a_0 = \sum_{k=0}^{0} a_k,$$

$$s_1 = a_0 + a_1 = \sum_{k=0}^{1} a_k,$$

$$s_2 = a_0 + a_1 + a_2 = \sum_{k=0}^{2} a_k,$$

$$s_3 = a_0 + a_1 + a_2 + a_3 = \sum_{k=0}^{3} a_k,$$

$$\vdots$$

$$s_n = a_0 + a_1 + a_2 + a_3 + \cdots + a_n = \sum_{k=0}^{n} a_k$$

$$\vdots$$

Continuing in this way, we are led to consider the "infinite sum" $\sum_{k=0}^{\infty} a_k$ which is called an *infinite series*. The corresponding sequence $\{s_n\}$ is called the *sequence of partial sums* of the series.

DEFINITION 11.2.1

Given the infinite series $\sum_{k=0}^{\infty} a_k$. If the sequence of partial sums $\{s_n\}$ converges to a finite limit L, then the series $\sum_{k=0}^{\infty} a_k$ is said to *converge* to L, written

$$\sum_{k=0}^{\infty} a_k = L.$$

The number L is called the *sum* of the series. If the sequence of partial sums diverges, then the series $\sum_{k=0}^{\infty} a_k$ *diverges*.

Remark   It is important to note that the sum of a series is not a sum in the ordinary sense.   It is a limit.   ❏

Here are some examples.

Example 1   We begin with the series

$$\sum_{k=0}^{\infty} \frac{1}{(k+1)(k+2)}.$$

To determine whether or not this series converges we must examine the partial sums. Since

$$\frac{1}{(k+1)(k+2)} = \frac{1}{k+1} - \frac{1}{k+2}, \qquad \text{(partial fraction decomposition, see Section 8.6)}$$

you can see that

$$\begin{aligned}
s_n &= \frac{1}{1 \cdot 2} + \frac{1}{2 \cdot 3} + \cdots + \frac{1}{n(n+1)} + \frac{1}{(n+1)(n+2)} \\
&= \left(\frac{1}{1} - \frac{1}{2}\right) + \left(\frac{1}{2} - \frac{1}{3}\right) + \cdots + \left(\frac{1}{n} - \frac{1}{n+1}\right) + \left(\frac{1}{n+1} - \frac{1}{n+2}\right) \\
&= 1 - \frac{1}{2} + \frac{1}{2} - \frac{1}{3} + \cdots + \frac{1}{n} - \frac{1}{n+1} + \frac{1}{n+1} - \frac{1}{n+2}.
\end{aligned}$$

Since all but the first and last terms occur in pairs with opposite signs, the sum collapses to give

$$s_n = 1 - \frac{1}{n+2}.$$

Obviously, as $n \to \infty$, $s_n \to 1$.   This means that the series converges to 1:

$$\sum_{k=0}^{\infty} \frac{1}{(k+1)(k+2)} = 1. \quad ❏$$

Remark   Infinite series with the special property illustrated in Example 1 (that is, except for the first and last term, the terms can be arranged in pairs with opposite signs) are called *telescoping series*.   In general,

$$\sum_{k=p}^{n} \{f(k) - f(k+1)\} = f(p) - f(n+1) \qquad \text{and}$$

$$\sum_{k=p}^{n} \{f(k) - f(k-1)\} = f(n) - f(p-1). \quad \text{(verify these)} \quad ❏$$

Example 2   Here we examine two divergent series

$$\sum_{k=0}^{\infty} 2^k \qquad \text{and} \qquad \sum_{k=1}^{\infty} (-1)^k.$$

The partial sums of the first series take the form

$$s_n = \sum_{k=0}^{n} 2^k = 1 + 2 + \cdots + 2^n.$$

The sequence $\{s_n\}$ is unbounded and therefore divergent (10.3.5). This means that the series diverges.

For the second series we have

$$s_n = -1 \quad \text{if } n \text{ is odd} \qquad \text{and} \qquad s_n = 0 \quad \text{if } n \text{ is even.}$$

The sequence of partial sums looks like this:

$$-1, 0, -1, 0, -1, 0, \ldots .$$

The series diverges since the sequence of partial sums diverges. ❏

Remark Example 2 illustrates two types of divergence. In the first case, $s_n \to \infty$ as $n \to \infty$. The notation $\Sigma_{k=0}^{\infty} a_k = \infty$ is sometimes used to denote this type of divergence. In the second case, s_n oscillates between -1 and 0. ❏

The Geometric Series

Fix a real number x. The sequence $1, x, x^2, x^3, \ldots = \{x^n\}$, $n = 0, 1, 2, 3, \ldots$, is called a *geometric progression*. Concerning convergence, we know that if $|x| < 1$, then $x^n \to 0$ ("special limit" (10.4.2)). If $x = 1$, then we have the constant sequence $1, 1, 1, \ldots$, which clearly converges to 1. Finally, it is easy to show that $\{x^n\}$ is divergent when $x = -1$ (the sequence oscillates between 1 and -1) and when $|x| > 1$ (the sequence is unbounded).

The sums

$$1, \quad 1 + x, \quad 1 + x + x^2, \quad 1 + x + x^2 + x^3, \ldots$$

generated by numbers in geometric progression are the partial sums of what is known as the *geometric series*:

$$\sum_{k=0}^{\infty} x^k.$$

The geometric series arises in so many contexts that it merits special attention. The following result is fundamental:

(11.2.2)

> (i) if $|x| < 1$, then $\displaystyle\sum_{k=0}^{\infty} x^k = \frac{1}{1-x}$;
>
> (ii) if $|x| \geq 1$, then $\displaystyle\sum_{k=0}^{\infty} x^k$ diverges.

PROOF The nth partial sum of the geometric series

$$\sum_{k=0}^{\infty} x^k$$

takes the form

(1)
$$s_n = 1 + x + \cdots + x^n.$$

Multiplication by x gives

$$xs_n = x + x^2 + \cdots + x^{n+1}.$$

Subtracting the second equation from the first, we find that

$$(1 - x)s_n = 1 - x^{n+1}.$$

For $x \neq 1$, this gives

(2)
$$s_n = \frac{1 - x^{n+1}}{1 - x}.$$

If $|x| < 1$, then $x^{n+1} \to 0$ as $n \to \infty$ and thus, by Equation (2),

$$s_n \to \frac{1}{1 - x}.$$

This proves (i).

Now let's prove (ii). For $x = 1$, we use Equation (1) and deduce that $s_n = n + 1$. Obviously, $\{s_n\}$ diverges. For $x \neq 1$ with $|x| \geq 1$, we use Equation (2). Since in this instance $\{x^{n+1}\}$ diverges, $\{s_n\}$ diverges. ❏

You may have seen (11.2.2) before, written as

$$a + ar + ar^2 + \cdots + ar^n + \cdots = \begin{cases} \dfrac{a}{1 - r}, & |r| < 1 \\ \text{diverges}, & |r| \geq 1, \end{cases} \qquad (a \neq 0).$$

Taking $a = 1$ and $r = \frac{1}{2}$, we have

$$\sum_{k=0}^{\infty} \frac{1}{2^k} = \frac{1}{1 - \frac{1}{2}} = 2.$$

Begin the summation at $k = 1$ instead of at $k = 0$, and you see that

(11.2.3)
$$\sum_{k=1}^{\infty} \frac{1}{2^k} = 1.$$

The partial sums of this series

$$s_1 = \tfrac{1}{2},$$
$$s_2 = \tfrac{1}{2} + \tfrac{1}{4} = \tfrac{3}{4},$$
$$s_3 = \tfrac{1}{2} + \tfrac{1}{4} + \tfrac{1}{8} = \tfrac{7}{8},$$
$$s_4 = \tfrac{1}{2} + \tfrac{1}{4} + \tfrac{1}{8} + \tfrac{1}{16} = \tfrac{15}{16},$$
$$s_5 = \tfrac{1}{2} + \tfrac{1}{4} + \tfrac{1}{8} + \tfrac{1}{16} + \tfrac{1}{32} = \tfrac{31}{32},$$

$$\vdots$$

are illustrated in Figure 11.2.1. Each new partial sum lies halfway between the previous partial sum and the number 1.

Figure 11.2.1

The convergence of the geometric series at $x = \frac{1}{10}$ enables us to assign a precise meaning to infinite decimals. Begin with the fact that

$$\sum_{k=0}^{\infty} \left(\frac{1}{10}\right)^k = \sum_{k=0}^{\infty} \frac{1}{10^k} = \frac{1}{1 - \frac{1}{10}} = \frac{10}{9}.$$

This gives

$$\sum_{k=1}^{\infty} \frac{1}{10^k} = \frac{1}{9}$$

and shows that the partial sums

$$s_n = \frac{1}{10} + \frac{1}{10^2} + \cdots + \frac{1}{10^n}$$

are all less than $\frac{1}{9}$. Now take a series of the form

$$\sum_{k=1}^{\infty} \frac{a_k}{10^k} \qquad \text{with} \qquad a_k = 0, 1, \ldots, \text{ or } 9.$$

Its partial sums

$$t_n = \frac{a_1}{10} + \frac{a_2}{10^2} + \cdots + \frac{a_n}{10^n}$$

are all less than 1:

$$t_n = \frac{a_1}{10} + \frac{a_2}{10^2} + \cdots + \frac{a_n}{10^n} \le 9\left(\frac{1}{10} + \frac{1}{10^2} + \cdots + \frac{1}{10^n}\right) = 9s_n < 9\left(\frac{1}{9}\right) = 1.$$

Since $\{t_n\}$ is nondecreasing, as well as bounded above, $\{t_n\}$ is convergent; this means that the series

$$\sum_{k=1}^{\infty} \frac{a_k}{10^k}$$

is convergent. The sum of this series is what we mean by the infinite decimal

$$0.a_1 a_2 a_3 \cdots a_n \cdots.$$

Following are two simple examples that lead naturally to geometric series. You will find more in the exercises.

Example 3 An electric fan is turned off, and the blades begin to lose speed. Given that the blades turn N times during the first second of no power and lose at least $\sigma\%$ of their speed with the passing of each ensuing second, show that the blades cannot turn more than $100N\sigma^{-1}$ times after power shutdown.

SOLUTION The number of turns during the first second of no power is

$$N.$$

The number of turns during the first 2 seconds is at most

$$N + \left(1 - \frac{1}{100}\sigma\right)N;$$

during the first 3 seconds, at most

$$N + \left(1 - \frac{1}{100}\sigma\right)N + \left(1 - \frac{1}{100}\sigma\right)^2 N;$$

and, during the first $n + 1$ seconds, at most

$$N \sum_{k=0}^{n} \left(1 - \frac{1}{100}\sigma\right)^k.$$

The total number of turns after power shutdown cannot exceed the limiting value

$$N \sum_{k=0}^{\infty} \left(1 - \frac{1}{100}\sigma\right)^k.$$

This is a geometric series with $x = 1 - \frac{1}{100}\sigma$. Thus

$$N \sum_{k=0}^{\infty} \left(1 - \frac{1}{100}\sigma\right)^k = N\left[\frac{1}{1 - (1 - \frac{1}{100}\sigma)}\right] = \frac{100N}{\sigma} = 100N\sigma^{-1}. \quad \square$$

Example 4 According to Figure 11.2.2, it is 2 o'clock. At what time between 2 and 3 o'clock will the two hands coincide?

Figure 11.2.2

SOLUTION We will solve the problem by setting up a geometric series. We will then confirm our answer by approaching the problem from a different perspective.

The hour hand travels one-twelfth as fast as the minute hand. At 2 o'clock the minute hand points to 12 and the hour hand points to 2. By the time the minute hand reaches 2, the hour hand points to $2 + \frac{1}{6}$. By the time the minute hand reaches $2 + \frac{1}{6}$, the hour hand points to

$$2 + \frac{1}{6} + \frac{1}{6 \cdot 12}.$$

By the time the minute hand reaches

$$2 + \frac{1}{6} + \frac{1}{6 \cdot 12},$$

the hour hand points to

$$2 + \frac{1}{6} + \frac{1}{6 \cdot 12} + \frac{1}{6 \cdot 12^2}$$

and so on. In general, by the time the minute hand reaches

$$2 + \frac{1}{6} + \frac{1}{6 \cdot 12} + \cdots + \frac{1}{6 \cdot 12^{n-1}} = 2 + \frac{1}{6} \sum_{k=0}^{n-1} \frac{1}{12^k},$$

the hour hand points to

$$2 + \frac{1}{6} + \frac{1}{6 \cdot 12} + \cdots + \frac{1}{6 \cdot 12^{n-1}} + \frac{1}{6 \cdot 12^n} = 2 + \frac{1}{6} \sum_{k=0}^{n} \frac{1}{12^k}.$$

The two hands coincide when they both point to the limiting value

$$2 + \frac{1}{6} \sum_{k=0}^{\infty} \frac{1}{12^k} = 2 + \frac{1}{6} \left(\frac{1}{1 - \frac{1}{12}} \right) = 2 + \frac{2}{11}.$$

This happens at $2 + \frac{2}{11}$ o'clock, approximately 10 minutes and 55 seconds after 2.

We can confirm this as follows. Suppose that the two hands meet at hour $2 + x$. The hour hand will have moved a distance x and the minute hand a distance $2 + x$. Since the minute hand moves 12 times as fast as the hour hand, we have

$$12x = 2 + x, \qquad 11x = 2, \qquad \text{and thus} \qquad x = \frac{2}{11}. \quad \square$$

We will return to the geometric series later. Right now we turn our attention to series in general.

Some Basic Results

THEOREM 11.2.4

1. If $\displaystyle\sum_{k=0}^{\infty} a_k$ converges and $\displaystyle\sum_{k=0}^{\infty} b_k$ converges, then $\displaystyle\sum_{k=0}^{\infty} (a_k + b_k)$ converges.

Moreover, if $\displaystyle\sum_{k=0}^{\infty} a_k = L$ and $\displaystyle\sum_{k=0}^{\infty} b_k = M$, then $\displaystyle\sum_{k=0}^{\infty} (a_k + b_k) = L + M.$

2. If $\displaystyle\sum_{k=0}^{\infty} a_k$ converges, then $\displaystyle\sum_{k=0}^{\infty} \alpha a_k$ converges for each real number α.

Moreover, if $\displaystyle\sum_{k=0}^{\infty} a_k = L$, then $\displaystyle\sum_{k=0}^{\infty} \alpha a_k = \alpha L.$

PROOF Let

$$s_n = \sum_{k=0}^{n} a_k, \qquad t_n = \sum_{k=0}^{n} b_k, \qquad u_n = \sum_{k=0}^{n} (a_k + b_k), \qquad v_n = \sum_{k=0}^{n} \alpha a_k.$$

Note that

$$u_n = s_n + t_n \qquad \text{and} \qquad v_n = \alpha s_n.$$

If $s_n \to L$ and $t_n \to M$, then

$$u_n \to L + M \qquad \text{and} \qquad v_n \to \alpha L. \qquad \text{(Theorem 10.3.7)} \quad \square$$

THEOREM 11.2.5

Let j be a positive integer. The series $\displaystyle\sum_{k=0}^{\infty} a_k$ converges iff the series $\displaystyle\sum_{k=j}^{\infty} a_k$ converges. Moreover, if $\displaystyle\sum_{k=0}^{\infty} a_k = L$, then $\displaystyle\sum_{k=j}^{\infty} a_k = L - (a_0 + a_1 + a_2 + \cdots + a_{j-1})$; or if $\displaystyle\sum_{k=j}^{\infty} a_k = M$, then $\displaystyle\sum_{k=0}^{\infty} a_k = M + (a_0 + a_1 + a_2 + \cdots + a_{j-1})$.

It is important to understand what this theorem means, namely, the convergence (or divergence) of an infinite series is not affected by where you start the summation. In the case of a convergent series, however, the limit (sum) does depend upon where you begin the summation. The proof of the theorem is left to you as an exercise.

THEOREM 11.2.6

The *kth term* of a convergent series tends to 0; namely,

$$\text{if } \sum_{k=0}^{\infty} a_k \text{ converges, } \quad \text{then } a_k \to 0 \text{ as } k \to \infty.$$

PROOF To say that the series converges is to say that the sequence of partial sums converges to some number L:

$$s_n = \sum_{k=0}^{n} a_k \to L.$$

Obviously, then, $s_{n-1} \to L$. Since $a_n = s_n - s_{n-1}$, we have $a_n \to L - L = 0$. A change in notation gives $a_k \to 0$. ❑

The next result is an obvious, but important, consequence of Theorem 11.2.6.

THEOREM 11.2.7 A DIVERGENCE TEST

$$\text{If } a_k \not\to 0 \text{ as } k \to \infty, \quad \text{then } \sum_{k=0}^{\infty} a_k \text{ diverges.}$$

Example 5

(a) Since $\dfrac{k}{k+1} \not\to 0$ as $k \to \infty$, the series

$$\sum_{k=0}^{\infty} \frac{k}{k+1} = 0 + \frac{1}{2} + \frac{2}{3} + \frac{3}{4} + \frac{4}{5} + \cdots \quad \text{diverges.}$$

(b) Since $\sin k \not\to 0$ as $k \to \infty$, the series

$$\sum_{k=0}^{\infty} \sin k = \sin 0 + \sin 1 + \sin 2 + \sin 3 + \cdots \quad \text{diverges.} \quad \square$$

CAUTION Theorem 11.2.6 does *not* say that, if $a_k \to 0$, then $\sum_{k=0}^{\infty} a_k$ converges. There are divergent series for which $a_k \to 0$. $\quad \square$

Example 6 In the case of

$$\sum_{k=1}^{\infty} \frac{1}{\sqrt{k}} = \frac{1}{\sqrt{1}} + \frac{1}{\sqrt{2}} + \frac{1}{\sqrt{3}} + \frac{1}{\sqrt{4}} + \cdots$$

we have

$$a_k = \frac{1}{\sqrt{k}} \to 0 \quad \text{as } k \to \infty,$$

but, since

$$s_n = \frac{1}{\sqrt{1}} + \frac{1}{\sqrt{2}} + \cdots + \frac{1}{\sqrt{n}} \geq \underbrace{\frac{1}{\sqrt{n}} + \frac{1}{\sqrt{n}} + \cdots + \frac{1}{\sqrt{n}}}_{n \text{ terms}} = \frac{n}{\sqrt{n}} = \sqrt{n},$$

the sequence of partial sums is unbounded, and therefore the series diverges. $\quad \square$

EXERCISES 11.2

In Exercises 1–18, find the sum of the series.

1. $\displaystyle\sum_{k=3}^{\infty} \frac{1}{(k+1)(k+2)}$.

2. $\displaystyle\sum_{k=0}^{\infty} \frac{1}{(k+3)(k+4)}$.

3. $\displaystyle\sum_{k=1}^{\infty} \frac{1}{2k(k+1)}$.

4. $\displaystyle\sum_{k=3}^{\infty} \frac{1}{k^2 - k}$.

5. $\displaystyle\sum_{k=1}^{\infty} \frac{1}{k(k+3)}$.

6. $\displaystyle\sum_{k=0}^{\infty} \frac{1}{(k+1)(k+3)}$.

7. $\displaystyle\sum_{k=0}^{\infty} \frac{3}{10^k}$.

8. $\displaystyle\sum_{k=0}^{\infty} \frac{12}{100^k}$.

9. $\displaystyle\sum_{k=0}^{\infty} \frac{67}{1000^k}$.

10. $\displaystyle\sum_{k=0}^{\infty} \frac{(-1)^k}{5^k}$.

11. $\displaystyle\sum_{k=0}^{\infty} \left(\frac{3}{4}\right)^k$.

12. $\displaystyle\sum_{k=0}^{\infty} \frac{3^k + 4^k}{5^k}$.

13. $\displaystyle\sum_{k=0}^{\infty} \frac{1 - 2^k}{3^k}$.

14. $\displaystyle\sum_{k=0}^{\infty} \left(\frac{25}{10^k} - \frac{6}{100^k}\right)$.

15. $\displaystyle\sum_{k=3}^{\infty} \frac{1}{2^{k-1}}$.

16. $\displaystyle\sum_{k=0}^{\infty} \frac{1}{2^{k+3}}$.

17. $\displaystyle\sum_{k=0}^{\infty} \frac{2^{k+3}}{3^k}$.

18. $\displaystyle\sum_{k=2}^{\infty} \frac{3^{k-1}}{4^{3k+1}}$.

In Exercises 19–26, write the decimal fraction as an infinite series and express the sum as the quotient of two integers.

19. $0.\overline{777}\ldots$

20. $0.\overline{999}\ldots$

21. $0.\overline{2424}\ldots$

22. $0.8\overline{989}\ldots$

23. $0.11\overline{2112112}\ldots$

24. $0.315\overline{315315}\ldots$

25. $0.62\overline{4545}\ldots$

26. $0.11\overline{2019019}\ldots$

27. Using series, show that every repeating decimal represents a rational number (the quotient of two integers).

28. Prove Theorem 11.2.5.

In Exercises 29 and 30, derive the indicated result from the geometric series.

29. $\displaystyle\sum_{k=0}^{\infty} (-1)^k x^k = \frac{1}{1+x}, \quad |x| < 1.$

30. $\displaystyle\sum_{k=0}^{\infty} (-1)^k x^{2k} = \frac{1}{1+x^2}, \quad |x| < 1.$

In Exercises 31–36, find a series expansion for the given expression.

31. $\displaystyle\frac{x}{1-x}$ for $|x| < 1$.

32. $\displaystyle\frac{x}{1+x}$ for $|x| < 1$.

33. $\dfrac{x}{1 + x^2}$ for $|x| < 1$. **34.** $\dfrac{1}{4 - x^2}$ for $|x| < 2$.

35. $\dfrac{1}{1 + 4x^2}$ for $|x| < \dfrac{1}{2}$.

36. $\dfrac{x^2}{1 - x}$ for $|x| < 1$.

In Exercises 37–40, show that the given series diverges.

37. $1 + \dfrac{3}{2} + \dfrac{9}{4} + \dfrac{27}{8} + \dfrac{81}{16} + \cdots$.

38. $\displaystyle\sum_{k=0}^{\infty} (-1)^k$. **39.** $\displaystyle\sum_{k=1}^{\infty} \left(\dfrac{k+1}{k}\right)^k$.

40. $\displaystyle\sum_{k=2}^{\infty} \dfrac{k^{k-2}}{3^k}$.

41. At some time between 4 and 5 o'clock the minute hand is directly above the hour hand. Express this time as a geometric series. What is the sum of this series?

42. Given that a ball dropped to the floor rebounds to a height proportional to the height from which it is dropped, find the total distance traveled by a ball dropped from a height of 6 feet if it rebounds initially to a height of 3 feet.

43. Exercise 42 under the supposition that the ball rebounds initially to a height of 2 feet.

44. In the setting of Exercise 42, to what height does the ball rebound initially if the total distance traveled by the ball is 21 feet?

45. How much money must you deposit at $r\%$ interest compounded annually to enable your descendants to withdraw n_1 dollars at the end of the first year, n_2 dollars at the end of the second year, n_3 dollars at the end of the third year, and so on in perpetuity? Assume that the sequence $\{n_k\}$ is bounded, $n_k \leq N$ for all k, and express your answer as an infinite series.

46. Sum the series you obtained in Exercise 45 setting
(a) $r = 5$, $n_k = 5000(\frac{1}{2})^{k-1}$.
(b) $r = 6$, $n_k = 1000(0.8)^{k-1}$.
(c) $r = 5$, $n_k = N$.

47. Suppose that 90% of each dollar is recirculated into the economy. That is, suppose that when a dollar is put into circulation, 90% of it is spent; then 90% of that is spent, and so on. What is the total economic value of the dollar?

48. Consider the following sequence of steps. First, take the unit interval $[0,1]$ and delete the open interval $(\frac{1}{3}, \frac{2}{3})$. Next delete the two open intervals $(\frac{1}{9}, \frac{2}{9})$ and $(\frac{7}{9}, \frac{8}{9})$ from the two intervals that remain after the first step. For the third step, delete the middle thirds from the four intervals that remain after the second step. Continue on in this

manner. What is the sum of the lengths of the intervals that have been deleted? The set that remains after all of the "middle thirds" have been deleted is called the *Cantor middle third set*. Can you name some points that are in the Cantor set?

49. Start with a square whose sides are four units long. Join the midpoints of the sides of the square to form a second square inside the first. Then join the midpoints of the sides of the second square to form a third square, and so on. See the figure. Find the sum of the areas of the squares.

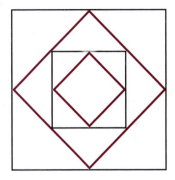

50. (a) Prove that if the series $\Sigma\, a_k$ converges and the series $\Sigma\, b_k$ diverges, then the series $\Sigma\, (a_k + b_k)$ diverges.
(b) Give examples to show that if $\Sigma\, a_k$ and $\Sigma\, b_k$ both diverge, then each of the series

$$\sum (a_k + b_k) \quad \text{and} \quad \sum (a_k - b_k)$$

may either converge or diverge.

51. Let $\Sigma_{k=0}^{\infty}\, a_k$ be a convergent series and let $R_n = \Sigma_{k=n+1}^{\infty}\, a_k$. Prove that $R_n \to 0$ as $n \to \infty$. Note that if s_n is the nth partial sum of the series, then $\Sigma_{k=0}^{\infty}\, a_k = s_n + R_n$; R_n is called the *remainder*.

52. (a) Prove that if $\Sigma_{k=0}^{\infty}\, a_k$ is a convergent series and $a_k \neq 0$ for all k, then $\Sigma_{k=0}^{\infty}\, (1/a_k)$ is divergent.
(b) Suppose that $a_k > 0$ for all k and $\Sigma_{k=0}^{\infty}\, a_k$ diverges. Show by means of examples that $\Sigma_{k=0}^{\infty}\, (1/a_k)$ may either converge or diverge.

53. Let $\{s_n\}$ be the sequence of partial sums of the series $\Sigma_{k=0}^{\infty}\, (-1)^k$. Find a formula for s_n. HINT: Find $s_0, s_1, s_2, \ldots$ and note the pattern.

54. Repeat Exercise 53 for the series $\displaystyle\sum_{k=1}^{\infty} \ln\left(\dfrac{k}{k+1}\right)$.

55. Show that

$$\sum_{k=1}^{\infty} \ln\left(\dfrac{k+1}{k}\right) \quad \text{diverges,} \quad \text{even though}$$

$$\lim_{k \to \infty} \ln\left(\dfrac{k+1}{k}\right) = 0.$$

56. Show that

$$\sum_{k=1}^{\infty} \left(\frac{k}{k+1}\right)^k \quad \text{diverges.}$$

57. (a) Let $\{d_k\}$ be a sequence of real numbers that converges to 0. Show that

$$\sum_{k=1}^{\infty} (d_k - d_{k+1}) = d_1.$$

(b) Sum the following series:

(i) $\displaystyle\sum_{k=1}^{\infty} \frac{\sqrt{k+1} - \sqrt{k}}{\sqrt{k(k+1)}}.$ (ii) $\displaystyle\sum_{k=1}^{\infty} \frac{2k+1}{2k^2(k+1)^2}.$

58. Show that

$$\sum_{k=1}^{\infty} kx^{k-1} = \frac{1}{(1-x)^2} \quad \text{for } |x| < 1.$$

HINT: Verify that s_n, the nth partial sum, satisfies the identity

$$(1-x)^2 s_n = 1 - (n+1)x^n + nx^{n+1}.$$

▶ **Speed of Convergence** Suppose that $\sum_{k=0}^{\infty} a_k$ is a convergent series with sum L and let $\{s_n\}$ be its sequence of partial sums. It follows from Exercise 51 that $|L - s_n| = |R_n|$. In Exercises 59–62, find the smallest integer N such that $|L - s_N| < 0.0001$.

59. $\displaystyle\sum_{k=0}^{\infty} \frac{1}{4^k}.$ **60.** $\displaystyle\sum_{k=0}^{\infty} (0.9)^k.$

61. $\displaystyle\sum_{k=1}^{\infty} \frac{1}{k(k+2)}.$ **62.** $\displaystyle\sum_{k=0}^{\infty} \left(\frac{2}{3}\right)^k.$

63. Given the geometric series $\sum_{k=0}^{\infty} x^k$, with $|x| < 1$, and a positive number ϵ. Determine the smallest positive integer N such that $|L - s_N| < \epsilon$, where L is the sum of the series and s_n is the nth partial sum.

■ **11.3 THE INTEGRAL TEST; COMPARISON THEOREMS**

Here and in the next section we direct our attention to *series with nonnegative terms*: $a_k \geq 0$ for all k. The significant feature of a series with nonnegative terms is the fact that its sequence of partial sums is nondecreasing:

$$s_{n+1} = \sum_{k=0}^{n+1} a_k = \sum_{k=0}^{n} a_k + a_{n+1} \geq \sum_{k=0}^{n} a_k = s_n, \qquad n = 0, 1, 2, \ldots.$$

The following theorem is fundamental.

THEOREM 11.3.1

A series with nonnegative terms converges iff the sequence of partial sums is bounded.

PROOF Assume that the series converges. Then the sequence of partial sums is convergent and therefore bounded (Theorem 10.3.4).

Suppose now that the sequence of partial sums is bounded. Since the terms are nonnegative, the sequence is nondecreasing. By being bounded and nondecreasing, the sequence of partial sums converges (Theorem 10.3.6). This means that the series converges. ❑

The convergence or divergence of a series can sometimes be deduced from the convergence or divergence of a closely related improper integral.

THEOREM 11.3.2 THE INTEGRAL TEST

If f is continuous, decreasing, and positive on $[1, \infty)$, then

$$\sum_{k=1}^{\infty} f(k) \quad \text{converges} \qquad \text{iff} \qquad \int_{1}^{\infty} f(x)\, dx \quad \text{converges.}$$

PROOF In Exercise 63, Section 10.7, you were asked to show that with f continuous, decreasing, and positive on $[1, \infty)$

$$\int_{1}^{\infty} f(x)\, dx \quad \text{converges} \qquad \text{iff} \qquad \text{the sequence} \quad \left\{ \int_{1}^{n} f(x)\, dx \right\} \quad \text{converges.}$$

We assume this result and base our proof on the behavior of the sequence of integrals. To visualize our argument see Figure 11.3.1.

Figure 11.3.1

Since f decreases on the interval $[1, n]$,

$$f(2) + \cdots + f(n) \quad \text{is a lower sum for } f \text{ on } [1, n]$$

and

$$f(1) + \cdots + f(n-1) \quad \text{is an upper sum for } f \text{ on } [1, n].$$

Consequently

$$(1) \quad f(2) + \cdots + f(n) \leq \int_{1}^{n} f(x)\, dx \quad \text{and} \quad \int_{1}^{n} f(x)\, dx \leq f(1) + \cdots + f(n-1).$$

If the sequence of integrals converges, it is bounded. By the first inequality the sequence of partial sums is bounded and the series is therefore convergent.

Suppose now that the sequence of integrals diverges. Since f is positive, the sequence of integrals increases:

$$\int_{1}^{n} f(x)\, dx < \int_{1}^{n+1} f(x)\, dx.$$

Since this sequence diverges, it must be unbounded. By the second inequality, the sequence of partial sums must be unbounded and the series diverges. ❏

Remark The inequalities established in the proof of Theorem 11.3.2 lead to bounds on the sum of the infinite series

$$\sum_{k=1}^{\infty} f(k),$$

where f is continuous, decreasing, and positive on $[1, \infty)$. In particular, it follows from the second inequality in (1) that

$$\int_1^\infty f(x)\, dx \le \sum_{k=1}^\infty f(k),$$

and from the first inequality in (1),

$$\sum_{k=1}^\infty f(k) \le f(1) + \int_1^\infty f(x)\, dx.$$

Combining these two inequalities, we have

$$\int_1^\infty f(x)\, dx \le \sum_{k=1}^\infty f(k) \le f(1) + \int_1^\infty f(x)\, dx.$$

These inequalities also make clear the relation between the convergence of the infinite series and the convergence of the corresponding improper integral. ❏

Applying the Integral Test

Example 1 (*The Harmonic Series*)

(11.3.3)
$$\sum_{k=1}^\infty \frac{1}{k} = 1 + \frac{1}{2} + \frac{1}{3} + \frac{1}{4} + \cdots \quad \text{diverges.}$$

PROOF The function $f(x) = 1/x$ is continuous, decreasing, and positive on $[1, \infty)$. We know that

$$\int_1^\infty \frac{dx}{x} \quad \text{diverges.} \tag{10.7.1}$$

By the integral test

$$\sum_{k=1}^\infty \frac{1}{k} \quad \text{diverges.} \quad ❏$$

The next example gives a more general result.

Example 2 (*The p-series*)

(11.3.4)
$$\sum_{k=1}^\infty \frac{1}{k^p} = 1 + \frac{1}{2^p} + \frac{1}{3^p} + \frac{1}{4^p} + \cdots \quad \text{converges} \quad \text{iff} \quad p > 1.$$

PROOF If $p \le 0$, then each term of the series is greater than or equal to 1. Therefore, the kth term does not have limit 0 and so, by the divergence test (11.2.7) the series cannot converge. (See, also, the remark following Example 2 in Section 10.7.) We assume, therefore, that $p > 0$. The function $f(x) = 1/x^p$ is then continuous, decreasing, and positive on $[1, \infty)$. Thus by the integral test

$$\sum_{k=1}^{\infty} \frac{1}{k^p} \quad \text{converges} \quad \text{iff} \quad \int_1^{\infty} \frac{dx}{x^p} \quad \text{converges.}$$

Earlier you saw that

$$\int_1^{\infty} \frac{dx}{x^p} \quad \text{converges} \quad \text{iff} \quad p > 1. \qquad\qquad (10.7.1)$$

It follows that

$$\sum_{k=1}^{\infty} \frac{1}{k^p} \quad \text{converges} \quad \text{iff} \quad p > 1. \quad \square$$

Example 3 Here we show that the series

$$\sum_{k=1}^{\infty} \frac{1}{k \ln (k + 1)} = \frac{1}{\ln 2} + \frac{1}{2 \ln 3} + \frac{1}{3 \ln 4} + \cdots$$

diverges. We begin by setting

$$f(x) = \frac{1}{x \ln (x + 1)}.$$

Since f is continuous, decreasing, and positive on $[1, \infty)$, we can use the integral test. Note that

$$\int_1^b \frac{dx}{x \ln (x + 1)} > \int_1^b \frac{dx}{(x + 1) \ln (x + 1)}$$

$$= \left[\ln (\ln (x + 1)) \right]_1^b = \ln (\ln (b + 1)) - \ln (\ln 2).$$

As $b \to \infty$, $\ln (\ln (b + 1)) \to \infty$. This shows that

$$\int_1^{\infty} \frac{dx}{x \ln (x + 1)} \quad \text{diverges.}$$

It follows that the series diverges. ❑

Remark on Notation You have seen that for each $j \geq 0$

$$\sum_{k=0}^{\infty} a_k \quad \text{converges} \quad \text{iff} \quad \sum_{k=j}^{\infty} a_k \quad \text{converges}$$

(Theorem 11.2.5). This tells you that, in determining whether or not a series converges, it does not matter where we begin the summation. Where detailed indexing would contribute nothing, we will omit it and write $\Sigma \, a_k$ without specifying where the summation begins. For instance, it makes sense to say that

$$\sum \frac{1}{k^2} \quad \text{converges} \quad \text{and} \quad \sum \frac{1}{k} \quad \text{diverges}$$

without specifying where we begin the summation. ❑

The convergence or divergence of a series with nonnegative terms can sometimes be deduced by comparison with a series of known behavior.

> ### THEOREM 11.3.5 THE BASIC COMPARISON TEST
>
> Let $\Sigma\, a_k$ be a series with nonnegative terms.
>
> **(i)** $\Sigma\, a_k$ converges if there exists a convergent series $\Sigma\, c_k$ with nonnegative terms such that $a_k \leq c_k$ for all k sufficiently large;
>
> **(ii)** $\Sigma\, a_k$ diverges if there exists a divergent series $\Sigma\, d_k$ with nonnegative terms such that $d_k \leq a_k$ for all k sufficiently large.

PROOF The proof is just a matter of noting that, in the first instance, the partial sums of $\Sigma\, a_k$ are bounded and, in the second instance, unbounded. The details are left to you. ❑

Example 4

(a) $\displaystyle\sum \frac{1}{2k^3 + 1}$ converges by comparison with $\displaystyle\sum \frac{1}{k^3}$:

$$\frac{1}{2k^3 + 1} < \frac{1}{k^3} \qquad \text{and} \qquad \sum \frac{1}{k^3} \text{ converges.}$$

(b) $\displaystyle\sum \frac{1}{3k + 1}$ diverges by comparison with $\displaystyle\sum \frac{1}{3(k + 1)}$:

$$\frac{1}{3(k + 1)} < \frac{1}{3k + 1} \qquad \text{and} \qquad \sum \frac{1}{3(k + 1)} = \frac{1}{3}\sum \frac{1}{k + 1} \qquad \text{diverges}$$

(it's the series $\Sigma 1/k$ with a change of index).

(c) $\displaystyle\sum \frac{k^3}{k^5 + 5k^4 + 7}$ converges by comparison with $\displaystyle\sum \frac{1}{k^2}$:

$$\frac{k^3}{k^5 + 5k^4 + 7} < \frac{k^3}{k^5} = \frac{1}{k^2} \qquad \text{and} \qquad \sum \frac{1}{k^2} \text{ converges.} \quad ❑$$

Example 5 Show that

$$\sum \frac{1}{\ln(k + 6)} \quad \text{diverges.}$$

SOLUTION We know that as $k \to \infty$

$$\frac{\ln k}{k} \to 0. \qquad\qquad \text{(see 10.4.5)}$$

It follows that

$$\frac{\ln(k + 6)}{k + 6} \to 0$$

and therefore that

$$\frac{\ln(k + 6)}{k} = \frac{\ln(k + 6)}{k + 6}\left(\frac{k + 6}{k}\right) \to 0.$$

Thus, for k sufficiently large,

$$\frac{\ln(k + 6)}{k} < 1,$$

and so

$$\ln(k + 6) < k \quad \text{and} \quad \frac{1}{k} < \frac{1}{\ln(k + 6)}.$$

Since

$$\sum \frac{1}{k} \quad \text{diverges,}$$

we can conclude that

$$\sum \frac{1}{\ln(k + 6)} \quad \text{diverges.} \quad ❏$$

Remark Another way to show that $\ln(k + 6) < k$ for large k is to examine the function $f(x) = x - \ln(x + 6)$. At $x = 3$ the function is positive:

$$f(3) = 3 - \ln 9 \cong 3 - 2.197 > 0.$$

Since

$$f'(x) = 1 - \frac{1}{x + 6} > 0 \quad \text{for all } x > 0,$$

$f(x) > 0$ for all $x \geq 3$. It follows that

$$\ln(x + 6) < x \quad \text{for all } x \geq 3. \quad ❏$$

The basic comparison test is algebraic in nature; it requires that certain inequalities hold. To apply the test to a series $\Sigma\, a_k$ you must show that the terms a_k are smaller than the corresponding terms c_k of a known convergent series to establish convergence, or larger than the terms d_k of a known divergent series to establish divergence. However, if the terms a_k are larger than the terms c_k, or smaller than the terms d_k, then the comparison test does not apply. For example, consider the series

$$\sum_{k=2}^{\infty} \frac{1}{k^3 - 1}.$$

It would be natural to compare this series with the convergent series

$$\sum_{k=2}^{\infty} \frac{1}{k^3},$$

but, unfortunately, the inequalities "go the wrong way:"

$$\frac{1}{k^3 - 1} > \frac{1}{k^3} \quad \text{for all } k \geq 2.$$

Our next theorem is a more sophisticated comparison test. It is analytic in the sense that it involves the evaluation of a limit.

THEOREM 11.3.6 THE LIMIT COMPARISON TEST

Let $\Sigma \, a_k$ and $\Sigma \, b_k$ be series with positive terms. If $a_k/b_k \to L$, where L is some *positive* number, then either both series converge or both series diverge.

PROOF   Choose ϵ between 0 and L. Since $a_k/b_k \to L$, we know that for all k sufficiently large (for all k greater than some k_0)

$$\left| \frac{a_k}{b_k} - L \right| < \epsilon.$$

For such k we have

$$L - \epsilon < \frac{a_k}{b_k} < L + \epsilon$$

and thus

$$(L - \epsilon)b_k < a_k < (L + \epsilon)b_k.$$

This last inequality is what we need:

if $\Sigma \, a_k$ converges, then $\Sigma \, (L - \epsilon)b_k$ converges, and thus $\Sigma \, b_k$ converges;
if $\Sigma \, b_k$ converges, then $\Sigma \, (L + \epsilon)b_k$ converges, and thus $\Sigma \, a_k$ converges.   ❏

To apply the limit comparison theorem to a series $\Sigma \, a_k$, we must first find a series $\Sigma \, b_k$ of known behavior for which a_k/b_k converges to a positive number. To complete the example started earlier,

$$\sum_{k=2}^{\infty} \frac{1}{k^3 - 1}$$

converges since

$$\sum_{k=2}^{\infty} \frac{1}{k^3} \quad \text{converges} \qquad \text{and} \qquad \left(\frac{1}{k^3 - 1} \right) \div \left(\frac{1}{k^3} \right) = \frac{k^3}{k^3 - 1} \to 1 \quad \text{as } k \to \infty.$$

Example 6   Determine whether the series

$$\sum \frac{3k^2 + 2k + 1}{k^3 + 1}$$

converges or diverges.

SOLUTION   For large k, the terms with the highest powers of k dominate. Here $3k^2$ dominates the numerator and k^3 dominates the denominator. Thus, for large k,

$$\frac{3k^2 + 2k + 1}{k^3 + 1} \quad \text{differs little from} \quad \frac{3k^2}{k^3} = \frac{3}{k}.$$

Since

$$\frac{3k^2 + 2k + 1}{k^3 + 1} \div \frac{3}{k} = \frac{3k^3 + 2k^2 + k}{3k^3 + 3} = \frac{1 + 2/(3k) + 1/(3k^2)}{1 + 1/k^3} \to 1$$

and

$$\sum \frac{3}{k} = 3 \sum \frac{1}{k} \quad \text{diverges,}$$

we know that the series diverges. ❏

Example 7 Determine whether the series

$$\sum \frac{5\sqrt{k} + 100}{2k^2\sqrt{k} + 9\sqrt{k}}$$

converges or diverges.

SOLUTION For large values of k, $5\sqrt{k}$ dominates the numerator and $2k^2\sqrt{k}$ dominates the denominator. Thus, for such k,

$$\frac{5\sqrt{k} + 100}{2k^2\sqrt{k} + 9\sqrt{k}} \quad \text{differs little from} \quad \frac{5\sqrt{k}}{2k^2\sqrt{k}} = \frac{5}{2k^2}.$$

Since

$$\frac{5\sqrt{k} + 100}{2k^2\sqrt{k} + 9\sqrt{k}} \div \frac{5}{2k^2} = \frac{10k^2\sqrt{k} + 200k^2}{10k^2\sqrt{k} + 45\sqrt{k}} = \frac{1 + 20/\sqrt{k}}{1 + 9/2k^2} \to 1 \quad \text{as } k \to \infty$$

and

$$\sum \frac{5}{2k^2} = \frac{5}{2} \sum \frac{1}{k^2} \quad \text{converges,}$$

the series converges. ❏

Example 8 Determine whether the series

$$\sum \sin \frac{\pi}{k}$$

converges or diverges.

SOLUTION Recall that

$$\text{as } x \to 0, \quad \frac{\sin x}{x} \to 1. \qquad\qquad (2.5.5)$$

As $k \to \infty$, $\pi/k \to 0$ and thus

$$\frac{\sin (\pi/k)}{\pi/k} \to 1.$$

Since $\sum \pi/k$ diverges, $\sum \sin (\pi/k)$ diverges. ❏

Remark The question of what we can and cannot conclude by limit comparison if $a_k/b_k \to 0$ or if $a_k/b_k \to \infty$ is taken up in Exercises 45 and 46. ❏

EXERCISES 11.3

In Exercises 1–34, determine whether the series converges or diverges.

1. $\Sigma \dfrac{k}{k^3 + 1}$.

2. $\Sigma \dfrac{1}{3k + 2}$.

3. $\Sigma \dfrac{1}{(2k + 1)^2}$.

4. $\Sigma \dfrac{\ln k}{k}$.

5. $\Sigma \dfrac{1}{\sqrt{k + 1}}$.

6. $\Sigma \dfrac{1}{k^2 + 1}$.

7. $\Sigma \dfrac{1}{\sqrt{2k^2 - k}}$.

8. $\Sigma \left(\dfrac{5}{2}\right)^{-k}$.

9. $\Sigma \dfrac{\tan^{-1}k}{1 + k^2}$.

10. $\Sigma \dfrac{\ln k}{k^3}$.

11. $\Sigma \dfrac{1}{k^{2/3}}$.

12. $\Sigma \dfrac{1}{(k + 1)(k + 2)(k + 3)}$.

13. $\Sigma \left(\dfrac{3}{4}\right)^{-k}$.

14. $\Sigma \dfrac{1}{1 + 2\ln k}$.

15. $\Sigma \dfrac{\ln \sqrt{k}}{k}$.

16. $\Sigma \dfrac{2}{k(\ln k)^2}$.

17. $\Sigma \dfrac{1}{2 + 3^{-k}}$.

18. $\Sigma \dfrac{7k + 2}{2k^5 + 7}$.

19. $\Sigma \dfrac{2k + 5}{5k^3 + 3k^2}$.

20. $\Sigma \dfrac{k^4 - 1}{3k^2 + 5}$.

21. $\Sigma \dfrac{1}{k \ln k}$.

22. $\Sigma \dfrac{1}{2^{k+1} - 1}$.

23. $\Sigma \dfrac{k^2}{k^4 - k^3 + 1}$.

24. $\Sigma \dfrac{k^{3/2}}{k^{5/2} + 2k - 1}$.

25. $\Sigma \dfrac{2k + 1}{\sqrt{k^4 + 1}}$.

26. $\Sigma \dfrac{2k + 1}{\sqrt{k^3 + 1}}$.

27. $\Sigma \dfrac{2k + 1}{\sqrt{k^5 + 1}}$.

28. $\Sigma \dfrac{1}{\sqrt{2k(k + 1)}}$.

29. $\Sigma k e^{-k^2}$.

30. $\Sigma k^2 2^{-k^3}$.

31. $\Sigma \dfrac{2 + \sin k}{k^2}$.

32. $\Sigma \dfrac{2 + \cos k}{\sqrt{k + 1}}$.

33. $\Sigma \dfrac{1}{1 + 2 + 3 + \cdots + k}$.

34. $\Sigma \dfrac{k}{1 + 2^2 + 3^2 + \cdots + k^2}$.

35. Find the values of p for which the series $\displaystyle\sum_{k=2}^{\infty} \dfrac{1}{k(\ln k)^p}$ converges.

36. Find the values of p for which the series $\displaystyle\sum_{k=2}^{\infty} \dfrac{\ln k}{k^p}$ converges.

37. (a) Prove that $\displaystyle\sum_{k=0}^{\infty} e^{-\alpha k}$ converges for any $\alpha > 0$.

(b) Prove that $\displaystyle\sum_{k=0}^{\infty} k e^{-\alpha k}$ converges for any $\alpha > 0$.

(c) In general, prove that $\displaystyle\sum_{k=0}^{\infty} k^n e^{-\alpha k}$ converges for any nonnegative integer n and any $\alpha > 0$.

38. Let $p > 1$. Use the integral test to show that

$$\dfrac{1}{(p - 1)(n + 1)^{p-1}} < \sum_{k=1}^{\infty} \dfrac{1}{k^p} - \sum_{k=1}^{n} \dfrac{1}{k^p} < \dfrac{1}{(p - 1)n^{p-1}}.$$

This result gives bounds on the *error* (or remainder) R_n that results from using s_n to approximate L the sum of the convergent p-series.

In Exercises 39–40, (a) compute the sum of the first four terms of the given series; use four decimal place accuracy. (b) Use the result in Exercise 38 to give upper and lower bounds on R_4. (c) Use parts (a) and (b) to estimate the sum of the series.

39. $\displaystyle\sum_{k=1}^{\infty} \dfrac{1}{k^3}$.

40. $\displaystyle\sum_{k=1}^{\infty} \dfrac{1}{k^4}$.

In Exercises 41–44, use the error bounds given in Exercise 38.

41. (a) If you were to use s_{100} to approximate $\displaystyle\sum_{k=1}^{\infty} \dfrac{1}{k^2}$ what would be the bounds on your error?

(b) How large would you have to choose n to ensure that R_n is less than 0.0001?

42. (a) If you were to use s_{100} to approximate $\displaystyle\sum_{k=1}^{\infty} \dfrac{1}{k^3}$ what would be the bounds on your error?

(b) How large would you have to choose n to ensure that R_n is less than 0.0001?

43. (a) How many terms of the series $\displaystyle\sum_{k=1}^{\infty} \dfrac{1}{k^4}$ should you use to ensure that R_n is less than 0.0001?

(b) Estimate $\displaystyle\sum_{k=1}^{\infty} \dfrac{1}{k^4}$ to three decimal places.

44. Repeat Exercise 43 for the series $\displaystyle\sum_{k=1}^{\infty} \dfrac{1}{k^5}$.

45. Let $\Sigma\, a_k$ and $\Sigma\, b_k$ be series with positive terms and suppose that $a_k/b_k \to 0$.
(a) Show that, if $\Sigma\, b_k$ converges, then $\Sigma\, a_k$ converges.
(b) Show that, if $\Sigma\, a_k$ diverges, then $\Sigma\, b_k$ diverges.

(c) Show by example that, if $\Sigma\, a_k$ converges, then $\Sigma\, b_k$ may converge or diverge.

(d) Show by example that, if $\Sigma\, b_k$ diverges, then $\Sigma\, a_k$ may converge or diverge.

[Parts (c) and (d) explain why we stipulated $L > 0$ in Theorem 11.3.6.]

46. Let $\Sigma\, a_k$ and $\Sigma\, b_k$ be series with positive terms and suppose that $a_k/b_k \to \infty$.

(a) Show that if $\Sigma\, b_k$ diverges, then $\Sigma\, a_k$ diverges.

(b) Show that if $\Sigma\, a_k$ converges, then $\Sigma\, b_k$ converges.

(c) Show by example that if $\Sigma\, a_k$ diverges, then $\Sigma\, b_k$ may converge or diverge.

(d) Show by example that if $\Sigma\, b_k$ converges, then $\Sigma\, a_k$ may converge or diverge.

47. Let $\Sigma\, a_k$ be a series with positive terms.

(a) Prove that if $\Sigma\, a_k$ converges, then $\Sigma\, a_k^2$ converges.

(b) Suppose that $\Sigma\, a_k^2$ converges. Does $\Sigma\, a_k$ converge or diverge? Prove or give a counterexample.

48. Let $\Sigma\, a_k$ be a series with positive terms. Prove that if $\Sigma\, a_k^2$ converges, then $\Sigma(a_k/k)$ converges.

49. Let f be a continuous, positive, decreasing function on $[1, \infty)$ such that $\int_1^\infty f(x)\, dx$ converges. Then the series $\Sigma_{k=1}^\infty f(k)$ also converges. Prove that

$$0 < L - s_n < \int_n^\infty f(x)\, dx,$$

where L is the sum of the series and s_n is the nth partial sum.

In Exercises 50 and 51, use the result of Exercise 49 to determine the smallest integer N such that the difference between the sum of the given series and the Nth partial sum is less than 0.001.

50. $\displaystyle\sum_{k=1}^\infty \frac{1}{k^2 + 1}$.

51. $\displaystyle\sum_{k=1}^\infty k e^{-k^2}$.

52. All the results of this section were stated for series with nonnegative terms. Corresponding results hold for *series with nonpositive terms*: $a_k \le 0$ for all k.

(a) State a comparison theorem analogous to Theorem 11.3.5, this time for series with nonpositive terms.

(b) As stated, the integral test (Theorem 11.3.2) applies only to series with positive terms. State the equivalent result for series with negative terms.

53. This exercise demonstrates that we cannot always use the same testing series for both the basic comparison test and the limit comparison test.

(a) Show that

$$\Sigma\, \frac{\ln n}{n\sqrt{n}} \quad \text{converges by comparison with} \quad \Sigma\, \frac{1}{n^{5/4}}.$$

(b) Show that the limit comparison test does not apply.

■ 11.4 THE ROOT TEST; THE RATIO TEST

We continue with our study of series with nonnegative terms. Comparison with geometric series

$$\Sigma\, x^k$$

and with the p-series

$$\Sigma\, \frac{1}{k^p}$$

leads to two important tests for convergence: the root test and the ratio test.

THEOREM 11.4.1 THE ROOT TEST

Let $\Sigma\, a_k$ be a series with nonnegative terms, and suppose that

$$(a_k)^{1/k} \to \rho \quad \text{as} \quad k \to \infty.$$

(a) If $\rho < 1$, then $\Sigma\, a_k$ converges.

(b) If $\rho > 1$, then $\Sigma\, a_k$ diverges.

(c) If $\rho = 1$, then the test is inconclusive; the series may either converge or diverge.

PROOF   We suppose first that $\rho < 1$ and choose μ so that

$$\rho < \mu < 1.$$

Since $(a_k)^{1/k} \to \rho$, we have

$$(a_k)^{1/k} < \mu \qquad \text{for all } k \text{ sufficiently large.} \qquad \text{(explain)}$$

Thus

$$a_k < \mu^k \qquad \text{for all } k \text{ sufficiently large.}$$

Since $\Sigma \, \mu^k$ converges (a geometric series with $0 < \mu < 1$), we know by the basic comparison theorem that $\Sigma \, a_k$ converges.

We suppose now that $\rho > 1$.   Since $(a_k)^{1/k} \to \rho$, we have

$$(a_k)^{1/k} > 1 \qquad \text{for all } k \text{ sufficiently large.} \qquad \text{(explain)}$$

Thus

$$a_k > 1 \qquad \text{for all } k \text{ sufficiently large.}$$

It now follows that $a_k \nrightarrow 0$ as $k \to \infty$.   Therefore $\Sigma \, a_k$ diverges by the divergence test (11.2.7).

To see the inconclusiveness of the root test when $\rho = 1$, consider the series $\Sigma(1/k^2)$ and $\Sigma(1/k)$.   The first series converges and the second series diverges.   However, in each case we have

$$(a_k)^{1/k} = \left(\frac{1}{k^2}\right)^{1/k} = \left(\frac{1}{k^{1/k}}\right)^2 \to 1^2 = 1 \quad \text{as } k \to \infty,$$

$$(a_k)^{1/k} = \left(\frac{1}{k}\right)^{1/k} = \frac{1}{k^{1/k}} \to 1 \qquad\qquad \text{as } k \to \infty.$$

(Recall that $k^{1/k} \to 1$ as $k \to \infty$, see (10.4.6).)   ❑

Applying the Root Test

Example 1   For the series

$$\Sigma \frac{1}{(\ln k)^k}$$

we have

$$(a_k)^{1/k} = \frac{1}{\ln k} \to 0.$$

The series converges.   ❑

Example 2   For the series

$$\Sigma \frac{2^k}{k^3}$$

we have

$$(a_k)^{1/k} = 2\left(\frac{1}{k}\right)^{3/k} = 2\left[\left(\frac{1}{k}\right)^{1/k}\right]^3 = 2\left[\frac{1}{k^{1/k}}\right]^3 \to 2 \cdot 1^3 = 2 \quad \text{as } k \to \infty.$$

The series diverges.   ❑

Example 3 In the case of

$$\Sigma \left(1 - \frac{1}{k}\right)^k,$$

we have

$$(a_k)^{1/k} = 1 - \frac{1}{k} \to 1.$$

Here the root test is inconclusive. It is also unnecessary: since $a_k = (1 - 1/k)^k$ converges to $1/e$ and not to 0 (10.4.7), the series diverges (11.2.7). ❏

THEOREM 11.4.2 THE RATIO TEST

Let Σa_k be a series with positive terms and suppose that

$$\frac{a_{k+1}}{a_k} \to \lambda \quad \text{as } k \to \infty$$

(a) If $\lambda < 1$, then Σa_k converges.
(b) If $\lambda > 1$, then Σa_k diverges.
(c) If $\lambda = 1$, then the test is inconclusive; the series may either converge or diverge.

PROOF We suppose first that $\lambda < 1$ and choose μ so that $\lambda < \mu < 1$. Since

$$\frac{a_{k+1}}{a_k} \to \lambda,$$

we know that there exists $k_0 > 0$ such that

$$\text{if } k \ge k_0, \quad \text{then} \quad \frac{a_{k+1}}{a_k} < \mu. \qquad \text{(explain)}$$

This gives

$$a_{k_0+1} < \mu a_{k_0}, \qquad a_{k_0+2} < \mu a_{k_0+1} < \mu^2 a_{k_0},$$

and more generally,

$$a_{k_0+j} < \mu^j a_{k_0}, \qquad j = 1, 2, \ldots .$$

For $k > k_0$ we have

(1) $$a_k < \mu^{k-k_0} a_{k_0} = \frac{a_{k_0}}{\mu^{k_0}} \mu^k.$$

$$\underset{\text{set } j = k - k_0}{\overset{\uparrow}{\rule{3cm}{0.4pt}}}$$

Since $\mu < 1$,

$$\Sigma \frac{a_{k_0}}{\mu^{k_0}} \mu^k = \frac{a_{k_0}}{\mu^{k_0}} \Sigma \mu^k \quad \text{converges.}$$

Recalling (1), you can see by the basic comparison theorem that Σa_k converges. The proof of the rest of the theorem is left to the exercises. ❏

Remark Contrary to some people's intuition the root and ratio tests are *not* equivalent. See Exercise 52. ❏

Applying the Ratio Test

Example 4   The ratio test shows that the series

$$\Sigma \frac{1}{k!}$$

converges:

$$\frac{a_{k+1}}{a_k} = \frac{1}{(k+1)!} \cdot \frac{k!}{1} = \frac{1}{k+1} \to 0 \qquad \text{as } k \to \infty. \quad \square$$

Example 5   For the series

$$\Sigma \frac{k}{10^k}$$

we have

$$\frac{a_{k+1}}{a_k} = \frac{k+1}{10^{k+1}} \cdot \frac{10^k}{k} = \frac{1}{10} \frac{k+1}{k} \to \frac{1}{10} \quad \text{as } k \to \infty.$$

The series converges.† $\square$

Example 6   For the series

$$\Sigma \frac{k^k}{k!}$$

we have

$$\frac{a_{k+1}}{a_k} = \frac{(k+1)^{k+1}}{(k+1)!} \cdot \frac{k!}{k^k} = \left(\frac{k+1}{k}\right)^k = \left(1 + \frac{1}{k}\right)^k \to e \quad \text{as } k \to \infty.$$

Since $e > 1$, the series diverges. $\square$

Example 7   For the series

$$\Sigma \frac{1}{2k+1}$$

the ratio test is inconclusive:

$$\frac{a_{k+1}}{a_k} = \frac{1}{2(k+1)+1} \cdot \frac{2k+1}{1} = \frac{2k+1}{2k+3} = \frac{2+1/k}{2+3/k} \to 1 \text{ as } k \to \infty.$$

Therefore, we have to look further.   Comparison with the harmonic series shows that the series diverges:

$$\frac{1}{2k+1} \div \frac{1}{k} = \frac{k}{2k+1} \to \frac{1}{2} \qquad \text{and} \qquad \Sigma \frac{1}{k} \text{ diverges.} \quad \square$$

---

† This series can be summed explicitly.   See Exercise 41.

Summary on Convergence Tests

In general, the root test is used only if powers are involved. The ratio test is particularly effective with factorials and with combinations of powers and factorials. If the terms are rational functions of k, the ratio test is inconclusive and the root test is difficult to apply. Rational terms are most easily handled by comparison or limit comparison with a p-series, $\Sigma \, 1/k^p$. If the terms have the configuration of a derivative, you may be able to apply the integral test. Finally, keep in mind that, if $a_k \nrightarrow 0$, then there is no reason to apply any special convergence test; the series diverges by Theorem 11.2.7.

EXERCISES 11.4

In Exercises 1–40, determine whether the series converges or diverges.

1. $\Sigma \dfrac{10^k}{k!}$.

2. $\Sigma \dfrac{1}{k \, 2^k}$.

3. $\Sigma \dfrac{1}{k^k}$.

4. $\Sigma \left(\dfrac{k}{2k+1} \right)^k$.

5. $\Sigma \dfrac{k!}{100^k}$.

6. $\Sigma \dfrac{(\ln k)^2}{k}$.

7. $\Sigma \dfrac{k^2+2}{k^3+6k}$.

8. $\Sigma \dfrac{1}{(\ln k)^k}$.

9. $\Sigma k \left(\dfrac{2}{3} \right)^k$.

10. $\Sigma \dfrac{1}{(\ln k)^{10}}$.

11. $\Sigma \dfrac{1}{1+\sqrt{k}}$.

12. $\Sigma \dfrac{2k+\sqrt{k}}{k^3+\sqrt{k}}$.

13. $\Sigma \dfrac{k!}{10^{4k}}$.

14. $\Sigma \dfrac{k^2}{e^k}$.

15. $\Sigma \dfrac{\sqrt{k}}{k^2+1}$.

16. $\Sigma \dfrac{2^k k!}{k^k}$.

17. $\Sigma \dfrac{k!}{(k+2)!}$.

18. $\Sigma \dfrac{1}{k} \left(\dfrac{1}{\ln k} \right)^{3/2}$.

19. $\Sigma \dfrac{1}{k} \left(\dfrac{1}{\ln k} \right)^{1/2}$.

20. $\Sigma \dfrac{1}{\sqrt{k^3-1}}$.

21. $\Sigma \left(\dfrac{k}{k+100} \right)^k$.

22. $\Sigma \dfrac{(k!)^2}{(2k)!}$.

23. $\Sigma k^{-(1+1/k)}$.

24. $\Sigma \dfrac{11}{1+100^{-k}}$.

25. $\Sigma \dfrac{\ln k}{e^k}$.

26. $\Sigma \dfrac{k!}{k^k}$.

27. $\Sigma \dfrac{\ln k}{k^2}$.

28. $\Sigma \dfrac{k!}{1 \cdot 3 \cdot \, \cdots \, \cdot (2k-1)}$.

29. $\Sigma \dfrac{2 \cdot 4 \cdot \, \cdots \, \cdot 2k}{(2k)!}$.

30. $\Sigma \dfrac{(2k+1)^{2k}}{(5k^2+1)^k}$.

31. $\Sigma \dfrac{k!(2k)!}{(3k)!}$.

32. $\Sigma \dfrac{\ln k}{k^{5/4}}$.

33. $\Sigma \dfrac{k^{k/2}}{k!}$.

34. $\Sigma \dfrac{k^k}{(3^k)^2}$.

35. $\Sigma \dfrac{k^k}{3^{(k^2)}}$.

36. $\Sigma (\sqrt{k} - \sqrt{k-1})^k$.

37. $\dfrac{1}{2} + \dfrac{2}{3^2} + \dfrac{4}{4^3} + \dfrac{8}{5^4} + \cdots$.

38. $1 + \dfrac{1 \cdot 2}{1 \cdot 3} + \dfrac{1 \cdot 2 \cdot 3}{1 \cdot 3 \cdot 5} + \dfrac{1 \cdot 2 \cdot 3 \cdot 4}{1 \cdot 3 \cdot 5 \cdot 7} + \cdots$.

39. $\dfrac{1}{4} + \dfrac{1 \cdot 3}{4 \cdot 7} + \dfrac{1 \cdot 3 \cdot 5}{4 \cdot 7 \cdot 10} + \dfrac{1 \cdot 3 \cdot 5 \cdot 7}{4 \cdot 7 \cdot 10 \cdot 13} + \cdots$.

40. $\dfrac{2}{3} + \dfrac{2 \cdot 4}{3 \cdot 7} + \dfrac{2 \cdot 4 \cdot 6}{3 \cdot 7 \cdot 11} + \dfrac{2 \cdot 4 \cdot 6 \cdot 8}{3 \cdot 7 \cdot 11 \cdot 15} + \cdots$.

41. Find the sum of the series $\frac{1}{10} + \frac{2}{100} + \frac{3}{1000} + \frac{4}{10000} + \cdots$. HINT: Exercise 58 of Section 11.2.

42. Complete the proof of the ratio test.
   (a) Prove that, if $\lambda > 1$, then $\Sigma \, a_k$ diverges.
   (b) Prove that, if $\lambda = 1$, the ratio test is inconclusive. HINT: Consider $\Sigma \, 1/k$ and $\Sigma \, 1/k^2$.

43. Prove that the sequence $\left\{ \dfrac{n!}{n^n} \right\}$ has limit 0. HINT: Consider the series $\Sigma \dfrac{k!}{k^k}$.

44. Let r be a positive number. Prove that the sequence $\left\{ \dfrac{r^n}{n!} \right\}$ has limit 0.

45. Let $p \geq 2$ be an integer. Find the values of p (if any) such that $\Sigma \dfrac{(k!)^2}{(pk)!}$ converges.

46. Let r be a positive number. For what values of r (if any) does $\Sigma \dfrac{r^k}{k^r}$ converge?

In Exercises 47–50, find the values of x for which the given series converges.

47. $\displaystyle\sum_{k=1}^{\infty} \frac{|x|^k}{k}$.

48. $\displaystyle\sum_{k=1}^{\infty} \frac{|x|^k}{2^k}$.

49. $\displaystyle\sum_{k=1}^{\infty} \frac{2^k |x|^k}{k!}$.

50. $\displaystyle\sum_{k=2}^{\infty} \frac{|x-2|^k}{k \, 3^k}$.

51. Let $\{a_k\}$ be a sequence of positive numbers and take $r > 0$. Use the root test to show that, if $(a_k)^{1/k} \to \rho$ and $\rho < 1/r$, then $\Sigma a_k r^k$ converges.

52. Consider the series
$\frac{1}{2} + 1 + \frac{1}{8} + \frac{1}{4} + \frac{1}{32} + \frac{1}{16} + \cdots$ formed by rearranging a convergent geometric series. (a) Use the root test to show that the series converges. (b) Show that the ratio test does not apply.

■ 11.5 ABSOLUTE AND CONDITIONAL CONVERGENCE; ALTERNATING SERIES

In this section we consider series that have both positive and negative terms.

Absolute and Conditional Convergence

Let Σa_k be a series with both positive and negative terms. One way to show that Σa_k converges is to show that the series of absolute values, $\Sigma |a_k|$, converges.

THEOREM 11.5.1

If $\Sigma |a_k|$ converges, then Σa_k converges.

PROOF For each k,

$$-|a_k| \le a_k \le |a_k| \quad \text{and therefore} \quad 0 \le a_k + |a_k| \le 2|a_k|.$$

If $\Sigma |a_k|$ converges, then $\Sigma 2|a_k| = 2 \Sigma |a_k|$ converges, and therefore, by the basic comparison theorem, $\Sigma (a_k + |a_k|)$ converges. Since

$$a_k = (a_k + |a_k|) - |a_k|,$$

we can conclude that $\Sigma |a_k|$ converges. ❏

DEFINITION 11.5.2 ABSOLUTE CONVERGENCE

A series Σa_k is *absolutely convergent* if the series of absolute values

$$|a_1| + |a_2| + |a_3| + \cdots = \Sigma |a_k|$$

is convergent

The theorem we have just proved says that *an absolutely convergent series is convergent*.

Example 1   Consider the series

$$1 - \frac{1}{2^2} + \frac{1}{3^2} - \frac{1}{4^2} + \frac{1}{5^2} - \frac{1}{6^2} + \cdots = \sum_{k=1}^{\infty} \frac{(-1)^{k+1}}{k^2}.$$

If we replace each term by its absolute value, we obtain the series

$$1 + \frac{1}{2^2} + \frac{1}{3^2} + \frac{1}{4^2} + \frac{1}{5^2} + \frac{1}{6^2} + \cdots = \sum_{k=1}^{\infty} \frac{1}{k^2}.$$

This is a *p*-series with $p = 2$. It is therefore convergent. This means that the initial series is absolutely convergent.   ❏

Example 2   Consider the series

$$1 - \frac{1}{2} - \frac{1}{2^2} + \frac{1}{2^3} - \frac{1}{2^4} + \frac{1}{2^5} + \frac{1}{2^6} - \frac{1}{2^7} - \frac{1}{2^8} + \cdots.$$

If we replace each term by its absolute value, we obtain the series

$$1 + \frac{1}{2} + \frac{1}{2^2} + \frac{1}{2^3} + \frac{1}{2^4} + \frac{1}{2^5} + \frac{1}{2^6} + \frac{1}{2^7} + \frac{1}{2^8} + \cdots = \sum_{k=0}^{\infty} \frac{1}{2^k}.$$

This is a convergent geometric series. The initial series is therefore absolutely convergent.   ❏

Example 3   As we will see after the next theorem, the series

$$1 - \frac{1}{2} + \frac{1}{3} - \frac{1}{4} + \frac{1}{5} - \frac{1}{6} + \cdots = \sum_{k=0}^{\infty} \frac{(-1)^k}{k+1}†$$

is convergent, but it is not absolutely convergent: if we replace each term by its absolute value, we get the divergent harmonic series

$$1 + \frac{1}{2} + \frac{1}{3} + \frac{1}{4} + \frac{1}{5} + \frac{1}{6} + \cdots = \sum_{k=0}^{\infty} \frac{1}{k+1}.$$ ❏

DEFINITION 11.5.3   CONDITIONAL CONVERGENCE

A series $\sum a_k$ is *conditionally convergent* if it converges but $\sum |a_k|$ diverges.

Thus, the series $\sum (-1)^k/(k+1)$ is conditionally convergent.

Alternating Series

Series in which the consecutive terms have opposite signs are called *alternating series*. For example, the series

$$1 - \frac{1}{2} + \frac{1}{3} - \frac{1}{4} + \frac{1}{5} - \frac{1}{6} + \cdots = \sum_{k=0}^{\infty} \frac{(-1)^k}{k+1}$$

---

† In Section 11.6 we show that the series $1 - \frac{1}{2} + \frac{1}{3} - \frac{1}{4} + \frac{1}{5} - \frac{1}{6} + \cdots$ converges to $\ln 2$.

and

$$-1 + \frac{1}{\sqrt{2}} - \frac{1}{\sqrt{3}} + \frac{1}{\sqrt{4}} - \frac{1}{\sqrt{5}} + \cdots = \sum_{k=1}^{\infty} \frac{(-1)^k}{\sqrt{k}}$$

are alternating series.

The series

$$1 - \frac{1}{2} - \frac{1}{3} + \frac{1}{4} - \frac{1}{5} - \frac{1}{6} + \cdots$$

is not an alternating series because there are consecutive terms with the same sign.

In general, an alternating series will either have the form

$$a_0 - a_1 + a_2 - a_3 + a_4 - \cdots = \sum_{k=0}^{\infty} (-1)^k a_k$$

or the form

$$-a_0 + a_1 - a_2 + a_3 - a_4 + \cdots = \sum_{k=0}^{\infty} (-1)^{k+1} a_k,$$

where $\{a_k\}$ is a sequence of positive numbers. Since the second form is simply the negative of the first, we will focus our attention on the first form.

THEOREM 11.5.4 ALTERNATING SERIES TEST†

Let $\{a_k\}$ be a sequence of positive numbers. If

(a) $a_{k+1} < a_k$ for all k; that is, if the sequence $\{a_k\}$ is decreasing, and

(b) $a_k \to 0$ as $k \to \infty$,

then $\displaystyle\sum_{k=0}^{\infty} (-1)^k a_k$ converges.

PROOF   First we look at the even partial sums, s_{2m}. Since

$$s_{2m} = (a_0 - a_1) + (a_2 - a_3) + \cdots + (a_{2m-2} - a_{2m-1}) + a_{2m}$$

is the sum of positive numbers, the even partial sums are all positive. Since

$$s_{2m+2} = s_{2m} - (a_{2m+1} - a_{2m+2}) \qquad \text{and} \qquad a_{2m+1} - a_{2m+2} > 0,$$

we have

$$s_{2m+2} < s_{2m}.$$

This means that the sequence of even partial sums is decreasing. Being bounded below by 0, it is convergent; say,

$$s_{2m} \to L \quad \text{as } m \to \infty.$$

Now

$$s_{2m+1} = s_{2m} - a_{2m+1}.$$

---

† This theorem dates back to Leibniz. He proved the result in 1705.

Since $a_{2m+1} \to 0$ as $m \to \infty$, we also have

$$s_{2m+1} \to L.$$

Since both the even and the odd partial sums tend to L, the sequence of all partial sums tends to L (Exercise 42, Section 10.3). ❏

From this theorem you can see that the following series all converge:

$$1 - \frac{1}{2} + \frac{1}{3} - \frac{1}{4} + \frac{1}{5} - \frac{1}{6} + \cdots, \qquad 1 - \frac{1}{\sqrt{2}} + \frac{1}{\sqrt{3}} - \frac{1}{\sqrt{4}} + \frac{1}{\sqrt{5}} - \frac{1}{\sqrt{6}} + \cdots,$$

$$1 - \frac{1}{2!} + \frac{1}{3!} - \frac{1}{4!} + \frac{1}{5!} - \frac{1}{6!} + \cdots.$$

The first two series converge only conditionally; the third is absolutely convergent.

Remark In the proof of Theorem 11.5.4, we showed that the sequence of even partial sums, $\{s_{2m}\}$, is decreasing and bounded below (by 0). In Exercise 48 you are asked to show that the sequence of odd partial sums $\{s_{2m+1}\}$ is increasing and bounded above. This provides an alternative way to show that the sequence of odd partial sums converges, and since

$$s_{2m+1} - s_{2m} = -a_{2m+1} \to 0 \quad \text{as } m \to \infty,$$

each sequence has the same limit L. This is illustrated in Figure 11.5.1. ❏

Figure 11.5.1

An Estimate for Alternating Series You have seen that if $\{a_k\}$ is a decreasing sequence of positive numbers that tends to 0, then

$$\sum_{k=0}^{\infty} (-1)^k a_k \quad \text{converges to a sum } L.$$

(11.5.5)

The sum L of a convergent alternating series lies between consecutive partial sums s_n, s_{n+1}, and thus s_n approximates L to within a_{n+1}:

$$|s_n - L| < a_{n+1}.$$

PROOF For all n

$$a_{n+1} > a_{n+2}.$$

If n is odd,

$$s_{n+2} = s_n + a_{n+1} - a_{n+2} > s_n;$$

if n is even,

$$s_{n+2} = s_n - a_{n+1} + a_{n+2} < s_n.$$

The odd partial sums increase toward L; the even partial sums decrease toward L. For odd n

$$s_n < L < s_{n+1} = s_n + a_{n+1},$$

and for even n

$$s_n - a_{n+1} = s_{n+1} < L < s_n.$$

Thus, for all n, L lies between s_n and s_{n+1}, and s_n approximates L to within a_{n+1}. ❏

Example 5 Both

$$1 - \frac{1}{2} + \frac{1}{3} - \frac{1}{4} + \frac{1}{5} - \frac{1}{6} + \cdots \qquad \text{and} \qquad 1 - \frac{1}{2^2} + \frac{1}{3^2} - \frac{1}{4^2} + \frac{1}{5^2} - \frac{1}{6^2} + \cdots$$

are convergent alternating series. The nth partial sum of the first series approximates the sum of that series within $1/(n+1)$; the nth partial sum of the second series approximates the sum of the second series within $1/(n+1)^2$. The second series converges more rapidly than the first series. ❏

Example 6 Approximate the sum L of the alternating series

$$1 - \frac{1}{3!} + \frac{1}{5!} - \frac{1}{7!} + \cdots = \sum_{k=0}^{\infty} \frac{(-1)^k}{(2k+1)!}$$

within 0.001.

SOLUTION It is easy to verify that the series converges. In fact, it is absolutely convergent. The fourth term of the series, $1/7! \cong 0.0002$, is the first term which is less than 0.001. Thus, we have

$$|L - s_3| < a_4 < 0.001.$$

Now,

$$s_3 = 1 - \frac{1}{3!} + \frac{1}{5!} = 1 - \frac{1}{6} + \frac{1}{120} \cong 0.8417$$

and so $L = 0.842$ with an error of less than 0.001.† ❏

---

† You will see in Section 11.6 that the actual sum of this series is $\sin 1 \cong 0.8415$.

Rearrangements

A *rearrangement* of a series $\Sigma\, a_k$ is a series that has exactly the same terms but in a different order. Thus, for example,

$$1 + \frac{1}{3^3} - \frac{1}{2^2} + \frac{1}{5^5} - \frac{1}{4^4} + \frac{1}{7^7} - \frac{1}{6^6} + \cdots$$

and

$$1 + \frac{1}{3^3} + \frac{1}{5^5} - \frac{1}{2^2} - \frac{1}{4^4} + \frac{1}{7^7} + \frac{1}{9^9} - \cdots$$

are both rearrangements of

$$1 - \frac{1}{2^2} + \frac{1}{3^3} - \frac{1}{4^4} + \frac{1}{5^5} - \frac{1}{6^6} + \frac{1}{7^7} - \cdots.$$

In 1867 Riemann published a theorem on rearrangements of series that underscores the importance of distinguishing between absolute convergence and conditional convergence. According to this theorem all rearrangements of an absolutely convergent series converge absolutely to the same sum. In sharp contrast, a series that is only conditionally convergent can be rearranged to converge to any number we please. It can also be arranged to diverge to $+\infty$, or to diverge to $-\infty$, or even to oscillate between any two bounds we choose.†

Example 7 We have shown that the series

$$\sum_{k=0}^{\infty} \frac{(-1)^k}{k+1}$$

is conditionally convergent, and in the next section you will see that its sum is $\ln 2$. Accepting this fact for now, we have

$$1 - \frac{1}{2} + \frac{1}{3} - \frac{1}{4} + \frac{1}{5} - \frac{1}{6} \cdots = \ln 2$$

and

$$\frac{1}{2} - \frac{1}{4} + \frac{1}{6} - \frac{1}{8} + \frac{1}{10} - \frac{1}{12} + \cdots = \tfrac{1}{2}\ln 2. \qquad \text{(multiply by } \tfrac{1}{2}\text{)}$$

Adding the two series, we get a rearrangement of the given series with a new sum:

$$1 + \frac{1}{3} - \frac{1}{2} + \frac{1}{5} + \frac{1}{7} - \frac{1}{4} + \cdots = \tfrac{3}{2}\ln 2. \quad \square$$

---

† For a complete proof see pp. 138–139, 318–320 in Konrad Knopp's *Theory and Applications of Infinite Series* (Second English Edition), Blackie & Son Limited, London, 1951.

EXERCISES 11.5

In Exercises 1–31, test these series for: (a) absolute convergence, (b) conditional convergence.

1. $1 + (-1) + 1 + \cdots + (-1)^k + \cdots$.

2. $\dfrac{1}{4} - \dfrac{1}{6} + \dfrac{1}{8} - \dfrac{1}{10} + \cdots + \dfrac{(-1)^k}{2k} + \cdots$.

3. $\dfrac{1}{2} - \dfrac{2}{3} + \dfrac{3}{4} - \dfrac{4}{5} + \cdots + (-1)^{k+1}\dfrac{k}{k+1} + \cdots$.

4. $\dfrac{1}{2 \ln 2} - \dfrac{1}{3 \ln 3} + \dfrac{1}{4 \ln 4} - \dfrac{1}{5 \ln 5} + \cdots +$
$(-1)^k\dfrac{1}{k \ln k} + \cdots$.

5. $\Sigma (-1)^k \dfrac{\ln k}{k}$.

6. $\Sigma (-1)^k \dfrac{k}{\ln k}$.

7. $\Sigma \left(\dfrac{1}{k} - \dfrac{1}{k!} \right)$.

8. $\Sigma \dfrac{k^3}{2^k}$.

9. $\Sigma (-1)^k \dfrac{1}{2k+1}$.

10. $\Sigma (-1)^k \dfrac{(k!)^2}{(2k)!}$.

11. $\Sigma \dfrac{k!}{(-2)^k}$.

12. $\Sigma \sin\left(\dfrac{k\pi}{4} \right)$.

13. $\Sigma (-1)^k (\sqrt{k+1} - \sqrt{k})$.

14. $\Sigma (-1)^k \dfrac{k}{k^2+1}$.

15. $\Sigma \sin\left(\dfrac{\pi}{4k^2} \right)$.

16. $\Sigma \dfrac{(-1)^k}{\sqrt{k(k+1)}}$.

17. $\Sigma (-1)^k \dfrac{k}{2^k}$.

18. $\Sigma \left(\dfrac{1}{\sqrt{k}} - \dfrac{1}{\sqrt{k+1}} \right)$.

19. $\Sigma \dfrac{(-1)^k}{k - 2\sqrt{k}}$.

20. $\Sigma (-1)^k \dfrac{k+2}{k^2+k}$.

21. $\Sigma (-1)^k \dfrac{4^{k-2}}{e^k}$.

22. $\Sigma (-1)^k \dfrac{k^2}{2^k}$.

23. $\Sigma (-1)^k k \sin(1/k)$.

24. $\Sigma (-1)^{k+1} \dfrac{k^k}{k!}$.

25. $\Sigma (-1)^k k\, e^{-k}$.

26. $\Sigma \dfrac{\cos \pi k}{k}$.

27. $\Sigma (-1)^k \dfrac{\cos \pi k}{k}$.

28. $\Sigma \dfrac{\sin(\pi k/2)}{k\sqrt{k}}$.

29. $\Sigma \dfrac{\sin(\pi k/4)}{k^2}$.

30. $\dfrac{1}{2} - \dfrac{1}{3} - \dfrac{1}{4} + \dfrac{1}{5} - \dfrac{1}{6} - \dfrac{1}{7} + \cdots + \dfrac{1}{3k+2} -$
$\dfrac{1}{3k+3} - \dfrac{1}{3k+4} + \cdots$.

31. $\dfrac{2 \cdot 3}{4 \cdot 5} - \dfrac{5 \cdot 6}{7 \cdot 8} + \cdots + (-1)^k\dfrac{(3k+2)(3k+3)}{(3k+4)(3k+5)} + \cdots$.

In Exercises 32–35, estimate the error if the partial sum s_n is used to approximate the sum of the given alternating series.

32. $\displaystyle\sum_{k=1}^{\infty} (-1)^{k+1}\dfrac{1}{k}$; s_{20}.

33. $\displaystyle\sum_{k=0}^{\infty} (-1)^k \dfrac{1}{\sqrt{k+1}}$; s_{80}.

34. $\displaystyle\sum_{k=0}^{\infty} (-1)^k \dfrac{1}{(10)^k}$; s_4.

35. $\displaystyle\sum_{k=1}^{\infty} (-1)^{k+1}\dfrac{1}{k^3}$; s_9.

36. Let s_n be the nth partial sum of the series

$$\sum_{k=0}^{\infty} (-1)^k \dfrac{1}{10^k}.$$

Find the least value of n for which s_n approximates the sum of the series within:
(a) 0.001; (b) 0.0001.

37. Find the sum of the series in Exercise 36.

In Exercises 38 and 39, find the smallest integer N such that s_N will approximate the sum of the given alternating series to within the indicated accuracy.

38. $\displaystyle\sum_{k=1}^{\infty} (-1)^k \dfrac{(0.9)^k}{k}$; 0.001.

39. $\displaystyle\sum_{k=0}^{\infty} (-1)^k \dfrac{1}{\sqrt{k+1}}$; 0.005.

40. Verify that the series

$$1 - \frac{1}{2} + \frac{1}{2} - \frac{1}{3} + \frac{1}{2} - \frac{1}{3} + \frac{1}{4} - \frac{1}{3} +$$
$$\frac{1}{4} + \frac{1}{3} - \frac{1}{4} + \cdots$$

diverges and explain how this does not violate the theorem on alternating series.

41. Let L be the sum of the series

$$\sum_{k=0}^{\infty} (-1)^k \dfrac{1}{k!}$$

and let s_n be the nth partial sum. Find the least value of n for which s_n approximates L within
(a) 0.01. (b) 0.001.

42. Let $\{a_k\}$ be a nonincreasing sequence of positive numbers that converges to 0. Does the alternating series $\Sigma(-1)^k a_k$ necessarily converge?

43. Can the hypothesis of Theorem 11.5.4 be relaxed to require only that $\{a_{2k}\}$ and $\{a_{2k+1}\}$ be decreasing sequences of positive numbers with limit zero?

44. Prove that if Σa_k is absolutely convergent and $|b_k| \le |a_k|$ for all k, then Σb_k is absolutely convergent.

45. (a) Prove that if Σa_k is absolutely convergent, then Σa_k^2 is convergent.
(b) Show by means of an example that the converse of the result in part (a) is false.

46. Indicate how a conditionally convergent series can be rearranged (a) to converge to an arbitrary real number L; (b) to diverge to $+\infty$; (c) to diverge to $-\infty$. HINT: Collect the positive terms $p_1, p_2, p_3, \ldots$ and also the negative terms $n_1, n_2, n_3, \ldots$ in the order in which they appear in the original series.

47. In Section 11.8 we prove that, if $\Sigma a_k x_1^k$ converges, then $\Sigma a_k x^k$ converges absolutely for $|x| < |x_1|$. Try to prove this now.

48. Let $\displaystyle\sum_{k=0}^{\infty} (-1)^k a_k$ be an alternating series with $\{a_k\}$ a decreasing sequence. Prove that the sequence of odd partial sums $\{s_{2m+1}\}$ is increasing and bounded above.

49. Let a and b be positive numbers and consider the series

$$a - \frac{b}{2} + \frac{a}{3} - \frac{b}{4} + \frac{a}{5} - \frac{b}{6} + \cdots.$$

(a) Express this series in Σ notation.
(b) For what values of a and b is this series absolutely convergent? Conditionally convergent?

■ 11.6 TAYLOR POLYNOMIALS IN x; TAYLOR SERIES IN x

Taylor Polynomials in x

We begin with a function f continuous at 0 and set $P_0(x) = f(0)$. If f is differentiable at 0, the linear function that best approximates f at points close to 0 is the linear function

$$P_1(x) = f(0) + f'(0)x;$$

P_1 has the same value as f at 0 and also the same first derivative (the same rate of change):

$$P_1(0) = f(0), \quad P_1'(0) = f'(0).$$

(See Section 3.9, Exercise 48.) If f is twice differentiable at 0, then we can get a better approximation to f by using the quadratic polynomial

$$P_2(x) = f(0) + f'(0)x + \frac{f''(0)}{2!}x^2;$$

P_2 has the same value as f at 0 and the same first two derivatives:

$$P_2(0) = f(0), \quad P_2'(0) = f'(0), \quad P_2''(0) = f''(0).$$

If f has three derivatives at 0, we can form the cubic polynomial

$$P_3(x) = f(0) + f'(0)x + \frac{f''(0)}{2!}x^2 + \frac{f'''(0)}{3!}x^3;$$

P_3 has the same value as f at 0 and the same first three derivatives:

$$P_3(0) = f(0), \quad P_3'(0) = f'(0), \quad P_3''(0) = f''(0), \quad P_3'''(0) = f'''(0).$$

More generally, if f has n derivatives at 0, we can form the polynomial

$$P_n(x) = f(0) + f'(0)x + \frac{f''(0)}{2!}x^2 + \cdots + \frac{f^{(n)}(0)}{n!}x^n;$$

P_n is the polynomial of degree n that has the same value as f at 0 and the same first n derivatives:

$$P_n(0) = f(0), \quad P'_n(0) = f'(0), \quad P''_n(0) = f''(0), \ldots, P_n^{(n)}(0) = f^{(n)}(0).$$

These approximating polynomials $P_0(x), P_1(x), P_2(x), \ldots, P_n(x)$ are called *Taylor polynomials* after the English mathematician Brook Taylor (1685–1731). In particular, for each nonnegative integer k, $P_k(x)$, is called the *Taylor polynomial of degree k for f* or the *kth Taylor polynomial of f*.

Example 1   The exponential function

$$f(x) = e^x$$

has derivatives

$$f'(x) = e^x, \quad f''(x) = e^x, \quad f'''(x) = e^x, \quad \text{and so on.}$$

Thus

$$f(0) = 1, \quad f'(0) = 1, \quad f''(0) = 1, \quad f'''(0) = 1, \ldots, \quad f^{(n)}(0) = 1.$$

The nth Taylor polynomial takes the form

$$P_n(x) = 1 + x + \frac{x^2}{2!} + \frac{x^3}{3!} + \cdots + \frac{x^n}{n!}.$$

The graphs of $f(x) = e^x$, $P_0(x)$, $P_1(x)$, $P_2(x)$, and $P_3(x)$ are indicated in Figure 11.6.1.   ❏

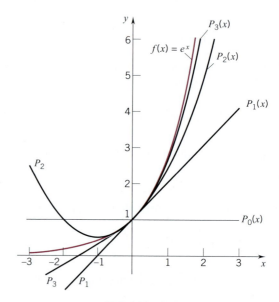

Figure 11.6.1

Example 2 To find the Taylor polynomials that approximate the sine function we write

$$f(x) = \sin x, \quad f'(x) = \cos x, \quad f''(x) = -\sin x, \quad f'''(x) = -\cos x.$$

The pattern now repeats itself:

$$f^{(4)}(x) = \sin x, \quad f^{(5)}(x) = \cos x, \quad f^{(6)}(x) = -\sin x, \quad f^{(7)}(x) = -\cos x.$$

At 0, the sine function and all its even derivatives are 0. The odd derivatives are alternately 1 and −1:

$$f'(0) = 1, \quad f'''(0) = -1, \quad f^{(5)}(0) = 1, \quad f^{(7)}(0) = -1, \quad \text{and so on.}$$

The Taylor polynomials are therefore as follows:

$$P_0(x) = 0$$

$$P_1(x) = P_2(x) = x$$

$$P_3(x) = P_4(x) = x - \frac{x^3}{3!}$$

$$P_5(x) = P_6(x) = x - \frac{x^3}{3!} + \frac{x^5}{5!}$$

$$P_7(x) = P_8(x) = x - \frac{x^3}{3!} + \frac{x^5}{5!} - \frac{x^7}{7!}, \quad \text{and so on.}$$

Only odd powers appear; you should relate this to the fact the $f(x) = \sin x$ is an *odd* function. The graphs of $f(x) = \sin x, P_1(x), P_3(x), P_5(x), P_7(x)$ are indicated in Figure 11.6.2. ❑

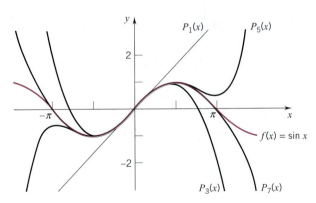

Figure 11.6.2

It is not enough to say that the Taylor polynomials

$$P_n(x) = f(0) + f'(0)x + \frac{f''(0)}{2!}x^2 + \cdots + \frac{f^{(n)}(0)}{n!}x^n$$

approximate $f(x)$. We must describe the accuracy of the approximation.

Our first step is to prove a result known as Taylor's theorem.

THEOREM 11.6.1 TAYLOR'S THEOREM

Suppose that f has $n + 1$ continuous derivatives on an open interval I containing 0. Then, for each $x \in I$,

$$f(x) = f(0) + f'(0)x + \frac{f''(0)}{2!}x^2 + \cdots + \frac{f^{(n)}(0)}{n!}x^n + R_{n+1}(x),$$

where the *remainder* $R_{n+1}(x)$ is given by the formula

$$R_{n+1}(x) = \frac{1}{n!}\int_0^x f^{(n+1)}(t)(x - t)^n \, dt.$$

PROOF   Fix x in the interval I.   Then

$$(1) \qquad \int_0^x f'(t) \, dt = f(x) - f(0).$$

On the other hand, if we evaluate the integral using integration by parts with

$$u = f'(t) \qquad \text{and} \quad dv = dt,$$

then

$$du = f''(t) \, dt \qquad \text{and} \qquad v = -(x - t) \qquad \text{(verify this expression for } v\text{)}$$

and

$$(2) \quad \int_0^x f'(t) \, dt = \left[-f'(t)(x - t) \right]_0^x$$
$$+ \int_0^x f''(t)(x - t) \, dt = f'(0)x + \int_0^x f''(t)(x - t) \, dt.$$

Thus, from Equations (1) and (2), we have

$$f(x) = f(0) + f'(0)x + \int_0^x f''(t)(x - t) \, dt.$$

Integrating by parts again [let $u = f''(t)$, $dv = (x - t) \, dt$, then $du = f'''(t) \, dt$, $v = -\frac{1}{2}(x - t)^2$], we get

$$f(x) = f(0) + f'(0)x + \frac{f''(0)}{2!}x^2 + \frac{1}{2!}\int_0^x f'''(t)(x - t)^2 \, dt.$$

If we continue integrating by parts (see Exercise 53), we will get, after n steps,

$$f(x) = f(0) + f'(0)x + \frac{f''(0)}{2!}x^2 + \frac{f'''(0)}{3!}x^3 + \cdots +$$

$$\frac{f^{(n)}(0)}{n!}x^n + \frac{1}{n!}\int_0^x f^{(n+1)}(t)(x - t)^n \, dt.$$

Thus,

$$f(x) = P_n(x) + \frac{1}{n!} \int_0^x f^{(n+1)}(t)(x-t)^n \, dt,$$

and

$$R_{n+1}(x) = \frac{1}{n!} \int_0^x f^{(n+1)}(t)(x-t)^n \, dt. \quad \Box$$

To see how closely

$$P_n(x) = f(0) + f'(0)x + \frac{f''(0)}{2!}x^2 + \cdots + \frac{f^{(n)}(0)}{n!}x^n$$

approximates $f(x)$ we need an estimate for the remainder term $R_{n+1}(x)$. The following corollary to Taylor's theorem gives a more convenient form of the remainder. It was established by Joseph Lagrange in 1797, and it is known as the Lagrange formula for the remainder. The proof is left to you as an exercise.

COROLLARY 11.6.2 LAGRANGE FORMULA FOR THE REMAINDER

Suppose that f has $n+1$ continuous derivatives on an open interval I containing 0. Let $x \in I$ and let $P_n(x)$ be the nth Taylor polynomial for f. Then

$$R_{n+1}(x) = \frac{f^{(n+1)}(c)}{(n+1)!}x^{n+1},$$

where c is some number between 0 and x.

Remark It is important to understand that the number c indicated in the corollary depends on x and, of course, on f and n. If we rewrite Taylor's theorem using the Lagrange form for the remainder, we have

$$f(x) = f(0) + f'(0)x + \frac{f''(0)}{2!}x^2 + \cdots + \frac{f^{(n)}}{n!}x^n + \frac{f^{(n+1)}(c)}{(n+1)!}x^{n+1}$$

where c is some number between 0 and x. This result is an extension of the mean-value theorem. Indeed, if $n = 0$, we get

$$f(x) = f(0) + f'(c)x \qquad \text{or} \qquad f(x) - f(0) = f'(c)(x - 0),$$

where c is between 0 and x. This is the mean-value theorem for f on the interval $[0, x]$. $\quad \Box$

The following estimate for $R_{n+1}(x)$ is an immediate consequence of Corollary 11.6.2. Let J be the interval joining 0 to x, $x \neq 0$. Then

(11.6.3)
$$|R_{n+1}(x)| \leq \left(\max_{t \in J} |f^{(n+1)}(t)| \right) \frac{|x|^{n+1}}{(n+1)!}.$$

Example 3 The Taylor polynomials of the exponential function

$$f(x) = e^x$$

take the form

$$P_n(x) = 1 + x + \frac{x^2}{2!} + \cdots + \frac{x^n}{n!}. \qquad \text{(Example 1)}$$

We will show with our remainder estimate that for all real x

$$R_{n+1}(x) \to 0 \quad \text{as } n \to \infty,$$

and therefore we can approximate e^x as closely as we wish by Taylor polynomials.

We begin by fixing x and letting M be the maximum value of the exponential function on the interval J that joins 0 to x. (If $x > 0$, then $M = e^x$; if $x < 0$, $M = e^0 = 1$.) Since

$$f^{(n+1)}(t) = e^t \qquad \text{for all } n,$$

we have

$$\max_{t \in J} |f^{(n+1)}(t)| = M \qquad \text{for all } n.$$

Thus by (11.6.3)

$$|R_{n+1}(x)| \leq M \frac{|x|^{n+1}}{(n+1)!}.$$

By (10.4.4) we know that

$$\frac{|x|^{n+1}}{(n+1)!} \to 0 \quad \text{as } n \to \infty.$$

It follows then that $R_{n+1}(x) \to 0$ as asserted. ❏

Example 4 We return to the sine function

$$f(x) = \sin x$$

and its Taylor polynomials

$$P_1(x) = P_2(x) = x$$

$$P_3(x) = P_4(x) = x - \frac{x^3}{3!}$$

$$P_5(x) = P_6(x) = x - \frac{x^3}{3!} + \frac{x^5}{5!}, \quad \text{and so on.}$$

The pattern of derivatives was established in Example 2; namely, for all k,

$$f^{(4k)}(x) = \sin x, \quad f^{(4k+1)}(x) = \cos x,$$

$$f^{(4k+2)}(x) = -\sin x, \quad f^{(4k+3)}(x) = -\cos x.$$

Thus, for all n and all real t,

$$|f^{(n+1)}(t)| \leq 1.$$

It follows from our remainder estimate (11.6.3) that

$$|R_{n+1}(x)| \leq \frac{|x|^{n+1}}{(n+1)!}.$$

Since

$$\frac{|x|^{n+1}}{(n+1)!} \to 0 \qquad \text{for all real } x,$$

we see that $R_{n+1}(x) \to 0$ for all real x. Thus the sequence of Taylor polynomials converges to the sine function and therefore can be used to approximate $\sin x$ for any real number x as closely as we may wish. ❑

Taylor Series in x

By definition $0! = 1$. By adopting the convention that $f^{(0)} = f$, we can write Taylor polynomials

$$P_n(x) = f(0) + f'(0)x + \frac{f''(0)}{2!}x^2 + \cdots + \frac{f^{(n)}(0)}{n!}x^n$$

in Σ notation:

$$P_n(x) = \sum_{k=0}^{n} \frac{f^{(k)}(0)}{k!}x^k.$$

If f is infinitely differentiable on an open interval I containing 0, then we have

$$f(x) = \sum_{k=0}^{n} \frac{f^{(k)}(0)}{k!}x^k + R_{n+1}(x), \qquad x \in I,$$

for all positive integers n. If, as in the case of the exponential function and the sine function, $R_{n+1}(x) \to 0$ as $n \to \infty$ for each $x \in I$, then

$$\sum_{k=0}^{n} \frac{f^{(k)}(0)}{k!}x^k \to f(x).$$

In this case, we say that $f(x)$ can be expanded as a *Taylor series in x* and write

(11.6.4)

$$f(x) = \sum_{k=0}^{\infty} \frac{f^{(k)}(0)}{k!}x^k.$$

Taylor series in x are sometimes called Maclaurin series after Colin Maclaurin, a Scottish mathematician (1698–1746). In some circles the name Maclaurin remains attached to these series, although Taylor considered them some twenty years before Maclaurin.

From Example 3 it is clear that

(11.6.5)

$$e^x = \sum_{k=0}^{\infty} \frac{x^k}{k!} = 1 + x + \frac{x^2}{2!} + \frac{x^3}{3!} + \cdots \qquad \text{for all real } x.$$

From Example 4 we have

$$
\textbf{(11.6.6)} \qquad \sin x = \sum_{k=0}^{\infty} \frac{(-1)^k}{(2k+1)!} x^{2k+1} = x - \frac{x^3}{3!} + \frac{x^5}{5!} - \frac{x^7}{7!} + \cdots \quad \text{for all real } x.
$$

Note that $\sin 1 = 1 - \frac{1}{3!} + \frac{1}{5!} - \frac{1}{7!} + \cdots$ as suggested in Section 11.5.
  We leave it to you as an exercise to show that

$$
\textbf{(11.6.7)} \qquad \cos x = \sum_{k=0}^{\infty} \frac{(-1)^k}{(2k)!} x^{2k} = 1 - \frac{x^2}{2!} + \frac{x^4}{4!} - \frac{x^6}{6!} + \cdots \quad \text{for all real } x.
$$

We come now to the logarithm function. Since $\ln x$ is not defined at $x = 0$, we cannot expand $\ln x$ in powers of x. We work instead with $\ln(1 + x)$.

$$
\textbf{(11.6.8)} \qquad \ln(1 + x) = \sum_{k=1}^{\infty} \frac{(-1)^{k+1}}{k} x^k = x - \frac{x^2}{2} + \frac{x^3}{3} - \cdots \quad \text{for } -1 < x \leq 1.
$$

PROOF† The function

$$
f(x) = \ln(1 + x)
$$

is defined on $(-1, \infty)$ and has derivatives

$$
f'(x) = \frac{1}{1+x}, \qquad f''(x) = -\frac{1}{(1+x)^2}, \qquad f'''(x) = \frac{2}{(1+x)^3},
$$

$$
f^{(4)}(x) = -\frac{3!}{(1+x)^4}, \qquad f^{(5)}(x) = \frac{4!}{(1+x)^5}, \qquad \text{and so on.}
$$

For $k \geq 1$

$$
f^{(k)}(x) = (-1)^{k+1} \frac{(k-1)!}{(1+x)^k}, \qquad f^{(k)}(0) = (-1)^{k+1}(k-1)!, \qquad \frac{f^{(k)}(0)}{k!} = \frac{(-1)^{k+1}}{k}.
$$

Since $f(0) = 0$, the nth Taylor polynomial takes the form

$$
P_n(x) = \sum_{k=1}^{n} (-1)^{k+1} \frac{x^k}{k} = x - \frac{x^2}{2} + \cdots + (-1)^{n+1} \frac{x^n}{n}.
$$

All we have to show therefore is that

$$
R_{n+1}(x) \to 0 \qquad \text{for } -1 < x \leq 1.
$$

---

† The proof we give here illustrates the methods of this section. A much simpler way of obtaining this series expansion is given in Section 11.9.

Instead of trying to apply our usual remainder estimate (in this case, that estimate is not delicate enough to show that $R_{n+1}(x) \to 0$ for $-1 < x < -\frac{1}{2}$), we write the remainder in its integral form. From Taylor's theorem

$$R_{n+1}(x) = \frac{1}{n!} \int_0^x f^{(n+1)}(t)(x - t)^n \, dt,$$

so that in this case

$$R_{n+1}(x) = \frac{1}{n!} \int_0^x (-1)^{n+2} \frac{n!}{(1 + t)^{n+1}} (x - t)^n \, dt = (-1)^n \int_0^x \frac{(x - t)^n}{(1 + t)^{n+1}} \, dt.$$

For $0 \le x \le 1$ we have

$$|R_{n+1}(x)| = \int_0^x \frac{(x - t)^n}{(1 + t)^{n+1}} \, dt \le \int_0^x (x - t)^n \, dt = \frac{x^{n+1}}{n + 1} \to 0.$$

$\underset{\text{explain}}{\uparrow\!\!\text{\textemdash\textemdash}}$

For $-1 < x < 0$ we have

$$|R_{n+1}(x)| = \left| \int_0^x \frac{(x - t)^n}{(1 + t)^{n+1}} \, dt \right| = \int_x^0 \left(\frac{t - x}{1 + t} \right)^n \frac{1}{1 + t} \, dt.$$

By the First Mean-Value Theorem for Integrals (5.9.1) there exists a number x_n between x and 0 such that

$$\int_x^0 \left(\frac{t - x}{1 + t} \right)^n \frac{1}{1 + t} \, dt = \left(\frac{x_n - x}{1 + x_n} \right)^n \left(\frac{1}{1 + x_n} \right) (-x).$$

Since $-x = |x|$ and $0 < 1 + x < 1 + x_n$, we can conclude that

$$|R_{n+1}(x)| < \left(\frac{x_n + |x|}{1 + x_n} \right)^n \left(\frac{|x|}{1 + x} \right).$$

Since $|x| < 1$ and $x_n < 0$, we have

$$x_n < |x| x_n, \qquad x_n + |x| < |x| x_n + |x| = |x|(1 + x_n)$$

and thus

$$\frac{x_n + |x|}{1 + x_n} < |x|.$$

It now follows that

$$|R_{n+1}(x)| < |x|^n \left(\frac{|x|}{1 + x} \right)$$

and, since $|x| < 1$, that $R_{n+1}(x) \to 0$ as $n \to \infty$. ❏

Remark The series expansion for $\ln(1 + x)$ that we have just verified for $-1 < x \le 1$ cannot be extended to other values of x. For $x \le -1$ neither side makes sense: $\ln(1 + x)$ is not defined, and the series on the right diverges. For $x > 1$, $\ln(1 + x)$ is defined, but the series on the right diverges and hence does not represent the function. At $x = 1$, the series gives the intriguing result that was mentioned in Section 11.5:

$$\ln 2 = 1 - \tfrac{1}{2} + \tfrac{1}{3} - \tfrac{1}{4} + \cdots.$$

The graphs of $f(x) = \ln(1 + x)$, $P_1(x)$, $P_2(x)$, and $P_3(x)$ are shown in Figure 11.6.3. Compare these approximations with the Taylor polynomial approximations of e^x and $\sin x$ shown in Figures 11.6.1 and 11.6.2. ❏

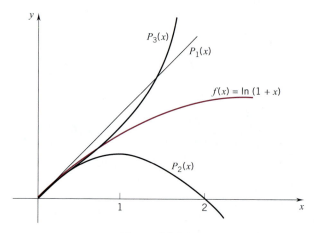

Figure 11.6.3

We want to emphasize again the role played by the remainder term $R_{n+1}(x)$. We can form a Taylor series

$$\sum_{k=0}^{\infty} \frac{f^{(k)}(0)}{k!} x^k$$

for any function f with derivatives of all orders at $x = 0$, but such a series need not converge at any number $x \neq 0$. Even if it does converge, the sum need not be $f(x)$. (See Exercise 55.) The Taylor series converges to $f(x)$ if and only if the remainder term $R_{n+1}(x)$ tends to 0.

Some Numerical Calculations

If the Taylor series converges to $f(x)$, we can use the partial sums (the Taylor polynomials) to calculate $f(x)$ as accurately as we wish. In what follows we show some sample calculations. For ready reference we list some values of $k!$ and $1/k!$ in Tables 11.6.1 and 11.6.2.

■ **Table 11.6.1**

$k!$
$2! = 2$
$3! = 6$
$4! = 24$
$5! = 120$
$6! = 720$
$7! = 5{,}040$
$8! = 40{,}320$

■ **Table 11.6.2**

$1/k!$	
$0.16666 < \dfrac{1}{3!} < 0.16667$	$0.00138 < \dfrac{1}{6!} < 0.00139$
$.04166 < \dfrac{1}{4!} < 0.04167$	$0.00019 < \dfrac{1}{7!} < 0.00020$
$0.00833 < \dfrac{1}{5!} < 0.00834$	$0.00002 < \dfrac{1}{8!} < 0.00003$

Example 5 Estimate e within 0.001.

SOLUTION For all x

$$e^x = 1 + x + \frac{x^2}{2!} + \cdots + \frac{x^n}{n!} + \cdots.$$

Taking $x = 1$ we have

$$e = 1 + 1 + \frac{1}{2!} + \cdots + \frac{1}{n!} + \cdots.$$

From Example 3 we know that the nth partial sum of this series, the nth Taylor polynomial of e^x evaluated at $x = 1$,

$$P_n(1) = 1 + 1 + \frac{1}{2!} + \cdots + \frac{1}{n!},$$

approximates e within

$$|R_{n+1}(1)| \le e \frac{|1|^{n+1}}{(n+1)!} < \frac{3}{(n+1)!}.$$

here $M = e^1 = e$ ⟶   ⟶ $e < 3$

Since

$$\frac{3}{7!} = \frac{3}{5040} = \frac{1}{1680} < 0.001,$$

we can take $n = 6$ and be sure that

$$P_6(1) = 1 + 1 + \frac{1}{2!} + \frac{1}{3!} + \frac{1}{4!} + \frac{1}{5!} + \frac{1}{6!} = \frac{1957}{720}$$

differs from e by less than 0.001.

Our calculator gives

$$\frac{1957}{720} \cong 2.7180556 \quad \text{and} \quad e \cong 2.7182818. \quad ❑$$

Example 6 Estimate $e^{0.2}$ within three decimal places (remainder less than 0.0005).

SOLUTION The exponential series at $x = 0.2$ gives

$$e^{0.2} = 1 + 0.2 + \frac{(0.2)^2}{2!} + \cdots + \frac{(0.2)^n}{n!} + \cdots.$$

From Example 3 we know that the nth partial sum of this series, the nth Taylor polynomial of e^x evaluated at $x = 0.2$

$$P_n(0.2) = 1 + 0.2 + \frac{(0.2)^2}{2!} + \cdots + \frac{(0.2)^n}{n!},$$

approximates $e^{0.2}$ within

$$|R_{n+1}(0.2)| \le e^{0.2} \frac{|0.2|^{n+1}}{(n+1)!} < 3^{0.2} \frac{(0.2)^{n+1}}{(n+1)!} < 1.25 \frac{(0.2)^{n+1}}{(n+1)!}.$$

here $M = e^{0.2}$

Since

$$1.25 \frac{(0.2)^4}{4!} = \frac{1.25(0.0016)}{24} = 0.0000833 \cdots < 0.00009,$$

we can take $n = 3$ and be sure that

$$P_3(0.2) = 1 + 0.2 + \frac{(0.2)^2}{2!} + \frac{(0.2)^3}{3!} = \frac{7.328}{6} \cong 1.22133 \cdots$$

differs from $e^{0.2}$ by less than 0.00009, which is a much better result than the one we asked for. You can verify that $n = 2$ will not yield the desired accuracy.

It now follows that $1.22124 < e^{0.2} < 1.22143$. Our calculator gives $e^{0.2} \cong 1.2214028$. ❏

Example 7 Estimate sin 0.5 within 0.001.

SOLUTION At $x = 0.5$ the sine series gives

$$\sin 0.5 = 0.5 - \frac{(0.5)^3}{3!} + \frac{(0.5)^5}{5!} - \frac{(0.5)^7}{7!} + \cdots .$$

From Example 4 we know that $P_n(0.5)$, where $P_n(x)$ is the nth Taylor polynomial of $\sin x$, approximates $\sin 0.5$ within

$$|R_{n+1}(0.5)| \le \frac{(0.5)^{n+1}}{(n+1)!}.$$

Since

$$\frac{(0.5)^5}{5!} = \frac{1}{(2^5)(5!)} = \frac{1}{(32)(120)} = \frac{1}{3840} < 0.001,$$

we can be sure that

the coefficient of x^4 is 0

$$P_4(0.5) = P_3(0.5) = 0.5 - \frac{(0.5)^3}{3!} = \frac{23}{48}$$

approximates $\sin 0.5$ within 0.001.

Our calculator gives

$$\frac{23}{48} \cong 0.4791666 \quad \text{and} \quad \sin 0.5 \cong 0.4794255. \quad ❏$$

Remark We could have solved the last problem without reference to the remainder estimate derived in Example 4. The series for $\sin 0.5$ is a convergent alternating series with decreasing terms. By (11.5.5) we can conclude immediately that $\sin 0.5$ lies between every two consecutive partial sums. In particular

$$0.5 - \frac{(0.5)^3}{3!} < \sin 0.5 < 0.5 - \frac{(0.5)^3}{3!} + \frac{(0.5)^5}{5!}. \quad ❏$$

Example 8 Estimate $\ln 1.4$ within 0.01.

SOLUTION By (11.6.8)

$$\ln 1.4 = \ln(1 + 0.4) = 0.4 - \tfrac{1}{2}(0.4)^2 + \tfrac{1}{3}(0.4)^3 - \tfrac{1}{4}(0.4)^4 + \cdots.$$

This is a convergent alternating series with decreasing terms. Therefore $\ln 1.4$ lies between every two consecutive partial sums.

The first term less than 0.01 is

$$\tfrac{1}{4}(0.4)^4 = \tfrac{1}{4}(0.0256) = 0.0064.$$

The relation

$$0.4 - \tfrac{1}{2}(0.4)^2 + \tfrac{1}{3}(0.4)^3 - \tfrac{1}{4}(0.4)^4 < \ln 1.4 < 0.4 - \tfrac{1}{2}(0.4)^2 + \tfrac{1}{3}(0.4)^3$$

gives

$$0.335 < \ln 1.4 < 0.341.$$

Within the prescribed limits of accuracy we can take $\ln 1.4 \cong 0.34$. ❏†

---

† A much more effective tool for computing logarithms is given in the exercises.

EXERCISES 11.6

In Exercises 1–4, find the Taylor polynomial $P_4(x)$ for the given function.

1. $f(x) = x - \cos x$.   **2.** $f(x) = \sqrt{1 + x}$.

3. $f(x) = \ln \cos x$.   **4.** $f(x) = \sec x$.

In Exercises 5–8, find the Taylor polynomial $P_5(x)$ for the given function.

5. $f(x) = (1 + x)^{-1}$.   **6.** $f(x) = e^x \sin x$.

7. $f(x) = \tan x$.   **8.** $f(x) = x \cos x^2$.

9. Determine $P_0(x)$, $P_1(x)$, $P_2(x)$, $P_3(x)$ for
$f(x) = 1 - x + 3x^2 + 5x^3$.

10. Determine $P_0(x)$, $P_1(x)$, $P_2(x)$, $P_3(x)$ for $f(x) = (x + 1)^3$.

In Exercises 11–16, determine the nth Taylor polynomial $P_n(x)$ for the given function.

11. $f(x) = e^{-x}$.   **12.** $f(x) = \sinh x$.

13. $f(x) = \cosh x$.   **14.** $f(x) = \ln(1 - x)$.

15. $f(x) = e^{rx}$, r a real number.

16. $f(x) = \cos bx$, b a real number.

In Exercises 17–24, use Taylor polynomials to estimate the following within 0.01.

17. $\sqrt{e}$.   **18.** $\sin 0.3$.

19. $\sin 1$.   **20.** $\ln 1.2$.

21. $\cos 1$.   **22.** $e^{0.8}$.

23. $\sin 10°$.   **24.** $\cos 6°$.

In Exercises 25–32, find the Lagrange form of the remainder R_{n+1} for the given function and the indicated integer n.

25. $f(x) = e^{2x}$; $n = 4$.

26. $f(x) = \ln(1 + x)$; $n = 5$.

27. $f(x) = \cos 2x$; $n = 4$.

28. $f(x) = \sqrt{x + 1}$; $n = 3$.

29. $f(x) = \tan x$; $n = 2$.

30. $f(x) = \sin x$; $n = 5$.

31. $f(x) = \tan^{-1} x$; $n = 2$.

32. $f(x) = \dfrac{1}{1 + x}$; $n = 4$.

In Exercises 33–36, find the Lagrange form of the remainder R_{n+1} for the given function.

33. $f(x) = e^{-x}$.   **34.** $f(x) = \sin 2x$.

35. $f(x) = \dfrac{1}{1 - x}$.   **36.** $f(x) = \ln(1 + x)$.

37. Let $P_n(x)$ be the nth Taylor polynomial of
$$f(x) = \ln(1 + x).$$

Find the least integer n for which: (a) $P_n(0.5)$ approximates ln 1.5 within 0.01; (b) $P_n(0.3)$ approximates ln 1.3 within 0.01; (c) $P_n(1)$ approximates ln 2 within 0.001.

38. Let $P_n(x)$ be the nth Taylor polynomial of

$$f(x) = \sin x.$$

Find the least integer n for which: (a) $P_n(1)$ approximates sin 1 within 0.001; (b) $P_n(2)$ approximates sin 2 within 0.001; (c) $P_n(3)$ approximates sin 3 within 0.001.

▷ **39.** Let $f(x) = e^x$.
   (a) Find the Taylor polynomial P_n of f of least degree that will approximate $\sqrt{e}$ with four decimal place accuracy. Then evaluate $P_n(1/2)$ to obtain your approximation of $\sqrt{e}$.
   (b) Find the Taylor polynomial P_n of f of least degree that will approximate $1/e$ with three decimal place accuracy. Then evaluate $P_n(-1)$ to obtain your approximation of $1/e$.

▷ **40.** Let $g(x) = \cos x$.
   (a) Find the Taylor polynomial P_n of g of least degree that will approximate $\cos(\pi/30)$ with three decimal place accuracy. Then evaluate $P_n(\pi/30)$ to obtain your approximation of $\cos(\pi/30)$.
   (b) Find the Taylor polynomial P_n of g of least degree that will approximate cos 9° with four decimal place accuracy. Then evaluate P_n to obtain your approximation of cos 9°. (Remember to convert to radian measure.)

41. Show that a polynomial $P(x) = a_0 + a_1 x + \cdots + a_n x^n$ is its own Taylor series.

42. Show that

$$\cos x = \sum_{k=0}^{\infty} \frac{(-1)^k}{(2k)!} x^{2k} \qquad \text{for all real } x.$$

43. Show that

$$\sinh x = \sum_{k=0}^{\infty} \frac{1}{(2k+1)!} x^{2k+1} \qquad \text{for all real } x.$$

44. Show that

$$\cosh x = \sum_{k=0}^{\infty} \frac{1}{(2k)!} x^{2k} \qquad \text{for all real } x.$$

In Exercises 45–49, derive a series expansion in x for the given function and specify the numbers x for which the expansion is valid. Take $a > 0$.

45. $f(x) = e^{ax}$.   HINT: Set $t = ax$ and expand e^t in powers of t.

46. $f(x) = \sin ax$.

47. $f(x) = \cos ax$.

48. $f(x) = \ln(1 - ax)$.

49. $f(x) = \ln(a + x)$.
   HINT: $\ln(a + x) = \ln\{a(1 + x/a)\}$.

50. The series we derived for $\ln(1 + x)$ converges too slowly to be of much practical use. The following logarithm series converges much more quickly:

(11.6.9)
$$\ln\left(\frac{1 + x}{1 - x}\right) = 2\left(x + \frac{x^3}{3} + \frac{x^5}{5} + \cdots\right)$$
$$\text{for } -1 < x < 1.$$

Derive this series expansion.

51. Set $x = \frac{1}{3}$ and use the first three nonzero terms of (11.6.9) to estimate ln 2.

52. Use the first two nonzero terms of (11.6.9) to estimate ln 1.4.

53. Verify the identity

$$\frac{f^{(k)}(0)}{k!} x^k = \frac{1}{(k-1)!} \int_0^x f^{(k)}(t)(x - t)^{k-1}\, dt$$
$$- \frac{1}{k!} \int_0^x f^{(k+1)}(t)(x - t)^k\, dt$$

by computing the second integral by parts.

54. Prove Corollary 11.6.2 and then derive the remainder estimate (11.6.3).

▷ **55.** (a) Use a graphing utility to draw the graph of the function

$$f(x) = \begin{cases} e^{-1/x^2}, & x \neq 0 \\ 0, & x = 0 \end{cases}.$$

   (b) Use L'Hospital's rule to show that for every positive integer n

$$\lim_{x \to 0} \frac{e^{-1/x^2}}{x^n} = 0.$$

   (c) Use mathematical induction to prove that $f^{(n)}(0) = 0$ for all $n \geq 1$.
   (d) What is the Taylor series of f?
   (e) For what values of x does the Taylor series of f actually represent f?

▷ **56.** Let $f(x) = \cos x$. Use a graphing utility to graph the Taylor polynomials $P_2(x)$, $P_4(x)$, $P_6(x)$, and $P_8(x)$ of f.

▷ **57.** Let $g(x) = \ln(1 + x)$. Use a graphing utility to graph the Taylor polynomials $P_2(x)$, $P_3(x)$, $P_4(x)$, and $P_5(x)$ of g.

58. (*Important*) Show that e is irrational by following these steps.

(1) Take the expansion

$$e = \sum_{k=0}^{\infty} \frac{1}{k!}$$

and show that the qth partial sum

$$s_q = \sum_{k=0}^{q} \frac{1}{k!}$$

satisfies the inequality

$$0 < q!(e - s_q) < \frac{1}{q}.$$

(2) Show that $q!s_q$ is an integer and argue that, if e were of the form p/q, then $q!(e - s_q)$ would be a positive integer less than 1.

■ 11.7 TAYLOR POLYNOMIALS AND TAYLOR SERIES IN $x - a$

So far we have considered series expansions only in powers of x. Here we generalize to expansions in powers of $x - a$, where a is an arbitrary real number. We begin with a more general version of Taylor's theorem.

THEOREM 11.7.1 TAYLOR'S THEOREM

Suppose that g has $n + 1$ continuous derivatives on an open interval I containing the point a. Then, for each $x \in I$,

$$g(x) = g(a) + g'(a)(x - a) + \frac{g''(a)}{2!}(x - a)^2 + \cdots$$

$$+ \frac{g^{(n)}(a)}{n!}(x - a)^n + R_{n+1}(x),$$

where

$$R_{n+1}(x) = \frac{1}{n!} \int_a^x g^{(n+1)}(t)(x - t)^n \, dt.$$

The polynomial

$$P_n(x) = g(a) + g'(a)(x - a) + \frac{g''(a)}{2!}(x - a)^2 + \cdots + \frac{g^{(n)}(a)}{n!}(x - a)^n$$

is call the *nth Taylor polynomial for g in powers of* $(x - a)$. In this more general setting, the Lagrange formula for the remainder $R_{n+1}(x)$ is given by the following Corollary.

COROLLARY 11.7.2 LAGRANGE FORMULA FOR THE REMAINDER

Suppose that g has $n + 1$ continuous derivatives on an open interval I containing a. Let $x \in I$ and let P_n be the nth Taylor polynomial for g in powers of $(x - a)$. Then

$$R_{n+1}(x) = \frac{g^{(n+1)}(c)}{(n + 1)!}(x - a)^{n+1},$$

where c is some number between a and x.

Now let $x \in I, x \neq a$, and let J be the interval joining a and x. Then an estimate for the remainder can be written:

(11.7.3)

$$|R_{n+1}(x)| \leq \left(\max_{t \in J} |g^{(n+1)}(t)| \right) \frac{|x-a|^{n+1}}{(n+1)!}.$$

If $R_{n+1}(x) \to 0$, then we have

$$g(x) = g(a) + g'(a)(x-a) + \frac{g''(a)}{2!}(x-a)^2 + \cdots + \frac{g^{(n)}(a)}{n!}(x-a)^n + \cdots.$$

In sigma notation we have

(11.7.4)

$$g(x) = \sum_{k=0}^{\infty} \frac{g^{(k)}(a)}{k!}(x-a)^k.$$

This is known as the Taylor expansion of $g(x)$ in powers of $x - a$. The series on the right is called a *Taylor series in $x - a$*.

All this differs from what you saw before only by a translation. Define

$$f(x) = g(x + a).$$

Then obviously

$$f^{(k)}(x) = g^{(k)}(x + a) \quad \text{and} \quad f^{(k)}(0) = g^{(k)}(a).$$

The results of this section as stated for g can be derived by applying the results of Section 11.6 to the function f.

Example 1 Expand $g(x) = 4x^3 - 3x^2 + 5x - 1$ in powers of $x - 2$.

SOLUTION We need to evaluate g and its derivatives at $x = 2$.

$$g(x) = 4x^3 - 3x^2 + 5x - 1$$
$$g'(x) = 12x^2 - 6x + 5$$
$$g''(x) = 24x - 6$$
$$g'''(x) = 24.$$

All higher derivatives are identically 0.

Substitution gives $g(2) = 29$, $g'(2) = 41$, $g''(2) = 42$, $g'''(2) = 24$, and $g^{(k)}(2) = 0$ for all $k \geq 4$. Thus from (11.7.4)

$$g(x) = 29 + 41(x-2) + \frac{42}{2!}(x-2)^2 + \frac{24}{3!}(x-2)^3$$
$$= 29 + 41(x-2) + 21(x-2)^2 + 4(x-2)^3. \quad \square$$

Example 2 Expand $g(x) = x^2 \ln x$ in powers of $x - 1$.

SOLUTION We need to evaluate g and its derivatives at $x = 1$.

$$g(x) = x^2 \ln x$$
$$g'(x) = x + 2x \ln x$$
$$g''(x) = 3 + 2 \ln x$$
$$g'''(x) = 2x^{-1}$$
$$g^{(4)}(x) = -2x^{-2}$$
$$g^{(5)}(x) = (2)(2)x^{-3}$$
$$g^{(6)}(x) = -(2)(2)(3)x^{-4} = -2(3!)x^{-4}$$
$$g^{(7)}(x) = (2)(2)(3)(4)x^{-5} = (2)(4!)x^{-5}, \quad \text{and so on.}$$

The pattern is now clear: for $k \geq 3$

$$g^{(k)}(x) = (-1)^{k+1}2(k-3)!x^{-k+2}.$$

Evaluation at $x = 1$ gives $g(1) = 0$, $g'(1) = 1$, $g''(1) = 3$ and, for $k \geq 3$,

$$g^{(k)}(1) = (-1)^{k+1}2(k-3)!.$$

The expansion in powers of $x - 1$ can be written

$$g(x) = (x - 1) + \frac{3}{2!}(x - 1)^2 + \sum_{k=3}^{\infty} \frac{(-1)^{k+1}(2)(k-3)!}{k!}(x - 1)^k$$
$$= (x - 1) + \frac{3}{2}(x - 1)^2 + 2\sum_{k=3}^{\infty} \frac{(-1)^{k+1}}{k(k-1)(k-2)}(x - 1)^k. \quad ❑$$

Another way to expand $g(x)$ in powers of $x - a$ is to expand $g(t + a)$ in powers of t and then set $t = x - a$. This is the approach we take when the expansion in t is either known to us, or is easily available.

Example 3 We can expand $g(x) = e^{x/2}$ in powers of $x - 3$ by expanding

$$g(t + 3) = e^{(t+3)/2} \qquad \text{in powers of } t$$

and then setting $t = x - 3$.

Note that

$$g(t + 3) = e^{3/2}e^{t/2} = e^{3/2}\underset{\underset{\text{exponential series}}{\uparrow}}{\sum_{k=0}^{\infty} \frac{(t/2)^k}{k!}} = e^{3/2}\sum_{k=0}^{\infty} \frac{1}{2^k k!}t^k.$$

Setting $t = x - 3$, we have

$$g(x) = e^{3/2}\sum_{k=0}^{\infty} \frac{1}{2^k k!}(x - 3)^k.$$

Since the expansion of $g(t + 3)$ is valid for all real t, the expansion of $g(x)$ is valid for all real x. ❑

Taking this same approach, we can prove:

(11.7.5)

> For $a > 0$ and $0 < x \le 2a$
>
> $$\ln x = \ln a + \frac{1}{a}(x - a) - \frac{1}{2a^2}(x - a)^2 + \frac{1}{3a^3}(x - a)^3 - \cdots$$
>
> $$= \ln a + \sum_{k=1}^{\infty} \frac{(-1)^{k+1}}{k\, a^k}(x - a)^k.$$

PROOF   We will expand $\ln(a + t)$ in powers of t and then set $t = x - a$.   In the first place

$$\ln(a + t) = \ln\left[a\left(1 + \frac{t}{a}\right)\right] = \ln a + \ln\left(1 + \frac{t}{a}\right).$$

From (11.6.8) it is clear that

$$\ln\left(1 + \frac{t}{a}\right) = \frac{t}{a} - \frac{1}{2}\left(\frac{t}{a}\right)^2 + \frac{1}{3}\left(\frac{t}{a}\right)^3 - \cdots \quad \text{for} \quad -1 < \frac{t}{a} \le 1 \quad \text{or} \quad -a < t \le a.$$

Adding $\ln a$ to both sides, we have

$$\ln(a + t) = \ln a + \frac{1}{a}t - \frac{1}{2a^2}t^2 + \frac{1}{3a^3}t^3 - \cdots \qquad \text{for} - a < t \le a.$$

Setting $t = x - a$, we find that

$$\ln x = \ln a + \frac{1}{a}(x - a) - \frac{1}{2a^2}(x - a)^2 + \frac{1}{3a^3}(x - a)^3 - \cdots$$

for all x such that $-a < x - a \le a$; that is, for all x such that $0 < x \le 2a$.   ❏

EXERCISES 11.7

In Exercises 1–6, find the Taylor polynomial of the function f for the given values of a and n, and give the Lagrange form of the remainder.

1. $f(x) = \sqrt{x}$; $a = 4$, $n = 3$.

2. $f(x) = \cos x$; $a = \pi/3$, $n = 4$.

3. $f(x) = \sin x$; $a = \pi/4$, $n = 4$.

4. $f(x) = \ln x$; $a = 1$, $n = 5$.

5. $f(x) = \tan^{-1} x$; $a = 1, n = 3$.

6. $f(x) = \cos \pi x$; $a = \frac{1}{2}, n = 4$.

In Exercises 7–22, expand $g(x)$ as indicated and specify the values of x for which the expansion is valid.

7. $g(x) = 3x^3 - 2x^2 + 4x + 1$ in powers of $x - 1$.

8. $g(x) = x^4 - x^3 + x^2 - x + 1$ in powers of $x - 2$.

9. $g(x) = 2x^5 + x^2 - 3x - 5$ in powers of $x + 1$.

10. $g(x) = x^{-1}$ in powers of $x - 1$.

11. $g(x) = (1 + x)^{-1}$ in powers of $x - 1$.

12. $g(x) = (b + x)^{-1}$ in powers of $x - a$, $a \ne -b$.

13. $g(x) = (1 - 2x)^{-1}$ in powers of $x + 2$.

14. $g(x) = e^{-4x}$ in powers of $x + 1$.

15. $g(x) = \sin x$ in powers of $x - \pi$.

16. $g(x) = \sin x$ in powers of $x - \frac{1}{2}\pi$.

17. $g(x) = \cos x$ in powers of $x - \pi$.

18. $g(x) = \cos x$ in powers of $x - \frac{1}{2}\pi$.

19. $g(x) = \sin \frac{1}{2}\pi x$ in powers of $x - 1$.

20. $g(x) = \sin \pi x$ in powers of $x - 1$.

21. $g(x) = \ln(1 + 2x)$ in powers of $x - 1$.

22. $g(x) = \ln(2 + 3x)$ in powers of $x - 4$.

In Exercises 23–32, expand $g(x)$ as indicated.

23. $g(x) = x \ln x$ in powers of $x - 2$.

24. $g(x) = x^2 + e^{3x}$ in powers of $x - 2$.

25. $g(x) = x \sin x$ in powers of x.

26. $g(x) = \ln (x^2)$ in powers of $x - 1$.

27. $g(x) = (1 - 2x)^{-3}$ in powers of $x + 2$.

28. $g(x) = \sin^2 x$ in powers of $x - \frac{1}{2}\pi$.

29. $g(x) = \cos^2 x$ in powers of $x - \pi$.

30. $g(x) = (1 + 2x)^{-4}$ in powers of $x - 2$.

31. $g(x) = x^n$ in powers of $x - 1$.

32. $g(x) = (x - 1)^n$ in powers of x.

33. (a) Expand e^x in powers of $x - a$.
(b) Use the expansion to show that $e^{x_1 + x_2} = e^{x_1}e^{x_2}$.
(c) Expand e^{-x} in powers of $x - a$.

34. (a) Expand $\sin x$ and $\cos x$ in powers of $x - a$.
(b) Show that both series are absolutely convergent for all real x.
(c) As noted earlier (Section 11.5), Riemann proved that the order of the terms of an absolutely convergent series may be changed without altering the sum of the series. Use Riemann's discovery and the Taylor expansions of part (a) to derive the addition formulas

$$\sin (x_1 + x_2) = \sin x_1 \cos x_2 + \cos x_1 \sin x_2,$$
$$\cos (x_1 + x_2) = \cos x_1 \cos x_2 - \sin x_1 \sin x_2.$$

▶ 35. (a) Determine the Taylor polynomial P_n in powers of $(x - \pi/6)$ and of least degree that will approximate $\sin 35°$ with four decimal place accuracy.
(b) Evaluate the polynomial that you found in part (a) to obtain your approximation of $\sin 35°$.

▶ 36. (a) Determine the Taylor polynomial P_n in powers of $(x - \pi/3)$ and of least degree that will approximate $\cos 57°$ with four decimal place accuracy.
(b) Evaluate the polynomial that you found in part (a) to obtain your approximation of $\cos 57°$.

▶ 37. Choose an appropriate Taylor polynomial of $f(x) = \sqrt{x}$ to approximate $\sqrt{38}$ with three decimal place accuracy.

▶ 38. Choose an appropriate Taylor polynomial for $f(x) = \sqrt{x}$ to approximate $\sqrt{61}$ with three decimal place accuracy.

■ 11.8 POWER SERIES

You have become familiar with Taylor series

$$\sum_{k=0}^{\infty} \frac{f^{(k)}(0)}{k!} x^k \quad \text{and} \quad \sum_{k=0}^{\infty} \frac{f^{(k)}(a)}{k!} (x - a)^k.$$

Here we study series of the form

$$\sum_{k=0}^{\infty} a_k x^k \quad \text{and} \quad \sum_{k=0}^{\infty} a_k(x - a)^k$$

without regard to how the coefficients a_k have been generated.   Such series are called *power series*. In particular, the first series is a *power series in powers of x*, and the second is a *power series in powers of $x - a$*.

Since a simple translation converts

$$\sum_{k=0}^{\infty} a_k (x - a)^k \quad \text{into} \quad \sum_{k=0}^{\infty} a_k x^k,$$

we can focus our attention on power series of the form

$$\sum_{k=0}^{\infty} a_k x^k.$$

When detailed indexing is unnecessary, we will omit it and write

$$\sum a_k x^k.$$

Note that if we replace the variable x by a real number c, then the result

$$\sum a_k c^k$$

is an infinite series, and so the ideas and methods of Sections 11.2–11.5 can be applied. In particular, our objective here is to determine the set of numbers c for which the resulting infinite series converges. We begin the discussion with a definition.

DEFINITION 11.8.1

A power series $\sum a_k x^k$ is said to converge

 (i) at c iff $\sum a_k c^k$ converges;

 (ii) on the set S iff $\sum a_k x^k$ converges for each $x \in S$.

The following result is fundamental.

THEOREM 11.8.2

If $\sum a_k x^k$ converges at $c \neq 0$, then it converges absolutely for $|x| < |c|$.

If $\sum a_k x^k$ diverges at c, then it diverges for $|x| > |c|$.

PROOF If $\sum a_k c^k$ converges, then $a_k c^k \to 0$ as $k \to \infty$. In particular, for k sufficiently large,

$$|a_k c^k| \leq 1$$

and thus

$$|a_k x^k| = |a_k c^k| \left| \frac{x}{c} \right|^k \leq \left| \frac{x}{c} \right|^k.$$

For $|x| < |c|$, we have

$$\left| \frac{x}{c} \right| < 1.$$

The convergence of $\sum |a_k x^k|$ follows by comparison with the geometric series. This proves the first statement.

Suppose now that $\sum a_k c^k$ diverges. By a similar argument, there cannot exist x with $|x| > |c|$ such that $\sum a_k x^k$ converges. The existence of such an x would imply the absolute convergence of $\sum a_k c^k$. This proves the second statement. ❑

From the theorem we just proved you can see that there are exactly three possibilities for a power series:

Case I. *The series converges only at $x = 0$.* This is what happens with

$$\sum k^k x^k.$$

For if $x \neq 0$, then $\lim\limits_{k \to \infty} k^k x^k \neq 0$, and so the series cannot converge (Theorem 11.2.7).

Case II. The series is absolutely convergent for all real numbers x. This is what happens with the exponential series

$$\sum \frac{x^k}{k!}.$$

Case III. *There exists a positive integer r such that the series converges for $|x| < r$ and diverges for $|x| > r$.* This is what happens with the geometric series

$$\sum x^k.$$

In this instance, there is absolute convergence for $|x| < 1$ and divergence for $|x| > 1$.

Associated with each case is a *radius of convergence*:

In Case I, we say that the radius of convergence is 0.
In Case II, we say that the radius of convergence is ∞.
In Case III, we say that the radius of convergence is r.

The three cases are pictured in Figure 11.8.1.

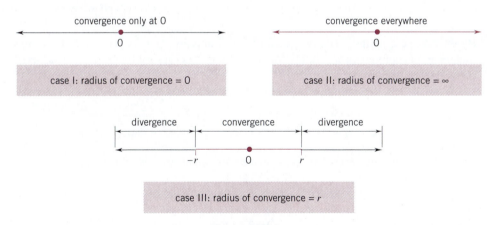

Figure 11.8.1

In general, the behavior of a power series at $-r$ and at r is not predictable. For example, the series

$$\sum x^k, \qquad \sum \frac{(-1)^k}{k} x^k, \qquad \sum \frac{1}{k} x^k, \qquad \sum \frac{1}{k^2} x^k$$

all have radius of convergence 1, but the first series converges only on $(-1, 1)$ (verify this) and, as we show next, the second series converges on $(-1, 1]$, the third on $[-1, 1)$, and the fourth on $[-1, 1]$.

The maximal interval on which a power series converges is called the *interval of convergence*. For a series with infinite radius of convergence, the interval of convergence is $(-\infty, \infty)$. For a series with radius of convergence r, the interval of convergence can be $[-r, r]$, $(-r, r]$, $[-r, r)$, or $(-r, r)$. For a series with radius of convergence 0, the interval of convergence reduces to a point, $\{0\}$.

Example 1 Verify that the series

$$(1) \qquad \sum \frac{(-1)^k}{k} x^k$$

has interval of convergence $(-1, 1]$.

SOLUTION First we show that the radius of convergence is 1 (that the series converges absolutely for $|x| < 1$ and diverges for $|x| > 1$). We do this by forming the series

$$(2) \qquad \sum \left| \frac{(-1)^k}{k} x^k \right| = \sum \frac{1}{k} |x|^k$$

and applying the ratio test.

We set

$$b_k = \frac{1}{k} |x|^k$$

and note that

$$\frac{b_{k+1}}{b_k} = \frac{|x|^{k+1}/(k+1)}{|x|^k/k} = \frac{k}{k+1} \frac{|x|^{k+1}}{|x|^k} = \frac{k}{k+1} |x| \to |x| \quad \text{as } k \to \infty.$$

By the ratio test, series (2) converges for $|x| < 1$ and diverges for $|x| > 1$.† It follows that series (1) converges absolutely for $|x| < 1$ and diverges for $|x| > 1$. The radius of convergence is therefore 1.

Now we test the endpoints $x = -1$ and $x = 1$. At $x = -1$

$$\sum \frac{(-1)^k}{k} x^k \quad \text{becomes} \quad \sum \frac{(-1)^k}{k} (-1)^k = \sum \frac{1}{k}.$$

This is the harmonic series which, as you know, diverges. At $x = 1$

$$\sum \frac{(-1)^k}{k} x^k \quad \text{becomes} \quad \sum \frac{(-1)^k}{k}.$$

This is a convergent alternating series.

We have shown that series (1) converges absolutely for $|x| < 1$, diverges at -1, and converges at 1. The interval of convergence is $(-1, 1]$. ❑

---

† We could also have used the root test:

$$(b_k)^{1/k} = \left| \frac{1}{k} \right|^{1/k} |x| = \frac{1}{k^{1/k}} |x| \to |x|.$$

Remark   The same arguments can be used to show that the series

$$\sum \frac{1}{k} x^k$$

converges on $[-1, 1)$.   ❏

Example 2   Verify that the series

(1) $$\sum \frac{1}{k^2} x^k$$

has interval of convergence $[-1, 1]$.

SOLUTION   First we examine the series

(2) $$\sum \left| \frac{1}{k^2} x^k \right| = \sum \frac{1}{k^2} |x|^k.$$

Here again we use the ratio test.   We set

$$b_k = \frac{1}{k^2} |x|^k$$

and note that

$$\frac{b_{k+1}}{b_k} = \frac{k^2}{(k+1)^2} \frac{|x|^{k+1}}{|x|^k} = \left(\frac{k}{k+1} \right)^2 |x| \to |x| \quad \text{as } k \to \infty.$$

By the ratio test, (2) converges for $|x| < 1$ and diverges for $|x| > 1$.†   This shows that (1) converges absolutely for $|x| < 1$ and diverges for $|x| > 1$.   The radius of convergence is therefore 1.

Now for the endpoints.   At $x = -1$,

$$\sum \frac{1}{k^2} x^k \quad \text{takes the form} \quad \sum \frac{(-1)^k}{k^2} = -1 + \tfrac{1}{4} - \tfrac{1}{9} + \tfrac{1}{16} - \cdots.$$

This is a convergent alternating series.   At $x = 1$,

$$\sum \frac{1}{k^2} x^k \quad \text{becomes} \quad \sum \frac{1}{k^2}.$$

This is a convergent *p*-series.   The interval of convergence is therefore the entire closed interval $[-1, 1]$.   ❏

Example 3   Find the interval of convergence of

(1) $$\sum \frac{k}{6^k} x^k.$$

---

† Once again we could have used the root test:

$$(b_k) = \frac{1}{k^{2/k}} |x| \to |x| \quad \text{as } k \to \infty.$$

SOLUTION   We begin by examining the series

(2) $$\sum \left| \frac{k}{6^k} x^k \right| = \sum \frac{k}{6^k} |x|^k.$$

We set

$$b_k = \frac{k}{6^k} |x|^k$$

and apply the root test.   (The ratio test will also work.)   Since

$$(b_k)^{1/k} = \tfrac{1}{6} k^{1/k} |x| \to \tfrac{1}{6} |x| \quad \text{as } k \to \infty \qquad \text{(recall } k^{1/k} \to 1)$$

you can see that (2) converges

$$\text{for} \quad \tfrac{1}{6} |x| < 1, \qquad \text{that is, for} \quad |x| < 6,$$

and diverges

$$\text{for} \quad \tfrac{1}{6} |x| > 1, \qquad \text{that is, for} \quad |x| > 6.$$

Testing the endpoints, we have:

$$\text{at} \quad x = 6, \qquad \sum \frac{k}{6^k} 6^k = \sum k, \qquad \text{which is divergent;}$$

$$\text{at} \quad x = -6, \qquad \sum \frac{k}{6^k} (-6)^k = \sum (-1)^k k, \qquad \text{which is also divergent.}$$

Thus, the interval of convergence is $(-6, 6)$.   ❏

Example 4   Find the interval of convergence of

(1) $$\sum \frac{(2k)!}{(3k)!} x^k.$$

SOLUTION   Again, we begin by examining the series

(2) $$\sum \left| \frac{(2k)!}{(3k)!} x^k \right| = \sum \frac{(2k)!}{(3k)!} |x|^k.$$

Set

$$b_k = \frac{(2k)!}{(3k)!} |x|^k.$$

Since factorials are involved, we will use the ratio test.   Note that

$$\frac{b_{k+1}}{b_k} = \frac{\{2(k+1)\}!}{\{3(k+1)\}!} \frac{(3k)!}{(2k)!} \frac{|x|^{k+1}}{|x|^k} = \frac{(2k+2)(2k+1)}{(3k+3)(3k+2)(3k+1)} |x|.$$

Since

$$\frac{(2k+2)(2k+1)}{(3k+3)(3k+2)(3k+1)} \to 0 \quad \text{as } k \to \infty,$$

(the numerator is a quadratic in k, the denominator is a cubic), the ratio b_{k+1}/b_k tends to 0 no matter what x is.   By the ratio test, (2) converges for all x and therefore (1) converges absolutely for all x.   The radius of convergence is ∞ and the interval of convergence is $(-\infty, \infty)$.   ❏

Example 5 Find the interval of convergence of

$$\sum (\tfrac{1}{2}k)^k x^k.$$

SOLUTION Since $(\tfrac{1}{2}k)^k x^k \to 0$ only if $x = 0$ (explain), there is no need to invoke the ratio test or the root test. By (11.2.7) the series can converge only at $x = 0$. There it converges trivially, that is, all the terms of the series are 0. ❑

Example 6 Find the interval of convergence of

$$\sum \frac{(-1)^k}{k^2 3^k} (x + 2)^k.$$

SOLUTION We consider the series

$$\sum \left| \frac{(-1)^k}{k^2 3^k} (x + 2)^k \right| = \sum \frac{1}{k^2 3^k} |x + 2|^k.$$

Set

$$b_k = \frac{1}{k^2 3^k} |x + 2|^k$$

and apply the ratio test (the root test will work equally as well):

$$\frac{b_{k+1}}{b_k} = \frac{|x + 2|^{k+1}/(k + 1)^2 3^{k+1}}{|x + 2|^k / k^2 3^k} = \frac{k^2}{3(k + 1)^2} |x + 2| \to \tfrac{1}{3}|x + 2| \quad \text{as } k \to \infty.$$

Thus, the series is absolutely convergent

$$\text{for} \quad \tfrac{1}{3}|x + 2| < 1 \qquad \text{or} \qquad |x + 2| < 3,$$

which is the same as $-5 < x < 1$.
    We now check the endpoints. At $x = -5$:

$$\sum \frac{(-1)^k}{k^2 3^k} (-3)^k = \sum \frac{1}{k^2}.$$

This is a convergent p-series. At $x = 1$:

$$\sum \frac{(-1)^k}{k^2 3^k} (3)^k = \sum \frac{(-1)^k}{k^2},$$

and this is a convergent alternating series. Therefore, the interval of convergence is $[-5, 1]$ ❑

EXERCISES 11.8

In Exercises 1–38, find the interval of convergence.

1. $\sum k x^k.$

2. $\sum \frac{1}{k} x^k.$

3. $\sum \frac{1}{(2k)!} x^k.$

4. $\sum \frac{2^k}{k^2} x^k.$

5. $\sum (-k)^{2k} x^{2k}.$

6. $\sum \frac{(-1)^k}{\sqrt{k}} x^k.$

7. $\sum \frac{1}{k 2^k} x^k.$

8. $\sum \frac{1}{k^2 2^k} x^k.$

9. $\sum \left(\frac{k}{100} \right)^k x^k.$

10. $\sum \frac{k^2}{1 + k^2} x^k.$

11. $\sum \frac{2^k}{\sqrt{k}} x^k.$

12. $\sum \frac{1}{\ln k} x^k.$

13. $\sum \frac{k - 1}{k} x^k.$

14. $\sum k a^k x^k.$

15. $\sum \dfrac{k}{10^k} x^k.$

16. $\sum \dfrac{3k^2}{e^k} x^k.$

17. $\sum \dfrac{x^k}{k^k}.$

18. $\sum \dfrac{7^k}{k!} x^k.$

19. $\sum \dfrac{(-1)^k}{k^k}(x - 2)^k.$

20. $\sum k! \, x^k.$

21. $\sum (-1)^k \dfrac{2^k}{3^{k+1}} x^k.$

22. $\sum \dfrac{2^k}{(2k)!} x^k.$

23. $\sum (-1)^k \dfrac{k!}{k^3}(x - 1)^k.$

24. $\sum \dfrac{(-e)^k}{k^2} x^k.$

25. $\sum \left(\dfrac{k}{k - 1} \right) \dfrac{(x + 2)^k}{2^k}.$

26. $\sum \dfrac{\ln k}{k}(x + 1)^k.$

27. $\sum (-1)^k \dfrac{k^2}{(k + 1)!}(x + 3)^k.$

28. $\sum \dfrac{k^3}{e^k}(x - 4)^k.$

29. $\sum \left(1 + \dfrac{1}{k} \right)^k x^k.$

30. $\sum \dfrac{(-1)^k a^k}{k^2}(x - a)^k.$

31. $\sum \dfrac{\ln k}{2^k}(x - 2)^k.$

32. $\sum \dfrac{1}{(\ln k)^k}(x - 1)^k.$

33. $\sum (-1)^k (\tfrac{2}{3})^k (x + 1)^k.$

34. $\sum \dfrac{2^{1/k} \pi^k}{k(k + 1)(k + 2)}(x - 2)^k.$

35. $1 - \dfrac{x}{2} + \dfrac{2x^2}{4} - \dfrac{3x^3}{8} + \dfrac{4x^4}{16} - \cdots.$

36. $\dfrac{(x - 1)}{5^2} + \dfrac{4}{5^4}(x - 1)^2 + \dfrac{9}{5^6}(x - 1)^3 +$

$\dfrac{16}{5^8}(x - 1)^4 + \cdots.$

37. $\dfrac{3x^2}{4} + \dfrac{9x^4}{9} + \dfrac{27x^6}{16} + \dfrac{81x^8}{25} + \cdots.$

38. $\dfrac{1}{16}(x + 1) - \dfrac{2}{25}(x + 1)^2 + \dfrac{3}{36}(x + 1)^3 -$

$\dfrac{4}{49}(x + 1)^4 + \cdots.$

39. Let $\sum a_k x^k$ be a power series, and let r be its radius of convergence.
 (a) Given that $|a_k|^{1/k} \to \rho$, show that, if $\rho \neq 0$, then $r = 1/\rho$ and, if $\rho = 0$, then $r = \infty$.
 (b) Given that $|a_{k+1}/a_k| \to \lambda$, show that, if $\lambda \neq 0$, then $r = 1/\lambda$ and, if $\lambda = 0$, then $r = \infty$.

40. Find the interval of convergence of the series $\sum s_k x^k$ where s_k is the kth partial sum of the series

$$\sum_{n=1}^{\infty} \dfrac{1}{n}.$$

41. Let $\sum a_k x^k$ be a power series and let r, $0 < r < \infty$, be its radius of convergence. Prove that if the series is absolutely convergent at one endpoint of its interval of convergence, then it is absolutely convergent at the other also.

42. Let $\sum a_k x^k$ be a power series and let r, $0 < r < \infty$, be its radius of convergence. Prove that the power series $\sum a_k x^{2k}$ has radius of convergence $\sqrt{r}$.

■ 11.9 DIFFERENTIATION AND INTEGRATION OF POWER SERIES

Suppose that the power series

$$\sum_{k=0}^{\infty} a_k x^k$$

converges on the interval $(-c, c)$. Then, for each $x \in (-c, c)$, the infinite series

$$\sum_{k=0}^{\infty} a_k x^k$$

converges to a number L_x; the sum L_x depends upon x. Let f be the function defined on $(-c, c)$ by $f(x) = L_x$, that is,

$$f(x) = \sum_{k=0}^{\infty} a_k x^k \qquad \text{for } x \in (-c, c).$$

In this section we show that this function is both infinitely differentiable and integrable on $(-c, c)$. We begin with a simple but important result.

THEOREM 11.9.1

If

$$\sum_{k=0}^{\infty} a_k x^k = a_0 + a_1 x + a_2 x^2 + \cdots + a_n x^n + \cdots$$

converges on $(-c, c)$, then

$$\sum_{k=0}^{\infty} \frac{d}{dx}(a_k x^k) = \sum_{k=1}^{\infty} k a_k x^{k-1}$$
$$= a_1 + 2a_2 x + 3a_3 x^2 + \cdots + n a_n x^{n-1} + \cdots$$

also converges on $(-c, c)$.

PROOF Assume that

$$\sum_{k=0}^{\infty} a_k x^k \quad \text{converges on } (-c, c).$$

By Theorem 11.8.2 the series is absolutely convergent on this interval.

Now let x be some fixed number in $(-c, c)$ and choose $\epsilon > 0$ such that

$$|x| < |x| + \epsilon < c.$$

Since $|x| + \epsilon$ lies within the interval of convergence,

$$\sum_{k=0}^{\infty} |a_k(|x| + \epsilon)^k| \quad \text{converges.}$$

In Exercise 48 you are asked to show that, for all k sufficiently large,

$$|kx^{k-1}| \leq (|x| + \epsilon)^k.$$

It follows that for all such k

$$|ka_k x^{k-1}| \leq |a_k(|x| + \epsilon)^k|.$$

Since

$$\sum_{k=0}^{\infty} |a_k(|x| + \epsilon)^k| \quad \text{converges,}$$

we can conclude that

$$\sum_{k=0}^{\infty} \left| \frac{d}{dx}(a_k x^k) \right| = \sum_{k=1}^{\infty} |ka_k x^{k-1}| \quad \text{converges,}$$

and thus that

$$\sum_{k=0}^{\infty} \frac{d}{dx}(a_k x^k) = \sum_{k=1}^{\infty} ka_k x^{k-1} \quad \text{converges.} \quad \square$$

Repeated application of the theorem shows that

$$\sum_{k=0}^{\infty} \frac{d^2}{dx^2}(a_k x^k), \qquad \sum_{k=0}^{\infty} \frac{d^3}{dx^3}(a_k x^k), \qquad \sum_{k=0}^{\infty} \frac{d^4}{dx^4}(a_k x^k), \qquad \text{and so on,}$$

all converge on $(-c, c)$.

Example 1 Since the geometric series

$$\sum_{k=0}^{\infty} x^k = 1 + x + x^2 + x^3 + x^4 + x^5 + x^6 + \cdots$$

converges on $(-1, 1)$, the series

$$\sum_{k=0}^{\infty} \frac{d}{dx}(x^k) = \sum_{k=1}^{\infty} kx^{k-1} = 1 + 2x + 3x^2 + 4x^3 + 5x^4 + 6x^5 + \cdots,$$

$$\sum_{k=0}^{\infty} \frac{d^2}{dx^2}(x^k) = \sum_{k=2}^{\infty} k(k-1)x^{k-2} = 2 + 6x + 12x^2 + 20x^3 + 30x^4 + \cdots,$$

$$\sum_{k=0}^{\infty} \frac{d^3}{dx^3}(x^k) = \sum_{k=3}^{\infty} k(k-1)(k-2)x^{k-3} = 6 + 24x + 60x^2 + 120x^3 + \cdots,$$

$$\vdots$$

all converge on $(-1, 1)$. ❑

COROLLARY 11.9.2

If

$$\sum_{k=0}^{\infty} a_k x^k$$

has a radius of convergence r, then each of the series

$$\sum_{k=0}^{\infty} \frac{d}{dx}(a_k x^k), \qquad \sum_{k=0}^{\infty} \frac{d^2}{dx^2}(a_k x^k), \qquad \sum_{k=0}^{\infty} \frac{d^3}{dx^3}(a_k x^k), \qquad \text{and so on,}$$

has radius of convergence r.

Remark Even though $\sum a_k x^k$ and its "derivative" $\sum ka_k x^{k-1}$ have the same radius of convergence, their intervals of convergence may be different. For example, the interval of convergence of the series

$$\sum_{k=1}^{\infty} \frac{1}{k^2} x^k$$

is $[-1, 1]$, whereas the interval of convergence of its derivative

$$\sum_{k=1}^{\infty} \frac{1}{k} x^{k-1}$$

is $[-1, 1)$. Endpoints must always be checked separately. Obviously, this remark does not apply in the cases where $r = 0$ or $r = \infty$. ❑

Suppose now that

$$\sum_{k=0}^{\infty} a_k x^k \quad \text{converges on } (-c, c).$$

Then, as we have seen,

$$\sum_{k=0}^{\infty} \frac{d}{dx}(a_k x^k) \quad \text{also converges on } (-c, c).$$

As we noted at the beginning of this section, we can define a function f on $(-c, c)$ by setting

$$f(x) = \sum_{k=0}^{\infty} a_k x^k.$$

Using the second series, we can define a function g on $(-c, c)$ by setting

$$g(x) = \sum_{k=0}^{\infty} \frac{d}{dx}(a_k x^k).$$

The crucial point is that

$$f'(x) = g(x).$$

THEOREM 11.9.3 THE DIFFERENTIABILITY THEOREM

If

$$f(x) = \sum_{k=0}^{\infty} a_k x^k \quad \text{for all } x \text{ in } (-c, c),$$

then f is differentiable on $(-c, c)$ and

$$f'(x) = \sum_{k=0}^{\infty} \frac{d}{dx}(a_k x^k) \quad \text{for all } x \text{ in } (-c, c).$$

By applying this theorem to f', you can see that f' is itself differentiable. This in turn implies that f'' is differentiable, and so on. In short, f has derivatives of all orders.

The discussion up to this point can be summarized as follows:

In the interior of its interval of convergence a power series defines an infinitely differentiable function, the derivatives of which can be obtained by differentiating term by term:

$$\frac{d^n}{dx^n}\left(\sum_{k=0}^{\infty} a_k x^k\right) = \sum_{k=0}^{\infty} \frac{d^n}{dx^n}(a_k x^k) \quad \text{for all } n.$$

For a detailed proof of the differentiability theorem see the supplement at the end of this section. We go on to examples.

Example 2 You know that

$$\frac{d}{dx}(e^x) = e^x.$$

You can see this directly by differentiating the exponential series:

$$\frac{d}{dx}(e^x) = \frac{d}{dx}\left(\sum_{k=0}^{\infty}\frac{x^k}{k!}\right) = \sum_{k=0}^{\infty}\frac{d}{dx}\left(\frac{x^k}{k!}\right) = \sum_{k=1}^{\infty}\frac{x^{k-1}}{(k-1)!} = \sum_{n=0}^{\infty}\frac{x^n}{n!} = e^x. \quad \square$$

set $n = k - 1$

Example 3 You have seen that

$$\sin x = x - \frac{x^3}{3!} + \frac{x^5}{5!} - \frac{x^7}{7!} + \frac{x^9}{9!} - \cdots$$

and

$$\cos x = 1 - \frac{x^2}{2!} + \frac{x^4}{4!} - \frac{x^6}{6!} + \frac{x^8}{8!} - \cdots.$$

The relations

$$\frac{d}{dx}(\sin x) = \cos x, \qquad \frac{d}{dx}(\cos x) = -\sin x$$

can be confirmed by differentiating the series term by term:

$$\frac{d}{dx}(\sin x) = 1 - \frac{3x^2}{3!} + \frac{5x^4}{5!} - \frac{7x^6}{7!} + \frac{9x^8}{9!} - \cdots$$

$$= 1 - \frac{x^2}{2!} + \frac{x^4}{4!} - \frac{x^6}{6!} + \frac{x^8}{8!} - \cdots = \cos x,$$

$$\frac{d}{dx}(\cos x) = -\frac{2x}{2!} + \frac{4x^3}{4!} - \frac{6x^5}{6!} + \frac{8x^7}{8!} - \cdots$$

$$= -x + \frac{x^3}{3!} - \frac{x^5}{5!} + \frac{x^7}{7!} - \cdots$$

$$= -\left(x - \frac{x^3}{3!} + \frac{x^5}{5!} - \frac{x^7}{7!} + \cdots\right) = -\sin x. \quad \square$$

Example 4 We can sum the series

$$\sum_{k=1}^{\infty}\frac{x^k}{k} \qquad \text{for all } x \text{ in } (-1, 1)$$

by setting

$$g(x) = \sum_{k=1}^{\infty}\frac{x^k}{k} \qquad \text{for all } x \text{ in } (-1, 1)$$

and noting that

$$g'(x) = \sum_{k=1}^{\infty} \frac{kx^{k-1}}{k} = \sum_{k=1}^{\infty} x^{k-1} = \sum_{n=0}^{\infty} x^n = \frac{1}{1-x}.$$

the geometric series

With

$$g'(x) = \frac{1}{1-x} \qquad \text{and} \qquad g(0) = 0,$$

we can conclude that

$$g(x) = -\ln(1-x) = \ln\left(\frac{1}{1-x}\right).$$

It follows that

$$\sum_{k=1}^{\infty} \frac{x^k}{k} = \ln\left(\frac{1}{1-x}\right) \qquad \text{for all } x \text{ in } (-1, 1). \quad \square$$

Power series can also be integrated term by term.

THEOREM 11.9.4 TERM-BY-TERM INTEGRATION

If $f(x) = \displaystyle\sum_{k=0}^{\infty} a_k x^k$ converges on $(-c, c)$, then

$$g(x) = \sum_{k=0}^{\infty} \frac{a_k}{k+1} x^{k+1} \quad \text{converges on } (-c, c) \text{ and } \int f(x)\, dx = g(x) + C.$$

PROOF If $\displaystyle\sum_{k=0}^{\infty} a_k x^k$ converges on $(-c, c)$, then $\displaystyle\sum_{k=0}^{\infty} |a_k x^k|$ converges on $(-c, c)$ by Theorem 11.8.2. Since

$$\left| \frac{a_k}{k+1} x^k \right| \le |a_k x^k| \qquad \text{for all } k,$$

we know by comparison that

$$\sum_{k=0}^{\infty} \left| \frac{a_k}{k+1} x^k \right| \quad \text{also converges on } (-c, c).$$

It follows that

$$x \sum_{k=0}^{\infty} \frac{a_k}{k+1} x^k = \sum_{k=0}^{\infty} \frac{a_k}{k+1} x^{k+1} \quad \text{converges on } (-c, c).$$

With

$$f(x) = \sum_{k=0}^{\infty} a_k x^k \qquad \text{and} \qquad g(x) = \sum_{k=0}^{\infty} \frac{a_k}{k+1} x^{k+1},$$

we know from the differentiability theorem that

$$g'(x) = f(x) \qquad \text{and therefore} \qquad \int f(x)\, dx = g(x) + C. \quad \square$$

Term-by-term integration can be expressed as follows:

(11.9.5)
$$\int \left(\sum_{k=0}^{\infty} a_k x^k \right) dx = \left(\sum_{k=0}^{\infty} \frac{a_k}{k+1} x^{k+1} \right) + C.$$

Remark It follows from Theorem 11.9.4 that if $\sum a_k x^k$ has radius of convergence r, then its "integral" $\sum [1/(k+1)]a_k x^{k+1}$ also has radius of convergence r. As in the case of differentiating power series, convergence at the endpoints (if any) has to be tested separately. $\square$

If a power series converges at c and converges at d, then it converges at all numbers in between and

(11.9.6)
$$\int_{c}^{d} \left(\sum_{k=0}^{\infty} a_k x^k \right) dx = \sum_{k=0}^{\infty} \left(\int_{c}^{d} a_k x^k\, dx \right) = \sum_{k=0}^{\infty} \frac{a_k}{k+1}(d^{k+1} - c^{k+1}).$$

Example 5 You are familiar with the series expansion

$$\frac{1}{1+x} = \frac{1}{1-(-x)} = \sum_{k=0}^{\infty} (-1)^k x^k.$$

It is valid for all x in $(-1, 1)$ and for no other x. Integrating term by term we have

$$\ln(1+x) = \int \left(\sum_{k=0}^{\infty} (-1)^k x^k \right) dx = \left(\sum_{k=0}^{\infty} \frac{(-1)^k}{k+1} x^{k+1} \right) + C$$

for all x in $(-1, 1)$. At $x = 0$ both $\ln(1+x)$ and the series on the right are 0. It follows that $C = 0$ and thus

$$\ln(1+x) = \sum_{k=0}^{\infty} \frac{(-1)^k}{k+1} x^{k+1} = x - \frac{x^2}{2} + \frac{x^3}{3} - \frac{x^4}{4} + \cdots$$

for all x in $(-1, 1)$. $\square$

In Section 11.6 we were able to prove that this expansion for $\ln(1+x)$ was valid on the half-closed interval $(-1, 1]$; this gave us an expansion for $\ln 2$. Term-by-term integration gives us only the open interval $(-1, 1)$. Well, you may say, it's easy to see that the logarithm series also converges at $x = 1$.† True enough, but why to $\ln 2$? This takes us back to consideration of the remainder term, the method of Section 11.6.

---

† An alternating series with $a_k \to 0$.

There is, however, another way to proceed. The great Norwegian mathematician Niels Henrik Abel (1802–1829) proved the following result: suppose that

$$\sum_{k=0}^{\infty} a_k x^k \quad \text{converges on } (-c, c) \text{ and there represents } f(x).$$

If f is continuous at one of the endpoints (c or $-c$) and the series converges there, then the series represents the function at that point. Using Abel's theorem it is evident that the series for $\ln(1 + x)$ does represent the function at $x = 1$.

We come now to another important series expansion:

(11.9.7) $$\tan^{-1} x = x - \frac{x^3}{3} + \frac{x^5}{5} - \frac{x^7}{7} + \cdots \qquad \text{for } -1 \le x \le 1.$$

PROOF For x in $(-1, 1)$

$$\frac{1}{1 + x^2} = \frac{1}{1 - (-x^2)} = \sum_{k=0}^{\infty} (-1)^k x^{2k}$$

so that, by integration,

$$\tan^{-1} x = \int \left(\sum_{k=0}^{\infty} (-1)^k x^{2k} \right) dx = \left(\sum_{k=0}^{\infty} \frac{(-1)^k}{2k + 1} x^{2k+1} \right) + C.$$

The constant C is 0 because the series on the right and the inverse tangent are both 0 at $x = 0$. Thus, for all x in $(-1, 1)$, we have

$$\tan^{-1} x = \sum_{k=0}^{\infty} \frac{(-1)^k}{2k + 1} x^{2k+1} = x - \frac{x^3}{3} + \frac{x^5}{5} - \frac{x^7}{7} + \cdots.$$

That the series also represents the function at $x = -1$ and $x = 1$ follows directly from Abel's theorem: at both these points $\tan^{-1} x$ is continuous, and at both of these points the series converges. ❑

Since $\tan^{-1} 1 = \frac{1}{4}\pi$, we have

$$\tfrac{1}{4}\pi = 1 - \tfrac{1}{3} + \tfrac{1}{5} - \tfrac{1}{7} + \tfrac{1}{9} - \cdots.$$

This series was known to the Scottish mathematician James Gregory in 1671. It is an elegant formula for π, but it converges too slowly for computational purposes. A much more effective way of computing π is outlined in the supplement at the end of this section.

Term-by-term integration provides a method of calculating some (otherwise rather intractable) definite integrals. Suppose that you are trying to evaluate

$$\int_a^b f(x)\, dx,$$

but cannot find an antiderivative. If you can expand $f(x)$ in a convergent power series, then you can estimate the integral by forming the series and integrating term by term.

EXAMPLE 6 We will estimate

$$\int_0^1 e^{-x^2} \, dx$$

by expanding the integral in a power series and integrating term by term. Our starting point is the expansion

$$e^x = 1 + x + \frac{x^2}{2!} + \frac{x^3}{3!} + \frac{x^4}{4!} + \frac{x^5}{5!} + \frac{x^6}{6!} + \cdots \qquad \text{for all } x.$$

From this we see that

$$e^{-x^2} = 1 - x^2 + \frac{x^4}{2!} - \frac{x^6}{3!} + \frac{x^8}{4!} - \frac{x^{10}}{5!} + \frac{x^{12}}{6!} - \cdots \qquad \text{for all } x,$$

and therefore

$$\int_0^1 e^{-x^2} \, dx = \left[x - \frac{x^3}{3} + \frac{x^5}{5(2!)} - \frac{x^7}{7(3!)} + \frac{x^9}{9(4!)} - \frac{x^{11}}{11(5!)} + \frac{x^{13}}{13(6!)} - \cdots \right]_0^1$$

$$= 1 - \frac{1}{3} + \frac{1}{5(2!)} - \frac{1}{7(3!)} + \frac{1}{9(4!)} - \frac{1}{11(5!)} + \frac{1}{13(6!)} - \cdots.$$

This is an alternating series with declining terms. Therefore we know that the integral lies between consecutive partial sums. In particular it lies between

$$1 - \frac{1}{3} + \frac{1}{5(2!)} - \frac{1}{7(3!)} + \frac{1}{9(4!)} - \frac{1}{11(5!)}$$

and

$$\left[1 - \frac{1}{3} + \frac{1}{5(2!)} - \frac{1}{7(3!)} + \frac{1}{9(4!)} - \frac{1}{11(5!)} \right] + \frac{1}{13(6!)}.$$

As you can check, the first sum is greater than 0.74673 and the second one is less than 0.74684. It follows that

$$0.74673 < \int_0^1 e^{-x^2} \, dx < 0.74684.$$

The estimate 0.7468 approximates the integral within 0.0001. ❏

The integral of Example 6 was easy to estimate numerically because it could be expressed as an alternating series with decreasing terms. The next example requires more subtlety and illustrates a method more general than that used in Example 6.

EXAMPLE 7 We want to estimate

$$\int_0^1 e^{x^2} \, dx.$$

If we proceed exactly as in Example 6, we find that

$$\int_0^1 e^{x^2} \, dx = 1 + \frac{1}{3} + \frac{1}{5(2!)} + \frac{1}{7(3!)} + \frac{1}{9(4!)} + \frac{1}{11(5!)} + \frac{1}{13(6!)} + \cdots.$$

We now have a series expansion for the integral, but that expansion does not guide us directly to a numerical estimate for the integral. We know that s_n, the nth partial sum of the series, approximates the integral, but we don't know the accuracy of the approximation. We have no handle on the remainder left by s_n.

We start again, this time keeping track of the remainder. For $x \in [0, 1]$

$$0 \leq e^x - \left(1 + x + \frac{x^2}{2!} + \cdots + \frac{x^n}{n!}\right) = R_{n+1}(x) \overset{(11.6.3)}{\leq} e\left[\frac{x^{n+1}}{(n+1)!}\right] \leq \frac{3}{(n+1)!}.$$

If $x \in [0, 1]$, then $x^2 \in [0, 1]$, and therefore

$$0 \leq e^{x^2} - \left(1 + x^2 + \frac{x^4}{2!} + \cdots + \frac{x^{2n}}{n!}\right) \leq \frac{3}{(n+1)!}$$

Integrating this inequality from $x = 0$ to $x = 1$, we have

$$0 \leq \int_0^1 \left[e^{x^2} - \left(1 + x^2 + \frac{x^4}{2!} + \cdots + \frac{x^{2n}}{n!}\right)\right] dx \leq \int_0^1 \frac{3}{(n+1)!} dx.$$

Carrying out the integration where possible, we see that

$$0 \leq \int_0^1 e^{x^2} dx - \left[1 + \frac{1}{3} + \frac{1}{5(2!)} + \cdots + \frac{1}{(2n+1)(n!)}\right] \leq \frac{3}{(n+1)!}.$$

We can use this inequality to estimate the integral as closely as we wish. Since

$$\frac{3}{7!} = \frac{1}{1680} < 0.0006,$$

we see that

$$\alpha = 1 + \frac{1}{3} + \frac{1}{5(2!)} + \frac{1}{7(3!)} + \frac{1}{9(4!)} + \frac{1}{11(5!)} + \frac{1}{13(6!)}$$

approximates the integral within 0.0006. Arithmetical computation shows that

$$1.4626 \leq \alpha \leq 1.4627.$$

It follows that

$$1.4626 \leq \int_0^1 e^{x^2} dx \leq 1.4627 + 0.0006 = 1.4633.$$

The estimate 1.463 approximates the integral within 0.0004. ❑

It is time to relate Taylor series

$$\sum_{k=0}^{\infty} \frac{f^{(k)}(0)}{k!} x^k$$

to power series in general. The relation is very simple:

> On its interval of convergence a power series is the Taylor series of its sum.

That is, if you have a power series representation of a function f, then the series must be the Taylor series for f. To see this, all you have to do is differentiate

$$f(x) = a_0 + a_1x + a_2x^2 + \cdots + a_kx^k + \cdots$$

term by term. Do this and you will find that $f^{(k)}(0) = k!\, a_k$, and therefore

$$a_k = \frac{f^{(k)}(0)}{k!}.$$

The a_k are the Taylor coefficients of f.

We end this section by carrying out a few simple expansions.

Example 8 Expand $\cosh x$ and $\sinh x$ in powers of x.

SOLUTION There is no need to go through the labor of computing the Taylor coefficients

$$\frac{f^{(k)}(0)}{k!}$$

by differentiation. We know that

$$\cosh x = \tfrac{1}{2}(e^x + e^{-x}) \qquad \text{and} \qquad \sinh x = \tfrac{1}{2}(e^x - e^{-x}). \qquad (7.9.1)$$

Since

$$e^x = 1 + x + \frac{x^2}{2!} + \frac{x^3}{3!} + \frac{x^4}{4!} + \frac{x^5}{5!} + \cdots,$$

we have

$$e^{-x} = 1 - x + \frac{x^2}{2!} - \frac{x^3}{3!} + \frac{x^4}{4!} - \frac{x^5}{5!} + \cdots.$$

Thus

$$\cosh x = \frac{1}{2}\left(2 + 2\frac{x^2}{2!} + 2\frac{x^4}{4!} + \cdots\right) = 1 + \frac{x^2}{2!} + \frac{x^4}{4!} + \cdots = \sum_{k=0}^{\infty} \frac{x^{2k}}{(2k)!}$$

and

$$\sinh x = \frac{1}{2}\left(2x + 2\frac{x^3}{3!} + 2\frac{x^5}{5!} + \cdots\right) = x + \frac{x^3}{3!} + \frac{x^5}{5!} + \cdots = \sum_{k=0}^{\infty} \frac{x^{2k+1}}{(2k+1)!}.$$

Both expansions are valid for all real x, since the exponential expansions are valid for all real x. ❑

Example 9 Expand $x^2 \cos x^3$ in powers of x.

SOLUTION

$$\cos x = 1 - \frac{x^2}{2!} + \frac{x^4}{4!} - \frac{x^6}{6!} + \cdots.$$

Thus

$$\cos x^3 = 1 - \frac{(x^3)^2}{2!} + \frac{(x^3)^4}{4!} - \frac{(x^3)^6}{6!} + \cdots = 1 - \frac{x^6}{2!} + \frac{x^{12}}{4!} - \frac{x^{18}}{6!} + \cdots,$$

and

$$x^2 \cos x^3 = x^2 - \frac{x^8}{2!} + \frac{x^{14}}{4!} - \frac{x^{20}}{6!} + \cdots.$$

This expansion is valid for all real x, since the expansion for $\cos x$ is valid for all real x. ❑

ALTERNATIVE SOLUTION Since

$$x^2 \cos x^3 = \frac{d}{dx}\left(\frac{1}{3}\sin x^3\right),$$

we can derive the expansion for $x^2 \cos x^3$ by expanding $\frac{1}{3}\sin x^3$ and then differentiating term by term. ❑

EXERCISES 11.9

In Exercises 1–6, expand f in powers of x, basing your calculations on the geometric series

$$\frac{1}{1-x} = 1 + x + x^2 + \cdots + x^n + \cdots.$$

1. $f(x) = \dfrac{1}{(1-x)^2}.$

2. $f(x) = \dfrac{1}{(1-x)^3}.$

3. $f(x) = \dfrac{1}{(1-x)^k}.$

4. $f(x) = \ln(1-x).$

5. $f(x) = \ln(1-x^2).$

6. $f(x) = \ln(2-3x).$

In Exercises 7 and 8, expand f in powers of x, basing your calculations on the tangent series:

$$\tan x = x + \tfrac{1}{3}x^3 + \tfrac{2}{15}x^5 + \tfrac{17}{315}x^7 + \cdots.$$

7. $f(x) = \sec^2 x.$

8. $f(x) = \ln \cos x.$

In Exercises 9 and 10, find $f^{(9)}(0)$.

9. $f(x) = x^2 \sin x.$

10. $f(x) = x \cos x^2.$

In Exercises 11–22, expand f in powers of x.

11. $f(x) = \sin x^2.$

12. $f(x) = x^2 \tan^{-1} x.$

13. $f(x) = e^{3x^3}.$

14. $f(x) = \dfrac{1-x}{1+x}.$

15. $f(x) = \dfrac{2x}{1-x^2}.$

16. $f(x) = x \sinh x^2.$

17. $f(x) = \dfrac{1}{1-x} + e^x.$

18. $f(x) = \cosh x \sinh x.$

19. $f(x) = x \ln(1+x^3).$

20. $f(x) = (x^2+x) \ln(1+x).$

21. $f(x) = x^3 e^{-x^3}.$

22. $f(x) = x^5 (\sin x + \cos 2x).$

In Exercises 23–26, evaluate the given limit in two ways: (a) using L'Hospital's rule, and (b) using power series.

23. $\displaystyle\lim_{x\to 0} \frac{1-\cos x}{x^2}.$

24. $\displaystyle\lim_{x\to 0} \frac{\sin x - x}{x^2}.$

25. $\displaystyle\lim_{x\to 0} \frac{\cos x - 1}{x \sin x}.$

26. $\displaystyle\lim_{x\to 0} \frac{e^x - 1 - x}{x \tan^{-1} x}.$

In Exercises 27–30, find a powers series representation of the improper integral.

27. $\displaystyle\int_0^x \frac{\ln(1+t)}{t}\,dt.$

28. $\displaystyle\int_0^x \frac{1-\cos t}{t^2}\,dt.$

29. $\displaystyle\int_0^x \frac{\tan^{-1} t}{t}\,dt.$

30. $\displaystyle\int_0^x \frac{\sinh t}{t}\,dt.$

▶ In Exercises 31–36, estimate within 0.01.

31. $\displaystyle\int_0^1 e^{-x^3}\,dx.$

32. $\displaystyle\int_0^1 \sin x^2\,dx.$

33. $\displaystyle\int_0^1 \sin \sqrt{x}\,dx.$

34. $\displaystyle\int_0^1 x^4 e^{-x^2}\,dx.$

35. $\displaystyle\int_0^1 \tan^{-1} x^2\,dx.$

36. $\displaystyle\int_1^2 \frac{1-\cos x}{x}\,dx.$

▶ In Exercises 37–40, use a power series to estimate the integral within 0.0001.

37. $\displaystyle\int_0^1 \frac{\sin x}{x}\,dx.$

38. $\displaystyle\int_0^{0.5} \frac{1-\cos x}{x^2}\,dx.$

39. $\displaystyle\int_0^{0.5} \frac{\ln(1+x)}{x}\,dx.$

40. $\displaystyle\int_0^{0.2} x \sin x\,dx.$

In Exercises 41–43, sum the series.

41. $\displaystyle\sum_{k=0}^{\infty} \frac{1}{k!}x^{3k}.$

42. $\displaystyle\sum_{k=0}^{\infty} \frac{1}{k!}x^{3k+1}.$

43. $\displaystyle\sum_{k=1}^{\infty} \frac{3k}{k!} x^{3k-1}$.

44. Let $f(x) = \dfrac{e^x - 1}{x}$.

   (a) Find a power series representation of f in powers of x.

   (b) Differentiate the power series in part (a) and show that

$$\sum_{n=1}^{\infty} \frac{n}{(n+1)!} = 1.$$

45. Let $f(x) = xe^x$.

   (a) Find a power series representation of f in powers of x.

   (b) Integrate the power series in part (a) and show that

$$\sum_{n=1}^{\infty} \frac{1}{n!(n+2)} = \frac{1}{2}.$$

46. Deduce the differentiation formulas

$$\frac{d}{dx}(\sinh x) = \cosh x, \qquad \frac{d}{dx}(\cosh x) = \sinh x$$

from the expansions of $\sinh x$ and $\cosh x$ in powers of x.

47. Show that, if $\sum a_k x^k$ and $\sum b_k x^k$ both converge to the same sum on some interval, then $a_k = b_k$ for each k.

48. Show that, if $\epsilon > 0$, then

$$|kx^{k-1}| < (|x| + \epsilon)^k \qquad \text{for all } k \text{ sufficiently large.}$$

HINT: Take the kth root of the left side and let $k \to \infty$.

49. Suppose that the function f has the power series representation $f(x) = \sum_{k=0}^{\infty} a_k x^k$.

   (a) Show that if f is an even function, then $a_{2k+1} = 0$ for all k.

   (b) Show that if f is an odd function, then $a_{2k} = 0$ for all k.

50. Suppose that the function f is infinitely differentiable on an interval containing 0, and suppose that $f'(x) = -2f(x)$ and $f(0) = 1$. Use these properties to find the power series representation of f in powers of x. Do you recognize this function?

In Exercises 51–53, estimate within 0.001 by the method of this section and check your result by carrying out the integration directly.

51. $\displaystyle\int_0^{1/2} x \ln(1+x)\, dx.$ **52.** $\displaystyle\int_0^1 x \sin x\, dx.$

53. $\displaystyle\int_0^1 xe^{-x}\, dx.$

54. Show that

$$0 \le \int_0^2 e^{x^2}\, dx - \left[2 + \frac{2^3}{3} + \frac{2^5}{5(2!)} + \cdots + \frac{2^{2n+1}}{(2n+1)n!}\right]$$
$$< \frac{e^4 2^{2n+3}}{(n+1)!}.$$

---

*SUPPLEMENT TO SECTION 11.9

PROOF OF THEOREM 11.9.3

Set

$$f(x) = \sum_{k=0}^{\infty} a_k x^k \qquad \text{and} \qquad g(x) = \sum_{k=0}^{\infty} \frac{d}{dx}(a_k x^k) = \sum_{k=1}^{\infty} k a_k x^{k-1}.$$

Select x from $(-c, c)$. We want to show that

$$\lim_{h \to 0} \frac{f(x+h) - f(x)}{h} = g(x).$$

For $x + h$ in $(-c, c)$, $h \ne 0$, we have

$$\left| g(x) - \frac{f(x+h) - f(x)}{h} \right| = \left| \sum_{k=1}^{\infty} k a_k x^{k-1} - \sum_{k=0}^{\infty} \frac{a_k(x+h)^k - a_k x^k}{h} \right|$$
$$= \left| \sum_{k=1}^{\infty} k a_k x^{k-1} - \sum_{k=1}^{\infty} a_k \left[\frac{(x+h)^k - x^k}{h}\right] \right|.$$

By the mean-value theorem

$$\frac{(x+h)^k - x^k}{h} = k(t_k)^{k-1}$$

for some number t_k between x and $x + h$. Thus we can write

$$\left| g(x) - \frac{f(x + h) - f(x)}{h} \right| = \left| \sum_{k=1}^{\infty} ka_k x^{k-1} - \sum_{k=1}^{\infty} ka_k(t_k)^{k-1} \right|$$

$$= \left| \sum_{k=1}^{\infty} ka_k[x^{k-1} - (t_k)^{k-1}] \right|$$

$$= \left| \sum_{k=2}^{\infty} ka_k[x^{k-1} - (t_k)^{k-1}] \right|.$$

By the mean-value theorem

$$\frac{x^{k-1} - (t_k)^{k-1}}{x - t_k} = (k - 1)(p_{k-1})^{k-2}$$

for some number p_{k-1} between x and t_k. Obviously, then,

$$|x^{k-1} - (t_k)^{k-1}| = |x - t_k||(k - 1)(p_{k-1})^{k-2}|.$$

Since $|x - t_k| < |h|$ and $|p_{k-1}| \le |\alpha|$ where $|\alpha| = \max\{|x|, |x + h|\}$,

$$|x^{k-1} - (t_k)^{k-1}| \le |h||(k - 1)\alpha^{k-2}|.$$

Thus

$$\left| g(x) - \frac{f(x + h) - f(x)}{h} \right| \le |h| \sum_{k=2}^{\infty} |k(k - 1)a_k\alpha^{k-2}|.$$

Since the series converges,

$$\lim_{h \to 0} \left(|h| \sum_{k=2}^{\infty} |k(k - 1)a_k\alpha^{k-2}| \right) = 0.$$

This gives

$$\lim_{h \to 0} \left| g(x) - \frac{f(x + h) - f(x)}{h} \right| = 0 \quad \text{and thus} \quad \lim_{h \to 0} \frac{f(x + h) - f(x)}{h} = g(x). \quad \square$$

Calculating π

We base our computation of π on the inverse tangent series (11.9.7)

$$\tan^{-1} x = x - \frac{x^3}{3} + \frac{x^5}{5} - \frac{x^7}{7} + \cdots \quad \text{for } -1 \le x \le 1$$

and the relation

(11.9.8)
$$\boxed{\tfrac{1}{4}\pi = 4 \tan^{-1} \tfrac{1}{5} - \tan^{-1} \tfrac{1}{239}.\dagger}$$

The inverse tangent series gives

$$\tan^{-1} \tfrac{1}{5} = \tfrac{1}{5} - \tfrac{1}{3}(\tfrac{1}{5})^3 + \tfrac{1}{5}(\tfrac{1}{5})^5 - \tfrac{1}{7}(\tfrac{1}{5})^7 + \cdots$$

---

† This relation was discovered in 1706 by John Machin, a Scotsman. It can be verified by repeated applications of the addition formula

$$\tan (A + B) = \frac{\tan A + \tan B}{1 - \tan A \tan B}.$$

First calculate $\tan (2 \tan^{-1} \tfrac{1}{5})$, then $\tan (4 \tan^{-1} \tfrac{1}{5})$, and finally $\tan (4 \tan^{-1} \tfrac{1}{5} - \tan^{-1} \tfrac{1}{239})$.

and

$$\tan^{-1} \tfrac{1}{239} = \tfrac{1}{239} - \tfrac{1}{3}(\tfrac{1}{239})^3 + \tfrac{1}{5}(\tfrac{1}{239})^5 - \tfrac{1}{7}(\tfrac{1}{239})^7 + \cdots.$$

These are alternating series $\Sigma\,(-1)^k a_k$ with a_k decreasing toward 0. Thus we know that

$$\tfrac{1}{5} - \tfrac{1}{3}(\tfrac{1}{5})^3 \le \tan^{-1} \tfrac{1}{5} \le \tfrac{1}{5} - \tfrac{1}{3}(\tfrac{1}{5})^3 + \tfrac{1}{5}(\tfrac{1}{5})^5$$

and

$$\tfrac{1}{239} - \tfrac{1}{3}(\tfrac{1}{239})^3 \le \tan^{-1} \tfrac{1}{239} \le \tfrac{1}{239}.$$

With these inequalities, together with relation (11.9.8), we can show that

$$3.14 < \pi < 3.147.$$

By using six terms of the series for $\tan^{-1} \tfrac{1}{5}$ and still only two of the series for $\tan^{-1} \tfrac{1}{239}$, we can show that

$$3.14159262 < \pi < 3.14159267.$$

Greater accuracy can be obtained by taking more terms into account. For instance, fifteen terms of the series for $\tan^{-1} \tfrac{1}{5}$ and just four terms of the series for $\tan^{-1} \tfrac{1}{239}$ determine π to twenty decimal places:

$$\pi \cong 3.14159\ 26535\ 89793\ 23846.$$

■ 11.10 THE BINOMIAL SERIES

Through a collection of problems we invite you to derive for yourself the basic properties of one of the most celebrated series of all—*the binomial series.*

Start with the binomial $1 + x$. Choose a real number $\alpha \ne 0$ and form the function

$$f(x) = (1 + x)^\alpha.$$

Note that if $\alpha = n$ is a positive integer, then

$$(1 + x)^n = 1 + nx + \frac{n(n-1)}{2!}x^2 + \cdots + nx^{n-1} + x^n$$

is the familiar binomial theorem. The binomial series is a generalization of the binomial theorem. You should also note that if $n = -1$, then $(1 + x)^{-1}$ is the sum of a geometric series

$$\frac{1}{1+x} = 1 - x + x^2 - x^3 + \cdots + (-1)^n x^n + \cdots$$

Problem 1 Show that

$$\frac{f^{(k)}(0)}{k!} = \frac{\alpha[\alpha - 1][\alpha - 2]\cdots[\alpha - (k-1)]}{k!}.$$

The number you just obtained is the coefficient of x^k in the expansion of $(1 + x)^\alpha$. It is called *the kth binomial coefficient* and is usually denoted by $\dbinom{\alpha}{k}$:

(11.10.1)

$$\binom{\alpha}{k} = \frac{\alpha[\alpha - 1][\alpha - 2]\cdots[\alpha - (k-1)]}{k!}.$$

For example, if $\alpha = 7$ and $k = 3$, then

$$\binom{7}{3} = \frac{7 \cdot 6 \cdot 5}{3!} = 35;$$

if $\alpha = 3/2$ and $k = 3$, then

$$\binom{3/2}{3} = \frac{(3/2)[(3/2) - 1][(3/2) - 2]}{3!} = \frac{(3/2)(1/2)(-1/2)}{6} = -\frac{1}{16}. \quad \square$$

Problem 2 Show that the binomial series

$$\sum \binom{\alpha}{k} x^k$$

has radius of convergence 1.   HINT: Use the ratio test.   ❑

From Problem 2 you know that the binomial series converges on the open interval $(-1, 1)$ and defines there an infinitely differentiable function. The next thing to show is that this function (the one defined by the series) is actually $(1 + x)^\alpha$. To do this, you first need some other results.

Problem 3   Verify the identity

$$(k + 1)\binom{\alpha}{k + 1} + k\binom{\alpha}{k} = \alpha\binom{\alpha}{k}. \quad \square$$

Problem 4   Use the identity of Problem 3 to show that the sum of the binomial series

$$\phi(x) = \sum_{k=0}^{\infty} \binom{\alpha}{k} x^k$$

satisfies the differential equation

$$(1 + x)\phi'(x) = \alpha\phi(x) \qquad \text{for all } x \text{ in } (-1, 1)$$

together with the side condition $\phi(0) = 1$.   ❑

You are now in a position to prove the main result.

Problem 5   Show that

(11.10.2)
$$(1 + x)^\alpha = \sum_{k=0}^{\infty} \binom{\alpha}{k} x^k \qquad \text{for all } x \text{ in } (-1, 1).$$

You can probably get a better feeling for the series by writing out the first few terms:

(11.10.3)
$$(1 + x)^\alpha = 1 + \alpha x + \frac{\alpha(\alpha - 1)}{2!}x^2 + \frac{\alpha(\alpha - 1)(\alpha - 2)}{3!}x^3 + \cdots. \quad \square$$

EXERCISES 11.10

In Exercises 1–10, expand f in powers of x up to x^4.

1. $f(x) = \sqrt{1 + x}$.

2. $f(x) = \sqrt{1 - x}$.

3. $f(x) = \sqrt{1 + x^2}$.

4. $f(x) = \sqrt{1 - x^2}$.

5. $f(x) = \dfrac{1}{\sqrt{1 + x}}$.

6. $f(x) = \dfrac{1}{\sqrt[3]{1 + x}}$.

7. $f(x) = \sqrt[4]{1 - x}$.

8. $f(x) = \dfrac{1}{\sqrt[4]{1 + x}}$.

9. $f(x) = (4 + x)^{3/2}$.

10. $f(x) = \sqrt{1 + x^4}$.

11. (a) Use a binomial series to find the Taylor series of $f(x) = 1/\sqrt{1 - x^2}$ in powers of x.
  (b) Use the series for f in part (a) to find the Taylor series for $F(x) = \sin^{-1} x$ and give the radius of convergence.

12. (a) Use a binomial series to find the Taylor series of $f(x) = 1/\sqrt{1 + x^2}$ in powers of x.

  (b) Use the series for f in part (a) to find the Taylor series for $F(x) = \sinh^{-1} x$ and give the radius of convergence.

▶ In Exercises 13–18, estimate by using the first three terms of a binomial expansion, rounding off your answer to four decimal places.

13. $\sqrt{98}$.   HINT: $\sqrt{98} = (100 - 2)^{1/2} = 10(1 - \frac{1}{50})^{1/2}$.

14. $\sqrt[5]{36}$.

15. $\sqrt[3]{9}$.

16. $\sqrt[4]{620}$.

17. $17^{-1/4}$.

18. $9^{-1/3}$.

▶ In Exercises 19–22, approximate each integral to within 0.001.

19. $\displaystyle\int_0^{1/3} \sqrt{1 + x^3}\, dx$.

20. $\displaystyle\int_0^{1/5} \sqrt{1 + x^4}\, dx$.

21. $\displaystyle\int_0^{1/2} \dfrac{1}{\sqrt{1 + x^2}}\, dx$.

22. $\displaystyle\int_0^{1/2} \dfrac{1}{\sqrt{1 - x^3}}\, dx$.

■ CHAPTER HIGHLIGHTS

11.1 Sigma Notation

11.2 Infinite Series

partial sums (p. 696)   convergence, divergence (p. 696)
sum of a series (p. 696)   a divergence test (p. 703)

$$\text{geometric series: } \sum_{k=0}^{\infty} x^k = \begin{cases} \dfrac{1}{1 - x}, & |x| < 1 \\ \text{diverges}, & |x| \geq 1 \end{cases}$$

If $\displaystyle\sum_{k=0}^{\infty} a_k$ converges, then $a_k \to 0$.   The converse is false.

11.3 The Integral Test; Comparison Theorems

integral test (p. 707)   basic comparison (p. 710)
limit comparison (p. 712)

$$\text{harmonic series: } \sum_{k=1}^{\infty} \frac{1}{k} \text{ diverges} \qquad p\text{-series: } \sum_{k=1}^{\infty} \frac{1}{k^p} \text{ converges iff } p > 1$$

11.4 The Root Test; The Ratio Test

root test (p. 715)   ratio test (p. 717)
summary on convergence tests (p. 719)

11.5 Absolute and Conditional Convergence; Alternating Series

absolutely convergent, conditionally convergent (pp. 720–721)
convergence theorem for alternating series (p. 722)
an estimate for alternating series (p. 723)
rearrangements (p. 725)

11.6 Taylor Polynomials in *x*; Taylor Series in *x*

Taylor polynomials in x (p. 728)   remainder term $R_{n+1}(x)$ (p. 730)
remainder estimate (p. 731)   Lagrange form of the remainder (p. 731)

Taylor series in x (Maclaurin series): $\displaystyle\sum_{k=0}^{\infty} \frac{f^{(k)}(0)}{k!} x^k$

$$e^x = \sum_{k=0}^{\infty} \frac{x^k}{k!}, \quad \text{all real } x \qquad\qquad \ln(1+x) = \sum_{k=1}^{\infty} \frac{(-1)^{k+1}}{k} x^k, \quad -1 < x \le 1$$

$$\sin x = \sum_{k=0}^{\infty} \frac{(-1)^k}{(2k+1)!} x^{2k+1}, \quad \text{all real } x \qquad \cos x = \sum_{k=0}^{\infty} \frac{(-1)^k}{(2k)!} x^{2k}, \quad \text{all real } x$$

11.7 Taylor Polynomials and Taylor Series in *x* − *a*

Taylor series in $x - a$: $\displaystyle\sum_{k=0}^{\infty} \frac{g^{(k)}(a)}{k!} (x-a)^k$

11.8 Power Series

power series (p. 745)   radius of convergence (p. 747)
interval of convergence (p. 748)

If a power series converges at $c \ne 0$, then it converges absolutely for $|x| < |c|$; if it diverges at c, then it diverges for $|x| > |c|$.

11.9 Differentiation and Integration of Power Series

$$\tan^{-1} x = \sum_{k=0}^{\infty} \frac{(-1)^k}{2k+1} x^{2k+1}, \quad -1 \le x \le 1$$

$$\cosh x = \sum_{k=0}^{\infty} \frac{x^{2k}}{(2k)!}, \quad \text{all real } x \qquad \sinh x = \sum_{k=0}^{\infty} \frac{x^{2k+1}}{(2k+1)!}, \quad \text{all real } x$$

On the interior of its interval of convergence, a power series can be differentiated and integrated term by term.

On its interval of convergence a power series is the Taylor series of its sum.

11.10 The Binomial Series

$$(1+x)^\alpha = \sum_{k=0}^{\infty} \binom{\alpha}{k} x^k = 1 + \alpha x + \frac{\alpha(\alpha-1)}{2!} x^2 + \cdots, \quad -1 < x < 1$$

■ PROJECTS AND EXPLORATIONS USING TECHNOLOGY

To do these exercises you will need a graphics calculator or a computer with graphing capability. The majority of these problems are open-ended so different approaches may be used to solve them. You should be aware that different approaches can result in slight variations in the answers. Round your numerical answers to at least four decimal places. The rounding method that your calculator or computer uses also may cause variations in answers.

11.1 The integral test is used to determine the convergence of a series of positive terms. In the case of convergent series, it can also be used to obtain reasonable approximations for the sum of the series. Let $\{a_n\}$ be the sequence defined by

$$a_n = \frac{[n^3 + \ln(n)] \ln(n)}{n^{4 + \ln(n)} + 3n^3 + 7}.$$

(a) Find j such that the sequence is decreasing for $n \geq j$.
(b) Use the integral test to show that the series

$$\sum_{n=1}^{\infty} \frac{[n^3 + \ln(n)] \ln(n)}{n^{4 + \ln(n)} + 3n^3 + 7}$$

converges. (You may need to start the series at a value $j > 1$.)
(c) Let

$$I_k = \int_j^k \frac{[x^3 + \ln(x)] \ln(x)}{x^{4 + \ln(x)} + 3x^3 + 7} \, dx,$$

and let

$$S_k = \sum_{n=j}^{k} \frac{[n^3 + \ln(n)] \ln(n)}{n^{4 + \ln(n)} + 3n^3 + 7}.$$

Show that $I_{k-1} < S_k < I_k$.
(d) How large must k be in order to get an estimate for the sum of the series that is accurate to three decimal places.
(e) Since you are approximating I_k, how can you be sure that you have enough accuracy for it so as to provide an accurate approximation of S_k?

11.2 The alternating series $\Sigma(-1)^n a_n$, $(a_n > 0$ for all $n)$, converges if the sequence $\{a_n\}$ is decreasing and has limit zero. Moreover, if the series converges, then the sum S is always between two consecutive partial sums, S_k and S_{k+1}. The following procedure allows even better approximations for the sum. Consider the alternating series

$$\sum_{n=0}^{\infty} \frac{(-1)^n(n^2 + 1)}{n^3 + 1}.$$

(a) Estimate the sum of the first 10, 20, 30, . . . , 100 terms of this series.
(b) Determine k such that the partial sum S_k approximates the sum S with three decimal place accuracy.
(c) Using the values of the partial sums from part (a), estimate the differences between the partial sums and S as a function of n. Add the value given by this function to the partial sum S_{100} to get a new approximation for S. This process is called *extrapolation*.
(d) Now consider the following weighted average of consecutive terms

$$T_n = \frac{a_n S_n + a_{n-1} S_{n-1}}{a_n + a_{n-1}}$$

and calculate T_n for $n = 10, 20, 30, . . . , 100$. Plot T_n versus n and estimate the error as a function of n.
(e) Generate the sequence $\{b_n\}$ where

$$b_n = \frac{(S - S_{n-1})}{(S_n - S_{n-1})}$$

and S is the sum of the series. For our example, what is $\lim_{n\to\infty} b_n$?
(f) Compare the convergence of $\{T_n\}$ with the convergence of the sequence $\{U_n\}$ given by

$$U_n = \frac{S_n + S_{n-1}}{2}.$$

11.3 Consider the function

$$F(z) = \int_0^1 [\cos zx^2 + \sin z^2x^3] \, dx.$$

(a) Using technology, graph the function F. Based on the graph, make conjectures about the properties of this function. For example, is it continuous? differentiable? Where is it increasing? decreasing? And so on.

(b) Determine the Taylor series in powers of x for each of the functions $\cos zx^2$ and $\sin z^2x^3$. Then approximate $F(z)$ using termwise integration. Can you verify the information that you found in part (a) from this expression for F?

(c) Now assume that the derivative of F with respect to z can be found by differentiating the integrand with respect to z, obtaining a series expansion in powers of x, and then integrating that series termwise. Do you get the same series?

VECTORS

■ 12.1 CARTESIAN SPACE COORDINATES

To introduce a Cartesian coordinate system in three-dimensional space we begin with a plane Cartesian coordinate system $O\text{-}xy$. Through the point O, which we continue to call the origin, we pass a third line, perpendicular to the other two. This third line we call the z-axis. We assign real number coordinates to the z-axis in the usual way, with the z-coordinate 0 being assigned to the origin O.

For later convenience we orient the z-axis so that $O\text{-}xyz$ forms a "right-handed" system. That is, if the index finger of the right hand points along the positive x-axis and the middle finger along the positive y-axis, then the thumb will point along the positive z-axis. (See Figure 12.1.1.)

Figure 12.1.1

The planes determined by the x- and y-axes, the x- and z-axes, and the y- and z-axes are called the *coordinate planes,* and are labeled the *xy-plane,* the *xz-plane,* and the *yz-plane,* respectively. If P is an arbitrary point in space, then the plane through P parallel to the yz-plane will intersect the x-axis at a point x_0, the plane through

P parallel to the xz-plane will intersect the y-axis at a point y_0, and the plane through P parallel to the xy-plane will intersect the z-axis at a point z_0. We associate P with the ordered triple of numbers (x_0, y_0, z_0) and call (x_0, y_0, z_0) the *Cartesian* (or *rectangular*) *space coordinates* of P, or simply the *space coordinates* of P. In particular, the point on the x-axis with x-coordinate x_0 has space coordinates $(x_0, 0, 0)$; the point on the y-axis with y-coordinate y_0 has space coordinates $(0, y_0, 0)$; and the point on the z-axis with z-coordinate z_0 has space coordinates $(0, 0, z_0)$. (See Figure 12.1.2.)

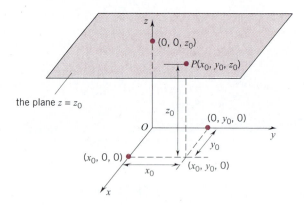

Figure 12.1.2

A point is in the xy-plane iff it is of the form $(x, y, 0)$. Thus, an equation for the xy-plane is $z = 0$. The equation

$$z = z_0$$

represents a plane parallel to the xy-plane, z_0 units above the xy-plane when z_0 is positive, $|z_0|$ units below the xy-plane when z_0 is negative. Similarly, equations of the form $x = x_0$, constant, represent planes parallel to the yz-plane, and equations of the form $y = y_0$, constant, represent planes parallel to xz-plane. (See, for example, Figure 12.1.3. There we have drawn the planes $x = 1$ and $y = 3$.)

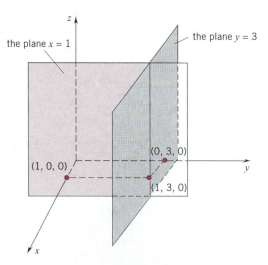

Figure 12.1.3

The Distance Formula

The distance $d(P_1, P_2)$ between two points $P_1(x_1, y_1, z_1)$ and $P_2(x_2, y_2, z_2)$ can be found by applying the Pythagorean theorem twice. With Q and R as in Figure 12.1.4, P_1P_2R and P_1RQ are both right triangles. From the first triangle

$$[d(P_1, P_2)]^2 = [d(P_1, R)]^2 + [d(R, P_2)]^2,$$

and from the second triangle

$$[d(P_1, R)]^2 = [d(Q, R)]^2 + [d(P_1, Q)]^2.$$

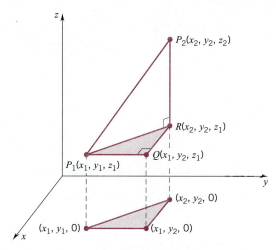

Figure 12.1.4

Combining equations,

$$[d(P_1, P_2)]^2 = [d(Q, R)]^2 + [d(P_1, Q)]^2 + [d(R, P_2)]^2$$
$$= (x_2 - x_1)^2 + (y_2 - y_1)^2 + (z_2 - z_1)^2.$$

Taking square roots, we have the distance formula:

(12.1.1)
$$d(P_1, P_2) = \sqrt{(x_2 - x_1)^2 + (y_2 - y_1)^2 + (z_2 - z_1)^2}.$$

The *sphere* of radius r centered at $P_0(a, b, c)$ is the set of all points $P(x, y, z)$ with $d(P, P_0) = r$. We can obtain an equation of this sphere by using (12.1.1);

(12.1.2)

Equation of a Sphere

$$(x - a)^2 + (y - b)^2 + (z - c)^2 = r^2.$$

See Figure 12.1.5. An equation for the sphere of radius r centered at the origin is

(12.1.3)
$$x^2 + y^2 + z^2 = r^2.$$

See Figure 12.1.6.

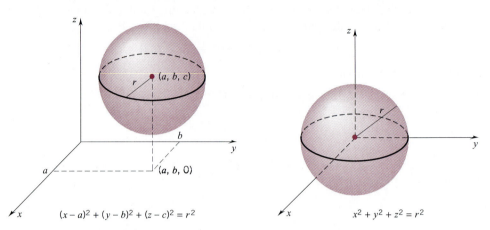

| Figure 12.1.5 | Figure 12.1.6 |

Example 1 An equation for the sphere with radius 3 and center $(5, -2, 0)$ is

$$(x - 5)^2 + (y + 2)^2 + z^2 = 9. \quad \square$$

Example 2 Show that

$$x^2 + y^2 + z^2 + 6x + 2y - 4z = 11$$

is an equation for a sphere. Find its center and radius.

SOLUTION We rewrite the equation as

$$(x^2 + 6x) + (y^2 + 2y) + (z^2 - 4z) = 11$$

and complete the squares. The result is

$$(x^2 + 6x + 9) + (y^2 + 2y + 1) + (z^2 - 4z + 4) = 11 + 9 + 1 + 4 = 25$$

or

$$(x + 3)^2 + (y + 1)^2 + (z - 2)^2 = 25.$$

The sphere has center $(-3, -1, 2)$ and radius $r = 5$. $\quad \square$

Symmetry

You are already familiar with two kinds of symmetry. Symmetry about a point: for example, the endpoints of a line segment are symmetric about the midpoint of the segment (Figure 12.1.7(a)); the graph of an odd function $y = f(x)$ is symmetric about the origin. Symmetry about a line (Figure 12.1.7(b)): the graph of an even function $y = f(x)$ is symmetric about the y-axis; if f and f^{-1} are inverse functions, then their graphs are symmetric about the line $y = x$. In space, we can also speak about symmetry about a plane. This idea is illustrated in Figure 12.1.7(c).

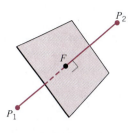

(a) $d(P_1, M) = d(M, P_2)$
point symmetry

(b) $d(P_1, R) = d(R, P_2)$
line symmetry

(c) $d(P_1, F) = d(F, P_2)$
plane symmetry

Figure 12.1.7

The endpoints $P_1(x_1, y_1, z_1)$ and $P_2(x_2, y_2, z_2)$ of the line segment $\overline{P_1P_2}$ are symmetric about the midpoint of the segment, which leads to:

(12.1.4)

> **Midpoint Formula**
> $$\left(\frac{x_1 + x_2}{2}, \frac{y_1 + y_2}{2}, \frac{z_1 + z_2}{2} \right).$$

EXERCISES 12.1

In Exercises 1–6, plot points A and B on a right-handed coordinate system. Then calculate the length of the line segment $\overline{AB}$ and find the midpoint.

1. $A(2, 0, 0)$, $B(0, 0, -4)$. **2.** $A(0, -2, 0)$, $B(0, 0, 6)$.

3. $A(0, -2, 5)$, $B(4, 1, 0)$. **4.** $A(4, 3, 0)$, $B(-2, 0, 6)$.

5. $A(2, 4, 7)$, $B(1, 6, 5)$. **6.** $A(-1, 2, 0)$, $B(3, -2, 4)$

In Exercises 7–12, find an equation for the plane through $(3, 1, -2)$ that satisfies the given condition.

7. Parallel to xy-plane. **8.** Parallel to xz-plane.

9. Perpendicular to y-axis. **10.** Perpendicular to z-axis.

11. Parallel to yz-plane. **12.** Perpendicular to x-axis.

In Exercises 13–20, find an equation for the sphere that satisfies the given conditions.

13. Centered at $(0, 2, -1)$ with radius 3.

14. Centered at $(1, 0, -2)$ with radius 4.

15. Centered at $(2, 4, -4)$ and passes through the origin.

16. Centered at the origin and passes through $(1, -2, 2)$.

17. The line segment joining $(0, 4, 2)$ and $(6, 0, 2)$ is a diameter.

18. Centered at $(2, 3, -4)$ and tangent to the xy-plane.

19. Centered at $(2, 3, -4)$ and tangent to the plane $x = 7$.

20. Centered at $(2, 3, -4)$ and tangent to the plane $y = 1$.

In Exercises 21–24, show that the given equation is an equation for a sphere, and find the center and radius.

21. $x^2 + y^2 + z^2 + 4x - 8y - 2z + 5 = 0$.

22. $x^2 + y^2 + z^2 - 2x + y - 4z = 4$.

23. $2x^2 + 2y^2 + 2z^2 + 8x - 4y = -1$.

24. $3x^2 + 3y^2 + 3z^2 - 12x - 6z + 3 = 0$.

The points $P(a, b, c)$ and $Q(2, 3, 5)$ are symmetric in the sense given in Exercises 25–36. Find a, b, c.

25. About the xy-plane. **26.** About the xz-plane.

27. About the yz-plane. **28.** About the x-axis.

29. About the y-axis. **30.** About the z-axis.

31. About the origin. **32.** About the plane $x = 1$.

33. About the plane $y = -1$.

34. About the plane $z = 4$.

35. About the point $(0, 2, 1)$.

36. About the point $(4, 0, 1)$.

37. Find an equation for each sphere that passes through the point $(5, 1, 4)$ and is tangent to all three coordinate planes.

38. Find an equation for the largest sphere that is centered at $(2, 1, -2)$ and intersects the sphere $x^2 + y^2 + z^2 = 1$.

39. Is the equation $x^2 + y^2 + z^2 - 4x + 4y + 6z + 20 = 0$ an equation for a sphere? If so, find its center and radius. If not, why isn't it?

40. Find conditions on A, B, C, D such that the equation

$$x^2 + y^2 + z^2 + Ax + By + Cz + D = 0$$

represents a sphere.

41. Show that the points $P(1, 2, 3)$, $Q(4, -5, 2)$, and $R(0, 0, 0)$ are the vertices of a right triangle.

42. The points $(5, -1, 3)$, $(4, 2, 1)$, and $(2, 1, 0)$ are the midpoints of the sides of a triangle PQR. Find the vertices P, Q, and R of the triangle.

In Exercises 43 and 44, the point R lies on the line segment that joins $P(a_1, a_2, a_3)$ and $Q(b_1, b_2, b_3)$.

43. (a) Find the coordinates of R given that

$$d(P, R) = t\, d(P, Q) \qquad \text{where } 0 \le t \le 1.$$

(b) Determine the value of t for which R is the midpoint of $\overline{PQ}$.

44. (a) Find the coordinates of R given that

$$d(P, R) = r\, d(R, Q) \qquad \text{where } r > 0.$$

(b) Determine the value of r for which R is the midpoint of $\overline{PQ}$.

■ 12.2 DISPLACEMENTS, FORCES AND VELOCITIES; VECTORS

Displacements

A *displacement* along a coordinate line can be specified by a real number and depicted by an arrow. For a displacement of a_1 units we use the number a_1 and an arrow that begins at any number x and ends at $x + a_1$. Following convention, if $a_1 > 0$, the arrow will point to the right, and if $a_1 < 0$, the arrow will point to the left. (Figure 12.2.1.) The *magnitude* of the displacement a_1 is defined to be the length of the arrow, which is the same as the distance between the points x and $x + a_1$. Thus, the magnitude of the displacement a_1 is $|a_1|$.

displacement a_1

Figure 12.2.1

Displacements in the plane are more interesting. Instead of having two possible directions, there are an infinite number of possible directions. Figure 12.2.2 shows several displacements beginning at the point (x, y). Displacements in the plane are specified by ordered pairs of real numbers. The magnitude of the displacement (a_1, a_2) is $\sqrt{a_1^2 + a_2^2}$, which equals the distance between the points (x, y) and $(x + a_1, y + a_2)$.

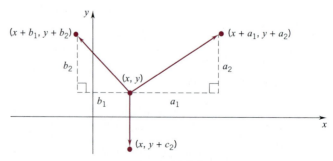

Figure 12.2.2

We now consider displacements in three-dimensional space. Here ordered triples of real numbers come into play. A displacement of a_1 units in the x-coordinate, a_2 units in the y-coordinate, a_3 units in the z-coordinate can be indicated by an arrow that begins at an arbitrary point (x, y, z) and ends at the point $(x + a_1, y + a_2, z + a_3)$. This displacement is represented by the ordered triple (a_1, a_2, a_3). (Figure 12.2.3.) The magnitude of the displacement (a_1, a_2, a_3) is $\sqrt{a_1^2 + a_2^2 + a_3^2}$.

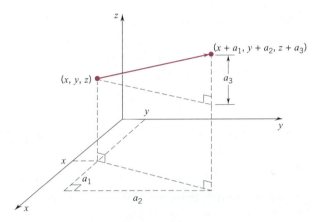

Figure 12.2.3

Remark You have probably noticed that we have ignored displacements of magnitude 0 in the preceding illustrations. In the case of three-dimensional space, the displacement of magnitude 0 is represented by the ordered triple $(0, 0, 0)$, which we will denote by $\mathbf{0}$. No direction is associated with this displacement, and consequently we will not use an arrow to depict it. As you will see, the ordered triple $\mathbf{0} = (0, 0, 0)$ arises naturally in the study of vectors, and we will have more to say about it later in this section. ❏

We can follow one displacement by another. A displacement (a_1, a_2, a_3) followed by (b_1, b_2, b_3) results in a total displacement $(a_1 + b_1, a_2 + b_2, a_3 + b_3)$. We can express this by writing

$$(a_1, a_2, a_3) + (b_1, b_2, b_3) = (a_1 + b_1, a_2 + b_2, a_3 + b_3).$$

We can picture the first displacement by an arrow from some point $P(x, y, z)$ to

$$Q(x + a_1, y + a_2, z + a_3),$$

the second displacement by an arrow from $Q(x + a_1, y + a_2, z + a_3)$ to

$$R(x + a_1 + b_1, y + a_2 + b_2, z + a_3 + b_3),$$

and then the resultant displacement by an arrow from $P(x, y, z)$ to

$$R(x + a_1 + b_1, y + a_2 + b_2, z + a_3 + b_3).$$

The three arrows then form a triangular pattern that is easy to remember (Figure 12.2.4).

From a displacement (a_1, a_2, a_3) and a real number α, we can form a new displacement $(\alpha a_1, \alpha a_2, \alpha a_3)$. We view this new displacement as α times the initial displacement and write

$$\alpha(a_1, a_2, a_3) = (\alpha a_1, \alpha a_2, \alpha a_3).$$

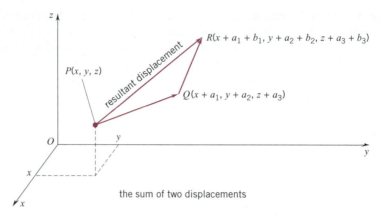

the sum of two displacements

Figure 12.2.4

The effect of multiplying a displacement by a number α is to change the magnitude of the displacement by a factor of $|\alpha|$:

$$\sqrt{(\alpha a_1)^2 + (\alpha a_2)^2 + (\alpha a_3)^2} = \sqrt{\alpha^2(a_1^2 + a_2^2 + a_3^2)} = |\alpha|\sqrt{a_1^2 + a_2^2 + a_3^2},$$

keeping the same direction if $\alpha > 0$, but reversing the direction if $\alpha < 0$. (If $\alpha = 0$, then $0(a_1, a_2, a_3) = (0, 0, 0) = \mathbf{0}$.) The displacement

$$2(a_1, a_2, a_3) = (2a_1, 2a_2, 2a_3)$$

is twice as long as (a_1, a_2, a_3) and has the same direction; the displacement

$$-(a_1, a_2, a_3) = (-a_1, -a_2, -a_3)$$

has the same length as (a_1, a_2, a_3) but the opposite direction; the displacement

$$-\tfrac{3}{2}(a_1, a_2, a_3) = (-\tfrac{3}{2}a_1, -\tfrac{3}{2}a_2, -\tfrac{3}{2}a_3)$$

is one and one-half times as long as (a_1, a_2, a_3) and has the opposite direction. (Figure 12.2.5.)

Forces and Velocities

As we have seen, a (nonzero) displacement is a quantity that has both direction and magnitude. Displacements arise naturally in many settings. For example, in tracking the motion of an object moving along a straight line, its *velocity* at time t can be represented by an arrow that points in the direction of motion. The length of the arrow represents the speed of the object.

Force and acceleration have both magnitude† and direction. If a force $\mathbf{F}$ is applied to an object of mass m, then the object will accelerate in the direction of the force (Figure 12.2.6). In this instance, force and acceleration are related by *Newton's Second Law of Motion*

$$\mathbf{F} = m\mathbf{a}.$$

A force $\mathbf{F}$ acting in three-dimensional space is completely determined by its components along the x-, y-, and z-axes. If these components are a_1, a_2, a_3, respectively, then $\mathbf{F}$ can be represented by the ordered triple (a_1, a_2, a_3). See Figure 12.2.7. In

---

† The standard units of measurement for the magnitude of a force $\mathbf{F}$ are: English—*pounds,* and metric—*newtons.*

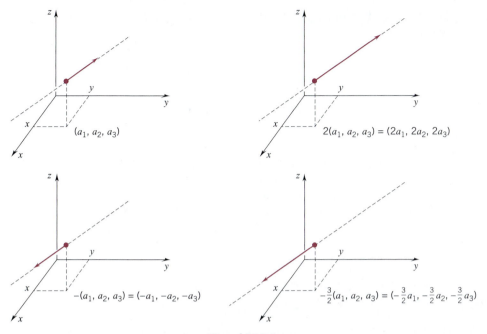

Figure 12.2.5

exactly the same way, the velocity **v** of an object moving in space can be represented by an ordered triple of real numbers.

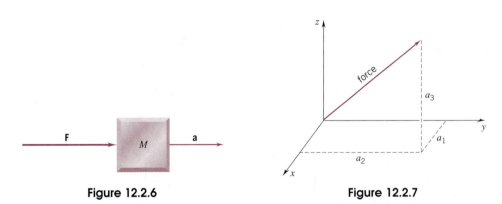

Figure 12.2.6   **Figure 12.2.7**

It can be verified by physical experiments that if two forces $\mathbf{F}_1 = (a_1, a_2, a_3)$ and $\mathbf{F}_2 = (b_1, b_2, b_3)$ are applied simultaneously at the same point, the effect is the same as that produced by the single force

$$\mathbf{F}_3 = (a_1 + b_1, a_2 + b_2, a_3 + b_3).$$

We call $\mathbf{F}_3$ the *resultant* or *total force* and write $\mathbf{F}_1 + \mathbf{F}_2 = \mathbf{F}_3$. For the ordered triples,

$$(a_1, a_2, a_3) + (b_1, b_2, b_3) = (a_1 + b_1, a_2 + b_2, a_3 + b_3).$$

Pictorially we have the usual force diagram, Figure 12.2.8. It is a parallelogram with the sides representing $\mathbf{F}_1 = (a_1, a_2, a_3)$ and $\mathbf{F}_2 = (b_1, b_2, b_3)$ and the diagonal representing

$$\mathbf{F}_3 = \mathbf{F}_1 + \mathbf{F}_2 = (a_1 + b_1, a_2 + b_2, a_3 + b_3).$$

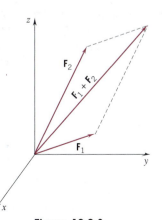

Figure 12.2.8

For any force $\mathbf{F} = (a_1, a_2, a_3)$ and any real number α, the force of $\alpha\mathbf{F}$ is defined by the equation

$$\alpha\mathbf{F} = (\alpha a_1, \alpha a_2, \alpha a_3).$$

Thus, once again we have

$$\alpha(a_1, a_2, a_3) = (\alpha a_1, \alpha a_2, \alpha a_3).$$

Vectors

The algebra of number triples that we have introduced to discuss displacements, forces, and velocities is so prodigiously rich in applications that it has found a firm place in the world of science and engineering and has generated much mathematics. It is to this mathematics that we now turn.

DEFINITION 12.2.1   VECTORS

Ordered triples of real numbers subject to addition defined by

$$(a_1, a_2, a_3) + (b_1, b_2, b_3) = (a_1 + b_1, a_2 + b_2, a_3 + b_3)$$

and multiplication by real numbers (*scalars*†) defined by

$$\alpha(a_1, a_2, a_3) = (\alpha a_1, \alpha a_2, \alpha a_3),$$

are called *vectors*.

Two vectors are called *equal* iff they have exactly the same *components*:

$$(a_1, a_2, a_3) = (b_1, b_2, b_3) \quad \text{iff} \quad a_1 = b_1, a_2 = b_2, a_3 = b_3.$$

Geometrically, we can depict vectors by arrows. To depict the vector (a_1, a_2, a_3) we can choose any initial point $Q = Q(x, y, z)$ and use the arrow

$$\overrightarrow{QR} \quad \text{with} \quad R = R(x + a_1, y + a_2, z + a_3). \qquad \text{(Figure 12.2.9)}$$

Usually we choose the origin as the initial point and use the arrow

$$\overrightarrow{OP} \quad \text{with} \quad P = P(a_1, a_2, a_3).\dagger\dagger$$

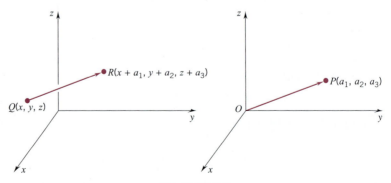

Figure 12.2.9

---

† In discussing vectors, *scalar* is synonymous with *real number*.

†† The ordered triple $(0, 0, 0)$ will not be represented by an arrow but simply by bold **0**.

Remark   Definition 12.2.1 specifically defines vectors in three-dimensional Euclidean space.   There are also two-dimensional vectors which are represented by ordered pairs of real numbers (a_1, a_2) and depicted by arrows in the xy-plane (Figure 12.2.10); four-dimensional vectors which are represented by ordered quadruples of real numbers; five-dimensional vectors which are represented by ordered quintuples, and so on.   Of course, we cannot draw figures to represent vectors in dimensions higher than three.   We will restrict our attention to three-dimensional vectors in the work which follows.   A brief discussion of two-dimensional vectors together with some examples involving forces and velocities is given at the end of the section.   ❏

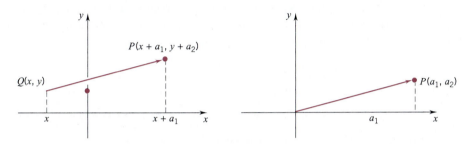

Figure 12.2.10

We use boldface letters to denote vectors.   Thus for

$$\mathbf{a} = (a_1, a_2, a_3) \qquad \text{and} \qquad \mathbf{b} = (b_1, b_2, b_3)$$

we have

$$\mathbf{a} + \mathbf{b} = (a_1 + b_1, a_2 + b_2, a_3 + b_3),$$

and, if α is a *scalar* (a real number),

$$\alpha\mathbf{a} = (\alpha a_1, \alpha a_2, \alpha a_3).$$

Addition of vectors satisfies the commutative and associative laws:

$$\mathbf{a} + \mathbf{b} = \mathbf{b} + \mathbf{a}$$

$$\mathbf{a} + (\mathbf{b} + \mathbf{c}) = (\mathbf{a} + \mathbf{b}) + \mathbf{c}.$$

These laws follow immediately from the definition of vector addition and the corresponding properties of real numbers.   For the *zero vector* $(0, 0, 0)$ we will use the symbol $\mathbf{0}$.   Obviously

$$0\mathbf{a} = \mathbf{0} \qquad \text{for all vectors } \mathbf{a}.$$

By the vector $-\mathbf{b}$ we mean $(-1)\mathbf{b}$; that is,

$$-(b_1, b_2, b_3) = (-b_1, -b_2, -b_3).$$

By $\mathbf{a} - \mathbf{b}$ we mean $\mathbf{a} + (-\mathbf{b})$; that is,

$$(a_1, a_2, a_3) - (b_1, b_2, b_3) = (a_1 - b_1, a_2 - b_2, a_3 - b_3).$$

Example 1   Given that

$$\mathbf{a} = (1, -1, 2), \qquad \mathbf{b} = (2, 3, -1), \qquad \mathbf{c} = (8, 7, 1),$$

find (a) $\mathbf{a} - \mathbf{b}$. (b) $2\mathbf{a} + \mathbf{b}$. (c) $3\mathbf{a} - 7\mathbf{b}$. (d) $2\mathbf{a} + 3\mathbf{b} - \mathbf{c}$.

SOLUTION

(a) $\mathbf{a} - \mathbf{b} = (1, -1, 2) - (2, 3, -1) = (1 - 2, -1 - 3, 2 + 1) = (-1, -4, 3)$.

(b) $2\mathbf{a} + \mathbf{b} = 2(1, -1, 2) + (2, 3, -1) = (2, -2, 4) + (2, 3, -1) = (4, 1, 3)$.

(c) $3\mathbf{a} - 7\mathbf{b} = 3(1, -1, 2) - 7(2, 3, -1)$
$$= (3, -3, 6) - (14, 21, -7) = (-11, -24, 13).$$

(d) $2\mathbf{a} + 3\mathbf{b} - \mathbf{c} = 2(1, -1, 2) + 3(2, 3, -1) - (8, 7, 1)$
$$= (2, -2, 4) + (6, 9, -3) - (8, 7, 1) = (0, 0, 0) = \mathbf{0}. \quad ❑$$

The addition of vectors can be visualized as the "addition" of arrows by the parallelogram law (as in the case of forces). If you picture $\mathbf{a}$, $\mathbf{b}$, and $\mathbf{a} + \mathbf{b}$ all as arrows emanating from the same point, say the origin, then $\mathbf{a} + \mathbf{b}$ acts as the diagonal of the parallelogram generated by $\mathbf{a}$ and $\mathbf{b}$. (See Figure 12.2.11.)

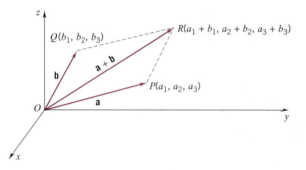

Figure 12.2.11   **Figure 12.2.12**

The addition of vectors can also be visualized as the "tail-to-head" addition of arrows (as in the case of displacements). If, instead of starting $\mathbf{b}$ at the origin, you start $\mathbf{b}$ at the tip of $\mathbf{a}$, then $\mathbf{a} + \mathbf{b}$ goes from the tail of $\mathbf{a}$ to the tip of $\mathbf{b}$. (Figure 12.2.12.) These two pictorial representations of vector addition lead to the same result.

DEFINITION 12.2.2 PARALLEL VECTORS

Two nonzero vectors $\mathbf{a}$ and $\mathbf{b}$ are said to be *parallel* provided that

$$\mathbf{a} = \alpha\mathbf{b} \qquad \text{for some real number } \alpha.$$

If $\alpha > 0$, $\mathbf{a}$ and $\mathbf{b}$ are said to have the *same direction*; if $\alpha < 0$, they are said to have *opposite directions*.

In the case of $\mathbf{a} = (2, -2, 6)$, $\mathbf{b} = (1, -1, 3)$, $\mathbf{c} = (-1, 1, -3)$ we have

$$\mathbf{a} = 2\mathbf{b} \qquad \text{and} \qquad \mathbf{a} = -2\mathbf{c}.$$

This tells us that $\mathbf{a}$ and $\mathbf{b}$ are parallel and have the same direction, whereas $\mathbf{a}$ and $\mathbf{c}$, though parallel, have opposite directions. (See Figure 12.2.13.)

SOLUTION

(a) $\|\mathbf{a}\| = \sqrt{1^2 + (-2)^2 + (3)^2} = \sqrt{1 + 4 + 9} = \sqrt{14}.$

(b) $\|\mathbf{b}\| = \sqrt{(-4)^2 + 1^2 + 0^2} = \sqrt{16 + 1 + 0} = \sqrt{17}.$

(c) $\|\mathbf{a} + \mathbf{b}\| = \|(-3, -1, 3)\| = \sqrt{(-3)^2 + (-1)^2 + (3)^2} = \sqrt{9 + 1 + 9} = \sqrt{19}.$

(d) $\|\mathbf{a} - \mathbf{b}\| = \|(5, -3, 3)\| = \sqrt{(5)^2 + (-3)^2 + (3)^2} = \sqrt{25 + 9 + 9} = \sqrt{43}.$

(e) $\|-7\mathbf{a}\| = |-7|\|\mathbf{a}\| = 7\sqrt{14}.$

(f) $\|2\mathbf{a} - 3\mathbf{b}\| = \|2(1, -2, 3) - 3(-4, 1, 0)\|$
$$= \|(14, -7, 6)\| = \sqrt{(14)^2 + (-7)^2 + (6)^2}$$
$$= \sqrt{196 + 49 + 36} = \sqrt{281}. \quad \square$$

To multiply a nonzero vector by a nonzero scalar α is to change its length by a factor of $|\alpha|$,

$$\|\alpha\mathbf{a}\| = |\alpha|\|\mathbf{a}\|,$$

keeping the same direction if $\alpha > 0$ and reversing the direction if $\alpha < 0$. The vector obtained from $\mathbf{a}$ simply by reversing its direction is the vector $(-1)\mathbf{a} = -\mathbf{a}$. (See Figure 12.2.16.)

Figure 12.2.16

Since $\mathbf{a} - \mathbf{b} = \mathbf{a} + (-\mathbf{b})$, we can draw the vector $\mathbf{a} - \mathbf{b}$ by drawing $-\mathbf{b}$ and adding it to the vector $\mathbf{a}$. (Figure 12.2.17.)

Figure 12.2.17   **Figure 12.2.18**

We can obtain the same result more easily by noting that $\mathbf{a} - \mathbf{b}$ is the vector that we must add to $\mathbf{b}$ to obtain $\mathbf{a}$. (Figure 12.2.18.)

Vectors of norm 1 are called *unit vectors*. If **b** is a nonzero vector, then there is a unit vector $\mathbf{u_b}$ that has the direction of **b**. To find $\mathbf{u_b}$, note that

$$\|\mathbf{u_b}\| = 1 \quad \text{and} \quad \mathbf{u_b} = \alpha\mathbf{b} \quad \text{for some } \alpha > 0.$$

It follows that

$$1 = \|\mathbf{u_b}\| = \|\alpha\mathbf{b}\| = |\alpha|\|\mathbf{b}\| = \alpha\|\mathbf{b}\|.$$

Thus

$$\alpha = \frac{1}{\|\mathbf{b}\|} \quad \text{and consequently} \quad \mathbf{u_b} = \frac{1}{\|\mathbf{b}\|}\mathbf{b} = \frac{\mathbf{b}}{\|\mathbf{b}\|}.$$

While

$$\mathbf{u_b} = \frac{\mathbf{b}}{\|\mathbf{b}\|}$$

is the unit vector in the direction of **b**,

$$-\mathbf{u_b} = -\frac{\mathbf{b}}{\|\mathbf{b}\|}$$

is the unit vector in the opposite direction.

We single out for special attention the vectors

$$\mathbf{i} = (1, 0, 0), \quad \mathbf{j} = (0, 1, 0), \quad \mathbf{k} = (0, 0, 1).$$

These vectors all have norm 1 and, if pictured as emanating from the origin, lie along the positive coordinate axes. They are called *the unit coordinate vectors*. (Figure 12.2.19.)

Every vector can be expressed as a linear combination of the unit coordinate vectors:

Figure 12.2.19

(12.2.5)

$$\text{for} \quad \mathbf{a} = (a_1, a_2, a_3) \quad \text{we have} \quad \mathbf{a} = a_1\mathbf{i} + a_2\mathbf{j} + a_3\mathbf{k}.$$

PROOF

$$\begin{aligned}(a_1, a_2, a_3) &= (a_1, 0, 0) + (0, a_2, 0) + (0, 0, a_3)\\ &= a_1(1, 0, 0) + a_2(0, 1, 0) + a_3(0, 0, 1)\\ &= a_1\mathbf{i} + a_2\mathbf{j} + a_3\mathbf{k}.\end{aligned}$$

The numbers a_1, a_2, a_3 are called the **i, j, k** components of the vector **a**.

Example 4 Given that

$$\mathbf{a} = 3\mathbf{i} - \mathbf{j} + \mathbf{k} \quad \text{and} \quad \mathbf{b} = 2\mathbf{i} + 3\mathbf{j} - \mathbf{k},$$

(1) Express $2\mathbf{a} - \mathbf{b}$ as a linear combination of **i, j, k**.

(2) Calculate $\|2\mathbf{a} - \mathbf{b}\|$.

(3) Find the unit vector $\mathbf{u_c}$ in the direction of $\mathbf{c} = 2\mathbf{a} - \mathbf{b}$.

SOLUTION

(1) $2\mathbf{a} - \mathbf{b} = 2(3\mathbf{i} - \mathbf{j} + \mathbf{k}) - (2\mathbf{i} + 3\mathbf{j} - \mathbf{k})$
$= 6\mathbf{i} - 2\mathbf{j} + 2\mathbf{k} - 2\mathbf{i} - 3\mathbf{j} + \mathbf{k} = 4\mathbf{i} - 5\mathbf{j} + 3\mathbf{k}.$

(2) $\|2\mathbf{a} - \mathbf{b}\| = \|4\mathbf{i} - 5\mathbf{j} + 3\mathbf{k}\| = \sqrt{16 + 25 + 9} = \sqrt{50} = 5\sqrt{2}.$

(3) $\mathbf{u_c} = \dfrac{2\mathbf{a} - \mathbf{b}}{\|2\mathbf{a} - \mathbf{b}\|} = \dfrac{1}{5\sqrt{2}}(4\mathbf{i} - 5\mathbf{j} + 3\mathbf{k}).$ ❑

Vectors in the Plane

A vector $\mathbf{a} = a_1\mathbf{i} + a_2\mathbf{j} + a_3\mathbf{k}$ for which $a_3 = 0$ can be written more simply as

(12.2.6)
$$\mathbf{a} = a_1\mathbf{i} + a_2\mathbf{j}$$

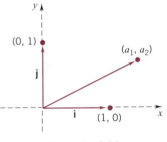

Figure 12.2.20

and identified with the vector (a_1, a_2) in the xy-plane. In particular, the vectors $\mathbf{i} = (1, 0)$ and $\mathbf{j} = (0, 1)$ are the unit coordinate vectors. See Figure 12.2.20.

It is easy to see that if $\mathbf{a}$ and $\mathbf{b}$ are vectors in the xy-plane and α is a scalar, then $\mathbf{a} + \mathbf{b}$ and $\alpha\mathbf{a}$ are also vectors in the plane. Thus, all of the definitions and results stated previously for vectors in space hold for vectors in the plane as well. In particular, if α is a scalar, and $\mathbf{a} = (a_1, a_2)$ and $\mathbf{b} = (b_1, b_2)$ are vectors in the xy-plane, then:

(1) $\mathbf{a} = \mathbf{b}$ iff $a_1 = b_1$ and $a_2 = b_2$.

(2) $\mathbf{a} + \mathbf{b} = (a_1 + b_1, a_2 + b_2) = (a_1 + b_1)\mathbf{i} + (a_2 + b_2)\mathbf{j}$.

(3) $\alpha\mathbf{a} = \alpha(a_1, a_2) = (\alpha a_1, \alpha a_2) = \alpha a_1\mathbf{i} + \alpha a_2\mathbf{j}$.

(4) $\mathbf{0} = (0, 0) = 0\mathbf{i} + 0\mathbf{j}$ is the zero vector.

(5) $\|\mathbf{a}\| = \sqrt{a_1^2 + a_2^2}$ is the norm, or magnitude, of $\mathbf{a}$.

A nonzero vector $\mathbf{a}$ can also be represented in terms of its magnitude and the angle θ that it makes with the positive x-axis (see Figure 12.2.21).

(12.2.7)
$$\mathbf{a} = \|\mathbf{a}\| \cos\theta\, \mathbf{i} + \|\mathbf{a}\| \sin\theta\, \mathbf{j}$$

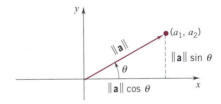

Figure 12.2.21

We will use either ordered pairs (a_1, a_2) or the form (12.2.6) to treat problems that are two-dimensional in nature. The following two examples illustrate analyses of forces and velocities in the plane.

Example 5 Two forces $\mathbf{F}_1$ and $\mathbf{F}_2$ with magnitudes 6 pounds and 4 pounds, respectively, are applied to an object at a point P as shown in Figure 12.2.22(a). Find the resultant force $\mathbf{F}$, give its magnitude, and give its direction in terms of an angle θ.

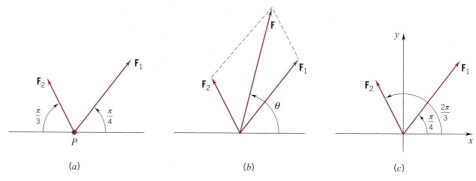

Figure 12.2.22

SOLUTION The resultant force **F** is shown in Figure 12.2.22(b). In Figure 12.2.22(c) we have introduced a coordinate system in order to find $\mathbf{F}_1$ and $\mathbf{F}_2$ in terms of their components:

$$\mathbf{F}_1 = \left(6 \cos \frac{\pi}{4} \right) \mathbf{i} + \left(6 \sin \frac{\pi}{4} \right) \mathbf{j} = 3\sqrt{2}\mathbf{i} + 3\sqrt{2}\mathbf{j}$$

$$\mathbf{F}_2 = \left(4 \cos \frac{2\pi}{3} \right) \mathbf{i} + \left(4 \sin \frac{2\pi}{3} \right) \mathbf{j} = -2\mathbf{i} + 2\sqrt{3}\mathbf{j}.$$

Thus, the resultant force $\mathbf{F} = \mathbf{F}_1 + \mathbf{F}_2$ is given by

$$\mathbf{F} = (3\sqrt{2} - 2)\mathbf{i} + (3\sqrt{2} + 2\sqrt{3})\mathbf{j} \cong 2.243\mathbf{i} + 7.707\mathbf{j}.$$

The magnitude of **F** is

$$\|\mathbf{F}\| = \sqrt{(3\sqrt{2} - 2)^2 + (3\sqrt{2} + 2\sqrt{3})^2} \cong \sqrt{(2.243)^2 + (7.707)^2} \cong 8.026.$$

To find the angle θ, note that

$$\cos \theta = \frac{3\sqrt{2} - 2}{\|\mathbf{F}\|} \cong \frac{2.243}{8.026} \cong 0.279 \qquad \text{so} \qquad \theta \cong 1.29 \text{ radians or } 73.8°. \quad \square$$

Example 6 A man wants to paddle his canoe across a river from a point X to the point Y on the opposite shore directly across from X. If he can paddle the canoe at the rate of 5 miles per hour and the current in the river is 3 miles per hour, in what direction θ should he steer his canoe in order to go straight across the river? See Figure 12.2.23a. Also, what is his resultant speed across the river?

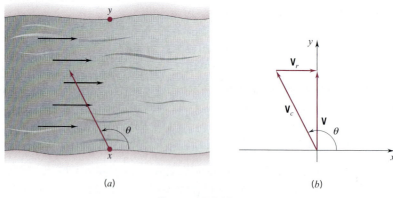

Figure 12.2.23

SOLUTION A velocity diagram is shown in Figure 12.2.23(b) where V_c denotes that velocity of the canoe, V_r denotes the velocity of the river, and V denotes the resultant velocity directed straight across the river. These velocities are given by

$$V_c = 5 \cos \theta \mathbf{i} + 5 \sin \theta \mathbf{j}, \qquad V_r = 3\mathbf{i} \qquad \text{and} \qquad V = \alpha\mathbf{j} \qquad \text{for some number } \alpha.$$

From the diagram, we have

$$V_c + V_r = V$$

so

$$5 \cos \theta \mathbf{i} + 5 \sin \theta \mathbf{j} + 3\mathbf{i} = \alpha\mathbf{j}$$

and

$$(5 \cos \theta + 3)\mathbf{i} + 5 \sin \theta \mathbf{j} = \alpha\mathbf{j}.$$

Thus, we must have

$$5 \cos \theta + 3 = 0 \qquad \text{or} \qquad \cos \theta = -\frac{3}{5} = -0.6.$$

It follows from this that $\theta \cong 2.214$ radians or $126.87°$.

The relative speed across the river is given by $\|V\| = \alpha$ and, from Figure 12.2.23(b),

$$\alpha^2 + 3^2 = 5^2 \qquad \text{and so} \qquad \alpha = 4;$$

therefore, the resultant speed is 4 miles per hour. ❏

EXERCISES 12.2

In Exercises 1–6, points P and Q are given. Find the vector $\overrightarrow{PQ}$ and determine its norm.

1. $P(1, -2, 5)$, $Q(4, 2, 3)$.

2. $P(-1, 0, 2)$, $Q(-3, -2, 4)$.

3. $P(4, -2)$, $Q(2, 4)$.

4. $P(\frac{3}{2}, -\frac{1}{4})$, $Q(-\frac{1}{2}, \frac{3}{4})$.

5. $P(3, 2, \frac{1}{2})$, $Q(-1, 4, \frac{5}{2})$.

6. $P(-4, 0, 7)$, $Q(0, 3, -1)$.

In Exercises 7–10, let $\mathbf{a} = (1, -2, 3)$, $\mathbf{b} = (3, 0, -1)$, and $\mathbf{c} = (-4, 2, 1)$, and find:

7. $2\mathbf{a} - \mathbf{b}$.

8. $2\mathbf{b} + 3\mathbf{c}$.

9. $-2\mathbf{a} + \mathbf{b} - \mathbf{c}$.

10. $\mathbf{a} + 3\mathbf{b} - 2\mathbf{c}$.

In Exercises 11–14, simplify the linear combinations.

11. $(2\mathbf{i} - \mathbf{j} + \mathbf{k}) + (\mathbf{i} - 3\mathbf{j} + 5\mathbf{k})$.

12. $(6\mathbf{j} - \mathbf{k}) + (3\mathbf{i} - \mathbf{j} + 2\mathbf{k})$.

13. $2(\mathbf{j} + \mathbf{k}) - 3(\mathbf{i} + \mathbf{j} - 2\mathbf{k})$.

14. $2(\mathbf{i} - \mathbf{j}) + 6(2\mathbf{i} + \mathbf{j} - 2\mathbf{k})$.

In Exercises 15–20, calculate the norm of the given vector.

15. $3\mathbf{i} + 4\mathbf{j}$.

16. $\mathbf{i} - \mathbf{j}$.

17. $2\mathbf{i} + \mathbf{j} - 2\mathbf{k}$.

18. $6\mathbf{i} + 2\mathbf{j} - \mathbf{k}$.

19. $\frac{1}{2}(\mathbf{i} + 4\mathbf{j}) - (\frac{3}{2}\mathbf{i} + \mathbf{k})$.

20. $(\mathbf{i} - \mathbf{j}) + 2(\mathbf{j} - \mathbf{i}) + (\mathbf{k} - \mathbf{j})$.

21. Let

$$\mathbf{a} = \mathbf{i} - \mathbf{j} + 2\mathbf{k}, \qquad \mathbf{b} = 2\mathbf{i} - \mathbf{j} + 2\mathbf{k},$$
$$\mathbf{c} = 3\mathbf{i} - 3\mathbf{j} + 6\mathbf{k}, \qquad \mathbf{d} = -2\mathbf{i} + 2\mathbf{j} - 4\mathbf{k}.$$

(a) Which vectors are parallel?

(b) Which vectors have the same direction?

(c) Which vectors have opposite directions?

22. (*Important*) Prove the following version of the triangle inequality:

$$\big|\|\mathbf{a}\| - \|\mathbf{b}\|\big| \le \|\mathbf{a} - \mathbf{b}\|.$$

HINT: Apply the previously stated triangle inequality to $\mathbf{a} = (\mathbf{a} - \mathbf{b}) + \mathbf{b}$.

In Exercises 23–28, find the unit vector in the direction of the given vector $\mathbf{a}$.

23. $\mathbf{a} = (3, -4)$.

24. $\mathbf{a} = (-2, 3)$.

25. $\mathbf{a} = (-4, 0, 3)$.

26. $\mathbf{a} = (2, 1, 2)$.

27. $\mathbf{a} = \mathbf{i} - 2\mathbf{j} + 2\mathbf{k}$.

28. $\mathbf{a} = 2\mathbf{i} + 3\mathbf{j} + \mathbf{k}$.

In Exercises 29–30, find the unit vector whose direction is opposite to the direction of **a**.

29. $\mathbf{a} = -\mathbf{i} + 3\mathbf{j} + 2\mathbf{k}$.   **30.** $\mathbf{a} = 2\mathbf{i} - \mathbf{k}$.

31. Label the vectors.

(i)

(ii)

(iii)

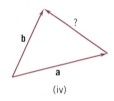

(iv)

32. Let $\mathbf{a} = (1, 1, 1)$, $\mathbf{b} = (-1, 3, 2)$, $\mathbf{c} = (-3, 0, 1)$, $\mathbf{d} = (4, -1, 1)$.
   (a) Express $\mathbf{a} + 2\mathbf{b} + 3\mathbf{c} + 4\mathbf{d}$ as a linear combination of $\mathbf{i}, \mathbf{j}, \mathbf{k}$.
   (b) Find scalars A, B, C such that $\mathbf{d} = A\mathbf{a} + B\mathbf{b} + C\mathbf{c}$.

33. Let $\mathbf{a} = (2, 0, -1)$, $\mathbf{b} = (1, 3, 5)$, $\mathbf{c} = (-1, 1, 1)$, $\mathbf{d} = (1, 1, 6)$.
   (a) Express $\mathbf{a} - 3\mathbf{b} + 2\mathbf{c} + 4\mathbf{d}$ as a linear combination of $\mathbf{i}, \mathbf{j}, \mathbf{k}$.
   (b) Find scalars A, B, C such that $\mathbf{d} = A\mathbf{a} + B\mathbf{b} + C\mathbf{c}$.

34. Find α given that $3\mathbf{i} + \mathbf{j} - \mathbf{k}$ and $\alpha\mathbf{i} - 4\mathbf{j} + 4\mathbf{k}$ are parallel.

35. Find α given that $3\mathbf{i} + \mathbf{j}$ and $\alpha\mathbf{j} - \mathbf{k}$ have the same length.

36. Find the unit vector in the direction of $\mathbf{i} - 2\mathbf{j} + 2\mathbf{k}$.

37. Find α given that $\| \alpha\mathbf{i} + (\alpha - 1)\mathbf{j} + (\alpha + 1)\mathbf{k} \| = 2$.

38. Find the vector of norm 2 in the direction of $\mathbf{i} + 2\mathbf{j} - \mathbf{k}$.

39. Find the vectors of norm 2 parallel to $3\mathbf{j} + 2\mathbf{k}$.

40. Express $\mathbf{c}$ in terms of $\mathbf{a}$ and $\mathbf{b}$, given that the tip of $\mathbf{c}$ bisects the line segment.

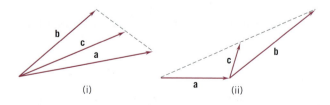

(i)   (ii)

In Exercises 41–46, find all vectors $\mathbf{v} = a\mathbf{i} + b\mathbf{j}$ that have the given properties.

41. Makes an angle of 30° with the positive x-axis and has norm 2.

42. Makes an angle of 135° with the positive x-axis and has norm 4.

43. Makes an angle of $\pi/4$ with the positive x-axis and has norm 1.

44. Makes an angle of $-5\pi/6$ with the positive x-axis and has norm 5.

45. Has norm 3 and the **i**-component is twice the **j**-component.

46. Has norm 5 and the **j**-component is three-fourths the **i**-component.

47. Let $\mathbf{a} = (2, -1)$ and $\mathbf{b} = (1, 2)$. Show that $\mathbf{a}$ and $\mathbf{b}$ are perpendicular.   HINT: Assume that $\mathbf{a}$ and $\mathbf{b}$ are the sides of a triangle.

48. Repeat Exercise 47 for $\mathbf{a} = (2, 0, 3)$ and $\mathbf{b} = (-6, 2, 4)$.

49. Let $\mathbf{a}$ and $\mathbf{b}$ be nonzero vectors such that $\|\mathbf{a} - \mathbf{b}\| = \|\mathbf{a} + \mathbf{b}\|$.
   (a) What can you conclude about the parallelogram generated by $\mathbf{a}$ and $\mathbf{b}$?
   (b) Show that, if $\mathbf{a} = a_1\mathbf{i} + a_2\mathbf{j} + a_3\mathbf{k}$ and $\mathbf{b} = b_1\mathbf{i} + b_2\mathbf{j} + b_3\mathbf{k}$, then
$$a_1 b_1 + a_2 b_2 + a_3 b_3 = 0.$$

50. (a) Show that, if $\mathbf{a}$ and $\mathbf{b}$ have the same direction, then $\|\mathbf{a} + \mathbf{b}\| = \|\mathbf{a}\| + \|\mathbf{b}\|$.
   (b) Does this equation necessarily hold if $\mathbf{a}$ and $\mathbf{b}$ are only parallel?

51. Let P and Q be two points in space and let M be the midpoint of the line segment $\overrightarrow{PQ}$. Let $\mathbf{p} = \overrightarrow{OP}$, $\mathbf{q} = \overrightarrow{OQ}$, and $\mathbf{m} = \overrightarrow{OM}$.
   (a) Show that $\mathbf{m} = \mathbf{p} + \frac{1}{2}(\mathbf{q} - \mathbf{p})$.
   (b) Derive the midpoint formula (12.1.4).

52. Let P and Q be two points in space and let R be the point on $\overrightarrow{PQ}$ which is twice as far from P as it is from Q. Let $\mathbf{p} = \overrightarrow{OP}$, $\mathbf{q} = \overrightarrow{OQ}$, and $\mathbf{r} = \overrightarrow{OR}$. Prove that $\mathbf{r} = \frac{1}{3}\mathbf{p} + \frac{2}{3}\mathbf{q}$.

▷53. A 200-pound weight is suspended by wires as shown in the following figure. Find the force exerted by each of the wires (called the *tension*) and determine the magnitude of each.   HINT: The resultant of the two tensions must counterbalance the weight.

54. The figure below illustrates two tugboats pulling a barge. If tugboat *A* exerts a force of 5000 pounds, what force must tugboat *B* exert if the barge is to move along the straight line *l*?

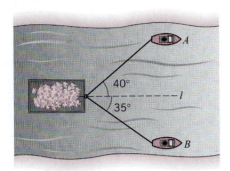

55. An airplane has an airspeed of 600 miles per hour. If the pilot is steering the plane in the direction N30°E and the wind is blowing from the northwest (N45°W) at the rate of 50 miles per hour, find the resultant velocity of the plane. The direction of the resultant is called the *true course* of the plane and the magnitude of the resultant is called the *ground speed*. Find the true course and ground speed of the plane.

56. The true course of an airplane is N40°W and its ground speed is 550 miles per hour (see Exercise 55). What is the airspeed of the plane if the wind is blowing from the east at 70 miles per hour?

A vector **r** emanating from the origin is called a *radius vector*. Each radius vector determines a unique point of space: the point at the tip of the vector. Conversely, each point of space determines a unique radius vector: the radius vector whose tip falls on that point. This one-to-one correspondence between the set of all radius vectors and the set of all points in three-dimensional space enables us to use radius vectors to specify sets in space. Thus, for example, the radius-vector equation $\|\mathbf{r}\| = 3$ can be used to represent the sphere of radius 3 centered at the origin: that sphere consists of the tips of all the radius vectors **r** that satisfy that equation.

57. Write a radius-vector equation or inequality for each of the following sets.
  (a) The sphere of radius 3 centered at $P(a_1, a_2, a_3)$.
  (b) The set of all points on or inside the sphere of radius 2 centered at the origin. (This set is called the *ball* of radius 2 about the origin.)
  (c) The ball of radius 1 about the point $P(a_1, a_2, a_3)$.
  (d) The set of all points equidistant from $P(a_1, a_2, a_3)$ and $Q(b_1, b_2, b_3)$. (This set forms a plane.)
  (e) The set of all points the sum of whose distances from $P(a_1, a_2, a_3)$ and $Q(b_1, b_2, b_3)$ is a constant $k > d(P, Q)$. [Such a set is called an *ellipsoid*. The ellipsoid is a three-dimensional analogue of the ellipse.] What happens if $k = d(P, Q)$? If $k < d(P, Q)$?

■ 12.3 THE DOT PRODUCT

In this section we introduce the first of two products that we define for vectors.

Introduction

We begin with two nonzero vectors

$$\mathbf{a} = a_1\mathbf{i} + a_2\mathbf{j} + a_3\mathbf{k}, \qquad \mathbf{b} = b_1\mathbf{i} + b_1\mathbf{j} + b_3\mathbf{k}.$$

How can we tell from the components of these vectors whether these vectors meet at right angles? To explore this question we draw Figure 12.3.1. By the Pythagorean Theorem **a** and **b** meet at right angles iff

$$\|\mathbf{a}\|^2 + \|\mathbf{b}\|^2 = \|\mathbf{b} - \mathbf{a}\|^2.$$

In terms of components this equation reads

$$(a_1^2 + a_2^2 + a_3^2) + (b_1^2 + b_2^2 + b_3^2) = (b_1 - a_1)^2 + (b_2 - a_2)^2 + (b_3 - a_3)^2,$$

which, as you can readily check, simplifies to

$$a_1b_1 + a_2b_2 + a_3b_3 = 0. \quad \square$$

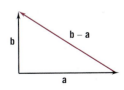

Figure 12.3.1

The expression $a_1 b_1 + a_2 b_2 + a_3 b_3$ is widely used in geometry and in physics. It has a name, the *dot product* of **a** and **b**, and there is a special notation for it, **a** · **b**. The notion is so important that it deserves a formal definition.

Definition of the Dot Product

DEFINITION 12.3.1

For any two vectors

$$\mathbf{a} = a_1\mathbf{i} + a_2\mathbf{j} + a_3\mathbf{k} \qquad \text{and} \qquad \mathbf{b} = b_1\mathbf{i} + b_2\mathbf{j} + b_3\mathbf{k}$$

we define the *dot product* **a** · **b** by setting

$$\mathbf{a} \cdot \mathbf{b} = a_1 b_1 + a_2 b_2 + a_3 b_3.$$

Remark Dot products for n-dimensional vectors, $n = 2, 4, 5, \ldots$, are defined in exactly the same way. In particular, for any 2 two-dimensional vectors $\mathbf{a} = a_1\mathbf{i} + a_2\mathbf{j}$ and $\mathbf{b} = b_1\mathbf{i} + b_2\mathbf{j}$

$$\mathbf{a} \cdot \mathbf{b} = a_1 b_1 + a_2 b_2.$$

The properties of the dot product discussed in this section for three-dimensional vectors also hold in the other dimensions. ❏

Example 1 For

$$\mathbf{a} = 2\mathbf{i} - \mathbf{j} + 3\mathbf{k}, \qquad \mathbf{b} = -3\mathbf{i} + \mathbf{j} + 4\mathbf{k}, \qquad \mathbf{c} = \mathbf{i} + 3\mathbf{j}$$

we have

$$\mathbf{a} \cdot \mathbf{b} = (2)(-3) + (-1)(1) + (3)(4) = -6 - 1 + 12 = 5,$$
$$\mathbf{a} \cdot \mathbf{c} = (2)(1) + (-1)(3) + (3)(0) = 2 - 3 = -1,$$
$$\mathbf{b} \cdot \mathbf{c} = (-3)(1) + (1)(3) + (4)(0) = -3 + 3 = 0.$$

The last equation tells us that **b** and **c** meet at right angles. (Verify this by drawing a figure.) ❏

Because **a** · **b** is not a vector, but a scalar, it is sometimes called the *scalar product* of **a** and **b**. We will continue to call it the dot product and speak of "dotting **a** with **b**."

Properties of the Dot Product

If we dot a vector with itself, we obtain the square of its norm:

(12.3.2)

$$\mathbf{a} \cdot \mathbf{a} = \|\mathbf{a}\|^2.$$

PROOF

$$\mathbf{a} \cdot \mathbf{a} = a_1 a_1 + a_2 a_2 + a_3 a_3 = a_1^2 + a_2^2 + a_3^2 = \|\mathbf{a}\|^2. \quad ❏$$

The dot product of any vector with the zero vector is zero:

(12.3.3)
$$\mathbf{a} \cdot \mathbf{0} = 0, \qquad \mathbf{0} \cdot \mathbf{a} = 0.$$

PROOF

$$(a_1)(0) + (a_2)(0) + (a_3)(0) = 0, \qquad (0)(a_1) + (0)(a_2) + (0)(a_3) = 0. \quad \square$$

The dot product is commutative:

(12.3.4)
$$\mathbf{a} \cdot \mathbf{b} = \mathbf{b} \cdot \mathbf{a},$$

and scalars can be factored:

(12.3.5)
$$\alpha \mathbf{a} \cdot \beta \mathbf{b} = \alpha\beta(\mathbf{a} \cdot \mathbf{b}).$$

PROOF

$$\mathbf{a} \cdot \mathbf{b} = a_1 b_1 + a_2 b_2 + a_3 b_3 = b_1 a_1 + b_2 a_2 + b_3 a_3 = \mathbf{b} \cdot \mathbf{a},$$

and

$$\alpha \mathbf{a} \cdot \beta \mathbf{b} = (\alpha a_1)(\beta b_1) + (\alpha a_2)(\beta b_2) + (\alpha a_3)(\beta b_3)$$
$$= \alpha\beta(a_1 b_1 + a_2 b_2 + a_3 b_3) = \alpha\beta(\mathbf{a} \cdot \mathbf{b}). \quad \square$$

The dot product satisfies the distributive laws:

(12.3.6)
$$\mathbf{a} \cdot (\mathbf{b} + \mathbf{c}) = \mathbf{a} \cdot \mathbf{b} + \mathbf{a} \cdot \mathbf{c}, \qquad (\mathbf{a} + \mathbf{b}) \cdot \mathbf{c} = \mathbf{a} \cdot \mathbf{c} + \mathbf{b} \cdot \mathbf{c}.$$

PROOF

$$\mathbf{a} \cdot (\mathbf{b} + \mathbf{c}) = a_1(b_1 + c_1) + a_2(b_2 + c_2) + a_3(b_3 + c_3)$$
$$= a_1 b_1 + a_1 c_1 + a_2 b_2 + a_2 c_2 + a_3 b_3 + a_3 c_3$$
$$= (a_1 b_1 + a_2 b_2 + a_3 b_3) + (a_1 c_1 + a_2 c_2 + a_3 c_3)$$
$$= \mathbf{a} \cdot \mathbf{b} + \mathbf{a} \cdot \mathbf{c}.$$

The second equation can be verified in a similar manner. ☐

Example 2 Given that

$$\|\mathbf{a}\| = 1, \qquad \|\mathbf{b}\| = 3, \qquad \|\mathbf{c}\| = 4, \qquad \mathbf{a} \cdot \mathbf{b} = 0, \qquad \mathbf{a} \cdot \mathbf{c} = 1, \qquad \mathbf{b} \cdot \mathbf{c} = -2,$$

find **(a)** $3\mathbf{a} \cdot (\mathbf{b} + 4\mathbf{c})$. **(b)** $(\mathbf{a} - \mathbf{b}) \cdot (2\mathbf{a} + \mathbf{b})$. **(c)** $[(\mathbf{b} \cdot \mathbf{c})\mathbf{a} - (\mathbf{a} \cdot \mathbf{c})\mathbf{b}] \cdot \mathbf{c}$.

SOLUTION

(a) $3\mathbf{a} \cdot (\mathbf{b} + 4\mathbf{c}) = (3\mathbf{a} \cdot \mathbf{b}) + (3\mathbf{a} \cdot 4\mathbf{c}) = 3(\mathbf{a} \cdot \mathbf{b}) + 12(\mathbf{a} \cdot \mathbf{c}) = 12.$

(b) $(\mathbf{a} - \mathbf{b}) \cdot (2\mathbf{a} + \mathbf{b}) = (\mathbf{a} \cdot 2\mathbf{a}) + (\mathbf{a} \cdot \mathbf{b}) + (-\mathbf{b} \cdot 2\mathbf{a}) + (-\mathbf{b} \cdot \mathbf{b})$
$$= 2(\mathbf{a} \cdot \mathbf{a}) + (\mathbf{a} \cdot \mathbf{b}) - 2(\mathbf{b} \cdot \mathbf{a}) - (\mathbf{b} \cdot \mathbf{b})$$
$$= 2\|\mathbf{a}\|^2 + (\mathbf{a} \cdot \mathbf{b}) - 2(\mathbf{a} \cdot \mathbf{b}) - \|\mathbf{b}\|^2$$
$$= 2 + 0 - 2(0) - 9 = -7.$$

(c) $[(\mathbf{b} \cdot \mathbf{c})\mathbf{a} - (\mathbf{a} \cdot \mathbf{c})\mathbf{b}] \cdot \mathbf{c} = [(\mathbf{b} \cdot \mathbf{c})\mathbf{a} \cdot \mathbf{c}] - [(\mathbf{a} \cdot \mathbf{c})\mathbf{b} \cdot \mathbf{c}]$
$$= (\mathbf{b} \cdot \mathbf{c})(\mathbf{a} \cdot \mathbf{c}) - (\mathbf{a} \cdot \mathbf{c})(\mathbf{b} \cdot \mathbf{c}) = 0. \quad ❑$$

Geometric Interpretation of the Dot Product

We begin with a triangle with sides a, b, c. (Figure 12.3.2.) If θ were $\frac{1}{2}\pi$, the Pythagorean Theorem would tell us that $c^2 = a^2 + b^2$. The Law of Cosines,

$$c^2 = a^2 + b^2 - 2ab \cos \theta,$$

is a generalization of the Pythagorean Theorem.

Figure 12.3.2

Figure 12.3.3

To derive the Law of Cosines we drop a perpendicular to side b. (Figure 12.3.3.) We then have

$$c^2 = z^2 + x^2 = (b - y)^2 + x^2 = b^2 - 2by + y^2 + x^2.$$

From the figure $y^2 + x^2 = a^2$ and $y = a \cos \theta$. Therefore

$$c^2 = a^2 + b^2 - 2ab \cos \theta,$$

as asserted. (What if the angle θ is obtuse? We leave that case to you.) ❑

Figure 12.3.4

Now back to dot products. If neither $\mathbf{a}$ nor $\mathbf{b}$ is zero, we can interpret $\mathbf{a} \cdot \mathbf{b}$ from the triangle of Figure 12.3.4. The lengths of the sides are $\|\mathbf{a}\|, \|\mathbf{b}\|, \|\mathbf{a} - \mathbf{b}\|$. By the law of cosines

$$\|\mathbf{a} - \mathbf{b}\|^2 = \|\mathbf{a}\|^2 + \|\mathbf{b}\|^2 - 2\|\mathbf{a}\| \|\mathbf{b}\| \cos \theta.$$

This gives

$$2\|\mathbf{a}\| \|\mathbf{b}\| \cos \theta = \|\mathbf{a}\|^2 + \|\mathbf{b}\|^2 - \|\mathbf{a} - \mathbf{b}\|^2 = a_1^2 + a_2^2 + a_3^2 + b_1^2 + b_2^2 + b_3^2$$
$$- (a_1 - b_1)^2 - (a_2 - b_2)^2 - (a_3 - b_3)^2 = 2(a_1 b_1 + a_2 b_2 + a_3 b_3) = 2(\mathbf{a} \cdot \mathbf{b}),$$

and thus

(12.3.7)

$$\boxed{\mathbf{a} \cdot \mathbf{b} = \|\mathbf{a}\| \|\mathbf{b}\| \cos \theta.}$$

(By convention, θ, the angle between $\mathbf{a}$ and $\mathbf{b}$, is measured in radians and taken from 0 to π, inclusive.)

From (12.3.7) you can see that the dot product of two vectors depends on the norms of the vectors and on the angle between them. For vectors of a given norm, the dot product measures the extent to which the vectors agree in direction. As the difference in direction increases, the dot product decreases:

If $\mathbf{a}$ and $\mathbf{b}$ have the same direction, then $\theta = 0$ and

$$\mathbf{a} \cdot \mathbf{b} = \|\mathbf{a}\| \|\mathbf{b}\|; \qquad (\cos 0 = 1)$$

this is the largest possible value for $\mathbf{a} \cdot \mathbf{b}$.

If $\mathbf{a}$ and $\mathbf{b}$ meet at right angles, then $\theta = \frac{1}{2}\pi$ and

$$\mathbf{a} \cdot \mathbf{b} = 0. \qquad (\cos \tfrac{1}{2}\pi = 0)$$

If $\mathbf{a}$ and $\mathbf{b}$ have opposite directions, then $\theta = \pi$ and

$$\mathbf{a} \cdot \mathbf{b} = -\|\mathbf{a}\| \|\mathbf{b}\|; \qquad (\cos \pi = -1)$$

this is the least possible value for $\mathbf{a} \cdot \mathbf{b}$

Two vectors are said to be *perpendicular* iff they lie at right angles or one of the vectors is the zero vector; in other words, two vectors are said to be perpendicular iff their dot product is zero.† In symbols

(12.3.8)
$$\boxed{\mathbf{a} \perp \mathbf{b} \quad \text{iff} \quad \mathbf{a} \cdot \mathbf{b} = 0.}$$
(Figure 12.3.5)

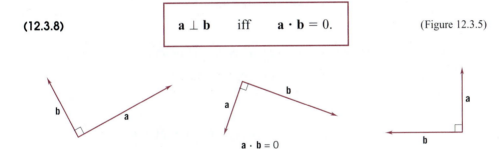

$$\mathbf{a} \cdot \mathbf{b} = 0$$

Figure 12.3.5

The unit coordinate vectors are obviously mutually perpendicular:

$$\mathbf{i} \cdot \mathbf{j} = 0, \qquad \mathbf{i} \cdot \mathbf{k} = 0, \qquad \mathbf{j} \cdot \mathbf{k} = 0.$$

Example 3 Verify that the vectors

$$\mathbf{a} = 2\mathbf{i} + \mathbf{j} + \mathbf{k} \qquad \text{and} \qquad \mathbf{b} = \mathbf{i} + \mathbf{j} - 3\mathbf{k}$$

are perpendicular.

SOLUTION

$$\mathbf{a} \cdot \mathbf{b} = (2)(1) + (1)(1) + (1)(-3) = 2 + 1 - 3 = 0. \quad \square$$

Example 4 Find the value of α for which

$$(3\mathbf{i} - \alpha\mathbf{j} + \mathbf{k}) \perp (\mathbf{i} + 2\mathbf{j}).$$

---

† This makes the zero vector both parallel and perpendicular to every vector. There is, however, no contradiction since we do not apply the notion of angle to the zero vector.

SOLUTION   For the two vectors to be perpendicular their dot product must be zero. Since

$$(3\mathbf{i} - \alpha\mathbf{j} + \mathbf{k}) \cdot (\mathbf{i} + 2\mathbf{j}) = (3)(1) + (-\alpha)2 + (1)(0) = 3 - 2\alpha,$$

α must be $\frac{3}{2}$.   ❏

With **a** and **b** both different from zero, we can divide the equation

$$\mathbf{a} \cdot \mathbf{b} = \|\mathbf{a}\| \|\mathbf{b}\| \cos\theta$$

by $\|\mathbf{a}\| \|\mathbf{b}\|$ to obtain

$$\cos\theta = \frac{\mathbf{a} \cdot \mathbf{b}}{\|\mathbf{a}\| \|\mathbf{b}\|} = \left(\frac{\mathbf{a}}{\|\mathbf{a}\|} \cdot \frac{\mathbf{b}}{\|\mathbf{b}\|} \right).$$

Writing

$$\mathbf{u_a} = \frac{\mathbf{a}}{\|\mathbf{a}\|} \quad \text{and} \quad \mathbf{u_b} = \frac{\mathbf{b}}{\|\mathbf{b}\|},$$

we have

(12.3.9)

$$\cos\theta = \mathbf{u_a} \cdot \mathbf{u_b}.$$

The cosine of the angle between two vectors is the dot product of the corresponding unit vectors.

Example 5   Calculate the angle between

$$\mathbf{a} = 2\mathbf{i} + 3\mathbf{j} + 2\mathbf{k} \quad \text{and} \quad \mathbf{b} = \mathbf{i} + 2\mathbf{j} - \mathbf{k}.$$

SOLUTION

$$\mathbf{u_a} = \frac{2\mathbf{i} + 3\mathbf{j} + 2\mathbf{k}}{\|2\mathbf{i} + 3\mathbf{j} + 2\mathbf{k}\|} = \frac{1}{\sqrt{17}}(2\mathbf{i} + 3\mathbf{j} + 2\mathbf{k}),$$

$$\mathbf{u_b} = \frac{\mathbf{i} + 2\mathbf{j} - \mathbf{k}}{\|\mathbf{i} + 2\mathbf{j} - \mathbf{k}\|} = \frac{1}{\sqrt{6}}(\mathbf{i} + 2\mathbf{j} - \mathbf{k}).$$

Therefore

$$\cos\theta = \mathbf{u_a} \cdot \mathbf{u_b} = \frac{1}{\sqrt{17}}\frac{1}{\sqrt{6}}[(2\mathbf{i} + 3\mathbf{j} + 2\mathbf{k}) \cdot (\mathbf{i} + 2\mathbf{j} - \mathbf{k})].$$

Since

$$(2\mathbf{i} + 3\mathbf{j} + 2\mathbf{k}) \cdot (\mathbf{i} + 2\mathbf{j} - \mathbf{k}) = (2)(1) + (3)(2) + (2)(-1) = 6,$$

we have

$$\cos\theta = \frac{6}{\sqrt{17}\sqrt{6}} = \frac{1}{17}\sqrt{102} \cong \frac{10.1}{17} \cong 0.594.$$

Thus, $\theta \cong 0.935$ radians, which is about 54 degrees.   ❏

Example 6 Show that, if **a** and **b** are both perpendicular to **c**, then every linear combination $\alpha\mathbf{a} + \beta\mathbf{b}$ is also perpendicular to **c**.

SOLUTION Suppose that **a** and **b** are both perpendicular to **c**. Then

$$\mathbf{a} \cdot \mathbf{c} = 0 \qquad \text{and} \qquad \mathbf{b} \cdot \mathbf{c} = 0.$$

It follows that

$$(\alpha\mathbf{a} + \beta\mathbf{b}) \cdot \mathbf{c} = \alpha\underbrace{(\mathbf{a} \cdot \mathbf{c})}_{0} + \beta\underbrace{(\mathbf{b} \cdot \mathbf{c})}_{0} = 0$$

and therefore $\alpha\mathbf{a} + \beta\mathbf{b}$ is perpendicular to **c**. ❑

Projections and Components

If $\mathbf{b} \neq \mathbf{0}$, then every vector **a** can be written in a unique manner as the sum of a vector $\mathbf{a}_{\|}$ parallel to **b** and a vector $\mathbf{a}_{\perp}$ perpendicular to **b**:

$$\mathbf{a} = \mathbf{a}_{\|} + \mathbf{a}_{\perp}. \qquad \text{(Exercise 39)}$$

The idea is illustrated in Figure 12.3.6. If **a** is parallel to **b**, then $\mathbf{a}_{\|} = \mathbf{a}$ and $\mathbf{a}_{\perp} = \mathbf{0}$. If **a** is perpendicular to **b**, then $\mathbf{a}_{\|} = \mathbf{0}$ and $\mathbf{a}_{\perp} = \mathbf{a}$.

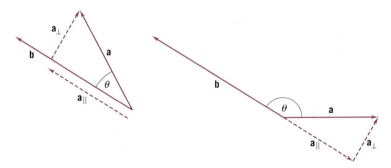

Figure 12.3.6

The vector $\mathbf{a}_{\|}$ is called *the projection of* **a** *on* **b** and is denoted by $\text{proj}_{\mathbf{b}}\mathbf{a}$. Since $\text{proj}_{\mathbf{b}}\mathbf{a}$ is parallel to **b**, it is a scalar multiple of the unit vector in the direction of **b**:

$$\text{proj}_{\mathbf{b}}\mathbf{a} = \lambda\mathbf{u}_{\mathbf{b}}.$$

The scalar λ is called *the component of* **a** *in the direction of* **b** (or more briefly, the **b**-*component of* **a**) and is denoted by $\text{comp}_{\mathbf{b}}\mathbf{a}$. In symbols,

(12.3.10)

$$\text{proj}_{\mathbf{b}}\mathbf{a} = (\text{comp}_{\mathbf{b}}\mathbf{a})\mathbf{u}_{\mathbf{b}}.$$

The component of **a** in the direction of **b** measures the "advance" of **a** in the direction of **b**. In Figure 12.3.6 we used θ to indicate the angle between **a** and **b**. If $0 \leq \theta < \frac{1}{2}\pi$, the projection and **b** have the same direction and the component is positive. If $\theta = \frac{1}{2}\pi$, the projection is **0** and the component is 0. If $\frac{1}{2}\pi < \theta \leq \pi$, the projection and **b** have opposite directions and, consequently, the component is negative.

Projections and components are closely related to dot products: if $\mathbf{b} \neq \mathbf{0}$, then

(12.3.11)

$$\text{proj}_b\mathbf{a} = (\mathbf{a} \cdot \mathbf{u_b})\mathbf{u_b} \qquad \text{and} \qquad \text{comp}_b\mathbf{a} = \mathbf{a} \cdot \mathbf{u_b}.$$

PROOF The second assertion follows trivially from the first. We will prove the first. We begin with the identity

$$\mathbf{a} = (\mathbf{a} \cdot \mathbf{u_b})\mathbf{u_b} + [\mathbf{a} - (\mathbf{a} \cdot \mathbf{u_b})\mathbf{u_b}].$$

Since the first vector $(\mathbf{a} \cdot \mathbf{u_b})\mathbf{u_b}$ is a scalar multiple of $\mathbf{b}$, it is parallel to $\mathbf{b}$. All we have to show now is that the second vector is perpendicular to $\mathbf{b}$. We do this by showing that its dot product with $\mathbf{u_b}$ is zero:

$$[\mathbf{a} - (\mathbf{a} \cdot \mathbf{u_b})\mathbf{u_b}] \cdot \mathbf{u_b} = (\mathbf{a} \cdot \mathbf{u_b}) - (\mathbf{a} \cdot \mathbf{u_b})(\mathbf{u_b} \cdot \mathbf{u_b}) = 0. \quad \square$$

$$\mathbf{u_b} \cdot \mathbf{u_b} = \|\mathbf{u_b}\|^2 = 1$$

Example 7 Find $\text{comp}_b\mathbf{a}$ and $\text{proj}_b\mathbf{a}$ given that

$$\mathbf{a} = -2\mathbf{i} + \mathbf{j} + \mathbf{k} \qquad \text{and} \qquad \mathbf{b} = 4\mathbf{i} - 3\mathbf{j} + \mathbf{k}.$$

SOLUTION Since

$$\|\mathbf{b}\| = \sqrt{(4)^2 + (-3)^2 + (1)^2} = \sqrt{26},$$

we have

$$\mathbf{u_b} = \frac{\mathbf{b}}{\|\mathbf{b}\|} = \frac{1}{\sqrt{26}}(4\mathbf{i} - 3\mathbf{j} + \mathbf{k}).$$

Thus

$$\text{comp}_b\mathbf{a} = \mathbf{a} \cdot \mathbf{u_b} = (-2\mathbf{i} + \mathbf{j} + \mathbf{k}) \cdot \frac{1}{\sqrt{26}}(4\mathbf{i} - 3\mathbf{j} + \mathbf{k})$$

$$= \frac{1}{\sqrt{26}}[(-2)(4) + (1)(-3) + (1)(1)] = -\frac{10}{\sqrt{26}} = -\frac{5}{13}\sqrt{26}$$

and

$$\text{proj}_b\mathbf{a} = (\text{comp}_b\mathbf{a})\,\mathbf{u_b} = -\frac{5}{13}(4\mathbf{i} - 3\mathbf{j} + \mathbf{k}). \quad \square$$

The following characterization of $\mathbf{a} \cdot \mathbf{b}$ is frequently used in physical applications:

(12.3.12)

$$\text{if } \mathbf{b} \neq \mathbf{0}, \quad \mathbf{a} \cdot \mathbf{b} = (\text{comp}_b\mathbf{a})\|\mathbf{b}\|.$$

PROOF

$$\mathbf{a} \cdot \mathbf{b} = \left(\mathbf{a} \cdot \frac{\mathbf{b}}{\|\mathbf{b}\|}\right)\|\mathbf{b}\| = (\mathbf{a} \cdot \mathbf{u_b})\|\mathbf{b}\| = (\text{comp}_b\mathbf{a})\|\mathbf{b}\|. \quad \square$$

For an arbitrary vector $\mathbf{a} = a_1\mathbf{i} + a_2\mathbf{j} + a_3\mathbf{k}$ we have

$$\text{comp}_\mathbf{i}\mathbf{a} = \mathbf{a} \cdot \mathbf{i} = a_1, \qquad \text{comp}_\mathbf{j}\mathbf{a} = \mathbf{a} \cdot \mathbf{j} = a_2, \qquad \text{comp}_\mathbf{k}\mathbf{a} = \mathbf{a} \cdot \mathbf{k} = a_3.$$

This agrees with our previous use of the term "component" (Section 12.2) and gives the identity

(12.3.13)
$$\mathbf{a} = (\mathbf{a} \cdot \mathbf{i})\mathbf{i} + (\mathbf{a} \cdot \mathbf{j})\mathbf{j} + (\mathbf{a} \cdot \mathbf{k})\mathbf{k}.$$

Direction Angles, Direction Cosines

In Figure 12.3.7 we have portrayed a nonzero vector $\mathbf{a}$. The angles α, β, γ that the vector makes with the unit coordinate vectors are called the *direction angles* of $\mathbf{a}$, and $\cos\alpha$, $\cos\beta$, $\cos\gamma$ are called the *direction cosines*. By (12.3.7)

$$\mathbf{a} \cdot \mathbf{i} = \|\mathbf{a}\| \cos\alpha, \qquad \mathbf{a} \cdot \mathbf{j} = \|\mathbf{a}\| \cos\beta, \qquad \mathbf{a} \cdot \mathbf{k} = \|\mathbf{a}\| \cos\gamma.$$

Thus by (12.3.13)

(12.3.14)
$$\mathbf{a} = \|\mathbf{a}\|(\cos\alpha\mathbf{i} + \cos\beta\mathbf{j} + \cos\gamma\mathbf{k}).$$

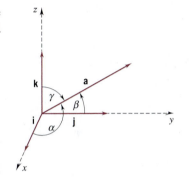

Figure 12.3.7

Taking the norm of both sides, we have

$$\|\mathbf{a}\| = \|\mathbf{a}\|\sqrt{\cos^2\alpha + \cos^2\beta + \cos^2\gamma}$$

and therefore

(12.3.15)
$$\cos^2\alpha + \cos^2\beta + \cos^2\gamma = 1.$$

The sum of the squares of the direction cosines is always 1.
For a unit vector $\mathbf{u}$, Equation 12.3.14 takes the form

(12.3.16)
$$\mathbf{u} = \cos\alpha\mathbf{i} + \cos\beta\mathbf{j} + \cos\gamma\mathbf{k}.$$

Thus, for a unit vector, the $\mathbf{i}, \mathbf{j}, \mathbf{k}$ components are simply the direction cosines.

Example 8 Find the unit vector with direction angles

$$\alpha = \tfrac{1}{4}\pi, \qquad \beta = \tfrac{2}{3}\pi, \qquad \gamma = \tfrac{1}{3}\pi.$$

What is the vector of norm 4 with these same direction angles?

SOLUTION The unit vector with these direction angles is

$$\cos\tfrac{1}{4}\pi\mathbf{i} + \cos\tfrac{2}{3}\pi\mathbf{j} + \cos\tfrac{1}{3}\pi\mathbf{k} = \tfrac{1}{2}\sqrt{2}\mathbf{i} - \tfrac{1}{2}\mathbf{j} + \tfrac{1}{2}\mathbf{k}.$$

The vector of norm 4 with these direction angles is

$$4(\tfrac{1}{2}\sqrt{2}\mathbf{i} - \tfrac{1}{2}\mathbf{j} + \tfrac{1}{2}\mathbf{k}) = 2\sqrt{2}\mathbf{i} - 2\mathbf{j} + 2\mathbf{k}. \quad \square$$

Example 9 Find the direction cosines of

$$\mathbf{a} = 2\mathbf{i} + 3\mathbf{j} - 6\mathbf{k}.$$

What are the direction angles?

SOLUTION Here

$$\|\mathbf{a}\| = \sqrt{2^2 + 3^2 + (-6)^2} = 7$$

so that

$$2 = 7 \cos \alpha, \qquad 3 = 7 \cos \beta, \qquad -6 = 7 \cos \gamma$$

and

$$\cos \alpha = \tfrac{2}{7}, \qquad \cos \beta = \tfrac{3}{7}, \qquad \cos \gamma = -\tfrac{6}{7}.$$

Since angles between vectors are measured in radians and taken from 0 to π,

$$\alpha = \cos^{-1} \tfrac{2}{7} \cong 1.28 \text{ radians}, \qquad \beta = \cos^{-1} \tfrac{3}{7} \cong 1.13 \text{ radians},$$

$$\gamma = \cos^{-1}(-\tfrac{6}{7}) \cong 2.60 \text{ radians}. \quad ❏$$

Proving the Triangle Inequality

By taking absolute values, the relation

$$\mathbf{a} \cdot \mathbf{b} = \|\mathbf{a}\| \, \|\mathbf{b}\| \cos \theta$$

gives

(12.3.17)

$$\boxed{|\mathbf{a} \cdot \mathbf{b}| \leq \|\mathbf{a}\| \, \|\mathbf{b}\|.}$$

$(|\cos \theta| \leq 1)$

This inequality, called *Schwarz's inequality*, enables us to give a simple proof of the triangle inequality

$$\|\mathbf{a} + \mathbf{b}\| \leq \|\mathbf{a}\| + \|\mathbf{b}\|.$$

PROOF

$$\begin{aligned}
\|\mathbf{a} + \mathbf{b}\|^2 &= (\mathbf{a} + \mathbf{b}) \cdot (\mathbf{a} + \mathbf{b}) \\
&= (\mathbf{a} \cdot \mathbf{a}) + (\mathbf{b} \cdot \mathbf{a}) + (\mathbf{a} \cdot \mathbf{b}) + (\mathbf{b} \cdot \mathbf{b}) \\
&= \|\mathbf{a}\|^2 + 2(\mathbf{a} \cdot \mathbf{b}) + \|\mathbf{b}\|^2 \\
&\leq \|\mathbf{a}\|^2 + 2|\mathbf{a} \cdot \mathbf{b}| + \|\mathbf{b}\|^2 \\
&\leq \|\mathbf{a}\|^2 + 2\|\mathbf{a}\| \, \|\mathbf{b}\| + \|\mathbf{b}\|^2 = (\|\mathbf{a}\| + \|\mathbf{b}\|)^2
\end{aligned}$$

$(\mathbf{a} \cdot \mathbf{b} \leq |\mathbf{a} \cdot \mathbf{b}|)$

by Schwarz's inequality ⟶

and therefore

$$\|\mathbf{a} + \mathbf{b}\| \leq \|\mathbf{a}\| + \|\mathbf{b}\|. \quad ❏$$

It is worth remarking that Schwarz's inequality, and hence the triangle inequality, can be proved purely algebraically. (See Exercise 50.)

EXERCISES 12.3

In Exercises 1–6, find $\mathbf{a} \cdot \mathbf{b}$.

1. $\mathbf{a} = (2, -3, 1)$, $\mathbf{b} = (-2, 0, 3)$.

2. $\mathbf{a} = (4, 2, -1)$, $\mathbf{b} = (-2, 2, 1)$.

3. $\mathbf{a} = (2, -4)$, $\mathbf{b} = (1, \frac{1}{2})$.

4. $\mathbf{a} = (-2, 5)$, $\mathbf{b} = (3, 1)$.

5. $\mathbf{a} = 2\mathbf{i} + \mathbf{j} - 2\mathbf{k}$, $\mathbf{b} = \mathbf{i} + \mathbf{j} + 2\mathbf{k}$.

6. $\mathbf{a} = 2\mathbf{i} + 3\mathbf{j} + \mathbf{k}$, $\mathbf{b} = \mathbf{i} + 4\mathbf{j}$.

In Exercises 7–10, simplify the given expression.

7. $(3\mathbf{a} \cdot \mathbf{b}) - (\mathbf{a} \cdot 2\mathbf{b})$.

8. $\mathbf{a} \cdot (\mathbf{a} - \mathbf{b}) + \mathbf{b} \cdot (\mathbf{b} + \mathbf{a})$.

9. $(\mathbf{a} - \mathbf{b}) \cdot \mathbf{c} + \mathbf{b} \cdot (\mathbf{c} + \mathbf{a})$.

10. $\mathbf{a} \cdot (\mathbf{a} + 2\mathbf{c}) + (2\mathbf{b} - \mathbf{a}) \cdot (\mathbf{a} + 2\mathbf{c}) - 2\mathbf{b} \cdot (\mathbf{a} + 2\mathbf{c})$.

11. Taking

$$\mathbf{a} = 2\mathbf{i} + \mathbf{j}, \qquad \mathbf{b} = 3\mathbf{i} - \mathbf{j} + 2\mathbf{k}, \qquad \mathbf{c} = 4\mathbf{i} + 3\mathbf{k}$$

calculate

(a) the three dot products $\mathbf{a} \cdot \mathbf{b}$, $\mathbf{a} \cdot \mathbf{c}$, $\mathbf{b} \cdot \mathbf{c}$;

(b) the cosines of the angles between these vectors;

(c) the component of $\mathbf{a}$ (i) in the $\mathbf{b}$ direction, (ii) in the $\mathbf{c}$ direction;

(d) the projection of $\mathbf{a}$ (i) in the $\mathbf{b}$ direction, (ii) in the $\mathbf{c}$ direction.

12. Exercise 11 with $\mathbf{a} = \mathbf{j} + 3\mathbf{k}$, $\mathbf{b} = 2\mathbf{i} - \mathbf{j} + 2\mathbf{k}$, $\mathbf{c} = 3\mathbf{i} - \mathbf{k}$.

13. Find the unit vector with direction angles $\frac{1}{3}\pi$, $\frac{1}{4}\pi$, $\frac{2}{3}\pi$.

14. Find the vector of norm 2 with direction angles $\frac{1}{4}\pi$, $\frac{1}{4}\pi$, $\frac{1}{2}\pi$.

15. Find the angle between the vectors $3\mathbf{i} - \mathbf{j} - 2\mathbf{k}$ and $\mathbf{i} + 2\mathbf{j} - 2\mathbf{k}$.

16. Find the angle between the vectors $2\mathbf{i} - 3\mathbf{j} + \mathbf{k}$ and $-3\mathbf{i} + \mathbf{j} + 9\mathbf{k}$.

17. Find the direction angles of the vector $\mathbf{i} - \mathbf{j} + \sqrt{2}\mathbf{k}$.

18. Find the direction angles of the vector $\mathbf{i} - \sqrt{3}\mathbf{k}$.

▶ In Exercises 19–22, estimate the angle between the given vectors. Express your answers in radians rounded to the nearest hundredth of a radian, and in degrees to the nearest tenth of a degree.

19. $\mathbf{a} = (3, 1, -1)$, $\mathbf{b} = (-2, 1, 4)$.

20. $\mathbf{a} = (-2, -3, 0)$, $\mathbf{b} = (-6, 0, 4)$.

21. $\mathbf{a} = -\mathbf{i} + 2\mathbf{k}$, $\mathbf{b} = 3\mathbf{i} + 4\mathbf{j} - 5\mathbf{k}$.

22. $\mathbf{a} = -3\mathbf{i} + \mathbf{j} - \mathbf{k}$, $\mathbf{b} = \mathbf{i} - \mathbf{j}$.

▶ In Exercises 23–26, find the direction cosines and direction angles of the given vector. Express the angles in degrees rounded to the nearest tenth of a degree.

23. $\mathbf{a} = (1, 2, 2)$.

24. $\mathbf{a} = (2, 6, -1)$.

25. $\mathbf{a} = 3\mathbf{i} + 12\mathbf{j} + 4\mathbf{k}$.

26. $\mathbf{a} = 3\mathbf{i} + 5\mathbf{j} - 4\mathbf{k}$.

27. Show that

(a) $\text{proj}_{\mathbf{b}} \alpha\mathbf{a} = \alpha \, \text{proj}_{\mathbf{b}}\mathbf{a}$ for all real α, and

(b) $\text{proj}_{\mathbf{b}}(\mathbf{a} + \mathbf{c}) = \text{proj}_{\mathbf{b}}\mathbf{a} + \text{proj}_{\mathbf{b}}\mathbf{c}$.

28. Show that

(a) $\text{proj}_{\beta\mathbf{b}}\mathbf{a} = \text{proj}_{\mathbf{b}}\mathbf{a}$ for all real $\beta \neq 0$, but

(b) $\text{comp}_{\beta\mathbf{b}}\mathbf{a} = \begin{cases} \text{comp}_{\mathbf{b}}\mathbf{a}, & \text{for } \beta > 0 \\ -\text{comp}_{\mathbf{b}}\mathbf{a}, & \text{for } \beta < 0. \end{cases}$

29. (a) (*Important*) Let $\mathbf{a} \neq \mathbf{0}$. Show that $\mathbf{a} \cdot \mathbf{b} = \mathbf{a} \cdot \mathbf{c}$ does not necessarily imply that $\mathbf{b} = \mathbf{c}$, but only that $\mathbf{b}$ and $\mathbf{c}$ have the same projection on $\mathbf{a}$. Draw a figure illustrating this for $\mathbf{b}$ and $\mathbf{c}$ different from $\mathbf{0}$.

(b) Show that if $\mathbf{u} \cdot \mathbf{b} = \mathbf{u} \cdot \mathbf{c}$ for all unit vectors $\mathbf{u}$, then $\mathbf{b} = \mathbf{c}$. HINT: Consider the unit coordinate vectors.

30. What can you conclude about $\mathbf{a}$ and $\mathbf{b}$ given that

(a) $\|\mathbf{a}\|^2 + \|\mathbf{b}\|^2 = \|\mathbf{a} + \mathbf{b}\|^2$?

(b) $\|\mathbf{a}\|^2 + \|\mathbf{b}\|^2 = \|\mathbf{a} - \mathbf{b}\|^2$?

HINT: Draw figures.

31. (a) Show that

$$4(\mathbf{a} \cdot \mathbf{b}) = \|\mathbf{a} + \mathbf{b}\|^2 - \|\mathbf{a} - \mathbf{b}\|^2.$$

(b) Use (a) to verify that

$$\mathbf{a} \perp \mathbf{b} \qquad \text{iff} \qquad \|\mathbf{a} + \mathbf{b}\| = \|\mathbf{a} - \mathbf{b}\|.$$

(c) Show that, if $\mathbf{a}$ and $\mathbf{b}$ are nonzero vectors such that

$$(\mathbf{a} + \mathbf{b}) \perp (\mathbf{a} - \mathbf{b}) \quad \text{and} \quad \|\mathbf{a} + \mathbf{b}\| = \|\mathbf{a} - \mathbf{b}\|,$$

then the parallelogram generated by $\mathbf{a}$ and $\mathbf{b}$ is a square.

32. Under what conditions does $|\mathbf{a} \cdot \mathbf{b}| = \|\mathbf{a}\| \, \|\mathbf{b}\|$?

33. Given two vectors $\mathbf{a}$ and $\mathbf{b}$, prove the *parallelogram law*:

$$\|\mathbf{a} + \mathbf{b}\|^2 + \|\mathbf{a} - \mathbf{b}\|^2 = 2\|\mathbf{a}\|^2 + 2\|\mathbf{b}\|^2.$$

Geometric interpretation: The sum of the squares of the lengths of the diagonals of a parallelogram determined by $\mathbf{a}$ and $\mathbf{b}$ equals the sum of the squares of the lengths of the four sides. See the following figure.

34. Suppose that the direction angles of a vector are equal. What are the angles?

35. What are the direction angles of $-\mathbf{a}$ if the direction angles of $\mathbf{a}$ are α, β, γ?

36. Let θ be the angle between $\mathbf{a}$ and $\mathbf{b}$, and let β be a negative number. Use the dot product to compute the angle between $\mathbf{a}$ and $\beta\mathbf{b}$ in terms of θ. Draw a figure to verify your answer geometrically.

37. Show that, if $\mathbf{a} \perp \mathbf{b}$ and $\mathbf{a} \perp \mathbf{c}$, then $\mathbf{a} \perp (\alpha\mathbf{b} + \beta\mathbf{c})$ for all real α, β.

38. Show that, if $\mathbf{a} \parallel \mathbf{b}$ and $\mathbf{a} \parallel \mathbf{c}$, then $\mathbf{a} \parallel (\alpha\mathbf{b} + \beta\mathbf{c})$ for all real α, β. ($\mathbf{a} \parallel \mathbf{b}$ means $\mathbf{a}$ is parallel to $\mathbf{b}$.)

39. (*Important*) Show that, if $\mathbf{b}$ is a nonzero vector, then every vector $\mathbf{a}$ can be written in a unique manner as the sum of a vector $\mathbf{a}_\parallel$ parallel to $\mathbf{b}$ and a vector $\mathbf{a}_\perp$ perpendicular to $\mathbf{b}$:

$$\mathbf{a} = \mathbf{a}_\parallel + \mathbf{a}_\perp.$$

40. Find all the numbers x for which
$$(x\mathbf{i} + 11\mathbf{j} - 3\mathbf{k}) \perp (2x\mathbf{i} - x\mathbf{j} - 5\mathbf{k}).$$

41. Find all the numbers x for which the angle between $\mathbf{c} = x\mathbf{i} + \mathbf{j} + \mathbf{k}$ and $\mathbf{d} = \mathbf{i} + x\mathbf{j} + \mathbf{k}$ is $\frac{1}{3}\pi$.

42. Set $\mathbf{a} = \mathbf{i} + x\mathbf{j} + \mathbf{k}$ and $\mathbf{b} = 2\mathbf{i} - \mathbf{j} + y\mathbf{k}$. Compute all values of x and y for which $\mathbf{a} \perp \mathbf{b}$ and $\|\mathbf{a}\| = \|\mathbf{b}\|$.

43. Find the angle between the diagonal of a cube and one of its edges.

44. Find the angle between the diagonal of a cube and the diagonal of one of its faces.

45. (a) Show that $\frac{1}{4}\pi$, $\frac{1}{6}\pi$, $\frac{2}{3}\pi$ cannot be the direction angles of a vector.
(b) Show that, if $\mathbf{a} = a_1\mathbf{i} + a_2\mathbf{j} + a_3\mathbf{k}$ has direction angles α, $\frac{1}{4}\pi$, $\frac{1}{4}\pi$, then $a_1 = 0$.

46. Let $r = f(\theta)$ be the polar equation of a curve in the plane and let

$$\mathbf{u}_r = (\cos\theta)\mathbf{i} + (\sin\theta)\mathbf{j} \qquad \mathbf{u}_\theta = (-\sin\theta)\mathbf{i} + (\cos\theta)\mathbf{j}$$

(a) Show that $\mathbf{u}_r$ and $\mathbf{u}_\theta$ are unit vectors and that they are perpendicular.
(b) Let $P(r, \theta)$ be a point on the curve. Show that $\mathbf{u}_r$ has the same direction as the vector $\overrightarrow{OP}$ and that the direction of $\mathbf{u}_\theta$ is 90° counterclockwise from $\mathbf{u}_r$.

47. Find the unit vectors $\mathbf{u}$ that are perpendicular to both $\mathbf{i} + 2\mathbf{j} + \mathbf{k}$ and $3\mathbf{i} - 4\mathbf{j} + 2\mathbf{k}$.

48. Find two mutually perpendicular unit vectors that are perpendicular to $2\mathbf{i} + 3\mathbf{j}$.

49. Two points on a sphere are called *antipodal* iff they are opposite endpoints of a diameter. Show that, if P_1 and P_2 are antipodal points on a sphere and Q is any other point on the sphere, then $\overrightarrow{P_1Q} \perp \overrightarrow{P_2Q}$.

50. Give an algebraic proof of Schwarz's inequality $|\mathbf{a} \cdot \mathbf{b}| \leq \|\mathbf{a}\| \|\mathbf{b}\|$. HINT: If $\mathbf{b} = \mathbf{0}$, the inequality is trivial, so assume $\mathbf{b} \neq \mathbf{0}$. Note that for any number λ we have $\|\mathbf{a} - \lambda\mathbf{b}\|^2 \geq \mathbf{0}$. First expand this inequality

using the fact that $\|\mathbf{a} - \lambda\mathbf{b}\|^2 = (\mathbf{a} - \lambda\mathbf{b}) \cdot (\mathbf{a} - \lambda\mathbf{b})$. After collecting terms, make the special choice $\lambda = (\mathbf{a} \cdot \mathbf{b})/\|\mathbf{b}\|^2$ and see what happens.

Work

In Section 6.5, we defined the *work W* done by a force F in moving an object along the x-axis from $x = a$ to $x = b$ by

$$W = F(b - a).$$

In words, *work equals force times displacement*. This definition can be extended as follows: If a constant force $\mathbf{F}$ is applied to an object moving in a straight line throughout a displacement $\mathbf{r}$ (see the following figure), then the work done by $\mathbf{F}$ is given by

$$W = (\text{comp}_\mathbf{r}\, \mathbf{F})\, \|\mathbf{r}\|.$$

If $\mathbf{F}$ is measured in pounds and distance in feet, then the work is measured in *foot-pounds*; if $\mathbf{F}$ is measured in newtons and distance in meters, then the work is measured in *newton-meters* or *joules*.

51. Let the force $\mathbf{F}$ be applied throughout a displacement $\mathbf{r}$.
(a) Express the work done by $\mathbf{F}$ as a dot product.
(b) What is the work done by $\mathbf{F}$ if $\mathbf{F} \perp \mathbf{r}$?
(c) Show that the work done by $\mathbf{F} = \|\mathbf{F}\|\mathbf{i}$ applied throughout the displacement $(b - a)\mathbf{i}$ reduces to the definition given in Section 6.5.

52. Find the work done by the force $\mathbf{F} = 2\mathbf{i} + \mathbf{j} + 3\mathbf{k}$ in moving an object from the point $P(1, 2, 0)$ to the point $Q(3, 5, 2)$.

▷53. (a) A sled is pulled along level ground by exerting a force of 15 Newtons on a rope that makes an angle of 35° with the ground. Find the work done by the force in pulling the sled 50 meters. See Figure A.

Figure A   Figure B

(b) Suppose the same sled is pulled 50 meters up a hill that makes an angle of 15° with level ground. Find the work done by the force in this case. See Figure B.

▷ 54. A wooden crate is pulled along a level floor by a rope that makes an angle of 40° with the floor. If the force of friction (which acts in a direction opposite to the motion) between the carton and the floor is 50 pounds, what is the minimum force that must be applied to the rope to move the crate?

55. Two forces of the same magnitude, $\mathbf{F}_1$ and $\mathbf{F}_2$, are applied throughout a displacement $\mathbf{r}$ at angles θ_1 and θ_2, respectively. Compare the work done by $\mathbf{F}_1$ to that done by $\mathbf{F}_2$ if
(a) $\theta_1 = -\theta_2$.
(b) $\theta_1 = \pi/3$ and $\theta_2 = \pi/6$.

56. What is the total work done by a constant force $\mathbf{F}$ if the object to which it is applied moves around a triangle? Justify your answer.

■ 12.4 THE CROSS PRODUCT

Although all of the focus in the preceding sections was on three-dimensional vectors, we did remark on several occasions that corresponding definitions and properties also held for vectors in other dimensions. In this section we study the *cross product* of two vectors, and we emphasize at the outset that this product is defined for three-dimensional vectors only. Throughout this section the term "vector" will mean "three-dimensional vector."

Definition of the Cross Product

While the dot product $\mathbf{a} \cdot \mathbf{b}$ is a scalar (and as such is sometimes called the scalar product of $\mathbf{a}$ and $\mathbf{b}$), the cross product $\mathbf{a} \times \mathbf{b}$ is a vector (sometimes called the *vector product* of $\mathbf{a}$ and $\mathbf{b}$). What is the vector $\mathbf{a} \times \mathbf{b}$? We could directly write down a formula that gives the components of $\mathbf{a} \times \mathbf{b}$ in terms of the components of $\mathbf{a}$ and $\mathbf{b}$, but at this stage that would reveal little. Instead we will begin geometrically. We will define $\mathbf{a} \times \mathbf{b}$ by giving its direction and its magnitude.

The Direction of a × b If the vectors $\mathbf{a}$ and $\mathbf{b}$ are not parallel, they determine a plane. The vector $\mathbf{a} \times \mathbf{b}$ is perpendicular to this plane and is directed in such a way that (like $\mathbf{i}, \mathbf{j}, \mathbf{k}$) the vectors $\mathbf{a}, \mathbf{b}, \mathbf{a} \times \mathbf{b}$ form a right-handed triple. (Figure 12.4.1.) (If the index finger of the right hand points along $\mathbf{a}$ and the middle finger points along $\mathbf{b}$, then the thumb will point in the direction of $\mathbf{a} \times \mathbf{b}$. NOTE: In saying that $\mathbf{a}, \mathbf{b}, \mathbf{a} \times \mathbf{b}$ form a right-handed triple, we do not require the vectors $\mathbf{a}$ and $\mathbf{b}$ to be perpendicular (like $\mathbf{i}$ and $\mathbf{j}$); the general notion of a "right-handed triple" does not require the vectors involved to be perpendicular to each other.)

Figure 12.4.1

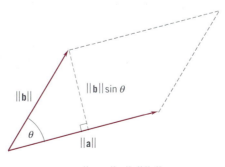

area $= \|\mathbf{a} \times \mathbf{b}\| = \|\mathbf{a}\| \|\mathbf{b}\| \sin \theta$

Figure 12.4.2

The Magnitude of a × b. If **a** and **b** are not parallel, they form the sides of a parallelogram. (Figure 12.4.2.) The magnitude of **a** × **b** is the area of this parallelogram: $\|\mathbf{a}\| \|\mathbf{b}\| \sin \theta$. (Recall that the area of a parallelogram with base b and height h is $A = bh$.)

One more point. What if **a** and **b** are parallel? Then there is no parallelogram and we define **a** × **b** = **0**.

We summarize all this below.

DEFINITION 12.4.1

If **a** and **b** are not parallel, then **a** × **b** is the vector with the following properties:

1. **a** × **b** is perpendicular to the plane of **a** and **b**.
2. **a**, **b**, **a** × **b** form a right-handed triple.
3. $\|\mathbf{a} \times \mathbf{b}\| = \|\mathbf{a}\| \|\mathbf{b}\| \sin \theta$, where θ is the angle between **a** and **b**, $0 < \theta < \pi$.

If **a** and **b** are parallel ($\theta = 0$ or $\theta = \pi$), then **a** × **b** = **0**.

Properties of Right-Handed Triples

I. Note first of all that if (**a**, **b**, **c**) form a right-handed triple, then (**b**, **c**, **a**) and (**c**, **a**, **b**) also form right-handed triples. (Figure 12.4.3.) To maintain right-handedness we don't have to keep the vectors in the same order, but we do have to keep them in the *same cyclic order:*

$$\mathbf{a} \rightarrow \mathbf{b} \rightarrow \mathbf{c}$$

Alter the cyclic order and you reverse the orientation.

Figure 12.4.3

II. A triple $(\mathbf{a}, \mathbf{b}, \mathbf{c})$ is right-handed iff $\mathbf{c}$ and $\mathbf{a} \times \mathbf{b}$ lie on the same side of the plane determined by $\mathbf{a}$ and $\mathbf{b}$. (Figure 12.4.4.) This means that $(\mathbf{a}, \mathbf{b}, \mathbf{c})$ is right-handed iff $(\mathbf{a} \times \mathbf{b}) \cdot \mathbf{c} > 0$. (Explain)

III. It follows from II that, if $(\mathbf{a}, \mathbf{b}, \mathbf{c})$ is right-handed, then $(\mathbf{a}, \mathbf{b}, -\mathbf{c})$ is not right-handed. Similarly $(-\mathbf{a}, \mathbf{b}, \mathbf{c})$ and $(\mathbf{a}, -\mathbf{b}, \mathbf{c})$ are also not right-handed. However, multiplication by positive scalars does maintain right-handedness: if $(\mathbf{a}, \mathbf{b}, \mathbf{c})$ is right-handed and α, β, γ are positive, then $(\alpha\mathbf{a}, \beta\mathbf{b}, \gamma\mathbf{c})$ is also right-handed.

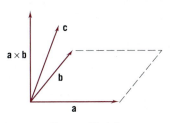

Figure 12.4.4

Properties of the Cross Product

The cross product is *anticommutative*:

(12.4.2)
$$\mathbf{b} \times \mathbf{a} = -(\mathbf{a} \times \mathbf{b}).$$

To see this, note that both vectors are perpendicular to the plane determined by $\mathbf{a}$ and $\mathbf{b}$ and both have the same norm. Thus $\mathbf{b} \times \mathbf{a} = \pm(\mathbf{a} \times \mathbf{b})$. That the minus sign holds and not the plus sign follows from observing that $\mathbf{b}, \mathbf{a}, \mathbf{b} \times \mathbf{a}$ is a right-handed triple and that $\mathbf{b}, \mathbf{a}, \mathbf{a} \times \mathbf{b}$ is not right-handed since $\mathbf{a}, \mathbf{b}, \mathbf{a} \times \mathbf{b}$ is right-handed. ❑

Scalars can be factored:

(12.4.3)
$$\alpha\mathbf{a} \times \beta\mathbf{b} = \alpha\beta(\mathbf{a} \times \mathbf{b}).$$

If α or β is zero, the result is obvious. We will assume that α and β are both nonzero. In this case the two vectors are perpendicular to $\mathbf{a}$, perpendicular to $\mathbf{b}$, and have the same norm. Thus $\alpha\mathbf{a} \times \beta\mathbf{b} = \pm\alpha\beta(\mathbf{a} \times \mathbf{b})$. That the positive sign holds comes from noting that $\alpha\mathbf{a}, \beta\mathbf{b}, \alpha\beta(\mathbf{a} \times \mathbf{b})$ is a right-handed triple. This is obvious if α and β are both positive. If not, two of the three coefficients $\alpha, \beta, \alpha\beta$ are negative and the other is positive. In this case, the first minus sign reverses the orientation but the second one restores it. ❑

Finally, there are two distributive laws, the verification of which we postpone for a moment.

(12.4.4)

$$\mathbf{a} \times (\mathbf{b} + \mathbf{c}) = (\mathbf{a} \times \mathbf{b}) + (\mathbf{a} \times \mathbf{c}), \qquad (\mathbf{a} + \mathbf{b}) \times \mathbf{c} = (\mathbf{a} \times \mathbf{c}) + (\mathbf{b} \times \mathbf{c}).$$

The Scalar Triple Product

Earlier we saw that $\mathbf{a}, \mathbf{b}, \mathbf{c}$ is a right-handed triple iff $(\mathbf{a} \times \mathbf{b}) \cdot \mathbf{c} > 0$. The expression $(\mathbf{a} \times \mathbf{b}) \cdot \mathbf{c}$ is called a *scalar triple product*. The absolute value of this number (it is a number and not a vector) has geometric significance. To describe it we refer to Figure

Figure 12.4.5

12.4.5. There you see a parallelepiped with edges **a**, **b**, **c**. The absolute value of the scalar triple product gives the volume of that parallelepiped:

(12.4.5)
$$V = |(\mathbf{a} \times \mathbf{b}) \cdot \mathbf{c}|.$$

PROOF The area of the base is $\|\mathbf{a} \times \mathbf{b}\|$. The height is $|\text{comp}_{\mathbf{a} \times \mathbf{b}} \, \mathbf{c}|$. Therefore

$$V = |\text{comp}_{\mathbf{a} \times \mathbf{b}} \, \mathbf{c}| \, \|\mathbf{a} \times \mathbf{b}\| = |(\mathbf{a} \times \mathbf{b}) \cdot \mathbf{c}|. \quad \square$$
$$\underset{(12.3.12)}{\uparrow}$$

Of course we could have formed the same parallelogram using a different base (for example, using the vectors **c** and **a**) with a correspondingly different height $(\text{comp}_{\mathbf{c} \times \mathbf{a}} \, \mathbf{b})$. Therefore

$$|(\mathbf{a} \times \mathbf{b}) \cdot \mathbf{c}| = |(\mathbf{c} \times \mathbf{a}) \cdot \mathbf{b}| = |(\mathbf{b} \times \mathbf{c}) \cdot \mathbf{a}|.$$

Since the **a**, **b**, **c** appear in the same cyclic order, the expressions inside the absolute value signs all have the same sign. (Property II of right-handed triples) Therefore

(12.4.6)
$$(\mathbf{a} \times \mathbf{b}) \cdot \mathbf{c} = (\mathbf{c} \times \mathbf{a}) \cdot \mathbf{b} = (\mathbf{b} \times \mathbf{c}) \cdot \mathbf{a}.$$

Verification of the Distributive Laws

We will verify the first distributive law,

$$\mathbf{a} \times (\mathbf{b} + \mathbf{c}) = (\mathbf{a} \times \mathbf{b}) + (\mathbf{a} \times \mathbf{c}).$$

The second follows readily from this one. The argument is left to you as an exercise.

Take an arbitrary vector **r** and form the dot product $[\mathbf{a} \times (\mathbf{b} + \mathbf{c})] \cdot \mathbf{r}$. We can then write

$$[\mathbf{a} \times (\mathbf{b} + \mathbf{c})] \cdot \mathbf{r} = (\mathbf{r} \times \mathbf{a}) \cdot (\mathbf{b} + \mathbf{c}) \qquad (12.4.6)$$
$$= [(\mathbf{r} \times \mathbf{a}) \cdot \mathbf{b}] + [(\mathbf{r} \times \mathbf{a}) \cdot \mathbf{c}] \qquad (12.3.6)$$
$$= [(\mathbf{a} \times \mathbf{b}) \cdot \mathbf{r}] + [(\mathbf{a} \times \mathbf{c}) \cdot \mathbf{r}] \qquad (12.4.6)$$
$$= [(\mathbf{a} \times \mathbf{b}) + (\mathbf{a} \times \mathbf{c})] \cdot \mathbf{r}. \qquad (12.3.6)$$

Since this holds true for all vectors **r**, it holds true for **i**, **j**, **k** and proves that

$$\mathbf{a} \times (\mathbf{b} + \mathbf{c}) = (\mathbf{a} \times \mathbf{b}) + (\mathbf{a} \times \mathbf{c}). \quad \square$$

The Components of a × b

You have learned a lot about cross products, but you still have not seen $\mathbf{a} \times \mathbf{b}$ expressed in terms of the components of $\mathbf{a}$ and $\mathbf{b}$. To derive the formula that does this, we need to observe one more fact, one which is obvious from the very definition of cross product:

(12.4.7)

$$\mathbf{i} \times \mathbf{j} = \mathbf{k}, \qquad \mathbf{j} \times \mathbf{k} = \mathbf{i}, \qquad \mathbf{k} \times \mathbf{i} = \mathbf{j}.$$

(Figure 12.4.6)

One way to remember these products is to arrange $\mathbf{i}$, $\mathbf{j}$, $\mathbf{k}$ in cyclic order, $\mathbf{i} \to \mathbf{j} \to \mathbf{k}$, and note that

[each coordinate unit vector] × [the next one] = [the third one].

Figure 12.4.6

THEOREM 12.4.8

For vectors $\mathbf{a} = a_1\mathbf{i} + a_2\mathbf{j} + a_3\mathbf{k}$ and $\mathbf{b} = b_1\mathbf{i} + b_2\mathbf{j} + b_3\mathbf{k}$

$$\mathbf{a} \times \mathbf{b} = (a_2 b_3 - a_3 b_2)\mathbf{i} - (a_1 b_3 - a_3 b_1)\mathbf{j} + (a_1 b_2 - a_2 b_1)\mathbf{k}.$$

Determinants provide an especially convenient notational device for representing this formula for $\mathbf{a} \times \mathbf{b}$. If you are not already familiar with determinants, a brief treatment is given in Appendix A-2. Theorem 12.4.8 stated in terms of determinants is:

THEOREM 12.4.8′

For vectors $\mathbf{a} = a_1\mathbf{i} + a_2\mathbf{j} + a_3\mathbf{k}$ and $\mathbf{b} = b_1\mathbf{i} + b_2\mathbf{j} + b_3\mathbf{k}$

$$\mathbf{a} \times \mathbf{b} = \begin{vmatrix} \mathbf{i} & \mathbf{j} & \mathbf{k} \\ a_1 & a_2 & a_3 \\ b_1 & b_2 & b_3 \end{vmatrix} = \begin{vmatrix} a_2 & a_3 \\ b_2 & b_3 \end{vmatrix} \mathbf{i} - \begin{vmatrix} a_1 & a_3 \\ b_1 & b_3 \end{vmatrix} \mathbf{j} + \begin{vmatrix} a_1 & a_2 \\ b_1 & b_2 \end{vmatrix} \mathbf{k}.$$

(The 3×3 determinant with $\mathbf{i}$, $\mathbf{j}$, $\mathbf{k}$ in the top row is there only as a mnemonic device.)

PROOF The hard work has all been done. With what you know about cross products now, the proof is just a matter of simple algebra:

$$\mathbf{a} \times \mathbf{b} = (a_1\mathbf{i} + a_2\mathbf{j} + a_3\mathbf{k}) \times (b_1\mathbf{i} + b_2\mathbf{j} + b_3\mathbf{k})$$
$$= a_1b_2(\mathbf{i} \times \mathbf{j}) + a_1b_3(\mathbf{i} \times \mathbf{k}) + a_2b_1(\mathbf{j} \times \mathbf{i}) + a_2b_3(\mathbf{j} \times \mathbf{k}) + a_3b_1(\mathbf{k} \times \mathbf{i}) + a_3b_2(\mathbf{k} \times \mathbf{j}$$

$$\overset{\displaystyle\longmapsto \mathbf{i} \times \mathbf{i} = \mathbf{j} \times \mathbf{j} = \mathbf{k} \times \mathbf{k} = \mathbf{0}}{}$$

$$= a_1b_2\mathbf{k} - a_1b_3\mathbf{j} - a_2b_1\mathbf{k} + a_2b_3\mathbf{i} + a_3b_1\mathbf{j} - a_3b_2\mathbf{i}$$
$$= (a_2b_3 - a_3b_2)\mathbf{i} - (a_1b_3 - a_3b_1)\mathbf{j} + (a_1b_2 - a_2b_1)\mathbf{k}$$
$$= \begin{vmatrix} a_2 & a_3 \\ b_2 & b_3 \end{vmatrix}\mathbf{i} - \begin{vmatrix} a_1 & a_3 \\ b_1 & b_3 \end{vmatrix}\mathbf{j} + \begin{vmatrix} a_1 & a_2 \\ b_1 & b_2 \end{vmatrix}\mathbf{k}. \quad \square$$

Examples

Example 1 Calculate $\mathbf{a} \times \mathbf{b}$ given that

$$\mathbf{a} = \mathbf{i} - 2\mathbf{j} + 3\mathbf{k} \quad \text{and} \quad \mathbf{b} = 2\mathbf{i} + \mathbf{j} - \mathbf{k}.$$

SOLUTION

$$\mathbf{a} \times \mathbf{b} = \begin{vmatrix} \mathbf{i} & \mathbf{j} & \mathbf{k} \\ 1 & -2 & 3 \\ 2 & 1 & -1 \end{vmatrix} = \begin{vmatrix} -2 & 3 \\ 1 & -1 \end{vmatrix}\mathbf{i} - \begin{vmatrix} 1 & 3 \\ 2 & -1 \end{vmatrix}\mathbf{j} + \begin{vmatrix} 1 & -2 \\ 2 & 1 \end{vmatrix}\mathbf{k}$$

$$= -\mathbf{i} + 7\mathbf{j} + 5\mathbf{k}. \quad \square$$

Example 2 Calculate $\mathbf{a} \times \mathbf{b}$ given that

$$\mathbf{a} = \mathbf{i} - \mathbf{j} \quad \text{and} \quad \mathbf{b} = \mathbf{i} + \mathbf{k}.$$

SOLUTION

$$\mathbf{a} \times \mathbf{b} = \begin{vmatrix} \mathbf{i} & \mathbf{j} & \mathbf{k} \\ 1 & -1 & 0 \\ 1 & 0 & 1 \end{vmatrix} = \begin{vmatrix} -1 & 0 \\ 0 & 1 \end{vmatrix}\mathbf{i} - \begin{vmatrix} 1 & 0 \\ 1 & 1 \end{vmatrix}\mathbf{j} + \begin{vmatrix} 1 & -1 \\ 1 & 0 \end{vmatrix}\mathbf{k}$$

$$= -\mathbf{i} - \mathbf{j} + \mathbf{k}. \quad \square$$

In Examples 1 and 2 we calculated some cross products using Theorem 12.4.8′. Of course, we can obtain the same results just by applying the distributive laws. For example, for $\mathbf{a} = \mathbf{i} - 2\mathbf{j} + 3\mathbf{k}$ and $\mathbf{b} = 2\mathbf{i} + \mathbf{j} - \mathbf{k}$, we have

$$\mathbf{a} \times \mathbf{b} = (\mathbf{i} - 2\mathbf{j} + 3\mathbf{k}) \times (2\mathbf{i} + \mathbf{j} - \mathbf{k})$$
$$= (\mathbf{i} \times \mathbf{j}) - (\mathbf{i} \times \mathbf{k}) - 4(\mathbf{j} \times \mathbf{i}) + 2(\mathbf{j} \times \mathbf{k}) + 6(\mathbf{k} \times \mathbf{i}) + 3(\mathbf{k} \times \mathbf{j})$$
$$= \mathbf{k} + \mathbf{j} + 4\mathbf{k} + 2\mathbf{i} + 6\mathbf{j} - 3\mathbf{i} = -\mathbf{i} + 7\mathbf{j} + 5\mathbf{k}.$$

Example 3 Show that the scalar triple product can be written as a determinant:

(12.4.9)
$$(\mathbf{a} \times \mathbf{b}) \cdot \mathbf{c} = \begin{vmatrix} a_1 & a_2 & a_3 \\ b_1 & b_2 & b_3 \\ c_1 & c_2 & c_3 \end{vmatrix}.$$

SOLUTION

$$(\mathbf{a} \times \mathbf{b}) \cdot \mathbf{c} = \mathbf{c} \cdot (\mathbf{a} \times \mathbf{b})$$

$$= (c_1\mathbf{i} + c_2\mathbf{j} + c_3\mathbf{k}) \cdot \left(\begin{vmatrix} a_2 & a_3 \\ b_2 & b_3 \end{vmatrix} \mathbf{i} - \begin{vmatrix} a_1 & a_3 \\ b_1 & b_3 \end{vmatrix} \mathbf{j} + \begin{vmatrix} a_1 & a_2 \\ b_1 & b_2 \end{vmatrix} \mathbf{k} \right)$$

$$= c_1 \begin{vmatrix} a_2 & a_3 \\ b_2 & b_3 \end{vmatrix} - c_2 \begin{vmatrix} a_1 & a_3 \\ b_1 & b_3 \end{vmatrix} + c_3 \begin{vmatrix} a_1 & a_2 \\ b_1 & b_2 \end{vmatrix}.$$

This is the expansion of

$$\begin{vmatrix} a_1 & a_2 & a_3 \\ b_1 & b_2 & b_3 \\ c_1 & c_2 & c_3 \end{vmatrix}$$

by the bottom row. (Exercise 45.) ❏

A SUGGESTION: Vectors were defined as ordered triples, and many of the early proofs were done by "breaking up" vectors into their components. This may give you the impression that the method of "breakup" and working with the components is the first thing to try when confronted with a problem that involves vectors. If it is a *computational* problem, this method may give good results. But if you have to *analyze* a situation involving vectors, particularly one in which geometry plays a role, then the "breakup" strategy is seldom the best. Think instead of using the *operations* we have defined on vectors: addition, subtraction, scalar multiplication, dot product, cross product. Being geometrically motivated, these operations are likely to provide greater understanding than breaking up everything in sight into components.

Example 4 Let $\mathbf{a}, \mathbf{b}, \mathbf{c}$ be nonzero vectors that do not lie in the same plane. Find all the vectors $\mathbf{d}$ for which

$$(*) \qquad\qquad \mathbf{d} \cdot \mathbf{a} = \mathbf{d} \cdot \mathbf{b} = \mathbf{d} \cdot \mathbf{c}.$$

SOLUTION We could begin by writing

$$\mathbf{d} = d_1\mathbf{i} + d_2\mathbf{j} + d_3\mathbf{k}, \qquad \mathbf{a} = a_1\mathbf{i} + a_2\mathbf{j} + a_3\mathbf{k}, \qquad \text{and so on.}$$

Equation $(*)$ would then take the form

$$d_1a_1 + d_2a_2 + d_3a_3 = d_1b_1 + d_2b_2 + d_3b_3 = d_1c_1 + d_2c_2 + d_3c_3,$$

and we would be faced with finding all d_1, d_2, d_3 that satisfied these equations. This is a messy task.

Here is a better approach. The vectors $\mathbf{d}$ that satisfy $(*)$ are the vectors $\mathbf{d}$ for which

$$\mathbf{d} \cdot (\mathbf{a} - \mathbf{b}) = 0 \qquad \text{and} \qquad \mathbf{d} \cdot (\mathbf{b} - \mathbf{c}) = 0.$$

These are the vectors $\mathbf{d}$ that are perpendicular to both $\mathbf{a} - \mathbf{b}$ and $\mathbf{b} - \mathbf{c}$. One such vector is $(\mathbf{a} - \mathbf{b}) \times (\mathbf{b} - \mathbf{c})$. The vectors $\mathbf{d}$ that satisfy $(*)$ are the scalar multiples of that cross product. ❏

Example 5 Verify that

$$\|\mathbf{a} \times \mathbf{b}\|^2 + (\mathbf{a} \cdot \mathbf{b})^2 = \|\mathbf{a}\|^2 \|\mathbf{b}\|^2.$$

SOLUTION We could begin by writing

$$\|\mathbf{a} \times \mathbf{b}\|^2 = (a_2b_3 - a_3b_2)^2 + (a_1b_3 - a_3b_1)^2 + (a_1b_2 - a_2b_1)^2$$

$$(\mathbf{a} \cdot \mathbf{b})^2 = (a_1b_1 + a_2b_2 + a_3b_3)^2$$

$$\|\mathbf{a}\|^2 \|\mathbf{b}\|^2 = (a_1^2 + a_2^2 + a_3^2)(b_1^2 + b_2^2 + b_3^2).$$

but this would take us into a morass of arithmetic. It is much more fruitful to proceed as follows:

$$\|\mathbf{a} \times \mathbf{b}\| = \|\mathbf{a}\| \, \|\mathbf{b}\| \sin \theta \qquad \text{and} \qquad \mathbf{a} \cdot \mathbf{b} = \|\mathbf{a}\| \, \|\mathbf{b}\| \cos \theta.$$

Therefore

$$\|\mathbf{a} \times \mathbf{b}\|^2 + (\mathbf{a} \cdot \mathbf{b})^2 = \|\mathbf{a}\|^2 \|\mathbf{b}\|^2 \sin^2 \theta + \|\mathbf{a}\|^2 \|\mathbf{b}\|^2 \cos^2 \theta$$

$$= \|\mathbf{a}\|^2 \|\mathbf{b}\|^2 (\sin^2 \theta + \cos^2 \theta) = \|\mathbf{a}\|^2 \|\mathbf{b}\|^2. \quad \square$$

Three Important Identities

It may be tempting to think that $\mathbf{a} \times (\mathbf{b} \times \mathbf{c}) = (\mathbf{a} \times \mathbf{b}) \times \mathbf{c}$. This is in general false:

$$\mathbf{i} \times (\mathbf{i} \times \mathbf{j}) = \mathbf{i} \times \mathbf{k} = -\mathbf{j} \qquad \text{but} \qquad (\mathbf{i} \times \mathbf{i}) \times \mathbf{j} = \mathbf{0} \times \mathbf{j} = \mathbf{0}.$$

What is true instead is that

(12.4.10)

$$\mathbf{a} \times (\mathbf{b} \times \mathbf{c}) = (\mathbf{a} \cdot \mathbf{c})\mathbf{b} - (\mathbf{a} \cdot \mathbf{b})\mathbf{c},$$

$$(\mathbf{a} \times \mathbf{b}) \times \mathbf{c} = (\mathbf{c} \cdot \mathbf{a})\mathbf{b} - (\mathbf{c} \cdot \mathbf{b})\mathbf{a}.$$

There is one more identity that we want to mention:

(12.4.11)

$$(\mathbf{a} \times \mathbf{b}) \cdot (\mathbf{c} \times \mathbf{d}) = (\mathbf{a} \cdot \mathbf{c})(\mathbf{b} \cdot \mathbf{d}) - (\mathbf{a} \cdot \mathbf{d})(\mathbf{b} \cdot \mathbf{c}).$$

The proof of this, as well as the proof of (12.4.10), is left to you in the exercises.

Remark Dot products and cross products are almost ubiquitous in physics and engineering. Work is a dot product. So is the power expended by a force. Every line integral is a dot product. Torque and angular momentum are cross products. Turn on a television set and watch the dots on the screen. How they move is determined by a cross product. Take a look at any text on electromagnetism. It is all based on Maxwell's four equations. Two of them are dot products, two of them are cross products. ❑

EXERCISES 12.4

In Exercises 1–12, find the indicated quantity.

1. $(\mathbf{i} + \mathbf{j}) \times (\mathbf{i} - \mathbf{j})$.

2. $(\mathbf{i} - \mathbf{j}) \times (\mathbf{j} - \mathbf{i})$.

3. $(\mathbf{i} - \mathbf{j}) \times (\mathbf{j} - \mathbf{k})$.

4. $\mathbf{j} \times (2\mathbf{i} - \mathbf{k})$.

5. $(2\mathbf{j} - \mathbf{k}) \times (\mathbf{i} - 3\mathbf{j})$.

6. $\mathbf{i} \cdot (\mathbf{j} \times \mathbf{k})$.

7. $\mathbf{j} \cdot (\mathbf{i} \times \mathbf{k})$.

8. $(\mathbf{j} \times \mathbf{i}) \cdot (\mathbf{i} \times \mathbf{k})$.

9. $(\mathbf{i} \times \mathbf{j}) \times \mathbf{k}$.

10. $\mathbf{k} \cdot (\mathbf{j} \times \mathbf{i})$.

11. $\mathbf{j} \cdot (\mathbf{k} \times \mathbf{i})$.　　　　　**12.** $\mathbf{j} \times (\mathbf{k} \times \mathbf{i})$.

In Exercises 13–20, find the indicated quantity.

13. $(\mathbf{i} + 3\mathbf{j} - \mathbf{k}) \times (\mathbf{i} + \mathbf{k})$.

14. $(3\mathbf{i} - 2\mathbf{j} + \mathbf{k}) \times (\mathbf{i} - \mathbf{j} + \mathbf{k})$.

15. $(\mathbf{i} + \mathbf{j} + \mathbf{k}) \times (2\mathbf{i} + \mathbf{k})$.

16. $(2\mathbf{i} - \mathbf{k}) \times (\mathbf{i} - 2\mathbf{j} + 2\mathbf{k})$.

17. $[2\mathbf{i} + \mathbf{j}] \cdot [(\mathbf{i} - 3\mathbf{j} + \mathbf{k}) \times (4\mathbf{i} + \mathbf{k})]$.

18. $[(-2\mathbf{i} + \mathbf{j} - 3\mathbf{k}) \times \mathbf{i}] \times [\mathbf{i} + \mathbf{j}]$.

19. $[(\mathbf{i} - \mathbf{j}) \times (\mathbf{j} - \mathbf{k})] \times [\mathbf{i} + 5\mathbf{k}]$.

20. $[\mathbf{i} - \mathbf{j}] \times [(\mathbf{j} - \mathbf{k}) \times (\mathbf{j} + 5\mathbf{k})]$.

21. Find two unit vectors which are perpendicular to the vectors $\mathbf{a} = (1, 3, -1)$ and $\mathbf{b} = (2, 0, 1)$.

22. Repeat Exercise 21 for the vectors $\mathbf{a} = (1, 2, 3)$ and $\mathbf{b} = (2, 1, 1)$.

In Exercises 23–26, find a vector $\mathbf{N}$ that is perpendicular to the plane determined by the points P, Q, R, and find the area of the triangle PQR.

23. $P(0, 1, 0), Q(-1, 1, 2), R(2, 1, -1)$.

24. $P(1, 2, 3), Q(-1, 3, 2), R(3, -1, 2)$.

25. $P(1, -1, 4), Q(2, 0, 1), R(0, 2, 3)$.

26. $P(2, -1, 3), Q(4, 1, -1), R(-3, 0, 5)$.

In Exercises 27 and 28, find the volume of the parallelepiped with the given edges.

27. $\mathbf{i} + \mathbf{j}, \quad 2\mathbf{i} - \mathbf{k}, \quad 3\mathbf{j} + \mathbf{k}$.

28. $\mathbf{i} - 3\mathbf{j} + \mathbf{k}, \quad 2\mathbf{j} - \mathbf{k}, \quad \mathbf{i} + \mathbf{j} - 2\mathbf{k}$.

29. Find the volume of the parallelepiped whose edges are $\overrightarrow{OP}, \overrightarrow{OQ}$, and $\overrightarrow{OR}$, where $O(0, 0, 0), P(1, 2, 3), Q(1, 1, 2)$, and $R(2, 1, 1)$.

30. Find the volume of the parallelepiped whose edges are $\overrightarrow{PQ}, \overrightarrow{PR}$, and $\overrightarrow{PS}$, where $P(1, -1, 4), \quad Q(2, 0, 1), R(0, 2, 3)$, and $S(3, 5, 7)$.

31. Express $(\mathbf{a} + \mathbf{b}) \times (\mathbf{a} - \mathbf{b})$ as a scalar multiple of $\mathbf{a} \times \mathbf{b}$.

32. Earlier we verified that $\mathbf{a} \times (\mathbf{b} + \mathbf{c}) = (\mathbf{a} \times \mathbf{b}) + (\mathbf{a} \times \mathbf{c})$. Show now that

$$(\mathbf{a} + \mathbf{b}) \times \mathbf{c} = (\mathbf{a} \times \mathbf{c}) + (\mathbf{b} \times \mathbf{c}).$$

33. Suppose that $\mathbf{a} \times \mathbf{i} = \mathbf{0}$ and $\mathbf{a} \times \mathbf{j} = \mathbf{0}$. What can you conclude about $\mathbf{a}$?

34. Let $\mathbf{a} = a_1\mathbf{i} + a_2\mathbf{j}$ and $\mathbf{b} = b_1\mathbf{i} + b_2\mathbf{j}$ be nonzero vectors in the xy-plane. Show that $\mathbf{a} \times \mathbf{b}$ is parallel to $\mathbf{k}$.

35. Express $(\alpha\mathbf{a} + \beta\mathbf{b}) \times (\gamma\mathbf{a} + \delta\mathbf{b})$ as a scalar multiple of $\mathbf{a} \times \mathbf{b}$.

36. (a) Let $\mathbf{a}, \mathbf{b}, \mathbf{c}$ be distinct nonzero vectors. Show that

$$\mathbf{a} \times \mathbf{b} = \mathbf{a} \times \mathbf{c} \quad \text{iff} \quad \mathbf{a} \text{ and } \mathbf{b} - \mathbf{c} \text{ are parallel.}$$

(b) Sketch a figure depicting all those vectors $\mathbf{c}$ that satisfy $\mathbf{a} \times \mathbf{b} = \mathbf{a} \times \mathbf{c}$.

37. Which of the following dot products are equal?

$$\mathbf{a} \cdot (\mathbf{b} \times \mathbf{c}) \quad \mathbf{a} \cdot (\mathbf{c} \times \mathbf{b}) \quad (\mathbf{a} \times \mathbf{b}) \cdot \mathbf{c} \quad (\mathbf{c} \times \mathbf{a}) \cdot \mathbf{b}$$

$$(\mathbf{b} \times \mathbf{c}) \cdot \mathbf{a} \quad \mathbf{c} \cdot (\mathbf{b} \times \mathbf{a}) \quad (-\mathbf{a} \times \mathbf{b}) \cdot \mathbf{c} \quad (\mathbf{a} \times -\mathbf{c}) \cdot \mathbf{b}.$$

38. Show that $(\mathbf{a} \times \mathbf{b}) \cdot \mathbf{b} = \mathbf{0}$ for all vectors $\mathbf{a}$ and $\mathbf{b}$.

39. Show that the vectors $\mathbf{a}, \mathbf{b}$, and $\mathbf{c}$ are coplanar iff $(\mathbf{a} \times \mathbf{b}) \cdot \mathbf{c} = 0$.

40. If $\mathbf{a}, \mathbf{b}$, and $\mathbf{c}$ are mutually perpendicular, show that $\mathbf{a} \times (\mathbf{b} \times \mathbf{c}) = \mathbf{0}$.

41. Let $\mathbf{a} \neq \mathbf{0}$. Show that if $\mathbf{a} \times \mathbf{b} = \mathbf{a} \times \mathbf{c}$ and $\mathbf{a} \cdot \mathbf{b} = \mathbf{a} \cdot \mathbf{c}$ then $\mathbf{b} = \mathbf{c}$.

42. (a) Show that

$$\mathbf{a} \times (\mathbf{b} \times \mathbf{c}) = (\mathbf{a} \cdot \mathbf{c})\mathbf{b} - (\mathbf{a} \cdot \mathbf{b})\mathbf{c}.$$

HINT: Verify that the $\mathbf{i}$ components of the two sides agree. By analogy, it will be clear that the same holds true for the $\mathbf{j}$ and $\mathbf{k}$ components.

(b) Show that

$$(\mathbf{a} \times \mathbf{b}) \times \mathbf{c} = (\mathbf{c} \cdot \mathbf{a})\mathbf{b} - (\mathbf{c} \cdot \mathbf{b})\mathbf{a}.$$

HINT: $(\mathbf{a} \times \mathbf{b}) \times \mathbf{c} = -[\mathbf{c} \times (\mathbf{a} \times \mathbf{b})]$.

(c) Finally, show that

$$(\mathbf{a} \times \mathbf{b}) \cdot (\mathbf{c} \times \mathbf{d}) = (\mathbf{a} \cdot \mathbf{c})(\mathbf{b} \cdot \mathbf{d}) - (\mathbf{a} \cdot \mathbf{d})(\mathbf{b} \cdot \mathbf{c}).$$

HINT: Set $\mathbf{c} \times \mathbf{d} = \mathbf{r}$ and use (12.4.6).

43. Let $\mathbf{a}, \mathbf{b}$ be nonzero vectors with $\mathbf{a} \perp \mathbf{b}$, and set $\mathbf{c} = \mathbf{a} \times \mathbf{b}$. Express $\mathbf{c} \times \mathbf{a}$ in terms of $\mathbf{b}$.

44. Given that $\mathbf{u}$ is a unit vector, show that each vector $\mathbf{a}$ can be decomposed as follows into a part parallel to $\mathbf{u}$ and a part perpendicular to $\mathbf{u}$:

$$\mathbf{a} = \underbrace{(\mathbf{a} \cdot \mathbf{u})\mathbf{u}}_{\substack{\text{parallel} \\ \text{to } \mathbf{u}}} + \underbrace{(\mathbf{u} \times \mathbf{a}) \times \mathbf{u}}_{\substack{\text{perpendicular} \\ \text{to } \mathbf{u}}}.$$

HINT: Use Exercise 42(b).

45. Verify the conclusion given in Example 3.

Torque. Suppose that a rigid body is free to rotate about a fixed point O. If a force $\mathbf{F}$ acts on the body at a point P, then the body will tend to rotate about an axis through O. This effect is measured by the *torque* vector $\boldsymbol{\tau}$ which is given by

$$\boldsymbol{\tau} = \mathbf{r} \times \mathbf{F},$$

where $\mathbf{r}$ is the position vector $\overrightarrow{OP}$. The straight line through O determined by $\boldsymbol{\tau}$ is the axis of rotation. By Definition 12.4.1, the vectors **r**, **F**, and **τ** form a right-handed system, and the magnitude of **τ** is

$$\|\boldsymbol{\tau}\| = \|\mathbf{r}\| \, \|\mathbf{F}\| \sin \theta$$

where θ is the angle between the position and force vectors. See the following figure.

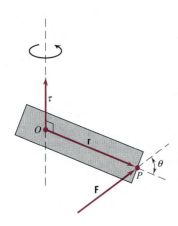

46. Find the magnitude of the torque exerted at the origin by the force $\mathbf{F} = \mathbf{i} + 2\mathbf{j} + \mathbf{k}$ at the point P(1, 1, 1).

▶**47.** A bolt is being tightened by a 20-pound force applied to a 10-inch wrench as shown in the figure. Find the magnitude of the torque. Assuming that the wrench and the force are in the plane of the paper, in what direction will the bolt move?

▶**48.** Repeat Exercise 47 if the 20-pound force is applied as shown in the figure.

■ 12.5 LINES

Vectors used to specify position are called *position vectors*. Position vectors that emanate from the origin are known as *radius vectors*. In this section we use radius vectors to characterize lines.

Vector Parametrizations

We begin with the idea that two distinct points determine a line. In Figure 12.5.1 we have marked two points, P and Q, and the line l that they determine. To obtain a vector characterization of l, we choose the vectors $\mathbf{r}_0$ and $\mathbf{d} = \overrightarrow{PQ}$ as in Figure 12.5.2. Since we began with two distinct points, the vector **d** is nonzero.

Figure 12.5.1

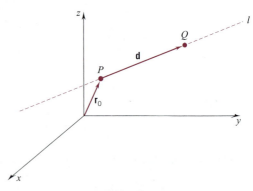

Figure 12.5.2

In Figure 12.5.3 we have drawn an additional vector $\mathbf{r}$. The vector that begins at the tip of $\mathbf{r}_0$ and ends at the tip of $\mathbf{r}$ is $\mathbf{r} - \mathbf{r}_0$. Therefore, the tip of $\mathbf{r}$ will fall on l iff

$$\mathbf{r} - \mathbf{r}_0 \qquad \text{and} \qquad \mathbf{d} \quad \text{are parallel;}$$

this in turn will happen iff

$$\mathbf{r} - \mathbf{r}_0 = t\mathbf{d} \qquad \text{for some real number } t,$$

or equivalently, iff

$$\mathbf{r} = \mathbf{r}_0 + t\mathbf{d} \qquad \text{for some real number } t.$$

The vector equation

(12.5.1)

$$\boxed{\mathbf{r}(t) = \mathbf{r}_0 + t\mathbf{d}, \qquad t \text{ real}}$$

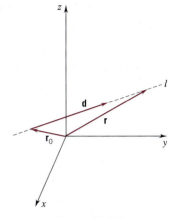

Figure 12.5.3

parametrizes the line l: by varying t, we vary the vector $\mathbf{r}(t)$, but its tip remains on l; as t ranges over the set of real numbers, the tip $\mathbf{r}(t)$ traces out the line l.

Now set

$$\mathbf{r}_0 = x_0\mathbf{i} + y_0\mathbf{j} + z_0\mathbf{k}, \qquad \mathbf{d} = d_1\mathbf{i} + d_2\mathbf{j} + d_3\mathbf{k}$$

and denote the tip of $\mathbf{r}_0$ by $P(x_0, y_0, z_0)$. The line l given by

(12.5.2)

$$\boxed{\mathbf{r}(t) = \mathbf{r}_0 + t\mathbf{d} = (x_0 + td_1)\mathbf{i} + (y_0 + td_2)\mathbf{j} + (z_0 + td_3)\mathbf{k}}$$

passes through the point $P(x_0, y_0, z_0)$ and is parallel to $\mathbf{d}$ (Figure 12.5.4). The vector $\mathbf{d}$, which by definition is not zero, is called a *direction vector* for l, and its components d_1, d_2, d_3 are called *direction numbers*.

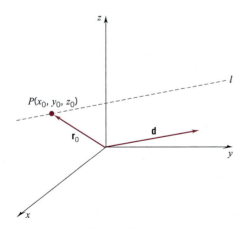

Figure 12.5.4

Remark A direction vector for the line l determined by two distinct points $P(x_0, y_0, z_0)$ and $Q(x_1, y_1, z_1)$ is $\mathbf{d} = \overrightarrow{PQ} = (x_1 - x_0)\mathbf{i} + (y_1 - y_0)\mathbf{j} + (z_1 - z_0)\mathbf{k}$; a set of direction numbers for l is: $(x_1 - x_0), (y_1 - y_0), (z_1 - z_0)$. ❏

Example 1 Find a vector equation that parametrizes the line that passes through the point $P(1, -1, 2)$ and is parallel to the vector $2\mathbf{i} - 3\mathbf{j} + \mathbf{k}$.

SOLUTION Here we can set

$$\mathbf{r}_0 = \mathbf{i} - \mathbf{j} + 2\mathbf{k} \qquad \text{and} \qquad \mathbf{d} = 2\mathbf{i} - 3\mathbf{j} + \mathbf{k}.$$

As a vector parametrization for the line we have

$$\mathbf{r}(t) = (\mathbf{i} - \mathbf{j} + 2\mathbf{k}) + t(2\mathbf{i} - 3\mathbf{j} + \mathbf{k}),$$

which we can rewrite as

$$\mathbf{r}(t) = (1 + 2t)\mathbf{i} - (1 + 3t)\mathbf{j} + (2 + t)\mathbf{k}. \qquad ❑$$

As a direction vector for a given line we can take any nonzero vector that is parallel to the line. Thus, if $\mathbf{d}$ is a direction vector for l, so is $\alpha\mathbf{d}$, provided that $\alpha \neq 0$. If d_1, d_2, d_3 are direction numbers for l, so are $\alpha d_1, \alpha d_2, \alpha d_3$, provided again that $\alpha \neq 0$.

The line that passes through the origin with direction vector $\mathbf{d}$ can be parametrized by the vector equation

(12.5.3)
$$\mathbf{r}(t) = t\mathbf{d} = t d_1 \mathbf{i} + t d_2 \mathbf{j} + t d_3 \mathbf{k}.$$
(Figure 12.5.5)

There are, however, other ways to parametrize this line.

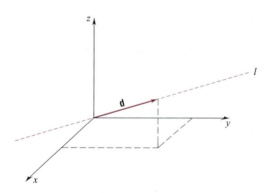

Figure 12.5.5

Example 2 Find all vector parametrizations (12.5.1) for the line through the origin with direction vector $\mathbf{d}$.

SOLUTION Since $\mathbf{d}$ is a direction vector, so is every vector $\alpha\mathbf{d}$ with $\alpha \neq 0$. We therefore write

(∗)
$$\mathbf{r}(t) = \mathbf{r}_0 + t\alpha\mathbf{d}.$$

Since the line passes through the origin, it may appear at first glance that $\mathbf{r}_0$ has to be $\mathbf{0}$, but that is not true. From the fact that the line passes through the origin we can conclude only that $\mathbf{r}_0 + t\alpha\mathbf{d}$ is $\mathbf{0}$ for some value of t. Call this value t_0. Then

$$\mathbf{r}_0 + t_0\alpha\mathbf{d} = \mathbf{0} \qquad \text{and thus} \qquad \mathbf{r}_0 = -t_0\alpha\mathbf{d}.$$

Substitution in ($*$) gives

$$\mathbf{r}(t) = -t_0\alpha\mathbf{d} + t\alpha\mathbf{d} = (t - t_0)\alpha\mathbf{d}.$$

All the desired parametrizations can be written

$$\mathbf{r}(t) = (t - t_0)\alpha\mathbf{d} \qquad \text{with } \alpha \text{ and } t_0 \text{ real, } \alpha \neq 0,$$

and all equations of this form parametrize that same line. ❑

Scalar Parametric Equations

It follows from equation (12.5.2) that the line that passes through the point $P(x_0, y_0, z_0)$ with direction numbers d_1, d_2, d_3 can be parametrized by three scalar equations:

(12.5.4)

$$x(t) = x_0 + d_1t, \qquad y(t) = y_0 + d_2t, \qquad z(t) = z_0 + d_3t.$$

These equations are the **i, j, k** components of the vector $\mathbf{r}(t) = \mathbf{r}_0 + t\mathbf{d}$.

Example 3 Write scalar parametric equations for the line that passes through the point $P(-1, 4, 2)$ with direction numbers $1, 2, 3$.

SOLUTION In this case the scalar equations

$$x(t) = x_0 + d_1t, \qquad y(t) = y_0 + d_2t, \qquad z(t) = z_0 + d_3t$$

take the form

$$x(t) = -1 + t, \qquad y(t) = 4 + 2t, \qquad z(t) = 2 + 3t. \quad ❑$$

Example 4 What direction numbers are displayed by the parametric equations

$$x(t) = 3 - t, \qquad y(t) = 2 + 4t, \qquad z(t) = 1 - 5t?$$

What other direction numbers could be used for the same line?

SOLUTION The direction numbers displayed are $-1, 4, -5$. Any triple of the form

$$-\alpha, \quad 4\alpha, \quad -5\alpha \qquad \text{with } \alpha \neq 0$$

could be used as a set of direction numbers for that same line. ❑

Symmetric Form

If the direction numbers are all nonzero, then each of the scalar parametric equations can be solved for t:

$$t = \frac{x(t) - x_0}{d_1}, \qquad t = \frac{y(t) - y_0}{d_2}, \qquad t = \frac{z(t) - z_0}{d_3}.$$

Eliminating the parameter t, we obtain three equations:

$$\frac{x - x_0}{d_1} = \frac{y - y_0}{d_2}, \qquad \frac{y - y_0}{d_2} = \frac{z - z_0}{d_3}, \qquad \frac{x - x_0}{d_1} = \frac{z - z_0}{d_3}.$$

Obviously, any two of these equations suffice; the third is redundant and can be discarded. Rather than decide which equation to discard, we simply write

(12.5.5)

$$\frac{x - x_0}{d_1} = \frac{y - y_0}{d_2} = \frac{z - z_0}{d_3}.$$

These are the equations of a line written in *symmetric form*. They can be used only if d_1, d_2, d_3 are all different from zero.

Example 5 Write equations in symmetric form for the line that passes through the points $P(x_0, y_0, z_0)$ and $Q(x_1, y_1, z_1)$. Under what conditions are the equations valid?

SOLUTION As direction numbers we can take the triple

$$x_1 - x_0, \qquad y_1 - y_0, \qquad z_1 - z_0.$$

We can base our calculations on $P(x_0, y_0, z_0)$ and write

$$\frac{x - x_0}{x_1 - x_0} = \frac{y - y_0}{y_1 - y_0} = \frac{z - z_0}{z_1 - z_0},$$

or we can base our calculations on $Q(x_1, y_1, z_1)$ and write

$$\frac{x - x_1}{x_1 - x_0} = \frac{y - y_1}{y_1 - y_0} = \frac{z - z_1}{z_1 - z_0}.$$

Both sets of equations are valid provided that $x_1 \neq x_0, y_1 \neq y_0, z_1 \neq z_0$. ❏

Equations 12.5.5 can be used only if the direction numbers are all different from zero. If one of the direction numbers is zero, then one of the coordinates is constant. As you will see, this simplifies the algebra. Geometrically, it means that the line lies on a plane that is parallel to one of the coordinate planes.

Suppose, for example, that $d_3 = 0$. Then the scalar parametric equations take the form

$$x(t) = x_0 + d_1 t, \qquad y(t) = y_0 + d_2 t, \qquad z(t) = z_0.$$

Eliminating t, we are left with two equations:

$$\frac{x - x_0}{d_1} = \frac{y - y_0}{d_2}, \qquad z = z_0.$$

The line lies on the horizontal plane $z = z_0$, and its projection onto the xy-plane (see Figure 12.5.6) is the line l' with equation

$$\frac{x - x_0}{d_1} = \frac{y - y_0}{d_2}.$$

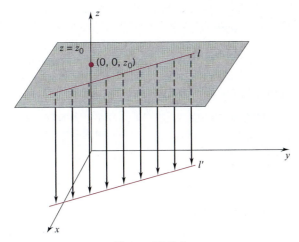

Figure 12.5.6

Intersecting Lines, Parallel Lines

Two distinct lines

$$l_1 : \mathbf{r}(t) = \mathbf{r}_0 + t\mathbf{d}, \qquad l_2 : \mathbf{R}(u) = \mathbf{R}_0 + u\mathbf{D}$$

intersect iff there are numbers t and u at which

$$\mathbf{r}(t) = \mathbf{R}(u).$$

Example 6 Find the point at which the lines

$$l_1 : \mathbf{r}(t) = (\mathbf{i} - 6\mathbf{j} + 2\mathbf{k}) + t(\mathbf{i} + 2\mathbf{j} + \mathbf{k}), \qquad l_2 : \mathbf{R}(u) = (4\mathbf{j} + \mathbf{k}) + u(2\mathbf{i} + \mathbf{j} + 2\mathbf{k})$$

intersect.

SOLUTION We set

$$\mathbf{r}(t) = \mathbf{R}(u)$$

and solve for t and u:

$$(\mathbf{i} - 6\mathbf{j} + 2\mathbf{k}) + t(\mathbf{i} + 2\mathbf{j} + \mathbf{k}) = (4\mathbf{j} + \mathbf{k}) + u(2\mathbf{i} + \mathbf{j} + 2\mathbf{k}),$$

$$\mathbf{i} - 6\mathbf{j} + 2\mathbf{k} + t\mathbf{i} + 2t\mathbf{j} + t\mathbf{k} = 4\mathbf{j} + \mathbf{k} + 2u\mathbf{i} + u\mathbf{j} + 2u\mathbf{k},$$

and therefore

$$(1 + t - 2u)\mathbf{i} + (-10 + 2t - u)\mathbf{j} + (1 + t - 2u)\mathbf{k} = 0.$$

This tells us that

$$1 + t - 2u = 0,$$
$$-10 + 2t - u = 0,$$
$$1 + t - 2u = 0.$$

Note that the first and third equations are the same. Solving the first two equations simultaneously, we obtain $t = 7$, $u = 4$. As you can verify,

$$\mathbf{r}(7) = 8\mathbf{i} + 8\mathbf{j} + 9\mathbf{k} = \mathbf{R}(4).$$

The two lines intersect at the tip of this vector, which is the point $P(8, 8, 9)$. ❏

Remark To give a physical interpretation of the result in Example 6, think of the parameters t and u as representing time, and think of particles moving along the lines l_1 and l_2. At time $t = u = 0$, the particle on l_1 is at the point $(1, -6, 2)$ and the particle on l_2 is at the point $(0, 4, 1)$. The particle on l_1 passes through the point $P(8, 8, 9)$ at time $t = 7$ while the particle on l_2 passes through P at time $u = 4$; both particles pass through the same point P, but at different times! ❏

In the setting of plane geometry we can think of two lines as parallel iff they do not intersect. This point of view is not satisfactory in three-dimensional space. (See Figure 12.5.7.)

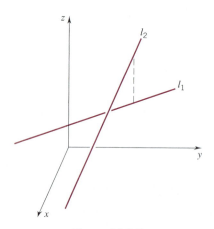

Figure 12.5.7

The lines l_1 and l_2 marked in Figure 12.5.7 do not intersect, and yet we would hesitate to call them parallel. We can avoid this difficulty by using direction vectors: we call two lines *parallel* iff their direction vectors are parallel. Nonparallel, nonintersecting lines are said to be *skew*.

If two lines l_1, l_2 intersect, we can find the angle between them by finding the angle between their direction vectors, **d** and **D**. Depending on our choice of direction vectors, there are two such angles, each the supplement of the other (Figure 12.5.8). We choose the smaller of the two angles, the one with nonnegative cosine:

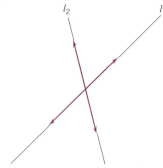

Figure 12.5.8

(12.5.6)
$$\cos \theta = |\mathbf{u_d} \cdot \mathbf{u_D}|.$$

Example 7 Earlier we verified that the lines

$$l_1: \mathbf{r}(t) = (\mathbf{i} - 6\mathbf{j} + 2\mathbf{k}) + t(\mathbf{i} + 2\mathbf{j} + \mathbf{k}), \qquad l_2: \mathbf{R}(u) = (4\mathbf{j} + \mathbf{k}) + u(2\mathbf{i} + \mathbf{j} + 2\mathbf{k})$$

intersect at $P(8, 8, 9)$. What is the angle between these lines?

SOLUTION As direction vectors we can take

$$\mathbf{d} = \mathbf{i} + 2\mathbf{j} + \mathbf{k} \qquad \text{and} \qquad \mathbf{D} = 2\mathbf{i} + \mathbf{j} + 2\mathbf{k}.$$

Then, as you can check,

$$\mathbf{u_d} = \tfrac{1}{6}\sqrt{6}(\mathbf{i} + 2\mathbf{j} + \mathbf{k}) \qquad \text{and} \qquad \mathbf{u_D} = \tfrac{1}{3}(2\mathbf{i} + \mathbf{j} + 2\mathbf{k}).$$

It follows that

$$\cos \theta = |\mathbf{u_d} \cdot \mathbf{u_D}| = \tfrac{1}{3}\sqrt{6} \cong 0.816 \qquad \text{and} \qquad \theta \cong 0.62 \text{ radian.} \quad \square$$

Two intersecting lines are said to be *perpendicular* iff their direction vectors are perpendicular.

Example 8 Let l_1 and l_2 be the lines of the last exercise. These lines intersect at $P(8, 8, 9)$. Find a vector parametrization for the line l_3 that passes through $P(8, 8, 9)$ and is perpendicular to both l_1 and l_2.

SOLUTION We are given that l_3 passes through $P(8, 8, 9)$. All we need to parametrize this line is a direction vector $\mathbf{d}_3$. We require that $\mathbf{d}_3$ be perpendicular to the direction vectors of l_1 and l_2; namely, we require that

$$\mathbf{d}_3 \perp \mathbf{d} \quad \text{and} \quad \mathbf{d}_3 \perp \mathbf{D}, \qquad \text{where} \quad \mathbf{d} = \mathbf{i} + 2\mathbf{j} + \mathbf{k} \quad \text{and} \quad \mathbf{D} = 2\mathbf{i} + \mathbf{j} + 2\mathbf{k}.$$

Since $\mathbf{d} \times \mathbf{D}$ is perpendicular to both $\mathbf{d}$ and $\mathbf{D}$, we can set

$$\mathbf{d}_3 = \mathbf{d} \times \mathbf{D} = \begin{vmatrix} \mathbf{i} & \mathbf{j} & \mathbf{k} \\ 1 & 2 & 1 \\ 2 & 1 & 2 \end{vmatrix} = \begin{vmatrix} 2 & 1 \\ 1 & 2 \end{vmatrix} \mathbf{i} - \begin{vmatrix} 1 & 1 \\ 2 & 2 \end{vmatrix} \mathbf{j} + \begin{vmatrix} 1 & 2 \\ 2 & 1 \end{vmatrix} \mathbf{k} = 3\mathbf{i} - 3\mathbf{k}.$$

As a parametrization for l_3 we can write

$$\gamma_3(t) = (8\mathbf{i} + 8\mathbf{j} + 9\mathbf{k}) + t(3\mathbf{i} - 3\mathbf{k}). \quad \square$$

Example 9

(a) Find a vector parametrization for the line

$$l: y = mx + b \qquad \text{in the } xy\text{-plane.}$$

(b) Show by vector methods that

$$l_1: y = m_1 x + b_1 \perp l_2: y = m_2 x + b_2 \qquad \text{iff} \qquad m_1 m_2 = -1.$$

SOLUTION

(a) We seek a parametrization of the form

$$\mathbf{r}(t) = \mathbf{r}_0 + t\mathbf{d}.$$

Since $P(0, b)$ lies on l, we can set

$$\mathbf{r}_0 = 0\mathbf{i} + b\mathbf{j} = b\mathbf{j}.$$

To find a direction vector, we take $x_1 \neq 0$ and note that the point $Q(x_1, mx_1 + b)$ also lies on l. (See Figure 12.5.9.) As direction numbers we can take

$$x_1 - 0 = x_1 \qquad \text{and} \qquad (mx_1 + b) - b = mx_1$$

or, more simply, 1 and m. This choice of direction numbers gives us the direction vector $\mathbf{d} = \mathbf{i} + m\mathbf{j}$. The vector equation

$$\mathbf{r}(t) = b\mathbf{j} + t(\mathbf{i} + m\mathbf{j})$$

parametrizes the line l.

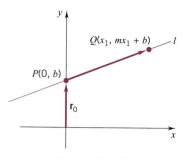

Figure 12.5.9

(b) As direction vectors for l_1 and l_2 we have

$$\mathbf{d}_1 = \mathbf{i} + m_1\mathbf{j} \qquad \text{and} \qquad \mathbf{d}_2 = \mathbf{i} + m_2\mathbf{j}.$$

Since

$$\mathbf{d}_1 \cdot \mathbf{d}_2 = (\mathbf{i} + m_1\mathbf{j}) \cdot (\mathbf{i} + m_2\mathbf{j}) = 1 + m_1 m_2,$$

you can see that

$$\mathbf{d}_1 \cdot \mathbf{d}_2 = 0 \qquad \text{iff} \qquad m_1 m_2 = -1. \qquad \square$$

Distance from a Point to a Line

In Figure 12.5.10 we have drawn a line l and a point P not on l. We are interested in finding the distance $d(P, l)$ between P and l.

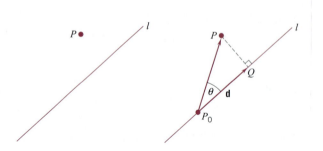

Figure 12.5.10

With P_0 and Q as in the figure, we can take $\mathbf{d} = \overrightarrow{P_0 Q}$ as a direction vector for l. From the figure you can see that

$$d(P, l) = \|\overrightarrow{QP}\| = \|\overrightarrow{P_0 P}\| \sin\theta.$$

Since $\|\overrightarrow{P_0 P} \times \mathbf{d}\| = \|\overrightarrow{P_0 P}\|\,\|\mathbf{d}\| \sin\theta$, we have

(12.5.7)

$$d(P, l) = \frac{\|\overrightarrow{P_0 P} \times \mathbf{d}\|}{\|\mathbf{d}\|}.$$

This elegant little formula gives the distance from a point P to any line l in terms of any point P_0 on l and any direction vector $\mathbf{d}$ for l. ❏

Computations based on this formula are left to the exercises.

EXERCISES 12.5

1. Which of the points $P(1, 2, 0)$, $Q(-5, 1, 5)$, $R(-4, 2, 5)$ lie on the line

$$l: \mathbf{r}(t) = (\mathbf{i} + 2\mathbf{j}) + t(6\mathbf{i} + \mathbf{j} - 5\mathbf{k})?$$

2. Determine which of the lines are parallel:

$$l_1: \mathbf{r}_1(t) = (\mathbf{i} + 2\mathbf{k}) + t(\mathbf{i} - 2\mathbf{j} + 3\mathbf{k}),$$

$$l_2: \mathbf{r}_2(u) = (\mathbf{i} + 2\mathbf{k}) + u(\mathbf{i} + 2\mathbf{j} - 3\mathbf{k}),$$

$$l_3: \mathbf{r}_3(v) = (6\mathbf{i} - \mathbf{j}) - v(2\mathbf{i} - 4\mathbf{j} + 6\mathbf{k}),$$

$$l_4: \mathbf{r}_4(w) = (\tfrac{1}{2} + \tfrac{1}{2}w)\mathbf{i} - w\mathbf{j} + (1 + \tfrac{3}{2}w)\mathbf{k}.$$

In Exercises 3–6, find a vector parametrization for the line that satisfies the given conditions.

3. Passes through $P(3, 1, 0)$ and is parallel to the line $\mathbf{r}(t) = (\mathbf{i} - \mathbf{j}) + t\mathbf{k}$.

4. Passes through $P(1, -1, 2)$ and is parallel to the line $\mathbf{r}(t) = t(3\mathbf{i} - \mathbf{j} + \mathbf{k})$.

5. Passes through the origin and $Q(x_1, y_1, z_1)$.

6. Passes through $P(x_0, y_0, z_0)$ and $Q(x_1, y_1, z_1)$.

In Exercises 7–10, find a set of scalar parametric equations for the line that satisfies the given conditions.

7. Passes through $P(1, 0, 3)$ and $Q(2, -1, 4)$.

8. Passes through $P(x_0, y_0, z_0)$ and $Q(x_1, y_1, z_1)$.

9. Passes through $P(2, -2, 3)$ and is perpendicular to the xz-plane.

10. Passes through $P(1, 4, -3)$ and is perpendicular to the yz-plane.

11. Give a vector parametrization for the line that passes through $P(-1, 2, -3)$ and is parallel to the line $2(x + 1) = 4(y - 3) = z$.

12. Write equations in symmetric form for the line that passes through the origin and the point $P(x_0, y_0, z_0)$.

In Exercises 13–16, find the point where l_1 and l_2 intersect and find the angle between l_1 and l_2.

13. $l_1: \mathbf{r}_1(t) = \mathbf{i} + t\mathbf{j}$, $l_2: \mathbf{r}_2(u) = \mathbf{j} + u(\mathbf{i} + \mathbf{j})$.

14. $l_1: \mathbf{r}_1(t) = (\mathbf{i} - 4\sqrt{3}\mathbf{j}) + t(\mathbf{i} + \sqrt{3}\mathbf{j})$,

    $l_2: \mathbf{r}_2(u) = (4\mathbf{i} + 3\sqrt{3}\mathbf{j}) + u(\mathbf{i} - \sqrt{3}\mathbf{j})$.

15. $l_1: \quad x_1(t) = 3 + t, \qquad y_1(t) = 1 - t, \qquad z_1(t) = 5 + 2t.$

    $l_2: \quad x_2(u) = 1, \qquad y_2(u) = 4 + u, \qquad z_2(u) = 2 + u.$

16. $l_1: \quad x_1(t) = 1 + t, \quad y_1(t) = -1 - t, \quad z_1(t) = -4 + 2t.$

    $l_2: \quad x_2(u) = 1 - u, \qquad y_2(u) = 1 + 3u, \qquad z_2(u) = 2u.$

17. Where does the line

$$\frac{x - x_0}{d_1} = \frac{y - y_0}{d_2} = \frac{z - z_0}{d_3}$$

intersect the xy-plane?

18. What can you conclude about the lines

$$\frac{x - x_0}{d_1} = \frac{y - y_0}{d_2} = \frac{z - z_0}{d_3}, \quad \frac{x - x_0}{D_1} = \frac{y - y_0}{D_2} = \frac{z - z_0}{D_3}$$

given that $d_1 D_1 + d_2 D_2 + d_3 D_3 = 0$?

19. What can you conclude about the lines

$$\frac{x - x_0}{d_1} = \frac{y - y_0}{d_2} = \frac{z - z_0}{d_3}, \quad \frac{x - x_1}{D_1} = \frac{y - y_1}{D_2} = \frac{z - z_1}{D_3}$$

given that $d_1/D_1 = d_2/D_2 = d_3/D_3$?

20. Let P_0, P_1 be two distinct points and let $\mathbf{r}_0$, $\mathbf{r}_1$ be the radius vectors that they determine:

$$\mathbf{r}_0 = \overrightarrow{OP_0}, \qquad \mathbf{r}_1 = \overrightarrow{OP_1}.$$

As t ranges over the set of real numbers, $\mathbf{r}(t) = \mathbf{r}_0 + t(\mathbf{r}_1 - \mathbf{r}_0)$ traces out the line determined by P_0 and P_1. Restrict t so that $\mathbf{r}(t)$ traces out only the line segment $\overline{P_0 P_1}$.

21. Find a vector parametrization for the line segment that begins at $(2, 7, -1)$ and ends at $(4, 2, 3)$.

22. Restrict t so that the equations

$$x(t) = 7 - 5t, \qquad y(t) = -3 + 2t, \qquad z(t) = 4 - t$$

parametrize the line segment that begins at $(12, -5, 5)$ and ends at $(-3, 1, 2)$.

23. Determine a unit vector $\mathbf{u}$ and the values of t for which the equation

$$\mathbf{r}(t) = (6\mathbf{i} - 5\mathbf{j} + \mathbf{k}) + t\mathbf{u}$$

parametrizes the line segment that begins at $P(0, -2, 7)$ and ends at $Q(-4, 0, 11)$.

24. Suppose that the lines

$$l_1: \mathbf{r}(t) = \mathbf{r}_0 + t\mathbf{d}, \qquad l_2: \mathbf{R}(u) = \mathbf{R}_0 + u\mathbf{D}$$

intersect at right angles. Show that the point of intersection is the origin iff $\mathbf{r}(t) \perp \mathbf{R}(u)$ for all real numbers t and u.

25. Find scalar parametric equations for all the lines that are perpendicular to the line

$$x(t) = 1 + 2t, \qquad y(t) = 3 - 4t, \qquad z(t) = 2 + 6t$$

and intersect that line at the point $P(3, -1, 8)$.

26. Suppose that $\mathbf{r}(t) = \mathbf{r}_0 + t\mathbf{d}$ and $\mathbf{R}(u) = \mathbf{R}_0 + u\mathbf{D}$ both parametrize the same line. (a) Show that $\mathbf{R}_0 = \mathbf{r}_0 + t_0\mathbf{d}$ for some real number t_0. (b) Then show that, for some real number α, $\mathbf{R}(u) = \mathbf{r}_0 + (t_0 + \alpha u)\mathbf{d}$ for all real u.

In Exercises 27 and 28, find the distance from $P(1, 0, 2)$ to the indicated line.

27. The line through the origin parallel to $2\mathbf{i} - \mathbf{j} + 2\mathbf{k}$.

28. The line through $P_0(1, -1, 1)$ parallel to $\mathbf{i} - 2\mathbf{j} - 2\mathbf{k}$.

In Exercises 29–31, find the distance from the point to the line.

29. $P(1, 2, 3)$, $l: \mathbf{r}(t) = \mathbf{i} + 2\mathbf{k} + t(\mathbf{i} - 2\mathbf{j} + 3\mathbf{k})$.

30. $P(0, 0, 0)$, $l: \mathbf{r}(t) = \mathbf{i} + t\mathbf{j}$.

31. $P(1, 0, 1)$, $l: \mathbf{r}(t) = 2\mathbf{i} - \mathbf{j} + t(\mathbf{i} + \mathbf{j})$.

32. Find the distance from the point $P(x_0, y_0, z_0)$ to the line $y = mx + b$ in the xy-plane.

33. What is the distance from the origin (a) to the line that joins $P(1, 1, 1)$ and $Q(2, 2, 1)$?   (b) To the line segment that joins these same points?   [For (b) find the point of the line segment $\overline{PQ}$ closest to the origin and calculate its distance from the origin.]

34. Let l be the line

$$\mathbf{r}(t) = \mathbf{r}_0 + t\mathbf{d}.$$

(a) Find the scalar t_0 for which $\mathbf{r}(t_0) \perp l$.
(b) Find the parametrizations $\mathbf{R}(t) = \mathbf{R}_0 + t\mathbf{D}$ for l in which $\mathbf{R}_0 \perp l$ and $\|\mathbf{D}\| = 1$. These are the *standard vector parametrizations* for l.

In Exercises 35 and 36, find the standard vector parametrizations (Exercise 34) for the specified line.

35. The line through $P(0, 1, -2)$ parallel to $\mathbf{i} - \mathbf{j} + 3\mathbf{k}$.

36. The line through $P(\sqrt{3}, 0, 0)$ parallel to $\mathbf{i} + \mathbf{j} + \mathbf{k}$.

37. Let A, B, C be the vertices of a triangle in the xy-plane. Given that $0 < s < 1$, determine the values of t for which the tip of the radius vector

$$\overrightarrow{OA} + s\,\overrightarrow{AB} + t\,\overrightarrow{BC}$$

lies inside the triangle.   HINT: Draw a diagram.

■ 12.6 PLANES

Ways of Specifying a Plane

How can we specify a plane?   There are a number of ways of doing so.   For example, by giving three distinct points on it, so long as they are not all on the same line; by giving two distinct lines on it; or by giving a line on it and a point on it, so long as the point does not lie on the line.   There is still another way to specify a plane, and that is to give a point on the plane and a nonzero vector perpendicular to the plane

Scalar Equation of a Plane

Figure 12.6.1 shows a plane.   On it we have marked a point $P(x_0, y_0, z_0)$ and starting at that point, a nonzero vector $\mathbf{N} = A\mathbf{i} + B\mathbf{j} + C\mathbf{k}$ perpendicular to the plane.   We call $\mathbf{N}$ a *normal vector*.   We can obtain an equation for the plane in terms of the coordinates of P and the components of $\mathbf{N}$.

To find such an equation we take an arbitrary point $Q(x, y, z)$ in space and form the vector

$$\overrightarrow{PQ} = (x - x_0)\mathbf{i} + (y - y_0)\mathbf{j} + (z - z_0)\mathbf{k}.$$

The point Q will lie on the given plane iff

$$\mathbf{N} \cdot \overrightarrow{PQ} = 0,$$

which is to say, iff

(12.6.1)
$$A(x - x_0) + B(y - y_0) + C(z - z_0) = 0.$$

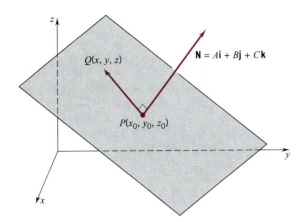

Figure 12.6.1

This is an equation in x, y, z for the plane that passes through $P(x_0, y_0, z_0)$ and has normal vector $\mathbf{N} = A\mathbf{i} + B\mathbf{j} + C\mathbf{k}$.

Remark If $\mathbf{N}$ is normal to a given plane, then so is every nonzero scalar multiple of $\mathbf{N}$. Suppose we had chosen $-2\mathbf{N}$ as our normal. Then Equation (12.6.1) would have read

$$-2A(x - x_0) - 2B(y - y_0) - 2C(z - z_0) = 0.$$

Cancelling the -2, we would have the same equation we had before. It does not matter which normal we choose. They all give equivalent equations. ❑

We can write Equation (12.6.1) in the form

$$Ax + By + Cz + D = 0$$

simply by setting $D = -Ax_0 - By_0 - Cz_0$.

Example 1 Write an equation for the plane that passes through the point $P(1, 0, 2)$ and has normal vector $\mathbf{N} = 3\mathbf{i} - 2\mathbf{j} + \mathbf{k}$.

SOLUTION The general equation

$$A(x - x_0) + B(y - y_0) + C(z - z_0) = 0$$

becomes

$$3(x - 1) + (-2)(y - 0) + (1)(z - 2) = 0,$$

which simplifies to

$$3x - 2y + z - 5 = 0. ❑$$

Example 2 Find an equation for the plane p that passes through $P(-2, 3, 5)$ and is perpendicular to the line l whose scalar parametric equations are $x = -2 + t$, $y = 1 + 2t$, $z = 4$.

SOLUTION We can take $\mathbf{N} = \mathbf{i} + 2\mathbf{j}$ as a direction vector for l. Since p and l are perpendicular, $\mathbf{N}$ is a normal vector for p. Thus, an equation for p is

$$1(x + 2) + 2(y - 3) + 0(z - 5) = 0,$$

which simplifies to

$$x + 2y - 4 = 0.$$

The last equation looks very much like the equation of a line in the xy-plane. If the context of our discussion were the xy-plane, then the equation $x + 2y - 4 = 0$ would represent a line. In this case, however, our context is three-dimensional space. Hence, the equation $x + 2y - 4 = 0$ represents the set of all points $Q(x, y, z)$, where $x + 2y - 4 = 0$ and z is unrestricted. This set forms a vertical plane that intersects the xy-plane in the line just mentioned (Figure 12.6.2). ❏

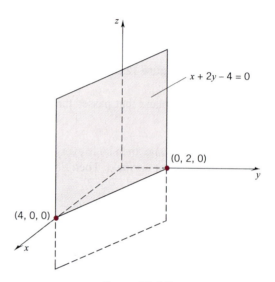

Figure 12.6.2

Example 3 Show that every equation of the form

$$ax + by + cz + d = 0 \quad \text{with} \quad \sqrt{a^2 + b^2 + c^2} \neq 0$$

represents a plane in space.

SOLUTION Since $\sqrt{a^2 + b^2 + c^2} \neq 0$, the numbers a, b, c are not all zero, and therefore there exist numbers x_0, y_0, z_0 such that

$$ax_0 + by_0 + cz_0 + d = 0.†$$

The equation

$$ax + by + cz + d = 0$$

can now be written

$$(ax + by + cz + d) - (ax_0 + by_0 + cz_0 + d) = 0,$$

and so, after factoring, we have

$$a(x - x_0) + b(y - y_0) + c(z - z_0) = 0.$$

This equation (and hence the initial equation) represents the plane through the point $P(x_0, y_0, z_0)$ with normal $\mathbf{N} = a\mathbf{i} + b\mathbf{j} + c\mathbf{k}$. The initial assumption that $\sqrt{a^2 + b^2 + c^2} \neq 0$ guarantees that $\mathbf{N} \neq \mathbf{0}$. ❏

---

† Would such numbers necessarily exist if $\sqrt{a^2 + b^2 + c^2}$ were zero?

Vector Equation of a Plane

We can write the equation of a plane entirely in vector notation. With

$$\mathbf{N} = A\mathbf{i} + B\mathbf{j} + C\mathbf{k}$$

and

$$\mathbf{r}_0 = x_0\mathbf{i} + y_0\mathbf{j} + z_0\mathbf{k}, \qquad \mathbf{r} = x\mathbf{i} + y\mathbf{j} + z\mathbf{k},$$

Equation (12.6.1) reads

(12.6.2)
$$\mathbf{N} \cdot (\mathbf{r} - \mathbf{r}_0) = 0.$$

This vector equation represents the plane that passes through the tip of $\mathbf{r}_0$ and has normal $\mathbf{N}$. (See Figure 12.6.3.) If the plane passes through the origin, we can take $\mathbf{r}_0 = \mathbf{0}$. Equation (12.6.2), then takes the form

(12.6.3)
$$\mathbf{N} \cdot \mathbf{r} = 0.$$

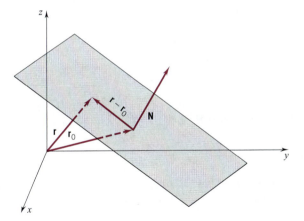

Figure 12.6.3

Collinear Vectors, Coplanar Vectors

Collinear points are points that lie on the same line; *coplanar points* are points that lie on the same plane. The terms "collinear" and "coplanar" are also applied to vectors: by definition, two vectors $\mathbf{a}$ and $\mathbf{b}$ are said to be *collinear* iff there exist scalars s and t not both 0 such that

$$s\mathbf{a} + t\mathbf{b} = \mathbf{0}.$$

If $s \neq 0$, then $\mathbf{a} = -(t/s)\mathbf{b}$; if $t \neq 0$, then $\mathbf{b} = -(s/t)\mathbf{a}$. Collinear vectors are thus parallel. If we set

$$\mathbf{a} = \overrightarrow{PA} \qquad \text{and} \qquad \mathbf{b} = \overrightarrow{PB},$$

then the points P, A, B all fall on the same line; hence the term "collinear vectors."

Three vectors **a**, **b**, **c** are said to be *coplanar* iff there exist scalars s, t, u not all zero such that

$$s\mathbf{a} + t\mathbf{b} + u\mathbf{c} = \mathbf{0}.$$

This term, too, is justified:

(12.6.4)

$\mathbf{a} = \overrightarrow{PA}$, $\mathbf{b} = \overrightarrow{PB}$, $\mathbf{c} = \overrightarrow{PC}$ are coplanar vectors iff the points P, A, B, C all lie on the same plane.

PROOF   Here we show that, if the three vectors are coplanar, then the four points all lie on the same plane. We leave the converse to you (Exercise 45).

Suppose that the three vectors are coplanar. Then we can write

$$s\,\overrightarrow{PA} + t\,\overrightarrow{PB} + u\,\overrightarrow{PC} = \mathbf{0} \qquad \text{with } s, t, u \text{ not all zero.}$$

Without loss of generality, we assume that $s \neq 0$. Then

$$\overrightarrow{PA} = -\frac{t}{s}\overrightarrow{PB} - \frac{u}{s}\overrightarrow{PC}.$$

Since $\overrightarrow{PB} \times \overrightarrow{PC}$ is perpendicular to both $\overrightarrow{PB}$ and $\overrightarrow{PC}$, we see that

$$(\overrightarrow{PB} \times \overrightarrow{PC}) \cdot \overrightarrow{PA} = (\overrightarrow{PB} \times \overrightarrow{PC}) \cdot \left(-\frac{t}{s}\overrightarrow{PB} - \frac{u}{s}\overrightarrow{PC} \right)$$

$$= -\frac{t}{s}(\overrightarrow{PB} \times \overrightarrow{PC}) \cdot \overrightarrow{PB} - \frac{u}{s}(\overrightarrow{PB} \times \overrightarrow{PC}) \cdot \overrightarrow{PC}$$

$$= 0.$$

But $|(\overrightarrow{PB} \times \overrightarrow{PC}) \cdot \overrightarrow{PA}|$ gives the volume of the parallelepiped with edges $\overrightarrow{PA}, \overrightarrow{PB}, \overrightarrow{PC}$. This volume can be zero only if P, A, B, C all lie on the same plane. ❑

Unit Normals

If **N** is normal to a given plane, then all other normals to that plane are parallel to **N** and hence scalar multiples of **N**. In particular there are only two normals of length 1:

$$\mathbf{u_N} = \frac{\mathbf{N}}{\|\mathbf{N}\|} \qquad \text{and} \qquad -\mathbf{u_N} = -\frac{\mathbf{N}}{\|\mathbf{N}\|}.$$

These are called the *unit normals*.

Example 4   Find the unit normals for the plane

$$3x - 4y + 12z + 8 = 0.$$

SOLUTION   We can take

$$\mathbf{N} = 3\mathbf{i} - 4\mathbf{j} + 12\mathbf{k}.$$

Since

$$\|\mathbf{N}\| = \sqrt{(3)^2 + (-4)^2 + (12)^2} = \sqrt{169} = 13,$$

we have

$$\mathbf{u_N} = \tfrac{1}{13}(3\mathbf{i} - 4\mathbf{j} + 12\mathbf{k}) \qquad \text{and} \qquad -\mathbf{u_N} = -\tfrac{1}{13}(3\mathbf{i} - 4\mathbf{j} + 12\mathbf{k}). ❑$$

Parallel Planes, Intersecting Planes

Two planes are called *parallel* iff their normals are parallel. If two planes, p_1, p_2 are not parallel, we can find the angle between them by finding the angle between their normals, $\mathbf{N}_1, \mathbf{N}_2$. (See Figure 12.6.4.) Depending on our choice of normals, there are two such angles, each the supplement of the other. We will choose the smaller angle, the one with the nonnegative cosine:

(12.6.5)

$$\cos \theta = |\mathbf{u}_{\mathbf{N}_1} \cdot \mathbf{u}_{\mathbf{N}_2}|.$$

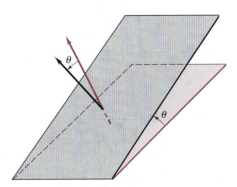

Figure 12.6.4

Example 5 Here are some planes:

$$p_1: 2(x - 1) - 3y + 5(z - 2) = 0, \qquad p_2: -4x + 6y + 10z + 24 = 0,$$
$$p_3: 4x - 6y - 10z + 1 = 0, \qquad p_4: 2x - 3y + 5z - 12 = 0.$$

(a) Indicate which planes are identical.

(b) Indicate which planes are distinct but parallel.

(c) Find the angle between p_1 and p_2.

SOLUTION

(a) p_1 and p_4 are identical, as you can verify by simplifying the equation of p_1.

(b) p_2 and p_3 are distinct but parallel. The planes are distinct since $P(0, 0, \frac{1}{10})$ lies on p_3 but not on p_2; they are parallel since the normals

$$-4\mathbf{i} + 6\mathbf{j} + 10\mathbf{k} \qquad \text{and} \qquad 4\mathbf{i} - 6\mathbf{j} - 10\mathbf{k}$$

are parallel.

(c) Taking

$$\mathbf{N}_1 = 2\mathbf{i} - 3\mathbf{j} + 5\mathbf{k} \qquad \text{and} \qquad \mathbf{N}_2 = -4\mathbf{i} + 6\mathbf{j} - 10\mathbf{k},$$

we have

$$\mathbf{u}_{\mathbf{N}_1} = \frac{1}{\sqrt{38}}(2\mathbf{i} - 3\mathbf{j} + 5\mathbf{k}) \qquad \text{and} \qquad \mathbf{u}_{\mathbf{N}_2} = \frac{1}{\sqrt{152}}(-4\mathbf{i} + 6\mathbf{j} + 10\mathbf{k}).$$

As you can check,

$$\cos \theta = |\mathbf{u}_{\mathbf{N}_1} \cdot \mathbf{u}_{\mathbf{N}_2}| = \tfrac{6}{19} \cong 0.316 \quad \text{and thus} \quad \theta \cong 1.25 \text{ radians.} \quad \square$$

Example 6 The planes

$$p_1 \colon A_1 x + B_1 y + C_1 z + D_1 = 0, \qquad p_2 \colon A_2 x + B_2 y + C_2 z + D_2 = 0$$

intersect to form a line l. Find a vector parametrization for l.

SOLUTION We need to find a point P_0 on l and a direction vector for l. Finding P_0 is a matter of finding numbers x, y, z that simultaneously satisfy the equations given above for p_1 and p_2. In concrete cases this is not hard, and we will not try to give a general formula for such a P_0. We will just assume that P_0 has been found and focus on finding a direction vector for l.

Since p_1 and p_2 intersect in a line, the normals

$$\mathbf{N}_1 = A_1 \mathbf{i} + B_1 \mathbf{j} + C_1 \mathbf{k}, \qquad \mathbf{N}_2 = A_2 \mathbf{i} + B_2 \mathbf{j} + C_2 \mathbf{k}$$

are not parallel. This guarantees that $\mathbf{N}_1 \times \mathbf{N}_2$ is not $\mathbf{0}$. Since l lies on both p_1 and p_2, the line l, like the vector $\mathbf{N}_1 \times \mathbf{N}_2$, is perpendicular to both $\mathbf{N}_1$ and $\mathbf{N}_2$. This makes l parallel to $\mathbf{N}_1 \times \mathbf{N}_2$. (See Figure 12.6.5.) We can therefore take $\mathbf{N}_1 \times \mathbf{N}_2$ as a direction vector for l and write

$$l \colon \mathbf{r}(t) = \overrightarrow{OP_0} + t(\mathbf{N}_1 \times \mathbf{N}_2). \quad \square$$

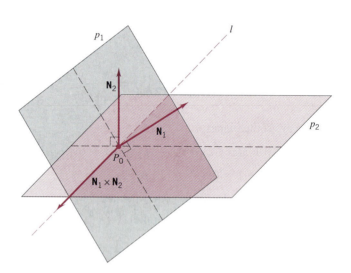

Figure 12.6.5

Example 7 Given the planes

$$p_1 \colon 2x - 3y + 2z = 9 \qquad \text{and} \qquad p_2 \colon x + 2y - z = -4.$$

Show that p_1 and p_2 are not parallel and find scalar parametric equations for the line of intersection l of the two planes.

SOLUTION Normal vectors for p_1 and p_2 are $\mathbf{N}_1 = 2\mathbf{i} - 3\mathbf{j} + 2\mathbf{k}$ and $\mathbf{N}_2 = \mathbf{i} + 2\mathbf{j} - \mathbf{k}$, respectively. Since neither vector is a scalar multiple of the other, the vectors are not parallel and therefore p_1 and p_2 are not parallel.

As shown in Example 6, a direction vector for l is given by

$$\mathbf{N}_1 \times \mathbf{N}_2 = \begin{vmatrix} \mathbf{i} & \mathbf{j} & \mathbf{k} \\ 2 & -3 & 2 \\ 1 & 2 & -1 \end{vmatrix} = -\mathbf{i} + 4\mathbf{j} + 7\mathbf{k}.$$

Now we need a point that lies on l. To find one, we solve the equations p_1 and p_2 simultaneously. If, for example, we set $x = 0$ in the two equations, we get

$$-3y + 2z = 9$$
$$2y - z = -4.$$

Solving this pair of equations for y and z, we find that $y = 1$ and $z = 6$. Thus, the point $(0, 1, 6)$ is on l and scalar parametric equations for l are

$$x = -t, \qquad y = 1 + 4t, \qquad z = 6 + 7t. \quad \square$$

The Plane Determined by Three Noncollinear Points

Suppose now that we are given three noncollinear points P_1, P_2, P_3. These points determine a plane. How can we find an equation for that plane?

First we form the vectors $\overrightarrow{P_1P_2}, \overrightarrow{P_1P_3}$. Since P_1, P_2, P_3 are noncollinear, the vectors are not parallel. Therefore their cross product $\overrightarrow{P_1P_2} \times \overrightarrow{P_1P_3}$ can be used as a normal for the plane. We are back in a familiar situation. We have a point of the plane, say P_1, and we have a normal vector, $\overrightarrow{P_1P_2} \times \overrightarrow{P_1P_3}$. A point P will lie on the plane iff

(12.6.6)
$$\boxed{\overrightarrow{P_1P} \cdot (\overrightarrow{P_1P_2} \times \overrightarrow{P_1P_3}) = 0.}$$
(Figure 12.6.6)

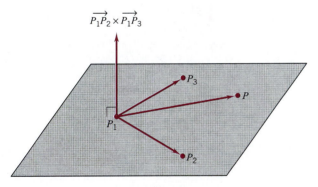

Figure 12.6.6

Example 8 Find an equation in x, y, z for the plane that passes through the points $P_1(0, 1, 1), P_2(1, 0, 1), P_3(1, 1, 0)$.

SOLUTION The point $P = P(x, y, z)$ will lie on this plane iff
$$\overrightarrow{P_1P} \cdot (\overrightarrow{P_1P_2} \times \overrightarrow{P_1P_3}) = 0.$$

Here
$$\overrightarrow{P_1P} = x\mathbf{i} + (y - 1)\mathbf{j} + (z - 1)\mathbf{k}, \qquad \overrightarrow{P_1P_2} = \mathbf{i} - \mathbf{j}, \quad \overrightarrow{P_1P_3} = \mathbf{i} - \mathbf{k}.$$

As you can check,
$$\overrightarrow{P_1P_2} \times \overrightarrow{P_1P_3} = \mathbf{i} + \mathbf{j} + \mathbf{k}.$$

and thus

$$\overrightarrow{P_1P} \cdot (\overrightarrow{P_1P_2} \times \overrightarrow{P_1P_3}) = [x\mathbf{i} + (y-1)\mathbf{j} + (z-1)\mathbf{k}] \cdot [\mathbf{i} + \mathbf{j} + \mathbf{k}]$$
$$= x + (y-1) + (z-1) = x + y + z - 2.$$

An equation for the plane reads

$$x + y + z - 2 = 0. \quad \square$$

The Distance from a Point to a Plane

In Figure 12.6.7, we have drawn a plane p: $Ax + By + Cz + D = 0$ and a point $P_1(x_1, y_1, z_1)$ not on p. The distance between the point P_1 and the plane p is given by the formula

(12.6.7)

$$d(P_1, p) = \frac{|Ax_1 + By_1 + Cz_1 + D|}{\sqrt{A^2 + B^2 + C^2}}.$$

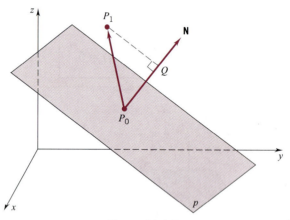

Figure 12.6.7

PROOF Pick any point $P_0(x_0, y_0, z_0)$ in the plane. As a normal to p we can take the vector

$$\mathbf{N} = A\mathbf{i} + B\mathbf{j} + C\mathbf{k}.$$

Then

$$\mathbf{u_N} = \frac{A\mathbf{i} + B\mathbf{j} + C\mathbf{k}}{\sqrt{A^2 + B^2 + C^2}}$$

is the corresponding unit normal. From Figure 12.6.7,

$$d(P_1, p) = d(P_0, Q) = |\text{comp}_\mathbf{N} \overrightarrow{P_0P_1}|$$
$$= |\overrightarrow{P_0P_1} \cdot \mathbf{u_N}|$$
$$= \frac{|(x_1 - x_0)A + (y_1 - y_0)B + (z_1 - z_0)C|}{\sqrt{A^2 + B^2 + C^2}}$$
$$= \frac{|Ax_1 + By_1 + Cz_1 - (Ax_0 + By_0 + Cz_0)|}{\sqrt{A^2 + B^2 + C^2}}.$$

Since $P_0(x_0, y_0, z_0)$ lies on p,

$$Ax_0 + By_0 + Cz_0 = -D$$

and

$$d(P_1, p) = \frac{|Ax_1 + By_1 + Cz_1 + D|}{\sqrt{A^2 + B^2 + C^2}}. \quad \square$$

EXERCISES 12.6

1. Which of the points $P(3, 2, 1)$, $Q(2, 3, 1)$, $R(1, 4, 1)$ lie on the plane

$$3(x - 1) + 4y - 5(z + 2) = 0?$$

2. Which of the points $P(2, 1, -2)$, $Q(2, 0, 0)$, $R(4, 1, -1)$, $S(0, -1, -3)$ lie on the plane $\mathbf{N} \cdot (\mathbf{r} - \mathbf{r}_0) = 0$ if $\mathbf{N} = \mathbf{i} - 3\mathbf{j} + \mathbf{k}$ and $\mathbf{r}_0 = 4\mathbf{i} + \mathbf{j} - \mathbf{k}$?

In Exercises 3–9, find an equation for the plane which satisfies the given conditions.

3. Passes through the point $P(2, 3, 4)$ and is perpendicular to $\mathbf{i} - 4\mathbf{j} + 3\mathbf{k}$.

4. Passes through the point $P(1, -2, 3)$ and is perpendicular to $\mathbf{j} + 2\mathbf{k}$.

5. Passes through the point $P(2, 1, 1)$ and is parallel to the plane $3x - 2y + 5z - 2 = 0$.

6. Passes through the point $P(3, -1, 5)$ and is parallel to the plane $4x + 2y - 7z + 5 = 0$.

7. Passes through the point $P(1, 3, 1)$ and contains the line l: $x = t$, $y = t$, $z = -2 + t$.

8. Passes through the point $P(2, 0, 1)$ and contains the line l: $x = 1 - 2t$, $y = 1 + 4t$, $z = 2 + t$.

9. Passes through the point $P_0(x_0, y_0, z_0)$ and is perpendicular to $\overrightarrow{OP_0}$.

In Exercises 10 and 11, find the unit normals to the given plane.

10. $2x - 3y + 7z - 3 = 0$.   11. $2x - y + 5z - 10 = 0$.

12. Show that the plane $x/a + y/b + z/c = 1$ intersects the coordinate axes at $x = a, y = b, z = c$. This is the equation of a plane in *intercept form.*

In Exercises 13 and 14, write the equation of the given plane in intercept form and find the points where it intersects the coordinate axes.

13. $4x + 5y - 6z = 60$.   14. $3x - y + 4z + 2 = 0$.

In Exercises 15–18, find the angle between the planes.

15. $5(x - 1) - 3(y + 2) + 2z = 0$,
$x + 3(y - 1) + 2(z + 4) = 0$.

16. $2x - y + 3z = 5$, $5x + 5y - z = 1$.

17. $x - y + z - 1 = 0$, $2x + y + 3z + 5 = 0$.

18. $4x + 4y - 2z = 3$, $2x + y + z = -1$.

In Exercises 19–22, determine whether or not the vectors are coplanar.

19. $4\mathbf{j} - \mathbf{k}$, $3\mathbf{i} + \mathbf{j} + 2\mathbf{k}$, $\mathbf{0}$.

20. $\mathbf{i}$, $\mathbf{i} - 2\mathbf{j}$, $3\mathbf{j} + \mathbf{k}$.

21. $\mathbf{i} + \mathbf{j} + \mathbf{k}$, $2\mathbf{i} - \mathbf{j}$, $3\mathbf{i} - \mathbf{j} - \mathbf{k}$.

22. $\mathbf{j} - \mathbf{k}$, $3\mathbf{i} - \mathbf{j} + 2\mathbf{k}$, $3\mathbf{i} - 2\mathbf{j} + 3\mathbf{k}$.

In Exercises 23–26, find the distance from the point P to the given plane.

23. $P(2, -1, 3)$; $2x + 4y - z + 1 = 0$.

24. $P(3, -5, 2)$; $8x - 2y + z = 5$.

25. $P(1, -3, 5)$; $-3x + 4z + 5 = 0$.

26. $P(1, 3, 4)$; $x + y - 2z = 0$.

In Exercises 27–30, find an equation in x, y, z for the plane that passes through the given points.

27. $P_1(1, 0, 1)$, $P_2(2, 1, 0)$, $P_3(1, 1, 1)$.

28. $P_1(1, 1, 1)$, $P_2(2, -2, -1)$, $P_3(0, 2, 1)$.

29. $P_1(3, -4, 1)$, $P_2(3, 2, 1)$, $P_3(-1, 1, -2)$.

30. $P_1(3, 2, -1)$, $P_2(3, -2, 4)$, $P_3(1, -1, 3)$.

31. Write equations in symmetric form for the line that passes through $P_0(x_0, y_0, z_0)$ and is perpendicular to the plane $Ax + By + Cz + D = 0$.

32. Find the distance between the parallel planes

$$Ax + By + Cz + D_1 = 0$$

and

$$Ax + By + Cz + D_2 = 0.$$

33. Show that the equations of a line in symmetric form

$$\frac{x - x_0}{d_1} = \frac{y - y_0}{d_2} = \frac{z - z_0}{d_3}$$

express the line as an intersection of two planes by finding equations for two such planes.

34. Suppose that we are given equations in x, y, z for two planes that intersect in a line. How can we determine a triple x_0, y_0, z_0 that places $P(x_0, y_0, z_0)$ on the line? One possible procedure is this: select a value for x, say x_0; substitute x_0 in the two equations given; we now have two

linear equations in y and z that we can solve simultaneously. Is this procedure foolproof? If not, identify the circumstances under which the procedure will fail and indicate how the procedure can be made airtight.

In Exercises 35 and 36, find a set of scalar parametric equations for the line formed by the two intersecting planes. HINT: To find a point P_0 on the line, see Exercise 34.

35. $p_1 : x + 2y + 3z = 0, \quad p_2 : -3x + 4y + z = 0.$

36. $p_1 : x + y + z + 1 = 0, \quad p_2 : x - y + z + 2 = 0.$

In Exercises 37 and 38, let l be the line determined by P_1, P_2, and let p be the plane determined by Q_1, Q_2, Q_3. Where, if anywhere, does l intersect p?

37. $P_1(1, -1, 2), \quad P_2(-2, 3, 1); \quad Q_1(2, 0, -4),$
$Q_2(1, 2, 3), \quad Q_3(-1, 2, 1).$

38. $P_1(4, -3, 1), \quad P_2(2, -2, 3); \quad Q_1(2, 0, -4),$
$Q_2(1, 2, 3), \quad Q_3(-1, 2, 1).$

39. Let l_1, l_2 be lines that pass through the origin and have direction vectors

$$\mathbf{d} = \mathbf{i} + 2\mathbf{j} + 4\mathbf{k}, \qquad \mathbf{D} = -\mathbf{i} - \mathbf{j} + 3\mathbf{k}.$$

Find an equation for the plane that contains l_1 and l_2.

40. Show that two nonparallel lines $\mathbf{r}(t) = \mathbf{r}_0 + t\mathbf{d}$ and $\mathbf{R}(t) = \mathbf{R}_0 + t\mathbf{D}$ intersect iff the vectors $\mathbf{r}_0 - \mathbf{R}_0$, $\mathbf{d}$, and $\mathbf{D}$ are coplanar.

41. Starting with the plane that contains the point P and has normal $\mathbf{N}$, describe the set of points Q on the plane for which $(\mathbf{N} + \overrightarrow{PQ}) \perp (\mathbf{N} - \overrightarrow{PQ})$. HINT: Draw a figure.

42. Let $\mathbf{a}, \mathbf{b}, \mathbf{c}$ be three nonzero vectors such that the angle between any pair of them is $\frac{1}{2}\pi$. Can these vectors be coplanar?

43. Suppose that $\mathbf{N} = A\mathbf{i} + B\mathbf{j} + C\mathbf{k}$ is a nonzero vector with its initial point on the plane $Ax + By + Cz + D = 0$. Take $P_1(x_1, y_1, z_1)$ as a point of space and set $\alpha = Ax_1 + By_1 + Cz_1 + D$. If $\alpha = 0$, then P_1 lies on the plane. What can you conclude about P_1 if α is positive? If α is negative?

44. Suppose that $\mathbf{a}, \mathbf{b}, \mathbf{c}$ are radius vectors the tips of which are not collinear. Give a geometric interpretation of the equation

$$\begin{vmatrix} x - a_1 & y - a_2 & z - a_3 \\ x - b_1 & y - b_2 & z - b_3 \\ x - c_1 & y - c_2 & z - c_3 \end{vmatrix} = 0.$$

45. Suppose that the points P, A, B, and C all lie on the same plane. Show that the vectors $\mathbf{a} = \overrightarrow{PA}$, $\mathbf{b} = \overrightarrow{PB}$, and $\mathbf{c} = \overrightarrow{PC}$ are coplanar vectors.

Graph Sketching

To sketch the graph of the plane $Ax + By + Cz + D = 0$, find, if possible, the *trace* of the graph in the coordinate

planes. The trace in the xy-plane is found by setting $z = 0$ in the given equation; the result will be the linear equation $Ax + By + D = 0$ which, in general, can be drawn as a line in the xy-plane. Similarly, to find the trace in the xz-plane, set $y = 0$ and sketch the line $Ax + Cz + D = 0$; and to find the trace in the yz-plane, set $x = 0$ and sketch the line $By + Cz + D = 0$. For example, the graph of the plane $3x + 4y + 2z - 12 = 0$ is shown in the following figure.

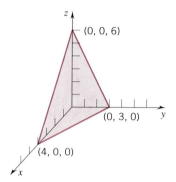

In Exercises 46–49, sketch the graph of the given plane.

46. $x + 2y + 3z - 6 = 0.$ **47.** $5x + 4y + 10z = 20.$

48. $3x + 2y - 6 = 0.$ **49.** $3x + 2z - 12 = 0.$

In Exercises 50–53, find an equation for the plane whose graph is shown in the figure.

50.

51.

52.

53.

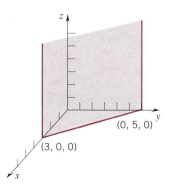

■ 12.7 SOME GEOMETRY BY VECTOR METHODS

Try your hand at proving the following theorems by vector methods. Follow the hints if you like, but you may find it more interesting to disregard them and come up with proofs that are entirely your own.

Proposition 1 The diagonals of a parallelogram are perpendicular iff the parallelogram is a rhombus.

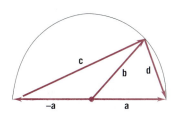

Figure 12.7.1

HINT FOR PROOF With $\mathbf{a}$ and $\mathbf{b}$ as in Figure 12.7.1, the diagonals are $\mathbf{a} + \mathbf{b}$ and $\mathbf{a} - \mathbf{b}$. Show that $(\mathbf{a} + \mathbf{b}) \cdot (\mathbf{a} - \mathbf{b}) = 0$ iff $\|\mathbf{a}\| = \|\mathbf{b}\|$. ❑

Proposition 2 Every angle inscribed in a semicircle is a right angle.

HINT FOR PROOF Take $\mathbf{c}$ and $\mathbf{d}$ as in Figure 12.7.2; express $\mathbf{c}$ and $\mathbf{d}$ in terms of $\mathbf{a}$ and $\mathbf{b}$; then show that $\mathbf{c} \cdot \mathbf{d} = 0$. ❑

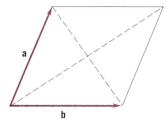

Figure 12.7.2

Proposition 3 In a parallelogram the sum of the squares of the lengths of the diagonals equals the sum of the squares of the lengths of the sides.

HINT FOR PROOF With $\mathbf{a}$ and $\mathbf{b}$ as in Figure 12.7.1, the diagonals are $\mathbf{a} + \mathbf{b}$ and $\mathbf{a} - \mathbf{b}$. Show that $\|\mathbf{a} + \mathbf{b}\|^2 + \|\mathbf{a} - \mathbf{b}\|^2 = 2\|\mathbf{a}\|^2 + 2\|\mathbf{b}\|^2$. ❑

Proposition 4 The three altitudes of a triangle meet at one point.

HINT FOR PROOF As in Figure 12.7.3 assume that the altitudes from P_1 and P_2 intersect at Q. Use the fact that $\overrightarrow{P_1Q} \perp \overrightarrow{OP_2}$ and $\overrightarrow{P_2Q} \perp \overrightarrow{OP_1}$ to show that $\overrightarrow{OQ} \perp \overrightarrow{P_1P_2}$. ❑

Figure 12.7.3

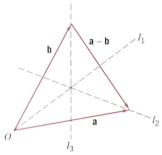

Figure 12.7.4

Proposition 5   The three medians of a triangle meet at one point.

HINT FOR PROOF   With l_1, l_2, l_3 as in Figure 12.7.4,

$$l_1: \mathbf{r}_1(t) = t(\mathbf{a} + \mathbf{b}), \qquad l_2: \mathbf{r}_2(u) = \tfrac{1}{2}\mathbf{b} + u(\mathbf{a} - \tfrac{1}{2}\mathbf{b}), \qquad l_3: \mathbf{r}_3(v) = \tfrac{1}{2}\mathbf{a} + v(\mathbf{b} - \tfrac{1}{2}\mathbf{a}).$$

Show that l_1 intersects both l_2 and l_3 at the same point.   ❏

Proposition 6 (The Law of Sines)   If a triangle has sides $\mathbf{a}$, $\mathbf{b}$, $\mathbf{c}$ and opposite angles A, B, C, then

$$\frac{\sin A}{\|\mathbf{a}\|} = \frac{\sin B}{\|\mathbf{b}\|} = \frac{\sin C}{\|\mathbf{c}\|}.$$

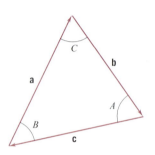

Figure 12.7.5

HINT FOR PROOF   With $\mathbf{a}$, $\mathbf{b}$, $\mathbf{c}$ as in Figure 12.7.5, $\mathbf{a} + \mathbf{b} + \mathbf{c} = \mathbf{0}$. Observe then that $\mathbf{a} \times [\mathbf{a} + \mathbf{b} + \mathbf{c}] = \mathbf{0}$ and $\mathbf{b} \times [\mathbf{a} + \mathbf{b} + \mathbf{c}] = \mathbf{0}$.   ❏

Proposition 7   If two planes have a point in common, then they have a line in common.

HINT FOR PROOF   As equations for the two planes, take $\mathbf{n} \cdot (\mathbf{r} - \mathbf{r}_0) = 0$ and $\mathbf{N} \cdot (\mathbf{R} - \mathbf{R}_0) = 0$. If the point $P(a_1, a_2, a_3)$ lies on both planes, then the vector $\mathbf{a} = a_1\mathbf{i} + a_2\mathbf{j} + a_3\mathbf{k}$ satisfies both equations. In that case we have $\mathbf{n} \cdot (\mathbf{a} - \mathbf{r}_0) = 0$ and $\mathbf{N} \cdot (\mathbf{a} - \mathbf{R}_0) = 0$. Consider the line $\mathbf{r}(t) = \mathbf{a} + t(\mathbf{n} \times \mathbf{N})$.   ❏

■ CHAPTER HIGHLIGHTS

12.1   Cartesian Space Coordinates

distance formula (p. 775)   equation of sphere (p. 775)
midpoint formula (p. 777)

12.2   Displacements, Forces and Velocities; Vectors

addition, multiplication by scalars (p. 782)   component (p. 782)
parallel vectors (p. 784)   norm (p. 785)   unit vector (p. 788)

The zero vector $\mathbf{0}$ is parallel to every vector.
Vectors in the plane.

12.3   The Dot Product

$\mathbf{a} \cdot \mathbf{b} = a_1b_1 + a_2b_2 + a_3b_3 = \|\mathbf{a}\| \|\mathbf{b}\| \cos \theta$ $\mathbf{a} \perp \mathbf{b}$ iff $\mathbf{a} \cdot \mathbf{b} = 0$ $\mathbf{a} = \mathbf{a}_{\parallel} + \mathbf{a}_{\perp}$
projections and components (p. 799)   direction angles, direction cosines
Schwarz's inequality (p. 802)   (p. 801)
work (p. 804)

12.4   The Cross Product

definition of $\mathbf{a} \times \mathbf{b}$ (p. 806)   properties of right-handed triples (p. 806)
properties of the cross product (p. 807)   distributive laws (p. 807)
scalar triple product (p. 807)   components of $\mathbf{a} \times \mathbf{b}$ (p. 809)
identities (pp. 812)   torque (p. 813)

12.5   Lines

position vector, radius vector (p. 814)
vector parametrization: $\mathbf{r}(t) = \mathbf{r}_0 + t\mathbf{d}$
direction vector, direction numbers (p. 815)
scalar parametric equations: $x(t) = x_0 + d_1t$, $y(t) = y_0 + d_2t$, $z(t) = z_0 + d_3t$

Two lines are parallel iff their direction vectors are parallel; two intersecting lines are perpendicular iff their direction vectors are perpendicular.

12.6 Planes

12.7 Some Geometry by Vector Methods

■ PROJECTS AND EXPLORATIONS USING TECHNOLOGY

To do these exercises you will need a graphics calculator or a computer with graphing capability. The majority of these problems are open-ended so different approaches may be used to solve them. You should be aware that different approaches can result in slight variations in the answers. Round your numerical answers to at least four decimal places. The rounding method that your calculator or computer uses also may cause variations in answers.

12.1 This exercise uses vectors to look at a geometric problem. To start, consider the triangle T with vertices $(0, 0)$, $(3, 0)$ and $(1, 2)$ in the plane.

(a) Show that the lengths of the medians of this triangle satisfy the triangle inequality. That is, show that the length of any one of the medians is less than or equal to the sum of the lengths of the other two.

(b) It follows from the result in part (a) that if we view the medians as vectors, then these vectors form the sides of a triangle M. How does the perimeter of M compare with the perimeter of T? How do the areas of the two triangles compare?

(c) Compare the angles of T and M.

(d) What can you say about the triangle formed by the medians of M?

(e) Repeat parts (a)–(c) for the triangle whose vertices are $(0, 0)$, $(3, 0)$, and (a, b), where (a, b) is an arbitrary point in the first quadrant. Based on these results, conjecture when the triangle formed by the medians is congruent to, or similar to, the original triangle.

12.2 Let P be a point on a circle of radius r that is rolling around the outside of the unit circle. See the figure. The curve traced out by P is called an *epicycloid*.

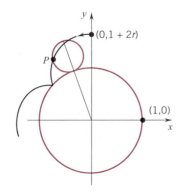

(a) Use vectors to describe the path of P, assuming that the initial position of P is $(0, 1 + 2r)$ and that the circle is rolling at a constant angular velocity ω.

(b) Use a graphing utility to draw the path of P in the case $r = \frac{1}{4}$. Have the graph drawn in such a way that you can watch the path being traced out.

(c) Now use your graphing utility to draw the path of P for several other values of r. Determine conditions on r that will ensure that the path is a closed curve.

(d) Using the graph drawn in part (b), does it appear that the angular velocity of the traced point is constant if the center of the rolling circle is moving at a constant rate.

(e) Repeat parts (a)–(d) for the case where $r < 1$ and the circle is rolling around the inside of the unit circle. The curve in this case is called a *hypocycloid*.

12.3 Consider the triangle R with vertices $A(0, 0, 0)$, $B(4, 0, 0)$, and $C(3, 2, 1)$.

(a) Show that if $0 \leq x \leq 1$ and $0 \leq y \leq 1$, then the point $P(4x(1 - y), 2y, y)$ is either inside or on the boundary of R.

(b) Let $f = f(x, y)$ be the sum of the lengths of the vectors from P to the vertices of R. Use a graphing utility to sketch the graph of f.

(c) Based on the graph of f, estimate the absolute minimum value and the absolute maximum value of f.

(d) Let P_m be a point where the absolute minimum value of f occurs. Find the angles between the vectors from P_m to the vertices of R.

(e) Repeat parts (a)–(d) for the triangle S with vertices $A(0, 0, 0)$, $B(4, 0, 0)$, and $C(3, 3, 3)$; and for the triangle T with vertices $A(0, 0, 0)$, $B(1, 0, 0)$, and $C(3, 3, 3)$.

(f) Suppose that the vertices A, B, and C are the locations of retail stores and P is the location of a distribution center. Then the points P_m found in parts (d) and (e) would be the locations of the distribution centers that would minimize transportation costs if the same amount is shipped to each store. Now suppose that twice as much is shipped to B as to A, and three times as much is shipped to C as to A. Use graphs to estimate the minimum transportation costs in the three cases. Find the angles between the vectors from P_m to the vertices of the respective triangles. Is there a relationship between the size of these angles and the corresponding transportation costs?

VECTOR CALCULUS

CHAPTER 13

■ 13.1 VECTOR FUNCTIONS

Introduction

If f_1, f_2, f_3 are real-valued functions defined on some interval I, then for each $t \in I$ we can form the vector

$$\mathbf{f}(t) = f_1(t)\mathbf{i} + f_2(t)\mathbf{j} + f_3(t)\mathbf{k}$$

and thereby create a *vector-valued function* $\mathbf{f}$. For short we will call such a function a *vector function*. The real-valued functions f_1, f_2, f_3 are called the *components* of $\mathbf{f}$. A point t is in the *domain* of a vector function $\mathbf{f}$ iff it is in the domain of each of its components. If a domain for $\mathbf{f}$ is not specified explicitly, then we will take the domain to be the common domain of its components.

For example, if we take

$$f_1(t) = x_0 + d_1 t, \qquad f_2(t) = y_0 + d_2 t, \qquad f_3(t) = z_0 + d_3 t$$

we can form the vector function

$$\mathbf{f}(t) = (x_0 + d_1 t)\mathbf{i} + (y_0 + d_2 t)\mathbf{j} + (z_0 + d_3 t)\mathbf{k}.$$

The domain of $\mathbf{f}$ is the set of all real numbers. If d_1, d_2, d_3 are not all 0, then, as we saw in Section 12.5, the radius vector $\mathbf{f}(t)$ traces out the line that passes through the point $P(x_0, y_0, z_0)$ and has direction numbers d_1, d_2, d_3. If d_1, d_2, d_3 are all 0, then we have the constant function

$$\mathbf{f}(t) = x_0\mathbf{i} + y_0\mathbf{j} + z_0\mathbf{k}.$$

Example 1 From the functions

$$f_1(t) = \cos t, \qquad f_2(t) = \sin t, \qquad f_3(t) = 0$$

we can form the vector function

$$\mathbf{f}(t) = \cos t\,\mathbf{i} + \sin t\,\mathbf{j}.$$

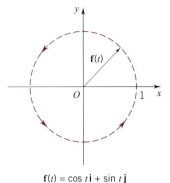

$f(t) = \cos t\,\mathbf{i} + \sin t\,\mathbf{j}$

Figure 13.1.1

For each t

$$\|\mathbf{f}(t)\| = \sqrt{\cos^2 t + \sin^2 t} = 1.$$

Since the third component is zero, the radius vector $\mathbf{f}(t)$ lies in the xy-plane. Note that the equations $x = \cos t$, $y = \sin t$ are parametric equations of the unit circle. As t increases, the tip of $\mathbf{f}(t)$ traces out the unit circle in a counterclockwise manner, effecting a complete revolution as t increases by 2π. (Figure 13.1.1.) ❑

Example 2 Each real-valued function f defined on some interval $[a, b]$ gives rise to a vector-valued function $\mathbf{f}$ in a natural way. Setting

$$f_1(t) = t, \qquad f_2(t) = f(t), \qquad f_3(t) = 0,$$

we obtain the vector function

$$\mathbf{f}(t) = t\mathbf{i} + f(t)\mathbf{j}.$$

As t ranges from a to b, the radius vector $\mathbf{f}(t)$ traces out the graph of f from left to right. (See Figure 13.1.2.) ❑

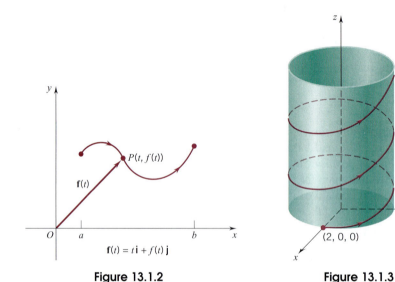

Figure 13.1.2

Figure 13.1.3

Example 3 From the functions

$$f_1(t) = 2\cos t, \qquad f_2(t) = 2\sin t, \qquad f_3(t) = t, \qquad t \geq 0,$$

we can form the vector function

$$\mathbf{f}(t) = 2\cos t\,\mathbf{i} + 2\sin t\,\mathbf{j} + t\mathbf{k}.$$

As indicated, the domain of $\mathbf{f}$ is the set of nonnegative real numbers. At $t = 0$, the radius vector $\mathbf{f}(0)$ has the value $(2, 0, 0)$. As t increases, the tip of $\mathbf{f}(t)$ winds around the circular cylinder $x^2 + y^2 = 4$ (z arbitrary) one revolution for each increase of 2π in t. This curve is called a *circular helix* (Figure 13.1.3). ❑

Example 4 Find $\lim\limits_{t \to 0} \mathbf{f}(t)$ given that

$$\mathbf{f}(t) = \cos(t + \pi)\mathbf{i} + \sin(t + \pi)\mathbf{j} + e^{-t^2}\mathbf{k}.$$

SOLUTION

$$\lim_{t \to 0} \mathbf{f}(t) = \lim_{t \to 0} [\cos(t + \pi)\mathbf{i} + \sin(t + \pi)\mathbf{j} + e^{-t^2}\mathbf{k}]$$

$$= \left[\lim_{t \to 0} \cos(t + \pi)\right]\mathbf{i} + \left[\lim_{t \to 0} \sin(t + \pi)\right]\mathbf{j} + \left[\lim_{t \to 0} e^{-t^2}\right]\mathbf{k}$$

$$= (-1)\mathbf{i} + (0)\mathbf{j} + (1)\mathbf{k} = -\mathbf{i} + \mathbf{k}. \quad \square$$

Continuity and Differentiability

As you would expect, **f** is said to be *continuous* at t_0 iff

$$\lim_{t \to t_0} \mathbf{f}(t) = \mathbf{f}(t_0).$$

Thus, by (13.1.4), **f** is continuous at t_0 iff each component of **f** is continuous at t_0.

The derivative of a vector function is defined as the limit of a *vector difference quotient*:

DEFINITION 13.1.5 DERIVATIVE OF A VECTOR FUNCTION

The vector function **f** is *differentiable* at t iff

$$\lim_{h \to 0} \frac{\mathbf{f}(t + h) - \mathbf{f}(t)}{h} \qquad \text{exists.}$$

If this limit exists, it is called the *derivative of* **f** *at t* and is denoted $\mathbf{f}'(t)$. The vector function **f** is *differentiable* if it is differentiable at each point of its domain.

Differentiation can be carried out component by component. If $\mathbf{f}(t) = f_1(t)\mathbf{i} + f_2(t)\mathbf{j} + f_3(t)\mathbf{k}$ is differentiable at t, then

$$\mathbf{f}'(t) = f_1'(t)\mathbf{i} + f_2'(t)\mathbf{j} + f_3'(t)\mathbf{k}.$$

PROOF

$$\mathbf{f}'(t) = \lim_{h \to 0} \frac{\mathbf{f}(t + h) - \mathbf{f}(t)}{h}$$

$$= \lim_{h \to 0} \left[\frac{f_1(t + h) - f_1(t)}{h}\mathbf{i} + \frac{f_2(t + h) - f_2(t)}{h}\mathbf{j} + \frac{f_3(t + h) - f_3(t)}{h}\mathbf{k}\right]$$

$$= \left[\lim_{h \to 0} \frac{f_1(t + h) - f_1(t)}{h}\right]\mathbf{i} + \left[\lim_{h \to 0} \frac{f_2(t + h) - f_2(t)}{h}\right]\mathbf{j}$$

$$+ \left[\lim_{h \to 0} \frac{f_3(t + h) - f_3(t)}{h}\right]\mathbf{k}$$

$$= f_1'(t)\mathbf{i} + f_2'(t)\mathbf{j} + f_3'(t)\mathbf{k}. \quad \square$$

As with real-valued functions, if **f** is differentiable at t, then **f** is continuous at t. (Exercise 53.)

Interpretations of the vector derivative and applications of vector differentiation are introduced later in the chapter. Here we limit ourselves to computation.

Example 5 Given that

$$\mathbf{f}(t) = t\mathbf{i} + \sqrt{t + 1}\,\mathbf{j} - e^t\mathbf{k},$$

find

(1) The domain of **f**. (2) $\mathbf{f}(0)$.
(3) $\mathbf{f}'(t)$. (4) $\mathbf{f}'(0)$.
(5) $\|\mathbf{f}(t)\|$. (6) $\mathbf{f}(t) \cdot \mathbf{f}'(t)$.

SOLUTION

(1) For a number to be in the domain of **f**, it is necessary only that it be in the domain of each of the components. The domain of **f** is $[-1, \infty)$.

(2) $\mathbf{f}(0) = 0\mathbf{i} + \sqrt{0 + 1}\,\mathbf{j} - e^0\mathbf{k} = \mathbf{j} - \mathbf{k}.$

(3) $\mathbf{f}'(t) = \mathbf{i} + \dfrac{1}{2\sqrt{t + 1}}\,\mathbf{j} - e^t\mathbf{k}.$

(4) $\mathbf{f}'(0) = \mathbf{i} + \dfrac{1}{2\sqrt{0 + 1}}\,\mathbf{j} - e^0\mathbf{k} = \mathbf{i} + \dfrac{1}{2}\mathbf{j} - \mathbf{k}.$

(5) $\|\mathbf{f}(t)\| = \sqrt{t^2 + (\sqrt{t + 1})^2 + (-e^t)^2} = \sqrt{t^2 + t + 1 + e^{2t}}.$

(6) $\mathbf{f}(t) \cdot \mathbf{f}'(t) = (t\mathbf{i} + \sqrt{t + 1}\,\mathbf{j} - e^t\mathbf{k}) \cdot \left(\mathbf{i} + \dfrac{1}{2\sqrt{t + 1}}\,\mathbf{j} - e^t\mathbf{k} \right)$

$$= (t)(1) + (\sqrt{t + 1})\left(\dfrac{1}{2\sqrt{t + 1}} \right) + (-e^t)(-e^t) = t + \dfrac{1}{2} + e^{2t}. \quad \square$$

If **f**$'$ is itself differentiable, we can form the second derivative **f**$''$ and so on.

Example 6 Find $\mathbf{f}''(t)$ for

$$\mathbf{f}(t) = t \sin t\,\mathbf{i} + e^{-t}\mathbf{j} + t\mathbf{k}.$$

SOLUTION

$$\mathbf{f}'(t) = (t \cos t + \sin t)\mathbf{i} - e^{-t}\mathbf{j} + \mathbf{k},$$

$$\mathbf{f}''(t) = (-t \sin t + \cos t + \cos t)\mathbf{i} + e^{-t}\mathbf{j} = (2 \cos t - t \sin t)\mathbf{i} + e^{-t}\mathbf{j}. \quad \square$$

Integration

Since we can differentiate vector functions component by component, we can define integration component by component. For $\mathbf{f}(t) = f_1(t)\mathbf{i} + f_2(t)\mathbf{j} + f_3(t)\mathbf{k}$ continuous on $[a, b]$, we set

(13.1.6)

$$\int_a^b \mathbf{f}(t)\,dt = \left(\int_a^b f_1(t)\,dt \right)\mathbf{i} + \left(\int_a^b f_2(t)\,dt \right)\mathbf{j} + \left(\int_a^b f_3(t)\,dt \right)\mathbf{k}.$$

Example 7 Find

$$\int_0^1 \mathbf{f}(t)\, dt \qquad \text{for} \qquad \mathbf{f}(t) = t\mathbf{i} + \sqrt{t+1}\,\mathbf{j} - e^t\mathbf{k}.$$

SOLUTION

$$\int_0^1 \mathbf{f}(t)\, dt = \left(\int_0^1 t\, dt\right)\mathbf{i} + \left(\int_0^1 \sqrt{t+1}\, dt\right)\mathbf{j} + \left(\int_0^1 (-e^t)\, dt\right)\mathbf{k}$$

$$= \left[\tfrac{1}{2}t^2\right]_0^1 \mathbf{i} + \left[\tfrac{2}{3}(t+1)^{3/2}\right]_0^1 \mathbf{j} + \left[-e^t\right]_0^1 \mathbf{k}$$

$$= \tfrac{1}{2}\mathbf{i} + \tfrac{2}{3}(2\sqrt{2} - 1)\mathbf{j} + (1 - e)\mathbf{k}. \quad \square$$

We can calculate indefinite integrals.

Example 8 Find $\mathbf{f}(t)$ given that

$$\mathbf{f}'(t) = 2\cos t\,\mathbf{i} - t\sin t^2\,\mathbf{j} + 2t\mathbf{k} \qquad \text{and} \qquad \mathbf{f}(0) = \mathbf{i} + 3\mathbf{k}.$$

SOLUTION By integrating $\mathbf{f}'(t)$, we find that

$$\mathbf{f}(t) = (2\sin t + C_1)\mathbf{i} + (\tfrac{1}{2}\cos t^2 + C_2)\mathbf{j} + (t^2 + C_3)\mathbf{k},$$

where C_1, C_2, C_3 are constants to be determined. Since

$$\mathbf{i} + 3\mathbf{k} = \mathbf{f}(0) = C_1\mathbf{i} + (\tfrac{1}{2} + C_2)\mathbf{j} + C_3\mathbf{k},$$

you can see that

$$C_1 = 1, \qquad C_2 = -\tfrac{1}{2}, \qquad C_3 = 3.$$

Thus

$$\mathbf{f}(t) = (2\sin t + 1)\mathbf{i} + (\tfrac{1}{2}\cos t^2 - \tfrac{1}{2})\mathbf{j} + (t^2 + 3)\mathbf{k}. \quad \square$$

(Integration can also be carried out without direct reference to components. See Exercise 54.)

Properties of the Integral

It is easy to see that

(13.1.7)

$$\int_a^b [\mathbf{f}(t) + \mathbf{g}(t)]\, dt = \int_a^b \mathbf{f}(t)\, dt + \int_a^b \mathbf{g}(t)\, dt$$

and

(13.1.8)

$$\int_a^b [\alpha\mathbf{f}(t)]\, dt = \alpha\int_a^b \mathbf{f}(t)\, dt \qquad \text{for every scalar } \alpha.$$

It is also true that

(13.1.9)
$$\int_a^b [\mathbf{c} \cdot \mathbf{f}(t)]\, dt = \mathbf{c} \cdot \left(\int_a^b \mathbf{f}(t)\, dt \right) \qquad \text{for every constant vector } \mathbf{c}$$

and

(13.1.10)
$$\left\| \int_a^b \mathbf{f}(t)\, dt \right\| \le \int_a^b \| \mathbf{f}(t) \|\, dt.$$

The proof of (13.1.9) is left as an exercise. (See Exercise 56.) Here we prove (13.1.10). It is an important inequality.

PROOF Set

$$\mathbf{r} = \int_a^b \mathbf{f}(t)\, dt$$

and note that

$$\| \mathbf{r} \|^2 = \mathbf{r} \cdot \mathbf{r} = \mathbf{r} \cdot \int_a^b \mathbf{f}(t)\, dt = \int_a^b [\mathbf{r} \cdot \mathbf{f}(t)]\, dt \le \int_a^b \| \mathbf{r} \|\, \| \mathbf{f}(t) \|\, dt = \| \mathbf{r} \| \int_a^b \| \mathbf{f}(t) \|\, dt.$$

$$\text{by (13.1.9)} \underline{\qquad} \uparrow \qquad\qquad \uparrow \underline{\qquad} \text{by Schwarz's inequality (12.3.17)}$$

If $\mathbf{r} \ne \mathbf{0}$, we can divide by $\| \mathbf{r} \|$ and conclude that

$$\| \mathbf{r} \| \le \int_a^b \| \mathbf{f}(t) \|\, dt.$$

If $\mathbf{r} = \mathbf{0}$, the result is obvious in the first place. ❑

EXERCISES 13.1

In Exercise 1–8, differentiate the given vector-valued function.

1. $\mathbf{f}(t) = (1 + 2t)\mathbf{i} + (3 - t)\mathbf{j} + (2 + 3t)\mathbf{k}.$
2. $\mathbf{f}(t) = 2\mathbf{i} - \cos t\mathbf{k}.$
3. $\mathbf{f}(t) = \sqrt{1 - t}\,\mathbf{i} + \sqrt{1 + t}\,\mathbf{j} + (1 - t)^{-1}\mathbf{k}.$
4. $\mathbf{f}(t) = e^t\mathbf{i} + \ln t\mathbf{j} + \tan^{-1} t\mathbf{k}.$
5. $\mathbf{f}(t) = \sin t\mathbf{i} + \cos t\mathbf{j} + \tan t\mathbf{k}.$
6. $\mathbf{f}(t) = e^t(\mathbf{i} + t\mathbf{j} + t^2\mathbf{k}).$
7. $\mathbf{f}(t) = \ln(1 - t)\mathbf{i} + \cos t\mathbf{j} + t^2\mathbf{k}.$
8. $\mathbf{f}(t) = \dfrac{t + 1}{t - 1}\mathbf{i} + te^{2t}\mathbf{j} + \sec t\mathbf{k}.$

In Exercises 9–12, calculate the second derivative of the given vector-valued function.

9. $\mathbf{f}(t) = 4t\mathbf{i} + 2t^3\mathbf{j} + (t^2 + 2t)\mathbf{k}.$
10. $\mathbf{f}(t) = t \sin t\mathbf{i} + t \cos t\mathbf{k}.$
11. $\mathbf{f}(t) = e^t \cos t\mathbf{i} + e^t \sin t\mathbf{j} + 4t\mathbf{k}.$
12. $\mathbf{f}(t) = \sqrt{t}\,\mathbf{i} + t\sqrt{t}\,\mathbf{j} + \ln t\mathbf{k}.$

In Exercises 13–18, Calculate the given integral.

13. $\displaystyle\int_1^2 \mathbf{f}(t)\, dt \quad \text{for} \quad \mathbf{f}(t) = \mathbf{i} + 2t\mathbf{j}.$
14. $\displaystyle\int_0^\pi \mathbf{r}(t)\, dt \quad \text{for} \quad \mathbf{r}(t) = \sin t\mathbf{i} + \cos t\mathbf{j} + t\mathbf{k}.$

15. $\int_0^1 \mathbf{g}(t)\,dt$ for $\mathbf{g}(t) = e^t\mathbf{i} + e^{-t}\mathbf{k}.$

16. $\int_0^1 \mathbf{h}(t)\,dt$ for $\mathbf{h}(t) = e^{-t}[t^2\mathbf{i} + \sqrt{2}\,t\mathbf{j} + \mathbf{k}].$

17. $\int_0^1 \mathbf{f}(t)\,dt$ for $\mathbf{f}(t) = \dfrac{1}{1+t^2}\mathbf{i} + \sec^2 t\mathbf{j}.$

18. $\int_1^3 \mathbf{F}(t)\,dt$ for $\mathbf{F}(t) = \dfrac{1}{t}\mathbf{i} + \dfrac{\ln t}{t}\mathbf{j} + e^{-2t}\mathbf{k}.$

In Exercises 19–24, find $\lim\limits_{t\to 0} \mathbf{f}(t)$ if it exists.

19. $\mathbf{f}(t) = \dfrac{\sin t}{2t}\mathbf{i} + e^{2t}\mathbf{j} + \dfrac{t^2}{e^t}\mathbf{k}.$

20. $\mathbf{f}(t) = 3(t^2 - 1)\mathbf{i} + \cos t\mathbf{j} + \dfrac{t}{|t|}\mathbf{k}.$

21. $\mathbf{f}(t) = t^2\mathbf{i} + \dfrac{1 - \cos t}{3t}\mathbf{j} + \dfrac{t}{t+1}\mathbf{k}.$

22. $\mathbf{f}(t) = 3t\mathbf{i} + (t^2 + 1)\mathbf{j} + e^{2t}\mathbf{k}.$

23. $\mathbf{f}(t) = \dfrac{1}{t}\mathbf{i} + \cos t\mathbf{j} + \dfrac{e^t}{t}\mathbf{k}.$

24. $\mathbf{f}(t) = (t^2 - 1)\mathbf{i} + t\sin\left(\dfrac{1}{t}\right)\mathbf{j} + \sqrt{t}\,\mathbf{k}.$

In Exercises 25–34, sketch the curve represented by the given vector-valued function and indicate the orientation.

25. $\mathbf{r}(t) = 2t\mathbf{i} + t^2\mathbf{j},$ $t \geq 0.$

26. $\mathbf{r}(t) = t^3\mathbf{i} + 2t\mathbf{j},$ $t \geq 0.$

27. $\mathbf{r}(t) = 2\sinh t\mathbf{i} + 2\cosh t\mathbf{j},$ $t \geq 0.$

28. $\mathbf{r}(t) = 3\cos t\mathbf{i} + 3\sin t\mathbf{k},$ $0 \leq t \leq 2\pi.$

29. $\mathbf{r}(t) = 2\cos t\mathbf{i} + 3\sin t\mathbf{j},$ $0 \leq t \leq 2\pi.$

30. $\mathbf{r}(t) = 2t\mathbf{i} + (5 - 2t)\mathbf{j} + 3t\mathbf{k},$ $t \geq 0.$

31. $\mathbf{r}(t) = (t^2 + 1)\mathbf{i} + t\mathbf{j} + 4\mathbf{k},$ $-2 \leq t \leq 2.$

32. $\mathbf{r}(t) = t\mathbf{i} + t^2\mathbf{j} + t^3\mathbf{k},$ $0 \leq t \leq 4.$

33. $\mathbf{r}(t) = 2\cos t\mathbf{i} + 2\sin t\mathbf{j} + (2\pi - t)\mathbf{k},$ $0 \leq t \leq 2\pi.$

34. $\mathbf{r}(t) = 3\sin t\mathbf{i} + 4\cos t\mathbf{j} + e^{-t}\mathbf{k},$ $t \geq 0.$

▶ In Exercises 35–37, use a graphing utility to sketch the curve represented by the given vector-valued function and indicate the orientation.

35. $\mathbf{r}(t) = 2\cos(t^2)\mathbf{i} + (2 - \sqrt{t})\mathbf{j}.$

36. $\mathbf{r}(t) = e^{\cos 2t}\mathbf{i} + e^{-\sin t}\mathbf{j}.$

37. $\mathbf{r}(t) = (2 - \sin 2t)\mathbf{i} + (3 + 2\cos t)\mathbf{j}.$

▶ **38.** The vector-valued function

$\mathbf{r}(t) = (A\cos at + B\cos bt)\mathbf{i} + A\sin at\mathbf{j} + B\sin bt\mathbf{k}$

where $A, B, a,$ and b are constants, describes the motion of an object on a torus. Use a graphing utility that can plot three-dimensional graphs to sketch the curve generated by letting:

(a) $A = 2, B = 1, a = 1, b = 1.$
(b) $A = 1, B = 2, a = 2, b = 1.$

In Exercises 39–44, find a vector-valued function $\mathbf{f}$ that traces out the given curve in the indicated direction.

39. $4x^2 + 9y^2 = 36$ (a) Counterclockwise.   (b) Clockwise.

40. $(x - 1)^2 + y^2 = 1$ (a) Counterclockwise.   (b) Clockwise.

41. $y = x^2$ (a) From left to right.   (b) From right to left.

42. $y = x^3$ (a) From left to right.   (b) From right to left.

43. The directed line segment from $(1, 4, -2)$ to $(3, 9, 6).$

44. The directed line segment from $(3, 2, -5)$ to $(7, 2, 9).$

45. Set $\mathbf{f}(t) = t\mathbf{i} + f(t)\mathbf{j}$ and calculate

$$\mathbf{f}'(t_0), \qquad \int_a^b \mathbf{f}(t)\,dt, \qquad \int_a^b \mathbf{f}'(t)\,dt$$

given that

$$f'(t_0) = m, \quad f(a) = c, \quad f(b) = d, \quad \int_a^b f(t)\,dt = A.$$

In Exercises 46–49, find $\mathbf{f}(t)$ from the following information.

46. $\mathbf{f}'(t) = t\mathbf{i} + t(1 + t^2)^{-1/2}\mathbf{j} + te^t\mathbf{k}$ and $\mathbf{f}(0) = \mathbf{i} + 2\mathbf{j} + 3\mathbf{k}.$

47. $\mathbf{f}'(t) = \mathbf{i} + t^2\mathbf{j}$ and $\mathbf{f}(0) = \mathbf{j} - \mathbf{k}.$

48. $\mathbf{f}'(t) = 2\mathbf{f}(t)$ and $\mathbf{f}(0) = \mathbf{i} - \mathbf{k}.$

49. $\mathbf{f}'(t) = \alpha\mathbf{f}(t)$ with α a real number and $\mathbf{f}(0) = \mathbf{c}.$

50. No $\epsilon,$ δ's have surfaced so far, but they are still there at the heart of the limit process. Give an $\epsilon,$ δ characterization of

$$\lim_{t\to t_0} \mathbf{f}(t) = \mathbf{L}.$$

51. (a) Show that, if $\mathbf{f}'(t) = \mathbf{0}$ for all t in an interval $I,$ then $\mathbf{f}$ is a constant vector on $I.$

(b) Show that, if $\mathbf{f}'(t) = \mathbf{g}'(t)$ for all t in an interval $I,$ then $\mathbf{f}$ and $\mathbf{g}$ differ by a constant vector on $I.$

52. Assume that, as $t \to t_0,$ $\mathbf{f}(t) \to \mathbf{L}$ and $\mathbf{g}(t) \to \mathbf{M}.$ Show that

$$\mathbf{f}(t) \cdot \mathbf{g}(t) \to \mathbf{L} \cdot \mathbf{M}.$$

53. Show that, if $\mathbf{f}$ is differentiable at $t,$ then $\mathbf{f}$ is continuous at $t.$

54. A vector-valued function $\mathbf{G}$ is called an *antiderivative* for $\mathbf{f}$ on $[a, b]$ iff (i) $\mathbf{G}$ is continuous on $[a, b]$ and (ii) $\mathbf{G}'(t) = \mathbf{f}(t)$ for all $t \in (a, b).$ Show that, if $\mathbf{f}$ is continuous on $[a, b]$ and $\mathbf{G}$ is an antiderivative for $\mathbf{f}$ on $[a, b],$ then

$$\int_a^b \mathbf{f}(t)\,dt = \mathbf{G}(b) - \mathbf{G}(a).$$

(This is a vector version of the fundamental theorem of integral calculus.)

55. Is it always true that

$$\int_a^b [\mathbf{f}(t) \cdot \mathbf{g}(t)] \, dt = \left[\int_a^b \mathbf{f}(t) \, dt \right] \cdot \left[\int_a^b \mathbf{g}(t) \, dt \right] ?$$

56. Prove that, if $\mathbf{f}$ is continuous on $[a, b]$, then for each vector $\mathbf{c}$

$$\int_a^b [\mathbf{c} \cdot \mathbf{f}(t)] \, dt = \mathbf{c} \cdot \int_a^b \mathbf{f}(t) \, dt \qquad \text{and}$$

$$\int_a^b [\mathbf{c} \times \mathbf{f}(t)] \, dt = \mathbf{c} \times \int_a^b \mathbf{f}(t) \, dt.$$

57. Let $\mathbf{f}$ be a differentiable vector-valued function. Show that if $\|\mathbf{f}(t)\| \neq 0$, then

$$\frac{d}{dt}(\|\mathbf{f}(t)\|) = \frac{\mathbf{f}(t) \cdot \mathbf{f}'(t)}{\|\mathbf{f}(t)\|}.$$

58. Let $\mathbf{f}$ be a differentiable vector-valued function. Show that if $\|\mathbf{f}(t)\| \neq 0$, then

$$\frac{d}{dt}\left(\frac{\mathbf{f}(t)}{\|\mathbf{f}(t)\|} \right) = \frac{\mathbf{f}'(t)}{\|\mathbf{f}(t)\|} - \frac{\mathbf{f}(t) \cdot \mathbf{f}'(t)}{\|\mathbf{f}(t)\|^3} \mathbf{f}(t).$$

■ 13.2 DIFFERENTIATION FORMULAS

Vector functions with a common domain can be combined in many ways to form new functions. From $\mathbf{f}$ and $\mathbf{g}$ we can form the sum $\mathbf{f} + \mathbf{g}$:

$$(\mathbf{f} + \mathbf{g})(t) = \mathbf{f}(t) + \mathbf{g}(t).$$

We can form scalar multiples $\alpha\mathbf{f}$ and thus linear combinations $\alpha\mathbf{f} + \beta\mathbf{g}$:

$$(\alpha\mathbf{f})(t) = \alpha\mathbf{f}(t), \qquad (\alpha\mathbf{f} + \beta\mathbf{g})(t) = \alpha\mathbf{f}(t) + \beta\mathbf{g}(t).$$

We can form the dot product $\mathbf{f} \cdot \mathbf{g}$:

$$(\mathbf{f} \cdot \mathbf{g})(t) = \mathbf{f}(t) \cdot \mathbf{g}(t).$$

We can also form the cross product $\mathbf{f} \times \mathbf{g}$:

$$(\mathbf{f} \times \mathbf{g})(t) = \mathbf{f}(t) \times \mathbf{g}(t).$$

Of course, these operations on vector functions are simply the pointwise application of the algebraic operations on vectors that we introduced in Chapter 12.

There are two ways of bringing *scalar functions* (real-valued functions) into play. If a scalar function u has the same domain as $\mathbf{f}$, we can form the scalar product $u\mathbf{f}$:

$$(u\mathbf{f})(t) = u(t)\mathbf{f}(t).$$

If $u(t)$ is in the domain of $\mathbf{f}$ for each t in some interval, then we can form the composition $\mathbf{f} \circ u$:

$$(\mathbf{f} \circ u)(t) = \mathbf{f}(u(t)).$$

Example 1 Let $u(t) = t^2$ and $\mathbf{f}(t) = e^t\mathbf{i} + \sin 2t\mathbf{j} + \sqrt{t^2 + 1}\mathbf{k}$. Then

$$(u\mathbf{f})(t) = u(t)\mathbf{f}(t) = t^2 e^t\mathbf{i} + t^2\sin 2t\mathbf{j} + t^2\sqrt{t^2 + 1}\mathbf{k}$$

and

$$(\mathbf{f} \circ u)(t) = \mathbf{f}(u(t)) = e^{t^2}\mathbf{i} + \sin 2t^2\mathbf{j} + \sqrt{t^4 + 1}\mathbf{k}. \qquad \square$$

From Theorem 13.1.3 it is clear that, if $\mathbf{f}$, $\mathbf{g}$, and u are continuous on a common domain, then $\mathbf{f} + \mathbf{g}$, $\alpha\mathbf{f}$, $\mathbf{f} \cdot \mathbf{g}$, $\mathbf{f} \times \mathbf{g}$, and $u\mathbf{f}$ are all continuous on that same set. We have yet to show the continuity of $\mathbf{f} \circ u$. The verification of that is left to you. (Exercise 36.) What interests us here is that, if $\mathbf{f}$, $\mathbf{g}$, and u are differentiable, then the newly constructed functions are also differentiable and their derivatives satisfy the following rules:

(13.2.1)

> (1) $(\mathbf{f} + \mathbf{g})'(t) = \mathbf{f}'(t) + \mathbf{g}'(t)$.
> (2) $(\alpha \mathbf{f})'(t) = \alpha \mathbf{f}'(t)$, ($\alpha$ constant).
> (3) $(u\mathbf{f})'(t) = u(t)\mathbf{f}'(t) + u'(t)\mathbf{f}(t)$.
> (4) $(\mathbf{f} \cdot \mathbf{g})'(t) = [\mathbf{f}(t) \cdot \mathbf{g}'(t)] + [\mathbf{f}'(t) \cdot \mathbf{g}(t)]$.
> (5) $(\mathbf{f} \times \mathbf{g})'(t) = [\mathbf{f}(t) \times \mathbf{g}'(t)] + [\mathbf{f}'(t) \times \mathbf{g}(t)]$.
> (6) $(\mathbf{f} \circ u)'(t) = \mathbf{f}'(u(t))u'(t) = u'(t)\mathbf{f}'(u(t))$ (the chain rule).

Rules (3), (4), (5) are all "product" rules and should remind you of the rule for differentiating the product of ordinary functions. Keep in mind, however, that the cross product is not commutative and therefore the order in Rule (5) is important.

In Rule (6) we first wrote the scalar part $u'(t)$ on the right so that the formula would look like the chain rule for ordinary functions. In general, $\mathbf{a}\alpha$ has the same meaning as $\alpha \mathbf{a}$.

Example 2 Taking

$$\mathbf{f}(t) = 2t^2\mathbf{i} - 3\mathbf{j}, \qquad \mathbf{g}(t) = \mathbf{i} + t\mathbf{j} + t^2\mathbf{k}, \qquad u(t) = \tfrac{1}{3}t^3$$

we have

$$\mathbf{f}'(t) = 4t\mathbf{i}, \qquad \mathbf{g}'(t) = \mathbf{j} + 2t\mathbf{k}, \qquad u'(t) = t^2.$$

Therefore

(a) $(\mathbf{f} + \mathbf{g})'(t) = \mathbf{f}'(t) + \mathbf{g}'(t) = 4t\mathbf{i} + (\mathbf{j} + 2t\mathbf{k}) = 4t\mathbf{i} + \mathbf{j} + 2t\mathbf{k}$;

(b) $(2\mathbf{f})'(t) = 2\mathbf{f}'(t) = 2(4t\mathbf{i}) = 8t\mathbf{i}$;

(c) $(u\mathbf{f})'(t) = u(t)\mathbf{f}'(t) + u'(t)\mathbf{f}(t) = \tfrac{1}{3}t^3(4t\mathbf{i}) + t^2(2t^2\mathbf{i} - 3\mathbf{j}) = \tfrac{10}{3}t^4\mathbf{i} - 3t^2\mathbf{j}$;

(d) $(\mathbf{f} \cdot \mathbf{g})'(t) = [\mathbf{f}(t) \cdot \mathbf{g}'(t)] + [\mathbf{f}'(t) \cdot \mathbf{g}(t)]$

$\qquad = [(2t^2\mathbf{i} - 3\mathbf{j}) \cdot (\mathbf{j} + 2t\mathbf{k})] + [4t\mathbf{i} \cdot (\mathbf{i} + t\mathbf{j} + t^2\mathbf{k})] = -3 + 4t$;

(e) $(\mathbf{f} \times \mathbf{g})'(t) = [\mathbf{f}(t) \times \mathbf{g}'(t)] + [\mathbf{f}'(t) \times \mathbf{g}(t)]$

$\qquad = [(2t^2\mathbf{i} - 3\mathbf{j}) \times (\mathbf{j} + 2t\mathbf{k})] + [4t\mathbf{i} \times (\mathbf{i} + t\mathbf{j} + t^2\mathbf{k})]$

$\qquad = (2t^2\mathbf{k} - 4t^3\mathbf{j} - 6t\mathbf{i}) + (4t^2\mathbf{k} - 4t^3\mathbf{j}) = -6t\mathbf{i} - 8t^3\mathbf{j} + 6t^2\mathbf{k}$

while

$(\mathbf{g} \times \mathbf{f})'(t) = [\mathbf{g}(t) \times \mathbf{f}'(t)] + [\mathbf{g}'(t) \times \mathbf{f}(t)]$

$\qquad = [(\mathbf{i} + t\mathbf{j} + t^2\mathbf{k}) \times 4t\mathbf{i}] + [(\mathbf{j} + 2t\mathbf{k}) \times (2t^2\mathbf{i} - 3\mathbf{j})]$

$\qquad = (-4t^2\mathbf{k} + 4t^3\mathbf{j}) + (-2t^2\mathbf{k} + 4t^3\mathbf{j} + 6t\mathbf{i})$

$\qquad = 6t\mathbf{i} + 8t^3\mathbf{j} - 6t^2\mathbf{k} = -(\mathbf{f} \times \mathbf{g})'(t)$;

(f) $(\mathbf{f} \circ u)'(t) = \mathbf{f}'(u(t))u'(t)$

$\qquad = [4u(t)\mathbf{i}]u'(t) = [4(\tfrac{1}{3}t^3)\mathbf{i}] t^2 = \tfrac{4}{3}t^5\mathbf{i}.$ ❑

The differentiation formulas that we have given can all be derived component by component, and they can all be derived in a component-free manner. Take for example formula (3):

$$(u\mathbf{f})'(t) = u(t)\mathbf{f}'(t) + u'(t)\mathbf{f}(t).$$

COMPONENT-BY-COMPONENT DERIVATION   Set

$$\mathbf{f}(t) = f_1(t)\mathbf{i} + f_2(t)\mathbf{j} + f_3(t)\mathbf{k}.$$

Then

$$(u\mathbf{f})(t) = u(t)\mathbf{f}(t) = u(t)f_1(t)\mathbf{i} + u(t)f_2(t)\mathbf{j} + u(t)f_3(t)\mathbf{k}$$

and

$$\begin{aligned}
(u\mathbf{f})'(t) &= [u(t)f_1'(t) + u'(t)f_1(t)]\mathbf{i} + [u(t)f_2'(t) + u'(t)f_2(t)]\mathbf{j} \\
&\quad + [u(t)f_3'(t) + u'(t)f_3(t)]\mathbf{k} \\
&= u(t)[f_1'(t)\mathbf{i} + f_2'(t)\mathbf{j} + f_3'(t)\mathbf{k}] + u'(t)[f_1(t)\mathbf{i} + f_2(t)\mathbf{j} + f_3(t)\mathbf{k}] \\
&= u(t)\mathbf{f}'(t) + u'(t)\mathbf{f}(t). \quad \square
\end{aligned}$$

COMPONENT-FREE DERIVATION   We find $(u\mathbf{f})'(t)$ by taking the limit as $h \to 0$ of the difference quotient

$$\frac{u(t+h)\mathbf{f}(t+h) - u(t)\mathbf{f}(t)}{h}.$$

By adding and subtracting $u(t+h)\mathbf{f}(t)$, we can rewrite this quotient as

$$\frac{u(t+h)\,\mathbf{f}(t+h) - u(t+h)\,\mathbf{f}(t) + u(t+h)\,\mathbf{f}(t) - u(t)\,\mathbf{f}(t)}{h}$$

which is equal to

$$u(t+h)\frac{\mathbf{f}(t+h) - \mathbf{f}(t)}{h} + \frac{u(t+h) - u(t)}{h}\mathbf{f}(t).$$

As $h \to 0$,

$$u(t+h) \to u(t), \qquad \text{(differentiable functions are continuous)}$$

$$\frac{\mathbf{f}(t+h) - \mathbf{f}(t)}{h} \to \mathbf{f}'(t), \qquad \text{(definition of derivative for vector functions)}$$

$$\frac{u(t+h) - u(t)}{h} \to u'(t). \qquad \text{(definition of derivative for scalar functions)}$$

It follows from the limit rules (Theorem 13.1.3) that

$$u(t+h)\frac{\mathbf{f}(t+h) - \mathbf{f}(t)}{h} \to u(t)\mathbf{f}'(t), \qquad \frac{u(t+h) - u(t)}{h}\mathbf{f}(t) \to u'(t)\mathbf{f}(t)$$

and therefore

$$u(t+h)\frac{\mathbf{f}(t+h) - \mathbf{f}(t)}{h} + \frac{u(t+h) - u(t)}{h}\mathbf{f}(t) \to u(t)\mathbf{f}'(t) + u'(t)\mathbf{f}(t). \quad \square$$

In Leibniz's notation the formulas take the following form:

(13.2.2)

$$(1) \quad \frac{d}{dt}(\mathbf{f} + \mathbf{g}) = \frac{d\mathbf{f}}{dt} + \frac{d\mathbf{g}}{dt}.$$

$$(2) \quad \frac{d}{dt}(\alpha\mathbf{f}) = \alpha\frac{d\mathbf{f}}{dt}. \qquad (\alpha \text{ constant})$$

$$(3) \quad \frac{d}{dt}(u\mathbf{f}) = u\frac{d\mathbf{f}}{dt} + \frac{du}{dt}\mathbf{f}. \qquad (u = u(t))$$

$$(4) \quad \frac{d}{dt}(\mathbf{f} \cdot \mathbf{g}) = \left(\mathbf{f} \cdot \frac{d\mathbf{g}}{dt}\right) + \left(\frac{d\mathbf{f}}{dt} \cdot \mathbf{g}\right).$$

$$(5) \quad \frac{d}{dt}(\mathbf{f} \times \mathbf{g}) = \left(\mathbf{f} \times \frac{d\mathbf{g}}{dt}\right) + \left(\frac{d\mathbf{f}}{dt} \times \mathbf{g}\right).$$

$$(6) \quad \frac{d\mathbf{f}}{dt} = \frac{d\mathbf{f}}{du}\frac{du}{dt}. \qquad (\mathbf{f} = \mathbf{f}(u(t)) \quad \text{(chain rule)}$$

We come now to a couple of results that will prove useful as we go on.

Example 3 Let $\mathbf{r}$ be a differentiable vector function of t and set $r = \|\mathbf{r}\|$. Show that r is differentiable wherever it is not 0 and

(13.2.3)

$$\mathbf{r} \cdot \frac{d\mathbf{r}}{dt} = r\frac{dr}{dt}.$$

SOLUTION If $\mathbf{r}$ is differentiable, then

$$\mathbf{r} \cdot \mathbf{r} = \|\mathbf{r}\|^2 = r^2$$

is differentiable. Let's assume now that $r \neq 0$. Since the square-root function is differentiable at all positive numbers and r^2 is positive, we can apply the square-root function to r^2 and conclude by the chain rule that r is itself differentiable.

To obtain the formula we differentiate the identity $\mathbf{r} \cdot \mathbf{r} = r^2$:

$$\mathbf{r} \cdot \frac{d\mathbf{r}}{dt} + \frac{d\mathbf{r}}{dt} \cdot \mathbf{r} = 2r\frac{dr}{dt}$$

$$2\mathbf{r} \cdot \frac{d\mathbf{r}}{dt} = 2r\frac{dr}{dt}$$

$$\mathbf{r} \cdot \frac{d\mathbf{r}}{dt} = r\frac{dr}{dt}. \quad \square$$

Example 4 Let $\mathbf{r}$ be a differentiable vector function of t and set $r = \|\mathbf{r}\|$. Show that where $r \neq 0$

(13.2.4)

$$\frac{d}{dt}\left(\frac{\mathbf{r}}{r}\right) = \frac{1}{r^3}\left[\left(\mathbf{r} \times \frac{d\mathbf{r}}{dt}\right) \times \mathbf{r}\right].$$

SOLUTION   This is a little tricky:

$$\frac{d}{dt}\left(\frac{\mathbf{r}}{r}\right) = \frac{1}{r}\frac{d\mathbf{r}}{dt} - \frac{1}{r^2}\frac{dr}{dt}\mathbf{r}$$

$$= \frac{1}{r^3}\left[r^2\frac{d\mathbf{r}}{dt} - r\frac{dr}{dt}\mathbf{r}\right]$$

$$= \frac{1}{r^3}\left[(\mathbf{r}\cdot\mathbf{r})\frac{d\mathbf{r}}{dt} - \left(\mathbf{r}\cdot\frac{d\mathbf{r}}{dt}\right)\mathbf{r}\right] = \frac{1}{r^3}\left[\left(\mathbf{r}\times\frac{d\mathbf{r}}{dt}\right)\times\mathbf{r}\right]. \quad \square$$

$$(\mathbf{a}\times\mathbf{b})\times\mathbf{c} = (\mathbf{c}\cdot\mathbf{a})\mathbf{b} - (\mathbf{c}\cdot\mathbf{b})\mathbf{a} \underline{\hspace{4cm}}$$

EXERCISES 13.2

In Exercises 1–12, find $\mathbf{f}'(t)$ and $\mathbf{f}''(t)$.

1. $\mathbf{f}(t) = \mathbf{a} + t\mathbf{b}$.

2. $\mathbf{f}(t) = \mathbf{a} + t\mathbf{b} + t^2\mathbf{c}$.

3. $\mathbf{f}(t) = e^{2t}\mathbf{i} - \sin t\mathbf{j}$.

4. $\mathbf{f}(t) = [(t^2\mathbf{i} - \mathbf{j})\cdot(\mathbf{i} - t^2\mathbf{j})]\,\mathbf{i}$.

5. $\mathbf{f}(t) = [(t^2\mathbf{i} - 2t\mathbf{j})\cdot(t\mathbf{i} + t^3\mathbf{j})]\,\mathbf{j}$.

6. $\mathbf{f}(t) = [(3t\mathbf{i} - t^2\mathbf{j} + \mathbf{k})\cdot(\mathbf{i} + t^3\mathbf{j} - 2t\mathbf{k})]\,\mathbf{k}$.

7. $\mathbf{f}(t) = (e^t\mathbf{i} + t\mathbf{k})\times(t\mathbf{j} + e^{-t}\mathbf{k})$.

8. $\mathbf{f}(t) = (t\mathbf{i} - t^2\mathbf{j} + \mathbf{k})\times(\mathbf{i} + t^3\mathbf{j} + 5t\mathbf{k})$.

9. $\mathbf{f}(t) = (\cos t\mathbf{i} + \sin t\mathbf{j} + \mathbf{k})\times(\sin 2t\mathbf{i} + \cos 2t\mathbf{j} + t\mathbf{k})$.

10. $\mathbf{f}(t) = t\mathbf{g}(t^2)$.

11. $\mathbf{f}(t) = t\mathbf{g}(\sqrt{t})$.

12. $\mathbf{f}(t) = (e^{2t}\mathbf{i} + e^{-2t}\mathbf{j} + \mathbf{k})\times(e^{2t}\mathbf{i} - e^{-2t}\mathbf{j} + \mathbf{k})$.

In Exercises 13–20, find the indicated derivative.

13. $\dfrac{d}{dt}[e^{\cos t}\mathbf{i} + e^{\sin t}\mathbf{j}]$.

14. $\dfrac{d^2}{dt^2}[e^t\cos t\mathbf{i} + e^t\sin t\mathbf{j}]$.

15. $\dfrac{d^2}{dt^2}\left[(e^t\mathbf{i} + e^{-t}\mathbf{j})\cdot(e^t\mathbf{i} - e^{-t}\mathbf{j})\right]$.

16. $\dfrac{d}{dt}\left[(\ln t\mathbf{i} + t\mathbf{j} - (t^2 + 1)\mathbf{k})\times\left(\dfrac{1}{t}\mathbf{i} + t^2\mathbf{j} - t\mathbf{k}\right)\right]$.

17. $\dfrac{d}{dt}[(\mathbf{a} + t\mathbf{b})\times(\mathbf{c} + t\mathbf{d})]$.

18. $\dfrac{d}{dt}[(\mathbf{a} + t\mathbf{b})\times(\mathbf{a} + t\mathbf{b} + t^2\mathbf{c})]$.

19. $\dfrac{d}{dt}[(\mathbf{a} + t\mathbf{b})\cdot(\mathbf{c} + t\mathbf{d})]$.

20. $\dfrac{d}{dt}[(\mathbf{a} + t\mathbf{b})\cdot(\mathbf{a} + t\mathbf{b} + t^2\mathbf{c})]$.

In Exercises 21–24, find $\mathbf{r}(t)$ given that:

21. $\mathbf{r}'(t) = \mathbf{b}$ for all real t, $\mathbf{r}(0) = \mathbf{a}$.

22. $\mathbf{r}''(t) = \mathbf{c}$ for all real t, $\mathbf{r}'(0) = \mathbf{b}$, $\mathbf{r}(0) = \mathbf{a}$.

23. $\mathbf{r}''(t) = \mathbf{a} + t\mathbf{b}$ for all real t, $\mathbf{r}'(0) = \mathbf{c}$, $\mathbf{r}(0) = \mathbf{d}$.

24. $\mathbf{r}''(t) = \cos 2t\mathbf{i} + \sin 2t\mathbf{j}$ for all real t, $\mathbf{r}'(0) = 2\mathbf{i} - \frac{1}{2}\mathbf{j}$, $\mathbf{r}(0) = \frac{3}{4}\mathbf{i} + \mathbf{j}$.

25. Show that, if $\mathbf{r}(t) = \sin t\mathbf{i} + \cos t\mathbf{j}$, then $\mathbf{r}(t)$ and $\mathbf{r}''(t)$ are parallel. Is there a value of t for which $\mathbf{r}(t)$ and $\mathbf{r}''(t)$ have the same direction?

26. Show that, if $\mathbf{r}(t) = e^{kt}\mathbf{i} + e^{-kt}\mathbf{j}$, then $\mathbf{r}(t)$ and $\mathbf{r}''(t)$ are parallel.

27. Calculate $\mathbf{r}(t)\cdot\mathbf{r}'(t)$ and $\mathbf{r}(t)\times\mathbf{r}'(t)$ given that $\mathbf{r}(t) = \cos t\mathbf{i} + \sin t\mathbf{j}$.

In Exercises 28–30, assume the rule for differentiating a cross product and show the following.

28. $(\mathbf{g}\times\mathbf{f})'(t) = -(\mathbf{f}\times\mathbf{g})'(t)$.

29. $\dfrac{d}{dt}[\mathbf{f}(t)\times\mathbf{f}'(t)] = \mathbf{f}(t)\times\mathbf{f}''(t)$.

30. $\dfrac{d}{dt}[u_1(t)\mathbf{r}_1(t)\times u_2(t)\mathbf{r}_2(t)] = u_1(t)u_2(t)\dfrac{d}{dt}[\mathbf{r}_1(t)\times\mathbf{r}_2(t)]$
$+ [\mathbf{r}_1(t)\times\mathbf{r}_2(t)]\dfrac{d}{dt}[u_1(t)u_2(t)]$.

31. Prove that $\{\mathbf{f}(t)\cdot[\mathbf{g}(t)\times\mathbf{h}(t)]\}' = \mathbf{f}'(t)\cdot[\mathbf{g}(t)\times\mathbf{h}(t)]$
$+ \mathbf{f}(t)\cdot[\mathbf{g}'(t)\times\mathbf{h}(t)] + \mathbf{f}(t)\cdot[\mathbf{g}(t)\times\mathbf{h}'(t)]$.

32. Suppose that $\mathbf{f}(t)$ is parallel to $\mathbf{f}''(t)$ for all t. Prove that $\mathbf{f}\times\mathbf{f}'$ is constant.   HINT: See Exercise 29.

33. Assume the rule for differentiating a dot product and show that

$\|\mathbf{r}(t)\|$ is constant   iff $\mathbf{r}(t)\cdot\mathbf{r}'(t) = 0$ identically.

34. Derive the formula

$$(\mathbf{f} \cdot \mathbf{g})'(t) = [\mathbf{f}(t) \cdot \mathbf{g}'(t)] + [\mathbf{f}'(t) \cdot \mathbf{g}(t)]$$

(a) by appealing to components;　(b) without appealing to components.

35. Derive the formula

$$(\mathbf{f} \times \mathbf{g})'(t) = [\mathbf{f}(t) \times \mathbf{g}'(t)] + [\mathbf{f}'(t) \times \mathbf{g}(t)]$$

without appealing to components.

36. (a) Show that, if u is continuous at t_0 and $\mathbf{f}$ is continuous at $u(t_0)$, then the composition $(\mathbf{f} \circ u)$ is continuous at t_0.

(b) Derive the chain rule for vector functions:

$$\frac{d\mathbf{f}}{dt} = \frac{d\mathbf{f}}{du}\frac{du}{dt}.$$

■ 13.3 CURVES

Introduction

Earlier we explained how every linear vector function

$$\mathbf{r}(t) = \mathbf{r}_0 + t\mathbf{d} \qquad \text{with} \qquad \mathbf{d} \neq \mathbf{0}$$

parametrizes a line. More generally, every differentiable vector function parametrizes a curve.

Suppose that

$$\mathbf{r}(t) = x(t)\mathbf{i} + y(t)\mathbf{j} + z(t)\mathbf{k}$$

is differentiable on some interval I. (At the endpoints, if there are any, we require only continuity.) For each number $t \in I$, the tip of the radius vector $\mathbf{r}(t)$ is the point $P(x(t), y(t), z(t))$. As t ranges over I, the point P traces out some path C. (Figure 13.3.1) We call C a *differentiable curve* and say that C is *parametrized* by $\mathbf{r}$. It is also important to understand that a parametrized curve C is an *oriented* curve in the sense that as t increases on I, the tip of the radius vector traces out C in a certain direction. For example, the unit circle parametrized by

$$\mathbf{r}(t) = \cos t\mathbf{i} + \sin t\mathbf{j}, \qquad t \in [0, 2\pi]$$

is traversed in the counterclockwise direction starting at the point $(1, 0)$ (Figure 13.3.2). The orientation of the elliptical helix parametrized by

$$\mathbf{r}(t) = t\mathbf{i} + 2\cos t\mathbf{j} + 3\sin t\mathbf{k}, \qquad t \geq 0,$$

is indicated by the arrows in Figure 13.3.3.

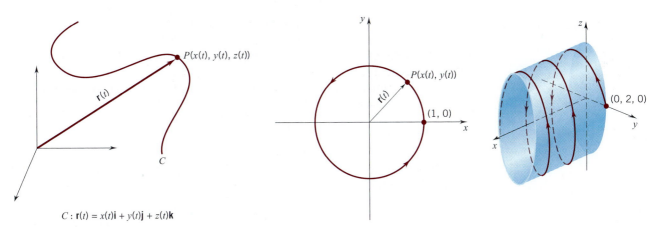

Figure 13.3.1　　　　Figure 13.3.2　　　　Figure 13.3.3

Tangent Vector, Tangent Line

Let's try to interpret the derivative

$$\mathbf{r}'(t) = x'(t)\mathbf{i} + y'(t)\mathbf{j} + z'(t)\mathbf{k}$$

geometrically. First of all, for fixed t

$$\mathbf{r}'(t) = \lim_{h \to 0} \frac{\mathbf{r}(t+h) - \mathbf{r}(t)}{h}.$$

If $\mathbf{r}'(t) \neq \mathbf{0}$, then we can be sure that for $t+h$ close enough to t, the vector

$$\mathbf{r}(t+h) - \mathbf{r}(t)$$

will not be $\mathbf{0}$. (Explain.) Consequently, we can think of the vector $\mathbf{r}(t+h) - \mathbf{r}(t)$ as pictured in Figure 13.3.4.

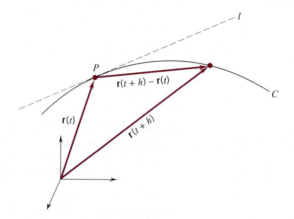

Figure 13.3.4

Let's agree that the line marked l in the figure corresponds to our intuitive notion of tangent line at the point P. As h tends to zero, the vector

$$\mathbf{r}(t+h) - \mathbf{r}(t)$$

comes increasingly closer to serving as a direction vector for that tangent line. It may be tempting therefore to take the limiting case

$$\lim_{h \to 0} [\mathbf{r}(t+h) - \mathbf{r}(t)]$$

and call that a direction vector for the tangent line. The trouble is that this limit vector is $\mathbf{0}$ and $\mathbf{0}$ has no direction.

We can circumvent this difficulty by replacing $\mathbf{r}(t+h) - \mathbf{r}(t)$ by a vector which, for small h, has greater length: the difference quotient

$$\frac{\mathbf{r}(t+h) - \mathbf{r}(t)}{h}.$$

For each real number $h \neq 0$, the vector $[\mathbf{r}(t+h) - \mathbf{r}(t)]/h$ is parallel to $\mathbf{r}(t+h) - \mathbf{r}(t)$, and therefore its limit,

$$\mathbf{r}'(t) = \lim_{h \to 0} \frac{\mathbf{r}(t+h) - \mathbf{r}(t)}{h},$$

which by assumption is not $\mathbf{0}$, can be taken as a direction vector for the tangent line.

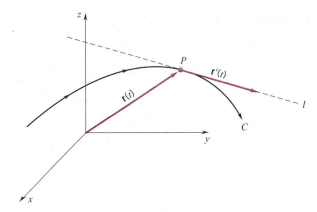

Figure 13.3.5

Finally, an obvious question is: In which of the two possible directions on l does $\mathbf{r}'(t)$ point? You are asked to show in Exercise 48 that $\mathbf{r}'(t)$ points in the direction of increasing t (see Figure 13.3.5). Hence the following definition:

DEFINITION 13.3.1 TANGENT VECTOR

Let

$$C: \quad \mathbf{r}(t) = x(t)\mathbf{i} + y(t)\mathbf{j} + z(t)\mathbf{k}$$

be a differentiable curve. The vector

$$\mathbf{r}'(t) = x'(t)\mathbf{i} + y'(t)\mathbf{j} + z'(t)\mathbf{k},$$

if not $\mathbf{0}$, is said to be *tangent* to the curve C at the point $P(x(t), y(t), z(t))$, and $\mathbf{r}'(t)$ points in the direction of increasing t.

At each point of a line

$$l: \quad \mathbf{r}(t) = \mathbf{r}_0 + t\mathbf{d}$$

the tangent vector $\mathbf{r}'(t)$ is parallel to the line itself:

$$\mathbf{r}'(t) = \mathbf{d} \qquad \text{and } \mathbf{d} \text{ is parallel to } l. \qquad \text{(Figure 13.3.6)}$$

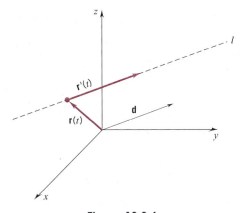

Figure 13.3.6

In the case of a circle

$$C: \quad \mathbf{r}(t) = a \cos t\,\mathbf{i} + a \sin t\,\mathbf{j}, \qquad a > 0,$$

the tangent vector $\mathbf{r}'(t)$ is perpendicular to the radius vector $\mathbf{r}(t)$:

$$\mathbf{r}'(t) \cdot \mathbf{r}(t) = (-a \sin t\,\mathbf{i} + a \cos t\,\mathbf{j}) \cdot (a \cos t\,\mathbf{i} + a \sin t\,\mathbf{j})$$

$$= -a^2 \sin t \cos t + a^2 \cos t \sin t = 0. \qquad \text{(Figure 13.3.7)}$$

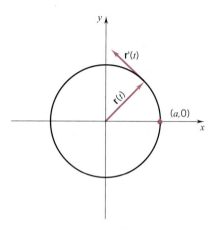

Figure 13.3.7

Example 1   Find the point P on the curve

$$\mathbf{r}(t) = (1 - 2t)\mathbf{i} + t^2\mathbf{j} + 2e^{2(t-1)}\mathbf{k}$$

at which the tangent vector $\mathbf{r}'(t)$ is parallel to the radius vector $\mathbf{r}(t)$.

SOLUTION

$$\mathbf{r}'(t) = -2\mathbf{i} + 2t\mathbf{j} + 4e^{2(t-1)}\mathbf{k}.$$

For $\mathbf{r}'(t)$ to be parallel to $\mathbf{r}(t)$ there must exist a scalar α such that

$$\mathbf{r}(t) = \alpha \mathbf{r}'(t).$$

This vector equation holds iff

$$1 - 2t = -2\alpha, \qquad t^2 = 2\alpha t, \qquad 2e^{2(t-1)} = 4\alpha e^{2(t-1)}.$$

The last scalar equation requires that $\alpha = \frac{1}{2}$. The only value of t that satisfies all three equations with $\alpha = \frac{1}{2}$ is $t = 1$. Therefore the only point at which $\mathbf{r}'(t)$ is parallel to $\mathbf{r}(t)$ is the tip of $\mathbf{r}(1)$. This is the point $P(-1, 1, 2)$. ❑

If $\mathbf{r}'(t_0) \neq \mathbf{0}$, then $\mathbf{r}'(t_0)$ is tangent to the curve at the tip of $\mathbf{r}(t_0)$. The *tangent line* at this point can be parametrized by setting

(13.3.2)
$$\mathbf{R}(u) = \mathbf{r}(t_0) + u\mathbf{r}'(t_0).$$

Example 2 Find a vector tangent to the *twisted cubic*

$$\mathbf{r}(t) = t\mathbf{i} + t^2\mathbf{j} + t^3\mathbf{k} \qquad \text{(Figure 13.3.8)}$$

at the point $P(2, 4, 8)$, and then parametrize the tangent line.

SOLUTION Here

$$\mathbf{r}'(t) = \mathbf{i} + 2t\mathbf{j} + 3t^2\mathbf{k}.$$

Since $P(2, 4, 8)$ is the tip of $\mathbf{r}(2)$, the vector

$$\mathbf{r}'(2) = \mathbf{i} + 4\mathbf{j} + 12\mathbf{k}$$

is tangent to the curve at the point $P(2, 4, 8)$. The vector function

$$\mathbf{R}(u) = (2\mathbf{i} + 4\mathbf{j} + 8\mathbf{k}) + u(\mathbf{i} + 4\mathbf{j} + 12\mathbf{k})$$

parametrizes the tangent line. ❑

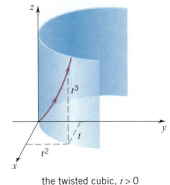

the twisted cubic, $t > 0$

Figure 13.3.8

Intersecting Curves

Two curves

$$C_1: \quad \mathbf{r}_1(t) = x_1(t)\mathbf{i} + y_1(t)\mathbf{j} + z_1(t)\mathbf{k}, \qquad C_2: \quad \mathbf{r}_2(u) = x_2(u)\mathbf{i} + y_2(u)\mathbf{j} + z_2(u)\mathbf{k}$$

intersect iff there are numbers t and u for which

$$\mathbf{r}_1(t) = \mathbf{r}_2(u).$$

The angle between two intersecting curves (which, by definition, is the angle between the corresponding tangent lines) can be obtained by examining the tangent vectors at the point of intersection.

Example 3 Show that the circles

$$C_1: \quad \mathbf{r}_1(t) = \cos t\,\mathbf{i} + \sin t\,\mathbf{j}, \qquad C_2: \quad \mathbf{r}_2(u) = \cos u\,\mathbf{j} + \sin u\,\mathbf{k}$$

intersect at right angles at $P(0, 1, 0)$ and $Q(0, -1, 0)$.

SOLUTION Since $\mathbf{r}_1(\pi/2) = \mathbf{j} = \mathbf{r}_2(0)$, the curves meet at the tip of $\mathbf{j}$, which is $P(0, 1, 0)$. Also, since $\mathbf{r}_1(3\pi/2) = -\mathbf{j} = \mathbf{r}_2(\pi)$, the curves meet at the tip of $-\mathbf{j}$, which is $Q(0, -1, 0)$. Differentiation gives

$$\mathbf{r}_1'(t) = -\sin t\,\mathbf{i} + \cos t\,\mathbf{j} \qquad \text{and} \qquad \mathbf{r}_2'(u) = -\sin u\,\mathbf{j} + \cos u\,\mathbf{k}.$$

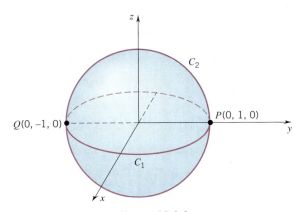

Figure 13.3.9

Since $\mathbf{r}_1'(\pi/2) = -\mathbf{i}$ and $\mathbf{r}_2'(0) = \mathbf{k}$, we have

$$\mathbf{r}_1'(\pi/2) \cdot \mathbf{r}_2'(0) = 0.$$

This tells us that the curves are perpendicular at $P(0, 1, 0)$. Since $\mathbf{r}_1'(3\pi/2) = \mathbf{i}$ and $\mathbf{r}_2'(\pi) = -\mathbf{k}$, we have

$$\mathbf{r}_1'(3\pi/2) \cdot \mathbf{r}_2'(\pi) = 0.$$

This tells us that the curves are perpendicular at $Q(0, -1, 0)$. The curves appear in Figure 13.3.9 on page 857. ❏

The Unit Tangent, the Principal Normal, the Osculating Plane

Suppose now that the curve

$$C: \quad \mathbf{r}(t) = x(t)\mathbf{i} + y(t)\mathbf{j} + z(t)\mathbf{k}$$

is twice differentiable and $\mathbf{r}'(t)$ is never zero. Then at each point $P(x(t), y(t), z(t))$ of the curve, there is a *unit tangent vector*:

(13.3.3)
$$\mathbf{T}(t) = \frac{\mathbf{r}'(t)}{\|\mathbf{r}'(t)\|}.$$

Since $\|\mathbf{r}'(t)\| > 0$, $\mathbf{T}(t)$ points in the same direction as $\mathbf{r}'(t)$, that is, in the direction of increasing t along the curve. Also, since $\|\mathbf{T}(t)\| = 1$, we have $\mathbf{T}(t) \cdot \mathbf{T}(t) = 1$ and differentiation gives

$$\mathbf{T}(t) \cdot \mathbf{T}'(t) + \mathbf{T}'(t) \cdot \mathbf{T}(t) = 0.$$

Since the dot product is commutative, we have

$$2[\mathbf{T}'(t) \cdot \mathbf{T}(t)] = 0 \qquad \text{and thus} \qquad \mathbf{T}'(t) \cdot \mathbf{T}(t) = 0.$$

At each point of the curve *the vector $\mathbf{T}'(t)$ is perpendicular to $\mathbf{T}(t)$*.

The vector $\mathbf{T}'(t)$ measures the rate of change of $\mathbf{T}(t)$ with respect to t. Since the norm of $\mathbf{T}(t)$ is constantly 1, $\mathbf{T}(t)$ can change only in direction. The vector $\mathbf{T}'(t)$ measures this change in direction.

If the unit tangent vector is not changing in direction (as in the case of a straight line), then $\mathbf{T}'(t) = \mathbf{0}$. If $\mathbf{T}'(t) \neq \mathbf{0}$, then we can form what is called the *principal normal vector*:

(13.3.4)
$$\mathbf{N}(t) = \frac{\mathbf{T}'(t)}{\|\mathbf{T}'(t)\|}.$$

This is the unit vector in the direction of $\mathbf{T}'(t)$. The *normal line* at P is the line through P parallel to the principal normal.

Figure 13.3.10 shows a curve on which we have marked several points. At each point we have drawn the unit tangent and the principal normal. The plane determined by these two vectors is called the *osculating plane* (literally, the "kissing plane"). This is the plane of greatest contact with the curve at the point in question.

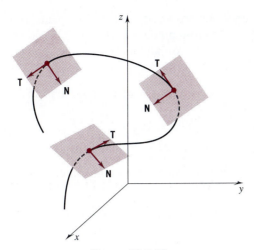

Figure 13.3.10

If a curve is a plane curve but not a straight line, then the osculating plane is simply the plane of the curve. A straight line does not have an osculating plane. There is no principal normal vector $[\mathbf{T}'(t) = \mathbf{0}]$ and there is no plane of greatest contact. Each straight line lies on an infinite number of planes.

Example 4 In Figure 13.3.11 you can see a curve spiraling up a circular cylinder. The curve is called a *circular helix* if the rate of climb is constant. The simplest parametrization for a circular helix takes the form

$$\mathbf{r}(t) = a\cos t\,\mathbf{i} + a\sin t\,\mathbf{j} + bt\,\mathbf{k} \qquad \text{with } a > 0, \quad b > 0.$$

The first two components produce the rotational effect; the third component gives the rate of climb.

We will find the unit tangent, the principal normal, and then an equation for the osculating plane. Since

$$\mathbf{r}'(t) = -a\sin t\,\mathbf{i} + a\cos t\,\mathbf{j} + b\,\mathbf{k},$$

we have

$$\|\mathbf{r}'(t)\| = \sqrt{a^2\sin^2 t + a^2\cos^2 t + b^2} = \sqrt{a^2 + b^2}$$

and therefore

$$\mathbf{T}(t) = \frac{\mathbf{r}'(t)}{\|\mathbf{r}'(t)\|} = \frac{1}{\sqrt{a^2 + b^2}}(-a\sin t\,\mathbf{i} + a\cos t\,\mathbf{j} + b\,\mathbf{k}).$$

This is the unit tangent vector.

The principal normal vector is the unit vector in the direction of $\mathbf{T}'(t)$. Since

$$\frac{d}{dt}(-a\sin t\,\mathbf{i} + a\cos t\,\mathbf{j} + b\,\mathbf{k}) = -a\cos t\,\mathbf{i} - a\sin t\,\mathbf{j} \qquad \text{and} \qquad a > 0,$$

you can see that

$$\mathbf{N}(t) = -\cos t\,\mathbf{i} - \sin t\,\mathbf{j}.$$

The principal normal is horizontal and points directly to the z-axis.

Circular Helix

Figure 13.3.11

Now let's find an equation for the osculating plane p at an arbitrary point $P(a \cos t, a \sin t, bt)$. The cross product $\mathbf{T}(t) \times \mathbf{N}(t)$ is perpendicular to p. Therefore, as a normal for p, we can take any nonzero scalar multiple of $\mathbf{T}(t) \times \mathbf{N}(t)$. In particular, we can take

$$(a \sin t\mathbf{i} - a \cos t\mathbf{j} - b\mathbf{k}) \times (\cos t\mathbf{i} + \sin t\mathbf{j}).$$

As you can check, this simplifies to

$$b \sin t\mathbf{i} - b \cos t\mathbf{j} + a\mathbf{k}.$$

The equation for the osculating plane at the point $P(a \cos t, a \sin t, bt)$ thus takes the form

$$b \sin t(x - a \cos t) - b \cos t(y - a \sin t) + a(z - bt) = 0. \qquad \text{(12.6.1)}$$

This simplifies to

$$(b \sin t)x - (b \cos t)y + az = abt.$$

To visualize how this osculating plane changes from point to point, think of a playground spiral slide or the threaded surface on a bolt. ❏

Reversing the Sense of a Curve

As noted at the beginning of this section, a parametrized curve is an *oriented* curve; not just a set of points, but rather a succession of points traversed in a certain order, a certain sense.

We make a distinction between the curve

$$\mathbf{r} = \mathbf{r}(t), \qquad t \in [a, b]$$

and the curve

$$\mathbf{R}(u) = \mathbf{r}(a + b - u), \qquad u \in [a, b].$$

Both vector functions trace out the same set of points (check that out), but the order has been reversed. Whereas the first curve starts at $\mathbf{r}(a)$ and ends at $\mathbf{r}(b)$, the second curve starts at $\mathbf{r}(b)$ and ends at $\mathbf{r}(a)$:

$$\mathbf{R}(a) = \mathbf{r}(a + b - a) = \mathbf{r}(b), \qquad \mathbf{R}(b) = \mathbf{r}(a + b - b) = \mathbf{r}(a).$$

Thus, for example,

$$\mathbf{r}(t) = \cos t\mathbf{i} + \sin t\mathbf{j}, \qquad t \in [0, 2\pi]$$

gives the unit circle traversed counterclockwise while

$$\mathbf{R}(u) = \cos (2\pi - u)\mathbf{i} + \sin (2\pi - u)\mathbf{j}, \qquad u \in [0, 2\pi]$$

gives the unit circle traversed clockwise.

What happens to the unit tangent $\mathbf{T}$, the principal normal $\mathbf{N}$, and the osculating plane when we reverse the sense of a curve? As you are asked to show in Exercise 43, $\mathbf{T}$ is replaced by $-\mathbf{T}$, but $\mathbf{N}$ remains the same. The osculating plane also remains the same.

Other Changes of Parameter

Not all changes of parameter change the succession of points. Suppose that

$$\mathbf{r} = \mathbf{r}(t), \qquad t \in I$$

is a differentiable curve, and let ϕ be a function that maps some interval J onto the interval I (domain J, range I). Now set

$$\mathbf{R}(u) = \mathbf{r}(\phi(u)) \qquad \text{for all } u \in J.$$

If $\phi'(u) > 0$ for all $u \in J$, then $\mathbf{r}$ and $\mathbf{R}$ are said to differ by a *sense-preserving change of parameter*. In this case, $\mathbf{r}$ and $\mathbf{R}$ take on exactly the same values in exactly the same order. In other words, they produce exactly the same oriented curve (Exercise 44).

If, on the other hand, $\phi'(u) < 0$ for all $u \in J$, then the change in parameter is said to be *sense-reversing*. In that case, $\mathbf{r}$ and $\mathbf{R}$ still take on exactly the same values but in opposite order. The paths are the same, but they are traversed in opposite directions. You have already seen one example of this:

$$\mathbf{R}(u) = \mathbf{r}(a + b - u).$$

Example 5 The parametrization

$$\mathbf{r}(t) = a \cos t\,\mathbf{i} + a \sin t\,\mathbf{j} + bt\,\mathbf{k}, \qquad t \in [0, 2\pi]$$

gives one "spiral" of the circular helix with the orientation indicated by the arrows (Figure 13.3.12). If we let $\varphi(u) = \pi u$, $u \in [0, 2]$, then φ maps the interval $J = [0, 2]$ onto the interval $I = [0, 2\pi]$, and $\varphi'(u) = \pi > 0$. Thus,

$$\mathbf{R}(u) = \mathbf{r}(\varphi(u)) = a \cos(\pi u)\mathbf{i} + a \sin(\pi u)\mathbf{j} + b\pi u\mathbf{k}, \qquad u \in [0, 2]$$

is precisely the same curve with the same orientation.

On the other hand, if we let $\psi(u) = (2 - u)\pi$, $u \in [0, 2]$, then ψ also maps the interval $J = [0, 2]$ onto the interval $I = [0, 2\pi]$, but $\psi'(u) = -\pi < 0$. Thus,

$$\mathbf{R}(u) = \mathbf{r}(\psi(u)) = a \cos(2 - u)\pi\mathbf{i} + a \sin(2 - u)\pi\mathbf{j} + b(2 - u)\pi\mathbf{k},$$

and this gives the same curve but with the opposite orientation. (See Figure 13.3.13.)

Figure 13.3.12

Figure 13.3.13

EXERCISES 13.3

In Exercises 1–8, find the tangent vector $\mathbf{r}'(t)$ and an equation for the tangent line at the indicated point.

1. $\mathbf{r}(t) = \cos \pi t \mathbf{i} + \sin \pi t \mathbf{j} + t \mathbf{k}$ at $t = 2$.

2. $\mathbf{r}(t) = e^t \mathbf{i} + e^{-t} \mathbf{j} - \ln t \mathbf{k}$ at $t = 1$.

3. $\mathbf{r}(t) = \mathbf{a} + t \mathbf{b} + t^2 \mathbf{c}$ at $t = -1$.

4. $\mathbf{r}(t) = (t + 1)\mathbf{i} + (t^2 + 1)\mathbf{j} + (t^3 + 1)\mathbf{k}$ at $P(1, 1, 1)$.

5. $\mathbf{r}(t) = 2t^2 \mathbf{i} + (1 - t)\mathbf{j} + (3 + 2t^2)\mathbf{k}$ at $P(2, 0, 5)$.

6. $\mathbf{r}(t) = 3t\mathbf{a} + \mathbf{b} - t^2 \mathbf{c}$ at $t = 2$.

7. $\mathbf{r}(t) = 2 \cos t \mathbf{i} + 3 \sin t \mathbf{j} + t \mathbf{k}; \ t = \pi/4$.

8. $\mathbf{r}(t) = t \sin t \mathbf{i} + t \cos t \mathbf{j} + 2t \mathbf{k}; \quad t = \pi/2$.

9. Show that $\mathbf{r}(t) = at\mathbf{i} + bt^2 \mathbf{j}$ parametrizes a parabola. Find an equation in x and y for this parabola.

10. Show that $\mathbf{r}(t) = \frac{1}{2}a(e^{\omega t} + e^{-\omega t})\mathbf{i} + \frac{1}{2}a(e^{\omega t} - e^{-\omega t})\mathbf{j}$ parametrizes the right branch $(x > 0)$ of the hyperbola $x^2 - y^2 = a^2$.

11. Find (a) the points on the curve $\mathbf{r}(t) = t\mathbf{i} + (1 + t^2)\mathbf{j}$ at which $\mathbf{r}(t)$ and $\mathbf{r}'(t)$ are perpendicular; (b) the points at which they have the same direction; (c) the points at which they have opposite directions.

12. Find the curve given that $\mathbf{r}'(t) = \alpha \mathbf{r}(t)$ for all real t and $\mathbf{r}(0) = \mathbf{i} + 2\mathbf{j} + 3\mathbf{k}$.

13. Suppose that $\mathbf{r}'(t)$ and $\mathbf{r}(t)$ are parallel for all t. Show that, if $\mathbf{r}'(t)$ is never $\mathbf{0}$, then the tangent line at each point passes through the origin.

▶ In Exercises 14–16, the given curves intersect at the indicated point. Find the angle of intersection. Express your answer in radians rounded to the nearest hundredth, and in degrees rounded to the nearest tenth.

14. $\mathbf{r}_1(t) = t\mathbf{i} + t^2 \mathbf{j} + t^3 \mathbf{k}$,
    $\mathbf{r}_2(u) = \sin 2u \mathbf{i} + u \cos u \mathbf{j} + u\mathbf{k}; \quad P(0, 0, 0)$.

15. $\mathbf{r}_1(t) = (e^t - 1)\mathbf{i} + 2 \sin t \mathbf{j} + \ln(t + 1)\mathbf{k}$,
    $\mathbf{r}_2(u) = (u + 1)\mathbf{i} + (u^2 - 1)\mathbf{j} + (u^3 + 1)\mathbf{k}; P(0, 0, 0)$.

16. $\mathbf{r}_1(t) = e^{-t}\mathbf{i} + \cos t \mathbf{j} + (t^2 + 4)\mathbf{k}$,
    $\mathbf{r}_2(u) = (2 + u)\mathbf{i} + u^4 \mathbf{j} + 4u^2 \mathbf{k}; \quad P(1, 1, 4)$.

17. Find the point at which the curves

$$\mathbf{r}_1(t) = e^t \mathbf{i} + 2 \sin(t + \tfrac{1}{2}\pi)\mathbf{j} + (t^2 - 2)\mathbf{k},$$

$$\mathbf{r}_2(u) = u\mathbf{i} + 2\mathbf{j} + (u^2 - 3)\mathbf{k}$$

intersect and find the angle of intersection.

18. Consider the vector function $\mathbf{f}(t) = t\mathbf{i} + f(t)\mathbf{j}$ formed from a differentiable real-valued function f. The vector function $\mathbf{f}$ parametrizes the graph of f.
    (a) Parametrize the tangent line at $P(t_0, f(t_0))$.

(b) Show that the parametrization obtained in (a) reduces to the usual equation for the tangent line:

$$y - f(t_0) = f'(t_0)(x - t_0) \quad \text{if } f'(t_0) \neq 0;$$

$$y = f(t_0) \quad \text{if } f'(t_0) = 0.$$

19. Define a vector function $\mathbf{r}$ on the interval $[0, 2\pi]$ that satisfies the initial condition $\mathbf{r}(0) = a\mathbf{i}$ and, as t increases to 2π, traces out the ellipse $b^2 x^2 + a^2 y^2 = a^2 b^2$.
    (a) Once in a counterclockwise manner.
    (b) Once in a clockwise manner.
    (c) Twice in a counterclockwise manner.
    (d) Three times in a clockwise manner.

20. Exercise 19 with $\mathbf{r}(0) = b\mathbf{j}$.

In Exercises 21–26, sketch the plane curve determined by the given vector-valued function $\mathbf{r}$ and indicate the orientation. Find $\mathbf{r}'(t)$ and draw the position vector and the tangent vector for the indicated value of t, placing the tangent vector at the tip of the position vector.

21. $\mathbf{r}(t) = \frac{1}{4}t^4 \mathbf{i} + t^2 \mathbf{j}; \quad t = 2$.

22. $\mathbf{r}(t) = 2t\mathbf{i} + (t^2 + 1)\mathbf{j}; \quad t = 4$.

23. $\mathbf{r}(t) = e^{2t}\mathbf{i} + e^{-4t}\mathbf{j}; \quad t = 0$.

24. $\mathbf{r}(t) = \sin t \mathbf{i} - 2 \cos t \mathbf{j}; \quad t = \pi/3$.

25. $\mathbf{r}(t) = 2 \cos t \mathbf{i} + 3 \sin t \mathbf{j}; \quad t = \pi/6$.

26. $\mathbf{r}(t) = \sec t \mathbf{i} + \tan t \mathbf{j}, \quad |t| < \pi/2; \quad t = \pi/4$.

In Exercises 27–30, find a vector parametrization for the curve.

27. $y^2 = x - 1, \quad y \geq 1$.

28. $r = 1 - \cos \theta, \quad \theta \in [0, 2\pi]$. (Polar coordinates.)

29. $r = \sin 3\theta, \quad \theta \in [0, \pi]$. (Polar coordinates.)

30. $y^4 = x^3, \quad y \leq 0$.

31. Find an equation in x and y for the curve $\mathbf{r}(t) = t^3 \mathbf{i} + t^2 \mathbf{j}$. Draw the curve. Does the curve have a tangent vector at the origin? If so, what is the unit tangent vector?

32. (a) Show that the curve

$$\mathbf{r}(t) = (t^2 - t + 1)\mathbf{i} + (t^3 - t + 2)\mathbf{j} + (\sin \pi t)\mathbf{k}$$

intersects itself at $P(1, 2, 0)$ by finding numbers $t_1 < t_2$ for which P is the tip of both $\mathbf{r}(t_1)$ and $\mathbf{r}(t_2)$.
    (b) Find the unit tangents at $P(1, 2, 0)$ first taking $t = t_1$, then taking $t = t_2$.

33. Find the point(s) at which the twisted cubic

$$\mathbf{r}(t) = t\mathbf{i} + t^2 \mathbf{j} + t^3 \mathbf{k}$$

intersects the plane $4x + 2y + z = 24$. What is the angle of intersection between the curve and the plane?

34. (a) Find the unit tangent and the principal normal at an arbitrary point of the ellipse

$$\mathbf{r}(t) = a \cos t\mathbf{i} + b \sin t\mathbf{j}.$$

(b) Write vector equations for the tangent line and the normal line at the tip of $\mathbf{r}(\frac{1}{4}\pi)$.

In Exercises 35–42, find the unit tangent vector, the principal normal vector, and an equation in x, y, z for the osculating plane at the point on the curve corresponding to the indicated value of t.

35. $\mathbf{r}(t) = \mathbf{i} + 2t\mathbf{j} + t^2\mathbf{k}$; $t = 1$.

36. $\mathbf{r}(t) = t\mathbf{i} + t^2\mathbf{j} + 2t^2\mathbf{k}$; $t = 1$.

37. $\mathbf{r}(t) = \cos 2t\mathbf{i} + \sin 2t\mathbf{j} + t\mathbf{k}$ at $t = \frac{1}{4}\pi$.

38. $\mathbf{r}(t) = t\mathbf{i} + 2t\mathbf{j} + t^2\mathbf{k}$ at $t = 2$.

39. $\mathbf{r}(t) = t\mathbf{i} + t^2\mathbf{j} + t^3\mathbf{k}$ at $t = 1$.

40. $\mathbf{r}(t) = \cos 3t\mathbf{i} + t\mathbf{j} - \sin 3t\mathbf{k}$ at $t = \frac{1}{3}\pi$.

41. $\mathbf{r}(t) = e^t\sin t\mathbf{i} + e^t\cos t\mathbf{j} + e^t\mathbf{k}$; $t = 0$.

42. $\mathbf{r}(t) = (\cos t + t \sin t)\mathbf{i} + (\sin t - t \cos t)\mathbf{j} + 2\mathbf{k}$; $t = \frac{1}{4}\pi$.

43. Let $\mathbf{r} = \mathbf{r}(t)$, $t \in [a, b]$ and set

$$\mathbf{R}(u) = \mathbf{r}(a + b - u), \qquad u \in [a, b].$$

Show that this change of parameter changes the sign of the unit tangent but does not alter the principal normal. HINT: Let P be the tip of $\mathbf{R}(u) = \mathbf{r}(a + b - u)$. At that point, $\mathbf{R}$ produces a unit tangent $\mathbf{T}_1(u)$ and a principal normal $\mathbf{N}_1(u)$. At that same point, $\mathbf{r}$ produces a unit tangent $\mathbf{T}(a + b - u)$ and a principal normal $\mathbf{N}(a + b - u)$.

44. Show that two vector functions that differ by a sense-preserving change of parameter take on exactly the same values in exactly the same order. That is, set

$$\mathbf{r} = \mathbf{r}(t), \qquad t \in I.$$

Assume that ϕ maps an interval J onto the interval I and that $\phi'(u) > 0$ for all $u \in J$. Set

$$\mathbf{R}(u) = \mathbf{r}(\phi(u)),$$

and show that $\mathbf{R}$ and $\mathbf{r}$ take on exactly the same values in exactly the same order.

45. Show that the unit tangent vector, the principal normal vector, and the osculating plane are left invariant (left unchanged) by every sense-preserving change of parameter.

In Exercises 46 and 47, let $\mathbf{r}$ be the vector-valued function defined by

$$\mathbf{r}(t) = 2 \cos t\mathbf{i} + 2 \sin t\mathbf{j} + 4t\mathbf{k} \quad \text{for } 0 \leq t \leq 2\pi.$$

The graph of $\mathbf{r}$ is one revolution on a circular helix, starting at the point $(2, 0, 0)$ and ending at the point $(2, 0, 8\pi)$.

46. Let $\varphi(u) = u^2$ for $0 \leq u \leq \sqrt{2\pi}$, and let

$$\mathbf{R}(u) = \mathbf{r}[\varphi(u)] = 2 \cos u^2\mathbf{i} + 2 \sin u^2\mathbf{j} + 4u^2\mathbf{k}.$$

(a) Show that φ determines a sense-preserving change of parameter on $[0, \sqrt{2\pi}]$.

(b) Show that the unit tangent and principal normal vectors for $\mathbf{r}$ at the point $t = \frac{1}{4}\pi$ are the same as the unit tangent and principal normal vectors for $\mathbf{R}$ at $u = \frac{1}{2}\sqrt{\pi}$.

47. Let $\psi(v) = (2\pi - v^2)$ for $0 \leq v \leq \sqrt{2\pi}$, and let

$$\mathbf{R}(v) = \mathbf{r}[\psi(v)] = 2 \cos(2\pi - v^2)\mathbf{i} + 2 \sin(2\pi - v^2)\mathbf{j} + 4(2\pi - v^2)\mathbf{k}.$$

(a) Show that ψ determines a sense-reversing change of parameter on $[0, \sqrt{2\pi}]$.

(b) Show that principal normal vector for $\mathbf{r}$ at $t = \pi/4$ is the same as the principal normal for $\mathbf{R}$ at $v = \frac{1}{2}\sqrt{7\pi}$, and show that the unit tangent vector for $\mathbf{r}$ at $\pi/4$ is the negative of the unit tangent vector for $\mathbf{R}$ at $v = \frac{1}{2}\sqrt{7\pi}$.

48. Let $\mathbf{r} = \mathbf{r}(t)$ be a differentiable vector-valued function. Prove that the tangent vector to the graph of $\mathbf{r}$ points in the direction of increasing t.

■ 13.4 ARC LENGTH

In Section 9.9, we treated arc length for plane curves in an intuitive manner. We showed that if a curve C is defined parametrically by the continuously differentiable functions

$$x = x(t), \qquad y = y(t), \qquad t \in [a, b],$$

then the length of C, $L(C)$, is given by the arc length formula

$$L(C) = \int_a^b \sqrt{[x'(t)]^2 + [y'(t)]^2} \, dt.$$

For a space curve C parametrized by the continuously differentiable functions

$$x = x(t), \qquad y = y(t), \qquad z = z(t), \qquad t \in [a, b],$$

the formula becomes

$$L(C) = \int_a^b \sqrt{[x'(t)]^2 + [y'(t)]^2 + [z'(t)]^2} \, dt.$$

In vector notation, both formulas can be written

$$L(C) = \int_a^b \|\mathbf{r}'(t)\| \, dt.$$

We will prove this result in this form, but first we give a precise definition of arc length.

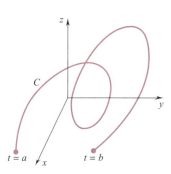

Figure 13.4.1

In Figure 13.4.1, we have sketched a space curve C which is parametrized by the continuously differentiable functions $x = x(t)$, $y = y(t)$, $z = z(t)$, $t \in [a, b]$. Before proceeding, we need to recall the underlying assumption about the nature of the parametrization of C that we made in Section 9.9. In particular, we assume throughout this discussion that the point $(x(t), y(t), z(t))$ traces out the curve C only once as t increases from a to b.

Now, to decide what should be meant by the length of C, we approximate C by the union of a finite number of line segments as we did in treating plane curves.

Choosing a finite number of points in $[a, b]$,

$$a = t_0 < t_1 < \cdots < t_{i-1} < t_i < \cdots < t_{n-1} < t_n = b,$$

we obtain a finite number of points $P_0, P_1, \ldots, P_{i-1}, P_i, \ldots, P_{n-1}, P_n$, on C, where for each k, $0 \le k \le n$, P_k denotes the point $P(x(t_k), y(t_k), z(t_k))$. Join these points consecutively by line segments and call the resulting path

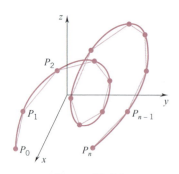

Figure 13.4.2

$$\gamma = \overline{P_0 P_1} \cup \cdots \cup \overline{P_{i-1} P_i} \cup \cdots \cup \overline{P_{n-1} P_n},$$

a *polygonal path* inscribed in C (Figure 13.4.2).

The length of such a polygonal path is the sum of the distances between consecutive vertices:

$$L(\gamma) = d(P_0, P_1) + \cdots + d(P_{i-1}, P_i) + \cdots + d(P_{n-1}, P_n).$$

The path γ serves as an approximation to the curve C, but obviously a better approximation can be obtained by adding more vertices to γ. We now ask ourselves exactly what should we require of the number that we shall call the length of C. Certainly we require that

$$L(\gamma) \le L(C) \qquad \text{for each } \gamma \text{ inscribed in } C.$$

But that is not enough. There is another requirement that seems reasonable. If we can choose γ to approximate C as closely as we wish, then we should be able to choose γ so that $L(\gamma)$ approximates the length of C as closely as we wish; namely, for each positive number ϵ there should exist a polygonal path γ such that

$$L(C) - \epsilon < L(\gamma) \le L(C).$$

In Section 10.1, we introduced the concept of least upper bound of a set of real numbers. Theorem 10.1.3 tells us that we can achieve the result we want by defining the length of the curve C to be the least upper bound of the set of all $L(\gamma)$. This is in fact what we do.

DEFINITION 13.4.1 ARC LENGTH

$$L(C) = \begin{cases} \text{the least upper bound of the set of all} \\ \text{lengths of polygonal paths inscribed in } C. \end{cases}$$

We are now ready to establish the arc length formula.

THEOREM 13.4.2 ARC LENGTH FORMULA

The length of a continuously differentiable curve

$$C: \quad \mathbf{r} = \mathbf{r}(t), \qquad t \in [a, b]$$

is given by the formula

$$L(C) = \int_a^b \|\mathbf{r}'(t)\| \, dt.$$

PROOF   First we show that

$$L(C) \le \int_a^b \|\mathbf{r}'(t)\| \, dt.$$

To do this, we begin with an arbitrary partition P of $[a, b]$:

$$P = \{a = t_0, \ldots, t_{i-1}, t_i, \ldots, t_n = b\}.$$

Such a partition gives rise to a finite number of points of C:

$$\mathbf{r}(a) = \mathbf{r}(t_0), \ldots, \mathbf{r}(t_{i-1}), \mathbf{r}(t_i), \ldots, \mathbf{r}(t_n) = \mathbf{r}(b)$$

and thus to an inscribed polygonal path of total length

$$L_P = \sum_{i=1}^{n} \|\mathbf{r}(t_i) - \mathbf{r}(t_{i-1})\|. \qquad \text{(Figure 13.4.3)}$$

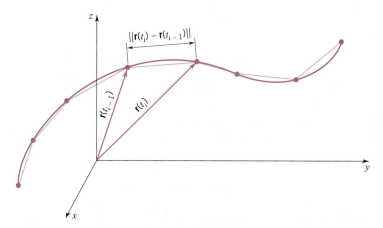

Figure 13.4.3

For each i,

$$\mathbf{r}(t_i) - \mathbf{r}(t_{i-1}) = \int_{t_{i-1}}^{t_i} \mathbf{r}'(t)\, dt.$$

This gives

by (13.1.10)

$$\|\mathbf{r}(t_i) - \mathbf{r}(t_{i-1})\| = \left\| \int_{t_{i-1}}^{t_i} \mathbf{r}'(t)\, dt \right\| \le \int_{t_{i-1}}^{t_i} \|\mathbf{r}'(t)\|\, dt$$

and thus

$$L_P = \sum_{i=1}^{n} \|\mathbf{r}(t_i) - \mathbf{r}(t_{i-1})\| \le \sum_{i=1}^{n} \int_{t_{i-1}}^{t_i} \|\mathbf{r}'(t)\|\, dt = \int_{a}^{b} \|\mathbf{r}'(t)\|\, dt.$$

From the arbitrariness of P, we know that the inequality

$$L_P \le \int_{a}^{b} \|\mathbf{r}'(t)\|\, dt$$

must hold for all the L_P. This makes the integral on the right an upper bound for all the L_P. Since $L(C)$ is the *least* upper bound of all the L_P, we can conclude right now that

$$L(C) \le \int_{a}^{b} \|\mathbf{r}'(t)\|\, dt.$$

To show equality, we need to know that arc length is additive: that is, with P, Q, R as in Figure 13.4.4, the arc length from P to Q, plus the arc length from Q to R, equals the arc length from P to R.† A proof of the additivity of arc length is given in the supplement at the end of this section. From this point on we will assume that arc length is additive.

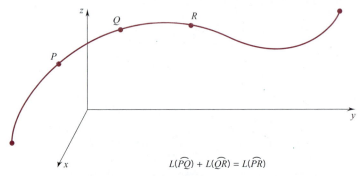

$$L(\widehat{PQ}) + L(\widehat{QR}) = L(\widehat{PR})$$

Figure 13.4.4

In Figure 13.4.5 we display the initial vector $\mathbf{r}(a)$, a general radius vector $\mathbf{r}(t)$, and a nearby vector $\mathbf{r}(t + h)$. Now set

$$s(t) = \text{length of the curve from } \mathbf{r}(a) \text{ to } \mathbf{r}(t),$$

$$s(t + h) = \text{length of the curve from } \mathbf{r}(a) \text{ to } \mathbf{r}(t + h).$$

---

† It is obvious that we want arc length to have this property. What we have to verify is that arc length *as we have defined it* has this property.

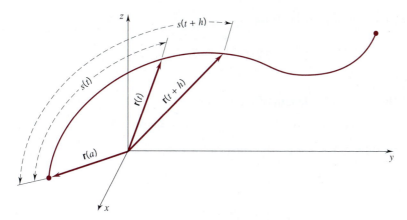

Figure 13.4.5

Then, by the additivity of arc length (remember we are assuming that),

$$s(t + h) - s(t) = \text{length of the curve from } \mathbf{r}(t) \text{ to } \mathbf{r}(t + h).$$

From what we have shown already, you can see that

$$\|\mathbf{r}(t + h) - \mathbf{r}(t)\| \le s(t + h) - s(t) \le \int_t^{t+h} \|\mathbf{r}'(u)\| \, du.$$

Dividing this inequality by h (which we are taking as positive), we get

$$\left\| \frac{\mathbf{r}(t + h) - \mathbf{r}(t)}{h} \right\| \le \frac{s(t + h) - s(t)}{h} \le \frac{1}{h} \int_t^{t+h} \|\mathbf{r}'(u)\| \, du.$$

As $h \to 0^+$, the left-hand side tends to $\|\mathbf{r}'(t)\|$ and, by the First Mean-Value Theorem for Integrals (Theorem 5.9.1), so does the right-hand side:

$$\frac{1}{h} \int_t^{t+h} \|\mathbf{r}'(u)\| \, du = \frac{1}{h} \|\mathbf{r}'(c_h)\|(t + h - t) = \|\mathbf{r}'(c_h)\| \to \|\mathbf{r}'(t)\|.$$

$$c_h \in (t, t+h) \underline{\qquad\qquad} \uparrow$$

Therefore

$$\lim_{h \to 0^+} \frac{s(t + h) - s(t)}{h} = \|\mathbf{r}'(t)\|.$$

By taking $h < 0$ and proceeding in a similar manner, one can show that

$$\lim_{h \to 0^-} \frac{s(t + h) - s(t)}{h} = \|\mathbf{r}'(t)\|.$$

It follows that

$$s'(t) = \lim_{h \to 0} \frac{s(t + h) - s(t)}{h} = \|\mathbf{r}'(t)\|.$$

Recall now that $s(t) = $ length of the curve from $\mathbf{r}(a)$ to $\mathbf{r}(t)$. Obviously, then, $s(a) = 0$. This, together with the fact that $s'(t) = \|\mathbf{r}'(t)\|$, implies that

$$s(t) = \text{length of the curve from } \mathbf{r}(a) \text{ to } \mathbf{r}(t) = \int_a^t \|\mathbf{r}'(u)\| \, du.$$

The total length of C is therefore

$$s(b) = \int_a^b \|\mathbf{r}'(t)\| \, dt. \quad \square$$

Example 1   Find the length of the curve

$$\mathbf{r}(t) = 2\cos t\mathbf{i} + 2\sin t\mathbf{j} + t^2\mathbf{k} \quad \text{from } t = 0 \text{ to } t = \pi/2$$

and compare it to the straight-line distance between the endpoints of the curve.   (See Figure 13.4.6.)

SOLUTION

$$\mathbf{r}'(t) = -2\sin t\mathbf{i} + 2\cos t\mathbf{j} + 2t\mathbf{k}.$$

$$\|\mathbf{r}'(t)\| = \sqrt{4\sin^2 t + 4\cos^2 t + 4t^2} = 2\sqrt{\sin^2 t + \cos^2 t + t^2} = 2\sqrt{1 + t^2}.$$

$$L(C) = \int_0^{\pi/2} \|\mathbf{r}'(t)\| \, dt = \int_0^{\pi/2} 2\sqrt{1 + t^2} \, dt = \left[t\sqrt{1 + t^2} + \ln\left(t + \sqrt{1 + t^2}\right) \right]_0^{\pi/2}$$

$$= \frac{\pi}{2}\sqrt{1 + \frac{\pi^2}{4}} + \ln\left[\frac{\pi}{2} + \sqrt{1 + \frac{\pi^2}{4}}\right] \cong 4.158.$$

$(0, 2, \frac{\pi^2}{4})$

$(2, 0, 0)$

Figure 13.4.6

The curve begins at $\mathbf{r}(0) = 2\mathbf{i}$ and ends at $\mathbf{r}(\pi/2) = 2\mathbf{j} + (\pi^2/4)\mathbf{k}$.   The distance between these two points is

$$\|\mathbf{r}(\pi/2) - \mathbf{r}(0)\| = \sqrt{2^2 + 2^2 + \frac{\pi^4}{16}} \cong 3.753.$$

The curve is about 11 percent longer than the straight-line distance between the endpoints of the curve.   □

EXERCISES 13.4

In Exercises 1–14, find the length of the given curve.

1. $\mathbf{r}(t) = t\mathbf{i} + \frac{2}{3}t^{3/2}\mathbf{j}$,   from $t = 0$ to $t = 8$.

2. $\mathbf{r}(t) = (\frac{1}{3}t^3 - t)\mathbf{i} + t^2\mathbf{j}$,   from $t = 0$ to $t = 2$.

3. $\mathbf{r}(t) = a\cos t\mathbf{i} + a\sin t\mathbf{j} + bt\mathbf{k}$ from $t = 0$ to $t = 2\pi$.

4. $\mathbf{r}(t) = t\mathbf{i} + \frac{2}{3}\sqrt{2}\, t^{3/2}\mathbf{j} + \frac{1}{2}t^2\mathbf{k}$ from $t = 0$ to $t = 2$.

5. $\mathbf{r}(t) = t\mathbf{i} + \ln(\sec t)\mathbf{j} + 3\mathbf{k}$ from $t = 0$ to $t = \frac{1}{4}\pi$.

6. $\mathbf{r}(t) = \tan^{-1}t\mathbf{i} + \frac{1}{2}\ln(1 + t^2)\mathbf{j}$ from $t = 0$ to $t = 1$.

7. $\mathbf{r}(t) = t^3\mathbf{i} + t^2\mathbf{j}$ from $t = 0$ to $t = 1$.

8. $\mathbf{r}(t) = t\mathbf{i} + \mathbf{j} + (\frac{1}{6}t^3 + \frac{1}{2}t^{-1})\mathbf{k}$ from $t = 1$ to $t = 3$.

9. $\mathbf{r}(t) = e^t[\cos t\mathbf{i} + \sin t\mathbf{j}]$ from $t = 0$ to $t = \pi$.

10. $\mathbf{r}(t) = 3t\cos t\mathbf{i} + 3t\sin t\mathbf{j} + 4t\mathbf{k}$ from $t = 0$ to $t = 4$.

11. $\mathbf{r}(t) = 2t\mathbf{i} + (t^2 - 2)\mathbf{j} + (1 - t^2)\mathbf{k}$ from $t = 0$ to $t = 2$.

12. $\mathbf{r}(t) = t^2\mathbf{i} + (t^2 - 2)\mathbf{j} + (1 - t^2)\mathbf{k}$ from $t = 0$ to $t = 2$.

13. $\mathbf{r}(t) = (\ln t)\mathbf{i} + 2t\mathbf{j} + t^2\mathbf{k}$,   from $t = 1$ to $t = e$.

14. $\mathbf{r}(t) = (t\sin t + \cos t)\mathbf{i} + (t\cos t - \sin t)\mathbf{j} + 2\mathbf{k}$,   from $t = 0$ to $t = 2$.

15. (*Important*) Let $\mathbf{r}(t) = x(t)\mathbf{i} + y(t)\mathbf{j} + z(t)\mathbf{k}$, $t \in [a, b]$ be a continuously differentiable curve.   Show that, if s is the length of the curve from the tip of $\mathbf{r}(a)$ to the tip of $\mathbf{r}(t)$, then

(13.4.3) $$\frac{ds}{dt} = \sqrt{\left(\frac{dx}{dt}\right)^2 + \left(\frac{dy}{dt}\right)^2 + \left(\frac{dz}{dt}\right)^2}.$$

16. Use vector methods to show that, if $y = f(x)$ has a continuous first derivative, then the length of the graph from $x = a$ to $x = b$ is given by the integral

$$\int_a^b \sqrt{1 + [f'(x)]^2} \, dx.$$

17. (*Important*) Let $y = f(x)$, $x \in [a, b]$ be a continuously differentiable function. Show that, if s is the length of the graph from $(a, f(a))$ to $(x, f(x))$, then

(13.4.4)
$$\frac{ds}{dx} = \sqrt{1 + \left(\frac{dy}{dx}\right)^2}.$$

18. Let C_1 be the curve

$$\mathbf{r}(t) = (t - \ln t)\mathbf{i} + (t + \ln t)\mathbf{j}, \qquad 1 \le t \le e$$

and let C_2 be the graph of

$$y = e^x, \qquad 0 \le x \le 1.$$

Find a relation between the length of C_1 and the length of C_2.

19. Show that the length of a continuously differentiable curve is left invariant (left unchanged) by a sense-preserving (or a sense-reversing) change of parameter. That is, set

$$\mathbf{r} = \mathbf{r}(t), \qquad t \in [a, b].$$

Assume that ϕ maps $[c, d]$ onto $[a, b]$ and that ϕ' is positive (or negative) and continuous on $[a, b]$. Set $\mathbf{R}(u) = \mathbf{r}(\phi(u))$ and show that the length of the curve as computed from $\mathbf{R}$ is the length of the curve as computed from $\mathbf{r}$.

20. Let $\mathbf{r}(t) = x(t)\mathbf{i} + y(t)\mathbf{j} + z(t)\mathbf{k}$ be a differentiable vector-valued function such that $\mathbf{r}'(t) \neq 0$ for all $t \ge 0$.
(a) Show that the arc length function s defined by

$$s(t) = \int_0^t \sqrt{\left(\frac{dx}{dt}\right)^2 + \left(\frac{dy}{dt}\right)^2 + \left(\frac{dz}{dt}\right)^2}\, dt, \quad t \ge 0$$

has an inverse, $t = \varphi(s)$.

(b) Let $\mathbf{R}(s) = \mathbf{r}[\varphi(s)]$. Show that $\|\mathbf{R}'(s)\| = 1$.

21. Consider the circular helix
$$\mathbf{r}(t) = 3\cos t\mathbf{i} + 3\sin t\mathbf{j} + 4t\mathbf{k} \text{ for } t \ge 0.$$
(a) Determine the arc length s as a function of t by evaluating the integral

$$s = \int_0^t \sqrt{\left(\frac{dx}{dt}\right)^2 + \left(\frac{dy}{dt}\right)^2 + \left(\frac{dz}{dt}\right)^2}\, dt.$$

(b) Use the relation found in (a) to express t as a function of s, $t = \varphi(s)$, and set

$$\mathbf{R}(s) = \mathbf{r}[\varphi(s)] = 3\cos\varphi(s)\mathbf{i} + 3\sin\varphi(s)\mathbf{j} + 4\varphi(s)\mathbf{k}.$$

(c) Find the coordinates of the point Q on the helix such that the arc length from $P(3, 0, 0)$ to Q is 5π.
(d) Show that $\|\mathbf{R}'(s)\| = 1$.

22. Repeat Exercise 21 (a), (b), (d) for the vector-valued function

$$\mathbf{r}(t) = (\sin t - t\cos t)\mathbf{i} + (\cos t + t\sin t)\mathbf{j} + \tfrac{1}{2}t^2\mathbf{k}$$

for $t \ge 0$.

▶ In Exercises 23–26, use a computer or calculator and Simpson's rule to estimate the length of the given curve. Obtain answers which are accurate to four decimal places.

23. $\mathbf{r}(t) = \frac{2}{3}t^{5/2}\mathbf{j} + t\mathbf{k}$ from $t = 0$ to $t = \frac{1}{2}$.
24. $\mathbf{r}(t) = t\mathbf{i} + \frac{1}{3}t^3\mathbf{j}$ from $t = 0$ to $t = 2$.
25. $\mathbf{r}(t) = 3\cos t\mathbf{i} + 4\sin t\mathbf{j} + 2\mathbf{k}$ from $t = 0$ to $t = 2\pi$.
26. $\mathbf{r}(t) = t\mathbf{i} + t^2\mathbf{j} + (\ln t)\mathbf{k}$ from $t = 1$ to $t = 4$.

---

*SUPPLEMENT TO SECTION 13.4

The Additivity of Arc Length

We wish to show that with P, Q, R as in Figure 13.4.3,

$$L(\widehat{PQ}) + L(\widehat{QR}) = L(\widehat{PR}).$$

Let γ_1 be an arbitrary polygonal path inscribed in $\widehat{PQ}$ and γ_2 an arbitrary polygonal path inscribed in $\widehat{QR}$. Then $\gamma_1 \cup \gamma_2$ is a polygonal path inscribed in $\widehat{PR}$. Since

$$L(\gamma_1) + L(\gamma_2) = L(\gamma_1 \cup \gamma_2) \qquad \text{and} \qquad L(\gamma_1 \cup \gamma_2) \le L(\widehat{PR}),$$

we have

$$L(\gamma_1) + L(\gamma_2) \le L(\widehat{PR}) \qquad \text{and thus} \qquad L(\gamma_1) \le L(\widehat{PR}) - L(\gamma_2).$$

From the arbitrariness of γ_1 we can conclude that $L(\widehat{PR}) - L(\gamma_2)$ is an upper bound for the set of all lengths of polygonal paths inscribed in $\widehat{PQ}$. Since $L(\widehat{PQ})$ is the *least* upper bound of this set, we have

$$L(\widehat{PQ}) \le L(\widehat{PR}) - L(\gamma_2).$$

It follows that

$$L(\gamma_2) \le L(\widehat{PR}) - L(\widehat{PQ}).$$

Arguing as we did with γ_1, we can conclude that

$$L(\widehat{QR}) \le L(\widehat{PR}) - L(\widehat{PQ}).$$

This gives

$$L(\widehat{PQ}) + L(\widehat{QR}) \le L(\widehat{PR}).$$

We now set out to prove that $L(\widehat{PR}) \le L(\widehat{PQ}) + L(\widehat{QR})$. To do this, we need only take $\gamma = \overline{T_0 T_1} \cup \cdots \cup \overline{T_{n-1} T_n}$ as an arbitrary polygonal path inscribed in $\widehat{PR}$ and show that

$$L(\gamma) \le L(\widehat{PQ}) + L(\widehat{QR}).$$

If Q is one of the T_i, say $Q = T_k$, then

$$\gamma_1 = \overline{T_0 T_1} \cup \cdots \cup \overline{T_{k-1} T_k} \quad \text{is inscribed in } \widehat{PQ}$$

and

$$\gamma_2 = \overline{T_k T_{k+1}} \cup \cdots \cup \overline{T_{n-1} T_n} \quad \text{is inscribed in } \widehat{QR}.$$

Moreover, $L(\gamma) = L(\gamma_1) + L(\gamma_2)$, so that

$$L(\gamma) \le L(\widehat{PQ}) + L(\widehat{QR}).$$

If Q is none of the T_i, then Q lies between two consecutive points T_k and T_{k+1}. Set

$$\gamma' = \overline{T_0 T_1} \cup \cdots \cup \overline{T_k Q} \cup \overline{Q T_{k+1}} \cup \cdots \cup \overline{T_{n-1} T_n}.$$

Since

$$d(T_k, T_{k+1}) \le d(T_k, Q) + d(Q, T_{k+1}),$$

we have

$$L(\gamma) \le L(\gamma').$$

Proceed as before and you will see that

$$L(\gamma') \le L(\widehat{PQ}) + L(\widehat{QR}),$$

and once again

$$L(\gamma) \le L(\widehat{PQ}) + L(\widehat{QR}). \quad \square$$

■ 13.5 CURVILINEAR MOTION; VECTOR CALCULUS IN MECHANICS

Introduction

The tools we have just developed find their premier application in Newtonian mechanics, the study of bodies in motion subject to Newton's laws. The heart of Newton's mechanics is his Second Law of Motion:

$$\text{force} = \text{mass} \times \text{acceleration}.$$

We have worked with Newton's second law, but only in a very restricted context: motion along a coordinate line under the influence of a force directed along that same line. In that special setting Newton's law was written as a scalar equation: $F = ma$. In general, objects do not move along straight lines (they move along curved paths) and the forces on them vary in direction. What happens to Newton's second law then? It becomes the vector equation

$$\mathbf{F} = m\mathbf{a}.$$

This is Newton's Second Law of Motion in its full glory. To understand this law and be able to use it effectively, you must learn to look at curvilinear motion from a vector viewpoint.

Curvilinear Motion from a Vector Viewpoint

We can describe the position of a moving object at time t by a radius vector $\mathbf{r}(t)$. As t ranges over a time interval I, the object traces out some path

$$C: \quad \mathbf{r}(t) = x(t)\mathbf{i} + y(t)\mathbf{j} + z(t)\mathbf{k}, \quad t \in I.$$

If $\mathbf{r}$ is twice differentiable, we can form $\mathbf{r}'(t)$ and $\mathbf{r}''(t)$. In this context these vectors have special names and special significance: $\mathbf{r}'(t)$ is called the *velocity* of the object at time t, and $\mathbf{r}''(t)$ is called the *acceleration*. In symbols, we have

(13.5.1)
$$\mathbf{r}'(t) = \mathbf{v}(t) \quad \text{and} \quad \mathbf{r}''(t) = \mathbf{v}'(t) = \mathbf{a}(t).$$

There should be nothing surprising about this. As before, velocity is the time rate of change of position and acceleration the time rate of change of velocity.

Since $\mathbf{v}(t) = \mathbf{r}'(t)$, the velocity vector, when not $\mathbf{0}$, is tangent to the path of the motion at the tip of $\mathbf{r}(t)$. (Section 13.3.) The direction of the velocity vector at time t thus gives the direction of the motion at time t. (Figure 13.5.1.)

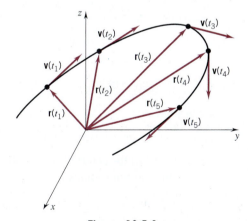

Figure 13.5.1

The magnitude of the velocity vector is called the *speed* of the object:

(13.5.2)
$$\|\mathbf{v}(t)\| = \text{the speed at time } t.$$

The reasoning is as follows: during a time interval $[t_0, t]$ the object moves along its path from $\mathbf{r}(t_0)$ to $\mathbf{r}(t)$ for a total distance

$$s(t) = \int_{t_0}^{t} \|\mathbf{r}'(u)\| \, du. \qquad \text{(Section 13.4.)}$$

Differentiating with respect to t, we have

$$s'(t) = \|\mathbf{r}'(t)\|.$$

The magnitude of the velocity vector is thus *the rate of change of arc distance with respect to time.* This is why we call it the speed of the object.

Speed is commonly denoted by the letter v. In symbols

(13.5.3)

$$s'(t) = \|\mathbf{v}(t)\| = v(t).$$

Motion Along a Straight Line

The position at time t is given by a function of the form

$$\mathbf{r}(t) = \mathbf{r}_0 + f(t)\mathbf{d}, \qquad \mathbf{d} \neq 0.$$

For convenience we take $\mathbf{d}$ as a unit vector.

The velocity and acceleration vectors are both directed along the line of the motion:

$$\mathbf{v}(t) = f'(t)\mathbf{d} \qquad \text{and} \qquad \mathbf{a}(t) = f''(t)\mathbf{d}.$$

The speed is $|f'(t)|$:

$$v(t) = \|f'(t)\mathbf{d}\| = |f'(t)| \, \|\mathbf{d}\| = |f'(t)|,$$

and the magnitude of the acceleration is $|f''(t)|$:

$$\|\mathbf{a}(t)\| = \|f''(t)\mathbf{d}\| = |f''(t)| \, \|\mathbf{d}\| = |f''(t)|.$$

Circular Motion About the Origin

The position function can be written

$$\mathbf{r}(t) = r[\cos \theta(t)\mathbf{i} + \sin \theta(t)\mathbf{j}].$$

Here $\theta'(t)$ gives the time rate of change of the central angle θ. If $\theta'(t) > 0$, the motion is counterclockwise; if $\theta'(t) < 0$, the motion is clockwise. We call $\theta'(t)$ the *angular velocity* and $|\theta'(t)|$ the *angular speed*.

Uniform circular motion is circular motion with constant angular speed ω ($\omega > 0$). The position function for uniform circular motion in the counterclockwise direction can be written

$$\mathbf{r}(t) = r(\cos \omega t \, \mathbf{i} + \sin \omega t \, \mathbf{j}).$$

Differentiation gives

$$\mathbf{v}(t) = r\omega(-\sin \omega t \, \mathbf{i} + \cos \omega t \, \mathbf{j}),$$

$$\mathbf{a}(t) = -r\omega^2(\cos \omega t \, \mathbf{i} + \sin \omega t \, \mathbf{j}) = -\omega^2 \mathbf{r}(t).$$

The acceleration is directed along the line of the radius vector toward the center of the circle and is therefore perpendicular to the velocity vector, which, as always, is tangential. As you can verify, the speed is $r\omega$ and the magnitude of acceleration is $r\omega^2$. ❑

Motion along a circular helix is a combination of circular motion and motion along a straight line.

Example 1 A particle moves along a circular helix (see Figure 13.3.11) with position at time t given by the function

$$\mathbf{r}(t) = a \cos \omega t\,\mathbf{i} + a \sin \omega t\,\mathbf{j} + b\omega t\,\mathbf{k}. \qquad (a > 0, b > 0, \omega > 0)$$

For each time t, find

(a) the velocity of the particle;   **(b)** the speed;   **(c)** the acceleration;
(d) the magnitude of acceleration;
(e) the angle between the velocity vector and the acceleration vector.

SOLUTION

(a) Velocity: $\mathbf{v}(t) = \mathbf{r}'(t) = -a\omega \sin \omega t\,\mathbf{i} + a\omega \cos \omega t\,\mathbf{j} + b\omega\,\mathbf{k}.$

(b) Speed $v(t) = \|\mathbf{v}(t)\| = \sqrt{a^2\omega^2\sin^2\omega t + a^2\omega^2\cos^2\omega t + b^2\omega^2}$
$$= \sqrt{a^2\omega^2 + b^2\omega^2} = \omega\sqrt{a^2 + b^2}.$$

(The speed is thus constant.)

(c) Acceleration: $\mathbf{a}(t) = \mathbf{v}'(t) = -a\omega^2\cos \omega t\,\mathbf{i} - a\omega^2\sin \omega t\,\mathbf{j}$
$$= -a\omega^2(\cos \omega t\,\mathbf{i} + \sin \omega t\,\mathbf{j}).$$

(Since the speed is constant, the acceleration comes entirely from the change in direction.)

(d) Magnitude of acceleration: $\|\mathbf{a}(t)\| = a\omega^2.$

(e) Angle between $\mathbf{v}(t)$ and $\mathbf{a}(t)$:

$$\cos \theta = \frac{\mathbf{v}(t)}{\|\mathbf{v}(t)\|} \cdot \frac{\mathbf{a}(t)}{\|\mathbf{a}(t)\|}$$

$$= \left[\frac{-a\omega \sin \omega t\,\mathbf{i} + a\omega \cos \omega t\,\mathbf{j} + b\omega\,\mathbf{k}}{\omega\sqrt{a^2 + b^2}}\right] \cdot \left[\frac{-a\omega^2(\cos \omega t\,\mathbf{i} + \sin \omega t\,\mathbf{j})}{a\omega^2}\right]$$

$$= \frac{a(\sin \omega t \cos \omega t - \cos \omega t \sin \omega t)}{\sqrt{a^2 + b^2}} = 0;$$

Therefore $\theta = \frac{1}{2}\pi.$ (At each point the acceleration vector is perpendicular to the velocity vector.) ❑

An Introduction to Vector Mechanics

We are now ready to work with Newton's Second Law of Motion in its vector form: $\mathbf{F} = m\mathbf{a}.$ Since at each time t we have $\mathbf{a}(t) = \mathbf{r}''(t)$, Newton's law can be written

(13.5.4)

$$\mathbf{F}(t) = m\mathbf{a}(t) = m\mathbf{r}''(t).$$

This is a second-order differential equation in t. In Chapter 18 we give an introduction to the general theory of differential equations. Second-order differential equations of the type that we encounter here are treated in Sections 18.5 and 18.6. Our approach in this section is intuitive; we will search for solutions of equation (13.5.4) in particular situations.

When objects are moving, certain quantities (positions, velocities, and so on) are continually changing. This can make a situation difficult to grasp. In these circumstances it is particularly satisfying to find quantities that do not change. Such quantities are said to be *conserved*. (These conserved quantities are called the *constants of the motion*.) Mathematically we can determine whether or not a quantity is conserved by looking at its derivative with respect to time (the time derivative): *The quantity is conserved (is constant) iff its time derivative remains zero*. A *conservation law* is the assertion that in such and such a context a certain quantity does not change.

Momentum

We start with the idea of momentum. The *momentum* $\mathbf{p}$ of an object is the mass of the object times the velocity of the object:

$$\mathbf{p} = m\mathbf{v}.$$

To indicate the time dependence we write

(13.5.5)
$$\mathbf{p}(t) = m\mathbf{v}(t) = m\mathbf{r}'(t).$$

Assume that the mass of the object is constant. Then differentiation gives

$$\mathbf{p}'(t) = m\mathbf{r}''(t) = \mathbf{F}(t).$$

Thus, *the time derivative of the momentum of an object is the net force on the object.* If the net force on an object is continually zero, the momentum $\mathbf{p}(t)$ is constant. This is the law of *conservation of momentum*:

(13.5.6)
> If the net force on an object is continually zero, then the momentum of the object is conserved.

Angular Momentum

The angular momentum of an object about any given point is a vector quantity that is intended to measure the extent to which the object is circling about that point. If the position of the object at time t is given by the radius vector $\mathbf{r}(t)$, then the object's *angular momentum about the origin* is defined by the formula

(13.5.7)
$$\mathbf{L}(t) = \mathbf{r}(t) \times \mathbf{p}(t) = \mathbf{r}(t) \times m\mathbf{v}(t).$$

At each time t of a motion, $\mathbf{L}(t)$ is perpendicular to $\mathbf{r}(t)$, perpendicular to $\mathbf{v}(t)$, and oriented so that $\mathbf{r}(t)$, $\mathbf{v}(t)$, $\mathbf{L}(t)$ form a right-handed triple. The magnitude of $\mathbf{L}(t)$ is given by the relation

$$\|\mathbf{L}(t)\| = \|\mathbf{r}(t)\|\|m\mathbf{v}(t)\| \sin \theta(t)$$

where $\theta(t)$ is the angle between $\mathbf{r}(t)$ and $\mathbf{v}(t)$. (All this, of course, comes simply from the definition of cross product.)

If $\mathbf{r}(t)$ and $\mathbf{v}(t)$ are not zero, then we can express $\mathbf{v}(t)$ as a vector parallel to $\mathbf{r}(t)$ plus a vector perpendicular to $\mathbf{r}(t)$:

$$\mathbf{v}(t) = \mathbf{v}_{\|}(t) + \mathbf{v}_{\perp}(t). \qquad \text{(See Figure 13.5.2.)}$$

Thus

$$
\begin{aligned}
\mathbf{L}(t) &= \mathbf{r}(t) \times m\mathbf{v}(t) \\
&= \mathbf{r}(t) \times m[\mathbf{v}_{\|}(t) + \mathbf{v}_{\perp}(t)] \\
&= \underbrace{[\mathbf{r}(t) \times m\mathbf{v}_{\|}(t)]}_{\mathbf{0}} + [\mathbf{r}(t) \times m\mathbf{v}_{\perp}(t)] \\
&= \mathbf{r}(t) \times m\mathbf{v}_{\perp}(t).
\end{aligned}
$$

Figure 13.5.2

The component of velocity that is parallel to the radius vector contributes nothing to angular momentum. *The angular momentum comes entirely from the component of velocity that is perpendicular to the radius vector.*

Example 2 In uniform circular motion about the origin,

$$\mathbf{r}(t) = r(\cos \omega t\,\mathbf{i} + \sin \omega t\,\mathbf{j}),$$

the velocity vector $\mathbf{v}(t)$ is always perpendicular to the radius vector $\mathbf{r}(t)$. In this case all of $\mathbf{v}(t)$ contributes to the angular momentum.

We can calculate $\mathbf{L}(t)$ as follows:

$$
\begin{aligned}
\mathbf{L}(t) &= \mathbf{r}(t) \times m\mathbf{v}(t) \\
&= [r(\cos \omega t\,\mathbf{i} + \sin \omega t\,\mathbf{j})] \times [mr(-\omega \sin \omega t\,\mathbf{i} + \omega \cos \omega t\,\mathbf{j})] \\
&= mr^2\omega(\cos^2 \omega t + \sin^2 \omega t)\mathbf{k} = mr^2\omega\mathbf{k}.
\end{aligned}
$$

The angular momentum is constant and is perpendicular to the xy-plane. If the motion is counterclockwise (if $\omega > 0$), then the angular momentum points up from the xy-plane. If the motion is clockwise (if $\omega < 0$), then the angular momentum points down from the xy-plane. (This is the right-handedness of the cross product coming in.) ❏

Example 3 In uniform straight-line motion with constant velocity $\mathbf{d}$,

$$\mathbf{r}(t) = \mathbf{r}_0 + t\mathbf{d},$$

we have

$$\mathbf{L}(t) = \mathbf{r}(t) \times m\mathbf{v}(t) = (\mathbf{r}_0 + t\mathbf{d}) \times m\mathbf{d} = m(\mathbf{r}_0 \times \mathbf{d}).$$

Figure 13.5.3

Here again the angular momentum is constant, but all is not quite so simple as it looks. In Figure 13.5.3 you can see the radius vector $\mathbf{r}(t)$, the velocity vector $\mathbf{v}(t) = \mathbf{d}$, and the part of the velocity vector that gives rise to angular momentum, the part perpendicular to $\mathbf{r}(t)$. As before we have called this $\mathbf{v}_{\perp}(t)$. While $\mathbf{v}(t)$ is constant,

$\mathbf{v}_\perp(t)$ is not constant. What happens here is that $\mathbf{r}(t)$ and $\mathbf{v}_\perp(t)$ vary in such a way that the cross product

$$\mathbf{L}(t) = \mathbf{r}(t) \times m\mathbf{v}_\perp(t)$$

remains constant. If, of course, the line of motion passes through the origin, then $\mathbf{v}(t)$ is parallel to $\mathbf{r}(t)$, $\mathbf{v}_\perp(t)$ is zero, and the angular momentum is zero. ❏

Torque

How the angular momentum of an object changes in time depends on the force acting on the object and on the position of the object relative to the origin: since $\mathbf{L}(t) = \mathbf{r}(t) \times m\mathbf{r}'(t)$,

$$\mathbf{L}'(t) = [\mathbf{r}(t) \times m\mathbf{r}''(t)] + \underbrace{[\mathbf{r}'(t) \times m\mathbf{r}'(t)]}_{\mathbf{0}} = \mathbf{r}(t) \times \mathbf{F}(t).$$

The cross product

(13.5.8)
$$\boxed{\boldsymbol{\tau}(t) = \mathbf{r}(t) \times \mathbf{F}(t)}$$

is called the *torque* about the origin.†

Since $\mathbf{L}'(t) = \boldsymbol{\tau}(t)$, we have the following conservation law:

(13.5.9)
> If the net torque on an object is continually zero,
> then the angular momentum of the object is conserved.

A force $\mathbf{F}(t)$ is called a *central force* (or a *radial force*) iff $\mathbf{F}(t)$ is always parallel to $\mathbf{r}(t)$. (Gravitational force, for example, is a central force.) For a central force, the cross product $\mathbf{r}(t) \times \mathbf{F}(t)$ is always zero. Thus a central force produces no torque about the origin. As you will see, this places severe restrictions on the kind of motion possible under a central force.

THEOREM 13.5.10

If an object moves under a central force and the angular momentum $\mathbf{L}$ is different from zero, then:

1. The object is confined to the plane that passes through the origin and is perpendicular to $\mathbf{L}$.
2. The radius vector of the object sweeps out equal areas in equal times.

(This theorem plays an important role in astronomy and is embodied in Kepler's three celebrated laws of planetary motion. We will study Kepler's laws in Section *13.6.)

---

† The symbol τ is the Greek letter tau. The word *torque* comes from the Latin word *torquere,* "to twist."

PROOF OF THEOREM 13.5.10 The first assertion is easy to verify: all the radius vectors pass through the origin and (by the very definition of angular momentum) they are all perpendicular to the constant vector **L**.

To verify the second assertion we introduce a right-handed coordinate system O-xyz setting the xy-plane as the plane of the motion, the positive z-axis pointing along **L**. On the xy-plane we introduce polar coordinates r, θ. Thus, at time t, the object has some position $[r(t), \theta(t)]$.

Let's denote by $A(t)$ the area swept out by the radius vector from some fixed time t_0 up to time t. Our task is to show that $A'(t)$ is constant.

The area swept out during the time interval $[t, t + h]$ is simply

$$A(t + h) - A(t).$$

Assuming that the motion takes place in the direction of increasing polar angle θ (see Figure 13.5.4), we have, with obvious notation,

$$\underbrace{\tfrac{1}{2} \min \, [r(t)]^2 \cdot [\theta(t + h) - \theta(t)]}_{\text{area of inner sector}} \leq A(t + h) - A(t) \leq \underbrace{\tfrac{1}{2} \max \, [r(t)]^2 \cdot [\theta(t + h) - \theta(t)]}_{\text{area of outer sector}}.$$

Divide through by h, take the limit as h tends to 0, and you will see that

(∗)
$$A'(t) = \tfrac{1}{2}[r(t)]^2 \theta'(t).$$

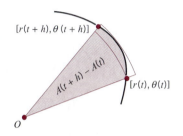

Figure 13.5.4

Now

$$\mathbf{r}(t) = r(t)[\cos \theta(t)\mathbf{i} + \sin \theta(t)\mathbf{j}],$$

$$\mathbf{v}(t) = r'(t)[\cos \theta(t)\mathbf{i} + \sin \theta(t)\mathbf{j}] + r(t)\theta'(t)[-\sin \theta(t)\mathbf{i} + \cos \theta(t)\mathbf{j}]$$
$$= [r'(t)\cos \theta(t) - r(t)\theta'(t)\sin \theta(t)]\mathbf{i} + [r'(t)\sin \theta(t) + r(t)\theta'(t)\cos \theta(t)]\mathbf{j}.$$

A calculation that you can carry out yourself shows that

$$\mathbf{L} = \mathbf{r}(t) \times m\mathbf{v}(t) = mr^2(t)\theta'(t)\mathbf{k}.$$

Since **L** is constant, $r^2(t)\theta'(t)$ is constant. Thus, by (∗), $A'(t)$ is constant:

$$A'(t) = L/2m \qquad \text{where} \quad L = \|\mathbf{L}\|. \quad \square$$

Initial-Value Problems

In physics one tries to make predictions about the future on the basis of current information and a knowledge of the forces at work. In the case of an object in motion, the task can be to determine $\mathbf{r}(t)$ for all t given the force and some "initial conditions." Frequently the initial conditions give the position and velocity of the object at some time t_0. The problem then is to solve the differential equation

$$\mathbf{F} = m\mathbf{r}''$$

subject to conditions of the form

$$\mathbf{r}(t_0) = \mathbf{r}_0, \qquad \mathbf{v}(t_0) = \mathbf{v}_0.$$

Such problems are known as initial-value problems. We have considered initial-value problems in several different contexts earlier in the text. For example, see Sections 5.6 and 7.6.

By far the simplest problem of this sort concerns a *free particle,* an object on which there is no net force.

Example 4   At time t_0 a free particle has position $\mathbf{r}(t_0) = \mathbf{r}_0$ and velocity $\mathbf{v}(t_0) = \mathbf{v}_0$. Find $\mathbf{r}(t)$ for all t.

SOLUTION   Since there is no net force on the object, the acceleration is zero and the velocity is constant.   Since $\mathbf{v}(t_0) = \mathbf{v}_0$,

$$\mathbf{v}(t) = \mathbf{v}_0 \qquad \text{for all } t.$$

Integration with respect to t gives

$$\mathbf{r}(t) = t\mathbf{v}_0 + \mathbf{c}$$

where $\mathbf{c}$, the constant of integration, is a vector that we can determine from the initial position.   The initial position $\mathbf{r}(t_0) = \mathbf{r}_0$ gives

$$\mathbf{r}_0 = t_0\mathbf{v}_0 + \mathbf{c} \qquad \text{and therefore} \qquad \mathbf{c} = \mathbf{r}_0 - t_0\mathbf{v}_0.$$

Using this value for $\mathbf{c}$ in our equation for $\mathbf{r}(t)$, we have

$$\mathbf{r}(t) = t\mathbf{v}_0 + (\mathbf{r}_0 - t_0\mathbf{v}_0),$$

which we write as

$$\mathbf{r}(t) = \mathbf{r}_0 + (t - t_0)\mathbf{v}_0.$$

This is the equation of a straight line with direction vector $\mathbf{v}_0$.   Free particles travel in straight lines with constant velocity.   (We have tacitly assumed that $\mathbf{v}_0 \neq \mathbf{0}$.   If $\mathbf{v}_0 = \mathbf{0}$, the particle remains at rest at $\mathbf{r}_0$.)   ❑

Example 5   An object of mass m is subject to a force of the form

$$\mathbf{F}(t) = -m\omega^2\mathbf{r}(t) \qquad \text{with } \omega > 0.$$

Find $\mathbf{r}(t)$ for all t given that

$$\mathbf{r}(0) = a\mathbf{i} \qquad \text{and} \qquad \mathbf{v}(0) = \omega a\mathbf{j} \qquad \text{with } a > 0.$$

SOLUTION   The force is a vector version of the restoring force exerted by a linear spring (Hooke's law).   Since the force is central, the angular momentum of the object, $\mathbf{L}(t) = \mathbf{r}(t) \times m\mathbf{v}(t)$, is conserved.   So $\mathbf{L}(t)$ is constantly equal to the value it had at time $t = 0$: for all t

$$\mathbf{L}(t) = \mathbf{L}(0) = \mathbf{r}(0) \times m\mathbf{v}(0) = a\mathbf{i} \times m\omega a\mathbf{j} = ma^2\omega\mathbf{k}.$$

From our earlier discussion (Theorem 13.5.10) we can conclude that the motion takes place in the plane that passes through the origin and is perpendicular to $\mathbf{k}$.   This is the xy-plane.   Thus we can write

$$\mathbf{r}(t) = x(t)\mathbf{i} + y(t)\mathbf{j}.$$

Since $\mathbf{F}(t) = m\mathbf{r}''(t)$, the force equation can be written $\mathbf{r}''(t) = -\omega^2\mathbf{r}(t)$.   In terms of components we have

$$x''(t) = -\omega^2 x(t), \qquad y''(t) = -\omega^2 y(t).$$

These are the equations of simple harmonic motion.   We have already seen that functions of the form

$$x(t) = A_1\sin(\omega t + \phi_1), \qquad y(t) = A_2\sin(\omega t + \phi_2)$$

are solutions of these equations (see Exercises 61 and 62, Section 3.6). In Section 18.5, we show that *all* solutions have this form. To evaluate the constants we use the initial conditions. That $\mathbf{r}(0) = a\mathbf{i}$ means that $x(0) = a$ and $y(0) = 0$. So

(*)
$$A_1 \sin \phi_1 = a, \qquad A_2 \sin \phi_2 = 0.$$

That $\mathbf{v}(0) = \omega a\mathbf{j}$ means that $x'(0) = 0$ and $y'(0) = \omega a$. Since

$$x'(t) = \omega A_1 \cos(\omega t + \phi_1) \qquad \text{and} \qquad y'(t) = \omega A_2 \cos(\omega t + \phi_2),$$

we have

(**)
$$\omega A_1 \cos \phi_1 = 0, \qquad \omega A_2 \cos \phi_2 = \omega a.$$

Conditions (*) and (**) are met by setting $A_1 = a$, $A_2 = a$, $\phi_1 = \frac{1}{2}\pi$, $\phi_2 = 0$. Thus

$$x(t) = a \sin(\omega t + \tfrac{1}{2}\pi) = a \cos \omega t, \qquad y(t) = a \sin \omega t.$$

The vector equation reads

$$\mathbf{r}(t) = a \cos \omega t\,\mathbf{i} + a \sin \omega t\,\mathbf{j}.$$

The object moves in a circle of radius a about the origin with constant angular velocity ω. ❑

Example 6 A particle of charge q in a magnetic field $\mathbf{B}$ is subject to the force

$$\mathbf{F}(t) = \frac{q}{c}\,[\mathbf{v}(t) \times \mathbf{B}(t)]$$

where c is the speed of light and $\mathbf{v}$ is the velocity of the particle. Given that $\mathbf{r}(0) = \mathbf{r}_0$ and $\mathbf{v}(0) = \mathbf{v}_0$, find the path of the particle in the constant magnetic field $\mathbf{B}(t) = B_0\mathbf{k}$, $B_0 \neq 0$.

SOLUTION There is no conservation law that we can conveniently appeal to here. Neither momentum nor angular momentum are conserved: the force is not zero and it is not central. We start directly with Newton's $\mathbf{F} = m\mathbf{r}''$.

Since $\mathbf{r}'' = \mathbf{v}'$, we have

$$m\mathbf{v}'(t) = \frac{q}{c}\,[\mathbf{v}(t) \times B_0\mathbf{k}],$$

which we can write as

$$\mathbf{v}'(t) = \frac{qB_0}{mc}\,[\mathbf{v}(t) \times \mathbf{k}].$$

To simplify notation we set $qB_0/mc = \omega$. We then have

$$\mathbf{v}'(t) = \omega[\mathbf{v}(t) \times \mathbf{k}].$$

Placing $\mathbf{v}(t) = v_1(t)\mathbf{i} + v_2(t)\mathbf{j} + v_3(t)\mathbf{k}$ in this last equation and working out the cross product, we find that

$$v_1'(t)\mathbf{i} + v_2'(t)\mathbf{j} + v_3'(t)\mathbf{k} = \omega[v_2(t)\mathbf{i} - v_1(t)\mathbf{j}].$$

This gives the scalar equations

$$v_1'(t) = \omega v_2(t), \qquad v_2'(t) = -\omega v_1(t), \qquad v_3'(t) = 0.$$

The last equation is trivial. It says that v_3 is constant:

$$v_3(t) = C.$$

The equations for v_1 and v_2 are linked together. We can get an equation that involves only v_1 by differentiating the first equation:

$$v_1''(t) = \omega v_2'(t) = -\omega^2 v_1(t).$$

As we know from our earlier work, this gives

$$v_1(t) = A_1 \sin(\omega t + \phi_1).$$

Since $v_1'(t) = \omega v_2(t)$, we have

$$v_2(t) = \frac{v_1'(t)}{\omega} = \frac{A_1 \omega}{\omega} \cos(\omega t + \phi_1) = A_1 \cos(\omega t + \phi_1).$$

Therefore

$$\mathbf{v}(t) = A_1 \sin(\omega t + \phi_1)\mathbf{i} + A_1 \cos(\omega t + \phi_1)\mathbf{j} + C\mathbf{k}.$$

A final integration with respect to t gives

$$\mathbf{r}(t) = \left[-\frac{A_1}{\omega} \cos(\omega t + \phi_1) + D_1 \right]\mathbf{i} + \left[\frac{A_1}{\omega} \sin(\omega t + \phi_1) + D_2 \right]\mathbf{j} + [Ct + D_3]\mathbf{k}$$

where D_1, D_2, D_3 are constants of integration. All six constants of integration—A_1, ϕ_1, C, D_1, D_2, D_3—can be evaluated from the initial conditions. We will not pursue this. What is important here is that the path of the particle is a circular helix with axis parallel to **B**, in this case parallel to **k**. You should be able to see this from the equation for $\mathbf{r}(t)$: the z-component of $\mathbf{r}$ varies linearly with t from the value D_3, while the x and y components represent uniform motion with angular velocity ω in a circle of radius $|A_1/\omega|$ around the center (D_1, D_2). ❑

(Physicists express the behavior just found by saying that charged particles *spiral around* the magnetic field lines. Qualitatively, this behavior still holds even if the magnetic field lines are "bent," as is the case with the earth's magnetic field. Many charged particles become trapped by the earth's magnetic field. They keep spiraling around the magnetic field lines that run from pole to pole.)

EXERCISES 13.5

1. A particle moves in a circle of radius r at constant speed v. Find the angular speed and the magnitude of the acceleration.

2. A particle moves so that

$$\mathbf{r}(t) = (a \cos \pi t + bt^2)\mathbf{i} + (a \sin \pi t - bt^2)\mathbf{j}.$$

Find the velocity, speed, acceleration, and the magnitude of the acceleration all at time $t = 1$.

3. A particle moves so that $\mathbf{r}(t) = at\mathbf{i} + b \sin at\mathbf{j}$. Show that the magnitude of the acceleration of the particle is proportional to the distance of the particle from the x-axis.

4. A particle moves so that $\mathbf{r}(t) = 2\mathbf{i} + t^2\mathbf{j} + (t - 1)^2\mathbf{k}$. At what time is the speed a minimum?

In Exercises 5–8, sketch the curve. Then compute and sketch the acceleration vector at the indicated points.

5. $\mathbf{r}(t) = (t/\pi)\mathbf{i} + \cos t\mathbf{j}$, $\quad t \in [0, 2\pi]$; at $t = \frac{1}{4}\pi, \frac{1}{2}\pi, \pi$.

6. $\mathbf{r}(t) = t^3\mathbf{i} + t\mathbf{j}$, $\quad t$ real; $\quad$ at $t = -\frac{1}{2}, \frac{1}{2}, 1$.

7. $\mathbf{r}(t) = \sec t\mathbf{i} + \tan t\mathbf{j}$, $\quad t \in [\frac{1}{4}\pi, \frac{1}{2}\pi)$; at $t = -\frac{1}{6}\pi, 0, \frac{1}{3}\pi$.

8. $\mathbf{r}(t) = \sin \pi t + t\mathbf{j}$, $\quad t \in [0, 2]$; $\quad$ at $t = \frac{1}{2}, 1, \frac{5}{4}$.

9. An object moves so that

$$\mathbf{r}(t) = x_0\mathbf{i} + [y_0 + (\alpha \cos \theta)t]\mathbf{j}$$
$$+ [z_0 + (\alpha \sin \theta)t - 16t^2]\mathbf{k}, \qquad t \geq 0.$$

Find (a) the initial position, (b) the initial velocity, (c) the initial speed, (d) the acceleration throughout the motion. Finally, (e) identify the curve.

10. A particle moves so that $\mathbf{r}(t) = 2 \cos 2t\mathbf{i} + 3 \cos t\mathbf{j}$.
(a) Show that the particle oscillates on an arc of the parabola $4y^2 - 9x = 18$. (b) Draw the path. (c) What are the acceleration vectors at the points of zero velocity?
(d) Draw these vectors at the points in question.

11. An object of mass m moves so that

$$\mathbf{r}(t) = \tfrac{1}{2}a(e^{\omega t} + e^{-\omega t})\mathbf{i} + \tfrac{1}{2}b(e^{\omega t} - e^{-\omega t})\mathbf{j}.$$

(a) What is the velocity at $t = 0$? (b) Show that the acceleration vector is a constant positive multiple of the radius vector. (This shows that the force is central and repelling.) (c) What does (b) imply about the angular momentum and the torque? Verify your answers by direct calculation.

12. (a) An object moves so that

$$\mathbf{r}(t) = a_1 e^{bt}\mathbf{i} + a_2 e^{bt}\mathbf{j} + a_3 e^{bt}\mathbf{k}.$$

Show that, if $b > 0$, the object experiences a repelling central force.
(b) An object moves so that

$$\mathbf{r}(t) = \sin t\mathbf{i} + \cos t\mathbf{j} + (\sin t + \cos t)\mathbf{k}.$$

Show that the object experiences an attracting central force.
(c) Compute the angular momentum $\mathbf{L}(t)$ for the motion in (b).

13. A constant force of magnitude α directed upward from the xy-plane is continually applied to an object of mass m. Given that the object starts at time 0 at the point $P(0, y_0, z_0)$ with initial velocity $2\mathbf{j}$, find: (a) the velocity of the object t seconds later; (b) the speed of the object t seconds later; (c) the momentum of the object t seconds later; (d) the path followed by the object, both in vector form and in Cartesian coordinates.

14. Show that, if the force on an object is always perpendicular to the velocity of the object, then the *speed* of the object is constant. (This tells us that the speed of a charged particle in a magnetic field is constant.)

15. Find the force required to propel a particle of mass m so that $\mathbf{r}(t) = t\mathbf{j} + t^2\mathbf{k}$.

16. Show that for an object of constant velocity the angular momentum is constant.

17. At each point $P(x(t), y(t), z(t))$ of its motion an object of mass m is subject to a force

$$\mathbf{F}(t) = m\pi^2[a \cos \pi t\mathbf{i} + b \sin \pi t\mathbf{j}]. \quad (a > 0, b > 0)$$

Given that $\mathbf{v}(0) = -\pi b\mathbf{j} + \mathbf{k}$ and $\mathbf{r}(0) = b\mathbf{j}$, find the following at time $t = 1$:
(a) The velocity.   (b) The speed.
(c) The acceleration.   (d) The momentum.
(e) The angular momentum.   (f) The torque.

18. If an object of mass m moves with velocity $\mathbf{v}(t)$ subject to a force $\mathbf{F}(t)$, the scalar product

(13.5.11)
$$\mathbf{F}(t) \cdot \mathbf{v}(t)$$

is called the *power* (expended by the force) and the number

(13.5.12)
$$\tfrac{1}{2}m[v(t)]^2$$

is called the *kinetic energy* of the object. Show that the time rate of change of the kinetic energy of an object is the power expended on it:

(13.5.13)
$$\frac{d}{dt}(\tfrac{1}{2}m[v(t)]^2) = \mathbf{F}(t) \cdot \mathbf{v}(t).$$

19. Two particles of equal mass m, one with constant velocity $\mathbf{v}$ and the other at rest, collide elastically (i.e., the kinetic energy of the system is preserved) and go off in different directions. Show that the two particles go off at right angles.

20. (*Elliptic harmonic motion*) Show that, if the force on a particle of mass m is of the form

$$\mathbf{F}(t) = -m\omega^2\mathbf{r}(t),$$

then the path of the particle may be written

$$\mathbf{r}(t) = \cos \omega t\mathbf{A} + \sin \omega t\mathbf{B}$$

with $\mathbf{A}$ and $\mathbf{B}$ constant vectors. Give the physical significance of $\mathbf{A}$ and $\mathbf{B}$ and specify conditions on $\mathbf{A}$ and $\mathbf{B}$ that restrict the particle to a circular path.
HINT: The solutions of the differential equation

$$x''(t) = -\omega^2 x(t)$$

can be written in the form

$$x(t) = A \cos \omega t + B \sin \omega t. \quad \text{(Exercise 61, Section 3.6)}$$

21. A particle moves with constant acceleration $\mathbf{a}$. Show that the path of the particle lies entirely in some plane. Find a vector equation for this plane.

22. In Example 6 we stated that the path

$$\mathbf{r}(t) = \left[-\frac{A_1}{\omega} \cos(\omega t + \phi_1) + D_1 \right] \mathbf{i}$$
$$+ \left[\frac{A_1}{\omega} \sin(\omega t + \phi_1) + D_2 \right] \mathbf{j} + [Ct + D_3]\mathbf{k}$$

was a circular helix. Set $\omega = -1$ and show that, if $\mathbf{r}(0) = a\mathbf{i}$ and $\mathbf{v}(0) = a\mathbf{j} + b\mathbf{k}$, then the path takes the form

$$\mathbf{r}(t) = a \cos t\mathbf{i} + a \sin t\mathbf{j} + bt\mathbf{k},$$

the circular helix described in Section 13.3.

23. A charged particle in a constant electric field $\mathbf{E}$ experiences the force $q\mathbf{E}$, where q is the charge of the particle. Assume that the field has the constant value $\mathbf{E} = E_0\mathbf{k}$ and find the path of the particle given that $\mathbf{r}(0) = \mathbf{i}$ and $\mathbf{v}(0) = \mathbf{j}$.

24. (*Important*) A wheel is rotating about an axle with angular speed ω. Let $\boldsymbol{\omega}$ be the *angular velocity vector*, the vector of length ω that points along the axis of the wheel in such a direction that, observed from the tip of $\boldsymbol{\omega}$, the wheel rotates counterclockwise. Take the origin as the center of the wheel and let $\mathbf{r}$ be the vector from the origin to a point P on the rim of the wheel. Express the velocity $\mathbf{v}$ of P in terms of $\boldsymbol{\omega}$ and $\mathbf{r}$.

25. Solve the initial-value problem

$$\mathbf{F}(t) = m\mathbf{r}''(t) = t\mathbf{i} + t^2\mathbf{j}, \qquad \mathbf{r}_0 = \mathbf{r}(0) = \mathbf{i},$$
$$\mathbf{v}_0 = \mathbf{v}(0) = \mathbf{k}.$$

26. Solve the initial-value problem

$$\mathbf{F}(t) = m\mathbf{r}''(t) = -m\beta^2 z(t)\mathbf{k}, \qquad \mathbf{r}_0 = \mathbf{r}(0) = \mathbf{k},$$
$$\mathbf{v}_0 = \mathbf{v}(0) = \mathbf{0}.$$

[Here $z(t)$ is the z-component of $\mathbf{r}(t)$.]

27. An object of mass m moves subject to the force

$$\mathbf{F}(\mathbf{r}) = 4r^2\mathbf{r}$$

where $\mathbf{r}$ is the position of the object. Suppose $\mathbf{r}(0) = \mathbf{0}$ and $\mathbf{v}(0) = 2\mathbf{u}$, where $\mathbf{u}$ is a unit vector. Show that at each time t the speed v of the object satisfies the relation

$$v = \sqrt{4 + \frac{2}{m}r^4}.$$

HINT: Examine the quantity $\frac{1}{2}mv^2 - r^4$. (This is the *energy* of the object, a notion we will take up in Chapter 17.)

28. A moving particle traces a path that at each point has a unit tangent vector $\mathbf{T}(t)$ and a principal normal $\mathbf{N}(t)$. In anticipation of Section 13.7, (a) write the velocity vector as a scalar multiple of $\mathbf{T}(t)$, and then (b) write the acceleration vector as a linear combination of $\mathbf{T}(t)$ and $\mathbf{N}(t)$. (Incidentally, this will show that the acceleration vector lies in the osculating plane.)

---

*■ **13.6 Planetary Motion**

Tycho Brahe, Johannes Kepler

In the middle of the sixteenth century the arguments on planetary motion persisted. Was Copernicus right? Did the planets move in circles about the sun? Obviously not. Was not the earth the center of the universe?

In 1576, with the generous support of his king, the Danish astronomer Tycho Brahe built an elaborate astronomical observatory on the isle of Hveen and began his painstaking observations. For more than twenty years he looked through his telescopes and recorded what he saw. He was a meticulous observer, but he could draw no definite conclusions.

In 1599 the German astronomer–mathematician Johannes Kepler began his study of Brahe's voluminous tables. For a year and a half Brahe and Kepler worked together. Then Brahe died and Kepler went on wrestling with the data. His persistence paid off. By 1619 Kepler had made three stupendous discoveries, known today as *Kepler's Laws of Planetary Motion:*

> **I.** Each planet moves in a plane, not in a circle, but in an elliptic orbit with the sun at one focus.
>
> **II.** The radius vector from sun to planet sweeps out equal areas in equal times.
>
> **III.** The square of the period of the motion varies directly as the cube of the major semiaxis, and the constant of proportionality is the same for all the planets.

What Kepler formulated, Newton was able to explain. Each of these laws, Newton showed, was deducible from his laws of motion and his law of gravitation.

Newton's Second Law of Motion for Extended Three-Dimensional Objects

Imagine an object that consists of n particles with masses $m_1, m_2, \ldots, m_n$ located at $\mathbf{r}_1, \mathbf{r}_2, \ldots, \mathbf{r}_n$.† The total mass M of the object is the sum of the masses of the constituent particles:

$$M = m_1 + \cdots + m_n.$$

The center of mass of the object is by definition the point $\mathbf{R}_M$ where

$$M\,\mathbf{R}_M = m_1\mathbf{r}_1 + \cdots + m_n\mathbf{r}_n.$$

The total force $\mathbf{F}_{\text{TOT}}$ on the object is by definition the sum of all the forces that act on the particles that constitute the object:

$$\mathbf{F}_{\text{TOT}} = \mathbf{F}_1 + \cdots + \mathbf{F}_n.$$

Since $\mathbf{F}_1 = m_1\mathbf{r}_1'', \ldots, \mathbf{F}_n = m_n\mathbf{r}_n''$, we have

$$\mathbf{F}_{\text{TOT}} = m_1\mathbf{r}_1'' + \cdots + m_n\mathbf{r}_n'',$$

which we can write as

$$\mathbf{F}_{\text{TOT}} = M\,\mathbf{R}_M''.$$

The total force on an extended object is thus the total mass of the object times the acceleration of the center of mass.

We can simplify this still further. The forces that act between the constituent particles, the so-called "internal forces," cancel in pairs: if particle 23 tugs at particle 71 in a certain direction with a certain strength, then particle 71 tugs at particle 23 in the opposite direction with the same strength. (Newton's third law: to every action there is an equal reaction.) Therefore, in calculating the total force on our object, we can disregard the internal forces and simply add up the external forces. $\mathbf{F}_{\text{TOT}} = \mathbf{F}_{\text{TOT}}^{(\text{Ext})}$ and Newton's second law takes the form

(13.6.1)

$$\mathbf{F}_{\text{TOT}}^{(\text{Ext})} = M\,\mathbf{R}_M''.$$

---

† The case of a continuously distributed mass is taken up in Chapter 16.

The total external force on an extended three-dimensional object is thus the total mass of the object times the acceleration of the center of mass.

When an external force is applied to an extended object, we cannot predict the reaction of all the constituent particles, but we can predict the reaction of the center of mass. The center of mass will react to the force as if it were a particle with all the mass concentrated there. Suppose, for example, that a bomb is dropped from an airplane. The center of mass, "feeling" only the force of gravity (we are neglecting air resistance), falls in a parabolic arc toward the ground even if the bomb explodes at a thousand meters and individual pieces fly every which way. The forces of explosion are internal and do not affect the center of mass.

Some Preliminary Comments About the Planets

Roughly speaking, a planet is a massive object in the form of a ball with the center of mass at the center. In what follows, when we refer to the position of a planet, you are to understand that we really mean the position of the center of mass of the planet. Two other points require comment. First, we will write our equations as if the sun affected the motion of the planet, but not vice versa: we will assume that the sun stays put and that the planet moves. Really each affects the other. Our viewpoint is justified by the immense difference in mass between the planets and the sun. In the case of the earth and the sun, for example, a reasonable analogy is to imagine a tug of war in space between someone who weighs three pounds and someone who weighs a million pounds: to a good approximation, the million-pound person does not move. The second point is that the planet is affected not only by the pull from the sun, but also by the gravitational pulls from the other planets and all the other celestial bodies. But these forces are much smaller, and they tend to cancel. We will ignore them. (Our results are only approximations, but they prove to be very good approximations.)

A Derivation of Kepler's Laws from Newton's Laws of Motion and His Law of Gravitation

The gravitational force exerted by the sun on a planet can be written in vector form as follows:

$$(*) \qquad\qquad \mathbf{F}(\mathbf{r}) = -G\frac{mM}{r^3}\,\mathbf{r}.$$

Here m is the mass of the planet, M is the mass of the sun, G is the gravitational constant, $\mathbf{r}$ is the vector from the center of the sun to the center of the planet, and r is the magnitude of $\mathbf{r}$. (Thus we are placing the sun at the origin of our coordinate system; namely, we are using what is known as a *heliocentric* coordinate system, from *Hēlios,* the Greek word for sun.)

Let's make sure that equation $(*)$ conforms to our earlier characterization of gravitational force. First of all, because of the minus sign, the direction is toward the origin, where the sun is located. Taking norms we have

$$\|\mathbf{F}(\mathbf{r})\| = \frac{GmM}{r^3}\,\|\mathbf{r}\| = \frac{GmM}{r^2}.$$

$$\underset{\Big\uparrow}{} \quad \|\mathbf{r}\| = r$$

Thus the magnitude of the force is as expected: it does vary directly as the product of the masses and inversely as the square of the distance between them. So we are back in familiar territory.

We will derive Kepler's laws in a somewhat piecemeal manner. Since the force on the planet is a central force, $\mathbf{L}$, the angular momentum of the planet, is conserved. (Section 13.5.) If $\mathbf{L}$ were zero, we would have

$$\|\mathbf{L}\| = m\|\mathbf{r}\|\,\|\mathbf{v}\| \sin\theta = 0.$$

This would mean that either $\mathbf{r} = \mathbf{0}$, $\mathbf{v} = \mathbf{0}$, or $\sin\theta = 0$. The first equation would place the planet at the center of the sun. The second equation could hold only if the planet stopped. The third equation could hold only if the planet moved directly toward the sun or directly away from the sun. We can be thankful that none of those things happen.

Since the planet moves under a central force and $\mathbf{L}$ is not zero, *the planet stays on the plane that passes through the center of the sun and is perpendicular to* $\mathbf{L}$, and the radius vector does sweep out equal areas in equal times. (We know all this from Theorem 13.5.10.)

Now we go on to show that the path is an ellipse with the sun at one focus.† The equation of motion of a planet of mass m can be written

$$m\mathbf{a}(t) = -GmM\frac{\mathbf{r}(t)}{[r(t)]^3}.$$

Clearly m drops out of the equation. Setting $GM = \rho$ and suppressing the explicit dependence on t, we have

$$\mathbf{a} = -\rho\frac{\mathbf{r}}{r^3},$$

which gives

(**)
$$\frac{\mathbf{r}}{r^3} = -\frac{1}{\rho}\mathbf{a} = -\frac{1}{\rho}\frac{d\mathbf{v}}{dt}.$$

From (13.2.4) we know that in general

$$\frac{d}{dt}\left(\frac{\mathbf{r}}{r}\right) = (\mathbf{r}\times\mathbf{v})\times\frac{\mathbf{r}}{r^3}.$$

Since $\mathbf{L} = \mathbf{r}\times m\mathbf{v}$, we have

$$\frac{d}{dt}\left(\frac{\mathbf{r}}{r}\right) = \frac{\mathbf{L}}{m}\times\frac{\mathbf{r}}{r^3}.$$

Inserting (**) we see that

$$\frac{d}{dt}\left(\frac{\mathbf{r}}{r}\right) = \frac{\mathbf{L}}{m}\times\left(-\frac{1}{\rho}\frac{d\mathbf{v}}{dt}\right) = \frac{d\mathbf{v}}{dt}\times\frac{\mathbf{L}}{m\rho} = \frac{d}{dt}\left(\mathbf{v}\times\frac{\mathbf{L}}{m\rho}\right)$$

$$\underset{\longleftarrow}{\text{L, } m, \rho \text{ are all constant}}$$

and therefore

$$\frac{d}{dt}\left[\left(\mathbf{v}\times\frac{\mathbf{L}}{m\rho}\right) - \frac{\mathbf{r}}{r}\right] = \mathbf{0}.$$

---

† If some of the steps seem unanticipated to you, you should realize that we are discussing one of the most celebrated physical problems in history and there have been three hundred years to think of ingenious ways to deal with it. The argument we give here follows the lines of the excellent discussion that appears in Harry Pollard's *Celestial Mechanics*, Mathematical Association of America (1976).

Integration with respect to t gives

(***)
$$\left(\mathbf{v} \times \frac{\mathbf{L}}{m\rho} \right) - \frac{\mathbf{r}}{r} = \mathbf{e}$$

where $\mathbf{e}$ is a constant vector that depends on the initial conditions. Dotting both sides with $\mathbf{r}$, we have

$$\mathbf{r} \cdot \left(\mathbf{v} \times \frac{\mathbf{L}}{m\rho} \right) - \frac{\mathbf{r} \cdot \mathbf{r}}{r} = \mathbf{r} \cdot \mathbf{e}.$$

Since

$$\mathbf{r} \cdot \left(\mathbf{v} \times \frac{\mathbf{L}}{m\rho} \right) = \frac{\mathbf{L}}{m\rho} \cdot (\mathbf{r} \times \mathbf{v}) = \frac{\mathbf{L}}{m\rho} \cdot \frac{\mathbf{L}}{m} = \frac{L^2}{m^2\rho} \qquad \text{and} \qquad \frac{\mathbf{r} \cdot \mathbf{r}}{r} = \frac{r^2}{r} = r,$$

(12.4.6) ——↑

we find that

$$\frac{L^2}{m^2\rho} - r = \mathbf{r} \cdot \mathbf{e},$$

which we write as

$$r + (\mathbf{r} \cdot \mathbf{e}) = \frac{L^2}{m^2\rho}. \qquad \text{(Orbit equation.)}$$

If $\mathbf{e} = \mathbf{0}$, the orbit is a circle:

$$r = \frac{L^2}{m^2\rho}.$$

This is a possibility that requires very special initial conditions, conditions not met by any of the planets in our solar system. In our solar system, at least, $\mathbf{e} \neq \mathbf{0}$.

Given that $\mathbf{e} \neq \mathbf{0}$, we can write the orbit equation as

(13.6.2)
$$r(1 + e \cos \theta) = \frac{L^2}{m^2\rho}$$

where $e = \|\mathbf{e}\|$ and θ is the angle between $\mathbf{r}$ and $\mathbf{e}$. We are almost through. Since $\mathbf{v} \times \mathbf{L}$ and $\mathbf{r}$ are both perpendicular to $\mathbf{L}$, we know from (***) that $\mathbf{e}$ is perpendicular to $\mathbf{L}$. Therefore, in the plane of $\mathbf{e}$ and $\mathbf{r}$ (see Figure 13.6.1) we can interpret r and θ as the usual polar coordinates. The orbit equation (13.6.2) is then a polar equation and, as such, represents a conic section with focus at the origin, which is to say, *focus at the sun*. According to 9.4.1, the conic section can be a parabola, a hyperbola, or an ellipse. The repetitiveness of planetary motion rules out the parabola and the hyperbola. *The orbit is therefore an ellipse.*

Finally we will verify Kepler's third law: that the square of the period of the motion varies directly as the cube of the major semi-axis and the constant of proportionality is the same for all planets.

The elliptic orbit has an equation of the form

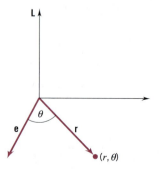

Figure 13.6.1

$$r(1 + e \cos \theta) = ed \qquad \text{with} \quad 0 < e < 1, \quad d > 0.$$

By Section 9.4 the major and minor semi-axes of the ellipse are given by

$$a = \frac{ed}{1 - e^2}, \qquad b = a\sqrt{1 - e^2}.$$

(See Theorem 9.4.2.) Denote the period of revolution by T. Since the radius vector sweeps out area at the constant rate of $L/2m$ (we know this from the proof of Theorem 13.5.10),

$$\left(\frac{L}{2m}\right) T = \text{area of the ellipse} = \pi ab = \pi a^2\sqrt{1 - e^2}.$$

Thus

$$T = \frac{2\pi m a^2\sqrt{1 - e^2}}{L} \qquad \text{and} \qquad T^2 = \frac{4\pi^2 m^2 a^4(1 - e^2)}{L^2}.$$

From (13.6.2) we know that

$$ed = \frac{L^2}{m^2\rho} = \frac{L^2}{m^2 GM} \qquad \text{and therefore} \qquad \frac{m^2}{L^2} = \frac{1}{edGM}.$$

It follows that

$$T^2 = \frac{4\pi^2 a^4(1 - e^2)}{edGM} = \frac{4\pi^2 a^4(1 - e^2)}{a(1 - e^2)GM} = \frac{4\pi^2}{GM}a^3.$$

$$\underleftarrow{\hspace{1cm}} ed = a(1 - e^2)$$

T^2 does vary directly with a^3, and the constant of proportionality $4\pi^2/GM$ is the same for all planets. ❏

EXERCISES *13.6

1. Kepler's third law can be stated as follows: for each planet the square of the period of revolution varies directly as the cube of the planet's average distance from the sun, and the constant of proportionality is the same for all planets. A *year* on a given planet is the time taken by the planet to make one circuit around the sun; thus a year on a planet is the period of revolution of that planet. Given that on average Venus is 0.72 times as far from the sun as the earth, how does the length of a "Venus year" compare with the length of an "earth year"?

2. Verify by differentiation with respect to time t that, if the acceleration of a planet is given by

$$\mathbf{a} = -\rho\frac{\mathbf{r}}{r^3},$$

then the energy $E = \frac{1}{2}mv^2 - m\rho r^{-1}$ is constant.

3. Given that a planet moves in a plane, its motion can be described by rectangular coordinates (x, y) or polar coordinates $[r, \theta]$, with origin at the sun. The kinetic energy of a planet is

$$\tfrac{1}{2}mv^2 = \tfrac{1}{2}m\left[\left(\frac{dx}{dt}\right)^2 + \left(\frac{dy}{dt}\right)^2\right].$$

Show that in polar coordinates

$$\tfrac{1}{2}mv^2 = \tfrac{1}{2}m\left[\left(\frac{dr}{dt}\right)^2 + r^2\left(\frac{d\theta}{dt}\right)^2\right].$$

4. We have seen that the energy of a planet

$$E = \tfrac{1}{2}mv^2 - \frac{m\rho}{r}$$

is constant. (Exercise 2.) Setting $dr/dt = \dot{r}$, $d\theta/dt = \dot{\theta}$ and using Exercise 3, we have

$$E = \tfrac{1}{2}m(\dot{r}^2 + r^2\,\dot{\theta}^2) - \frac{m\rho}{r}.$$

We also know that the angular momentum $\mathbf{L}$ is constant and that $L = mr^2\dot{\theta}$. Use this fact to verify that

$$E = \frac{L^2}{2m}\left\{\frac{1}{r^2} + \frac{1}{r^4}\left(\frac{dr}{d\theta}\right)^2\right\} - \frac{m\rho}{r}.$$

Since E is a constant, this is a differential equation for r as a function of θ.

5. Show that the function

$$r = \frac{a}{1 + e\cos\theta} \quad \text{with} \quad a = \frac{L^2}{m^2\rho} \quad \text{and} \quad e^2 = \frac{2Ea}{m\rho} + 1$$

satisfies the equation derived in Exercise 4.

■ 13.7 CURVATURE

The Curvature of a Plane Curve

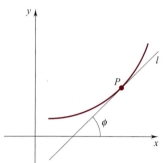

Figure 13.7.1 shows a plane curve which we assume to be twice differentiable. At the point P we have drawn the tangent line l. The angle that l makes with the x-axis has been labeled ϕ. As P moves along the curve, l changes and ϕ changes also. The magnitude k of the change in ϕ per unit of arc length is called the *curvature*:

(13.7.1)

$$k = \left| \frac{d\phi}{ds} \right|.$$

Figure 13.7.1

Calculating Curvature

If the curve is the graph of a twice differentiable function

$$y = y(x),$$

then the curvature can be calculated from the formula

(13.7.2)

$$k = \frac{|y''|}{[1 + (y')^2]^{3/2}}$$

where the primes indicate differentiation with respect to x.

DERIVATION OF FORMULA 13.7.2 We know that $\tan \phi = y'$. Therefore,

$$\phi = \tan^{-1}(y').$$

Differentiating with respect to x, we have

$$\frac{d\phi}{dx} = \frac{1}{1 + (y')^2} \cdot \frac{d}{dx}(y') = \frac{y''}{1 + (y')^2}.$$

Since

$$\frac{d\phi}{dx} = \frac{d\phi}{ds}\frac{ds}{dx} = \frac{d\phi}{ds}\sqrt{1 + (y')^2},$$

chain rule ——↑   ↑—— (13.4.4)

we have

$$\frac{d\phi}{ds}\sqrt{1 + (y')^2} = \frac{y''}{1 + (y')^2} \qquad \text{and therefore} \qquad \left| \frac{d\phi}{ds} \right| = \frac{|y''|}{[1 + (y')^2]^{3/2}}. \quad \square$$

If the curve is given parametrically by a twice differentiable vector function

$$\mathbf{r}(t) = x(t)\mathbf{i} + y(t)\mathbf{j},$$

then

(13.7.3)
$$k = \frac{|x'y'' - y'x''|}{[(x')^2 + (y')^2]^{3/2}}$$

where the primes now indicate differentiation with respect to t.

We will derive the formula under the assumption that $x' \neq 0$. Actually, the formula holds provided that $(x')^2 + (y')^2 \neq 0$.

DERIVATION OF FORMULA 13.7.3

$$\frac{dy}{dx} = \frac{dy/dt}{dx/dt} = \frac{y'}{x'}.$$

Therefore, as you can verify,

$$\frac{d^2y}{dx^2} = \frac{(dx/dt)(d^2y/dt^2) - (dy/dt)(d^2x/dt^2)}{(dx/dt)^3} = \frac{x'y'' - y'x''}{(x')^3}.$$

$(9.8.5)$ ⟶

Thus

$$k = \frac{|d^2y/dx^2|}{[1 + (dy/dx)^2]^{3/2}} = \left|\frac{x'y'' - y'x''}{(x')^3}\right| \frac{1}{[1 + (y'/x')^2]^{3/2}} = \frac{|x'y'' - y'x''|}{[(x')^2 + (y')^2]^{3/2}}. \quad ❏$$

⟶ $(13.7.2)$

Example 1 Since a straight line has a constant angle of inclination, we have

$$\frac{d\phi}{ds} = 0 \quad \text{and thus} \quad k = 0.$$

(13.7.4) Along a straight line the curvature is constantly zero. ❏

Example 2 For a circle of radius r,

$$\mathbf{r}(t) = r(\cos t\,\mathbf{i} + \sin t\,\mathbf{j}),$$

we have

$$x = r \cos t, \qquad y = r \sin t.$$

Differentiation with respect to t gives

$$x' = -r \sin t, \quad x'' = -r \cos t; \qquad y' = r \cos t, \quad y'' = -r \sin t.$$

Thus

$$k = \frac{|x'y'' - y'x''|}{[(x')^2 + (y')^2]^{3/2}} = \frac{|(-r \sin t)(-r \sin t) - (r \cos t)(-r \cos t)|}{[(-r \sin t)^2 + (r \cos t)^2]^{3/2}} = \frac{r^2}{r^3} = \frac{1}{r}.$$

(13.7.5) Along a circle of radius r the curvature is constantly $1/r$.

Hardly surprising. It is geometrically evident that along a circular path the change in direction takes place at a constant rate. Since a complete revolution entails a change of direction of 2π radians and this change is effected on a path of length $2\pi r$, the change in direction per unit of arc length is $2\pi/2\pi r = 1/r$. ❑

Figure 13.7.2

To say that a curve $y = f(x)$ has slope m at a point P is to say that at the point P the curve is rising or falling at the rate of a line of slope m. To say that a plane curve C has curvature $1/r$ at a point P is to say that at the point P the curve is turning at the rate of a circle of radius r. (Figure 13.7.2.) The smaller the circle, the tighter the turn and, thus, the greater the curvature.

The reciprocal of the curvature,

$$\rho = \frac{1}{k}, \qquad \text{(for } k \neq 0)$$

is called the *radius of curvature*. The point at a distance ρ from the curve in the direction of the principal normal is called the *center of curvature*.

Example 3 Take $a > 0$. Show that the point of maximal curvature on the catenary

$$y = \frac{a}{2}(e^{x/a} + e^{-x/a})$$

is the lowest point on the curve. (See Exercise 46, Section 9.9.) What is the radius of curvature at that point?

SOLUTION We will use the formula

$$k = \frac{|y''|}{[1 + (y')^2]^{3/2}}.$$

Here

$$y' = \frac{1}{2}(e^{x/a} - e^{-x/a}) \qquad \text{and} \qquad y'' = \frac{1}{2a}(e^{x/a} + e^{-x/a}) = \frac{y}{a^2} > 0.$$

Observe that

$$[1 + (y')^2]^{3/2} = [1 + \tfrac{1}{4}(e^{x/a} - e^{-x/a})^2]^{3/2},$$

which, as you can check, reduces to give

$$[1 + (y')^2]^{3/2} = \frac{y^3}{a^3}.$$

Thus

$$k = \left(\frac{y}{a^2}\right)\left(\frac{a^3}{y^3}\right) = \frac{a}{y^2}.$$

Since $y > 0$, the curvature is maximal where y is minimal; that is, at the lowest point of the curve.

The value of y at the lowest point of the curve is a. The curvature at that point is $1/a$; the radius of curvature ρ at that point is a. ❑

Example 4 Find the curvature at an arbitrary point (x, y) of the ellipse

$$\frac{x^2}{a^2} + \frac{y^2}{b^2} = 1. \qquad\qquad (a > b > 0)$$

Determine the points of maximal curvature and the points of minimal curvature. What is the radius of curvature at each of these points?

SOLUTION We parametrize the ellipse by setting

$$\mathbf{r}(t) = a \cos t\mathbf{i} + b \sin t\mathbf{j}$$

and use the fact that

$$k = \frac{|x'y'' - y'x''|}{[(x')^2 + (y')^2]^{3/2}}.$$

Here

$$x = a \cos t, \qquad y = b \sin t$$

and therefore

$$x' = -a \sin t, \quad x'' = -a \cos t; \qquad y' = b \cos t, \quad y'' = -b \sin t.$$

Thus

$$k = \frac{|(-a \sin t)(-b \sin t) - (b \cos t)(-a \cos t)|}{[(-a \sin t)^2 + (b \cos t)^2]^{3/2}} = \frac{ab}{[a^2 \sin^2 t + b^2 \cos^2 t]^{3/2}}.$$

As you can check, the curvature at the point (x, y) can be written

$$\frac{a^4b^4}{(b^4x^2 + a^4y^2)^{3/2}}.$$

To find the points of maximal and minimal curvature, we go back to the parameter t. Observe that

$$a^2\sin^2t + b^2\cos^2t = (a^2 - b^2)\sin^2t + b^2(\sin^2t + \cos^2t)$$
$$= (a^2 - b^2)\sin^2t + b^2.$$

Thus we have

$$k = \frac{ab}{[(a^2 - b^2)\sin^2t + b^2]^{3/2}}.$$

Since we have assumed that $a > b > 0$, the curvature is maximal when $\sin^2t = 0$; that is, when $t = 0$ and when $t = \pi$. Thus the points of maximal curvature are the points $P(\pm a, 0)$, the ends of the major axis. The curvature at these points is a/b^2 and the radius of curvature is b^2/a. The curvature is minimal when $\sin^2t = 1$; that is, when $t = \frac{1}{2}\pi$ and when $t = \frac{3}{2}\pi$. The points of minimal curvature are the points $P(0, \pm b)$, the ends of the minor axis. The curvature at these points is b/a^2 and the radius of curvature is a^2/b. ❑

In Figure 13.7.3 you can see a plane curve. At the point P we have affixed the unit tangent vector $\mathbf{T}$. As P moves along the curve, $\mathbf{T}$ changes, not in length, but in

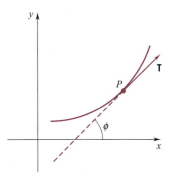

Figure 13.7.3

direction. The curvature of the curve is the magnitude of the change in **T** per unit of arc length:

(13.7.6)

$$k = \left\| \frac{d\mathbf{T}}{ds} \right\|.$$

PROOF Since **T** has length 1, we can write

$$\mathbf{T} = \cos \phi \mathbf{i} + \sin \phi \mathbf{j}.$$

Note that ϕ represents the angle between the tangent line and the x-axis. Differentiation with respect to s gives

$$\frac{d\mathbf{T}}{ds} = -\sin \phi \frac{d\phi}{ds} \mathbf{i} + \cos \phi \frac{d\phi}{ds} \mathbf{j} = \frac{d\phi}{ds}(-\sin \phi \mathbf{i} + \cos \phi \mathbf{j}).$$

Taking norms we have

$$\left\| \frac{d\mathbf{T}}{ds} \right\| = \left| \frac{d\phi}{ds} \right| \sqrt{\sin^2 \theta + \cos^2 \theta} = \left| \frac{d\phi}{ds} \right| = k$$

as asserted. ❏

This characterization of curvature generalizes to space curves.

The Curvature of a Space Curve

A space curve bends in two ways. It bends in the osculating plane (the plane of the unit tangent **T** and the principal normal **N**) and it bends away from that plane. The first form of bending is measured by the rate at which the unit tangent **T** changes direction. The second form of bending is measured by the rate at which the vector **T** × **N** changes direction. We will concentrate here on the first form of bending, the bending in the osculating plane. The measure of this is called *curvature*.

What was a theorem on curvature in the case of a plane curve becomes a definition of curvature in the case of a space curve; namely, in the case of a space curve, we *define* the curvature k by setting

$$k = \left\| \frac{d\mathbf{T}}{ds} \right\|.$$

If the space curve is given in terms of a parameter t, say

$$C: \quad \mathbf{r}(t) = x(t)\mathbf{i} + y(t)\mathbf{j} + z(t)\mathbf{k}, \quad t \in [a, b],$$

then the curvature can be calculated from the formula

(13.7.7)

$$k = \frac{\|d\mathbf{T}/dt\|}{ds/dt}.$$

We arrive at this by noting that

$$\frac{d\mathbf{T}}{ds}\frac{ds}{dt} = \frac{d\mathbf{T}}{dt},$$

dividing through by ds/dt and taking norms. The assumption here is that $ds/dt = \|\mathbf{r}'(t)\|$ remains nonzero.

Example 5 Calculate the curvature of the circular helix

$$\mathbf{r}(t) = r\sin t\,\mathbf{i} + r\cos t\,\mathbf{j} + t\,\mathbf{k}. \qquad\qquad (r > 0)$$

SOLUTION We will use the Leibniz notation.

$$\frac{d\mathbf{r}}{dt} = r\cos t\,\mathbf{i} - r\sin t\,\mathbf{j} + \mathbf{k}, \qquad \frac{ds}{dt} = \left\|\frac{d\mathbf{r}}{dt}\right\| = \sqrt{r^2 + 1},$$

$$\mathbf{T} = \frac{d\mathbf{r}/dt}{\|d\mathbf{r}/dt\|} = \frac{r\cos t\,\mathbf{i} - r\sin t\,\mathbf{j} + \mathbf{k}}{\sqrt{r^2 + 1}}, \qquad \frac{d\mathbf{T}}{dt} = \frac{-r\sin t\,\mathbf{i} - r\cos t\,\mathbf{j}}{\sqrt{r^2 + 1}},$$

$$k = \frac{\|d\mathbf{T}/dt\|}{ds/dt} = \frac{1}{\sqrt{r^2 + 1}} \cdot \frac{r}{\sqrt{r^2 + 1}} = \frac{r}{r^2 + 1}. \qquad □$$

Components of Acceleration

In straight-line motion, acceleration is purely tangential; that is, the acceleration vector points in the direction of the motion. (Section 13.5.) In uniform circular motion, the acceleration is normal; the acceleration vector is perpendicular to the tangent vector and points along the line of the normal vector toward the center of the circle. (Also Section 13.5.) In general, acceleration has two components, a tangential component and a normal component. To see this, let's suppose that the position of an object at time t is given by the vector function

$$\mathbf{r}(t) = x(t)\mathbf{i} + y(t)\mathbf{j} + z(t)\mathbf{k}.$$

Since

$$\mathbf{T} = \frac{d\mathbf{r}/dt}{\|d\mathbf{r}/dt\|} = \frac{\mathbf{v}}{ds/dt},$$

we have

$$\mathbf{v} = \frac{ds}{dt}\mathbf{T}.$$

Differentiation gives

$$\mathbf{a} = \frac{d^2s}{dt^2}\mathbf{T} + \frac{ds}{dt}\frac{d\mathbf{T}}{dt}.$$

Observe now that

$$\frac{d\mathbf{T}}{dt} = \left\|\frac{d\mathbf{T}}{dt}\right\|\mathbf{N} = k\frac{ds}{dt}\mathbf{N}.$$

$$(13.3.4) \xrightarrow{\hspace{1.5cm}} \qquad \xleftarrow{\hspace{1.5cm}} (13.7.7)$$

Substitution in the previous display gives

(13.7.8)

$$a = \frac{d^2s}{dt^2}T + k\left(\frac{ds}{dt}\right)^2 N.$$

The acceleration vector lies in the osculating plane, the plane of **T** and **N**. The tangential component of acceleration

$$a_T = \frac{d^2s}{dt^2}$$

depends only on the change of speed of the object; if the speed is constant, the tangential component of acceleration is zero and the acceleration is directed entirely toward the center of curvature of the path. On the other hand, the normal component of acceleration

$$a_N = k\left(\frac{ds}{dt}\right)^2$$

depends both on the speed of the object and the curvature of the path. At a point where the curvature is zero, the normal component of acceleration is zero and the acceleration is directed entirely along the path of motion. If the curvature is not zero, then the normal component of acceleration is a multiple of the *square* of the speed. This means, for example, that if you are in a car going around a curve at 50 miles per hour, you will feel *four times* the normal component of acceleration that you would feel going around the same curve at 25 miles per hour.

EXERCISES 13.7

In Exercises 1–10, find the curvature of the given curve.

1. $y = e^{-x}$.

2. $y = x^3$.

3. $y = \sqrt{x}$.

4. $y = 1/x$.

5. $y = \ln \sec x$.

6. $y = x - x^2$.

7. $y = \sin x$.

8. $y = \tan x$.

9. $y = x^{-3/2}$.

10. $x^2 - y^2 = a^2$.

In Exercises 11–18, find the radius of curvature at the indicated point.

11. $6y = x^3$; $(2, \frac{4}{3})$.

12. $2y = x^2$; $(0, 0)$.

13. $y^2 = 2x$; $(2, 2)$.

14. $y = 2 \sin 2x$; $(\frac{1}{4}\pi, 2)$.

15. $y = \ln(x + 1)$; $(2, \ln 3)$.

16. $y = \sec x$; $(\frac{1}{4}\pi, \sqrt{2})$.

17. $4x^2 + 9y^2 = 36$; $(3, 0)$.

18. $x^2 - 4y^2 = 9$; $(5, 2)$.

19. Find the point of maximal curvature on the curve $y = \ln x$.

20. Find the curvature of the graph of $y = 3x - x^3$ at the point where the function takes on its local maximum value.

In Exercises 21–28, express the curvature in terms of t.

21. $\mathbf{r}(t) = t\mathbf{i} + \frac{1}{2}t^2\mathbf{j}$.

22. $\mathbf{r}(t) = e^t\mathbf{i} + e^{-t}\mathbf{j}$.

23. $\mathbf{r}(t) = 2t\mathbf{i} + t^3\mathbf{j}$.

24. $\mathbf{r}(t) = t^2\mathbf{i} + t^3\mathbf{j}$.

25. $\mathbf{r}(t) = e^t(\cos t\mathbf{i} + \sin t\mathbf{j})$.

26. $\mathbf{r}(t) = 2\cos t\mathbf{i} + 3\sin t\mathbf{j}$.

27. $\mathbf{r}(t) = (t\cos t)\mathbf{i} + (t\sin t)\mathbf{j}$.

28. $\mathbf{r}(t) = (\cos t + t\sin t)\mathbf{i} + (\sin t - t\cos t)\mathbf{j}$, $t > 0$.

29. Find the radius of curvature at the vertices of the hyperbola $xy = 1$.

30. Find the radius of curvature at the vertices of the hyperbola $x^2 - y^2 = 1$.

31. Find the curvature at each (x, y) of the hyperbola $b^2x^2 - a^2y^2 = a^2b^2$.

HINT: Parametrize the hyperbola by setting $\mathbf{r}(t) = a\cosh t\mathbf{i} + b\sinh t\mathbf{j}$.

32. Find the curvature at the highest point of an arch of the cycloid

$$x(t) = r(t - \sin t), \qquad y(t) = r(1 - \cos t).$$

33. Show that the curvature of the path of a moving object can be written

(13.7.9)

$$k = \frac{\|\mathbf{v} \times \mathbf{a}\|}{(ds/dt)^3}.$$

HINT: $\mathbf{v} = \dfrac{ds}{dt}\mathbf{T}$ and $\mathbf{a} = \dfrac{d^2s}{dt^2}\mathbf{T} + k\left(\dfrac{ds}{dt}\right)^2\mathbf{N}$.

In Exercises 34–39, interpret $\mathbf{r}(t)$ as the position of a moving object at time t. Find the curvature of the path and determine the tangential and normal components of acceleration.

34. $\mathbf{r}(t) = (1 - 2t)\mathbf{i} + (3 + 4t)\mathbf{j} + (2 - 3t)\mathbf{k}$.

35. $\mathbf{r}(t) = e^t\cos t\mathbf{i} + e^t\sin t\mathbf{j} + e^t\mathbf{k}$.

36. $\mathbf{r}(t) = 2\cos t\mathbf{i} + t\mathbf{j} + \sin t\mathbf{k}$

37. $\mathbf{r}(t) = \cos 2t\mathbf{i} + \sin 2t\mathbf{j} + \mathbf{k}$.

38. $\mathbf{r}(t) = t\mathbf{i} + t^2\mathbf{j} + (\ln t)\mathbf{k}$.

39. $\mathbf{r}(t) = t\mathbf{i} + \frac{1}{2}t^2\mathbf{j} + \frac{1}{3}t^3\mathbf{k}$.

40. Show that the curvature of a polar curve $r = f(\theta)$ is given by

$$k = \frac{|[f(\theta)]^2 + 2[f'(\theta)]^2 - f(\theta)f''(\theta)|}{([f(\theta)]^2 + [f'(\theta)]^2)^{3/2}}.$$

41. Find the curvature of the logarithmic spiral $r = e^{a\theta}$, $a > 0$.

42. Find the curvature of the spiral of Archimedes $r = a\theta$, $a > 0$.

43. Find the curvature of the cardioid $r = a(1 - \cos\theta)$ in terms of r.

44. (*Transition Curves*) When engineers lay railroad track, they cannot allow abrupt changes in curvature. To join a straightaway that ends at a point P to a curved track that begins at a point Q, they need to lay some transitional track that has curvature zero at P and the curvature of the second piece at Q. Find an arc of the form

$$y = Cx^n, \qquad x \in [0, 1]$$

that joins the arcs C_1 and C_2 of Figure A without any discontinuities in curvature.

45. Let $s(\theta)$ be the arc distance from the highest point of the cycloidal arch

$$x(\theta) = R(\theta - \sin\theta), \quad y(\theta) = R(1 - \cos\theta), \quad \theta \in [0, 2\pi]$$

to the point $(x(\theta), y(\theta))$ of that same arch. (The arch is pictured in Figure 9.11.1.) Let $\rho(\theta)$ be the radius of curvature at the point $(x(\theta), y(\theta))$. (a) Calculate $s(\theta)$. (b) Calculate $\rho(\theta)$. (c) Then find an equation in s and ρ for that arch. (Such an equation is called a *natural equation* for the curve.)

46. Let $s(\theta)$ be the arc distance from the origin to the point $(x(\theta), y(\theta))$ along the exponential spiral $r = ae^{c\theta}$. (Take $a > 0, c > 0$.) Let $\rho(\theta)$ be the radius of curvature at that same point. Find an equation in s and ρ for that curve.

47. Let $s(\theta)$ be the arc distance from the point $(-2a, 0)$ to the point $(x(\theta), y(\theta))$ along the cardioid $r = a(1 - \cos\theta)$. (Take $a > 0$.) Let $\rho(\theta)$ be the radius of curvature at that same point. Find an equation in s and ρ for that curve.

(*The Frenet Formulas*) Figure B shows a space curve. At a point of the curve we have drawn the unit tangent $\mathbf{T}$, the principal normal $\mathbf{N}$, and the vector $\mathbf{B} = \mathbf{T} \times \mathbf{N}$, which, being normal to both $\mathbf{T}$ and $\mathbf{N}$, is called the *binormal*. At each point of the curve, the vectors $\mathbf{T}$, $\mathbf{N}$, $\mathbf{B}$ form what is called the *Frenet trihedral*, a set of mutually perpendicular unit vectors that, in the order given, form a local right-handed coordinate system.

Figure A

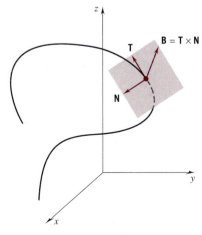

Figure B

48. (a) Show that $d\mathbf{B}/ds$ is parallel to $\mathbf{N}$, and therefore there is a scalar τ for which

$$\frac{d\mathbf{B}}{ds} = \tau\mathbf{N}.$$

HINT: Since $\mathbf{B}$ has constant length one, $d\mathbf{B}/ds \perp \mathbf{B}$. Show that $d\mathbf{B}/ds \perp \mathbf{T}$ by carrying out the differentiation

$$\frac{d\mathbf{B}}{ds} = \frac{d}{ds}(\mathbf{T} \times \mathbf{N}).$$

(b) Now show that

$$\frac{d\mathbf{N}}{ds} = -k\mathbf{T} - \tau\mathbf{B}.$$

HINT: Since $\mathbf{T}, \mathbf{N}, \mathbf{B}$ form a right-handed system of mutually perpendicular unit vectors, we can show that

$$\mathbf{N} \times \mathbf{B} = \mathbf{T} \quad \text{and} \quad \mathbf{B} \times \mathbf{T} = \mathbf{N}.$$

You can assume these relations.

(c) The scalar τ is called the *torsion* of the curve. Give a geometric interpretation to $|\tau|$.

The three formulas

$$(13.7.10) \quad \boxed{\frac{d\mathbf{T}}{ds} = k\mathbf{N}, \quad \frac{d\mathbf{N}}{ds} = -k\mathbf{T} - \tau\mathbf{B}, \quad \frac{d\mathbf{B}}{ds} = \tau\mathbf{N}}$$

were developed in 1847 by the French geometer Frederic Frenet (1816–1900). Another Frenchman, Alfred Serret (1819–1885), discovered these same relations independently in 1850. Accordingly, the formulas are often called the Frenet–Serret formulas.

■ CHAPTER HIGHLIGHTS

13.1 Vector Functions

basic definition: $\lim_{t \to t_0} \mathbf{f}(t) = \mathbf{L}$ iff $\lim_{t \to t_0} \|\mathbf{f}(t) - \mathbf{L}\| = 0$

theorem: if $\lim_{t \to t_0} \mathbf{f}(t) = \mathbf{L}$ then $\lim_{t \to t_0} \|\mathbf{f}(t)\| = \|\mathbf{L}\|$

limit rules (p. 842)
limits can be taken component by component (p. 842)
continuity and differentiability (p. 843)   integration (p. 844)
properties of the integral (p. 845)

13.2 Differentiation Formulas

$$\frac{d}{dt}(\mathbf{f} + \mathbf{g}) = \frac{d\mathbf{f}}{dt} + \frac{d\mathbf{g}}{dt}. \qquad \frac{d}{dt}(\alpha\mathbf{f}) = \alpha\frac{d\mathbf{f}}{dt} \quad (\alpha \text{ constant}).$$

$$\frac{d}{dt}(u\mathbf{f}) = u\frac{d\mathbf{f}}{dt} + \frac{du}{dt}\mathbf{f} \quad (u = u(t)). \qquad \frac{d}{dt}(\mathbf{f} \cdot \mathbf{g}) = \left(\mathbf{f} \cdot \frac{d\mathbf{g}}{dt}\right) + \left(\frac{d\mathbf{f}}{dt} \cdot \mathbf{g}\right).$$

$$\frac{d}{dt}(\mathbf{f} \times \mathbf{g}) = \left(\mathbf{f} \times \frac{d\mathbf{g}}{dt}\right) + \left(\frac{d\mathbf{f}}{dt} \times \mathbf{g}\right). \qquad \frac{d\mathbf{f}}{dt} = \frac{d\mathbf{f}}{du}\frac{du}{dt} \quad (\text{chain rule}).$$

13.3 Curves

tangent vector (p. 854)   tangent line (p. 856)
unit tangent (p. 858)   principal normal (p. 858)
normal line (p. 858)   osculating plane (p. 858)
sense-preserving (sense-reversing) change of parameter (p. 861)

13.4 Arc Length

$$L(C) = \int_a^b \|\mathbf{r}'(t)\| \, dt \qquad \frac{ds}{dt} = \sqrt{\left(\frac{dx}{dt}\right)^2 + \left(\frac{dy}{dt}\right)^2 + \left(\frac{dz}{dt}\right)^2}$$

13.5 Curvilinear Motion; Vector Calculus in Mechanics

*13.6 Planetary Motion

13.7 Curvature

■ PROJECTS AND EXPLORATIONS USING TECHNOLOGY

To do these exercises you will need a graphics calculator or a computer with graphing capability. The majority of these problems are open-ended so different approaches may be used to solve them. You should be aware that different appraoches can result in slight variations in the answers. Round your numerical answers to at least four decimal places. The rounding method that your calculator or computer uses also may cause variations in answers.

13.1 (a) Use your technology to calculate the derivatives of the following vector functions:

   (i) $\mathbf{f}(t) = t^3 \mathbf{i} - \sin t \mathbf{j} + \ln t \mathbf{k}$ at $t = 1$.

   (ii) $\mathbf{f}(t) = \sec t \mathbf{i} + e^{3t} \mathbf{j} - (3 + \tan t^3)^6 \mathbf{k}$ at $t = 0.75$.

   If your technology can do this directly, do so. Otherwise, find an efficient way to use it component-wise.

(b) Use the difference quotient

$$\frac{\mathbf{f}(t + h) - \mathbf{f}(t)}{h}$$

with $h = 0.1, 0.01, 0.001, 0.0001$ to find approximations for the derivatives of the vector functions in part (a).

(c) We are interested in "how good" our approximations are. Use the standard vector norm $\| \cdot \|$, called the *Euclidean norm,* to measure the difference between the approximate derivative and the analytic derivative for each of the values found in part (b). Does the norm approach zero in the same way as it would for a scalar function?

(d) Mathematicians often use norms other than the Euclidean norm to find the lengths of vectors. One such norm, called the "sup norm" and denoted $\| \cdot \|_\infty$, is given by

$$\|a \mathbf{i} + b \mathbf{j} + c \mathbf{k}\|_\infty = \max(|a|, |b|, |c|).$$

For example, $\|2 \mathbf{i} - 3 \mathbf{j} + \mathbf{k}\|_\infty = 3$. Discuss the possible advantages of this norm over the Euclidean norm.

(e) Repeat part (c) using the sup norm.

(f) Use the symmetric difference quotient

$$\frac{\mathbf{f}(t + h) - \mathbf{f}(t - h)}{2h}$$

with $h = 0.1, 0.01, 0.001, 0.0001$ to find approximations for the derivatives in part (a).

(g) Use the Euclidean norm to measure the difference between the approximate derivative and the analytic derivative for each of the values calculated in part (f). Repeat using the sup norm.

13.2 (a) Use your technology to calculate the integrals of the following vector functions:

$$\text{(i) } \mathbf{f}(t) = t^3 \,\mathbf{i} - \sin t \,\mathbf{j} + \ln t \,\mathbf{k} \quad \text{from } t = 1 \text{ to } t = 3.$$

$$\text{(ii) } \mathbf{f}(t) = \sec t \,\mathbf{i} + e^{3t} \,\mathbf{j} - (3 + \tan t^3)^6 \,\mathbf{k} \quad \text{from } t = 0.75 \text{ to } t = 1.$$

If your technology can do this directly, do so. Otherwise, determine an efficient way to use it component-wise.

(b) Approximate the integrals in part (a) by using a variation of the trapezoidal rule modified to deal with vector functions. Let $n = 2, 4, 8, 16, 32$.

(c) Using the Euclidean norm, the trapezoidal rule, and the number of subintervals $n = 2$, 4, 8, 16, 32, compare the rate of convergence here with that for the one-dimensional case.

(d) Repeat part (c) using the sup norm introduced in Exercise 1.

(e) Repeat parts (b), (c), and (d) using a vector version of Simpson's rule.

13.3 Let $\mathbf{r}(t) = x(t) \,\mathbf{i} + y(t) \,\mathbf{j} + z(t) \,\mathbf{k}$, $t \in [a, b]$, be a continuously differentiable curve. Consider using the summation

$$\sum_{i=0}^{n-1} \sqrt{(x_{i+1} - x_i)^2 + (y_{i+1} - y_i)^2 + (z_{i+1} - z_i)^2}$$

as an approximation for arc length, where the points (x_i, y_i, z_i) are spaced along the curve. In particular, let $h = (b - a)/n$, $t_i = a + ih$, $x_i = x(t_i)$, $y_i = y(t_i)$, and $z_i = z(t_i)$ for $i = 0, 1, 2, \ldots, n$. Not only is this method intuitively clear, it may have advantages over approximations of the arc length integral.

(a) Test this method on the curve $\mathbf{r}(t) = t^5 \,\mathbf{i} + t \,\mathbf{j} - \sin t \,\mathbf{k}$, $t \in [0, 1]$, with $n = 1, 2, 4$, 8, 16, 32.

(b) Use arc length software or technology to calculate the arc length. Plot n versus error and try to fit a curve through these points.

(c) Using the same number of intervals, approximate the definite integral that represents this arc length by applying either the trapezoidal rule or Simpson's rule. How do these approximations compare with those found in part (a)?

(d) Our theory claims that the summations should increase toward the arc length. Does this seem to be true?

(f) The arc length of a curve should be independent of the particular parametrization used for the curve. Find another parametrization for the curve in part (a) and check this result.

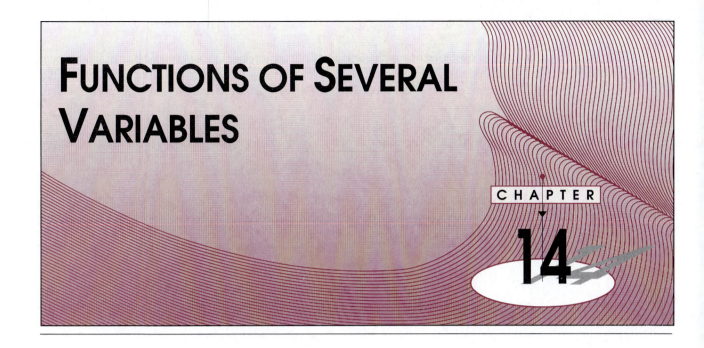

FUNCTIONS OF SEVERAL VARIABLES

CHAPTER

14

■ 14.1 ELEMENTARY EXAMPLES

First a remark on notation. Points $P(x, y)$ of the xy-plane will be written (x, y) and points $P(x, y, z)$ of three-space will be written (x, y, z).

Let D be a nonempty subset of the xy-plane. A rule f that assigns a real number $f(x, y)$ to each point (x, y) in D is called a *real-valued function of two variables.* The set D is called the *domain of f* and the set of all values $f(x, y)$ is called the *range of f.*

Example 1   Take D as the entire xy-plane and to each point (x, y) assign the number

$$f(x, y) = xy. \quad \square$$

Example 2   Take D as the set of all points (x, y) with $y \neq 0$. To each such point assign the number

$$f(x, y) = \tan^{-1}\left(\frac{x}{y}\right). \quad \square$$

Example 3   Take D as the *open unit disc*:

$$D = \{(x, y): x^2 + y^2 < 1\}.$$

This set consists of all points inside the *unit circle* $x^2 + y^2 = 1$; the circle itself is omitted.† To each point (x, y) in D assign the number

$$f(x, y) = \frac{1}{\sqrt{1 - (x^2 + y^2)}}. \quad \square$$

---

† The *closed unit disc* $D = \{(x, y): x^2 + y^2 \leq 1\}$ consists of all points on or inside the unit circle; the circle is included.

899

Now let D be a nonempty subset of three-space. A rule f that assigns a real number $f(x, y, z)$ to each point (x, y, z) in D is called a *real-valued function of three variables*. The set D is called the *domain of f* and the set of all values $f(x, y, z)$ is called the *range of f*.

Example 4 Take D as all of three-space and to each point (x, y, z) assign the number

$$f(x, y, z) = xyz. \quad ❑$$

Example 5 Take D as the set of all points (x, y, z) with $z \neq x + y$. (Thus D consists of all points not on the plane $x + y - z = 0$.) To each point of D assign the number

$$f(x, y, z) = \cos\left(\frac{1}{x + y - z}\right). \quad ❑$$

Example 6 Take D as the *open unit ball*:

$$D = \{(x, y, z): x^2 + y^2 + z^2 < 1\}.$$

This set consists of all points inside the *unit sphere* $x^2 + y^2 + z^2 = 1$; the sphere itself is omitted.† To each point (x, y, z) in D assign the number

$$f(x, y, z) = \frac{1}{\sqrt{1 - (x^2 + y^2 + z^2)}}. \quad ❑$$

Functions of several variables arise naturally in very elementary settings.

$f(x, y) = \sqrt{x^2 + y^2}$ gives the distance between $P(x, y)$ and the origin;

$f(x, y) = xy$ gives the area of a rectangle of dimensions x, y, and

$f(x, y) = 2(x + y)$ gives the perimeter.

$f(x, y, z) = \sqrt{x^2 + y^2 + z^2}$ gives the distance between $P(x, y, z)$ and the origin;

$f(x, y, z) = xyz$ gives the volume of of rectangular solid of dimensions x, y, z, and

$f(x, y, z) = 2(xy + xz + yz)$ gives the total surface area.

Functions of several variables are also key to many important problems in science, engineering, economics, and so on. Indeed, the mathematical models for "real" problems are much more likely to involve functions of several variables than functions of a single variable. Here are some simple examples.

If interest is compounded continuously, then the accumulation of principal A is a function of the interest rate r and the time t given by

$$A(t, r) = A_0 e^{rt},$$

where A_0 is the initial investment (Section 7.6).

---

† The *closed unit ball* $D = \{(x, y, z): x^2 + y^2 + z^2 \leq 1\}$ consists of all points on or inside the unit sphere; the sphere is included.

The magnitude of the gravitational force exerted by a body of mass M situated at the origin on a body of mass m at the point (x, y, z) is given by

$$F(x, y, z) = \frac{GmM}{x^2 + y^2 + z^2}.$$ (Section 13.6.)

The *ideal gas law* states that the pressure P of a gas is a function of the volume V and the temperature T according to the equation

$$P = \frac{cT}{V},$$

where c is a constant.

The deflection S at the midpoint of a rectangular beam which is supported at both ends and subjected to a uniform load is given by

$$S(L, w, h) = \frac{CL^3}{wh^3},$$

where L is the length, w the width, h the height, and C is a constant.

If the domain of a function of several variables is not explicitly given, it is to be understood that the domain is the set of all points for which the definition makes sense. Thus, in the case of

$$f(x, y) = \frac{1}{x - y},$$

the domain is understood to be all points (x, y) with $y \neq x$. That is, all points of the plane not on the line $y = x$. In the case of

$$g(x, y, z) = \sin^{-1}(x + y + z),$$

the domain is understood to be all points (x, y, z) with $-1 \leq x + y + z \leq 1$. This set is the slab bounded by the parallel planes

$$x + y + z = -1 \quad \text{and} \quad x + y + z = 1.$$

To say that a function is *bounded* is to say that its range is bounded. Since the function

$$f(x, y) = \frac{1}{x - y}$$

takes on all values other than 0, its range is $(-\infty, 0) \cup (0, \infty)$. The function is unbounded. In the case of

$$g(x, y, z) = \sin^{-1}(x + y + z)$$

the range is the closed interval $[-\frac{1}{2}\pi, \frac{1}{2}\pi]$. This function is bounded, below by $-\frac{1}{2}\pi$ and above by $\frac{1}{2}\pi$.

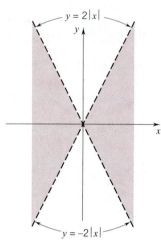

Figure 14.1.1

Example 7 Find the domain and range of

$$f(x, y) = \frac{1}{\sqrt{4x^2 - y^2}}$$

SOLUTION   A point (x, y) is in the domain of f iff

$$4x^2 - y^2 > 0.$$

This occurs iff

$$y^2 < 4x^2$$

and thus iff

$$-2|x| < y < 2|x|.$$

The domain of f is shaded in Figure 14.1.1. It consists of all the points of the xy-plane that lie between the graph of $y = -2|x|$ and the graph of $y = 2|x|$. On this set $\sqrt{4x^2 - y^2}$ takes on all positive values, and so does its reciprocal $f(x, y)$. The range of f is $(0, \infty)$. ❏

EXERCISES 14.1

In Exercises 1–22, find the domain and range of the given function.

1. $f(x, y) = \sqrt{xy}$.

2. $f(x, y) = \sqrt{1 - xy}$.

3. $f(x, y) = \dfrac{1}{x + y}$.

4. $f(x, y) = \dfrac{1}{x^2 + y^2}$.

5. $f(x, y) = \dfrac{e^x - e^y}{e^x + e^y}$.

6. $f(x, y) = \dfrac{x^2}{x^2 + y^2}$.

7. $f(x, y) = \ln(xy)$.

8. $f(x, y) = \ln(1 - xy)$.

9. $f(x, y) = \dfrac{1}{\sqrt{y - x^2}}$.

10. $f(x, y) = \dfrac{\sqrt{9 - x^2}}{1 + \sqrt{1 - y^2}}$.

11. $f(x, y) = \sqrt{9 - x^2} - \sqrt{4 - y^2}$.

12. $f(x, y, z) = \cos x + \cos y + \cos z$.

13. $f(x, y, z) = \dfrac{x + y + z}{|x + y + z|}$.

14. $f(x, y, z) = \dfrac{z^2}{x^2 - y^2}$.

15. $f(x, y, z) = -\dfrac{z^2}{\sqrt{x^2 - y^2}}$.

16. $f(x, y, z) = \dfrac{z}{x - y}$.

17. $f(x, y) = \dfrac{2}{\sqrt{9 - (x^2 + y^2)}}$.

18. $f(x, y, z) = \ln(|x + 2y + 3z| + 1)$.

19. $f(x, y, z) = \ln(x + 2y + 3z)$.

20. $f(x, y, z) = e^{\sqrt{4 - (x^2 + y^2 + z^2)}}$.

21. $f(x, y, z) = e^{-(x^2 + y^2 + z^2)}$.

22. $f(x, y, z) = \dfrac{\sqrt{1 - x^2} + \sqrt{4 - y^2}}{1 + \sqrt{9 - z^2}}$.

23. Let $f(x) = \sqrt{x}$, $g(x, y) = \sqrt{x}$, and $h(x, y, z) = \sqrt{x}$. Determine the domain and range of each function and compare your results. Sketch the graphs of f and g.

24. Let $f(x, y) = \cos \pi x \sin \pi y$ and $g(x, y, z) = \cos \pi x \sin \pi y$. Determine the domain and range of each of these functions and compare your results.

In Exercises 25–30, form the difference quotients

$$\frac{f(x + h, y) - f(x, y)}{h} \quad \text{and} \quad \frac{f(x, y + h) - f(x, y)}{h}, \quad (h \neq 0).$$

Then, assuming that x and y are fixed, calculate the limit as $h \to 0$. Can you make a connection between your results and derivatives?

25. $f(x, y) = 2x^2 - y$.

26. $f(x, y) = xy + 2y$.

27. $f(x, y) = 3x - xy + 2y^2$.

28. $f(x, y) = x \sin y$.

29. $f(x, y) = \cos(xy)$.

30. $f(x, y) = x^2 e^y$.

31. Determine a function f of two variables whose value at (x, y) is:
(a) The volume of a box with a square base of side length x and height y.
(b) The volume of a right circular cylinder whose radius is x and whose height is y.
(c) The area of the parallelogram whose sides are the vectors $2\mathbf{i}$ and $x\mathbf{i} + y\mathbf{j}$.

32. Determine a function f of three variables whose value at (x, y, z) is:
(a) The surface area of a box with no top whose sides have lengths x, y, and z.

(b) The angle between the vectors $\mathbf{i} + \mathbf{j}$ and $x\mathbf{i} + y\mathbf{j} + z\mathbf{k}$.

(c) The volume of the parallelepiped whose sides are the vectors $\mathbf{i}$, $\mathbf{i} + \mathbf{j}$, and $x\mathbf{i} + y\mathbf{j} + z\mathbf{k}$.

33. A closed box is to have a total surface area of 20 square feet. Express the volume V of the box as a function of its length l and height h.

34. An open box is to contain a volume of 12 cubic meters. If the material for the sides of the box costs \$2 per square meter and the material for the bottom costs \$4 per square meter, express the total cost C of the box as a function of the length l and width w.

35. A petrochemical company is designing a cylindrical tank with hemispherical ends to be used in the transportation of its products. See the figure. Express the volume of the tank as a function of its radius r and the length h of the cylindrical portion.

36. A 10-foot section of gutter is to be made from a 12-inch-wide strip of metal by folding up strips of length x on each side so that they make an angle θ with the bottom of the gutter. See the figure. Express the cross-sectional area of the gutter as a function of x and θ.

■ 14.2 A BRIEF CATALOG OF THE QUADRIC SURFACES; PROJECTIONS

As you know, the graph of an equation in two variables, say x and y, is typically a curve in the xy-plane. We will see in this section, and in the sections which follow, that the graph of an equation in three variables is, in general, a *surface* in three-space.

In this section we examine in a systematic manner the surfaces defined by equations of the form

$$Ax^2 + By^2 + Cz^2 + Dxy + Exz + Fyz + Hx + Iy + Jz + K = 0$$

where $A, B, C, \ldots, K$ are constants. Such surfaces are called *quadric surfaces*.

By suitable translations and rotations of the coordinate axes (see Section 1.4 and Appendix A.1) we can simplify such equations and thereby show that the nondegenerate† quadrics fall into nine distinct types:

1. The ellipsoid.

2. The hyperboloid of one sheet.

3. The hyperboloid of two sheets.

4. The quadric cone.

5. The elliptic paraboloid.

6. The hyperbolic paraboloid.

7. The parabolic cylinder.

8. The elliptic cylinder.

9. The hyperbolic cylinder.

---

† We are excluding such degenerate quadrics as

$$x^2 + y^2 + z^2 + 1 = 0 \qquad \text{and} \qquad x^2 + y^2 + z^2 = 0.$$

The first one has no points and the second consists of only one point, the origin.

As you go on with calculus, you will encounter these surfaces time and time again. Here we give you a picture of each one, together with its equation in standard form and some information about its special properties. These are some of the things to look for:

(a) The *intercepts* (the points at which the surface intersects the coordinate axes).

(b) The *traces* (the intersections with the coordinate planes).

(c) The *sections* (the intersections with planes in general).

(d) The *center* (some quadrics have a center; some do not).

(e) *Symmetry.*

(f) *Boundedness, unboundedness.*

1. The Ellipsoid

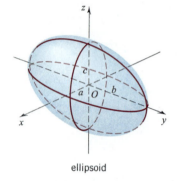

$$\frac{x^2}{a^2} + \frac{y^2}{b^2} + \frac{z^2}{c^2} = 1.†$$
(Figure 14.2.1.)

This ellipsoid is centered at the origin and is symmetric about the three coordinate planes. It intersects the coordinate axes at six points: $(\pm a, 0, 0)$, $(0, \pm b, 0)$, $(0, 0, \pm c)$. These points are called the *vertices*. The surface is bounded, being contained in the rectangular solid: $|x| \le a$, $|y| \le b$, $|z| \le c$. All three traces are ellipses; thus, for example, the trace in the xy-plane (set $z = 0$) is the ellipse

$$\frac{x^2}{a^2} + \frac{y^2}{b^2} = 1.$$

ellipsoid

Figure 14.2.1

All sections parallel to the coordinate planes are also ellipses; for example, taking $y = y_0$ we have

$$\frac{x^2}{a^2} + \frac{z^2}{c^2} = 1 - \frac{y_0^2}{b^2}.$$

This ellipse is the intersection of the ellipsoid with the plane $y = y_0$. The numbers a, b, c are called the *semi-axes* of the ellipsoid. If two of the semi-axes are equal, then we have an *ellipsoid of revolution*. (If, for example, $a = c$, then all sections parallel to the xz-plane are circles and the surface can be obtained by revolving the trace in the xy-plane about the y-axis.) If all three semi-axes are equal, the surface is a *sphere*. ❏

2. The Hyperboloid of One Sheet

$$\frac{x^2}{a^2} + \frac{y^2}{b^2} - \frac{z^2}{c^2} = 1.$$
(Figure 14.2.2.)

The surface is unbounded. It is centered at the origin and is symmetric about the three coordinate planes. The surface intersects the coordinate axes at four points: $(\pm a, 0, 0)$, $(0, \pm b, 0)$. The trace in the xy-plane (set $z = 0$) is the ellipse

$$\frac{x^2}{a^2} + \frac{y^2}{b^2} = 1.$$

hyperboloid of one sheet

Figure 14.2.2

† Throughout this section we take a, b, c as positive constants.

All sections parallel to the xy-plane are ellipses. The trace in the xz-plane (set $y = 0$) is the hyperbola

$$\frac{x^2}{a^2} - \frac{z^2}{c^2} = 1,$$

and the trace in the yz-plane (set $x = 0$) is the hyperbola

$$\frac{y^2}{b^2} - \frac{z^2}{c^2} = 1.$$

All sections parallel to the xz-plane or yz-plane are hyperbolas. If $a = b$, then all sections parallel to the xy-plane are circles and we have a *hyperboloid of revolution*. ❏

3. The Hyperboloid of Two Sheets

$$\frac{x^2}{a^2} + \frac{y^2}{b^2} - \frac{z^2}{c^2} = -1. \qquad \text{(Figure 14.2.3.)}$$

The surface intersects the coordinate axes only at the two vertices $(0, 0, \pm c)$. The surface consists of two parts: one for which $z \geq c$, another for which $z \leq -c$. We can see this by rewriting the equation as

$$\frac{x^2}{a^2} + \frac{y^2}{b^2} = \frac{z^2}{c^2} - 1$$

and noting that we must have

$$\frac{z^2}{c^2} - 1 \geq 0$$

in order for there to be solutions for x and y. It follows from this that $z^2 \geq c^2$, and so $|z| \geq c$. Each of the parts is unbounded. All sections parallel to the xy-plane are ellipses: for $z = z_0$ with $|z_0| \geq c$, we have

$$\frac{x^2}{a^2} + \frac{y^2}{b^2} = \frac{z_0^2}{c^2} - 1.$$

Sections parallel to the other coordinate planes are hyperbolas; for example, for $y = y_0$ we have

$$\frac{z^2}{c^2} - \frac{x^2}{a^2} = 1 + \frac{y_0^2}{b^2}.$$

The surface is symmetric about the three coordinate planes and is centered at the origin. ❏

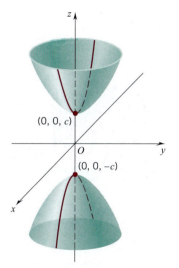

hyperboloid of two sheets

Figure 14.2.3

4. The Quadric Cone

$$\frac{x^2}{a^2} + \frac{y^2}{b^2} = z^2. \qquad \text{(Figure 14.2.4.)}$$

The surface intersects the coordinate axes only at the origin. The surface is unbounded. Once again there is symmetry about the three coordinate planes. The trace in the xz-plane is a pair of intersecting lines: $z = \pm x/a$. The trace in the yz-plane is

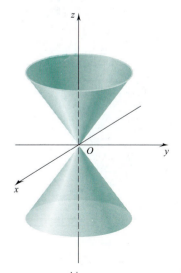

quadric cone

Figure 14.2.4

also a pair of intersecting lines: $z = \pm y/b$. The trace in the xy-plane is just the origin. Sections parallel to the xy-plane are ellipses. If $a = b$, these sections are circles and we have what is commonly called a *double circular cone* or simply a *cone*. The upper and lower portions of the cone are called *nappes*. ❏

We come now to the *paraboloids*. The equations in standard form will involve x^2 and y^2, but then z instead of z^2.

5. The Elliptic Paraboloid

$$\frac{x^2}{a^2} + \frac{y^2}{b^2} = z. \qquad \text{(Figure 14.2.5.)}$$

The surface does not extend below the xy-plane; it is unbounded above. The origin is called the *vertex*. Sections parallel to the xy-plane are ellipses; sections parallel to the other coordinate planes are parabolas. Hence the term "elliptic paraboloid." The surface is symmetric about the xz-plane and about the yz-plane. It is also symmetric about the z-axis. If $a = b$, then the surface is a *paraboloid of revolution*. ❏

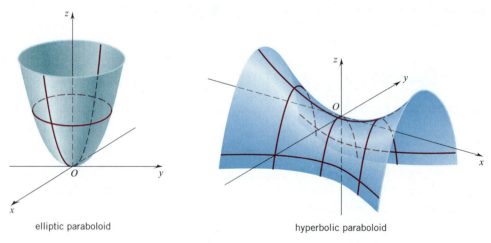

elliptic paraboloid

Figure 14.2.5

hyperbolic paraboloid

Figure 14.2.6

6. The Hyperbolic Paraboloid

$$\frac{x^2}{a^2} - \frac{y^2}{b^2} = z. \qquad \text{(Figure 14.2.6.)}$$

Here there is symmetry about the xz-plane and yz-plane. Sections parallel to the xy-plane are hyperbolas; sections parallel to the other coordinate planes are parabolas. Hence the term "hyperbolic paraboloid." The origin is a minimum point for the trace in the xz-plane, but a maximum point for the trace in the yz-plane. The origin is called a *minimax* or *saddle point* of the surface. *Note:* The orientation of the coordinate axes was chosen to enhance the view of the surface. ❏

The rest of the quadric surfaces are *cylinders*. The term deserves definition. Take any plane curve C. All the lines through C that are perpendicular to the plane of C form a surface. Such a surface is called a *cylinder*, *the cylinder with base curve C*. The perpendicular lines are known as the *generators* of the cylinder.

If the base curve lies in the xy-plane (or in a plane parallel to the xy-plane), then the generators of the cylinder are parallel to the z-axis. In such a case the equation of the cylinder involves only x and y. The z-coordinate is left unrestricted; it can take on all values.

There are three basic types of quadric cylinders. We give you their equations in standard form: base curve in the xy-plane, generators parallel to the z-axis.

7. The Parabolic Cylinder

$$x^2 = 4cy.$$ (Figure 14.2.7.)

This surface is formed by all lines that pass through the parabola $x^2 = 4cy$ and are perpendicular to the xy-plane. ❏

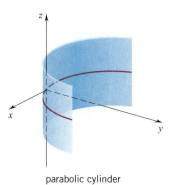

parabolic cylinder

Figure 14.2.7

8. The Elliptic Cylinder

$$\frac{x^2}{a^2} + \frac{y^2}{b^2} = 1.$$ (Figure 14.2.8.)

The surface is formed by all lines that pass through the ellipse

$$\frac{x^2}{a^2} + \frac{y^2}{b^2} = 1$$

and are perpendicular to the xy-plane. If $a = b$, we have the common *right circular cylinder*. ❏

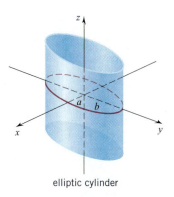

elliptic cylinder

Figure 14.2.8

9. The Hyperbolic Cylinder

$$\frac{x^2}{a^2} - \frac{y^2}{b^2} = 1.$$ (Figure 14.2.9.)

The surface has two parts, each generated by a branch of the hyperbola

$$\frac{x^2}{a^2} - \frac{y^2}{b^2} = 1. \quad ❏$$

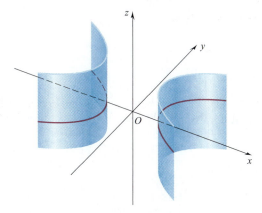

hyperbolic cylinder

Figure 14.2.9

Projections

Suppose that $S_1 : z = f(x, y)$ and $S_2 : z = g(x, y)$ are surfaces in three-space that intersect in a space curve C. See Figure 14.2.10.

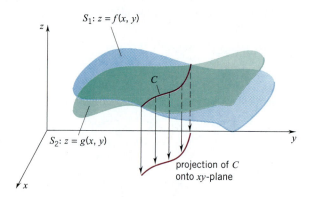

Figure 14.2.10

The curve C is the set of all points (x, y, z) with

$$z = f(x, y) \qquad \text{and} \qquad z = g(x, y).$$

The set of all points (x, y, z) with

$$f(x, y) = g(x, y) \qquad \text{(here } z \text{ is unrestricted)}$$

is the vertical cylinder that passes through C.

The set of all points $(x, y, 0)$ with

$$f(x, y) = g(x, y) \qquad \text{(here } z = 0)$$

is called the *projection of C onto the xy-plane*. In Figure 14.2.10 it appears as the curve in the xy-plane that lies directly below C.

Example 1 The paraboloid of revolution

$$z = x^2 + y^2$$

and the plane

$$z = 2y + 3$$

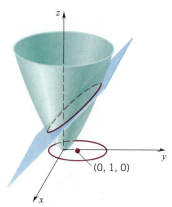

Figure 14.2.11

intersect in a curve C. See Figure 14.2.11. The projection of this curve onto the xy-plane is the set of all points $(x, y, 0)$ with

$$x^2 + y^2 = 2y + 3.$$

This equation can be written

$$x^2 + (y - 1)^2 = 4.$$

The projection of C onto the xy-plane is the circle of radius 2 centered at $(0, 1, 0)$. ❑

EXERCISES 14.2

In Exercises 1–12, identify the given surface.

1. $x^2 + 4y^2 - 16z^2 = 0$.

2. $x^2 + 4y^2 + 16z^2 - 12 = 0$.

3. $x - 4y^2 = 0$.   **4.** $x^2 - 4y^2 - 2z = 0$.

5. $5x^2 + 2y^2 - 6z^2 - 10 = 0$.

6. $2x^2 + 4y^2 - 1 = 0$.   **7.** $x^2 + y^2 + z^2 - 4 = 0$.

8. $5x^2 + 2y^2 - 6z^2 + 10 = 0$.

9. $x^2 + 2y^2 - 4z = 0$.   **10.** $2x^2 - 3y^2 - 6 = 0$.

11. $x - y^2 + 2z^2 = 0$.   **12.** $x - y^2 - 6z^2 = 0$.

In Exercises 13–24, sketch the given cylinder.

13. $25y^2 + 4z^2 - 100 = 0$.   **14.** $25x^2 + 4y^2 - 100 = 0$.

15. $y^2 - z = 0$.   **16.** $x^2 - y + 1 = 0$.

17. $y^2 + z = 0$.   **18.** $25x^2 - 9y^2 - 225 = 0$.

19. $x^2 + y^2 = 9$.   **20.** $\dfrac{x^2}{4} + \dfrac{y^2}{9} = 1$.

21. $y^2 - 4x^2 = 4$.   **22.** $z = x^2$.

23. $y = x^2 + 1$.

24. $(x - 1)^2 + (y - 1)^2 = 1$.

In Exercises 25–38, identify the surface and find the traces. Then sketch the surface.

25. $9x^2 + 4y^2 - 36z = 0$.

26. $9x^2 + 4y^2 + 36z^2 - 36 = 0$.

27. $9x^2 + 4y^2 - 36z^2 = 0$.

28. $9x^2 + 4y^2 - 36z^2 - 36 = 0$.

29. $9x^2 + 4y^2 - 36z^2 + 36 = 0$.

30. $9x^2 - 4y^2 - 36z = 0$.

31. $9x^2 - 4y^2 - 36z^2 = 36$.

32. $9x^2 + 4z^2 - 36y^2 - 36 = 0$.

33. $9x^2 + 4y^2 - 36z = 0$.

34. $9x^2 + 4z^2 - 36y^2 = 0$.

35. $9y^2 - 4x^2 - 36z^2 - 36 = 0$.

36. $9y^2 + 4z^2 - 36x = 0$.

37. $x^2 + y^2 - 4z = 0$.

38. $36x^2 + 9y^2 + 4z^2 - 36 = 0$.

39. Identify all possibilities for the surface

$$z = Ax^2 + By^2$$

taking (a) $AB > 0$.   (b) $AB < 0$.   (c) $AB = 0$.

40. Find the planes of symmetry for the cylinder $x - 4y^2 = 0$.

41. Write an equation for the surface obtained by revolving the parabola $4z - y^2 = 0$ about the z-axis.

42. The hyperbola $c^2y^2 - b^2z^2 - b^2c^2 = 0$ is revolved about the z-axis. Find an equation for the resulting surface.

43. (a) The equation

$$\sqrt{x^2 + y^2} = kz \qquad \text{with } k > 0$$

represents the upper nappe of a cone, with vertex at the origin and the positive z-axis as the axis of symmetry. Describe the section in the plane $z = z_0$, $z_0 > 0$.

(b) Let S be one nappe of a cone, with vertex at the origin. Write an equation for S given that

(i) The negative z-axis is the axis of symmetry and the section in the plane $z = -2$ is a circle of radius 6.

(ii) The positive y-axis is the axis of symmetry and the section in the plane $y = 3$ is a circle of radius 1.

44. Form the elliptic paraboloid

$$x^2 + \frac{y^2}{b^2} = z.$$

(a) Describe the section in the plane $z = 1$.

(b) What happens to this section as b tends to infinity?

(c) What happens to the paraboloid as b tends to infinity?

In Exercises 45–52, the given surfaces intersect in a space curve C. Determine the projection of C onto the xy-plane.

45. The planes $x + 2y + 3z = 6$ and $x + y - 2z = 6$.

46. The planes $x - 2y + z = 4$ and $3x + y - 2z = 1$.

47. The sphere $x^2 + y^2 + (z - 1)^2 = \frac{3}{2}$ and the hyperboloid $x^2 + y^2 - z^2 = 1$.

48. The sphere $x^2 + y^2 + (z - 2)^2 = 2$ and the cone $x^2 + y^2 = z^2$.

49. The paraboloids $x^2 + y^2 + z = 4$ and $x^2 + 3y^2 = z$.

50. The cylinder $y^2 + z - 4 = 0$ and the paraboloid $x^2 + 3y^2 = z$.

51. The cone $x^2 + y^2 = z^2$ and the plane $y + z = 2$.

52. The cone $x^2 + y^2 = z^2$ and the plane $y + 2z = 2$.

■ 14.3 GRAPHS; LEVEL CURVES AND LEVEL SURFACES

We begin with a function f of two variables defined on a subset D of the xy-plane. By the *graph of f* we mean the graph of the equation

$$z = f(x, y) \qquad (x, y) \in D.$$

Figure 14.3.1

Example 1 In the case of

$$f(x, y) = x^2 + y^2$$

the domain is the entire plane. The graph of f is a paraboloid of revolution:

$$z = x^2 + y^2.$$

This surface can be generated by revolving the parabola

$$z = x^2 \text{ (in the } xz\text{-plane)}$$

about the z-axis. See Figure 14.3.1. ❑

Example 2 Let a, b, and c be positive constants. The domain of the function

$$g(x, y) = c - ax - by$$

is also the entire xy-plane. The graph of g is the plane

$$z = c - ax - by \qquad \text{(Figure 14.3.2.)}$$

with intercepts $x = c/a$, $y = c/b$, $z = c$. ❑

Figure 14.3.2

Figure 14.3.3

Example 3 The function

$$f(x, y) = \sqrt{r^2 - (x^2 + y^2)}, \qquad r \text{ a positive constant,}$$

is defined only on the closed disc $x^2 + y^2 \le r^2$. The graph is the surface

$$z = \sqrt{r^2 - (x^2 + y^2)}, \qquad \text{(Figure 14.3.3.)}$$

which is the upper half of the sphere

$$x^2 + y^2 + z^2 = r^2. \quad ❑$$

Example 4 The function

$$f(x, y) = xy$$

is simple enough, but its graph, the surface $z = xy$, is quite difficult to draw. It is a "saddle-shaped" surface, a hyperbolic paraboloid: rotate the x and y axes by $\frac{1}{4}\pi$ radians and the equation takes the form

$$\frac{X^2}{2} - \frac{Y^2}{2} = z.†$$

A computer-generated graph of this surface is shown in Figure 14.3.9. ❏

Level Curves

In practice the graph of a function of two variables is usually difficult to visualize and draw. Moreover, if drawn, it is often difficult to interpret. Of course, the current techniques of computer graphics do make it possible to sketch and analyze more complicated surfaces. We will provide some illustrations of this later in the section.

Here we discuss an approach to sketching surfaces in space which we take from the map maker. In mapping mountainous terrain it is a common practice to sketch curves joining points of constant elevation. A collection of such curves, properly labeled, gives a good idea of the altitude variation in a region and suggests the shape of mountains and valleys. (See Figure 14.3.4.)

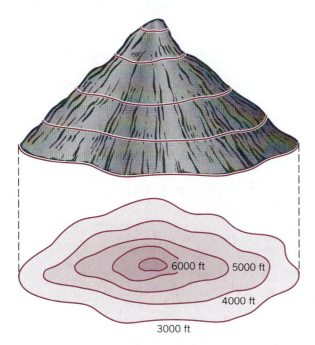

Figure 14.3.4

We can do the same thing to portray functions of two variables. Suppose that f is a nonconstant function defined on some portion of the xy-plane. If c is some value in the range of f, then we can sketch the curve $f(x, y) = c$. Such a curve is called a *level*

---

† To see this, set $\alpha = \frac{1}{4}\pi$ in Formula A.1.2, Appendix A-1.

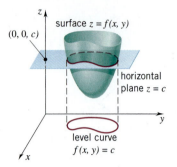

Figure 14.3.5

curve for f. It can be obtained by intersecting the graph of f with the horizontal plane $z = c$ and then projecting this intersection onto the xy-plane. (See Figure 14.3.5.)

The level curve $f(x, y) = c$ lies entirely in the domain of f, and on this curve f is constantly c.

A collection of level curves, properly drawn and labeled, can lead to a good understanding of the overall behavior of a function.

Example 5 We begin with the function $f(x, y) = x^2 + y^2$ (see Figure 14.3.1). The level curves are circles centered at the origin:

$$x^2 + y^2 = c, \qquad c \geq 0 \qquad \text{(Figure 14.3.6.)}$$

The function has the value c on the circle of radius $\sqrt{c}$ centered at the origin. At the origin, the function has the value 0. ❑

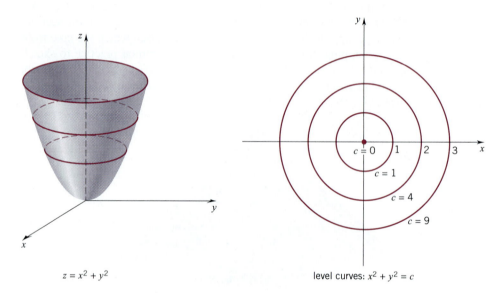

$z = x^2 + y^2$

level curves: $x^2 + y^2 = c$

Figure 14.3.6

Example 6 The graph of the function $g(x, y) = 4 - x - y$ is a plane. The level curves are parallel lines of the form

$$4 - x - y = c.$$

The surface and the level curves are indicated in Figure 14.3.7. ❑

Example 7 Now we consider the function

$$h(x, y) = \begin{cases} \sqrt{x^2 + y^2}, & x \geq 0 \\ |y|, & x < 0. \end{cases}$$

For $x \geq 0$, $h(x, y)$ is the distance from (x, y) to the origin. For $x < 0$, $h(x, y)$ is the distance from (x, y) to the x-axis. The level curves are pictured in Figure 14.3.8. The 0-level curve is the nonpositive x-axis. The other level curves are horseshoe-shaped: pairs of horizontal rays capped on the right by semicircles. ❑

Figure 14.3.7

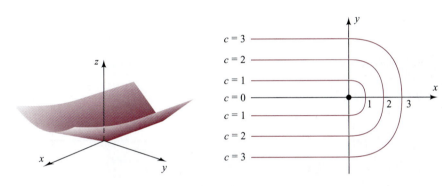

Figure 14.3.8

Example 8 Let's return to the function $f(x, y) = xy$. Earlier we noted that the graph is a "saddled-shaped" surface. You can visualize the surface from the few level curves sketched in Figure 14.3.9. The 0-level curve, $xy = 0$, consists of the two coordinate axes. The other level curves, $xy = c$ with $c \neq 0$, are hyperbolas. ❑

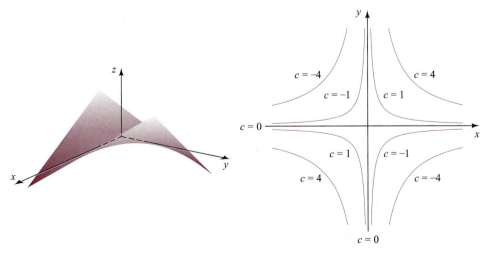

level curves: $xy = c$

Figure 14.3.9

Computer-Generated Graphs

The preceding examples illustrate how difficult it is to sketch an accurate graph of a function of two variables. But powerful help is at hand. Three-dimensional graphing programs for modern computers make it possible to visualize even quite complicated surfaces in three-space. These programs allow the user to view a surface from different perspectives, and they show level curves and the traces of various planes. Examples of computer-generated graphs are shown in Figure 14.3.10 and in the Exercises.

Level Surfaces

While drawing graphs for functions of two variables is quite difficult, drawing graphs for functions of three variables is actually impossible. To draw such figures we would need four dimensions at our disposal; the domain itself is a portion of three-space.

One can try to visualize the behavior of a function of three variables by examining the *level surfaces*. These are the subsets of the domain with equations of the form

$$f(x, y, z) = c.$$

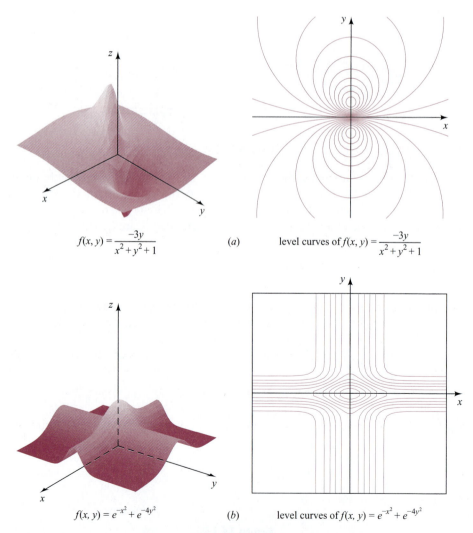

$$f(x, y) = \frac{-3y}{x^2 + y^2 + 1}$$

(a) level curves of $f(x, y) = \dfrac{-3y}{x^2 + y^2 + 1}$

$$f(x, y) = e^{-x^2} + e^{-4y^2}$$

(b) level curves of $f(x, y) = e^{-x^2} + e^{-4y^2}$

Figure 14.3.10

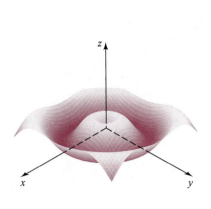

$f(x, y) = \sin \sqrt{x^2 + y^2}$

(c)   level curves of $f(x, y) = \sin \sqrt{x^2 + y^2}$

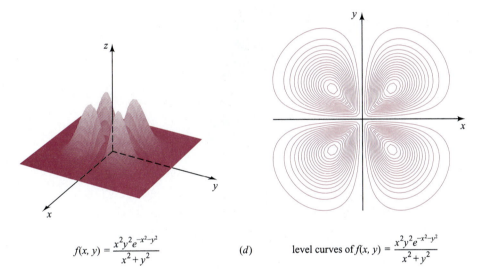

$f(x, y) = \dfrac{x^2 y^2 e^{-x^2 - y^2}}{x^2 + y^2}$

(d)   level curves of $f(x, y) = \dfrac{x^2 y^2 e^{-x^2 - y^2}}{x^2 + y^2}$

Figure 14.3.10 *Continued*

Level surfaces are usually difficult to draw. Nevertheless, a knowledge of what they are can be helpful. Here we restrict ourselves to a few simple examples.

Example 9   For the function $f(x, y, z) = Ax + By + Cz$ the level surfaces are parallel planes

$$Ax + By + Cz = c. \quad \square$$

Example 10   For the function $g(x, y, z) = \sqrt{x^2 + y^2 + z^2}$ the level surfaces are concentric spheres

$$x^2 + y^2 + z^2 = c^2. \quad \square$$

Example 11   As our final example we take the function

$$f(x, y, z) = \frac{|z|}{x^2 + y^2}.$$

We extend this function to the origin by defining it to be zero there. At other points of the z-axis we leave f undefined.

In the first place note that f takes on only nonnegative values. Since f is zero only when $z = 0$, the 0-level surface is the xy-plane. To find the other level surfaces, we take $c > 0$ and set $f(x, y, z) = c$. This gives

$$\frac{|z|}{x^2 + y^2} = c \quad \text{and thus} \quad |z| = c(x^2 + y^2).$$

(Figure 14.3.11). Each of these surfaces is a double-paraboloid of revolution.† ❑

level surface: $|z| = c(x^2 + y^2)$, $(c > 0)$

Figure 14.3.11

---

† Surface 5 of the last section together with its mirror image below the xy-plane.

EXERCISES 14.3

In Exercises 1–18, identify the level curves $f(x, y) = c$ and sketch the curves corresponding to the indicated values of c.

1. $f(x, y) = x - y$; $c = -2, 0, 2$.

2. $f(x, y) = 2x - y$; $c = -2, 0, 2$.

3. $f(x, y) = x^2 - y$; $c = -1, 0, 1, 2$.

4. $f(x, y) = \dfrac{1}{x - y^2}$; $c = -2, -1, 1, 2$.

5. $f(x, y) = \dfrac{x}{x + y}$; $c = -1, 0, 1, 2$.

6. $f(x, y) = \dfrac{y}{x^2}$; $c = -1, 0, 1, 2$.

7. $f(x, y) = x^3 - y$; $c = -1, 0, 1, 2$.

8. $f(x, y) = e^{xy}$; $c = \frac{1}{2}, 1, 2, 3$.

9. $f(x, y) = x^2 - y^2$; $c = -2, -1, 0, 1, 2$.

10. $f(x, y) = x^2$; $c = 0, 1, 4, 9$.

11. $f(x, y) = y^2$; $c = 0, 1, 4, 9$.

12. $f(x, y) = x(y - 1)$; $c = -2, -1, 0, 1, 2$.

13. $f(x, y) = \ln(x^2 + y^2)$; $c = -1, 0, 1$.

14. $f(x, y) = \ln\left(\dfrac{y}{x^2}\right)$; $c = -2, -1, 0, 1, 2$.

15. $f(x, y) = \dfrac{\ln y}{x^2}$; $c = -2, -1, 0, 1, 2$.

16. $f(x, y) = x^2 y^2$; $c = -4, -1, 0, 1, 4$.

17. $f(x, y) = \dfrac{x^2}{x^2 + y^2}$, $c = 0, \frac{1}{2}, \frac{1}{4}$.

18. $f(x, y) = \dfrac{\ln y}{x}$; $c = -2, -1, 0, 1, 2$.

In Exercises 19–24, identify the c-level surface and sketch it.

19. $f(x, y, z) = x + 2y + 3z, \quad c = 0$.

20. $f(x, y, z) = x^2 + y^2, \quad c = 4$.

21. $f(x, y, z) = z(x^2 + y^2)^{-1/2}, \quad c = 1$.

22. $f(x, y, z) = x^2/4 + y^2/6 + z^2/9, \quad c = 1$.

23. $f(x, y, z) = 4x^2 + 9y^2 - 72z, \quad c = 0$.

24. $f(x, y, z) = z^2 - 36x^2 - 9y^2, \quad c = 1$.

25. Identify the c-level surfaces of

$$f(x, y, z) = x^2 + y^2 - z^2$$

taking (i) $c < 0$, (ii) $c = 0$, (iii) $c > 0$.

26. Identify the c-level surfaces of

$$f(x, y, z) = 9x^2 - 4y^2 + 36z^2$$

taking (i) $c < 0$, (ii) $c = 0$, (iii) $c > 0$.

In Exercises 27–30, find an equation for the level curve of f that contains the given point P.

27. $f(x, y) = 1 - 4x^2 - y^2; \quad P(0, 1)$.

28. $f(x, y) = (x^2 + y^2)e^{xy}; \quad P(1, 0)$.

29. $f(x, y) = y^2 \tan^{-1}x; \quad P(1, 2)$.

30. $f(x, y) = (x^2 + y)\ln[2 - x + e^y]; \quad P(2, 1)$.

In Exercises 31 and 32, find an equation for the level surface of f that contains the given point P.

31. $f(x, y, z) = x^2 + 2y^2 - 2xyz; \quad P(-1, 2, 1)$.

32. $f(x, y, z) = \sqrt{x^2 + y^2} - \ln z; \quad P(3, 4, e)$.

33. The magnitude of the gravitational force exerted by a body of mass M situated at the origin on a body of mass m located at the point (x, y, z) is given by

$$F(x, y, z) = \frac{GmM}{x^2 + y^2 + z^2}$$

where G is the universal gravitational constant. If m and M are constants, describe the level surfaces of F. What is the physical significance of these surfaces?

34. The strength E of an electric field at a point (x, y, z) due to an infinitely long charged wire lying along the y-axis is given by

$$E(x, y, z) = \frac{k}{\sqrt{x^2 + z^2}}$$

where k is a positive constant. Describe the level surfaces of E.

35. A thin metal plate is situated in the xy-plane. The temperature T (in C°) at the point (x, y) is inversely proportional to the square of the distance from the origin.
(a) Express T as a function of x and y.

(b) Describe the level curves and sketch a representative set. NOTE: The level curves of T are called *isothermals*; all points on an isothermal have the same temperature.
(c) Suppose the temperature at the point $(1, 2)$ is 50°. What is the temperature at the point $(4, 3)$?

36. The formula

$$V(x, y) = \frac{k}{\sqrt{r^2 - x^2 - y^2}},$$

where k and r are positive constants, gives the electric potential (in volts) at a point (x, y) in the xy-plane. Describe the level curves of V and sketch a representative set. NOTE: The level curves of V are called the *equipotential curves*; all points on an equipotential curve have the same electric potential.

In Exercises 37–42, a function f, together with a set of level curves for f, is given. Figures $A–F$ are the surfaces $z = f(x, y)$ (in some order). Match f and its system of level curves with its graph $z = f(x, y)$.

37. $f(x, y) = y^2 - y^3$.

38. $f(x, y) = \sin x, \ 0 \le x \le 2\pi$.

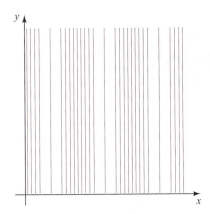

39. $f(x, y) = \cos \sqrt{x^2 + y^2}$, $-10 \le x \le 10$, $-10 \le y \le 10$.

41. $f(x, y) = xy\, e^{-\frac{1}{2}(x^2 + y^2)}$.

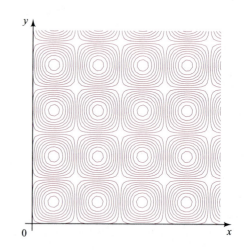

40. $f(x, y) = 2x^2 + 4y^2$.

42. $f(x, y) = \sin x \sin y$.

A.

B.

C.

D.

E.

F.

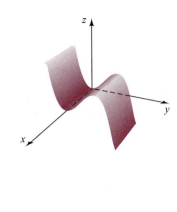

■ 14.4 PARTIAL DERIVATIVES

Functions of Two Variables

Let f be a function of x and y; for example

$$f(x, y) = 3x^2y - 5x \cos \pi y.$$

The *partial derivative of f with respect to x* is the function f_x obtained by differentiating f with respect to x, treating y as a constant; in this case

$$f_x(x, y) = 6xy - 5 \cos \pi y.$$

The *partial derivative of f with respect to y* is the function f_y obtained by differentiating f with respect to y, treating x as a constant; in this case

$$f_y(x, y) = 3x^2 + 5\pi x \sin \pi y.$$

These partial derivatives are formally defined as limits:

DEFINITION 14.4.1 PARTIAL DERIVATIVES OF $f(x, y)$

Let f be a function of two variables. The partial derivatives of f with respect to x and y are the functions f_x and f_y defined by

$$f_x(x, y) = \lim_{h \to 0} \frac{f(x + h, y) - f(x, y)}{h}$$

$$f_y(x, y) = \lim_{h \to 0} \frac{f(x, y + h) - f(x, y)}{h}$$

provided these limits exist.

Example 1 For

$$f(x, y) = x \tan^{-1} xy$$

we have

$$f_x(x, y) = x\frac{y}{1 + (xy)^2} + \tan^{-1} xy = \frac{xy}{1 + x^2y^2} + \tan^{-1} xy$$

and

$$f_y(x, y) = x\frac{x}{1 + (xy)^2} = \frac{x^2}{1 + x^2y^2}. \quad ❑$$

In the one-variable case, $f'(x_0)$ gives the rate of change with respect to x of $f(x)$ at $x = x_0$. In the two-variable case, $f_x(x_0, y_0)$ *gives the rate of change with respect to x of* $f(x, y_0)$ *at* $x = x_0$, *and* $f_y(x_0, y_0)$ *gives the rate of change with respect to y of* $f(x_0, y)$ *at* $y = y_0$.

Example 2 For the function

$$f(x, y) = e^{xy} + \ln (x^2 + y)$$

we have

$$f_x(x, y) = ye^{xy} + \frac{2x}{x^2 + y} \quad \text{and} \quad f_y(x, y) = xe^{xy} + \frac{1}{x^2 + y}.$$

The number

$$f_x(2, 1) = e^2 + \frac{4}{4 + 1} = e^2 + \frac{4}{5}$$

gives the rate of change with respect to x of the function

$$f(x, 1) = e^x + \ln (x^2 + 1) \qquad \text{at } x = 2;$$

the number

$$f_y(2, 1) = 2e^2 + \frac{1}{4 + 1} = 2e^2 + \frac{1}{5}$$

gives the rate of change with respect to y of the function

$$f(2, y) = e^{2y} + \ln (4 + y) \qquad \text{at } y = 1. \quad ❑$$

A Geometric Interpretation

In Figure 14.4.1 we have sketched a surface $z = f(x, y)$, which you can take as everywhere defined. Through the surface we have passed a plane $y = y_0$ parallel to the xz-plane. The plane $y = y_0$ intersects the surface in a curve, the y_0-section of the surface.

The y_0-section of the surface is the graph of the function

$$g(x) = f(x, y_0).$$

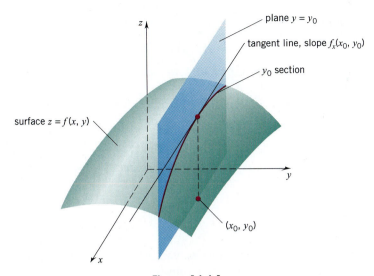

Figure 14.4.1

Differentiating with respect to x, we have

$$g'(x) = f_x(x, y_0)$$

and in particular

$$g'(x_0) = f_x(x_0, y_0).$$

The number $f_x(x_0, y_0)$ is thus the slope of the y_0-section of the surface $z = f(x, y)$ at the point $P(x_0, y_0, f(x_0, y_0))$.

The other partial derivative f_y can be given a similar interpretation. In Figure 14.4.2 you can see the same surface $z = f(x, y)$, this time sliced by a plane $x = x_0$ parallel to the yz-plane. The plane $x = x_0$ intersects the surface in a curve, the x_0-section of the surface.

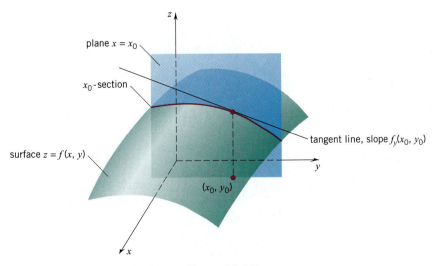

Figure 14.4.2

The x_0-section of the surface is the graph of the function

$$h(y) = f(x_0, y).$$

Differentiating, this time with respect to y, we have

$$h'(y) = f_y(x_0, y)$$

and thus

$$h'(y_0) = f_y(x_0, y_0).$$

The number $f_y(x_0, y_0)$ is the slope of the x_0-section of the surface $z = f(x, y)$ at the point $P(x_0, y_0, f(x_0, y_0))$.

Functions of Three Variables

In the case of a function of three variables, you can look for three partial derivatives: the partial with respect to x, the partial with respect to y, and also the partial with respect to z. These partials

$$f_x(x, y, z), \qquad f_y(x, y, z), \qquad f_z(x, y, z)$$

are defined as follows:

DEFINITION 14.4.2 PARTIAL DERIVATIVES OF $f(x, y, z)$

Let f be a function of three variables. The partial derivatives of f with respect to x, y, and z are the functions f_x, f_y, and f_z defined by

$$f_x(x, y, z) = \lim_{h \to 0} \frac{f(x + h, y, z) - f(x, y, z)}{h},$$

$$f_y(x, y, z) = \lim_{h \to 0} \frac{f(x, y + h, z) - f(x, y, z)}{h},$$

$$f_z(x, y, z) = \lim_{h \to 0} \frac{f(x, y, z + h) - f(x, y, z)}{h},$$

provided these limits exist.

Each partial can be found by differentiating with respect to the subscript variable, treating the other two variables as constants.

Example 3 For the function

$$f(x, y, z) = xy^2z^3$$

the partial derivatives are

$$f_x(x, y, z) = y^2z^3, \qquad f_y(x, y, z) = 2xyz^3, \qquad f_z(x, y, z) = 3xy^2z^2.$$

In particular

$$f_x(1, -2, -1) = -4, \qquad f_y(1, -2, -1) = 4, \qquad f_z(1, -2, -1) = 12. \quad \square$$

Example 4 For

$$g(x, y, z) = x^2 e^{y/z}$$

we have

$$g_x(x, y, z) = 2xe^{y/z}, \qquad g_y(x, y, z) = \frac{x^2}{z} e^{y/z}, \qquad g_z(x, y, z) = -\frac{x^2 y}{z^2} e^{y/z}. \quad \square$$

Example 5 For a function of the form

$$f(x, y, z) = F(x, y)G(y, z)$$

we can write

$$f_x(x, y, z) = F_x(x, y)G(y, z),$$

$$f_y(x, y, z) = F(x, y)G_y(y, z) + F_y(x, y)G(y, z),$$

$$f_z(x, y, z) = F(x, y)G_z(y, z). \quad \square$$

The number $f_x(x_0, y_0, z_0)$ gives the rate of change with respect to x of $f(x, y_0, z_0)$ at $x = x_0$; $f_y(x_0, y_0, z_0)$ gives the rate of change with respect to y of $f(x_0, y, z_0)$ at $y = y_0$; and $f_z(x_0, y_0, z_0)$ gives the rate of change with respect to z of $f(x_0, y_0, z)$ at $z = z_0$.

Example 6 The function

$$f(x, y, z) = xy^2 - yz^2$$

has partial derivatives

$$f_x(x, y, z) = y^2, \qquad f_y(x, y, z) = 2xy - z^2, \qquad f_z(x, y, z) = -2yz.$$

The number $f_x(1, 2, 3) = 4$ gives the rate of change with respect to x of the function

$$f(x, 2, 3) = 4x - 18 \qquad \text{at } x = 1;$$

$f_y(1, 2, 3) = -5$ gives the rate of change with respect to y of the function

$$f(1, y, 3) = y^2 - 9y \qquad \text{at } y = 2;$$

$f_z(1, 2, 3) = -12$ gives the rate of change with respect to z of the function

$$f(1, 2, z) = 4 - 2z^2 \qquad \text{at } z = 3. \quad \square$$

The geometric significance that we attached to the partials of a function of two variables has no analog in the case of three variables. The interpretations shown in Figures 14.4.1 and 14.4.2 do not generalize to three variables.

Other Notations

There is obviously no need to restrict ourselves to the variables x, y, z. Where more convenient we can use other letters.

Example 7 The volume of the frustum of a cone (Figure 14.4.3) is given by the function

$$V(R, r, h) = \tfrac{1}{3} \pi h (R^2 + Rr + r^2).$$

Find the rate of change of the volume with respect to each of its dimensions if the other dimensions are held constant. Determine the values of these rates of change when $R = 8$, $r = 4$, and $h = 6$.

frustum of a cone

Figure 14.4.3

SOLUTION The partial derivatives of V with respect to R, r, and h are as follows:

$$V_R(R, r, h) = \tfrac{1}{3}\pi h(2R + r),$$

$$V_r(R, r, h) = \tfrac{1}{3}\pi h(R + 2r),$$

$$V_h(R, r, h) = \tfrac{1}{3}\pi(R^2 + Rr + r^2).$$

When $R = 8$, $r = 4$, and $h = 6$,

the rate of change of V with respect to R is $V_R(8, 4, 6) = 40\pi$,
the rate of change of V with respect to r is $V_r(8, 4, 6) = 32\pi$,
the rate of change of V with respect to h is $V_h(8, 4, 6) = \tfrac{112}{3}\pi$. ❏

The subscript notation is not the only one used for partial differentiation. A variant of Leibniz's double-d notation is also commonly used. In this notation the partials f_x, f_y, f_z are denoted by

$$\frac{\partial f}{\partial x}, \quad \frac{\partial f}{\partial y}, \quad \frac{\partial f}{\partial z}.$$

Thus, for

$$f(x, y, z) = x^3 y^2 z + \sin xy$$

we have

$$\frac{\partial f}{\partial x}(x, y, z) = 3x^2 y^2 z + y \cos xy,$$

$$\frac{\partial f}{\partial y}(x, y, z) = 2x^3 yz + x \cos xy,$$

$$\frac{\partial f}{\partial z}(x, y, z) = x^3 y^2,$$

or more simply,

$$\frac{\partial f}{\partial x} = 3x^2 y^2 z + y \cos xy,$$

$$\frac{\partial f}{\partial y} = 2x^3 yz + x \cos xy,$$

$$\frac{\partial f}{\partial z} = x^3 y^2.$$

We can also write

$$\frac{\partial}{\partial x}(x^3 y^2 z + \sin xy) = 3x^2 y^2 z + y \cos xy,$$

$$\frac{\partial}{\partial y}(x^3 y^2 z + \sin xy) = 2x^3 yz + x \cos xy,$$

$$\frac{\partial}{\partial z}(x^3 y^2 z + \sin xy) = x^3 y^2.$$

Of course, this notation is not restricted to the letters x, y, z. For instance, we can write

$$\frac{\partial}{\partial r}(r^2\cos\theta + e^{\theta r}) = 2r\cos\theta + \theta e^{\theta r},$$

$$\frac{\partial}{\partial\theta}(r^2\cos\theta + e^{\theta r}) = -r^2\sin\theta + re^{\theta r}.$$

For the function

$$\rho = \sin 2\theta \cos 3\phi$$

we have

$$\frac{\partial\rho}{\partial\theta} = 2\cos 2\theta\cos 3\phi \quad\text{and}\quad \frac{\partial\rho}{\partial\phi} = -3\sin 2\theta\sin 3\phi.$$

EXERCISES 14.4

In Exercises 1–28, calculate the partial derivatives of the given function

1. $f(x, y) = 3x^2 - xy + y$.

2. $g(x, y) = x^2 e^{-y}$.

3. $\rho = \sin\phi\cos\theta$.

4. $\rho = \sin^2(\theta - \phi)$.

5. $f(x, y) = e^{x-y} - e^{y-x}$.

6. $z = \sqrt{x^2 - 3y}$.

7. $g(x, y) = \dfrac{Ax + By}{Cx + Dy}$.

8. $u = \dfrac{e^z}{xy^2}$.

9. $u = xy + yz + zx$.

10. $z = Ax^2 + Bxy + Cy^2$.

11. $f(x, y, z) = z\sin(x - y)$.

12. $g(u, v, w) = \ln(u^2 + vw - w^2)$.

13. $\rho = e^{\theta+\phi}\cos(\theta - \phi)$.

14. $f(x, y) = (x + y)\sin(x - y)$.

15. $f(x, y) = x^2 y\sec xy$.

16. $g(x, y) = \tan^{-1}(2x + y)$.

17. $h(x, y) = \dfrac{x}{x^2 + y^2}$.

18. $z = \ln\sqrt{x^2 + y^2}$.

19. $f(x, y) = \dfrac{x\sin y}{y\cos x}$.

20. $f(x, y, z) = e^{xy}\sin xz$.

21. $h(x, y) = [f(x)]^2 g(y)$.

22. $h(x, y) = e^{f(x)g(y)}$.

23. $f(x, y, z) = z^{xy^2}$.

24. $h(x, y, z) = [f(x, y)]^3 [g(x, z)]^2$.

25. $h(r, \theta, t) = r^2 e^{2t}\cos(\theta - t)$.

26. $u = \ln(x/y) - ye^{xz}$.

27. $f(x, y, z) = z\tan^{-1}(y/x)$.

28. $w = xy\sin z - yz\sin x$.

29. Find $f_x(0, e)$ and $f_y(0, e)$ given that $f(x, y) = e^x\ln y$.

30. Find $g_x(0, \frac{1}{4}\pi)$ and $g_y(0, \frac{1}{4}\pi)$ given that $g(x, y) = e^{-x}\sin(x + 2y)$.

31. Find $f_x(1, 2)$ and $f_y(1, 2)$ given that $f(x, y) = \dfrac{x}{x + y}$.

32. Find $g_x(1, 2)$ and $g_y(1, 2)$ given that $g(x, y) = \dfrac{y}{x + y^2}$.

In Exercises 33–38, find $f_x(x, y)$ and $f_y(x, y)$ by forming the appropriate difference quotient and taking the limit as h tends to zero (Definition 14.4.1).

33. $f(x\ y) = x^2 y$.

34. $f(x, y) = y^2$.

35. $f(x, y) = \ln(x^2 y)$.

36. $f(x, y) = \dfrac{1}{x + 4y}$.

37. $f(x\ y) = \dfrac{1}{x - y}$.

38. $f(x, y) = e^{2x+3y}$.

In Exercises 39 and 40, find $f_x(x, y, z), f_y(x, y, z)$, and $f_z(x, y, z)$ by forming the appropriate difference quotient and taking the limit as h tends to zero (Definition 14.4.2).

39. $f(x, y, z) = xy^2 z$.

40. $f(x, y, z) = \dfrac{x^2 y}{z}$.

41. The intersection of a surface $z = f(x, y)$ with a plane $y = y_0$ is a curve C in space. The slope of the tangent line to C at the point $P(x_0, y_0, f(x_0, y_0))$ is $f_x(x_0, y_0)$. (See Figure 14.4.1.)

(a) Show that equations for the tangent line can be written in the form:

$$y = y_0, \qquad z - z_0 = f_x(x_0, y_0)(x - x_0).$$

(b) Now let C be the curve formed by intersecting the surface $z = f(x, y)$ with the plane $x = x_0$. Derive equations for the tangent line to C at the point $P(x_0, y_0, f(x_0, y_0))$. (See Figure 14.4.2.)

In Exercises 42 and 43, let $z = x^2 + y^2$ and let C be the curve of intersection of the surface with the given plane. Find equations for the tangent line to the graph of C at the given point.

42. Plane: $y = 3$; point $(1, 3, 10)$.

43. Plane: $x = 2$; point $(2, 1, 5)$.

In Exercises 44 and 45, let

$$z = \frac{x^2}{y^2 - 3}$$

and let C be the curve of intersection of the surface with the given plane. Find equations for the tangent line to C at the given point.

44. Plane: $x = 3$; point $(3, 2, 9)$.

45. Plane: $y = 2$; point $(3, 2, 9)$.

46. The surface $z = \sqrt{4 - x^2 - y^2}$ is a hemisphere of radius 2 centered at the origin.
  (a) Find equations for the tangent line l_1 to the curve of intersection of the hemisphere with the plane $x = 1$ at the point $(1, 1, \sqrt{2})$.
  (b) Find equations for the tangent line l_2 to the curve of intersection of the hemisphere with the plane $y = 1$ at the point $(1, 1, \sqrt{2})$.
  (c) The tangent lines l_1 and l_2 determine a plane. Find an equation for this plane. As you might expect, this plane is tangent to the surface at the point $(1, 1, \sqrt{2})$

In Exercises 47–50, show that the functions u and v satisfy

$$u_x(x, y) = v_y(x, y) \qquad \text{and} \qquad u_y(x, y) = -v_x(x, y).$$

These equations are called the *Cauchy–Riemann equations*. They arise in the study of functions of a complex variable and are of fundamental importance.

47. $u(x, y) = x^2 - y^2;$ $\quad v(x, y) = 2xy.$

48. $u(x, y) = e^x \cos y;$ $\quad v(x, y) = e^x \sin y.$

49. $u(x, y) = \frac{1}{2} \ln(x^2 + y^2);$ $\quad v(x, y) = \tan^{-1}\frac{y}{x}.$

50. $u(x, y) = \dfrac{x}{x^2 + y^2};$ $\quad v(x, y) = \dfrac{-y}{x^2 + y^2}.$

51. Assume that f is a function which is defined on a domain D in the xy-plane, and assume that its partial derivatives exist throughout D.
  (a) Suppose that $f_x(x, y) = 0$ for all $(x, y) \in D$. What can you conclude about f?
  (b) Suppose that $f_y(x, y) = 0$ for all $(x, y) \in D$. What can you conclude about f?

52. The law of cosines for a triangle can be written

$$a^2 = b^2 + c^2 - 2bc \cos \theta.$$

At time t_0 we have $b_0 = 10$ inches, $c_0 = 15$ inches, $\theta_0 = \frac{1}{3}\pi$ radians.

(a) Find a_0.
(b) Find the rate of change of a with respect to b at time t_0 if c and θ remain constant.
(c) Using the rate found in (b), calculate (by differentials) the approximate change in a if b is decreased by 1 inch.
(d) Find the rate of change of a with respect to θ at time t_0 if b and c remain constant.
(e) Find the rate of change of c with respect to θ at time t_0 if a and b remain constant.

53. The area of a triangle is given by the formula

$$A = \tfrac{1}{2}bc \sin \theta.$$

At time t_0 we have $b_0 = 10$ inches, $c_0 = 20$ inches, $\theta_0 = \frac{1}{3}\pi$ radians.
(a) Find the area of the triangle at time t_0.
(b) Find the rate of change of the area with respect to b at time t_0 if c and θ remain constant.
(c) Find the rate of change of the area with respect to θ at time t_0 if b and c remain constant.
(d) Using the rate found in (c), calculate (by differentials) the approximate change in area if the angle is increased by one degree.
(e) Find the rate of change of c with respect to b at time t_0 if the area and the angle are to remain constant.

54. Let f be a function of x and y that satisfies a relation of the form

$$\frac{\partial f}{\partial x} = kf, \qquad k \text{ a constant.}$$

Show that

$$f(x, y) = g(y)e^{kx},$$

where g is some function of y.

55. Let $z = f(x, y)$ be a surface everywhere defined.
  (a) Find a vector function that parametrizes the y_0-section of the surface. (See Figure 14.4.1.) Find a vector function that parametrizes the line tangent to this section at the point $P(x_0, y_0, f(x_0, y_0))$. See Exercise 41.
  (b) Find a vector function that parametrizes the x_0-section of the surface. (See Figure 14.4.2.) Find a vector function that parametrizes the line tangent to this section at the point $P(x_0, y_0, f(x_0, y_0))$.
  (c) Show that the equation of the plane determined by the tangent lines found in (a) and (b) can be written

$$z - f(x_0, y_0) = (x - x_0)\frac{\partial f}{\partial x}(x_0, y_0)$$

$$+ (y - y_0)\frac{\partial f}{\partial y}(x_0, y_0).$$

(We show in Chapter 15 that, for certain functions of two variables, this plane plays a role similar to the role played by the tangent line for differentiable functions of a single variable.)

56. *(A chain rule)* Let f be a function of x and y, and g be a function of a single variable. Form the composition $h(x, y) = g(f(x, y))$ and show that

$$h_x(x, y) = g'(f(x, y))f_x(x, y) \quad \text{and}$$
$$h_y(x, y) = g'(f(x, y))f_y(x, y).$$

In Leibniz's notation, setting $u = f(x, y)$, we have

$$\frac{\partial h}{\partial x} = \frac{dg}{du}\frac{\partial u}{\partial x} \quad \text{and} \quad \frac{\partial h}{\partial y} = \frac{dg}{du}\frac{\partial u}{\partial y}.$$

57. Let g be a differentiable function of a single variable. Use Exercise 56 to verify the following results.
(a) If $w = g(ax + by)$, then

$$b\frac{\partial w}{\partial x} = a\frac{\partial w}{\partial y}.$$

(b) If m and n are nonzero integers and $w = g(x^m y^n)$, then

$$nx\frac{\partial w}{\partial x} = my\frac{\partial w}{\partial y}.$$

Partial Differential Equations

Differential equations were introduced in Section 7.6. At that stage, the only type of derivative that we had available was an "ordinary" derivative, that is, the derivative of a function of one variable. The differential equations we considered are called *ordinary differential equations.* This term is used to distinguish them from *partial differential equations,* which we introduce here. A *partial differential equation* is a differential equation in which the unknown function is a function of more than one variable. Since the unknown is a function of several variables, the derivatives of the unknown that appear in the equation must be partial derivatives. Partial differential equations play a significant role in the science and engi-neering disciplines since the descriptions of most natural phenomena have mathematical models that involve functions of several variables.

58. Show that $u = \dfrac{x^2 y^2}{x + y}$ satisfies the partial differential equation

$$x\frac{\partial u}{\partial x} + y\frac{\partial u}{\partial y} = 3u.$$

59. Show that $u = Ax^4 + 2Bx^2y^2 + Cy^4$ satisfies the partial differential equation

$$x\frac{\partial u}{\partial x} + y\frac{\partial u}{\partial y} = 4u.$$

60. Show that $u = x^2y + y^2z + z^2x$ satisfies the partial differential equation

$$\frac{\partial u}{\partial x} + \frac{\partial u}{\partial y} + \frac{\partial u}{\partial z} = (x + y + z)^2.$$

61. Given that $x = r\cos\theta$ and $y = r\sin\theta$, find

$$\frac{\partial x}{\partial r}\frac{\partial y}{\partial \theta} - \frac{\partial x}{\partial \theta}\frac{\partial y}{\partial r}.$$

62. For a gas confined in a container, the ideal gas law states that the pressure P is related to the volume V and the temperature T by an equation of the form

$$P = k\frac{T}{V}$$

where k is a positive constant. Show that

$$V\frac{\partial P}{\partial V} = -P \quad \text{and} \quad V\frac{\partial P}{\partial V} + T\frac{\partial P}{\partial T} = 0.$$

■ 14.5 OPEN SETS AND CLOSED SETS

Recall that the fundamental concepts of calculus for a function f of one variable (limit, continuity, derivative) were so-called point concepts, and in the definitions we required f to be defined on some open interval of the form $(x - p, x + p)$ containing x (except possibly at x itself in the case of the limit). Open intervals played a basic role in that development. In this section we extend the notion of open interval to sets in the plane and sets in space. In the next section we consider limits, continuity, and differentiability for functions of several variables.

Points in the domain of a function of several variables can be written in vector notation. In the two-variable case, set

$$\mathbf{x} = (x, y),$$

and, in the three-variable case, set

$$\mathbf{x} = (x, y, z).$$

The vector notation enables us to treat the two cases together.

In this section we introduce five important notions:

(1) Neighborhood of a point.

(2) Interior of a set.

(3) Boundary of a set.

(4) Open set.

(5) Closed set.

For our purposes, the fundamental notion here is "neighborhood of a point." The other four notions can be derived from it.

DEFINITION 14.5.1. NEIGHBORHOOD OF A POINT

A *neighborhood* of a point $\mathbf{x}_0$ is a set of the form

$$\{\mathbf{x}\colon \|\mathbf{x} - \mathbf{x}_0\| < \delta\}$$

where δ is some number greater than zero.

In the plane, a neighborhood of $\mathbf{x}_0 = (x_0, y_0)$ consists of all the points inside a disc centered at (x_0, y_0). In three-space, a neighborhood of $\mathbf{x}_0 = (x_0, y_0, z_0)$ consists of all the points inside a ball (sphere) centered at (x_0, y_0, z_0). See Figure 14.5.1.

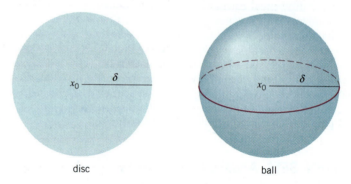

disc   ball

Figure 14.5.1

Remark On the real line, the norm of a number x is simply its absolute value, $\|x\| = |x|$. Thus a neighborhood of a point x_0 on the real line is a set of the form $\{x\colon |x - x_0| < \delta\}$ for some number $\delta > 0$, which is the same as the open interval $(x_0 - \delta, x_0 + \delta)$. ❑

DEFINITION 14.5.2 THE INTERIOR OF A SET

A point $\mathbf{x}_0$ is said to be an *interior point* of the set S iff the set S contains some neighborhood of $\mathbf{x}_0$. The set of all interior points of S is called the *interior* of S.

Example 1 Let Ω be the plane set shown in Figure 14.5.2. The point marked $\mathbf{x}_1$ is an interior point of Ω because Ω contains a neighborhood of $\mathbf{x}_1$. The point $\mathbf{x}_2$ is not an interior point of Ω because *no* neighborhood of $\mathbf{x}_2$ is completely contained in Ω. (Every neighborhood of $\mathbf{x}_2$ has points that lie outside of Ω.) ❏

Figure 14.5.2

DEFINITION 14.5.3 THE BOUNDARY OF A SET

A point $\mathbf{x}_0$ is said to be a *boundary point* of the set S iff every neighborhood of $\mathbf{x}_0$ contains points that are in S and points that are not in S. The set of all boundary points of S is called the *boundary* of S.

Example 2 The point marked $\mathbf{x}_2$ in Figure 15.4.2 is a boundary point of Ω: each neighborhood of $\mathbf{x}_2$ contains points in Ω and points not in Ω. ❏

DEFINITION 14.5.4 OPEN SET

A set S is said to be *open* iff it contains a neighborhood of each of its points.

Statements which are equivalent to this definition of open set are:

(a) A set S is open iff each of its points is an interior point.

(b) A set S is open iff it contains no boundary points.

DEFINITION 14.5.5 CLOSED SET

A set S is said to be *closed* iff it contains its boundary.

Here are some examples of sets that are open, sets that are closed, and sets that are neither open nor closed:

Two-Dimensional Examples

The sets

$$S_1 = \{(x, y): 1 < x < 2, 1 < y < 2\},$$
$$S_2 = \{(x, y): 3 \le x \le 4, 1 \le y \le 2\},$$
$$S_3 = \{(x, y): 5 \le x \le 6, 1 < y < 2\}$$

Figure 14.5.3

are displayed in Figure 14.5.3. S_1 is the inside of the first square. S_1 is open because it contains a neighborhood of each of its points. S_2 is the inside of the second square together with the four bounding line segments. S_2 is closed because it contains its entire boundary. S_3 is the inside of the last square together with the two vertical bounding line segments. S_3 is not open because it contains part of its boundary, and it is not closed because it does not contain all of its boundary. ❏

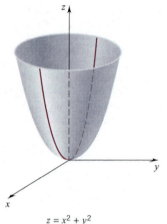

$z = x^2 + y^2$

Figure 14.5.4

Three-Dimensional Examples

We now examine some three-dimensional sets:

$$S_1 = \{(x, y, z): z > x^2 + y^2\},$$

$$S_2 = \{(x, y, z): z \geq x^2 + y^2\},$$

$$S_3 = \left\{(x, y, z): 1 \geq \frac{x^2 + y^2}{z}\right\}.$$

The boundary of each of these sets is the paraboloid of revolution

$$z = x^2 + y^2. \qquad \text{(Figure 14.5.4)}$$

The first set consists of all points above this surface. This set is open because, if a point is above this surface, then all points sufficiently close to it are also above this surface. Thus the set contains a neighborhood of each of its points. The second set is closed because it contains all of its boundary. The third set is neither open nor closed. It is not open because it contains some boundary points; for example, it contains the point $(1, 1, 2)$. It is not closed because it fails to contain the boundary point $(0, 0, 0)$. ❏

A Final Remark A neighborhood of $\mathbf{x}_0$ is a set of the form

$$\{\mathbf{x}: \|\mathbf{x} - \mathbf{x}_0\| < \delta\}.$$

If we remove $\mathbf{x}_0$ from the set, we have the set

$$\{\mathbf{x}: 0 < \|\mathbf{x} - \mathbf{x}_0\| < \delta\}.$$

Such a set is called a *deleted neighborhood* of $\mathbf{x}_0$. (We use deleted neighborhoods in the next section.) ❏

EXERCISES 14.5

In Exercises 1–10, specify the interior and the boundary of the set. State whether the set is open, closed, or neither. Then sketch the set.

1. $\{(x, y): 2 \le x \le 4, 1 \le y \le 3\}$.

2. $\{(x, y): 2 < x < 4, 1 < y < 3\}$.

3. $\{(x, y): 1 < x^2 + y^2 < 4\}$.

4. $\{(x, y): 1 \le x^2 \le 4\}$.

5. $\{(x, y): 1 < x^2 \le 4\}$. **6.** $\{(x, y): y < x^2\}$.

7. $\{(x, y): y \le x^2\}$.

8. $\{(x, y, z): 1 \le x \le 2, 1 \le y \le 2, 1 \le z < 2\}$.

9. $\{(x, y, z): x^2 + y^2 \le 1, 0 \le z \le 4\}$.

10. $\{(x, y, z): (x - 1)^2 + (y - 1)^2 + (z - 1)^2 < \frac{1}{4}\}$.

11. Let $S = \{\mathbf{x}_1, \mathbf{x}_2, \ldots, \mathbf{x}_n\}$ be a non-empty, finite set of points. (a) What is the interior of S? (b) What is the boundary of S? (c) Is S open, closed, or neither?

All the notions introduced in this section can be applied to sets of real numbers: write x for $\mathbf{x}$ and $|x|$ for $\|\mathbf{x}\|$. A *neighborhood* of a number x_0 is then a set of the form

$$\{x: |x - x_0| < \delta\} \qquad \text{with } \delta > 0.$$

This is just an open interval $(x_0 - \delta, x_0 + \delta)$. In Exercises 12–19, specify the interior and boundary of the given set of real numbers. State whether the set is open, closed, or neither.

12. $\{x: 1 < x < 3\}$. **13.** $\{x: 1 \le x \le 3\}$.

14. $\{x: 1 \le x < 3\}$. **15.** $\{x: x > 1\}$.

16. $\{x: x \le -1\}$. **17.** $\{x: x < -1 \text{ or } x \ge 1\}$.

18. The set of positive integers: $\{1, 2, 3, \ldots, n, \ldots\}$.

19. The set of reciprocals: $\{1, 1/2, 1/3, \ldots, 1/n, \ldots\}$.

20. Let $\varnothing$ be the empty set. Let X be the real line, the entire plane, or, in the three-dimensional case, all of three-space. For each subset A of X, let $X - A$ be the set of all points $\mathbf{x} \in X$ such that $\mathbf{x} \notin A$.

(a) Show that $\varnothing$ is both open and closed.

(b) Show that X is both open and closed. (It can be shown that $\varnothing$ and X are the only subsets of X that are both open and closed.)

(c) Let U be a subset of X. Show that U is open iff $X - U$ is closed.

(d) Let F be a subset of X. Show that F is closed iff $X - F$ is open.

■ 14.6 LIMITS AND CONTINUITY; EQUALITY OF MIXED PARTIALS

The Basic Notions

The limit process used in taking partial derivatives involved nothing new because in each instance all but one of the variables remained fixed. In this section we take up limits of the form

$$\lim_{(x,y) \to (x_0, y_0)} f(x, y) \qquad \text{and} \qquad \lim_{(x,y,z) \to (x_0, y_0, z_0)} f(x, y, z).$$

To avoid having to treat the two- and three-variable cases separately, we will write instead

$$\lim_{\mathbf{x} \to \mathbf{x}_0} f(\mathbf{x}).$$

This gives us both the two-variable case [set $\mathbf{x} = (x, y)$ and $\mathbf{x}_0 = (x_0, y_0)$] and the three-variable case [set $\mathbf{x} = (x, y, z)$ and $\mathbf{x}_0 = (x_0, y_0, z_0)$].

To take the limit of $f(\mathbf{x})$ as $\mathbf{x}$ tends to $\mathbf{x}_0$ we do not need f to be defined at $\mathbf{x}_0$ itself, but we do need f to be defined at points $\mathbf{x}$ close to $\mathbf{x}_0$. At this stage, we will assume that f is defined at all points $\mathbf{x}$ in some deleted neighborhood of $\mathbf{x}_0$ (f may or may not be defined at $\mathbf{x}_0$). This will guarantee that we can form $f(\mathbf{x})$ for all $\mathbf{x} \ne \mathbf{x}_0$ that are "sufficiently close" to $\mathbf{x}_0$. This approach is consistent with our approach to limits of functions of one variable in Chapter 2.

To say that

$$\lim_{\mathbf{x} \to \mathbf{x}_0} f(\mathbf{x}) = L$$

is to say that for **x** sufficiently close to $\mathbf{x}_0$ but different from $\mathbf{x}_0$, the number $f(\mathbf{x})$ is close to L; or, to put it another way, as $\|\mathbf{x} - \mathbf{x}_0\|$ tends to zero but remains different from zero, $|f(\mathbf{x}) - L|$ tends to zero. The ϵ-δ definition is a direct generalization of the ϵ-δ definition in the single-variable case.

DEFINITION 14.6.1 THE LIMIT OF A FUNCTION OF SEVERAL VARIABLES

Let f be a function defined at least on some deleted neighborhood of $\mathbf{x}_0$.

$$\lim_{\mathbf{x} \to \mathbf{x}_0} f(\mathbf{x}) = L \qquad \text{iff}$$

for each $\epsilon > 0$ there exists $\delta > 0$ such that

$$\text{if} \quad 0 < \|\mathbf{x} - \mathbf{x}_0\| < \delta \qquad \text{then} \qquad |f(\mathbf{x}) - L| < \epsilon$$

Example 1 We will show that the function

$$f(x, y) = \frac{xy + y^3}{x^2 + y^2}$$

does not have a limit at $(0, 0)$. Note that f is not defined at $(0, 0)$, but is defined for all $(x, y) \neq (0, 0)$.

Along the obvious paths to $(0, 0)$, the coordinate axes, the limiting value is 0:

Along the x-axis, $y = 0$; thus, $f(x, y) = f(x, 0) = 0$ and $\lim_{x \to 0} f(x, 0) = \lim_{x \to 0} 0 = 0$.

Along the y-axis, $x = 0$; thus, $f(x, y) = f(0, y) = y$ and $\lim_{y \to 0} f(0, y) = \lim_{y \to 0} y = 0$.

Along the line $y = 2x$, however, the limiting value is $\frac{2}{5}$:

$$f(x, y) = f(x, 2x) = \frac{2x^2 + 8x^3}{x^2 + 4x^2} = \frac{2}{5} + \frac{8}{5}x \to \frac{2}{5} \qquad \text{as } x \to 0.$$

There is nothing special about the line $y = 2x$ here. For example, you can verify that $f(x, y) \to -\frac{1}{2}$ as $(x, y) \to (0, 0)$ along the line $y = -x$. See Figure 14.6.1.

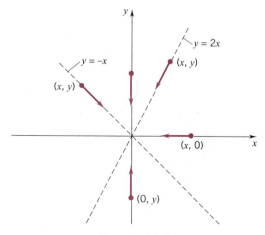

Figure 14.6.1

We have shown that not all paths to (0, 0) yield the same limiting value. It follows that f does not have a limit at (0, 0). ❑

Example 2 Show that the function

$$g(x, y) = \frac{x^2 y}{x^4 + y^2}$$

has limiting value 0 as $(x, y) \to (0, 0)$ along *any* line through the origin, but

$$\lim_{(x,y) \to (0,0)} g(x, y)$$

still does not exist. Note that the domain of g is all $(x, y) \neq (0, 0)$.

SOLUTION As in Example 1, it is easy to verify that $g(x, y) \to 0$ when $(x, y) \to (0, 0)$ along the coordinate axes. If we let $y = mx$, then

$$g(x, y) = g(x, mx) = \frac{mx^3}{x^4 + m^2 x^2} = \frac{mx}{x^2 + m^2} \qquad (x \neq 0)$$

and

$$\lim_{x \to 0} g(x, mx) = \lim_{x \to 0} \frac{mx}{x^2 + m^2} = 0.$$

Therefore, $g(x, y) \to 0$ as $(x, y) \to (0, 0)$ along any line through the origin.

Now suppose that $(x, y) \to (0, 0)$ along the parabola $y = x^2$. Then we have

$$g(x, y) = g(x, x^2) = \frac{x^4}{x^4 + x^4} = \frac{1}{2}$$

and

$$\lim_{x \to 0} g(x, x^2) = \lim_{x \to 0} \frac{1}{2} = \frac{1}{2}.$$

Thus, $g(x, y) \to \dfrac{1}{2}$ as $(x, y) \to (0, 0)$ along $y = x^2$ (Figure 14.6.2).

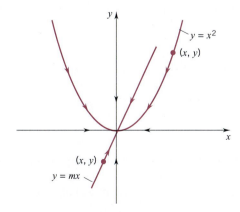

Figure 14.6.2

Since not all paths to (0, 0) yield the same limiting value, we conclude that g does not have a limit at (0, 0). ❑

As in the one-variable case, the limit (if it exists) is unique. Moreover, if

$$\lim_{\mathbf{x} \to \mathbf{x}_0} f(\mathbf{x}) = L \qquad \text{and} \qquad \lim_{\mathbf{x} \to \mathbf{x}_0} g(\mathbf{x}) = M,$$

then

$$\lim_{\mathbf{x} \to \mathbf{x}_0} [f(\mathbf{x}) + g(\mathbf{x})] = L + M, \qquad \lim_{\mathbf{x} \to \mathbf{x}_0} [\alpha f(\mathbf{x})] = \alpha L, \quad \alpha \text{ a real number,}$$

$$\lim_{\mathbf{x} \to \mathbf{x}_0} [f(\mathbf{x})g(\mathbf{x})] = LM, \qquad \text{and} \qquad \lim_{\mathbf{x} \to \mathbf{x}_0} [f(\mathbf{x})/g(\mathbf{x})] = L/M \qquad \text{provided } M \neq 0.$$

These results are not hard to derive. You can do it simply by imitating the comparable arguments in the one-variable case.

Suppose now that $\mathbf{x}_0$ is an interior point of the domain of f. To say that f is *continuous* at $\mathbf{x}_0$ is to say that

(14.6.2)
$$\lim_{\mathbf{x} \to \mathbf{x}_0} f(\mathbf{x}) = f(\mathbf{x}_0)$$

or, equivalently, that

(14.6.3)
$$\lim_{\mathbf{h} \to \mathbf{0}} f(\mathbf{x}_0 + \mathbf{h}) = f(\mathbf{x}_0).$$

For two variables we can write

$$\lim_{(x,y) \to (x_0, y_0)} f(x, y) = f(x_0, y_0)$$

and for three variables

$$\lim_{(x,y,z) \to (x_0, y_0, z_0)} f(x, y, z) = f(x_0, y_0, z_0).$$

To say that f is *continuous on an open set S* is to say that f is continuous at all points of S.

Some Examples of Continuous Functions

Polynomials in several variables, for example,

$$P(x, y) = x^2 y + 3x^3 y^4 - x + 2y \qquad \text{and} \qquad Q(x, y, z) = 6x^3 z - z^3 y + 2xyz,$$

are everywhere continuous. In the two-variable case, that means continuity at each point of the xy-plane; in the three-variable case, continuity at each point of three-space.

Rational functions (quotients of polynomials) are continuous everywhere except where the denominator is zero. Thus

$$f(x, y) = \frac{2x - y}{x^2 + y^2}$$

is continuous at each point of the xy-plane other than the origin $(0, 0)$;

$$g(x, y) = \frac{x^4}{x - y}$$

is continuous except on the line $y = x$;

$$h(x, y) = \frac{1}{x^2 - y}$$

is continuous except on the parabola $y = x^2$;

$$F(x, y, z) = \frac{2x}{x^2 + y^2 + z^2}$$

is continuous at each point of three-space other than the origin $(0, 0, 0)$;

$$G(x, y, z) = \frac{x^5 - y}{ax + by + cz},$$

where a, b, c are constants, is continuous except on the plane $ax + by + cz = 0$.

You can construct more elaborate continuous functions by forming composites: take, for example,

$$f(x, y, z) = \tan^{-1}\left(\frac{xz^2}{x + y}\right), \qquad g(x, y, z) = \sqrt{x^2 + y^4 + z^6}, \qquad h(x, y, z) = \sin xyz.$$

The first function is continuous except along the vertical plane $x + y = 0$. The other two functions are continuous at each point of space. The continuity of such composites follows from a simple theorem that we now state and prove. In the theorem, g is a function of several variables, but f is a function of a single variable.

THEOREM 14.6.4 THE CONTINUITY OF COMPOSITE FUNCTIONS

If g is continuous at the point $\mathbf{x}_0$ and f is continuous at the number $g(\mathbf{x}_0)$, then the composition $f \circ g$ is continuous at the point $\mathbf{x}_0$.

PROOF   We begin with $\epsilon > 0$. We must show that there exists $\delta > 0$ such that

$$\text{if} \quad \|\mathbf{x} - \mathbf{x}_0\| < \delta \quad \text{then} \quad |f(g(\mathbf{x})) - f(g(\mathbf{x}_0))| < \epsilon.$$

From the continuity of f at $g(\mathbf{x}_0)$, we know that there exists $\delta_1 > 0$ such that

$$\text{if} \quad |u - g(\mathbf{x}_0)| < \delta_1 \quad \text{then} \quad |f(u) - f(g(\mathbf{x}_0))| < \epsilon.$$

From the continuity of g at $\mathbf{x}_0$, we know that there exists $\delta > 0$ such that

$$\text{if} \quad \|\mathbf{x} - \mathbf{x}_0\| < \delta \quad \text{then} \quad |g(\mathbf{x}) - g(\mathbf{x}_0)| < \delta_1.$$

This last δ obviously works; namely,

$$\text{if} \quad \|\mathbf{x} - \mathbf{x}_0\| < \delta \quad \text{then} \quad |g(\mathbf{x}) - g(\mathbf{x}_0)| < \delta_1$$

$$\text{and} \quad |f(g(\mathbf{x})) - f(g(\mathbf{x}_0))| < \epsilon \quad \square$$

Continuity in Each Variable Separately

A *continuous function of several variables is continuous in each of its variables separately.* In the two-variable case, this means that, if

$$\lim_{(x,y)\to(x_0,y_0)} f(x, y) = f(x_0, y_0),$$

then

$$\lim_{x\to x_0} f(x, y_0) = f(x_0, y_0) \qquad \text{and} \qquad \lim_{y\to y_0} f(x_0, y) = f(x_0, y_0).$$

(This is not hard to prove.) The converse is false. *It is possible for a function to be continuous in each variable separately and yet fail to be continuous as a function of several variables.* You can see this in the next example.

Example 3 We set

$$f(x, y) = \begin{cases} \dfrac{2xy}{x^2 + y^2}, & (x, y) \neq (0, 0) \\ 0, & (x, y) = (0, 0). \end{cases}$$

Since

$$f(x, 0) = 0 \quad \text{for all } x \qquad \text{and} \qquad f(0, y) = 0 \quad \text{for all } y,$$

we have

$$\lim_{x\to 0} f(x, 0) = 0 = f(0, 0) \qquad \text{and} \qquad \lim_{y\to 0} f(0, y) = 0 = f(0, 0).$$

Thus, at the point $(0, 0)$, f is continuous in x and continuous in y. As a function of two variables, however, f is not continuous at $(0, 0)$. One way to see this is to note that we can approach $(0, 0)$ as closely as we wish by points of the form (t, t) with $t \neq 0$, (that is, along the line $y = x$). (See Figure 14.6.3.) At such points f takes on the value 1:

$$f(t, t) = \frac{2t^2}{t^2 + t^2} = 1.$$

Hence, f cannot tend to $f(0, 0) = 0$ as required. ❑

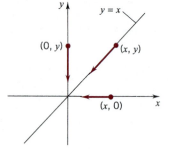

Figure 14.6.3

Continuity and Partial Differentiability

For functions of a single variable the existence of the derivative guarantees continuity (Theorem 3.1.4). *For functions of several variables the existence of partial derivatives fails to guarantee continuity.*†

To show this, we can use the same function

$$f(x, y) = \begin{cases} \dfrac{2xy}{x^2 + y^2}, & (x, y) \neq (0, 0) \\ 0, & (x, y) = (0, 0). \end{cases}$$

Since both $f(x, 0)$ and $f(0, y)$ are constantly zero, both partials exist (and are zero) at $(0, 0)$, and yet, as you saw, the function is discontinuous at $(0, 0)$.

---

† See, however, Exercise 34.

It is not hard to understand how a function can have partial derivatives and yet fail to be continuous. The existence of $\partial f/\partial x$ at (x_0, y_0) depends on the behavior of f only at points of the form $(x_0 + h, y_0)$. Similarly, the existence of $\partial f/\partial y$ at (x_0, y_0) depends on the behavior of f only at points of the form $(x_0, y_0 + k)$. On the other hand, continuity at (x_0, y_0) depends on the behavior of f at points of the more general form $(x_0 + h, y_0 + k)$. More briefly, we can put it this way: *the existence of a partial derivative depends on the behavior of the function along a line segment (two directions), whereas continuity depends on the behavior of the function in all directions.*

Derivatives of Higher Order; Equality of Mixed Partials

Suppose that f is a function of x and y with first partials f_x and f_y. These are again functions of x and y and may themselves possess partial derivatives such as $(f_x)_x$, $(f_x)_y$, $(f_y)_x$, and $(f_y)_y$. These functions are called the *second-order partials*. If $z = f(x, y)$, we use the following notations for second-order partials

$$(f_x)_x = f_{xx} = \frac{\partial}{\partial x}\left(\frac{\partial f}{\partial x}\right) = \frac{\partial^2 f}{\partial x^2} = \frac{\partial^2 z}{\partial x^2},$$

$$(f_x)_y = f_{xy} = \frac{\partial}{\partial y}\left(\frac{\partial f}{\partial x}\right) = \frac{\partial^2 f}{\partial y \partial x} = \frac{\partial^2 z}{\partial y \partial x},$$

$$(f_y)_x = f_{yx} = \frac{\partial}{\partial x}\left(\frac{\partial f}{\partial y}\right) = \frac{\partial^2 f}{\partial x \partial y} = \frac{\partial^2 z}{\partial x \partial y},$$

$$(f_y)_y = f_{yy} = \frac{\partial}{\partial y}\left(\frac{\partial f}{\partial y}\right) = \frac{\partial^2 f}{\partial y^2} = \frac{\partial^2 z}{\partial y^2}.$$

Note that there are two "mixed" partials: f_{xy} (or $\partial^2 f/\partial y \partial x$) and f_{yx} (or $\partial^2 f/\partial x \partial y$). The first of these is obtained by differentiating first with respect to x and then with respect to y. The second is obtained by differentiating first with respect to y and then with respect to x.

Example 4 The function $f(x, y) = \sin x^2 y$ has first partials

$$\frac{\partial f}{\partial x} = 2xy \cos x^2 y \qquad \text{and} \qquad \frac{\partial f}{\partial y} = x^2 \cos x^2 y.$$

The second-order partials are

$$\frac{\partial^2 f}{\partial x^2} = -4x^2 y^2 \sin x^2 y + 2y \cos x^2 y, \qquad \frac{\partial^2 f}{\partial y \partial x} = -2x^3 y \sin x^2 y + 2x \cos x^2 y,$$

$$\frac{\partial^2 f}{\partial x \partial y} = -2x^3 y \sin x^2 y + 2x \cos x^2 y, \qquad \frac{\partial^2 f}{\partial y^2} = -x^4 \sin x^2 y. \quad \square$$

Example 5 Setting $f(x, y) = \ln(x^2 + y^3)$, we have

$$\frac{\partial f}{\partial x} = \frac{2x}{x^2 + y^3} \qquad \text{and} \qquad \frac{\partial f}{\partial y} = \frac{3y^2}{x^2 + y^3}.$$

The second-order partials are

$$\frac{\partial^2 f}{\partial x^2} = \frac{(x^2 + y^3)2 - 2x(2x)}{(x^2 + y^3)^2} = \frac{2(y^3 - x^2)}{(x^2 + y^3)^2},$$

$$\frac{\partial^2 f}{\partial y \partial x} = \frac{-2x(3y^2)}{(x^2 + y^3)^2} = -\frac{6xy^2}{(x^2 + y^3)^2},$$

$$\frac{\partial^2 f}{\partial x \partial y} = \frac{-3y^2(2x)}{(x^2 + y^3)^2} = -\frac{6xy^2}{(x^2 + y^3)^2},$$

$$\frac{\partial^2 f}{\partial y^2} = \frac{(x^2 + y^3)6y - 3y^2(3y^2)}{(x^2 + y^3)^2} = \frac{3y(2x^2 - y^3)}{(x^2 + y^3)^2}. \quad \square$$

Perhaps you noticed that in both examples we had

$$\frac{\partial^2 f}{\partial y \partial x} = \frac{\partial^2 f}{\partial x \partial y}.$$

Since in neither case was f symmetric in x and y, this equality of the mixed partials was not due to symmetry. Actually it was due to continuity. It can be proved that

(14.6.5)

$$\frac{\partial^2 f}{\partial y \partial x} = \frac{\partial^2 f}{\partial x \partial y}$$

on every open set U on which f and its partials

$$\frac{\partial f}{\partial x}, \quad \frac{\partial f}{\partial y}, \quad \frac{\partial^2 f}{\partial y \partial x}, \quad \frac{\partial^2 f}{\partial x \partial y}$$

are all continuous.†

In the case of a function of three variables you can look for three first partials

$$\frac{\partial f}{\partial x}, \quad \frac{\partial f}{\partial y}, \quad \frac{\partial f}{\partial z},$$

and nine second partials

$$\frac{\partial^2 f}{\partial x^2}, \quad \frac{\partial^2 f}{\partial x \partial y}, \quad \frac{\partial^2 f}{\partial x \partial z} \qquad \frac{\partial^2 f}{\partial y \partial x}, \quad \frac{\partial^2 f}{\partial y^2}, \quad \frac{\partial^2 f}{\partial y \partial z} \qquad \frac{\partial^2 f}{\partial z \partial x}, \quad \frac{\partial^2 f}{\partial z \partial y}, \quad \frac{\partial^2 f}{\partial z^2}.$$

Here again, there is equality of the mixed partials

$$\frac{\partial^2 f}{\partial y \partial x} = \frac{\partial^2 f}{\partial x \partial y}, \qquad \frac{\partial^2 f}{\partial z \partial x} = \frac{\partial^2 f}{\partial x \partial z}, \qquad \frac{\partial^2 f}{\partial y \partial z} = \frac{\partial^2 f}{\partial z \partial y}$$

provided that f and its first and second partials are continuous.

---

† For a proof consult a text on advanced calculus.

Example 6 For

$$f(x, y, z) = xe^y \sin \pi z$$

we have

$$\frac{\partial f}{\partial x} = e^y \sin \pi z, \qquad \frac{\partial f}{\partial y} = xe^y \sin \pi z, \qquad \frac{\partial f}{\partial z} = \pi xe^y \cos \pi z,$$

$$\frac{\partial^2 f}{\partial x^2} = 0, \qquad \frac{\partial^2 f}{\partial y^2} = xe^y \sin \pi z, \qquad \frac{\partial^2 f}{\partial z^2} = -\pi^2 xe^y \sin \pi z,$$

$$\frac{\partial^2 f}{\partial y \partial x} = e^y \sin \pi z = \frac{\partial^2 f}{\partial x \partial y},$$

$$\frac{\partial^2 f}{\partial z \partial x} = \pi e^y \cos \pi z = \frac{\partial^2 f}{\partial x \partial z},$$

$$\frac{\partial^2 f}{\partial y \partial z} = \pi xe^y \cos \pi z = \frac{\partial^2 f}{\partial z \partial y}. \qquad \square$$

EXERCISES 14.6

In Exercises 1–20, calculate the second-order partial derivatives of the given function. (NOTE: Treat A, B, C, and D as constants.)

1. $f(x, y) = Ax^2 + 2Bxy + Cy^2$.

2. $f(x, y) = Ax^3 + Bx^2y + Cxy^2$.

3. $f(x, y) = Ax + By + Ce^{xy}$

4. $f(x, y) = x^2\cos y + y^2\sin x$.

5. $f(x, y, z) = (x + y^2 + z^3)^2$.

6. $f(x, y) = \sqrt{x + y^2}$.

7. $f(x, y) = \ln\left(\dfrac{x}{x + y}\right)$.

8. $f(x, y) = \dfrac{Ax + By}{Cx + Dy}$.

9. $f(x, y, z) = (x + y)(y + z)(z + x)$.

10. $f(x, y, z) = \tan^{-1}xyz$.

11. $f(x, y) = x^y$.

12. $f(x, y, z) = \sin(x + z^y)$.

13. $f(x, y) = xe^y + ye^x$.

14. $f(x, y) = \tan^{-1}(y/x)$.

15. $f(x, y) = \ln\sqrt{x^2 + y^2}$.

16. $f(x, y) = \sin(x^3y^2)$.

17. $f(x, y) = \cos^2(xy)$.

18. $f(x, y) = e^{xy^2}$.

19. $f(x, y, z) = xy\sin z - xz\sin y$.

20. $f(x, y, z) = xe^y + ye^z + ze^x$.

21. Show that

$$\text{if} \quad u = \frac{xy}{x + y}$$

$$\text{then} \quad x^2\frac{\partial^2 u}{\partial x^2} + 2xy\frac{\partial^2 u}{\partial x \partial y} + y^2\frac{\partial^2 u}{\partial y^2} = 0.$$

Laplace's Equation† The partial differential equation

$$\frac{\partial^2 f}{\partial x^2} + \frac{\partial^2 f}{\partial y^2} = 0$$

is known as *Laplace's equation in two dimensions.* It is used to describe potentials and steady-state temperature distributions in the plane. In three dimensions, Laplace's equation is

$$\frac{\partial^2 f}{\partial x^2} + \frac{\partial^2 f}{\partial y^2} + \frac{\partial^2 f}{\partial z^2} = 0$$

and it is satisfied by gravitational and electrostatic potentials, and by steady-state temperature distributions in space. Functions that satisfy Laplace's equation are called *harmonic* functions.

---

† Named after the French mathematician Pierre-Simon Laplace (1749–1827). Laplace wrote two monumental works: one on celestial mechanics, the other on probability theory. He also made major contributions to the theory of differential equations.

In Exercises 22–25, show that the given function is harmonic.

22. $f(x, y) = x^3 - 3xy^2$.　　　**23.** $f(x, y) = e^x \sin y$.

24. $f(x, y) = \cos x \sinh y + \sin x \cosh y$.

25. $f(x, y) = \ln \sqrt{x^2 + y^2}$.

In Exercises 26 and 27, show that the given function satisfies the three-dimensional Laplace equation.

26. $f(x, y, z) = \dfrac{1}{\sqrt{x^2 + y^2 + z^2}}$.

27. $f(x, y, z) = e^{x+y} \cos \sqrt{2}z$.

The Wave Equation　The partial differential equation

$$\frac{\partial^2 f}{\partial t^2} - c^2 \frac{\partial^2 f}{\partial x^2} = 0,$$

where c is a positive constant, is known as the *wave equation*. It arises in the study of phenomena involving the propagation of waves in a continuous medium. For example, studies of water waves, sound waves, and light waves are all based on this equation. The wave equation is also used in the study of mechanical vibrations such as a vibrating string.

In Exercises 28–32, show that the given function is a solution of the wave equation.

28. $f(x, t) = (Ax + B)(Ct + D)$, where A, B, C, and D are constants.

29. $f(x, t) = \sin(x + ct) \cos(2x + 2ct)$.

30. $f(x, t) = \ln(x + ct)$.

31. $f(x, t) = (Ae^{kx} + Be^{-kx})(Ce^{ckt} + De^{-ckt})$, where c and k are constants.

32. $f(x, t) = g(x + ct) + h(x - ct)$, where g and h are any two, twice differentiable functions. (This is the most general form of a solution of the wave equation.)

33. Verify that

$$\frac{\partial^2 f}{\partial y \partial x} = \frac{\partial^2 f}{\partial x \partial y}$$

given that
(a) $f(x, y) = g(x) + h(y)$ with g and h differentiable.
(b) $f(x, y) = g(x)h(y)$ with g and h differentiable.
(c) $f(x, y)$ is a polynomial in x and y.
　　HINT: Check each term $x^m y^n$ separately.

34. If a function of several variables has all first partials at a point, then it is continuous in each variable separately at that point. Show, for example, that
if $\dfrac{\partial f}{\partial x}$ exists at (x_0, y_0), then f is continuous in x at (x_0, y_0).

35. Let f be a function of x and y with everywhere continuous second partials. Is it possible that

(a) $\dfrac{\partial f}{\partial x} = x + y$ and $\dfrac{\partial f}{\partial y} = y - x$?

(b) $\dfrac{\partial f}{\partial x} = xy$ and $\dfrac{\partial f}{\partial y} = xy$?

36. Let g be a twice-differentiable function of one variable and set

$$h(x, y) = g(x + y) + g(x - y).$$

Show that

$$\frac{\partial^2 h}{\partial x^2} = \frac{\partial^2 h}{\partial y^2}.$$

HINT: The chain rule (Exercise 56, Section 14.4).

37. Let f be a function of x and y with third-order partials

$$\frac{\partial^3 f}{\partial x^2 \partial y} = \frac{\partial}{\partial x}\left(\frac{\partial^2 f}{\partial x \partial y}\right) \quad \text{and} \quad \frac{\partial^3 f}{\partial y \partial x^2} = \frac{\partial}{\partial y}\left(\frac{\partial^2 f}{\partial x^2}\right).$$

Show that, if all the partials are continuous, then

$$\frac{\partial^3 f}{\partial x^2 \partial y} = \frac{\partial^3 f}{\partial y \partial x^2}.$$

38. Show that the following functions do not have a limit at $(0, 0)$:

(a) $f(x, y) = \dfrac{x^2 - y^2}{x^2 + y^2}$.　　(b) $f(x, y) = \dfrac{y^2}{x^2 + y^2}$.

In Exercises 39 and 40, evaluate the limit as (x, y) approaches the origin along:

(a) The x-axis.　　(b) The y-axis.
(c) The line $y = mx$.　　(d) The spiral $r = \theta$, $\theta > 0$.
(e) The differentiable curve $y = f(x)$, given that $f(0) = 0$.
(f) The arc $r = \sin 3\theta$, $\frac{1}{6}\pi < \theta < \frac{1}{3}\pi$.
(g) The path $\mathbf{r}(t) = \dfrac{1}{t}\mathbf{i} + \dfrac{\sin t}{t}\mathbf{j}$, $t > 0$.

39. $\displaystyle\lim_{(x,y)\to(0,0)} \frac{xy}{x^2 + y^2}$.　　**40.** $\displaystyle\lim_{(x,y)\to(0,0)} \frac{xy^2}{(x^2 + y^2)^{3/2}}$.

41. Set

$$g(x, y) = \begin{cases} \dfrac{x^2 y^2}{x^4 + y^4}, & (x, y) \neq (0, 0) \\ 0, & (x, y) = (0, 0). \end{cases}$$

(a) Show that $\partial g/\partial x$ and $\partial g/\partial y$ both exist at $(0, 0)$. What are their values at $(0, 0)$?
(b) Show that $\displaystyle\lim_{(x,y)\to(0,0)} g(x, y)$ does not exist.

42. Set

$$f(x, y) = \frac{x - y^4}{x^3 - y^4}.$$

Determine whether or not f has a limit at $(1, 1)$.
HINT: Let (x, y) tend to $(1, 1)$ along the lines $x = 1$ and $y = 1$.

43. Set

$$f(x, y) = \begin{cases} \dfrac{xy(y^2 - x^2)}{x^2 + y^2}, & (x, y) \neq (0, 0) \\ 0, & (x, y) = (0, 0). \end{cases}$$

It can be shown that some of the second partials are discontinuous at $(0, 0)$. Show that

$$\frac{\partial^2 f}{\partial y \partial x}(0, 0) \neq \frac{\partial^2 f}{\partial x \partial y}(0, 0).$$

44. Let f be a function of x and y which has continuous first and second partial derivatives throughout some set D in the plane. Suppose that $f_{xy}(x, y) = 0$ for all $(x, y) \in D$. What can you conclude about f?

■ CHAPTER HIGHLIGHTS

14.1   Elementary Examples

14.2   A Brief Catalog of the Quadric Surfaces; Projections

14.3   Graphs; Level Curves and Level Surfaces

14.4   Partial Derivatives

14.5   Open Sets and Closed Sets

Some sets are neither open nor closed.

14.6   Limits and Continuity; Equality of Mixed Partials

Points in the domain of a function of several variables can be written in vector notation. In the two-variable case, set $\mathbf{x} = (x, y)$, and, in the three-variable case, set $\mathbf{x} = (x, y, z)$. The vector notation enables us to treat the two cases together simply by writing $f(\mathbf{x})$.

A continuous function of several variables is continuous in each of its variables. The converse is false: it is possible for a function to be continuous in each variable separately and yet fail to be continuous as a function of several variables. (p. 936)

For functions of several variables the existence of partial derivatives fails to guarantee continuity. (p. 936)

The existence of a partial derivative depends on the behavior of the function along a line segment (two directions), whereas continuity depends on the behavior of the function in all directions. (p. 937)

equality of mixed partials (p. 938)

■ PROJECTS AND EXPLORATIONS USING TECHNOLOGY

To do these exercises you will need a graphics calculator or a computer with graphing capability. The majority of these problems are open-ended so different approaches may be used to solve them. You should be aware that different approaches can result in slight variations in the answers. Round your numerical answers to at least four decimal places. The rounding method that your calculator or computer uses also may cause variations in answers.

14.1 A problem about surface area known as Schwarz's paradox involves computing the following limit for the surface area of a cylinder

$$\lim_{n,m \to \infty} 2\pi r \, \frac{\sin(\pi/m)}{\pi/m} \left[h^2 + \pi^2 r^2 \left(\frac{n^2}{m^2} \right) \left(\frac{1 - \cos(\pi/m)}{\pi/m} \right)^2 \right]^{1/2},$$

where r and h are positive constants.
- (a) Suppose that a cylinder of height h and radius r is subdivided into n small cylindrical bands and each band is divided into m equilateral triangles (see the figure). Show that the sum of the areas of these triangles is the expression just given.

- (b) Try to compute the limit by holding n fixed and letting $m \to \infty$, and then letting $n \to \infty$.
- (c) Now fix m and let $n \to \infty$, and then let $m \to \infty$.
- (d) Set $n = m$ and let $n \to \infty$.
- (e) Set $n = m^2$ and let $m \to \infty$.

14.2 One version of the Newton-Raphson iteration method for finding a zero of a function of two variables is

$$x_{n+1} = x_n - \frac{f(x_n, y_n)}{\dfrac{\partial f}{\partial x}(x_n, y_n)}$$

$$y_{n+1} = y_n - \frac{f(x_n, y_n)}{\dfrac{\partial f}{\partial y}(x_n, y_n)}.$$

(a) Let $f(x, y) = \sqrt{x^2 + y^2}$. If this version of Newton's method is used with an initial guess of $(1, 1)$, it does not converge. Does it converge for other initial guesses? Can you explain? Now try the method using the function

$$g(x, y) = e^{xy^2} - \ln(x + y).$$

(b) As a variation of the method, consider

$$x_{n+1} = x_n - \frac{f(x_n, y_n)}{\frac{\partial f}{\partial x}(x_n, y_n)}$$

$$y_{n+1} = y_n - \frac{f(x_{n+1}, y_n)}{\frac{\partial f}{\partial y}(x_{n+1}, y_n)}.$$

Apply this method to the functions f and g in part (a) using the same initial guesses. Compare the convergence of this method with the method in part (a).

(c) A modification to find the intersecton of two surfaces is given by the solution to the following system:

$$0 = f(x_n, y_n) + (x_{n+1} - x_n)\frac{\partial f}{\partial x}(x_n, y_n) + (y_{n+1} - y_n)\frac{\partial f}{\partial y}(x_n, y_n)$$

$$0 = g(x_n, y_n) + (x_{n+1} - x_n)\frac{\partial g}{\partial x}(x_n, y_n) + (y_{n+1} - y_n)\frac{\partial g}{\partial y}(x_n, y_n).$$

Apply this method to the pair of functions

$$f(x, y) = 1 - \sqrt{x^2 + 2y^4},$$

$$g(x, y) = 1 + x^6 - 5y^2,$$

with an initial guess of $(\frac{1}{2}, \frac{1}{2})$. Does this method seem to converge as quickly as the Newton-Raphson method for functions of one variable?

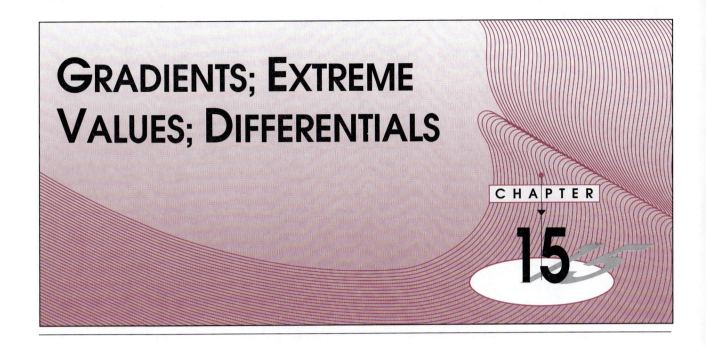

GRADIENTS; EXTREME VALUES; DIFFERENTIALS

■ 15.1 **DIFFERENTIABILITY AND GRADIENT**

The Notion of Differentiability

Our object here is to extend the notion of differentiability from real-valued functions of one variable to real-valued functions of several variables. Partial derivatives alone do not fulfill this role because they reflect behavior only in particular directions (the coordinate directions).

In the one-variable case we formed the difference quotient

$$\frac{f(x + h) - f(x)}{h}$$

and called f differentiable at x provided that this quotient had a limit as h tended to zero. In the multivariable case we can still form the difference

$$f(\mathbf{x} + \mathbf{h}) - f(\mathbf{x}),$$

but the "quotient"

$$\frac{f(\mathbf{x} + \mathbf{h}) - f(\mathbf{x})}{\mathbf{h}}$$

is not defined because it makes no sense to divide a real number $[f(\mathbf{x} + \mathbf{h}) - f(\mathbf{x})]$ by a vector $\mathbf{h}$.

We can get around this difficulty by going back to an idea introduced in the exercises to Section 3.9. We review the idea here.

Let g be a real-valued function of a single variable which is defined on some open interval containing 0. We say that $g(h)$ is *little-o(h)* (read "little oh of h") and write $g(h) = o(h)$ iff

$$\lim_{h \to 0} \frac{g(h)}{|h|} = 0.$$

For a function of one variable the following statements are equivalent:

$$\lim_{h \to 0} \frac{f(x + h) - f(x)}{h} = f'(x),$$

$$\lim_{h \to 0} \frac{[f(x + h) - f(x)] - f'(x)h}{h} = 0,$$

$$\lim_{h \to 0} \frac{[f(x + h) - f(x)] - f'(x)h}{|h|} = 0,$$

$$[f(x + h) - f(x)] - f'(x)h = o(h),$$

$$[f(x + h) - f(x)] = f'(x)h + o(h).$$

Thus, for a function of one variable, the derivative of f at x is the unique number $f'(x)$ such that

$$f(x + h) - f(x) = f'(x)h + o(h).$$

It is this view of the derivative that inspires the notion of differentiability in the multivariable case.

Paralleling the definition for a function of a single variable, let g be a function of several variables which is defined in some neighborhood of **0**. We will say that $g(\mathbf{h})$ is $o(\mathbf{h})$ iff

$$\lim_{\mathbf{h} \to 0} \frac{g(\mathbf{h})}{\|\mathbf{h}\|} = 0.$$

We will denote by $o(\mathbf{h})$ any function $g(\mathbf{h})$ that has this property.

Now let f be a function of several variables *defined at least in some neighborhood of* **x**. [In the three-variable case, $\mathbf{x} = (x, y, z)$; in the two-variable case, $\mathbf{x} = (x, y)$.]

DEFINITION 15.1.1 DIFFERENTIABILITY

We say that f is *differentiable* at **x** iff there exists a vector **y** such that

$$f(\mathbf{x} + \mathbf{h}) - f(\mathbf{x}) = \mathbf{y} \cdot \mathbf{h} + o(\mathbf{h}).$$

It is not hard to show that, if such a vector **y** exists, it is unique (Exercise 43). We call this unique vector *the gradient of f at* **x** and denote it by $\nabla f(\mathbf{x})$:†

DEFINITION 15.1.2 GRADIENT

Let f be differentiable at **x**. The *gradient of f at* **x** is the unique vector $\nabla f(\mathbf{x})$ such that

$$f(\mathbf{x} + \mathbf{h}) - f(\mathbf{x}) = \nabla f(\mathbf{x}) \cdot \mathbf{h} + o(\mathbf{h}).$$

---

† The notation grad f is sometimes used to denote the gradient of f.

The similarities between the one-variable case,

$$f(x + h) - f(x) = f'(x)h + o(h),$$

and the multivariable case,

$$f(\mathbf{x} + \mathbf{h}) - f(\mathbf{x}) = \nabla f(\mathbf{x}) \cdot \mathbf{h} + o(\mathbf{h}),$$

are obvious. We point to the differences. There are essentially two of them:

(1) While the derivative $f'(x)$ is a number, the gradient $\nabla f(\mathbf{x})$ is a vector.

(2) While $f'(x)h$ is the ordinary product of two real numbers, $\nabla f(\mathbf{x}) \cdot \mathbf{h}$ is the dot product of two vectors.

Calculating Gradients

First we calculate some gradients by applying the definition directly. Then we give a theorem that makes such calculations much easier. Finally, we calculate some gradients with the aid of the theorem. As for notation, in the two-variable case we write

$$\nabla f(\mathbf{x}) = \nabla f(x, y) \quad \text{and} \quad \mathbf{h} = (h_1, h_2),$$

and in the three-variable case,

$$\nabla f(\mathbf{x}) = \nabla f(x, y, z) \quad \text{and} \quad \mathbf{h} = (h_1, h_2, h_3).$$

Example 1 For the function

$$f(x, y) = x^2 + y^2$$

we have

$$
\begin{aligned}
f(\mathbf{x} + \mathbf{h}) - f(\mathbf{x}) &= f(x + h_1, y + h_2) - f(x, y) \\
&= [(x + h_1)^2 + (y + h_2)^2] - [x^2 + y^2] \\
&= [2xh_1 + 2yh_2] + [h_1^2 + h_2^2] \\
&= [2x\mathbf{i} + 2y\mathbf{j}] \cdot \mathbf{h} + \|\mathbf{h}\|^2.
\end{aligned}
$$

The remainder $\|\mathbf{h}\|^2$ is $o(\mathbf{h})$:

$$\frac{\|\mathbf{h}\|^2}{\|\mathbf{h}\|} = \|\mathbf{h}\| \to 0 \quad \text{as} \quad \mathbf{h} \to \mathbf{0}.$$

Thus

$$\nabla f(\mathbf{x}) = \nabla f(x, y) = 2x\mathbf{i} + 2y\mathbf{j}. \quad \square$$

Example 2 For the function

$$f(x, y, z) = 2xy - 3z^2$$

we have

$$
\begin{aligned}
f(\mathbf{x} + \mathbf{h}) - f(\mathbf{x}) &= f(x + h_1, y + h_2, z + h_3) - f(x, y, z) \\
&= 2(x + h_1)(y + h_2) - 3(z + h_3)^2 - [2xy - 3z^2] \\
&= 2xh_2 + 2yh_1 + 2h_1h_2 - 6zh_3 - 3h_3^2 \\
&= (2y\mathbf{i} + 2x\mathbf{j} - 6z\mathbf{k}) \cdot (h_1\mathbf{i} + h_2\mathbf{j} + h_3\mathbf{k}) + 2h_1h_2 - 3h_3^2 \\
&= (2y\mathbf{i} + 2x\mathbf{j} - 6z\mathbf{k}) \cdot \mathbf{h} + 2h_1h_2 - 3h_3^2.
\end{aligned}
$$

It remains to show that the remainder $g(\mathbf{h}) = 2h_1h_2 - 3h_3^2$ is $o(\mathbf{h})$. We can rewrite g as

$$g(\mathbf{h}) = (2h_2\mathbf{i} - 3h_3\mathbf{k}) \cdot (h_1\mathbf{i} + h_2\mathbf{j} + h_3\mathbf{k}) = (2h_2\mathbf{i} - 3h_3\mathbf{k}) \cdot \mathbf{h}.$$

Then

$$\frac{|g(\mathbf{h})|}{\|\mathbf{h}\|} = \frac{\|2h_2\mathbf{i} - 3h_3\mathbf{k}\| \cdot \|\mathbf{h}\| \, |\cos\theta|}{\|\mathbf{h}\|} \leq \frac{\|2h_2\mathbf{i} - 3h_3\mathbf{k}\| \cdot \|\mathbf{h}\|}{\|\mathbf{h}\|} = \|2h_2\mathbf{i} - 3h_3\mathbf{k}\|.$$

Now, since $\mathbf{h} \to \mathbf{0}$ iff $h_1 \to 0$, $h_2 \to 0$, and $h_3 \to 0$, it follows that $\|2h_2\mathbf{i} - 3h_3\mathbf{k}\| \to 0$ as $\mathbf{h} \to \mathbf{0}$. Therefore, $g(\mathbf{h})/\|\mathbf{h}\| \to 0$ as $\mathbf{h} \to \mathbf{0}$, and

$$\nabla f(\mathbf{x}) = 2y\mathbf{i} + 2x\mathbf{j} - 6z\mathbf{k}. \quad \square$$

Examples 1 and 2 illustrate the use of the definition to calculate gradients. The following theorem provides an alternative method which you will find to be much easier to apply.

THEOREM 15.1.3

If f has continuous first partials in a neighborhood of $\mathbf{x}$, then f is differentiable at $\mathbf{x}$ and

$$\nabla f(\mathbf{x}) = \frac{\partial f}{\partial x}(\mathbf{x})\mathbf{i} + \frac{\partial f}{\partial y}(\mathbf{x})\mathbf{j} \qquad (f \text{ a function of two variables})$$

or

$$\nabla f(\mathbf{x}) = \frac{\partial f}{\partial x}(\mathbf{x})\mathbf{i} + \frac{\partial f}{\partial y}(\mathbf{x})\mathbf{j} + \frac{\partial f}{\partial z}(\mathbf{x})\mathbf{k}. \qquad (f \text{ a function of three variables})$$

The proof is somewhat difficult. A proof of the two-variable case is given in a supplement at the end of this section. $\square$

Returning to Examples 1 and 2: for $f(x, y) = x^2 + y^2$, we have $\partial f/\partial x = 2x$ and $\partial f/\partial y = 2y$, and since these functions are continuous, we have

$$\nabla f(\mathbf{x}) = 2x\mathbf{i} + 2y\mathbf{j}.$$

For $f(x, y, z) = 2xy - 3z^2$, we have $\partial f/\partial x = 2y$, $\partial f/\partial y = 2x$ and $\partial f/\partial z = -6z$, and each of these functions is continuous. Therefore,

$$\nabla f(\mathbf{x}) = 2y\mathbf{i} + 2x\mathbf{j} - 6z\mathbf{k}.$$

Example 3 For

$$f(x, y) = xe^y - ye^x$$

we have

$$\frac{\partial f}{\partial x}(x, y) = e^y - ye^x, \qquad \frac{\partial f}{\partial y}(x, y) = xe^y - e^x$$

and therefore

$$\nabla f(x, y) = (e^y - ye^x)\mathbf{i} + (xe^y - e^x)\mathbf{j}. \quad \square$$

When there is no reason to emphasize the point of evaluation, we don't write

$$\nabla f(\mathbf{x}) \quad \text{or} \quad \nabla f(x, y) \quad \text{or} \quad \nabla f(x, y, z)$$

but simply ∇f. Thus for the function

$$f(x, y) = xe^y - ye^x$$

we write

$$\frac{\partial f}{\partial x} = e^y - ye^x, \qquad \frac{\partial f}{\partial y} = xe^y - e^x$$

and

$$\nabla f = (e^y - ye^x)\mathbf{i} + (xe^y - e^x)\mathbf{j}.$$

Example 4 For

$$f(x, y, z) = \sin(xy^2z^3)$$

we have

$$\frac{\partial f}{\partial x} = y^2z^3\cos(xy^2z^3), \qquad \frac{\partial f}{\partial y} = 2xyz^3\cos(xy^2z^3), \qquad \frac{\partial f}{\partial z} = 3xy^2z^2\cos(xy^2z^3)$$

and

$$\nabla f = yz^2\cos(xy^2z^3)[yz\mathbf{i} + 2xz\mathbf{j} + 3xy\mathbf{k}]. \quad ❑$$

Example 5 Let f be the function defined by

$$f(x, y, z) = x \sin \pi y + y \cos \pi z.$$

Evaluate ∇f at $(0, 1, 2)$.

SOLUTION Here

$$\frac{\partial f}{\partial x} = \sin \pi y, \qquad \frac{\partial f}{\partial y} = \pi x \cos \pi y + \cos \pi z, \qquad \frac{\partial f}{\partial z} = -\pi y \sin \pi z.$$

At $(0, 1, 2)$

$$\frac{\partial f}{\partial x} = 0, \qquad \frac{\partial f}{\partial y} = 1, \qquad \frac{\partial f}{\partial z} = 0 \qquad \text{and thus} \qquad \nabla f(0, 1, 2) = \mathbf{j}. \quad ❑$$

Of special interest for later work are the powers of r where, as usual,

$$r = \|\mathbf{r}\| \qquad \text{and} \qquad \mathbf{r} = x\mathbf{i} + y\mathbf{j} + z\mathbf{k}.$$

We begin by showing that, for $r \neq 0$,

(15.1.4)

$$\nabla r = \frac{\mathbf{r}}{r} \qquad \text{and} \qquad \nabla\left(\frac{1}{r}\right) = -\frac{\mathbf{r}}{r^3}.$$

PROOF

$$\nabla r = \nabla(x^2 + y^2 + z^2)^{1/2}$$

$$= \frac{\partial}{\partial x}(x^2 + y^2 + z^2)^{1/2}\mathbf{i} + \frac{\partial}{\partial y}(x^2 + y^2 + z^2)^{1/2}\mathbf{j} + \frac{\partial}{\partial z}(x^2 + y^2 + z^2)^{1/2}\mathbf{k}$$

$$= \frac{x}{(x^2 + y^2 + z^2)^{1/2}}\mathbf{i} + \frac{y}{(x^2 + y^2 + z^2)^{1/2}}\mathbf{j} + \frac{z}{(x^2 + y^2 + z^2)^{1/2}}\mathbf{k}$$

$$= \frac{1}{(x^2 + y^2 + z^2)^{1/2}}(x\mathbf{i} + y\mathbf{j} + z\mathbf{k}) = \frac{\mathbf{r}}{r}.$$

$$\nabla\left(\frac{1}{r}\right) = \nabla(x^2 + y^2 + z^2)^{-1/2}$$

$$= \frac{\partial}{\partial x}(x^2 + y^2 + z^2)^{-1/2}\mathbf{i} + \frac{\partial}{\partial y}(x^2 + y^2 + z^2)^{-1/2}\mathbf{j} + \frac{\partial}{\partial z}(x^2 + y^2 + z^2)^{-1/2}\mathbf{k}$$

$$= -\frac{x}{(x^2 + y^2 + z^2)^{3/2}}\mathbf{i} - \frac{y}{(x^2 + y^2 + z^2)^{3/2}}\mathbf{j} - \frac{z}{(x^2 + y^2 + z^2)^{3/2}}\mathbf{k}$$

$$= -\frac{1}{(x^2 + y^2 + z^2)^{3/2}}(x\mathbf{i} + y\mathbf{j} + z\mathbf{k}) = -\frac{\mathbf{r}}{r^3}. \quad ❑$$

The formulas we just derived can be generalized. As you are asked to show in the exercises, for each integer n and all $\mathbf{r} \neq \mathbf{0}$,

(15.1.5)

$$\nabla r^n = nr^{n-2}\mathbf{r}.$$

(If n is positive and even, the result also holds at $\mathbf{r} = \mathbf{0}$.)

Differentiability Implies Continuity

As in the one-variable case, differentiability implies continuity; namely,

(15.1.6)

if f is differentiable at $\mathbf{x}$, then f is continuous at $\mathbf{x}$.

To see this, write

$$f(\mathbf{x} + \mathbf{h}) - f(\mathbf{x}) = \nabla f(\mathbf{x}) \cdot \mathbf{h} + o(\mathbf{h})$$

and note that

$$|f(\mathbf{x} + \mathbf{h}) - f(\mathbf{x})| = |\nabla f(\mathbf{x}) \cdot \mathbf{h} + o(\mathbf{h})| \leq |\nabla f(\mathbf{x}) \cdot \mathbf{h}| + |o(\mathbf{h})|. \qquad \text{(triangle inequality)}$$

As $\mathbf{h} \to \mathbf{0}$,

$$|\nabla f(\mathbf{x}) \cdot \mathbf{h}| \leq \|\nabla f(\mathbf{x})\| \, \|\mathbf{h}\| \to 0 \qquad \text{and} \qquad |o(\mathbf{h})| \to 0.$$

Schwarz's Inequality ———↑   ↑—— Exercise 44

It follows that

$$[f(\mathbf{x} + \mathbf{h}) - f(\mathbf{x})] \to 0 \qquad \text{and therefore} \quad f(\mathbf{x} + \mathbf{h}) \to f(\mathbf{x}). \quad ❑$$

EXERCISES 15.1

In Exercises 1–20, find the gradient.

1. $f(x, y) = xe^{xy}$.

2. $f(x, y) = x^3 + y^2$.

3. $f(x, y) = 3x^2 - xy + y$.

4. $f(x, y) = Ax^2 + Bxy + Cy^2$.

5. $f(x, y) = x^2 y^{-2}$.

6. $f(x, y) = e^{x-y} - e^{y-x}$.

7. $f(x, y, z) = z \sin(x - y)$.

8. $f(x, y, z) = xy^2 e^{-z}$.

9. $f(x, y, z) = xy + yz + zx$.

10. $f(x, y, z) = \sqrt{x + y + z}$.

11. $f(x, y) = (x + y)e^{x-y}$.

12. $f(x, y) = (x + y) \sin(x - y)$.

13. $f(x, y) = e^x \ln y$.

14. $f(x, y, z) = x^2 y + y^2 z + z^2 x$.

15. $f(x, y) = \dfrac{Ax + By}{Cx + Dy}$.

16. $f(x, y) = \dfrac{x - y}{x^2 + y^2}$.

17. $f(x, y, z) = xye^x + ye^z - e^y \sin xz$.

18. $f(x, y, z) = e^{yz^2/x^3}$.

19. $f(x, y, z) = e^{x+2y} \cos(z^2 + 1)$.

20. $f(x, y, z) = \sin(xy) + e^{xyz} + \ln(x^2 z)$.

In Exercises 21–30, find the gradient vector at the indicated point.

21. $f(x, y) = 2x^2 - 3xy + 4y^2$ at $(2, 3)$.

22. $f(x, y) = 2x(x - y)^{-1}$ at $(3, 1)$.

23. $f(x, y) = \ln(x^2 + y^2)$ at $(2, 1)$.

24. $f(x, y) = x \tan^{-1}\left(\frac{y}{x}\right)$ at $(1, 1)$.

25. $f(x, y) = x \sin(xy)$ at $(1, \pi/2)$.

26. $f(x, y) = xye^{-(x^2+y^2)}$ at $(1, -1)$.

27. $f(x, y, z) = e^{-x} \sin(z + 2y)$ at $(0, \frac{1}{4}\pi, \frac{1}{4}\pi)$.

28. $f(x, y, z) = (x - y)\cos \pi z$ at $(1, 0, \frac{1}{2})$.

29. $f(x, y, z) = x - \sqrt{y^2 + z^2}$ at $(2, -3, 4)$.

30. $f(x, y, z) = \cos(xyz^2)$ at $(\pi, \frac{1}{4}, -1)$.

In Exercises 31–34, use Definition 15.1.2 to find the gradient of the function f.

31. $f(x, y) = 3x^2 - xy + y$.

32. $f(x, y) = \frac{1}{2}x^2 + 2xy + y^2$.

33. $f(x, y, z) = x^2 y + y^2 z + z^2 x$.

34. $f(x, y, z) = 2x^2 y - \frac{1}{z}$.

In Exercises 35–38, find a function f whose gradient is the given vector-valued function.

35. $\mathbf{F}(x, y) = 2xy\mathbf{i} + (1 + x^2)\mathbf{j}$.

36. $\mathbf{F}(x, y) = (2xy + x)\mathbf{i} + (x^2 + y)\mathbf{j}$.

37. $\mathbf{F}(x, y) = (x + \sin y)\mathbf{i} + (x \cos y - 2y)\mathbf{j}$.

38. $\mathbf{F}(x, y, z) = yz\mathbf{i} + (xz + 2yz)\mathbf{j} + (xy + y^2)\mathbf{k}$.

39. Calculate: (a) $\nabla(\ln r)$. (b) $\nabla(\sin r)$. (c) $\nabla(e^r)$, where $r = \sqrt{x^2 + y^2 + z^2}$.

40. Derive (15.1.5).

41. Let $f(x, y) = 1 + x^2 + y^2$.
   (a) Find the points (x, y), if any, at which $\nabla f(x, y) = \mathbf{0}$.
   (b) Sketch the graph of the surface $z = f(x, y)$.
   (c) What can you say about the surface at the point(s) found in part (a)?

42. Repeat Exercise 41 for the function $f(x, y) = \sqrt{4 - x^2 - y^2}$.

43. (a) Show that, if $\mathbf{c} \cdot \mathbf{h}$ is $o(\mathbf{h})$, then $\mathbf{c} = \mathbf{0}$.   HINT: First set $\mathbf{h} = h\mathbf{i}$, then set $\mathbf{h} = h\mathbf{j}$, then $\mathbf{h} = h\mathbf{k}$.
   (b) Show that, if

   $$f(\mathbf{x} + \mathbf{h}) - f(\mathbf{x}) = \mathbf{y} \cdot \mathbf{h} + o(\mathbf{h})$$

   and

   $$f(\mathbf{x} + \mathbf{h}) - f(\mathbf{x}) = \mathbf{z} \cdot \mathbf{h} + o(\mathbf{h}),$$

   then

   $$\mathbf{y} = \mathbf{z}.$$

   HINT: Verify that $\mathbf{y} \cdot \mathbf{h} - \mathbf{z} \cdot \mathbf{h} = (\mathbf{y} - \mathbf{z}) \cdot \mathbf{h}$ is $o(\mathbf{h})$.

44. Show that, if g is $o(\mathbf{h})$, then

   $$\lim_{\mathbf{h} \to \mathbf{0}} g(\mathbf{h}) = 0.$$

45. Set

   $$f(x, y) = \begin{cases} \dfrac{2xy}{x^2 + y^2}, & (x, y) \neq (0, 0) \\ 0, & (x, y) = (0, 0). \end{cases}$$

   (a) Show that f is not differentiable at $(0, 0)$.
   (b) In Section 14.6 you saw that the first partials $\partial f/\partial x$ and $\partial f/\partial y$ exist at $(0, 0)$. Since these partials obviously exist at every other point of the plane, we can conclude from Theorem 15.1.3 that at least one of these partials is not continuous in a neighborhood of $(0, 0)$. Show that $\partial f/\partial x$ is discontinuous at $(0, 0)$.

▶ 46. Let $f(x, y) = 2x^2 - y^2 - x^4 + 2$.
   (a) Find the points (x, y), if any, at which $\nabla f(x, y) = \mathbf{0}$.
   (b) Use a graphing utility to graph the surface $z = f(x, y)$ and investigate the behavior of the surface at the points found in part (a).

▶ 47. Repeat Exercise 46 for the function $f(x, y) = 4xye^{-(x^2+y^2)}$.

▶ 48. Repeat Exercise 46 for the function $f(x, y) = (x^2 + 4y^2)e^{1-(x^2+y^2)}$.

*SUPPLEMENT TO SECTION 15.1

PROOF OF THEOREM 15.1.3

We prove the theorem in the two-variable case. A similar argument yields a proof in the three-variable case, but there the details are more burdensome.

In the first place,

$$f(\mathbf{x} + \mathbf{h}) - f(\mathbf{x}) = f(x + h_1, y + h_2) - f(x, y).$$

Adding and subtracting $f(x, y + h_2)$, we have

(1) $f(\mathbf{x} + \mathbf{h}) - f(\mathbf{x}) = [f(x + h_1, y + h_2) - f(x, y + h_2)] + [f(x, y + h_2) - f(x, y)].$

By the mean-value theorem for functions of one variable, we know that there are numbers

$$0 < \theta_1 < 1 \quad \text{and} \quad 0 < \theta_2 < 1$$

such that

$$f(x + h_1, y + h_2) - f(x, y + h_2) = \frac{\partial f}{\partial x}(x + \theta_1 h_1, y + h_2)h_1$$

and

$$f(x, y + h_2) - f(x, y) = \frac{\partial f}{\partial y}(x, y + \theta_2 h_2)h_2. \quad \text{(Exercise 36, Section 4.1)}$$

By the continuity of $\partial f/\partial x$,

$$\frac{\partial f}{\partial x}(x + \theta_1 h_1, y + h_2) = \frac{\partial f}{\partial x}(x, y) + \epsilon_1(\mathbf{h})$$

where

$$\epsilon_1(\mathbf{h}) \to 0 \quad \text{as} \quad \mathbf{h} \to \mathbf{0}.†$$

By the continuity of $\partial f/\partial y$,

$$\frac{\partial f}{\partial y}(x, y + \theta_2 h_2) = \frac{\partial f}{\partial y}(x, y) + \epsilon_2(\mathbf{h})$$

where

$$\epsilon_2(\mathbf{h}) \to 0 \quad \text{as} \quad \mathbf{h} \to \mathbf{0}.$$

Substituting these expressions in equation (1), we find that

$$f(\mathbf{x} + \mathbf{h}) - f(\mathbf{x}) = \left[\frac{\partial f}{\partial x}(x, y) + \epsilon_1(\mathbf{h})\right]h_1 + \left[\frac{\partial f}{\partial y}(x, y) + \epsilon_2(\mathbf{h})\right]h_2$$

$$= \left[\frac{\partial f}{\partial x}(x, y) + \epsilon_1(\mathbf{h})\right](\mathbf{i} \cdot \mathbf{h}) + \left[\frac{\partial f}{\partial y}(x, y) + \epsilon_2(\mathbf{h})\right](\mathbf{j} \cdot \mathbf{h})$$

$$= \left[\frac{\partial f}{\partial x}(x, y)\mathbf{i} + \epsilon_1(\mathbf{h})\mathbf{i}\right] \cdot \mathbf{h} + \left[\frac{\partial f}{\partial y}(x, y)\mathbf{j} + \epsilon_2(\mathbf{h})\mathbf{j}\right] \cdot \mathbf{h}$$

$$= \left[\frac{\partial f}{\partial x}(x, y)\mathbf{i} + \frac{\partial f}{\partial y}(x, y)\mathbf{j}\right] \cdot \mathbf{h} + \left[\epsilon_1(\mathbf{h})\mathbf{i} + \epsilon_2(\mathbf{h})\mathbf{j}\right] \cdot \mathbf{h}.$$

---

†
$$\epsilon_1(\mathbf{h}) = \frac{\partial f}{\partial x}(x + \theta_1 h_1, y + h_2) - \frac{\partial f}{\partial x}(x, y) \to 0$$

since, by the continuity of $\partial f/\partial x$,

$$\frac{\partial f}{\partial x}(x + \theta_1 h_1, y + h_2) \to \frac{\partial f}{\partial x}(x, y).$$

To complete the proof of the theorem we need only show that

(2) $$[\epsilon_1(\mathbf{h})\mathbf{i} + \epsilon_2(\mathbf{h})\mathbf{j}] \cdot \mathbf{h} = o(\mathbf{h}).$$

From Schwarz's inequality, $|\mathbf{a} \cdot \mathbf{b}| \leq \|\mathbf{a}\|\,\|\mathbf{b}\|$, we know that

$$|[\epsilon_1(\mathbf{h})\mathbf{i} + \epsilon_2(\mathbf{h})\mathbf{j}] \cdot \mathbf{h}| \leq \|\epsilon_1(\mathbf{h})\mathbf{i} + \epsilon_2(\mathbf{h})\mathbf{j}\|\,\|\mathbf{h}\|.$$

It follows that

$$\frac{|[\epsilon_1(\mathbf{h})\mathbf{i} + \epsilon_2(\mathbf{h})\mathbf{j}] \cdot \mathbf{h}|}{\|\mathbf{h}\|} \leq \|\epsilon_1(\mathbf{h})\mathbf{i} + \epsilon_2(\mathbf{h})\mathbf{j}\| \leq \|\epsilon_1(\mathbf{h})\mathbf{i}\| + \|\epsilon_2(\mathbf{h})\mathbf{j}\| = |\epsilon_1(\mathbf{h})| + |\epsilon_2(\mathbf{h})|.$$

by the triangle inequality

As $\mathbf{h} \to \mathbf{0}$, the expression on the right tends to 0. This shows that (2) holds and completes the proof of the theorem. ❏

■ 15.2 GRADIENTS AND DIRECTIONAL DERIVATIVES

Some Elementary Formulas

In many respects gradients behave just as derivatives do in the one-variable case. In particular, if $\nabla f(\mathbf{x})$ and $\nabla g(\mathbf{x})$ exist, then $\nabla[f(\mathbf{x}) + g(\mathbf{x})]$, $\nabla[\alpha f(\mathbf{x})]$, and $\nabla[f(\mathbf{x})g(\mathbf{x})]$ all exist, and

(15.2.1)
$$\nabla[f(\mathbf{x}) + g(\mathbf{x})] = \nabla f(\mathbf{x}) + \nabla g(\mathbf{x}),$$
$$\nabla[\alpha f(\mathbf{x})] = \alpha\,\nabla f(\mathbf{x}),$$
$$\nabla[f(\mathbf{x})g(\mathbf{x})] = f(\mathbf{x})\,\nabla g(\mathbf{x}) + g(\mathbf{x})\,\nabla f(\mathbf{x}).$$

The first two formulas are easy to derive. To derive the third formula, let's assume that $\nabla f(\mathbf{x})$ and $\nabla g(\mathbf{x})$ both exist. Our task is to show that

$$f(\mathbf{x} + \mathbf{h})g(\mathbf{x} + \mathbf{h}) - f(\mathbf{x})g(\mathbf{x}) = [f(\mathbf{x})\,\nabla g(\mathbf{x}) + g(\mathbf{x})\,\nabla f(\mathbf{x})] \cdot \mathbf{h} + o(\mathbf{h}).$$

We now sketch how this can be done. We leave it to you to justify each step. The "trick" in proving a product rule is to add and subtract an appropriate expression (for example, see the proof of the product rule, Theorem 3.2.5). Starting from

$$f(\mathbf{x} + \mathbf{h})g(\mathbf{x} + \mathbf{h}) - f(\mathbf{x})g(\mathbf{x})$$

we add and subtract the term $f(\mathbf{x})g(\mathbf{x} + \mathbf{h})$. This gives:

$$[f(\mathbf{x} + \mathbf{h})g(\mathbf{x} + \mathbf{h}) - f(\mathbf{x})g(\mathbf{x} + \mathbf{h})] + [f(\mathbf{x})g(\mathbf{x} + \mathbf{h}) - f(\mathbf{x})g(\mathbf{x})]$$
$$= [f(\mathbf{x} + \mathbf{h}) - f(\mathbf{x})]g(\mathbf{x} + \mathbf{h}) + f(\mathbf{x})[g(\mathbf{x} + \mathbf{h}) - g(\mathbf{x})]$$
$$= [\nabla f(\mathbf{x}) \cdot \mathbf{h} + o(\mathbf{h})]g(\mathbf{x} + \mathbf{h}) + f(\mathbf{x})[\nabla g(\mathbf{x}) \cdot \mathbf{h} + o(\mathbf{h})]$$
$$= g(\mathbf{x} + \mathbf{h})\,\nabla f(\mathbf{x}) \cdot \mathbf{h} + f(\mathbf{x})\,\nabla g(\mathbf{x}) \cdot \mathbf{h} + o(\mathbf{h}) \qquad \text{(Exercise 30)}$$
$$= g(\mathbf{x})\,\nabla f(\mathbf{x}) \cdot \mathbf{h} + f(\mathbf{x})\,\nabla g(\mathbf{x}) \cdot \mathbf{h} + [g(\mathbf{x} + \mathbf{h}) - g(\mathbf{x})]\,\nabla f(\mathbf{x}) \cdot \mathbf{h} + o(\mathbf{h})$$
$$= [g(\mathbf{x})\,\nabla f(\mathbf{x}) + f(\mathbf{x})\,\nabla g(\mathbf{x})] \cdot \mathbf{h} + o(\mathbf{h}). \qquad \text{(Exercise 30)} \quad ❏$$

In Exercise 43, you are asked to verify this formula using Theorem 15.1.3.

Directional Derivatives

Here we take up an idea that generalizes the notion of partial derivative. Its connection with gradients will be made clear as we go on. First, recall the definitions of the first partial derivatives:

(two variables)   (three variables)

$$\frac{\partial f}{\partial x}(x, y) = \lim_{h \to 0} \frac{f(x + h, \, y) - f(x, y)}{h}, \qquad \frac{\partial f}{\partial x}(x, y, z) = \lim_{h \to 0} \frac{f(x + h, y, z) - f(x, y, z)}{h},$$

$$\frac{\partial f}{\partial y}(x, y) = \lim_{h \to 0} \frac{f(x, y + h) - f(x, y)}{h}, \qquad \frac{\partial f}{\partial y}(x, y, z) = \lim_{h \to 0} \frac{f(x, y + h, z) - f(x, y, \, z)}{h},$$

$$\frac{\partial f}{\partial z}(x, y, z) = \lim_{h \to 0} \frac{f(x, y, z + h) - f(x, y, z)}{h}.$$

Expressed in vector notation, these definitions take the form

$(\mathbf{x} = (x, y))$ $(\mathbf{x} = (x, y, z))$

$$\frac{\partial f}{\partial x}(\mathbf{x}) = \lim_{h \to 0} \frac{f(\mathbf{x} + h\mathbf{i}) - f(\mathbf{x})}{h}, \qquad \frac{\partial f}{\partial x}(\mathbf{x}) = \lim_{h \to 0} \frac{f(\mathbf{x} + h\mathbf{i}) - f(\mathbf{x})}{h},$$

$$\frac{\partial f}{\partial y}(\mathbf{x}) = \lim_{h \to 0} \frac{f(\mathbf{x} + h\mathbf{j}) - f(\mathbf{x})}{h}, \qquad \frac{\partial f}{\partial y}(\mathbf{x}) = \lim_{h \to 0} \frac{f(\mathbf{x} + h\mathbf{j}) - f(\mathbf{x})}{h},$$

$$\frac{\partial f}{\partial z}(\mathbf{x}) = \lim_{h \to 0} \frac{f(\mathbf{x} + h\mathbf{k}) - f(\mathbf{x})}{h}.$$

Each partial is thus the limit of a quotient

$$\frac{f(\mathbf{x} + h\mathbf{u}) - f(\mathbf{x})}{h},$$

where $\mathbf{u}$ is one of the unit coordinate vectors $\mathbf{i}$, $\mathbf{j}$, or $\mathbf{k}$. There is no reason to be so restrictive on $\mathbf{u}$. If f is defined in a neighborhood of $\mathbf{x}$, then, for small h, the difference quotient

$$\frac{f(\mathbf{x} + h\mathbf{u}) - f(\mathbf{x})}{h}$$

makes sense for any unit vector $\mathbf{u}$.

DEFINITION 15.2.2 DIRECTIONAL DERIVATIVE

For each unit vector $\mathbf{u}$, the limit

$$f'_{\mathbf{u}}(\mathbf{x}) = \lim_{h \to 0} \frac{f(\mathbf{x} + h\mathbf{u}) - f(\mathbf{x})}{h},$$

if it exists, is called the *directional derivative of f at* $\mathbf{x}$ *in the direction* $\mathbf{u}$.

It is important to realize that each of the partial derivatives $(\partial f/\partial x)$, $(\partial f/\partial y)$, $(\partial f/\partial z)$ is itself a directional derivative:

(15.2.3)
$$\frac{\partial f}{\partial x}(\mathbf{x}) = f'_{\mathbf{i}}(\mathbf{x}), \qquad \frac{\partial f}{\partial y}(\mathbf{x}) = f'_{\mathbf{j}}(\mathbf{x}), \qquad \frac{\partial f}{\partial z}(\mathbf{x}) = f'_{\mathbf{k}}(\mathbf{x}).$$

As you know, the partials of f give the rates of change of f in the $\mathbf{i}$, $\mathbf{j}$, $\mathbf{k}$ directions. The directional derivative $f'_{\mathbf{u}}$ gives *the rate of change of f in the direction* $\mathbf{u}$.

The geometric interpretation of the directional derivative for a function f of two variables is shown in Figure 15.2.1. Fix a point (x, y) in the domain of f and let $\mathbf{u}$ be a unit vector with initial point (x, y) in the xy-plane. Let C be the curve of intersection of the surface $z = f(x, y)$ and the plane p which contains $\mathbf{u}$ and is perpendicular to the xy-plane. Then $f'_{\mathbf{u}}(\mathbf{x})$ is the slope of the tangent line to C at the point $(x, y, f(x, y))$.

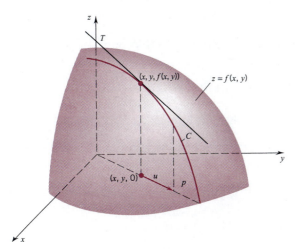

Figure 15.2.1

Remark The definition of the directional derivative of f in the direction $\mathbf{u}$ requires $\mathbf{u}$ to be a unit vector. However, we can extend the definition to arbitrary nonzero vectors as follows: The directional derivative of f at $\mathbf{x}$ in the direction of a nonzero vector $\mathbf{a}$ is $f'_{\mathbf{u}}(\mathbf{x})$ where $\mathbf{u} = \mathbf{a}/\|\mathbf{a}\|$ is the unit vector having the same direction as $\mathbf{a}$. ❑

There is an important connection between the gradient at $\mathbf{x}$ and the directional derivatives at $\mathbf{x}$.

THEOREM 15.2.4

If f is differentiable at $\mathbf{x}$, then f has a directional derivative at $\mathbf{x}$ in every direction $\mathbf{u}$, where $\mathbf{u}$ is a unit vector, and

$$f'_{\mathbf{u}}(\mathbf{x}) = \nabla f(\mathbf{x}) \cdot \mathbf{u}.$$

PROOF We take **u** as a unit vector and assume that f is differentiable at **x**. The differentiability at **x** tells us that $\nabla f(\mathbf{x})$ exists and

$$f(\mathbf{x} + h\mathbf{u}) - f(\mathbf{x}) = \nabla f(\mathbf{x}) \cdot h\mathbf{u} + o(h\mathbf{u}).$$

Division by h gives

$$\frac{f(\mathbf{x} + h\mathbf{u}) - f(\mathbf{x})}{h} = \nabla f(\mathbf{x}) \cdot \mathbf{u} + \frac{o(h\mathbf{u})}{h}.$$

Since

$$\left| \frac{o(h\mathbf{u})}{h} \right| = \frac{|o(h\mathbf{u})|}{|h|} = \frac{|o(h\mathbf{u})|}{\|h\mathbf{u}\|} \to 0,$$

you can see that

$$\frac{o(h\mathbf{u})}{h} \to 0 \quad \text{and thus} \quad \frac{f(\mathbf{x} + h\mathbf{u}) - f(\mathbf{x})}{h} \to \nabla f(\mathbf{x}) \cdot \mathbf{u}. \quad \square$$

Earlier (Theorem 15.1.3) you saw that, *if f has continuous first partials in a neighborhood of* **x**, *then f is differentiable at* **x** *and*

$$\nabla f(\mathbf{x}) = \frac{\partial f}{\partial x}(\mathbf{x})\mathbf{i} + \frac{\partial f}{\partial y}(\mathbf{x})\mathbf{j}. \qquad (\mathbf{x} = (x, y))$$

$$\nabla f(\mathbf{x}) = \frac{\partial f}{\partial x}(\mathbf{x})\mathbf{i} + \frac{\partial f}{\partial y}(\mathbf{x})\mathbf{j} + \frac{\partial f}{\partial z}(\mathbf{x})\mathbf{k}. \qquad (\mathbf{x} = (x, y, z))$$

The next theorem shows that this formula for $\nabla f(\mathbf{x})$ holds wherever f is differentiable.

THEOREM 15.2.5

If f is differentiable at **x**, then all the first partial derivatives of f exist at **x** and

$$\nabla f(\mathbf{x}) = \frac{\partial f}{\partial x}(\mathbf{x})\mathbf{i} + \frac{\partial f}{\partial y}(\mathbf{x})\mathbf{j} \qquad (\mathbf{x} = (x, y)),$$

$$\nabla f(\mathbf{x}) = \frac{\partial f}{\partial x}(\mathbf{x})\mathbf{i} + \frac{\partial f}{\partial y}(\mathbf{x})\mathbf{j} + \frac{\partial f}{\partial z}(\mathbf{x})\mathbf{k} \qquad (\mathbf{x} = (x, y, z)).$$

PROOF It is sufficient to prove the theorem for the case $\mathbf{x} = (x, y, z)$. Assume that f is differentiable at **x**. Then $\nabla f(\mathbf{x})$ exists and we can write

$$\nabla f(\mathbf{x}) = [\nabla f(\mathbf{x}) \cdot \mathbf{i}]\mathbf{i} + [\nabla f(\mathbf{x}) \cdot \mathbf{j}]\mathbf{j} + [\nabla f(\mathbf{x}) \cdot \mathbf{k}]\mathbf{k}. \qquad (12.3.13)$$

The result follows from observing that

$$\overset{(15.2.4)}{\downarrow} \quad \overset{(15.2.3)}{\downarrow}$$

$$\nabla f(\mathbf{x}) \cdot \mathbf{i} = f'_{\mathbf{i}}(\mathbf{x}) = \frac{\partial f}{\partial x}(\mathbf{x}),$$

$$\nabla f(\mathbf{x}) \cdot \mathbf{j} = f'_{\mathbf{j}}(\mathbf{x}) = \frac{\partial f}{\partial y}(\mathbf{x}),$$

$$\nabla f(\mathbf{x}) \cdot \mathbf{k} = f'_{\mathbf{k}}(\mathbf{x}) = \frac{\partial f}{\partial z}(\mathbf{x}). \quad \square$$

Example 1 Find the directional derivative of the function

$$f(x, y) = x^2 + y^2$$

at the point $(1, 2)$ in the direction of the vector $2\mathbf{i} - 3\mathbf{j}$.

SOLUTION In the first place, $2\mathbf{i} - 3\mathbf{j}$ is not a unit vector; its norm is $\sqrt{13}$. The unit vector in the direction of $2\mathbf{i} - 3\mathbf{j}$ is the vector

$$\mathbf{u} = \frac{1}{\sqrt{13}}[2\mathbf{i} - 3\mathbf{j}].$$

Next

$$\nabla f = 2x\mathbf{i} + 2y\mathbf{j},$$

and therefore

$$\nabla f(1, 2) = 2\mathbf{i} + 4\mathbf{j}.$$

By (15.2.4) we have

$$f'_{\mathbf{u}}(1, 2) = \nabla f(1, 2) \cdot \mathbf{u}$$

$$= (2\mathbf{i} + 4\mathbf{j}) \cdot \frac{1}{\sqrt{13}}[2\mathbf{i} - 3\mathbf{j}] = \frac{-8}{\sqrt{13}} \cong -2.219. \quad \square$$

Example 2 Find the directional derivative of the function

$$f(x, y, z) = x \cos y \sin z$$

at the point $(1, \pi, \frac{1}{4}\pi)$ in the direction of the vector $2\mathbf{i} - \mathbf{j} + 4\mathbf{k}$.

SOLUTION The unit vector in the direction of $2\mathbf{i} - \mathbf{j} + 4\mathbf{k}$ is the vector

$$\mathbf{u} = \frac{1}{\sqrt{21}}[2\mathbf{i} - \mathbf{j} + 4\mathbf{k}].$$

Here

$$\frac{\partial f}{\partial x}(x, y, z) = \cos y \sin z, \qquad \frac{\partial f}{\partial y}(x, y, z) = -x \sin y \sin z,$$

$$\frac{\partial f}{\partial z}(x, y, z) = x \cos y \cos z$$

so that

$$\frac{\partial f}{\partial x}\left(1, \pi, \frac{1}{4}\pi\right) = -\frac{1}{2}\sqrt{2}, \qquad \frac{\partial f}{\partial y}\left(1, \pi, \frac{1}{4}\pi\right) = 0, \qquad \frac{\partial f}{\partial z}\left(1, \pi, \frac{1}{4}\pi\right) = -\frac{1}{2}\sqrt{2}.$$

Therefore

$$\nabla f\left(1, \pi, \frac{1}{4}\pi\right) = -\frac{1}{2}\sqrt{2}\,\mathbf{i} - \frac{1}{2}\sqrt{2}\,\mathbf{k}$$

and

$$f'_{\mathbf{u}}\left(1, \pi, \frac{1}{4}\pi\right) = \nabla f\left(1, \pi, \frac{1}{4}\pi\right) \cdot \mathbf{u}$$

$$= -\frac{1}{2}\sqrt{2}[\mathbf{i} + \mathbf{k}] \cdot \frac{1}{\sqrt{21}}[2\mathbf{i} - \mathbf{j} + 4\mathbf{k}] = -\frac{3\sqrt{2}}{\sqrt{21}} \cong -0.926. \quad \square$$

You know that for each unit vector **u**

$$f'_{\mathbf{u}}(\mathbf{x}) = \nabla f(\mathbf{x}) \cdot \mathbf{u}.$$

Since

$$\nabla f(\mathbf{x}) \cdot \mathbf{u} = \operatorname{comp}_{\mathbf{u}} \nabla f(\mathbf{x}), \qquad (12.3.11)$$

we have

(15.2.6)

$$\boxed{f'_{\mathbf{u}}(\mathbf{x}) = \operatorname{comp}_{\mathbf{u}} \nabla f(\mathbf{x}).}$$

*Namely, the directional derivative in a direction **u** is the component of the gradient vector in that direction.* (Figure 15.2.2)

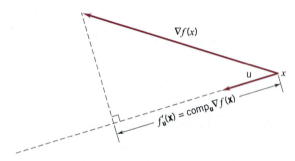

Figure 15.2.2

If $\nabla f(\mathbf{x}) \neq \mathbf{0}$, then

$$f'_{\mathbf{u}}(\mathbf{x}) = \nabla f(\mathbf{x}) \cdot \mathbf{u} = \|\nabla f(\mathbf{x})\| \, \|\mathbf{u}\| \cos \theta = \|\nabla f(\mathbf{x})\| \cos \theta$$

$$\underset{\text{(12.3.7)}}{\uparrow} \qquad \underset{\|\mathbf{u}\| = 1}{\uparrow}$$

where θ is the angle between $\nabla f(\mathbf{x})$ and **u**. Since $-1 \leq \cos \theta \leq 1$, we have

$$-\|\nabla f(\mathbf{x})\| \leq f'_{\mathbf{u}}(\mathbf{x}) \leq \|\nabla f(\mathbf{x})\| \qquad \text{for all directions } \mathbf{u}.$$

If **u** points in the direction of $\nabla f(\mathbf{x})$, then

$$f'_{\mathbf{u}}(\mathbf{x}) = \|\nabla f(\mathbf{x})\|, \qquad (\theta = 0, \cos \theta = 1)$$

and, if **u** points in the direction of $-\nabla f(\mathbf{x})$, then

$$f'_{\mathbf{u}}(\mathbf{x}) = -\|\nabla f(\mathbf{x})\|. \qquad (\theta = \pi, \cos \theta = -1)$$

Since the directional derivative gives the rate of change of the function in that direction, it is clear that

(15.2.7)

a differentiable function f increases most rapidly in the direction of the gradient (the rate of change is then $\|\nabla f(\mathbf{x})\|$) and it decreases most rapidly in the opposite direction (the rate of change is then $-\|\nabla f(\mathbf{x})\|$).

Example 3 The graph of the function

$$f(x, y) = \sqrt{1 - (x^2 + y^2)}$$

is the upper half of the unit sphere $x^2 + y^2 + z^2 = 1$. The function is defined on the closed unit disc $x^2 + y^2 \le 1$, but differentiable only on the open unit disc.

In Figure 15.2.3 we have marked a point (x, y) and drawn the corresponding radius vector $\mathbf{r} = x\mathbf{i} + y\mathbf{j}$. The gradient

$$\nabla f(x, y) = \frac{-x}{\sqrt{1 - (x^2 + y^2)}}\mathbf{i} + \frac{-y}{\sqrt{1 - (x^2 + y^2)}}\mathbf{j}$$

is a negative multiple of $\mathbf{r}$:

$$\nabla f(x, y) = -\frac{1}{\sqrt{1 - (x^2 + y^2)}}(x\mathbf{i} + y\mathbf{j}) = -\frac{1}{\sqrt{1 - (x^2 + y^2)}}\mathbf{r}.$$

Since $\mathbf{r}$ points from the origin to (x, y), the gradient points from (x, y) to the origin. This means that f increases most rapidly toward the origin. This is borne out by the observation that along the hemispherical surface the path of steepest ascent from the point $P(x, y, f(x, y))$ is the "great circle route to the north pole." ❏

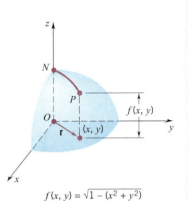

$f(x, y) = \sqrt{1 - (x^2 + y^2)}$

Figure 15.2.3

Example 4 Suppose that the temperature at each point of a metal plate is given by the function

$$T(x, y) = e^x \cos y + e^y \cos x.$$

(a) In what direction does the temperature increase most rapidly at the point $(0, 0)$? What is this rate of increase?

(b) In what direction does the temperature decrease most rapidly at $(0, 0)$?

SOLUTION

$$\nabla T(x, y) = \frac{\partial T}{\partial x}(x, y)\mathbf{i} + \frac{\partial T}{\partial y}(x, y)\mathbf{j}$$
$$= (e^x \cos y - e^y \sin x)\mathbf{i} + (e^y \cos x - e^x \sin y)\mathbf{j}.$$

(a) At $(0, 0)$ the temperature increases most rapidly in the direction of the gradient

$$\nabla T(0, 0) = \mathbf{i} + \mathbf{j}.$$

This rate of increase is

$$\|\nabla T(0, 0)\| = \|\mathbf{i} + \mathbf{j}\| = \sqrt{2}.$$

(b) The temperature decreases most rapidly in the direction of

$$-\nabla T(0, 0) = -\mathbf{i} - \mathbf{j}. \quad ❏$$

Example 5 The mass density (mass per unit volume) of a metal ball centered at the origin is given by the function

$$\lambda(x, y, z) = ke^{-(x^2 + y^2 + z^2)}, \qquad k \text{ a positive constant.}$$

(a) In what direction does the density increase most rapidly at the point (x, y, z)? What is this rate of density increase?

(b) In what direction does the density decrease most rapidly?

(c) What are the rates of density change at (x, y, z) in the $\mathbf{i}, \mathbf{j}, \mathbf{k}$ directions?

SOLUTION The gradient

$$\nabla\lambda(x, y, z) = \frac{\partial\lambda}{\partial x}(x, y, z)\mathbf{i} + \frac{\partial\lambda}{\partial y}(x, y, z)\mathbf{j} + \frac{\partial\lambda}{\partial z}(x, y, z)\mathbf{k}$$

$$= -2ke^{-(x^2+y^2+z^2)}[x\mathbf{i} + y\mathbf{j} + z\mathbf{k}],$$

and since $\lambda(x, y, z) = ke^{-(x^2+y^2+z^2)}$, we have

$$\nabla\lambda(x, y, z) = -2\lambda(x, y, z)\mathbf{r}.$$

From this, we see that the gradient points from (x, y, z) in the direction opposite to that of the radius vector.

(a) The density increases most rapidly toward the origin. The rate of increase is

$$\|\nabla\lambda(x, y, z)\| = 2\lambda(x, y, z)\|\mathbf{r}\| = 2\lambda(x, y, z)\sqrt{x^2 + y^2 + z^2}.$$

(b) The density decreases most rapidly directly away from the origin.

(c) The rates of density change in the $\mathbf{i}$, $\mathbf{j}$, $\mathbf{k}$ directions are given by the directional derivatives

$$\lambda_\mathbf{i}'(x, y, z) = \nabla\lambda(x, y, z) \cdot \mathbf{i} = -2x\,\lambda(x, y, z),$$

$$\lambda_\mathbf{j}'(x, y, z) = \nabla\lambda(x, y, z) \cdot \mathbf{j} = -2y\,\lambda(x, y, z),$$

$$\lambda_\mathbf{k}'(x, y, z) = \nabla\lambda(x, y, z) \cdot \mathbf{k} = -2z\,\lambda(x, y, z).$$

These are just the first partials of λ. ❏

Example 6 Suppose that the temperature at each point of a metal plate is given by the function

$$T(x, y) = 1 + x^2 - y^2.$$

Find the path followed by a heat-seeking particle that originates at $(-2, 1)$.

SOLUTION The particle moves in the direction of the gradient vector

$$\nabla T = 2x\mathbf{i} - 2y\mathbf{j}.$$

We want the curve

$$C: \mathbf{r}(t) = x(t)\mathbf{i} + y(t)\mathbf{j}$$

which begins at $(-2, 1)$ and at each point has its tangent vector in the direction of ∇T. We can satisfy the first condition by setting

$$x(0) = -2, \qquad y(0) = 1.$$

We can satisfy the second condition by setting

$$x'(t) = 2x(t), \qquad y'(t) = -2y(t). \qquad\qquad \text{(explain)}$$

These differential equations, together with initial conditions at $t = 0$, imply that

$$x(t) = -2e^{2t}, \qquad y(t) = e^{-2t}. \qquad\qquad \text{(Section 7.6)}$$

We can eliminate the parameter t by noting that

$$x(t)y(t) = (-2e^{2t})(e^{-2t}) = -2.$$

In terms of just x and y we have

$$xy = -2.$$

The particle moves from the point $(-2, 1)$ along the left branch of the hyperbola $xy = -2$ in the direction of decreasing x (see Figure 15.2.4).

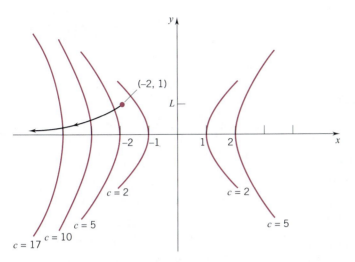

Figure 15.2.4

The level curves (isothermals) of the temperature distribution T are also hyperbolas. You can verify that the path of the particle is perpendicular to each of the level curves $x^2 - y^2 = c - 1$. ❏

Remark The pair of differential equations

$$x'(t) = 2x(t), \qquad y'(t) = -2y(t)$$

can be set as a single differential equation in x and y: the relation

$$\frac{y'(t)}{x'(t)} = -\frac{y(t)}{x(t)}$$

gives

$$\frac{dy}{dx} = -\frac{y}{x}.$$

This equation is readily solved directly (see Section 7.6):

$$\frac{1}{y}\frac{dy}{dx} = -\frac{1}{x}$$

$$\ln|y| = -\ln|x| + C$$

$$\ln|x| + \ln|y| = C$$

$$\ln|xy| = C.$$

Thus xy is constant:

$$xy = k.$$

Since the curve passes through the point $(-2, 1)$, $k = -2$ and once again we have the curve

$$xy = -2.$$

You will be called upon to use this method of solution in some of the exercises. ❏

EXERCISES 15.2

In Exercises 1–14, find the directional derivative at the given point in the direction indicated.

1. $f(x, y) = x^2 + 3y^2$ at $(1, 1)$ in the direction of $\mathbf{i} - \mathbf{j}$.

2. $f(x, y) = x + \sin(x + y)$ at $(0, 0)$ in the direction of $2\mathbf{i} + \mathbf{j}$.

3. $f(x, y) = xe^y - ye^x$ at $(1, 0)$ in the direction of $3\mathbf{i} + 4\mathbf{j}$.

4. $f(x, y) = \dfrac{2x}{x - y}$ at $(1, 0)$ in the direction of $\mathbf{i} - \sqrt{3}\mathbf{j}$.

5. $f(x, y) = \dfrac{ax + by}{x + y}$ at $(1, 1)$ in the direction of $\mathbf{i} - \mathbf{j}$.

6. $f(x, y) = \dfrac{x + y}{cx + dy}$ at $(1, 1)$ in the direction of $c\mathbf{i} - d\mathbf{j}$.

7. $f(x, y) = \ln(x^2 + y^2)$ at $(0, 1)$ in the direction of $8\mathbf{i} + \mathbf{j}$.

8. $f(x, y) = x^2y + \tan y$ at $(-1, \pi/4)$ in the direction of $\mathbf{i} - 2\mathbf{j}$.

9. $f(x, y, z) = xy + yz + zx$ at $(1, -1, 1)$ in the direction of $\mathbf{i} + 2\mathbf{j} + \mathbf{k}$.

10. $f(x, y, z) = x^2y + y^2z + z^2x$ at $(1, 0, 1)$ in the direction of $3\mathbf{j} - \mathbf{k}$.

11. $f(x, y, z) = (x + y^2 + z^3)^2$ at $(1, -1, 1)$ in the direction of $\mathbf{i} + \mathbf{j}$.

12. $f(x, y, z) = Ax^2 + Bxyz + Cy^2$ at $(1, 2, 1)$ in the direction of $A\mathbf{i} + B\mathbf{j} + C\mathbf{k}$.

13. $f(x, y, z) = x\tan^{-1}(y + z)$ at $(1, 0, 1)$ in the direction of $\mathbf{i} + \mathbf{j} - \mathbf{k}$.

14. $f(x, y, z) = xy^2\cos z - 2yz^2\sin \pi x + 3zx^2$ at $(0, -1, \pi)$ in the direction of $2\mathbf{i} - \mathbf{j} + 2\mathbf{k}$.

15. Find the directional derivative of $f(x, y) = \ln \sqrt{x^2 + y^2}$ at $(x, y) \neq (0, 0)$ toward the origin.

16. Find the directional derivative of $f(x, y) = (x - 1)y^2e^{xy}$ at $(0, 1)$ toward the point $(-1, 3)$.

17. Find the directional derivative of $f(x, y) = Ax^2 + 2Bxy + Cy^2$ at (a, b) toward (b, a) (a) if $a > b$; (b) if $a < b$.

18. Find the directional derivative of $f(x, y, z) = z \ln\left(\dfrac{x}{y}\right)$ at $(1, 1, 2)$ toward the point $(2, 2, 1)$.

19. Find the directional derivative of $f(x, y, z) = xe^{y^2-z^2}$ at $(1, 2, -2)$ in the direction of the path $\mathbf{r}(t) = t\mathbf{i} + 2\cos(t - 1)\mathbf{j} - 2e^{t-1}\mathbf{k}$.

20. Find the directional derivative of $f(x, y, z) = x^2 + yz$ at $(1, -3, 2)$ in the direction of the path $\mathbf{r}(t) = t^2\mathbf{i} + 3t\mathbf{j} + (1 - t^3)\mathbf{k}$.

21. Find the directional derivative of $f(x, y, z) = x^2 + 2xyz - yz^2$ at $(1, 1, 2)$ in a direction parallel to the straight line

$$\frac{x - 1}{2} = y - 1 = \frac{z - 2}{-3}.$$

22. Find the directional derivative of $f(x, y, z) = e^x\cos \pi yz$ at $(0, 1, \frac{1}{2})$ in a direction parallel to the line of intersection of the planes $x + y - z - 5 = 0$ and $4x - y - z + 2 = 0$.

In Exercises 23–26, find a unit vector in the direction in which f increases most rapidly at P and give the rate of change of f in that direction; find a unit vector in the direction in which f decreases most rapidly at P and give the rate of change of f in that direction.

23. $f(x, y) = y^2e^{2x}$; $P(0, 1)$.

24. $f(x, y) = x + \sin(x + 2y)$; $P(0, 0)$.

25. $f(x, y, z) = \sqrt{x^2 + y^2 + z^2}$; $P(1, -2, 1)$.

26. $f(x, y, z) = x^2ze^y + xz^2$; $P(1, \ln 2, 2)$.

27. Let $f = f(x)$ be a differentiable function of one variable. What is the gradient of f at x_0? What is the geometric significance of the direction of this gradient?

28. Suppose that f is differentiable at (x_0, y_0) and $\nabla f(x_0, y_0) \neq \mathbf{0}$. Compute the rate of change of f in the direction of the vector

$$\frac{\partial f}{\partial y}(x_0, y_0)\mathbf{i} - \frac{\partial f}{\partial x}(x_0, y_0)\mathbf{j}.$$

Give a geometric interpretation to your answer.

29. Let

$$f(x, y) = \sqrt{x^2 + y^2}.$$

(a) Show that $\partial f/\partial x$ is not defined at $(0, 0)$.
(b) Is f differentiable at $(0, 0)$?

30. Verify that, if g is continuous at $\mathbf{x}$, then
   (a) $g(\mathbf{x} + \mathbf{h})o(\mathbf{h}) = o(\mathbf{h})$ and
   (b) $[g(\mathbf{x} + \mathbf{h}) - g(\mathbf{x})] \nabla f(\mathbf{x}) \cdot \mathbf{h} = o(\mathbf{h})$.

31. Given the density function $\lambda(x, y) = 48 - \frac{4}{3}x^2 - 3y^2$, find the rate of density change (a) at $(1, -1)$ in the direction of the most rapid density decrease; (b) at $(1, 2)$ in the **i** direction; (c) at $(2, 2)$ away from the origin.

32. The intensity of light in a neighborhood of the point $(-2, 1)$ is given by a function of the form $I(x, y) = A - 2x^2 - y^2$. Find the path followed by a light-seeking particle that originates at the center of the neighborhood.

33. Determine the path of steepest descent along the surface $z = x^2 + 3y^2$ from each of the following points: (a) $(1, 1, 4)$; (b) $(1, -2, 13)$.

34. Determine the path of steepest ascent along the hyperbolic paraboloid $z = \frac{1}{2}x^2 - y^2$ from each of the following points: (a) $(-1, 1, -\frac{1}{2})$; (b) $(1, 0, \frac{1}{2})$.

35. Determine the path of steepest descent along the surface $z = a^2x^2 + b^2y^2$ from the point $(a^2, b^2, a^4 + b^4)$.

36. The temperature in a neighborhood of the origin is given by a function of the form

$$T(x, y) = T_0 + e^y \sin x.$$

Find the path followed by a heat-fleeing particle that originates at the origin.

37. The temperature in a neighborhood of the point $(\frac{1}{4}\pi, 0)$ is given by the function

$$T(x, y) = \sqrt{2}\, e^{-y} \cos x.$$

Find the path followed by a heat-seeking particle that originates at the center of the neighborhood.

38. Determine the path of steepest descent along the surface $z = A + x + 2y - x^2 - 3y^2$ from the point $(0, 0, A)$.

39. Set $f(x, y) = 3x^2 + y$.
   (a) Find

$$\lim_{h \to 0} \frac{f(x(2 + h), y(2 + h)) - f(2, 4)}{h}$$

given that $x(t) = t$ and $y(t) = t^2$. (These functions parametrize the parabola $y = x^2$.)
   (b) Find

$$\lim_{h \to 0} \frac{f(x(4 + h), y(4 + h)) - f(2, 4)}{h}$$

given that $x(t) = \frac{1}{4}(t + 4)$ and $y(t) = t$. (These functions parametrize the line $y = 4x - 4$.)
   (c) Compute the directional derivative of f at $(2, 4)$ in the direction of $\mathbf{i} + 4\mathbf{j}$.
   (d) Notice that $\mathbf{i} + 4\mathbf{j}$ is a direction vector for the line $y = 4x - 4$ and this line is tangent to the parabola $y = x^2$ at $(2, 4)$. Explain then why the computations in (a), (b), and (c) yield different values.

40. According to Newton's law of gravitation, the force exerted on a particle of mass m located at the point (x, y, z) by a particle of mass M located at the origin is given by

$$\mathbf{F}(x, y, z) = \frac{-GMm}{\|\mathbf{r}\|^3}\mathbf{r}$$

where $\mathbf{r} = x\mathbf{i} + y\mathbf{j} + z\mathbf{k}$ and G is the gravitational constant. Show that $\mathbf{F}$ is the gradient of

$$f(x, y, z) = \frac{GMm}{\sqrt{x^2 + y^2 + z^2}}.$$

41. Let $\mathbf{u}$ be a unit position vector (initial point at the origin) in the plane and let θ be the angle measured in the counterclockwise direction from the positive x-axis to $\mathbf{u}$. Let f be a function of two variables.
   (a) Show that $f'_{\mathbf{u}}(x, y) = \dfrac{\partial f}{\partial x}\cos\theta + \dfrac{\partial f}{\partial y}\sin\theta$.
   (b) Let $f(x, y) = x^3 + 2xy - xy^2$ and $\theta = 2\pi/3$. Find $f'_{\mathbf{u}}(-1, 2)$.

42. Refer to Exercise 41. Let $f(x, y) = x^2 e^{2y}$ and $\theta = 5\pi/4$. Find $f'_{\mathbf{u}}(x, y)$ and $f'_{\mathbf{u}}(2, \ln 2)$.

43. Assume that $\nabla f(\mathbf{x})$ and $\nabla g(\mathbf{x})$ exist. Use Theorem 15.1.3 to derive the product rule

$$\nabla[f(\mathbf{x})g(\mathbf{x})] = f(\mathbf{x})\nabla g(\mathbf{x}) + g(\mathbf{x})\nabla f(\mathbf{x}).$$

44. Assume that $\nabla f(\mathbf{x})$ and $\nabla g(\mathbf{x})$ exist, and that $g(\mathbf{x}) \neq 0$. Derive the quotient rule

$$\nabla\left[\frac{f(\mathbf{x})}{g(\mathbf{x})}\right] = \frac{g(\mathbf{x})\nabla f(\mathbf{x}) - f(\mathbf{x})\nabla g(\mathbf{x})}{g^2(\mathbf{x})}.$$

45. Assume that $\nabla f(\mathbf{x})$ exists. Prove that, for any integer n,

$$\nabla f^n(\mathbf{x}) = nf^{n-1}(\mathbf{x})\nabla f(\mathbf{x}).$$

Does this result hold if n is replaced by any real number?

■ 15.3 THE MEAN-VALUE THEOREM; TWO INTERMEDIATE-VALUE THEOREMS

You have seen the important role played by the mean-value theorem in the calculus of functions of one variable. Here we take up the analogous result for functions of several variables. Let $\mathbf{a}$ and $\mathbf{b}$ be points (either in the plane or in three space); by $\overline{\mathbf{a}\mathbf{b}}$ we mean the line segment that joins point $\mathbf{a}$ to point $\mathbf{b}$.

THEOREM 15.3.1 **THE MEAN-VALUE THEOREM (SEVERAL VARIABLES)**

If f is differentiable at each point of the line segment $\overline{\mathbf{ab}}$, then there exists on that line segment a point $\mathbf{c}$ between $\mathbf{a}$ and $\mathbf{b}$ such that

$$f(\mathbf{b}) - f(\mathbf{a}) = \nabla f(\mathbf{c}) \cdot (\mathbf{b} - \mathbf{a}).$$

PROOF As t ranges from 0 to 1, $\mathbf{a} + t(\mathbf{b} - \mathbf{a})$ traces out the line segment $\overline{\mathbf{ab}}$. The idea of the proof is to apply the one-variable mean-value theorem to the function

$$g(t) = f(\mathbf{a} + t[\mathbf{b} - \mathbf{a}]), \qquad t \in [0, 1].$$

To show that g is differentiable on the open interval $(0, 1)$, we take $t \in (0, 1)$ and form

$$\begin{aligned}
g(t + h) - g(t) &= f(\mathbf{a} + (t + h)[\mathbf{b} - \mathbf{a}]) - f(\mathbf{a} + t[\mathbf{b} - \mathbf{a}]) \\
&= f(\mathbf{a} + t[\mathbf{b} - \mathbf{a}] + h[\mathbf{b} - \mathbf{a}]) - f(\mathbf{a} + t[\mathbf{b} - \mathbf{a}]) \\
&= \nabla f(\mathbf{a} + t[\mathbf{b} - \mathbf{a}]) \cdot h[\mathbf{b} - \mathbf{a}] + o(h[\mathbf{b} - \mathbf{a}]).
\end{aligned}$$

Since

$$\nabla f(\mathbf{a} + t[\mathbf{b} + \mathbf{a}]) \cdot h(\mathbf{b} - \mathbf{a}) = [\nabla f(\mathbf{a} + t[\mathbf{b} - \mathbf{a}]) \cdot (\mathbf{b} - \mathbf{a})]h$$

and the $o(h[\mathbf{b} - \mathbf{a}])$ term is obviously $o(h)$, we can write

$$g(t + h) - g(t) = [\nabla f(\mathbf{a} + t[\mathbf{b} - \mathbf{a}]) \cdot (\mathbf{b} - \mathbf{a})]h + o(h).$$

Dividing both sides by h, we see that g is differentiable and

$$g'(t) = \nabla f(\mathbf{a} + t[\mathbf{b} - \mathbf{a}]) \cdot (\mathbf{b} - \mathbf{a}).$$

The function g is clearly continuous at 0 and at 1. Applying the one-variable mean-value theorem to g, we can conclude that there exists a number t_0 between 0 and 1 such that

(1) $$g(1) - g(0) = g'(t_0)(1 - 0).$$

Since $g(1) = f(\mathbf{b})$, $g(0) = f(\mathbf{a})$, and $g'(t_0) = \nabla f(\mathbf{a} + t_0[\mathbf{b} - \mathbf{a}]) \cdot (\mathbf{b} - \mathbf{a})$, condition (1) gives

$$f(\mathbf{b}) - f(\mathbf{a}) = \nabla f(\mathbf{a} + t_0[\mathbf{b} - \mathbf{a}]) \cdot (\mathbf{b} - \mathbf{a}).$$

Setting $\mathbf{c} = \mathbf{a} + t_0[\mathbf{b} - \mathbf{a}]$, we have

$$f(\mathbf{b}) - f(\mathbf{a}) = \nabla f(\mathbf{c}) \cdot (\mathbf{b} - \mathbf{a}). \quad \Box$$

U is connected

Figure 15.3.1

A nonempty open set U (in the plane or in three-space) is said to be *connected* iff any two points of U can be joined by a polygonal path that lies entirely in U. You can see such a set in Figure 15.3.1.

The set shown in Figure 15.3.2 is the union of two disjoint open sets. The set is open but not connected: it is impossible to join $\mathbf{a}$ and $\mathbf{b}$ by a polygonal path that lies within the set.

In Chapter 4 you saw that, if $f'(x) = 0$ for all x in an open interval I, then f is constant on I. We have a similar result for functions of several variables.

$U = A \cup B$ is not connected

Figure 15.3.2

THEOREM 15.3.2

Let U be an open connected set and let f be a differentiable function on U.

If $\nabla f(\mathbf{x}) = \mathbf{0}$ for all $\mathbf{x}$ in U, then f is constant on U.

PROOF Let $\mathbf{a}$ and $\mathbf{b}$ be any two points in U. Since U is connected, we can join these points by a polygonal path with vertices $\mathbf{a} = \mathbf{c}_0, \mathbf{c}_1, \mathbf{c}_2, \ldots, \mathbf{c}_{n-1}, \mathbf{c}_n = \mathbf{b}$ (see Figure 15.3.3). By the mean-value theorem (15.3.1) there exist points

$\mathbf{c}_1^*$ between $\mathbf{c}_0$ and $\mathbf{c}_1$ $\quad$ such that $\quad f(\mathbf{c}_1) - f(\mathbf{c}_0) = \nabla f(\mathbf{c}_1^*) \cdot (\mathbf{c}_1 - \mathbf{c}_0),$

$\mathbf{c}_2^*$ between $\mathbf{c}_1$ and $\mathbf{c}_2$ $\quad$ such that $\quad f(\mathbf{c}_2) - f(\mathbf{c}_1) = \nabla f(\mathbf{c}_2^*) \cdot (\mathbf{c}_2 - \mathbf{c}_1),$

$$\vdots \qquad\qquad\qquad\qquad\qquad\qquad\qquad \vdots$$

$\mathbf{c}_n^*$ between $\mathbf{c}_{n-1}$ and $\mathbf{c}_n$ $\quad$ such that $\quad f(\mathbf{c}_n) - f(\mathbf{c}_{n-1}) = \nabla f(\mathbf{c}_n^*) \cdot (\mathbf{c}_n - \mathbf{c}_{n-1}).$

Figure 15.3.3

If $\nabla f(\mathbf{x}) = \mathbf{0}$ for all $\mathbf{x}$ in U, then

$$f(\mathbf{c}_1) - f(\mathbf{c}_0) = 0, \quad f(\mathbf{c}_2) - f(\mathbf{c}_1) = 0, \ldots, f(\mathbf{c}_n) - f(\mathbf{c}_{n-1}) = 0.$$

This shows that

$$f(\mathbf{a}) = f(\mathbf{c}_0) = f(\mathbf{c}_1) = f(\mathbf{c}_2) = \cdots = f(\mathbf{c}_{n-1}) = f(\mathbf{c}_n) = f(\mathbf{b}).$$

Since $\mathbf{a}$ and $\mathbf{b}$ are arbitrary points of U, f must be constant on U. ❑

THEOREM 15.3.3

Let U be an open connected set, and let f and g be differentiable functions on U.

If $\nabla f(\mathbf{x}) = \nabla g(\mathbf{x})$ for all $\mathbf{x}$ in U, then f and g differ by a constant on U.

PROOF   If $\nabla f(\mathbf{x}) = \nabla g(\mathbf{x})$ for all $\mathbf{x}$ in U, then

$$\nabla[f(\mathbf{x}) - g(\mathbf{x})] = \nabla f(\mathbf{x}) - \nabla g(\mathbf{x}) = \mathbf{0} \qquad \text{for all } \mathbf{x} \text{ in } U.$$

By Theorem 15.3.2, $f - g$ must be constant on U.   ❑

You have seen that a function continuous on an interval skips no values (Theorem 2.6.1). There are analogous results for functions of several variables. Here is one of them.

THEOREM 15.3.4   AN INTERMEDIATE-VALUE THEOREM (SEVERAL VARIABLES)

Suppose that f is continuous on an open connected set U and A, B, C are real numbers such that $A < C < B$. If, somewhere in U, f takes on the value A and, somewhere in U, f takes on the value B, then, somewhere in U, f takes on the value C.

PROOF   Let $\mathbf{a}$ and $\mathbf{b}$ be points of U for which

$$f(\mathbf{a}) = A \qquad \text{and} \qquad f(\mathbf{b}) = B.$$

We must show that there exists a point $\mathbf{c}$ in U for which $f(\mathbf{c}) = C$.

Since U is connected, there is a polygonal path γ in U that joins $\mathbf{a}$ to $\mathbf{b}$. Let $\mathbf{r} = \mathbf{r}(t),\ a \le t \le b$, be a continuous parametrization of the path γ with $\mathbf{r}(a) = \mathbf{a}$ and $\mathbf{r}(b) = \mathbf{b}$. Since $\mathbf{r}$ is continuous, the composition

$$g(t) = f(\mathbf{r}(t))$$

is also continuous on $[a, b]$. Since

$$g(a) = f(\mathbf{r}(a)) = f(\mathbf{a}) = A \qquad \text{and} \qquad g(b) = f(\mathbf{r}(b)) = f(\mathbf{b}) = B,$$

we know from Theorem 2.6.1 that there is a number c in $[a, b]$ for which $g(c) = C$. Setting $\mathbf{c} = \mathbf{r}(c)$ we have $f(\mathbf{c}) = C$.   ❑

Figure 15.3.4

Figure 15.3.5

Continuity on a Closed Region

An open connected set is called an *open region*. If we start with an open region and adjoin to it the boundary, then we have what is called a *closed region*. (A closed region is therefore a closed set, the interior of which is an open region.)

Continuity on a closed region Ω requires continuity at the boundary points of Ω as well as the interior points.

Let the function f be defined on a closed region Ω, and let $\mathbf{x}_0 \in \Omega$.

If $\mathbf{x}_0$ is an interior point of Ω, then all points $\mathbf{x}$ sufficiently close to $\mathbf{x}_0$ are in Ω and, by definition, f is continuous at $\mathbf{x}_0$ iff

$$\text{as } \mathbf{x} \text{ approaches } \mathbf{x}_0, f(\mathbf{x}) \text{ approaches } f(\mathbf{x}_0). \qquad \text{(Figure 15.3.4)}$$

If $\mathbf{x}_0$ is a boundary point of Ω, then we have to modify the definition and say: f is continuous at $\mathbf{x}_0$ iff

$$\text{as } \mathbf{x} \text{ approaches } \mathbf{x}_0 \text{ within } \Omega, f(\mathbf{x}) \text{ approaches } f(\mathbf{x}_0). \qquad \text{(Figure 15.3.5)}$$

In terms of ϵ-δ, f is continuous at a boundary point $\mathbf{x}_0$ iff for each $\epsilon > 0$ there exists $\delta > 0$ such that

$$\text{if } \|\mathbf{x} - \mathbf{x}_0\| < \delta \text{ and } \mathbf{x} \in \Omega, \quad \text{then } |f(\mathbf{x}) - f(\mathbf{x}_0)| < \epsilon.$$

(This is completely analogous to one-sided continuity at an endpoint of a closed interval $[a, b]$ as defined in Section 2.4.)

The intermediate-value result that we just proved for open connected sets can be extended to closed regions.

THEOREM 15.3.5 A SECOND INTERMEDIATE-VALUE THEOREM (SEVERAL VARIABLES)

Suppose that f is continuous on a closed region Ω and A, B, C are real numbers such that $A < C < B$. If, somewhere in Ω, f takes on the value A and, somewhere in Ω, f takes on the value B, then, somewhere in Ω, f takes on the value C.

PROOF Let $\mathbf{a}$ and $\mathbf{b}$ be points of Ω for which

$$f(\mathbf{a}) = A \quad \text{and} \quad f(\mathbf{b}) = B.$$

If $\mathbf{a}$ and $\mathbf{b}$ are both in the interior of Ω, then the result follows from the previous theorem. But one or both of these points could lie on the boundary of Ω. To take care of that possibility, we can proceed as follows. Take ϵ small enough that

$$A + \epsilon < C < B - \epsilon.$$

By continuity there exist points $\mathbf{x}_1$, $\mathbf{x}_2$ in the interior of Ω for which

$$f(\mathbf{x}_1) < A + \epsilon \quad \text{and} \quad B - \epsilon < f(\mathbf{x}_2). \qquad \text{(Exercise 7)}$$

Then

$$f(\mathbf{x}_1) < C < f(\mathbf{x}_2)$$

and the result follows from the previous theorem. ❑

EXERCISES 15.3

1. The function $f(x, y) = x^3 - xy$ is differentiable for all (x, y). Let $\mathbf{a} = (0, 1)$ and $\mathbf{b} = (1, 3)$. Find a point $\mathbf{c}$ on the line segment joining $\mathbf{a}$ and $\mathbf{b}$ such that

$$f(\mathbf{b}) - f(\mathbf{a}) = \nabla f(\mathbf{c}) \cdot (\mathbf{b} - \mathbf{a}).$$

2. The function $f(x, y, z) = 4xz - y^2 + z^2$ is differentiable for all (x, y, z). Let $\mathbf{a} = (0, 1, 1)$ and $\mathbf{b} = (1, 3, 2)$. Find a point $\mathbf{c}$ on the line segment joining $\mathbf{a}$ and $\mathbf{b}$ such that

$$f(\mathbf{b}) - f(\mathbf{a}) = \nabla f(\mathbf{c}) \cdot (\mathbf{b} - \mathbf{a}).$$

3. (a) Find f if $\nabla f(x, y, z) = a_1 \mathbf{i} + a_2 \mathbf{j} + a_3 \mathbf{k}$ for all (x, y, z).

(b) What can you conclude about f and g if $\nabla f(x, y, z) - \nabla g(x, y, z) = a_1 \mathbf{i} + a_2 \mathbf{j} + a_3 \mathbf{k}$ for all (x, y, z)?

4. (*Rolle's theorem for functions of several variables*) Show that, if f is differentiable at each point of the line segment $\overline{\mathbf{ab}}$ and $f(\mathbf{a}) = f(\mathbf{b})$, then there exists a point $\mathbf{c}$ between $\mathbf{a}$ and $\mathbf{b}$ for which $\nabla f(\mathbf{c}) \perp (\mathbf{b} - \mathbf{a})$.

5. Let $U = \{\mathbf{x} : \|\mathbf{x}\| \neq 1\}$. Define f on U by setting

$$f(\mathbf{x}) = \begin{cases} 0, & \|\mathbf{x}\| < 1 \\ 1, & \|\mathbf{x}\| > 1. \end{cases}$$

(a) Note that $\nabla f(\mathbf{x}) = \mathbf{0}$ for all $\mathbf{x}$ in U, but f is not constant on U. Explain how this does not contradict Theorem 15.3.2.

(b) Define a function g on U different from f such that $\nabla f(\mathbf{x}) = \nabla g(\mathbf{x})$ for all $\mathbf{x}$ in U and $f - g$ is (i) constant on U, (ii) not constant on U.

(c) The function f takes on the value 0, takes on the value 1, but takes on no value in between. Explain how this does not contradict Theorem 15.3.4.

6. A set of points is said to be *convex* provided that every pair of points in the set can be joined by a line segment that lies entirely within the set. Show that, if $\|\nabla f(\mathbf{x})\| \le M$ for all $\mathbf{x}$ in some convex set Ω, then

$$|f(\mathbf{x}_1) - f(\mathbf{x}_2)| \le M\|\mathbf{x}_1 - \mathbf{x}_2\| \qquad \text{for all } \mathbf{x}_1 \text{ and } \mathbf{x}_2 \text{ in } \Omega.$$

7. Justify the assertion made in the proof of Theorem 15.3.5 that there exist points $\mathbf{x}_1, \mathbf{x}_2$ in the interior of Ω for which

$$f(\mathbf{x}_1) < A + \epsilon \qquad \text{and} \qquad B - \epsilon < f(\mathbf{x}_2).$$

■ **15.4 CHAIN RULES**

For functions of a single variable there is basically only one chain rule: Theorem 3.5.7. For functions of several variables there are many chain rules.

A vector-valued function is said to be *continuous* provided that its components are continuous. If $f = f(x, y, z)$ is a scalar-valued function (a real-valued function), then its gradient ∇f is a vector-valued function. We say that f is *continuously differentiable* on an *open set* U iff f is differentiable on U and ∇f is continuous on U.

If a curve $\mathbf{r}$ lies in the domain of f, then we can form the composition

$$(f \circ \mathbf{r})(t) = f(\mathbf{r}(t)).$$

The composition $f \circ \mathbf{r}$ is a real-valued function of a real variable t. The numbers $f(\mathbf{r}(t))$ are the values taken on by f *along the curve* $\mathbf{r}$.

THEOREM 15.4.1   CHAIN RULE (ALONG A CURVE)

If f is continuously differentiable on an open set U and $\mathbf{r} = \mathbf{r}(t)$ is a differentiable curve that lies entirely in U, then the composition $f \circ \mathbf{r}$ is differentiable and

$$\frac{d}{dt}[f(\mathbf{r}(t))] = \nabla f(\mathbf{r}(t)) \cdot \mathbf{r}'(t).$$

PROOF   We will show that

$$\lim_{h \to 0} \frac{f(\mathbf{r}(t + h)) - f(\mathbf{r}(t))}{h} = \nabla f(\mathbf{r}(t)) \cdot \mathbf{r}'(t).$$

For $h \neq 0$ and sufficiently small, the line segment that joins $\mathbf{r}(t)$ to $\mathbf{r}(t + h)$ lies entirely in U. This we know because U is open and $\mathbf{r}$ is continuous. (See Figure 15.4.1.) For such h, the mean-value theorem we proved in Section 15.3 assures us that there exists a point $\mathbf{c}(h)$ between $\mathbf{r}(t)$ and $\mathbf{r}(t + h)$ such that

$$f(\mathbf{r}(t + h)) - f(\mathbf{r}(t)) = \nabla f(\mathbf{c}(h)) \cdot [\mathbf{r}(t + h) - \mathbf{r}(t)].$$

Dividing both sides by h, we have

$$\frac{f(\mathbf{r}(t + h)) - f(\mathbf{r}(t))}{h} = \nabla f(\mathbf{c}(h)) \cdot \left[\frac{\mathbf{r}(t + h) - \mathbf{r}(t)}{h}\right].$$

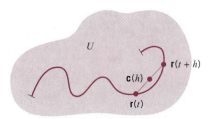

Figure 15.4.1

As h tends to zero, $\mathbf{c}(h)$ tends to $\mathbf{r}(t)$ and by the continuity of ∇f,

$$\nabla f(\mathbf{c}(h)) \rightarrow \nabla f(\mathbf{r}(t)).$$

Since

$$\frac{\mathbf{r}(t+h) - \mathbf{r}(t)}{h} \rightarrow \mathbf{r}'(t),$$

the result follows. ❏

Example 1 Use the chain rule to find the rate of change of

$$f(x, y) = \tfrac{1}{3}(x^3 + y^3)$$

with respect to t along the ellipse $\mathbf{r}(t) = a \cos t\,\mathbf{i} + b \sin t\,\mathbf{j}$.

SOLUTION The rate of change of f with respect to t along the curve $\mathbf{r}$ is the derivative

$$\frac{d}{dt}[f(\mathbf{r}(t))].$$

By the chain rule (Theorem 15.4.1)

$$\frac{d}{dt}[f(\mathbf{r}(t))] = \nabla f(\mathbf{r}(t)) \cdot \mathbf{r}'(t).$$

Here

$$\nabla f = x^2\mathbf{i} + y^2\mathbf{j}.$$

With $x(t) = a \cos t$ and $y(t) = b \sin t$, we have

$$\nabla f(\mathbf{r}(t)) = a^2\cos^2 t\,\mathbf{i} + b^2\sin^2 t\,\mathbf{j}.$$

Since $\mathbf{r}'(t) = -a \sin t\,\mathbf{i} + b \cos t\,\mathbf{j}$, you can see that

$$\begin{aligned}
\frac{d}{dt}[f(\mathbf{r}(t))] &= \nabla f(\mathbf{r}(t)) \cdot \mathbf{r}'(t) \\
&= (a^2\cos^2 t\,\mathbf{i} + b^2\sin^2 t\,\mathbf{j}) \cdot (-a \sin t\,\mathbf{i} + b \cos t\,\mathbf{j}) \\
&= -a^3\sin t \cos^2 t + b^3\sin^2 t \cos t \\
&= \sin t \cos t(b^3\sin t - a^3\cos t). \quad ❏
\end{aligned}$$

Remark Note that we could have obtained the same result without invoking Theorem 15.4.1 by first forming $f(\mathbf{r}(t))$ and then differentiating with respect to t:

$$f(\mathbf{r}(t)) = f(x(t), y(t)) = \tfrac{1}{3}([x(t)]^3 + [y(t)]^3) = \tfrac{1}{3}(a^3\cos^3 t + b^3\sin^3 t)$$

so that

$$\frac{d}{dt}[f(\mathbf{r}(t))] = \tfrac{1}{3}[3a^3(\cos^2 t)(-\sin t) + 3b^3\sin^2 t \cos t]$$

$$= \sin t \cos t \,(b^3\sin t - a^3\cos t). \quad \square$$

Example 2 Use the chain rule to find the rate of change of

$$f(x, y, z) = x^2 y + z \cos x$$

with respect to t along the twisted cubic $\mathbf{r}(t) = t\mathbf{i} + t^2\mathbf{j} + t^3\mathbf{k}$.

SOLUTION Once again we use the relation

$$\frac{d}{dt}[f(\mathbf{r}(t))] = \nabla f(\mathbf{r}(t)) \cdot \mathbf{r}'(t).$$

This time

$$\nabla f = (2xy - z \sin x)\mathbf{i} + x^2\mathbf{j} + \cos x \, \mathbf{k}.$$

With $x(t) = t$, $y(t) = t^2$, $z(t) = t^3$, we have

$$\nabla f(\mathbf{r}(t)) = (2t^3 - t^3\sin t)\mathbf{i} + t^2\mathbf{j} + \cos t\,\mathbf{k}.$$

Since $\mathbf{r}'(t) = \mathbf{i} + 2t\mathbf{j} + 3t^2\mathbf{k}$, we have

$$\frac{d}{dt}[f(\mathbf{r}(t))] = \nabla f(\mathbf{r}(t)) \cdot \mathbf{r}'(t)$$

$$= [(2t^3 - t^3\sin t)\mathbf{i} + t^2\mathbf{j} + \cos t\,\mathbf{k}] \cdot [\mathbf{i} + 2t\mathbf{j} + 3t^2\mathbf{k}]$$
$$= 2t^3 - t^3\sin t + 2t^3 + 3t^2\cos t$$
$$= 4t^3 - t^3\sin t + 3t^2\cos t.$$

You can check this answer by first forming $f(\mathbf{r}(t))$ and then differentiating. $\quad \square$

Another Formulation of (15.4.1)

The chain rule for functions of one variable,

$$\frac{d}{dt}[u(x(t))] = u'(x(t))x'(t),$$

can be written

$$\frac{du}{dt} = \frac{du}{dx}\frac{dx}{dt}.$$

In a similar manner, the relation

$$\frac{d}{dt}[u(\mathbf{r}(t))] = \nabla u(\mathbf{r}(t)) \cdot \mathbf{r}'(t)$$

can be written

(1) $$\frac{du}{dt} = \nabla u \cdot \frac{d\mathbf{r}}{dt}.$$

With

$$\nabla u = \frac{\partial u}{\partial x}\mathbf{i} + \frac{\partial u}{\partial y}\mathbf{j} + \frac{\partial u}{\partial z}\mathbf{k} \quad \text{and} \quad \frac{d\mathbf{r}}{dt} = \frac{dx}{dt}\mathbf{i} + \frac{dy}{dt}\mathbf{j} + \frac{dz}{dt}\mathbf{k},$$

equation (1) takes the form

(15.4.2)
$$\frac{du}{dt} = \frac{\partial u}{\partial x}\frac{dx}{dt} + \frac{\partial u}{\partial y}\frac{dy}{dt} + \frac{\partial u}{\partial z}\frac{dz}{dt}.$$

In the two-variable case, the z-term drops out and we have

(15.4.3)
$$\frac{du}{dt} = \frac{\partial u}{\partial x}\frac{dx}{dt} + \frac{\partial u}{\partial y}\frac{dy}{dt}.$$

Example 3 Find du/dt if
$$u = x^2 - y^2 \quad \text{and} \quad x = t^2 - 1, \quad y = 3 \sin \pi t.$$

SOLUTION Here we are in the two-variable case
$$\frac{du}{dt} = \frac{\partial u}{\partial x}\frac{dx}{dt} + \frac{\partial u}{\partial y}\frac{dy}{dt}.$$

Since
$$\frac{\partial u}{\partial x} = 2x, \quad \frac{\partial u}{\partial y} = -2y \quad \text{and} \quad \frac{dx}{dt} = 2t, \quad \frac{dy}{dt} = 3\pi \cos \pi t,$$

we have

$$\begin{aligned}
\frac{du}{dt} &= (2x)(2t) + (-2y)(3\pi \cos \pi t) \\
&= 2(t^2 - 1)(2t) + (-2)(3 \sin \pi t)(3\pi \cos \pi t) \\
&= 4t^3 - 4t - 18\pi \sin \pi t \cos \pi t.
\end{aligned}$$

You can obtain this same result by first writing u directly as a function of t and then differentiating:
$$u = x^2 - y^2 = (t^2 - 1)^2 - (3 \sin \pi t)^2$$

so that

$$\frac{du}{dt} = 2(t^2 - 1)2t - 2(3 \sin \pi t)3\pi \cos \pi t = 4t^3 - 4t - 18\pi \sin \pi t \cos \pi t. \quad \square$$

Example 4 A solid is in the shape of a frustum of a right circular cone (see Figure 15.4.2). If the upper radius x decreases at the rate of 2 inches per minute, the lower radius y increases at the rate of 3 inches per minute, and the height z decreases at the

Figure 15.4.2

rate of 4 inches per minute, at what rate is the volume V changing at the instant the upper radius is 10 inches, the lower radius is 12 inches, and the height is 18 inches.

SOLUTION The formula for the volume of a frustum of a right circular cone is

$$V = \frac{1}{3}\pi z(x^2 + xy + y^2) \qquad \text{(Exercise 35, Section 6.2)}$$

so that

$$\frac{\partial V}{\partial x} = \frac{1}{3}\pi z(2x + y), \qquad \frac{\partial V}{\partial y} = \frac{1}{3}\pi z(x + 2y), \qquad \frac{\partial V}{\partial z} = \frac{1}{3}\pi(x^2 + xy + y^2).$$

Since

$$\frac{dV}{dt} = \frac{\partial V}{\partial x}\frac{dx}{dt} + \frac{\partial V}{\partial y}\frac{dy}{dt} + \frac{\partial V}{\partial z}\frac{dz}{dt},$$

we have

$$\frac{dV}{dt} = \frac{1}{3}\pi z(2x + y)\frac{dx}{dt} + \frac{1}{3}\pi z(x + 2y)\frac{dy}{dt} + \frac{1}{3}\pi(x^2 + xy + y^2)\frac{dz}{dt}.$$

Set

$$x = 10, \quad y = 12, \quad z = 18, \quad \frac{dx}{dt} = -2, \quad \frac{dy}{dt} = 3, \quad \frac{dz}{dt} = -4,$$

and you will find that

$$\frac{dV}{dt} = -\frac{772}{3}\pi \cong -808.4.$$

The volume decreases at the rate of approximately 808 cubic inches per minute. ❏

Other Chain Rules

In the setting of functions of several variables there are numerous chain rules. Some are stated here, others in the exercises. They can all be deduced from Theorem 15.4.1 and its corollaries, (15.4.2) and (15.4.3).

If, for example,

$$u = u(x, y) \qquad \text{where} \quad x = x(s, t) \qquad \text{and} \qquad y = y(s, t),$$

then

(15.4.4)
$$\frac{\partial u}{\partial s} = \frac{\partial u}{\partial x}\frac{\partial x}{\partial s} + \frac{\partial u}{\partial y}\frac{\partial y}{\partial s} \qquad \text{and} \qquad \frac{\partial u}{\partial t} = \frac{\partial u}{\partial x}\frac{\partial x}{\partial t} + \frac{\partial u}{\partial y}\frac{\partial y}{\partial t}.$$

To obtain the first equation, keep t fixed and differentiate u with respect to s according to Formula (15.4.3); to obtain the second equation, keep s fixed and differentiate u with respect to t. These formulas can be viewed as a chain rule on a surface, extending the chain rule along a curve.

In Figure 15.4.3 we have drawn a *tree diagram* for Formula (15.4.4). We construct such a tree by branching at each stage from a function to all the variables that directly determine it. Each path starting at u and ending at a variable determines a product of (partial) derivatives. The partial derivative of u with respect to each variable is the sum of the products generated by all the direct paths to that variable.

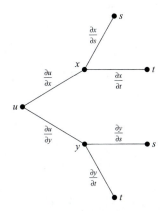

Figure 15.4.3

Example 5 Let $u = x^2 - 2xy + 2y^3$, where $x = s^2 \ln t$ and $y = 2st^3$. Find $\partial u/\partial s$ and $\partial u/\partial t$.

SOLUTION Here u is a function of two variables, x and y, where each of x and y is itself a function of two variables. Thus (15.4.4) applies. Since

$$\frac{\partial u}{\partial x} = (2x - 2y), \qquad \frac{\partial u}{\partial y} = (-2x + 6y^2)$$

and

$$\frac{\partial x}{\partial s} = 2s \ln t, \qquad \frac{\partial y}{\partial s} = 2t^3, \qquad \frac{\partial x}{\partial t} = \frac{s^2}{t}, \qquad \frac{\partial y}{\partial t} = 6st^2,$$

we have

$$\frac{\partial u}{\partial s} = (2x - 2y)(2s \ln t) + (-2x + 6y^2)(2t^3),$$

and

$$\frac{\partial u}{\partial t} = (2x - 2y)\left(\frac{s^2}{t}\right) + (-2x + 6y^2)(6st^2).$$

These results can be expressed entirely in terms of s and t by replacing x by $s^2 \ln t$ and y by $2st^3$:

$$\frac{\partial u}{\partial s} = (2s^2 \ln t - 4st^3)(2s \ln t) + (-2s^2 \ln t + 24s^2t^6)(2t^3),$$

$$\frac{\partial u}{\partial t} = (2s^2 \ln t - 4st^3)\left(\frac{s^2}{t}\right) + (-2s^2 \ln t + 24s^2t^6)(6st^2). \qquad \square$$

Now suppose that u is a function of three variables:

$$u = u(x, y, z) \qquad \text{where} \quad x = x(s, t), \quad y = y(s, t), \quad z = z(s, t).$$

A tree diagram for the partials of u appears in Figure 15.4.4. The partials of u with respect to s and t can be read from the diagram:

(15.4.5)
$$\frac{\partial u}{\partial s} = \frac{\partial u}{\partial x}\frac{\partial x}{\partial s} + \frac{\partial u}{\partial y}\frac{\partial y}{\partial s} + \frac{\partial u}{\partial z}\frac{\partial z}{\partial s}, \qquad \frac{\partial u}{\partial t} = \frac{\partial u}{\partial x}\frac{\partial x}{\partial t} + \frac{\partial u}{\partial y}\frac{\partial y}{\partial t} + \frac{\partial u}{\partial z}\frac{\partial z}{\partial t}.$$

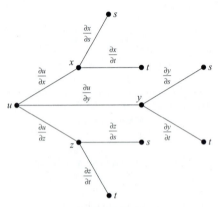

Figure 15.4.4

Example 6 Let $u = x^2 y^3 e^{xz}$, where $x = s^2 + t^2$, $y = 2st$ and $z = s \ln t$. Find $\partial u/\partial s$.

SOLUTION In this case, u is a function of three variables, x, y, z, where each of x, y, and z is a function of s and t. To find $\partial u/\partial s$, we use chain rule (15.4.5):

$$\frac{\partial u}{\partial s} = \frac{\partial u}{\partial x}\frac{\partial x}{\partial s} + \frac{\partial u}{\partial y}\frac{\partial y}{\partial s} + \frac{\partial u}{\partial z}\frac{\partial z}{\partial s}.$$

Now,

$$\frac{\partial u}{\partial x} = 2xy^3 e^{xz} + x^2 y^3 z e^{xz}, \qquad \frac{\partial u}{\partial y} = 3x^2 y^2 e^{xz}, \qquad \frac{\partial u}{\partial z} = x^3 y^3 e^{xz}$$

and

$$\frac{\partial x}{\partial s} = 2s, \qquad \frac{\partial y}{\partial s} = 2t, \qquad \frac{\partial z}{\partial s} = \ln t.$$

Therefore,

$$\frac{\partial u}{\partial s} = (2xy^3 e^{xz} + x^2 y^3 z e^{xz})(2s) + (3x^2 y^2 e^{xz})(2t) + (x^3 y^3 e^{xz})(\ln t).$$

This result can be expressed entirely in terms of s and t by substituting $s^2 + t^2$ for x, $2st$ for y, and $s \ln t$ for z. ❏

Implicit Differentiation

In Section 3.7 we presented a method for finding the derivative of a function $y = y(x)$ defined implicitly by an equation $u(x, y) = 0$. Here we show how the chain rule (15.4.1) can be used to derive a simpler method for finding the derivative of an implicitly defined function.

Let $u = u(x, y)$ be a continuously differentiable function and suppose that the equation

$$u(x, y) = 0$$

defines y implicitly as a differentiable function of x. We introduce a variable t by setting $x = t$. Then we have

$$u = u(x, y) \qquad \text{with} \quad x = t \quad \text{and} \quad y = y(t).$$

By (15.4.3),

$$\frac{du}{dt} = \frac{\partial u}{\partial x} \frac{dx}{dt} + \frac{\partial u}{\partial y} \frac{dy}{dt}$$

Now, $u(t, y(t)) = 0$ for all t, so $du/dt = 0$. Also, since $x = t$, we have $dx/dt = 1$ and $dy/dt = dy/dx$. Therefore,

$$0 = \frac{\partial u}{\partial x} + \frac{\partial u}{\partial y} \frac{dy}{dx}.$$

If $\partial u/\partial y \neq 0$, we can solve for dy/dx to obtain

$$\frac{dy}{dx} = -\frac{\partial u/\partial x}{\partial u/\partial y}.$$

This result can be summarized as follows:

(15.4.6)

> Let $u = u(x, y)$ be differentiable and suppose that the equation $u(x, y) = 0$ defines y implicitly as a differentiable function of x. If $\partial u/\partial y \neq 0$, then
> $$\frac{dy}{dx} = -\frac{\partial u/\partial x}{\partial u/\partial y}.$$

Example 7 The equation

$$u(x, y) = 2xy - y^3 + 1 - x - 2y = 0$$

appeared in Example 1(a), Section 3.7. Using the method we have just derived,

$$\frac{\partial u}{\partial x} = 2y - 1, \qquad \frac{\partial u}{\partial y} = 2x - 3y^2 - 2,$$

and

$$\frac{dy}{dx} = -\frac{2y - 1}{2x - 3y^2 - 2} = \frac{1 - 2y}{2x - 3y^2 - 2},$$

which is the same as our previous result. ❏

As illustrated in the next example, this method can be extended to expressions involving more than two variables.

Example 8 Let $u = u(x, y, z)$ be a continuously differentiable function, and suppose that the equation

$$u(x, y, z) = 0$$

defines z implicitly as a differentiable function of x and y. Show that

(15.4.7) if $\dfrac{\partial u}{\partial z} \neq 0$, then $\dfrac{\partial z}{\partial x} = -\dfrac{\partial u/\partial x}{\partial u/\partial z}$ and $\dfrac{\partial z}{\partial y} = -\dfrac{\partial u/\partial y}{\partial u/\partial z}$.

SOLUTION To be able to apply (15.4.5), we write

$$u = u(x, y, z) \quad \text{with} \quad x = s, \quad y = t, \quad z = z(s, t).$$

Then

$$\frac{\partial u}{\partial s} = \frac{\partial u}{\partial x}\frac{\partial x}{\partial s} + \frac{\partial u}{\partial y}\frac{\partial y}{\partial s} + \frac{\partial u}{\partial z}\frac{\partial z}{\partial s}.$$

Since $u(s, t, z(s, t)) = 0$ for all s and t,

$$\frac{\partial u}{\partial s} = 0.$$

Since

$$\frac{\partial x}{\partial s} = 1 \quad \text{and} \quad \frac{\partial y}{\partial s} = 0,$$

we have

$$0 = \frac{\partial u}{\partial x} \cdot 1 + \frac{\partial u}{\partial y} \cdot 0 + \frac{\partial u}{\partial z}\frac{\partial z}{\partial s} = \frac{\partial u}{\partial x} + \frac{\partial u}{\partial z}\frac{\partial z}{\partial s} = \underbrace{\frac{\partial u}{\partial x} + \frac{\partial u}{\partial z}\frac{\partial z}{\partial x}}_{x = s}.$$

If $\partial u/\partial z \neq 0$, then

$$\frac{\partial z}{\partial x} = -\frac{\partial u/\partial x}{\partial u/\partial z}.$$

The formula for $\partial z/\partial y$ can be obtained in a similar manner. ❏

EXERCISES 15.4

In Exercises 1–10, find the rate of change of f with respect to t along the given curve.

1. $f(x, y) = x^2 y$, $\mathbf{r}(t) = e^t\mathbf{i} + e^{-t}\mathbf{j}$.

2. $f(x, y) = x - y$, $\mathbf{r}(t) = at\mathbf{i} + b\cos at\mathbf{j}$.

3. $f(x, y) = \tan^{-1}(y^2 - x^2)$, $\mathbf{r}(t) = \sin t\mathbf{i} + \cos t\mathbf{j}$.

4. $f(x, y) = \ln(2x^2 + y^3)$, $\mathbf{r}(t) = e^{2t}\mathbf{i} + t^{1/3}\mathbf{j}$.

5. $f(x, y) = xe^y + ye^{-x}$, $\mathbf{r}(t) = (\ln t)\mathbf{i} + t(\ln t)\mathbf{j}$.

6. $f(x, y, z) = \ln(x^2 + y^2 + z^2)$,
$\mathbf{r}(t) = \sin t\mathbf{i} + \cos t\mathbf{j} + e^{2t}\mathbf{k}$.

7. $f(x, y, z) = xy - yz$, $\mathbf{r}(t) = t\mathbf{i} + t^2\mathbf{j} + t^3\mathbf{k}$.

8. $f(x, y, z) = x^2 + y^2$,
$\mathbf{r}(t) = a\cos \omega t\mathbf{i} + b\sin \omega t\mathbf{j} + b\omega t\mathbf{k}$.

9. $f(x, y, z) = x^2 + y^2 + z,$
  $\mathbf{r}(t) = a \cos \omega t \mathbf{i} + b \sin \omega t \mathbf{j} + b \omega t \mathbf{k}.$

10. $f(x, y, z) = y^2 \sin(x + z),$ $\mathbf{r}(t) = 2t\mathbf{i} + \cos t\mathbf{j} + t^3\mathbf{k}.$

In Exercises 11–18, find du/dt by applying (15.4.2) or (15.4.3).

11. $u = x^2 - 3xy + 2y^2;$ $x = \cos t,$ $y = \sin t.$
12. $u = x + 4\sqrt{xy} - 3y;$ $x = t^3,$ $y = t^{-1}$ $(t > 0).$
13. $u = e^x \sin y + e^y \sin x;$ $x = \frac{1}{2}t,$ $y = 2t.$
14. $u = 2x^2 - xy + y^2;$ $x = \cos 2t,$ $y = \sin t.$
15. $u = e^x \sin y;$ $x = t^2,$ $y = \pi t.$
16. $u = z \ln\left(\dfrac{y}{x}\right);$ $x = t^2 + 1, y = \sqrt{t}, z = te^t.$
17. $u = xy + yz + zx;$ $x = t^2, y = t(1 - t),$ $z = (1 - t)^2.$
18. $u = x \sin \pi y - z \cos \pi x;$ $x = t^2, y = 1 - t,$ $z = 1 - t^2.$

19. The radius of a right circular cone is increasing at the rate of 3 inches per second and the height is decreasing at the rate of 2 inches per second. At what rate is the volume of the cone changing at the instant the height is 20 inches and the radius is 14 inches?

20. The radius of a right circular cylinder is decreasing at the rate of 2 centimeters per second and the height is increasing at the rate of 3 centimeters per second. At what rate is the volume of the cylinder changing at the instant the radius is 13 centimeters and the height is 18 centimeters?

21. If the lengths of two sides of a triangle are x and y, and θ is the angle between the two sides, then the area A of the triangle is given by $A = \frac{1}{2}xy \sin \theta.$ See the figure. If the sides are each increasing at the rate of 3 inches per second and θ is decreasing at the rate of 0.10 radian per second, how fast is the area changing at the instant $x = 1.5$ feet, $y = 2$ feet, and $\theta = 1$ radian?

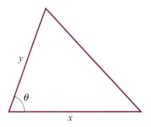

22. An object is moving along the curve of intersection of the paraboloid $z = x^2 + \frac{1}{4}y^2$ and the circular cylinder $x^2 + y^2 = 13.$ If the x-coordinate is increasing at the rate of 5 centimeters per second, how fast is the z-coordinate changing at the instant when $x = 2$ centimeters and $y = 3$ centimeters?

In Exercises 23–28, find $\partial u/\partial s$ and $\partial u/\partial t.$

23. $u = x^2 - xy$ where $x = s \cos t,$ $y = t \sin s.$
24. $u = \sin(x - y) + \cos(x + y)$ where $x = st,$ $y = s^2 - t^2.$
25. $u = x^2 \tan y$ where $x = s^2 t,$ $y = s + t^2.$
26. $u = z^2 \sec xy$ where $x = 2st,$ $y = s - t^2,$ $z = s^2 t.$
27. $u = x^2 - xy + z^2$ where $x = s \cos t,$ $y = \sin(t - s),$ $z = t \sin s.$
28. $u = xe^{yz^2}$ where $x = \ln st,$ $y = t^3,$ $z = s^2 + t^2.$

29. An object moves so that at time t it has position $\mathbf{r}(t) = x(t)\mathbf{i} + y(t)\mathbf{j} + z(t)\mathbf{k}.$ Show that

$$\frac{d}{dt}[f(\mathbf{r}(t))]$$

is the directional derivative of f in the direction of the motion times the speed of the object.

30. (*Important*) Set $r = \|\mathbf{r}\|$ where $\mathbf{r} = x\mathbf{i} + y\mathbf{j} + z\mathbf{k}.$ If f is a continuously differentiable function of r, then

(15.4.8) $$\nabla[f(r)] = f'(r)\frac{\mathbf{r}}{r}.$$

Derive this formula.

In Exercises 31–33, calculate the following gradients $(r = \sqrt{x^2 + y^2 + z^2}).$

31. (a) $\nabla(\sin r).$ (b) $\nabla(r \sin r).$
32. (a) $\nabla(r \ln r).$ (b) $\nabla(e^{1-r^2}).$
33. (a) $\nabla\left(\dfrac{\sin r}{r}\right).$ (b) $\nabla\left(\dfrac{r}{\sin r}\right).$

34. (a) Draw a tree diagram for du/dt given that
  $u = u(x, y)$ where $x = x(s),$ $y = y(s),$ and $s = s(t).$
  (b) Compute $du/dt.$

35. Set $u = u(x, y, z)$ where
  $$x = x(w, t),\quad y = y(w, t),\ z = z(w, t)$$
  and $$w = w(r, s),\ t = t(r, s).$$
  (a) Draw a tree diagram for the partials of $u.$
  (b) Compute $\partial u/\partial r$ and $\partial u/\partial s.$

36. Set $u = u(x, y, z, w)$ where
  $$x = x(r, s, t),\quad y = y(s, t, v),\ z = z(r, t)$$
  and $$w = w(r, s, t, v).$$
  (a) Draw a tree diagram for the partials of $u.$
  (b) Compute $\partial u/\partial r$ and $\partial u/\partial v.$

Higher Derivatives

37. Let $u = u(x, y)$, where $x = x(t)$ and $y = y(t)$, and assume that these functions have continuous second derivatives. Show that

$$\frac{d^2u}{dt^2} = \frac{\partial^2 u}{\partial x^2}\left(\frac{dx}{dt}\right)^2 + 2\frac{\partial^2 u}{\partial x \partial y}\frac{dx}{dt}\frac{dy}{dt} + \frac{\partial^2 u}{\partial y^2}\left(\frac{dy}{dt}\right)^2$$
$$+ \frac{\partial u}{\partial x}\frac{d^2x}{dt^2} + \frac{\partial u}{\partial y}\frac{d^2y}{dt^2}.$$

38. Let $u = u(x, y)$, where $x = x(s, t)$ and $y = y(s, t)$, and assume that these functions have continuous second derivatives. Show that

$$\frac{\partial^2 u}{\partial s^2} = \frac{\partial^2 u}{\partial x^2}\left(\frac{\partial x}{\partial s}\right)^2 + 2\frac{\partial^2 u}{\partial x \partial y}\frac{\partial x}{\partial s}\frac{\partial y}{\partial s} + \frac{\partial^2 u}{\partial y^2}\left(\frac{\partial y}{\partial s}\right)^2$$
$$+ \frac{\partial u}{\partial x}\frac{\partial^2 x}{\partial s^2} + \frac{\partial u}{\partial y}\frac{\partial^2 y}{\partial s^2}.$$

Polar Coordinates

39. Let $u = u(x, y)$ and assume that u is differentiable.
(a) Show that the change of variables to polar coordinates $x = r \cos \theta$ and $y = r \sin \theta$ gives

$$\frac{\partial u}{\partial r} = \frac{\partial u}{\partial x}\cos \theta + \frac{\partial u}{\partial y}\sin \theta,$$

$$\frac{\partial u}{\partial \theta} = -\frac{\partial u}{\partial x}r \sin \theta + \frac{\partial u}{\partial y}r \cos \theta.$$

(b) Express

$$\left(\frac{\partial u}{\partial r}\right)^2 + \frac{1}{r^2}\left(\frac{\partial u}{\partial \theta}\right)^2$$

entirely in terms of $\partial u/\partial x$ and $\partial u/\partial y$.

40. Let w be a function of polar coordinates r and θ. Then w is also a function of rectangular coordinates

$$x = r \cos \theta \quad \text{and} \quad y = r \sin \theta.$$

(a) Using the first part of Exercise 39, verify that

$$\frac{\partial w}{\partial x} = \frac{\partial w}{\partial r}\cos \theta - \frac{1}{r}\frac{\partial w}{\partial \theta}\sin \theta,$$

$$\frac{\partial w}{\partial y} = \frac{\partial w}{\partial r}\sin \theta + \frac{1}{r}\frac{\partial w}{\partial \theta}\cos \theta.$$

(b) Deduce from (a) that

$$\frac{\partial r}{\partial x} = \cos \theta, \qquad \frac{\partial r}{\partial y} = \sin \theta;$$

$$\frac{\partial \theta}{\partial x} = -\frac{1}{r}\sin \theta, \qquad \frac{\partial \theta}{\partial y} = \frac{1}{r}\cos \theta.$$

(c) Find the fallacy:

$$x = r \cos \theta, \qquad r = \frac{x}{\cos \theta}, \qquad \frac{\partial r}{\partial x} = \frac{1}{\cos \theta}.$$

41. *The gradient in polar coordinates*: Let $u = u(x, y)$ be differentiable. Show that if u is written in terms of polar coordinates, then

$$\nabla u = \frac{\partial u}{\partial r}\mathbf{e_r} + \frac{1}{r}\frac{\partial u}{\partial \theta}\mathbf{e_\theta},$$

where

$$\mathbf{e_r} = \cos \theta \mathbf{i} + \sin \theta \mathbf{j} \quad \text{and} \quad \mathbf{e_\theta} = -\sin \theta \mathbf{i} + \cos \theta \mathbf{j}.$$

In Exercises 42 and 43, use the formula in Exercise 41 to express the gradient of the given function in polar coordinates.

42. $u(x, y) = x^2 + y^2$.   **43.** $u(x, y) = x^2 - xy + y^2$.

44. Let $u = u(x, y)$, where $x = r \cos \theta$ and $y = r \sin \theta$, and assume that u has continuous second partial derivatives. Derive a formula for $\partial^2 u/\partial r \partial \theta$.

45. *Laplace's equation in polar coordinates*: Let $u = u(x, y)$ have continuous second partial derivatives. Show that

$$\frac{\partial^2 u}{\partial x^2} + \frac{\partial^2 u}{\partial y^2} = \frac{\partial^2 u}{\partial r^2} + \frac{1}{r^2}\frac{\partial^2 u}{\partial \theta^2} + \frac{1}{r}\frac{\partial u}{\partial r}.$$

In Exercises 46–49, y is defined implicitly as a differentiable function of x by the given equation. Find dy/dx.

46. $x^2 - 2xy + y^4 = 4$.   **47.** $xe^y + ye^x - 2x^2y = 0$.

48. $x^{2/3} + y^{2/3} = a^{2/3}$ (*a* a constant).

49. $x \cos xy + y \cos x = 2$.

In Exercises 50–53, z is defined implicitly as a differentiable function of x and y by the given equation. Find $\partial z/\partial x$ and $\partial z/\partial y$.

50. $z^4 + x^2z^3 + y^2 + xy = 2$.

51. $\cos xyz + \ln(x^2 + y^2 + z^2) = 0$.

52. $xe^y + ye^z + ze^x = 1$.

53. $z \tan x - xy^2z^3 = 2xyz$.

54. (*A chain rule for vector-valued functions*) Suppose that

$$\mathbf{u}(x, y) = u_1(x, y)\mathbf{i} + u_2(x, y)\mathbf{j}$$
$$\text{where} \quad x = x(t), \quad y = y(t).$$

(a) Show that

(15.4.9) $\dfrac{d\mathbf{u}}{dt} = \dfrac{\partial \mathbf{u}}{\partial x}\dfrac{dx}{dt} + \dfrac{\partial \mathbf{u}}{\partial y}\dfrac{dy}{dt},$

where

$$\frac{\partial \mathbf{u}}{\partial x} = \frac{\partial u_1}{\partial x}\mathbf{i} + \frac{\partial u_2}{\partial x}\mathbf{j} \quad \text{and} \quad \frac{\partial \mathbf{u}}{\partial y} = \frac{\partial u_1}{\partial y}\mathbf{i} + \frac{\partial u_2}{\partial y}\mathbf{j}.$$

(b) Let

$$\mathbf{u} = e^x\cos y\,\mathbf{i} + e^x\sin y\,\mathbf{j}$$

where $x = \frac{1}{2}t^2, \quad y = \pi t.$

Compute $d\mathbf{u}/dt$ (i) by applying (15.4.9), (ii) by forming $\mathbf{u}(t)$ directly.

55. Set

$$\mathbf{u}(x, y) = u_1(x, y)\mathbf{i} + u_2(x, y)\mathbf{j}$$

where $x = x(s, t), \quad y = y(s, t).$

Find $\partial \mathbf{u}/\partial s$ and $\partial \mathbf{u}/\partial t$.

56. Set

$$\mathbf{u}(x, y, z) = u_1(x, y, z)\mathbf{i} + u_2(x, y, z)\mathbf{j} + u_3(x, y, z)\mathbf{k},$$

where

$$x = x(t), \quad y = y(t), \quad z = z(t).$$

Derive a formula for $d\mathbf{u}/dt$ analogous to (15.4.9).

57. Homogeneous Functions. A function $f = f(x, y)$ is *homogeneous of degree n*, where n is a positive integer, if $f(tx, ty) = t^n f(x, y)$. Assume that f has continuous second partial derivatives, and is homogeneous of degree n.

(a) Show that f satisfies the equation

$$x\frac{\partial f}{\partial x} + y\frac{\partial f}{\partial y} = nf(x, y).$$

(b) Show that f satisfies the equation

$$x^2\frac{\partial^2 f}{\partial x^2} + 2xy\frac{\partial^2 f}{\partial x\partial y} + y^2\frac{\partial^2 f}{\partial y^2} = n(n-1)f(x, y).$$

■ 15.5 THE GRADIENT AS A NORMAL; TANGENT LINES AND TANGENT PLANES

Functions of Two Variables

We begin with a nonconstant function $f = f(x, y)$ that is continuously differentiable. (*Remember:* That means f is differentiable and its gradient ∇f is continuous.) You have seen that at each point of the domain the gradient vector, if not $\mathbf{0}$, points in the direction of the most rapid increase of f. Here we show that

(15.5.1) | at each point of the domain, the gradient vector, if not $\mathbf{0}$, is perpendicular to the level curve that passes through that point.

PROOF We choose a point (x_0, y_0) in the domain and assume that $\nabla f(x_0, y_0) \neq \mathbf{0}$. The level curve through this point has equation

$$f(x, y) = c \quad \text{where } c = f(x_0, y_0).$$

Under our assumptions on f, this curve can be parametrized in a neighborhood of (x_0, y_0) by a continuously differentiable vector function

$$\mathbf{r}(t) = x(t)\mathbf{i} + y(t)\mathbf{j}, \quad t \in I$$

with nonzero tangent vector $\mathbf{r}'(t).$†

Now take t_0 such that

$$\mathbf{r}(t_0) = x_0\mathbf{i} + y_0\mathbf{j} = (x_0, y_0).$$

---

† This follows from a result of advanced calculus known as the *implicit function theorem*.

We will show that

$$\nabla f(\mathbf{r}(t_0)) \perp \mathbf{r}'(t_0).$$

Since f is constantly c on the curve, we have

$$f(\mathbf{r}(t)) = c \qquad \text{for all } t \in I.$$

For such t

$$\frac{d}{dt}[f(\mathbf{r}(t))] = \nabla f(\mathbf{r}(t)) \cdot \mathbf{r}'(t) = 0.$$

$\longmapsto$ (Theorem 15.4.1)

In particular

$$\nabla f(\mathbf{r}(t_0)) \cdot \mathbf{r}'(t_0) = 0,$$

and thus

$$\nabla f(\mathbf{r}(t_0)) \perp \mathbf{r}'(t_0). \quad \square$$

Figure 15.5.1 illustrates the result.

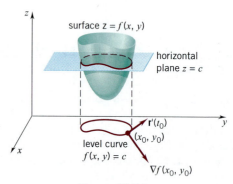

Figure 15.5.1

Example 1 For the function

$$f(x, y) = x^2 + y^2$$

the level curves are concentric circles:

$$x^2 + y^2 = c.$$

At each point $(x, y) \neq (0, 0)$ the gradient vector

$$\nabla f(x, y) = 2x\mathbf{i} + 2y\mathbf{j} = 2\mathbf{r}$$

points away from the origin along the line of the radius vector and is thus perpendicular to the circle in question.   At the origin the level curve is reduced to a point and the gradient is simply $\mathbf{0}$.   See Figure 15.5.2.   ❑

Consider now a curve in the xy-plane

$$C: \quad f(x, y) = c.$$

As before we assume that f is nonconstant and continuously differentiable.   Let's suppose that (x_0, y_0) lies on the curve and $\nabla f(x_0, y_0) \neq \mathbf{0}$.

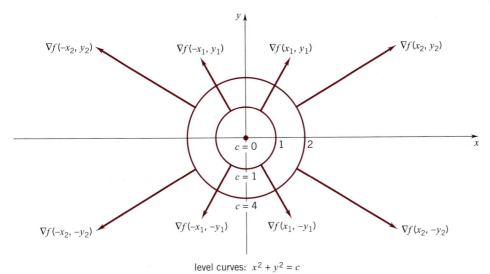

level curves: $x^2 + y^2 = c$

Figure 15.5.2

We can view C as the c-level curve of f and conclude from (15.5.1) that the gradient

(15.5.2)
$$\nabla f(x_0, y_0) = \frac{\partial f}{\partial x}(x_0, y_0)\mathbf{i} + \frac{\partial f}{\partial y}(x_0, y_0)\mathbf{j}$$

is perpendicular to C at (x_0, y_0). We call it a *normal vector*.

The vector

(15.5.3)
$$\mathbf{t}(x_0, y_0) = \frac{\partial f}{\partial y}(x_0, y_0)\mathbf{i} - \frac{\partial f}{\partial x}(x_0, y_0)\mathbf{j}$$

is perpendicular to the gradient:

$$\nabla f(x_0, y_0) \cdot \mathbf{t}(x_0, y_0) = \frac{\partial f}{\partial x}(x_0, y_0)\frac{\partial f}{\partial y}(x_0, y_0) - \frac{\partial f}{\partial y}(x_0, y_0)\frac{\partial f}{\partial x}(x_0, y_0) = 0.$$

It is therefore a *tangent vector*.

The line through (x_0, y_0) perpendicular to the gradient is the tangent line. To obtain an equation for the tangent line we refer to Figure 15.5.3. A point (x, y) will lie on the tangent line iff

$$[(x - x_0)\mathbf{i} + (y - y_0)\mathbf{j}] \cdot \nabla f(x_0, y_0) = 0,$$

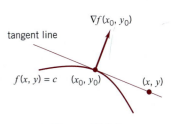

Figure 15.5.3

that is, iff

(15.5.4)

$$\frac{\partial f}{\partial x}(x_0, y_0)(x - x_0) + \frac{\partial f}{\partial y}(x_0, y_0)(y - y_0) = 0.$$

This is an equation for the *tangent line*.

The line through (x_0, y_0) perpendicular to the tangent vector $\mathbf{t}(x_0, y_0)$ is the normal line (Figure 15.5.4). A point (x, y) will lie on the normal line iff

$$[(x - x_0)\mathbf{i} + (y - y_0)\mathbf{j}] \cdot \mathbf{t}(x_0, y_0) = 0,$$

that is, iff

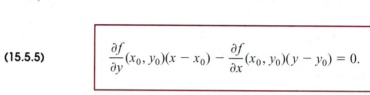

normal line

(x, y)

(x_0, y_0)

$f(x, y) = c$

$\mathbf{t}(x_0, y_0)$

Figure 15.5.4

(15.5.5)

$$\frac{\partial f}{\partial y}(x_0, y_0)(x - x_0) - \frac{\partial f}{\partial x}(x_0, y_0)(y - y_0) = 0.$$

This is an equation for the *normal line*.

Example 2 Choose a point (x_0, y_0) on the hyperbola

$$\frac{x^2}{a^2} - \frac{y^2}{b^2} = 1.$$

To avoid denominators, we write

$$b^2 x^2 - a^2 y^2 = a^2 b^2.$$

This equation is of the form

$$f(x, y) = c \quad \text{with} \quad f(x, y) = b^2 x^2 - a^2 y^2 \quad \text{and} \quad c = a^2 b^2.$$

Partial differentiation gives

$$\frac{\partial f}{\partial x}(x, y) = 2b^2 x, \qquad \frac{\partial f}{\partial y}(x, y) = -2a^2 y.$$

At (x_0, y_0) the gradient

$$\frac{\partial f}{\partial x}(x_0, y_0)\mathbf{i} + \frac{\partial f}{\partial y}(x_0, y_0)\mathbf{j} = 2b^2 x_0 \mathbf{i} - 2a^2 y_0 \mathbf{j}$$

is normal to the curve and the vector

$$\frac{\partial f}{\partial y}(x_0, y_0)\mathbf{i} - \frac{\partial f}{\partial x}(x_0, y_0)\mathbf{j} = -2a^2 y_0 \mathbf{i} - 2b^2 x_0 \mathbf{j}$$

is tangent to the curve.

The equation of the tangent line can be written

$$2b^2 x_0(x - x_0) + (-2a^2 y_0)(y - y_0) = 0.$$

Dividing by 2 and noting that $b^2 x_0^2 - a^2 y_0^2 = a^2 b^2$,† we can simplify the equation to

$$(b^2 x_0)x - (a^2 y_0)y = a^2 b^2.$$

The equation of the normal line takes the form

$$(-2a^2 y_0)(x - x_0) + (-2b^2 x_0)(y - y_0) = 0.$$

This simplifies to

$$(a^2 y_0)x + (b^2 x_0)y = (a^2 + b^2)x_0 y_0. \quad \square$$

Functions of Three Variables

Here, instead of level curves, we have level surfaces, but the results are similar. If $f = f(x, y, z)$ is nonconstant and continuously differentiable, then

(15.5.6)

> at each point of the domain, the gradient vector, if not **0**, is perpendicular to the level surface that passes through that point.

PROOF We choose a point $\mathbf{x}_0 = (x_0, y_0, z_0)$ in the domain and assume that $\nabla f(x_0, y_0, z_0) \neq \mathbf{0}$. The level surface through this point has equation

$$f(x, y, z) = c \qquad \text{where } c = f(x_0, y_0, z_0).$$

We suppose now that

$$\mathbf{r}(t) = x(t)\mathbf{i} + y(t)\mathbf{j} + z(t)\mathbf{k}, \qquad t \in I$$

is a differentiable curve that lies on this surface and passes through the point $\mathbf{x}_0 = (x_0, y_0, z_0)$. We choose t_0 so that

$$\mathbf{r}(t_0) = \mathbf{x}_0 = (x_0, y_0, z_0)$$

and suppose that $\mathbf{r}'(t_0) \neq \mathbf{0}$.

Since the curve lies on the given surface, we have

$$f(\mathbf{r}(t)) = c \qquad \text{for all } t \in I.$$

For such t

$$\frac{d}{dt}[f(\mathbf{r}(t))] = \nabla f(\mathbf{r}(t)) \cdot \mathbf{r}'(t) = 0.$$

In particular,

$$\nabla f(\mathbf{r}(t_0)) \cdot \mathbf{r}'(t_0) = 0.$$

The gradient vector

$$\nabla f(\mathbf{r}(t_0)) = \nabla f(\mathbf{x}_0) = \nabla f(x_0, y_0, z_0)$$

is thus perpendicular to the curve in question.

---

† The point (x_0, y_0) lies on the hyperbola.

This same argument applies to *every* differentiable curve that lies on this level surface and passes through the point $\mathbf{x}_0 = (x_0, y_0, z_0)$ with nonzero tangent vector. (See Figure 15.5.5.) Consequently, $\nabla f(\mathbf{x}_0)$ must be perpendicular to the surface itself. ❑

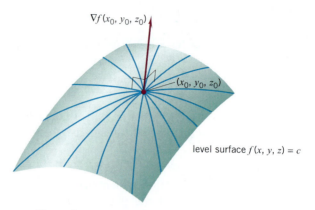

The gradient vector $\nabla f(x_0, y_0, z_0)$ is perpendicular
to the level surface at (x_0, y_0, z_0)

Figure 15.5.5

Example 3   For the function

$$f(x, y, z) = x^2 + y^2 + z^2$$

the level surfaces are concentric spheres:

$$x^2 + y^2 + z^2 = c.$$

At each point $(x, y, z) \neq (0, 0, 0)$ the gradient vector

$$\nabla f(x, y, z) = 2x\mathbf{i} + 2y\mathbf{j} + 2z\mathbf{k} = 2\mathbf{r}$$

points away from the origin along the line of the radius vector and is thus perpendicular to the sphere in question.   At the origin the level surface is reduced to a point and the gradient is $\mathbf{0}$. ❑

The *tangent plane* to a surface

$$f(x, y, z) = c$$

at a point $\mathbf{x}_0 = (x_0, y_0, z_0)$ is the plane through $\mathbf{x}_0$ with normal $\nabla f(\mathbf{x}_0)$.   See Figure 15.5.6.

The tangent plane at a point $\mathbf{x}_0$ is the plane through $\mathbf{x}_0$ that best approximates the surface in a neighborhood of $\mathbf{x}_0$.   (We return to this later.)

A point $\mathbf{x}$ lies on the tangent plane through $\mathbf{x}_0$ iff

(15.5.7)

$$\nabla f(\mathbf{x}_0) \cdot (\mathbf{x} - \mathbf{x}_0) = 0.$$

(Figure 15.5.6)

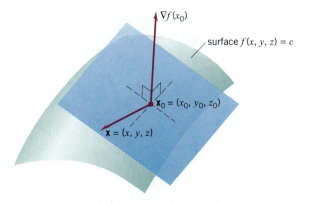

tangent plane to a surface $f(x, y, z) = c$

Figure 15.5.6

This is an equation for the tangent plane in vector notation. In Cartesian coordinates the equation takes the form

(15.5.8)
$$\frac{\partial f}{\partial x}(x_0, y_0, z_0)(x - x_0) + \frac{\partial f}{\partial y}(x_0, y_0, z_0)(y - y_0) + \frac{\partial f}{\partial z}(x_0, y_0, z_0)(z - z_0) = 0.$$

The *normal line* to the surface $f(x, y, z) = c$ at a point $\mathbf{x}_0 = (x_0, y_0, z_0)$ on the surface is the line which passes through (x_0, y_0, z_0) parallel to $\nabla f(\mathbf{x}_0)$. Thus, $\nabla f(\mathbf{x}_0)$ is a direction vector for the normal line and

(15.5.9)
$$\mathbf{r}(t) = \mathbf{r}_0 + \nabla f(\mathbf{x}_0)t \qquad (\mathbf{r}_0 = x_0\mathbf{i} + y_0\mathbf{j} + z_0\mathbf{k})$$

is a vector equation for the line. In scalar parametric form, equations for the normal line are

(15.5.10)
$$x = x_0 + \frac{\partial f}{\partial x}(x_0, y_0, z_0)t,$$
$$y = y_0 + \frac{\partial f}{\partial y}(x_0, y_0, z_0)t,$$
$$z = z_0 + \frac{\partial f}{\partial z}(x_0, y_0, z_0)t.$$

Example 4 Find an equation in x, y, z for the plane tangent to the surface
$$xy + yz + zx = 11 \qquad \text{at the point } (1, 2, 3).$$

SOLUTION   The surface is of the form

$$f(x, y, z) = c \qquad \text{with} \quad f(x, y, z) = xy + yz + zx \quad \text{and} \quad c = 11.$$

Observe that

$$\frac{\partial f}{\partial x} = y + z, \qquad \frac{\partial f}{\partial y} = x + z, \qquad \frac{\partial f}{\partial z} = x + y.$$

At the point $(1, 2, 3)$

$$\frac{\partial f}{\partial x} = 5, \qquad \frac{\partial f}{\partial y} = 4, \qquad \frac{\partial f}{\partial z} = 3.$$

The equation for the tangent plane can therefore be written

$$5(x - 1) + 4(y - 2) + 3(z - 3) = 0.$$

This simplifies to

$$5x + 4y + 3z - 22 = 0. \quad ❑$$

Example 5   Find an equation for the tangent plane and scalar parametric equations for the normal line to the quadric cone

$$x^2 + 4y^2 = z^2$$

at the point $(3, 2, 5)$ on the cone (Figure 15.5.7).

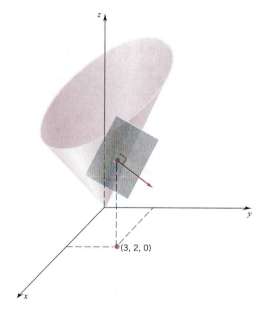

Figure 15.5.7

SOLUTION   The surface is of the form $f(x, y, z) = c$ with

$$f(x, y, z) = x^2 + 4y^2 - z^2 \quad \text{and} \quad c = 0.$$

The partial derivatives of f are

$$\frac{\partial f}{\partial x} = 2x, \qquad \frac{\partial f}{\partial y} = 8y, \qquad \frac{\partial f}{\partial z} = -2z.$$

and

$$\nabla f = 2x\mathbf{i} + 8y\mathbf{j} - 2z\mathbf{k}.$$

Now, $\nabla f(3, 2, 5) = 6\mathbf{i} + 16\mathbf{j} - 10\mathbf{k}$ is normal to the cone at the point $(3, 2, 5)$. Note that $\frac{1}{2}\nabla f(3, 2, 5) = 3\mathbf{i} + 8\mathbf{j} - 5\mathbf{k}$ is also normal to the cone and is a little simpler. An equation for the tangent plane is

$$3(x - 3) + 8(y - 2) - 5(z - 5) = 0,$$

which simplifies to

$$3x + 8y - 5z = 0.$$

Note that this plane passes through the origin, as we would expect. Scalar parametric equations for the normal line are

$$x = 3 + 3t, \qquad y = 2 + 8t, \qquad z = 5 - 5t. \quad \square$$

Example 6 The curve

$$\mathbf{r}(t) = \tfrac{1}{2}t^2\mathbf{i} + 4t^{-1}\mathbf{j} + (\tfrac{1}{2}t - t^2)\mathbf{k}$$

intersects the hyperbolic paraboloid $x^2 - 4y^2 - 4z = 0$ at the point $(2, 2, -3)$. What is the angle of intersection?

SOLUTION We want the angle φ between the tangent vector of the curve and the tangent plane of the surface at the point of intersection (Figure 15.5.8).

A simple calculation shows that the curve passes through the point $(2, 2, -3)$ at $t = 2$. Since

$$\mathbf{r}'(t) = t\mathbf{i} - 4t^{-2}\mathbf{j} + (\tfrac{1}{2} - 2t)\mathbf{k},$$

we have

$$\mathbf{r}'(2) = 2\mathbf{i} - \mathbf{j} - \tfrac{7}{2}\mathbf{k}.$$

Now set

$$f(x, y, z) = x^2 - 4y^2 - 4z.$$

This function has gradient $2x\mathbf{i} - 8y\mathbf{j} - 4\mathbf{k}$. At the point $(2, 2, -3)$

$$\nabla f = 4\mathbf{i} - 16\mathbf{j} - 4\mathbf{k}.$$

Now let θ be the angle between $\mathbf{r}'(2)$ and this gradient. By (12.3.9)

$$\cos\theta = \frac{\mathbf{r}'(2)}{\|\mathbf{r}'(2)\|} \cdot \frac{\nabla f}{\|\nabla f\|} = \frac{19}{414}\sqrt{138} \cong 0.539$$

so that $\theta \cong 1.00$ radian. Since the gradient is normal to the tangent plane, the angle ϕ we want is

$$\phi = \tfrac{1}{2}\pi - \theta \cong 1.57 - 1.00 = 0.57 \text{ radian.} \quad \square$$

A surface of the form

$$z = g(x, y)$$

can be written in the form

$$f(x, y, z) = 0$$

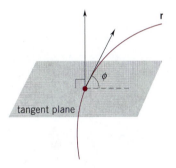

Figure 15.5.8

by setting

$$f(x, y, z) = g(x, y) - z.$$

If g is differentiable, so is f. Moreover,

$$\frac{\partial f}{\partial x}(x, y, z) = \frac{\partial g}{\partial x}(x, y), \qquad \frac{\partial f}{\partial y}(x, y, z) = \frac{\partial g}{\partial y}(x, y), \qquad \frac{\partial f}{\partial z}(x, y, z) = -1.$$

By (15.5.8), the tangent plane at (x_0, y_0, z_0) has equation

$$\frac{\partial g}{\partial x}(x_0, y_0)(x - x_0) + \frac{\partial g}{\partial y}(x_0, y_0)(y - y_0) + (-1)(z - z_0) = 0,$$

which we can rewrite as

(15.5.11)
$$(z - z_0) = \frac{\partial g}{\partial x}(x_0, y_0)(x - x_0) + \frac{\partial g}{\partial y}(x_0, y_0)(y - y_0).$$

If $\nabla g(x_0, y_0) = \mathbf{0}$, then both partials of g are zero at (x_0, y_0) and the equation reduces to

$$z = z_0.$$

In this case the tangent plane is *horizontal*.

Scalar parametric equations for the normal line to the surface $z = g(x, y)$ at the point (x_0, y_0, z_0) are

(15.5.12)
$$x = x_0 + \frac{\partial g}{\partial x}(x_0, y_0)t, \qquad y = y_0 + \frac{\partial g}{\partial y}(x_0, y_0)t, \qquad z = z_0 + (-1)t.$$

Example 7 Find an equation for the tangent plane and symmetric equations for the normal line to the surface

$$z = \ln(x^2 + y^2)$$

at the point $(-2, 1, \ln 5)$ on the surface.

SOLUTION Set

$$g(x, y) = \ln(x^2 + y^2).$$

The partial derivatives of g are

$$\frac{\partial g}{\partial x}(x, y) = \frac{2x}{x^2 + y^2}, \qquad \frac{\partial g}{\partial y}(x, y) = \frac{2y}{x^2 + y^2},$$

and when $x = -2$ and $y = 1$,

$$\frac{\partial g}{\partial x} = -\frac{4}{5}, \qquad \frac{\partial g}{\partial y} = \frac{2}{5}.$$

Therefore, at the point $(-2, 1, \ln 5)$, the tangent plane has equation

$$z - \ln 5 = -\frac{4}{5}(x + 2) + \frac{2}{5}(y - 1).$$

Symmetric equations for the normal line are

$$\frac{x + 2}{-\frac{4}{5}} = \frac{y - 1}{\frac{2}{5}} = \frac{z - \ln 5}{-1}.$$

A graph of this surface is shown in Figure 15.5.9. ❏

Figure 15.5.9

Example 8   At what points of the surface

$$z = 3xy - x^3 - y^3$$

is the tangent plane horizontal?

SOLUTION   The function

$$g(x, y) = 3xy - x^3 - y^3$$

has first partials

$$\frac{\partial g}{\partial x}(x, y) = 3y - 3x^2, \qquad \frac{\partial g}{\partial y}(x, y) = 3x - 3y^2.$$

We set these partial derivatives equal to 0 and solve the resulting system of equations:

$$3y - 3x^2 = 0 \qquad\qquad y - x^2 = 0$$
$$\qquad\qquad\qquad \text{or}$$
$$3x - 3y^2 = 0 \qquad\qquad x - y^2 = 0.$$

From the first equation, we get $y = x^2$.   Substituting this into the second equation, it follows that

$$x - x^4 = 0$$

or

$$x(1 - x^3) = 0.$$

Therefore, $x = 0$ (which implies $y = 0$) or $x = 1$ (which implies $y = 1$). Thus, these partials are both zero only at $(0, 0)$ and $(1, 1)$. The surface has a horizontal tangent plane only at $(0, 0, 0)$ and $(1, 1, 1)$. A graph of this surface is shown in Figure 15.5.10. ❑

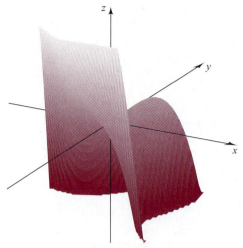

Figure 15.5.10

EXERCISES 15.5

In Exercises 1–8, find a normal vector and a tangent vector at the point indicated. Write an equation for the tangent line and an equation for the normal line.

1. $x^2 + xy + y^2 = 3$; $(-1, -1)$.

2. $(y - x)^2 = 2x$; $(2, 4)$.

3. $(x^2 + y^2)^2 = 9(x^2 - y^2)$; $(\sqrt{2}, 1)$.

4. $x^3 + y^3 = 9$; $(1, 2)$.

5. $xy^2 - 2x^2 + y + 5x = 6$; $(4, 2)$.

6. $x^5 + y^5 = 2x^3$; $(1, 1)$.

7. $2x^3 - x^2y^2 = 3x - y - 7$; $(1, -2)$.

8. $x^3 + y^2 + 2x = 6$; $(-1, 3)$.

9. Find the slope of the curve $x^2y = a^2(a - y)$ at the point $(0, a)$.

In Exercises 10–18, find an equation for the tangent plane and scalar parametric equations for the normal line at the point indicated.

10. $z = (x^2 + y^2)^2$; $(1, 1, 4)$.

11. $x^3 + y^3 = 3xyz$; $(1, 2, \frac{3}{2})$.

12. $xy^2 + 2z^2 = 12$; $(1, 2, 2)$.

13. $z = axy$; $(1, 1/a, 1)$.

14. $\sqrt{x} + \sqrt{y} + \sqrt{z} = 4$; $(1, 4, 1)$.

15. $z = \sin x + \sin y + \sin(x + y)$; $(0, 0, 0)$.

16. $z = x^2 + xy + y^2 - 6x + 2$; $(4, -2, -10)$.

17. $b^2c^2x^2 - a^2c^2y^2 - a^2b^2z^2 = a^2b^2c^2$; (x_0, y_0, z_0).

18. $z = \sin(x \cos y)$; $(0, \frac{1}{2}\pi, 0)$.

In Exercises 19–23, find the point(s) on the surface at which the tangent plane is horizontal.

19. $xy + a^3x^{-1} + b^3y^{-1} - z = 0$.

20. $z = 4x + 2y - x^2 + xy - y^2$.

21. $z = xy$.

22. $x + y + z + xy - x^2 - y^2 = 0$.

23. $z - 2x^2 - 2xy + y^2 + 5x - 3y + 2 = 0$.

24. (a) Find the *upper unit normal* (the unit normal with positive $\mathbf{k}$ component) for the surface $z = xy$ at the point $(1, 1, 1)$.

(b) Find the *lower unit normal* (the unit normal with negative $\mathbf{k}$ component) for the surface $z = 1/x - 1/y$ at the point $(1, 1, 0)$.

25. Let $f = f(x, y, z)$ be continuously differentiable. Write equations in symmetric form for the line normal to the surface $f(x, y, z) = c$ at the point (x_0, y_0, z_0).

26. Show that in the case of a surface of the form $z = xf(x/y)$ with f continuously differentiable, all the tangent planes share a point in common.

27. Given that the surfaces $F(x, y, z) = 0$ and $G(x, y, z) = 0$ intersect at right angles in a curve γ, what condition must be satisifed by the partial derivatives of F and G on γ?

28. Show that, for all planes tangent to the surface $\sqrt{x} + \sqrt{y} + \sqrt{z} = \sqrt{a}$, the sum of the intercepts is the same.

29. Show that all pyramids formed by the coordinate planes and a plane tangent to the surface $xyz = a^3$ have the same volume. What is this volume?

30. Show that, for all planes tangent to the surface $x^{2/3} + y^{2/3} + z^{2/3} = a^{2/3}$, the sum of the squares of the intercepts is the same.

31. The curve $\mathbf{r}(t) = 2t\mathbf{i} + 3t^{-1}\mathbf{j} - 2t^2\mathbf{k}$ and the ellipsoid $x^2 + y^2 + 3z^2 = 25$ intersect at $(2, 3, -2)$. What is the angle of intersection?

32. Show that the curve $\mathbf{r}(t) = \frac{3}{2}(t^2 + 1)\mathbf{i} + (t^4 + 1)\mathbf{j} + t^3\mathbf{k}$ is perpendicular to the ellipsoid $x^2 + 2y^2 + 3z^2 = 20$ at the point $(3, 2, 1)$.

33. The surfaces $x^2y^2 + 2x + z^3 = 16$ and $3x^2 + y^2 - 2z = 9$ intersect in a curve that passes through the point $(2, 1, 2)$. What are the equations of the respective tangent planes to the two surfaces at this point?

34. Show that the sphere $x^2 + y^2 + z^2 - 8x - 8y - 6z + 24 = 0$ is tangent to the ellipsoid $x^2 + 3y^2 + 2z^2 = 9$ at the point $(2, 1, 1)$.

35. Show that the sphere $x^2 + y^2 + z^2 - 4y - 2z + 2 = 0$ is perpendicular to the paraboloid $3x^2 + 2y^2 - 2z = 1$ at the point $(1, 1, 2)$.

36. Show that the following surfaces are mutually perpendicular: $xy = az^2$, $x^2 + y^2 + z^2 = b$, $z^2 + 2x^2 = c(z^2 + 2y^2)$.

37. The surface S: $z = x^2 + 3y^2 + 2$ intersects the vertical plane p: $3x + 4y + 6 = 0$ in a space curve C.
(a) Let C_1 be the projection of C in the xy-plane. Find an equation for C_1.
(b) Find a parametrization $\mathbf{r}(t) = x(t)\mathbf{i} + y(t)\mathbf{j} + z(t)\mathbf{k}$ for C setting $x(t) = 4t - 2$.
(c) Find a parametrization $\mathbf{R}(s) = \mathbf{R}_0 + s\mathbf{d}$ for the line l tangent to C at $(2, -3, 33)$.
(d) Find an equation for the plane p_1 tangent to S at $(2, -3, 33)$.
(e) Find a parametrization $\mathbf{r}(t) = x(t)\mathbf{i} + y(t)\mathbf{j} + z(t)\mathbf{k}$ for the line l' formed by the intersection of p with p_1 taking $x(t) = t$. What is the relation between l and l'?

■ 15.6 MAXIMUM AND MINIMUM VALUES

In Chapter 4 we discussed the local extreme values of a function of one variable. Here we take up the same subject for functions of several variables. The ideas are very similar.

DEFINITION 15.6.1 LOCAL MAXIMUM AND LOCAL MINIMUM

Let f be a function of several variables and let $\mathbf{x}_0$ be an interior point of the domain:
f is said to have a *local maximum* at $\mathbf{x}_0$ iff

$$f(\mathbf{x}_0) \geq f(\mathbf{x}) \qquad \text{for all } \mathbf{x} \text{ in some neighborhood of } \mathbf{x}_0;$$

f is said to have a *local minimum* at $\mathbf{x}_0$ iff

$$f(\mathbf{x}_0) \leq f(\mathbf{x}) \qquad \text{for all } \mathbf{x} \text{ in some neighborhood of } \mathbf{x}_0.$$

As in the one-variable case, the local maxima and local minima together comprise the *local extreme values*.

In the one-variable case we know that, if f has a local extreme value at x_0, then

$$\text{either} \quad f'(x_0) = 0 \quad \text{or} \quad f'(x_0) \text{ does not exist.}$$

We have a similar result for functions of several variables.

THEOREM 15.6.2

If f has a local extreme value at $\mathbf{x}_0$, then

$$\text{either} \quad \nabla f(\mathbf{x}_0) = \mathbf{0} \quad \text{or} \quad \nabla f(\mathbf{x}_0) \text{ does not exist.}$$

PROOF We assume that f has a local extreme value at $\mathbf{x}_0$ and that f is differentiable at $\mathbf{x}_0$ [that $\nabla f(\mathbf{x}_0)$ exists]. We need to show that $\nabla f(\mathbf{x}_0) = \mathbf{0}$. For simplicity we set $\mathbf{x}_0 = (x_0, y_0)$. The three-variable case is similar.

Since f has a local extreme value at (x_0, y_0), the function $g(x) = f(x, y_0)$ has a local extreme value at x_0. Since f is differentiable at (x_0, y_0), g is differentiable at x_0 and therefore

$$g'(x_0) = \frac{\partial f}{\partial x}(x_0, y_0) = 0.$$

Similarly, the function $h(y) = f(x_0, y)$ has a local extreme value at y_0 and, being differentiable there, satisfies the relation

$$h'(y_0) = \frac{\partial f}{\partial y}(x_0, y_0) = 0.$$

The gradient is $\mathbf{0}$ since both partials are 0. ❏

Interior points of the domain at which the gradient is zero or the gradient does not exist are called *critical points*. By Theorem 15.6.2 these are the only points at which a local extreme value can occur.

Although the ideas introduced so far are completely general, their application to functions of more than two variables is generally laborious. We restrict ourselves mostly to functions of two variables. Not only are the computations less formidable, but also we can make use of our geometric intuition.

Two Variables

We suppose for the moment that $f = f(x, y)$ is defined on an open set and is continuously differentiable there. The graph of f is a surface

$$z = f(x, y).$$

Where f has a local maximum, the surface has a local high point. Where f has a local minimum, the surface has a local low point. Where f has either a local maximum or a local minimum, the gradient is $\mathbf{0}$ and therefore the tangent plane is horizontal. See Figure 15.6.1.

A zero gradient signals the possibility of a local extreme value; it does not guarantee it. For example, in the case of the saddle-shaped surface of Figure 15.6.2, there is a horizontal tangent plane at the origin and therefore the gradient is zero there, yet the origin gives neither a local maximum nor a local minimum.

Critical points at which the gradient is zero are called *stationary points*. The stationary points that do not give rise to extreme values are called *saddle points*.

Below we test some differentiable functions for extreme values. In each case, our first step is to seek out the stationary points.

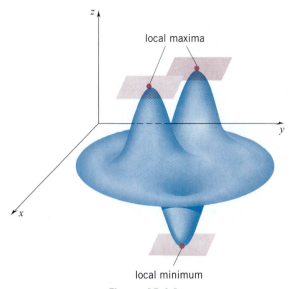

local maxima

local minimum

Figure 15.6.1

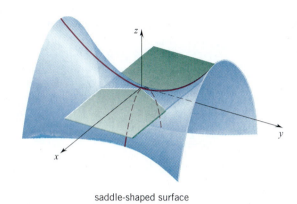

saddle-shaped surface

Figure 15.6.2

Example 1 For the function

$$f(x, y) = 2x^2 + y^2 - xy - 7y,$$

we have

$$\nabla f(x, y) = (4x - y)\mathbf{i} + (2y - x - 7)\mathbf{j}.$$

To find the stationary points, we set $\nabla f(x, y) = \mathbf{0}$. This gives

$$4x - y = 0 \quad \text{and} \quad 2y - x - 7 = 0.$$

The only simultaneous solution to these equations is $x = 1, y = 4$. The point $(1, 4)$ is therefore the only stationary point.

We now compare the value of f at $(1, 4)$ with the values of f at nearby points $(1 + h, 4 + k)$:

$$f(1, 4) = 2 + 16 - 4 - 28 = -14,$$

$$\begin{aligned} f(1 + h, 4 + k) &= 2(1 + h)^2 + (4 + k)^2 - (1 + h)(4 + k) - 7(4 + k) \\ &= 2 + 4h + 2h^2 + 16 + 8k + k^2 - 4 - 4h - k - hk - 28 - 7k \\ &= 2h^2 + k^2 - hk - 14. \end{aligned}$$

The difference

$$\begin{aligned} f(1 + h, 4 + k) - f(1, 4) &= 2h^2 + k^2 - hk \\ &= h^2 + (h^2 - hk + k^2) \\ &\geq h^2 + (h^2 - 2|h||k| + k^2) \\ &= h^2 + (|h| - |k|)^2 \geq 0. \end{aligned}$$

Thus $f(1 + h, 4 + k) \geq f(1, 4)$ for all small h and k (in fact for all real h and k).† It follows that f has a local minimum at $(1, 4)$. This local minimum is -14. ❏

---

† Another way to see that $2h^2 + k^2 - hk$ is nonnegative is to complete the square:

$$2h^2 + k^2 - hk = \tfrac{1}{4}h^2 - hk + k^2 + \tfrac{7}{4}h^2 = (\tfrac{1}{2}h - k)^2 + \tfrac{7}{4}h^2 \geq 0.$$

Example 2 In the case of

$$f(x, y) = y^2 - xy + 2x + y + 1$$

we have

$$\nabla f(x, y) = (2 - y)\mathbf{i} + (2y - x + 1)\mathbf{j}.$$

The gradient is **0** where

$$2 - y = 0 \quad \text{and} \quad 2y - x + 1 = 0.$$

The only simultaneous solution to these equations is $x = 5$, $y = 2$. The point $(5, 2)$ is the only stationary point.

We now compare the value of f at $(5, 2)$ with the values of f at nearby points $(5 + h, 2 + k)$:

$$f(5, 2) = 4 - 10 + 10 + 2 + 1 = 7,$$

$$\begin{aligned} f(5 + h, 2 + k) &= (2 + k)^2 - (5 + h)(2 + k) + 2(5 + h) + (2 + k) + 1 \\ &= 4 + 4k + k^2 - 10 - 2h - 5k - hk + 10 + 2h + 2 + k + 1 \\ &= k^2 - hk + 7. \end{aligned}$$

The difference

$$d = f(5 + h, 2 + k) - f(5, 2) = k^2 - hk = k(k - h)$$

does not keep a constant sign for small h and k. See Figure 15.6.3: $d < 0$ in region I $(k > 0, k < h)$; $d > 0$ in region II $(k > 0, k > h)$; and so on. Therefore, it follows that $(5, 2)$ is a saddle point. ❑

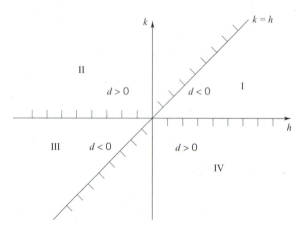

Figure 15.6.3

Example 3 For the function

$$f(x, y) = -xye^{-(x^2+y^2)/2}$$

we have

$$\frac{\partial f}{\partial x} = -ye^{-(x^2+y^2)/2} + x^2ye^{-(x^2+y^2)/2} = y(x^2 - 1)e^{-(x^2+y^2)/2},$$

$$\frac{\partial f}{\partial y} = -xe^{-(x^2+y^2)/2} + xy^2e^{-(x^2+y^2)/2} = x(y^2 - 1)e^{-(x^2+y^2)/2},$$

and

$$\nabla f(x, y) = e^{-(x^2+y^2)/2}[y(x^2 - 1)\mathbf{i} + x(y^2 - 1)\mathbf{j}].$$

Since $e^{-(x^2+y^2)/2} \neq \mathbf{0}$ for all (x, y), $\nabla f(x, y) = \mathbf{0}$ iff

$$y(1 - x^2) = 0 \qquad \text{and} \qquad x(y^2 - 1) = 0.$$

The simultaneous solutions to these equations are $x = y = 0$; $x = 1$, $y = \pm 1$; $x = -1$, $y = \pm 1$. Thus, $(0, 0)$, $(1, 1)$, $(1, -1)$, $(-1, 1)$, and $(-1, -1)$ are the stationary points.

A computer-generated graph of this function is shown in Figure 15.6.4. From the graph, we see that f has local minima at $(1, 1)$ and $(-1, -1)$, and local maxima at $(1, -1)$ and $(-1, 1)$. The point $(0, 0)$ is a saddle point since

$$f(h, k) - f(0, 0) = f(h, k) = -hke^{-(h^2+k^2)/2}$$

does not keep constant sign for small h and k. ❏

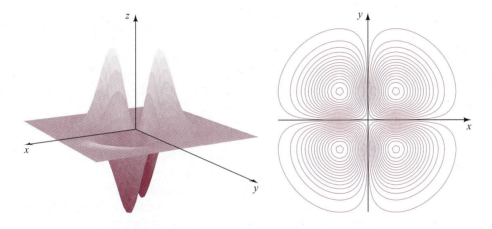

level curves of $f(x, y) = -xye^{-(x^2+y^2)^2}$

Figure 15.6.4

The function in the next example has an infinite number of stationary points.

Example 4 For all points (x, y) with $x + y \neq 0$, set

$$f(x, y) = -\frac{x^2 + 2y^2}{(x + y)^2}.$$

Then

$$\nabla f(x, y) = 2y\frac{2y - x}{(x + y)^3}\mathbf{i} + 2x\frac{x - 2y}{(x + y)^3}\mathbf{j}.$$

The gradient is $\mathbf{0}$ where

$$2y\frac{2y - x}{(x + y)^3} = 0 \qquad \text{and} \qquad 2x\frac{x - 2y}{(x + y)^3} = 0.$$

Solving these two equations simultaneously, we get

$$x = 2y.$$

The stationary points are all the points of the line $x = 2y$ other than the point $(0, 0)$, which is not in the domain of f. At each stationary point (x_0, y_0) the function takes on the value $-\frac{2}{3}$.

$$f(x_0, y_0) = f(2y_0, y_0) = -\frac{(2y_0)^2 + 2y_0^2}{(2y_0 + y_0)^2} = -\frac{6y_0^2}{9y_0^2} = -\frac{2}{3}.$$

In a moment we will show that for all (x, y) in the domain,

$$f(x, y) = -\frac{x^2 + 2y^2}{(x + y)^2} \leq -\frac{2}{3}.$$

It will then follow that the number $-\frac{2}{3}$ is a local maximum value.

The inequality

$$-\frac{x^2 + 2y^2}{(x + y)^2} \leq -\frac{2}{3}$$

can be justified by the following sequence of equivalent inequalities, the last of which is obvious:

$$-3(x^2 + 2y^2) \leq -2(x + y)^2,$$
$$3(x^2 + 2y^2) \geq 2(x + y)^2,$$
$$3x^2 + 6y^2 \geq 2x^2 + 4xy + 2y^2,$$
$$x^2 - 4xy + 4y^2 \geq 0,$$
$$(x - 2y)^2 \geq 0. \quad \square$$

Each function we have considered so far was differentiable on its entire domain. The only critical points were thus the stationary points. Our next example exhibits a function that is everywhere defined but is not differentiable at the origin. The origin is therefore a critical point, though not a stationary point.

Example 5 The function

$$f(x, y) = 1 + \sqrt{x^2 + y^2}$$

is everywhere defined and everywhere continuous. The graph is the upper nappe of a right circular cone. (See Figure 15.6.5.) The number $f(0, 0) = 1$ is obviously a local minimum.

Since the partials

$$\frac{\partial f}{\partial x} = \frac{x}{\sqrt{x^2 + y^2}}, \qquad \frac{\partial f}{\partial y} = \frac{y}{\sqrt{x^2 + y^2}}$$

are not defined at $(0, 0)$, the gradient is not defined at $(0, 0)$ (Theorem 15.2.5). The point $(0, 0)$ is thus a critical point, but not a stationary point. At $(0, 0, 1)$ the surface comes to a sharp point and there is no tangent plane. $\quad \square$

Absolute Extreme Values

In Chapter 2 we saw that a function of one variable that is continuous on a bounded closed interval takes on both an absolute maximum and an absolute minimum on that interval. (Theorem 2.6.2.) More generally, it can be proved that, if f, now a function of one or several variables, is continuous on a *bounded closed set* (a closed set that can

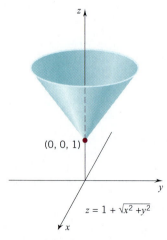

$z = 1 + \sqrt{x^2 + y^2}$

Figure 15.6.5

be contained in a ball of finite radius), then f takes on both an absolute maximum and an absolute minimum on that set.

In the search for local extreme values the critical points are interior points of the domain: the stationary points and the interior points at which the gradient does not exist. In the search for absolute extreme values we must also test the boundary points. This usually requires special methods. One approach is to try to parametrize the boundary by some vector function $\mathbf{r} = \mathbf{r}(t)$ and then work with $f(\mathbf{r}(t))$. This is the approach we take in the example below. Another, altogether different approach is outlined in Section 15.8.

Example 6 The triangular region

$$\Omega = \{(x, y): 0 \leq x \leq 4, 0 \leq y \leq 2x\} \qquad \text{(Figure 15.6.6)}$$

is a closed, bounded set. The function

$$f(x, y) = xy - 2x - 3y,$$

being continuous everywhere, is continuous on Ω. Therefore, we know that f takes on an absolute maximum and an absolute minimum on Ω.

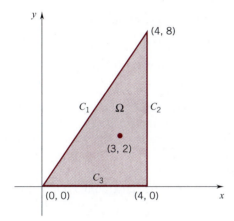

Figure 15.6.6

To see whether either of these values is taken on in the interior of the set, we form the gradient

$$\nabla f(x, y) = (y - 2)\mathbf{i} + (x - 3)\mathbf{j}.$$

The gradient, defined everywhere, is $\mathbf{0}$ only at the point $(3, 2)$, and $(3, 2) \in \Omega$. As you can check,

$$f(3 + h, 2 + k) - f(3, 2) = hk.$$

Since this difference does not keep a constant sign for small h and k, the point $(3, 2)$ does not give rise to an extreme value. It is a saddle point.

Now we will look for extreme values on the boundary by writing each side of the triangle in the form $\mathbf{r} = \mathbf{r}(t)$ and then analyzing $f(\mathbf{r}(t))$. With C_1, C_2, C_3 as in the figure, we have

$$C_1: \quad \mathbf{r}_1(t) = t\mathbf{i} + 2t\mathbf{j}, \quad t \in [0, 4],$$
$$C_2: \quad \mathbf{r}_2(t) = 4\mathbf{i} + t\mathbf{j}, \quad t \in [0, 8],$$
$$C_3: \quad \mathbf{r}_3(t) = t\mathbf{i}, \quad t \in [0, 4].$$

The values of f on these line segments are given by the functions

$$f_1(t) = f(\mathbf{r}_1(t)) = 2t^2 - 8t, \quad t, \in [0, 4],$$

$$f_2(t) = f(\mathbf{r}_2(t)) = t - 8, \qquad t \in [0, 8],$$

$$f_3(t) = f(\mathbf{r}_3(t)) = -2t, \qquad t \in [0, 4].$$

The maximum value of f is the maximum of the maxima of f_1, f_2, f_3. And, the minimum value of f is the minimum of the minima of f_1, f_2, f_3. In this case, it happens that each of the f_i has a maximum of 0 and a minimum of -8. (Check this out.) It follows that the maximum value of f is 0 and the minimum value of f is -8. ❑

In some cases, physical or geometric considerations may allow us to conclude that an absolute maximum or an absolute minimum exists even if the function is not continuous, or the domain is not bounded or not closed.

Example 7 The rectangle

$$\{(x, y): 0 \le x \le a, \ -b \le y \le b\}$$

is a bounded closed subset of the plane. The function

$$f(x, y) = 1 + \sqrt{x^2 + y^2}$$

being everywhere continuous, is continuous on this rectangle. Thus we can be sure that f takes on both an absolute maximum and an absolute minimum on this set. The absolute maximum is taken on at the points $(a, -b)$ and (a, b), the points of the rectangle furthest away from the origin. (This should be clear from Figure 15.6.5.) The value at these points is $1 + \sqrt{a^2 + b^2}$. The absolute minimum is taken on at the origin $(0, 0)$. The value there is 1.

Now let's continue with the same function but apply it instead to the rectangle

$$\{(x, y): 0 < x \le a, \ -b \le y \le b\}.$$

This rectangle is bounded but not closed. On this set f takes on an absolute maximum (the same maximum as before and at the same points), but it takes on no absolute minimum (the origin is not in the set).

Finally, on the entire plane (which is closed but not bounded), f takes on an absolute minimum (1 at the origin) but no maximum. ❑

EXERCISES 15.6

In Exercises 1–14, find the stationary points and determine the local extreme values.

1. $f(x, y) = 2x - x^2 - y^2$.

2. $f(x, y) = x^2 + 2y^2 - 4y$.

3. $f(x, y) = 2x + 2y - x^2 + y^2 + 5$.

4. $f(x, y) = x^2 - xy + y^2 + 1$.

5. $f(x, y) = x^2 + xy + y^2 + 3x + 1$.

6. $f(x, y) = x^3 - y^2 + 2y$.

7. $f(x, y) = x^3 - 3x + y$.

8. $f(x, y) = x^5 - 3x^2 - 2y$.

9. $f(x, y) = x^2 + xy + y^2 - 3x - 3y$.

10. $f(x, y) = x^2 - xy + y^2 + 2x + 2y - 6$.

11. $f(x, y) = x^2 + xy + 2x + 2y + 1$.

12. $f(x, y) = 5xy - 7x^2 - y^2 + 3x - 6y$.

13. $f(x, y) = 4x^3y - 4xy^3$.

14. $f(x, y) = x \sin y, \ -\pi < y < \pi$.

In Exercises 15–24, find the absolute extreme values taken on by the function on the set D.

15. $f(x, y) = \dfrac{1}{\sqrt{x^2 + y^2}}$,
$D = \{(x, y): 1 \leq x \leq 3, 1 \leq y \leq 4\}$.

16. $f(x, y) = |x| + |y|$,
$D = \{(x, y): -1 \leq x \leq 1, -1 \leq y \leq 1\}$.

17. $f(x, y) = 4x^2 - 9y^2$,
$D = \{(x, y): -1 \leq x \leq 1\}$.

18. $f(x, y) = (x^2 + y^2)e^{-(x^2+y^2)}$,
$D = \{(x, y): x^2 + y^2 \leq 25\}$. HINT: Consider te^{-t}.

19. $f(x, y) = (x - 1)^2 + (y - 1)^2$,
$D = \{(x, y): x^2 + y^2 \leq 4\}$.

20. $f(x, y) = y(x - 3)$, $D = \{(x, y): x^2 + y^2 \leq 9\}$.

21. $f(x, y) = (x - y)^2$,
$D = \{(x, y): 0 \leq x \leq 6, 0 \leq y \leq 12 - 2x\}$.

22. $f(x, y) = (x - 3)^2 + y^2$,
$D = \{(x, y): 0 \leq x \leq 4, x^2 \leq y \leq 4x\}$.

23. $f(x, y) = (x - 4)^2 + y^2$,
$D = \{(x, y): 0 \leq x \leq 2, x^3 \leq y \leq 4x\}$.

24. $f(x, y) = (x + y - 2)^2$,
$D = \{(x, y): 0 \leq x \leq 3, x \leq y \leq 3\}$.

25. Find the point in the plane $2x - y + 2z = 16$ that is closest to the origin, and calculate the distance from the origin to the plane. Check your answer by using Formula (12.6.7).

26. Find the point in the plane $3x - 4y + 2z + 32 = 0$ that is closest to the point $P(-1, 2, 4)$ and calculate the distance from P to the plane. Check your answer by using Formula (12.6.7).

27. Find positive numbers x, y, z such that $x + y + z = 18$ and xyz is a maximum. HINT: Maximize $f(x, y) = 18xy - x^2y - xy^2$ on the triangle formed by the positive x- and y-axes and the line $x + y = 18$.

28. Find positive numbers x, y, z such that $x + y + z = 30$ and xyz^2 is a maximum. HINT: Maximize $f(y, z) = 30yz^2 - y^2z^2 - yz^3$ over the triangle formed by the positive y- and z-axes and the line $y + z = 30$.

29. A rectangular box without a top is to have a volume of 12 cubic feet. Find the dimensions for the box that will yield the minimum surface area.

30. Suppose that the material to be used to construct the box in Exercise 29 costs $3 per square foot for the sides and

$4 per square foot for the bottom. What dimensions will yield the minimum cost?

31. Define $f(x, y) = \frac{1}{4}x^2 - \frac{1}{9}y^2$ on the closed unit disc. Find (a) the stationary points, (b) the local extreme values, (c) the absolute extreme values.

32. Find the dimensions of the rectangular solid of maximum volume given that the sum of these dimensions is b.

33. Find the point with the property that the sum of the squares of its distances from $P_1(x_1, y_1)$, $P_2(x_2, y_2)$, $P_3(x_3, y_3)$ is an absolute minimum.

34. A manufacturer finds that it can sell in two distinct markets. If it sells Q_1 units in the first market and Q_2 units in the second market, the revenue functions are

$$R_1 = A_1Q_1 - B_1Q_1^2, \qquad R_2 = A_2Q_2 - B_2Q_2^2$$

and the total cost function is

$$C = A_3 + B_3(Q_1 + Q_2).$$

Given that the A's and B's are known positive constants, find the quantities Q_1 and Q_2 that will maximize profit.

35. Find the distance from the point $(\frac{1}{2}, \frac{1}{2}, \frac{1}{2})$ to the sphere $x^2 + y^2 + z^2 = 1$.

36. Given that $0 < a < b$, find the absolute maximum value taken on by the function

$$f(x, y) = \frac{xy}{(a + x)(x + y)(b + y)}$$

on the open square $\{(x, y): a < x < b, a < y < b\}$.

37. A manufacturer produces razors and blades at a constant average cost of 80 cents per razor and 60 cents per dozen blades. If the razors are sold at x cents each and the blades at y cents per dozen, the demand of the market each week is

$1000(320 - x)$ razors and $1000(140 - y)$ dozen blades.

Find the selling prices for maximum profit.

▶ In Exercises 38–41, use a graphing utility to graph the surface and locate any stationary points and saddle points.

38. $f(x, y) = 4xy - x^4 - y^4 + 1$.

39. $f(x, y) = 3xy - x^3 - y^3 + 2$.

40. $f(x, y) = (x^2 + 2y^2)e^{-(x^2+y^2)}$.

41. $f(x, y) = \dfrac{-2x}{x^2 + y^2 + 1}$.

■ 15.7 SECOND-PARTIALS TEST

We start with a function g of one variable and suppose that $g'(x_0) = 0$. According to the second-derivative test, (Theorem 4.3.5), g has

a local minimum at x_0 if $g''(x_0) > 0$,

a local maximum at x_0 if $g''(x_0) < 0$.

We have a similar test for functions of two variables. As one might expect, the test is somewhat more complicated to state and definitely more difficult to prove. We will omit the proof.†

THEOREM 15.7.1 THE SECOND-PARTIALS TEST

Suppose that f has continuous second-order partial derivatives in a neighborhood of (x_0, y_0) and that $\nabla f(x_0, y_0) = \mathbf{0}$. Set

$$A = \frac{\partial^2 f}{\partial x^2}(x_0, y_0), \qquad B = \frac{\partial^2 f}{\partial y \partial x}(x_0, y_0), \qquad C = \frac{\partial^2 f}{\partial y^2}(x_0, y_0)$$

and form the *discriminant $D = B^2 - AC$.*

1. If $D > 0$, then (x_0, y_0) is a saddle point.
2. If $D < 0$, then f has

$$\begin{aligned}&\text{a local minimum at } (x_0, y_0) \quad \text{if} \quad A > 0,\\&\text{a local maximum at } (x_0, y_0) \quad \text{if} \quad A < 0.\end{aligned}$$

The test is geometrically evident for functions of the form

$$f(x, y) = \tfrac{1}{2}ax^2 + \tfrac{1}{2}cy^2. \qquad\qquad (a \neq 0, c \neq 0)$$

The graph of such a function is a paraboloid:

$$z = \tfrac{1}{2}ax^2 + \tfrac{1}{2}cy^2.$$

The gradient is $\mathbf{0}$ at the origin $(0, 0)$. Moreover

$$A = \frac{\partial^2 f}{\partial x^2}(0, 0) = a, \qquad B = \frac{\partial^2 f}{\partial y \partial x}(0, 0) = 0, \qquad C = \frac{\partial^2 f}{\partial y^2}(0, 0) = c.$$

and $D = B^2 - AC = -ac$. If $D > 0$, then a and c have opposite signs and the surface has a saddle point. (Figure 15.7.1.)

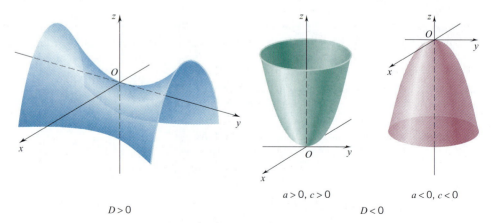

$D > 0$ $a > 0, c > 0$ $a < 0, c < 0$ $D < 0$

Figure 15.7.1

---

† You can find a proof in most texts on advanced calculus.

Suppose now that $D < 0$. If $a > 0$, then $c > 0$ and the surface has a minimum point; if $a < 0$, then $c < 0$ and the surface has a maximum point.

In the examples that follow we apply the second-partials test to a variety of functions.

Example 1 For the function

$$f(x, y) = 2x^2 + y^2 - xy - 7y$$

we have

$$\frac{\partial f}{\partial x} = 4x - y, \qquad \frac{\partial f}{\partial y} = 2y - x - 7.$$

Setting both partials equal to zero, we have

$$4x - y = 0, \qquad 2y - x - 7 = 0.$$

The only simultaneous solution to these equations is $x = 1$, $y = 4$. The point $(1, 4)$ is thus the only stationary point.

The second partials are constant:

$$\frac{\partial^2 f}{\partial x^2} = 4, \qquad \frac{\partial^2 f}{\partial y \partial x} = -1, \qquad \frac{\partial^2 f}{\partial y^2} = 2.$$

Thus $A = 4$, $B = -1$, $C = 2$, and

$$D = B^2 - AC = -7 < 0.$$

Since $A > 0$, it follows from the second-partials test that

$$f(1, 4) = 2 + 16 - 4 - 28 = -14$$

is a local minimum. (We obtained this result in Section 15.6 without benefit of the second-partials test; see Example 1.) ❑

Example 2 The function

$$f(x, y) = \frac{x}{y^2} + xy$$

has partial derivatives

$$\frac{\partial f}{\partial x} = \frac{1}{y^2} + y, \qquad \frac{\partial f}{\partial y} = -\frac{2x}{y^3} + x.$$

Setting both of these partials equal to zero, we have

$$\frac{1}{y^2} + y = 0, \qquad x\left(-\frac{2}{y^3} + 1\right) = 0.$$

As you can check, the only simultaneous solution to these equations is $x = 0$, $y = -1$. The point $(0, -1)$ is thus the only stationary point.

The second partials are

$$\frac{\partial^2 f}{\partial x^2} = 0, \qquad \frac{\partial^2 f}{\partial y \partial x} = -\frac{2}{y^3} + 1, \qquad \frac{\partial^2 f}{\partial y^2} = \frac{6x}{y^4}.$$

Evaluating these partials at the point $(0, -1)$, we find that $A = 0$, $B = 3$, $C = 0$, and thus

$$D = B^2 - AC = 9 > 0.$$

By the second-partials test, $(0, -1)$ is a saddle point.

We could have determined that $(0, -1)$ is a saddle point by comparing the value of f at $(0, -1)$ with its value at nearby points $(h, -1 + k)$:

$$f(h, -1 + k) - f(0, -1) = \frac{h}{(k - 1)^2} + h(k - 1) - 0 = h \left[\frac{1 + (k - 1)^3}{(k - 1)^2} \right].$$

For small $k > 0$, this expression is positive if $h > 0$ and negative if $h < 0$. ❑

Example 3 The function

$$f(x, y) = x^3 + y^3 - 3xy$$

has partial derivatives

$$\frac{\partial f}{\partial x} = 3x^2 - 3y, \qquad \frac{\partial f}{\partial y} = 3y^2 - 3x.$$

Setting both partials equal to zero, we have

$$x^2 - y = 0, \qquad y^2 - x = 0.$$

The only simultaneous solutions are $x = 1, y = 1$ and $x = 0, y = 0$. The points $(1, 1)$ and $(0, 0)$ are the only stationary points.

The second partials are

$$\frac{\partial^2 f}{\partial x^2} = 6x, \qquad \frac{\partial^2 f}{\partial y \partial x} = -3, \qquad \frac{\partial^2 f}{\partial y^2} = 6y.$$

At $(1, 1)$, we have $A = 6$, $B = -3$, $C = 6$, and thus

$$D = B^2 - AC = 9 - 36 = -27 < 0.$$

Since $A > 0$, we know that

$$f(1, 1) = 1 + 1 - 3 = -1$$

is a local minimum. At $(0, 0)$, we have $A = 0$, $B = -3$, $C = 0$. Thus

$$D = B^2 - AC = 9 > 0.$$

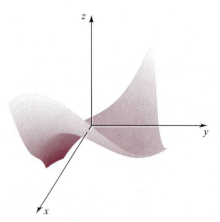

Figure 15.7.2

The origin is a saddle point. A computer-generated graph of this function is shown in Figure 15.7.2. ❏

Example 4 We considered the function

$$f(x, y) = -xye^{-(x^2+y^2)/2}$$

in Example 3 of the preceding section and found the stationary points $(0, 0)$, $(1, 1)$, $(1, -1)$, $(-1, 1)$, and $(-1, -1)$. Here we apply the second-partials test to confirm the results that we found there. You can verify that the second-partial derivatives of f are

$$\frac{\partial^2 f}{\partial x^2} = xy(3 - x^2)e^{-(x^2+y^2)/2}, \qquad \frac{\partial^2 f}{\partial y^2} = xy(3 - y^2)e^{-(x^2+y^2)/2}$$

and

$$\frac{\partial^2 f}{\partial y \partial x} = (x^2 - 1)(1 - y^2)e^{-(x^2+y^2)/2}.$$

The data for the second partials test are recorded in the following table.

Point	A	B	C	D	Result
$(0, 0)$	0	-1	0	1	Saddle point
$(1, 1)$	$2e^{-1}$	0	$2e^{-1}$	$-4e^{-2}$	Loc. min.
$(1, -1)$	$-2e^{-1}$	0	$-2e^{-1}$	$-4e^{-2}$	Loc. max.
$(-1, 1)$	$-2e^{-1}$	0	$-2e^{-1}$	$-4e^{-2}$	Loc. max.
$(-1, -1)$	$2e^{-1}$	0	$2e^{-1}$	$-4e^{-2}$	Loc. min.

❏

Example 5 Here we test the function

$$f(x, y) = \sin x + \sin y + \sin(x + y)$$

on the open square $0 < x < \pi, 0 < y < \pi$.

In the first place

$$\frac{\partial f}{\partial x} = \cos x + \cos(x + y), \qquad \frac{\partial f}{\partial y} = \cos y + \cos(x + y).$$

Setting both of these partials equal to zero, we have

$$\cos x + \cos(x + y) = 0, \qquad \cos y + \cos(x + y) = 0.$$

Together these equations imply that $\cos x = \cos y$. Since both x and y lie between 0 and π, and since the cosine function is one-to-one on $(0, \pi)$, we can conclude that $x = y$. The condition

$$\cos x + \cos(x + y) = 0$$

now gives

$$\cos x + \cos 2x = 0.$$

Into this last equation we substitute the identity $\cos 2x = 2 \cos^2 x - 1$ and thereby obtain the relation

$$2 \cos^2 x + \cos x - 1 = 0.$$

This is a quadratic in $\cos x$, which can be written in factored form

$$(2 \cos x - 1)(\cos x + 1) = 0$$

Thus

$$\cos x = -1 \quad \text{or} \quad \tfrac{1}{2}$$

We throw out the solution -1 because x has to lie strictly between 0 and π. The other possibility, $\cos x = \tfrac{1}{2}$, gives $x = \tfrac{1}{3}\pi$. Since $x = y$, we also have $y = \tfrac{1}{3}\pi$. This shows that $(\tfrac{1}{3}\pi, \tfrac{1}{3}\pi)$ is the only stationary point.

The second partials are

$$\frac{\partial^2 f}{\partial x^2} = -\sin x - \sin(x + y), \qquad \frac{\partial^2 f}{\partial y \partial x} = -\sin(x + y),$$

$$\frac{\partial^2 f}{\partial y^2} = -\sin y - \sin(x + y).$$

At $(\tfrac{1}{3}\pi, \tfrac{1}{3}\pi)$ we have

$$A = -\sin \tfrac{1}{3}\pi - \sin \tfrac{2}{3}\pi = -\tfrac{1}{2}\sqrt{3} - \tfrac{1}{2}\sqrt{3} = -\sqrt{3}, \qquad B = -\sin \tfrac{2}{3}\pi = -\tfrac{1}{2}\sqrt{3},$$

$$C = -\sin \tfrac{1}{3}\pi - \sin \tfrac{2}{3}\pi = -\tfrac{1}{2}\sqrt{3} - \tfrac{1}{2}\sqrt{3} = -\sqrt{3},$$

and thus

$$D = B^2 - AC = \tfrac{3}{4} - 3 < 0.$$

Since $A < 0$, we can conclude that

$$f(\tfrac{1}{3}\pi, \tfrac{1}{3}\pi) = \sin \tfrac{1}{3}\pi + \sin \tfrac{1}{3}\pi + \sin \tfrac{2}{3}\pi = \tfrac{3}{2}\sqrt{3}$$

is a local maximum. A computer-generated drawing of the graph of $f(x, y) = \sin x + \sin y + \sin(x + y)$ is shown in Figure 15.7.3. ❑

Figure 15.7.3

The second-derivative test for a function of one variable applies to points x_0 where $g'(x_0) = 0$ but $g''(x_0) \neq 0$. If $g''(x_0) = 0$, the second-derivative test provides no conclusive information. The second-partials test suffers from a similar limitation. It applies to points (x_0, y_0) where $\nabla f(x_0, y_0) = \mathbf{0}$ but $D \neq 0$. If $D = 0$, the second-partials test provides no information.

Consider, for example, the functions

$$f(x, y) = x^4 + y^4, \qquad g(x, y) = -(x^4 + y^4), \qquad h(x, y) = x^4 - y^4.$$

Each of these functions has zero gradient at the origin, and, as you can check, in each case $D = 0$. Yet,

(1) for f, $(0, 0)$ gives a local minimum;
(2) for g, $(0, 0)$ gives a local maximum;
(3) for h, $(0, 0)$ is a saddle point.

Statements (1) and (2) are obvious. To confirm (3), note that $h(0, 0) = 0$, but in every neighborhood of $(0, 0)$ the function h takes on both positive and negative values:

$$h(x, 0) > 0 \quad \text{for } x \neq 0 \qquad \text{while} \qquad h(0, y) < 0 \quad \text{for } y \neq 0. \quad \square$$

EXERCISES 15.7

In Exercises 1–24, find the stationary points and the local extreme values.

1. $f(x, y) = x^2 + xy + y^2 - 6x + 2$.

2. $f(x, y) = x^2 + 2xy + 3y^2 + 2x + 10y + 1$.

3. $f(x, y) = 4x + 2y - x^2 + xy - y^2$.

4. $f(x, y) = x^2 - xy + y^2 + 2x + 2y - 1$.

5. $f(x, y) = x^3 - 6xy + y^3$.

6. $f(x, y) = 3x^2 + xy - y^2 + 5x - 5y + 4$.

7. $f(x, y) = x^3 + y^2 - 6xy + 6x + 3y - 2$.

8. $f(x, y) = x^2 + xy + y^2 - 6y + 5$.

9. $f(x, y) = e^x \cos y$.

10. $f(x, y) = x^2 - 2xy + 2y^2 - 3x + 5y$.

11. $f(x, y) = x \sin y$.   **12.** $f(x, y) = y + x \sin y$.

13. $f(x, y) = (x + y)(xy + 1)$.

14. $f(x, y) = xy^{-1} - yx^{-1}$.

15. $f(x, y) = xy + x^{-1} + 8y^{-1}$.

16. $f(x, y) = x^2 - 2xy - y^2 + 1$.

17. $f(x, y) = xy + x^{-1} + y^{-1}$.

18. $f(x, y) = (x - y)(xy - 1)$.

19. $f(x, y) = \dfrac{-2x}{x^2 + y^2 + 1}$.   **20.** $f(x, y) = (x - 3) \ln xy$.

21. $f(x, y) = x^4 - 2x^2 + y^2 - 2$.

22. $f(x, y) = (x^2 + y^2)e^{x^2 - y^2}$.

23. $f(x, y) = \cos x + \cos y - \cos(x + y)$ where $-\pi/2 < x < \pi/2, -\pi/2 < y < \pi/2$.

24. $f(x, y) = 8xye^{-(x^2 + y^2)}$.

25. Let $f(x, y) = x^2 + kxy + y^2$, k a constant.
  (a) Show that f has a stationary point at $(0, 0)$ independent of the value of k.
  (b) For what values of k will f have a saddle point at $(0, 0)$?
  (c) For what values of k will f have a local minimum at $(0, 0)$?

(d) For what values of k is the second-derivative test inconclusive?

26. Repeat Exercise 25 for the function

$$f(x, y) = x^2 + kxy + 4y^2, k \text{ a constant}.$$

27. Find the point in the plane $2x - 3y + 6z = 14$ that is closest to the point $(-1, 1, 2)$. Justify your answer using the second-partials test.

28. Use the methods of this section to derive the formula for the distance from the origin to the plane
$$Ax + By + Cz + D = 0.$$

29. Find the maximum volume for a rectangular solid in the first octant $(x \geq 0, y \geq 0, z \geq 0)$ with one vertex at the origin and opposite vertex on the plane $x + y + z = 1$.

30. Find the maximum volume for a rectangular solid in the first octant with one vertex at the origin and the opposite vertex on the plane

$$\frac{x}{a} + \frac{y}{b} + \frac{z}{c} = 1.$$

31. Find the shortest distance from the point $(1, 2, 0)$ to the elliptic cone $z = \sqrt{x^2 + 2y^2}$.   HINT: Minimize the square of the distance.

32. Let n be an integer greater than 2 and set

$$f(x, y) = ax^n + cy^n, \text{ taking } ac \neq 0.$$

  (a) Find the stationary points.
  (b) Find the discriminant at each stationary point.
  (c) Find the local and absolute extreme values given that
    (i) $a > 0, c > 0$.
    (ii) $a < 0, c < 0$.
    (iii) $a > 0, c < 0$.

33. Describe the behavior of the function at the origin.
  (a) $f(x, y) = x^2 - y^3$.
  (b) $f(x, y) = 2 \cos(x + y) + e^{xy}$.

34. Find the maximum volume for a rectangular solid inscribed in the sphere

$$x^2 + y^2 + z^2 = a^2$$

35. A pentagon is composed of a rectangle surmounted by an isosceles triangle (see the figure). Given that the perimeter of the pentagon has a fixed value P, find the dimensions for maximum area.

36. A bakery produces two types of bread, one at a cost of 50 cents per loaf, the other at a cost of 60 cents per loaf. Assume that, if the first bread is sold at x cents a loaf and the second at y cents a loaf, then the number of loaves that can be sold each week is given by the formulas

$$N_1 = 250(y - x), \qquad N_2 = 32{,}000 + 250(x - 2y).$$

Determine x and y for maximum profit.

37. Find the distance between the lines $x = \frac{1}{2}y = \frac{1}{3}z$ and $x = y - 2 = z$.

38. Find the absolute maximum value of

$$f(x, y) = \frac{(ax + by + c)^2}{x^2 + y^2 + 1}.$$

39. Find the dimensions of the most economical open-top rectangular crate 96 cubic meters in volume given that the base costs 30 cents per square meter and the sides cost 10 cents per square meter.

40. Let $f(x, y) = ax^2 + bxy + cy^2$, taking $abc \neq 0$.
   (a) Find the discriminant D.
   (b) Find the stationary points and local extreme values if $D \neq 0$.
   (c) Suppose that $D = 0$. Find the stationary points and the local and absolute extreme values given that (i) $a > 0, c > 0$. (ii) $a < 0, c < 0$.

41. Show that a closed rectangular box of maximum volume having a prescribed surface area S is a cube.

42. If an open rectangular box has a prescribed surface area S, what dimensions yield the maximum volume?

43. (*The method of least squares*) In this exercise we illustrate an important method of fitting a curve to a collection of points. Consider three points

$$(x_1, y_1) = (0, 2), \quad (x_2, y_2) = (1, -5), \quad (x_3, y_3) = (2, 4).$$

   (a) Find the line $y = mx + b$ that minimizes the sum of the squares of the vertical distances $d_i = |y_i - (mx_i + b)|$ from these points to the line.

   (b) Find the parabola $y = \alpha x^2 + \beta$ that minimizes the sum of the squares of the vertical distances $d_i = |y_i - (\alpha x_i^2 + \beta)|$ from the points to the parabola.

44. Repeat Exercise 43 taking $(x_1, y_1) = (-1, 2)$, $(x_2, y_2) = (0, -1)$, $(x_3, y_3) = (1, 1)$.

45. According to U.S. Postal Service regulations the length plus the girth (the perimeter of a cross section) of a package cannot exceed 108 inches. (See the figure.)

   (a) Find the dimensions of the rectangular box of maximum volume that is acceptable for mailing.
   (b) Find the dimensions of the cylindrical tube of maximum volume that is acceptable for mailing.

46. A petrochemical company is designing a cylindrical tank with hemispherical ends to be used in transporting its products. (See Exercise 35, Section 14.1.) If the volume of the tank is to be 10,000 cubic meters, what dimensions should be used to minimize the amount of metal required?

47. A 10-foot section of gutter is to be made from a 12-inch-wide strip of metal by folding up strips of length x on each side so that they make an angle θ with the bottom of the gutter. (See the figure.) Determine values for x and θ that will maximize the cross-sectional area of the gutter.

48. Find the volume of the largest rectangular box with edges parallel to the coordinate axes that can be inscribed in the ellipsoid

$$\frac{x^2}{a^2} + \frac{y^2}{b^2} + \frac{z^2}{c^2} = 1.$$

■ 15.8 MAXIMA AND MINIMA WITH SIDE CONDITIONS

When we ask for the distance from a point $P(x_0, y_0)$ to a line $l: Ax + By + C = 0$, we are asking for the minimum value of

$$f(x, y) = \sqrt{(x - x_0)^2 + (y - y_0)^2}$$

with (x, y) subject to the side condition† $Ax + By + C = 0$. When we ask for the distance from a point $P(x_0, y_0, z_0)$ to a plane $p: Ax + By + Cz + D = 0$, we are asking for the minimum value of

$$f(x, y, z) = \sqrt{(x - x_0)^2 + (y - y_0)^2 + (z - z_0)^2}$$

with (x, y, z) subject to the side condition $Ax + By + Cz + D = 0$.

We have already treated these particular problems by special techniques. Our interest here is to present techniques for handling problems of this sort in general. In the two-variable case, the problems will take the form of maximizing (or minimizing) some expression $f(x, y)$ subject to a side condition $g(x, y) = 0$. In the three-variable case, we will seek to maximize (or minimize) some expression $f(x, y, z)$ subject to a side condition $g(x, y, z) = 0$. We begin with two simple examples.

Example 1 Maximize the product xy subject to the side condition $x + y - 1 = 0$.

SOLUTION The condition $x + y - 1 = 0$ gives $y = 1 - x$. Our initial problem can therefore be solved simply by maximizing the product $h(x) = x(1 - x)$. The derivative $h'(x) = 1 - 2x$ is 0 only at $x = \frac{1}{2}$. Since $h''(x) = -2 < 0$, we know from the second-derivative test that $h(\frac{1}{2}) = \frac{1}{2}(1 - \frac{1}{2}) = \frac{1}{4}$ is the desired maximum. ❏

Example 2 Find the maximum volume of a rectangular solid given that the sum of the lengths of its edges is $12a$.

SOLUTION We denote the dimensions of the solid by x, y, z. (Figure 15.8.1.) The volume is given by

$$V = xyz.$$

The stipulation on the edges requires that

$$4(x + y + z) = 12a.$$

Solving this last equation for z, we find that

$$z = 3a - (x + y).$$

Substituting this expression for z in the volume formula, we have

$$V = xy[3a - (x + y)].$$

Since $x, y,$ and $z = 3a - (x + y)$ must all remain positive, our problem is to find the maximum value of V on the interior of the triangle shown in Figure 15.8.2.

The first partials are

$$\frac{\partial V}{\partial x} = xy(-1) + y[3a - (x + y)] = 3ay - 2xy - y^2,$$

$$\frac{\partial V}{\partial y} = xy(-1) + x[3a - (x + y)] = 3ax - x^2 - 2xy.$$

Figure 15.8.1

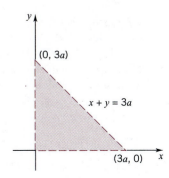

Figure 15.8.2

---

† Side conditions are often called *constraints*.

Setting both partials equal to zero, we have

$$(3a - 2x - y)y = 0 \qquad \text{and} \qquad (3a - x - 2y)x = 0.$$

Since x and y are assumed positive, we can divide by x and y and get

$$3a - 2x - y = 0 \qquad \text{and} \qquad 3a - x - 2y = 0.$$

Solving these equations simultaneously, we find that $x = y = a$. The point (a, a), which does lie within the triangle, is the only stationary point. The value of V at that point is a^3. The conditions of the problem make it clear that this is a maximum. (If you are skeptical, you can confirm this by appealing to the second-partials test.) ❑

The last two problems were easy. They were easy in part because the side conditions were such that we could solve for one of the variables in terms of the other(s). In general this is not possible and a more sophisticated approach is required.

The Method of Lagrange

We begin with what looks like a detour. To avoid having to make separate statements for the two- and three-variable cases, we will use vector notation.

Throughout the discussion f will be a function of two or three variables which is continuously differentiable on some open set U. We take

$$C: \quad \mathbf{r} = \mathbf{r}(t), \quad t \in I$$

as a curve that lies entirely in U and has at each point a nonzero tangent vector $\mathbf{r}'(t)$. The basic result is this:

(15.8.1)

> if $\mathbf{x}_0$ maximizes (or minimizes) $f(\mathbf{x})$ on C,
> then $\nabla f(\mathbf{x}_0)$ is perpendicular to C at $\mathbf{x}_0$.

PROOF Choose t_0 so that

$$\mathbf{r}(t_0) = \mathbf{x}_0.$$

The composition $f(\mathbf{r}(t))$ has a maximum (or minimum) at t_0. Consequently, its derivative,

$$\frac{d}{dt}[f(\mathbf{r}(t))] = \nabla f(\mathbf{r}(t)) \cdot \mathbf{r}'(t).$$

must be zero at t_0:

$$0 = \nabla f(\mathbf{r}(t_0)) \cdot \mathbf{r}'(t_0) = \nabla f(\mathbf{x}_0) \cdot \mathbf{r}'(t_0).$$

This shows that

$$\nabla f(\mathbf{x}_0) \perp \mathbf{r}'(t_0).$$

Since $\mathbf{r}'(t_0)$ is tangent to C at $\mathbf{x}_0$, $\nabla f(\mathbf{x}_0)$ is perpendicular to C at $\mathbf{x}_0$. ❑

We are now ready for the side condition problems. Suppose that g is a continuously differentiable function of two or three variables defined on a subset of the domain of f. Lagrange made the following observation:†

(15.8.2)

> if $\mathbf{x}_0$ maximizes (or minimizes) $f(\mathbf{x})$ subject to the side condition $g(\mathbf{x}) = 0$, then $\nabla f(\mathbf{x}_0)$ and $\nabla g(\mathbf{x}_0)$ are parallel. Thus, if $\nabla g(\mathbf{x}_0) \neq \mathbf{0}$, then there exists a scalar λ such that
> $$\nabla f(\mathbf{x}_0) = \lambda \nabla g(\mathbf{x}_0).$$

Such a scalar λ has come to be called a *Lagrange multiplier.*

PROOF OF (15.8.2) Let's suppose that $\mathbf{x}_0$ maximizes (or minimizes) $f(\mathbf{x})$ subject to the side condition $g(\mathbf{x}) = 0$. If $\nabla g(\mathbf{x}_0) = \mathbf{0}$, the result is trivially true: every vector is parallel to the zero vector. We suppose therefore that $\nabla g(\mathbf{x}_0) \neq \mathbf{0}$.

In the two-variable case we have

$$\mathbf{x}_0 = (x_0, y_0) \quad \text{and} \quad \text{the side condition} \quad g(x, y) = 0.$$

The side condition defines a curve C that has a nonzero tangent vector at (x_0, y_0).†† Since (x_0, y_0) maximizes (or minimizes) $f(x, y)$ on C, we know from (15.8.1) that $\nabla f(x_0, y_0)$ is perpendicular to C at (x_0, y_0). By (15.5.2), $\nabla g(x_0, y_0)$ is also perpendicular to C at (x_0, y_0). The two gradients are therefore parallel.

In the three-variable case we have

$$\mathbf{x}_0 = (x_0, y_0, z_0) \quad \text{and} \quad \text{the side condition} \quad g(x, y, z) = 0.$$

The side condition defines a surface Γ that lies in the domain of f. Now let C be a curve that lies on Γ and passes through (x_0, y_0, z_0) with nonzero tangent vector. We know that (x_0, y_0, z_0) maximizes (or minimizes) $f(x, y, z)$ on C. Consequently, $\nabla f(x_0, y_0, z_0)$ is perpendicular to C at (x_0, y_0, z_0). Since this is true for each such curve C, $\nabla f(x_0, y_0, z_0)$ must be perpendicular to Γ itself. But $\nabla g(x_0, y_0, z_0)$ is also perpendicular to Γ at (x_0, y_0, z_0) (15.5.6). It follows that $\nabla f(x_0, y_0, z_0)$ and $\nabla g(x_0, y_0, z_0)$ are parallel. ❑

We come now to some problems that are susceptible to Lagrange's method. In each case ∇g is not $\mathbf{0}$ where g is 0 and therefore we can focus entirely on those points $\mathbf{x}$ that satisfy the Lagrange condition

$$\nabla f(\mathbf{x}) = \lambda \nabla g(\mathbf{x}). \tag{15.8.2}$$

Example 3 Maximize and minimize

$$f(x, y) = xy \quad \text{on the unit circle} \quad x^2 + y^2 = 1.$$

SOLUTION Since f is continuous and the unit circle is closed and bounded, it is clear that both a maximum and a minimum exist (see Section 15.6).

---

† Another contribution of the French mathematician Joseph Louis Lagrange.

†† $\mathbf{t}(x_0, y_0) = \dfrac{\partial g}{\partial y}(x_0, y_0)\mathbf{i} - \dfrac{\partial g}{\partial x}(x_0, y_0)\mathbf{j} \neq \mathbf{0}.$

To apply Lagrange's method we set

$$g(x, y) = x^2 + y^2 - 1.$$

We want to maximize and minimize

$$f(x, y) = xy \quad \text{subject to the side condition} \quad g(x, y) = 0.$$

The gradients are

$$\nabla f(x, y) = y\mathbf{i} + x\mathbf{j}, \qquad \nabla g(x, y) = 2x\mathbf{i} + 2y\mathbf{j}.$$

Setting

$$\nabla f(x, y) = \lambda \nabla g(x, y),$$

we obtain

$$y = 2\lambda x, \qquad x = 2\lambda y.$$

Multiplying the first equation by y and the second equation by x, we find that

$$y^2 = 2\lambda xy, \qquad x^2 = 2\lambda xy$$

and thus

$$y^2 = x^2.$$

The side condition $x^2 + y^2 = 1$ now implies that $2x^2 = 1$ and therefore that $x = \pm\frac{1}{2}\sqrt{2}$. The only points that can give rise to an extreme value are

$$(\tfrac{1}{2}\sqrt{2}, \tfrac{1}{2}\sqrt{2}), \quad (\tfrac{1}{2}\sqrt{2}, -\tfrac{1}{2}\sqrt{2}), \quad (-\tfrac{1}{2}\sqrt{2}, \tfrac{1}{2}\sqrt{2}), \quad (-\tfrac{1}{2}\sqrt{2}, -\tfrac{1}{2}\sqrt{2}).$$

At the first and fourth points f takes on the value $\frac{1}{2}$. At the second and third points f takes on the value $-\frac{1}{2}$. Clearly then, $\frac{1}{2}$ is the maximum value and $-\frac{1}{2}$ the minimum value. ❑

Example 4 Find the minimum value taken on by the function

$$f(x, y) = x^2 + (y - 2)^2 \quad \text{on the hyperbola} \quad x^2 - y^2 = 1.$$

SOLUTION Note that the expression $f(x, y) = x^2 + (y - 2)^2$ gives the square of the distance between the points $(0, 2)$ and (x, y). Therefore, the problem asks us to minimize the square of the distance from the point $(0, 2)$ to the hyperbola, and this minimum value clearly exists. Note, also, that there is no maximum value. See Figure 15.8.3.

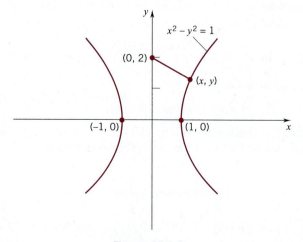

Figure 15.8.3

Now set

$$g(x, y) = x^2 - y^2 - 1.$$

We want to minimize

$$f(x, y) = x^2 + (y - 2)^2 \quad \text{subject to the side condition} \quad g(x, y) = 0.$$

Here

$$\nabla f(x, y) = 2x\mathbf{i} + 2(y - 2)\mathbf{j}, \qquad \nabla g(x, y) = 2x\mathbf{i} - 2y\mathbf{j}.$$

The Lagrange condition $\nabla f(x, y) = \lambda \nabla g(x, y)$ gives

$$2x = 2\lambda x, \qquad 2(y - 2) = -2\lambda y,$$

which we can simplify to

$$x = \lambda x, \qquad y - 2 = -\lambda y.$$

The side condition $x^2 - y^2 = 1$ shows that x cannot be zero. Dividing $x = \lambda x$ by x, we get $\lambda = 1$. This means that $y - 2 = -y$ and therefore $y = 1$. With $y = 1$, the side condition gives $x = \pm\sqrt{2}$. The points to be checked are therefore $(-\sqrt{2}, 1)$ and $(\sqrt{2}, 1)$. At each of these points f takes on the value 3. This is the desired minimum. ❏

Remark The last problem could have been solved more simply by rewriting the side condition as $x^2 = 1 + y^2$ and eliminating x from $f(x, y)$ by substitution. It would then have been only a matter of minimizing the function $h(y) = 1 + y^2 + (y - 2)^2 = 2y^2 - 4y + 5$. ❏

Example 5 Maximize

$$f(x, y, z) = xyz \quad \text{subject to the side condition} \quad x^3 + y^3 + z^3 = 1$$

with $x \geq 0, y \geq 0, z \geq 0$.

SOLUTION The set of all points (x, y, z) that satisfy the side condition can be shown to be closed and bounded. Since f is continuous, we can be sure that the desired maximum exists.

We begin by setting

$$g(x, y, z) = x^3 + y^3 + z^3 - 1$$

so that the side condition becomes $g(x, y, z) = 0$. We seek those triples (x, y, z) that simultaneously satisfy the Lagrange condition

$$\nabla f(x, y, z) = \lambda \nabla g(x, y, z) \quad \text{and the side condition} \quad g(x, y, z) = 0.$$

The gradients are

$$\nabla f(x, y, z) = yz\mathbf{i} + xz\mathbf{j} + xy\mathbf{k}, \qquad \nabla g(x, y, z) = 3x^2\mathbf{i} + 3y^2\mathbf{j} + 3z^2\mathbf{k}.$$

The Lagrange condition $\nabla f(x, y, z) = \lambda \nabla g(x, y, z)$ gives

$$yz = 3\lambda x^2, \qquad xz = 3\lambda y^2, \qquad xy = 3\lambda z^2.$$

Multiplying the first equation by x, the second by y, and the third by z, we get

$$xyz = 3\lambda x^3, \qquad xyz = 3\lambda y^3, \qquad xyz = 3\lambda z^3$$

and consequently

$$\lambda x^3 = \lambda y^3 = \lambda z^3.$$

We can exclude $\lambda = 0$ because, if $\lambda = 0$, then x, y, or z would have to be zero. That would force xyz to be 0, and 0 is obviously not a maximum. Having excluded $\lambda = 0$, we can divide by λ and get $x^3 = y^3 = z^3$ and thus $x = y = z$. The side condition $x^3 + y^3 + z^3 = 1$ now gives

$$x = (\tfrac{1}{3})^{1/3}, \qquad y = (\tfrac{1}{3})^{1/3}, \qquad z = (\tfrac{1}{3})^{1/3}.$$

The desired maximum is $\tfrac{1}{3}$. ❏

Example 6 Show that, of all the triangles inscribed in a fixed circle of radius R, the equilateral triangle has the largest perimeter.

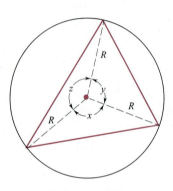

Figure 15.8.4

SOLUTION It is intuitively clear that this maximum exists and geometrically clear that the triangle that offers this maximum contains the center of the circle in its interior or on its boundary. As in Figure 15.8.4 we denote by x, y, z the central angles that subtend the three sides. As you can verify by trigonometry, the perimeter of the triangle is given by the function

$$f(x, y, z) = 2R(\sin \tfrac{1}{2}x + \sin \tfrac{1}{2}y + \sin \tfrac{1}{2}z).$$

As a side condition we have

$$g(x, y, z) = x + y + z - 2\pi = 0.$$

To maximize the perimeter we form the gradients

$$\nabla f(x, y, z) = R[\cos \tfrac{1}{2}x\,\mathbf{i} + \cos \tfrac{1}{2}y\,\mathbf{j} + \cos \tfrac{1}{2}z\,\mathbf{k}], \qquad \nabla g(x, y, z) = \mathbf{i} + \mathbf{j} + \mathbf{k}.$$

The Lagrange condition $\nabla f(x, y, z) = \lambda \nabla g(x, y, z)$ gives

$$\lambda = R \cos \tfrac{1}{2}x, \qquad \lambda = R \cos \tfrac{1}{2}y, \qquad \lambda = R \cos \tfrac{1}{2}z$$

and therefore

$$\cos \tfrac{1}{2}x = \cos \tfrac{1}{2}y = \cos \tfrac{1}{2}z.$$

With x, y, z all in $(0, \pi]$, we can conclude that $x = y = z$. Since the central angles are equal, the sides are equal. The triangle is therefore equilateral. ❏

An Application of the Cross Product

The Lagrange condition can be replaced by a cross-product equation: points that satisfy $\nabla f = \lambda \nabla g$ satisfy

(15.8.3)
$$\boxed{\nabla f \times \nabla g = \mathbf{0}.}$$

If f and g are functions of two variables, then $\nabla f \times \nabla g = 0$ has a simpler form:

$$\nabla f \times \nabla g = \begin{vmatrix} \mathbf{i} & \mathbf{j} & \mathbf{k} \\ f_x & f_y & 0 \\ g_x & g_y & 0 \end{vmatrix} = (f_x g_y - f_y g_x)\mathbf{k}$$

which implies that

(15.8.4)

$$\frac{\partial f}{\partial x}\frac{\partial g}{\partial y} - \frac{\partial f}{\partial y}\frac{\partial g}{\partial x} = 0.$$

Example 7 Maximize and minimize $f(x, y) = xy$ on the unit circle $x^2 + y^2 = 1$.

SOLUTION This problem was solved earlier by means of the Lagrange equation. This time we will use (15.8.4) instead.
    As before we set

$$g(x, y) = x^2 + y^2 - 1$$

so that the side condition takes the form $g(x, y) = 0$. Since

$$\frac{\partial f}{\partial x} = y, \quad \frac{\partial f}{\partial y} = x \quad \text{and} \quad \frac{\partial g}{\partial x} = 2x, \quad \frac{\partial g}{\partial y} = 2y,$$

(15.8.4) takes the form

$$y(2y) - x(2x) = 0.$$

This gives $x^2 = y^2$.
    As before, the side condition $x^2 + y^2 = 1$ implies that $2x^2 = 1$ and therefore that $x = \pm\frac{1}{2}\sqrt{2}$. The points under consideration are

$$(\tfrac{1}{2}\sqrt{2}, \tfrac{1}{2}\sqrt{2}), \quad (\tfrac{1}{2}\sqrt{2}, -\tfrac{1}{2}\sqrt{2}), \quad (-\tfrac{1}{2}\sqrt{2}, \tfrac{1}{2}\sqrt{2}), \quad (-\tfrac{1}{2}\sqrt{2}, -\tfrac{1}{2}\sqrt{2}).$$

At the first and fourth points f takes on the value $\frac{1}{2}$. At the second and third points f takes on the value $-\frac{1}{2}$. Thus, $\frac{1}{2}$ is the maximum value of f and $-\frac{1}{2}$ is the minimum value of f on the unit circle. ❑

    In three variables the computations demanded by the cross-product equation are often quite complicated, and it is usually easier to follow the method of Lagrange.

EXERCISES 15.8

1. Minimize $x^2 + y^2$ on the hyperbola $xy = 1$.

2. Maximize xy on the ellipse $b^2x^2 + a^2y^2 = a^2b^2$.

3. Minimize xy on the ellipse $b^2x^2 + a^2y^2 = a^2b^2$.

4. Minimize xy^2 on the unit circle $x^2 + y^2 = 1$.

5. Maximize xy^2 on the ellipse $b^2x^2 + a^2y^2 = a^2b^2$.

6. Maximize $x + y$ on the curve $x^4 + y^4 = 1$.

7. Maximize $x^2 + y^2$ on the curve $x^4 + 7x^2y^2 + y^4 = 1$.

8. Minimize xyz on the unit sphere $x^2 + y^2 + z^2 = 1$.

9. Maximize xyz on the ellipsoid $x^2/a^2 + y^2/b^2 + z^2/c^2 = 1$.

10. Minimize $x + 2y + 4z$ on the sphere $x^2 + y^2 + z^2 = 7$.

11. Maximize $2x + 3y + 5z$ on the sphere $x^2 + y^2 + z^2 = 19$.

12. Minimize $x^4 + y^4 + z^4$ on the plane $x + y + z = 1$.

13. Maximize the volume of a rectangular solid in the first octant with one vertex at the origin and opposite vertex on the plane $x/a + y/b + z/c = 1$. (Take $a > 0, b > 0, c > 0$.)

14. Show that the square has the largest area of all the rectangles with a given perimeter.

15. Find the distance from the point $(0, 1)$ to the parabola $x^2 = 4y$.

16. Find the distance from the point $(p, 4p)$ to the parabola $y^2 = 2px$.

17. Find the points on the sphere $x^2 + y^2 + z^2 = 1$ that are closest to and furthest from the point $(2, 1, 2)$.

18. Let x, y, and z be the angles of a triangle. Determine the maximum value of $f(x, y, z) = \sin x \sin y \sin z$.

19. Determine the maximum value of $f(x, y, z) = 3x - 2y + z$ on the sphere $x^2 + y^2 + z^2 = 14$.

20. A rectangular box has three of its faces on the coordinate planes and one vertex in the first octant on the paraboloid $z = 4 - x^2 - y^2$. Determine the maximum volume of the box.

21. Use the method of Lagrange to find the distance from the origin to the plane with equation $Ax + By + Cz + D = 0$.

22. Maximize the volume of a rectangular solid given that the sum of the areas of the six faces is $6a^2$.

23. Within a triangle there is a point P such that the sum of the squares of its distances from the sides of the triangle is a minimum. Find this minimum.

24. Show that of all the triangles inscribed in a fixed circle the equilateral one has the largest: (a) product of the lengths of the sides; (b) sum of the squares of the lengths of the sides.

25. The curve $x^3 - y^3 = 1$ is asymptotic to the line $y = x$. Find the point(s) on the curve $x^3 - y^3 = 1$ furthest from the line $y = x$.

26. A plane passes through the point (a, b, c). Find its intercepts with the coordinate axes if the volume of the solid bounded by the plane and the coordinate planes is to be a minimum.

27. Show that, of all the triangles with a given perimeter, the equilateral triangle has the largest area. HINT: Area $= \sqrt{s(s-a)(s-b)(s-c)}$, where s represents the semiperimeter $s = \frac{1}{2}(a + b + c)$.

28. Show that the rectangular box of maximum volume that can be inscribed in the sphere $x^2 + y^2 + z^2 = a^2$ is a cube.

29. (a) Determine the maximum value of $f(x, y) = (xy)^{1/2}$ given that x and y are nonnegative numbers and $x + y = k$, k a constant.
   (b) Use the result in (a) to show that if x and y are nonnegative numbers, then

$$(xy)^{1/2} \le \frac{x + y}{2}.$$

   (See Exercise 76, Section 1.3.)

30. (a) Determine the maximum value of $f(x, y, z) = (xyz)^{1/3}$ given that x, y, and z are nonnegative numbers and $x + y + z = k$, k a constant.
   (b) Use the result in (a) to show that if x, y, and z are nonnegative numbers, then

$$(xyz)^{1/3} \le \frac{x + y + z}{3}.$$

   Note: $(xyz)^{1/3}$ is the *geometric mean* of x, y, z.

31. Let $x_1, x_2, \ldots, x_n$ be nonnegative numbers such that $x_1 + x_2 + \cdots + x_n = k$, k a constant. Prove that

$$(x_1 x_2 \cdots x_n)^{1/n} \le \frac{x_1 + x_2 + \cdots + x_n}{n}.$$

In words, the geometric mean of n nonnegative numbers cannot exceed the arithmetic mean of the numbers.

32. The Celsius temperature T at a point (x, y, z) on the sphere $x^2 + y^2 + z^2 = 1$ is given by

$$T(x, y, z) = 10xy^2z.$$

Find the point(s) on the sphere at which the temperature is greatest and the point(s) at which it is least. Give the temperature at each of these points.

33. A soft drink manufacturer wants to design an aluminum can in the shape of a right circular cylinder to hold a given volume V (measured in cubic inches). If the objective is to minimize the amount of aluminum needed (top, sides, and bottom), what dimensions should be used?

In Exercises 34–38, use the Lagrange method to give alternative solutions to the indicated exercises in Sections 15.6 and 15.7.

34. Exercise 28, Section 15.6.

35. Exercise 29, Section 15.6.

36. Exercise 29, Section 15.7.

37. Exercise 45, Section 15.7.

38. Exercise 46, Section 15.7.

39. A manufacturer can produce three distinct products in quantities Q_1, Q_2, Q_3, respectively, and thereby derive a profit $P(Q_1, Q_2, Q_3) = 2Q_1 + 8Q_2 + 24Q_3$. Find the values of Q_1, Q_2, Q_3 that maximize profit if production is subject to the constraint $Q_1^2 + 2Q_2^2 + 4Q_3^2 = 4.5 \times 10^9$.

40. Find the volume of the largest rectangular box that can be inscribed in the ellipsoid

$$4x^2 + 9y^2 + 36z^2 = 36$$

if the edges of the box are parallel to the coordinate axes.

Problems with Two Side Conditions

The Lagrange method can be extended to problems with two side conditions as follows: If $\mathbf{x}_0$ is a maximum (or minimum) of $f(\mathbf{x})$ subject to the two side conditions $g(\mathbf{x}) = 0$ and $h(\mathbf{x}) = 0$, then there exist scalars λ and μ such that

$$\nabla f(\mathbf{x}_0) = \lambda \nabla g(\mathbf{x}_0) + \mu \nabla h(\mathbf{x}_0).$$

41. Find the maximum value of $f(x, y, z) = xy^2z - x^2yz$ subject to the side conditions $x^2 + y^2 = 1$ and $z = \sqrt{x^2 + y^2}$.   HINT: Set $\nabla f = \lambda \nabla g + \mu \nabla h$ and equate components.

42. The planes $x + 2y + 3z = 0$ and $2x + 3y + z = 4$ intersect in a straight line.   Find the point on the line that is closest to the origin.   HINT: Minimize $D(x, y, z) = $

$x^2 + y^2 + z^2$ subject to the side conditions $x + 2y + 3z = 0$ and $2x + 3y + z = 4$; set $\nabla f = \lambda \nabla g + \mu \nabla h$ and equate components.

43. The plane $x + y - z + 1 = 0$ intersects the upper nappe of the cone $z^2 = x^2 + y^2$ in an ellipse.   Find the points on this ellipse that are closest to and furthest from the origin.

■ 15.9 DIFFERENTIALS

We begin by reviewing the one-variable case.   If f is differentiable at x, then, for small h, the increment

$$\Delta f = f(x + h) - f(x)$$

can be approximated by the differential

$$df = f'(x)h.$$

The geometric interpretations of Δf and df are shown in Figure 15.9.1.   We write

$$\Delta f \cong df.$$

How good is this approximation?   Good enough that the ratio

$$\frac{\Delta f - df}{|h|}$$

tends to 0 as h tends to 0.

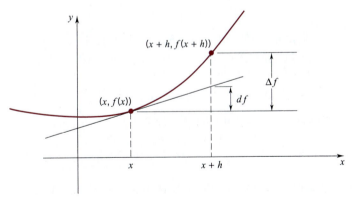

Figure 15.9.1

The differential of a function of several variables, defined in an analogous manner, plays a similar approximating role.   Let's suppose that f, now a function of several variables, is differentiable at $\mathbf{x}$.   The difference

(15.9.1)

$$\Delta f = f(\mathbf{x} + \mathbf{h}) - f(\mathbf{x})$$

is called the *increment* of *f*, and the dot product

(15.9.2)
$$df = \nabla f(\mathbf{x}) \cdot \mathbf{h}$$

is called the *differential* (more formally, the *total differential*). As in the one-variable case, for small **h**, the differential and the increment are approximately equal:

(15.9.3)
$$\Delta f \cong df.$$

How approximately equal are they? Enough so that the ratio

$$\frac{\Delta f - df}{\|\mathbf{h}\|}$$

tends to 0 as **h** tends to **0**. How do we know this? We know that

$$f(\mathbf{x} + \mathbf{h}) - f(\mathbf{x}) = \nabla f(\mathbf{x}) \cdot \mathbf{h} + o(\mathbf{h}).$$

Therefore,

$$[f(\mathbf{x} + \mathbf{h}) - f(\mathbf{x})] - \nabla f(\mathbf{x}) \cdot \mathbf{h} = o(\mathbf{h}).$$

Now, it follows that

$$\frac{\overbrace{[f(\mathbf{x} + \mathbf{h}) - f(\mathbf{x})]}^{\Delta f} - \overbrace{\nabla f(\mathbf{x}) \cdot \mathbf{h}}^{df}}{\|\mathbf{h}\|} \to 0$$

as $\mathbf{h} \to \mathbf{0}$ since $(o(\mathbf{h}))/\|\mathbf{h}\| \to 0$ as $\mathbf{h} \to \mathbf{0}$. (See Section 15.1.)

In the two-variable case we set $\mathbf{x} = (x, y)$ and $\mathbf{h} = (\Delta x, \Delta y)$. The increment $\Delta f = f(\mathbf{x} + \mathbf{h}) - f(\mathbf{x})$ then takes the form

$$\Delta f = f(x + \Delta x, y + \Delta y) - f(x, y),$$

and the differential $df = \nabla f(\mathbf{x}) \cdot \mathbf{h}$ takes the form

$$df = \frac{\partial f}{\partial x}(x, y)\, \Delta x + \frac{\partial f}{\partial y}(x, y)\, \Delta y.$$

By suppressing the point of evaluation, we can write

(15.9.4)
$$df = \frac{\partial f}{\partial x}\, \Delta x + \frac{\partial f}{\partial y}\, \Delta y.$$

The approximation $\Delta f \cong df$ is illustrated in Figure 15.9.2. There we have represented *f* as a surface $z = f(x, y)$. Through a point $P(x_0, y_0, f(x_0, y_0))$ we have drawn the tangent plane. *The difference $df - \Delta f$ is the vertical separation between this tangent plane and the surface as measured at the point $(x_0 + \Delta x, y_0 + \Delta y)$.*

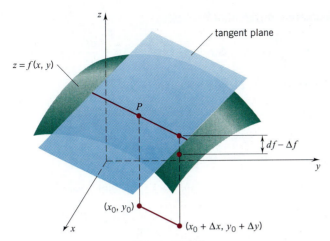

Figure 15.9.2

PROOF The tangent plane at P has equation

$$z - f(x_0, y_0) = \frac{\partial f}{\partial x}(x_0, y_0)(x - x_0) + \frac{\partial f}{\partial y}(x_0, y_0)(y - y_0).$$

The z-coordinate of this plane at the point $(x_0 + \Delta x, y_0 + \Delta y)$ is

$$f(x_0, y_0) + \frac{\partial f}{\partial x}(x_0, y_0)\, \Delta x + \frac{\partial f}{\partial y}(x_0, y_0)\, \Delta y. \qquad \text{(check this)}$$

The z-coordinate of the surface at this same point is

$$f(x_0 + \Delta x, y_0 + \Delta y).$$

The difference between these two:

$$\left[f(x_0, y_0) + \frac{\partial f}{\partial x}(x_0, y_0)\, \Delta x + \frac{\partial f}{\partial y}(x_0, y_0)\, \Delta y \right] - \left[f(x_0 + \Delta x, y_0 + \Delta y) \right]$$

can be written as

$$\left[\frac{\partial f}{\partial x}(x_0, y_0)\, \Delta x + \frac{\partial f}{\partial y}(x_0, y_0)\, \Delta y \right] - \left[f(x_0 + \Delta x, y_0 + \Delta y) - f(x_0, y_0) \right].$$

This is just $df - \Delta f$. ❑†

For the three-variable case we set $\mathbf{x} = (x, y, z)$ and $\mathbf{h} = (\Delta x, \Delta y, \Delta z)$. The increment becomes

$$\Delta f = f(x + \Delta x, y + \Delta y, z + \Delta z) - f(x, y, z),$$

---

† As in the figure, we have been assuming that the tangent plane lies above the surface. If the tangent plane lies below the surface, then $df - \Delta f$ is negative. The vertical separation between the tangent plane and the surface is then $\Delta f - df$.

and the approximating differential becomes

$$df = \frac{\partial f}{\partial x}(x, y, z) \, \Delta x + \frac{\partial f}{\partial y}(x, y, z) \, \Delta y + \frac{\partial f}{\partial z}(x, y, z) \, \Delta z.$$

Suppressing the point of evaluation we have

(15.9.5)

$$df = \frac{\partial f}{\partial x} \Delta x + \frac{\partial f}{\partial y} \Delta y + \frac{\partial f}{\partial z} \Delta z.$$

To illustrate the use of differentials we begin with a rectangle of sides x and y. The area is given by

$$A(x, y) = xy.$$

An increase in the dimensions of the rectangle to $x + \Delta x$ and $y + \Delta y$ produces a change in area

$$\begin{aligned}
\Delta A &= (x + \Delta x)(y + \Delta y) - xy \\
&= (xy + x \, \Delta y + y \, \Delta x + \Delta x \, \Delta y) - xy \\
&= x \, \Delta y + y \, \Delta x + \Delta x \, \Delta y.
\end{aligned}$$

The differential estimate for this change in area is

$$dA = \frac{\partial A}{\partial x} \Delta x + \frac{\partial A}{\partial y} \Delta y = y \, \Delta x + x \, \Delta y. \qquad \text{(Figure 15.9.3)}$$

The error of our estimate, the difference between the actual change and the estimated change, is the difference $\Delta A - dA = \Delta x \, \Delta y.$ ❑

Figure 15.9.3

Example 1 Given that

$$f(x, y) = yx^{2/5} + x\sqrt{y},$$

estimate by a differential the change in f from (32, 16) to (35, 18).

SOLUTION Since

$$\frac{\partial f}{\partial x} = \frac{2}{5}\left(\frac{1}{x}\right)^{3/5} y + \sqrt{y} \qquad \text{and} \qquad \frac{\partial f}{\partial y} = x^{2/5} + \frac{x}{2\sqrt{y}},$$

we have

$$df = \left[\frac{2}{5}\left(\frac{1}{x}\right)^{3/5} y + \sqrt{y} \right] \Delta x + \left[x^{2/5} + \frac{x}{2\sqrt{y}} \right] \Delta y.$$

At $x = 32$, $y = 16$, $\Delta x = 3$, $\Delta y = 2$,

$$df = \left[\frac{2}{5}\left(\frac{1}{32}\right)^{3/5} 16 + \sqrt{16} \right] 3 + \left[32^{2/5} + \frac{32}{2\sqrt{16}} \right] 2 = 30.4.$$

The change increases the value of f by approximately 30.4. ❏

Example 2 Use differentials to estimate $\sqrt{27}\sqrt[3]{1021}$.

SOLUTION We know $\sqrt{25}$ and $\sqrt[3]{1000}$. What we need is an estimate for the increase of

$$f(x, y) = \sqrt{x}\sqrt[3]{y} = x^{1/2}y^{1/3}$$

from $x = 25$, $y = 1000$ to $x = 27$, $y = 1021$. The differential is

$$df = \tfrac{1}{2}x^{-1/2}y^{1/3}\,\Delta x + \tfrac{1}{3}x^{1/2}y^{-2/3}\,\Delta y.$$

With $x = 25$, $y = 1000$, $\Delta x = 2$, $\Delta y = 21$, df becomes

$$(\tfrac{1}{2} \cdot 25^{-1/2} \cdot 1000^{1/3})2 + (\tfrac{1}{3} \cdot 25^{1/2} \cdot 1000^{-2/3})21 = 2.35.$$

The change increases the value of the function by about 2.35. It follows that

$$\sqrt{27}\sqrt[3]{1021} \cong \sqrt{25}\sqrt[3]{1000} + 2.35 = 52.35.$$

(Our calculator gives $\sqrt{27}\sqrt[3]{1021} \cong 52.323$.) ❏

Example 3 Estimate by a differential the change in the volume of the frustum of a right circular cone if the upper radius r is decreased from 3 to 2.7 centimeters, the base radius R is increased from 8 to 8.1 centimeters, and the height h is increased from 6 to 6.3 centimeters.

SOLUTION Since

$$V(r, R, h) = \tfrac{1}{3}\pi h(R^2 + Rr + r^2),$$

we have

$$dV = \tfrac{1}{3}\pi h(R + 2r)\,\Delta r + \tfrac{1}{3}\pi h(2R + r)\,\Delta R + \tfrac{1}{3}\pi(R^2 + Rr + r^2)\,\Delta h.$$

At $r = 3$, $R = 8$, $h = 6$, $\Delta r = -0.3$, $\Delta R = 0.1$, and $\Delta h = 0.3$,

$$dV = (28\pi)(-0.3) + (38\pi)(0.1) + \tfrac{1}{3}(97\pi)(0.3) = 5.1\pi \cong 16.02.$$

According to our differential estimate, the volume increases by about 16 cubic centimeters. (As you can check, the exact change in volume is 5.015π cubic centimeters, $5.015\pi \cong 15.76$.) ❏

EXERCISES 15.9

In Exercises 1–12, find the differential df.

1. $f(x, y) = x^3y - x^2y^2$.

2. $f(x, y, z) = xy + yz + xz$.

3. $f(x, y) = x\cos y - y\cos x$.

4. $f(x, y, z) = x^2ye^{2z}$.

5. $f(x, y, z) = x - y\tan z$.

6. $f(x, y) = (x - y) \ln(x + y)$.

7. $f(x, y, z) = \dfrac{xy}{x^2 + y^2 + z^2}$.

8. $f(x, y) = \ln(x^2 + y^2) + xe^{xy}$.

9. $f(x, y) = \sin(x + y) + \sin(x - y)$.

10. $f(x, y) = x \ln\left[\dfrac{1 + y}{1 - y} \right]$.

11. $f(x, y, z) = y^2 e^{xz} + x \ln z$.

12. $f(x, y) = xye^{-(x^2 + y^2)}$.

13. Compute Δu and du for $u = x^2 - 3xy + 2y^2$ at $x = 2$, $y = -3$, $\Delta x = -0.3$, $\Delta y = 0.2$.

14. Compute du for $u = (x + y)\sqrt{x - y}$ at $x = 6$, $y = 2$, $\Delta x = \frac{1}{4}$, $\Delta y = -\frac{1}{2}$.

15. Compute Δu and du for $u = x^2 z - 2yz^2 + 3xyz$ at $x = 2$, $y = 1$, $z = 3$, $\Delta x = 0.1$, $\Delta y = 0.3$, $\Delta z = -0.2$.

16. Calculate du for

$$ u = \frac{xy}{\sqrt{x^2 + y^2 + z^2}} $$

at $x = 1$, $y = 3$, $z = -2$, $\Delta x = \frac{1}{2}$, $\Delta y = \frac{1}{4}$, $\Delta z = -\frac{1}{4}$.

In Exercises 17–20, use differentials to find the approximate value.

17. $\sqrt{125}\sqrt[4]{17}$.

18. $(1 - \sqrt{10})(1 + \sqrt{24})$.

19. $\sin \frac{6}{7}\pi \cos \frac{1}{5}\pi$.

20. $\sqrt{8} \tan \frac{5}{16}\pi$.

In Exercises 21–24, use differentials to approximate the value of f at the given point.

21. $f(x, y) = x^2 e^{xy}$; $(2.9, 0.01)$.

22. $f(x, y, z) = x^2 y \cos \pi z$; $(2.12, 2.92, 3.02)$.

23. $f(x, y, z) = x \tan^{-1} yz$; $(2.94, 1.1, 0.92)$.

24. $f(x, y) = \sqrt{x^2 + y^2}$; $(3.06, 3.88)$.

25. Given that $z = (x - y)(x + y)^{-1}$, use dz to find the approximate change in z if x is increased from 4 to $4\frac{1}{10}$ and y is increased from 2 to $2\frac{1}{10}$.   What is the exact change?

26. Estimate by a differential the change in the volume of a right circular cylinder if the height is increased from 12 to 12.2 inches and the radius is decreased from 8 to 7.7 inches.

27. Estimate the change in total surface area for the cylinder of Exercise 26.

28. Use a differential to estimate the change in $T = x^2 \cos \pi z - y^2 \sin \pi z$ from $x = 2, y = 2, z = 2$ to $x = 2.1$, $y = 1.9$, $z = 2.2$.

29. Estimate the surface area of a closed rectangular box whose dimensions are length = 9.98 inches; width = 5.88 inches; height = 4.08 inches.

30. Estimate the volume of a right circular cone whose dimensions are radius = 7.2 centimeters and height = 10.15 centimeters.

31. The dimensions of a closed rectangular box change from length = 12; width = 8; height = 6, to length = 12.02; width = 7.95; height = 6.03.
   (a) Use a differential to approximate the change in volume.
   (b) Calculate the exact change in volume.

32. Use the dimensions of the rectangular box in Exercise 31.
   (a) Approximate the change in surface area using a differential.
   (b) Calculate the exact change in the surface area.

33. The function $T(x, y, z) = 100 - x^2 - y^2 - z^2 + 2xyz$ gives the temperature T at the point $P(x, y, z)$ in space. Use differentials to approximate the temperature difference between the points $P(1, 3, 4)$ and $Q(1.15, 2.90, 4.10)$.

34. According to the ideal gas law, the relation between the pressure P, the temperature T, and the volume V of a confined gas is given by the equation $PV = kT$, where k is a constant.   If $P = 4$ pounds per square inch when $V = 81$ cubic inches and $T = 300$ K, approximate the change in pressure if the volume is decreased to 75 cubic inches and the temperature is increased to 325 K.

35. As illustrated in the following figure, the side x in the right triangle is increased by Δx and the angle θ is increased by $\Delta \theta$.   Use a differential to approximate the change in the area of the triangle.   Is the area more sensitive to a change in x or to a change in θ?

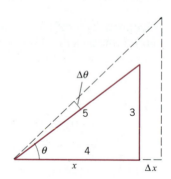

36. Use a differential to approximate the change in the area of the isosceles triangle shown in the following figure if x changes by Δx and θ changes by $\Delta \theta$.   Is the area more sensitive to changes in x or to changes in θ? HINT: The area of a triangle with sides a and b and included angle θ is $A = \frac{1}{2}ab \sin \theta$.

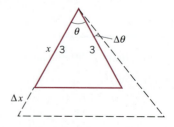

37. The radius of a right circular cylinder of height h is increased from r to $r + \Delta r$.
   (a) Determine the exact change in h that will keep the volume constant. Then estimate this change in h by using a differential.
   (b) Determine the exact change in h that will keep the total surface area constant. Then estimate this change in h by using a differential.

38. The dimensions of a rectangular box with a top are length = 4 feet; width = 2 feet; height = 3 feet. It has a coat of paint $\frac{1}{16}$ inch thick. Estimate the amount of paint (cubic inches) on the box.

Error Estimates

Let $u = u(x, y)$ be differentiable. If the variables x and y are known to be $x_0 \pm \Delta x$ and $y_0 \pm \Delta y$, then the maximum possible error in the calculated value $u(x_0, y_0)$ is

$$\frac{\partial u}{\partial x}(x_0, y_0)(\pm \Delta x) + \frac{\partial u}{\partial y}(x_0, y_0)(\pm \Delta y)$$

39. The legs of a right triangle are measured to be 5 and 12 centimeters with a possible error of ± 15 millimeters in each measurement. What is the maximum possible error in the calculated value of (a) the hypotenuse and (b) the area of the triangle?

40. The radius of a right circular cone is measured to be 5 inches with a possible error of ± 0.2 inch and the height is measured to be 12 inches with a possible error of ± 0.3 inch. What is the maximum possible error in the calculated values of (a) the volume and (b) the lateral surface area of the cone?

41. The specific gravity of a solid is given by the formula $s = A(A - W)^{-1}$ where A is the weight in air and W is the weight in water. What is the maximum possible error in the calculated value of s if A is measured to be 9 pounds (within a tolerance of 0.01 pound) and W is measured to be 5 pounds (within a tolerance of 0.02 pound)?

42. The measurements of a closed rectangular box are length = 5 feet; width = 3 feet; height = 3.5 feet, with a possible error of $\pm \frac{1}{12}$ inch in each measurement. What is the maximum possible error in the calculated value of (a) the volume and (b) the surface area of the box? HINT: Extend the error estimate to three dimensions.

■ 15.10 RECONSTRUCTING A FUNCTION FROM ITS GRADIENT

This section has three parts. In Part 1 we show how to find $f(x, y)$ given its gradient

$$\nabla f(x, y) = \frac{\partial f}{\partial x}(x, y)\mathbf{i} + \frac{\partial f}{\partial y}(x, y)\mathbf{j}.$$

In Part 2 we show that, although all gradients $\nabla f(x, y)$ are expressions of the form

$$P(x, y)\mathbf{i} + Q(x, y)\mathbf{j}$$

(set $P = \partial f/\partial x$ and $Q = \partial f/\partial y$), not all such expressions are gradients. In Part 3 we tackle the problem of recognizing which expressions $P(x, y)\mathbf{i} + Q(x, y)\mathbf{j}$ are actually gradients.

Part 1

Example 1 Find f given that

$$\nabla f(x, y) = (4x^3y^3 - 3x^2)\mathbf{i} + (3x^4y^2 + \cos 2y)\mathbf{j}.$$

SOLUTION The first partial derivatives of f are

$$\frac{\partial f}{\partial x}(x, y) = 4x^3y^3 - 3x^2, \qquad \frac{\partial f}{\partial y}(x, y) = 3x^4y^2 + \cos 2y.$$

Integrating $\partial f/\partial x$ with respect to x, treating y as a constant, we find that

$$f(x, y) = x^4y^3 - x^3 + \phi(y)$$

where ϕ is an unknown function of y. The function ϕ plays the same role as the arbitrary constant C that arises when you integrate a function of one variable. Now, differentiation with respect to y gives

$$\frac{\partial f}{\partial y}(x, y) = 3x^4y^2 + \phi'(y).$$

Equating the two expressions for $\partial f/\partial y$, we have

$$3x^4y^2 + \phi'(y) = 3x^4y^2 + \cos 2y,$$

which implies

$$\phi'(y) = \cos 2y \quad \text{and thus} \quad \phi(y) = \tfrac{1}{2}\sin 2y + C \quad (C \text{ a constant}).$$

This means that

$$f(x, y) = x^4y^3 - x^3 + \tfrac{1}{2}\sin 2y + C. \quad \square$$

Remark The procedure for finding f just illustrated is symmetric in x and y. That is, rather than integrating $\partial f/\partial x$ with respect to x, we could have started by integrating $\partial f/\partial y$ with respect to y, with x held constant, followed by differentiating the result with respect to x:

$$\text{if} \quad \frac{\partial f}{\partial y}(x, y) = 3x^4y^2 + \cos 2y \quad \text{then} \quad f(x, y) = x^4y^3 + \tfrac{1}{2}\sin 2y + \psi(x),$$

where ψ is an unknown function of x. Now, differentiating with respect to x, we have

$$\frac{\partial f}{\partial x}(x, y) = 4x^3y^3 + \psi'(x).$$

Equating the two expressions for $\partial f/\partial x$ gives

$$4x^3y^3 + \psi'(x) = 4x^3y^3 - 3x^2.$$

Therefore,

$$\psi'(x) = -3x^2 \quad \text{which implies} \quad \psi(x) = -x^3 + C \quad (C \text{ a constant}).$$

Thus,

$$f(x, y) = x^4y^3 + \tfrac{1}{2}\sin 2y - x^3 + C$$

as we found before. $\quad \square$

Example 2 Find f given that

$$\nabla f(x, y) = \left(\sqrt{y} - \frac{y}{2\sqrt{x}} + 2x\right)\mathbf{i} + \left(\frac{x}{2\sqrt{y}} - \sqrt{x} + 1\right)\mathbf{j}.$$

SOLUTION Here we have

$$\frac{\partial f}{\partial x}(x, y) = \sqrt{y} - \frac{y}{2\sqrt{x}} + 2x, \qquad \frac{\partial f}{\partial y}(x, y) = \frac{x}{2\sqrt{y}} - \sqrt{x} + 1.$$

We will proceed as in the first example. Integrating $\partial f/\partial x$ with respect to x, we have

$$f(x, y) = x\sqrt{y} - y\sqrt{x} + x^2 + \phi(y)$$

with $\phi(y)$ independent of x. Differentiation with respect to y gives

$$\frac{\partial f}{\partial y}(x, y) = \frac{x}{2\sqrt{y}} - \sqrt{x} + \phi'(y).$$

The two equations for $\partial f/\partial y$ can be reconciled only by having

$$\phi'(y) = 1 \quad \text{and thus} \quad \phi(y) = y + C.$$

This means that

$$f(x, y) = x\sqrt{y} - y\sqrt{x} + x^2 + y + C. \quad \square$$

The function

$$f(x, y) = x\sqrt{y} - y\sqrt{x} + x^2 + y + C$$

is the *general solution* of the vector differential equation

$$\nabla f(x, y) = \left(\sqrt{y} - \frac{y}{2\sqrt{x}} + 2x\right)\mathbf{i} + \left(\frac{x}{2\sqrt{y}} - \sqrt{x} + 1\right)\mathbf{j}.$$

Each *particular* solution can be obtained by assigning a particular value to the constant of integration C.

Part 2

The next example shows that not all linear combinations $P(x, y)\mathbf{i} + Q(x, y)\mathbf{j}$ are gradients.

Example 3 Show that $y\mathbf{i} - x\mathbf{j}$ is not a gradient.

SOLUTION Suppose on the contrary that it is a gradient. Then there exists a function f such that

$$\nabla f(x, y) = y\mathbf{i} - x\mathbf{j}.$$

Obviously

$$\frac{\partial f}{\partial x}(x, y) = y, \qquad \frac{\partial f}{\partial y}(x, y) = -x$$

$$\frac{\partial^2 f}{\partial y \partial x}(x, y) = 1, \qquad \frac{\partial^2 f}{\partial x \partial y}(x, y) = -1$$

and thus

$$\frac{\partial^2 f}{\partial y \partial x}(x, y) \neq \frac{\partial^2 f}{\partial x \partial y}(x, y).$$

This contradicts (14.6.5): the four partial derivatives under consideration are everywhere continuous and thus, according to (14.6.5), we must have

$$\frac{\partial^2 f}{\partial y \partial x}(x, y) = \frac{\partial^2 f}{\partial x \partial y}(x, y).$$

This contradiction shows that $y\mathbf{i} - x\mathbf{j}$ is not a gradient. $\square$

Part 3

We come now to the problem of recognizing which linear combinations

$$P(x, y)\mathbf{i} + Q(x, y)\mathbf{j}$$

are actually gradients. But first we need to review some ideas and establish some new terminology.

As indicated earlier (Section 15.3), an open set (in the plane or in three-space) is said to be *connected* iff any two points of the set can be joined by a polygonal path that lies entirely within the set. An open connected set is called an *open region*. A curve

$$C: \quad \mathbf{r} = \mathbf{r}(t), \qquad t \in [a, b]$$

is said to be *closed* iff it begins and ends at the same point:

$$\mathbf{r}(a) = \mathbf{r}(b).$$

It is said to be *simple* iff it does not intersect itself:

$$a < t_1 < t_2 < b \qquad \text{implies} \qquad \mathbf{r}(t_1) \neq \mathbf{r}(t_2).$$

These notions are illustrated in Figure 15.10.1.

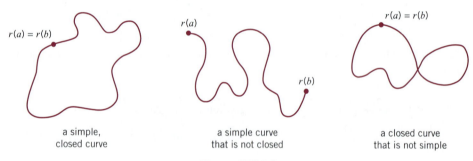

a simple,
closed curve

a simple curve
that is not closed

a closed curve
that is not simple

Figure 15.10.1

Figure 15.10.2

As is intuitively clear (Figure 15.10.2), a simple closed curve in the plane separates the plane into two disjoint open connected sets: a bounded inner region consisting of all points surrounded by the curve and an unbounded outer region consisting of all points not surrounded by the curve.†

Finally, we come to the notion we need in our work with gradients:

(15.10.1)

> Let Ω be an open region of the plane, and let C be an arbitrary simple closed curve. Ω is said to be *simply connected* iff
>
> C is in Ω implies   the inner region of C is in Ω.

The first two regions in Figure 15.10.3 are simply connected. The annular region is not; the annular region contains the simple closed curve drawn there, but it does not contain all of the inner region of that curve.

open rectangle   open disc   open annular region

Figure 15.10.3

THEOREM 15.10.2

Let P and Q be functions of two variables, each continuously differentiable on a simply connected open region Ω. The linear combination $P(x, y)\mathbf{i} + Q(x, y)\mathbf{j}$ is a gradient on Ω iff

$$\frac{\partial P}{\partial y}(x, y) = \frac{\partial Q}{\partial x}(x, y) \qquad \text{for all } (x, y) \in \Omega.$$

A complete proof of this theorem for a general region Ω is complicated. We will prove the result under the additional assumption that Ω has the form of an open rectangle with sides parallel to the coordinate axes.

Suppose that $P(x, y)\mathbf{i} + Q(x, y)\mathbf{j}$ is a gradient on this open rectangle Ω, say

$$\nabla f(x, y) = P(x, y)\mathbf{i} + Q(x, y)\mathbf{j}.$$

---

† Add to this the assertion that the curve in question constitutes the total boundary of both regions and you have what is called the Jordan curve theorem, named after the French mathematician Camille Jordan (1838–1922). Jordan was the first to point out that, although all this is apparently obvious, it nevertheless requires proof. In recognition of Jordan, plane simple closed curves are now commonly known as *Jordan curves*.

Since

$$\nabla f(x, y) = \frac{\partial f}{\partial x}(x, y)\mathbf{i} + \frac{\partial f}{\partial y}(x, y)\mathbf{j},$$

we have

$$P = \frac{\partial f}{\partial x} \quad \text{and} \quad Q = \frac{\partial f}{\partial y}.$$

Since P and Q have continuous first partials, f has continuous second partials. Thus, according to (14.6.5), the mixed partials are equal and we have

$$\frac{\partial P}{\partial y} = \frac{\partial^2 f}{\partial y \partial x} = \frac{\partial^2 f}{\partial x \partial y} = \frac{\partial Q}{\partial x}.$$

Conversely, suppose that

$$\frac{\partial P}{\partial y}(x, y) = \frac{\partial Q}{\partial x}(x, y) \quad \text{for all } (x, y) \in \Omega.$$

We must show that $P(x, y)\mathbf{i} + Q(x, y)\mathbf{j}$ is a gradient on Ω. To do this, we choose a point (x_0, y_0) from Ω and form the function

$$f(x, y) = \int_{x_0}^{x} P(u, y_0)\, du + \int_{y_0}^{y} Q(x, v)\, dv, \quad (x, y) \in \Omega.$$

[If you want to visualize f, you can refer to Figure 15.10.4. The function P is being integrated along the horizontal line segment joining (x_0, y_0) to (x, y_0), and Q is being integrated along the vertical line segment joining (x, y_0) to (x, y). Our assumptions on Ω guarantee that these line segments remain in Ω.]

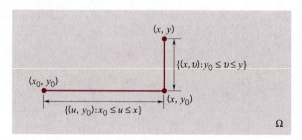

Figure 15.10.4

The first integral is independent of y. Hence

$$\frac{\partial f}{\partial y}(x, y) = \frac{\partial}{\partial y}\left(\int_{y_0}^{y} Q(x, v)\, dv\right) = Q(x, y).$$

The last equality holds because we are differentiating an integral with respect to its upper limit. (See Theorem 5.3.5.) Differentiating f with respect to x we have

$$\frac{\partial f}{\partial x}(x, y) = \frac{\partial}{\partial x}\left(\int_{x_0}^{x} P(u, y_0)\, du\right) + \frac{\partial}{\partial x}\left(\int_{y_0}^{y} Q(x, v)\, dv\right).$$

The first term is $P(x, y_0)$ since once again we are differentiating with respect to the upper limit. In the second term the variable x appears under the sign of integration. It can be shown that, since Q and $\partial Q/\partial x$ are continuous,

$$\frac{\partial}{\partial x}\left(\int_{y_0}^{y} Q(x, v)\, dv\right) = \int_{y_0}^{y} \frac{\partial Q}{\partial x}(x, v)\, dv.\dagger$$

Anticipating this result, we have

$$\frac{\partial f}{\partial x}(x, y) = P(x, y_0) + \int_{y_0}^{y} \frac{\partial Q}{\partial x}(x, v)\, dv = P(x, y_0) + \int_{y_0}^{y} \frac{\partial P}{\partial y}(x, v)\, dv$$

$$\text{explain} \longrightarrow$$

$$= P(x, y_0) + P(x, y) - P(x, y_0) = P(x, y).$$

We have now shown that

$$P(x, y) = \frac{\partial f}{\partial x}(x, y) \qquad \text{and} \qquad Q(x, y) = \frac{\partial f}{\partial y}(x, y) \qquad \text{for all } (x, y) \in \Omega.$$

It follows that $P(x, y)\mathbf{i} + Q(x, y)\mathbf{j}$ is the gradient of f on Ω. ❏

Example 4 The vector functions

$$\mathbf{F}(x, y) = 2x \sin y\,\mathbf{i} + x^2 \cos y\,\mathbf{j} \qquad \text{and} \qquad \mathbf{G}(x, y) = xy\mathbf{i} + \tfrac{1}{2}(x + 1)^2 y^2\,\mathbf{j}$$

are both defined everywhere. The first vector function is the gradient of a function that is defined everywhere since for $P(x, y) = 2x \sin y$ and $Q(x, y) = x^2 \cos y$, we have

$$\frac{\partial P}{\partial y}(x, y) = 2x \cos y = \frac{\partial Q}{\partial x} \qquad \text{for all } (x, y).$$

It is easy to verify that if $f(x, y) = x^2 \sin y + C$, where C is a constant, then $\nabla f(x, y) = \mathbf{F}(x, y)$.

The vector function $\mathbf{G}$ is not a gradient: $P(x, y) = xy$, $Q(x, y) = \tfrac{1}{2}(x + 1)^2 y^2$, and

$$\frac{\partial P}{\partial y}(x, y) = x \qquad \text{and} \qquad \frac{\partial Q}{\partial x}(x, y) = (x + 1)y^2.$$

Thus,

$$\frac{\partial P}{\partial y}(x, y) \neq \frac{\partial Q}{\partial x}(x, y). \qquad ❏$$

Example 5 The vector function $\mathbf{F}$ defined on the *punctured disc* $0 < x^2 + y^2 < 1$ by setting

$$\mathbf{F}(x, y) = \frac{y}{x^2 + y^2}\,\mathbf{i} - \frac{x}{x^2 + y^2}\,\mathbf{j}$$

satisfies

$$\frac{\partial P}{\partial y}(x, y) = \frac{\partial Q}{\partial x}(x, y) \qquad \text{on the punctured disc.}$$

---

† See Exercise 60, Section 16.4.

(Check this out.) Nevertheless, as you will see in Chapter 17, **F** is not a gradient on that set. The punctured disc is not simply connected and therefore Theorem 15.10.2 does not apply. ❑

EXERCISES 15.10

In Exercises 1–16, determine whether the vector function is the gradient $\nabla f(x, y)$ of a function everywhere defined. If so, find such a function f.

1. $xy^2\mathbf{i} + x^2y\mathbf{j}$

2. $x\mathbf{i} + y\mathbf{j}$.

3. $y\mathbf{i} + x\mathbf{j}$.

4. $(x^2 + y)\mathbf{i} + (y^3 + x)\mathbf{j}$.

5. $(y^3 + x)\mathbf{i} + (x^2 + y)\mathbf{j}$.

6. $(y^2e^x - y)\mathbf{i} + (2ye^x - x)\mathbf{j}$.

7. $(\cos x - y \sin x)\mathbf{i} + \cos x\mathbf{j}$.

8. $(1 + e^y)\mathbf{i} + (xe^y + y^2)\mathbf{j}$.

9. $e^x\cos y^2\mathbf{i} - 2ye^x\sin y^2\mathbf{j}$.

10. $e^x\cos y\mathbf{i} + e^x\sin y\mathbf{j}$.

11. $ye^x(1 + x)\mathbf{i} + (xe^x - e^{-y})\mathbf{j}$.

12. $(e^x + 2xy)\mathbf{i} + (x^2 + \sin y)\mathbf{j}$.

13. $(xe^{xy} + x^2)\mathbf{i} + (ye^{xy} - 2y)\mathbf{j}$.

14. $(y \sin x + xy \cos x)\mathbf{i} + (x \sin x + 2y + 1)\mathbf{j}$.

15. $(1 + y^2 + xy^2)\mathbf{i} + (x^2y + y + 2xy + 1)\mathbf{j}$.

16. $\left[2 \ln(3y) + \dfrac{1}{x} \right]\mathbf{i} + \left[\dfrac{2x}{y} + y^2 \right]\mathbf{j}$.

In Exercises 17–20, find the most general function with the given gradient.

17. $\dfrac{x}{\sqrt{x^2 + y^2}}\mathbf{i} + \dfrac{y}{\sqrt{x^2 + y^2}}\mathbf{j}$.

18. $(x \tan y + \sec^2 x)\mathbf{i} + (\frac{1}{2}x^2\sec^2 y + \pi y)\mathbf{j}$.

19. $(x^2\sin^{-1}y)\mathbf{i} + \left(\dfrac{x^3}{3\sqrt{1 - y^2}} - \ln y \right)\mathbf{j}$.

20. $\left(\dfrac{\tan^{-1}y}{\sqrt{1 - x^2}} + \dfrac{x}{y} \right)\mathbf{i} + \left(\dfrac{\sin^{-1}x}{1 + y^2} - \dfrac{x^2}{2y^2} + 1 \right)\mathbf{j}$.

21. Find the general solution of the differential equation $\nabla f(x, y) = f(x, y)\mathbf{i} + f(x, y)\mathbf{j}$.

22. Given that g and its first and second partials are everywhere continuous, find the general solution of the differential equation $\nabla f(x, y) = e^{g(x, y)}[g_x(x, y)\mathbf{i} + g_y(x, y)\mathbf{j}]$.

Theorem 15.10.2 has a three-dimensional analog. In particular we can show that, if P, Q, R are continuously differentiable on an open rectangular box S, then the vector function

$$P(x, y, z)\mathbf{i} + Q(x, y, z)\mathbf{j} + R(x, y, z)\mathbf{k}$$

is a gradient on S iff

$$\dfrac{\partial P}{\partial y} = \dfrac{\partial Q}{\partial x}, \qquad \dfrac{\partial P}{\partial z} = \dfrac{\partial R}{\partial x}, \qquad \dfrac{\partial Q}{\partial z} = \dfrac{\partial R}{\partial y} \qquad \text{throughout } S.$$

23. (a) Verify that $2x\mathbf{i} + z\mathbf{j} + y\mathbf{k}$ is the gradient of a function f that is everywhere defined.
 (b) Deduce from the relation $\partial f/\partial x = 2x$ that $f(x, y, z) = x^2 + g(y, z)$.
 (c) Verify then that $\partial f/\partial y = z$ gives $g(y, z) = zy + h(z)$ and finally that $\partial f/\partial z = y$ gives $h(z) = C$.
 (d) What is $f(x, y, z)$?

In Exercises 24–28, determine whether the function is a gradient $\nabla f(x, y, z)$ and, if so, find such a function f.

24. $yz\mathbf{i} + xz\mathbf{j} + xy\mathbf{k}$.

25. $(2x + y)\mathbf{i} + (2y + x + z)\mathbf{j} + (y - 2z)\mathbf{k}$.

26. $(2x \sin 2y \cos z)\mathbf{i} + (2x^2 \cos 2y \cos z)\mathbf{j} - (x^2 \sin 2y \sin z)\mathbf{k}$.

27. $(y^2z^3 + 1)\mathbf{i} + (2xyz^3 + y)\mathbf{j} + (3xy^2z^2 + 1)\mathbf{k}$.

28. $\left[\dfrac{y}{z} - e^z \right]\mathbf{i} + \left[\dfrac{x}{z} + 1 \right]\mathbf{j} - \left[xe^z + \dfrac{xy}{z^2} \right]\mathbf{k}$.

29. Verify that the gravitational force function

$$\mathbf{F}(\mathbf{r}) = -G\dfrac{mM}{r^3}\mathbf{r} \qquad (\mathbf{r} = x\mathbf{i} + y\mathbf{j} + z\mathbf{k})$$

is a gradient.

30. Verify that every vector function of the form

$$\mathbf{h}(\mathbf{r}) = kr^n\mathbf{r} \qquad (k \text{ constant}, n \text{ an integer})$$

is a gradient.

■ CHAPTER HIGHLIGHTS

15.1 Differentiability and Gradient

gradient of f at $\mathbf{x}$: $\nabla f(\mathbf{x})$ (p. 946)

$$\nabla f(x, y) = \dfrac{\partial f}{\partial x}\mathbf{i} + \dfrac{\partial f}{\partial y}\mathbf{j} \qquad \nabla f(x, y, z) = \dfrac{\partial f}{\partial x}\mathbf{i} + \dfrac{\partial f}{\partial y}\mathbf{j} + \dfrac{\partial f}{\partial z}\mathbf{k}$$

If f is differentiable at $\mathbf{x}$, then f is continuous at $\mathbf{x}$.

$$\nabla r^n = n r^{n-2} \mathbf{r}$$

15.2 Gradients and Directional Derivatives

directional derivative: $f_{\mathbf{u}}'(\mathbf{x}) = \nabla f(\mathbf{x}) \cdot \mathbf{u}$ (p. 955)

The directional derivative $f_{\mathbf{u}}'$ gives the rate of change of f in the direction of the unit vector $\mathbf{u}$.

The directional derivative in a direction $\mathbf{u}$ is the component of the gradient vector in that direction. (p. 958)

A differentiable function f increases most rapidly in the direction of the gradient (the rate of change is then $\|\nabla f(\mathbf{x})\|$) and it decreases most rapidly in the opposite direction (the rate of change is then $-\|\nabla f(\mathbf{x})\|$).

15.3 The Mean-Value Theorem; Two Intermediate-Value Theorems

The Mean-Value Theorem (p. 964)   open, connected set (p. 964)
intermediate-value theorems (p. 966, 967)   open, closed regions (p. 966)

15.4 Chain Rules

chain rule along a curve (p. 968)   tree diagram (p. 973)

In the setting of functions of several variables, there are numerous chain rules. They can all be deduced from the chain rule along a curve. If, for example, $u = u(x, y)$ where $x = x(s, t)$ and $y = y(s, t)$, then

$$\frac{\partial u}{\partial s} = \frac{\partial u}{\partial x}\frac{\partial x}{\partial s} + \frac{\partial u}{\partial y}\frac{\partial y}{\partial s} \quad \text{and} \quad \frac{\partial u}{\partial t} = \frac{\partial u}{\partial x}\frac{\partial x}{\partial t} + \frac{\partial u}{\partial y}\frac{\partial y}{\partial t}.$$

If f is a continuously differentiable function of $r = \|\mathbf{r}\|$, then

$$\nabla[f(r)] = f'(r)\frac{\mathbf{r}}{r}.$$

15.5 The Gradient as a Normal; Tangent Lines and Tangent Planes

tangent and normal lines to a curve $f(x, y) = c$ (p. 982)
tangent plane to a surface $f(x, y, z) = c$ (p. 984)
upper and lower unit normals (p. 990)

At each point of the domain of a function, the gradient vector, if not $\mathbf{0}$, is perpendicular to the level curve (level surface) that passes through that point.

15.6 Maximum and Minimum Values

local maximum and local minimum (p. 991)   critical points (p. 992)
stationary points, saddle points (p. 992)   absolute extreme values (p. 996)

If f has a local extreme value at $\mathbf{x}_0$, then either $\nabla f(\mathbf{x}_0) = \mathbf{0}$ or $\nabla f(\mathbf{x}_0)$ does not exist.

15.7 Second-Partials Test

second-partials test, discriminant (p. 1000)   method of least squares (p. 1006)

15.8 Maxima and Minima with Side Conditions

If $\mathbf{x}_0$ maximizes (or minimizes) $f(\mathbf{x})$ subject to the side condition $g(\mathbf{x}) = 0$, then $\nabla f(\mathbf{x}_0)$ and $\nabla g(\mathbf{x}_0)$ are parallel. Thus, if $\nabla g(\mathbf{x}_0) \neq \mathbf{0}$, then there exists a scalar λ, called a Lagrange multiplier, such that $\nabla f(\mathbf{x}_0) = \lambda \nabla g(\mathbf{x}_0)$.

15.9 Differentials

increment: $\Delta f = f(\mathbf{x} + \mathbf{h}) - f(\mathbf{x})$ differential: $df = \nabla f(\mathbf{x}) \cdot \mathbf{h}$

$\Delta f \cong df$ in the sense that $\dfrac{\Delta f - df}{\|\mathbf{h}\|} \to 0$ as $\mathbf{h} \to \mathbf{0}$

two variables: $df = \dfrac{\partial f}{\partial x} \Delta x + \dfrac{\partial f}{\partial y} \Delta y$

three variables: $df = \dfrac{\partial f}{\partial x} \Delta x + \dfrac{\partial f}{\partial y} \Delta y + \dfrac{\partial f}{\partial z} \Delta z$

15.10 Reconstructing a Function from its Gradient

finding f from ∇f (pp. 1021–1023)   closed curve, simple curve (p. 1024)
simply-connected open region (p. 1025)

necessary and sufficient conditions for a vector-valued function to be a gradient:

two variables (p. 1025)   three variables (p. 1028)

■ PROJECTS AND EXPLORATIONS USING TECHNOLOGY

To do these exercises you will need a graphics calculator or a computer with graphing capability. The majority of these problems are open-ended so different approaches may be used to solve them. You should be aware that different approaches can result in slight variations in the answers. Round your numerical answers to at least four decimal places. The rounding method that your calculator or computer uses also may cause variations in answers.

15.1 (a) A fence of height h is located w feet in front of a tall building. We want to find the length of the shortest ladder that can be leaned over the fence, against the side of the building. Determine the function that is to be minimized and the side condition.

(b) Suppose that the fence is 6 feet tall and 4 feet from the building. Find the length of the shortest ladder using the methods of single-variable calculus.

(c) Solve the same problem using Lagrange multipliers.

(d) Suppose that a moat of width w has walls of heights h_1 and h_2 on either side. Ladders of lengths L_1 and L_2 lean over the walls and meet at a point which is at a height y above the moat. Let x be the distance from the first wall to the point directly below the point where the ladders meet. (See the figure.) We want to minimize the sum of the lengths of the two ladders. Determine the function that is to be minimized and the side conditions. Explain why this problem should have a solution.

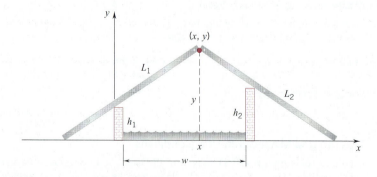

(e) If x and y are given, solve for L_1 and L_2 in terms of x, y, h_1, h_2, and w.

(f) Let $w = 20$, $h_1 = 5$, and $h_2 = 6$. Find L_1, L_2 and $L_1 + L_2$ for the following pairs (x, y): $(5, 8)$, $(7, 10)$, $(10, 12)$, $(12, 15)$.

(g) Now solve the problem using Lagrange multipliers.

15.2 In this Exercise and the next, we are going to use Lagrange multipliers to study municipal recycling. In this problem we will determine where to locate the recycling center. We assume that the garbage is picked up from the homes and businesses, transported to the recycling center, and then sorted at the center. A percentage γ of the garbage is recycled and the remaining $1 - \gamma$ is trucked to a landfill. Let c_1 be the cost per mile of trucking garbage from homes and business to the recycling center and c_2 be the cost per mile of trucking garbage from the recycling center to the landfill. For simplicity, assume that the city is rectangular with vertices $(0, 0)$, $(a, 0)$, (a, b), and $(0, b)$, that the recycling center is at the point (p, q) inside the city limits, and that the landfill is at the point (r, s) outside the city limits. (See the figure.)

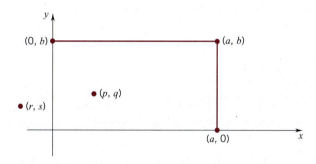

(a) Using the usual distance function for points in the plane
$$d[(x_1, y_1), (x_2, y_2)] = \sqrt{(x_2 - x_1)^2 + (y_2 - y_1)^2},$$
give the integrals that represent the average distance from a home or business to the recycling center. Find the location (p, q) for the recycling center which will minimize the average distance.

(b) Next, assume that the distance between points in the plane is given by
$$d[(x_1, y_1), (x_2, y_2)] = |x_2 - x_1| + |y_2 - y_1|.$$
Give the integrals that represent the average distance from a home to the recycling center and find the location (p, q) for the recycling center which will minimize this average distance.

(c) Using the distance function from part (b), find an expression for the total cost of transporting the garbage to the center and transporting the nonrecycled material to the landfill.

(d) Using the cost function from part (c), find the point (p, q) which minimizes the total cost (as a function of γ, c_1, and c_2) for each of the following cases (i) $r = 0$, $0 < s < b$; (ii) $r < 0$, $0 < s < b$; (iii) $r < 0$, $s < 0$.

15.3 Now we will determine how much sorting to do in order to minimize total costs. In addition to the transportation costs, c_1 and c_2, the costs that we consider are (i) the unit cost c_3 of dumping at the landfill; (ii) p, the unit profit of selling recycled material; and (iii) $c_4\gamma/(1 - \gamma)$, the labor cost of sorting.

(a) Discuss the labor cost function. Determine the total cost function.

(b) Let $a = 2$, $b = 6$, $r = 0$, $s = 1$, and $c_1 = 1$. Find the value of γ that minimizes the total cost in each of the following situations:

(i) $c_2 = 0.5$, $c_3 = 1$, $c_4 = 1$.   (ii) $c_2 = 0.5$, $c_3 = 0.5$, $c_4 = 1$.
(iii) $c_2 = 0.5$, $c_3 = 1$, $c_4 = 0.5$.   (iv) $c_2 = 0.5$, $c_3 = 0.5$, $c_4 = 0.5$.
(v) $c_2 = 0.25$, $c_3 = 1$, $c_4 = 1$.

(c) Suppose that the recycling center is built at the point (p, q) found in Exercise 2, part (b). How much is the total cost above the minimum in each of the cases in part (b).

(d) Use the values from part (c) to approximate the partials of γ with respect to c_2, c_3, and c_4. What is the importance of these partials?

15.4 Here we are going to reexamine a familiar problem—the problem of minimizing the cost of manufacturing a can in the shape of a right circular cylinder that has a given volume. Suppose the cost per unit area of the top and bottom is k times the cost per unit area of the sides. Let r be the radius of the can, h the height, and V the given volume.

(a) Find the ratio of the height to the radius of the cheapest can as a function of k. Also give the minimal cost as a function of k. Treat this as a single-variable problem.

(b) Now do the problem using Lagrange multipliers.

(c) Find the partial derivatives of the minimal cost with respect to k and with respect to V. Is the minimal cost an increasing function of V for all k?

(d) Suppose that for aesthetic reasons we add the constraint that the height of the can must be at least equal to the diameter of the base, but at most twice the diameter. Use Lagrange multipliers to solve this problem.

DOUBLE AND TRIPLE INTEGRALS

■ 16.1 MULTIPLE-SIGMA NOTATION

In Chapter 5 we studied ordinary integrals

$$\int_J f(x) \, dx$$

where $J = [a, b]$ is an interval on the real line and

$$\int_J f(x) \, dx = \int_a^b f(x) \, dx.$$

Here we study double integrals

$$\iint_\Omega f(x, y) \, dx \, dy,$$

where Ω is some region in the x,y-plane, and a little later we will study triple integrals

$$\iiint_T f(x, y, z) \, dx \, dy \, dz,$$

where T is a solid in three-dimensional space.

Our first step is to introduce some new notation. In an ordinary sequence $\{a_i\}$, each term a_i is indexed by a single integer. The sum of all the a_i from $i = 1$ to $i = m$ is then denoted by

$$\sum_{i=1}^m a_i.$$

When two indices are involved, say

$$a_{ij} = 2^i 5^j, \qquad a_{ij} = \frac{2i}{5+j}, \quad \text{or} \quad a_{ij} = (1 + i)^j,$$

then we use double-sigma notation to denote the sum of all the doubly indexed terms. By

(16.1.1)

$$\sum_{i=1}^{m} \sum_{j=1}^{n} a_{ij}$$

we mean *the sum of all the a_{ij} where i ranges from 1 to m and j ranges from 1 to n.* For example,

$$\sum_{i=1}^{3} \sum_{j=1}^{2} 2^i 5^j = 2 \cdot 5 + 2 \cdot 5^2 + 2^2 \cdot 5 + 2^2 \cdot 5^2 + 2^3 \cdot 5 + 2^3 \cdot 5^2 = 420.$$

Since addition is associative and commutative, we can add the terms of (16.1.1) in any order we choose. Usually we set

(16.1.2)

$$\sum_{i=1}^{m} \sum_{j=1}^{n} a_{ij} = \sum_{i=1}^{m} \left(\sum_{j=1}^{n} a_{ij} \right).$$

We can expand the expression on the right by expanding first with respect to i and then with respect to j:

$$\sum_{i=1}^{m} \left(\sum_{j=1}^{n} a_{ij} \right) = \sum_{j=1}^{n} a_{1j} + \sum_{j=1}^{n} a_{2j} + \cdots + \sum_{j=1}^{n} a_{mj}$$
$$= (a_{11} + a_{12} + \cdots + a_{1n}) + (a_{21} + a_{22} + \cdots + a_{2n})$$
$$+ \cdots + (a_{m1} + a_{m2} + \cdots + a_{mn}),$$

or we can expand first with respect to j and then with respect to i:

$$\sum_{i=1}^{m} \left(\sum_{j=1}^{n} a_{ij} \right) = \sum_{i=1}^{m} (a_{i1} + a_{i2} + \cdots + a_{in})$$
$$= (a_{11} + a_{12} + \cdots + a_{1n}) + (a_{21} + a_{22} + \cdots + a_{2n})$$
$$+ \cdots + (a_{m1} + a_{m2} + \cdots + a_{mn}).$$

The results are the same.

For example, we can write

$$\sum_{i=1}^{3} \left(\sum_{j=1}^{2} a_{ij} \right) = \sum_{j=1}^{2} a_{1j} + \sum_{j=1}^{2} a_{2j} + \sum_{j=1}^{2} a_{3j}$$
$$= (a_{11} + a_{12}) + (a_{21} + a_{22}) + (a_{31} + a_{32}),$$

or we can write

$$\sum_{i=1}^{3} \left(\sum_{j=1}^{2} a_{ij} \right) = \sum_{i=1}^{3} (a_{i1} + a_{i2}) = (a_{11} + a_{12}) + (a_{21} + a_{22}) + (a_{31} + a_{32}).$$

Since constants can be factored through single sums, they can also be factored through double sums; namely,

(16.1.3)

$$\sum_{i=1}^{m} \sum_{j=1}^{n} \alpha a_{ij} = \alpha \sum_{i=1}^{m} \sum_{j=1}^{n} a_{ij}.$$

Also,

(16.1.4)

$$\sum_{i=1}^{m} \sum_{j=1}^{n} (a_{ij} + b_{ij}) = \sum_{i=1}^{m} \sum_{j=1}^{n} a_{ij} + \sum_{i=1}^{m} \sum_{j=1}^{n} b_{ij}.$$

The easiest double sums to handle are those where each term a_{ij} appears as a product $b_i c_j$ in which each of the factors bears only one index. In that case, we can express the double sum as the product of two single sums:

(16.1.5)

$$\sum_{i=1}^{m} \sum_{j=1}^{n} b_i c_j = \left(\sum_{i=1}^{m} b_i \right) \left(\sum_{j=1}^{n} c_j \right).$$

PROOF Set

$$B = \sum_{i=1}^{m} b_i, \qquad C = \sum_{j=1}^{n} c_j.$$

Then

$$\sum_{i=1}^{m} \sum_{j=1}^{n} b_i c_j = \sum_{i=1}^{m} \left(\sum_{j=1}^{n} b_i c_j \right) \overset{\dagger}{=} \sum_{i=1}^{m} b_i \left(\sum_{j=1}^{n} c_j \right) = \sum_{i=1}^{m} b_i C$$

$$= C \sum_{i=1}^{m} b_i = CB = BC. \quad \square$$

For example,

$$\sum_{i=1}^{4} \sum_{j=1}^{2} 2^i 5^j = \left(\sum_{i=1}^{4} 2^i \right) \left(\sum_{j=1}^{2} 5^j \right) = (2 + 2^2 + 2^3 + 2^4)(5 + 5^2) = (30)(30) = 900.$$

Triple-sigma notation is used when three indices are involved. The sum of all the a_{ijk} where i ranges from 1 to m, j from 1 to n, and k from 1 to q can be written

(16.1.6)

$$\sum_{i=1}^{m} \sum_{j=1}^{n} \sum_{k=1}^{q} a_{ijk}.$$

---

† Since b_i is independent of j, it can be factored through the j-summation.

Multiple sums appear in the following sections. We introduced them here so as to avoid lengthy asides later.

EXERCISES 16.1

In Exercises 1–4, evaluate the given sum.

1. $\displaystyle\sum_{i=1}^{3}\sum_{j=1}^{3} 2^{i-1}3^{j+1}.$

2. $\displaystyle\sum_{i=1}^{4}\sum_{j=1}^{2} (1 + i)^{j}.$

3. $\displaystyle\sum_{i=1}^{4}\sum_{j=1}^{3} (i^2 + 3i)(j - 2).$

4. $\displaystyle\sum_{i=1}^{3}\sum_{j=1}^{2}\sum_{k=1}^{3} \frac{2i}{k + j^2}.$

In Exercises 5–16, let

$P_1 = \{x_0, x_1, \ldots, x_m\}$ be a partition of $[a_1, a_2]$,

$P_2 = \{y_0, y_1, \ldots, y_n\}$ be a partition of $[b_1, b_2]$,

$P_3 = \{z_0, z_1, \ldots, z_q\}$ be a partition of $[c_1, c_2]$,

and let

$$\Delta x_i = x_i - x_{i-1}, \quad \Delta y_j = y_j - y_{j-1}, \quad \Delta z_k = z_k - z_{k-1}.$$

Evaluate the following sums.

5. $\displaystyle\sum_{i=1}^{m} \Delta x_i.$

6. $\displaystyle\sum_{j=1}^{n} \Delta y_j.$

7. $\displaystyle\sum_{i=1}^{m}\sum_{j=1}^{n} \Delta x_i\,\Delta y_j.$

8. $\displaystyle\sum_{j=1}^{n}\sum_{k=1}^{q} \Delta y_j\,\Delta z_k.$

9. $\displaystyle\sum_{i=1}^{m} (x_i + x_{i-1})\Delta x_i.$

10. $\displaystyle\sum_{j=1}^{n} \tfrac{1}{2}(y_j^2 + y_j y_{j-1} + y_{j-1}^2)\Delta y_j.$

11. $\displaystyle\sum_{i=1}^{m}\sum_{j=1}^{n} (x_i + x_{i-1})\Delta x_i\,\Delta y_j.$

12. $\displaystyle\sum_{i=1}^{m}\sum_{j=1}^{n} (y_j + y_{j-1})\Delta x_i\,\Delta y_j.$

13. $\displaystyle\sum_{i=1}^{m}\sum_{j=1}^{n} (2\Delta x_i - 3\Delta y_j).$

14. $\displaystyle\sum_{i=1}^{m}\sum_{j=1}^{n} (3\Delta x_i - 2\Delta y_j).$

15. $\displaystyle\sum_{i=1}^{m}\sum_{j=1}^{n}\sum_{k=1}^{q} \Delta x_i\,\Delta y_j\,\Delta z_k.$

16. $\displaystyle\sum_{i=1}^{m}\sum_{j=1}^{n}\sum_{k=1}^{q} (x_i + x_{i-1})\Delta x_i\,\Delta y_j\,\Delta z_k.$

17. Evaluate

$$\sum_{i=1}^{n}\sum_{j=1}^{n}\sum_{k=1}^{n} \delta_{ijk}a_{ijk} \quad \text{where} \quad \delta_{ijk} = \begin{cases} 1, & \text{if } i = j = k \\ 0, & \text{otherwise.} \end{cases}$$

18. Show that any sum written in double-sigma notation can also be expressed in ordinary (that is, single) sigma notation.

19. Verify property (16.1.3).

20. Verify property (16.1.4).

■ 16.2 THE DOUBLE INTEGRAL OVER A RECTANGLE

We start with a function f continuous on a rectangle

$$R: \quad a \le x \le b, \quad c \le y \le d.$$

See Figure 16.2.1. Our object is to define the double integral of f over R:

$$\iint_R f(x, y)\,dx\,dy.$$

Recall that our approach to defining the definite integral of a function f of one variable,

$$\int_a^b f(x)\,dx,$$

involved some auxiliary notions, namely: partition P of $[a, b]$, upper sum $U_f(P)$, and lower sum $L_f(P)$ (see Section 5.2). We were then able to define

$$\int_a^b f(x)\,dx$$

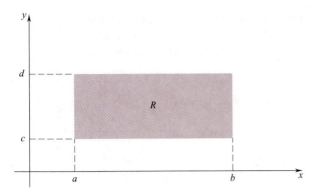

Figure 16.2.1

as the unique number I that satisfies the inequality

$$L_f(P) \le I \le U_f(P) \qquad \text{for all partitions } P \text{ of } [a, b].$$

We will follow exactly the same procedure to define the double integral

$$\iint_R f(x, y)\, dx\, dy.$$

First we explain what we mean by a partition of the rectangle R. To do this, we begin with a partition

$$P_1 = \{x_0, x_1, \ldots, x_m\} \quad \text{of} \quad [a, b]$$

and a partition

$$P_2 = \{y_0, y_1, \ldots, y_n\} \quad \text{of} \quad [c, d].$$

The set

$$P = P_1 \times P_2 = \{(x_i, y_j): x_i \in P_1, y_j \in P_2\}\dagger$$

is called a *partition of R* (see Figure 16.2.2); P consists of all the grid points (x_i, y_j).

Figure 16.2.2

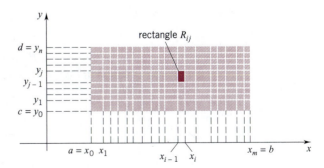

Figure 16.2.3

The partition P breaks up R into $m \times n$ nonoverlapping rectangles

$$R_{ij}: \quad x_{i-1} \le x \le x_i, \quad y_{j-1} \le y \le y_j, \qquad \text{(Figure 16.2.3)}$$

$\dagger$ $P_1 \times P_2$ is the Cartesian product of P_1 and P_2; if A and B are sets, then the *Cartesian product* of A and B is the set $A \times B = \{(a, b): a \in A \text{ and } b \in B\}$.

where $1 \leq i \leq m$, $1 \leq j \leq n$. On each rectangle R_{ij}, the function f takes on a maximum value M_{ij} and a minimum value m_{ij}. We know this because f is continuous and R_{ij} is closed and bounded (Section 15.6). The sum of all the products

$$M_{ij}(\text{area of } R_{ij}) = M_{ij}(x_i - x_{i-1})(y_j - y_{j-1}) = M_{ij}\, \Delta x_i\, \Delta y_j$$

is called the P *upper sum* for f:

(16.2.1)
$$U_f(P) = \sum_{i=1}^{m} \sum_{j=1}^{n} M_{ij}(\text{area of } R_{ij}) = \sum_{i=1}^{m} \sum_{j=1}^{n} M_{ij}\, \Delta x_i\, \Delta y_j.$$

The sum of all the products

$$m_{ij}(\text{area of } R_{ij}) = m_{ij}(x_i - x_{i-1})(y_j - y_{j-1}) = m_{ij}\, \Delta x_i\, \Delta y_j$$

is called the P *lower sum* for f:

(16.2.2)
$$L_f(P) = \sum_{i=1}^{m} \sum_{j=1}^{n} m_{ij}(\text{area of } R_{ij}) = \sum_{i=1}^{m} \sum_{j=1}^{n} m_{ij}\, \Delta x_i\, \Delta y_j.$$

Example 1 Consider the function

$$f(x, y) = x + y - 2$$

on the rectangle

$$R: \quad 1 \leq x \leq 4, \quad 1 \leq y \leq 3.$$

As a partition of $[1, 4]$ take

$$P_1 = \{1, 2, 3, 4\},$$

and as a partition of $[1, 3]$ take

$$P_2 = \{1, \tfrac{3}{2}, 3\}.$$

The partition $P = P_1 \times P_2$ then breaks up the initial rectangle into the six rectangles marked in Figure 16.2.4. On each rectangle R_{ij}, the function f takes on its maximum value M_{ij} at the point (x_i, y_j), the corner farthest from the origin:

$$M_{ij} = f(x_i, y_j) = x_i + y_j - 2.$$

Thus

$$U_f(P) = M_{11}(\text{area of } R_{11}) + M_{12}(\text{area of } R_{12}) + M_{21}(\text{area of } R_{21})$$
$$+ M_{22}(\text{area of } R_{22}) + M_{31}(\text{area of } R_{31}) + M_{32}(\text{area of } R_{32})$$
$$= \tfrac{3}{2}(\tfrac{1}{2}) + 3(\tfrac{3}{2}) + \tfrac{5}{2}(\tfrac{1}{2}) + 4(\tfrac{3}{2}) + \tfrac{7}{2}(\tfrac{1}{2}) + 5(\tfrac{3}{2}) = \tfrac{87}{4}.$$

On each rectangle R_{ij}, f takes on its minimum value m_{ij} at the point (x_{i-1}, y_{j-1}), the corner closest to the origin:

$$m_{ij} = f(x_{i-1}, y_{j-1}) = x_{i-1} + y_{j-1} - 2.$$

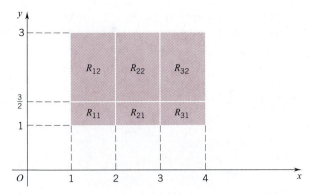

Figure 16.2.4

Thus

$$L_f(P) = m_{11}(\text{area of } R_{11}) + m_{12}(\text{area of } R_{12}) + m_{21}(\text{area of } R_{21})$$
$$+ m_{22}(\text{area of } R_{22}) + m_{31}(\text{area of } R_{31}) + m_{32}(\text{area of } R_{32})$$
$$= 0(\tfrac{1}{2}) + \tfrac{1}{2}(\tfrac{3}{2}) + 1(\tfrac{1}{2}) + \tfrac{3}{2}(\tfrac{3}{2}) + 2(\tfrac{1}{2}) + \tfrac{5}{2}(\tfrac{3}{2}) = \tfrac{33}{4}. \quad ❑$$

We return now to the general situation. As in the one-variable case, it can be shown that, if f is continuous, then there exists one and only one number I that satisfies the inequality

$$L_f(P) \le I \le U_f(P) \qquad \text{for all partitions } P \text{ of } R.$$

DEFINITION 16.2.3 THE DOUBLE INTEGRAL OVER A RECTANGLE R

Let f be continuous on a closed rectangle R. The unique number I that satisfies the inequality

$$L_f(P) \le I \le U_f(P) \qquad \text{for all partitions } P \text{ of } R$$

is called the *double integral* of f over R and is denoted by

$$\iint\limits_R f(x, y) \, dx \, dy.$$

The Double Integral as a Volume

If f is continuous and nonnegative on the rectangle R, the equation

$$z = f(x, y)$$

represents a surface that lies above R. (See Figure 16.2.5.) In this case the double integral

$$\iint\limits_R f(x, y) \, dx \, dy$$

Figure 16.2.5

gives the volume of the solid that is bounded below by R and bounded above by the surface $z = f(x, y)$.

Figure 16.2.6

To see this, consider a partition P of R. P breaks up R into subrectangles R_{ij} and thus the entire solid T into parts T_{ij}. Since T_{ij} contains a rectangular solid with base R_{ij} and height m_{ij} (Figure 16.2.6), we must have

$$m_{ij}(\text{area of } R_{ij}) \leq \text{volume of } T_{ij}.$$

Since T_{ij} is contained in a rectangular solid with base R_{ij} and height M_{ij} (Figure 16.2.7), we must have

$$\text{volume of } T_{ij} \leq M_{ij}(\text{area of } R_{ij}).$$

In short, for each pair of indices i and j, we must have

$$m_{ij}(\text{area of } R_{ij}) \leq \text{volume of } T_{ij} \leq M_{ij}(\text{area of } R_{ij}).$$

Adding up these inequalities, we can conclude that

$$L_f(P) \leq \text{volume of } T \leq U_f(P).$$

Figure 16.2.7

Since P is arbitrary, the volume of T must be the double integral:

(16.2.4)

$$\text{volume of } T = \iint_R f(x, y) \, dx \, dy.$$

The double integral

$$\iint_R 1 \, dx \, dy = \iint_R dx \, dy$$

gives the volume of a solid of constant height 1 erected over R. In square units this is just the area of R:

(16.2.5)

$$\text{area of } R = \iint_R dx \, dy.$$

Some Computations

Double integrals are generally computed by techniques that we will take up later. It is possible, however, to evaluate simple double integrals directly from the definition.

Example 2 Evaluate

$$\iint_R \alpha \, dx \, dy$$

where α is a constant and R is the rectangle

$$R: \quad a \le x \le b, \quad c \le y \le d.$$

SOLUTION Here

$$f(x, y) = \alpha \qquad \text{for all } (x, y) \in R.$$

We begin with $P_1 = \{x_0, x_1, \ldots, x_m\}$ as an arbitrary partition of $[a, b]$ and $P_2 = \{y_0, y_1, \ldots, y_n\}$ as an arbitrary partition of $[c, d]$. This gives

$$P = P_1 \times P_2 = \{(x_i, y_j) : x_i \in P_1, y_j \in P_2\}$$

as an arbitrary partition of R. On each rectangle R_{ij}, f has constant value α. This gives $M_{ij} = \alpha$ and $m_{ij} = \alpha$ throughout. Thus

$$U_f(P) = \sum_{i=1}^{m} \sum_{j=1}^{n} \alpha \, \Delta x_i \, \Delta y_j = \alpha \left(\sum_{i=1}^{m} \Delta x_i \right) \left(\sum_{j=1}^{n} \Delta y_j \right) = \alpha(b - a)(d - c).$$

Similarly

$$L_f(P) = \alpha(b - a)(d - c).$$

Since

$$L_f(P) \le \alpha(b - a)(d - c) \le U_f(P)$$

and P was chosen arbitrarily, the inequality must hold for all partitions P of R. This means that

$$\iint_R f(x, y) \, dx \, dy = \alpha(b - a)(d - c). \quad \square$$

Remark If $\alpha > 0$,

$$\iint_R \alpha \, dx \, dy = \alpha(b - a)(d - c)$$

gives the volume of the rectangular solid of constant height α erected over the rectangle R (Figure 16.2.8). ❑

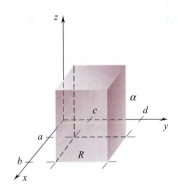

Figure 16.2.8

Example 3 Evaluate

$$\iint_R (y - 2x) \, dx \, dy$$

where R is the rectangle

$$R: \quad 1 \le x \le 2, \quad 3 \le y \le 5.$$

SOLUTION With

$$P_1 = \{x_0, x_1, \ldots, x_m\} \text{ as an arbitrary partition of } [1, 2]$$

and

$$P_2 = \{y_0, y_1, \ldots, y_n\} \text{ as an arbitrary partition of } [3, 5]$$

we have

$$P = P_1 \times P_2 = \{(x_i, y_j): x_i \in P_1, y_j \in P_2\}$$

as an arbitrary partition of R. On each rectangle

$$R_{ij}: \quad x_{i-1} \le x \le x_i, \quad y_{j-1} \le y \le y_j$$

the function

$$f(x, y) = y - 2x$$

has a maximum

$$M_{ij} = y_j - 2x_{i-1} \qquad (y \text{ is maximized and } x \text{ is minimized})$$

and a minimum

$$m_{ij} = y_{j-1} - 2x_i. \qquad (y \text{ is minimized and } x \text{ is maximized})$$

Thus,

$$U_f(P) = \sum_{i=1}^{m} \sum_{j=1}^{n} (y_j - 2x_{i-1}) \, \Delta x_i \, \Delta y_j \quad \text{and} \quad L_f(P) = \sum_{i=1}^{m} \sum_{j=1}^{n} (y_{j-1} - 2x_i) \, \Delta x_i \, \Delta y_j.$$

For each pair of indices i and j,

$$y_{j-1} - 2x_i \le \tfrac{1}{2}(y_j + y_{j-1}) - (x_i + x_{i-1}) \le y_j - 2x_{i-1}. \qquad \text{(explain)}$$

This means that for arbitrary P we have

$$L_f(P) \le \sum_{i=1}^{m} \sum_{j=1}^{n} [\tfrac{1}{2}(y_j + y_{j-1}) - (x_i + x_{i-1})] \, \Delta x_i \, \Delta y_j \le U_f(P).$$

The double sum in the middle of this set of inequalities can be written

$$\sum_{i=1}^{m} \sum_{j=1}^{n} \tfrac{1}{2}(y_j + y_{j-1}) \, \Delta x_i \, \Delta y_j - \sum_{i=1}^{m} \sum_{j=1}^{n} (x_i + x_{i-1}) \, \Delta x_i \, \Delta y_j.$$

The first double sum reduces to

$$\sum_{i=1}^{m} \sum_{j=1}^{n} \tfrac{1}{2} \Delta x_i (y_j^2 - y_{j-1}^2) = \tfrac{1}{2} \left(\sum_{i=1}^{m} \Delta x_i \right) \left(\sum_{j=1}^{n} (y_j^2 - y_{j-1}^2) \right)$$

$$= \tfrac{1}{2}(2 - 1)(25 - 9) = 8.$$

The second double sum reduces to

$$-\sum_{i=1}^{m} \sum_{j=1}^{n} (x_i^2 - x_{i-1}^2) \, \Delta y_j = -\left(\sum_{i=1}^{m} (x_i^2 - x_{i-1}^2) \right) \left(\sum_{j=1}^{n} \Delta y_j \right)$$

$$= -(4 - 1)(5 - 3) = -6.$$

The sum of these two numbers

$$I = 8 + (-6) = 2$$

satisfies the inequality

$$L_f(P) \le 2 \le U_f(P) \qquad \text{for arbitrary } P.$$

The integral is therefore 2:

$$\iint_R (y - 2x) \, dx \, dy = 2. \quad \square$$

Remark This last integral should not be interpreted as a volume. The expression $y - 2x$ does not keep a constant sign on

$$R: \quad 1 \le x \le 2, \quad 3 \le y \le 5. \quad \square$$

EXERCISES 16.2

For Exercises 1–3, take

$$f(x, y) = x + 2y \qquad \text{on} \qquad R: \quad 0 \le x \le 2, \quad 0 \le y \le 1,$$

and P as the partition $P = P_1 \times P_2$.

1. Find $L_f(P)$ and $U_f(P)$ if $P_1 = \{0, 1, \frac{3}{2}, 2\}$ and $P_2 = \{0, \frac{1}{2}, 1\}$.
2. Find $L_f(P)$ and $U_f(P)$ if $P_1 = \{0, \frac{1}{2}, 1, \frac{3}{2}, 2\}$ and $P_2 = \{0, \frac{1}{4}, \frac{1}{2}, \frac{3}{4}, 1\}$.
3. (a) Find $L_f(P)$ and $U_f(P)$ if

$$P_1 = \{x_0, x_1, \ldots, x_m\}$$

   is an arbitrary partition of $[0, 2]$,

$$P_2 = \{y_0, y_1, \ldots, y_n\}$$

   is an arbitrary partition of $[0, 1]$.
   (b) Use (a) to evaluate the double integral

$$\iint_R (x + 2y) \, dx \, dy$$

   and give a geometric interpretation to your answer.

For Exercises 4–6, take

$$f(x, y) = x - y \qquad \text{on} \qquad R: \quad 0 \le x \le 1, \quad 0 \le y \le 1,$$

and P as the partition $P = P_1 \times P_2$.

4. Find $L_f(P)$ and $U_f(P)$ if $P_1 = \{0, \frac{1}{2}, \frac{3}{4}, 1\}$ and $P_2 = \{0, \frac{1}{2}, 1\}$.
5. Find $L_f(P)$ and $U_f(P)$ if $P_1 = \{0, \frac{1}{4}, \frac{1}{2}, \frac{3}{4}, 1\}$ and $P_2 = \{0, \frac{1}{3}, \frac{2}{3}, 1\}$.

6. (a) Find $L_f(P)$ and $U_f(P)$ if

$$P_1 = \{x_0, x_1, \ldots, x_m\} \text{ and } P_2 = \{y_0, y_1, \ldots, y_n\}$$

   are arbitrary partitions of $[0, 1]$.
   (b) Use (a) to evaluate the double integral

$$\iint_R (x - y) \, dx \, dy.$$

For Exercises 7–9, take $R: 0 \le x \le b, 0 \le y \le d$ and let $P = P_1 \times P_2$ where

$$P_1 = \{x_0, x_1, \ldots, x_m\} \text{ is an arbitrary partition of } [0, b],$$

$$P_2 = \{y_0, y_1, \ldots, y_n\} \text{ is an arbitrary partition of } [0, d].$$

7. (a) Find $L_f(P)$ and $U_f(P)$ for $f(x, y) = 4xy$.
   (b) Calculate

$$\iint_R 4xy \, dx \, dy.$$

   HINT: $4x_{i-1}y_{j-1} \le (x_i + x_{i-1})(y_j + y_{j-1}) \le 4x_i y_j.$
8. (a) Find $L_f(P)$ and $U_f(P)$ for $f(x, y) = 3(x^2 + y^2)$.
   (b) Calculate

$$\iint_R 3(x^2 + y^2) \, dx \, dy.$$

   HINT: If $0 \le s \le t$, then $3s^2 \le t^2 + ts + s^2 \le 3t^2$.
9. (a) Find $L_f(P)$ and $U_f(P)$ for $f(x, y) = 3(x^2 - y^2)$.
   (b) Calculate

$$\iint_R 3(x^2 - y^2) \, dx \, dy.$$

10. Let $f = f(x, y)$ be continuous on the rectangle R: $a \le x \le b$, $c \le y \le d$. Suppose that $L_f(P) = U_f(P)$ for some partition P of R. What can you conclude about f? What is

$$\iint_R f(x, y) \, dx \, dy?$$

11. Let $f(x, y) = \sin(x + y)$ on R: $0 \le x \le 1$, $0 \le y \le 1$. Show that

$$0 \le \iint_R \sin(x + y) \, dx \, dy \le 1.$$

■ 16.3 THE DOUBLE INTEGRAL OVER A REGION

We start with a closed bounded region Ω in the xy-plane (Figure 16.3.1). We assume that Ω is a *basic region*. A *basic region* is a region whose boundary consists of a finite number of continuous arcs $y = \phi(x)$, $x = \psi(y)$. See Figure 16.3.2.

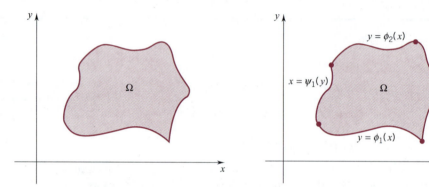

Figure 16.3.1   **Figure 16.3.2**

Now let f be a continuous function on Ω. We want to define the double integral

$$\iint_\Omega f(x, y) \, dx \, dy.$$

To do this, we enclose Ω by a rectangle R as in Figure 16.3.3. We now extend f to all of R by setting f equal to 0 outside Ω. This extended function, which we continue to call f, is bounded on R, and it is continuous on all of R except possibly at the boundary of Ω. In spite of these possible discontinuities, it can be shown that f is still integrable on R; that is, there still exists a unique number I such that

$$L_f(P) \le I \le U_f(P) \qquad \text{for all partitions } P \text{ of } R.$$

(We will not attempt to prove this, but an analogous result is given in Exercises 29 and 30, Section 5.2.) This number I is by definition the double integral

$$\iint_R f(x, y) \, dx \, dy.$$

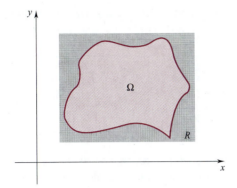

Figure 16.3.3

As you probably guessed by now, we define the double integral over Ω by setting

(16.3.1)

$$\iint\limits_{\Omega} f(x, y)\, dx\, dy = \iint\limits_{R} f(x, y)\, dx\, dy.$$

If f is continuous and nonnegative over Ω, the extended f is nonnegative on all of R. (Figure 16.3.4.) The double integral gives the volume of the solid whose upper boundary is the surface $z = f(x, y)$ and whose lower boundary is the rectangle R. But since the surface has height 0 outside of Ω, the volume outside Ω is 0. Therefore, it follows that

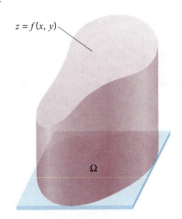

$$\iint\limits_{\Omega} f(x, y)\, dx\, dy$$

gives the volume of the solid T bounded above by $z = f(x, y)$ and below by Ω:

(16.3.2)

$$\text{volume of } T = \iint\limits_{\Omega} f(x, y)\, dx\, dy.$$

Figure 16.3.4

The double integral

$$\iint\limits_{\Omega} 1\, dx\, dy = \iint\limits_{\Omega} dx\, dy$$

gives the volume of a solid of constant height 1 over Ω. In square units this is the area of Ω:

(16.3.3)
$$\text{area of } \Omega = \iint_{\Omega} dx\, dy.$$

Below we list four elementary properties of the double integral. They are all analogous to what you saw in the one-variable case. The Ω referred to is a basic region. The functions f and g are assumed to be continuous on Ω.

I. Linearity: the double integral of a linear combination is the linear combination of the double integrals

$$\iint_{\Omega} [\alpha f(x, y) + \beta g(x, y)]\, dx\, dy = \alpha \iint_{\Omega} f(x, y)\, dx\, dy + \beta \iint_{\Omega} g(x, y)\, dx\, dy.$$

II. Order:

$$\text{if } f \geq 0 \text{ on } \Omega, \text{ then } \iint_{\Omega} f(x, y)\, dx\, dy \geq 0;$$

$$\text{if } f \leq g \text{ on } \Omega, \text{ then } \iint_{\Omega} f(x, y)\, dx\, dy \leq \iint_{\Omega} g(x, y)\, dx\, dy.$$

III. Additivity: if Ω is broken up into a finite number of non-overlapping basic regions $\Omega_1, \ldots \Omega_n$, then

$$\iint_{\Omega} f(x, y)\, dx\, dy = \iint_{\Omega_1} f(x, y)\, dx\, dy + \cdots + \iint_{\Omega_n} f(x, y)\, dx\, dy.$$

See, for example, Figure 16.3.5.

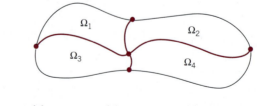

$$\iint_{\Omega} f(x, y)\, dxdy = \iint_{\Omega_1} f(x, y)\, dxdy + \iint_{\Omega_2} f(x, y)\, dxdy + \iint_{\Omega_3} f(x, y)\, dxdy + \iint_{\Omega_4} f(x, y)\, dxdy$$

Figure 16.3.5

IV. Mean-value condition: There is a point (x_0, y_0) in Ω for which

$$\iint_{\Omega} f(x, y)\, dx\, dy = f(x_0, y_0) \cdot (\text{area of } \Omega).$$

We call $f(x_0, y_0)$ *the average value of f on Ω.*

The notion of average given in (IV) enables us to write

(16.3.4)

$$\iint_\Omega f(x, y)\, dxdy = \left(\begin{array}{c}\text{the average value} \\ \text{of } f \text{ on } \Omega\end{array}\right) \cdot (\text{area of } \Omega).$$

This is a powerful, intuitive way of viewing the double integral. We will capitalize on it as we go on.

THEOREM 16.3.5 **MEAN-VALUE THEOREM FOR DOUBLE INTEGRALS**

Let f and g be continuous functions on the basic region Ω. If g is nonnegative on Ω, then there exists a point (x_0, y_0) in Ω for which

$$\iint_\Omega f(x, y)g(x, y)\, dx\, dy = f(x_0, y_0) \iint_\Omega g(x, y)\, dx\, dy\dagger$$

NOTE: The value $f(x_0, y_0)$ specified in this theorem is called the *g-weighted average of f on* Ω.

PROOF Since f is continuous on Ω and Ω is closed and bounded, we know that f takes on a minimum value m and a maximum value M. Since g is nonnegative on Ω,

$$mg(x, y) \le f(x, y)g(x, y) \le Mg(x, y) \qquad \text{for all } (x, y) \text{ in } \Omega.$$

Therefore (by property II)

$$\iint_\Omega mg(x, y)\, dx\, dy \le \iint_\Omega f(x, y)g(x, y)\, dx\, dy \le \iint_\Omega Mg(x, y)\, dx\, dy$$

and (by property I)

$$(*) \qquad m \iint_\Omega g(x, y)\, dx\, dy \le \iint_\Omega f(x, y)g(x, y)\, dx\, dy \le M \iint_\Omega g(x, y)\, dx\, dy.$$

We know that $\iint_\Omega g(x, y)\, dx\, dy \ge 0$ (again, by Property II). If $\iint_\Omega g(x, y)\, dx\, dy = 0$, then by (*) we have $\iint_\Omega f(x, y)g(x, y)\, dx\, dy = 0$ and the theorem holds for all choices of (x_0, y_0) in Ω. If $\iint_\Omega g(x, y)\, dx\, dy > 0$, then

$$m \le \frac{\iint_\Omega f(x, y)g(x, y)\, dx\, dy}{\iint_\Omega g(x, y)\, dx\, dy} \le M$$

---

† Property IV is this equation with g constantly 1.

and by the intermediate-value theorem (Theorem 15.3.5) there exists (x_0, y_0) in Ω for which

$$f(x_0, y_0) = \frac{\iint_\Omega f(x, y)g(x, y)\, dx\, dy}{\iint_\Omega g(x, y)\, dx\, dy}.$$

Obviously then

$$f(x_0, y_0) \iint_\Omega g(x, y)\, dx\, dy = \iint_\Omega f(x, y)g(x, y)\, dx\, dy. \quad \square$$

EXERCISES 16.3

1. Let $\phi = \phi(x)$ be a continuous, nonnegative function on $[a, b]$, and set

$$\Omega = \{(x, y): \ a \leq x \leq b, \ 0 \leq y \leq \phi(x)\}.$$

Compare

$$\iint_\Omega dx\, dy \quad \text{to} \quad \int_a^b \phi(x)\, dx.$$

2. Begin with a function f continuous on a closed bounded region Ω. Now surround Ω by a rectangle R as in Figure 16.3.3 and extend f to all of R by defining f to be 0 outside of Ω. Explain how the extended f can fail to be continuous on the boundary of Ω although the original function f, being continuous on all of Ω, was continuous on the boundary of Ω.

3. Suppose that f is continuous on a disc Ω with center at (x_0, y_0) and that

$$\iint_R f(x, y)\, dx\, dy = 0$$

for every rectangle R contained in Ω. Show that $f(x_0, y_0) = 0$.

4. Calculate the average value of $f(x, y) = x + 2y$ on the rectangle $R: 0 \leq x \leq 2, 0 \leq y \leq 1$. HINT: See Exercise 3, Section 16.2.

5. Calculate the average value of $f(x, y) = 4xy$ on the rectangle $R: 0 \leq x \leq 2, 0 \leq y \leq 3$. HINT: See Exercise 7, Section 16.2.

6. Calculate the average value of $f(x, y) = x^2 + y^2$ on the rectangle $R: 0 \leq x \leq b, 0 \leq y \leq d$. Hint: See Exercise 8, Section 16.2.

7. Let f be continuous on a closed bounded region Ω and let (x_0, y_0) be a point in the interior of Ω. Let D_r be a closed disc with center (x_0, y_0) and radius r. Show that

$$\lim_{r \to 0} \frac{1}{\pi r^2} \iint_{D_r} f(x, y)\, dx\, dy = f(x_0, y_0).$$

HINT: Use the mean-value theorem.

■ 16.4 THE EVALUATION OF DOUBLE INTEGRALS BY REPEATED INTEGRALS

The Reduction Formulas

If an integral

$$\int_a^b f(x)\, dx$$

proves difficult to evaluate, it is not because of the interval $[a, b]$ but because of the integrand f. Difficulty in evaluating a double integral

$$\iint_\Omega f(x, y)\, dx\, dy$$

can come from two sources: from the integrand f or from the base region Ω. Even such a simple looking integral as $\iint_\Omega 1 \, dx \, dy$ is difficult to evaluate if Ω is complicated.

In this section we introduce a technique for evaluating double integrals of continuous functions over regions that have special shapes, the shapes shown in Figure 16.4.1. In each case, the region Ω is a basic region and so we know that the double integral exists. The fundamental idea of this section is that double integrals over sets of these shapes can be reduced to a pair of ordinary integrals.

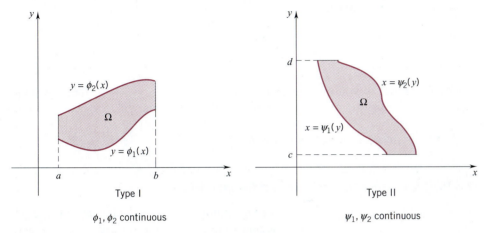

Figure 16.4.1

Type I Region. The *projection* of Ω onto the x-axis is a closed interval $[a, b]$ and Ω consists of all points (x, y) with

$$a \le x \le b \quad \text{and} \quad \phi_1(x) \le y \le \phi_2(x).$$

Then

(16.4.1)
$$\iint_\Omega f(x, y) \, dx \, dy = \int_a^b \left(\int_{\phi_1(x)}^{\phi_2(x)} f(x, y) \, dy \right) dx.$$

Here we first calculate

$$\int_{\phi_1(x)}^{\phi_2(x)} f(x, y) \, dy$$

by integrating $f(x, y)$ with respect to y from $y = \phi_1(x)$ to $y = \phi_2(x)$. The resulting expression is a function of x alone, which we then integrate with respect to x from $x = a$ to $x = b$.

Type II Region. The *projection* of Ω onto the y-axis is a closed interval $[c, d]$ and Ω consists of all points (x, y) with

$$c \le y \le d \quad \text{and} \quad \psi_1(y) \le x \le \psi_2(y).$$

Then

(16.4.2)
$$\iint_{\Omega} f(x, y)\, dx\, dy = \int_{c}^{d} \left(\int_{\psi_1(y)}^{\psi_2(y)} f(x, y)\, dx \right) dy.$$

This time we first calculate

$$\int_{\psi_1(y)}^{\psi_2(y)} f(x, y)\, dx$$

by integrating $f(x, y)$ with respect to x from $x = \psi_1(y)$ to $x = \psi_2(y)$. The resulting expression is a function of y alone, which we then integrate with respect to y from $y = c$ to $y = d$.

These formulas are easy to understand geometrically.

The Reduction Formulas Viewed Geometrically

Suppose that f is nonnegative and Ω is a region of type I. The double integral over Ω gives the volume of the solid T whose upper boundary is the surface $z = f(x, y)$ and whose lower boundary is the region Ω:

(1) $$\iint_{\Omega} f(x, y)\, dx\, dy = \text{volume of } T.$$ (Figure 16.4.2)

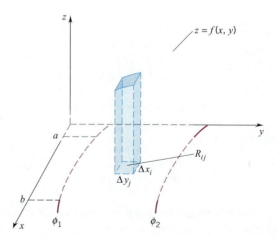

volume by double integration

$$V = \iint_{\Omega} f(x, y)\, dx\, dy$$

Figure 16.4.2

We can also compute the volume of T by the method of parallel cross sections (see Section 6.2). As in Figure 16.4.3, let $A(x)$ be the area of that cross section of T that has first coordinate x. Then by (6.2.1)

$$\int_a^b A(x)\, dx = \text{volume of } T.$$

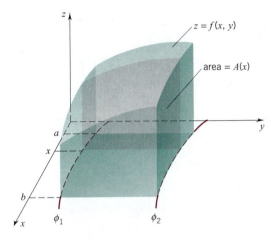

volume by parallel cross sections

$$V = \int_a^b A(x)\, dx = \int_a^b \left(\int_{\phi_1(x)}^{\phi_2(x)} f(x, y)\, dy \right) dx$$

Figure 16.4.3

Since

$$A(x) = \int_{\phi_1(x)}^{\phi_2(x)} f(x, y)\, dy,$$

we have

(2)
$$\int_a^b \left(\int_{\phi_1(x)}^{\phi_2(x)} f(x, y)\, dy \right) dx = \text{volume of } T.$$

Combining (1) with (2), we have the first reduction formula

$$\iint_\Omega f(x, y)\, dx\, dy = \int_a^b \left(\int_{\phi_1(x)}^{\phi_2(x)} f(x, y)\, dy \right) dx.$$

The other reduction formula can be obtained in a similar manner. ❏

Remark Note that our argument was a very loose one and certainly not a proof. How do we know, for example, that the "volume" obtained by double integration is the same as the "volume" obtained by the method of parallel cross sections? Intuitively it seems evident, but actually it is quite difficult to prove. ❏

Computations

Example 1 Evaluate

$$\iint_\Omega (x^4 - 2y)\, dx\, dy$$

with Ω as in Figure 16.4.4.

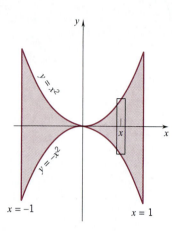

Figure 16.4.4

SOLUTION By projecting Ω onto the x-axis we obtain the interval $[-1, 1]$. The region Ω consists of all points (x, y) with

$$-1 \le x \le 1 \quad \text{and} \quad -x^2 \le y \le x^2.$$

This is a region of type I. By (16.4.1)

$$\iint_\Omega (x^4 - 2y)\, dx\, dy = \int_{-1}^1 \left(\int_{-x^2}^{x^2} [x^4 - 2y]\, dy \right) dx$$

$$= \int_{-1}^1 \left[x^4 y - y^2 \right]_{-x^2}^{x^2} dx$$

$$= \int_{-1}^1 [(x^6 - x^4) - (-x^6 - x^4)]\, dx$$

$$= \int_{-1}^1 2x^6\, dx = \left[\tfrac{2}{7} x^7 \right]_{-1}^1 = \tfrac{4}{7}. \quad \square$$

Example 2 Evaluate

$$\iint_\Omega (xy - y^3)\, dx\, dy$$

with Ω as in Figure 16.4.5.

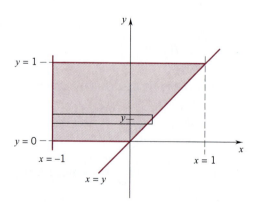

Figure 16.4.5

SOLUTION By projecting Ω onto the y-axis we obtain the interval $[0, 1]$. The region Ω consists of all points (x, y) with

$$0 \le y \le 1 \quad \text{and} \quad -1 \le x \le y.$$

This is a region of type II. By (16.4.2)

$$\iint_{\Omega} (xy - y^3)\, dx\, dy = \int_0^1 \left(\int_{-1}^y (xy - y^3)\, dx \right) dy$$

$$= \int_0^1 \left[\tfrac{1}{2}x^2 y - xy^3 \right]_{-1}^y dy$$

$$= \int_0^1 \left[(\tfrac{1}{2}y^3 - y^4) - (\tfrac{1}{2}y + y^3) \right] dy$$

$$= \int_0^1 (-\tfrac{1}{2}y^3 - y^4 - \tfrac{1}{2}y)\, dy$$

$$= \left[-\tfrac{1}{8}y^4 - \tfrac{1}{5}y^5 - \tfrac{1}{4}y^2 \right]_0^1 = -\tfrac{23}{40}.$$

We can also project Ω onto the x-axis and express Ω as a region of type I, but then the lower boundary is defined piecewise (see figure) and the calculations are somewhat more complicated: setting

$$\phi(x) = \begin{cases} 0, & -1 \le x \le 0 \\ x, & 0 \le x \le 1, \end{cases}$$

we have Ω as the set of all points (x, y) with

$$-1 \le x \le 1 \quad \text{and} \quad \phi(x) \le y \le 1;$$

thus

$$\iint_{\Omega} (xy - y^3)\, dx\, dy = \int_{-1}^1 \left(\int_{\phi(x)}^1 (xy - y^3)\, dy \right) dx$$

$$= \int_{-1}^0 \left(\int_{\phi(x)}^1 (xy - y^3)\, dy \right) dx$$

$$+ \int_0^1 \left(\int_{\phi(x)}^1 (xy - y^3)\, dy \right) dx$$

$$= \int_{-1}^0 \left(\int_0^1 (xy - y^3)\, dy \right) dx + \int_0^1 \left(\int_x^1 (xy - y^3)\, dy \right) dx$$

as you can check $\overset{\cdot}{=} (-\tfrac{1}{2}) + (-\tfrac{3}{40}) = -\tfrac{23}{40}.$ ❑

Repeated integrals

$$\int_a^b \left(\int_{\phi_1(x)}^{\phi_2(x)} f(x, y)\, dy \right) dx \quad \text{and} \quad \int_c^d \left(\int_{\psi_1(y)}^{\psi_2(y)} f(x, y)\, dx \right) dy$$

can be written in more compact form by omitting the large parentheses. From now on we will simply write

$$\int_a^b \int_{\phi_1(x)}^{\phi_2(x)} f(x, y)\, dy\, dx \quad \text{and} \quad \int_c^d \int_{\psi_1(y)}^{\psi_2(y)} f(x, y)\, dx\, dy.$$

Sometimes a region can be expressed either as a type I region: $a \leq x \leq b$, $\phi_1(x) \leq y \leq \phi_2(x)$, or as a type II region: $c \leq y \leq d$, $\psi_1(y) \leq x \leq \psi_2(y)$. Since

$$\iint\limits_{\Omega} f(x, y)\, dx\, dy = \int_a^b \int_{\phi_1(x)}^{\phi_2(x)} f(x, y)\, dy\, dx$$

and

$$\iint\limits_{\Omega} f(x, y)\, dx\, dy = \int_c^d \int_{\psi_1(y)}^{\psi_2(y)} f(x, y)\, dx\, dy,$$

it follows that

$$\int_a^b \int_{\phi_1(x)}^{\phi_2(x)} f(x, y)\, dy\, dx = \int_c^d \int_{\psi_1(y)}^{\psi_2(y)} f(x, y)\, dx\, dy.$$

Therefore we can, at least in theory, perform the integration in either order. However, there are instances where one order is preferable over the other. This is illustrated at the end of the section.

Example 3   Evaluate

$$\iint\limits_{\Omega} (x^{1/2} - y^2)\, dx\, dy$$

with Ω as in Figure 16.4.6.

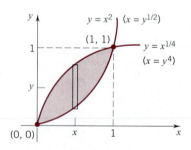

Figure 16.4.6

SOLUTION   The projection of Ω onto the x-axis is the closed interval $[0, 1]$, and Ω can be characterized as the set of all (x, y) with

$$0 \leq x \leq 1 \quad \text{and} \quad x^2 \leq y \leq x^{1/4}.$$

Thus

$$\iint\limits_{\Omega} (x^{1/2} - y^2)\, dx\, dy = \int_0^1 \int_{x^2}^{x^{1/4}} (x^{1/2} - y^2)\, dy\, dx$$

$$= \int_0^1 \left[x^{1/2} y - \tfrac{1}{3} y^3 \right]_{x^2}^{x^{1/4}} dx$$

$$= \int_0^1 (\tfrac{2}{3} x^{3/4} - x^{5/2} + \tfrac{1}{3} x^6)\, dx$$

$$= \left[\tfrac{8}{21} x^{7/4} - \tfrac{2}{7} x^{7/2} + \tfrac{1}{21} x^7 \right]_0^1 = \tfrac{8}{21} - \tfrac{2}{7} + \tfrac{1}{21} = \tfrac{1}{7}.$$

We can also integrate in the other order. The projection of Ω onto the y-axis is the closed interval $[0, 1]$, and Ω can be characterized as the set of all (x, y) with

$$0 \le y \le 1 \quad \text{and} \quad y^4 \le x \le y^{1/2}.$$

This gives the same result:

$$\iint_\Omega (x^{1/2} - y^2)\, dx\, dy = \int_0^1 \int_{y^4}^{y^{1/2}} (x^{1/2} - y^2)\, dx\, dy$$

$$= \int_0^1 \left[\tfrac{2}{3}x^{3/2} - y^2 x \right]_{y^4}^{y^{1/2}} dy$$

$$= \int_0^1 (\tfrac{2}{3}y^{3/4} - y^{5/2} + \tfrac{1}{3}y^6)\, dy$$

$$= \left[\tfrac{8}{21}y^{7/4} - \tfrac{2}{7}y^{7/2} + \tfrac{1}{21}y^7 \right]_0^1 = \tfrac{8}{21} - \tfrac{2}{7} + \tfrac{1}{21} = \tfrac{1}{7}. \quad \square$$

Example 4 Calculate by double integration the area of the region Ω that lies between

$$\sqrt{x} + \sqrt{y} = \sqrt{a} \quad \text{and} \quad x + y = a, \qquad a > 0.$$

SOLUTION The region Ω is pictured in Figure 16.4.7. Its area is given by the double integral

$$\iint_\Omega dx\, dy. \qquad \text{(by 16.3.3)}$$

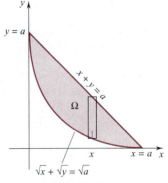

Figure 16.4.7

Here again we can integrate in either order. We can project Ω onto the x-axis and write the boundaries as functions of x:

$$y = (\sqrt{a} - \sqrt{x})^2 \quad \text{and} \quad y = a - x;$$

Ω is then the set of all (x, y) with

$$0 \le x \le a \quad \text{and} \quad (\sqrt{a} - \sqrt{x})^2 \le y \le a - x.$$

This gives

$$\iint_\Omega dx\, dy = \int_0^a \int_{(\sqrt{a}-\sqrt{x})^2}^{a-x} dy\, dx$$

$$= \int_0^a [(a - x) - (\sqrt{a} - \sqrt{x})^2]\, dx$$

$$= \int_0^a (-2x + 2\sqrt{a}\sqrt{x})\, dx = \left[-x^2 + \tfrac{4}{3}\sqrt{a}\, x^{3/2} \right]_0^a = -a^2 + \tfrac{4}{3}a^2 = \tfrac{1}{3}a^2.$$

We can also project Ω onto the y-axis and write the boundaries as functions of y:

$$x = (\sqrt{a} - \sqrt{y})^2 \quad \text{and} \quad x = a - y;$$

Ω is then the set of all (x, y) with

$$0 \leq y \leq a \quad \text{and} \quad (\sqrt{a} - \sqrt{y})^2 \leq x \leq a - y.$$

This gives

$$\iint_\Omega dx\, dy = \int_0^a \int_{(\sqrt{a} - \sqrt{y})^2}^{a-y} dx\, dy,$$

which is also $\frac{1}{3}a^2$. ❑

Symmetry in Double Integration

First we go back to the one-variable case (see Section 5.7). Let's suppose that g is continuous on an interval that is symmetric about the origin, say $[-a, a]$.

If g is odd, then $\displaystyle\int_{-a}^a g(x)\, dx = 0.$

If g is even, then $\displaystyle\int_{-a}^a g(x)\, dx = 2\int_0^a g(x)\, dx.$

We have similar results for double integrals.

Suppose that Ω is symmetric about the y-axis. If f is odd in x [if $f(-x, y) = -f(x, y)$], then

$$\iint_\Omega f(x, y)\, dx\, dy = 0.$$

If f is even in x [if $f(-x, y) = f(x, y)$], then

$$\iint_\Omega f(x, y)\, dx\, dy = 2 \iint_{\substack{\text{right half} \\ \text{of } \Omega}} f(x, y)\, dx\, dy.$$

Suppose that Ω is symmetric about the x-axis. If f is odd in y [if $f(x, -y) = -f(x, y)$], then

$$\iint_\Omega f(x, y)\, dx\, dy = 0.$$

If f is even in y [if $f(x, -y) = f(x, y)$], then

$$\iint_\Omega f(x, y)\, dx\, dy = 2 \iint_{\substack{\text{upper half} \\ \text{of } \Omega}} f(x, y)\, dx\, dy.$$

As an example take the region Ω of Figure 16.4.8. Suppose we wanted to calculate

$$\iint_\Omega (2x - \sin x^2 y)\, dx\, dy.$$

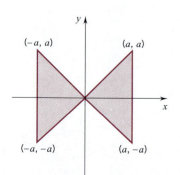

Figure 16.4.8

First,

$$\iint_{\Omega} (2x - \sin x^2 y)\, dx\, dy = \iint_{\Omega} 2x\, dx\, dy - \iint_{\Omega} \sin x^2 y\, dx\, dy.$$

Now, since $f(x, y) = 2x$ is odd in x, the symmetry of Ω about the y-axis gives

$$\iint_{\Omega} 2x\, dx\, dy = 0.$$

Since $g(x, y) = \sin x^2 y$ is odd in y, the symmetry of Ω about the x-axis gives

$$\iint_{\Omega} \sin x^2 y\, dx\, dy = 0.$$

Therefore

$$\iint_{\Omega} (2x - \sin x^2 y)\, dx\, dy = 0. \quad \square$$

As another example we go to Example 1 and reevaluate

$$\iint_{\Omega} (x^4 - 2y)\, dx\, dy,$$

this time capitalizing on the symmetry of Ω. Note that

$$\underbrace{\iint_{\Omega} 2y\, dx\, dy = 0}_{\text{symmetry about }x\text{-axis}} \quad \text{and} \quad \underbrace{\iint_{\Omega} x^4\, dx\, dy = 2}_{\text{symmetry about }y\text{-axis}} \underbrace{\iint_{\Omega} x^4\, dx\, dy}_{\substack{\text{right half}\\\text{of }\Omega}} = 4 \underbrace{\iint_{\Omega} x^4\, dx\, dy.}_{\substack{\text{upper part}\\\text{of right half}\\\text{of }\Omega}}$$

symmetry about x-axis   symmetry about y-axis   symmetry about x-axis

Therefore

$$\iint_{\Omega} (x^4 - 2y)\, dx\, dy = 4 \int_0^1 \int_0^{x^2} x^4\, dy\, dx = 4 \int_0^1 x^6\, dx = \tfrac{4}{7}. \quad \square$$

Example 5 Calculate the volume within the cylinder $x^2 + y^2 = b^2$ between the planes $y + z = a$ and $z = 0$ given that $a \geq b > 0$.

SOLUTION See Figure 16.4.9. The solid in question is bounded below by the disc

$$\Omega: \quad 0 \leq x^2 + y^2 \leq b^2$$

and above by the plane

$$z = a - y.$$

The volume is given by the double integral

$$\iint_{\Omega} (a - y)\, dx\, dy.$$

Figure 16.4.9

Since Ω is symmetric about the x-axis,

$$\iint_{\Omega} y \, dx \, dy = 0.$$

Thus

$$\iint_{\Omega} (a - y) \, dx \, dy = \iint_{\Omega} a \, dx \, dy = a \iint_{\Omega} dx \, dy = a \, (\text{area of } \Omega) = \pi a b^2. \quad \square$$

Concluding Remarks

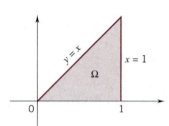

Figure 16.4.10

When two orders of integration are possible, one order may be easy to carry out, while the other may present serious difficulties. Take as an example the double integral

$$\iint_{\Omega} \cos \tfrac{1}{2} \pi x^2 \, dx \, dy$$

with Ω as in Figure 16.4.10. Projection onto the x-axis leads to

$$\int_0^1 \int_0^x \cos \tfrac{1}{2} \pi x^2 \, dy \, dx.$$

Projection onto the y-axis leads to

$$\int_0^1 \int_y^1 \cos \tfrac{1}{2} \pi x^2 \, dx \, dy.$$

The first double integral is easy to calculate:

$$\int_0^1 \int_0^x \cos \tfrac{1}{2} \pi x^2 \, dy \, dx = \int_0^1 \left(\int_0^x \cos \tfrac{1}{2} \pi x^2 \, dy \right) dx$$

$$= \int_0^1 \left[y \cos \tfrac{1}{2} \pi x^2 \right]_0^x dx = \int_0^1 x \cos \tfrac{1}{2} \pi x^2 \, dx$$

$$= \left[\tfrac{1}{\pi} \sin \tfrac{1}{2} \pi x^2 \right]_0^1 = \tfrac{1}{\pi},$$

but the second is not:

$$\int_0^1 \int_y^1 \cos \tfrac{1}{2} \pi x^2 \, dx \, dy = \int_0^1 \left(\int_y^1 \cos \tfrac{1}{2} \pi x^2 \, dx \right) dy,$$

and $\cos \tfrac{1}{2} \pi x^2$ does not have an elementary antiderivative. One possible way to handle the integral

$$\int_y^1 \cos \tfrac{1}{2} \pi x^2 \, dx$$

is to expand $\cos \tfrac{1}{2} \pi x^2$ as a series in x and then integrate term by term.

Finally, if Ω, the region of integration, is neither of type I nor of type II, it is often possible to break it up into a finite number of regions $\Omega_1, \ldots, \Omega_n$, each of type I or type II. (See Figure 16.4.11.) Since the double integral is additive,

$$\iint_{\Omega_1} f(x, y) \, dx \, dy + \cdots + \iint_{\Omega_n} f(x, y) \, dx \, dy = \iint_{\Omega} f(x, y) \, dx \, dy.$$

Each of the integrals on the left can be evaluated by the methods of this section.

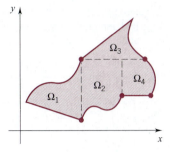

Figure 16.4.11

EXERCISES 16.4

In Exercises 1–3, evaluate taking Ω: $0 \le x \le 1$, $0 \le y \le 3$.

1. $\displaystyle\iint_{\Omega} x^2 \, dx \, dy.$

2. $\displaystyle\iint_{\Omega} e^{x+y} \, dx \, dy.$

3. $\displaystyle\iint_{\Omega} xy^2 \, dx \, dy.$

In Exercises 4–6, evaluate taking Ω: $0 \le x \le 1$, $0 \le y \le x$.

4. $\displaystyle\iint_{\Omega} x^3 y \, dx \, dy.$

5. $\displaystyle\iint_{\Omega} xy^3 \, dx \, dy.$

6. $\displaystyle\iint_{\Omega} x^2 y^2 \, dx \, dy.$

In Exercises 7–9, evaluate taking Ω: $0 \le x \le \frac{1}{2}\pi$, $0 \le y \le \frac{1}{2}\pi$.

7. $\displaystyle\iint_{\Omega} \sin(x + y) \, dx \, dy.$

8. $\displaystyle\iint_{\Omega} \cos(x + y) \, dx \, dy.$

9. $\displaystyle\iint_{\Omega} (1 + xy) \, dx \, dy.$

In Exercises 10–18, evaluate the double integral.

10. $\displaystyle\iint_{\Omega} (x + 3y^3) \, dx \, dy, \quad \Omega: \ 0 \le x^2 + y^2 \le 1.$

11. $\displaystyle\iint_{\Omega} \sqrt{xy} \, dx \, dy, \quad \Omega: \ 0 \le y \le 1, \ y^2 \le x \le y.$

12. $\displaystyle\iint_{\Omega} ye^x \, dx \, dy, \quad \Omega: \ 0 \le y \le 1, \ 0 \le x \le y^2.$

13. $\displaystyle\iint_{\Omega} (4 - y^2) \, dx \, dy, \quad \Omega$ the bounded region between $y^2 = 2x$ and $y^2 = 8 - 2x.$

14. $\displaystyle\iint_{\Omega} (x^4 + y^2) \, dx \, dy, \quad \Omega$ the bounded region between $y = x^3$ and $y = x^2.$

15. $\displaystyle\iint_{\Omega} (3xy^2 - y) \, dx \, dy, \quad \Omega$ the region between $y = |x|$ and $y = -|x|, \ x \in [-1, 1].$

16. $\displaystyle\iint_{\Omega} e^{-y^2/2} \, dx \, dy, \quad \Omega$ the triangle formed by the y-axis, $2y = x$, $y = 1.$

17. $\displaystyle\iint_{\Omega} e^{x^2} \, dx \, dy, \quad \Omega$ the triangle formed by the x-axis, $2y = x$, $x = 2.$

18. $\displaystyle\iint_{\Omega} (x + y) \, dx \, dy, \quad \Omega$ the region between $y = x^3$ and $y = x^4, \ x \in [-1, 1].$

In Exercises 19–24, sketch the region Ω that gives rise to the repeated integral and change the order of integration.

19. $\displaystyle\int_0^1 \int_{x^4}^{x^2} f(x, y) \, dy \, dx.$

20. $\displaystyle\int_0^1 \int_0^{y^2} f(x, y) \, dx \, dy.$

21. $\displaystyle\int_0^1 \int_{-y}^{y} f(x, y) \, dx \, dy.$

22. $\displaystyle\int_{1/2}^1 \int_{x^3}^{x} f(x, y) \, dy \, dx.$

23. $\displaystyle\int_1^4 \int_x^{2x} f(x, y) \, dy \, dx.$

24. $\displaystyle\int_1^3 \int_{-x}^{x^2} f(x, y) \, dy \, dx.$

In Exercises 25–28, calculate by double integration the area of the bounded region determined by the given pairs of curves.

25. $x^2 = 4y, \quad 2y - x - 4 = 0.$

26. $y = x, \quad x = 4y - y^2.$

27. $y = x, \quad 4y^3 = x^2.$

28. $x + y = 5, \quad xy = 6.$

In Exercises 29–32, sketch the region Ω that gives rise to the repeated integral, change the order of integration, and then evaluate.

29. $\displaystyle\int_0^1 \int_{\sqrt{x}}^1 \sin\left(\frac{y^3 + 1}{2}\right) dy\, dx.$

30. $\displaystyle\int_{-1}^0 \int_{-\sqrt{y+1}}^{\sqrt{y+1}} x^2\, dx\, dy.$ **31.** $\displaystyle\int_1^2 \int_0^{\ln y} e^{-x}\, dx\, dy.$

32. $\displaystyle\int_0^1 \int_{x^2}^1 \frac{x^3}{\sqrt{x^4 + y^2}} dy\, dx.$

33. Find the area of the first quadrant region bounded by $xy = 2, y = 1, y = x + 1$.

34. Find the volume of the solid bounded above by $z = x + y$ and below by the triangle with vertices $(0, 0, 0), (0, 1, 0), (1, 0, 0)$.

35. Find the volume of the solid bounded by $\frac{1}{2}x + \frac{1}{3}y + \frac{1}{4}z = 1$ and the coordinate planes.

36. Find the volume of the solid bounded above by the plane $z = 2x + 3y$ and below by the unit square $0 \le x \le 1, 0 \le y \le 1$.

37. Find the volume of the solid bounded above by $z = x^3 y$ and below by the triangle with vertices $(0, 0, 0), (2, 0, 0), (0, 1, 0)$.

38. Find the volume under the paraboloid $z = x^2 + y^2$ within the cylinder $x^2 + y^2 \le 1$.

39. Find the volume of the solid bounded above by the plane $z = 2x + 1$ and below by the disc $(x - 1)^2 + y^2 \le 1$.

40. Find the volume of the solid bounded above by $z = 4 - y^2 - \frac{1}{4}x^2$ and below by the disc $(y - 1)^2 + x^2 \le 1$.

41. Find the volume of the solid in the first octant bounded by $z = x^2 + y^2$, the plane $x + y = 1$ and the coordinate planes.

42. Find the volume of the solid bounded by the circular cylinder $x^2 + y^2 = 1$, the plane $z = 0$ and the plane $x + z = 1$.

43. Find the volume of the solid in the first octant bounded above by $z = x^2 + 3y^2$, below by the xy-plane, and on the sides by the cylinder $y = x^2$ and the plane $y = x$.

44. Find the volume of the solid bounded above by the surface $z = 1 + xy$ and below by the triangle with vertices $(1, 1), (4, 1)$, and $(3, 2)$.

45. Find the volume of the solid in the first octant bounded by the two cylinders $x^2 + y^2 = a^2, x^2 + z^2 = a^2$.

46. Find the volume of the tetrahedron bounded by the co-ordinate planes and the plane

$$\frac{x}{a} + \frac{y}{b} + \frac{z}{c} = 1, \qquad a, b, c > 0.$$

In Exercises 47–50, evaluate the double integral.

47. $\displaystyle\int_0^1 \int_y^1 e^{y/x}\, dx\, dy.$ **48.** $\displaystyle\int_0^1 \int_0^{\cos^{-1} y} e^{\sin x}\, dx\, dy.$

49. $\displaystyle\int_0^1 \int_x^1 x^2 e^{y^4}\, dy\, dx.$ **50.** $\displaystyle\int_0^1 \int_x^1 e^{y^2}\, dy\, dx.$

In Exercises 51–54, calculate the average value of f over the region Ω.

51. $f(x, y) = x^2 y; \qquad \Omega: \ -1 \le x \le 1, \quad 0 \le y \le 4.$

52. $f(x, y) = xy; \qquad \Omega: \ 0 \le x \le 1, \quad 0 \le y \le \sqrt{1 - x^2}.$

53. $f(x, y) = \dfrac{1}{xy}; \qquad \Omega: \ \ln 2 \le x \le 2 \ln 2,$

$\ln 2 \le y \le 2 \ln 2.$

54. $f(x, y) = e^{x+y}; \qquad \Omega: \ 0 \le x \le 1,$

$x - 1 \le y \le x + 1.$

55. (*Separated variables over a rectangle*) Let R be the rectangle $a \le x \le b, c \le y \le d$. Show that, if f is continuous on $[a, b]$ and g is continuous on $[c, d]$, then

(16.4.3)

$$\iint_R f(x)g(y)\, dx\, dy$$

$$= \left[\int_a^b f(x)\, dx\right] \cdot \left[\int_c^d g(y)\, dy\right].$$

56. Let R be a rectangle symmetric about the x-axis, sides parallel to the coordinate axes. Show that, if f is odd with respect to y, then the double integral of f over R is 0.

57. Let R be a rectangle symmetric about the y-axis, sides parallel to the coordinate axes. Show that, if f is odd with respect to x, then the double integral of f over R is 0.

58. Given that $f(-x, -y) = -f(x, y)$ for all (x, y) in Ω, what form of symmetry in Ω will ensure that the double integral of f over Ω is zero?

59. Let Ω be the triangle with vertices $(0, 0), (0, 1), (1, 1)$. Show that

$$\text{if } \int_0^1 f(x)\, dx = 0 \qquad \text{then} \qquad \iint_\Omega f(x)f(y)\, dx\, dy = 0.$$

60. (*Differentiation under the integral sign*) If f and $\partial f/\partial x$ are continuous, then the function

$$H(t) = \int_a^b \frac{\partial f}{\partial x}(t, y)\, dy$$

can be shown to be continuous. Use the identity

$$\int_0^x \int_a^b \frac{\partial f}{\partial x}(t, y) \, dy \, dt = \int_a^b \int_0^x \frac{\partial f}{\partial x}(t, y) \, dt \, dy$$

to verify that

$$\frac{d}{dx} \left[\int_a^b f(x, y) \, dy \right] = \int_a^b \frac{\partial f}{\partial x}(x, y) \, dy.$$

61. We integrate over regions of type I by means of the formula

$$\iint_\Omega f(x, y) \, dx \, dy = \int_a^b \int_{\phi_1(x)}^{\phi_2(x)} f(x, y) \, dy \, dx.$$

Here f is assumed to be continuous on Ω and ϕ_1, ϕ_2 are assumed to be continuous on $[a, b]$. Show that the function

$$F(x) = \int_{\phi_1(x)}^{\phi_2(x)} f(x, y) \, dy$$

is continuous on $[a, b]$.

■ 16.5 THE DOUBLE INTEGRAL AS THE LIMIT OF RIEMANN SUMS; POLAR COORDINATES

The Double Integral as the Limit of Riemann Sums

In the one-variable case we can write the integral as the limit of Riemann sums:

$$\int_a^b f(x) \, dx = \lim_{\max \Delta x_i \to 0} \sum_{i=1}^n f(x_i^*) \, \Delta x_i.$$

The same approach works with double integrals. To explain it we need to explain what we mean by the *diameter of a set*.

Suppose that S is a bounded closed set (on the line, in the plane, or in three-space). For any two points P and Q of S we can measure their separation, $d(P, Q)$ (the distance between P and Q). The maximum separation between points of S is called the *diameter of S*:

$$\text{diam } S = \max_{P, Q \in S} d(P, Q).$$

NOTE: The distance function $d(P, Q)$ is continuous on the bounded closed set S, and so it has an absolute maximum value.

For a circle, a circular disc, a sphere, or a ball, this sense of diameter agrees with the usual one. Figure 16.5.1 gives some other examples.

diam $= \Delta x_i$

(Length of the interval)

diam $= \sqrt{(\Delta x_i)^2 + (\Delta y_j)^2}$

(Length of the diagonal)

diam $= \sqrt{(\Delta x_i)^2 + (\Delta y_j)^2 + (\Delta z_k)^2}$

(Length of the diagonal)

Figure 16.5.1

Now let's start with a basic region Ω and decompose it into a finite number of basic subregions $\Omega_1, \ldots, \Omega_N$. (See Figure 16.5.2.) If f is continuous on Ω, then f is continuous on each Ω_i. Now from each Ω_i we pick an arbitrary point (x_i^*, y_i^*) and form the *Riemann sum*

$$\sum_{i=1}^{N} f(x_i^*, y_i^*)(\text{area of } \Omega_i).$$

As you would expect, the double integral over Ω can be obtained as the limit of such sums; namely, given any $\epsilon > 0$, there exists $\delta > 0$ such that, if the diameters of the Ω_i are all less than δ, then

$$\left| \sum_{i=1}^{N} f(x_i^*, y_i^*)(\text{area of } \Omega_i) - \iint_{\Omega} f(x, y) \, dx \, dy \right| < \epsilon$$

no matter how the (x_i^*, y_i^*) are chosen within the Ω_i. We express this by writing

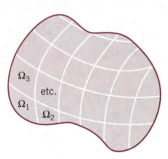

Figure 16.5.2

(16.5.1)
$$\iint_{\Omega} f(x, y) \, dx \, dy = \lim_{\text{diam } \Omega_i \to 0} \sum_{i=1}^{N} f(x_i^*, y_i^*)(\text{area of } \Omega_i).$$

Evaluating Double Integrals Using Polar Coordinates

Here we explain how to calculate double integrals

$$\iint_{\Omega} f(x, y) \, dx \, dy$$

using polar coordinates (r, θ). Throughout we take $r \geq 0$.

We will work with the type of region shown in Figure 16.5.3. The region Ω is then the set of all points (x, y) that have polar coordinates (r, θ) in the set

$$\Gamma: \quad \alpha \leq \theta \leq \beta, \quad \rho_1(\theta) \leq r \leq \rho_2(\theta),$$

where $\beta \leq \alpha + 2\pi$.

You already know how to calculate the area of Ω. By (9.6.2)

$$\text{area of } \Omega = \int_{\alpha}^{\beta} \tfrac{1}{2}[\rho_2^2(\theta) - \rho_1^2(\theta)] \, d\theta.$$

We can write this as a double integral over Γ:

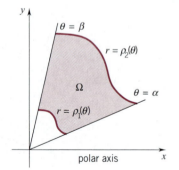

Figure 16.5.3

(16.5.2)
$$\text{area of } \Omega = \iint_{\Gamma} r \, dr \, d\theta.$$

PROOF Simply note that

$$\tfrac{1}{2}[\rho_2^2(\theta) - \rho_1^2(\theta)] = \int_{\rho_1(\theta)}^{\rho_2(\theta)} r \, dr$$

and therefore

$$\text{area of } \Omega = \int_\alpha^\beta \int_{\rho_1(\theta)}^{\rho_2(\theta)} r \, dr \, d\theta = \int\int_\Gamma r \, dr \, d\theta. \quad \square$$

Now let's suppose that f is some function continuous at each point (x, y) of Ω. Then the composition

$$F(r, \theta) = f(r \cos \theta, r \sin \theta)$$

is continuous at each point (r, θ) of Γ. We will show that

(16.5.3)
$$\int\int_\Omega f(x, y) \, dx \, dy = \int\int_\Gamma f(r \cos \theta, r \sin \theta) r \, dr \, d\theta. \qquad \text{(note the extra } r\text{)}$$

PROOF Our first step is to place a grid on Ω by using a finite number of rays $\theta = \theta_j$ and a finite number of continuous curves $r = \rho_k(\theta)$ in the manner of Figure 16.5.4. This grid decomposes Ω into a finite number of regions

$$\Omega_1, \ldots \Omega_N$$

with polar coordinates in sets $\Gamma_1, \ldots, \Gamma_N$. Note that by (16.5.2)

$$\text{area of each } \Omega_i = \int\int_{\Gamma_i} r \, dr \, d\theta.$$

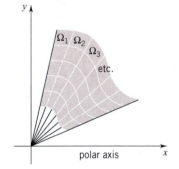

Figure 16.5.4

Writing $F(r, \theta)$ for $f(r \cos \theta, r \sin \theta)$, we have

$$\int\int_\Gamma F(r, \theta) r \, dr \, d\theta = \sum_{i=1}^N \int\int_{\Gamma_i} F(r, \theta) r \, dr \, d\theta$$

additivity ⎯⎯⎯⎯⎯⎯⎯⎯⎯⎯⎯⎯⎯⎯⎯↑

for some $(r_i^*, \theta_i^*) \in \Gamma_i$
(Theorem 16.3.5) ⎯⎯⎯⎯⎯⎯↑
$$= \sum_{i=1}^N F(r_i^*, \theta_i^*) \int\int_{\Gamma_i} r \, dr \, d\theta$$

$$= \sum_{i=1}^N F(r_i^*, \theta_i^*)(\text{area of } \Omega_i)$$

(with $x_i^* = r_i^* \cos \theta_i^*, y_i^* = r_i^* \sin \theta_i^*$) ⎯⎯⎯↓
$$= \sum_{i=1}^N f(x_i^*, y_i^*)(\text{area of } \Omega_i).$$

This last expression is a Riemann sum for the double integral

$$\int\int_\Omega f(x, y) \, dx \, dy$$

and, as such, by (16.5.1), differs from that integral by less than any preassigned positive ϵ provided only that the diameters of all the Ω_i are sufficiently small. This we can guarantee by making our grid sufficiently fine. ❑

Example 1 Use polar coordinates to evaluate

$$\iint_{\Omega} xy \, dx \, dy,$$

where Ω is the portion of the unit disc that lies in the first quadrant.

SOLUTION Here

$$\Gamma: \quad 0 \le \theta \le \tfrac{1}{2}\pi, \quad 0 \le r \le 1.$$

Therefore

$$\iint_{\Omega} xy \, dx \, dy = \iint_{\Gamma} (r \cos \theta)(r \sin \theta) r \, dr \, d\theta$$

$$= \int_{0}^{\pi/2} \int_{0}^{1} r^3 \cos \theta \sin \theta \, dr \, d\theta = \tfrac{1}{8}. \quad \square$$

check this ⟶

Example 2 Use polar coordinates to calculate the volume of a sphere of radius R.

SOLUTION In rectangular coordinates

$$V = 2 \iint_{\Omega} \sqrt{R^2 - (x^2 + y^2)} \, dx \, dy$$

where Ω is the disc of radius R centered at the origin. (Verify.) In polar coordinates, the disc of radius R centered at the origin is given by

$$\Gamma: \quad 0 \le \theta \le 2\pi, \quad 0 \le r \le R.$$

Therefore

$$V = 2 \iint_{\Gamma} \sqrt{R^2 - r^2} \, r \, dr \, d\theta = 2 \int_{0}^{2\pi} \int_{0}^{R} \sqrt{R^2 - r^2} \, r \, dr \, d\theta = \tfrac{4}{3}\pi R^3. \quad \square$$

check this ⟶

Example 3 Calculate the volume of the solid bounded above by the cone $z = 2 - \sqrt{x^2 + y^2}$ and bounded below by the disc $\Omega: (x - 1)^2 + y^2 \le 1$. (See Figure 16.5.5.)

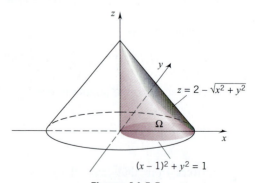

$z = 2 - \sqrt{x^2 + y^2}$

Ω

$(x - 1)^2 + y^2 = 1$

Figure 16.5.5

SOLUTION

$$V = \iint_\Omega (2 - \sqrt{x^2 + y^2}) \, dx \, dy$$

$$= 2 \iint_\Omega dx \, dy - \iint_\Omega \sqrt{x^2 + y^2} \, dx \, dy.$$

The first integral is $2\times$ (area of Ω) $= 2\pi$. We evaluate the second integral by changing to polar coordinates.

The equation $(x - 1)^2 + y^2 = 1$ simplifies to $x^2 + y^2 = 2x$. In polar coordinates this becomes $r^2 = 2r \cos\theta$, which simplifies to $r = 2 \cos\theta$. The disc Ω is the set of all points with polar coordinates in the set

$$\Gamma: \quad -\tfrac{1}{2}\pi \le \theta \le \tfrac{1}{2}\pi, \quad 0 \le r \le 2 \cos\theta.$$

Therefore

$$\iint_\Omega \sqrt{x^2 + y^2} \, dx \, dy = \iint_\Gamma r^2 \, dr \, d\theta = \int_{-\pi/2}^{\pi/2} \int_0^{2\cos\theta} r^2 \, dr \, d\theta = \tfrac{32}{9}.$$

check this ⟶

We then have $V = 2\pi - \tfrac{32}{9} \cong 2.73$. ❏

Example 4 Evaluate

$$\iint_\Omega \frac{1}{(1 + x^2 + y^2)^{3/2}} \, dx \, dy$$

where Ω is the triangle of Figure 16.5.6.

SOLUTION The vertical side of the triangle is part of the line $x = 1$. In polar coordinates this is $r \cos\theta = 1$, which can be written $r = \sec\theta$. Therefore

$$\iint_\Omega \frac{1}{(1 + x^2 + y^2)^{3/2}} \, dx \, dy = \iint_\Gamma \frac{r}{(1 + r^2)^{3/2}} \, dr \, d\theta$$

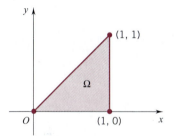

Figure 16.5.6

where

$$\Gamma: \quad 0 \le \theta \le \pi/4, \quad 0 \le r \le \sec\theta. \qquad \text{(Figure 16.5.7)}$$

The double integral over Γ reduces to

$$\int_0^{\pi/4} \int_0^{\sec\theta} \frac{r}{(1 + r^2)^{3/2}} \, dr \, d\theta = \int_0^{\pi/4} \left[\frac{1}{\sqrt{1 + r^2}} \right]_{\sec\theta}^0 \, d\theta$$

$$= \int_0^{\pi/4} \left(1 - \frac{1}{\sqrt{1 + \sec^2\theta}} \right) d\theta.$$

For $\theta \in [0, \pi/4]$

$$\frac{1}{\sqrt{1 + \sec^2\theta}} = \frac{\cos\theta}{\sqrt{\cos^2\theta + 1}} = \frac{\cos\theta}{\sqrt{2 - \sin^2\theta}}.$$

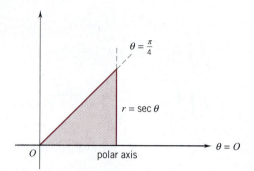

Figure 16.5.7

Therefore the integral can be written

$$\int_0^{\pi/4} \left(1 - \frac{\cos \theta}{\sqrt{2 - \sin^2 \theta}} \right) d\theta = \left[\theta - \sin^{-1} \left(\frac{\sin \theta}{\sqrt{2}} \right) \right]_0^{\pi/4} = \frac{\pi}{4} - \frac{\pi}{6} = \frac{\pi}{12}. \quad \square$$

(7.8.4) ──────↑

The function $f(x) = e^{-x^2}$ has no elementary antiderivative. Nevertheless, by taking a circuitous route and then using polar coordinates, we can show that

(16.5.4)
$$\int_{-\infty}^{\infty} e^{-x^2} \, dx = \sqrt{\pi}. \quad †$$

PROOF The circular disc $D_b: x^2 + y^2 \leq b^2$ is the set of all (x, y) with polar coordinates (r, θ) in the set $\Gamma: 0 \leq \theta \leq 2\pi, 0 \leq r \leq b$. Therefore

$$\iint_{D_b} e^{-(x^2 + y^2)} \, dx \, dy = \iint_{\Gamma} e^{-r^2} r \, dr \, d\theta = \int_0^{2\pi} \int_0^b e^{-r^2} r \, dr \, d\theta$$

$$= \int_0^{2\pi} \tfrac{1}{2}(1 - e^{-b^2}) \, d\theta = \pi(1 - e^{-b^2}).$$

Let S_a be the square $-a \leq x \leq a, -a \leq y \leq a$. Since $D_a \subseteq S_a \subseteq D_{2a}$ and $e^{-(x^2 + y^2)}$ is positive,

$$\iint_{D_a} e^{-(x^2 + y^2)} \, dx \, dy \leq \iint_{S_a} e^{-(x^2 + y^2)} \, dx \, dy \leq \iint_{D_{2a}} e^{-(x^2 + y^2)} \, dx \, dy.$$

It follows that

$$\pi(1 - e^{-a^2}) \leq \iint_{S_a} e^{-(x^2 + y^2)} \, dx \, dy \leq \pi(1 - e^{-4a^2}).$$

---

† This integral comes up frequently in probability theory and plays an important role in the branch of physics called "statistical mechanics."

As $a \to \infty$, $\pi(1 - e^{-a^2}) \to \pi$ and $\pi(1 - e^{-4a^2}) \to \pi$. Therefore

$$\lim_{a \to \infty} \iint_{S_a} e^{-(x^2 + y^2)} \, dx \, dy = \pi.$$

But

$$\iint_{S_a} e^{-(x^2 + y^2)} \, dx \, dy = \int_{-a}^{a} \int_{-a}^{a} e^{-(x^2 + y^2)} \, dx \, dy$$

$$= \int_{-a}^{a} \int_{-a}^{a} e^{-x^2} \cdot e^{-y^2} \, dx \, dy$$

$$= \left(\int_{-a}^{a} e^{-x^2} \, dx \right) \left(\int_{-a}^{a} e^{-y^2} \, dy \right) = \left(\int_{-a}^{a} e^{-x^2} \, dx \right)^2.$$

Therefore

$$\lim_{a \to \infty} \int_{-a}^{a} e^{-x^2} \, dx = \lim_{a \to \infty} \left(\iint_{S_a} e^{-(x^2 + y^2)} \, dx \, dy \right)^{1/2} = \sqrt{\pi}. \quad \square$$

EXERCISES 16.5

In Exercises 1–4, evaluate the iterated integral.

1. $\displaystyle \int_{0}^{\pi/2} \int_{0}^{\sin \theta} r \cos \theta \, dr \, d\theta.$

2. $\displaystyle \int_{0}^{\pi/4} \int_{0}^{\cos 2\theta} r \, dr \, d\theta.$

3. $\displaystyle \int_{0}^{\pi/2} \int_{0}^{3 \sin \theta} r^2 \, dr \, d\theta.$

4. $\displaystyle \int_{-\pi/3}^{2\pi/3} \int_{0}^{2 \cos \theta} r \sin \theta \, dr \, d\theta.$

5. Integrate $f(x, y) = \cos(x^2 + y^2)$ over:
(a) the closed unit disc;
(b) the annular region $1 \le x^2 + y^2 \le 4$.

6. Integrate $f(x, y) = \sin(\sqrt{x^2 + y^2})$ over:
(a) the closed unit disc;
(b) the annular region $1 \le x^2 + y^2 \le 4$.

7. Integrate $f(x, y) = x + y$ over:
(a) $0 \le x^2 + y^2 \le 1, x \ge 0, y \ge 0$;
(b) $1 \le x^2 + y^2 \le 4, x \ge 0, y \ge 0$.

8. Integrate $f(x, y) = \sqrt{x^2 + y^2}$ over the triangle with vertices $(0, 0)$, $(1, 0)$, $(1, \sqrt{3})$.

In Exercises 9–16, calculate by changing to polar coordinates.

9. $\displaystyle \int_{-1}^{1} \int_{0}^{\sqrt{1 - y^2}} \sqrt{x^2 + y^2} \, dx \, dy.$

10. $\displaystyle \int_{0}^{2} \int_{0}^{\sqrt{4 - x^2}} \sqrt{x^2 + y^2} \, dy \, dx.$

11. $\displaystyle \int_{1/2}^{1} \int_{0}^{\sqrt{1 - x^2}} (x^2 + y^2)^{3/2} \, dy \, dx.$

12. $\displaystyle \int_{0}^{1/2} \int_{0}^{\sqrt{1 - x^2}} xy\sqrt{x^2 + y^2} \, dy \, dx.$

13. $\displaystyle \int_{0}^{1} \int_{0}^{\sqrt{1 - x^2}} \sin(x^2 + y^2) \, dy \, dx.$

14. $\displaystyle \int_{-1}^{1} \int_{-\sqrt{1 - y^2}}^{\sqrt{1 - y^2}} e^{-(x^2 + y^2)} \, dx \, dy.$

15. $\displaystyle \int_{0}^{2} \int_{0}^{\sqrt{2x - x^2}} x \, dy \, dx.$

16. $\displaystyle \int_{0}^{1} \int_{-\sqrt{x - x^2}}^{\sqrt{x - x^2}} (x^2 + y^2) \, dy \, dx.$

In Exercises 17–22, use double integrals to find the area of the given region.

17. One leaf of the petal curve $r = 3 \sin 3\theta$.

18. The region enclosed by the cardioid $r = 2(1 - \cos \theta)$.

19. The region inside the circle $r = 4 \cos \theta$ and outside the circle $r = 2$.

20. The region inside the large loop and outside the small loop of the limaçon $r = 1 + 2 \cos \theta$.

21. The region enclosed by the lemniscate $r^2 = 4 \cos 2\theta$.

22. The region inside the circle $r = 3 \cos \theta$ and outside the cardioid $r = 1 + \cos \theta$.

23. Find the volume of the solid bounded above by the plane $z = y + b$, below by the xy-plane, and on the sides by the circular cylinder $x^2 + y^2 = b^2$.

24. Find the volume of the solid bounded below by the xy-plane and above by the paraboloid $z = 1 - (x^2 + y^2)$.

25. Find the volume of the ellipsoid $x^2/4 + y^2/4 + z^2/3 = 1$.

26. Find the volume of the solid bounded below by the xy-plane and above by the surface $x^2 + y^2 + z^6 = 5$.

27. Find the volume of the solid bounded below by the xy-plane, above by the spherical surface $x^2 + y^2 + z^2 = 4$, and on the sides by the cylinder $x^2 + y^2 = 1$.

28. Find the volume of the solid bounded above by $z = 1 - (x^2 + y^2)$, below by the xy-plane, and on the sides by the cylinder $x^2 + y^2 - x = 0$.

29. Find the volume of the solid bounded above by the plane $z = 2x$ and below by the disc $(x - 1)^2 + y^2 \le 1$.

30. Find the volume of the solid bounded above by the cone $z^2 = x^2 + y^2$ and below by the region Ω which lies inside the curve $x^2 + y^2 = 2ax$.

31. Find the volume of the solid bounded below by the xy-plane, above by the ellipsoid of revolution $b^2x^2 + b^2y^2 + a^2z^2 = a^2b^2$, and on the sides by the cylinder $x^2 + y^2 - ay = 0$.

32. A cylindrical hole of radius r is drilled through the center of a sphere of radius R, $0 < r < R$.
    (a) Determine the volume of the material that has been removed from the sphere.
    (b) Determine the volume of the ring-shaped solid that remains.

■ 16.6 SOME APPLICATIONS OF DOUBLE INTEGRATION

The Mass of a Plate

Suppose that a thin distribution of matter, called a *plate*, is laid out in the xy-plane in the form of a basic region Ω. If the mass density of the plate (the mass per unit area) is a constant λ, then the total mass M of the plate is simply the density λ times the area of the plate:

$$M = \lambda \times \text{the area of } \Omega.$$

If the density varies continuously from point to point, say $\lambda = \lambda(x, y)$, then the mass of the plate is the average density of the plate times the area of the plate:

$$M = \text{average density} \times \text{the area of } \Omega.$$

This is an integral:

(16.6.1)
$$M = \iint_\Omega \lambda(x, y) \, dx \, dy.$$

The Center of Mass of a Plate

The center of mass of a rod x_M is a density-weighted average taken over the interval occupied by the rod:

$$x_M M = \int_a^b x\lambda(x) \, dx. \qquad \text{[This you have seen: (5.9.5).]}$$

The center of mass of a plate (x_M, y_M) is determined by two density-weighted averages, each taken over the region occupied by the plate:

(16.6.2)
$$x_M M = \iint\limits_{\Omega} x\lambda(x, y)\, dx\, dy, \qquad y_M M = \iint\limits_{\Omega} y\lambda(x, y)\, dx\, dy.$$

Example 1 A plate is in the form of a half-disc of radius a. Find the mass of the plate and the center of mass given that the mass density of the plate is directly proportional to the distance from the midpoint of the straight edge of the plate.

SOLUTION Place the plate over the region Ω: $-a \le x \le a,\ 0 \le y \le \sqrt{a^2 - x^2}$. See Figure 16.6.1. The mass density can then be written $\lambda(x, y) = k\sqrt{x^2 + y^2}$, where $k > 0$ is the constant of proportionality. Now

$$M = \iint\limits_{\Omega} k\sqrt{x^2 + y^2}\, dx\, dy = \int_0^\pi \int_0^a (kr) r\, dr\, d\theta = k\left(\int_0^\pi 1\, d\theta\right)\left(\int_0^a r^2\, dr\right)$$

change to polar coordinates

$$= k(\pi)(\tfrac{1}{3} a^3) = \tfrac{1}{3} ka^3 \pi,$$

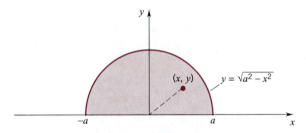

Figure 16.6.1

and

$$x_M M = \iint\limits_{\Omega} x(k\sqrt{x^2 + y^2})\, dx\, dy = 0$$

(Ω is symmetric with respect to the y-axis and the integrand is odd with respect to x). Thus $x_M = 0$. (This result is also clear from the symmetry of the plate.) Also,

$$y_M M = \iint\limits_{\Omega} y(k\sqrt{x^2 + y^2})\, dx\, dy = \int_0^\pi \int_0^a (r\sin\theta)(kr) r\, dr\, d\theta$$

$$= k\left(\int_0^\pi \sin\theta\, d\theta\right)\left(\int_0^a r^3\, dr\right)$$

$$= k(2)(\tfrac{1}{4} a^4) = \tfrac{1}{2} ka^4.$$

Since $M = \tfrac{1}{3} ka^3 \pi$, we have $y_M = (\tfrac{1}{2} ka^4)/(\tfrac{1}{3} ka^3 \pi) = 3a/2\pi$ Thus, the center of mass of the plate is the point $(0, 3a/2\pi) \cong (0, 0.48a)$. ❑

Centroids

If the plate is homogeneous, then the mass density λ is constantly M/A where A is the area of the base region Ω. In this case the center of mass of the plate falls on the *centroid* of the base region (a notion with which you are already familiar). The centroid $(\bar{x}, \bar{y})$ depends only on the geometry of Ω:

$$\bar{x}M = \iint\limits_{\Omega} x(M/A)\, dx\, dy = (M/A) \iint\limits_{\Omega} x\, dx\, dy,$$

$$\bar{y}M = \iint\limits_{\Omega} y(M/A)\, dx\, dy = (M/A) \iint\limits_{\Omega} y\, dx\, dy.$$

Dividing by M and multiplying through by A we have

(16.6.3)
$$\bar{x}A = \iint\limits_{\Omega} x\, dx\, dy, \qquad \bar{y}A = \iint\limits_{\Omega} y\, dx\, dy.$$

Thus $\bar{x}$ is the average x-coordinate on Ω and $\bar{y}$ is the average y-coordinate on Ω. The mass of the plate does not enter into this at all.

Example 2 Find the centroid of the region

$$\Omega: \quad a \leq x \leq b, \quad \phi_1(x) \leq y \leq \phi_2(x). \qquad \text{(Figure 16.6.2.)}$$

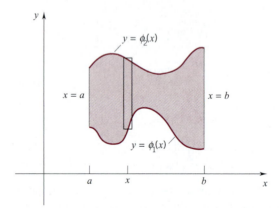

Figure 16.6.2

SOLUTION

$$\bar{x}A = \iint\limits_{\Omega} x\, dx\, dy = \int_a^b \int_{\phi_1(x)}^{\phi_2(x)} x\, dy\, dx = \int_a^b x[\phi_2(x) - \phi_1(x)]\, dx;$$

$$\bar{y}A = \iint\limits_{\Omega} y\, dx\, dy = \int_a^b \int_{\phi_1(x)}^{\phi_2(x)} y\, dy\, dx = \int_a^b \tfrac{1}{2}([\phi_2(x)]^2 - [\phi_1(x)]^2)\, dx.$$

These are the formulas for the centroid that we developed in Section 6.4. Having calculated many centroids at that time, we won't do so here. ❏

Kinetic Energy and Moment of Inertia

A particle of mass m at a distance r from a given line rotates about that line (called the *axis of rotation*) with angular speed ω. The speed v of the particle is then $r\omega$ and the kinetic energy is given by the formula

$$\text{KE} = \tfrac{1}{2}mv^2 = \tfrac{1}{2}mr^2\omega^2.$$

Imagine now a rigid body composed of a finite number of point masses m_i located at distances r_i from some fixed line. If the rigid body rotates about that line with angular speed ω, then all the point masses rotate about that same line with that same angular speed ω. The kinetic energy of the body can be obtained by adding up the kinetic energies of all the individual particles:

$$\text{KE} = \sum_i \tfrac{1}{2}m_i r_i^2 \omega^2 = \tfrac{1}{2}\left(\sum_i m_i r_i^2\right)\omega^2.$$

The expression in parentheses is called the *moment of inertia* (or *rotational inertia*) of the body and is denoted by the letter I:

(16.6.4)

$$I = \sum_i m_i r_i^2.$$

For a rigid body in straight-line motion

$$\text{KE} = \tfrac{1}{2}Mv^2, \qquad \text{where } v \text{ is the speed of the body.}$$

For a rigid body in rotational motion

$$\text{KE} = \tfrac{1}{2}I\omega^2, \qquad \text{where } \omega \text{ is the angular speed of the body.}$$

The Moment of Inertia of a Plate

Suppose that a plate in the shape of a basic region Ω rotates about a line. The moment of inertia of the plate about that axis of rotation is given by the formula

(16.6.5)

$$I = \iint\limits_{\Omega} \lambda(x, y)[r(x, y)]^2 \, dx \, dy$$

where $\lambda = \lambda(x, y)$ is the mass density function and $r(x, y)$ is the distance from the axis to the point (x, y).

Derivation of (16.6.5) Decompose the plate into N pieces in the form of basic regions $\Omega_1, \ldots, \Omega_N$. From each Ω_i choose a point (x_i^*, y_i^*) and view all the mass of the ith piece as concentrated there. The moment of inertia of this piece is then approximately

$$\underbrace{[\lambda(x_i^*, y_i^*)(\text{area of } \Omega_i)]}_{\text{approx. mass of piece}}\underbrace{[r(x_i^*, y_i^*)]^2}_{\text{(approx. distance)}^2} = \lambda(x_i^*, y_i^*)[r(x_i^*, y_i^*)]^2 \,(\text{area of } \Omega_i).$$

The sum of these approximations,

$$\sum_{i=1}^{N} \lambda(x_i^*, y_i^*)[r(x_i^*, y_i^*)]^2 (\text{area of } \Omega_i),$$

is a Riemann sum for the double integral

$$\iint_{\Omega} \lambda(x, y)[r(x, y)]^2 \, dx \, dy.$$

As the maximum diameter of the Ω_i tends to zero, the Riemann sum tends to this integral. ❏

Figure 16.6.3

Example 3 A rectangular plate of mass M, length L, and width W rotates about the line shown in Figure 16.6.3. Find the moment of inertia of the plate about that line: **(a)** given that the plate has uniform mass density; **(b)** given that the mass density of the plate is directly proportional to the square of the distance from the rightmost side.

SOLUTION Coordinatize the plate as in Figure 16.6.4 and call the base region R.
(a) Here $\lambda(x, y) = M/LW$ and $r(x, y) = x$. Thus

$$I = \iint_{R} \frac{M}{LW} x^2 \, dx \, dy = \frac{M}{LW} \int_0^W \int_0^L x^2 \, dx \, dy$$

$$= \frac{M}{LW} W \int_0^L x^2 \, dx = \frac{M}{L}(\tfrac{1}{3}L^3) = \tfrac{1}{3}ML^2.$$

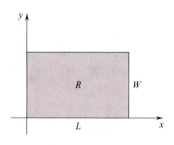

Figure 16.6.4

(b) In this case $\lambda(x, y) = k(L - x)^2$, but we still have $r(x, y) = x$. Therefore

$$I = \iint_{R} k(L - x)^2 x^2 \, dx \, dy = k \int_0^W \int_0^L (L - x)^2 x^2 \, dx \, dy$$

$$= kW \int_0^L (L^2 x^2 - 2Lx^3 + x^4) \, dx = \tfrac{1}{30} kL^5 W.$$

We can eliminate the constant of proportionality k by noting that

$$M = \iint_{R} k(L - x)^2 \, dx \, dy = k \int_0^W \int_0^L (L - x)^2 \, dx \, dy$$

$$= kW \left[-\tfrac{1}{3}(L - x)^3 \right]_0^L = \tfrac{1}{3} kWL^3.$$

Therefore,

$$k = \frac{3M}{WL^3} \quad \text{and} \quad I = \tfrac{1}{30}\left(\frac{3M}{WL^3}\right) L^5 W = \tfrac{1}{10} ML^2. \quad ❏$$

Radius of Gyration

If the mass M of an object is all concentrated at a distance r from a given line, then the moment of inertia about that line is given by the product Mr^2.

Suppose now that we have a plate of mass M (actually any object of mass M will do here), and suppose that l is some line. The object has some moment of inertia I about l. Its *radius of gyration* about l is the distance K for which

$$I = MK^2.$$

Namely, the radius of gyration about l is the distance from l at which all the mass of the object would have to be concentrated to effect the same moment of inertia. The formula for radius of gyration is usually written

(16.6.6)
$$K = \sqrt{I/M}.$$

Example 4 A homogeneous circular plate of mass M and radius R rotates about an axle that passes through the center of the plate and is perpendicular to the plate. Calculate the moment of inertia and the radius of gyration.

SOLUTION Take the axle as the z-axis and let the plate rest on the circular region Ω: $x^2 + y^2 \le R^2$ (Figure 16.6.5). The density of the plate is $M/A = M/\pi R^2$ and $r(x, y) = \sqrt{x^2 + y^2}$. Hence

$$I = \iint_\Omega \frac{M}{\pi R^2}(x^2 + y^2)\, dx\, dy = \frac{M}{\pi R^2} \int_0^{2\pi} \int_0^R r^3\, dr\, d\theta = \tfrac{1}{2}MR^2.$$

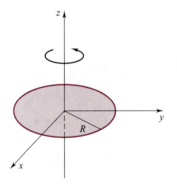

Figure 16.6.5

The radius of gyration K is $\sqrt{I/M} = R/\sqrt{2}$.

The circular plate of radius R has the same moment of inertia about the central axle as a circular wire of the same mass with radius $R/\sqrt{2}$. The circular wire is a more efficient carrier of moment of inertia than the circular plate. ❑

The Parallel Axis Theorem

Suppose we have an object of mass M and a line l_M that passes through the center of mass of the object. The object has some moment of inertia about that line; call it I_M. If l is any line parallel to l_M, then the object has a certain moment of inertia about l; call that I. The parallel axis theorem states that

(16.6.7)
$$I = I_M + d^2M$$

where d is the distance between the axes.

We prove the theorem under somewhat restrictive assumptions. Assume that the object is a plate of mass M in the shape of a basic region Ω, and assume that l_M is perpendicular to the plate. Call l the z-axis. Call the plane of the plate the xy-plane.

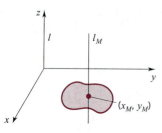

Figure 16.6.6

(See Figure 16.6.6.) Denoting the points of Ω by (x, y) we have

$$I - I_M = \iint_\Omega \lambda(x, y)(x^2 + y^2) \, dx \, dy - \iint_\Omega \lambda(x, y)[(x - x_M)^2 + (y - y_M)^2] \, dx \, dy$$

$$= \iint_\Omega \lambda(x, y)[2x_M x + 2y_M y - (x_M^2 + y_M^2)] \, dx \, dy$$

$$= 2x_M \iint_\Omega x\lambda(x, y) \, dx \, dy + 2y_M \iint_\Omega y\lambda(x, y) \, dx \, dy$$

$$- (x_M^2 + y_M^2) \iint_\Omega \lambda(x, y) \, dx \, dy$$

$$= 2x_M^2 M + 2y_M^2 M - (x_M^2 + y_M^2)M = (x_M^2 + y_M^2)M = d^2 M. \quad \square$$

An obvious consequence of the parallel axis theorem is that $I_M \leq I$ for all lines l parallel to l_M. To minimize the moment of inertia we must pass our axis of rotation through the center of mass.

EXERCISES 16.6

In Exercises 1–10, find the mass and center of mass of the plate that occupies the given region Ω and has the given density function λ.

1. Ω: $-1 \leq x \leq 1$, $0 \leq y \leq 1$, $\lambda(x, y) = x^2$.

2. Ω: $0 \leq x \leq 1$, $0 \leq y \leq \sqrt{x}$, $\lambda(x, y) = x + y$.

3. Ω: $0 \leq x \leq 1$, $x^2 \leq y \leq 1$, $\lambda(x, y) = xy$.

4. Ω: $0 \leq x \leq \pi$, $0 \leq y \leq \sin x$, $\lambda(x, y) = y$.

5. Ω: $0 \leq x \leq 8$, $0 \leq y \leq \sqrt[3]{x}$, $\lambda(x, y) = y^2$.

6. Ω: $0 \leq x \leq a$, $0 \leq y \leq \sqrt{a^2 - x^2}$, $\lambda(x, y) = xy$.

7. Ω is the triangle with vertices $(0, 0)$, $(1, 2)$, and $(1, 3)$; $\lambda(x, y) = xy$.

8. Ω is the triangular region in the first quadrant bounded by the lines $x = 0$, $y = 0$, and $3x + 2y = 6$; $\lambda(x, y) = x + y$.

9. Ω is the region bounded by the cardioid $r = 1 + \cos\theta$; λ is the distance to the pole.

10. Ω is the region inside the circle $r = 2\sin\theta$ and outside the circle $r = 1$; $\lambda(x, y) = y$.

In the exercises that follow, I_x, I_y, and I_z denote the moments of inertia about the x, y, and z axes.

11. A rectangular plate of mass M, length L, and width W is placed on the xy-plane with center at the origin, long sides parallel to the x-axis. (We assume here that $L \geq W$.) Find I_x, I_y, I_z if the plate is homogeneous. Determine the corresponding radii of gyration K_x, K_y, K_z.

12. Verify that I_x, I_y, I_z are unchanged if the mass density of

the plate of Exercise 11 varies directly as the distance from the leftmost side.

13. Determine the center of mass of the plate of Exercise 11 if the mass density varies as in Exercise 12.

14. Show that for any plate in the xy-plane

$$I_z = I_x + I_y.$$

How are the corresponding radii of gyration K_x, K_y, K_z related?

15. A homogeneous plate of mass M in the form of a quarter disc of radius R is placed in the xy-plane as in the figure. Find I_x, I_y, I_z and the corresponding radii of gyration.

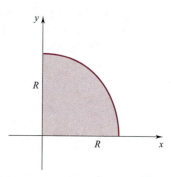

16. A plate in the xy-plane undergoes a rotation in that plane about its center of mass. Show that I_z remains unchanged.

17. A homogeneous disc of mass M and radius R is to be placed on the xy-plane so that it has moment of inertia I_0 about the z-axis. Where should the disc be placed?

18. A homogeneous plate of mass density λ occupies the region under the curve $y = f(x)$ from $x = a$ to $x = b$. Show that

$$I_x = \tfrac{1}{3}\lambda \int_a^b [f(x)]^3\, dx \qquad \text{and} \qquad I_y = \lambda \int_a^b x^2 f(x)\, dx.$$

19. A homogeneous plate of mass M in the form of an elliptical quadrant is placed on the xy-plane. (See the figure below.) Find I_x, I_y, I_z.

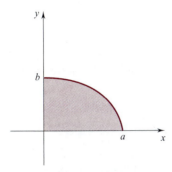

20. Find I_x, I_y, I_z for the plate in Exercise 2.

21. Find I_x, I_y, I_z for the plate in Exercise 3.

22. Find I_x, I_y, I_z for the plate in Exercise 5.

23. Find I_x, I_y, I_z for the plate in Exercise 9.

24. A plate of varying density occupies the region $\Omega = \Omega_1 \cup \Omega_2$ shown in the figure below. Find the center of mass of the plate given that the Ω_1 piece has mass M_1 and center of mass (x_1, y_1), and the Ω_2 piece has mass M_2 and center of mass (x_2, y_2).

25. A homogeneous plate of mass M is in the form of a

ring. (See the figure below.) Calculate the moment of inertia of the plate:
(a) about a diameter;
(b) about a tangent to the inner circle;
(c) about a tangent to the outer circle.

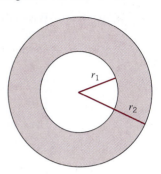

26. Find the moment of inertia of a homogeneous circular wire of mass M and radius r:
(a) about a diameter;
(b) about a tangent. HINT: Use the previous exercise.

27. The plate of Exercise 25 rotates about the axis that is perpendicular to the plate and passes through the center. Find the moment of inertia.

28. Prove the parallel axis theorem for the case where the line through the center of mass lies in the plane of the plate.

29. A plate of mass M has the form of a half disc Ω: $-R \leq x \leq R$, $0 \leq y \leq \sqrt{R^2 - x^2}$. Find the center of mass given that the mass density varies directly as the distance from the curved boundary.

30. Find I_x, I_y, I_z for the plate of Exercise 29.

31. A plate of mass M is in the form of a disc of radius R. Given that the mass density of the plate varies directly as the distance from a point P on the boundary of the plate, locate the center of mass.

32. A plate of mass M is in the form of a right triangle of base b and height h. Given that the mass density of the plate varies directly as the square of the distance from the vertex of the right angle, locate the center of mass of the plate.

33. Use double integrals to justify an assumption we made about centroids in Chapter 6: Formula (6.4.1).

■ 16.7 TRIPLE INTEGRALS

Now that you are familiar with double integrals

$$\iint_\Omega f(x, y)\, dx\, dy,$$

you will find it easy to understand triple integrals

$$\iiint_T f(x, y, z) \, dx \, dy \, dz.$$

Basically the only difference is this: instead of working with functions of two variables continuous on a plane region Ω, we will be working with functions of three variables continuous on some portion T of three-space.

The Triple Integral over a Box

For double integration we began with a rectangle

$$R: \quad a_1 \leq x \leq a_2, \quad b_1 \leq y \leq b_2.$$

For triple integration we begin with a *box* (a rectangular solid)

$$\Pi: \quad a_1 \leq x \leq a_2, \quad b_1 \leq y \leq b_2, \quad c_1 \leq z \leq c_2. \qquad \text{(Figure 16.7.1)}$$

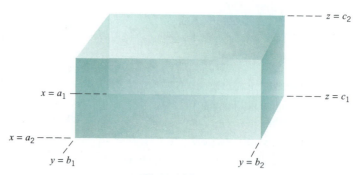

Figure 16.7.1

To partition this box, we first partition the edges. Taking

$$\begin{aligned}
P_1 &= \{x_0, \ldots, x_m\} \quad \text{as a partition of } [a_1, a_2], \\
P_2 &= \{y_0, \ldots, y_n\} \quad \text{as a partition of } [b_1, b_2], \\
P_3 &= \{z_0, \ldots, z_q\} \quad \text{as a partition of } [c_1, c_2],
\end{aligned}$$

we form the set

$$P = P_1 \times P_2 \times P_3 = \{(x_i, y_j, z_k): x_i \in P_1, y_j \in P_2, z_k \in P_3\}\dagger$$

and call this a *partition of* Π. The partition P breaks up Π into $m \times n \times q$ nonoverlapping boxes

$$\Pi_{ijk}: \quad x_{i-1} \leq x \leq x_i, \quad y_{j-1} \leq y \leq y_j, \quad z_{k-1} \leq z \leq z_k.$$

$\dagger P_1 \times P_2 \times P_3$ is the Cartesian product of $P_1, P_2,$ and P_3.

A typical such box is pictured in Figure 16.7.2.

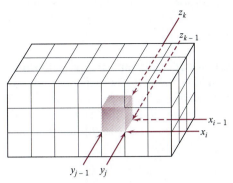

Figure 16.7.2

Let f be continuous on Π. Then, taking

$$M_{ijk} \text{ as the maximum value of } f \text{ on } \Pi_{ijk}$$

and

$$m_{ijk} \text{ as the minimum value of } f \text{ on } \Pi_{ijk},†$$

we form the *upper sum*

$$U_f(P) = \sum_{i=1}^{m} \sum_{j=1}^{n} \sum_{k=1}^{q} M_{ijk}(\text{volume of } \Pi_{ijk}) = \sum_{i=1}^{m} \sum_{j=1}^{n} \sum_{k=1}^{q} M_{ijk} \, \Delta x_i \, \Delta y_j \, \Delta z_k$$

and the *lower sum*

$$L_f(P) = \sum_{i=1}^{m} \sum_{j=1}^{n} \sum_{k=1}^{q} m_{ijk}(\text{volume of } \Pi_{ijk}) = \sum_{i=1}^{m} \sum_{j=1}^{n} \sum_{k=1}^{q} m_{ijk} \, \Delta x_i \, \Delta y_j \, \Delta z_k.$$

As in the case of functions of one and two variables, it turns out that, with f continuous on Π, there is one and only one number I that satisfies the inequality

$$L_f(P) \le I \le U_f(P) \qquad \text{for all partitions } P \text{ of } \Pi.$$

DEFINITION 16.7.1 THE TRIPLE INTEGRAL OVER A BOX Π

Let f be continuous on the closed box Π. The unique number I that satisfies the inequality

$$L_f(P) \le I \le U_f(P) \qquad \text{for all partitions } P \text{ of } \Pi$$

is called the *triple integral* of f over Π and is denoted by

$$\iiint_{\Pi} f(x, y, z) \, dx \, dy \, dz.$$

---

† A continuous function on a closed bounded set has an absolute maximum and an absolute minimum.

The Triple Integral over a More General Solid

We start with a three-dimensional, bounded, closed, connected set T. We assume that T is a *basic solid*; that is, we assume that the boundary of T consists of a finite number of continuous surfaces $z = \alpha(x, y)$, $y = \beta(x, z)$, $x = \gamma(y, z)$. See, for example, Figure 16.7.3.

Now let's suppose that f is some function continuous on T. To define the triple integral of f over T we first encase T in a rectangular box Π. (Figure 16.7.4.) We then extend f to all of Π by defining f to be zero outside of T. This extended function f is bounded on Π, and it is continuous on all of Π except possibly at the boundary of T. In spite of these possible discontinuities, f is still integrable over Π; that is, there still exists a unique number I such that

$$L_f(P) \le I \le U_f(P) \qquad \text{for all partitions } P \text{ of } \Pi.$$

Figure 16.7.3

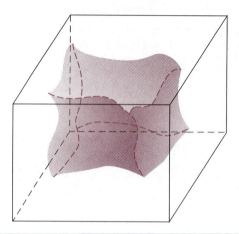

Figure 16.7.4

(We will not attempt to prove this.) The number I is by definition the triple integral

$$\iiint_{\Pi} f(x, y, z) \, dx \, dy \, dz.$$

We define the triple integral over T by setting

(16.7.2)
$$\iiint_{T} f(x, y, z) \, dx \, dy \, dz = \iiint_{\Pi} f(x, y, z) \, dx \, dy \, dz.$$

Volume as a Triple Integral

The simplest triple integral of interest is the triple integral of the function that is constantly one on T. This gives the volume of T:

(16.7.3)
$$\text{volume of } T = \iiint_{T} dx \, dy \, dz.$$

PROOF Set $f(x, y, z) = 1$ for all (x, y, z) in T. Encase T in a box Π. Define f to be zero outside of T. An arbitrary partition P of Π breaks up T into little boxes Π_{ijk}. Note that

$L_f(P) =$ the sum of the volumes of all the Π_{ijk} that are contained in T

$U_f(P) =$ the sum of the volumes of all the Π_{ijk} that intersect T.

It follows that

$$L_f(P) \le \text{the volume of } T \le U_f(P).$$

The arbitrariness of P gives the formula. ❏

Some Properties of the Triple Integral

Below we give without proof the salient elementary properties of the triple integral. They are all analogous to what you saw in the one- and two-variable cases. Assume throughout that T is a basic solid. The functions f and g are assumed to be continuous on T.

I. Linearity:

$$\iiint_T [\alpha f(x, y, z) + \beta g(x, y, z)] \, dx \, dy \, dz =$$

$$\alpha \iiint_T f(x, y, z) \, dx \, dy \, dz + \beta \iiint_T g(x, y, z) \, dx \, dy \, dz.$$

II. Order:

if $f \ge 0$ on T, then $\iiint_T f(x, y, z) \, dx \, dy \, dz \ge 0$;

if $f \le g$ on T, then $\iiint_T f(x, y, z) \, dx \, dy \, dz \le \iiint_T g(x, y, z) \, dx \, dy \, dz.$

III. Additivity: if T is broken up into a finite number of basic solids $T_1, \ldots, T_n$, then

$$\iiint_T f(x, y, z) \, dx \, dy \, dz = \iiint_{T_1} f(x, y, z) \, dx \, dy \, dz +$$

$$\cdots + \iiint_{T_n} f(x, y, z) \, dx \, dy \, dz.$$

IV. Mean-value condition: there is a point (x_0, y_0, z_0) in T for which

$$\iiint_T f(x, y, z) \, dx \, dy \, dz = f(x_0, y_0, z_0) \cdot (\text{volume of } T).$$

We call $f(x_0, y_0, z_0)$ *the average value of f on T.*

The notion of average given above enables us to write

(16.7.4)

$$\iiint_T f(x, y, z) \, dx \, dy \, dz = \left(\begin{array}{c} \text{the average value} \\ \text{of } f \text{ on } T \end{array} \right) \cdot (\text{volume of } T).$$

We can also take weighted averages: if f and g are continuous and g is nonnegative on T, then there is a point (x_0, y_0, z_0) in T for which

(16.7.5)

$$\iiint_T f(x, y, z)g(x, y, z) \, dx \, dy \, dz = f(x_0, y_0, z_0) \iiint_T g(x, y, z) \, dx \, dy \, dz.$$

As you would expect, we call $f(x_0, y_0, z_0)$ *the g-weighted average of f on T.*

The formulas for mass, center of mass, and moments of inertia derived in the previous section for two-dimensional plates are easily extended to three-dimensional objects.

Suppose that T is an object in the form of a basic solid. If T has constant mass density λ (here density is mass per unit volume), then the mass of T is the density λ times the volume of T:

$$M = \lambda V.$$

If the mass density varies continuously over T, say $\lambda = \lambda(x, y, z)$, then *the mass of T is the average density of T times the volume of T.* Thus, by (16.7.4) the mass of T is given by the triple integral

(16.7.6)

$$M = \iiint_T \lambda(x, y, z) \, dx \, dy \, dz.$$

The coordinates of the center of mass (x_M, y_M, z_M) are density-weighted averages:

(16.7.7)

$$x_M M = \iiint_T x\lambda(x, y, z) \, dx \, dy \, dz, \qquad \text{etc.}$$

If the object T is homogeneous (constant mass density M/V), then the center of mass of T depends only on the geometry of T and falls on the centroid $(\bar{x}, \bar{y}, \bar{z})$ of the space occupied by T.   The density is irrelevant.   The coordinates of the centroid are simple averages over T:

(16.7.8)

$$\bar{x} V = \iiint_T x \, dx \, dy \, dz, \qquad \text{etc.}$$

The moment of inertia of T about a line is given by the formula

(16.7.9)

$$I = \iiint_T \lambda(x, y, z)[r(x, y, z)]^2 \, dx \, dy \, dz.$$

Here $\lambda(x, y, z)$ is the mass density of T at (x, y, z) and $r(x, y, z)$ is the distance of (x, y, z) from the line in question.   The moments of inertia about the x, y, z axes are again denoted by I_x, I_y, I_z.

All of this should be readily understandable.   Techniques for evaluating triple integrals are considered in the next three sections.

EXERCISES 16.7

1. Let $f = f(x, y)$ be a function continuous and nonnegative on a basic region Ω and set

$$T = \{(x, y, z): \quad (x, y) \in \Omega, \quad 0 \le z \le f(x, y)\}.$$

Compare

$$\iiint_T dx \, dy \, dz \qquad \text{to} \qquad \iint_\Omega f(x, y) \, dx \, dy.$$

2. Set $f(x, y, z) = xyz$ on $\Pi: 0 \le x \le 1, 0 \le y \le 1, 0 \le z \le 1$ and take P as the partition $P_1 \times P_2 \times P_3$.
  (a) Find $L_f(P)$ and $U_f(P)$ given that

$$P_1 = \{x_0, \ldots, x_m\}, \qquad P_2 = \{y_0, \ldots, y_n\},$$
$$P_3 = \{z_0, \ldots, z_q\}$$

are all arbitrary partitions of $[0, 1]$.
  (b) Use your answer to (a) to calculate

$$\iiint_\Pi xyz \, dx \, dy \, dz.$$

3. Let $f(x, y, z) = \alpha$, constant, over the rectangular solid Π: $a_1 \le x \le a_2, b_1 \le y \le b_2, c_1 \le z \le c_2$.   Show that

$$\iiint_\Pi \alpha \, dx \, dy \, dz = \alpha(a_2 - a_1)(b_2 - b_1)(c_2 - c_1).$$

4. Find the average value of $f(x, y, z) = xyz$ over the region Π in Exercise 2.

5. Calculate

$$\iiint_\Pi xy \, dx \, dy \, dz \quad \text{where} \quad \Pi: 0 \le x \le a, \quad 0 \le y \le b,$$
$$0 \le z \le c.$$

6. Let T be a basic solid of varying mass density $\lambda = \lambda(x, y, z)$.   The moment of inertia of T about the xy-plane is defined by setting

$$I_{xy} = \iiint_T \lambda(x, y, z)z^2 \, dx \, dy \, dz.$$

The other plane moments of inertia, I_{xz} and I_{yz}, have comparable definitions.   Express I_x, I_y, I_z in terms of the plane moments of inertia.

7. A box Π_1: $0 \le x \le 2a, \ 0 \le y \le 2b, \ 0 \le z \le 2c$ is cut away from a larger box $\Pi_0: 0 \le x \le 2A, 0 \le y \le 2B, 0 \le z \le 2C$.   Locate the centroid of the remaining solid.

8. Show that, if f is continuous and nonnegative on a basic solid T, then the triple integral of f over T is nonnegative.

9. Calculate the mass M of the cube Π: $0 \leq x \leq a$, $0 \leq y \leq a$, $0 \leq z \leq a$ given that the mass density varies directly with the distance from the face on the xy-plane.

10. Locate the center of mass of the cube of Exercise 9.

11. Find the moment of inertia I_z of the cube of Exercise 9.

■ 16.8 REDUCTION TO REPEATED INTEGRALS

In this section we give no proofs. You can assume that all the solids that appear are basic solids and all the functions that you encounter are continuous.

In Figure 16.8.1 we have sketched a solid T. The projection of T onto the xy-plane has been labeled Ω_{xy}. The solid T is then the set of all (x, y, z) with

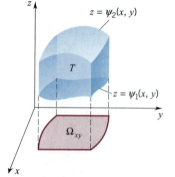

Figure 16.8.1

$$(x, y) \text{ in } \Omega_{xy} \quad \text{and} \quad \psi_1(x, y) \leq z \leq \psi_2(x, y).$$

The triple integral over T can be evaluated by setting

$$(*) \qquad \iiint_T f(x, y, z) \, dx \, dy \, dz = \iint_{\Omega_{xy}} \left(\int_{\psi_1(x,y)}^{\psi_2(x,y)} f(x, y, z) \, dz \right) dx \, dy.$$

Moving over to Figure 16.8.2 we see that in this case Ω_{xy} is the region

$$a_1 \leq x \leq a_2, \quad \phi_1(x) \leq y \leq \phi_2(x),$$

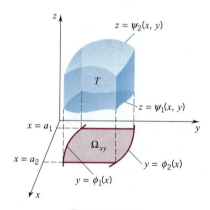

Figure 16.8.2

and T itself is the set of all (x, y, z) with

$$a_1 \leq x \leq a_2, \quad \phi_1(x) \leq y \leq \phi_2(x), \quad \psi_1(x, y) \leq z \leq \psi_2(x, y).$$

The triple integral over T can then be expressed by three ordinary integrals:

$$\iiint_T f(x, y, z) \, dx \, dy \, dz = \int_{a_1}^{a_2} \left[\int_{\phi_1(x)}^{\phi_2(x)} \left(\int_{\psi_1(x,y)}^{\psi_2(x,y)} f(x, y, z) \, dz \right) dy \right] dx.$$

It is customary to omit the brackets and parentheses and write

(16.8.1)

$$\iiint\limits_{T} f(x, y, z)\, dx\, dy\, dz = \int_{a_1}^{a_2} \int_{\phi_1(x)}^{\phi_2(x)} \int_{\psi_1(x,y)}^{\psi_2(x,y)} f(x, y, z)\, dz\, dy\, dx. \qquad \dagger$$

Here we first integrate with respect to z [from $z = \psi_1(x, y)$ to $z = \psi_2(x, y)$], then with respect to y [from $y = \phi_1(x)$ to $y = \phi_2(x)$], and finally with respect to x [from $x = a_1$ to $x = a_2$].

There is nothing special about this order of integration. Other orders of integration are possible and in some cases more convenient. Suppose, for example, that the projection of T onto the xz-plane is a region of the form

$$\Omega_{xz}: \quad a_1 \le z \le a_2, \quad \phi_1(z) \le x \le \phi_2(z).$$

If T is the set of all (x, y, z) with

$$a_1 \le z \le a_2, \quad \phi_1(z) \le x \le \phi_2(z), \quad \psi_1(x, z) \le y \le \psi_2(x, z)$$

then

$$\iiint\limits_{T} f(x, y, z)\, dx\, dy\, dz = \int_{a_1}^{a_2} \int_{\phi_1(z)}^{\phi_2(z)} \int_{\psi_1(x,z)}^{\psi_2(x,z)} f(x, y, z)\, dy\, dx\, dz.$$

In this case we integrate first with respect to y, then with respect to x, and finally with respect to z. Still four other orders of integration are possible.

Example 1 Evaluate the triple integral

$$\int_0^2 \int_0^x \int_0^{4-x^2} xyz\, dz\, dy\, dx.$$

SOLUTION

$$\int_0^2 \int_0^x \int_0^{4-x^2} xyz\, dz\, dy\, dx = \int_0^2 \int_0^x \left(\int_0^{4-x^2} xyz\, dz \right) dy\, dx$$

$$= \int_0^2 \int_0^x \left(\left[\tfrac{1}{2} xyz^2 \right]_0^{4-x^2} \right) dy\, dx$$

$$= \tfrac{1}{2} \int_0^2 \int_0^x x(4 - x^2)^2 y\, dy\, dx$$

$$= \tfrac{1}{2} \int_0^2 \left(\int_0^x x(4 - x^2)^2 y\, dy \right) dx$$

$$= \tfrac{1}{2} \int_0^2 \left(\left[\tfrac{1}{2} x(4 - x^2)^2 y^2 \right]_0^x \right) dx$$

$$= \tfrac{1}{4} \int_0^2 x^3(4 - x^2)^2\, dx = \tfrac{1}{4} \int_0^2 x^3(16 - 8x^2 + x^4)\, dx$$

$$= \tfrac{1}{4} \left[4x^4 - \tfrac{8}{6} x^6 + \tfrac{1}{8} x^8 \right]_0^2 = \tfrac{8}{3}. \quad \square$$

$\dagger$ This formula is formula $(*)$ taken one step further. Usually we skip the double-integral stage and go directly to three integrals.

Remark The solid determined by the limits of integration in Example 1 is the region T in the first octant bounded by the parabolic cylinder $z = 4 - x^2$, the plane $z = 0$, the plane $y = x$, and the plane $y = 0$. This solid is shown in Figure 16.8.3. ❑

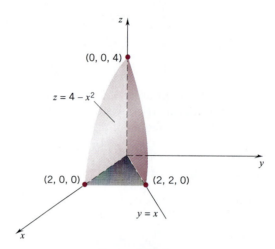

Figure 16.8.3

Example 2 Use triple integration to find the volume of the tetrahedron T shown in Figure 16.8.4, and find the coordinates of the centroid.

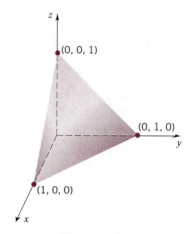

Figure 16.8.4

SOLUTION The volume of T is given by the triple integral

$$V = \iiint_T dx \, dy \, dz.$$

To evaluate this triple integral we can project T onto any one of the three coordinate planes. We will project onto the xy-plane. The base region is then the triangle

$$\Omega_{xy}: \quad 0 \le x \le 1, \quad 0 \le y \le 1 - x. \qquad \text{(Figure 16.8.5)}$$

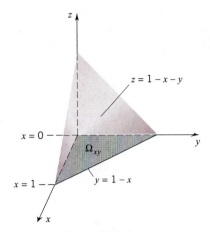

Figure 16.8.5

Since the inclined face is part of the plane $z = 1 - x - y$, we have T as the set of all (x, y, z) with

$$0 \le x \le 1, \quad 0 \le y \le 1 - x, \quad 0 \le z \le 1 - x - y.$$

It follows that

$$V = \iiint_T dx\, dy\, dz = \int_0^1 \int_0^{1-x} \int_0^{1-x-y} dz\, dy\, dx$$

$$= \int_0^1 \int_0^{1-x} (1 - x - y)\, dy\, dx$$

$$= \int_0^1 \left[(1 - x)y - \tfrac{1}{2}y^2 \right]_0^{1-x} dx$$

$$= \int_0^1 \tfrac{1}{2}(1 - x)^2\, dx = \left[-\tfrac{1}{6}(1 - x)^3 \right]_0^1 = \tfrac{1}{6}.$$

By symmetry $\bar{x} = \bar{y} = \bar{z}$. We can calculate $\bar{x}$ as follows:

$$\bar{x}\, V = \iiint_T x\, dx\, dy\, dz = \int_0^1 \int_0^{1-x} \int_0^{1-x-y} x\, dz\, dy\, dx = \tfrac{1}{24}.$$

check this ⟶

Since $V = \tfrac{1}{6}$, we have $\bar{x} = \tfrac{1}{4}$. The centroid is the point $(\tfrac{1}{4}, \tfrac{1}{4}, \tfrac{1}{4})$. See Exercise 37 for a general result concerning the volume and centroid of a tetrahedron. ❑

Example 3 Find the mass of a solid right circular cylinder of radius r and height h given that the mass density is directly proportional to the distance from the lower base.

SOLUTION Call the solid T. In the setup of Figure 16.8.6 we can characterize T by the following inequalities:

$$-r \le x \le r, \quad -\sqrt{r^2 - x^2} \le y \le \sqrt{r^2 - x^2}, \quad 0 \le z \le h.$$

Figure 16.8.6

The first two inequalities define the base region Ω_{xy}. Since the density varies directly with the distance from the lower base, we have $\lambda(x, y, z) = kz$, where $k > 0$ is the constant of proportionality. Then

$$M = \iiint_T kz \, dx \, dy \, dz$$

$$= \int_{-r}^{r} \int_{-\sqrt{r^2 - x^2}}^{\sqrt{r^2 - x^2}} \int_0^h kz \, dz \, dy \, dx$$

$$= \int_{-r}^{r} \int_{-\sqrt{r^2 - x^2}}^{\sqrt{r^2 - x^2}} \tfrac{1}{2} kh^2 \, dy \, dx$$

$$= 4 \int_0^r \int_0^{\sqrt{r^2 - x^2}} \tfrac{1}{2} kh^2 \, dy \, dx \qquad \text{(using the symmetry)}$$

$$= 2kh^2 \int_0^r \sqrt{r^2 - x^2} \, dx.$$

This integral can be evaluated either by using a trigonometric substitution (Section 8.5), or by using Formula 87 in the Table of Integrals. Either way,

$$\int_0^r \sqrt{r^2 - x^2} \, dx = \tfrac{1}{4} \pi r^2.$$

Thus,

$$M = 2kh^2(\tfrac{1}{4}\pi r^2) = \tfrac{1}{2} kh^2 r^2 \pi. \qquad \square$$

Remark In Example 3 we would have profited by not skipping the double integral stage; namely, we could have written

$$M = \iint_{\Omega_{xy}} \left(\int_0^h kz \, dz \right) dx \, dy = \iint_{\Omega_{xy}} \tfrac{1}{2} kh^2 \, dx \, dy$$

$$= \tfrac{1}{2} kh^2(\text{area of } \Omega_{xy}) = \tfrac{1}{2} kh^2 r^2 \pi. \qquad \square$$

Example 4 Integrate $f(x, y, z) = yz$ over that part of the first octant $x \geq 0$, $y \geq 0$, $z \geq 0$ that is cut off by the ellipsoid

$$\frac{x^2}{a^2} + \frac{y^2}{b^2} + \frac{z^2}{c^2} = 1.$$

SOLUTION Call the solid T. The upper boundary of T has equation

$$z = \psi(x, y) = \frac{c}{ab} \sqrt{a^2 b^2 - b^2 x^2 - a^2 y^2}.$$

This surface intersects the xy-plane in the curve

$$y = \phi(x) = \frac{b}{a} \sqrt{a^2 - x^2}.$$

We can take

$$\Omega_{xy}: \quad 0 \leq x \leq a, \quad 0 \leq y \leq \phi(x)$$

as the base region (see Figure 16.8.7) and characterize T as the set of all (x, y, z) with

$$0 \le x \le a, \quad 0 \le y \le \phi(x), \quad 0 \le z \le \psi(x, y).$$

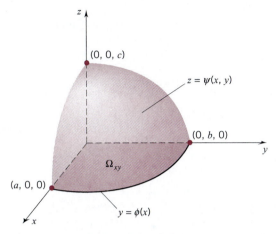

Figure 16.8.7

We can therefore calculate the triple integral by evaluating

$$\int_0^a \int_0^{\phi(x)} \int_0^{\psi(x,y)} yz \, dz \, dy \, dx.$$

A straightforward (but somewhat lengthy) computation that you can carry out yourself gives an answer of $\frac{1}{15} ab^2 c^2$. ❑

ANOTHER SOLUTION We return to Example 4 but this time carry out the integration in a different order. In Figure 16.8.8 you can see the same solid projected this time onto the yz-plane. In terms of y and z the curved surface has equation

$$x = \Psi(y, z) = \frac{a}{bc} \sqrt{b^2 c^2 - c^2 y^2 - b^2 z^2}.$$

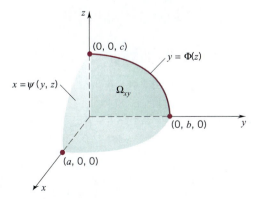

Figure 16.8.8

This surface intersects the yz-plane in the curve

$$y = \Phi(z) = \frac{b}{c}\sqrt{c^2 - z^2}.$$

We can take

$$\Omega_{yz}: \quad 0 \le z \le c, \quad 0 \le y \le \Phi(z)$$

as the base region and characterize T as the set of all (x, y, z) with

$$0 \le z \le c, \quad 0 \le y \le \Phi(z), \quad 0 \le x \le \Psi(y, z).$$

This leads to the repeated integral

$$\int_0^c \int_0^{\Phi(z)} \int_0^{\Psi(y,z)} yz \, dx \, dy \, dz$$

which, as you can check, also gives $\frac{1}{15} ab^2c^2$. ❑

Example 5 Use triple integration to find the volume of the solid T bounded above by the parabolic cylinder $z = 4 - y^2$ and bounded below by the elliptic paraboloid $z = x^2 + 3y^2$.

SOLUTION Solving the two equations simultaneously, we have

$$4 - y^2 = x^2 + 3y^2 \quad \text{and thus} \quad x^2 + 4y^2 = 4.$$

This tells us that the two surfaces intersect in a space curve that lies along the elliptic cylinder $x^2 + 4y^2 = 4$. The projection of this intersection onto the xy-plane is the ellipse $x^2 + 4y^2 = 4$. (See Figure 16.8.9.)

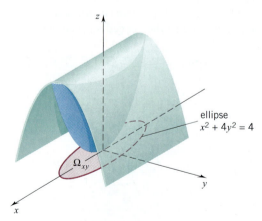

ellipse
$x^2 + 4y^2 = 4$

Figure 16.8.9

The projection of T onto the xy-plane is the region

$$\Omega_{xy}: \quad -2 \le x \le 2, \quad -\tfrac{1}{2}\sqrt{4 - x^2} \le y \le \tfrac{1}{2}\sqrt{4 - x^2}.$$

The solid T is then the set of all (x, y, z) with

$$-2 \leq x \leq 2, \quad -\tfrac{1}{2}\sqrt{4 - x^2} \leq y \leq \tfrac{1}{2}\sqrt{4 - x^2}, \quad x^2 + 3y^2 \leq z \leq 4 - y^2.$$

Its volume is given by

$$V = \int_{-2}^{2} \int_{-\sqrt{4-x^2}/2}^{\sqrt{4-x^2}/2} \int_{x^2+3y^2}^{4-y^2} dz\, dy\, dx$$

$$= 4 \int_{0}^{2} \int_{0}^{\sqrt{4-x^2}/2} \int_{x^2+3y^2}^{4-y^2} dz\, dy\, dx = 4\pi. \quad \square$$

$\underset{\text{check this}}{\uparrow}$

EXERCISES 16.8

In Exercises 1–10, evaluate the given repeated integral.

1. $\displaystyle\int_{0}^{a} \int_{0}^{b} \int_{0}^{c} dx\, dy\, dz.$

2. $\displaystyle\int_{0}^{1} \int_{0}^{x} \int_{0}^{y} y\, dz\, dy\, dx.$

3. $\displaystyle\int_{0}^{1} \int_{1}^{2y} \int_{0}^{x} (x + 2z)\, dz\, dx\, dy.$

4. $\displaystyle\int_{0}^{1} \int_{1-x}^{1+x} \int_{0}^{xy} 4z\, dz\, dy\, dx.$

5. $\displaystyle\int_{0}^{2} \int_{-1}^{1} \int_{1}^{3} (z - xy)\, dz\, dy\, dx.$

6. $\displaystyle\int_{0}^{2} \int_{-1}^{1} \int_{1}^{3} (z - xy)\, dy\, dx\, dz.$

7. $\displaystyle\int_{0}^{\pi/2} \int_{0}^{1} \int_{0}^{\sqrt{1-x^2}} x \cos z\, dy\, dx\, dz.$

8. $\displaystyle\int_{-1}^{2} \int_{1}^{y+2} \int_{e}^{e^2} \frac{x + y}{z}\, dz\, dx\, dy.$

9. $\displaystyle\int_{1}^{2} \int_{y}^{y^2} \int_{0}^{\ln x} y e^z\, dz\, dx\, dy.$

10. $\displaystyle\int_{0}^{\pi/2} \int_{0}^{\pi/2} \int_{0}^{1} e^z \cos x \sin y\, dz\, dy\, dx.$

11. (*Separated variables over a box*) Set Π: $a_1 \leq x \leq a_2$, $b_1 \leq y \leq b_2$, $c_1 \leq z \leq c_2$. Show that, if f is continuous

on $[a_1, a_2]$, g is continuous on $[b_1, b_2]$ and h is continuous on $[c_1, c_2]$, then

(16.8.2)
$$\iiint_{\Pi} f(x) g(y) h(z)\, dx\, dy\, dx =$$
$$\left(\int_{a_1}^{a_2} f(x)\, dx \right) \left(\int_{b_1}^{b_2} g(y)\, dy \right) \left(\int_{c_1}^{c_2} h(z)\, dz \right).$$

In Exercises 12 and 13, evaluate taking Π: $0 \leq x \leq 1$, $0 \leq y \leq 2$, $0 \leq z \leq 3$.

12. $\displaystyle\iiint_{\Pi} x^3 y^2 z\, dx\, dy\, dz.$

13. $\displaystyle\iiint_{\Pi} x^2 y^2 z^2\, dx\, dy\, dz.$

In Exercises 14–16, the mass density of a box Π: $0 \leq x \leq a$, $0 \leq y \leq b$, $0 \leq z \leq c$ varies directly with the product xyz.

14. Calculate the mass of Π.

15. Locate the center of mass.

16. Determine the moment of inertia of Π about: (a) the vertical line that passes through the point (a, b, c); (b) the vertical line that passes through the center of mass.

In Exercises 17–20, a homogeneous solid T of mass M consists of all points (x, y, z) with $0 \leq x \leq 1$, $0 \leq y \leq 1$, $0 \leq z \leq 1 - y$.

17. Sketch T.

18. Find the volume of T.

19. Locate the center of mass.

20. Find the moments of inertia of T about the coordinate axes.

In Exercises 21–26, express the indicated quantity by repeated integrals. Do not evaluate.

21. The mass of a ball $x^2 + y^2 + z^2 \leq r^2$ given that the density varies directly with the distance from the outer shell.

22. The mass of the solid bounded above by $z = 1$ and bounded below by $z = \sqrt{x^2 + y^2}$ given that the density varies directly with the distance from the origin. Identify the solid.

23. The volume of the solid bounded above by the parabolic cylinder $z = 1 - y^2$, below by the plane $2x + 3y + z + 10 = 0$, and on the sides by the circular cylinder $x^2 + y^2 - x = 0$.

24. The volume of the solid bounded above by the paraboloid $z = 4 - x^2 - y^2$ and bounded below by the parabolic cylinder $z = 2 + y^2$.

25. The mass of the solid bounded by the elliptic paraboloids $z = 4 - x^2 - \frac{1}{4}y^2$ and $z = 3x^2 + \frac{1}{4}y^2$ given that the density varies directly with the vertical distance from the lower surface.

26. The mass of the solid bounded by the paraboloid $x = z^2 + 2y^2$ and the parabolic cylinder $x = 4 - z^2$ given that the density varies directly with the distance from the z-axis.

In Exercises 27–32, evaluate the given triple integral.

27. $\iiint\limits_T (x^2z + y) \, dx \, dy \, dz$, where T is the solid bounded by the planes $x = 0$, $x = 1$, $y = 1$, $y = 3$, $z = 0$, $z = 2$.

28. $\iiint\limits_T 2ye^x \, dx \, dy \, dz$, where T is the solid given by $0 \leq y \leq 1$, $0 \leq x \leq y$, $0 \leq z \leq x + y$.

29. $\iiint\limits_T x^2y^2z^2 \, dx \, dy \, dz$, where T is the solid bounded by the planes $z = y + 1$, $y + z = 1$, $x = 0$, $x = 1$, $z = 0$.

30. $\iiint\limits_T xy \, dx \, dy \, dz$, where T is the solid in the first octant bounded by the coordinate planes and the hemisphere $z = \sqrt{4 - x^2 - y^2}$.

31. $\iiint\limits_T y^2 \, dx \, dy \, dz$, where T is the tetrahedron in the first octant bounded by the coordinate planes and the plane $2x + 3y + z = 6$.

32. $\iiint\limits_T y^2 \, dx \, dy \, dz$, where T is the solid in the first octant bounded by the cylinders $x^2 + y = 1$, $z^2 + y = 1$.

33. Find the volume of the portion of the first octant bounded by the planes $z = x$, $y - x = 2$, and the cylinder $y = x^2$. Where is the centroid?

34. Find the mass of a block in the shape of a unit cube given that the density varies directly with: (a) the distance from one of the faces; (b) the square of the distance from one of the vertices.

35. Find the volume and the centroid of the solid bounded above by the cylindrical surface $x^2 + z = 4$, below by the plane $x + z = 2$, and on the sides by the planes $y = 0$ and $y = 3$.

36. Show that, if $(\bar{x}, \bar{y}, \bar{z})$ is the centroid of a solid T, then

$$\iiint\limits_T (x - \bar{x}) \, dx \, dy \, dz = 0,$$

$$\iiint\limits_T (y - \bar{y}) \, dx \, dy \, dz = 0,$$

$$\iiint\limits_T (z - \bar{z}) \, dx \, dy \, dz = 0.$$

37. Taking a, b, c as positive, find the volume of the tetrahedron with vertices $(0, 0, 0)$, $(a, 0, 0)$, $(0, b, 0)$, and $(0, 0, c)$. Where is the centroid?

38. A homogeneous solid of mass M in the form and position of the tetrahedron of Figure 16.8.4 rotates about the z-axis with moment of inertia I_z. Find I_z.

39. A homogeneous box of mass M has edges a, b, c. Calculate the moment of inertia about:
(a) the edge c;
(b) the line that passes through the center of the box and is parallel to the edge c;
(c) the line that passes through the center of the face bc and is parallel to the edge c.

40. Where is the centroid of the solid bounded above by the plane $z = 1 + x + y$, below by the plane $z = -2$, and on the sides by the planes $x = 1$, $x = 2$, $y = 1$, $y = 2$?

41. Let T be the solid bounded above by the plane $z = y$, below by the xy-plane, and on the sides by the planes $x = 0$, $x = 1$, $y = 0$, $y = 1$. Find the mass of T given that the density varies directly with the square of the distance from the origin. Where is the center of mass?

42. What can you conclude about T given that

$$\iiint\limits_T f(x, y, z) \, dx \, dy \, dz = 0$$

(a) for every continuous function f that is odd in x? (b) for every continuous function f that is odd in y? (c) for every continuous function f that is odd in z? (d) for every continuous function f that satisfies the relation $f(-x, -y, -z) = -f(x, y, z)$?

43. (a) Integrate $f(x, y, z) = x + y^3 + z$ over the unit ball centered at the origin.

(b) Integrate $f(x, y, z) = a_1x + a_2y + a_3z + a_4$ over the unit ball centered at the origin.

44. Integrate $f(x, y, z) = x^2y^2$ over the solid bounded above by the cylinder $y^2 + z = 4$, below by the plane $y + z = 2$, and on the sides by the planes $x = 0$ and $x = 2$.

45. Use triple integrals to find the volume enclosed by the sphere $x^2 + y^2 + z^2 = a^2$.

46. Use triple integrals to find the volume enclosed by the ellipsoid

$$\frac{x^2}{a^2} + \frac{y^2}{b^2} + \frac{z^2}{c^2} = 1.$$

47. Find the mass of the solid Example 5 given that the density varies directly with $|x|$.

48. Find the volume of the solid bounded by the paraboloids $z = 2 - x^2 - y^2$ and $z = x^2 + y^2$.

49. Find the mass and the center of mass of the solid of Exercise 35 given that the density varies directly with $1 + y$.

50. Let T be a solid with volume

$$V = \iiint_T dx\, dy\, dz = \int_0^2 \int_0^{9-x^2} \int_0^{2-x} dz\, dy\, dx.$$

Sketch T and fill in the blanks.

(a) $V = \int_\square^\square \int_\square^\square \int_\square^\square dy\, dx\, dz.$

(b) $V = \int_\square^\square \int_\square^\square \int_\square^\square dy\, dz\, dx.$

(c) $V = \int_0^5 \int_\square^\square \int_\square^\square dz\, dx\, dy + \int_5^9 \int_\square^\square \int_\square^\square dz\, dx\, dy.$

51. Let T be a solid with volume

$$V = \iiint_T dx\, dy\, dz = \int_0^3 \int_x^{6-x} \int_0^{2x} dz\, dy\, dx.$$

Sketch T and fill in the blanks.

(a) $V = \int_\square^\square \int_\square^\square \int_\square^\square dy\, dx\, dz.$

(b) $V = \int_\square^\square \int_\square^\square \int_\square^\square dy\, dz\, dx.$

(c) $V = \int_\square^\square \int_\square^\square \int_\square^\square dx\, dy\, dz + \int_\square^\square \int_\square^\square \int_\square^\square dx\, dy\, dz.$

For the remaining exercises, let V be the volume of the solid T enclosed by the parabolic cylinder $y = 4 - z^2$ and the V-shaped cylinder $y = |x|$. Let Ω_{xy}, Ω_{yz}, Ω_{xz} be the projections of T onto the xy, yz, xz planes, respectively. Fill in the blanks.

52. (a) $V = \iint_{\Omega_{xy}} \square\, dx\, dy.$

(b) $V = \iint_{\Omega_{xy}} \left(\int_\square^\square dz \right) dx\, dy.$

(c) $V = \int_\square^\square \int_\square^\square \int_\square^\square dz\, dy\, dx.$

(d) $V = \int_\square^\square \int_\square^\square \int_\square^\square dz\, dx\, dy.$

53. (a) $V = \iint_{\Omega_{yz}} \square\, dy\, dz.$

(b) $V = \iint_{\Omega_{yz}} \left(\int_\square^\square dx \right) dy\, dz.$

(c) $V = \int_\square^\square \int_\square^\square \int_\square^\square dx\, dz\, dy.$

(d) $V = \int_\square^\square \int_\square^\square \int_\square^\square dx\, dy\, dz.$

54. (a) $V = \iint_{\Omega_{xz}} \square\, dx\, dz.$

(b) $V = \iint_{\Omega_{xz}} \left(\int_\square^\square dy \right) dx\, dz.$

(c) $V = \int_\square^\square \int_\square^\square \int_\square^\square dy\, dx\, dz.$

(d) $V = \int_{-2}^0 \int_\square^\square \int_\square^\square dy\, dz\, dx + \int_0^2 \int_\square^\square \int_\square^\square dy\, dz\, dx.$

■ 16.9 CYLINDRICAL COORDINATES

Introduction to Cylindrical Coordinates

The cylindrical coordinates (r, θ, z) of a point P in xyz-space are shown geometrically in Figure 16.9.1. The first two coordinates, r and θ, are the usual plane polar coordinates, except that r is taken as nonnegative and θ is restricted to the interval $[0, 2\pi)$. The third coordinate is the third rectangular coordinate z. Thus, rectangular coordinates (x, y, z) are related to cylindrical coordinates (r, θ, z) by the equations

$$x = r \cos\theta, \quad y = r \sin\theta, \quad z = z.$$

Conversely, with the obvious exclusions, cylindrical coordinates are related to rectangular coordinates by

$$r = \sqrt{x^2 + y^2}, \quad \tan\theta = \frac{y}{x}, \quad z = z.$$

In rectangular coordinates, the coordinate surfaces

$$x = x_0, \quad y = y_0, \quad z = z_0$$

are three mutually perpendicular planes. In cylindrical coordinates the coordinate surfaces take the form

$$r = r_0, \quad \theta = \theta_0, \quad z = z_0.$$

These surfaces are drawn in Figure 16.9.2.

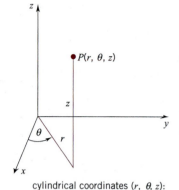

cylindrical coordinates (r, θ, z):
$r \geq 0$, $0 \leq \theta < 2\pi$, z real

Figure 16.9.1

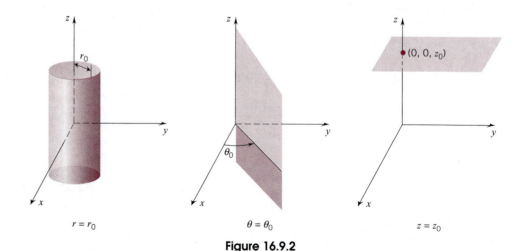

| $r = r_0$ | $\theta = \theta_0$ | $z = z_0$ |

Figure 16.9.2

The first surface is a right circular cylinder of radius r_0. The central axis of the cylinder is the z-axis. The surface $\theta = \theta_0$ is a vertical half-plane hinged at the z-axis. The plane stands at an angle of θ_0 radians from the positive x-axis. The last coordinate surface is the plane $z = z_0$.

The xyz-solids easiest to describe in cylindrical coordinates are the *cylindrical* wedges. One is pictured in Figure 16.9.3. That wedge consists of all points (x, y, z) that have cylindrical coordinates (r, θ, z) in the box

$$\Pi: \quad a_1 \leq r \leq a_2, \quad b_1 \leq \theta \leq b_2, \quad c_1 \leq z \leq c_2.$$

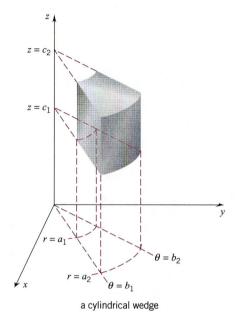

a cylindrical wedge

Figure 16.9.3

Evaluating Triple Integrals Using Cylindrical Coordinates

Suppose that T is some basic solid in xyz-space, not necessarily a wedge. If T is the set of all (x, y, z) with cylindrical coordinates in some basic solid S in $r\theta z$-space, then

(16.9.1)
$$\iiint_T f(x, y, z)\, dx\, dy\, dz = \iiint_S f(r \cos \theta, r \sin \theta, z)\, r\, dr\, d\theta\, dz.$$

Derivation of (16.9.1) We will carry out the argument on the assumption that T is projectable onto some basic region Ω_{xy} of the xy-plane. (It is for such solids that the formula is most useful.) T has some lower boundary $z = \psi_1(x, y)$ and some upper boundary $z = \psi_2(x, y)$. T is then the set of all (x, y, z) with

$$(x, y) \in \Omega_{xy} \quad \text{and} \quad \psi_1(x, y) \leq z \leq \psi_2(x, y).$$

The region Ω_{xy} has polar coordinates in some set $\Omega_{r\theta}$ (which we assume is a basic region). Then S is the set of all (r, θ, z) with

$$(r, \theta) \in \Omega_{r\theta} \quad \text{and} \quad \psi_1(r \cos \theta, r \sin \theta) \leq z \leq \psi_2(r \cos \theta, r \sin \theta).$$

Therefore

$$\iiint_T f(x, y, z)\, dx\, dy\, dz = \iint_{\Omega_{xy}} \left(\int_{\psi_1(x, y)}^{\psi_2(x, y)} f(x, y, z)\, dz \right) dx\, dy$$

(16.5.3) —— $$= \iint_{\Omega_{r\theta}} \left(\int_{\psi_1(r\cos\theta,\, r\sin\theta)}^{\psi_2(r\cos\theta,\, r\sin\theta)} f(r\cos\theta,\, r\sin\theta,\, z)\, dz \right) r\, dr\, d\theta$$

$$= \iiint_S f(r\cos\theta,\, r\sin\theta,\, z) r\, dr\, d\theta\, dz. \quad \square$$

Volume Formula

If $f(x, y, z) = 1$ for all (x, y, z) in T, then (16.9.1) reduces to

$$\iiint_T dx\, dy\, dz = \iiint_S r\, dr\, d\theta\, dz.$$

The triple integral on the left is the volume of T. In summary, if T is a basic solid in xyz-space and the cylindrical coordinates of T constitute a basic solid S in $r\theta z$-space, then the volume of T is given by the formula

(16.9.2)

$$V = \iiint_S r\, dr\, d\theta\, dz.$$

Calculations

Cylindrical coordinates are particularly useful when an axis of symmetry is present. The axis of symmetry is then taken as the z-axis.

Example 1 Evaluate

$$\int_{-2}^{2} \int_{-\sqrt{4-x^2}}^{\sqrt{4-x^2}} \int_{0}^{4-x^2-y^2} (x^2 + y^2)\, dz\, dy\, dx.$$

SOLUTION This is a triple integral over the solid T given by

$$T: \quad -2 \le x \le 2, \quad -\sqrt{4-x^2} \le y \le \sqrt{4-x^2}, \quad 0 \le z \le 4 - x^2 - y^2.$$

This solid is bounded above by the paraboloid of revolution $z = 4 - x^2 - y^2$ and below by the xy-plane (see Figure 16.9.4). Since the solid is symmetric about the z-axis, the integral has a simpler representation in cylindrical coordinates:

$$S: \quad 0 \le r \le 2, \quad 0 \le \theta \le 2\pi, \quad 0 \le z \le 4 - r^2.$$

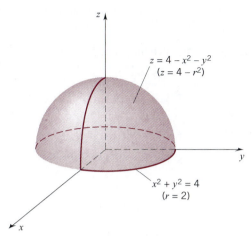

$z = 4 - x^2 - y^2$
$(z = 4 - r^2)$

$x^2 + y^2 = 4$
$(r = 2)$

Figure 16.9.4

Now,

$$\iiint_T (x^2 + y^2)\, dx\, dy\, dz = \iiint_S r^2\, r\, dr\, d\theta\, dz = \int_0^{2\pi} \int_0^2 \int_0^{4-r^2} r^3\, dz\, dr\, d\theta$$

$$= \int_0^{2\pi} \int_0^2 \left[r^3 z \right]_0^{4-r^2} dr\, d\theta = \int_0^{2\pi} \int_0^2 (4r^3 - r^5)\, dr\, d\theta$$

$$= \int_0^{2\pi} \left[r^4 - \tfrac{1}{6} r^6 \right]_0^2 d\theta$$

$$= \tfrac{16}{3} \int_0^{2\pi} d\theta = \frac{32\pi}{3}. \quad \square$$

Example 2 Find the mass of a solid right circular cylinder T of radius R and height h given that the density varies directly with the distance from the axis of the cylinder.

SOLUTION Place the cylinder T on the xy-plane so that the axis of T coincides with the z-axis. The density function then takes the form $\lambda(x, y, z) = k\sqrt{x^2 + y^2}$ and T consists of all points (x, y, z) with cylindrical coordinates (r, θ, z) in the set

$$S: \quad 0 \le r \le R, \quad 0 \le \theta \le 2\pi, \quad 0 \le z \le h.$$

Therefore

$$M = \iiint_T k\sqrt{x^2 + y^2}\, dx\, dy\, dz = \iiint_S (kr)r\, dr\, d\theta\, dz$$

$$= k \int_0^R \int_0^{2\pi} \int_0^h r^2\, dz\, d\theta\, dr = \tfrac{2}{3} k\pi R^3 h. \quad \square$$

└─ check this

Example 3 Use cylindrical coordinates to find the volume of the solid T bounded above by the plane $z = y$ and below by the paraboloid $z = x^2 + y^2$.

SOLUTION In cylindrical coordinates the plane has equation $z = r \sin \theta$ and the paraboloid has equation $z = r^2$. Solving these two equations simultaneously, we have $r = \sin \theta$. This tells us that the two surfaces intersect in a space curve that lies along the circular cylinder $r = \sin \theta$. The projection of this intersection onto the xy-plane is the circle with polar equation $r = \sin \theta$. (See Figure 16.9.5.) The base region Ω_{xy} is thus the set of all (x, y) with polar coordinates in the set

$$0 \le \theta \le \pi, \quad 0 \le r \le \sin \theta.$$

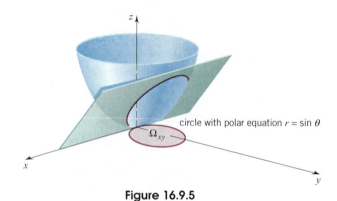

circle with polar equation $r = \sin \theta$

Ω_{xy}

Figure 16.9.5

T itself is the set of all (x, y, z) with cylindrical coordinates in the set

$$S: \quad 0 \le \theta \le \pi, \quad 0 \le r \le \sin \theta, \quad r^2 \le z \le r \sin \theta. \qquad \text{(check this)}$$

Therefore,

$$V = \iiint_T dx\, dy\, dz = \iiint_S r\, dr\, d\theta\, dz$$

$$= \int_0^\pi \int_0^{\sin \theta} \int_{r^2}^{r \sin \theta} r\, dz\, dr\, d\theta$$

$$= \int_0^\pi \int_0^{\sin \theta} (r^2 \sin \theta - r^3)\, dr\, d\theta$$

$$= \int_0^\pi \left[\tfrac{1}{3} r^3 \sin \theta - \tfrac{1}{4} r^4 \right]_0^{\sin \theta} d\theta$$

$$= \tfrac{1}{12} \int_0^\pi \sin^4 \theta\, d\theta = \tfrac{1}{12} (\tfrac{3}{8} \pi) = \tfrac{1}{32} \pi. \quad \square$$

└── (verify this; Exercise 11, Section 8.3)

Example 4 Locate the centroid of the solid in Example 3.

SOLUTION Since T is symmetric about the yz-plane, we see that $\bar{x} = 0$. To get $\bar{y}$ we begin as usual:

$$\bar{y}V = \iiint_T y\, dx\, dy\, dz = \iiint_S (r\sin\theta)\, r\, dr\, d\theta\, dz$$

$$= \int_0^\pi \int_0^{\sin\theta} \int_{r^2}^{r\sin\theta} r^2 \sin\theta\, dz\, dr\, d\theta$$

$$= \int_0^\pi \int_0^{\sin\theta} (r^3\sin^2\theta - r^4\sin\theta)\, dr\, d\theta$$

$$= \int_0^\pi \left[\tfrac{1}{4}r^4\sin^2\theta - \tfrac{1}{5}r^5\sin\theta \right]_0^{\sin\theta} d\theta$$

$$= \tfrac{1}{20} \int_0^\pi \sin^6\theta\, d\theta \; = \; \tfrac{1}{20}\left(\tfrac{5}{16}\pi\right) = \tfrac{1}{64}\pi.$$

$\qquad\qquad\qquad\qquad\qquad$ (verify this; Exercise 19, Section 8.3)

Since $V = \tfrac{1}{32}\pi$, we have $\bar{y} = \tfrac{1}{2}$. Now for $\bar{z}$:

$$\bar{z}V = \iiint_T z\, dx\, dy\, dz$$

$$= \iiint_S zr\, dr\, d\theta\, dz$$

$$= \int_0^\pi \int_0^{\sin\theta} \int_{r^2}^{r\sin\theta} zr\, dz\, dr\, d\theta = \cdots = \tfrac{1}{24} \int_0^\pi \sin^6\theta\, d\theta = \tfrac{1}{24}\left(\tfrac{5}{16}\pi\right) = \tfrac{5}{384}\pi.$$

$\qquad\qquad\qquad\qquad$ details are left to you

Division by $V = \tfrac{1}{32}\pi$ gives $\bar{z} = \tfrac{5}{12}$. The centroid is thus the point $(0, \tfrac{1}{2}, \tfrac{5}{12})$. ❑

EXERCISES 16.9

In Exercises 1–6, write the given equation in cylindrical coordinates and sketch the graph of the surface.

1. $x^2 + y^2 + z^2 = 9$.   **2.** $x^2 + y^2 = 4$.

3. $z = 2\sqrt{x^2 + y^2}$.   **4.** $x = 4z$.

5. $4x^2 + 4y^2 - z^2 = 0$.   **6.** $y^2 + z^2 = 8$.

In Exercises 7–10, evaluate the repeated integral.

7. $\displaystyle \int_0^\pi \int_0^2 \int_0^{4-r^2} r\, dz\, dr\, d\theta.$

8. $\displaystyle \int_0^{\pi/4} \int_0^1 \int_0^{\sqrt{1-r^2}} r\cos\theta\, dz\, dr\, d\theta.$

9. $\displaystyle \int_0^{\pi/4} \int_0^1 \int_0^{r\cos\theta} r\sec^3\theta\, dz\, dr\, d\theta.$

10. $\displaystyle \int_0^\pi \int_0^{4\cos\theta} \int_0^{\sqrt{16-r^2}} r\, dz\, dr\, d\theta.$

In Exercises 11–13, let T be a solid right circular cylinder of base radius R and height h. Assume that the mass density varies directly with the distance from one of the bases.

11. Use cylindrical coordinates to find the mass M of T.

12. Locate the center of mass of T.

13. Find the moment of inertia of T about the axis of the cylinder.

14. Let T be a homogeneous right circular cylinder of mass M, base radius R, and height h. Find the moment of inertia of the cylinder about: (a) the central axis; (b) a line that lies in the plane of one of the bases and passes through the center of that base; (c) a line that passes through the center of the cylinder and is parallel to the bases.

In Exercises 15–18, let T be a homogeneous solid right circular cone of mass M, base radius R, and height h.

15. Use cylindrical coordinates to verify that the volume of the cone is given by the formula $V = \frac{1}{3}\pi R^2 h$.

16. Locate the center of mass.

17. Find the moment of inertia about the axis of the cone.

18. Find the moment of inertia about a line that passes through the vertex and is parallel to the base.

In Exercises 19–21, let T be the solid bounded above by the paraboloid $z = 1 - (x^2 + y^2)$ and bounded below by the xy-plane.

19. Use cylindrical coordinates to find the volume of T.

20. Find the mass of T if the density varies directly with the distance to the xy-plane.

21. Find the mass of T if the density varies directly with the square of the distance from the origin.

In Exercises 22–27, evaluate using cylindrical coordinates.

22. $\displaystyle\int_{-1}^{1}\int_{0}^{\sqrt{1-x^2}}\int_{\sqrt{x^2+y^2}}^{1} z^3 \, dz \, dy \, dx.$

23. $\displaystyle\int_{0}^{1}\int_{0}^{\sqrt{1-x^2}}\int_{0}^{\sqrt{4-(x^2+y^2)}} dz \, dy \, dx.$

24. $\displaystyle\int_{0}^{1}\int_{0}^{\sqrt{1-x^2}}\int_{0}^{\sqrt{1-x^2-y^2}} z \, dz \, dy \, dx.$

25. $\displaystyle\int_{0}^{3}\int_{0}^{\sqrt{9-y^2}}\int_{0}^{\sqrt{9-x^2-y^2}} \frac{1}{\sqrt{x^2+y^2}} \, dz \, dx \, dy.$

26. $\displaystyle\int_{-1}^{1}\int_{-\sqrt{1-x^2}}^{\sqrt{1-x^2}}\int_{x^2+y^2}^{2-x^2-y^2} \sqrt{x^2+y^2} \, dz \, dy \, dx.$

27. $\displaystyle\int_{0}^{1}\int_{0}^{\sqrt{1-x^2}}\int_{0}^{2} \sin(x^2 + y^2) \, dz \, dy \, dx.$

28. Find the volume of the solid bounded by the paraboloid of revolution $x^2 + y^2 = az$, the xy-plane, and the cylinder $x^2 + y^2 = 2ax$.

29. Find the volume of the solid bounded above by the cone $z^2 = x^2 + y^2$, below by the xy-plane, and on the sides by the cylinder $x^2 + y^2 = 2ax$.

30. Find the volume of the solid bounded above by the plane $2z = 4 + x$, below by the xy-plane, and on the sides by the cylinder $x^2 + y^2 = 2x$.

31. Find the volume of the solid bounded above by $z = a - \sqrt{x^2 + y^2}$, below by the xy-plane, and on the sides by the cylinder $x^2 + y^2 = ax$.

32. Find the volume of the solid that is bounded above by $x^2 + y^2 + z^2 = 25$ and below by $z = \sqrt{x^2 + y^2} + 1$.

33. Find the volume of the solid bounded by the paraboloid $z = x^2 + y^2$ and the plane $z = x$.

34. Find the volume of the solid bounded by the hyperboloid $z^2 = a^2 + x^2 + y^2$ and the upper nappe of the cone $z^2 = 2(x^2 + y^2)$.

35. Find the volume of the "ice cream cone" bounded below by the half-cone $z = \sqrt{3(x^2 + y^2)}$ and above by the unit sphere $x^2 + y^2 + z^2 = 1$.

36. Find the volume of the solid that lies between the cylinders $x^2 + y^2 = 1$ and $x^2 + y^2 = 4$, and is bounded above by the ellipsoid $x^2 + y^2 + 4z^2 = 36$ and below by the xy-plane.

■ **16.10 THE TRIPLE INTEGRAL AS THE LIMIT OF RIEMANN SUMS; SPHERICAL COORDINATES**

The Triple Integral as the Limit of Riemann Sums

You have seen how single integrals and double integrals can be obtained as limits of Riemann sums. The same holds true for triple integrals.

Start with a basic solid T in xyz-space and decompose it into a finite number of basic solids $T_1, \ldots, T_N$. If f is continuous on T, then f is continuous on each T_i. From each T_i pick an arbitrary point (x_i^*, y_i^*, z_i^*) and form the *Riemann sum*

$$\sum_{i=1}^{N} f(x_i^*, y_i^*, z_i^*)(\text{volume of } T_i).$$

As you would expect, the triple integral over T is the limit of such sums; namely, given any $\epsilon > 0$, there exists $\delta > 0$ such that, if the diameters of the T_i are all less than δ, then

$$\left| \sum_{i=1}^{N} f(x_i^*, y_i^*, z_i^*)(\text{volume of } T_i) - \iiint_{T} f(x, y, z) \, dx \, dy \, dz \right| < \epsilon$$

no matter how the (x_i^*, y_i^*, z_i^*) are chosen within the T_i. We express this by writing

(16.10.1)
$$\iiint_T f(x, y, z)\, dx\, dy\, dz = \lim_{\text{diam } T_i \to 0} \sum_{i=1}^{N} f(x_i^*, y_i^*, z_i^*)(\text{volume of } T_i).$$

Introduction to Spherical Coordinates

The spherical coordinates (ρ, θ, ϕ) of a point P in xyz-space are shown geometrically in Figure 16.10.1. The first coordinate, ρ, is the distance from P to the origin; thus $\rho \geq 0$. The second coordinate, the angle marked θ, is the second coordinate of cylindrical coordinates; θ ranges from 0 to 2π. We call θ the *longitude*. The third coordinate, the angle marked ϕ, ranges only from 0 to π. We call ϕ the *colatitude*, or more simply the *polar angle*. (The complement of ϕ would be the *latitude* on a globe.)

The coordinate surfaces

$$\rho = \rho_0, \quad \theta = \theta_0, \quad \phi = \phi_0$$

are shown in Figure 16.10.2. The surface $\rho = \rho_0$ is a sphere; the radius is ρ_0 and the center is the origin. The second surface, $\theta = \theta_0$, is the same as in cylindrical coordinates: the vertical half-plane hinged at the z-axis and standing at an angle of θ_0 radians from the positive x-axis. The surface $\phi = \phi_0$ requires detailed explanation. If $0 < \phi_0 < \frac{1}{2}\pi$ or $\frac{1}{2}\pi < \phi_0 < \pi$, the surface is a nappe of a cone; it is generated by revolving about the z-axis any ray that emerges from the origin at an angle of ϕ_0 radians from the positive z-axis. The surface $\phi = \frac{1}{2}\pi$ is the xy-plane. (The nappe of the cone has opened up completely.) The equation $\phi = 0$ gives the nonnegative z-axis, and the equation $\phi = \pi$ gives the nonpositive z-axis. (When $\phi = 0$ or $\phi = \pi$, the nappe of the cone has closed up completely.)

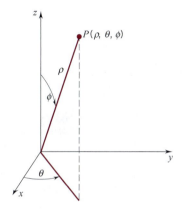

spherical coordinates (ρ, θ, ϕ):
$\rho \geq 0,\ 0 \leq \theta < 2\pi,\ 0 \leq \phi \leq \pi$

Figure 16.10.1

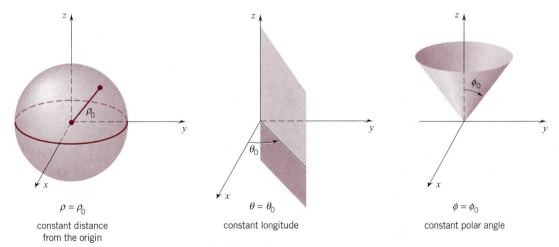

$\rho = \rho_0$
constant distance
from the origin

$\theta = \theta_0$
constant longitude

$\phi = \phi_0$
constant polar angle

Figure 16.10.2

Rectangular coordinates (x, y, z) are related to spherical coordinates (ρ, θ, ϕ) by the following equations:

$$x = \rho \sin \phi \cos \theta, \qquad y = \rho \sin \phi \sin \theta, \qquad z = \rho \cos \phi.$$

You can verify these relations by referring to Figure 16.10.3. (Note that the factor $\rho \sin \phi$ appearing in the first two equations is the r of cylindrical coordinates: $r = \rho \sin \phi$.) Conversely, with obvious exclusions, we have

$$\rho = \sqrt{x^2 + y^2 + z^2}, \qquad \tan \theta = \frac{y}{x}, \qquad \cos \phi = \frac{z}{\sqrt{x^2 + y^2 + z^2}}.$$

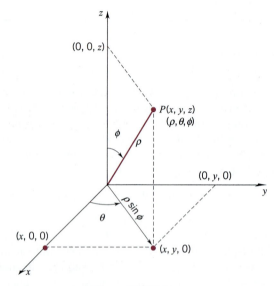

Figure 16.10.3

The Volume of a Spherical Wedge

Figure 16.10.4 shows a *spherical wedge* W in xyz-space. The wedge W consists of all points (x, y, z) that have spherical coordinates in the box

$$\Pi: \quad a_1 \leq \rho \leq a_2, \quad b_1 \leq \theta \leq b_2, \quad c_1 \leq \phi \leq c_2.$$

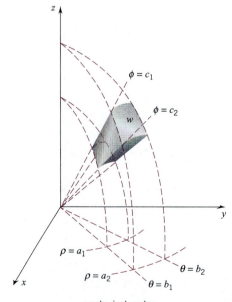

a spherical wedge

Figure 16.10.4

The volume of this wedge is given by the formula

(16.10.2)

$$V = \iiint_{\Pi} \rho^2 \sin\phi \, d\rho \, d\theta \, d\phi.$$

PROOF Note first that W is a solid of revolution. One way to obtain W is to rotate the $\theta = b_1$ face of W, call it Ω, about the z-axis for $b_2 - b_1$ radians. (See Figure 16.10.4.) On that face ρ and $\alpha = \frac{1}{2}\pi - \phi$ play the role of polar coordinates. (See Figure 16.10.5.) In the setup of Figure 16.10.5 the face Ω is the set of all (z, X) with polar coordinates (ρ, α) in the set $\Gamma: a_1 \le \rho \le a_2, \frac{1}{2}\pi - c_2 \le \alpha \le \frac{1}{2}\pi - c_1$. The centroid of Ω is at a distance $\overline{X}$ from the z-axis where

$$\overline{X}(\text{area of } \Omega) = \iint_{\Omega} X \, dX \, dz = \iint_{\Gamma} \rho^2 \cos\alpha \, d\rho \, d\alpha$$

this follows from (16.5.3) together with the fact that here (ρ, α) play the role of polar coordinates

$$= \left(\int_{a_1}^{a_2} \rho^2 \, d\rho \right) \left(\int_{\frac{1}{2}\pi - c_2}^{\frac{1}{2}\pi - c_1} \cos\alpha \, d\alpha \right)$$

$\phi = \frac{1}{2}\pi - \alpha$

$$= \left(\int_{a_1}^{a_2} \rho^2 \, d\rho \right) \left(\int_{c_1}^{c_2} \sin\phi \, d\phi \right).$$

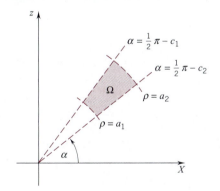

Figure 16.10.5

As the face Ω is rotated from $\theta = b_1$ to $\theta = b_2$, the centroid travels through a circular arc of length

$$s = (b_2 - b_1)\overline{X} = (b_2 - b_1)\frac{1}{\text{area of } \Omega}\left(\int_{a_1}^{a_2} \rho^2 \, d\rho \right) \left(\int_{c_1}^{c_2} \sin\phi \, d\phi \right).$$

From Pappus (see the remark on p. 375) we know that

$$\text{the volume of } W = s(\text{area of } \Omega) = (b_2 - b_1)\left(\int_{a_1}^{a_2} \rho^2 \, d\rho\right)\left(\int_{c_1}^{c_2} \sin\phi \, d\phi\right)$$

$$= \left(\int_{b_1}^{b_2} d\theta\right)\left(\int_{a_1}^{a_2} \rho^2 \, d\rho\right)\left(\int_{c_1}^{c_2} \sin\phi \, d\phi\right)$$

$$= \int_{a_1}^{a_2}\int_{b_1}^{b_2}\int_{c_1}^{c_2} \rho^2 \sin\phi \, d\phi \, d\theta \, d\rho$$

$$= \iiint_{\Pi} \rho^2 \sin\phi \, d\rho \, d\theta \, d\phi. \quad \square$$

Evaluating Triple Integrals Using Spherical Coordinates

Suppose that T is a basic solid in xyz-space with spherical coordinates in some basic solid S of $\rho\theta\phi$-space. Then

(16.10.3)

$$\iiint_{T} f(x, y, z) \, dx \, dy \, dz =$$
$$\iiint_{S} f(\rho \sin\phi \cos\theta, \rho \sin\phi \sin\theta, \rho \cos\phi) \, \rho^2 \sin\phi \, d\rho \, d\theta \, d\phi.$$

DERIVATION OF (16.10.3) Assume first that T is a spherical wedge W. The solid S is then a box Π. Now decompose Π into N boxes $\Pi_1, \ldots, \Pi_N$. This induces a subdivision of W into N spherical wedges $W_1, \ldots, W_N$.

Writing $F(\rho, \theta, \phi)$ for $f(\rho \sin\phi \cos\theta, \rho \sin\phi \sin\theta, \rho \cos\phi)$ to save space, we have

$$\iiint_{\Pi} F(\rho, \theta, \phi) \, \rho^2 \sin\phi \, d\rho \, d\theta \, d\phi = \sum_{i=1}^{N} \iiint_{\Pi_i} F(\rho, \theta, \phi) \, \rho^2 \sin\phi \, d\rho \, d\theta \, d\phi$$

additivity

$$= \sum_{i=1}^{N} F(\rho_i^*, \theta_i^*, \phi_i^*) \iiint_{\Pi_i} \rho^2 \sin\phi \, d\rho \, d\theta \, d\phi$$

for some $(\rho_i^*, \theta_i^*, \phi_i^*) \in \Pi_i$
(by 16.7.5)

$$= \sum_{i=1}^{N} F(\rho_i^*, \theta_i^*, \phi_i^*)(\text{volume of } W_i)$$

(16.10.2) applied to Π_i

$$x_i^* = \rho_i^* \sin\phi_i^* \cos\theta_i^*, \qquad = \sum_{i=1}^{N} f(x_i^*, y_i^*, z_i^*)(\text{volume of } W_i).$$
$$y_i^* = \rho_i^* \sin\phi_i^* \sin\theta_i^*,$$
$$z_i^* = \rho_i^* \cos\phi_i^*$$

This last expression is a Riemann sum for

$$\iiint_{W} f(x, y, z) \, dx \, dy \, dz$$

and, as such, by (16.10.1), will differ from that integral by less than any preassigned positive number ϵ provided only that the diameters of all the W_i are sufficiently small. This we can guarantee by making the diameters of all the Π_i sufficiently small.

This verifies the formula for the case where T is a spherical wedge. The more general case is left to you.   HINT: Encase T in a wedge W and define f to be zero outside of T.   ❏

Volume Formula

If $f(x, y, z) = 1$ for all (x, y, z) in T, then the change of variables formula reduces to

$$\iiint_T dx \, dy \, dz = \iiint_S \rho^2 \sin \phi \, d\rho \, d\theta \, d\phi.$$

The integral on the left is the volume of T. It follows that the volume of T is given by the formula

(16.10.4)
$$V = \iiint_S \rho^2 \sin \phi \, d\rho \, d\theta \, d\phi.$$

Calculations

Spherical coordinates are commonly used in applications where there is a center of symmetry. The center of symmetry is then taken as the origin.

Example 1   Calculate the mass M of a solid ball of radius 1 given that the density varies directly with the square of the distance from the center of the ball.

SOLUTION   Center the ball at the origin. The ball, call it T, is now the set of all (x, y, z) with spherical coordinates (ρ, θ, ϕ) in the box

$$S: \quad 0 \leq \rho \leq 1, \quad 0 \leq \theta \leq 2\pi, \quad 0 \leq \phi \leq \pi. \qquad \text{(Figure 16.10.6)}$$

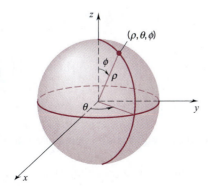

Figure 16.10.6

Therefore

$$M = \iiint_T k(x^2 + y^2 + z^2)\, dx\, dy\, dz$$

$$= \iiint_S (k\rho^2)\rho^2\sin\phi\, d\rho\, d\theta\, d\phi = k\int_0^\pi \int_0^{2\pi} \int_0^1 \rho^4\sin\phi\, d\rho\, d\theta\, d\phi$$

$$= k\left(\int_0^\pi \sin\phi\, d\phi\right)\left(\int_0^{2\pi} d\theta\right)\left(\int_0^1 \rho^4\, d\rho\right) = k(2)(2\pi)(\tfrac{1}{5}) = \tfrac{4}{5}k\pi. \quad ❑$$

Example 2 Find the volume of the solid T that is bounded above by the cone $z^2 = x^2 + y^2$, below by the xy-plane, and on the sides by the hemisphere $z = \sqrt{4 - x^2 - y^2}$ (see Figure 16.10.7).

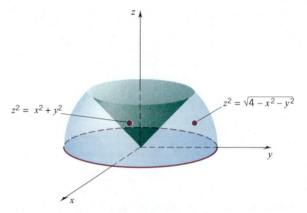

Figure 16.10.7

SOLUTION The hemisphere is given by $\rho = 2$, $0 \le \phi \le \pi/2$, in spherical coordinates. You can verify that the hemisphere and the cone intersect in a circle which lies in the plane $z = \sqrt{2}$ and has its center on the z-axis. For points on this circle, the angle $\phi = \pi/4$ (verify this). Therefore, it follows that the solid T is the set of all (x, y, z) with spherical coordinates (ρ, θ, ϕ) in the set

$$S: \quad 0 \le \rho \le 2, \quad 0 \le \theta \le 2\pi, \quad \pi/4 \le \phi \le \pi/2.$$

Thus,

$$V = \iiint_T dx\, dy\, dz = \iiint_S \rho^2\sin\phi\, d\rho\, d\phi\, d\theta$$

$$= \int_{\pi/4}^{\pi/2} \int_0^{2\pi} \int_0^2 \rho^2\sin\phi\, d\rho\, d\theta\, d\phi = \left(\int_{\pi/4}^{\pi/2} \sin\phi\, d\phi\right)\left(\int_0^{2\pi} d\theta\right)\left(\int_0^2 \rho^2\, d\rho\right)$$

$$= (\sqrt{2}/2)(2\pi)(\tfrac{8}{3}) = \frac{8\pi\sqrt{2}}{3} \cong 11.85.$$

This is about 70.7 percent of the volume of a hemisphere of radius 2. ❑

Example 3 Find the volume of the solid T enclosed by the surface

$$(x^2 + y^2 + z^2)^2 = 2z(x^2 + y^2).$$

SOLUTION In spherical coordinates the bounding surface takes the form

$$\rho = 2 \sin^2 \phi \cos \phi. \qquad \text{(check this out)}$$

This equation places no restriction on θ; thus θ can range from 0 to 2π. Since ρ remains nonnegative, ϕ can range only from 0 to $\frac{1}{2}\pi$. Thus the solid T is the set of all (x, y, z) with spherical coordinates (ρ, θ, ϕ) in the set

$$S: \quad 0 \le \theta \le 2\pi, \quad 0 \le \phi \le \tfrac{1}{2}\pi, \quad 0 \le \rho \le 2 \sin^2 \phi \cos \phi.$$

The rest is straightforward:

$$V = \iiint_T dx\, dy\, dz = \iiint_S \rho^2 \sin \phi \, d\rho\, d\theta\, d\phi$$

$$= \int_0^{2\pi} \int_0^{\pi/2} \int_0^{2\sin^2\phi\cos\phi} \rho^2 \sin \phi \, d\rho\, d\phi\, d\theta = \int_0^{2\pi} \int_0^{\pi/2} \tfrac{8}{3} \sin^7 \phi \cos^3 \phi \, d\phi\, d\theta$$

$$= \tfrac{8}{3} \left(\int_0^{2\pi} d\theta \right) \left(\int_0^{\pi/2} (\sin^7 \phi \cos \phi - \sin^9 \phi \cos \phi) \, d\phi \right)$$

$$= \tfrac{8}{3}(2\pi)(\tfrac{1}{40}) = \tfrac{2}{15}\pi \cong 0.42. \quad \square$$

EXERCISES 16.10

1. Find the spherical coordinates (ρ, θ, ϕ) of the point with rectangular coordinates $(1, 1, 1)$.

2. Find the rectangular coordinates of the point with spherical coordinates $(2, \frac{1}{6}\pi, \frac{1}{4}\pi)$.

3. Find the rectangular coordinates of the point with spherical coordinates $(3, \frac{1}{3}\pi, \frac{1}{6}\pi)$.

4. Find the spherical coordinates of the point with cylindrical coordinates $(2, \frac{2}{3}\pi, 6)$.

5. Find the spherical coordinates of the point with rectangular coordinates $(2, 2, \frac{2}{3}\sqrt{6})$.

6. Find the spherical coordinates of the point with rectangular coordinates $(2\sqrt{2}, -2\sqrt{2}, -4\sqrt{3})$.

7. Find the rectangular coordinates of the point with spherical coordinates $(3, \pi/2, 0)$.

8. Find the spherical coordinates of the point with cylindrical coordinates $(1/\sqrt{2}, \pi/4, 1/\sqrt{2})$.

In Exercises 9–14, equations are given in spherical coordinates. Interpret each one geometrically.

9. $\rho \sin \phi = 1$.

10. $\sin \phi = 1$.

11. $\cos \phi = -\frac{1}{2}\sqrt{2}$.

12. $\tan \theta = 1$.

13. $\rho \cos \phi = 1$.

14. $\rho = \cos \phi$.

In Exercises 15–18, evaluate the repeated integral.

15. $\displaystyle\int_0^{\pi/3} \int_0^{2\pi} \int_0^1 \rho^2 \sin \phi \, d\rho\, d\theta\, d\phi$.

16. $\displaystyle\int_0^{2\pi} \int_0^{\pi} \int_0^2 \rho^3 \sin \phi \, d\rho\, d\phi\, d\theta$.

17. $\displaystyle\int_0^{\pi/4} \int_0^{\pi} \int_0^{2\cos\phi} \rho^2 \sin \phi \, d\rho\, d\theta\, d\phi$.

18. $\displaystyle\int_0^{\pi/4} \int_0^{2\pi} \int_0^{\sec\phi} \rho^3 \sin \phi \cos \phi \, d\rho\, d\theta\, d\phi$.

In Exercises 19–22, evaluate the repeated integral by changing to spherical coordinates.

19. $\displaystyle\int_0^1 \int_0^{\sqrt{1-x^2}} \int_{\sqrt{x^2+y^2}}^{\sqrt{2-x^2-y^2}} dz\, dy\, dx$.

20. $\displaystyle\int_0^2 \int_0^{\sqrt{4-y^2}} \int_{\sqrt{x^2+y^2}}^{\sqrt{4-x^2-y^2}} (x^2 + y^2 + z^2)\, dz\, dx\, dy$.

21. $\displaystyle\int_0^3 \int_0^{\sqrt{9-y^2}} \int_0^{\sqrt{9-x^2-y^2}} z\sqrt{x^2+y^2+z^2} \, dz\, dx\, dy$.

22. $\displaystyle\int_0^1 \int_0^{\sqrt{1-x^2}} \int_0^{\sqrt{1-x^2-y^2}} \frac{1}{x^2+y^2+z^2} dz\, dy\, dx$.

23. Derive the formula for the volume of a sphere of radius R using spherical coordinates.

24. Express cylindrical coordinates in terms of spherical coordinates.

25. A wedge is cut from a ball of radius R by two planes that meet in a diameter. Find the volume of the wedge if the angle between the planes is α radians.

26. Find the mass of a ball of radius R given that the density varies directly with the distance from the boundary.

27. Find the mass of a right circular cone of base radius r and height h given that the density varies directly with the distance from the vertex.

28. Use spherical coordinates to derive the formula for the volume of a right circular cone of base radius r and height h.

In Exercises 29 and 30, let T be a homogeneous ball of mass M and radius R.

29. Calculate the moment of inertia about: (a) a diameter; (b) a tangent line.

30. Locate the center of mass of the upper half given that the center of the ball is at the origin.

In Exercises 31 and 32, let T be a homogeneous solid bounded by two concentric spherical shells, an outer shell of radius R_2 and an inner shell of radius R_1.

31. (a) Calculate the moment of inertia about a diameter.
    (b) Use your result in (a) to determine the moment of inertia of a spherical shell of radius R and mass M about a diameter.
    (c) What is the moment of inertia of that same shell about a tangent line?

32. (a) Locate the center of mass of the upper half of T given that the center of T is at the origin.
    (b) Use your result in (a) to locate the center of mass of a homogeneous hemispherical shell of radius R.

33. Find the volume of the solid common to the sphere $\rho = a$ and the cone $\phi = \alpha$. Take $\alpha \in (0, \frac{1}{2}\pi)$.

34. Let T be the solid bounded below by the half-cone $z = \sqrt{x^2 + y^2}$ and above by the spherical surface $x^2 + y^2 + z^2 = 1$. Use spherical coordinates to evaluate

$$\iiint_T e^{(x^2 + y^2 + z^2)^{3/2}} \, dx \, dy \, dz.$$

35. (a) Find an equation for the sphere
    $x^2 + y^2 + (z - R)^2 = R^2$ in spherical coordinates.

(b) Express the upper half of the ball
$x^2 + y^2 + (z - R)^2 \le R^2$ by inequalities in spherical coordinates.

36. Find the mass of the ball $\rho \le 2R \cos \phi$ given that the density varies directly with:
    (a) ρ. (b) $\rho \sin \phi$.
    (c) $\rho \cos^2\theta \sin \phi$.

37. Find the volume of the solid common to the spheres $\rho = 2\sqrt{2} \cos \phi$ and $\rho = 2$.

38. Find the volume of the solid enclosed by the surface $\rho = 1 - \cos \phi$.

39. Finish the argument in (16.10.3).

40. (*Gravitational Attraction*) Let T be a basic solid and let (a, b, c) be a point not in T. Show that, if T has continuously varying mass density $\lambda = \lambda(x, y, z)$, then T attracts a point mass m at (a, b, c) with a force

$$\mathbf{F} = \iiint_T Gm\lambda(x, y, z)\mathbf{f}(x, y, z) \, dx \, dy \, dz, \text{ where}$$

$$\mathbf{f}(x, y, z) = \frac{[(x - a)\mathbf{i} + (y - b)\mathbf{j} + (z - c)\mathbf{k}]}{[(x - a)^2 + (y - b)^2 + (z - c)^2]^{3/2}}.$$

(Assume that a point mass m_1 at P_1 attracts a point mass m_2 at P_2 with a force $\mathbf{F} = -(Gm_1m_2/r^3)\mathbf{r}$, where $\mathbf{r}$ is the vector $\overrightarrow{P_1P_2}$. Interpret the triple integral component by component.)

41. Let T be the upper half of the ball $x^2 + y^2 + (z - R)^2 \le R^2$. Given that T is homogeneous and has mass M, find the gravitational force exerted by T on a point mass m located at the origin. (Note Exercise 40.)

42. A point mass m is placed on the axis of a homogeneous solid right circular cylinder at a distance α from the nearest base of the cylinder. Find the gravitational force exerted by the cylinder on the point mass given that the cylinder has base radius R, height h, and mass M. (Note Exercise 40.)

■ 16.11 JACOBIANS; CHANGING VARIABLES IN MULTIPLE INTEGRATION

During the course of the last few sections you have met several formulas for changing variables in multiple integration: to polar coordinates, to cylindrical coordinates, to spherical coordinates. The purpose of this section is to bring some unity into that material and provide a general description for other changes of variable.

We begin with a consideration of area. Figure 16.11.1 shows a basic region Γ in a plane that we are calling the uv-plane. (In this plane we denote the abscissa of a point by u and the ordinate by v.) Suppose that

$$x = x(u, v), \qquad y = y(u, v)$$

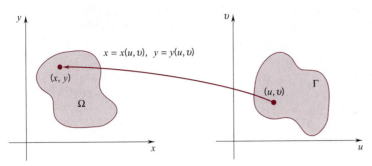

Figure 16.11.1

are functions continuously differentiable on Γ. As (u, v) ranges over Γ, the point $(x, y) = (x(u, v), y(u, v))$ generates a region Ω in the xy-plane. If the mapping

$$(u, v) \longrightarrow (x, y)$$

is one-to-one on the interior of Γ and the *Jacobian J* given by the 2×2 determinant

$$J(u, v) = \begin{vmatrix} \dfrac{\partial x}{\partial u} & \dfrac{\partial y}{\partial u} \\ \dfrac{\partial x}{\partial v} & \dfrac{\partial y}{\partial v} \end{vmatrix} = \frac{\partial x}{\partial u}\frac{\partial y}{\partial v} - \frac{\partial x}{\partial v}\frac{\partial y}{\partial u}$$

is never zero on the interior of Γ, then

(16.11.1)
$$\text{area of } \Omega = \iint\limits_{\Gamma} |J(u, v)| \, du \, dv.$$

It is very difficult to prove this assertion without making additional assumptions. A proof valid for all cases of practical interest is given in the supplement to Section 17.5. At this point we simply assume this area formula and go on from there.

Suppose now that we want to integrate some continuous function $f = f(x, y)$ over Ω. If this proves difficult to do directly, then we can change variables to u, v and try to integrate over Γ instead. It follows from (16.11.1) that

(16.11.2)
$$\iint\limits_{\Omega} f(x, y) \, dx \, dy = \iint\limits_{\Gamma} f(x(u, v), y(u, v))|J(u, v)| \, du \, dv.$$

The derivation of this formula from (16.11.1) follows the usual lines. Break up Γ into N little basic regions $\Gamma_1, \ldots, \Gamma_N$. These induce a decomposition of Ω into N little basic regions $\Omega_1, \ldots, \Omega_N$. We can then write

$$\iint_{\Gamma} f(x(u, v), y(u, v))|J(u, v)| \, du \, dv = \sum_{i=1}^{N} \iint_{\Gamma_i} f(x(u, v), y(u, v))|J(u, v)| \, du \, dv$$

additivity ⌐

$$= \sum_{i=1}^{N} f(x(u_i^*, v_i^*), y(u_i^*, v_i^*)) \iint_{\Gamma_i} |J(u, v)| \, du \, dv$$

Theorem (16.3.5) applied to Γ_i ⌐

$$= \sum_{i=1}^{N} f(x_i^*, y_i^*) \iint_{\Gamma_i} |J(u, v)| \, du \, dv$$

set $x_i^* = x(u_i^*, v_i^*), y_i^* = y(u_i^*, v_i^*)$ ⌐

$$= \sum_{i=1}^{N} f(x_i^*, y_i^*)(\text{area of } \Omega_i).$$

(16.11.1) applied to Γ_i ⌐

This last expression is a Riemann sum for

$$\iint_{\Omega} f(x, y) \, dx \, dy$$

and tends to that integral as the maximum diameter of the Ω_i tends to zero. This we can ensure by letting the maximum diameter of the Γ_i tend to zero. ❑

Example 1 Evaluate

$$\iint_{\Omega} (x + y)^2 \, dx \, dy$$

where Ω is the parallelogram bounded by the lines

$$x + y = 0, \qquad x + y = 1, \qquad 2x - y = 0, \qquad 2x - y = 3.$$

SOLUTION The parallelogram is shown in Figure 16.11.2. The boundaries suggest that we set

$$u = x + y, \qquad v = 2x - y.$$

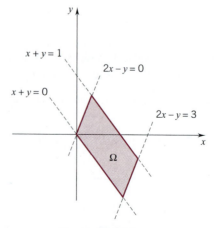

Figure 16.11.2

We want x and y in terms of u and v. Since

$$u + v = (x + y) + (2x - y) = 3x \qquad \text{and} \qquad 2u - v = (2x + 2y) - (2x - y) = 3y,$$

we have

$$x = \frac{u + v}{3}, \qquad y = \frac{2u - v}{3}.$$

This transformation maps the rectangle Γ of Figure 16.11.3 onto Ω. The Jacobian is given by

$$J(u, v) = \begin{vmatrix} \dfrac{\partial}{\partial u}\left(\dfrac{u + v}{3}\right) & \dfrac{\partial}{\partial u}\left(\dfrac{2u - v}{3}\right) \\[2mm] \dfrac{\partial}{\partial v}\left(\dfrac{u + v}{3}\right) & \dfrac{\partial}{\partial v}\left(\dfrac{2u - v}{3}\right) \end{vmatrix} = \begin{vmatrix} \frac{1}{3} & \frac{2}{3} \\[1mm] \frac{1}{3} & -\frac{1}{3} \end{vmatrix} = -\frac{1}{3}.$$

Figure 16.11.3

Therefore

$$\iint\limits_{\Omega} (x + y)^2 \, dx \, dy = \iint\limits_{\Gamma} u^2 \, |J(u, v)| \, du \, dv$$

$$= \tfrac{1}{3} \int_0^3 \int_0^1 u^2 \, du \, dv$$

$$= \tfrac{1}{3}\left(\int_0^3 dv\right)\left(\int_0^1 u^2 \, du\right) = \tfrac{1}{3}(3)\tfrac{1}{3} = \tfrac{1}{3}. \quad \square$$

Example 2 Evaluate

$$\iint\limits_{\Omega} xy \, dx \, dy$$

where Ω is the first-quadrant region bounded by the curves

$$x^2 + y^2 = 4, \qquad x^2 + y^2 = 9, \qquad x^2 - y^2 = 1, \qquad x^2 - y^2 = 4.$$

SOLUTION The region is shown in Figure 16.11.4. The boundaries suggest that we set

$$u = x^2 + y^2, \qquad v = x^2 - y^2.$$

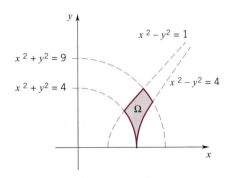

Figure 16.11.4

We want x and y in terms of u and v. Since

$$u + v = 2x^2 \quad \text{and} \quad u - v = 2y^2,$$

we have

$$x = \sqrt{\frac{u+v}{2}}, \qquad y = \sqrt{\frac{u-v}{2}}.$$

The transformation maps the rectangle Γ of Figure 16.11.5 onto Ω with Jacobian

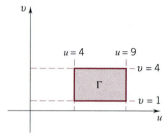

$u = 4$ $u = 9$

$v = 4$

Γ

$v = 1$

Figure 16.11.5

$$J(u, v) = \begin{vmatrix} \dfrac{\partial}{\partial u}\left(\sqrt{\dfrac{u+v}{2}}\right) & \dfrac{\partial}{\partial u}\left(\sqrt{\dfrac{u-v}{2}}\right) \\[2mm] \dfrac{\partial}{\partial v}\left(\sqrt{\dfrac{u+v}{2}}\right) & \dfrac{\partial}{\partial v}\left(\sqrt{\dfrac{u-v}{2}}\right) \end{vmatrix} \overset{\text{check this}}{\underset{}{=}} -\frac{1}{4\sqrt{u^2 - v^2}}.$$

Therefore

$$\iint_{\Omega} xy \, dx \, dy = \iint_{\Gamma} \left(\sqrt{\frac{u+v}{2}}\right)\left(\sqrt{\frac{u-v}{2}}\right)\left(\frac{1}{4\sqrt{u^2 - v^2}}\right) du \, dv$$

$$= \iint_{\Gamma} \tfrac{1}{8} \, du \, dv = \tfrac{1}{8}(\text{area of } \Gamma) = \tfrac{15}{8}. \quad \square$$

In Section 16.5 you saw the formula for changing variables from rectangular coordinates (x, y) to polar coordinates (r, θ). The formula reads

$$\iint_{\Omega} f(x, y) \, dx \, dy = \iint_{\Gamma} f(r \cos \theta, r \sin \theta) \, r \, dr \, d\theta. \qquad (16.5.3)$$

The factor r in the double integral over Γ is the Jacobian of the transformation $x = r \cos \theta, \, y = r \sin \theta$:

$$J(r, \theta) = \begin{vmatrix} \dfrac{\partial}{\partial r}(r \cos \theta) & \dfrac{\partial}{\partial r}(r \sin \theta) \\[2mm] \dfrac{\partial}{\partial \theta}(r \cos \theta) & \dfrac{\partial}{\partial \theta}(r \sin \theta) \end{vmatrix} = \begin{vmatrix} \cos \theta & \sin \theta \\ -r \sin \theta & r \cos \theta \end{vmatrix} = r(\cos^2 \theta + \sin^2 \theta) = r.$$

As you can see, (16.5.3) is a special case of (16.11.2).

When changing variables in a triple integral we make three coordinate changes:

$$x = x(u, v, w), \qquad y = y(u, v, w), \qquad z = z(u, v, w).$$

If these functions carry a basic solid Γ onto a solid T, then, under conditions analogous to the two-dimensional case,

$$\text{volume of } T = \iiint_{\Gamma} |J(u, v, w)| \, du \, dv \, dw$$

where now the Jacobian† is a three-by-three determinant:

$$
J(u, v, w) = \begin{vmatrix} \dfrac{\partial x}{\partial u} & \dfrac{\partial y}{\partial u} & \dfrac{\partial z}{\partial u} \\[2mm] \dfrac{\partial x}{\partial v} & \dfrac{\partial y}{\partial v} & \dfrac{\partial z}{\partial v} \\[2mm] \dfrac{\partial x}{\partial w} & \dfrac{\partial y}{\partial w} & \dfrac{\partial z}{\partial w} \end{vmatrix}.
$$

In this case the change of variables formula reads

$$
\iiint_T f(x, y, z)\, dx\, dy\, dz = \iiint_\Gamma f(x(u, v, w), y(u, v, w), z(u, v, w))\, |J(u, v, w)|\, du\, dv\, dw.
$$

---

†A historical note on Jacobians: the study of these functional determinants goes back to a memoir by the German mathematician C. G. Jacobi, 1804–1851.

EXERCISES 16.11

In Exercises 1–11, find the Jacobian of the transformation.

1. $x = au + bv$, $y = cu + dv$. (linear transformation)

2. $x = u \cos \theta - v \sin \theta$, $y = u \sin \theta + v \cos \theta$. (rotation by θ)

3. $x = uv$, $y = u^2 + v^2$.   **4.** $x = u \ln v$, $y = uv$.

5. $x = uv^2$, $y = u^2 v$.

6. $x = u - \ln v$, $y = \ln u + v$.

7. $x = au$, $y = bv$, $z = cw$.

8. $x = v + w$, $y = u + w$, $z = u + v$.

9. $x = r \cos \theta$, $y = r \sin \theta$, $z = z$. (cylindrical coordinates)

10. $x = \rho \sin \phi \cos \theta$, $y = \rho \sin \phi \sin \theta$, $z = \rho \cos \phi$. (spherical coordinates)

11. $x = (1 + w \cos v) \cos u$, $y = (1 + w \cos v) \sin u$, $z = w \sin v$.

12. Every linear transformation

$$x = au + bv, \quad y = cu + dv \quad \text{with} \quad ad - bc \neq 0$$

maps lines of the uv-plane onto lines of the xy-plane. Find the image of:
(a) a vertical line $u = u_0$;
(b) a horizontal line $v = v_0$.

For Exercises 13–15, take Ω as the parallelogram bounded by

$$x + y = 0, \quad x + y = 1, \quad x - y = 0, \quad x - y = 2.$$

Evaluate.

13. $\displaystyle\iint_\Omega (x^2 - y^2)\, dx\, dy.$ **14.** $\displaystyle\iint_\Omega 4xy\, dx\, dy.$

15. $\displaystyle\iint_\Omega (x + y) \cos [\pi(x - y)]\, dx\, dy.$

For Exercises 16–18, take Ω as the parallelogram bounded by $x - y = 0$, $x - y = \pi$, $x + 2y = 0$, $x + 2y = \frac{1}{2}\pi$.

Evaluate.

16. $\displaystyle\iint_\Omega (x + y)\, dx\, dy.$

17. $\displaystyle\iint_\Omega \sin (x - y) \cos (x + 2y)\, dx\, dy.$

18. $\displaystyle\iint_\Omega \sin 3x\, dx\, dy.$

19. Let Ω be the first-quadrant region bounded by the curves $xy = 1$, $xy = 4$, $y = x$, $y = 4x$.
(a) Determine the area of Ω and (b) locate the centroid.

20. Show that the ellipse $b^2 x^2 + a^2 y^2 = a^2 b^2$ has area πab setting $x = ar \cos \theta$, $y = br \cos \theta$.

21. A homogeneous plate in the xy-plane is in the form of a parallelogram. The parallelogram is bounded by the lines $x + y = 0$, $x + y = 1$, $3x - 2y = 0$, $3x - 2y = 2$. Calculate the moments of inertia of the plate about the three coordinate axes. Express your answers in terms of the mass of the plate.

22. Calculate the area of the region Ω bounded by the curves

$$x^2 - 2xy + y^2 + x + y = 0, \qquad x + y + 4 = 0.$$

HINT: Set $u = x - y$, $v = x + y$.

23. Calculate the area of the region Ω bounded by the curves

$$x^2 - 4xy + 4y^2 - 2x - y - 1 = 0, \qquad y = \tfrac{2}{5}.$$

24. Locate the centroid of the region in Exercise 22.

25. Calculate the area of the region Ω enclosed by the curve

$$11x^2 + 4\sqrt{3}xy + 7y^2 - 1 = 0.$$

HINT: Use a rotation $x = u \cos\theta - v \sin\theta$, $y = u \sin\theta + v \cos\theta$ such that the resulting uv-equation has no uv-term.

26. *(Generalized Polar Coordinates)* For a, b, α fixed positive numbers, define

$$x = ar(\cos\theta)^\alpha, \qquad y = br(\sin\theta)^\alpha.$$

(a) Show that the mapping carries the strip

$$0 \le r < \infty, \quad 0 \le \theta \le \tfrac{1}{2}\pi$$

onto the first quadrant of the xy-plane. [HINT: Find a point (r, θ) in the strip that maps onto (x, y) given that $x \ge 0$, $y \ge 0$.]
(b) Show that the mapping is one-to-one on the interior of the strip.

27. Determine the Jacobian of the mapping of Exercise 26.

28. Calculate the area enclosed by the curve $(x/a)^{1/4} + (y/b)^{1/4} = 1$ in the first quadrant by setting $x = ar \cos^8\theta$, $y = br \sin^8\theta$.

For Exercises 29–32, let T be the solid ellipsoid $x^2/a^2 + y^2/b^2 + z^2/c^2 \le 1$.

29. Calculate the volume of T by setting

$$x = a\rho \sin\phi \cos\theta, \quad y = b\rho \sin\phi \sin\theta, \quad z = c\rho \cos\phi.$$

30. Locate the centroid of the upper half of T.

31. View the upper half of T as a homogeneous solid of mass M. Find the moments of inertia of this solid about the coordinate axes.

32. Evaluate

$$\iiint_T \left(\frac{x^2}{a^2} + \frac{y^2}{b^2} + \frac{z^2}{c^2} \right) dx\, dy\, dz.$$

33. Evaluate

$$\int_{-\infty}^{\infty} \int_{-\infty}^{\infty} \frac{e^{-(x-y)^2}}{1 + (x + y)^2} dx\, dy$$

by integrating over the square S_a: $-a \le x \le a$, $-a \le y \le a$ and taking the limit as $a \to \infty$.
HINT: Set $u = x - y$, $v = x + y$ and see (16.5.4).

■ **CHAPTER HIGHLIGHTS**

16.1 Multiple-Sigma Notation

16.2 The Double Integral Over a Rectangle

partition (p. 1037)   upper sum, lower sum (p. 1038)
double integral over a rectangle (p. 1039)

16.3 The Double Integral Over a Region

basic region (p. 1044)   double integral over a region (p. 1045)
double integral as a volume (p. 1045)   area as a double integral (p. 1046)
properties of the double integral (pp. 1046–1047)
average value over a region (p. 1046)   weighted average over a region (p. 1047)

16.4 The Evaluation of Double Integrals by Repeated Integrals

evaluation formulas (pp. 1049–1050)   symmetry (p. 1056)
separated variables over a rectangle (p. 1060)

16.5 The Double Integral as the Limit of Riemann Sums; Polar Coordinates

diameter of a set (p. 1061)   Riemann sum (p. 1062)

$$\iint_\Omega f(x, y)\, dx\, dy = \lim_{\operatorname{diam} \Omega_i \to 0} \sum_{i=1}^{N} f(x_i^*, y_i^*)(\text{area of } \Omega_i)$$

evaluating double integrals using polar coordinates (p. 1063)
area formula (p. 1062)

16.6 Some Applications of Double Integration

16.7 Triple Integrals

16.8 Reduction to Repeated Integrals

16.9 Cylindrical Coordinates

16.10 The Triple Integral as the Limit of Riemann Sums; Spherical Coordinates

$$\iiint\limits_{T} f(x, y, z)\, dx\, dy\, dz = \lim_{\text{diam } T_i \to 0} \sum_{i=1}^{N} f(x_i^*, y_i^*, z_i^*)(\text{volume of } T_i)$$

16.11 Jacobians; Changing Variables in Multiple Integration

■ PROJECTS AND EXPLORATIONS USING TECHNOLOGY

To do these exercises you will need a graphics calculator or a computer with graphing capability. The majority of these problems are open-ended so different approaches may be used to solve them. You should be aware that different approaches can result in slight variations in the answers. Round your numerical answers to at least four decimal places. The rounding method that your calculator or computer uses also may cause variations in answers.

The methods used to approximate ordinary definite integrals can be extended to multiple integrals, but due to the large number of calculations involved, numerical methods for multiple integrals are much more time-consuming. This is illustrated in the following exercises for integrals of the form

$$I = \int_a^b \int_{g(x)}^{h(x)} f(x, y)\, dy\, dx.$$

16.1 (a) Write a program for your calculator or computer to implement the trapezoidal rule for the integral I.
   (b) Test your program on the integral

$$\int_1^5 \int_{\sin x}^{e^x} \cos(e^x + y)\, dy\, dx.$$

   (c) Explain why it takes much longer for this program to run than the trapezoidal rule for the definite integral of a function of one variable. Your program might run slow enough that you will be able to hand time it. If so, find an approximate relationship between n, the number of subintervals, and the length of time for your program to run.
   (d) If the number of subintervals is doubled when using the trapezoidal rule for a function of one variable, then the error is approximately quartered. See Section 8.8. Show that if J is the value of the integral of a function of one variable and J_k is the value of the trapezoidal approximation for k subintervals, then

$$J \cong \frac{4 J_{2k} - J_k}{3}.$$

   (e) Use the values from part (b) and the formula in part (d) to test whether doubling the number of subintervals in the trapezoidal approximation of a double integral also quarters the error.

16.2 (a) Write a program for your calculator or computer to implement Simpson's rule for the integral I.
   (b) Test your program on the integral

$$\int_1^5 \int_{\sin x}^{e^x} \cos(e^x + y)\, dy\, dx.$$

   (c) Explain why it takes much longer for this program to run than Simpson's rule for a function of one variable. Your program might run slow enough that you will be able to hand time it. If so, find the approximate relationship between n, the number of subintervals, and the length of time for your program to run.
   (d) If the number of subintervals is doubled when using Simpson's rule for a function of one variable, then the error is approximately divided by 16. See Section 8.8. Show that if J is the value of the integral of a function of one variable and J_k is the value of Simpson's rule approximation for k subintervals, then

$$J \cong \frac{16 J_{2k} - J_k}{15}.$$

   (e) Use the values from part (b) and the formula in part (d) to test whether doubling the number of subintervals in Simpson's rule approximation of a double integral also divides the error by 16.

16.3 (a) Write a program for your calculator or computer to approximate I using random numbers.
   (b) Run your program 10 times on the integral

$$\int_1^5 \int_{\sin x}^{e^x} \cos(e^x + y)\, dy\, dx.$$

   (c) Compute the mean and standard deviation of the numbers found in part (b).

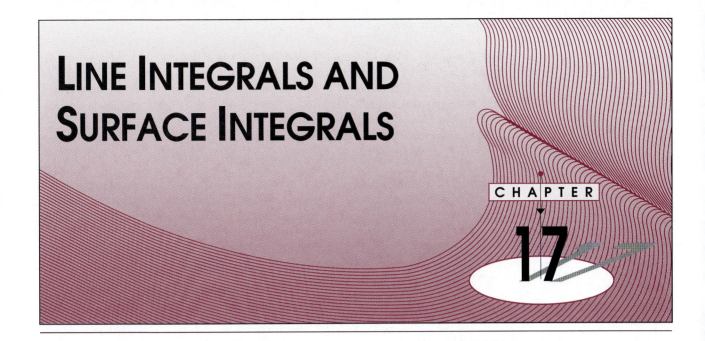

LINE INTEGRALS AND SURFACE INTEGRALS

CHAPTER 17

In this chapter we will study integration along curves and integration over surfaces. At the heart of this subject lie three great integration theorems: *Green's Theorem, Gauss' Theorem* (commonly known as *The Divergence Theorem*), and *Stokes's Theorem.*

All three theorems are ultimately based on *The Fundamental Theorem of Integral Calculus,* and all can be cast in the same general form:

An integral over a set S = a related integral over the boundary of S.

A word about terminology. Suppose that S is some subset of the plane or of three-dimensional space. A function that assigns a scalar to each point of S (say, the temperature at that point or the mass density at that point) is known in science as a *scalar field. A function that assigns a vector to each point of S (say, the wind-velocity at that point) is called a *vector field.* We will be using this "field" language throughout.

■ 17.1 LINE INTEGRALS

We are led to the definition of *line integral* by the notion of work.

The Work Done by a Varying Force Over a Curved Path

The work done by a constant force $\mathbf{F}$ on an object that moves along a straight line is, by definition, the component of $\mathbf{F}$ in the direction of the displacement multiplied by the length of the displacement vector $\mathbf{d}$ (see Exercises 12.3):

$$W = (\text{comp}_{\mathbf{d}} \, \mathbf{F}) \, \|\mathbf{d}\|$$

We can write this more briefly as a dot product:

(17.1.1)

$$W = \mathbf{F} \cdot \mathbf{d}$$

This elementary notion of work is useful, but it is not sufficient. Consider, for example, an object that moves through a magnetic field or a gravitational field. The path of the motion is then usually not a straight line but a curve, and the force, rather than remaining constant, tends to vary from point to point. What we want now is a notion of work that applies to this more general situation.

Let's suppose that an object moves along a curve

$$C: \quad \mathbf{r}(u) = x(u)\mathbf{i} + y(u)\mathbf{j} + z(u)\mathbf{k}, \qquad u \in [a, b]$$

subject to a continuous force $\mathbf{F}$. (The vector field $\mathbf{F}$ may vary from point to point, not only in magnitude but also in direction.) We will suppose that the curve is *smooth;* namely, we will suppose that the tangent vector $\mathbf{r}'$ is continuous and never zero. What we want to do here is define the total work done by $\mathbf{F}$ along the curve C.

To decide how to do this, we begin by focusing our attention on what happens over a short parameter interval $[u, u + h]$. As an estimate for the work done over this interval we can use the dot product

$$\mathbf{F}(\mathbf{r}(u)) \cdot [\mathbf{r}(u + h) - \mathbf{r}(u)].$$

In making this estimate we are using the force vector at $\mathbf{r}(u)$ and we are replacing the curved path from $\mathbf{r}(u)$ to $\mathbf{r}(u + h)$ by the line segment from $\mathbf{r}(u)$ to $\mathbf{r}(u + h)$. (See Figure 17.1.1.) If we set

$$W(u) = \text{total work done by } \mathbf{F} \text{ from } \mathbf{r}(a) \text{ to } \mathbf{r}(u), \text{ and}$$

$$W(u + h) = \text{total work done by } \mathbf{F} \text{ from } \mathbf{r}(a) \text{ to } \mathbf{r}(u + h),$$

then the work done by $\mathbf{F}$ from $\mathbf{r}(u)$ to $\mathbf{r}(u + h)$ must be the difference

$$W(u + h) - W(u).$$

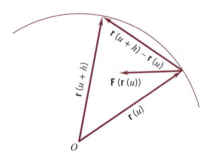

Figure 17.1.1

Bringing our estimate into play, we are led to the approximate equation

$$W(u + h) - W(u) \cong \mathbf{F}(\mathbf{r}(u)) \cdot [\mathbf{r}(u + h) - \mathbf{r}(u)],$$

which, upon division by h, becomes

$$\frac{W(u + h) - W(u)}{h} \cong \mathbf{F}(\mathbf{r}(u)) \cdot \frac{[\mathbf{r}(u + h) - \mathbf{r}(u)]}{h}.$$

The quotients here are average rates of change, and the equation is only an approximate one. The notion of work is made precise by requiring that both sides have exactly the same limit as h tends to zero; in other words, by requiring that

$$W'(u) = \mathbf{F}(\mathbf{r}(u)) \cdot \mathbf{r}'(u).$$

The rest is now determined. Since

$$W(a) = 0 \qquad \text{and} \qquad W(b) = \text{total work done by } \mathbf{F} \text{ on } C,$$

we have

total work done by $\mathbf{F}$ on $C = W(b) - W(a) = \int_a^b W'(u)\, du = \int_a^b [\mathbf{F}(\mathbf{r}(u)) \cdot \mathbf{r}'(u)]\, du.$

In short, we have arrived at the following definition of work:

(17.1.2)

$$W = \int_a^b [\mathbf{F}(\mathbf{r}(u)) \cdot \mathbf{r}'(u)]\, du.$$

Example 1 Determine the work done by the vector field

$$\mathbf{F}(x, y, z) = xy\,\mathbf{i} + 2\mathbf{j} + 4z\,\mathbf{k}$$

along the circular helix C: $\quad r(u) = \cos u\,\mathbf{i} + \sin u\,\mathbf{j} + u\,\mathbf{k}, \qquad 0 \le u \le 2\pi.$

SOLUTION First we calculate $\mathbf{F}(\mathbf{r}(u))$ and $\mathbf{r}'(u)$:

$$\mathbf{F}(\mathbf{r}(u)) = \cos u \sin u\,\mathbf{i} + 2\mathbf{j} + 4u\,\mathbf{k},$$

$$\mathbf{r}'(u) = -\sin u\,\mathbf{i} + \cos u\,\mathbf{j} + \mathbf{k}.$$

Now,

$$\mathbf{F}(\mathbf{r}(u)) \cdot \mathbf{r}'(u) = -\sin^2 u \cos u + 2 \cos u + 4u$$

and

$$W = \int_0^{2\pi} \mathbf{F}(\mathbf{r}(u)) \cdot \mathbf{r}'(u)\, du = \int_0^{2\pi} (-\sin^2 u \cos u + 2 \cos u + 4u)\, du$$

$$= \left[-\frac{\sin^3 u}{3} + 2 \sin u + 2u^2 \right]_0^{2\pi} = 8\pi^2. \quad \square$$

Line Integrals

The integral on the right of (17.1.2) can be calculated not only for a force function $\mathbf{F}$ but for any vector field $\mathbf{h}$ continuous on C.

DEFINITION 17.1.3 LINE INTEGRAL

Let $\mathbf{h}(x, y, z) = h_1(x, y, z)\mathbf{i} + h_2(x, y, z)\mathbf{j} + h_3(x, y, z)\mathbf{k}$ be a vector field continuous on a smooth curve

$$C: \quad \mathbf{r}(u) = x(u)\mathbf{i} + y(u)\mathbf{j} + z(u)\mathbf{k}, \qquad u \in [a, b].$$

The *line integral* of $\mathbf{h}$ over C is the number

$$\int_C \mathbf{h}(\mathbf{r}) \cdot d\mathbf{r} = \int_a^b [\mathbf{h}(\mathbf{r}(u)) \cdot \mathbf{r}'(u)]\, du.$$

Although we stated this definition in terms of three-dimensional vector fields and curves in space, it also includes the two-dimensional case: $\mathbf{h}(x, y) = h_1(x, y)\mathbf{i} + h_2(x, y)\mathbf{j}$ is a two-dimensional vector field and C: $\mathbf{r}(u) = x(u)\mathbf{i} + y(u)\mathbf{j}$ is a smooth plane curve. Some of the examples which follow are two-dimensional.

Note also that, while we speak of integrating over C, we actually carry out the calculations over the parameter set $[a, b]$. If our definition of line integral is to make sense, the line integral as defined must be independent of the particular parametrization chosen for C. Within the limitations spelled out in the following this is indeed the case:

THEOREM 17.1.4

Let $\mathbf{h}$ be a vector field continuous on a smooth curve C. The line integral

$$\int_C \mathbf{h}(\mathbf{r}) \cdot d\mathbf{r} = \int_a^b [\mathbf{h}(\mathbf{r}(u)) \cdot \mathbf{r}'(u)] \, du$$

is left invariant by every *sense-preserving* change of parameter.

NOTE: Sense-preserving and sense-reversing changes of parameter were discussed in Section 13.3.

PROOF Suppose that ϕ maps $[c, d]$ onto $[a, b]$ and that ϕ' is positive and continuous on $[c, d]$. We must show that the line integral over C as parametrized by

$$\mathbf{R}(w) = \mathbf{r}(\phi(w)), \qquad w \in [c, d]$$

equals the line integral over C as parametrized by $\mathbf{r}$. The argument is straightforward:

$$\int_C \mathbf{h}(\mathbf{R}) \cdot d\mathbf{R} = \int_c^d [\mathbf{h}(\mathbf{R}(w)) \cdot \mathbf{R}'(w)] \, dw$$

$$= \int_c^d [\mathbf{h}(\mathbf{r}(\phi(w))) \cdot \mathbf{r}'(\phi(w)) \, \phi'(w)] \, dw$$

$$\left. \begin{array}{l} \text{Set } u = \phi(w), \, du = \phi'(w) \, dw. \\ \text{At } w = c, \, u = a; \text{ at } w = d, \, u = b. \end{array} \right\} \longrightarrow \quad = \int_c^d [\mathbf{h}(\mathbf{r}(\phi(w))) \cdot \mathbf{r}'(\phi(w))]\phi'(w) \, dw$$

$$= \int_a^b [\mathbf{h}(\mathbf{r}(u)) \cdot \mathbf{r}'(u)] \, du = \int_C \mathbf{h}(\mathbf{r}) \cdot d\mathbf{r}. \quad \square$$

Example 2 Calculate

$$\int_C \mathbf{h}(\mathbf{r}) \cdot d\mathbf{r}$$

given that

$$\mathbf{h}(x, y) = xy\mathbf{i} + y^2\mathbf{j} \qquad \text{and} \qquad C: \ \mathbf{r}(u) = u\mathbf{i} + u^2\mathbf{j}, \qquad u \in [0, 1].$$

SOLUTION Here

$$x(u) = u, \qquad y(u) = u^2$$

and

$$\begin{aligned}
\mathbf{h}(\mathbf{r}(u)) \cdot \mathbf{r}'(u) &= [x(u)y(u)\mathbf{i} + [y(u)]^2\mathbf{j}] \cdot [x'(u)\mathbf{i} + y'(u)\mathbf{j}] \\
&= x(u)y(u)x'(u) + [y(u)]^2 y'(u) \\
&= u(u^2)1 + u^4(2u) = u^3 + 2u^5.
\end{aligned}$$

If follows that

$$\int_C \mathbf{h}(\mathbf{r}) \cdot d\mathbf{r} = \int_0^1 (u^3 + 2u^5)\, du = \left[\tfrac{1}{4}u^4 + \tfrac{1}{3}u^6 \right]_0^1 = \tfrac{7}{12}. \quad \square$$

Example 3 Integrate the vector field $\mathbf{h}(x, y, z) = xy\,\mathbf{i} + yz\,\mathbf{j} + xz\,\mathbf{k}$ over the twisted cubic

$$\mathbf{r}(u) = u\,\mathbf{i} + u^2\,\mathbf{j} + u^3\,\mathbf{k}$$

from $(-1, 1, -1)$ to $(1, 1, 1)$.

SOLUTION The path of integration begins at $u = -1$ and ends at $u = 1$. In this case

$$x(u) = u, \qquad y(u) = u^2, \qquad z(u) = u^3.$$

Therefore

$$\begin{aligned}
\mathbf{h}(\mathbf{r}(u)) \cdot \mathbf{r}'(u) &= [x(u)y(u)\mathbf{i} + y(u)z(u)\mathbf{j} + x(u)z(u)\mathbf{k}] \cdot [x'(u)\mathbf{i} + y'(u)\mathbf{j} + z'(u)\mathbf{k}] \\
&= x(u)y(u)x'(u) + y(u)z(u)y'(u) + x(u)z(u)z'(u) \\
&= u(u^2)1 + u^2(u^3)2u + u(u^3)3u^2 \\
&= u^3 + 5u^6
\end{aligned}$$

and

$$\int_C \mathbf{h}(\mathbf{r}) \cdot d\mathbf{r} = \int_{-1}^1 (u^3 + 5u^6)\, du = \left[\tfrac{1}{4}u^4 + \tfrac{5}{7}u^7 \right]_{-1}^1 = \tfrac{10}{7}. \quad \square$$

If a curve C is not smooth but is made up of a finite number of adjoining smooth pieces $C_1, C_2, \ldots, C_n$, then we define the integral over C as the sum of the integrals over the C_i:

(17.1.5)

$$\int_C = \int_{C_1} + \int_{C_2} + \cdots + \int_{C_n}$$

A curve of this type is said to be *piecewise smooth*. See Figure 17.1.2.

All polygonal paths are piecewise-smooth curves. In the next problem we integrate over a triangle. We do this by integrating over each of the sides and then adding up the results. Observe that the directed line segment that begins at $\mathbf{a} = (a_1, a_2)$ and ends at $\mathbf{b} = (b_1, b_2)$ can be parametrized by setting

$$\mathbf{r}(u) = (1 - u)\mathbf{a} + u\mathbf{b}, \qquad u \in [0, 1].$$

Figure 17.1.2

Example 4 Evaluate the line integral

$$\int_C \mathbf{h}(\mathbf{r}) \cdot d\mathbf{r}$$

if $\mathbf{h}(x, y) = e^y \mathbf{i} - \sin \pi x \, \mathbf{j}$ and C is the triangle with vertices $(1, 0)$, $(0, 1)$, $(-1, 0)$ traversed counterclockwise.

SOLUTION The path C is made up of three line segments:

$$C_1: \quad \mathbf{r}(u) = (1 - u)\mathbf{i} + u\mathbf{j}, \quad u \in [0, 1],$$

$$C_2: \quad \mathbf{r}(u) = (1 - u)\mathbf{j} + u(-\mathbf{i}) = -u\mathbf{i} + (1 - u)\mathbf{j}, \quad u \in [0, 1],$$

$$C_3: \quad \mathbf{r}(u) = (1 - u)(-\mathbf{i}) + u\mathbf{i} = (2u - 1)\mathbf{i}, \quad u \in [0, 1].$$

See Figure 17.1.3.

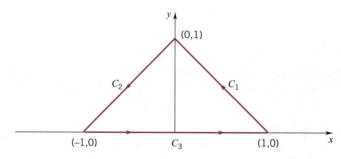

Figure 17.1.3

As you can verify,

$$\int_{C_1} \mathbf{h}(\mathbf{r}) \cdot d\mathbf{r} = \int_0^1 \left[e^{y(u)} x'(u) - \sin[\pi x(u)] y'(u) \right] du$$

$$= \int_0^1 \left[-e^u - \sin[\pi(1 - u)] \right] du = 1 - e - \frac{2}{\pi};$$

$$\int_{C_2} \mathbf{h}(\mathbf{r}) \cdot d\mathbf{r} = \int_0^1 \left[e^{y(u)} x'(u) - \sin\left[\pi x(u)\right] y'(u) \right] du$$

$$= \int_0^1 \left[-e^{1-u} + \sin\left(-\pi u\right) \right] du = 1 - e - \frac{2}{\pi};$$

$$\int_{C_1} \mathbf{h}(\mathbf{r}) \cdot d\mathbf{r} = \int_0^1 \left[e^{y(u)} x'(u) - \sin\left[\pi x(u)\right] y'(u) \right] du = \int_0^1 2 \, du = 2.$$

The integral over the entire triangle is the sum of these integrals:

$$\int_C \mathbf{h}(\mathbf{r}) \cdot d\mathbf{r} = \left(1 - e - \frac{2}{\pi}\right) + \left(1 - e - \frac{2}{\pi}\right) + 2 = 4 - 2e - \frac{4}{\pi} \cong -2.71. \quad \square$$

When we integrate over a parametrized curve, we integrate in the direction determined by the parametrization. If we integrate in the opposite direction, our answer is altered by a factor of -1. To be precise, let C be a piecewise-smooth curve and let $-C$ denote the same path traversed in the *opposite direction*. (See Figure 17.1.4.) If C is parameterized by a vector function $\mathbf{r}$ defined on $[a, b]$, then $-C$ can be parameterized by setting

$$R(w) = \mathbf{r}(a + b - w), \qquad w \in [a, b]. \qquad \text{(Section 13.3)}$$

Figure 17.1.4

Our assertion is that

(17.1.6)
$$\int_{-C} \mathbf{h}(\mathbf{R}) \cdot d\mathbf{R} = -\int_C \mathbf{h}(\mathbf{r}) \cdot d\mathbf{r}$$

or more briefly that

(17.1.7)
$$\int_{-C} = -\int_C$$

We leave the proof of this to you. ❏

We were led to the definition of line integral by the notion of work. From (17.1.2) it is clear that, if a force **F** is continually applied to an object that moves over a piecewise-smooth curve C, then the work done by **F** is the line integral of **F** over C:

(17.1.8)

$$W = \int_C \mathbf{F}(\mathbf{r}) \cdot d\mathbf{r}.$$

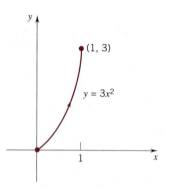

Figure 17.1.5

Example 5 An object, acted upon by various forces, moves along the parabola $y = 3x^2$ from the origin to the point $(1, 3)$. (Figure 17.1.5.) One of the forces acting on the object is $\mathbf{F}(x, y) = x^3\,\mathbf{i} + y\,\mathbf{j}$. Calculate the work done by **F**.

SOLUTION We can parameterize the path by setting

$$C: \quad \mathbf{r}(u) = u\,\mathbf{i} + 3u^2\,\mathbf{j}, \quad u \in [0, 1].$$

Here

$$x(u) = u, \qquad y(u) = 3u^2$$

and

$$\mathbf{F}(\mathbf{r}(u)) \cdot \mathbf{r}'(u) = [x(u)]^3 x'(u) + y(u)y'(u) = u^3(1) + 3u^2(6u) = 19u^3.$$

If follows that

$$W = \int_C \mathbf{F}(\mathbf{r}) \cdot d\mathbf{r} = \int_0^1 19u^3\,du = \tfrac{19}{4}. \quad \square$$

If an object of mass m moves so that at time t it has position $\mathbf{r}(t)$, then, from Newton's second law $\mathbf{F} = m\mathbf{a}$, the total force acting on the object at time t must be $m\mathbf{r}''(t)$.

Example 6 An object of mass m moves from time $t = 0$ to $t = 1$ so that its position at time t is given by the vector function

$$\mathbf{r}(t) = \alpha t^2\,\mathbf{i} + \sin\,\beta t\,\mathbf{j} + \cos\,\beta t\,\mathbf{k}, \qquad \alpha, \beta \text{ constant.}$$

Find the total force acting on the object at time t and calculate the total work done by this force.

SOLUTION Differentiation gives

$$\mathbf{r}'(t) = 2\alpha t\,\mathbf{i} + \beta\,\cos\,\beta t\,\mathbf{j} - \beta\,\sin\,\beta t\,\mathbf{k}, \qquad \mathbf{r}''(t) = 2\alpha\,\mathbf{i} - \beta^2\sin\,\beta t\,\mathbf{j} - \beta^2\cos\,\beta t\,\mathbf{k}.$$

The total force on the object at time t is therefore

$$m\mathbf{r}''(t) = m(2\alpha\,\mathbf{i} - \beta^2\sin\,\beta t\,\mathbf{j} - \beta^2\cos\,\beta t\,\mathbf{k}).$$

We can calculate the total work done by this force by integrating the force over the curve

$$C: \quad \mathbf{r}(t) = \alpha t^2\,\mathbf{i} + \sin\,\beta t\,\mathbf{j} + \cos\,\beta t\,\mathbf{k}, \qquad t \in [0, 1].$$

We leave it to you to verify that

$$W = \int_0^1 [m\mathbf{r}''(t) \cdot \mathbf{r}'(t)]\,dt = m\int_0^1 4\alpha^2 t\,dt = 2\alpha^2 m. \quad \square$$

EXERCISES 17.1

1. Integrate $\mathbf{h}(x, y) = y\mathbf{i} + x\mathbf{j}$ over the indicated path:
   (a) $\mathbf{r}(u) = u\mathbf{i} + u^2\mathbf{j}, \quad u \in [0, 1]$.
   (b) $\mathbf{r}(u) = u^3\mathbf{i} - 2u\mathbf{j}, \quad u \in [0, 1]$.

2. Integrate $\mathbf{h}(x, y) = x\mathbf{i} + y\mathbf{j}$ over the paths of Exercise 1.

3. Integrate $\mathbf{h}(x, y) = y\mathbf{i} + x\mathbf{j}$ over the unit circle traversed clockwise.

4. Integrate $\mathbf{h}(x, y) = y^2x\mathbf{i} + 2\mathbf{j}$ over the indicated path:
   (a) $\mathbf{r}(u) = e^u\mathbf{i} + e^{-u}\mathbf{j}, \quad u \in [0, 1]$.
   (b) $\mathbf{r}(u) = (1 - u)\mathbf{i}, \quad u \in [0, 2]$.

5. Integrate $\mathbf{h}(x, y) = (x - y)\mathbf{i} + xy\mathbf{j}$ over the indicated path:
   (a) The line segment from $(2, 3)$ to $(1, 2)$.
   (b) The line segment from $(1, 2)$ to $(2, 3)$.

6. Integrate $\mathbf{h}(x, y) = x^{-1}y^{-2}\mathbf{i} + x^{-2}y^{-1}\mathbf{j}$ over the indicated path:
   (a) $\mathbf{r}(u) = \sqrt{u}\mathbf{i} + \sqrt{1 + u}\mathbf{j}, \quad u \in [1, 4]$.
   (b) The line segment from $(1, 1)$ to $(2, 2)$.

7. Integrate $\mathbf{h}(x, y) = y\mathbf{i} - x\mathbf{j}$ over the triangle with vertices $(-2, 0), (2, 0), (0, 2)$ traversed counterclockwise.

8. Integrate $\mathbf{h}(x, y) = e^{x-y}\mathbf{i} + e^{x+y}\mathbf{j}$ over the line segment from $(-1, 1)$ to $(1, 2)$.

9. Integrate $\mathbf{h}(x, y) = (x + y)\mathbf{i} + (y^2 - x)\mathbf{j}$ over the closed curve that begins at $(-1, 0)$, goes along the x-axis to $(1, 0)$, and returns to $(-1, 0)$ by the upper part of the unit circle.

10. Integrate $\mathbf{h}(x, y) = 3x^2y\mathbf{i} + (x^3 + 2y)\mathbf{j}$ over the square with vertices $(0, 0), (1, 0), (1, 1), (0, 1)$ traversed counterclockwise.

11. Integrate $\mathbf{h}(x, y, z) = yz\mathbf{i} + x^2\mathbf{j} + xz\mathbf{k}$ over the indicated path:
    (a) The line segment from $(0, 0, 0)$ to $(1, 1, 1)$.
    (b) $\mathbf{r}(u) = u\mathbf{i} + u^2\mathbf{j} + u^3\mathbf{k}, \quad u \in [0, 1]$.

12. Integrate $\mathbf{h}(x, y, z) = e^x\mathbf{i} + e^y\mathbf{j} + e^z\mathbf{k}$ over the paths of Exercise 11.

13. Integrate $\mathbf{h}(x, y, z) = \cos x\mathbf{i} + \sin y\mathbf{j} + yz\mathbf{k}$ over the indicated path:
    (a) The line segment from $(0, 0, 0)$ to $(2, 3, -1)$.
    (b) $\mathbf{r}(u) = u^2\mathbf{i} - u^3\mathbf{j} + u\mathbf{k}, \quad u \in [0, 1]$.

14. Integrate $\mathbf{h}(x, y, z) = xy\mathbf{i} + x^2z\mathbf{j} + xyz\mathbf{k}$ over the indicated path:
    (a) The line segment from $(0, 0, 0)$ to $(2, -1, 1)$.
    (b) $\mathbf{r}(u) = e^u\mathbf{i} + e^{-u}\mathbf{j} + u\mathbf{k}, \quad u \in [0, 1]$.

15. Calculate the work done by the force $\mathbf{F}(x, y, z) = x\mathbf{i} + xy\mathbf{j} + xyz\mathbf{k}$ applied to an object that moves in a straight line from $(0, 1, 4)$ to $(1, 0, -4)$.

16. A mass m, moving in a force field, traces out a circular arc at constant speed. Show that the force field does no work. Give a physical explanation for this.

17. Find the work done by the force $\mathbf{F}(x, y) = (x + 2)y\mathbf{i} + (2x + y)\mathbf{j}$ applied to an object that moves along the indicated path:
    (a) The parabola from $y = x^2$ from $(0, 0)$ to $(2, 4)$.
    (b) The quarter circle $\mathbf{r}(u) = \cos u\mathbf{i} + \sin u\mathbf{j}, \quad u \in [0, \pi/2]$.

18. Find the work done by the force $\mathbf{F}(x, y) = x \cos y\mathbf{i} - y \sin x\mathbf{j}$ applied to an object that moves along the line segments that connect $(0, 0), (1, 0), (1, 1)$, and $(0, 1)$ in that order.

19. Find the work done by the force $\mathbf{F}(x, y, z) = x^2\mathbf{i} + xy\mathbf{j} + z^2\mathbf{k}$ applied to an object that moves along the circular helix $\mathbf{r}(u) = \cos u\mathbf{i} + \sin u\mathbf{j} + u\mathbf{k}, \quad u \in [0, 2\pi]$.

20. Find the work done by the force $\mathbf{F}(x, y, z) = yz\mathbf{i} + xz\mathbf{j} + xy\mathbf{k}$ applied to an object that moves along the line segments that connect $(0, 0, 0), (1, 0, 0), (1, 1, 0)$, and $(1, 1, 1)$ in that order.

21. Let $C: \mathbf{r} = \mathbf{r}(u), u \in [a, b]$ be a smooth curve and $\mathbf{q}$ a fixed vector. Show that
$$\int_C \mathbf{q} \cdot d\mathbf{r} = \mathbf{q} \cdot [\mathbf{r}(b) - \mathbf{r}(a)] \quad \text{and}$$
$$\int_C \mathbf{r} \cdot d\mathbf{r} = \frac{\|\mathbf{r}(b)\|^2 - \|\mathbf{r}(a)\|^2}{2}.$$

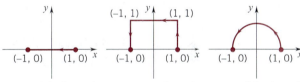

C_1: straight-line path C_2: rectangular path C_3: semicircular path

22. The preceding figure shows three paths from $(1, 0)$ to $(-1, 0)$. Calculate the line integral of
$$\mathbf{h}(x, y) = x^2\mathbf{i} + y\mathbf{j}$$
over (a) the straight-line path; (b) the rectangular path; (c) the semicircular path.

23. Let f be a continuous real-valued function of a real variable. Show that, if
$$\mathbf{f}(x, y, z) = f(x)\mathbf{i} \quad \text{and} \quad C: \mathbf{r}(u) = u\mathbf{i}, \quad u \in [a, b],$$
then
$$\int_C \mathbf{f}(\mathbf{r}) \cdot d\mathbf{r} = \int_a^b f(u)\, du.$$

24. (*Linearity*) Show that, if **f** and **g** are continuous vector fields and C is piecewise smooth, then

(17.1.9)
$$\int_C [\alpha \mathbf{f}(\mathbf{r}) + \beta \mathbf{g}(\mathbf{r})] \cdot d\mathbf{r}$$
$$= \alpha \int_C \mathbf{f}(\mathbf{r}) \cdot d\mathbf{r} + \beta \int_C \mathbf{g}(\mathbf{r}) \cdot d\mathbf{r}$$

for all real α, β.

25. The force $\mathbf{F}(x, y) = -\frac{1}{2}[y\mathbf{i} - x\mathbf{j}]$ is continually applied to an object that orbits an ellipse in standard position. Find a relation between the work done and the area of the ellipse.

26. An object of mass m moves from time $t = 0$ to $t = 1$ so that its position at time t is given by the vector function

$$\mathbf{r}(t) = \alpha t \mathbf{i} + \beta t^2 \mathbf{j}, \quad \alpha, \beta \text{ constant}$$

Find the total force acting on the object at time t and calculate the work done by that force during the time interval $[0, 1]$.

27. Exercise 26 with $\mathbf{r}(t) = \alpha t \mathbf{i} + \beta t^2 \mathbf{j} + \gamma t^3 \mathbf{k}$.

28. (*Important*) The *circulation* of a vector field **v** around an oriented closed curve C is by definition the line integral

$$\int_C \mathbf{v}(\mathbf{r}) \cdot d\mathbf{r}.$$

Let **v** be the velocity field of a fluid in counterclockwise circular motion about the z-axis with constant angular speed ω.
(a) Verify that $\mathbf{v}(\mathbf{r}) = \omega \mathbf{k} \times \mathbf{r}$.
(b) Show that the circulation of **v** around any circle C in the xy-plane with center at the origin is $\pm 2\omega$ times the area of the circle.

29. Let **v** be the velocity field of a fluid that moves radially from the origin: $\mathbf{v} = f(x, y)\mathbf{r}$. What is the circulation of **v** around a circle C centered at the origin?

30. The force exerted on a charged particle at the point $(x, y) \neq (0, 0)$ in the xy-plane by an infinitely long uniformly charged wire lying along the z-axis is given by

$$\mathbf{F}(x, y) = \frac{k(x\mathbf{i} + y\mathbf{j})}{x^2 + y^2}, \quad k > 0, \quad \text{constant.}$$

Find the work done by **F** in moving the particle along the indicated path:
(a) The line segment from $(1, 0)$ to $(1, 2)$.
(b) The line segment from $(0, 1)$ to $(1, 1)$.

31. An inverse-square force field is given by

$$\mathbf{F}(x, y, z) = \frac{k\mathbf{r}}{\|\mathbf{r}\|^3} = \frac{k(x\mathbf{i} + y\mathbf{j} + z\mathbf{k})}{(x^2 + y^2 + z^2)^{3/2}}, \quad k > 0, \text{ constant.}$$

Find the work done by **F** in moving an object along the indicated path:
(a) The line segment from $(1, 0, 2)$ to $(1, 3, 2)$.
(b) From $(1, 0, 0)$ to $(0, \frac{5}{2}\sqrt{2}, \frac{5}{2}\sqrt{2})$ by the line segment from $(1, 0, 0)$ to $(5, 0, 0)$ and then to $(0, \frac{5}{2}\sqrt{2}, \frac{5}{2}\sqrt{2})$ along a path on the surface of the sphere $x^2 + y^2 + z^2 = 25$.

32. Assume that the earth is located at the origin of a rectangular coordinate system. The gravitational force on an object at the point (x, y) is given by

$$\mathbf{F}(x, y) = \frac{-k\mathbf{r}}{\|\mathbf{r}\|^3} = \frac{-k(x\mathbf{i} + y\mathbf{j})}{(x^2 + y^2)^{3/2}}, \quad k > 0, \quad \text{constant.}$$

Find the work done by **F** in moving an object from $(3, 0)$ to $(0, 4)$ along the indicated path:
(a) The first quadrant part of the ellipse $\mathbf{r}(u) = 3\cos u \mathbf{i} + 4\sin u \mathbf{j}, \quad u \in [0, \pi/2]$.
(b) The line segment connecting the two points.

33. An object moves from the point $(0, 0)$ to the point $(1, 0)$ along the curve $y = \alpha x(1 - x)$ in the force field $\mathbf{F}(x, y) = (y^2 + 1)\mathbf{i} + (x + y)\mathbf{j}$. Find α so that the work done by **F** is a minimum.

■ 17.2 THE FUNDAMENTAL THEOREM FOR LINE INTEGRALS

In general, if we integrate a vector field **h** from one point to another, the value of the line integral depends upon the path chosen. There is, however, an important exception. If the vector field is a *gradient field*,

$$\mathbf{h} = \nabla f,$$

then the value of the line integral depends only on the endpoints of the path and not on the path itself. The details are spelled out in the following theorem.

THEOREM 17.2.1   THE FUNDAMENTAL THEOREM FOR LINE INTEGRALS

Let $C: \mathbf{r} = \mathbf{r}(u)$, $u \in [a, b]$, be a piecewise-smooth curve that begins at $\mathbf{a} = \mathbf{r}(a)$ and ends at $\mathbf{b} = \mathbf{r}(b)$. If the scalar field f is continuously differentiable on an open set that contains the curve C, then

$$\int_C \nabla f(\mathbf{r}) \cdot d\mathbf{r} = f(\mathbf{b}) - f(\mathbf{a}).$$

NOTE: It is important to see that this result is an extension of the fundamental theorem of integral calculus: $\int_a^b f'(x)\, dx = f(b) - f(a)$. (See Theorem 5.4.2.)

PROOF   If C is smooth,

$$\int_C \nabla f(\mathbf{r}) \cdot d\mathbf{r} = \int_a^b [\nabla f(\mathbf{r}(u)) \cdot \mathbf{r}'(u)]\, du$$

$$= \int_a^b \frac{d}{du}[f(\mathbf{r}(u))]\, du$$

chain rule (15.4.1) ⟶

$$= f(\mathbf{r}(b)) - f(\mathbf{r}(a))$$
$$= f(\mathbf{b}) - f(\mathbf{a}).$$

If C is not smooth but only piecewise smooth, then we break up C into smooth pieces

$$C = C_1 \cup C_2 \cup \cdots \cup C_n.$$

With obvious notation,

$$\int_C \nabla f(\mathbf{r}) \cdot d\mathbf{r} = \int_{C_1} \nabla f(\mathbf{r}) \cdot d\mathbf{r} + \int_{C_2} \nabla f(\mathbf{r}) \cdot d\mathbf{r} + \cdots + \int_{C_n} \nabla f(\mathbf{r}) \cdot d\mathbf{r}$$

$$= [f(\mathbf{a}_1) - f(\mathbf{a}_0)] + [f(\mathbf{a}_2) - f(\mathbf{a}_1)] + \cdots + [f(\mathbf{a}_n) - f(\mathbf{a}_{n-1})]$$
$$= f(\mathbf{a}_n) - f(\mathbf{a}_0)$$
$$= f(\mathbf{b}) - f(\mathbf{a}). \quad ❑$$

The theorem we just proved has an important corollary:

(17.2.2)

If the curve C is *closed* (that is, if $\mathbf{a} = \mathbf{b}$), then $f(\mathbf{b}) = f(\mathbf{a})$ and

$$\int_C \nabla f(\mathbf{r}) \cdot d\mathbf{r} = 0.$$

Example 1   Integrate the vector field $\mathbf{h}(x, y) = y^2 \mathbf{i} + (2xy - e^y)\mathbf{j}$ over the circular arc

$$C: \quad \mathbf{r}(u) = \cos u\, \mathbf{i} + \sin u\, \mathbf{j}, \qquad u \in [0, \tfrac{1}{2}\pi].$$

SOLUTION First we try to determine whether **h** is a gradient. We do this by applying Theorem 15.10.2.

Note that $\mathbf{h}(x, y)$ has the form $P(x, y)\mathbf{i} + Q(x, y)\mathbf{j}$ with

$$P(x, y) = y^2 \quad \text{and} \quad Q(x, y) = 2xy - e^y.$$

Since P and Q are continuously differentiable everywhere and

$$\frac{\partial P}{\partial y}(x, y) = 2y = \frac{\partial Q}{\partial x}(x, y),$$

we can conclude that **h** is a gradient. Since the integral depends then only on the endpoints of C and not on C itself, we can simplify the computations by integrating over the line segment C' that joins these same endpoints. (See Figure 17.2.1.)

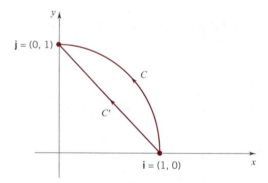

Figure 17.2.1

We parametrize C' by setting

$$\mathbf{r}(u) = (1 - u)\mathbf{i} + u\mathbf{j}, \quad u \in [0, 1].$$

We then have

$$\int_C \mathbf{h}(\mathbf{r}) \cdot d\mathbf{r} = \int_{C'} \mathbf{h}(\mathbf{r}) \cdot d\mathbf{r}$$

$$= \int_0^1 [\mathbf{h}(\mathbf{r}(u)) \cdot \mathbf{r}'(u)] \, du$$

$$= \int_0^1 \left[[y(u)]^2 x'(u) + [2x(u)y(u) - e^{y(u)}]y'(u) \right] du$$

$$= \int_0^1 [2u - 3u^2 - e^u] \, du = \left[u^2 - u^3 - e^u \right]_0^1$$

$$= 1 - e \approx -1.718 \quad \square$$

ALTERNATIVE SOLUTION Once we have recognized that $\mathbf{h}(x, y) = y^2\mathbf{i} + (2xy - e^y)\mathbf{j}$ is a gradient $\nabla f(x, y)$, we can determine $f(x, y)$ by the methods of

Section 15.10. Since

$$\frac{\partial f}{\partial x}(x, y) = y^2 \quad \text{and} \quad \frac{\partial f}{\partial y}(x, y) = 2xy - e^y,$$

we have

$$f(x, y) = xy^2 + \phi(y) \quad \text{and therefore} \quad \frac{\partial f}{\partial y}(x, y) = 2xy + \phi'(y).$$

The two expressions for $\partial f/\partial y$ can be reconciled only if

$$\phi'(y) = -e^y \quad \text{and thus} \quad \phi(y) = -e^y + c.$$

This means that

$$f(x, y) = xy^2 - e^y + c.$$

Since the curve C begins at $(1, 0)$ and ends at $(0, 1)$, we see that

$$\int_C \mathbf{h}(\mathbf{r}) \cdot d\mathbf{r} = f(0, 1) - f(1, 0) = (-e + c) - (-1 + c) = 1 - e. \quad \square$$

Example 2 Evaluate the line integral

$$\int_C \mathbf{h}(\mathbf{r}) \cdot d\mathbf{r}$$

where C is the unit circle

$$\mathbf{r}(u) = \cos u\, \mathbf{i} + \sin u\, \mathbf{j}, \quad u \in [0, 2\pi].$$

and

$$\mathbf{h}(x, y) = (y^2 + y)\mathbf{i} + (2xy - e^y)\mathbf{j}.$$

SOLUTION Although $(y^2 + y)\mathbf{i} + (2xy - e^y)\mathbf{j}$ is not a gradient, part of it,

$$y^2\mathbf{i} + (2xy - e^y)\mathbf{j},$$

is a gradient. (See Example 1.) Thus, we can write $\mathbf{h}$ as

$$\mathbf{h}(x, y) = (y^2 + y)\mathbf{i} + (2xy - e^y)\mathbf{j} = \nabla f(x, y) + \mathbf{g}(x, y),$$

where $\mathbf{g}(x, y) = y\mathbf{i}$, and

$$\int_C \mathbf{h}(\mathbf{r}) \cdot d\mathbf{r} = \int_C \nabla f(\mathbf{r}) \cdot d\mathbf{r} + \int_C \mathbf{g}(\mathbf{r}) \cdot d\mathbf{r}.$$

Now, since we are integrating over a closed curve, the contribution of the gradient part is 0. The contribution of the remaining part is easy to evaluate:

$$\int_C \mathbf{g}(\mathbf{r}) \cdot d\mathbf{r} = \int_0^{2\pi} [\mathbf{g}(\mathbf{r}(u)) \cdot \mathbf{r}'(u)]\, du = \int_0^{2\pi} y(u)x'(u)\, du$$

$$= \int_0^{2\pi} -\sin^2 u\, du = -\pi. \quad \square$$

EXERCISES 17.2

In Exercises 1–11, calculate the line integral of **h** over the indicated curve.

1. $\mathbf{h}(x, y) = x\mathbf{i} + y\mathbf{j}$; $\mathbf{r}(u) = a \cos u\mathbf{i} + b \sin u\mathbf{j}$, $u \in [0, 2\pi]$.

2. $\mathbf{h}(x, y) = (x + y)\mathbf{i} + y\mathbf{j}$; the curve of Exercise 1.

3. $\mathbf{h}(x, y) = \cos \pi y\mathbf{i} - \pi x \sin \pi y\mathbf{j}$; $\mathbf{r}(u) = u^2\mathbf{i} - u^3\mathbf{j}$, $u \in [0, 1]$.

4. $\mathbf{h}(x, y) = (x^2 - y)\mathbf{i} + (y^2 - x)\mathbf{j}$; the curve of Exercise 1.

5. $\mathbf{h}(x, y) = xy^2\mathbf{i} + x^2y\mathbf{j}$; $\mathbf{r}(u) = u \sin \pi u\mathbf{i} + \cos \pi u^2\mathbf{j}$, $u \in [0, 1]$.

6. $\mathbf{h}(x, y) = (1 + e^y)\mathbf{i} + (xe^y - x)\mathbf{j}$; the square with vertices $(-1, -1), (1, -1), (1, 1), (-1, 1)$ traversed counterclockwise.

7. $\mathbf{h}(x, y) = (2xy - y^2)\mathbf{i} + (x^2 - 2xy)\mathbf{j}$; $\mathbf{r}(u) = \cos u\mathbf{i} + \sin u\mathbf{j}$, $u \in [0, \pi]$.

8. $\mathbf{h}(x, y) = 3x(x^2 + y^4)^{1/2}\mathbf{i} + 6y^3(x^2 + y^4)^{1/2}\mathbf{j}$; the circular arc $y = (1 - x^2)^{1/2}$ from $(1,0)$ to $(-1, 0)$.

9. $\mathbf{h}(x, y) = 3x(x^2 + y^4)^{1/2}\mathbf{i} + 6y^3(x^2 + y^4)^{1/2}\mathbf{j}$; the arc $y = -(1 - x^2)^{1/2}$ from $(-1, 0)$ to $(1, 0)$.

10. $\mathbf{h}(x, y) = 2xy \sinh x^2y\mathbf{i} + x^2\sinh x^2y\mathbf{j}$; the curve of Exercise 1.

11. $\mathbf{h}(x, y) = (2x \cosh y - y)\mathbf{i} + (x^2\sinh y - y)\mathbf{j}$; the square of Exercise 6.

In Exercises 12–15, verify that the given vector field **h** is a gradient. Then calculate the line integral of **h** over the indicated curve C in two ways: (a) by calculating $\int_C \mathbf{h}(\mathbf{r}) \cdot d\mathbf{r}$ and (b) by finding f such that $\nabla f = \mathbf{h}$ and evaluating f at the endpoints of C.

12. $\mathbf{h}(x, y) = xy^2\mathbf{i} + yx^2\mathbf{j}$; $\mathbf{r}(u) = u\mathbf{i} + u^2\mathbf{j}$, $u \in [0, 2]$.

13. $\mathbf{h}(x, y) = (3x^2y^3 + 2x)\mathbf{i} + (3x^3y^2 - 4y)\mathbf{j}$; $\mathbf{r}(u) = u\mathbf{i} + e^u\mathbf{j}$, $u \in [0, 1]$.

14. $(2x \sin y - e^x)\mathbf{i} + (x^2 \cos y)\mathbf{j}$; $\mathbf{r}(u) = \cos u\mathbf{i} + u\mathbf{j}$, $u \in [0, \pi]$.

15. $\mathbf{h}(x, y) = (e^{2y} - 2xy)\mathbf{i} + (2xe^{2y} - x^2 + 1)\mathbf{j}$; $\mathbf{r}(u) = ue^u\mathbf{i} + (1 + u)\mathbf{j}$, $u \in [0, 1]$.

In Exercises 16–18, use the three dimensional analog of Theorem 15.10.2 given in Exercises 15.10 to show that the vector function **h** is a gradient. Then evaluate the line integral of **h** over the indicated curve.

16. $\mathbf{h}(x, y, z) = y^2z^3\mathbf{i} + 2xyz^3\mathbf{j} + 3xy^2z^2\mathbf{k}$; $\mathbf{r}(u) = u^2\mathbf{i} + u^4\mathbf{j} + u^6\mathbf{k}$, $u \in [0, 1]$.

17. $\mathbf{h}(x, y, z) = (2xz + \sin y)\mathbf{i} + x \cos y\mathbf{j} + x^2\mathbf{k}$; $\mathbf{r}(u) = \cos u\mathbf{i} + \sin u\mathbf{j} + u\mathbf{k}$, $u \in [0, 2\pi]$.

18. $\mathbf{h}(x, y, z) = e^{-x} \ln y\mathbf{i} - [\frac{e^{-x}}{y}]\mathbf{j} + 3z^2\mathbf{k}$; $\mathbf{r}(u) = (u + 1)\mathbf{i} + e^{2u}\mathbf{j} + (u^2 + 1)\mathbf{k}$, $u \in [0, 1]$.

19. Calculate the work done by the force $\mathbf{F}(x, y) = (x + e^{2y})\mathbf{i} + (2y + 2xe^{2y})\mathbf{j}$ applied to an object that moves along the curve $\mathbf{r}(u) = 3 \cos u\mathbf{i} + 4 \sin u\mathbf{j}$, $u \in [0, 2\pi]$.

20. Calculate the work done by the force $\mathbf{F}(x, y, z) = (2x \ln y - yz)\mathbf{i} + [(x^2/y) - xz]\mathbf{j} - xy\mathbf{k}$ applied to an object that moves from the point $(1, 2, 1)$ to the point $(3, 2, 2)$.

21. If g is a continuously differentiable real-valued function defined on $[a, b]$, then by the fundamental theorem of integral calculus

$$\int_a^b g'(u) \, du = g(b) - g(a).$$

Show how this result is included in Theorem 17.2.1.

22. Let $\mathbf{r} = x\mathbf{i} + y\mathbf{j} + z\mathbf{k}$ and set $r = \|\mathbf{r}\|$. The central force field

$$\mathbf{F}(\mathbf{r}) = \frac{K}{r^n}\mathbf{r}, \qquad n \text{ a positive integer}$$

is a gradient field. Find f such that $\nabla f(\mathbf{r}) = \mathbf{F}(\mathbf{r})$ if: (a) $n = 2$; (b) $n \neq 2$.

23. Let $\mathbf{r} = x\mathbf{i} + y\mathbf{j} + z\mathbf{k}$ and set $r = \|\mathbf{r}\|$. The function

$$\mathbf{F}(\mathbf{r}) = -\frac{mG}{r^3}\mathbf{r} \qquad (G \text{ is a gravitational constant})$$

gives the gravitational force exerted by a unit mass at the origin on a mass m located at $\mathbf{r}$. What is the work done by $\mathbf{F}$ if m moves from $\mathbf{r}_1$ to $\mathbf{r}_2$?

24. Set

$$P(x, y) = \frac{y}{x^2 + y^2} \qquad \text{and} \qquad Q(x, y) = -\frac{x}{x^2 + y^2}$$

on *the punctured unit disc* $\Omega: 0 < x^2 + y^2 < 1$.
(a) Verify that P and Q are continuously differentiable on Ω and that

$$\frac{\partial P}{\partial y}(x, y) = \frac{\partial Q}{\partial x}(x, y) \qquad \text{for all } (x, y) \in \Omega.$$

(b) Verify that, in spite of (a), the vector field $\mathbf{h}(x, y) = P(x, y)\mathbf{i} + Q(x, y)\mathbf{j}$ is not a gradient on Ω.
   HINT: Integrate **h** over a circle of radius less than one centered at the origin.
(c) Explain how (b) does not contradict Theorem 15.10.2.

25. The gravitational force acting on an object of mass m at a height z above the surface of the earth is given by

$$\mathbf{F}(x, y, z) = \frac{-mGr_0^2}{(r_0 + z)^2}\mathbf{k},$$

where G is the gravitational constant and r_0 is the radius of the earth. Show that $\mathbf{F}$ is a gradient field and find f such that $\nabla f = \mathbf{F}$.

▷ 26. A 4-ton rocket falls to earth from a height of 300 miles. How much work is done by the gravitational force? Use Exercise 25 and assume that the radius of the earth is 4000 miles.

▷ 27. An elevator carries 20 people to the fiftieth floor of a building. If the elevator weighs 5000 pounds, the people weigh 3000 lbs, and the fiftieth floor is 500 feet above the ground, find the work done by the elevator against the force of gravity. Use Exercise 25 and assume that the radius of the earth is 4000 miles.

---

*■ 17.3 WORK–ENERGY FORMULA; CONSERVATION OF MECHANICAL ENERGY

Suppose that a continuous force field $\mathbf{F} = \mathbf{F}(\mathbf{r})$ accelerates a mass m from $\mathbf{r} = \mathbf{a}$ to $\mathbf{r} = \mathbf{b}$ along some smooth curve C. The object undergoes a change in kinetic energy:

$$\tfrac{1}{2}m[v(\beta)]^2 - \tfrac{1}{2}m[v(\alpha)]^2.$$

The force does a certain amount of work W. How are these quantities related? They are equal:

(17.3.1)
$$W = \tfrac{1}{2}m[v(\beta)]^2 - \tfrac{1}{2}m[v(\alpha)]^2.$$

This relation is called the *Work–Energy Formula.*

Derivation of the Work–Energy Formula

We parametrize the path of the motion by the time parameter t:

$$C: \quad \mathbf{r} = \mathbf{r}(t), \quad t \in [\alpha, \beta],$$

Where $\mathbf{r}(\alpha) = \mathbf{a}$ and $\mathbf{r}(\beta) = \mathbf{b}$. The work done by $\mathbf{F}$ is given by the formula

$$W = \int_C \mathbf{F}(\mathbf{r}) \cdot d\mathbf{r} = \int_\alpha^\beta [\mathbf{F}(\mathbf{r}(t)) \cdot \mathbf{r}'(t)]\, dt.$$

From Newton's second law of motion, we know that at time t,

$$\mathbf{F}(\mathbf{r}(t)) = m\, \mathbf{a}(t) = m\, \mathbf{r}''(t).$$

It follows that

$$\mathbf{F}(\mathbf{r}(t)) \cdot \mathbf{r}'(t) = m\, \mathbf{r}''(t) \cdot \mathbf{r}'(t)$$
$$= \frac{d}{dt}\left[\frac{1}{2}m(\mathbf{r}'(t) \cdot \mathbf{r}'(t))\right] = \frac{d}{dt}\left[\frac{1}{2}m\|\mathbf{r}'(t)\|^2\right] = \frac{d}{dt}\left[\frac{1}{2}m[v(t)]^2\right].$$

Substituting this last expression into the work integral, we see that

$$W = \int_\alpha^\beta \frac{d}{dt}\left(\frac{1}{2}m[v(t)]^2\right) dt = \frac{1}{2}m[v(\beta)]^2 - \frac{1}{2}m[v(\alpha)]^2.$$

as asserted. ❑

Conservative Force Fields

In general, if an object moves from one point to another, the work done (and hence the change in kinetic energy) depends on the path of the motion. There is, however, an important exception: if the force field is a gradient field,

$$\mathbf{F} = \nabla f,$$

then the work done (and hence the change in kinetic energy) depends only on the endpoints of the path and not on the path itself. (This follows directly from the Fundamental Theorem for Line Integrals.) A force field that is a gradient field is called a *conservative field.*

Since the line integral over a closed path is zero, *the work done by a conservative field over a closed path is always zero. An object that passes through a given point with a certain kinetic energy returns to that same point with exactly the same kinetic energy.*

Potential Energy Functions

Suppose that $\mathbf{F}$ is a conservative force field. It is then a gradient field. Then $-\mathbf{F}$ is a gradient field. The functions U for which $\nabla U = -\mathbf{F}$ are called *potential energy functions* for $\mathbf{F}$.

The Conservation of Mechanical Energy

Suppose that $\mathbf{F}$ is a conservative force field: $\mathbf{F} = -\nabla U$. In our derivation of the work–energy formula we showed that

$$\frac{d}{dt}(\tfrac{1}{2}m[v(t)]^2) = \mathbf{F}(\mathbf{r}(t)) \cdot \mathbf{r}'(t).$$

Since

$$\frac{d}{dt}[U(\mathbf{r}(t))] = \nabla U(\mathbf{r}(t)) \cdot \mathbf{r}'(t) = -\mathbf{F}(\mathbf{r}(t)) \cdot \mathbf{r}'(t),$$

we have

$$\frac{d}{dt}[\tfrac{1}{2}m[v(t)]^2 + U(\mathbf{r}(t))] = 0$$

and therefore

$$\underbrace{\tfrac{1}{2}m[v(t)]^2}_{\text{KE}} + \underbrace{U(\mathbf{r}(t))}_{\text{PE}} = \text{a constant.}$$

As an object moves in a conservative force field, its kinetic energy can vary and its potential energy can vary, but the sum of these two quantities remains constant. We call this constant *the total mechanical energy.*

The total mechanical energy is usually denoted by the letter E. The law of conservation of mechanical energy can then be written

(17.3.2.)

$$\tfrac{1}{2}mv^2 + U = E.$$

The conservation of energy is one of the cornerstones of physics. Here we have been talking about mechanical energy. There are other forms of energy and other energy conservation laws.

Differences in Potential Energy

Potential energy at a particular point has no physical significance. Only differences in potential energy are significant:

$$U(\mathbf{b}) - U(\mathbf{a}) = \int_C -\mathbf{F}(\mathbf{r}) \cdot d\mathbf{r}$$

is the work required to move from $\mathbf{r} = \mathbf{a}$ to $\mathbf{r} = \mathbf{b}$ *against* the force field $\mathbf{F}$.

Example 1 A planet moves in the gravitational field of the sun:

$$\mathbf{F}(\mathbf{r}) = -m\rho \frac{\mathbf{r}}{r^3}. \qquad \text{(\textit{m} is the mass of the planet)}$$

Show that the force field is conservative, find a potential energy function, and determine the total energy of the planet. How does the planet's speed vary with the planet's distance from the sun?

SOLUTION The field is conservative since

$$\mathbf{F}(\mathbf{r}) = -m\rho \frac{\mathbf{r}}{r^3} = \nabla\left(\frac{\rho m}{r}\right). \qquad \text{(check this out)}$$

As a potential energy function we can use

$$U(\mathbf{r}) = -\frac{\rho m}{r}.$$

The total energy of the planet is the constant

$$E = \tfrac{1}{2}mv^2 - \frac{\rho m}{r}.$$

(You met this quantity before: Exercises 2 and 4 of Section 13.6.)
Solving the energy equation for v we have

$$v = \sqrt{\frac{2E}{m} + \frac{2\rho}{r}}.$$

As r decreases, $2\rho/r$ increases, and v increases; as r increases, $2\rho/r$ decreases, and v decreases. Thus every planet speeds up as it comes near the sun and slows down as it moves away. (The same holds true for Halley's comet. The fact that it slows down as it gets farther away helps explain why it comes back. The simplicity of all this is a testimony to the power of the principle of energy conservation.) ❑

EXERCISES 17.3

1. Let f be a continuous real-valued function of the real variable x. Show that the force field $F(x, y, z) = f(x)i$ is conservative and the potential functions for F are (except for notation) the antiderivatives of $-f$.

2. A particle with electric charge e and velocity v moves in a magnetic field B, experiencing the force

$$F = \frac{e}{c}[v \times B]. \quad (c \text{ is the velocity of light})$$

F is not a gradient—it can't be, depending as it does on the *velocity* of the particle. Still, we can find a conserved quantity: the *kinetic energy* $\frac{1}{2}mv^2$. Show by differentiation with respect to t that this quantity is constant. (Assume Newton's second law.)

3. An object is subject to a constant force in the direction of $-k$: $F = -ck$ with $c > 0$. Find a potential energy function for F, and use energy conservation to show that the speed of the object at time t_2 is related to that at time t_1 by the equation

$$v(t_2) = \sqrt{[v(t_1)]^2 + \frac{2c}{m}[z(t_1) - z(t_2)]}$$

where $z(t_1)$ and $z(t_2)$ are the z-coordinates of the object at times t_1 and t_2. (This analysis is sometimes used to model the behavior of an object in the gravitational field near the surface of the earth.)

4. (*Escape velocity*) An object is to be fired straight up from the surface of the earth. Assume that the only force acting on the object is the gravitational pull of the earth and determine the initial speed v_0 necessary to send the object off to infinity. HINT: Appeal to conservation of energy and use the idea that the object is to arrive at infinity with zero speed.

5. (a) Justify the statement that a conservative force field F always acts so as to encourage motion toward regions of lower potential energy U.
   (b) Evaluate F at a point where U has a minimum.

6. A harmonic oscillator has a restoring force $F = -\lambda x i$. The associated potential is $U(x, y, z) = -\frac{1}{2}\lambda x^2$, and the constant total energy is

$$E = \frac{1}{2}mv^2 + U(x, y, z) = \frac{1}{2}mv^2 + \frac{1}{2}\lambda x^2.$$

Given that $x(0) = 2$ and $x'(0) = 1$, calculate the maximum speed of the oscillator and the maximum value of x.

7. The *equipotential surfaces* of a conservative field F are the surfaces where the potential energy is constant. Show that: (a) the speed of an object in such a field is constant on every equipotential surface; and (b) at each point of such a surface the force field is perpendicular to the surface.

8. Suppose a force field F is directed away from the origin with a magnitude that is inversely proportional to the distance from the origin. Show that F is a conservative field.

9. Let F be the inverse-square force field:

$$F(x, y, z) = \frac{k(xi + yj + zk)}{(x^2 + y^2 + z^2)^{3/2}}$$

and let C be any curve on the unit sphere $x^2 + y^2 + z^2 = 1$. Show that the work done by F in moving an object along C is 0. Explain this result.

■ 17.4 ANOTHER NOTATION FOR LINE INTEGRALS; LINE INTEGRALS WITH RESPECT TO ARC LENGTH

If $h(x, y, z) = P(x, y, z)i + Q(x, y, z)j + R(x, y, z)k$, the line integral

$\int_C h(r) \cdot dr$ can be written $\int_C P(x, y, z)\, dx + Q(x, y, z)\, dy + R(x, y, z)\, dz.$

The notation arises as follows. With

$$C: \quad r(u) = x(u)i + y(u)j + z(u)k, \quad u \in [a, b]$$

the line integral

$$\int_C h(r) \cdot dr = \int_a^b [h(r(t)) \cdot r'(t)]\, dt$$

expands to

$$\int_a^b \{P[x(u), y(u), z(u)]x'(u) + Q[x(u), y(u), z(u)]y'(u) + R[x(u), y(u), z(u)]z'(u)\}\, du.$$

Now set

$$\int_C P(x, y, z)\, dx = \int_a^b P[x(u), y(u), z(u)]x'(u)\, du,$$

$$\int_C Q(x, y, z)\, dy = \int_a^b Q[x(u), y(u), z(u)]y'(u)\, du,$$

$$\int_C R(x, y, z)\, dz = \int_a^b R[x(u), y(u), z(u)]z'(u)\, du.$$

Writing the sum of these integrals as

$$\int_C P(x, y, z)\, dx + Q(x, y, z)\, dy + R(x, y, z)\, dz,$$

we have

$$\int_C P(x, y, z)\, dx + Q(x, y, z)\, dy + R(x, y, z)\, dz = \int_C \mathbf{h}(\mathbf{r}) \cdot d\mathbf{r}. \quad \square$$

If C lies in the xy-plane and $\mathbf{h}(x, y) = P(x, y)\mathbf{i} + Q(x, y)\mathbf{j}$, then the line integral reduces to

$$\int_C P(x, y)\, dx + Q(x, y)\, dy.$$

The notation is easy to use; for example, with C as before,

$$\int_C y\, dx + 2z\, dy - x\, dz = \int_a^b [y(u)x'(u) + 2z(u)y'(u) - x(u)z'(u)]\, du,$$

$$\int_C \sin(xy)\, dx - \cos(xy)\, dy = \int_a^b \{\sin[x(u)y(u)]x'(u) - \cos[x(u)y(u)]y'(u)\}\, du.$$

Line Integrals with Respect to Arc Length

Suppose that f is a scalar field continuous on a piecewise-smooth curve

$$C: \quad \mathbf{r}(u) = x(u)\mathbf{i} + y(u)\mathbf{j} + z(u)\mathbf{k}, \quad u \in [a, b].$$

If $s(u)$ is the length of the curve from the tip of $\mathbf{r}(a)$ to the tip of $\mathbf{r}(u)$, then, as you have seen,

$$s'(u) = \|\mathbf{r}'(u)\| = \sqrt{[x'(u)]^2 + [y'(u)]^2 + [z'(u)]^2}.$$

The line integral of f over C *with respect to arc length s* is defined by setting

(17.4.1)

$$\int_C f(\mathbf{r})\, ds = \int_a^b f(\mathbf{r}(u))\, s'(u)\, du.$$

In the xyz-notation we have

$$\int_C f(x, y, z)\, ds = \int_a^b f(x(u), y(u), z(u))\, s'(u)\, du,$$

which, in the two-dimensional case, becomes

$$\int_C f(x, y)\, ds = \int_a^b f(x(u), y(u))\, s'(u)\, du.$$

Suppose now that C represents a thin wire (a material curve) of varying mass density $\lambda = \lambda(\mathbf{r})$. [Here mass density is mass per unit length.] The *length* of the wire can be written

(17.4.2)
$$L = \int_C ds.$$

The *mass* of the wire is given by

(17.4.3)
$$M = \int_C \lambda(\mathbf{r})\, ds,$$

and the *center of mass* $\mathbf{r}_M$ can be obtained from the vector equation

(17.4.4)
$$\mathbf{r}_M M = \int_C \mathbf{r}\lambda(\mathbf{r})\, ds.$$

The equivalent scalar equations read

$$x_M M = \int_C x\lambda(\mathbf{r})\, ds, \qquad y_M M = \int_C y\lambda(\mathbf{r})\, ds, \qquad z_M M = \int_C z\lambda(\mathbf{r})\, ds.$$

Finally, the *moment of inertia* about an axis is given by the formula

(17.4.5)
$$I = \int_C \lambda(\mathbf{r})[R(\mathbf{r})]^2\, ds$$

where $R(\mathbf{r})$ is the distance from the axis to the tip of $\mathbf{r}$.

Example 1 The mass density of a semicircular wire of radius a varies directly as the distance from the diameter that joins the two endpoints of the wire. (a) Find the mass of the wire. (b) Locate the center of mass. (c) Determine the moment of inertia of the wire about the diameter.

SOLUTION Placed as in Figure 17.4.1 the wire can be parameterized by

$$\mathbf{r}(u) = a \cos u\, \mathbf{i} + a \sin u\, \mathbf{j}, \qquad u \in [0, \pi]$$

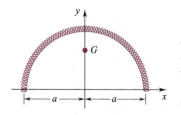

Figure 17.4.1

and the mass density function can be written $\lambda(x, y) = ky$. Since $\mathbf{r}'(u) = -a \sin u\,\mathbf{i} + a \cos u\,\mathbf{j}$, we have

$$s'(u) = \|\mathbf{r}'(u)\| = a.$$

Therefore

$$
\begin{aligned}
M &= \int_C \lambda(x, y)\,ds = \int_C ky\,ds \\
&= \int_0^\pi ky(u)s'(u)\,du \\
&= \int_0^\pi k(a \sin u)a\,du = ka^2 \int_0^\pi \sin u\,du = 2ka^2.
\end{aligned}
$$

By the symmetry of the configuration, $x_M = 0$. To find y_M we have to integrate:

$$
\begin{aligned}
y_M M &= \int_C y\lambda(x, y)\,ds = \int_C ky^2\,ds \\
&= \int_0^\pi k[y(u)]^2 s'(u)\,du \\
&= \int_0^\pi k(a \sin u)^2 a\,du = ka^3 \int_0^\pi \sin^2 u\,du = \tfrac{1}{2}ka^3\pi.
\end{aligned}
$$

Since $M = 2ka^2$, we have $y_M = (\tfrac{1}{2}ka^3\pi)/(2ka^2) = \tfrac{1}{4}a\pi$. The center of mass G lies on the perpendicular bisector of the wire at a distance $\tfrac{1}{4}a\pi$ from the diameter. (See Figure 17.4.1.) Note that in this instance, the center of mass does not lie on the wire.

Now let's find the moment of inertia about the diameter:

$$
\begin{aligned}
I &= \int_C \lambda(x, y)[R(x, y)]^2\,ds = \int_C (ky)y^2\,ds \\
&= \int_C k[y(u)]^3 s'(u)\,du \\
&= \int_0^\pi k(a \sin u)^3 a\,du \\
&= ka^4 \int_0^\pi \sin^3 u\,du = \tfrac{4}{3}ka^4.
\end{aligned}
$$

It is customary to express I in terms of M. With $M = 2ka^2$, we have

$$I = \tfrac{2}{3}(2ka^2)a^2 = \tfrac{2}{3}Ma^2. \quad \square$$

EXERCISES 17.4

In Exercises 1–4, evaluate

$$\int_C (x - 2y)\,dx + 2x\,dy$$

along the given path C from $(0, 0)$ to $(1, 2)$.

1. The straight-line path.
2. The parabolic path $y = 2x^2$.
3. The polygonal path $(0, 0)$, $(1, 0)$, $(1, 2)$.
4. The polygonal path $(0, 0)$, $(0, 2)$, $(1, 2)$.

In Exercises 5–8, evaluate

$$\int_C y\,dx + xy\,dy$$

along the given path C from $(0, 0)$ to $(2, 1)$.

5. The parabolic path $x = 2y^2$.
6. The straight-line path.
7. The polygonal path $(0, 0)$, $(0, 1)$, $(2, 1)$.
8. The cubic path $x = 2y^3$.

In Exercises 9–12, evaluate

$$\int_C y^2 \, dx + (xy - x^2) \, dy$$

along the given path C from $(0, 0)$ to $(2, 4)$.

9. The straight-line path.

10. The parabolic path $y = x^2$.

11. The parabolic path $y^2 = 8x$.

12. The polygonal path $(0, 0) \, (2, 0) \, (2, 4)$.

In Exercises 13–16, evaluate

$$\int_C x^2 y \, dx + xy \, dy$$

along the given path C from $(1, 0)$ to $(0, 1)$.

13. The straight-line path.

14. The circular path $y = \sqrt{1 - x^2}, 0 \le x \le 1$.

15. $\mathbf{r}(u) = \cos u \, \mathbf{i} + \sin u \, \mathbf{j}, u \in [0, \pi/2]$.

16. The polygonal path $(1, 0), (1, 1), (0, 1)$.

In Exercises 17–20, evaluate

$$\int_C (y^2 + 2x + 1) \, dx + (2xy + 4y - 1) \, dy$$

along the given path C from $(0, 0)$ to $(1, 1)$.

17. The straight-line path.

18. The parabolic path $y = x^2$.

19. The cubic path $y = x^3$.

20. The polygonal path $(0, 0), (4, 0), (4, 2), (1, 1)$.

In Exercises 21–24, evaluate

$$\int_C y \, dx + 2z \, dy + x \, dz$$

along the given path C from $(0, 0, 0)$ to $(1, 1, 1)$.

21. The straight-line path.

22. $\mathbf{r}(u) = u \mathbf{i} + u^2 \mathbf{j} + u^3 \mathbf{k}$.

23. The polygonal path $(0, 0, 0), (0, 0, 1), (0, 1, 1), (1, 1, 1)$.

24. The polygonal path $(0, 0, 0), (1, 0, 0), (1, 1, 0), (1, 1, 1)$.

In Exercises 25–28, evaluate

$$\int_C xy \, dx + 2z \, dy + (y + z) \, dz$$

along the given path C from $(0, 0, 0)$ to $(2, 2, 8)$.

25. The straight-line path.

26. The polygonal path $(0, 0, 0), (2, 0, 0), (2, 2, 0), (2, 2, 8)$.

27. The parabolic path $\mathbf{r}(u) = t \mathbf{i} + t \mathbf{j} + 2t^2 \mathbf{k}$.

28. The polygonal path $(0, 0, 0), (2, 2, 2) \, (2, 2, 8)$.

29. Evaluate $\displaystyle\int_C x^2 y \, dx + y \, dy + xz \, dz$, where C is the curve of intersection of the cylinder $y - 2z^2 = 1$ and the plane $z = x + 1$ from $(0, 3, 1)$ to $(1, 9, 2)$.

30. Evaluate $\displaystyle\int_C y \, dx + yz \, dy + z(x - 1) \, dz$, where C is the curve of intersection of the sphere $x^2 + y^2 + z^2 = 4$ and the cylinder $(x - 1)^2 + y^2 = 1$ from $(2, 0, 0)$ to $(0, 0, 2)$.

31. Given the vector field
$\mathbf{F}(x, y) = (x^2 + 6xy - 2y^2)\mathbf{i} + (3x^2 - 4xy + 2y)\mathbf{j}$.
 (a) Show that $\mathbf{F}$ is a gradient field.
 (b) Evaluate

$$\int_C (x^2 + 6xy - 2y^2) \, dx + (3x^2 - 4xy + 2y) \, dy$$

along any piecewise smooth curve from $(3, 0)$ to $(0, 4)$.
 (c) What is the value of

$$\int_{C'} (x^2 + 6xy - 2y^2) \, dx + (3x^2 - 4xy + 2y) \, dy$$

where C' is any piecewise smooth curve from $(4, 0)$ to $(0, 3)$.

32. Given the vector field
$\mathbf{F}(x, y, z) = (2xy + z^2)\mathbf{i} + (x^2 - 2yz)\mathbf{j} + (2xz - y^2)\mathbf{k}$.
 (a) Show that $\mathbf{F}$ is a gradient field.
 (b) Evaluate

$$\int_C (2xy + z^2) \, dx + (x^2 - 2yz) \, dy + (2xz - y^2) \, dz$$

along any piecewise smooth curve from $(1, 0, 1)$ to $(3, 2, -1)$.
 (c) What is the value of

$$\int_{C'} (2xy + z^2) \, dx + (x^2 - 2yz) \, dy + (2xz - y^2) \, dz$$

where C' is any piecewise smooth curve from $(3, 2, -1)$ to $(1, 0, 1)$?

33. A wire in the shape of the quarter-circle

$$C: \quad \mathbf{r}(u) = a(\cos u \, \mathbf{i} + \sin u \, \mathbf{j}), \quad u \in [0, \tfrac{1}{2}\pi]$$

has varying mass density $\lambda(x, y) = k(x + y)$ where k is a positive constant.
 (a) Find the total mass of the wire and locate the center of mass.
 (b) What is the moment of inertia of the wire about the x-axis?

34. Find the moment of inertia of a homogeneous circular wire of radius a and mass M about: (a) a diameter; (b) the axis through the center that is perpendicular to the plane of the wire.

35. Find the moment of inertia of the wire of Exercise 33 about: (a) the z-axis; (b) the line $y = x$.

36. A wire of constant mass density k has the form

$$\mathbf{r}(u) = (1 - \cos u)\mathbf{i} + (u - \sin u)\mathbf{j}, \qquad u \in [0, 2\pi].$$

(a) Determine the mass of the wire.
(b) Locate the center of mass.

37. A homogeneous wire of mass M winds around the z-axis as

$$C: \quad \mathbf{r}(u) = a \cos u\,\mathbf{i} + a \sin u\,\mathbf{j} + bu\,\mathbf{k}, \qquad u \in [0, 2\pi].$$

(a) Find the length of the wire.
(b) Locate the center of mass.
(c) Determine the moments of inertia of the wire about the coordinate axes.

38. A homogeneous wire of mass M is of the form

$$C: \quad \mathbf{r}(u) = u\mathbf{i} + u^2\mathbf{j} + \tfrac{2}{3}u^3\mathbf{k}, \quad u \in [0, a].$$

(a) Find the length of the wire.
(b) Locate the center of mass.
(c) Determine the moment of inertia of the wire about the z-axis.

39. Calculate the mass of the wire of Exercise 37 if the mass density varies directly as the square of the distance from the origin.

40. Show that

(17.4.6)

$$\int_C \mathbf{h}(\mathbf{r}) \cdot d\mathbf{r} = \int_C [\mathbf{h}(\mathbf{r}) \cdot \mathbf{T}(\mathbf{r})]\, ds$$

where $\mathbf{T}$ is the unit tangent vector.

■ 17.5 GREEN'S THEOREM

(Green's theorem is the first of the three integration theorems heralded at the beginning of this chapter.)

Recall that a Jordan curve is a plane curve that is both closed and simple. Thus circles, ellipses, and triangles are Jordan curves; figure eights are not. (See Section 15.10, page 1025.)

Figure 17.5.1 depicts a closed region Ω, the total boundary of which is a Jordan curve C. Such a region is called a *Jordan region*.

We know how to integrate over Ω, and, if the boundary C is piecewise smooth, we know how to integrate over C. Green's theorem expresses a double integral over Ω as a line integral over C.

Figure 17.5.1

THEOREM 17.5.1 GREEN'S THEOREM†

Let Ω be a Jordan region with a piecewise-smooth boundary C. If P and Q are scalar fields continuously differentiable on an open set that contains Ω, then

$$\iint_\Omega \left[\frac{\partial Q}{\partial x}(x, y) - \frac{\partial P}{\partial y}(x, y) \right] dx\, dy = \oint_C P(x, y)\, dx + Q(x, y)\, dy$$

where the integral on the right is the line integral over C taken in the counterclockwise direction.

---

† The result was established in 1828 by the English mathematician George Green (1793–1841).

NOTE: As indicated, the symbol $\displaystyle\oint_C$ is used to denote the line integral over a simple closed curve C taken in the counterclockwise direction. A line integral over C taken in the clockwise direction would be written $\displaystyle\oint_C$.

We will prove the theorem only for special cases. First of all let's assume that Ω is an *elementary region*, a region that is both of type I and type II as defined in Section 16.4. For simplicity we take Ω as in Figure 17.5.2.

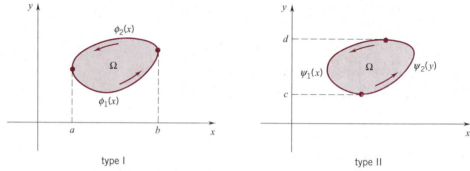

type I   type II

Ω is an elementary region: it is both of type I and type II

Figure 17.5.2

With Ω being of type I, we can show that

(1)
$$\oint_C P(x, y)\, dx = \iint_\Omega -\frac{\partial P}{\partial y}(x, y)\, dx\, dy.$$

In the first place

$$\iint_\Omega -\frac{\partial P}{\partial y}(x, y)\, dx\, dy = -\int_a^b \int_{\phi_1(x)}^{\phi_2(x)} \frac{\partial P}{\partial y}(x, y)\, dy\, dx$$

$$= -\int_a^b \{P[x, \phi_2(x)] - P[x, \phi_1(x)]\}\, dx$$

by The Fundamental Theorem
of Integral Calculus ⟶

(*)
$$= \int_a^b P[x, \phi_1(x)]\, dx - \int_a^b P[x, \phi_2(x)]\, dx.$$

The graph of ϕ_1 parametrized from left to right is the curve

$$C_1: \quad \mathbf{r}_1(u) = u\,\mathbf{i} + \phi_1(u)\mathbf{j}, \qquad u \in [a, b];$$

the graph of ϕ_2, also parametrized from left to right, is the curve

$$C_2: \quad \mathbf{r}_2(u) = u\,\mathbf{i} + \phi_2(u)\mathbf{j}, \qquad u \in [a, b].$$

Since C traversed counterclockwise consists of C_1 followed by $-C_2$ (C_2 traversed from right to left), you can see that

$$\oint_C P(x, y)\, dx = \int_{C_1} P(x, y)\, dx - \int_{C_2} P(x, y)\, dx$$

$$= \int_a^b P[u, \phi_1(u)]\, du - \int_a^b P[u, \phi_2(u)]\, du.$$

Since u is a dummy variable, it can be replaced by x. Comparison with $(*)$ proves (1).

We leave it to you to verify that

(2)
$$\oint_C Q(x, y)\, dy = \iint_\Omega \frac{\partial Q}{\partial x}(x, y)\, dx\, dy$$

by using the fact that Ω is of type II. This completes the proof of the theorem for Ω as in Figure 17.5.2.

A slight modification of this argument applies to elementary regions which are bordered entirely or in part by line segments parallel to the coordinate axes.

Figure 17.5.3 shows a Jordan region that is not elementary but can be broken up into two elementary regions. (Figure 17.5.4.) Green's theorem applied to the elementary parts tells us that.

$$\iint_{\Omega_1} \left(\frac{\partial Q}{\partial x} - \frac{\partial P}{\partial y} \right) dx\, dy = \oint_{\text{bdry of } \Omega_1} P\, dx + Q\, dy,$$

$$\iint_{\Omega_2} \left(\frac{\partial Q}{\partial x} - \frac{\partial P}{\partial y} \right) dx\, dy = \oint_{\text{bdry of } \Omega_2} P\, dx + Q\, dy.$$

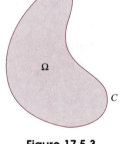

Figure 17.5.3

We now add these equations. The sum of the double integrals is, by additivity, the double integral over Ω. The sum of the line integrals is the integral over C (see the figure) plus the integrals over the crosscut. Since the crosscut is traversed twice and in opposite directions, the total contribution of the crosscut is zero and therefore Green's theorem holds:

$$\iint_\Omega \left(\frac{\partial Q}{\partial x} - \frac{\partial P}{\partial y} \right) dx\, dy = \oint_C P\, dx + Q\, dy.$$

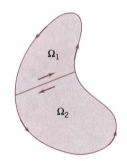

This same argument can be extended to a Jordan region Ω that breaks up into n elementary regions $\Omega_1, \ldots, \Omega_n$. (Figure 17.5.5 gives an example with $n = 4$.)

Figure 17.5.4

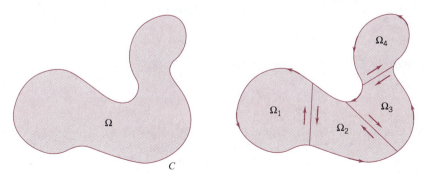

Figure 17.5.5

The double integrals over the Ω_i add up to the double integral over Ω, and, since the line integrals over the crosscuts cancel, the line integrals over the boundaries of the Ω_i add up to the line integral over C. (This is as far as we will carry the proof of Green's theorem. It is far enough to cover all the Jordan regions encountered in practice.) ❏

$x^2 + y^2 = a^2$

$(a, 0)$

Figure 17.5.6

Example 1 Use Green's theorem to evaluate

$$\oint_C^C (3x^2 + y)\, dx + (2x + y^3)\, dy$$

where C is the circle $x^2 + y^2 = a^2$. (Figure 17.5.6.)

SOLUTION Let Ω be the closed disc $0 \leq x^2 + y^2 \leq a^2$. With

$$P(x, y) = 3x^2 + y \qquad \text{and} \qquad Q(x, y) = 2x + y^3,$$

we have

$$\frac{\partial P}{\partial y} = 1, \qquad \frac{\partial Q}{\partial x} = 2, \qquad \text{and} \qquad \frac{\partial Q}{\partial x} - \frac{\partial P}{\partial y} = 2 - 1 = 1.$$

By Green's theorem

$$\oint_C (3x^2 + y)\, dx + (2x + y^3)\, dy = \iint_\Omega 1\, dx\, dy = \text{area of } \Omega = \pi a^2. \quad ❏$$

Remark The line integral in Example 1 could have been calculated directly as follows: The circle $x^2 + y^2 = a^2$ is parametrized by

$$x = a \cos u, \qquad y = a \sin u, \qquad 0 \leq u \leq 2\pi.$$

Thus,

$$\oint_C (3x^2 + y)\, dx + (2x + y^3)\, dy$$

$$= \int_0^{2\pi} [(3a^2\cos^2 u + a \sin u)(-a \sin u) + (2a \cos u + a^3\sin^3 u)(a \cos u)]\, du$$

$$= \int_0^{2\pi} [-3a^3\cos^2 u \sin u - a^2\sin^2 u + 2a^2\cos^2 u + a^4\sin^3 u \cos u]\, du$$

which is obviously a much more difficult integral to evaluate. You can verify, however, that the result is πa^2. ❏

Example 2 Use Green's theorem to evaluate

$$\oint_C (1 + 10xy + y^2)\, dx + (6xy + 5x^2)\, dy$$

where C is the square with vertices $(0, 0)$, $(a, 0)$, (a, a), $(0, a)$. (Figure 17.5.7.)

$(0, a)$ (a, a)

$(a, 0)$

Figure 17.5.7

SOLUTION Let Ω be the square region enclosed by C. With

$$P(x, y) = 1 + 10xy + y^2 \qquad \text{and} \qquad Q(x, y) = 6xy + 5x^2,$$

we have

$$\frac{\partial P}{\partial y} = 10x + 2y, \qquad \frac{\partial Q}{\partial x} = 6y + 10x, \qquad \text{and} \qquad \frac{\partial Q}{\partial x} - \frac{\partial P}{\partial y} = 4y.$$

By Green's theorem

$$\oint_C (1 + 10xy + y^2)\, dx + (6xy + 5x^2)\, dy = \iint_\Omega 4y\, dx\, dy$$

$$= \int_0^a \int_0^a 4y\, dx\, dy$$

$$= \left(\int_0^a dx \right) \left(\int_0^a 4y\, dy \right)$$

$$= (a)(2a^2) = 2a^3. \quad \square$$

ALTERNATE SOLUTION   By 16.6.3,

$$\iint_\Omega y\, dx\, dy = \bar{y}\, (\text{area of } \Omega)$$

where $\bar{y}$ is the y-coordinate of the centroid of the region Ω. Since $\bar{y} = \frac{1}{2}a$, we have,

$$\iint_\Omega 4y\, dx\, dy = 4\,\bar{y}\,(\text{area of } \Omega) = 4\,(\tfrac{1}{2}a)\,a^2 = 2a^3. \quad \square$$

Example 3   Use Green's theorem to evaluate

$$\oint_C e^x \sin y\, dx + e^x \cos y\, dy$$

where C is the closed curve consisting of the semicircle $y = \sqrt{1 - x^2}$ and the interval $[-1, 1]$. (Figure 17.5.8.)

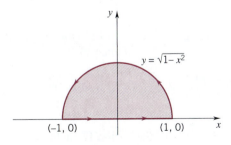

$y = \sqrt{1 - x^2}$

$(-1, 0)$ $(1, 0)$

Figure 17.5.8

SOLUTION   The curve bounds the closed semicircular disc $\Omega: x^2 + y^2 \leq 1,\ y \geq 0$.
Here

$$\frac{\partial P}{\partial y} = e^x \cos y, \qquad \frac{\partial Q}{\partial x} = e^x \cos y \qquad \text{and} \qquad \frac{\partial Q}{\partial x} - \frac{\partial P}{\partial y} = 0.$$

By Green's theorem,

$$\oint_C e^x \sin y \; dx + e^x \cos y \; dy = \iint_\Omega 0 \; dx \; dy = 0.$$

To see the power of Green's theorem, try to evaluate this line integral directly. ❑

The preceding examples illustrate the use of Green's theorem to convert a line integral along the boundary of a Jordan region Ω into a double integral over Ω. Green's theorem can also be used in reverse. That is, the value of a double integral over a Jordan region can be found by evaluating a line integral along its boundary. A useful application of this involves the computation of the area of a region enclosed by a piecewise-smooth Jordan curve C.

(17.5.2)

$$\text{The area enclosed by } C = \oint_C -y \; dx = \oint_C x \; dy$$
$$= \tfrac{1}{2} \oint_C -y \; dx + x \; dy.$$

PROOF Let Ω be the region enclosed by C. In the first integral

$$P(x, y) = -y, \qquad Q(x, y) = 0.$$

Therefore

$$\frac{\partial P}{\partial y} = -1, \qquad \frac{\partial Q}{\partial x} = 0, \qquad \text{and} \qquad \frac{\partial Q}{\partial x} - \frac{\partial P}{\partial y} = 1.$$

Thus by Green's theorem

$$\oint_C -y \; dx = \iint_\Omega 1 \; dx \; dy = \text{area of } \Omega.$$

That the second integral also gives the area of Ω can be verified in a similar manner. Finally, the third integral is a consequence of

$$\oint_C -y \; dx + \oint_C x \; dy = \text{twice the area enclosed by } C. \quad ❑$$

Example 4 Show that the area of the region Ω enclosed by the ellipse

$$\frac{x^2}{a^2} + \frac{y^2}{b^2} = 1 \qquad \text{(Figure 17.5.9)}$$

is πab.

SOLUTION The ellipse is oriented counterclockwise by the parametrization

$$x = a \cos u, \qquad y = b \sin u, \qquad 0 \le u \le 2\pi.$$

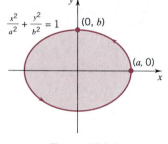

$$\frac{x^2}{a^2} + \frac{y^2}{b^2} = 1$$

(0, b)

(a, 0)

Figure 17.5.9

Although the third integral in (17.5.2) appears to be more complicated than either of the other two integrals, it is actually simpler to use in this case:

$$\text{area of } \Omega = \tfrac{1}{2}\oint_C - y\,dx + x\,dy$$

$$= \tfrac{1}{2}\int_0^{2\pi}[-(b\sin u)(-a\sin u) + (a\cos u)(b\cos u)]\,du$$

$$= \tfrac{1}{2}ab\int_0^{2\pi}du = \pi ab. \quad \square$$

Example 5 Let Ω be a Jordan region of area A with a piecewise-smooth boundary C. Show that the coordinates of the centroid of Ω are given by

$$\bar{x}A = \tfrac{1}{2}\oint_C x^2\,dy, \qquad \bar{y}A = -\tfrac{1}{2}\oint_C y^2\,dx.$$

SOLUTION

$$\tfrac{1}{2}\oint_C x^2\,dy = \tfrac{1}{2}\iint_\Omega 2x\,dx\,dy = \iint_\Omega x\,dx\,dy = \bar{x}A,$$

by Green's theorem

$$-\tfrac{1}{2}\oint_C y^2\,dx = -\tfrac{1}{2}\iint_\Omega(-2y)\,dx\,dy = \iint_\Omega y\,dx\,dy = \bar{y}A. \quad \square$$

Regions Bounded by Two or More Jordan Curves

(All the curves appearing in the following are assumed to be piecewise smooth.)

Figure 17.5.10 shows an annular region Ω. The region is not a Jordan region: the boundary consists of two Jordan curves C_1 and C_2. We cannot apply Green's theorem to Ω directly, but we can break up Ω into two Jordan regions as in Figure 17.5.11 and then apply Green's theorem to each piece. With Ω_1 and Ω_2 as in Figure 17.5.11,

$$\iint_{\Omega_1}\left(\frac{\partial Q}{\partial x} - \frac{\partial P}{\partial y}\right)dx\,dy = \oint_{\text{bdry of }\Omega_1} P\,dx + Q\,dy,$$

$$\iint_{\Omega_2}\left(\frac{\partial Q}{\partial x} - \frac{\partial P}{\partial y}\right)dx\,dy = \oint_{\text{bdry of }\Omega_2} P\,dx + Q\,dy.$$

Figure 17.5.10

Figure 17.5.11

When we add the double integrals, we get the double integral over Ω. When we add the line integrals, the integrals over the crosscuts cancel and we are left with the *counterclockwise integral* over C_1 and the *clockwise integral* over C_2. (See the figure.) Thus, for the annular region,

(17.5.3)
$$\iint_\Omega\left(\frac{\partial Q}{\partial x} - \frac{\partial P}{\partial y}\right)dx\,dy = \oint_{C_1} P\,dx + Q\,dy + \oint_{C_2} P\,dx + Q\,dy.$$

As a corollary to this we see that, if $\partial Q/\partial x = \partial P/\partial y$ throughout Ω, then the double integral on the left is 0, the sum of the integrals on the right is 0. Therefore

(17.5.4)

$$\text{if } \partial Q/\partial x = \partial P/\partial y \text{ throughout } \Omega, \text{ then}$$
$$\oint_{C_1} P \, dx + Q \, dy = \oint_{C_2} P \, dx + Q \, dy.$$

Example 6 Let C_1 be a Jordan curve that does not pass through the origin $(0, 0)$. Show that

$$\oint_{C_1} -\frac{y}{x^2 + y^2} \, dx + \frac{x}{x^2 + y^2} \, dy = \begin{cases} 0, \text{ if } C_1 \text{ does not enclose the origin} \\ 2\pi, \text{ if } C_1 \text{ does enclose the origin.} \end{cases}$$

SOLUTION In this case

$$\frac{\partial P}{\partial y} = \frac{\partial}{\partial y}\left(-\frac{y}{x^2 + y^2}\right) = -\left[\frac{(x^2 + y^2)1 - 2y^2}{(x^2 + y^2)^2}\right] = \frac{y^2 - x^2}{(x^2 + y^2)^2},$$

$$\frac{\partial Q}{\partial x} = \frac{\partial}{\partial x}\left(\frac{x}{x^2 + y^2}\right) = \frac{(x^2 + y^2)1 - 2x^2}{(x^2 + y^2)^2} = \frac{y^2 - x^2}{(x^2 + y^2)^2}.$$

Thus

$$\frac{\partial Q}{\partial x} = \frac{\partial P}{\partial y} \qquad \text{except at the origin.}$$

If C_1 does not enclose the origin, then $\partial Q/\partial x - \partial P/\partial y = 0$ throughout the region enclosed by C_1, and, by Green's theorem, the line integral is 0.

If C_1 does enclose the origin, we draw within the inner region of C_1 a small circle centered at the origin

$$C_2: \quad x^2 + y^2 = a^2. \qquad\qquad \text{(See Figure 17.5.12)}$$

Since $\partial Q/\partial x - \partial P/\partial y = 0$ on the annular region bounded by C_1 and C_2, we know from (17.5.4) that the line integral over C_1 equals the line integral over C_2. All we have to show now is that the line integral over C_2 is 2π. This is straightforward. Parametrizing the circle by

$$\mathbf{r}(u) = a \cos u \, \mathbf{i} + a \sin u \, \mathbf{j} \qquad \text{with} \quad u \in [0, 2\pi],$$

we have

$$\oint_{C_2} -\frac{y}{x^2 + y^2} \, dx + \frac{x}{x^2 + y^2} \, dy = \int_0^{2\pi} (\sin^2 u + \cos^2 u) \, du = \int_0^{2\pi} du = 2\pi. \quad \square$$

check this ⟶

Figure 17.5.13 shows a region bounded by three Jordan curves: C_2 and C_3 both exterior to one another, both within C_1. For such a region Green's theorem gives

$$\iint_\Omega \left(\frac{\partial Q}{\partial x} - \frac{\partial P}{\partial y}\right) dx \, dy =$$

$$\oint_{C_1} P \, dx + Q \, dy + \oint_{C_2} P \, dx + Q \, dy + \oint_{C_3} P \, dx + Q \, dy.$$

Figure 17.5.12

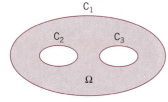

Figure 17.5.13

To see this, break up Ω into two regions by making the crosscuts shown in Figure 17.5.14.

The general formula for configurations of this type reads

$$\iint_\Omega \left(\frac{\partial Q}{\partial x} - \frac{\partial P}{\partial y} \right) dx\, dy = \oint_{C_1} P\, dx + Q\, dy + \sum_{i=2}^n \oint_{C_i} P\, dx + Q\, dy.$$

Figure 17.5.14

EXERCISES 17.5

In Exercises 1–4, evaluate the line integral two ways: (a) directly, and (b) using Green's theorem.

1. $\oint_C xy\, dx + x^2\, dy$; where C is the triangle with vertices $(0, 0)$, $(0, 1)$, $(1, 1)$.

2. $\oint_C x^2 y\, dx + 2y^2\, dy$; where C is the square with vertices $(0, 0)$, $(1, 0)$, $(1, 1)$, $(0, 1)$.

3. $\oint_C (3x^2 + y)\, dx + (2x + y^3)\, dy$; $C: x^2 + y^2 = 1$.

4. $\oint_C y^2\, dx + x^2\, dy$; where C is the boundary of the region bounded by the curves $y = x$ and $y = x^2$.

In Exercises 5–16, evaluate by Green's theorem.

5. $\oint_C 3y\, dx + 5x\, dy$; $C: x^2 + y^2 = 1$.

6. $\oint_C 5x\, dx + 3y\, dy$; $C: (x - 1)^2 + (y + 1)^2 = 1$.

7. $\oint_C x^2\, dy$; where C is the rectangle with vertices $(0, 0)$, $(a, 0)$, (a, b), $(0, b)$.

8. $\oint_C y^2\, dx$; where C is the rectangle of Exercise 7.

9. $\oint_C (3xy + y^2)\, dx + (2xy + 5x^2)\, dy$; $C: (x - 1)^2 + (y + 2)^2 = 1$.

10. $\oint_C (xy + 3y^2)\, dx + (5xy + 2x^2)\, dy$; $C: (x - 1)^2 + (y + 2)^2 = 1$.

11. $\oint_C (2x^2 + xy - y^2)\, dx + (3x^2 - xy + 2y^2)\, dy$; $C:$ $(x - a)^2 + y^2 = r^2$.

12. $\oint_C (x^2 - 2xy + 3y^2)\, dx + (5x + 1)\, dy$; $C: x^2 + (y - b)^2 = r^2$.

13. $\oint_C e^x \sin y\, dx + e^x \cos y\, dy$; $C: (x - a)^2 + (y - b)^2 = r^2$.

14. $\oint_C e^x \cos y\, dx + e^x \sin y\, dy$; C the rectangle with vertices $(0, 0)$, $(1, 0)$, $(1, \pi)$, $(0, \pi)$.

15. $\int_C 2xy\, dx + x^2\, dy$; where C is the cardioid $r = 1 - \cos \theta$, $\theta \in [0, 2\pi]$.

16. $\int_C y^2\, dx + 2xy\, dy$; where C is the first quadrant loop of the petal curve $r = 2 \sin 2\theta$.

In Exercises 17 and 18, find the area enclosed by the curve by integrating over the curve.

17. The circle $x^2 + y^2 = a^2$.

18. The astroid $x^{2/3} + y^{2/3} = a^{2/3}$.

19. Find the area of the region in the first quadrant bounded by the coordinate axes and the line $x/a + y/b = 1$.

20. Find the area of the region enclosed by the parabola $y = x^2$ and the line $y = 4$.

21. Let C be a piecewise-smooth Jordan curve. Calculate

$$\oint_C (ay + b)\, dx + (cx + d)\, dy$$

given that C encloses a region of area A.

22. Calculate

$$\oint_C \mathbf{F}(\mathbf{r}) \cdot d\mathbf{r}$$

given that $\mathbf{F}(x, y) = 2y\,\mathbf{i} - 3x\,\mathbf{j}$ and C is the astroid $x^{2/3} + y^{2/3} = a^{2/3}$.

23. Use Green's theorem to find the area under one arch of the cycloid

$$x(\theta) = R(\theta - \sin \theta), \qquad y(\theta) = R(1 - \cos \theta).$$

24. Find the Jordan curve C that maximizes the line integral

$$\oint_C y^3 \, dx + (3x - x^3) \, dy.$$

25. Find the work done by the force $\mathbf{F}(x, y) = (x^2 - y^3)\mathbf{i} + (x^2 + y^2)\mathbf{j}$ in moving an object around the circle $x^2 + y^2 = 1$ in the counterclockwise direction.

26. Find the work done by the force $\mathbf{F}(x, y) = (x^3 - x^2y)\mathbf{i} + xy^2\mathbf{j}$ in moving an object in the counterclockwise direction around the boundary of the region determined by the parabolas $y = x^2$ and $x = y^2$.

27. Complete the proof of Green's theorem for the elementary region of Figure 17.5.2 by showing that

$$\oint_C Q(x, y) \, dy = \iint_\Omega \frac{\partial Q}{\partial x}(x, y) \, dx \, dy.$$

28. Let Ω be a plate of constant mass density λ in the form of a Jordan region with a piecewise-smooth boundary C. Show that the moments of inertia of the plate about the coordinate axes are given by the formulas

(17.5.5) $\qquad I_x = -\dfrac{\lambda}{3} \oint_C y^3 \, dx, \qquad I_y = \dfrac{\lambda}{3} \oint_C x^3 \, dy.$

29. Let P and Q be continuously differentiable functions on the region Ω of Figure 17.5.13. Given that $\partial P/\partial y = \partial Q/\partial x$ on Ω, find a relation between the line integrals

$$\oint_{C_1} P \, dx + Q \, dy, \qquad \oint_{C_2} P \, dx + Q \, dy,$$

$$\oint_{C_3} P \, dx + Q \, dy.$$

30. Show that, if $f = f(x)$ and $g = g(y)$ are continuously differentiable, then

$$\int_C f(x) \, dx + g(y) \, dy = 0$$

for all piecewise-smooth Jordan curves C.

31. Let C be a piecewise-smooth Jordan curve that does not pass through the origin. Evaluate

$$\oint_C \frac{x}{x^2 + y^2} \, dx + \frac{y}{x^2 + y^2} \, dy.$$

(a) If C does not enclose the origin.
(b) If C does enclose the origin.

32. Let C be a piecewise-smooth Jordan curve that does not pass through the origin. Evaluate

$$\oint_C -\frac{y^3}{(x^2 + y^2)^2} \, dx + \frac{xy^2}{(x^2 + y^2)^2} \, dy.$$

(a) If C does not enclose the origin.
(b) If C does enclose the origin.

33. Let $\mathbf{v}$ be a continuously differentiable vector field on the entire plane. Use Green's theorem to verify that if $\mathbf{v}$ is a gradient field $[\mathbf{v} = \nabla\phi]$, then

$$\oint_C \mathbf{v} \cdot d\mathbf{r} = 0$$

for every piecewise-smooth Jordan curve C.

34. Let C be the line segment from the point (x_1, y_1) to the point (x_2, y_2). Show that

$$\int_C -y \, dx + x \, dy = x_1 y_2 - x_2 y_1.$$

35. Let $(x_1, y_1), (x_2, y_2), \ldots, (x_n, y_n)$ be the vertices of a polygon in counterclockwise order. Show that the area of the polygon is

$$A = \tfrac{1}{2}[(x_1 y_2 - x_2 y_1) + (x_2 y_3 - x_3 y_2) + \cdots$$
$$+ (x_{n-1} y_n - x_n y_{n-1}) + (x_n y_1 - x_1 y_n)].$$

36. Use the result in Exercise 35 to find the area of:
(a) The triangle with vertices $(0, 0), (2, 1), (1, 4)$.
(b) The pentagon with vertices $(0, 0), (3, 1), (2, 4), (0, 6), (-1, 2)$.

---

*SUPPLEMENT TO SECTION 17.5

A JUSTIFICATION OF THE JACOBIAN AREA FORMULA

We based the change of variables for double integrals on the Jacobian area formula (Formula 16.10.1). Green's theorem enables us to derive this formula under the conditions spelled out in the following.

> Let Γ be a Jordan region in the uv-plane with a piecewise-smooth boundary C_Γ. A vector function $\mathbf{r}(u, v) = x(u, v)\mathbf{i} + y(u, v)\mathbf{j}$ with continuous second partials maps Γ onto a region Ω of the xy-plane. If $\mathbf{r}$ is one-to-one on the interior of Γ and the Jacobian J of the components of $\mathbf{r}$ is different from zero on the interior of Γ, then
>
> $$\text{area of } \Omega = \iint_\Gamma |J(u, v)| \, du \, dv.$$

PROOF Suppose that C_Γ is parametrized $u = u(t)$, $v = v(t)$ with $t \in [a, b]$. Then the boundary of Ω is a piecewise-smooth curve C given by

$$\mathbf{r}[u(t), v(t)] = x[u(t), v(t)]\mathbf{i} + y[u(t), v(t)]\mathbf{j}, \qquad t \in [a, b].$$

By Green's theorem, the area of Ω is

$$\oint_C x \, dy = \left| \int_a^b x[u(t), v(t)] \frac{d}{dt}(y[u(t), v(t)]) \, dt \right|$$

$$= \left| \int_a^b x[u(t), v(t)]\left(\frac{\partial y}{\partial u}[u(t), v(t)]u'(t) + \frac{\partial y}{\partial v}[u(t), v(t)]v'(t) \right) dt \right|$$

$$= \left| \int_a^b \left(x[u(t), v(t)]\frac{\partial y}{\partial u}[u(t), v(t)]u'(t) + x[u(t), v(t)]\frac{\partial y}{\partial v}[u(t), v(t)]v'(t) \right) dt \right|$$

$$= \left| \int_{C_\Gamma} x \frac{\partial y}{\partial u} \, du + x \frac{\partial y}{\partial v} \, dv \right|$$

↓——— again by Green's theorem

$$= \left| \iint_\Gamma \left[\frac{\partial}{\partial u}\left(x \frac{\partial y}{\partial v} \right) - \frac{\partial}{\partial v}\left(x \frac{\partial y}{\partial u} \right) \right] du \, dv \right|.$$

Now

$$\frac{\partial}{\partial u}\left(x \frac{\partial y}{\partial v} \right) - \frac{\partial}{\partial v}\left(x \frac{\partial y}{\partial u} \right) = \frac{\partial x}{\partial u}\frac{\partial y}{\partial v} + x \frac{\partial^2 y}{\partial u \, \partial v} - \frac{\partial x}{\partial v}\frac{\partial y}{\partial u} - x \frac{\partial^2 y}{\partial v \, \partial u}$$

$$= \frac{\partial x}{\partial u}\frac{\partial y}{\partial v} - \frac{\partial x}{\partial v}\frac{\partial y}{\partial u} = J(u, v).$$

Therefore

$$\text{area of } \Omega = \left| \iint_\Gamma J(u, v) \, du \, dv \right| = \iint_\Gamma |J(u, v)| \, du \, dv,$$

the final equality holding because $J(u, v)$ cannot change sign on Γ. ❑

■ 17.6 PARAMETRIZED SURFACES; SURFACE AREA

You have seen that a space curve C can be parametrized by a vector function $\mathbf{r} = \mathbf{r}(u)$ where u ranges over some interval I of the u-axis (Figure 17.6.1). In an analogous manner we can parametrize a surface S in space by a vector function $\mathbf{r} = \mathbf{r}(u, v)$ where (u, v) ranges over some region Ω of the uv-plane (Figure 17.6.2).

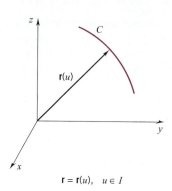

$r = r(u), \quad u \in I$

Figure 17.6.1

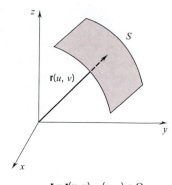

$r = r(u, v), \quad (u, v) \in \Omega$

Figure 17.6.2

Example 1 (*The graph of a function*) As you have seen, the graph of a function

$$y = f(x), \qquad x \in [a, b]$$

can be parametrized by setting

$$\mathbf{r}(u) = u\mathbf{i} + f(u)\mathbf{j}, \qquad u \in [a, b].$$

In the same vein the graph of a function

$$z = f(x, y), \qquad (x, y) \in \Omega$$

is a surface which can be parametrized by setting

$$\mathbf{r}(u, v) = u\mathbf{i} + v\mathbf{j} + f(u, v)\mathbf{k}, \qquad (u, v) \in \Omega.$$

As (u, v) ranges over Ω, the tip of $\mathbf{r}(u, v)$ traces out the graph of f. ❑

Example 2 (*A plane*) If two vectors $\mathbf{a}$ and $\mathbf{b}$ are not parallel, then the set of all linear combinations $u\mathbf{a} + v\mathbf{b}$ generate a plane p_0 that passes through the origin. We can parametrize this plane by setting

$$\mathbf{r}(u, v) = u\mathbf{a} + v\mathbf{b}, \qquad u, v \text{ real numbers.}$$

The plane p that is parallel to p_0 and passes through the tip of $\mathbf{c}$ can be parametrized by setting

$$\mathbf{r}(u, v) = u\mathbf{a} + v\mathbf{b} + \mathbf{c}, \qquad u, v \text{ real numbers.}$$

Note that the plane contains the lines

$$l_1: \quad \mathbf{r}(u, 0) = u\mathbf{a} + \mathbf{c} \qquad \text{and} \qquad l_2: \quad \mathbf{r}(0, v) = v\mathbf{b} + \mathbf{c}. \quad ❑$$

Example 3 (*A sphere*) The sphere of radius a centered at the origin can be parametrized by

$$\mathbf{r}(u, v) = a \cos u \cos v \, \mathbf{i} + a \sin u \cos v \, \mathbf{j} + a \sin v \, \mathbf{k}$$

with (u, v) ranging over the rectangle $R: 0 \le u \le 2\pi, \ -\frac{1}{2}\pi \le v \le \frac{1}{2}\pi$.

To derive this parametrization we refer to Figure 17.6.3. The points of latitude v (see the figure) form a circle of radius $a \cos v$ on the horizontal plane $z = a \sin v$. This circle can be parametrized by

$$\mathbf{R}(u) = a \cos v(\cos u \, \mathbf{i} + \sin u \, \mathbf{j}) + a \sin v \, \mathbf{k}, \qquad u \in [0, 2\pi].$$

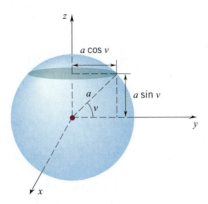

Figure 17.6.3

This expands to give

$$\mathbf{R}(u) = a \cos u \cos v \, \mathbf{i} + a \sin u \cos v \, \mathbf{j} + a \sin v \, \mathbf{k}, \qquad u \in [0, 2\pi].$$

Letting v range from $-\frac{1}{2}\pi$ to $\frac{1}{2}\pi$, we obtain the entire sphere.

The xyz-equation for this same sphere is $x^2 + y^2 + z^2 = a^2$. It is easy to verify that the parametrization satisfies this equation:

$$x^2 + y^2 + z^2 = a^2\cos^2u \cos^2v + a^2\sin^2u \cos^2v + a^2\sin^2u$$
$$= a^2(\cos^2u + \sin^2u)\cos^2u + a^2\sin^2v$$
$$= a^2(\cos^2v + \sin^2v) = a^2. \quad \square$$

Example 4 (*A cone*) Figure 17.6.4 shows a right circular cone with apex semiangle α and slant height s. The points of slant height v (see the figure) form a circle of radius $v \sin \alpha$ on the horizontal plane $z = v \cos \alpha$. This circle can be parametrized by

$$R(u) = v \sin \alpha \, (\cos u \, \mathbf{i} + \sin u \, \mathbf{j}) + v \cos \alpha \, \mathbf{k}$$
$$= v \cos u \sin \alpha \, \mathbf{i} + v \sin u \sin \alpha \, \mathbf{j} + v \cos \alpha \, \mathbf{k}, \qquad u \in [0, 2\pi].$$

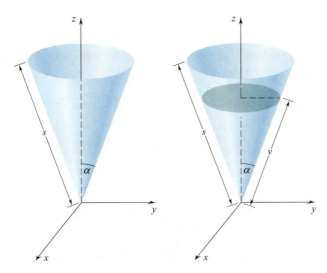

Figure 17.6.4

Since we can obtain the entire cone by letting v range from 0 to s, the cone is parametrized by

$$\mathbf{r}(u, v) = v \cos u \sin \alpha \, \mathbf{i} + v \sin u \sin \alpha \, \mathbf{j} + v \cos \alpha \, \mathbf{k},$$

with $0 \leq u \leq 2\pi$, $0 \leq v \leq s$. ❏

Example 5 (*A spiral ramp*) A rod of length l initially resting on the x-axis and attached at one end to the z-axis sweeps out a surface by rotating about the z-axis at constant rate ω while climbing at a constant rate b. The surface is pictured in Figure 17.6.5.

To parametrize this surface we mark the point of the rod at a distance u from the z-axis ($0 \leq u \leq l$) and ask for the position of this point at time v. At time v the rod will have climbed a distance bv and rotated through an angle ωv. Thus the point will be found at the tip of the vector

$$u(\cos \omega v \, \mathbf{i} + \sin \omega v \, \mathbf{j}) + bv \, \mathbf{k} = u \cos \omega v \, \mathbf{i} + u \sin \omega v \, \mathbf{j} + bv \, \mathbf{k}.$$

The entire surface can be parametrized by

$$\mathbf{r}(u, v) = u \cos \omega v \, \mathbf{i} + u \sin \omega v \, \mathbf{j} + bv \, \mathbf{k} \qquad \text{with } 0 \leq u \leq l, \quad v \geq 0. \quad ❏$$

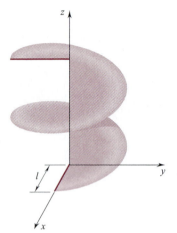

Figure 17.6.5

The Fundamental Vector Product

Let S be a surface parametrized by a differentiable vector function

$$\mathbf{r} = \mathbf{r}(u, v) = x(u, v)\mathbf{i} + y(u, v)\mathbf{j} + z(u, v)\mathbf{k}.$$

For simplicity, let us suppose that (u, v) varies over the open rectangle R: $a < u < b$, $c < v < d$. Since $\mathbf{r}$ is a function of two variables, we can take two partial derivatives:

$$\mathbf{r}'_u = \frac{\partial x}{\partial u} \mathbf{i} + \frac{\partial y}{\partial u} \mathbf{j} + \frac{\partial z}{\partial u} \mathbf{k},$$

the partial of $\mathbf{r}$ with respect to u, and

$$\mathbf{r}'_v = \frac{\partial x}{\partial v} \mathbf{i} + \frac{\partial y}{\partial v} \mathbf{j} + \frac{\partial z}{\partial v} \mathbf{k},$$

the partial of $\mathbf{r}$ with respect to v. Now let (u_0, v_0) be a point of R for which

$$\mathbf{r}'_u(u_0, v_0) \times \mathbf{r}'_v(u_0, v_0) \neq \mathbf{0}.$$

(The reason for this condition will be apparent as we go on.) The vector function

$$\mathbf{r}_1(u) = \mathbf{r}(u, v_0), \qquad u \in (a, b)$$

(here we are keeping v fixed at v_0) traces a differentiable curve C_1 that lies on S (Figure 17.6.6). The vector function

$$\mathbf{r}_2(v) = \mathbf{r}(u_0, v), \qquad v \in (c, d)$$

(this time we are keeping u fixed at u_0) traces a differentiable curve C_2 that also lies on S. Both curves pass through the tip of $\mathbf{r}(u_0, v_0)$:

$$C_1 \text{ with tangent vector } \mathbf{r}'_1(u_0) = \mathbf{r}'_u(u_0, v_0)$$

$$C_2 \text{ with tangent vector } \mathbf{r}'_2(v_0) = \mathbf{r}'_v(u_0, v_0).$$

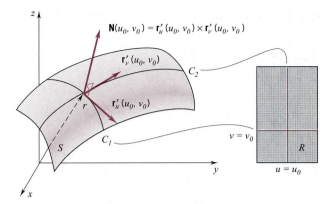

Figure 17.6.6

The cross product $N(u_0, v_0) = r_u'(u_0, v_0) \times r_v'(u_0, v_0)$, which we have assumed to be different from zero, is thus perpendicular to both curves at the tip of $r(u_0, v_0)$ and can be taken as a normal to the surface at that point. We record the result as follows:

(17.6.1)
> If S is a surface given by a differentiable function $\mathbf{r} = \mathbf{r}(u, v)$, then the vector $N(u, v) = r_u'(u, v) \times r_v'(u, v)$ is perpendicular to the surface at the point with position vector $\mathbf{r}(u, v)$ and, if different from zero, can be taken as a normal to the surface at that point.

The cross product

$$N = r_u' \times r_v' = \begin{vmatrix} \mathbf{i} & \mathbf{j} & \mathbf{k} \\ \dfrac{\partial x}{\partial u} & \dfrac{\partial y}{\partial u} & \dfrac{\partial z}{\partial u} \\ \dfrac{\partial x}{\partial v} & \dfrac{\partial y}{\partial v} & \dfrac{\partial z}{\partial v} \end{vmatrix}$$

is called the *fundamental vector product* of the surface.

Example 6 For the plane $\mathbf{r}(u, v) = u\mathbf{a} + v\mathbf{b} + \mathbf{c}$ we have

$$r_u'(u, v) = \mathbf{a}, \qquad r_v'(u, v) = \mathbf{b} \qquad \text{and therefore} \qquad N(u, v) = \mathbf{a} \times \mathbf{b}.$$

The vector $\mathbf{a} \times \mathbf{b}$ is normal to the plane. ❑

Example 7 We parametrized the sphere $x^2 + y^2 + z^2 = a^2$ by setting

$$\mathbf{r}(u, v) = a \cos u \cos v\,\mathbf{i} + a \sin u \cos v\,\mathbf{j} + a \sin v\,\mathbf{k}$$

with $0 \le u \le 2\pi$, $-\frac{1}{2}\pi \le v \le \frac{1}{2}\pi$. In this case

$$r_u'(u, v) = -a \sin u \cos v\,\mathbf{i} + a \cos u \cos v\,\mathbf{j}$$

and

$$r_v'(u, v) = -a \cos u \sin v\,\mathbf{i} - a \sin u \sin v\,\mathbf{j} + a \cos v\,\mathbf{k}.$$

Thus

$$\mathbf{N}(u, v) = \begin{vmatrix} \mathbf{i} & \mathbf{j} & \mathbf{k} \\ -a \sin u \cos v & a \cos u \cos v & 0 \\ -a \cos u \sin v & -a \sin u \sin v & a \cos v \end{vmatrix}$$

check this

$$= a \cos v \, (a \cos u \cos v \, \mathbf{i} + a \sin u \cos v \, \mathbf{j} + a \sin v \, \mathbf{k})$$
$$= a \cos v \, \mathbf{r}(u, v).$$

As was to be expected, the fundamental vector product of a sphere is parallel to the radius vector $\mathbf{r}(u, v)$. ❏

The Area of a Parametrized Surface

A linear function

$$\mathbf{r}(u, v) = u\mathbf{a} + v\mathbf{b} + \mathbf{c} \qquad\qquad \text{(\textbf{a} and \textbf{b} not parallel)}$$

parametrizes a plane p. Horizontal lines from the uv-plane, lines with equations of the form $v = v_0$, are mapped onto lines parallel to $\mathbf{a}$, and vertical lines, $u = u_0$, are carried onto lines parallel to $\mathbf{b}$:

$$\mathbf{r}(u, v_0) = u\mathbf{a} \; + \underbrace{v_0\mathbf{b} + \mathbf{c}}, \qquad \mathbf{r}(u_0, v) = v\mathbf{b} \; + \underbrace{u_0\mathbf{a} + \mathbf{c}}.$$

direction vector ⏐ constant   direction vector   constant

Thus a rectangle R in the uv-plane with sides parallel to the u and v axes,

$$R: \quad u_1 \leq u \leq u_2, \quad v_1 \leq v \leq v_2,$$

is mapped onto a parallelogram on p with sides parallel to $\mathbf{a}$ and $\mathbf{b}$. What is important to us here is that

the area of the parallelogram $= \|\mathbf{a} \times \mathbf{b}\| \cdot$ (the area of R).

To show this, we refer to Figure 17.6.7. The parallelogram is generated by the vectors

$$\mathbf{r}(u_2, v_1) - \mathbf{r}(u_1, v_1) = (u_2\mathbf{a} + v_1\mathbf{b} + \mathbf{c}) - (u_1\mathbf{a} + v_1\mathbf{b} + \mathbf{c}) = (u_2 - u_1)\mathbf{a},$$
$$\mathbf{r}(u_1, v_2) - \mathbf{r}(u_1, v_1) = (u_1\mathbf{a} + v_2\mathbf{b} + \mathbf{c}) - (u_1\mathbf{a} + v_1\mathbf{b} + \mathbf{c}) = (v_2 - v_1)\mathbf{b}.$$

Figure 17.6.7

The area of the parallelogram is thus

$$\|(u_2 - u_1)\mathbf{a} \times (v_2 - v_1)\mathbf{b}\| = \|\mathbf{a} \times \mathbf{b}\|(u_2 - u_1)(v_2 - v_1)$$
$$= \|\mathbf{a} \times \mathbf{b}\| \cdot (\text{area of } R). \quad ❏$$

We can summarize as follows:

(17.6.2)

Let R be a rectangle in the uv-plane with sides parallel to the coordinate axes. If $\mathbf{a}$ and $\mathbf{b}$ are not parallel, the linear function

$$\mathbf{r}(u, v) = u\mathbf{a} + v\mathbf{b} + \mathbf{c}, \qquad (u, v) \in R$$

parametrizes a parallelogram with sides parallel to $\mathbf{a}$ and $\mathbf{b}$ and

the area of the parallelogram $= \|\mathbf{a} \times \mathbf{b}\| \cdot$ (the area of R).

More generally, let's suppose that we have a surface S parametrized by a continuously differentiable function

$$\mathbf{r} = \mathbf{r}(u, v), \qquad (u, v) \in \Omega.$$

We assume that Ω is a basic region in the uv-plane and that $\mathbf{r}$ is one-to-one on the interior of Ω. (We don't want $\mathbf{r}$ to cover parts of S more than once.) Also we assume that the fundamental vector product $\mathbf{N} = \mathbf{r}'_u \times \mathbf{r}'_v$ is never zero on the interior of Ω. (We can then use it as a normal.) Under these conditions we call S a *continuously differentiable surface* and define

(17.6.3)

$$\text{area of } S = \iint_\Omega \|\mathbf{N}(u, v)\| \, du \, dv.$$

We show the reasoning behind this definition in the case that Ω is a rectangle R with sides parallel to the coordinate axes. We begin by breaking up R into N little rectangles $R_1, \ldots, R_N$. This induces a decomposition of S into little pieces $S_1, \ldots, S_N$. Taking (u_i^*, v_i^*) as the center of R_i, we have the tip of $\mathbf{r}(u_i^*, v_i^*)$ in S_i. Since the vector $\mathbf{r}'_u(u_i^*, v_i^*) \times \mathbf{r}'_v(u_i^*, v_i^*)$ is normal to the surface at the tip of $\mathbf{r}(u_i^*, v_i^*)$, we can parametrize the tangent plane at this point by the linear function

$$\mathbf{f}(u, v) = u\,\mathbf{r}'_u(u_i^*, v_i^*) + v\,\mathbf{r}'_v(u_i^*, v_i^*) + [\mathbf{r}(u_i^*, v_i^*) - u_i^*\mathbf{r}'_u(u_i^*, v_i^*) - v_i^*\mathbf{r}'_v(u_i^*, v_i^*)].$$

(Check that this linear function gives the right plane.) S_i is the portion of S that corresponds to R_i. The portion of the tangent plane that corresponds to this same R_i is a parallelogram with area

$$\|\mathbf{r}'_u(u_i^*, v_i^*) \times \mathbf{r}'_v(u_i^*, v_i^*)\| \cdot (\text{area of } R_i) = \|\mathbf{N}(u_i^*, v_i^*)\| \cdot (\text{area of } R_i). \quad \text{(by (17.6.2))}$$

Taking this as our estimate for the area of S_i, we have

$$\text{area of } S = \sum_{i=1}^{N} \text{area of } S_i \cong \sum_{i=1}^{N} \|\mathbf{N}(u_i^*, v_i^*)\| \cdot (\text{area of } R_i).$$

This is a Riemann sum for

$$\iint_R \|\mathbf{N}(u, v)\| \, du \, dv$$

and tends to this integral as the maximum diameter of the R_i tends to zero. ❑

To make sure that Formula 17.6.3 does not violate our previously established notion of area, we must verify that it gives the expected result both for plane regions and surfaces of revolution. This is done in Examples 9 and 10. By way of introduction we begin with the sphere.

Example 8 (*The surface area of a sphere*) The function

$$\mathbf{r}(u, v) = a \cos u \cos v \mathbf{i} + a \sin u \cos v \mathbf{j} + a \sin v \mathbf{k}$$

with (u, v) ranging over the set $\Omega: 0 \leq u \leq 2\pi, -\frac{1}{2}\pi \leq v \leq \frac{1}{2}\pi$ parametrizes a sphere of radius a. For this parametrization

$$\mathbf{N}(u, v) = a \cos v \, \mathbf{r}(u, v) \qquad \text{and} \qquad \|\mathbf{N}(u, v)\| = a^2|\cos v| = a^2\cos v.$$

Example 7 ——⌐ $-\frac{1}{2}\pi \leq v \leq \frac{1}{2}\pi$ ——⌐

According to the new formula

$$\text{area of the sphere} = \iint_\Omega a^2\cos v \; du \; dv$$

$$= \int_0^{2\pi} \left(\int_{-\frac{1}{2}\pi}^{\frac{1}{2}\pi} a^2\cos v \; dv \right) du = 2\pi a^2 \int_{-\frac{1}{2}\pi}^{\frac{1}{2}\pi} \cos v \; dv = 4\pi a^2,$$

which, as you know, is correct. ❑

Example 9 (*The area of a region*) If S is a plane region Ω, then S can be parametrized by setting

$$\mathbf{r}(u, v) = u\mathbf{i} + v\mathbf{j}, \qquad (u, v) \in \Omega.$$

Here $\mathbf{N}(u, v) = \mathbf{r}'_u(u, v) \times \mathbf{r}'_v(u, v) = \mathbf{i} \times \mathbf{j} = \mathbf{k}$ and $\|\mathbf{N}(u, v)\| = 1$. In this case (17.6.3) reduces to the familiar formula

$$A = \iint_\Omega du \; dv. \quad ❑$$

Example 10 (*The area of a surface of revolution*) Let S be the surface generated by revolving the graph of a function

$$y = f(x), \qquad x \in [a, b] \qquad\qquad \text{(Figure 17.6.8)}$$

about the x-axis. We assume that f is positive and continuously differentiable. We can parametrize S by setting

$$\mathbf{r}(u, v) = v\mathbf{i} + f(v)\cos u \mathbf{j} + f(v)\sin u \mathbf{k}$$

with (u, v) ranging over the set $\Omega: 0 \leq u \leq 2\pi, a \leq v \leq b$. (We leave it to you to verify that this is right.) In this case

$$\mathbf{N}(u, v) = \mathbf{r}'_u(u, v) \times \mathbf{r}'_v(u, v) = \begin{vmatrix} \mathbf{i} & \mathbf{j} & \mathbf{k} \\ 0 & -f(v)\sin u & f(v)\cos u \\ 1 & f'(v)\cos u & f'(v)\sin u \end{vmatrix}$$

$$= -f(v)f'(v)\mathbf{i} + f(v)\cos u \mathbf{j} + f(v)\sin u \mathbf{k}.$$

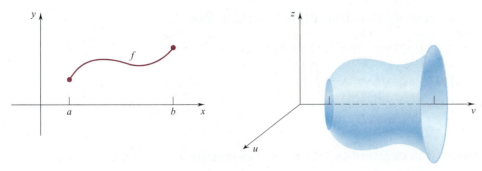

Figure 17.6.8

Therefore $\|\mathbf{N}(u, v)\| = f(v)\sqrt{[f'(v)]^2 + 1}$ and

$$\text{area of } S = \iint_\Omega f(v)\sqrt{[f'(v)]^2 + 1}\, du\, dv$$

$$= \int_0^{2\pi} \left(\int_a^b f(v)\sqrt{[f'(v)]^2 + 1}\, dv \right) du = \int_a^b 2\pi f(v)\sqrt{[f'(v)]^2 + 1}\, dv.$$

This is in agreement with Formula 9.10.3. ❏

Example 11 (*Spiral ramp*)   One turn of the spiral ramp of Example 5 is the surface

$$S: \quad \mathbf{r}(u, v) = u \cos \omega v\, \mathbf{i} + u \sin \omega v\, \mathbf{j} + bv\, \mathbf{k}$$

with (u, v) ranging over the set $\Omega: 0 \le u \le l, 0 \le v \le 2\pi/\omega$. In this case

$$\mathbf{r}'_u(u, v) = \cos \omega v\, \mathbf{i} + \sin \omega v\, \mathbf{j}, \qquad \mathbf{r}'_v(u, v) = -\omega u \sin \omega v\, \mathbf{i} + \omega u \cos \omega v\, \mathbf{j} + b\, \mathbf{k}.$$

Therefore

$$\mathbf{N}(u, v) = \begin{vmatrix} \mathbf{i} & \mathbf{j} & \mathbf{k} \\ \cos \omega v & \sin \omega v & 0 \\ -\omega u \sin \omega v & \omega u \cos \omega v & b \end{vmatrix} = b \sin \omega v\, \mathbf{i} - b \cos \omega v\, \mathbf{j} + \omega u\, \mathbf{k}$$

and

$$\|\mathbf{N}(u, v)\| = \sqrt{b^2 + \omega^2 u^2}.$$

Thus

$$\text{area of } S = \iint_\Omega \sqrt{b^2 + \omega^2 u^2}\, du\, dv$$

$$= \int_0^{2\pi/\omega} \left(\int_0^1 \sqrt{b^2 + \omega^2 u^2}\, du \right) dv = \frac{2\pi}{\omega} \int_0^1 \sqrt{b^2 + \omega^2 u^2}\, du.$$

The integral can be evaluated by setting $u = (b/\omega) \tan x$. ❏

The Area of a Surface $z = f(x, y)$

Figure 17.6.9 shows a surface that projects onto a basic region Ω of the xy-plane. Above each point (x, y) of Ω there is one and only one point of S. The surface S is then the graph of a function

$$z = f(x, y), \qquad (x, y) \in \Omega.$$

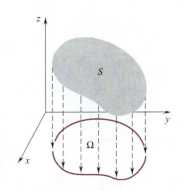

Figure 17.6.9

As we show, if f is continuously differentiable, then

(17.6.4)

$$\text{area of } S = \iint_\Omega \sqrt{[f_x(x, y)]^2 + [f_y(x, y)]^2 + 1} \; dx \, dy.$$

DERIVATION OF FORMULA 17.6.4   We can parametrize S by setting

$$\mathbf{r}(u, v) = u\mathbf{i} + v\mathbf{j} + f(u, v)\mathbf{k}, \qquad (u, v) \in \Omega.$$

We may just as well use x and y and write

$$\mathbf{r}(x, y) = x\mathbf{i} + y\mathbf{j} + f(x, y)\mathbf{k}, \qquad (x, y) \in \Omega.$$

Clearly

$$\mathbf{r}'_x(x, y) = \mathbf{i} + f_x(x, y)\mathbf{k} \qquad \text{and} \qquad \mathbf{r}'_y(x, y) = \mathbf{j} + f_y(x, y)\mathbf{k}.$$

Thus

$$\mathbf{N}(x, y) = \begin{vmatrix} \mathbf{i} & \mathbf{j} & \mathbf{k} \\ 1 & 0 & f_x(x, y) \\ 0 & 1 & f_y(x, y) \end{vmatrix} = -f_x(x, y)\mathbf{i} - f_y(x, y)\mathbf{j} + \mathbf{k}.$$

Therefore $\|\mathbf{N}(x, y)\| = \sqrt{[f_x(x, y)]^2 + [f_y(x, y)]^2 + 1}$ and the formula is verified. ❏

Example 12   Find the surface area of that part of the parabolic cylinder $z = y^2$ that lies over the triangle with vertices $(0, 0)$, $(0, 1)$, $(1, 1)$ in the xy-plane.

SOLUTION   Here $f(x, y) = y^2$ so that

$$f_x(x, y) = 0, \qquad f_y(x, y) = 2y.$$

The base triangle can be expressed by writing

$$\Omega: \quad 0 \le y \le 1, \quad 0 \le x \le y.$$

The surface has area

$$A = \iint_\Omega \sqrt{[f_x(x, y)]^2 + [f_y(x, y)]^2 + 1} \; dx \, dy$$

$$= \int_0^1 \int_0^y \sqrt{4y^2 + 1} \; dx \, dy$$

$$= \int_0^1 y\sqrt{4y^2 + 1} \; dy = \left[\tfrac{1}{12}(4y^2 + 1)^{3/2} \right]_0^1 = \tfrac{1}{12}(5\sqrt{5} - 1) \cong 0.85. \quad ❏$$

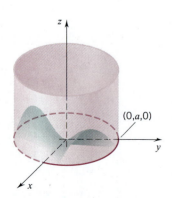

Figure 17.6.10

Example 13   Find the surface area of that part of the hyperbolic paraboloid $z = xy$ that lies inside the cylinder $x^2 + y^2 = a^2$. See Figure 17.6.10.

SOLUTION   Here $f(x, y) = xy$ so that

$$f_x(x, y) = y, \qquad f_y(x, y) = x.$$

The formula gives

$$A = \int\int_{\Omega} \sqrt{y^2 + x^2 + 1} \; dx \; dy.$$

In polar coordinates the base region takes the form

$$\Gamma: \quad 0 \le r \le a, \quad 0 \le \theta \le 2\pi.$$

Thus we have

$$A = \int\int_{\Gamma} \sqrt{r^2 + 1} \; r \; dr \; d\theta = \int_0^{2\pi} \int_0^a \sqrt{r^2 + 1} \; r \; dr \; d\theta = \tfrac{2}{3}\pi[(a^2 + 1)^{3/2} - 1]. \quad \square$$

There is an elegant version of this last area formula (Formula 17.6.4) that is geometrically vivid. We know that the vector

$$\mathbf{r}'_x(x, y) \times \mathbf{r}'_y(x, y) = -f_x(x, y)\mathbf{i} - f_y(x, y)\mathbf{j} + \mathbf{k}$$

is normal to the surface at the point $(x, y, f(x, y))$. The unit vector in that direction, the vector

$$\mathbf{n}(x, y) = \frac{-f_x(x, y)\mathbf{i} - f_y(x, y)\mathbf{j} + \mathbf{k}}{\sqrt{[f_x(x, y)]^2 + [f_y(x, y)]^2 + 1}},$$

is called the *upper unit normal.* (It is the unit normal with a nonnegative **k**-component.)

Now let $\gamma(x, y)$ be the angle between $\mathbf{n}(x, y)$ and $\mathbf{k}$ (Figure 17.6.11). Since $\mathbf{n}(x, y)$ and $\mathbf{k}$ are both unit vectors,

$$\cos[\gamma(x, y)] = \mathbf{n}(x, y) \cdot \mathbf{k} = \frac{1}{\sqrt{[f_x(x, y)]^2 + [f_y(x, y)]^2 + 1}}.$$

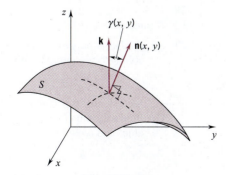

Figure 17.6.11

Taking reciprocals we have

$$\sec[\gamma(x, y)] = \sqrt{[f_x(x, y)]^2 + [f_y(x, y)]^2 + 1}.$$

The area formula can therefore be written

(17.6.5)

$$A = \iint_\Omega \sec[\gamma(x, y)] \, dx \, dy.$$

EXERCISES 17.6

In Exercises 1–4, calculate the fundamental vector product.

1. $\mathbf{r}(u, v) = (u^2 - v^2)\mathbf{i} + (u^2 + v^2)\mathbf{j} + 2uv\,\mathbf{k}$.

2. $\mathbf{r}(u, v) = u \cos v\,\mathbf{i} + u \sin v\,\mathbf{j} + \mathbf{k}$.

3. $\mathbf{r}(u, v) = (u + v)\mathbf{i} + (u + v)\mathbf{j} + (u - v)\mathbf{k}$.

4. $\mathbf{r}(u, v) = \cos u \sin v\,\mathbf{i} + \sin u \cos v\,\mathbf{j} + u\,\mathbf{k}$.

In Exercises 5–10, find a parametric representation for the given surface.

5. The upper half of the ellipsoid $4x^2 + 9y^2 + z^2 = 36$.

6. The part of the cylinder $x^2 + y^2 = 4$ that lies between the planes $z = 1$ and $z = 4$.

7. The part of the sphere $x^2 + y^2 + z^2 = 4$ that lies above the plane $z = \sqrt{2}$.

8. The part of the plane $z = x + 2$ that lies inside the cylinder $x^2 + y^2 = 1$.

9. $y = g(x, z), (x, z) \in \Omega$.   10. $x = h(y, z), (y, z) \in \Gamma$.

In Exercises 11–13, find an equation in x, y, z for the given surface and identify the surface.

11. $\mathbf{r}(u, v) = a \cos u \cos v\,\mathbf{i} + b \sin u \cos v\,\mathbf{j} + c \sin v\,\mathbf{k}$;
$0 \le u \le 2\pi, -\frac{1}{2}\pi \le v \le \frac{1}{2}\pi$.

12. $\mathbf{r}(u, v) = au \cos v\,\mathbf{i} + bu \sin v\,\mathbf{j} + u^2\,\mathbf{k}$; $0 \le u$,
$0 \le v \le 2\pi$.

13. $\mathbf{r}(u, v) = au \cosh v\,\mathbf{i} + bu \sinh v\,\mathbf{j} + u^2\,\mathbf{k}$; u real, v real.

14. Show that the hyperboloid of one sheet

$$\frac{x^2}{a^2} + \frac{y^2}{b^2} - \frac{z^2}{c^2} = 1 \qquad \text{(Figure 14.2.2)}$$

can be parametrized by the vector function

$\mathbf{r}(u, v) = a \cos u \cosh v\,\mathbf{i} + b \sin u \cosh v\,\mathbf{j}$
$\qquad + c \sinh v\,\mathbf{k};\quad 0 \le u \le 2\pi, v$ real.

15. The graph of a continuously differentiable function $y = f(x), x \in [a, b]$ is revolved about the y-axis. Parametrize the surface given that $a \ge 0$.

16. Show that the area of the surface of Exercise 15 is given by the formula

$$A = \int_a^b 2\pi x\sqrt{1 + [f'(x)]^2} \, dx.$$

17. A plane p intersects the xy-plane at an angle γ. (Draw a figure.) Find the area of a region Γ on p given that the projection of Γ onto the xy-plane is a region Ω of area A_Ω.

18. Determine the area of the portion of the plane $x + y + z = a$ that lies within the cylinder $x^2 + y^2 = b^2$.

19. Find the area of the part of the plane $bcx + acy + abz = abc$ that lies within the first octant.

20. Find the area of the surface $z^2 = x^2 + y^2$ from $z = 0$ to $z = 1$.

21. Find the area of the surface $z = x^2 + y^2$ from $z = 0$ to $z = 4$.

In Exercises 22–28, calculate the area of the surface.

22. $z^2 = 2xy$ with $0 \le x \le a,\ 0 \le y \le b,\ z \ge 0$.

23. $z = a^2 - (x^2 + y^2)$ with $\frac{1}{4}a^2 \le x^2 + y^2 \le a^2$.

24. $3z^2 = (x + y)^3$ with $x + y \le 2,\ x \ge 0,\ y \ge 0$.

25. $3z = x^{3/2} + y^{3/2}$ with $0 \le x \le 1,\ 0 \le y \le x$.

26. $z = y^2$ with $0 \le x \le 1,\ 0 \le y \le 1$.

27. $x^2 + y^2 + z^2 - 4z = 0$ with $0 \le 3(x^2 + y^2) \le z^2$,
$z \ge 2$.

28. $x^2 + y^2 + z^2 - 2az = 0$ with $0 \le x^2 + y^2 \le bz$.
Assume $a > b > 0$.

29. (a) Find a formula for the area of a surface that is projectable onto a region Ω of the yz-plane; say,

$$S: \quad x = g(y, z),\quad (y, z) \in \Omega.$$

Assume that g is continuously differentiable.
(b) Find a formula for the area of a surface that is projectable onto a region Ω of the xz-plane; say,

$$S: \quad y = h(x, z),\quad (x, z) \in \Omega.$$

Assume that h is continuously differentiable.

30. (a) Determine the fundamental vector product for the cylindrical surface

$\mathbf{r}(u, v) = a \cos u\,\mathbf{i} + a \sin u\,\mathbf{j} + v\,\mathbf{k}$;
$0 \le u \le 2\pi,\ 0 \le v \le l$.

(b) Use your answer to (a) to find the area of the surface.

31. (a) Determine the fundamental vector product for the cone of Example 4:

$$\mathbf{r}(u, v) = v \cos u \sin \alpha \mathbf{i} + v \sin u \sin \alpha \mathbf{j} + v \cos \alpha \mathbf{k};$$
$$0 \leq u \leq 2\pi, \quad 0 \leq v \leq s.$$

(b) Use your answer to (a) to calculate the area of the cone.

32. Verify that the ellipsoid of revolution

$$\frac{x^2}{a^2} + \frac{y^2}{a^2} + \frac{z^2}{b^2} = 1$$

can be parametrized by

$$\mathbf{r}(u, v) = a \cos u \sin v \mathbf{i} + a \sin u \sin v \mathbf{j} + b \cos v \mathbf{k};$$
$$0 \leq u \leq 2\pi, \quad 0 \leq v \leq \pi.$$

Then show that the surface area of the ellipsoid is given by the formula

$$A = 2\pi a \int_0^\pi \sin v \sqrt{b^2 \sin^2 v + a^2 \cos^2 v}\, dv.$$

33. Let Ω be a plane region in space and let A_1, A_2, A_3 be the areas of the projections of Ω onto the three coordinate planes. Express the area of Ω in terms of A_1, A_2, A_3.

34. Let S be a surface given in cylindrical coordinates by $z = f(r, \theta)$, $(r, \theta) \in \Omega$. Show that, if f is continuously differentiable, then

$$\text{area of } S = \iint_\Omega \sqrt{r^2[f_r(r, \theta)]^2 + [f_\theta(r, \theta)]^2 + r^2}\, dr\, d\theta$$

provided the integrand is never zero on the interior of Ω.

35. The following surfaces are given in cylindrical coordinates. Find the surface area.
  (a) $z = r + \theta$; $\quad 0 \leq r \leq 1, \quad 0 \leq \theta \leq \pi.$
  (b) $z = re^\theta$; $\quad 0 \leq r \leq a, \quad 0 \leq \theta \leq 2\pi.$

36. Show that, for a flat surface S that is part of the xy-plane, (17.6.3) gives

$$\text{area of } S = \iint_\Omega |J(u, v)|\, du\, dv$$

where J is the Jacobian of the components of the vector function that parametrizes S. Except for notation this is Formula 16.11.1.

■ 17.7 SURFACE INTEGRALS

The Mass of a Material Surface

Imagine a thin distribution of matter spread out over a surface S. We call this a *material surface*.

If the mass density (the mass per unit area) is a constant λ throughout, then the total mass of the material surface is the density λ times the area of S:

$$M = \lambda(\text{area of } S).$$

If, however, the mass density varies continuously from point to point, $\lambda = \lambda(x, y, z)$, then the total mass must be calculated by integration.

To develop the appropriate integral we suppose that

$$S: \quad \mathbf{r} = \mathbf{r}(u, v) = x(u, v)\mathbf{i} + y(u, v)\mathbf{j} + z(u, v)\mathbf{k}, \quad (u, v) \in \Omega$$

is a continuously differentiable surface, a surface that meets the conditions for area formula (17.6.3).† Our first step is to break up Ω into N little basic regions $\Omega_1, \ldots, \Omega_N$. This decomposes the surface into N little pieces $S_1, \ldots, S_N$. The area of S_i is given by the integral

$$\iint_{\Omega_i} \|\mathbf{N}(u, v)\|\, du\, dv. \qquad \text{(Formula 17.6.3)}$$

---

† We repeat the conditions here: $\mathbf{r}$ is continuously differentiable; Ω is a basic region in the uv-plane; $\mathbf{r}$ is one-to-one on the interior of Ω; $\mathbf{N} = \mathbf{r}_u' \times \mathbf{r}_v'$ is never zero on the interior of Ω.

By the mean-value theorem for double integrals, there exists a point (u_i^*, v_i^*) in Ω_i for which

$$\iint\limits_{\Omega_i} \|\mathbf{N}(u, v)\| \, du \, dv = \|\mathbf{N}(u_i^*, v_i^*)\|(\text{area of } \Omega_i).$$

It follows that

$$\text{area of } S_i = \|\mathbf{N}(u_i^*, v_i^*)\|(\text{area of } \Omega_i).$$

Since the point (u_i^*, v_i^*) is in Ω_i, the tip of $\mathbf{r}(u_i^*, v_i^*)$ is on S_i. The mass density at this point is $\lambda[\mathbf{r}(u_i^*, v_i^*)]$. If S_i is small (which we can guarantee by choosing Ω_i small), then the mass density on S_i is approximately the same throughout. Thus we can estimate M_i, the mass contribution of S_i, by writing

$$M_i \cong \lambda[\mathbf{r}(u_i^*, v_i^*)](\text{area of } S_i) = \lambda[\mathbf{r}(u_i^*, v_i^*)]\|\mathbf{N}(u_i^*, v_i^*)\|(\text{area of } \Omega_i).$$

Adding up these estimates we have an estimate for the total mass of the surface:

$$M \cong \sum_{i=1}^{N} \lambda[\mathbf{r}(u_i^*, v_i^*)]\|\mathbf{N}(u_i^*, v_i^*)\|(\text{area of } \Omega_i)$$
$$= \sum_{i=1}^{N} \lambda[x(u_i^*, v_i^*), y(u_i^*, v_i^*), z(u_i^*, v_i^*)]\|\mathbf{N}(u_i^*, v_i^*)\|(\text{area of } \Omega_i).$$

This last expression is a Riemann sum for

$$\iint\limits_{\Omega} \lambda[x(u, v), y(u, v), z(u, v)]\|\mathbf{N}(u, v)\| \, du \, dv$$

and tends to this integral as the maximum diameter of the Ω_i tends to zero. We can therefore conclude that

(17.7.1)
$$M = \iint\limits_{\Omega} \lambda[x(u, v), y(u, v), z(u, v)]\|\mathbf{N}(u, v)\| \, du \, dv.$$

Surface Integrals

The double integral in (17.7.1) can be calculated not only for a mass density function λ but for any scalar field H continuous over S. We call this integral *the surface integral of H over S* and write

(17.7.2)
$$\iint\limits_{S} H(x, y, z) \, d\sigma = \iint\limits_{\Omega} H[x(u, v), y(u, v), z(u, v)]\|\mathbf{N}(u, v)\| \, du \, dv.$$

Note that, if $H(x, y, z)$ is identically 1, then the right-hand side of (17.7.2) gives the area of S. Thus

(17.7.3)
$$\iint_S d\sigma = \text{area of } S.$$

Example 1 Let $\mathbf{a} = a_1\mathbf{i} + a_2\mathbf{j} + a_3\mathbf{k}$ and $\mathbf{b} = b_1\mathbf{i} + b_2\mathbf{j} + b_3\mathbf{k}$ be vectors, and calculate

$$\iint_S xy \, d\sigma \quad \text{where} \quad S: \quad \mathbf{r}(u, v) = u\mathbf{a} + v\mathbf{b}; \quad 0 \le u \le 1, \quad 0 \le v \le 1.$$

SOLUTION Call the parameter set Ω. Then

$$\iint_S xy \, d\sigma = \iint_\Omega x(u, v)\, y(u, v)\|\mathbf{N}(u, v)\| \, du \, dv.$$

By Example 6, Section 17.6, $\|\mathbf{N}(u, v)\| = \|\mathbf{a} \times \mathbf{b}\|$. Thus

$$\iint_S xy \, d\sigma = \|\mathbf{a} \times \mathbf{b}\| \iint_\Omega x(u, v)\, y(u, v) \, du \, dv.$$

To find $x(u, v)$ and $y(u, v)$ we need the $\mathbf{i}$ and $\mathbf{j}$ components of $\mathbf{r}(u, v)$. We can get these as follows:

$$\mathbf{r}(u, v) = u\mathbf{a} + v\mathbf{b} = u(a_1\mathbf{i} + a_2\mathbf{j} + a_3\mathbf{k}) + v(b_1\mathbf{i} + b_2\mathbf{j} + b_3\mathbf{k})$$
$$= (a_1u + b_1v)\mathbf{i} + (a_2u + b_2v)\mathbf{j} + (a_3u + b_3v)\mathbf{k}.$$

Therefore $x(u, v) = a_1u + b_1v$ and $y(u, v) = a_2u + b_2v$. We can now write

$$\iint_S xy \, d\sigma = \|\mathbf{a} \times \mathbf{b}\| \iint_\Omega (a_1u + b_1v)(a_2u + b_2v) \, du \, dv$$

$$= \|\mathbf{a} \times \mathbf{b}\| \int_0^1 \left(\int_0^1 [a_1a_2u^2 + (a_1b_2 + b_1a_2)uv + b_1b_2v^2] \, du \right) dv$$

$$= \|\mathbf{a} \times \mathbf{b}\|[\tfrac{1}{3}a_1a_2 + \tfrac{1}{4}(a_1b_2 + b_1a_2) + \tfrac{1}{3}b_1b_2]. \quad \square$$

check this ⟶

Example 2 Calculate

$$\iint_S \sqrt{x^2 + y^2} \, d\sigma$$

where S is the spiral ramp of Example 11, Section 17.6:

$$S: \quad \mathbf{r}(u, v) = u \cos \omega v\, \mathbf{i} + u \sin \omega v\, \mathbf{j} + bv\, \mathbf{k}; \quad 0 \le u \le l, \quad 0 \le v \le 2\pi/\omega.$$

SOLUTION   Call the parameter set Ω.   As we saw in Example 11, Section 17.6,

$$\|\mathbf{N}(u, v)\| = \sqrt{b^2 + \omega^2 u^2}.$$

Therefore

$$
\begin{aligned}
\iint_S \sqrt{x^2 + y^2}\, d\sigma &= \iint_\Omega \sqrt{[x(u, v)]^2 + [y(u, v)]^2}\,\|\mathbf{N}(u, v)\|\, du\, dv \\[2mm]
&= \iint_\Omega \sqrt{u^2\cos^2\omega v + u^2\sin^2\omega v}\,\sqrt{b^2 + \omega^2 u^2}\, du\, dv \\[2mm]
&= \iint_\Omega u\sqrt{b^2 + \omega^2 u^2}\, du\, dv \\[2mm]
&= \int_0^{2\pi/\omega} \left(\int_0^l u\sqrt{b^2 + \omega^2 u^2}\, du \right) dv \\[2mm]
&= \frac{2\pi}{\omega} \int_0^l u\sqrt{b^2 + \omega^2 u^2}\, du = \frac{2\pi}{3\omega^3}[(b^2 + \omega^2 l^2)^{3/2} - b^3]. \quad \square
\end{aligned}
$$

$u \geq 0$ on Ω ⟶ (annotation pointing to third line)

Like the other integrals you have studied, the surface integral satisfies a mean-value condition; namely, if the scalar field H is continuous, then there is a point (x_0, y_0, z_0) on S for which

$$\iint_S H(x, y, z)\, d\sigma = H(x_0, y_0, z_0)(\text{area of } S).$$

We call $H(x_0, y_0, z_0)$ *the average value of H on s.*   We can thus write

(17.7.4)
$$\iint_S H(x, y, z)\, d\sigma = \left(\begin{array}{c} \text{average value} \\ \text{of } H \text{ on } S \end{array} \right) \cdot (\text{area of } S).$$

We can also take weighted averages: if H and G are continuous on S, then there is a point (x_0, y_0, z_0) on S for which

(17.7.5)
$$\iint_S H(x, y, z)G(x, y, z)\, d\sigma = H(x_0, y_0, z_0) \iint_S G(x, y, z)\, d\sigma.$$

As you would expect, we call $H(x_0, y_0, z_0)$ *the G-weighted average of H on S.*

The coordinates of the centroid $(\bar{x}, \bar{y}, \bar{z})$ of a surface are simply averages over the surface: for a surface S of area A

$$\bar{x}A = \iint_S x\, d\sigma, \qquad \bar{y}A = \iint_S y\, d\sigma, \qquad \bar{z}A = \iint_S z\, d\sigma.$$

In the case of a material surface of mass density $\lambda = \lambda(x, y, z)$, the coordinates of the center of mass (x_M, y_M, z_M) are density-weighted averages: for a surface S of total mass M

$$x_M M = \iint_S x\lambda(x, y, z)\, d\sigma, \quad y_M M = \iint_S y\lambda(x, y, z)\, d\sigma, \quad z_M M = \iint_S z\lambda(x, y, z)\, d\sigma.$$

Example 3 Locate the center of mass of a material surface in the form of a hemispherical shell $x^2 + y^2 + z^2 = a^2$ with $z \geq 0$ given that the mass density is directly proportional to the distance from the xy-plane. See Figure 17.7.1.

SOLUTION The surface S can be parametrized by

$$\mathbf{r}(u, v) = a \cos u \cos v\,\mathbf{i} + a \sin u \cos v\,\mathbf{j} + a \sin v\,\mathbf{k}; \quad 0 \leq u \leq 2\pi, 0 \leq v \leq \tfrac{1}{2}\pi.$$

Call the parameter set Ω and recall that $\|\mathbf{N}(u, v)\| = a^2 \cos v$ (Example 8, Section 17.6). The density function can be written $\lambda(x, y, z) = kz$, where k is the constant of proportionality. We can calculate the mass as follows:

$$M = \iint_S \lambda(x, y, z)\, d\sigma = k \iint_S z\, d\sigma = k \iint_\Omega z(u, v)\|\mathbf{N}(u, v)\|\, du\, dv$$

$$= k \int_0^{2\pi} \left(\int_0^{\pi/2} (a \sin v)(a^2 \cos v)\, dv \right) du$$

$$= 2\pi k a^3 \int_0^{\pi/2} \sin v \cos v\, dv = \pi k a^3.$$

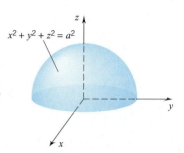

$x^2 + y^2 + z^2 = a^2$

Figure 17.7.1

By symmetry $x_M = 0$ and $y_M = 0$. To find z_M we write

$$z_M M = \iint_S z\lambda(x, y, z)\, d\sigma = k \iint_S z^2\, d\sigma$$

$$= k \iint_\Omega [z(u, v)]^2 \|\mathbf{N}(u, v)\|\, du\, dv$$

$$= k \int_0^{2\pi} \left(\int_0^{\pi/2} (a^2 \sin^2 v)(a^2 \cos v)\, dv \right) du$$

$$= 2\pi k a^4 \int_0^{\pi/2} \sin^2 v \cos v\, dv = \tfrac{2}{3}\pi k a^4.$$

Since $M = \pi k a^3$, we see that $z_M = \tfrac{2}{3}\pi k a^4 / M = \tfrac{2}{3}a$. The center of mass is the point $(0, 0, \tfrac{2}{3}a)$. ❏

Suppose that a material surface S rotates about an axis. The moment of inertia of the surface about that axis is given by the formula

(17.7.6)

$$I = \iint_S \lambda(x, y, z)[R(x, y, z)]^2\, d\sigma$$

where $\lambda = \lambda(x, y, z)$ is the mass density function and $R(x, y, z)$ is the distance from the axis to the point (x, y, z). (As with other configurations the moments of inertia about the x, y, z axes are denoted by I_x, I_y, I_z.)

Example 4 A homogeneous material surface with mass density 1 is in the shape of a spherical shell

$$S: \quad x^2 + y^2 + z^2 = a^2. \qquad \text{(Figure 17.7.2)}$$

Calculate the moment of inertia about the z-axis.

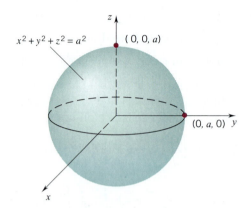

Figure 17.7.2

SOLUTION We parametrize S by setting

$$\mathbf{r}(u, v) = a \cos u \cos v \, \mathbf{i} + a \sin u \cos v \, \mathbf{j} + a \sin v \, \mathbf{k}; \quad 0 \le u \le 2\pi, \ -\tfrac{1}{2}\pi \le v \le \tfrac{1}{2}\pi.$$

Call the parameter set Ω and recall that $\|\mathbf{N}(u, v)\| = a^2 \cos v$. We can calculate the moment of inertia as follows:

$$I_z = \iint_S 1(x^2 + y^2)\, d\sigma = \iint_\Omega ([x(u, v)]^2 + [y(u, v)]^2)\|\mathbf{N}(u, v)\|\, d\sigma$$

$$= \iint_\Omega (a^2 \cos^2 v)(a^2 \cos v)\, du\, dv$$

$$= a^4 \int_0^{2\pi} \left(\int_{-\pi/2}^{\pi/2} \cos^3 v\, dv \right) du$$

$$= 2\pi a^4 \int_{-\pi/2}^{\pi/2} \cos^3 v\, dv = \tfrac{8}{3}\pi a^4.$$
$$\underset{\text{check this}}{\uparrow\underline{\qquad}}$$

Since the surface has mass $M = A = 4\pi a^2$, it follows that $I_z = \tfrac{2}{3}Ma^2$. ❏

A surface

$$S: \quad z = f(x, y), \quad (x, y) \in \Omega$$

can be parametrized by

$$\mathbf{r}(x, y) = x\mathbf{i} + y\mathbf{j} + f(x, y)\mathbf{k}, \qquad (x, y) \in \Omega.$$

As you saw in Section 17.6, $\|\mathbf{N}(x, y)\| = \sec[\gamma(x, y)]$ where $\gamma(x, y)$ is the angle between $\mathbf{k}$ and the upper unit normal. Therefore, for any continuous scalar field H on S,

$$(17.7.7) \qquad \iint\limits_S H(x, y, z) \, d\sigma = \iint\limits_\Omega H(x, y, z) \sec[\gamma(x, y)] \, dx \, dy.$$

In evaluating this last integral we use the fact that

$$\sec[\gamma(x, y)] = \sqrt{[f_x(x, y)]^2 + [f_y(x, y)]^2 + 1}. \qquad \text{(Section 17.6)}$$

Example 5 Calculate

$$\iint\limits_S \sqrt{x^2 + y^2} \, d\sigma \qquad \text{with} \qquad S: \quad z = xy, \quad 0 \le x^2 + y^2 \le 1.$$

SOLUTION The base region Ω is the unit disc. The function $z = f(x, y) = xy$ has partial derivatives $f_x(x, y) = y$, $f_y(x, y) = x$. Therefore

$$\sec[\gamma(x, y)] = \sqrt{y^2 + x^2 + 1} = \sqrt{x^2 + y^2 + 1}$$

and

$$\iint\limits_S \sqrt{x^2 + y^2} \, d\sigma = \iint\limits_\Omega \sqrt{x^2 + y^2} \sqrt{x^2 + y^2 + 1} \, dx \, dy.$$

We evaluate this last integral by changing to polar coordinates. The region Ω is the set of all (x, y) with polar coordinates (r, θ) in the set

$$\Gamma: \quad 0 \le \theta \le 2\pi, \quad 0 \le r \le 1.$$

Therefore

$$\iint\limits_S \sqrt{x^2 + y^2} \, d\sigma = \iint\limits_\Gamma r\sqrt{r^2 + 1} \, r \, dr \, d\theta = \int_0^{2\pi} \left(\int_0^1 r^2 \sqrt{r^2 + 1} \, dr \right) d\theta$$

$$= 2\pi \int_0^1 r^2 \sqrt{r^2 + 1} \, dr$$

$$= 2\pi \int_0^{\pi/4} \tan^2\phi \, \sec^3\phi \, d\phi$$

$$= 2\pi \int_0^{\pi/4} [\sec^5\phi - \sec^3\phi] \, d\phi$$

$$= \tfrac{1}{4}\pi[3\sqrt{2} - \ln(\sqrt{2} + 1)] \cong 2.64. \quad \square$$

The Flux of a Vector Field

Suppose that

$$S: \quad \mathbf{r} = \mathbf{r}(u, v), \quad (u, v) \in \Omega$$

is a continuously differentiable surface with a unit normal $\mathbf{n} = \mathbf{n}(x, y, z)$ that is continuous on all of S. Such a surface is called a *smooth surface*. If $\mathbf{v} = \mathbf{v}(x, y, z)$ is a vector field continuous on S, then we can form the surface integral

(17.7.8) $$\iint\limits_{S} (\mathbf{v} \cdot \mathbf{n}) \, d\sigma = \iint\limits_{S} [\mathbf{v}(x, y, z) \cdot \mathbf{n}(x, y, z)] \, d\sigma.$$

This surface integral is called *the flux of* $\mathbf{v}$ *across S in the direction of* $\mathbf{n}$.

Note that the flux across a surface depends on the choice of unit normal. If $-\mathbf{n}$ is chosen instead of $\mathbf{n}$, the sign of the flux is reversed:

$$\iint\limits_{S} (\mathbf{v} \cdot -\mathbf{n}) \, d\sigma = \iint\limits_{S} -(\mathbf{v} \cdot \mathbf{n}) \, d\sigma = -\iint\limits_{S} (\mathbf{v} \cdot \mathbf{n}) \, d\sigma.$$

Example 6 Calculate the flux of the vector field $\mathbf{v}(x, y, z) = x\mathbf{i} + y\mathbf{j}$ across the sphere S: $x^2 + y^2 + z^2 = a^2$ in the outward direction (Figure 17.7.3).

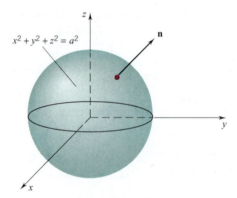

Figure 17.7.3

SOLUTION The outward unit normal is the vector

$$\mathbf{n}(x, y, z) = \frac{1}{a}(x\mathbf{i} + y\mathbf{j} + z\mathbf{k}).$$

Here

$$\mathbf{v} \cdot \mathbf{n} = (x\mathbf{i} + y\mathbf{j}) \cdot \frac{1}{a}(x\mathbf{i} + y\mathbf{j} + z\mathbf{k}) = \frac{1}{a}(x^2 + y^2).$$

Therefore

$$\text{flux out of } S = \frac{1}{a} \iint\limits_{S} (x^2 + y^2) \, d\sigma.$$

To evaluate this integral we use the usual parametrization

$$\mathbf{r}(u, v) = a \cos u \cos v \, \mathbf{i} + a \sin u \cos v \, \mathbf{j} + a \sin v \, \mathbf{k}; \quad 0 \le u \le 2\pi, \ -\tfrac{1}{2}\pi \le v \le \tfrac{1}{2}\pi.$$

Also, recall that $\|\mathbf{N}(u, v)\| = a^2 \cos v$. Thus

$$\text{flux out of } S = \frac{1}{a} \iint\limits_{\Omega} ([x(u, v)]^2 + [y(u, v)]^2) \|\mathbf{N}(u, v)\| \, du \, dv$$

$$= \frac{1}{a} \iint\limits_{\Omega} (a^2 \cos^2 u \cos^2 v + a^2 \sin^2 u \cos^2 v)(a^2 \cos v) \, du \, dv$$

$$= a^3 \iint\limits_{\Omega} \cos^3 v \, du \, dv$$

$$= a^3 \int_0^{2\pi} \left(\int_{-\pi/2}^{\pi/2} \cos^3 v \, dv \right) du = 2\pi a^3 \int_{-\pi/2}^{\pi/2} \cos^3 v \, dv = \tfrac{8}{3} \pi a^3. \quad \square$$

If S is the graph of a function $z = f(x, y), (x, y) \in \Omega$ and $\mathbf{n}$ is the upper unit normal, then the flux of the vector field $\mathbf{v} = v_1 \mathbf{i} + v_2 \mathbf{j} + v_3 \mathbf{k}$ across S in the direction of $\mathbf{n}$ is

(17.7.9)
$$\iint\limits_{S} (\mathbf{v} \cdot \mathbf{n}) \, d\sigma = \iint\limits_{\Omega} (-v_1 f_x - v_2 f_y + v_3) \, dx \, dy.$$

PROOF From Section 17.6 we know that

$$\mathbf{n} = \frac{-f_x \mathbf{i} - f_y \mathbf{j} + \mathbf{k}}{\sqrt{(f_x)^2 + (f_y)^2 + 1}} = (-f_x \mathbf{i} - f_y \mathbf{j} + \mathbf{k}) \cos \gamma,$$

where γ is the angle between $\mathbf{n}$ and $\mathbf{k}$. Therefore $\mathbf{v} \cdot \mathbf{n} = (-v_1 f_x - v_2 f_y + v_3) \cos \gamma$ and

$$\iint\limits_{S} (\mathbf{v} \cdot \mathbf{n}) \, d\sigma = \iint\limits_{\Omega} (\mathbf{v} \cdot \mathbf{n}) \sec \gamma \, dx \, dy = \iint\limits_{\Omega} (-v_1 f_x - v_2 f_y + v_3) \, dx \, dy. \quad \square$$

Example 7 Let S be the portion of the paraboloid $z = 1 - (x^2 + y^2)$ that lies above the unit disc Ω. Calculate the flux of $\mathbf{v} = x\mathbf{i} + y\mathbf{j} + z\mathbf{k}$ across this surface in the direction of the upper unit normal (Figure 17.7.4).

SOLUTION

$$f(x, y) = 1 - (x^2 + y^2); \qquad f_x = -2x, \quad f_y = -2y.$$

$$\text{Flux} = \iint\limits_{\Omega} (-v_1 f_x - v_2 f_y + v_3) \, dx \, dy$$

$$= \iint\limits_{\Omega} [(-x)(-2x) - y(-2y) + 1 - (x^2 + y^2)] \, dx \, dy$$

$$= \iint\limits_{\Omega} (1 + x^2 + y^2) \, dx \, dy = \int_0^{2\pi} \left(\int_0^1 (1 + r^2) r \, dr \right) d\theta = \tfrac{3}{2} \pi. \quad \square$$

in polar coordinates

Figure 17.7.4

The flux through a closed *piecewise-smooth surface* (a closed surface that consists of a finite number of smooth pieces joined together at the boundaries) can be evaluated by integrating over each smooth piece and adding up the results.

Example 8   Calculate the total flux of $\mathbf{v}(x, y, z) = xy\,\mathbf{i} + 4yz^2\mathbf{j} + yz\,\mathbf{k}$ out of the unit cube: $0 \le x \le 1, 0 \le y \le 1, 0 \le z \le 1$.   (Figure 17.7.5)

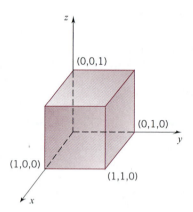

Figure 17.7.5

SOLUTION   The total flux is the sum of the fluxes across the faces of the cube:

Face	$\mathbf{n}$	$\mathbf{v} \cdot \mathbf{n}$	Flux
$x = 0$	$-\mathbf{i}$	$-xy = 0$	0
$x = 1$	$\mathbf{i}$	$xy = y$	$\frac{1}{2}$
$y = 0$	$-\mathbf{j}$	$-4yz^2 = 0$	0
$y = 1$	$\mathbf{j}$	$4yz^2 = 4z^2$	$\frac{4}{3}$
$z = 0$	$-\mathbf{k}$	$-yz = 0$	0
$z = 1$	$\mathbf{k}$	$yz = y$	$\frac{1}{2}$

The total flux is $\frac{1}{2} + \frac{4}{3} + \frac{1}{2} = \frac{7}{3}$.   ❑

Flux plays an important role in the study of fluid motion.   Imagine a surface S within a fluid and choose a unit normal $\mathbf{n}$.   Take $\mathbf{v}$ as the velocity of the fluid that passes through S.   You can expect $\mathbf{v}$ to vary from point to point, but, for simplicity, let's assume that $\mathbf{v}$ does not change with time.   (Such a time-independent flow is said to be steady state.)   The flux of $\mathbf{v}$ across S is the average component of $\mathbf{v}$ in the direction of $\mathbf{n}$ times the area of the surface S.   This is just the volume of fluid that passes through S in unit time, from the $-\mathbf{n}$ side of S to the $\mathbf{n}$ side of S.   If S is a closed surface (such as a cube, a sphere, or an ellipsoid) and $\mathbf{n}$ is chosen as the *outer unit normal*, then the flux across S gives the volume of fluid that flows *out* through S in unit time; if $\mathbf{n}$ is chosen as the *inner unit normal,* then the flux gives the volume of liquid that flows *in* through S in unit time.

EXERCISES 17.7

In Exercises 1–6, evaluate these integrals over the surface S: $z = \frac{1}{2}y^2$; $0 \leq x \leq 1$, $0 \leq y \leq 1$.

1. $\displaystyle\iint_S d\sigma.$

2. $\displaystyle\iint_S x^2 \, d\sigma.$

3. $\displaystyle\iint_S 3y \, d\sigma.$

4. $\displaystyle\iint_S (x - y) \, d\sigma.$

5. $\displaystyle\iint_S \sqrt{2z} \, d\sigma.$

6. $\displaystyle\iint_S \sqrt{1 + y^2} \, d\sigma.$

In Exercises 7–12, evaluate the given surface integral.

7. $\displaystyle\iint_S xy \, d\sigma$; S is the first octant part of the plane

$x + 2y + 3z = 6.$

8. $\displaystyle\iint_S xyz \, d\sigma$; S is the first octant part of the plane

$x + y + z = 1.$

9. $\displaystyle\iint_S x^2 z \, d\sigma$; S is that part of the cylinder $x^2 + z^2 = 1$,

which is between the planes $y = 0$ and $y = 2$, and above the xy-plane.

10. $\displaystyle\iint_S (x^2 + y^2 + z^2) \, d\sigma$; S is that part of the plane $z = $

$x + 2$ that lies inside the cylinder $x^2 + y^2 = 1.$

11. $\displaystyle\iint_S (x^2 + y^2) \, d\sigma$; S is the hemisphere

$z = \sqrt{1 - x^2 + y^2}.$

12. $\displaystyle\iint_S (x^2 + y^2) \, d\sigma$; S is that part of the paraboloid $z = $

$1 - x^2 - y^2$ that lies above the xy-plane.

In Exercises 13–15, find the mass of a material surface in the shape of a triangle $(a, 0, 0)$, $(0, a, 0)$, $(0, 0, a)$ with the given density function.

13. $\lambda(x, y, z) = k.$

14. $\lambda(x, y, z) = k(x + y).$

15. $\lambda(x, y, z) = kx^2.$

16. Locate the centroid of the triangle $(a, 0, 0)$, $(0, a, 0)$, $(0, 0, a)$, $a > 0.$

17. Locate the centroid of the hemisphere $x^2 + y^2 + z^2 = a^2$, $z \geq 0.$

In Exercises 18–19, let S be the parallelogram given by $\mathbf{r}(u, v) = (u + v)\mathbf{i} + (u - v)\mathbf{j} + 2u\mathbf{k}$; $0 \leq u \leq 1$, $0 \leq v \leq 1.$

18. Find the area of S.

19. Determine the flux of $\mathbf{v} = x\mathbf{i} - y\mathbf{j}$ across S in the direction of the fundamental vector product.

20. Find the mass of the material surface S: $z = 1 - \frac{1}{2}(x^2 + y^2)$ with $0 \leq x \leq 1$, $0 \leq y \leq 1$ if the density at each point (x, y, z) is proportional to xy.

In Exercises 21–23, calculate the flux out of the sphere $x^2 + y^2 + z^2 = a^2.$

21. $\mathbf{v} = z\mathbf{k}.$ **22.** $\mathbf{v} = x\mathbf{i} + y\mathbf{j} + z\mathbf{k}.$ **23.** $\mathbf{v} = y\mathbf{i} - x\mathbf{j}.$

24. A homogeneous plate of mass density 1 is in the form of the parallelogram of Exercises 18 and 19. Determine the moments of inertia about the coordinate axes: (a) I_x. (b) I_y. (c) I_z.

In Exercises 25–27, determine the flux across the triangle $(a, 0, 0)$, $(0, a, 0)$, $(0, 0, a)$, $a > 0$, in the direction of the upper unit normal.

25. $\mathbf{v} = x\mathbf{i} + y\mathbf{j} + z\mathbf{k}.$ **26.** $\mathbf{v} = (x + z)\mathbf{k}.$

27. $\mathbf{v} = x^2\mathbf{i} - y^2\mathbf{j}.$

In Exercises 28–30, determine the flux across S: $z = xy$ with $0 \leq x \leq 1$, $0 \leq y \leq 2$ in the direction of the upper unit normal.

28. $\mathbf{v} = -xy^2\mathbf{i} + z\mathbf{j}.$ **29.** $\mathbf{v} = xz\mathbf{j} - xy\mathbf{k}.$

30. $\mathbf{v} = x^2y\mathbf{i} + z^2\mathbf{k}.$

31. Calculate the flux of $\mathbf{v} = x\mathbf{i} + y\mathbf{j} + z\mathbf{k}$ out of the cylindrical surface

$S: \quad \mathbf{r}(u, v) = a \cos u\mathbf{i} + a \sin u\mathbf{j} + v\mathbf{k};$
$0 \leq u \leq 2\pi, 0 \leq v \leq l.$

32. Calculate the flux of the gravitational force function

$$\mathbf{F}(\mathbf{r}) = G\frac{mM}{r^3}\mathbf{r} \quad \text{out of the sphere } x^2 + y^2 + z^2 = a^2.$$

In Exercises 33–36, find the flux across S: $z = \frac{2}{3}(x^{3/2} + y^{3/2})$ with $0 \leq x \leq 1$, $0 \leq y \leq 1 - x$ in the direction of the upper unit normal.

33. $\mathbf{v} = x\mathbf{i} - y\mathbf{j} + \frac{3}{2}z\mathbf{k}.$ **34.** $\mathbf{v} = x^2\mathbf{i}.$

35. $\mathbf{v} = y^2\mathbf{j}.$ **36.** $\mathbf{v} = y\mathbf{i} - \sqrt{xy}\mathbf{j}.$

37. The cone

$$\mathbf{r}(u, v) = v \cos u \sin \alpha\mathbf{i} + v \sin u \sin \alpha\mathbf{j} + v \cos \alpha\mathbf{k};$$

$$0 \leq u \leq 2\pi, \quad 0 \leq v \leq s$$

has area $A = \pi s^2 \sin \alpha$. Locate the centroid.

In Exercises 38–40, the mass density of a material cone $z = \sqrt{x^2 + y^2}$ with $0 \leq z \leq 1$ varies directly as the distance from the z-axis.

38. Find the mass of the cone.

39. Locate the center of mass.

40. Determine the moments of inertia about the coordinate axes: (a) I_x.　(b) I_y.　(c) I_z.

41. You have seen that, if S is a smooth surface immersed in a fluid of velocity $\mathbf{v}$, then the flux

$$\iint_S (\mathbf{v} \cdot \mathbf{n}) \, d\sigma$$

is the volume of fluid that passes through S in unit time from the $-\mathbf{n}$ side of S to the $\mathbf{n}$ side of S. This requires that S be a two-sided surface. There are, however, one-sided surfaces: for example, the Möbius band. To construct a material Möbius band start with a piece of paper in the form of the rectangle in the figure above. Now give the piece of paper a single twist and join the two far edges together so that C coincides with A and D coincides with B. (a) Convince yourself that this surface is one-

sided and therefore the notion of flux cannot be applied to it. (b) The surface is not smooth because it is impossible to erect a unit normal $\mathbf{n}$ that varies continuously over the entire surface. Convince yourself of this as follows: erect a unit normal $\mathbf{n}$ at some point P and make a circuit of the surface with $\mathbf{n}$. Note that, as $\mathbf{n}$ returns to P, the direction of $\mathbf{n}$ has been reversed.

In Exercises 42–44, assume that the parallelogram of Exercises 18 and 19 is a material surface with a mass density that varies directly as the square of the distance from the x-axis.

42. Determine the mass.

43. Find the x-coordinate of the center of mass.

44. Find the moment of inertia about the z-axis.

45. Calculate the total flux of $\mathbf{v}(x, y, z) = y\mathbf{i} - x\mathbf{j}$ out of the solid bounded on the sides by the cylinder $x^2 + y^2 = 1$ and above and below by the planes $z = 1$ and $z = 0$.　HINT: Draw a figure.

46. Calculate the total flux of $\mathbf{v}(x, y, z) = y\mathbf{i} - x\mathbf{j}$ out of the solid bounded above by $z = 4$ and below by $z = x^2 + y^2$.

47. Calculate the total flux of $\mathbf{v}(x, y, z) = x\mathbf{i} + y\mathbf{j} + z\mathbf{k}$ out of the solid bounded above by $z = \sqrt{2 - (x^2 + y^2)}$ and below by $z = x^2 + y^2$.

48. Calculate the total flux of $\mathbf{v}(x, y, z) = xz\mathbf{i} + 4xyz^2\mathbf{j} + 2z\mathbf{k}$ out of the unit cube: $0 \leq x \leq 1$, $0 \leq y \leq 1, 0 \leq z \leq 1$.

■ 17.8 THE VECTOR DIFFERENTIAL OPERATOR ∇

Divergence ∇ · v, Curl ∇ × v

The *vector differential operator* ∇ (this is an inverted delta and it's read *del*) is defined formally by setting

(17.8.1)
$$\nabla = \frac{\partial}{\partial x}\mathbf{i} + \frac{\partial}{\partial y}\mathbf{j} + \frac{\partial}{\partial z}\mathbf{k}.$$

By "formally" we mean that this is not an ordinary vector. Its "components" are differentiation symbols. As the term "operator" suggests, ∇ is to be thought of as something that "operates" on things. What sort of things? Scalar fields and vector fields.

Suppose that f is a differentiable scalar field. Then ∇ operates on f as follows:

$$\nabla f = \left(\frac{\partial}{\partial x}\mathbf{i} + \frac{\partial}{\partial y}\mathbf{j} + \frac{\partial}{\partial z}\mathbf{k} \right) f = \frac{\partial f}{\partial x}\mathbf{i} + \frac{\partial f}{\partial y}\mathbf{j} + \frac{\partial f}{\partial z}\mathbf{k}.$$

This is just the *gradient* of f, with which you are already familiar.

How does ∇ operate on vector fields? In two ways. If $\mathbf{v} = v_1\mathbf{i} + v_2\mathbf{j} + v_3\mathbf{k}$ is a differentiable vector field, then by definition

(17.8.2)
$$\nabla \cdot \mathbf{v} = \frac{\partial v_1}{\partial x} + \frac{\partial v_2}{\partial y} + \frac{\partial v_3}{\partial z}$$

and

(17.8.3)
$$\nabla \times \mathbf{v} = \begin{vmatrix} \mathbf{i} & \mathbf{j} & \mathbf{k} \\ \dfrac{\partial}{\partial x} & \dfrac{\partial}{\partial y} & \dfrac{\partial}{\partial z} \\ v_1 & v_2 & v_3 \end{vmatrix}$$
$$= \left(\frac{\partial v_3}{\partial y} - \frac{\partial v_2}{\partial z}\right)\mathbf{i} + \left(\frac{\partial v_1}{\partial z} - \frac{\partial v_3}{\partial x}\right)\mathbf{j} + \left(\frac{\partial v_2}{\partial x} - \frac{\partial v_1}{\partial y}\right)\mathbf{k}.$$

The first "product," $\nabla \cdot \mathbf{v}$, defined in imitation of the ordinary dot product, is called the *divergence* of $\mathbf{v}$:

$$\nabla \cdot \mathbf{v} = \text{div } \mathbf{v}.$$

The second "product," $\nabla \times \mathbf{v}$, defined in imitation of the ordinary cross product, is called the *curl* of $\mathbf{v}$:

$$\nabla \times \mathbf{v} = \text{curl } \mathbf{v}.$$

Interpretation of Divergence and Curl

Suppose we know the divergence of a field and also the curl. What does that tell us? For definitive answers we must wait for the divergence theorem and Stokes's theorem, but, in a preliminary way, we can give you some rough answers right now.

View $\mathbf{v}$ as the velocity field of some fluid. The divergence of $\mathbf{v}$ at a point P gives us an indication of whether the fluid tends to accumulate near P (negative divergence) or tends to move away from P (positive divergence). In the first case, P is sometimes called a *sink*, and in the second case, it is called a *source*. The curl at P measures the rotational tendency of the fluid.

Example 1 Set

$$\mathbf{v}(x, y, z) = \alpha x\,\mathbf{i} + \alpha y\,\mathbf{j} + \alpha z\,\mathbf{k}. \qquad (\alpha \text{ a constant})$$

Divergence:

$$\nabla \cdot \mathbf{v} = \alpha \frac{\partial x}{\partial x} + \alpha \frac{\partial y}{\partial y} + \alpha \frac{\partial z}{\partial z} = 3\alpha.$$

Curl:

$$\nabla \times \mathbf{v} = \begin{vmatrix} \mathbf{i} & \mathbf{j} & \mathbf{k} \\ \dfrac{\partial}{\partial x} & \dfrac{\partial}{\partial y} & \dfrac{\partial}{\partial z} \\ \alpha x & \alpha y & \alpha z \end{vmatrix} = \alpha \begin{vmatrix} \mathbf{i} & \mathbf{j} & \mathbf{k} \\ \dfrac{\partial}{\partial x} & \dfrac{\partial}{\partial y} & \dfrac{\partial}{\partial z} \\ x & y & z \end{vmatrix} = \mathbf{0}$$

because the partial derivatives that appear in the expanded determinant

$$\frac{\partial y}{\partial x}, \qquad \frac{\partial x}{\partial y}, \qquad \text{etc.,}$$

are all zero. ❑

The field of Example 1

$$\mathbf{v}(x, y, z) = \alpha(x\mathbf{i} + y\mathbf{j} + z\mathbf{k}) = \alpha\mathbf{r}$$

can be viewed as the velocity field of a fluid in radial motion—toward the origin if $\alpha < 0$, away from the origin if $\alpha > 0$. Figure 17.8.1 shows a point (x, y, z), a spherical neighborhood of that point, and a cone emanating from the origin that is tangent to the boundary of the neighborhood.

Note two things: all the fluid in the cone stays in the cone, and the speed of the fluid is proportional to its distance from the origin. Therefore, if the divergence 3α is negative, then α is negative, the motion is toward the origin, and the neighborhood *gains fluid* because the fluid coming in is moving more quickly than the fluid going out. (Also the entry area is larger than the exit area.) If, however, the divergence 3α is positive, then α is positive, the motion is away from the origin, and the neighborhood *loses fluid* because the fluid coming in is moving more slowly than the fluid going out. (Also the entry area is smaller than the exit area.)

Since the motion is radial, the fluid has no rotational tendency whatsoever, and we would expect the curl to be identically zero. It is. ❑

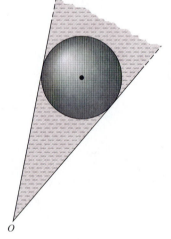

O

Figure 17.8.1

Example 2   Set

$$\mathbf{v}(x, y, z) = -\omega y\mathbf{i} + \omega x\mathbf{j}. \qquad (\omega \text{ a positive constant})$$

Divergence:

$$\nabla \cdot \mathbf{v} = -\omega\frac{\partial y}{\partial x} + \omega\frac{\partial x}{\partial y} = 0 + 0 = 0.$$

Curl:

$$\nabla \times \mathbf{v} = \begin{vmatrix} \mathbf{i} & \mathbf{j} & \mathbf{k} \\ \dfrac{\partial}{\partial x} & \dfrac{\partial}{\partial y} & \dfrac{\partial}{\partial z} \\ -\omega y & \omega x & 0 \end{vmatrix} = \left(\omega\frac{\partial x}{\partial x} + \omega\frac{\partial y}{\partial y}\right)\mathbf{k} = 2\omega\mathbf{k}. ❑$$

The field of Example 2

$$\mathbf{v}(x, y, z) = -\omega y\mathbf{i} + \omega x\mathbf{j}$$

is the velocity field of uniform counterclockwise rotation about the z-axis with angular speed ω. You can see this by noting that $\mathbf{v}$ is perpendicular to $\mathbf{r} = x\mathbf{i} + y\mathbf{j} + z\mathbf{k}$:

$$\mathbf{v} \cdot \mathbf{r} = (-\omega y\mathbf{i} + x\mathbf{j}) \cdot (x\mathbf{i} + y\mathbf{j} + z\mathbf{k}) = -\omega yx + \omega xy = 0$$

and the speed at each point is ωR where R is the radius of rotation:

$$v = \sqrt{\omega^2 y^2 + \omega^2 x^2} = \omega \sqrt{x^2 + y^2} = \omega R.$$

How is the curl, $2\omega\mathbf{k}$, related to the rotation? The angular velocity vector (see Exercise 24, Section 13.5) is the vector $\boldsymbol{\omega} = \omega\mathbf{k}$. In this case, then, the curl of $\mathbf{v}$ is twice the angular velocity vector.

With this rotation no neighborhood gains any fluid and no neighborhood loses any fluid. As we saw, the divergence is identically zero. ❏

Basic Identities

For vectors we have $\mathbf{a} \times \mathbf{a} = \mathbf{0}$. Is it true that $\nabla \times \nabla = \mathbf{0}$? Define $(\nabla \times \nabla)f$ by

$$(\nabla \times \nabla)f = \nabla \times (\nabla f).$$

THEOREM 17.8.4 (THE CURL OF A GRADIENT IS ZERO)

If f is a scalar field with continuous second partials, then

$$\nabla \times (\nabla f) = \mathbf{0}.$$

PROOF

$$\nabla \times (\nabla f) = \begin{vmatrix} \mathbf{i} & \mathbf{j} & \mathbf{k} \\ \dfrac{\partial}{\partial x} & \dfrac{\partial}{\partial y} & \dfrac{\partial}{\partial z} \\ \dfrac{\partial f}{\partial x} & \dfrac{\partial f}{\partial y} & \dfrac{\partial f}{\partial z} \end{vmatrix} =$$

$$\left(\frac{\partial^2 f}{\partial y \partial z} - \frac{\partial^2 f}{\partial z \partial y} \right)\mathbf{i} + \left(\frac{\partial^2 f}{\partial z \partial x} - \frac{\partial^2 f}{\partial x \partial z} \right)\mathbf{j} + \left(\frac{\partial^2 f}{\partial x \partial y} - \frac{\partial^2 f}{\partial y \partial x} \right)\mathbf{k} = \mathbf{0}. \quad ❏$$

For vectors we have $\mathbf{a} \cdot (\mathbf{a} \times \mathbf{c}) = 0$. The analogous operator formula, $\nabla \cdot (\nabla \times \mathbf{v}) = 0$, is also valid.

THEOREM 17.8.5 (THE DIVERGENCE OF A CURL IS ZERO)

If the components of the vector field $\mathbf{v} = v_1\mathbf{i} + v_2\mathbf{j} + v_3\mathbf{k}$ have continuous second partials, then

$$\nabla \cdot (\nabla \times \mathbf{v}) = 0.$$

PROOF Again the key is the equality of the mixed partials:

$$\nabla \cdot (\nabla \times \mathbf{v}) = \frac{\partial}{\partial x}\left(\frac{\partial v_3}{\partial y} - \frac{\partial v_2}{\partial z} \right) + \frac{\partial}{\partial y}\left(\frac{\partial v_1}{\partial z} - \frac{\partial v_3}{\partial x} \right) + \frac{\partial}{\partial z}\left(\frac{\partial v_2}{\partial x} - \frac{\partial v_1}{\partial y} \right) = 0,$$

since for each component v_i the mixed partials cancel. Try it for v_1. ❏

The next two identities are product rules. Here f is a scalar field and $\mathbf{v}$ is a vector field.

(17.8.6)

$$\nabla \cdot (f\mathbf{v}) = (\nabla f) \cdot \mathbf{v} + f(\nabla \cdot \mathbf{v}). \quad [\text{div}(f\mathbf{v}) = (\text{grad}\, f) \cdot \mathbf{v} + f(\text{div } \mathbf{v})]$$

(17.8.7)

$$\nabla \times (f\mathbf{v}) = (\nabla f) \times \mathbf{v} + f(\nabla \times \mathbf{v}). \quad [\text{curl}(f\mathbf{v}) = (\text{grad}\, f) \times \mathbf{v} + f(\text{curl } \mathbf{v})]$$

The verification of these identities is left to you in the exercises.

We know from Example 1 that $\nabla \cdot \mathbf{r} = 3$ and $\nabla \times \mathbf{r} = \mathbf{0}$ at all points of space. Now we can show that, if n is an integer, then, for all $\mathbf{r} \neq \mathbf{0}$,

(17.8.8)

$$\nabla \cdot (r^n \mathbf{r}) = (n + 3)r^n \qquad \text{and} \qquad \nabla \times (r^n \mathbf{r}) = \mathbf{0}. \quad †$$

PROOF Recall that $\nabla r^n = nr^{n-2}\mathbf{r}$. (See 15.1.5.) Using (17.8.6), we have

$$\nabla \cdot (r^n \mathbf{r}) = (\nabla r^n) \cdot \mathbf{r} + r^n(\nabla \cdot \mathbf{r})$$
$$= (nr^{n-2}\mathbf{r}) \cdot \mathbf{r} + r^n(3)$$
$$= nr^{n-2}(\mathbf{r} \cdot \mathbf{r}) + 3r^n = nr^n + 3r^n = (n + 3)r^n.$$

From (15.1.5) you can see that $r^n \mathbf{r}$ is a gradient. Its curl is therefore $\mathbf{0}$ (17.8.4). ❏

The Laplacian

From the operator ∇ we can construct other operators. The most important of these is the Laplacian $\nabla^2 = \nabla \cdot \nabla$. The Laplacian (named after the French mathematician Pierre-Simon Laplace) operates on scalar fields according to the following rule:

(17.8.9)

$$\nabla^2 f = \nabla \cdot (\nabla f) = \frac{\partial^2 f}{\partial x^2} + \frac{\partial^2 f}{\partial y^2} + \frac{\partial^2 f}{\partial z^2}. \quad ††$$

Example 3 If $f(x, y, z) = x^2 + y^2 + z^2$, then

$$\nabla^2 f = \frac{\partial^2}{\partial x^2}(x^2 + y^2 + z^2) + \frac{\partial^2}{\partial y^2}(x^2 + y^2 + z^2) + \frac{\partial^2}{\partial z^2}(x^2 + y^2 + z^2)$$
$$= 2 + 2 + 2 = 6. \quad ❏$$

---

† If n is positive and even, these formulas also hold at $\mathbf{r} = \mathbf{0}$.

†† In some texts, you will see the Laplacian of f written Δf. Unfortunately this can be misread as the increment of f.

Example 4   If $f(x, y, z) = e^{xyz}$, then

$$\nabla^2 f = \frac{\partial}{\partial x^2}(e^{xyz}) + \frac{\partial^2}{\partial y^2}(e^{xyz}) + \frac{\partial^2}{\partial z^2}(e^{xyz})$$

$$= \frac{\partial}{\partial x}(yze^{xyz}) + \frac{\partial}{\partial y}(xze^{xyz}) + \frac{\partial}{\partial z}(xye^{xyz})$$

$$= y^2 z^2 e^{xyz} + x^2 z^2 e^{xyz} + x^2 y^2 e^{xyz}$$

$$= (y^2 z^2 + x^2 z^2 + x^2 y^2)e^{xyz}. \quad ❏$$

Example 5   To calculate $\nabla^2(\sin r) = \nabla^2(\sin\sqrt{x^2 + y^2 + z^2})$ we could write

$$\frac{\partial^2}{\partial x^2}(\sin\sqrt{x^2 + y^2 + z^2}) + \frac{\partial^2}{\partial y^2}(\sin\sqrt{x^2 + y^2 + z^2}) + \frac{\partial^2}{\partial z^2}(\sin\sqrt{x^2 + y^2 + z^2})$$

and proceed from there.   The calculations are straightforward but lengthy.   We will do it in a different way.

Recall that

$$\nabla^2 f = \nabla \cdot \nabla f \quad \text{(definition)} \qquad \nabla \cdot (f\mathbf{v}) = (\nabla f \cdot \mathbf{v}) + f(\nabla \cdot \mathbf{v}) \quad (17.8.6)$$

$$\nabla f(r) = f'(r)r^{-1}\mathbf{r} \quad (15.4.8) \qquad \nabla \cdot (r^n\mathbf{r}) = (n + 3)r^n \quad (17.8.8)$$

Using these relations, we have

$$\nabla^2(\sin r) = \nabla \cdot (\nabla \sin r) = \nabla \cdot [(\cos r)r^{-1}\mathbf{r}]$$
$$= [(\nabla \cos r) \cdot r^{-1}\mathbf{r}] + \cos r(\nabla \cdot r^{-1}\mathbf{r})$$
$$= \{[(-\sin r)r^{-1}\mathbf{r}] \cdot r^{-1}\mathbf{r}\} + (\cos r)(2r^{-1})$$
$$= -\sin r + 2r^{-1} \cos r.$$

We leave it to you to justify each step.   ❏

EXERCISES 17.8

In Exercises 1–12, find $\nabla \cdot \mathbf{v}$ and $\nabla \times \mathbf{v}$.

1. $\mathbf{v}(x, y) = x\mathbf{i} + y\mathbf{j}$.   **2.** $\mathbf{v}(x, y) = y\mathbf{i} + x\mathbf{j}$.

3. $\mathbf{v}(x, y) = \dfrac{x}{x^2 + y^2}\mathbf{i} + \dfrac{y}{x^2 + y^2}\mathbf{j}$.

4. $\mathbf{v}(x, y) = \dfrac{y}{x^2 + y^2}\mathbf{i} + \dfrac{x}{x^2 + y^2}\mathbf{j}$.

5. $\mathbf{v}(x, y, z) = x\mathbf{i} + 2y\mathbf{j} + 3z\mathbf{k}$.

6. $\mathbf{v}(x, y, z) = yz\mathbf{i} + xz\mathbf{j} + xy\mathbf{k}$.

7. $\mathbf{v}(x, y, z) = xyz\mathbf{i} + xz\mathbf{j} + z\mathbf{k}$.

8. $\mathbf{v}(x, y, z) = x^2 y\mathbf{i} + y^2 z\mathbf{j} + xy^2\mathbf{k}$.

9. $\mathbf{v}(\mathbf{r}) = r^{-2}\mathbf{r}$.   **10.** $\mathbf{v}(\mathbf{r}) = e^x\mathbf{r}$.

11. $\mathbf{v}(\mathbf{r}) = e^{r^2}(\mathbf{i} + \mathbf{j} + \mathbf{k})$.

12. $\mathbf{v}(\mathbf{r}) = e^{y^2}\mathbf{i} + e^{z^2}\mathbf{j} + e^{x^2}\mathbf{k}$.

13. Suppose that f is a differentiable function of one variable and $\mathbf{v}(x, y, z) = f(x)\mathbf{i}$.   Determine $\nabla \cdot \mathbf{v}$ and $\nabla \times \mathbf{v}$.

14. Show that, if $\mathbf{v}$ is a differentiable field of the form $\mathbf{v}(\mathbf{r}) = f(x)\mathbf{i} + g(y)\mathbf{j} + h(z)\mathbf{k}$, then $\nabla \times \mathbf{v} = 0$.

15. Show that divergence and curl are *linear* operators:

$$\nabla \cdot (\alpha\mathbf{u} + \beta\mathbf{v}) = \alpha(\nabla \cdot \mathbf{u}) + \beta(\nabla \cdot \mathbf{v}) \quad \text{and}$$

$$\nabla \times (\alpha\mathbf{u} + \beta\mathbf{v}) = \alpha(\nabla \times \mathbf{u}) + \beta(\nabla \times \mathbf{v}).$$

16. (*Important*) Show that the gravitational field

$$\mathbf{F}(\mathbf{r}) = -\frac{GmM}{r^3}\mathbf{r}$$

has zero divergence and zero curl at each $\mathbf{r} \neq \mathbf{0}$.

A vector field $\mathbf{F}$ with the property that $\nabla \cdot \mathbf{F} = 0$ is said to be *solenoidal*.   Exercise 16 shows that the gravitational field is solenoidal.   If $\mathbf{F}$ is the velocity field of some fluid and $\nabla \cdot \mathbf{F} = 0$ in a solid T in three-dimensional space, then $\mathbf{F}$ has no sources or sinks within T.

17. Show that the vector field $\mathbf{F}(x, y, z) = (2x + y + 2z)\mathbf{i} + (x + 4y - 3z)\mathbf{j} + (2x - 3y - 6z)\mathbf{k}$ is solenoidal.

18. Show that the vector field $\mathbf{F}(x, y, z) = 3x^2\mathbf{i} - y^2\mathbf{j} + (2yz - 6xz)\mathbf{k}$ is solenoidal.

A vector field **F** with the property that $\nabla \times \mathbf{F} = \mathbf{0}$ is said to be *irrotational*. Exercise 16 shows that the gravitational field is irrotational. If **F** is the velocity field of some fluid, then $\nabla \times \mathbf{F}$ measures the tendency of the fluid to "curl" or rotate about an axis. Thus $\nabla \times \mathbf{F} = \mathbf{0}$ in some solid T in three-dimensional space can be interpreted to mean that the fluid is tending to move in a straight line.

19. Show that the vector field $\mathbf{F}(x, y, z) = x\mathbf{i} + y\mathbf{j} - 2z\mathbf{k}$ is irrotational.

20. Show that the vector field of Exercise 17 is irrotational.

In Exercises 21–26, calculate the Laplacian $\nabla^2 f$.

21. $f(x, y, z) = x^4 + y^4 + z^4$.

22. $f(x, y, z) = xyz$.

23. $f(x, y, z) = x^2 y^3 z^4$.   **24.** $f(\mathbf{r}) = \cos r$.

25. $f(\mathbf{r}) = e^r$.   **26.** $f(\mathbf{r}) = \ln r$.

27. Given a vector field **u**, the operator $\mathbf{u} \cdot \nabla$ is defined by setting

$$(\mathbf{u} \cdot \nabla)f = \mathbf{u} \cdot \nabla f = u_1 \frac{\partial f}{\partial x} + u_2 \frac{\partial f}{\partial y} + u_3 \frac{\partial f}{\partial z}.$$

Calculate $(\mathbf{r} \cdot \nabla)f$ for: (a) $f(\mathbf{r}) = r^2$;   (b) $f(\mathbf{r}) = 1/r$.

28. (Based on Exercise 27) We can also apply $\mathbf{u} \cdot \nabla$ to a vector field **v** by applying it to each component. By definition

$$(\mathbf{u} \cdot \nabla)\mathbf{v} = (\mathbf{u} \cdot \nabla v_1)\mathbf{i} + (\mathbf{u} \cdot \nabla v_2)\mathbf{j} + (\mathbf{u} \cdot \nabla v_3)\mathbf{k}.$$

  (a) Calculate $(\mathbf{u} \cdot \nabla)\mathbf{r}$ for an arbitrary vector field **u**.
  (b) Calculate $(\mathbf{r} \cdot \nabla)\mathbf{u}$ given that $\mathbf{u} = yz\mathbf{i} + xz\mathbf{j} + xy\mathbf{k}$.

29. Show that, if $f(\mathbf{r}) = g(r)$ and g is twice differentiable, then

$$\nabla^2 f = g''(r) + 2r^{-1} g'(r).$$

30. Verify the following identities.
  (a) $\nabla \cdot (f\mathbf{v}) = (\nabla f) \cdot \mathbf{v} + f(\nabla \cdot \mathbf{v})$.
  (b) $\nabla \times (f\mathbf{v}) = (\nabla f) \times \mathbf{v} + f(\nabla \times \mathbf{v})$.
  (c) $\nabla \times (\nabla \times \mathbf{v}) = \nabla(\nabla \cdot \mathbf{v}) - \nabla^2 \mathbf{v}$ where
$$\nabla^2 \mathbf{v} = (\nabla^2 v_1)\mathbf{i} + (\nabla^2 v_2)\mathbf{j} + (\nabla^2 v_3)\mathbf{k}.$$
   HINT: Begin (c) by writing out the ith-component of each side.

As you saw in Exercises 14.6, the equation

$$\nabla^2 f = \frac{\partial^2 f}{\partial x^2} + \frac{\partial^2 f}{\partial y^2} + \frac{\partial^2 f}{\partial z^2} = 0$$

is called *Laplace's equation in three dimensions*. A scalar field $f = f(x, y, z)$ with continuous second partial derivatives is said to be *harmonic* on a solid T if it is a solution of Laplace's equation.

31. Show that the scalar field

$$f(x, y, z) = x^2 + 2y^2 - 3z^2 + xy + 2xz - 3yz$$

is harmonic.

32. Show that the scalar field

$$f(x, y, z) = \frac{1}{\sqrt{x^2 + y^2 + z^2}}$$

is harmonic on every solid T that excludes the origin. Except for a constant multiplier, f is a potential function for the gravitational field.

33. For what nonzero integers n is $f(\mathbf{r}) = r^n$ harmonic on every solid T that excludes the origin? Here $\mathbf{r} = x\mathbf{i} + y\mathbf{j} + z\mathbf{k}$ and $r = \|\mathbf{r}\| = \sqrt{x^2 + y^2 + z^2}$.

34. Show that if $f = f(x, y, z)$ satisfies Laplace's equation, then its gradient field is both solenoidal and irrotational.

■ 17.9 THE DIVERGENCE THEOREM

Let Ω be a Jordan region with a piecewise smooth boundary C, and let P and Q be continuously differentiable scalar fields on an open set containing Ω. Green's theorem allows us to express a double integral over Ω as a line integral over C:

$$\iint_\Omega \left(\frac{\partial Q}{\partial x} - \frac{\partial P}{\partial y} \right) dx\, dy = \oint_C P\, dx + Q\, dy.$$

In vector terms Green's theorem can be written:

(17.9.1)
$$\iint_\Omega (\nabla \cdot \mathbf{v})\, dx\, dy = \oint_C (\mathbf{v} \cdot \mathbf{n})\, ds.$$

Here $\mathbf{n}$ is the outer unit normal and the integral on the right is taken with respect to arc length. (Section 17.4)

PROOF Set $\mathbf{v} = Q\mathbf{i} - P\mathbf{j}$. Then

$$\iint_{\Omega} (\boldsymbol{\nabla} \cdot \mathbf{v})\, dx\, dy = \iint_{\Omega} \left(\frac{\partial Q}{\partial x} - \frac{\partial P}{\partial y} \right) dx\, dy.$$

All we have to show then is that

$$\oint_{C} (\mathbf{v} \cdot \mathbf{n})\, ds = \oint_{C} P\, dx + Q\, dy.$$

For C traversed counterclockwise, $\mathbf{n} = \mathbf{T} \times \mathbf{k}$ where $\mathbf{T}$ is the unit tangent vector. (Draw a figure.) Thus

$$\mathbf{v} \cdot \mathbf{n} = \mathbf{v} \cdot (\mathbf{T} \times \mathbf{k}) = (-\mathbf{v}) \cdot (\mathbf{k} \times \mathbf{T}) = (-\mathbf{v} \times \mathbf{k}) \cdot \mathbf{T}.$$
$$\underset{\mathbf{a} \cdot (\mathbf{b} \times \mathbf{c}) = (\mathbf{a} \times \mathbf{b}) \cdot \mathbf{c}}{\uparrow}$$

Since $-\mathbf{v} \times \mathbf{k} = (P\mathbf{j} - Q\mathbf{i}) \times \mathbf{k} = P\mathbf{i} + Q\mathbf{j}$, we have $\mathbf{v} \cdot \mathbf{n} = (P\mathbf{i} + Q\mathbf{j}) \cdot \mathbf{T}$. Therefore

$$\oint_{C} (\mathbf{v} \cdot \mathbf{n})\, ds = \oint_{C} [(P\mathbf{i} + Q\mathbf{j}) \cdot \mathbf{T}]\, ds = \oint_{C} (P\mathbf{i} + Q\mathbf{j}) \cdot d\mathbf{r} = \oint_{C} P\, dx + Q\, dy.$$
$$\underset{(17.4.6)}{\uparrow}$$

❑

Green's theorem expressed as (17.9.1) has a higher dimensional analog that is known as the divergence theorem.†

THEOREM 17.9.2 THE DIVERGENCE THEOREM

Let T be a solid bounded by a closed surface S which, if not smooth, is piecewise smooth. If the vector field $\mathbf{v} = \mathbf{v}(x, y, z)$ is continuously differentiable throughout T, then

$$\iiint_{T} (\boldsymbol{\nabla} \cdot \mathbf{v})\, dx\, dy\, dz = \iint_{S} (\mathbf{v} \cdot \mathbf{n})\, d\sigma$$

where $\mathbf{n}$ is the outer unit normal.

PROOF We will carry out the proof under the assumption that S is smooth and that any line parallel to a coordinate axis intersects S at most twice.

Our first step is to express the outer unit normal $\mathbf{n}$ in terms of its direction cosines:

$$\mathbf{n} = \cos \alpha_1 \mathbf{i} + \cos \alpha_2 \mathbf{j} + \cos \alpha_3 \mathbf{k}.$$

Then, for $\mathbf{v} = v_1 \mathbf{i} + v_2 \mathbf{j} + v_3 \mathbf{k}$,

$$\mathbf{v} \cdot \mathbf{n} = v_1 \cos \alpha_1 + v_2 \cos \alpha_2 + v_3 \cos \alpha_3.$$

---

† Also called Gauss' theorem after the German mathematician Karl Friedrich Gauss (1777–1855). Often referred to as "The Prince of Mathematicians," Gauss is regarded by many as one of the greatest geniuses of all time.

The idea of the proof is to show that

(1)
$$\iint_S v_1 \cos \alpha_1 \, d\sigma = \iiint_T \frac{\partial v_1}{\partial x} \, dx \, dy \, dz,$$

(2)
$$\iint_S v_2 \cos \alpha_2 \, d\sigma = \iiint_T \frac{\partial v_2}{\partial y} \, dx \, dy \, dz,$$

(3)
$$\iint_S v_3 \cos \alpha_3 \, d\sigma = \iiint_T \frac{\partial v_3}{\partial z} \, dx \, dy \, dz.$$

All three equations can be verified in much the same manner. We will carry out the details only for the third equation.

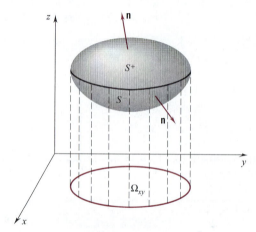

Figure 17.9.1

Let Ω_{xy} be the projection of T onto the xy-plane. (See Figure 17.9.1.) If $(x, y) \in \Omega_{xy}$, then, by assumption, the vertical line through (x, y) intersects S in at most two points, an upper point P^+ and a lower point P^-. (If the vertical line intersects S at only one point P, we set $P = P^+ = P^-$.) As (x, y) ranges over Ω_{xy}, the upper point P^+ describes a surface

$$S^+: \quad z = f^+(x, y), \quad (x, y) \in \Omega_{xy} \qquad \text{(see the figure)}$$

and the lower point describes a surface

$$S^-: \quad z = f^-(x, y), \quad (x, y) \in \Omega_{xy}.$$

By our assumptions, f^+ and f^- are continuously differentiable, $S = S^+ \cup S^-$, and the solid T is the set of all points (x, y, z) with

$$f^-(x, y) \leq z \leq f^+(x, y), \quad (x, y) \in \Omega_{xy}.$$

Now let γ be the angle between the positive z-axis and the upper unit normal. On S^+ the outer unit normal $\mathbf{n}$ is the upper unit normal. Thus on S^+

$$\gamma = \alpha_3 \qquad \text{and} \qquad \cos \alpha_3 \sec \gamma = 1.$$

On S^- the outer unit normal $\mathbf{n}$ is the lower unit normal. In this case

$$\gamma = \pi - \alpha_3 \qquad \text{and} \qquad \cos \alpha_3 \sec \gamma = -1.$$

Thus,

$$\iint_{S^-} v_3 \cos \alpha_3 \, d\sigma = \iint_{\Omega_{xy}} v_3 \cos \alpha_3 \sec \gamma \, dx \, dy = \iint_{\Omega_{xy}} v_3[x, y, f^+(x, y)] \, dx \, dy$$

and

$$\iint_{S^-} v_3 \cos \alpha_3 \, d\sigma = \iint_{\Omega_{xy}} v_3 \cos \alpha_3 \sec \gamma \, dx \, dy = -\iint_{\Omega_{xy}} v_3[x, y, f^-(x, y)] \, dx \, dy.$$

It follows that

$$\iint_{S} v_3 \cos \alpha_3 \, d\sigma = \iint_{S^+} v_3 \cos \alpha_3 \, d\sigma + \iint_{S^-} v_3 \cos \alpha_3 \, d\sigma$$

$$= \iint_{\Omega_{xy}} (v_3[x, y, f^+(x, y)] - v_3[x, y, f^-(x, y)]) \, dx \, dy$$

$$= \iint_{\Omega_{xy}} \left(\int_{f^-(x, y)}^{f^+(x, y)} \frac{\partial v_3}{\partial z}(x, y, z) \, dz \right) dx \, dy$$

$$= \iiint_{T} \frac{\partial v_3}{\partial z}(x, y, z) \, dx \, dy \, dz.$$

This confirms Equation (3). Equation (2) can be confirmed by projection onto the xz plane; Equation (1) can be confirmed by projection onto the yz-plane. ❑

Divergence as Outward Flux per Unit Volume

Choose a point P and surround it by a closed ball N_ϵ of radius ϵ. According to the divergence theorem

$$\iiint_{N_\epsilon} (\nabla \cdot \mathbf{v}) \, dx \, dy \, dz = \text{flux of } \mathbf{v} \text{ out of } N_\epsilon.$$

Thus

$$(\text{average divergence of } \mathbf{v} \text{ on } N_\epsilon)(\text{volume of } N_\epsilon) = \text{flux of } \mathbf{v} \text{ out of } N_\epsilon$$

and

$$\text{average divergence of } \mathbf{v} \text{ on } N_\epsilon = \frac{\text{flux of } \mathbf{v} \text{ out of } N_\epsilon}{\text{volume of } N_\epsilon}.$$

Taking the limit of both sides as ϵ shrinks to 0, we have

$$\text{divergence of } \mathbf{v} \text{ at } P = \lim_{\epsilon \to 0^+} \frac{\text{flux of } \mathbf{v} \text{ out of } N_\epsilon}{\text{volume of } N_\epsilon}.$$

In this sense *divergence is outward flux per unit volume.*

Think of $\mathbf{v}$ as the velocity of a fluid. As suggested in Section 17.8, negative divergence at P signals an accumulation of fluid near P:

$$\nabla \cdot \mathbf{v} < 0 \text{ at } P \Rightarrow \text{flux out of } N_\epsilon < 0 \Rightarrow \text{net flow into } N_\epsilon.$$

Positive divergence at P signals a flow of liquid away from P:

$$\nabla \cdot \mathbf{v} > 0 \text{ at } P \Rightarrow \text{flux out of } N_\epsilon > 0 \Rightarrow \text{net flow out of } N_\epsilon.$$

Points at which the divergence is negative are called *sinks*; points at which the divergence is positive are called *sources*. If the divergence of $\mathbf{v}$ is 0 throughout, then the flow has no sinks and no sources and $\mathbf{v}$ is called *solenoidal* (see Exercises 17.8).

Solids Bounded by Two or More Closed Surfaces

The divergence theorem, stated for solids bounded by a single closed surface, can be extended to solids bounded by several closed surfaces. Suppose, for example, that we start with a solid bounded by a closed surface S_1 and extract from the interior of that solid a solid bounded by a closed surface S_2. The remaining solid, call it T and see Figure 17.9.2, has a boundary S that consists of two pieces: an outer piece S_1 and an inner piece S_2. The key here is to note that the *outer* normal for T points *out of* S_1 but *into* S_2. The divergence theorem can be proved for T by slicing T into two pieces T_1 and T_2 as in Figure 17.9.3 and applying the divergence theorem to each piece:

$$\iiint_{T_1} (\nabla \cdot \mathbf{v}) \, dx \, dy \, dz = \iint_{\text{bdry of } T_1} (\mathbf{v} \cdot \mathbf{n}) \, d\sigma,$$

$$\iiint_{T_2} (\nabla \cdot \mathbf{v}) \, dx \, dy \, dz = \iint_{\text{bdry of } T_2} (\mathbf{v} \cdot \mathbf{n}) \, d\sigma.$$

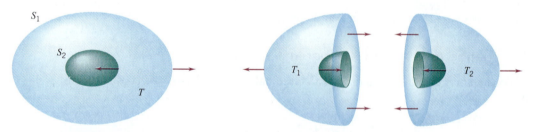

Figure 17.9.2   Figure 17.9.3

The triple integrals over T_1 and T_2 add up to the triple integral over T. When the surface integrals are added together, the integrals along the common cut cancel (because the normals are in opposite directions), and therefore only the integrals over S_1 and S_2 remain. Thus the surface integrals add up to the surface integral over $S = S_1 \cup S_2$ and the divergence theorem still holds:

$$\iiint_{T} (\nabla \cdot \mathbf{v}) \, dx \, dy \, dz = \iint_{S} (\mathbf{v} \cdot \mathbf{n}) \, d\sigma.$$

An Application to Static Charges

Consider a point charge q somewhere in space. This charge creates around itself an *electric field* $\mathbf{E}$, which in turn exerts an electric force on every other nearby charge. If we center our coordinate system at q, then the electric field at $\mathbf{r}$ can be written

$$\mathbf{E}(\mathbf{r}) = q \, \frac{\mathbf{r}}{r^3}.$$

(This is found experimentally.) Note that this field has exactly the same form as the gravitational field: a constant multiple of r^{-3} **r**. It follows from (17.8.8) that

$$\nabla \cdot \mathbf{E} = 0 \qquad \text{for all } \mathbf{r} \neq \mathbf{0}.$$

We are interested in the flux of **E** out of a closed surface S. We assume that S does not pass through q.

If the charge q is outside of S, then **E** is continuously differentiable on the region T bounded by S, and, by the divergence theorem.

$$\text{flux of } \mathbf{E} \text{ out of } S = \iiint_T (\nabla \cdot \mathbf{E}) \, dx \, dy \, dz = \iiint_T 0 \, dx \, dy \, dz = 0.$$

If q is inside of S, then the divergence theorem does not apply to T directly because **E** is not differentiable on all of T. We can circumvent this difficulty by surrounding q by a small sphere S_a of radius a and applying the divergence theorem to the region T' bounded on the outside by S and on the inside by S_a. (Figure 17.9.4)

Since **E** is continuously differentiable on T',

$$\iiint_{T'} (\nabla \cdot \mathbf{E}) \, dx \, dy \, dz = \text{flux of } \mathbf{E} \text{ out of } S + \text{flux of } \mathbf{E} \text{ into } S_a$$

$$= \text{flux of } \mathbf{E} \text{ out of } S - \text{flux of } \mathbf{E} \text{ out of } S_a.$$

Since $\nabla \cdot \mathbf{E} = 0$ on T', the triple integral on the left is zero and therefore

$$\text{flux of } \mathbf{E} \text{ out of } S = \text{flux of } \mathbf{E} \text{ out of } S_a.$$

The quantity on the right is easy to calculate: on S_a, $\mathbf{n} = \mathbf{r}/r$ and therefore

$$\mathbf{E} \cdot \mathbf{n} = q \frac{\mathbf{r}}{r^3} \cdot \frac{\mathbf{r}}{r} = \frac{q}{r^2} = \frac{q}{a^2}.$$

Figure 17.9.4

Thus

$$\text{flux of } \mathbf{E} \text{ out of } S_a = \iint_{S_a} (\mathbf{E} \cdot \mathbf{n}) \, d\sigma = \iint_{S_a} \frac{q}{a^2} \, d\sigma = \frac{q}{a^2} (\text{area of } S_a) = 4\pi q.$$

$$\text{area of } S_a = 4\pi a^2 \underline{\qquad} \uparrow$$

It follows that the flux of **E** out of $S = 4\pi q$.

In summary, if **E** is the electric field of a point charge q and S is a closed surface that does not pass through q, then

$$\text{flux of } \mathbf{E} \text{ out of } S = \begin{cases} 0, & \text{if } q \text{ is outside of } S \\ 4\pi q, & \text{if } q \text{ is inside of } S. \end{cases}$$

EXERCISES 17.9

In Exercises 1–4, calculate the flux **v** out of the unit ball $x^2 + y^2 + z^2 \leq 1$ by applying the divergence theorem.

1. $\mathbf{v}(x, y, z) = x\mathbf{i} + y\mathbf{j} + z\mathbf{k}$.
2. $\mathbf{v}(x, y, z) = (1 - x)\mathbf{i} + (2 - y)\mathbf{j} + (3 - z)\mathbf{k}$.
3. $\mathbf{v}(x, y, z) = x^2\mathbf{i} + y^2\mathbf{j} + z^2\mathbf{k}$.
4. $\mathbf{v}(x, y, z) = (1 - x^2)\mathbf{i} - y^2\mathbf{j} + z\mathbf{k}$.

In Exercises 5–8, verify the divergence theorem on the unit cube $0 \leq x \leq 1$, $0 \leq y \leq 1$, $0 \leq z \leq 1$ for the following vector fields.

5. $\mathbf{v}(x, y, z) = x\mathbf{i} + y\mathbf{j} + z\mathbf{k}$.
6. $\mathbf{v}(x, y, z) = xy\mathbf{i} + yz\mathbf{j} + xz\mathbf{k}$.
7. $\mathbf{v}(x, y, z) = x^2\mathbf{i} - xz\mathbf{j} + z^2\mathbf{k}$.

8. $\mathbf{v}(x, y, z) = x\mathbf{i} + xy\mathbf{j} + xyz\mathbf{k}$.

In Exercises 9–14, use the divergence theorem to find the total flux out of the given solid.

9. $\mathbf{v}(x, y, z) = x\mathbf{i} + 2y^2\mathbf{j} + 3z^2\mathbf{k}$; $0 \leq x^2 + y^2 \leq 9$, $0 \leq z \leq 1$.

10. $\mathbf{v}(x, y, z) = xy\mathbf{i} + yz\mathbf{j} + xz\mathbf{k}$; $0 \leq x \leq 1$, $0 \leq y \leq 1 - x$, $0 \leq z \leq 1 - x - y$.

11. $\mathbf{v}(x, y, z) = x^2\mathbf{i} + xy\mathbf{j} - 2xz\mathbf{k}$; $0 \leq x \leq 1$, $0 \leq y \leq 1 - x$, $0 \leq z \leq 1 - x - y$.

12. $\mathbf{v}(x, y, z) = (2xy + 2z)\mathbf{i} + (y^2 + 1)\mathbf{j} - (x + y)\mathbf{k}$; $0 \leq x \leq 4$, $0 \leq y \leq 4 - x$, $0 \leq z \leq 4 - x - y$.

13. $\mathbf{v}(x, y, z) = x^2\mathbf{i} + y^2\mathbf{j} + z^2\mathbf{k}$; the cylinder $0 \leq x^2 + y^2 \leq 4$, $0 \leq z \leq 4$, including the top and base.

14. $\mathbf{v}(x, y, z) = 2x\mathbf{i} + xy\mathbf{j} + xz\mathbf{k}$; the ball $x^2 + y^2 + z^2 \leq 4$.

In Exercises 15 and 16 calculate the total flux of $\mathbf{v}(x, y, z) = 2xy\mathbf{i} + y^2\mathbf{j} + 3yz\mathbf{k}$ out of the given solid.

15. The ball: $0 \leq x^2 + y^2 + z^2 \leq a^2$.

16. The cube: $0 \leq x \leq a, 0 \leq y \leq a, 0 \leq z \leq a$.

17. What is the flux of $\mathbf{v}(x, y, z) = Ax\mathbf{i} + By\mathbf{j} + Cz\mathbf{k}$ out of a solid of volume V?

18. Let T be a basic solid with a piecewise-smooth boundary. Show that, if f is harmonic on T (defined in Exercises 17.8), then the flux of ∇f out of T is zero.

19. Let S be a closed smooth surface with continuous unit normal $\mathbf{n} = \mathbf{n}(x, y, z)$. Show that

$$\iint_S \mathbf{n}\, d\sigma = \left(\iint_S n_1\, d\sigma \right)\mathbf{i} + \left(\iint_S n_2\, d\sigma \right)\mathbf{j}$$
$$+ \left(\iint_S n_3\, d\sigma \right)\mathbf{k} = \mathbf{0}.$$

20. Let T be a solid with a piecewise-smooth boundary S and let $\mathbf{n}$ be the outer unit normal.
(a) Verify the identity $\nabla \cdot (f\nabla f) = \|\nabla f\|^2 + f(\nabla^2 f)$ and show that, if f is harmonic on T, then

$$\iint_S (ff_\mathbf{n}')\, d\sigma = \iiint_T \|\nabla f\|^2\, dx\, dy\, dz$$

where $f_\mathbf{n}'$ is the directional derivative $\nabla f \cdot \mathbf{n}$.
(b) Show that, if g is continuously differentiable on T, then

$$\iint_S (gf_\mathbf{n}')\, d\sigma =$$
$$\iiint_T [(\nabla g \cdot \nabla f) + g(\nabla^2 f)]\, dx\, dy\, dz.$$

21. Let T be a solid with a piecewise-smooth boundary. Show that, if f and g have continuous second partials, then the flux of $\nabla f \times \nabla g$ out of T is zero.

22. Let T be a solid with a piecewise-smooth boundary S. Express the volume of T as a surface integral over S.

23. Suppose that a solid T (boundary S, outer unit normal $\mathbf{n}$) is immersed in a fluid. The fluid exerts a pressure $p = p(x, y, z)$ at each point of S, and therefore the solid T experiences a force. The total force on the solid due to the pressure distribution is given by the surface integral

$$\mathbf{F} = -\iint_S p\mathbf{n}\, d\sigma.$$

(The formula says that the force on the solid is the average pressure against S times the area of S.) Now choose a coordinate system with the z-axis vertical and assume that the fluid fills a region of space to the level $z = c$. The depth of a point (x, y, z) is then $c - z$ and we have $p(x, y, z) = \rho(c - z)$, where ρ is the weight density of the fluid (the weight per unit volume). Apply the divergence theorem to each component of $\mathbf{F}$ to show that $\mathbf{F} = W\mathbf{k}$, where W is the weight of the fluid displaced by the solid. We call this the *buoyant force* on the solid. (This shows that the object is not pushed from side to side by the pressure and verifies the *principle of Archimedes*: that the buoyant force on an object at rest in a fluid equals the weight of the fluid displaced.)

24. If $\mathbf{F}$ is a force applied at the tip of a radius vector $\mathbf{r}$, then the *torque*, or twisting strength, of $\mathbf{F}$ about the origin is given by the cross product $\boldsymbol{\tau} = \mathbf{r} \times \mathbf{F}$. From physics we learn that the total torque on the solid T of Exercise 23 is given by the formula

$$\boldsymbol{\tau}_{\text{Tot}} = -\iint_S [\mathbf{r} \times \rho(c - z)\mathbf{n}]\, d\sigma.$$

Use the divergence theorem to find the components of $\boldsymbol{\tau}_{\text{Tot}}$ and show that $\boldsymbol{\tau}_{\text{Tot}} = \bar{\mathbf{r}} \times \mathbf{F}$ where $\mathbf{F}$ is the buoyant force $W\mathbf{k}$ and $\bar{\mathbf{r}}$ is the centroid of T.

(This indicates that for calculating the twisting effect of the buoyant force we can view this force as being applied at the centroid. This is very important in ship design. Imagine, for example, a totally submerged submarine. While the buoyant force acts upward through the centroid of the submarine, gravity acts downward through the center of mass. Suppose the submarine should tilt a bit to one side as depicted in the figure. If the centroid lies above the center of mass, then the buoyant force acts to restore the submarine to an upright position. If, however, the centroid lies below the center of mass, then disaster. Once the submarine has tilted a bit, the buoyant

buoyant force

buoyant force

Centroid above the center of mass: restoration

gravity

Centroid below the center of mass: disaster

gravity

force will make it tilt further. This kind of analysis also applies to surface ships, but in this case the buoyant force acts upward through *the centroid of the portion of the ship that is submerged.* This point is called the *center of flotation.* One must design and load a ship to keep the center of flotation above the center of mass. Putting a lot of heavy cargo on the deck, for instance, tends to raise the center of mass and destabilize the ship.)

■ 17.10 STOKES'S THEOREM

We return to Green's theorem

$$\iint_{\Omega} \left(\frac{\partial Q}{\partial x} - \frac{\partial P}{\partial y} \right) dx\, dy = \oint_C P\, dx + Q\, dy.$$

Setting $\mathbf{v} = P\mathbf{i} + Q\mathbf{j} + R\mathbf{k}$, we have

$$(\nabla \times \mathbf{v}) \cdot \mathbf{k} = \begin{vmatrix} \mathbf{i} & \mathbf{j} & \mathbf{k} \\ \dfrac{\partial}{\partial x} & \dfrac{\partial}{\partial y} & \dfrac{\partial}{\partial z} \\ P & Q & R \end{vmatrix} \cdot \mathbf{k} = \frac{\partial Q}{\partial x} - \frac{\partial P}{\partial y}.$$

Thus in terms of $\mathbf{v}$, Green's theorem can be written

$$\iint_{\Omega} [(\nabla \times \mathbf{v}) \cdot \mathbf{k}]\, dx\, dy = \oint_C \mathbf{v}(\mathbf{r}) \cdot d\mathbf{r}.$$

Since any plane can be coordinatized as the xy-plane, this result can be phrased as follows: Let S be a flat surface in space bounded by a Jordan curve C: If $\mathbf{v}$ is continuously differentiable on S, then

$$\iint_{S} [(\nabla \times \mathbf{v}) \cdot \mathbf{n}]\, d\sigma = \oint_C \mathbf{v}(\mathbf{r}) \cdot d\mathbf{r}$$

where $\mathbf{n}$ is a unit normal for S and the line integral is taken in the *positive sense,* meaning, in the direction of the unit tangent $\mathbf{T}$ for which $\mathbf{T} \times \mathbf{n}$ points away from the surface. See Figure 17.10.1. (An observer marching along C with the same

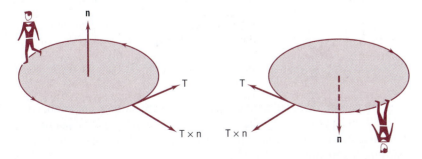

Figure 17.10.1

orientation as **n** keeps the surface to his left.) The symbol $\oint_C$ denotes this line integral.

Figure 17.10.2 shows a *polyhedral surface S* bounded by a closed polygonal path C. The surface S consists of a finite number of flat faces $S_1, \ldots, S_N$ with polygonal boundaries $C_1, \ldots, C_N$ and unit normals $\mathbf{n}_1, \ldots, \mathbf{n}_N$. We choose these unit

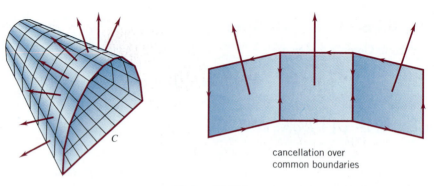

cancellation over
common boundaries

Figure 17.10.2

normals in a consistent manner; that is, they emanate from the same side of the surface. Now let $\mathbf{n} = \mathbf{n}(x, y, z)$ be a vector function of norm 1 which is $\mathbf{n}_1$ on S_1, $\mathbf{n}_2$ on S_2, $\mathbf{n}_3$ on S_3, etc. It is immaterial how $\mathbf{n}$ is defined on the line segments that join the different faces. Suppose now that $\mathbf{v} = \mathbf{v}(x, y, z)$ is a vector function continuously differentiable on an open set that contains S. Then

$$\iint_S [(\nabla \times \mathbf{v}) \cdot \mathbf{n}]\, d\sigma = \sum_{i=1}^{N} \iint_{S_i} [(\nabla \times \mathbf{v}) \cdot \mathbf{n}_i]\, d\sigma = \sum_{i=1}^{N} \oint_{C_i} \mathbf{v}(\mathbf{r}) \cdot d\mathbf{r},$$

the integral over C_i being taken in the positive sense with respect to $\mathbf{n}_i$. Now when we add these line integrals, we find that all the line segments that make up the C_i but are not part of C are traversed twice and in opposite directions. (See the figure.) Thus these line segments contribute nothing to the sum of the line integrals and we are left with the integral around C. It follows that for a polyhedral surface S with boundary C

$$\iint_S [(\nabla \times \mathbf{v}) \cdot \mathbf{n}]\, d\sigma = \oint_C \mathbf{v}(\mathbf{r}) \cdot d\mathbf{r}.$$

This result can be extended to smooth surfaces with smooth bounding curves (see Figure 17.10.3) by approximating these configurations by polyhedral configurations of the type considered and using a limit process. In an admittedly informal way we have arrived at Stokes's theorem.

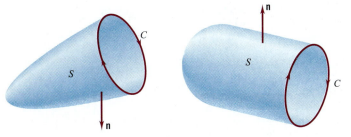

Figure 17.10.3

THEOREM 17.10.1 STOKES'S THEOREM†

Let S be a smooth surface with a smooth bounding curve C. If $\mathbf{v} = \mathbf{v}(x, y, z)$ is a continuously differentiable vector field on an open set that contains S, then

$$\iint_S [(\nabla \times \mathbf{v}) \cdot \mathbf{n}]\, d\sigma = \oint_C \mathbf{v}(\mathbf{r}) \cdot d\mathbf{r}$$

where $\mathbf{n} = \mathbf{n}(x, y, z)$ is a unit normal that varies continuously on S and the line integral is taken in the positive sense with respect to $\mathbf{n}$.

Example 1 Verify Stokes's theorem for

$$\mathbf{v} = -3y\mathbf{i} + 3x\mathbf{j} + z^4\mathbf{k}$$

taking S as the portion of the ellipsoid $2x^2 + 2y^2 + z^2 = 1$ that lies above the plane $z = 1/\sqrt{2}$ (Figure 17.10.4).

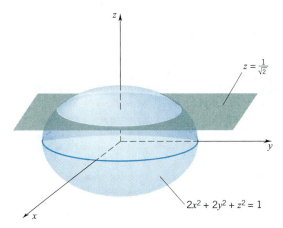

Figure 17.10.4

---

† The result was announced publicly for the first time by George Gabriel Stokes (1819–1903), an Irish mathematician and physicist who, like Green, was a Cambridge professor.

SOLUTION A little algebra shows that S is the graph of

$$f(x, y) = \sqrt{1 - 2(x^2 + y^2)}$$

with (x, y) restricted to the disc $\Omega : x^2 + y^2 \leq \frac{1}{4}$. Now

$$\nabla \times \mathbf{v} = \begin{vmatrix} \mathbf{i} & \mathbf{j} & \mathbf{k} \\ \dfrac{\partial}{\partial x} & \dfrac{\partial}{\partial y} & \dfrac{\partial}{\partial z} \\ -3y & 3x & z^4 \end{vmatrix} = \left[\dfrac{\partial}{\partial x}(3x) + \dfrac{\partial}{\partial y}(3y) \right] \mathbf{k} = 6\mathbf{k}.$$

Taking $\mathbf{n}$ as the upper unit normal (i.e., $\mathbf{n} = -f_x \mathbf{i} - f_y \mathbf{j} + \mathbf{k}$), we have

$$\iint_S [(\nabla \times \mathbf{v}) \cdot \mathbf{n}] \, d\sigma = \iint_S (6\mathbf{k} \cdot \mathbf{n}) \, d\sigma$$

$$(17.7.9) \underline{\quad\quad} \uparrow \qquad = \iint_\Omega (-(0)f_x - (0)f_y + 6) \, dx \, dy$$

$$= \iint_\Omega 6 \, dx \, dy = 6(\text{area of } \Omega) = 6(\tfrac{1}{4}\pi) = \tfrac{3}{2}\pi.$$

The bounding curve C is the set of all (x, y, z) with $x^2 + y^2 = \frac{1}{4}$ and $z = 1/\sqrt{2}$. We can parametrize C by setting

$$\mathbf{r}(u) = \tfrac{1}{2} \cos u \, \mathbf{i} + \tfrac{1}{2} \sin u \, \mathbf{j} + \frac{1}{\sqrt{2}} \mathbf{k}, \qquad u \in [0, 2\pi].$$

Since $\mathbf{n}$ is the upper unit normal, this parametrization gives C in the positive sense. Thus

$$\oint_C \mathbf{v}(\mathbf{r}) \cdot d\mathbf{r} = \int_0^{2\pi} (-\tfrac{3}{2} \sin u \, \mathbf{i} + \tfrac{3}{2} \cos u \, \mathbf{j} + \tfrac{1}{4}\mathbf{k}) \cdot (-\tfrac{1}{2} \sin u \, \mathbf{i} + \tfrac{1}{2} \cos u \, \mathbf{j}) \, du$$

$$= \int_0^{2\pi} (\tfrac{3}{4} \sin^2 u + \tfrac{3}{4} \cos^2 u) \, du = \int_0^{2\pi} \tfrac{3}{4} \, du = \tfrac{3}{2}\pi$$

This is the value we obtained for the surface integral. ❏

Example 2 Verify Stokes's theorem for

$$\mathbf{v} = z^2 \mathbf{i} - 2x\mathbf{j} + y^3 \mathbf{k}$$

taking S as the upper half of the unit sphere $x^2 + y^2 + z^2 = 1$.

SOLUTION We use the upper unit normal $\mathbf{n} = x\mathbf{i} + y\mathbf{j} + z\mathbf{k}$. Now

$$\nabla \times \mathbf{v} = \begin{vmatrix} \mathbf{i} & \mathbf{j} & \mathbf{k} \\ \dfrac{\partial}{\partial x} & \dfrac{\partial}{\partial y} & \dfrac{\partial}{\partial z} \\ z^2 & -2x & y^3 \end{vmatrix} = 3y^2 \mathbf{i} + 2z\mathbf{j} - 2\mathbf{k}.$$

Therefore

$$\iint_S [(\nabla \times \mathbf{v}) \cdot \mathbf{n}] \, d\sigma = \iint_S [(3y^2\mathbf{i} + 2z\mathbf{j} - 2\mathbf{k}) \cdot (x\mathbf{i} + y\mathbf{j} + z\mathbf{k}) \, d\sigma$$

$$= \iint_S (3xy^2 + 2yz - 2z) \, d\sigma$$

$$= \iint_S 3xy^2 \, d\sigma + \iint_S 2yz \, d\sigma - \iint_S 2z \, d\sigma.$$

The first integral is zero because S is symmetric about the yz-plane and the integrand is odd with respect to x. The second integral is zero because S is symmetric about the xz-plane and the integrand is odd with respect to y. Thus

$$\iint_S [(\nabla \times \mathbf{v}) \cdot \mathbf{n}] \, d\sigma = -\iint_S 2z \, d\sigma = -2\bar{z} \text{ (area of } S) = -2(\tfrac{1}{2})2\,\pi = -2\,\pi.$$

$$S \quad \text{Exercise 17, Section 17.7} \underline{}\uparrow$$

This is also the value of the integral along the bounding base circle taken in the positive sense: $\mathbf{r}(u) = \cos u\,\mathbf{i} + \sin u\,\mathbf{j}, \quad u \in [2, \pi]$, and

$$\oint_C \mathbf{v}(\mathbf{r}) \cdot d\mathbf{r} = \oint_C z^2 \, dx - 2x \, dy = -2 \oint_C x \, dy$$

$$= -2 \int_0^{2\pi} \cos^2 u \, du = -2\,\pi. \quad \square$$

Earlier you saw that the curl of a gradient is zero. Using Stokes's theorem we can prove a partial converse.

(17.10.2) If a vector field $\mathbf{v} = \mathbf{v}(x, y, z)$ is continuously differentiable on an open convex† set U and $\nabla \times \mathbf{v} = 0$ on all of U, then $\mathbf{v}$ is the gradient of some scalar field ϕ defined on U.

PROOF Choose a point $\mathbf{a}$ in U and for each point $\mathbf{x}$ in U define

$$\phi(\mathbf{x}) = \int_{\mathbf{a}}^{\mathbf{x}} \mathbf{v}(\mathbf{r}) \cdot d\mathbf{r}.$$

(This is the line integral from $\mathbf{a}$ to $\mathbf{x}$ taken along the line segment that joins these two points. We know that this line segment lies in U because U is convex.)

Since U is open, $\mathbf{x} + \mathbf{h}$ is in U for all $\mathbf{h}$ sufficiently small. Assume then that $\mathbf{h}$ is sufficiently small for $\mathbf{x} + \mathbf{h}$ to be in U. Since U is convex, the triangular region with vertices at $\mathbf{a}, \mathbf{x}, \mathbf{x} + \mathbf{h}$ lies in U. (See Figure 17.10.5.) Since $\nabla \times \mathbf{v} = 0$ on U, we can conclude from Stokes's theorem that

$$\int_{\mathbf{a}}^{\mathbf{x}} \mathbf{v}(\mathbf{r}) \cdot d\mathbf{r} + \int_{\mathbf{x}}^{\mathbf{x}+\mathbf{h}} \mathbf{v}(\mathbf{r}) \cdot d\mathbf{r} + \int_{\mathbf{x}+\mathbf{h}}^{\mathbf{a}} \mathbf{v}(\mathbf{r}) \cdot d\mathbf{r} = 0.$$

Figure 17.10.5

----

† Recall that the set U is *convex* if for each pair of points $p, q \in U$, the line segment $\overline{pq}$ lies entirely in U.

Therefore

$$\int_{\mathbf{x}}^{\mathbf{x}+\mathbf{h}} \mathbf{v}(\mathbf{r}) \cdot d\mathbf{r} = -\int_{\mathbf{x}+\mathbf{h}}^{\mathbf{a}} \mathbf{v}(\mathbf{r}) \cdot d\mathbf{r} - \int_{\mathbf{a}}^{\mathbf{x}} \mathbf{v}(\mathbf{r}) \cdot d\mathbf{r}$$

$$= \int_{\mathbf{a}}^{\mathbf{x}+\mathbf{h}} \mathbf{v}(\mathbf{r}) \cdot d\mathbf{r} - \int_{\mathbf{a}}^{\mathbf{x}} \mathbf{v}(\mathbf{r}) \cdot d\mathbf{r}.$$

By our definition of ϕ we have

$$\phi(\mathbf{x}+\mathbf{h}) - \phi(\mathbf{x}) = \int_{\mathbf{x}}^{\mathbf{x}+\mathbf{h}} \mathbf{v}(\mathbf{r}) \cdot d\mathbf{r}.$$

We can parametrize the line segment from $\mathbf{x}$ to $\mathbf{x} + \mathbf{h}$ by $\mathbf{r}(u) = \mathbf{x} + u\mathbf{h}$ with $u \in [0, 1]$. Therefore

$$\phi(\mathbf{x}+\mathbf{h}) - \phi(\mathbf{x}) = \int_{0}^{1} [\mathbf{v}(\mathbf{r}(u)) \cdot \mathbf{r}'(u)]\, du$$

$$= \int_{0}^{1} [\mathbf{v}(\mathbf{r}(u)) \cdot \mathbf{h}]\, du$$

Theorem 5.9.1 $\longrightarrow = \mathbf{v}(\mathbf{r}(u_0)) \cdot \mathbf{h}$ for some u_0 in $[0, 1]$

$$= \mathbf{v}(\mathbf{x} + u_0\mathbf{h}) \cdot \mathbf{h} = \mathbf{v}(\mathbf{x}) \cdot \mathbf{h} + [\mathbf{v}(\mathbf{x} + u_0\mathbf{h}) - \mathbf{v}(\mathbf{x})] \cdot \mathbf{h}.$$

The fact that $\mathbf{v} = \nabla\phi$ follows from observing that $[\mathbf{v}(\mathbf{x} + u_0\mathbf{h}) - \mathbf{v}(\mathbf{x})] \cdot \mathbf{h}$ is $o(\mathbf{h})$:

$$\frac{|[\mathbf{v}(\mathbf{x} + u_0\mathbf{h}) - \mathbf{v}(\mathbf{x})] \cdot \mathbf{h}|}{\|\mathbf{h}\|} \leq \frac{\|\mathbf{v}(\mathbf{x} + u_0\mathbf{h}) - \mathbf{v}(\mathbf{x})\| \, \|\mathbf{h}\|}{\|\mathbf{h}\|}$$

$$= \|\mathbf{v}(\mathbf{x} + u_0\mathbf{h}) - \mathbf{v}(\mathbf{x})\| \to 0$$

as $\mathbf{h} \to \mathbf{0}$. ❑

The Normal Component of $\nabla \times \mathbf{v}$ as Circulation per Unit Area; Irrotational Flow

Interpret $\mathbf{v} = \mathbf{v}(x, y, z)$ as the velocity of a fluid flow. In Section 17.8 we stated that $\nabla \times \mathbf{v}$ measures the rotational tendency of the fluid. Now we can be more precise.

Take a point P within the flow and choose a unit vector $\mathbf{n}$. Let D_ϵ be the ϵ-disc that is centered at P and is perpendicular to $\mathbf{n}$. Let C_ϵ be the circular boundary of D_ϵ directed in the positive sense with respect to $\mathbf{n}$. (See Figure 17.10.6) By Stokes's theorem

$$\iint_{D_\epsilon} [(\nabla \times \mathbf{v}) \cdot \mathbf{n}]\, d\sigma = \oint_{C_\epsilon} \mathbf{v}(\mathbf{r}) \cdot d\mathbf{r}.$$

Figure 17.10.6

The line integral on the right is called *the circulation* of $\mathbf{v}$ around C_ϵ. Thus we can say that

$$\left(\begin{matrix} \text{The average } \mathbf{n}\text{-component} \\ \text{of } \nabla \times \mathbf{v} \text{ on } D_\epsilon \end{matrix}\right) (\text{the area of } D_\epsilon) = \text{the circulation of } \mathbf{v} \text{ around } C_\epsilon.$$

It follows that

$$\text{The average } \mathbf{n}\text{-component of } \nabla \times \mathbf{n} \text{ on } D_\epsilon = \frac{\text{the circulation of } \mathbf{v} \text{ around } C_\epsilon}{\text{the area of } D_\epsilon}$$

Taking the limit as ϵ shrinks to 0, you can see that

$$\text{The } \mathbf{n}\text{-component of } \nabla \times \mathbf{v} \text{ at } P = \lim_{\epsilon \to 0^+} \frac{\text{the circulation of } \mathbf{v} \text{ around } C_\epsilon}{\text{the area of } D_\epsilon}.$$

At each point P the component of $\nabla \times \mathbf{v}$ in any direction $\mathbf{n}$ is the circulation of $\mathbf{v}$ per unit area in the plane normal to $\mathbf{n}$. If $\nabla \times \mathbf{v} = \mathbf{0}$ identically, the fluid has no rotational tendency and the flow is called irrotational.

EXERCISES 17.10

In Exercises 1–4, let S be the upper half of the unit sphere $x^2 + y^2 + z^2 = 1$ and take $\mathbf{n}$ as the upper unit normal. Find

$$\iint_S [(\nabla \times \mathbf{v}) \cdot \mathbf{n}] \, d\sigma$$

(a) By direct calculation.   (b) By Stokes's theorem.

1. $\mathbf{v}(x, y, z) = x\mathbf{i} + y\mathbf{j} + z\mathbf{k}$.

2. $\mathbf{v}(x, y, z) = y\mathbf{i} - x\mathbf{j} + z\mathbf{k}$.

3. $\mathbf{v}(x, y, z) = z^2\mathbf{i} + 2x\mathbf{j} - y^3\mathbf{k}$.

4. $\mathbf{v}(x, y, z) = 6xz\mathbf{i} - x^2\mathbf{j} - 3y^2\mathbf{k}$.

In Exercises 5–7, let S be the triangular surface with vertices $(2, 0, 0)$, $(0, 2, 0)$, $(0, 0, 2)$ and take $\mathbf{n}$ as the upper unit normal. Find

$$\iint_S [(\nabla \times \mathbf{v}) \cdot \mathbf{n}] \, d\sigma$$

(a) By direct calculation.   (b) By Stokes's theorem.

5. $\mathbf{v}(x, y, z) = 2z\mathbf{i} - y\mathbf{j} + x\mathbf{k}$.

6. $\mathbf{v}(x, y, z) = (x^2 + y^2)\mathbf{i} + y^2\mathbf{j} + (x^2 + z^2)\mathbf{k}$.

7. $\mathbf{v}(x, y, z) = x^4\mathbf{i} + xy\mathbf{j} + z^4\mathbf{k}$.

8. Show that, if $\mathbf{v} = \mathbf{v}(x, y, z)$ is continuously differentiable everywhere and its curl is identically zero, then

$$\int_C \mathbf{v}(\mathbf{r}) \cdot d\mathbf{r} = 0 \qquad \text{for every smooth closed curve } C.$$

9. Let $\mathbf{v} = y\mathbf{i} + z\mathbf{j} + x^2y^2\mathbf{k}$ and let S be the surface $z = x^2 + y^2$ from $z = 0$ to $z = 4$. Calculate the flux of $\nabla \times \mathbf{v}$ in the direction of the lower unit normal $\mathbf{n}$.

10. Let $\mathbf{v} = \frac{1}{2}y\mathbf{i} + 2xz\mathbf{j} - 3x\mathbf{k}$ and let S be the surface $y = 1 - (x^2 + z^2)$ from $y = -8$ to $y = 1$. Calculate the flux of $\nabla \times \mathbf{v}$ in the direction of the unit normal $\mathbf{n}$ with positive $\mathbf{j}$-component.

11. Let $\mathbf{v} = 2x\mathbf{i} + 2y\mathbf{j} + x^2y^2z^2\mathbf{k}$ and let S be the lower half of the ellipsoid

$$\frac{x^2}{4} + \frac{y^2}{9} + \frac{z^2}{27} = 1.$$

Calculate the flux of $\nabla \times \mathbf{v}$ in the direction of the upper unit normal $\mathbf{n}$.

12. Let S be a smooth closed surface and let $\mathbf{v} = \mathbf{v}(x, y, z)$ be a vector field with second partials continuous on an open convex set that contains S. Show that

$$\iint_S [(\nabla \times \mathbf{v}) \cdot \mathbf{n}] \, d\sigma = 0$$

where $\mathbf{n} = \mathbf{n}(x, y, z)$ is the outer unit normal.

13. The upper half of the ellipsoid $\frac{1}{2}x^2 + \frac{1}{2}y^2 + z^2 = 1$ intersects the cylinder $x^2 + y^2 - y = 0$ in a curve C. Calculate the circulation of $\mathbf{v} = y^3\mathbf{i} + (xy + 3xy^2)\mathbf{j} + z^4\mathbf{k}$ around C by using Stokes's theorem.

14. The sphere $x^2 + y^2 + z^2 = a^2$ intersects the plane $x + 2y + z = 0$ in a curve C. Calculate the circulation of $\mathbf{v} = 2y\mathbf{i} - z\mathbf{j} + 2x\mathbf{k}$ about C by using Stokes's theorem.

15. The paraboloid $z = x^2 + y^2$ intersects the plane $z = y$ in a curve C. Calculate the circulation of $\mathbf{v} = 2z\mathbf{i} + x\mathbf{j} + y\mathbf{k}$ about C using Stokes's theorem.

16. The cylinder $x^2 + y^2 = b^2$ intersects the plane $y + z = a^2$ in a curve C. Assume $a^2 \geq b > 0$. Calculate the circulation of $\mathbf{v} = xy\mathbf{i} + yz\mathbf{j} + xz\mathbf{k}$ about C using Stokes's theorem.

17. Let S be a smooth surface with a smooth bounding curve C and let $\mathbf{a}$ be a fixed vector. Show that

$$\iint_S (2\mathbf{a} \cdot \mathbf{n}) \, d\sigma = \oint_C (\mathbf{a} \times \mathbf{r}) \cdot d\mathbf{r}$$

where $\mathbf{n} = \mathbf{n}(x, y, z)$ is a unit vector that varies continuously over S and the line integral is taken in the positive sense with respect to $\mathbf{n}$.

18. Let S be a smooth surface with smooth bounding curve C. Show that, if ϕ and ψ are sufficiently differentiable scalar fields, then

$$\iint_S [(\nabla \phi \times \nabla \psi) \cdot \mathbf{n}] \, d\sigma = \oint_C (\phi \nabla \psi) \cdot d\mathbf{r}$$

where $\mathbf{n} = \mathbf{n}(x, y, z)$ is a unit normal that varies continuously on S and the line integral is taken in the positive sense with respect to $\mathbf{n}$.

19. Let S be a smooth surface with a smooth plane bounding curve C and let $\mathbf{v} = \mathbf{v}(x, y, z)$ be a vector field with second partials continuous on an open convex set that contains S. If S does not cross the plane of C, then Stokes's theorem for S follows readily from the divergence theorem and Green's theorem. Carry out the argument.

20. Our derivation of Stokes's theorem was admittedly non-rigorous. The following version of Stokes's theorem lends itself more readily to rigorous proof.

THEOREM 17.10.3

Let Γ be a Jordan region in the uv-plane with a piecewise-smooth boundary C_Γ given in a counterclockwise sense by a pair of functions $u = u(t)$, $v = v(t)$ with $t \in [a, b]$. Let $\mathbf{R}(u, v) = x(u, v)\mathbf{i} + y(u, v)\mathbf{j} + z(u, v)\mathbf{k}$ be a vector function with continuous second partials on Γ. Assume that $\mathbf{R}$ is one-to-one on Γ and that the fundamental vector product $\mathbf{N} = \mathbf{R}'_u \times \mathbf{R}'_v$ is never zero. The surface S: $\mathbf{R} = \mathbf{R}(u, v)$, $(u, v) \in \Gamma$ is a smooth surface bounded by the oriented space curve C: $\mathbf{r}(t) = \mathbf{R}(u(t), v(t))$, $t \in [a, b]$. If $\mathbf{v} = \mathbf{v}(x, y, z)$ is a vector field continuously differentiable on S, then

$$\iint_S [(\nabla \times \mathbf{v}) \cdot \mathbf{n}] \, d\sigma = \oint_C \mathbf{v}(\mathbf{r}) \cdot d\mathbf{r}$$

where $\mathbf{n}$ is the unit normal in the direction of the fundamental vector product.

Give a detailed proof of the theorem.   HINT: Set $\mathbf{v} = v_1\mathbf{i} + v_2\mathbf{j} + v_3\mathbf{k}$. Then

$$\iint_S [(\nabla \times \mathbf{v}) \cdot \mathbf{n}] \, d\sigma = \iint_S [(\nabla \times v_1\mathbf{i}) \cdot \mathbf{n}] \, d\sigma$$

$$+ \iint_S [(\nabla \times v_2\mathbf{j}) \cdot \mathbf{n}] \, d\sigma + \iint_S [(\nabla \times v_3\mathbf{k}) \cdot \mathbf{n}] \, d\sigma$$

and

$$\oint_C \mathbf{v}(\mathbf{r}) \cdot d\mathbf{r} = \int_C v_1 \, dx + \int_C v_2 \, dy + \int_C v_3 \, dz.$$

Show that

$$\iint_S [(\nabla \times v_1\mathbf{i}) \cdot \mathbf{n}] \, d\sigma = \int_C v_1 \, dx$$

by showing that both integrals can be written

$$\iint_\Gamma \left[\frac{\partial v_1}{\partial u} \frac{\partial x}{\partial v} - \frac{\partial v_1}{\partial v} \frac{\partial x}{\partial u} \right] du \, dv.$$

A similar argument (no need to carry it out) equates the integrals for v_2 and v_3 and proves the theorem.

■ CHAPTER HIGHLIGHTS

17.1   Line Integrals

17.2   The Fundamental Theorem for Line Integrals

*17.3   Work–Energy Formula; Conservation of Mechanical Energy

17.4   Another Notation for Line Integrals; Line Integrals with Respect to Arc Length

$$\int_C P\,dx + Q\,dy + R\,dz \qquad \int_C f(\mathbf{r})\,ds$$

mass, center of mass, moment of inertia of a wire (pp. 1133–1134)

17.5   Green's Theorem

Jordan region (p. 1137)   Green's theorem for a Jordan region (p. 1137)
area enclosed by a plane curve (p. 1142)
Green's theorem for regions bounded by several Jordan curves (pp. 1143–1145)
a justification of the Jacobian area formula (supplement) (p. 1146)

17.6   Parametrized Surfaces; Surface Area

parametrized surface: $\mathbf{r} = \mathbf{r}(u, v)$, $(u, v) \in \Omega$
fundamental vector product: $\mathbf{N} = \mathbf{r}_u' \times \mathbf{r}_v'$
$\mathbf{N}$ as a normal (p. 1151)   area of a parametrized surface (p. 1152)
area of a surface $z = f(x, y)$ (p. 1155)   upper unit normal (p. 1157)
secant area formula (p. 1158)

17.7   Surface Integrals

mass of a material surface (p. 1159)   surface integral (p. 1160)
average value on a surface, weighted average (p. 1162)   centroid (p. 1162)
center of mass of a material surface (p. 1163)   moment of inertia (p. 1163)
smooth surface (p. 1166)   flux of a vector field (p. 1166)
closed, piecewise-smooth surface (p. 1168)
one sided surface, Möbius band (p. 1170)

17.8   The Vector Differential Operator ∇

the operator del: $\nabla = \dfrac{\partial}{\partial x}\mathbf{i} + \dfrac{\partial}{\partial y}\mathbf{j} + \dfrac{\partial}{\partial z}\mathbf{k}$ gradient of f: ∇f

divergence of $\mathbf{v}$: $\nabla \cdot \mathbf{v}$ curl of $\mathbf{v}$: $\nabla \times \mathbf{v}$
curl of a gradient is zero (p. 1173)   divergence of a curl is zero (p. 1173)
$\nabla \cdot (f\mathbf{v}) = (\nabla f) \cdot \mathbf{v} + f(\nabla \cdot \mathbf{v})$ $\nabla \times (f\mathbf{v}) = (\nabla f) \times \mathbf{v} + f(\nabla \times \mathbf{v})$
$\nabla \cdot (r^n \mathbf{r}) = (n + 3)r^n$ $\nabla \times (r^n \mathbf{r}) = \mathbf{0}$

Laplacian: $\nabla^2 f = \nabla \cdot \nabla f = \dfrac{\partial^2 f}{\partial x^2} + \dfrac{\partial^2 f}{\partial y^2} + \dfrac{\partial^2 f}{\partial z^2}$

17.9   The Divergence Theorem

divergence theorem for a solid bounded by a single closed surface (p. 1177)
divergence as outward flux per unit volume (p. 1179)   sinks and sources (p. 1180)
divergence theorem for solids bounded by two or more closed surfaces (p. 1180)
an application to static charges (p. 1180)
buoyant force, principle of Archimedes (p. 1182)

17.10   Stokes's Theorem

positive sense along a curve bounding an open surface (p. 1183)
Stokes's theorem (p. 1185)
conditions under which $\nabla \times \mathbf{v} = \mathbf{0}$ implies $\mathbf{v} = \nabla f$
normal component of $\nabla \times \mathbf{v}$ as circulation per unit area
irrotational flow (p. 1189)

■ PROJECTS AND EXPLORATIONS USING TECHNOLOGY

To do these exercises you will need a graphics calculator or a computer with graphing capability. The majority of these problems are open-ended so different approaches may be used to solve them. You should be aware that different approaches can result in slight variations in the answers. Round your numerical answers to at least four decimal places. The rounding method that your calculator or computer uses also may cause variations in answers.

17.1 This exercise uses technology to approximate line integrals.

(a) A line integral is defined in terms of a single variable definite integral. Thus, any existing technology for approximating definite integrals can be used. Set up a procedure for manipulating the integrand of a line integral to use definite integral software. Test your procedure on the following examples from the text

(i) $\displaystyle\int_C \mathbf{f(r) \cdot dr}$ where $\mathbf{f}(x, y) = y\,\mathbf{i} + x\,\mathbf{j}$; $\mathbf{r}(u) = u\,\mathbf{i} + u^2\,\mathbf{j}$, $u \in [0, 1]$.

(ii) $\displaystyle\int_C \mathbf{f(r) \cdot dr}$ where $\mathbf{f}(x, y, z) = y^2 z^3\,\mathbf{i} + 2xyz^3\,\mathbf{j} + 3xy^2 z^2\,\mathbf{k}$;

$$\mathbf{r}(u) = u^2\,\mathbf{i} + u^4\,\mathbf{j} + u^6\,\mathbf{k}, \quad u \in [0, 1].$$

(b) Prepare a trapezoidal rule procedure for approximating line integrals and test it on the two examples in part (a).

(c) We have approximated definite integrals by averaging random function values over the domain (see, for example, Technology Exercise 2 at the end of Chapter 8). Prepare such a procedure for approximating line integrals and test it on the two examples in part (a).

(d) Theorem 17.1.4 states that, under certain conditions, line integrals are independent of the parametrization used for the curve. Check this result by calculating the length of a spiral on a sphere of radius π using the following two parametrizations of the spiral.

(i) $\mathbf{r}(u) = u \sin u\,\mathbf{i} + u \cos u\,\mathbf{j} + \sqrt{\pi^2 - u^2}\,\mathbf{k}, \quad u \in [0, \pi]$,

(ii) $\mathbf{R}(u) = u^2 \cos\left(\dfrac{\pi}{2} - u^2\right)\mathbf{i} + u^2 \sin\left(\dfrac{\pi}{2} - u^2\right)\mathbf{j} + \sqrt{\pi^2 - u^4}\,\mathbf{k}, u \in [0, \sqrt{\pi}]$.

Use any one of the approximation methods developed above.

(e) Using the paths in part (d), calculate

$$\int_C \mathbf{f(r) \cdot dr} \text{ and } \int_C \mathbf{f(R) \cdot dR}, \quad \text{where } \mathbf{f}(x, y, z) = x^3\,\mathbf{i} - xz\,\mathbf{j} + yz\,\mathbf{k}.$$

(f) Approximating line integrals using random function values over the domain [part (c)] may seem to give different results for different parametrizations of the curve. Explain why an arc length parametrization should give the best result.

17.2 This exercise uses technology to approximate a surface integral.

(a) Surface integrals can be expressed in terms of double integrals. If you have technology that evaluates double integrals, determine a method for entering a surface integral. Test your method on the following integral:

$$\iint_S H(x, y, z)\, d\sigma, \quad \text{where } H(x, y, z) + x^2 + y^3 + z^4 \quad \text{and}$$

S: $\mathbf{r}(u, v) = \cos u \cosh v\,\mathbf{i} + \sin u \cosh v\,\mathbf{j} + \sinh v\,\mathbf{k}, \quad u \in [0, 2\pi], \quad v \in [0, 1].$

(b) Exercise 17.1, part (b), involved a trapezoidal rule for approximating a line integral. Modify that procedure to allow surface integrals to be the input. Then test it on the integral in part (a).

(c) Prepare a procedure for approximating a surface integral using random points on the surface and test it on the integral in part (a). What parametrization of the surface do you think you should use?

(d) Use the procedures from this exercise and Exercise 17.1 to verify Green's theorem for the following problem: Find the work done by the force $\mathbf{F}(x, y) = (x^3 - x^2y)\,\mathbf{i} + xy^2\,\mathbf{j}$ in moving an object in a counterclockwise direction around the boundary of the region enclosed by the parabolas $y = x^2$ and $x = y^2$.

17.3 This exercise uses technology to simplify calculations with the vector differential operator ∇.

(a) Test the capabilities of your technology by calculating the divergence and curl of the vector functions

$$\text{(i)}\quad \mathbf{v}(x, y, z) = xyz\,\mathbf{i} + \ln(xyz)\,\mathbf{j} - \sin(2x + 3y - 4z)\,\mathbf{k},$$
$$\text{(ii)}\quad \mathbf{v}(x, y, z) = (x^2 + y^3 - 3z^7)\,\mathbf{i} - \tan^{-1}(x/y)\,\mathbf{j} + e^{x-z}\,\mathbf{k},$$

and the gradient of the function

$$f(x, y, z) = \cos(\sqrt{x^2 + 3\sin y + 5z^4}).$$

If you use symbolic manipulation software, use it on these functions directly. If your technology has only numeric capabilities, then evaluate your results at $x = 2, y = -3$, and $z = 5$.

(b) As you have seen, round off error is a problem in computation. Use the following identities to test the accuracy of the technology that you used in part (a):

$$\nabla \times (\nabla f) = 0 \quad \text{and} \quad \nabla \cdot (\nabla \times \mathbf{v}) = 0.$$

(c) Use the results of the preceding exercises to rework Exercise 13, Section 17.10: The upper half of the ellipsoid $\frac{1}{2}x^2 + \frac{1}{2}y^2 + z^2 = 1$ intersects the cylinder $x^2 + y^2 - y = 0$ in a curve C. Calculate the circulation of $\mathbf{v}(x, y, z) = y^3\,\mathbf{i} + (xy + 3xy^2)\mathbf{j} + z^4\,\mathbf{k}$ around C using Stokes's theorem.

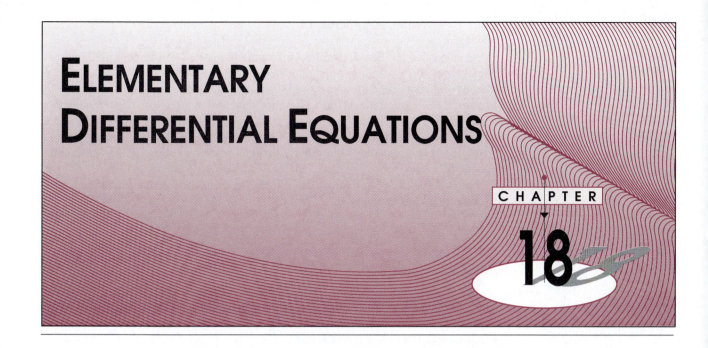

ELEMENTARY DIFFERENTIAL EQUATIONS

CHAPTER

18

■ 18.1 INTRODUCTION

From time to time throughout this text we have spoken about differential equations. In Chapter 7, for instance, we considered the differential equations which modeled exponential growth and decay, and saw that the only functions that satisfy the equation

$$y' = ky \qquad (k \text{ a constant})$$

have the form $y = Ce^{kx}$, where C is an arbitrary constant. Differential equations involving Newton's second law of motion

$$F = ma = m\frac{d^2r}{dt^2},$$

were studied in several different contexts, including the motion of freely falling bodies and planetary motion (see Sections 3.4 and 13.5).

Many problems in the physical and life sciences, engineering, and the social sciences involve "change" (change in size, shape, motion, and so forth). Since a derivative is a "rate of change," the mathematical formulations of such problems are usually described by differential equations. This chapter is intended as an elementary introduction to a study of differential equations. We begin with some terminology.

First, recall that a *differential equation* is an equation which contains an unknown function together with one or more of its derivatives. Some examples are

$$x\frac{dy}{dx} - y = e^x, \qquad y''' + xy'' - yy' = y \sin x,$$

$$\frac{\partial^2 u}{\partial x^2} + \frac{\partial^2 u}{\partial y^2} = 0 \qquad \text{(Laplace's equation, see Exercises 14.6).}$$

In the first two equations, x is the independent variable and $y = y(x)$ is the unknown function. In the third equation, x and y are independent variables and $u = u(x, y)$ is the unknown function.

As suggested by these examples, differential equations can be classified into two general types corresponding to the number of independent variables involved. If the unknown function depends on a single variable, then the equation is an *ordinary differential equation*; if the unknown function depends on more than one variable, then the equation is a *partial differential equation* since the derivatives that appear in the equation must be partial derivatives. Thus, for example, Laplace's equation is a partial differential equation; the other equations just given are ordinary differential equations.

The *order* of a differential equation is the order of the highest derivative (of the unknown function) that appears in the equation

$$y' + ay = b, \qquad y' = x^2 y^2, \qquad \frac{\partial u}{\partial x} = k \frac{\partial u}{\partial y}$$

are first-order equations;

$$x^2 y'' + xy' + x^2 y = 0, \qquad y'' + \omega^2 y = \sin 2x, \qquad \frac{\partial^2 u}{\partial x^2} - \frac{\partial^2 u}{\partial y^2} = 0$$

are equations of the second order; and

$$y''' + ay'' + by' + cy = d$$

is a third-order differential equation.

The obvious question that we want to consider in this chapter is that of "solving" a given differential equation. By a *solution* of a differential equation we mean a function f defined on some interval (in the case of an ordinary differential equation), or on some region in n-dimensional space, $n \geq 2$ (in the case of a partial differential equation), with the property that f and its derivatives satisfy the equation. To determine whether a given function satisfies a certain differential equation, we simply substitute the function and its derivatives into the equation. If the result is an identity, then the function is a solution; otherwise it is not.

Example 1 In each of the following, determine whether the given function satisfies the corresponding differential equation:

(a) $y(x) = x^2 - 5x$; $\quad x\dfrac{dy}{dx} - y = x^2$.

(b) $u(x, y) = (\sin x)(\sinh y)$; $\quad \dfrac{\partial^2 u}{\partial x^2} + \dfrac{\partial^2 u}{\partial y^2} = 0$.

(c) $y(x) = e^x - x^2$; $\quad (x - 1)y'' - xy' + y = (x - 1)^2$.

SOLUTION **(a)** Substituting $y(x) = x^2 - 5x$, $y'(x) = 2x - 5$ into the differential equation

$$x\frac{dy}{dx} - y = x^2,$$

we have

$$x(2x - 5) - (x^2 - 5x) \overset{?}{=} x^2,$$

$$2x^2 - 5x - x^2 + 5x \overset{?}{=} x^2,$$

$$x^2 = x^2.$$

Thus, $y(x) = x^2 - 5x$ is a solution of the differential equation. It can be verified in the same manner that $y = x^2 - Cx$ is a solution of the differential equation for *any* constant C.

(b) Let $u(x, y) = (\sin x)(\sinh y)$. Then

$$\frac{\partial u}{\partial x} = (\cos x)(\sinh y) \qquad \text{and} \qquad \frac{\partial u}{\partial y} = (\sin x)(\cosh y),$$

$$\frac{\partial^2 u}{\partial x^2} = -(\sin x)(\sinh y) \qquad \text{and} \qquad \frac{\partial^2 u}{\partial y^2} = (\sin x)(\sinh y).$$

Since $-(\sin x)(\sinh y) + (\sin x)(\sinh y) = 0$, it follows that $u(x, y) = (\sin x)(\sinh y)$ is a solution of the differential equation $\partial^2 u/\partial x^2 + \partial^2 u/\partial y^2 = 0$.

(c) Substituting $y(x) = e^x - x^2$ and its derivatives $y'(x) = e^x - 2x$, $y''(x) = e^x - 2$, into the differential equation

$$(x - 1)y'' - xy' + y = (x - 1)^2,$$

we have

$$(x - 1)(e^x - 2) - x(e^x - 2x) + (e^x - x^2) \stackrel{?}{=} (x - 1)^2,$$

$$xe^x - e^x - 2x + 2 - xe^x + 2x^2 + e^x - x^2 \stackrel{?}{=} (x - 1)^2,$$

$$x^2 - 2x + 2 \neq (x - 1)^2.$$

Therefore, $y(x) = e^x - x^2$ is not a solution of the differential equation. ❏

From your experience in calculus, you know that the study of functions of several variables (graphing, differentiation, integration) is more complicated than the corresponding study of functions of a single variable. You would probably suspect, therefore, that partial differential equations are typically more complicated than ordinary differential equations. This is indeed the case. Since the intent of this chapter is simply to introduce some of the basic theory and methods for differential equations, we shall confine ourselves to ordinary differential equations. Hereafter, the term *differential equation* will mean *ordinary differential equation.*

In Chapter 5 we saw that the set of all solutions of the first-order differential equation

$$(1) \qquad \frac{dy}{dx} = f(x)$$

is given by $y(x) = F(x) + C$, where F is an antiderivative of f and C is an arbitrary constant. Also, as we noted at the beginning of this section, the set of functions $y(x) = Ce^{kx}$, C an arbitrary constant, represents all solutions of

$$(2) \qquad y' = ky \qquad (k \text{ a constant}).$$

The set of solutions of the second-order differential equation

$$(3) \qquad \frac{d^2 y}{dt^2} = -32$$

governing the motion of a freely falling body is given by

$$y(t) = -16t^2 + C_1 t + C_2,$$

where C_1 and C_2 are independent arbitrary constants (in Section 3.4 these constants were denoted v_0, the initial velocity, and y_0, the initial height). The arbitrary constants involved in these sets of solutions are often called *parameters*; Equations (1) and (2) are said to have *one-parameter families of solutions* while Equation (3) has a *two-parameter family of solutions*.

Intuitively, to find a set of solutions of an *n*th-order differential equation, we would expect to integrate *n* times, with each integration step producing (independently) an arbitrary constant. Thus, we would expect an *n*th-order equation to have an *n-parameter family of solutions*. To *solve* a differential equation of order *n* means to find an *n*-parameter family of solutions. In this chapter we study equations of orders one and two, and so we will be looking for one- and two-parameter families of solutions.

In the examples just given, the families of solutions actually included *all* solutions of the corresponding differential equations. In such cases, the term *general solution* is often used in place of *n-parameter family of solutions*. We comment further on this in the sections which follow.

If specific values are assigned to each of the arbitrary constants in an *n*-parameter family of solutions, then the resulting solution is called a *particular solution*. For example, the function

(4) $$y(t) = -16t^2 + 40t + 144$$

is a particular solution of Equation (3) (it represents the path of an object which is projected upward with an initial velocity of 40 ft/sec from an initial height of 144 ft). Specific values for the constants are usually determined by imposing additional conditions, called *side conditions* or *initial conditions*. The function given in (4) is the solution of (3) which satisfies the side conditions

$$y(0) = 144, \qquad y'(0) = 40.$$

Example 2 Show that $y(x) = -\frac{3}{2} + Ce^{x^2}$ is a one-parameter family of solutions of

$$\frac{dy}{dx} - 2xy = 3x$$

and determine a solution which satisfies the side condition $y(1) = 0$.

SOLUTION First, we substitute

$$y = -\tfrac{3}{2} + Ce^{x^2}, \qquad y' = 2Cxe^{x^2}$$

into the differential equation:

$$2Cxe^{x^2} - 2x\left[-\frac{3}{2} + Ce^{x^2}\right] = 2Cxe^{x^2} + 3x - 2Cxe^{x^2} = 3x.$$

Thus, $y(x) = -\frac{3}{2} + Ce^{x^2}$ is a one-parameter family of solutions. Imposing the side condition $y(1) = 0$, we have

$$y(1) = -\tfrac{3}{2} + Ce = 0,$$

which gives

$$C = \frac{3}{2e}.$$

Therefore, $y(x) = -\dfrac{3}{2} + \dfrac{3}{2e}e^{x^2}$ is a particular solution which satisfies the given side condition. ❏

Example 3 Show that $y = C_1e^{2x} + C_2e^{-x}$ is a two-parameter family of solutions of

$$y'' - y' - 2y = 0$$

and determine a particular solution that satisfies the side conditions

$$y(0) = 2, \quad y'(0) = 1.$$

SOLUTION Substituting

$$y = C_1e^{2x} + C_2e^{-x}, \qquad y' = 2C_1e^{2x} - C_2e^{-x}, \qquad y'' = 4C_1e^{2x} + C_2e^{-x}$$

into the differential equation, we have

$$4C_1e^{2x} + C_2e^{-x} - (2C_1e^{2x} - C_2e^{-x}) - 2(C_1e^{2x} + C_2e^{-x})$$
$$= (4C_1 - 2C_1 - 2C_1)e^{2x} + (C_2 + C_2 - 2C_2)e^{-x} = 0,$$

which shows that $y(x) = C_1e^{2x} + C_2e^{-x}$ is a two-parameter family of solutions. Now, imposing the side conditions leads to the pair of equations

$$y(0) = C_1 + C_2 = 2$$
$$y'(0) = 2C_1 - C_2 = 1.$$

This system has the solution $C_1 = 1$, $C_2 = 1$. Thus $y(x) = e^{2x} + e^{-x}$ is a particular solution that satisfies the side conditions. ❏

Remark Since an nth order differential equation will have an n-parameter family of solutions, we will need to impose n side conditions to evaluate the n arbitrary constants. Typically, the n side conditions lead to a system of n equations in the n "unknowns" $C_1, C_2, \ldots, C_n$. ❏

EXERCISES 18.1

In Exercises 1–8, classify the given differential equation with respect to type (ordinary or partial) and give the order.

1. $xy' - y = e^{2x}$.

2. $(y')^3 + xy'' = x^2 - 1$.

3. $\dfrac{\partial u}{\partial x} = k\dfrac{\partial u}{\partial y}$, k constant.

4. $y''' + xy' + (\sin x)y = \cos x$.

5. $y'' + e^{xy} = \tan x$.

6. $\alpha^2\dfrac{\partial^2 u}{\partial x^2} = \dfrac{\partial u}{\partial t}$, α constant.

7. $\dfrac{\partial^2 u}{\partial x^2} + \dfrac{\partial^2 u}{\partial y^2} = 0$.

8. $y^{(4)} + 3y''' + 3y'' + y' = xe^{-x}$.

In Exercises 9–18, determine whether the given differential equation has the indicated functions as solutions.

9. $2y' - y = 0$; $y_1(x) = e^{x/2}$, $y_2(x) = x^2 + 2e^{x/2}$.

10. $y' + xy = x$; $y_1(x) = e^{-x^2/2}$, $y_2(x) = 1 + Ce^{-x^2/2}$, C any constant.

11. $y' + y = y^2$; $y_1(x) = \dfrac{1}{e^x + 1}$, $y_2(x) = \dfrac{1}{Ce^x + 1}$, C any constant.

12. $y'' + 4y = 0$; $y_1(x) = 2\sin 2x$, $y_2(x) = 2\cos x$.

13. $y'' - 4y = 0$; $y_1(x) = e^{2x}$, $y_2(x) = C\sinh 2x$, C any constant.

14. $y'' - 2y' - 3y = 7e^{3x}$; $y_1(x) = e^{-x} + 2e^{3x}$, $y_2(x) = \frac{7}{4}xe^{3x}$.

15. $a^2\dfrac{\partial^2 u}{\partial x^2} = \dfrac{\partial^2 u}{\partial t^2}$, $u_1(x, t) = \cos \lambda x \sin \lambda at$, λ any constant, $u_2(x, t) = \sin(x - at)$.

16. $\dfrac{\partial^2 u}{\partial x^2} + \dfrac{\partial^2 u}{\partial y^2} = 0$; $u_1(x, y) = \sin x \cosh y$, $u_2(x, y) = \ln (x^2 + y^2)$.

17. $4x^2y'' - 12xy' - 9y = 0;$ $y_1(x) = x^{-1/2},$
$y_2(x) = C_1 x^{-1/2} + C_2 x^{9/2}.$

18. $a^2 \dfrac{\partial^2 u}{\partial x^2} = \dfrac{\partial u}{\partial t},$ $u_1(x, t) = e^{-a^2 t} \sin x,$
$u_2(x, t) = e^{-a^2 t} \sin \lambda x.$

In Exercises 19–26, an n-parameter family of functions is given. Show that the members of the family are solutions of the differential equation and find a member of the family that satisfies the side conditions.

19. $y = Ce^{5x};$ $y' = 5y,$ $y(0) = 2.$

20. $y = \dfrac{x^2}{3} + \dfrac{C}{x};$ $xy' + y = x^2,$ $y(3) = 2.$

21. $y = \dfrac{1}{Ce^x + 1};$ $y' + y = y^2,$ $y(1) = -1.$

22. $y = x \ln \dfrac{C}{x};$ $y' = \dfrac{y - x}{x},$ $y(2) = 4.$

23. $y = C_1 x + C_2 x^{1/2};$ $2x^2 y'' - xy' + y = 0,$
$y(4) = 1,$ $y'(4) = -2.$

24. $y = C_1 \sin 3x + C_2 \cos 3x;$ $y'' + 9y = 0,$
$y(\pi/2) = y'(\pi/2) = 1.$

25. $y = C_1 x^2 + C_2 x^2 \ln x;$ $x^2 y'' - 3xy' + 4y = 0,$
$y(1) = 0,$ $y'(1) = 1.$

26. $y = C_1 + C_2 e^x + C_3 e^{2x} + \frac{1}{4}x^2 + \frac{3}{4}x - xe^x;$
$y''' - 3y'' + 2y' = x + e^x,$
$y(0) = 1,$ $y'(0) = -\frac{1}{4},$ $y''(0) = -\frac{3}{2}.$

In Exercises 27–30, determine the values of r, if any, such that $y = e^{rx}$ is a solution of the given differential equation.

27. $y' + 3y = 0.$ **28.** $y'' - 5y' + 6y = 0.$

29. $y'' + 6y' + 9y = 0.$ **30.** $y''' - 3y' + 2y = 0.$

In Exercises 31–34, determine values of r, if any, such that $y = x^r$ is a solution of the given differential equation.

31. $xy'' + y' = 0.$ **32.** $x^2 y'' + xy' - y = 0.$

33. $4x^2 y'' - 4xy' + 3y = 0.$

34. $x^3 y''' - 2x^2 y'' - 2xy' + 8y = 0.$

An nth-order differential equation together with conditions that are specified at two, or more, points on an interval I is usually called a *boundary-value problem*. In particular, if conditions are specified at two points, the problem is called a *two-point boundary-value problem*.

35. Each member of the two-parameter family $y = C_1 \sin 4x + C_2 \cos 4x$ is a solution of the differential equation $y'' + 16y = 0.$
(a) Find all the members of this family that satisfy the boundary conditions:
$$y(0) = 0, \quad y(\pi/2) = 0.$$
(b) Find all the members of this family that satisfy the boundary conditions:
$$y(0) = 0, \quad y(\pi/8) = 0.$$

36. For each real number r, each member of the two-parameter family $y = C_1 \sin rx + C_2 \cos rx$ is a solution of the differential equation $y'' + r^2 y = 0.$
(a) Determine the numbers r such that the two-point boundary-value problem
$$y'' + r^2 y = 0, \quad y(0) = 0, \quad y(\pi) = 0$$
has a nonzero solution.
(b) Determine the numbers r such that the two-point boundary-value problem
$$y'' + r^2 y = 0, \quad y(0) = 0, \quad y(\pi/2) = 0$$
has a nonzero solution.

■ 18.2 FIRST ORDER LINEAR DIFFERENTIAL EQUATIONS

A first-order differential equation is *linear* if it can be written in the form

(18.2.1)
$$y' + p(x)y = q(x),$$

where p and q are continuous functions on some interval I.

To solve (18.2.1), we first calculate

$$H(x) = \int p(x)\, dx$$

(let the constant of integration equal 0) and form $e^{H(x)}$. Note that

$$\frac{d}{dx}[e^{H(x)}] = e^{H(x)} \frac{d}{dx}[H(x)] = p(x)\, e^{H(x)}.$$

Now multiply the differential equation by $e^{H(x)}$ to obtain the equivalent equation (since $e^{H(x)} \neq 0$)

$$e^{H(x)} y' + e^{H(x)} p(x) y = e^{H(x)} q(x).$$

The left side of this equation has the form

$$\frac{d}{dx}[e^{H(x)} y],$$

and this was our motivation for multiplying by $e^{H(x)}$; it converted the left-hand side of the equation into the derivative of a product. Now the equation can be written

$$\frac{d}{dx}[e^{H(x)} y] = e^{H(x)} q(x),$$

and integration gives

$$e^{H(x)} y = \int e^{H(x)} q(x)\, dx + C.$$

Thus

(18.2.2)
$$y = e^{-H(x)} \left\{ \int e^{H(x)} q(x)\, dx + C \right\}$$

is a one-parameter family of solutions. In Exercise 37 you are asked to show that *every* solution can be written in this form, and so (18.2.2) is actually the *general solution* of Equation (18.2.1). Different choices of C give different *particular solutions*.

Remark The key step in solving the first-order linear differential equation

$$y' + p(x) y = q(x)$$

is multiplication by $e^{H(x)}$, where

$$H(x) = \int p(x)\, dx.$$

It is multiplication by $e^{H(x)}$ that enables us to write the left side of the equation in a form that we can integrate directly. The expression $e^{H(x)}$ is called an *integrating factor*. Integrating factors are discussed again in Section 18.4. ❑

Example 1 Find the general solution of the linear differential equation

$$y' + ay = b, \qquad a, b \text{ constants and } a \neq 0.$$

SOLUTION Here $p(x) = a$ and $q(x) = b$ are continuous for all x, and so we can take the interval I to be $(-\infty, \infty)$. An integrating factor is

$$e^{\int a\,dx} = e^{ax}.$$

Multiplication by e^{ax} gives

$$e^{ax}y' + e^{ax}ay = e^{ax}b.$$

Since the left-hand side is the derivative of $e^{ax}y$, we have

$$\frac{d}{dx}[e^{ax}y] = e^{ax}b.$$

Integrating this equation, we get

$$e^{ax}y = \frac{b}{a}e^{ax} + C$$

and so

$$y = \frac{b}{a} + Ce^{-ax}. \quad \square$$

Example 2 Find the general solution of the linear differential equation

$$y' + 2xy = x$$

and determine a particular solution that satisfies the side condition $y(0) = 2$.

SOLUTION The functions $p(x) = 2x$ and $q(x) = x$ are continuous for all x, and so we can take $I = (-\infty, \infty)$. An integrating factor is

$$e^{\int (2x)\,dx} = e^{x^2}.$$

Multiplication by e^{x^2} gives

$$e^{x^2}y' + 2xe^{x^2}y = xe^{x^2},$$

which is the same as

$$\frac{d}{dx}[e^{x^2}y] = xe^{x^2}.$$

Integrating both sides of this equation, we get

$$e^{x^2}y = \tfrac{1}{2}e^{x^2} + C.$$

Thus, the general solution of the equation is

$$y = \tfrac{1}{2} + Ce^{-x^2}.$$

To determine a solution that satisfies the given side condition, we set $x = 0$ and $y = 2$ in the general solution and solve for C:

$$2 = \tfrac{1}{2} + C \qquad \text{and so} \qquad C = \tfrac{3}{2}.$$

Therefore

$$y = \tfrac{1}{2} + \tfrac{3}{2}e^{-x^2}$$

is a particular solution that satisfies $y(0) = 2$. $\square$

Remark Note that the particular solution in Example 2 was uniquely determined by the side condition $y(0) = 2$. In Exercise 42 you are asked to show that this situation holds in general. That is, if p and q are continuous functions on the interval I, and if $x_0 \in I$ and y_0 is any real number, then the linear differential equation

$$y' + p(x)y = q(x) \qquad \text{with the side condition} \qquad y(x_0) = y_0$$

has a unique solution. ❏

Example 3 Find the general solution of

$$xy' - 2y = 2x^2 + x.$$

SOLUTION This equation is not in the form of (18.2.1), but it can be put in that form by dividing through by x:

$$y' - \frac{2}{x}y = 2x + 1.$$

The functions $p(x) = -2/x$ and $q(x) = 2x + 1$ are continuous on $(0, \infty)$ and on $(-\infty, 0)$. We take $I = (0, \infty)$. Now,

$$H(x) = \int -\frac{2}{x}dx = -2 \ln x = \ln x^{-2},$$

and so

$$e^{\ln x^{-2}} = x^{-2}$$

is an integrating factor. Multiplication by x^{-2} gives

$$x^{-2}y' - 2x^{-3}y = 2x^{-1} + x^{-2}.$$

Therefore,

$$\frac{d}{dx}[x^{-2}y] = 2x^{-1} + x^{-2},$$

and

$$x^{-2}y = 2 \ln x - x^{-1} + C.$$

Solving for y, we get

$$y = 2x^2 \ln x - x + Cx^2, \qquad (x > 0)$$

and this is the general solution. ❏

Applications

Example 4 A new company is being formed. A market analysis projects that the rate of growth of the company's income at any future time t will be proportional to the difference between the actual income at that time and an upper limit of $10 million. Further, it is estimated that the income after 3 years will be $4 million; the income is $0 initially. Find an expression for the company's income at any time t, and determine how long it will take for the income to reach $8 million.

SOLUTION Let $A(t)$ denote the company's income at time t. According to the market analysis, the rate of growth of A is proportional to the difference between A and $10 million. That is,

$$\frac{dA}{dt} = k(10 - A),$$

where k is the constant of proportionality and A is measured in millions. Rewriting this equation, we have

$$\frac{dA}{dt} + kA = 10k,$$

a first-order linear equation in which $p(x) = k$ and $q(x) = 10k$ are constants. By the result in Example 1,

$$A(t) = \frac{10k}{k} + Ce^{-kt} = 10 + Ce^{-kt}.$$

Since $A(0) = 0$, we have

$$0 = 10 + Ce^0 = 10 + C \quad \text{and} \quad C = -10.$$

Therefore,

$$A(t) = 10 - 10\, e^{-kt}.$$

To determine k, we use the fact that $A = 4$ when $t = 3$:

$$4 = 10 - 10e^{-3k},$$

$$e^{-3k} = \tfrac{3}{5} = 0.6,$$

$$-3k = \ln(0.6) \cong -0.511,$$

$$k \cong 0.17.$$

Thus, the company's income at any time t is projected to be (approximately)

$$A(t) = 10 - 10\, e^{-0.17t}.$$

To find out how long it will take for the income to reach $8 million, we solve

$$8 = 10 - 10e^{-0.17t}$$

for t:

$$e^{-0.17t} = \tfrac{1}{5} = 0.2,$$

$$-0.17t = \ln(0.2) \cong -1.609,$$

$$t \cong 9.47.$$

Thus, it will take approximately $9\frac{1}{2}$ years for the company's income to reach $8 million. ❏

Example 5 A chemical manufacturing company has a 1000-gallon holding tank which it uses to control the release of pollutants into a sewage system. Initially the tank has 360 gallons of water containing 2 pounds of pollutant per gallon. Water containing 3 pounds of pollutant per gallon enters the tank at the rate of 80 gallons per hour and is uniformly mixed with the water already in the tank. Simultaneously, water is released from the tank at the rate of 40 gallons per hour. Determine how many pounds of pollutant are in the tank at any time t. Also, determine the rate (lbs/gal) at which the pollutant is being released after 10 hours.

SOLUTION Let $P(t)$ be the amount of pollutant (in pounds) in the tank at time t. The rate of change of pollutant in the tank, dP/dt, is given by

$$\frac{dP}{dt} = (\text{rate in}) - (\text{rate out}).$$

The pollutant is entering the tank at the rate of $3 \times 80 = 240$ pounds per hour ($=$ rate in).

Water is entering the tank at the rate of 80 gallons per hour and is leaving at the rate of 40 gallons per hour. Therefore the amount of water in the tank is increasing at the rate of 40 gallons per hour, and so there are $360 + 40t$ gallons of water in the tank at time t, where $0 \leq t \leq 16$ (the tank will be full at time $t = 16$ hours). We can now conclude that the amount of pollutant per gallon in the tank at time t is

$$\frac{P(t)}{360 + 40t},$$

and the rate at which pollutant is leaving the tank is

$$\frac{40P(t)}{360 + 40t} = \frac{P(t)}{9 + t}(= \text{rate out}).$$

Therefore, our differential equation is

$$\frac{dP}{dt} = 240 - \frac{P}{9 + t}$$

or

$$\frac{dP}{dt} + \frac{1}{9 + t}P = 240,$$

a first-order, linear differential equation. Here we have

$$p(t) = \frac{1}{9 + t} \quad \text{and} \quad H(t) = \int \frac{1}{9 + t}dt = \ln|9 + t| = \ln(9 + t) \quad (9 + t > 0).$$

Thus, an integrating factor is

$$e^{H(t)} = e^{\ln(9 + t)} = 9 + t.$$

Multiplying the equation by $9 + t$ gives

$$(9 + t)\frac{dP}{dt} + P = 240(9 + t),$$

$$\frac{d}{dt}[(9 + t)P] = 240(9 + t),$$

$$(9 + t)P = 120(9 + t)^2 + C,$$

and

$$P(t) = 120(9 + t) + \frac{C}{9 + t}.$$

Since the amount of pollutant in the tank initially is $2 \times 360 = 720$ (pounds), we have

$$P(0) = 120(9) + \frac{C}{9} = 720, \quad \text{which implies} \quad C = -3240.$$

Thus, the amount of pollutant in the tank at any time t is

$$P(t) = 120(9 + t) - \frac{3240}{9 + t} \quad \text{(pounds)}.$$

After 10 hours there are $360 + 40(10) = 760$ gallons of water in the tank, and there are

$$P(10) = 120(19) - \frac{3240}{19} \cong 2109$$

pounds of pollutant. Therefore, the rate at which pollutant is being released into the sewage system after 10 hours is

$$\frac{2109}{760} \cong 2.78 \quad \text{(pounds per gallon)}. \quad \square$$

EXERCISES 18.2

In Exercises 1–16, find the general solution.

1. $y' - 2y = 1$.

2. $xy' - 2y = -x$.

3. $2y' + 5y = 2$.

4. $y' - y = -2e^{-x}$.

5. $y' - 2y = 1 - 2x$.

6. $xy' - 2y = -3x$.

7. $xy' - 4y = -2nx$.

8. $y' + y = 2 + 2x$.

9. $y' - e^x y = 0$.

10. $y' - y = e^x$.

11. $(1 + e^x)y' + y = 1$.

12. $xy' + y = (1 + x)e^x$.

13. $y' + 2xy = xe^{-x^2}$.

14. $xy' - y = 2x \ln x$.

15. $y' + \dfrac{2}{x + 1}y = 0$.

16. $y' + \dfrac{2}{x + 1}y = (x + 1)^{5/2}$.

In Exercises 17–24, find the particular solution determined by the given side condition.

17. $y' + y = x, \quad y(0) = 1$.

18. $y' - y = e^{2x}, \quad y(1) = 1$.

19. $y' + y = \dfrac{1}{1 + e^x}, \quad y(0) = e$.

20. $y' + y = \dfrac{1}{1 + 2e^x}, \quad y(0) = e$.

21. $xy' - 2y = x^3 e^x, \quad y(1) = 0$.

22. $x^2 y' + 2xy = 1, \quad y(1) = 2$.

23. $y' + 2xy = 0, \quad y(x_0) = y_0$.

24. $xy' + 2y = xe^{-x}, \quad y(1) = -1$.

25. A 200-liter tank initially full of water develops a leak at the bottom. Given that 20% of the water leaks out in the first 5 minutes, find the amount of water left in the tank t minutes after the leak develops

(a) if the water drains off at a rate that is proportional to the amount of water present;

(b) if the water drains off at a rate that is proportional to the product of the time elapsed and the amount of water present.

26. An object falling in air is subject not only to gravitational force but also to air resistance. Find $v(t)$, the velocity of the object at time t, given that

$$v'(t) + Kv(t) = 32 \quad \text{and} \quad v(0) = 0 \quad \text{(take } K > 0\text{)}.$$

Show that $v(t)$ cannot exceed $32/K$ ($32/K$ is called the *terminal velocity*).

27. At a certain moment a 100-gallon mixing tank is full of brine containing 0.25 pounds of salt per gallon. Find the amount of salt present t minutes later if the brine is being continuously drawn off at the rate of 3 gallons per minute and replaced by brine containing 0.2 pound of salt per gallon.

28. The current i in an electric circuit varies with time t according to the formula

$$L\frac{di}{dt} + Ri = E$$

where E (the voltage), L (the inductance), and R (the resistance) are positive constants. Measure time in seconds, current in amperes, and suppose that the initial current is 0.

(a) Find a formula for the current at each subsequent time t.

(b) What upper limit does the current approach as t increases?

(c) In how many seconds will the current reach 90% of its upper limit?

29. *Newton's Law of Cooling* states that the rate of change of the temperature T of an object is proportional to the difference between T and the temperature τ of the surrounding medium:

$$\frac{dT}{dt} = -k(T - \tau)$$

where k is a positive constant.

(a) Solve this equation for T. (Find the general solution.)

(b) Solve this equation for T given that $T(0) = T_0$.

(c) Take $T_0 > \tau$. What is the limiting temperature to which the object cools as t increases? What happens if $T_0 < \tau$?

(d) A cup of coffee is served to you at 185°F in a room where the temperature is 65°F. Two minutes later the temperature of the coffee has dropped to 155°F. How many more minutes would you expect to wait for the coffee to cool to 105°F?

30. A drug is fed intravenously into a patient's blood stream at the constant rate r. Simultaneously, the drug diffuses into the patient's body at a rate proportional to the amount of the drug present.

(a) Determine the differential equation that describes the amount $Q(t)$ of the drug in the patient's bloodstream at time t.

(b) Determine the solution $Q = Q(t)$ of the differential equation from part (a) that satisfies the initial condition $Q(0) = 0$.

(c) Find $\lim_{t \to \infty} Q(t)$.

31. An advertising company is trying to expose a new product in a certain metropolitan area by advertising on television. Suppose that the rate of exposure to new people is proportional to the number of people who have not seen the product out of a total population of M viewers. Let $P(t)$ denote the number of viewers who have been exposed to the product at time t. The company has determined that no one was aware of the product at the start of the advertising campaign ($P(0) = 0$) and that 30% of the viewers were aware of the product after 10 days.

(a) Determine the differential equation that describes the number of viewers who are aware of the product at time t.

(b) Determine the solution of the differential equation from part (a) that satisfies the side condition $P(0) = 0$.

(c) How long will it take for 90% of the population to be aware of the product.

32. Let k be an integer different from 0 and 1. Transform *Bernoulli's equation*

$$y' + a(x)y = b(x)y^k$$

into the linear equation

$$v' + (1 - k)a(x)v = (1 - k)b(x)$$

by setting $v = y^{1-k}$.

In Exercises 33–36, use the result in Exercise 32 to solve the given differential equation.

33. $y' + y = 3e^x y^3$.

34. $y' - y = xy^2$.

35. $y' + \dfrac{y}{x} = x^3 y^2$.

36. $y' + xy = \dfrac{x}{y}$.

37. Show that if $y = y(x)$ is a solution of the differential equation (18.2.1), then y is a member of the one-parameter family (18.2.2). Thus, the one-parameter family of solutions (18.2.2) is the general solution of (18.2.1). HINT: If $y(x)$ is a solution of (18.2.1), then

$$y'(x) + p(x)y(x) = q(x).$$

Multiply by $e^{H(x)}$, where $H(x) = \int p(x)\,dx$, and continue as in the derivation of (18.2.2).

First Order, Linear, Homogeneous Differential Equations

If the function q in equation (18.2.1) is identically zero on I, then the equation is said to be *homogeneous*; if q is not identically zero on I, then equation (18.2.1) is said to be *nonhomogeneous*. The first order, linear homogeneous differential equation is:

(18.2.3) $$y' + p(x)y = 0$$

where p is continuous on I.

38. Show that the one-parameter family of solutions of equation (18.2.3) is given by

$$y(x) = Ce^{-H(x)}, \qquad \text{where} \qquad H(x) = \int p(x)\,dx.$$

39. Let $y = y_1(x)$ and $y = y_2(x)$ be any two solutions of (18.2.3), and let c be any real number.

(a) Prove that $y = y_1(x) + y_2(x)$ is a solution of (18.2.3).

(b) Prove that $y = cy_1(x)$ is a solution of (18.2.3).

40. Suppose that the coefficient function p in equation (18.2.3) is continuously differentiable. Show that each solution of (18.2.3) has a continuous second derivative.

41. (a) Prove that if $y = y(x)$ is a solution of equation (18.2.3) and $y(a) = 0$ for some $a \in I$, then $y(x) = 0$ for all $x \in I$.

(b) Suppose that $y = y_1(x)$ and $y = y_2(x)$ are solutions of (18.2.3). Prove that if $y_1(a) = y_2(a)$ at some $a \in I$, then $y_1(x) = y_2(x)$ for all $x \in I$.

42. Let $y = u_1(x)$ and $y = u_2(x)$ be solutions of equation (18.2.1).

(a) Prove that $y = u_1(x) - u_2(x)$ is a solution of the homogeneous equation (18.2.3).

(b) Prove that if $u_1(a) = u_2(a)$ at some $a \in I$, then $u_1(x) = u_2(x)$ for all $x \in I$. Note that this result implies that equation (18.2.1) with a side condition $y(x_0) = y_0$ has a unique solution.

43. Consider the linear homogeneous equation $y' + ry = 0$, where r is a real number, and let $y = y(x)$ be any solution.

(a) Show that if $r > 0$, then $\lim\limits_{x \to \infty} y(x) = 0$.

(b) Show that if $r < 0$, then the solution $y = y(x)$ is unbounded.

44. Suppose that the coefficient function p in equation (18.2.3) is positive on $[0, \infty)$.

(a) Show that all solutions of (18.2.3) are bounded on $[0, \infty)$.

(b) Under what conditions will every solution $y = y(x)$ have the property $\lim\limits_{x \to \infty} y(x) = 0$?

■ 18.3 SEPARABLE EQUATIONS; HOMOGENEOUS EQUATIONS

A first-order differential equation can always be written in the form

(18.3.1)
$$F(x, y, y') = 0$$

by gathering all of the terms on the left side of the equation. For example, the first-order linear equation (18.2.1) written in this form is

$$y' + p(x)y - q(x) = 0.$$

Of course, in order for (18.3.1) to be a differential equation, y' must appear among the terms of F. We assume, in addition, that the equation can be solved for y'. This means that the equation can be expressed as

(18.3.2)
$$P(x, y) + Q(x, y)y' = 0.$$

It will also be convenient to write this equation in the equivalent differential form

$$P(x, y)\,dx + Q(x, y)\,dy = 0.$$

The first-order equation (18.3.2) is *separable* iff it can be put in the form

(18.3.3)
$$p(x) + q(y)y' = 0$$

or in the equivalent differential form

(18.3.4)
$$p(x)\,dx + q(y)\,dy = 0.$$

Separable equations were introduced in Section 7.6. We provide the details of the solution method here.

To solve such an equation, we begin by writing (18.3.3) as

$$q(y)y' = -p(x).$$

This separates the variables: y's on the left, x's on the right. If p and q are continuous, they have antiderivatives. Setting

$$P(x) = \int p(x) \, dx \quad \text{and} \quad Q(y) = \int q(y) \, dy,$$

we have

$$Q'(y)y' = -P'(x).$$

Since, by the chain rule,

$$Q'(y)y' = \frac{d}{dx}[Q(y)],$$

we have

$$\frac{d}{dx}[Q(y)] = -P'(x).$$

Integration with respect to x gives

$$Q(y) = -P(x) + C.$$

Back in terms of p and q, we have

$$\int q(y) \, dy = -\int p(x) \, dx + C,$$

which is usually written

(18.3.5)
$$\int p(x) \, dx + \int q(y) \, dy = C.$$

This is a one-parameter family of solutions of (18.3.3). It may or may not be the general solution of the equation, as we illustrate below. Now that we have established the form of the solutions, note that (18.3.5) can be obtained simply by integrating the differential form (18.3.4) term-by-term.

If

$$Q(y) = \int q(y) \, dy$$

is one-to-one, then Q has an inverse and (18.3.5) can be solved for y:

$$y = Q^{-1}\left(C - \int p(x) \, dx\right).$$

Usually this is not the case and we have to be satisfied with (18.3.5) as it stands. For each choice of C, (18.3.5) gives a curve in the xy-plane. Such a curve is called an *integral curve* of the differential equation. An integral curve is a *solution* of the differential equation in the sense that at each point (x, y) of the curve, the slope y' is related to x and y by the equation

$$p(x) + q(y)y' = 0.$$

Example 1 The differential equation

$$x + yy' = 0$$

is separable. We can solve the equation (that is, we can obtain the integral curves) by writing the equation in differential form and integrating term-by-term:

$$x \, dx + y \, dy = 0,$$

$$\int x \, dx + \int y \, dy = C.$$

Carrying out the integrations, we have

$$\tfrac{1}{2}x^2 + \tfrac{1}{2}y^2 = C,$$

and thus

$$x^2 + y^2 = 2C.$$

Since C is an arbitrary constant, $2C$ is also arbitrary, and so we simply write the one-parameter family of solutions as

$$x^2 + y^2 = C.$$

This equation defines y implicitly as a function of x; the integral curves are circles centered at the origin. See Figure 18.3.1.

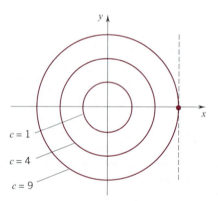

Figure 18.3.1

We can find the integral curve that passes through a specified point by adjusting the constant C. The integral curve that passes through the point (x_0, y_0) is the circle

$$x^2 + y^2 = x_0^2 + y_0^2.$$

It can be obtained by setting

$$C = x_0^2 + y_0^2. \quad \square$$

Let's return briefly to the general form

$$p(x) + q(y)y' = 0.$$

At a point (x, y) where $q(y) = 0$ but $p(x) \neq 0$, y' is infinite, and thus the tangent line, if it exists, must be vertical. In the case of

$$x + yy' = 0$$

there are two points on each circle where $y = 0$ and $x \neq 0$ (see Figure 18.3.1). At each of these points the tangent line is vertical.

Example 2 Show that the differential equation

$$y' = -2x(y - 1)^2$$

is separable and find a one-parameter family of solutions.

SOLUTION We try to write the equation in the form of (18.3.3):

$$2x(y - 1)^2 + y' = 0$$

and for $y \neq 1$,

$$2x + \frac{1}{(y - 1)^2} y' = 0.$$

Thus, the equation is separable. To solve, we write the equation in differential form and integrate term-by-term:

$$2x \, dx + \frac{1}{(y - 1)^2} dy = 0,$$

$$\int 2x \, dx + \int (y - 1)^{-2} \, dy = C,$$

and

$$x^2 - (y - 1)^{-1} = C$$

is a one-parameter family of solutions. In this case, we can solve for y explicitly:

$$\frac{1}{(y - 1)} = x^2 - C \quad \text{and} \quad y = 1 + \frac{1}{x^2 - C}.$$

Since y cannot be 1, no integral curve can cross the line $y = 1$. We can, however, take any point (x_0, y_0) with $y_0 \neq 1$ and adjust the constant C so that the curve passes through that point: the integral curve

$$y = 1 + \frac{1}{x^2 - C}$$

passes through the point (x_0, y_0) iff $C = x_0^2 - (y_0 - 1)^{-1}$. Representative graphs of the integral curves are shown in Figure 18.3.2. ❏

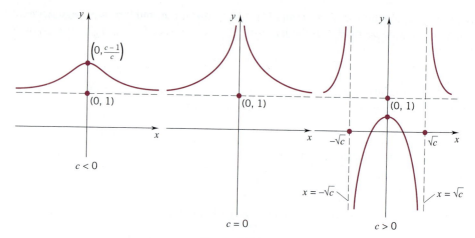

Figure 18.3.2

Remark It is easy to verify that the constant function $y = 1$ is a solution of the differential equation in Example 2 which is not a member of the one-parameter family of solutions (we had to exclude this function when we divided through by $(y - 1)^2$). Thus, the one-parameter family is not the *general solution* of the equation. A solution of a first-order differential equation which is not included in a one-parameter family of solutions is called a *singular solution*. ❏

Example 3 Find a solution of the differential equation

$$y' = \frac{xy - y}{y + 1},$$

which satisfies the side condition $y(2) = 1$.

SOLUTION We show first that the equation is separable:

$$y - xy + (y + 1)y' = 0,$$

$$1 - x + \frac{y + 1}{y}y' = 0 \qquad (y \neq 0)$$

or

$$1 - x + \left(1 + \frac{1}{y}\right)y' = 0.$$

Now,

$$(1 - x)\,dx + \left(1 + \frac{1}{y}\right)dy = 0$$

$$\int (1 - x)\,dx + \int \left(1 + \frac{1}{y}\right)dy = C,$$

and

$$x - \tfrac{1}{2}x^2 + y + \ln|y| = C$$

is a one-parameter family of solutions which defines y implicitly as a function of x. The constant function $y = 0$ is not included in the one-parameter family; it is a singular solution of the equation.

To find a solution which satisfies the side condition, we set $x = 2$ and $y = 1$ in the one-parameter family:

$$2 - \tfrac{1}{2}(2)^2 + 1 + \ln(1) = C \qquad \text{and} \qquad C = 1.$$

Thus, the particular solution

$$x - \tfrac{1}{2}x^2 + y + \ln|y| = 1$$

satisfies the side condition $y(2) = 1$. ❏

An Application

In the mid-nineteenth century the Belgian biologist P. F. Verhulst used the differential equation

(18.3.6)
$$\frac{dy}{dt} = ky(M - y),$$

where k and M are positive constants, to study the population growth of various countries. This equation is now known as the *logistic equation* and its solutions are called *logistic functions*. Life scientists have used this equation to model the spread of a disease through a population, and social scientists have used it to study the flow of information. In the case of a disease, if M denotes the total number of people in the population and $y(t)$ is the number of infected people at time t, then the differential equation states that the rate of growth of the disease is proportional to the product of the number of people who are infected and the number who are not.

The differential equation is separable since it can be written in the form

$$\frac{1}{y(M - y)}y' - k = 0.$$

Writing the equation in differential form and integrating term-by-term, we have

$$\frac{1}{y(M - y)}dy - k\,dt = 0,$$

$$\int \frac{1}{y(M - y)}dy - \int k\,dt = C_1,$$

$$\int \left[\frac{1/M}{y} + \frac{1/M}{M - y}\right]dy - kt = C_1, \qquad \text{(partial fraction decomposition)}$$

and

$$\frac{1}{M}\ln|y| - \frac{1}{M}\ln|M - y| = kt + C_1$$

is a one-parameter family of solutions. We can solve this equation for y as follows:

$$\frac{1}{M} \ln \left| \frac{y}{M-y} \right| = kt + C_1$$

$$\ln \left| \frac{y}{M-y} \right| = Mkt + MC_1 = Mkt + C_2, \qquad (C_2 = MC_1)$$

$$\left| \frac{y}{M-y} \right| = e^{(C_2+Mkt)} = e^{C_2}e^{Mkt} = Ce^{Mkt}. \qquad (C = e^{C_2})$$

Now, in the context of this application, $y = y(t)$ satisfies $0 < y < M$. Therefore $y/(M-y) > 0$ and we have

$$\frac{y}{M-y} = Ce^{Mkt}.$$

Solving this equation for y we get

$$y(t) = \frac{CM}{C + e^{-Mkt}}.$$

If $y(t)$ satisfies the side condition $y(0) = R$, $R < M$, then

$$R = \frac{CM}{C+1}, \qquad \text{which implies} \quad C = \frac{R}{M-R}$$

and

(18.3.7)

$$y(t) = \frac{MR}{R + (M-R)e^{-Mkt}}.$$

The graph of this particular solution is shown in Figure 18.3.3. Note that y is increasing for all $t \geq 0$. In Exercise 38 you are asked to show that the graph is concave up on $[0, t_1]$ and concave down on $[t_1, \infty)$. This means that the disease is spreading at an increasing rate up to time $t = t_1$; after t_1, the disease is still spreading, but at a decreasing rate. Note, also, that $y(t) \rightarrow M$ as $t \rightarrow \infty$.

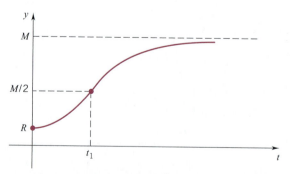

Figure 18.3.3

Example 4 A rumor spreads through a population of 5000 people at a rate proportional to the product of the number of people who have heard it and the number who have not. Suppose that 100 people initiate the rumor and that a total of 500 people know the rumor after two days. How long will it take for half the population to hear the rumor?

SOLUTION Let $y(t)$ denote the number of people who know the rumor at time t. Then y satisfies the differential equation (18.3.6) with $M = 5000$ and the side condition $R = y(0) = 100$. Thus, by (18.3.7),

$$y(t) = \frac{100(5000)}{100 + 4900e^{-5000kt}} = \frac{5000}{1 + 49e^{-5000kt}},$$

where the constant of proportionality k can be determined from the condition $y(2) = 500$. We have

$$500 = \frac{5000}{1 + 49e^{-10,000k}},$$

$$1 + 49e^{-10,000k} = 10,$$

$$e^{-10,000k} = \frac{9}{49},$$

$$-10,000k = \ln\left[\frac{9}{49}\right] \quad \text{and} \quad k \cong 0.00017.$$

Therefore,

$$y(t) = \frac{5000}{1 + 49e^{-0.85t}}.$$

To determine how long it will take for half the population to hear the rumor, we solve the equation

$$2500 = \frac{5000}{1 + 49e^{-0.85t}}$$

for t. We get

$$1 + 49e^{-0.85t} = 2, \quad e^{-0.85t} = \frac{1}{49}, \quad \text{and} \quad t = \frac{\ln(1/49)}{-0.85} \cong 4.58.$$

Thus, it will take slightly more than $4\frac{1}{2}$ days for half the population to hear the rumor. ❑

Homogeneous Equations

If we solve equation (18.3.2) for y', we get

$$y' = \frac{-P(x, y)}{Q(x, y)},$$

which we can write as

(18.3.8)

$$y' = f(x, y).$$

The first-order differential equation (18.3.8) is *homogeneous* iff the function f has the property

$$f(tx, ty) = f(x, y)$$

for all $t \neq 0$. This condition is equivalent to saying that f can be expressed as a function of y/x. For if f is homogeneous, then

$$f(x, y) = f\left(x, x\left(\frac{y}{x}\right)\right) = f\left(1, \frac{y}{x}\right) = g\left(\frac{y}{x}\right) \qquad (x \neq 0).$$

A homogeneous equation can be transformed into a separable equation by setting

$$v = \frac{y}{x}.$$

To see this, write

$$xv = y$$

and differentiate with respect to x:

$$v + xv' = y'.$$

Substituting y and y' into (18.3.8), gives

$$v + xv' = f(x, xv) = f(1, v) = g(v).$$

Thus, we have

$$v + xv' = g(v)$$

and

(18.3.9)

$$\boxed{\frac{1}{x} + \left[\frac{1}{v - g(v)}\right]v' = 0.}$$

This equation is separable.

If Equation (18.3.8) is homogeneous, then we solve it by solving the transformed equation, Equation (18.3.9), and then substitute y/x back in for v.

Example 5 Show that the differential equation

$$y' = \frac{3x^2 + y^2}{xy}$$

is homogeneous and find a one-parameter family of solutions.

SOLUTION The equation is homogeneous since

$$f(tx, ty) = \frac{3(tx)^2 + (ty)^2}{(tx)(ty)} = \frac{t^2(3x^2 + y^2)}{t^2(xy)} = \frac{3x^2 + y^2}{xy} = f(x, y).$$

Note that f can be written as

$$\frac{3 + (y/x)^2}{(y/x)} \left[= g\left(\frac{y}{x}\right)\right],$$

which is a function of y/x as previously indicated.

Now, set $vx = y$. Then, $v + xv' = y'$ and

$$v + xv' = \frac{3x^2 + v^2x^2}{vx^2} = \frac{3 + v^2}{v} \, [= g(v)].$$

Thus,

$$v - \frac{3 + v^2}{v} + xv' = 0,$$

$$-3 + xvv' = 0,$$

$$-\frac{3}{x} \, dx + v \, dv = 0,$$

$$-\int \frac{3}{x} \, dx + \int v \, dv = C_1$$

$$-3 \ln|x| + \frac{v^2}{2} = C_1 \quad \text{or} \quad \ln|x| - \tfrac{1}{6}v^2 = C. \quad (C = -\tfrac{1}{3}C_1)$$

Substituting y/x back in for v, we have

$$\ln|x| - \frac{1}{6}\left(\frac{y}{x}\right)^2 = C.$$

The integral curves take the form

$$y^2 = 6x^2 \, (\ln|x| - C). \quad \square$$

EXERCISES 18.3

In Exercises 1–10, solve the differential equation.

1. $y' = y \sin(2x + 3)$.

2. $y' = (x^2 + 1)(y^2 + y)$.

3. $y' = (xy)^3$.

4. $y' = 3x^2(1 + y^2)$.

5. $y' = -\dfrac{\sin 1/x}{x^2 y \cos y}$.

6. $y' = \dfrac{y^2 + 1}{y + yx}$.

7. $y' = xe^{y-x}$.

8. $y' = xy^2 - x - y^2 + 1$.

9. $(y \ln x)y' = \dfrac{(y + 1)^2}{x}$.

10. $e^y \sin 2x \, dx + \cos x(e^{2y} - y) \, dy = 0$.

In Exercises 11–16, find the particular solution determined by the side condition.

11. $y' = x\sqrt{\dfrac{1 - y^2}{1 - x^2}}, \quad y(0) = 0$.

12. $y' = \dfrac{e^{x-y}}{1 + e^x}, \quad y(1) = 0$.

13. $y' = \dfrac{x^2 y - y}{y + 1}, \quad y(3) = 1$.

14. $x^2 y' = y - xy, \quad y(-1) = -1$.

15. $(xy^2 + y^2 + x + 1) \, dx + (y - 1) \, dy = 0, \quad y(2) = 0$.

16. $\cos y \, dx + (1 + e^{-x})\sin y \, dy = 0, \quad y(0) = \pi/4$.

In Exercises 17–24, verify that the equation is homogeneous, and then solve the equation.

17. $y' = \dfrac{x^2 + y^2}{2xy}$.

18. $y' = \dfrac{y^2}{xy + x^2}$.

19. $y' = \dfrac{x - y}{x + y}$.

20. $y' = \dfrac{x + y}{x - y}$.

21. $y' = \dfrac{x^2(e^y)^{1/x} + y^2}{xy}$.

22. $y' = \dfrac{x^2 + 3y^2}{4xy}$.

23. $y' = \dfrac{y}{x} + \sin\left(\dfrac{y}{x}\right)$.

24. $x \, dy = y\left[1 + \ln\left(\dfrac{y}{x}\right)\right] dx$.

In Exercises 25 and 26, find the particular solution determined by the side condition

25. $y' = \dfrac{y^3 - x^3}{xy^2}, \quad y(1) = 2$.

26. $x \sin\left(\dfrac{y}{x}\right) dy = \left[x + y \sin\left(\dfrac{y}{x}\right)\right] dx, \qquad y(1) = 0.$

27. (a) Given that

$$x'(t) = a_1 x - a_2 xy \quad \text{and} \quad y'(t) = -a_3 y + a_4 xy,$$

find a differential equation that links y to x independently of the parameter t.
HINT: $dy/dx = y'(t)/x'(t)$.
(b) Solve the equation obtained in (a).

28. Given that

$$x'(t) = -Ax \quad \text{and} \quad y'(t) = -Bxy,$$

(a) Express y as a function of x.
(b) Express x as a function of t with $x(0) = x_0$.
(c) Express y as a function of t with $y(0) = y_0$.

Orthogonal trajectories If two curves intersect at right angles, one with slope m_1 and the other with slope m_2, then $m_1 m_2 = -1$. A curve that intersects every member of a family of curves at right angles is called an *orthogonal trajectory* for that family of curves. A differential equation of the form

$$y' = f(x, y)$$

generates a family of curves: all the integral curves of that particular equation. The orthogonal trajectories of this family are the integral curves of the differential equation

$$y' = -\frac{1}{f(x, y)}.$$

The figure shows the family of parabolas $y = Cx^2$. The orthogonal trajectories of this family of parabolas are ellipses.

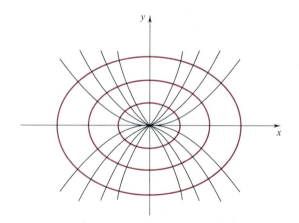

We can establish this by starting with the equation

$$y = Cx^2$$

and differentiating. Differentiation gives

$$y' = 2Cx.$$

Since $y = Cx^2$, we have $C = y/x^2$, and thus

$$y' = 2\left(\frac{y}{x^2}\right)x = \frac{2y}{x}.$$

The orthogonal trajectories are the integral curves of the differential equation

$$y' = -\frac{x}{2y}.$$

As you can check, the solutions of this equation are all of the form

$$x^2 + 2y^2 = K. \quad \square$$

In Exercises 29–34, find the orthogonal trajectories for the following families of curves. In each case draw several of the curves and several of the orthogonal trajectories.

29. $2x + 3y = C.$ **30.** $y = Cx.$

31. $xy = C.$ **32.** $y = Cx^3.$

33. $y = Ce^x.$ **34.** $x = Cy^4.$

▶ In Exercises 35 and 36, show that the given family is *self-orthogonal*. Use a graphing utility to graph at least four members of the family.

35. $y^2 = 4C(x + C).$ **36.** $\dfrac{x^2}{C^2} + \dfrac{y^2}{C^2 - 4} = 1.$

▶ **37.** A flu virus is spreading rapidly through a small town with a population of 25,000. The disease is spreading at a rate proportional to the product of the number of people who have it and the number who don't. Suppose that 100 people had the disease initially, and 400 had it after 10 days.
(a) How many people will have the disease at an arbitrary time t? How many will have it after 20 days?
(b) How long it will take for half the population to have the flu?
(c) Use a graphing utility to sketch the graph of your solution in part (a).

38. Consider the logistic equation (18.3.6). Show that dy/dt is increasing if $y < M/2$ and is decreasing if $y > M/2$. What can you conclude about dy/dt at the instant $y = M/2$? Explain.

▶ **39.** A rescue package whose mass is 100 kilograms is dropped from an airplane at a height of 4000 meters. As it falls, the air resistance is equal to twice its velocity. After 10 seconds, the package's parachute opens and the air resistance is now four times the square of its velocity.
(a) What is the package's velocity at the instant the parachute opens?
(b) Determine an expression for the velocity of the package at time t after the parachute opens.
(c) What is the terminal velocity of the package?

HINT: There are two differential equations that govern the package's velocity and position; one for the free-fall period and one for the period after the parachute opens.

40. It is known that m parts of chemical A combine with n parts of chemical B to produce a compound C. Suppose that the rate at which C is produced varies directly with the product of the amounts of A and B present at that instant. Find the amount of C produced in t minutes from an initial mixing of A_0 pounds of A with B_0 pounds of B, given that:
   (a) $n = m$, $A_0 = B_0$, and A_0 pounds of C are produced in the first minute.
   (b) $n = m$, $A_0 = \frac{1}{2}B_0$, and A_0 pounds of C are produced in the first minute.

(c) $n \neq m$, $A_0 = B_0$, and A_0 pounds of C are produced in the first minute.

HINT: Denote by $A(t)$, $B(t)$, and $C(t)$ the amounts of A, B, and C present at time t. Observe that $C'(t) = kA(t)B(t)$. Then note that

$$A_0 - A(t) = \frac{m}{m + n}C(t) \quad \text{and} \quad B_0 - B(t) = \frac{n}{m + n}C(t)$$

and thus

$$C'(t) = k\left[A_0 - \frac{m}{m + n}C(t)\right]\left[B_0 - \frac{n}{m + n}C(t)\right].$$

■ 18.4 EXACT EQUATIONS; INTEGRATING FACTORS

We continue to consider the first-order differential equation

(18.4.1)
$$P(x, y) + Q(x, y)y' = 0$$

and its equivalent differential form

(18.4.2)
$$P(x, y)\,dx + Q(x, y)\,dy = 0.$$

Throughout this section, we assume that the functions $P = P(x, y)$ and $Q = Q(x, y)$ are continuously differentiable on a simply connected region Ω in the xy-plane.

Equation (18.4.1) (or (18.4.2)) is said to be exact on Ω iff

(18.4.3)
$$\frac{\partial P}{\partial y}(x, y) = \frac{\partial Q}{\partial x}(x, y) \qquad \text{for all } (x, y) \in \Omega.$$

The reason for this terminology is as follows. If the equation

$$P(x, y) + Q(x, y)y' = 0$$

is exact, then (by Theorem 15.10.2) the vector-valued function

$$P(x, y)\mathbf{i} + Q(x, y)\mathbf{j}$$

is "exactly" in the form of a gradient and there is a function f defined on Ω such that

$$\frac{\partial f}{\partial x} = P \quad \text{and} \quad \frac{\partial f}{\partial y} = Q.$$

We can therefore write (18.4.1) as

(1)
$$\frac{\partial f}{\partial x}(x, y) + \frac{\partial f}{\partial y}(x, y)y' = 0.$$

Now, if we regard y as a differentiable function of x, then, by the chain rule,

$$\frac{d}{dx}[f(x, y)] = \frac{\partial f}{\partial x}(x, y) + \frac{\partial f}{\partial y}(x, y)y'$$

and equation (1) can be written

$$\frac{d}{dx}[f(x, y)] = 0.$$

Integrating with respect to x, we have

$$f(x, y) = C.$$

The level curves of f are thus the integral curves of equation (18.4.1).

A procedure for calculating f from P and Q is shown in the following example. This procedure was explained in detail in Section 15.10.

Example 1   Show that the differential equation

$$(xy^2 - x^3 + y) + (x^2y + x - 2y + 3)y' = 0$$

is exact on $\Omega = $ the xy-plane, and find a one-parameter family of solutions.

SOLUTION   The coefficients

$$P(x, y) = xy^2 - x^3 + y \qquad \text{and} \qquad Q(x, y) = (x^2y + x - 2y + 3)$$

are continuously differentiable on Ω, and for all (x, y) we have

$$\frac{\partial P}{\partial y} = 2xy + 1 = \frac{\partial Q}{\partial x}.$$

To find the integral curves

$$f(x, y) = C$$

we set

$$\frac{\partial f}{\partial x}(x, y) = xy^2 - x^3 + y \qquad \text{and} \qquad \frac{\partial f}{\partial y}(x, y) = x^2y + x - 2y + 3.$$

Integrating $\partial f/\partial x$ with respect to x, holding y constant, we have

$$f(x, y) = \tfrac{1}{2}x^2y^2 - \tfrac{1}{4}x^4 + xy + \phi(y),$$

where $\phi(y)$ is a function of y which is to be determined.   Differentiation with respect to y gives

$$\frac{\partial f}{\partial y}(x, y) = x^2y + x + \phi'(y).$$

It now follows that

$$x^2y + x - 2y + 3 = x^2y + x + \phi'(y),$$

and so

$$\phi'(y) = -2y + 3.$$

Thus,

$$\phi(y) = -y^2 + 3y, \qquad \text{(we omit the arbitrary constant)}$$

and

$$f(x, y) = \tfrac{1}{2}x^2y^2 - \tfrac{1}{4}x^4 + xy - y^2 + 3y.$$

Finally, a one-parameter family of solutions is

$$\tfrac{1}{2}x^2y^2 - \tfrac{1}{4}x^4 + xy - y^2 + 3y = C.$$

You can verify this by differentiating implicitly. ❏

If the equation

$$P(x, y) + Q(x, y)y' = 0$$

is not exact, it may be possible to find a function $\mu = \mu(x, y)$ not identically zero such that the equation

$$\mu(x, y)\, P(x, y) + \mu(x, y)\, Q(x, y)y' = 0$$

is exact. If μ is never zero or is zero only where P and Q are both zero, then any solution of this second equation gives a solution of the first equation. We call $\mu(x, y)$ an *integrating factor*.

Example 2 The equation

(∗)
$$(e^x - \sin y) + (\cos y)y' = 0$$

is not exact:

$$\frac{\partial}{\partial y}(e^x - \sin y) = -\cos y \quad \text{but} \quad \frac{\partial}{\partial x}(\cos y) = 0.$$

Multiplication by e^{-x} gives

(∗∗)
$$(1 - e^{-x}\sin y) + (e^{-x}\cos y)y' = 0.$$

This equation is exact:

$$\frac{\partial}{\partial y}(1 - e^{-x}\sin y) = -e^{-x}\cos y = \frac{\partial}{\partial x}(e^{-x}\cos y).$$

Thus $\mu(x, y) = e^{-x}$ is an integrating factor, and we can solve equation (∗) by solving (∗∗).
To solve (∗∗) we set

$$\frac{\partial f}{\partial x} = 1 - e^{-x}\sin y, \qquad \frac{\partial f}{\partial y} = e^{-x}\cos y.$$

Thus

$$f(x, y) = \int (1 - e^{-x}\sin y)\, dx = x + e^{-x}\sin y + \phi(y)$$

so that

$$\frac{\partial f}{\partial y} = e^{-x} \cos y + \phi'(y).$$

This gives $\phi'(y) = 0$ and shows that we can set $\phi(y) = 0$. The integral curves take the form

$$x + e^{-x} \sin y = C.$$

You can verify this by differentiation. ❏

Let's return to the general situation and write our equation in the form

(1) $$P + Qy' = 0.$$

How can we find an integrating factor if this equation is not exact?

Observe first of all that the equation

$$\mu P + \mu Q y' = 0$$

is exact iff

$$\frac{\partial}{\partial y}(\mu P) = \frac{\partial}{\partial x}(\mu Q)$$

and this occurs iff

(2) $$\mu \frac{\partial P}{\partial y} + \frac{\partial \mu}{\partial y} P = \mu \frac{\partial Q}{\partial x} + \frac{\partial \mu}{\partial x} Q.$$

Thus μ is an integrating factor for (1) iff μ satisfies (2).

In theory, all we have to do now to find an integrating factor for (1) is to solve equation (2) for μ. Unfortunately, Equation (2) is a partial differential equation which is, in general, much more difficult to solve than Equation (1). However, there are two special cases where an integrating factor can be determined.

(18.4.4)

> If $v = \dfrac{1}{Q}\left(\dfrac{\partial P}{\partial y} - \dfrac{\partial Q}{\partial x}\right)$ is a function of x only, then
>
> $\mu(x) = e^{\int v(x)\,dx}$ is an integrating factor for equation (1);
>
> if $w = \dfrac{1}{P}\left(\dfrac{\partial P}{\partial y} - \dfrac{\partial Q}{\partial x}\right)$ is a function of y only, then
>
> $\mu(y) = e^{\int -w(y)\,dy}$ is an integrating factor for equation (1).

PROOF We prove the first part and leave the proof of the second part as an exercise. Assume that

$$v = \frac{1}{Q}\left(\frac{\partial P}{\partial y} - \frac{\partial Q}{\partial x}\right)$$

is a function of x only and multiply (1) by $e^{\int v}$. This gives

$$P e^{\int v(x)\,dx} + Q e^{\int v(x)\,dx} y' = 0.$$

All we have to do is show that

$$\frac{\partial}{\partial y}\left(Pe^{\int v(x)\,dx}\right) = \frac{\partial}{\partial x}\left(Qe^{\int v(x)\,dx}\right).$$

This can be seen as follows:

$$\frac{\partial}{\partial x}\left(Qe^{\int v(x)\,dx}\right) = Qve^{\int v(x)\,dx} + \frac{\partial Q}{\partial x}e^{\int v(x)\,dx}$$

$$= \left[Qv + \frac{\partial Q}{\partial x}\right]e^{\int v(x)\,dx}$$

$$= \left[\left(\frac{\partial P}{\partial y} - \frac{\partial Q}{\partial x}\right) + \frac{\partial Q}{\partial x}\right]e^{\int v(x)\,dx}$$

$$= \frac{\partial P}{\partial y}e^{\int v(x)\,dx} = \frac{\partial}{\partial y}\left(Pe^{\int v(x)\,dx}\right). \quad ❏$$

Example 3 Earlier we showed that the equation

$$(e^x - \sin y) + (\cos y)y' = 0$$

is not exact, but that e^{-x} is an integrating factor. We can discover this integrating factor by using (18.4.4). In this case,

$$\frac{1}{Q}\left(\frac{\partial P}{\partial y} - \frac{\partial Q}{\partial x}\right) = \frac{1}{\cos y}(-\cos y - 0) = -1,$$

and

$$e^{\int v(x)\,dx} = e^{\int(-1)\,dx} = e^{-x}. \quad ❏$$

EXERCISES 18.4

In Exercises 1–10, verify that the equation is exact and solve the equation.

1. $(xy^2 - y) + (x^2y - x)y' = 0.$

2. $e^x \sin y + (e^x \cos y)y' = 0.$

3. $(e^y - ye^x) + (xe^y - e^x)y' = 0.$

4. $\sin y + (x \cos y + 1)y' = 0.$

5. $(\ln y + 2xy) + \left(\dfrac{x}{y} + x^2\right)y' = 0.$

6. $(2x \tan^{-1} y) + \left(\dfrac{x^2}{1 + y^2}\right)y' = 0.$

7. $\left(\dfrac{y}{x} + 6x\right)dx + (\ln x - 2)\,dy = 0.$

8. $e^x + \ln y + \dfrac{y}{x} + \left(\dfrac{x}{y} + \ln x + \sin y\right)y' = 0.$

9. $(y^3 - y^2 \sin x - x)\,dx + $
$\qquad\qquad (3xy^2 + 2y \cos x + e^{2y})\,dy = 0.$

10. $(e^{2y} - y \cos xy) + (2x\,e^{2y} - x \cos xy + 2y)\,y' = 0.$

11. Let p and q be functions of one variable everywhere continuously differentiable.
   (a) Is the equation $p(x) + q(y)y' = 0$ necessarily exact?
   (b) Show that the equation $p(y) + q(x)y' = 0$ is not necessarily exact. Then find an integrating factor.

12. Prove the second part of (18.4.4).

In Exercises 13–18, solve each of the following equations using an integrating factor if necessary.

13. $(e^{y-x} - y) + (xe^{y-x} - 1)y' = 0.$

14. $(x + e^y) - \frac{1}{2}x^2y' = 0.$

15. $(3x^2y^2 + x + e^y) + (2x^3y + y + xe^y)y' = 0.$

16. $(\sin 2x \cos y) - (\sin^2 x \sin y)y' = 0.$

17. $(y^3 + x + 1) + (3y^2)y' = 0.$

18. $(e^{2x+y} - 2y) + (xe^{2x+y} + 1)y' = 0.$

In Exercises 19–25, find the integral curve that passes through (x_0, y_0). Use an integrating factor if necessary.

19. $(x^2 + y) + (x + e^y)y' = 0;\quad (x_0, y_0) = (1, 0).$

20. $(3x^2 - 2xy + y^3) + (3xy^2 - x^2)y' = 0$;
    $(x_0, y_0) = (1, -1)$.

21. $(2y^2 + x^2 + 2) + (2xy)y' = 0$; $(x_0, y_0) = (1, 0)$.

22. $(x^2 + y) + (3x^2y^2 - x)y' = 0$; $(x_0, y_0) = (1, 1)$.

23. $y^3 + (1 + xy^2)y' = 0$; $(x_0, y_0) = (-2, -1)$.

24. $(x + y)^2 + (2xy + x^2 - 1)y' = 0$; $(x_0, y_0) = (1, 1)$.

25. $[\cosh(x - y^2) + e^{2x}]\, dx + y[1 - 2\cosh(x - y^2)]$
    $dy = 0$; $(x_0, y_0) = (2, \sqrt{2})$.

26. In Section 18.2 we solved the linear differential equation

    $$y' + p(x)y = q(x)$$

    by using the integrating factor

    $$e^{\int p(x)\, dx}.$$

    Show that this integrating factor is obtainable by the methods of this section.

27. (a) Find a value of k, if possible, such that the differential equation

    $$(xy^2 + k\, x^2y + x^3)\, dx + (x^3 + x^2y + y^2)\, dy = 0$$

    is exact.

    (b) Find a value of k, if possible, such that the differential equation

    $$ye^{2xy} + 2x + (k\, xe^{2xy} - 2y)y' = 0$$

    is exact.

28. (a) Find functions f and g, not both identically zero, such that the differential equation

    $$g(y)\sin x\, dx + y^2 f(x)\, dy = 0$$

    is exact.

    (b) Find all functions g such that the differential equation $g(y)\, e^y + xy\, y' = 0$ is exact.

In Exercises 29–34, solve the given differential equation by using a suitable method from Sections 18.2–18.4.

29. $y' = y^2x^3$.

30. $y' = \dfrac{y}{x + \sqrt{xy}}$.

31. $y' + \dfrac{4y}{x} = x^4$.

32. $y' + 2xy - 2x^3 = 0$.

33. $(y\, e^{xy} - 2x)\, dx + \left(\dfrac{2}{y} + xe^{xy}\right) dy = 0$.

34. $y\, dx + (2xy - e^{-2y})\, dy = 0$.

■ **18.5 THE EQUATION** $y'' + ay' + by = 0$

A differential equation of the form

$$y'' + ay' + by = \phi(x),$$

where a and b are real numbers and ϕ is a continuous function on some interval I, is a *second-order linear differential equation with constant coefficients.* By a solution of such an equation we mean a function $y = y(x)$ that satisfies the equation for all x in I. In this section we set $\phi = 0$ and consider the *reduced equation*

(18.5.1)
$$\boxed{y'' + ay' + by = 0} \;\dagger$$

Linear Combinations of Solutions; The Characteristic Equation

Observe, first of all, that *if y_1 and y_2 are both solutions of the reduced equation, then every linear combination*

$$C_1y_1 + C_2y_2$$

is also a solution.

---

† Often called the *homogeneous equation;* see Exercises 18.2.

PROOF Set

$$u = C_1 y_1 + C_2 y_2$$

and observe that

$$u' = C_1 y_1' + C_2 y_2' \quad \text{and} \quad u'' = C_1 y_1'' + C_2 y_2''.$$

If

$$y_1'' + ay_1' + by_1 = 0 \quad \text{and} \quad y_2'' + ay_2' + by_2 = 0,$$

then

$$\begin{aligned}
u'' + au' + bu &= (C_1 y_1'' + C_2 y_2'') + a(C_1 y_1' + C_2 y_2') + b(C_1 y_1 + C_2 y_2) \\
&= C_1(y_1'' + ay_1' + by_1) + C_2(y_2'' + ay_2' + by_2) \\
&= C_1(0) + C_2(0) = 0. \quad \square
\end{aligned}$$

Earlier you saw that the function $y = e^{-ax}$ satisfies the differential equation

$$y' + ay = 0.$$

This suggests that the differential equation

$$y'' + ay' + by = 0$$

might also have solutions of the form $y = e^{rx}$.

If $y = e^{rx}$, then

$$y' = re^{rx} \quad \text{and} \quad y'' = r^2 e^{rx}.$$

Substitution into the differential equation gives

$$r^2 e^{rx} + are^{rx} + be^{rx} = 0$$

and, since $e^{rx} \neq 0$,

$$r^2 + ar + b = 0.$$

This shows that *the function $y = e^{rx}$ satisfies the differential equation iff*

$$r^2 + ar + b = 0.$$

This quadratic equation in r is called the *characteristic equation* of (18.5.1).†
By the quadratic formula, the roots of the characteristic equation are

$$r = \frac{-a \pm \sqrt{a^2 - 4b}}{2}.$$

The nature of the solutions of the differential equation

$$y'' + ay' + by = 0$$

depends on the nature of these roots.

Case 1. $a^2 > 4b$

If $a^2 > 4b$, the characteristic equation has two distinct real roots:

$$r_1 = -\tfrac{1}{2}a + \tfrac{1}{2}\sqrt{a^2 - 4b}, \qquad r_2 = -\tfrac{1}{2}a - \tfrac{1}{2}\sqrt{a^2 - 4b}.$$

---

† Sometimes called the *auxiliary equation*.

Both

$$y_1 = e^{r_1 x} \qquad \text{and} \qquad y_2 = e^{r_2 x}$$

are solutions of the differential equation. ❑

Case 2. $a^2 = 4b$

If $a^2 = 4b$, the characteristic equation has only one root

$$r_1 = -\tfrac{1}{2}a.$$

In this case, the characteristic equation must be

$$(r - r_1)^2 = r^2 - 2r_1 r + r_1^2 = 0 \qquad \text{(verify this)}$$

and the corresponding differential equation is

(1) $$y'' - 2r_1 y' + r_1^2 y = 0.$$

We know that $y_1 = e^{r_1 x}$ is one solution of this equation; we need to find a second solution. To motivate our choice for the second solution, consider the special case $r_1 = 0$. Setting $r_1 = 0$ in equation (1), we get

$$y'' = 0,$$

which has

$$y = C_1 + C_2 x$$

as a two-parameter family of solutions. Since $e^{0x} = 1$, we can write this family of solutions in the form

$$y = C_1 e^{0x} + C_2 x e^{0x},$$

and $y_1 = e^{0x}$ and $y_2 = x e^{0x}$ are particular solutions. Thus, we are led to consider $y_2 = x e^{r_1 x}$ as a possible second solution of equation (1). The function y_2 and its first two derivatives are

$$y_2 = x e^{r_1 x}$$
$$y_2' = e^{r_1 x} + r_1 x e^{r_1 x}$$
$$y_2'' = 2r_1 e^{r_1 x} + r_1^2 x e^{r_1 x}.$$

Substituting y_2 and its derivatives into equation (1), we have

$$y_2'' - 2r_1 y_2' + r_1^2 y_2 =$$
$$(2 r_1 e^{r_1 x} + r_1^2 x e^{r_1 x}) - 2r_1(e^{r_1 x} + r_1 x e^{r_1 x}) + r_1^2 x e^{r_1 x} = 0,$$

and so $y_2(x) = x e^{r_1 x}$ is also a solution of (1). ❑

Case 3. $a^2 < 4b$

If $a^2 < 4b$, the characteristic equation has two complex conjugate roots:

$$r_1 = -\tfrac{1}{2}a + i\tfrac{1}{2}\sqrt{4b - a^2}, \qquad r_2 = -\tfrac{1}{2}a - i\tfrac{1}{2}\sqrt{4b - a^2} \qquad (i^2 = -1).$$

Setting

$$\alpha = -\tfrac{1}{2}a \qquad \text{and} \qquad \beta = \tfrac{1}{2}\sqrt{4b - a^2},$$

we can write

$$r_1 = \alpha + i\beta, \qquad r_2 = \alpha - i\beta.$$

The complex-valued functions

$$z_1 = e^{(\alpha + i\beta)x}, \qquad z_2 = e^{(\alpha - i\beta)x}$$

both satisfy the differential equation.

We can obtain real solutions by forming suitable linear combinations of z_1 and z_2. In Exercise 41 you are asked to verify Euler's formulas

$$e^{i\theta} = \cos\theta + i\sin\theta \qquad \text{and} \qquad e^{-i\theta} = \cos\theta - i\sin\theta.$$

Now, using these formulas, we have

$$z_1 = e^{\alpha x + i\beta x} = e^{\alpha x}e^{i\beta x} = e^{\alpha x}[\cos\beta x + i\sin\beta x]$$

and

$$z_2 = e^{\alpha x - i\beta x} = e^{\alpha x}e^{-i\beta x} = e^{\alpha x}[\cos\beta x - i\sin\beta x].$$

Adding and subtracting these equations, we have

$$z_1 + z_2 = 2e^{\alpha x}\cos\beta x, \qquad z_1 - z_2 = 2ie^{\alpha x}\sin\beta x$$

and thus

$$\frac{1}{2}(z_1 + z_2) = e^{\alpha x}\cos\beta x, \qquad \frac{1}{2i}(z_1 - z_2) = e^{\alpha x}\sin\beta x.$$

The functions

$$y_1 = e^{\alpha x}\cos\beta x, \qquad y_2 = e^{\alpha x}\sin\beta x$$

are real-valued and, being linear combinations of solutions, are themselves solutions. ❑

Remark In our manipulations of complex exponential functions we have implicitly assumed that the basic properties of real exponents (for example, $e^a e^b = e^{a+b}$) carry over to complex exponents. This is indeed true, but the treatment needed to verify these properties is beyond the scope of our discussion here. ❑

Existence and Uniqueness of Solutions; Wronskians, Fundamental Solutions

You have seen how to obtain solutions of the differential equation

$$y'' + ay' + by = 0$$

from the characteristic equation

$$r^2 + ar + b = 0.$$

We can form more solutions by taking linear combinations of these solutions. Question: Are there still other solutions or do all solutions arise in this manner? Answer: All solutions of the reduced equation are linear combinations of the solutions that we have already found.

To show this, we have to go a little deeper into the theory. Our point of departure is a result proved in more advanced texts.

THEOREM 18.5.2

Let x_0, y_0, y_1 be arbitrary real numbers. The reduced equation

$$y'' + ay' + by = 0$$

has a unique solution $y = y(x)$ that satisfies the side conditions

$$y(x_0) = y_0, \qquad y'(x_0) = y_1.$$

Geometrically, the theorem says that there is one and only one solution whose graph passes through a prescribed point (x_0, y_0) with a prescribed slope y_1. We assume this result and go on from there.

DEFINITION 18.5.3

Let y_1 and y_2 be two solutions of

$$y'' + ay' + by = 0.$$

The *Wronskian*† of y_1 and y_2 is the function W defined for all real x by

$$W(x) = y_1(x)y_2'(x) - y_2(x)y_1'(x).$$

It follows from this definition that the Wronskian can be written as the 2×2 determinant (see Appendix A.2).

$$W(x) = \begin{vmatrix} y_1(x) & y_2(x) \\ y_1'(x) & y_2'(x) \end{vmatrix}.$$

THEOREM 18.5.4

If both y_1 and y_2 are solutions of the reduced equation

$$y'' + ay' + by = 0,$$

then their Wronskian W is either identically zero or never zero.

PROOF Assume that both y_1 and y_2 are solutions of the reduced equation and set

$$W = y_1 y_2' - y_1' y_2.$$

Differentiation gives

$$W' = y_1 y_2'' + y_1' y_2' - y_1' y_2' - y_1'' y_2 = y_1 y_2'' - y_1'' y_2.$$

---

† Named in honor of Count Hoëne Wronski, a Polish mathematician (1778–1853).

Since y_1 and y_2 are solutions, we know that

$$y_1'' + ay_1' + by_1 = 0$$

and

$$y_2'' + ay_2' + by_2 = 0.$$

Multiplying the first equation by $-y_2$ and the second equation by y_1, we have

$$-y_1''y_2 - ay_1'y_2 - by_1y_2 = 0$$

$$y_2''y_1 + ay_2'y_1 + by_2y_1 = 0.$$

We now add these two equations and obtain

$$(y_1y_2'' - y_2y_1'') + a(y_1y_2' - y_2y_1') = 0.$$

In terms of the Wronskian, we have

$$W' + aW = 0.$$

This is a first-order linear differential equation with general solution

$$W = Ce^{-ax}.$$

If $C = 0$, then W is identically 0; if $C \neq 0$, then W is never zero. ❏

Two solutions y_1, y_2 of the reduced equation

$$y'' + ay' + by = 0$$

are said to be *fundamental* iff their Wronskian is not identically zero (in which case, by Theorem 18.5.4, the Wronskian is never zero).

THEOREM 18.5.5

Every solution of the reduced equation

$$y'' + ay' + by = 0$$

can be expressed in a unique manner as the linear combination of any two fundamental solutions.

PROOF Let u be any solution of the reduced equation and let y_1, y_2 be any pair of fundamental solutions. Choose a number x_0 and form the equations

(1)
$$C_1y_1(x_0) + C_2y_2(x_0) = u(x_0)$$

$$C_1y_1'(x_0) + C_2y_2'(x_0) = u'(x_0).$$

Since y_1, y_2 are fundamental solutions, the Wronskian at x_0,

$$W(x_0) = y_1(x_0)y_2'(x_0) - y_1'(x_0)y_2(x_0),$$

is different from zero. This guarantees that equations (1) can be solved for C_1, C_2. By elementary algebra

$$C_1 = \frac{u(x_0)y_2'(x_0) - u'(x_0)y_2(x_0)}{y_1(x_0)y_2'(x_0) - y_1'(x_0)y_2(x_0)}, \qquad C_2 = \frac{y_1(x_0)u'(x_0) - y_1'(x_0)u(x_0)}{y_1(x_0)y_2'(x_0) - y_1'(x_0)y_2(x_0)},$$

The denominator in both cases is $W(x_0)$, which, since y_1 and y_2 are fundamental, is different from zero.

Our work is finished. The function $C_1y_1 + C_2y_2$ is a solution of the reduced equation which by (1) has the same value as u at x_0 and the same derivative. Thus, by the uniqueness part of Theorem 18.5.2, $C_1y_1 + C_2y_2$ and u cannot be different functions. It shows that

$$u = C_1y_1 + C_2y_2$$

and proves the theorem. ❏

The arbitrary linear combination

$$y = C_1y_1 + C_2y_2$$

of any two fundamental solutions is the *general solution*. By Theorem 18.5.5 we can obtain any *particular solution* by adjusting C_1 and C_2.

We now return to the solutions obtained earlier and prove the final result.

THEOREM 18.5.6

Given the reduced equation

$$(1) \qquad\qquad y'' + ay' + by = 0,$$

we form the characteristic equation

$$r^2 + ar + b = 0.$$

I. If the characteristic equation has two distinct real roots r_1 and r_2, then the general solution of (1) takes the form

$$y = C_1e^{r_1 x} + C_2e^{r_2 x}.$$

II. If the characteristic equation has only one real root r_1, then the general solution takes the form

$$y = C_1e^{r_1 x} + C_2xe^{r_1 x} = (C_1 + C_2x)e^{r_1 x}.$$

III. If the characteristic equation has two complex roots

$$r_1 = \alpha + i\beta \qquad \text{and} \qquad r_2 = \alpha - i\beta, \quad \beta \neq 0,$$

then the general solution takes the form

$$y = C_1e^{\alpha x} \cos \beta x + C_2e^{\alpha x} \sin \beta x = e^{\alpha x}(C_1 \cos \beta x + C_2 \sin \beta x).$$

PROOF To prove this theorem it is enough to show that the three solution pairs

$$e^{r_1 x}, e^{r_2 x} \qquad e^{r_1 x}, xe^{r_1 x} \qquad e^{\alpha x} \cos \beta x, e^{\alpha x} \sin \beta x$$

all have nonzero Wronskians. The Wronskian of the first pair is the function

$$W(x) = e^{r_1 x}\frac{d}{dx}(e^{r_2 x}) - \frac{d}{dx}(e^{r_1 x})e^{r_2 x}$$

$$= e^{r_1 x}r_2e^{r_2 x} - r_1e^{r_1 x}e^{r_2 x} = (r_2 - r_1)e^{(r_1 + r_2)x}.$$

$W(x)$ is different from zero since by assumption $r_2 \neq r_1$.

We leave it to you to verify that the other two pairs also have nonzero Wronskians. ❏

It's time to look at specific examples.

Example 1   Find the general solution of the equation
$$y'' + 2y' - 15y = 0.$$
Then find the particular solution that satisfies the side conditions
$$y(0) = 0, \qquad y'(0) = -1.$$

SOLUTION   The characteristic equation is the quadratic
$$r^2 + 2r - 15 = 0.$$
Factoring the left side, we have
$$(r + 5)(r - 3) = 0.$$
There are two real roots: -5 and 3.   The general solution takes the form
$$y = C_1 e^{-5x} + C_2 e^{3x}.$$

Differentiating the general solution, we have
$$y' = -5C_1 e^{-5x} + 3C_2 e^{3x}.$$
The conditions
$$y(0) = 0, \qquad y'(0) = -1$$
are satisfied iff
$$C_1 + C_2 = 0 \qquad \text{and} \qquad -5C_1 + 3C_2 = -1.$$
Solving these two equations simultaneously, we find that
$$C_1 = \tfrac{1}{8}, \qquad C_2 = -\tfrac{1}{8}.$$
The solution that satisfies the prescribed side conditions is the function
$$y = \tfrac{1}{8} e^{-5x} - \tfrac{1}{8} e^{3x}.$$
You should verify this.   ❑

Example 2   Find the general solution of the equation
$$y'' + 4y' + 4y = 0.$$
Then find the particular solution that satisfies the side conditions
$$y(1) = 1, \qquad y'(1) = 2.$$

SOLUTION   The characteristic equation takes the form
$$r^2 + 4r + 4 = 0,$$
which can be written
$$(r + 2)^2 = 0.$$
The number -2 is the only root.   The general solution can be written
$$y = C_1 e^{-2x} + C_2 x e^{-2x}.$$

Differentiating the general solution, we have
$$y' = -2C_1 e^{-2x} - 2C_2 x e^{-2x} + C_2 e^{-2x} = -2C_1 e^{-2x} + C_2(1 - 2x)e^{-2x}.$$

The conditions

$$y(1) = 1, \qquad y'(1) = 2$$

require that

$$C_1 e^{-2} + C_2 e^{-2} = 1 \qquad \text{and} \qquad -2C_1 e^{-2} - C_2 e^{-2} = 2.$$

Solve these equations simultaneously for C_1, C_2 and you will find that

$$C_1 = -3e^2, \qquad C_2 = 4e^2.$$

The solution that satisfies the prescribed side conditions is the function

$$y = -3e^2 e^{-2x} + 4e^2 x e^{-2x} = -3e^{2(1-x)} + 4x e^{2(1-x)}.$$

We leave the checking to you. ❏

Example 3 Find the general solution of the equation

$$y'' + y' + 3y = 0.$$

SOLUTION The characteristic equation is

$$r^2 + r + 3 = 0.$$

The quadratic formula shows that there are two complex roots:

$$r_1 = -\tfrac{1}{2} + i\tfrac{1}{2}\sqrt{11}, \qquad r_2 = -\tfrac{1}{2} - i\tfrac{1}{2}\sqrt{11}.$$

The general solution takes the form

$$y = e^{-\frac{1}{2}x}[C_1 \cos(\tfrac{1}{2}\sqrt{11}\,x) + C_2 \sin(\tfrac{1}{2}\sqrt{11}\,x)] \quad ❏$$

Example 4 Find the general solution of the equation

$$y'' + \omega^2 y = 0, \qquad (\omega \neq 0).$$

SOLUTION The characteristic equation is

$$r^2 + \omega^2 = 0$$

and the roots are

$$r_1 = \omega i, \qquad r_2 = -\omega i.$$

Thus, the general solution is

$$y = C_1 \cos \omega x + C_2 \sin \omega x. \quad ❏$$

Remark The equation in Example 4 describes the oscillatory motion of an object suspended by a spring under the assumption that there are no forces acting on the spring–mass system other than the restoring force of the spring. This application and some generalizations are considered in Section 18.7. In the Exercises you are asked to show that the general solution can also be written as

$$y = A \sin(\omega x + \phi_0),$$

where A and ϕ_0 are constants with $A > 0$ and $\phi_0 \in [0, 2\pi)$.

EXERCISES 18.5

In Exercises 1–18, find the general solution of the given equation.

1. $y'' + 2y' - 15y = 0.$

2. $y'' - 13y' + 42y = 0.$

3. $y'' + 8y' + 16y = 0.$

4. $y'' + 7y' + 3y = 0.$

5. $y'' + 2y' + 5y = 0.$

6. $y'' - 3y' + 8y = 0.$

7. $2y'' + 5y' - 3y = 0.$

8. $y'' - 12y = 0.$

9. $y'' + 12y = 0.$

10. $y'' - 3y' + \frac{9}{4}y = 0.$

11. $5y'' + \frac{11}{4}y' - \frac{3}{4}y = 0.$

12. $2y'' + 3y' = 0.$

13. $y'' + 9y = 0.$

14. $y'' - y' - 30y = 0.$

15. $2y'' + 2y' + y = 0.$

16. $y'' - 4y' + 4y = 0.$

17. $8y'' + 2y' - y = 0.$

18. $5y'' - 2y' + y = 0.$

In Exercises 19–24, solve the given initial value problems.

19. $y'' - 5y' + 6y = 0, \quad y(0) = 1, \quad y'(0) = 1.$

20. $y'' + 2y' + y = 0, \quad y(2) = 1, \quad y'(2) = 2.$

21. $y'' + \frac{1}{4}y = 0, \quad y(\pi) = 1, \quad y'(\pi) = -1.$

22. $y'' - 2y' + 2y = 0, \quad y(0) = -1, \quad y'(0) = -1.$

23. $y'' + 4y' + 4y = 0, \quad y(-1) = 2, \quad y'(-1) = 1.$

24. $y'' - 2y' + 5y = 0, \quad y(\pi/2) = 0, \quad y'(\pi/2) = 2.$

25. Find all solutions of the equation $y'' - y' - 2y = 0$ that satisfy the given side conditions:
(a) $y(0) = 1.$ (b) $y'(0) = 1.$
(c) $y(0) = 1, \quad y'(0) = 1.$

26. Prove that the general solution of the differential equation

$$y'' - \omega^2 y = 0 \qquad (\omega > 0)$$

can be written

$$y = C_1 \cosh \omega x + C_2 \sinh \omega x.$$

27. Suppose that the roots r_1 and r_2 of the characteristic equation of (18.5.1) are real and distinct. Then they can be written as $r_1 = \alpha + \beta$ and $r_2 = \alpha - \beta$, where α and β are real. Show that the general solution of the equation (18.5.1) can be expressed in the form

$$y = e^{\alpha x}(C_1 \cosh \beta x + C_2 \sinh \beta x).$$

28. Show that the general solution of the differential equation

$$y'' + \omega^2 y = 0$$

can be written

$$y = A \sin(\omega x + \phi_0)$$

where A and ϕ_0 are constants with $A > 0$ and $\phi_0 \in [0, 2\pi)$.

29. Complete the proof of Theorem 18.5.6 by showing that the following solutions are fundamental:
(a) $y_1 = e^{r_1 x}, \quad y_2 = xe^{r_1 x}.$ (one root case)
(b) $y_1 = e^{\alpha x}\cos \beta x, \quad y_2 = e^{\alpha x}\sin \beta x.$ (complex root case)

30. In the absence of any external electromotive force, the current i in a simple electrical circuit varies with time t according to the formula

$$L\frac{d^2 i}{dt^2} + R\frac{di}{dt} + \frac{1}{C}i = 0. \qquad (L, R, C \text{ constants})†$$

Find the general solution of this equation given that $L = 1, R = 10^3$, and
(a) $C = 5 \times 10^{-6}.$ (b) $C = 4 \times 10^{-6}.$
(c) $C = 2 \times 10^{-6}.$

31. Determine a differential equation $y'' + ay' + by = 0$ that has the given functions as solutions:
(a) $y_1 = e^{2x}, \quad y_2 = e^{-4x}.$
(b) $y_1 = 3e^{-x}, \quad y_2 = 4e^{5x}.$
(c) $y_1 = 2e^{3x}, \quad y_2 = xe^{3x}.$

32. Determine a differential equation $y'' + ay' + by = 0$ that has the given functions as solutions:
(a) $y_1 = 2\cos 2x, \quad y_2 = -\sin 2x.$
(b) $y_1 = e^{-2x}\cos 3x, \quad y_2 = 2e^{-2x}\sin 3x.$

33. Let y_1, y_2 be two solutions of the reduced equation with y_2 never zero. Show that the Wronskian of y_1, y_2 is zero iff y_1 is a constant multiple of y_2.

Exercises 34 and 35 concern the differential equation $y'' + ay' + by = 0$ where a and b are nonnegative constants.

34. Prove that if a and b are both positive, then $y(x) \to 0$ as $x \to \infty$ for all solutions y of the equation.

35. (a) Prove that if $a = 0$ and $b > 0$, then all solutions of the equation are bounded.
(b) Suppose that $a > 0$ and $b = 0$, and let $y = y(x)$ be a solution of the equation. Prove that

$$\lim_{x \to \infty} y(x) = k$$

for some constant k. Determine k for the solution that satisfies the side conditions: $y(0) = y_0, \quad y'(0) = y_1.$

Euler Equations   An equation of the form

$$(*) \qquad x^2 y'' + \alpha xy' + \beta y = 0,$$

where α and β are real numbers, is called an *Euler equation*.

36. Show that the Euler equation $(*)$ can be transformed into an equation of the form

$$\frac{d^2 y}{dz^2} + a\frac{dy}{dz} + by = 0,$$

---

† L is inductance, R is resistance, and C is capacitance. If L is given in henrys, R in ohms, C in farads, and t in seconds, then the current i is given in amperes.

where a and b are real numbers, by means of the change of variable $z = \ln x$.   HINT: If $z = \ln x$, then, by the chain rule,

$$\frac{dy}{dx} = \frac{dy}{dz}\frac{dz}{dx} = \frac{dy}{dz}\frac{1}{x}.$$

Now calculate $\dfrac{d^2y}{dx^2}$ and substitute into the differential equation.

In Exercises 37–40, use the change of variable indicated in Exercise 36 to transform the given equation into an equation with constant coefficients. Find the general solution of that equation, and then express it in terms of x.

37. $x^2y'' - xy' - 8y = 0.$ **38.** $x^2y'' - 2xy' + 2y = 0.$
39. $x^2y'' - 3xy' + 4y = 0.$ **40.** $x^2y'' - xy' + 5y = 0.$

41. (a) Expand the function $f(x) = e^x$ in a Taylor series in powers of x (see Section 11.6).
  (b) Assuming that the manipulations can be justified, replace x in the series for e^x by $i\theta$, where $i^2 = -1$, and derive Euler's formula

$$e^{i\theta} = \cos\theta + i\sin\theta.$$

  (c) Now replace x by $-i\theta$ and conclude that

$$e^{-i\theta} = \cos\theta - i\sin\theta.$$

■ 18.6 THE EQUATION $y'' + ay' + by = \phi(x)$

In Section 18.5, we solved the reduced equation

$$y'' + ay' + by = 0.$$

In this section we go on to the *complete equation*†

(18.6.1)

$$\boxed{y'' + ay' + by = \phi(x).}$$

The function $\phi = \phi(x)$ appearing on the right side of this equation is often called the *nonhomogeneous term* or the *forcing function*. We assume that ϕ is continuous on some interval I.
    We begin by proving three simple but important results.

(18.6.2)

> If y_1 and y_2 are solutions of the complete equation, then their difference $u = y_1 - y_2$ is a solution of the reduced equation.

PROOF   Suppose the y_1 and y_2 are solutions of the complete equation.   Then

$$y_1'' + ay_1' + by_1 = \phi(x) \qquad \text{and} \qquad y_2'' + ay_2' + by_2 = \phi(x).$$

Now let $u = y_1 - y_2$.   Then we have

$$
\begin{aligned}
u'' + au' + bu &= (y_1'' - y_2'') + a(y_1' - y_2') + b(y_1 - y_2) \\
&= (y_1'' + ay_1' + by_1) - (y_2'' + ay_2' + by_2) \\
&= \phi(x) - \phi(x) = 0. \quad \square
\end{aligned}
$$

(18.6.3)

> If y_p is a particular solution of the complete equation, then every solution of the complete equation can be written as a solution of the reduced equation plus y_p.

_____
† Sometimes called the *nonhomogeneous equation*.

PROOF Let y_p be a particular solution of the complete equation. If y is also a solution of the complete equation, then, by (18.6.2), $y - y_p$ is a solution of the reduced equation. Obviously

$$y = (y - y_p) + y_p. \quad \square$$

It follows from (18.6.3) that we can obtain *the general solution* of the complete equation by starting with the general solution of the reduced equation and adding a particular solution of the complete equation. The general solution of the complete equation can thus be written

(18.6.4)
$$\boxed{y = C_1 u_1 + C_2 u_2 + y_p,}$$

where u_1, u_2 are any two fundamental solutions of the reduced equation and y_p is any particular solution of the complete equation.

The third result is called *the superposition principle*. It allows us to reduce a given complete equation to a set of simpler complete equations whose forcing functions have only one term.

(18.6.5)

> Given the complete equation
>
> (∗) $\quad y'' + ay' + by = \phi_1(x) + \phi_2(x)$.
>
> If y_1 is a solution of
>
> $\quad y'' + ay' + by = \phi_1(x)$
>
> and y_2 is a solution of
>
> $\quad y'' + ay' + by = \phi_2(x)$,
>
> then $y_p = y_1 + y_2$ is a solution of (∗).

The proof of the superposition principle is left as an exercise.

We now describe two methods for finding a particular solution of the complete equation.

Variation of Parameters

The method that we present here gives (at least in theory) particular solutions to all equations of the form

(1)
$$y'' + ay' + by = \phi(x).$$

The general solution of the reduced equation

$$y'' + ay' + by = 0$$

can be written

$$y = C_1 u_1(x) + C_2 u_2(x),$$

where u_1, u_2 are any fundamental solutions and the coefficients C_1 and C_2 are arbitrary constants. In the method called *variation of parameters* we replace the constants C_1, C_2 by functions

$$z_1 = z_1(x), \qquad z_2 = z_2(x),$$

which are to be determined so that

(2) $$y_p = z_1 u_1 + z_2 u_2$$

is a solution of Equation (1). Since we have two "unknowns," $z_1(x)$ and $z_2(x)$, we can impose two conditions on them. One condition is that (2) satisfies (1). The second condition is still at our disposal, and we shall choose it in a way that will simplify our calculations.

Differentiating (2), we have

$$y'_p = z_1 u'_1 + z'_1 u_1 + z_2 u'_2 + z'_2 u_2 = (z_1 u'_1 + z_2 u'_2) + (z'_1 u_1 + z'_2 u_2).$$

We now impose the second condition: we require that

(3) $$z'_1 u_1 + z'_2 u_2 = 0.$$

With this additional restriction, we have

$$y'_p = z_1 u'_1 + z_2 u'_2,$$

and, differentiating once more,

$$y''_p = z_1 u''_1 + z'_1 u'_1 + z_2 u''_2 + z'_2 u'_2.$$

Substituting y_p and its derivatives into the differential equation (1), we have

$$y''_p + ay'_p + by_p = (z_1 u''_1 + z'_1 u'_1 + z_2 u''_2 + z'_2 u'_2) + a(z_1 u'_1 + z_2 u'_2) + b(z_1 u_1 + z_2 u_2)$$
$$= \phi(x).$$

Rearranging the terms, we obtain

$$(u''_1 + au'_1 + bu_1)z_1 + (u''_2 + au'_2 + bu_2)z_2 + (z'_1 u'_1 + z'_2 u'_2) = \phi(x).$$

Since u_1 and u_2 are solutions of the reduced equation, this equation reduces to

(4) $$z'_1 u'_1 + z'_2 u'_2 = \phi(x).$$

Equations (3) and (4) can now be solved simultaneously for z'_1 and z'_2. As you can verify yourself, the unique solutions are

(5) $$z'_1 = -\frac{u_2 \phi}{W} \qquad \text{and} \qquad z'_2 = \frac{u_1 \phi}{W},$$

where the denominator W is the Wronskian of u_1, u_2. The functions z_1, z_2 are now determined up to an additive constant:

$$z_1 = -\int \frac{u_2 \phi}{W} \, dx, \qquad z_2 = \int \frac{u_1 \phi}{W} \, dx.$$

It follows then that the function

(18.6.6) $$y_p(x) = \left(-\int \frac{u_2(x)\phi(x)}{W} \, dx \right) u_1(x) + \left(\int \frac{u_1(x)\phi(x)}{W} \, dx \right) u_2(x)$$

is a particular solution of the complete equation

$$y'' + ay' + by = \phi(x). \quad \square$$

Example 1 Use variation of parameters to find a solution of the equation

$$y'' + y = \tan x, \qquad -\frac{1}{2}\pi < x < \frac{1}{2}\pi.$$

Also give the general solution of this equation.

SOLUTION The reduced equation

$$y'' + y = 0$$

has fundamental solutions

$$u_1 = \cos x, \qquad u_2 = \sin x.$$

In this instance the Wronskian W is 1:

$$W = u_1 u_2' - u_1' u_2 = (\cos x)(\cos x) - (-\sin x)\sin x = \cos^2 x + \sin^2 x = 1.$$

Since $\phi(x) = \tan x$, we can set

$$\begin{aligned}
z_1 &= -\int \frac{u_2 \phi}{W}\, dx \\
&= -\int \frac{\sin x \tan x}{1}\, dx \\
&= -\int \frac{\sin^2 x}{\cos x}\, dx \\
&= \int \frac{\cos x^2 - 1}{\cos x}\, dx = \int (\cos x - \sec x)\, dx = \sin x - \ln|\sec x + \tan x|
\end{aligned}$$

and

$$z_2 = \int \frac{u_1 \phi}{W}\, dx = \int \frac{\cos x \tan x}{1}\, dx = \int \sin x\, dx = -\cos x.$$

By (18.6.6) the function

$$\begin{aligned}
y_p &= (\sin x - \ln|\sec x + \tan x|)\cos x + (-\cos x)\sin x \\
&= -(\ln|\sec x + \tan x|)\cos x
\end{aligned}$$

is a particular solution of the complete equation.

The general solution of the complete equation is

$$y = C_1 \cos x + C_2 \sin x - (\ln|\sec x + \tan x|)\cos x \quad \square$$

Example 2 Find the general solution of

$$y'' - 5y' + 6y = 4e^{2x}.$$

SOLUTION The reduced equation

$$y'' - 5y' + 6y = 0$$

has characteristic equation

$$r^2 - 5r + 6 = (r - 2)(r - 3) = 0.$$

Thus, $u_1(x) = e^{2x}$, $u_2(x) = e^{3x}$ are fundamental solutions. Their Wronskian W is given by

$$W = u_1u_2' - u_2u_1' = e^{2x}3e^{3x} - e^{3x}2e^{2x} = e^{5x}.$$

Since $\phi(x) = 4e^{2x}$, we have

$$z_1 = -\int \frac{u_2\phi}{W}\,dx = -\int \frac{e^{3x}\,4e^{2x}}{e^{5x}}\,dx = -\int 4\,dx = -4x$$

and

$$z_2 = \int \frac{u_1\phi}{W}\,dx = \int \frac{e^{2x}\,4e^{2x}}{e^{5x}}\,dx = \int 4e^{-x}\,dx = -4e^{-x}.$$

Therefore, by (18.6.6),

$$y_p = -4xe^{2x} - 4e^{-x}e^{3x} = -4xe^{2x} - 4e^{2x}$$

is a particular solution of the complete equation.

The general solution of the complete equation is given by

$$\begin{aligned}
y &= A_1u_1 + A_2u_2 + y_p = A_1e^{2x} + A_2e^{3x} - 4xe^{2x} - 4e^{2x} \\
&= (A_1 - 4)e^{2x} + A_2e^{3x} - 4xe^{2x} \\
&= C_1e^{2x} + C_2e^{3x} - 4xe^{2x}, \qquad (C_1 = A_1 - 4, C_2 = A_2). \quad \square
\end{aligned}$$

The Method of Undetermined Coefficients

In contrast to variation of parameters, the method of undetermined coefficients only works for equations

(1) $$y'' + ay' + by = \phi(x),$$

where the forcing function ϕ has a special form. To motivate the method, consider the expression on the left side of (1):

(2) $$y'' + ay' + by.$$

If we calculate (2) for an exponential function $y = Ae^{rx}$, we have

$$y = Ae^{rx}, \qquad y' = Are^{rx}, \qquad y'' = Ar^2e^{rx}$$

and

$$\begin{aligned}
y'' + ay' + by &= Ar^2e^{rx} + aAre^{rx} + bAe^{rx} = (Ar^2 + aAr + bA)e^{rx} \\
&= Ke^{rx}, \qquad \text{where } K = Ar^2 + aAr + bA.
\end{aligned}$$

Thus, the expression (2) "transforms" $y = Ae^{rx}$ into a constant multiple of itself.

Similarly, if we calculate (2) for $y = A\cos\beta x$, $y = A\sin\beta x$, or $y = A\cos\beta x + B\sin\beta x$, we will get

$$y'' + ay' + by = K\cos\beta x + M\sin\beta x,$$

where K and M are constants which are given in terms of A, B, β, a, and b. For example, if we let $y = A\cos\beta x + B\sin\beta x$ and substitute into (2), we will get

$$\begin{aligned}
y'' + ay' + by &= (-A\beta^2 + aB\beta + bA)\cos\beta x + (-B\beta^2 - aA\beta + bB)\sin\beta x \\
&= K\cos\beta x + M\sin\beta x,
\end{aligned}$$

where $K = -A\beta^2 + aB\beta + bA$ and $M = -B\beta^2 - aA\beta + bB$.

Finally, if we substitute $y = Ae^{\alpha x} \cos \beta x$, $y = Ae^{\alpha x} \sin \beta x$, or $y = Ae^{\alpha x} \cos \beta x + Be^{\alpha x} \sin \beta x$ into (2), we will obtain

$$y'' + ay' + by = Ke^{\alpha x} \cos \beta x + Me^{\alpha x} \sin \beta x,$$

where K and M are constants which are given in terms of A, B, α, β, a, and b.

The point of these observations is that if the forcing function ϕ in Equation (1) has one of the forms

$$\phi(x) = \begin{cases} ce^{rx}, \\ c \cos \beta x, \quad c \sin \beta x, \quad \text{or} \quad c_1 \cos \beta x + c_2 \sin \beta x \\ ce^{\alpha x} \cos \beta x, \quad ce^{\alpha x} \sin \beta x, \quad \text{or} \quad c_1 e^{\alpha x} \cos \beta x + c_2 e^{\alpha x} \sin \beta x, \end{cases}$$

then we can assume that a particular solution of equation (1) will have the corresponding form

$$y_p = \begin{cases} Ae^{rx}, \\ A \cos \beta x + B \sin \beta x, \\ Ae^{\alpha x} \cos \beta x + Be^{\alpha x} \sin \beta x, \end{cases}$$

where the coefficient A, or the coefficients A and B, are to be determined. We then substitute y_p into equation (1) and solve for the *undetermined coefficients*. The following example illustrates the method.

Example 3 Use the method of undetermined coefficients to find a particular solution of each of the following differential equations:

(a) $y'' + 2y' + 5y = 10e^{-2x}$.   **(b)** $y'' + 2y' + y = 10 \cos 3x + 6 \sin 3x$.

SOLUTION **(a)** We assume that the equation has a solution of the form $y_p = Ae^{-2x}$. Then

$$y_p' = -2Ae^{-2x} \quad \text{and} \quad y_p'' = 4Ae^{-2x}.$$

Substituting y_p and its derivatives into the equation, we have

$$4Ae^{-2x} + 2(-2Ae^{-2x}) + 5Ae^{-2x} = 10e^{-2x}$$

$$5Ae^{-2x} = 10e^{-2x}.$$

Therefore,

$$5A = 10 \quad \text{and} \quad A = 2.$$

Thus, a particular solution of the equation is $y_p = 2e^{-2x}$. You can verify that the general solution of the complete equation is

$$y = C_1 e^{-x} \cos 2x + C_2 e^{-x} \sin 2x + 2e^{-2x}.$$

(b) We assume that the equation has a solution of the form $y_p = A \cos 3x + B \sin 3x$. Then

$$y_p' = -3A \sin 3x + 3B \cos 3x \quad \text{and} \quad y_p'' = -9A \cos 3x - 9B \sin 3x.$$

Substituting y_p and its derivatives into the equation, we have

$$(-9A \cos 3x - 9B \sin 3x) + 2(-3A \sin 3x + 3B \cos 3x)$$
$$+ (A \cos 3x + B \sin 3x) = 10 \cos 3x + 6 \sin 3x$$

and

$$(-8A + 6B) \cos 3x + (-6A - 8B)\sin 3x = 10 \cos 3x + 6 \sin 3x.$$

This equation will be satisfied for all x iff

$$\begin{array}{ll} -8A + 6B = 10 & \quad -4A + 3B = 5 \\ -6A - 8B = 6 & \quad \text{or} \quad 3A + 4B = -3. \end{array}$$

The solution of this pair of equations is $A = -\frac{29}{25}$, $B = \frac{3}{25}$. Thus

$$y_p = -\frac{29}{25}\cos 3x + \frac{3}{25}\sin 3x$$

is a particular solution. You can verify that

$$y = C_1 e^{-x} + C_2 x e^{-x} - \frac{29}{25}\cos 3x + \frac{3}{25}\sin 3x$$

is the general solution of the complete equation. ❏

There is a difficulty that can arise in trying to use the method of undetermined coefficients. Consider the differential equation in Example 2:

$$y'' - 5y' + 6y = 4e^{2x}.$$

To use the method of undetermined coefficients, we would assume that a particular solution has the form $y = Ae^{2x}$ since $\phi(x) = 4e^{2x}$. Substituting y and its derivatives

$$y' = 2Ae^{2x}, \qquad y'' = 4Ae^{2x}$$

into the equation, we have

$$4Ae^{2x} - 5(2Ae^{2x}) + 6Ae^{2x} = 4e^{2x}$$

$$0 \cdot Ae^{2x} = 4e^{2x}$$

and

$$0 \cdot A = 4.$$

Clearly, this equation has no solutions. The problem is that our assumed solution $y = Ae^{2x}$ is a solution of the reduced equation

$$y'' - 5y' + 6y = 0.$$

From Example 2 we note that the equation has the particular solution $y = -4xe^{2x}$. This suggests that we should try $y_p = Axe^{2x}$ instead of $y = Ae^{2x}$. The derivatives of y_p are

$$y_p' = Ae^{2x} + 2Axe^{2x}, \qquad y_p'' = 4Ae^{2x} + 4Axe^{2x}.$$

Substituting into the differential equation, we have

$$4Ae^{2x} + 4Axe^{2x} - 5(Ae^{2x} + 2Axe^{2x}) + 6Axe^{2x} = 4e^{2x}$$

$$-Ae^{2x} = 4e^{2x}$$

and

$$A = -4.$$

Thus $y_p = -4xe^{2x}$ is a particular solution as we saw in Example 2.

In a similar manner, you can verify that the form of a particular solution of

$$y'' - 4y' + 4y = e^{2x}$$

is $y_p = Ax^2 e^{2x}$ since $u_1 = e^{2x}$ and $u_2 = xe^{2x}$ are fundamental solutions of the reduced equation:

$$y'' - 4y' + 4y = 0.$$

The form of a particular solution of

$$y'' + 9y = \cos 3x$$

is $y_p = Ax \cos 3x + Bx \sin 3x$ since $u_1 = \cos 3x$ and $u_2 = \sin 3x$ are fundamental solutions of

$$y'' + 9y = 0.$$

Table 18.6.1 summarizes our discussion up to this point.

■ **Table 18.6.1** A Particular Solution of $y'' + ay' + by = \phi(x)$

If $\phi(x)$ is	Try $y_p =$
ce^{rx}	Ae^{rx}
$c_1 \cos \beta x + c_2 \sin \beta x,$	$A \cos \beta x + B \sin \beta x$
$c_1 e^{\alpha x} \cos \beta x + c_2 e^{\alpha x} \sin \beta x,$	$Ae^{\alpha x} \cos \beta x + Be^{\alpha x} \sin \beta x$

Note: If y_p satisfies the reduced equation $y'' + ay' + by = 0$, try xy_p; if xy_p also satisfies the reduced equation, then $x^2 y_p$ will give a particular solution.

Remark In practice it is a good idea to solve the reduced equation before selecting the trial particular solution of the complete equation. In this way, you will not make the mistake of selecting a y_p which satisfies the reduced equation. ❏

Example 4 Find the general solution of

$$y'' + 3y' + 2y = 2e^{-x} + 4e^{-x} \cos x.$$

SOLUTION First consider the reduced equation

$$y'' + 3y' + 2y = 0.$$

The characteristic equation is

$$r^2 + 3r + 2 = (r + 2)(r + 1) = 0,$$

and it follows that $u_1 = e^{-2x}$ and $u_2 = e^{-x}$ are fundamental solutions.

Now we need to find a particular solution of the complete equation. By the superposition principle (18.6.5), if y_1 is a particular solution of

$$(*) \qquad\qquad y'' + 3y' + 2y = 2e^{-x}$$

and y_2 is a particular solution of

$$(**) \qquad\qquad y'' + 3y' + 2y = 4e^{-x} \cos x,$$

then $y_p = y_1 + y_2$ is a particular solution of the complete equation.

Since $y = e^{-x}$ is a solution of the reduced equation, a particular solution of $(*)$ has the form $y_1 = Axe^{-x}$. We then have

$$y_1' = Ae^{-x} - Axe^{-x} \qquad \text{and} \qquad y_1'' = -2Ae^{-x} + Axe^{-x}.$$

Substituting y_1 and its derivatives into $(*)$ gives

$$y_1'' + 3y_1' + 2y_1 = -2Ae^{-x} + Axe^{-x} + 3(Ae^{-x} - Axe^{-x}) + 2Axe^{-x} = 2e^{-x},$$

$$Ae^{-x} = 2e^{-x}.$$

Thus, $A = 2$ and $y_1 = 2xe^{-x}$.

A particular solution of (∗∗) has the form $y_2 = Be^{-x} \cos x + Ce^{-x} \sin x$. The derivatives of y_2 are

$$y_2' = (-B + C)e^{-x} \cos x - (B + C)e^{-x} \sin x,$$

$$y_2'' = -2Ce^{-x} \cos x + 2Be^{-x} \sin x. \qquad \text{(verify these).}$$

Substituting y_2 and its derivatives into equation (∗∗), we get

$$[-2C + 3(-B + C) + 2B]e^{-x} \cos x + [2B - 3(B + C) + 2C]e^{-x} \sin x$$
$$= 4e^{-x} \cos x,$$

$$(-B + C)e^{-x} \cos x - (B + C)e^{-x} \sin x = 4e^{-x} \cos x.$$

It follows that

$$-B + C = 4$$
$$B + C = 0$$

and so $B = -2, C = 2$. Thus, $y_2 = -2e^{-x} \cos x + 2e^{-x} \sin x$ is a particular solution of (∗∗).

Now, a particular solution of the complete equation is

$$y_p = 2xe^{-x} - 2e^{-x} \cos x + 2e^{-x} \sin x$$

and

$$y = C_1 e^{-2x} + C_2 e^{-x} + 2xe^{-x} - 2e^{-x} \cos x + 2e^{-x} \sin x$$

is the general solution. ❑

Remark It was not actually necessary to solve for the two particular solutions y_1 and y_2 separately. We could have substituted

$$y_p = Axe^{-x} + Be^{-x} \sin x + Ce^{-x} \cos x$$

and its derivatives directly into the complete equation and solved for the three coefficients A, B, C, simultaneously. ❑

We have illustrated how the method of undetermined coefficients can be applied to the complete equation

$$y'' + ay' + by = \phi(x),$$

where the forcing function ϕ has one of the forms ce^{rx}, $c \cos \beta x$, $c \sin \beta x$, $ce^{\alpha x} \cos \beta x$, $ce^{\alpha x} \sin \beta x$, or is a sum of such forms. However, the method actually applies in the more general situation where ϕ has one of the forms:

$$p(x)e^{rx}, \quad p(x) \cos \beta x, \quad p(x) \sin \beta x, \quad p(x)e^{\alpha x} \cos \beta x, \quad p(x)e^{\alpha x} \sin \beta x,$$

where $p(x)$ is a polynomial, or is a sum of such forms. This follows from the fact that the expression

$$y'' + ay' + by$$

will transform $y_p = (A_0 + A_1 x + A_2 x^2 + \cdots + A_n x^n)e^{rx}$ into $P(x)e^{rx}$, where P is a polynomial of degree n (or less); it will transform

$$y_p = (A_0 + A_1 x + A_2 x^2 + \cdots + A_n x^n) \cos \beta x$$

into

$$P(x) \cos \beta x + Q(x) \sin \beta x,$$

where P and Q are polynomials of degree n (or less); and so on. For example, if we calculate

$$y_p'' + 3y_p$$

for $y_p = (A + Bx + Cx^2)e^{2x}$, we get

$$y_p' = 2(A + Bx + Cx^2)e^{2x} + (B + 2Cx)e^{2x},$$

$$y_p'' = 4(A + Bx + Cx^2)e^{2x} + 4(B + 2Cx)e^{2x} + 2Ce^{2x},$$

and

$$y_p'' + 3y_p = 4(A + Bx + Cx^2)e^{2x} + 4(B + 2Cx)e^{2x} + 2Ce^{2x} + 3(A + Bx + Cx^2)e^{2x}$$
$$= [(7A + 4B + 2C) + (7B + 8C)x + 7Cx^2]e^{2x}.$$

The general version of the method of undetermined coefficients can be summarized as follows:

(1) If $\phi(x) = p(x)\,e^{rx}$, where p is a polynomial of degree n, then try

$$y_p(x) = (A_0 + A_1x + \cdots + A_nx^n)\,e^{rx}.$$

(2) If $\phi(x) = p_1(x) \cos \beta x + p_2(x) \sin \beta x$, where p_1 and p_2 are polynomials with maximum degree $(p_1, p_2) = n$, then try

$$y_p(x) = (A_0 + A_1x + \cdots + A_nx^n) \cos \beta x + (B_0 + B_1x + \cdots + B_nx^n) \sin \beta x.$$

(3) If $\phi(x) = p_1(x)\,e^{rx} \cos \beta x + p_2(x)\,e^{rx} \sin \beta x$, where p_1 and p_2 are polynomials with maximum degree $(p_1, p_2) = n$, then try

$$y_p(x)$$
$$= (A_0 + A_1x + \cdots + A_nx^n)\,e^{rx} \cos \beta x + (B_0 + B_1x + \cdots + B_nx^n)\,e^{rx} \sin \beta x.$$

Note: If any term in y_p satisfies the reduced equation $y'' + ay' + by = 0$, then try $x \cdot y_p$; if any term in $x \cdot y_p$ satisfies the reduced equation then $x^2 \cdot y_p$ will give a particular solution.

Example 5 Find a particular solution of

$$y'' + 4y = (3 + 2x)e^{-x}.$$

SOLUTION You can verify that $u_1 = \cos 2x$ and $u_2 = \sin 2x$ are fundamental solutions of the reduced equation. Therefore we should try to find a particular solution of the form

$$y_p = (A + Bx)e^{-x}.$$

The derivatives of y_p are

$$y_p' = -(A + Bx)e^{-x} + Be^{-x}, \qquad y_p'' = (A + Bx)e^{-x} - 2Be^{-x}.$$

Substituting y_p and its derivatives into the differential equation, we have

$$y_p'' + 4y_p = (A + Bx)e^{-x} - 2Be^{-x} + 4(A + Bx)e^{-x}$$
$$= (3 + 2x)e^{-x},$$

$$[(5A - 2B) + 5Bx]e^{-x} = (3 + 2x)e^{-x},$$

and it follows that

$$5B = 2$$

$$5A - 2B = 3.$$

The solution of this pair of equations is $A = \frac{19}{25}$, $B = \frac{2}{5}$. Thus,

$$y_p = (\tfrac{19}{25} + \tfrac{2}{5}x)e^{-x}$$

is a particular solution of the complete equation. ❏

EXERCISES 18.6

In Exercises 1–16, find a particular solution.

1. $y'' + 5y' + 6y = 3x + 4$.

2. $y'' - 3y' - 10y = 5$.

3. $y'' + 2y' + 5y = x^2 - 1$.

4. $y'' + y' - 2y = x^3 + x$.

5. $y'' + 6y' + 9y = e^{3x}$. **6.** $y'' + 6y' + 9y = e^{-3x}$.

7. $y'' + 2y' + 2y = e^x$. **8.** $y'' + 4y' + 4y = xe^{-x}$.

9. $y'' - y' - 12y = \cos x$. **10.** $y'' - y' - 12y = \sin x$.

11. $y'' + 7y' + 6y = 3 \cos 2x$.

12. $y'' + y' + 3y = \sin 3x$.

13. $y'' - 2y' + 5y = e^{-x} \sin 2x$.

14. $y'' + 4y' + 5y = e^{2x} \cos x$.

15. $y'' + 6y' + 8y = 3e^{-2x}$.

16. $y'' - 2y' + 5y = e^x \sin x$.

In Exercises 17–24, find the general solution.

17. $y'' + y = e^x$.

18. $y'' - 2y' + y = -25 \sin 2x$.

19. $y'' - 3y' - 10y = -x - 1$.

20. $y'' + 4y = x \cos 2x$. **21.** $y'' + 3y' - 4y = e^{-4x}$.

22. $y'' + 2y' = 4 \sin 2x$. **23.** $y'' + y' - 2y = 3xe^x$.

24. $y'' + 4y' + 4y = x e^{-2x}$.

25. Prove the superposition principle (18.6.5).

26. Use (18.6.5) to find a particular solution.
(a) $y'' + 2y' - 15y = x + e^{2x}$.
(b) $y'' - 7y' + 12y = e^{-x} + \sin 2x$.

27. Find the general solution of the equation

$$y'' + 4y' + 3y = \cosh x.$$

In Exercises 28–35, use variation of parameters to find a particular solution.

28. $y'' + y = 3 \sin x \sin 2x$.

29. $y'' - 2y' + y = xe^x \cos x$.

30. $y'' + y = \csc x$, $\quad 0 < x < \pi$.

31. $y'' - 4y' + 4y = \frac{1}{3}x^{-1}e^{2x}$, $\quad x > 0$.

32. $y'' + 4y = \sec^2 2x$. **33.** $y'' + 4y' + 4y = \dfrac{e^{-2x}}{x^2}$.

34. $y'' + 2y' + y = e^{-x} \ln x$.

35. $y'' - 2y' + 2y = e^x \sec x$.

36. Show that the change of variable $y = ve^{kx}$ transforms the equation

$$y'' + ay' + by = (c_n x^n + \cdots + c_1 x + c_0)e^{kx}$$

into

$$v'' + (2k + a)v' + (k^2 + ak + b)v$$
$$= c_n x^n + \cdots + c_1 x + c_0.$$

37. In Exercise 30 of Section 18.5 we introduced a differential equation for electrical current in a simple circuit. In the presence of an external electromotive force $F(t)$, the equation takes the form

$$L\frac{d^2i}{dt^2} + R\frac{di}{dt} + \frac{1}{C}i = F(t).$$

Find the current i given that $F(t) = F_0$, $i(0) = 0$, $i'(0) = F_0/L$ and
(a) $CR^2 = 4L$. (b) $CR^2 < 4L$.

38. (a) Show that $y_1 = x$, $y_2 = x \ln x$ are solutions of the Euler equation (see Exercises 18.5)

$$x^2y'' - xy' + y = 0$$

and that their Wronskian is nonzero on $(0, \infty)$.
(b) Use variation of parameters to find a particular solution of

$$x^2y'' - xy' + y = 4x \ln x.$$

39. (a) Show that $y_1 = \sin(\ln x^2)$, $y_2 = \cos(\ln x^2)$ are solutions of the Euler equation

$$x^2y'' + xy' + 4y = 0$$

and that their Wronskian is nonzero on $(0, \infty)$.
(b) Use variation of parameters to find a particular solution of

$$x^2y'' + xy' + 4y = \sin(\ln x).$$

■ 18.7 MECHANICAL VIBRATIONS

An object moves along a straight line. Instead of continuing in one direction, it moves back and forth, oscillating about a central point. Call the central point $x = 0$ and denote by $x(t)$ the displacement of the object at time t. If the acceleration is a constant negative multiple of the displacement,

$$a(t) = -kx(t), \qquad k > 0,$$

then the object is said to be in *simple harmonic motion*.

Since, by definition,

$$a(t) = x''(t),$$

in simple harmonic motion, we have

$$x''(t) = -kx(t),$$

which is the same as

$$x''(t) + kx(t) = 0.$$

To emphasize that k is positive, we set $k = \omega^2$, where $\omega = \sqrt{k} > 0$. The equation of motion then takes the form

(18.7.1)
$$x''(t) + \omega^2 x(t) = 0.$$

This is a second-order, linear differential equation with constant coefficients. The characteristic equation is

$$r^2 + \omega^2 = 0,$$

and the roots are $\pm \omega i$. Therefore, the general solution of Equation (18.7.1) is

$$x(t) = C_1 \cos \omega t + C_2 \sin \omega t.$$

A routine calculation shows that the general solution can be written (Exercise 28, Section 18.5)

(18.7.2)
$$x(t) = A \sin(\omega t + \phi_0),$$

where A and ϕ_0 are constants with $A > 0$ and $\phi_0 \in [0, 2\pi)$.

Now let's analyze the motion measuring t in seconds. By adding $2\pi/\omega$ to t we increase $\omega t + \phi_0$ by 2π:

$$\omega \left(t + \frac{2\pi}{\omega} \right) + \phi_0 = \omega t + \phi_0 + 2\pi.$$

This means that the motion is *periodic* with *period T* given by:

$$T = \frac{2\pi}{\omega}.$$

A complete oscillation takes $2\pi/\omega$ seconds. The reciprocal of the period gives the number of complete oscillations per second. This is called the *frequency f*:

$$f = \frac{\omega}{2\pi}.$$

The number ω is called the *angular frequency*. Since $\sin(\omega t + \phi_0)$ oscillates between -1 and 1,

$$x(t) = A \sin(\omega t + \phi_0)$$

oscillates between $-A$ and A. The number A is called the *amplitude* of the motion.

In Figure 18.7.1 we have plotted x against t. The oscillations along the x-axis are now waves in the xt-plane. The period of the motion, $2\pi/\omega$, is the t distance (the time separation) between consecutive wave crests. The amplitude of the motion, A, is the height of the waves measured in x units from $x = 0$. The number ϕ_0 is known as the *phase constant,* or *phase shift*. The phase constant determines the initial displacement (in the xt-plane the height of the wave at time $t = 0$). If $\phi_0 = 0$, the object starts at the center of the interval of motion (the wave starts at the origin of the xt-plane).

simple harmonic motion

Figure 18.7.1

Example 1 Find an equation for the oscillatory motion of an object, given that the period is $2\pi/3$ and, at time $t = 0$, $x(0) = 1$, $v(0) = x'(0) = 3$.

SOLUTION We begin by setting

$$x(t) = A \sin(\omega t + \phi_0).$$

In general the period is $2\pi/\omega$, so that here

$$\frac{2\pi}{\omega} = \frac{2\pi}{3} \quad \text{and thus} \quad \omega = 3.$$

The equation of motion takes the form

$$x(t) = A \sin(3t + \phi_0).$$

By differentiation

$$v(t) = 3A \cos(3t + \phi_0).$$

The conditions at $t = 0$ give

$$1 = x(0) = A \sin \phi_0, \qquad 3 = v(0) = 3A \cos \phi_0$$

and therefore

$$1 = A \sin \phi_0, \qquad 1 = A \cos \phi_0.$$

Adding the squares, we have

$$2 = A^2 \sin^2 \phi_0 + A^2 \cos^2 \phi_0 = A^2.$$

Since $A > 0$, $A = \sqrt{2}$.

To find ϕ_0 we note that

$$1 = \sqrt{2} \sin \phi_0, \qquad 1 = \sqrt{2} \cos \phi_0.$$

These equations are satisfied by setting $\phi_0 = \frac{1}{4}\pi$. The equation of motion can be written

$$x(t) = \sqrt{2} \sin(3t + \frac{1}{4}\pi). \quad \square$$

Undamped Vibrations

A coil spring hangs naturally to a length l_0. When a bob of mass m is attached to it, the spring stretches l_1 inches. The bob is later pulled down an additional x_0 inches and then released. What is the resulting motion? Throughout we refer to Figure 18.7.2, taking the downward direction as positive.

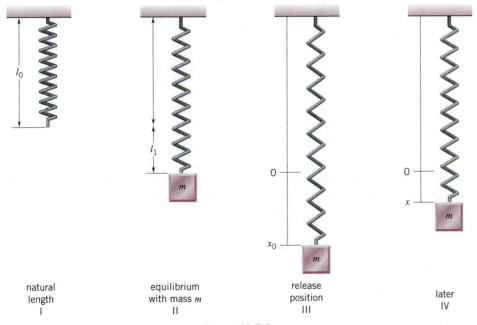

| natural length I | equilibrium with mass m II | release position III | later IV |

Figure 18.7.2

We begin by analyzing the forces acting on the bob at general position x (stage IV). First there is the weight of the bob:

$$F_1 = mg.$$

This is a downward force, and by our choice of coordinate system, positive. Then there is the restoring force of the spring. This force, by Hooke's law, is proportional to the total displacement $l_1 + x$ and acts in the opposite direction:

$$F_2 = -k(l_1 + x) \qquad \text{with } k > 0.$$

If we neglect resistance, then these are the only forces acting on the bob. Under these conditions the total force is

$$F = F_1 + F_2 = mg - k(l_1 + x),$$

which we rewrite as

$$(1) \qquad\qquad F = (mg - kl_1) - kx.$$

At stage II (Figure 18.7.2) there was equilibrium. The force of gravity, mg, plus the force of the spring, $-kl_1$, must have been 0:

$$mg - kl_1 = 0.$$

Equation (1) can therefore be simplified to

$$F = -kx.$$

Using Newton's second law

$$F = ma \qquad\qquad \text{(force = mass × acceleration)}$$

we have

$$ma = -kx \quad \text{and} \quad \text{thus} \quad a = -\frac{k}{m}x.$$

At any time t,

$$x''(t) = -\frac{k}{m}x(t).$$

Since $k/m > 0$, we can set $\omega = \sqrt{k/m}$ and write

$$x''(t) = -\omega^2 x(t).$$

The motion of the bob is simple harmonic motion with period $T = 2\pi/\omega$. ❑

There is something remarkable about harmonic motion that we have not yet specifically pointed out; namely, that the frequency $f = \omega/2\pi$ is completely independent of the amplitude of the motion. The oscillations of the bob occur with frequency

$$f = \frac{\sqrt{k/m}}{2\pi}, \qquad\qquad \text{(here } \omega = \sqrt{k/m}\text{)}.$$

By adjusting the spring constant k and the mass of the bob m, we can calibrate the spring–bob system so that the oscillations take place exactly once a second (at least almost exactly). We then have a primitive timepiece (a first cousin of the windup clock). With the passing of time, friction and air resistance reduce the amplitude of the oscillations but not their frequency. By giving the bob a little push or pull once in a while (by rewinding our clock), we can restore the amplitude of the oscillations and thus maintain the steady "ticking."

Damped Vibrations

We derived the equation of motion

$$x'' + \frac{k}{m}x = 0$$

from the force equation

$$F = -kx.$$

Unless the spring is frictionless and the motion takes place in a vacuum, there is a resistance to the motion that tends to dampen the vibrations. Experiment shows that the resistance force R is approximately proportional to the velocity x':

$$R = -cx'. \qquad\qquad (c > 0)$$

Taking this resistance term into account, the force equation reads

$$F = -kx - cx'.$$

Newton's law $F = ma = mx''$ then gives

$$mx'' = -cx' - kx,$$

which we can write as

(18.7.3)
$$\boxed{x'' + \frac{c}{m}x' + \frac{k}{m}x = 0.}$$

This is the equation of motion in the presence of a *damping* factor. To study the motion we analyze this equation.

The characteristic equation

$$r^2 + \frac{c}{m}r + \frac{k}{m} = 0$$

has roots

$$r = \frac{-c \pm \sqrt{c^2 - 4km}}{2m}.$$

There are three possibilities:

$$c^2 - 4km < 0, \qquad c^2 - 4km > 0, \qquad c^2 - 4km = 0.$$

Case 1. $c^2 - 4km < 0$

In this case the characteristic equation has two complex conjugate roots:

$$r_1 = -\frac{c}{2m} + i\omega, \qquad r_2 = -\frac{c}{2m} - i\omega \qquad \text{where} \quad \omega = \frac{\sqrt{4km - c^2}}{2m}.$$

The general solution of (18.7.3),

$$x = e^{-(c/2m)t}(c_1 \cos \omega t + c_2 \sin \omega t),$$

can be written

(18.7.4)

$$x(t) = Ae^{(-c/2m)t} \sin(\omega t + \phi_0),$$

where, as before, A and ϕ_0 are constants, $A > 0$, $\phi_0 \in [0, 2\pi)$. This is called the *underdamped case*. The motion is similar to simple harmonic motion, except that the damping term $e^{(-c/2m)t}$ ensures that $x \to 0$ as $t \to \infty$. The vibrations continue indefinitely with constant frequency $\omega/2\pi$ but diminishing amplitude $Ae^{(-c/2m)t}$. As $t \to \infty$, the amplitude of the vibrations tends to zero; the vibrations die down. The motion is illustrated in Figure 18.7.3. ❏

damped harmonic motion

Figure 18.7.3

Case 2. $c^2 - 4km > 0$

In this case the characteristic equation has two distinct real roots:

$$r_1 = \frac{-c + \sqrt{c^2 - 4km}}{2m}, \qquad r_2 = \frac{-c - \sqrt{c^2 - 4km}}{2m}.$$

The general solution takes the form

(18.7.5)

$$x = C_1 e^{r_1 t} + C_2 e^{r_2 t}.$$

This is called the *overdamped case*. The motion is nonoscillatory. Since $\sqrt{c^2 - 4km} < \sqrt{c^2} = c$, both r_1 and r_2 are negative. As $t \to \infty$, $x \to 0$. ❏

Case 3. $c^2 - 4km = 0$

In this case the characteristic equation has only one root

$$r_1 = -\frac{c}{2m}$$

and the general solution takes the form

(18.7.6)

$$x = C_1 e^{-(c/2m)t} + C_2 t e^{-(c/2m)t}.$$

This is called the *critically damped case*. Once again the motion is nonoscillatory. Moreover, as $t \to \infty$, $x \to 0$. ❏

In both the overdamped and critically damped cases, the mass moves slowly back to its equilibrium position ($x \to 0$ as $t \to \infty$). Depending upon the side conditions, the mass may move through the equilibrium once, but only once; there is no oscillatory motion. Two typical examples of the motion are shown in Figure 18.7.4.

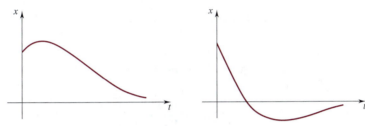

Figure 18.7.4

Forced Vibrations

The vibrations that we have been considering result from the interplay of three forces: the force of gravity, the elastic force of the spring, and the retarding force of the surrounding medium. Such vibrations are called *free vibrations*.

The application of an external force to a freely vibrating system modifies the vibrations and results in what are called *forced vibrations*. In what follows we examine the effect of a pulsating force $F_0 \cos \gamma t$. Without loss of generality we can take both F_0 and γ as positive.

In an undamped system the force equation reads

$$F = -kx + F_0 \cos \gamma t$$

and the equation of motion takes the form

(18.7.7)

$$x'' + \frac{k}{m} x = \frac{F_0}{m} \cos \gamma t.$$

As usual we set $\omega = \sqrt{k/m}$ and write

(18.7.8)

$$x'' + \omega^2 x = \frac{F_0}{m} \cos \gamma t.$$

As you'll see, the nature of the vibrations depends on the relation between the *applied frequency*, $\gamma/2\pi$, and the *natural frequency* of the system, $\omega/2\pi$.

Case 1. $\gamma \neq \omega$

In this case the method of undetermined coefficients gives the particular solution

$$x_p = \frac{F_0/m}{\omega^2 - \gamma^2} \cos \gamma t.$$

The general equation of motion can thus be written

(18.7.9)
$$x = A \sin(\omega t + \phi_0) + \frac{F_0/m}{\omega^2 - \gamma^2} \cos \gamma t.$$

If ω/γ is rational, the vibrations are periodic. If, on the other hand, ω/γ is not rational, then the vibrations are not periodic and the motion, though bounded by

$$|A| + \left| \frac{F_0/m}{\omega^2 - \gamma^2} \right|,$$

can be highly irregular. ❏

Case 2. $\gamma = \omega$

In this case the method of undetermined coefficients gives

$$x_p = \frac{F_0}{2\omega m} t \sin \omega t$$

and the general solution takes the form

(18.7.10)
$$x = A \sin(\omega t + \phi_0) + \frac{F_0}{2\omega m} t \sin \omega t.$$

The undamped system is said to be in *resonance*. The motion is oscillatory, but, because of the extra t present in the second summand, it is far from periodic. As $t \to \infty$, the amplitude of vibration increases without bound. The motion is illustrated in Figure 18.7.5. ❏

Undamped systems and unbounded vibrations are mathematical fictions. No real mechanical system is totally undamped, and unbounded vibrations do not occur in nature. Nevertheless a form of resonance can occur in a real mechanical system. (See Exercises 24–28.) A periodic external force applied to a mechanical system that is insufficiently damped can set up vibrations of very large amplitude. Such vibrations have caused the destruction of some formidable man-made structures. In 1850 the suspension bridge at Angers, France, was destroyed by vibrations set up by the unified step of a column of marching soldiers. More than two hundred French soldiers were killed in that catastrophe. (Soldiers today are told to break ranks before crossing a

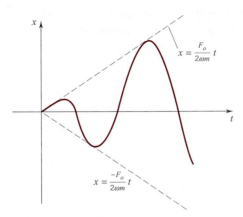

Figure 18.7.5

bridge.) The collapse of the bridge at Tacoma, Washington, is a more recent event. Slender in construction and graceful in design, the Tacoma bridge was opened to traffic on July 1, 1940. The third longest suspension bridge in the world, with a main span of 2800 feet, the bridge attracted many admirers. On November 1 of that same year, after less than five months of service, the main span of the bridge broke loose from its cables and crashed into the water below. (Luckily only one person was on the bridge at the time, and he was able to crawl to safety.) A driving wind had set up vibrations in resonance with the natural vibrations of the roadway, and the stiffening girders of the bridge had not provided sufficient damping to keep the vibrations from reaching destructive magnitude.

EXERCISES 18.7

1. An object is in simple harmonic motion. Find an equation for the motion given that the period is $\frac{1}{4}\pi$ and, at time $t = 0, x = 1$ and $v = 0$. What is the amplitude? What is the frequency?

2. An object is in simple harmonic motion. Find an equation for the motion given that the frequency is $1/\pi$ and, at time $t = 0$, $x = 0$ and $v = -2$. What is the amplitude? What is the period?

3. An object is in simple harmonic motion with period T and amplitude A. What is the velocity at the central point $x = 0$?

4. An object is in simple harmonic motion with period T. Find the amplitude given that $v = \pm v_0$ at $x = x_0$.

5. An object in simple harmonic motion passes through the central point $x = 0$ at time $t = 0$ and every 3 seconds thereafter. Find the equation of motion given that $v(0) = 5$.

6. Show that simple harmonic motion $x(t) = A \sin(\omega t + \phi_0)$ can just as well be written: (a) $x(t) = A \cos(\omega t + \phi_1)$; (b) $x(t) = B \sin \omega t + C \cos \omega t$.

Exercises 7–12 are concerned with the motion of the bob depicted in Figure 18.7.2.

7. What is $x(t)$ for the bob of mass m?

8. Find the positions of the bob where the bob attains: (a) maximum speed; (b) zero speed; (c) maximum acceleration; (d) zero acceleration.

9. Where does the bob take on half of its maximum speed?

10. Find the maximal kinetic energy obtained by the bob. (Remember: KE $= \frac{1}{2}mv^2$ where m is the mass of the object and v is the speed.)

11. Find the time average of the kinetic energy of the bob during one period T.

12. Express the velocity of the bob in terms of k, m, x_0, and $x(t)$.

13. Given that $x''(t) = 9 - 4x(t)$ with $x(0) = 0$ and $x'(0) = 0$, show that the motion is simple harmonic motion centered at $x = 2$. Find the amplitude and the period.

14. The figure shows a pendulum of mass m swinging on an arm of length L. The angle θ is measured counterclockwise. Neglecting friction and the weight of the arm, we can describe the motion by the equation

$$mL\theta''(t) = -mg \sin \theta(t)$$

which reduces to

$$\theta''(t) = -\frac{g}{L}\sin \theta(t).$$

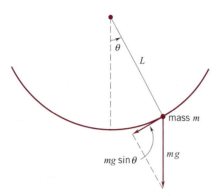

mass m

$mg \sin\theta$ mg

(a) For small angles we replace $\sin \theta$ by θ and write

$$\theta''(t) \cong -\frac{g}{L}\,\theta(t).$$

Justify this step.
(b) Solve the approximate equation in (a) given that the pendulum
   (i) is held at an angle $\theta_0 > 0$ and released at time $t = 0$.
   (ii) passes through the vertical position at time $t = 0$ and $\theta'(0) = -\sqrt{g/L}\;\theta_0$.
(c) Find L given that the motion repeats itself every 2 seconds.

15. A cylindrical buoy of mass m and radius r centimeters floats with its axis vertical in a liquid of density ρ kilograms per cubic centimeter. Suppose that the buoy is pushed x_0 centimeters down into the liquid. See the figure.

x_0

(a) Neglecting friction and given that the buoyancy force is equal to the weight of the liquid displaced, show that the buoy bobs up and down in simple harmonic motion by finding the equation of motion.
(b) Solve the equation obtained in (a). Specify the amplitude and the period.

16. Explain in detail the connection between uniform circular motion and simple harmonic motion.

17. What is the effect of an increase in the resistance constant c on the amplitude and frequency of the vibrations given by Equation (18.7.4).

18. Prove that the motion given by (18.7.5) can pass through the equilibrium point at most once. How many times can the motion change directions?

19. Prove that the motion given by (18.7.6) can pass through the equilibrium point at most once. How many times can the motion change directions?

20. Show that, if $\gamma \neq \omega$, then the method of undetermined coefficients applied to (18.7.8) gives

$$x_p = \frac{F_0/m}{\omega^2 - \gamma^2}\cos \gamma t.$$

21. Show that, if ω/γ, is rational, then the vibrations given by (18.7.9) are periodic.

22. Show that, if $\gamma = \omega$, then the method of undetermined coefficients applied to (18.7.8) gives

$$x_p = \frac{F_0}{2\omega m}t \sin \omega t.$$

Forced vibrations in a damped system Write the equation

$$x'' + \frac{c}{m}x' + \frac{k}{m}x = \frac{F_0}{m}\cos \gamma t$$

as

(∗) $$x'' + 2\alpha x' + \omega^2 x = \frac{F_0}{m}\cos \gamma t.$$

We will assume throughout that $0 < \alpha < \omega$. (For large α the resistance is large and the motion is not as interesting.)

23. Find the general solution of the reduced equation $x'' + 2\alpha x' + \omega^2 x = 0$.

24. Verify that the function

$$x_p = \frac{F_0/m}{(\omega^2 - \gamma^2)^2 + 4\alpha^2\gamma^2}[(\omega^2 - \gamma^2)\cos \gamma t + 2\alpha\gamma \sin \gamma t]$$

is a particular solution of (∗).

25. Determine x_p if $\omega = \gamma$. Show that the amplitude of the vibrations is very large if the resistance constant c is very small.

26. Show that the solution x_p in Exercise 24 can be written

$$x_p = \frac{F_0/m}{\sqrt{(\omega^2 - \gamma^2)^2 + 4\alpha^2\gamma^2}}\sin(\gamma t + \phi).$$

27. Show that, if $2\alpha^2 \geq \omega^2$, then the amplitude of vibration of the solution x_p in Exercise 26 decreases as γ increases.

28. Suppose now that $2\alpha^2 \leq \omega^2$.
   (a) Find the value of γ that maximizes the amplitude of the solution x_p in Exercise 26.
   (b) Determine the frequency that corresponds to this value of γ. (This is called the *resonant frequency* of the damped system.)
   (c) What is the *resonant amplitude* of the system? (In other words, what is the amplitude of vibration when the applied force is at resonant frequency?)
   (d) Show that, if c, the constant of resistance, is very small, then the resonant amplitude is very large.

■ CHAPTER HIGHLIGHTS

18.1 Introduction

18.2 First Order Linear Differential Equations

A first order differential equation is *linear* if it can be written in the form

$$y' + p(x)y = q(x), \quad \text{where } p \text{ and } q \text{ are continuous on some interval } I.$$

The general solution is $y = e^{-H(x)} \left\{ \displaystyle\int e^{H(x)} q(x) \, dx + C \right\}$ (p. 1201).

Integrating factor (p. 1201)

18.3 Separable Equations; Homogeneous Equations

A first order differential equation is *separable* if it can be written in the form

$$p(x) + q(y)y' = 0 \quad \text{or} \quad p(x) \, dx + q(y) \, dy = 0.$$

A one-parameter family of solutions is $\displaystyle\int p(x) \, dx + \int q(y) \, dy = C$ (p. 1209).

The logistic equation (p. 1213)
The first order differential equation $y' = f(x, y)$ is *homogeneous* if $f(tx, ty) = f(x, y)$ for all $t \neq 0$ (p. 1216).
The transformation $v = y/x$ transforms a homogeneous equation into a separable equation (p. 1216).

18.4 Exact Equations; Integrating Factors

The first order differential equation

$$P(x, y) + Q(x, y)y' = 0, \text{ where } P \text{ and } Q \text{ are continuously differentiable in a region } \Omega,$$

is *exact* iff $\partial P/\partial y = \partial Q/\partial x$ for all $(x, y) \in \Omega$.
The function $\mu = \mu(x, y)$ is an *integrating factor* for $P(x, y) + Q(x, y)y' = 0$ if

$$\mu(x, y)P(x, y) + \mu(x, y)Q(x, y)y' = 0$$

is exact (p. 1221).

18.5 The Equation $y'' + ay' + by = 0$

Reduced equation (homogeneous equation) (p. 1224)
Linear combinations of solutions (p. 1224)   The characteristic equation (p. 1225)
Existence and uniqueness of solutions (p. 1228)
The Wronskian (p. 1228)   Fundamental solutions (p. 1229)
Let r_1, r_2 be the roots of the characteristic equation. The general solution of

$$y'' + ay' + by = 0 \text{ is}$$

  I. $y = C_1 e^{r_1 x} + C_2 e^{r_2 x}$ if r_1 and r_2 are real and unequal.
  II. $y = C_1 e^{rx} + C_2 x e^{rx}$ if $r_1 = r_2 = r$ are real and equal.
  III. $y = e^{\alpha x}(C_1 \cos \beta x + C_2 \sin \beta x)$ if $r_1 = \alpha + i\beta$ and $r_2 = \alpha - i\beta$ are complex.

18.6 The Equation $y'' + ay' + by = \phi(x)$

Complete equation (nonhomogeneous equation) (p. 1234)
Nonhomogeneous term (forcing function) (p. 1234)
The general solution of the complete equation is $y = C_1 u_1(x) + C_2 u_2(x) + y_p(x)$, where u_1 and u_2 are fundamental solutions of the reduced equation and y_p is a particular solution of the complete equation.

If u_1 and u_2 are fundamental solutions of the reduced equation and W is their Wronskian, then

$$y_p = \left(-\int \frac{u_2(x)\phi(x)}{W} \, dx \right) u_1 + \left(\int \frac{u_1(x)\alpha(x)}{W} \, dx \right) u_2$$

is a particular solution of the complete equation.

Undetermined coefficients, see Table 18.6.1 (p. 1241).

18.7 Mechanical Vibrations

■ PROJECTS AND EXPLORATIONS USING TECHNOLOGY

To do these exercises you will need a graphics calculator or a computer with graphing capability. The majority of these problems are open-ended so different appraoches may be used to solve them. You should be aware that different approaches can result in slight variations in the answers. Round your numerical answers to at least four decimal places. The rounding method that your calculator or computer uses also may cause variations in answers.

18.1 The problem of solving a first order differential equation $y' = f(x, y)$ subject to an initial condition $y(a) = \alpha$ is called an *initial value problem*. Recall that a function $y = g(x)$ defined on an interval I is a solution of the initial value problem

$$(1) \qquad\qquad y' = f(x, y), \quad y(a) = \alpha, \quad \text{where } a \in I,$$

if $g(a) = \alpha$ and $g'(x) = f(x, g(x))$ for all $x \in I$. For our purposes in these exercises, we assume that (1) has a unique solution. For example, it can be shown that if f and $\partial f / \partial y$ are continuous functions on a region Ω in the x,y-plane and $(a,\alpha) \in \Omega$, then (1) has a unique solution that is defined on some interval I containing a.

There are no general methods for solving initial value problems such as (1). Consequently, there is a long history of methods for approximating solutions. Numerical methods for constructing approximate solutions of differential equations were among the first applications of digital computers.

Euler's method is a relatively simple method for constructing an approximate solution of (1) on a closed interval $[a, b]$. It is based on the first order Taylor (or tangent line) approximation. If $y = g(x)$ is the solution of (1) and $h \neq 0$ is fixed, then

$$g(x + h) \cong g(x) + g'(x)h = g(x) + f(x, g(x))h.$$

Euler's method uses this approximation to calculate a sequence of points (x_i, w_i), $i = 0, 1, 2, \ldots, n$, where $x_i = a + ih$, $h = (b - a)/n$ and $w_i \cong g(x_i)$. The values w_i are calculated recursively by

$$w_0 = \alpha, \quad \text{and} \quad w_{i+1} = w_i + h f(x_i, w_i), \quad i = 0, 1, 2, \ldots, n - 1.$$

These ideas are illustrated in the figure below.

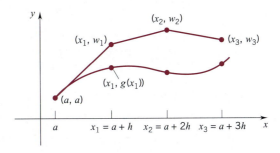

(a) Write a program for your calculator or computer to implement Euler's method. Test your program on the following problems:

$$\text{(i) } y' = y, \quad y(0) = 1; \quad x \in [0, 2].$$

$$\text{(ii) } y' = x + 2y, \quad y(0) = 1; \quad x \in [0, 1].$$

$$\text{(iii) } y' = 2xy, \quad y(1) = 1; \quad x \in [1, 2].$$

(b) The initial value problems in part (a) can be solved exactly. Using your results in (a), plot the points (x_j, w_i), connect them with straight line segments, and compare with the graph of the exact solution for each of the problems.

(c) We would expect the computed sequence (x_j, w_j) to get closer to the exact solution as n increases. We want to determine how quickly this happens. For each of the problems in part (a) calculate $w_n \cong g(b)$ for $n = 1, 2, 4, 8, 16, 32, 64, 128$, and plot $|w_n - g(b)|$ versus n. Then try to find curves that fit each of the data sets.

(d) We would expect to find better approximations if we used a second order Taylor approximation instead of the first order (tangent line) approximation. Fix $h \neq 0$. Show that if the function f in (1) is differentiable, then the solution $y = g(x)$ of the initial value problem (1) is twice differentiable and

$$g(x + h) \cong g(x) + g'(x)h + g''(x)\frac{h^2}{2}$$

$$= g(x) + f(x, g(x))h + \left[\frac{\partial f}{\partial x}(x, g(x)) + \frac{\partial f}{\partial y}(x, g(x))f(x, g(x)) \right]\frac{h^2}{2}.$$

(e) The approximation in (d) leads to the recurrence relation

$$w_0 = \alpha, \text{ and } w_{i+1} = w_i + h f(x_i, w_i) + \frac{h^2}{2}\left[\frac{\partial f}{\partial x}(x_i, w_i) + \frac{\partial f}{\partial y}(x_i, w_i)f(x_i, w_i) \right],$$

$$i = 0, 1, \ldots, n - 1.$$

Modify the procedures written for part (a) to use this recurrence relation. Then repeat part (c) and compare your results with the results of Euler's method.

18.2 A geometric method for studying the first order differential equation

$$(2) \qquad\qquad y' = f(x, y)$$

involves the use of *slope fields* (also called *direction fields*). Slope fields are used to give qualitative information about the solutions (are solutions increasing? decreasing? concave up? and so forth).

According to the differential equation (2), if y is a solution, then its slope y' at a point (x, y) is simply $f(x, y)$. A slope field for (2) is constructed as follows. Select a set of grid points (x_i, y_i), $i = 1, 2, \ldots, n$ in the plane, calculate $f(x_i, y_i)$ at each grid point, and then draw a short line segment through (x_i, y_i) with slope $f(x_i, y_i)$. For example, a slope field for

$$y' = x - y$$

on the rectangle $-3 \leq x \leq 3$, $-3 \leq y \leq 3$, is shown in Figure A. To sketch the graph of a solution of an initial value problem, start at the initial point and follow the line segments. A sketch of the solution to $y' = x - y$, $y(0) = 0$, is shown in Figure B.

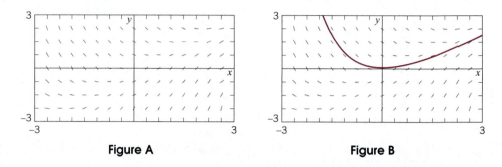

Figure A   **Figure B**

(a) Use technology to draw slope fields for the differential equations in Exercise 18.1 (a). Then use the slope fields to sketch the graphs of the solutions with the given initial conditions. (Some software has these capabilities programmed in, your instructor may provide special software, or you may need to write your own programs.)

(b) Use your technology from part (a) to draw a slope field for the differential equation $y' = x^4 - y^5$. Select grid points in the rectangle $-10 \leq x \leq 10$, $-10 \leq y \leq 10$.

(c) The initial value problem $y' = x^4 - y^5$, $y(0) = \alpha$ has a unique solution. Using your slope field in part (b), sketch the solution curves for the initial value problems corresponding to $\alpha = -1$, $\alpha = 0$, $\alpha = 1$, $\alpha = 5$, and $\alpha = 10$. If possible, compare these sketches with the solution curves drawn by technology.

(d) Explain why each solution is decreasing to a minimum, and then increasing.

(e) Show that every solution is eventually less than $y = \sqrt{x}$. Explain why the solutions get close together as $x \to \infty$.

(f) Let $y = g(x)$ be the solution of the initial value problem $y' = x^4 - y^5$, $y(0) = 10$, on the interval $[0, 2]$. Use Euler's method from Exercise 18.1 to see how well it does with this problem.

SOME ADDITIONAL TOPICS

APPENDIX

A

■ A.1 ROTATION OF AXES; EQUATIONS OF SECOND DEGREE

Rotation of Axes

We begin by referring to Figure A.1.1. From the figure,

$$\cos \theta = \frac{x}{r}, \qquad \sin \theta = \frac{y}{r}.$$

Thus

(A.1.1)
$$x = r \cos \theta, \qquad y = r \sin \theta.$$

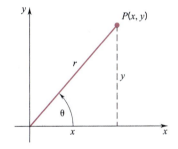

Figure A.1.1

Equations (A.1.1) come up repeatedly in calculus. In particular, these are the equations that you use to convert polar coordinates to rectangular coordinates (see Section 9.2).

Consider now a rectangular coordinate system Oxy. If we rotate this system counterclockwise α radians about the origin, we obtain a new coordinate system OXY. See Figure A.1.2.

A point P will now have two pairs of rectangular coordinates:

$$(x, y) \text{ in the } Oxy \text{ system} \qquad \text{and} \qquad (X, Y) \text{ in the } OXY \text{ system.}$$

Here we investigate the relation between (x, y) and (X, Y). With P as in Figure A.1.3,

$$x = r \cos (\alpha + \beta), \qquad y = r \sin (\alpha + \beta)$$

and

$$X = r \cos \beta, \qquad Y = r \sin \beta.$$

Figure A.1.2

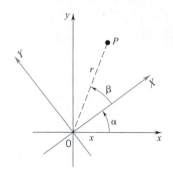

Figure A.1.3

Since

$$\cos(\alpha + \beta) = \cos\alpha\cos\beta - \sin\alpha\sin\beta,$$
$$\sin(\alpha + \beta) = \sin\alpha\cos\beta + \cos\alpha\sin\beta,$$

we have

$$x = r\cos(\alpha + \beta) = (\cos\alpha)r\cos\beta - (\sin\alpha)r\sin\beta,$$
$$y = r\sin(\alpha + \beta) = (\sin\alpha)r\cos\beta + (\cos\alpha)r\sin\beta,$$

and therefore

(A.1.2)
$$x = (\cos\alpha)X - (\sin\alpha)Y, \qquad y = (\sin\alpha)X + (\cos\alpha)Y.$$

These formulas give the algebraic consequences of a counterclockwise rotation of α radians.

Equations of Second Degree

Equations of the form

(1) $$ax^2 + cy^2 + dx + ey + f = 0,$$

where a, c, d, e, f are constants and a and c are not both zero, are studied in Sections 1.4 and 9.1, and they occur throughout the text in the examples and exercises. Except for degenerate cases (for example, $x^2 + y^2 + 1 = 0$ or $x^2 - y^2 = 0$), the graph of (1) is a conic section: circle, ellipse, parabola, or hyperbola.

The *general equation of second degree in x and y* is an equation of the form

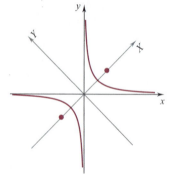

Figure A.1.4

(A.1.3)
$$ax^2 + bxy + cy^2 + dx + ey + f = 0,$$

where a, b, c, d, e, f are constants and a, b, c are not all zero. The graph of such an equation is still a conic section (again, except for degenerate cases). For example, the graph of the equation

$$xy - 2 = 0$$

is the hyperbola shown in Figure A.1.4.

Eliminating the xy-Term

Rotations of the coordinate system enable us to simplify equations of the second degree by eliminating the xy-term. That is, if in the Oxy coordinate system, a curve S has an equation of the form

(2) $$ax^2 + bxy + cy^2 + dx + ey + f = 0 \qquad \text{with} \qquad b \neq 0,$$

then there exists a coordinate system OXY, differing from Oxy by a rotation α, where $0 < \alpha < \pi/2$, such that in the OXY system S has an equation of the form

(3) $$AX^2 + CY^2 + DX + EY + F = 0,$$

where A and C are not both zero. To see this, substitute

$$x = (\cos \alpha)X - (\sin \alpha)Y, \qquad y = (\sin \alpha)X + (\cos \alpha)Y$$

in equation (2). This will give you a second-degree equation in X and Y in which the coefficient of XY is

$$-2a \cos \alpha \sin \alpha + b(\cos^2 \alpha - \sin^2 \alpha) + 2c \cos \alpha \sin \alpha.$$

This can be simplified to

$$(c - a) \sin 2\alpha + b \cos 2\alpha.$$

To eliminate the XY term we must have this coefficient equal to zero, that is, we must have

$$b \cos 2\alpha = (a - c)\sin 2\alpha$$

or

$$\cot 2\alpha = \frac{a - c}{b} \qquad\qquad \text{(recall } b \neq 0\text{)}.$$

Therefore,

$$2\alpha = \cot^{-1}\left(\frac{a - c}{b}\right)$$

and

$$\alpha = \frac{1}{2}\cot^{-1}\left(\frac{a - c}{b}\right).$$

Since the range of the inverse cotangent function is $(0, \pi)$, it follows that $0 < \alpha < \pi/2$.

We have shown that an equation of the form (2) can be transformed into an equation of the form (3) by rotating the axes through the angle α given by

(A.1.5)

$$\boxed{\alpha = \frac{1}{2}\cot^{-1}\left(\frac{a - c}{b}\right)}$$

We leave it as an exercise to show that the coefficients A and C in (3) are not both zero.

Example 1 In the case of

$$xy - 2 = 0,$$

we have $a = c = 0$, $b = 1$, and $\alpha = \frac{1}{2}\cot^{-1}(0) = \frac{1}{4}\pi$. Setting

$$x = (\cos \tfrac{1}{4}\pi)X - (\sin \tfrac{1}{4}\pi)Y = \tfrac{1}{2}\sqrt{2}\,(X - Y),$$
$$y = (\sin \tfrac{1}{4}\pi)X + (\cos \tfrac{1}{4}\pi)Y = \tfrac{1}{2}\sqrt{2}\,(X + Y),$$

we find that $xy - 2 = 0$ becomes

$$\tfrac{1}{2}(X^2 - Y^2) - 2 = 0,$$

which can be written

$$\frac{X^2}{4} - \frac{Y^2}{4} = 1.$$

This is the equation of a hyperbola in standard position in the OXY system. The hyperbola is shown in Figure A.1.4. ❏

Example 2 In the case of

$$11x^2 + 4\sqrt{3}xy + 7y^2 - 1 = 0,$$

we have $a = 11$, $b = 4\sqrt{3}$, and $c = 7$. Thus we can choose

$$\alpha = \tfrac{1}{2}\cot^{-1}\left(\frac{11-7}{4\sqrt{3}}\right) = \tfrac{1}{2}\cot^{-1}\left(\frac{1}{\sqrt{3}}\right) = \tfrac{1}{6}\pi$$

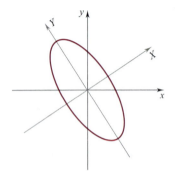

Setting

$$x = (\cos \tfrac{1}{6}\pi)X - (\sin \tfrac{1}{6}\pi)Y = \tfrac{1}{2}(\sqrt{3}X - Y),$$

$$y = (\sin \tfrac{1}{6}\pi)X + (\cos \tfrac{1}{6}\pi)Y = \tfrac{1}{2}(X + \sqrt{3}Y),$$

we find that our initial equation simplifies to $13X^2 + 5Y^2 - 1 = 0$, which we can write as

$$\frac{X^2}{(1/\sqrt{13})^2} + \frac{Y^2}{(1/\sqrt{5})^2} = 1.$$

Figure A.1.5

This is the equation of an ellipse. The ellipse is pictured in Figure A.1.5. ❏

The Discriminant

It is possible to draw general conclusions about the graph of a second-degree equation

$$ax^2 + bxy + cy^2 + dx + ey + f = 0, \qquad a, b, c \text{ not all } 0,$$

just from the *discriminant* $\Delta = b^2 - 4ac$. There are three cases:

Case 1. If $\Delta < 0$, the graph is an ellipse, a circle, a point, or empty.

Case 2. If $\Delta > 0$, the graph is a hyperbola or a pair of intersecting lines.

Case 3. If $\Delta = 0$, the graph is a parabola, a line, a pair of lines, or empty.

Below we outline how these assertions can be verified. A useful first step is to rotate the coordinate system so that the equation takes the form

(4) $$AX^2 + CY^2 + DX + EY + F = 0.$$

An elementary but time-consuming computation shows that the discriminant is unchanged by a rotation, so that in this instance we have

$$\Delta = b^2 - 4ac = -4AC.$$

Moreover, A and C cannot both be zero. If $\Delta < 0$, then $AC > 0$ and we can rewrite (4) as

$$\frac{X^2}{C} + \frac{D}{AC}X + \frac{Y^2}{A} + \frac{E}{AC}Y + \frac{F}{AC} = 0.$$

By completing the squares, we obtain an equation of the form

$$\frac{(X - \alpha)^2}{(\sqrt{|C|})^2} + \frac{(Y - \beta)^2}{(\sqrt{|A|})^2} = K.$$

If $K > 0$, we have an ellipse or a circle. If $K = 0$, we have the point (α, β). If $K < 0$, the set is empty.

If $\Delta > 0$, then $AC < 0$. Proceeding as before, we obtain an equation of the form

$$\frac{(X - \alpha)^2}{(\sqrt{|C|})^2} - \frac{(Y - \beta)^2}{(\sqrt{|A|})^2} = K.$$

If $K \neq 0$, we have a hyperbola. If $K = 0$, the equation becomes

$$\left(\frac{X - \alpha}{\sqrt{|C|}} - \frac{Y - \beta}{\sqrt{|A|}}\right)\left(\frac{X - \alpha}{\sqrt{|C|}} + \frac{Y - \beta}{\sqrt{|A|}}\right) = 0,$$

so that we have a pair of lines intersecting at the point (α, β).

If $\Delta = 0$, then $AC = 0$, so that either $A = 0$ or $C = 0$. Since A and C are not both zero, there is no loss in generality in assuming that $A \neq 0$ and $C = 0$. In this case equation (4) reduces to

$$AX^2 + DX + EY + F = 0.$$

Dividing by A and completing the square we have an equation of the form

$$(X - \alpha)^2 = \beta Y + K.$$

If $\beta \neq 0$, we have a parabola. If $\beta = 0$ and $K = 0$, we have a line. If $\beta = 0$ and $K > 0$, we have a pair of parallel lines. If $\beta = 0$ and $K < 0$, the set is empty. ❏

EXERCISES A.1

In Exercises 1–8, (a) use the discriminant to give a possible identification of the graph of the equation; (b) find a rotation $\alpha \in (0, \pi/2)$ that eliminates the xy-term; (c) rewrite the equation in terms of the new coordinate system; (d) sketch the graph displaying both coordinate systems.

1. $xy = 1.$ **2.** $xy - y + x = 1.$

3. $11x^2 + 10\sqrt{3}xy + y^2 - 4 = 0.$

4. $52x^2 - 72xy + 73y^2 - 100 = 0.$

5. $x^2 - 2xy + y^2 + x + y = 0.$

6. $3x^2 + 2\sqrt{3}xy + y^2 - 2x + 2\sqrt{3}y = 0.$

7. $x^2 + 2\sqrt{3}xy + 3y^2 + 2\sqrt{3}x - 2y = 0.$

8. $2x^2 + 4\sqrt{3}xy + 6y^2 + (8 - \sqrt{3})x + (8\sqrt{3} + 1)y + 8 = 0.$

In Exercises 9 and 10, find a rotation $\alpha \in (0, \pi/2)$ that eliminates the xy-term. Then find $\cos \alpha$ and $\sin \alpha$.

9. $x^2 + xy + Kx + Ly + M = 0.$

10. $5x^2 + 24xy + 12y^2 + Kx + Ly + M = 0.$

11. Show that after a rotation of axes through an angle α, the coefficients in the equation

$$AX^2 + BXY + CY^2 + DX + EY + F = 0$$

are related to the coefficients in the equation

$$ax^2 + bxy + cy^2 + dx + ey + f = 0, \quad a, b, c \text{ not all } 0,$$

as follows:

$A = a \cos^2\alpha + b \cos \alpha \sin \alpha + c \sin^2\alpha,$

$B = 2(c - a)\cos \alpha \sin \alpha + b(\cos^2\alpha - \sin^2\alpha),$

$C = a \sin^2\alpha - b \cos \alpha \sin \alpha + c \cos^2\alpha,$

$D = d \cos \alpha + e \sin \alpha,$

$E = e \cos \alpha - d \sin \alpha,$

$F = f.$

12. Use the results of Exercise 11 to show:
   (a) $B^2 - 4AC = b^2 - 4ac.$
   (b) If $B = 0$, then A and C cannot both be 0.

■ A.2 DETERMINANTS

By a *matrix* we mean a rectangular arrangement of numbers enclosed in parentheses. For example,

$$\begin{pmatrix} 2 & 4 \\ 3 & 1 \end{pmatrix} \quad \begin{pmatrix} 1 & 6 & 3 \\ 5 & 2 & 2 \end{pmatrix} \quad \begin{pmatrix} 2 & 4 & 0 \\ 4 & 7 & 1 \\ 0 & 1 & 1 \end{pmatrix}$$

are all matrices. The numbers occurring in a matrix are called the *entries*.

Each matrix has a certain number of rows and a certain number of columns. A matrix with m rows and n columns is called an $m \times n$ *matrix*. Thus the first matrix above is a 2×2 matrix, the second a 2×3 matrix, the third a 3×3 matrix. The first and third matrices are called *square*; they have the same number of rows as columns. Here we will be working with square matrices as these are the only ones that have determinants.

We could give a definition of determinant that is applicable to all square matrices, but the definition is complicated and would serve little purpose at this point. Our interest here is in the 2×2 case and the 3×3 case. We begin with the 2×2 case.

(A.2.1)

> The *determinant* of the matrix
> $$\begin{pmatrix} a_1 & a_2 \\ b_1 & b_2 \end{pmatrix}$$
> is the number $a_1 b_2 - a_2 b_1$.

We have a special notation for the determinant. We change the parentheses of the matrix to vertical bars:

$$\text{Determinant of } \begin{pmatrix} a_1 & a_2 \\ b_1 & b_2 \end{pmatrix} = \begin{vmatrix} a_1 & a_2 \\ b_1 & b_2 \end{vmatrix} = a_1 b_2 - a_2 b_1.$$

Thus, for example,

$$\begin{vmatrix} 5 & 8 \\ 4 & 2 \end{vmatrix} = (5 \cdot 2) - (8 \cdot 4) = 10 - 32 = -22$$

$$\text{and} \quad \begin{vmatrix} 4 & 0 \\ 0 & \frac{1}{4} \end{vmatrix} = (4 \cdot \tfrac{1}{4}) - (0 \cdot 0) = 1.$$

We remark on three properties of 2×2 determinants:

1. If the rows or columns of a 2×2 determinant are interchanged, the determinant changes sign:

$$\begin{vmatrix} b_1 & b_2 \\ a_1 & a_2 \end{vmatrix} = - \begin{vmatrix} a_1 & a_2 \\ b_1 & b_2 \end{vmatrix}, \quad \begin{vmatrix} a_2 & a_1 \\ b_2 & b_1 \end{vmatrix} = - \begin{vmatrix} a_1 & a_2 \\ b_1 & b_2 \end{vmatrix}.$$

PROOF Just note that

$$b_1 a_2 - b_2 a_1 = -(a_1 b_2 - a_2 b_1) \quad \text{and} \quad a_2 b_1 - a_1 b_2 = -(a_1 b_2 - a_2 b_1). \quad \square$$

2. A common factor can be removed from any row or column and placed as a multiplier in front of the determinant:

$$\begin{vmatrix} \lambda a_1 & \lambda a_2 \\ b_1 & b_2 \end{vmatrix} = \lambda \begin{vmatrix} a_1 & a_2 \\ b_1 & b_2 \end{vmatrix}, \qquad \begin{vmatrix} \lambda a_1 & a_2 \\ \lambda b_1 & b_2 \end{vmatrix} = \lambda \begin{vmatrix} a_1 & a_2 \\ b_1 & b_2 \end{vmatrix}.$$

PROOF   Just note that

$$(\lambda a_1)b_2 - (\lambda a_2)b_1 = \lambda(a_1 b_2 - a_2 b_1)$$

and $(\lambda a_1)b_2 - a_2(\lambda b_1) = \lambda(a_1 b_2 - a_2 b_1).$ ❏

3. If the rows or columns of a 2×2 determinant are the same, the determinant is 0.

PROOF

$$\begin{vmatrix} a_1 & a_2 \\ a_1 & a_2 \end{vmatrix} = a_1 a_2 - a_2 a_1 = 0, \qquad \begin{vmatrix} a_1 & a_1 \\ b_1 & b_1 \end{vmatrix} = a_1 b_1 - a_1 b_1 = 0.$$ ❏

The determinant of a 3×3 matrix is harder to define.   One definition is this:

$$\begin{vmatrix} a_1 & a_2 & a_3 \\ b_1 & b_2 & b_3 \\ c_1 & c_2 & c_3 \end{vmatrix} = a_1 b_2 c_3 - a_1 b_3 c_2 + a_2 b_3 c_1 - a_2 b_1 c_3 + a_3 b_1 c_2 - a_3 b_2 c_1.$$

The problem with this definition is that it is hard to remember.   What saves us is that the expansion on the right can be conveniently written in terms of 2×2 determinants; namely, the expression on the right can be written

$$a_1(b_2 c_3 - b_3 c_2) - a_2(b_1 c_3 - b_3 c_1) + a_3(b_1 c_2 - b_2 c_1),$$

which turns into

$$a_1 \begin{vmatrix} b_2 & b_3 \\ c_2 & c_3 \end{vmatrix} - a_2 \begin{vmatrix} b_1 & b_3 \\ c_1 & c_3 \end{vmatrix} + a_3 \begin{vmatrix} b_1 & b_2 \\ c_1 & c_2 \end{vmatrix}.$$

We then have

(A.2.2)
$$\begin{vmatrix} a_1 & a_2 & a_3 \\ b_1 & b_2 & b_3 \\ c_1 & c_2 & c_3 \end{vmatrix} = a_1 \begin{vmatrix} b_2 & b_3 \\ c_2 & c_3 \end{vmatrix} - a_2 \begin{vmatrix} b_1 & b_3 \\ c_1 & c_3 \end{vmatrix} + a_3 \begin{vmatrix} b_1 & b_2 \\ c_1 & c_2 \end{vmatrix}.$$

We will take this as our definition.   It is called the *expansion of the determinant along the first row*.   Note that the coefficients are the entries a_1, a_2, a_3 of the first row, that they occur alternately with $+$ and $-$ signs, and that each is multiplied by a determinant.   You can remember which determinant goes with which entry a_i as follows: in the original matrix, mentally cross out the row and column in which the entry a_i is found, and take the determinant of the remaining 2×2 matrix.   For example, the determinant that goes with a_3 is

$$\begin{vmatrix} \cancel{a_1} & \cancel{a_2} & \cancel{a_3} \\ b_1 & b_2 & \cancel{b_3} \\ c_1 & c_2 & \cancel{c_3} \end{vmatrix} = \begin{vmatrix} b_1 & b_2 \\ c_1 & c_2 \end{vmatrix}.$$

When first starting to work with specific 3×3 determinants, it is a good idea to set up the formula with blank 2×2 determinants:

$$\begin{vmatrix} a_1 & a_2 & a_3 \\ b_1 & b_2 & b_3 \\ c_1 & c_2 & c_3 \end{vmatrix} = a_1 \begin{vmatrix} & \\ & \end{vmatrix} - a_2 \begin{vmatrix} & \\ & \end{vmatrix} + a_3 \begin{vmatrix} & \\ & \end{vmatrix}$$

and then fill in the 2×2 determinants by the "crossing out" rule explained above.

Example 1

$$\begin{vmatrix} 1 & 2 & 1 \\ 0 & 3 & 4 \\ 6 & 2 & 5 \end{vmatrix} = 1 \begin{vmatrix} 3 & 4 \\ 2 & 5 \end{vmatrix} - 2 \begin{vmatrix} 0 & 4 \\ 6 & 5 \end{vmatrix} + 1 \begin{vmatrix} 0 & 3 \\ 6 & 2 \end{vmatrix}$$

$$= 1(15 - 8) - 2(0 - 24) + 1(0 - 18)$$
$$= 7 + 48 - 18 = 37. \quad \square$$

A straightforward (but somewhat laborious) calculation shows that 3×3 determinants have the same three properties we proved earlier for 2×2 determinants:

1. If two rows or columns are interchanged, the determinant changes sign.

2. A common factor can be removed from any row or column and placed as a multiplier in front of the determinant.

3. If two rows or columns are the same, the determinant is 0.

EXERCISES A.2

Evaluate the following determinants.

1. $\begin{vmatrix} 1 & 2 \\ 3 & 4 \end{vmatrix}$.

2. $\begin{vmatrix} 1 & -1 \\ -1 & 1 \end{vmatrix}$.

3. $\begin{vmatrix} 1 & 1 \\ a & a \end{vmatrix}$.

4. $\begin{vmatrix} a & b \\ b & d \end{vmatrix}$.

5. $\begin{vmatrix} 1 & 0 & 3 \\ 2 & 4 & 1 \\ 0 & 1 & 0 \end{vmatrix}$.

6. $\begin{vmatrix} 1 & 0 & 0 \\ 0 & 2 & 0 \\ 0 & 0 & 3 \end{vmatrix}$.

7. $\begin{vmatrix} 0 & 0 & 1 \\ 0 & 2 & 0 \\ 3 & 0 & 0 \end{vmatrix}$.

8. $\begin{vmatrix} a & 0 & 0 \\ b & c & 0 \\ d & e & f \end{vmatrix}$.

9. If A is a matrix, its *transpose* A^T is obtained by interchanging the rows and columns. Thus

$$\begin{pmatrix} a_1 & a_2 \\ b_1 & b_2 \end{pmatrix}^T = \begin{pmatrix} a_1 & b_1 \\ a_2 & b_2 \end{pmatrix}$$

and

$$\begin{pmatrix} a_1 & a_2 & a_3 \\ b_1 & b_2 & b_3 \\ c_1 & c_2 & c_3 \end{pmatrix}^T = \begin{pmatrix} a_1 & b_1 & c_1 \\ a_2 & b_2 & c_2 \\ a_3 & b_3 & c_3 \end{pmatrix}.$$

Show that the determinant of a matrix equals the determinant of its transpose: (a) the 2×2 case. (b) the 3×3 case.

Verify the assertions in Exercises 10–14.

10. $\begin{vmatrix} 1 & 2 & 3 \\ 4 & 5 & 6 \\ 7 & 8 & 9 \end{vmatrix} + \begin{vmatrix} 4 & 5 & 6 \\ 1 & 2 & 3 \\ 7 & 8 & 9 \end{vmatrix} = 0$.

11. $\begin{vmatrix} 1 & 2 & 3 \\ 4 & 5 & 6 \\ 7 & 8 & 9 \end{vmatrix} = \begin{vmatrix} 4 & 5 & 6 \\ 7 & 8 & 9 \\ 1 & 2 & 3 \end{vmatrix}$.

12. $\begin{vmatrix} 1 & 2 & 3 \\ 4 & 5 & 6 \\ 7 & 8 & 9 \end{vmatrix} + \begin{vmatrix} 1 & 2 & 3 \\ 1 & 2 & 3 \\ 7 & 8 & 9 \end{vmatrix} = \begin{vmatrix} 1 & 2 & 3 \\ 4 & 5 & 6 \\ 7 & 8 & 9 \end{vmatrix}$.

13. $\frac{1}{2} \begin{vmatrix} 1 & 0 & 7 \\ 3 & 4 & 5 \\ 2 & 4 & 6 \end{vmatrix} = \begin{vmatrix} 1 & 0 & 7 \\ 3 & 4 & 5 \\ 1 & 2 & 3 \end{vmatrix}$.

14. $\begin{vmatrix} 1 & 2 & 3 \\ x & 2x & 3x \\ 4 & 5 & 6 \end{vmatrix} = 0$.

15. (a) Verify that the equations

$$3x + 4y = 6$$
$$2x - 3y = 7$$

are solved by the prescription

$$x = \frac{\begin{vmatrix} 6 & 4 \\ 7 & -3 \end{vmatrix}}{\begin{vmatrix} 3 & 4 \\ 2 & -3 \end{vmatrix}}, \qquad y = \frac{\begin{vmatrix} 3 & 6 \\ 2 & 7 \end{vmatrix}}{\begin{vmatrix} 3 & 4 \\ 2 & -3 \end{vmatrix}}.$$

(b) More generally, verify that the equations

$$a_1 x + a_2 y = d$$
$$b_1 x + b_2 y = e$$

are solved by the prescription

$$x = \frac{\begin{vmatrix} d & a_2 \\ e & b_2 \end{vmatrix}}{\begin{vmatrix} a_1 & a_2 \\ b_1 & b_2 \end{vmatrix}}, \qquad y = \frac{\begin{vmatrix} a_1 & d \\ b_1 & e \end{vmatrix}}{\begin{vmatrix} a_1 & a_2 \\ b_1 & b_2 \end{vmatrix}}$$

provided that the determinant in the denominator is different from 0.

(c) Conjecture an analogous rule for solving three linear equations in three unknowns.

16. Show that a 3×3 determinant can be "expanded along the bottom row" as follows:

$$\begin{vmatrix} a_1 & a_2 & a_3 \\ b_1 & b_2 & b_3 \\ c_1 & c_2 & c_3 \end{vmatrix} = c_1 \begin{vmatrix} a_2 & a_3 \\ b_2 & b_3 \end{vmatrix} - c_2 \begin{vmatrix} a_1 & a_3 \\ b_1 & b_3 \end{vmatrix} + c_3 \begin{vmatrix} a_1 & a_2 \\ b_1 & b_2 \end{vmatrix}$$

HINT: You can check this directly by writing out the values of the determinants on the right, or you can interchange rows twice to bring the bottom row to the top and then use expansion along the top row.

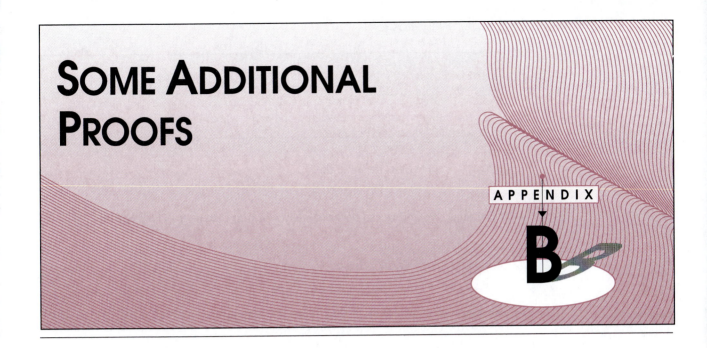

SOME ADDITIONAL PROOFS

In this appendix we present some proofs that many would consider too advanced for the main body of the text. Some details are omitted. These are left to you.

The arguments presented in Sections B.1, B.2, and B.4 require some familiarity with the *least upper bound axiom.* This is discussed in Section 10.1. In addition, Section B.4 requires some understanding of *sequences,* for which we refer you to Sections 10.2 and 10.3.

■ B.1 THE INTERMEDIATE-VALUE THEOREM

> **LEMMA B.1.1**
>
> Let f be continuous on $[a, b]$. If $f(a) < 0 < f(b)$ or $f(b) < 0 < f(a)$, then there is a number c between a and b for which $f(c) = 0$.

PROOF Suppose that $f(a) < 0 < f(b)$. (The other case can be treated in a similar manner.) Since $f(a) < 0$, we know from the continuity of f that there exists a number ξ such that f is negative on $[a, \xi)$. Let

$$c = \text{lub } \{\xi : f \text{ is negative on } [a, \xi)\}.$$

Clearly, $c \le b$. We cannot have $f(c) > 0$, for then f would be positive on some interval extending to the left of c, and we know that, to the left of c, f is negative. Incidentally this argument excludes the possibility $c = b$ and means that $c < b$. We cannot have $f(c) < 0$, for then there would be an interval $[a, t)$, with $t > c$, on which f is negative, and this would contradict the definition of c. It follows that $f(c) = 0$. ❑

THEOREM B.1.2   THE INTERMEDIATE-VALUE THEOREM

If f is continuous on $[a, b]$ and C is a number between $f(a)$ and $f(b)$, then there is at least one number c between a and b for which $f(c) = C$.

PROOF   Suppose for example that

$$f(a) < C < f(b).$$

(The other possibility can be handled in a similar manner.)   The function

$$g(x) = f(x) - C$$

is continuous on $[a, b]$.   Since

$$g(a) = f(a) - C < 0 \quad \text{and} \quad g(b) = f(b) - C > 0,$$

we know from the lemma that there is a number c between a and b for which $g(c) = 0$.   Obviously, then, $f(c) = C$.   ❑

■ B.2 THE MAXIMUM-MINIMUM THEOREM

LEMMA B.2.1

If f is continuous on $[a, b]$, then f is bounded on $[a, b]$.

PROOF   Consider

$$\{x: x \in [a, b] \text{ and } f \text{ is bounded on } [a, x]\}.$$

It is easy to see that this set is nonempty and bounded above by b.   Thus we can set

$$c = \text{lub } \{x: f \text{ is bounded on } [a, x]\}.$$

Now we argue that $c = b$.   To do so, we suppose that $c < b$.   From the continuity of f at c, it is easy to see that f is bounded on $[c - \epsilon, c + \epsilon]$ for some $\epsilon > 0$.   Being bounded on $[a, c - \epsilon]$ and on $[c - \epsilon, c + \epsilon]$, it is obviously bounded on $[a, c + \epsilon]$. This contradicts our choice of c.   We can therefore conclude that $c = b$.   This tells us that f is bounded on $[a, x]$ for all $x < b$.   We are now almost through.   From the continuity of f, we know that f is bounded on some interval of the form $[b - \epsilon, b]$. Since $b - \epsilon < b$, we know from what we have just proved that f is bounded on $[a, b - \epsilon]$.   Being bounded on $[a, b - \epsilon]$ and bounded on $[b - \epsilon, b]$, it is bounded on $[a, b]$.   ❑

THEOREM B.2.2   THE MAXIMUM-MINIMUM THEOREM

If f is continuous on $[a, b]$, then f takes on both a maximum value M and a minimum value m on $[a, b]$.

PROOF By the lemma, f is bounded on $[a, b]$. Set

$$M = \text{lub } \{f(x): x \in [a, b]\}.$$

We must show that there exists c in $[a, b]$ such that $f(c) = M$. To do this, we set

$$g(x) = \frac{1}{M - f(x)}.$$

If f does not take on the value M, then g is continuous on $[a, b]$ and thus, by the lemma, bounded on $[a, b]$. A look at the definition of g makes it clear that g cannot be bounded on $[a, b]$. The assumption that f does not take on the value M has led to a contradiction. (That f takes on a minimum value m can be proved in a similar manner.) ❑

■ B.3 INVERSES

THEOREM B.3.1 CONTINUITY OF THE INVERSE

Let f be a one-to-one function defined on an interval (a, b). If f is continuous, then its inverse f^{-1} is also continuous.

PROOF If f is continuous, then, being one-to-one, f either increases throughout (a, b) or it decreases throughout (a, b). The proof of this assertion we leave to you.

Let's suppose now that f increases throughout (a, b). Let's take c in the domain of f^{-1} and show that f^{-1} is continuous at c.

We first observe that $f^{-1}(c)$ lies in (a, b) and choose $\epsilon > 0$ sufficiently small so that $f^{-1}(c) - \epsilon$ and $f^{-1}(c) + \epsilon$ also lie in (a, b). We seek $\delta > 0$ such that

$$\text{if } c - \delta < x < c + \delta, \quad \text{then} \quad f^{-1}(c) - \epsilon < f^{-1}(x) < f^{-1}(c) + \epsilon.$$

This condition can be met by choosing δ to satisfy

$$f(f^{-1}(c) - \epsilon) < c - \delta \quad \text{and} \quad c + \delta < f(f^{-1}(c) + \epsilon)$$

for then, if $c - \delta < x < c + \delta$, then

$$f(f^{-1}(c) - \epsilon) < x < f(f^{-1}(c) + \epsilon),$$

and, since f^{-1} also increases,

$$f^{-1}(c) - \epsilon < f^{-1}(x) < f^{-1}(c) + \epsilon.$$

The case where f decreases throughout (a, b) can be handled in a similar manner. ❑

THEOREM B.3.2 DIFFERENTIABILITY OF THE INVERSE

Let f be a one-to-one function defined on an interval (a, b). If f is differentiable and its derivative does not take on the value 0, then f^{-1} is differentiable and

$$(f^{-1})'(x) = \frac{1}{f'(f^{-1}(x))}.$$

PROOF (Here we use the characterization of derivative spelled out in Theorem 3.5.8.) Let x be in the domain of f^{-1}. We take $\epsilon > 0$ and show that there exists $\delta > 0$ such that

$$\text{if } 0 < |t - x| < \delta, \quad \text{then} \quad \left| \frac{f^{-1}(t) - f^{-1}(x)}{t - x} - \frac{1}{f'(f^{-1}(x))} \right| < \epsilon.$$

Since f is differentiable at $f^{-1}(x)$ and $f'(f^{-1}(x)) \neq 0$, there exists $\delta_1 > 0$ such that

$$\text{if } 0 < |y - f^{-1}(x)| < \delta_1, \quad \text{then} \quad \left| \frac{1}{\dfrac{f(y) - f(f^{-1}(x))}{y - f^{-1}(x)}} - \frac{1}{f'(f^{-1}(x))} \right| < \epsilon$$

and therefore

$$\left| \frac{y - f^{-1}(x)}{f(y) - f(f^{-1}(x))} - \frac{1}{f'(f^{-1}(x))} \right| < \epsilon.$$

By the previous theorem, f^{-1} is continuous at x and therefore there exists $\delta > 0$ such that

$$\text{if } 0 < |t - x| < \delta, \quad \text{then} \quad 0 < |f^{-1}(t) - f^{-1}(x)| < \delta_1.$$

It follows from the special property of δ_1 that

$$\left| \frac{f^{-1}(t) - f^{-1}(x)}{t - x} - \frac{1}{f'(f^{-1}(x))} \right| < \epsilon. \quad \square$$

■ B.4 THE INTEGRABILITY OF CONTINUOUS FUNCTIONS

The aim here is to prove that, if f is continuous on $[a, b]$, then there is one and only one number I that satisfies the inequality

$$L_f(P) \leq I \leq U_f(P) \qquad \text{for all partitions } P \text{ of } [a, b].$$

DEFINITION B.4.1

A function f is said to be *uniformly continuous* on $[a, b]$ iff for each $\epsilon > 0$ there exists $\delta > 0$ such that

$$\text{if } x, y \in [a, b] \text{ and } |x - y| < \delta, \quad \text{then } |f(x) - f(y)| < \epsilon.$$

For convenience, let's agree to say that *the interval $[a, b]$ has the property P_ϵ* iff there exist sequences $\{x_n\}, \{y_n\}$ satisfying

$$x_n, y_n \in [a, b], \qquad |x_n - y_n| < 1/n, \qquad |f(x_n) - f(y_n)| \geq \epsilon.$$

LEMMA B.4.2

If f is not uniformly continuous on $[a, b]$, then $[a, b]$ has the property P_ϵ for some $\epsilon > 0$.

PROOF If f is not uniformly continuous on $[a, b]$, then there is no $\delta > 0$ such that

$$\text{if } x, y \in [a, b] \text{ and } |x - y| < \delta, \qquad \text{then } |f(x) - f(y)| < \epsilon.$$

The interval $[a, b]$ has the property P_ϵ for that choice of ϵ. The details of the argument are left to you. ❏

LEMMA B.4.3

Let f be continuous on $[a, b]$. If $[a, b]$ has the property P_ϵ, then at least one of the subintervals $[a, \frac{1}{2}(a + b)]$, $[\frac{1}{2}(a + b), b]$ has the property P_ϵ.

PROOF Let's suppose that the lemma is false. For convenience, we let $c = \frac{1}{2}(a + b)$, so that the halves become $[a, c]$ and $[c, b]$. Since $[a, c]$ fails to have the property P_ϵ, there exists an integer p such that

$$\text{if } x, y \in [a, c] \text{ and } |x - y| < 1/p, \qquad \text{then } |f(x) - f(y)| < \epsilon.$$

Since $[c, b]$ fails to have the property P_ϵ, there exists an integer q such that

$$\text{if } x, y \in [c, b] \text{ and } |x - y| < 1/q, \qquad \text{then} \qquad |f(x) - f(y)| < \epsilon.$$

Since f is continuous at c, there exists an integer r such that, if $|x - c| < 1/r$, then $|f(x) - f(c)| < \frac{1}{2}\epsilon$. Set $s = \max\{p, q, r\}$ and suppose that

$$x, y \in [a, b], \qquad |x - y| < 1/s.$$

If x, y are both in $[a, c]$ or both in $[c, b]$, then

$$|f(x) - f(y)| < \epsilon.$$

The only other possibility is that $x \in [a, c]$ and $y \in [c, b]$. In this case we have

$$|x - c| < 1/r, \qquad |y - c| < 1/r,$$

and thus

$$|f(x) - f(c)| < \tfrac{1}{2}\epsilon, \qquad |f(y) - f(c)| < \tfrac{1}{2}\epsilon.$$

By the triangle inequality, we again have

$$|f(x) - f(y)| < \epsilon.$$

In summary, we have obtained the existence of an integer s with the property that

$$x, y \in [a, b], |x - y| < 1/s \qquad \text{implies} \qquad |f(x) - f(y)| < \epsilon.$$

Hence $[a, b]$ does not have the property P_ϵ. This is a contradiction and proves the lemma. ❏

THEOREM B.4.4

If f is continuous on $[a, b]$, then f is uniformly continuous on $[a, b]$.

PROOF   We suppose that f is not uniformly continuous on $[a, b]$ and base our argument on a mathematical version of "bisection."

By the first lemma of this section, we know that $[a, b]$ has the property P_ϵ for some $\epsilon > 0$. We bisect $[a, b]$ and note by the second lemma that one of the halves, say $[a_1, b_1]$, has the property P_ϵ. We then bisect $[a_1, b_1]$ and note that one of the halves, say $[a_2, b_2]$, has the property P_ϵ. Continuing in this manner, we obtain a sequence of intervals $[a_n, b_n]$, each with the property P_ϵ. Then for each n, we can choose x_n, $y_n \in [a_n, b_n]$ such that

$$|x_n - y_n| < 1/n \qquad \text{and} \qquad |f(x_n) - f(y_n)| \geq \epsilon.$$

Since

$$a \leq a_n \leq a_{n+1} < b_{n+1} \leq b_n \leq b,$$

we see that the sequences $\{a_n\}$ and $\{b_n\}$ are both bounded and monotonic. Thus they are convergent. Since $b_n - a_n \to 0$, we see that $\{a_n\}$ and $\{b_n\}$ both converge to the same limit, say L. From the inequality

$$a_n \leq x_n \leq y_n \leq b_n,$$

we conclude that

$$x_n \to L \quad \text{and} \quad y_n \to L.$$

This tells us that

$$|f(x_n) - f(y_n)| \to |f(L) - f(L)| = 0,$$

which contradicts the statement that $|f(x_n) - f(y_n)| \geq \epsilon$ for all n.   ❏

LEMMA B.4.5

If P and Q are partitions of $[a, b]$, then $L_f(P) \leq U_f(Q)$.

PROOF $P \cup Q$ is a partition of $[a, b]$ that contains both P and Q. It is obvious then that

$$L_f(P) \leq L_f(P \cup Q) \leq U_f(P \cup Q) \leq U_f(Q).   ❏$$

From the last lemma it follows that the set of all lower sums is bounded above and has a least upper bound L. The number L satisfies the inequality

$$L_f(P) \leq L \leq U_f(P) \qquad \text{for all partitions } P$$

and is clearly the least of such numbers. Similarly, we find that the set of all upper sums is bounded below and has a greatest lower bound U. The number U satisfies the inequality

$$L_f(P) \leq U \leq U_f(P) \qquad \text{for all partitions } P$$

and is clearly the greatest of such numbers.

We are now ready to prove the basic theorem.

> **THEOREM B.4.6 THE INTEGRABILITY THEOREM**
>
> If f is continuous on $[a, b]$, then there exists one and only one number I that satisfies the inequality
>
> $$L_f(P) \leq I \leq U_f(P) \qquad \text{for all partitions } P \text{ of } [a, b].$$

PROOF We know that

$$L_f(P) \leq L \leq U \leq U_f(P) \qquad \text{for all } P,$$

so that existence is no problem. We will have uniqueness if we can prove that

$$L = U.$$

To do this, we take $\epsilon > 0$ and note that f, being continuous on $[a, b]$, is uniformly continuous on $[a, b]$. Thus there exists $\delta > 0$ such that, if

$$x, y \in [a, b] \text{ and } |x - y| < \delta, \qquad \text{then} \qquad |f(x) - f(y)| < \frac{\epsilon}{b - a}.$$

We now choose a partition $P = \{x_0, x_1, \ldots, x_n\}$ for which max $\Delta x_i < \delta$. For this partition P, we have

$$U_f(P) - L_f(P) = \sum_{i=1}^{n} M_i \, \Delta x_i - \sum_{i=1}^{n} m_i \, \Delta x_i$$

$$= \sum_{i=1}^{n} (M_i - m_i) \, \Delta x_i$$

$$< \sum_{i=1}^{n} \frac{\epsilon}{b - a} \Delta x_i = \frac{\epsilon}{b - a} \sum_{i=1}^{n} \Delta x_i = \frac{\epsilon}{b - a}(b - a) = \epsilon.$$

Since

$$U_f(P) - L_f(P) < \epsilon \qquad \text{and} \qquad 0 \leq U - L \leq U_f(P) - L_f(P),$$

you can see that

$$0 \leq U - L < \epsilon.$$

Since ϵ was chosen arbitrarily, we must have $U - L = 0$ and $L = U$. ❑

■ B.5 THE INTEGRAL AS THE LIMIT OF RIEMANN SUMS

For the notation we refer to Section 5.10.

> **THEOREM B.5.1**
>
> If f is continuous on $[a, b]$, then
>
> $$\int_a^b f(x) \, dx = \lim_{\|P\| \to 0} S^*(P).$$

PROOF Let $\epsilon > 0$. We must show that there exists $\delta > 0$ such that

$$\text{if } \|P\| < \delta, \quad \text{then} \quad \left| S^*(P) - \int_a^b f(x)\, dx \right| < \epsilon.$$

From the proof of Theorem B.4.6 we know that there exists $\delta > 0$ such that

$$\text{if } \|P\| < \delta, \quad \text{then} \quad U_f(P) - L_f(P) < \epsilon.$$

For such P we have

$$U_f(P) - \epsilon < L_f(P) \le S^*(P) \le U_f(P) < L_f(P) + \epsilon.$$

This gives

$$\int_a^b f(x)\, dx - \epsilon < S^*(P) < \int_a^b f(x)\, dx + \epsilon,$$

and therefore

$$\left| S^*(P) - \int_a^b f(x)\, dx \right| < \epsilon. \quad \square$$

CHAPTER 1
SECTION 1.2

1. rational   **3.** rational   **5.** integer, rational   **7.** integer, rational   **9.** integer, rational   **11.** irrational   **13.** $\dfrac{13188}{999}$

15. $\dfrac{23}{99}$ **17.** $\dfrac{5247}{999}$ **19.** $=$ **21.** $>$ **23.** $<$ **25.** 6   **27.** 4   **29.** 13   **31.** $5 - \sqrt{5}$

33.   **35.**   **37.**

39.   **41.**   **43.**

45.   **47.**   **49.**

51. bounded; lower bound 0, upper bound 4   **53.** not bounded   **55.** not bounded   **57.** bounded above, $\sqrt{2}$ is an upper bound.

59. $(x - 5)^2$ **61.** $8(x^2 + 2)(x^4 - 2x^2 + 4)$ **63.** $(2x + 3)^2$ **65.** $2, -1$ **67.** 3   **69.** none   **71.** $x = 0; R(1) = 0$

73. $R(x)$ is real for all x; $R(-1) = R(1) = 0$ **75.** $R(x)$ is real for all x; $R(-\frac{3}{2}) = 0$ **77.** $\dfrac{1}{336}$ **79.** 84

81. If r and $r + s$ are rational, then $s = (r + s) - r$ is rational.

83. The product could be either rational or irrational; $0 \cdot \sqrt{2} = 0$ is rational, $1 \cdot \sqrt{2} = \sqrt{2}$ is irrational.

85. Suppose $\sqrt{2} = \dfrac{p}{q}$, where p, q are integers with no common divisor (other than ± 1). Then $p^2 = 2q^2$, which implies p^2 is even, which, in turn, implies $p = 2r$ is even. Thus $4r^2 = 2q^2$, which implies q^2 and hence q are even. This is a contradiction.

87. If the length of a rectangle with perimeter P is x, $0 < x < \frac{1}{2}P$, then the width is $\frac{1}{2}P - x$ and the area A is $A(x) = x(\frac{1}{2}P - x) = \left(\dfrac{P}{4}\right)^2 - \left(x - \dfrac{P}{4}\right)^2$. Clearly A is a maximum when $x = P/4$. The width in this case is also $P/4$, and the rectangle is a square.

SECTION 1.3

1. $(-\infty, 1)$ **3.** $(-\infty, -3]$ **5.** $(-\infty, -\frac{1}{5})$ **7.** $(-1, 1)$ **9.** $(0, 1) \cup (2, \infty)$ **11.** $[0, \infty)$ **13.** $(-\infty, 2 - \sqrt{3}) \cup (2 + \sqrt{3}, \infty)$

15. $(-\infty, 0) \cup (0, \infty)$ **17.** $(-1, 0) \cup (1, \infty)$ **19.** $(-\infty, 0] \cup (5, \infty)$ **21.** $(-\infty, -\frac{5}{3}) \cup (5, \infty)$ **23.** $(-3, -1) \cup (3, \infty)$

25. $(0, 2)$ **27.** $(-\infty, -6) \cup (2, \infty)$ **29.** $(-2, 0) \cup (2, \infty)$ **31.** $(1, 2) \cup (6, \infty)$ **33.** $(-\infty, 1) \cup (3, 5)$

35. $(-\infty, 0) \cup (0, 1) \cup (3, \infty)$ **37.** $(-2, 2)$ **39.** $(-\infty, -3) \cup (3, \infty)$ **41.** $(\frac{3}{2}, \frac{5}{2})$ **43.** $(-1, 0) \cup (0, 1)$ **45.** $(\frac{3}{2}, 2) \cup (2, \frac{5}{2})$

47. $(-5, 3) \cup (3, 11)$ **49.** $(-\frac{5}{8}, -\frac{3}{8})$ **51.** $(-\infty, -4) \cup (-1, \infty)$ **53.** $(-\infty, -\frac{8}{3}) \cup (2, \infty)$ **55.** $|x| < 3$ **57.** $|x - 2| < 5$

59. $|x + 2| < 5$ **61.** $0 < A \leq \frac{3}{2}$ **63.** $A \geq 6$ **65.** $x < \sqrt{x} < 1 < \dfrac{1}{\sqrt{x}} < \dfrac{1}{x}$

67. $a < b$ and $ab > 0$ implies   **69.** $b - a = (\sqrt{b} + \sqrt{a})(\sqrt{b} - \sqrt{a})$; since $\sqrt{b} + \sqrt{a} \geq 0$, $b - a$ and $\sqrt{b} - \sqrt{a}$ have the same sign.

$\dfrac{1}{b} - \dfrac{1}{a} = \dfrac{a - b}{ab} < 0.$

The result follows.

71. $\big||a| - |b|\big|^2 = (|a| - |b|)^2 = |a|^2 - 2|a||b| + |b|^2 = a^2 - 2|a||b| + b^2 \leq a^2 - 2ab + b^2 = (a - b)^2$

$\qquad\qquad\qquad\qquad\qquad\qquad\qquad\qquad\qquad\qquad\uparrow\_\_\_\_\_ (ab \leq |ab|)$

Thus, $\big||a| - |b|\big| \leq \sqrt{(a - b)^2} = |a - b|$.

73. $0 \leq a \leq b$ implies

$$\frac{a}{1+a} - \frac{b}{1+b} = \frac{a-b}{(1+a)(1+b)} < 0.$$

The result follows.

75. $a < b$ implies

$$a - \frac{a+b}{2} = \frac{a-b}{2} < 0$$

and

$$\frac{a+b}{2} - b = \frac{a-b}{2} < 0.$$

Thus

$$a < \frac{a+b}{2} < b;$$

$(a+b)/2$ is the midpoint of the line segment $\overline{ab}$.

SECTION 1.4

1. 10   **3.** $4\sqrt{5}$ **5.** 6   **7.** (4, 6)   **9.** $(\frac{9}{2}, -3)$ **11.** $\left(\frac{\sqrt{3}}{2}, \frac{\sqrt{3}}{2}\right)$ **13.** $-\frac{2}{3}$ **15.** 0   **17.** -1 **19.** $-y_0/x_0$

21. slope 2, y-intercept -4 **23.** slope $\frac{1}{3}$, y-intercept 2   **25.** slope undefined, no y-intercept   **27.** slope $\frac{7}{3}$, y-intercept $\frac{4}{3}$

29. $y = 5x + 2$ **31.** $y = -5x + 2$ **33.** $y = 3$ **35.** $x = -3$ **37.** $y = 7$ **39.** $3y - 2x - 17 = 0$ **41.** $2y + 3x - 20 = 0$

43. $45°$ **45.** $90°$ **47.** approx. $143°$ **49.** $y = \frac{1}{3}\sqrt{3}\,x + 2$ **51.** $y = -\sqrt{3}\,x + 3$ **53.** $(\frac{1}{2}\sqrt{2}, \frac{1}{2}\sqrt{2}), (-\frac{1}{2}\sqrt{2}, -\frac{1}{2}\sqrt{2})$

55. (3, 4)   **57.** (1, 1); approx. $39°$ **59.** $(-\frac{2}{23}, \frac{38}{23})$; approx. $17°$ **61.** (a) $\frac{2}{13}$ (b) $\frac{29}{13}$ **63.** (0, 1) is closest; $(-1, 1)$ is farthest away

65. $\frac{17}{2}$ **67.** parabola, vertex $(-1, -1)$ **69.** ellipse, center $(2, -1)$ **71.** hyperbola, center $(1, 2)$ **73.** hyperbola, center $(3, -2)$

75. $-\frac{5}{12}$ **77.** $x - 2y - 3 = 0$ **79.** $3x + 13y - 40 = 0$ **81.** isosceles right triangle   **83.** isosceles right triangle

85. The midpoint of the hypotenuse is $M(\frac{a}{2}, \frac{b}{2})$. The distance between M and the origin, M and $(0, b)$, and M and $(a, 0)$ is $\frac{1}{2}\sqrt{a^2 + b^2}$

87. $(1, \frac{10}{3})$ **89.** If A (0, 0) and B (a, 0) are two vertices of a parallelogram, and C (b, c) is the vertex opposite B, then D (a + b, c) is the vertex opposite A. The midpoint of the diagonal AD = midpoint of the diagonal $BC = (\frac{1}{2}(a + b), \frac{1}{2}c)$.

91. $F = \frac{9}{5}C + 32; -40°$

SECTION 1.5

1. $f(0) = 2, f(1) = 1, f(-2) = 16, f(\frac{3}{2}) = 2$ **3.** $f(0) = 0, f(1) = \sqrt{3}, \quad f(-2) = 0, f(\frac{3}{2}) = \dfrac{\sqrt{21}}{2}$

5. $f(0) = 0, f(1) = \frac{1}{2}, f(-2) = -1, f(\frac{3}{2}) = \frac{12}{23}$ **7.** $f(-x) = x^2 + 2x, f\left(\dfrac{1}{x}\right) = \dfrac{1}{x^2} - \dfrac{2}{x}, f(a + b) = (a + b)^2 - 2(a + b)$

9. $f(-x) = \sqrt{1 + x^2}, f\left(\dfrac{1}{x}\right) = \dfrac{|x|}{\sqrt{1 + x^2}}, f(a + b) = \sqrt{a^2 + 2ab + b^2 + 1}$ **11.** $2a^2 + 4ah + 2h^2 - 3a - 3h; 4a - 3 + 2h$

13. 1, 3   **15.** -2 **17.** 3, -3 **19.** dom $(f) = (-\infty, \infty)$; range $(f) = [0, \infty)$ **21.** dom $(f) = (-\infty, \infty)$; range $(f) = (-\infty, \infty)$

23. dom $(f) = (-\infty, 0) \cup (0, \infty)$; range $(f) = (0, \infty)$ **25.** dom $(f) = (-\infty, 1]$; range $(f) = [0, \infty)$

27. dom $(f) = (-\infty, 7]$; range $(f) = [-1, \infty)$ **29.** dom $(f) = (-\infty, 2)$; range $(f) = (0, \infty)$

31. horizontal line one unit above x-axis   **33.** line through the origin with slope 2   **35.** line through the origin with slope $\frac{1}{2}$

37. line through (0, 2) with slope $\frac{1}{2}$ **39.** upper semicircle of radius 2 centered at the origin

41.

43.

45. dom $(f) = (-\infty, 0) \cup (0, \infty)$; range $(f) = \{-1, 1\}$

47. dom $(f) = [0, \infty)$; range $(f) = [1, \infty)$

49. yes, dom$(f) = [-2, 2]$; range $(f) = [-2, 2]$ **51.** no   **53.** odd   **55.** neither   **57.** even

59. (a)

61.

(b) $x_1 = -6.566$, $x_2 = -0.493$, $x_3 = 5.559$
(c) $A(-4, 28.667)$, $B(3, 28.500)$

$-5 \le x \le 8$, $0 \le y \le 100$

63. $A = \dfrac{C^2}{4\pi}$, where C is the circumference; dom $(A) = [0, \infty)$ **65.** $V = s^{3/2}$, where s is the area of a face; dom $V = [0, \infty)$

67. $S = 3d^2$, where d is the diagonal of a face; dom $(S) = [0, \infty)$ **69.** $A = \dfrac{\sqrt{3}}{4}x^2$, where x is the length of a side; dom $(A) = [0, \infty)$

71. $V = 108x^2 - 4x^3$, $0 < x \le 27$ **73.** $A = \dfrac{15x}{2} - \dfrac{x^2}{2} + \dfrac{\pi x^2}{8}$, $0 < x < \dfrac{30}{\pi + 2}$ **75.** $A = x\sqrt{d^2 - x^2}$, $0 < x < d$

77. $A = bx - \dfrac{b}{a}x^2$, $0 \le x \le a$ **79.** $A = \dfrac{P^2}{16} + \dfrac{(28 - P)^2}{4\pi}$, $0 \le P \le 28$

SECTION 1.6

1. polynomial, degree 0   **3.** rational function   **5.** neither   **7.** neither   **9.** neither

11. dom $(f) = (-\infty, \infty)$ **13.** dom $(f) = (-\infty, \infty)$ **15.** dom $(f) = (-\infty, -2) \cup (-2, 2) \cup (2, \infty)$

       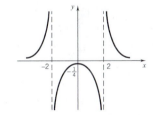

17. $\dfrac{5\pi}{4}$ **19.** $-\dfrac{5\pi}{3}$ **21.** $\dfrac{\pi}{12}$ **23.** $-270°$ **25.** $300°$ **27.** $114.59°$ **29.** $\dfrac{\pi}{6}, \dfrac{5\pi}{6}$ **31.** $\dfrac{\pi}{2}$ **33.** $\dfrac{\pi}{4}, \dfrac{7\pi}{4}$

35. $\dfrac{\pi}{4}, \dfrac{3\pi}{4}, \dfrac{5\pi}{4}, \dfrac{7\pi}{4}$ **37.** 0.7772 **39.** 0.7101 **41.** 3.1524 **43.** -3.8611 **45.** -2.8974 **47.** 0.5505 **49.** 1.4231

51. 1.7997 **53.** dom$(f) = (-\infty, \infty)$; range $(f) = [0, 1]$ **55.** dom$(f) = (-\infty, \infty)$; range $(f) = [-2, 2]$

57. dom $(f) = \left(k\pi - \dfrac{\pi}{2}, k\pi + \dfrac{\pi}{2}\right)$, $k = 0, \pm 1, \pm 2, \ldots$; range $(f) = [1, \infty)$

59.

61.

63.

65. odd   **67.** even   **69.** odd   **71.** Let $m_1 = \tan \theta_1$, $m_2 = \tan \theta_2$, $\alpha = |\theta_2 - \theta_1|$

$$\tan \alpha = |\tan (\theta_2 - \theta_1)| = \left| \frac{\tan \theta_2 - \tan \theta_1}{1 + \tan \theta_2 \tan \theta_1} \right| = \left| \frac{m_2 - m_1}{1 + m_2 m_1} \right|$$

73. $b \sin A = a \sin B = h$, so $\dfrac{\sin A}{a} = \dfrac{\sin B}{b}$; similarly, $\dfrac{\sin A}{a} = \dfrac{\sin C}{c}$

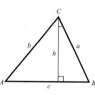

75. $A = \frac{1}{2}ah = \frac{1}{2}a^2 \sin \theta$

77.

(b)

79.

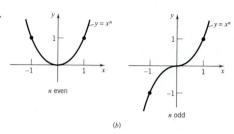

n even

n odd

(b)

(c) $f_k(x) \geq f_{k+1}(x)$ on $[0, 1]$; $f_{k+1}(x) \geq f_k(x)$ on $[1, \infty)$

SECTION 1.7

1. $\frac{15}{2}$ **3.** $\frac{105}{2}$ **5.** $\frac{-27}{4}$ **7.** 3

9. $(f + g)(x) = x - 1$; domain $(-\infty, \infty)$
  $(f - g)(x) = 3x - 5$; domain $(-\infty, \infty)$
  $(f \cdot g)(x) = -2x^2 + 7x - 6$; domain $(-\infty, \infty)$
  $\left(\dfrac{f}{g}\right)(x) = \dfrac{2x - 3}{2 - x}$; domain: all real numbers except $x = 2$

11. $(f + g)(x) = x + \sqrt{x - 1} - \sqrt{x + 1}$; domain $[1, \infty)$
  $(f - g)(x) = \sqrt{x - 1} + \sqrt{x + 1} - x$; domain $[1, \infty)$
  $(f \cdot g)(x) = \sqrt{x - 1}\,(x - \sqrt{x + 1}) = x\sqrt{x - 1} - \sqrt{x^2 - 1}$; domain $[1, \infty)$
  $\left(\dfrac{f}{g}\right)(x) = \dfrac{\sqrt{x - 1}}{x - \sqrt{x + 1}}$; domain $\left[1, \dfrac{1 + \sqrt{5}}{2}\right) \cup \left(\dfrac{1 + \sqrt{5}}{2}, \infty\right)$

13. (a) $(6f + 3g)(x) = 6x - \dfrac{6}{\sqrt{x}}$, $x > 0$ (b) $(f - g)(x) = x + \dfrac{3}{\sqrt{x}} - \sqrt{x}$, $x > 0$ (c) $(f/g)(x) = \dfrac{x\sqrt{x} + 1}{x - 2}$, $x > 0$, $x \neq 2$.

15.

17.

19.

21.

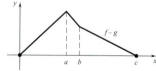

23. $(f \circ g)(x) = 2x^2 + 5$; domain $(-\infty, \infty)$ **25.** $(f \circ g)(x) = \sqrt{x^3 + 5}$; domain $(-\infty, \infty)$

27. $(f \circ g)(x) = \dfrac{x}{x - 2}$; domain: all real numbers except $x = 0$, $x = 2$ **29.** $(f \circ g)(x) = \dfrac{1}{|x^2 - 1| - 3}$; domain: all real numbers except $x = \pm 2$

31. $(f \circ g)(x) = |\sin 2x|$; domain $(-\infty, \infty)$ **33.** $(f \circ g \circ h)(x) = 4(x^2 - 1)$; domain $(-\infty, \infty)$ **35.** $(f \circ g \circ h)(x) = 2x^2 + 1$; domain $(-\infty, \infty)$

37. $f(x) = \dfrac{1}{x}$ **39.** $f(x) = 2 \sin x$ **41.** $g(x) = \left(1 - \dfrac{1}{x^4}\right)^{2/3}$ **43.** $g(x) = 2x^3 - 1$ **45.** $(f \circ g)(x) = |x|$; $(g \circ f)(x) = x$

47. $(f \circ g)(x) = \cos^2 x$; $(g \circ f)(x) = \sin (1 - x^2)$ **49.** $(f \circ g)(x) = x$; $(g \circ f)(x) = x$

51. fg is an even function since $(fg)(-x) = f(-x)g(-x) = f(x)g(x) = (fg)(x)$ **53.** (a) $f(x) = \begin{cases} -x, & -1 \leq x < 0 \\ 1, & x < -1 \end{cases}$ (b) $f(x) = \begin{cases} x, & -1 \leq x < 0 \\ -1, & x < -1 \end{cases}$

55. $g(-x) = f(-x) + f(x) = f(x) + f(-x) = g(x)$

57.

	f_1	f_2	f_3	f_4	f_5	f_6
f_1	f_1	f_2	f_3	f_4	f_5	f_6
f_2	f_2	f_1	f_4	f_3	f_6	f_5
f_3	f_3	f_5	f_1	f_6	f_2	f_4
f_4	f_4	f_6	f_2	f_5	f_1	f_3
f_5	f_5	f_3	f_6	f_1	f_4	f_2
f_6	f_6	f_4	f_5	f_2	f_3	f_1

59. $f(g(x)) = x$ and $g(f(x)) = x$ **61.** $f(g(x)) = x$ and $g(f(x)) = x$

63. (a) Varying a varies the x-coordinate of the vertex (a, b) of the parabola.
  (b) Varying b varies the y-coordinate of the vertex (a, b) of the parabola.

65. (a) For $a > 0$ ($a < 0$), the graph of $f(x - a)$ is the graph of $f(x)$ shifted horizontally $|a|$ units to the right (left).
  (b) For $b > 1$, the graph of $f(bx)$ is compressed horizontally.
   For $0 < b < 1$, the graph of $f(bx)$ is stretched horizontally.
   For $-1 < b < 0$, the graph of $f(bx)$ is stretched horizontally and reflected in the y-axis.
   For $b < -1$, the graph of $f(bx)$ is compressed horizontally and reflected in the y-axis.
  (c) The graph of $f(x) + c$ is the graph of $f(x)$ shifted $|c|$ units up if $c > 0$ and shifted $|c|$ units down if $c < 0$.

67. (a) For $A > 0$, the graph of Af is the graph of f scaled vertically by the factor A.
   For $A < 0$, the graph of Af is the graph of f scaled vertically by the factor $|A|$ and then reflected in the x-axis.
  (b) See 65(b).

SECTION 1.8

1. Let S be the set of integers for which the statement is true. Since $2(1) \leq 2^1$, S contains 1. Assume now that $k \in S$. This tells us that $2k \leq 2^k$, and thus

$$2(k + 1) = 2k + 2 \leq 2^k + 2 \leq 2^k + 2^k = 2(2^k) = 2^{k+1}.$$
$$\uparrow \!\!\!\!\!\!\_\_\_\_\_ (k \geq 1)$$

This places $k + 1$ in S.
  We have shown that

$$1 \in S \quad \text{and that} \quad k \in S \quad \text{implies} \quad k + 1 \in S.$$

It follows that S contains all the positive integers.

3. Let S be the set of integers for which the statement is true. Since $(1)(2) = 2$ is divisible by 2, $1 \in S$. Assume now that $k \in S$. This tells us that $k(k + 1)$ is divisible by 2 and therefore

$$(k + 1)(k + 2) = k(k + 1) + 2(k + 1)$$

is also divisible by 2. This places $k + 1 \in S$.
  We have shown that

$$1 \in S \quad \text{and that} \quad k \in S \quad \text{implies} \quad k + 1 \in S.$$

It follows that S contains all the positive integers.

5. Use $1^2 + 2^2 + \cdots + k^2 + (k + 1)^2 = \frac{1}{6}k(k + 1)(2k + 1) + (k + 1)^2$
$= \frac{1}{6}(k + 1)[k(2k + 1) + 6(k + 1)]$
$= \frac{1}{6}(k + 1)(2k^2 + 7k + 6)$
$= \frac{1}{6}(k + 1)(k + 2)(2k + 3) = \frac{1}{6}(k + 1)[(k + 1) + 1][2(k + 1) + 1].$

7. By Exercise 6 and Example 1.

$$1^3 + 2^3 + \cdots + (n - 1)^3 = [\tfrac{1}{2}(n - 1)n]^2 = \tfrac{1}{4}(n - 1)^2 n^2 < \tfrac{1}{4}n^4$$

and

$$1^3 + 2^3 + \cdots + n^3 = [\tfrac{1}{2}n(n + 1)]^2 = \tfrac{1}{4}n^2(n + 1)^2 > \tfrac{1}{4}n^4.$$

9. Use $\dfrac{1}{\sqrt{1}} + \dfrac{1}{\sqrt{2}} + \dfrac{1}{\sqrt{3}} + \cdots + \dfrac{1}{\sqrt{n}} + \dfrac{1}{\sqrt{n + 1}} > \sqrt{n} + \dfrac{1}{\sqrt{n + 1}} > \sqrt{n} + \dfrac{1}{\sqrt{n + 1} + \sqrt{n}} \left(\dfrac{\sqrt{n + 1} - \sqrt{n}}{\sqrt{n + 1} - \sqrt{n}} \right) = \sqrt{n + 1}.$

11. Let S be the set of integers for which the statement is true. Since

$$3^{2(1)+1} + 2^{1+2} = 27 + 8 = 35$$

is divisible by 7, $1 \in S$.
  Assume now that $k \in S$. This tells us that

$$3^{2k+1} + 2^{k+2} \text{ is divisible by 7}.$$

It follows that

$$32^{(k+1)+1} + 2^{(k+1)+2} = 3^2 \cdot 3^{2k+1} + 2 \cdot 2^{k+2} = 9 \cdot 3^{2k+1} + 2 \cdot 2^{k+2} = 7 \cdot 3^{2k+1} + 2(3^{2k+1} + 2^{k+2})$$

, is also divisible by 7. This places $k + 1 \in S$.

We have shown that

$$1 \in S \qquad \text{and that} \qquad k \in S \quad \text{implies} \quad k + 1 \in S.$$

It follows that S contains all the positive integers.

13. For all positive integers $n \geq 2$,

$$\left(1 - \frac{1}{2}\right)\left(1 - \frac{1}{3}\right) \cdots \left(1 - \frac{1}{n}\right) = \frac{1}{n}.$$

To see this let S be the set of integers n for which the formula holds. Since $1 - \frac{1}{2} = \frac{1}{2}, 2 \in S$. Suppose now that $k \in S$. This tells us that

$$\left(1 - \frac{1}{2}\right)\left(1 - \frac{1}{3}\right) \cdots \left(1 - \frac{1}{k}\right) = \frac{1}{k}$$

and therefore that

$$\left(1 - \frac{1}{2}\right)\left(1 - \frac{1}{3}\right) \cdots \left(1 - \frac{1}{k}\right)\left(1 - \frac{1}{k+1}\right) = \frac{1}{k}\left(1 - \frac{1}{k+1}\right) = \frac{1}{k}\left(\frac{k}{k+1}\right) = \frac{1}{k+1}.$$

This places $k + 1$ in S and verifies the formula for $n \geq 2$.

15. From the figure, observe that adding a vertex V_{N+1} to an N-sided polygon increases the number of diagonals by $(N - 2) + 1 = N - 1$.

Then use the identity

$$\tfrac{1}{2}N(N - 3) + (N - 1) = \tfrac{1}{2}(N + 1)(N + 1 - 3).$$

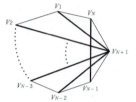

17. To go from k to $k + 1$, take $A = \{a_1, \ldots, a_{k+1}\}$ and $B = \{a_1, \ldots, a_k\}$. Assume that B has 2^k subsets: $B_1, B_2, \ldots, B_{2^k}$. The subsets of A are then $B_1, B_2, \ldots, B_{2^k}$ together with

$$B_1 \cup \{a_{k+1}\}, B_2 \cup \{a_{k+1}\}, \ldots, B_{2^k} \cup \{a_{k+1}\}.$$

This gives $2(2^k) = 2^{k+1}$ subsets for A.

CHAPTER 2

SECTION 2.1

1. (a) 2 (b) -1 (c) does not exist (d) -3 **3.** (a) does not exist (b) -3 (c) does not exist (d) -3

5. (a) does not exist (b) does not exist (c) does not exist (d) 1   **7.** (a) 2 (b) 2 (c) 2 (d) -1

9. (a) -1 (b) -1 (c) -1 (d) undefined   **11.** (a) 0 (b) 0 (c) 0 (d) 0   **13.** $c = 0, 6$ **15.** -1 **17.** 4   **19.** 1

21. $\frac{3}{2}$ **23.** does not exist   **25.** 2   **27.** does not exist   **29.** 1   **31.** does not exist   **33.** 2   **35.** 2   **37.** 0   **39.** 1

41. 16   **43.** does not exist   **45.** 4   **47.** 4   **49.** $1/\sqrt{2}$ **51.** 4; $y = 4x - 4$ **53.** -4; $y = -4x$

55. $\frac{1}{2}$; $y = \frac{1}{2}x + \frac{1}{2}$ **57.** does not exist   **59.** (b) the limits do not exist (c)

61. 2   **63.** $\frac{3}{2}$ **65.** 2.71828 . . .

SECTION 2.2

1. $\frac{1}{2}$ **3.** does not exist   **5.** $\frac{4}{5}\sqrt{5}$ **7.** 4   **9.** does not exist   **11.** -1 **13.** does not exist   **15.** 0   **17.** 2   **19.** 1

21. 1   **23.** δ_1 and δ_2 **25.** $\frac{1}{2}\epsilon$ **27.** 2ϵ **29.** Take $\delta = \frac{1}{2}\epsilon$. If $0 < |x - 4| < \frac{1}{2}\epsilon$, then $|(2x - 5) - 3| = 2|x - 4| < \epsilon$.

31. Take $\delta = \frac{1}{6}\epsilon$. If $0 < |x - 3| < \frac{1}{6}\epsilon$, then $|(6x - 7) - 11| = 6|x - 3| < \epsilon$.

33. Take $\delta = \frac{1}{3}\epsilon$. If $0 < |x - 2| < \frac{1}{3}\epsilon$, then $\left| |1 - 3x| - 5 \right| \leq 3|x - 2| < \epsilon$.   **35.** Statements (b), (e), (g), and (i) are necessarily true.

37. (i) $\lim\limits_{x \to 3} \dfrac{1}{x - 1} = \dfrac{1}{2}$ (ii) $\lim\limits_{h \to 0} \dfrac{1}{(3 + h) - 1} = \dfrac{1}{2}$ (iii) $\lim\limits_{x \to 3}\left(\dfrac{1}{x - 1} - \dfrac{1}{2}\right) = 0$ (iv) $\lim\limits_{x \to 3}\left|\dfrac{1}{x - 1} - \dfrac{1}{2}\right| = 0$

39. 5 **41.** $4 + 10x$ **43.** $-\dfrac{1}{(x + 1)^2}$ **45.** (i) and (iv) of (2.2.5) with $L = 0$ **47.** $\delta = 0.001$ **49.** $\delta = 0.04$

51. Let $\epsilon > 0$. If $\lim\limits_{x \to c} f(x) = L$, then there must exist $\delta > 0$ such that

(*) if $0 < |x - c| < \delta$ then $|f(x) - L| < \epsilon$.

Suppose now that $0 < |h| < \delta$. Then $0 < |(c + h) - c| < \delta$, and thus by (*), $|f(c + h) - L| < \epsilon$. This proves that, if $\lim\limits_{x \to c} f(x) = L$, then $\lim\limits_{h \to 0} f(c + h) = L$.

If, on the other hand, $\lim\limits_{h \to 0} f(c + h) = L$, then there must exist $\delta > 0$ such that

(**) if $0 < |h| < \delta$ then $|f(c + h) - L| < \epsilon$.

Suppose now that $0 < |x - c| < \delta$. Then by (**), $|f(c + (x - c)) - L| < \epsilon$. More simply stated, $|f(x) - L| < \epsilon$. This proves that, if $\lim\limits_{h \to 0} f(c + h) = L$, then $\lim\limits_{x \to c} f(x) = L$.

53. (a) Set $\delta = \epsilon\sqrt{c}$. By the hint

$$\text{if} \quad 0 < |x - c| < \epsilon\sqrt{c}, \quad \text{then} \quad |\sqrt{x} - \sqrt{c}| < \dfrac{1}{\sqrt{c}}|x - c| < \epsilon.$$

(b) Set $\delta = \epsilon^2$. If $0 < x < \epsilon^2$, then $|\sqrt{x} - 0| = \sqrt{x} < \epsilon$.

55. Take $\delta = $ minimum of 1 and $\epsilon/7$. If $0 < |x - 1| < \delta$, then $0 < x < 2$ and $|x - 1| < \epsilon/7$. Therefore $|x^3 - 1| = |x^2 + x + 1||x - 1| < 7|x - 1| < 7(\epsilon/7) = \epsilon$.

57. Set $\delta = \epsilon^2$. If $3 - \epsilon^2 < x < 3$, then $-\epsilon^2 < x - 3$, $0 < 3 - x < \epsilon^2$ and therefore $|\sqrt{3 - x} - 0| < \epsilon$.

59. Suppose, on the contrary, that $\lim\limits_{x \to c} f(x) = L$ for some particular c. Taking $\epsilon = \frac{1}{2}$, there must exist $\delta > 0$ such that,

if $0 < |x - c| < \delta$, then $|f(x) - L| < \frac{1}{2}$.

Let x_1 be a rational number satisfying $0 < |x_1 - c| < \delta$, and x_2 an irrational number satisfying $0 < |x_2 - c| < \delta$. (That such numbers exist follows from the fact that every interval contains both rational and irrational numbers.) Now $f(x_1) = 1$ and $f(x_2) = 0$. Thus we must have both $|1 - L| < \frac{1}{2}$ and $|0 - L| < \frac{1}{2}$. From the first inequality we conclude that $L > \frac{1}{2}$. From the second, we conclude that $L < \frac{1}{2}$. Clearly no such number L exists.

61. We begin by assuming that $\lim\limits_{x \to c^+} f(x) = L$ and showing that $\lim\limits_{h \to 0} f(c + |h|) = L$.

Let $\epsilon > 0$. Since $\lim\limits_{x \to c^+} f(x) = L$, there exists $\delta > 0$ such that

(*) if $c < x < c + \delta$ then $|f(x) - L| < \epsilon$.

Suppose now that $0 < |h| < \delta$. Then $c < c + |h| < c + \delta$ and, by (*), $|f(c + |h|) - L| < \epsilon$. Thus $\lim\limits_{h \to 0} f(c + |h|) = L$.

Conversely let's assume that $\lim\limits_{h \to 0} f(c + |h|) = L$ and again take $\epsilon > 0$. Then there exists $\delta > 0$ such that

(**) if $0 < |h| < \delta$ then $|f(c + |h|) - L| < \epsilon$.

Suppose now that $c < x < c + \delta$. Then $0 < x - c < \delta$ so that, by (**),

$$|f(x) - L| = |f(c + (x - c)) - L| < \epsilon.$$

Thus $\lim\limits_{x \to c^+} f(x) = L$.

63. (a) Let $\epsilon = L$. There exists $\gamma > 0$ such that if $0 < |x - c| < \gamma$, then $|f(x) - L| < L$, which is equivalent to $0 < f(x) < 2L$. Thus $f(x) > 0$ for all $x \in (c - \gamma, c + \gamma)$, $x \neq c$.

(b) Use the same argument.

65. (a) Suppose $\lim\limits_{x \to c} f(x) = L$ and $\lim\limits_{x \to c} g(x) = M$. Then $\lim\limits_{x \to c} [g(x) - f(x)] = M - L$ exists. If $M - L < 0$, then, by Exercise 63(b), there exists $\gamma > 0$ such that $g(x) - f(x) < 0$ on $(c - \gamma, c + \gamma)$, $x \neq c$, which contradicts the fact that $f(x) \leq g(x)$ on $(c - p, c + p)$, $x \neq c$. Thus $M - L \geq 0$.

(b) No. Consider $f(x) = 1 - x^2$ and $g(x) = 1 + x^2$ on $(-1, 1)$, and let $c = 0$.

67. $5; y = 5x - 8$ **69.** $\frac{1}{4}; y = \frac{1}{4}x + 1$ **71.** -2.7207 **73.** 1.0986

SECTION 2.3

1. (a) 3 (b) 4 (c) -2 (d) 0 (e) does not exist (f) $\frac{1}{3}$

3. $\lim\limits_{x \to 4}\left[\left(\dfrac{1}{x} - \dfrac{1}{4}\right)\left(\dfrac{1}{x - 4}\right)\right] = \lim\limits_{x \to 4}\left[\left(\dfrac{4 - x}{4x}\right)\left(\dfrac{1}{x - 4}\right)\right] = \lim\limits_{x \to 4}\dfrac{-1}{4x} = -\dfrac{1}{16}$; Theorem 2.3.2 does not apply since $\lim\limits_{x \to 4}\dfrac{1}{x - 4}$ does not exist.

5. 3 **7.** -3 **9.** 5 **11.** does not exist **13.** -1 **15.** does not exist **17.** 1 **19.** 4 **21.** $\frac{1}{4}$ **23.** $-\frac{2}{3}$

25. does not exist   **27.** -1 **29.** 4   **31.** a/b **33.** $5/4$ **35.** does not exist   **37.** 2

39. (a) 0 (b) $-\frac{1}{16}$ (c) 0 (d) does not exist   **41.** (a) 4 (b) -2 (c) 2 (d) does not exist   **43.** $f(x) = 1/x$, $g(x) = -1/x$, $c = 0$

45. True. Let $\lim_{x \to c} [f(x) + g(x)] = L$. If $\lim_{x \to c} g(x) = M$ exists, then $\lim_{x \to c} f(x) = \lim_{x \to c} [f(x) + g(x) - g(x)] = L - M$ also exists. Thus, $\lim_{x \to c} g(x)$ cannot exist.

47. True. If $\lim_{x \to c} \sqrt{f(x)} = L$ exists, then $\lim_{x \to c} f(x) = \lim_{x \to c} \sqrt{f(x)} \cdot \sqrt{f(x)} = L^2$ exists.   **49.** False. Let $f(x) = x$, $c = 0$

51. False. Let $f(x) = 1 - x^2$, $g(x) = 1 + x^2$, and $c = 0$ **53.** If $\lim_{x \to c} f(x) = L$ and $\lim_{x \to c} g(x) = L$, then

$$\lim_{x \to c} h(x) = \lim_{x \to c} \tfrac{1}{2}\{[f(x) + g(x)] - |f(x) - g(x)|\}$$
$$= \lim_{x \to c} \tfrac{1}{2}[f(x) + g(x)] - \lim_{x \to c} \tfrac{1}{2}|f(x) - g(x)| = \tfrac{1}{2}(L + L) - \tfrac{1}{2}|L - L| = L$$

A similar argument works for H.

55. (a) Suppose $\lim_{x \to c} g(x) = k$ exists. Then $\lim_{x \to c} f(x) \cdot g(x) = 0 \cdot k = 0$. Thus $\lim_{x \to c} g(x)$ cannot exist. (b) $\lim_{x \to c} g(x) = \dfrac{1}{L}$

57. (a) 1 (b) $2x$ (c) $3x^2$ (d) $4x^3$ (e) nx^{n-1}

SECTION 2.4

1. (a) $x = -3, x = 0, x = 2, x = 6$ (b) At -3, neither; at 0, continuous from the right; at 2, neither; at 6, neither
(c) removable discontinuity at $x = 2$; jump discontinuity at $x = 0$

3. continuous   **5.** continuous   **7.** continuous   **9.** removable discontinuity   **11.** jump discontinuity   **13.** continuous

15. jump discontinuity   **17.** removable discontinuity at 2   **19.** no discontinuities   **21.** jump discontinuity at 1

23. no discontinuities   **25.** no discontinuities   **27.** jump discontinuities at 0 and 2

29. removable discontinuity at -2; jump discontinuity at 3   **31.** $f(1) = 2$ **33.** impossible   **35.** 4   **37.** $A - B = 3$ with $B \neq 3$

39. $f(5) = \frac{1}{6}$ **41.** $f(5) = \frac{1}{3}$ **43.** nowhere   **45.** $x = 0, x = 2$, and all nonintegral values of x

47. Refer to (2.2.5). Use the equivalence of (i) to (ii) setting $L = f(c)$.

49. Let $A = \{x_1, x_2, \ldots, x_n\}$ and let $\epsilon > 0$. In $A - \{c\}$ there is one point closest to c. Call it d and set $\delta_1 = |c - d|$. Note that if $0 < |x - c| < \delta_1$, then $f(x) = g(x)$.
　　　Suppose now that g is continuous at c. Then there exists a positive number δ less than δ_1 such that

$$\text{if} \quad 0 < |x - c| < \delta, \quad \text{then} \quad |g(x) - g(c)| < \epsilon.$$

But for such x, $f(x) = g(x)$. Therefore we see that

$$\text{if} \quad 0 < |x - c| < \delta, \quad \text{then} \quad |f(x) - g(c)| < \epsilon.$$

This means that

$$\lim_{x \to c} f(x) = g(c)$$

and contradicts the assumption that f has a nonremovable discontinuity at c.

51. The hypothesis implies that f is defined at c and $B > 0$. Let $\epsilon > 0$ and choose $\delta = \min\,(p, B/\epsilon)$. Then $|x - c| < \delta$ implies

$$|f(x) - f(c)| < B\,|x - c| < B(\epsilon/B) = \epsilon$$

It now follows that f is continuous at c.

53. $\lim_{h \to 0} f(c + h) - f(c) = \lim_{h \to 0} \left[\dfrac{f(c + h) - f(c)}{h} \cdot h \right] = \lim_{h \to 0} \left[\dfrac{f(c + h) - f(c)}{h} \right] \cdot \lim_{h \to 0} h = L \cdot 0 = 0$. Therefore f is continuous at c by Exercise 47.

SECTION 2.5

1. 3   **3.** $\frac{3}{5}$ **5.** 2   **7.** 0   **9.** does not exist   **11.** $\frac{9}{5}$ **13.** $\frac{2}{3}$ **15.** 1   **17.** $\frac{1}{2}$ **19.** -4 **21.** 1   **23.** $\frac{3}{5}$

25. 0   **27.** $\dfrac{2}{\pi}\sqrt{2}$ **29.** -1 **31.** 0

33. We will show that

$$\lim_{x \to c} \cos x = \cos c \quad \text{by showing that} \quad \lim_{h \to 0} \cos\,(c + h) = \cos c.$$

Note that $\cos\,(c + h) = \cos c \cos h - \sin c \sin h$. We know that

$$\lim_{h \to 0} \cos h = 1 \quad \text{and} \quad \lim_{h \to 0} \sin h = 0.$$

Therefore

$$\lim_{h \to 0} \cos(c + h) = (\cos c)(\lim_{h \to 0} \cos h) - (\sin c)(\lim_{h \to 0} \sin h) = (\cos c)(1) - (\sin c)(0) = \cos c.$$

Here is a different proof. The addition formula for the sine gives $\cos x = \sin(\tfrac{1}{2}\pi + x)$. Being the composition of functions that are everywhere continuous, the cosine function is itself everywhere continuous. Therefore

$$\lim_{x \to c} \cos x = \cos c \qquad \text{for all real } c.$$

35. $\dfrac{\sqrt{2}}{2}; \; y - \dfrac{\sqrt{2}}{2} = \dfrac{\sqrt{2}}{2}\left(x - \dfrac{\pi}{4}\right)$ **37.** $-\sqrt{3}; \; y - \tfrac{1}{2} = -\sqrt{3}\left(x - \dfrac{\pi}{6}\right)$ **39.** $0 \le |x \sin(1/x)| \le |x|$ for $x \ne 0$

41. $0 \le |x - 1| \, |\sin x| \le |x - 1|$ for all x **43.** $0 \le |xf(x)| \le B|x|$ for $x \ne 0$ **45.** $0 \le |f(x) - L| \le B|x - c|$ for $x \ne c$

SECTION 2.6

1. **3.** **5.** **7.**

9. impossible by the intermediate-value theorem

11. **13.** **15.**

17. $f(1) = -1 < 0$ and $f(2) = 6 > 0$ **19.** $\displaystyle\lim_{x \to 1^+} f(x) = \infty$ and $\displaystyle\lim_{x \to 4^-} f(x) = -\infty$

21. Set $g(x) = x - f(x)$. Since g is continuous on $[0, 1]$ and $g(0) \le 0 \le g(1)$, there exists c in $[0, 1]$ such that $g(c) = c - f(c) = 0$.

23. By Exercise 43, Section 2.5, $\displaystyle\lim_{x \to 0} x \, f(x) = 0$. Thus, $\displaystyle\lim_{x \to 0} |g(x) - g(0)| = \lim_{x \to 0} |x \, f(x)| = 0$ and g is continuous at 0.

25. The cubic polynomial $P(x) = x^3 + ax^2 + bx + c$ is continuous on $(-\infty, \infty)$. Writing P as

$$P(x) = x^3\left(1 + \frac{a}{x} + \frac{b}{x^2} + \frac{c}{x^3}\right), \; x \ne 0,$$

it follows that $P(x) < 0$ for large negative values of x and $P(x) > 0$ for large positive values of x. Thus, there exists a negative number N such that $P(x) < 0$ for $x < N$, and a positive number M such that $P(x) > 0$ for $x > M$. By the intermediate-value theorem, P has a zero on $[N, M]$.

27. A circle of radius r has area πr^2. Let $A(r) = \pi r^2, \; r \in [0, 10]$. Then $A(0) = 0$ and $A(10) = 100\pi \cong 314$. Since $0 < 250 < 314$, it follows from the intermediate-value theorem that there exists a number $c \in (0, 10)$ such that $A(c) = 250$.

29. Inscribe a rectangle in a circle of radius R and then introduce a coordinate system as shown in the figure. Then the area of the rectangle is given by

$$A(x) = 4x\sqrt{R^2 - x^2}, \; x \in [0, R]$$

Since A is continuous on $[0, R]$, A has a maximum value. (A also has a minimum value, namely 0.)

31. $m_{10} = 1.7314$ **33.** $m_{10} = 0.7392$

35. f has a zero on $(-3, -2)$, $(0, 1)$ and $(1, 2)$. The zeros of f are: $r_1 = -2.4909$, $r_2 = 0.6566$, $r_3 = 1.8343$

37. f has a zero on $(-2, -1)$, $(0, 1)$, $(1, 2)$. The zeros of f are: $r_1 = -1.3482$, $r_2 = 0.2620$, $r_3 = 1.0816$

39. f is bounded; the maximum value of f is 1, the minimum value is -1

41. f is bounded; the maximum value of f is 0.5, the minimum value is approximately 0.3540

CHAPTER 3

SECTION 3.1

1. 0   **3.** −3   **5.** $5 - 2x$ **7.** $4x^3$ **9.** $\frac{1}{2}(x-1)^{-1/2}$ **11.** $-2x^{-3}$ **13.** −6   **15.** $-\frac{1}{4}$ **17.** $\frac{3}{2}$

19. tangent $y - 4x + 4 = 0$; normal $4y + x - 18 = 0$ **21.** tangent $y + 3x - 16 = 0$; normal $3y - x - 8 = 0$

23. tangent $y + x + 4 = 0$; normal $y - x - 2 = 0$

25. (a) Removable discontinuity at $c = -1$; jump discontinuity at $c = 1$ (b) f is continuous but not differentiable at $c = 0$ and $c = 3$

27. $x = -1$ **29.** $x = 0$ **31.** $x = 1$ **33.** 4   **35.** does not exist

37.

39.

41.

43. $f(x) = x^2$; $c = 1$ **45.** $f(x) = \sqrt{x}$; $c = 4$ **47.** $f(x) = \cos x$; $c = \pi$

49. Since $f(1) = 1$ and $\lim\limits_{x \to 1^+} f(x) = 2$, f is not continuous at 1 and thus, by (3.1.4), is not differentiable at 1.

51. (a) $f'(x) = \begin{cases} 2(x+1), & x < 0 \\ 2(x-1), & x > 0 \end{cases}$ (b) $\lim\limits_{h \to 0^-} \dfrac{f(0+h) - f(0)}{h} = \lim\limits_{h \to 0^-} \dfrac{(h+1)^2 - 1}{h} = \lim\limits_{h \to 0^-} (h+2) = 2$,

$\lim\limits_{h \to 0^+} \dfrac{f(0+h) - f(0)}{h} = \lim\limits_{h \to 0^+} \dfrac{(h-1)^2 - 1}{h} = \lim\limits_{h \to 0^+} (h-2) = -2$

In 53–57, there are many possible answers. Here are some.

53. $f(x) = c$, c any constant   **55.** $f(x) = |x+1|$; $f(x) = \begin{cases} 0, & x \neq -1 \\ 1, & x = -1 \end{cases}$ **57.** $f(x) = 2x + 5$

59. $f'(-x) = \lim\limits_{h \to 0} \dfrac{f(-x+h) - f(-x)}{h} = \lim\limits_{h \to 0} \dfrac{-f(x-h) + f(x)}{h} = \lim\limits_{h \to 0} \dfrac{f[x + (-h)] - f(x)}{-h} = f'(x)$

61. (a) $\lim\limits_{x \to 2^-} (x^2 - x) = \lim\limits_{x \to 2^+} (2x - 2) = 2 = f(2)$ (b) $f'_-(2) = 3, f'_+(2) = 2$ (c) f is not differentiable at 2

63. (a) $f'(x) = \dfrac{-1}{2\sqrt{1-x}}$ (b) $f'_+(0) = -\frac{1}{2}$ (c) $f'_-(1)$ does not exist

65. Let $L = f'_-(c) = f'_+(c)$ and let $\epsilon > 0$. There exists $\delta_1 > 0$ such that

$$\left| \frac{f(c+h) - f(c)}{h} - L \right| < \epsilon$$

whenever $h \in (-\delta_1, 0)$. There exists $\delta_2 > 0$ such that

$$\left| \frac{f(c+h) - f(c)}{h} - L \right| < \epsilon$$

whenever $h \in (0, \delta_2)$. Let $\delta = \min (\delta_1, \delta_2)$. Then

$$\left| \frac{f(c+h) - f(c)}{h} - L \right| < \epsilon$$

whenever $|h| < \delta, h \neq 0$. Thus f is differentiable at c and $f'(c) = L$.

67. (a) $\lim\limits_{x \to 0} x \sin\dfrac{1}{x} = 0 = f(0)$; $\lim\limits_{x \to 0} x^2 \sin\dfrac{1}{x} = 0 = g(0)$

(b) $\lim\limits_{h \to 0} \dfrac{h \sin (1/h) - 0}{h} = \lim\limits_{h \to 0} \sin\dfrac{1}{h}$ does not exist

(c) $\lim\limits_{h \to 0} \dfrac{h^2 \sin (1/h) - 0}{h} = \lim\limits_{h \to 0} h \sin\dfrac{1}{h} = 0$; $g'(0) = 0$

69. $f'(1) = -1$ **71.** $f'(-1) = \frac{1}{3}$ **73.** (b) 7.071   (c) $D(0.001) \approx 7.074$; $D(-0.001) \approx 7.068$

75. (a) $T(x) = -\frac{11}{4}(x - \frac{3}{2}) + \frac{21}{8}$ (c) $(1.453, 1.547)$

SECTION 3.2

1. -1 **3.** $55x^4 - 18x^2$ **5.** $2ax + b$ **7.** $\dfrac{2}{x^3}$ **9.** $3x^2 - 6x - 1$ **11.** $\dfrac{3x^2 - 2x^3}{(1 - x)^2}$ **13.** $\dfrac{2(x^2 + 3x + 1)}{(2x + 3)^2}$ **15.** $2x - 3$

17. $-\dfrac{2(3x^2 - x + 1)}{x^2(x - 2)^2}$ **19.** $-80x^9 + 81x^8 - 64x^7 + 63x^6$ **21.** $f'(0) = -\frac{1}{4}; f'(1) = -1$ **23.** $f'(0) = 0; f'(1) = -1$

25. $f'(0) = \dfrac{ad - bc}{d^2}; f'(1) = \dfrac{ad - bc}{(c + d)^2}$ **27.** $f'(0) = 3$ **29.** $f'(0) = \frac{20}{9}$ **31.** $2y - x - 8 = 0$ **33.** $y - 4x + 12 = 0$

35. $9y + 4x - 18 = 0$ **37.** $(-1, 27), (3, -5)$ **39.** $(-1, -\frac{5}{2}), (1, \frac{5}{2})$ **41.** $(2, 3)$ **43.** $(-2, -10)$ **45.** $(-1, -2), (\frac{5}{3}, \frac{50}{27})$

47. $\frac{425}{8}$ **49.** $A = -1, B = 0, C = 4$ **51.** $x = -\dfrac{b}{2a}$ **53.** $c = -1, 1$

55. Let $a > 0$. An equation for the tangent line to the graph of $f(x) = 1/x$ at $x = a$ is $y = (-1/a^2)x + 2/a$. The y-intercept is $2/a$ and the x-intercept is $2a$. Thus, the area of the triangle formed by this line and the coordinate axes is $A = \frac{1}{2}(2/a)(2a) = 2$ square units.

57. Since f and $f + g$ are differentiable, $g = (f + g) - f$ is differentiable. The functions $f(x) = |x|$ and $g(x) = -|x|$ are not differentiable at 0; their sum $h(x) = 0$ is differentiable everywhere.

59. Since

$$\left(\frac{f}{g}\right)(x) = \frac{f(x)}{g(x)} = f(x) \cdot \frac{1}{g(x)},$$

it follows from the product and reciprocal rules that

$$\left(\frac{f}{g}\right)'(x) = \left(f \cdot \frac{1}{g}\right)'(x) = f(x)\left(-\frac{g'(x)}{[g(x)]^2}\right) + f'(x) \cdot \frac{1}{g(x)} = \frac{g(x)f'(x) - f(x)g'(x)}{[g(x)]^2}.$$

61. $F'(x) = 2x\left(1 + \dfrac{1}{x}\right)(2x^3 - x + 1) + (x^2 + 1)\left(-\dfrac{1}{x^2}\right)(2x^3 - x + 1) + (x^2 + 1)\left(1 + \dfrac{1}{x}\right)(6x^2 - 1)$

63. $g(x) = [f(x)]^2 = f(x) \cdot f(x); g'(x) = f(x)f'(x) + f(x)f'(x) = 2f(x)f'(x)$ **65.** $g'(x) = 3(x^3 - 2x^2 + x + 2)^2(3x^2 - 4x + 1)$

67. (a) $f'_+(-1) = -6; f'_-(3) = 2$ (b) $g'(x) = 2x - 4; g'(-1) = f'_+(-1); g'(3) = f'_-(3)$ **69.** $A = -2, B = -8$

71. (a) Let $D(h) = \dfrac{\sin(x + h) - \sin x}{h}$.

At $x = 0$, $D(0.001) \cong 0.99999$, $D(-0.001) \cong 0.99999$; at $x = \pi/6$, $D(0.001) \cong 0.86578$, $D(-0.001) \cong 0.86628$; at $x = \pi/4$, $D(0.001) \cong 0.70675$, $D(-0.001) \cong 0.70746$; at $x = \pi/3$, $D(0.001) \cong 0.49957$, $D(-0.001) \cong 0.50043$; at $x = \pi/2$, $D(0.001) \cong 0.0005$, $D(-0.001) \cong 0.0005$

(b) $\cos(0) = 1$, $\cos(\pi/6) \cong 0.866025$, $\cos(\pi/4) \cong 0.707107$, $\cos(\pi/3) = 0.5$, $\cos(\pi/2) = 1$ (c) $f'(x) = \cos x$.

73. (a) Let $D(h) = \dfrac{2^{x+h} - 2^x}{h}$.

At $x = 0$, $D(0.001) \cong 0.69339$, $D(-0.001) \cong 0.69291$; at $x = 1$, $D(0.001) \cong 1.38678$, $D(-0.001) \cong 1.38581$; at $x = 2$, $D(0.001) \cong 2.77355$, $D(-0.001) \cong 2.77163$; at $x = 3$, $D(0.001) \cong 5.54710$, $D(-0.001) \cong 5.54326$

(b) $\dfrac{f'(x)}{f(x)} \cong 0.693$ (c) $f'(x) \cong (0.693)\, 2^x$

SECTION 3.3

1. $\dfrac{dy}{dx} = 12x^3 - 2x$ **3.** $\dfrac{dy}{dx} = 1 + \dfrac{1}{x^2}$ **5.** $\dfrac{dy}{dx} = \dfrac{1 - x^2}{(1 + x^2)^2}$ **7.** $\dfrac{dy}{dx} = \dfrac{2x - x^2}{(1 - x)^2}$ **9.** $\dfrac{dy}{dx} = \dfrac{-6x^2}{(x^3 - 1)^2}$ **11.** 2

13. $18x^2 + 30x + 5x^{-2}$ **15.** $\dfrac{-4t}{(t^2 - 1)^2}$ **17.** $\dfrac{2t^3(t^3 - 2)}{(2t^3 - 1)^2}$ **19.** $\dfrac{2}{(1 - 2u)^2}$ **21.** $-\left[\dfrac{1}{(u - 1)^2} + \dfrac{1}{(u + 1)^2}\right] = -2\left[\dfrac{u^2 + 1}{(u^2 - 1)^2}\right]$

23. $2x\left[\dfrac{1}{(1 - x^2)^2} + \dfrac{1}{x^4}\right]$ **25.** $\dfrac{-2}{(x - 1)^2}$ **27.** 47 **29.** $\frac{1}{4}$ **31.** $42x - 120x^3$ **33.** $-6x^{-3}$ **35.** $4 - 12x^{-4}$ **37.** 2

39. 0 **41.** $6 + 60x^{-6}$ **43.** $1 - 4x$ **45.** -24 **47.** -24 **49.** $p(x) = 2x^2 - 6x + 7$

51. (a) $n!$ (b) 0 (c) $f^{(k)}(x) = n(n - 1) \cdots (n - k + 1) x^{n-k}$

53. (a) $f'_+(0) = 0$ and $f'_-(0) = 0$. Thus $f'(0) = 0$. (d)

(b) $f'(x) = \begin{cases} 2x, & x \geq 0 \\ 0, & x < 0 \end{cases}$

(c) $f''_+(0) = 2$ and $f''_-(0) = 0$. Thus $f''(0)$ does not exist.

55. If $f(x) = g(x) = x$, then $(fg)(x) = x^2$, so that $(fg)''(x) = 2$, but $f(x)g''(x) + f''(x)g(x) = x \cdot 0 + 0 \cdot x = 0$.

57. (a) $x = 0$ (b) $x > 0$ (c) $x < 0$ **59.** (a) $x = -2, x = 1$ (b) $x < -2, x > 1$ (c) $-2 < x < 1$

61. The result is true for $n = 1$:

$$\frac{d^1 y}{dx^1} = \frac{dy}{dx} = -x^{-2} = (-1)^1 1! \, x^{-1-1}.$$

If the result is true for $n = k$:

$$\frac{d^k y}{dx^k} = (-1)^k k! \, x^{-k-1},$$

then the result is true for $n = k + 1$:

$$\frac{d^{k+1} y}{dx^{k+1}} = \frac{d}{dx}\left[\frac{d^k y}{dx^k}\right] = (-1)^k k! \, (-k - 1)x^{-k-2} = (-1)^{k+1}(k + 1)! \, x^{-(k+1)-1}.$$

63. (a) $(f \cdot g)'' = (f'g + fg')' = f''g + f'g' + f'g' + fg''$
$= f''g + 2f'g' + fg''$

(b) $(f \cdot g)''' = (f''g + 2f'g' + fg'')'$
$= f'''g + f''g' + 2f''g' + 2f'g'' + f'g'' + fg'''$
$= f'''g + 3f''g' + 3f'g'' + fg'''$

65. $\dfrac{d}{dx}(uvw) = vw\dfrac{du}{dx} + uw\dfrac{dv}{dx} + uv\dfrac{dw}{dx}$

67. (a) $f'(x) = 3x^2 + 2x - 4$ (c) The graph is "falling" when $f'(x) < 0$; the graph is rising when $f'(x) > 0$. **69.** (a) $y = 4x + 1$ (c) $(6, 25)$

SECTION 3.4

1. $\dfrac{dA}{dr} = 2\pi r, 4\pi$ **3.** $\dfrac{dA}{dz} = z, 4$ **5.** $-\frac{5}{36}$ **7.** $\dfrac{dV}{dr} = 4\pi r^2$ **9.** $x_0 = \frac{3}{4}$ **11.** (a) $\dfrac{3\sqrt{2}}{4} w^2$ (b) $\dfrac{\sqrt{3}}{3} z^2$

13. (a) $\frac{1}{2}r^2$ (b) $r\theta$ (c) $-4Ar^{-3} = -2\theta/r$ **15.** $x = \frac{1}{2}$ **17.** $x(5) = -6, v(5) = -7, a(5) = -2$, speed $= 7$

19. $x(2) = -4, v(2) = 6, a(2) = 12$, speed $= 6$ **21.** $x(1) = 6, v(1) = -2, a(1) = \frac{4}{3}$, speed $= 2$

23. $x(1) = 0, v(1) = 18, a(1) = 54$, speed $= 18$ **25.** never **27.** at $-2 + \sqrt{5}$ **29.** at $2\sqrt{2}$ **31.** A **33.** A **35.** A and B

37. A **39.** A and C **41.** $(0, 2), (7, \infty)$ **43.** $(0, 3), (4, \infty)$ **45.** $(2, 5)$ **47.** $(0, 2 - \frac{2}{3}\sqrt{3}), (4, \infty)$ **49.** 576 ft **51.** $v_0^2/2g$ meters

53. $y(t) = \frac{1}{2}gt^2 + v_0 t + y_0$. If $y(t_1) = y(t_2)$, $t_1 \neq t_2$, then $v(t_1) = gt_1 + v_0 = -gt_2 - v_0 = -v(t_2)$. Thus $|v(t_1)| = |v(t_2)|$.

55. 9 ft/sec **57.** (a) 2 sec (b) 16 ft (c) 48 ft/sec **59.** (a) $\frac{1625}{16}$ ft (b) $\frac{6475}{64}$ ft (c) 100 ft

61. $C'(100) = 0.04$
$C(101) - C(100) = 0.0401$

63. $C'(100) = 0$
$C(101) - C(100) = 0$

65. $C'(10 = 23$
$C(11) - C(10) = 22.90$

67. (a) $\overline{C}'(x) = \dfrac{-200}{x^2} + 0.0001$ (b) $\overline{C}'(x) = \dfrac{-200}{x^2} - \dfrac{200}{x^3}$

69. (a) $v(t) = 3t^2 - 14t + 10$
(b) moving to the right when $0 < t < 0.88$ and when $3.79 < t < 5$; moving to the left when $0.88 < t < 3.79$
(c) the object stops at $t \cong 0.88$ and $t \cong 3.79$; the maximum speed is approximately 6.33
(d) $a(t) = 6t - 14$; slowing down when $0 < t < 0.88$ and when $2.33 < t < 3.79$; speeding up when $0.88 < t < 2.33$ and when $3.79 < t < 5$

SECTION 3.5

1. $f(x) = x^4 + 2x^2 + 1; f'(x) = 4x^3 + 4x = 4x(x^2 + 1)$
$f(x) = (x^2 + 1)^2; f'(x) = 2(x^2 + 1)(2x) = 4x(x^2 + 1)$

3. $f(x) = 8x^3 + 12x^2 + 6x + 1; f'(x) = 24x^2 + 24x + 6 = 6(2x + 1)^2$
$f(x) = (2x + 1)^3; f'(x) = 3(2x + 1)^2(2) = 6(2x + 1)^2$

5. $f(x) = x^2 + 2 + x^{-2}; f'(x) = 2x - 2x^{-3} = 2x(1 - x^{-4})$
$f(x) = (x + x^{-1})^2; f'(x) = 2(x + x^{-1})(1 - x^{-2}) = 2x(1 + x^{-2})(1 - x^{-2}) = 2x(1 - x^{-4})$

7. $2(1 - 2x)^{-2}$ **9.** $20(x^5 - x^{10})^{19}(5x^4 - 10x^9)$ **11.** $4\left(x - \dfrac{1}{x}\right)^3\left(1 + \dfrac{1}{x^2}\right)$ **13.** $4(x - x^3 - x^5)^3(1 - 3x^2 - 5x^4)$

15. $200t\,(t^2 - 1)^{99}$ **17.** $-4(t^{-1} + t^{-2})^3(t^{-2} + 2t^{-3})$ **19.** $324x^3\left[\dfrac{1 - x^2}{(x^2 + 1)^5}\right]$ **21.** $2(x^4 + x^2 + x)(4x^3 + 2x + 1)$

23. $-\left(\dfrac{x^3}{3} + \dfrac{x^2}{2} + \dfrac{x}{1}\right)^{-2}(x^2 + x + 1)$ **25.** $3[(x + x^{-1})^2 - (x^2 + x^{-2})^{-1}]^2 \, [2(x + x^{-1})(1 - x^{-2}) + (x^2 + x^{-2})^{-2}(2x - 2x^{-3})]$

27. -1 **29.** 0 **31.** $\dfrac{dy}{dt} = \dfrac{dy}{du}\dfrac{du}{dx}\dfrac{dx}{dt} = \dfrac{7(2t - 5)^4 + 12(2t + 5)^2 - 2}{[(2t - 5)^4 + 2(2t - 5)^2 + 2]^2}[4(2t - 5)]$ **33.** 16 **35.** 1 **37.** 1 **39.** 1 **41.** 2

43. 0 **45.** $4(x^3 + x)^2\,[3(3x^2 + 1)^2 + 6x(x^3 + x)]$ **47.** $\dfrac{6x(1 + x)}{(1 - x)^5}$ **49.** $2xf'(x^2 + 1)$ **51.** $2f(x)f'(x)$

53. (a) $x = 0$ (b) $x < 0$ (c) $x > 0$ **55.** (a) $x = -1, x = 1$ (b) $-1 < x < 1$ (c) $x < -1, x > 1$ **57.** at 3

59. at 2 and $2\sqrt{3}$ **61.** $L'(x^2 + 1) = \dfrac{2x}{x^2 + 1}$ **63.** $T'(x) = 0$

65. If $p(x) = (x - a)^2 q(x)$, where $q(a) \neq 0$, then $p(a) = p'(a) = 0$ and $p''(a) \neq 0$. Now suppose that $p(a) = p'(a) = 0$ and $p''(a) \neq 0$. Then $p(x) = (x - a)g(x)$ for some polynomial g. Since $p'(x) = (x - a)g'(x) + g(x)$ and $p'(a) = 0$, it follows that $g(a) = 0$. Therefore $g(x) = (x - a)q(x)$ for some polynomial q, and $p(x) = (x - a)^2 q(x)$. Finally $p''(a) \neq 0$ implies $q(a) \neq 0$.

67. Let $p(x)$ be a polynomial function of degree n. The number a is a zero of multiplicity k for p, $k < n$, if and only if $p(a) = p'(a) = \cdots = p^{(k-1)}(a) = 0$ and $p^{(k)}(a) \neq 0$.

69. 132.3   **71.** $800\,\pi\,\text{cm}^3/\text{sec}$ **73.** $\dfrac{d(KE)}{dt} = mv\dfrac{dv}{dt}$

SECTION 3.6

1. $\dfrac{dy}{dx} = -3\sin x - 4\sec x \tan x$ **3.** $\dfrac{dy}{dx} = 3x^2\csc x - x^3\csc x \cot x$ **5.** $\dfrac{dy}{dt} = -2\cos t \sin t$ **7.** $\dfrac{dy}{du} = 2u^{-1/2}\sin^3\sqrt{u}\,\cos\sqrt{u}$

9. $\dfrac{dy}{dx} = 2x\sec^2 x^2$ **11.** $\dfrac{dy}{dx} = 4(1 - \pi\csc^2\pi x)(x + \cot\pi x)^3$ **13.** $\dfrac{d^2y}{dx^2} = -\sin x$ **15.** $\dfrac{d^2y}{dx^2} = \cos x\,(1 + \sin x)^{-2}$

17. $\dfrac{d^2y}{du^2} = 12\cos 2u\,(2\sin^2 2u - \cos^2 2u)$ **19.** $\dfrac{d^2y}{dt^2} = 8\sec^2 2t \tan 2t$ **21.** $\dfrac{d^2y}{dx^2} = (2 - 9x^2)\sin 3x + 12x\cos 3x$ **23.** $\dfrac{d^2y}{dx^2} = 0$

25. $\sin x$ **27.** $(27t^3 - 12t)\sin 3t - 45t^2\cos 3t$ **29.** $3\cos 3x\,f'(\sin 3x)$ **31.** $y = x$ **33.** $y - \sqrt{3} = -4(x - \tfrac16\pi)$

35. $y - \sqrt{2} = \sqrt{2}(x - \tfrac14\pi)$ **37.** at π **39.** at $\tfrac16\pi, \tfrac76\pi$ **41.** at $\tfrac12\pi, \pi, \tfrac32\pi$ **43.** at $\tfrac14\pi, \tfrac34\pi, \tfrac54\pi, \tfrac74\pi$ **45.** at $\tfrac76\pi, \tfrac{11}{6}\pi$

47. $(\tfrac12\pi, \tfrac23\pi), (\tfrac76\pi, \tfrac43\pi), (\tfrac{11}{6}\pi, 2\pi)$ **49.** $(0, \tfrac14\pi), (\tfrac74\pi, 2\pi)$ **51.** $(\tfrac56\pi, \tfrac32\pi)$

53. (a) $\dfrac{dy}{dt} = \dfrac{dy}{du}\dfrac{du}{dx}\dfrac{dx}{dt} = (2u)(\sec x \tan x)(\pi) = 2\pi\sec^2\pi t \tan\pi t$ (b) $y = \sec^2\pi t - 1;\ \dfrac{dy}{dt} = 2\sec\pi t\,(\sec\pi t \tan\pi t)\pi = 2\pi\sec^2\pi t \tan\pi t$

55. (a) $\dfrac{dy}{dt} = \dfrac{dy}{du}\dfrac{du}{dx}\dfrac{dx}{dt} = 4[\tfrac12(1 - u)]^3(-\tfrac12)\cdot(-\sin x)\cdot 2 = 4[\tfrac12(1 - \cos 2t)]^3\sin 2t = (4\sin^6 t)(2\sin t\cos t) = 8\sin^7 t\cos t$

(b) $y = [\tfrac12(1 - \cos 2t)]^4 = \sin^8 t;\ \dfrac{dy}{dt} = 8\sin^7 t\cos t$

57. $\dfrac{d^n}{dx^n}(\cos x) = \begin{cases}(-1)^{(n+1)/2}\sin x, & n\ \text{odd}\\ (-1)^{n/2}\cos x, & n\ \text{even}\end{cases}$

59. (a) $f'(x) = \sin\dfrac1x - \dfrac1x\cos\dfrac1x;\ g'(x) = 2x\sin\dfrac1x - \cos\dfrac1x$ (b) $\displaystyle\lim_{x\to 0}g'(x) = \lim_{x\to 0}\left(2x\sin\dfrac1x - \cos\dfrac1x\right) = \lim_{x\to 0}\cos\dfrac1x$ does not exist

61. $y'' = -A\omega^2\sin\omega t - B\omega^2\cos\omega t = -\omega^2 y$. Thus $y'' + \omega^2 y = 0$.   **63.** $A(x) = \tfrac12 c^2\sin x;\ A'(x) = \tfrac12 c^2\cos x$ **65.** $-\tfrac16$ rad/sec

67. 0.74 rad/sec   **69.** by numerical work $f'(0) \cong 0$;   by the chain rule $f'(x) = -2x\sin x^2, f'(0) = 0$

71. (b) $f(x) = 0$ at $x = 0$ and $x \cong 0.81$ (c) $f'(x) = 0$ at $x \cong -1.25, x \cong 0.68, x \cong 0.43$

73.

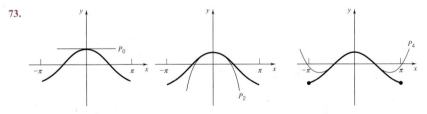

SECTION 3.7

1. $-\dfrac{x}{y}$ **3.** $-\dfrac{4x}{9y}$ **5.** $-\dfrac{x^2(x + 3y)}{x^3 + y^3}$ **7.** $\dfrac{2(x - y)}{2(x - y) + 1}$ **9.** $\dfrac{y - \cos(x + y)}{\cos(x + y) - x}$ **11.** $\dfrac{16}{(x + y)^3}$ **13.** $\dfrac{90}{(2y + x)^3}$

15. $\dfrac{d^2y}{dx^2} = \tfrac32 x\cos^2 y - \tfrac98 x^4\sin y\cos^3 y$ **17.** $\dfrac{dy}{dx} = \tfrac58, \dfrac{d^2y}{dx^2} = -\dfrac{9}{128}$ **19.** $\dfrac{dy}{dx} = -\tfrac12, \dfrac{d^2y}{dx^2} = 0$

21. tangent $2x + 3y - 5 = 0$; normal $3x - 2y + 12 = 0$ **23.** tangent $x + 2y + 8 = 0$; normal $2x - y + 1 = 0$

25. tangent: $y = \dfrac{-2}{\sqrt{3}}x + \left(\dfrac{1}{\sqrt{3}} + \dfrac{\pi}{3}\right)$; normal: $y = \dfrac{\sqrt{3}}{2}x + \left(\dfrac{1}{\sqrt{3}} + \dfrac{\pi}{3}\right)$ **27.** $\tfrac32 x^2(x^3 + 1)^{-1/2}$ **29.** $\dfrac{2x^2 + 1}{\sqrt{x^2 + 1}}$ **31.** $\dfrac{x}{(\sqrt[4]{2x^2 + 1})^3}$

33. $\dfrac{x(2x^2 - 5)}{\sqrt{2 - x^2}\sqrt{3 - x^2}}$ **35.** $\dfrac12\left(\dfrac{1}{\sqrt{x}} - \dfrac{1}{x\sqrt{x}}\right)$ **37.** $\dfrac{1}{(\sqrt{x^2 + 1})^3}$ **39.** $\dfrac13\left[\dfrac{1}{(\sqrt[3]{x})^2} - \dfrac{1}{(\sqrt[3]{x})^4}\right]$

41. (a) 　(b) 　(c)

43. $-\dfrac{2b^2}{9(\sqrt[3]{a+bx})^5}$

45. $\dfrac{\sqrt{x}\,\sec^2\sqrt{x}-\tan\sqrt{x}+2x\sec^2\sqrt{x}\,\tan\sqrt{x}}{4x\sqrt{x}}$

47. Let (x_0, y_0) be a point of the circle. If $x_0 = 0$, the normal line is the y-axis; if $y_0 = 0$, the normal line is the x-axis. For all other choices of (x_0, y_0) the normal line takes the form

$$y - y_0 = \frac{y_0}{x_0}(x - x_0) \qquad \text{which reduces to} \qquad y = \frac{y_0}{x_0}x.$$

In each case the normal line passes through the origin.

49. at right angles　**51.** at $(1, 1)$, $\alpha = \pi/4$; at $(0, 0)$, $\alpha = \pi/2$

53. The hyperbola and the ellipse intersect at the four points $(\pm 3, \pm 2)$. For the hyperbola, $\dfrac{dy}{dx} = \dfrac{x}{y}$. For the ellipse, $\dfrac{dy}{dx} = -\dfrac{4x}{9y}$. The product of these slopes is therefore $-\dfrac{4x^2}{9y^2}$. At each of the points of intersection this product is -1.

55. For the circles, $\dfrac{dy}{dx} = -\dfrac{x}{y}$, $y \neq 0$. For the straight lines $\dfrac{dy}{dx} = m = \dfrac{y}{x}$, $x \neq 0$. Thus, at a point of intersection of a circle $x^2 + y^2 = r^2$ and a line $y = mx$, we have $-\dfrac{x}{y} \cdot \dfrac{y}{x} = -1$ $(x \neq 0, y \neq 0)$.

57. $y - 2x + 12 = 0$, $y - 2x - 12 = 0$　**59.** $\left(\dfrac{\sqrt{6}}{4}, \pm\dfrac{\sqrt{2}}{4}\right)$, $\left(-\dfrac{\sqrt{6}}{4}, \pm\dfrac{\sqrt{2}}{4}\right)$

61. Let (x_0, y_0) be a point on the graph. An equation for the tangent line at (x_0, y_0) is

$$y - y_0 = -\left(\frac{y_0}{x_0}\right)^{1/2}(x - x_0)$$

The y-intercept is: $(x_0y_0)^{1/2} + y_0$; the x-intercept is: $(x_0, y_0)^{1/2} + x_0$; and $(x_0y_0)^{1/2} + y_0 + (x_0y_0)^{1/2} + x_0 = (x_0^{1/2} + y_0^{1/2})^2 = (c^{1/2})^2 = c$

63. (a)

(b) $\sqrt{3}, -\sqrt{3}, 0$, respectively　(c) $-\dfrac{1}{\sqrt{3}}, \dfrac{1}{\sqrt{3}}$, respectively

65. from numerical work, $f'(16) \cong 0.375$; from (3.7.1) $f'(x) = \frac{3}{4}x^{-1/4}$, $f'(16) = \frac{3}{8} = 0.375$

67.
$x = t, y = \sqrt{4 - t^2}$

$x = t, y = -\sqrt{4 - t^2}$

69. (a) 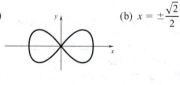　(b) $x = \pm\dfrac{\sqrt{2}}{2}$

SECTION 3.8

1. (a) -2 units/sec　(b) 4 units/sec　**3.** 9 units/sec　**5.** $\dfrac{-12\sin t\cos t}{\sqrt{16\cos^2 t + 4\sin^2 t}}$; $-\frac{3}{5}\sqrt{10}$　**7.** $-\frac{2}{27}$ m/min, $-\frac{8}{3}$ m²/min

9. $1/(50\pi)$ ft/min, $\frac{8}{3}$ ft²/min   **11.** (a) $A(\theta) = 50 \sin \theta$ (b) $\dfrac{25\pi}{8} \cong 4.36$ cm²/min   (c) $\theta = \dfrac{\pi}{2}$

13. 6 cm   **15.** decreasing 7 in.²/min   **17.** boat A **19.** 10 ft³/hr   **21.** $\frac{1600}{3}$ ft/min   **23.** 0.5634 lb/sec

25. dropping $1/2\pi$ in./min   **27.** decreasing 0.04 rad/min   **29.** 5π mi/min   **31.** decreasing 0.12 rad/sec   **33.** increasing $\frac{4}{101}$ rad/min

SECTION 3.9

1.
$$\Delta V = (x + h)^3 - x^3$$
$$= (x^3 + 3x^2h + 3xh^2 + h^3) - x^3$$
$$= 3x^2h + 3xh^2 + h^3$$
$$dV = 3x^2h$$
$$\Delta V - dV = 3xh^2 + h^3 \quad \text{(see figure)}$$

3. $10\frac{1}{30}$ taking $x = 1000$; 10.0332   **5.** $1\frac{31}{32}$ taking $x = 16$; 1.9680   **7.** 1.975 taking $x = 32$; 1.9744   **9.** 8.15 taking $x = 32$; 8.1491

11. 0.719; 0.7193   **13.** 0.531; 0.5317   **15.** 1.6   **17.** $2\pi rht$ **19.** error ≤ 0.01 ft   **21.** 98 gallons

23. $P^2 = 4\pi^2 \dfrac{L}{g}$ **25.** 0.00307 sec   **27.** within $\frac{1}{2}\%$ **29.** (a) $x_{n+1} = \frac{1}{2}x_n + 12\left(\dfrac{1}{x_n}\right)$ (b) $x_4 \cong 4.89898$

$2P\dfrac{dP}{dL} = \dfrac{4\pi^2}{g} = \dfrac{P^2}{L}$

Thus $\dfrac{dP}{P} = \dfrac{1}{2}\dfrac{dL}{L}$

31. (a) $x_{n+1} = \frac{2}{3}x_n + \frac{25}{3}\left(\dfrac{1}{x_n}\right)^2$ (b) $x_4 \cong 2.92402$ **33.** (a) $x_{n+1} = \dfrac{x_n \sin x_n + \cos x_n}{\sin x_n + 1}$ (b) $x_4 \cong 0.73909$

35. (a) $x_{n+1} = \dfrac{2x_n \cos x_n - 2 \sin x_n}{2 \cos x_n - 1}$ (b) $x_4 \cong 1.89549$ **37.** $x_2 = -2x_1, x_3 = 4x_1, x_4 = -8x_1, \ldots, x_n = (-2)^{n-1}x_1, \ldots$

39. (a) $x_1 = \frac{1}{2}, x_2 = -\frac{1}{2}, x_3 = \frac{1}{2}, \ldots, x_n = (-1)^{n-1}\frac{1}{2}, \ldots$ (b) $x_4 \cong 1.56165$ **41.** (b) $x_4 \cong 2.84382$; $f(x_4) \cong -0.00114$

43. (a) Let $F(x) = \frac{1}{2}\cos x - x$. Then $F(0) = \frac{1}{2}$ and $F\left(\dfrac{\pi}{2}\right) = -\pi/2$. Thus, F has a zero in $(0, \pi/2)$.   (b) $x_4 \cong 0.4502$; $f(x_4) \cong 0.45018$

45. (a) and (b)   **47.** $\displaystyle\lim_{h\to 0} \dfrac{g_1(h) + g_2(h)}{h} = \lim_{h\to 0} \dfrac{g_1(h)}{h} + \lim_{h\to 0} \dfrac{g_2(h)}{h} = 0 + 0 = 0$

$\displaystyle\lim_{h\to 0} \dfrac{g_1(h)g_2(h)}{h} = \lim_{h\to 0} h\dfrac{g_1(h)g_2(h)}{h^2} = \left(\lim_{h\to 0} h\right)\left(\lim_{h\to 0} \dfrac{g_1(h)}{h}\right)\left(\lim_{h\to 0} \dfrac{g_2(h)}{h}\right) = (0)(0)(0) = 0$

CHAPTER 4

SECTION 4.1

1. $c = \dfrac{\sqrt{3}}{3} \cong 0.577$ **3.** $c = \dfrac{\pi}{4}, \dfrac{3\pi}{4}, \dfrac{5\pi}{4}, \dfrac{7\pi}{4}$ **5.** $c = \frac{3}{2}$ **7.** $c = \frac{1}{3}\sqrt{39}$ **9.** $c = \frac{1}{2}\sqrt{2}$ **11.** $c = 0$

13. No. By mean-value theorem there exists at least one number $c \in (0, 2)$ such that $f'(c) = \dfrac{f(2) - f(0)}{2 - 0} = \dfrac{3}{2}$.

15. $f'(x) = \begin{cases} 2, & x \leq -1 \\ 3x^2 - 1, & x > -1; \end{cases} \quad -3 < c \leq -1$ and $c = 1$

17. $\dfrac{f(b) - f(a)}{b - a} = A(b + a) + B; f'(c) = 2Ac + B$. Equating and solving for c gives $c = \dfrac{a + b}{2}$.

19. $\dfrac{f(1) - f(-1)}{(1) - (-1)} = 0$ and $f'(x)$ is never zero; f is not differentiable at 0

21. Set $P(x) = 6x^4 - 7x + 1$. If there existed three numbers $a < b < c$ at which $P(x) = 0$, then by Rolle's theorem $P'(x)$ would have to be zero for some x in (a, b) and also for some x in (b, c). This is not the case: $P'(x) = 24x^3 - 7$ is zero only at $x = (\frac{7}{24})^{1/3}$.

23. Set $P(x) = x^3 + 9x^2 + 33x - 8$. Note that $P(0) < 0$ and $P(1) > 0$. Thus by the intermediate-value theorem there exists some number c between 0 and 1 at which $P(x) = 0$. If the equation $P(x) = 0$ had an additional real root, then by Rolle's theorem there would have to be some real number at which $P'(x) = 0$. This is not the case: $P'(x) = 3x^2 + 18x + 33$ is never 0 since the discriminant $b^2 - 4ac = (18)^2 - 12(33) < 0$.

25. Let c and d be two consecutive roots of the equation $P'(x) = 0$. The equation $P(x) = 0$ cannot have two or more roots between c and d, for then, by Rolle's theorem, $P'(x)$ would have to be zero somewhere between these two roots and thus between c and d. In this case, c and d would no longer be consecutive roots of $P'(x) = 0$.

27. If $x_1, x_2 \in I$ are fixed points of f, then $g(x_1) = g(x_2) = 0$. By Rolle's theorem there exists a number $c \in (x_1, x_2)$ such that $g'(c) = 1 - f'(c) = 0$, which implies $f'(c) = 1$, contradicting the hypothesis.

29. If $x_1 = x_2$, then $|f(x_1) - f(x_2)|$ and $|x_1 - x_2|$ are both 0 and the inequality holds. If $x_1 \neq x_2$, then you know by the mean-value theorem that

$$\frac{f(x_1) - f(x_2)}{x_1 - x_2} = f'(c)$$

for some number c between x_1 and x_2. Since $|f'(c)| \leq 1$, you can conclude that

$$\left| \frac{f(x_1) - f(x_2)}{x_1 - x_2} \right| \leq 1 \qquad \text{and thus that} \qquad |f(x_1) - f(x_2)| \leq |x_1 - x_2|.$$

31. set, for instance, $f(x) = \begin{cases} 1, & a < x < b \\ 0, & x = a, b \end{cases}$

33. (a) By the mean-value theorem, there exists $c \in (a, b)$ such that $f(b) - f(a) = f'(c)(b - a)$. Since $f'(x) \leq M$ for all $x \in (a, b)$, it follows that $f(b) \leq f(a) + M(b - a)$. (b) Same argument as in (a). (c) $|f'(x)| \leq L$ implies $-L \leq f'(x) \leq L$.

35. $f(x)g'(x) - g(x)f'(x) = \cos^2 x + \sin^2 x = 1$ for all $x \in I$. The result follows from Exercise 34.

37. $f'(x_0) = \lim_{y \to 0} \dfrac{f(x_0 + y) - f(x_0)}{y} = \lim_{y \to 0} \dfrac{f'(x_0 + \theta y)y}{y} = \lim_{y \to 0} f'(x_0 + \theta y) = \lim_{x \to x_0} f'(x) = L$

                                   └── by the hint   └── by (2.2.5)

39. (a) Between any two times that the object is at the origin there is at least one instant when the velocity is zero. (b) On any time interval there is at least one instant when the instantaneous velocity equals the average velocity over the interval.

41. Yes. Since $\dfrac{f(6) - f(0)}{6 - 0} = \dfrac{280}{6} \cong 46.67 \text{ ft/sec}$, the driver's speed at some time $c \in (0, 6)$ was $46.67 \text{ ft/sec} \cong 32 \text{ mph}$, by the mean-value theorem. The driver's speed must have been greater than 32 mph at the instant the brakes were applied.

43. $\sqrt{65} \cong 8.0625$ **45.** (a) $f'(x) = 4 + 2 \sin x > 0$ for all x. (b) f has a zero in $(0, 1)$. (c) $x_3 = 0.2361$.   **47.** $c = 0.676$

49. $c = 2.205$

SECTION 4.2

1. increases on $(-\infty, -1]$ and $[1, \infty)$, decreases on $[-1, 1]$ **3.** increases on $(-\infty, -1]$ and $[1, \infty)$, decreases on $[-1, 0)$ and $(0, 1]$

5. increases on $[-\frac{3}{4}, \infty)$, decreases on $(-\infty, -\frac{3}{4}]$ **7.** increases on $[-1, \infty)$, decreases on $(-\infty, -1]$

9. increases on $(-\infty, 2)$, decreases on $(2, \infty)$ **11.** increases on $(-\infty, -1)$ and $(-1, 0]$, decreases on $[0, 1)$ and $(1, \infty)$

13. increases on $[-\sqrt{5}, 0]$ and $[\sqrt{5}, \infty)$, decreases on $(-\infty, -\sqrt{5}]$ and $[0, \sqrt{5}]$ **15.** increases on $(-\infty, -1)$ and $(-1, \infty)$

17. decreases on $[0, \infty)$ **19.** increases on $[0, \infty)$, decreases on $(-\infty, 0]$ **21.** decreases on $(0, 3]$ **23.** increases on $[0, 2\pi]$

25. increases on $[\frac{2}{3}\pi, \pi]$, decreases on $[0, \frac{2}{3}\pi]$ **27.** increases on $[0, \frac{2}{3}\pi]$ and $[\frac{5}{6}\pi, \pi]$, decreases on $[\frac{2}{3}\pi, \frac{5}{6}\pi]$

29. $f(x) = \frac{1}{3}x^3 - x + \frac{8}{3}$ **31.** $f(x) = x^5 + x^4 + x^3 + x^2 + x + 5$ **33.** $f(x) = \frac{3}{4}x^{4/3} - \frac{2}{3}x^{3/2} + 1, x \geq 0$ **35.** $f(x) = 2x - \cos x + 4$

37. increases on $(-\infty, -3)$ and $[-1, 1]$, decreases on $[-3, -1]$ and $[1, \infty)$ **39.** increases on $(-\infty, 0]$ and $[3, \infty)$, decreases on $[0, 1)$ and $[1, 3]$

41.   **43.**   **45.**   **47.**

49. Not possible; f is increasing, so $f(2)$ must be greater than $f(-1)$.

51. $v(t) = 3t^2 - 12t + 9 = 3(t - 1)(t - 3)$ **53.** $v(t) = 6 \cos 3t$

sign of v: sign of v:

$a(t) = 6t - 12$ $a(t) = -18 \sin 3t$

sign of a: sign of a:

55. (a) $M \leq L \leq N$ (b) none (c) $M = L = N$ **57.** set, for instance, $f(x) = \begin{cases} 1, & x \text{ rational} \\ 0, & x \text{ irrational} \end{cases}$

59. (a) $f'(x) = 2 \sec x(\sec x \tan x) = 2 \sec^2 x \tan x$ $g'(x) = 2 \tan x (\sec^2 x) = 2 \sec^2 x \tan x$ (b) $C = 1$

61. (a) Let $h(x) = f^2(x) + g^2(x)$. Then $h'(x) = 0$; thus $h(x) = C$, constant.
(b) $f^2(a) + g^2(a) = C$ implies $C = 1$; $f(x) = \cos(x - a)$, $g(x) = \sin(x - a)$.

63. (a) $f'(x) = 1 - \cos x > 0$ except for $x = \dfrac{\pi}{2} + 2n\pi$, $n = 0, \pm 1, \pm 2, \ldots$. Thus f is increasing on $(-\infty, \infty)$. (b) Since f is increasing on
$(-\infty, \infty)$ and $f(0) = 0$, $f(x) < 0$ on $(-\infty, 0)$ and $f(x) > 0$ on $(0, \infty)$. Thus $\sin x < x$ on $(-\infty, 0)$ and $\sin x > x$ on $(0, \infty)$.

65. Let $f(x) = \tan x$ and $g(x) = x$. Then $f(0) = g(0) = 0$ and $f'(x) = \sec^2 x > g'(x) = 1$ on $\left(0, \dfrac{\pi}{2}\right)$. The result follows from Exercise 64.

67. Choose an integer $n > 1$, and let $f(x) = (1 + x)^n$, $g(x) = 1 + nx$. Then $f(0) = g(0) = 1$ and $f'(x) = n(1 + x)^{n-1} > g'(x) = n$
since $(1 + x)^{n-1} > 1$. The result follows from Exercise 64.

69. $0.06975 < \sin 4° < 0.06981$

71. $f'(x) = 0$ at $x = -0.633, 0.5, 2.633$
f is decreasing on $[-2, -0.633]$ and $[0.5, 2.633]$
f is increasing on $[-0.633, 0.5]$ and $[2.633, 5]$

73. $f'(x) = 0$ at $x = 0.770, 2.155, 3.798, 5.812$
f is decreasing on $[0, 0.770], [2.155, 3.798]$, and $[5.812, 6]$
f is increasing on $[0.770, 2.155]$ and $[3.798, 5.812]$

$-2 \leq x \leq 5, -40 \leq y \leq 20$

$0 \leq x \leq 6, -10 \leq y \leq 12$

75. $mgy_0 = mgy + \frac{1}{2}mv^2$; $|v| = \sqrt{2g(y_0 - y)}$.   **77.** 54.22 m/sec

SECTION 4.3

1. no critical nos.; no local extreme values   **3.** critical nos. ± 1; local max $f(-1) = -2$, local min $f(1) = 2$
5. critical nos. $0, \frac{2}{3}$; $f(0) = 0$ local min, $f(\frac{2}{3}) = \frac{4}{27}$ local max   **7.** no critical nos.; no local extreme values
9. critical no. $-\frac{1}{2}$; local max $f(-\frac{1}{2}) = -8$ **11.** critical nos. $0, \frac{3}{5}, 1$; local max $f(\frac{3}{5}) = 2^2 3^3/5^5$, local min $f(1) = 0$
13. critical nos. $\frac{5}{8}, 1$; local max $f(\frac{5}{8}) = \frac{27}{2048}$ **15.** critical nos. $-2, 0$; local max $f(-2) = 4$, local min $f(0) = 0$
17. critical nos. $-2, -\frac{1}{2}, 1$; local max $f(-\frac{1}{2}) = \frac{9}{4}$, local min $f(-2) = f(1) = 0$
19. critical nos. $-2, -\frac{12}{7}, 0$; local max $f(-\frac{12}{7}) = \frac{144}{49}(\frac{2}{7})^{1/3}$, local min $f(0) = 0$ **21.** critical nos. $-\frac{1}{2}, 3$; local min $f(-\frac{1}{2}) = \frac{7}{2}$
23. critical no. 1; local min $f(1) = 3$ **25.** critical nos. $\frac{1}{4}\pi, \frac{3}{4}\pi$; local max $f(\frac{1}{4}\pi) = \sqrt{2}$, local min $f(\frac{3}{4}\pi) = -\sqrt{2}$
27. critical nos. $\frac{1}{3}\pi, \frac{1}{2}\pi, \frac{2}{3}\pi$; local max $f(\frac{1}{2}\pi) = 1 - \sqrt{3}$, local min $f(\frac{1}{3}\pi) = f(\frac{2}{3}\pi) = -\frac{3}{4}$
29. critical nos. $\frac{1}{3}\pi, \frac{5}{3}\pi$; local max $f(\frac{1}{3}\pi) = \frac{4}{3}\sqrt{3} + \frac{10}{3}\pi$, local min $f(\frac{5}{3}\pi) = -\frac{5}{4}\sqrt{3} + \frac{2}{3}\pi$

31. (i) f increases on $(c - \delta, c]$ and decreases on $[c, c + \delta)$. (ii) f decreases on $(c - \delta, c]$ and increases on $[c, c + \delta)$. (iii) If $f'(x) > 0$ on
$(c - \delta, c) \cup (c, c + \delta)$, then, since f is continuous at c, f increases on $(c - \delta, c]$ and also on $[c, c + \delta)$. Therefore, in this case, f increases on
$(c - \delta, c + \delta)$. A similar argument shows that, if $f'(x) < 0$ on $(c - \delta, c) \cup (c, c + \delta)$, then f decreases on $(c - \delta, c + \delta)$.

33. critical pts 1, 2, 3; local max $P(2) = -4$, local min $P(1) = P(3) = -5$
Since $P'(x) < 0$ for $x < 0$, P decreases on $(-\infty, 0]$. Since $P(0) > 0$, P does not take on the value 0 on $(-\infty, 0]$.
Since $P(0) > 0$ and $P(1) < 0$, P takes on the value 0 at least once on $(0, 1)$. Since $P'(x) < 0$ on $(0, 1)$, P decreases on $[0, 1]$. It follows that
P takes on the value zero only once on $[0, 1]$.
Since $P'(x) > 0$ on $(1, 2)$ and $P'(x) < 0$ on $(2, 3)$, P increases on $[1, 2]$ and decreases on $[2, 3]$. Since $P(1), P(2), P(3)$ are all negative, P
cannot take on the value 0 between 1 and 3.
Since $P(3) < 0$ and $P(100) > 0$, P takes on the value 0 at least once on $(3, 100)$. Since $P'(x) > 0$ on $(3, 100)$, P increases on $[3, 100]$. It
follows that P takes on the value zero only once on $[3, 100]$.
Since $P'(x) > 0$ on $(100, \infty)$, P increases on $[100, \infty)$. Since $P(100) > 0$, P does not take on the value 0 on $[100, \infty)$.

35.

n odd

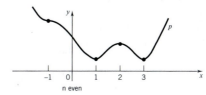

n even

37. $a = 4, b = \pm 2$ **39.** The function f takes on both positive and negative values in every open interval containing $x = 0$.

41. The line through $(0, 0)$ and $(c, f(c))$ has equation $y = \dfrac{f(c)}{c} x$; the tangent line to the graph of f at c has equation $y - f(c) = f'(c)(x - c)$.   If D has

an extreme value at c, then $\dfrac{f(c)}{c} = -\dfrac{1}{f'(c)}$.

43. (a) $f'(x) = 4x^3 - 4x - 3; f'(1) = -3, f'(2) = 21.$ Therefore f' has a zero in $(1, 2)$.   Since $f''(x) = 12x^2 - 4 > 0$ on $[1, 2], f'$ has exactly one zero in $(1, 2)$.   (b) $c \cong 1.3125; f$ has a local minimum at c.

45. (a) $f'(x) = 4x^3 - 14x - 8; f'(2) = -4, f'(3) = 58.$ Therefore f' has a zero in $(2, 3)$.   Since $f''(x) = 12x^2 - 14 > 0$ on $[2, 3], f'$ has exactly one zero in $(2, 3)$.   (b) $c \cong 2.1091; f$ has a local minimum at c.

47. (a) $f'(x) = \cos x + x - 2; f'(2) = -0.4161, f'(3) = 0.01.$ Therefore f' has a zero in $(2, 3)$.   Since $f''(x) = -\sin x + 1 \neq 0$ on $[2, 3], f'$ has exactly one zero in $(2, 3)$.   (b) $c \cong 2.9883; f$ has a local minimum at c.

49. (a) critical nos.: $-2.085, -1, 0.207, 1.096, 1.544$
   local extreme values: $f(-2.085) \cong -6.255, f(-1) = 7, f(0.207) \cong 0.621, f(1.096) \cong 7.097, f(1.544) \cong 4.635$
   (b) f is increasing on $[-2.085, -1], [0.207, 1.096], [1.544, 4]$; f is decreasing on $[-4, -2.085], [-1, 0.207], [1.096, 1.544]$

51. (a) critical nos.: $-2.204, -0.654, 0.654, 2.204$
   local extreme values: $f(-2.204) \cong 2.226, f(-0.654) \cong -6.364, f(0.654) \cong 6.364, f(2.204) \cong -2.226$
   (b) f is increasing on $[-3, -2.204], [-0.654, 0.654], [2.204, 3]$; f is decreasing on $[-2.204, -0.654], [0.654, 2.204]$

53. critical nos.: $-1.326, 0, 1.816$ local maxima at -1.326 and 1.816, local minimum at 0

SECTION 4.4

1. critical no. $-2; f(-2) = 0$ endpt min and absolute min

3. critical nos. $0, 2, 3; f(0) = 1$ endpt max and absolute max, $f(2) = -3$ local min and absolute min, $f(3) = -2$ endpt max

5. critical no. $2^{-1/3}; f(2^{-1/3}) = 3 \cdot 2^{-2/3}$ local min

7. critical nos. $\frac{1}{10}, 2^{-1/3}, 2; f(\frac{1}{10}) = 10 \frac{1}{100}$ endpt max and absolute max; $f(2^{-1/3}) = 3 \cdot 2^{-2/3}$ local min and absolute min, $f(2) = 4\frac{1}{2}$ endpt max

9. critical nos. $0, \frac{3}{2}, 2; f(0) = 2$ endpt max and absolute max, $f(\frac{3}{2}) = -\frac{1}{4}$ local min and absolute min, $f(2) = 0$ endpt max

11. critical nos. $-3, -2, 1; f(-3) = -\frac{3}{13}$ endpt max, $f(-2) = -\frac{1}{4}$ local min and absolute min, $f(1) = \frac{1}{5}$ endpt max and absolute max

13. critical nos. $0, \frac{1}{4}, 1; f(0) = 0$ endpt min and absolute min, $f(\frac{1}{4}) = \frac{1}{16}$ local max, $f(1) = 0$ local min and absolute min

15. critical nos. $2, 3; f(2) = 2$ local max and absolute max, $f(3) = 0$ endpt min

17. critical no. 1; no extreme values

19. critical nos. $0, \frac{5}{6}\pi, \pi; f(0) = -\sqrt{3}$ endpt min and absolute min, $f(\frac{5}{6}\pi) = \frac{7}{4}$ local max and absolute max, $f(\pi) = \sqrt{3}$ endpt min

21. critical nos. $0, \pi; f(0) = 5$ endpt max and absolute max, $f(\pi) = -5$ endpt min and absolute min

23. critical nos. $-\frac{1}{3}\pi, 0; f(-\frac{1}{3}\pi) = \frac{1}{3}\pi - \sqrt{3}$ endpt min and absolute min, no absolute max

25. critical nos. $0, 1, 4, 7; f(0) = 0$ endpt min, $f(1) = -2$ local min and absolute min, $f(4) = 1$ local max and absolute max, $f(7) = -2$ endpt min and absolute min

27. critical nos. $-2, -1, 1, 3; f(-2) = 5$ endpt max, $f(-1) = 2$ local min and absolute min, $f(1) = 6$ local max and absolute max, $f(3) = 2$ local min and absolute min

29. critical nos. $-3, -1, 0, 2; f(-3) = 2$ endpt max and absolute max, $f(-1) = 0$ local min, $f(0) = 2$ local max and absolute max, $f(2) = -2$ local min and absolute min

31.

33. Not possible: $f(1) > 0$ and f is increasing on $(1, 3)$.   Therefore $f(3)$ cannot be 0.

35. The discriminant of the quadratic polynomial $p'(x) = 3x^2 + 2ax + b$ is $4a^2 - 12b = 4(a^2 - 3b)$.   If $a^2 < 3b$, then p' has no real zeros.

37. By contradiction.   If f is continuous at c, then $f(c)$ is not a local maximum by the first-derivative test (4.3.4).

39. If f is not differentiable on (a, b), then f has a critical number at each point c in (a, b) where $f'(c)$ does not exist.   If f is differentiable on (a, b), then there exists c in (a, b) where $f'(c) = (f(b) - f(a))/(b - a)$ (mean-value theorem).   With $f(b) = f(a)$, we have $f'(c) = 0$ and thus c is a critical point of f.

41.
$$P(x) - M \geq a_0 x^n - (|a_1| x^{n-1} + \cdots + |a_{n-1}| x + |a_n| + M)$$
for $x > 0$
$$\geq a_0 x^n - (|a_1| + \cdots + |a_{n-1}| + |a_n| + M) \geq 0 \quad \text{for } x \geq \left(\frac{|a_1| + \cdots + |a_{n-1}| + |a_n| + M}{a_0} \right)^{1/n}$$
for $x > 1$

43. Let x be the length and y the width of a rectangle with diagnonal of length c. Then $y = \sqrt{c^2 - x^2}$ and the area $A = x\sqrt{c^2 - x^2}$. The maximum value of A occurs when $x = y = \dfrac{c}{\sqrt{2}}$.

45. (a) Use the wire to form a square.   (b) Use $\dfrac{9L}{(4\sqrt{3} + 9)} \cong 0.57L$ for the triangle and the remainder for the square.

47. critical nos.: -1.452, local max; 0.727, local min.

49. critical nos.: -1.683, local max; -0.284, local min; 0.645, local max; 1.760, local min. absolute min at $-\pi$, absolute max at π.

SECTION 4.5

1. 400   **3.** 20 by 10 ft   **5.** 32   **7.** 100 by 150 ft with the divider 100 ft long

9. radius of semicircle $\dfrac{90}{12 + 5\pi} \cong 3.25$ ft; height of rectangle $\dfrac{90 + 30\pi}{12 + 5\pi} \cong 6.65$ ft   **11.** $x = 2, y = \frac{3}{2}$ **13.** $-\frac{5}{2}$ **15.** $\frac{10}{3}\sqrt{3}$ in. by $\frac{5}{3}\sqrt{3}$ in.

17. equilateral triangle with side 4   **19.** $(1, 1)$ **21.** height of rectangle $\frac{15}{11}(5 - \sqrt{3}) \cong 4.46$ in.; side of triangle $\frac{10}{11}(6 + \sqrt{3}) \cong 7.03$ in.

23. $\frac{5}{3}$ **25.** $(0, \sqrt{3})$ **27.** $5\sqrt{5}$ ft   **29.** 54 by 72 in.   **31.** (a) use it all for the circle   (b) use $28\pi/(4 + \pi) \cong 12.32$ in. for the circle

33. base radius $\frac{10}{3}$ and height $\frac{8}{3}$ **35.** 10 by 10 by 12.5 ft   **37.** equilateral triangle with side $2r\sqrt{3}$ **39.** base radius $\frac{1}{3}R\sqrt{6}$ and height $\frac{2}{3}R\sqrt{3}$

41. base radius $\frac{2}{3}R\sqrt{2}$ and height $\frac{4}{3}R$ **43.** \$160,000   **45.** $\tan \theta = m$ **47.** $6\sqrt{6}$ ft   **49.** 125 customers

SECTION 4.6

1. (a) increasing on $[a, b]$, $[d, n]$, decreasing on $[b, d]$, $[n, p]$
  (b) concave up on (c, k), (l, m), concave down on (a, c), (k, l), (m, p); points of inflection at $x = c, k, l,$ and m.

3. concave down on $(-\infty, 0)$, concave up on $(0, \infty)$ **5.** concave down on $(-\infty, 0)$, concave up on $(0, \infty)$; pt of inflection $(0, 2)$

7. concave up on $(-\infty, -\frac{1}{3}\sqrt{3})$, concave down on $(-\frac{1}{3}\sqrt{3}, \frac{1}{3}\sqrt{3})$, concave up on $(\frac{1}{3}\sqrt{3}, \infty)$; pts of inflection $(-\frac{1}{3}\sqrt{3} -\frac{5}{36})$, $(\frac{1}{3}\sqrt{3}, -\frac{5}{36})$

9. concave down on $(-\infty, -1)$ and on $(0, 1)$, concave up on $(-1, 0)$ and on $(1, \infty)$; pt of inflection $(0, 0)$

11. concave up on $(-\infty, -\frac{1}{3}\sqrt{3})$, concave down on $(-\frac{1}{3}\sqrt{3}, \frac{1}{3}\sqrt{3})$, concave up on $(\frac{1}{3}\sqrt{3}, \infty)$; pts of inflection $(-\frac{1}{3}\sqrt{3}, \frac{4}{9})$, $(\frac{1}{3}\sqrt{3}, \frac{4}{9})$

13. concave up on $(0, \infty)$ **15.** concave down on $(-\infty, -2)$, concave up on $(-2, \infty)$; pt of inflection $(-2, 0)$

17. concave up on $(0, \frac{1}{4}\pi)$, concave down on $(\frac{1}{4}\pi, \frac{3}{4}\pi)$, concave up on $(\frac{3}{4}\pi, \pi)$; pts of inflection $(\frac{1}{4}\pi, \frac{1}{2})$ and $(\frac{3}{4}\pi, \frac{1}{2})$

19. concave up on $(0, \frac{1}{12}\pi)$, concave down on $(\frac{1}{12}\pi, \frac{5}{12}\pi)$, concave up on $(\frac{5}{12}\pi, \pi)$; pts of inflection $(\frac{1}{12}\pi, \frac{1}{2} + \frac{1}{144}\pi^2)$ and $(\frac{5}{12}\pi, \frac{1}{2} + \frac{25}{144}\pi^2)$

21. (a) f increases on $(-\infty, -\sqrt{3}]$ and $[\sqrt{3}, \infty)$, decreases on $[-\sqrt{3}, \sqrt{3}]$
  (b) $f(-\sqrt{3}) \cong 10.39$ local max; $f(\sqrt{3}) \cong -10.39$ local min
  (c) concave down on $(-\infty, 0)$, concave up on $(0, \infty)$
  (d) $(0, 0)$ is a point of inflection

23. (a) f decreases on $(-\infty, -1]$ and $[1, \infty)$, increases on $[-1, 1]$
  (b) $f(-1) = -1$ local min; $f(1) = 1$ local max
  (c) concave down on $(-\infty, -\sqrt{3})$ and $(0, \sqrt{3})$, concave up on $(-\sqrt{3}, 0)$ and $(\sqrt{3}, \infty)$
  (d) points of inflection $\left(-\sqrt{3}, -\dfrac{\sqrt{3}}{2}\right)$, $(0, 0)$, $\left(\sqrt{3}, \dfrac{\sqrt{3}}{2}\right)$

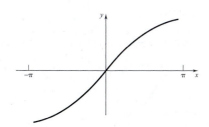

25. (a) increasing on $[-\pi, \pi]$
  (b) no local max or min
  (c) concave up on $(-\pi, 0)$, concave down on $(0, \pi)$
  (d) point of inflection $(0, 0)$

27. (a) increasing on $(-\infty, \infty)$
  (b) no local max or min
  (c) concave down on $(-\infty, 0)$, concave up on $(0, 1)$, no concavity on $(1, \infty)$
  (d) point of inflection $(0, 0)$

29.

31.

33. $d = \frac{1}{3}(a + b + c)$ **35.** $a = -\frac{1}{2}, b = \frac{1}{2}$ **37.** $A = 18, B = -4$ **39.** $f(x) = x^3 - 3x^2 + 3x - 3$

41. (a) $p''(x) = 6x - 2a$ has exactly one zero
(b) $p'(x) = 3x^2 + 2ax + b$ has real zeros iff $a^2 \geq 3b$.

43. (a) concave up on $(-\infty, -0.913)$ and $(0.913, \infty)$; concave down on $(-0.913, 0.913)$
(b) points of inflection at $x \cong -0.913, 0.913$

45. (a) concave up on $(-\pi, -1.996)$ and $(-0.345, 2.550)$; concave down on $(-1.996, -0.345), (2.550, \pi)$
(b) points of inflection at $x \cong -1.996, -0.345, 2.550$

47. (a) concave up on $(-2.726, 0.402)$ and $(1.823, 3)$; concave down on $(-3, -2.726), (0.402, 1.823)$
(b) points of inflection at $x \cong -2.726, 0.402, 1.823$

SECTION 4.7

1. (a) ∞ (b) $-\infty$ (c) ∞ (d) 1 (e) 0 (f) $x = -1, x = 1$ (g) $y = 0, y = 1$

3. vertical: $x = \frac{1}{3}$; horizontal: $y = \frac{1}{3}$ **5.** vertical: $x = 2$; horizontal: none   **7.** vertical: $x = \pm 3$; horizontal: $y = 0$

9. vertical: $x = -\frac{4}{3}$; horizontal: $y = \frac{4}{9}$ **11.** vertical: $x = \frac{5}{2}$; horizontal: $y = 0$ **13.** vertical: none; horizontal: $y = \pm\frac{3}{2}$

15. vertical: $x = 1$; horizontal: $y = 0$ **17.** vertical: none; horizontal: $y = 0$ **19.** vertical: $x = (2n + \frac{1}{2})\pi$; horizontal: none

21. neither   **23.** cusp   **25.** tangent   **27.** neither   **29.** cusp   **31.** cusp   **33.** neither

35.

37.

39. (a) increasing on $(-\infty, -1], [1, \infty)$;
decreasing on $[-1, 1]$
(b) concave down on $(-\infty, 0)$;
concave up on $(0, \infty)$
vertical tangent at $x = 0$

41. (a) decreasing on $[0, 2]$;
increasing on $(-\infty, 0], [2, \infty)$
(b) concave up on $(-1, 0), (0, \infty)$;
concave down on $(-\infty, -1)$
vertical cusp at $x = 0$

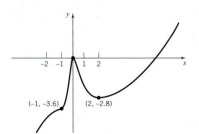

43. (a) decreasing on $[0, 1)$;
increasing on $(1, \infty)$
(b) concave up on $(0, 1)$;
concave down on $(1, \infty)$
vertical tangent line at $x = 0$

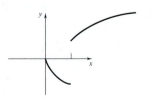

45. $x = 1$ vertical asymptote
$y = 0$ and $y = 2$ horizontal asymptotes

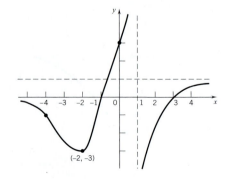

47. vertical cusp at $x = 0$

49. vertical tangent line at $x = 0$
$y = 1$ and $y = -1$ horizontal asymptotes

51. (a) $p < q$ and p odd
(b) $p < q$ and p even

53.

55.

SECTION 4.8

1.

3.

5.

7.

9.

11.

13.

15.

17.

19.
(2,1)

21.
$y = \frac{1}{2}$
$x = \frac{3}{4}$

23.
3 6
$x = -3$

25.
$y = 1$
$x = -2$ $x = 2$

27.
$\left(\frac{2}{3}, \frac{2}{9}\sqrt{3}\right)$
(1,0)

29.
(π, π)
$\left(\frac{1}{3}\pi, \frac{1}{3}\pi + \frac{1}{2}\sqrt{3}\right)$
$\left(\frac{1}{2}\pi, \frac{1}{2}\pi\right)$
$\left(\frac{2\pi}{3}, \frac{2}{3}\pi - \frac{1}{2}\sqrt{3}\right)$

31.
1
$\frac{9}{16}$
$\frac{1}{6}\pi$ $\frac{1}{2}\pi$ $\frac{5}{6}\pi$ π

33.
5
$2\sqrt{2}$
$\frac{1}{4}\pi$ $\frac{1}{2}\pi$ $\frac{3}{4}\pi$ π

35.
(0, 1)

37.
(2, 108)
$2 - \frac{1}{2}\sqrt{6}$ $2 + \frac{1}{2}\sqrt{6}$ 5

39.
3
2
-2 -1 1 2

41.
$\frac{5}{6}$ 1 $\frac{5}{3}$

43.
$\left(-\frac{2}{3}\right)^{3/5}$
(1, −5)

45.
$y = 1$
$x = 2$

47.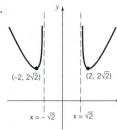
$(-2, 2\sqrt{2})$ $(2, 2\sqrt{2})$
$x = -\sqrt{2}$ $x = \sqrt{2}$

49.
2
$\frac{1}{3}\pi$ $\frac{2}{3}\pi$ π
-2

51.
$\frac{1}{4}\pi$
-1
$x = \frac{1}{2}\pi$

53.
$-\pi$ $-\frac{\pi}{2}$ $\frac{\pi}{2}$

55. (a) increasing on $(-\infty, -1], (0, 1], [3, \infty)$; decreasing on $[-1, 0), [1, 3]$ critical nos. $x = -1, 1, 3$
   (b) concave up on $(-\infty, -3), (2, \infty)$ concave down on $(-3, 0), (0, 2)$ (c)

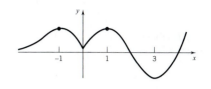

57. Solve for y: $y = \pm\dfrac{b}{a}x\sqrt{1 - \dfrac{a^2}{x^2}}$.   For $|x|$ large, $y \cong \pm\dfrac{b}{a}x$.

CHAPTER 5
SECTION 5.2

1. $L_f(P) = \frac{5}{8}$, $U_f(P) = \frac{11}{8}$ **3.** $L_f(P) = \frac{9}{64}$, $U_f(P) = \frac{37}{64}$ **5.** $L_f(P) = \frac{17}{16}$, $U_f(P) = \frac{25}{16}$ **7.** $L_f(P) = \frac{7}{16}$, $U_f(P) = \frac{25}{16}$

9. $L_f(P) = \frac{3}{16}$, $U_f(P) = \frac{43}{32}$ **11.** $L_f(P) = \frac{1}{6}\pi$, $U_f(P) = \frac{11}{12}\pi$ **13.** (a) $L_f(P) \le U_f(P)$ but $3 \nleq 2$

(b) $L_f(P) \le \displaystyle\int_{-1}^{1} f(x)\,dx \le U_f(P)$ but $3 \nleq 2 \le 6$

(c) $L_f(P) \le \displaystyle\int_{-1}^{1} f(x)\,dx \le U_f(P)$ but $3 \le 10 \nleq 6$

15. (a) $L_f(P) = -3x_1(x_1 - x_0) - 3x_2(x_2 - x_1) - \cdots - 3x_n(x_n - x_{n-1})$ (b) $-\frac{3}{2}(b^2 - a^2)$
   $U_f(P) = -3x_0(x_1 - x_0) - 3x_1(x_2 - x_1) - \cdots - 3x_{n-1}(x_n - x_{n-1})$

17. $L_f(P) = x_0^3(x_1 - x_0) + x_1^3(x_2 - x_1) + \cdots + x_{n-1}^3(x_n - x_{n-1})$
   $U_f(P) = x_1^3(x_1 - x_0) + x_2^3(x_2 - x_1) + \cdots + x_n^3(x_n - x_{n-1})$
   For each index i, $x_{i-1}^3 \le \frac{1}{4}(x_i^3 + x_i^2 x_{i-1} + x_i x_{i-1}^2 + x_{i-1}^3) \le x_i^3$ and thus by the hint $x_{i-1}^3(x_i - x_{i-1}) \le \frac{1}{4}(x_i^4 - x_{i-1}^4) \le x_i^3(x_i - x_{i-1})$. Adding up
   these inequalities, we find that $L_f(P) \le \frac{1}{4}(x_n^4 - x_0^4) \le U_f(P)$. Since $x_n = 1$ and $x_0 = 0$, the middle term is $\frac{1}{4}$. Thus the integral is $\frac{1}{4}$.

19. Let f be continuous and increasing on $[a, b]$, and let P be a regular partition. Then

$$U_f(P) = [f(x_1) + f(x_2) + \cdots + f(x_n)]\Delta x$$
$$L_f(P) = [f(x_0) + f(x_1) + \cdots + f(x_{n-1})]\Delta x$$

and $U_f(P) - L_f(P) = [f(x_n) - f(x_0)]\Delta x = [f(b) - f(a)]\Delta x$.

21. (a) From Definition 5.2.3, $0 \le I - L_f(P) \le U_f(P) - L_f(P)$ **23.** (a) Let $x_1, x_2 \in [0, 2]$, $x_2 \ge x_1$. Then
   (b) Use Definition 5.2.3

$$f(x_2) - f(x_1) = \frac{(x_2 + x_1)(x_2 - x_1)}{\sqrt{1 + x_2^2} + \sqrt{1 + x_1^2}} \ge 0.$$

(b) $n = 25$

25. Let S be the set of positive integers for which the statement is true. Since $1 = \dfrac{1(2)}{2}$, $1 \in S$. Assume $k \in S$. Then

$$1 + 2 + \cdots + k + k + 1 = (1 + 2 + \cdots + k) + k + 1$$
$$= \frac{k(k + 1)}{2} + k + 1 = \frac{(k + 1)(k + 2)}{2}.$$

Thus $k + 1 \in S$ and so S is the set of positive integers.

27. Let $f(x) = x$ on $[0, b]$ and let $P = \{x_0, x_1, \ldots, x_n\}$ be a regular partition.

(a) $L_f(P) = \left[0 \cdot \dfrac{b}{n} + 1 \cdot \dfrac{b}{n} + \cdots + (n - 1) \cdot \dfrac{b}{n}\right]\dfrac{b}{n}$ (c) $L_f(P) = \dfrac{b^2}{n^2} \cdot \dfrac{(n - 1)n}{2} \to \dfrac{b^2}{2}$ as $n \to \infty$

$= \dfrac{b^2}{n^2}[1 + 2 + \cdots + (n - 1)]$ $U_f(P) = \dfrac{b^2}{n^2} \cdot \dfrac{n(n + 1)}{2} \to \dfrac{b^2}{2}$ as $n \to \infty$

(b) $U_f(P) = \left[1 \cdot \dfrac{b}{n} + 2 \cdot \dfrac{b}{n} + \cdots + n \cdot \dfrac{b}{n}\right]\dfrac{b}{n}$ Thus $\displaystyle\int_0^1 x\,dx = \dfrac{b^2}{2}$.

$= \dfrac{b^2}{n^2}[1 + 2 + \cdots + n]$

29. Let P be an arbitrary partition of $[0, 4]$. Since each $m_i = 2$ and each $M_i \geq 2$.

$$L_g(P) = 2\Delta x_i + \cdots + 2\Delta x_n = 2(\Delta x_1 + \cdots + \Delta x_n) = 2 \cdot 4 = 8,$$
$$U_g(P) \geq 2\Delta x_1 + \cdots + 2\Delta x_n = 2(\Delta x_1 + \cdots + \Delta x_n) = 2 \cdot 4 = 8.$$

Thus $L_g(P) \leq 8 \leq U_g(P)$ for all partitions P of $[0, 4]$.
Uniqueness: Suppose that

(∗) $\qquad\qquad\qquad\qquad\qquad L_g(P) \leq I \leq U_g(P) \qquad$ for all partitions P of $[0, 4]$.

Since $L_g(P) = 8$ for all P, I is at least 8. Suppose now that $I > 8$ and choose a partition P of $[0, 4]$ with max $\Delta x_i < \frac{1}{5}(I - 8)$ and $0 = x_1 < \cdots < x_{i-1} < 3 < x_i < \cdots < x_n = 4$. Then

$$U_g(P) = 2\Delta x_1 + \cdots + 2\Delta x_{i-1} + 7\Delta x_i + 2\Delta x_{i+1} + \cdots + 2\Delta x_n$$
$$= 2(\Delta x_1 + \cdots + \Delta x_n) + 5\Delta x_i = 8 + 5\Delta x_i < 8 + \tfrac{5}{5}(I - 8) = I$$

and I does not satisfy (∗). This contradiction proves that I is not greater than 8 and therefore $I = 8$.

31. Let $P = \{x_0, x_1, \ldots, x_n\}$ be any partition of $[2, 10]$.
   (a) Since each subinterval of $[2, 10]$ contains both rational and irrational numbers,

$$L_f(P) = 4\Delta x_1 + 4\Delta x_2 + \cdots + 4\Delta x_n = 4(10 - 2) = 32$$
$$U_f(P) = 7\Delta x_1 + 7\Delta x_2 + \cdots + 7\Delta x_n = 7(10 - 2) = 56.$$

   (b) There is more than one number I that satisfies $L_f(P) \leq I \leq U_f(P)$ for all partitions P.
   (c) See (a).

SECTION 5.3

1. (a) 5 (b) -2 (c) -1 (d) 0 (e) -4 (f) 1

3. With $P = \{1, \frac{3}{2}, 2\}$ and $f(x) = \dfrac{1}{x}$, we have $0.5 < \frac{7}{12} = L_f(P) \leq \displaystyle\int_1^2 \dfrac{dx}{x} \leq U_f(P) = \frac{5}{6} < 1$.

5. (a) $F(0) = 0$ (b) $F'(x) = x\sqrt{x+1}$ (c) $F'(2) = 2\sqrt{3}$ (d) $F(2) = \displaystyle\int_0^2 t\sqrt{t+1}\, dt$ (e) $-F(x) = \displaystyle\int_x^0 t\sqrt{t+1}\, dt$

7. (a) $F'(x) = \dfrac{1}{x}, x > 0$; F is increasing on $(0, \infty)$.   (c)

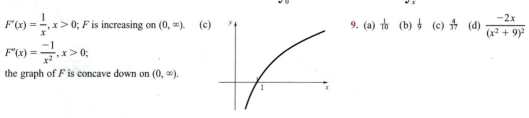

   (b) $F''(x) = \dfrac{-1}{x^2}, x > 0$;

   the graph of F is concave down on $(0, \infty)$.

9. (a) $\frac{1}{10}$ (b) $\frac{1}{9}$ (c) $\frac{4}{37}$ (d) $\dfrac{-2x}{(x^2+9)^2}$

11. $F'(x) = -\sqrt{x^2 + 1}$ (a) $-\sqrt{2}$ (b) -1 (c) $-\frac{1}{2}\sqrt{5}$ (d) $F''(x) = \dfrac{-x}{\sqrt{x^2+1}}$ 13. (a) -1 (b) 1 (c) 0 (d) $-\pi \sin \pi x$

15. (a) Since $P_1 \subseteq P_2$, $U_f(P_2) \leq U_f(P_1)$ but $5 \not\leq 4$. (b) Since $P_1 \subseteq P_2$, $L_f(P_1) \leq L_f(P_2)$ but $5 \not\leq 4$.   17. $F'(x) = 3x^5 \cos(x^3)$.

19. $F'(x) = 2x[\sin^2(x^2) - x^3]$.   21. $F'(x) = 4x^3 \sin(x^2) - 2x \sin x$ (assuming $x > 0$).   23. (a) $f(0) = \frac{1}{2}$ (b) $2, -2$

25. By the hint $F(b) - F(a) = F'(c)(b - a)$ for some c in (a, b). The desired result follows by observing that

$$F(b) = \int_a^b f(t)\, dt, \qquad F(a) = 0, \qquad \text{and} \qquad F'(c) = f(c).$$

27. Set $G(x) = \displaystyle\int_a^x f(t)\, dt$. Then $F(x) = \displaystyle\int_c^a f(t)\, dt + G(x)$. By (5.3.5) G, and thus F, is continuous on $[a, b]$, is differentiable on (a, b), and $F'(x) = G'(x) = f(x)$ for all x in (a, b).

SECTION 5.4

1. -2 3. 1   5. $\frac{28}{3}$ 7. $\frac{32}{3}$ 9. $\frac{2}{3}$ 11. 0   13. $\frac{5}{72}$ 15. $\frac{13}{2}$ 17. $-\frac{4}{15}$ 19. $\frac{1}{18}(2^{18} - 1)$ 21. $\frac{1}{6}a^2$ 23. $\frac{7}{4}$

25. $-\frac{1}{12}$ 27. $\frac{21}{2}$ 29. 1   31. 2   33. $2 - \sqrt{2}$ 35. 0   37. $\dfrac{\pi}{9} - 2\sqrt{3}$ 39. $\sqrt{13} - 2$

41. (a) $\displaystyle\int_2^x \dfrac{dt}{t}$ (b) $-3 + \displaystyle\int_2^x \dfrac{dt}{t}$ 43. $\dfrac{32}{3}$ 45. $2 + \sqrt{2}$ 47. (a) $3/2$ (b) $5/2$ 49. (a) $4/3$ (b) 4

51. (a) $x(t) = 5t^2 - \frac{1}{3}t^3, 0 \le t \le 10$ (b) At $t = 5$; $x(5) = \frac{250}{3}$ **53.** $\frac{13}{2}$ **55.** $\frac{2 + \sqrt{3}}{2} + 2\pi$

57. (a)
$$g(x) = \begin{cases} x^2 + 2x + 2, & -2 \le x \le 0 \\ 2x + 2, & 0 < x \le 1 \\ -x^2 + 4x + 1, & 1 < x \le 2 \end{cases}$$

(b)

(c) f is differentiable on $(-2, 0)$, $(0, 1)$, $(1, 2)$; g is differentiable on $(-2, 2)$.

59. $f(x)$ and $f(x) - f(a)$, respectively

SECTION 5.5

1. $\frac{9}{4}$ **3.** $\frac{38}{3}$ **5.** $\frac{47}{15}$ **7.** $\frac{5}{3}$ **9.** $\frac{1}{2}$

11. area $= \frac{1}{3}$

13. area $= \frac{9}{2}$

15. area $= \frac{64}{3}$

17. area $= 10$

19. area $= \frac{32}{3}$

21. area $= 4$

23. area $= 2 + \frac{2}{3}\pi^3$

25. area $= \frac{1}{8}\pi^2 - 1$

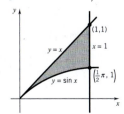

27. (a) $-\frac{91}{6}$, the area of the region bounded by f and the x-axis for $x \in [-3, -2] \cup [3, 4]$ minus the area of the region bounded by f and the x-axis for $x \in [-2, 3]$.

(b) $\frac{53}{2}$ (c) $\frac{125}{6}$

29. (a) 0 (b) 5

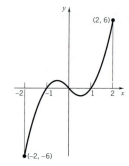

31. (a) $\dfrac{65}{4}$ (b) 17.87   **33.** $\dfrac{10}{3}$ **35.** area $= 2 - \sqrt{2}$

Area $= 2 - \sqrt{2}$

37. 2.86

SECTION 5.6

1. $-\dfrac{1}{3x^3} + C$ **3.** $\frac{1}{2}ax^2 + bx + C$ **5.** $2\sqrt{1+x} + C$ **7.** $\dfrac{1}{2}x^2 + \dfrac{1}{x} + C$ **9.** $\frac{1}{3}t^3 - \frac{1}{2}(a + b)t^2 + abt + C$

11. $\frac{2}{9}t^{9/2} - \frac{2}{5}(a + b)t^{5/2} + 2abt^{1/2} + C$ **13.** $\frac{1}{2}[g(x)]^2 + C$ **15.** $\frac{1}{2}\sec^2 x + C$ **17.** $-\dfrac{1}{4x + 1} + C$ **19.** $x^2 - x - 2$

21. $\frac{1}{2}ax^2 + bx - 2a - 2b$ **23.** $3 - \cos x$ **25.** $x^3 - x^2 + x + 2$ **27.** $\frac{1}{12}(x^4 - 2x^3 + 2x + 23)$ **29.** $x - \cos x + 3$

31. $\frac{1}{3}x^3 - \frac{3}{2}x^2 - \frac{1}{3}x + 3$ **33.** $\dfrac{d}{dx}\left(\displaystyle\int f(x)\,dx\right) = f(x), \quad \displaystyle\int \dfrac{d}{dx}[f(x)]\,dx = f(x) + C$

35. (a) 34 units to the right of the origin (b) 44 units   **37.** (a) $v(t) = 2(t + 1)^{1/2} - 1$ (b) $x(t) = \frac{4}{3}(t + 1)^{3/2} - t - \frac{4}{3}$

39. (a) 4.4 sec (b) 193.6 ft   **41.** $[v(t)]^2 = (at + v_0)^2 = a^2t^2 + 2av_0t + v_0^2 = v_0^2 + 2a(\frac{1}{2}at^2 + v_0t) = v_0^2 + 2a[x(t) - x_0]$ **43.** 42 sec
$$x(t) = \tfrac{1}{2}at^2 + v_0t + x_0$$

45. $x(t) = x_0 + v_0t + At^2 + Bt^3$ **47.** at $(\frac{160}{3}, 50)$ **49.** $A = -\frac{5}{2}, B = 2$ **51.** (a) at $t = \frac{11}{6}\pi$ sec (b) at $t = \frac{13}{6}\pi$ sec

53. mean-value theorem   **55.** $v(t) = v_0(1 - 2tv_0)^{-1}$

SECTION 5.7

1. $\dfrac{1}{3(2 - 3x)} + C$ **3.** $\frac{1}{3}(2x + 1)^{3/2} + C$ **5.** $\dfrac{4}{7a}(ax + b)^{7/4} + C$ **7.** $-\dfrac{1}{8(4t^2 + 9)} + C$ **9.** $\frac{1}{75}(5t^3 + 9)^5 + C$

11. $\frac{4}{15}(1 + x^3)^{5/4} + C$ **13.** $-\dfrac{1}{4(1 + s^2)^2} + C$ **15.** $\sqrt{x^2 + 1} + C$ **17.** $-\frac{1}{5}(1 - x^3)^{5/3} + C$ **19.** $-\frac{5}{4}(x^2 + 1)^{-2} + C$

21. $-4(x^{1/4} + 1)^{-1} + C$ **23.** $-\dfrac{b^3}{2a^4}\sqrt{1 - a^4x^4} + C$ **25.** $\frac{15}{8}$ **27.** $\frac{31}{2}$ **29.** 0   **31.** $\frac{1}{3}|a|^3$ **33.** $\frac{2}{5}(x + 1)^{5/2} - \frac{2}{3}(x + 1)^{3/2} + C$

35. $\frac{1}{10}(2x - 1)^{5/2} + \frac{1}{6}(2x - 1)^{3/2} + C$ **37.** $\frac{1}{14}(y + 1)^{14} - \frac{1}{13}(y + 1)^{13} + C$ **39.** $-\frac{1}{2}(t - 2)^{-2} - \frac{4}{3}(t - 2)^{-3} - (t - 2)^{-4} + C$

41. $\frac{16}{3}\sqrt{2} - \frac{14}{3}$ **43.** $-\frac{769}{112}$ **45.** $\frac{1}{3}\sin(3x - 1) + C$ **47.** $-(\cot \pi x)/\pi + C$ **49.** $\frac{1}{2}\cos(3 - 2x) + C$ **51.** $-\frac{1}{5}\cos^5 x + C$

53. $-2\cos x^{1/2} + C$ **55.** $\frac{2}{3}(1 + \sin x)^{3/2} + C$ **57.** $\tan x + C$ **59.** $\frac{1}{8}\sin^4 x^2 + C$ **61.** $-\cot x - \frac{1}{3}\cot^3 x + C$

63. $2(1 + \tan x)^{1/2} + C$ **65.** 0   **67.** $(\sqrt{3} - 1)/\pi$ **69.** $\frac{1}{4}$ **71.** $2 - \sqrt{2}$

73. $\displaystyle\int \sin^2 x\,dx = \int \dfrac{1 - \cos 2x}{2}\,dx = \frac{1}{2}\int (1 - \cos 2x)\,dx = \frac{1}{2}x - \frac{1}{4}\sin 2x + C$ **75.** $\frac{1}{2}x + \frac{1}{20}\sin 10x + C$ **77.** $\dfrac{\pi}{4}$ **79.** $1/2\,\pi$

81. $(4\sqrt{3} - 6)/3\,\pi$ **83.** (a) $\frac{1}{2}\sec^2 x + C$ (b) $\frac{1}{2}\tan^2 x + C'$
(c) $\frac{1}{2}\sec^2 x + C = \frac{1}{2}(1 + \tan^2 x) + C = \frac{1}{2}\tan^2 x + (C + \frac{1}{2}) = \frac{1}{2}\tan^2 x + C'; C + \frac{1}{2}$ and C' each represent an arbitrary constant

85. πab

SECTION 5.8

1. yes; $\displaystyle\int_a^b [f(x) - g(x)]\,dx = \int_a^b f(x)\,dx - \int_a^b g(x)\,dx > 0$ **3.** yes; otherwise we would have $f(x) \leq g(x)$ for all $x \in [a, b]$, and it would follow that
$$\int_a^b f(x)\,dx \leq \int_a^b g(x)\,dx$$

5. no; take $f(x) = 0$, $g(x) = -1$ on $[0, 1]$ **7.** no; take, for example, any odd function on an interval of the form $[-c, c]$

9. no; $\int_{-1}^{1} x\,dx = 0$ but $\int_{-1}^{1} |x|\,dx \neq 0$ **11.** yes; $U_f(P) \geq \int_a^b f(x)\,dx = 0$ **13.** no; $L_f(P) \leq \int_a^b f(x)\,dx = 0$

15. yes; $\int_a^b [f(x) + 1]\,dx = \int_a^b f(x)\,dx + \int_a^b 1\,dx = 0 + b - a = b - a$ **17.** $\dfrac{2x}{\sqrt{2x^2 + 7}}$ **19.** $-f(x)$ **21.** $-\dfrac{2\sin(x^2)}{x}$

23. $\dfrac{\sqrt{x}}{2(1 + x)}$ **25.** $\dfrac{1}{x}$ **27.** $\dfrac{3}{1 + (2 + 3x)^{3/2}} - \dfrac{1}{3x^{2/3}(1 + x^{1/2})}$ **29.** $2x^3 \cos(x^4) - \tfrac{1}{2}\cos x$ **31.** $4x\sqrt{1 + 4x^2} - \tan x \sec^2 x\,|\sec x|$

33. (a) With P a partition of $[a, b]$

$$L_f(P) \leq \int_a^b f(x)\,dx.$$

If f is nonnegative on $[a, b]$, then $L_f(P$ is nonnegative and, consequently, so is the integral. If f is positive on $[a, b]$, then $L_f(P$ is positive and, consequently, so is the integral.

(b) Take F as an antiderivative of f on $[a, b]$. Observe that

$$F'(x) = f(x) \quad \text{on } (a, b) \qquad \text{and} \qquad \int_a^b f(x)\,dx = F(b) - F(a).$$

If $f(x) \geq 0$ on $[a, b]$, then F is nondecreasing on $[a, b]$ and $F(b) - F(a) \geq 0$.
If $f(x) > 0$ on $[a, b]$, then F is increasing on $[a, b]$ and $F(b) - F(a) > 0$.

35. For all $x \in [a, b]$

$$-f(x) \leq |f(x)| \qquad \text{and} \qquad f(x) \leq |f(x)|.$$

It follows from II that

$$\int_a^b -f(x)\,dx \leq \int_a^b |f(x)|\,dx \qquad \text{and} \qquad \int_a^b f(x)\,dx \leq \int_a^b |f(x)|\,dx,$$

and, consequently, that

$$\left| \int_a^b f(x)\,dx \right| \leq \int_a^b |f(x)|\,dx.$$

37. $H(x) = \int_{2x}^{x^3-4} \dfrac{x\,dt}{1 + \sqrt{t}} = x \int_{2x}^{x^3-4} \dfrac{dt}{1 + \sqrt{t}}$

$H'(x) = x \cdot \left[\dfrac{3x^2}{1 + \sqrt{x^3 - 4}} - \dfrac{2}{1 + \sqrt{2x}} \right] + 1 \cdot \int_{2x}^{x^3-4} \dfrac{dt}{1 + \sqrt{t}}$

$H'(2) = 2\left[\dfrac{12}{3} - \dfrac{2}{3} \right] + \underbrace{\int_4^4 \dfrac{dt}{1 + \sqrt{t}}}_{= 0} = \dfrac{20}{3}$

39. (a) Let $u = -x$ (b) $\int_{-a}^a f(x)\,dx = \int_{-a}^0 f(x)\,dx + \int_0^a f(x)\,dx = \int_0^a f(-x)\,dx + \int_0^a f(x)\,dx = \int_0^a [f(x) + f(-x)]\,dx$

41. 0   **43.** $\dfrac{2}{3}\pi + \dfrac{2}{81}\pi^3 - \sqrt{3}$

SECTION 5.9
1. A.V. $= \frac{1}{2}mc + b$, $x = \frac{1}{2}c$ **3.** A.V. $= 0$, $x = 0$ **5.** A.V. $= 1$, $x = \pm 1$ **7.** A.V. $= \frac{2}{3}$, $x = 1 \pm \frac{1}{3}\sqrt{3}$

9. A.V. $= 2$, $x = 4$ **11.** A.V. $= 0$, $x = 0, \pi, 2\pi$ **13.** (a) $\dfrac{13}{6}$ (b) 4.694   **15.** (a) $\dfrac{2}{\pi} \approx 0.637$ (b) 0.691

17. A = average value of f on $[a, b] = \dfrac{1}{b - a} \int_a^b f'(x)\,dx$ **19.** average of f' on $[a, b] = \dfrac{1}{b - a} \int_a^b f'(x)\,dx = \dfrac{1}{b - a}\left[f(x) \right]_a^b = \dfrac{f(b) - f(a)}{b - a}$

21. (a)1 (b) $\frac{2}{3}\sqrt{3}$ (c) $\frac{7}{9}\sqrt{3}$ **23.** (a) The terminal velocity is twice the average velocity.
(b) The average velocity during the first $\frac{1}{2}x$ seconds is one-third of the average velocity during the next $\frac{1}{2}x$ seconds.

25. (a) $v(t) = at$, $x(t) = \frac{1}{2}at^2 + x_0$ (b) $v_{\text{avg}} = \dfrac{1}{t_2 - t_1} \int_{t_1}^{t_2} at\,dt = \dfrac{at_1 + at_2}{2} = \dfrac{v(t_1) + v(t_2)}{2}$

27. (a) $M = \frac{2}{3}kL^{3/2}$, $x_M = \frac{3}{5}L$ (b) $M = \frac{1}{3}kL^3$, $x_M = \frac{1}{4}L$ **29.** $x_{M_2} = (2M - M_1)L/8M_2$ **31.** see answer to Exercise 25, Section 5.3

33. If f and g take on the same average value on every interval $[a, x]$, then

$$\dfrac{1}{x - a} \int_a^x f(t)\,dt = \dfrac{1}{x - a} \int_a^x g(t)\,dt.$$

Multiplication by $(x - a)$ gives

$$\int_a^x f(t)\,dt = \int_a^x g(t)\,dt.$$

Differentiation with respect to x gives $f(x) = g(x)$. This shows that, if the averages are the same on every interval, then the functions are everywhere the same.

SECTION 5.10

1.

3. (a) $\Delta x_1 = \Delta x_2 = \frac{1}{8}, \Delta x_3 = \Delta x_4 = \Delta x_5 = \frac{1}{4}$
(b) $\|P\| = \frac{1}{4}$
(c) $m_1 = 0, m_2 = \frac{1}{4}, m_3 = \frac{1}{2}, m_4 = 1, m_5 = \frac{3}{2}$
(d) $f(x_1^*) = \frac{1}{8}, f(x_2^*) = \frac{3}{8}, f(x_3^*) = \frac{3}{4}, f(x_4^*) = \frac{5}{4}, f(x_5^*) = \frac{3}{2}$
(e) $M_1 = \frac{1}{4}, M_2 = \frac{1}{2}, M_3 = 1, M_4 = \frac{3}{2}, M_5 = 2$
(f) $L_f(P) = \frac{25}{32}$
(g) $S^*(P) = \frac{15}{16}$
(h) $U_f(P) = \frac{39}{32}$
(i) $\displaystyle\int_a^b f(x)\, dx = 1$

5. (a) $\dfrac{1}{n^2}(1 + 2 + \cdots + n) = \dfrac{1}{n^2}\left[\dfrac{n(n+1)}{2}\right] = \dfrac{1}{2} + \dfrac{1}{2n}$

(b) $S_n^* = \dfrac{1}{2} + \dfrac{1}{2n}, \quad \displaystyle\int_0^1 x\, dx = \left[\dfrac{1}{2}x^2\right]_0^1 = \dfrac{1}{2}$

$\left| S_n^* - \displaystyle\int_0^1 x\, dx \right| = \dfrac{1}{2n} < \dfrac{1}{n} < \epsilon \quad \text{if} \quad n > \dfrac{1}{\epsilon}$

7. (a) $\dfrac{1}{n^4}(1^3 + 2^3 + \cdots + n^3) = \dfrac{1}{n^4}\left[\dfrac{n^2(n+1)^2}{4}\right] = \dfrac{1}{4} + \dfrac{1}{2n} + \dfrac{1}{4n^2}$

(b) $S_n^* = \dfrac{1}{4} + \dfrac{1}{2n} + \dfrac{1}{4n^2}, \quad \displaystyle\int_0^1 x^3\, dx = \left[\dfrac{1}{4}x^4\right]_0^1 = \dfrac{1}{4}$

$\left| S_n^* - \displaystyle\int_0^1 x^3\, dx \right| = \dfrac{1}{2n} + \dfrac{1}{4n^2} < \dfrac{1}{n} < \epsilon \quad \text{if} \quad n > \dfrac{1}{\epsilon}$

9. $S^*(P) = \frac{1}{3}[\frac{1}{6}\cos(\frac{1}{6})^2 + \frac{3}{6}\cos(\frac{3}{6})^2 + \frac{5}{6}\cos(\frac{5}{6})^2 + \frac{7}{6}\cos(\frac{7}{6})^2 + \frac{9}{6}\cos(\frac{9}{6})^2 + \frac{11}{6}\cos(\frac{11}{6})^2] \cong -0.3991$

$\displaystyle\int_0^2 x \cos x^2\, dx = \frac{1}{2}\sin 4 \cong -0.3784$

CHAPTER 6
SECTION 6.1

1.

(a) $\displaystyle\int_{-1}^{2} [(x+2) - x^2]\, dx$

(b) $\displaystyle\int_0^1 [\sqrt{y} - (-\sqrt{y})]\, dy + \int_1^4 [\sqrt{y} - (y-2)]\, dy$

3.

(a) $\displaystyle\int_0^2 [2x^2 - x^3]\, dx$

(b) $\displaystyle\int_0^8 \left[y^{1/3} - \left(\frac{1}{2}y\right)^{1/2} \right] dy$

5.

(a) $\displaystyle\int_0^4 [0 - (-\sqrt{x})]\, dx + \int_4^6 [0 - (x-6)]\, dx$

(b) $\displaystyle\int_{-2}^0 [(y+6) - y^2]\, dy$

7.

(a) $\displaystyle\int_{-2}^0 \left[\frac{8+x}{3} - (-x)\right] dx + \int_0^4 \left[\frac{8+x}{3} - x\right] dx$

(b) $\displaystyle\int_0^2 [y - (-y)]\, dy + \int_2^4 [y - (3y-8)]\, dy$

9.

(a) $\displaystyle\int_{-4}^5 [\sqrt{4+x} - (-\sqrt{4+x})]\, dx$

(b) $\displaystyle\int_{-3}^3 [5 - (y^2 - 4)]\, dy$

11.

(a) $\displaystyle\int_{-1}^{3} [2x - (x - 1)]\, dx + \int_{3}^{5} [(9 - x) - (x - 1)]\, dx$

(b) $\displaystyle\int_{-2}^{4} \left[(y + 1) - \frac{1}{2}\, y \right] dy + \int_{4}^{6} \left[(9 - y) - \frac{1}{2}\, y \right] dy$

13.

(a) $\displaystyle\int_{-1}^{1} [x^{1/3} - (x^2 + x - 1)]\, dx$

(b) $\displaystyle\int_{-5/4}^{-1} \left[\left(-\frac{1}{2} + \frac{1}{2}\sqrt{4y + 5}\right) - \left(-\frac{1}{2} - \frac{1}{2}\sqrt{4y + 5}\right) \right] dy + \int_{-1}^{1} \left[\left(-\frac{1}{2} + \frac{1}{2}\sqrt{4y + 5}\right) - y^3 \right] dy$

15. area $= \frac{9}{8}$ **17.** area $= 4$ **19.** area $= \frac{5}{12}$ **21.** area $= 2 - \sqrt{2}$

23. area $= 8$ **25.** 4   **27.** $\frac{39}{2}$ **29.** area $= 27$

31.

$C = 4^{2/3}$

33.

Area $\cong 7.93$

35. 1536 cu in. $\cong 0.89$ cu ft   **37.** (a) 35.4% (b) $\dfrac{7}{36} \cong 0.194$

SECTION 6.2

1. $\frac{1}{3}\pi$ **3.** $\frac{1944}{5}\pi$ **5.** $\frac{5}{14}\pi$ **7.** $\frac{3790}{21}\pi$ **9.** $\frac{72}{5}\pi$ **11.** $\frac{32}{3}\pi$ **13.** π **15.** $\dfrac{\pi^2}{24}(\pi^2 + 6\pi + 6)$ **17.** $\frac{16}{3}\pi$ **19.** $\frac{768}{7}\pi$

21. $\frac{2}{5}\pi$ **23.** $\frac{128}{3}\pi$ **25.** $\frac{16}{3}\pi$ **27.** (a) $\frac{16}{3}r^3$ (b) $\frac{4}{3}\sqrt{3}\,r^3$ **29.** (a) $\frac{512}{15}$ (b) $\frac{64}{15}\pi$ (c) $\frac{128}{15}\sqrt{3}$

31. (a) 32 (b) 4π (c) $8\sqrt{3}$ **33.** $\frac{4}{3}\pi ab^2$ **35.** $\frac{1}{3}\pi h(R^2 + rR + r^2)$ **37.** (a) $31\frac{1}{4}\%$ (b) $14\frac{22}{27}\%$

39. $\dfrac{1}{\pi}$ ft/min when the depth is 2 feet; $\dfrac{2}{3\pi}$ ft/min when the depth is 2 feet.

41. The cross section with coordinate x is a washer with outer radius k, inner radius $k - f(x)$, and area

$$A(x) = \pi k^2 - \pi[k - f(x)]^2 = 2\pi kf(x) - \pi[f(x)]^2$$

Thus,

$$V(x) = \int_{a}^{b} \pi(2kf(x) - [f(x)]^2)\, dx.$$

43. $\dfrac{40\pi}{3}$ **45.** $4\pi - \dfrac{1}{2}\pi^2$ **47.** 250π **49.** (a) $\dfrac{32\pi}{3}$ (b) $\dfrac{64\pi}{5}$ **51.** (a) 64π (b) $\dfrac{1024}{35}\pi$ (c) $\dfrac{704}{5}\pi$ (d) $\dfrac{512}{7}\pi$

SECTION 6.3

1. $\frac{2}{3}\pi$ **3.** $\frac{128}{5}\pi$ **5.** $\frac{2}{3}\pi$ **7.** 16π **9.** $\frac{72}{5}\pi$ **11.** 36π **13.** 8π **15.** $\frac{1944}{5}\pi$ **17.** $\frac{5}{14}\pi$ **19.** $\frac{72}{5}\pi$ **21.** 64π

23. $\frac{1}{3}\pi$ **25.** (a) $V = \displaystyle\int_0^1 2\pi x(1 - \sqrt{x})\,dx$ (b) $V = \displaystyle\int_0^1 \pi y^4\,dy$; $V = \dfrac{1}{5}\pi$

27. (a) $V = \displaystyle\int_0^1 \pi(x - x^4)\,dx$ (b) $V = \displaystyle\int_0^1 2\pi y(\sqrt{y} - y^2)\,dy$; $V = \dfrac{3}{10}\pi$

29. (a) $V = \displaystyle\int_0^1 2\pi x^3\,dx$ (b) $V = \displaystyle\int_0^1 \pi(1 - y)\,dy$; $V = \dfrac{\pi}{2}$ **31.** $\frac{4}{3}\pi ba^2$ **33.** $\frac{1}{4}\pi a^3\sqrt{3}$

35. (a) 64π (b) $\frac{1024}{35}\pi$ (c) $\frac{704}{5}\pi$ (d) $\frac{512}{7}\pi$ **37.** (a) $F'(x) = x\cos x$ (b) $V = \pi^2 - 2\pi$

39. (a) $V = \displaystyle\int_0^1 2\sqrt{3}\,\pi x^2\,dx + \int_1^2 2\pi x\sqrt{4 - x^2}\,dx$ (b) $V = \displaystyle\int_0^{\sqrt{3}} \pi\left(4 - \frac{4}{3}y^2\right)dy$ (c) $V = \dfrac{8\pi\sqrt{3}}{3}$

41. (a) $V = \displaystyle\int_0^1 2\sqrt{3}\,\pi x(2 - x)\,dx + \int_1^2 2\pi(2 - x)\sqrt{4 - x^2}\,dx$ (b) $V = \displaystyle\int_0^{\sqrt{3}} \pi\left[\left(2 - \frac{y}{\sqrt{3}}\right)^2 - (2 - \sqrt{4 - y^2})^2\right]dy$

43. (a) $V = 2\displaystyle\int_{b-a}^{b+a} 2\pi x\sqrt{a^2 - (x - b)^2}\,dx$ (b) $V = \displaystyle\int_{-a}^a \pi[(b + \sqrt{a^2 - y^2})^2 - (b - \sqrt{a^2 - y^2})^2]\,dy$

SECTION 6.4

1. $(\frac{12}{5}, \frac{3}{4})$, $V_x = 8\pi$, $V_y = \frac{128}{5}\pi$ **3.** $(\frac{3}{7}, \frac{12}{25})$, $V_x = \frac{2}{3}\pi$, $V_y = \frac{5}{14}\pi$ **5.** $(\frac{7}{3}, \frac{10}{3})$, $V_x = \frac{80}{3}\pi$, $V_y = \frac{56}{3}\pi$ **7.** $(\frac{3}{4}, \frac{22}{5})$, $V_x = \frac{704}{15}\pi$, $V_y = 8\pi$

9. $(\frac{2}{5}, \frac{2}{5})$, $V_x = \frac{4}{15}\pi$, $V_y = \frac{4}{15}\pi$ **11.** $(\frac{45}{28}, \frac{93}{70})$, $V_x = \frac{31}{5}\pi$, $V_y = \frac{15}{2}\pi$ **13.** $(3, \frac{3}{5})$, $V_x = \frac{40}{3}\pi$, $V_y = 24\pi$ **15.** $(\frac{5}{2}, 5)$ **17.** $(1, \frac{8}{5})$

19. $(\frac{10}{3}, \frac{40}{21})$ **21.** $(2, 4)$ **23.** $(-\frac{3}{5}, 0)$ **25.** (a) $(0, 0)$ (b) $\left(\dfrac{14}{5\pi}, \dfrac{14}{5\pi}\right)$ (c) $\left(0, \dfrac{14}{5\pi}\right)$ **27.** $V = \pi ab(2c + \sqrt{a^2 + b^2})$

29. (a) $(\frac{2}{3}a, \frac{1}{3}h)$ (b) $(\frac{2}{3}a + \frac{1}{3}b, \frac{1}{3}h)$ (c) $(\frac{1}{3}a + \frac{1}{3}b, \frac{1}{3}h)$ **31.** (a) $\frac{1}{3}\pi R^3\sin^2\theta\,(2\sin\theta + \cos\theta)$ (b) $\dfrac{2R\sin\theta\,(2\sin\theta + \cos\theta)}{3(\pi\sin\theta + 2\cos\theta)}$

33. (a) The mass contributed by $[x_{i-1}, x_i]$ is approximately $\lambda(x_i^*)\Delta x_i$ where x_i^* is the midpoint of $[x_{i-1}, x_i]$. The sum of these contributions,

$$\lambda(x_1^*)\Delta x_1 + \cdots + \lambda(x_n^*)\Delta x_n,$$

is a Riemann sum, which as $\|P\| \to 0$, tends to the given integral.
(b) Take M_i as the mass contributed by $[x_{i-1}, x_i]$. Then $x_{M_i}M_i \cong x_i^*\lambda(x_i^*)\Delta x_i$ where x_i^* is the midpoint of $[x_{i-1}, x_i]$. Therefore

$$x_M M = x_{M_1}M_1 + \cdots + x_{M_n}M_n \cong x_1^*\lambda(x_1^*)\Delta x_1 + \cdots + x_n^*\lambda(x_n^*)\Delta x_n.$$

As $\|P\| \to 0$, the sum on the right converges to the given integral.

35. on the axis of the cone at distance $\frac{3}{4}h$ from the vertex   **37.** (a) $(\frac{2}{3}, 0)$ (b) $(0, \frac{5}{12})$ **39.** $(\frac{3}{8}a, 0)$

SECTION 6.5

1. 817.5 ft-lb   **3.** $\dfrac{35\pi^2}{72} - \dfrac{1}{4}$ newton-meters   **5.** 625 ft-lb   **7.** (a) 25 ft-lb (b) $\frac{225}{4}$ ft-lb   **9.** 1.95 ft

11. (a) $(6480\pi + 8640)$ ft-lb (b) $(15{,}120\pi + 8640)$ ft-lb   **13.** (a) $\frac{11}{192}\pi r^2 h^2\sigma$ ft-lb (b) $(\frac{11}{192}\pi r^2 h^2\sigma + \frac{7}{24}\pi r^2 hk\sigma)$ ft-lb

15. $GmM\left(\dfrac{1}{r_2} - \dfrac{1}{r_1}\right)$ **17.** 48,000 ft-lb   **19.** (a) 20,000 ft-lb (b) 30,000 ft-lb   **21.** 788 ft-lb   **23.** (a) $\frac{1}{2}\sigma l^2$ ft-lb (b) $\frac{3}{2}l^2\sigma$ ft-lb

25. 20,800 ft-lb   **27.** $W = \displaystyle\int_a^b ma\,dx = \int_a^b mv\,dv = \frac{1}{2}mv_b^2 - \frac{1}{2}mv_a^2$ **29.** (a) 670 sec or 11 min, 10 sec (b) 1116 sec or 18 min, 36 sec

SECTION 6.6

1. 9000 lb   **3.** 1.437×10^8 newtons   **5.** 1.7502×10^6 newtons   **7.** 2160 lb   **9.** $\frac{8000}{3}\sqrt{2}$ lb   **11.** 333.33 lb   **13.** 2560 lb

15. (a) 41,250 lb (b) 41,250 lb   **17.** (a) 297,267 newtons (b) 39,200 newtons at the shallow end; 352,800 newtons at the deep end

19. 2.21749×10^6 newtons   **21.** $F = \sigma\bar{x}A$ where A is the area of the submerged surface and $\bar{x}$ is the depth of its centroid

CHAPTER 7

SECTION 7.1

1. $f^{-1}(x) = \frac{1}{5}(x-3)$ **3.** $f^{-1}(x) = \frac{1}{4}(x+7)$ **5.** not one-to-one **7.** $f^{-1}(x) = (x-1)^{1/5}$ **9.** $f^{-1}(x) = [\frac{1}{3}(x-1)]^{1/3}$

11. $f^{-1}(x) = 1 - x^{1/3}$ **13.** $f^{-1}(x) = (x-2)^{1/3} - 1$ **15.** $f^{-1}(x) = x^{5/3}$ **17.** $f^{-1}(x) = \frac{1}{3}(2-x^{1/3})$ **19.** $f^{-1}(x) = 1/x$

21. not one-to-one **23.** $f^{-1}(x) = (1/x - 1)^{1/3}$ **25.** $f^{-1}(x) = (2-x)/(x-1)$ **27.** they are equal

29. **31.** **33.**

35. $f'(x) = 3x^2 \geq 0$ on $(-\infty, \infty)$, $f'(x) = 0$ only at $x = 0$; $(f^{-1})'(9) = \frac{1}{12}$ **37.** $f'(x) = 1 + \dfrac{1}{\sqrt{x}} > 0$ on $(0, \infty)$; $(f^{-1})'(8) = \frac{2}{3}$

39. $f'(x) = 2 - \sin x > 0$ on $(-\infty, \infty)$; $(f^{-1})'(\pi) = 1$ **41.** $f'(x) = \sec^2 x > 0$ on $(-\pi/2, \pi/2)$; $(f^{-1})'(\sqrt{3}) = \frac{1}{4}$

43. $(f^{-1})'(x) = \dfrac{1}{x}$ **45.** $(f^{-1})'(x) = \dfrac{1}{\sqrt{1-x^2}}$ **47.** (a) $f'(x) = \dfrac{ad - bc}{(cx+d)^2} \neq 0$ iff $ad - bc \neq 0$ (b) $f^{-1}(x) = \dfrac{dx - b}{a - cx}$

49. (a) $f'(x) = 8\sqrt{1 + x^4} > 0$ (b) $(f^{-1})'(0) = \dfrac{\sqrt{17}}{34}$ **51.** (a) $g'(x) = \dfrac{1}{f'(g(x))}$; $g''(x) = -\dfrac{f''(g(x))g'(x)}{[f'(g(x))]^2} = -\dfrac{f''(g(x))}{[f'(g(x))]^3}$

(b) If the graph of f is concave up (down), then the graph of $g = f^{-1}$ is concave down (up).

53. $(f^{-1})'(x) = \dfrac{1}{\sqrt{1-x^2}}$. **55.** **57.**

SECTION 7.2

1. $\ln 2 + \ln 10 \cong 2.99$

3. $2 \ln 4 - \ln 10 \cong 0.48$

5. $-\ln 10 \cong -2.30$

7. $\ln 8 + \ln 9 - \ln 10 \cong 1.98$

9. $\frac{1}{2} \ln 2 \cong 0.35$ **11.** $5 \ln 2 \cong 3.45$

15. 0.406 **17.** (a) 1.65 (b) 1.57 (c) 1.71

13.

19. $x = e^2$ **21.** $x = 1, e^2$ **23.** $x = 1$ **25.** $\lim\limits_{x \to 1} \dfrac{\ln x}{x-1} = \dfrac{d(\ln x)}{dx}\Big|_{x=1} = 1$

27. (a) $P = \{1, 2, \ldots, n\}$ is a regular partition of $[1, n]$; $L_f(P) = \dfrac{1}{2} + \dfrac{1}{3} + \cdots + \dfrac{1}{n} < \displaystyle\int_1^n \dfrac{1}{t}\,dt = \ln n < 1 + \dfrac{1}{2} + \cdots + \dfrac{1}{n-1} = U_f(P)$

(b) Sum of shaded areas $= U_f(P) - \displaystyle\int_1^n \dfrac{1}{t}\,dt = 1 + \dfrac{1}{2} + \cdots + \dfrac{1}{n-1} - \ln n$.

(c) Connect the points $(1, 1), (2, \frac{1}{2}), \ldots, \left(n, \dfrac{1}{n}\right)$ by straight line segments. The sum of the areas of the triangles that are formed is

$$\frac{1}{2} \cdot 1 \left[\left(1 - \frac{1}{2}\right) + \left(\frac{1}{2} - \frac{1}{3}\right) + \cdots + \left(\frac{1}{n-1} - \frac{1}{n}\right) \right] = \frac{1}{2}\left(1 - \frac{1}{n}\right),$$

so

$$\frac{1}{2}\left(1 - \frac{1}{n}\right) < \gamma.$$

The sum of the areas of the indicated rectangles is

$$1\left[\left(1 - \frac{1}{2}\right) + \left(\frac{1}{2} - \frac{1}{3}\right) + \cdots + \left(\frac{1}{n-1} - \frac{1}{n}\right)\right] = 1 - \frac{1}{n},$$

so

$$\gamma < 1 - \frac{1}{n}.$$

Letting $n \to \infty$, we have $\frac{1}{2} < \gamma < 1$.

29. (a) $\ln 3 - \sin 3 \cong 0.96 > 0$; $\ln 2 - \sin 2 \cong -0.22 < 0$ (b) $r \cong 2.2191$ **31.** 1   **33.** 0

SECTION 7.3

1. domain $(0, \infty)$, $f'(x) = \dfrac{1}{x}$ **3.** domain $(-1, \infty)$, $f'(x) = \dfrac{3x^2}{x^3 + 1}$ **5.** domain $(-\infty, \infty)$, $f'(x) = \dfrac{x}{1 + x^2}$

7. domain all $x \neq \pm 1$, $f'(x) = \dfrac{4x^3}{x^4 - 1}$ **9.** domain $(0, \infty)$, $f'(x) = x + 2x \ln x$ **11.** domain $(0, 1) \cup (1, \infty)$, $f'(x) = -\dfrac{1}{x(\ln x)^2}$

13. domain $(-1, \infty)$, $f'(x) = \dfrac{1 - \ln(x + 1)}{(x + 1)^2}$ **15.** domain $(0, \infty)$; $f'(x) = \dfrac{1}{x}\cos(\ln x)$ **17.** domain all $x \neq (2k + 1)\dfrac{\pi}{2}$; $f'(x) = -\tan x$

19. domain: the union of all intervals $\left(\dfrac{[2k - 1]\pi}{2}, \dfrac{[2k + 1]\pi}{2}\right)$, $k = 0, \pm 1, \pm 2, \ldots$; $f'(x) = \sec x$ **21.** $\ln|x + 1| + C$

23. $-\frac{1}{2}\ln|3 - x^2| + C$ **25.** $\dfrac{1}{2(3 - x^2)} + C$ **27.** $-\ln(2 + \cos x) + C$ **29.** $\ln|\sec x| + C$ **31.** $\ln\left|\dfrac{x + 2}{x - 2}\right| + C$

33. $\dfrac{-1}{\ln x} + C$ **35.** $-\ln|\sin x + \cos x| + C$ **37.** $\frac{2}{3}\ln|1 + x\sqrt{x}| + C$ **39.** 1   **41.** 1   **43.** $\frac{1}{2}\ln\frac{8}{3}$ **45.** $\frac{1}{2}(\ln 2)^2$

47. $g'(x) = (x^2 + 1)^2(x - 1)^5 x^3\left(\dfrac{4x}{x^2 + 1} + \dfrac{5}{x - 1} + \dfrac{3}{x}\right)$ **49.** $g'(x) = \dfrac{x^4(x - 1)}{(x + 2)(x^2 + 1)}\left(\dfrac{4}{x} + \dfrac{1}{x - 1} - \dfrac{1}{x + 2} - \dfrac{2x}{x^2 + 1}\right)$

51. $g'(x) = \dfrac{1}{2}\sqrt{\dfrac{(x - 1)(x - 2)}{(x - 3)(x - 4)}}\left(\dfrac{1}{x - 1} + \dfrac{1}{x - 2} - \dfrac{1}{x - 3} - \dfrac{1}{x - 4}\right)$ **53.** $\pi \ln 9$ **55.** $\frac{15}{8} - \ln 4$

57. $\ln 5$ ft   **59.** $(-1)^{n-1}\dfrac{(n - 1)!}{x^n}$ **61.** $(-1)^{n-1}\dfrac{(n - 1)!}{x^n}$

63. (a) for $t \in (1, x)$.   (b) for $x > 1$ (c) for $0 < x < 1$ $0 < \ln\dfrac{1}{x} < 2\left(\sqrt{\dfrac{1}{x}} - 1\right)$ by (b)

$$\dfrac{1}{t} < \dfrac{1}{\sqrt{t}}$$ $$\ln x = \int_1^x \dfrac{dt}{t} < \int_1^x \dfrac{dt}{\sqrt{t}} = \left[2\sqrt{t}\right]_1^x = 2(\sqrt{x} - 1)$$ $$0 < -\ln x < 2\left(\dfrac{1}{\sqrt{x}} - 1\right)$$

$$2\left(1 - \dfrac{1}{\sqrt{x}}\right) < \ln x < 0$$

$$2x\left(1 - \dfrac{1}{\sqrt{x}}\right) < x \ln x < 0$$

(d) Use (c) and the pinching theorem for one-sided limits.

65. (i) domain $(0, \infty)$
   (ii) increases on $(0, \infty)$
   (iii) no extreme values
   (iv) concave down on $(0, \infty)$;
      no pts of inflection

(v)

vertical asymptote $x = 0$

67. (i) domain $(-\infty, 4)$
   (ii) decreases throughout
   (iii) no extreme values
   (iv) concave down throughout;
      no pts of inflection

(v)

$x = 4$, $(3, 0)$

69. (i) domain $(0, \infty)$
   (ii) decreases on $(0, e^{-1/2}]$,
      increases on $[e^{-1/2}, \infty)$
   (iii) $f(e^{-1/2}) = -\frac{1}{2}e$ local and
      absolute min.
   (iv) concave up throughout;
      no pts of inflection

(v)

$(e^{-\frac{1}{2}}, -\frac{1}{2}e)$

71. (i) domain $(0, \infty)$
   (ii) increases on $(0, 1]$,
      decreases on $[1, \infty)$
   (iii) $f(1) = -\ln 2$ local and absolute
      max
   (iv) concave down on $(0, \sqrt{2 + \sqrt{5}})$,
      concave up on $(\sqrt{2 + \sqrt{5}}, \infty)$;
      pt of inflection at $x = \sqrt{2 + \sqrt{5}}$

(v)

$(1, -\ln 2)$

vertical asymptote $x = 0$

73. x-intercept: 1; absolute min at $x = \frac{1}{4}$, absolute max at $x = 10$

75. x-intercepts: 1, 23.1407; absolute max at $x \cong 4.8105$, absolute min at $x = 100$

77. (a) $v(t) = 2 + 2t - t^2 + 3\ln(t + 1)$ (c) max velocity at $t \cong 1.5811$; min velocity at $t = 0$

SECTION 7.4

1. $\frac{dy}{dx} = -2e^{-2x}$ **3.** $\frac{dy}{dx} = 2xe^{x^2-1}$ **5.** $\frac{dy}{dx} = e^x\left(\frac{1}{x} + \ln x\right)$ **7.** $\frac{dy}{dx} = -(x^{-1} + x^{-2})e^{-x}$ **9.** $\frac{dy}{dx} = \frac{1}{2}(e^x - e^{-x})$

11. $\frac{dy}{dx} = \frac{1}{2}e^{\sqrt{x}}\left(\frac{1}{x} + \frac{\ln\sqrt{x}}{\sqrt{x}}\right)$ **13.** $\frac{dy}{dx} = 2(e^{2x} - e^{-2x})$ **15.** $\frac{dy}{dx} = 4xe^{x^2}(e^{x^2} + 1)$ **17.** $\frac{dy}{dx} = x^2 e^x$ **19.** $\frac{dy}{dx} = \frac{2e^x}{(e^x + 1)^2}$

21. $\frac{dy}{dx} = 2(a - b)\frac{e^{(a+b)x}}{(e^{ax} + e^{bx})^2}$ **23.** $\frac{dy}{dx} = 4x^3$ **25.** $f'(x) = 2\cos(e^{2x})e^{2x}$ **27.** $f'(x) = -e^{-2x}(2\cos x + \sin x)$

29. $f'(x) = -\frac{3}{2}\sqrt{e^{-3x}}\sec^2\sqrt{e^{-3x}}$ **31.** $\frac{1}{2}e^{2x} + C$ **33.** $\frac{1}{k}e^{kx} + C$ **35.** $\frac{1}{2}e^{x^2} + C$ **37.** $-e^{1/x} + C$ **39.** $\frac{1}{2}e^{2x} - \frac{1}{2}e^{-2x} + 2x + C$

41. $\frac{1}{2}x^2 + C$ **43.** $-8e^{-x/2} + C$ **45.** $2\sqrt{e^x + 1} + C$ **47.** $\frac{1}{4}\ln(2e^{2x} + 3) + C$ **49.** $e^{\sin x} + C$ **51.** $-\sin(e^{-x}) - e^{-x} + C$

53. $e - 1$ **55.** $\frac{1}{6}(1 - \pi^{-6})$ **57.** $2 - \frac{1}{e}$ **59.** $\ln\frac{3}{2}$ **61.** $\frac{185}{72} + \ln\frac{4}{9}$ **63.** $\frac{1}{2}e + \frac{1}{2}$ **65.** $e^{-0.4} = \frac{1}{e^{0.4}} \cong \frac{1}{1.49} \cong 0.67$

67. $e^{2.8} = (e^2)(e^{0.8}) \cong (7.39)(2.23) \cong 16.48$ **69.** $x''(t) = Ac^2 e^{ct} + Bc^2 e^{-ct} = c^2(Ae^{ct} + Be^{-ct}) = c^2 x(t)$ **71.** $\frac{1}{2}(3e^4 + 1)$ **73.** $e^2 - e - 2$

75. (i) domain $(-\infty, \infty)$ (v)
(ii) decreases on $(-\infty, 0]$,
increases on $[0, \infty)$
(iii) $f(0) = 1$ local and absolute
min
(iv) concave up everywhere

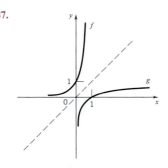

77. (i) domain $(-\infty, 0) \cup (0, \infty)$ (v)
(ii) increases on $(-\infty, 0)$,
decreases on $(0, \infty)$
(iii) no extreme values
(iv) concave up on $(-\infty, 0)$
and on $(0, \infty)$

79. (a) $\left(\pm\frac{1}{a}, e\right)$ (b) $\frac{1}{a}(e - 2)$ (c) $\frac{1 + 2a^2 e}{a^3 e}$ **81.** for $x > (n + 1)!$

$$e^x > 1 + x + \cdots + \frac{x^{n+1}}{(n + 1)!} > \frac{x^{n+1}}{(n + 1)!} = x^n \left[\frac{x}{(n + 1)!}\right] > x^n$$

83. Numerically, $8.15 \leq L \leq 8.16$. The limit is the derivative of $f(x) = e^{x^3}$ at $x = 1$; note that $f'(x) = 3x^2 e^{x^3}$ and $f'(1) = 3e \cong 8.15485$.

85. (a)

87.

89.

(b) $x = -1.9646$; $x = 1.0580$
(c) 6.4240

SECTION 7.5

1. 6 **3.** $-\frac{1}{6}$ **5.** 0 **7.** 3 **9.** $\frac{3}{5}$ **11.** $-\frac{8}{5}$ **13.** $\log_p xy = \frac{\ln xy}{\ln p} = \frac{\ln x + \ln y}{\ln p} = \frac{\ln x}{\ln p} + \frac{\ln y}{\ln p} = \log_p x + \log_p y$

15. $\log_p x^y = \frac{\ln x^y}{\ln p} = y\frac{\ln x}{\ln p} = y\log_p x$ **17.** 0 **19.** 2 **21.** e^c, where $c = \frac{(\ln 2)^2}{\ln 2 - 1}$ **23.** $t_1 < \ln a < t_2$

25. $f'(x) = 2(\ln 3)3^{2x}$ **27.** $f'(x) = \left(5\ln 2 + \frac{\ln 3}{x}\right)2^{5x}3^{\ln x}$ **29.** $g'(x) = \frac{1}{2\ln 3} \cdot \frac{1}{x\sqrt{\log_3 x}}$ **31.** $f'(x) = \frac{\sec^2(\log_5 x)}{x\ln 5}$

33. $F'(x) = \ln 2\,(2^{-x} - 2^x)\sin(2^x + 2^{-x})$ **35.** $g'(x) = \frac{2}{\ln 2\,\log_4(2x + 1)\ln 4\,(2x + 1)}$ **37.** $\frac{3^x}{\ln 3} + C$

39. $\frac{1}{4}x^4 - \frac{3^{-x}}{\ln 3} + C$ **41.** $\log_5 |x| + C$ **43.** $\frac{3}{\ln 4}(\ln x)^2 + C$ **45.** $\frac{10^{x^2}}{2\ln 10} + C$ **47.** $\frac{1}{e\ln 3}$ **49.** $\frac{1}{e}$

51. $f(x) = p^x$
$\ln f(x) = x \ln p$
$\dfrac{f'(x)}{f(x)} = \ln p$
$f'(x) = p^x \ln p$

53. $(x + 1)^x \left[\dfrac{x}{x + 1} + \ln (x + 1) \right]$

55. $(\ln x)^{\ln x} \left[\dfrac{1 + \ln (\ln x)}{x} \right]$

57. $(x^2 + 2)^{\ln x} \left[\dfrac{2x \ln x}{x^2 + 2} + \dfrac{\ln (x^2 + 2)}{x} \right]$

59. $x^{\sin x} \left(\cos x \ln x + \dfrac{\sin x}{x} \right)$

61. $(\sin x)^{\cos x} \left[\dfrac{\cos^2 x}{\sin x} - \sin x \ln (\sin x) \right]$

63. $x^{2^x} \left[\dfrac{2^x}{x} + 2^x (\ln x)(\ln 2) \right]$

65.

67.

69.

71.

73. domain $(-\infty, \infty)$; increasing on $(-\infty, 0]$, decreasing on $[0, \infty)$; $f(0) = 10$ local and absolute max

75. domain $[-1, 1]$; increasing on $[-1, 0]$, decreasing on $[0, 1]$; $f(0) = 10$ local and absolute max, $f(-1) = f(1) = 1$ endpt and absolute min

77. $\dfrac{1}{4 \ln 2}$ **79.** 2   **81.** $\dfrac{45}{\ln 10}$ **83.** $(\ln 2)(\ln 10)$ **85.** $\dfrac{1}{3} + \dfrac{1}{\ln 2}$ **87.** $\log_{10} 7 = \dfrac{\ln 7}{\ln 10} \cong \dfrac{1.95}{2.30} \cong 0.85$

89. $\log_{10} 45 = \dfrac{\ln 45}{\ln 10} = \dfrac{\ln 9 + \ln 5}{\ln 10} \cong \dfrac{2.20 + 1.61}{2.30} \cong 1.66$ **91.** approx. 16.999999; $5^{(\ln 17)/(\ln 5)} = (e^{\ln 5})^{(\ln 17)/(\ln 5)} = e^{\ln 17} = 17$

93. approx. 54.59815; $16^{1/\ln 2} = (e^{\ln 16})^{1/\ln 2} = e^{(\ln 16)/(\ln 2)} = e^{4(\ln 2)/(\ln 2)} = e^4 \cong 54.59815$

SECTION 7.6

1. (a)$411.06   (b) $612.77   (c)$859.14   **3.** about $5\frac{1}{2}\%$: $(\ln 3)/20 \cong 0.0549$ **5.** 16,000   **7.** $20 + 5e^{-0.03t}$ lb

9. $200(\frac{2}{3})^{t^2/25}$ liters   **11.** a little more than $9\frac{1}{2}$ years: $\dfrac{4 \ln \frac{1}{2}}{\ln \frac{3}{4}} \cong 9.64$ **13.** $5(\frac{4}{3})^{5/2} \cong 2.86$ gms   **15.** $100[1 - (\frac{1}{2})^{1/n}]\%$

17. (a) $x_1(t) = 10^6 t$, $x_2(t) = e^t - 1$

(b) $\dfrac{d}{dt} [x_1(t) - x_2(t)] = \dfrac{d}{dt} [10^6 t - (e^t - 1)] = 10^6 - e^t$.
    This derivative is zero at $t = 6 \ln 10 \cong 13.8$. After that the derivative is negative.

(c) $x_2(15) < e^{15} = (e^3)^5 \cong 20^5 = 2^5(10^5) = 3.2(10^6) < 15(10^6) = x_1(15)$
    $x_2(18) = e^{18} - 1 = (e^3)^6 - 1 \cong 20^6 - 1 = 64(10^6) - 1 > 18(10^6) = x_1(18)$
    $x_2(18) - x_1(18) \cong 64(10^6) - 1 - 18(10^6) \cong 46(10^6)$

(d) If by time t_1 EXP has passed LIN, then $t_1 > 6 \ln 10$. For all $t \geq t_1$ the speed of EXP is greater than the speed of LIN: for $t \geq t_1 > 6 \ln 10$, $v_2(t) = e^t > 10^6 = v_1(t)$.

19. (a) $15(\frac{2}{3})^{1/2} \cong 12.25$ lb/in.^{2} (b) $15(\frac{2}{3})^{3/2} \cong 8.16$ lb/in.^{2} **21.** 6.4%   **23.** (a) $18,589.35   (b) $20,339.99   (c) $22,933.27

25. $176/\ln 2 \cong 254$ ft   **27.** about 284 million: $203(\frac{227}{203})^3 \cong 283.85$ **29.** 11,400 years   **31.** about $16\frac{1}{2}$ years: $100 (\ln \frac{3}{2})^2 \cong 16.44$

33. 640 lb   **35.** Proceeding from the hint, we know from Theorem 7.6.1 that

$$f(t) + k_2/k_1 = Ae^{k_1(t + k_2/k_1)} = Ae^{k_1 t + k_2} = (Ae^{k_2})e^{k_1 t} = Ce^{k_1 t}.$$

where A is an arbitrary constant   set $Ae^{k_2} = C$

Therefore $f(t) = Ce^{k_1 t} - k_2/k_1$, C an arbitrary constant.

37. (a) From Exercise 35 you can determine that
$$v(t) = \frac{32}{K}(1 - e^{-Kt}).$$

(b) At each time t, $1 - e^{-Kt} < 1$. With $K > 0$,
$$v(t) = \frac{32}{K}(1 - e^{-Kt}) < \frac{32}{K}.$$

39. $\dfrac{2 \ln 3}{\ln \frac{4}{3}} - 2 \cong 5.64$ min   **41.** (a) 46.48°C (b) 9.2 minutes   **43.** $y = cx$ (c constant)

45. $y = -\ln |c - \sin x|$ (c constant)

47. $\ln |y - a| = rx + C$
$|y - a| = e^C e^{rx}$
$y = a + Be^{rx}$ $(B = \pm e^C)$

SECTION 7.7

1. $\frac{1}{3}\ln|\sec 3x| + C$ **3.** $(1/\pi)\ln|\csc \pi x - \cot \pi x| + C$ **5.** $\ln|\sin e^x| + C$ **7.** $\frac{1}{4}\ln|3 - 2\cos 2x| + C$ **9.** $\frac{1}{3}e^{\tan 3x} + C$

11. $\frac{1}{2}\ln|\sec x^2 + \tan x^2| + C$ **13.** $\frac{1}{2}(\ln|\sin x|)^2 + C$ **15.** $x + 2\ln|\sec x + \tan x| + \tan x + C$ **17.** $(1 + \cot x)^{-1} + C$

19. $2\sqrt{\ln|\sec x + \tan x|} + C$ **21.** $\ln \frac{4}{3}$ **23.** $\frac{1}{2}\ln 2$ **25.** $\ln[(1 + \sqrt{2})/(\sec 1 + \tan 1)] \cong -0.345$ **27.** $\frac{1}{3}\pi - \frac{1}{2}\ln 3$ **29.** $\frac{1}{4}\pi - \frac{1}{2}\ln 2$

31. $\frac{3}{4}\ln 2$ **33.** $2\pi\ln(2 + \sqrt{3})$ **35.** $\frac{\ln 3}{\pi}$ **37.**
$$\int \csc x\, dx = \int \frac{\csc x(\csc x - \cot x)}{\csc x - \cot x}\, dx$$

Let $u = \csc x - \cot x$, $du = \csc x(\csc x - \cot x)\, dx$. Then
$$\int \csc x\, dx = \int \frac{1}{u}\, du = \ln|u| + C = \ln|\csc x - \cot x| + C.$$

SECTION 7.8

1. 0   **3.** $\frac{\pi}{3}$ **5.** $\frac{2\pi}{3}$ **7.** $-\frac{\pi}{4}$ **9.** $-\frac{2}{\sqrt{3}}$ **11.** $\frac{1}{2}$ **13.** 1.1630   **15.** -0.4580 **17.** 1.2002   **19.** $1/\sqrt{1 + x^2}$

21. x **23.** $\sqrt{1 + x^2}$ **25.** $\sqrt{1 - x^2}$ **27.** $\frac{1}{x}\sqrt{1 + x^2}$ **29.** $\frac{x}{\sqrt{1 - x^2}}$ **31.** $\frac{1}{x^2 + 2x + 2}$ **33.** $\frac{2}{x\sqrt{4x^4 - 1}}$

35. $\frac{2x}{\sqrt{1 - 4x^2}} + \sin^{-1} 2x$ **37.** $\frac{2\sin^{-1}x}{\sqrt{1 - x^2}}$ **39.** $\frac{x - (1 + x^2)\tan^{-1}x}{x^2(1 + x^2)}$ **41.** $\frac{1}{(1 + 4x^2)\sqrt{\tan^{-1} 2x}}$ **43.** $\frac{1}{x[1 + (\ln x)^2]}$

45. $-\frac{r}{|r|\sqrt{1 - r^2}}$ **47.** $2x\sec^{-1}\left(\frac{1}{x}\right) - \frac{x^2}{\sqrt{1 - x^2}}$ **49.** $\cos[\sec^{-1}(\ln x)] \cdot \frac{1}{x|\ln x|\sqrt{(\ln x)^2 - 1}}$ **51.** $\sqrt{\frac{c - x}{c + x}}$ **53.** $\frac{x^2}{(c^2 - x^2)^{3/2}}$

55. Set $au = x + b$, $a\, du = dx$.
$$\int \frac{dx}{\sqrt{a^2 - (x + b)^2}} = \int \frac{a\, du}{\sqrt{a^2 - a^2u^2}} = \int \frac{du}{\sqrt{1 - u^2}} = \sin^{-1}u + C = \sin^{-1}\left(\frac{x + b}{a}\right) + C$$

57. Set $au = x + b$, $a\, du = dx$.
$$\int \frac{dx}{(x + b)\sqrt{(x + b)^2 - a^2}} = \int \frac{a\, du}{au\sqrt{a^2u^2 - a^2}} = \frac{1}{a}\int \frac{du}{u\sqrt{u^2 - 1}} = \frac{1}{a}\sec^{-1}\frac{|x + b|}{a} + C.$$

59. domain $(-\infty, \infty)$, range $(0, \pi)$ **61.** $\frac{1}{4}\pi$ **63.** $\frac{1}{4}\pi$ **65.** $\frac{1}{20}\pi$ **67.** $\frac{1}{24}\pi$ **69.** $\frac{1}{3}\sec^{-1} 4 \cong 0.439$ **71.** $\frac{1}{6}\pi$

73. $\tan^{-1} 2 - \frac{1}{4}\pi \cong 0.322$ **75.** $\frac{1}{2}\sin^{-1}x^2 + C$ **77.** $\frac{1}{2}\tan^{-1}x^2 + C$ **79.** $\tan^{-1}(\frac{1}{3}\tan x) + C$ **81.** $\frac{1}{2}(\sin^{-1}x)^2 + C$

83. $\sin^{-1}(\ln x) + C$ **85.** $\frac{\pi}{3}$ **87.** $2\pi - \frac{4}{3}$ **89.** $\sqrt{s^2 + sk}$ feet from the point where the line of the sign intersects the road

91. (a) There exists constants C_1, C_2 such that
$$f(x) = g(x) + C_1 \quad \text{for } x < 1/2; \qquad f(x) = g(x) + C_2 \quad \text{for } x > 1/2.$$
(b) $\displaystyle\lim_{x\to 1/2^+} f(x) = -\frac{\pi}{2}; \quad \lim_{x\to 1/2^-} f(x) = \frac{\pi}{2}$ (e) $C_1 = \tan^{-1} 2; \quad C_2 = \tan^{-1} 3 - \frac{\pi}{4}$

93. (a) There exist constants C_1, C_2 such that
$$f(x) + g(x) = C_1 \quad \text{for } x < 0; \qquad f(x) + g(x) = C_2 \quad \text{for } x > 0.$$
(b) $\displaystyle\lim_{x\to 0^+} f(x) = \frac{\pi}{2}; \quad \lim_{x\to 0^-} f(x) = -\frac{\pi}{2}$ (e) $C_1 = \frac{\pi}{2}; \quad C_2 = -\frac{\pi}{2}$

95. estimate $\cong 0.523$, $\sin 0.523 \cong 0.499$ explanation: the integral $= \sin^{-1} 0.5$; therefore $\sin$ (integral) $= 0.5$

SECTION 7.9

1. $2x\cosh x^2$ **3.** $\frac{a\sinh ax}{2\sqrt{\cosh ax}}$ **5.** $\frac{1}{1 - \cosh x}$ **7.** $ab(\cosh bx - \sinh ax)$ **9.** $\frac{a\cosh ax}{\sinh ax}$ **11.** $2e^{2x}\cosh(e^{2x})$

13. $-e^{-x}\cosh 2x + 2e^{-x}\sinh 2x$ **15.** $\tanh x$ **17.** $(\sinh x)^x[\ln(\sinh x) + x\coth x]$

19. $\cosh^2 t - \sinh^2 t = \left(\frac{e^t + e^{-t}}{2}\right)^2 - \left(\frac{e^t - e^{-t}}{2}\right)^2 = \frac{e^{2t} + 2 + e^{-2t}}{4} - \frac{e^{2t} - 2 + e^{-2t}}{4} = 1$

21. $\cosh t\cosh s + \sinh t\sinh s = \left(\frac{e^t + e^{-t}}{2}\right)\left(\frac{e^s + e^{-s}}{2}\right) + \left(\frac{e^t - e^{-t}}{2}\right)\left(\frac{e^s - e^{-s}}{2}\right)$
$$= \tfrac{1}{4}(e^{t+s} + e^{s-t} + e^{t-s} + e^{-t-s} + e^{t+s} - e^{s-t} - e^{t-s} + e^{-t-s})$$
$$= \tfrac{1}{2}(e^{t+s} + e^{-(t+s)}) = \cosh(t + s)$$

23. $\cosh^2 t + \sinh^2 t = \left(\dfrac{e^t + e^{-t}}{2}\right)^2 + \left(\dfrac{e^t - e^{-t}}{2}\right)^2 = \tfrac{1}{4}(e^{2t} + 2 + e^{-2t} + e^{2t} - 2 - e^{-2t}) = \dfrac{e^{2t} + e^{-2t}}{2} = \cosh 2t$

25. $\sinh(-t) = \dfrac{e^{-t} - e^{-(-t)}}{2} = \dfrac{e^{-t} - e^t}{2} = -\sinh t$ **27.** absolute max -3

29. $[\cosh x + \sinh x]^n = \left[\dfrac{e^x + e^{-x}}{2} + \dfrac{e^x - e^{-x}}{2}\right]^n = [e^x]^n = e^{nx} = \dfrac{e^{nx} + e^{-nx}}{2} + \dfrac{e^{nx} - e^{-nx}}{2} = \cosh nx + \sinh nx$

31. $A = 2,\ B = \tfrac{1}{3},\ c = 3$ **33.** $\dfrac{1}{a}\sinh ax + C$ **35.** $\dfrac{1}{3a}\sinh^3 ax + C$ **37.** $\dfrac{1}{a}\ln(\cosh ax) + C$ **39.** $-\dfrac{1}{a\cosh ax} + C$

41. $\tfrac{1}{2}(\sinh x \cosh x + x) + C$ **43.** $2\cosh\sqrt{x} + C$ **45.** $\dfrac{e^2 - 1}{2e} \cong 1.175$ **47.** π

SECTION 7.10

1. $2\tanh x \operatorname{sech}^2 x$ **3.** $\operatorname{sech} x \operatorname{csch} x$ **5.** $\dfrac{2e^{2x}\cosh(\tan^{-1} e^{2x})}{1 + e^{4x}}$ **7.** $\dfrac{-x\operatorname{csch}^2(\sqrt{x^2+1})}{\sqrt{x^2+1}}$ **9.** $\dfrac{-\operatorname{sech} x(\tanh x + 2\sinh x)}{(1 + \cosh x)^2}$

11. $\dfrac{d}{dx}(\coth x) = \dfrac{d}{dx}\left(\dfrac{\cosh x}{\sinh x}\right) = \dfrac{\sinh^2 x - \cosh^2 x}{\sinh^2 x} = \dfrac{-1}{\sinh^2 x} = -\operatorname{csch}^2 x$

13. $\dfrac{d}{dx}(\operatorname{csch} x) = \dfrac{d}{dx}\left(\dfrac{1}{\sinh x}\right) = -\dfrac{\cosh x}{\sinh^2 x} = -\operatorname{csch} x \coth x$ **15.** (a) $\tfrac{3}{5}$ (b) $\tfrac{5}{3}$ (c) $\tfrac{4}{3}$ (d) $\tfrac{4}{4}$ (e) $\tfrac{3}{4}$

17. If $x \le 0$, the result is obvious. Suppose then that $x > 0$. Since $x^2 \ge 1$, we have $x \ge 1$. Consequently,
$$x - 1 = \sqrt{x-1}\,\sqrt{x-1} \le \sqrt{x-1}\,\sqrt{x+1} = \sqrt{x^2-1} \quad \text{and therefore} \quad x - \sqrt{x^2-1} \le 1.$$

19. $\dfrac{d}{dx}(\sinh^{-1} x) = \dfrac{d}{dx}[\ln(x + \sqrt{x^2+1})] = \dfrac{1 + \dfrac{x}{\sqrt{x^2+1}}}{x + \sqrt{x^2+1}} = \dfrac{1}{\sqrt{x^2+1}}$

21. $\dfrac{d}{dx}(\tanh^{-1} x) = \dfrac{1}{2}\dfrac{d}{dx}\left[\ln\left(\dfrac{1+x}{1-x}\right)\right] = \dfrac{1}{2}\cdot\dfrac{1}{\left(\dfrac{1+x}{1-x}\right)}\cdot\dfrac{2}{(1-x)^2} = \dfrac{1}{1-x^2}$

23. Let $y = \operatorname{csch}^{-1} x$. Then $\operatorname{csch} y = x$ and $\sinh y = \dfrac{1}{x}$. Thus $\cosh y \cdot y' = -\dfrac{1}{x^2}$ and
$$y' = -\dfrac{1}{x^2\cosh y} = -\dfrac{1}{x^2\sqrt{1 + \left(\dfrac{1}{x}\right)^2}} = -\dfrac{1}{|x|\sqrt{1 + x^2}}.$$

25. (a) absolute max $(0, 1)$
(b) points of inflection at $x = \ln(1 + \sqrt{2}) \cong 0.881$,
    $x = -\ln(1 + \sqrt{2}) \cong -0.881$
(c) concave up on $(-\infty, -\ln(1 + \sqrt{2})) \cup (\ln(1 + \sqrt{2}), \infty)$;
    concave down on $(-\ln(1 + \sqrt{2}), \ln(1 + \sqrt{2}))$

(d)

27.

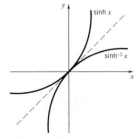

(0, 0) is a point of inflection for both graphs

29. (a) $\tan\phi = \sinh x$
$\phi = \tan^{-1}(\sinh x)$
$\dfrac{d\phi}{dx} = \dfrac{\cosh x}{1 + \sinh^2 x} = \dfrac{\cosh x}{\cosh^2 x} = \dfrac{1}{\cosh x} = \operatorname{sech} x$
(b) $\sinh x = \tan\phi$
    $x = \sinh^{-1}(\tan\phi)$
    $= \ln(\tan\phi + \sqrt{\tan^2\phi + 1})$
    $= \ln(\tan\phi + \sec\phi) = \ln(\sec\phi + \tan\phi)$
(c) $x = \ln(\sec\phi + \tan\phi)$
    $\dfrac{dx}{d\phi} = \dfrac{\sec\phi\tan\phi + \sec^2\phi}{\tan\phi + \sec\phi} = \sec\phi$

31. $\ln|\cosh x| + C$ **33.** $2\tan^{-1}(e^x) + C$ **35.** $-\tfrac{1}{3}\operatorname{sech}^3 x + C$ **37.** $\tfrac{1}{2}[\ln(\cosh x)]^2 + C$ **39.** $\ln|1 + \tanh x| + C$

41. Let $x = a\sinh u$, $dx = a\cosh u\, du$. Then
$$\int \dfrac{dx}{\sqrt{a^2 + x^2}} = \int \dfrac{a\cosh u}{\sqrt{a^2 + a^2\sinh^2 u}}\, du = \int du = \sinh^{-1}\left(\dfrac{x}{a}\right) + C.$$

43. Suppose $|x| < a$. Let $x = a \tanh u$, $dx = a \operatorname{sech}^2 u \, du$. Then

$$\int \frac{dx}{a^2 - x^2} = \int \frac{a \operatorname{sech}^2 u}{a^2 - a^2 \tanh^2 u} \, du = \frac{1}{a} \int du = \frac{1}{a} \tanh^{-1}\left(\frac{x}{a}\right) + C.$$

The other case is done in the same way.

CHAPTER 8

SECTION 8.1

1. $-e^{2-x} + C$ **3.** $2/\pi$ **5.** $-\tan(1-x) + C$ **7.** $\frac{1}{2}\ln 3$ **9.** $-\sqrt{1-x^2} + C$ **11.** 0 **13.** $e - \sqrt{e}$ **15.** $\frac{1}{2}\ln(x^2+1) + C$

17. $\pi/4c$ **19.** $\frac{2}{3}\sqrt{3\tan\theta + 1} + C$ **21.** $(1/a)\ln|ae^x - b| + C$ **23.** $-\frac{1}{2}(\cot 2x + \csc 2x) + C$

25. $\frac{1}{2}\ln[(x+1)^2 + 4] - \frac{1}{2}\tan^{-1}(\frac{1}{2}[x+1]) + C$ **27.** $\frac{1}{2}\sin^{-1}x^2 + C$ **29.** $\tan^{-1}(x+3) + C$ **31.** $-\frac{1}{2}\cos x^2 + C$ **33.** $\tan x - x + C$

35. $\frac{1}{2}\ln(\frac{8}{5})$ **37.** $3/2$ **39.** $\frac{1}{2}(\sin^{-1}x)^2 + C$ **41.** $\frac{1}{2}\sin^{-1}(e^{2x}) + C$ **43.** $\ln|\ln x| + C$ **45.** $\frac{2}{3}(\ln 4)^{3/2}$ **47.** $\sqrt{2}$ **49.** $2\sqrt{2}$

51. (a) $\displaystyle\int_0^\pi \sin^2 nx \, dx = \int_0^\pi \left(\frac{1}{2} - \frac{\cos 2nx}{2}\right) dx = \left[\frac{1}{2}x - \frac{\sin 2nx}{4n}\right]_0^\pi = \frac{\pi}{2}$

(b) $\displaystyle\int_0^\pi \sin nx \cos nx \, dx = \frac{1}{n}\int_0^0 u \, du = 0$ $\left(u = \sin nx, \, du = \frac{1}{n}\cos nx \, dx\right)$ (c) $\displaystyle\int_0^{\pi/n} \sin nx \cos nx \, dx = \frac{1}{n}\int_0^0 u \, du = 0.$

53. (a) $\frac{1}{2}\tan^2 x - \ln|\sec x| + C$ (b) $\frac{1}{4}\tan^4 x - \frac{1}{2}\tan^2 x + \ln|\sec x| + C$ (c) $\frac{1}{6}\tan^6 x - \frac{1}{4}\tan^4 x + \frac{1}{2}\tan^2 x - \ln|\sec x| + C$

(d) $\displaystyle\int \tan^{2k+1}x \, dx = \frac{1}{2k}\tan^{2k}x - \int \tan^{2k-1}x \, dx$

55. (b) $A = \sqrt{2}$, $B = \frac{\pi}{4}$ (c) $\frac{\sqrt{2}}{2}\ln\left(\frac{\sqrt{2}+1}{\sqrt{2}-1}\right)$ **57.** (b) -0.80, 5.80 (c) 27.60

SECTION 8.2

1. $-xe^{-x} - e^{-x} + C$ **3.** $\frac{2}{9}e^3 - \frac{1}{9}$ **5.** $-\frac{1}{3}e^{-x^3} + C$ **7.** $2 - 5e^{-1}$ **9.** $-2x^2(1-x)^{1/2} - \frac{8}{3}x(1-x)^{3/2} - \frac{16}{15}(1-x)^{5/2} + C$

11. $\frac{3}{8}e^4 - \frac{1}{8}$ **13.** $2\sqrt{x+1}\ln(x+1) - 4\sqrt{x+1} + C$ **15.** $x(\ln x)^2 - 2x\ln x + 2x + C$

17. $3x\left(\dfrac{x^3}{\ln 3} - \dfrac{3x^2}{(\ln 3)^2} + \dfrac{6x}{(\ln 3)^3} - \dfrac{6}{(\ln 3)^4}\right) + C$ **19.** $\frac{1}{15}x(x+5)^{15} - \frac{1}{240}(x+5)^{16} + C$ **21.** $\dfrac{1}{2\pi} - \dfrac{1}{\pi^2}$

23. $\frac{1}{10}x^2(x+1)^{10} - \frac{1}{55}x(x+1)^{11} + \frac{1}{660}(x+1)^{12} + C$ **25.** $\frac{1}{2}e^x(\sin x - \cos x) + C$ **27.** $\ln 2 + \frac{\pi}{2} - 2$ **29.** $\dfrac{x^{n+1}}{n+1}\ln x - \dfrac{x^{n+1}}{(n+1)^2} + C$

31. $-\frac{1}{2}x^2\cos x^2 + \frac{1}{2}\sin x^2 + C$ **33.** $e^x(x^4 - 4x^3 + 12x^2 - 24x + 24) + C$ **35.** $\dfrac{\pi}{24} + \dfrac{\sqrt{3}-2}{4}$ **37.** $x\sec^{-1}x - \ln|x + \sqrt{x^2-1}| + C$

39. $\dfrac{\pi}{8} - \dfrac{1}{4}\ln 2$ **41.** $\frac{1}{2}x^2\sinh 2x - \frac{1}{2}x\cosh 2x + \frac{1}{4}\sinh 2x + C$ **43.** $\dfrac{x}{2}[\sin(\ln x) - \cos(\ln x)] + C$ **45.** $\ln x\sin^{-1}(\ln x) + \sqrt{1-(\ln x)^2} + C$

47. $\dfrac{\pi}{12} + \dfrac{\sqrt{3}-2}{2}$ **49.** (a) 1 (b) $\bar{x} = \dfrac{e^2}{4} + \dfrac{1}{4}$, $\bar{y} = e - 2$ (c) x-axis: $\pi(e-2)$, y-axis: $\dfrac{\pi}{2}(e^2+1)$ **51.** $\bar{x} = 1/(e-1)$, $\bar{y} = (e+1)/4$

53. $\bar{x} = \frac{1}{2}\pi$, $\bar{y} = \frac{1}{8}\pi$ **55.** (a) $M = (e^k - 1)/k$ (b) $x_M = [(k-1)e^k + 1]/[k(e^k - 1)]$ **57.** $V = (2\pi/\alpha^2)(\alpha e^\alpha - e^\alpha + 1)$

59. $V = 4 - 8/\pi$ **61.** $V = 2\pi(e-2)$ **63.** $\bar{x} = (e^2+1)/[2(e^2-1)]$ **65.** area $= \sinh 1 = \dfrac{e^2-1}{2e}$; $\bar{x} = \dfrac{2}{e+1}$, $\bar{y} = \dfrac{e^4 + 4e^2 - 1}{8e(e^2-1)}$

67. Let $u = x^n$, $dv = e^{ax}dx$. Then $du = nx^{n-1}\,dx$, $v = \dfrac{1}{a}e^{ax}$.   **69.** $(\frac{1}{2}x^3 - \frac{3}{4}x^2 + \frac{3}{4}x - \frac{3}{8})e^{2x} + C$ **71.** $x[(\ln x)^3 - 3(\ln x)^2 + 6\ln x - 6] + C$

73. Let $u = f(x)$, $dv = g''(x)\,dx$. Then $du = f'(x)\,dx$, $v = g'(x)$, and

$$\int_a^b f(x)g''(x)\,dx = [f(x)g'(x)]_a^b - \int_a^b f'(x)g'(x)\,dx.$$

Now let $u = f'(x)$, $dv = g'(x)\,dx$ and integrate by parts again.   The result follows.

75. The present value of R dollars at the interest rate r on the interval $[t, t + \Delta t]$ is approximately $Re^{-rt}\,\Delta t$.   **77.** (a) about \$2016   (b) about \$1918

SECTION 8.3

1. $\frac{1}{3}\cos^3 x - \cos x + C$ **3.** $\dfrac{\pi}{12}$ **5.** $-\frac{1}{5}\cos^5 x + \frac{1}{7}\cos^7 x + C$ **7.** $\frac{1}{4}\sin^4 x - \frac{1}{6}\sin^6 x + C$ **9.** $\frac{1}{3}\sin^3 x - \frac{1}{5}\sin^5 x + C$

11. $\frac{3}{8}\pi$ **13.** $\frac{1}{2}\cos x - \frac{1}{10}\cos 5x + C$ **15.** $\frac{1}{2}\sin^4 x + C$ **17.** $\frac{3}{128}x - \frac{1}{128}\sin 4x + \frac{1}{1024}\sin 8x + C$

19. $\frac{5}{16}x - \frac{1}{4}\sin 2x + \frac{3}{64}\sin 4x + \frac{1}{48}\sin^3 2x + C$ **21.** $\sin x - \sin^3 x + \frac{3}{5}\sin^5 x - \frac{1}{7}\sin^7 x + C$

23. $\frac{3}{5}$ **25.** $\frac{1}{6}\sin 3x - \frac{1}{14}\sin 7x + C$ **27.** $\dfrac{2}{105\pi}$ **29.** $-1/6$ **31.** $\frac{1}{3}\cos(\frac{3}{2}x) - \frac{1}{5}\cos(\frac{5}{2}x) + C$

33. $\frac{3}{8} \tan^{-1} x + \frac{x}{2(x^2 + 1)} + \frac{x(1 - x^2)}{8(x^2 + 1)^2} + C$ **35.** $\frac{1}{2}\left[\tan^{-1} (x + 1) + \frac{x + 1}{x^2 + 2x + 2} \right] + C$ **37.** $\pi/2$ **39.** $\frac{3\pi^2}{8}$

41. $\sin mx \sin nx = \frac{1}{2}[\cos (m - n)x - \cos (m + n)x], \quad m \neq n$

$\sin mx \sin nx = \sin^2 mx = \frac{1 - \cos 2mx}{2}, \quad m = n$

45. Let $u = \cos^{n-1}x$, $dv = \cos x \, dx$. Then $du = (n - 1)\cos^{n-2}x(-\sin x) \, dx$, $\quad v = \sin x$.   **47.** $\frac{16}{35}$

$$\int \cos^n x \, dx = \int \cos^{n-1}x \cos x \, dx = \cos^{n-1}x \sin x + (n - 1) \int \cos^{n-2}x \sin^2 x \, dx$$

$$= \cos^{n-1}x \sin x + (n - 1) \int (\cos^{n-2}x - \cos^n x) \, dx$$

Now solve for $\int \cos^n x \, dx$.

SECTION 8.4

1. $\frac{1}{3} \tan 3x - x + C$ **3.** $(1/\pi) \tan \pi x + C$ **5.** $\frac{1}{2} \tan^2 x + \ln |\cos x| + C$ **7.** $\frac{1}{3} \tan^3 x + C$

9. $-\frac{1}{2} \csc x \cot x + \frac{1}{2} \ln |\csc x - \cot x| + C$ **11.** $-\frac{1}{3} \cot^3 x + \cot x + x + C$ **13.** $-\frac{1}{5} \csc^5 x + \frac{1}{3} \csc^3 x + C$

15. $-\frac{1}{6} \cot^3 2x - \frac{1}{2} \cot 2x + C$ **17.** $-\frac{1}{2} \cot x \csc x - \frac{1}{2} \ln |\csc x - \cot x| + C$ **19.** $\frac{1}{12} \tan^4 3x - \frac{1}{6} \tan^2 3x + \frac{1}{3} \ln |\sec 3x| + C$

21. $\frac{1}{4} \sec^3 x \tan x + \frac{3}{8} \sec x \tan x + \frac{3}{8} \ln |\sec x + \tan x| + C$ **23.** $\frac{1}{7} \tan^7 x + \frac{1}{5} \tan^5 x + C$ **25.** $\frac{1}{6} \tan^3 (e^{2x}) + C$ **27.** $\frac{1}{4}$ **29.** $\frac{\sqrt{3}}{2} - \frac{\pi}{6}$

31. $\frac{8\sqrt{3} + 504}{135}$ **33.** $\frac{\pi^2}{2} - \pi$ **35.** $\pi\left[1 - \frac{\pi}{4} + \ln 2 \right]$ **37.** $\int \cot^n x \, dx = \int \cot^{n-2} x \, (\csc^2 x - 1) \, dx = -\frac{\cot^{n-1} x}{n - 1} - \int \cot^{n-2} x \, dx$

39. $\int \csc^n x \, dx = \int \csc^{n-2} x \csc^2 x \, dx$. Now let $u = \csc^{n-2} x$, $dv = \csc^2 x \, dx$ and use integration by parts.

SECTION 8.5

1. $\sin^{-1}\left(\frac{x}{a} \right) + C$ **3.** $\frac{1}{10}$ **5.** $\frac{1}{2}x\sqrt{x^2 - 1} - \frac{1}{2} \ln |x + \sqrt{x^2 - 1}| + C$ **7.** $2 \sin^{-1}\left(\frac{x}{2} \right) - \frac{1}{2}x\sqrt{4 - x^2} + C$

9. $\frac{1}{\sqrt{1 - x^2}} + C$ **11.** $\frac{2\sqrt{3} - \pi}{6}$ **13.** $-\frac{1}{3}(4 - x^2)^{3/2} + C$ **15.** $\frac{625\pi}{16}$ **17.** $\ln (\sqrt{8 + x^2} + x) - \frac{x}{\sqrt{8 + x^2}} + C$

19. $\frac{1}{a} \ln \left| \frac{a - \sqrt{a^2 - x^2}}{x} \right| + C$ **21.** $\ln |x + \sqrt{x^2 - a^2}| + C$ **23.** $\frac{1}{2}e^x\sqrt{e^{2x} - 1} - \frac{1}{2} \ln (e^x + \sqrt{e^{2x} - 1}) + C$ **25.** $18 - 9\sqrt{2}$

27. $-\frac{1}{a^2 x}\sqrt{a^2 + x^2} + C$ **29.** $\frac{1}{a^2 x}\sqrt{x^2 - a^2} + C$ **31.** $2 - \frac{\pi}{2}$ **33.** $\frac{1}{9}e^{-x}\sqrt{e^{2x} - 9} + C$ **35.** $-\frac{1}{2(x - 2)^2} + C$

37. $-\frac{1}{3}(6x - x^2 - 8)^{3/2} + \frac{3}{2} \sin^{-1} (x - 3) + \frac{3}{2}(x - 3) \sqrt{6x - x^2 - 8} + C$ **39.** $\frac{x^2 + x}{8(x^2 + 2x + 5)} - \frac{1}{16} \tan^{-1}\left(\frac{x + 1}{2} \right) + C$

41. $\sqrt{x^2 + 4x + 13} + \ln (x + 2 + \sqrt{x^2 + 4x + 13}) + C$ **43.** $\frac{1}{2}(x - 3)\sqrt{6x - x^2 - 8} + \frac{1}{2} \sin^{-1} (x - 3) + C$

45. Let $x = a \tan u$, $dx = a \sec^2 u \, du$; $\sqrt{x^2 + a^2} = a \sec u$. Then $(x^2 + a^2)^n = a^{2n} \sec^{2n} u$ and the result follows by substitution.

47. $\frac{3}{8} \tan^{-1}x + \frac{x}{2(x^2 + 1)} + \frac{x(1 - x^2)}{8(x^2 + 1)^2} + C$

49. (a) Let $x = a \tan u$. Then $dx = a \sec^2 u \, du$ and $\sqrt{x^2 + a^2} = a \sec u$. Thus $\int \frac{1}{\sqrt{x^2 + a^2}} \, dx = \int \sec u \, du$ and the result follows.

(b) Let $x = a \sinh u$. Then $dx = a \cosh u \, du$ and $\sqrt{x^2 + a^2} = a \cosh u$. Thus $\int \frac{1}{\sqrt{x^2 + a^2}} \, dx = \int du$ and the result follows.

(c) See Theorem 7.10.2

51. $\frac{\pi^2}{8} + \frac{\pi}{4}$ **53.** $A = \frac{1}{2} r^2\sin \theta \cos \theta + \int_{r\cos \theta}^{r} \sqrt{r^2 - x^2} \, dx = \frac{1}{2}r^2\theta$ **55.** $M = \ln (1 + \sqrt{2})$, $x_M = \frac{(\sqrt{2} - 1)a}{\ln (1 + \sqrt{2})}$

57. $A = \frac{1}{2}a^2[\sqrt{2} - \ln (\sqrt{2} + 1)]$; $\bar{x} = \frac{2a}{3[\sqrt{2} - \ln (\sqrt{2} + 1)]}$, $\bar{y} = \frac{(2 - \sqrt{2})a}{3[\sqrt{2} - \ln (\sqrt{2} + 1)]}$ **59.** $V_y = \frac{2}{3}\pi a^3$, $\bar{y} = \frac{3}{8}a$

61. (b) $\ln (2 + \sqrt{3}) - \frac{\sqrt{3}}{2}$ (c) $\bar{x} = \frac{2(3\sqrt{3} - \pi)}{2 \ln (2 + \sqrt{3}) - \sqrt{3}}$, $\bar{y} = \frac{5}{72[\ln (2 + \sqrt{3}) - \sqrt{3}]}$

SECTION 8.6

1. $\dfrac{1/5}{x+1} - \dfrac{1/5}{x+6}$ **3.** $\dfrac{1/4}{x-1} + \dfrac{1/4}{x+1} - \dfrac{x/2}{x^2+1}$ **5.** $\dfrac{1/2}{x} + \dfrac{3/2}{x+2} - \dfrac{1}{x-1}$ **7.** $\dfrac{3/2}{x-1} - \dfrac{9}{x-2} + \dfrac{19/2}{x-3}$ **9.** $\ln\left|\dfrac{x-2}{x+5}\right| + C$

11. $5\ln|x-1| - 3\ln|x| + \dfrac{3}{x} + C$ **13.** $\dfrac{1}{4}x^4 + \dfrac{4}{3}x^3 + 6x^2 + 32x - \dfrac{32}{x-2} + 80\ln|x-2| + C$ **15.** $5\ln|x-2| - 4\ln|x-1| + C$

17. $\dfrac{-1}{2(x-1)^2} + C$ **19.** $\dfrac{3}{4}\ln|x-1| - \dfrac{1}{2(x-1)} + \dfrac{1}{4}\ln|x+1| + C$ **21.** $\dfrac{1}{32}\ln\left|\dfrac{x-2}{x+2}\right| - \dfrac{1}{16}\tan^{-1}\dfrac{x}{2} + C$

23. $\dfrac{1}{2}\ln(x^2+1) + \dfrac{3}{2}\tan^{-1}x + \dfrac{5(1-x)}{2(x^2+1)} + C$ **25.** $\dfrac{1}{16}\ln\left[\dfrac{x^2+2x+2}{x^2-2x+2}\right] + \dfrac{1}{8}\tan^{-1}(x+1) + \dfrac{1}{8}\tan^{-1}(x-1) + C$

27. $\dfrac{3}{x} + 4\ln\left|\dfrac{x}{x+1}\right| + C$ **29.** $-\dfrac{1}{6}\ln|x| + \dfrac{3}{10}\ln|x-2| - \dfrac{2}{15}\ln|x+3| + C$ **31.** $\ln\left(\dfrac{125}{108}\right)$ **33.** $\dfrac{1}{4}\ln\left(\dfrac{5}{2}\right) + \dfrac{3}{4}$ **35.** $\ln\left(\dfrac{27}{4}\right) - 2$

37. Note that

$$y = \frac{1}{x^2-1} = \frac{1}{2}\left[\frac{1}{x-1} - \frac{1}{x+1}\right] \qquad \text{and thus} \qquad \frac{d^0y}{dx^0} = \left(\frac{1}{2}\right)(-1)^0\,0!\left[\frac{1}{(x-1)^{0+1}} - \frac{1}{(x+1)^{0+1}}\right].$$

The rest is a routine induction.

39. $\bar{x} = (2\ln 2)/\pi$, $\bar{y} = (\pi+2)/\pi$ **41.** (b) $3\ln 7 - 5\ln 3$ **43.** (b) $11 - \ln 12$

45. (a) $C(t) = \dfrac{kA_0^2 t}{1 + kA_0 t}$ (b) $C(t) = \dfrac{A_0 B_0(e^{kA_0 t} - e^{kB_0 t})}{B_0 e^{kA_0 t} - A_0 e^{kB_0 t}}$ **47.** (a) $v(t) = \dfrac{\alpha}{Ce^{(\alpha/m)t} - \beta}$, where C is an arbitrary constant.

(b) $v(t) = \dfrac{\alpha v_0}{(\alpha + \beta v_0)e^{(\alpha/m)t} - \beta v_0} = \dfrac{\alpha v_0 e^{-(\alpha/m)t}}{\alpha + \beta v_0 - \beta v_0 e^{-(\alpha/m)t}}$

(c) $\lim\limits_{t\to\infty} v(t) = 0$

SECTION 8.7

1. $-2(\sqrt{x} + \ln|1 - \sqrt{x}|) + C$ **3.** $2\ln(\sqrt{1+e^x} - 1) - x + 2\sqrt{1+e^x} + C$ **5.** $\dfrac{2}{5}(1+x)^{5/2} - \dfrac{2}{3}(1+x)^{3/2} + C$

7. $\dfrac{2}{5}(x-1)^{5/2} + 2(x-1)^{3/2} + C$ **9.** $-\dfrac{1+2x^2}{4(1+x^2)^2} + C$ **11.** $x + 2\sqrt{x} + 2\ln|\sqrt{x} - 1| + C$ **13.** $x + 4\sqrt{x-1} + 4\ln|\sqrt{x-1} - 1| + C$

15. $2\ln(\sqrt{1+e^x} - 1) - x + C$ **17.** $\dfrac{2}{3}(x-8)\sqrt{x+4} + C$ **19.** $\dfrac{1}{16}(4x+1)^{1/2} + \dfrac{1}{8}(4x+1)^{-1/2} - \dfrac{1}{48}(4x+1)^{-3/2} + C$ **21.** $\dfrac{4b+2ax}{a^2\sqrt{ax+b}} + C$

23. $-\ln\left|1 - \tan\dfrac{x}{2}\right| + C$ **25.** $\dfrac{2}{\sqrt{3}}\tan^{-1}\left[\dfrac{1}{\sqrt{3}}\left(2\tan\dfrac{x}{2} + 1\right)\right] + C$ **27.** $\dfrac{1}{2}\ln\left|\tan\dfrac{x}{2}\right| - \dfrac{1}{4}\tan^2\dfrac{x}{2} + C$

29. $\ln\left|\dfrac{1}{1+\sin x}\right| - \dfrac{2}{1 + \tan(x/2)} + C$ **31.** $\dfrac{4}{3} + 2\tan^{-1}2$ **33.** $2 + 4\ln\dfrac{2}{3}$ **35.** $\ln\left(\dfrac{\sqrt{3}-1}{\sqrt{3}}\right)$

37. Let $u = \tan\dfrac{x}{2}$. Then $\displaystyle\int \dfrac{1}{\cos x}\,dx = 2\int \dfrac{du}{1-u^2}$ and the result follows.

39. $\displaystyle\int \csc x\,dx = \int \dfrac{\sin x}{\sin^2 x}\,dx = \int \dfrac{\sin x}{1 - \cos^2 x}\,dx = -\int \dfrac{du}{1-u^2}$ where $u = \cos x$.   The result follows.   **41.** $2\tan^{-1}\left(\tanh\dfrac{x}{2}\right) + C$

43. $\dfrac{-2}{\tanh\dfrac{x}{2} + 1} + C$

SECTION 8.8

1. (a) 506 (b) 650 (c) 575 (d) 578.8 (e) 576   **3.** (a) 1.394 (b) 0.9122 (c) 1.1852 (d) 1.1533 (e) 1.1614

5. (a) $\frac{1}{4}\pi \cong 0.7828$ (b) $\frac{1}{4}\pi \cong 0.7854$ **7.** (a) 1.8440 (b) 1.7915 (c) 1.8090   **9.** (a) 0.8818 (b) 0.8821

11. Such a curve passes through the three points $(a_1, b_1), (a_2, b_2), (a_3, b_3)$ iff

$$b_1 = a_1^2 A + a_1 B + C, \qquad b_2 = a_2^2 A + a_2 B + C, \qquad b_3 = a_3^2 A + a_3 B + C,$$

which happens iff

$$A = \frac{b_1(a_2 - a_3) - b_2(a_1 - a_3) + b_3(a_1 - a_2)}{(a_1 - a_3)(a_1 - a_2)(a_2 - a_3)}, \qquad B = -\frac{b_1(a_2^2 - a_3^2) - b_2(a_1^2 - a_3^2) + b_3(a_1^2 - a_2^2)}{(a_1 - a_3)(a_1 - a_2)(a_2 - a_3)},$$

$$C = \frac{a_1^2(a_2 b_3 - a_3 b_2) - a_2^2(a_1 b_3 - a_3 b_1) + a_3^2(a_1 b_2 - a_2 b_1)}{(a_1 - a_3)(a_1 - a_2)(a_2 - a_3)}.$$

13. (a) $n \geq 8$ (b) $n \geq 4$ **15.** (a) $n \geq 238$ (b) $n \geq 19$ **17.** (a) $n \geq 51$ (b) $n \geq 7$ **19.** (a) $n \geq 37$ (b) $n \geq 5$

21. (a) 78 (b) 7   **23.** $f^{(4)}(x) = 0$ for all x; therefore by (8.7.2) the theoretical error is zero

25. (a) $\left| T_2 - \int_0^1 x^2 \, dx \right| = \dfrac{3}{8} - \dfrac{1}{3} = \dfrac{1}{24} = E_2^T$ (b) $\left| S_1 - \int_0^1 x^4 \, dx \right| = \dfrac{5}{24} - \dfrac{1}{5} = \dfrac{1}{120} = E_1^S$

27. Using the hint, $M_n = $ area $ABCD = $ area $AEFD \leq \displaystyle\int_a^b f(x) \, dx \leq T_n$.

CHAPTER 9
SECTION 9.1

1. $y^2 = 8x$

3. $(x + 1)^2 = -12(y - 3)$

5. $4y = (x - 1)^2$

7. $(y - 1)^2 = -2(x - \frac{3}{2})$

9. vertex $(0, 0)$
focus $(\frac{1}{2}, 0)$
axis $y = 0$
directrix $x = -\frac{1}{2}$

11. vertex $(0, -\frac{1}{2})$
focus $(0, -\frac{3}{8})$
axis $x = 0$
directrix $y = -\frac{5}{8}$

13. vertex $(-2, \frac{3}{2})$
focus $(-2, -\frac{1}{2})$
axis $x = -2$
directrix $y = \frac{7}{2}$

15. vertex $(\frac{3}{4}, -\frac{1}{2})$
focus $(1, -\frac{1}{2})$
axis $y = -\frac{1}{2}$
directrix $x = \frac{1}{2}$

17. center $(0, 0)$
foci $(\pm\sqrt{5}, 0)$
length of major axis 6
length of minor axis 4

19. center $(0, 0)$
foci $(0, \pm\sqrt{2})$
length of major axis $2\sqrt{6}$
length of minor axis 4

21. center $(0, 1)$
foci $(\pm\sqrt{5}, 1)$
length of major axis 6
length of minor axis 4

23. center $(1, 0)$
foci $(1, \pm 4\sqrt{3})$
length of major axis 16
length of minor axis 8

25. $\dfrac{x^2}{9} + \dfrac{y^2}{8} = 1$ **27.** $\dfrac{(x - 1)^2}{16} + \dfrac{(y - 6)^2}{25} = 1$ **29.** $\dfrac{(x - 1)^2}{21} + \dfrac{(y - 3)^2}{25} = 1$ **31.** $\dfrac{(x - 3)^2}{25} + \dfrac{(y + 1)^2}{9} = 1$ **33.** $\dfrac{x^2}{9} - \dfrac{y^2}{16} = 1$

35. $\dfrac{y^2}{25} - \dfrac{x^2}{144} = 1$ **37.** $\dfrac{x^2}{9} - \dfrac{(y - 1)^2}{16} = 1$ **39.** $16y^2 - \frac{16}{15}(x + 1)^2 = 1$

41. center $(0, 0)$
transverse axis 2
vertices $(\pm 1, 0)$
foci $(\pm\sqrt{2}, 0)$
asymptotes $y = \pm x$

43. center $(0, 0)$
transverse axis 6
vertices $(\pm 3, 0)$
foci $(\pm 5, 0)$
asymptotes $y = \pm\frac{4}{3}x$

45. center $(0, 0)$
transverse axis 8
vertices $(0, \pm 4)$
foci $(0, \pm 5)$
asymptotes $y = \pm\frac{4}{3}x$

47. center $(1, 3)$
transverse axis 6
vertices $(4, 3)$ and $(-2, 3)$
foci $(6, 3)$ and $(-4, 3)$
asymptotes $y = \pm\frac{4}{3}(x - 1) + 3$

49. center $(1, 3)$
transverse axis 4
vertices $(1, 5)$ and $(1, 1)$
foci $(1, 3 \pm \sqrt{5})$
asymptotes $y = 2x + 1, y = -2x + 5$

51. $(x - y)^2 = 6x + 10y - 9$ **53.** $(x + y)^2 = -12x + 20y + 28$

55. $P(x, y)$ is on the parabola with directrix l: $Ax + By + C = 0$ and focus $F(a, b)$ iff

$$d(P, l) = d(P, F) \quad \text{which happens iff} \quad \frac{|Ax + By + C|}{\sqrt{A^2 + B^2}} = \sqrt{(x - a)^2 + (y - b)^2}.$$

Square this last equation and simplify.

57. We can choose the coordinate system so that the parabola has an equation of the form $y = \alpha x^2$, $\alpha > 0$. One of the points of intersection is then the origin and the other is of the form $(c, \alpha c^2)$. We will assume that $c > 0$.

$$\text{area of } R_1 = \int_0^c \alpha x^2\, dx = \tfrac{1}{3}\alpha c^3 = \tfrac{1}{3}A$$

$$\text{area of } R_2 = A - \tfrac{1}{3}A = \tfrac{2}{3}A.$$

59. $2y = x^2 - 4x + 7$, $\quad 18y = x^2 - 4x + 103$ $\qquad$ **61.** $2\sqrt{\pi^2 a^4 - A^2}/\pi a$ $\qquad$ **63.** $(5 \pm \tfrac{5}{21}\sqrt{5}, 0)$ $\qquad$ **65.** $\tfrac{3}{5}$ $\qquad$ **67.** $\tfrac{4}{5}$

69. E_1 is fatter than E_2, more like a circle $\qquad$ **71.** the ellipse tends to a line segment of length $2a$ $\qquad$ **73.** $x^2/9 + y^2 = 1$

75. center $(0, 0)$, vertices $(1, 1)$ and $(-1, 1)$, foci $(\sqrt{2}, \sqrt{2})$ and $(-\sqrt{2}, \sqrt{2})$, asymptotes $x = 0$ and $y = 0$, transverse axis $2\sqrt{2}$

77. $[2\sqrt{3} - \ln(2 + \sqrt{3})]ab$ $\qquad$ **79.** $\tfrac{5}{3}$ $\qquad$ **81.** $\sqrt{2}$ $\qquad$ **83.** the branches of H_1 open up less quickly than the branches of H_2

85. the hyperbola tends to a pair of parallel lines separated by the transverse axis $\qquad$ **87.** $4c$ $\qquad$ **89.** $A = \tfrac{8}{3}c^3$; $\bar{x} = 0$, $\bar{y} = \tfrac{3}{5}c$

91. $\dfrac{kx}{p(0)} = \tan\theta = \dfrac{dy}{dx}$, $\quad y = \dfrac{k}{2p(0)}x^2 + C$

In our figure, $C = y(0) = 0$. Thus the equation of the cable is $y = kx^2/2p(0)$, the equation of a parabola.

93. Start with any two parabolas γ_1, γ_2. By moving them we can see to it that they have equations of the following form:

$$\gamma_1: \quad x^2 = 4c_1 y, \quad c_1 > 0; \qquad \gamma_2: \quad x^2 = 4c_2 y, \quad c_2 > 0.$$

Now we change the scale for γ_2 so that the equation for γ_2 will look exactly like the equation for γ_1. Set $X = (c_1/c_2)x$, $Y = (c_1/c_2)y$. Then

$$x^2 = 4c_2 y \Longrightarrow (c_2/c_1)^2 X^2 = 4c_2(c_2/c_1)Y \Longrightarrow X^2 = 4c_1 Y.$$

Now γ_2 has exactly the same equation as γ_1; only the scale, the units by which we measure distance, has changed.

95. about 0.25 mile west and 1.5 miles north of point A.

SECTION 9.2

1–7. See figure to the right. $\qquad$ **9.** $(0, 3)$ $\qquad$ **11.** $(1, 0)$ $\qquad$ **13.** $(-\tfrac{3}{2}, \tfrac{3}{2}\sqrt{3})$ $\qquad$ **15.** $(0, -3)$

17. $[1, \tfrac{1}{2}\pi + 2n\pi]$, $[-1, \tfrac{3}{2}\pi + 2n\pi]$ $\qquad$ **19.** $[3, \pi + 2n\pi]$, $[-3, 2n\pi]$

21. $[2\sqrt{2}, \tfrac{7}{4}\pi + 2n\pi]$, $[-2\sqrt{2}, \tfrac{3}{4}\pi + 2n\pi]$ $\qquad$ **23.** $[8, \tfrac{1}{6}\pi + 2n\pi]$, $[-8, \tfrac{7}{6}\pi + 2n\pi]$

25. $\sqrt{r_1^2 + r_2^2 - 2r_1 r_2 \cos(\theta_1 - \theta_2)}$ $\qquad$ **27.** (a) $[\tfrac{1}{2}, \tfrac{11}{6}\pi]$ (b) $[\tfrac{1}{2}, \tfrac{5}{6}\pi]$ (c) $[\tfrac{1}{2}, \tfrac{7}{6}\pi]$

29. (a) $[2, \tfrac{2}{3}\pi]$ (b) $[2, \tfrac{5}{3}\pi]$ (c) $[2, \tfrac{1}{3}\pi]$ $\qquad$ **31.** symmetry about the x-axis

33. no symmetry about the coordinate axes; no symmetry about the origin

35. symmetry about the origin $\qquad$ **37.** $r \cos\theta = 2$ $\qquad$ **39.** $r^2 \sin 2\theta = 1$ $\qquad$ **41.** $r = 4 \sin\theta$

43. $\theta = \pi/4$ $\qquad$ **45.** $r = 1 - \cos\theta$ $\qquad$ **47.** $r^2 = \sin 2\theta$ $\qquad$ **49.** the horizontal line $y = 4$ $\qquad$ **51.** the line $y = \sqrt{3}x$

53. the parabola $y^2 = 4(x + 1)$ $\qquad$ **55.** the circle $x^2 + y^2 = 3x$ $\qquad$ **57.** the line $y = 2x$ $\qquad$ **59.** $3x^2 + 4y^2 - 8x = 16$, ellipse

61. $y^2 = 8x + 16$, parabola $\qquad$ **63.** $\left(x - \dfrac{b}{2}\right)^2 + \left(y - \dfrac{a}{2}\right)^2 = \dfrac{a^2 + b^2}{4}$; center: $\left(\dfrac{b}{2}, \dfrac{a}{2}\right)$, radius: $\dfrac{\sqrt{a^2 + b^2}}{2}$ $\qquad$ **65.** $r = \dfrac{d}{2 - \cos\theta}$

SECTION 9.3

1.

3.

5.

7.

9.

11.

13.

15.

17.

19.

21.

23.

25.

27.

29.

31.

33.

37. (a) center: $\left(\dfrac{e^2c}{1-e^2}, 0\right)$ length of major axis: $\dfrac{2ec}{1-e^2}$; length of minor axis: $\dfrac{2ec}{\sqrt{1-e^2}}$

(b) (i) the ellipse in (a) rotated $\pi/2$ radians in the counterclockwise direction
(ii) the ellipse in (a) rotated π radians in the counterclockwise direction
(iii) the ellipse in (a) rotated $\pi/2$ radians in the clockwise direction

39. (a) a point, the origin (b) parabola (c) ellipse, center at $\left(\dfrac{2d}{3}, 0\right)$, length of major axis $\dfrac{8}{3}d$ **41.** Butterfly

SECTION *9.4

1. (a) $\frac{1}{2}$ unit to the right of the pole (b) 2 units (c) Figure A. **3.** the parabola of Figure A rotated by π radians

5. (a) $e = \frac{2}{3}$ (b) 2 units to the left of the pole and $\frac{2}{3}$ units to the right of the pole (c) $\frac{4}{5}$ units to the left of the pole
(d) $\frac{8}{5}$ units to the left of the pole (e) $\frac{4}{5}\sqrt{5}$ units (about 1.79 units) (f) $\frac{4}{3}$ units (g) Figure B

(A)

(B)

7. the ellipse of Figure B rotated by $\frac{1}{2}\pi$ radians

9. (a) $e = 2$
(b) 2 units to the right of the pole and 6 units to the right of the pole
(c) 4 units to the right of the pole
(d) 8 units to the right of the pole
(e) 12 units

(f)

11. the hyperbola of Exercise 9(f) rotated by $\frac{3}{2}\pi$ radians   **13.** ellipse: $x^2/48 + (y+4)^2/64 = 1$ **15.** ellipse: $x^2/9 + (y-4)^2/25 = 1$

SECTION 9.5

1. yes; $[1, \pi] = [-1, 0]$ and the pair $r = -1$, $\theta = 0$ satisfies the equation   **3.** yes; the pair $r = \frac{1}{2}$, $\theta = \frac{1}{2}\pi$ satisfies the equation

5. $[2, \pi] = [-2, 0]$. The coordinates of $[-2, 0]$ satisfy the equation $r^2 = 4\cos\theta$, and the coordinates of $[2, \pi]$ satisfy the equation $r = 3 + \cos\theta$.

7. $(0, 0)$, $(-\frac{1}{2}, \frac{1}{2})$ **9.** $(-1, 0)$, $(1, 0)$ **11.** $(0, 0)$, $(\frac{1}{4}, \pm\frac{1}{4}\sqrt{3})$ **13.** $(\frac{3}{2}, 2)$ **15.** $(0, 0)$, $(1, 0)$

17. (b) The curves intersect at the pole and at $(1.172, 0.173)$,   **19.** (b) The curves intersect at the pole and at:
     $(1.86, 1.036)$, $(0.90, 3.245)$.

$r = 1 - 3\sin\theta$	$r = 2 - 5\sin\theta$
$(-2, 0)$	$(2, \pi)$
$(3.800, 3.510)$	$(3.800, 3.510)$
$(2.412, 4.223)$	$(-2.412, 1.081)$
$(-1.267, 0.713)$	$(-1.267, 0.713)$

SECTION 9.6

1. $\frac{1}{4}\pi a^2$ **3.** $\frac{1}{2}a^2$ **5.** $\frac{1}{2}\pi a^2$ **7.** $\frac{1}{4} - \frac{1}{16}\pi$ **9.** $\frac{3}{16}\pi + \frac{3}{8}$ **11.** $\frac{5}{2}a^2$ **13.** $\frac{1}{12}(3e^{2\pi} - 3 - 2\pi^3)$ **15.** $\frac{1}{4}(e^{2\pi} + 1 - 2e^\pi)$

17. $\displaystyle\int_{\pi/6}^{5\pi/6} \frac{1}{2}([4\sin\theta]^2 - [2]^2)\, d\theta$ **19.** $\displaystyle\int_{-\pi/3}^{\pi/3} \frac{1}{2}([4]^2 - [2\sec\theta]^2)\, d\theta$ **21.** $2\left[\displaystyle\int_{0}^{\pi/3} \frac{1}{2}(2\sec\theta)^2\, d\theta + \int_{\pi/3}^{\pi/2} \frac{1}{2}(4)^2\, d\theta\right]$

23. $\displaystyle\int_{0}^{\pi/3} \frac{1}{2}(2\sin 3\theta)^2\, d\theta$ **25.** $2\left[\displaystyle\int_{0}^{\pi/6} \frac{1}{2}(\sin\theta)^2\, d\theta + \int_{\pi/6}^{\pi/2} \frac{1}{2}(1 - \sin\theta)^2\, d\theta\right]$ **27.** $\pi - 8\displaystyle\int_{0}^{\pi/4} \frac{1}{2}(\cos 2\theta)^2\, d\theta$

29. For $r = a\cos 2n\theta$, area of one petal is $a^2\displaystyle\int_{0}^{\pi/4n} \cos^2 2n\theta\, d\theta = \frac{\pi a^2}{8n}$. For $r = a\sin 2n\theta$, area of one petal is $a^2\displaystyle\int_{0}^{\pi/4n} \sin^2 2n\theta\, d\theta = \frac{\pi a^2}{8n}$.

Total area $= \dfrac{\pi a^2}{2}$.

33. $(5/6, 0)$ **35.** (a) Substitute $x = r\cos\theta$, $y = r\sin\theta$ into the equation and solve for r.   (c) $8 - 2\pi$

SECTION 9.7

1. $4x = (y-1)^2$ **3.** $y = 4x^2 + 1$, $x \geq 0$ **5.** $9x^2 + 4y^2 = 36$ **7.** $1 + x^2 = y^2$ **9.** $y = 2 - x^2$, $-1 \leq x \leq 1$

11. $2y - 6 = x$, $-4 \leq x \leq 4$

13. $y = x - 1$

15. $xy = 1$

17. $y + 2x = 11$

19. $x = \sin\frac{1}{2}\pi y$

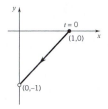

21. $y^2 = x^2 + 1$

23. (a) $x(t) = -\sin 2\pi t$, $y(t) = \cos 2\pi t$ (b) $x(t) = \sin 4\pi t$, $y(t) = \cos 4\pi t$ (c) $x(t) = \cos\frac{1}{2}\pi t$, $y(t) = \sin\frac{1}{2}\pi t$
     (d) $x(t) = \cos\frac{3}{2}\pi t$, $y(t) = -\sin\frac{3}{2}\pi t$

25. $x(t) = \tan\frac{1}{2}\pi t$, $y(t) = 2$ **27.** $x(t) = 3 + 5t$, $y(t) = 7 - 2t$ **29.** $x(t) = \sin^2 \pi t$, $y(t) = -\cos \pi t$

31. $x(t) = (2-t)^2$, $y(t) = (2-t)^3$ **33.** $x(t) = t(b-a) + a$, $y(t) = f(t(b-a) + a)$ **35.** $A = 2\pi a^2$

37. (a) $V_x = 3\pi^2 r^3$ (b) $V_y = 4\pi^3 r^3$ **39.** $x(t) = -a\cos t$, $y(t) = b\sin t$; $t \in [0, \pi]$

41. (a) paths intersect at $(6, 5)$ and $(8, 1)$ (b) particles collide at $(8, 1)$ **43.** curve intersects itself at $(0, 0)$

45. curve intersects itself at $(0, 0)$ and $(0, \frac{3}{4})$

47. (a) The coefficient a determines the amplitude and the period.   (b) $\dfrac{dy}{dx} \to -\infty$ as $\theta \to 2\pi^-$; $\dfrac{dy}{dx} \to \infty$ as $\theta \to 2\pi^+$

49.

  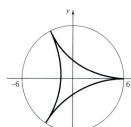

51. $y = -\dfrac{16}{v_0^2}(\sec^2\theta)x^2 + (\tan\theta)x$ **53.** $\dfrac{1}{16}v_0^2\cos\theta\sin\theta$ feet   **55.** $\pi/4$

SECTION 9.8

1. $3x - y - 3 = 0$ **3.** $y = 1$ **5.** $3x + y - 3 = 0$ **7.** $2x + 2y - \sqrt{2} = 0$ **9.** $2x + y - 8 = 0$ **11.** $x - 5y + 4 = 0$

13. $x + 2y + 1 = 0$ **15.** $x(t) = t$, $y(t) = t^3$;   **17.** $x(t) = t^{5/3}$, $y(t) = t$;   **19.** (a) none;
tangent line $y = 0$ tangent line $x = 0$ (b) at $(2, 2)$ and $(-2, 0)$

21. (a) at $(3, 7)$ and $(3, 1)$;   **23.** (a) at $(-\frac{2}{3}, \pm\frac{2}{9}\sqrt{3})$;   **25.** (a) at $(\pm\frac{1}{2}\sqrt{2}, \pm 1)$;
(b) at $(-1, 4)$ and $(7, 4)$ (b) at $(-1, 0)$ (b) at $(\pm 1, 0)$

27. $y = 0$, $(\pi - 2)y + 32x - 64 = 0$

29. The slope of $\overline{OP}$ is $\tan\theta_1$. The curve $r = f(\theta)$ can be parametrized by setting

$$x(\theta) = f(\theta)\cos\theta, \qquad y(\theta) = f(\theta)\sin\theta.$$

Differentiation gives

$$x'(\theta) = -f(\theta)\sin\theta + f'(\theta)\cos\theta, \qquad y'(\theta) = f(\theta)\cos\theta + f'(\theta)\sin\theta.$$

If $f'(\theta_1) = 0$, then

$$x'(\theta_1) = -f(\theta_1)\sin\theta_1, \qquad y'(\theta_1) = f(\theta_1)\cos\theta_1.$$

Since $f(\theta_1) \neq 0$, we have

$$m = \frac{y'(\theta_1)}{x'(\theta_1)} = -\cot\theta_1 = -\frac{1}{\text{slope of }\overline{OP}}.$$

31.

33.

35.

37. -8 **39.** 2

SECTION 9.9

1. $\sqrt{5}$ **3.** 7   **5.** $2\sqrt{3}$ **7.** $\frac{4}{3}$ **9.** $6 + \frac{1}{2}\ln 5$ **11.** $\frac{63}{8}$ **13.** $\ln(1 + \sqrt{2})$ **15.** $\frac{3}{2}$ **17.** $\frac{1}{3}\pi + \frac{1}{2}\sqrt{3}$

19. initial speed 2, terminal speed 4; $s = 2\sqrt{3} + \ln(2 + \sqrt{3})$ **21.** initial speed 0, terminal speed $\sqrt{13}$; $x = \frac{1}{27}(13\sqrt{13} - 8)$

23. initial speed $\sqrt{2}$, terminal speed $\sqrt{2}e^{\pi}$; $s = \sqrt{2}(e^{\pi} - 1)$ **25.** $8a$

27. (a) $24a$ (b) use the identities $\cos 3\theta = 4\cos^3\theta - 3\cos\theta$, $\sin 3\theta = 3\sin\theta - 4\sin^3\theta$ **29.** 2π **31.** $\sqrt{2}(e^{4\pi} - 1)$

33. $\frac{1}{2}\sqrt{5}(e^{4\pi} - 1)$ **35.** $4 - 2\sqrt{2}$ **37.** $\ln(1 + \sqrt{2}$ **39.** $c = 1$ **41.** $L = \int_a^b \sqrt{1 + \sinh^2 x}\, dx = \int_a^b \sqrt{\cosh^2 x}\, dx = \int_a^b \cosh x\, dx = A$

43. (a) Express GPE + KE as a function of t and verify that the derivative with respect to t is zero.

(b) From (a) we know that throughout the motion

$$32my + \tfrac{1}{2}mv^2 = C.$$

At the time of firing $y = 0$ and $v = |v_0| = v_0$.   Therefore

$$32my + \tfrac{1}{2}mv^2 = \tfrac{1}{2}mv_0^2.$$

At impact $y = 0$, $\tfrac{1}{2}mv^2 = \tfrac{1}{2}mv_0^2$, and $v = v_0$.

45. $\sqrt{1 + [f'(x)]^2} = \sqrt{1 + \tan^2[\alpha(x)]} = |\sec[\alpha(x)]|$ **47.** (a)

(b) 2.7156   **49.** (b) 28.3617

SECTION 9.10

1. $L = 1$, $(\bar{x}, \bar{y}) = (\frac{1}{2}, 4)$, $A_x = 16\pi$ **3.** $L = 5$, $(\bar{x}, \bar{y}) = (\frac{3}{2}, 2)$, $A_x = 20\pi$ **5.** $L = 10$, $(\bar{x}, \bar{y}) = (3, 4)$, $A_x = 80\pi$

7. $L = \frac{1}{3}\pi$; $\bar{x} = 6/\pi$, $\bar{y} = 6(2 - \sqrt{3})/\pi$; $A_x = 4\pi(2 - \sqrt{3})$ **9.** $L = \frac{1}{3}\pi a$; $\bar{x} = 0$, $\bar{y} = 3a/\pi$; $A_x = 2\pi a^2$ **11.** $\frac{1}{9}\pi(17\sqrt{7} - 1)$

13. $\frac{61}{432}\pi$ **15.** $\pi[\sqrt{2} + \ln(1 + \sqrt{2})]$ **17.** $\frac{2}{3}\sqrt{2}\pi(2e^n + 1)$ **19.** (a) $3\pi a^2$ (b) $\dfrac{64\pi a^2}{3}$

21. See the figure.

$A = \frac{1}{2}\theta s_2^2 - \frac{1}{2}\theta s_1^2$
$\quad = \frac{1}{2}(\theta s_2 + \theta s_1)(s_2 - s_1)$
$\quad = \frac{1}{2}(2\pi R + 2\pi r)s = \pi(R + r)s$

23. (a) the 3, 4, 5 sides have centroids   **25.** $4\pi^2 ab$
$\qquad (\frac{3}{2}, 0), (4, 2), (\frac{3}{2}, 2)$
(b) $\bar{x} = 2$, $\bar{y} = \frac{3}{2}$
(c) $\bar{x} = 2$, $\bar{y} = \frac{4}{3}$
(d) $\bar{x} = \frac{13}{6}$, $\bar{y} = 2$
(e) $A = 20\pi$
(f) $A = 36\pi$

27. The band can be obtained by revolving about the x-axis the graph of a function

$$f(x) = \sqrt{r^2 - x^2}, \quad x \in [a, b].$$

A straightforward calculation shows that the surface area of the band is $2\pi r(b - a)$.

29. (a) $2\pi b^2 + \dfrac{2\pi ab}{e}\sin^{-1}e$ (b) $2\pi a^2 + \dfrac{\pi b^2}{e}\ln\left|\dfrac{1 + e}{1 - e}\right|$, where e is the eccentricity $c/a = \sqrt{a^2 - b^2}/a$

31. at the midpoint of the axis of the hemisphere   **33.** on the axis of the cone $\left(\dfrac{2R + r}{R + r}\right)\dfrac{h}{3}$ units from the base of radius r

SECTION 9.11

1. $x(\theta) = \overline{OB} - \overline{AB} = R\theta - R\sin\theta = R(\theta - \sin\theta),\quad y(\theta) = \overline{BQ} - \overline{QC} = R - R\cos\theta = R(1 - \cos\theta)$

3. (a) The slope at P is

$$m = \frac{y'(\theta)}{x'(\theta)} = \frac{\sin\theta}{1 - \cos\theta}.$$

The line tangent to the cycloid at P has equation

$$y - R(1 - \cos\theta) = \frac{\sin\theta}{1 - \cos\theta}[x - R(\theta - \sin\theta)].$$

The top of the circle is the point $(R\theta, 2R)$. Its coordinates satisfy the equation for the tangent:

$$2R - R(1 - \cos\theta) \overset{?}{=} \frac{\sin\theta}{1 - \cos\theta}[R\theta - R(\theta - \sin\theta)]$$

$$R(1 + \cos\theta) \overset{?}{=} \frac{R\sin^2\theta}{1 - \cos\theta}$$

$$1 - \cos^2\theta \overset{\checkmark}{=} \sin^2\theta.$$

(b) In view of the symmetry and repetitiveness of the curve we can assume that $\theta \in (0, \pi)$. Then

$$\tan\alpha = \frac{\sin\theta}{1 - \cos\theta} = \frac{\sin\theta}{2\sin^2\frac{1}{2}\theta} = \frac{\sin\frac{1}{2}\theta\cos\frac{1}{2}\theta}{\sin^2\frac{1}{2}\theta} = \cot\frac{1}{2}\theta$$

and $\alpha = \frac{1}{2}\pi - \frac{1}{2}\theta = \frac{1}{2}(\pi - \theta)$.

5. $8R$ **7.** $\bar{x} = \pi R,\quad \bar{y} = \frac{5}{6}R$ **9.** $V_y = 6\pi^3R^3$ **11.** $A = \frac{64}{3}\pi R^2$ **13.** $\alpha = \frac{1}{2}\phi$ (radian measure)

15. (a) already shown more generally in Example 6, Section 9.9
(b) Combining $d^2s/dt^2 = -g\sin\alpha$ with $s = 4R\sin\alpha$, we have

$$\frac{d^2s}{dt^2} = -\frac{g}{4R}s.$$

This is simple harmonic motion with angular frequency $\omega = \frac{1}{2}\sqrt{g/R}$ and period $T = 2\pi/\omega = 4\pi\sqrt{R/g}$.

CHAPTER 10

SECTION 10.1

1. lub = 2; glb = 0   **3.** no lub; glb = 0   **5.** lub = 2; glb = -2 **7.** no lub; glb = 2   **9.** lub = $2\frac{1}{2}$; glb = 2

11. lub = 1; glb = 0.9   **13.** lub = e; glb = 0   **15.** lub = $\frac{1}{2}(-1 + \sqrt{5})$; glb = $\frac{1}{2}(-1 - \sqrt{5})$ **17.** no lub; no glb

19. no lub; no glb   **21.** glb $S = 0,\quad 0 \le (\frac{1}{11})^3 < 0 + 0.001$ **23.** glb $S = 0,\quad 0 \le (\frac{1}{10})^{2n-1} < 0 + (\frac{1}{10})^k,\quad n > \frac{1}{2}(k + 1)$

25. Let $\epsilon > 0$. The condition $m \le s$ is satisfied by all numbers s in S. All we have to show therefore is that there is some number s in S such that $s < m + \epsilon$. Suppose on the contrary that there is no such number in S. We then have $m + \epsilon \le x$ for all $x \in S$, so that $m + \epsilon$ becomes a lower bound for S. But this cannot happen, for it makes $m + \epsilon$ a lower bound that is *greater* than m, and by assumption, m is the *greatest* lower bound.

27. Let $c = $ lub S. Since $b \in S, b \le c$. Since b is an upper bound for $S, c \le b$. Thus $b = c$.

29. (a) Any upper bound for S is an upper bound for T; any lower bound for S is a lower bound for T.
(b) Let $a = $ glb S. Then $a \le t$ for all $t \in T$. Therefore $a \le$ glb T. Similarly, if $b = $ lub S, then $t \le b$ for all $t \in T$, so lub $T \le b$. It now follows that glb $S \le$ glb $T \le$ lub $T \le$ lub S.

31. Let M be any positive number and consider M/c. Since the set of positive integers is not bounded above, there exists a positive integer k such that $k \ge M/c$. This implies $kc \ge M$. Since $kc \in S$, it follows that S is not bounded above.

33. (a)

a_1	a_2	a_3	a_4	a_5	a_6	a_7	a_8	a_9	a_{10}
1.4142	1.6818	1.8340	1.9152	1.9571	1.9785	1.9892	1.9946	1.9973	1.9986

(b) Let S be the set of positive integers for which $a_n < 2$. Then $1 \in S$ since $a_1 = \sqrt{2} \cong 1.4142 < 2$. Assume that $k \in S$. Now $a_{k+1}^2 = 2a_k < 4$, which implies $a_{k+1} < 2$. Thus $k + 1 \in S$ and S is the set of positive integers.
(c) yes   (d) The number you chose is the least upper bound of the corresponding set S.

SECTION 10.2

1. $a_n = 2 + 3(n - 1),\quad n = 1, 2, 3, \ldots$ **3.** $a_n = \frac{(-1)^{n-1}}{2n - 1},\quad n = 1, 2, 3, \ldots$ **5.** $a_n = \frac{n^2 + 1}{n},\quad n = 1, 2, 3, \ldots$

7. $a_n = \begin{cases} n & \text{if } n = 2k - 1, \\ \frac{1}{n} & \text{if } n = 2k, \end{cases}$ where $k = 1, 2, 3, \ldots$ **9.** decreasing; bounded below by 0 and above by 2

11. increasing; bounded below by 1 but not bounded above **13.** not monotonic; bounded below by 0 and above by $\frac{3}{2}$

15. decreasing; bounded below by 0 and above by 0.9 **17.** increasing; bounded below by $\frac{1}{2}$ but not bounded above

19. increasing; bounded below by $\frac{4}{5}\sqrt{5}$ and above by 2 **21.** increasing; bounded below by $\frac{2}{51}$ but not bounded above

23. decreasing; bounded below by 0 and above by $\frac{1}{2}(10^{10})$ **25.** increasing; bounded below by 0 and above by $\ln 2$

27. decreasing; bounded below by 1 and above by 4 **29.** increasing; bounded below by $\sqrt{3}$ and above by 2

31. decreasing; bounded above by -1 but not bounded below **33.** increasing; bounded below by $\frac{1}{2}$ and above by 1

35. decreasing; bounded below by 0 and above by 1 **37.** decreasing; bounded below by 0 and above by $\frac{5}{6}$

39. decreasing; bounded below by 0 and above by $\frac{1}{2}$ **41.** decreasing; bounded below by 0 and above by $\frac{1}{3}\ln 3$

43. increasing; bounded below by $\frac{3}{4}$ but not bounded above **45.** For $n \geq 5$

$$\frac{a_{n+1}}{a_n} = \frac{5^{n+1}}{(n+1)!} \cdot \frac{n!}{5^n} = \frac{5}{n+1} < 1 \qquad \text{and thus} \qquad a_{n+1} < a_n.$$

Sequence is not nonincreasing: $a_1 = 5 < \frac{25}{2} = a_2$.

47. boundedness: $0 < (c^n + d^n)^{1/n} < (2d^n)1/n = 2^{1/n} d \leq 2d.$

monotonicity: $a_{n+1}^{n+1} = c^{n+1} + d^{n+1} = cc^n + dd^n < (c^n + d^n)^{1/n} c^n + (c^n + d^n)^{1/n} d^n$

$= (c^n + d^n)^{1 + (1/n)} = (c^n + d^n)^{(n+1)/n} = a_n^{n+1}.$

Taking the $n + 1$-th root of each side we have $a_{n+1} < a_n$. The sequence is monotonic decreasing.

49. $a_1 = 1, a_2 = \frac{1}{2}, a_3 = \frac{1}{6}, a_4 = \frac{1}{24}, a_5 = \frac{1}{120}, a_6 = \frac{1}{720}$; $a_n = 1/n!$ **51.** $a_1 = a_2 = a_3 = a_4 = a_5 = a_6 = 1$; $a_n = 1$

53. $a_1 = 1, a_2 = 3, a_3 = 5, a_4 = 7, a_5 = 9, a_6 = 11$; $a_n = 2n - 1$ **55.** $a_1 = 1, a_2 = 4, a_3 = 13, a_4 = 40, a_5 = 121, a_6 = 364$; $a_n = \frac{1}{2}(3^n - 1)$

57. $a_1 = 1, a_2 = 4, a_3 = 9, a_4 = 16, a_5 = 25, a_6 = 36$; $a_n = n^2$ **59.** $a_1 = 1, a_2 = 3, a_3 = 4, a_4 = 8, a_5 = 16, a_6 = 32$; $a_n = 2^{n-1}$ $(n \geq 3)$

61. $a_1 = 2, a_2 = 1, a_3 = 2, a_4 = 1, a_5 = 2, a_6 = 1$; $a_n = \frac{1}{2}[3 - (-1)^n]$ **63.** $a_1 = 1, a_2 = 3, a_3 = 5, a_4 = 7, a_5 = 9, a_6 = 11$; $a_n = 2n - 1$

65. First $a_1 = 2^1 - 1 = 1$. Next suppose $a_k = 2^k - 1$ for some $k \geq 1$. Then

$$a_{k+1} = 2a_k + 1 = 2(2^k - 1) + 1 = 2^{k+1} - 1.$$

67. First $a_1 = \frac{1}{2^0} = 1$. Next suppose $a_k = \frac{k}{2^{k-1}}$ for some $k \geq 1$. Then $a_{k+1} = \frac{k+1}{2k} a_k = \frac{k+1}{2k} \frac{k}{2^{k-1}} = \frac{k+1}{2^k}.$

69. (a) n (b) $\frac{1 - r^n}{1 - r}$ **71.** (a) $150(\frac{3}{4})^{n-1}$ (b) $\frac{5\sqrt{3}}{2}\left(\frac{3}{4}\right)^{\frac{n-1}{2}}$

73. (a) $a_2 = 1 + \sqrt{a_1} = 2 > 1 = a_1$. Assume $a_k = 1 + \sqrt{a_{k-1}} > a_{k-1}$. Then $a_{k+1} = 1 + \sqrt{a_k} > 1 + \sqrt{a_{k-1}} = a_k$. Thus $\{a_n\}$ is an increasing
sequence.

(b) $a_n = 1 + \sqrt{a_{n-1}} < 1 + \sqrt{a_n}$, since $a_{n-1} < a_n$.

$a_n - \sqrt{a_n} - 1 < 0$, or $(\sqrt{a_n})^2 - \sqrt{a_n} - 1 < 0$, which implies (solve the inequality) that $\sqrt{a_n} < \frac{1 + \sqrt{5}}{2}$, hence $a_n < \frac{3 + \sqrt{5}}{2}$ for all n.

(c) lub $\{a_n\} \cong 2.6180.$

SECTION 10.3

1. diverges **3.** converges to 0 **5.** converges to 1 **7.** converges to 0 **9.** converges to 0 **11.** diverges **13.** converges to 0

15. converges to 1 **17.** converges to $\frac{4}{9}$ **19.** converges to $\frac{1}{2}\sqrt{2}$ **21.** diverges **23.** converges to 1 **25.** diverges

27. converges to 0 **29.** converges to $\frac{1}{2}$ **31.** converges to e^2 **33.** diverges **35.** converges to 0

37. Use $|(a_n + b_n) - (L + M)| \leq |a_n - L| + |b_n - M|.$ **39.** $\left(1 + \frac{1}{n}\right)^{n+1} = \left(1 + \frac{1}{n}\right)^n\left(1 + \frac{1}{n}\right)$. Note that $\left(1 + \frac{1}{n}\right)^n \to e$

and $\left(1 + \frac{1}{n}\right) \to 1.$

41. Imitate the proof given for the nondecreasing case in Theorem 10.3.6. **43.** Let $\epsilon > 0$. Choose k so that, for $n \geq k$,

$$L - \epsilon < a_n < L + \epsilon, \quad L - \epsilon < c_n < L + \epsilon \quad \text{and} \quad a_n \leq b_n \leq c_n.$$

For such n,

$$L - \epsilon < b_n < L + \epsilon.$$

45. Let $\epsilon > 0$. Since $a_n \to L$, there exists a positive integer N such that $L - \epsilon < a_n < L + \epsilon$ for all $n \geq N$. Now $a_n \leq M$ for all n, so $L - \epsilon < M$, or
$L < M + \epsilon$. Since ϵ is arbitrary, $L \leq M.$

47. By the continuity of f, $f(L) = f(\lim_{n\to\infty} a_n) = \lim_{n\to\infty} f(a_n) = \lim_{n\to\infty} a_{n+1} = L.$ **49.** Use Theorem 10.3.12 with $f(x) = x^{1/p}.$ **51.** converges to 0

53. converges to 0   **55.** diverges   **57.** converges to 2   **59.** $L = 0$, $n = 32$ **61.** $L = 0$, $n = 4$ **63.** $L = 0$, $n = 7$

65. $L = 0$, $n = 65$ **67.** (a) $\dfrac{3 + \sqrt{5}}{2}$ (b) 3

69. (a)

a_2	a_3	a_4	a_5	a_6	a_7	a_8	a_9	a_{10}
0.540302	0.857553	0.654290	0.793480	0.701369	0.763960	0.722102	0.750418	0.731404

(b) 0.739085; it is the fixed point of $f(x) = \cos x$.

71. (a)

a_2	a_3	a_4	a_5	a_6	a_7	a_8
2.000000	1.750000	1.732143	1.732051	1.732051	1.732051	1.732051

(b) $L = \dfrac{1}{2}\left(L + \dfrac{3}{L}\right)$, which implies $L^2 = 3$, or $L = \sqrt{3}$.

SECTION 10.4

1. converges to 1   **3.** converges to 0   **5.** converges to 0   **7.** converges to 0   **9.** converges to 1   **11.** converges to 0

13. converges to 1   **15.** converges to 1   **17.** converges to π **19.** converges to 1   **21.** converges to 0   **23.** diverges

25. converges to 0   **27.** converges to e^{-1} **29.** converges to 0   **31.** converges to 0   **33.** converges to e^x **35.** converges to 0

37. $\sqrt{n+1} - \sqrt{n} = \dfrac{\sqrt{n+1} - \sqrt{n}}{\sqrt{n+1} + \sqrt{n}}(\sqrt{n+1} + \sqrt{n}) = \dfrac{1}{\sqrt{n+1} + \sqrt{n}} \to 0$

39. (b) $2\pi r$. As $n \to \infty$, the perimeter of the polygon tends to the circumference of the circle.   **41.** $\frac{1}{2}$ **43.** $\frac{1}{8}$

45. (a) $m_{n+1} - m_n = \dfrac{1}{n+1}(a_1 + \cdots + a_n + a_{n+1}) - \dfrac{1}{n}(a_1 + \cdots + a_n)$

$= \dfrac{1}{n(n+1)}\left[na_{n+1} - (a_1 + \cdots + a_n)\right] > 0$ since $\{a_n\}$ is increasing.

(b) We begin with the hint $m_n < \dfrac{|a_1 + \cdots + a_j|}{n} + \dfrac{\epsilon}{2}\left(\dfrac{n-j}{n}\right)$. Since j is fixed, $\dfrac{|a_1 + \cdots + a_j|}{n} \to 0$, and therefore for n sufficiently

large $\dfrac{|a_1 + \cdots + a_j|}{n} < \dfrac{\epsilon}{2}$. Since $\dfrac{\epsilon}{2}\left(\dfrac{n-j}{n}\right) < \dfrac{\epsilon}{2}$, we see that, for n sufficiently large, $|m_n| < \epsilon$. This shows that $m_n \to 0$.

47. (a) Let S be the set of positive integers n ($n \geq 2$) for which the inequalities hold. Since $(\sqrt{b})^2 - 2\sqrt{ab} + (\sqrt{a})^2 = (\sqrt{b} - \sqrt{a})^2 > 0$, it follows that

$\dfrac{a+b}{2} > \sqrt{ab}$ and $a_1 > b_1$. Now $a_2 = \dfrac{a_1 + b_1}{2} < a_1$ and $b_2 = \sqrt{a_1 b_1} < b_1$. Also, by the argument above, $a_2 = \dfrac{a_1 + b_1}{2} > \sqrt{a_1 b_1} = b_2$, and

so $a_1 > a_2 > b_2 > b_1$. Thus $2 \in S$. Assume that $k \in S$. Then $a_{k+1} = \dfrac{a_k + b_k}{2} < \dfrac{a_k + a_k}{2} = a_k$, $b_{k+1} = \sqrt{a_k b_k} > \sqrt{b_k^2} = b_k$, and $a_{k+1} =$

$\dfrac{a_k + b_k}{2} > \sqrt{a_k b_k} = b_{k+1}$. Thus $k + 1 \in S$. Therefore the inequalities hold for all $n \geq 2$.

(b) $\{a_n\}$ is a decreasing sequence which is bounded below.
$\{b_n\}$ is an increasing sequence which is bounded above.
Let $L_a = \lim\limits_{n \to \infty} a_n$, $L_b = \lim\limits_{n \to \infty} b_n$. Then $a_n = \dfrac{a_{n-1} + b_{n-1}}{2}$ implies $L_a = \dfrac{L_a + L_b}{2}$ and $L_a = L_b$.

49. The numerical work suggests $L \cong 1$. Justification: Set $f(x) = \sin x - x^2$. Note that $f(0) = 0$ and for x close to 0, $f'(x) = \cos x - 2x > 0$.
Therefore $\sin x - x^2 > 0$ for x close to 0 and $\sin(1/n) - 1/n^2 > 0$ for n large. Thus, for n large,

$$\frac{1}{n^2} < \sin\frac{1}{n} < \frac{1}{n}$$

$|\sin x| \leq |x|$ for all x

$$\left(\frac{1}{n^2}\right)^{1/n} < \left(\sin\frac{1}{n}\right)^{1/n} < \left(\frac{1}{n}\right)^{1/n}$$

$$\left(\frac{1}{n^{1/n}}\right)^2 < \left(\sin\frac{1}{n}\right)^{1/n} < \frac{1}{n^{1/n}}.$$

As $n \to \infty$ both bounds tend to 1 and therefore the middle term also tends to 1.

51. (a)

a_3	a_4	a_5	a_6	a_7	a_8	a_9	a_{10}
2	3	5	8	13	21	34	55

(b)

r_1	r_2	r_3	r_4	r_5	r_6
1	2	1.5	1.6667	1.6000	1.625

(c) $L = \dfrac{1 + \sqrt{5}}{2} \cong 1.618033989$

SECTION 10.5

1. 0   **3.** 1   **5.** $\frac{1}{2}$ **7.** $\ln 2$ **9.** $\frac{1}{4}$ **11.** 2   **13.** $\dfrac{1 + \pi}{1 - \pi}$ **15.** $\frac{1}{2}$ **17.** π **19.** $-\frac{1}{2}$ **21.** -2 **23.** $\frac{1}{3}\sqrt{6}$

25. $-\frac{1}{8}$ **27.** 4   **29.** $\frac{1}{2}$ **31.** $\frac{1}{2}$ **33.** 1   **35.** 1   **37.** $\lim\limits_{x \to 0} (2 + x + \sin x) \neq 0$, $\lim\limits_{x \to 0} (x^3 + x - \cos x) \neq 0$

39. $a = \pm 4, \;\; b = 1$ **41.** $f(0)$ **43.** $\frac{3}{4}$ **45.** (a) $f(x) \to \infty$ as $x \to \pm\infty$ (b) 10   **47.** (b) $\ln 2 \cong 0.6931$

SECTION 10.6

1. ∞ **3.** -1 **5.** ∞ **7.** $\frac{1}{5}$ **9.** 1   **11.** 0   **13.** ∞ **15.** $\frac{1}{3}$ **17.** e **19.** 1   **21.** $\frac{1}{2}$ **23.** 0   **25.** 1

27. e^3 **29.** e **31.** e^2 **33.** 0   **35.** $-\frac{1}{2}$ **37.** 0   **39.** 1   **41.** 1   **43.** 0

45. y-axis vertical asymptote   **47.** x-axis horizontal asymptote   **49.** x-axis horizontal asymptote

51. $\dfrac{b}{a}\sqrt{x^2 - a^2} - \dfrac{b}{a}x = \dfrac{\sqrt{x^2 - a^2} + x}{\sqrt{x^2 - a^2} + x}\left(\dfrac{b}{a}\right)(\sqrt{x^2 - a^2} - x) = \dfrac{-ab}{\sqrt{x^2 - a^2} + x} \to 0$ as $x \to \infty$ **53.** Example: $f(x) = x^2 + \dfrac{(x - 1)(x - 2)}{x^3}$

55. $\lim\limits_{x \to 0^+} \cos x \neq 0$

57. (a) Let S be the set of positive integers for which the statement is true. Since $\lim\limits_{x \to \infty} \dfrac{\ln x}{x} = 0$, $1 \in S$. Assume that $k \in S$. By L'Hospital's rule,

$$\lim_{x \to \infty} \frac{(\ln x)^{k+1}}{x} \;\overset{\ast}{=}\; \lim_{x \to \infty} \frac{(k + 1)(\ln x)^k}{x} = 0 \quad \text{(since } k \in S\text{)}.$$

Thus $k + 1 \in S$, and S is the set of positive integers.

(b) Choose any positive number α. Let $k - 1$ and k be positive integers such that $k - 1 \leq \alpha \leq k$. Then, for $x > e$,

$$\frac{(\ln x)^{k-1}}{x} \leq \frac{(\ln x)^\alpha}{x} \leq \frac{(\ln x)^k}{x}$$

and the result follows by the pinching theorem.

59. (a) L'Hospital's rule applied to the given limit results in $\lim\limits_{x \to 0} \dfrac{2e^{-1/x^2}}{x^3}$. Rewrite the quotient as $\dfrac{1/x}{e^{1/x^2}} \to 0$ as $x \to 0$.   (b) $f'(0) = 0$.

61. $\lim\limits_{x \to 0^+} (1 + x^2)^{1/x} = 1$.   **63.** $\lim\limits_{x \to \infty} g(x) = -5/3$.

SECTION 10.7

1. 1   **3.** $\frac{1}{4}\pi$ **5.** diverges   **7.** 6   **9.** $\frac{1}{2}\pi$ **11.** 2   **13.** diverges   **15.** $-\frac{1}{4}$ **17.** π **19.** diverges   **21.** $\ln 2$

23. 4   **25.** diverges   **27.** diverges   **29.** diverges   **31.** $\frac{1}{2}$ **33.** $2e - 2$ **35.** $\dfrac{\pi}{2} - 1$ **37.** π

39. (a)   (b) 2   (c) $V = \displaystyle\int_0^1 \pi\left(\dfrac{1}{\sqrt{x}}\right)^2 dx = \pi \int_0^1 \dfrac{1}{x}\, dx$,   diverges   **41.** (b) 1   (a)

(c) $\frac{1}{2}\pi$

(d) 2π

(e) $\pi[\sqrt{2} + \ln(1 + \sqrt{2})]$

43. (a) The interval $[0, 1]$ causes no problem. For $x \geq 1$, $e^{-x^2} \leq e^{-x}$ and $\displaystyle\int_1^\infty e^{-x} \, dx$ is finite.

(b) $V_y = \displaystyle\int_0^\infty 2\pi x e^{-x^2} \, dx = \pi$

45. (a)

(b) $\frac{4}{3}$

(c) 2π

(d) $\frac{8}{7}\pi$

47. converges by comparison with $\displaystyle\int_0^\infty \frac{dx}{x^{3/2}}$

49. diverges since for x large the integrand is greater than $\dfrac{1}{x}$ and $\displaystyle\int_1^\infty \frac{1}{x} \, dx$ diverges

51. converges by comparison with $\displaystyle\int_1^\infty \frac{dx}{x^{3/2}}$

53. $L = (a\sqrt{1 + c^2}/c) e^{c\theta_1}$ **55.** $\dfrac{1}{s}$; $\text{dom}(F) = (0, \infty)$ **57.** $\dfrac{s}{s^2 + 4}$; $\text{dom}(F) = (0, \infty)$

59. $f(x) \geq 0$ for all x and $\displaystyle\int_{-\infty}^\infty f(x) \, dx = \int_0^\infty \frac{6x}{(1 + 3x^2)^2} \, dx = 1$ **61.** $\dfrac{1}{k}$

CHAPTER 11
SECTION 11.1

1. 12   **3.** 15   **5.** -10 **7.** $\frac{15}{16}$ **9.** $-\frac{2}{15}$ **11.** $\frac{85}{64}$ **13.** $\displaystyle\sum_{n=1}^{11} 2n - 1$ **15.** $\displaystyle\sum_{n=1}^{25} 2n$ **17.** $\displaystyle\sum_{n=1}^{81} (-1)^{n-1} \sqrt{n}$ **19.** $\displaystyle\sum_{k=1}^{n} m_k \Delta x_k$

21. $\displaystyle\sum_{k=1}^{n} f(x_k^*) \Delta x_k$ **23.** $\displaystyle\sum_{k=0}^{5} (-1)^k a^{5-k} b_k$ **25.** $\displaystyle\sum_{k=0}^{4} a_k x^{4-k}$ **27.** $\displaystyle\sum_{k=0}^{4} (-1)^k (k + 1) x^k$ **29.** $\displaystyle\sum_{k=3}^{10} \frac{1}{2^k}$, $\displaystyle\sum_{i=0}^{7} \frac{1}{2^{i+3}}$

31. $\displaystyle\sum_{k=3}^{10} (-1)^{k+1} \frac{k}{k + 1}$, $\displaystyle\sum_{i=0}^{7} (-1)^i \frac{i + 3}{i + 4}$ **33.** let $k = n + 3$ **35.** let $k = n - 3$

37. (a) $(1 - x) \displaystyle\sum_{k=0}^{n} x^k = \sum_{k=0}^{n} (x^k - x^{k+1})$ (b) converges to $\frac{3}{2}$

$$= (1 - x) + (x - x^2) + (x^2 - x^3) + \cdots + (x^n - x^{n+1}) = 1 - x^{n+1}.$$

39. $|a_n - L| < \epsilon$ for $n \geq k$ iff $|a_{n-p} - L| < \epsilon$ for $n \geq k + p$.

41. True for $n = 1$: $\displaystyle\sum_{k=1}^{1} k = 1 = \frac{1}{2}(1)(2)$. Suppose true for $n = p$. Then

$$\sum_{k=1}^{p+1} k = \sum_{k=1}^{p} k + (p + 1) = \frac{1}{2}(p)(p + 1) + (p + 1) = \frac{1}{2}(p + 1)(p + 2) = \frac{1}{2}(p + 1)[(p + 1) + 1]$$

and thus true for $n = p + 1$.

43. True for $n = 1$: $\displaystyle\sum_{k=1}^{1} k^2 = 1 = \frac{1}{6}(1)(2)(3)$. Suppose true for $n = p$. Then   **45.** 140   **47.** 680

$$\sum_{k=1}^{p+1} k^2 = \sum_{k=1}^{p} k^2 + (p + 1)^2 = \frac{1}{6}(p)(p + 1)(2p + 1) + (p + 1)^2$$

$$= \frac{1}{6}(p + 1)(2p^2 + 7p + 6) = \frac{1}{6}(p + 1)(p + 2)(2p + 3) = \frac{1}{6}(p + 1)[(p + 1) + 1][2(p + 1) + 1]$$

and thus true for $n = p + 1$.

SECTION 11.2

1. $\frac{1}{4}$ **3.** $\frac{1}{2}$ **5.** $\frac{11}{18}$ **7.** $\frac{10}{3}$ **9.** $\frac{67000}{999}$ **11.** 4   **13.** $-\frac{3}{2}$ **15.** $\frac{1}{2}$ **17.** 24   **19.** $\displaystyle\sum_{k=1}^{\infty} \frac{7}{10^k} = \frac{7}{9}$ **21.** $\displaystyle\sum_{k=1}^{\infty} \frac{24}{100^k} = \frac{8}{33}$

23. $\displaystyle\sum_{k=1}^{\infty} \frac{112}{1000^k} = \frac{112}{999}$ **25.** $\dfrac{62}{100} + \dfrac{1}{100} \displaystyle\sum_{k=1}^{\infty} \frac{45}{100^k} = \frac{687}{1100}$

27. Let $x = .\overline{a_1 a_2 \cdots a_n} a_1 a_2 \cdots a_n \cdots$. Then

$$x = \sum_{k=1}^{\infty} \frac{a_1 a_2 \cdots a_n}{(10^n)^k} = a_1 a_2 \cdots a_n \sum_{k=1}^{\infty} \left(\frac{1}{10^n}\right)^k = a_1 a_2 \cdots a_n \left[\frac{1}{1 - \dfrac{1}{10^n}} - 1\right] = \frac{a_1 a_2 \cdots a_n}{10^n - 1}.$$

29. $\dfrac{1}{1+x} = \dfrac{1}{1-(-x)} = \sum\limits_{k=0}^{\infty} (-x)^k = \sum\limits_{k=0}^{\infty} (-1)^k x^k$ **31.** $\sum\limits_{k=0}^{\infty} x^{k+1}$ **33.** $\sum\limits_{k=0}^{\infty} (-1)^k x^{2k+1}$ **35.** $\sum\limits_{k=0}^{\infty} (-1)^k (2x)^{2k}$

37. $\sum\limits_{n=0}^{\infty} \left(\dfrac{3}{2}\right)^n$;   geometric series with $r = \dfrac{3}{2} > 1$ **39.** $\lim\limits_{k\to\infty} \left(\dfrac{k}{k+1}\right)^k = \dfrac{1}{e} \neq 0$ **41.** $4 + \dfrac{1}{3} \sum\limits_{k=0}^{\infty} \left(\dfrac{1}{12}\right)^k = 4 + \dfrac{4}{11}$ o'clock   **43.** 12 ft

45. $\sum\limits_{k=1}^{\infty} n_k \left(1 + \dfrac{r}{100}\right)^{-k}$ **47.** $9   **49.** 32   **51.** $\lim\limits_{n\to\infty} s_n = L = \sum\limits_{k=0}^{\infty} a_k.$ Thus $\lim\limits_{n\to\infty} R_n = \lim\limits_{n\to\infty} \left(s_n - \sum\limits_{k=0}^{\infty} a_k\right) = \lim\limits_{n\to\infty} s_n - L = 0.$

53. $\dfrac{1 + (-1)^n}{2}, \qquad n = 0, 1, 2, \ldots$ **55.** $s_n = \sum\limits_{k=1}^{n} \ln\left(\dfrac{k+1}{k}\right) = \sum\limits_{k=1}^{n} [\ln(k+1) - \ln k] = \ln(n+1) \to \infty$

57. (a) $s_n = \sum\limits_{k=1}^{n} (d_k - d_{k+1}) = d_1 - d_{n+1} \to d_1$

     (b) (i) $\sum\limits_{k=1}^{\infty} \dfrac{\sqrt{k+1} - \sqrt{k}}{\sqrt{k(k+1)}} = \sum\limits_{k=1}^{\infty} \left(\dfrac{1}{\sqrt{k}} - \dfrac{1}{\sqrt{k+1}}\right) = 1$ (ii) $\sum\limits_{k=1}^{\infty} \dfrac{2k+1}{2k^2(k+1)^2} = \sum\limits_{k=1}^{\infty} \dfrac{1}{2}\left(\dfrac{1}{k^2} - \dfrac{1}{(k+1)^2}\right) = \dfrac{1}{2}$

59. $N = 6$ **61.** $N = 9999$ **63.** $N = \left[\!\left[\dfrac{\ln(\epsilon[1 - x])}{\ln|x|} \right]\!\right]$, where $[\![\quad]\!]$ denotes the greatest integer function.

SECTION 11.3

1. converges; comparison $\Sigma\, 1/k^2$ **3.** converges; comparison $\Sigma\, 1/k^2$ **5.** diverges; comparison $\Sigma\, 1/(k+1)$ **7.** diverges; limit comparison $\Sigma\, 1/k$

9. converges; integral test   **11.** diverges; p-series with $p = \frac{2}{3} \leq 1$ **13.** diverges; $a_k \nrightarrow 0$ **15.** diverges; comparison $\Sigma\, 1/k$

17. diverges; $a_k \nrightarrow 0$ **19.** converges; limit comparison $\Sigma\, 1/k^2$ **21.** diverges; integral test   **23.** converges; limit comparison $\Sigma\, 1/k^2$

25. diverges; limit comparison $\Sigma\, 1/k$ **27.** converges; limit comparison $\Sigma\, 1/k^{3/2}$ **29.** converges; integral test

31. converges; comparison $\Sigma\, 3/k^2$ **33.** converges; comparison $\Sigma\, 2/k^2$ **35.** $p > 1$

37. (a) The improper integral $\displaystyle\int_0^\infty e^{-\alpha x}\,dx = \dfrac{1}{\alpha}$ converges.   (b) The improper integral $\displaystyle\int_0^\infty x e^{-\alpha x}\,dx = \dfrac{1}{\alpha^2}$ converges.

     (c) The improper integral $\displaystyle\int_0^\infty x^n e^{-\alpha x}\,dx = \dfrac{1}{\alpha^{n+1}}$ converges.

39. (a) 1.1777   (b) $0.02 < R_4 < 0.0313$ (c) $1.1977 < \sum\limits_{k=1}^{\infty} \dfrac{1}{k^3} < 1.209$ **41.** (a) $1/101 < R_{100} < 1/100$ (b) 10,001   **43.** (a) 15   (b) 1.082

45. (a) If $a_k/b_k \to 0$, then $a_k/b_k < 1$ for all $k \geq K$ for some K. But then $a_k < b_k$ for all $k \geq K$ and, since Σb_k converges, Σa_k converges.
     [The basic comparison test, Theorem 11.3.5.]
     (b) Similar to (a) except that this time we appeal to part (ii) of Theorem 11.3.5.

     (c) $\Sigma a_k = \Sigma \dfrac{1}{k^2}$ converges, $\Sigma b_k = \Sigma \dfrac{1}{k^{3/2}}$ converges, $\dfrac{1/k^2}{1/k^{3/2}} = \dfrac{1}{\sqrt{k}} \to 0$

     $\Sigma a_k = \Sigma \dfrac{1}{k^2}$ converges, $\Sigma b_k = \Sigma \dfrac{1}{\sqrt{k}}$ diverges, $\dfrac{1/k^2}{1/\sqrt{k}} = \dfrac{1}{k^{3/2}} \to 0$

     (d) $\Sigma b_k = \Sigma \dfrac{1}{\sqrt{k}}$ diverges, $\Sigma a_k = \Sigma \dfrac{1}{k^2}$ converges, $\dfrac{1/k^2}{1/\sqrt{k}} = \dfrac{1}{k^{3/2}} \to 0$

     $\Sigma b_k = \Sigma \dfrac{1}{\sqrt{k}}$ diverges, $\Sigma a_k = \Sigma \dfrac{1}{k}$ diverges, $\dfrac{1/k}{1/\sqrt{k}} = \dfrac{1}{\sqrt{k}} \to 0$

47. (a) Since Σa_k converges, $\lim\limits_{k\to\infty} a_k = 0$. Therefore there exists a positive integer N such that $0 < a_k < 1$ for $k \geq N$.
     Thus, for $k \geq N$, $a_k^2 < a_k$ and so Σa_k^2 converges by the comparison test.
     (b) Σa_k may either converge or diverge.
     $\Sigma\, 1/k^4$ converges and $\Sigma\, 1/k^2$ converges; $\Sigma\, 1/k^2$ converges and $\Sigma\, 1/k$ diverges.

49. $0 < L - \sum\limits_{k=1}^{n} f(k) = L - s_n = \sum\limits_{k=n+1}^{\infty} f(k) < \displaystyle\int_n^\infty f(x)\,dx$ **51.** $N = 3$.

53. (a) Set $f(x) = x^{1/4} - \ln x$. Then $f'(x) = \dfrac{1}{4}x^{-3/4} - \dfrac{1}{x} = \dfrac{1}{4x}(x^{1/4} - 4)$. Since $f(e^{12}) = e^3 - 12 > 0$ and $f'(x) > 0$ for $x > e^{12}$, we know that $n^{1/4} >$

     $\ln n$ and therefore $\dfrac{1}{n^{5/4}} > \dfrac{\ln n}{n^{3/2}}$ for sufficiently large n. Since $\Sigma \dfrac{1}{n^{5/4}}$ is a convergent p-series, $\Sigma \dfrac{\ln n}{n^{3/2}}$ converges by the basic comparison test.

     (b) By L'Hospital's rule $\lim\limits_{x\to\infty} \left[\left(\dfrac{\ln x}{x^{3/2}}\right) \Big/ \left(\dfrac{1}{x^{5/4}}\right)\right] = 0$

SECTION 11.4

1. converges; ratio test **3.** converges; root test **5.** diverges; $a_k \nrightarrow 0$ **7.** diverges; limit comparison $\Sigma\, 1/k$ **9.** converges; root test

11. diverges; limit comparison $\Sigma\, 1/\sqrt{k}$ **13.** diverges; ratio test **15.** converges; comparison $\Sigma\, 1/k^{3/2}$ **17.** converges; comparison $\Sigma\, 1/k^2$

19. diverges; integral test **21.** diverges; $a_k \to e^{-100} \neq 0$ **23.** diverges; limit comparison $\Sigma\, 1/k$ **25.** converges; ratio test

27. converges; comparison $\Sigma\, 1/k^{3/2}$ **29.** converges; ratio test **31.** converges; ratio test: $a_{k+1}/a_k \to \frac{4}{27}$ **33.** converges; ratio test

35. converges; root test **37.** converges; root test **39.** converges; ratio test **41.** $\frac{10}{81}$

43. The series $\Sigma\, \dfrac{k!}{k^k}$ converges. Therefore $\lim\limits_{k\to\infty} \dfrac{k!}{k^k} = 0$ by Theorem 11.2.6. **45.** $p \geq 2$ **47.** $|x| < 1$ **49.** converges for all x

51. Set $b_k = a_k r^k$. If $(a_k)^{1/k} \to \rho$ and $\rho < 1/r$, then

$$(b_k)^{1/k} = (a_k r^k)^{1/k} = (a_k)^{1/k} r \to \rho r < 1$$

and thus, by the root test, $\Sigma b_k = \Sigma a_k r^k$ converges.

SECTION 11.5

1. diverges; $a_k \nrightarrow 0$ **3.** diverges; $a_k \nrightarrow 0$ **5.** (a) does not converge absolutely; integral test (b) converges conditionally; Theorem 11.5.4

7. diverges; limit comparison $\Sigma\, 1/k$ **9.** (a) does not converge absolutely; limit comparison $\Sigma\, 1/k$ (b) converges conditionally; Theorem 11.5.4

11. diverges; $a_k \nrightarrow 0$ **13.** (a) does not converge absolutely; comparison $2\,\Sigma\, 1/\sqrt{k+1}$ (b) converges conditionally; Theorem 11.5.4

15. converges absolutely (terms already positive); $\Sigma \sin\left(\dfrac{\pi}{4k^2}\right) \leq \Sigma \dfrac{\pi}{4k^2} = \dfrac{\pi}{4} \Sigma \dfrac{1}{k^2}$ $(|\sin x| \leq |x|)$ **17.** converges absolutely; ratio test

19. (a) does not converge absolutely; limit comparison $\Sigma\, 1/k$ (b) converges conditionally; Theorem 11.5.4 **21.** diverges; $a_k \nrightarrow 0$

23. diverges; $a_k \nrightarrow 0$ **25.** converges absolutely; ratio test **27.** diverges; $a_k = \dfrac{1}{k}$ for all k **29.** converges absolutely; comparison $\Sigma\, 1/k^2$

31. diverges; $a_k \nrightarrow 0$ **33.** 0.1104 **35.** 0.001 **37.** $\frac{10}{11}$ **39.** $N = 39{,}998$ **41.** (a) 4 (b) 6

43. No. For instance, set $a_{2k} = 2/k$ and $a_{2k+1} = 1/k$.

45. (a) Since $\Sigma |a_k|$ converges, $\Sigma |a_k|^2 = \Sigma a_k^2$ converges (Exercise 47, Section 11.3). (b) $\Sigma\, 1/k^2$ is convergent, $\Sigma \dfrac{(-1)^k}{k}$ is not absolutely convergent.

47. See the proof of Theorem 11.8.2. **49.** (a) $\displaystyle\sum_{k=1}^{\infty} \dfrac{(-1)^{k-1}(a+b) + (a-b)}{2k}$ (b) if $a = b = 0$, absolutely convergent (vacuously)

if $a = b \neq 0$, conditionally convergent

if $a \neq b$, divergent

SECTION 11.6

1. $-1 + x + \frac{1}{2}x^2 - \frac{1}{24}x^4$ **3.** $-\frac{1}{2}x^2 - \frac{1}{12}x^4$ **5.** $1 - x + x^2 - x^3 + x^4 - x^5$ **7.** $x + \frac{1}{3}x^3 + \frac{2}{15}x^5$

9. $P_0(x) = 1$, $P_1(x) = 1 - x$, $P_2(x) = 1 - x + 3x^2$, $P_3(x) = 1 - x + 3x^2 + 5x^3$

11. $\displaystyle\sum_{k=0}^{n} (-1)^k \dfrac{x^k}{k!}$ **13.** $\displaystyle\sum_{k=0}^{m} \dfrac{x^{2k}}{(2k)!}$ where $m = \dfrac{n}{2}$ and n is even **15.** $\displaystyle\sum_{k=0}^{n} \dfrac{r^k}{k!} x^k$ **17.** $79/48$ $(79/48 \cong 1.646)$

19. $5/6$ $(5/6 \cong 0.833)$ **21.** $13/24$ $(13/24 \cong 0.542)$ **23.** 0.17 **25.** $\dfrac{4e^{2c}}{15} x^5$, $|c| < |x|$ **27.** $\dfrac{-4\sin 2c}{15} x^5$, $|c| < |x|$

29. $\dfrac{3\sec^4 c - 2\sec^2 c}{3} x^3$, $|c| < |x|$ **31.** $\dfrac{3c^2 - 1}{3(1 + c^2)^3} x^3$, $|c| < |x|$ **33.** $\dfrac{(-1)^{n+1} e^{-c}}{(n+1)!} x^{n+1}$, $|c| < |x|$ **35.** $\dfrac{1}{(1-c)^{n+2}} x^{n+1}$, $|c| < |x|$

37. (a) 4 (b) 2 (c) 999 **39.** (a) 1.649 (b) 0.368 **41.** For $0 \leq k \leq n$, $P^{(k)}(0) = k! a_k$; for $k > n$, $P^{(k)}(0) = 0$. Thus $P(x) = \displaystyle\sum_{k=0}^{\infty} P^{(k)}(0) \dfrac{x^k}{k!}$.

43. $\dfrac{d^{2k}(\sinh x)}{dx^{2k}}\bigg|_{x=0} = \sinh(0) = 0;$ $\dfrac{d^{2k+1}(\sinh x)}{dx^{2k+1}}\bigg|_{x=0} = \cosh(0) = 1$

Therefore $\sinh x = x + \dfrac{x^3}{3!} + \dfrac{x^5}{5!} + \cdots = \displaystyle\sum_{k=0}^{\infty} \dfrac{1}{(2k+1)!} x^{2k+1}$

45. $\displaystyle\sum_{k=0}^{\infty} \dfrac{a^k}{k!} x^k$, $(-\infty, \infty)$ **47.** $\displaystyle\sum_{k=0}^{\infty} \dfrac{(-1)^k a^{2k}}{(2k)!} x^{2k}$, $(-\infty, \infty)$ **49.** $\ln a + \displaystyle\sum_{k=1}^{\infty} \dfrac{(-1)^{k-1}}{ka^k} x^k$, $(-a, a]$

51. $\ln 2 = \ln\left(\dfrac{1 + \frac{1}{3}}{1 - \frac{1}{3}}\right) \cong 2\left[\dfrac{1}{3} + \dfrac{1}{3}\left(\dfrac{1}{3}\right)^3 + \dfrac{1}{5}\left(\dfrac{1}{3}\right)^5\right] = \dfrac{842}{1215}$ $\left(\dfrac{842}{1215} \cong 0.693\right)$ **53.** routine; use $u = (x-t)^k$ and $dv = f^{(k+1)}(t)\, dt$

55. (b) $f(x) = \dfrac{x^{-n}}{e^{1/x^2}}$ and $\lim\limits_{x\to 0} f(x)$ has the form ∞/∞. Successive applications of L'Hospital's rule will finally produce a quotient of the form

$\dfrac{cx^k}{e^{1/x^2}}$, where k is a nonnegative integer and c is a constant. It follows that $\lim\limits_{x\to 0} f(x) = 0$.

(c) $f'(0) = \lim\limits_{x \to 0} \dfrac{e^{-1/x^2} - 0}{x} = 0$ by part (b). Assume that $f^{(k)}(0) = 0$. Then

$$f^{(k+1)}(0) = \lim_{x \to 0} \frac{f^{(k)}(x) - 0}{x} = \lim_{x \to 0} \frac{f^{(k)}(x)}{x}.$$

Now, $\dfrac{f^{(k)}(x)}{x}$ is a sum of terms of the form $\dfrac{ce^{-1/x^2}}{x^n}$, where n is a positive integer and c is a constant. By part (b), $f^{(k+1)}(0) = 0$.

Therefore $f^{(n)}(0) = 0$ for all n.

(d) 0 (e) $x = 0$

57.

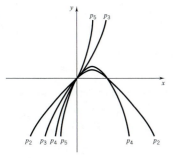

SECTION 11.7

1. $P_3(x) = 2 + \frac{1}{4}(x - 4) - \frac{1}{64}(x - 4)^2 + \frac{1}{512}(x - 4)^3$

$R_4(x) = \dfrac{-5}{128c^{7/2}}x^4, \quad |c| < |x|$

3. $P_4(x) = \dfrac{\sqrt{2}}{2} + \dfrac{\sqrt{2}}{2}\left(x - \dfrac{\pi}{4}\right) - \dfrac{\sqrt{2}}{4}\left(x - \dfrac{\pi}{4}\right)^2 - \dfrac{\sqrt{2}}{12}\left(x - \dfrac{\pi}{4}\right)^3 + \dfrac{\sqrt{2}}{48}\left(x - \dfrac{\pi}{4}\right)^4$

$R_5(x) = \dfrac{\cos c}{120}\left(x - \dfrac{\pi}{4}\right)^5, \quad \left|c - \dfrac{\pi}{4}\right| < \left|x - \dfrac{\pi}{4}\right|$

5. $P_3(x) = \dfrac{\pi}{4} + \dfrac{1}{2}(x - 1) - \dfrac{1}{4}(x - 1)^2 + \dfrac{1}{12}(x - 1)^3$ $R_4(x) = \dfrac{c(1 - c^2)}{(1 + c^2)^3}(x - 1)^4, \quad |c - 1| < |x - 1|$

7. $6 + 9(x - 1) + 7(x - 1)^2 + 3(x - 1)^3, \quad (-\infty, \infty)$ **9.** $-3 + 5(x + 1) - 19(x + 1)^2 + 20(x + 1)^3 - 10(x + 1)^4 + 2(x + 1)^5, \quad (-\infty, \infty)$

11. $\displaystyle\sum_{k=0}^{\infty} (-1)^k \left(\frac{1}{2}\right)^{k+1}(x - 1)^k, \quad (-1, 3)$ **13.** $\dfrac{1}{5}\displaystyle\sum_{k=0}^{\infty} \left(\frac{2}{5}\right)^k(x + 2)^k, \quad \left(-\dfrac{9}{2}, \dfrac{1}{2}\right)$ **15.** $\displaystyle\sum_{k=0}^{\infty} \dfrac{(-1)^{k+1}}{(2k + 1)!}(x - \pi)^{2k+1}, \quad (-\infty, \infty)$

17. $\displaystyle\sum_{k=0}^{\infty} \dfrac{(-1)^{k+1}}{(2k)!}(x - \pi)^{2k}, \quad (-\infty, \infty)$ **19.** $\displaystyle\sum_{k=0}^{\infty} \dfrac{(-1)^k}{(2k)!}\left(\frac{\pi}{2}\right)^{2k}(x - 1)^{2k}, \quad (-\infty, \infty)$ **21.** $\ln 3 + \displaystyle\sum_{k=1}^{\infty} \dfrac{(-1)^{k+1}}{k}\left(\frac{2}{3}\right)^k(x - 1)^k, \quad \left(-\dfrac{1}{2}, \dfrac{5}{2}\right]$

23. $2 \ln 2 + (1 + \ln 2)(x - 2) + \displaystyle\sum_{k=2}^{\infty} \dfrac{(-1)^k}{k(k - 1)2^{k-1}}(x - 2)^k$ **25.** $\displaystyle\sum_{k=0}^{\infty} \dfrac{(-1)^k}{(2k + 1)!}x^{2k+2}$ **27.** $\displaystyle\sum_{k=0}^{\infty} (k + 2)(k + 1)\dfrac{2^{k-1}}{5^{k+3}}(x + 2)^k$

29. $1 + \displaystyle\sum_{k=1}^{\infty} \dfrac{(-1)^k 2^{2k-1}}{(2k)!}(x - \pi)^{2k}$ **31.** $\displaystyle\sum_{k=0}^{\infty} \dfrac{n!}{(n - k)! \, k!}(x - 1)^k$

33. (a) $\dfrac{e^x}{e^a} = e^{x-a} = \displaystyle\sum_{k=0}^{\infty} \dfrac{(x - a)^k}{k!}, \quad e^x = e^a \displaystyle\sum_{k=0}^{\infty} \dfrac{(x - a)^k}{k!}$ (b) $e^{a+(x-a)} = e^x = e^a \displaystyle\sum_{k=0}^{\infty} \dfrac{(x - a)^k}{k!}, \quad e^{x_1+x_2} = e^{x_1} \displaystyle\sum_{k=0}^{\infty} \dfrac{x_2^k}{k!} = e^{x_1}e^{x_2}$

(c) $e^{-a} \displaystyle\sum_{k=0}^{\infty} (-1)^k \dfrac{(x - a)^k}{k!}$

35. (a) $P_2(x) = \dfrac{1}{2} + \dfrac{\sqrt{3}}{2}\left(x - \dfrac{\pi}{6}\right) - \dfrac{1}{4}\left(x - \dfrac{\pi}{6}\right)^2 - \dfrac{\sqrt{3}}{12}\left(x - \dfrac{\pi}{6}\right)^3$ (b) 0.5736

37. $P_2(x) = 6 + \frac{1}{12}(x - 36) - \frac{1}{1728}(x - 36)^2; \quad \sqrt{38} \cong 6.164$

SECTION 11.8

1. $(-1, 1)$ **3.** $(-\infty, \infty)$ **5.** $\{0\}$ **7.** $[-2, 2)$ **9.** $\{0\}$ **11.** $[-\frac{1}{2}, \frac{1}{2})$ **13.** $(-1, 1)$ **15.** $(-10, 10)$ **17.** $(-\infty, \infty)$

19. $(-\infty, \infty)$ **21.** $(-3/2, 3/2)$ **23.** converges only at $x = 1$ **25.** $(-4, 0)$ **27.** $(-\infty, \infty)$ **29.** $(-1, 1)$ **31.** $(0, 4)$

33. $(-\frac{5}{2}, \frac{1}{2})$ **35.** $(-2, 2)$ **37.** $\left[-\dfrac{1}{\sqrt{3}}, \dfrac{1}{\sqrt{3}}\right]$

39. Examine the convergence of $\Sigma \, |a_k x^k|$; for (a) use the root test and for (b) use the ratio test. **41.** $\Sigma \, |a_k(-r)^k| = \Sigma \, |a_k r^k|$

SECTION 11.9

1. $1 + 2x + 3x^2 + \cdots + nx^{n-1} + \cdots$ **3.** $1 + kx + \dfrac{(k + 1)k}{2!}x^2 + \cdots + \dfrac{(n + k - 1)!}{n!(k - 1)!}x^n + \cdots$

5. $\ln(1 - x^2) = -x^2 - \dfrac{1}{2}x^4 - \dfrac{1}{3}x^6 - \cdots - \dfrac{1}{n + 1}x^{2n+2} - \cdots$ **7.** $1 + x^2 + \frac{2}{3}x^4 + \frac{17}{45}x^6 + \cdots$ **9.** -72 **11.** $\displaystyle\sum_{k=0}^{\infty} \dfrac{(-1)^k}{(2k + 1)!}x^{4k+2}$

13. $\displaystyle\sum_{k=0}^{\infty} \frac{3^k}{k!} x^{3k}$ **15.** $\displaystyle 2\sum_{k=0}^{\infty} x^{2k+1}$ **17.** $\displaystyle\sum_{k=0}^{\infty} \frac{(k!+1)}{k!} x^k$ **19.** $\displaystyle\sum_{k=1}^{\infty} \frac{(-1)^{k+1}}{k} x^{3k+1}$ **21.** $\displaystyle\sum_{k=0}^{\infty} \frac{(-1)^k}{k!} x^{3k+3}$ **23.** $\frac{1}{2}$ **25.** $-\frac{1}{2}$

27. $\displaystyle\sum_{k=1}^{\infty} \frac{(-1)^{k-1}}{k^2} x^k, \quad -1 \le x \le 1$ **29.** $\displaystyle\sum_{k=1}^{\infty} \frac{(-1)^{k-1}}{(2k-1)^2} x^{2k-1}$ **31.** $0.804 \le I \le 0.808$ **33.** $0.600 \le I \le 0.603$ **35.** $0.294 \le I \le 0.304$

37. 0.9461 **39.** 0.4485 **41.** e^{x^3} **43.** $3x^2 e^{x^3}$ **45.** (a) $\displaystyle\sum_{n=0}^{\infty} \frac{1}{n!} x^{n+1}$ (b) $\displaystyle\int_0^1 xe^x \, dx = 1 = \int_0^1 \left(\sum_{n=0}^{\infty} \frac{1}{n!} x^{n+1}\right) dx = \sum_{n=0}^{\infty} \frac{1}{n!(n+2)}$

47. Let $f(x)$ be the sum of these series; a_k and b_k are both $f^{(k)}(0)/k!$.

49. (a) If f is even, then $f^{(2k-1)}$ is odd for $k = 1, 2, \ldots$. This implies that $f^{(2k-1)}(0) = (0)$, and so $a_{2k-1} = \dfrac{f^{(2k-1)}(0)}{(2k-1)!} = 0$ for all k.

   (b) If f is odd, then $f^{(2k)}$ is odd for $k = 1, 2, \ldots$, which implies $a_{2k} = 0$ for all k.

51. $0.0352 \le I \le 0.0359; \quad I = \frac{3}{6} - \frac{3}{8} \ln 1.5 \cong 0.0354505$ **53.** $0.2640 \le I \le 0.2643; \quad I = 1 - 2/e \cong 0.2642411$

SECTION 11.10
1. $1 + \frac{1}{2}x - \frac{1}{8}x^2 + \frac{1}{16}x^3 - \frac{5}{128}x^4$ **3.** $1 + \frac{1}{2}x^2 - \frac{1}{8}x^4$ **5.** $1 - \frac{1}{2}x + \frac{3}{8}x^2 - \frac{5}{16}x^3 + \frac{35}{128}x^4$ **7.** $1 - \frac{1}{4}x - \frac{3}{32}x^2 - \frac{7}{128}x^3 - \frac{77}{2048}x^4$

9. $8 + 3x + \frac{3}{16}x^2 - \frac{1}{128}x^3 + \frac{3}{4096}x^4$ **11.** (a) $\displaystyle\sum_{k=0}^{\infty} (-1)^k \binom{-1/2}{k} x^{2k}$ (b) $\displaystyle\sum_{k=0}^{\infty} (-1)^k \binom{-1/2}{k} \frac{1}{2k+1} x^{2k+1}, \quad R = 1$ **13.** 9.8995

15. 2.0799 **17.** 0.4925 **19.** 0.3349 **21.** 0.4815

CHAPTER 12
SECTION 12.1

1. length $\overline{AB}$: $2\sqrt{5}$ mid point: $(1,0,-2)$ **3.** length $\overline{AB}$: $5\sqrt{2}$ mid point: $(2, -\frac{1}{2}, \frac{5}{2})$ **5.** length $\overline{AB}$: 3 mid point: $(\frac{1}{2}, 5, 6)$

7. $z = -2$ **9.** $y = 1$ **11.** $x = 3$ **13.** $x^2 + (y-2)^2 + (z+1)^2 = 9$ **15.** $(x-2)^2 + (y-4)^2 + (z+4)^2 = 36$

17. $(x-3)^2 + (y-2)^2 + (z-2)^2 = 13$ **19.** $(x-2)^2 + (y-3)^2 + (z+4)^2 = 25$ **21.** center $(-2, 4, 1)$, radius 4

23. center $(-2, 1, 0)$, radius $\frac{3}{2}\sqrt{2}$ **25.** $(2, 3, -5)$ **27.** $(-2, 3, 5)$ **29.** $(-2, 3, -5)$ **31.** $(-2, -3, -5)$ **33.** $(2, -5, 5)$

35. $(-2, 1, -3)$ **37.** $(x-3)^2 + (y-3)^2 + (z-3)^2 = 9, \quad (x-7)^2 + (y-7)^2 + (z-7)^2 = 49$

39. not a sphere; the equation is equivalent to $(x-2)^2 + (y+2)^2 + (z+3)^2 = -3$

41. $d(P, R) = \sqrt{14}, \quad d(Q, R) = \sqrt{45}, \quad d(P, Q) = \sqrt{59}, \quad [d(P, R)]^2 + [d(Q, R)]^2 = [d(P, Q)]^2$

43. (a) $x = a_1 + t(b_1 - a_1), \quad y = a_2 + t(b_2 - a_2), \quad z = a_3 + t(b_3 - a_3)$ (b) $t = \frac{1}{2}$

SECTION 12.2
1. $(3, 4, -2), \quad \sqrt{29}$ **3.** $(-2, 6), \quad 2\sqrt{10}$ **5.** $(-4, 2, 2), \quad 2\sqrt{6}$ **7.** $(-1, -4, 7)$ **9.** $(5, 2, -8)$ **11.** $3\mathbf{i} - 4\mathbf{j} + 6\mathbf{k}$

13. $-3\mathbf{i} - \mathbf{j} + 8\mathbf{k}$ **15.** 5 **17.** 3 **19.** $\sqrt{6}$ **21.** (a) $\mathbf{a}, \mathbf{c}, \mathbf{d}$ (b) $\mathbf{a}, \mathbf{c}$ (c) $\mathbf{a}$ and $\mathbf{c}$ both have direction opposite to $\mathbf{d}$

23. $(\frac{3}{5}, -\frac{4}{5})$ **25.** $(-\frac{4}{5}, 0, \frac{3}{5})$ **27.** $\frac{1}{3}\mathbf{i} - \frac{2}{3}\mathbf{j} + \frac{2}{3}\mathbf{k}$ **29.** $\dfrac{1}{\sqrt{14}}\mathbf{i} - \dfrac{3}{\sqrt{14}}\mathbf{j} - \dfrac{2}{\sqrt{14}}\mathbf{k}$

31. (i) $\mathbf{a} + \mathbf{b}$ (ii) $-(\mathbf{a} + \mathbf{b})$ (iii) $\mathbf{a} - \mathbf{b}$ (iv) $\mathbf{b} - \mathbf{a}$ **33.** (a) $\mathbf{i} - 3\mathbf{j} + 10\mathbf{k}$ (b) $A = -2, B = \frac{3}{2}, C = -\frac{7}{2}$

35. $\alpha = \pm 3$ **37.** $\alpha = \pm\frac{1}{3}\sqrt{6}$ **39.** $\pm\frac{2}{13}\sqrt{13}(3\mathbf{j} + 2\mathbf{k})$ **41.** $\sqrt{3}\mathbf{i} + \mathbf{j}$ **43.** $\dfrac{\sqrt{2}}{2}\mathbf{i} + \dfrac{\sqrt{2}}{2}\mathbf{j}$

45. $\dfrac{6\sqrt{5}}{5}\mathbf{i} + \dfrac{3\sqrt{5}}{5}\mathbf{j}$ or $-\dfrac{6\sqrt{5}}{5}\mathbf{i} - \dfrac{3\sqrt{5}}{5}\mathbf{j}$ **47.** $\|\mathbf{a}\|^2 + \|\mathbf{b}\|^2 = \|\mathbf{b} - \mathbf{a}\|^2$

49. (a) the parallelogram is a rectangle

   (b) simplify $\sqrt{(a_1 - b_1)^2 + (a_2 - b_2)^2 + (a_3 - b_3)^2} = \sqrt{(a_1 + b_1)^2 + (a_2 + b_2)^2 + (a_3 + b_3)^2}$

51. (a)

$$m = p + \tfrac{1}{2}(q - p)$$

(b) Let $P = (x_1, y_1, z_1)$, $Q = (x_2, y_2, z_2)$, and $M = (x_m, y_m, z_m)$. Then

$$(x_m, y_m, z_m) = (x_1, y_1, z_1) + \frac{1}{2}(x_2 - x_1, y_2 - y_1, z_2 - z_1)$$

$$= \left(\frac{x_1 + x_2}{2}, \frac{y_1 + y_2}{2}, \frac{z_1 + z_2}{2} \right).$$

53. $\mathbf{F}_1 \cong -153.21\mathbf{i} - 128.56\mathbf{j}$; $\|\mathbf{F}_1\| = 200.02$; $\mathbf{F}_2 \cong 153.21\mathbf{i} + 71.44\mathbf{j}$; $\|\mathbf{F}_2\| = 169.05$

55. $\mathbf{R} = (300 + 25\sqrt{2})\mathbf{i} + (300\sqrt{3} - 25\sqrt{2})\mathbf{j} \cong 335.36\mathbf{i} + 484.26\mathbf{j}$. True course: N34.70°E; ground speed 589.05 miles/hr.

57. (a) $\|\mathbf{r} - \mathbf{a}\| = 3$ where $\mathbf{a} = a_1\mathbf{i} + a_2\mathbf{j} + a_3\mathbf{k}$ (b) $\|\mathbf{r}\| \le 2$ (c) $\|\mathbf{r} - \mathbf{a}\| \le 1$ where $\mathbf{a} = a_1\mathbf{i} + a_2\mathbf{j} + a_3\mathbf{k}$
(d) $\|\mathbf{r} - \mathbf{a}\| = \|\mathbf{r} - \mathbf{b}\|$ (e) $\|\mathbf{r} - \mathbf{a}\| + \|\mathbf{r} - \mathbf{b}\| = k$

SECTION 12.3

1. -1 **3.** 0 **5.** -1 **7.** $\mathbf{a} \cdot \mathbf{b}$ **9.** $\mathbf{a} \cdot (\mathbf{b} + \mathbf{c})$

11. (a) $\mathbf{a} \cdot \mathbf{b} = 5$, $\mathbf{a} \cdot \mathbf{c} = 8$, $\mathbf{b} \cdot \mathbf{c} = 18$ (b) $\cos \sphericalangle (\mathbf{a}, \mathbf{b}) = \tfrac{1}{14}\sqrt{70}$, $\cos \sphericalangle (\mathbf{a}, \mathbf{c}) = \tfrac{8}{25}\sqrt{5}$, $\cos \sphericalangle (\mathbf{b}, \mathbf{c}) = \tfrac{9}{35}\sqrt{14}$
(c) $\text{comp}_\mathbf{b}\, \mathbf{a} = \tfrac{5}{14}\sqrt{14}$, $\text{comp}_\mathbf{c}\, \mathbf{a} = \tfrac{8}{5}$ (d) $\text{proj}_\mathbf{b}\, \mathbf{a} = \tfrac{5}{14}(3\mathbf{i} - \mathbf{j} + 2\mathbf{k})$, $\text{proj}_\mathbf{c}\, \mathbf{a} = \tfrac{8}{25}(4\mathbf{i} + 3\mathbf{k})$

13. $\tfrac{1}{2}\mathbf{i} + \tfrac{1}{2}\sqrt{2}\mathbf{j} - \tfrac{1}{2}\mathbf{k}$ **15.** $\tfrac{1}{3}\pi$ **17.** $\tfrac{1}{3}\pi, \tfrac{2}{3}\pi, \tfrac{1}{4}\pi$ **19.** 2.2 radians, or 126.3° **21.** 2.5 radians, or 145.3°

23. $\cos \alpha = \tfrac{1}{3}$, $\cos \beta = \tfrac{2}{3}$, $\cos \gamma = \tfrac{2}{3}$; $\alpha \cong 70.5°$, $\beta \cong 48.2°$, $\gamma \cong 48.2°$

25. $\cos \alpha = \tfrac{3}{13}$, $\cos \beta = \tfrac{12}{13}$, $\cos \gamma = \tfrac{4}{13}$; $\alpha \cong 76.7°$, $\beta \cong 22.6°$, $\gamma \cong 72.1°$

27. (a) $\text{proj}_\mathbf{b}\, \alpha\mathbf{a} = (\alpha\mathbf{a} \cdot \mathbf{u}_\mathbf{b})\mathbf{u}_\mathbf{b} = \alpha(\mathbf{a} \cdot \mathbf{u}_\mathbf{b})\mathbf{u}_\mathbf{b} = \alpha\text{proj}_\mathbf{b}\, \mathbf{a}$
(b) $\text{proj}_\mathbf{b}(\mathbf{a} + \mathbf{c}) = [(\mathbf{a} + \mathbf{c}) \cdot \mathbf{u}_\mathbf{b}]\mathbf{u}_\mathbf{b}$
$= (\mathbf{a} \cdot \mathbf{u}_\mathbf{b} + \mathbf{c} \cdot \mathbf{u}_\mathbf{b})\mathbf{u}_\mathbf{b}$
$= (\mathbf{a} \cdot \mathbf{u}_\mathbf{b})\mathbf{u}_\mathbf{b} + (\mathbf{c} \cdot \mathbf{u}_\mathbf{b})\mathbf{u}_\mathbf{b} = \text{proj}_\mathbf{b}\, \mathbf{a} + \text{proj}_\mathbf{b}\, \mathbf{c}$

29. (a) for $\mathbf{a} \ne 0$ the following statements are equivalent:

$$\mathbf{a} \cdot \mathbf{b} = \mathbf{a} \cdot \mathbf{c}, \qquad \mathbf{b} \cdot \mathbf{a} = \mathbf{c} \cdot \mathbf{a},$$

$$\mathbf{b} \cdot \frac{\mathbf{a}}{\|\mathbf{a}\|} = \mathbf{c} \cdot \frac{\mathbf{a}}{\|\mathbf{a}\|}, \qquad \mathbf{b} \cdot \mathbf{u}_\mathbf{a} = \mathbf{c} \cdot \mathbf{u}_\mathbf{a},$$

$$(\mathbf{b} \cdot \mathbf{u}_\mathbf{a})\mathbf{u}_\mathbf{a} = (\mathbf{c} \cdot \mathbf{u}_\mathbf{a})\mathbf{u}_\mathbf{a}, \qquad \text{proj}_\mathbf{a}\, \mathbf{b} = \text{proj}_\mathbf{a}\, \mathbf{c}. \qquad \mathbf{a} \cdot \mathbf{b} = \mathbf{a} \cdot \mathbf{c} \text{ but } \mathbf{b} \ne \mathbf{c}.$$

(b) $\mathbf{b} = (\mathbf{b} \cdot \mathbf{i})\mathbf{i} + (\mathbf{b} \cdot \mathbf{j})\mathbf{j} + (\mathbf{b} \cdot \mathbf{k})\mathbf{k} = (\mathbf{c} \cdot \mathbf{i})\mathbf{i} + (\mathbf{c} \cdot \mathbf{j})\mathbf{j} + (\mathbf{c} \cdot \mathbf{k})\mathbf{k} = \mathbf{c}$.

31. (a) Express the norms as dot products.
(b) The following statements are equivalent:

$$\mathbf{a} \perp \mathbf{b}, \qquad \mathbf{a} \cdot \mathbf{b} = 0, \qquad \|\mathbf{a} + \mathbf{b}\|^2 - \|\mathbf{a} - \mathbf{b}\|^2 = 0, \qquad \|\mathbf{a} + \mathbf{b}\| = \|\mathbf{a} - \mathbf{b}\|.$$

(c) By (b), the relation $\|\mathbf{a} + \mathbf{b}\| = \|\mathbf{a} - \mathbf{b}\|$ gives $\mathbf{a} \perp \mathbf{b}$. The relation $\mathbf{a} + \mathbf{b} \perp \mathbf{a} - \mathbf{b}$ gives

$$0 = (\mathbf{a} + \mathbf{b}) \cdot (\mathbf{a} - \mathbf{b}) = \|\mathbf{a}\|^2 - \|\mathbf{b}\|^2 \qquad \text{and thus} \qquad \|\mathbf{a}\| = \|\mathbf{b}\|.$$

The parallelogram is a square since it has two adjacent sides of equal length that meet at right angles.

33. $\|\mathbf{a} + \mathbf{b}\|^2 = (\mathbf{a} + \mathbf{b}) \cdot (\mathbf{a} + \mathbf{b}) = \mathbf{a} \cdot \mathbf{a} + 2\mathbf{a} \cdot \mathbf{b} + \mathbf{b} \cdot \mathbf{b}$ **35.** $\pi - \alpha, \pi - \beta, \pi - \gamma$
$\|\mathbf{a} - \mathbf{b}\|^2 = (\mathbf{a} - \mathbf{b}) \cdot (\mathbf{a} - \mathbf{b}) = \mathbf{a} \cdot \mathbf{a} - 2\mathbf{a} \cdot \mathbf{b} + \mathbf{b} \cdot \mathbf{b}$
and the result follows.

37. If $\mathbf{a} \perp \mathbf{b}$ and $\mathbf{a} \perp \mathbf{c}$, then $\mathbf{a} \cdot \mathbf{b} = 0$ and $\mathbf{a} \cdot \mathbf{c} = 0$, so that

$$\mathbf{a} \cdot (\alpha\mathbf{b} + \beta\mathbf{c}) = \alpha(\mathbf{a} \cdot \mathbf{b}) + \beta(\mathbf{a} \cdot \mathbf{c}) = 0.$$

Thus $\mathbf{a} \perp (\alpha\mathbf{b} + \beta\mathbf{c})$.

39. Existence of decomposition: $\mathbf{a} = (\mathbf{a} \cdot \mathbf{u}_\mathbf{b})\mathbf{u}_\mathbf{b} + [\mathbf{a} - (\mathbf{a} \cdot \mathbf{u}_\mathbf{b})\mathbf{u}_\mathbf{b}]$. Uniqueness of decomposition: suppose that $\mathbf{a} = \mathbf{a}_\| + \mathbf{a}_\perp = \mathbf{A}_\| + \mathbf{A}_\perp$. Then the vector $\mathbf{a}_\| - \mathbf{A}_\| = \mathbf{A}_\perp - \mathbf{a}_\perp$ is both parallel to $\mathbf{b}$ and perpendicular to $\mathbf{b}$. (Exercises 37 and 38.) Therefore it is zero. Consequently $\mathbf{A}_\| = \mathbf{a}_\|$ and $\mathbf{A}_\perp = \mathbf{a}_\perp$.

41. $x = 0$, $x = 4$ **43.** $\theta = \cos^{-1}(\tfrac{1}{3}\sqrt{3}) \cong 0.96$ radians

45. (a) The direction angles of a vector always satisfy $\cos^2 \alpha + \cos^2 \beta + \cos^2 \gamma = 1$, and, as you can check, $\cos^2 \tfrac{1}{4}\pi + \cos^2 \tfrac{1}{6}\pi + \cos^2 \tfrac{2}{3}\pi \ne 1$.
(b) The relation $\cos^2 \alpha + \cos^2 \tfrac{1}{4}\pi + \cos^2 \tfrac{1}{4}\pi = 1$ gives

$$\cos^2 \alpha + \tfrac{1}{2} + \tfrac{1}{2} = 1, \quad \cos \alpha = 0, \quad a_1 = \|\mathbf{a}\| \cos \alpha = 0.$$

47. $u = \pm\frac{1}{165}\sqrt{165}\,(8i + j - 10k)$ **49.** Place center of sphere at the origin.

$$\overrightarrow{P_1Q} \cdot \overrightarrow{P_2Q} = (-a + b) \cdot (a + b)$$
$$= -\|a\|^2 + \|b\|^2$$
$$= 0.$$

51. (a) $W = F \cdot r$ (b) 0 (c) $\|F\|i \cdot (b - a)i = \|F\|(b - a)$ **53.** (a) 614.5 joules (b) 614.29 joules

55. (a) $W_1 = W_2$ (b) $W_1 = \frac{1}{2}\|F\|\,\|r\|$, $W_2 = \frac{\sqrt{3}}{2}\|F\|\,\|r\|$, $W_2 = \sqrt{3}\,W_1$

SECTION 12.4

1. $-2k$ **3.** $i + j + k$ **5.** $-3i - j - 2k$ **7.** -1 **9.** 0   **11.** 1   **13.** $3i - 2j - 3k$ **15.** $i + j - 2k$ **17.** -3

19. $5i - 4j - k$ **21.** $\left(\dfrac{1}{\sqrt{6}}, \dfrac{-1}{\sqrt{6}}, \dfrac{-2}{\sqrt{6}}\right), \left(\dfrac{-1}{\sqrt{6}}, \dfrac{1}{\sqrt{6}}, \dfrac{2}{\sqrt{6}}\right)$ **23.** $N = 3j$; area $= \frac{3}{2}$ **25.** $N = 8i + 4j + 4k$; area $= 2\sqrt{6}$ **27.** 1

29. 2   **31.** $-2(a \times b)$ **33.** $a = 0$ **35.** $\begin{vmatrix} \alpha & \beta \\ \gamma & \delta \end{vmatrix}(a \times b)$ **37.** $a \cdot (b \times c) = (a \times b) \cdot c = (c \times a) \cdot b = (b \times c) \cdot a = (a \times -c) \cdot b$
$a \cdot (c \times b) = c \cdot (b \times a) = (-a \times b) \cdot c$

39. $a \times b$ is perpendicular to the plane determined by a and b;   **41.** $a \cdot b = a \cdot c$ implies $a \cdot (b - c) = 0$; a is perpendicular to $b - c$.
c is in this plane iff $a \times b \cdot c = 0$. $a \times b = a \times c$ implies $a \times (b - c) = 0$; a is parallel to $b - c$.
Since $a \neq 0$, it follows that $b - c = 0$, or $b = c$.

43. $c \times a = \|a\|^2 b$ **47.** 12.77 ft-lb; bolt moves into the plane of the paper.

SECTION 12.5

1. P and Q **3.** $r(t) = (3i + j) + tk$ **5.** $r(t) = t(x_1 i + y_1 j + z_1 k)$ **7.** $x(t) = 1 + t$, $y(t) = -t$, $z(t) = 3 + t$

9. $x(t) = 2$, $y(t) = t$, $z(t) = 3$ **11.** $r(t) = (-i + 2j - 3k) + t(2i + j + 4k)$ **13.** $P(1, 2, 0)$, $\frac{1}{4}\pi$ rad

15. $P(1, 3, 1)$, $\cos^{-1}\left(\frac{1}{6}\sqrt{3}\right) \cong 1.28$ rad   **17.** $(x_0 - [d_1/d_3]z_0, y_0 - [d_2/d_3]z_0, 0)$ **19.** The lines are parallel.

21. $r(t) = (2i + 7j - k) + t(2i - 5j + 4k)$, $0 \le t \le 1$ **23.** $u = -\frac{2}{3}i + \frac{1}{3}j + \frac{2}{3}k$, $9 \le t \le 15$

25. triples of the form $X(u) = 3 + au$, $Y(u) = -1 + bu$, $Z(u) = 8 + cu$ with $2a - 4b + 6c = 0$ **27.** 1   **29.** $\sqrt{69/14} \cong 2.22$

31. $\sqrt{3} \cong 1.73$ **33.** (a) 1 (b) $\sqrt{3}$ **35.** $r(t) = \frac{1}{11}(7i + 4j - k) \pm t[\frac{1}{11}\sqrt{11}(i - j + 3k)]$ **37.** $0 < t < s$

SECTION 12.6

1. Q **3.** $x - 4y + 3z - 2 = 0$ **5.** $3x - 2y + 5z - 9 = 0$ **7.** $y - z - 2 = 0$ **9.** $x_0(x - x_0) + y_0(y - y_0) + z_0(z - z_0) = 0$

11. $\dfrac{1}{\sqrt{30}}(2i - j + 5k)$, $-\dfrac{1}{\sqrt{30}}(2i - j + 5k)$ **13.** $\dfrac{1}{15}x + \dfrac{1}{12}y - \dfrac{1}{10}z = 1$ **15.** $\frac{1}{2}\pi$ **17.** $\cos\theta = \frac{2}{21}\sqrt{42} \cong 0.617$, $\theta \cong 0.91$ rad

19. coplanar   **21.** not coplanar   **23.** $\dfrac{2}{\sqrt{21}}$ **25.** $\frac{22}{5}$ **27.** $x + z = 2$ **29.** $3x - 4z - 5 = 0$ **31.** $\dfrac{x - x_0}{A} = \dfrac{y - y_0}{B} = \dfrac{z - z_0}{C}$

33. $(x - x_0)/d_1 = (y - y_0)/d_2$, $(y - y_0)/d_2 = (z - z_0)/d_3$ **35.** $x(t) = t$, $y(t) = t$, $z(t) = -t$ **37.** $P(-\frac{19}{14}, \frac{15}{7}, \frac{17}{14})$

39. $10x - 7y + z = 0$ **41.** circle centered at P with radius $\|N\|$ **43.** if $\alpha > 0$, then P_1 lies on the same side of the plane as the tip of N; if $\alpha < 0$, then P_1 and the tip of N lie on opposite sides of the plane

45. $a \cdot b \times c = 0$ **47.**

49.

51. $10x + 4y + 5z = 20$ **53.** $5x + 3y = 15$

CHAPTER 13

SECTION 13.1

1. $\mathbf{f}'(t) = 2\mathbf{i} - \mathbf{j} + 3\mathbf{k}$ **3.** $\mathbf{f}'(t) = -\dfrac{1}{2\sqrt{1-t}}\mathbf{i} + \dfrac{1}{2\sqrt{1+t}}\mathbf{j} + \dfrac{1}{(1-t)^2}\mathbf{k}$ **5.** $\mathbf{f}'(t) = \cos t\,\mathbf{i} - \sin t\,\mathbf{j} + \sec^2 t\,\mathbf{k}$ **7.** $-\dfrac{1}{1-t}\mathbf{i} - \sin t\mathbf{j} + 2t\mathbf{k}$

9. $12t\mathbf{j} + 2\mathbf{k}$ **11.** $-2e^t\sin t\mathbf{i} + 2e^t\cos t\mathbf{j}$ **13.** $\mathbf{i} + 3\mathbf{j}$ **15.** $(e-1)\mathbf{i} + (1 - 1/e)\mathbf{k}$ **17.** $\dfrac{\pi}{4}\mathbf{i} + \tan(1)\mathbf{j}$ **19.** $\frac{1}{2}\mathbf{i} + \mathbf{j}$

21. $0\mathbf{i} + 0\mathbf{j} + 0\mathbf{k} = \mathbf{0}$ **23.** does not exist

25. **27.** **29.** **31.**

33. **35.** **37.**

39. (a) $\mathbf{f}(t) = 3\cos t\mathbf{i} + 2\sin t\mathbf{j}$ (b) $\mathbf{f}(t) = 3\cos t\mathbf{i} - 2\sin t\mathbf{j}$ **41.** (a) $\mathbf{f}(t) = t\mathbf{i} + t^2\mathbf{j}$ (b) $\mathbf{f}(t) = -t\mathbf{i} + t^2\mathbf{j}$

43. $\mathbf{f}(t) = (1+2t)\mathbf{i} + (4+5t)\mathbf{j} + (-2+8t)\mathbf{k},\ 0 \le t \le 1$ **45.** $\mathbf{f}'(t_0) = \mathbf{i} + m\mathbf{j},\ \displaystyle\int_a^b \mathbf{f}(t)\,dt = \frac{1}{2}(b^2-a^2)\mathbf{i} + A\mathbf{j},\ \displaystyle\int_a^b \mathbf{f}'(t)\,dt = (b-a)\mathbf{i} + (d-c)\mathbf{j}$

47. $\mathbf{f}(t) = t\mathbf{i} + (\frac{1}{3}t^3 + 1)\mathbf{j} - \mathbf{k}$ **49.** $\mathbf{f}(t) = e^{\alpha t}\mathbf{c}$

51. (a) if $\mathbf{f}'(t) = \mathbf{0}$ on an interval, then the derivative of each component is 0 on that interval, each component is constant on that interval, and therefore $\mathbf{f}$ itself is constant on that interval
(b) set $\mathbf{h}(t) = \mathbf{f}(t) - \mathbf{g}(t)$ and apply part (a)

53. set $\mathbf{f}(t) = f_1(t)\mathbf{i} + f_2(t)\mathbf{j} + f_3(t)\mathbf{k}$, and apply (3.1.4) to f_1, f_2, f_3 **55.** no; as a counterexample set $\mathbf{f}(t) = \mathbf{i} = \mathbf{g}(t)$

57. $\|\mathbf{f}(t)\|^2 = \mathbf{f}(t) \cdot \mathbf{f}(t)$

$2\|\mathbf{f}(t)\|\dfrac{d\|\mathbf{f}(t)\|}{dt} = 2\mathbf{f}(t) \cdot \mathbf{f}'(t)$

$\dfrac{d\|\mathbf{f}(t)\|}{dt} = \dfrac{\mathbf{f}(t) \cdot \mathbf{f}'(t)}{\|\mathbf{f}(t)\|}$

SECTION 13.2

1. $\mathbf{f}'(t) = \mathbf{b},\ \mathbf{f}''(t) = \mathbf{0}$ **3.** $\mathbf{f}'(t) = 2e^{2t}\mathbf{i} - \cos t\mathbf{j},\ \mathbf{f}''(t) = 4e^{2t}\mathbf{i} + \sin t\mathbf{j}$ **5.** $\mathbf{f}'(t) = (3t^2 - 8t^3)\mathbf{j},\ \mathbf{f}''(t) = (6t - 24t^2)\mathbf{j}$

7. $\mathbf{f}'(t) = -2t\mathbf{i} + e^t(t+1)\mathbf{k},\ \mathbf{f}''(t) = -2\mathbf{i} + e^t(t+2)\mathbf{k}$

9. $\mathbf{f}'(t) = (\sin t + t\cos t + 2\sin 2t)\mathbf{i} + (2\cos 2t - \cos t + t\sin t)\mathbf{j} - 3\sin 3t\mathbf{k}$
$\mathbf{f}''(t) = (2\cos t - t\sin t + 4\cos 2t)\mathbf{i} + (-4\sin 2t + 2\sin t + t\cos t)\mathbf{j} - 9\cos 3t\mathbf{k}$

11. $\mathbf{f}'(t) = \frac{1}{2}\sqrt{t}\,\mathbf{g}'(\sqrt{t}) + \mathbf{g}(\sqrt{t}),\ \mathbf{f}''(t) = \frac{1}{4}\mathbf{g}''(\sqrt{t}) + \frac{3}{4}(1/\sqrt{t})\mathbf{g}'(\sqrt{t})$ **13.** $-\sin t\,e^{\cos t}\mathbf{i} + \cos t\,e^{\sin t}\mathbf{j}$ **15.** $4e^{2t} - 4e^{-2t}$

17. $(\mathbf{a} \times \mathbf{d}) + (\mathbf{b} \times \mathbf{c}) + 2t(\mathbf{b} \times \mathbf{d})$ **19.** $(\mathbf{a} \cdot \mathbf{d}) + (\mathbf{b} \cdot \mathbf{c}) + 2t(\mathbf{b} \cdot \mathbf{d})$ **21.** $\mathbf{r}(t) = \mathbf{a} + t\mathbf{b}$ **23.** $\mathbf{r}(t) = \frac{1}{2}t^2\mathbf{a} + \frac{1}{6}t^3\mathbf{b} + t\mathbf{c} + \mathbf{d}$

25. $\mathbf{r}''(t) = -\sin t\mathbf{i} - \cos t\mathbf{j} = -\mathbf{r}(t)$; no. **27.** $\mathbf{r}(t) \cdot \mathbf{r}'(t) = 0,\ \mathbf{r}(t) \times \mathbf{r}'(t) = \mathbf{k}$

29. $\dfrac{d}{dt}[\mathbf{f}(t) \times \mathbf{f}'(t)] = [\mathbf{f}(t) \times \mathbf{f}''(t)] + \underbrace{[\mathbf{f}'(t) \times \mathbf{f}'(t)]}_{\mathbf{0}} = \mathbf{f}(t) \times \mathbf{f}''(t)$

31. $[\mathbf{f} \cdot \mathbf{g} \times \mathbf{h}]' = \mathbf{f}' \cdot (\mathbf{g} \times \mathbf{h}) + \mathbf{f} \cdot (\mathbf{g} \times \mathbf{h})' = \mathbf{f}' \cdot (\mathbf{g} \times \mathbf{h}) + \mathbf{f} \cdot [\mathbf{g}' \times \mathbf{h} + \mathbf{g} \times \mathbf{h}']$ and the result follows.

33. The following four statements are equivalent: $\|\mathbf{r}(t)\| = \sqrt{\mathbf{r}(t) \cdot \mathbf{r}(t)}$ is constant, $\mathbf{r}(t) \cdot \mathbf{r}(t)$ is constant, $d/dt\,[\mathbf{r}(t) \cdot \mathbf{r}(t)] = 2[\mathbf{r}(t) \cdot \mathbf{r}'(t)] = 0$ identically, $\mathbf{r}(t) \cdot \mathbf{r}'(t) = 0$ identically.

35. $\dfrac{[\mathbf{f}(t+h) \times \mathbf{g}(t+h)] - [\mathbf{f}(t) \times \mathbf{g}(t)]}{h} = \left(\mathbf{f}(t+h) \times \left[\dfrac{\mathbf{g}(t+h) - \mathbf{g}(t)}{h} \right] \right) + \left(\left[\dfrac{\mathbf{f}(t+h) - \mathbf{f}(t)}{h} \right] \times \mathbf{g}(t) \right).$ Now take the limit as $h \to 0$.

(Appeal to Theorem 13.1.3.)

SECTION 13.3

1. $\pi \mathbf{j} + \mathbf{k}$, $\mathbf{R}(u) = (\mathbf{i} + 2\mathbf{k}) + u(\pi \mathbf{j} + \mathbf{k})$ **3.** $\mathbf{b} - 2\mathbf{c}$, $\mathbf{R}(u) = (\mathbf{a} - \mathbf{b} + \mathbf{c}) + u(\mathbf{b} - 2\mathbf{c})$

5. $4\mathbf{i} - \mathbf{j} + 4\mathbf{k}$, $\mathbf{R}(u) = (2\mathbf{i} + 5\mathbf{k}) + u(4\mathbf{i} - \mathbf{j} + 4\mathbf{k})$ **7.** $-\sqrt{2}\mathbf{i} + \dfrac{3\sqrt{2}}{2}\mathbf{j} + \mathbf{k}$, $\mathbf{R}(u) = \left(\sqrt{2}\mathbf{i} + \dfrac{3\sqrt{2}}{2}\mathbf{j} + \dfrac{\pi}{4}\mathbf{k} \right) + u\left(-\sqrt{2}\mathbf{i} + \dfrac{3\sqrt{2}}{2}\mathbf{j} + \mathbf{k} \right)$

9. The scalar components $x(t) = at$ and $y(t) = bt^2$ satisfy the equation $a^2 y(t) = a^2(bt^2) = b(at)^2 = b[x(t)]^2$ and generate the parabola $a^2 y = bx^2$.

11. (a) $P(0, 1)$ (b) $P(1, 2)$ (c) $P(-1, 2)$

13. The tangent line at $t = t_0$ has the form $\mathbf{R}(u) = \mathbf{r}(t_0) + u\mathbf{r}'(t_0)$. If $\mathbf{r}'(t_0) = a\mathbf{r}(t_0)$, then

$$\mathbf{R}(u) = \mathbf{r}(t_0) + ua\mathbf{r}(t_0) = (1 + ua)\mathbf{r}(t_0).$$

The tangent line passes through the origin at $u = -1/a$.

15. $\pi/2 \cong 1.57$, or $90°$

17. $P(1, 2, -2)$; $\cos^{-1}(\tfrac{1}{3}\sqrt{5}) \cong 1.11$ rad   **19.** (a) $\mathbf{r}(t) = a \cos t\mathbf{i} + b \sin t\mathbf{j}$ (b) $\mathbf{r}(t) = a \cos t\mathbf{i} - b \sin t\mathbf{j}$
                                                       (c) $\mathbf{r}(t) = a \cos 2t\mathbf{i} + b \sin 2t\mathbf{j}$ (d) $\mathbf{r}(t) = a \cos 3t\mathbf{i} - b \sin 3t\mathbf{j}$

21. $\mathbf{r}'(t) = t^3\mathbf{i} + 2t\mathbf{j}$ **23.** $\mathbf{r}'(t) = 2e^{2t}\mathbf{i} - 4e^{-4t}\mathbf{j}$ **25.** $\mathbf{r}'(t) = -2 \sin t\mathbf{i} + 3 \cos t\mathbf{j}$

27. $\mathbf{r}(t) = (t^2 + 1)\mathbf{i} + t\mathbf{j}$, $t \geq 1$;   or, $\mathbf{r}(t) = \sec^2 t\mathbf{i} + \tan t\mathbf{j}$, $t \in [\tfrac{1}{4}\pi, \tfrac{1}{2}\pi)$ **29.** $\mathbf{r}(t) = \cos t \sin 3t\mathbf{i} + \sin t \sin 3t\mathbf{j}$, $t \in [0, \pi]$

31.
$y^3 = x^2$

33. $(2, 4, 8)$; $\cos^{-1}\left(\dfrac{24}{\sqrt{21}\,\sqrt{161}} \right) \cong 1.15$ rad

There is no tangent vector at the origin.

35. $\mathbf{T}(1) = \dfrac{1}{\sqrt{2}}\mathbf{j} + \dfrac{1}{\sqrt{2}}\mathbf{k}$, $\mathbf{N}(1) = \dfrac{1}{\sqrt{2}}\mathbf{j} + \dfrac{1}{\sqrt{2}}\mathbf{k}$, $x - 1 = 0$ **37.** $\tfrac{1}{5}\sqrt{5}\,(-2\mathbf{i} + \mathbf{k})$, $-\mathbf{j}$, $x + 2z = \tfrac{1}{2}\pi$

39. $\tfrac{1}{14}\sqrt{14}\,(\mathbf{i} + 2\mathbf{j} + 3\mathbf{k})$, $\tfrac{1}{266}\sqrt{266}\,(-11\mathbf{i} - 8\mathbf{j} + 9\mathbf{k})$, $3x - 3y + z = 1$

41. $\mathbf{T}(0) = \dfrac{1}{\sqrt{3}}\mathbf{i} + \dfrac{1}{\sqrt{3}}\mathbf{j} + \dfrac{1}{\sqrt{3}}\mathbf{k}$, $\mathbf{N}(0) = \dfrac{1}{\sqrt{2}}\mathbf{i} - \dfrac{1}{\sqrt{2}}\mathbf{j}$, $x + y - 2z + 1 = 0$

43. $\mathbf{T}_1 = \dfrac{\mathbf{R}'(u)}{\|\mathbf{R}'(u)\|} = -\dfrac{\mathbf{r}'(a + b - u)}{\|\mathbf{r}'(a + b - u)\|} = -\mathbf{T}$. Then $\mathbf{T}_1'(u) = \mathbf{T}'(a + b - u)$ and thus $\mathbf{N}_1 = \mathbf{N}$.

45. Let $\mathbf{T}$ be the unit tangent at the tip of $\mathbf{R}(u) = \mathbf{r}(\phi(u))$ as calculated from the parametrization $\mathbf{r}$ and let $\mathbf{T}_1$ be the unit tangent at the same point as calculated from the parametrization $\mathbf{R}$. Then

$$\mathbf{T}_1 = \frac{\mathbf{R}'(u)}{\|\mathbf{R}'(u)\|} = \frac{\mathbf{r}'(\phi(u))\,\phi'(u)}{\|\mathbf{r}'(\phi(u))\,\phi'(u)\|} = \frac{\mathbf{r}'(\phi(u))}{\|\mathbf{r}'(\phi(u))\|} = \mathbf{T}.$$

$$\phi'(u) > \underset{\uparrow}{\rule{1.5cm}{0.4pt}}$$

This shows the invariance of the unit tangent. The invariance of the principal normal and the osculating plane follows directly from the invariance of the unit tangent.

47. (a) Let $t = \Psi(v) = 2\pi - v^2$. When t increases from 0 to 2π, v decreases from $\sqrt{2\pi}$ to 0.

(b) $T_r\left(\dfrac{\pi}{4}\right) = -\dfrac{1}{\sqrt{10}}\,\mathbf{i} + \dfrac{1}{\sqrt{10}}\,\mathbf{j} + \dfrac{2}{\sqrt{5}}\,\mathbf{k}$, $\quad T_R\left(\dfrac{\sqrt{7\pi}}{2}\right) = \dfrac{1}{\sqrt{10}}\,\mathbf{i} - \dfrac{1}{\sqrt{10}}\,\mathbf{j} - \dfrac{2}{\sqrt{5}}\,\mathbf{k}$

$N_r\left(\dfrac{\pi}{4}\right) = -\dfrac{\sqrt{2}}{2}\,\mathbf{i} - \dfrac{\sqrt{2}}{2}\,\mathbf{j}$; $\quad N_R\left(\dfrac{\sqrt{7\pi}}{2}\right) = -\dfrac{\sqrt{2}}{2}\,\mathbf{i} - \dfrac{\sqrt{2}}{2}\,\mathbf{j}$

SECTION 13.4

1. $\frac{52}{3}$ **3.** $2\pi\sqrt{a^2 + b^2}$ **5.** $\ln(1 + \sqrt{2})$ **7.** $\frac{1}{27}(13\sqrt{13} - 8)$ **9.** $\sqrt{2}(e^\pi - 1)$ **11.** $6 + \frac{1}{2}\sqrt{2}\ln(2\sqrt{2} + 3)$ **13.** e^2

15. differentiate $s(t) = \displaystyle\int_a^t \sqrt{[x'(u)]^2 + [y'(u)]^2 + [z'(u)]^2}\ du$ **17.** see Exercise 16, differentiate $s(x) = \displaystyle\int_a^x \sqrt{1 + [f'(t)]^2}\ dt$

19. Let L be the length as computed from $\mathbf{r}$ and L^* the length as computed from $\mathbf{R}$. Then

$$L^* = \int_c^d \|\mathbf{R}'(u)\|\ du = \int_c^d \|\mathbf{r}'(\phi(u))\|\ \phi'(u)\ du = \int_a^b \|\mathbf{r}'(t)\|\ dt = L.$$

$$t = \phi(u) \underrightarrow{\qquad}$$

21. (a) $s = 5t$ (b) $\mathbf{R}(s) = 3\cos\left(\dfrac{s}{5}\right)\mathbf{i} + 3\sin\left(\dfrac{s}{5}\right)\mathbf{j} + \dfrac{4s}{5}\,\mathbf{k}$ (c) $Q(-3, 0, 4\pi)$

(d) $\mathbf{R}'(s) = \dfrac{-3}{5}\sin\left(\dfrac{s}{5}\right)\mathbf{i} + \dfrac{3}{5}\cos\left(\dfrac{s}{5}\right)\mathbf{j} + \dfrac{4}{5}\,\mathbf{k}$, $\|\mathbf{R}'(s)\| = 1$

23. 0.5077 **25.** 22.0939

SECTION 13.5

1. $v/r,\ v^2/r$ **3.** $\|\mathbf{r}''(t)\| = a^2|b\sin at| = a^2|y(t)|$ **5.** $y = \cos\pi x,\quad 0 \le x \le 2$ **7.** $x = \sqrt{1 + y^2},\quad y \ge -1$

9. (a) (x_0, y_0, z_0) (b) $\alpha\cos\theta\mathbf{j} + \alpha\sin\theta\mathbf{k}$ (c) $|\alpha|$ (d) $-32\mathbf{k}$
(e) arc from parabola $z = z_0 + (\tan\theta)(y - y_0) - 16(y - y_0)^2/(\alpha^2\cos^2\theta)$ in the plane $x = x_0$

11. (a) $\mathbf{r}'(0) = b\omega\mathbf{j}$ (b) $\mathbf{r}''(t) = \omega^2\mathbf{r}(t)$ (c) The torque is $\mathbf{0}$ and the angular momentum is constant.

13. (a) $\mathbf{v}(t) = 2\mathbf{j} + (\alpha/m)t\mathbf{k}$ (b) $v(t) = (1/m)\sqrt{4m^2 + \alpha^2 t^2}$ (c) $\mathbf{p}(t) = 2m\mathbf{j} + \alpha t\mathbf{k}$
(d) $\mathbf{r}(t_1) = [2t + y_0]\mathbf{j} + [(\alpha/2m)t^2 + z_0]\mathbf{k}$, $t \ge 0$ $z = (\alpha/8m)(y - y_0)^2 + z_0$, $y \ge y_0$, $x = 0$

15. $\mathbf{F}(t) = 2m\mathbf{k}$ **17.** (a) $\pi b\mathbf{j} + \mathbf{k}$ (b) $\sqrt{\pi^2 b^2 + 1}$ (c) $-\pi^2 a\mathbf{i}$ (d) $m(\pi b\mathbf{j} + \mathbf{k})$
(e) $m[b(1 - \pi)\mathbf{i} - 2a\mathbf{j} + 2\pi ab\mathbf{k}]$ (f) $-m\pi^2 a[\mathbf{j} - b\mathbf{k}]$

19. We have $m\mathbf{v} = m\mathbf{v}_1 + m\mathbf{v}_2$ and $\frac{1}{2}mv^2 = \frac{1}{2}mv_1^2 + \frac{1}{2}mv_2^2$. Therefore $\mathbf{v} = \mathbf{v}_1 + \mathbf{v}_2$ and $v^2 = v_1^2 + v_2^2$. Since

$$v^2 = \mathbf{v} \cdot \mathbf{v} = (\mathbf{v}_1 + \mathbf{v}_2) \cdot (\mathbf{v}_1 + \mathbf{v}_2) = v_1^2 + v_2^2 + 2(\mathbf{v}_1 \cdot \mathbf{v}_2),$$

we have $\mathbf{v}_1 \cdot \mathbf{v}_2 = 0$ and $\mathbf{v}_1 \perp \mathbf{v}_2$.

21. Here $\mathbf{r}''(t) = \mathbf{a}$, $\mathbf{r}'(t) = \mathbf{v}(0) + t\mathbf{a}$, $\mathbf{r}(t) = \mathbf{r}(0) + t\mathbf{v}(0) + \frac{1}{2}t^2\mathbf{a}$. If neither $\mathbf{v}(0)$ nor $\mathbf{a}$ is zero, the displacement $\mathbf{r}(t) - \mathbf{r}(0)$ is a linear combination of $\mathbf{v}(0)$ and $\mathbf{a}$ and thus remains on the plane determined by these vectors. The equation of this plane can be written

$$[\mathbf{a} \times \mathbf{v}(0)] \cdot [\mathbf{r} - \mathbf{r}(0)] = 0.$$

(If either $\mathbf{v}(0)$ or $\mathbf{a}$ is zero, the motion is restricted to a straight line; if both of these vectors are zero, the particle remains at its initial position $\mathbf{r}(0)$.)

23. $\mathbf{r}(t) = \mathbf{i} + t\mathbf{j} + (qE_0/2m)t^2\mathbf{k}$ **25.** $\mathbf{r}(t) = (1 + t^3/6m)\mathbf{i} + (t^4/12m)\mathbf{j} + \mathbf{k}$.

27. $\dfrac{d}{dt}(\frac{1}{2}mv^2) = mv\,\dfrac{dv}{dt} = m\left(\mathbf{v} \cdot \dfrac{d\mathbf{v}}{dt}\right) = m\dfrac{d\mathbf{v}}{dt} \cdot \mathbf{v} = \mathbf{F} \cdot \dfrac{d\mathbf{r}}{dt} = 4r^2\left(\mathbf{r} \cdot \dfrac{d\mathbf{r}}{dt}\right) = 4r^2\left(r\dfrac{dr}{dt}\right) = 4r^3\dfrac{dr}{dt} = \dfrac{d}{dt}(r^4)$. Therefore $d/dt\,(\frac{1}{2}mv^2 - r^4) = 0$
and $\frac{1}{2}mv^2 - r^4$ is a constant E. Evaluating E from $t = 0$, we find that $E = 2m$. Thus $\frac{1}{2}mv^2 - r^4 = 2m$ and $v = \sqrt{4 + (2/m)r^4}$.

SECTION *13.6

1. about 61.5% of an earth year **3.** set $x = r\cos\theta,\quad y = r\sin\theta$

5. Substitute

$$r = \frac{a}{1 + e\cos\theta}, \qquad \left(\frac{dr}{d\theta}\right)^2 = \frac{(ae\sin\theta)^2}{(1 + e\cos\theta)^4}$$

into the right side of the equation and you will see that, with a and e^2 as given, the expression reduces to E.

SECTION 13.7

1. $\dfrac{e^{-x}}{(1 + e^{-2x})^{3/2}}$ **3.** $\dfrac{2}{(1 + 4x)^{3/2}}$ **5.** $|\cos x|$ **7.** $\dfrac{|\sin x|}{(1 + \cos^2 x)^{3/2}}$ **9.** $\dfrac{30x^4}{(9 + 4x^5)^{3/2}}$ **11.** $\frac{5}{2}\sqrt{5}$ **13.** $5\sqrt{5}$ **15.** $\dfrac{10\sqrt{10}}{3}$

17. $\frac{4}{3}$ **19.** $(\frac{1}{2}\sqrt{2}, \frac{1}{2}\ln\frac{1}{2})$ **21.** $\dfrac{1}{(1 + t^2)^{3/2}}$ **23.** $\dfrac{12|t|}{(4 + 9t^4)^{3/2}}$ **25.** $\frac{1}{2}\sqrt{2}\,e^{-t}$ **27.** $\dfrac{2 + t^2}{(1 + t^2)^{3/2}}$ **29.** $\sqrt{2}$ **31.** $\dfrac{a^4b^4}{(b^4x^2 + a^4y^2)^{3/2}}$

33. follow the hint   **35.** $k = \frac{1}{3}\sqrt{2}\,e^{-t}$, $a_{\mathbf{T}} = \sqrt{3}\,e^t$, $a_{\mathbf{N}} = \sqrt{2}\,e^t$ **37.** $k = 2$, $a_{\mathbf{T}} = 0$, $a_{\mathbf{N}} = 8$

39. $k = \dfrac{\sqrt{t^4 + 4t^2 + 1}}{(t^4 + t^2 + 1)^{3/2}}$, $a_{\mathbf{T}} = \dfrac{2t^3 + t}{\sqrt{t^4 + t^2 + 1}}$, $a_{\mathbf{N}} = \dfrac{\sqrt{t^4 + 4t^2 + 1}}{\sqrt{t^4 + t^2 + 1}}$ **41.** $\dfrac{e^{-a\theta}}{\sqrt{1 + a^2}}$ **43.** $\dfrac{3}{2\sqrt{2a^2(1 - \cos\theta)}} = \dfrac{3}{2\sqrt{2ar}}$

45. (a) $s(\theta) = 4R|\cos\frac{1}{2}\theta|$ (b) $\rho(\theta) = 4R\sin\frac{1}{2}\theta$ (c) $\rho^2 + s^2 = 16R^2$ **47.** $9\rho^2 + s^2 = 16a^2$

CHAPTER 14

SECTION 14.1

1. dom (f) = the first and third quadrants, including the axes; ran $(f) = [0, \infty)$

3. dom (f) = the set of all points (x, y) not on the line $y = -x$; ran $(f) = (-\infty, 0) \cup (0, \infty)$ **5.** dom (f) = the entire plane; ran $(f) = (-1, 1)$

7. dom (f) = the first and third quadrants, excluding the axes; ran $(f) = (-\infty, \infty)$

9. dom (f) = the set of all points (x, y) with $x^2 < y$; in other words, the set of all points of the plane above the parabola $y = x^2$; ran $(f) = (0, \infty)$

11. dom (f) = the set of all points (x, y) with $-3 \le x \le 3$, $-2 \le y \le 2$ (a rectangle); ran $(f) = [-2, 3]$

13. dom (f) = the set of all points (x, y, z) not on the plane $x + y + z = 0$; ran $(f) = \{-1, 1\}$

15. dom (f) = the set of all points (x, y, z) with $|y| < |x|$; ran $(f) = (-\infty, 0]$

17. dom (f) = the set of all points (x, y) such that $x^2 + y^2 < 9$; in other words, the set of all points of the plane inside the circle $x^2 + y^2 = 9$; ran $(f) = [\frac{2}{3}, \infty)$

19. dom (f) = the set of all points (x, y, z) with $x + 2y + 3z > 0$; in other words, the set of all points in space that lie on the same side of the plane $x + 2y + 3z = 0$ as the point $(1, 1, 1)$; ran $(f) = (-\infty, \infty)$

21. dom (f) = all of space; ran $(f) = (0, \infty)$

23. dom $(f) = \{x: x \ge 0\}$; range $(f) = [0, \infty)$
dom $(g) = \{(x, y): x \ge 0, y \text{ real}\}$; range $(g) = [0, \infty)$
dom $(h) = \{(x, y, z): x \ge 0, y, z \text{ real}\}$; range $(h) = [0, \infty)$

25. $\displaystyle\lim_{h\to 0}\frac{f(x + h, y) - f(x, y)}{h} = 4x;$ $\displaystyle\lim_{h\to 0}\frac{f(x, y + h) - f(x, y)}{h} = -1$

27. $\displaystyle\lim_{h\to 0}\frac{f(x + h, y) - f(x, y)}{h} = 3 - y;$ $\displaystyle\lim_{h\to 0}\frac{f(x, y + h) - f(x, y)}{h} = -x + 4y$

29. $\displaystyle\lim_{h\to 0}\frac{f(x + h, y) - f(x, y)}{h} = -y\sin(xy);$ $\displaystyle\lim_{h\to 0}\frac{f(x, y + h) - f(x, y)}{h} = -x\sin(xy)$

31. (a) $f(x, y) = x^2y$ (b) $f(x, y) = \pi x^2 y$ (c) $f(x, y) = 2|y|$ **33.** $V = \dfrac{lw(10 - lw)}{l + w}$ **35.** $V = \pi r^2 h + \frac{4}{3}\pi r^3$

SECTION 14.2

1. a quadric cone   **3.** a parabolic cylinder   **5.** a hyperboloid of one sheet   **7.** sphere of radius 2 centered at the origin

9. an elliptic paraboloid   **11.** a hyperbolic paraboloid

13.

15.

17.

19.

21.

23.

25. elliptic paraboloid, xy-trace: the origin, xz-trace: the parabola $x^2 = 4z$, yz-trace: the parabola $y^2 = 9z$, surface has the form of Figure 14.2.5

27. a quadric cone, xy-trace: the origin, xz-trace: the lines $x = \pm 2z$, yz-trace: the lines $y = \pm 3z$, surface has the form of Figure 14.2.4

29. a hyperboloid of two sheets, xy-trace: none, xz-trace: the hyperbola $4z^2 - x^2 = 4$, yz-trace: the hyperbola $9z^2 - y^2 = 9$, surface has the form of Figure 14.2.3

31. hyperboloid of two sheets, xy-trace: the hyperbola $\dfrac{x^2}{4} - \dfrac{y^2}{9} = 1$, xz-trace: the hyperbola $\dfrac{x^2}{4} - z^2 = 1$, yz-trace: none, see Figure 14.2.3 for an example

33. elliptic paraboloid, xy-trace: the origin, xz-trace: the parabola $x^2 = 4z$, yz-trace: the parabola $y^2 = 9z$, surface has the form of Figure 14.2.5

35. hyperboloid of two sheets, xy-trace: the hyperbola $\dfrac{y^2}{4} - \dfrac{x^2}{9} = 1$, xz-trace: none, yz-trace: the hyperbola $\dfrac{y^2}{4} - z^2 = 1$, see Figure 14.2.3 for an example

37. paraboloid of revolution, xy-trace: the origin, xz-trace: the parabola $x^2 = 4z$, yz-trace: the parabola $y^2 = 4z$, surface has the form of Figure 14.2.5.

39. (a) an elliptic paraboloid (vertex down if A and B are both positive, vertex up if A and B are both negative)
(b) a hyperbolic paraboloid   (c) the xy-plane if A and B are both zero; otherwise, a parabolic cylinder

41. $x^2 + y^2 - 4z = 0$ (paraboloid of revolution)   **43.** (a) a circle   (b) (i) $\sqrt{x^2 + y^2} = -3z$ (ii) $\sqrt{x^2 + z^2} = \frac{1}{3}y$

45. the line $5x + 7y = 30$ **47.** the circle $x^2 + y^2 = \frac{5}{4}$ **49.** the ellipse $x^2 + 2y^2 = 2$ **51.** the parabola $x^2 = -4(y - 1)$

SECTION 14.3

1. lines of slope 1: $y = x - c$

3. parabolas, $y = x^2 - c$

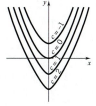

5. the y-axis and the lines $y = [(1 - c)/c]x$, the origin omitted throughout

7. the cubics $y = x^3 - c$

9. the lines $y = \pm x$ and the hyperbolas $x^2 - y^2 = c$

11. pairs of horizontal lines $y = \pm\sqrt{c}$ and the x-axis

13. the circles $x^2 + y^2 = e^c$ with c real

15. the curves $y = e^{cx^2}$ with the point $(0, 1)$ omitted

17. the coordinate axes and pairs of lines $y = \pm(\sqrt{1 - c}/\sqrt{c})x$, the origin omitted throughout

19. $x + 2y + 3z = 0$, plane through the origin **21.** $z = \sqrt{x^2 + y^2}$, the upper nappe of the circular cone $z^2 = x^2 + y^2$; Figure 14.2.4

23. the elliptic paraboloid $\dfrac{x^2}{(3\sqrt{2})^2} + \dfrac{y^2}{(2\sqrt{2})^2} = z$; Figure 14.2.5

25. (i) hyperboloid of two sheets; Figure 14.2.3   (ii) circular cone; Figure 14.2.4   (iii) hyperboloid of one sheet; Figure 14.2.2

27. $4x^2 + y^2 = 1$ **29.** $y^2\tan^{-1}x = \pi$ **31.** $x^2 + 2y^2 - 2xyz = 13$

33. $x^2 + y^2 + z^2 = \dfrac{GmM}{c}$. The surfaces of constant gravitational force are concentric spheres.

35. (a) $T(x, y) = \dfrac{k}{x^2 + y^2}$ (b) $x^2 + y^2 = \dfrac{k}{c}$, concentric circles   (c) $10°$ **37.** F   **39.** A   **41.** E

SECTION 14.4

1. $\dfrac{\partial f}{\partial x} = 6x - y$, $\dfrac{\partial f}{\partial y} = 1 - x$ **3.** $\dfrac{\partial \rho}{\partial \phi} = \cos\phi\cos\theta$, $\dfrac{\partial \rho}{\partial \theta} = -\sin\phi\sin\theta$ **5.** $\dfrac{\partial f}{\partial x} = e^{x-y} + e^{y-x}$, $\dfrac{\partial f}{\partial y} = -e^{x-y} - e^{y-x}$

7. $\dfrac{\partial g}{\partial x} = \dfrac{(AD - BC)y}{(Cx + Dy)^2}$, $\dfrac{\partial g}{\partial y} = \dfrac{(BC - AD)x}{(Cx + Dy)^2}$ **9.** $\dfrac{\partial u}{\partial x} = y + z$, $\dfrac{\partial u}{\partial y} = x + z$, $\dfrac{\partial u}{\partial z} = x + y$

11. $\dfrac{\partial f}{\partial x} = z \cos (x - y)$, $\dfrac{\partial f}{\partial y} = -z \cos (x - y)$, $\dfrac{\partial f}{\partial z} = \sin (x - y)$

13. $\dfrac{\partial \rho}{\partial \theta} = e^{\theta + \phi} [\cos (\theta - \phi) - \sin(\theta - \phi)]$, $\dfrac{\partial \rho}{\partial \phi} = e^{\theta + \phi} [\cos (\theta - \phi) + \sin (\theta - \phi)]$

15. $\dfrac{\partial f}{\partial x} = 2xy \sec(xy) + x^2 y^2 \sec(xy)\tan (xy)$, $\dfrac{\partial f}{\partial y} = x^2 \sec(xy) + x^3 y \sec(xy)\tan (xy)$ **17.** $\dfrac{\partial h}{\partial x} = \dfrac{y^2 - x^2}{(x^2 + y^2)^2}$, $\dfrac{\partial h}{\partial y} = -\dfrac{2xy}{(x^2 + y^2)^2}$

19. $\dfrac{\partial f}{\partial x} = \dfrac{\sin y \,(\cos x + x \sin x)}{y \cos^2 x}$, $\dfrac{\partial f}{\partial y} = \dfrac{x \,(y \cos y - \sin y)}{y^2 \cos x}$ **21.** $\dfrac{\partial h}{\partial x} = 2f(x)f'(x) \, g(y)$, $\dfrac{\partial h}{\partial y} = [f(x)]^2 \, g'(y)$

23. $\dfrac{\partial f}{\partial x} = (y^2 \ln z)z^{xy^2}$, $\dfrac{\partial f}{\partial y} = (2xy \ln z)z^{xy^2}$, $\dfrac{\partial f}{\partial z} = xy^2 z^{xy^2 - 1}$

25. $\dfrac{\partial h}{\partial r} = 2re^{2t}\cos (\theta - t)$, $\dfrac{\partial h}{\partial \theta} = -r^2 e^{2t}\sin (\theta - t)$, $\dfrac{\partial h}{\partial t} = r^2 e^{2t}[2 \cos (\theta - t) + \sin (\theta - t)]$

27. $\dfrac{\partial f}{\partial x} = -\dfrac{yz}{x^2 + y^2}$, $\dfrac{\partial f}{\partial y} = \dfrac{xz}{x^2 + y^2}$, $\dfrac{\partial f}{\partial z} = \tan^{-1}(y/x)$ **29.** $f_x(0, e) = 1$, $f_y(0, e) = e^{-1}$ **31.** $f_x(1, 2) = \frac{2}{9}$, $f_y(1, 2) = -\frac{1}{9}$

33. $f_x(x, y) = 2xy$, $f_y(x, y) = x^2$ **35.** $f_x(x, y) = \dfrac{2}{x}$, $f_y(x, y) = \dfrac{1}{y}$

37. $f_x(x, y) = -\dfrac{1}{(x - y)^2}$, $f_y(x, y) = \dfrac{1}{(x - y)^2}$ **39.** $f_x(x, y, z) = y^2 z$, $f_y(x, y, z) = 2xyz$, $f_z(x, y, z) = xy^2$

41. (b) $x = x_0$, $z - z_0 = f_y(x_0, y_0)(y - y_0)$ **43.** $x = 2$, $z - 5 = 2(y - 1)$ **45.** $y = 2$, $z - 9 = 6(x - 3)$

47. $u_x = v_y = 2x$, $u_y = -v_x = -2y$ **49.** $u_x = v_y = \dfrac{x}{x^2 + y^2}$, $u_y = -v_x = \dfrac{y}{x^2 + y^2}$

51. (a) f depends only on y (b) f depends only on x **53.** (a) $50\sqrt{3}$ in.^{2} (b) $5\sqrt{3}$ (c) 50 (d) $\frac{5}{18} \pi$ in.^{2} (e) -2

55. (a) y_0-section: $\mathbf{r}(x) = x\mathbf{i} + y_0\mathbf{j} + f(x, y_0)\mathbf{k}$

tangent line: $\mathbf{R}(t) = [x_0\mathbf{i} + y_0\mathbf{j} + f(x_0, y_0)\mathbf{k}] + t\left[\mathbf{i} + \dfrac{\partial f}{\partial x}(x_0, y_0)\mathbf{k}\right]$

(b) x_0-section: $\mathbf{r}(y) = x_0\mathbf{i} + y\mathbf{j} + f(x_0, y)\mathbf{k}$

tangent line: $\mathbf{R}(t) = [x_0\mathbf{i} + y_0\mathbf{j} + f(x_0, y_0)\mathbf{k}] + t\left[\mathbf{j} + \dfrac{\partial f}{\partial y}(x_0, y_0)\mathbf{k}\right]$

(c) For (x, y, z) in the plane

$[(x - x_0)\mathbf{i} + (y - y_0)\mathbf{j} + (z - f(x_0, y_0))\mathbf{k}] \cdot \left[\left(\mathbf{i} + \dfrac{\partial f}{\partial x}(x_0, y_0)\mathbf{k}\right) \times \left(\mathbf{j} + \dfrac{\partial f}{\partial y}(x_0, y_0)\mathbf{k}\right)\right] = 0.$

From this it follows that

$z - f(x_0, y_0) = (x - x_0)\dfrac{\partial f}{\partial x}(x_0, y_0) + (y - y_0)\dfrac{\partial f}{\partial y}(x_0, y_0).$

57. (a) Set $u = ax + by$. Then $\dfrac{\partial w}{\partial x} = ag'(u)$ and $\dfrac{\partial w}{\partial y} = bg'(u)$. (b) Set $u = x^m y^n$. Then $\dfrac{\partial w}{\partial x} = mx^{m-1}y^n \, g'(u)$ and $\dfrac{\partial w}{\partial y} = nx^m y^{n-1} \, g'(u)$.

59. $x\dfrac{\partial u}{\partial x} + y\dfrac{\partial u}{\partial y} = x(4Ax^3 + 4Bxy^2) + y(4Bx^2 y + 4Cy^3) = 4(Ax^4 + 2Bx^2 y^2 + Cy^4) = 4u$ **61.** r

SECTION 14.5

1. interior $= \{(x, y): 2 < x < 4, 1 < y < 3\}$ (the inside of the rectangle)
boundary $=$ the union of the four line segments that bound the rectangle
set is closed

3. interior $=$ the entire set (region between two concentric circles)
boundary $= \{(x, y): x^2 + y^2 = 1 \text{ or } x^2 + y^2 = 4\}$ (the two circles)
set is open

5. interior $= \{(x, y): 1 < x^2 < 4\} = \{(x, y): -2 < x < 1\} \cup \{(x, y): 1 < x < 2\}$

(two vertical strips without the boundary lines)

boundary $= \{(x, y): x = -2, x = -1, x = 1, \text{ or } x = 2\}$ (four vertical lines)

set is neither open nor closed

7. interior $= \{(x, y): y < x^2\}$ (region below the parabola)

boundary $= \{(x, y): y = x^2\}$ (the parabola)

set is closed

9. interior $= \{(x, y, z): x^2 + y^2 < 1, 0 < z < 4\}$

(the inside of a cylinder)

boundary = the total surface of the cylinder

(the curved part, the top, the bottom)

set is closed

11. (a) ϕ (b) S (c) closed   13. interior $= \{x: 1 < x < 3\}$, boundary $= \{1, 3\}$; set is closed

15. interior = the entire set, boundary $= \{1\}$; set is open   17. interior $= \{x: |x| > 1\}$, boundary $= \{1, -1\}$; set is neither open nor closed

19. interior $= \phi$, boundary $= \{\text{the entire set}\} \cup \{0\}$; the set is neither open nor closed

SECTION 14.6

1. $\dfrac{\partial^2 f}{\partial x^2} = 2A, \quad \dfrac{\partial^2 f}{\partial y^2} = 2C, \quad \dfrac{\partial^2 f}{\partial y \partial x} = \dfrac{\partial^2 f}{\partial x \partial y} = 2B$ 3. $\dfrac{\partial^2 f}{\partial x^2} = Cy^2 e^{xy}, \quad \dfrac{\partial^2 f}{\partial y^2} = Cx^2 e^{xy}, \quad \dfrac{\partial^2 f}{\partial y \partial x} = \dfrac{\partial^2 f}{\partial x \partial y} = Ce^{xy}(xy + 1)$

5. $\dfrac{\partial^2 f}{\partial x^2} = 2, \quad \dfrac{\partial^2 f}{\partial y^2} = 4(x + 3y^2 + z^3), \quad \dfrac{\partial^2 f}{\partial z^2} = 6z(2x + 2y^2 + 5z^3)$

$\dfrac{\partial^2 f}{\partial x \partial y} = \dfrac{\partial^2 f}{\partial y \partial x} = 4y, \quad \dfrac{\partial^2 f}{\partial z \partial x} = \dfrac{\partial^2 f}{\partial x \partial z} = 6z^2, \quad \dfrac{\partial^2 f}{\partial z \partial y} = \dfrac{\partial^2 f}{\partial y \partial z} = 12yz^2$

7. $\dfrac{\partial^2 f}{\partial x^2} = \dfrac{1}{(x + y)^2} - \dfrac{1}{x^2}, \quad \dfrac{\partial^2 f}{\partial y^2} = \dfrac{1}{(x + y)^2}, \quad \dfrac{\partial^2 f}{\partial y \partial x} = \dfrac{\partial^2 f}{\partial x \partial y} = \dfrac{1}{(x + y)^2}$

9. $\dfrac{\partial^2 f}{\partial x^2} = 2(y + z), \quad \dfrac{\partial^2 f}{\partial y^2} = 2(x + z), \quad \dfrac{\partial^2 f}{\partial z^2} = 2(x + y)$; the second mixed partials are all $2(x + y + z)$

11. $\dfrac{\partial^2 f}{\partial x^2} = y(y - 1)x^{y-2}, \quad \dfrac{\partial^2 f}{\partial y^2} = (\ln x)^2 x^y, \quad \dfrac{\partial^2 f}{\partial y \partial x} = \dfrac{\partial^2 f}{\partial x \partial y} = x^{y-1}(1 + y \ln x)$

13. $\dfrac{\partial^2 f}{\partial x^2} = ye^x, \quad \dfrac{\partial^2 f}{\partial y^2} = xe^y, \quad \dfrac{\partial^2 f}{\partial y \partial x} = e^y + e^x = \dfrac{\partial^2 f}{\partial x \partial y}$ 15. $\dfrac{\partial^2 f}{\partial x^2} = \dfrac{y^2 - x^2}{(x^2 + y^2)^2}, \quad \dfrac{\partial^2 f}{\partial y^2} = \dfrac{x^2 - y^2}{(x^2 + y^2)^2}, \quad \dfrac{\partial^2 f}{\partial y \partial x} = -\dfrac{2xy}{(x^2 + y^2)^2} = \dfrac{\partial^2 f}{\partial x \partial y}$

17. $\dfrac{\partial^2 f}{\partial x^2} = -2y^2 \cos 2xy, \quad \dfrac{\partial^2 f}{\partial y^2} = -2x^2 \cos 2xy, \quad \dfrac{\partial^2 f}{\partial y \partial x} = -[\sin 2xy + 2xy \cos 2xy] = \dfrac{\partial^2 f}{\partial x \partial y}$

19. $\dfrac{\partial^2 f}{\partial x^2} = 0, \quad \dfrac{\partial^2 f}{\partial y^2} = xz \sin y, \quad \dfrac{\partial^2 f}{\partial z^2} = -xy \sin z$

$\dfrac{\partial^2 f}{\partial y \partial x} = \sin z - z \cos y = \dfrac{\partial^2 f}{\partial x \partial y}$

$\dfrac{\partial^2 f}{\partial z \partial x} = y \cos z - \sin y = \dfrac{\partial^2 f}{\partial x \partial z}$

$\dfrac{\partial^2 f}{\partial z \partial y} = x \cos z - x \cos y = \dfrac{\partial^2 f}{\partial y \partial z}$

21. $x^2 \dfrac{\partial^2 u}{\partial x^2} + 2xy \dfrac{\partial^2 u}{\partial x \partial y} + y^2 \dfrac{\partial^2 u}{\partial y^2} = x^2 \left(\dfrac{-2y^2}{(x + y)^3} \right) + 2xy \left(\dfrac{2xy}{(x + y)^3} \right) + y^2 \left(\dfrac{-2x^2}{(x + y)^3} \right) = 0$ 23. $\dfrac{\partial^2 f}{\partial x^2} + \dfrac{\partial^2 f}{\partial y^2} = e^x \sin y - e^x \sin y = 0$

25. $\dfrac{\partial^2 f}{\partial x^2} + \dfrac{\partial^2 f}{\partial y^2} = \dfrac{y^2 - x^2}{x^2 + y^2} + \dfrac{x^2 - y^2}{x^2 + y^2} = 0$ 27. $\dfrac{\partial^2 f}{\partial x^2} = e^{x+y} \cos \sqrt{2}z, \quad \dfrac{\partial^2 f}{\partial y^2} = e^{x+y} \cos \sqrt{2}z, \quad \dfrac{\partial^2 f}{\partial z^2} = -2e^{x+y} \cos \sqrt{2}z$

29. $\dfrac{\partial^2 f}{\partial t^2} = -5c^2\sin(x+ct)\cos(2x+2ct) - 4c^2\cos(x+ct)\sin(2x+2ct)$

$\dfrac{\partial^2 f}{\partial x^2} = -5\sin(x+ct)\cos(2x+2ct) - 4\cos(x+ct)\sin(2x+2ct)$, so $\dfrac{\partial^2 f}{\partial t^2} - c^2\dfrac{\partial^2 f}{\partial x^2} = 0$

31. $\dfrac{\partial^2 f}{\partial t^2} = c^2 k^2(Ae^{kx} + Be^{-kx})(Ce^{ckt} + De^{-ckt})$, $\dfrac{\partial^2 f}{\partial x^2} = k^2(Ae^{kx} + Be^{-kx})(Ce^{ckt} + De^{-ckt})$, so $\dfrac{\partial^2 f}{\partial t^2} - c^2\dfrac{\partial^2 f}{\partial x^2} = 0$

33. (a) mixed partials are 0   (b) mixed partials are $g'(x)\, h'(y)$ (c) by the hint mixed partials for each term $x^m y^n$ are $mnx^{m-1}y^{n-1}$

35. (a) no, since $\dfrac{\partial^2 f}{\partial x\,\partial y} \neq \dfrac{\partial^2 f}{\partial y\,\partial x}$ (b) no, since $\dfrac{\partial^2 f}{\partial x\,\partial y} \neq \dfrac{\partial^2 f}{\partial y\,\partial x}$ for $x \neq y$

37.
$$\underset{\substack{\uparrow \\ \text{by definition}}}{\dfrac{\partial^3 f}{\partial x^2 \partial y}} = \dfrac{\partial}{\partial x}\left(\dfrac{\partial^2 f}{\partial x\,\partial y}\right)$$

$$= \underset{\substack{\uparrow \\ (14.6.5)}}{\dfrac{\partial}{\partial x}\left(\dfrac{\partial^2 f}{\partial y\,\partial x}\right)} = \underset{\substack{\uparrow \\ \text{by definition}}}{\dfrac{\partial^2}{\partial x\,\partial y}\left(\dfrac{\partial f}{\partial x}\right)} = \underset{\substack{\uparrow \\ (14.6.5)}}{\dfrac{\partial^2}{\partial y\,\partial x}\left(\dfrac{\partial f}{\partial x}\right)} = \underset{\substack{\uparrow \\ \text{by definition}}}{\dfrac{\partial}{\partial y}\left(\dfrac{\partial^2 f}{\partial x^2}\right)} = \dfrac{\partial^3 f}{\partial y\,\partial x^2}$$

39. (a) 0   (b) 0   (c) $\dfrac{m}{1+m^2}$ (d) 0   (e) $\dfrac{f'(0)}{1+[f'(0)]^2}$ (f) $\tfrac{1}{4}\sqrt{3}$ (g) does not exist

41. (a) $\dfrac{\partial g}{\partial x}(0,0) = \lim_{h\to 0}\dfrac{g(h,0) - g(0,0)}{h} = \lim_{h\to 0} 0 = 0$, $\dfrac{\partial g}{\partial y}(0,0) = \lim_{h\to 0}\dfrac{g(0,h) - g(0,0)}{h} = \lim_{h\to 0} 0 = 0$

(b) as (x,y) tends to $(0,0)$ along the x-axis, $g(x,y) = g(x,0) = 0$ tends to 0;
as (x,y) tends to $(0,0)$ along the line $y = x$, $g(x,y) = g(x,x) = \tfrac{1}{2}$ tends to $\tfrac{1}{2}$

43. For $y \neq 0$, $\dfrac{\partial f}{\partial x}(0,y) = \lim_{h\to 0}\dfrac{f(h,y) - f(0,y)}{h} = \lim_{h\to 0}\dfrac{y(y^2 - h^2)}{h^2 + y^2} = y$.   Since $\dfrac{\partial f}{\partial x}(0,0) = \lim_{h\to 0}\dfrac{f(h,0) - f(0,0)}{h} = \lim_{h\to 0} 0 = 0$,

we have $\dfrac{\partial f}{\partial x}(0,y) = y$ for all y.   For $x \neq 0$, $\dfrac{\partial f}{\partial y}(x,0) = \lim_{h\to 0}\dfrac{f(x,h) - f(x,0)}{h} = \lim_{h\to 0}\dfrac{x(h^2 - x^2)}{x^2 + h^2} = -x$.

Since $\dfrac{\partial f}{\partial y}(0,0) = \lim_{h\to 0}\dfrac{f(0,h) - f(0,0)}{h} = \lim_{h\to 0} 0 = 0$, we have $\dfrac{\partial f}{\partial y}(x,0) = -x$ for all x.

Therefore $\dfrac{\partial^2 f}{\partial y\,\partial x}(0,y) = 1$ for all y and $\dfrac{\partial^2 f}{\partial x\,\partial y}(x,0) = -1$ for all x.   In particular, $\dfrac{\partial^2 f}{\partial y\,\partial x}(0,0) = 1$, while $\dfrac{\partial^2 f}{\partial x\,\partial y}(0,0) = -1$.

CHAPTER 15

SECTION 15.1

1. $\nabla f = e^{xy}[(xy+1)\mathbf{i} + x^2\mathbf{j}]$ **3.** $\nabla f = (6x-y)\mathbf{i} + (1-x)\mathbf{j}$ **5.** $\nabla f = 2xy^{-2}\mathbf{i} - 2x^2 y^{-3}\mathbf{j}$

7. $\nabla f = z\cos(x-y)\mathbf{i} - z\cos(x-y)\mathbf{j} + \sin(x-y)\mathbf{k}$ **9.** $\nabla f = (y+z)\mathbf{i} + (x+z)\mathbf{j} + (x+y)\mathbf{k}$

11. $\nabla f = e^{x-y}[(1+x+y)\mathbf{i} + (1-x-y)\mathbf{j}]$ **13.** $\nabla f = e^x[\ln y\,\mathbf{i} + y^{-1}\mathbf{j}]$ **15.** $\nabla f = \dfrac{AD - BC}{(Cx+Dy)^2}[y\mathbf{i} - x\mathbf{j}]$

17. $\nabla f = (ye^x + xye^x - ze^y\cos xz)\mathbf{i} + (xe^x + e^z - e^y\sin xz)\mathbf{j} + (ye^z - xe^y\cos xz)\mathbf{k}$

19. $\nabla f = e^{x+2y}\cos(z^2+1)\mathbf{i} + 2e^{x+2y}\cos(z^2+1)\mathbf{j} - 2ze^{x+2y}\sin(z^2+)\mathbf{k}$ **21.** $\nabla f = -\mathbf{i} + 18\mathbf{j}$ **23.** $\tfrac{4}{5}\mathbf{i} + \tfrac{2}{5}\mathbf{j}$ **25.** $\mathbf{i}$

27. $\nabla f = -\tfrac{1}{2}\sqrt{2}(\mathbf{i} + 2\mathbf{j} + \mathbf{k})$ **29.** $\mathbf{i} + \tfrac{3}{5}\mathbf{j} - \tfrac{4}{5}\mathbf{k}$ **31.** $(6x-y)\mathbf{i} + (1-x)\mathbf{j}$ **33.** $(2xy+z^2)\mathbf{i} + (2yz+x^2)\mathbf{j} + (2xz+y^2)\mathbf{k}$

35. $f(x,y) = x^2 y + y$ **37.** $f(x,y) = \dfrac{x^2}{2} + x\sin y - y^2$ **39.** (a) $(1/r^2)\mathbf{r}$ (b) $[(\cos r)/r]\mathbf{r}$ (c) $(e^r/r)\mathbf{r}$

41. (a) $(0,0)$ (b)   (c) f has an absolute minimum at $(0,0)$

$(0,0,1)$

43. (a) Let $\mathbf{c} = c_1\mathbf{i} + c_2\mathbf{j} + c_3\mathbf{k}$. First, we take $\mathbf{h} = h\mathbf{i}$. Since $\mathbf{c} \cdot \mathbf{h}$ is $o(\mathbf{h})$,

$$0 = \lim_{h\to 0} \frac{\mathbf{c} \cdot \mathbf{h}}{\|\mathbf{h}\|} = \lim_{h\to 0} \frac{c_1 h}{h} = c_1.$$

Similarly, $c_2 = 0$ and $c_3 = 0$.

(b) $(\mathbf{y} - \mathbf{z}) \cdot \mathbf{h} = [f(\mathbf{x} + \mathbf{h}) - f(\mathbf{x}) - \mathbf{z} \cdot \mathbf{h}] + [\mathbf{y} \cdot \mathbf{h} - f(\mathbf{x} + \mathbf{h}) + f(\mathbf{x})] = o(\mathbf{h}) + o(\mathbf{h}) = o(\mathbf{h})$, so that, by part (a), $\mathbf{y} - \mathbf{z} = \mathbf{0}$.

45. (a) In Section 14.6 we showed that f was not continuous at $(0, 0)$. It is therefore not differentiable at $(0, 0)$.

(b) For $(x, y) \neq (0, 0)$, $\dfrac{\partial f}{\partial x} = \dfrac{2y(y^2 - x^2)}{(x^2 + y^2)^2}$. As (x, y) tends to $(0, 0)$ along the y-axis, $\partial f/\partial x = 2/y$ tends to ∞.

47. (a), (b) $\left(\dfrac{\sqrt{2}}{2}, \dfrac{\sqrt{2}}{2}\right)$ maximum, $\left(-\dfrac{\sqrt{2}}{2}, \dfrac{\sqrt{2}}{2}\right)$ minimum, $\left(\dfrac{\sqrt{2}}{2}, -\dfrac{\sqrt{2}}{2}\right)$ minimum, $\left(-\dfrac{\sqrt{2}}{2}, -\dfrac{\sqrt{2}}{2}\right)$ maximum

SECTION 15.2

1. $-2\sqrt{2}$ **3.** $\frac{1}{3}(7 - 4e)$ **5.** $\frac{1}{4}\sqrt{2}(a - b)$ **7.** $\dfrac{2}{\sqrt{65}}$ **9.** $\frac{2}{3}\sqrt{6}$ **11.** $-3\sqrt{2}$ **13.** $\dfrac{\sqrt{3}\pi}{12}$ **15.** $-(x^2 + y^2)^{-1/2}$

17. (a) $\sqrt{2}[a(B - A) + b(C - B)]$ (b) $\sqrt{2}[a(A - B) + b(B - C)]$ **19.** $-\frac{7}{3}\sqrt{5}$ **21.** $\dfrac{18}{\sqrt{14}}$ or $\dfrac{-18}{\sqrt{14}}$

23. increases most rapidly in the direction of $\dfrac{1}{\sqrt{2}}\mathbf{i} + \dfrac{1}{\sqrt{2}}\mathbf{j}$, rate of change $2\sqrt{2}$; decreases most rapidly in the direction of $-\dfrac{1}{\sqrt{2}}\mathbf{i} - \dfrac{1}{\sqrt{2}}\mathbf{j}$, rate of change $-2\sqrt{2}$

25. increases most rapidly in the direction of $\dfrac{1}{\sqrt{6}}\mathbf{i} - \dfrac{2}{\sqrt{6}}\mathbf{j} + \dfrac{1}{\sqrt{6}}\mathbf{k}$, rate of change 1; decreases most rapidly in the direction of $-\dfrac{1}{\sqrt{6}}\mathbf{i} + \dfrac{2}{\sqrt{6}}\mathbf{j} - \dfrac{1}{\sqrt{6}}\mathbf{k}$, rate of change -1

27. $\nabla f = f'(x_0)\mathbf{i}$. If $f'(x_0) \neq 0$, the gradient points in the direction in which f increases: to the right if $f'(x_0) > 0$, to the left if $f'(x_0) < 0$.

29. (a) $\lim\limits_{h\to 0} \dfrac{f(h, 0) - f(0, 0)}{h} = \lim\limits_{h\to 0} \dfrac{\sqrt{h^2}}{h} = \lim\limits_{h\to 0} \dfrac{|h|}{h}$ does not exist (b) no; by Theorem 15.2.5 f cannot be differentiable at $(0, 0)$

31. (a) $-\frac{2}{3}\sqrt{97}$ (b) $-\frac{8}{3}$ (c) $-\frac{26}{3}\sqrt{2}$ **33.** (a) its projection onto the xy-plane is the curve $y = x^3$ from $(1, 1)$ to $(0, 0)$

(b) its projection onto the xy-plane is the curve $y = -2x^3$ from $(1, -2)$ to $(0, 0)$

35. its projection onto the xy-plane is the curve $(b^2)^{a^2} x^{b^2} = (a^2)^{b^2} y^{a^2}$ from (a^2, b^2) to $(0, 0)$

37. the curve $y = \ln |\sqrt{2} \sin x|$ in the direction of decreasing x

39. (a) 16 (b) 4 (c) $\frac{16}{17}\sqrt{17}$

(d) The limits computed in (a) and (b) are not directional derivatives. In (a) and (b) we have, in essence, computed $\nabla f(2, 4) \cdot \mathbf{r}_0$ taking $\mathbf{r}_0 = \mathbf{i} + 4\mathbf{j}$ in (a) and $\mathbf{r}_0 = \frac{1}{4}\mathbf{i} + \mathbf{j}$ in (b). In neither case is $\mathbf{r}_0$ a unit vector.

41. (b) $\dfrac{2\sqrt{3} - 3}{2}$ **43.** $\nabla(fg) = \left(f\dfrac{\partial g}{\partial x} + g\dfrac{\partial f}{\partial x}\right)\mathbf{i} + \left(f\dfrac{\partial g}{\partial y} + g\dfrac{\partial f}{\partial y}\right)\mathbf{j} = f\nabla g + g\nabla f$

45. $\nabla f^n = \dfrac{\partial f^n}{\partial x}\mathbf{i} + \dfrac{\partial f^n}{\partial y}\mathbf{j} = nf^{n-1}\dfrac{\partial f}{\partial x}\mathbf{i} + nf^{n-1}\dfrac{\partial f}{\partial y}\mathbf{j} = nf^{n-1}\nabla f$

SECTION 15.3

1. $C = (\frac{1}{3}, \frac{5}{3})$ **3.** (a) $f(x, y, z) = a_1 x + a_2 y + a_3 z + C$ (b) $f(x, y, z) = g(x, y, z) + a_1 x + a_2 y + a_3 z + C$

5. (a) U is not connected (b) (i) $g(\mathbf{x}) = f(\mathbf{x}) - 1$ (ii) $g(\mathbf{x}) = -f(\mathbf{x})$ (c) U is not connected

7. Since f is continuous at $\mathbf{a}$ and $f(\mathbf{a}) = A$, there exists $\delta > 0$ such that

if $\|\mathbf{x} - \mathbf{a}\| < \delta$ and $\mathbf{x} \in \Omega$, then $|f(\mathbf{x}) - A| < \epsilon$.

Whether $\mathbf{a}$ is on the boundary of Ω or in the interior of Ω, there exists $\mathbf{x}_1$ in the interior of Ω within δ of $\mathbf{a}$. That implies that $|f(\mathbf{x}_1) - A| < \epsilon$. A similar argument shows the existence of $\mathbf{x}_2$ with the desired property.

SECTION 15.4

1. e^t **3.** $\dfrac{-2\sin 2t}{1 + \cos^2 2t}$ **5.** $t^t\left[\dfrac{1}{t} + \ln t + (\ln t)^2\right] + \dfrac{1}{t}$ **7.** $3t^2 - 5t^4$ **9.** $2\omega(b^2 - a^2)\sin \omega t \cos \omega t + b\omega$ **11.** $\sin 2t - 3\cos 2t$

13. $e^{t/2}(\frac{1}{2}\sin 2t + 2\cos 2t) + e^{2t}(2\sin \frac{1}{2}t + \frac{1}{2}\cos \frac{1}{2}t)$ **15.** $e^{t^2}\left[2t\sin \pi t + \pi \cos \pi t\right]$ **17.** $1 - 4t + 6t^2 - 4t^3$

19. increasing $\frac{1288}{3}\pi$ in.^{3}/sec   **21.** 41.34 sq in./sec   **23.** $\dfrac{\partial u}{\partial s} = 2s\cos^2 t - t\sin s\cos t - st\cos s\cos t$

$$\frac{\partial u}{\partial t} = -2s^2\sin t\cos t + st\sin s\sin t - s\sin s\cos t$$

25. $\dfrac{\partial u}{\partial s} = 4s^3 t^2\tan(s + t^2) + s^4 t^2\sec^2(s + t^2);$ **27.** $\dfrac{\partial u}{\partial s} = 2s\cos^2 t - \sin(t - s)\cos t + s\cos t\cos(t - s) + 2t^2\sin s\cos s$

$\dfrac{\partial u}{\partial t} = 2s^4 t\tan(s + t^2) + 2s^4 t^3\sec^2(s + t^2)$ $\dfrac{\partial u}{\partial t} = -2s^2\sin t\cos t + s\sin(t - s)\sin t - s\cos t\cos(t - s) + 2t\sin^2 s$

29. $\dfrac{d}{dt}[f(\mathbf{r}(t))] = \left[\nabla f(\mathbf{r}(t)) \cdot \dfrac{\mathbf{r}'(t)}{\|\mathbf{r}'(t)\|}\right]\|\mathbf{r}'(t)\| = f'_{\mathbf{u}(t)}(\mathbf{r}(t))\,\|\mathbf{r}'(t)\|$ where $\mathbf{u}(t) = \dfrac{\mathbf{r}'(t)}{\|\mathbf{r}'(t)\|}$

31. (a) $(\cos r)\dfrac{\mathbf{r}}{r}$ (b) $(r\cos r + \sin r)\dfrac{\mathbf{r}}{r}$ **33.** (a) $(r\cos r - \sin r)\dfrac{\mathbf{r}}{r^3}$ (b) $\left(\dfrac{\sin r - r\cos r}{\sin^2 r}\right)\dfrac{\mathbf{r}}{r}$

35. (a) See the figure

(b) $\dfrac{\partial u}{\partial r} = \dfrac{\partial u}{\partial x}\left(\dfrac{\partial x}{\partial w}\dfrac{\partial w}{\partial r} + \dfrac{\partial x}{\partial t}\dfrac{\partial t}{\partial r}\right) + \dfrac{\partial u}{\partial y}\left(\dfrac{\partial y}{\partial w}\dfrac{\partial w}{\partial r} + \dfrac{\partial y}{\partial t}\dfrac{\partial t}{\partial r}\right) + \dfrac{\partial u}{\partial z}\left(\dfrac{\partial z}{\partial w}\dfrac{\partial w}{\partial r} + \dfrac{\partial z}{\partial t}\dfrac{\partial t}{\partial r}\right)$

To obtain $\dfrac{\partial u}{\partial s}$, replace each r by s.

37. $\dfrac{du}{dt} = \dfrac{\partial u}{\partial x}\dfrac{dx}{dt} + \dfrac{\partial u}{\partial y}\dfrac{dy}{dt}$

$\dfrac{d^2 u}{dt^2} = \dfrac{\partial u}{\partial x}\dfrac{d^2 x}{dt^2} + \dfrac{dx}{dt}\left[\dfrac{\partial^2 u}{\partial x^2}\dfrac{dx}{dt} + \dfrac{\partial^2 u}{\partial y\partial x}\dfrac{dy}{dt}\right] + \dfrac{\partial u}{\partial y}\dfrac{d^2 y}{dt^2} + \dfrac{dy}{dt}\left[\dfrac{\partial^2 u}{\partial x\partial y}\dfrac{dx}{dt} + \dfrac{\partial^2 u}{\partial y^2}\dfrac{dy}{dt}\right]$ and the result follows.

39. (a) $\dfrac{\partial u}{\partial r} = \dfrac{\partial u}{\partial x}\dfrac{\partial x}{\partial r} + \dfrac{\partial u}{\partial y}\dfrac{\partial y}{\partial r} = \dfrac{\partial u}{\partial x}\cos\theta + \dfrac{\partial u}{\partial y}\sin\theta,\quad \dfrac{\partial u}{\partial\theta} = \dfrac{\partial u}{\partial x}\dfrac{\partial x}{\partial\theta} + \dfrac{\partial u}{\partial y}\dfrac{\partial y}{\partial\theta} = \dfrac{\partial u}{\partial x}(-r\sin\theta) + \dfrac{\partial u}{\partial y}(r\cos\theta)$

(b) $\left(\dfrac{\partial u}{\partial r}\right)^2 = \left(\dfrac{\partial u}{\partial x}\right)^2\cos^2\theta + 2\dfrac{\partial u}{\partial x}\dfrac{\partial u}{\partial y}\cos\theta\sin\theta + \left(\dfrac{\partial u}{\partial y}\right)^2\sin^2\theta,\quad \dfrac{1}{r^2}\left(\dfrac{\partial u}{\partial\theta}\right)^2 = \left(\dfrac{\partial u}{\partial x}\right)^2\sin^2\theta - 2\dfrac{\partial u}{\partial x}\dfrac{\partial u}{\partial y}\cos\theta\sin\theta + \left(\dfrac{\partial u}{\partial y}\right)^2\cos^2\theta,$

$\left(\dfrac{\partial u}{\partial r}\right)^2 + \dfrac{1}{r^2}\left(\dfrac{\partial u}{\partial\theta}\right)^2 = \left(\dfrac{\partial u}{\partial x}\right)^2(\cos^2\theta + \sin^2\theta) + \left(\dfrac{\partial u}{\partial y}\right)^2(\sin^2\theta + \cos^2\theta) = \left(\dfrac{\partial u}{\partial x}\right)^2 + \left(\dfrac{\partial u}{\partial y}\right)^2$

41. Solve the equations in Exercise 39 (a) for $\dfrac{\partial u}{\partial x}, \dfrac{\partial u}{\partial y}$:   **43.** $\nabla u = r(2 - \sin 2\theta)\mathbf{e}_r + r^2\cos 2\theta\,\mathbf{e}_\theta$

$\dfrac{\partial u}{\partial x} = \dfrac{\partial u}{\partial r}\cos\theta - \dfrac{1}{r}\dfrac{\partial u}{\partial\theta}\sin\theta;\quad \dfrac{\partial u}{\partial y} = \dfrac{\partial u}{\partial r}\sin\theta + \dfrac{1}{r}\dfrac{\partial u}{\partial\theta}\cos\theta$

Then $\nabla f = \dfrac{\partial u}{\partial x}\mathbf{i} + \dfrac{\partial u}{\partial y}\mathbf{j} = \dfrac{\partial u}{\partial r}(\cos\theta\mathbf{i} + \sin\theta\mathbf{j}) + \dfrac{1}{r}\dfrac{\partial u}{\partial\theta}(-\sin\theta\mathbf{i} + \cos\theta\mathbf{j})$

45. From Exercise 39 (a),   **47.** $\dfrac{dy}{dx} = -\dfrac{e^y + ye^x - 4xy}{xe^y + e^x - 2x^2}$

$\dfrac{\partial^2 u}{\partial r^2} = \dfrac{\partial^2 u}{\partial x^2}\cos^2\theta + 2\dfrac{\partial^2 u}{\partial y\partial x}\sin\theta\cos\theta + \dfrac{\partial^2 u}{\partial y^2}\sin^2\theta$

$\dfrac{\partial^2 u}{\partial\theta^2} = \dfrac{\partial^2 u}{\partial x^2}r^2\sin^2\theta - 2\dfrac{\partial^2 u}{\partial y\partial x}r^2\sin\theta\cos\theta + \dfrac{\partial^2 u}{\partial y^2}r^2\cos^2\theta - r\left(\dfrac{\partial u}{\partial x}\cos\theta + \dfrac{\partial u}{\partial y}\sin\theta\right).$

The term in parentheses is just $\dfrac{\partial u}{\partial r}$, and the result follows.

49. $\dfrac{dy}{dx} = \dfrac{\cos xy - xy\sin xy - y\sin x}{x^2\sin xy - \cos x}$ **51.** $\dfrac{\partial z}{\partial x} = -\dfrac{2x - yz(x^2 + y^2 + z^2)\sin xyz}{2z - xy(x^2 + y^2 + z^2)\sin xyz};\quad \dfrac{\partial z}{\partial y} = -\dfrac{2y - xz(x^2 + y^2 + z^2)\sin xyz}{2z - xy(x^2 + y^2 + z^2)\sin xyz}$

53. $\dfrac{\partial z}{\partial x} = -\dfrac{z\sec^2 x - y^2 z^3 - 2yz}{\tan x - 3xy^2 z^2 - 2xy}$; $\dfrac{\partial z}{\partial y} = \dfrac{2xyz^3 + 2xz}{\tan x - 3xy^2 z^2 - 2xy}$

55. $\dfrac{\partial \mathbf{u}}{\partial s} = \dfrac{\partial \mathbf{u}}{\partial x}\dfrac{\partial x}{\partial s} + \dfrac{\partial \mathbf{u}}{\partial y}\dfrac{\partial y}{\partial s}$, $\dfrac{\partial \mathbf{u}}{\partial t} = \dfrac{\partial \mathbf{u}}{\partial x}\dfrac{\partial x}{\partial t} + \dfrac{\partial \mathbf{u}}{\partial y}\dfrac{\partial y}{\partial t}$

where

$$\dfrac{\partial \mathbf{u}}{\partial x} = \dfrac{\partial u_1}{\partial x}\mathbf{i} + \dfrac{\partial u_2}{\partial x}\mathbf{j}, \quad \dfrac{\partial \mathbf{u}}{\partial y} = \dfrac{\partial u_1}{\partial y}\mathbf{i} + \dfrac{\partial u_2}{\partial y}\mathbf{j}$$

57. (a) $f(x, y) = f\left(x, x\dfrac{y}{x}\right) = x^n f(1, \nu)$, where $\nu = y/x$

$$\dfrac{\partial f}{\partial x} = nx^{n-1}f(1, \nu) + x^n\dfrac{df}{d\nu}\left(\dfrac{-y}{x^2}\right)$$

Multiplying through by x and noting that $x^{n-1}\dfrac{df}{d\nu} = \dfrac{\partial f}{\partial y}$, we get

$$x\dfrac{\partial f}{\partial x} = nx^n f(1, \nu) - yx^{n-1}\dfrac{df}{d\nu} = nf(x, y) - y\dfrac{\partial f}{\partial y}$$

and the result follows

(b) Differentiate the result in part (a) with respect to x and multiply the resulting equation by x; then do the same with y. Add the two results.

SECTION 15.5

1. normal vector $\mathbf{i} + \mathbf{j}$; tangent vector $\mathbf{i} - \mathbf{j}$
tangent line $x + y + 2 = 0$; normal line $x - y = 0$

3. normal vector $\sqrt{2}\mathbf{i} - 5\mathbf{j}$; tangent vector $5\mathbf{i} + \sqrt{2}\mathbf{j}$
tangent line $\sqrt{2}x - 5y + 3 = 0$; normal line $5x + \sqrt{2}y - 6\sqrt{2} = 0$

5. normal vector $7\mathbf{i} - 17\mathbf{j}$; tangent vector $17\mathbf{i} + 7\mathbf{j}$
tangent line $7x - 17y + 6 = 0$; normal line $17x + 7y - 82 = 0$

7. normal vector $\mathbf{i} - \mathbf{j}$; tangent vector $\mathbf{i} + \mathbf{j}$
tangent line $x - y - 3 = 0$; normal line $x + y + 1 = 0$

9. 0

11. $4x - 5y + 4z = 0$ **13.** $x + ay - z - 1 = 0$ **15.** $2x + 2y - z = 0$ **17.** $b^2 c^2 x_0 x - a^2 c^2 y_0 y - a^2 b^2 z_0 z - a^2 b^2 c^2 = 0$

19. $(a^2/b, b^2/a, 3ab)$ **21.** $(0, 0, 0)$ **23.** $(\frac{1}{3}, \frac{11}{6}, -\frac{1}{12})$ **25.** $\dfrac{x - x_0}{\partial f/\partial x\,(x_0, y_0, z_0)} = \dfrac{y - y_0}{\partial f/\partial y\,(x_0, y_0, z_0)} = \dfrac{z - z_0}{\partial f/\partial z\,(x_0, y_0, z_0)}$

27. the tangent planes meet at right angles and therefore the normals ∇F and ∇G must meet at right angles:

$$\dfrac{\partial F}{\partial x}\dfrac{\partial G}{\partial x} + \dfrac{\partial F}{\partial y}\dfrac{\partial G}{\partial y} + \dfrac{\partial F}{\partial z}\dfrac{\partial G}{\partial z} = 0$$

29. $\frac{9}{2}a^3$ $(V = \frac{1}{3}Bh)$ **31.** approx. 0.528 rad **33.** $3x + 4y + 6z = 22$, $6x + y - z = 11$

35. $(1, 1, 2)$ lies on both surfaces and the normals at this point are perpendicular

37. (a) $3x - 4y + 6 = 0$ (b) $\mathbf{r}(t) = (4t - 2)\mathbf{i} - 3t\mathbf{j} + (43t^2 - 16t + 6)\mathbf{k}$ (c) $\mathbf{R}(s) = (2\mathbf{i} - 3\mathbf{j} + 33\mathbf{k}) + s(4\mathbf{i} - 3\mathbf{j} + 70\mathbf{k})$
(d) $4x - 18y - z = 29$ (e) $\mathbf{r}(t) = t\mathbf{i} - (\frac{3}{4}t + \frac{3}{2})\mathbf{j} + (\frac{35}{2}t - 2)\mathbf{k}$; $l = l'$

SECTION 15.6

1. $(1, 0)$ gives a local max of 1 **3.** $(1, -1)$ is a saddle point **5.** $(-2, 1)$ gives a local min of -2

7. no stationary points and no local extreme values **9.** $(1, 1)$ gives a local min of -3 **11.** $(-2, 2)$ is a saddle point

13. $(0, 0)$ is a saddle point **15.** $(3, 4)$ gives absolute min of $\frac{1}{3}$; $(1, 1)$ gives absolute max of $\frac{1}{2}\sqrt{2}$

17. absolute max of 4 occurs at $(-1, 0)$ and $(1, 0)$; there is no absolute min

19. $(1, 1)$ gives absolute min of 0; $(-\sqrt{2}, -\sqrt{2})$ gives absolute max of $6 + 4\sqrt{2}$

21. absolute min of 0 occurs at each point (x, x) with $0 \le x \le 4$; absolute max of 144 occurs at $(0, 12)$

23. $(1, 1)$ gives absolute min of 10; $(2, 8)$ gives absolute max of 68 **25.** $(\frac{32}{9}, -\frac{16}{9}, \frac{32}{9})$; $\frac{16}{3}$ **27.** $x = 6, y = 6, z = 6$; maximum $= 216$

29. length = width = $2\sqrt[3]{3}$ feet; height = $\sqrt[3]{3}$ feet

31. (a) $(0, 0)$ (b) no local extremes as $(0, 0)$ is a saddle point
(c) $(1, 0)$ and $(-1, 0)$ give absolute max of $\frac{1}{4}$; $(0, 1)$ and $(0, -1)$ give absolute min of $-\frac{1}{9}$

33. $\left(\dfrac{x_1 + x_2 + x_3}{3}, \dfrac{y_1 + y_2 + y_3}{3}\right)$

35. $1 - \frac{1}{2}\sqrt{3}$ **37.** \$1 per dozen blades, \$2 per razor **39.** $(0, 0)$ saddle point; $(1, 1)$ local maximum

41. $(1, 0)$ local minimum; $(-1, 0)$ local maximum

SECTION 15.7

1. $(4, -2)$ gives a local min of -10 **3.** $(\frac{10}{3}, \frac{8}{3})$ gives a local max of $\frac{28}{3}$ **5.** $(0, 0)$ is a saddle point; $(2, 2)$ gives a local min of -8

7. $(1, \frac{3}{2})$ is a saddle point; $(5, \frac{27}{2})$ gives a local min of $-\frac{117}{4}$ **9.** no stationary points and no local extreme values

11. $(0, n\pi)$ for integral n are saddle points; no local extreme values **13.** $(1, -1)$ and $(-1, 1)$ are saddle points; no local extreme values

15. $(\frac{1}{2}, 4)$ gives a local min of 6   **17.** $(1, 1)$ gives a local min of 3   **19.** $(1, 0)$ gives a local min of -1; $(-1, 0)$ gives a local max of 1

21. $(0, 0)$ is a saddle point; $(1, 0)$ and $(-1, 0)$ give a local min of -3 **23.** $(0, 0)$ is a saddle point; $(\frac{\pi}{3}, \frac{\pi}{3})$ and $(-\frac{\pi}{3}, -\frac{\pi}{3})$ give a local max of $\frac{3}{2}$

25. (a) $f_x = 2x + ky, f_y = 2y + kx$; $f_x(0, 0) = f_y(0, 0) = 0$ independent of k. (b) $|k| > 2$ (c) $|k| < 2$ (d) $|k| = 2$

27. $(-\frac{5}{7}, \frac{4}{7}, \frac{20}{7})$ **29.** $\frac{1}{27}$ **31.** $\dfrac{\sqrt{114}}{36}$ **33.** (a) saddle point, $f(x, y) = 0$ along the plane curve $y = x^{2/3}$ (see figure)

(b) $(0, 0)$ gives a local max of 3

35. $\theta = \frac{1}{6}\pi, x = (2 - \sqrt{3})P, y = \frac{1}{6}(3 - \sqrt{3})P$ **37.** $\frac{2}{3}\sqrt{6}$ **39.** $4 \times 4 \times 6$ m

41. $V = \dfrac{xy(S - 2xy)}{2(x + y)}$ has a maximum value when $x = y = z = \sqrt{\dfrac{S}{6}}$.   **43.** (a) $y = x - \frac{2}{3}$ (b) $y = \frac{14}{13}x^2 - \frac{19}{13}$

45. (a) cross section: 18×18 inches; length: 36 inches   **47.** $x = 4$ in., $\theta = \dfrac{\pi}{3}$
(b) radius of cross section: $36/\pi$ inches; length: 36 inches.

SECTION 15.8

1. 2   **3.** $-\frac{1}{2}ab$ **5.** $\frac{2}{9}\sqrt{3}ab^2$ **7.** 1   **9.** $\frac{1}{9}\sqrt{3}abc$ **11.** $19\sqrt{2}$ **13.** $\frac{1}{27}abc$ **15.** 1

17. closest point $(\frac{2}{3}, \frac{1}{3}, \frac{2}{3})$; furthest point $(-\frac{2}{3}, -\frac{1}{3}, -\frac{2}{3})$ **19.** $f(3, -2, 1) = 14$ **21.** $|D|(A^2 + B^2 + C^2)^{-1/2}$

23. $4A^2(a^2 + b^2 + c^2)^{-1}$, where A is the area of the triangle and a, b, c, are the sides   **25.** $(2^{-1/3}, -2^{-1/3})$ **27.** hint is given

29. (a) $f(\frac{k}{2}, \frac{k}{2}) = \frac{k}{2}$ is the maximum value   (b) $(xy)^{1/2} = f(x, y) \le f(\frac{k}{2}, \frac{k}{2}) = \frac{k}{2} = \dfrac{x + y}{2}$

31. Same argument as Exercises 29 and 30: $f\left(\dfrac{k}{n}, \dfrac{k}{n}, \ldots, \dfrac{k}{n}\right) = \dfrac{k}{n}$ is the maximum value of $f(x_1, x_2, \ldots, x_n) = (x_1 x_2 \cdots x_n)^{1/n}$

33. radius: $\sqrt[3]{\dfrac{V}{2\pi}}$; height: $\sqrt[3]{\dfrac{4V}{\pi}}$ **35.** length = width = $2\sqrt[3]{3}$ feet; height = $\sqrt[3]{3}$ feet

37. (a) cross section: 18×18 inches; length: 36 inches   (b) radius of cross-section: $36/\pi$ inches; length: 36 inches

39. $Q_1 = 10,000, Q_2 = 20,000, Q_3 = 30,000$ **41.** $\dfrac{\sqrt{2}}{2}$

43. closest point $(-1 + \dfrac{\sqrt{2}}{2}, -1 + \dfrac{\sqrt{2}}{2}, -1 + \sqrt{2})$; furthest point $(-1 - \dfrac{\sqrt{2}}{2}, -1 - \dfrac{\sqrt{2}}{2}, -1 - \sqrt{2})$.

SECTION 15.9

1. $df = (3x^2y - 2xy^2)\,\Delta x + (x^3 - 2x^2y)\,\Delta y$ **3.** $df = (\cos y + y \sin x)\,\Delta x - (x \sin y + \cos x)\,\Delta y$ **5.** $df = \Delta x - (\tan z)\,\Delta y - (y \sec^2 z)\,\Delta z$

7. $df = \dfrac{y(y^2 + z^2 - x^2)}{(x^2 + y^2 + z^2)^2}\,\Delta x + \dfrac{x(x^2 + z^2 - y^2)}{(x^2 + y^2 + z^2)^2}\,\Delta y - \dfrac{2xyz}{(x^2 + y^2 + z^2)^2}\,\Delta z$ **9.** $df = [\cos(x + y) + \cos(x - y)]\,\Delta x + [\cos(x + y) - \cos(x - y)]\,\Delta y$

11. $df = (y^2 z e^{xz} + \ln z)\,\Delta x + 2y e^{xz}\,\Delta y + \left(xy^2 e^{xz} + \dfrac{x}{z}\right)\Delta z$ **13.** $\Delta u = -7.15, du = -7.5$ **15.** $\Delta u = 2.896; du = 2.5$

17. $22\frac{249}{352}$ taking $u = x^{1/2}y^{1/4}$, $x = 121$, $y = 16$, $\Delta x = 4$, $\Delta y = 1$

19. $\frac{1}{14}\sqrt{2}\pi$ taking $u = \sin x \cos y$, $x = \pi$, $y = \frac{1}{4}\pi$, $\Delta x = -\frac{1}{7}\pi$, $\Delta y = -\frac{1}{20}\pi$ **21.** $f(2.9, 0.01) \cong 8.67$

23. $f(2.94, 1.1, 0.92) \cong 23.392$ **25.** $dz = -\frac{1}{90}$, $\Delta z = -\frac{1}{93}$ **27.** decreases about 13.6π in.^{2} **29.** $S \cong 246.8$

31. (a) $dV = 0.24$ (b) $\Delta V = 0.22077$ **33.** $dT = 2.9$ **35.** $dA = x \tan \theta\,\Delta x + \frac{1}{2}x^2\sec^2\theta\,\Delta\theta$; more sensitive to a change in θ

37. (a) $\Delta h = -\dfrac{(2r + \Delta r)h}{(r + \Delta r)^2}\,\Delta r$, $\Delta h \cong -\left(\dfrac{2h}{r}\right)\Delta r$ (b) $\Delta h = -\dfrac{(2r + h + \Delta r)}{r + \Delta r}\,\Delta r$, $\Delta h \cong -\left(\dfrac{2r + h}{r}\right)\Delta r$

39. (a) 1.962 mm (b) 1.275 cm^2 **41.** $2.23 \le s \pm |\Delta s| \le 2.27$

SECTION 15.10

1. $f(x, y) = \frac{1}{2}x^2y^2 + C$ **3.** $f(x, y) = xy + C$ **5.** not a gradient   **7.** $f(x, y) = \sin x + y \cos x + C$ **9.** $f(x, y) = e^x \cos y^2 + C$

11. $f(x, y) = xye^x + e^{-y} + C$ **13.** not a gradient   **15.** $f(x, y) = x + xy^2 + \frac{1}{2}x^2y^2 + \frac{1}{2}y^2 + y + C$ **17.** $f(x, y) = \sqrt{x^2 + y^2} + C$

19. $f(x, y) = \frac{1}{3}x^3 \sin^{-1} y + y - y \ln y + C$ **21.** $f(x, y) = Ce^{x+y}$ **23.** (a), (b), (c) routine; (d) $f(x, y, z) = x^2 + yz + C$

25. $f(x, y, z) = x^2 + y^2 - z^2 + xy + yz + C$ **27.** $f(x, y, z) = xy^2z^3 + x + \frac{1}{2}y^2 + z + C$ **29.** $\mathbf{F}(\mathbf{r}) = \nabla\left(G\dfrac{mM}{r}\mathbf{r}\right)$

CHAPTER 16

SECTION 16.1

1. 819   **3.** 0   **5.** $a_2 - a_1$ **7.** $(a_2 - a_1)(b_2 - b_1)$ **9.** $a_2^2 - a_1^2$ **11.** $(a_2^2 - a_1^2)(b_2 - b_1)$

13. $2n(a_2 - a_1) - 3m(b_2 - b_1)$ **15.** $(a_2 - a_1)(b_2 - b_1)(c_2 - c_1)$ **17.** $a_{111} + a_{222} + \cdots + a_{nnn} = \sum_{p=1}^{n} a_{ppp}$

19. $\sum_{i=1}^{m} \sum_{j=1}^{n} \alpha a_{ij} = \sum_{i=1}^{m} \alpha \left(\sum_{j=1}^{n} a_{ij} \right) = \alpha \sum_{i=1}^{m} \sum_{j=1}^{n} a_{ij}$ **21.** $\sum_{i=1}^{m} \sum_{j=1}^{n} b_i c_j = \sum_{i=1}^{m} b_i \left(\sum_{j=1}^{n} c_j \right) = \left(\sum_{j=1}^{n} c_j \right) \sum_{i=1}^{m} b_i = \left(\sum_{i=1}^{m} b_i \right) \left(\sum_{j=1}^{n} c_j \right)$

SECTION 16.2

1. $L_f(P) = 2\frac{1}{4}$, $U_f(P) = 5\frac{3}{4}$ **3.** (a) $L_f(P) = \sum_{i=1}^{m} \sum_{j=1}^{n} (x_{i-1} + 2y_{i-1}) \, \Delta x_i \, \Delta y_j$, $U_f(P) = \sum_{i=1}^{m} \sum_{j=1}^{n} (x_i + 2y_j) \, \Delta x_i \, \Delta y_j$

(b) $I = 4$; the volume of the prism bounded above by the plane $z = x + 2y$ and below by R

5. $L_f(P) = -\frac{7}{24}$, $U_f(P) = \frac{7}{24}$ **7.** (a) $L_f(P) = \sum_{i=1}^{m} \sum_{j=1}^{n} 4x_{i-1} y_{j-1} \, \Delta x_i \, \Delta y_j$, $U_f(P) = \sum_{i=1}^{m} \sum_{j=1}^{n} 4x_i y_j \Delta x_i \Delta y_j$ (b) $I = b^2 d^2$

9. (a) $L_f(P) = \sum_{i=1}^{m} \sum_{j=1}^{n} 3(x_{i-1}^2 - y_j^2) \, \Delta x_i \, \Delta y_j$, $U_f(P) = \sum_{i=1}^{m} \sum_{j=1}^{n} 3(x_i^2 - y_{j-1}^2) \, \Delta x_i \, \Delta y_j$ (b) $I = bd(b^2 - d^2)$

11. $0 \le \sin(x + y) \le 1$ for all $(x, y) \in R$. Thus $0 \le \iint_R \sin(x + y) \, dx \, dy \le \iint_R dx \, dy = 1$.

SECTION 16.3

1. $\iint_\Omega dx \, dy = \int_a^b \phi(x) \, dx$

3. Suppose $f(x_0, y_0) \ne 0$. Assume $f(x_0, y_0) > 0$. Since f is continuous, there exists a disc Ω_ϵ with radius ϵ centered at (x_0, y_0) such that $f(x, y) > 0$ on Ω_ϵ. Let R be a rectangle contained in Ω_ϵ. Then $\iint_R f(x, y) \, dx \, dy > 0$, which contradicts the hypothesis.

5. 6

7. By Theorem 16.3.5, there exists a point (x_1, y_1) in D_r such that

$$\iint_{D_r} f(x, y) \, dx \, dy = f(x_1, y_1) \iint_{D_r} dx \, dy = f(x_1, y_1)\pi r^2$$

Thus $f(x_1, y_1) = \dfrac{1}{\pi r^2} \iint_{D_r} f(x, y) \, dx \, dy$. As $r \to 0$, $(x_1, y_1) \to (x_0, y_0)$ and $f(x_1, y_1) \to f(x_0, y_0)$ since f is continuous. The result follows.

SECTION 16.4

1. 1   **3.** $\frac{9}{2}$ **5.** $\frac{1}{24}$ **7.** -2 **9.** $\frac{1}{4}\pi^2 + \frac{1}{64}\pi^4$ **11.** $\frac{2}{27}$ **13.** $\frac{512}{15}$ **15.** 0   **17.** $\frac{1}{4}(e^4 - 1)$

19. $\int_0^1 \int_{y^{1/2}}^{y^{1/4}} f(x, y) \, dx \, dy$ **21.** $\int_{-1}^0 \int_{-x}^1 f(x, y) \, dy \, dx + \int_0^1 \int_x^1 f(x, y) \, dy \, dx$

23. $\int_1^2 \int_1^y f(x, y) \, dx \, dy + \int_2^4 \int_{y/2}^y f(x, y) \, dx \, dy + \int_4^8 \int_{y/2}^4 f(x, y) \, dx \, dy$ **25.** 9   **27.** $\frac{1}{160}$

29. $\frac{2}{3}(\cos\frac{1}{2} - \cos 1)$ **31.** $1 - \ln 2$ **33.** $\ln 4 - \frac{1}{2}$ **35.** 4   **37.** $\frac{2}{15}$

39. 3π **41.** $\frac{1}{6}$ **43.** $\frac{11}{70}$ **45.** $\frac{2}{3}a^3$ **47.** $\frac{1}{2}(e - 1)$ **49.** $\frac{1}{12}(e - 1)$ **51.** $\frac{2}{3}$ **53.** 1

55. $\displaystyle\iint_R f(x)g(y)\,dx\,dy = \int_c^d \int_a^b f(x)g(y)\,dx\,dy = \int_c^d \left(\int_a^b f(x)g(y)\,dx\right)dy = \int_c^d g(y)\left(\int_a^b f(x)\,dx\right)dy$

$$= \left(\int_a^b f(x)\,dx\right)\left(\int_c^d g(y)\,dy\right)$$

57. We have R: $-a \le x \le a,\ c \le y \le d$. Set $f(x, y) = g_y(x)$. For each fixed $y \in [c, d]$, g_y is an odd function. Thus

$$\int_{-a}^a g_y(x)\,dx = 0$$

Therefore

$$\iint_R f(x, y)\,dx\,dy = \int_c^d \int_{-a}^a f(x, y)\,dx\,dy = \int_c^d \int_{-a}^a g_y(x)\,dx\,dy = \int_c^d 0\,dy = 0.$$

59. Note that $\Omega = \{(x, y): 0 \le x \le y,\ 0 \le y \le 1\}$. Set $\Omega' = \{(x, y): 0 \le y \le x,\ 0 \le x \le 1\}$.

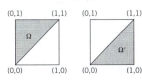

$$\iint_\Omega f(x)f(y)\,dx\,dy = \int_0^1 \int_0^y f(x)f(y)\,dx\,dy$$

$$= \int_0^1 \int_0^x f(y)f(x)\,dy\,dx$$

x and y are dummy variables ⟶

$$= \iint_{\Omega'} f(x)f(y)\,dx\,dy.$$

Note that Ω and Ω' don't overlap and their union is the unit square R: $\{(x, y): 0 \le x \le 1, 0 \le y \le 1\}$. If $\displaystyle\int_0^1 f(x)\,dx = 0$, then

$$0 = \left(\int_0^1 f(x)\,dx\right)\left(\int_0^1 f(y)\,dy\right) = \iint_R f(x)f(y)\,dx\,dy$$

by Exercise 55 ⟶

$$= \iint_\Omega f(x)f(y)\,dx\,dy + \iint_{\Omega'} f(x)f(y)\,dx\,dy$$

$$= 2\iint_\Omega f(x)f(y)\,dx\,dy$$

and therefore $\displaystyle\iint_\Omega f(x)f(y)\,dx\,dy = 0.$

61. Let M be the maximum value of $|f(x, y)|$ on Ω.

$$\int_{\phi_1(x+h)}^{\phi_2(x+h)} = \int_{\phi_1(x+h)}^{\phi_1(x)} + \int_{\phi_1(x)}^{\phi_2(x)} + \int_{\phi_2(x)}^{\phi_2(x+h)}$$

$$|F(x + h) - F(x)| = \left|\int_{\phi_1(x+h)}^{\phi_2(x+h)} f(x, y)\,dy - \int_{\phi_1(x)}^{\phi_2(x)} f(x, y)\,dy\right|$$

$$= \left|\int_{\phi_1(x+h)}^{\phi_1(x)} f(x, y)\,dy + \int_{\phi_2(x)}^{\phi_2(x+h)} f(x, y)\,dy\right| \le \left|\int_{\phi_1(x+h)}^{\phi_1(x)} f(x, y)\,dy\right| + \left|\int_{\phi_2(x)}^{\phi_2(x+h)} f(x, y)\,dy\right|$$

$$\le |\phi_1(x) - \phi_1(x + h)|M + |\phi_2(x + h) - \phi_2(x)|M.$$

The expression on the right tends to 0 as h tends to 0 since ϕ_1 and ϕ_2 are continuous.

SECTION 16.5

1. $\frac{1}{6}$ **3.** 6   **5.** (a) $\pi \sin 1 \cong 0.84\pi$ (b) $\pi(\sin 2 - \sin 1) \cong 0.07\pi$ **7.** (a) $\frac{2}{3}$ (b) $\frac{14}{3}$ **9.** $\frac{1}{3}\pi$ **11.** $\frac{1}{15}\pi$ **13.** $\dfrac{\pi}{4}(1 - \cos 1)$

15. $\frac{\pi}{2}$ **17.** $\frac{3\pi}{4}$ **19.** $\frac{4\pi}{3} + 2\sqrt{3}$ **21.** 4   **23.** $b^3\pi$ **25.** $\frac{16}{3}\sqrt{3}\pi$ **27.** $\frac{2}{3}(8 - 3\sqrt{3})\pi$ **29.** 2π **31.** $\frac{1}{3}\pi a^2 b$

SECTION 16.6

1. $M = \frac{2}{3}$, $x_M = 0$, $y_M = \frac{1}{2}$ **3.** $M = \frac{1}{6}$, $x_M = \frac{12}{21}$, $y_M = \frac{3}{4}$ **5.** $M = \frac{32}{3}$, $x_M = \frac{16}{3}$, $y_M = \frac{9}{7}$ **7.** $M = \frac{5}{8}$, $x_M = \frac{4}{5}$, $y_M = \frac{152}{75}$

9. $M = \frac{5\pi}{3}$, $x_M = \frac{21\pi}{20}$, $y_M = \frac{48}{25\pi}$

11. $I_x = \frac{1}{12}MW^2$, $I_y = \frac{1}{12}ML^2$, $I_z = \frac{1}{12}M(L^2 + W^2)$; $K_x = \frac{1}{6}\sqrt{3}W$, $K_y = \frac{1}{6}\sqrt{3}L$, $K_z = \frac{1}{6}\sqrt{3}\sqrt{L^2 + W^2}$ **13.** $x_M = \frac{1}{12}L$, $y_M = 0$

15. $I_x = I_y = \frac{1}{4}MR^2$, $I_z = \frac{1}{2}MR^2$; $K_x = K_y = \frac{1}{2}R$, $K_z = \frac{1}{2}\sqrt{2}R$ **17.** center the disc at a distance $\sqrt{I_0 - \frac{1}{2}MR^2}/\sqrt{M}$ from the origin

19. $I_x = \frac{1}{4}Mb^2$, $I_y = \frac{1}{4}Ma^2$, $I_z = \frac{1}{4}M(a^2 + b^2)$ **21.** $I_x = \frac{1}{10}$, $I_y = \frac{1}{16}$, $I_z = \frac{13}{80}$ **23.** $I_x = \frac{33\pi}{40}$, $I_y = \frac{93\pi}{40}$, $I_z = \frac{63\pi}{20}$

25. (a) $\frac{1}{4}M(r_2^2 + r_1^2)$ (b) $\frac{1}{4}M(r_2^2 + 5r_1^2)$ (c) $\frac{1}{4}M(5r_2^2 + r_1^2)$ **27.** $\frac{1}{2}M(r_2^2 + r_1^2)$ **29.** $x_M = 0$, $y_M = R/\pi$

31. on the diameter through P at a distance $\frac{6}{5}R$ from P

33. Suppose Ω, a basic region of area A, is broken up into n basic regions $\Omega_1, \ldots, \Omega_n$ with areas $A_1, \ldots, A_n$. Then

$$\bar{x}A = \iint_\Omega x \, dx \, dy = \sum_{i=1}^n \left(\iint_{\Omega_i} x \, dx \, dy \right) = \sum_{i=1}^n \bar{x}_i A_i = \bar{x}_1 A_1 + \cdots + \bar{x}_n A_n.$$

The second formula follows in the same manner.

SECTION 16.7

1. they are equal   **3.** $\iiint_\Pi \alpha \, dx \, dy \, dz = \alpha \iiint_\Pi dx \, dy \, dz = \alpha(\text{volume of } \Pi) = \alpha(a_2 - a_1)(b_2 - b_1)(c_2 - c_1)$ **5.** $\frac{1}{8}a^2 b^2 c$

7. $\bar{x} = \dfrac{A^2 BC - a^2 bc}{ABC - abc}$, $\bar{y} = \dfrac{AB^2 C - ab^2 c}{ABC - abc}$, $\bar{z} = \dfrac{ABC^2 - abc^2}{ABC - abc}$

9. $M = \frac{1}{2}ka^4$ where k is the constant of proportionality for the density function   **11.** $I_z = \frac{2}{3}Ma^2$

SECTION 16.8

1. abc **3.** $\frac{2}{3}$ **5.** 16   **7.** $\frac{1}{3}$ **9.** $\frac{47}{24}$ **11.** $\displaystyle\iiint_\Pi f(x)g(y)h(z) \, dx \, dy \, dz = \int_{c_1}^{c_2} \left[\int_{b_1}^{b_2} \left(\int_{a_1}^{a_2} f(x)g(y)h(z) \, dx \right) dy \right] dz$ **13.** 8

$$= \int_{c_1}^{c_2} \left[\int_{b_1}^{b_2} g(y)h(z) \left(\int_{a_1}^{a_2} f(x) \, dx \right) dy \right] dz$$

$$= \int_{c_1}^{c_2} \left[h(z) \left(\int_{a_1}^{a_2} f(x) \, dx \right) \left(\int_{b_1}^{b_2} g(y) \, dy \right) dz \right]$$

$$= \left(\int_{a_1}^{a_2} f(x) \, dx \right) \left(\int_{b_1}^{b_2} g(y) \, dy \right) \left(\int_{c_1}^{c_2} h(z) \, dz \right)$$

15. $(\frac{2}{3}a, \frac{2}{3}b, \frac{2}{3}c)$ **17.**   **19.** $(\frac{1}{2}, \frac{1}{3}, \frac{1}{3})$

21. $\displaystyle\int_{-r}^r \int_{-\sqrt{r^2 - x^2}}^{\sqrt{r^2 - x^2}} \int_{-\sqrt{r^2 - (x^2 + y^2)}}^{\sqrt{r^2 - (x^2 + y^2)}} k(r - \sqrt{x^2 + y^2 + z^2}) \, dz \, dy \, dx$ **23.** $\displaystyle\int_0^1 \int_{-\sqrt{x - x^2}}^{\sqrt{x - x^2}} \int_{-2x - 3y - 10}^{1 - y^2} dz \, dy \, dx$

25. $\displaystyle\int_{-1}^1 \int_{-2\sqrt{2 - 2x^2}}^{2\sqrt{2 - 2x^2}} \int_{3x^2 + y^2/4}^{4 - x^2 - y^2/4} k(z - 3x^2 - \frac{1}{4}y^2) \, dz \, dy \, dx$ **27.** $\frac{28}{3}$ **29.** $\frac{1}{270}$ **31.** $\frac{12}{5}$ **33.** $V = \frac{8}{3}$, $(\frac{11}{10}, \frac{9}{4}, \frac{11}{20})$

35. $V = \frac{27}{2}$, $(\frac{1}{2}, \frac{3}{2}, \frac{12}{5})$ **37.** $V = \frac{1}{6}abc$, $(\frac{1}{4}a, \frac{1}{4}b, \frac{1}{4}c)$ **39.** (a) $\frac{1}{3}M(a^2 + b^2)$ (b) $\frac{1}{12}M(a^2 + b^2)$ (c) $\frac{1}{3}Ma^2 + \frac{1}{12}Mb^2$

41. $M = \frac{1}{2}k$, $(\frac{7}{12}, \frac{34}{45}, \frac{37}{90})$ **43.** (a) 0 by symmetry   (b) $\frac{4}{3}\pi a_4$ **45.** $8 \displaystyle\int_0^a \int_0^{\sqrt{a^2 - x^2}} \int_0^{\sqrt{a^2 - x^2 - y^2}} dz \, dy \, dz = \frac{4}{3}\pi a^3$ **47.** $M = \frac{128}{15}k$

49. $M = \frac{135}{4}k$, $(\frac{1}{2}, \frac{9}{5}, \frac{12}{5})$ **51.** (a) $V = \int_0^6 \int_{z/2}^3 \int_x^{6-x} dy\, dx\, dz$

(b) $V = \int_0^3 \int_0^{2x} \int_x^{6-x} dy\, dz\, dx$

(c) $V = \int_0^6 \int_{z/2}^3 \int_{z/2}^y dx\, dy\, dz + \int_0^6 \int_3^{(12-z)/2} \int_{z/2}^{6-y} dx\, dy\, dz$

53. (a) $V = \iint_{\Omega_{yz}} 2y\, dy\, dz$ (b) $V = \iint_{\Omega_{yz}} \left(\int_{-y}^y dx \right) dy\, dz$ (c) $V = \int_0^4 \int_{-\sqrt{4-y}}^{\sqrt{4-y}} \int_{-y}^y dx\, dz\, dy$ (d) $V = \int_{-2}^2 \int_0^{4-z^2} \int_{-y}^y dx\, dy\, dz$

SECTION 16.9

1. $r^2 + z^2 = 9$ **3.** $z = 2r$ **5.** $4r^2 = z^2$ **7.** 4π **9.** $\frac{1}{3}$

11. $M = \frac{1}{2}k\pi R^2 h^2$ **13.** $\frac{1}{2}MR^2$ **15.** Inverting the cone and placing the vertex at the origin, we have
$$V = \int_0^h \int_0^{2\pi} \int_0^{(R/h)z} r\, dr\, d\theta\, dz = \frac{1}{3}\pi R^2 h.$$

17. $\frac{3}{10}MR^2$ **19.** $\frac{1}{2}\pi$ **21.** $\frac{1}{4}k\pi$ **23.** $\frac{1}{6}(8 - 3\sqrt{3})\pi$ **25.** $\dfrac{9\pi^2}{8}$ **27.** $\dfrac{\pi}{2}(1 - \cos 1) \cong 0.7221$ **29.** $\frac{32}{9}a^3$

31. $\frac{1}{36}a^2(9\pi - 16)$ **33.** $\frac{1}{32}\pi$ **35.** $\frac{1}{3}\pi(2 - \sqrt{3})$

SECTION 16.10

1. $(\sqrt{3}, \frac{1}{4}\pi, \cos^{-1}[\frac{1}{3}\sqrt{3}])$ **3.** $(\frac{3}{4}, \frac{3}{4}\sqrt{3}, \frac{3}{2}\sqrt{3})$ **5.** $(\rho, \theta, \phi) = \left(\dfrac{4\sqrt{6}}{3}, \dfrac{\pi}{4}, \dfrac{\pi}{3} \right)$ **7.** $(x, y, z) = (0, 0, 3)$

9. the circular cylinder $x^2 + y^2 = 1$; the radius of the cylinder is 1 and the axis is the z-axis   **11.** the lower nappe of the cone $z^2 = x^2 + y^2$

13. horizontal plane one unit above the xy-plane   **15.** $\dfrac{\pi}{3}$ **17.** $\dfrac{\pi}{2}$ **19.** $\dfrac{\pi}{3}(\sqrt{2} - 1)$ **21.** $\dfrac{243\pi}{20}$ **23.** $V = \frac{4}{3}\pi R^3$ **25.** $V = \frac{2}{3}\alpha R^3$

27. $M = \frac{1}{6}k\pi h[(r^2 + h^2)^{3/2} - h^3]$ **29.** (a) $\frac{2}{5}MR^2$ (b) $\frac{7}{5}MR^2$ **31.** (a) $\frac{2}{5}M\left(\dfrac{R_2^5 - R_1^5}{R_2^3 - R_1^3} \right)$ (b) $\frac{2}{3}MR^2$ (c) $\frac{5}{3}MR^2$

33. $V = \frac{2}{3}\pi(1 - \cos \alpha)a^3$ **35.** (a) $\rho = 2R \cos\phi$ (b) $0 \le \theta \le 2\pi$, $0 \le \phi \le \frac{1}{4}\pi$, $R \sec\phi \le \rho \le 2R \cos\phi$ **37.** $V = \frac{1}{3}(16 - 6\sqrt{2})\pi$

39. Encase T in a spherical wedge W. W has spherical coordinates in a box Π that contains S. Define f to be zero outside of T. Then $F(\rho, \theta, \phi) = f(\rho \sin\phi \cos\theta, \rho \sin\phi \sin\theta, \rho \cos\phi)$ is zero outside of S and
$$\iiint_T f(x, y, z)\, dx\, dy\, dz = \iiint_W f(x, y, z)\, dx\, dy\, dz$$
$$= \iiint_\Pi F(\rho, \theta, \phi)\rho^2 \sin\phi\, d\rho\, d\theta\, d\phi = \iiint_S F(\rho, \theta, \phi)\rho^2 \sin\phi\, d\rho\, d\theta\, d\phi.$$

41. $\mathbf{F} = \dfrac{GmM}{R^2}(\sqrt{2} - 1)\mathbf{k}$

SECTION 16.11

1. $ad - bc$ **3.** $2(v^2 - u^2)$ **5.** $u^2v^2 - 4uv$ **7.** abc **9.** r **11.** $w(1 + w \cos v)$ **13.** $\frac{1}{2}$ **15.** 0   **17.** $\frac{2}{3}$

19. (a) $A = 3 \ln 2$ (b) $\bar{x} = \dfrac{7}{9 \ln 2}$, $\bar{y} = \dfrac{14}{9 \ln 2}$ **21.** $I_x = \frac{4}{75}M$, $I_y = \frac{14}{75}M$, $I_z = \frac{18}{75}M$ **23.** $A = \frac{32}{15}$ **25.** $A = \pi/\sqrt{65}$

27. $J = ab\alpha r \cos^{\alpha-1}\theta \sin^{\alpha-1}\theta$ **29.** $V = \frac{4}{3}\pi abc$ **31.** $I_x = \frac{1}{5}M(b^2 + c^2)$, $I_y = \frac{1}{5}M(a^2 + c^2)$, $I_z = \frac{1}{5}M(a^2 + b^2)$ **33.** $\frac{1}{2}\pi^{3/2}$

CHAPTER 17

SECTION 17.1

1. (a) 1   (b) -2 3. 0   5. (a) $-\frac{17}{6}$ (b) $\frac{17}{6}$ 7. -8 9. $-\pi$ 11. (a) 1   (b) $\frac{23}{21}$

13. (a) $2 + \sin 2 - \cos 3$ (b) $\frac{4}{5} + \sin 1 - \cos 1$ 15. $\frac{1}{3}$ 17. (a) 28   (b) $\frac{1}{6}$ 19. $\dfrac{8\pi^3}{3}$

21. $\displaystyle\int_C \mathbf{q} \cdot d\mathbf{r} = \int_a^b [\mathbf{q} \cdot \mathbf{r}'(u)] \, du + \int_a^b \frac{d}{du}[\mathbf{q} \cdot \mathbf{r}(u)] \, du = [\mathbf{q} \cdot \mathbf{r}(b)] - [\mathbf{q} \cdot \mathbf{r}(a)] = \mathbf{q} \cdot [\mathbf{r}(b) - \mathbf{r}(a)]$

$\displaystyle\int_C \mathbf{r} \cdot d\mathbf{r} = \int_a^b [\mathbf{r}(u) \cdot \mathbf{r}'(u)] \, du = \frac{1}{2}\int_a^b \frac{d}{du}[\mathbf{r}(u) \cdot \mathbf{r}(u)] \, du = \frac{1}{2}\int_a^b \frac{d}{du}(\|\mathbf{r}(u)\|^2) \, du = \frac{1}{2}(\|\mathbf{r}(b)\|^2 - \|\mathbf{r}(a)\|^2)$

23. $\displaystyle\int_C \mathbf{f}(\mathbf{r}) \cdot d\mathbf{r} = \int_a^b [\mathbf{f}(\mathbf{r}(u)) \cdot \mathbf{r}'(u)] \, du = \int_a^b [f(u)\mathbf{i} \cdot \mathbf{i}] \, du = \int_a^b f(u) \, du$ 25. $|W| = $ area of ellipse

27. force at time $t = m\mathbf{r}''(t) = m(2\beta\mathbf{j} + 6\gamma t\mathbf{k})$; $W = (2\beta^2 + \frac{9}{2}\gamma^2)m$ 29. 0   31. (a) $\left(\dfrac{1}{\sqrt{5}} - \dfrac{1}{\sqrt{14}}\right)\mathbf{k}$ (b) $\dfrac{4}{5}\mathbf{k}$ 33. $\alpha = \dfrac{15}{6}$

SECTION 17.2

1. 0   3. -1 5. 0   7. 0   9. 0   11. 4   13. $e^3 - 2e^2 + 3$ 15. $e^5 - 2e^2 + 1$ 17. 2π 19. 0

21. Set $f(x, y, z) = g(x)$ and $C: \mathbf{r}(u) = u\mathbf{i}, \quad u \in [a, b]$. In this case, $\nabla f(\mathbf{r}(u)) = g'(x(u))\mathbf{i} = g'(u)\mathbf{i}$ and $\mathbf{r}'(u) = \mathbf{i}$, so that

$$\int_C \nabla f(\mathbf{r}) \cdot d\mathbf{r} = \int_a^b [\nabla f(\mathbf{r}(u)) \cdot \mathbf{r}'(u)] \, du = \int_a^b g'(u) \, du$$

and

$$f(\mathbf{r}(b)) - f(\mathbf{r}(a)) = g(b) - g(a).$$

The statement $\displaystyle\int_C \nabla f(\mathbf{r}) \cdot d\mathbf{r} = f(\mathbf{r}(b)) - f(\mathbf{r}(a))$ reduces to $\displaystyle\int_a^b g'(u) \, du = g(b) - g(a)$.

23. $W = mG\left(\dfrac{1}{r_2} - \dfrac{1}{r_1}\right)$ 25. $f(x, y, z) = \dfrac{mGr_0^2}{r_0 + z}$ 27. $7575.6G$

SECTION 17.3

1. If f is continuous, then $-f$ is continuous and has antiderivatives u. The scalar fields $U(x, y, z) = u(x)$ are potential functions for $\mathbf{F}$:

$$\nabla U = \frac{\partial U}{\partial x}\mathbf{i} + \frac{\partial U}{\partial y}\mathbf{j} + \frac{\partial U}{\partial z}\mathbf{k} = \frac{du}{dx}\mathbf{i} = -f\mathbf{i} = -\mathbf{F}.$$

3. The scalar field $U(x, y, z) = cz + d$ is a potential energy function for $\mathbf{F}$. We know that the total mechanical energy remains constant. Thus, for any times t_1 and t_2,

$$\tfrac{1}{2}m[v(t_1)]^2 + U(\mathbf{r}(t_1)) = \tfrac{1}{2}m[v(t_2)]^2 + U(\mathbf{r}(t_2)).$$

This gives

$$\tfrac{1}{2}m[v(t_1)]^2 + cz(t_1) + d = \tfrac{1}{2}m[v(t_2)]^2 + cz(t_2) + d.$$

Solve this equation for $v(t_2)$ and you have the desired formula.

5. (a) We know that $-\nabla U$ points in the direction of maximum decrease of U. Thus $\mathbf{F} = -\nabla U$ attempts to drive objects toward a region where U has lower values.   (b) At a point where U has a minimum, $\nabla U = \mathbf{0}$ and therefore $\mathbf{F} = \mathbf{0}$.

7. (a) By conservation of energy $\frac{1}{2}mv^2 + U = E$. Since E is constant and U is constant, v is constant.
   (b) ∇U is perpendicular to any surface where U is constant. Obviously so is $\mathbf{F} = -\nabla U$.

9. $f(x, y, z) = -\dfrac{k}{\sqrt{x^2 + y^2 + z^2}}$ is a potential function for $\mathbf{F}$. The work done by $\mathbf{F}$ moving an object along C is $W = \displaystyle\int_C \mathbf{F}(\mathbf{r}) \cdot d\mathbf{r} = \int_a^b \nabla f \cdot d\mathbf{r} = f(\mathbf{r}(b)) - f(\mathbf{r}(a))$. Since $\mathbf{r}(a) = (x_0, y_0, z_0)$ and $\mathbf{r}(b) = (x_1, y_1, z_1)$ are points on the unit sphere, $f(\mathbf{r}(b)) = f(\mathbf{r}(a)) = -k$, and so $W = 0$.

SECTION 17.4

1. $\frac{1}{2}$ 3. $\frac{9}{2}$ 5. $\frac{11}{6}$ 7. 2   9. 16   11. $\frac{104}{5}$ 13. $\frac{1}{12}$ 15. $\dfrac{1}{3} - \dfrac{\pi}{16}$ 17. 4   19. 4   21. 2   23. 3   25. $\frac{176}{3}$

27. 32   29. $\frac{1177}{30}$ 31. (a) $\dfrac{\partial P}{\partial y} = 6x - 4y = \dfrac{\partial Q}{\partial x}$ (b) 7   (c) $\dfrac{-37}{3}$

33. (a) $M = 2ka^2$, $x_M = y_M = \frac{1}{8}a(\pi + 2)$ (b) $I_x = ka^4 = \frac{1}{2}Ma^2$ 35. (a) $I_z = 2ka^4 = Ma^2$ (b) $I = \frac{1}{3}ka^4 = \frac{1}{6}Ma^2$

37. (a) $L = 2\pi\sqrt{a^2 + b^2}$ (b) $x_M = y_M = 0$, $z_M = \pi b$ (c) $I_x = I_y = \frac{1}{6}M(3a^2\pi + 8b^2\pi^2)$, $I_z = Ma^2$ 39. $M = \frac{2}{3}\pi k\sqrt{a^2 + b^2}(3a^2 + 4\pi^2b^2)$

SECTION 17.5

1. $\frac{1}{6}$ **3.** π **5.** 2π **7.** a^2b **9.** 7π **11.** $5a\pi r^2$ **13.** 0   **15.** 0   **17.** πa^2 **19.** $\dfrac{ab}{2}$ **21.** $(c-a)A$

23. $3\pi r^2$ **25.** $\dfrac{3}{4}\pi$

27. Taking Ω to be of type II, we have

$$\iint_\Omega \frac{\partial Q}{\partial x}(x,y)\,dx\,dy = \int_c^d \int_{\phi_3(y)}^{\phi_4(y)} \frac{\partial Q}{\partial x}(x,y)\,dx\,dy = \int_c^d \{Q[\phi_4(y),y] - Q[\phi_3(y),y]\}\,dy$$

$$* = \int_c^d Q[\phi_4(y),y]\,dy - \int_c^d Q[\phi_3(y),y]\,dy.$$

Set $C_3:\mathbf{r}_3(u) = \phi_3(u)\mathbf{i} + u\mathbf{j},\ \ u\in[c,d]$ and $C_4:\mathbf{r}_4(u) = \phi_4(u)\mathbf{i} + u\mathbf{j},\ \ u\in[c,d].$ Then

$$\oint_C Q(x,y)\,dy = \int_{C_4} Q(x,y)\,dy - \int_{C_3} Q(x,y)\,dy = \int_c^d Q[\phi_4(u),u]\,du - \int_c^d Q[\phi_3(u),u]\,du.$$

Comparison with * proves the result.

29. $\displaystyle\oint_{C_1} = \oint_{C_2} + \oint_{C_3}$ **31.** (a) 0 (b) 0   **33.** If Ω is the region bounded by C, then

$$\oint_C \mathbf{v}\cdot d\mathbf{r} = \oint_C \frac{\partial\phi}{\partial x}\,dx + \frac{\partial\phi}{\partial y}\,dy = \iint_\Omega \left\{ \frac{\partial}{\partial x}\left(\frac{\partial\phi}{\partial y}\right) - \frac{\partial}{\partial y}\left(\frac{\partial\phi}{\partial x}\right) \right\}\,dx\,dy$$

is zero by equality of mixed partials.

35. $A = \dfrac{1}{2}\displaystyle\oint_C (-y\,dx + x\,dy)$

$= \dfrac{1}{2}\left[\displaystyle\int_{C_1} + \int_{C_2} + \cdots + \int_{C_n}\right]$

Now

$$\int_{C_i} (-y\,dx + x\,dy) = \int_0^1 [(y_i + u(y_{i+1} - y_i))(x_{i+1} - x_i) + (x_i + u(x_{i+1} - x_i))(y_{i+1} - y_i)]\,du$$

$$= x_i y_{i+1} - x_{i+1}y_i, \qquad i = 1, 2, \ldots, n \quad (x_{n+1} = x_1, y_{n+1} = y_1).$$

Thus $A = \dfrac{1}{2}[(x_1 y_2 - x_2 y_1) + (x_2 y_3 - x_3 y_2) + \cdots + (x_n y_1 - x_1 y_n)].$

SECTION 17.6

1. $4[(u^2 - v^2)\mathbf{i} - (u^2 + v^2)\mathbf{j} + 2uv\mathbf{k}]$ **3.** $2(\mathbf{j} - \mathbf{i})$ **5.** $\mathbf{r}(u,v) = 3\cos u\cos v\mathbf{i} + 2\sin u\cos v\mathbf{j} + 6\sin v\mathbf{k},\qquad u\in[0,2\pi],\quad v\in[0,\pi/2]$

7. $\mathbf{r}(u,v) = 2\cos u\cos v\mathbf{i} + 2\sin u\cos v\mathbf{j} + 2\sin v\mathbf{k},\qquad u\in[0,2\pi],\quad v\in(\pi/4,\pi/2]$ **9.** $\mathbf{r}(u,v) = u\mathbf{i} + g(u,v)\mathbf{j} + v\mathbf{k},\qquad (u,v)\in\Omega$

11. $x^2/a^2 + y^2/b^2 + z^2/c^2 = 1$; ellipsoid   **13.** $x^2/a^2 - y^2/b^2 = z$; hyperbolic paraboloid

15. $\mathbf{r}(u,v) = v\cos u\mathbf{i} + v\sin u\mathbf{j} + f(v)\mathbf{k}; 0\le u\le 2\pi, a\le v\le b$ **17.** area of $\Gamma = A_\Omega\sec\gamma$ **19.** $\frac{1}{2}\sqrt{a^2b^2 + a^2c^2 + b^2c^2}$

21. $\frac{1}{6}\pi(17\sqrt{17} - 1)$ **23.** $\frac{1}{6}\pi[(4a^2 + 1)^{3/2} - (a^2 + 1)^{3/2}]$ **25.** $\frac{1}{15}(36\sqrt{6} - 50\sqrt{5} + 32)$ **27.** 4π

29. (a) $\displaystyle\iint_\Omega \sqrt{\left[\frac{\partial g}{\partial y}(y,z)\right]^2 + \left[\frac{\partial g}{\partial z}(y,z)\right]^2 + 1}\,dy\,dz = \iint_\Omega \sec[\alpha(y,z)]\,dy\,dz$

where α is the angle between the unit normal with positive $\mathbf{i}$ component and the x-axis

(b) $\displaystyle\iint_\Omega \sqrt{\left[\frac{\partial h}{\partial x}(x,z)\right]^2 + \left[\frac{\partial h}{\partial z}(x,z)\right]^2 + 1}\,dx\,dz = \iint_\Omega \sec[\beta(x,z)]\,dx\,dz$

where β is the angle between the unit normal with positive $\mathbf{j}$ component and the y-axis

31. (a) $\mathbf{N}(u,v) = v\cos u\sin\alpha\cos\alpha\mathbf{i} + v\sin u\sin\alpha\cos\alpha\mathbf{j} - v\sin^2\alpha\mathbf{k}$ (b) $A = \pi s^2\sin\alpha$

33. $A = \sqrt{A_1^2 + A_2^2 + A_3^2}$ **35.** (a) $\frac{1}{4}\sqrt{2}\pi[\sqrt{6} + \ln(\sqrt{2} + \sqrt{3})]$ (b) $\frac{1}{2}a^2[\sqrt{2e^{4\pi} + 1} - \sqrt{3} + 2\pi + \ln(1 + \sqrt{3}) - \ln(1 + \sqrt{2e^{4x} + 1})]$

SECTION 17.7

1. $\frac{1}{2}[\sqrt{2} + \ln(1 + \sqrt{2})]$ **3.** $2\sqrt{2} - 1$ **5.** $\frac{1}{3}[2\sqrt{2} - 1]$ **7.** $\dfrac{9\sqrt{14}}{2}$ **9.** $\dfrac{4}{3}$ **11.** $\dfrac{4\pi}{3}$ **13.** $\frac{1}{2}\sqrt{3}a^2k$ **15.** $\frac{1}{12}\sqrt{3}a^4k$

17. $(0, 0, \frac{1}{2}a)$ **19.** 2   **21.** $\frac{4}{3}\pi a^3$ **23.** 0   **25.** $\frac{1}{2}a^3$ **27.** 0   **29.** $-\frac{3}{2}$ **31.** $2\pi la^2$ **33.** $\frac{8}{35}$ **35.** $-\frac{4}{63}$

37. $\bar{x} = \bar{y} = 0$, $\bar{z} = \frac{2}{3}s \cos \alpha$ **39.** $x_M = y_M = 0$, $z_M = \frac{3}{4}$ **41.** no answer required   **43.** $x_M = \frac{11}{9}$

45. Total flux out of the solid is 0. It is clear from a diagram that the outer unit normal to the cylindrical side of the solid is given by $\mathbf{n} = x\mathbf{i} + y\mathbf{j}$, in which case $\mathbf{v} \cdot \mathbf{n} = 0$. The outer unit normals to the top and bottom of the solid are $\mathbf{k}$ and $-\mathbf{k}$, respectively. So, here as well, $\mathbf{v} \cdot \mathbf{n} = 0$ and the total flux is 0.

47. $(4\sqrt{2} - \frac{7}{2})\pi$

SECTION 17.8

1. $\nabla \cdot \mathbf{v} = 2$, $\nabla \times \mathbf{v} = \mathbf{0}$ **3.** $\nabla \cdot \mathbf{v} = 0$, $\nabla \times \mathbf{v} = \mathbf{0}$ **5.** $\nabla \cdot \mathbf{v} = 6$, $\nabla \times \mathbf{v} = \mathbf{0}$

7. $\nabla \cdot \mathbf{v} = yz + 1$, $\nabla \times \mathbf{v} = -x\mathbf{i} + xy\mathbf{i} + (1 - x)z\mathbf{k}$ **9.** $\nabla \cdot \mathbf{v} = 1/r^2$, $\nabla \times \mathbf{v} = \mathbf{0}$

11. $\nabla \cdot \mathbf{v} = 2(x + y + z)e^{r^2}$, $\nabla \times \mathbf{v} = 2e^{r^2}[(y - z)\mathbf{i} - (x - z)\mathbf{j} + (x - y)\mathbf{k}]$ **13.** $\nabla \cdot \mathbf{v} = f'(x)$, $\nabla \times \mathbf{v} = \mathbf{0}$ **15.** use components

17. $\nabla \cdot \mathbf{F} = \dfrac{\partial P}{\partial x} + \dfrac{\partial Q}{\partial y} + \dfrac{\partial R}{\partial z} = 2 + 4 - 6 = 0$ **19.** $\nabla \times \mathbf{F} = \begin{vmatrix} \mathbf{i} & \mathbf{j} & \mathbf{k} \\ \frac{\partial}{\partial x} & \frac{\partial}{\partial y} & \frac{\partial}{\partial z} \\ x & y & -2z \end{vmatrix} = \mathbf{0}$ **21.** $\nabla^2 f = 12(x^2 + y^2 + z^2)$

23. $\nabla^2 f = 2y^3 z^4 + 6x^2 yz^4 + 12x^2 y^3 z^2$ **25.** $\nabla^2 f = e^r(1 + 2r^{-1})$ **27.** (a) $2r^2$ (b) $-1/r$

29. $\nabla^2 f = \nabla^2 g(r) = \nabla \cdot (\nabla g(r)) = \nabla \cdot (g'(r)r^{-1}\mathbf{r})$
$= [(\nabla g'(r)) \cdot r^{-1}\mathbf{r}] + g'(r)(\nabla \cdot r^{-1}\mathbf{r})$
$= \{[g''(r)r^{-1}\mathbf{r}] \cdot r^{-1}\mathbf{r}\} + g'(r)(2r^{-1}) = g''(r) + 2r^{-1}g'(r)$

33. $n = -1$

SECTION 17.9

1. $\displaystyle\iint_S (\mathbf{v} \cdot \mathbf{n}) \, d\sigma = \iiint_T (\nabla \cdot \mathbf{v}) \, dx \, dy \, dz = \iiint_T 3 \, dx \, dy \, dz = 3V = 4\pi$

3. $\displaystyle\iint_S (\mathbf{v} \cdot \mathbf{n}) \, d\sigma = \iiint_T (\nabla \cdot \mathbf{v}) \, dx \, dy \, dz = \iiint_T 2(x + y + z) \, dx \, dy \, dz.$

The flux is zero since the function $f(x, y, z) = 2(x + y + z)$ satisfies the relation $f(-x, -y, -z) = -f(x, y, z)$ and T is symmetric about the origin.

5.

Face	$\mathbf{n}$	$\mathbf{v} \cdot \mathbf{n}$	Flux	
$x = 0$	$-\mathbf{i}$	0	0	
$x = 1$	$\mathbf{i}$	1	1	
$y = 0$	$-\mathbf{j}$	0	0	
$y = 1$	$\mathbf{j}$	1	1	total flux = 3
$z = 0$	$-\mathbf{k}$	0	0	
$z = 1$	$\mathbf{k}$	1	1	

7.

Face	$\mathbf{n}$	$\mathbf{v} \cdot \mathbf{n}$	Flux	
$x = 0$	$-\mathbf{i}$	0	0	
$x = 1$	$\mathbf{i}$	1	1	
$y = 0$	$-\mathbf{j}$	xz		fluxes added up to 0 total flux = 2
$y = 1$	$\mathbf{j}$	$-xz$		
$z = 0$	$-\mathbf{k}$	0	0	
$z = 1$	$\mathbf{k}$	1	1	

$\displaystyle\iiint_T (\nabla \cdot \mathbf{v}) \, dx \, dy \, dz = \iiint_T 3 \, dx \, dy \, dz = 3V = 3$

$\displaystyle\iiint_T (\nabla \cdot \mathbf{v}) \, dx \, dy \, dz = \iiint_T 2(x + z) \, dx \, dy \, dz = 2(\bar{x} + \bar{z})V = 2(\frac{1}{2} + \frac{1}{2})1$

9. flux $= \displaystyle\iiint_T (1 + 4y + 6z) \, dx \, dy \, dz = (1 + 4\bar{y} + 6\bar{z})V = (1 + 0 + 3)9\pi = 36\pi$ **11.** $\frac{1}{24}$ **13.** 64π **15.** 0   **17.** $(A + B + C)V$

19. Let T be the solid enclosed by S and set $n = n_1\mathbf{i} + n_2\mathbf{j} + n_3\mathbf{k}$.

$\displaystyle\iint_S n_1 d\sigma = \iint_S (\mathbf{i} \cdot \mathbf{n}) \, d\sigma = \iiint_T (\nabla \cdot \mathbf{i}) \, dx \, dy \, dz = \iiint_T 0 \, dx \, dy \, dz = 0.$

Similarly $\displaystyle\iint_S n_2 d\sigma = 0$ and $\displaystyle\iint_S n_3 d\sigma = 0.$

21. A routine computation shows that $\nabla \cdot (\nabla f \times \nabla g) = 0$. Therefore

$\displaystyle\iint_S [(\nabla f \times \nabla g) \cdot \mathbf{n}] \, d\sigma = \iiint_T [\nabla \cdot (\nabla f \times \nabla g)] \, dx \, dy \, dz = 0.$

23. Set $\mathbf{F} = F_1\mathbf{i} + F_2\mathbf{j} + F_3\mathbf{k}$.

$$F_1 = \iint_S [\rho(z - c)\mathbf{i} \cdot \mathbf{n}] \, d\sigma = \iiint_T [\nabla \cdot \rho(z - c)\mathbf{i}] \, dx \, dy \, dz = \iiint_T \underbrace{\frac{\partial}{\partial x}[\rho(z - c)]}_{0} \, dx \, dy \, dz = 0.$$

Similarly $F_2 = 0$.

$$F_3 = \iint_S [\rho(z - c)\mathbf{k} \cdot \mathbf{n}] \, d\sigma = \iiint_T [\nabla \cdot \rho(z - c)\mathbf{k}] \, dx \, dy \, dz$$

$$= \iiint_T \frac{\partial}{\partial z}[(\rho(z - c)] \, dx \, dy \, dz = \iiint_T \rho \, dx \, dy \, dz = W.$$

SECTION 17.10

For Exercises 1 and 3: $\mathbf{n} = x\mathbf{i} + y\mathbf{j} + z\mathbf{k}$ and C: $\mathbf{r}(u) = \cos u\mathbf{i} + \sin u\mathbf{j}$, $u \in [0, 2\pi]$.

1. (a) $\displaystyle\iint_S [(\nabla \times \mathbf{v}) \cdot \mathbf{n}] \, d\sigma = \iint_S (\mathbf{0} \cdot \mathbf{n}) \, d\sigma = 0$.

(b) S is bounded by the unit circle C: $\mathbf{r}(u) = \cos u\mathbf{i} + \sin u\mathbf{j}$, $u \in [0, 2\pi]$.

$\displaystyle\oint_C \mathbf{v}(\mathbf{r}) \cdot d\mathbf{r} = 0$ since $\mathbf{v}$ is a gradient.

3. (a) $\displaystyle\iint_S [(\nabla \times \mathbf{v}) \cdot \mathbf{n}] \, d\sigma = \iint_S [(-3y^2\mathbf{i} + 2z\mathbf{j} + 2\mathbf{k}) \cdot (x\mathbf{i} + y\mathbf{j} + z\mathbf{k})] \, d\sigma$

$$= \iint_S (-3xy^2 + 2yz + 2z) \, d\sigma$$

$$= \underbrace{\iint_S (-3xy^2) \, d\sigma}_{\substack{\text{0 by symmetry}}} + \underbrace{\iint_S 2yz \, d\sigma}_{\substack{\text{0 by}\\\text{symmetry}}} + \underbrace{\iint_S 2z \, d\sigma}_{\substack{2\bar{z}A = 2\pi\\\text{by Ex. 17,}\\\text{Section 17.7}}} = 2\pi$$

(b) $\displaystyle\oint_C \mathbf{v}(\mathbf{r}) \cdot d\mathbf{r} = \oint_C z^2 \, dx + 2x \, dy = \oint_C 2x \, dy = \int_0^{2\pi} 2\cos^2 u \, du = 2\pi$

For Exercises 5 and 7 take S: $z = 2 - x - y$ with $0 \le x \le 2$, $0 \le y \le 2 - x$ and C as the triangle $(2, 0, 0)$, $(0, 2, 0)$, $(0, 0, 2)$. Then $C = C_1 \cup C_2 \cup C_3$ with

$$C_1: \mathbf{r}_1(u) = 2(1 - u)\mathbf{i} + 2u\mathbf{j}, \ u \in [0, 1],$$
$$C_2: \mathbf{r}_2(u) = 2(1 - u)\mathbf{j} + 2u\mathbf{k}, \ u \in [0, 1],$$
$$C_3: \mathbf{r}_3(u) = 2(1 - u)\mathbf{k} + 2u\mathbf{i}, \ u \in [0, 1].$$

$\mathbf{n} = \frac{1}{3}\sqrt{3}(\mathbf{i} + \mathbf{j} + \mathbf{k})$.

5. (a) $\displaystyle\iint_S [(\nabla \times \mathbf{v}) \cdot \mathbf{n}] \, d\sigma = \iint_S \tfrac{1}{3}\sqrt{3} \, d\sigma = \tfrac{1}{3}\sqrt{3}A = 2$

(b) $\displaystyle\oint_C \mathbf{v}(\mathbf{r}) \cdot d\mathbf{r} = \left(\int_{C_1} + \int_{C_2} + \int_{C_3}\right)\mathbf{v}(\mathbf{r}) \cdot d\mathbf{r} = -2 + 2 + 2 = 2$

7. (a) $\displaystyle\iint_S [(\nabla \times \mathbf{v}) \cdot \mathbf{n}] \, d\sigma = \iint_S [y\mathbf{k} \cdot \tfrac{1}{3}\sqrt{3}(\mathbf{i} + \mathbf{j} + \mathbf{k})] \, d\sigma = \tfrac{1}{3}\sqrt{3} \iint_S y \, d\sigma = \tfrac{1}{3}\sqrt{3}\bar{y}A = \tfrac{4}{3}$ **9.** 4π **11.** 0   **13.** $\pm\tfrac{1}{8}\pi$ **15.** $\pm\tfrac{1}{4}\pi$

(b) $\displaystyle\oint_C \mathbf{v}(\mathbf{r}) \cdot d\mathbf{r} = \left(\int_{C_1} + \int_{C_2} + \int_{C_3}\right)\mathbf{v}(\mathbf{r}) \cdot d\mathbf{r} = (\tfrac{4}{3} - \tfrac{32}{5}) + \tfrac{32}{5} + 0 = \tfrac{4}{3}$

17. Straightforward calculation shows that

$$\nabla \times (\mathbf{a} \times \mathbf{r}) = \nabla \times [(a_2z - a_3y)\mathbf{i} + (a_3x - a_1z)\mathbf{j} + (a_1y - a_2x)\mathbf{k}] = 2\mathbf{a}.$$

19. In the plane of C, the curve C bounds some Jordan region that we call Ω. The surface $S \cup \Omega$ is a piecewise-smooth surface that bounds a solid T. Note that $\nabla \times \mathbf{v}$ is continuously differentiable on T. Thus, by the divergence theorem,

$$\iiint_T [\nabla \cdot (\nabla \times \mathbf{v})] \, dx \, dy \, dz = \iint_{S \cup \Omega} [(\nabla \times \mathbf{v}) \cdot \mathbf{n}] \, d\sigma$$

where **n** is the outer unit normal. Since the divergence of a curl is identically zero, we have

$$\iint_{S \cup \Omega} [(\nabla \times \mathbf{v}) \cdot \mathbf{n}] \, d\sigma = 0.$$

Now **n** is $\mathbf{n}_1$ on S and $\mathbf{n}_2$ on Ω. Thus

$$\iint_{S} [(\nabla \times \mathbf{v}) \cdot \mathbf{n}_1] \, d\sigma + \iint_{\Omega} [(\nabla \times \mathbf{v}) \cdot n_2] \, d\sigma = 0.$$

This gives

$$\iint_{S} [(\nabla \times \mathbf{v}) \cdot \mathbf{n}_1] \, d\sigma = \iint_{\Omega} [(\nabla \times \mathbf{v}) \cdot (-\mathbf{n}_2)] \, d\sigma = \oint_{C} \mathbf{v}(\mathbf{r}) \cdot \mathbf{dr}$$

where C is traversed in a positive sense with respect to $-\mathbf{n}_2$ and therefore in a positive sense with respect to $\mathbf{n}_1$. ($-\mathbf{n}_2$ points toward S)

CHAPTER 18
SECTION 18.1
1. first-order ordinary **3.** first-order partial **5.** second-order ordinary **7.** second-order partial **9.** y_1 is; y_2 is not
11. both y_1 and y_2 are solutions **13.** both y_1 and y_2 are solutions **15.** both u_1 and u_2 are solutions **17.** both y_1 and y_2 are solutions
19. $y = 2e^{5x}$ **21.** $y = \dfrac{1}{-2e^{x-1} + 1}$ **23.** $y = -\frac{17}{4}x + 9x^{1/2}$ **25.** $y = x^2 \ln x$ **27.** $r = -3$ **29.** $r = -3$ **31.** $r = 0$
33. $r = \frac{1}{2}$ or $r = \frac{3}{2}$ **35.** (a) $y = C_1 \sin 4x$ (b) $y = 0$

SECTION 18.2
1. $y = -\frac{1}{2} + Ce^{2x}$ **3.** $y = \frac{2}{5} + Ce^{-(5/2)x}$ **5.** $y = x + Ce^{2x}$ **7.** $y = \frac{2}{3}nx + Cx^4$ **9.** $y = Ce^{e^x}$ **11.** $y = 1 + C(e^{-x} + 1)$
13. $y = e^{-x^2}(\frac{1}{2}x^2 + C)$ **15.** $y = C(x + 1)^{-2}$ **17.** $y = 2e^{-x} + x - 1$ **19.** $y = e^{-x}[\ln(1 + e^x) + e - \ln 2]$ **21.** $y = x^2(e^x - e)$
23. $y = y_0 e^{x_0^2 - x^2}$ **25.** (a) $200(\frac{4}{5})^{t/5}$ (b) $200(\frac{4}{5})^{t^2/25}$ **27.** $20 + 5e^{-0.03t}$ lb

29. (a) $T(t) = \tau + Ce^{-kt}$ (b) $T(t) = \tau + (T_0 - \tau)e^{-kt}$ (c) $\lim\limits_{t \to \infty} T(t) = \tau$ (d) $\dfrac{2 \ln 3}{\ln(\frac{4}{3})} - 2 \cong 5.64$ min

31. (a) $\dfrac{dP}{dt} = k(M - P)$ (b) $P(t) = M(1 - e^{-0.0357t})$ (c) 65 days **33.** $y^2 = \dfrac{1}{6e^x + Ce^{2x}}$ **35.** $y = \dfrac{1}{Cx - \frac{1}{3}x^4}$

37. If $y = y(x)$ is a solution of (18.2.1), then

$$y'(x) + p(x)y(x) = q(x)$$

and

$$e^{h(x)}y'(x) + p(x)e^{h(x)}y(x) = e^{h(x)}q(x),$$

where $h(x) = \displaystyle\int p(x) \, dx$. This implies that

$$e^{h(x)}y(x) = \int e^{h(x)}q(x) \, dx + C$$

for some constant C. It now follows that $y = y(x)$ is a member of the one-parameter family (18.2.2).

39. Let y_1 and y_2 be solutions of (18.2.3) and let $z = y_1 + y_2$. Then

$$z' = (y_1 + y_2)' = y_1' + y_2' = -py_1 - py_2 = -pz. \text{ Thus } z' + p(x)z = 0 \text{ and } z \text{ is a solution of (18.2.3).}$$

41. (a) From Exercise 38, $y(x) = Ce^{-h(x)}$. If $y(a) = 0$, then $0 = Ce^{-h(a)}$, which implies $C = 0$ since $e^{-h(a)} \neq 0$. Thus $y(x) = 0$ for all $x \in I$.
(b) Put $u(x) = y_1(x) - y_2(x)$. Then u is a solution of (18.2.3) by Exercise 39. Since $u(a) = y_1(a) - y_2(a) = 0$, $u(x) = 0$ for all $x \in I$ by part (a). Thus $y_1(x) = y_2(x)$ for all $x \in I$.

43. $y(x) = Ce^{-rx}$
(a) If $r > 0$, then $\lim\limits_{x \to \infty} y(x) = \lim\limits_{x \to \infty} Ce^{-rx} = 0$ (b) If $r < 0$, then $\lim\limits_{x \to \infty} y(x) = \lim\limits_{x \to \infty} Ce^{-rx} = \infty$

SECTION 18.3
1. $y = Ce^{-(1/2)\cos(2x+3)}$ **3.** $x^4 + \dfrac{2}{y^2} = C$ **5.** $y \sin y + \cos y = \cos\left(\dfrac{1}{x}\right) + C$ **7.** $e^{-y} = xe^{-x} + e^{-x} + C$

9. $\ln|y + 1| + \dfrac{1}{y + 1} = \ln|\ln x| + C$ **11.** $\sin^{-1}y = 1 - \sqrt{1 - x^2}$ **13.** $y + \ln|y| = \dfrac{x^3}{3} - x - 5$

15. $\dfrac{x^2}{2} + x + \dfrac{1}{2}\ln(y^2+1) - \tan^{-1}y = 4$ **17.** $y^2 - x^2 = Cx$ **19.** $x^2 - 2xy - y^2 = C$ **21.** $y + x = xe^{y/x}(C - \ln x)$

23. $x\left(1 + \cos\dfrac{y}{x}\right) = C\sin\dfrac{y}{x}$ **25.** $y^3 + 3x^3\ln|x| = 8x^3$ **27.** (a) $y' = \dfrac{y(a_4x - a_3)}{x(a_1 - a_2y)}$ (b) $x^{a_3}y^{a_1} = Ce^{a_4x + a_2y}$

29. $y = \frac{3}{2}x + C$ **31.** $x^2 - y^2 = C$ **33.** $y^2 = -2x + C$

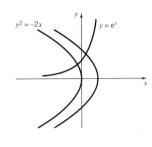

35. A differential equation for the given family is $y^2 = 2xyy' + y^2(y')^2$. Now replace y' by $-\dfrac{1}{y'}$. Then the resulting differential equation is

$y^2 = -\dfrac{2xy}{y'} + \dfrac{y^2}{(y')^2}$, which simplifies to $y^2 = 2xyy' + y^2(y')^2$. Thus the given family is self-orthogonal.

37. (a) $y(t) = \dfrac{25,000}{1 + 249e^{-0.1382t}}$, $y(20) \cong 1498$ (b) 40 days   **39.** (a) 88.82 m/sec   (b) $v(t) = \dfrac{15.65(1 + 0.48e^{-1.25t})}{1 - 0.48e^{-1.25t}}$ (c) 15.65 m/sec

SECTION 18.4

1. $\dfrac{x^2y^2}{2} - xy = C$ **3.** $xe^y - ye^x = C$ **5.** $x\ln y + x^2y = C$ **7.** $y\ln x - 3x^2 - 2y = C$ **9.** $xy^3 + y^2\cos x - \frac{1}{2}x^2 + \frac{1}{2}e^{2y} = C$

11. (a) yes   (b) $\dfrac{1}{p(y)q(x)}$ $(p(y)q(x) \neq 0)$ **13.** $xe^y - ye^x = C$ **15.** $x^3y^2 + \frac{1}{2}x^2 + xe^y + \frac{1}{2}y^2 = C$ **17.** $y^3e^x + xe^x = C$

19. $x^3 + 3xy + 3e^y = 4$ **21.** $4x^2y^2 + x^4 + 4x^2 = 5$ **23.** $xy - \dfrac{1}{y} = 3$ **25.** $\sinh(x - y^2) + \frac{1}{2}e^{2x} + \frac{1}{2}y^2 = \frac{1}{2}e^4 + 1$

27. (a) $k = 3$ (b) $k = 1$ **29.** $y = \dfrac{-4}{x^4 + C}$ **31.** $y = \frac{1}{9}x^5 + Cx^{-4}$ **33.** $e^{xy} - x^2 + 2\ln|y| = C$

SECTION 18.5

1. $y = C_1e^{3x} + C_2e^{-5x}$ **3.** $y = C_1e^{-4x} + C_2xe^{-4x}$ **5.** $y = e^{-x}(C_1\cos 2x + C_2\sin 2x)$ **7.** $y = C_1e^{(1/2)x} + C_2e^{-3x}$

9. $y = C_1\cos 2\sqrt{3}x + C_2\sin 2\sqrt{3}x$ **11.** $y = C_1e^{(1/5)x} + C_2e^{-(3/4)x}$ **13.** $y = C_1\cos 3x + C_2\sin 3x$ **15.** $y = e^{-(1/2)x}(C_1\cos\frac{1}{2}x + C_2\sin\frac{1}{2}x)$

17. $y = C_1e^{(1/4)x} + C_2e^{-(1/2)x}$ **19.** $y = 2e^{2x} - e^{3x}$ **21.** $y = 2\cos\dfrac{x}{2} + \sin\dfrac{x}{2}$ **23.** $y = 7e^{-2(x+1)} + 5xe^{-2(x+1)}$

25. (a) $y = Ce^{2x} + (1 - C)e^{-x}$ (b) $y = Ce^{2x} + (2C - 1)e^{-x}$ (c) $y = \frac{2}{3}e^{2x} + \frac{1}{3}e^{-x}$

27. $\alpha = \dfrac{r_1 + r_2}{2}$; $\beta = \dfrac{r_1 - r_2}{2}$

$y = k_1e^{r_1x} + k_2e^{r_2x} = e^{\alpha x}(C_1\cosh\beta x + C_2\sinh\beta x)$, where $k_1 = \dfrac{C_1 + C_2}{2}$, $k_2 = \dfrac{C_1 - C_2}{2}$.

29. (a) The Wronskian of $y_1 = e^{r_1x}$, $y_2 = xe^{r_1x}$ is

$W(x) = e^{r_1x}(e^{r_1x} + r_1xe^{r_1x}) - xe^{r_1x}(r_1e^{r_1x}) = e^{2r_1x} \neq 0$.

(b) The Wronskian of $y_1 = e^{\alpha x}\cos\beta x$, $y_2 = e^{\alpha x}\sin\beta x$, $\beta \neq 0$, is

$W(x) = e^{\alpha x}\cos\beta x[\alpha e^{\alpha x}\sin\beta x + \beta e^{\alpha x}\cos\beta x] - e^{\alpha x}\sin\beta x[\alpha e^{\alpha x}\cos\beta x - \beta e^{\alpha x}\sin\beta x] = \beta e^{2\alpha x} \neq 0$.

31. (a) $y'' + 2y' - 8y = 0$ (b) $y'' - 4y' - 5y = 0$ (c) $y'' - 6y' + 9y = 0$

33. If $y_1(x) = ky_2(x)$, then $y_2(x)y_1'(x) - y_1(x)y_2'(x) = ky_2(x)y_2'(x) - ky_2(x)y_2'(x) = 0$. If $y_2(x)y_1'(x) - y_1(x)y_2'(x) = 0$, then $\left[\dfrac{y_1(x)}{y_2(x)}\right]' = 0$, which implies

$y_1(x) = ky_2(x)$ for some constant k.

35. (a) If $a = 0$, $b > 0$, then the general solution is

$y = C_1\cos\sqrt{b}\,x + C_2\sin\sqrt{b}\,x = A\cos(\sqrt{b}\,x + \phi)$

where A and ϕ are constants. Clearly $|y(x)| \leq |A|$ for all x.

(b) If $a > 0$, $b = 0$, then the general solution is $y = C_1 + C_2 e^{-ax}$ and $\lim_{x \to \infty} y(x) = C_1$. The solution which satisfies the conditions $y(0) = y_0$, $y'(0) = y_1$ is

$$y = y_0 + \frac{y_1}{a} - \frac{y_1}{a} e^{-ax};$$

$$k = y_0 + \frac{y_1}{a}.$$

37. $y = C_1 x^4 + C_2 x^{-2}$ **39.** $y = C_1 x^2 + C_2 x^2 \ln x$

41. (a) $e^x = 1 + x + \frac{x^2}{2!} + \frac{x^3}{3!} + \cdots + \frac{x^n}{n!} + \cdots = \sum_{k=0}^{\infty} \frac{1}{k!} x^k$ (b) $e^{i\theta} = 1 + i\theta - \frac{1}{2!} \theta^2 - \frac{i}{3!} \theta^3 + \cdots + \frac{(i)^n}{n!} \theta^n + \cdots = \cos\theta + i\sin\theta$

(c) $e^{-i\theta} = 1 - i\theta - \frac{1}{2!} \theta^2 + \frac{i}{3!} \theta^3 + \cdots + \frac{(-i)^n}{n!} \theta^n + \cdots = \cos\theta - i\sin\theta$

SECTION 18.6

1. $y = \frac{1}{2} x + \frac{1}{4}$ **3.** $y = \frac{1}{5} x^2 - \frac{4}{25} x - \frac{27}{125}$ **5.** $y = \frac{1}{36} e^{3x}$ **7.** $y = \frac{1}{5} e^x$ **9.** $y = -\frac{13}{170} \cos x - \frac{1}{170} \sin x$ **11.** $y = \frac{3}{100} \cos 2x + \frac{21}{100} \sin 2x$

13. $y = \frac{1}{20} e^{-x} \sin 2x + \frac{1}{10} e^{-x} \cos 2x$ **15.** $y = \frac{3}{2} x e^{-2x}$ **17.** $y = C_1 \cos x + C_2 \sin x + \frac{1}{2} e^x$ **19.** $y = C_1 e^{5x} + C_2 e^{-2x} + \frac{1}{10} x + \frac{7}{100}$

21. $y = C_1 e^x + C_2 e^{-4x} - \frac{1}{5} x e^{-4x}$ **23.** $y = C_1 e^{-2x} + C_2 e^x + \frac{1}{2} x^2 e^x - \frac{1}{3} x e^x$

25. Let $z = y_1 + y_2$. Then   **27.** $y = C_1 e^{-3x} + C_2 e^{-x} + \frac{1}{4} x e^{-x} + \frac{1}{16} e^x$

$$z'' + az' + bz = (y_1'' + y_2'') + a(y_1' + y_2') + b(y_1 + y_2)$$
$$= (y_1'' + ay_1' + by_1) + (y_2'' + ay_2' + by_2) = \phi_1 + \phi_2$$

29. $y = 2 e^x \sin x - x e^x \cos x$ **31.** $y = \frac{1}{3} x \ln|x| e^{2x}$ **33.** $y = -\ln|x| e^{-2x}$ **35.** $y = e^x (x \sin x + \cos x \ln|\cos x|)$

37. (a) $i(t) = -CF_0 e^{-(R/2L)t} + \frac{F_0}{2L} (2 - RC) t e^{-(R/2L)t} + CF_0$

(b) $i(t) = e^{-(R/2L)t} \left[\frac{F_0(2 - RC)}{2L\beta} \sin\beta t - CF_0 \cos\beta t \right] + CF_0$,   where $\beta = \sqrt{\dfrac{4L - CR^2}{C}}$

39. (a) $y_1 y_2' - y_2 y_1' = -\dfrac{2}{x} \neq 0$ (b) $y = \frac{1}{3} \sin(\ln x)$

SECTION 18.7

1. $x(t) = \sin(8t + \frac{1}{2}\pi)$; $A = 1$, $f = 4/\pi$ **3.** $\pm 2\pi A/T$ **5.** $x(t) = (15/\pi) \sin \frac{1}{3}\pi t$ **7.** $x(t) = x_0 \sin(\sqrt{k/m}\, t + \frac{1}{2}\pi)$

9. at $x = \pm \frac{1}{2}\sqrt{3}\, x_0$ **11.** $\frac{1}{4} k x_0^2$

13. Set $y(t) = x(t) - 2$. Equation $x''(t) = 8 - 4x(t)$ can be written $y''(t) + 4y(t) = 0$. This is simple harmonic motion centered at $y = 0$, which is $x = 2$.

$$y(t) = A \sin(2t + \phi_0).$$

The condition $x(0) = 0$ gives $y(0) = -2$ and thus

$$A \sin \phi_0 = -2. \tag{$*$}$$

Since $y'(t) = x'(t)$ and $y'(t) = 2A \cos(2t + \phi_0)$, the condition $x'(0) = 0$ gives $y'(0) = 0$, and thus

$$2A \cos \phi_0 = 0. \tag{$**$}$$

Equations $(*)$ and $(**)$ are satisfied by $A = 2$, $\phi_0 = \frac{3}{2}\pi$. The equation of motion can therefore be written

$$y(t) = 2 \sin(2t + \frac{3}{2}\pi).$$

The amplitude is 2 and the period is π.

15. (a) $x''(t) + \omega^2 x(t) = 0$ with $\omega = r\sqrt{\pi\rho/m}$ (b) $x(t) = x_0 \sin(r\sqrt{\pi\rho/m}\, t + \frac{1}{2}\pi)$, taking downward as positive; $A = x_0$, $T = (2/r)\sqrt{m\pi/\rho}$

17. amplitude and frequency both decrease   **19.** at most once; at most once   **21.** if $\omega/\gamma = m/n$, then $m/\omega = n/\gamma$ is a period

23. $x = e^{-\alpha t}[c_1 \cos(\sqrt{\omega^2 - \alpha^2}\, t) + c_2 \sin(\sqrt{\omega^2 - \alpha^2}\, t)]$ or equivalently $x = C_1 e^{-\alpha t}[\sin(\sqrt{\omega^2 - \alpha^2}\, t) + C_2]$

25. $x_P = \dfrac{F_0}{2\alpha\gamma m} \sin\gamma t$;   as $c = 2\alpha m \to 0^+$, the amplitude $\dfrac{F_0}{2\alpha\gamma m} = \dfrac{F_0}{cm} \to \infty$

27. $(\omega^2 - \gamma^2)^2 + 4\alpha^2\gamma^2 = \omega^2 + \gamma^4 + 2\gamma^2(2\alpha^2 - \omega^2)$ increases as γ increases

APPENDIX A

SECTION A.1

1. hyperbola, $\alpha = \dfrac{\pi}{4}$, $\frac{1}{2}(X^2 - Y^2) = 1$ **3.** hyperbola, $\alpha = \dfrac{\pi}{6}$, $4X^2 - Y^2 = 1$ **5.** parabola, $\alpha = \dfrac{\pi}{4}$, $2Y^2 + \sqrt{2} X = 0$

7. parabola, $\alpha = \dfrac{\pi}{3}$, $X^2 - Y = 0$ **9.** $\alpha = \dfrac{\pi}{8}$, $\cos\alpha = \dfrac{1}{2}\sqrt{2 + \sqrt{2}}$, $\sin\alpha = \dfrac{1}{2}\sqrt{2 - \sqrt{2}}$

SECTION A.2

1. -2 **3.** 0 **5.** 5 **7.** -6 **9.** $\begin{vmatrix} 4 & 5 & 6 \\ 1 & 2 & 3 \\ 7 & 8 & 9 \end{vmatrix} = -\begin{vmatrix} 1 & 2 & 3 \\ 4 & 5 & 6 \\ 7 & 8 & 9 \end{vmatrix}$ (interchanged two rows)

11. The second determinant is zero because two rows are the same. **13.** $\begin{vmatrix} 1 & 2 & 3 \\ x & 2x & 3x \\ 4 & 5 & 6 \end{vmatrix} = x\begin{vmatrix} 1 & 2 & 3 \\ 1 & 2 & 3 \\ 4 & 5 & 6 \end{vmatrix} = 0$ ⌐— two rows are the same

15. The expression on the left can be written

$$\lambda^2 \begin{vmatrix} 1 & 1 \\ a & b \end{vmatrix} - 2\lambda \begin{vmatrix} 1 & 1 \\ a & b \end{vmatrix} + 1 \begin{vmatrix} 1 & 1 \\ a & b \end{vmatrix} = (\lambda^2 - 2\lambda + 1)\begin{vmatrix} 1 & 1 \\ a & b \end{vmatrix} = (\lambda - 1)^2 \begin{vmatrix} 1 & 1 \\ a & b \end{vmatrix}.$$

17. $\begin{vmatrix} a_1 & a_2 & a_3 \\ b_1 & b_2 & b_3 \\ c_1 & c_2 & c_3 \end{vmatrix} = -\begin{vmatrix} a_1 & a_2 & a_3 \\ c_1 & c_2 & c_3 \\ b_1 & b_2 & b_3 \end{vmatrix} = \begin{vmatrix} c_1 & c_2 & c_3 \\ a_1 & a_2 & a_3 \\ b_1 & b_2 & b_3 \end{vmatrix} = c_1 \begin{vmatrix} a_2 & a_3 \\ b_2 & b_3 \end{vmatrix} - c_2 \begin{vmatrix} a_1 & a_3 \\ b_1 & b_3 \end{vmatrix} + c_3 \begin{vmatrix} a_1 & a_2 \\ b_1 & b_2 \end{vmatrix}$

■ INDEX

(continued from the front)

INVERSE TRIGONOMETRIC FUNCTIONS

64. $\displaystyle\int \sin^{-1}u\,du = u\sin^{-1}u + \sqrt{1-u^2} + C$

65. $\displaystyle\int \cos^{-1}u\,du = u\cos^{-1}u - \sqrt{1-u^2} + C$

66. $\displaystyle\int \tan^{-1}u\,du = u\tan^{-1}u - \tfrac{1}{2}\ln(1+u^2) + C$

67. $\displaystyle\int \cot^{-1}u\,du = u\cot^{-1}u + \tfrac{1}{2}\ln(1+u^2) + C$

68. $\displaystyle\int \sec^{-1}u\,du = u\sec^{-1}u - \ln|u + \sqrt{u^2-1}| + C$

69. $\displaystyle\int \csc^{-1}u\,du = u\csc^{-1}u + \ln|u + \sqrt{u^2-1}| + C$

70. $\displaystyle\int u\sin^{-1}u\,du = \tfrac{1}{4}(2u^2-1)\sin^{-1}u + u\sqrt{1-u^2} + C$

71. $\displaystyle\int u\tan^{-1}u\,du = \tfrac{1}{2}(u^2+1)\tan^{-1}u - \tfrac{1}{2}u + C$

72. $\displaystyle\int u\cos^{-1}u\,du = \tfrac{1}{4}(2u^2-1)\cos^{-1}u - u\sqrt{1-u^2} + C$

73. $\displaystyle\int u^n\sin^{-1}u\,du = \frac{1}{n+1}\left[u^{n+1}\sin^{-1}u - \int\frac{u^{n+1}\,du}{\sqrt{1-u^2}}\right], n \ne -1$

74. $\displaystyle\int u^n\cos^{-1}u\,du = \frac{1}{n+1}\left[u^{n+1}\cos^{-1}u + \int\frac{u^{n+1}\,du}{\sqrt{1-u^2}}\right], n \ne -1$

75. $\displaystyle\int u^n\tan^{-1}u\,du = \frac{1}{n+1}\left[u^{n+1}\tan^{-1}u - \int\frac{u^{n+1}\,du}{1+u^2}\right], n \ne -1$

$\sqrt{a^2 + u^2},\ a > 0$

76. $\displaystyle\int \frac{du}{a^2+u^2} = \frac{1}{a}\tan^{-1}\frac{u}{a} + C$

77. $\displaystyle\int \frac{du}{\sqrt{a^2+u^2}} = \ln|u + \sqrt{a^2+u^2}| + C$

78. $\displaystyle\int \sqrt{a^2+u^2}\,du = \frac{u}{2}\sqrt{a^2+u^2} + \frac{a^2}{2}\ln|u + \sqrt{a^2+u^2}| + C$

79. $\displaystyle\int u^2\sqrt{a^2+u^2}\,du = \frac{u}{8}(a^2+2u^2)\sqrt{a^2+u^2} - \frac{a^4}{8}\ln|u + \sqrt{a^2+u^2}| + C$

80. $\displaystyle\int \frac{\sqrt{a^2+u^2}}{u}\,du = \sqrt{a^2+u^2} - a\ln\left|\frac{a+\sqrt{a^2+u^2}}{u}\right| + C$

81. $\displaystyle\int \frac{\sqrt{a^2+u^2}}{u^2}\,du = -\frac{\sqrt{a^2+u^2}}{u} + \ln|u + \sqrt{a^2+u^2}| + C$

82. $\displaystyle\int \frac{u^2\,du}{\sqrt{a^2+u^2}} = \frac{u}{2}\sqrt{a^2+u^2} - \frac{a^2}{2}\ln|u + \sqrt{a^2+u^2}| + C$

83. $\displaystyle\int \frac{du}{u\sqrt{a^2+u^2}} = -\frac{1}{a}\ln\left|\frac{\sqrt{a^2+u^2}+a}{u}\right| + C$

84. $\displaystyle\int \frac{du}{u^2\sqrt{a^2+u^2}} = -\frac{\sqrt{a^2+u^2}}{a^2 u} + C$

85. $\displaystyle\int \frac{du}{(a^2+u^2)^{3/2}} = \frac{u}{a^2\sqrt{a^2+u^2}} + C$

$\sqrt{a^2 - u^2},\ a > 0$

86. $\displaystyle\int \frac{du}{\sqrt{a^2-u^2}} = \sin^{-1}\frac{u}{a} + C$

87. $\displaystyle\int \sqrt{a^2-u^2}\,du = \frac{u}{2}\sqrt{a^2-u^2} + \frac{a^2}{2}\sin^{-1}\frac{u}{a} + C$

88. $\displaystyle\int u^2\sqrt{a^2-u^2}\,du = \frac{u}{8}(2u^2-a^2)\sqrt{a^2-u^2} + \frac{a^4}{8}\sin^{-1}\frac{u}{a} + C$